Protostars and Planets V

Protostars and Planets V

Bo Reipurth
David Jewitt
Klaus Keil

Editors

With 249 collaborating authors

THE UNIVERSITY OF ARIZONA PRESS
Tucson

in collaboration with

LUNAR AND PLANETARY INSTITUTE
Houston

About the front cover:

This view of part of the Orion Nebula was obtained by combining a large number of optical images taken by the Advanced Camera System onboard the Hubble Space Telescope (HST). The area shows Messier 43 (NGC 1982), a small circular HII region centered on NU Orionis (HR 2159, HD 41753), a bright (V = 4.4 mag) spectroscopic binary of spectral type B3V. The Orion Trapezium is seen in the lower right corner. Numerous young low-mass stars and brown dwarfs are scattered throughout the image. The high resolution offered by HST shows the dark lanes in this region in exquisite detail, and vividly portrays the effect of the newborn massive stars on the surrounding molecular clouds. *Courtesy of the Space Telescope Science Institute.*

About the back cover:

The image shows a snapshot taken from a simulation of five mutually interacting protoplanets embedded in the gaseous protoplanetary disk out of which they formed. Such a scenario is predicted by the oligarchic growth model of planet formation. The protoplanets are each 10 $M_\oplus$, and are located at the disk midplane. The image clearly shows the spiral waves generated by the protoplanets as they interact with the disk, causing them ultimately to migrate in toward the central star. *Courtesy of Richard P. Nelson and Paul Cresswell, Queen Mary, University of London.*

The University of Arizona Press
in collaboration with the Lunar and Planetary Institute
© 2007 The Arizona Board of Regents
All rights reserved
⊗ This book is printed on acid-free, archival-quality paper.
Manufactured in the United States of America

12 11 10 09 08 07   6 5 4 3 2 1

Library of Congress Cataloging-in-Publication Data

Protostars and planets V / Bo Reipurth, David Jewitt, and
Klaus Keil, editors.
       p. cm. — (The University of Arizona space science series)
    Includes bibliographical references and index.
    ISBN-13: 978-0-8165-2654-3 (hardcover : alk. paper)
    ISBN-10: 0-8165-2654-0 (hardcover : alk. paper)
    1. Protostars—Congresses. 2. Planetology—Congresses.
3. Meteorites—Congresses. I. Reipurth, Bo. II. Jewitt, D.
(David) III. Keil, Klaus. IV. Title: Protostars and planets 5.
V. Title: Protostars and planets five.
QB806.P78 2007
523.8'8—dc22   2006037043

Dedicated to

George H. Herbig

in recognition of his fundamental
contributions to early stellar evolution

Contents

PART I: MOLECULAR CLOUDS

PART II: STAR FORMATION

PART III: OUTFLOWS

PART IV: YOUNG STARS AND CLUSTERS

PART VI: CIRCUMSTELLAR DISKS

PART VII: PLANET FORMATION AND EXTRASOLAR PLANETS

PART VIII: DUST, METEORITES, AND THE EARLY SOLAR SYSTEM

Scientific Advisory Committee

List of Collaborating Authors

Adams F. C.	361	Ceccarelli C.	47	Goodwin S. P.	133
Aikawa Y.	17, 751	Cesaroni R.	197	Gounelle M.	835
Akeson R.	539	Chabrier G.	623	Greene T. P.	117
Alencar S. H. P.	479	Charbonneau D.	701	Grimm R.	863
Alexander C. M. O'D.	801	Chiang E.	895	Grundy W.	879, 895
Alibert Y.	623	Chizmadia L.	863	Güdel M.	313, 329
Allen L.	361	Chrysostomou A.	231	Gueth F.	245
Allen P.	427	Churchwell E. B.	165	Guilloteau S.	495
Alves J. F.	3	Clarke C. J.	395	Gutermuth R.	361
Amelin Y.	835	Close L.	427	Haisch K. E. Jr.	379
André P.	33	Covey K. R.	117	Harker D.	815
Arce H. G.	245	Cruikshank D. P.	879	Harries T. J.	479
Artymowicz P.	655	Crutcher R.	33	Hennebelle P.	623
Bacciotti F.	231	Cuzzi J. N.	783, 849	Henning Th.	767
Bachiller R.	245	D'Alessio P.	555	Herbst E.	47
Backman D.	573, 669	Davis A. M.	835	Herbst W.	297
Ballesteros-Paredes J.	63	Davis C. J.	215	Hillenbrand L. A.	117
Bally J.	215	Delgado-Donate E.	379	Hirano N.	261
Baraffe I.	411, 623	Desch S.	815, 835	Ho P.	495
Barman T.	623, 733	Di Francesco J.	17	Hoare M. G.	181
Barucci M. A.	879	Dominik C.	783	Hofner P.	181
Bate M. R.	459	Doppmann G. W.	117	Hollenbach D.	555
Beichman C. A.	915	Dougados C.	231, 329	Holman M.	895
Bergin E. A.	751	Duchêne G.	379	Inutsuka S.-I.	99
Beuther H.	165, 245	Dullemond C. P.	555	Jewitt D.	863
Beuzit J.-L.	717	Durisen R. H.	607	Joergens V.	443
Blake G. A.	751	Dutrey A.	495	Johns-Krull C. M.	479
Blitz L.	81	Eislöffel J.	231, 297	Johnstone D.	33
Blum J.	783	Emery J. P.	879	Kamp I.	555
Bonnell I. A.	149	Evans N. J. II	17	Kawamura A.	81
Boss A. P.	607, 801	Feigelson E.	313	Keller L.	801, 815
Bouvier J.	479	Fendt Ch.	277	Kenyon S. J.	639
Brandenburg A.	277	Fernández Y. R.	879	Keto E.	181
Briceño C.	345	Fischer D.	685	Klein R. I.	99
Bromley B. C.	639	Fortney J.	733	Klessen R. S.	63
Brown T. M.	701	Fridlund M.	915	Kley W.	655
Buie M.	895	Fukui Y.	81	Krot A. N.	849
Burgasser A. J.	427	Gaidos E.	929	Kroupa P.	133
Burkert A.	133	Gail H.-P.	815	Kuramoto K.	849
Burrows A.	701	Galli D.	197	Kurtz S. E.	181
Calvet N.	767	Gizis J.	427	Lada C.	3, 443
Carr J. S.	507	Glassgold A. E.	507	Larson R. B.	149
Caselli P.	17, 47	Gomes R.	669	Laughlin G.	701
Caux E.	47	Goodman A.	133	Lee C.-F.	245

List of Collaborating Authors

(continued)

Leinert C.	539	Najita J. R.	507	Simon M.	411
Leroy A.	81	Natta A.	767	Stansberry J. A.	879
Levison H. F.	669	Nelson A. F.	607	Stapelfeldt K. R.	523, 915
Li Z.-Y.	261	Nelson R. P.	655	Stassun K.	313, 411
Lin D. N. C.	639	Noll K. S.	879	Stevenson D. J.	591
Lissauer J. J.	591	Nordlund Å.	459	Tafalla M.	17
Lithwick Y.	895	Nuth J. A.	801	Tan J. C.	165
Lizano S.	181	Onishi T.	33	Testi L.	767
Lodato G.	197	Oppenheimer B. R.	717	Thiemens M. H.	849
Loinard L.	379	Ouyed R.	277	Thommes E. W.	639
Lombardi M.	3	Padgett D. L.	329	Tielens A. G. G. M.	47
Lowrance P.	427	Padoan P.	99	Tomisaka K.	99
Lugmair G. W.	835	Papaloizou J. C. B.	655	Townsley L.	313
Luhman K. L.	443	Pascucci I.	443	Traub W. A.	915
Lyons J. R.	849	Pipher J. L.	361	Udry S.	685
Mac Low M.-M.	63	Prato L.	395	Valenti J. A.	507
Malbet F.	539	Preibisch T.	345	van Dishoeck E. F.	751
Mamajek E. A.	345	Prialnik D.	863	Vázquez-Semadeni E.	63
Marley M. S.	733	Pudritz R. E.	277	Wadhwa M.	835
Masset F. S.	655	Queloz D.	685	Walmsley C. M.	197
Mathieu R. D.	345, 411	Quinn T.	607	Walter F. M.	345
Mayer L.	607	Quirrenbach A.	915	Ward-Thompson D.	33
McCabe C.	395	Ray T.	231	Waters R.	539, 767
McKee C. F.	165	Reid I. N.	427	Watson A. M.	523
Megeath S. T.	361	Reipurth B.	215, 459	Weinberger A.	573, 801
Ménard F.	523	Rice W. K. M.	607	White R.	117, 411, 443
Meyer B.	835	Rodríguez L. F.	379	Whitworth A.	459
Meyer M. R.	573	Romanova M. M.	479	Wilner D.	767
Millan-Gabet R.	539	Rosen A.	245	Wilson C.	33
Mizuno N.	81	Rosolowsky E.	81	Wolk S.	361
Monin J.-L.	395	Scholz A.	297	Wood K.	523
Monnier J.	539, 717	Scott E. R. D.	849	Wooden D.	815
Morbidelli A.	669	Seager S.	733, 915	Wurm G.	783
Mouillet D.	717	Selsis F.	623, 929	Wyatt M. C.	573
Mundt R.	297	Shang H.	261	Young E.	361
Murray-Clay R.	895	Shepherd D.	245	Yurimoto H.	849
Muzerolle J.	361, 443	Sherry W. H.	345	Zhang Q.	197
Myers P. C.	17, 361	Shirley Y.	17	Zinnecker H.	149, 345, 459
Nagasawa M.	639	Siegler N.	427		

Preface

The Protostars and Planets project was initiated by Tom Gehrels, and commenced on January 3, 1978, when the first Protostars and Planets conference began in Tucson, Arizona. The goal was the then-novel one of bridging the gap between the fields of star formation and planetary science. Or, in the words of Gehrels in the preface to the first *Protostars and Planets* book: "Cross-fertilization of information and understanding is bound to occur when investigators who are familiar with the stellar and interstellar phases meet with those who study the early phases of solar system formation." This goal remained the same for the subsequent meetings — Protostars and Planets II in January 1984, Protostars and Planets III in March 1990, and Protostars and Planets IV in July 1998 — as well as for the resulting books. It is amusing today to note that Gehrels had a wager with several people that a conference on protostars and planets could not be successful. They wagered on the basis of two similar attempts at Kitt Peak in the 1970s, and on the belief that the stellar and planetary populations were too far apart to even speak to each other or use the same language. The obvious success of the Protostars and Planets conferences and resulting books has since vindicated Gehrels' vision and put such doubts to shame. The original concept was again the foundation on which Protostars and Planets V was organized.

The Protostars and Planets V conference was held at the Hilton Waikoloa Village on the Big Island of Hawaii on October 24–28, 2005. The meeting attracted 805 participants, half of which were from 30 countries outside the United States. The present book contains 58 chapters, corresponding to the review talks given at the conference, and together they cover in detail the key aspects of our present understanding of star and planet formation, young stars, and the early solar system. The average time between Protostars and Planets meetings is seven years, a period long enough that very substantial progress in a field can be made. Indeed, every preface for the earlier Protostars and Planets books has exulted in the major strides made since the previous conference, and this book is no exception. As documented in this volume, we have since the previous conference seen the field of extrasolar planets burgeon, with the discovery of new planets around others stars growing by more than a factor of 10, with the detection of multiple planets around stars that allow us to compare empirically the architectures of planetary systems, and with the development of detailed theoretical work on the formation of both terrestrial planets and gas giants. The earliest stages of protostars are being routinely studied with new advanced submillimeter detectors, and the collapse process can be followed with sophisticated numerical simulations on ever-more-powerful computers. While we are gaining confidence that fundamental aspects of low-mass star formation are becoming understood, the formation of massive stars is emerging as a challenging and fruitful area of investigation. X-ray studies of star-forming regions have opened entirely new vistas into the active lives of young stars. High-spatial-resolution techniques have developed in recent years to offer amazingly detailed views of circumstellar disks, allowing meaningful comparisons with detailed theoretical models. Most stars are born in binary systems, and both observations and theory have pointed to the importance for binary formation of dynamical evolution and of decay in small multiple systems. The study of the nature and formation of brown dwarfs has now become part of mainstream astrophysics, and continues to offer exciting results and new challenges. More locally, a whole new field of research has opened up concerning the contents and structure of the Kuiper belt. New surveys and physical observations with large telescopes show dynami-cal and compositional complexity that challenges our understanding of the accretion process. Developments in understanding the dynamics of the early solar system have also been dramatic, with firm evidence for the radial migration of the planets setting new and unexpected constraints on early conditions. Refined dating techniques of meteorites have resulted in a more precise age for the solar system of 4567 million years. An even greater emphasis on the cosmochemical connection between meteorites and studies of dust and disks could have been made in this volume, but the recent publication of the *Meteorites and the Early Solar System II* volume, also in the Space Science Series, provides a current in-depth overview of the links between meteoritics and astrophysics that we see no reason to duplicate.

The five volumes published so far in the Protostars and Planets series offer a vivid portrait of the development of the field of early stellar evolution and its many associated aspects. For younger readers in particular it may be instructive to peruse the earlier volumes as a reminder of the long winding process, the hard work, the enduring successes, and the inevitable blind alleys that all were part of the path to our current insights. Numerous researchers have participated in this process, but none has contributed more to shaping the foundation for our understanding of early stellar evolution than George H. Herbig, to whom we dedicate this book.

The volume that you hold in your hands is a product of care and effort by numerous people and organizations. First and foremost we want to extend our thanks to the 249 contributing authors of this book; it is through their work and expertise that we have been able to prepare a compendium having both depth and breadth. The Protostars and Planets V conference, which formed the basis of this volume, was a major undertaking that would not have been possible without the support and sponsorship by numerous entities (listed in alphabetical order): Apple Computer, Inc., Caltech Submillimeter Observatory, Canada-France-Hawaii Telescope, Gemini Observatory, Hawaii Institute of Geophysics and Planetology, ITT Industries, Jet Propulsion Laboratory, Joint Astronomy Centre, Lockheed Martin Corporation, Lunar and Planetary Institute, NASA Astrobiology Institute, NASA Headquarters, NASA Infrared Telescope Facility, National Science Foundation, Northrop Grumman Corporation, Smithsonian Submillimeter Array, Stratospheric Observatory for Infrared Astronomy, Subaru Telescope, Terrestrial Planet Finder, UH Astrobiology Institute, UH Institute for Astronomy, UH School of Ocean and Earth Science and Technology, University of Hawaii, Watumull Foundation, and W. M. Keck Observatory. To all, we express our gratitude for their support. We further want to thank the members of the Scientific Advisory Committee, listed elsewhere, who selected the subjects and author teams from four times more proposals than could be accommodated, a difficult task that they executed with insight and sensitivity. Similarly, we are grateful to the 58, mostly anonymous, referees, who offered their expertise to ensure that all details of this book have been examined and approved. We owe a great mahalo to Judith Fox-Goldstein of the University of Hawaii at Hilo Conference Center, who with her expert team handled the long preparations and difficult logistics with grace and élan, and resolved a seemingly never-ending series of difficulties. Likewise, thanks are due to Karen Meech and the members of the Local Organizing Committee, whose efforts ensured a successful conference. We also wish to thank Richard P. Binzel, General Editor of the Space Science Series, for his unwavering support during the lengthy and at times stressful process leading to this book. Finally, but certainly not least, special thanks to the editorial staff of the Lunar and Planetary Institute, particularly Renée Dotson, who handled the detailed processing of all manuscripts and the production of the book with her legendary expertise and competence, and to the University of Arizona Press for speedy publication.

Bo Reipurth, David Jewitt, and Klaus Keil
October 2006

Part I:
Molecular Clouds

Near-Infrared Extinction and Molecular Cloud Structure

Charles J. Lada

Harvard-Smithsonian Center for Astrophysics

João F. Alves and Marco Lombardi

European Southern Observatory

A little more than a decade has passed since the advent of large-format infrared array cameras opened a new window on molecular cloud research. This powerful observational tool has enabled dust extinction and column density maps of molecular clouds to be constructed with unprecedented precision, depth, and angular resolution. Near-infrared extinction studies can achieve column density dynamic ranges of $0.3 < A_V < 40$ mag ($6 \times 10^{20} < N < 10^{23}$ cm^{-2}), allowing with one simple tracer a nearly complete description of the density structure of a cloud free from the uncertainties that typically plague measurements derived from radio spectroscopy and dust emission. This has led in recent years to an empirical characterization of the evolutionary status of dense cores based on the shapes of their radial column density profiles and revealed the best examples in nature of Bonnor-Ebert spheres. Widefield infrared extinction mapping of large cloud complexes provides the most complete inventory of cloudy material that can be derived from observations. Such studies enable the measurement of the mass function of dense cores within a cloud, a critical piece of information for developing an understanding of the origin of the stellar IMF. Comparison with radio spectroscopic data has allowed detailed chemical structure studies of starless cores and provided some of the clearest evidence for differential depletion of molecular species in cold gaseous configurations. Recent studies have demonstrated the feasibility of infrared extinction mapping of galactic molecular clouds (GMCs) in external galaxies, enabling the fundamental measurement of the GMC mass function in these systems. In this contribution we review recent results arising from this powerful technique, ranging from studies of Bok globules to local GMCs, to GMCs in external galaxies.

1. INTRODUCTION

Although first discovered in the late eighteenth century with visual telescopic observations by William and Caroline Herschel (*Herschel,* 1785), well over a century passed before photographic observations of *Barnard* (1919) and *Wolf* (1923) established dark nebulae as discrete, optically opaque interstellar clouds. Wolf's photographic photometry further demonstrated that the opaque material in these clouds consisted of solid, small particles now known as interstellar dust grains. For nearly three decades astronomers could not discern whether such nebulae contained gaseous matter or were solely made up of dust. Indeed, in the 1950s the discovery of the 21-cm line of H I led to the measurement of a general correlation between dust absorption and H I emission at low extinctions (*Lilley,* 1955). However, the first searches for H I gas toward the centers of dark nebulae found H I emission to be either weak or entirely absent (*Bok et al.,* 1955; *Heiles,* 1969). This fact led *Bok et al.* (1955) to the prescient suggestion that if any gas existed within these nebulae, it had to be molecular in form.

The discovery of a cold molecular component of the interstellar medium primarily via CO observations in the early 1970s (e.g., *Wilson et al.,* 1970) quickly lead to the realization that dark clouds were molecular clouds, consisting almost entirely of molecular hydrogen mixed with small amounts of interstellar dust and trace amounts of more complex molecular species. Over the last 30 years infrared observations from space and the ground (e.g., *Becklin and Neugebauer,* 1967; *Strom et al.,* 1975; *Wilking et al.,* 1989; *Yun and Clemens,* 1990; *Lada,* 1992) have established the true astrophysical importance of molecular clouds as the sites of all star formation in the Milky Way.

Knowledge of the structure and physical conditions in molecular clouds is of critical importance for understanding the process of star formation. Such observations can set the initial conditions for star formation and enable the direct investigation of the evolution of cloudy material as it is transformed into stellar form. However, although almost entirely composed of molecular hydrogen, two factors render H$_2$ generally unobservable in molecular clouds. First, because it is a homonuclear molecule, it lacks a permanent dipole moment and its rotational transitions are extremely weak. Second, being the lightest molecule, its lowest energy rotational transitions are at mid-infrared wavelengths, which are simultaneously inaccessible to observation from Earth and too energetic to be collisionally excited at the cold temperatures (e.g., 10 K) that are characteristic of dark molecular clouds. Thus the structure and physical conditions of dark clouds must be derived from H_2 surrogates, namely dust and trace molecular gases such as CO, NH_3, and HCN.

Despite early promise, use of molecular lines from trace molecular species for this purpose has turned out to be severely hampered by the presence of significant variations in opacities, chemical abundances, and excitation conditions within dark nebulae. Yet most of what we know of the physical conditions in molecular clouds derives from such studies. Observations of dust may be the most direct and reliable way to trace the hydrogen content of a molecular cloud. This is because of the constancy of the gas-to-dust ratio in interstellar clouds, which is observationally well established (e.g., *Lilley,* 1955; *Ryter et al.,* 1975; *Bohlin et al.,* 1978; *Predehl and Schmitt,* 1995). In principle, the dust content of a cloud can be traced through measurements of either dust emission or extinction (i.e., absorption plus scattering of background starlight). Interpretation of measurements of dust emission is complicated by the fact that the observed emission is a product of dust column density and dust temperature. For the range of temperatures characteristic of dark clouds (8–50 K), dust emission is brightest and most readily detected at far-infrared wavelengths. This emission is near the peak and often on the Wien side of the corresponding Planck curve and thus nonlinearly sensitive to the temperature of dust along the line-of-sight. Moreover, because of the opacity of our atmosphere at these wavelengths such observations are best carried out on space platforms. On the other hand, *dust extinction* is independent of *dust temperature,* and measurements of dust extinction are directly proportional to dust optical depth and column density and thus measurement of dust extinction is the most direct and straightforward way to detect and map the dust content of a cloud. There are two observational techniques used to make such extinction measurements: the classical method of star counts (e.g., *Bok,* 1956) and the method of measuring stellar color excess (e.g., *Lada et al.,* 1994).

2. MEASURING DUST EXTINCTION

2.1. Star Counts

Throughout the last century the general method to derive and map the extinction through a dark cloud has been the use of optical star counts (e.g., *Wolf,* 1923; *Bok,* 1956; *Dickman,* 1978; *Cernicharo and Bachiller,* 1984). In this method typically a rectilinear grid is overlaid on an image of the target cloud, and the number of stars to a fixed limiting magnitude is counted in each box of the grid. These counts are compared to those in a nearby, unobscured region. The extinction is then given by

$$A_\lambda = \log(N_{off}/N_{on})/b_\lambda \qquad (1)$$

Here N is the surface density of stars either on or off the cloud and b is the slope of the logarithmic cumulative luminosity function of stars in the control field. One of the most impressive studies of this type is the recent work of *Dobashi et al.* (2005). They used the Digitized Sky Survey to obtain star counts and a map of extinction encompassing the entire galactic plane. In the process they were able to identify 2448 dark clouds. However, such measurements suffer from a number of limitations, the most significant of which is the uncertainty due to Poisson statistics. For example, at optical wavelengths (using the POSS) it takes only about 4–5 mag of visual extinction to reduce the number of stars in a single counting pixel to unity in a typical arc-minute-sized region of the galactic plane. Somewhat higher values of extinction ($A_V \sim 5$–9 mag) can be measured, but at the significant expense of angular resolution and thus the loss of structural information (*Cambresy,* 1999). For example, the widefield maps of *Dobashi et al.* (2005) were limited to resolutions of 6 and 18 arcminutes. One can take advantage of the extinction law and instead perform star counts at near-infrared wavelengths where extinction is significantly reduced compared to the optical regime. Therefore, significantly larger numbers of background stars can be observed through a dark cloud in the infrared than at traditional optical wavelengths. Such measurements can attain higher values (10–20 mag) of visual extinction without sacrificing too much angular resolution (e.g., *Lada et al.,* 1994; *Harvey et al.,* 2001, 2003; *Pagani et al.,* 2004; *Kandori et al.,* 2005). However, multiwavelength infrared observations can produce even deeper and much more direct extinction determinations with greatly improved angular resolution and smaller overall uncertainties by using measurements of near-infrared color-excess to derive the extinction. With large-format infrared arrays it is now possible to use moderate-sized telescopes to detect and simultaneously determine the colors of thousands of stars behind a dark molecular cloud.

2.2. Near-Infrared Color Excess: NICE and NICER

Because of the wavelength dependence of extinction (see Fig. 1) it is advantageous to observe at the longest wavelength possible to penetrate the most heavily obscured regions of a dark cloud. The opacity of Earth's atmosphere severely limits observations at wavelengths much longer than 2 µm. The standard near-infrared H (1.65 µm) and K (2.2 µm) bands are closely matched to windows of atmospheric transmission and represent the longest wavelengths for optimized groundbased observation with infrared array cameras. Moreover, typical field stars are relatively bright in the near-infrared since these wavelengths are near the peak of their stellar energy distributions, and thus these stars are readily detected with existing infrared cameras on modest-sized telescopes. Because of the wavelength dependence of extinction, light from a star suffers a change in color as well as a general diminution. The change in a star's color is much more easily measured than the change in a star's brightness. This is because the intrinsic variation in unextincted stellar colors is considerably smaller than the magnitude of the change in stellar colors induced by extinction,

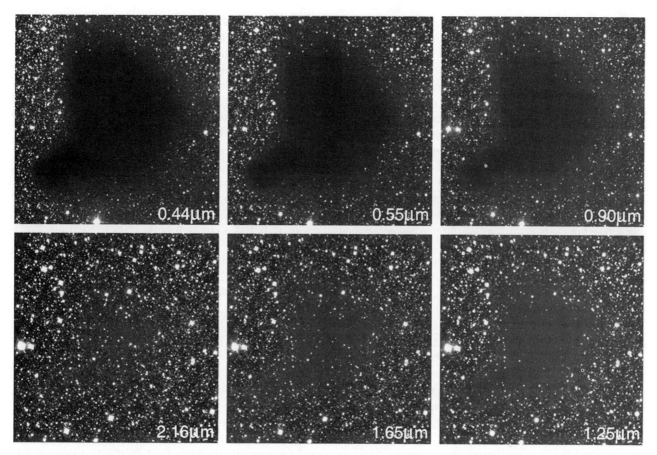

Fig. 1. Optical (BVI) and infrared (JHKs) images of the globule Barnard 68 obtained with ESO's VLT and NTT (*Alves et al.,* 2001a). This sequence shows the increasing transparency of interstellar dust with increasing wavelength. At the optical bands B (0.44 µm) and V (0.55 µm) the globule is completely opaque, while in the infrared K_s (2.16 µm) band it is almost completely transparent.

while the intrinsic variation in unextincted stellar brightnesses of background stars is comparable to and often significantly exceeds the magnitude of the diminution in stellar brightnesses caused by extinction.

The line-of-sight extinction to an individual star can be directly determined from knowledge of its color excess and the extinction law. The color excess, E(H–K), can be directly derived from observations provided the intrinsic color of the star is known

$$E(H–K) = (H–K)_{observed} - (H–K)_{intrinsic} \qquad (2)$$

The intrinsic (H–K) colors of normal main-sequence and giant stars are small and span a narrow range in magnitude (0.0–0.3 mag for stars with spectral types between A0 and late M).

One expects the range in intrinsic colors of background field stars to span an even smaller range since such stars span a relatively narrow range in spectral type (typically K and M). Observations of field stars in nearby (unobscured) control regions off the target cloud can be used very effectively to determine the intrinsic colors of background stars provided that the control field stars themselves do not suffer significant additional extinction in comparison to the stars

in the target field (e.g., from unrelated background clouds along the same line-of-sight).

With the assumption that all the background field stars observed through a target molecular cloud are identical in nature to those in the control field, we can use the mean H–K color of the control field stars to approximate the intrinsic H–K color of all stars background to the cloud

$$(H–K)_{intrinsic} \equiv <(H–K)>_{control} \qquad (3)$$

The infrared color excess is directly related to the extinction (at any wavelength) via the extinction law

$$A_\lambda = r_\lambda^{H,K} E(H–K) \qquad (4)$$

where $r_\lambda^{H,K}$ is the appropriate constant of proportionality given by the adopted extinction law (e.g., *Rieke and Lebosky,* 1985). Use of infrared colors to measure extinction is known as the Near-Infrared Color Excess (NICE) method of extinction determination.

As shown by *Lombardi and Alves* (2001), one can take full advantage of observations carried out in multiple (>2) bands to obtain even more accurate column density measurements. The improved technique, called NICER (NICE

Revisited), optimally balances the information from different bands and different stars. As a byproduct of the analysis, NICER also allows one to evaluate the expected error on the column density map, which is useful to estimate the significance for the detection of substructures and cores. The NICER technique is better described by noting that, in principle, equations (2) and (4) can be equally well applied to any combination of infrared bands (e.g., the color H–K usually used in NICE, or the alternative JH color, if the J-band photometry is available): Hence, this way we can have several different estimates of the cloud column density. The NICER technique finds an optimal linear combination of the column density estimates from the different bands. For example, in the case of three bands J, H, and K, the extinction is estimated as

$$A_\lambda = b_\lambda^{H,K} E(H\text{–}K) + b_\lambda^{J,H} E(J\text{–}H) \qquad (5)$$

and the coefficients $b_\lambda^{H,K}$ and $b_\lambda^{J,H}$ are found by requiring that (1) the column density estimate is unbiased and (2) it has minimum variance. The first condition, by inspection of equation (4), implies $b_\lambda^{H,K} r_\lambda^{H,K} + b_\lambda^{J,H} r_\lambda^{J,H} = 1$, while the second depends on the intrinsic scatter in the infrared colors and on the individual errors on the photometric measurements of each star (see *Lombardi and Alves*, 2001, for details). When applied to typical NIR observations, the NICER method is able to reduce by a factor of 2 the average variance of the NICE column density measurements, thus improving the maximum resolution achievable. More significantly, use of NICER greatly improves the overall sensitivity of the extinction measurements at low extinctions ($0.5 < A_V < 2.0$ mag). Its implementation thus enables the mapping of the outer envelopes of molecular clouds, regions that cannot be traced by CO observations because of the dissociation of CO by the interstellar radiation field.

A further enhancement on the accuracy of column density measurements can be gained by using a maximum likelihood (ML) technique (*Lombardi*, 2005). The method takes advantage of both the multicolor distribution of reddened stars and of their spatial distribution. In other words, the ML method optimally integrates the color excess and star counts methods. The ML technique is especially convenient in the high-density cores of molecular clouds, and in the presence of contamination by foreground stars.

Although the exact number of background stars that can be detected behind a molecular cloud depends on its galactic latitude and longitude, its angular size, and the sensitivity of the observations, a modest depth (i.e., K ≤ 16) infrared imaging survey can easily yield colors for thousands of background stars behind a typical nearby dark cloud. The resulting database of infrared color excesses and source positions would represent a map of the distribution of color excess (and extinction) through the cloud obtained with extraordinarily high (pencil-beam) angular resolution. However, we can expect this map to be randomly and heavily undersampled in space because of various observational selection effects expected in a magnitude-limited survey (e.g.,

the observed surface density of background stars is a function of detector sensitivity, galactic longitude and latitude, and extinction). The data in such a survey can also be used to construct an ordered and uniformly sampled map of the distribution of color excess through the cloud by smoothing the angular resolution of the observations. The smoothing functions or kernels that have been employed are typically of fixed angular resolution and either in the shape of a square (e.g., *Lada et al.,* 1994) or a two-dimensional gaussian [better for comparison with radio data (e.g., *Alves et al.,* 1999, 2001a)]. Cambresy and co-workers have advocated the use of an adaptive smoothing kernel that constantly adjusts the size of the kernel to maintain a constant number of stars and a constant noise level in each pixel of a map (*Cambresy et al.,* 1997, 2002; *Cambresy,* 1999). This method produces lower angular resolution at the highest extinctions, but eliminates empty pixels or pixels with only a single star that might otherwise occur in maps with fixed angular resolution.

3. EXTINCTION MAPPING OF CLOUD CORES

Low-mass dense cores within large molecular complexes and isolated dense cores (also known as Bok globules) are the simplest configurations of dense molecular gas and dust known to form stars (e.g., *Benson and Myers,* 1989; *Yun and Clemens,* 1990). They have been long recognized as important laboratories for investigating the physical processes that lead to the formation of stars and planets (e.g., *Bok,* 1948). These cores come in two varieties: protostellar and starless. Detailed knowledge of the structure of both types of cores is essential for obtaining an empirical picture of the evolution of dense gas to form stars. Such information can provide critical constraints for developing an overall theory of star formation. Early work by *Tomita et al.* (1979) attempted to probe the structure of dark globules using extinction measurements derived from optical star counts of the Palomar Sky Survey. They found core structure to be characterized by very steep density gradients with power-law exponents ranging between –3 to –5. However, the high extinctions that characterized the globules limited their measurements to only the very outer regions or atmospheres of these objects and were not useful for constraining the overall structure of these cores.

It has been long understood that infrared observations could probe much larger extinctions in such objects and initial studies of globules using single-channel infrared detectors showed much promise for probing higher extinctions and obtaining more complete measurements of the structure of globules and dense cores (*Jones et al.,* 1980; *Casali,* 1986). But these studies were very limited by sensitivity and the small numbers of background stars that could be measured in the regions. The development of sensitive infrared imaging array detectors radically altered this situation. Such arrays enabled the detection of thousands of background stars behind clouds and resulted in the first detailed extinc-

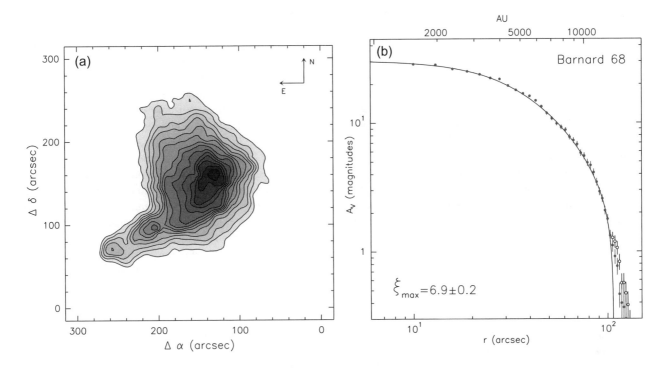

Fig. 2. **(a)** Distribution of extinction (dust column density) for the B68 cloud (see Fig. 1) derived from deep near-infrared extinction measurements sampled with a gaussian spatial filter of 10 arcsec (1000 AU) width (FWHP). **(b)** Azimuthally averaged dust column density profile for the cloud constructed with a spatial resolution of 5 arcsec (500 AU). With this resolution the radial structure of the cloud is extremely well resolved. (Open circles include the contribution of the tongue or protrusion of material at the southeast edge of the cloud; filled circles exclude this material.) Plotted for comparison is the corresponding best-fit column density profile (solid line) for a pressure-truncated, isothermal sphere (Bonnor-Ebert sphere).

tion maps of dark nebulae with high angular resolution and sufficient depth to probe the structure at relatively high extinction (*Lada et al.,* 1994, 1999; *Alves et al.,* 1998). Indeed, coupled with larger-aperture telescopes, such arrays also allowed deep imaging of nearby dark clouds and the ability to measure extinctions in regions of very high opacity, that is, $20 < A_V < 35$ mag (*Alves et al.,* 2001a). This is the level of extinction produced by the dense cores of dark clouds, the very places in which star formation takes place.

3.1. Starless Cores

Two of the least-understood aspects of the star-formation process are the initial conditions that describe dense cores that ultimately form stars and the origin of such dense cores from more diffuse atomic and molecular material. Deep infrared observations of starless cores or globules offer among the best opportunities to quantitatively investigate these issues. The first starless globule or core to be mapped at high resolution in dust extinction was the prominent globule Barnard 68 (B68) (*Alves et al.,* 2001a). Optical and infrared images of this globule are shown in Fig. 1. This globule is situated in front of the rich star field of the galactic bulge where the homogeneous nature of the background stars resulted in a very accurate determination of

their intrinsic colors and very accurate measurements of extinction. Moreover, the high density of background stars enabled the cloud to be mapped with relatively high angular resolution.

Figure 2 shows the resulting extinction map of B68 smoothed with a 10-arcsec (~1000 AU) gaussian spatial filter. Also plotted is the radial column density profile constructed by azimuthally averaging the data into annuli of 5-arcsec (500 AU)-wide bins. The radial column density profile of this cloud is very well resolved by the observations. *Alves et al.* (2001a) showed that this profile could be extremely well fit by the predicted profile of a Bonnor-Ebert (BE) sphere.

A BE sphere is a pressure-truncated isothermal ball of gas within which internal pressure everywhere precisely balances the inward push of self-gravity and external surface pressure. The fluid equation that describes such a self-gravitating, isothermal sphere in hydrostatic equilibrium is the following well-known variant of the Lane-Emden equation

$$\frac{1}{\xi^2}\frac{d}{d\xi}\left(\xi^2\frac{d\psi}{d\xi}\right) = e^{-\psi} \qquad (6)$$

where ξ is the dimensionless radius

$$\xi = r/r_c \tag{7}$$

and r_c is the characteristic or scale radius

$$r_c = c_s/(4\pi G\rho_0)^{1/2} \tag{8}$$

where c_s is the sound speed in the cloud and ρ_0 is the density at the origin. Equation (6) is Poisson's equation in dimensionless form where $\psi(\xi)$ is the dimensionless potential and is set by the requirement of hydrostatic equilibrium to be $\psi(\xi) = -\ln(\rho/\rho_0)$. The equation can be solved using the boundary conditions that the function ψ and its first derivative are zero at the origin. Equation (6) has an infinite family of solutions that are characterized by a single parameter, the dimensionless radius at outer edge (R) of the sphere

$$\xi_{max} = R/r_c \tag{9}$$

Each solution thus corresponds to a truncation of the infinite isothermal sphere at a different outer radius, R. The external pressure at a given R must then be equal to that which would be produced by the weight of material that otherwise would extend from R to infinity in an infinite isothermal sphere. The shape of the BE density profile for a pressure-truncated isothermal cloud therefore depends on the single parameter ξ_{max}. As it turns out, the higher the value of ξ_{max}, the more centrally concentrated the cloud is. The stability of such pressure truncated clouds was investigated by *Bonnor* (1956) and *Ebert* (1955), who showed that when $\xi_{max} > 6.5$ the clouds are in a state of unstable equilibrium, susceptible to gravitational collapse.

For B68, *Alves et al.* (2001a) found the radial density profile to be extremely well fit by a Bonnor-Ebert model with $\xi_{max} = 6.9$ very close to the critically stable value. The close correspondence of the data with the BE prediction strongly suggests that this globule is a relatively stable configuration of gas in which thermal pressure is a significant source of support against gravity. This close agreement of theory and observation has very important implications for the physical and evolutionary nature of this source since it implies that the globule is a coherent dynamical unit and not a transient structure.

The nature of B68 as a thermally dominated, quasistable cloud has been decisively confirmed by molecular-line observations of *Hotzel et al.* (2002) and *Lada et al.* (2003), who found the molecular-line profiles to be characterized by thermal line widths. Indeed, the latter study showed the thermal pressure to be a factor of 5 greater than the nonthermal or turbulent pressure in the center of the cloud. Moreover, these observations demonstrated that the cloud is undergoing global oscillation around a state of overall dynamical equilibrium. In a more recent survey of globules, *Kandori et al.* (2005) found that most of the starless globules they studied are dominated by thermal pressure, similar to B68. *Tafalla et al.* (2004) also reported thermally dominated, quasistable states for two more massive cores in the Taurus dark cloud complex. These findings are in clear contradic-

tion to and rule out the suggestion of *Ballesteros-Paredes et al.* (2003), who posited B68 (and similar cores) to be a transient dynamical fluctuation in a turbulent velocity field.

Equations (8) and (9) coupled with the extinction measurements enable the fundamental physical conditions of the globule, i.e., its temperature, central density, external pressure, size, etc., to be exquisitely specified. Indeed, if the temperature of the cloud can be independently determined from molecular line observations (e.g., NH_3), then the terms in equations (8) and (9) are overspecified and "reverse engineering" is possible, resulting in precise determination of either the cloud distance or its gas-to-dust ratio (*Alves et al.*, 2001b; *Hotzel et al.*, 2002; *Lai et al.*, 2003). Integration of the extinction map (i.e., Fig. 2) also produces a very precise estimate of the total dust mass of the cloud. With the assumption of distance and gas-to-dust ratio, a relatively accurate determination of the total cloud mass can be made with an uncertainty completely dominated by the uncertainties in these two assumed parameters.

Infrared extinction maps of a number of starless globules and cores have now been published and the cores are typically found to exhibit Bonnor-Ebert column density profiles with shapes close to the critical stability limit, similar to B68 (*Lada et al.*, 2004; *Teixeira et al.*, 2005; *Kandori et al.*, 2005). Figure 3 shows the distribution of the fit parameter ξ_{max} for a sample of mostly starless globules studied by *Kandori et al.* (2005). The concentration near the critical value is evident. Similar findings have resulted from surveys and mapping of dust continuum emission from starless cores (e.g., *Kirk et al.*, 2005). Such gaseous configurations likely represent the initial conditions for star formation in dense cores.

3.2. Protostellar Cores

Infrared extinction studies have also been carried out toward a few star-forming, protostellar cores (e.g., *Harvey et al.*, 2001; *Teixeira et al.*, 2005; *Kandori et al.*, 2005). These studies show that cores containing protostars tend to have much steeper radial column and volume density profiles than starless cores. Indeed, the central extinctions rise to such high values ($A_V > 40$ mag) in these cores that even the deepest ground- and spacebased imaging cannot detect background stars in their innermost regions. Surprisingly perhaps, the radial density profiles, where measured, can still be fit by Bonnor-Ebert curves, but these are characterized by large values of ξ_{max} (12–25) consistent with large center-to-edge density contrasts and indicative of highly unstable equilibrium configurations (see Fig. 3).

It therefore has been suggested that in the process of forming a star the shape of a core's radial density profile evolves from a relatively flat, equilibrium structure to a highly condensed collapsing structure in a systematic manner that can be quantified using a Bonnor-Ebert analysis (e.g., *Lada et al.*, 2004; *Teixeira et al.*, 2005; *Kandori et al.*, 2005). A quasistable globule or core could become unstable by either an increase in the external pressure (an external

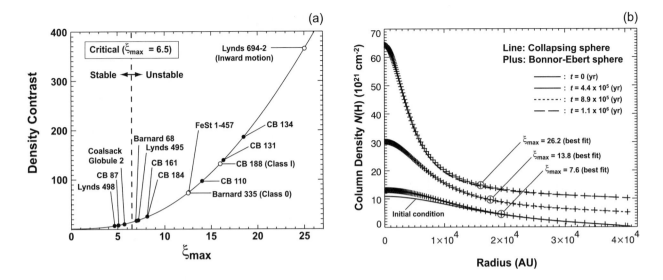

Fig. 3. **(a)** The relation of ξ_{max} to a cloud's center-to-edge density contrast for Bonnor-Ebert configurations. The distributions of the values of ξ_{max} derived from extinction observations of starless (closed symbols) and protostellar (open symbols) cores are also plotted (*Kandori et al., 2005*). **(b)** The predicted radial column density profiles for a collapsing spherical cloud at different times (solid lines) starting from an initial state characterized by a critically stable Bonnor-Ebert sphere. The fitted Bonnor-Ebert relations are plotted as plus signs and the corresponding values of ξ_{max} are indicated. From *Kandori et al.* (2005).

perturbation) or a decrease in the internal pressure. Passage of external shocks, such as produced by supernovae or in cloud-cloud collisions, could result in abrupt increases in external pressure. Although many of these globules and cores are thermally dominated, residual magnetic fields and/or turbulence could provide significant sources of internal pressure. The dissipation of even small amounts of residual internal turbulence (*Nakano*, 1998) or ambipolar diffusion (*Shu et al.*, 1987; *Mouschovias and Ciolek*, 1998) could result in a decrease of internal pressure and instability. Recently *Kandori et al.* (2005) and *Myers* (2005) have calculated the evolution of a spherical cloud assuming that the initial condition is that of a critically stable Bonnor-Ebert configuration. These calculations demonstrate that as the cloud collapses its radial density profile maintains the shape of a Bonnor-Ebert profile with the density contrast (i.e., ξ_{max}) systematically increasing as the collapse proceeds. This effect is nicely illustrated in Fig. 3, which shows the theoretical prediction for the structure of a collapsing core at four different times (*Kandori et al.*, 2005). Also plotted are the best fit Bonnor-Ebert relations, which suggest that the evolutionary status of a collapsing core can be estimated from the instantaneous shape (parameterized by ξ_{max}) of its column density profile. Thus, *even when a dense core is out of equilibrium and collapsing*, its density structure maintains a Bonnor-Ebert like configuration, provided the initial condition was given by a critically stable Bonnor-Ebert state. Moreover, the calculations of Kandori et al. predict that the frequency distribution of ξ_{max} derived from observations should peak near the critically stable value since the timescale for evolution decreases with increasing density. This is in fact evident in the data (Fig. 3). Indeed, according

to the calculations more than 60% of the total collapse time passes before the center-to-edge density contrast increases by as much as a factor of 4 over the initial critical state (*Kandori et al.*, 2005). Finally, the fact that the structure of both stable and collapsing cores can be fit by Bonnor-Ebert models indicates that the initial conditions for core collapse and star formation must be similar to those described by a critically stable Bonnor-Ebert configuration, such as B68.

The observations and calculations described above demonstrate that extinction measurements of cores and globules can be a powerful tool for constructing a quantitative empirical description of their structural evolution to form stars. As a result a more complete theoretical understanding of this process may now be closer to realization. The Bonnor-Ebert formalism appears to provide an initial theoretical framework that can describe quite well the physical process of core evolution to form stars. However, this framework will need to be refined to account for more realistic cloud geometries (e.g., *Myers*, 2005) and equations of state (e.g., *McKee*, 1999; *McKee and Holliman*, 1999). Moreover, extinction properties of a larger sample of cores and globules as well as improved characterizations of their dynamical states from molecular-line observations are clearly warranted.

4. EXTINCTION MAPPING OF CLOUD COMPLEXES

Infrared extinction mapping has also provided new information about the larger-scale structure of nearby molecular cloud complexes within which dense cores are usually embedded. Initial studies focused on nearby filamentary

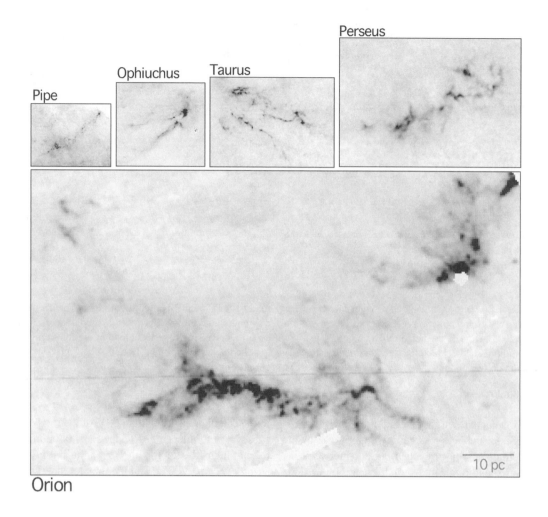

Fig. 4. NICER extinction maps of local molecular cloud complexes. The maps are shown on the same physical scale but are characterized by different angular resolutions.

clouds including IC 5146 (B168) (*Lada et al.,* 1994, 1999) and L 977 (*Alves et al.,* 1998, 1999). These studies showed that the filamentary structures exhibited smooth radial density gradients orthogonal to their main axis that are well modeled by pressure-truncated isothermal cylinders (Huard et al., in preparation). The extinction observations produced detailed measurements of the frequency distribution of extinction (i.e., the cloud mass distribution function), showing that most of the mass of a cloud is at low extinction. Only about 25–30% of the material in these dark clouds is characterized by extinctions in excess of 10 visual magnitudes (*Alves et al.,* 1999). The form of the cloud mass distribution function was also found to be consistent with that predicted in simulations of turbulent clouds (*Ostriker et al.,* 2001). Moreover, integration of such maps yields very precise estimates (to a few percent) of the masses of these clouds. However, the accuracy of such mass determinations is entirely dominated and limited by the uncertainty in the assumed distance determinations.

The availability of the 2MASS near-infrared all sky survey has enabled the construction of widefield extinction maps that can encompass the entire extent of a galactic molecular cloud (GMC) (e.g., *Cambresy et al.,* 2002; *Lombardi and Alves,* 2001). Using NICER, complete maps of a number of prominent local cloud complexes have been made. These maps are sufficiently sensitive to trace relatively low levels of extinction ($A_V \sim 0.5$ mag) and have extended the boundaries of the clouds beyond the level that can be traced by CO emission. Figure 4 shows NICER extinction maps of five prominent cloud complexes drawn to the same physical scale (*Lombardi and Alves,* 2001; *Lombardi et al.,* 2006; Lombardi and Alves, in preparation). These maps can reveal interesting structural information about the cloudy material. For example, the cumulative frequency distribution of extinction for the Pipe Nebula (*Lombardi et al.,* 2006) shows the cloud to contain only about 1% of its mass at extinctions in excess of $A_V > 10$ mag. In this respect the Pipe cloud differs from the IC 5146 and L 977 clouds having a considerably smaller fraction of its mass at these levels of extinction that are typically characteristic of star-forming material (e.g., *Johnstone et al.,* 2004). The Pipe cloud also contains much lower levels of star-formation activity than clouds like Ophiuchus, Orion, and Taurus and may be at a much earlier stage of overall development. Further examination of its cumulative extinction distribution shows that about 30% of the cloud's mass exists at extinctions (A_V) < 2 mag, the threshold for the CO dissociation. This fact indicates that even under perfect circumstances, CO emission is insensitive to a significant fraction of a cloud's total mass and that dust extinction measurements provide a more complete inventory of a cloud's total mass content.

Widefield extinction maps can also provide a relatively complete inventory of dense cores within cloud complexes. For example, 170 distinct cores were identified within the Pipe Nebula from an analysis of its extinction map (*Lombardi et al.*, 2006). The sizes and masses of this population of cores are readily obtained from the extinction data. Since dense cores are the progenitors of stars, the core mass function of a cloud is of particular interest for developing a theory of star formation. Figure 5 shows the core mass function (CMF) for the Pipe Nebula (Alves et al., in preparation) and the stellar IMF for the Trapezium cluster (*Muench et al.*, 2001) plotted for comparison. The shape of the Pipe CMF is surprisingly very similar to that of the stellar IMF. This similarity has potentially profound implications for star formation.

Although molecular-line studies (e.g., *Blitz*, 1993) have long suggested that the forms of the stellar and core mass functions were fundamentally different, observations of dust emission in a few regions hinted at a similarity of the shapes,

at least for relatively high (>2–3 $M_\odot$) mass (e.g., *Motte et al.*, 1998; *Testi and Sargent*, 1998). Such a similarity would imply a 1-to-1 mapping of cores to stars in the star-formation process and indicate that the origin of the stellar IMF has its direct roots in the origin of the CMF. The extinction derived CMF of the Pipe Nebula is the first to reveal a break in the power-law form of the CMF at low mass (~2 $M_\odot$). This break is similar to that in the stellar IMF but occurs at a mass a factor of ~3 higher than the corresponding break in the stellar IMF (see Fig. 5). However, if the stellar IMF is shifted to higher mass by a multiplicative factor of about 3, the shapes of the two mass functions agree extremely well over all masses. The close correspondence in shape between the two mass functions indicates that there is indeed a direct transformation of the CMF to the stellar IMF. However, the factor of 3 difference in the mass scale is fundamentally significant since it further indicates that this transformation is governed by a universal efficiency of ~30% across the entire stellar mass range. The important implication of this observation is that understanding the origin of the stellar IMF requires understanding the origin of the mass spectrum of their progenitor cores. It is interesting that the Pipe CMF and the stellar IMF are so similar. The stellar IMF plotted is for stars formed in the Trapezium cluster where the environment and physical conditions differ significantly from those found in the Pipe Nebula, although the former are probably typical of those characterizing most star formation, since most stars are believed to be formed in clusters (*Lada and Lada*, 2003). This similarity may be another indication that the stellar IMF is a very robust product of the star-formation process.

5. COMPARISON WITH MOLECULAR DATA

Infrared extinction measurements provide the most straightforward way to make direct determinations of molecular abundances in regions of high extinction (i.e, A_V > 5 mag). Such measurements are therefore critical to quantitative exploration of cloud chemistry and chemical evolution. In Fig. 6 we show the map of $C^{18}O$ emission of the B68 cloud along with a detailed point-to-point comparison of the CO integrated intensity with extinction across the entire cloud (*Bergin et al.*, 2002). The spatial map clearly shows the CO emission to peak in a ring around the cloud center. Comparison with the extinction map in Fig. 2 shows that CO and extinction are essentially anticorrelated in the sense that CO is apparently absent where the extinction is the highest.

The relation between integrated intensity and dust optical depth is linearly correlated (but with significant scatter) until a visual opacity of about 10 mag, at which point there is a sharp break followed by a flattening and then a decrease in the integrated intensity with increasing cloud extinction. Calculations of the non-LTE excitation of CO that directly incorporate the observed Bonnor-Ebert density distribution derived from the extinction indicate that the observed emission is largely optically thin. This, in turn, is confirmed by

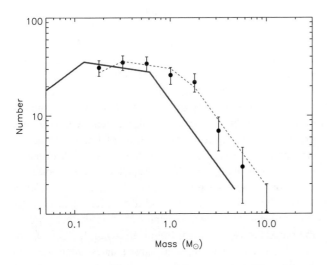

Fig. 5. The mass function of dense cores in the Pipe Nebula (Alves et al., in preparation). The points with error bars represent the observed core mass function. The solid trace on the left of the diagram is the stellar IMF derived for the Trapezium cluster (*Muench et al.*, 2002), which is very similar to the galactic field star IMF. The dotted trace passing through the points on the right is the stellar IMF scaled by a factor of 3 in mass and binned to the same mass intervals as the core mass function. The two mass functions have very similar shapes (apart from the factor of 3 shift in mass scale).

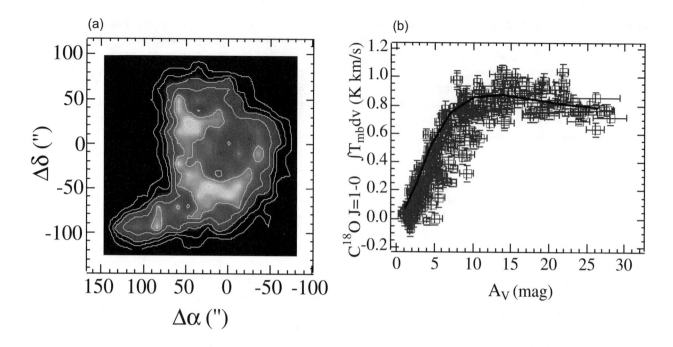

Fig. 6. (a) Map of $C^{18}O$ (J = 1–0) emission from the B68 cloud together with (b) a plot of the correlation between CO integrated intensity and extinction for the cloud. The break from a linear correlation at 10 mag in the latter plot indicates a sharp decrease of CO abundance in the high-extinction regions. This drop in abundance in the central regions is also manifest in the spatial map, where a bright ring of emission surrounds a central depression. The solid line in the righthand plot represents the prediction of a chemical model in which the CO abundance is decreased by 2 orders of magnitude in the central regions of the cloud due to depletion onto grains (from *Bergin et al., 2002*).

observations of the rarer $C^{17}O$ isotope at numerous locations (*Bergin et al., 2002*). Thus, the flattening of the relation is due to a sharp decline in the CO abundance in the center of the cloud. Indeed the inner regions may be completely devoid of gaseous CO. In a similar fashion Bergin et al. also measured the abundances of CS and N_2H^+ across the cloud and demonstrated a pattern of differential depletion for these species in which CS is most heavily depleted and N_2H^+ is least depleted of the species observed. This pattern of depletion agrees with predictions of models of time-dependent chemical evolution. This is illustrated by the chemical model plotted in Fig. 6, which matches the data very well and requires a reduction in the CO abundance of over 2 orders of magnitude. These observations and calculations suggest that in the very center of the cloud all observable molecules may be frozen onto grains, except perhaps for a species such as H_2D^+. Very similar results have been derived from extinction studies of other clouds (*Kramer et al., 1999; Bergin et al., 2001*) and from studies comparing molecular and dust emission from clouds (*Caselli et al., 1999; Tafalla et al., 2002*). Together these observations show that it may not be possible to probe the conditions in the inner regions of prestellar cores with molecular-line tracers and that maps of the distribution of dust are the most reliable tracers of cloudy material at the highest extinctions.

Combining infrared extinction data with molecular-line observations of the more abundant CO isotopes, particularly ^{12}CO and ^{13}CO, can provide important constraints for determining masses of GMCs in the Milky Way and in other galaxies where measurements of dust may be difficult or impractical. One of the standard methods to determine cloud masses of galactic and extragalactic GMCs is to scale CO integrated intensities by the so-called "X-factor," which is defined to be the ratio of total column density to CO integrated intensity. Direct comparisons of CO observations with infrared extinction measurements at thousands of points within a molecular cloud have greatly improved the determination of the X-factor (e.g., *Lada et al., 1994; Lombardi et al., 2006*). These studies find a definite linear correlation between CO integrated intensity and extinction over the range $2 < A_V < 6$ mag. Below an A_V of 2, CO is dissociated and undetectable, while above an A_V of 6, the CO lines are optically thick and the correlation saturates. However, both studies find a large intrinsic scatter in the correlated quantities. The correlation coefficient is low (~0.5) for both the observed ^{12}CO- (*Lombardi et al., 2006*) and ^{13}CO-extinction correlations.

These CO-extinction comparisons indicate that there is an intrinsic and irreducible uncertainty of about a factor of ~1.4–2.0 in the X-factor. The likely reason for this circumstance is the volatile chemistry that must characterize the low extinction regions of GMCs where most of the mass is found (*Lada et al.,* 1994).

6. EXTINCTION MAPPING OF GALAXIES

The physics of the formation of GMCs is one of the major unsolved problems of the interstellar medium. Although many papers have been written on the subject (for a review, see *Elmegreen,* 1993, and references therein) it is not yet known what the dominant formation mechanism is, nor even what the relative importance of gravity, shocks, and magnetic fields are in the cloud-formation process. One avenue for testing these theories is to study molecular clouds in a wide range of environments, ideally from a view point outside the galaxy's plane, and to determine which aspects of the environment set the cloud properties.

A study of GMCs in external galaxies can address the fundamental questions of whether the molecular ISM in external galaxies is organized differently than in the Milky Way and whether GMCs play the same central role in massive star formation as in the Milky Way, and are then responsible for galaxy evolution. Moreover, in an external galaxy we can easily disentangle molecular clouds and assess their basic properties and star forming status, as opposed to the confused "inside view" of GMCs. Ironically, GMCs are perhaps the most overlooked parameter in extragalactic star formation studies, and only recently, with the development of millimeter (CO) interferometry, they are beginning to be consistently incorporated into the picture.

In the previous sections we presented the merits of deep NIR imaging to trace column density not only in dense cores but also entire molecular cloud complexes where these dense cores are embedded (see Fig. 4). Recently, we have been trying to extend this successful idea to extragalactic GMCs. Instead of measuring the color of thousands of background stars to molecular clouds, we will measure the average color of the unresolved thousands of stars that will fall on a pixel of a NIR detector. We will measure the NIR diffuse galaxy light seen through GMCs, in a simple analogy to our galactic work. Dark dust lanes (as judged by available deep optical imaging), and a bright diffuse background, are necessary conditions to make the method work. It is easy to understand the impact this approach might have when one realizes that we will be able to map the distribution of dust column density at subarcsecond resolutions better than present-day interferometers (with typical resolutions of a few arcseconds; see Fig. 7). A caveat that does not affect our galactic work on the nearby complexes is the unknown contribution of foreground light along the line-of-sight. This contribution will dilute the signal, and without modeling, any derived column density along a specific line-of-sight will be a lower limit to the true column density. Still, one can minimize this problem by selecting targets with conspicuous dark lanes as seen in deep optical images.

Bialetsky et al. (in preparation) and Alves et al. (in preparation) present the first successful application of this extension of the technique to the radio-galaxy Centaurus A (NGC 5128) (Fig. 7). We have successfully mapped dust extinction in this galaxy down to a 3σ level of 0.7 mag (visual) and a physical resolution of 10 pc, i.e., the seeing in the final NIR image (0.6"). In our map we can resolve and identify individual GMCs. Figure 8 shows an optical (HST) image of a massive GMC associated with what appears to be an emerging or recently emerged OB association or young cluster. Next to the optical image is the derived infrared extinction map. This map of dust column density closely traces the optically obscured material that defines the cloud. We were able to identify more than 400 GMCs from our map of NGC 5128 and derived, for the first time, a GMC mass spectrum ($dN/dM \propto M^{-\alpha}$) that does not rely on CO

Fig. 7. Optical image and corresponding dust extinction map derived from near-infrared imaging of Cen A obtained with ESO's New Technology Telescope.

HST-Optical

Dust extinction

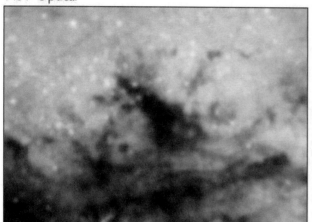
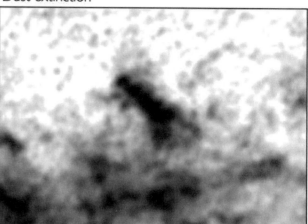

Fig. 8.   Optical HST image and infrared extinction map zooming in on a single GMC with an associated, recently emerged young OB association/cluster within the Cen A dusty disk (see Fig. 7). The extinction map of dust column density clearly resolves the molecular cloud.

observations. The result, a Salpeter-like law ($\alpha \sim 2.31$), is puzzling as the CO-derived GMC mass spectrum for our galaxy and many others is $\alpha \sim 1.5$ (*Solomon et al.,* 1987; *Scoville et al.,* 1987). Overlap of GMCs along the line-of-sight could influence the mass spectrum that is derived for Centaurus A, but such an effect would lead to an apparent excess of more massive clouds and a flatter slope and smaller spectral index than derived. There is, nevertheless, a CO-derived GMC mass spectrum for M33 (*Engargiola et al.,* 2003) that is similar to our result, and we might be seeing an intrinsic and physical difference of GMCs in galaxies. There is also the possibility that because we are much more sensitive to column density, and have exquisite resolution when compared to CO data [the *Engargiola et al.* (2003) result uses an interferometer], we might be determining the true GMC mass spectrum in galaxies. This issue remains open and more observations of other galaxies are needed. Clearly, NIR dust column tracing in external galaxies is possible and promises to bring a new and complementary look to the morphology of the dense ISM in nearby galaxies.

7. CONCLUDING REMARKS

The development of infrared arrays has enabled the measurement and mapping of dust extinction in molecular clouds with unprecedented detail, depth, and precision. Such studies are providing new insights into the structure and structural evolution of dense cores. A clearer empirical and theoretical picture of core evolution from an initial state of quasistable equilibrium to the development of a protostellar embryo is now beginning to emerge. Mapping of large cloud complexes can produce very reliable estimates of total cloud mass and enable their structure to be quantified, facilitating comparison with theories of cloud evolution and star formation. Measurement of the core mass

function is now possible for an interesting range of core mass and can be put on a reliable footing. Comparison with molecular-line observations yields improved determinations of molecular abundances and investigation of cloud chemistry. For example, such comparisons have produced some of the most compelling evidence for differential depletion in cold starless cores. Although not discussed here, comparison with observations of dust emission can produce important constraints on the nature of dust and on the thermal structure in cloud cores (*Kramer et al.,* 1998, 1999; *Bianchi et al.,* 2003). Finally, infrared extinction mapping techniques can be applied to investigate the star-forming clouds in external galaxies with angular resolution comparable to that of the best radio interferometers, including ALMA, and in doing so, provide an unexpected new window on the nature of extragalactic GMCs.

Acknowledgments.   We thank the referee, R. Kandori, for a careful reading of this chapter and suggestions that improved its content. This work was partially supported by the NASA Origins Program under grants NAG 5-9520 and NAG 5-13041.

REFERENCES

Alves J. F., Lada C. J., Lada E. A., Kenyon S., and Phelps R. (1998) *Astrophys. J., 506,* 292–305.

Alves J. F., Lada C. J., and Lada E. A. (1999) *Astrophys. J., 515,* 265–274.

Alves J. F., Lada C. J., and Lada E. A. (2001a) *Nature, 409,* 159–161.

Alves J. F., Lada C. J., and Lada E. A. (2001b) *ESO Messenger, 103,* 1–20.

Ballesteros-Paredes J., Klessen R. S., and Vázquez-Semadeni E. (2003) *Astrophys. J., 592,* 188–202.

Barnard E. E. (1919) *Astrophys. J., 49,* 1–24.

Becklin E. E. and Neugebauer G. (1967) *Astrophys. J., 147,* 799–802.

Benson P. and Myers P. C. (1989) *Astrophys. J. Suppl., 71,* 89–108.

Bergin E. A., Ciardi D. R., Lada C. J., Alves J. F., and Lada E. A. (2001) *Astrophys. J., 557,* 209–225.

Bergin E. A., Alves J. F., Huard T. L., and Lada C. J. (2002) *Astrophys. J., 570,* L101–L104.

Bianchi S., Goncalves J., Albrecht M., et al. (2003) *Astron. Astrophys., 339,* L43–L46.

Blitz L. (1993) In *Protostars and Planets III* (E. H. Levy and J. I. Lunine, eds.), pp. 125–161. Univ. of Arizona, Tucson.

Bok B. J. (1948) In *Centennial Symposia,* pp. 53–72. Harvard College Observatory Monographs, Vol. 7, Cambridge.

Bok B. J. (1956) *Astron. J., 61,* 309–316.

Bok B. J., Lawrence R. S., and Menon T. K. (1955) *Publ. Astron. Soc. Pac., 67,* 108–112.

Bohlin R. C., Savage B. D., and Drake J. F. (1978) *Astrophys. J., 224,* 132–142.

Bonnor W. (1956) *Mon. Not. R. Astron. Soc., 116,* 351–359.

Casali M. M. (1986) *Mon. Not. R. Astron. Soc., 223,* 341–352.

Caselli P., Walmsley C. M., Tafalla M., Dore L., and Myers P. (1999) *Astrophys. J., 523,* L165–L169.

Cambresy L. (1999) *Astron. Astrophys., 345,* 965–976.

Cambresy L., Epchtein N., Copet E., de Batz B., Kimeswenger S., et al. (1997) *Astron. Astrophys., 324,* L5–L8.

Cambresy L., Beichman C. A., Jarrett T. H., and Cutri R. M. (2002) *Astron. J., 123,* 2559–2573.

Cernicharo J. and Bachiller R. (1984) *Astron. Astrophys. Suppl., 58,* 327–350.

Dickman R. (1978) *Astron. J., 83,* 363–372.

Dobashi K., Uehara H., Kandori R., Sakurai T., Kaiden M., Umemoto T., and Sato F. (2005) *Publ. Astr. Soc. Japan, 57,* 1–368.

Ebert R. (1955) *Z. Astrophys., 37,* 217–223.

Elmegreen B. G. (1993) *Astrophys. J., 419,* L29–33.

Engargiola G., Plambeck R. L., Rosolowsky E., and Blitz L. (2003) *Astrophys. J. Suppl., 149,* 343–363.

Harvey D. W., Wilner D. J., Lada C. J., Myers P. C., Alves J. F., and Chen H. (2001) *Astrophys. J., 563,* 903–918.

Harvey D. W., Wilner D. J., Lada C. J., Myers P. C., and Alves J. (2003) *Astrophys. J., 598,* 1112–1126.

Heiles C. (1969) *Astrophys. J., 156,* 493–499.

Herschel W. (1785) *Philos. Trans. LXXV,* 213–224.

Hotzel S., Harju J., and Juvela M. (2002) *Astron. Astrophys., 395,* L5–L8.

Johnstone D., Di Francesco J., and Kirk H. (2004) *Astrophys. J., 611,* L45–L49.

Jones T. J., Hyland A. R., Robinson G., Smith R., and Thomas J. (1980) *Astrophys. J., 242,* 132–140.

Kandori R., Nakajima M., Tamura K., Tatematsu K., et al. (2005) *Astron. J., 130,* 2166–2184.

Kirk J. M., Ward-Thompson D., and André P. (2005) *Mon. Not. R. Astron. Soc., 360,* 1506–1526.

Kramer C., Alves J., Lada C. J., Lada E. A., et al. (1998) *Astron. Astrophys., 329,* L33–L36.

Kramer C., Alves J., Lada C. J., Lada E. A., et al. (1999) *Astron. Astrophys., 342,* 257–270.

Lada C. J. and Lada E. A. (2003) *Ann. Rev. Astron. Astrophys., 41,* 57–115.

Lada C. J., Lada E. A., Clemens D. P., and Bally J. (1994) *Astrophys. J., 429,* 694–709.

Lada C. J., Alves J., and Lada E. A. (1999) *Astrophys. J., 521,* 250–259.

Lada C. J., Bergin E. A., Alves J. F., and Huard T. L. (2003) *Astrophys. J., 586,* 286–295.

Lada C. J., Huard T. L., Crews L. J., and Alves J. (2004) *Astrophys. J., 610,* 303–312.

Lada E. A. (1992) *Astrophys. J., 393,* L25–L28.

Lai S-P., Velusamy T., Langer W. D., and Kuiper T. B. H. (2003) *Astrophys. J., 126,* 311–318.

Lilley A. E. (1955) *Astrophys. J., 121,* 559–568.

Lombardi M. (2005) *Astron. Astrophys., 438,* 169–185.

Lombardi M. and Alves J. F. (2001) *Astron. Astrophys., 377,* 1023.

Lombardi M., Alves J. F., and Lada C. J. (2006) *Astron. Astrophys., 454,* 781–796.

McKee C. F. (1999) In *The Origin of Stars and Planetary Systems* (C. J. Lada and N. D. Kylafis, eds.), pp. 29–66. Kluwer, Dordrecht.

McKee C. F. and Holliman J. H. (1999) *Astrophys. J., 522,* 313–337.

Motte F., André P., and Neri R. (1998) *Astron. Astrophys., 336,* 150–172.

Mouschovias T. and Ciolek G. E. (1998) In *The Origins of Stars and Planetary Systems* (C. J. Lada and N. Kylafis, eds.), pp. 305–339. Kluwer, Dordrecht.

Muench A. A., Lada E. A., Lada C. J., and Alves J. F. (2002) *Astrophys. J., 573,* 366–393.

Myers P. C. (2005) *Astrophys. J., 623,* 280–290.

Nakano T. (1998) *Astrophys. J., 494,* 587–604.

Ostriker E., Stone J., and Gammie C. (2001) *Astrophys. J., 546,* 980–1005.

Pagani L., Bacmann A., Motte F., et al. (2004) *Astron. Astrophys., 417,* 605–613.

Predehl P. and Schmitt J. H. M. M. (1995) *Astron. Astrophys., 293,* 889–905.

Rieke G. and Lebosky M. (1985) *Astrophys. J., 288,* 618–621.

Ryter C., Cesarsky C. J., and Audouze J. (1975) *Astrophys. J., 198,* 103–109.

Scoville N. Z., Yun M. S., Sanders D. B., Clemens D. P., and Waller W. H. (1987) *Astrophys. J. Suppl., 63,* 821–915.

Solomon P. M., Rivolo A. R., Barrett J., and Yahil A. (1987) *Astrophys. J., 319,* 730–741.

Shu F. H., Adams F. C., and Lizano S. (1987) *Ann. Rev. Astron. Astrophys., 25,* 23–81.

Strom S. E., Strom K. M., and Grasdalen G. (1975) *Ann. Rev. Astron. Astrophys., 13,* 187–216.

Tafalla M., Myers P. C., Caselli P., Walmsley C. M., and Comito C. (2002) *Astrophys. J., 569,* 815–835.

Tafalla M., Myers P. C., Caselli P., and Walmsley C. M. (2004) *Astron. Astrophys., 416,* 191–212.

Teixeira P. S., Lada C. J., and Alves J. F. (2005) *Astrophys. J., 629,* 276–287.

Testi L. and Sargent A. I. (1998) *Astrophys. J., 508,* L91–L94.

Tomita Y., Saito T., and Ohtani H. (1979) *Publ. Astron. Soc. Japan, 31,* 407–416.

Yun J. L. and Clemens D. P. (1990) *Astrophys. J., 365,* L73–L76.

Wilking B. A., Lada C. J., and Young E. (1989) *Astrophys. J., 340,* 823–852.

Wilson R. W., Jefferts K. B., and Penzias A. A. (1970) *Astrophys. J., 161,* L43–L46.

Wolf M. (1923) *Astron. Nachr., 219,* 109–114.

An Observational Perspective of Low-Mass Dense Cores I: Internal Physical and Chemical Properties

James Di Francesco
National Research Council of Canada

Neal J. Evans II
The University of Texas at Austin

Paola Caselli
Arcetri Observatory

Philip C. Myers
Harvard-Smithsonian Center for Astrophysics

Yancy Shirley
University of Arizona

Yuri Aikawa
Kobe University

Mario Tafalla
National Astronomical Observatory of Spain

Low-mass dense cores represent the state of molecular gas associated with the earliest phases of low-mass star formation. Such cores are called "protostellar" or "starless," depending on whether they do or do not contain compact sources of luminosity. In this chapter, the first half of the review of low-mass dense cores, we describe the numerous inferences made about the nature of starless cores as a result of recent observations, since these reveal the initial conditions of star formation. We focus on the identification of isolated starless cores and their internal physical and chemical properties, including morphologies, densities, temperatures, kinematics, and molecular abundances. These objects display a wide range of properties since they are each at different points on evolutionary paths from ambient molecular cloud material to cold, contracting, and centrally concentrated configurations with significant molecular depletions and, in rare cases, enhancements.

1. INTRODUCTION AND SCOPE OF REVIEW

Over the last decade, dedicated observations have revealed much about the earliest phases of low-mass star formation, primarily through studies of low-mass dense cores with and without internal protostellar sources, e.g., protostellar and starless cores. Such cores are dense zones of molecular gas of relatively high density, and represent the physical and chemical conditions of interstellar gas just after or prior to localized gravitational collapse. These objects are essentially the seeds from which young stars may spring, and define the fundamental starting point of stellar evolution. Since Protostars and Planets IV (PPIV) (see *Langer et al.,* 2000; *André et al,* 2000), enormous strides have been made in the observational characterization of

such cores, thanks to the increased interest stemming from the promising early results reported at PPIV and to new instrumentation.

This review summarizes the advances made since PPIV, and, given the large number, it has been divided into two parts. In Part I (i.e., this chapter), we summarize major recent observational studies of the initial conditions of star formation, i.e., the individual physical and chemical properties of starless cores, with emphasis on cores not found in crowded regions. This subject includes their identification, morphologies, densities, temperatures, molecular abundances, and kinematics. In recent years, the internal density and temperature structures of starless cores have been resolved, regions of chemical depletion (or enhancement) have been studied, and inward motions have been measured.

Part II of this review (see next chapter by Ward-Thompson et al.) summarizes what observed individual and group characteristics, including magnetic field properties, mass distributions, and apparent lifetimes, have taught us about the evolution of low-mass dense cores through protostellar formation. Despite significant progress, the observational picture of the earliest stages of star formation remains incomplete, and new observational and experimental data and theoretical work will be necessary to make further advances.

2.   STARLESS CORES

2.1.   Definition

Stars form within molecular gas behind large amounts of extinction from dust. This extinction has been used to locate molecular clouds (*Lynds*, 1962), isolated smaller clouds (*Clemens and Barvainis*, 1988), and particularly opaque regions within larger clouds (*Myers et al.*, 1983; *Lee and Myers*, 1999). Locations within these clouds of relatively high density or column density, although identifiable from optical or infrared absorption (e.g., see chapter by Alves et al.), can also be detected using submillimeter, millimeter, or radio emission. For instance, spectroscopic studies of clouds found via extinction studies further selected those with evidence for dense gas (e.g., see *Myers et al.*, 1983; *Myers and Benson*, 1983). These became known collectively as "dense cores," based primarily on whether or not they showed emission from NH_3, indicative of densities above about 10^4 cm^{-3} (*Benson and Myers*, 1989; see also *Jijina et al.*, 1999). Using data from the Infrared Astronomical Satellite (IRAS), *Beichman et al.* (1986) found that roughly half the known dense cores in clouds contained IRAS sources, while *Yun and Clemens* (1990) found that about 23% of the isolated smaller clouds contained infrared sources within; the remainder became known as "starless cores." Follow-up studies of clouds with sensitive submillimeter continuum detection systems allowed further discrimination within the class of starless cores. Those with detected emission, indicating relatively high densities of 10^{5-6} cm^{-3} (*Ward-Thompson et al.*, 1994), were called "pre-protostellar" cores or later "prestellar cores." Molecular line studies showed that prestellar cores were indeed more likely to show evidence for inflowing material than were the merely starless cores (e.g., see *Gregersen and Evans*, 2000).

In this review, we describe the recent observational characterization of "starless cores," low-mass dense cores without compact luminous sources of any mass. (We call such sources "young stellar objects," regardless of whether or not they are stellar or substellar in mass.) By "low-mass," we mean cores with masses $M < 10$ $M_\odot$, such as those found in the nearby Taurus or Ophiuchus clouds. Starless cores are important because they represent best the physical conditions of dense gas prior to star formation. Conceptually, we distinguish starless cores that are gravitationally bound as "prestellar cores" although it is very difficult to determine

observationally at present whether a given core is bound or not (see below). In addition, we consider such cores as either "isolated" or "clustered" depending on the local surface density of other nearby cores and young stellar objects. Observations of isolated cores (e.g., *Clemens and Barvainis*, 1988) are easier to interpret since they generally inhabit simpler, less confused environments, i.e., without nearby cores or young stellar objects. Observations of clustered cores (e.g., *Motte et al.*, 1998) are important also because they are more representative of star formation within embedded clusters, i.e., where most stars form within the galaxy (*Lada and Lada*, 2003). Finally, we describe cores as either "shielded" from or "exposed" to the interstellar radiation field (ISRF), depending on whether they do [e.g., many of those described by *Myers et al.* (1983)] or do not [e.g., those cataloged by *Clemens and Barvainis* (1988)] reside within a larger molecular cloud.

Observational limitations strongly influence how dense cores are defined. For example, low-resolution data may reveal cores over a wide range of mass, but higher-resolution data may reveal substructures within each, with their own range of (lower) mass, i.e., cores within cores. In addition, different tracers may probe different kinds of structures within clouds, due to varying density, temperature, or chemistry. Also, whether or not a given dense core contains a compact infrared-luminous source is ultimately a matter of observational sensitivity (e.g., L1014; see below). Finally, learning whether or not a given core is prestellar (i.e., gravitationally bound) requires accurate determination of its physical state (mass, temperature, and internal magnetic field). Masses alone can be uncertain by factors of ~3 due to uncertainties in distances, dust opacities, molecular abundances, or calibration. Various new instrumentation of higher resolution and sensitivity will soon be available, making such observational characterization more tenable.

2.2.   Identification

Observations of starless cores have focused mostly on examples within 1 kpc, where the highest mass sensitivities and linear resolutions are possible. Many starless cores have been detected serendipitously, adjacent to previously known protostellar cores or young stellar objects (e.g., see *Wilson et al.*, 1999; *Williams and Myers*, 1999; *Swift et al.*, 2005). Only recently, more systematic mapping has begun to identify larger numbers of starless cores in less-biased surveys.

Recent large line maps across significant portions of nearby molecular clouds using transitions that trace relatively high densities have been useful in identifying dense cores. For example, 179 cores in clouds within 200 pc were identified from substructures detected in $C^{18}O$ (1–0) maps of L1333 in Cassiopeia (*Obayashi et al.*, 1998), Taurus (*Onishi et al.*, 1998), the Southern Coalsack region (*Kato et al.*, 1999), the Pipe Nebula region (*Onishi et al.*, 1999), Chamaeleon (*Mizuno et al.*, 1999), Lupus (*Hara et al.*, 1999), Corona Australis (*Yonekura et al.*, 1999), and Ophiuchus (*Tachihara et al., 2000*). *Tachihara et al.* (2002) identified 76% of these cores as "starless" (i.e., without associated IRAS sources; see below) with masses of 1–100 $M_\odot$ (on average, 15 $M_\odot$). Using a still higher density molecular tracer, $H^{13}CO^+$ (1–0), *Onishi et al.* (2002) identified 55 "condensations" in Taurus, 80% of which were similarly deemed "starless" with estimated masses of 0.4–20 $M_\odot$. Using line maps to identify cores can be problematic, however, because of line optical depths and chemical evolution within the cloud. For example, lines like $C^{18}O$ (1–0) can be too opaque to sample most of the material in dense cores. Furthermore, this line and $H^{13}CO^+$ (1–0) can be affected by the significant depletion of the source molecules, again making it difficult to sample their column densities. In contrast, N-bearing molecules, in particular NH_3 and N_2H^+, appear to survive at higher densities than C-bearing species, and in fact they have been used to probe core nuclei (see *Caselli et al., 2002c*) (see also section 4).

Submillimeter or millimeter continuum emission maps of molecular clouds can trace the high column densities associated with starless cores without the potentially confusing effects of line opacity or chemical evolution. These studies trace the dust component, which emits radiation at long wavelengths even at the very low dust temperatures (T_D) in dense cores (~10 K), although the emission is sensitive to T_D. The optical properties of grains are also likely to evolve at relatively high densities due to ice mantle growth and coagulation, but these effects are much less severe than those for gas phase tracers (see section 4). Recent single-dish telescope studies include the inner regions of Ophiuchus (*Motte et al., 1998; Johnstone et al., 2000, 2004; Stanke et al., 2006*), northern Orion A (*Chini et al., 1997; Lis et al., 1998; Johnstone and Bally, 1999*), NGC 2068/2071 in Orion B (*Mitchell et al., 2001; Johnstone et al., 2001; Motte et al., 2001*), NGC 1333 in Perseus (*Sandell and Knee, 2001*), and R Corona Australis (*Chini et al., 2003; Nutter et al., 2005*). For example, *Motte et al.* (1998) found 58 "starless clumps" in their map of Ophiuchus with masses of 0.05–3 $M_\odot$. Even larger regions (e.g., 8–10 deg²) have now been mapped (*Enoch et al., 2006; Young et al., 2006; Hatchell et al., 2005; Kirk et al., 2006*), providing expansive (although shallow) coverage of the Perseus and Ophiuchus clouds. Collectively these studies have revealed hundreds of compact continuum objects. In those cases where comparisons with near- to far-infrared data were made, more than ~60% were deemed "starless." The fraction of these that are prestellar is not known, however, and determining this (through appropriate molecular tracers; see section 4) is a key goal of future research.

In both line and continuum maps, the methods by which objects are identified have varied widely. Some studies have identified objects by eye, while others have used more objective automated structure identification algorithms [e.g., *Clumpfind* by *Williams et al.* (1994)], multiscale wavelet analyses, and even hybrid techniques. In addition, objects identified in continuum studies are significantly more compact in scale than those found in the line maps described

above. First, angular resolutions of the data have substantially differed, with line and continuum maps typically having >1' and <1' resolutions respectively. Second, the spatial sampling of the data also have substantially differed, although single-dish telescopes were used for both. Line data are obtained through frequency or position switching to line-free locations, allowing large-scale information to be retained, but continuum data are obtained through chopping (or sky background removal for instruments that do not chop), filtering out large-scale information (e.g., over 2–10' depending on the instrument.) Indeed, many continuum objects can be found embedded within objects identified via line surveys. Continuum cores have been identified, however, using relatively low-resolution data, e.g., from IRAS (*Jessop and Ward-Thompson, 2000*) or the Infrared Space Observatory (ISO) (*Tóth et al., 2004*).

Identification of objects found through line or continuum observations as starless cores is ultimately dependent upon the availability and sensitivity of ancillary data at the time of analysis. For example, Fig. 1 shows the starless core L1551-MC in Taurus recently discovered by *Swift et al.* (2005) in NH_3 and CCS emission, with the positions of nearby, noncoincident objects listed in the SIMBAD database (as of late 2004). Far-infrared data from IRAS were used traditionally to find evidence for embedded protostars associated with detected cores, but more recently wide-field near-infrared data, including those from 2MASS, have also been used. Such wide-field data can be insufficiently sensitive to detect the youngest, most-embedded protostars, however, given high extinctions (e.g., $A_V > 30$). Near-infrared observations targeting specific cores can attain much higher sensitivities but these have not been done extensively. Some targeted studies, e.g., *Allen et al.* (2002) or *Murphy and Myers* (2003), however, did not detect low-luminosity protostars in various cores despite relatively high sensitivity. The extraordinarily high sensitivity of the Spitzer Space Telescope in the mid-infrared has revealed about 12 low-luminosity protostars within cores previously identified as starless, e.g., in L1014 by *C. H. Young et al.* (2004) and in L1148 by *Kauffmann et al.* (2005; see also Bourke et al., in preparation; Huard et al., in preparation). These are candidates to join the class of very low luminosity objects (VeLLOs), sources with $L_{int} < 0.1 L_\odot$ that are located within dense cores. (Note that L_{int} is the luminosity of the object, in excess of that supplied by the ISRF.) Based on this definition, the previously identified, extremely young, low-luminosity Class 0 protostar IRAM 04191 (*André et al., 1999*) is also a VeLLO (Dunham et al., in preparation).

In addition to near- to far-infrared imaging, deeply embedded protostars may be identified using other signposts of star formation not affected by high extinction, including compact molecular outflows, "hot core" molecular line emission, masers, or compact thermal radio emission. For example, *Yun et al.* (1996) and *Harvey et al.* (2002) used the latter method to limit luminosities of protostellar sources within several starless cores using limits on radio emission expected from shock-ionized gas, i.e., from outflows.

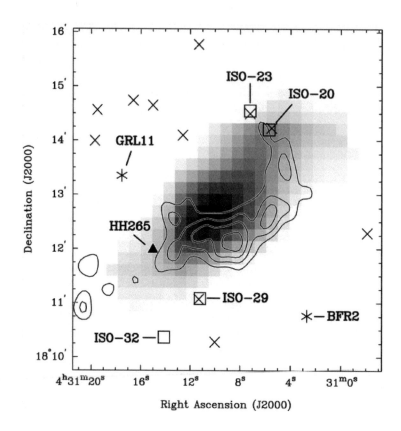

Fig. 1.   The prestellar core L1551-MC shown in NH_3 (1,1) integrated intensity (grayscale) and CCS (3_2–2_1) integrated intensity (contours). Positions of nearby objects are shown as symbols (see *Swift et al., 2005*).

3. DENSITY AND TEMPERATURE STRUCTURES

3.1. Probing Physical Structure

In this section, we describe the morphologies and internal configurations of density and temperature of starless cores as derived in recent studies. Much progress in understanding dense core structure has come from analyses of far-infrared to millimeter continuum emission observations, typically of relatively isolated and exposed examples. Continuum data have been used preferentially over molecular line data because interpretation of the latter can be complicated by large internal molecular abundance variations (see section 4). Near- or mid-infrared absorption of background emission by cores can be also a powerful probe of the density structure since extinction is independent of temperature and can trace those locations where dust column densities are too low to be easily detected in emission at present (e.g., see *Bacmann et al., 2000*). Results about core structure derived from "extinction mapping" have agreed generally with those from emission studies, with some exceptions (see below and the chapter by Alves et al.).

The observed emission from cores depends on the entangled effects of temperature, density, properties of the tracer being used, and the observational technique. For molecular lines, the abundance and excitation of the tracer varies within the core. For continuum emission from dust, the interpretation depends on dust properties and the characteristics of the ISRF. Two general approaches are taken: In the first, simplifying assumptions are made and the observations are inverted to derive core properties, usually averages over the line of sight and the beam; in the second, specific physical models are assumed and a self-consistent calculation of predicted observations is compared to the actual observations, allowing the best model to be chosen. Both approaches have their strengths and weaknesses.

Self-consistent models must compute the temperature distribution for a given density distribution, dust properties, and radiation field. For starless cores, only the ISRF heats the dust. In their models, *Evans et al.* (2001) included the "Black-Draine" (BD) ISRF, one based on the Cosmic Background Explorer (COBE) results of *Black* (1994), which include the cosmic microwave background (CMB) blackbody and the ISRF in the ultraviolet of *Draine* (1978; cf. *van Dishoeck* 1988), rather than the older, pre-COBE model of *Mathis et al.* (1983; MMP). The BD and MMP models can differ significantly, e.g., up to a factor of 13 between 5 μm and 400 μm. For dust opacities, those currently in use have been theoretically derived from optical constants of grain and ice materials and include features such as mid-infrared vibrational bands (e.g., 9.7 μm Si-O stretch) and

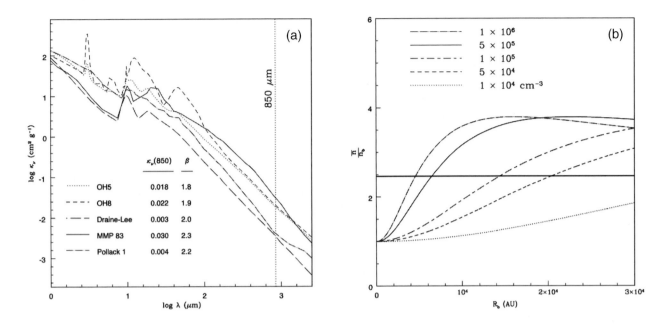

Fig. 2. **(a)** Opacity vs. wavelength from selected dust models (see section 3.3) (*Shirley et al.,* 2005). Values (κ_ν) and spectral indices (β) of opacities at 850 μm are tabulated. Units of κ_ν are cm² g⁻¹ and the β at 850 μm were obtained from linear regressions of respective model values between 350 μm and 1.3 mm. **(b)** The ratio of average density ($\bar{n}$) to density at the BES boundary radius (n_b) vs. BES boundary radius (R_b) for families of BES of given n_c (Shirley and Jorgensen, in preparation). The solid horizontal line corresponds to the stability criterion of *Lombardi and Bertin* (2001).

tend toward a power-law decrease of opacity at submillimeter wavelengths ($\kappa_\nu \propto \lambda^{-\beta}$). Figure 2a shows five dust opacity models: OH5 and OH8 are models of grains that have coagulated for 10⁵ yr and acquired varied depths of ice mantles (*Ossenkopf and Henning,* 1994); "Draine-Lee" are the "standard" interstellar medium dust opacities derived from silicate and graphite grains by *Draine and Lee* (1984); MMP opacities are derived from an empirical fit to observed dust properties in star-forming regions; and Pollack opacities are derived from an alternative grain composition based on silicate grains (olivine and orthopyroxene), iron compounds (troilite and metallic iron), and various organic C compounds (*Pollack et al.,* 1994). The mass opacity (κ_ν) at long wavelengths can vary by 1 order of magnitude between the opacity models.

These ingredients then provide inputs to radiative transport codes (e.g., *Egan et al.,* 1988; *Ivezić et al.,* 1999), which attempt to fit simultaneously the observed intensity profiles, $I_\nu(r)$ [most sensitive to variation in n(r)], and spectral energy distribution (SED, most sensitive to the total mass and the assumed dust opacity). To test various ISRF and dust opacity models for the L1498 starless core, *Shirley et al.* (2005) varied the BD ISRF by a scaling factor, s_{isrf}, and tested eight dust opacity models, including those shown in Fig. 2a. In their models, the resultant SED was more sensitive to the ISRF than the resultant normalized continuum intensity profiles, and they found best fits with $s_{isrf} = 0.5–1.0$. Furthermore, only the OH5 and OH8 opacity models were consistent with the observed SED.

Chemical differentiation observed in starless cores (see section 4) should lead to variations in ice compositions, and hence dust opacities. Indeed, *Kramer et al.* (2003) found in IC 5146 an increase in dust emissivity by a factor of 4 between 20 K and 12 K. To date, variations of dust opacity with radius have not been self-consistently included in radiative transfer models. Moreover, most radiative transfer models are one-dimensional and are constrained by one-dimensional intensity profiles, although starless cores can be quite ellipsoidal in three dimensions (see section 3.2). If magnetic fields are dynamically important, they require core geometries that are inherently multidimensional since the magnetic field only provides support perpendicular to the direction of the field lines (e.g., *Li and Shu,* 1996). Promising work on modeling cores with three-dimensional radiative transfer codes has begun (e.g., *Doty et al.,* 2005a,b; *Steinacker et al.,* 2005). A reanalysis of L1544 core data by *Doty et al.* (2005a) using a three-dimensional code yielded results that were overall similar to those found from one-dimensional models but with reduced size and mass.

3.2. Morphology and Density Structure

Starless cores vary widely in morphology or shape, ranging between filamentary to compact and round. Cores are seen as two-dimensional projections against the sky, making it difficult to constrain their three-dimensional shape. Furthermore, core morphologies can depend on the wavelength and the angular resolution of observation (e.g., see

Stamatellos et al., 2004). Observed morphologies also depend upon the intensity level chosen as the boundary between cores and their surroundings, and this can be affected strongly by observational sensitivity. In general, cores that are round (in projection) have been typically interpreted as being spherical in shape, while cores that are elongated (in projection) have been typically interpreted as being either oblate or prolate spheroids. Statistical analyses of core map aspect ratios indicate that most isolated cores are prolate rather than oblate (*Myers et al.,* 1991; *Ryden,* 1996), and many cores in Taurus share the projected alignment of the filamentary clouds in which they reside (*Hartmann,* 2002). *Jones et al.* (2001), using a sample of cores identified from the extensive NH_3 catalog of *Jijina et al.* (1999), posited that elongated cores actually have intrinsically triaxial shapes that are neither purely oblate nor prolate. *Goodwin et al.* (2002), using the same catalog, found that starless cores have more extreme axial ratios than protostellar cores.

Column densities within starless cores can be traced using the optically thin millimeter or submillimeter continuum specific intensity. Derivation of absolute column densities and conversion to volume densities requires knowledge or assumptions about dust temperatures, opacities, gas-to-dust ratios, geometry, and telescope beam pattern. Azimuthally averaged (or elliptically averaged in cases of extreme axial ratio), one-dimensional intensity profiles of starless cores are found to have "flat" slopes at radii <3000–7500 AU and "steep" slopes at larger radii (see *Ward-Thompson et al.,* 1994, 1999; *André et al.,* 1996; *Shirley et al.,* 2000; *Kirk et al.,* 2005). Such profiles depart from the $n(r) \propto r^{-2}$ distribution of the "singular isothermal sphere" postulated by *Shu* (1977) as the initial state of isolated dense cores prior to gravitational collapse. Similar profiles have been seen also in studies of dense core structure utilizing the absorption of background emission by core dust (e.g., *Alves et al.,* 2001; see chapter by Alves et al.) The observed configurations are better described by "Bonnor-Ebert spheres" (BES) (*Emden,* 1907; *Ebert,* 1955; *Bonnor,* 1956; *Chandrasekhar,* 1957), nonsingular solutions to the equations of hydrostatic equilibrium that can be critically stable in the presence of external pressure. In the one-dimensional isothermal approximation, the family of BES solutions are parameterized by the central density, n_c, and are characterized by density profiles with two regimes: a "plateau" of slowly decreasing density at small radii and a power-law decrease ($\sim r^{-2}$) at large radii (*Chandrasekhar and Wares,* 1949). The size of the plateau scales inversely with the central density, i.e., $R\left(\frac{1}{2}n_c\right) \propto \sqrt{T/n_c}$. For reference, Fig. 2b shows the average density contrast, $\frac{\bar{n}}{n_b}$, where $\bar{n}$ is the average density and n_b is the density at the BES (outer) boundary radius, vs. BES boundary radius, R_b, for families of BES of given n_c (Shirley and Jorgensen, in preparation). Thermally supported density configurations, regardless of geometry, are unstable if they have $\frac{\bar{n}}{n_b} \geq 2.46457$ (*Lombardi and Bertin,* 2001). Such configurations may remain stable if nonthermal internal pressures (e.g., from magnetic fields or turbulence) are relevant, however; see section 5, as well as the next chapter by Ward-Thompson et al.

Detailed modeling of continuum intensity profiles have confirmed that BES are a good approximation to the density structures of starless cores (e.g., *Ward-Thompson et al.,* 1999; *Evans et al.,* 2001; *Kirk et al.,* 2005; *Schnee and Goodman,* 2005, but see *Harvey et al.,* 2003). Recently, Shirley and Jorgensen (in preparation) modeled submillimeter continuum emission from 33 isolated starless cores, the largest sample so far, and found a log-normal distribution of central densities with median log $n_c = 5.3$ cm^{-3}. This large median central density indicates that nonthermal support (e.g., magnetic pressure) has stabilized the cores, that the density scale is inaccurate due to the dust opacity assumed (see below), or that these cores are not in equilibrium. Unfortunately, it is not currently possible to distinguish between collapsing and static BES with dust emission or absorption profiles because the density structure, for a given n_c, does not significantly vary until late in the collapse history (*Myers,* 2005). Molecular probes of the kinematics are much better suited to resolving this issue (e.g., see section 5). Nevertheless, increasing central concentration within starless cores could indicate an evolutionary path. Note, however, that while BES are consistent with current observations, this does not prove that given physical structures *are* BES (cf. *Tafalla et al.,* 2002; *Ballesteros-Paredes et al.,* 2003) (but see section 5.2).

3.3. Temperature Structures

Early calculations of dust temperature (T_D) in cores heated only by the ISRF indicated $T_D \sim 10$ K (*Leung,* 1975), cooler on the inside and warmer on the outside. Such a gradient in T_D is unavoidable theoretically because the cores are optically thin to their own radiation ($\lambda_{peak} \approx 200$ μm) and are primarily heated externally by a local ISRF for which short wavelength radiation is significantly obnubilated in the center of the core. Alternative heating mechanisms for dust, including cosmic rays (i.e., direct heating of grains, secondary UV heating from excitation of H_2, and heating of gas followed by gas-grain heating), are not significant compared to heating from the ISRF. More recent calculations using current ideas for the ISRF and dust opacities confirm the earlier work (e.g., *Zucconi et al.,* 2001; *Evans et al.,* 2001), finding T_D falling from about 12 K on the exposed surface to as low as 7 K at very small radii. Such low temperatures would suppress radiation even at millimeter wavelengths from the center, possibly affecting interpretation of the central densities. Gradients of T_D found in starless cores, however, do not have a strong effect on overall core equilibria, and density profiles very different from isothermal BES are not expected (*Evans et al.,* 2001; *Galli et al.,* 2002).

The temperature gradient within a core depends on the strength and spectral properties of the ISRF, which are affected by surrounding extinction, and the density structure in the core. Shielded cores have lower temperatures and shallower gradients than exposed cores, because the ISRF is attenuated and reddened (*Stamatellos and Whitworth,* 2003; *André et al.,* 2003; *K. E. Young et al.,* 2004). The reddening leaves more of the heating to the longer wavelength photons, which "cook" more evenly. Cores of relatively low central concentration or low central density should have shallower temperature gradients due to lower extinction (*Shirley et al., 2005*). Calculations in higher dimensions also show smaller gradients. For clumpy cores, the heating is dominated by unobscured lines of sight (*Doty et al., 2005b*). Such clumpy cores can have cold patches ($T_D < 6$ K) within an overall flatter distribution of T_D.

Observational tests do indicate warmer T_D on the outside of cores. *Langer and Willacy* (2001), *Ward-Thompson et al.* (2002), and *Pagani et al.* (2004, see also 2003), each using ISO data, found evidence for T_D gradients in exposed cores, with innermost $T_D \leq 10$ K. Limited spatial resolution at far-infrared wavelengths, however, does not allow direct tests of the predictions of very low T_D at very small radii. One indirect test comes from studies of starless cores in extinction (e.g., *Bacmann et al., 2000*), which are temperature independent. These studies also show "flat" inner density gradients, and hence these gradients are not due to decreased T_D in the centers. Another possible indirect test is the high-resolution study of molecular lines (see below).

Examining gas kinetic temperature (T_K), *Goldreich and Kwan* (1974) showed it is forced to be very close to T_D at high densities ($n \geq 10^5$ cm^{-3}). This coupling fails for less dense cores and for the outer parts of even denser cores (*Goldsmith, 2001; Galli et al., 2002*). At first, T_K may drop below T_D, but cosmic-ray heating prevents T_K from dropping too low. If the core is not very deeply shielded, however, photoelectric heating causes $T_K > T_D$ in the outer layers (e.g., *K. E. Young et al., 2004*).

Reductions in the ISRF affect the chemistry in the outer layers (*J.-E. Lee et al., 2004*) and change the gas kinetic temperature (T_K) profile as well by decreasing photoelectric heating. Conversely, cores in regions of high ISRF will have warmer surfaces in both dust and gas (*Lis et al., 2001; Jørgensen et al., 2006*). Since this heating is caused by ultraviolet light, it is very sensitive to small amounts of shielding. CO is a particularly good constraint on the short-wavelength end of the ISRF (e.g., *Evans et al., 2005*), but with chemical models (section 4) other species can also be excellent probes (e.g., *K. E. Young et al., 2004; J.-E. Lee et al., 2004*).

Observationally, T_K can be determined directly from CO transitions, which probe the outer layers, and by inversion transitions of NH$_3$, a good tracer of dense core material (e.g., see *Jijina et al., 1999*, and references therein), which probe deeper layers. Analyses of such lines from starless cores have found nearly constant values of $T_K \approx 10$ K [e.g., *Hotzel et al.* (2002) for B68 and *Tafalla et al.* (2004) for L1498 and L1517B]. If $T_K = T_D$, and T_D decreases toward the center, one might expect to see changes in T_K. For L1498 and L1517B, however, *Galli et al.* (2002) found T_K was ~2 K higher than T_D for cores with central densities ~10^5 cm^{-3} (i.e., like L1498 and L1517B), and that neither T_K nor T_D varied significantly within the inner core radii. *Shirley et al.* (2005), however, found a much lower central density (i.e., $n_c = 1-3 \times 10^4$ cm^{-3}) for L1498, which would

explain the difference of T_K and T_D. Also, recent VLA observations of NH$_3$ (1,1) and (2,2) across the L1544 prestellar core by Crapsi et al. (in preparation) have revealed a clear T_K gradient that is consistent with the dust temperature profile derived from models of submillimeter continuum emission with a central density of 3×10^6 cm^{-3}.

4. CHEMICAL CHARACTERISTICS

Structures of starless cores in principle can be probed using molecular emission lines, typically rotational transitions excited at the relatively low temperatures and high densities of such objects. Indeed, line data can complement continuum data to provide new constraints to models (e.g., *Jessop and Ward-Thompson, 2001*). Line studies also have the advantage of tracing kinematic behavior of starless cores. In practice, however, line observations are affected by the abundance variations within the cores and by the optical depths of the specific lines used; unlike submillimeter or millimeter continuum emission, the optical depths of molecular lines are often quite large. To circumvent this problem, lines from rare isotopologues are observed and interpreted using non-LTE radiative transfer codes, which take into account the core chemical structure.

Although the basic chemical processes in clouds have been known for some time [e.g., formation of H$_2$ on dust grains (*Gould and Salpeter, 1963; Jura, 1974*), ionization of H$_2$ by cosmic rays (*Solomon and Werner, 1971*), ion-molecule reactions (*Herbst and Klemperer, 1973*), and grain-surface chemistry (*Watson and Salpeter, 1972; Allen and Robinson, 1977*)], there are still several uncertainties for denser regions, one of which is grain-surface processes. Theoretical studies predicted molecular depletion onto grain surfaces (e.g., *Léger, 1983*). While adsorbed molecules have long been observed in infrared absorption bands (e.g., *Gillett and Forrest, 1973*; see also *Pontoppidan et al., 2005*), only in the past few years have molecular line observations found clear and quantitative evidence for gas-phase depletion (e.g., *Kuiper et al., 1996; Willacy et al., 1998; Kramer et al., 1998*) (see section 4.1). Enhanced deuterium fractionation, which should accompany molecular depletion (e.g., *Brown and Millar, 1989*), is also found in both starless and protostellar cores (see section 4.2). Uncertainties still remain, however, and further updates of chemical models and detailed comparisons with new observational data are needed.

4.1. Molecular Freeze-Out

The dominant gas phase constituent in molecular clouds is H$_2$ but its line emission is not used to trace the interiors of molecular clouds and cores. H$_2$, as a homonuclear molecule, has no electric dipole moment and thus has no dipole transitions between rotational states. Quadrupole emission between states is possible but the upper level of the first allowed transition is 512 K above ground (since H$_2$ is a light molecule) and is only excited where gas is suitably hot, e.g., on cloud surfaces and not in cloud interiors.

CO and its isotopologues ^{13}CO, $C^{18}O$, or $C^{17}O$ are commonly used as surrogate tracers to H_2 since they are relatively abundant and intermixed with H_2, and have rotational transitions excited at the ambient densities in molecular clouds, i.e., 10^{2-3} cm^{-3}. Recent studies, however, have found significant depletions of CO abundance within the innermost regions of starless cores by comparing line observations of CO isotopologues with the submillimeter dust continuum or extinction maps of cores. For example, *Bacmann et al.* (2002) found evidence for CO depletions by factors of 4–15 in a sample of 7 cores. Furthermore, *Caselli et al.* (1999) and *Bergin et al.* (2002), using $C^{17}O$ and $C^{18}O$ respectively, found CO abundance depletions by factors of ~1000 and ~100 in the centers of the L1544 and B68 cores respectively, i.e., where n $\geq 10^5$ cm^{-3}.

The most common explanation for these abundance variations is that CO and its isotopologues adsorb, i.e., "freeze-out," easily onto grain mantles at high densities and dust temperatures <20 K. Other molecules are also similarly depleted onto dust grains within low-mass cores. For example, *Bergin et al.* (2001), *K. E. Young et al.* (2004), *Lai et al.* (2003), and *Ohashi et al.* (1999) found observational evidence for depletions of CS, H_2CO, C_3H_2, and CCS respectively toward various dense cores. More recently, Tafalla et al. (in preparation) carried out a multimolecular study of the L1498 and L1517B cores and found evidence for the depletion of 11 species, including all the aforementioned species plus DCO^+, HCN, HC_3N, HCO^+, CH_3OH, and SO.

Water is also clearly frozen onto dust grains. First, solid water is seen (with abundances ~10^{-4} relative to H_2) in absorption within the near-infrared spectra of stars located behind molecular clouds and those of embedded protostars (e.g., *van Dishoeck*, 2004, and references therein). Indeed, water ice is the major compound of grain mantles, followed by CO (e.g., *Chiar et al.*, 1995; *Ehrenfreund and Charnley*, 2000). Second, the recent SWAS mission furnished stringent upper limits to abundances of water vapor [<10^{-8} relative to H_2 (*Bergin and Snell*, 2002)] that are orders of magnitude lower than values predicted by gas-phase chemical models that did not account for the gas-grain interaction and the consequent molecular freeze-out. Other ices on grain surfaces recently detected through absorption spectra of background stars include CO_2 (*Bergin et al.*, 2005), and probably HCOOH and NH_4^+ (*Knez et al.*, 2005).

The degree to which a given molecule depletes onto grains is likely a function of its respective binding energy onto grain surfaces, but other chemical factors may be also important. For example, carbon-chain molecules such as CCS and C_3H_2, which are called "early-time" species because in gas-phase chemical models they form before atomic carbon is mainly locked in CO (*Herbst and Leung*, 1989), can decrease in abundance due to gas-phase reactions, especially with atomic oxygen, which tends to convert all C-bearing species into CO. When CO and the reactive O start to freeze-out, carbon-chain species increase their abundance again, showing a "late-time" secondary peak, limited in time

by their own adsorption onto grains (*Ruffle et al.*, 1997, 1999; *Li et al.*, 2002; *Lee et al.*, 2003).

Nitrogen-bearing molecules, e.g., NH_3 and N_2H^+, have been shown to be quite resilient tracers of dense cores, relative to carbon-bearing molecules. They are called "late-time" species; they both have N_2 as a parent molecule, which forms slowly (relative to CO) in the gas phase via neutral-neutral reactions. As described in section 2, NH_3 transitions have long been tracers of dense cores since these can be excited at densities $\approx 10^4$ cm^{-3}. In recent years, however, N_2H^+ transitions have grown in popularity for tracing dense cores with surveys by *Benson et al.* (1998), *Caselli et al.* (2002c), and *Tatematsu et al.* (2004). The intensity distribution of N_2H^+ 1–0 in a given core closely matches that of the millimeter or submillimeter continuum (e.g., *Caselli et al.*, 2002a) and its hyperfine structure can be fit to determine excitation temperatures and line opacities [under the assumption that all the hyperfine components have the same excitation temperature (see *Womack et al.*, 1992; *Caselli et al.*, 1995)]. Within five cores, *Tafalla et al.* (2002; see also *Tafalla et al.*, 2004) found inner zones of CO and CS depletion by factors ≥ 100 but NH_3 abundances that increased by factors of ~1–10 toward the core centers and constant N_2H^+ abundances.

The relatively high abundances of NH_3 and N_2H^+ in starless cores were thought to be due to the low binding energy of their parent N_2 relative to CO (*Bergin and Langer*, 1997), but recent laboratory experiments by *Öberg et al.* (2005; see also *Bisschop et al.*, 2006) have found the binding energies to be quite similar. Another possible explanation is a slow formation of the N_2 molecule in the gas phase (Caselli and Aikawa, in preparation). In the initial formation of molecular clouds, the N atom would be the dominant form of nitrogen. The N atom has a lower adsorption energy than N_2 and CO, and thus is more resilient to adsorption. Formation of N_2 is much slower than the CO formation in the gas phase, and hence many N atoms are still available when the adsorption starts (i.e., when the core age exceeds the collisional timescale between gas and dust). Adsorption of N_2 is compensated by the formation in the gas phase. The high abundance of their mother molecule helps N_2H^+ and NH_3 to maintain their abundances. Two additional factors lead to high N_2H^+ abundance. First, N_2H^+ does not directly adsorb onto grains; it recombines on charged grain surfaces and returns to the gas-phase as N_2. Second, the main reactants of N_2H^+, e.g., CO and electrons, decrease as the core density rises. Indeed, in Model A of *Aikawa et al.* (2001), depletion of N_2H^+ and NH_3 is much less significant than that of CO, even though the adsorption energies of CO and N_2 are assumed to be equal. The central NH_3 enhancement relative to N_2H^+ can be caused by CO depletion. In the outer regions with high CO abundance, N_2H^+ mainly reacts with CO to produce N_2, while in the central regions with heavy CO depletion, it recombines to produce NH as well as N_2 (*Geppert et al.*, 2004), which is transformed to NH_3 (*Aikawa et al.*, 2005).

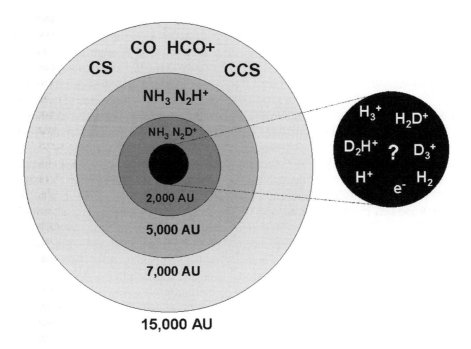

Fig. 3. A schematic representation of eventual molecular differentiation within a starless core. The external shell of the core (where $n(H_2) \simeq 10^4$ cm^{-3}) can be traced by CO, CS, and other carbon-bearing species. At radii <7000 AU, where $n(H_2) \simeq 10^5$ cm^{-3}, CO and CS disappears from the gas phase and the best gas tracers are NH$_3$ and N$_2$H$^+$. At higher densities, deuterated species become quite abundant and, when the density exceeds $\sim 10^6$ cm^{-3}, the chemistry will be dominated by light molecular ions, in particular H$_3^+$ and its deuterated forms, as well as H$^+$ (e.g., *Walmsley et al.*, 2004).

Nitrogen-bearing species are not immune to depletion and even N$_2$ will be eventually adsorbed onto grains at high densities in starless cores (e.g., *Bergin and Langer,* 1997; *Aikawa et al.,* 2001). *Bergin et al.* (2002) found evidence for small zones of N$_2$H$^+$ depletion (by factors of ~2) coincident with the centers of the exposed cores in B68 and L1521F, respectively. In addition, *Di Francesco et al.* (2004) and *Pagani et al.* (2005) found evidence for N$_2$H$^+$ depletion in the Oph A and L183 cores, respectively. *Belloche and André* (2004) and *Jørgensen et al.* (2004) found N$_2$H$^+$ depletion within cold IRAM 04191 and NGC 1333 IRAS 2 protostellar cores respectively, but outflows may have also partially affected the local chemistry if CO has been released from grain mantles near the protostars. Furthermore, *Caselli et al.* (2003) and *Walmsley et al.* (2004) have suggested that all molecules with C, N, or O can be utterly depleted in the innermost regions of denser cores based on interpretation of the H$_2$D$^+$ emission in the L1544 core. Indeed, molecules without heavy elements, e.g., H$_2$D$^+$, may be the only remaining molecular tracers of such regions.

Figure 3 schematically summarizes how the eventual molecular differentiation within a starless core may look. At radii of ~7000–15000 AU, CO and CS are still mainly in the gas phase and the main molecular ion is HCO$^+$, with which one can deduce a stringent lower limit of the electron density (e.g., *Caselli et al.,* 1998) (see section 4.3). The fraction of atomic carbon in the gas is still large, as testi-

fied by the large observed abundances of carbon-chain molecules, such as CCS (e.g., *Ohashi et al.,* 1999). Deeper in the core (~5000–7000 AU), where the density approaches values of the order of 10^5 cm^{-3}, CO and CS disappear from the gas-phase because of the freeze-out onto dust grains. The physical and chemical properties (as well as kinematics) are better traced by N-bearing species, in particular NH$_3$. Within the central 5000 AU, deuterium fractionation takes over (see section 4.2), and N$_2$D$^+$ becomes the best probe (*Caselli et al.,* 2002a). NH$_3$ is still abundant in these regions, however, as suggested by the observed increase of the NH$_3$ abundance toward core centers (*Tafalla et al.,* 2002). At r $\leq$ 2000 AU, where $n(H_2) \geq 10^6$ cm^{-3}, all neutral species are expected to freeze-out in short timescales ($\leq$1000 yr) and light species, such as H$_3^+$ and its deuterated forms, are thought to dominate the chemistry and the degree of ionization (*Caselli et al.,* 2003; *Vastel et al.,* 2004).

The schematic picture of Fig. 3 of course depends on the time spent by a starless core in this condensed phase, so that one expects to find less significant depletion and more typical cloud chemistry in those objects that just entered this phase [e.g., possibly the L1521E core (see *Tafalla and Santiago,* 2004)]. Moreover, recent VLA observations of NH$_3$ (1,1) and (2,2) by Crapsi et al. (in preparation) have shown that NH$_3$ is still present in the central 800 AU of the L1544 core, where the gas density is a few times 10^6 cm^{-3}. If no efficient desorption mechanisms are at work, the time spent by the L1544 core nucleus in its high-density phase may be <500 yr (see section 4.4). More examples are needed to refine chemical models of starless cores.

4.2. Deuterium Fractionation

Another important chemical process recently identified within starless cores is deuterium fractionation, i.e., the enhancement of deuterated isotopologues beyond levels expected from the elemental D/H ratio of ~1.5×10^{-5} (*Oliveira et al.,* 2003). For example, in a sample of dense cores, *Bacmann et al.* (2003) found a D_2CO/H_2CO column density ratio between 0.01 and 0.1. In addition, *Crapsi et al.* (2005) found N_2D^+/N_2H^+ ratios between 0.05 and 0.4 in a similar sample of cores. Deuterium fractionation is related to core temperature and CO depletion (e.g., *Dalgarno and Lepp,* 1984). Species such as H_3^+ and CH_3^+ are enriched in deuterium in cold clouds because of the difference in zero-point energies between deuterated and nondeuterated species and rapid exchange reactions such as $H_3^+ + HD \rightarrow H_2D^+ + H_2$ (e.g., *Millar et al.,* 1989). The enrichments are propagated to other molecules by chemical reactions. At high densities, heavy element molecules like CO will also deplete onto grains, reducing the destruction rate of H_2D^+ and further increasing the H_2D^+/H_3^+ ratio. For example, the N_2D^+/N_2H^+ ratio is higher than the DCO^+/HCO^+ ratio by a factor of about 5 in the L1544 core (*Caselli et al.,* 2002b) because N_2H^+ can trace the central region with CO depletion, while HCO^+, which is produced by the protonation of CO, is not abundant at the core center.

Multiply deuterated species, such as HD_2^+ and D_3^+, can be similarly increased in abundance (*Roberts et al.,* 2003; see chapter by Ceccarelli et al.). For example, the L1544 and IRAS 16293E core centers are also rich in H_2D^+ and HD_2^+, as confirmed respectively by *Caselli et al.* (2003) and *Vastel et al.* (2004).

The sensitivity of the D/H ratio to physical conditions and molecular depletion could make it a potentially useful probe of chemical or dynamic evolution of cores. For example, *Crapsi et al.* (2005) observed 14 cores and found that deuterium enhancement was higher in those that are more centrally concentrated and have larger peak H_2 and N_2H^+ column densities, i.e., arguably those more likely to collapse into stars.

4.3. Fractional Ionization

Observation of molecular ions, combined with chemical models, are used frequently to derive ionization degree, x_e. This value is important for dynamics since the amounts of free charge within cores determine the relative influence on core evolution of ambipolar diffusion, the gravitational inward motion of neutral species retarded by interactions with ionic species bound to a strong magnetic field (e.g., see *Ciolek and Mouschovias,* 1995). For example, *Caselli et al.* (1998; see also *Williams et al.,* 1998) found, for dense cores observed in DCO^+ and HCO^+, x_e values between 10^{-8} and 10^{-6}, with associated ambipolar-diffusion/free-fall time-

scale (AD/FF) ratios of 3–200 and no significant differences between starless and protostellar cores. In the L1544 core, using N_2D^+ and N_2H^+, *Caselli et al.* (2002b) found a significantly smaller x_e, ~10^{-9}, and an associated AD/FF ratio of ~1, suggesting this particular core is near dynamical collapse.

Estimates of x_e, however, depend on many assumptions, e.g., the cosmic-ray ionization rate and molecular depletion degrees, as well as the accuracy of the chemical models used. In particular, the cosmic-ray ionization rate, ζ, typically assumed to be around $1-3 \times 10^{-17}$ s^{-1} in molecular clouds (e.g., *Herbst and Klemperer,* 1973; *van der Tak and van Dishoeck,* 2000), seems to be closer to 10^{-15} s^{-1} in diffuse clouds (*McCall et al.,* 2003). Therefore, a large variation of ζ is expected in the transition between diffuse and molecular clouds (e.g., *Padoan and Scalo,* 2005) and, maybe, between dense cores with different amounts of shielding. Such variation may, in turn, cause significant variations of the ionization degree between cores (e.g., *Caselli et al.,* 1998) with consequent differences in AD/FF timescales. *Padoan et al.* (2004) have further argued that the ISRF strengths (i.e., A_V) and core age variations may account for at least some of the large range of x_e seen in dense cores. *Walmsley et al.* (2004) and *Flower et al.* (2005) have also suggested that the charge and size distributions of dust grains impact ionic abundances, and thus the local AD/FF ratio, in the centers of cores depleted of heavy-element species.

4.4. "Chemodynamical" Evolution

Depletions or enhancements of various molecular species in starless cores can be used as probes of dynamical evolution, since chemical and dynamical timescales can significantly differ. For example, if the dynamical timescale of a core is shorter than the adsorption timescale of CO

$$\left[\sim 10^4 \left(\frac{10^5 [\mathrm{cm}^{-3}]}{n(\mathrm{H}_2)} \right) \left(\frac{10[\mathrm{K}]}{\mathrm{T}_{\mathrm{gas}}} \right)^{0.5} \mathrm{yr} \right]$$

it will not experience CO depletion. Recent theoretical studies have investigated the evolution of molecular abundance distributions in cores (e.g., in L1544) using models including Larson-Penston collapse, ambipolar diffusion of magnetized cores, and contraction of a critical BES (*Aikawa et al.,* 2001, 2003, 2005; *Li et al.,* 2002; *Shematovich et al.,* 2003; *J.-E. Lee et al.,* 2004), showing that chemistry can be used to probe, although not uniquely, how the core contracts.

It is important to note that not all starless cores display similar chemical differentiation. *Hirota et al.* (2002) found that CCS is centrally peaked in the L1521E core, while it is depleted at the center of the L1498 core (*Kuiper et al.,* 1996), although the central density of the L1521E core (~3×10^5 cm^{-3}) is *higher* than that of the L1498 core (~$1 \times 10^{4-5}$ cm^{-3}) (see *Shirley et al.,* 2005; Shirley and Jorgensen, in preparation; *Tafalla et al.,* 2002, 2004; Tafalla et al., in preparation). The L1521E core also has low NH_3 and N_2H^+ abundance, low DNC/HNC ratio, and no CO depletion (*Hirota et al.*, 2001, 2002; *Tafalla and Santiago*, 2004). The core is apparently "chemically young"; these molecular abundances are reproduced either at early stages of chemical evolution in a pseudo-time-dependent model (i.e., assuming that the density does not vary with time), or in a dynamical model where the contraction or formation timescale of a core is smaller than the chemical timescale (e.g., *Lee et al.*, 2003; *Aikawa et al.*, 2001, 2005). Other candidates of such "chemically young" cores are those in L1689B, L1495B, and L1521B (*Lee et al.*, 2003; *Hirota et al.*, 2004). Further determinations of kinematics and physical conditions in these cores and "chemically old" cores (e.g., that in L1498) will be very useful toward understanding chemical evolution in starless cores.

5. BULK AND TURBULENT MOTIONS

5.1. Bulk Motions

Molecular line profiles can be effective tracers of starless core kinematics, provided appropriate lines are observed and modeled. For example, internal velocity gradients in dense cores can be derived from the variation of line profiles across cores. Recent interpretations of such gradients as rotational in origin include those made by *Barranco and Goodman* (1998) of NH_3 (1,1) emission over three isolated starless cores, finding velocity gradients of ~0.3 to 1.40 km s^{-1} pc^{-1} with projected rotational axes not aligned along projected core axes. In addition, *Swift et al.* (2005) found a 1.2 km s^{-1} pc^{-1} gradient of NH_3 emission aligned with the minor axis of the isolated starless core L1551-MC (see Fig. 1). Furthermore, *Lada et al.* (2003) found evidence for differential rotation and angular momentum evolution in the B68 core, finding velocity gradients of 3.4 km s^{-1} pc^{-1} and 4.8 km s^{-1} pc^{-1} from $C^{18}O$ (1–0) and N_2H^+ (1–0) emission respectively, which in turn likely trace outer and inner core material respectively.

Further investigations of velocity gradients include a study of 12 starless and 14 protostellar cores mapped in N_2H^+ (1–0) by *Caselli et al.* (2002c), who derived a typical velocity gradient of 2 km s^{-1} pc^{-1} for both types of cores, and found that the spatial variations of the gradients indicate complex motions not consistent with simple solid-body rotation. As found in the earlier study of cores by *Goodman et al.* (1993), these motions appear dynamically insignificant, since the derived rotational energies are at most a few percent of the respectively derived gravitational potential energies. Even relatively small rotational energies, however, could influence protostellar formation during core collapse. For example, angular momentum evolution due to contraction (see below) may induce disk formation or fragmentation into multiple objects (e.g., see *Hennebelle et al.*, 2004).

In addition to plane-of-sky motions, molecular line profiles can be used to trace the gas kinematics along the line-of-sight. For example, *Walsh et al.* (2004) determined that the relative motion between 42 isolated cores and their surrounding gas was relatively low (e.g., ≤0.1 km s^{-1}) by comparing profiles of a core tracer like N_2H^+ with profiles of cloud tracers like $C^{18}O$ and ^{13}CO. This result implies that cores do not move ballistically with respect to their surrounding gas.

Molecular line profiles can also be used to detect inward motions in cores. Such motions can be detected using optically thick lines, which will appear asymmetrically blue relative to symmetric, optically thin lines in a centrally contracting system. Indeed, inward motions may distinguish prestellar cores from starless ones, since they may indicate gravitational binding (see section 2.1) (*Gregersen and Evans*, 2000; *Crapsi et al.*, 2005). Asymmetrically blue profiles were reproduced very well in collapse models of the protostellar envelope of B335 by *Zhou et al.* (1993), *Choi et al.* (1995), and *Evans et al.* (2005). In those protostellar cases, the models indicated that the self-absorption is due to the static envelope expected in the inside-out collapse (*Shu*, 1977) and the blue peak is stronger because of radiative transfer effects (see *Zhou et al.*, 1993). Although such profiles may in principle also arise from rotational or outflow motions, two-dimensional mapping and detailed modeling of the emission can distinguish between the different cases (e.g., *Narayanan and Walker*, 1998; *Redman et al.*, 2004). In addition, explanations other than inward motions would not favor blue-skewed over red-skewed profiles in a statistical sample.

Figure 4 shows two examples of such profiles, seen in the $F_1,F = 2,3 \rightarrow 1,2$ hyperfine component of the N_2H^+ (1–0) transition, observed toward the central positions of the L694-2 and L1544 starless cores from *Williams et al.* (2006). Such a blue-skewed profile results from redward self-absorption, i.e., absorption along the line-of-sight by foreground, inward-moving gas of lower excitation. The statistics and the extended nature of many of the blue profiles suggest that inward motions exist before formation of the central luminosity source, in contrast to predictions of the simplest inside-out collapse models (e.g., *Shu*, 1977).

Such blue-skewed line profiles can be modeled with radiative transfer codes to estimate inward motion velocities. A simplified solution to the problem has been presented by *Myers et al.* (1996) with a "two-layer" model, i.e., with two isothermal gas layers along the line-of-sight approaching each other, that can be used to determine rapidly the essentials of inward velocity. Figure 4 shows how well such modeling can reproduce observed asymmetrically blue profiles, with infall velocities of ~0.1 km s^{-1} for both cores. With the same goal, *De Vries and Myers* (2005) have recently provided a more realistic "hill" model, i.e., where the excitation increases with optical depth along the line-of-sight. Full radiative transfer solutions, on the other hand, can now be obtained relatively rapidly, and can be used to derive velocities (as well as densities, temperatures, and molecular abundances) of starless cores (e.g., see *Keto et al.*, 2004; *J.-E. Lee et al.*, 2004).

Surveys of starless cores have revealed many candidates with profiles suggestive of inward motions. For example, *Lee et al.* (1999), using CS (2–1) as an optically thick tracer

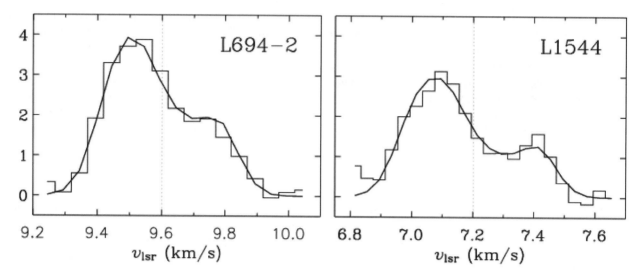

Fig. 4. Spectra of the $F_1,F = 2,3 \rightarrow 1,2$ hyperfine component of $N_2H^+(1-0)$ (histograms) from the starless cores in L694-2 (left) and L1544 (right) from *Williams et al.* (2006). Each spectrum is an average of emission within the central 10″ of each core. Each profile, asymmetrically blue with respect to the systemic velocity of each core (dotted lines), is indicative of inward motions, and can be reproduced well with a two-layer radiative transfer model (solid curves).

and N_2H^+ (1–0) as an optically thin counterpart in single-pointing observations, found 17 of 220 optically selected "starless" cores with asymmetrically blue profiles. Subsequent modeling of these profiles suggested inward velocities of 0.04–0.1 km s^{-1}, i.e., subsonic. (The cores here are generally isolated and not found in turbulent molecular clouds.) *C. W. Lee et al.* (2004) also observed 94 cores in the higher excitation CS (3–2) and DCO$^+$ (2–1) lines, finding 18 strong inward motion candidates. The average inward velocity of 10 cores derived from CS (3–2) was found to be slightly larger than the average velocity derived from CS (2–1), i.e., 0.07 km s^{-1} vs. 0.04 km s^{-1} respectively, suggesting more dense, inner gas moves faster than less dense, outer gas in these cores, although the statistical sample was small. Indeed, CS, being depleted in the central and denser regions of the cores (see section 4), only traces the external (radii ≥ 7000 AU) envelope of the "chemically old" cores, so that the extended infall may reflect the (gravity driven) motion of the material surrounding these cores.

The nature of the inward motions observed in starless cores can be probed in detail through multiline analyses of individual objects. For example, the isolated core in L1544 was found by *Tafalla et al.* (1998; see also *Williams et al.,* 1999) to have CS and H_2CO line profiles suggesting inward motions over a 0.2 pc extent. This scale is inconsistent with that expected from thermal "inside-out" collapse of a singular isothermal sphere since a central protostar of easily detectable luminosity would have likely formed in the time required for such a collapse wave to have propagated so far. [A CS/N_2H^+ mapping survey of 53 cores by *Lee et al.* (2001) also found 19 with infall that was very extended.] In addition, the inward velocities determined via two-layer modeling were up to ~0.1 km s^{-1}.

Although inward velocities derived from observations are larger than expected from some models of ambipolar diffusion, *Ciolek and Basu* (2000; CB00) argued that such motions could indeed result from magnetically retarded contraction through a background magnetic field of relatively low strength, e.g., ~10 μG, and tailored a specific ambipolar diffusion model for the L1544 core. Multiple lines of N_2H^+ and N_2D^+ observed from L1544 by *Caselli et al.* (2002a) indeed showed some consistency with features of this model, including the velocity gradient observed across the minor axis and a slight broadening of line width and double-peaked line profiles at the core center [possibly due to central depletion, but see *Myers* (2005)]. Indeed, the ionization fraction of the L1544 core found by *Caselli et al.* (2002b) suggests its AD/FF timescale ratio ≈1 (see section 4.3). The observed lines at central locations were wider than predicted from CB00, however, suggesting possible modifications such as turbulence and, in particular, an initial quasistatic layer contraction due to Alfvénic turbulence dissipation (e.g., *Myers and Zweibel*, 2001). Still more recent data have revealed further deviations from the CB00 model, however. For example, *van der Tak et al.* (2005) found evidence in double-peaked H_2D^+ 1_{10}–1_{11} line emission from the L1544 core for velocities (and densities) that are higher in its innermost regions than predicted by the CB00 model. (The CB00 model could fit N_2H^+ and $H^{13}CO^+$ data of the core but they sampled larger radii than H_2D^+.) Such velocities could occur if the central region of the L1544 core, where thermal pressure dominates, was smaller than predicted by CB00 model, or if collapse in these regions has become slightly more dynamic (but still subsonic).

Other isolated cores have shown inward velocities similar to those seen in L1544. For example, *Tafalla et al.*

(2004) found evidence in N_2H^+ (1–0) line emission toward the L1498 and L1517B cores for relatively high velocity gas (i.e., ~0.1 km s^{-1} offset from systemic velocities) that was coincident with regions of relatively high CS abundance. Such material may have been accreted onto the core only recently, suggesting a more stochastic growth of cores over time. In addition, *Williams et al.* (2006), comparing N_2H^+ 1–0 of the L694-2 and L1544 cores, found similar inward velocity increases with smaller projected radii, although the L694-2 core has a shallower gradient but a similar central value to that of the L1544 core (see Fig. 4).

Not all starless cores show kinematic behavior similar to L1544, however. For example, *Lada et al.* (2003) found both blue- and red-skewed profiles of CS (2–1) across the B68 core grouped in patterns reminiscent of a low-order acoustic oscillation mode. Since the B68 core may be more in structural equilibrium than the L1544 core (e.g., see *Alves et al.,* 2001), simultaneous inward and outward velocities of ~0.1 km s^{-1} magnitude in the B68 core may not have to do with core contraction and expansion. Instead, such oscillations may have been excited by external influences, e.g., recent passage of a shock from expansion of the Loop I superbubble.

Ambipolar diffusion models like that of CB00 may not be required to explain inward motions within starless cores. For example, *Myers* (2005) has recently proposed that the pressure-free, gravitational collapse of nonsingular, centrally concentrated gas configurations alone is sufficient to explain observed velocity features of starless cores; i.e., no magnetic retardation is required.

5.2. Turbulent Motions

Pressure support against gravity within starless cores is provided by both thermal and turbulent motions of the gas, in addition to what magnetic field pressure can impart to neutral species through ions. Observationally, widths of lines from undepleted molecules in starless cores can be used to probe the relative influence of thermal and non-thermal (turbulent) motions, and thus pressures in such cores, provided gas temperatures can be well determined [and magnetic field strengths (see chapter by Ward-Thompson et al.)]. Early NH_3 observations showed that low-mass dense cores have typically subsonic levels of gas turbulence (*Myers,* 1983). More recently, *Goodman et al.* (1998; see also *Barranco and Goodman,* 1998) found, in the cases they examined, that the width of the NH_3 lines within ~0.1 pc of the core center does not follow the line width-size relations seen on large scales [e.g., $\Delta V \propto R^{0.5}$ (see *Larson,* 1981)], suggesting that starless cores may constitute localized islands of "coherence." *Goodman et al.* (1998) further speculated that this reduction of turbulent motions may be the result of dense cores at 0.1 pc providing a threshold for turbulent dissipation, e.g., through lower ion-neutral coupling from reduced internal ionization at high density and extinction.

Regions of quiescent line width in other starless cores were analyzed by *Caselli et al.* (2002c), who found a lack of general correlation between the N_2H^+ (1–0) line width and the core impact parameter in a sample of 26 cores, in agreement with earlier results. Furthermore, *Tafalla et al.* (2004; see also *Tafalla et al.,* 2002) found a close to constant dependence between the NH_3 or N_2H^+ line widths and impact parameter in the L1498 and L1517B cores. This breakdown of the line width-size relation has been replicated in recent numerical MHD models of dense cores, described elsewhere in this volume (e.g., see chapter by Ballesteros-Paredes et al.). Large numbers of the "cores" produced in these models, although even resembling BES in density configuration, have larger internal velocities than the subsonic cores seen in Taurus. For example, *Klessen et al.* (2005) found only <23% of their model cores were subsonic. Comparisons between observed internal velocities of cores in other clouds and those predicted by turbulent models would be very interesting. Regions of quiescence within clustered cores have also been noted by *Williams and Myers* (2000) and *Di Francesco et al.* (2004). (For further discussion of the evolution of turbulent motions in cores, see the following chapter by Ward-Thompson et al.)

6. CONCLUSIONS

This review described the numerous recent observations and resulting inferences about the nature of starless cores. These objects represent the state of molecular gas preceding gravitational collapse that defines the initial conditions of star formation. Recent observations, together with recent advances in theory and experimental chemistry, have shown that these objects reside at different points on evolutionary paths from ambient molecular material to centrally concentrated, slowly contracting configurations with significant molecular differentiation. Although significant observational strides have been made, sensitivity and resolution limits on past instrumentation have in turn limited our ability to probe their internal physical and chemical properties. Fortunately, new instruments such as the wide-field line and continuum focal-plane arrays on single-dish telescopes like the JCMT, LMT, and the IRAM 30 m, Nobeyama 45 m, and APEX telescopes, and the new sensitive interferometers like the SMA, CARMA, and ALMA will be available soon. These will provide the means to characterize numerous cores to high degrees, moving our understanding of their internal physical and chemical properties from an anecdotal to a statistical regime.

Acknowledgments. We thank an anonymous referee whose comments improved this review. N.J.E. and P.C.M. acknowledge support for this work as part of the Spitzer Legacy Science Program, provided by NASA through contract 1224608 issued by JPL, California Institute of Technology, under NASA contract 1407. In addition, N.J.E. acknowledges support from NASA OSS Grant NNG04GG24G and NSF Grant AST-0307250 to the University of Texas at Austin. P.C. acknowledges support from the MIUR grant "Dust and Molecules in Astrophysical Environments." P.C.M. also acknowledges the support of NASA OSS Grant NAG5-13050. Y.A. acknowledges support from the COE Program at Kobe University and Grant-in-Aid for Scientific Research (14740130, 16036205) of MEXT, Japan.

REFERENCES

Aikawa Y., Ohashi N., Inutsuka S., Herbst E., and Takakuwa S. (2001) *Astrophys. J., 552*, 639–653.

Aikawa Y., Ohashi N., and Herbst E. (2003) *Astrophys. J., 593*, 906–924.

Aikawa Y., Herbst E., Roberts H., and Caselli P. (2005) *Astrophys. J., 620*, 330–346.

Allen L. E., Myers P. C., Di Francesco J., Mathieu R., Chen H., et al. (2002) *Astrophys. J., 566*, 993–1004.

Allen M. and Robinson G. W. (1977) *Astrophys. J., 212*, 396–415.

Alves J. F., Lada C. J., and Lada E. A. (2001) *Nature, 409*, 159–161.

André P., Ward-Thompson D. A., and Motte F. (1996) *Astron. Astrophys., 314*, 625–635.

André P., Motte F., and Bacmann A. (1999) *Astrophys. J., 513*, L57–L60.

André P., Ward-Thompson D. A., and Barsony M. (2000) In *Protostars and Planets IV* (V. Mannings et al., eds.), pp. 59–96. Univ. of Arizona, Tucson.

André P., Bouwman J., Belloche A., and Hennebelle P. (2003) In *Chemistry as a Diagnostic of Star Formation* (C. L. Curry and M. Fich, eds.), pp. 127–138. NRC, Ottawa.

Bacmann A., André P., Puget J.-L., Abergel A., Bontemps S., et al. (2000) *Astron. Astrophys., 361*, 555–580.

Bacmann A., Lefloch B., Ceccarelli C., Castets A., Steinacker J., et al. (2002) *Astron. Astrophys., 389*, L6–L10.

Bacmann A., Lefloch B., Ceccarelli C., Steinacker J., Castets A., et al. (2003) *Astrophys. J., 585*, L55–L58.

Ballesteros-Paredes J., Klessen R. S., and Vázquez-Semadeni E. (2003) *Astrophys. J., 592*, 188–202.

Barranco J. A. and Goodman A. A. (1998) *Astrophys. J., 504*, 207–222.

Beichman C. A., Myers P. C., Emerson J. P., Harris S., Mathieu R., et al. (1986) *Astrophys. J., 307*, 337–339.

Belloche A. and André P. (2004) *Astron. Astrophys., 419*, L35–L38.

Benson P. J. and Myers P. C. (1989) *Astrophys. J. Suppl., 71*, 89–108.

Benson P. J., Caselli P., and Myers P. C. (1998) *Astrophys. J., 506*, 743–757.

Bergin E. A. and Langer W. D. (1997) *Astrophys. J., 486*, 316–328.

Bergin E. A. and Snell R. L. (2002) *Astrophys. J., 581*, L105–L108.

Bergin E. A., Plume R., Williams J. P., and Myers P. C. (1999) *Astrophys. J., 512*, 724–739.

Bergin E. A., Ciardi D. R., Lada C. J., Alves J., and Lada E. A. (2001) *Astrophys. J., 557*, 209–225.

Bergin E. A., Alves J., Huard T., and Lada C. J. (2002) *Astrophys. J., 570*, L101–L104.

Bergin E. A., Melnick G. J., Gerakines P. A., Neufeld D. A., and Whittet D. C. B. (2005) *Astrophys. J., 627*, L33–L36.

Bisschop S. E., Fraser H. J., Öberg K. I., van Dishoeck E. F., and Schlemmer S. (2006) *Astron. Astrophys., 449*, 1297.

Black J. H. (1994) In *The First Symposium on the Infrared Cirrus and Diffuse Interstellar Clouds* (R. M. Cutri and W. B. Latter, eds.), pp. 355–367. ASP Conference Series 58, Astronomical Society of the Pacific, San Francisco.

Bonnor W. B. (1956) *Mon. Not. R. Astron. Soc., 116*, 351–359.

Brown P. D. and Millar T. J. (1989) *Mon. Not. R. Astron. Soc., 240*, 25–29.

Caselli P., Myers P. C., and Thaddeus P. (1995) *Astrophys. J., 455*, L77–L80.

Caselli P., Walmsley C. M., Terzieva R., and Herbst E. (1998) *Astrophys. J., 499*, 234–249.

Caselli P., Walmsley C. M., Tafalla M., Dore L., and Myers P. C. (1999) *Astrophys. J., 523*, L165–L169.

Caselli P., Walmsley C. M., Zucconi A., Tafalla M., Dore L., et al. (2002a) *Astrophys. J., 565*, 331–343.

Caselli P., Walmsley C. M., Zucconi A., Tafalla M., Dore L., et al. (2002b) *Astrophys. J., 565*, 344–358.

Caselli P., Benson P. J., Myers P. C., and Tafalla M. (2002c) *Astrophys. J., 572*, 238–263.

Caselli P., van der Tak F. F. S., Ceccarelli C., and Bacmann A. (2003) *Astron. Astrophys., 403*, L37–L41.

Chandrasekhar S. (1957) *An Introduction to the Study of Stellar Structure.* Dover, Chicago. 155 pp.

Chandrasekhar S. and Wares G. W. (1949) *Astrophys. J., 109*, 551–554.

Chiar J. E., Adamson A. J., Kerr T. H., and Whittet D. C. B. (1995) *Astrophys. J., 455*, 234–243.

Chini R., Reipurth B., Ward-Thompson D., Bally J., Nyman L.-Å., et al. (1997) *Astrophys. J., 474*, L135–L138.

Chini R., Kämpgen K., Reipurth B., Albrecht M., Kreysa E., et al. (2003) *Astron. Astrophys., 409*, 235–244.

Choi M., Evans N. J. II, Gregersen E. M., and Wang Y. (1995) *Astrophys. J., 448*, 742–747.

Ciolek G. E. and Basu S. (2000) *Astrophys. J., 529*, 925–931.

Ciolek G. E. and Mouschovias T. Ch. (1995) *Astrophys. J., 454*, 194–216.

Clemens D. P. and Barvainis R. (1988) *Astrophys. J. Suppl., 68*, 257–286.

Crapsi A., Caselli P., Walmsley C. M., Myers P. C., Tafalla M., et al. (2005) *Astrophys. J., 619*, 379–406.

Dalgarno A. and Lepp S. (1984) *Astrophys. J., 287*, L47–L50.

De Vries C. H. and Myers P. C. (2005) *Astrophys. J., 620*, 800–815.

Di Francesco J., Myers P. C., and André P. (2004) *Astrophys. J., 617*, 425–438.

Doty S. D., Everett S. E., Shirley Y. L., Evans N. J. II, and Palotti M. L. (2005a) *Mon. Not. R. Astron. Soc., 359*, 228–236.

Doty S. D., Metzler R. A., and Palotti M. L. (2005b) *Mon. Not. R. Astron. Soc., 362*, 737–747.

Draine B. T. (1978) *Astrophys. J. Suppl., 36*, 595–619.

Draine B. T. and Lee H. M. (1984) *Astrophys. J., 285*, 89–108.

Ebert R. (1955) *Z. Ap., 37*, 217–232.

Egan M. P., Leung C. M., and Spagna G. F. (1988) *Comp. Phys. Comm., 48*, 271–292.

Ehrenfreund P. and Charnley S. B. (2000) *Ann. Rev. Astron. Astrophys., 38*, 427–483.

Emden R. (1907) *Gaskugeln*, Leipzig and Berlin, Table 14.

Enoch M. L., Young K. E., Glenn J., Evans N. J. II, Golwala S., et al. (2006) *Astrophys. J., 638*, 293.

Evans N. J. II, Rawlings J. M. C., Shirley Y. L., and Mundy L. G. (2001) *Astrophys. J., 557*, 193–208.

Evans N. J. II, Lee J.-E., Rawlings J. M. C., and Choi M. (2005) *Astrophys. J., 626*, 919–932.

Flower D. R., Pineau des Forêts G., and Walmsley C. M. (2005) *Astron. Astrophys., 436*, 933–943.

Galli D., Walmsley M., and Gonçalves J. (2002) *Astron. Astrophys., 394*, 275–284.

Geppert W. D., Thomas R., Semaniak J., Ehlerding A., Millar T. J., et al. (2004) *Astrophys. J., 609*, 459–464.

Gillett F. C. and Forrest W. J. (1973) *Astrophys. J., 179*, 483

Goldreich P. and Kwan J. (1974) *Astrophys. J., 189*, 441–453.

Goldsmith P. F. (2001) *Astrophys. J., 557*, 736–746.

Goodman A. A., Benson P. J., Fuller G. A., and Myers P. C. (1993) *Astrophys. J., 406*, 528–547.

Goodman A. A., Barranco J. A., Wilner D. J., and Heyer M. H. (1998) *Astrophys. J., 504*, 223–246.

Goodwin S. P., Ward-Thompson D., and Whitworth A. P. (2002) *Mon. Not. R. Astron. Soc., 30*, 769–771.

Gould R. J. and Salpeter E. E. (1963) *Astrophys. J., 138*, 393–407.

Gregersen E. M. and Evans N. J. (2000) *Astrophys. J., 538*, 260–267.

Hara A., Tachihara K., Mizuno A., Onishi T., Kawamura A., et al. (1999) *Publ. Astron. Soc. Japan, 51*, 895–910.

Hartmann L. (2002) *Astrophys. J., 578*, 914–924.

Harvey D. W. A., Wilner D. J., Di Francesco J., Lee C. W., Myers P. C., et al. (2002) *Astron. J., 123*, 3325–3328.

Harvey D. W. A., Wilner D. J., Myers P. C., and Tafalla M. (2003) *Astrophys. J., 597*, 424–433.

Hatchell J., Richer J. S., Fuller G. A., Qualtrough C. J., Ladd E. F., et al. (2005) *Astron. Astrophys., 440*, 151–161.

Hennebelle P., Whitworth A. P., Cha S.-H., and Goodwin S. P. (2004) *Mon. Not. R. Astron. Soc., 348*, 687–701.

Herbst E. and Klemperer W. (1973) *Astrophys. J., 185*, 505–534.

Herbst E. and Leung C. M. (1989) *Astrophys. J. Suppl., 69*, 271–300.

Hirota T., Ikeda M., and Yamamoto S. (2001) *Astrophys. J., 547*, 814–828.

Hirota T., Ito T., and Yamamoto S. (2002) *Astrophys. J., 565*, 359–363.

Hirota T., Maezawa H., and Yamamoto S. (2004) *Astrophys. J., 617*, 399–405.

Hotzel S., Harju J., and Juvela M. (2002) *Astron. Astrophys., 395*, L5–L8.

Ivezić Ž., Nenkova M., and Elitzur M. (1999) *User Manual for DUSTY.* Univ. of Kentucky Internal Report.

Jessop N. E. and Ward-Thompson D. (2000) *Mon. Not. R. Astron. Soc., 311*, 63–74.

Jessop N. E. and Ward-Thompson D. (2001) *Mon. Not. R. Astron. Soc., 323*, 1025–1034.

Jijina J., Myers P. C., and Adams F. C. (1999) *Astrophys. J. Suppl., 125*, 161–236.

Johnstone D. and Bally J. (1999) *Astrophys. J., 510*, L49–L53.

Johnstone D., Wilson C. D., Moriarty-Schieven G., Joncas G., Smith G., et al. (2000) *Astrophys. J., 545*, 327–339.

Johnstone D., Fich M., Mitchell G. F., and Moriarty-Schieven G. (2001) *Astrophys. J., 559*, 307–317.

Johnstone D., Di Francesco J., and Kirk H. M. (2004) *Astrophys. J., 611*, L45–L48.

Jones C. E., Basu S., and Dubinski J. (2001) *Astrophys. J., 551*, 387–393.

Jørgensen J. K., Hogerheijde M. R., van Dishoeck E. F., Blake G. A., and Schöier F. L. (2004) *Astron. Astrophys., 413*, 993–1007.

Jørgensen J. K., Johnstone D., van Dishoeck E. F., and Doty S. D. (2006) *Astron. Astrophys., 449*, 609.

Jura M. (1974) *Astrophys. J., 191*, 375–379.

Kato S., Mizuno N., Asayama S., Mizuno A., Ogawa H., et al. (1999) *Publ. Astron. Soc. Japan, 51*, 883–893.

Kauffmann J., Bertoldi F., Evans N. J. II, et al. (2005) *Astron. Nach., 326*, 878–881.

Keto E., Rybicki G. B., Bergin E. A., and Plume R. (2004) *Astrophys. J., 613*, 355–373.

Kirk H. M., Johnstone D., and Di Francesco J. (2006) *Astrophys. J.,* in press.

Kirk J. M., Ward-Thompson D., and André P. (2005) *Mon. Not. R. Astron. Soc., 360*, 1506–1526.

Klessen R. S., Ballesteros-Paredes J., Vázquez-Semadeni E., and Durán-Rojas C. (2005) *Astrophys. J., 620*, 786–794.

Knez C., Boogert A. C. A., Pontoppidan K. M., Kessler-Silacci J. E., van Dishoeck E. F., et al. (2005) *Astrophys. J., 635*, L145–L148.

Kramer C., Alves J., Lada C. J., Lada E. A., Sievers A., et al. (1998) *Astron. Astrophys., 329*, L33–L36.

Kramer C., Richer J., Mookerjea B., Alves J., and Lada C. (2003) *Astron. Astrophys., 399*, 1073–1082.

Kuiper T. B. H., Langer W. D., and Velusamy T. (1996) *Astrophys. J., 468*, 761–773.

Lada C. J. and Lada E. A. (2003) *Ann. Rev. Astron. Astrophys., 41*, 57–115.

Lada C. J., Bergin E. A., Alves J. F., and Huard T. L. (2003) *Astrophys. J., 586*, 286–295.

Lai S.-P., Velusamy T., Langer W. D., and Kuiper T. B. H. (2003) *Astron. J., 126*, 311–318.

Larson R. B. (1981) *Mon. Not. R. Astron. Soc., 194*, 809–826.

Langer W. D. and Willacy K. (2001) *Astrophys. J., 557*, 714–726.

Langer W. D., van Dishoeck E. F., Bergin E. A., Blake G. A., Tielens A. G. G. M., et al. (2000) In *Protostars and Planets IV* (V. Mannings et al., eds.), pp. 29–57. Univ. of Arizona, Tucson.

Lee C. W. and Myers P. C. (1999) *Astrophys. J. Suppl., 123*, 233–250.

Lee C. W., Myers P. C., and Tafalla M. (1999) *Astrophys. J., 526*, 788–805.

Lee C. W., Myers P. C., and Tafalla M. (2001) *Astrophys. J. Suppl., 136*, 703–734.

Lee C. W., Myers P. C., and Plume R. (2004) *Astrophys. J. Suppl., 153*, 523–543.

Lee J.-E., Evans N. J. II, Shirley Y. L., and Tatematsu K. (2003) *Astrophys. J., 583*, 789–808.

Lee J.-E., Bergin E. A., and Evans N. J. II (2004) *Astrophys. J., 617*, 360–383.

Léger A. (1983) *Astron. Astrophys., 123*, 271–278.

Leung C. M. (1975) *Astrophys. J., 199*, 340–360.

Li Z.-Y. and Shu F. H. (1996) *Astrophys. J., 472*, 211–224.

Li Z.-Y., Shematovich V. I., Wiebe D. S., and Shustov B. M. (2002) *Astrophys. J., 569*, 792–802.

Lis D. C., Serabyn E., Keene J., Dowell C. D., Benford D. J., et al. (1998) *Astrophys. J., 509*, 299–308.

Lis D. C., Serabyn E., Zylka R., and Li Y. (2001) *Astrophys. J., 550*, 761–777.

Lombardi M. and Bertin G. (2001) *Astron. Astrophys., 375*, 1091–1099.

Lynds B. T. (1962) *Astrophys. J. Suppl., 7*, 1–52.

Mathis J. S., Mezger P. G., and Panagia N. (1983) *Astron. Astrophys., 128*, 212–229.

McCall B. J., Huneycutt A. J., Saykally R. J., Geballe T. R., Djuric N., et al. (2003) *Nature, 422*, 500–502.

Millar T. J., Bennett A., and Herbst E. (1989) *Astrophys. J., 340*, 906–920.

Mitchell G. F., Johnstone D., Moriarty-Schieven G., Fich M., and Tothill N. F. H. (2001) *Astrophys. J., 556*, 215–229.

Mizuno A., Hayakawa T., Tachihara K., Onishi T., Yonekura Y., et al. (1999) *Publ. Astron. Soc. Japan, 51*, 859–870.

Motte F., André P., and Neri R. (1998) *Astron. Astrophys., 336*, 150–172.

Motte F., André P., Ward-Thompson D., and Bontemps S. (2001) *Astron. Astrophys., 372*, L41–L44.

Murphy D. C. and Myers P. C. (2003) *Astrophys. J., 591*, 1034–1048.

Myers P. C. (1983) *Astrophys. J., 270*, 105–118.

Myers P. C. (2005) *Astrophys. J., 623*, 280–290.

Myers P. C. and Benson P. J. (1983) *Astrophys. J., 266*, 309–320.

Myers P. C. and Zweibel E. (2001) *Bull. Am. Astron. Soc., 33*, 915.

Myers P. C., Linke R. A., and Benson P. J. (1983) *Astrophys. J., 264*, 517–537.

Myers P. C., Fuller G. A., Goodman A. A., and Benson P. J. (1991) *Astrophys. J., 376*, 561–572.

Myers P. C., Mardones D., Tafalla M., Williams J. P., and Wilner D. J. (1996) *Astrophys. J., 465*, L133–L136.

Narayanan G. and Walker C. K. (1998) *Astrophys. J., 508*, 780–790.

Nutter D. J., Ward-Thompson D., and André P. (2005) *Mon. Not. R. Astron. Soc., 357*, 975–982.

Obayashi A., Kun M., Sato F., Yonekura Y., and Fukui Y. (1998) *Astron. J., 115*, 274–285.

Öberg K. I., van Broekhuizen F., Fraser H. J., Bisschop S. E., van Dishoeck E. F. et al. (2005) *Astrophys. J., 621*, L33–L36.

Ohashi N., Lee S. W., Wilner D. J., and Hayashi M. (1999) *Astrophys. J., 518*, L41–L44.

Oliveira C. M., Hébrard G., Howk C. J., Kruk J. W., Chayer P., et al. (2003) *Astrophys. J., 587*, 235–255.

Onishi T., Mizuno A., Kawamura A., Ogawa H., and Fukui Y. (1998) *Astrophys. J., 502*, 296–314.

Onishi T., Kawamura A., Abe R., Yamaguchi N., Saito H., et al. (1999) *Publ. Astron. Soc. Japan, 51*, 871–881.

Onishi T., Mizuno A., Kawamura A., Tachihara K., and Fukui Y. (2002) *Astrophys. J., 575*, 950–973.

Ossenkopf V. and Henning T. (1994) *Astron. Astrophys., 291*, 943–959.

Padoan P. and Scalo J. (2005) *Astrophys. J., 624*, L97–L100.

Padoan P., Willacy K., Langer W. D., and Juvela M. (2004) *Astrophys. J., 614*, 203–210.

Pagani L., Lagache G., Bacmann A., Motte F., Cambrésy L., et al. (2003)

Astron. Astrophys., 406, L59–L62.

Pagani L., Bacmann A., Motte F., Cambrésy L., Fich M., et al. (2004) *Astron. Astrophys., 417,* 605–613.

Pagani L., Pardo J.-R., Apponi A. J., Bacmann A., and Cabrit S. (2005) *Astron. Astrophys., 429,* 181–192.

Pollack J. B., Hollenbach D., Beckwith S., Simonelli D. P., Roush T., et al. (1994) *Astrophys. J., 421,* 615–639.

Pontoppidan K. M., Dullemond C. P., van Dishoeck E. F., Blake G. A., Boogert A. C. A., Evans N. J. II, et al. (2005) *Astrophys. J., 622,* 463–481.

Redman M. P., Keto E., Rawlings J. M. C., and Williams D. A. (2004) *Mon. Not. R. Astron. Soc., 352,* 1365–1371.

Roberts H., Herbst E., and Millar T. J. (2003) *Astrophys. J., 591,* L41–L44.

Ruffle D. P., Hartquist T. W., Taylor S. D., and Williams D. A. (1997) *Mon. Not. R. Astron. Soc., 291,* 235–240.

Ruffle D. P., Hartquist T. W., Caselli P., and Williams D. A. (1999) *Mon. Not. R. Astron. Soc., 306,* 691–695.

Ryden B. S. (1996) *Astrophys. J., 471,* 822–831.

Sandell G. and Knee L. B. G. (2001) *Astrophys. J., 546,* L49–L52.

Schnee S. and Goodman A. A. (2005) *Astrophys. J., 624,* 254–266.

Shematovich V. I., Wiebe D. S., Shustov B. M., and Li Z.-Y. (2003) *Astrophys. J., 588,* 894–909.

Shirley Y. L., Evans N. J. II, Rawlings J. M. C., and Gregersen E. M. (2000) *Astrophys. J. Suppl., 131,* 249–271.

Shirley Y. L., Nordhaus M. K., Grcevich J. M., Evans N. J. II, Rawlings J. M. C., et al. (2005) *Astrophys, J., 632,* 982–1000.

Shu F. H. (1977) *Astrophys. J., 214,* 488–497.

Solomon P. M. and Werner M. W. (1971) *Astrophys. J., 165,* 41–49.

Stamatellos D. and Whitworth A. P. (2003) *Astron. Astrophys., 407,* 941–955.

Stamatellos D., Whitworth A. P., André P., and Ward-Thompson D. (2004) *Astron. Astrophys., 420,* 1009–1023.

Stanke T., Smith M. D., Gredel R., and Khanzadyan T. (2006) *Astron. Astrophys., 447,* 609.

Steinacker J., Bacmann A., Henning Th., Klessen R., and Stickel M. (2005) *Astron. Astrophys., 434,* 167–180.

Swift J., Welch W. J., and Di Francesco J. (2005) *Astrophys. J., 620,* 823–834.

Tachihara K., Mizuno A., and Fukui Y. (2000) *Astrophys. J., 528,* 817–840.

Tachihara K., Onishi T., Mizuno A., and Fukui Y. (2002) *Astron. Astrophys., 385,* 909–920.

Tafalla M. and Santiago J. (2004) *Astron. Astrophys., 414,* L53–L56.

Tafalla M., Mardones D., Myers P. C., Caselli P., Bachiller R., et al. (1998) *Astrophys. J., 504,* 900–914.

Tafalla M., Myers P. C., Caselli P., Walmsley C. M., and Comito C. (2002) *Astrophys. J., 569,* 815–835.

Tafalla M., Myers P. C., Caselli P., and Walmsley C. M. (2004) *Astron. Astrophys., 416,* 191–212.

Tatematsu K., Umemoto T., Kandori R., and Sekimoto Y. (2004) *Astrophys. J., 606,* 333–340.

Tóth L. V., Haas M., Lemke D., Mattila K., and Onishi T. (2004) *Astron. Astrophys., 420,* 533–546.

van der Tak F. F. S. and van Dishoeck E. F. (2000) *Astron. Astrophys., 358,* L79–L82.

van der Tak F. F. S., Caselli P., and Ceccarelli C. (2005) *Astron. Astrophys., 439,* 195–203.

van Dishoeck E. F. (1988) In *Rate Coefficients in Astrochemistry* (T. L. Millar and D. A. Williams, eds.), pp. 49–62. Kluwer, Dordrecht.

van Dishoeck E. F. (2004) *Ann. Rev. Astron. Astrophys., 42,* 119–167.

Vastel C., Phillips T. G., and Yoshida H. (2004) *Astrophys. J., 606,* L127–L130.

Walmsley C. M., Flower D. R., and Pineau des Forêts G. (2004) *Astron. Astrophys., 418,* 1035–1043.

Walsh A. J., Myers P. C., and Burton M. G. (2004) *Astrophys. J., 614,* 194–202.

Ward-Thompson D., Scott P. F., Hills R. E., and André P. (1994) *Mon. Not. R. Astron. Soc., 268,* 276–290.

Ward-Thompson D., Motte F., and André P. (1999) *Mon. Not. R. Astron. Soc., 305,* 143–150.

Ward-Thompson D., André P., and Kirk J. M. (2002) *Mon. Not. R. Astron. Soc., 329,* 257–276.

Watson W. D. and Salpeter E. E. (1972) *Astrophys. J., 174,* 321–340.

Williams J. P. and Myers P. C. (1999) *Astrophys. J., 518,* L37–L40.

Williams J. P. and Myers P. C. (2000) *Astrophys J., 537,* 891–903.

Williams J. P., de Geus E. J., and Blitz L. (1994) *Astrophys. J., 428,* 693–712.

Williams J. P., Bergin E. A., Caselli P., Myers P. C., and Plume R. (1998) *Astrophys. J., 503,* 689–699.

Williams J. P., Myers P. C., Wilner D. J., and Di Francesco J. (1999) *Astrophys. J., 513,* L61–L64.

Williams J. P., Lee C. W., and Myers P. C. (2006) *Astrophys. J., 636,* 952–958.

Willacy K., Langer W. D., and Velusamy T. (1998) *Astrophys. J., 507,* L171–L175.

Wilson C. D., Avery L. W., Fich M., Johnstone D., Joncas G., et al. (1999) *Astrophys. J., 513,* L139–L142.

Womack M., Ziurys L. M., and Wyckoff S. (1992) *Astrophys. J., 387,* 417–429.

Yonekura Y., Mizuno N., Saito H., Mizuno A., Ogawa H., et al. (1999) *Publ. Astron. Soc. Japan, 51,* 911–918.

Young C. H., Jørgensen J. K., Shirley Y. L., Kauffmann J., Huard T., et al. (2004) *Astrophys. J. Suppl., 154,* 396–401.

Young K. E., Lee J.-E., Evans N. J., Goldsmith P. F., and Doty S. D. (2004) *Astrophys. J., 614,* 252–266.

Young K. E., Enoch M. L., Evans N. J. II, Glenn J., Sargent A. I., et al. (2006) *Astrophys. J., 644,* 326–343.

Yun J. L. and Clemens D. P. (1990) *Astrophys. J., 365,* L73–L76.

Yun J. L., Moreira M. C., Torrelles J. M., Afonso J. M., and Santos N. C. (1996) *Astron. J., 111,* 841–845.

Zhou S., Evans N. J. II, Kömpe C., and Walmsley C. M. (1993) *Astrophys. J., 404,* 232–246.

Zucconi A., Walmsley C. M., and Galli D. (2001) *Astron. Astrophys., 376,* 650–662.

An Observational Perspective of Low-Mass Dense Cores II: Evolution Toward the Initial Mass Function

Derek Ward-Thompson
Cardiff University

Philippe André
Service d'Astrophysique de Saclay

Richard Crutcher
University of Illinois

Doug Johnstone
National Research Council of Canada

Toshikazu Onishi
Nagoya University

Christine Wilson
McMaster University

We review the properties of low-mass dense molecular cloud cores, including starless, prestellar, and Class 0 protostellar cores, as derived from observations. In particular we discuss them in the context of the current debate surrounding the formation and evolution of cores. There exist several families of model scenarios to explain this evolution (with many variations of each) that can be thought of as a continuum of models lying between two extreme paradigms for the star and core formation process. At one extreme there is the dynamic, turbulent picture, while at the other extreme there is a slow, quasistatic vision of core evolution. In the latter view the magnetic field plays a dominant role, and it may also play some role in the former picture. Polarization and Zeeman measurements indicate that some, if not all, cores contain a significant magnetic field. Wide-field surveys constrain the timescales of the core formation and evolution processes, as well as the statistical distribution of core masses. The former indicates that prestellar cores typically live for 2–5 freefall times, while the latter seems to determine the stellar initial mass function. In addition, multiple surveys allow one to compare core properties in different regions. From this it appears that aspects of different models may be relevant to different star-forming regions, depending on the environment. Prestellar cores in cluster-forming regions are smaller in radius and have higher column densities, by up to an order of magnitude, than isolated prestellar cores. This is probably due to the fact that in cluster-forming regions the prestellar cores are formed by fragmentation of larger, more turbulent cluster-forming cores, which in turn form as a result of strong external compression. It is then the fragmentation of the cluster-forming core (or cores) that forms a stellar cluster. In more isolated, more quiescent, star-forming regions the lower ambient pressure can only support lower-density cores, which go on to form only a single star or a binary/multiple star system. Hence the evolution of cluster-forming cores appears to differ from the evolution of more isolated cores. Furthermore, for the isolated prestellar cores studied in detail, the magnetic field and turbulence appear to be playing a roughly equal role.

1. INTRODUCTION

A great deal is now known about dense cores in molecular clouds that are the progenitors of protostars — see the previous chapter by Di Francesco et al., which details many of the observational constraints that have been placed upon their physical parameters (this chapter and the preceding chapter should be read in conjunction). What is less clear is the manner in which the cores are formed and subsequently evolve. In this chapter we discuss what the observations can tell us about the formation and evolution of cores. Clearly the evolution depends heavily upon the formation mecha-

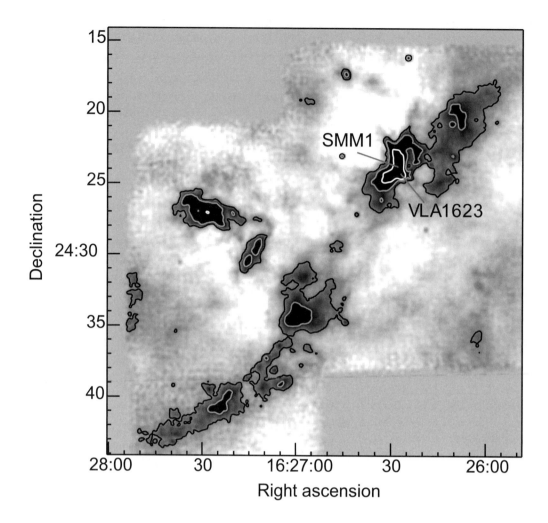

Fig. 1. SCUBA image of the ρ Oph molecular cloud region seen in dust continuum at 850 μm (adapted from *Johnstone et al.,* 2000). Prestellar, protostellar, and cluster-forming cores can all be seen in this molecular cloud region. For example, the cluster-forming core ρ Oph A (extended region in the upper right of this image) contains within it (inside the white contour) the prestellar core SM1 and Class 0 protostellar core VLA1623 (cf. *André et al.,* 1993). Note also that large areas of the cloud contain no dense cores, leading to the idea of a threshold criterion discussed in section 6.

nism, and upon the dominant physics of that formation. Several model scenarios have been proposed for this mechanism.

These models can be thought of as a small number of families of models, each of which contains many variations, representing a continuum lying between two extremes. At one extreme there is a school of thought that proposes a slow, quasistatic evolution, in which a core gradually becomes more centrally condensed. This evolution may be moderated by the magnetic field (e.g., *Mouschovias and Ciolek,* 1999) or else by the gradual dissipation of low-level turbulent velocity fields (e.g., *Myers,* 1998, 2000). At the other extreme is a very dynamic picture (e.g., *Ballesteros-Paredes et al.,* 2003), in which highly turbulent gas creates large density inhomogeneities, some of which become gravitationally unstable and collapse to form stars (for a review, see *Ward-Thompson,* 2002). Once again the magnetic field

may play a role in this latter picture, in which magneto-hydrodynamic (MHD) waves may be responsible for carrying away excess turbulent energy (e.g., *Ostriker et al.,* 1999).

What we find from the observations is that some aspects of each of these different model scenarios may be relevant in different regions of star formation, depending on the local environment. No two regions are the same, and the effects of local density, pressure and magnetic field strength, and the presence or absence of other nearby stars and protostars all play an important role in determining what dominates the formation and evolution of dense molecular cloud cores.

Throughout this chapter we define a dense core as any region in a molecular cloud that is observed to be significantly overdense relative to its surroundings. We define a starless core as any dense core that does not contain any evidence that it harbors a protostar, young stellar object or young star (*Beichman et al.,* 1986). Such evidence would include an embedded infrared source, centimeter radio source, or bipolar outflow, for example (cf. *André et al.,* 1993, 2000).

Any core that does contain such evidence we define as a protostellar core. This might be a Class 0 protostellar core (*André et al.,* 1993, 2000) or a Class I protostellar core (*Lada,* 1987; *Wilking et al.,* 1989) depending upon its evolutionary status.

We here define prestellar cores [formerly pre-protostellar cores (*Ward-Thompson et al.,* 1994)] as that subset of starless cores that are gravitationally bound and hence are expected to participate in the star-formation process. We further define cluster-forming cores as those cores that have significant observed structure within them, such that they appear to be forming a small cluster or group of stars rather than a single star or star system. Examples of the various types of cores can be seen in Fig. 1.

We note that the resolution of current single-dish telescopes is insufficient in more distant regions to differentiate between cluster-forming cores and other types of cores. Hence we restrict most of our discussion to nearby molecular clouds — typically we restrict our discussion to d < 0.5 kpc.

2. EVOLUTIONARY MODELS

We begin by summarizing some of the key model parameters and predictions. One such discriminator between the extreme pictures mentioned above is the timescale of core evolution. Therefore we first discuss some predictions of the models regarding core lifetimes.

If turbulent dissipation in a quasistatic scenario is the relevant physics, then the timescale of the dissipation of turbulence could be several times the freefall time (e.g., *Nakano,* 1998). However, if highly turbulent processes dominate molecular cloud evolution, then detailed modeling yields results that suggest that cores only live for approximately one or two freefall times (e.g., *Ballesteros-Paredes et al.,* 2003; *Vazquez-Semadeni et al.,* 2005).

In the magnetically dominated paradigm, molecular clouds may form by accumulation of matter along flux tubes, by (for example) the Parker instability (*Parker,* 1966). Furthermore, if magnetic fields dominate the evolution, then a key parameter is the ratio of core mass to magnetic flux (M/Φ). A critical cloud or core is defined as one in which the energy density of the magnetic field exactly balances the gravitational potential energy.

For clouds with magnetic fields stronger than is necessary for support against gravitational collapse, M/Φ is subcritical; for fields too weak to support clouds, M/Φ is supercritical. Consequently, two possible extreme-case scenarios arise: one in which low-mass stars form in originally highly magnetically subcritical clouds, with ambipolar diffusion leading to core formation and quasistatic contraction of the cores (e.g., *Mouschovias,* 1991; *Shu et al.,* 1987); and the other in which clouds are originally supercritical (e.g., *Na-*

kano, 1998). In the absence of turbulent support, highly supercritical collapse occurs on essentially the freefall time.

Since magnetic fields can be frozen into only the ionized component of clouds, neutral matter can be driven by gravity through the field. Hence, if a star is formed in an originally very magnetically subcritical cloud, the relevant timescale is the ambipolar diffusion timescale, τ_{AD}, which is proportional to the ionization fraction X_e. This is normally taken to have a power-law dependence on density: $\tau_{AD} \propto X_e \propto n(H_2)^{-0.5}$ for $A_V > 4$, where cosmic-ray ionization dominates (*McKee,* 1989; *Mouschovias,* 1991). For $A_V < 4$, UV ionization dominates, leading to a steeper dependence (*McKee,* 1989), but this regime is not believed to be significant for prestellar cores.

Since τ_{AD} is shorter in denser regions, the process of ambipolar diffusion increases M/Φ in overdense regions of the cloud, leading to the formation of cores. Eventually, M/Φ is increased from subcritical to supercritical and the core collapses. For highly subcritical clouds τ_{AD} is roughly 10× the freefall time (*Nakano,* 1998), although *Ciolek and Basu* (2001) point out that the ambipolar diffusion timescale of marginally subcritical cores within clouds can be as little as a few times the freefall time. Only observations can establish the original M/Φ in clouds; it is a free parameter in the theory.

Magnetic fields may also play another crucial role in star formation — transferring angular momentum outward from collapsing, rotating cores; resolving the angular momentum problem; and allowing collapse to protostellar densities (*Mouschovias,* 1991). Although supersonic motions are allowed in this paradigm, they do not dominate.

This picture has been challenged by interpretations of observations of the ratio of numbers of starless cores to cores with protostars and young stars that suggest that molecular clouds are short-lived compared with the ambipolar diffusion timescale, and that star formation takes place on a cloud-crossing time (e.g., *Elmegreen,* 2000; *Hartmann et al.,* 2001).

This alternative paradigm is that molecular clouds are intermittent phenomena in an interstellar medium dominated by compressible turbulence (e.g., *MacLow and Klessen,* 2004). Turbulent flows form density enhancements that may or may not be self-gravitating. If self-gravitating, they may be supported against collapse for a short time by the turbulent energy. However, the supersonic turbulence will decay on a short (freefall) timescale (e.g., *MacLow et al.,* 1998), and collapse will ensue. This would mean that molecular clouds were transient objects, forming and either dissolving quickly or rapidly collapsing to form stars.

One way to distinguish between the theories is to determine the lifetimes of the large-scale molecular clouds. *Hartmann et al.* (2001) argue for short lifetimes, whereas *Tassis and Mouschovias* (2004) and *Mouschovias et al.* (2006) suggest that all available observational data are consistent with lifetimes of molecular clouds as a whole being $\sim10^7$ yr (see also *Goldsmith and Li,* 2005). Another way is to determine the lifetimes of individual cloud cores. We attempt to do this in the next section. The role of magnetic fields can be assessed by measuring the M/Φ values in cores. We discuss the current data on this in section 4.

3. OBSERVED CORE LIFETIMES

It was shown in the previous section that it is of vital importance to estimate observationally the timescale of cores with various densities if we are to distinguish between the different model pictures. The numbers of cores detected can be used to determine typical statistical timescales for particular evolutionary stages. This method was first used by *Beichman et al.* (1986), who extrapolated from the typical T Tauri star lifetime and estimated the starless core lifetime to be roughly a few times 10^6 yr.

This estimate was subsequently refined by *Lee and Myers* (1999), using an optically selected sample, to ~0.3–1.6 × 10^6 yr for a mean density of ~6–8 × 10^3 cm^{-3}. This age is based upon an estimated range in lifetimes for Class I sources of ~1–5 × 10^5 yr. Within this range the best estimate for the Class I lifetime is ~2 ± 1 × 10^5 yr (e.g., *Greene et al.*, 1994; *Kenyon and Hartmann*, 1995). This corresponds to a starless core lifetime of ~6 ± 3 × 10^5 yr.

Toward some molecular cloud complexes optical selection can miss deeply embedded cores in the complex. In these cases, observations in the millimeter/submillimeter regime are the best way to observe cores and to carry out the statistical study. For example, *Onishi et al.* (1998, 2002) estimated the timescale of cores with a density of ~10^5 cm^{-3} to be ~4 × 10^5 yr, based on a large-scale molecular line study of cores in Taurus.

Kirk et al. (2005) carried out a similar exercise using submillimeter continuum observations of dust in molecular cloud cores. They found a timescale for prestellar cores of ~3 × 10^5 yr with a minimum central density of ~5 × 10^4 cm^{-3}. At this density the freefall time is ~10^5 yr. They made a similar calculation for the cores they classified as "bright," and derived a timescale of ~1.5 × 10^5 yr for cores with a minimum central volume density of ~2 × 10^5 cm^{-3}. At this density the freefall timescale is ~7 × 10^4 yr.

Kandori et al. (2005) derived detailed radial column density profiles for Bok globules and, by comparison with the theoretical calculations of *Aikawa et al.* (2005), estimated their timescale to be ~10^6 yr for a density of ~2 × 10^4 cm^{-3}. A number of other chemical models have been used to carry out a similar exercise in estimating the "chemical age" of cores (see previous chapter by Di Francesco et al.). In many cases this leads to values much longer than a freefall time.

A similar comparison to that discussed in this section was carried out for a number of different datasets in the literature by *Jessop and Ward-Thompson* (2000), who plotted the calculated statistical lifetime against the mean volume density of each sample of cores. We reproduce those data here in Fig. 2, along with other, more recent data. For example, we include the "bright" and "intermediate" cores from *Kirk et al.* (2005), labeled K1 and K2 respectively.

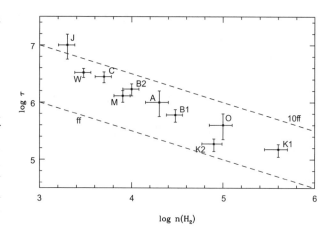

Fig. 2. A "JWT plot" (after *Jessop and Ward-Thompson*, 2000) — plot of inferred starless core lifetime against mean volume density (see also *Kirk et al.*, 2005). The dashed lines correspond to models discussed in the text. The symbols refer to literature data as follows: J — *Jessop and Ward-Thompson* (2000); W — *Wood et al.* (1994); C — *Clemens and Barvainis* (1988); B1, B2 — *Bourke et al.* (1995a,b); M — *Myers et al.* (1983); A — *Aikawa et al.* (2005), *Kandori et al.* (2005); O — *Onishi et al.* (2002); K1, K2 — *Kirk et al.* (2005).

We also plot on Fig. 2 some of the model predictions discussed above (following *Kirk et al.*, 2005). The lower dashed line is the freefall time, relevant to models such as the highly magnetically supercritical models and the highly dynamic, turbulent models (e.g., *Vazquez-Semadeni et al.*, 2005). The upper dashed line is the power-law formulation of *Mouschovias* (1991) discussed in section 2 above for a quasistatic, magnetically subcritical core evolving on the ambipolar diffusion timescale at 10× the freefall time (*Nakano*, 1998).

In summary, almost all the literature estimates lie between the two dashed lines on Fig. 2. All the observed timescales are longer than the freefall time by a factor of ~2–5 in the density range of 10^4–10^5 cm^{-3}. Hence we see quite clearly that prestellar and starless cores cannot generally all be in freefall collapse. Their timescales also appear to be too short for them all to be in a highly magnetically subcritical state. They are all roughly consistent both with mildly subcritical magnetized cores and with models invoking low levels of turbulent support. Hence we must look to observations of magnetic fields to help differentiate between models. In the next section we summarize some of the key observations of magnetic fields.

4. OBSERVATIONS OF MAGNETIC FIELDS

Given that the relative importance of the magnetic field is a key way in which to choose between models, one must try to determine observationally the role of magnetic fields in the star-formation process.

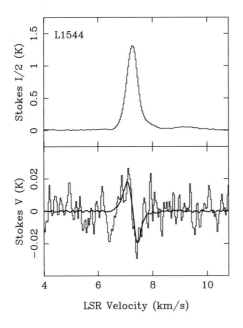
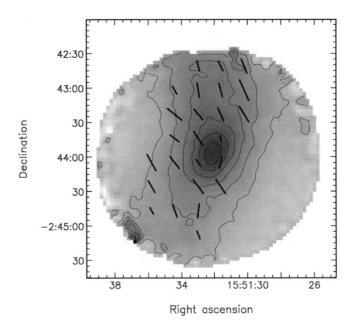

Fig. 3. Plot of Arecibo OH Zeeman spectra (left) in the prestellar core L1544, from *Crutcher and Troland* (2000), and SCUBA submillimeter dust intensity and polarized intensity data of prestellar core L183 (right), from *Crutcher et al.* (2004). The polarized intensity half vectors have been rotated by 90° to show the plane-of-sky magnetic field direction.

Many observations of magnetic fields in regions of low-mass star formation have attempted to test the various paradigms. The observations have utilized the Zeeman effect, mainly in the 18-cm lines of OH (e.g., *Crutcher et al.*, 1993; *Crutcher and Troland*, 2000) and linearly polarized emission of dust at submillimeter wavelengths (e.g., *Ward-Thompson et al.*, 2000; *Matthews et al.*, 2001; *Crutcher et al.*, 2004; *J. Kirk et al.*, 2006).

Unfortunately, the observations are difficult and the results remain somewhat sparse (see, e.g., *Crutcher*, 1999; *Heiles and Crutcher*, 2005). One of the best-studied prestellar cores is L1544, a relatively isolated core in Taurus that has been studied by single-dish (*Tafalla et al.*, 1998) and interferometer spectroscopy (*Williams et al.*, 1999). These studies have suggested that L1544 is contracting.

Information on the magnetic field in L1544 includes OH Zeeman observations with the Arecibo telescope (*Crutcher and Troland*, 2000) and dust polarization mapping with the JCMT SCUBA polarimeter (*Ward-Thompson et al.*, 2000). This cloud is therefore a good example of observational results in low-mass star-formation regions.

Figure 3 shows the Arecibo OH Zeeman spectra, which imply a line-of-sight (los) magnetic field strength of $B_{los} = 10.8 \pm 1.7$ µG, with column density $N(H_2) \approx 4.8 \times 10^{21}$ cm^{-2}, mean radius $\bar{r}(OH) \approx 0.08$ pc, and volume density $n(H_2) \approx 1 \times 10^4$ cm^{-3}. Because all three components of the magnetic field vector are not generally observed, and because the inferred column densities are not generally along the direction of the magnetic field vector, the directly observed M/Φ is typically an overestimate of the true value. A statisti-

cal correction for this is possible (see *Heiles and Crutcher*, 2005). For a large randomly oriented sample, the observed M/Φ average should be divided by 3 to obtain the statistically correct result. This correction may be applied to each cloud individually, but it must be kept in mind that this correction is only strictly valid for a large sample of measurements. The directly observed M/Φ for L1544 is ≈3.4. *Crutcher and Troland* (2000) corrected this value statistically for geometrical bias, finding M/Φ ≈ 1.1, or roughly critical.

Figure 3 also shows the SCUBA dust intensity and polarized intensity map of L183 (*Crutcher et al.*, 2004). *Ward-Thompson et al.* (2000) had previously mapped L183 and two other prestellar cores: L1544 and L43. *Crutcher et al.* (2004) used the Chandrasekhar-Fermi (CF) method (*Chandrasekhar and Fermi*, 1953) to measure the magnetic field strengths and hence the relative criticality of all three cores.

In Fig. 3 the polarization half vectors have been rotated by 90° to indicate the plane-of-sky (pos) magnetic field direction (a half vector is a vector with a 180° bidirectional ambiguity, such as we have here). The field is seen to be fairly uniform in direction in L183, as it was in L1544 and L43, with a position angle dispersion δφ ≈ 14°, but the direction of the field in the plane of the sky is at an angle of 34° ± 7° to the minor axis.

This difference between the magnetic field direction and the minor axis of the core is in conflict with symmetric models that rely only on thermal and static magnetic pressure to balance gravity, since the minor axis projected onto the sky should lie along B_{pos}. However, projection effects can produce the observed position angle difference if the core has a more complicated shape, such as a triaxial geometry (*Basu*, 2000). The initial conditions of cloud formation and turbulence may produce the more complicated shapes (*Gammie et al.*, 2003).

The physical parameters of the L1544 prestellar core inferred from the SCUBA data (*Crutcher et al.*, 2004) are $\bar{r}(dust) \approx 0.021$ pc, $N(H_2) \approx 4.2 \times 10^{22}$ cm^{-2}, $n(H_2) \approx 4.9 \times 10^5$ cm^{-3}, and total mass $M \approx 1.3$ $M_\odot$. With the velocity dispersion $\Delta V_{NT} \approx 0.28$ km s^{-1}, as measured from N_2H^+ data (*Caselli et al.*, 2002), the CF method yielded $B_{pos} \approx 140$ µG. Then $M/\Phi \approx 2.3$ (*Crutcher et al.*, 2004) and the statistically corrected value is then $M/\Phi \approx 0.8$. Hence L1544 is approximately critical or mildly supercritical.

The M/Φ for L1544 found from the OH Zeeman and the dust polarization techniques are essentially in agreement, but very different regions are sampled by the two methods. The region sampled by OH has 4× the radius, 0.1× the column density, and 0.02× the volume density of the region sampled by the dust emission. Therefore, the data probe separately the envelope and the core regions of the cloud.

We have argued that the two M/Φ values are consistent to within the errors, but if the difference between them were real, then M/Φ decreases from envelope to core, the opposite of the ambipolar diffusion prediction. However, the Zeeman effect measures B_{los} and dust emission measures B_{pos}, and we do not know the inclination of **B** to the line of sight. Direct measurement of an increase in M/Φ from envelope to core would strongly support the ambipolar diffusion model. However, present data do not allow one to do this.

Other prestellar cores have recently been mapped in submillimeter polarization. *J. Kirk et al.* (2006) mapped two cores, L1498 and L1517B. They measured the magnetic field strength by the CF method and estimated both the (nonmagnetic) virial mass and the magnetic critical mass. In both cases they found the prestellar cores to be supercritical by a factor of ~2–3. For comparison the three cores of *Crutcher et al.* (2004) were also seen to be mildly supercritical, as predicted by the ambipolar diffusion model.

However, when *J. Kirk et al.* (2006) calculated the magnetic virial mass (i.e., including the effects of both magnetic fields and turbulent line-widths) they found the cores to be roughly virialized, with the magnetic field providing less than half the support [less than was the case for the three cores studied by *Crutcher et al.* (2004)].

Hence we see that for the five prestellar cores whose magnetic fields have been studied in detail, both turbulence and magnetic fields are seen to be playing a role in the support against gravitational collapse, and thus in the evolution of the cores (*J. Kirk et al.*, 2006). These cores are all relatively isolated and moderately quiescent cases, and all give a similar result. Thus we may perhaps conclude that for isolated star formation in fairly quiescent molecular clouds, one must consider the influences of mildly turbulent motions and magnetic fields together. Finally, we note that similar submillimeter dust polarization results have been ob-

tained for a number of Class 0 protostellar cores (e.g., *Matthews and Wilson*, 2002; *Wolf et al.*, 2003).

5. CORE MASS FUNCTION

The last seven years since Protostars and Planets IV have seen some progress in measuring the mass function of cold cores in molecular clouds with a wide range of intrinsic mass scales. This progress has been made possible by the availability of new, sensitive cameras at millimeter and submillimeter wavelengths (e.g., *Kreysa et al.*, 1999; *Holland et al.*, 1999).

Wide-area millimeter continuum mapping of the Ophiuchus molecular cloud (*Motte et al.*, 1998) first revealed a core mass function that bears a striking similarity to the stellar initial mass function (IMF) (e.g., *Kroupa*, 2002; *Chabrier*, 2003). This was subsequently confirmed by others: e.g., in Serpens (*Testi and Sargent*, 1998), ρ Oph (*Johnstone et al.*, 2000; *Reid and Wilson*, 2006; *Stanke et al.*, 2006), and Orion (*Motte et al.*, 2001; *Johnstone et al.*, 2001, 2006; *Nutter*, 2004; *Reid and Wilson*, 2006). Figure 4 shows the core mass functions for ρ Oph and Orion (*Reid and Wilson*, 2006), based on the original results of *Motte et al.* (1998) and *Johnstone et al.* (2000).

In all these regions the slope of the cumulative core mass function above 0.5–1 $M_\odot$ is –1.0 to –1.5 (see Fig. 4), in good agreement with the high-mass slope of –1.35 for the stellar IMF (*Salpeter*, 1955). The core mass function is observed to have a shallower slope at smaller masses, although it has been questioned as to whether the change in slope at lower masses is an intrinsic property of the clump mass function or is caused by some kind of incompleteness in the observations (e.g., *Johnstone et al.*, 2000, 2001).

In addition, the peak of the core mass function in each of these cluster-forming regions lies in the range of 0.2–1 $M_\odot$ (in dN/dlogM format), only slightly larger than the peaks of the mass functions of ~0.08 $M_\odot$ for single stars and ~0.2 $M_\odot$ for multiple systems (*Chabrier*, 2003). In short, both the shape and the intrinsic scale of the core mass function in these regions appear to be well-matched to the observed properties of the stellar IMF.

Some similar work on more distant, higher-mass regions has also been carried out (e.g., *Tothill et al.*, 2002; *Motte et al.*, 2003; *Mookerjea et al.*, 2004; *Beuther and Schilke*, 2004; *Reid and Wilson*, 2005, 2006), although this is strictly beyond the scope of this chapter on low-mass cores. In addition, these studies suffer from problems: e.g., a cluster of low-mass cores can appear, in these more distant regions, to merge into a single higher-mass core; any incompleteness in the mass function will set in at relatively higher masses; and most of these studies make no distinction between starless cores and those with protostars.

It has been suggested that the fact that the shape of the core mass function does not appear to vary from region to region, even as its intrinsic scale is changing, appears to be consistent with the core mass function being determined primarily by turbulent fragmentation (e.g., *Reid*, 2005). How-

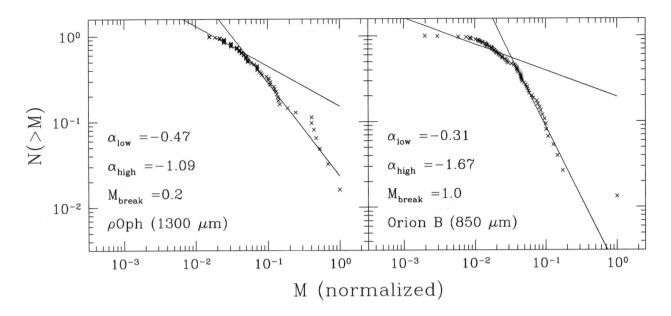

Fig. 4. Plot of normalized core mass function for the ρ Oph and Orion molecular clouds [adapted from *Reid* (2005); based on original data from *Motte et al.* (1998) and *Johnstone et al.* (2000)]. The two slopes and break-point mass of double power-law fits are given in each panel. The break-point masses are quoted in $M_{\odot}$. Note the similarity to the stellar initial mass function (e.g., *Kroupa*, 2002).

ever, this result is subject to the caveats mentioned above. Nonetheless, numerical simulations by several groups have shown that turbulent fragmentation can produce clump mass functions whose shape does not depend strongly on the intrinsic mass scale of the region (*Klessen et al.*, 1998; *Klessen and Burkert*, 2000; *Klessen*, 2001a; *Padoan and Nordlund*, 2002; *Gammie et al.*, 2003; *Tilley and Pudritz*, 2004).

6. CLUSTER-FORMING VS. ISOLATED CORES

The environment in which a core forms is crucial to its subsequent evolution. This has been known for some time. The sequential model of star formation (*Lada*, 1987) predicts that where young stars have already formed, their combined effects will cause further star formation in the remainder of the molecular cloud. This was seen, for example, in the ρ Oph molecular cloud, where *Loren* (1989) hypothesized that the upper Sco OB association was triggering star formation in L1688. Further evidence in support of this hypothesis was provided by a comparison of the relative star-formation activity in L1688 and L1689 (*Nutter et al.*, 2006), wherein these two adjacent clouds were seen to have very different levels of star formation due to L1689 being further from the OB association.

Furthermore, the dense cores that are observed on an ~0.1 pc scale in nearby cluster-forming clouds using classical high-density tracers, such as NH_3, N_2H^+, $H^{13}CO^+$, DCO^+, $C^{18}O$, and dust continuum emission, tend to have higher masses and column densities than isolated prestellar cores (e.g., *Jijina et al.*, 1999). Figure 5 shows a comparison between the Taurus and central Ophiuchus star-form-

ing regions. It can be seen that the region occupied by a typical single prestellar core in Taurus plays host to a small cluster in Ophiuchus. Moreover, the level of cluster-forming activity in a core clearly correlates with core mass and column density (e.g., *Aoyama et al.*, 2001).

High column-density, cluster-forming cores are typically fragmented and show a great deal of substructure (see Fig. 5a). Submillimeter dust continuum mapping of the ρ Oph, Serpens, and Orion B cluster-forming cores has revealed a wealth of compact starless and prestellar cores (e.g., *Motte et al.*, 1998, 2001; *Johnstone et al.*, 2000, 2001; *Kaas et al.*, 2004; *Testi and Sargent*, 1998), which appear to be the direct precursors of individual stars or systems. In particular, their mass distribution is remarkably similar to the stellar IMF (see section 5 above).

These prestellar cores in clusters are denser ($\langle n \rangle \gtrsim 10^6$–$10^7$ cm^{-3}), more compact (diameter D ~ 0.02–0.03 pc), and more closely spaced (L ~ 0.03 pc) than isolated prestellar cores, such as those seen in Taurus, which typically have $\langle n \rangle \gtrsim 10^5$ cm^{-3}, D ~ 0.1 pc, and L ~ 0.25 pc (e.g., *Onishi et al.*, 2002; previous chapter by Di Francesco et al.).

We define the local star-forming efficiency (SFE$_{pre}$) associated with a prestellar core as

$$\mathrm{SFE}_{pre} = \frac{M_*}{M_{pre}}$$

where M_{pre} is the initial mass of a prestellar core that forms a star of mass M_*. We find that the star-formation efficiency within prestellar cores in cluster-forming regions is high. Most of their initial mass at the onset of collapse appears to

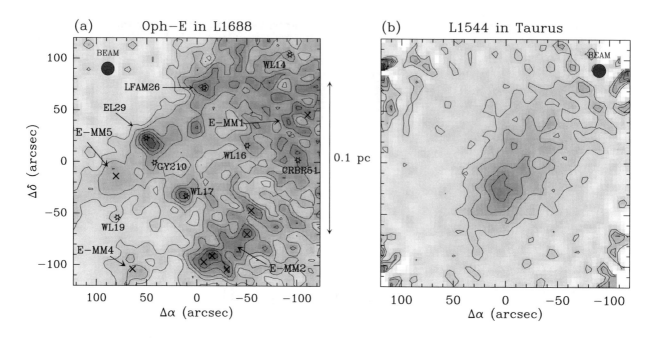

Fig. 5. Millimeter dust continuum images of **(a)** ρ Oph E and **(b)** Taurus taken at the same resolution with the IRAM 30-m telescope (from *Motte et al.*, 1998, and *Ward-Thompson et al.*, 1999, respectively). Note how the cluster forming the ρ Oph core shows far more substructure than the more isolated Taurus prestellar core at the same linear scale. In **(a)** the crosses mark starless cores and the stars mark protostellar objects.

end up in a star or stellar system: $SFE_{pre} \geq 50\%$ (cf. *Motte et al.*, 1998; *Bontemps et al.*, 2001). This contrasts with the lower (~15%) local star-formation efficiency associated with the isolated prestellar cores in Taurus (*Onishi et al.*, 2002).

Interestingly, extensive searches for cores in the Ophiuchus and Pipe Nebula complexes (e.g., *Onishi et al.*, 1999; *Tachihara et al.*, 2000; *Johnstone et al.*, 2004; *Nutter et al.*, 2006) suggest that cluster-forming cores and prestellar cores can only form in a very small fraction of the volume of any given molecular cloud complex, typically at a compressed extremity.

Recent analysis of the physical conditions within cluster-forming molecular clouds has revealed an apparent extinction threshold criterion. *Johnstone et al.* (2004) noted that in Ophiuchus almost all cores were located in high extinction regions ($A_V > 10$), despite the fact that most of the cloud mass was found at much lower extinctions (cf. Fig. 1).

Analysis of the Perseus cloud (*Kirk*, 2005; *H. Kirk et al.*, 2006; *Enoch et al.*, 2006) reveals a similar extinction threshold, although at a somewhat lower value ($A_V > 5$). These results are in agreement with the analysis of Taurus using $C^{18}O$ by *Onishi et al.* (1998), who found that only regions with column densities above $N(H_2) = 8 \times 10^{21}$ cm^{-2} ($A_V \sim 8$) contained IRAS sources, indicating that high column density is necessary for star formation.

These observations are in fact consistent with the idea that magnetic fields play an important role in supporting molecular clouds (*McKee*, 1989). The outer region of the cloud is maintained at a higher fractional ionization level by ultraviolet photons from the interstellar radiation field.

However, the ultraviolet photons cannot penetrate deep into the cloud due to extinction.

The ionization fraction thus drops in the inner region, shortening the ambipolar diffusion timescale. According to this scenario, one expects small-scale structure and star formation to proceed only in the inner, denser regions of the cloud. McKee estimates the column density depth required for sufficient ultraviolet attenuation to be in the region of $A_V \sim 4$–8.

Column density and line width are among the key parameters for a core in determining its evolution. Figure 6 plots the column density $N(H_2)$ vs. the nonthermal velocity dispersion σ_{NT}, measured in the N_2H^+ line, for a large number of cores. The open circles are isolated prestellar cores. The open triangles and pentagons are cluster-forming cores. The filled triangles are prestellar cores in cluster-forming regions.

It is seen that the cluster-forming cores of, for example, Ophiuchus, Serpens, Perseus, and Orion have line widths dominated by nonthermal motions (e.g., *Jijina et al.*, 1999; *Aso et al.*, 2000). They are significantly more turbulent than the more isolated prestellar cores of Taurus, whose line widths are dominated by thermal motions (e.g., *Tatematsu et al.*, 2004; *Benson and Myers*, 1989). In plotting Fig. 6 we have removed the thermal velocities and plot only the nonthermal velocity dispersions.

However, the more compact (~0.03 pc) prestellar cores observed within cluster-forming regions are characterized by fairly narrow $N_2H^+(1-0)$ line widths ($\Delta V_{FWHM} \leq 0.5$ km s^{-1}), more reminiscent of the isolated prestellar cores of Tau-

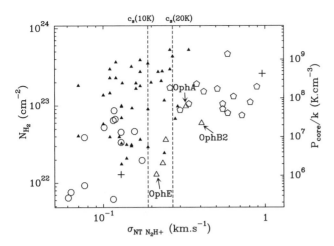

Fig. 6. Plot of column density against nonthermal velocity dispersion in low-mass dense cores (following *André et al., 2006*). The two vertical dashed lines represent the sound speeds for 10 K and 20 K gas respectively. The open circles represent the isolated prestellar cores in Taurus. The large, open triangles are the cluster-forming cores in ρ Oph. The small, filled triangles are the prestellar cores in ρ Oph (*Motte et al., 1998; André et al., 2006*). The large, open pentagons are the cluster-forming cores in NGC 2264D (*Peretto et al., 2006*). The two crosses are two equilibrium models representative of quasistatic scenarios for low-mass (lower left) and high-mass (upper right) star formation (cf. *Shu et al., 1987,* and *McKee and Tan, 2003,* respectively). Note the broad trend seen in the open symbols, and that cluster-forming cores lie on the right-hand (supersonic) side of the figure, while prestellar cores lie on the lefthand (subsonic) side. Note also that most of the prestellar cores in the cluster-forming region of ρ Oph lie above the "sequence" observed in the open polygons, at up to an order of magnitude higher densities than isolated prestellar cores.

rus (*Belloche et al., 2001*). This indicates subsonic or at most transonic levels of internal turbulence and suggests that, even in cluster-forming clouds, the initial conditions for individual protostellar collapse are relatively free of supersonic turbulence. It can be seen from Fig. 6 that the nonthermal velocity dispersion, measured toward the prestellar cores of the ρ Oph cluster, is only a fraction (typically 0.5–1) of the isothermal sound speed (*Belloche et al., 2001; André et al., 2006*).

The narrow N_2H^+ line widths measured for the prestellar cores in these clusters imply virial masses that generally agree well with the mass estimates derived from the dust continuum. This confirms that most of the starless cores identified in the submillimeter dust continuum are self-gravitating and very likely prestellar in nature.

Furthermore, Fig. 6 appears to show that the prestellar cores in a cluster-forming region such as ρ Oph (filled triangles) largely occupy a different parameter space from the isolated prestellar cores of Taurus (open circles). This tends to imply a different formation mechanism for prestellar cores in isolated and clustered regions, with the latter forming by fragmentation of higher-mass, more turbulent, clus-

ter-forming cores (cf. *Myers, 1998*). For this reason (following *Motte et al., 2001*), we suggest that prestellar cores in clustered regions could perhaps be called prestellar condensations to indicate this difference.

In addition, there appears to be a broad trend of increasing velocity dispersion with increasing column density from isolated prestellar cores to cluster-forming cores (cf. *Larson, 1981*). This perhaps reflects the fact that all these cores are self-gravitating, hence characterized by virial mass ratios close to unity. Higher-density cores form in clustered regions, where the ambient pressure is higher, and subsequently fragment before forming stars.

One of the clear signposts of star formation is direct detection of infall. Detection of blue infall profiles (see previous chapter by Di Francesco et al.) in optically thick line tracers such as $HCO^+(3–2)$ toward a number of starless cores in cluster-forming regions [e.g., OphE-MM2 in ρ Oph (*Belloche et al., 2001*)] suggests that some of them are in fact already collapsing and on the verge of forming protostars.

Prestellar cores in low-mass protoclusters also appear to be characterized by small core-core relative motions (e.g., *Walsh et al., 2004*). For instance, based on the observed distribution of $N_2H^+(1–0)$ line-of-sight velocities, a global, one-dimensional velocity dispersion σ_{1D} of <0.4 km s^{-1} was found for the cores of the ρ Oph cluster (*Belloche et al., 2001; André et al., 2006*).

7. FROM CORES TO PROTOSTARS

Most of the starless cores and all the prestellar cores that we have been discussing are expected to evolve into Class 0 protostars (*André et al., 1993, 2000*) and subsequently into Class I protostars (*Lada, 1987; Wilking et al., 1989*). Therefore, another approach to constraining the initial conditions for protostellar collapse consists of studying the structure of young Class 0 protostars. These objects are observed early enough after point mass formation that they still retain some memory of their initial conditions (cf. *André et al., 2000*).

By comparing the properties of prestellar cores with those of Class 0 cores, we can hope to bracket the physical conditions at point mass formation. Furthermore, since Class 0 objects have already begun to form stars at their centers, we can be sure that they are participating in the star-formation process (which is not certain for all starless cores). In fact, in some cases it is difficult to differentiate between the most centrally condensed prestellar cores and the youngest Class 0 protostars. Examples of low-luminosity, very young Class 0 protostars that look like prestellar cores include L1014 (*Young et al., 2004; Bourke et al., 2005*) and MC27/L1521F (*Onishi et al., 1999*) (see also previous chapter by Di Francesco et al.).

When discussing the density and velocity structure of Class 0 envelopes, we here again contrast isolated and clustered objects. In terms of their density profiles, protostellar envelopes are found to be more strongly centrally con-

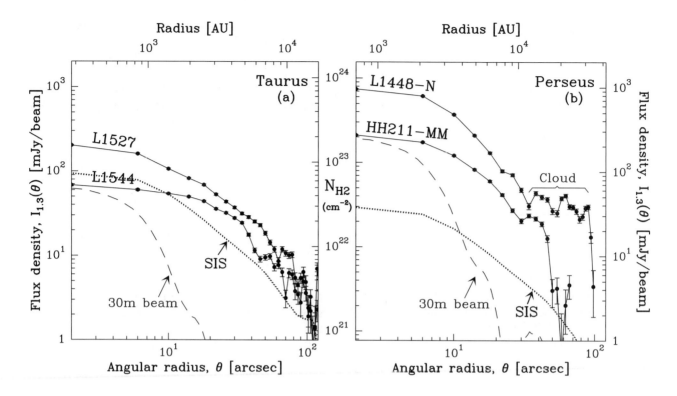

Fig. 7. Radial density profiles in Taurus (left) and Perseus (right), from *Motte and André* (2001). In Taurus we show a prestellar core, L1544, and a Class 0 protostar, L1527. In Perseus we show two Class 0 protostars, L1448-N and HH211-MM. The dotted line marked SIS shows the initial conditions for spontaneous collapse (e.g., *Shu et al.,* 1987) convolved with the beam (dashed line). Note that the prestellar core has a flatter profile than the protostellar cores and that the column densities in Perseus are an order of magnitude higher than both the model and the Taurus cores.

densed than prestellar cores, and do not exhibit any marked inner flattening in their radial column density profiles (see Fig. 7), unlike prestellar cores (*Kirk et al.,* 2005).

In regions of isolated star formation such as Taurus, protostellar envelopes have radial density gradients consistent with $\rho(r) \propto r^{-p}$ with $p \sim 1.5$–2 over ~10,000–15,000 AU in radius (e.g., *Chandler and Richer,* 2000; *Hogerheijde and Sandell,* 2000; *Shirley et al.,* 2000, 2003; *Motte and André,* 2001). Furthermore, the absolute level of the density distributions observed toward Taurus Class 0 sources is roughly consistent with the predictions of spontaneous collapse models (see Fig. 7) starting from quasi-equilibrium, thermally dominated prestellar cores (e.g., *Hennebelle et al.,* 2003).

By contrast, in cluster-forming regions such as Serpens, Perseus, or ρ Oph, Class 0 envelopes are clearly not scale-free. They merge either with other cores or other protostellar envelopes, or the ambient cloud, at a finite radius $R_{out} \lesssim$ 5000 AU (*Motte et al.,* 1998; *Looney et al.,* 2003). They are also typically an order of magnitude more dense than models of the spontaneous collapse of isothermal (e.g., Bonnor-Ebert) spheres predict immediately after point mass formation (cf. *Motte and André,* 2001) (see Fig. 7).

Turning to velocity profiles, the surrounding environment can play an important role in the mass infall rate, since in a clustered environment this can vary strongly even for proto-

stars with similar final masses (*Klessen,* 2001b). Models suggest that the mass infall rate may be a strongly varying function of time, with a peak infall rate occurring in the Class 0 stage (e.g., *Henriksen et al.,* 1997; *Whitworth and Ward-Thompson,* 2001; *Schmeja and Klessen,* 2004). The mean mass infall rate is also predicted to decrease as the Mach number of the turbulence increases (*Schmeja and Klessen,* 2004). In addition, the relative importance of turbulent and gravitational energy can change the number of binary systems that are formed as well as their properties, such as semimajor axis and mass ratio (*Goodwin et al.,* 2004).

There have been few detailed studies of velocity profiles, but one example of an isolated Class 0 object is IRAM 04191 in Taurus (see Fig. 8) (*Belloche et al.,* 2002). In this case, the inner part of the envelope ($r \lesssim$ 2000–4000 AU) is rapidly collapsing and rotating, while the outer part (4000 $\lesssim$ $r \lesssim$ 11,000 AU) undergoes only moderate infall and slower rotation. This dramatic drop in rotational velocity beyond $r \sim$ 4000 AU, combined with the flat infall velocity profile, suggests that angular momentum is conserved in the collapsing inner envelope but efficiently dissipated, perhaps due to magnetic braking, in the slowly contracting outer envelope.

The mass infall rate of IRAM 04191 is estimated to be $\dot{M}_{inf} \sim 3 \times 10^{-6}$ $M_\odot$ yr^{-1}, which is ~2–3 times the canonical a_s^3/G value often used (where $a_s \sim$ 0.15–0.2 km s^{-1} is the iso-

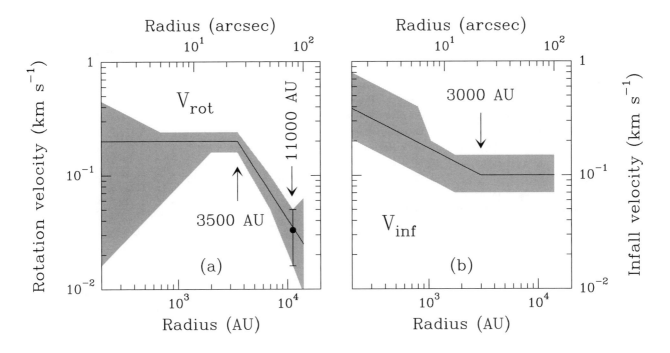

Fig. 8. Velocity profiles in the Class 0 protostellar core IRAM04191. Rotational velocity (left) and infall velocity (right) are plotted as a function of radius (from *Belloche et al.*, 2002). The inner part of the envelope is rapidly collapsing and rotating, while the outer part undergoes only moderate infall and slower rotation.

thermal sound speed). Similar $\dot{M}_{inf}$ values have been reported for several other bona-fide Class 0 and Class I protostars in Taurus (e.g., *Ohashi*, 1999; *Hirano et al.*, 2002).

A very different example in a clustered region is IRAS 4A in the NGC 1333 protocluster. *Di Francesco et al.* (2001) observed inverse P-Cygni profiles toward IRAS 4A, from which they derived a very large mass infall rate ~1.1 × 10⁻⁴ $M_\odot$ yr⁻¹ at r ~ 2000 AU. A similar infall rate was independently found by *Maret et al.* (2002).

This value of $\dot{M}_{inf}$ corresponds to more than ~15× the canonical a_{eff}^3/G value (where $a_{eff} \lesssim 0.3$ km s⁻¹ is the effective sound speed). This high infall rate results both from a very dense envelope and a large, supersonic infall velocity: ~0.68 km s⁻¹ at ~2000 AU (*Di Francesco et al.*, 2001).

Other examples of Class 0 protostars in cluster-forming regions with quantitative estimates of the mass infall rate include NGC 1333-IRAS 2, Serpens-SMM4, and IRAS 16293. In all these objects, high $\dot{M}_{inf}$ values ≳3 × 10⁻⁵ $M_\odot$ yr⁻¹ are found (e.g., *Ceccarelli et al.*, 2000; *Ward-Thompson and Buckley*, 2001).

The velocity structures of prestellar cores have also been studied in some cases. The isolated prestellar core L1544 has been seen to have a "flat" velocity profile over a wide range of radii, with no evidence for velocity increasing toward the center (*Tafalla et al.*, 1998; *Williams et al.*, 1999). Infall profiles have also been observed in a number of other prestellar cores at large radii (e.g., *Lee et al.*, 1999; *Gregersen and Evans*, 2000) and it seems that a significant number may already be contracting (see previous chapter by Di Francesco et al.).

The observational constraints summarized above have strong implications for collapse models. First, the extended infall velocity profiles observed in prestellar cores and young Class 0 objects are inconsistent with pure inside-out collapse and in better agreement with isothermal collapse models starting from Bonnor-Ebert spheres (e.g., *Whitworth and Summers*, 1985; *Foster and Chevalier*, 1993), or similar density profiles (e.g., *Whitworth and Ward-Thompson*, 2001).

For isolated cores, the fact that the measured infall velocities are subsonic and that there is indirect evidence of magnetic braking (*Belloche et al.*, 2002) (see above) suggests that the collapse is spontaneous and moderated by magnetic effects in mildly magnetized versions of Bonnor-Ebert cloudlets (cf. *Basu and Mouschovias*, 1994). In Taurus, the measured infall rates seem to rule out models based on competitive accretion (e.g., *Bonnell et al.*, 2001) or gravo-turbulent fragmentation (e.g., *Schmeja and Klessen*, 2004) that predict large time and spatial variations of $\dot{M}_{inf}$.

By contrast, in protoclusters such as NGC 1333 or ρ Oph, the large overdensity factors measured in Class 0 envelopes compared to hydrostatic isothermal structures, as well as the supersonic infall velocities and very high infall rates observed in some cases, are inconsistent with self-initiated forms of collapse and require strong external compression.

This point is supported by recent numerical simulations of the collapse of Bonnor-Ebert spheres (*Hennebelle et al.*, 2003, 2004), which show that large overdensity factors, together with supersonic infall velocities, and high infall rates (≳10 a_s^3/G) are produced near point mass formation when, and only when, the collapse is induced by a strong and rapid increase in external pressure (see also *Motoyama and Yoshida,* 2003).

The high infall rates at the Class 0 stage, as well as the strong decline of $\dot{M}_{inf}$ observed between the Class 0 and the Class I stage in clusters (e.g., *Henriksen et al.,* 1997; *Whitworth and Ward-Thompson,* 2001), can also be reproduced in the context of the turbulent fragmentation picture (cf. *Schmeja and Klessen,* 2004), according to which dense cores form by strong turbulent compression (e.g., *Padoan and Nordlund,* 2002).

8. DISCUSSION AND CONCLUSIONS

We have presented observational results that bear on the evolution of dense low-mass cores in an endeavor to estimate which aspects of the continuum of models discussed in section 2 relate to the different environments of star formation that we observe. The formation and evolution of cores is crucial to an understanding of the star-formation process, not least because the results presented in section 5 indicate that the core mass function has a very strong bearing on the stellar IMF. The results summarized in section 6 help to discriminate between possible theoretical scenarios for the formation and evolution of isolated cores compared to cluster-forming cores.

The narrow line widths observed in prestellar cores in cluster-forming regions are in qualitative agreement with the picture according to which such cores form by dissipation of internal MHD turbulence (e.g., *Nakano,* 1998) (cf. Fig. 6). These cores may correspond to self-gravitating "kernels," of size comparable to the cutoff wavelength for MHD waves (e.g., *Kulsrud and Pearce,* 1969), that can develop only in turbulent cloud cores (e.g., *Myers,* 1998).

However, at variance with this picture, we see that some cluster-forming cores such as ρ Oph E (*Belloche et al.,* 2001) also exhibit narrow line widths (see Fig. 6), similar to those of the prestellar cores within them. This tends to suggest that spontaneous dissipation of internal MHD turbulence may not be the only mechanism responsible for core fragmentation. In an alternative view, the formation of cluster-forming cores may primarily reflect the action of a strong external trigger at the head of elongated, head-tail cloud structures (e.g., *Tachihara et al.,* 2002; *Nutter et al.,* 2006).

A marked increase in external pressure resulting from the propagation of neighboring stellar winds and/or supernova shells into a cloud can indeed significantly reduce the critical Bonnor-Ebert mass and the corresponding Jeans fragmentation length scale (cf. *Nakano,* 1998). It may also trigger protostellar collapse (e.g., *Boss,* 1995) and account for the enhanced infall rates observed at the Class 0 stage in cluster-forming clouds (see section 7).

Furthermore, the small velocity dispersions measured for the prestellar cores in the ρ Oph cluster imply a crossing time, $\sim 2 \times 10^6$ yr (*Belloche et al.,* 2001; *André et al.,* 2006), that is larger than the estimated core lifetime ($< 2.5 \times 10^5$ yr; see section 3). This suggests that typical prestellar cores in

clusters do not have time to interact with one another before collapsing to protostars. Taken at face value, this seems inconsistent with models that resort to dynamical interactions to build up a mass spectrum comparable to the IMF (e.g., *Bate et al.,* 2003; *Bonnell et al.,* 2003). Nonetheless, these models may still be relevant in higher-mass star-forming regions (cf. *Peretto et al.,* 2006).

Therefore, it appears that the influence of the external environment plays a crucial role in the formation and evolution of low-mass dense cores. An isolated, low-density, quiescent environment will most likely lead to a more quasi-static evolution. A clustered, dense environment in which the external pressure is increased by the action of nearby, newly formed stars, will probably yield a more dynamic evolutionary scenario.

The fact that most isolated prestellar cores appear to be within a factor of a few of magnetic criticality suggests that the magnetic field is playing an important role and is consistent with the ambipolar diffusion picture (see section 4). However, whether or not this role is dominant depends on the balance between the field strength and the other environmental factors.

9. FUTURE DEVELOPMENTS

There are many exciting developments in telescopes and instrumentation planned in the next few years that will impact this field. These include new, more sensitive cameras for single-dish telescopes, such as SCUBA-2 on the James Clerk Maxwell Telescope (JCMT) as well as SPIRE and PACS on the Herschel Space Observatory, and new interferometers such as the Combined Array for Research in Millimeter-wave Astronomy (CARMA) and the Atacama Large Millimeter Array (ALMA).

The new submillimeter cameras on JCMT and Herschel will increase our wide-area mapping coverage, so that, for example, SCUBA-2 and SPIRE will map almost all star-forming regions within 0.5 kpc. These observations will produce a flux-limited, multiwavelength snapshot of star formation near the Sun, providing a legacy of images, as well as point-source and extended-source catalogs, covering up to 700 square degrees of sky.

On small scales, the Herschel observations will, for the first time, resolve the detailed dust temperature structure of the nearest isolated prestellar cores. On global molecular cloud scales, the large spatial dynamic range of the Herschel images will provide a unique view of the formation of both isolated prestellar cores and cluster-forming cores.

CARMA will bring improved angular resolution (~ 0.13 arcsec) and sensitivity — $\sim 25\times$ better than the Berkeley Illinois Maryland Array (BIMA) — to millimeter-wave polarization studies of the dust and molecular line emission in dense clouds and lead to routine high-resolution polarization mapping. These instrumental gains will enable Zeeman mapping of the CN J = (1 → 0) transition and measurement of line-of-sight magnetic field strengths at densities $n(H_2) \sim 10^{5-6}$ cm^{-3} and mapping of dust and CO linearly polarized emission toward both high-mass and low-mass star-formation regions.

ALMA will have more than an order of magnitude greater sensitivity and resolution compared to existing millimeter arrays. Observations with ALMA will allow us to probe the nearby star-forming regions discussed in this chapter on spatial scales of a few tens of AU, while allowing these types of analyses to be extended to the more distant regions of high-mass star formation.

ALMA's broad wavelength coverage and flexible spectrometer will allow detailed studies of cores throughout the submillimeter windows, while its dual polarization receivers will allow sensitive high-resolution observations of magnetic field signatures, both with polarization and with Zeeman observations.

Perhaps these instrumental advances will have helped to answer some of the questions raised in this chapter in time for Protostars and Planets VI.

Acknowledgments. This work was carried out while D.W.-T. was on sabbatical at the Observatoire de Bordeaux and CEA Saclay, and he would like to thank both institutions for their hospitality. R.M.C. received partial research support under grants NSF AST 02-28953 and NSF AST 02-05810. The research of D.J. is supported through a grant from the Natural Sciences and Engineering Council of Canada, held at the University of Victoria. This work was partially supported by a Discovery grant to C.W. from the Natural Sciences and Engineering Research Council of Canada. H. Kirk, D. Nutter, and M. Reid are thanked for assistance in producing some of the figures for this chapter. The referee is thanked for helpful comments on the manuscript.

REFERENCES

Aikawa Y., Herbst E., Roberts H., and Caselli P. (2005) *Astrophys. J., 620,* 330–346.

André P., Ward-Thompson D., and Barsony M. (1993) *Astrophys. J., 406,* 122–141.

André P., Ward-Thompson D., and Barsony M. (2000) In *Protostars and Planets IV* (V. Mannings et al., eds.), pp. 59–96. Univ. of Arizona, Tucson.

André P., Belloche A., Motte F., and Peretto N. (2006) *Astron. Astrophys.,* in press.

Aoyama H., Mizuno N., Yamamoto H., Onishi T., Mizuno A., and Fukui Y. (2001) *Publ. Astron. Soc. Japan, 53,* 1053–1062.

Aso Y., Tatematsu K., Sekimoto Y., Nakano T., Umemoto T., et al. (2000) *Astrophys. J. Suppl., 131,* 465–482.

Ballesteros-Paredes J., Klessen R. S., and Vazquez-Semadeni E. (2003) *Astrophys. J., 592,* 188–202.

Basu S. (2000) *Astrophys. J., 540,* L103–L106.

Basu S. and Mouschovias T. Ch. (1994) *Astrophys. J., 432,* 720–741.

Bate M. R., Bonnell I. A., and Bromm V. (2003) *Mon. Not. R. Astron. Soc., 339,* 577–599.

Beichman C. A., Myers P. C., Emerson J. P., Harris S., Mathieu R., Benson P. J., and Jennings R. E. (1986) *Astrophys. J., 307,* 337–349.

Belloche A., André P., and Motte F. (2001) In *From Darkness to Light: Origin and Evolution of Young Stellar Clusters* (T. Montmerle and P. André, eds.), pp. 313–318. ASP Conf. Series 243, San Francisco.

Belloche A., André P., Despois D., and Blinder S. (2002) *Astron. Astrophys., 393,* 927–947.

Benson P. J. and Myers P. C. (1989) *Astrophys. J. Suppl., 71,* 89–108.

Beuther H. and Schilke P. (2004) *Science, 303,* 1167–1169.

Bonnell I. A., Bate M. R., Clarke C. J., and Pringle J. E. (2001) *Mon. Not. R. Astron. Soc., 323,* 785–794.

Bonnell I. A., Bate M. R., and Vine S. G. (2003) *Mon. Not. R. Astron. Soc., 343,* 413–418.

Bontemps S., André P., Kaas A. A., Nordh L., Olofsson G., et al. (2001) *Astron. Astrophys., 372,* 173–194.

Boss A. P. (1995) *Astrophys. J., 439,* 224–236.

Bourke T. L., Hyland A. R., and Robinson G. (1995a) *Mon. Not. R. Astron. Soc., 276,* 1052–1066.

Bourke T. L., Hyland A. R., Robinson G., James S. D., and Wright C. M. (1995b) *Mon. Not. R. Astron. Soc., 276,* 1067–1084.

Bourke T. L., Crapsi A., Myers P. C., Evans N. J., Wilner D. J., et al. (2005) *Astrophys. J., 633,* L129–L132.

Caselli P., Benson P. J., Myers P. C., and Tafalla M. (2002) *Astrophys. J., 572,* 238–263.

Ceccarelli C., Castets A., Caux E., Hollenbach D., Loinard L., et al. (2000) *Astron. Astrophys., 355,* 1129–1137.

Chabrier G. (2003) *Publ. Astron. Soc. Pac., 115,* 763–795.

Chandler C. J. and Richer J. S. (2000) *Astrophys. J., 530,* 851–866.

Chandrasekhar S. and Fermi E. (1953) *Astrophys. J., 118,* 113–115.

Ciolek G. E. and Basu S. (2001) *Astrophys. J., 547,* 272–279.

Clemens D. P. and Barvainis R. (1988) *Astrophys. J. Suppl., 68,* 257–286.

Crutcher R. M. (1999) *Astrophys. J., 520,* 706–713.

Crutcher R.M. and Troland T. H. (2000) *Astrophys. J., 537,* L139–L142.

Crutcher R. M., Troland T. H., Goodman A. A., Heiles C., Kazés I., and Myers P. C. (1993) *Astrophys. J., 407,* 175–184.

Crutcher R. M., Nutter D., Ward-Thompson D., and Kirk J. M. (2004) *Astrophys. J., 600,* 279–285.

Di Francesco J., Myers P. C., Wilner D. J., and Ohashi N. (2001) *Astrophys. J., 562,* 770–789.

Elmegreen B. G. (2000) *Astrophys. J., 530,* 277–281.

Enoch M. L., Young K. E., Glenn J., Evans N. J., Golwala S., et al. (2006) *Astrophys. J., 638,* 293–313.

Foster P. N. and Chevalier R. A. (1993) *Astrophys. J., 416,* 303–311.

Gammie C. F., Lin Y., Stone J. M., and Ostriker E. C. (2003) *Astrophys. J., 592,* 203–216.

Goldsmith P. F. and Li D. (2005) *Astrophys. J., 622,* 938–958.

Goodwin S. P., Whitworth A. P., and Ward-Thompson D. (2004) *Astron. Astrophys., 423,* 169–182.

Greene T. P., Wilking B. A., André P., Young E. T., and Lada C. J. (1994) *Astrophys. J., 434,* 614–626.

Gregersen E. M. and Evans N. J. (2000) *Astrophys. J., 538,* 260–267.

Hartmann L., Ballesteros-Paredes J., and Bergin E. A. (2001) *Astrophys. J., 562,* 852–868.

Heiles C. and Crutcher R. M. (2005) In *Cosmic Magnetic Fields* (R. Wielebinski and R. Beck, eds.), pp. 137–183. Lecture Notes in Physics Vol. 664, Springer, Heidelberg.

Hennebelle P.,Whitworth A. P., Gladwin P. P., and André P. (2003) *Mon. Not. R. Astron. Soc., 340,* 870–882.

Hennebelle P., Whitworth A. P., Cha S., and Goodwin S. (2004) *Mon. Not. R. Astron. Soc., 348,* 687–701.

Henriksen R., André P., and Bontemps S. (1997) *Astron. Astrophys., 323,* 549–565.

Hirano N., Ohashi N., and Dobashi K. (2002) In *8th Asian-Pacific Regional Meeting* (S. Ikeuchi et al., eds.), pp. 141–142. Astronomical Society of Japan, Tokyo.

Hogerheijde M. R. and Sandell G. (2000) *Astrophys. J., 534,* 880–893.

Holland W. S., Robson E. I., Gear W. K., Cunningham C. R., Lightfoot J. F., et al. (1999) *Mon. Not. R. Astron. Soc., 303,* 659–672.

Jessop N. E. and Ward-Thompson D. (2000) *Mon. Not. R. Astron. Soc., 311,* 63–74.

Jijina J., Myers P. C., and Adams F. C. (1999) *Astrophys. J. Suppl., 125,* 161–236.

Johnstone D., Wilson C. D. Moriarty-Schieven G., Joncas G., Smith G., Gregersen E., and Fich M. (2000) *Astrophys. J., 545,* 327–339.

Johnstone D., Fich M., Mitchell G. F., and Moriarty-Schieven G. (2001) *Astrophys. J., 559,* 307–317.

Johnstone D., Di Francesco J., and Kirk H. (2004) *Astrophys. J., 611,* L45–L48.

Johnstone D., Matthews H., and Mitchell G. F. (2006) *Astrophys. J., 639,* 259–274.

Kaas A. A., Olofsson G., Bontemps S., André P., Nordh L., et al. (2004) *Astron. Astrophys., 421,* 623–642.

Kandori R., Nakajima Y., Tamura M., Tatematsu K., Aikawa Y., et al. (2005) *Astron. J., 130,* 2166–2184.

Kenyon S. J. and Hartmann L. (1995) *Astrophys. J. Suppl., 101,* 117–171.

Kirk H. (2005) M.Sc. thesis, Univ. of Victoria, Canada.

Kirk H., Johnstone D., and Di Francesco J. (2006) *Astrophys. J., 646,* 1009–1023.

Kirk J. M., Ward-Thompson D., and André P. (2005) *Mon. Not. R. Astron. Soc., 360,* 1506–1526.

Kirk J. M., Ward-Thompson D., and Crutcher R. M. (2006) *Mon. Not. R. Astron. Soc., 369,* 1445–1450.

Klessen R. S. (2001a) *Astrophys. J., 550,* L77–L80.

Klessen R. S. (2001b) *Astrophys. J., 556,* 837–846.

Klessen R. S. and Burkert A. (2000) *Astrophys. J. Suppl., 128,* 287–319.

Klessen R. S., Burkert A., and Bate M. R. (1998) *Astrophys. J., 501,* L205–L208.

Kreysa E., Gemuend H. P., Gromke J., Haslam C. G., Reichertz L., et al. (1999) *Soc. Phot. Inst. Eng., 3357,* 319–325.

Kroupa P. (2002) *Science, 295,* 82–91.

Kulsrud R. and Pearce W. P. (1969) *Astrophys. J., 156,* 445–469.

Lada C. (1987) In *Star Forming Regions* (M. Peimbert and J. Jugaku, eds.), pp. 1–17. Reidel, Dordrecht.

Larson R. B. (1981) *Mon. Not. R. Astron. Soc., 194,* 809–826.

Lee C. W. and Myers P. C. (1999) *Astrophys. J. Suppl., 123,* 233–250.

Lee C. W., Myers P. C., and Tafalla M. (1999) *Astrophys. J., 526,* 788–805.

Looney L. W., Mundy L. G., and Welch W. J. (2003) *Astrophys. J., 592,* 255–265.

Loren R. B. (1989) *Astrophys. J., 338,* 902–924.

MacLow M.-M. and Klessen R. S. (2004) *Rev. Mod. Phys., 76,* 125–194.

MacLow M.-M., Klessen R. S., Burkert A., and Smith M. D. (1998) *Phys. Rev. Lett., 80,* 2754–2757.

Maret S., Ceccarelli C., Caux E., Tielens A. G. G. M., and Castets A. (2002) *Astron. Astrophys., 395,* 573–585.

Matthews B. C. and Wilson C. D. (2002) *Astrophys. J., 574,* 822–833.

Matthews B. C., Wilson C. D., and Fiege J. D. (2001) *Astrophys. J., 562,* 400–423.

McKee C. F. (1989) *Astrophys. J., 345,* 782–801.

McKee C. F. and Tan J. (2003) *Astrophys. J., 585,* 850–871.

Mookerjea B., Kramer C., Nielbock M., and Nyman L. A. (2004) *Astron. Astrophys., 426,* 119–129.

Motoyama K. and Yoshida T. (2003) *Mon. Not. R. Astron. Soc., 344,* 461–467.

Motte F. and André P. (2001) *Astron. Astrophys., 365,* 440–464.

Motte F., André P., and Neri R. (1998) *Astron. Astrophys., 336,* 150–172.

Motte F., André P., Ward-Thompson D., and Bontemps S. (2001) *Astron. Astrophys., 372,* L41–L44.

Motte F., Schilke P., and Lis D. C. (2003) *Astrophys. J., 582,* 277–291.

Mouschovias T. Ch. (1991) *Astrophys. J., 373,* 169–186.

Mouschovias T. Ch. and Ciolek G. E. (1999) In *The Origin of Stars and Planetary Systems* (C. J. Lada and N. D. Kylafis, eds.), pp. 305–339. Kluwer, Dordrecht.

Mouschovias T. Ch., Tassis K., and Kunz M. W. (2006) *Astrophys. J., 646,* 1043–1049.

Myers P. C. (1998) *Astrophys. J., 496,* L109–L112.

Myers P. C. (2000) *Astrophys. J., 530,* L119–L122.

Myers P. C., Linke R. A., and Benson P. J. (1983) *Astrophys. J., 264,* 517–537.

Nakano T. (1998) *Astrophys. J., 494,* 587–604.

Nutter D. J. (2004) Ph.D. thesis, Cardiff University.

Nutter D. J., Ward-Thompson D., and André P. (2006) *Mon. Not. R. Astron. Soc., 368,* 1833–1842.

Ohashi N. (1999) In *Star Formation 1999* (T. Nakamoto, ed.), pp. 129–135. Nobeyama Radio Observatory, Nobeyama.

Onishi T., Mizuno A., Kawamura A., Ogawa H., and Fukui Y. (1998) *Astrophys. J., 502,* 296–314.

Onishi T., Kawamura A., Abe R., Yamaguchi N., Saito H., et al. (1999) *Publ. Astron. Soc. Japan, 51,* 871–881.

Onishi T., Mizuno A., Kawamura A., Tachihara K., and Fukui Y. (2002) *Astrophys. J., 575,* 950–973.

Ostriker E. C., Gammie C. F., and Stone J. M. (1999) *Astrophys. J., 513,* 259–274.

Padoan P. and Nordlund A. (2002) *Astrophys. J., 576,* 870–879.

Parker E. N. (1966) *Astrophys. J., 145,* 811–833.

Peretto N., André P., and Belloche A. (2006) *Astron. Astrophys., 445,* 979–998.

Reid M. A. (2005) Ph.D. thesis, McMaster University, Canada.

Reid M. A. and Wilson C. D. (2005) *Astrophys. J., 625,* 891–905.

Reid M. A. and Wilson C. D. (2006) *Astrophys. J., 644,* 990–1005.

Salpeter E. E. (1955) *Astrophys. J., 121,* 161–167.

Schmeja S. and Klessen R. S. (2004) *Astron. Astrophys., 419,* 405–417.

Shirley Y. L., Evans N. J., Rawlings J. M. C., and Gregersen E. M. (2000) *Astrophys. J. Suppl., 131,* 249–271.

Shirley Y. L., Evans N. J., Young K. E., Knez C., and Jaffe D. T. (2003) *Astrophys. J. Suppl., 149,* 375–403.

Shu F. H., Adams F. C., and Lizano S. (1987) *Ann. Rev. Astron. Astrophys., 25,* 23–81.

Stanke T., Smith M. D., Gredel R., and Khanzadyan T. (2006) *Astron. Astrophys., 447,* 609–622.

Tachihara K., Mizuno A., and Fukui Y. (2000) *Astrophys. J., 528,* 817–840.

Tachihara K., Onishi T., Mizuno A., and Fukui Y. (2002) *Astron. Astrophys., 385,* 909–920.

Tafalla M., Mardones D., Myers P. C., Caselli P., Bachiller R., and Benson P. J. (1998) *Astrophys. J., 504,* 900–914.

Tassis K. and Mouschovias T. Ch. (2004) *Astrophys. J., 616,* 283–287.

Tatematsu K., Umemoto T., Kandori R., and Sekimoto Y. (2004) *Astrophys. J., 606,* 333–340.

Testi L. and Sargent A. I. (1998) *Astrophys. J., 508,* L91–L94.

Tilley D. and Pudritz R. E. (2004) *Mon. Not. R. Astron. Soc., 353,* 769–788.

Tothill N. F. H., White G. J., Matthews H. E., McCutcheon W. H., McCaughrean M. J., and Kenworthy M. A. (2002) *Astrophys. J., 580,* 285–304.

Vazquez-Semadeni E., Kim J., Shadmehri M., and Ballesteros-Paredes J. (2005) *Astrophys. J., 618,* 344–359.

Walsh A. J., Myers P. C., and Burton M. G. (2004) *Astrophys. J., 614,* 194–202.

Ward-Thompson D. (2002) *Science, 295,* 76–81.

Ward-Thompson D. and Buckley H. D. (2001) *Mon. Not. R. Astron. Soc., 327,* 955–983.

Ward-Thompson D., Scott P. F., Hills R. E., and André P. (1994) *Mon. Not. R. Astron. Soc., 268,* 276–290.

Ward-Thompson D., Motte F., and André P. (1999) *Mon. Not. R. Astron. Soc., 305,* 143–150.

Ward-Thompson D., Kirk J. M., Crutcher R. M., Greaves J. S., Holland W. S., and André P. (2000) *Astrophys. J., 537,* L135–L138.

Whitworth A. P. and Summers D. (1985) *Mon. Not. R. Astron. Soc., 214,* 1–25.

Whitworth A. P. and Ward-Thompson D. (2001) *Astrophys. J., 547,* 317–322.

Wilking B. A., Lada C. J., and Young E. T. (1989) *Astrophys. J., 340,* 823–852.

Williams J. P., Myers P. C., Wilner D. J., and Di Francesco J. (1999) *Astrophys. J., 513,* L61–L64.

Wolf S., Launhardt R., and Henning T. (2003) *Astrophys. J., 592,* 233–244.

Wood D. O. S., Myers P. C., and Daugherty D. A. (1994) *Astrophys. J. Suppl., 95,* 457–501.

Young C. H., Jorgensen J. K., Shirley Y. L., Kauffmann J., Huard T., et al. (2004) *Astrophys. J. Suppl., 154,* 396–401.

Extreme Deuteration and Hot Corinos: The Earliest Chemical Signatures of Low-Mass Star Formation

Cecilia Ceccarelli
Laboratoire d'Astrophysique de Grenoble

Paola Caselli
INAF–Osservatorio Astrofisico di Arcetri

Eric Herbst
The Ohio State University

Alexander G. G. M. Tielens
Space Research Organization of the Netherlands

Emmanuel Caux
Centre d'Etude Spatiale des Rayonnements

Low-mass protostars form from condensations inside molecular clouds when gravity overwhelms thermal and magnetic supporting forces. The first phases of the formation of a solar-type star are characterized by dramatic changes not only in the physical structure but also in the chemical composition. Since Protostars and Planets IV (e.g., *Langer et al., 2000*), exciting new developments have occurred in our understanding of the processes driving this chemical evolution. These developments include two new discoveries: (1) extremely enhanced molecular deuteration, which is caused by the freeze-out of heavy-element-bearing molecules onto grain mantles during the prestellar core and Class 0 source phases; and (2) hot corinos, which are warm and dense regions at the center of Class 0 source envelopes and which are characterized by a multitude of complex organic molecules. In this chapter we will review these two new topics, and will show how they contribute to our understanding of the first phases of solar-type stars.

1. INTRODUCTION

Molecular clouds are the placentas inside which matter evolves from embryos to stars and planetary systems. The material in the molecular cloud feeds the newly forming system. During this formation, the material undergoes several dramatic changes. This chapter focuses on the changes of the chemical composition during the first phases of star formation.

There are several reasons why the study of chemistry in the first phases of star formation is fascinating and important. Among them, two stand out. First, chemistry is a very powerful diagnostic, both of the current and the past physical conditions of the forming protostar. Much of this chapter will be spent in illustrating that point in detail. In addition, chemistry in the first phases of star formation may affect the chemical composition of the objects that will eventually form the planetary system: planets, comets, and asteroids.

In this sense, the study of chemistry during the first phases of star formation is far-reaching. Indeed, one of the major and more fascinating questions linked to the process of the formation of a star and its planetary system, especially if similar to our solar system, is "What is the chemical budget acquired during the protostellar phase and inherited by the forming planets?" Answering this question might even shed some light on our own origins. The above ultimate question implies answering several linked questions: What molecules are formed during the protostellar phase? Do molecules exist that are formed prevalently during the protostellar phase, and that are therefore a hallmark of this period? What is their fate? Do they condense onto the grain mantles during the protoplanetary phase? Are they incorporated into planetesimals, which eventually form comets, meteorites, and planets? Are these pristine molecules released into nascent planetary atmospheres during early, intense cometary bombardment? And what is the ultimate molecular complexity reached during star formation? In this contribution we will focus on the first questions, those related to the first phase of star formation.

The story starts with the first step toward collapse: the prestellar core phase (see chapters by Di Francesco et al. and Ward-Thompson et al.). These objects are cold (≤ 10 K) and dense ($\geq 10^5$ cm^{-3}) condensations inside molecular clouds. What makes them particularly important is that they are believed to be on the verge of collapse (*Tafalla et al.*, 1998; *Caselli et al.*, 2002b; *Young et al.*, 2004; *Crapsi et al.*, 2005b) and, in this sense, they are considered to be representative of the initial conditions of star formation. During this phase, matter slowly accumulates toward the center under the gravitational force, which counteracts the thermal and/or magnetic pressure. As the density increases, gaseous molecules start to freeze out onto the cold dust grains, forming H$_2$O-dominated ice mantles "dirtied" with several other molecules. The process is so efficient that in the innermost and densest regions of the condensation, heavy-element-bearing molecules are thought to be virtually all frozen out. The low temperatures and the disappearance of most molecules, and particularly of CO, from the gas phase trigger a peculiar chemistry: an extreme molecular deuteration.

Once gravitational contraction finally takes over, a protostar is born, consisting of a central object, which eventually will become a star, surrounded by an envelope from which the future star accretes matter. In the beginning, the envelope is so thick that it obscures the central object, and the spectral energy distribution (SED) is totally dominated by the cold outer regions of the envelope, with a temperature lower than ~30 K. This phase is represented by so-called Class 0 sources (*André et al.*, 2000). Most of the envelope is cold and depleted of heavy-element-bearing molecules, frozen onto the grain mantles, exactly as in prestellar cores. However, the presence of a central source, powered by gravitational energy, causes the heating of the innermost regions of the envelope. In these regions the dust temperature reaches 100 K, causing the evaporation of the grain mantles formed during the prestellar-core phase. The molecules trapped in the ices are injected back into the gas phase, giving rise to a rich and peculiar chemistry. Complex organic molecules are found in these regions, called "hot corinos."

In this chapter, we will show that extreme deuteration and hot corinos are specific signatures of the first phases of low-mass star formation. The chapter is organized as follows. We first briefly describe the physical and chemical structures of prestellar cores and Class 0 source (section 2). We then discuss in detail the deuteration phenomenon (section 3), hot corinos (section 4), and the chemical models that have been developed to explain these observations (section 5). The chapter ends with our conclusions.

We conclude this section by listing a few reviews that appeared in the literature related to the subjects treated in this chapter. Previous reviews of the chemistry in general are reported in *Herbst* (2005) and during the star-formation period in *van Dishoeck and Blake* (1998), *Langer et al.* (2000), and *Caselli* (2006). Reviews of the observations of multi-

ply deuterated molecules as well as the related chemical models can be found in *Roueff and Gerin* (2003), *Ceccarelli* (2004a), and *Millar* (2005), and on the hot corinos in *Ceccarelli* (2004b).

2. THE PHYSICAL AND CHEMICAL STRUCTURE IN THE FIRST PHASES OF THE COLLAPSE

Since Protostars and Planets IV, our understanding of the physical (temperature and density profiles) and chemical (molecular and atomic abundance profiles) structure of matter in the first phases of low-mass star formation has improved considerably. This is due both to the increased sensitivity of groundbased instruments and to the development of sophisticated models to interpret the data. In this section, we review the observations and their interpretations, which have led to the reconstruction of the gas density, temperature, and molecular composition in both prestellar cores and Class 0 sources.

2.1. Prestellar Cores

Several authors have studied the physical and chemical structure of starless cores; detailed reviews are presented in the chapters by Di Francesco et al. and Ward-Thompson et al. Here we briefly recall the basic properties characterizing prestellar cores, and discuss some aspects complementary to those presented by the aforementioned authors. First, the density profiles of the studied prestellar cores are reasonably well represented by Bonnor-Ebert spheres. In practice, the density profile can be approximated by a power law r^{-2} in the outer regions and a plateau in the center regions of the prestellar core. In the following we will refer to the density in the plateau as the "central" density. Typically, the radius of the plateau is 3000–6000 AU (*Crapsi et al.*, 2005a). Figure 1 shows the example of L1544, one of the best-studied prestellar cores (*Tafalla et al.*, 1998; *Caselli et al.*, 2002b,c), but other prestellar cores present basically the same structure (*Alves et al.*, 2001; *Crapsi et al.*, 2005a). In the same figure, the predicted temperature profile is also shown. Note the temperature drop in the inner 3000 AU region, confirmed by observations of NH$_3$ (Craspi et al., in preparation). A similar drop in the dust temperature has been previously observed in other prestellar cores (*Evans et al.*, 2001; *Pagani et al.*, 2003, 2004, 2005), as predicted by radiative transfer modeling of centrally concentrated and externally heated dense cloud cores (*Zucconi et al.*, 2001). These large central densities and low temperatures are accompanied by the depletion of CO molecules in the central regions, as they condense out onto the grain mantles (*Evans et al.*, 2001; *Caselli et al.*, 1999; *Galli et al.*, 2002). Indeed, if one had to give a short list of properties for the prestellar cores they would be large central densities: $\geq 10^5$ cm^{-3}; very low central temperatures: ≤ 10 K; depletion of molecular species, including CO; and enhanced molecular deuteration. Table 1 summarizes these properties in a sample of prestellar cores. In Fig. 2 we show the observed CO depletion fac-

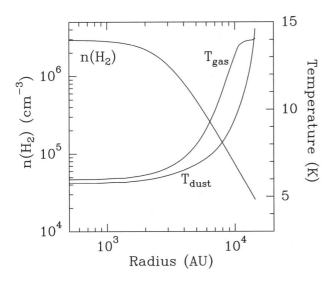

Fig. 1.   The density and temperature profiles derived for the prestellar core L1544. Adapted from Crapsi et al. (in preparation).

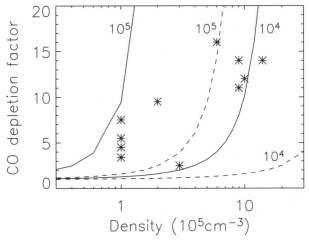

Fig. 2.   CO depletion factor vs. the H_2 density. The asterisks represent the observations in the prestellar cores quoted in Table 1. The solid lines show the theoretical curves for an age of 1×10^4 and 1×10^5 yr, respectively, assuming an average grain radius of 0.1 μm. The dashed lines are computed taking an average grain radius of 0.5 μm.

TABLE 1.   Summary of the properties of a sample of prestellar cores.

Name	$n_c(H_2)$ (10^5 cm^{-3})	$f_D(CO)$	Mol. D/H	Reference
L1521F	3	2.5	0.01	[1,2,3]
B68	1	3.4	0.03	[3,4]
L1689B	1	4.5	0.09	[3,6,7,8]
L183	10	12	0.22	[3,9]
L1544	14	14	0.23	[3,6,10]
L694-2	9	11	0.26	[3]
L429	6	16	0.28	[3,6,7]
OphD	3	14	0.44	[3,6,7]
L1709A	1	5.5	0.17	[6,7]
L1498	1	7.5	0.04	[3,5]
L1517B	2	9.5	0.06	[3,5]

$n_c(H_2)$ is the central density; $f_D(CO)$ is the CO depletion factor; the molecular D/H ratio column contains the observed N_2D^+/N_2H^+ and/or $(D_2CO/H_2CO)^{1/2}$ ratio.

References: [1] *Tafalla and Santiago* (2004); [2] *Hirota et al.* (2001); [3] *Crapsi et al.* (2005a); [4] *Bergin et al.* (2001); [5] *Tafalla et al.* (2004); [6] *Bacmann et al.* (2002); [7] *Bacmann et al.* (2003); [8] *Redman et al.* (2002); [9] *Pagani et al.* (2005); [10] *Tafalla et al.* (2002).

tors vs. the central densities, where the depletion factor is computed with respect to the standard CO abundance in molecular clouds, 9.5×10^{-5} with respect to H_2 (*Frerking et al.,* 1982). As can easily be seen, there is a clear correlation between the CO depletion factors and the densities. This observationally confirms the basic idea that CO molecules disappear from the gas phase because they freeze out onto the grains, and that the condensation rate is proportional to

the density (see below). Figure 2 also reports theoretical curves, obtained assuming that the CO molecules condense onto the grains at a rate k_{freeze} and are released in the gas phase because of the evaporation caused by cosmic rays at a rate $k_{cr} = 9.8 \times 10^{-15}$ s^{-1} (*Hasegawa and Herbst,* 1993). The freezing rate can be written as

$$k_{freeze} = S \langle \pi a_{gr}^2 v_{CO} \rangle n_g \qquad (1)$$

where we adopted a sticking coefficient S = 1 (*Burke and Hollenbach,* 1983) and a mean grain radius of a_{gr}. The grain number density n_g is given by the (mass) dust-to-gas ratio (0.01) multiplied by the gas density, and divided by the grain mass (computed assuming a grain density of 2.5 g cm^{-3}). The figure shows curves assuming an average grain radius of 0.1 (solid lines) and 0.5 μm (dashed lines) respectively. In the first case, which is the typical value assumed in chemical models for the ISM, the denser prestellar cores lie around the curve with an age of 1×10^4 yr, which is definitively too short an age for these objects, based on several arguments. However, if the average grain radius is larger and equal to 0.5 μm, for example, the observed points lie around the curve at 1×10^5 yr, which is a more realistic estimate of the age of these objects. Larger grains would shift the curve even more to larger ages. Note that we used the same value for the cosmic-ray evaporation rate, although increasing the grain sizes decreases it because of the larger volume to heat (*Bringa and Johnson,* 2004; *Shen et al.,* 2004), enhancing the effect. Since the surface πa_{gr}^2 is dominated by the small grains, Fig. 2 suggests that the small grains are efficiently removed by coagulation with large grains in the innermost regions of prestellar cores (*Ossenkopf,* 1993; *Bianchi et al.,* 2003; *Flower et al.,* 2005). Other explanations are also possible. For example, the cosmic-ray evaporation rate can be enhanced if the ice mantles contain a substantial (1%) fraction of radicals due to UV photolysis (*D'Hendecourt et al., 1982, 1986*). However, in order to explain Fig. 2, the mantle evaporation rate should be increased by more than a factor of 30 with respect to the *Hasegawa and Herbst* (1993) rate. Therefore, our preliminary conclusion is that, most likely, the average sizes of the grains in prestellar cores are larger than in the ISM, and that these objects are older than about 1×10^5 yr. It is also interesting to note that less-dense prestellar cores point to greater ages and/or smaller average grain sizes. It is not clear at this stage whether this means that those cores will not evolve into denser ones, or whether prestellar cores spend more time in this first "low-density" phase than in the "high-density" phase.

2.2. Class 0 Sources

The physical and chemical structure of the envelopes of Class 0 sources is affected by the presence of a central object, i.e., a heating source, which implies gradients in the density and temperature. This, in turn, implies a gradient in the chemical composition of the gas across the envelope. Two basic methods have been used in the literature to derive the chemical and physical structure of Class 0 envelopes: (1) modeling of multifrequency observations of several molecules, and (2) modeling of the continuum SED and map, coupled with molecular multifrequency observations. The first method makes use of the intrinsic property that each line from any molecule probes the specific region where the line is excited, namely a region with a specific density and temperature. Using several lines from the same molecule can, therefore, be used to reconstruct the *gas* density and temperature profiles of the envelope. The observations of several molecules are then used to derive the abundance profile of each molecule. Theoretical examples of this method, applied to the problem of collapsing envelopes, are discussed in *Ceccarelli et al.* (2000a, 2003), *Maret et al.* (2002a), and *Parise et al.* (2005b). In these models the density is assumed to follow the evolution of a singular isothermal sphere contracting under the hypothesis of isothermal collapse — the so-called "inside-out" framework, developed by Shu and collaborators (*Shu,* 1977). In this framework, the density follows a two-slope power law: In the regions not reached by the collapse the density has an r^{-2} dependence, whereas in the collapsing regions the density follows an $r^{-3/2}$ law. In addition, in using molecular line observations, one has to also take into account the abundance profile of the molecule used. The simplest profile is that of a single-step function, which assumes that the relevant molecule has an almost constant abundance in the cold outer envelope and a jump in the inner region when the grain mantles sublimate. Other profiles have been used though, in which the abundance also has a drop in the outer envelope, to take into account the molecular depletion as in prestellar cores (*Ceccarelli et al.,* 2001; *Schöier et al.,* 2004; *Jørgensen et al.,* 2004).

The second method uses the dust continuum SED simultaneously with maps to trace the H_2 distribution, and the observations are best-fitted with assumed single power law density distributions (*Shirley et al.,* 2002; *Jørgensen et al.,* 2002; *Williams et al.,* 2005; *Young and Evans,* 2005). Then, in order to derive the molecular abundance profiles, single-jump and/or drop models are used to interpret multifrequency observations, as in *Schöier et al.* (2002, 2004), *Jørgensen et al.* (2004), and *Evans et al.* (2005). Both methods give approximatively the same structure (temperature and density distribution) when applied to the same object, as in the case of IRAS16293-2422 (*Ceccarelli et al.,* 2000a; *Schöier et al.,* 2002). Figure 3 shows the derived structure of the envelope of this source, obtained by applying the first method to several molecules (*Ceccarelli et al.,* 2000a,b).

Both classes of models predict the existence of regions with dust temperatures larger than 100 K, the sublimation temperature of the grain mantles (*Ceccarelli et al.,* 1996). In these regions, therefore, the components of the grain mantles are injected into the gas phase, and the abundances of the corresponding molecules should increase with respect to the abundance in the cold ($T_{dust} \leq 100$ K) envelope. Water is the major component of the ice mantles, but unfortunately water lines are not observable from groundbased telescopes. The spectrometer LWS onboard ISO provided observations of water lines toward a few Class 0 sources (*Ceccarelli et al.,* 1999), but with relatively poor spatial and spectral resolution, so that the interpretation of the data is not unique. Two interpretations have been advanced. The first one assumes that the water lines originate in the shocks at the interface between the outflows and the envelopes of these

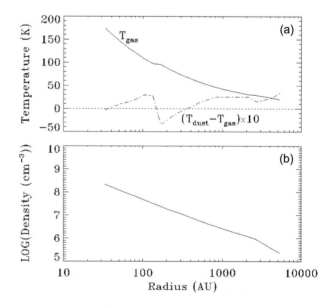

Fig. 3. The **(a)** temperature and **(b)** density profile of the envelope surrounding the Class 0 protostar IRAS16293-2422. From *Ceccarelli et al.* (2000a).

sources (*Giannini et al.,* 2001; *Nisini et al.,* 2002). The second one assumes that the water lines originate mostly in the envelope. Based only on the ISO data, it is impossible to discriminate observationally between the two hypotheses. However, if the water line spectra are interpreted as being emitted in the envelope, allowing a jump in the water abundance in the $T_{dust} \geq 100$ K region, the agreement among the model predictions and the observed line fluxes is rather good (*Ceccarelli et al.,* 2000a; *Maret et al.,* 2002b). The observed data are reproduced if the water abundance is ~3 × 10^{-7} in the outer envelope, and jumps by about a factor of 10 in the inner region. In support of this interpretation, the physical structure derived by this analysis is substantially confirmed by observations of other molecules obtained with groundbased telescopes (see below).

Formaldehyde is also an important grain mantle component, and this molecule has the advantage that it has several transitions in the millimeter to submillimeter wavelength range observable with groundbased telescopes (*Mangum and Wootten,* 1993; *Ceccarelli et al.,* 2003). These observations have a much better spatial and spectral resolution than ISO-LWS and provide stronger constraints for the models. A survey of the H_2CO line emission toward a bit fewer than a dozen Class 0 sources has shown that all the targeted sources, except VLA16293, have a region where the formaldehyde abundance jumps by more than one order of magnitude (*Ceccarelli et al.,* 2000b; *Maret et al.,* 2004). Several of these sources also show jumps in the methanol abundance (*Maret et al.,* 2005). The *predicted* sizes of the jump regions range from 10 to 150 AU, and are therefore comparable to the sizes of the solar system. The densities are predicted to be larger than 10^8 cm^{-3}. However, one should be aware that the above observations have been obtained with *single-dish* telescopes, which, at best, have spatial resolutions corresponding to about 1000 AU in radius. Therefore, both the estimates of the abundances and the warm region sizes suffer from a relatively large uncertainty. Besides, the details of the model adopted in the analysis of the data give rise to additional systematic uncertainties so that the very existence of the H_2CO abundance jumps is contested by some authors (*Schöier et al.,* 2004; *Jørgensen et al.,* 2005a,c). In addition, the situation is further confused by the presence at small scales of cavities and outflows, which some authors also consider a major component in the line emission attributed to the abundance jumps of the above analysis (*Chandler et al.,* 2005; *Jørgensen et al.,* 2005b).

Finally, several simple molecules (HCO^+, N_2H^+, CS, SO, SO_2, HCN, HNC, HC_3N, and CN) have been observed in a sample of Class 0 sources (*Jørgensen et al.,* 2004). Given the observed transitions, this study was able to trace the abundance of the observed species in the outer envelope and probe some formation routes, such as for N_2H^+ and HCO^+. One remarkable result is the discovery of large regions in the outer envelopes where the molecular abundances are depleted, because of condensation onto grain mantles (*Maret et al.,* 2004; *Jørgensen et al.,* 2005d). However, even more

remarkable is the discovery of extreme deuterium fractionation in the Class 0 sources, which will be discussed in section 3, and which is indeed linked to the molecular depletion.

In summary, from a chemical point of view, the similarity to prestellar cores is the most prominent characteristic of the outer envelopes of Class 0 sources. With respect to molecular depletion and extreme deuteration, they are indeed virtually indistinguishable, suggesting that prestellar cores are the likely precursors of Class 0 sources (*Maret et al.,* 2004; *Jørgensen et al.,* 2005c).

3. EXTREME DEUTERATION

The previous section has anticipated a major characteristic shared by prestellar cores and the envelopes of Class 0 sources: an extreme molecular deuteration. Although the deuterium abundance is 1.5 × 10^{-5} relative to hydrogen (*Linsky,* 2003), singly deuterated molecules have been observed in both types of objects with abundances relative to their hydrogenated counterparts between 10% and 50% (*Guélin et al.,* 1977; *Loren et al.,* 1990; *van Dishoeck et al.,* 1995; *Tiné et al.,* 2000; *Shah and Wootten,* 2001; *Roberts et al.,* 2002; *Crapsi et al.,* 2004, 2005a). Even more extreme, doubly and triply deuterated molecules have been observed (Table 2), with D/H ratios reaching 30% for D_2CO (*Loinard et al.,* 2002) and 3% for CD_3OH (*Parise et al.,* 2004). This enhances the molecular D/H ratio by up to 13 orders of magnitude with respect to the elemental D/H ratio. In this section, we review these spectacular observations,

TABLE 2.   Summary of the multiply deuterated molecules observed in prestellar cores and Class 0 sources.

Species	Deuterium Fractionation	Reference
NHD_2	0.03	[1,2,3]
ND_3	0.006	[3,4,5]
D_2CO	0.15	[6,7,8]
D_2S	0.12	[9]
HD_2^+	1	[10]
CHD_2OH	0.06	[11,8]
CD_3OH	0.03	[12]
D_2CS	0.10	[13]

Typical deuterium fractionation ratios are given with respect to the fully hydrogenated counterparts except for D_2S and HD_2^+, where the ratio is with respect to the singly deuterated isotopologue. Note that variations among different sources amount to about a factor of 3 or so.

References: [1] *Roueff et al.* (2000); [2] *Loinard et al.* (2001); [3] *Roueff et al.* (2005); [4] *Lis et al.* (2002); [5] *van der Tak et al.* (2002); [6] *Ceccarelli et al.* (1998); [7] *Bacmann et al.* (2003); [8] *Parise et al.* (2006); [9] *Vastel et al.* (2003); [10] *Vastel et al.* (2004); [11] *Parise et al.* (2002); [12] *Parise et al.* (2004); [13] *Marcelino et al.* (2005).

which have led to the development of a new class of models for molecular deuteration, described in section 5. It is worth emphasizing that extreme fractionations have only been observed in *low-mass* prestellar cores (section 3.1) and protostars (section 3.2) so far. Massive protostars do not show the same deuterium enrichment, probably because the chemistry occurs in warmer environments.

3.1. Prestellar Cores

In the last few years, it has become increasingly clear that molecular deuteration reaches extreme values in prestellar cores, where multiply deuterated molecules have been detected with ratios larger than 1%. The list of multiply deuterated molecules detected in prestellar cores (and Class 0 sources) is reported in Table 2, as well as the typical measured D/H ratio for each species. To the best of our knowledge, two surveys of multiply deuterated molecules — D_2CO (*Bacmann et al., 2003*) and ND_3 (*Roueff et al., 2005*) — have been obtained in (small) samples of prestellar cores. *Bacmann et al.* (2003) found that the D_2CO/H_2CO ratios correlate with the CO depletion factors, proving observationally the key role of CO depletion in the deuterium-enhanced fractionation. This trend is confirmed now by the larger sample of Table 2, as shown in Fig. 4: *the larger the CO depletion the larger the molecular D/H ratio* (see also *Crapsi et al., 2005a*). This is a key point whose importance will be further discussed in section 5. Another aspect shown in the figure is the similarity of the D/H ratio of the three observed molecules: N_2H^+, H_2CO, and NH_3. Note that the three molecules refer to the D/H ratios of singly, doubly, and triply deuterated isotopologues respectively, and the ratios shown in the figure are scaled to account for the number of D atoms. While N_2H^+ is certainly a gas-phase product

(formed by the reaction of N_2 with H_3^+), there is considerable debate as to whether formaldehyde and ammonia are formed in the gas phase or on grain surfaces (*Tielens, 1983; Ceccarelli et al., 2001; Roueff et al., 2005*) (section 5). From a purely observational point of view, Fig. 4 points to a similar origin for the three molecules, but this is very likely an oversimplification, and may be simply due to the common origin of the molecular deuteration, the molecular ion H_2D^+ (section 5).

Indeed, it has been long believed that the deuterium fractionation is started by the fast ion-neutral reaction $H_3^+ + HD$ leading to H_2D^+ (section 5.1). For this reason, H_2D^+ has long been searched for in regions of massive star formation, where, however, it has never been detected (*van Dishoeck et al., 1992; Pagani et al., 1992*). The first detection of H_2D^+ was finally obtained by *Stark et al.* (1999) toward the low-mass Class 0 source NGC1333-IRAS4. However, the breakthrough detection is considered to be that obtained in the prestellar core L1544 (*Caselli et al., 2003*), where the ground ortho-H_2D^+ line at 372 GHz is 10× brighter than in NGC1333-IRAS4. The observed flux implies an abundance of H_2D^+ similar to that of electrons, and therefore to that of H_3^+. This was unexpected and unpredicted by the chemical models at the time of the discovery. It is now accepted that the H_2D^+/H_3^+ ratio can reach unity in cold gas depleted by heavy-element-bearing molecules, as occurs at the centers of prestellar cores (section 2.1). Actually, in the extreme conditions of the centers of prestellar cores, even the doubly deuterated form of H_3^+ has been detected (*Vastel et al., 2004*). The map of H_2D^+ emission in L1544 shows that it is very abundant in a zone with diameter ~6000 AU (*Vastel et al., 2006*). This zone coincides with the central density plateau, where CO is highly depleted (*Caselli et al., 1999*). This further proves the key role played by the freezing of heavy-element-bearing molecules in the enhancement of molecular deuteration. Finally, one has to note that L1544 is not a peculiarity. An ongoing project shows that prestellar cores are in general strong emitters of the ground ortho-H_2D^+ line [$T_{MB} > 0.5$ K (Caselli et al., in preparation)].

Thanks to the above observations, the very high diagnostic power of the ground-state ortho-H_2D^+ line is now realized. This is true not only for chemistry, but also for the study of the evolutionary and dynamical status of prestellar cores. For example, the ortho-to-para ratio of H_2D^+ is predicted to critically depend on the density, temperature, and grain sizes (*Flower et al., 2005, 2006*). Comparing these model predictions with observations, *Vastel et al.* (2006) found that the average grain sizes in L1544 are larger than in the ISM (in agreement with the discussion in section 2.1). Finally, since very likely most of the heavy-element-bearing molecules are frozen out onto the grain mantles in the centers of the prestellar cores, H_2D^+ is the best, if not the only, way to probe those regions. Of particular interest is knowing the kinematical structure and whether and when the collapse starts at the center. *van der Tak et al.* (2005) studied the line profiles of the fundamental ortho-H_2D^+ line toward L1544

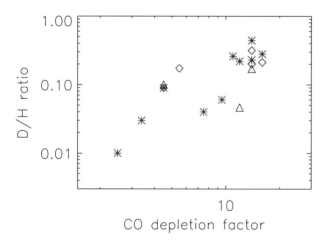

Fig. 4. The measured N_2D^+/N_2H^+ [stars (from *Crapsi et al., 2005a*)], $(D_2CO/H_2CO)^{1/2}$ [diamonds (*Bacmann et al., 2003*)], and $(ND_3/NH_3)^{1/3}$ [triangles (*Roueff et al., 2005*)] in the sample of prestellar cores of Table 2, as function of the CO depletion factor.

with this scope. They found that the infall likely started at the center of L1544. We predict that similar studies with submillimeter interferometers will be extremely useful to unveil the first instants of collapse.

3.2. Class 0 Sources

As for prestellar cores, recent observations have revealed a zoo of molecules with enhanced deuterium fractionation. Actually, the discovery of abundant D_2CO in a Class 0 source started the hunt for multiply deuterated molecules. Although doubly deuterated formaldehyde had been detected 15 years ago in Orion by *Turner* (1990), the measured abundance (~0.2% H_2CO) did not seem to draw particular attention (probably because Orion has always been considered a peculiar region). The discovery that D_2CO has an abundance about 5% that of H_2CO in a low-mass protostar, IRAS16293-2422 (*Ceccarelli et al.,* 1998), on the contrary, started a flurry of activity in this field, and initiated the hunt for multiply deuterated molecules. In a few years, the deuterated species quoted in Table 2 were detected.

The best-studied Class 0 source is IRAS16293-2422, where multifrequency observations of singly and doubly deuterated formaldehyde constrain the abundance ratios of these molecules quite well (*Loinard et al.,* 2000) to be $HDCO/H_2CO = 0.15$ and $D_2CO/H_2CO = 0.05$. Besides, D_2CO is very abundant across the entire envelope, which extends up to a radius of more than 5000 AU (*Ceccarelli et al.,* 2001). Remarkably, the D_2CO/H_2CO ratio reaches a peak of 0.16 at about 2000 AU from the center. Note that the dust at this distance is warm enough to prevent the condensation of CO, and the gas is too warm, ~35 K, to favor the deuterium fractionation in the gas phase (*Ceccarelli et al.,* 2000a; *Schöier et al.,* 2002). In a few words, the two key factors for the enhanced fractionation, cold and CO-depleted gas, are absent in the largest fraction of the IRAS16293-2422 envelope, where D_2CO is abundant. Therefore, the observed enhanced deuterium fractionation is unlikely to be a present-day product (*Ceccarelli et al.,* 2001). In support of this view, the formaldehyde abundance profile across the IRAS16293-2422 envelope increases by about a factor of 10 where the dust temperature exceeds 50 K, suggesting that gaseous formaldehyde is due to the partial, but continuous, sublimation of the grain mantles across the entire envelope. Layered like onions, the components of the mantles sublimate at different distances, corresponding to different dust temperatures. In the regions where the dust temperature is larger than about 25 K, CO-rich mantles sublimate, injecting into the gas phase the CO and the trace molecules trapped in the ice, like formaldehyde. When the dust temperature reaches 50 K the H_2CO-rich ices sublimate and so on (*Ceccarelli et al.,* 2001). The observed deuteration is therefore a heritage of the precollapse phase (section 3.1).

This can also be seen in Fig. 5, where the measured D/H ratios of formaldehyde and methanol are reported as function of CO depletion in a sample of Class 0 sources (*Parise*

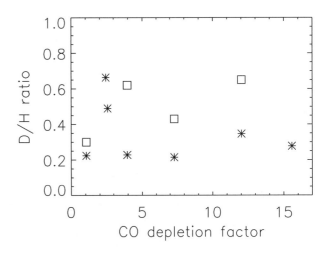

Fig. 5. The measured $(D_2CO/H_2CO)^{1/2}$ (stars) and CH_2DOH/CH_3OH (squares) in a sample of Class 0 sources (*Parise et al.,* 2006) as a function of the CO depletion factor.

et al., 2006). Contrary to the prestellar cores, there is no clear correlation between the molecular deuteration and the *present* CO depletion. The most plausible explanation is that the deuterated formaldehyde and methanol have been built up during the previous phase and that Class 0 sources are young enough for it not being substantially modified, likely younger than ~10^5 yr based on modeling (*Charnley et al.,* 1997). Figure 5 also shows that methanol is systematically more enriched with deuterium than is formaldehyde. This may reflect a different epoch of formation of these two molecules on the ices: The formaldehyde perhaps formed in an earlier stage than methanol, when CO depletion was less severe. Alternatively, present-day gas-phase reactions may have changed the initial mantle composition already, or, finally, formaldehyde and methanol follow different routes of formation (gas phase vs. grain surface?) (*Parise et al.,* 2006).

Although our comprehension of the mechanisms leading to molecular deuteration in the ISM has undoubtedly improved in these last few years, there still remain a number of unresolved questions. First, the relative role of gas phase vs. grain surface chemistry is uncertain: It is not completely clear yet what molecules and to what extent are formed in the gas phase vs. on the grain surfaces. Deuterium fractionation promises to be a key aspect in this issue. Another open question is represented by deuterated water. Searches for solid HDO have so far been in vain, giving some stringent upper limits to the water deuteration: $\leq 2\%$ (*Dartois et al.,* 2003; *Parise et al.,* 2003), considerably less than the deuteration observed in the other molecules. One possible explanation is that observations of solid H_2O and HDO may be "contaminated" by the contribution of the molecular cloud in front of the observed protostar [see, for example, the discussion in *Boogert et al.* (2002)]. A more stringent con-

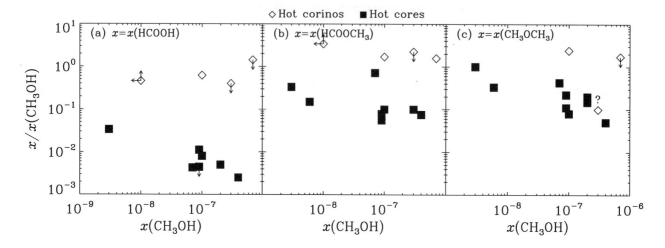

Fig. 6. Abundance ratios of complex O-bearing molecules to methanol, plotted as a function of the methanol abundance. Diamonds and filled squares represent hot corinos and hot cores respectively (from *Bottinelli et al., 2006*).

straint is set by observations of gas-phase HDO in the outer envelope and hot corino, where the icy mantles sublimate. Observations toward IRAS16293-2422 set a stringent constraint to the deuteration of water; in the sublimated ices it is ~3%, whereas in the cold envelope it is ≤0.2% (*Parise et al., 2005a*). Therefore, water deuteration seems to follow a different route from formaldehyde, methanol, and the other molecules.

4. HOT CORINOS

While the outer envelopes of Class 0 sources are so cold that heavy-element-bearing molecules are at least partially frozen onto the grain mantles, the innermost regions are characterized by warm dust, so warm that the mantles sublimate totally (section 2.2). These regions share several aspects of the hot cores, discovered in the 1980s around massive protostars (*Kurtz et al., 2000*). Being "smaller" in size and mass, they have been called "hot corinos" (*Ceccarelli, 2004b; Bottinelli et al., 2004b*). Based on the analysis discussed in section 2.2, the radii of the hot corinos are predicted to be around 50 AU (IRAS16293-2422 represents here an exception with a radius of 150 AU), comparable to the sizes of the solar system. The key question, and the most relevant for the chemical structure of Class 0 sources, is which molecules are found in the hot corinos, and, in particular, whether complex organic molecules are formed, in analogy with massive hot cores. More generally, are the hot corinos a scaled version of the hot cores, or does a peculiar chemistry take place? And how reliable are the hot corino sizes derived by the analysis of single-dish data?

Once again, the benchmark for these studies has been IRAS16293-2422. In the hot corino of this source, complex organic molecules were detected for the first time (*Cazaux et al., 2003*), and this hot corino was the first to be imaged (*Kuan et al., 2004; Bottinelli et al., 2004b*). In practice, all

the complex molecules typical of massive hot cores that have been searched for were detected, with abundances similar to if not larger than those measured in hot cores (*Cazaux et al., 2003*). These species include O- and N-bearing molecules such as formic acid, HCOOH; acetaldehyde, CH₃CHO; methyl formate, HCOOCH₃; dimethyl ether, CH₃OCH₃; acetic acid, CH₃COOH; methyl cyanide, CH₃CN; ethyl cyanide, C₂H₅CN; and propyne, CH₃CCH. Two years after the first detection in IRAS16293-2422, complex organic molecules were also detected toward the other Class 0 sources NGC1333-IRAS4A and B, and NGC1333-IRAS2 (*Bottinelli et al., 2004a, 2006; Jørgensen et al., 2005a*). The abundances of the O-bearing complex molecules detected in the four hot corinos are shown in Fig. 6 as functions of the methanol abundance (*Bottinelli et al., 2006*). Note that the abundances have been normalized to the methanol abundance so that they do not depend on the uncertain sizes of the hot corinos. The number of hot corinos where complex organic molecules have been detected is evidently much too small to try to draw firm conclusions based on statistical considerations. Nevertheless, it is obvious that complex organic molecules are rather common in hot corinos. In this respect, the prototypical Class 0 source, IRAS16293-2422, may actually be representative for Class 0 sources. Furthermore, the abundances of HCOOH, HCOOCH₃, and CH₃OCH₃ in hot corinos are comparable to the methanol abundance and to formaldehyde (*Bottinelli et al., 2006*), and do not depend appreciably on the source luminosity, nor on the methanol and formaldehyde abundance. If the O-bearing complex molecules are formed in the gas phase from methanol and formaldehyde, as some theories predict (section 5.3), then they "burn" the majority of their "parent" molecules, and this burned fraction is independent of the amount of H₂CO and/or CH₃OH sublimated from the grain mantles. Alternatively, if the O-bearing complex molecules are synthesized on the grain surfaces, they are an important ice component.

So far, however, only HCOOH has been claimed to be found in its solid form (*Schutte et al.,* 1999; *Gibb et al.,* 2000). The other evident message of Fig. 6 is that hot corinos are not a simply scaled version of hot cores. The abundance of the O-bearing complex molecules in hot corinos is larger than that in hot cores, by more than a factor of 10: HCOOH is even 100× more abundant.

So far, we have focused on results based on single-dish observations. These observations encompass regions of at least 1000 AU, whereas the predicted sizes of the hot corinos do not exceed 150 AU. Recently, interferometric observations have resolved the emission of complex organic molecules in the hot corinos toward IRAS16293-2422 (*Kuan et al.,* 2004; *Bottinelli et al.,* 2004b; *Chandler et al.,* 2005; *Remijan and Hollis,* 2006), NGC1333-IRAS2 (*Jørgensen et al.,* 2005a), and NGC1333-IRAS4B (Bottinelli et al., in preparation). From these data, a few general conclusions can be drawn. First, for IRAS16293-2422 and NGC1333-IRAS4B, the line emission from the complex molecules is concentrated in two spots, centered on the two objects forming in each case a protobinary system (*Wootten,* 1989; *Looney et al.,* 2003; *Sandell and Knee,* 2001; *Di Francesco et al.,* 2001). The full line flux of the observed complex molecules, detected with the single dish, is recovered in these two spots, which means that most of the single-dish emission originates in those regions. In addition, there is no sign of outflows in the line emission of complex molecules; the emission is compact and the line profile is relatively narrow. The brightest — and so far only resolved — spot in IRAS16293-2422 has a radius of ~150 AU, close to the size expected for warm (T > 100 K) gas (*Ceccarelli et al.,* 2000a). As emphasized above, each of these two objects is indeed a protobinary system. In each case, the two objects in the protobinary have different characteristics; e.g., one spot is bright in the continuum but barely detected in the complex molecule line emission, while this is reversed for the other spot. This difference does not necessarily imply that the abundances in the individual spots of these protobinaries are truly different. The spot with the weaker line emission is also unresolved and the line likely optically thick (*Bottinelli et al.,* 2004b). Evidently, high-resolution observations of complex organic molecules are a valuable tool to study the chemistry of the hot corinos and how it depends on physical conditions, rather than the mantle composition and age, which presumably are the same for the two objects of the binary system.

5. CHEMICAL MODELS

5.1. Molecular Deuteration in the Gas Phase

Although the basic gas-phase mechanism of molecular deuteration in the cold interstellar medium was elucidated almost 30 years ago (*Guélin et al.,* 1977; *Watson et al.,* 1978) and incorporated into detailed models over a decade ago (*Millar et al.,* 1989), it is only recently that an understanding

of the effect in star-forming regions has been achieved by astrochemists. Since the reservoir of deuterium lies in the molecule HD, deuteration proceeds by removing deuterium atoms from this molecule onto the many other trace constituents of the gas. The principal mode of transfer at low temperatures has been thought to be the reaction system

$$H_3^+ + HD \rightleftharpoons H_2D^+ + H_2 \qquad (2)$$

in which the left-to-right reaction is exothermic by approximately 230 K. If no other reactions need be considered to determine the abundance ratio of H_2D^+ to H_3^+, the slowness of the backward reaction leads to the stunning prediction that the abundance of the deuterated ion, now referred to as an isotopologue, can exceed that of H_3^+. For cold cores at the standard density of 10^4 cm^{-3}, the prediction is in error because H_2D^+ reacts quickly with both electrons and heavy species such as CO. The reactions with heavy species serve to spread the enhanced deuteration around so that typical ratios between singly deuterated isotopologues (e.g., DCO$^+$) and normal species (e.g. HCO$^+$) are predicted to lie in the range 0.01–0.10, in agreement with observation. Two other exchange reactions are of lesser importance: these involve the ions CH_3^+ and $C_2H_2^+$. Their deuterium exchange reactions with HD leading to the ions CH_2D^+ and C_2HD^+ are more exothermic than the H_3^+ case, so that the backward reactions are slow up to higher temperatures, and subsequent reactions involving the deuterated ions can maintain deuterium fractionation when H_2D^+ is no longer enhanced. Other reactions are also non-negligible, including some involving deuterium atoms (*Brown and Millar,* 1989).

As the density increases, the picture changes dramatically. Near the center of a cold prestellar core, for example, the gas density increases to 10^{5-6} cm^{-3} and the timescale for accretion of species onto dust particles becomes so short that most heavy species are strongly depleted onto dust particles. In addition, the fractional ionization in the gas is thought to become quite low if one takes account of gas-grain interactions to deplete the chemically rather inert atomic ion H$^+$ (*Liszt,* 2003). Since seemingly all destruction mechanisms of H_2D^+ are slow, a large abundance of H_2D^+ does indeed build up, in agreement with observations (section 3.1). But the story does not end here: In analogy with the reaction above, further reactions with HD convert a significant portion of the H_2D^+ to the more deuterated isotopologues HD_2^+ and even D_3^+ (*Roberts et al.,* 2003; *Walmsley et al.,* 2004; *Ceccarelli and Dominik,* 2005). Indeed, simple chemical models of the densest portions show the triply deuterated ion to be the most abundant of the four ions in the series, and possibly the most abundant ion in the gas (*Roberts et al.,* 2003). Figure 7 shows the abundances of the deuterated forms of H_3^+, H$^+$, and e$^-$ as functions of the CO depletion factor respectively. Since the age and any other parameter is constant, increasing the depletion factor corresponds to increasing the density. For "intermediate" (≤30) CO depletion factors, H_2D^+ is the most abundant deuterated form of

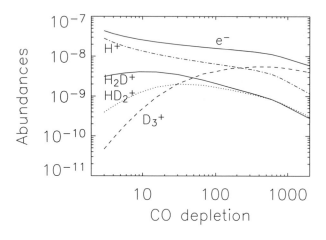

Fig. 7. Abundance of the deuterated forms of H_3^+, H^+, and e^- as functions of the CO depletion factor: thick solid line, e^-; thin solid line, H_2D^+; dashed line, D_3^+; dotted line, HD_2^+; and dotted-dashed line, H^+.

H_3^+. However, for larger depletions D_3^+ takes over, and becomes the dominant positive charge carrier. A similar effect leads to the ions CD_3^+ and $C_2D_2^+$.

Although a tiny region in the center may be completely devoid of heavy species (*Walmsley et al.*, 2004), such molecules do exist, albeit with depleted abundances, near the center (*Bergin et al.*, 2002; *Pagani et al.*, 2005; *Lis et al.*, 2006). The remaining heavy species will react with all three of the deuterating ions, especially D_3^+, leading to highly deuterated isotopologues such as D_2CO and ND_3. Critical here is the high efficiency of D_3^+ since, unlike the partially deuterated analogs, it can only deuterate and not protonate heavy species. Contrast the reaction

$$D_3^+ + CO \rightarrow DCO^+ + D_2 \qquad (3)$$

with

$$H_2D^+ + CO \rightarrow DCO^+ + H_2; \; HCO^+ + HD \qquad (4)$$

In the latter reaction, deuteration occurs on only one of three collisions. The increase in efficiency with D_3^+ as the deuterating agent is especially important in the formation of multiply deuterated neutral isotopologues, which occurs via a cycle of deuteration followed by dissociative recombination reactions. For example, starting with normal ammonia, deuteration leads to NH_3D^+, followed by dissociative recombination to form NH_2D, followed by deuteration to form $NH_2D_2^+$, followed by dissociative recombination to form NHD_2, etc. (*Roueff et al.*, 2005).

Gas-phase chemical models of prestellar cores including deuteration range from simple homogeneous treatments relevant to the center of the core (*Roberts et al.*, 2003) to shell models that can be static (*Roberts et al.*, 2004) or in a state of collapse (*Aikawa et al.*, 2003, 2005). Accretion onto grains must be included in all approaches. The collapse

can be treated via analytical approaches (*Aikawa et al.*, 2003) or hydrodynamic simulations (*Aikawa et al.*, 2005). The chemistry occurring on grain surfaces can also be included (*Aikawa et al.*, 2003, 2005). A relatively simple static approach by *Roberts et al.* (2004) illustrates the power of the chemical models. In terms of column densities, *Roberts et al.* (2004) obtain a value for D_2CO higher by only a factor of 3 than that observed by *Bacmann et al.* (2003), and a CO depletion of a factor of 20 compared with the standard value, in good agreement with the observed depletion of 14 (*Bacmann et al.*, 2002) (see also sections 2 and 3). The overall computed D_2CO/H_2CO ratio is 0.3, also in good agreement with observation. In general, models with collapse are not in as good agreement with observation because it is not facile to vary the details of the collapse to obtain optimal agreement.

5.2. Mantle Formation and Surface Chemistry

As the gas-phase chemistry occurs and mantles accrete in cold sources, a surface chemistry is also active. A proper mathematical treatment of diffusive surface reactions on interstellar grains is not simple. The first such treatments used rate equations analogous to those used for gas-phase processes (*Pickles and Williams*, 1977), and the method is still the only practical approach to the construction of large gas-grain models (*Ruffle and Herbst*, 2000). The method is only a rough approximation to reality for a number of reasons, and a variety of more accurate but more complex methods have been advocated. If we imagine grains to be rather smooth surfaces on which binding parameters for adsorbates do not differ strongly from one site to another, there is the problem that only small (fractional) numbers of reactive species such as atomic hydrogen are present on a given grain on average. With such small numbers, it is more reasonable to consider discrete rather than average quantities. Both the discreteness of the problem and the large fluctuations possible from grain to grain then argue for a stochastic approach. Following the pioneering work of *Tielens and Hagen* (1982), a truly stochastic method was utilized by *Charnley* (2001) based on Monte Carlo methods, while a master equation approach was advocated by *Biham et al.* (2001) and by *Green et al.* (2001). The latter approach has recently been incorporated into moderately sized gas-grain networks by *Biham and Lipshtat* (2002), *Lipshtat and Biham* (2004), and *Stantcheva and Herbst* (2004).

To add to the complexity of the problem, interstellar surfaces are likely to be irregular, porous, and even amorphous, so that binding sites for adsorbates are likely to be diverse, including some that contain much deeper wells than others. A new extension of the Monte Carlo method for irregular grains with silicate, carbonaceous, and icy surfaces has been introduced by *Chang et al.* (2005) and *Cuppen and Herbst* (2005) for the specific case of the formation of hydrogen molecules. Here, the fraction of binding sites with deep potential wells totally changes the character of the problem since numerous H atoms are present over wide temperature ranges on each grain. Whether or not an approach of this complexity can be extended to large models is unclear. A simpler approach to different binding sites is the physisorption-chemisorption model of *Cazaux and Tielens* (2004), where only two types of binding sites are considered for H atoms, shallow and deep. In this approximation, rate equations can be used to a reasonable degree of accuracy.

At later stages, the surface chemistry includes deuteration via atomic deuterium, which is produced in the gas by dissociative recombination of deuterated ions; viz., $D_3^+ + e^- \to D + D + D$, and can lead in very dense regions to an abundance of atomic deuterium nearly equal to the abundance of atomic hydrogen (*Roberts et al.*, 2003, 2004). The H and D atoms land on grains and, because of their rapid diffusion and high surface reactivity, both hydrogenate and deuterate heavy atoms and reactive molecules to form more saturated forms via sequential reactions. For example, the formation of water ice can occur by the addition of two hydrogen atoms to an oxygen atom that lands on a grain surface: $O + H \to OH$; $OH + H \to H_2O$ or can start from OH, which can be formed on grain surfaces by successive reactions involving O_2 and O_3 (*Tielens and Hagen*, 1982). Note that even if the gaseous O_2 is low, an appreciable abundance of O_2 can built up on the surface through reactions of O plus O, or O_2 can be formed in the reaction of O_3 with H. There is even laboratory evidence that hydrogenation of CO into formaldehyde (H_2CO) and methanol (CH_3OH) occurs (*Watanabe et al.*, 2003; *Hidaka et al.*, 2004): $CO \to HCO \to H_2CO \to H_3CO \to CH_3OH$, although two of the hydrogenation reactions possess small activation energy barriers. Following a similar path, not only H_2CO and CH_3OH are formed but also their deuterated isotopologues (*Charnley et al.*, 1997). It is interesting to note that there is currently no viable formation mechanism for methanol in the gas phase, so that all methanol detected in the gas has its genesis on grain surfaces. This statement is true even for the small amount of methanol detected in cold sources. Here surface formation is followed by inefficient desorption. The results of a number of models — which differ in the sophistication of the gas-grain interaction, their assumptions on initial conditions including gas-phase abundances, and the chemical routes included — have been reported in the literature (*Tielens and Hagen*, 1982; *Hasegawa et al.*, 1992; *Aikawa et al.*, 2003; *Stantcheva and Herbst*, 2003; *Aikawa et al.*, 2005). Deuteration on cold surfaces was first explored by *Tielens* (1983) and later studied by *Brown and Millar* (1989), *Charnley et al.* (1997), *Caselli et al.* (2002a), and *Stantcheva and Herbst* (2003) among others. The basic picture adopted by astrochemists is that D atoms can deuterate atoms and radicals in a similar manner to the reactions of H atoms. If we, for example, consider the competing deuteration and hydrogenation processes starting with surface CO, all the different isotopologues and isotopomers of methanol can be produced: CH_3OD, CH_2DOH, CH_2DOD, CHD_2OH, CHD_2OD, CD_3OH, and CD_3OD (*Charnley et al.*, 1997). Their relative abundances on grain surfaces depend on the

flux ratio between H and D atoms landing on the grains. That the deuteration is more complex than envisioned by astrochemists, however, is shown by some recent experimental evidence provided by *Nagaoka et al.* (2005), in which it seems that even normal methanol can be deuterated by reactions with atomic D on grain surfaces.

All model calculations confirm that ices are produced by surface reactions, and that these ices are, at least on average, dominated by water, as observed. The large amount of CO ice detected by observers is probably mainly the result of accretion from the gas, where this molecule is produced copiously. The production of surface methanol occurs once there is sufficient CO to be hydrogenated. The production of the other very abundant surface species — carbon dioxide (CO_2) — is not well understood.

5.3. Mantle Evaporation and Complex Organic Molecule Formation

A variety of processes can return ice mantle species to the gas phase, even during the cold prestellar core phase. The most important one here is cosmic-ray-driven desorption where an Fe-cosmic-ray ($E \sim 10$–100 MeV/nucleon) temporarily heats the grain to a high enough temperature that some mantle species evaporate (*Leger*, 1983; *Leger et al.*, 1985). By itself, this process will only return weakly bound species (eg., CO, N_2) to the gas phase. However, this process becomes very effective, even for more tightly bound species such as H_2O, if the ices contain chemical energy in the form of stored radicals produced by UV photolysis (*D'Hendecourt et al.*, 1982, 1986; *Tielens*, 2005). It should be emphasized that the thermal shock driven into the grain by the penetrating cosmic ray can also "lift" off a small quantity of larger species (*Johnson et al.*, 1991).

As the protostellar stage commences and temperatures begin to increase, the "dirty" grain mantles will be lost partially or totally via evaporation in a rather complex process stemming from the heterogeneous nature of the mantles (*Viti et al.*, 2004). Nearer to the bipolar outflows associated with protostellar regions, shock waves can also lead to the removal of the mantles. In particular, heating of ice mantles will lead to evaporation and an injection into the warm dense gas of large quantities of grain surface produced species such as water, methanol, and formaldehyde. Reactions among these simple molecules can then lead to more complex species (*Millar et al.*, 1991; *Charnley et al.*, 1992; *Caselli et al.*, 1993; *Rodgers and Charnley*, 2003; *Nomura and Millar*, 2004). The process begins with the production of H_3^+ via cosmic-ray bombardment of H_2, followed by the reaction of H_2^+ and H_2. This triatomic species, possibly through some intermediaries, transfers its proton to methanol. It has long been believed that the resulting protonated methanol, $CH_3OH_2^+$, will rapidly undergo alkyl cation transfer reactions with other species. With CH_3OH and H_2CO, this might lead to the formation of dimethyl ether and methyl formate, which are abundant complex molecules in hot corinos (section 4). However, the above reactions advocated to produce methyl formate do not occur, based on laboratory experiments (*Horn et al.,* 2004). A more successful synthesis seems to be the one leading to dimethyl ether

$$CH_3OH + CH_3OH_2^+ \rightarrow CH_3OHCH_3^+ + H_2O \qquad (5)$$

$$CH_3OHCH_3^+ + e^- \rightarrow CH_3OCH_3 + H \qquad (6)$$

but it is to be noted that experimental measurements of the neutral products of dissociative recombination reactions show that two-body products such as shown here are often minor channels. Eventually, these complex species are destroyed on a timescale of 10^{4-5} yr, depending on the temperature and density of the gas, to reform CO. Indeed, if the chemistry were understood well enough, abundances of these species have the potential to be a chemical "clock," timing the formation of the central object, which has heated up and evaporated the ices. Other determinants of age are abundance ratios of sulfur-bearing species, although the use of these indicators requires sophisticated modeling of both the chemical and the physical structure of the studied object (*Wakelam et al.,* 2004, 2005). The need for such a sophisticated approach arises because the abundance ratios of sulfur-bearing species, used as clocks, turn out to also be dependent on initial abundances following evaporation as well as on physical conditions.

Deuterium fractionation in the ices arises mainly from reactions involving deuterium atoms accreted from the highly fractionated gas. Such fractionation occurs in the abundant surface species (H_2O, H_2CO, CH_3OH, NH_3, CH_4) acting as signposts of a cold prestellar-core history in which gaseous D atoms are abundant. When these species are returned to the gas, the fractionation becomes more apparent. Although it is sometimes difficult to disentangle purely gas-phase fractionation from surface fractionation followed by evaporation, fractionation in gaseous molecules formed mainly on surfaces (e.g., methanol) is due at least initially to surface reactions. This fractionation will also be passed on to daughter products such as dimethyl ether and methyl formate via gas-phase processes. Thus, the deuteratium fractionation can be used as a tracer of the chemical routes involved in the formation of complex molecules in hot cores. Indeed, the fractionation pattern of dimethyl ether and methyl formate should reflect directly that of methanol and formaldehyde. Eventually, reactions in the warm dense gas of the hot core will reset this deuterium fractionation to that appropriate for the gas temperature (*Charnley et al.,* 1997). This chemical kinetic evolution may run differently for different species or even for different chemical groups in one species. Specifically, for the fractionation pattern of methanol, it has been suggested that protonation of methanol isotopomers followed by dissociative electron recombination reforming methanol will preferentially remove the deuteration from the OH group but not affect the CH_3 group (*Osamura et al.,* 2004). The analysis may be incorrect, however, because the most recent storage ring experiments (*Geppert et al.,* 2004) show that protonated and deuterated methanol ions undergo dissociative recombination reactions that lead to virtually no neutral methanol or its isotopologues at all. If all the gas-phase isotopologues of methanol are depleted at the same rate, then the abundance ratios are indeed determined by the surface chemistry during the previous cooler eon.

6. CONCLUSIONS: CLOSED AND OPEN QUESTIONS

In this contribution, we have summarized new insights into the physical and chemical composition and evolution of prestellar cores and Class 0 low-mass protostars:

1. On the physical structure, prestellar cores can be described by the density distribution of a Bonnor-Ebert sphere, whereas Class 0 sources are consistent with the Shu "inside-out" recipe.

2. At the center of prestellar core condensations, densities are so high and temperatures so low that possibly all heavy-element-bearing molecules freeze out onto the dust grains, forming mantles of ices. We discussed the possibility that small grains coagulate into larger grains during this phase.

3. A similar molecular depletion is observed in large regions of the outer envelopes of Class 0 sources. These regions are indistinguishable from prestellar cores on this basis, arguing that the latter are precursors of Class 0 sources.

4. The molecular depletion is accompanied by a dramatic enhancement of the molecular deuteration, observed in both prestellar cores and Class 0 sources. In the innermost regions of prestellar cores, the only observable molecules may be H_2D^+ and HD_2^+, which are likely as abundant as H_3^+ or even more so.

5. In Class 0 sources, many multiply deuterated molecules have been observed. Even triply deuterated molecules have been detected, with abundances enhanced by up to 13 orders of magnitude with respect to the D/H elemental abundance.

6. Triggered by the observations of multiply deuterated molecules in prestellar cores and Class 0 sources, a new class of models for deuteration has been developed in the last few years. These models predict that the low temperatures coupled with the disappearance of heavy-element-bearing molecules from the gas phase cause the formation of abundant H_2D^+, HD_2^+, and D_3^+, which are extremely efficient in passing their deuterium atoms on to other molecules. In extreme conditions, D_3^+ is predicted to be the most abundant molecular ion in the gas, surpassing by orders of magnitude the number density of H_3^+.

7. During the cold phases of low-mass star formation, grain mantles are formed, consisting mainly of water ice, but with many other molecules in quantities that can be important. Extreme molecular deuteration is a clear hallmark of this phase and is recorded in the mantles. Hydrogenated molecules, like formaldehyde and methanol, are also believed to be formed on the grain surfaces during the cold precollapse phase.

8. Class 0 envelopes consist of two chemically distinct regions: the outer envelopes and the inner regions, called hot corinos. The border between the two is where the dust temperature reaches the sublimation temperature of the grain ices. This has a major impact on the gas chemistry. In the hot corinos, the chemistry is dominated by the evaporation of the mantles, built up during the precollapse phase. Once the mantle components are in the gas phase, they undergo successive reactions leading to the formation of many complex organic molecules, similar to those observed in the hot cores.

Although important progress has been achieved in these last few years, it has also been accompanied by the rise of numerous new questions. For example, while molecular depletion is now an accepted "fact," it is still not totally clear whether in the innermost regions of the prestellar cores the density increases so much that not even a trace of heavy-element species remains in the gas phase. Also, it still remains unclear when and exactly how the collapse starts. It is clear now that observations of heavy-element-bearing molecules may not be the best way to proceed. In contrast, as we now realize, the ground-state transition of ortho-H_2D^+ (and, to a lesser extent, para-HD_2^+) provides a new and exciting probe with which to search for the long-sought collapse in the innermost regions of prestellar cores (*van der Tak et al.*, 2005). However, in order to fully exploit this new diagnostic, we have to first fully understand the ortho-para ratio of H_2D^+ (*Flower et al.*, 2006), which is not measured at all so far. It is also clear that models for the prestellar core phase have improved dramatically with the inclusion of the extreme depletions of the gas phase and the formation of the multiply deuterated forms of H_3^+. However, these models still struggle to reproduce the extreme observed deuteration, especially that of formaldehyde and methanol. While this is likely due to grain surface chemistry, it remains the case that, unless there is an enhancement in the reactivity of D over H, the gas landing on the grain surfaces at the moment of the mantle formation must have an atomic D/H ratio higher than accounted for by the current gas-phase chemistry models. We also have little knowledge as to whether the prediction that D_3^+ is the most abundant molecular ion in extreme conditions is correct. The coming years will likely see many of these questions answered, with the advent of the satellite Herschel and, foremost, the interferometer ALMA.

If many questions remain about the cold precollapse phase, even more questions are unanswered for the protostellar phase, and many have likely not even been asked. For example, the very nature of the hot corinos is still very much debated. Several authors question how much of the observed warm gas, with complex organic molecules, is due to the interaction of an outflow with the inner envelope (*Chandler et al.*, 2005). Another possibility invoked is that this warm gas enriched with complex organic molecules resides in a disk-like atmosphere (*Jørgensen et al.*, 2005a). The reality is indeed that very few observations have resolved hot corinos; actually, in just one source has this so far occurred (*Bottinelli et al.*, 2004b). Until more sensitive observations are available with SMA and ALMA, the question will remain open for debate. Another open question regards the chemistry leading to the observed molecular complexity in hot corinos. No clear observations are able so far to distinguish the roles of gas-phase and grain-surface chemistry in the formation of formic acid or methyl formate, just to mention two molecules frequently detected. The answer to this question will likely need not only more and more-sensitive observations toward the hot corinos, but also laboratory experiments coupled with modeling in order to elucidate the role of the grains in the story. Last, but not least, we suspect that we are just viewing the tip of the iceberg: What is still hidden from our eyes? What is the ultimate molecular complexity reached during the hot corino phase? Do biotic molecules form? Do they play a role in the trigger and/or diffusion of life in the universe (see chapter by Gaidos and Selsis)?

Finally, many more questions exist regarding the fate of the molecules formed through the prestellar and protostellar phases. It is likely that some condense again onto the grain mantles during the phase of the protoplanetary disk, at least in the outer zones of the disk that are shielded by FUV photons (see chapters by Bergin et al., Dutrey et al., and Dullemond et al.). The same grains are likely to coagulate into larger aggregates, which are the bricks from which the planets eventually form (see chapters by Dominik et al. and Wadhwa et al.). How much of the mantle formed in the prestellar core phase is preserved? Do the molecules, and particularly the complex organic molecules, formed during the hot corino phase freeze out onto the new mantles without other alterations? Or what kind of alterations do they undergo? How pristine is the material forming the comets, asteroids, and planets?

To reconstruct the full story, we need many more pieces of the puzzle. One strong motivation for pursuing the story of protoplanetary disks is that their aftermath may have been part of the history of our own solar system. The water of the oceans is an emblematic example. The ratio between HDO and H_2O (known as SMOW) is ~10^{-4} (*De Witt et al.*, 1980), fully 10× larger than the D/H elemental abundance of the solar nebula (*Geiss*, 1993). The origin of the oceans has been long debated and new theories keep coming. The deuterium enhancement is in line with the theory that much of the ocean water comes from the bombardment of the early Earth by comets (*Owen and Bar-Nun*, 1995), as comets show a similarly enhanced HDO/H_2O ratio (*Meier et al.*, 1998; *Bockelée-Morvan et al.*, 1998). Other possibilities include water brought by smaller bodies of the solar system (meteorites and asteroids) (*Raymond et al.*, 2004; *Gomes et al.*, 2005). These bodies, as well as comets, have the imprint of the precollapse/protostellar phase in their ices (see chapters by Jewitt et al., Wooden et al., and Alexander et al.). Regarding deuterium fractionation, the question remains as to what extent the enhanced HDO/H_2O ratio is a legacy from the protostellar/protoplanetary phase and to what extent it has been modified by subsequent chemistry.

And in addition to water, what other molecules were inherited by the early Earth and how did they arrive? As with other molecular indicators, how much of the terrestrial HDO/H$_2$O ratio is a legacy depends on how it evolved during the various phases of Earth's formation. So far, we have constraints on the evolution of deuterium fractionation from the precollapse, protostellar, and protoplanetary phases for only two molecules, HCO$^+$ and H$_2$O, and in the outer zones of the disk only, which may not necessarily be connected with cometary formation. The case of DCO$^+$/HCO$^+$, representing deuteration in the gas phase, seems to keep a rather constant value, around 10% (*van Dishoeck et al.*, 2003). The HDO/H$_2$O ratio, very likely a product of the grain mantles, seems to remain quite similar too, at a level of ~1%, with just a relatively small decrease in the protoplanetary phase (*Ceccarelli et al.*, 2005; *Dominik et al.*, 2005). Evidently, this information alone is far too little to resolve the problem of the oceanic origin, but shows how important the study of chemistry in the prestellar and protostellar phases of star formation is and how much it will eventually aid our understanding of later phases up through the present eon. The coming decade will likely see many of these questions answered with the advent of the satellite Herschel and the submillimeter interferometer ALMA.

Acknowledgments. We wish to thank the referee for valuable comments on the manuscript. This study was supported in part by the European Community's Human Potential Programme under contract MCRTN 512302,Molecular Universe, the French "Projet National PCMI," and the NASA Planetary Geology and Geophysics Program under grant NAG 5-10201.

REFERENCES

Aikawa Y., Ohashi N., and Herbst E. (2003) *Astrophys. J., 593*, 906–924.

Aikawa Y., Herbst E., Roberts H., and Caselli P. (2005) *Astrophys. J., 620*, 330–346.

Alves J. F., Lada C. J., and Lada E. A. (2001) *Nature, 409*, 159–161.

André P., Ward-Thompson D., and Barsony M. (2000) In *Protostars and Planets IV* (V. Mannings et al., eds.), pp. 59–63. Univ. of Arizona, Tucson.

Bacmann A., Lefloch B., Ceccarelli C., Castets A., Steinacker J., and Loinard L. (2002) *Astron. Astrophys., 389*, L6–L10.

Bacmann A., Lefloch B., Ceccarelli C., Steinacker J., Castets A., and Loinard L. (2003) *Astrophys. J., 585*, L55–L58.

Bergin E. A., Ciardi D. R., Lada C. J., Alves J., and Lada E. A. (2001) *Astrophys. J., 557*, 209–225.

Bergin E. A., Alves J., Huard T., and Lada C. J. (2002) *Astrophys. J., 570*, L101–L104.

Bianchi S., Gonçalves J., Albrecht M., Caselli P., Chini R., Galli D., and Walmsley M. (2003) *Astron. Astrophys., 399*, L43–L46.

Biham O. and Lipshtat A. (2002) *Phys. Rev. E, 66*, 100–112.

Biham O., Furman I., Pirronello V., and Vidali G. (2001) *Astrophys. J., 553*, 595–603.

Bockelée-Morvan D., Gautier D., Lis D. C., et al. (1998) *Icarus, 133*, 147–162.

Boogert A. C. A., Hogerheijde M. R., Ceccarelli C., et al. (2002) *Astrophys. J., 570*, 708–723.

Bottinelli S., Ceccarelli C., Lefloch B., et al. (2004a) *Astrophys. J., 615*, 354–363.

Bottinelli S., Ceccarelli C., Neri R., Williams J. P., Caux E., Cazaux S., Lefloch B., Maret S., and Tielens A. G. G. M. (2004b) *Astrophys. J., 617*, L69–L72.

Bottinelli S., Ceccarelli C., Williams J. P., and Lefloch B. (2006) *Astrophys. J.*, in press.

Bringa E. M. and Johnson R. E. (2004) *Astrophys. J., 603*, 159–164.

Brown P. D. and Millar T. J. (1989) *Mon. Not. R. Astron. Soc., 237*, 661–671.

Burke J. R. and Hollenbach D. J. (1983) *Astrophys. J., 265*, 223–234.

Caselli P. (2006) *Mem. Soc. Astron. Ital. Suppl., 10*, 146.

Caselli P., Hasegawa T. I., and Herbst E. (1993) *Astrophys. J., 408*, 548–558.

Caselli P., Walmsley C. M., Tafalla M., Dore L., and Myers P. C. (1999) *Astrophys. J., 523*, L165–L169.

Caselli P., Stantcheva T., Shalabiea O., Shematovich V. I., and Herbst E. (2002a) *Planet. Space Sci., 50*, 1257–1266.

Caselli P., Walmsley C. M., Zucconi A., Tafalla M., Dore L., and Myers P. C. (2002b) *Astrophys. J., 565*, 331–343.

Caselli P., Walmsley C. M., Zucconi A., Tafalla M., Dore L., and Myers P. C. (2002c) *Astrophys. J., 565*, 344–358.

Caselli P., van der Tak F. F. S., Ceccarelli C., and Bacmann A. (2003) *Astron. Astrophys., 403*, L37–L41.

Cazaux S. and Tielens A. G. G. M. (2004) *Astrophys. J., 604*, 222–237.

Cazaux S., Tielens A. G. G. M., Ceccarelli C., Castets A., Wakelam V., Caux E., Parise B., and Teyssier D. (2003) *Astrophys. J., 593*, L51–L55.

Ceccarelli C. (2004a) In *The Dense Interstellar Medium in Galaxies* (S. Pfalzner et al., eds.), pp. 473–481. Springer Proceedings in Physics, Berlin.

Ceccarelli C. (2004b) In *Star Formation in the Interstellar Medium* (D. Johnstone et al., eds.), pp. 195–207. ASP Conf. Series 323, San Francisco.

Ceccarelli C. and Dominik C. (2005) *Astron. Astrophys., 440*, 583–593.

Ceccarelli C., Hollenbach D. J., and Tielens A. G. G. M. (1996) *Astrophys. J., 471*, 400–432.

Ceccarelli C., Castets A., Loinard L., Caux E., and Tielens A. G. G. M. (1998) *Astron. Astrophys., 338*, L43–L46.

Ceccarelli C., Caux E., Loinard L., et al. (1999) *Astron. Astrophys., 342*, L21–L24.

Ceccarelli C., Castets A., Caux E., Hollenbach D., Loinard L., Molinari S., and Tielens A. G. G. M. (2000a) *Astron. Astrophys., 355*, 1129–1137.

Ceccarelli C., Loinard L., Castets A., Tielens A. G. G. M., and Caux E. (2000b) *Astron. Astrophys., 357*, L9–L12.

Ceccarelli C., Loinard L., Castets A., Tielens A. G. G. M., Caux E., Lefloch B., and Vastel C. (2001) *Astron. Astrophys., 372*, 998–1004.

Ceccarelli C., Maret S., Tielens A. G. G. M., Castets A., and Caux E. (2003) *Astron. Astrophys., 410*, 587–595.

Ceccarelli C., Dominik C., Caux E., Lefloch B., and Caselli P. (2005) *Astrophys. J., 631*, L81–L84.

Chandler C. J., Brogan C. L., Shirley Y. L., and Loinard L. (2005) *Astrophys. J., 632*, 371–396.

Chang Q., Cuppen H. M., and Herbst E. (2005) *Astron. Astrophys., 434*, 599–611.

Charnley S. B. (2001) *Astrophys. J., 562*, L99–L102.

Charnley S. B., Tielens A. G. G. M., and Millar T. J. (1992) *Astrophys. J., 399*, L71–L74.

Charnley S. B., Tielens A. G. G. M., and Rodgers S. D. (1997) *Astrophys. J., 482*, L203–L207.

Crapsi A., Caselli P., Walmsley C. M., Tafalla M., Lee C. W., Bourke T. L., and Myers P. C. (2004) *Astron. Astrophys., 420*, 957–974.

Crapsi A., Caselli P., Walmsley C. M., Myers P. C., Tafalla M., Lee C. W., and Bourke T. L. (2005a) *Astrophys. J., 619*, 379–406.

Crapsi A., Devries C. H., Huard T. L., et al. (2005b) *Astron. Astrophys., 439*, 1023–1032.

Cuppen H. M. and Herbst E. (2005) *Mon. Not. R. Astron. Soc., 361*, 565–576.

Dartois E., Thi W.-F., Geballe T. R., Deboffle D., d'Hendecourt L., and van Dishoeck E. (2003) *Astron. Astrophys., 399*, 1009–1020.

De Witt J., van der Straaten C., and Mook W. (1980) *Geostand. Newslett., 4*, 33–36.

D'Hendecourt L. B., Allamandola L. J., Baas F., and Greenberg J. M. (1982) *Astron. Astrophys., 109*, L12–L14.

D'Hendecourt L. B., Allamandola L. J., Grim R. J. A., and Greenberg J. M. (1986) *Astron. Astrophys., 158,* 119–134.

Di Francesco J., Myers P. C., Wilner D. J., Ohashi N., and Mardones D. (2001) *Astrophys. J., 562,* 770–789.

Dominik C., Ceccarelli C., Hollenbach D., and Kaufman M. (2005) *Astrophys. J., 635,* L85–L88.

Evans N. J., Rawlings J. M. C., Shirley Y. L., and Mundy L. G. (2001) *Astrophys. J., 557,* 193–208.

Evans N. J., Lee J.-E., Rawlings J. M. C., and Choi M. (2005) *Astrophys. J., 626,* 919–932.

Flower D. R., Pineau des Forêts G., and Walmsley C. M. (2005) *Astron. Astrophys., 436,* 933–943.

Flower D. R., Pineau des Forêts G., and Walmsley C. M. (2006) *Astron. Astrophys., 449,* 621.

Frerking M. A., Langer W. D., and Wilson R. W. (1982) *Astrophys. J., 262,* 590–605.

Galli D., Walmsley M., and Gonçalves J. (2002) *Astron. Astrophys., 394,* 275–284.

Geiss J. (1993) In *Origin and Evolution of Elements* (N. Pranzos et al., eds.), pp. 89–106. Cambridge Univ., Cambridge.

Geppert W. D., Ehlerding A., Hellberg F., et al. (2004) *Astrophys. J., 613,* 1302–1309.

Giannini T., Nisini B., and Lorenzetti D. (2001) *Astrophys. J., 555,* 40–57.

Gibb E. L.,Whittet D. C. B., Schutte W. A., et al. (2000) *Astrophys. J., 536,* 347–356.

Gomes R., Levison H. F., Tsiganis K., and Morbidelli A. (2005) *Nature, 435,* 466–469.

Green N. J. B., Toniazzo T., Pilling M. J., Ruffle D. P., Bell N., and Hartquist T. W. (2001) *Astron. Astrophys., 375,* 1111–1119.

Guélin M., Langer W. D., Snell R. L., and Wootten H. A. (1977) *Astrophys. J., 217,* L165–L168.

Hasegawa T. I. and Herbst E. (1993) *Mon. Not. R. Astron. Soc., 261,* 83–102.

Hasegawa T. I., Herbst E., and Leung C. M. (1992) *Astrophys. J. Suppl., 82,* 167–195.

Herbst E. (2005) In *The Dusty and Molecular Universe: A Prelude to Herschel and ALMA* (A.Wilson, ed.), pp. 205–210. ESA SP-577, Noordwijk, Netherlands.

Hidaka H., Watanabe N., Shiraki T., Nagaoka A., and Kouchi A. (2004) *Astrophys. J., 614,* 1124–1131.

Hirota T., Ikeda M., and Yamamoto S. (2001) *Astrophys. J., 547,* 814–828.

Horn A., Møllendal H., Sekiguchi O., et al. (2004) *Astrophys. J., 611,* 605–614.

Johnson R. E., Donn B., Pirronello V., and Sundqvist B. (1991) *Astrophys. J., 379,* L75–L77.

Jørgensen J. K., Schöier F. L., and van Dishoeck E. F. (2002) *Astron. Astrophys., 389,* 908–930.

Jørgensen J. K., Schöier F. L., and van Dishoeck E. F. (2004) *Astron. Astrophys., 416,* 603–622.

Jørgensen J. K., Bourke T. L., Myers P. C., Schöier F. L., van Dishoeck E. F., and Wilner D. J. (2005a) *Astrophys. J., 632,* 973–981.

Jørgensen J. K., Lahuis F., Schöier F. L., et al. (2005b) *Astrophys. J., 631,* L77–L80.

Jørgensen J. K., Schöier F. L., and van Dishoeck E. F. (2005c) *Astron. Astrophys., 437,* 501–515.

Jørgensen J. K., Schöier F. L., and van Dishoeck E. F. (2005d) *Astron. Astrophys., 435,* 177–182.

Kuan Y.-J., Huang H.-C., Charnley S. B., et al. (2004) *Astrophys. J., 616,* L27–L30.

Kurtz S., Cesaroni R., Churchwell E., Hofner P., and Walmsley C. M. (2000) In *Protostars and Planets IV* (V. Mannings et al., eds.), pp. 299–315. Univ. of Arizona, Tucson.

Langer W. D., van Dishoeck E. F., Bergin E. A., Blake G. A., Tielens A. G. G. M., Velusamy T., and Whittet D. C. B. (2000) In *Protostars and Planets IV* (V. Mannings et al., eds.), pp. 29–43. Univ. of Arizona, Tucson.

Leger A. (1983) *Astron. Astrophys., 123,* 271–278.

Leger A., Jura M., and Omont A. (1985) *Astron. Astrophys., 144,* 147–160.

Linsky J. L. (2003) *Space Sci. Rev., 106,* 49–60.

Lipshtat A. and Biham O. (2004) *Phys. Rev. Lett., 93,* 1706–1710.

Lis D. C., Roueff E., Gerin M., Phillips T. G., Coudert L. H., van der Tak F. F. S., and Schilke P. (2002) *Astrophys. J., 571,* L55–L58.

Lis D. C., Gerin M., Roueff E., Vastel C., and Phillips T. G. (2006) *Astrophys. J., 636,* 916–922.

Liszt H. (2003) *Astron. Astrophys., 398,* 621–630.

Loinard L., Castets A., Ceccarelli C., Tielens A. G. G. M., Faure A., Caux E., and Duvert G. (2000) *Astron. Astrophys., 359,* 1169–1174.

Loinard L., Castets A., Ceccarelli C., Caux E., and Tielens A. G. G. M. (2001) *Astrophys. J., 552,* L163–L166.

Loinard L., Castets A., Ceccarelli C., et al. (2002) *Planet. Space Sci., 50,* 1205–1213.

Looney L. W., Mundy L. G., and Welch W. J. (2003) *Astrophys. J., 592,* 255–265.

Loren R. B., Wootten A., and Wilking B. A. (1990) *Astrophys. J., 365,* 269–286.

Mangum J. G. and Wootten A. (1993) *Astrophys. J. Suppl., 89,* 123–153.

Marcelino N., Cernicharo J., Roueff E., Gerin M., and Mauersberger R. (2005) *Astrophys. J., 620,* 308–320.

Maret S., Caux E., and Ceccarelli C. (2002a) *Astrophys. Space Sci., 281,* 139–140.

Maret S., Ceccarelli C., Caux E., Tielens A. G. G. M., and Castets A. (2002b) *Astron. Astrophys., 395,* 573–585.

Maret S., Ceccarelli C., Caux E., et al. (2004) *Astron. Astrophys., 416,* 577–594.

Maret S., Ceccarelli C., Tielens A. G. G. M., et al. (2005) *Astron. Astrophys., 442,* 527–538.

Meier R., Owen T. C., Matthews H. E., et al. (1998) *Science, 279,* 842–846.

Millar T. J. (2005) *Astron. Geophys., 46,* 29–32.

Millar T. J., Bennett A., and Herbst E. (1989) *Astrophys. J., 340,* 906–920.

Millar T. J., Herbst E., and Charnley S. B. (1991) *Astrophys. J., 369,* 147–156.

Nagaoka A., Watanabe N., and Kouchi A. (2005) *Astrophys. J., 624,* L29–L32.

Nisini B., Giannini T., and Lorenzetti D. (2002) *Astrophys. J., 574,* 246–257.

Nomura H. and Millar T. J. (2004) *Astron. Astrophys., 414,* 409–423.

Osamura Y., Roberts H., and Herbst E. (2004) *Astron. Astrophys., 421,* 1101–1111.

Ossenkopf V. (1993) *Astron. Astrophys., 280,* 617–646.

Owen T. and Bar-Nun A. (1995) *Icarus, 116,* 215–226.

Pagani L., Wannier P. G., Frerking M. A., et al. (1992) *Astron. Astrophys., 258,* 472–478.

Pagani L., Lagache G., Bacmann A., et al. (2003) *Astron. Astrophys., 406,* L59–L62.

Pagani L., Bacmann A., Motte F., et al. (2004) *Astron. Astrophys., 417,* 605–613.

Pagani L., Pardo J.-R., Apponi A. J., Bacmann A., and Cabrit S. (2005) *Astron. Astrophys., 429,* 181–192.

Parise B., Ceccarelli C., Tielens A. G. G. M., Herbst E., Lefloch B., Caux E., Castets A., Mukhopadhyay I., Pagani L., and Loinard L. (2002) *Astron. Astrophys., 393,* L49–L53.

Parise B., Simon T., Caux E., Dartois E., Ceccarelli C., Rayner J., and Tielens A. G. G. M. (2003) *Astron. Astrophys., 410,* 897–904.

Parise B., Castets A., Herbst E., Caux E., Ceccarelli C., Mukhopadhyay I., and Tielens A. G. G. M. (2004) *Astron. Astrophys., 416,* 159–163.

Parise B., Caux E., Castets A., et al. (2005a) *Astron. Astrophys., 431,* 547–554.

Parise B., Ceccarelli C., and Maret S. (2005b) *Astron. Astrophys., 441,* 171–179.

Parise B., Ceccarelli C., Tielens A., Castets A., Caux E., et al. (2006) *Astron. Astrophys., 453,* 949.

Pickles J. B. and Williams D. A. (1977) *Astrophys. Space Sci., 52,* 443–452.

Raymond S. N., Quinn T., and Lunine J. I. (2004) *Icarus, 168,* 1–17.

Redman M. P., Rawlings J. M. C., Nutter D. J., Ward-Thompson D., and Williams D. A. (2002) *Mon. Not. R. Astron. Soc., 337,* L17–L21.

Remijan A. J. and Hollis J. M. (2006) *Astrophys J., 640,* 842.

Roberts H., Fuller G. A., Millar T. J., Hatchell J., and Buckle J. V. (2002) *Planet. Space Sci., 50,* 1173–1178.

Roberts H., Herbst E., and Millar T. J. (2003) *Astrophys. J., 591,* L41–L44.

Roberts H., Herbst E., and Millar T. J. (2004) *Astron. Astrophys., 424,* 905–917.

Rodgers S. D. and Charnley S. B. (2003) *Astrophys. J., 585,* 355–371.

Roueff E. and Gerin M. (2003) *Space Sci. Rev., 106,* 61–72.

Roueff E., Tiné S., Coudert L. H., Pineau des Forêts G., Falgarone E., and Gerin M. (2000) *Astron. Astrophys., 354,* L63–L66.

Roueff E., Lis D. C., van der Tak F. F. S., Gerin M., and Goldsmith P. F. (2005) *Astron. Astrophys., 438,* 585–598.

Ruffle D. P. and Herbst E. (2000) *Mon. Not. R. Astron. Soc., 319,* 837–850.

Sandell G. and Knee L. B. G. (2001) *Astrophys. J., 546,* L49–L52.

Schöier F. L., Jørgensen J. K., van Dishoeck E. F., and Blake G. A. (2002) *Astron. Astrophys., 390,* 1001–1021.

Schöier F. L., Jørgensen J. K., van Dishoeck E. F., and Blake G. A. (2004) *Astron. Astrophys., 418,* 185–202.

Schutte W. A., Boogert A. C. A., Tielens A. G. G. M., et al. (1999) *Astron. Astrophys., 343,* 966–976.

Shah R. Y. and Wootten A. (2001) *Astrophys. J., 554,* 933–947.

Shen C. J., Greenberg J. M., Schutte W. A., and van Dishoeck E. F. (2004) *Astron. Astrophys., 415,* 203–215.

Shirley Y. L., Evans N. J., and Rawlings J. M. C. (2002) *Astrophys. J., 575,* 337–353.

Shu F. H. (1977) *Astrophys. J., 214,* 488–497.

Stantcheva T. and Herbst E. (2003) *Mon. Not. R. Astron. Soc., 340,* 983–988.

Stantcheva T. and Herbst E. (2004) *Astron. Astrophys., 423,* 241–251.

Stark R., van der Tak F. F. S., and van Dishoeck E. F. (1999) *Astrophys. J., 521,* L67–L70.

Tafalla M. and Santiago J. (2004) *Astron. Astrophys., 414,* L53–L56.

Tafalla M., Mardones D., Myers P. C., Caselli P., Bachiller R., and Benson P. J. (1998) *Astrophys. J., 504,* 900–908.

Tafalla M., Myers P. C., Caselli P., Walmsley C. M., and Comito C. (2002) *Astrophys. J., 569,* 815–835.

Tafalla M., Myers P. C., Caselli P., and Walmsley C. M. (2004) *Astron. Astrophys., 416,* 191–212.

Tielens A. G. G. M. (1983) *Astron. Astrophys., 119,* 177–184.

Tielens A. G. G. M. (2005) *The Physics and Chemistry of the Interstellar Medium.* Cambridge Univ., Cambridge.

Tielens A. G. G. M. and Hagen W. (1982) *Astron. Astrophys., 114,* 245–260.

Tiné S., Roueff E., Falgarone E., Gerin M., and Pineau des Forêts G. (2000) *Astron. Astrophys., 356,* 1039–1049.

Turner B. E. (1990) *Astrophys. J., 362,* L29–L33.

van der Tak F. F. S., Schilke P., Müller H. S. P., Lis D. C., Phillips T. G., Gerin M., and Roueff E. (2002) *Astron. Astrophys., 388,* L53–L56.

van der Tak F. F. S., Caselli P., and Ceccarelli C. (2005) *Astron. Astrophys., 439,* 195–203.

van Dishoeck E. F. and Blake G. A. (1998) *Ann. Rev. Astron. Astrophys., 36,* 317–368.

van Dishoeck E. F., Phillips T. G., Keene J., and Blake G. A. (1992) *Astron. Astrophys., 261,* L13–L16.

van Dishoeck E. F., Blake G. A., Jansen D. J., and Groesbeck T. D. (1995) *Astrophys. J., 447,* 760–774.

van Dishoeck E. F., Thi W.-F., and van Zadelhoff G.-J. (2003) *Astron. Astrophys., 400,* L1–L4.

Vastel C., Phillips T. G., Ceccarelli C., and Pearson J. (2003) *Astrophys. J., 593,* L97–L100.

Vastel C., Phillips T. G., and Yoshida H. (2004) *Astrophys. J., 606,* L127–L130.

Vastel C., Caselli P., Ceccarelli C., et al. (2006) *Astrophys. J., 645,* 1198.

Viti S., Collings M. P., Dever J. W., McCoustra M. R. S., and Williams D. A. (2004) *Mon. Not. R. Astron. Soc., 354,* 1141–1145.

Wakelam V., Caselli P., Ceccarelli C., Herbst E., and Castets A. (2004) *Astron. Astrophys., 422,* 159–169.

Wakelam V., Ceccarelli C., Castets A., et al. (2005) *Astron. Astrophys., 437,* 149–158.

Walmsley C. M., Flower D. R., and Pineau des Forêts G. (2004) *Astron. Astrophys., 418,* 1035–1043.

Watanabe N., Shiraki T., and Kouchi A. (2003) *Astrophys. J., 588,* L121–L124.

Watson W. D., Snyder L. E., and Hollis J. M. (1978) *Astrophys. J., 222,* L145–L147.

Williams S. J., Fuller G. A., and Sridharan T. K. (2005) *Astron. Astrophys., 434,* 257–274.

Wootten A. (1989) *Astrophys. J., 337,* 858–866.

Young C. H. and Evans N. J. (2005) *Astrophys. J., 627,* 293–309.

Young K. E., Lee J.-E., Evans N. J., Goldsmith P. F., and Doty S. D. (2004) *Astrophys. J., 614,* 252–266.

Zucconi A., Walmsley C. M., and Galli D. (2001) *Astron. Astrophys., 376,* 650–662.

Molecular Cloud Turbulence and Star Formation

Javier Ballesteros-Paredes
Universidad Nacional Autónoma de México

Ralf S. Klessen
Astrophysikalisches Institut Potsdam

Mordecai-Mark Mac Low
American Museum of Natural History at New York

Enrique Vázquez-Semadeni
Universidad Nacional Autónoma de México

We review the properties of turbulent molecular clouds (MCs), focusing on the physical processes that influence star formation (SF). Molecular cloud formation appears to occur during large-scale compression of the diffuse interstellar medium (ISM) driven by supernovae, magnetorotational instability, or gravitational instability in galactic disks of stars and gas. The compressions generate turbulence that can accelerate molecule production and produce the observed morphology. We then review the properties of MC turbulence, including density enhancements observed as clumps and cores, magnetic field structure, driving scales, the relation to observed scaling relations, and the interaction with gas thermodynamics. We argue that MC cores are dynamical, not quasistatic, objects with relatively short lifetimes not exceeding a few megayears. We review their morphology, magnetic fields, density and velocity profiles, and virial budget. Next, we discuss how MC turbulence controls SF. On global scales turbulence prevents monolithic collapse of the clouds; on small scales it promotes local collapse. We discuss its effects on the SF efficiency, and critically examine the possible relation between the clump mass distribution and the initial mass function, and then turn to the redistribution of angular momentum during collapse and how it determines the multiplicity of stellar systems. Finally, we discuss the importance of dynamical interactions between protostars in dense clusters, and the effect of the ionization and winds from those protostars on the surrounding cloud. We conclude that the interaction of self-gravity and turbulence controls MC formation and behavior, as well as the core- and star-formation processes within them.

1. INTRODUCTION

Star formation occurs within molecular clouds (MCs). These exhibit supersonic linewidths, which are interpreted as evidence for supersonic turbulence (*Zuckerman and Evans,* 1974). Early studies considered this property mainly as a mechanism of MC support against gravity. In more recent years, however, it has been realized that turbulence is a fundamental ingredient of MCs, determining properties such as their morphology, lifetimes, rate of star formation, etc.

Turbulence is a multiscale phenomenon in which kinetic energy cascades from large scales to small scales. The bulk of the specific kinetic energy remains at large scales. Turbulence appears to be dynamically important from scales of whole MCs down to cores (e.g., *Larson,* 1981; *Ballesteros-Paredes et al.,* 1999a; *Mac Low and Klessen,* 2004). Thus, early microturbulent descriptions postulating that turbulence *only* acts on small scales did not capture major effects at large scales such as cloud and core formation by the turbulence.

Turbulence in the warm diffuse interstellar medium (ISM) is transonic, with both the sound speed and the nonthermal motions being ~10 km s^{-1} (*Kulkarni and Heiles,* 1987; *Heiles and Troland,* 2003), while within MCs it is highly supersonic, with Mach numbers $\mathcal{M} \approx 5$–20 (*Zuckerman and Palmer,* 1974). Both media are highly compressible. Hypersonic velocity fluctuations in the roughly isothermal molecular gas produce large density enhancements at shocks. The velocity fluctuations in the warm diffuse medium, despite being only transonic, can still drive large density enhancements because they can push the medium into a thermally unstable regime in which the gas cools rapidly into a cold, dense regime (*Hennebelle and Pérault,* 1999). In general, the atomic gas responds close to isobarically to dynamic compressions for densities $0.5 \le n/cm^{-3} \le 40$ (*Gazol et al.,* 2005). We argue in this review that MCs form from dynamically evolving, high-density features in the diffuse ISM. Similarly, their internal substructure of clumps and cores are also transient density enhancements continually changing their shape and even the material contained in the turbulent flow, behaving as something between discrete objects and waves (*Vázquez-Semadeni et al.*, 1996). Because of their higher density, some clumps and cores become gravitationally unstable and collapse to form stars.

We here discuss the main advances in our understanding of the turbulent properties of MCs and their implications for star formation since the reviews by *Vázquez-Semadeni et al.* (2000) and *Mac Low and Klessen* (2004), proceeding from large (giant MCs) to small (core) scales.

2. MOLECULAR CLOUD FORMATION

The questions of how MCs form and what determines their physical properties have remained unanswered until recently (e.g., *Elmegreen*, 1991; *Blitz and Williams*, 1999). Giant molecular clouds (GMCs) have gravitational energy far exceeding their thermal energy (*Zuckerman and Palmer*, 1974), although comparable to their turbulent (*Larson*, 1981) and magnetic energies (*Myers and Goodman*, 1988; *Crutcher*, 1999; *Bourke et al.*, 2001; *Crutcher et al.*, 2004). This near equipartition of energies has traditionally been interpreted as indicative of approximate virial equilibrium, and thus of general stability and longevity of the clouds (e.g., *McKee et al.*, 1993; *Blitz and Williams*, 1999). In this picture, the fact that MCs have thermal pressures exceeding that of the general ISM by roughly one order of magnitude (e.g., *Blitz and Williams*, 1999) was interpreted as a consequence of their being strongly self-gravitating (e.g., *McKee*, 1995), while the magnetic and turbulent energies were interpreted as support against gravity. Because of the interpretation that their overpressures were due to self-gravity, MCs could not be incorporated into global ISM models based on thermal pressure equilibrium, such as those by *Field et al.* (1969), *McKee and Ostriker* (1977), and *Wolfire et al.* (1995).

Recent work suggests instead that MCs are likely to be *transient*, dynamically evolving features produced by compressive motions of either gravitational or turbulent origin, or some combination thereof. In what follows we first discuss these two formation mechanisms, and then discuss how they can give rise to the observed physical and chemical properties of MCs.

2.1. Formation Mechanisms

Large-scale gravitational instability in the combined medium of the collisionless stars and the collisional gas appears likely to be the main driver of GMC formation in galaxies. In the Milky Way, MCs, and particularly GMCs, are observed to be concentrated toward the spiral arms (*Lee et al.*, 2001; *Blitz and Rosolowsky*, 2005; *Stark and Lee*, 2005), which are the first manifestation of gravitational instability in the combined medium. We refer to the instability parame-

ter for stars and gas combined (*Rafikov*, 2001) as the Toomre parameter for stars and gas Q_{sg}. As Q_{sg} drops below unity, gravitational instability drives spiral density waves (*Lin and Shu*, 1964; *Lin et al.*, 1969). High-resolution numerical simulations of the process then show the appearance of regions of local gravitational collapse in the arms, with the timescale for collapse depending exponentially on the value of Q_{sg} (*Li et al.*, 2005). In more strongly gravitationally unstable regions toward the centers of galaxies, collapse occurs more generally, with molecular gas dominating over atomic gas, as observed by *Wong and Blitz* (2002).

The multikiloparsec scale of spiral arms driven by gravitational instability suggests that this mechanism should preferentially form GMCs, consistent with the fact that GMCs are almost exclusively confined to spiral arms in our galaxy (*Stark and Lee*, 2005). Also, stronger gravitational instability causes collapse to occur closer to the disk midplane, producing a smaller scale height for GMCs (*Li et al.*, 2005), as observed by *Dalcanton et al.* (2004) and *Stark and Lee* (2005). In gas-poor galaxies like our own, the self-gravity of the gas in the unperturbed state is negligible with respect to that of the stars, so this mechanism reduces to the standard scenario of the gas falling into the potential well of the stellar spiral and shocking there (*Roberts*, 1969), with its own self-gravity only becoming important as the gas is shocked and cooled (section 2.2).

A second mechanism of MC formation operating at somewhat smaller scales (tens to hundreds of parsecs) is the ram pressure from supersonic flows, which can be produced by a number of different sources. Supernova explosions drive blast waves and superbubbles into the ISM. Compression and gravitational collapse can occur in discrete superbubble shells (*McCray and Kafatos*, 1987). The ensemble of supernova remnants, superbubbles, and expanding HII regions in the ISM drives a turbulent flow (*Vázquez-Semadeni et al.*, 1995; *Passot et al.*, 1995; *Rosen and Bregman*, 1995; *Korpi et al.*, 1999; *Gazol-Patiño and Passot*, 1999; *de Avillez*, 2000; *Avila-Reese and Vázquez-Semadeni*, 2001; *Slyz et al.*, 2005; *Mac Low et al.*, 2005; *Dib et al.*, 2006a; *Joung and Mac Low*, 2006). Finally, the magnetorotational instability (*Balbus and Hawley*, 1998) in galaxies can drive turbulence with velocity dispersion close to that of supernova-driven turbulence even in regions far from recent star formation (*Sellwood and Balbus*, 1999; *Kim et al.*, 2003; *Dziourkevitch et al.*, 2004; *Piontek and Ostriker*, 2005). Both of these sources of turbulence drive flows with velocity dispersion ~10 km s^{-1}, similar to that observed (*Kulkarni and Heiles*, 1987; *Dickey and Lockman*, 1990; *Heiles and Troland*, 2003). The high-$\mathcal{M}$ tail of the shock distribution in the warm medium involved in this turbulent flow can drive MC formation.

Small-scale, scattered, ram-pressure compressions in a globally gravitationally stable medium can produce clouds that are globally supported against collapse, but that still undergo local collapse in their densest substructures. These events involve a small fraction of the total cloud mass and are characterized by a shorter freefall time than that of the parent cloud because of the enhanced local density (*Sasao*, 1973; *Elmegreen*, 1993; *Padoan*, 1995; *Vázquez-Semadeni et al.*, 1996; *Ballesteros-Paredes et al.*, 1999a,b; *Klessen et al.*, 2000; *Heitsch et al.*, 2001a; *Vázquez-Semadeni et al.*, 2003, 2005a; *Li and Nakamura*, 2004; *Nakamura and Li*, 2005; *Clark et al.*, 2005). Large-scale gravitational contraction, on the other hand, can produce strongly bound GMCs in which gravitational collapse proceeds efficiently. In either case, mass that does not collapse sees its density reduced, and may possibly remain unbound throughout the evolution (*Bonnell et al.*, 2006). The duration of the entire gas accumulation process to form a GMC by either kind of compressions may be ~10–20 m.y., with the duration of the mainly *molecular* phase probably constituting only the last 3–5 m.y. (*Ballesteros-Paredes et al.*, 1999b; *Hartmann*, 2001, 2003; *Hartmann et al.*, 2001).

Thus, GMCs and smaller MCs may represent two distinct populations: one formed by the large-scale gravitational instability in the spiral arms, and concentrated toward the midplane because of the extra gravity in the arms; the other formed by more scattered turbulent compression events, probably driven by supernovae, and distributed similarly to the turbulent atomic gas.

2.2. Origin of Molecular Cloud Properties

Large-scale compressions can account for the physical and chemical conditions characteristic of MCs, whether the compression comes from gravity or ram pressure. *Hartmann et al.* (2001) estimated that the column density thresholds for becoming self-gravitating, molecular, and magnetically supercritical are very similar (see also *Franco and Cox*, 1986), $N \sim 1.5 \times 10^{21} (P_e/k)_4^{1/2}$ cm^{-2}, where $(P_e/k)_4$ is the pressure external to the cloud in units of 10^4 K cm^{-3}.

Molecular clouds are observed to be magnetically critical (*Crutcher et al.*, 2003), yet the diffuse atomic ISM is reported to be subcritical (e.g., *Crutcher et al.*, 2003; *Heiles and Troland*, 2005), apparently violating mass-to-magnetic flux conservation. However, the formation of a GMC of 10^6 M$_\odot$ out of diffuse gas at 1 cm^{-3} requires accumulation lengths ~400 pc, a scale over which the diffuse ISM is magnetically critical (*Hartmann et al.*, 2001). Thus, reports of magnetic subcriticality need to be accompanied by an estimate of the applicable length scale.

Large-scale compressions also appear capable of producing the observed internal turbulence in MCs. *Vishniac* (1994) showed analytically that bending modes in isothermal shock-bounded layers are *nonlinearly* unstable. This analysis has been confirmed numerically (*Hunter et al.*, 1986; *Stevens et al.*, 1992; *Blondin and Marks*, 1996; *Walder and Folini*, 1998, 2000), demonstrating that shock-bounded, radiatively cooled layers can become unstable and develop turbulence. Detailed models of the compression of the diffuse ISM show that the compressed postshock gas undergoes both thermal and dynamical instability, fragmenting into cold, dense clumps with clump-to-clump velocity dispersions of a few kilometers per second, supersonic with

respect to their internal sound speeds (*Koyama and Inutsuka*, 2002; *Inutsuka and Koyama*, 2004; *Audit and Hennebelle*, 2005; *Heitsch et al.*, 2005; *Vázquez-Semadeni et al.*, 2006a). *Inutsuka and Koyama* (2004) included magnetic fields, with similar results. *Vázquez-Semadeni et al.* (2006a) further showed that gas with MC-like densities occurs in regions overpressured by dynamic compression to factors of as much as 5 above the mean thermal pressure (see Fig. 1). These regions are the density peaks in a turbulent flow, rather than ballistic objects in a diffuse substrate. Very mildly supersonic compressions can form thin, cold atomic sheets like those observed by *Heiles and Troland* (2003) rather than turbulent, thick objects like MCs. Depending on the inflow Mach number, turbulence can take 5–100 m.y. to develop.

A crucial question is whether molecules can form on the short timescales (3–5 m.y.) implied by the ages of the stellar populations in nearby star-forming regions (*Ballesteros-Paredes et al.*, 1999b; *Hartmann*, 2001; *Hartmann et al.*, 2001). H$_2$ molecule formation on grains occurs on timescales $t_f = 10^9$ yr/n (*Hollenbach and Salpeter*, 1971; *Jura*, 1975). *Pavlovski et al.* (2002) showed that, in a turbulent flow, H$_2$ formation proceeds fastest in the highest density enhancements, yet the bulk of the mass passes through those enhancements quickly. *Glover and Mac Low* (2005) confirm that this mechanism produces widespread molecular gas at average densities on the order of 10^2 cm^{-3} in compressed, turbulent regions. This may explain recent observations of diffuse ($n_H \le 30$ cm^{-3}) cirrus clouds that exhibit significant fractions of H$_2$ (~1–30%) (*Gillmon and Shull*, 2006). Rapid H$_2$ formation suggests that a more relevant MC formation timescale is that required to accumulate a sufficient column density for extinction to allow CO for-

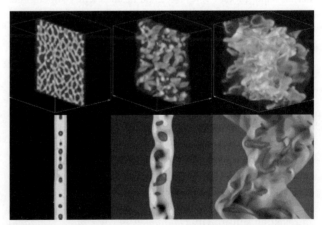

Fig. 1. *Top:* Projections of the density field in a simulation of MC formation by the collision of convergent streams, each at a Mach number $\mathcal{M} = 2.4$. From left to right, times t = 2, 4, and 7.33 m.y. are shown, illustrating how the collision first forms a thin sheet that then fragments, becomes turbulent, and thickens, until it becomes a fully three-dimensional cloud. *Bottom:* Cross sections through the pressure field at the xy plane at half the z extension of the simulation. The color code ranges from P/k = 525 (black) to 7000 K cm^{-3} (white). From *Vázquez-Semadeni et al.* (2006a).

mation. Starting from HI with n ~ 1 cm^{-3}, and δv ~ 10 km/s, this is ~10–20 m.y. (*Bergin et al.*, 2004).

In the scenario we have described, MCs are generally not in equilibrium, but rather evolving secularly. They start as atomic gas that is compressed, increasing its mean density. The atomic gas may or may not be initially self-gravitating, but in either case the compression causes the gas to cool via thermal instability and to develop turbulence via dynamical instabilities. Thermal instability and turbulence promote fragmentation and, even though the self-gravity of the cloud as a whole is increased due to cooling and the compression, the freefall time in the fragments is shorter, so collapse proceeds locally, preventing monolithic collapse of the cloud and thus reducing the star-formation efficiency (section 5.1). The near equipartition between their gravitational, magnetic, and turbulent kinetic energies is not necessarily a condition of equilibrium (*Maloney*, 1990; *Vázquez-Semadeni et al.*, 1995, 2006b; *Ballesteros-Paredes and Vázquez-Semadeni*, 1997; *Ballesteros-Paredes et al.*, 1999a; *Ballesteros-Paredes*, 2006; *Dib et al.*, 2006b), and instead may simply indicate that self-gravity has become important either due to large-scale gravitational instability or local cooling.

3. PROPERTIES OF MOLECULAR CLOUD TURBULENCE

In a turbulent flow with velocity power spectrum having negative slope, the typical velocity difference between two points decreases as their separation decreases. Current observations give $\Delta v \approx 1$ km s^{-1} [R/(1 pc)]$^{\beta}$ with $\beta \simeq 0.38$–0.5 (e.g., *Larson*, 1981; *Blitz*, 1993; *Heyer and Brunt*, 2004). This scaling law implies that the typical velocity difference across separations $\ell > \lambda_s \sim 0.05$ pc is statistically supersonic, while it is statistically subsonic for $\ell < \lambda_s$, where we define λ_s as the sonic scale. Note that dropping to subsonic velocity is unrelated to reaching the dissipation scale of the turbulence, although there may be a change in slope of the velocity power spectrum at λ_s.

The roughly isothermal supersonic turbulence in MCs at scales $\ell > \lambda_s$ produces density enhancements that appear to constitute the clumps and cores of MCs (*von Weizsäcker*, 1951; *Sasao*, 1973; *Elmegreen*, 1993; *Padoan*, 1995; *Ballesteros-Paredes et al.*, 1999a), since the density jump associated with isothermal shocks is $\mathcal{M}^2$. Conversely, the subsonic turbulence at $\ell < \lambda_s$ within cores does not drive further fragmentation, and is less important than thermal pressure in resisting self-gravity (*Padoan*, 1995; *Vázquez-Semadeni et al.*, 2003).

3.1. Density Fluctuations and Magnetic Fields

The typical amplitudes and sizes of density fluctuations driven by supersonic turbulence determine if they will become gravitationally unstable and collapse to form one or more stars. One of the simplest statistical indicators is the density probability distribution function (PDF). For nearly isothermal flows subject to random accelerations, the density PDF develops a log-normal shape (*Vázquez-Semadeni*, 1994; *Padoan et al.*, 1997a; *Passot and Vázquez-Semadeni*, 1998; *Klessen*, 2000). This is the expected shape if the density fluctuations are built by successive passages of shock waves at any given location and the jump amplitude is independent of the local density (*Passot and Vázquez-Semadeni*, 1998; *Nordlund and Padoan*, 1999). Numerical studies in three dimensions suggest that the standard deviation of the logarithm of the density fluctuations $\sigma_{\log\rho}$ scales with the logarithm of the Mach number (*Padoan et al.*, 1997a; *Mac Low et al.*, 2005).

Similar features appear to persist in the isothermal, ideal MHD case (*Ostriker et al.*, 1999, 2001; *Falle and Hartquist*, 2002), with the presence of the field not significantly affecting the shape of the density PDF except for particular angles between the field and the direction of the MHD wave propagation (*Passot and Vázquez-Semadeni*, 2003). In general, $\sigma_{\log\rho}$ tends to decrease with increasing field strength B, although it exhibits slight deviations from monotonicity. Some studies find that intermediate values of B inhibit the production of large density fluctuations better than larger values (*Passot et al.*, 1995; *Ostriker et al.*, 2001; *Ballesteros-Paredes and Mac Low*, 2002; *Vázquez-Semadeni et al.*, 2005a). This may be due to a more isotropic behavior of the magnetic pressure at intermediate values of B (see also *Heitsch et al.*, 2001a).

An important observational (*Crutcher et al.*, 2003) and numerical (*Padoan and Nordlund*, 1999; *Ostriker et al.*, 2001; *Passot and Vázquez-Semadeni*, 2003) result is that the magnetic field strength appears to be uncorrelated with the density at low-to-moderate densities or strong fields, while a correlation appears at high densities or weak fields. *Passot and Vázquez-Semadeni* (2003) suggest this occurs because the slow mode dominates for weak fields, while the fast mode dominates for strong fields.

Comparison of observations to numerical models has led *Padoan and Nordlund* (1999) and *Padoan et al.* (2003, 2004a) to strongly argue that MC turbulence is not just supersonic but also super-Alfvénic with respect to the initial mean field. *Padoan and Nordlund* (1999) note that such turbulence can easily produce trans- or even sub-Alfvénic cores. Also, at advanced stages, the total magnetic energy including fluctuations in driven turbulence approaches the equipartition value (*Padoan et al.*, 2003), due to turbulent amplification of the magnetic energy (*Ostriker et al.*, 1999; *Schekochihin et al.*, 2002). Observations, on the other hand, are generally interpreted as showing that MC cores are magnetically critical or moderately supercritical and therefore trans- or super-Alfvénic if the turbulence is in equipartition with self-gravity (*Bertoldi and McKee*, 1992; *Crutcher*, 1999; *Bourke et al.*, 2001; *Crutcher et al.*, 2003; see also the chapter by di Francesco et al.). However, the observational determination of the relative importance of the magnetic field strength in MCs depends on their assumed geometry, and the clouds could even be strongly supercritical (*Bourke et al.*, 2001). Unfortunately, the lack of self-gravity, which could substitute for stronger turbulence in the production of denser cores in the simulations by Padoan and co-workers, implies that the evidence in favor of super-Alfvénic MCs remains inconclusive.

3.2. Driving and Decay of the Turbulence

Supersonic turbulence should decay in a crossing time of the driving scale (*Goldreich and Kwan*, 1974; *Field*, 1978). For a number of years it was thought that magnetic fields might modify this result (*Arons and Max*, 1975), but it has been numerically confirmed that both hydrodynamical and MHD turbulence decay in less than a freefall time, whether or not an isothermal equation of state is assumed (*Stone et al.*, 1998; *Mac Low et al.*, 1998; *Padoan and Nordlund*, 1999; *Biskamp and Müller*, 2000; *Avila-Reese and Vázquez-Semadeni*, 2001; *Pavlovski et al.*, 2002). Decay does proceed more slowly if there are order-of-magnitude imbalances in motions in opposite directions (*Maron and Goldreich*, 2001; *Cho et al.*, 2002), although this seems unlikely to occur in MCs.

Traditionally, MCs have been thought to have lifetimes much larger than their freefall times (e.g., *Blitz and Shu*, 1980). However, recent work suggests that MC lifetimes may be actually comparable or shorter than their freefall times (*Ballesteros-Paredes et al.*, 1999b; *Hartmann et al.*, 2001; *Hartmann*, 2003), and that star formation occurs within a single crossing time at all scales (*Elmegreen*, 2000b). If true, and if clouds are destroyed promptly after star formation has occurred (*Fukui et al.*, 1999; *Yamaguchi et al.*, 2001), this suggests that turbulence is ubiquitously observed simply because it is produced together with the cloud itself (see section 2.2) and does not have time to decay afterward.

Observations of nearby MCs show them to be self-similar up to the largest scales traced by molecules (*Mac Low and Ossenkopf*, 2000; *Ossenkopf et al.*, 2001; *Ossenkopf and Mac Low*, 2002; *Brunt*, 2003; *Heyer and Brunt*, 2004), suggesting that they are driven from even larger scales (see also section 4.2). If the turbulent or gravitational compressions that form MCs drive the turbulence, then the driving scale would indeed be larger than the clouds.

3.3. Spectra and Scaling Relations

Scaling relations between the mean density and size, and between velocity dispersion and size (*Larson*, 1981), have been discussed in several reviews (e.g., *Vázquez-Semadeni et al.*, 2000; *Mac Low and Klessen*, 2004; *Elmegreen and Scalo*, 2004). In order to avoid repetition, we comment in detail only on the caveats not discussed there.

The Larson relation $\langle \rho \rangle \sim R^{-1}$ implies a constant column density. However, observations tend to have a limited dynamic range, and thus to select clouds with similar column densities (*Kegel*, 1989; *Scalo*, 1990; *Schneider and Brooks*, 2004), as already noted by Larson himself. Numerical simulations of both the diffuse ISM (*Vázquez-Semadeni et al.*, 1997) and of MCs (*Ballesteros-Paredes and Mac Low*, 2002)

suggest that clouds of lower column density exist. However, when the observational procedure is simulated, then the mean density-size relation appears, showing that it is likely to be an observational artifact.

Elmegreen and Scalo (2004) suggested that the mass-size relation found in the numerical simulations of *Ostriker et al.* (2001) supports the reality of the $\langle \rho \rangle$-R^{-1} relation. However, those authors defined their clumps using a procedure inspired by observations, focusing on regions with higher-than-average column density. As in noise-limited observations, this introduces an artificial cutoff in column density, which again produces the appearance of a $\langle \rho \rangle$-R^{-1} relation.

Nevertheless, the fact that the column density threshold for a cloud to become molecular is similar to that for becoming self-gravitating (*Franco and Cox*, 1986; *Hartmann et al.*, 2001) may truly imply a limited column density range for *molecular* gas, since clouds of lower column density are not molecular, while clouds with higher column densities collapse quickly. Molecular clouds constitute only the tip of the iceberg of the neutral gas distribution.

The Δv-R relation is better established. However, it does not appear to depend on self-gravitation, as has often been argued. Rather, it appears to be a measurement of the second-order structure function of the turbulence (*Elmegreen and Scalo*, 2004), which has a close connection with the energy spectrum $E(k) \propto k^n$. The relation between the spectral index n and the exponent in the velocity dispersion-size relation $\Delta v \propto R^\beta$, is $\beta = -(n + 1)/2$. Observationally, the value originally found by *Larson* (1981), $\beta \approx 0.38$, is close to the value expected for incompressible, Kolmogorov turbulence ($\beta = 1/3$, $n = -5/3$). More recent surveys of whole GMCs tend to give values $\beta \sim 0.5$–0.65 (*Blitz*, 1993; *Mizuno et al.*, 2001; *Heyer and Brunt*, 2004), with smaller values arising in low surface brightness regions of clouds ($\beta \sim 0.4$) (*Falgarone et al.*, 1992) and massive core surveys ($\beta \sim 0$–0.2, and very poor correlations) (*Caselli and Myers*, 1995; *Plume et al.*, 1997). The latter may be affected by intermittency effects, stellar feedback, or strong gravitational contraction. Numerically, simulations of MCs and the diffuse ISM (*Vázquez-Semadeni et al.*, 1997; *Ostriker et al.*, 2001; *Ballesteros-Paredes and Mac Low*, 2002) generally exhibit very noisy Δv-R relations both in physical and observational projected space, with slopes $\beta \sim 0.2$–0.4. The scatter is generally larger at small clump sizes, both in physical space, perhaps because of intermittency, and in projection, probably because of feature superposition along the line of sight.

There is currently no consensus on the spectral index for compressible flows. A value $n = -2$ ($\beta = 1/2$) is expected for shock-dominated flows (e.g., *Elsässer and Schamel*, 1976), while *Boldyrev* (2002) has presented a theoretical model suggesting that $n \approx -1.74$, and thus $\beta \approx 0.37$. The predictions of this model for the higher-order structure function scalings appear to be confirmed numerically (*Boldyrev et al.*, 2002; *Padoan et al.*, 2004b; *Joung and Mac Low*, 2006; *de Avillez and Breitschwerdt*, 2006). However, the model assumes incompressible Kolmogorov scaling at large scales on the basis of simulations driven with purely solenoidal motions (*Boldyrev et al.*, 2002), which does not appear to agree with the behavior seen by ISM simulations (e.g., *de Avillez*, 2000; *Joung and Mac Low*, 2006).

Numerically, *Passot et al.* (1995) reported n = –2 in their two-dimensional simulations of magnetized, self-gravitating turbulence in the diffuse ISM. More recently, *Boldyrev et al.* (2002) have reported n = –1.74 in isothermal, non-self-gravitating MHD simulations, while other studies find that n appears to depend on the rms Mach number of the flow in both magnetic and nonmagnetic cases (*Cho and Lazarian*, 2003; *Ballesteros-Paredes et al.*, 2006), approaching n = –2 at high Mach numbers.

3.4. Thermodynamic Properties of the Star-forming Gas

While the atomic gas is tenuous and warm, with densities $1 < n < 100$ cm^{-3} and temperatures 100 K $< T < 5000$ K, MCs have $10^2 < n < 10^3$ cm^{-3}, and locally $n > 10^6$ cm^{-3} in dense cores. The kinetic temperature inferred from molecular line ratios is typically about 10 K for dark, quiescent clouds and dense cores, but can reach 50–100 K in regions heated by UV radiation from high-mass stars (*Kurtz et al.*, 2000).

MC temperatures are determined by the balance between heating and cooling, which in turn both depend on the chemistry and dust content. Early studies predicted that the equilibrium temperatures in MC cores should be 10–20 K, tending to be lower at the higher densities (e.g., *Hayashi*, 1966; *Larson*, 1969, 1973; *Goldsmith and Langer*, 1978). Observations generally agree with these values (*Jijina et al.*, 1999), prompting most theoretical and numerical star-formation studies to adopt a simple isothermal description, with T ~ 10 K.

In reality, however, the temperature varies by factors of 2 or 3 above and below the assumed constant value (*Goldsmith*, 2001; *Larson*, 2005). This variation can be described by an effective equation of state (EOS) derived from detailed balance between heating and cooling when cooling is fast (*Vázquez-Semadeni et al.*, 1996; *Passot and Vázquez-Semadeni*, 1998; *Scalo et al.*, 1998; *Spaans and Silk*, 2000; *Li et al.*, 2003; *Spaans and Silk*, 2005). This EOS can often be approximated by a polytropic law of the form P ∝ ρ^γ, where γ is a parameter. For molecular clouds, there are three main regimes for which different polytropic EOS can be defined: (1) For $n \lesssim 2.5 \times 10^5$ cm^{-3}, MCs are externally heated by cosmic rays or photoelectric heating, and cooled mainly through the emission of molecular (e.g., CO, H_2O) or atomic (CII, O, etc.) lines, depending on ionization level and chemical composition (e.g., *Genzel*, 1991). The strong dependence of the cooling rate on density yields an equilibrium temperature that decreases with increasing density. In this regime, the EOS can be approximated by a polytrope with $\gamma = 0.75$ (*Koyama and Inutsuka*, 2000). Small, dense cores indeed are observed to have T ~ 8.5 K at a density n ~ 10^5 cm^{-3} (*Evans*, 1999). (2) At $n > 10^6$ cm^{-3}, the gas becomes thermally coupled to dust grains, which then con-

trol the temperature by their far-infrared thermal emission. In this regime, the effective polytropic index is found to be $1 < \gamma < 1.075$ (*Scalo et al.*, 1998; *Spaans and Silk*, 2000; *Larson*, 2005). The temperature is predicted to reach a minimum of 5 K (*Larson*, 2005). Such cold gas is difficult to observe, but observations have been interpreted to suggest central temperatures between 6 K and 10 K for the densest observed prestellar cores. (3) Finally, at $n > 10^{11}$ cm^{-3}, the dust opacity increases, causing the temperature to rise rapidly. This results in an opacity limit to fragmentation that is somewhat below 0.01 M$_\odot$ (*Low and Lynden-Bell*, 1976; *Masunaga and Inutsuka*, 2000). The transition from $\gamma < 1$ to $\gamma \geq 1$ in molecular gas may determine the characteristic stellar mass resulting from gravoturbulent fragmentation (section 5.2).

4. PROPERTIES OF MOLECULAR CLOUD CORES

4.1. Dynamical Evolution

4.1.1. Dynamic or hydrostatic cores? The first successful models of MCs and their cores used hydrostatic equilibrium configurations as the starting point (e.g., *de Jong et al.*, 1980; *Shu et al.*, 1987; *Mouschovias*, 1991) (see also section 5.1). However, supersonic turbulence generates the initial density enhancements from which cores develop (e.g., *Sasao*, 1973; *Passot et al.*, 1988; *Vázquez-Semadeni*, 1994; *Padoan et al.*, 1997b, 2001b; *Passot and Vázquez-Semadeni*, 1998, 2003; *Padoan and Nordlund*, 1999; *Klessen et al.*, 2000, 2005; *Heitsch et al.*, 2001a; *Falle and Hartquist*, 2002; *Ballesteros-Paredes et al.*, 2003; *Hartquist et al.*, 2003; *Kim and Ryu*, 2005; *Beresnyak et al.*, 2005). Therefore, they do not necessarily approach hydrostatic equilibrium at any point in their evolution (*Ballesteros-Paredes et al.*, 1999a). For the dynamically formed cores to settle into equilibrium configurations, it is necessary that the equilibrium be *stable*. In this subsection we discuss the conditions necessary for the production of stable self-gravitating equilibria, and whether they are likely to be satisfied in MCs.

Known hydrostatic stable equilibria in nonmagnetic, self-gravitating media require confinement by an external pressure to prevent expansion (*Bertoldi and McKee*, 1992; *Hartmann*, 1998; *Hartmann et al.*, 2001; *Vázquez-Semadeni et al.*, 2005b). The most common assumption is a discontinuous phase transition to a warm, tenuous phase that provides pressure without adding weight (i.e., mass) beyond the core's boundary. The requirement of a confining medium implies that the medium must be two-phase, as in the model of *Field et al.* (1969) for the diffuse ISM. Turbulent pressure is probably not a suitable confining agent because it is large-scale, transient, and anisotropic, thus being more likely to cause distortion than confinement of the cores (*Ballesteros-Paredes et al.*, 1999a; *Klessen et al.*, 2005).

Bonnor-Ebert spheres are stable for a ratio of central to external density <14. Unstable equilibrium solutions exist that dispense with the confining medium by pushing the boundaries to infinity, with vanishing density there. Other, nonspherical, equilibrium solutions exist (e.g., *Curry*, 2000) that do not require an external confining medium, although their stability remains an open question. A sufficiently large pressure increase, such as might occur in a turbulent flow, can drive a stable, equilibrium object to become unstable.

For mass-to-flux ratios below a critical value, the equations of ideal MHD without ambipolar diffusion (AD) predict unconditionally stable equilibrium solutions, in the sense that no amount of external compression can cause collapse (*Shu et al.*, 1987). However, these configurations still require a confining thermal or magnetic pressure, applied at the boundaries parallel to the mean field, to prevent expansion in the perpendicular direction (*Nakano*, 1998; *Hartmann et al.*, 2001). Treatments avoiding such a confining medium (e.g., *Lizano and Shu*, 1989) have used periodic boundary conditions, in which the pressure from the neighboring boxes effectively confines the system. When boundary conditions do not restrict the cores to remain subcritical, as in globally supercritical simulations, subcritical cores do not appear (*Li et al.*, 2004; *Vázquez-Semadeni et al.*, 2005b).

Thus it appears that the boundary conditions of the cores determine whether they can find stable equilibria. Some specific key questions are (1) whether MC interiors can be considered two-phase media, (2) whether the clouds can really be considered magnetically subcritical when the masses of their atomic envelopes are taken into account (cf. section 2), and (3) whether the magnetic field inside the clouds is strong and uniform enough to prevent expansion of the field lines at the cores. If any of these questions is answered in the positive, the possibility of nearly hydrostatic cores cannot be ruled out. At this point, however, they do not appear likely to dominate MCs.

4.1.2. Core lifetimes. Core lifetimes are determined using either chemical or statistical methods. Statistical studies use the observed number ratio of starless to stellar cores in surveys n_{sl}/n_s (*Beichman et al.*, 1986; *Ward-Thompson et al.*, 1994; *Lee and Myers*, 1999; *Jijina et al.*, 1999; *Jessop and Ward-Thompson*, 2000; see also the chapter by di Francesco et al.) as a measure of the time spent by a core in the pre- and protostellar stages, respectively. This method gives estimates of prestellar lifetimes τ of a few hundred thousand to a few million years. Although *Jessop and Ward-Thompson* (2000) report $6 \leq \tau \leq 17$ m.y., they likely overestimate the number of starless cores because to find central stars they rely on IRAS measurements of relatively distant clouds (~350 pc), and so probably miss many of the lower-mass stars (see *Ballesteros-Paredes and Hartmann*, 2006). This is a general problem of the statistical method (*Jørgensen et al.*, 2005). Indeed, Spitzer observations have begun to reveal embedded sources in cores previously thought to be starless (e.g., *Young et al.*, 2004). Another problem is that if not all present-day starless cores eventually form stars, then the ratio n_{sl}/n_s overestimates the lifetime ratio. Numerical simulations of isothermal turbulent MCs indeed show that a significant fraction of the cores (dependent on the exact

criterion used to define them) end up reexpanding rather than collapsing (*Vázquez-Semadeni et al.*, 2005b; *Nakamura and Li*, 2005).

Chemical methods rely on matching the observed chemistry in cores to evolutionary models (e.g., *Taylor et al.*, 1996, 1998; *Morata et al.*, 2005). They suggest that not all cores will proceed to form stars. *Langer et al.* (2000) summarize comparisons of observed molecular abundances to time-dependent chemical models, suggesting typical ages for cores of $\sim 10^5$ yr. Using a similar technique, *Jørgensen et al.* (2005) report a timescale of $10^{5 \pm 0.5}$ yr for the heavy depletion, dense prestellar stage.

These timescales have been claimed to pose a problem for the model of AD-mediated, quasistatic, core contraction (*Lee and Myers*, 1999; *Jijina et al.*, 1999), as they amount to only a few local freefall times ($t_{ff} \equiv L_J/c_s$, ~0.5 m.y. for typical cores with $\langle n \rangle \sim 5 \times 10^4$ cm^{-3}, where L_J is the local Jeans length, and c_s is the sound speed), rather than the 10–20 t_{ff} predicted by the theory of AD-controlled star formation under the assumption of low initial values (~0.25) of the mass-to-magnetic flux ratio (in units of the critical value) μ_0 (e.g., *Ciolek and Mouschovias*, 1995). However, *Ciolek and Basu* (2001) have pointed out that the lifetimes of the ammonia-observable cores predicted by the AD theory are much lower, reaching values as small as ~2t_{ff} when $\mu_0 \sim 0.8$. Moreover, nonlinear effects can enhance the AD rates over the linear estimates (*Fatuzzo and Adams*, 2002; *Heitsch et al.*, 2004).

The lifetimes and evolution of dense cores have recently been investigated numerically by two groups. *Vázquez-Semadeni et al.* (2005a,b) considered driven, ideal MHD, magnetically supercritical turbulence with sustained rms Mach number $\mathcal{M} \sim 10$, comparable to Taurus, while *Nakamura and Li* (2005) considered subcritical, decaying turbulence including AD, in which the flow spent most of the time at low $\mathcal{M} \sim 2$–3. Thus, the two setups probably bracket the conditions in actual MCs. Nevertheless, the lifetimes found in both studies agree within a factor of ~1.5, ranging from 0.5–10 m.y., with a median value ~2–3 t_{ff} (~1–1.5 m.y.), with the longer timescales corresponding to the decaying cases. Furthermore, a substantial fraction of quiescent, trans- or subsonic cores, and low star-formation efficiencies were found (*Klessen et al.*, 2005) (see also section 4.2), comparable to those observed in the Taurus molecular cloud (TMC). In all cases, the simulations reported a substantial fraction of reexpanding ("failed") cores.

This suggests that the AD timescale, when taking into account the nearly magnetically critical nature of the clouds as well as nonlinear effects and turbulence-accelerated core formation, is comparable to the dynamical timescale. The notion of quasistatic, long-lived cores as the general rule is therefore unfounded, except in cases where a warm confining medium is clearly available (B68 may be an example) (*Alves et al.*, 2001). Both observational and numerical evidence point toward short typical core lifetimes of a few t_{ff}, or a few hundred thousand to a million years, although the scatter around these values can be large.

4.2. Core Structure

4.2.1. Morphology and magnetic field. Molecular cloud cores in general, and starless cores in particular, have a median projected aspect ratio ~1.5 (e.g., *Jijina et al., 1999*, and references therein). It is not clear whether the observed aspect ratios imply that MC cores are (nearly) oblate or prolate. *Myers et al.* (1991) and *Ryden* (1996) found that cores are more likely to be prolate than oblate, while *Jones et al.* (2001) found them to be more likely oblate, as favored by AD-mediated collapse. Magnetohydrodynamic numerical simulations of turbulent MCs agree with generally triaxial shapes, with a slight preference for prolateness (*Gammie et al., 2003; Li et al., 2004*).

The importance of magnetic fields to the morphology and structure of MCs remains controversial. Observationally, although some cores exhibit alignment (e.g., *Wolf et al., 2003*), it is common to find misalignments between the polarization angles and the minor axis of MC cores (e.g., *Heiles et al., 1993; Ward-Thompson et al., 2000; Crutcher et al., 2004*). On the other hand, in modern models of AD-controlled contraction, triaxial distributions may exist, but the cores are expected to still be intrinsically flattened along the field (*Basu, 2000; Jones et al., 2001; Basu and Ciolek, 2004*). However, numerical simulations of turbulent magnetized clouds tend to produce cores with little or no alignment with the magnetic field and a broad distribution of shapes (*Ballesteros-Paredes and Mac Low, 2002; Gammie et al., 2003; Li et al., 2004*).

Finally, a number of theoretical and observational works have studied the polarized emission of dust from MC cores. Their main conclusions are (1) cores in magnetized simulations exhibit degrees of polarization between 1% and 10%, regardless of whether the turbulence is sub- or super-Alfvénic (*Padoan et al., 2001a*); (2) submillimeter polarization maps of quiescent cores do not map the magnetic fields inside the cores for visual extinctions larger than $A_V \sim 3$ (*Padoan et al., 2001a*), in agreement with observations (e.g., *Ward-Thompson et al., 2000; Wolf et al., 2003; Crutcher et al., 2004; Bourke and Goodman, 2004*); (3) although the simplest Chandrasekhar-Fermi method of estimating the mean magnetic field in a turbulent medium underestimates very weak magnetic fields by as much as a factor of 150, it can be readily modified to yield estimates good to a factor of 2.5 for all field strengths (*Heitsch et al., 2001b*); and (4) limited telescope resolution leads to a systematic overestimation of field strength, as well as to an impression of more regular, structureless magnetic fields (*Heitsch et al., 2001b*).

4.2.2. Density and velocity fields. Low-mass MC cores often exhibit column density profiles resembling those of Bonnor-Ebert spheres (e.g., *Alves et al., 2001*; see also the chapter by Lada et al.), and trans- or even subsonic nonthermal linewidths (e.g., *Jijina et al., 1999; Tafalla et al., 2004*; see also the chapter by di Francesco et al.). These results have been traditionally interpreted as evidence that MC

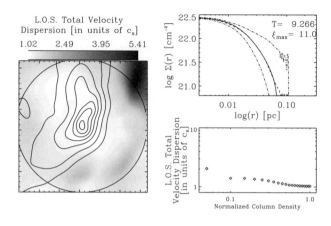

Fig. 2. Properties of a typical quiescent core formed by turbulence in nonmagnetic simulations of molecular clouds. *Left:* Column-density contours on the projected plane of the sky, superimposed on a grayscale image of the *total* (thermal plus nonthermal) velocity dispersion along each line of sight, showing that the maximum velocity dispersion occurs near the periphery of the core. *Upper right:* A fit of a Bonnor-Ebert profile to the angle-averaged column density radial profile (solid line) within the circle on the map. The dash-dotted lines give the range of variation of individual radial profiles before averaging. *Lower right:* Plot of averaged *total* velocity dispersion vs. column density, showing that the core is subsonic. Adapted from *Ballesteros-Paredes et al.* (2003) and *Klessen et al.* (2005).

cores are in quasistatic equilibrium, and are therefore long-lived. This is in apparent contradiction with the discussion from section 4.1 (see, e.g., discussion in *Keto and Field, 2006*), which presented evidence favoring the view that cores are formed by supersonic compressions in MCs, and that their lifetimes are on the order of one dynamical timescale. It is thus necessary to understand how those observational properties are arrived at, and whether dynamically evolving cores reproduce them.

Radial column density profiles of MC cores are obtained by averaging the actual three-dimensional structure of the density field twice: first, along the line of sight during the observation itself, and second, during the angle-averaging (*Ballesteros-Paredes et al., 2003*). These authors (see also Fig. 2) further showed that nonmagnetic numerical simulations of turbulent MCs may form rapidly evolving cores that nevertheless exhibit Bonnor-Ebert-like angle-averaged column density profiles, even though they are formed by a supersonic turbulent compression and are transient structures. Furthermore, *Klessen et al.* (2005) have shown that approximately one-fourth of the cores in their simulations have subsonic internal velocity dispersions, while about one-half of them are transonic, with velocity dispersions $c_s < \sigma_{turb} \leq 2c_s$, where c_s is the sound speed (Fig. 2). These fractions are intermediate between those reported by *Jijina et al.* (1999) for the Taurus and the Perseus MCs. These authors also showed that the ratio of virial to actual core masses in simulations with large-scale driving (LSD) (one-

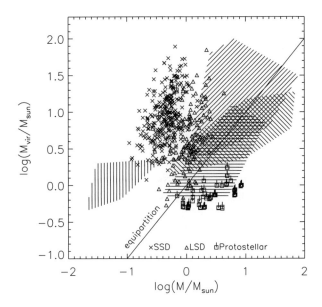

Fig. 3. Virial mass M_{vir} vs. actual mass M for cores in turbulent models driven at small scales (crosses; "SSD") and at large scales (triangles; "LSD"). Tailed squares indicate lower limits on the estimates on M_{vir} for cores in the LSD model containing "protostellar objects" (sink particles). The solid line denotes the identity $M_{vir} =$ M. Also shown are core survey data by *Morata et al.* (2005) (vertical hatching), *Onishi et al.* (2002) (horizontal hatching), *Caselli et al.* (2002) (–45° hatching), and *Tachihara et al.* (2002) (+45° hatching). From *Klessen et al.* (2005).

half the simulation size) is reasonably consistent with observations, while that in simulations with small-scale driving (SSD) is not (Fig. 3), in agreement with the discussion in section 3.2. Improved agreement with observations may be expected for models spanning a larger range of cloud masses.

Quiescent cores in turbulence simulations are not a consequence of (quasi)hydrostatic equilibrium, but a natural consequence of the Δv-R relation, which implies that the typical velocity differences across small enough regions ($l_s \lesssim 0.05$ pc) are subsonic, with $\sigma_{turb} \lesssim c_s$. Thus, it can be expected that small cores within turbulent environments may have trans- or even subsonic nonthermal linewidths. Moreover, the cores are assembled by random ram-pressure compressions in the large-scale flow, and as they are compressed, they trade kinetic compressive energy for internal energy, so that when the compression is at its maximum, the velocity is at its minimum (*Klessen et al.,* 2005). Thus, the observed properties of quiescent low-mass protostellar cores in MCs do not necessarily imply the existence of strong magnetic fields providing support against collapse, nor of nearly hydrostatic states.

4.2.3. Energy and virial budget. The notion that MCs and their cores are in near virial equilibrium is almost universally encountered in the literature (e.g., *Larson,* 1981; *Myers and Goodman,* 1988; *McKee,* 1999; *Krumholz et al.,*

2005a, and references therein). However, what is actually observed is energy equipartition between self-gravity and one or more forms of their internal energy, as *Myers and Goodman* (1988), for example, explicitly recognize. This by no means implies equilibrium, as the full virial theorem also contains hard to observe surface- and time-derivative terms (*Spitzer,* 1968; *Parker,* 1979; *McKee and Zweibel,* 1992). Even the energy estimates suffer from important observational uncertainties (e.g., *Blitz,* 1994). On the other hand, a core driven into gravitational collapse by an external compression naturally has an increasing ratio of its gravitational to the kinetic plus magnetic energies, and passes through equipartition at the verge of collapse, even if the process is fully dynamic. This is consistent with the observational fact that, in general, a substantial fraction of clumps in MCs are gravitationally unbound (*Maloney,* 1990; *Falgarone et al.,* 1992; *Bertoldi and McKee,* 1992), a trend also observed in numerical simulations (*Vázquez-Semadeni et al.,* 1997).

Indeed, in numerical simulations of the turbulent ISM, the cores formed as turbulent density enhancements are continually changing their shapes and exchanging mass, momentum, and energy with their surroundings, giving large values of the surface- and time-derivative terms in the virial theorem (*Ballesteros-Paredes and Vázquez-Semadeni,* 1995, 1997; *Tilley and Pudritz,* 2004; see *Ballesteros-Paredes,* 2004, for a review). That the surface terms are large indicates that there are fluxes of mass, momentum, and energy across cloud boundaries. Thus, we conclude that observations of equipartition do not necessarily support the notion of equilibrium, and that a dynamical picture is consistent with the observations.

5. CONTROL OF STAR FORMATION

5.1. Star-Formation Efficiency as Function of Turbulence Parameters

The fraction of molecular gas mass converted into stars, called the star-formation efficiency (SFE), is known to be small, ranging from a few percent for entire MC complexes (e.g., *Myers et al.,* 1986) to 10–30% for cluster-forming cores (e.g., *Lada and Lada,* 2003) and 30% for starburst galaxies whose gas is primarily molecular (*Kennicutt,* 1998). *Li et al.* (2006) suggest that large-scale gravitational instabilities in galaxies may directly determine the global SFE in galaxies. They find a well-defined, nonlinear relationship between the gravitational instability parameter Q_{sg} (see section 2.1) and the SFE.

Stellar feedback reduces the SFE in MCs both by destroying them, and by driving turbulence within them (e.g., *Franco et al.,* 1994; *Williams and McKee,* 1997; *Matzner and McKee,* 2000; *Matzner,* 2001, 2002; *Nakamura and Li,* 2005). Supersonic turbulence has two countervailing effects that must be accounted for. If it has energy comparable to the gravitational energy, it disrupts monolithic collapse, and if driven can prevent large-scale collapse. However, these same motions produce strong density enhancements that can collapse locally, but only involve a small fraction of the total mass. Its net effect is to delay collapse compared to the same situation without turbulence (*Vázquez-Semadeni and Passot,* 1999; *Mac Low and Klessen,* 2004). The magnetic field is only an additional contribution to the support, rather than the fundamental mechanism regulating the SFE.

As discussed in section 3, the existence of a scaling relation between the velocity dispersion and the size of a cloud or clump implies that within structures sizes smaller than the sonic scale $\lambda_s \sim 0.05$ pc turbulence is subsonic. Within structures with scale $\ell > \lambda_s$, turbulence drives subfragmentation and prevents monolithic collapse hierarchically (smaller fragments form within larger ones), while on scales $\ell < \lambda_s$ turbulence ceases to be a dominant form of support, and cannot drive further subfragmentation (*Padoan,* 1995; *Vázquez-Semadeni et al.,* 2003) [subsonic turbulence, however, can still produce weakly nonlinear density fluctuations that can act as seeds for gravitational fragmentation of the core during its collapse (*Goodwin et al.,* 2004a)]. Cores with $\ell < \lambda_s$ collapse if gravity overwhelms their thermal and magnetic pressures, perhaps helped by ram pressure acting across their boundaries (*Hunter and Fleck,* 1982).

Numerical simulations neglecting magnetic fields showed that the SFE decreases monotonically with increasing turbulent energy or with decreasing energy injection scale (*Léorat et al.,* 1990; *Klessen et al.,* 2000; *Clark and Bonnell,* 2004). A remarkable fact is that the SFE was larger than zero even for simulations in which the turbulent Jeans mass [the Jeans mass including the turbulent RMS velocity (*Chandrasekhar,* 1951)] was larger than the simulation mass, showing that turbulence can induce local collapse even in globally unbound clouds (*Vázquez-Semadeni et al.,* 1996; *Klessen et al.,* 2000; *Clark and Bonnell,* 2004; *Clark et al.,* 2005).

A correlation between the SFE and λ_s, $SFE(\lambda_s) = \exp(-\lambda_0/\lambda_s)$, with $\lambda_0 \sim 0.05$ pc, was found in the numerical simulations of *Vázquez-Semadeni et al.* (2003). This result provided support to the idea that the sonic scale helps determine the SFE. *Krumholz and McKee* (2005) have further suggested that the other crucial parameter is the Jeans length L_J, and computed, from the probability distribution of the gas density, the fraction of the gas mass that has $L_J \lesssim \lambda_s$, which then gives the star-formation rate per freefall time. A recent observational study by *Heyer et al.* (2006) only finds a correlation between the sonic scale and the SFE for a subset of clouds with particularly high SFE, while no correlation is observed in general. This result might be a consequence of *Heyer et al.* (2006) not taking into account the clouds' Jeans lengths, as required by *Krumholz and McKee* (2005).

The influence of magnetic fields in controlling the SFE is a central question in the turbulent model, given their fundamental role in the AD-mediated model. Numerical studies using ideal MHD, neglecting AD, have shown that the field can only prevent gravitational collapse if it provides *magne-*

tostatic support, while MHD waves can only delay the collapse, but not prevent it (*Ostriker et al.,* 1999; *Heitsch et al.,* 2001a; *Li et al.,* 2004). *Vázquez-Semadeni et al.* (2005a) found a trend of decreasing SFE with increasing mean field strength in *driven* simulations ranging from nonmagnetic to moderately magnetically supercritical. In addition, a trend to form fewer but more massive objects in more strongly magnetized cases was also observed (see Fig. 4). Whole-cloud efficiencies $\lesssim 5\%$ were obtained for moderately supercritical regimes at Mach numbers $\mathcal{M} \sim 10$. For comparison, *Nakamura and Li* (2005) found that comparable efficiencies in decaying simulations at *effective* Mach numbers $\mathcal{M} \sim 2$–3 required moderately subcritical clouds in the presence of AD. These results suggest that whether MC turbulence is driven or decaying is a crucial question for quantifying the role of the magnetic field in limiting the SFE in turbulent MCs.

The relationship between the global and local SFE in galaxies still needs to be elucidated. The existence of the Kennicutt-Schmidt law shows that global SFEs vary as a function of galactic properties, but there is little indication that the local SFE in individual MCs varies strongly from galaxy to galaxy. Are they in fact independent?

5.2. Distribution of Clump Masses and Relation to the Initial Mass Function

The stellar IMF is a fundamental diagnostic of the star-formation process, and understanding its physical origin is one of the main goals of any star-formation theory. Recently, the possibility that the IMF is a direct consequence of the protostellar core mass distribution (CMD) has gained considerable attention. Observers of dense cores report that the slope at the high-mass end of the CMD resembles the

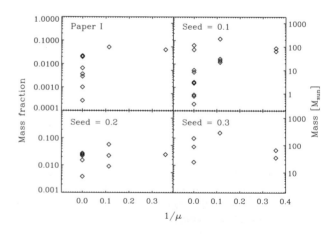

Fig. 4. Masses of the collapsed objects formed in four sets of simulations with rms Mach number $\mathcal{M} \sim 10$ and 64 Jeans masses. Each set contains one nonmagnetic run and two MHD runs each. The latter with have initial mass-to-flux ratio $\mu_0 = 8.8$ and $\mu_0 = 2.8$, showing that the magnetic runs tend to form fewer but more massive objects. From *Vázquez-Semadeni et al.* (2005a).

Salpeter (1955) slope of the IMF (e.g., *Motte et al.,* 1998; *Testi and Sargent,* 1998). This has been interpreted as evidence that those cores are the direct progenitors of single stars.

Padoan and Nordlund (2002) computed the mass spectrum of self-gravitating density fluctuations produced by isothermal supersonic turbulence with a log-normal density PDF and a power-law turbulent energy spectrum with slope n, and proposed to identify this with the IMF. In this theory, the predicted slope of the IMF is given by $-3/(4 - n)$. For n = 1.74, this gives -1.33, in agreement with the Salpeter IMF, although this value of n may not be universal (see section 3.3). However, this theory suffers from a problem common to any approach that identifies the core mass function with the IMF: It implicitly assumes that the final stellar mass is proportional to the core mass, at least on average. Instead, high-mass cores have a larger probability of building up multiple stellar systems than low-mass ones.

Moreover, the number of physical processes that may play an important role during cloud fragmentation and protostellar collapse is large. In particular, simulations show: (1) The mass distribution of cores changes with time as cores merge with each other (e.g., *Klessen,* 2001a). (2) Cores generally produce not a single star but several with the number stochastically dependent on the global parameters (e.g., *Klessen et al.,* 1998; *Bate and Bonnell,* 2005; *Dobbs et al.,* 2005), even for low levels of turbulence (*Goodwin et al.,* 2004b). (3) The shape of the clump mass spectrum appears to depend on parameters of the turbulent flow, such as the scale of energy injection (*Schmeja and Klessen,* 2004) and the rms Mach number of the flow (*Ballesteros-Paredes et al.,* 2006), perhaps because the density power spectrum of isothermal, turbulent flows becomes shallower when the Mach number increases (*Cho and Lazarian,* 2003; *Beresnyak et al.,* 2005; *Kim and Ryu,* 2005). Convergence of this result as numerical resolution increases sufficiently to resolve the turbulent inertial range needs to be confirmed. A physical explanation may be that stronger shocks produce denser, and thus thinner, density enhancements. Finally, there may be other processes, such as (4) competitive accretion influencing the mass-growth history of individual stars (see, e.g., *Bate and Bonnell,* 2005; but for an opposing view see *Krumholz et al.,* 2005b), (5) stellar feedback through winds and outflows (e.g., *Shu et al.,* 1987), or (6) changes in the equation of state introducing preferred mass scales (e.g., *Scalo et al.,* 1998; *Li et al.,* 2003; *Jappsen et al.,* 2005; *Larson,* 2005). Furthermore, even though some observational and theoretical studies of dense, compact cores fit power laws to the high-mass tail of the CMD, the actual shape of those CMDs is often not a single-slope power law, but rather a function with a continuously varying slope, frequently similar to a log-normal distribution (*Reid and Wilson,* 2006; *Stanke et al.,* 2006; *Ballesteros-Paredes et al.,* 2006; see also Fig. 6 in the chapter by Bonnell et al.). So, the dynamic range in which a power law with slope -1.3 can be fitted is often smaller than one order of magnitude.

All these facts call into question the existence of a simple one-to-one mapping between the purely turbulent clump-mass spectrum and the observed IMF.

Finally, it is important to mention that, although it is generally accepted that the IMF has a slope of -1.3 at the high-mass range, the universal character of the IMF is still a debated issue, with strong arguments being given both in favor of a universal IMF (*Elmegreen,* 2000a; *Kroupa,* 2001, 2002; *Chabrier,* 2005) and against it in the Arches cluster (*Stolte et al.,* 2005), the galactic center region (*Nayakshin and Sunyaev,* 2005), and near the central black hole of M31 (*Bender et al.,* 2005).

We conclude that both the uniqueness of the IMF and the relationship between the IMF and the CMD are presently uncertain. The definition of core boundaries usually involves taking a density threshold. For a sufficiently high threshold, the CMD must reflect the IMF, while for lower thresholds subfragmentation is bound to occur. Thus, future studies may find it more appropriate to ask *at what density threshold* does a one-to-one relation between the IMF and the CMD finally occur.

5.3. Angular Momentum and Multiplicity

The specific angular momentum and the magnetic field of a collapsing core determines the multiplicity of the stellar system that it forms. We first review how core angular momentum is determined, and then how it in turn influences multiplicity.

5.3.1. Core angular momentum. Galactic differential rotation corresponds to a specific angular momentum $j \approx 10^{23}$ cm^2 s^{-1}, while on scales of cloud cores, below 0.1 pc, $j \approx 10^{21}$ cm^2 s^{-1}. A binary G star with an orbital period of 3 d has $j \approx 10^{19}$ cm^2 s^{-1}, while the spin of a typical T Tauri star is a few $\sim 10^{17}$ cm^2 s^{-1}. Our own Sun rotates only with $j \approx 10^{15}$ cm^2 s^{-1}. Angular momentum must be lost at all stages (for a review, see *Bodenheimer,* 1995).

Specific angular momentum has been measured in clumps and cores at various densities with different tracers: (1) $n \sim 10^3$ cm^{-3} observed in ^{13}CO (*Arquilla and Goldsmith,* 1986); (2) $n \sim 10^4$ cm^{-3} observed in NH$_3$ (e.g., *Goodman et al.,* 1993; *Barranco and Goodman,* 1998; see also *Jijina et al.,* 1999); (3) $n \sim 10^4$–10^5 cm^{-3} observed in N$_2$H$^+$ in both low-mass (*Caselli et al.,* 2002) and high-mass (*Pirogov et al.,* 2003) star-forming regions. In all these objects, the rotational energy is considerably lower than required for support; in the densest cores the differences is an order of magnitude. *Jappsen and Klessen* (2004) showed that these clumps form an evolutionary sequence with j declining with decreasing scale. Main-sequence binaries have j just below that of the densest cores.

Magnetic braking has long been thought to be the primary mechanism for the redistribution of angular momentum out of collapsing objects (*Ebert et al.,* 1960; *Mouschovias and Paleologou,* 1979, 1980). Recent results suggest that magnetic braking may be less important than was thought. Two groups have now performed MHD models of self-gravitating, decaying (*Gammie et al.,* 2003), and driven (*Li et al.,* 2004) turbulence, finding distributions of specific angular momentum consistent with observed cores and with each other. Surprisingly, however, *Jappsen and Klessen* (2004) found similar results with *hydrodynamic* models (see also *Tilley and Pudritz,* 2004), suggesting that accretion and gravitational interactions may dominate over magnetic braking in determining how angular momentum is lost during the collapse of cores. *Fisher* (2004) reached similar conclusions based on semi-analytical models.

5.3.2. Multiplicity. Young stars frequently have a higher multiplicity than main-sequence stars in the solar neighborhood (*Duchêne,* 1999; *Mathieu et al.,* 2000), suggesting that stars are born with high multiplicity, but binaries are then destroyed dynamically during their early lifetime (*Simon et al.,* 1993; *Kroupa,* 1995; *Patience et al.,* 1998). In addition, multiplicity depends on mass. Main-sequence stars with lower masses have lower multiplicity (*Lada,* 2006).

Fragmentation during collapse is considered likely to be the dominant mode of binary formation (*Bodenheimer et al.,* 2000). Hydrodynamic models showed that isothermal spheres with a ratio of thermal to gravitational energy α < 0.3 fragment (*Miyama et al.,* 1984; *Boss,* 1993; *Boss and Myhill,* 1995; *Tsuribe and Inutsuka,* 1999a,b), and that even clumps with larger α can still fragment if they rotate fast enough to form a disk (*Matsumoto and Hanawa,* 2003).

The presence of protostellar jets driven by magnetized accretion disks (e.g., *Königl and Ruden,* 1993) strongly suggests that magnetic braking must play a significant role in the final stages of collapse when multiplicity is determined. Low-resolution MHD models showed strong braking and no fragmentation (*Dorfi,* 1982; *Benz,* 1984; *Phillips,* 1986a,b; *Hosking and Whitworth,* 2004). *Boss* (2000, 2001, 2002, 2005) found that fields enhanced fragmentation, but neglected magnetic braking by only treating the radial component of the magnetic tension force. *Ziegler* (2005) showed that field strengths small enough to allow for binary formation cannot provide support against collapse, offering additional support for a dynamic picture of star formation.

High-resolution models using up to 17 levels of nested grids have now been done for a large number of cores (*Machida et al.,* 2005a,b). These clouds were initially cylindrical, in hydrostatic equilibrium, threaded by a magnetic field aligned with the rotation axis having uniform plasma β_p (the ratio of magnetic field to thermal pressure) (*Machida et al.,* 2004). At least for these initial conditions, fragmentation and binary formation happens during collapse if the initial ratio

$$\frac{\Omega_0}{B_0} > \frac{G^{1/2}}{2^{1/2}c_s} \sim 3 \times 10^7 \text{ yr}^{-1} \, \mu \, G^{-1} \quad (1)$$

Whether this important result generalizes to other geometries clearly needs to be confirmed.

If prestellar cores form by collapse and fragmentation in a turbulent flow, then nearby protostars might be expected to have aligned angular momentum vectors as seen in models (*Jappsen and Klessen,* 2004). This appears to be ob-

served in binaries (*Monin et al.,* 2006) and stars with ages less than 10^6 yr. However, the effect fades with later ages (e.g., *Ménard and Duchêne,* 2004). Ménard and Duchêne also make the intriguing suggestion that jets may only appear if disks are aligned with the local magnetic field.

5.4. Importance of Dynamical Interactions During Star Formation

Rich compact clusters can have large numbers of protostellar objects in small volumes. Thus, dynamical interactions between protostars may become important, introducing a further degree of stochasticity to the star-formation process besides the statistical chaos associated with the gravoturbulent fragmentation process.

Numerical simulations have shown that when a MC region of a few hundred solar masses or more coherently becomes gravitationally unstable, it contracts and builds up a dense clump highly structured in a hierarchical way, containing compact protostellar cores (e.g., *Klessen and Burkert,* 2000, 2001; *Bate et al.,* 2003; *Clark et al.,* 2005). Those cores may contain multiple collapsed objects that will compete with each other for further mass accretion in a common limited and rapidly changing reservoir of contracting gas. The relative importance of these competitive processes depends on the initial ratio between gravitational vs. turbulent and magnetic energies of the cluster-forming core (*Krumholz et al.,* 2005b; see also the chapter by Bonnell et al.). If gravitational contraction strongly dominates the dynamical evolution, as seems probable for the formation of massive, bound clusters, then the following effects need to be considered.

1. Dynamical interactions between collapsed objects during the embedded phase of a nascent stellar cluster evolve toward energy equipartition. As a consequence, massive objects will have, on average, smaller velocities than low-mass objects. They sink toward the cluster center, while low-mass stars predominantly populate large cluster radii. This effect already holds for the embedded phase (e.g., *Bonnell and Davies,* 1998; *Klessen and Burkert,* 2000; *Bonnell et al.,* 2004), and agrees with observations of young clusters [e.g., *Sirianni et al.* (2002) for NGC 330 in the small Magellanic cloud].

2. The most massive cores within a large clump have the largest density contrast and collapse first. In there, the more massive protostars begin to form first and continue to accrete at a high rate throughout the entire cluster evolution ($\sim 10^5$ yr). As their parental cores merge with others, more gas is fed into their "sphere of influence." They are able to maintain or even increase the accretion rate when competing with lower-mass objects (e.g., *Schmeja and Klessen,* 2004; *Bonnell et al.,* 2004).

3. The previous processes lead to highly time-variable protostellar mass growth rates (e.g., *Bonnell et al.,* 2001a; *Klessen,* 2001b; *Schmeja and Klessen,* 2004). As a consequence, the resulting stellar mass spectrum can be modified (*Field and Saslaw,* 1965; *Silk and Takahashi,* 1979; *Lejeune and Bastien,* 1986; *Murray and Lin,* 1996; *Bonnell et al.,*

2001b; *Klessen,* 2001a; *Schmeja and Klessen,* 2004; *Bate and Bonnell,* 2005; see also the chapters by Bonnell et al. and Whitworth et al.). *Krumholz et al.* (2005b) argue that these works typically do not resolve the Bondi-Hoyle radius well enough to derive quantitatively correct accretion rates, but the qualitative statement appears likely to remain correct.

4. Stellar systems with more than two members are in general unstable, with the lowest-mass member having the highest probability of being expelled (e.g., *van Albada,* 1968).

5. Although stellar collisions have been proposed as a mechanism for the formation of high-mass stars (*Bonnell et al.,* 1998; *Stahler et al.,* 2000), detailed two- and three-dimensional calculations (*Yorke and Sonnhalter,* 2002; *Krumholz et al.,* 2005a) show that mass can be accreted from a protostellar disk onto the star. Thus, collisional processes need not be invoked for the formation of high-mass stars (see also *Krumholz et al.,* 2005b).

6. Close encounters in nascent star clusters can truncate and/or disrupt the accretion disk expected to surround (and feed) every protostar. This influences mass accretion through the disk, modifies the ability to subfragment and form a binary star, and this affects the probability of planet formation (e.g., *Clarke and Pringle,* 1991; *McDonald and Clarke,* 1995; *Scally and Clarke,* 2001; *Kroupa and Burkert,* 2001; *Bonnell et al.,* 2001c). In particular, *Ida et al.* (2000) note that an early stellar encounter may explain features of our own solar system, namely the high eccentricities and inclinations observed in the outer part of the Edgeworth-Kuiper belt at distances larger than 42 AU.

7. While competitive coagulation and accretion is a viable mechanism in very massive and dense star-forming regions, protostellar cores in low-mass, low-density clouds such as Taurus or ρ Ophiuchus are less likely to strongly interact with each other.

5.5. Effects of Ionization and Winds

When gravitational collapse proceeds to star formation, ionizing radiation begins to act on the surrounding MC. This can drive compressive motions that accelerate collapse in surrounding gas, or raise the energy of the gas sufficiently to dissociate molecular gas and even drive it out of the potential well of the cloud, ultimately destroying it. Protostellar outflows can have similar effects, but are less powerful, and probably do not dominate cloud energy (*Matzner,* 2002; *Nakamura and Li,* 2005; see also the chapter by Arce et al.).

The idea that the expansion of HII regions can compress the surrounding gas sufficiently to trigger star formation was proposed by *Elmegreen and Lada* (1977). Observational and theoretical work supporting it was reviewed by *Elmegreen* (1998). Since then, it has become reasonably clear that star formation is more likely to occur close to already formed stars than elsewhere in a dark cloud (e.g., *Ogura et al.,* 2002; *Stanke et al.,* 2002; *Karr and Martin,* 2003; *Barbá et al.,* 2003; *Clark and Porter,* 2004; *Healy et al.,* 2004; *Deharveng et al.,* 2005). However, the question remains

whether this seemingly triggered star formation would have occurred anyway in the absence of the trigger. Massive star formation occurs in gravitationally collapsing regions where low-mass star formation is already abundant. The triggering shocks produced by the first massive stars may determine in detail the configuration of newly formed stars without necessarily changing the overall star-formation efficiency of the region (*Elmegreen,* 2002; *Hester et al.,* 2004). Furthermore, the energy input by the shock fronts has the net effect of inhibiting collapse by driving turbulence (*Matzner,* 2002; *Vázquez-Semadeni et al.,* 2003; *Mac Low and Klessen,* 2004), even if it cannot prevent local collapse (see section 5.1).

The study of the dissociation of clouds by ionizing radiation also has a long history, stretching back to the one-dimensional analytic models of *Whitworth* (1979) and *Bodenheimer et al.* (1979). The latter group first noted the champagne effect, in which ionized gas in a density gradient can blow out toward vacuum at supersonic velocities. This mechanism was also invoked by *Blitz and Shu* (1980), and further studied by *Franco et al.* (1994) and *Williams and McKee* (1997). *Matzner* (2002) has performed a detailed analytic study recently, which suggests that the energy injected by HII regions is sufficient to support GMCs for as long as 2×10^7 yr. One major uncertainty remaining is whether expanding HII regions can couple to the clumpy, turbulent gas as well as is assumed.

6. CONCLUSIONS

We have reviewed recent work on MC turbulence and discussed its implications for our understanding of star formation. There are several aspects of MC and star formation where our understanding of the underlying physical processes has developed considerably, but there are also a fair number of questions that remain unanswered. The following gives a brief overview.

1. Molecular clouds appear to be transient objects, representing the brief molecular phase during the evolution of dense regions formed by compressions in the diffuse gas, rather than long-lived, equilibrium structures.

2. Several competing mechanisms for MC formation remain viable. At the largest scales, the gravitational instability of the combined system of stars plus gas drives the formation of spiral density waves in which GMCs can form either by direct gravitational instability of the diffuse gas (in gas-rich systems), or by compression in the stellar potential followed by cooling (in gas-poor systems). At smaller scales, supersonic flows driven by the ensemble of SN explosions have a similar effect.

3. Similar times (<10–15 m.y.) are needed both to assemble enough gas for a MC and to form molecules in it. This is less than earlier estimates. Turbulence-triggered thermal instability can further produce denser (10^3 cm^{-3}), compact HI clouds, where the production of molecular gas takes ≤1 m.y. Turbulent flow through the dense regions results in broad regions of molecular gas at lower average densities of ~100 cm^{-3}.

4. Interstellar turbulence seems to be driven from scales substantially larger than MCs. Internal MC turbulence may well be a byproduct of the cloud-formation mechanism itself, explaining why turbulence is ubiquitously observed on all scales.

5. Molecular cloud turbulence appears to produce a well-defined velocity-dispersion to size relation at the level of entire MC complexes, with an index ~0.5–0.6, but becoming less well defined as smaller objects are considered. The apparent density-size relation, on the other hand, is probably an observational artifact.

6. Molecular cloud cores are produced by turbulent compression and may, or may not, undergo gravitational instability. They evolve dynamically on timescales of ~10^6 yr, but still can mimic certain observational properties of quasi-static objects, such as sub- or transonic nonthermal central linewidth, or Bonnor-Ebert density profiles. These results agree with cloud statistics of nearby, well-studied regions.

7. Magnetic fields appear unlikely to be of qualitative importance in structuring MC cores, although they probably do make a quantitative difference.

8. Comparison of magnetized and unmagnetized simulations of self-gravitating, turbulent gas reveals similar angular momentum distributions. The angular momentum distribution of MC cores and protostars may be determined by turbulent accretion rather than magnetic braking.

9. The interplay between supersonic turbulence and gravity on galactic scales can explain the global efficiency of converting warm atomic gas into cold molecular clouds. The local efficiency of conversion of MCs into stars, on the other hand, seems to depend primarily on stellar feedback evaporating and dispersing the cloud, and only secondarily on the presence of magnetic fields.

10. There are multiple reasons to doubt the existence of a simple one-to-one mapping between the purely turbulent clump mass spectrum and the observed IMF, despite observations showing that the dense self-gravitating core mass spectrum may resemble the IMF. A comprehensive understanding of the physical origin of the IMF remains elusive.

11. In dense clusters, star formation might be affected by dynamical interactions. Close encounters between newborn stars and protostellar cores have the potential to modify their accretion history, and consequently the final stellar mass, as well as the size of protostellar disks, and possibly their ability to form planets and planetary systems.

Acknowledgments. We acknowledge useful comments and suggestions from our referee, P. Padoan. J.B-P. and E.V-S. were supported by CONACYT grant 36571-E. R.S.K. was supported by the Emmy Noether Program of the DFG under grant KL1358/1. M-M.M.L. was supported by NSF grants AST99-85392, AST03-07793, and AST03-07854, and by NASA grant NAG5-10103.

REFERENCES

Alves J. F., Lada C. J., and Lada E. A. (2001) *Nature, 409,* 159–161.
Arons J. and Max C. E. (1975) *Astrophys. J., 196,* L77–L81.
Arquilla R. and Goldsmith P. F. (1986) *Astrophys. J., 303,* 356–374.

Audit E. and Hennebelle P. (2005) *Astron. Astrophys., 433,* 1–13.
Avila-Reese V. and Vázquez-Semadeni E. (2001) *Astrophys. J., 553,* 645–660.
Balbus S. A. and Hawley J. F. (1998) *Rev. Mod. Phys., 70,* 1–53.
Ballesteros-Paredes J. (2004) *Astrophys. Space Sci., 292,* 193–205.
Ballesteros-Paredes J. (2006) *Mon. Not. R. Astron. Soc., 372,* 443–449.
Ballesteros-Paredes J. and Hartmann L. (2006) *Rev. Mex. Astron. Astrofis.,* in press (astro-ph/0605268).
Ballesteros-Paredes J. and Mac Low M.-M. (2002) *Astrophys. J., 570,* 734–748.
Ballesteros-Paredes J. and Vázquez-Semadeni E. (1995) *Rev. Mex. Astron. Astrofís. Conf. Ser.,* pp. 105–108.
Ballesteros-Paredes J. and Vázquez-Semadeni E. (1997) In *Star Formation Near and Far* (S. Holt and L. G. Mundy, eds.), pp. 81–84. AIP Conf. Series 393, New York.
Ballesteros-Paredes J., Vázquez-Semadeni E., and Scalo J. (1999a) *Astrophys. J., 515,* 286–303.
Ballesteros-Paredes J., Hartmann L., and Vázquez-Semadeni E. (1999b) *Astrophys. J., 527,* 285–297.
Ballesteros-Paredes J., Klessen R. S., and Vázquez-Semadeni E. (2003) *Astrophys. J., 592,* 188–202.
Ballesteros-Paredes J., Gazol A., Kim J., Klessen R. S., Jappsen A.-K., et al. (2006) *Astrophys. J., 637,* 384–391.
Barbá R. H., Rubio M., Roth M. R., and García J. (2003) *Astron. J., 125,* 1940–1957.
Barranco J. A. and Goodman A. A. (1998) *Astrophys. J., 504,* 207–222.
Basu S. (2000) *Astrophys. J., 540,* L103–L106.
Basu S. and Ciolek G. E. (2004) *Astrophys. J., 607,* L39–L42.
Bate M. R. and Bonnell I. A. (2005) *Mon. Not. R. Astron. Soc., 356,* 1201–1221.
Bate M. R., Bonnell I. A., and Bromm V. (2003) *Mon. Not. R. Astron. Soc., 339,* 577–599.
Beichman C. A., Myers P. C., Emerson J. P., Harris S., Mathieu R., et al. (1986) *Astrophys. J., 307,* 337–349.
Bender R., Kormendy J., Bower G., Green R., Thomas J., et al. (2005) *Astrophys. J., 631,* 280–300.
Benz W. (1984) *Astron. Astrophys., 139,* 378–388.
Beresnyak A., Lazarian A., and Cho J. (2005) *Astrophys. J., 624,* L93–L96.
Bergin E. A., Hartmann L. W., Raymond J. C., and Ballesteros-Paredes J. (2004) *Astrophys. J., 612,* 921–939.
Bertoldi F. and McKee C. F. (1992) *Astrophys. J., 395,* 140–157.
Biskamp D. and Müller W.-C. (2000) *Physics of Plasmas, 7,* 4889–4900.
Blitz L. (1993) In *Protostars and Planets III* (E. H. Levy and J. I. Lunine, eds.), pp. 125–161. Univ. of Arizona, Tucson.
Blitz L. (1994) In *The Cold Universe* (T. Montmerle et al., eds.), pp. 99–106. Editions Frontieres, Cedex, France.
Blitz L. and Rosolowsky E. (2005) In *The Initial Mass Function 50 Years Later* (E. Corbelli et al., eds.), pp. 287–295. Springer, Dordrecht.
Blitz L. and Shu F. H. (1980) *Astrophys. J., 238,* 148–157.
Blitz L. and Williams J. P. (1999) In *The Origin of Stars and Planetary Systems* (C. J. Lada and N. D. Kylafis, eds.), pp. 3–28. NATO ASIC Proc. 540, Kluwer, Dordrecht.
Blondin J. M. and Marks B. S. (1996) *New Astron., 1,* 235–244.
Bodenheimer P. (1995) *Ann. Rev. Astron. Astrophys., 33,* 199–238.
Bodenheimer P., Tenorio-Tagle G., and Yorke H. W. (1979) *Astrophys. J., 233,* 85–96.
Bodenheimer P., Burkert A., Klein R. I., and Boss A. P. (2000) In *Protostars and Planets IV* (V. Mannings et al., eds.) pp. 675–701. Univ. of Arizona, Tucson.
Boldyrev S. (2002) *Astrophys. J., 569,* 841–845.
Boldyrev S., Nordlund Å., and Padoan P. (2002) *Astrophys. J., 573,* 678–684.
Bonnell I. A. and Davies M. B. (1998) *Mon. Not. R. Astron. Soc., 295,* 691–698.
Bonnell I. A., Bate M. R., and Zinnecker H. (1998) *Mon. Not. R. Astron. Soc., 298,* 93–102.
Bonnell I. A., Bate M. R., Clarke C. J., and Pringle J. E. (2001a) *Mon. Not. R. Astron. Soc., 323,* 785–794.

Bonnell I. A., Clarke C. J., Bate M. R., and Pringle J. E. (2001b) *Mon. Not. R. Astron. Soc., 324,* 573–579.

Bonnell I. A., Smith K. W., Davies M. B., and Horne K. (2001c) *Mon. Not. R. Astron. Soc., 322,* 859–865.

Bonnell I. A., Vine S. G., and Bate M. R. (2004) *Mon. Not. R. Astron. Soc., 349,* 735–741.

Bonnell I. A., Dobbs C. L., Robitaille T. P., and Pringle J. E. (2006) *Mon. Not. R. Astron. Soc., 365,* 37–45.

Boss A. P. (1993) *Astrophys. J., 410,* 157–167.

Boss A. P. (2000) *Astrophys. J., 545,* L61–L64.

Boss A. P. (2001) *Astrophys. J., 551,* L167–L170.

Boss A. P. (2002) *Astrophys. J., 568,* 743–753.

Boss A. P. (2005) *Astrophys. J., 622,* 393–403.

Boss A. P. and Myhill E. A. (1995) *Astrophys. J., 451,* 218–224.

Bourke T. L. and Goodman A. A. (2004) In *Star Formation at High Angular Resolution* (M. Burton et al., eds.), pp. 83–94. IAU Symposium 221, ASP, San Francisco.

Bourke T. L., Myers P. C., Robinson G., and Hyland A. R. (2001) *Astrophys. J., 554,* 916–932.

Brunt C. M. (2003) *Astrophys. J., 583,* 280–295.

Caselli P. and Myers P. C. (1995) *Astrophys. J., 446,* 665–686.

Caselli P., Benson P. J., Myers P. C., and Tafalla M. (2002) *Astrophys. J., 572,* 238–263.

Chabrier G. (2005) In *The Initial Mass Function 50 Years Later* (E. Corbelli et al., eds.), pp. 41–52. Springer, Dordrecht.

Chandrasekhar S. (1951) *Proc. R. Soc. London, A210,* 18.

Cho J. and Lazarian A. (2003) *Mon. Not. R. Astron. Soc., 345,* 325–339.

Cho J., Lazarian A., and Vishniac E. T. (2002) *Astrophys. J., 564,* 291–301.

Ciolek G. E. and Basu S. (2001) *Astrophys. J., 547,* 272–279.

Ciolek G. E. and Mouschovias T. C. (1995) *Astrophys. J., 454,* 194–216.

Clark P. C. and Bonnell I. A. (2004) *Mon. Not. R. Astron. Soc., 347,* L36–L40.

Clark J. S. and Porter J. M. (2004) *Astron. Astrophys., 427,* 839–847.

Clarke C. J. and Pringle J. E. (1991) *Mon. Not. R. Astron. Soc., 249,* 584–587.

Clark P. C., Bonnell I. A., Zinnecker H., and Bate M. R. (2005) *Mon. Not. R. Astron. Soc., 359,* 809–818.

Crutcher R. M. (1999) *Astrophys. J., 520,* 706–713.

Crutcher R., Heiles C., and Troland T. (2003) In *Turbulence and Magnetic Fields in Astrophysics* (E. Falgarone and T. Passot, eds.), pp. 155–181. LNP Vol. 614, Springer, Dordrecht.

Crutcher R. M., Nutter D. J., Ward-Thompson D., and Kirk J. M. (2004) *Astrophys. J., 600,* 279–285.

Curry C. L. (2000) *Astrophys. J., 541,* 831–840.

Dalcanton J. J., Yoachim P., and Bernstein R. A. (2004) *Astrophys. J., 608,* 189–207.

de Avillez M. A. (2000) *Mon. Not. R. Astron. Soc., 315,* 479–497.

de Avillez M. A. and Breitschwerdt D. (2006) Paper presented at the Portuguese National Astronomical & Astrophysics Meeting, held in Lisbon (Portugal) on 28–29 July 2005 (astro-ph/0601228).

Deharveng L., Zavagno A., and Caplan J. (2005) *Astron. Astrophys., 433,* 565–577.

de Jong T., Boland W., and Dalgarno A. (1980) *Astron. Astrophys., 91,* 68–84.

Dib S., Bell E., and Burkert A. (2006a) *Astrophys. J., 638,* 797–810.

Dib S., Vázquez-Semadeni E., Kim J., Burkert A., and Shadmehri M. (2006b) *Astrophys. J.,* in press (astro-ph/0607362).

Dickey J. M. and Lockman F. J. (1990) *Ann. Rev. Astron. Astrophys., 28,* 215–261.

Dobbs C. L., Bonnell I. A., and Clark P. C. (2005) *Mon. Not. R. Astron. Soc., 360,* 2–8.

Dorfi E. (1982) *Astron. Astrophys., 114,* 151–164.

Duchêne G. (1999) *Astron. Astrophys., 341,* 547–552.

Dziourkevitch N., Elstner D., and Rüdiger G. (2004) *Astron. Astrophys., 423,* L29–L32.

Ebert R., von Hoerner S., and Temesváry S. (1960) *Die Entstehung von Sternen durch Kondensation diffuser Materie,* Springer-Verlag, Berlin.

Elmegreen B. G. (1991) In *The Physics of Star Formation and Early Stel-*

lar Evolution (C. J. Lada and N. D. Kylafis, eds.), pp. 35–59. NATO ASIC Proc. 342, Kluwer, Dordrecht.

Elmegreen B. G. (1993) *Astrophys. J., 419,* L29–L32.

Elmegreen B. G. (1998) In *Origins* (C. E. Woodward et al., eds.), pp. 150–183. ASP Conf. Series 148, San Francisco.

Elmegreen B. G. (2000a) *Astrophys. J., 539,* 342–351.

Elmegreen B. G. (2000b) *Astrophys. J., 530,* 277–281.

Elmegreen B. G. (2002) *Astrophys. J., 577,* 206–220.

Elmegreen B. G. and Lada C. J. (1977) *Astrophys. J., 214,* 725–741.

Elmegreen B. G. and Scalo J. (2004) *Ann. Rev. Astron. Astrophys., 42,* 211–273.

Elsässer K. and Schamel H. (1976) *Z. Physik B, 23,* 89.

Evans N. J. (1999) *Ann. Rev. Astron. Astrophys., 37,* 311–362.

Falgarone E., Puget J.-L., and Perault M. (1992) *Astron. Astrophys., 257,* 715–730.

Falle S. A. E. G. and Hartquist T. W. (2002) *Mon. Not. R. Astron. Soc., 329,* 195–203.

Fatuzzo M. and Adams F. C. (2002) *Astrophys. J., 570,* 210–221.

Field G. B. (1978) In *Protostars and Planets* (T. Gehrels, ed.), pp. 243–264. Univ. of Arizona, Tucson.

Field G. B. and Saslaw W. C. (1965) *Astrophys. J., 142,* 568–583.

Field G. B., Goldsmith D. W., and Habing H. J. (1969) *Astrophys. J., 155,* L149–L154.

Fisher R. T. (2004) *Astrophys. J., 600,* 769–780.

Franco J. and Cox D. P. (1986) *Publ. Astron. Soc. Pac., 98,* 1076–1079.

Franco J., Shore S. N., and Tenorio-Tagle G. (1994) *Astrophys. J., 436,* 795–799.

Fukui Y., Mizuno N., Yamaguchi R., Mizuno A., Onishi T., et al. (1999) *Publ. Astron. Soc. Japan, 51,* 745–749.

Gammie C. F., Lin Y., Stone J. M., and Ostriker E. C. (2003) *Astrophys. J., 592,* 203–216.

Gazol A., Vázquez-Semadeni E., and Kim J. (2005) *Astrophys. J., 630,* 911–924.

Gazol-Patiño A. and Passot T. (1999) *Astrophys. J., 518,* 748–759.

Genzel R. (1991) In *The Physics of Star Formation and Early Stellar Evolution* (C. J. Lada and N. D. Kylafis), pp. 155–219. NATO ASIC Proc. 342, Kluwer, Dordrecht.

Gillmon K. and Shull J. M. (2006) *Astrophys. J., 636,* 908.

Glover S. C. O. and Mac Low M.-M. (2005) In *PPV Poster Proceedings,* www.lpi.usra.edu/meetings/ppv2005/pdf/8577.pdf.

Goldreich P. and Kwan J. (1974) *Astrophys. J., 189,* 441–454.

Goldsmith P. F. (2001) *Astrophys. J., 557,* 736–746.

Goldsmith P. F. and Langer W. D. (1978) *Astrophys. J., 222,* 881–895.

Goodman A. A., Benson P. J., Fuller G. A., and Myers P. C. (1993) *Astrophys. J., 406,* 528–547.

Goodwin S. P., Whitworth A. P., and Ward-Thompson D. (2004a) *Astron. Astrophys., 414,* 633–650.

Goodwin S. P., Whitworth A. P., and Ward-Thompson D. (2004b) *Astron. Astrophys., 423,* 169–182.

Hartmann L. (1998) *Accretion Processes in Star Formation.* Cambridge Univ., Cambridge.

Hartmann L. (2001) *Astron. J., 121,* 1030–1039.

Hartmann L. (2003) *Astrophys. J., 585,* 398–405.

Hartmann L., Ballesteros-Paredes J., and Bergin E. A. (2001) *Astrophys. J., 562,* 852–868.

Hartquist T. W., Falle S. A. E. G., and Williams D. A. (2003) *Astrophys. Space Sci., 288,* 369–375.

Hayashi C. (1966) *Ann. Rev. Astron. Astrophys., 4,* 171–192.

Healy K. R., Hester J. J., and Claussen M. J. (2004) *Astrophys. J., 610,* 835–850.

Heiles C. and Troland T. H. (2003) *Astrophys. J., 586,* 1067–1093.

Heiles C. and Troland T. H. (2005) *Astrophys. J., 624,* 773–793.

Heiles C., Goodman A. A., McKee C. F., and Zweibel E. G. (1993) In *Protostars and Planets III* (E. H. Levy and J. I. Lunine, eds.), pp. 279–326. Univ. of Arizona, Tucson.

Heitsch F., Mac Low M.-M., and Klessen R. S. (2001a) *Astrophys. J., 547,* 280–291.

Heitsch F., Zweibel E. G., Mac Low M.-M., Li P., and Norman M. L. (2001b) *Astrophys. J., 561,* 800–814.

Heitsch F., Zweibel E. G., Slyz A. D., and Devriendt J. E. G. (2004) *Astrophys. J., 603,* 165–179.

Heitsch F., Burkert A., Hartmann L. W., Slyz A. D., and Devriendt J. E. G. (2005) *Astrophys. J., 633,* L113–L116.

Hennebelle P. and Pérault M. (1999) *Astron. Astrophys., 351,* 309–322.

Hester J. J., Desch S. J., Healy K. R., and Leshin L. A. (2004) *Science, 304,* 1116–1117.

Heyer M. H. and Brunt C. M. (2004) *Astrophys. J., 615,* L45–L48.

Heyer M., Williams J., and Brunt C. (2006) *Astrophys. J., 643,* 956.

Hollenbach D. and Salpeter E. E. (1971) *Astrophys. J., 163,* 155–164.

Hosking J. G. and Whitworth A. P. (2004) *Mon. Not. R. Astron. Soc., 347,* 1001–1010.

Hunter J. H. and Fleck R. C. (1982) *Astrophys. J., 256,* 505–513.

Hunter J. H., Sandford M. T., Whitaker R. W., and Klein R. I. (1986) *Astrophys. J., 305,* 309–332.

Ida S., Larwood J., and Burkert A. (2000) *Astrophys. J., 528,* 351–356.

Inutsuka S. and Koyama H. (2004) *Rev. Mex. Astron. Astrofís. Conf. Ser.,* pp. 26–29.

Jappsen A.-K. and Klessen R. S. (2004) *Astron. Astrophys., 423,* 1–12.

Jappsen A.-K., Klessen R. S., Larson R. B., Li Y., and Mac Low M.-M. (2005) *Astron. Astrophys., 435,* 611–623.

Jessop N. E. and Ward-Thompson D. (2000) *Mon. Not. R. Astron. Soc., 311,* 63–74.

Jijina J., Myers P. C., and Adams F. C. (1999) *Astrophys. J. Suppl., 125,* 161–236.

Jones C. E., Basu S., and Dubinski J. (2001) *Astrophys. J., 551,* 387–393.

Jørgensen J. K., Schöier F. L., and van Dishoeck E. F. (2005) *Astron. Astrophys., 437,* 501–515.

Joung M. K. R. and Mac Low M.-M. (2006) *Astrophys. J.,* in press (astro-ph/0601005).

Jura M. (1975) *Astrophys. J., 197,* 575–580.

Karr J. L. and Martin P. G. (2003) *Astrophys. J., 595,* 900–912.

Kegel W. H. (1989) *Astron. Astrophys., 225,* 517–520.

Kennicutt R. C. (1998) *Astrophys. J., 498,* 541–552.

Keto E. and Field G. (2006) *Astrophys. J., 635,* 1151.

Kim J. and Ryu D. (2005) *Astrophys. J., 630,* L45–L48.

Kim W.-T., Ostriker E. C., and Stone J. M. (2003) *Astrophys. J., 599,* 1157–1172.

Klessen R. S. (2000) *Astrophys. J., 535,* 869–886.

Klessen R. S. (2001a) *Astrophys. J., 556,* 837–846.

Klessen R. S. (2001b) *Astrophys. J., 550,* L77–L80.

Klessen R. S. and Burkert A. (2000) *Astrophys. J. Suppl., 128,* 287–319.

Klessen R. S. and Burkert A. (2001) *Astrophys. J., 549,* 386–401.

Klessen R. S., Burkert A., and Bate M. R. (1998) *Astrophys. J., 501,* L205–L208.

Klessen R. S., Heitsch F., and Mac Low M.-M. (2000) *Astrophys. J., 535,* 887–906.

Klessen R. S., Ballesteros-Paredes J., Vázquez-Semadeni E., and Durán-Rojas C. (2005) *Astrophys. J., 620,* 786–794.

Königl A. and Ruden S. P. (1993) In *Protostars and Planets III* (E. H. Levy and J. I. Lunine, eds.), pp. 641–687. Univ. of Arizona, Tucson.

Korpi M. J., Brandenburg A., Shukurov A., Tuominen I., and Nordlund Å. (1999) *Astrophys. J., 514,* L99–L102.

Koyama H. and Inutsuka S. (2000) *Astrophys. J., 532,* 980–993.

Koyama H. and Inutsuka S.-i. (2002) *Astrophys. J., 564,* L97–L100.

Kroupa P. (1995) *Mon. Not. R. Astron. Soc., 277,* 1507–1521.

Kroupa P. (2001) *Mon. Not. R. Astron. Soc., 322,* 231–246.

Kroupa P. (2002) *Science, 295,* 82–91.

Kroupa P. and Burkert A. (2001) *Astrophys. J., 555,* 945–949.

Krumholz M. R. and McKee C. F. (2005) *Astrophys. J., 630,* 250–268.

Krumholz M. R., McKee C. F., and Klein R. I. (2005a) *Astrophys. J., 618,* L33–L36.

Krumholz M. R., McKee C. F., and Klein R. I. (2005b) *Nature, 438,* 332–334.

Kulkarni S. R. and Heiles C. (1987) In *Interstellar Processes* (D. J. Hollenbach and H. A. Thronson Jr., eds.), pp. 87–122. ASSL Vol. 134, Reidel, Dordrecht.

Kurtz S., Cesaroni R., Churchwell E., Hofner P., and Walmsley C. M. (2000) In *Protostars and Planets IV* (V. Mannings et al., eds.), pp. 299–326. Univ. of Arizona, Tucson.

Lada C. J. (2006) *Astrophys. J., 640,* L63.

Lada C. J. and Lada E. A. (2003) *Ann. Rev. Astron. Astrophys., 41,* 57–115.

Langer W. D., van Dishoeck E. F., Bergin E. A., Blake G. A., Tielens A. G. G. M., et al. (2000) In *Protostars and Planets IV* (V. Mannings et al., eds.), pp. 29–57. Univ. of Arizona, Tucson.

Larson R. B. (1969) *Mon. Not. R. Astron. Soc., 145,* 271–295.

Larson R. B. (1973) *Fund. Cosmic Phys., 1,* 1–70.

Larson R. B. (1981) *Mon. Not. R. Astron. Soc., 194,* 809–826.

Larson R. B. (2005) *Mon. Not. R. Astron. Soc., 359,* 211–222.

Lee C. W. and Myers P. C. (1999) *Astrophys. J., Suppl., 123,* 233–250.

Lee Y., Stark A. A., Kim H.-G., and Moon D.-S. (2001) *Astrophys. J. Suppl., 136,* 137–187.

Lejeune C. and Bastien P. (1986) *Astrophys. J., 309,* 167–175.

Léorat J., Passot T., and Pouquet A. (1990) *Mon. Not. R. Astron. Soc., 243,* 293–311.

Li P. S., Norman M. L., Mac Low M.-M., and Heitsch F. (2004) *Astrophys. J., 605,* 800–818.

Li Y., Klessen R. S., and Mac Low M.-M. (2003) *Astrophys. J., 592,* 975–985.

Li Y., Mac Low M.-M., and Klessen R. S. (2005) *Astrophys. J., 626,* 823–843.

Li Y., Mac Low M.-M., and Klessen R. S. (2006) *Astrophys. J., 639,* 879.

Li Z.-Y. and Nakamura F. (2004) *Astrophys. J., 609,* L83–L86.

Lin C. C. and Shu F. H. (1964) *Astrophys. J., 140,* 646–655.

Lin C. C., Yuan C., and Shu F. H. (1969) *Astrophys. J., 155,* 721–746.

Lizano S. and Shu F. H. (1989) *Astrophys. J., 342,* 834–854.

Low C. and Lynden-Bell D. (1976) *Mon. Not. R. Astron. Soc., 176,* 367–390.

Machida M. N., Tomisaka K., and Matsumoto T. (2004) *Mon. Not. R. Astron. Soc., 348,* L1–L5.

Machida M. N., Matsumoto T., Tomisaka K., and Hanawa T. (2005a) *Mon. Not. R. Astron. Soc., 362,* 369–381.

Machida M. N., Matsumoto T., Hanawa T., and Tomisaka K. (2005b) *Mon. Not. R. Astron. Soc., 362,* 382–402.

Mac Low M.-M. and Klessen R. S. (2004) *Rev. Mod. Phys. 76,* 125–194.

Mac Low M.-M. and Ossenkopf V. (2000) *Astron. Astrophys., 353,* 339–348.

Mac Low M.-M., Klessen R. S., Burkert A., and Smith M. D. (1998) *Phys. Rev. Lett., 80,* 2754–2757.

Mac Low M.-M., Balsara D. S., Kim J., and de Avillez M. A. (2005) *Astrophys. J., 626,* 864–876.

Maloney P. (1990) *Astrophys. J., 348,* L9–L12.

Maron J. and Goldreich P. (2001) *Astrophys. J., 554,* 1175–1196.

Masunaga H. and Inutsuka S. (2000) *Astrophys. J., 531,* 350–365.

Mathieu R. D., Ghez A. M., Jensen E. L. N., and Simon M. (2000) In *Protostars and Planets IV* (V. Mannings et al., eds.), pp. 703–730. Univ. of Arizona, Tucson.

Matsumoto T. and Hanawa T. (2003) *Astrophys. J., 595,* 913–934.

Matzner C. D. (2001) In *From Darkness to Light: Origin and Evolution of Young Stellar Clusters* (T. Montmerle and P. André, eds.), pp. 757–768. ASP Conf. Series 243, San Francisco.

Matzner C. D. (2002) *Astrophys. J., 566,* 302–314.

Matzner C. D. and McKee C. F. (2000) *Astrophys. J., 545,* 364–378.

McCray R. and Kafatos M. (1987) *Astrophys. J., 317,* 190–196.

McDonald J. M. and Clarke C. J. (1995) *Mon. Not. R. Astron. Soc., 275,* 671–684.

McKee C. F. (1995) In *The Physics of the Interstellar Medium and Intergalactic Medium* (A. Ferrara et al., eds.), pp. 292–316. ASP Conf. Series 80, San Francisco.

McKee C. F. (1999) In *The Origin of Stars and Planetary Systems* (C. J. Lada and N. D. Kylafis, eds.), pp. 29–66. NATO ASIC Proc. 540, Kluwer Dordrecht.

McKee C. F. and Ostriker J. P. (1977) *Astrophys. J., 218,* 148–169.

McKee C. F. and Zweibel E. G. (1992) *Astrophys. J., 399,* 551–562.

McKee C. F., Zweibel E. G., Goodman A. A., and Heiles C. (1993) In *Protostars and Planets III* (E. H. Levy and J. I. Lunine, eds.), pp. 327–366. Univ. of Arizona, Tucson.

Ménard F. and Duchêne G. (2004) *Astron. Astrophys., 425,* 973–980.

Miyama S. M., Hayashi C., and Narita S. (1984) *Astrophys. J., 279,* 621–632.

Mizuno N., Yamaguchi R., Mizuno A., Rubio M., Abe R., et al. (2001) *Publ. Astron. Soc. Japan, 53,* 971–984.

Monin J.-L., Ménard F., and Peretto N. (2006) *Astron. Astrophys., 446,* 201–210.

Morata O., Girart J. M., and Estalella R. (2005) *Astron. Astrophys., 435,* 113–124.

Motte F., André P., and Neri R. (1998) *Astron. Astrophys., 336,* 150–172.

Mouschovias T. C. (1991) In *The Physics of Star Formation and Early Stellar Evolution* (C. J. Lada and N. D. Kylafis, eds.), pp. 61–122. NATO ASIC Proc. 342, Kluwer, Dordrecht.

Mouschovias T. C. and Paleologou E. V. (1979) *Astrophys. J., 230,* 204–222.

Mouschovias T. C. and Paleologou E. V. (1980) *Astrophys. J., 237,* 877–899.

Murray S. D. and Lin D. N. C. (1996) *Astrophys. J., 467,* 728–748.

Myers P. C. and Goodman A. A. (1988) *Astrophys. J., 326,* L27–L30.

Myers P. C., Dame T. M., Thaddeus P., Cohen R. S., Silverberg R. F., et al. (1986) *Astrophys. J., 301,* 398–422.

Myers P. C., Fuller G. A., Goodman A. A., and Benson P. J. (1991) *Astrophys. J., 376,* 561–572.

Nakamura F. and Li Z.-Y. (2005) *Astrophys. J., 631,* 411–428.

Nakano T. (1998) *Astrophys. J., 494,* 587–604.

Nayakshin S. and Sunyaev R. (2005) *Mon. Not. R. Astron. Soc., 364,* L23–L27.

Nordlund Å. K. and Padoan P. (1999) In *Interstellar Turbulence* (J. Franco and A. Carramiñana, eds.), pp. 218–222. Cambridge Univ., Cambridge.

Ogura K., Sugitani K., and Pickles A. (2002) *Astron. J., 123,* 2597–2626.

Onishi T., Mizuno A., Kawamura A., Tachihara K., and Fukui Y. (2002) *Astrophys. J., 575,* 950–973.

Ossenkopf V. and Mac Low M.-M. (2002) *Astron. Astrophys., 390,* 307–326.

Ossenkopf V., Klessen R. S., and Heitsch F. (2001) *Astron. Astrophys., 379,* 1005–1016.

Ostriker E. C., Gammie C. F., and Stone J. M. (1999) *Astrophys. J., 513,* 259–274.

Ostriker E. C., Stone J. M., and Gammie C. F. (2001) *Astrophys. J., 546,* 980–1005.

Padoan P. (1995) *Mon. Not. R. Astron. Soc., 277,* 377–388.

Padoan P. and Nordlund Å. (1999) *Astrophys. J., 526,* 279–294.

Padoan P. and Nordlund Å. (2002) *Astrophys. J., 576,* 870–879.

Padoan P., Jones B. J. T., and Nordlund Å. (1997a) *Astrophys. J., 474,* 730–734.

Padoan P., Nordlund Å., and Jones B. J. T. (1997b) *Mon. Not. R. Astron. Soc., 288,* 145–152.

Padoan P., Goodman A., Draine B. T., Juvela M., Nordlund Å., et al. (2001a) *Astrophys. J., 559,* 1005–1018.

Padoan P., Juvela M., Goodman A. A., and Nordlund Å. (2001b) *Astrophys. J., 553,* 227–234.

Padoan P., Goodman A. A., and Juvela M. (2003) *Astrophys. J., 588,* 881–893.

Padoan P., Jimenez R., Juvela M., and Nordlund Å. (2004a) *Astrophys. J., 604,* L49–L52.

Padoan P., Jimenez R., Nordlund Å., and Boldyrev S. (2004b) *Phys. Rev. Lett., 92,* 191102-1 to 191102-4.

Parker E. N. (1979) *Cosmical Magnetic Fields: Their Origin and Their Activity.* Oxford Univ., New York.

Passot T. and Vázquez-Semadeni E. (1998) *Phys. Rev. E, 58,* 4501–4510.

Passot T. and Vázquez-Semadeni E. (2003) *Astron. Astrophys., 398,* 845–855.

Passot T., Pouquet A., and Woodward P. (1988) *Astron. Astrophys., 197,* 228–234.

Passot T., Vázquez-Semadeni E., and Pouquet A. (1995) *Astrophys. J., 455,* 536–555.

Patience J., Ghez A. M., Reid I. N., Weinberger A. J., and Matthews K. (1998) *Astron. J., 115,* 1972–1988.

Pavlovski G., Smith M. D., Mac Low M.-M., and Rosen A. (2002) *Mon. Not. R. Astron. Soc., 337,* 477–487.

Phillips G. J. (1986a) *Mon. Not. R. Astron. Soc., 221,* 571–587.

Phillips G. J. (1986b) *Mon. Not. R. Astron. Soc., 222,* 111–119.

Piontek R. A. and Ostriker E. C. (2005) *Astrophys. J., 629,* 849–864.

Pirogov L., Zinchenko I., Caselli P., Johansson L. E. B., and Myers P. C. (2003) *Astron. Astrophys., 405,* 639–654.

Plume R., Jaffe D. T., Evans N. J., Martin-Pintado J., and Gomez-Gonzalez J. (1997) *Astrophys. J., 476,* 730–749.

Rafikov R. R. (2001) *Mon. Not. R. Astron. Soc., 323,* 445–452.

Reid M. A. and Wilson C. D. (2006) *Astrophys. J., 644,* 990–1005.

Roberts W. W. (1969) *Astrophys. J., 158,* 123–144.

Rosen A. and Bregman J. N. (1995) *Astrophys. J., 440,* 634–665.

Ryden B. S. (1996) *Astrophys. J., 471,* 822–831.

Salpeter E. E. (1955) *Astrophys. J., 121,* 161–167.

Sasao T. (1973) *Publ. Astron. Soc. Japan, 25,* 1–33.

Scally A. and Clarke C. (2001) *Mon. Not. R. Astron. Soc., 325,* 449–456.

Scalo J. (1990) In *Physical Processes in Fragmentation and Star Formation* (R. Capuzzo-Dolcetta et al., eds.), pp. 151–176. ASSL Vol. 162, Kluwer, Dordrecht.

Scalo J., Vázquez-Semadeni E., Chappell D., and Passot T. (1998) *Astrophys. J., 504,* 835–853.

Schekochihin A., Cowley S., Maron J., and Malyshkin L. (2002) *Phys. Rev. E, 65,* 016305-1 to 016305-18.

Schmeja S. and Klessen R. S. (2004) *Astron. Astrophys., 419,* 405–417.

Schneider N. and Brooks K. (2004) *Publ. Astron. Soc. Austral., 21,* 290–301.

Sellwood J. A. and Balbus S. A. (1999) *Astrophys. J., 511,* 660–665.

Shu F. H., Adams F. C., and Lizano S. (1987) *Ann. Rev. Astron. Astrophys., 25,* 23–81.

Silk J. and Takahashi T. (1979) *Astrophys. J., 229,* 242–256.

Simon M., Ghez A. M., and Leinert C. (1993) *Astrophys. J., 408,* L33–L36.

Sirianni M., Nota A., De Marchi G., Leitherer C., and Clampin M. (2002) *Astrophys. J., 579,* 275–288.

Slyz A. D., Devriendt J. E. G., Bryan G., and Silk J. (2005) *Mon. Not. R. Astron. Soc., 356,* 737–752.

Spaans M. and Silk J. (2000) *Astrophys. J., 538,* 115–120.

Spaans M. and Silk J. (2005) *Astrophys. J., 626,* 644–648.

Spitzer L. (1968) *Diffuse Matter in Space.* Interscience, New York.

Stahler S. W., Palla F., and Ho P. T. P. (2000) In *Protostars and Planets IV* (V. Mannings et al., eds.), pp. 327–351 Univ. of Arizona, Tucson.

Stanke T., Smith M. D., Gredel R., and Szokoly G. (2002) *Astron. Astrophys., 393,* 251–258.

Stanke T., Smith M. D., Gredel R., and Khanzadyan T. (2006) *Astron. Astrophys., 447,* 609–622.

Stark A. A. and Lee Y. (2005) *Astrophys. J., 619,* L159–L162.

Stevens I. R., Blondin J. M., and Pollock A. M. T. (1992) *Astrophys. J., 386,* 265–287.

Stolte A., Brandner W., Grebel E. K., Lenzen R., and Lagrange A.-M. (2005) *Astrophys. J., 628,* L113–L117.

Stone J. M., Ostriker E. C., and Gammie C. F. (1998) *Astrophys. J., 508,* L99–L102.

Tachihara K., Onishi T., Mizuno A., and Fukui Y. (2002) *Astron. Astrophys., 385,* 909.

Tafalla M., Myers P. C., Caselli P., and Walmsley C. M. (2004) *Astron. Astrophys., 416,* 191–212.

Taylor S. D., Morata O., and Williams D. A. (1996) *Astron. Astrophys., 313,* 269–276.

Taylor S. D., Morata O., and Williams D. A. (1998) *Astron. Astrophys., 336,* 309–314.

Testi L. and Sargent A. I. (1998) *Astrophys. J., 508,* L91–L94.

Tilley D. A. and Pudritz R. E. (2004) *Mon. Not. R. Astron. Soc., 353,* 769–788.

Tsuribe T. and Inutsuka S.-I. (1999b) *Astrophys. J., 523,* L155–L158.

Tsuribe T. and Inutsuka S.-I. (1999a) *Astrophys. J., 526,* 307–313.

van Albada T. S. (1968) *Bull. Astron. Inst. Netherlands, 19,* 479–499.

Vázquez-Semadeni E. (1994) *Astrophys. J., 423,* 681–692.

Vázquez-Semadeni E. and Passot T. (1999) In *Interstellar Turbulence* (J. Franco and A. Carramiñana, eds.), pp. 223–231. Cambridge Univ., Cambridge.

Vázquez-Semadeni E., Passot T., and Pouquet A. (1995) *Astrophys. J., 441,* 702–725.

Vázquez-Semadeni E., Passot T., and Pouquet A. (1996) *Astrophys. J., 473,* 881–893.

Vázquez-Semadeni E., Ballesteros-Paredes J., and Rodríguez L. F. (1997) *Astrophys. J., 474,* 292–307.

Vázquez-Semadeni E., Ostriker E. C., Passot T., Gammie C. F., and Stone J. M. (2000) *Protostars and Planets IV* (V. Mannings et al., eds.), pp. 3–28. Univ. of Arizona, Tucson.

Vázquez-Semadeni E., Ballesteros-Paredes J., and Klessen R. S. (2003) *Astrophys. J., 585,* L131–L134.

Vázquez-Semadeni E., Kim J., and Ballesteros-Paredes J. (2005a) *Astrophys. J., 630,* L49–L52.

Vázquez-Semadeni E., Kim J., Shadmehri M., and Ballesteros-Paredes J. (2005b) *Astrophys. J., 618,* 344–359.

Vázquez-Semadeni E., Ryu D., Passot T., Gonzalez R. F., and Gazol A. (2006a) *Astrophys. J., 643,* 245.

Vázquez-Semadeni E., Gómez G. C., Jappsen A. K., Ballesteros-Paredes J., González R. F., and Klessen R. S. (2006b) *Astrophys. J.,* in press (astro-ph/0608375).

Vishniac E. T. (1994) *Astrophys. J., 428,* 186–208.

von Weizsäcker C. F. (1951) *Astrophys. J., 114,* 165–186.

Walder R. and Folini D. (1998) *Astron. Astrophys., 330,* L21–L24.

Walder R. and Folini D. (2000) *Astrophys. Space Sci., 274,* 343–352.

Ward-Thompson D., Scott P. F., Hills R. E., and Andre P. (1994) *Mon. Not. R. Astron. Soc., 268,* 276–290.

Ward-Thompson D., Kirk J. M., Crutcher R. M., Greaves J. S., Holland W. S., et al. (2000) *Astrophys. J., 537,* L135–L138.

Whitworth A. (1979) *Mon. Not. R. Astron. Soc., 186,* 59–67.

Williams J. P. and McKee C. F. (1997) *Astrophys. J., 476,* 166–183.

Wolf S., Launhardt R., and Henning T. (2003) *Astrophys. J., 592,* 233–244.

Wolfire M. G., Hollenbach D., McKee C. F., Tielens A. G. G. M., and Bakes E. L. O. (1995) *Astrophys. J., 443,* 152–168.

Wong T. and Blitz L. (2002) *Astrophys. J., 569,* 157–183.

Yamaguchi R., Mizuno N., Mizuno A., Rubio M., Abe R., et al. (2001) *Publ. Astron. Soc. Japan, 53,* 985–1001.

Yorke H. W. and Sonnhalter C. (2002) *Astrophys. J., 569,* 846–862.

Young C. H., Jørgensen J. K., Shirley Y. L., Kauffmann J., Huard T., et al. (2004) *Astrophys. J. Suppl., 154,* 396–401.

Ziegler U. (2005) *Astron. Astrophys., 435,* 385–395.

Zuckerman B. and Evans N. J. (1974) *Astrophys. J., 192,* L149–L152.

Zuckerman B. and Palmer P. (1974) *Ann. Rev. Astron. Astrophys., 12,* 279–313.

Giant Molecular Clouds in Local Group Galaxies

Leo Blitz
University of California, Berkeley

Yasuo Fukui and Akiko Kawamura
Nagoya University

Adam Leroy
University of California, Berkeley

Norikazu Mizuno
Nagoya University

Erik Rosolowsky
Harvard-Smithsonian Center for Astrophysics

We present the first comparative study of extragalactic Giant Molecular Clouds (GMCs) using complete datasets for entire galaxies and a uniform set of reduction and analysis techniques. We present results based on CO observations for the Large Magellanic Cloud (LMC), Small Magellanic Cloud (SMC), M33, M31, IC 10, and the nucleus of M64, and make comparisons with archival Milky Way observations. Our sample includes large spirals and dwarf irregulars with metallicities that vary by an order of magnitude. GMCs in HI-rich galaxies are seen to be well-correlated with HI filaments that pervade the galactic disks, suggesting that they form from preexisting HI structures. Virial estimates of the ratio of CO line strength to H_2 column density, X_{CO}, suggest that a value of 4×10^{20} cm^{-2} (K km s^{-1})$^{-1}$ is a good value to use in most galaxies (except the SMC) if the GMCs are virialized. However, if the clouds are only marginally self-gravitating, as appears to be the case judging from their appearance, half the virial value may be more appropriate. There is no clear trend of X_{CO} with metallicity. The clouds within a galaxy are shown to have the about the same H_2 surface density, and differences between galaxies seem to be no more than a factor of ~2. We show that hydrostatic pressure appears to be the main factor in determining what fraction of atomic gas is turned into molecules. In the high-pressure regions often found in galactic centers, the observed properties of GMCs appear to be different from those in the found in the Local Group. From the association of tracers of star formation with GMCs in the LMC, we find that about one-fourth of the GMCs exhibit no evidence of star formation, and we estimate that the lifetime of a typical GMC in these galaxies is 20–30 m.y.

1. INTRODUCTION

Although a great deal of progress has been made on the topic of star and planet formation since the Protostars and Planets IV conference in Santa Barbara, little work has been done to connect what we know about star formation in the Milky Way to star formation in the universe as a whole. Fundamental limitations include only a weak understanding of how the massive stars form, how clusters and associations form, and the constancy of the initial mass function (IMF). After all, in external galaxies, we generally observe only the effects of massive star formation and the formation of star clusters. Furthermore, knowledge of the initial conditions for star formation at all masses remains elusive both within and outside of the Milky Way.

Since nearly all stars form in GMCs, one way to make progress is to examine the properties of GMCs in a number of different extragalactic environments to see how they differ. From the similarities and differences, it might be possible to make some general conclusions about how star formation varies throughout the universe. Although individual, extragalactic GMCs had been observed previously at high enough resolution to at least marginally resolve them (e.g., *Vogel et al.,* 1987; *Lada et al.,* 1988), the first attempts to do this in a systematic way were by C. Wilson (*Wilson and Scoville,* 1990; *Wilson and Reid,* 1991; *Wilson and Rudolph,*

TABLE 1.   Local Group GMC data.

Galaxy	Telescope	Metallicity	Spatial Resolution	Reference
LMC	NANTEN	$0.33\ Z_\odot$	40 pc	[1]
SMC	NANTEN	$0.1\ Z_\odot$	48 pc	[2]
IC 10	OVRO/BIMA	$0.25\ Z_\odot$	14–20 pc	[3]
M33	BIMA	0.1–$1.0\ Z_\odot$	20–30 pc	[4]
M31	BIMA	$0.5\ Z_\odot$	26–36 pc	[5]

References: [1] Fukui et al. (in preparation); [2] Mizuno et al. (in preparation); [3] *Leroy et al.* (2006); [4] *Engargiola et al.* (2003); [5] *Rosolowsky* (2006).

1993; *Wilson,* 1994) using the OVRO and BIMA interferometers. Her efforts were hampered by small survey areas in a few galaxies, so general conclusions could only be made by extrapolation. Numerous other authors subsequently studied one or a few extragalactic GMCs, both in the Local Group and beyond. An exhaustive list of their efforts is beyond the scope of the present article.

The situation has changed in the last five years as a result of the construction of the NANTEN telescope in the southern hemisphere and the completion of the 10-element BIMA array. The former made it possible to map the Magellanic Clouds completely with high enough spatial resolution and signal-to-noise to identify all the GMCs with masses $>3 \times 10^4\ M_\odot$; the completion of the BIMA interferometer made it possible to identify GMCs in other, more distant galaxies in the Local Group. Because of their relatively large fields of view, these two telescopes could completely survey nearby galaxies. Thus, the first complete survey of GMCs in *any* galaxy was of the LMC (*Fukui et al.,* 1999; *Mizuno et al.,* 2001b) and not the Milky Way (MW). Although the molecular gas in the MW has been essentially completely mapped, velocity crowding in many directions makes it impossible to generate a full catalog of GMCs. Similarly, the first complete CO surveys of the Magellanic Clouds were by *Cohen et al.* (1988) and *Rubio et al.* (1991), but the resolution was too poor to determine the properties of individual molecular clouds.

In this paper, we review the recent surveys of CO in Local Group galaxies that (1) have sufficient resolution to study individual molecular clouds and (2) span all or most of the target galaxy. We compare the results of observations of GMCs in the four external Local Group galaxies that have been mapped in their entirety in CO: the LMC (*Fukui et al.,* 2001; Fukui et al., in preparation), the SMC (*Mizuno et al.,* 2001a; Mizuno et al., in preparation), IC 10 (*Leroy et al.,* 2006), and M33 (*Engargiola et al.,* 2003). We have also made observations in a small strip in M31 (*Rosolowsky,* 2006), and we compare the properties of the GMCs in all these galaxies to clouds in the outer MW (from *Dame et al.,* 2001) using a uniform set of analytic techniques. The LMC and SMC observations were made with the single-dish NANTEN telescope in Chile; the remaining galaxies were observed with the BIMA millimeter-wave interferometer at Hat Creek, California (combined with obsevations from the Caltech OVRO millimeter interferometer for IC 10). A tabu-

lation of the galaxies we observed, their metallicities, and the resolution used to observe them is given in Table 1.

2.   THE GALAXIES

In this section, we examine the distribution of CO emission in the surveyed galaxies and we compare the CO to emission in other wavebands.

2.1.   The Large Magellanic Cloud

Figure 1 shows the molecular clouds detected with the NANTEN Survey (*Fukui et al.,* 2001; Fukui et al., in preparation) on an optical image of the LMC. Except for a region near the eastern edge of the galaxy (left side of Fig. 1) below 30 Doradus, the clouds appear to be spatially well separated and it is possible to pick them out individually

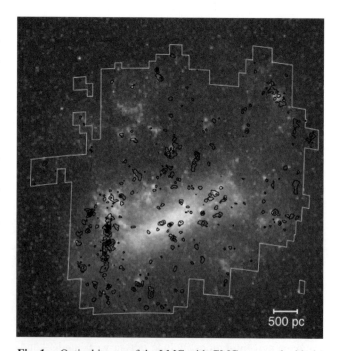

500 pc

Fig. 1.   Optical image of the LMC with GMCs mapped with the NANTEN telescope indicated within the boundary of the survey area. The CO is well correlated with H II regions. The GMCs are easily identified by eye except for the region south of 30 Doradus, where they appear as a vertical line of clouds, and the individual GMCs may be overlapping in this region.

by eye. The long string of bright CO emission along the eastern edge of the galaxy is likely composed of several clouds that cannot be separated at this resolution. Some have speculated that this feature is due to hydrodynamical collision between the LMC and SMC (*Fujimoto and Noguchi,* 1990) or ram pressure pileup of gas due to the motion of the LMC through a halo of hot, diffuse gas (*de Boer et al.,* 1998; *Kim et al.,* 1998). Supershells may also be playing a role in the formation of GMCs as in the case LMC4 (*Yamaguchi et al.,* 2001a). A comprehensive comparison between supergiant shells and GMCs shows that only about one-third of the GMCs are located toward supershells, suggesting the effects of supershells are not predominant (*Yamaguchi et al.,* 2001b). There is neither an excess nor a deficit of CO associated with the stellar bar, but the bright HII regions are all clearly associated with molecular clouds. Individual clouds are frequently associated with young clusters of stars. Not every cluster of young stars is associated with a cloud nor does every cloud show evidence of massive star formation. Using this association and the ages of the stellar clusters, we can establish the evolutionary timescale for GMCs (section 6).

2.2. The Small Magellanic Cloud

Figure 2 shows the GMCs superimposed on a grayscale image made using the 3.6-, 4.5-, and 8.0-μm bands from the IRAC instrument on the Spitzer Space Telescope (Bolatto et al., in preparation). The CO map is from the NANTEN telescope (Mizuno et al., in preparation). As in the LMC, the GMCs in the SMC are easily identified by eye. Unlike the LMC, they are not spread throughout the galaxy but appear preferentially on the northern and southern ends of the galaxy. Another grouping is located to the east (left) of

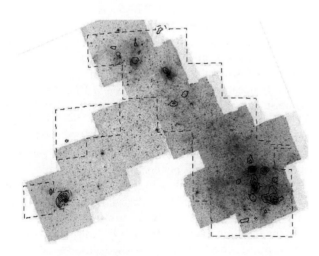

Fig. 2. GMCs in the SMC observed with the NANTEN telescope overlayed on an a near-infrared image of the galaxy from the Spitzer Space Telescope (Bolatto et al., in preparation). The lines indicate the survey boundary. The CO clouds are clearly associated with regions of transiently heated small grains or PAHs that appear as dark, nebulous regions in the image.

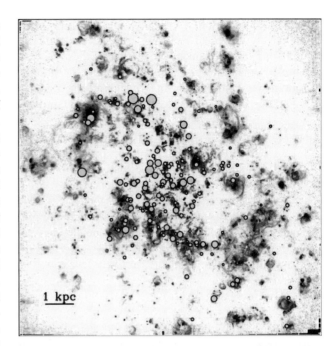

Fig. 3. The locations of GMCs in M33 as derived from the 759-field BIMA mosaic of *Engargiola et al.* (2003). Since sources of CO emission in a map would be too small to identify in the figure, the locations of GMCs are instead indicated by light gray circles. The area of the circles is scaled to the CO luminosity, which should be proportional to the H_2 mass. The GMC locations are overlayed on a continuum subtracted Hα image of the galaxy (*Massey et al.,* 2001). There is significant correlation between the GMCs and massive star formation as traced by Hα.

the SMC along the HI bridge that connects the LMC and SMC, apparently outside the stellar confines of the galaxy. The Spitzer image traces the stellar continuum as well as warm dust and PAH emission. The 8.0-μm emission is associated with the molecular gas traced by CO, but appears to be more extended than the CO emission. The SMC has the lowest metallicity in our sample and provides an opportunity to explore the behavior of molecular gas in chemically primitive environments.

2.3. M33

Figure 3 shows the locations of GMCs in M33 from the BIMA telescope (*Engargiola et al.,* 2003) superimposed on an Hα image of the galaxy (*Massey et al.,* 2001). The two low-contrast spiral arms (*Regan and Vogel,* 1994) are well-traced by GMCs, but the GMCs are not confined to these arms, as is evident in the center of the galaxy. There is good spatial correlation between the GMCs and the HII regions. Once again, the correlation is not perfect and there are GMCs without HII regions and vice versa. Unlike the other images, we show the locations of the GMCs as circles with areas proportional to the CO luminosity of each GMC; the CO luminosity is expected to be proportional to the H_2 mass of each GMC. Note that the most massive GMCs (~10^6 $M_\odot$) are not found toward the center of the galaxy but along spiral arms north of the galactic nucleus. These massive clouds are relatively devoid of Hα emission. The completeness limit of this survey is about 1.5×10^5 M$_\odot$; thus there are presumably many lower-mass clouds below the limit of sensitivity. Many of these low-mass clouds are likely associated with the unaccompanied HII regions in the figure.

2.4. IC 10

Figure 4 is an image of the GMCs in IC 10 from a 50-field CO mosaic with the BIMA telescope (*Leroy et al.,* 2006) superimposed on a 2-μm image of the galaxy made from 2MASS data (*Jarrett et al.,* 2003). As with the Magellanic Clouds and M33, the GMCs show no obvious spatial correlation with old stellar population — some massive clouds are found where there are relatively few stars.

2.5. The Correlation with HI

The distribution of GMCs in these four galaxies shows little correlation with old stars (see Figs. 1 and 4). The obvious correlations with Hα (Figs. 1 and 3) and young stellar clusters (Fig. 1) are expected since these trace the star formation that occurs within GMCs. That the correlation is not perfect can be used to deduce information about the evolution of the clouds (section 6). To examine the relationship of GMCs to the remainder of the neutral ISM, we plot the locations of CO emission on top of HI maps of these four galaxies in Fig. 5. A strong correlation between the atomic and molecular gas is immediately apparent. Every

GMC in each of the galaxies is found on a bright filament or clump of HI, but the reverse is not true: There are many bright filaments of HI without molecular gas. In M33, the largest of the fully mapped galaxies, the ratio of HI to CO in the filaments in the center of the galaxy is smaller than in the outer parts. In the LMC, the CO is generally found at peaks of the HI, but most of the short filaments have no associated CO. In the SMC, the HI is so widespread that the CO clouds appear as small, isolated clouds in a vast sea of HI. Apparently, HI is a necessary but not a sufficient condition for the formation of GMCs in these galaxies.

Figures 1–4 show that the molecular gas forms from the HI, rather than the HI being a dissociation product of the molecular clouds as some have advocated (e.g., *Allen,* 2001). First, in all four galaxies the HI is much more widespread than the detected CO emission. Thus, most of the HI cannot be dissociated H$_2$ without violating mass conservation if the GMC lifetimes are as short as we derive in section 6. Second, there is no CO associated with most of the filaments in the LMC, M33, and IC 10, and the column density of these CO-free filaments is about the same as the column density of filaments that have CO emission. Because there is no transition in HI properties at radii where one observes CO, and the radii where it is absent, it is difficult to imagine that two separate origins for the HI would produce a seamless transition. Finally, the HI in the filaments between GMCs has the wrong geometry to be a dissociation product; there is too much gas strung out along the filaments to have come from dissociation of the molecular gas.

2.6. Implications for Giant Molecular Cloud Formation

What can the morphology of the atomic gas tell us about GMCs and their formation? All of the HI images are characterized by filamentary structures that demarcate holes in the atomic distribution. In IC 10, there is good evidence that some of the holes are evacuated by the action of supernovae or stellar winds that sweep up the atomic gas into the observed filamentary structure (*Wilcots and Miller,* 1998). In contrast, most of the large HI holes observed in the M33 HI are *not* likely to be caused by supernovae. The large holes require about 10^{53} ergs to evacuate, but there are no obvious stellar clusters remaining at the center of the holes. Furthermore, X-ray emission is not concentrated in the holes. The large holes in M33 are thus likely to have a gravitational or density-wave origin. Small holes with D < 200 pc, on the other hand, are found to be well correlated with OB associations (*Deul and van der Hulst,* 1987); these tend to be concentrated toward the center of the galaxy.

This leads to some qualitative conclusions about the formation of GMCs and ultimately the star formation that occurs within them. Because the CO forms from HI filaments and not the other way around, it is the filaments in a galaxy that must form first as precursors to the GMCs. In some of the galaxies, such as M33 and apparently in the

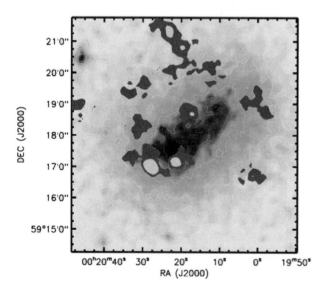

Fig. 4. The GMCs in IC 10 made from a 50-field mosaic of the galaxy with the BIMA telescope overlayed on a 2-μm image of the galaxy from the 2MASS survey. The dark gray area corresponds to CO brightness above 1 K km s^{-1}; the light gray area corresponds to CO brightness above 10 K km s^{-1}. The black region in the center has the highest stellar surface density. The rms noise of the CO data is ~0.3 K km s^{-1} in each channel map; the peak value in the integrated intensity map is 48 K km s^{-1} (*Leroy et al.,* 2006).

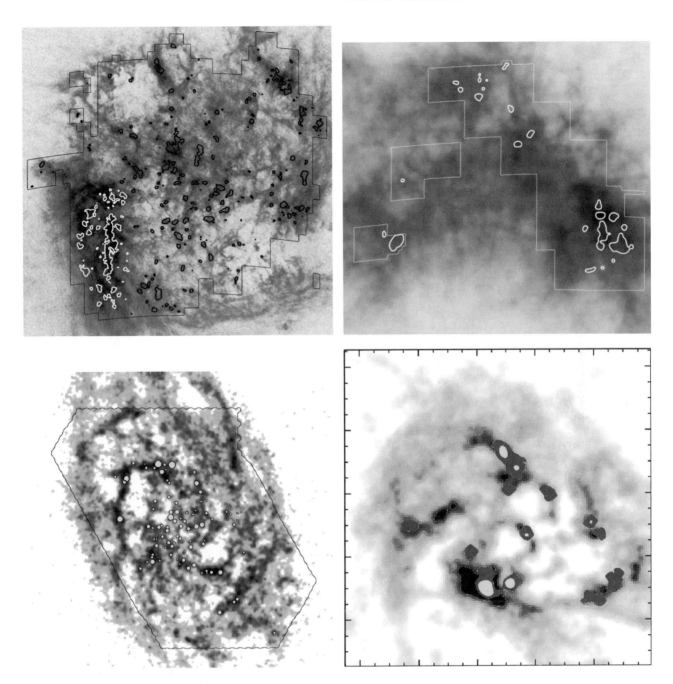

Fig. 5. CO emission overlayed on maps of Hɪ emission for the LMC (top left), the SMC (top right), M33 (bottom left), and IC 10 (bottom right). The Hɪ maps are the work of *Kim et al.* (2003) (LMC), *Stanimirović et al.* (1999) (SMC), *Deul and van der Hulst* (1987) (M33), and *Wilcots and Miller* (1998) (IC 10). Contours of the CO emission are shown in each case except for M33, where the emission is indicated as circles with area proportional to the flux. Where appropriate, the boundaries of the surveys are indicated. CO emission is found exclusively on bright filaments of atomic gas, although not every bright Hɪ filament has CO emission.

LMC and the SMC, most of the filaments are not associated with energetic phenomena. This clearly rules out the self-propagating star formation picture that was promoted some years back by *Gerola and Seiden* (1978) for most of our galaxies. In their picture, GMC formation and thus star formation propagates by means of supernovae that explode in regions of a galaxy adjacent to a previous episode of star formation. However, in IC 10, because there is evidence that some of the Hɪ morphology may be the result of energetic

events from previous generations of stars, self-propagating star formation may be a viable mechanism. *The critical element of GMC formation across all these systems appears to be the assembly of Hɪ filaments, although the mechanism that collects the atomic gas appears to vary across the systems.*

But why, then, do some filaments form GMCs and not others? We argue in section 5 that it is the result of the pressure to which filaments are subjected.

3. MOLECULAR CLOUD PROPERTIES

Our main goal in this section is to compare the properties of GMCs made with different telescopes, resolutions, and sensitivities. We use GMC catalogs from the studies of the four galaxies listed above, and we supplement our work with a sample of GMCs in M31 (*Rosolowsky*, 2006) as well as a compilation of molecular clouds in the outer Milky Way as observed by *Dame et al.* (2001) and cataloged in *Rosolowsky and Leroy* (2006).

To aid in the systematic comparison of cloud properties, *Rosolowsky and Leroy* (2006, hereafter *RL06*) have recently published a method for minimizing the biases that plague such comparisons. For example, measurement of the cloud radius depends on the sensitivity of the measurements, and *RL06* suggest a robust method to extrapolate to the expected radius in the limit of infinite sensitivity. They also suggest a method to correct cloud sizes for beam convolution, which has been ignored in many previous studies of extragalactic clouds. We use the *RL06* extrapolated moment method on all the data used in this paper since it is least affected by relatively poor signal-to-noise and resolution effects. We have also applied the *RL06* methodology to the outer Milky Way data of *Dame et al.* (2001) rather than relying on published properties (e.g., *Heyer et al.*, 2001). It is for this reason that we have not included the cloud properties of *Solomon et al.* (1987) in our plots, but we do make comparisons to their work at the end of this section. Except where noted, we consider only clouds that are well resolved by the telescope beam; the GMCs must have angular diameters at least twice that of the beam used to observe them.

Are we seeing single or multiple objects in the beam? The issue of velocity blending of multiple clouds in the beam is much less of an issue in extragalactic observations than in the galactic case, where the overwhelming majority of GMCs are observed only in the galactic plane. Extragalactic observations of all but the most highly inclined galaxies do not suffer from this problem and as can be seen in Figs. 1–4, the clouds are, in general, spatially well separated, ensuring that we are almost always seeing only a single GMC along the line of sight.

One of the long-debated questions related to GMCs is: How does metallicity affect the value of X_{CO}, the conversion factor from CO line strength to H_2 column density? Figure 6 is a plot of the virial mass of the GMCs as a function of CO luminosity. Diagonal lines indicate constant X_{CO}. A compilation of X_{CO} values is given in Table 2. We note first that most of the points lie above the dashed line that indicates the value determined from γ-rays in the Milky Way (*Strong and Mattox*, 1996). A value of $X_{CO} = 4 \times 10^{20}$ cm^{-2} (K km s^{-1})$^{-1}$ would allow virial masses to be derived to within about a factor of 2 for all the GMCs in our sample, with the clouds in the SMC and the outer galaxy requiring a somewhat higher value.

Note, however, that the SMC clouds are systematically higher in this plot than the GMCs for any other galaxy, and that the GMCs in IC 10 are systematically a bit lower. Solv-

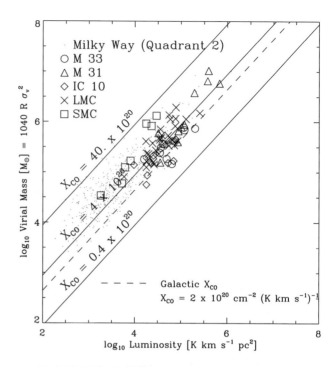

Fig. 6. Plot of the virial mass of the GMCs in our sample as a function of luminosity. The value of X_{CO} from γ-ray investigations in the Milky Way (*Strong and Mattox*, 1996) is shown by the dashed line. The plot shows that while there are some differences in X_{CO} from galaxy to galaxy, except for the SMC, a value of $X_{CO} = 4 \times 10^{20}$ cm^{-2} (K km s^{-1})$^{-1}$ can be used for all the other galaxies to a reasonable degree of approximation.

TABLE 2. X_{CO} across the Local Group.

Galaxy	Mean X_{CO} $\times 10^{20}$ cm^{-2} (K km s^{-1})$^{-1}$	Scatter in X_{CO}*
LMC	5.4 ± 0.5	1.7
SMC	13.5 ± 2.6	2.2
M33	3.0 ± 0.4	1.5
IC 10	2.6 ± 0.5	2.2
M31	5.6 ± 1.1	2.7
Quad 2†	6.6 ± 0.6	2.0
Local Group‡	5.4 ± 0.5	2.0

*Scatter is a factor based on median absolute deviation of the log.
†Clouds with luminosities corresponding to $M_{Lum} \geq 5 \times 10^4$ M$_\odot$ (for $X_{CO} = 2 \times 10^{20}$).
‡Excluding the Milky Way.

ing for X_{CO} in the SMC gives a value of 13.5×10^{20} cm^{-2} (K km s^{-1})$^{-1}$, more than a factor of 3 above the mean. In contrast, IC 10 yields $X_{CO} = 2 \times 10^{20}$ cm^{-2} (K km s^{-1})$^{-1}$. Surprisingly, the galaxies differ in metallicity from one another only by a factor of two, and both are much less than solar. In M33, the metallicity decreases by almost an order of magnitude from the center out (*Henry and Howard*, 1995), but *Rosolowsky et al.* (2003) find no change in X_{CO} with radius. Although metallicity may be a factor in determining

X_{CO} in different galaxies, there is no clear trend with metallicity alone — other factors appear to be as important as the metallicity in determining X_{CO}.

The discrepancy between the galactic γ-ray value of 2×10^{20} cm^{-2} (K km s^{-1})$^{-1}$ and the virial value we derive here is not necessarily a problem. Taken at face value, it may be telling us is that the GMCs are not in virial equilibrium, but are nearly gravitationally neutral: The overall potential energy is equal to the kinetic energy. The γ-ray value of X_{CO} is independent of the dynamical state of the cloud, thus uncertainties about the self-gravity of GMCs do not come into play. Since GMCs do not look as if they are in virial equilibrium (they are highly filamentary structures and do not appear to be strongly centrally concentrated), these two different values of X_{CO} are consistent if the clouds are only marginally self-gravitating.

Figure 7a is a plot of the CO luminosity of GMCs as a function of line width. It may be thought of as a plot of H$_2$ mass vs. line width for a single, but undetermined, value of X_{CO}. The dashed line, which is the relation $L_{CO} \propto \sigma_v^4$, is not a fit, but is a good representation of the data for the five external galaxies in our sample as well as for the outer Milky Way. The scatter in the relationship is 0.5 dex, or a factor of 3 over 3 orders of magnitude in luminosity. If the GMCs are self-gravitating, then they obey

$$M = 5R\sigma_v^2/(\alpha G) \qquad (1)$$

where α is a constant of order unity. Provided the CO luminosity is proportional to the mass of a GMC, the plot shows that $M(H_2) \propto \sigma_v^4$; thus

$$\sigma_v \propto R^{0.5} \text{ and } M/R^2 = \text{constant} \qquad (2)$$

These two relations are shown in Figs. 8a and 8b respectively.

The advantage of a luminosity-line width plot, especially for extragalactic studies, is that one need not resolve the individual clouds, since the luminosity, and by implication, the mass, is independent of resolution. One need only be sure that individual GMCs are isolated in the beam. Figure 7b shows all the individual clouds identified in the galaxy surveys, most of which are unresolved. We see that the clouds populate the same $L_{CO} \propto \sigma_v^4$ line as in Fig. 7a. This plot demonstrates, probably better than any other, that the GMCs in our sample are much more alike than they are different.

Figure 8a shows the size-line width relation for the GMCs in our sample. The dashed line is the size-line width relation for GMCs in the inner region of the Milky Way from *Solomon et al.* (1987). First, we note that the correlation for the extragalactic clouds is very weak. However, if we add the outer galaxy clouds, the correlation does seem to be consistent with a power-law relation $\sigma_v \propto R^{0.5}$. However, there is a clear offset from the relation determined for the inner galaxy [dashed line (*Solomon et al.,* 1987)]. At least part of this offset can be attributed to differences in the methods used to measure cloud properties. The sense of the

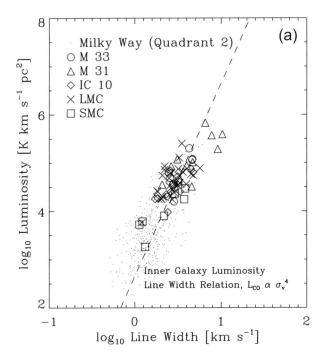
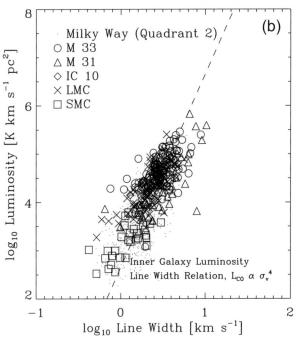

Fig. 7. (a) Luminosity vs. line width plot for all the resolved clouds in our survey. The dashed line, $L_{CO} \propto \sigma_v^4$, with a single constant of proportionality, is a good representation of the data. (b) The same as the lefthand panel but including the unresolved clouds in our sample. The dashed line remains a good representation of the data even with much more data included.

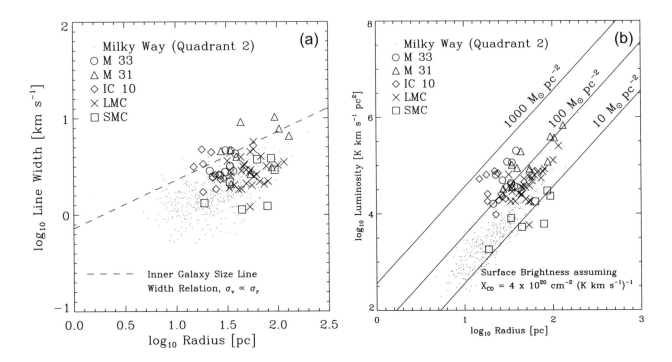

Fig. 8. (a) Line width-size relation for the GMCs in our sample. The dashed line is the relation found for the GMCs in the inner Milky Way, showing a clear offset from the extragalactic GMCs. (b) Luminosity vs. radius relation for the GMCs in our sample. Solid lines are lines of constant surface density assuming $X_{CO} = 4 \times 10^{20}$ cm^{-2} (K km s^{-1})$^{-1}$. The galaxies show clear differences in CO luminosity for a given cloud radius.

offset is that for a given cloud radius, inner Milky Way clouds have larger line widths. This may be partially due to the relatively high value of T_A used by *Solomon et al.* (1987) to define the cloud radius, implying that the clouds might be inferred to be smaller for a given value of σ_v.

But part of the offset may also be real. We see that there is a clear separation of the clouds by galaxy in the plot. The IC 10 clouds lie to the left of the diagram, while the LMC clouds lie to the right. The SMC clouds tend to lie at the bottom of the group. The apparently weak correlation of extragalactic clouds is probably due to the small dynamic range in the plot compared to the measurement error in the cloud properties; the rms scatter in Fig. 8a is only 0.2 dex, or less than a factor of 2. We therefore conclude that the GMCs in our sample are consistent with a power law relation $\sigma_v \propto R^{0.5}$. There are, however, real differences in the coefficient of proportionality, and this gives rise to some of the scatter in the relationship. The size-line width relationship arises from the turbulent nature of the molecular gas motions. Differences in the constant of proportionality imply variations in the normalization of the turbulent motions of GMCs in different galaxies, independent of cloud luminosity.

These conclusions help to explain Fig. 8b, which is a plot of luminosity vs. radius. Assuming that luminosity is proportional to mass, at least within a single galaxy, we can plot lines of constant surface brightness. After all, Fig. 6 suggests that the clouds have a nearly constant surface brightness. In fact, it appears that for a given galaxy, the individual GMCs

are strung out along lines of constant surface density, but with each galaxy lying on a different line. The SMC clouds, for example, have a mean surface density of 10 M$_\odot$ pc^{-2}, but the IC 10 clouds have a mean surface density >100 M$_\odot$ pc^{-2}. A direct interpretation of Fig. 8b implies that for a given radius, the SMC clouds are less luminous than the rest, and the IC 10 clouds are more luminous. Another way of saying this is that for a given cloud luminosity, the SMC clouds are larger, as are the LMC clouds, only less so. This difference disappears, for the most part, if we consider the mass surface density rather than the surface brightness. In that case one must multiply the luminosity of the GMCs in each galaxy by its appropriate value of X_{CO}. When that is done, the difference in the mean surface density of clouds from galaxy to galaxy is less than a factor of 2.

In Fig. 7 we see that the GMCs in the SMC are well separated from the GMCs in M31, implying that the median luminosity of the two sets of clouds is different by nearly two orders of magnitude. The differences due to X_{CO} are only a factor of about 4; but is the distribution of GMC masses in the two systems really different? There are not enough clouds to measure a mass spectrum in the SMC, but Fig. 9 shows the mass distribution of molecular clouds normalized to the survey area for the other five galaxies. Power-law fits to the masses of all cataloged molecular clouds above the completeness limit give the index of the mass distributions listed in Table 3. All the galaxies have remarkably similar mass distributions except for M33, which is much steeper than the others. In addition, the mass distribu-

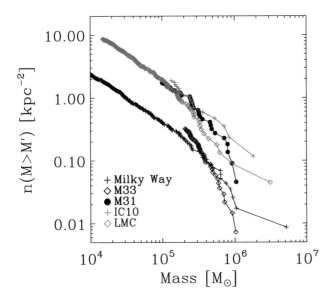

Fig. 9. Cumulative mass distribution for the galaxies in our sample. The mass distributions have been normalized by the area surveyed in each galaxy. In this figure, we use all clouds above the completeness limits in each survey, whether or not the clouds are resolved. All the galaxies look similar except for M33, which has a steeper mass spectrum than the others.

TABLE 3.   Mass distributions of the five galaxies.

Galaxy	Index
LMC	−1.74 ± 0.08
IC 10	−1.74 ± 0.41
M33	−2.49 ± 0.48
M31	−1.55 ± 0.20
Outer MW	−1.71 ± 0.06

tions in M31 and the LMC show a truncation at high mass similar to that found in the inner Milky Way (e.g., *Williams and McKee,* 1997), suggesting that there is a characteristic cloud mass in these systems. In addition, *Engargiola et al.* (2003) also argue for a characteristic cloud mass in M33 but it is not at the high-mass end, as it is for the LMC and M31; rather it has a value of 4–6 × 10⁴ M⊙. The variation in the mass distributions is unexplained and may offer an avenue to understanding differences in star formation rates between galaxies.

4.   THE ROLE OF HYDROSTATIC PRESSURE

A number of authors have speculated on the role that hydrostatic pressure plays in the formation of molecular clouds in the centers of galaxies (*Helfer and Blitz,* 1993; *Spergel and Blitz,* 1992) and galactic disks (e.g., *Elmegreen,* 1993; *Wong and Blitz,* 2002). *Blitz and Rosolowsky* (2004) showed that if hydrostatic pressure is the only parameter governing the molecular gas fraction in galaxies, then one

predicts that the location where the ratio of molecular to atomic gas is unity occurs at constant *stellar* surface density. They probed this prediction and found that the constancy holds to within 40% for 30 nearby galaxies.

The functional form of the relationship between hydrostatic pressure and molecular gas fraction has recently been investigated by *Blitz and Rosolowsky* (2006) for 14 galaxies covering 3 orders of magnitude in pressure. Hydrostatic pressure is determined by

$$P_{hydro} = 0.84(G\Sigma_*)^{0.5}\Sigma_g \frac{v_g}{(h_*)^{0.5}} \qquad (3)$$

The quantities v_g, the gas velocity dispersion, and h_*, the stellar scale height, vary by less than a factor of 2 both within and among galaxies (*van der Kruit and Searle,* 1981a,b; *Kregel et al.,* 2002). The quantities Σ_*, the stellar surface density, and Σ_g, the gas surface density, can be obtained from observations. The results for 14 galaxies is given in Fig. 10.

The figure shows that the galaxies all have similar slopes for the relationship: $\Sigma_{H_2}/\Sigma_{HI} \propto P^{0.92}$, very nearly linear. Moreover, except for three galaxies, NGC 3627, NGC 4321, and NGC 4501, all have the same constant of proportionality. The three exceptional galaxies all are interacting with their environments and may be subject to additional pressure forces. It is important to point out that we expect this pressure relation to break down at some lower scale no smaller than the scale of a typical GMC, ~50 pc. However, on the scale of the pressure scale height, typically a few hundred parsecs, the pressure should be more or less constant both vertically and in the plane of a galaxy.

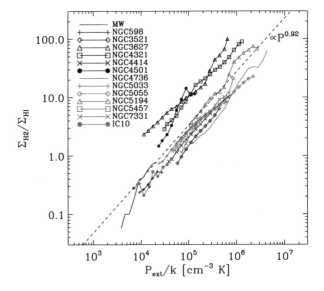

Fig. 10. Plot of the ratio of molecular to atomic surface density as a function of hydrostatic pressure for 14 galaxies. The plot covers 3 mag in pressure and molecular fraction.

The two axes in Fig. 10 are not completely independent; both are proportional to Σ_{H_2}. However, each axis is also dependent on other quantities such as Σ_{HI} and Σ_*. Since Σ_* varies by a larger amount in a given galaxy than Σ_{H_2}, because Σ_{HI} dominates at low pressures (P/k < 10^5 cm^{-3} K) and because both axes have different dependencies on Σ_{H_2}, the constancy of the slopes and the agreement of the intercepts cannot be driven by the common appearance of Σ_{H_2} on each axis. A more detailed discussion of this point is given in *Blitz and Rosolowsky* (2006).

As of this writing we do not know how the LMC and the SMC fit into this picture; no good map giving the stellar surface density for these objects is currently available. Although we do not know the stellar scale heights for these galaxies, because of the weak dependence on h_* in equation (3), this ignorance should not be much of a difficulty. The results for the SMC are particularly interesting because of its low metallicity and low dust-to-gas ratio (*Koorneef,* 1984; *Stanimirović et al.,* 2000). Since the extinction in the UV is significantly smaller than in other galaxies, one might expect higher pressures to be necessary to achieve the same fraction of molecular gas in the SMC, although care must be taken since CO may be compromised as a mass tracer in such environments.

The following picture for the formation of molecular clouds in galaxies is therefore suggested by the observations. Density waves or some other process collects the atomic gas into filamentary structures. This process may be the result of energetic events, as is thought to be the case for IC 10, or dynamical processes, as is primarily the case for M33. Depending on how much gas is collected, and where in the gravitational potential of the galaxy the gas is located, a fraction of the atomic hydrogen is turned in molecular gas. In very gas-rich, high-pressure regions near galactic centers, this conversion is nearly complete. But some other process, perhaps instabilities, collects the gas into clouds. Whether this is done prior to the formation of molecules or after is not clear.

5. BEYOND THE LOCAL GROUP

5.1. Giant Molecular Clouds in Starburst Galaxies

In many galaxies the average surface density of molecular gas is much greater than the surface densities of individual GMCs shown in Fig. 8b (*Helfer et al.,* 2003). These regions of high surface density can be as much as a kiloparsec in extent. Indeed, about half of the galaxies in the BIMA SONG survey (*Helfer et al.,* 2003) have central surface densities in excess of 100 M$_\odot$ pc^{-2}. Moreover, regions with high surface densities of molecular gas are invariably associated with dramatically enhanced star formation rates (*Kennicutt,* 1998). In regions of such high surface density, are there even individual, identifiable GMCs? If so, do they obey the same relationships shown in Figs. 6–8?

Several recent studies have begun to attack these questions. The only such molecule-rich region in the Local Group is in the vicinity of the galactic center, where cloud properties were analyzed by *Oka et al.* (2001). They found that clouds in the galactic center were smaller, denser and had larger line widths than the GMCs in the galactic disk. For targets beyond the Local Group, achieving the requisite spatial resolution to study individual GMCs requires significant effort. To date, only a few extragalactic, molecule-rich regions have been studied. *Keto et al.* (2005) show clouds in M82 to be roughly in virial equilibrium. At the high surface densities of molecular gas observed in M82, this requires clouds to be smaller and denser than those found in the galactic disk. Similarly, *Rosolowsky and Blitz* (2005) observed the inner region of the galaxy M64, which has a surface density of ~100 M$_\odot$ pc^{-2} over the inner 300 pc of the galaxy. They found 25 GMCs with densities 2.5× higher, and are 10× more massive, on average, than typical disk GMCs. This conclusion is quite robust against differences in cloud decomposition because if some of the clouds they identify are in fact blends of smaller clouds, then the derived densities are lower limits, reinforcing their conclusions. In M64, *Rosolowsky and Blitz* (2005) examine many of the relationships shown in Figs. 7 and 8 and find that all are significantly different.

5.2. Giant Molecular Cloud Formation in Galactic Centers

The peak H$_2$ surface density in the central 1 kpc of M64 is about 20× the HI surface density (*Braun et al.,* 1994; *Rosolowsky and Blitz,* 2005), which is typical of many galaxies (*Helfer et al.,* 2003). In such regions, the formation of GMCs cannot take place by first collecting atomic hydrogen into filaments and then turning the gas molecular. If the gas is cycled between the atomic and molecular phases, as is required by the presence of HII regions in the central regions of M64, then continuity requires that the amount of time that the gas remains in each phase is roughly equal to the ratio of surface densities at each particular radius. Thus, gas ionized by the O stars must quickly return to the molecular phase, which is catalyzed by the very large pressures in the central region (section 4). More than likely, the GMCs are formed and destroyed without substantially leaving the molecular phase, unlike what happens in the disks. Indeed, *Rosolowsky and Blitz* (2005) present evidence for a diffuse molecular component that is not bound into GMCs. Thus it seems likely that, as in galactic disks, the formation of structure (filaments?) in galactic nuclei occurs before the formation of the GMCs. The gas, though, is largely molecular prior to the formation of the clouds.

Measuring the properties of individual GMCs in more distant molecule-rich galaxies will rely upon future improvements in angular resolution and sensitivity. At present, some information can be gleaned from single-dish spectra of the regions in multiple tracers of molecular gas. The observations of *Gao and Solomon* (2004) and *Narayanan et al.* (2005) show that the star formation rate is linearly proportional to the mass of molecular gas found at high densities ($\geq 10^5$ cm^{-3}), and that the fraction of dense gas increases with the amount of molecular mass in the system. Since the fraction of molecular mass found at high densities is relatively small in galactic GMCs, this implies there are substantial differences in GMC properties in these starburst systems, compared to those in the Local Group.

6. STAR FORMATION IN EXTRAGALACTIC GIANT MOLECULAR CLOUDS

The evolution of GMCs substantially influences the evolution of galaxies. In particular, star formation in GMCs is a central event that affects galactic structure, energetics, and chemistry. A detailed understanding of star formation is therefore an important step for a better understanding of galaxy evolution.

6.1. Identification of Star Formation

In galactic molecular clouds, we are able to study the formation of stars from high mass to low mass including even brown dwarfs. In all external galaxies, even those in the Local Group, such studies are limited to only the highest-mass stars as a result of limited sensitivity. It is nonetheless worthwhile to learn how high-mass stars form in extragalactic GMCs because high-mass stars impart the highest energies to the ISM via UV photons, stellar winds, and supernova explosions.

Young, high-mass stars are apparent at optical/radio wavelengths as the brightest members of stellar clusters or associations or by the Hα and radio continuum emission from HII regions. The positional coincidence between these signposts of star formation and GMCs is the most common method of identifying the star formation associated with individual clouds. Such associations can be made with reasonable confidence when the source density is small enough that confusion is not important. When confusion becomes significant, however, conclusions can only be drawn by either making more careful comparisons at higher angular resolution, or by adopting a statistical approach.

6.2. The Large Magellanic Cloud

The most complete datasets for young stars are available for the LMC, which has a distance of 50 kpc. They include catalogs of clusters and associations (e.g., *Bica et al.,* 1996) and of optical and radio HII regions (*Henize,* 1956; *Davies et al.,* 1976; *Kennicutt and Hodge,* 1986; *Filipovic et al.,* 1998). The colors of the stellar clusters are studied in detail at four optical wavelengths and are classified into an age sequence from SWB 0 to SWB VII, where SWB 0 is the youngest with an age of less than 10 m.y., SWB I in a range 10–30 m.y., and so on (*Bica et al.,* 1996). The sensitivity limit of the published catalogs of star clusters is 14.5 mag (V); it is not straightforward to convert this into the number of stars since a stellar mass function must be assumed. The datasets of HII regions have a detection limit in Hα flux

of 10^{-12} ergs cm^{-2} s^{-1}, and the radio sensitivity limit at 5-GHz thermal emission corresponds to 20 mJy. The faintest detectable HII regions correspond to the ionization by an O5 star if a single ionizing source is assumed. We note that the detection limit of HII regions is quite good, L(Hα) = 2 × 10^{36} ergs s^{-1}, corresponding to one-fourth the luminosity of the Orion Nebula.

Using the first NANTEN CO survey (*Fukui et al.,* 1999; *Mizuno et al.,* 2001b), the GMCs in the LMC were classified into the three categories according to their associated young objects (*Fukui et al.,* 1999; *Yamaguchi et al.,* 2001c): (1) starless GMCs (no early O stars) — "starless" here indicates no associated early O star capable of ionizing an HII region, which does not exclude the possibility of associated young stars later than B-type; (2) GMCs with HII regions only — those with small HII regions whose Hα luminosity is less than 10^{37} erg s^{-1}; (3) GMCs with HII regions and stellar clusters — those with stellar clusters and large HII regions of Hα luminosity greater than 10^{37} erg s^{-1}.

The new NANTEN GMC catalog (Fukui et al., in preparation) is used to improve and confirm the statistics of these three classes (Kawamura et al., in preparation). For the updated sample of 181 GMCs in Fig. 1, Fig. 11 shows the frequency distribution of the apparent separation of young objects, i.e., optical HII regions and stellar clusters, measured from the lowest contour of the nearest GMC. Obviously, the youngest stellar clusters, SWB 0 and HII regions, exhibit marked peaks within 50–100 pc, indicating their strong concentrations toward GMCs. Comparisons of these distributions with a purely random distribution is shown by lines. The differences between these peaks are significant. The correlation with young clusters establishes the physical association of the young objects with the GMCs. On the other hand, clusters older than SWB I show almost no correlation with GMCs.

In order to look for any optically obscured HII regions we have also used the Parkes/ATNF radio continuum survey carried out at five frequencies: 1.4, 2.45, 4.75, 4.8, and 8.55 GHz (*Filipovic et al.,* 1995, 1998). The typical sensitivity limits of these new datasets are quite good, allowing us to reach flux limits equivalent to those in Hα. The radio continuum results are summarized in a catalog of 483 sources, and the spectral information makes it possible to select HII regions and to eliminate background sources not related the LMC. By comparing these data with the GMCs, we found that all the starless GMCs have no embedded HII regions that are detectable at radio wavelengths (Kawamura et al., in preparation).

Table 4 summarizes the results of the present comparison between GMCs and young objects, SWB 0 clusters, and the HII regions including radio sources. It shows that ~25% of the GMCs are starless in the sense that they are not associated with HII regions or young clusters. Figure 12 shows mass histograms of the three classes, I, II and III. These indicate that the mass range of the three is from $10^{4.5}$ M$_\odot$ to a few times 10^6 M$_\odot$. It is also noteworthy that class I GMCs tend to be less massive than the other two in the sense that

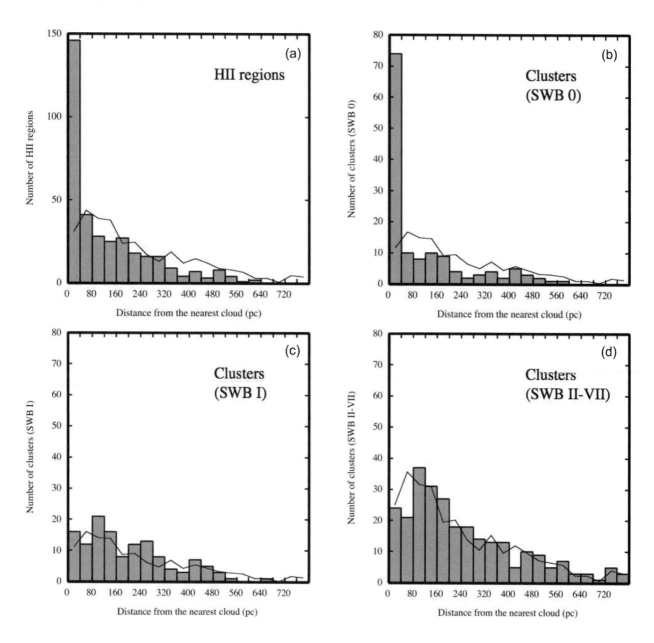

Fig. 11. Histograms of the projected separation from **(a)** the HII regions (*Davies et al., 1976*) and clusters cataloged by *Bica et al.* (1996) to the nearest CO emission; **(b)** clusters with τ < 10 m.y. (SWB 0), **(c)** clusters with 10 m.y. < τ < 30 m.y. (SWB I), and **(d)** clusters with 30 m.y. < τ (SWB II–VII), respectively. The lines represent the frequency distribution expected if the same number of the clusters are distributed at random in the observed area.

TABLE 4.   Association of the young objects with GMCs.

Class of GMC	Number of GMCs*	Timescale (m.y.)[†]	Association
Class I	44 (25.7%)	7	Starless
Class II	88 (51.5%)	14	HII regions
Class III	39 (22.8%)	6	HII regions and clusters[‡]

*GMCs with $M > 10^{4.5}\,M_\odot$; mass is derived by using $X_{CO} = 5.4 \times 10^{20}$ cm^{-2} (K km s^{-1})$^{-1}$ (Table 2).
[†]A steady-state evolution is assumed. The absolute timescale is based on the age of stellar clusters; the age of SWB 0 clusters, half of which are associated with the GMCs, is taken to be 10 m.y.
[‡]Young clusters or associations, SWB 0, by *Bica et al.* (1996).

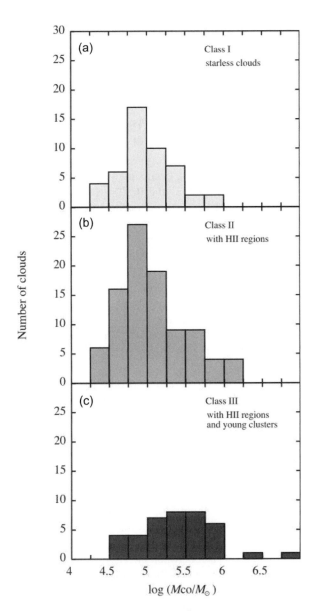

Fig. 12. Histograms of the mass of **(a)** class I, **(b)** class II, and **(c)** class III, respectively. Mass is derived by using $X_{CO} = 5.4 \times 10^{20}$ cm^{-2} (K km s^{-1})$^{-1}$ (Table 2).

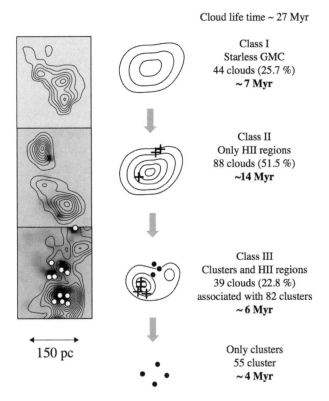

Fig. 13. Evolutionary sequence of the GMCs in the LMC. An example of the GMCs and illustration at each class are shown in the left panels and the middle column, respectively. The images and contours in the left panels are Hα (*Kim et al., 1999*) and CO integrated intensity by NANTEN (*Fukui et al., 2001*; Fukui et al., in preparation); contour levels are from K km s^{-1} with 1.2 K km s^{-1} intervals. Crosses and filled circles indicate the position of the H II regions and young clusters, SWB 0 (*Bica et al., 1996*), respectively. The number of the objects and the timescale at each class are also presented on the right.

the number of GMCs more massive than 10^5 M$_\odot$ is about half of those of class II and class III GMCs, respectively.

6.3. The Evolution of Giant Molecular Clouds in the Large Magellanic Cloud

The completeness of the present GMC sample covering the whole LMC enables us to infer the evolutionary timescales of the GMCs. We assume a steady-state evolution and therefore time spent in each phase is proportional to the number of objects in Table 4. Figure 13 is a scheme representing the evolution suggested from Table 4. The absolute timescale is based on the age of stellar clusters: The age of SWB 0 clusters is taken to be 10 m.y. The first stage corresponds to starless GMCs, having a long timescale of 7 m.y. This is followed by a phase with small H II regions,

implying the formation of a few to several O stars. The subsequent phase indicates the most active formation of rich clusters including many early O stars (one such example is N 159N). In the final phase, the GMC has been more or less dissipated under the strong ionization and stellar winds from O stars. The lifetime of a typical GMC in the LMC is then estimated as the total of the timescales in Table 4: ~27 m.y., assuming that the GMC is completely disrupted by the star formation. As noted earlier (section 6), the mass of class I GMCs tends to be smaller than the rest. We may speculate that class I GMCs, and possibly some of the class II GMCs, are still growing in mass via mass accretion from their surrounding lower-density atomic gas. In addition, the lifetime of GMCs likely varies with cloud mass, so 27 m.y. is only a characteristic value and is probably uncertain by about 50%.

6.4. Star Formation in M33

None of the other galaxies in our sample has as complete a record of interstellar gas and star formation as does the LMC, which is due, in part, to its proximity. Nevertheless, it is possible to draw some conclusions about the star formation in M33. *Engargiola et al.* (2003) correlated the HII regions cataloged by *Hodge et al.* (1999) with the 149 GMCs in the M33 catalog. For reference, the completeness limit of the *Hodge et al.* (1999) catalog is $L(H\alpha) = 3 \times 10^{35}$ erg s^{-1}; a similar range of HII region luminosities is cataloged in the LMC and M33. *Engargiola et al.* (2003) assumed that an HII region is associated with a GMC if its boundary lies either within or tangent to a GMC; 36% of the flux from HII regions can be associated with the cataloged GMCs. Correcting for the incompleteness of the GMCs cataloged below their sensitivity limit suggests that >90% of the total flux of ionized gas from M33 originates from GMCs. Within the uncertainties, essentially all the flux from HII regions is consistent with an origin in GMCs. Apparently, about half the star formation in M33 originates in GMCs below the the sensitivity limit of our survey.

A related question is, what fraction of GMCs in M33 is actively forming stars? *Engargiola et al.* (2003) counted the fraction of GMCs with at least one HII region having a separation Δr. They defined the correlation length, such that half the GMCs have at least one HII region within this distance. The correlation length for the GMCs and HII regions is 35 pc; a random distribution of GMCs and HII regions would return a correlation length of 80 pc. They assumed that a GMC is actively forming stars if there is an HII region within 50 pc of the centroid of a GMC. With this assumption, as many as 100 GMCs (67%) are forming massive stars. Of the 75 GMCs with masses above the median cataloged mass, the fraction of clouds actively forming stars rises to 85%. They estimate that the number of totally obscured HII regions affect these results by at most 5%.

Thus the fraction of GMCs without star formation is estimated to be about one-third, a fraction similar to that in the LMC. The M33 study estimated the lifetime of GMCs to be ~20 m.y., also similar to that found for the LMC. The fraction of clouds without active star formation is much higher than that found in the vicinity of the Sun, where only one of all the GMCs within 2 kpc is found to be devoid of star formation. It is unclear whether this difference is significant. Neither the LMC analysis nor the M33 analysis would detect the low-mass star formation that it is proceeding in the Taurus molecular clouds. In any event, both the LMC and M33 studies suggest that the fraction of clouds without star formation is small. Thus the onset of star formation in GMCs is rather rapid not only in the Milky Way, but in at least some lower-mass spiral and irregular galaxies.

7. FUTURE PROSPECTS

Studying GMCs in galaxies using CO emission requires spatial resolutions higher than 30–40 pc. It will be possible to extend studies such as ours to a few tens of Mpc soon, with the advent of ALMA and CARMA arrays in the southern and northern hemispheres respectively. These instruments will provide angular resolutions of 0.1–1 arcsec in millimeter and submillimeter CO emission, corresponding 5–50 pc at 10 Mpc, and will provide unprecedented details of physical conditions in GMCs in galaxies. The work described in this chapter should be just the beginning of extragalactic GMC studies.

8. SUMMARY AND CONCLUSIONS

We have compared the properties of GMCs in five galaxies, four of which have been surveyed in their entirety: the LMC, the SMC, M33, and IC 10. M31 was observed over a very limited area. The interstellar medium of all five galaxies is dominated by the atomic phase.

1. The GMCs do not, in general, show any relationship to the stellar content of the galaxies except for the O stars born in the GMCs.

2. There is a very good correlation between the locations of the GMCs and filaments of HI. Many filaments contain little of no molecular gas even though they have similar surface densities compared to those that are rich in GMCs. This suggests that clouds form from the HI rather than vice-versa.

3. There appears to be a clear evolutionary trend going from filament formation → molecule formation → GMC formation. It is not clear, however, whether the condensations that form GMCs are first formed in the atomic filaments, or only after the molecules have formed.

4. We derive X_{CO} for all the galaxies assuming that the GMCs are virialized. Although there is some variation, a value of $X_{CO} = 4 \times 10^{20}$ cm^{-2} (K kms^{-1})$^{-1}$ is a representative value to within about 50% except for the SMC, which has a value more than three times higher. There is no clear trend of X_{CO} with metallicity.

5. The discrepancy between the virial value and the value determined from γ-ray observations in the MilkyWay suggests that the GMCs are not virialized, if the γ-ray value is applicable to other galaxies in the Local Group. In that case, a value of $X_{CO} = 2 \times 10^{20}$ cm^{-2} (K km s^{-1})$^{-1}$ may be more appropriate.

6. The GMCs in our sample appear to satisfy the line width-size relation for the Milky Way, but with an offset in the constant of proportionality. This offset may be due, at least in part, to the different data analysis techniques for the MW and extragalactic data sets. For a given line width, the extragalactic clouds appear to be about 50% larger. Despite the systematic offset, there are small but significant differences in the line width-size relationship among GMCs in different galaxies.

7. The GMCs within a particular galaxy have a roughly constant surface density. If the value of X_{CO} we derive for each galaxy is applied, the surface densities of the sample as a whole have a scatter of less than a factor of 2.

8. The mass spectra for the GMCs in all the galaxies can be characterized as a power law with a slope of ~–1.7, with the exception of M33, which has a slope of –2.5.

9. The ratio of H_2 to HI on a pixel-by-pixel basis in gal-

axies appears to be determined by the hydrostatic pressure in the disk.

10. About one-fourth to one-third of the GMCs in the LMC and M33 appear to be devoid of high-mass star formation.

11. The association of stars and HII regions in the LMC suggests a lifetime for the GMCs of about 27 m.y., with a quiescent phase that is about 25% of the age of the GMCs. In M33, a lifetime of ~20 m.y. is measured. For GMCs in these galaxies we estimate that typical lifetimes are roughly 20–30 m.y. Both lifetimes are uncertain by about 50%.

Acknowledgments.   This work is partially supported by the U.S. National Science Foundation under grants AST-0228963 and AST-0540567 and a Grant-in-Aid for Scientific Research from the Ministry of Education, Culture, Sports, Science and Technology of Japan (No. 15071203) and from JSPS (No. 14102003). The NANTEN project is based on a mutual agreement between Nagoya University and the Carnegie Institution of Washington (CIW). We greatly appreciate the hospitality of all the staff members of the Las Campanas Observatory of CIW. We are thankful to many Japanese public donors and companies who contributed to the realization of the project. We would like to acknowledge Drs. L. Stavely-Smith and M. Filipovic for the kind use of their radio continuum data prior to publication.

REFERENCES

Allen R. J. (2001) In *Gas and Galaxy Evolution* (J. Hibbard and J. van Gorkom, eds.), pp. 331–337. ASP Conf. Series 240, San Francisco.

Bica E., Claria J. J., Dottori H., Santos J. F. C. Jr., and Piatti A. E. (1996) *Astrophys. J. Suppl., 102,* 57–73.

Blitz L. and Rosolowsky E. (2004) *Astrophys. J., 612,* L29–L32.

Blitz L. and Rosolowsky E. (2006) *Astrophys. J.,* in press.

Braun R., Walterbos R. A. M., Kennicutt R. C., and Tacconi L. J. (1994) *Astrophys. J., 420,* 558–569.

Cohen R. S., Dame T. M., Garay G., Montani J., Rubio M., and Thaddeus P. (1988) *Astrophys. J., 331,* L95–L99.

Dame T. M., Hartmann D., and Thaddeus P. (2001) *Astrophys. J., 547,* 792–813.

Davies R. D., Elliott K. H., and Meaburn J. (1976) *Mem. R. Astron. Soc., 81,* 89–128.

de Boer K. S., Braun J. M., Vallenari A., and Mebold U. (1998) *Astron. Astrophys., 329,* L49–L52.

Deul E. R. and van der Hulst J. M. (1987) *Astron. Astrophys. Suppl., 67,* 509–539.

Elmegreen B. G. (1993) *Astrophys. J., 411,* 170–177.

Engargiola G., Plambeck R. L, Rosolowsky E., and Blitz L. (2003) *Astrophys. J. Suppl., 149,* 343–363.

Filipovic M. D., Haynes R. F., White G. L., Jones P. A., Klein U., and Wielebinski R. (1995) *Astron. Astrophys. Suppl., 111,* 311–332.

Filipovic M. D., Haynes R. F., White G. L., and Jones P. A. (1998) *Astron. Astrophys. Suppl., 130,* 421–440.

Fujimoto M. and Noguchi M. (1990) *Publ. Astron. Soc. Japan, 42,* 505–516.

Fukui Y., Mizuno N., Yamaguchi R., Mizuno A., Onishi T., et al. (1999) *Publ. Astron. Soc. Japan, 51,* 745–749.

Fukui Y., Mizuno N., Yamaguchi R., Mizuno A., and Onishi T. (2001) *Publ. Astron. Soc. Japan Lett., 53,* L41–L44.

Gao Y. and Solomon P. M. (2004) *Astrophys. J., 606,* 271–290.

Gerola H. and Seiden P. E. (1978) *Astrophys. J., 223,* 129–135.

Helfer T. T. and Blitz L. (1993) *Astrophys. J., 419,* 86–93.

Helfer T. T., Thornley M. D., Regan M. W., Wong T., Sheth K., Vogel S. N., Blitz L., and Bock D. C.-J. (2003) *Astrophys. J. Suppl., 145,* 259–327.

Henize K. G. (1956) *Astrophys. J. Suppl., 2,* 315–344.

Henry R. B. C. and Howard J. W. (1995) *Astrophys. J., 438,* 170–180.

Heyer M. H., Carpenter J. M., and Snell R. L. (2001) *Astrophys. J., 551,* 852–866.

Hodge P. W., Balsley J., Wyder T. K., and Skelton B. P. (1999) *Publ. Astron. Soc. Pac., 111,* 685–690.

Jarrett T. H., Chester T., Cutri R., Schneider S. E., and Huchra J. P. (2003) *Astron. J., 125,* 525–554.

Kennicutt R. C. (1998) *Astrophys. J., 498,* 541–552.

Kennicutt R. C. Jr. and Hodge P. W. (1986) *Astrophys. J., 306,* 130–141.

Keto E., Ho L., and Lo K. Y. (2005) *Astrophys. J., 635,* 1062–1076.

Kim S., Staveley-Smith L., Dopita M. A., Freeman K. C., Sault R. J., Kesteven M. J., and McConnell D. (1998) *Astrophys. J., 503,* 674–688.

Kim S., Staveley-Smith L., Dopita M. A., Sault R. J., Freeman K. C., Lee Y., and Chu Y.-H. (2003) *Astrophys. J. Suppl., 148,* 473–486.

Kregel M., van der Kruit P. C., and de Grijs R. (2002) *Mon. Not. R. Astron. Soc., 334,* 646–668.

Koornneef J. (1984) In *Structure and Evolution of the Magellanic Clouds* (S. van den Bergh and K. S. de Boer, eds.), pp. 333–339. IAU Symposium 108, Reidel, Dordrecht.

Lada C. J., Margulis M., Sofue Y., Nakai N., and Handa T. (1988) *Astrophys. J., 328,* 143–160.

Leroy A., Bolatto A., Walter F., and Blitz L. (2006) *Astrophys. J., 643,* 825–843.

Massey P., Hodge P. W., Holmes S., Jacoby G., King N. L., Olsen K., Saha A., and Smith C. (2001) *Bull. Am. Astron. Soc., 33,* 1496.

Mizuno N., Rubio M., Mizuno A., Yamaguchi R., Onishi T., and Fukui Y. (2001a) *Publ. Astron. Soc. Japan, 53,* L45–L49.

Mizuno N., Yamaguchi R., Mizuno A., Rubio M., Abe R., Saito H., Onishi T., Yonekura Y., Yamaguchi N., Ogawa H., and Fukui Y. (2001b) *Publ. Astron. Soc. Japan, 53,* 971–984.

Narayanan D., Groppi C. E., Kulesa C. A., and Walker C. K. (2005) *Astrophys. J., 630,* 269–279.

Oka T., Hasegawa T., Sato F., Tsuboi M., Miyazaki A., and Sugimoto M. (2001) *Astrophys. J., 562,* 348–362.

Regan M. and Vogel S. (1994) *Astrophys. J., 434,* 536–545.

Rosolowsky E. (2006) *Astrophys. J.,* in press.

Rosolowsky E. and Blitz L. (2005) *Astrophys. J., 623,* 826–845.

Rosolowsky E. and Leroy A. (RL06) (2006) *Publ. Astron. Soc. Pac., 118,* 590–610.

Rosolowsky E., Plambeck R., Engargiola G., and Blitz L. (2003) *Astrophys. J., 599,* 258–274.

Rubio M., Garay G., Montani J., and Thaddeus P. (1991) *Astrophys. J., 368,* 173–177.

Solomon P. M., Rivolo A. R., Barrett J. and Yahil A. (1987) *Astrophys. J., 319,* 730–741.

Spergel D. N. and Blitz L. (1992) *Nature, 357,* 665–667.

Stanimirović S., Staveley-Smith L., Dickey J. M., Sault R. J., and Snowden S. L. (1999) *Mon. Not. R. Astron. Soc., 302,* 417–436.

Stanimirović S., Staveley-Smith L., van der Hulst J. M., Bontekoe T. R., Kester D. J. M., and Jones P. A. (2000) *Mon. Not. R. Astron. Soc., 315,* 791–807.

Strong A. W. and Mattox J. R. (1996) *Astron. Astrophys., 308,* L21–L24.

van der Kruit P. C. and Searle L. (1981a) *Astron. Astrophys., 95,* 105–115.

van der Kruit P. C. and Searle L. (1981b) *Astron. Astrophys., 95,* 116–126.

Vogel S. N., Boulanger F., and Ball R. (1987) *Astrophys. J., 321,* L145–L149.

Wilcots E. M. and Miller B. W. (1998) *Astron. J., 116,* 2363–2394.

Williams J. P. and McKee C. F. (1997) *Astrophys. J., 476,* 166–183.

Wilson C. D. (1994) *Astrophys. J., 434,* L11–L14.

Wilson C. D. and Reid I. N. (1991) *Astrophys. J., 366,* L11–L14.

Wilson C. D. and Rudolph A. L. (1993) *Astrophys. J., 406,* 477–481.

Wilson C. D. and Scoville N. (1990) *Astrophys. J., 363,* 435–450.

Wong T. and Blitz L. (2002) *Astrophys. J., 569,* 157–183.

Yamaguchi R., Mizuno N., Onishi T., Mizuno A., and Fukui Y. (2001a) *Astrophys. J., 553,* L185–L188.

Yamaguchi R., Mizuno N., Onishi T., Mizuno A., and Fukui Y. (2001b) *Publ. Astron. Soc. Japan, 53,* 959–969.

Yamaguchi R., Mizuno N., Mizuno A., Rubio M., Abe R., Saito H., Moriguchi Y., Matsunaga L., Onishi T., Yonekura Y., and Fukui Y. (2001c) *Publ. Astron. Soc. Japan, 53,* 985–1001.

Part II:

Star Formation

Current Advances in the Methodology and Computational Simulation of the Formation of Low-Mass Stars

Richard I. Klein
University of California, Berkeley,
and Lawrence Livermore National Laboratory

Shu-ichiro Inutsuka
Kyoto University

Paolo Padoan
University of California, San Diego

Kohji Tomisaka
National Astronomical Observatory, Japan

Developing a theory of low-mass star formation (~0.1 to 3 $M_\odot$) remains one of the most elusive and important goals of theoretical astrophysics. The star-formation process is the outcome of the complex dynamics of interstellar gas involving nonlinear interactions of turbulence, gravity, magnetic field, and radiation. The evolution of protostellar condensations, from the moment they are assembled by turbulent flows to the time they reach stellar densities, spans an enormous range of scales, resulting in a major computational challenge for simulations. Since the previous Protostars and Planets conference, dramatic advances in the development of new numerical algorithmic techniques have been successfully implemented on large-scale parallel supercomputers. Among such techniques, adaptive mesh refinement and smoothed particle hydrodynamics have provided frameworks to simulate the process of low-mass star formation with a very large dynamic range. It is now feasible to explore the turbulent fragmentation of molecular clouds and the gravitational collapse of cores into stars self-consistently within the same calculation. The increased sophistication of these powerful methods comes with substantial caveats associated with the use of the techniques and the interpretation of the numerical results. In this review, we examine what has been accomplished in the field and present a critique of both numerical methods and scientific results. We stress that computational simulations should obey the available observational constraints and demonstrate numerical convergence. Failing this, results of numerical simulations do not advance our understanding of low-mass star formation.

1. INTRODUCTION

Most of the stars in the galaxy exist in gravitationally bound binary and low-order multiple systems. Although several mechanisms have been put forth to account for binary star formation, fragmentation has emerged as the leading mechanism in the past decade (*Bodenheimer et al.,* 2000). This point of view has been strengthened by observations that have shown the binary frequency among pre-main-sequence stars is comparable to or greater than that among nearby main-sequence stars (*Duchene et al.,* 1999). This suggests that most binary stars are formed during the protostellar collapse phase. Developing a theory for low-mass star formation (0.1 to 3 $M_\odot$) that explains the physical properties of the formation of binary and multiple stellar systems remains one of the most elusive and important goals of theoretical astrophysics.

Until very recently, the extreme variations in length scale inherent in the star-formation process have made it difficult to perform accurate calculations of fragmentation and collapse, which are intrinsically three-dimensional in nature. Since the last review in *Protostars and Planets IV* by *Bodenheimer et al.* (2000), dramatic advances in the development of new numerical algorithmic techniques, including adaptive mesh refinement (AMR) and smoothed particle hydrodynamics (SPH), as well as advances in large-scale parallel machines, have allowed a significant increase in the dynamic range of simulations of low-mass star formation. It is now feasible to explore the turbulent fragmentation of molecular clouds and the gravitational collapse of cores into stars self-consistently within the same calculation. In this chapter we examine what has been recently accomplished in the field of numerical simulation of low-mass star formation, and we critically review both numerical methods and scientific results.

1.1. Key Questions Posed by the Observations

Observational surveys present us with a basic picture of star-forming regions, including the structure and dynamics of star-forming clouds and the properties of protostellar cores. A theory of star formation should explain both the large-scale environment and the properties of protostellar cores self-consistently.

As shown first by *Larson* (1981), and later confirmed by many other studies, star-forming regions are characterized by a correlation between internal velocity dispersion and size, $\delta V \approx 1$ km/s$(L = 1$ pc$)^{0.4}$. This scaling law has been interpreted as evidence of supersonic turbulence on a wide range of scales. The turbulence can provide support against the gravitational collapse, but can also create gravitationally unstable compressed regions through shocks. A theory of star formation should elucidate whether turbulence controls the star-formation rate and efficiency, or whether those properties are controlled by stellar outflows and winds. Are the scaling laws of turbulent flows related to scaling laws of core and stellar properties? Is the turbulence setting the initial density perturbations that collapse into stars? What is the effect of turbulence on the accretion of mass onto protostars?

On smaller scales, observational surveys have shown that prestellar cores are elongated. Their density profiles are flat near the center, steeper at larger radii, may show very sharp edges, and are sometimes consistent with Bonnor-Ebert profiles (e.g., *Bacmann et al.*, 2000; *Alves et al.*, 2001). Cores have rotational energies on the average only a few percent of their binding energies (e.g., *Goodman et al.*, 1993). They are marginally supercritical (*Crutcher*, 1999) and their mass distribution is very similar to the stellar mass distribution (e.g., *Motte et al.*, 1998, 2001). The large majority of cores are found to contain stars (e.g., *Jijina et al.*, 1999), and individual cores produce at most two to three protostars (e.g., *Goodwin and Kroupa*, 2005). A theory of low-mass star formation should be consistent with these observations and address the following questions: Why are prestellar cores so shortlived? Why do they have Bonnor-Ebert profiles? How does the observed core angular momentum affect the formation of binary and multiple systems? What is the role played by magnetic fields in their evolution? Why is the mass distribution of prestellar cores so similar to the stellar initial mass function? Why are cores barely fragmenting into binaries or low-multiplicity systems?

Observations of young stellar populations provide important constraints as well. We know that young stars are always found in association with dense gas with an efficiency ~10–20%, much higher than the overall star-formation efficiency in GMCs, ~1–3% (e.g., *Myers et al.*, 1986). Stars are often found in clusters ranging from 10 to 1000 members (e.g., *Lada and Lada*, 2003). The stellar initial mass function peaks around a fraction of a solar mass and its lognormal shape around the peak is roughly the same in open clusters, globular clusters, and field stars (*Chabrier*, 2003). What determines the efficiency of star formation? Why is the stellar multiplicity higher in younger populations? What determines the typical stellar mass and the initial mass function?

Although answering all these questions is outside the scope of this review, we pose them because these are the questions to be addressed by the computational simulations of the formation of low-mass stars.

1.2. Generation of Initial Conditions Consistent with the Observations

Simulations of low-mass star formation should generate initial conditions for the collapse of protostars consistent with the observed physical properties of star-forming clouds. This can be achieved if a relatively large scale is simulated (>1 pc) with numerical methods that can accurately reproduce fundamental statistics measured in molecular clouds. Such statistics include (1) scaling laws of velocity, density, and magnetic fields; and (2) mean relative values of turbulent, thermal, magnetic, and gravitational energies (the normalization of the scaling laws).

1.2.1. Scaling laws. *Larson* (1981) found that velocity and size of interstellar clouds are correlated over many orders of magnitude in size. This correlation has been confirmed by many more recent studies (e.g., *Fuller and Myers*, 1992; *Falgarone et al.*, 1992). The most accepted interpretation is that the scaling law reflects the presence of supersonic turbulence in the ISM (e.g., *Larson*, 1981; *Ossenkopf and Mac Low*, 2002; *Heyer and Brunt*, 2004). Large-scale velocity-column density correlations from molecular line surveys of giant molecular clouds also suggest a turbulent origin of the observed density enhancements (*Padoan et al.*, 2001b). Starting with the work of *Troland and Heiles* (1986), a correlation between magnetic field strength and gas density, $B \propto n^{1/2}$, has been reported for mean densities larger than $n \sim 100$ cm^{-3}. However, density and magnetic field scalings are very uncertain because both quantities are difficult to measure.

1.2.2. Mean energies. Assuming an average temperature of $T = 10$ K, Larson's velocity-size correlation corresponds to an rms sonic Mach number $M_s \approx 5$ on the scale of 1 pc, and $M_s \approx 1$ at 0.02 pc. So, on the average, the turbulent kinetic energy is larger than the thermal energy. Indirect evidence of super-Alfvénic dynamics in giant molecular clouds has been presented by *Padoan and Nordlund* (1999) and *Padoan et al.* (2004a). They have shown that the magnetic energy averaged over a large scale has an intermediate value between the thermal and the kinetic energies, even if it can be significantly larger than this average value within dense prestellar cores. Observations suggest that in dense cores gravitational, kinetic, thermal, and magnetic energies are all comparable. However, the magnetic energy is very difficult to estimate. Accounting for both detections and upper limits, there is a large dispersion in the ratio of magnetic to gravitational energy of dense cores (*Crutcher et al.*, 1993, 1999; *Bourke et al.*, 2001; *Nutter et al.*, 2004). In the case of turbulence that is super-Alfvénic on the large scale, this dispersion and the B-n relation are predicted to be real (*Padoan and Nordlund*, 1999).

In summary, large-scale simulations may provide realistic initial and boundary conditions for protostellar collapse, but they must be consistent with the turbulent nature of the ISM. On the scale of giant molecular clouds, observations suggest the turbulence is on the average supersonic and super-Alfvénic and its kinetic energy is roughly equal to the cloud gravitational energy.

2. A BRIEF SURVEY OF LOW-MASS STAR-FORMATION MECHANISMS

Although much progress in numerical simulations of collapse and fragmentation has been made in the intervening six years since Protostars and Planets IV, a self-consistent theory of binary and multiple star formation that addresses the key questions posed by observations is still not at hand. As discussed by *Bodenheimer et al.* (2000), binary and multiple formation can occur through the processes of (1) capture, (2) fission, (3) prompt initial fragmentation, (4) disk fragmentation, and (5) fragmentation during the protostellar collapse phase.

A recent mechanism for multiple star formation has been put forth by *Shu et al.* (2000) and *Galli et al.* (2001). They develop equilibrium models of strongly magnetized isopedic disks and explored their bifurcation to nonaxisymmetric, multilobed structures of increasing rotation rates. Possible problems with this mechanism include the observed low rotational energies, the observed random alignment of disks with the ambient magnetic fields, the complexity of star-forming regions relative to the two-dimensional geometry, and the absence of turbulence in the model.

Disk fragmentation from gravitational instability can result in multiple systems in an equilibrium disk if the minimum Toomre Q parameter falls below ≈1. However, *Bodenheimer et al.* (2000) have pointed out that the required initial conditions to obtain Q < 1 may not be easily realized since the mass-accretion timescale is significantly longer than the dynamical timescale throughout most of the evolution of the protostar. Disk fragmentation plays a key role in one of the theories of the formation of brown dwarfs (BDs). This scenario, known as the "failed embryo" scenario, begins with a gravitationally unstable disk surrounding a protostar. The disk fragments into a number of substellar objects. If the crossing time of the cluster of embryos is much less than the freefall time of the collapsing core, one or more of the members will be rapidly ejected, resulting in a BD (*Reipurth and Clarke*, 2001). Problems with this model include observational evidence of BD clustering (*Duchêne et al.*, 2004), Ly-α signatures of BD accretion (e.g., *Jayawardhana*

et al., 2002; *Natta et al.*, 2004; *Barrado y Navascués et al.*, 2004; *Mohanty et al.*, 2005), and evidence that individual cores produce only two or three stars (*Goodwin and Kroupa*, 2005).

Currently, there are two dominant models to explain what determines the mass of stars. The direct gravitational collapse theory suggests that star-forming turbulent clouds fragment into cores that eventually collapse to make individual stars or small multiple systems (*Shu et al.*, 1987; *Padoan and Nordlund*, 2002, 2004). In contrast, the competitive accretion theory suggests that at birth all stars are much smaller than the typical stellar mass and that the final stellar mass is determined by the subsequent accretion of unbound gas from the clump (*Bonnell et al.*, 1998, 2001a; *Bate and Bonnell*, 2005). Significant problems with competitive accretion include the large value of the observed virial parameter relative to the one required by competitive accretion (*Krumholz et al.*, 2005b). We discuss this problem with competitive accretion in detail in section 5.2.

3. PHYSICAL PROCESSES NECESSARY FOR DETAILED SIMULATION OF LOW-MASS STAR FORMATION

3.1. Turbulence

The Reynolds number estimates the relative importance of the nonlinear advection term and the viscosity term in the Navier-Stokes equation, $Re = V_0 L_0/\nu$. V_0 is the flow rms velocity, L_0 is the integral scale of the turbulence (say the energy injection scale), and ν is the kinematic viscosity that we can approximate as $\nu \approx v_{th} = (\sigma n)$. v_{th} is the gas thermal velocity, n is the gas mean number density, and $\sigma \sim 10^{-15}$ cm^2 is the typical gas collisional cross section. For typical molecular cloud values, $Re \sim 10^7$–10^8, which implies flows are highly unstable to turbulence. It is important to recognize the significance of turbulent gas dynamics in astrophysical processes, as turbulence is a dominant transport mechanism. In molecular clouds, the turbulence is supersonic and the postshock gas cooling time is very short. This results in the highly fragmented structure of molecular clouds.

There has been significant progress in our understanding of supersonic turbulence in recent years [progress on the scaling properties of subsonic and sub-Alfvénic turbulence is reviewed in *Cho et al.* (2003)]. Phenomenological models of the intermittency of incompressible turbulence (e.g., *She and Leveque*, 1994; *Dubrulle*, 1994) have been extended to supersonic turbulence by *Boldyrev* (2002) and the predictions of the model have been confirmed by numerical simulations (*Boldyrev et al.*, 2002a; *Padoan et al.*, 2004b). The intermittency correction is small for the exponent of the velocity power spectrum (corresponding to the second-order velocity structure function) and large only at high order. However, *Boldyrev et al.* (2002b) have shown that low-order density correlators depend on high-order velocity statistics, so intermittency is likely to play a significant role in turbulent fragmentation, despite being only a small effect in the velocity power spectrum.

Because supersonic turbulence can naturally generate, at very small scale, strong density enhancements of mass comparable to a low-mass stars or even a BD, its correct description is of paramount importance for simulations of molecular cloud fragmentation into low-mass stars and BDs. At present, the largest simulations of supersonic turbulence may achieve a Reynolds number Re ~ 10^4. The scale of turbulence dissipation is therefore much larger in numerical simulations (on the order of the computational mesh size) than in nature (~10^{14} cm). However, the ratio of the Kolmogorov dissipation scale, η_K, and the Jeans length, λ_J, is very small and remarkably independent of temperature and density, $\eta_K/\lambda_J \approx 10^{-4}(T = 10$ K$)^{-1/8}(n = 10^3$ cm$^{-3})^{-1/4}$. One may hope to successfully simulate the process of turbulent fragmentation by numerically resolving the turbulence to scales smaller than λ_J, but not as small as η_K, unless the nature of turbulent flows varies dramatically between Re ~ 10^3 and Re ~ 10^7. Experimental results seem to indicate that the asymptotic behavior of turbulence is recovered in the approximate range Re = 10,000–20,000 (*Dimotakis*, 2000), which can be achieved with PPM simulations on a 2048^3 mesh.

In order to (1) generate a sizable inertial range (a power-law power spectrum of the turbulent velocity over an extended range of scales) and (2) resolve the turbulence just below the Jeans length, a minimum computational box size of at least 1000^3 zones is required for a grid code. This accounts for the fact that the velocity power spectrum starts to decay with increasing wave number faster than a power law already at approximately 30× the Nyquist frequency. It is still an optimistic estimate, because at this resolution the velocity power spectrum is also affected by the bottleneck effect (e.g., *Falkovich*, 1994; *Dobler et al.*, 2003; *Haugen and Brandenburg*, 2004). Assuming the standard SPH kernel of 50 particles, this corresponds to at least 50 × 1000^3 particles to describe the density field, and at least few × 1000^3 particles to describe the velocity field, if a Godunov SPH method is used (see below).

Grid-code simulations have started to achieve this dynamical range only recently, while particle codes appear unsuitable to the task. The calculation of *Bate et al.* (2003) has $3.5 × 100^3$ particles, more than 4 orders of magnitude below the above estimate and therefore inadequate to describe the process of turbulent fragmentation (the formation of small-scale density enhancements by the supersonic turbulence). Studies proposing to directly test the effect of turbulence on star formation, based on numerical simulations with resolution well below the above estimate, should be regarded with caution.

3.2. Gravity

3.2.1. The Jeans condition. *Jeans* (1902) analyzed the linearized equations of one-dimensional isothermal self-gravitational hydrodynamics (GHD) for a medium of infinite extent and found that perturbations on scales larger than the Jeans length

$$\lambda_J \equiv \left(\frac{\pi c_s^2}{G\rho} \right)^{1/2}$$

are unstable. Thermal pressure cannot resist the self-gravity of a perturbation larger than λ_J, resulting in runaway collapse. *Truelove et al.* (1997) showed that the errors generated by numerical GHD solvers can act as unstable perturbations to the flow. In a simulation with variable resolution, cell-scale errors introduced in regions of coarser resolution can be advected to regions of finer resolution, allowing these errors to grow. The unstable collapse of numerical perturbations can lead to artificial fragmentation. The strategy to avoid artificial fragmentation is to maintain a sufficient resolution of λ_J. Defining the Jeans number

$$J \equiv \frac{\Delta x}{\lambda_J}$$

Truelove et al. (1997) found that keeping $J \leq 0.25$ avoided artificial fragmentation in the isothermal evolution of a collapse spanning seven decades of density, the approximate range separating typical molecular cloud cores from nonisothermal protostellar fragments. This Jeans condition arises because perturbations on scales above λ_J are physically unstable, and discretization of the GHD PDEs introduces perturbations on all scales above Δx. It is essential to keep λ_J as resolved as possible in order to diminish the initial amplitude of perturbations that exceed this scale. Although it has been shown to hold only for isothermal evolution, it is reasonable to expect that it is necessary (although not necessarily sufficient) for nonisothermal collapse as well where the transition to nonisothermal evolution may produce structure on smaller scales than the local Jeans length.

3.2.2. Runaway collapse. The self-gravitational collapse in nearly spherical geometry tends to show a so-called "runaway collapse," where the denser central region collapses much faster than the less-dense surrounding region. The mass of the central fast-collapsing region is on the order of the Jeans mass, $M_J = \rho \lambda_J^3 \sim G^{-3/2} C_s^{3/2} \rho^{-1/2}$, which decreases monotonically in this runaway stage. The description of this process requires increasingly higher resolution, not only on the spatial scale but also on the mass scale. Therefore, an accurate description is not guaranteed even with Lagrangian particle methods such as SPH, if the number of particles is conserved. The end of the runaway stage corresponds to the deceleration of the gravitational collapse. If the effective ratio of specific heats, $\gamma(P \propto \rho^\gamma)$, becomes larger than $\gamma_{crit} = 4/3$, the increased pressure can decelerate the gravitational collapse. For example, the question of how and when the isothermal evolution terminates was explored in *Masunaga and Inutsuka* (1999).

3.2.3. Thermal budget. In the low-density regime the gas temperature is affected by various heating and cooling processes (e.g., *Wolfire et al.*, 1995; *Koyama and Inutsuka*, 2000; *Juvela et al.*, 2001). However, above a gas density of 10^4–10^5 cm^{-3}, depending on the timescale of interest, the gas is thermally well coupled with the dust grains that maintain a temperature on the order of 10 K. During the dynamical collapse, gas and dust are isothermal until a density of 10^{10}–10^{11} cm^{-3}, when the compressional heating rate becomes larger than the cooling rate (*Inutsuka and Miyama*, 1997; *Masunaga and Inutsuka*, 1999). The further evolution of a collapsing core and the formation of a protostar are radiation-hydrodynamical (RHD) processes that should be modeled by solving the radiation transfer and the hydrodynamics simultaneously and in three dimensions. Presently, the most sophisticated multidimensional models are based on the (flux-limited) diffusion approximation (*Bodenheimer et al.*, 1990; *Krumholz et al.*, 2005c).

Once the compressional heating dominates the radiative cooling, the central temperature increases gradually from the initial value of ~10 K. The initial slope of the temperature as a function of gas density corresponds to an effective ratio of specific heats $\gamma = 5/3$: $T(\rho) \propto \rho^{2/3}$ for 10 K < T < 100 K. This monatomic gas property is due to the fact that the rotational degree of freedom of molecular hydrogen is not excited in this low-temperature regime [e.g., E(J = 2–0)/k_B = 512 K]. When the temperature becomes larger than ~10^2 K, the slope corresponds to $\gamma = 7/5$, as for diatomic molecules. This value of γ is larger than the minimum required for thermal pressure support against gravitational collapse: $\gamma > \gamma_{crit} \equiv 4/3$. The collapse is therefore decelerated and a shock is formed at the surface of a quasiadiabatic core, called "the first core." Its radius is about 1 AU in spherically symmetric models, but can be significantly larger in more realistic multidimensional models. It consists mainly of H_2.

The increase of density and temperature inside the first core is slow but monotonic. When the temperature becomes >10^3 K, the dissociation of H_2 starts. The dissociation of H_2 acts as an efficient cooling of the gas, which makes $\gamma < 4/3$, triggering the second dynamical collapse. In this second collapse phase, the collapsing velocity becomes very large and engulfs the first core. As a result, the first core lasts for only ~10^3 yr. In the course of the second collapse, the central density attains the stellar value, $\rho \sim 1$ g/cm^3, and the second adiabatic core, or "protostar," is formed. The time evolution of the SED obtained from the self-consistent RHD calculation can be found in *Masunaga et al.* (1998) and *Masunaga and Inutsuka* (2000a,b).

3.2.4. Core fragmentation. *Tsuribe and Inutsuka* (1999a,b) have shown that the fragmentation of a rotating collapsing core into a multiple system is difficult in the isothermal stage. *Matsumoto and Hanawa* (2003) have extended the collapse calculation by using a nested-grid hydrocode and a simplified barotropic equation of state that mimics the thermal evolution, and have shown that the first-core disk increases the rotation-to-gravitational energy ratio (T/|W|) by mass accretion. A stability analysis of a rotating polytropic gas shows that gas with T/|W| > 0.27 is unstable for nonaxisymmetric perturbations (e.g., *Imamura et al.*, 2000). If the first-core disk rotates fast enough that the angular speed × the freefall time satisfies $\Omega_c(4\pi G\rho_c)^{-1/2} \geq (0.2$–$0.3)$, fragments appear and grow into binaries and multiples in the first core phase. The nonaxisymmetric nonlinear spi-

ral pattern can transfer the angular momentum of the accreting gas.

3.3. Magnetic Fields

Detailed self-consistent calculations accounting for both thermal and magnetic support (*Mouschovias and Spitzer*, 1976; *Tomisaka et al.*, 1988) show that the maximum stable mass can be expressed as $M_{mag,max} \sim M_{BE} \{1 - [0.17/(G^{1/2} M/\Phi)_c]^2\}^{-3/2}$, where $(M/\Phi)_c$ is the central mass-to-flux ratio and $M_{BE} = 1.18c_s^4/G^{3/2}p_{ext}^{1/2}$ is the Bonnor-Ebert mass (*Bonnor*, 1956, 1957; *Ebert*, 1957). A similar formula was proposed by *McKee* (1989), $M_{mag,max} \sim M_{BE} + \Phi_B/2\pi G^{1/2}$.

Further support is provided by rotation. For a core with specific angular momentum j, the maximum stable mass is given by $M_{max} \sim [M_{mag,max}^2 + (4.8c_s j/G)^2]^{1/2}$ (*Tomisaka et al.*, 1989). The dynamical runaway collapse begins when the core mass exceeds this maximum stable mass (magnetically supercritical cloud). Quasistatic equilibrium configurations exist for cores less massive than the maximum stable mass. The evolution of these subcritical cores is controlled by the processes of ambipolar diffusion and magnetic braking, both of which have longer timescales than the gravitational freefall. As the core contracts, the density grows and, when $n \geq 10^{12}$ cm^{-3}, the magnetic field is effectively decoupled from the gas. At these densities, Joule dissipation becomes important and particle drifts are qualitatively different from ambipolar diffusion (*Nakano et al.*, 2002).

The magnetic field is also responsible for the transfer of angular momentum in magnetized rotating cores, by a process called magnetic braking. Magnetic braking is caused by the azimuthal component of the Lorentz force $(\vec{j} \times \vec{B})_\phi$. In the evolution of subcritical cores, the magnetic braking is important during the quasistatic contraction phase controlled by the ambipolar diffusion (*Basu and Mouschovias*, 1994). In the dynamical runaway collapse, the rotational speed is smaller than the inflow speed (*Tomisaka*, 2000).

3.4. Outflows

The magnetic field generates an outflow, by which star-forming gas loses its angular momentum and accretes onto a protostar. Magnetohydrodynamical simulations of the contraction of molecular cores (*Tomisaka*, 1998, 2000, 2002; *Allen et al.*, 2003; *Banerjee and Pudritz*, 2006) have shown that after the formation of the first core, the gas rotates around the core, a toroidal magnetic field is induced, and magnetic torques transfer angular momentum from the disk midplane to the surface. Outflows are accelerated in two ways: (1) The gas that has received enough angular momentum compared with the gravity is ejected by the excess centrifugal force [magnetocentrifugal wind mechanism (*Blandford and Payne*, 1982)]. (2) In a core with a weak magnetic field, the magnetic pressure gradient of the toroidal magnetic field accelerates the gas and an outflow is formed in the direction perpendicular to the disk.

Axisymmetric two-dimensional simulations show that (1) at least 10% of the accreted mass is ejected and (2) the angular momentum is reduced to a factor of 10^{-4} of the value of the parent cloud at the age of ≈ 7000 yr from the core formation. This resolves the angular momentum problem (*Tomisaka*, 2000). After a period of 7000 yr from the first core formation, the mass of the core reaches ~ 0.1 $M_\odot$ and the outflow extends to a distance from the core of ≈ 2000 AU with a speed of ~ 2 km s^{-1} (*Tomisaka*, 2002). If the accretion continues and the core mass grows to 1 $M_\odot$, the outflow expands and its speed is further increased. It should be noted that the outflow refers to the physics of the first core collapse only; the energetics of outflows during the second core collapse phase are yet to be determined.

3.5. Radiative Transfer in Multidimensions

Radiation transport has been shown to play a significant role in the outcome of fragmentation into binary and multiple systems (*Boss et al.*, 2000; *Whitehouse and Bate*, 2005) and in limiting the largest stellar mass in two-dimensional (*Yorke and Sonnhalter*, 2002) and three-dimensional simulations (*Edgar and Clarke*, 2004; *Krumholz et al.*, 2005c). The strong dependence of the evolution of isothermal and nonisothermal cloud models on the handling of the cloud's thermodynamics implies that collapse calculations must treat the thermodynamics accurately in order to obtain the correct solution (*Boss et al.*, 2000). Because of the great computational burden imposed by solving the mean intensity equation in the Eddington approximation (the computational time is increased by a factor of 10 or more), it is tempting to sidestep the Eddington approximation solution altogether and employ a simple barotropic prescription (e.g., *Boss*, 1981; *Bonnell*, 1994; *Bonnell and Bate*, 1994a; *Burkert et al.*, 1997; *Klein et al.*, 1999). However, *Boss et al.* (2000) showed that a simple barotropic approximation is insufficient and radiative transfer must be used. We discuss the various methods of radiation transport in section 4.6.

4. METHODOLOGY OF NUMERICAL SIMULATIONS

4.1. Complexity of the Problem of Low-Mass Star Formation

The computational challenge for simulations of low-mass star formation is that star formation occurs in clouds over a huge dynamic range of spatial scales, with different physical mechanisms being important on different scales. The gas densities in these clouds also varies over many orders of magnitude as a result of supersonic turbulence and gravitational collapse. Gravity, turbulence, radiation, and magnetic fields all contribute to the star-formation process. Thus the numerical problem is multiscale, multiphysics, and highly nonlinear. To develop a feel for the range of scale a simulation must cover, we can consider the internal structure of GMCs as hierarchical, consisting of smaller subunits within larger ones (*Elmegreen and Falgarone*, 1996). GMCs vary in

size from 20 to 100 pc, in density from 50 to 100 H_2 cm^{-3}, and in mass from 10^4 to 10^6 $M_\odot$.

Self-gravity and turbulence are equally important in controlling the structure and evolution of these clouds. Magnetic fields are likely to play an important role as well (*Heiles et al.*, 1993; *McKee et al.*, 1993). Embedded within the GMCs are dense clumps that may form clusters of stars. These clumps are few pc in size, have masses of a few thousand $M_\odot$, and mean densities $\sim 10^3$ H_2 cm^{-3}. The clumps contain dense cores with radii ~ 0.1 pc, densities 10^4-10^6 H_2 cm^{-3}, and masses ranging from 1 to several $M_\odot$. These cores likely form individual stars or low-order multiple systems. The role of turbulence and magnetic fields in the fragmentation of molecular clouds has been investigated by three-dimensional numerical simulations (e.g., *Padoan and Nordlund*, 1999; *Ostriker et al.*, 1999; *Ballesteros-Paredes et al.*, 2003; *Mac Low and Klessen*, 2004; *Nordlund and Padoan*, 2003).

A simulation that starts from a region of a turbulent molecular clouds (R ~ few pc) and evolves through the isothermal core collapse into the formation of the first hydrostatic core at densities of 10^{13} H_2 cm^{-3} requires an accurate calculation across 10 orders of magnitude in density and 4–5 orders of magnitude in spatial scale. To resolve 100-AU separation binaries, one needs a resolution of about 10 AU. To follow the collapse all the way to an actual star would require a further 10 orders of magnitude increase in density and 2–3 more orders of magnitude in spatial scale. Such extraordinary computational demands rule out fixed-grid simulations entirely and can be addressed only with accurate AMR or SPH approaches.

4.2. Smoothed Particle Hydrodynamics

The description of the gravitational collapse requires a large dynamic range of spatial resolution. An efficient way to achieve this is to use Lagrangian methods. Smoothed particle hydrodynamics (SPH) is a fully Lagrangian particle method designed to describe compressible fluid dynamics. This method is economical in handling hydrodynamic problems that have large, almost empty regions. A variety of astrophysical problems including star formation have been studied with SPH, because of its simplicity in programming two- and three-dimensional codes and its versatility to incorporate self-gravity. A broad discussion of the method can be found in a review by *Monaghan* (1994). An advantage of SPH is its conservation property; SPH is Galilean invariant and, in contrast to grid-based methods, conserves both linear and angular momentum simultaneously. The method to conserve the total energy within a computer roundoff error is explained in *Inutsuka* (2002). In order to further increase the dynamic range of spatial resolution, *Kitsionas and Whitworth* (2002) introduced particle splitting, which is an adaptive approach in SPH.

The "standard" SPH formalism adopts artificial viscosity that mimics the classical von-Neumann Richtmeyer viscosity. This tends to give poor performance in the description of strong shocks. In two- or three-dimensional calculations of colliding streams, standard SPH particles often penetrate into the opposite side. This unphysical effect can be partially eliminated by the so-called XSPH prescription (*Monaghan*, 1989), which does not introduce the (required) additional *dissipation*, but results in additional *dispersion* of the waves. As a more efficient method for handling strong shocks in the SPH framework, the so-called "Godunov SPH," was proposed by *Inutsuka* (2002), who implemented the exact Riemann solver in the strictly conservative particle method. This was used in the simulation of the collapse and fragmentation of self-gravitating cores (*Tsuribe and Inutsuka*, 1999a; *Cha and Whitworth*, 2003a,b).

The implementation of self-gravity in SPH is relatively easy and one can use various acceleration methods, such as *Tree-Codes*, and special purpose processors (e.g., GRAPE board). The flux-limited diffusion radiative transfer was recently incorporated in SPH by *Whitehouse and Bate* (2004), *Whitehouse et al.* (2005), and *Bastien et al.* (2004).

Several groups are now using "sink particles" to follow the subsequent evolution even after protostars are formed (*Bate et al.*, 1995). This is a prescription to continue the calculations without resolving processes of extremely short timescale around stellar objects. *Krumholz et al.* (2004) have introduced sink particles for the first time into Eulerian grid-based methods and in particular for AMR.

4.3. Fixed-Grid Hydrodynamics

Since the time of its introduction, the numerical code of choice for supersonic hydrodynamic turbulence has been the Piecewise Parabolic Method (PPM) (*www.lcse.umn.edu*) of *Colella and Woodward* (1984). PPM is based on a Rieman solver (the discretized approximation to the solution is locally advanced analytically) with a third-order accurate reconstruction scheme, which allows an accurate and stable treatment of strong shocks, while maintaining numerical viscosity to a minimum away from discontinuities. Because the physical viscosity is not explicitly computed (PPM solves the Euler equations), large-scale PPM flows are characterized by a very large effective Reynolds number (*Porter and Woodward*, 1994). Direct numerical simulations (DNS) of the Navier-Stokes equation, where the physical viscosity is explicitly computed, require a linear numerical resolution four times larger than PPM to achieve the same wave-number extension of the inertial range of turbulence as PPM (*Sytine et al.*, 2000). From this point of view, therefore, PPM has a significant advantage over DNS codes. Furthermore, DNS codes are generally designed for incompressible turbulence, and hence of limited use for simulations of the ISM.

Codes based on straightforward staggered mesh finite-difference methods, rather than Rieman solvers, have also been used in applications to star formation and interstellar turbulence, such as the Zeus code (*cosmos.ucsd.edu*) (*Stone and Norman*, 1992a,b) and the Stagger Code (*www.astro. ku.dk/StaggerCode/*) (*Galsgaard*, 1995; *Gudiksen and Nord-*

lund, 2005). Finite-difference codes address the problem of supersonic turbulence with the introduction of localized numerical viscosity to stabilize the shocks while keeping viscosity as low as possible away from shocks. The main advantage of this type of code, compared with Riemann solvers, is their flexibility in incorporating new physics and their computational efficiency.

Fixed-grid codes cannot achieve the dynamical range required by problems involving the gravitational collapse of protostellar cores. Such problems are better addressed with particle methods such as SPH, or by generalizing the methods used for fixed-grid codes into AMR schemes. The main advantage of large fixed-grid experiments is their ability to simulate the physics of turbulent flows. As supersonic turbulence is believed to play a crucial role in the initial fragmentation of star-forming clouds, fixed-grid codes may still be the method of choice to generate realistic large-scale initial conditions for the collapse of protostellar cores. Recent attempts of simulating supersonic turbulence with AMR methods are promising (*Kritsuk et al.*, 2006), but may be truly advantageous only at a resolution above $\sim 1000^3$. SPH simulations to date have resolution far too small for the task, as commented above, and have not been used so far as an alternative method to investigate the physics of turbulence.

4.4. Adaptive Mesh Refinement Hydrodynamics and Nested Grids

The adaptive mesh refinement (AMR) scheme utilizes underlying rectangular grids at different levels of resolution. Linear resolution varies by integral refinement factors between levels, and a given grid is always fully contained within one at the next coarser level (excluding the coarsest grid). The origin of the method stems from the seminal work of *Berger and Oliger* (1984) and *Berger and Collela* (1989). The AMR method dynamically resizes and repositions these grids and inserts new, finer ones within them according to adjustable refinement criteria, such as the numerical Jean's condition (*Truelove et al.*, 1997). Fine grids are automatically removed as flow conditions require less resolution. During the course of the calculation, some pointwise measure of the error is computed at frequent intervals, typically every other timestep. At those times, the cells that are identified are covered by a relatively small number of rectangular patches, which are refined by some even integer factor.

Refinement is in both time and space, so that the calculation on the refined grids is computed at the same Courant number as that on the coarse grid. The finite-difference approximations on each level of refinement are in conservation form, as is the coupling at the interface between grids at different levels of refinement. AMR has three substantial advantages over standard SPH. Combined with high-order Godunov methods, AMR achieves a much higher resolution of shocks. This is important in obtaining accuracy in supersonic turbulent flows in star-forming clumps and cores and in accretion shocks onto forming protostars. AMR allows high resolution at *all* points in the flow as dictated by the physics. Unlike SPH, where particles are taken away from low-density regions, where accuracy is lost, and concentrated into high-density regions, AMR maintains high accuracy everywhere. An important consequence of this is that if SPH were to maintain the same comparable resolution as AMR everywhere in the flow, it would be prohibitively expensive. AMR is based on fixed Eulerian grids and thus can take advantage of sophisticated algorithms to incorporate magnetic fields and radiative transfer. This is far more difficult in a particle-based scheme. AMR was first introduced into astrophysics by *Klein et al.* (1990, 1994) and has been used extensively both in low-mass and high-mass star-formation simulations (*Truelove et al.,* 1998; *Klein,* 1999; *Klein et al.,* 2001, 2003, 2004a; *Krumholz et al.,* 2005c).

An advantage of SPH over Cartesian grid-based AMR is that for pure hydrodynamics it can conserve both linear and angular momentum simultaneously to within roundoff errors, whereas Cartesian grid-based AMR cannot. However, if one uses a cylindrical or spherical coordinate system for the simulation of protostellar disks, for instance, then grid-based AMR conserves total angular momentum to roundoff. We point out, however, that these statements apply only to pure hydrodynamics. Once forces such as gravity are included, the situation becomes worse and both grid codes that solve the Poisson equation or SPH codes that use tree-type acceleration or grid-based methods for gravity lose the conservation property for total linear momentum as well.

There are several ways to implement AMR. They can be broadly divided into two categories: meshes with fixed number of cells, such as in Lagrangian or rezoning approaches, and meshes with variable number of cells, such as unstructured finite elements, structured cell-by-cell and structured subgrid blocks. For various reasons the most widely adopted approaches in astrophysics are structured subgrid blocks and cell-by-cell. The first was developed by *Berger and Oliger* (1984) and *Berger and Collela* (1989). It is used in the AMR code ORION developed by Klein and collaborators (*Klein,* 1999; *Crockett et al.,* 2005) and in the community code ENZO (*Norman and Bryan,* 1999). The cell-by-cell approach such as in PARAMESH (*MacNeice et al.,* 2000) is used in the community code Flash (*Banerjee et al.,* 2004). A hybrid approach is used in the code NIRVANA (*Ziegler,* 2005). The cell-by-cell method has the advantages of flexible and efficient refinement patterns and low memory overhead and the disadvantages of expensive interpolation and derivation formulas and large tree data structures. The subgrid block method is more efficient and more suitable for shock capturing schemes than the cell-by-cell method, at the price of some memory overhead.

Finally, nested grids consisting of concentric hierarchical rectangular subgrids can also be very effective for problems of well-defined geometry (*Yorke et al.,* 1993). These methods are advantageous for tracing the nonhomologous runaway collapse of an initially symmetrical cloud in which the coordinates of a future dense region are known in ad-

vance (*Tomisaka,* 1998). The finest subgrid is added dynamically when spatial resolution is needed as in AMR methods.

4.5. Approaches for Magnetohydrodynamics

Since strong shocks often appear in the astrophysical phenomena, a shock-capturing scheme is needed also in MHD. Upwind schemes based on the Riemann solver are used as the MHD engine. Schemes well known in hydrosimulations, such as Roe's approximate Riemann solver (*Brio and Wu,* 1988; *Ryu and Jones,* 1995; *Nakajima and Hanawa,* 1996) and PPM (*Dai and Woodward,* 1994), are also applicable to MHD.

Special attention should be paid to guarantee $\mathrm{div}\vec{B} = 0$ in MHD simulations. To ensure that the divergence of Maxwell stress tensor $T_{ij} = -(1/4\pi)B_iB_j + (1/8\pi)B^2\delta_{ij}$ gives the Lorentz force; the first term on the righthand side $\partial_j(B_iB_j)$ must be equal to $B_j\partial_jB_i$. This requires $B_i\partial_jB_j = 0$ and means that a fictitious force appears along the magnetic field if the divergence-free condition is broken. The divergence of the magnetic field amplifies the instability of the solution even for a linear wave. Thus, it is necessary for the MHD scheme to keep the divergence of the magnetic field zero within a roundoff error or at least small enough. This divergence-free nature should be satisfied for the boundaries of subgrids in AMR and nested grid schemes.

One solution is based on "constrained transport" (CT) (*Evans and Hawley,* 1988), in which the staggered collocation of the components of magnetic field on the cell faces makes the numerical divergence vanish exactly. In the staggered collocation, the electric field $-v \times \vec{B}$ of the induction equation $\partial_t\vec{B} = \nabla \times (v \times \vec{B})$ is evaluated on the edge of the cell face and the line integral along the edge gives the time difference of a component of the magnetic field. Note that the electric field on one edge appears twice to complete the induction equation. To guarantee a vanishing divergence of the magnetic field, CT requires the two evaluations to coincide with each other.

To utilize the Godunov-type Riemann solver in the context of CT, *Balsara and Spicer* (1999) proposed the following scheme: (1) The face-centered magnetic field is interpolated to the cell center; (2) from the cell-centered variables, the numerical flux at the cell face is obtained using a Riemann solver; (3) the flux is interpolated to the edge of the cell face and the electric field in the induction equation is obtained; (4) a new face-centered magnetic field is obtained from the induction equation. Variants of this method are widely used (see also *Ryu et al.,* 1998; *Ziegler,* 2005).

Avoiding staggered collocation of the magnetic field requires divergence cleaning. In this case, divergence cleaning is realized by replacing the magnetic field every step as $\vec{B}^{\mathrm{new}} = \vec{B} - \vec{\nabla}\Phi$, where $\vec{\nabla}^2\Phi = \mathrm{div}\vec{B}$ (Hodge projection), or by solving a diffusion equation for $\mathrm{div}\vec{B}$ as $\partial t\vec{B} = \eta\nabla(\nabla \cdot \vec{B})$. The former is combined with pure Godunov-type Riemann solvers using only cell-centered variables (*Ryu et al.,* 1995). *Crockett et al.* (2005) reported that the divergence cleaning of the face-centered magnetic field appearing in the numerical flux based on an unsplit, cell-centered Godunov scheme improves its accuracy and stability.

Powell et al. (1999) proposed a different formalism, in which the div$\vec{B}$ term is kept in the MHD equations as a source [e.g., the Lorentz force $(\nabla \times \vec{B}) \times \vec{B}/4\pi$ gives an extra term related to div$\vec{B}$ as $-\vec{B}\nabla \cdot \vec{B}$ besides the Maxwell stress tensor term $-T^{ij}$, in the equation for momentum density]. In this formalism, div$\vec{B}$ is not amplified but advected along the flow. Comparison between various methods is found in *Tóth* (2000), *Dedner et al.* (2002), *Balsara and Kim* (2004), and *Crockett et al.* (2005).

There have been attempts to solve the induction equation with SPH methods (e.g., *Stellingwerf and Peterkin*, 1994; *Byleveld and Pongracic*, 1996; *Price and Monaghan*, 2004a,b,c, 2005). A major obstacle is an instability that develops when the momentum and energy equations are written in conservation form. As a result, the equations must be written in a way that does not conserve momentum (*Phillips and Monaghan*, 1985; *Morris*, 1996), which is a major concern for the accurate treatment of shocks. Results of recent tests of the state-of-the-art SPH MHD code (*Price and Monaghan*, 2004c) appear to be rather poor even for very mild shocks, and we conclude that MHD with SPH is not yet viable for simulations.

4.6. Approaches for Radiation Transport

Several levels of approximation of the radiation transport in star-formation simulations can be used and details of various methods can be found in *Mihalas and Mihalas* (1984) and *Castor* (2004). Here we briefly describe these methods and point out their strengths and weaknesses. Although modern simulations using radiative transfer are still at an early stage, we include methods that hold promise for the future that will circumvent the weaknesses of more approximate approaches currently in place.

The simplest improvement beyond a barotropic-stiffened EOS is the diffusion approximation, which pertains to the limit in which radiation can be treated as an ideal fluid with small corrections. The approximation holds when the photon mean free path is small compared with other length scales. The combined energy equation for the gas and radiation results in an implicit nonlinear diffusion equation for the temperature. The weakness of the diffusion approximation is that it is strictly applicable to optically thick regimes and performs poorly in optically thin regions. This can be severe in optically thin regions of an inhomogeneous turbulent core or in the optically thin atmosphere surrounding a developing protostar.

The next level of approximation is the Eddington approximation (*Boss and Myhill*, 1992; *Boss et al.*, 2000). It can be shown that the diffusion approximation leads directly to Eddington's approximation $P_\nu = \frac{1}{3}E_\nu I$, where P_ν is the pressure tensor moment of the specific intensity of radiation, E_ν is the scalar energy density of radiation, and I_ν is the isotropic identity tensor. This approximation, coupled with dropping the time-dependent term in the second-mo-

ment equation of transfer results in a combined parabolic second-order time-dependent diffusion equation for the energy density of the radiation field. This formulation of the Eddington approximation is used in *Boss et al.* (2000). The approximation results in a loss of the finite propagation speed of light c and a loss of the radiation momentum density, thus there is an error in the total momentum budget. In optically thin regions, the radiation flux can increase without limit. As with all diffusion-like methods, this approach also suffers from shadow effects whereby radiation will tend to fill in behind optically thick structures and may lead to unphysical heating.

An alternative approach is *Flux Limited Diffusion* (FLD), which modifies the Eddington approximation and compensates for the errors made in dropping the time-dependent flux term by including a correction factor in the diffusion coefficient for the radiation flux. This correction factor, called a flux limiter, is in general a tensor and has the property that the flux goes to the diffusion limit at large optical depth and it correctly limits the flux to be no larger than cE in the optically thin regime. This improvement over the Eddington approximation has been used by *Klein et al.* (2004c) for the simulation of both low-mass and high-mass star formation. The resulting sparse matrices introduced by the diffusion-like terms are solved by multigrid iterative methods in an AMR framework. The flux-limiting correction can cause errors on the order of 20% in the flux-limiter (or the flux), similar to the errors of the Eddington approximation of 20% in the Eddington factor at $\tau = 0$ in the Milne problem (*Castor*, 2004). FLD methods also suffer from shadow effects, which can be severe.

The next level of approximation, the variable Eddington tensor method, removes many of the inaccuracies of the Eddington approximation and the flux limiter modification. It was first formulated in multidimensions for astrophysical problems by *Dykema et al.* (1996). In essence, if the precise ratio of the pressure tensor to the energy density were included as an *ad hoc* multiplier in the Eddington approximation equations, they would represent an exact closure of the system. The tensor ratio is obtained iteratively from either an auxiliary solution of the exact transport equation for the specific intensity or using an approximate analytic representation of the tensor. This method holds promise for future simulations, but has yet to be used in star formation.

The final two approaches, which are highly accurate and deal with the angle-dependent transport equation directly, are S_N methods and Monte Carlo methods. They have not yet been developed for simulations in star formation because the cost in three dimensions is prohibitive. The S_N method is a short characteristic method in which a bundle of rays is created at every mesh point and are extended in the upwind direction only as far as the next spatial cell. The main problem is in finding the efficient angle set to represent the radiation field in two or three dimensions (*Castor*, 2004). Finally, one might consider Monte Carlo methods to solve the transport equation. Although simple to implement (its great advantage), this method suffers from needing a vast number of operations per timestep to get accurate statistics in following the particles used to track the radiation field. Both of these methods will avoid shadow effects and may be necessary to accurately treat optically thick inhomogeneous structures that form in accretion flows onto protostars.

Radiative transfer implementations have recently been developed also for SPH methods, based on the flux-limited diffusion (*Whitehouse et al.,* 2005) or the Monte Carlo method (*Stamatello and Whitworth,* 2003, 2005).

4.7. Comparison of Computational Methods

Based on the physical processes and numerical methodologies discussed in the previous sections, we can compare numerical schemes according to their ability to handle the following problems both accurately and efficiently: (a) turbulence, (b) strong shocks, (c) self-gravity, (d) magnetic fields, and (e) radiative transfer.

The standard SPH method has been successful with (c) and implementations of (e) have been recently developed in the flux-limited diffusion approximation and with a Monte Carlo method. It does not include (d), it is well known to be inadequate for (b), and has had virtually no applications to (a) to date. As with any Lagrangian particle method, SPH provides good resolution in high-density regions, but very poor in low-density ones. The Godunov SPH method improves the standard SPH codes because of its ability to address (b), but does not provide a solution to (d) and is untested for (a) as well. Although MHD is currently under development in SPH, results of standard MHD tests with a state-of-the-art code show the need for significant improvements even in the case of very mild shocks. Current applications of SPH should therefore be limited to non-MHD problems and the accuracy and performance of SPH with turbulent flows must be thoroughly tested.

In hydrodynamical problems, grid-based methods such as MUSCL (*van Leer,* 1979) and PPM (*Colella and Woodward,* 1984) use exact Riemann solvers to construct the numerical fluxes and provide very accurate description in astrophysical flows with strong shocks (b). They have also been thoroughly tested with compressible turbulent flows, and MHD versions have been developed that can address both (d) and (e). Traditional finite-difference grid-based schemes may still be useful, because the best of them can also accurately address (a), (b), (d), and (e), at a lower cost of code development and computer resources. Point (c) can also be efficiently dealt with by grid-based codes thanks to AMR methods. However, the development of AMR schemes that satisfy (c) and at the same time (d) has begun only recently. These schemes exist and have been successfully tested, but it is unclear at present which approach will provide the best tradeoff between accuracy and performance.

The constrained transport method appears to be the ideal one to guarantee the $\nabla \cdot B = 0$ condition. Various schemes have been proposed even in the category of Godunov-type methods with a linearized Riemann solver. An exact MHD

Riemann solver would be more adequate to handle strong shocks, but it is not available yet. In MHD we have to solve seven characteristics even in one-dimensional problems, which hinders an efficient construction of numerical fluxes based on the nonlinear waves. Furthermore, the discovery of the existence of the MHD intermediate shocks (*Brio and Wu,* 1988) introduces an additional difficulty in the categorization and prediction of the emerging nonlinear waves. Among possible solutions, a linearized Riemann solver with artificial viscosity may still be a useful option.

The Godunov MHD code of *Crockett et al.* (2005) has been merged with the AMR RHD code of *Klein et al.* (2004a,b) into the first fully developed AMR magnetoradiation-hydrodynamic code (ORION) to be used in simulations of star formation capable of addressing (a) through (e).

5. RECENT SIMULATIONS AND CONFRONTATION WITH THE OBSERVATIONS

5.1. Turbulent Fragmentation of Molecular Clouds

The fragmentation of molecular clouds is the result of a complex interaction of supersonic turbulence with gravity and magnetic fields. Supersonic turbulent flows generate nonlinear density enhancements through a complex network of interacting shocks. Some density enhancements are massive and dense enough to undergo gravitational instability and collapse into stars. A fundamental problem with our understanding of star formation is that the physics of turbulence is not fully understood. In order to investigate the process of turbulent fragmentation we must rely on large numerical simulations that barely resolve the scale-free nature of interstellar turbulent flows. If the scaling laws of turbulence play a role in the star-formation process, we cannot accurately test their effects numerically, unless those scaling laws are well reproduced in the simulations. *Ballesteros-Paredes et al.* (2006) have tested the idea that the power spectrum of turbulence determines the Salpeter IMF (*Padoan and Nordlund,* 2002). However, their simulations do not generate a turbulence inertial range, due to the combined effect of low resolution and large numerical diffusivity. Their velocity power spectra do not show any extended power laws, and they even differ between their grid and SPH simulations. As a consequence, such numerical simulations fail to reproduce a scale-free mass distribution of unstable cores.

Other recent SPH simulations have been used to compute the stellar mass distribution (e.g., *Bonnell et al.,* 2003; *Klessen,* 2001) and to test the effect of turbulence on star formation (*Delgado-Donate et al.,* 2004). Although such SPH simulations are ideally suited to follow the collapse of individual objects due to their Lagrangian nature, their size is far too small to generate an inertial range of turbulence, as discussed in section 3.1.

Despite their limitations, numerical simulations carried out over the last few years have given us a good statistical picture of the process of fragmentation of magnetized supersonic flows. The following are the most important results: (1) The dissipation time of turbulence is almost independent of the magnetic field strength. The turbulence decays in approximately one dynamical time in both equipartition and super-Alfvénic flows (*Padoan and Nordlund,* 1997, 1999; *Mac Low et al.,* 1998; *Stone et al.,* 1998; *Muller and Biskamp,* 2000). (2) The velocity power spectrum and structure functions are power laws over an inertial range of scales (*Boldyrev,* 2002; *Boldyrev et al.,* 2002a,b; *Padoan et al.,* 2004b). In the limit of very large rms Mach number, the turbulent velocity power spectrum scales approximately as $u_k^2 \propto k^{-1.8}$ and the velocity structure functions follow the relative scaling predicted by *Boldyrev* (2002). For intermediate levels of compressibility, the scaling exponents depend on the rms Mach number (*Padoan et al.,* 2004b). (3) The power spectrum of the gas density is a power law over an inertial range of scales. Its slope is a function of the rms Mach number of the flow and of the average magnetic field strength (*Padoan et al.,* 2004b; *Beresnyak et al.,* 2005). (4) With an isothermal equation of state, the probability density function of gas density is well approximated by a log-normal distribution (*Vázquez-Semadeni,* 1994; *Padoan et al.,* 1997; *Scalo et al.,* 1998; *Passot and Vázquez-Semadeni,* 1998; *Nordlund and Padoan,* 1999; *Ostriker et al.,* 1999; *Wada and Norman,* 2001), with the dispersion of linear density proportional to the rms Mach number of the flow (*Padoan et al.,* 1997; *Nordlund and Padoan,* 1999; *Ostriker et al.,* 1999; *Li et al.,* 2004). (5) If the kinetic energy exceeds the magnetic energy, the distribution of the magnetic field strength, B, is very intermittent and is correlated with the gas density, n. The scatter plot of B vs. n shows a very large dispersion and a well-defined power-law upper envelope (*Padoan and Nordlund,* 1999; *Ostriker et al.,* 2001). If the kinetic energy is comparable to the magnetic energy, strong density enhancements are still possible in the direction of the magnetic field, but fluctuations of B are small and independent of n. (6) The flow velocity is correlated to the gas density. Because density is increased by shocks, where the velocity is dissipated, dense filaments and cores have lower velocity than the low-density gas (*Padoan et al.,* 2001b). (7) The mass distribution of gravitationally unstable turbulent density peaks is very close to the stellar IMF and follows the analytical model of *Padoan and Nordlund* (2002) (*Li et al.,* 2004; *Tilley and Pudritz,* 2004).

There is now substantial observational evidence indicating that these main properties of supersonic MHD turbulence are indeed found in molecular clouds. The comparison of numerical simulations of turbulence with observational data was pioneered by *Falgarone et al.* (1994) and continued by many others (e.g., *Padoan et al.,* 1998, 1999, 2001a,b, 2004a; *Ostriker et al.,* 2001; *Ballesteros-Paredes and Mac Low,* 2002; *Ossenkopf,* 2002; *Gammie et al.,* 2003; *Klessen et al.,* 2005; *Esquivel and Lazarian,* 2005).

However, with the exception of several works by Padoan et al. and the work of *Ossenkopf* (2002), where postprocessed three-dimensional non-LTE radiative transfer calculations were carried out, all other studies are based on a simple comparison of density and velocity fields in the simulations with the observed quantities. Some recent works addressing the comparison of turbulence simulations and observational data include studies of velocity scaling showing that molecular cloud turbulence is driven on large-scale (e.g., *Ossenkopf and Mac Low,* 2002; *Heyer and Brunt,* 2004) studies of core properties, showing that turbulent flows naturally generate dense cores with shapes, internal turbulence, rotation velocity, and magnetic field strength consistent with the observations (e.g., *Padoan and Nordlund,* 1999; *Gammie et al.,* 2003; *Tilley and Pudritz,* 2004; Tilley and Pudritz, in preparation; *Li et al.,* 2004).

5.2. Collapse and Fragmentation of Molecular Cloud Cores into Low-Mass Stars

Over the past several years two dominant models of how stars acquire their final mass have emerged: direct gravitational collapse and competitive accretion. In both theories, a star initially forms when gravitational bound gas collapses. In the gravitational collapse scenario, after a protostar has consumed or expelled all the gas in its initial core, it may continue accreting from the parent clump, but it will not significantly alter its mass (*McKee and Tan,* 2003; *Padoan et al.,* 2005; *Krumholz et al.,* 2005b). Competitive accretion, in contrast, requires that the amount accreted after consuming the initial core be substantially larger than the protostellar mass.

Krumholz et al. (2005a) define $f_m \equiv \dot{m}_* t_{dyn} = m_*$ as the fractional change in mass that a protostar of mass m_* undergoes each dynamical time t_{dyn} of its parent clump, starting after the initial core has been consumed by the accreting protostar. Gravitational collapse theory suggests that $f_m \ll 1$, while competitive accretion *requires* $f_m \gg 1$. In recent work examining the plausibility of competitive accretion, *Krumholz et al.* (2005a) considered two possible scenarios. In the first scenario, the gas the protostar is accreting is not accumulated into bound structures on scales smaller than the entire clump. For unbound gas, self-gravity may be neglected and the entire problem can be treated as Bondi-Hoyle accretion in a turbulent medium of non-self-gravitating gas onto a point particle. In a companion paper (*Krumholz et al.,* 2006) they develop the theory for Bondi-Hoyle accretion in a turbulent medium. Using this theory they derive the accretion rate for such a turbulent medium and they confirm their theory with detailed, high-resolution, converged AMR simulations. By using this accretion rate and the definition of the virial parameter, $\alpha_{vir} \equiv M_{vir}/M$, and the dynamical time, $t_{dyn} \equiv R/\sigma$, where σ is the velocity dispersion in the gas, they show that the accretion of unbound gas gives

$$f_{m-BH} = \left(14.4, 3.08\frac{L}{R}\right)\phi_{BH}\alpha_{vir}^{-2}\left(\frac{m_*}{M}\right)$$

for a (spherical, filamentary) star-forming region, where ϕ_{BH} represents the effects of turbulence (*Krumholz et al.,* 2005a).

From this it is clear that competitive accretion is most effective in low-mass clumps with virial parameters $\alpha_{vir} \ll 1$. They then examined the observed properties of a large range of star-forming regions spanning both low-mass and high-mass stars and computed the properties for each region yielding α_{vir}, ϕ_{BH}, and f_{m-BH}. In virtually every region examined, the virial parameter $\alpha_{vir} \sim 1$ and $f_{m-BH} \ll 1$. Thus *none* of the star-forming regions are consistent with competitive accretion, but all are consistent with direct gravitational collapse. *Edgar and Clarke* (2004) examined Bondi-Hoyle accretion onto stars including radiation pressure effects and found that radiation pressure halts further accretion around stars more massive than $\sim 10\, M_\odot$. It is important to point out that while this result may be true for accretion of unbound gas onto a point particle, it is not true for global collapse and accretion from a bound core as shown in more realistic full three-dimensional radiation hydrodynamics simulations by *Krumholz et al.* (2005c), who form massive stars with $M \sim 40\, M_\odot$. If *Edgar and Clarke*'s (2004) results do hold for Bondi-Hoyle accretion (but not accretion from a core), then the only way for massive stars to grow in competitive accretion is by direct collisions. This requires densities of $10^8\, pc^{-3}$, ~ 3 orders of magnitude larger than any observed in the galactic plane. Furthermore, no competitive accretion model to date has included the effects of radiation pressure, a glaring omission if the model is attempting to explain high-mass stars. It follows that competitive accretion is not a viable mechanism for producing the stellar IMF.

In a second possible competitive accretion scenario, *Krumholz et al.* (2005a) examined another way that a star could increase its mass by capturing and accreting other gravitationally bound cores. Their theory results in the calculation of f_{m-cap}, the fractional change in mass that a protostar undergoes by capturing bound cores. As found with f_{m-BH}, all the values are estimated to be 3 more orders of magnitude below unity and again, competitive accretion is found not to work.

If competitive accretion is clearly not supported by observations in any known star-forming region, why do the simulations (*Bonnell et al.*, 1998, 2001a,b; *Bate and Bonnell*, 2005) almost invariably find competitive accretion to work? Is there a fundamental flaw in the methodology used in competitive accretion scenarios (SPH), or is the problem one of physics and initial conditions? As *Krumholz et al.* (2005a) point out, all competitive accretion have virial parameters $\alpha_{vir} \ll 1$. Some of the simulations start with $\alpha_{vir} \approx 0.01$ as a typical choice (*Bonnell et al.*, 2001a,b; *Klessen and Burkert*, 2000, 2001). For other simulations the virial parameter is initially on the order of unity but decreases to $\ll 1$ in a crossing time as turbulence decays (*Bonnell et al.*, 2004; *Bate et al.*, 2002a,b, 2003). It is also noteworthy that many of the simulations begin with clumps of mass considerably smaller ($M \leq 100\, M_\odot$) than that typically found in star-forming regions, $\sim 5000\, M_\odot$ (*Plume et al.*, 1997). *Krumholz et al.* (2005a) show that for competitive accretion to work, $\alpha_{vir}^2 M < 10\, M_\odot$, but for typical star-forming regions $\alpha_{vir} \approx 1$ and $M \approx 10^2$–$10^4\, M_\odot$ and the inequality is almost never satisfied.

One reason why simulations evolve to $\alpha_{vir} \ll 1$ is that they omit feedback from star formation. Observations by *Quillen et al.* (2005) show that outflows inject enough energy on the scale of a clump to sustain turbulence, thereby keeping the virial parameter from declining to values much less than unity. Another possible reason is that the simulations consider isolated clumps with too little material. Real clumps are embedded in larger molecular clouds, where larger-scale turbulent motions can cascade down to the clump scale, preventing the rapid decay of the turbulence.

5.3. Three-Dimensional Collapse with Radiation

Radiation transfer is an important element in both low-mass and high-mass star formation. *Boss et al.* (2000) carried out the first three-dimensional simulations including radiation transfer to study the effect of radiation on the formation of stars in collapsing molecular cloud cores. Starting from cores with Gaussian initial density profiles, this work compared collapse calculations based on the isothermal and barotropic approximations with the more realistic case of detailed radiative transfer in the Eddington approximation. The use of the isothermal equation of state resulted in a collapse leading to a thin isothermal filament (*Truelove et al.*, 1997). In the more realistic case with nonisothermal heating using radiative transfer in the Eddington approximation, they showed that thermal retardation of the collapse caused the formation of a binary protostar system at the same maximum density where the isothermal collapse yielded a thin filament. Eventually, the binary clumps evolved into a central protostar surrounded by spiral arms containing two more clumps. The corresponding collapse using the barotropic approximation allowed a transition from an isothermal optically thin to an optically thick flow. It resulted in a transient binary merging into a central object surrounded by spiral arms with no evidence of further fragmentation. The barotropic result differs significantly from the Eddington result at the same maximum density, indicating the importance of detailed radiative transfer effects.

Boss et al. (2000) examined the differences in the use of a barotropic-stiffened EOS approximation and radiative transfer. In the former, the thermal properties of the gas are specified solely as a function of density. This implies that a single global value of a critical density represents the entire calculation and its value depends weakly on the assumed geometry of the cloud (*Inutsuka and Miyama*, 1997). In three dimensions the effective value of this critical density depends on the local geometry surrounding a fluid element. In addition, whereas the specific entropy of a gas parcel generally depends on the thermal history of the parcel, the specific entropy of the gas using the barotropic approximation depends solely on the density. Thus the derived pressures used in the momentum equations will differ between a calculation using a stiffened EOS approximation and a fully consistent calculation using radiative transfer. As a result, the temperature is determined not simply by adiabatic compression, but by compressional heating in a three-dimensional volume with highly variable optical depth. Thus the dependence of the temperature on the density cannot be represented with a simple barotropic approximation with any great accuracy. This causes concern for the validity of current simulations of multiple star formation and cluster formation, since essentially all use either the isothermal or the barotropic approximation. Recent work by *Whitehouse and Bate* (2005) examined core collapse with radiation and the adequacy of the barotropic approximation as well.

5.4. The Debate Over Disk Fragmentation

As pointed out in section 2, most simulations performed with SPH to date find circumstellar disk fragmentation and rapid ejection of BDs in most cases (*Bate et al.*, 2003; *Delgato-Donate et al.*, 2004; *Goodwin et al.*, 2004a,b), even possibly to an excess of BD and low-mass companions (*Goodwin et al.*, 2004b). Furthermore, most simulations are terminated at an arbitrary time, when much of the gas is still present; hence the simulations may provide only a lower bound to the number of companions that would be produced. As discussed in section 2, these simulations are increasingly contradicted by recent observations. *Goodwin and Kroupa* (2005) have recently pointed out that observations suggest that individual cores produce at most two to three protostars. A possible reason for this problem is the absence of a magnetic field and inaccurate thermodynamics (i.e., barotropic equation rather than radiative transfer) in the SPH simulations.

However, there is an alternative explanation: The SPH simulations are likely not converged. Indeed, recent high-resolution AMR simulations of the collapse and fragmentation of turbulent cores (*Klein et al.*, 2004a,b) show that a magnetic field and radiative transfer is not required to explain the observational results of *Goodwin and Kroupa* (2005). *Klein et al.* (2004a,b) find that over a broad range of turbulent Mach numbers (M ~ 1–3) and rotational energies (β ~ 10^{-4}–10^{-1}), low-order single or binary stars are formed through fragmentation of the core, not the ensuing disk. Is it then a matter of faith in one numerical method or the other? SPH vs. AMR? Not really; it is rather a matter of testing the convergence of the numerical simulations, irrespective of the method.

Fisher et al. (in preparation) have performed high-resolution, full-convergence studies using the SPH code Gadget with the same model and initial conditions as *Goodwin et al.* (2004a), except for the initial turbulent seed. They confirmed the results of *Goodwin et al.* (2004a) at the same low resolution (only one smoothing kernel per minimum Jeans mass). But they have also found *no convergence* of either the multiple number of companion protostars produced or the time for multiple fragmentation to occur in the disk, with up to 32× the resolution of *Goodwin et al.* (2004a). This suggests that the current SPH simulations showing high-order multiple disk fragmentation may be

grossly underresolved. If so, the disagreement with the observations may be due not only to the absence of the magnetic field and radiation, but also to unconverged numerics. Recent grid-based simulations attempting to study fragmentation in isolated disks (cf. chapter by Durisen et al.) indicate that for two-dimensional axisymmetric disk studies, very-high-resolution, 125 cells in the radial direction alone (Ostriker, personal communication) are required to demonstrate that disks do not suffer from numerical instability. This level of resolution is far beyond any SPH simulation of cores or clusters to date showing disk fragmentation, and would be difficult to achieve also with AMR simulations starting from globally collapsing cores.

At the time of this writing, the debate between SPH and AMR with respect to the issue of disk fragmentation continues, but it is our strong opinion that all simulations using SPH or AMR must demonstrate adequate convergence to be credible. This should apply also to physical systems that may display chaotic behavior, such as the gas dynamics in a molecular cloud core. A high sensitivity to initial conditions may result in statistical distributions of the measured quantities (e.g., number of collapsing objects and their formation time) around some mean value. In this case, an expensive approach to test numerical convergence would consist of running a large number of experiments at each resolution and test for the convergence of the statistical distributions of the quantities of interest. A less-expensive approach is to measure quantities with a weak sensitivity to initial conditions because they already represent the average of some statistical distribution, or just because they are even more sensitive to the numerical resolution than to the initial conditions. For example, Fisher et al. (in preparation) find that as they increase the resolution of their SPH simulations, the time of disk fragmentation (when the first object is formed) increases monotonically with resolution, a sign of lack of convergence in the SPH simulations, rather than a signature of chaotic behavior.

5.5. Star Formation in a Cloud with Magnetic Field and Rotation

5.5.1. Fragmentation of the first core. A nonrotating cloud core without magnetic field contracts in a self-similar manner to form a first core that is composed of molecular hydrogen (*Larson*, 1969; *Masunaga et al.*, 1998). Recently, nested-grid MHD simulations (*Machida et al.*, 2004, 2005a) have revealed that (1) a rotating magnetized core evolves maintaining a ratio of angular speed to magnetic field strength at the center $\Omega_c = B_c \simeq$ const; and (2) Ω_c and B_c are well correlated at the first core phase and satisfy the "magnetic flux-spin relation" as $\Omega_c^2 \simeq 0.2^2 4\pi G\rho_c + B_c^2 = 0.36^2 8\pi c_s^2 \rho_c \simeq 1$, using a central density ρ_c and isothermal sound speed c_s. This is regarded as an appearance of self-similarity in magnetized rotating cores.

The fragmentation of the first core is regarded as one of the mechanisms to explain close binary systems (*Bonnell and Bate*, 1994b; *Bate*, 1998). The magnetic field affects the rotational motion (magnetic braking) and thus the fragmentation. Whether or not the magnetic field stabilizes the first core against the fragmentation (*Machida et al.,* 2005b; *Ziegler,* 2005) is attracting attention in relation to the binary formation. As well as nonmagnetic cores, a magnetized first core can fragment if it is rotating sufficiently fast, $\Omega_c \simeq 0.2(4\pi G\rho_c)^{1/2}$ (*Machida et al.,* 2005b) and has only a weak magnetic field, $B_c \lesssim 0.3(8\pi c_s^2\rho_c)^{1/2}$. Simulations show that increasing the magnetic field strength, fragmentation is stabilized by the suppression of rotational motion by magnetic braking (see also *Hosking and Whitworth,* 2004). This is not found by *Boss* (2002), but his model equation is not fully consistent with MHD and does not account for magnetic braking. In order to achieve enough rotation to cause fragmentation, the initial Ω-to-B ratio must satisfy the condition $(\Omega/B)_{init} > 0.39G^{1/2}c_s \sim 17 \times 10^{-7}(c_s/0.19 \text{ km s}^{-1})^{-1}\mu G^{-1}$ (*Machida et al.,* 2005b).

When $\vec{B}$ and angular momentum $\vec{J}$ are not parallel to each other, the magnetic braking works more efficiently for the component of $\vec{J}$ perpendicular to $\vec{B}$. A magnetically dominated cloud core whose Ω-to-B ratio is less than the above critical value forms a disk perpendicular to the magnetic field (*Matsumoto and Tomisaka,* 2004) and an outflow is ejected in the direction of the local magnetic field, even if $\vec{B}$ is not parallel to $\vec{J}$. The difference between the local field in the vicinity of a protostar and on the cloud scale is restricted to $\lesssim 30°$ for this case.

If the first core is fragmented into binary or multiple cores, each fragment spins and multiple outflows (Fig. 1) are ejected (*Machida et al.,* 2004, 2005b; *Ziegler,* 2005; *Banerjee and Pudritz,* 2006), which explains binary outflows (*Liseau et al.,* 2005).

5.5.2. Second collapse with magnetic field. Once the dissociation of molecular hydrogen starts at the central region of the first core, the second collapse begins. Further calculation of the evolution to form the second core (i.e., a protostar) requires the inclusion of resistivity in the MHD description, as the high density reduces the degree of ionization and the conductivity of the medium. Machida et al. (in preparation) have adopted parameterized resistivity as a function of density in their nested-grid resistive MHD code, and extended the calculations of the collapse. During the isothermal phase, the magnetic Reynolds number is a decreasing function of density. If the magnetic Reynolds number decreases below unity, the magnetic field is effectively decoupled from the collapsing gas. However, the temperature of the gas becomes sufficiently high ($\sim10^3$ K) that the magnetic field is recoupled again with the collapsing gas. This relatively sudden grab of the field lines tends to make a very collimated fast outflow around the second core. Figure 2 shows a snapshot of the propagation of a fast jet from the protostar. The density distribution on the cross section is also shown on the left and right walls. Note that bow shocks are clearly seen in the density plot. The results of this calculation indicate that a realistic modeling of the evolution of temperature and resistivity as a function of density is required for a precise description of the jet.

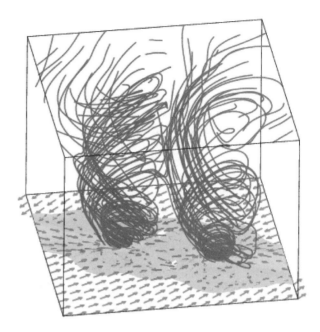

Fig. 1. Binary outflows. Magnetic field lines, velocity field vectors, and density distribution on the z = 0 plane. Taken from model DL of *Machida et al.* (2005b).

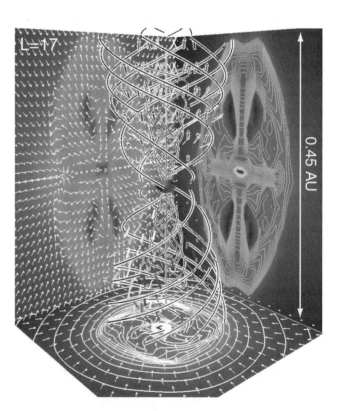

Fig. 2. Bird's-eye view of the magnetic field lines in the fast jet emanating from the central region around a protostar. Only the seventeenth grid in the nested-grid resistive MHD calculations by Machida et al. (in preparation) is shown. The density contours and velocity vectors are projected on the walls on both sides.

6. SUMMARY AND FUTURE DIRECTIONS

6.1. Summary

Observations of molecular clouds and cores provide a wealth of data that are important constraints for initial conditions of numerical simulations. Large-scale simulations should be consistent with the turbulent nature of the ISM and the observed properties of prestellar cores should emerge self-consistently from the simulations. All calculations must adhere to the Jeans condition for grid-based schemes or a well-established equivalent for particle-based schemes. It is important to stress that the Jeans condition is a necessary but not sufficient condition to guarantee avoidance of artificial fragmentation. Simulations must be tested at more resolved Jeans numbers to establish convergence.

The importance of turbulence in star formation is now well accepted. A very large spatial resolution is required to simulate the turbulent fragmentation, barely achieved by the largest grid-based simulations. Present SPH simulations fall several orders of magnitude below the required spatial resolution. It is possible that almost no available simulations have yet accurately tested the effect of turbulent fragmentation.

Magnetic fields play a crucial role in the star-formation process. At this time there are no three-dimensional, self-gravitational MHD simulations that have evolved stars from turbulent clouds. Grid-based schemes such as AMR have developed high-order accurate PPM and Godunov-based MHD that can provide accurate solutions across several orders of magnitude of collapse. SPH has developed cruder approaches to MHD that appear to be rather poor even for very mild shocks, but this will hopefully improve. Godunov approaches to MHD for SPH may provide increased accuracy.

Radiation transport has been shown to play a significant role in the outcome of fragmentation to binary and multiple systems. It has been shown that sidestepping the issue of radiation transport with a barotropic approximation can lead to incorrect results. Current AMR calculations have implemented radiation transfer in the flux-limited approximation and have used this for both low-mass and high-mass star formation. Radiation transfer schemes have recently been developed for SPH as well.

Current star-formation simulations are not yet adequate to accurately span the many orders of magnitude of density and spatial range necessary to account for stars from initially turbulent clouds and encompassing all the relevant physics. In our opinion, AMR approaches with recent developments coupling radiation transfer and MHD hold out the best promise for achieving that goal. SPH, while making significant strides in the last few years, is still faced with difficult challenges of accurate handling of turbulence, strong shocks, MHD, and radiation transfer. However, we anticipate further progress with SPH.

Current calculations are still not adequate to explain the stellar IMF. Of the two dominant theories for the origin of stellar masses described as direct gravitational collapse and competitive accretion, observations appear to favor the first. Recent theoretical work by *Krumholz et al.* (2005a) has now demonstrated that seed protostars cannot gain mass efficiently by competitive accretion processes in observed star-forming regions that are approximately in virial balance. There is no observational evidence for the existence of regions that are far from virial balance, as required by competitive accretion models. This suggests that competitive accretion is not a viable mechanism for producing the stellar IMF and that current simulations resulting in competitive accretion must have initial or boundary conditions inconsistent with the observations, or rather neglect some crucial physics. Theoretical efforts directed toward the picture of direct gravitational collapse set up by turbulent fragmentation appear to be more promising.

6.2. Future Directions for Numerical Simulation

The lack of demonstrated convergence for most simulations in the field presents us with uncalibrated results. We strongly emphasize that future simulations should demonstrate numerical convergence before detailed comparison with observations can be credible. Otherwise there is no way to normalize the accuracy of large-scale simulations, and the results of such simulations will not advance our understanding of low-mass star formation. Convergence tests should be always carried out irrespective of the numerical approach. A better understanding of the numerical treatment of disk fragmentation must occur to clear up the current discrepancies between AMR and SPH.

Future simulations of low-mass star formation must endeavor to include MHD and radiation transfer. With the development of accurate approaches to these processes, we can expect to see simulations become more relevant to addressing the observations. For simulations to make a real connection to the observations, detailed line profiles and continuum submillimeter and millimeter maps should be calculated with three-dimensional radiative transfer codes. Approaches that go beyond the Eddington approximation and flux-limited diffusion such as SN transport and Monte Carlo need further development to work efficiently in full three-dimensional simulations. They will become increasingly important in treating the flow of radiation in highly inhomogeneous regions. As future simulations encompass multicoupled physics, significant progress must be made in algorithms that improve the parallel scalability. That is necessary in order to simulate the full dynamic range of collapse and fragmentation from clouds to stars, while capturing all the relevant physics.

REFERENCES

Allen A., Li Z.-Y., and Shu F. H. (2003) *Astrophys. J., 599*, 363–379.
Alves J. F., Lada C. J., and Lada E. A. (2001) *Nature, 409*, 159–161.
Bacmann A., André P., Puget J.-L., Abergel A., Bontemps S., and Ward-Thompson D. (2000) *Astron. Astrophys., 361*, 555–580.
Ballesteros-Paredes J. and Mac Low M.-M. (2002) *Astrophys. J., 570*, 734–748.

Ballesteros-Paredes J., Klessen R. S., and Vázquez-Semadeni E. (2003) *Astrophys. J., 592,* 188–202.

Ballesteros-Paredes J., Gazol A., Kim J., Klessen R. S., Jappsen A.-K., and Tejero E. (2006) *Astrophys. J., 637,* 384–391.

Balsara D. S. and Kim J. (2004) *Astrophys. J., 602,* 1079–1090.

Balsara D. S. and Spicer D. S. (1999) *J. Comp. Phys., 149,* 270–292.

Banerjee R. and Pudritz R. E. (2006) *Astrophys. J., 641,* 949–960.

Banerjee R., Pudritz R. E., and Holmes L. (2004) *Mon. Not. R. Astron. Soc., 355,* 248–272.

Barrado y Navascués D., Mohanty S., and Jayawardhana R. (2004) *Astrophys. J., 604,* 284–296.

Bastien P., Cha S.-H., and Viau S. (2004) *Rev. Mex. Astron. Astrofis., 22,* 144–147.

Basu S. and Mouschovias T. C. (1994) *Astrophys. J., 432,* 720–741.

Bate M. R. (1998) *Astrophys. J., 508,* L95–L98.

Bate M. R. and Bonnell I. A. (2005) *Mon. Not. R. Astron. Soc., 356,* 1201–1221.

Bate M. R., Bonnell I. A., and Price N. M. (1995) *Mon. Not. R. Astron. Soc., 277,* 362–376.

Bate M. R., Bonnell I. A., and Bromm V. (2002a) *Mon. Not. R. Astron. Soc., 332,* L65–L68.

Bate M. R., Bonnell I. A., and Bromm V. (2002b) *Mon. Not. R. Astron. Soc., 336,* 705–713.

Bate M. R., Bonnell I. A., and Bromm V. (2003) *Mon. Not. R. Astron. Soc., 339,* 577–599.

Beresnyak A., Lazarian A., and Cho J. (2005) *Astrophys. J., 624,* L93–L96.

Berger M. J. and Colella P. (1989) *J. Comp. Phys., 82,* 64–84.

Berger M. J. and Oliger S. (1984) *J. Comp. Phys., 53,* 484.

Blandford R. D. and Payne D. G. (1982) *Mon. Not. R. Astron. Soc., 199,* 883–903.

Bodenheimer P., Yorke H. W., Rożycka M., and Tohline J. E. (1990) *Astrophys. J., 355,* 651–660.

Bodenheimer P., Burkert A., Klein R. I., and Boss A. P. (2000) In *Protostars and Planets IV* (V. Mannings et al., eds.), pp. 675–690. Univ. of Arizona, Tucson.

Boldyrev S. (2002) *Astrophys. J., 569,* 841–845.

Boldyrev S., Nordlund Å., and Padoan P. (2002a) *Astrophys. J., 573,* 678–684.

Boldyrev S., Nordlund Å., and Padoan P. (2002b) *Phys. Rev. Lett., 89,* 031102–031105.

Bonnell I. A. (1994) *Mon. Not. R. Astron. Soc., 269,* 837–848.

Bonnell I. A. and Bate M. R. (1994a) *Mon. Not. R. Astron. Soc., 269,* L45–L48.

Bonnell I. A. and Bate M. R. (1994b) *Mon. Not. R. Astron. Soc., 271,* 999–1004.

Bonnell I. A., Bate M. R., and Zinnecker H. (1998) *Mon. Not. R. Astron. Soc., 298,* 93–102.

Bonnell I. A., Bate M. R., Clarke C. J., and Pringle J. E. (2001a) *Mon. Not. R. Astron. Soc., 323,* 785–794.

Bonnell I. A., Clarke C. J., Bate M. R., and Pringle J. E. (2001b) *Mon. Not. R. Astron. Soc., 324,* 573–579.

Bonnell I. A., Bate M. R., and Vine S. G. (2003) *Mon. Not. R. Astron. Soc., 343,* 413–418.

Bonnell I. A., Vine S. G., and Bate M. R. (2004) *Mon. Not. R. Astron. Soc., 349,* 735–741.

Bonnor W. B. (1956) *Mon. Not. R. Astron. Soc., 116,* 351–359.

Bonnor W. B. (1957) *Mon. Not. R. Astron. Soc., 117,* 104–117.

Boss A. P. (1981) *Astrophys. J., 250,* 636–644.

Boss A. P. (2002) *Astrophys. J., 568,* 743–753.

Boss A. P. and Myhill E. A. (1992) *Astrophys. J. Suppl., 83,* 311–327.

Boss A. P., Fisher R. T., Klein R. I., and McKee C. F. (2000) *Astrophys. J., 528,* 325–335.

Bourke T. L., Myers P. C., Robinson G., and Hyland A. R. (2001) *Astrophys. J., 554,* 916–932.

Brio M. and Wu C. C. (1988) *J. Comp. Phys., 75,* 400–422.

Burkert A., Bate M. R., and Bodenheimer P. (1997) *Mon. Not. R. Astron. Soc., 289,* 497–504.

Byleveld S. E. and Pongracic H. (1996) *Publ. Astron. Soc. Pac., 13,* 71–74.

Castor J. I. (2004) *Radiation Hydrodynamics.* Cambridge Univ., Cambridge.

Cha S.-H. and Whitworth A. P. (2003a) *Mon. Not. R. Astron. Soc., 340,* 73–90.

Cha S.-H. and Whitworth A. P. (2003b) *Mon. Not. R. Astron. Soc., 340,* 91–104.

Chabrier G. (2003) *Publ. Astron. Soc. Pac., 115,* 763–795.

Cho J., Lazarian A., and Vishniac E. T. (2003) *Astrophys. J., 595,* 812–823.

Colella P. and Woodward P. R. (1984) *J. Comp. Phys., 54,* 174–201.

Crockett R. K., Colella P., Fisher R. T., Klein R. I., and McKee C. F. (2005) *J. Comp. Phys., 203,* 422–448.

Crutcher R. M. (1999) *Astrophys. J., 520,* 706–713.

Crutcher R. M., Troland T. H., Goodman A. A., Heiles C., Kazes I., and Myers P. C. (1993) *Astrophys. J., 407,* 175–184.

Crutcher R. M., Troland T. H., Lazareff B., Paubert G., and Kazès I. (1999) *Astrophys. J., 514,* L121–L124.

Dai W. and Woodward P. R. (1994) *J. Comp. Phys., 115,* 485–514.

Dedner A., Kemm F., Kröner D., Munz C.-D., Schnitzer T., and Wesenberg M. (2002) *J. Comp. Phys., 175,* 645–673.

Delgado-Donate E. J., Clarke C. J., and Bate M. R. (2004) *Mon. Not. R. Astron. Soc., 347,* 759–770.

Dimotakis P. E. (2000) *J. Fluid Mech., 409,* 69–98.

Dobler W., Haugen N. E., Yousef T. A., and Brandenburg A. (2003) *Phys. Rev. E, 68,* 026304–026311.

Dubrulle B. (1994) *Phys. Rev. Lett., 73,* 959–962.

Duchêne G., Bouvier J., and Simon T. (1999) *Astron. Astrophys., 343,* 831–840.

Duchêne G., Bouvier J., Bontemps S., André P., and Motte F. (2004) *Astron. Astrophys., 427,* 651–665.

Dykema P. G., Klein R. I., and Castor J. I. (1996) *Astrophys. J., 457,* 892–921.

Ebert R. (1957) *Z. Astrophys., 42,* 263–272.

Edgar R. and Clarke C. (2004) *Mon. Not. R. Astron. Soc., 349,* 678–686.

Elmegreen B. G. and Falgarone E. (1996) *Astrophys. J., 471,* 816–821.

Esquivel A. and Lazarian A. (2005) *Astrophys. J., 631,* 320–350.

Evans C. R. and Hawley J. F. (1988) *Astrophys. J., 332,* 659–677.

Falgarone E., Puget J.-L., and Perault M. (1992) *Astron. Astrophys., 257,* 715–730.

Falgarone E., Lis D. C., Phillips T. G., Pouquet A., Porter D. H., and Woodward P. R. (1994) *Astrophys. J., 436,* 728–740.

Falkovich G. (1994) *Phys. Fluids, 6,* 1411–1414.

Fuller G. A. and Myers P. C. (1992) *Astrophys. J., 384,* 523–527.

Galli D., Shu F. H., Laughlin G., and Lizano S. (2001) *Astrophys. J., 551,* 367–386.

Galsgaard K. (1995) Ph.D. thesis, Univ. of Copenhagen, *www.astro.ku.dk/~kg/.*

Gammie C. F., Lin Y.-T., Stone J. M., and Ostriker E. C. (2003) *Astrophys. J., 592,* 203–216.

Goodman A. A., Benson P. J., Fuller G. A., and Myers P. C. (1993) *Astrophys. J., 406,* 528–547.

Goodwin S. P. and Kroupa P. (2005) *Astron. Astrophys., 439,* 565–569.

Goodwin S. P., Whitworth A. P., and Ward-Thompson D. (2004a) *Astron. Astrophys., 414,* 633–650.

Goodwin S. P., Whitworth A. P., and Ward-Thompson D. (2004b) *Astron. Astrophys., 423,* 169–182.

Gudiksen B. V. and Nordlund Å. (2005) *Astrophys. J., 618,* 1020–1030.

Haugen N. E. and Brandenburg A. (2004) *Phys. Rev. E, 70,* 026405–026411.

Heiles C., Goodman A. A., McKee C. F., and Zweibel E. G. (1993) In *Protostars and Planets III* (E. H. Levy and J. I. Lunine, eds.), pp. 279–326. Univ. of Arizona, Tucson.

Heyer M. H. and Brunt C. M. (2004) *Astrophys. J., 615,* L45–L48.

Hosking J. G. and Whitworth A. P. (2004) *Mon. Not. R. Astron. Soc., 347,* 1001–1010.

Imamura J. N., Durisen R. H., and Pickett B. K. (2000) *Astrophys. J., 528,* 946–964.

Inutsuka S. (2002) *J. Comp. Phys., 179,* 238–267.

Inutsuka S.-I. and Miyama S. M. (1997) *Astrophys. J., 480,* 681–693.

Jayawardhana R., Mohanty S., and Basri G. (2002) *Astrophys. J., 578,* L141–L144.

Jeans J. H. (1902) *Philos. Trans. A, 199,* 1–53.

Jijina J., Myers P. C., and Adams F. C. (1999) *Astrophys. J. Suppl., 125,* 161–236.

Juvela M., Padoan P., and Nordlund Å. (2001) *Astrophys. J., 563,* 853–866.

Kitsionas S. and Whitworth A. P. (2002) *Mon. Not. R. Astron. Soc., 330,* 129–136.

Klein R. I. (1999) *J. Comp. Appl. Math., 109,* 123–152.

Klein R. I., Colella P., and McKee C. F. (1990) *Publ. Astron. Soc. Pac., 12,* 117–136.

Klein R. I., McKee C. F., and Colella P. (1994) *Astrophys. J., 420,* 213–236.

Klein R. I., Fisher R. T., McKee C. F., and Truelove J. K. (1999) In *Numerical Astrophysics* (S. M. Miyama et al., eds.), pp. 131–140. Proceedings of the International Conference on Numerical Astrophysics, Vol. 240, Kluwer, Dordrecht.

Klein R. I., Fisher R., and McKee C. F. (2001) In *The Formation of Binary Stars* (H. Zinnecker and R. D. Mathieu, eds.), pp. 361–370. IAU Symposium 200, ASP, San Francisco.

Klein R. I., Fisher R. T., Krumholz M. R., and McKee C. F. (2003) *Rev. Mex. Astron. Astrofis., 15,* 92–96.

Klein R. I., Fisher R. T., McKee C. F., and Krumholz M. R. (2004a) *Publ. Astron. Soc. Pac., 323,* 227–234.

Klein R. I., Fisher R., and McKee C. F. (2004b) *Rev. Mex. Astron. Astrofis., 22,* 3–7.

Klein R. I., Fisher R. T., McKee C. F., and Krumholz M. R. (2004c) In *Adaptive Mesh Refinement* (T. Plewa, ed.), pp. 112–118. Univ. of Chicago, Chicago.

Klessen R. S. (2001) *Astrophys. J., 556,* 837–846.

Klessen R. S. and Burkert A. (2000) *Astrophys. J. Suppl., 128,* 287–319.

Klessen R. S. and Burkert A. (2001) *Astrophys. J., 549,* 386–401.

Klessen R. S., Ballesteros-Paredes J., Vázquez-Semadeni E., and Durán-Rojas C. (2005) *Astrophys. J., 620,* 786–794.

Koyama H. and Inutsuka S.-I. (2000) *Astrophys. J., 532,* 980–993.

Kritsuk A. G., Norman M. L., and Padoan P. (2006) *Astrophys. J., 638,* L25–L28.

Krumholz M. R., McKee C. F., and Klein R. I. (2004) *Astrophys. J., 611,* 399–412.

Krumholz M. R., McKee C. F., and Klein R. I. (2005a) *Nature, 438,* 332–334.

Krumholz M. R., McKee C. F., and Klein R. I. (2005b) *Astrophys. J., 618,* 757–768.

Krumholz M. R., Klein R. I., and McKee C. F. (2005c) In *Massive Star Birth: A Crossroads of Astrophysics* (R. Cesaroni et al., eds.), pp. 231–236. Cambridge Univ., Cambridge.

Krumholz M. R., McKee C. F., and Klein R. I. (2006) *Astrophys. J., 638,* 369–381.

Lada C. J. and Lada E. A. (2003) *Ann. Rev. Astron. Astrophys., 41,* 57–115.

Larson R. B. (1969) *Mon. Not. R. Astron. Soc., 145,* 271–295.

Larson R. B. (1981) *Mon. Not. R. Astron. Soc., 194,* 809–826.

Li P. S., Norman M. L., Mac Low M.-M., and Heitsch F. (2004) *Astrophys. J., 605,* 800–818.

Liseau R., Fridlund C. V. M., and Larsson B. (2005) *Astrophys. J., 619,* 959–967.

Mac Low M.-M. and Klessen R. S. (2004) *Rev. Mod. Phys., 76,* 125–194.

Mac Low M.-M., Smith M. D., Klessen R. S., and Burkert A. (1998)

Astron. Astrophys. Suppl., 261, 195–196.

Machida M. N., Tomisaka K., and Matsumoto T. (2004) *Mon. Not. R. Astron. Soc., 348,* L1–L5.

Machida M. N., Matsumoto T., Tomisaka K., and Hanawa T. (2005a) *Mon. Not. R. Astron. Soc., 362,* 369–381.

Machida M. N., Matsumoto T., Hanawa T., and Tomisaka K. (2005b) *Mon. Not. R. Astron. Soc., 362,* 382–402.

MacNeice P., Olson K. M., Mobarry C., de Fainchtein R., and Packer C. (2000) *Comp. Phys. Comm., 126,* 330–354.

Masunaga H. and Inutsuka S.-I. (1999) *Astrophys. J., 510,* 822–827.

Masunaga H. and Inutsuka S.-I. (2000a) *Astrophys. J., 531,* 350–365.

Masunaga H. and Inutsuka S.-I. (2000b) *Astrophys. J., 536,* 406–415.

Masunaga H., Miyama S. M., and Inutsuka S.-I. (1998) *Astrophys. J., 495,* 346–369.

Matsumoto T. and Hanawa T. (2003) *Astrophys. J., 595,* 913–934.

Matsumoto T. and Tomisaka K. (2004) *Astrophys. J., 616,* 266–282.

McKee C. F. (1989) *Astrophys. J., 345,* 782–801.

McKee C. F. and Tan J. C. (2003) *Astrophys. J., 585,* 850–871.

McKee C. F., Zweibel E. G., Goodman A. A., and Heiles C. (1993) In *Protostars and Planets III* (E. H. Levy and J. I. Lunine, eds.), pp. 327–342. Univ. of Arizona, Tucson.

Mihalas D. and Mihalas B. W. (1984) *Foundations of Radiation Hydrodynamics.* Oxford Univ., Oxford.

Mohanty S., Jayawardhana R., and Basri G. (2005) *Astrophys. J., 626,* 498–522.

Monaghan J. J. (1989) *J. Comp. Phys., 82,* 1–15.

Monaghan J. J. (1994) *J. Comp. Phys., 110,* 399–406.

Morris J. P. (1996) *Publ. Astron. Soc. Aus., 13,* 97–102.

Motte F., Andre P., and Neri R. (1998) *Astron. Astrophys., 336,* 150–172.

Motte F., André P., Ward-Thompson D., and Bontemps S. (2001) *Astron. Astrophys., 372,* L41–L44.

Mouschovias T. C. and Spitzer L. (1976) *Astrophys. J., 210,* 326–327.

Muller W. C. and Biskamp D. (2000) *Phys. Rev. Lett., 84,* 475–478.

Myers P. C., Dame T. M., Thaddeus P., Cohen R. S., Silverberg R. F., Dwek E., and Hauser M. G. (1986) *Astrophys. J., 301,* 398–422.

Nakajima Y. and Hanawa T. (1996) *Astrophys. J., 467,* 321–333.

Nakano T., Nishi R., and Umebayashi T. (2002) *Astrophys. J., 573,* 199–214.

Natta A., Testi L., Muzerolle J., Randich S., Comerón F., and Persi P. (2004) *Astron. Astrophys., 424,* 603–612.

Nordlund Å. and Padoan P. (1999) In *Interstellar Turbulence* (J. Franco and A. Carraminana, eds.), pp. 218–231. Cambridge Univ., Cambridge.

Nordlund Å. and Padoan P. (2003) In *Turbulence and Magnetic Fields in Astrophysics* (E. Falgarone and T. Passot, eds.), pp. 271–298. Springer-Verlag, Berlin.

Norman M. L. and Bryan G. L. (1999) In *Numerical Astrophysics* (S. M. Miyama et al., eds.), pp. 19–33. Kluwer, Dordrecht.

Nutter D. J., Ward-Thompson D., Crutcher R. M., and Kirk J. M. (2004) *Astron. Astrophys. Suppl., 292,* 179–184.

Ossenkopf V. (2002) *Astron. Astrophys., 391,* 295–315.

Ossenkopf V. and Mac Low M.-M. (2002) *Astron. Astrophys., 390,* 307–326.

Ostriker E. C., Gammie C. F., and Stone J. M. (1999) *Astrophys. J., 513,* 259–274.

Ostriker E. C., Stone J. M., and Gammie C. F. (2001) *Astrophys. J., 546,* 980–1005.

Padoan P. and Nordlund Å. (1997) *arXiv e-prints,* astro-ph/9706177.

Padoan P. and Nordlund Å. (1999) *Astrophys. J., 526,* 279–294.

Padoan P. and Nordlund Å. (2002) *Astrophys. J., 576,* 870–879.

Padoan P. and Nordlund Å. (2004) *Astrophys. J., 617,* 559–564.

Padoan P., Jones B. J. T., and Nordlund Å. P. (1997) *Astrophys. J., 474,* 730–734.

Padoan P., Juvela M., Bally J., and Nordlund Å. (1998) *Astrophys. J., 504,* 300–313.

Padoan P., Bally J., Billawala Y., Juvela M., and Nordlund Å. (1999)

Astrophys. J., 525, 318–329.

Padoan P., Goodman A., Draine B. T., Juvela M., and Nordlund Å., and Rögnvaldsson Ö. E. (2001a) *Astrophys. J., 559,* 1005–1018.

Padoan P., Juvela M., Goodman A. A., and Nordlund Å. (2001b) *Astrophys. J., 553,* 227–234.

Padoan P., Jimenez R., Juvela M., and Nordlund Å. (2004a) *Astrophys. J., 604,* L49–L52.

Padoan P., Jimenez R., Nordlund Å., and Boldyrev S. (2004b) *Phys. Rev. Lett., 92,* 191102–191105.

Padoan P., Kritsuk A., Norman M. L., and Nordlund Å. (2005) *Astrophys. J., 622,* L61–L64.

Passot T. and Vázquez-Semadeni E. (1998) *Phys. Rev. E, 58,* 4501–4510.

Phillips G. J. and Monaghan J. J. (1985) *Mon. Not. R. Astron. Soc., 216,* 883–895.

Plume R., Jaffe D. T., Evans N. J., Martin-Pintado J., and Gomez-Gonzalez J. (1997) *Astrophys. J., 476,* 730–749.

Porter D. H. and Woodward P. R. (1994) *Astrophys. J. Suppl., 93,* 309–349.

Powell K. G., Roe P. L., Linde T. J., Gombosi T. I., and de Zeeuw D. L. (1999) *J. Comp. Phys., 154,* 284–309.

Price D. J. and Monaghan J. J. (2004a) *Mon. Not. R. Astron. Soc., 348,* 123–138.

Price D. J. and Monaghan J. J. (2004b) *Mon. Not. R. Astron. Soc., 348,* 139–152.

Price D. J. and Monaghan J. J. (2004c) *Astron. Astrophys. Suppl., 292,* 279–283.

Price D. J. and Monaghan J. J. (2005) *Mon. Not. R. Astron. Soc., 364,* 384–406.

Quillen A. C., Thorndike S. L., Cunningham A., Frank A., Gutermuth R. A., Blackman E. G., Pipher J. L., and Ridge N. (2005) *Astrophys. J., 632,* 941–955.

Reipurth B. and Clarke C. (2001) *Astron. J., 122,* 432–439.

Ryu D. and Jones T. W. (1995) *Astrophys. J., 442,* 228–258.

Ryu D., Jones T. W., and Frank A. (1995) *Astrophys. J., 452,* 785–796.

Ryu D., Miniati F., Jones T. W., and Frank A. (1998) *Astrophys. J., 509,* 244–255.

Scalo J., Vázquez-Semadeni E., Chappell D., and Passot T. (1998) *Astrophys. J., 504,* 835–853.

She Z.-S. and Leveque E. (1994) *Phys. Rev. Lett., 72,* 336–339.

Shu F. H., Adams F. C., and Lizano S. (1987) *Ann. Rev. Astron. Astrophys., 25,* 23–81.

Shu F. H., Laughlin G., Lizano S., and Galli D. (2000) *Astrophys. J., 535,* 190–210.

Stamatellos D. and Whitworth A. P. (2003) *Astron. Astrophys., 407,* 941–955.

Stamatellos D. and Whitworth A. P. (2005) *Astron. Astrophys., 439,* 153–158.

Stellingwerf R. F. and Peterkin R. E. (1994) *Mem. Soc. Astron. Ital., 65,* 991–1011.

Stone J. M. and Norman M. L. (1992a) *Astrophys. J. Suppl., 80,* 753–790.

Stone J. M. and Norman M. L. (1992b) *Astrophys. J. Suppl., 80,* 791–818.

Stone J. M., Ostriker E. C., and Gammie C. F. (1998) *Astrophys. J., 508,* L99–L102.

Sytine I. V., Porter D. H., Woodward P. R., Hodson S. W., and Winkler K.-H. (2000) *J. Comp. Phys., 158,* 225–238.

Tilley D. A. and Pudritz R. E. (2004) *Mon. Not. R. Astron. Soc., 353,* 769–788.

Tomisaka K. (1998) *Astrophys. J., 502,* L163–L167.

Tomisaka K. (2000) *Astrophys. J., 528,* L41–L44.

Tomisaka K. (2002) *Astrophys. J., 575,* 306–326.

Tomisaka K., Ikeuchi S., and Nakamura T. (1988) *Astrophys. J., 335,* 239–262.

Tomisaka K., Ikeuchi S., and Nakamura T. (1989) *Astrophys. J., 341,* 220–237.

Tóth G. (2000) *J. Comp. Phys., 161,* 605–652.

Troland T. H. and Heiles C. (1986) *Astrophys. J., 301,* 339–345.

Truelove J. K., Klein R. I., McKee C. F., Holliman J. H., Howell L. H., and Greenough J. A. (1997) *Astrophys. J., 489,* L179–L183.

Truelove J. K., Klein R. I., McKee C. F., Holliman J. H., Howell L. H., Greenough J. A., and Woods D. T. (1998) *Astrophys. J., 495,* 821–852.

Tsuribe T. and Inutsuka S.-I. (1999a) *Astrophys. J., 523,* L155–L158.

Tsuribe T. and Inutsuka S.-I. (1999b) *Astrophys. J., 526,* 307–313.

van Leer B. (1979) *J. Comp. Phys., 32,* 101–136.

Vázquez-Semadeni E. (1994) *Astrophys. J., 423,* 681–692.

Wada K. and Norman C. A. (2001) *Astrophys. J., 547,* 172–186.

Whitehouse S. C. and Bate M. R. (2004) *Mon. Not. R. Astron. Soc., 353,* 1078–1094.

Whitehouse S. C., Bate M. R., and Monaghan J. J. (2005) *Mon. Not. R. Astron. Soc., 364,* 1367–1377.

Wolfire M. G., Hollenbach D., McKee C. F., Tielens A. G. G. M., and Bakes E. L. O. (1995) *Astrophys. J., 443,* 152–168.

Yorke H. W. and Sonnhalter C. (2002) *Astrophys. J., 569,* 846–862.

Yorke H. W., Bodenheimer P., and Laughlin G. (1993) *Astrophys. J., 411,* 274–284.

Ziegler U. (2005) *Astron. Astrophys., 435,* 385–395.

Stellar Properties of Embedded Protostars

R. J. White
University of Alabama in Huntsville

T. P. Greene
NASA Ames Research Center

G. W. Doppmann
Gemini Observatory

K. R. Covey
University of Washington

L. A. Hillenbrand
California Institute of Technology

Protostars are precursors to the nearly fully assembled T Tauri and Herbig Ae/Be type stars undergoing quasistatic contraction toward the zero-age main sequence; they are in the process of acquiring the majority of their stellar mass. Although numerous young stars with spatially extended envelope-like structures appear to fit this description, their high extinction has inhibited observers from directly measuring their stellar and accretion properties and confirming that they are in fact in the main phase of mass accretion (i.e., true protostars). Recently, however, high-dispersion spectrographs on large-aperture telescopes have allowed observers to begin studying the stellar and accretion properties of a subset of these stars, commonly referred to as Class I stars. In this chapter, we summarize the newly determined properties of Class I stars and compare them with observations of Class II stars, which are the more optically revealed T Tauri stars, to better understand the relative evolutionary state of the two classes. Class I stars have distributions of spectral types and stellar luminosities that are similar to those of Class II stars, suggesting similar masses and ages. The stellar luminosity and resulting age estimates, however, are especially uncertain given the difficulty in accounting for the large extinctions, scattered light emission, and continuum excesses typical of Class I stars. Several candidate Class I brown dwarfs are identified. Class I stars appear to rotate somewhat more rapidly than T Tauri stars, by roughly a factor of 2 in the mean. Likewise, the disk-accretion rates inferred from optical excesses and Brγ luminosities are similar to, but larger in the mean by a factor of a few than, the disk-accretion rates of T Tauri stars. There is some evidence that the disk-accretion rates of Class I stars are more distinct from T Tauri stars within the ρ Ophiuchi star-forming region than in others (e.g., Taurus-Auriga), suggesting a possible environmental influence. The determined disk-accretion rates are nevertheless 1–2 orders of magnitude less than the mass-infall rates predicted by envelope models. In at least a few cases the discrepancy appears to be caused by T Tauri stars being misclassified as Class I stars because of their edge-on disk orientation. In cases where the envelope density and infall velocity have been determined directly and unambiguously, the discrepancy suggests that the stellar mass is not acquired in a steady-state fashion, but instead through brief outbursts of enhanced accretion. If the ages of some Class I stars are in fact as old as T Tauri stars, replenishment may be necessary to sustain the long-lived envelopes, possibly via continued dynamical interactions with cloud material.

1. THE DISCOVERY AND CLASSIFICATION OF PROTOSTARS

The early phases of star and planet formation are difficult to observe because this process occurs while the protostar is buried within its natal molecular cloud material. Nevertheless, infrared and submillimeter observations, which are able to penetrate this high extinction material, have revealed much about the bolometric luminosities, spectral energy distributions (SEDs), and circumstellar material of embedded young stars (e.g., *Lada and Wilking,* 1984; *Myers et al.,* 1987; *Wilking et al.,* 1989; *Kenyon et al.,* 1990; *André and Montmerle,* 1994; *Motte and André,* 2001; *Onishi et al.,* 2002; *Andrews and Williams,* 2005). The earliest of these observations spurred development of the theory of isolated low-mass star formation, advancing initial considerations of the collapse of a singular isothermal sphere (e.g., *Shu,* 1977) to include circumstellar disks and envelopes

(*Cassen and Moosman*, 1981; *Terebey et al.*, 1984; *Adams et al.*, 1987).

An easy marriage of observation and theory was found by equating different stages of this theoretical evolutionary process with observed differences in the spectral energy distributions of very young stars. Four classes have been proposed (Class 0, I, II, and III), and are now commonly used to classify young stars. In this proposed scheme, Class 0 stars are cloud cores that are just beginning their protostellar collapse; Class I stars are embedded within an "envelope" of circumstellar material that is infalling, accumulating in a disk, and being channeled onto the star; Class II stars are nearly fully assembled stars undergoing pure disk accretion with perhaps some evidence for tenuous amounts of envelope material; and, finally, Class III stars are post-accretion but still pre-main-sequence stars. The Class II and Class III stars are also known as classical T Tauri stars and weak-lined T Tauri stars, respectively. It is believed that the majority of the stellar mass is acquired prior to the Class II phase; these younger stars are thus considered to be the true "protostars."

Despite the discretization of the class classification scheme, there is a continuum of circumstellar evolutionary states and thus a continuum of observational properties exhibited by young stars. Figure 1 illustrates two popular criteria used to segregate the classes, bolometric temperature

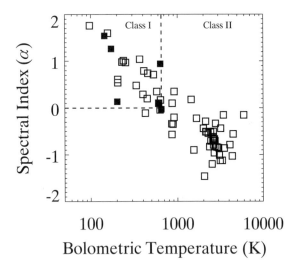

Fig. 1.   Spectral index vs. bolometric temperature for young stars observed spectroscopically (*White and Hillenbrand*, 2004; *Doppmann et al.*, 2005) in Taurus and ρ Ophiuchi that have both evolutionary diagnostics determined. Open symbols correspond to stars with detected photospheric features from which stellar properties can be extracted; filled symbols are too heavily veiled to extract these features. Class I stars are bolometrically cold ($T_{bol} < 650$ K) and have rising mid-infrared energy distributions ($\alpha > 0.0$). The new spectroscopic observations extend well into the Class I regime.

[T_{bol}, defined as the temperature of a blackbody with the same mean frequency as the observed SED (*Myers and Ladd*, 1993)] and infrared spectral slope [$\alpha = d\log(\lambda F_\lambda)/(d\log\lambda)$, typically determined over the wavelength interval 2–25 μm (*Lada and Wilking*, 1984; *Lada*, 1987)], plotted against one another. Class I stars are distinguished from Class II stars as having $\alpha > 0.0$ or $T_{bol} < 650$ K; their SEDs rise into the infrared. A subsample of "flat spectrum" or "transitional Class I/II" stars are often distinguished as those with $-0.3 < \alpha < 0.3$ or $650 < T_{bol} < 1000$ K. However, since these criteria are based on observations that typically do not spatially resolve the circumstellar structures, it is not clear that the observed SED differences truly correspond to distinct evolutionary stages. Line of sight orientation or unresolved companions, as examples, can significantly alter the observed SED.

Studies of the emergent SEDs at wavelengths ≥10 μm have provided important, albeit ambiguous, constraints on the circumstellar dust distributions for Class I stars. Although existing data are based on relatively low spatial resolution observations from IRAS and ISO, with the promise of the Spitzer Space Telescope (*Werner et al.*, 2004) currently being realized, a single generic representation of Class I and some I/II stars has been developed. Models incorporating infalling, rotating envelopes with mass-infall rates on the order of 10^{-6} M$_\odot$/yr predict SEDs that are consistent with observations (*Adams et al.*, 1987; *Kenyon et al.*, 1993a; *Whitney et al.*, 1997, 2003). However, only in a few cases are these mass-infall rates supported by kinematic measurements of spatially resolved envelope structures (e.g., *Gregerson et al.*, 1997). For some young stars whose SEDs can be explained by spherically symmetric dust distributions, it has been suggested that nearly edge-on flared disk models may also be able to reproduce the SEDs (e.g., *Chiang and Goldreich*, 1999; *Hogerheijde and Sandell*, 2000).

Whether the observable diagnostics trace distinct evolutionary states bears directly on the issue of whether Class I stars are in fact younger than Class II stars, as is often assumed, or whether they are simply less environmentally developed; they could be T Tauri age stars still (or perhaps just currently) embedded within circumstellar material. What is needed is an understanding of the *stellar* properties of these systems. To date, stellar properties such as mass and age have been derived for Class I stars predicated on the assumption that they are in the main stage of infall (see, e.g., *Evans*, 1999), that this material is accumulating in a circumstellar disk, and then accretes onto the star at a rate sufficient to match the bolometric luminosity (defined as the luminosity of a star's entire energy distribution). Given an assumed mass, the age of the star is then simply the mass divided by the mass-infall rate (0.6 M$_\odot$/3 × 10^{-6} M$_\odot$/yr = 2 × 10^5 yr). Buttressing the argument for the extreme youth of Class I stars is the relative number of Class I, II, and III stars in clouds such as Tau-Aur. As discussed by *Benson and Myers* (1989) and *Kenyon et al.* (1990), the relative ages of stars in different stages can be inferred from their relative numbers, assuming a constant star formation rate. For Taurus-Auriga, there are 10× fewer Class I stars than Class II and Class III stars, implying the Class I phase must be 10× shorter, leading to age estimates of ~2 × 10^5 yr assuming typical ages of ~2 × 10^6 yr for the Class II/III population, as inferred from the Hertzsprung-Russell diagram (e.g., *Kenyon and Hartmann,* 1995).

A more robust confirmation that Class I stars are bona-fide protostars would be an unambiguous demonstration that they are acquiring mass at a much higher rate than Class II stars. Although the mass-infall rates inferred (indirectly, in most cases) for Class I stars are roughly 2 orders of magnitude larger than the disk-accretion rates determined for Class II stars [~10^{-8} M$_\odot$/yr (e.g., *Gullbring et al.,* 1998)], it has not yet been shown that the infalling material is channeled through the disk and onto the star at this same prodigious rate. Under the assumption that these two rates are the same leads to a historical difficulty with the Class I paradigm — the so-called "luminosity problem." As first pointed out by *Kenyon et al.* (1990), if the material infalling from the envelope is channeled through the disk via steady state accretion and onto the star, the accretion luminosity would be dominant at roughly 10× the luminosity emitted from the photosphere. However, Class I stars, at least in Tau-Aur, do not have integrated luminosities substantially different from those of neighboring T Tauri stars. Several reconciliations have been proposed, including disk accretion that is not steady-state, very-low-mass (i.e., substellar) central masses, or simply erroneously large mass-infall rates. Direct measurement, rather than indirect inference, of both the stellar and the accretion luminosities of Class I stars is needed to distinguish between these.

The most straightforward way to unambiguously determine the stellar and accretion properties of young stars at any age is to observe their spectra at wavelengths shorter than ~3 μm where the peak flux from the stellar photosphere is emitted. While this has been possible for over five decades for Class II stars, the faintness of Class I stars at optical and near-infrared wavelengths have made it difficult to obtain high-resolution, high-signal-to-noise observations necessary for such measurements. The development of sensitive spectrographs mounted on moderate- to large-aperture telescopes now allow direct observations of Class I and I/II photospheres via light scattered through circumstellar envelopes. These observational windows provide an opportunity to study Class I stars with the same tools and techniques developed for the study of Class II stars.

2. PHOTOSPHERES AND ACCRETION

Detailed spectroscopic studies of young stars less embedded than protostars (e.g., T Tauri stars) have provided much of the observational basis for theories of how stars are assembled and how they interact with their environment. The spectrum of the canonical young, accreting, low-mass star consists of a late-type photosphere with strong emission lines and excess continuum emission (i.e., veiling) at optical and infrared wavelengths. At optical wavelengths, measurement of this excess emission, which is attributed to high-temperature regions generated in the accretion flow, provides a direct estimate of the mass-accretion rate and constrains physical conditions of accretion shock models (see chapter by Bouvier et al.). At infrared wavelengths, measurement of the excess thermal emission from warm circumstellar dust reveals structural information of the inner accretion disk (*Najita,* 2004; *Muzerolle et al.,* 2004; *Johns-Krull et al.,* 2003). Additionally, the strengths and profile shapes of permitted emission lines delimit how circumstellar material is channeled onto the stellar surface (*Calvet and Hartmann,* 1992; *Muzerolle et al.,* 1998b, 2001), while density-sensitive forbidden emission lines trace how and how much mass is lost in powerful stellar jets (e.g., *Hartigan et al.,* 1995). Perhaps most importantly, extraction of the underlying photospheric features permit the determination of precise stellar properties [T$_{eff}$, log g, (Fe/H)], which can be compared to evolutionary models to determine stellar masses and ages. Doppler broadening of these features also provides a measure of the stellar rotation rate (v sin i), which is important for tracing the evolution of angular momentum. Spectroscopic observations at visible and near-infrared wavelengths are two powerful tools for studying a young star's photospheric properties and its circumstellar accretion, if realizable.

2.1. Visible Light

Although observations at visible or optical wavelengths (≲1 μm) are especially challenging for highly extincted stars, there are nevertheless two motivations for pursuing this. First, visible light is dominated by emission from both the photosphere and high-temperature accretion shocks; it therefore offers the most direct view of stellar properties and accretion luminosity. Second, for small dust grains (≲1 μm), visible light scatters more efficiently than infrared light. Thus, even if the direct line-of-sight extinction is too large for an embedded star to be observed directly, the cavities commonly seen in the envelopes surrounding Class I stars (e.g., *Padgett et al.,* 1999) may permit observations of the photosphere and inner accretion processes through scattered light. This is only feasible in low-column-density star-forming environments like Taurus-Auriga, where the young stars are not deeply embedded within the large-scale molecular cloud.

Recognition of faint but nevertheless detectable emission from these embedded stars inspired several low-resolution spectroscopic studies with the aim of putting the first solid constraints on the stellar and accretion properties of suspected protostars. This work began even prior to the now established "class" classification scheme; some of the first embedded young stars were identified by the strong stellar jets that they power (e.g., *Cohen and Schwartz,* 1983; *Graham,* 1991). These stars typically have nearly featureless continua with strong emission lines superimposed. *Mundt et al.* (1985) obtained an optical spectra of the Class I star L1551 IRS 5 in Taurus-Auriga and identified the star as a G or K spectral type (but see *Osorio et al.,* 2003); the emission line features show P Cygni-like profiles suggestive of a strong outflowing wind. More recently, *Kenyon et al.* (1998) reported spectroscopic observations for 10 of the Class I stars in Taurus-Auriga, detecting M-spectral-type features (i.e., TiO bands) in several and strong emission line features in all. These initial spectroscopic studies suggested that at least some Class I stars resemble their more evolved T Tauri star counterparts (Class II stars), but with heavily veiled spectra and strong emission lines. Unfortunately, the limited numbers of stars with revealed spectroscopic features, due in part to the low spectral resolution of the observations, precluded accurate determination of stellar properties and specific mass accretion and mass-outflow rates for unbiased comparisons with the more optically revealed T Tauri stars.

2.2. Near-Infrared Light

The development of infrared detector technology during the 1980s and 1990s has provided another valuable tool for the study of protostars. Since many stars form in high extinction clouds that block nearly all visible light (e.g., ρ Ophiuchi, Serpens), they are not amenable to study at visible wavelengths. It has been recognized for some time that late-type stellar photospheres exhibit a number of atomic and molecular features in the 2–2.4-μm wavelength region (K band) that are diagnostic of effective temperatures and surface gravities (*Kleinmann and Hall,* 1986; *Wallace and Hinkle,* 1996), and can be used to measure stellar projected rotational and radial velocities. Interstellar dust is also relatively transparent in this wavelength region, $A_K \simeq 0.1 A_V$ (in magnitudes), permitting spectroscopic observations of even highly extinguished young stars in nearby dark clouds to be obtained. However, the near-infrared spectra of embedded young stars are frequently complicated by the presence of thermal emission from warm dust grains in their inner circumstellar disks or inner envelope regions. This excess circumstellar emission can be several times greater than the photospheric flux of an embedded young star in the K-band wavelength region, causing an increased continuum level that veils photospheric features.

Initial near-infrared observations at low resolution found that the CO absorption features at 2.3 μm could be identified less often for Class I stars than Class II stars (*Casali and Matthews,* 1992; *Casali and Eiroa,* 1996). This was interpreted as Class I stars having larger near-infrared excess emission than Class II stars, possibly because of more luminous circumstellar disks caused by larger mass-accretion rates or alternatively, envelope emission (*Greene and Lada,* 1996; *Calvet et al.,* 1997). *Muzerolle et al.* (1998c) demonstrated that the Brγ (2.166 μm) luminosity correlates well with the total accretion luminosity, and used this relation to measure the first accretion luminosities for Class I

stars. The determined accretion luminosities were only a small fraction (~one-tenth) of the bolometric luminosity; assuming a typical T Tauri star mass and a radius, these accretion luminosities correspond to mass-accretion rates that are overall similar to those of T Tauri stars (~10^{-8} $M_\odot$/ yr). With regard to stellar features, *Greene and Lada* (1996) and *Luhman and Rieke* (1999) showed that at least ~25% of Class I and flat-spectrum stars exhibited temperature-sensitive photospheric features, suggesting that stellar properties could potentially be determined directly (see also *Ishii et al.,* 2004). As with early optical observations, however, low spectral resolution and large infrared excesses prevented accurate extraction of these properties. More recently, *Nisini et al.* (2005) presented spectra of three Class I stars in R CrA at moderate resolution (R ~ 9000), sufficient to measure the amount of continuum excess and assign spectral types (i.e., temperatures), but (in this case) insufficient to measure radial and rotational velocities.

2.3. The Promise of High-Resolution Spectra

Fortunately, high-dispersion spectrographs on large-aperture telescopes have allowed observers to begin studying the stellar and accretion properties of embedded low-mass protostars in detail, at both optical and near-infrared wavelengths. Initial measurements demonstrated that the key to spectroscopically resolving faint photospheric features, given the large continuum excess emission, is high-signal-to-noise, high-dispersion spectroscopy (*Greene and Lada,* 1997, 2000, 2002; *Doppmann et al.,* 2003). This pioneering work showed that fundamental photospheric diagnostics (temperatures, surface gravities, rotational velocities) and circumstellar features (continuum excesses, emission line luminosities) could be measured nearly as precisely for Class I stars as for Class II stars. The small number of Class I stars "revealed" however, inhibited statistically meaningful comparisons with T Tauri stars to search for evolutionary differences.

Very recently the situation changed dramatically with two large surveys of embedded stars. *White and Hillenbrand* (2004; hereafter *WH04*) conducted a high-resolution (R ≃ 34,000) optical spectroscopic study of 36 "environmentally young" stars in Taurus-Auriga (Tau-Aur). *WH04* classify stars as "environmentally young" if they are either Class I stars or power a Herbig-Haro flow. Their sample consisted of 15 Class I stars and 21 Class II stars; they detected photospheric features in 11 of the Class I stars and all the Class II stars. Figure 2 shows three optical spectra from this survey. *Doppmann et al.* (2005; hereafter *D05*) conducted a complementary high-resolution (R ≃ 18,000) K-band study of 52 Class I and flat-SED stars, selected from 5 nearby star-forming regions — Taurus-Auriga (Tau-Aur), ρ Ophiuchi (ρ Oph), Serpens, Perseus, and R Corona Australi (R CrA). Forty-one of the 52 stars were found to have photospheric absorption features from which stellar properties and excess emission could be measured. Figure 3 shows three near-infrared spectra from this survey.

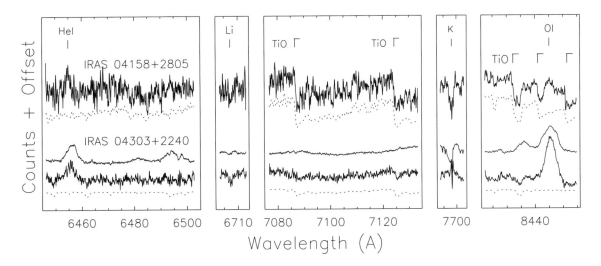

Fig. 2. Portions of the Keck/HIRES spectra from *White and Hillenbrand* (2004). IRAS 04158 + 2805 (α = +0.71) has a very cool spectral type (~M6) and possibly a substellar mass. The two epochs of IRAS 04303 + 2240 (α = –0.35) show dramatic variations in the veiling and the inferred mass-accretion rate; the heavily veiled spectrum is less noisy because the star was also much brighter. Spectra of the best-fit dwarf stars, veiled and rotationally broadened, are shown as dotted lines.

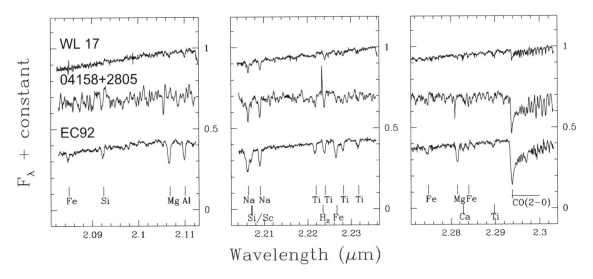

Fig. 3. High-resolution near-infrared spectra of the embedded protostars from *Doppmann et al.* (2005). WL 17 (α = +0.42) is heavily veiled, IRAS 04158 + 2805 (α = +0.71) has a very cool spectral type (~M6) and possibly a substellar mass, and EW 92 is moderately rapidly rotating (v sin i = 47 km/s).

3. SPECTROSCOPIC PROPERTIES OF PROTOSTARS REVEALED

In this section, we present a combined assessment of the stellar and accretion properties of Class I and transitional Class I/II stars as inferred in the *WH04* and *D05* studies, including other results when applicable. We note that the two primary studies were able to determine astrophysical properties for six of the same stars, permitting a direct comparison of the two techniques. Agreement is good for five of the six overlapping stars (in v sin i and effective tempera-

ture); the one discrepancy occurs in a heavily veiled, very low signal-to-noise (optical) observation; the infrared properties are adopted in this case.

To identify how stellar and circumstellar properties change as a star evolves through the proposed evolutionary scheme, we present the inferred properties as a function of the evolutionary diagnostic α, the infrared spectral index. We adopt this diagnostic simply because it is available for most of the stars observed. In addition to the primary samples of *WH04* and *D05*, we include a sample of accreting Class II stars from Tau-Aur (as assembled in *WH04*) and ρ Ophiuchi (assembled in *Greene and Lada*, 1997, and *Doppmann et al.*, 2003), whose properties have been determined from high-dispersion spectra as well. When available, we selected values of α calculated from observations at 2 and 25 μm; when such measurements are not available, we use α values calculated over a smaller wavelength interval (typically based on ISO observations extending to 14 μm). Specifically, stars in Tau-Aur, NGC 1333, and R CrA have 2–25-μm α values determined from IRAS observations by *Kenyon and Hartmann* (1995), *Ladd et al.* (1993), and *Wilking et al.* (1992), respectively. Serpens and ρ Oph stars have 2–14-μm α values from the work of *Kaas et al.* (2004) and *Bontemps et al.* (2001). As emphasized in the introduction, however, all evolutionary diagnostics are subject to significant biases, which can mask subtle trends. Thus, we will primarily make ensemble comparisons between stars classified as Class I stars (α > 0.0) and stars classified as Class II stars.

3.1. Stellar Masses

Historically, the masses of embedded young stars have been poorly determined by observations, since in most cases, the only measurable property was the bolometric luminosity from the (often poorly determined) SED. IRAS surveys of the Tau-Aur, ρ Oph, R CrA, and Chamaeleon I dark clouds revealed populations of Class I stars in each region with bolometric luminosities spanning from below 0.1 $L_\odot$ to approximately 50 $L_\odot$, with a median value near 1 $L_\odot$ (*Kenyon et al.*, 1990; *Wilking et al.*, 1989, 1992; *Prusti et al.*, 1992). Converting these luminosities to mass estimates requires an assumed mass-luminosity relation, which strongly depends upon age, and an assumed accretion luminosity. If the embedded Class I stars in these regions have luminosities dominated by accretion, then their masses can be approximated by applying the spherical accretion luminosity relation $L_{bol} = L_{acc} = GM_*\dot{M}/R_*$. Adopting an infall rate of 2×10^{-6} $M_\odot$/yr and an protostellar mass-radius relation (e.g., *Adams et al.*, 1987; *Hartmann*, 1998) leads to a mass of 0.5 $M_\odot$ (at a radius of 3 $R_\odot$) for a 10 $L_\odot$ star. Thus, only the most luminous Class I stars would have inferred masses consistent with those of T Tauri stars [0.1– a few $M_\odot$ (*Kenyon and Hartmann*, 1995; *Luhman and Rieke*, 1999)]; the majority would have masses ≲0.1 $M_\odot$. Although there remain considerable uncertainties in the calculated bolometric luminosities and the prescription for accretion for Class I stars, the emerging census suggests a "luminosity problem" as described in section 1; the typical Class I star is underluminous relative to what is expected for a canonical T Tauri-sized star (in mass and radius) accreting at the predicted envelope infall rates. The luminosity problem is most severe in the Tau-Aur star-forming region (*Kenyon et al.*, 1990); there is tentative evidence for a regional dependence upon the distribution of bolometric luminosities of Class I stars. One proposed solution to the luminosity problem is that Class I stars are in fact much lower in mass than T Tauri stars (i.e., brown dwarfs), either because they are forming from

less massive cores/envelopes or because they have yet accreted only a small fraction of their final mass. These possible solutions introduce yet additional problems, however. If almost all Class I stars are producing brown dwarfs, then "star" formation in most regions must have already ceased, implying an unexpected mass dependent formation timescale. Alternatively, the hypothesis that Class I stars have accreted only a small fraction of their final mass is inconsistent with their relatively low envelope mas-ses (~0.1 $M_\odot$), estimated from millimeter wavelength observations (e.g., *Motte and André*, 2001). Accurately determined stellar mass estimates are needed to test this proposed yet problematic solution to the luminosity problem.

One direct way to estimate the mass of a young star is to observationally determine its stellar effective temperature and luminosity and then compare them with the predictions of pre-main-sequence (PMS) evolutionary models. The recent optical and near-IR spectroscopic studies of *WH04* and *D05* have been able to achieve this for the first time for several dozen embedded young stars in nearby dark clouds. Since low-mass, fully convective stars primarily evolve in luminosity while young (e.g., *Baraffe et al.*, 1998), temperature is especially important in determining a young star's mass. In most cases, the temperature estimates for the Class I stars are as precisely determined as those for T Tauri stars (~150 K), which translates into similar uncertainties in the inferred stellar masses (a few tens of percent), but large systematic uncertainties remain [e.g., temperature scale (see chapter by Mathieu et al.); effects of accretion (*Siess et al.*, 1999; *Tout et al.*, 1999)]. The uncertainties in the stellar luminosities of embedded stars are, on the other hand, typically much larger than those for T Tauri stars. *WH04* estimated stellar luminosities by performing a bolometric correction from near-infrared J-band (λ ≈ 1.25 μm) photometric data, which is expected to be least contaminated by circumstellar excesses (see, however, *Cieza et al.*, 2005); extinctions were determined by comparing the observed J–H colors to that expected for a dwarf-like photosphere. *D05* estimated luminosities by performing a bolometric correction to near-infrared K-band (λ ≈ 2.3 μm) photometric data, after accounting for K-band veilings determined from their spectra; extinctions were determined by comparing the H–K colors to a typical value for a T Tauri star. However, much of the flux detected from embedded young stars at visible and near-infrared wavelengths has been scattered from their circumstellar environments. The physical nature of circumstellar dust grains (sizes, shape, composition), distribution of material in disks and envelopes, and system inclination all change how the photospheric flux is scattered into our line of sight, changing both a star's brightness and color. Comparisons of luminosities determined via different techniques differ by factors of 2–3, and we suggest this as a typical uncertainty. In addition to this, stars with edge-on disk orientations often have calculated luminosities that can be low by factors of 10 to 100; the preferential short-wavelength scattering leads to artificially low extinction estimates. With these uncertainties and possible systematic errors in mind, all the Class I and flat-spectrum stars ($\alpha > -0.3$) in Tau-Aur, ρ Oph, and Serpens observed in the *D05* and the *WH04* surveys are shown on a Hertzprung-Russell diagram in Fig. 4. The PMS evolutionary models of *Baraffe et al.* (1998) are shown for comparison.

Several points can be extracted from Fig. 4 regarding the masses of Class I and flat-spectrum stars. First, the stars generally span a similar range of effective temperatures and stellar luminosities in all regions, although there is a slightly narrower range of temperatures in ρ Oph. Second, based on comparisons with the *Baraffe et al.* (1998) PMS evolutionary models, the combined distribution of masses span from substellar to several solar masses. This is similar to the distributions of Class II star masses in Tau-Aur and ρ Oph (*Kenyon and Hartmann*, 1995; *Luhman and Rieke*, 1999), while little has been reported on the masses of Class II stars in the Serpens clouds. Other studies corroborate these findings. *Nisini et al.* (2005) determine masses spanning from 0.3 to 1.2 $M_\odot$ for three Class I stars in R CrA, based on spectral types determined from moderate-resolution spectra. *Brown and Chandler* (1999) determine masses of 0.2–0.7 $M_\odot$ for two Class I stars in Tau-Aur based on disk kinematics under the assumption of Keplerian rotation.

Apparently the majority of Class I stars have masses that are similar to those of T Tauri stars. The present spectroscopic data does not support the notion that the majority of the low-luminosity ($L < 1\ L_\odot$) Class I stars are substellar. This proposed resolution to the luminosity problem can now be excluded. Nevertheless, it is particularly notable that

there are several Class I stars that have masses that are close to or below the substellar boundary [0.075 $M_\odot$ (*Baraffe et al.*, 1998)]. *WH04* identified three Class I stars with spectral types of M5.5 or M6, and considered them candidate Class I brown dwarfs. Two of these stars, IRAS 04158 + 2805 and IRAS 04489 + 3042, were also observed and analyzed by *D05*. The *D05* infrared spectra also indicate a M6 spectral type for IRAS 04158 + 2805, but yield a slightly earlier M4 spectral type for IRAS 04489 + 3042. It is encouraging that both optical and infrared spectra are yielding very similar results for these stars, and strengthens the case for the existence of a Class I object at or below the substellar boundary.

On the other hand, some Class I stars that had been previously interpreted as accreting brown dwarfs from photometric data are now revealed to be low-mass stars instead. For example, *Young et al.* (2003) interpreted the very complete photometric data on the low-luminosity Class I star IRAS 04385 + 2250 as evidence that it is a brown dwarf of only 0.01 $M_\odot$ (~10 M_{Jup}) by assuming an accretion rate of $2 \times 10^{-6}\ M_\odot$/yr. As pointed out by *Kenyon et al.* (1990, 1994), such an assumption implies that all low-luminosity ($L < 1\ L_\odot$) Class I stars are actually substellar. However, *WH04* find that IRAS 04385 + 2250 (also known as Haro 6-33) has a spectral type of M0, placing it squarely in the regime of low-mass stars and not brown dwarfs.

3.2. Stellar Ages

In addition to providing stellar masses, comparisons of observationally determined stellar properties with the predictions of evolutionary models can provide useful age estimates. Comparisons of Class II stars in nearby dark clouds consistently yield ages spanning from less than 1 to a few million years (e.g., see *Kenyon and Hartmann*, 1995; *Luhman and Rieke*, 1999). If Class I stars are really the precursors to Class II stars, as their less-evolved circumstellar environments suggest, then they should have younger ages.

As emphasized above, the calculated luminosities of Class I stars are especially uncertain given the large extinctions, uncertain scattered light contributions, and continuum excesses; they are also occasionally subject to large systematic errors caused by orientation effects. These large uncertainties bear directly upon how well the stellar ages can be determined, since low-mass stars evolve primarily along vertical evolutionary tracks at ages less than a few × 10^7 yr. Nevertheless, comparisons of the observed luminosities and temperatures with the predictions of pre-main-sequence evolutionary models, as shown in Fig. 4, provide a large ensemble of ages estimates. The range of ages is broadest for Tau-Aur and narrowest for Serpens, although these regions have the largest and smallest samples measured, respectively. Several stars in Tau-Aur appear to have unrealistically old ages (below the main sequence), likely a consequence of the stellar luminosity being severely underestimated because of an edge-on disk orientation (see *WH04*). The absence of low-luminosity stars in ρ Oph and Serpens

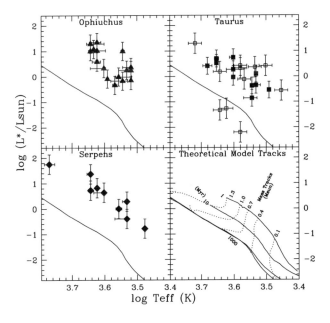

Fig. 4. Stellar luminosities and effective temperatures of Class I and flat-SED stars ($\alpha > -0.3$) in ρ Oph, Tau-Aur, and Serpens are shown on an H-R diagram. Filled symbols are from *D05* and unfilled symbols are from *WH04*. The evolutionary models of *Baraffe et al.* (1998) are also shown for comparison.

suggests they may be more difficult to identify in regions of high extinction.

Despite these possible regional differences and large uncertainties, calculating a median age in all three regions yields a consistent value of ~1 m.y. This is remarkably similar to the average age of Class II stars in ρ Oph and Tau-Aur; few are known in Serpens. This suggests that most Class I stars are not systematically younger than Class II stars. Unfortunately, the large uncertainties in the luminosity estimates of Class I stars, as well as current evolutionary models at early ages (*Baraffe et al.*, 2002), limits the robustness of this comparison at this time.

Finally, we note that a comparison of the inferred stellar luminosities with calculated bolometric luminosities for Class I stars suggests that in most cases studied here, the stellar luminosity is the dominant source of luminosity in the system ($L_{star}/L_{bol} > 0.5$). Most Class I stars with detected photospheric features do not have accretion dominated luminosities as had been initially proposed. We caution, however, that the observational biases in the sample studied here (revealed at <3 μm, with moderate veiling or less) prevent extrapolation of this finding to Class I stars in general. The most bolometrically luminous stars for which photospheric features are detected are IRS 43 ($L_{bol} = 7.2\ L_\odot$) and YLW 16A ($L_{bol} = 8.9\ L_\odot$) (see *D05*). Thus it is not yet known if the most luminous Class I stars ($L_{bol} > 10\ L_\odot$) have accretion-dominated luminosities or are simply more massive stars.

3.3. Stellar Rotation

Studies of stellar rotation at very young ages have revealed clues regarding the evolution of angular momentum from the epoch of formation to the zero age main sequence. Conservation of angular momentum during the collapse of a molecular core to form a low-mass star should lead to rotation velocities near breakup

$$v_{breakup} = \sqrt{(GM_{star}/R_{star})} \sim 200\ \text{km/s}$$

However, the small projected rotation velocities of Class II stars (v sin i ≲ 20 km/s) (*Bouvier et al.*, 1986, 1993; *Hartmann et al.*, 1986; *Stassun et al.*, 1999; *Rhode et al.*, 2001; *Rebull et al.*, 2002; see chapter by Herbst et al.) show that angular momentum must be extracted quickly, on timescales of <1–10 m.y. A number of theories have been proposed to rotationally "brake" young stars. One favored model involves magnetic linkage between the star and slowly rotating disk material at a distance of several stellar radii (*Königl*, 1991; *Collier Cameron and Campbell*, 1993; *Shu et al.*, 1994; *Armitage and Clarke*, 1996). Initial observational evidence supported this picture. T Tauri stars without disks were found to rotate somewhat more rapidly than stars with disks (*Edwards et al.*, 1993; *Bouvier et al.*, 1993, 1995), which was interpreted as evidence that disk presence keeps stars rotating at fixed angular velocity, while disk absence allows stars to conserve angular momentum and spin up as they contract toward their main-sequence radii.

Since then, the observational case for disk locking has become less clear-cut (e.g., *Stassun et al.*, 2001; but see *Rebull et al.*, 2004), while detailed theoretical and magnetohydrodynamical considerations suggest that disk locking in and of itself is unable to extract sufficient amounts of angular momentum (*Safier*, 1998). Strong stellar winds are one possible alternative (e.g., *Matt and Pudritz*, 2005).

The uncertainties in our understanding of *how* angular momentum is extracted from young stars provide motivation for determining *when* it is extracted, since knowing the appropriate timescale could help distinguish between proposed models. The rotation velocities of Class I stars as revealed by spectroscopic studies provide the earliest measurements of stellar angular momentum; Fig. 5 shows the distribution of v sin i values for Class I and Class II stars in Tau-Aur, ρ Oph, and R CrA vs. the evolutionary diagnostic α. The largest v sin i value observed for a Class I star is 77 km/s, while the remainder have v sin i ≤ 56 km/s. These values are only a few tenths of the typical breakup velocity. Comparing Class I (α > 0.0) to Class II (α < 0.0) stars, Class I stars have slightly higher rotation rates. Although the distributions of rotation rates are statistically different (*Covey et al.*, 2005), the difference in the mean is only a factor of 2. The distributions are less distinct for any particular region [e.g., Tau-Aur (*WH04*)], likely from smaller number statistics, although global properties of a region could lead to correlated biases (e.g., age). Although the distributions of Class II rotational velocities in some star-forming regions have been shown to be statistically differ-

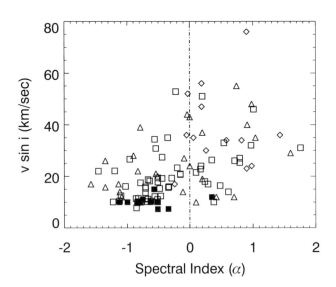

Fig. 5. Projected rotational velocity (v sin i) vs. spectral index. Triangles are stars in ρ Oph, squares are stars in Tau-Aur, and diamonds are stars in Serpens. v sin i measurements are shown as open symbols while upper limits are shown as filled symbols. The dashed vertical line separates Class I stars from Class II stars. Data originally presented in *WH04*, *Covey et al.* (2005), and references therein.

ent [e.g., Orion vs. Tau-Aur (*Clarke and Bouvier*, 2000; *White and Basri*, 2003)], the evidence for this at the Class I stage is still tentative (~2σ); *Covey et al.* (2005) found Tau-Aur to have the lowest mean observed rotation velocity for the three regions in their study (30.1 km/s vs. 31.1 km/s in ρ Oph and 36.8 km/s in Serpens). These comparisons are likewise limited by small number statistics. Overall, the observational evidence demonstrates that Class I stars are rotating somewhat more rapidly than Class II stars, but at rates that are well below breakup velocities. If Class I stars are indeed in the main phase of mass accretion (section 4.3), this implies that angular momentum is removed concurrently with this process.

3.4. Circumstellar Accretion

If Class I stars are to acquire the majority of their mass (e.g., 0.6 $M_\odot$) on a timescale of ~2 × 10^5 yr, they must have time-averaged mass-accretion rates that are ~3 × 10^{-6} $M_\odot$/yr, assuming a simple spherical infall model (e.g., *Hartmann*, 1998). For comparison, this mass-accretion rate is at least 2 orders of magnitude larger than what is typically observed for T Tauri stars (e.g., *Gullbring et al.*, 1998). The newly available high dispersion spectra of Class I stars permit measurements of the mass-accretion rate (from the disk onto the star) by two independent methods. The first of these comes from measurements of optical excess emission in high-dispersion optical spectra under the assumption that the liberated energy is gravitational potential energy (see chapter by Bouvier et al.). Unfortunately, there remain considerable uncertainties in measuring the total liberated energy, which typically requires a large bolometric correction from an optical measurement; the majority of the accretion luminosity is emitted at ultraviolet wavelengths. Additionally, estimating the potential energy requires estimates of stellar and inner disk properties, which have large uncertainties themselves. Nevertheless, by calculating mass-accretion rates for Class I stars following the same assumptions used for T Tauri stars, many of these systematic uncertainties can be removed, thereby permitting a more robust comparison if the same accretion mechanism applies.

WH04 have measured optical excess emission at 6500 Å for 11 Class I stars, and several borderline Class I/II stars, all within the Tau-Aur star-forming region. These measurements, along with a sample of excess measurements of accreting T Tauri stars from *Hartigan et al.* (1995; as compiled in *WH04*), are shown in Fig. 6. Similar to the T Tauri stars, the Class I stars have continuum excesses that range from not detected (<0.1) to several times the photosphere; in the general case, their optical emission is not dominated by accretion luminosity. Class I stars have veiling values that are only modestly larger in the mean (×1.3) than those of Class II stars. *WH04* proceed to convert these continuum excesses to mass-accretion rates, and find values of a few × 10^{-8} $M_\odot$/yr, which are again similar to those of Class II stars. Furthermore, by accounting for other components to

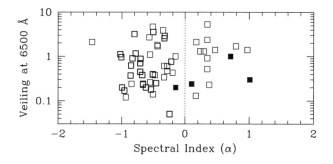

Fig. 6. Optical veiling vs. spectral index for stars in Tau-Aur. Measurements are from *WH04* and *Hartigan et al.* (1995). Veiling measurements are shown as open symbols while upper limits are shown as filled symbols. The dotted vertical line separates Class I stars from Class II stars.

the star's luminosity, they find that the accretion luminosity only accounts for ~25% of the bolometric luminosity, on average.

A second measure of the mass-accretion rate comes from emission-line luminosities. Emission-line studies, in combination with radiative transfer models of circumstellar accretion, suggest that many of the permitted lines originate in the infalling magnetospheric flow (*Hartmann et al.*, 1994; *Muzerolle et al.*, 1998a), and that the line strengths are proportional to the amount of infalling mass. *Muzerolle et al.* (1998b,c) demonstrated this to be true for the Ca II infrared triplet and Brγ by correlating these emission-line luminosities with mass-accretion rates determined from blue excess emission. As emphasized by the authors, accurate corrections for extinction and scattered light are critical for this. The near-infrared emission-line Brγ is of special interest in the study for Class I stars since the high extinction often inhibits observations at shorter wavelengths. Using the Brγ correlation, *Muzerolle et al.* (1998c) found that the Brγ luminosities of Class I stars, with assumed stellar properties, are similar to those of Class II stars. The implication is that they have similar mass-accretion rates.

Figure 7 shows a compilation of logarithmic Brγ luminosities from *Muzerolle et al.* (1998c) and *D05* [also includes measurements from *Luhman and Rieke* (1999); *Folha and Emerson* (1999); and *Doppmann et al.* (2003)], plotted vs. spectral index. As with optical excess emission, the Brγ luminosities of Class I stars span a similar, though slightly broader range than the Class II stars, but are larger in the average by a factor of a few; the distributions are different at approximately the 3σ level according to a K-S test. Much of the difference between the Class I and Class II stars appears to be driven by stars in the ρ Oph region, where the Brγ luminosities of Class I stars are systematically larger than those of Class II stars by a factor of ~5 in the mean; the distributions are different at the ~2σ level, or ≥3σ if the low Brγ luminosity (−4.58), α = 0.0 star GY21 is considered a Class II. Stars in Tau-Aur show no difference between the two classes (<1σ).

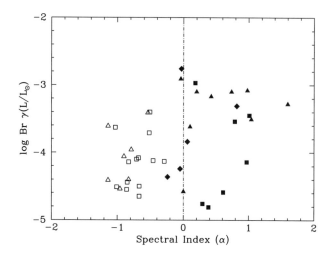

Fig. 7. Brγ luminosity vs. spectral index. Triangles are stars in ρ Oph, squares are stars in Tau-Aur, and diamonds are stars in Serpens. Filled symbols are from *D05* while open symbols are from *Muzerolle et al.* (1998c). The dashed vertical line separates Class I stars from Class II stars.

Conversion of these luminosities to mass-accretion rates leads to values for Class I stars that are similar to Class II stars, and corroborates the initial study of *Muzerolle et al.* (1998c). The largest mass-accretion rates are ~10^{-7} M$_\odot$/yr, and many of these are in the ρ Oph star-forming region. The larger mean accretion luminosities in ρ Oph is consistent with its larger mean near-infrared excess for Class I stars relative to Class II stars ($<r_K> = 2.2$ vs. 0.94), compared with other regions. Overall it appears that the mass-accretion rate during the majority of the Class I phase is similar to that of T Tauri stars, and 1–2 orders of magnitude less than the envelope infall rates inferred from SED modeling (few × 10^{-6} M$_\odot$/yr). We note that there is tentative evidence that the mass-accretion rate is extremely time variable during the embedded phase. As one example, the borderline Class I/II star IRAS 04303 + 2240 changed its mass-accretion rate dramatically (>4×) during two observational epochs (Fig. 2). Little observational work has been done to characterize the amplitudes or timescales of candidate protostar variability.

3.5. Jet Emission

Optically thin forbidden emission lines are believed to originate in an outflowing jet or wind. Their intensity is expected to be directly proportional to the amount of material being funneled along the jet, as viewed through the slit of the spectrograph. The luminosity of these emission lines can therefore be used to estimate the mass-outflow rate in a young stellar jet (see chapter by Bally et al.). Since jets are believed to be powered by circumstellar accretion, the mass-outflow rate should correlate with the mass-accretion rate.

Equivalent width measurements of the forbidden line [S II] 6731 Å for stars in Tau-Aur vs. spectral index are shown in Fig. 8. Measurements are from *WH04* and *Harti-*

gan et al. (1995; as compiled in *WH04*). Unlike the optical excess and Brγ luminosity accretion diagnostics, which are only slightly enhanced among Class I stars relative to Class II stars, Class I stars systematically have larger [S II] equivalent widths by roughly a factor of 20 in the mean. The implication is that Class I stars power much more energetic outflows than Class II stars. *Kenyon et al.* (1998) found similar results based on some of the same stars presented here.

However, as emphasized by *WH04*, many of the Class I stars they observed show signatures of having an edge-on disk orientation. In such a case, the emission-line region may be more directly observable than the partially embedded central star is. The preferentially attenuated continuum flux will consequently produce artificially large equivalent width values and biased mass-outflow rates. Without accurate geometric information for these stars, however, it is difficult to tell the significance of this bias in the average case. If these larger forbidden line equivalent widths indeed correspond to larger mass-outflow rates, perhaps it is because their spatially extended location make them a better tracer of the time-averaged mass-outflow rate. As is considered below, the mass-accretion rate (and corresponding mass-outflow rate) for Class I stars could be T Tauri-like for the majority of the time, with occasional large outbursts. In such a case, Class I stars would then have larger time-averaged mass-accretion and mass-outflow rates. The alternative to this, in the absence of any significant continuum attenuation bias, is that the ratio of mass-loss to mass-accretion rate is dramatically different between Class I and Class II stars (>10×), possibly suggesting a different accretion mechanism.

4. IMPLICATIONS FOR CLASSIFICATION SCHEMES AND FORMATION THEORY

A broad comparison of the properties derived for Class I stars and Class II stars from spectroscopy reveals some surprising similarities and differences. Class I stars appear to occupy the same ranges of effective temperature and

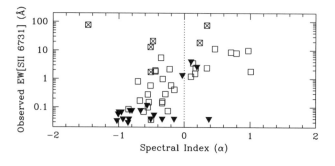

Fig. 8. Equivalent width measurements of [S II] 6731 Å vs. spectral index for stars in Tau-Aur. Measurements are from *WH04* and *Hartigan et al.* (1995). Detections are shown as open squares while upper limits are shown as filled triangles; stars with an x have a known or suspected edge-on orientation. The dashed vertical line separates Class I stars from Class II stars.

(photospheric) luminosity as Class II stars; applying our current theoretical understanding of PMS evolution to these observations implies that Class I stars have similar stellar masses as Class II stars. If Class I stars are indeed precursors to T Tauri stars, this means that by the Class I stage a young star has accreted the majority of its final stellar mass. Although the inferred stellar luminosities of Class I stars are, on average, similar to those of Class II stars, implying similar ages, the large uncertainties and systematic biases in these estimates prevent strict age comparisons at this time. Class I stars appear to be more rapidly rotating, on average, than Class II stars, although there are many Class I stars with low projected rotational velocities. Spectroscopic indicators of mass accretion, such as optical veiling and Brγ luminosity, do appear slightly elevated in Class I stars relative to Class II stars, but still well below predicted mass-infall rates ($\sim 10^{-6}$ M$_\odot$/yr). Although Class I stars have larger forbidden line emission strengths, implying larger mass-outflow rates, there is a yet unaccounted for continuum attenuation bias in these measurements. Here we use the combined results to investigate possible regional differences among Class I stars, to assess whether Class I stars are being properly classified, and to improve our understanding of how mass is acquired during the main phase of mass accretion.

4.1. Are There Regional Differences Among Class I Stars?

Initial studies of Class I stars suggested that their properties may differ in different regions. *Kenyon et al.* (1990) noted that although the Tau-Aur and ρ Oph clouds both contain similar numbers of Class I stars, ρ Oph contains many more with $L_{bol} > 10$ L$_\odot$. If stars in both regions have similar stellar masses, then this luminosity difference should translate into ρ Oph stars having mass-accretion rates 3–10× higher than those in Tau-Aur (and correspondingly larger mass-infall rates, if the accretion is steady-state). This is expected according to classical star-formation theory (*Shu*, 1977), which predicts that the infall rate should scale as the cube of the isothermal sound speed. The warmer gas in ρ Oph, relative to Tau-Aur (*Myers and Benson*, 1983), should consequently yield large mass-infall rates, and larger time-averaged mass-accretion rates. However, more recent work suggests that cloud turbulence may primarily set the initial infall rates (e.g., *Mac Low and Klessen*, 2004), and possibly even the initial mass function (e.g., *Goodwin et al.*, 2004), and the resulting binary fraction and rotational distribution (*Jappsen and Klessen*, 2004). Searching for possible differences in the stellar and accretion properties of stars produced in regions with different global properties (temperature, turbulence, density) can therefore help distinguish between proposed scenarios for mass assembly and early evolution.

The distributions of effective temperatures shown in Fig. 4 indicate that there are no significant differences in the masses of Class I stars in either the Tau-Aur, ρ Oph, or Serpens star-forming regions ($<1\sigma$, according to K-S tests). While the stellar luminosities and ages are also similar

among the three regions (given their large uncertainties), Serpens is somewhat distinct in that all its Class I and flat-SED members appear coeval at an age younger than 1 m.y., as would be expected for bona fide protostars. The larger scatter in stellar luminosity in ρ Oph and Tau-Aur is not well understood, although some apparently low-luminosity stars are a consequence of their edge-on disk orientation (section 3.2).

The analysis of accretion diagnostics in section 3.4 partially supports a scenario in which Class I stars in ρ Oph are accreting at higher rates than those in Tau-Aur. The Brγ luminosities of Class I stars in ρ Oph are systematically larger than those in Tau-Aur, implying larger mass-accretion rates. This is also supported, although less directly, with the larger near-IR veiling and higher bolometric luminosities of Class I stars in ρ Oph relative to Tau-Aur (*D05*). In a case study of the luminous ($L_{bol} = 10$ L$_\odot$) protostar YLW 15, *Greene and Lada* (2002) determine that 70% of the star's luminosity is due to mass accretion and infer a rate of 2×10^{-6} M$_\odot$/yr. At least in this one case, the disk-accretion rate appears consistent with the mass-infall rate inferred from envelope models (although these infall rates are derived primarily from Class I stars in Tau-Aur, because of less confusion with cloud material in that region). Given the small number and broad range of Brγ luminosities for Class I stars in Serpens, this distribution is consistent with the distributions of either ρ Oph or Tau-Aur Class I stars.

Finally, the present data do not reveal any notable differences in the distributions of rotation velocities, or angular momenta, of embedded protostars in different regions. *Covey et al.* (2005) found that Class I and flat-spectrum stars in Serpens had a somewhat larger mean v sin i rotation velocity than those in Tau-Aur or ρ Oph, but this difference is not statistically significant ($<2\sigma$). We caution that any orientation bias present in the samples studied (e.g., edge-on disk systems), will also bias the distribution of projected rotational velocities.

4.2. Are Class I Stars Properly Classified?

In the traditional classification scheme, Class I stars are true protostars — stellar embryos surrounded by an infalling envelope — while Class II stars are pre-main-sequence stars surrounded by circumstellar disks only. Here we consider the ability of popular evolutionary diagnostics to unambiguously distinguish between these two classes. Radiative transfer models of still-forming stars find that the SED shape typically used to distinguish Class I and Class II stars (as parameterized by T_{bol} and α) has an important dependence on the orientation of the disk and envelope relative to the observer's line of sight (*Kenyon et al.*, 1993a,b; *Yorke et al.*, 1993; *Sonnhalter et al.*, 1995; *Whitney et al.*, 2003, 2004). As an example, the models of *Whitney et al.* (2003) show that mid-latitude (i ~ 40°) Class I stars have optical, near-, and mid-infrared characteristics similar to those of more edge-on disk (i ~ 75°) Class II stars. The effects of edge-on disk orientation are most severe for evolutionary diagnostics determined in the near- and mid-infrared such as the 2–25-μm spectral index. Bolometric temperatures are also biased, but less so, while diagnostics based at much longer wavelengths, such as the ratio of submillimeter to bolometric luminosity (*André et al.*, 1993), are the least affected. Unfortunately, longer-wavelength SEDs are not yet available for Class I stars in many star-forming regions. Consequently, we conclude that current samples of Class I stars defined by either spectral index or bolometric temperature are contaminated with at least a few edge-on disk Class II stars.

Other observable characteristics, however, can be helpful in identifying Class II stars that have been mistakenly classified as Class I stars due to orientation effects. The radiative transfer models of *Whitney et al.* (2003) show that edge-on Class II stars are nearly 5× fainter than "true" Class I stars with the same value of α, suggesting that misclassified edge-on systems should appear significantly lower in an H-R diagram. *WH04* identified several likely disk edge-on systems in their sample of optically revealed Class I stars in Tau-Aur, several of which (but not all) appear underluminous relative to other cluster members. Unfortunately, the large luminosity spread of Class I stars inhibit identifying edge-on disk systems based on this criterion alone, unless the system is almost precisely edge-on (e.g., HH 30). Column-density sensitive spectral features [e.g., Si at 9.7 μm (*Kessler-Silacci et al.*, 2005)] or high-spatial-resolution imaging may provide less ambiguous orientation information.

The presence of spatially extended envelope material, as determined from image morphology at infrared and millimeter wavelengths, has been proposed as a more direct way to constrain the evolutionary class. Such features are only expected during the main accretion phase. Based on criteria put forth by *Motte and André* (2001), only 58% (15/26) of the Class I stars in Tau-Aur are true protostars. The remaining 42% (11 stars) have envelope masses ≤0.1 $M_\odot$ and are spatially unresolved at 1.3-mm wavelengths [referred to as "unresolved Class I sources" in *Motte and André* (2001)]. *Motte and André* (2001) suggest that these stars are more likely transitional Class I/II stars or highly reddened Class II stars (e.g., edge-on disk systems). The complementary near-infrared morphology survey by *Park and Kenyon* (2002) supports the claim that these stars are not bona fide Class I stars. However, we note that the morphological criteria used in these studies do not account for the luminosity and mass of the central star. For example, IRAS 04158 + 2805 may appear more evolved and point-like because it is a lower-luminosity Class I brown dwarf with a smaller disk and envelope.

André and Montmerle (1994) present a similar morphological study based on 1.3-mm continuum observations of Class I and Class II stars in the ρ Oph star-forming region. They found that Class I and Class II stars, as classified by the 2.2–10-μm spectral index, have similar 1.3-mm flux densities. Class I stars, however, were more often spatially extended, consistent with a significant envelope component, although of relatively low mass (≤0.1 $M_\odot$). Thus, it appears that a much smaller fraction of Class I stars in ρ Oph, rela-

tive to Tau-Aur, are candidate misclassified Class II stars. Nevertheless, their low envelope masses imply that they have already acquired the majority of their stellar mass (discussed below), like Class II stars. Complementary comparisons of Class I and Class II stars in the Serpens and R CrA starforming region have not yet been carried out.

Based on this mostly indirect evidence, we conclude that between one-third and one-half of the Class I stars in Tau-Aur are candidate misclassified Class II stars; the emission line profiles and image morphology suggests that in some cases the misclassification is caused by a nearly edge-on orientation. There is less evidence for significant misclassification in other regions (e.g., ρ Oph).

4.3. Are Class I Stars in the Main Accretion Phase?

Although the absolute values of the circumstellar diskaccretion rates have large systematic uncertainties, the rates inferred for Class I stars and Class II stars, under the same assumptions, are similar. However, these values are typically 1–2 orders of magnitude less than both the envelope infall rates inferred from SED modeling of Class I stars (e.g., few × 10^{-8} $M_\odot$/yr vs. few × 10^{-6} $M_\odot$/yr) and the time-averaged accretion rate necessary to assemble a solar mass star in a few × 10^5 yr. Here we explore possible ways to reconcile this apparent discrepancy.

The first possibility to consider is that either the diskaccretion rates or the mass-infall rates are wrong, or both. Given the large uncertainty in determining the total accretion luminosity from an observed excess, which is roughly an order of magnitude (see e.g., *Gullbring et al.*, 1998; *WH04*), the average disk-accretion rate could be as large as 10^{-7} $M_\odot$/yr. Much larger disk-accretion rates would invoke statistical problems since classical T Tauri stars are accreting at this rate as well, for 1–10 m.y., and would consequently produce a much more massive population than what is observed. Larger rates would also be inconsistent with emission-line profile analyses (*Hartmann et al.*, 1994; *Muzerolle et al.*, 1998c). Assessing possible errors in the mass-infall rates is more challenging since most are not determined directly from kinematic infall signatures. Instead, they are primarily set by the density of the envelope material; denser envelopes yield higher mass-infall rates and redder SEDs. However, effects such as orientation (*Whitney et al.*, 2003) and disk emission (*Kenyon et al.*, 1993a; *Wolf et al.*, 2003) can also shift the SED toward redder wavelengths, if unaccounted for. Using sophisticated envelope plus disk models combined with spatially resolved images to constrain orientation, *Eisner et al.* (2005) and *Terebey et al.* (2006) nevertheless find that mass-infall rates of a few × 10^{-6} $M_\odot$/yr still provide the best fits to the SED and image morphology. How low these infall rates could be and still provide reasonable fits is unclear; a factor of ~10 decrease in the assumed infall rate could potentially reconcile the discrepancy, if disk-accretion rates are correspondingly increased by a factor of 10. It is important to keep in mind, as highlighted by *Terebey et al.* (2006), that the amount of envelope material that actually reaches the star may be only one-fourth of the infalling mass because of mass lost to stellar jets/winds and companions. With all this in mind, we conclude that it is possible to reconcile the infall/disk-accretion rate discrepancy based on systematic errors and model assumptions alone. However, since the current best estimates strongly favor values that are ~2 orders of magnitude discrepant, we will also consider other possibilities for reconciling these rates.

One possibility, as first suggested by *Kenyon et al.* (1990), is that the infalling envelope material is not transferred to the star via disk accretion in a steady-state fashion. Instead, the accreting envelope mass accumulates in the circumstellar disk until it becomes gravitationally unstable (e.g., *Larson*, 1984) and then briefly accretes at a prodigious rate [~10^{-5} $M_\odot$/yr (see *Calvet et al.*, 2000)]. This scenario is consistent with the small population of young, often embedded stars that dramatically increase their luminosity for a few years to a few centuries [e.g., FU Ori (*Hartmann and Kenyon*, 1987); V1647 Ori (*Briceño et al.*, 2004)]. If Class I stars intermittently accrete at this rate, they must spend 5–10% of their lifetime in the high accretion state to achieve typical T Tauri masses within 1 m.y. Statistically, 5–10% of Class I stars should then be accreting at this rate. The sample of Class I stars with mass-accretion rates is now becoming large enough to suggest a possible problem with these expected percentages; none appear to accrete at a rate this high (section 3.4). However, there is a strong observational bias in that stars accreting at this rate are likely to be too heavily veiled, at both optical and infrared wavelengths, to identify photospheric features from which the amount of excess can be measured. Indeed, several stars observed by *WH04* and *D05* are too veiled to measure mass-accretion rates. L1551 IRS 5, for example, which is the most luminous Class I star in Tau-Aur, has been proposed to be a young star experiencing an FU Ori-like outburst (*Hartmann and Kenyon*, 1996; *Osorio et al.*, 2003). Without a more accurate measure of its stellar properties this is difficult to confirm; its larger luminosity could be a consequence of it being a somewhat more massive star.

An independent test of the episodic accretion hypothesis is the relative masses of Class I disks compared to Class II disks. If the envelope material of Class I stars is accumulating in their circumstellar disks, they should be more massive than Class II stars. *WH04* investigated this using 1.3-mm continuum observations from *Beckwith et al.* (1990), *Osterloh and Beckwith* (1995), and *Motte and André* (2001), restricted to beam sizes of 11–12" to avoid contamination from envelope emission of Class I stars. This comparison showed that the 1.3-mm flux densities of Class I and Class II stars in Tau-Aur are indistinguishable, implying similar disk masses if the Class I and Class II disks have similar dust opacity and dust temperature (*Henning et al.*, 1995). However, *Andrews and Williams* (2005) drew a different conclusion based on submillimeter observations at 450 μm and 850 μm (with beam sizes of 9" and 15", respectively). They showed that the distribution of submillimeter

flux densities and disk masses of Class I stars are statistically different from those of Class II stars (being more massive), although Class I and Class II samples nevertheless span the same range of disk masses. Unfortunately, biases introduced by stellar mass, multiplicity, envelope emission, and low-spatial-resolution evolutionary diagnostics inhibit robust comparisons of these samples. We conclude there is at most marginal evidence for Class I stars having more massive disks than Class II stars, as would be expected if they undergo FU Ori-like outbursts more often than Class II stars.

Given the overall similarities of Class I and Class II stars, *WH04* put forth the still controversial suggestion that many (but not all) Class I stars are no longer in the main-accretion phase and are much older than traditionally assumed; *WH04* focus their study on Class I stars in Tau-Aur, where the case for this is most compelling. This proposal does not eliminate the luminosity problem for bona fide Class I stars, but minimizes the statistical significance of it in general. Support for this idea originates in the known biases introduced by current classification criteria that are inadequate to unambiguously identify young stars with infalling envelopes. The two largest biases are the low-spatial-resolution mid-infrared measurements upon which most SEDs are based and the effects of an unknown orientation on the SED. These biases likely explain why ~42% of stars classified as Class I stars in Tau-Aur do not appear to be bona fide protostars (section 4.2). Indeed, some authors have claimed that Class I stars like IRAS 04016 + 2610 and IRAS 04302 + 2247 have morphologies and kinematics that are better described by a rotating disk-like structure (*Hogerheijde and Sandell*, 2000; *Boogert et al.*, 2002; *Wolf*, 2003) than a collapsing envelope model (*Kenyon et al.*, 1993b; *Whitney et al.*, 1997), although more recent work still favors massive envelopes (e.g., *Eisner et al.*, 2005). However, the limitation of all these models is that they only account for the spatial distribution of circumstellar material, which can be confused with diffuse cloud emission (*Motte and André*, 2001) or companion stars with ~10^3 AU separations (*Haisch et al.*, 2004; *Duchêne et al.*, 2004). A convincing case for a massive infalling envelope can only be established by spatially mapping molecular line profiles and accounting for the effects of outflows and rotations (*Evans*, 1999). Currently the Class 0 star IRAS 04368 + 2557 (L1527) is the only star in Tau-Aur that has been shown to retain a massive extended envelope with unambiguous evidence for infall (*Gregersen et al.*, 1997).

In regions outside Tau-Aur, there is less evidence as well as less motivation for Class I stars being older than presumed and past the main phase of mass accretion. As discussed in section 3.4, the higher disk-accretion rates of many Class I stars in ρ Oph, for example, are within a factor of ~10 of predicted mass-infall rates, and thus easier to reconcile given current uncertainties in observations and models assumptions. Additionally, there is less evidence that these Class I stars are misclassified Class II stars, compared with Tau-Aur Class I stars. However, we strongly caution that it is not yet possible to tell if the apparent differences between the Class I population in Tau-Aur and other regions reflects real differences in their evolutionary state or is simply a consequence of Tau-Aur being a lower-density environment and its members being more optically revealed. One important similarity of Class I stars in all star-forming regions is their relatively low-mass envelopes (e.g. *André and Montmerle*, 1994; *Motte and André*, 2001), suggesting that at this phase they have already acquired the majority of their stellar mass.

If the ages of Class I stars are indeed as old as T Tauri stars ($\geq$1 m.y.) as the comparisons tentatively suggest (section 3.2), there is a potential dynamical timescale problem. In such a case the envelope is surviving nearly a factor of 10 longer than its dynamical collapse timescale, which seems unlikely. However, it is well known that there is nearly an order of magnitude spread in the *disk* dispersal timescale of Class II stars (e.g., *Hillenbrand et al.*, 1998); a similar spread in the envelope dispersal timescale seems plausible. One possibility for generating a large spread in the envelope dispersal timescale is that in some cases the envelopes are replenished. Recent simulations of cluster formation (e.g., *Bate et al.*, 2003) suggest that even after the initial phase of mass accretion, a young star continues to dynamically interact with the cloud from which it formed, and in some cases even significantly increase its mass. Thus some embedded stars could in fact come from an older population. These would be difficult to distinguish from younger stars in their initial main-accretion phase based on circumstellar properties alone. More accurately determined age estimates is likely the best way to test this intriguing hypothesis.

Summarizing, we find that in most cases the disk-accretion rates of Class I stars are well below predicted envelope infall rates. In some cases this may be a consequence of misclassification. In the more general case, it implies that if the envelope material is indeed infalling, it is not transferred to the star efficiently (e.g., *Terebey et al.*, 2006) or at a steady rate (e.g., *Kenyon et al.*, 1990), or both. While it is known that some young stars dramatically increase in brightness, presumably due to enhanced accretion (e.g., FU Ori), the idea that this is the process by which stars acquire the majority of their mass is still unconfirmed. If the ages of some Class I stars are indeed as old as T Tauri stars, the long-lived envelope lifetimes may stem from envelope replenishment, possibly caused by continued interactions with the cloud after formation. Overall, it appears that most of Class I stars, as currently defined, have already acquired the majority of their final stellar mass.

5. FUTURE PROSPECTS

The ensemble of newly determined stellar and disk accretion properties of Class I stars offer powerful constraints on how and when young stars (and brown dwarfs) are assembled. However, many unknowns still remain. Here we highlight seven key areas of research that would help re-

solve the remaining uncertainties and advance our understanding of the earliest stages of star formation.

1. *More accurately determined circumstellar properties.* Much of the suspected misclassification of Class I stars could be confirmed or refuted with more accurately determined SEDs based on observations over a broad wavelength range that spatially resolve structures (e.g., edge-on disks) and nearby neighbors. In concert with this, more accurate and less orientation dependent criteria for identifying Class I stars needs to be established.

2. *Extensive surveys for Class I stars.* Larger, more complete (and less flux limited) surveys for Class I stars in multiple star-forming regions will help confirm or refute tentative trends identified with the small samples studied so far, and may likewise reveal real environmental and/or (sub)stellar mass dependencies upon the formation process.

3. *Improved models of the circumstellar environment.* With larger, more accurately determined samples of Class I stars, there will be a need for more sophisticated envelope-plus disk models of embedded stars that can fit the observed SED and scattered light and polarization images (e.g., *Osorio et al.*, 2003; *Whitney et al.*, 2005; *Eisner et al.*, 2005; *Terebey et al.*, 2006). This work is important for directly determining the density of the envelope material (which constrains the mass-infall rate) and estimating the extinction to the central star (which is often in error because of scattered light), and can potentially determine the system orientation.

4. *Detailed kinematic mapping.* The case for massive infalling envelopes can be unambiguously resolved using interferometric techniques that kinematically map the surrounding envelope-like material. In addition, these techniques offer the most direct and accurate way to determine the envelope infall rate, which can be compared to the newly determined disk-accretion rates.

5. *Improved models of disk accretion.* Unfortunately, there remain considerable uncertainties in observationally determining disk-accretion rates (e.g., bolometric correction), and these uncertainties are magnified for embedded stars with high extinction and scattered light. Consequently the absolute value of the disk-accretion rates and their agreement with infall model predictions are difficult to assess. Observations of emission-line profiles and continuum excess measurements over a broad range of wavelengths are promising methods to help resolve this. Additionally, observational monitoring to determine the timescale and magnitude of variations in the mass-accretion rate may yield important constraints on how and how quickly mass is acquire during this stage.

6. *More sensitive spectroscopic surveys.* While the recent spectroscopic observations focused upon here have revealed much about the pre-T Tauri evolutionary stage, the data suffer from rather severe observational biases. These include biases in flux, extinction, and mass-accretion rate. More sensitive spectrographs and/or larger-aperture telescopes may be needed to address the first two biases, while higher signal-to-noise observations of "featureless" Class I

stars may help reveal their stellar and accretion properties. Specifically, particular attention should be paid to the most luminous Class I stars in a given region, to establish better if they are more luminous because they have accretion-dominated luminosities, or if they are simply more massive stars (e.g., L1551 IRS 5).

7. *Comparisons with synthetically generated spectra.* Fortunately, in the last decade there has been considerable progress in the area of synthetically generated spectra of stars. The implication is that stellar properties (e.g., temperature and surface gravity) can be directly extracted from the spectra (e.g., *Johns-Krull et al.,* 1999, 2004; *Doppmann et al.,* 2003, 2005), as opposed to indirectly determined by historic spectral-comparison techniques. The most exciting application is the determination of surface gravities, as the inferred stellar radii from these measurements can be used to establish more precise age estimates, and perhaps unambiguously determine the age of Class I stars, even if only relative to T Tauri stars.

Acknowledgments. We thank C. Lada for his substantial contributions to many of the papers that constitute much of the work discussed in this chapter. This work was partially supported by the NASA Origins of Solar Systems program. We have appreciated the privilege to observe on the revered summit of Mauna Kea.

REFERENCES

Adams F. C., Lada C. J., and Shu F. J. (1987) *Astrophys. J., 312,* 788–806.

André P. and Montmerle T. (1994) *Astrophys. J., 420,* 837–862.

André P., Ward-Thompson D., and Barsony M. (1993) *Astrophys. J., 406,* 122–141.

Andrews S. and Williams J. (2005) *Astrophys. J., 631,* 1134–1160.

Armitage P. J. and Clarke C. J. (1996) *Mon. Not. R. Astron. Soc., 280,* 458–468.

Baraffe I., Chabrier G., Allard F., and Hauschildt P. H. (1998) *Astron. Astrophys., 337,* 403–412.

Baraffe I., Chabrier G., Allard F., and Hauschildt P. H. (2002) *Astron. Astrophys., 382,* 563–572.

Bate M. R., Bonnell I. A., and Volker B. (2003) *Mon. Not. R. Astron. Soc., 339,* 577–599.

Beckwith S. V. W., Sargent A. I., Chini R. S., and Güsten R. (1990) *Astrophys. J., 99,* 924–945.

Benson P. J. and Myers P. C. (1989) *Astrophys. J. Suppl., 71,* 89–108.

Bontemps S., André P., Kaas A. A., Nordh L., Olofsson G., et al. (2001) *Astron. Astrophys., 372,* 173–194.

Boogert A. C. A., Blake G. A., and Tielens A. G. G. M. (2002) *Astrophys. J., 577,* 271–280.

Bouvier J., Bertout C., Benz W., and Mayor M. (1986) *Astron. Astrophys., 165,* 110–119.

Bouvier J., Cabrit S., Fernández M., Martin E. L., and Matthews J. M. (1993) *Astron. Astrophys., 272,* 176–206.

Bouvier J., Covino E., Kovo O., Martin E. L., Matthews J. M., et al. (1995) *Astron. Astrophys., 299,* 89–107.

Briceño C., Vivas A., Hernández J., Calvet N., Hartmann L., et al. (2004) *Astrophys. J., 606,* L123–L126.

Brown D. W. and Chandler C. J. (1999) *Mon. Not. R. Astron. Soc., 303,* 855–863.

Calvet N. and Hartmann L. (1992) *Astrophys. J., 386,* 239–247.

Calvet N., Hartmann L., and Strom S. E. (1997) *Astrophys. J., 481,* 912–917.

Calvet N., Hartmann L., and Strom S. E. (2000) In *Protostars and Planets IV* (V. Mannings et al., eds.), pp. 377–399. Univ. of Arizona, Tucson.

Casali M. M. and Eiroa C. (1996) *Astron. Astrophys., 306,* 427–435.

Casali M. M. and Matthews H. E. (1992) *Mon. Not. R. Astron. Soc., 258,* 399–403.

Cassen P. and Moosman A. (1981) *Icarus, 48,* 353–376.

Chiang E. I. and Goldreich P. (1999) *Astrophys. J., 519,* 279–284.

Cieza L. A., Kessler-Silacci J. E., Jaffe D. T., Harvey P. M., and Evans N. J. II (2005) *Astrophys. J., 635,* 422–441.

Clarke C. J. and Bouvier J. (2000) *Mon. Not. R. Astron. Soc., 319,* 457–466.

Cohen M. and Schwartz R. D. (1983) *Astrophys. J., 265,* 877–900.

Collier Cameron A. and Campbell C. G. (1993) *Astron. Astrophys., 274,* 309–318.

Covey K. R., Greene T. P., Doppmann G. W., and Lada C. J. (2005) *Astron. J., 129,* 2765–2776.

Doppmann G. W., Jaffe D. T., and White R. J. (2003) *Astron. J., 126,* 3043–3057.

Doppmann G. W., Greene T. P., Covey K. R., and Lada C. J. (2005) *Astron. J., 130,* 1145–1170.

Duchêne G., Bouvier J., Bontemps S., André P., and Motte F. (2004) *Astron. Astrophys., 427,* 651–665.

Edwards S., Strom S., Hartigan P., Strom K., Hillenbrand L., et al. (1993) *Astron. J., 106,* 372–382.

Eisner J. A., Hillenbrand L. A., Carpenter J. M., and Wolf S. (2005) *Astrophys. J., 635,* 396–421.

Evans N. J. (1999) *Ann. Rev. Astron. Astrophys., 37,* 311–362.

Folha D. F. M. and Emerson J. P. (1999) *Astron. Astrophys., 352,* 517–531.

Goodwin S. P., Whitworth A. P., and Ward-Thompson D. (2004) *Astron. Astrophys., 423,* 169–182.

Graham J. A. (1991) *Publ. Astron. Soc. Pac., 103,* 79–84.

Greene T. P. and Lada C. J. (1996) *Astron. J., 112,* 2184–2221.

Greene T. P. and Lada C. J. (1997) *Astron. J., 114,* 2157–2165.

Greene T. P. and Lada C. J. (2000) *Astron. J., 120,* 430–436.

Greene T. P. and Lada C. J. (2002) *Astron. J., 124,* 2185–2193.

Gregersen E. M., Evans N. J., Zhou S., and Choi M. (1997) *Astrophys. J., 484,* 256–276.

Gullbring E., Hartmann L., Briceño C., and Calvet N. (1998) *Astrophys. J., 492,* 323–341.

Haisch K. E. Jr., Greene T. P., Barsony M., and Stahler S. W. (2004) *Astrophys. J., 127,* 1747–1754.

Hartigan P., Edwards S., and Ghandour L. (1995) *Astrophys. J., 452,* 736–768.

Hartmann L. (1998) *Accretion Processes in Star Formation,* Chapter 4. Cambridge Univ., Cambridge.

Hartmann L. and Kenyon S. J. (1987) *Astrophys. J., 312,* 243–253.

Hartmann L. and Kenyon S. J. (1996) *Ann. Rev. Astron. Astrophys., 34,* 207–240.

Hartmann L., Hewett R., Stahler S., and Mathieu R. D. (1986) *Astrophys. J., 309,* 275–293.

Hartmann L., Hewett R., and Calvet N. (1994) *Astrophys. J., 426,* 669–687.

Henning T., Begemann B., Mutschke H., and Dorshner J. (1995) *Astron. Astrophys. Suppl., 112,* 143.

Hillenbrand L. A., Strom S. E., Calvet N., Merrill K. M., Gatley I., et al. (1998) *Astron. J., 116,* 1816–1841.

Hogerheijde M. R. and Sandell G. (2000) *Astrophys. J., 534,* 880–893.

Ishii M., Tamura M., and Itoh Y. (2004) *Astrophys. J., 612,* 956–965.

Jappsen A.-K. and Klessen R. S. (2004) *Astron. Astrophys., 423,* 1.

Johns-Krull C. M., Valenti J. A., and Koresko C. (1999) *Astrophys. J., 516,* 900–915.

Johns-Krull C. M., Valenti J. A., and Gafford A. D. (2003) *Rev. Mex. Astron. Astrofys., 18,* 38–44.

Johns-Krull C. M., Valenti J. A., and Saar S. H. (2004) *Astrophys. J., 617,* 1204–1215.

Kaas A. A., Olofsson G., Bontemps S., André P., Nordh L., et al. (2004) *Astron. Astrophys., 421,* 623–642.

Kenyon S. J. and Hartmann L. (1995) *Astrophys. J. Suppl., 101,* 117–171.

Kenyon S. J., Hartmann L. W., Strom K. M., and Strom S. E. (1990) *Astron. J., 99,* 869–887.

Kenyon S. J., Calvet N., and Hartmann L. (1993a) *Astrophys. J., 414,* 676–694.

Kenyon S. J., Whitney B. A., Gómez M., and Hartmann L. (1993b) *Astrophys. J., 414,* 773–792.

Kenyon S. J., Gómez M., Marzke R. O., and Hartmann L. (1994) *Astron. J., 108,* 251–261.

Kenyon S. J., Brown D. I., Tout C. A., and Berlind P. (1998) *Astron. J., 115,* 2491–2503.

Kessler-Silacci J. E., Hillenbrand L. A., Blake G. A., and Meyer M. R. (2005) *Astrophys. J., 622,* 404–429.

Kleinmann S. G. and Hall D. N. B. (1986) *Astrophys. J. Suppl., 62,* 501–517.

Königl A. (1991) *Astrophys. J., 370,* L39–L43.

Lada C. J. (1987) In *Star Forming Regions* (M. Peimbert and J. Jugaku, eds.), pp. 1–18. IAU Symposium 115, Reidel, Dordrecht.

Lada C. J. and Wilking B. A. (1984) *Astrophys. J., 287,* 610–621.

Ladd E. F., Lada E. A., and Myers P. C. (1993) *Astrophys. J., 410,* 168–178.

Larson R. B. (1984) *Mon. Not. R. Astron. Soc., 206,* 197–207.

Luhman K. L. and Rieke G. H. (1999) *Astrophys. J., 525,* 440–465.

Mac Low M. and Klessen R. S. (2004) *Rev. Mod. Phys., 76,* 125–194.

Matt S. and Pudritz R. (2005) *Astrophys. J., 632,* L135–L138.

Motte F. and André P. (2001) *Astron. Astrophys., 365,* 440–464.

Mundt R., Stocke J., Strom S. E., Strom K. M., and Anderson E. R. (1985) *Astrophys. J., 297,* L41–L45.

Muzerolle J., Calvet N., and Hartmann L. (1998a) *Astrophys. J., 492,* 743–753.

Muzerolle J., Hartmann L., and Calvet N. (1998b) *Astron. J., 116,* 455–468.

Muzerolle J., Hartmann L., and Calvet N. (1998c) *Astron. J., 116,* 2965–2974.

Muzerolle J., Calvet N., and Hartmann L. (2001) *Astrophys. J., 550,* 944–961.

Muzerolle J., D'Alessio P., Calvet N., and Hartmann L. (2004) *Astrophys. J., 617,* 406.

Myers P. C. and Benson P. (1983) *Astrophys. J., 266,* 309–320.

Myers P. C. and Ladd E. F. (1993) *Astrophys. J., 413,* L47–L50.

Myers P. C., Fuller G. A., Mathieu R. D., Beichman C. A., Benson P. J., et al. (1987) *Astrophys. J., 319,* 340–357.

Najita J. (2004) In *Star Formation in the Interstellar Medium: In Honor of David Hollenbach, Chris McKee and Frank Shu* (D. Johnstone et al., eds.), pp. 271–277. ASP Conf. Series 323, San Francisco.

Nisini B., Antoniucci S., Giannini T., and Lorenzetti D. (2005) *Astron. Astrophys., 429,* 543–557.

Onishi R., Mizuno A., Kawamura A., Tachihara K., and Fukui Y. (2002) *Astrophys. J, 575,* 950–973.

Osorio M., D'Alessio P., Muzerolle J., Calvet N., and Hartmann L. (2003) *Astrophys. J., 586,* 1148–1161.

Osterloh M. and Beckwith S. V. W. (1995) *Astrophys. J., 439,* 288–302.

Padgett D. L., Brandner W., Stapelfeldt K. R., Strom S. E., Terebey S., et al. (1999) *Astron. J., 117,* 1490–1504.

Park S. and Kenyon S. J. (2002) *Astron. J., 123,* 3370–3379.

Prusti T., Whittet D. C. B., and Wesselius P. R. (1992) *Mon. Not. R. Astron. Soc., 254,* 361–368.

Rebull L. M.,Wolff S. C., Strom S. E., and Makidon R. B. (2002) *Astron. J., 124,* 546–559.

Rebull L. M., Wolff S. C., and Strom S. E. (2004) *Astron. J., 127,* 1029–1051.

Rhode K. L., Herbst W., and Mathieu R. D. (2001) *Astron. J., 122,* 3258–3279.

Safier P. N. (1998) *Astrophys. J., 494,* 336–341.

Shu F. H. (1977) *Astrophys. J., 214,* 488–497.

Shu F., Najita J., Ostriker E., Wilkin F., Ruden S., et al. (1994) *Astrophys. J., 429,* 781–796.

Siess L., Forestini M., and Bertout C. (1999) *Astron. Astrophys., 342,* 480–491.

Sonnhalter C., Preibisch T., and Yorke H. W. (1995) *Astron. Astrophys., 299,* 545–556.

Stassun K. G., Mathieu R. D., Mazeh T., and Vrba F. (1999) *Astron. J., 117,* 2941–2979.

Stassun K., Mathieu R., Vrba J., Mazeh T., and Henden A. (2001) *Astron. J., 121,* 1003–1012.

Terebey S., Shu F. H., and Cassen P. (1984) *Astrophys. J., 286,* 529–551.

Terebey S., Van Buren D., and Hancock T. (2006) *Astrophys. J., 637,* 811–822

Tout C. A., Livio M., and Bonnell I. A. (1999) *Mon. Not. R. Astron. Soc., 310,* 360–376.

Wallace L. and Hinkle K. (1996) *Astrophys. J. Suppl., 107,* 312–390.

Werner M. W., Roellig T. L., Low F. J., Rieke G. H., Rieke M., et al. (2004) *Astrophys. J. Suppl., 154,* 154–162.

White R. J. and Basri G. (2003) *Astrophys. J., 582,* 1109–1122.

White R. J. and Hillenbrand L. A. (2004) *Astrophys. J., 616,* 998–1032.

Whitney B. A., Kenyon S. J., and Gómez M. (1997) *Astrophys. J., 485,* 703–734.

Whitney B. A., Wood K., Bjorkman J. E., and Cohen M. (2003) *Astrophys. J., 598,* 1079–1099.

Whitney B. A., Indeebetouw R., Bjorkman J. E., and Wood K. (2004) *Astrophys. J., 617,* 1177–1190.

Whitney B. A., Robitaille T. P., Wood K., Denzmore P., and Bjorkman J. E. (2005) In *PPV Poster Proceedings,* www.lpi.usra.edu/meetings/ppv2005/pdf/8460.pdf.

Wilking B. A., Lada C. J., and Young E. T. (1989) *Astrophys. J., 340,* 823–852.

Wilking B. A., Greene T. P., Lada C. J., Meyer M. R., and Young E. T. (1992) *Astrophys. J., 397,* 520–533.

Wolf S., Padgett D. L., and Stapelfeldt K. R. (2003) *Astrophys. J., 588,* 373–386.

Yorke H.W., Bodenheimer P., and Laughlin G. (1993) *Astrophys. J., 411,* 274–284.

Young C. H., Shirley Y. L., Evans N. J., and Rawlings J. M. C. (2003) *Astrophys. J. Suppl., 145,* 111–145.

The Fragmentation of Cores and the Initial Binary Population

Simon P. Goodwin
University of Sheffield

Pavel Kroupa
Rheinische Fredrich-Wilhelms-Universität Bonn

Alyssa Goodman
Harvard-Smithsonian Center for Astrophysics

Andreas Burkert
Ludwig-Maximillian-Universität Munich

Almost all young stars are found in multiple systems. This suggests that protostellar cores almost always fragment into multiple objects. The observed properties of multiple systems such as their separation distribution and mass ratios provide strong constraints on star-formation theories. We review the observed properties of young and old multiple systems and find that the multiplicity of stars changes. Such an evolution is probably due to (a) the dynamical decay of small-N systems and/or (b) the destruction of multiple systems within dense clusters. We review simulations of the fragmentation of rotating and turbulent molecular cores. Such models almost always produce multiple systems; however, the properties of those systems do not match observations at all well. Magnetic fields appear to supress fragmentation, perhaps suggesting that they are not dynamically important in the formation of multiple systems. We finish by discussing possible reasons why theory fails to match observation, and the future prospects for this field.

1. INTRODUCTION

Correctly predicting the properties of young multiple systems is one of the most challenging tests of any theory of star formation. In this chapter we discuss the current understanding of how dense prestellar cores fragment into multiple stars (by "stars" we generally mean both stars and brown dwarfs).

First, we will discuss the important observed properties of both young and old binary systems and the differences between them. Then we will describe the possible origins of the differences between the young and old systems and hopefully convince the reader that almost all stars must form in multiple systems. This initial binary population must form from the fragmentation of star-forming dense molecular cores and so we discuss the observed properties of these cores that may influence their ability to form multiple systems. We will then review the current models of core fragmentation, with an emphasis on turbulence as the mecha-

nism that promotes fragmentation. Finally we will examine why theory currently fails to correctly predict binary properties.

2. THE PROPERTIES OF MULTIPLE SYSTEMS

There has been an extensive study of binary properties over the past two decades with the modern study often marked as beginning with the detailed survey by *Duquennoy and Mayor* (1991) (hereafter *DM91*). Multiple systems in the field are by far the best studied due to the availability of local samples whose completeness is easier to estimate, and the properties of the field provide the benchmark against which younger samples are measured.

2.1. Multiple Systems in the Field

2.1.1. Multiplicity fraction. The fraction of field stars in multiple systems is found to be high and increases with increasing primary mass. Many different measures are used to quantify multiplicity, which can become very confusing. Most important is the "multiplicity frequency" $f_{mult} = (B + T + Q + ...)/(S + B + T + Q + ...)$ where S, B, T, and Q are the numbers of single, binary, triple, and quadruple systems respectively, thus f_{mult} represents the probability that any system is a multiple system (see *Reipurth and Zinnecker,* 1993).

The raw value of the field G-dwarf multiplicity frequency found by *DM91* is 0.49; when corrected for incompleteness this rises to 0.58. However, recent studies using Hipparcos data have shown that it may even be higher than this (*Quist and Lindegren,* 2000; *Söderhjelm,* 2000). It should be noted that the brown dwarf (BD) binary fraction has generally been considered to be much lower than that of stars at ~0.1–0.2 (*Bouy et al.,* 2003; *Close et al.,* 2003; *Gizis et al.,* 2003; *Martín et al.,* 2003). However, recent studies have suggested that the BD multiplicity frequency may be significantly higher, possibly exceeding 0.5 (*Pinfield et al.,* 2003; *Maxted and Jeffries,* 2005), provided multiple BDs reside in very tight BD-BD pairs.

2.1.2. Separation distribution. The binary separation distribution is very wide and flat, usually modeled as a log-normal with mean ~30 AU and variance $\sigma_{\log d} \sim 1.5$ (*DM91* for G dwarfs). This is illustrated in Fig. 1 where the field period distribution is compared to that of young stars (note that $a^3/P^2 = m_{sys}$, where a is in AU, P is in years, and m_{sys} is the system mass in $M_\odot$).

A similar distribution is found for M dwarfs by *Fischer and Marcy* (1992), and generally seems to hold for all stars, although the maximum separations do appear to decrease somewhat, but not substantially, for stars with decreasing mass (*Close et al.,* 2003). Very-low-mass stars (VLMSs) and BDs seem strongly biased toward very close companions

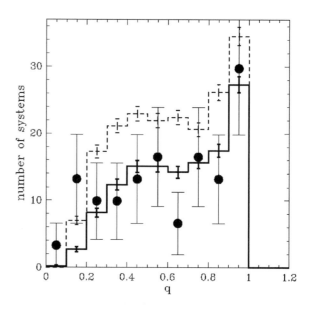

Fig. 2. The mass-ratio distribution of all late-type primary stars. The solid dots are observational data by *Reid and Gizis* (1997), while the expected initial distribution is shown as the dashed histogram. In a typical star cluster it evolves to the solid histogram, which reproduces the observed data quite well (from *Kroupa et al.,* 2003).

with semimajor axes $a \leq a_{max} \approx 15$ AU (*Close et al.,* 2003; *Gizis et al.,* 2003; *Pinfield et al.,* 2003; *Maxted and Jeffries,* 2005), in contrast to those of stars that have $a_{max} \geq 1000$ AU (*DM91*; *Fischer and Marcy,* 1992; *Mayor et al.,* 1992). It is this unusual separation distribution that may have led to the underestimate of the BD multiplicity fraction. The much smaller a_{max} for VLMSs and BDs compared to the other stars cannot be a result of disruption in a cluster environment but must be due to the inherent physics of their formation (*Kroupa et al.,* 2003).

2.1.3. Mass ratio distribution. *DM91* found that galactic-field systems with a G-dwarf primary have a mass-ratio distribution biased toward small values such that it does not follow the stellar IMF, which would predict a far larger number of companions with masses $m_2 \lesssim 0.3 M_\odot$ (*Kroupa,* 1995b). For short-period binaries, the mass-ratio distribution is biased toward similar-mass pairs (*Mazeh et al.,* 1992). Integrating over all periods, for a sample of nearby systems with primary masses in the range $0.1 \leq m_1/M_\odot \leq 1$, *Reid and Gizis* (1997) find the mass-ratio distribution to be approximately flat and consistent with the IMF (Fig. 2).

2.1.4. Eccentricity distribution. Binary systems have a thermalized eccentricity distribution (equation (3) below) for periods $P \geq 10^{3-4}$ d, with tidally circularized binaries dominating at low separations (*DM91*; *Fischer and Marcy,* 1992).

2.1.5. Higher-order systems. *DM91* find the uncorrected ratio of systems of different multiplicity to be S:B:T:Q = 1.28:1:0.175:0.05 (see also *Tokovinin and Smekhov,* 2002), suggesting that roughly 20% of multiple systems are high-order systems. Concerning the origin of high-order multiple

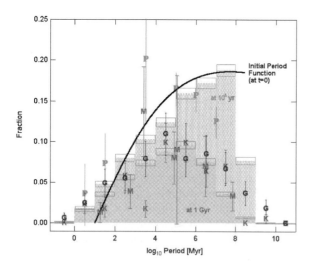

Fig. 1. The period distribution function. Letters show the observed fraction of field G, K, and M stars and PMS stars (P). The solid curve shows the model initial period distribution (see equation (2)). The light histogram is the initial binary population in the simulations of *Kroupa* (1995b), which evolves through dynamical interactions in a cluster into a field-like distribution shown by the heavy histogram.

systems in the galactic field, we note that many, and perhaps most, of these may be the remnants of star clusters (*Goodwin and Kroupa, 2005*).

2.2. Pre-Main-Sequence Multiple Systems

The properties of pre-main-sequence (PMS) multiple systems are much harder to determine than those in the field. We refer the reader to the chapter by Duchêne et al. for a detailed review of the observations of PMS multiple systems and the inherent problems.

Probably the most important difference between the PMS and field populations is that young stars have a significantly higher *multiplicity fraction* than the field (see the chapter by Duchêne et al.) (see also Fig. 1).

The *separation distribution* of PMS stars also appears different to that in the field with an overabundance of binaries with separations of a few hundred AU (*Mathieu, 1994; Patience et al., 2002*) (Fig. 1). More specifically, the binary frequency in the separation range ~100–1000 AU is a factor of ~2 higher than in the field (*Mathieu, 1994; Patience et al., 2002; Duchêne et al., 2004*). Extrapolating this increase across the whole separation range implies that f_{mult} for PMS stars could be as high as 100%. [It appears that in Taurus the binary frequency is ~100% for stars >0.3 $M_\odot$ (*Leinert et al., 1993; Köhler and Leinert, 1998*).]

The *mass-ratio distribution* of PMS stars is similar to the field population. A detailed comparison is not yet possible because low-mass companions to pre-main-sequence primaries are very difficult to observe as the available results depend mostly on direct imaging or speckle interferometry, while for main-sequence systems radial-velocity surveys have been done over decades (*DM91*). Using near-infrared speckle interferometry observations to obtain resolved JHK photometry for the components of 58 young binary systems, *Woitas et al.* (2001) found that the mass-ratio distribution is flat for mass ratios q ≥ 0.2, which is consistent with random pairing from the IMF, i.e., fragmentation processes rather than common-accretion (Fig. 2).

The *eccentricity distribution* of PMS stars also is similar to the field, with a thermalized distribution except at low separations where tidal circularization has occurred rapidly (e.g., *Kroupa, 1995b; White and Ghez, 2001*). Finally, the ages of components in young multiple systems appear to be very similar (*White and Ghez, 2001*).

It is currently unclear what the proportion of higher-order multiples is in young systems (see the chapter by Duchêne et al.). We will revisit this question in the next section.

2.3. The Evolution of Binary Properties

The observations described in sections 2.1 and 2.2 clearly show that at least the binary fraction and separation distributions evolve significantly between young stellar populations and the field.

Indeed, binary properties are seen to change even within populations in star-forming regions. The binary fraction is found to vary between embedded and (older) nonembedded

sources in both Taurus and ρ Oph (*Duchêne et al., 2004; Haisch et al., 2004*). Also, the mass-ratio distributions of massive stars appear to depend on the age of the cluster, with those in young clusters being consistent with random sampling from the IMF and those in dynamically evolved populations favoring equal-mass companions (section 2.4).

The evolution of binary properties has been ascribed to two mechanisms: (1) the rapid dynamical decay of young small-N clusters within cluster cores (e.g., *Reipurth and Clarke, 2001; Sterzik and Durisen, 1998, 2003; Durisen et al., 2001; Hubber and Whitworth, 2005; Goodwin and Kroupa, 2005; Umbreit et al., 2005*); and (2) the dynamical destruction of multiples by interactions in a clustered environment, which can modify an initial PMS-like distribution into a field-like distribution (*Kroupa, 1995a,b; Kroupa et al., 2003*) (Figs. 1, 3).

2.3.1. Small-N decay. Multiple systems containing N ≥ 3 stars are unstable to dynamical decay unless they form in a strongly hierarchical configuration [stability criteria for N > 2 systems are provided by *Eggleton and Kiseleva* (1995)]. Generally, a triple system is unstable to decay with a half-life of

$$t_{decay} = 14 \left(\frac{R}{AU} \right)^{3/2} \left(\frac{M_{stars}}{M_\odot} \right)^{-1/2} yr \qquad (1)$$

where R is the size of the system, and M_{stars} is the mass of the components (*Anosova, 1986*). The decay time for R =

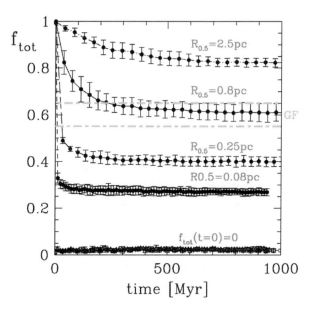

Fig. 3. Evolution of the total binary fraction as a function of time for four cluster models with decreasing half-mass radii $R_{0.5}$ from top to bottom. Each cluster initially contains 200 late-type binaries (with periods, eccentricities, and mass ratios consistent with the initial distribution functions described in section 2.4) except the bottom simulation, which contains no binaries. The area marked "GF" shows the observed field binary fraction (*DM91*). The cluster with $R_{0.5}$ = 0.8 pc evolves into a field-like distribution. With no initial binaries capture is unable to produce a significant number of binaries. From *Kroupa* (1995a).

250 AU and $M_{stars} = 1\ M_{\odot}$ is ~55 k.y., which is on the order of the duration of the embedded phases of young stars, thus ejections should mainly occur during the main Class 0 accretion phase of PMS objects (e.g., *Reipurth*, 2000). Indeed, one such early dynamical decay appears to have been observed by *Gómez et al.* (2005), and this is probably the process at work to reduce the binary fraction between the embedded and nonembedded stars seen by *Duchêne et al.* (2004) and *Haisch et al.* (2004). These early dynamical processes cause embedded protostars to be ejected from their natal envelopes, possibly causing abrupt transitions of objects from Class 0/I to Class II/III (*Reipurth*, 2000; *Goodwin and Kroupa*, 2005).

Significant numbers of small-N decays will dilute any initial high-multiplicity fraction very rapidly to a small binary fraction for the whole population (e.g., N = 5 systems would lead to a population with a binary fraction f = 1/4 within <10^5 yr). The observed high binary fraction in ~1-m.y.-old populations thus suggests that the formation of N > 2 systems is the exception rather than the rule (*Goodwin and Kroupa*, 2005). Ejections would occur mostly during the very early Class 0 stage such that ejected embryos later appear as free-floating single very-low-mass stars and BDs (*Reipurth and Clarke*, 2001). However, the small ratio of the number of BDs per star, ≈0.25 (*Munech et al.*, 2002; *Kroupa et al.*, 2003; *Kroupa and Bouvier*, 2003b; *Luhman*, 2004), again suggests this not to be a very common process even if *all* BDs form from ejections.

Ejections have two main consequences: a significant reduction in the semimajor axis of the remaining stars (*Anosova*, 1986; *Reipurth*, 2000; *Umbreit et al.*, 2005), and the preferential ejection of the lowest-mass component (*Anosova*, 1986; *Sterzik and Durisen*, 2003). The early ejection of the lowest-mass component forms the basis of the embryo ejection scenario of BD formation (*Reipurth and Clarke*, 2001; *Bate et al.*, 2002).

The N-body statistics of the decay of small-N systems has been studied by a number of authors (*Anosova*, 1986; *Sterzik and Durisen*, 1998, 2003; *Durisen et al.*, 2001; *Goodwin et al.*, 2005; *Hubber and Whitworth*, 2005). However, only *Umbreit et al.* (2005) have attempted to include the effects of accretion on the N-body dynamics, which appear to have a significant effect — especially on the degree of hardening of the binary after ejection. *Goodwin et al.* (2004a,b) and *Delgado Donate et al.* (2004a,b) have simulated ensembles of cores, including the full hydrodynamics of star formation; however, proper statistical conclusions about the effects of ejections are difficult to draw due to the different numbers of stars forming in each ensemble (which there is no way of controlling *a priori*) and the smaller number of ensembles that may be run in a fully hydrodynamic context. However, some conclusions appear from these and other studies (*Whitworth et al.*, 1995; *Bate and Bonnell*, 1997; *Bate et al.*, 2003; *Delgado Donate et al.*, 2003): First, that early ejections are very effective at hardening the remaining stars (cf. *Umbreit et al.*, 2005). Second, this early

hardening tends to push the mass ratios of close binaries toward unity. This occurs as the low-mass component has a higher specific angular momentum than the primary and so is more able to accrete mass from the high angular momentum circumstellar material (see *Whitworth et al.*, 1995; *Bate and Bonnell*, 1997). However, *Ochi et al.* (2005) find in detailed two-dimensional simulations of accretion onto binaries that the gas accretes mainly onto the primary due to shocks removing angular momentum.

One significant caveat to the previous discussion is that the gradual formation (over ~0.1 m.y.) of stars allows far more stable triples and higher-order multiples to form than expected observationally, which are stable for at least 10 m.y. (*Delgado Donate et al.*, 2004a,b). Indeed, simulations that form a large number of stars often form very hierarchical higher-order multiples (often quadruples and quintuples formed when even larger systems decay), which are not observed (*Delgado Donate et al.*, 2003, 2004a,b; *Goodwin et al.*, 2004a,b). Such systems would probably be destroyed during the cluster destruction phase (see below), but not dilute the binary fraction on very short timescales.

2.3.2. Dynamical destruction in clusters. In the highly clustered environments in which most stars are thought to form (e.g., *Lada and Lada*, 2003) dynamical interactions will be common and may disrupt many initial binary systems.

Binaries can be subdivided into three dynamical groups: (1) the wide, or soft, binaries; (2) the dynamically active binaries; and (3) the tight or hard binaries.

Wide binaries have orbital velocities much smaller than the velocity dispersion, σ, in a cluster and are easily disrupted. This is best seen by a gedanken experiment, where we construct a reduced-mass particle (a test particle orbiting in a fixed potential with total mass, eccentricity, and orbital period equal to that of the binary in question) in a heat bath of field particles (the cluster stars). If the orbital velocity of the test particle is smaller than the typical velocities of the field particles ($v_{orb} \ll \sigma$), then the test particle will gain kinetic energy by encounters, i.e., by the principle of energy equipartition, until its orbital velocity surpasses the binding energy of the binary. The binary consequently gets disrupted. Energy conservation requires the heat bath to cool; cluster cooling has been seen in N-body computations by *Kroupa et al.* (1999), but the effect is not significant for cluster dynamics. The general effect of this process is that binaries with weak binding energies are disrupted (i.e., binaries with long periods and/or small mass ratios).

Hard binaries, on the other hand, can be represented by a reduced-mass binary in which the test particle has $v_{orb} \gg \sigma$, so that energy equipartition leads to a reduction of v_{orb} and to an increase of σ (cluster heating). This increases the binding energy of the binary, which heats up further (v_{orb} increases as the test particle falls toward the potential minimum). This runaway process only stops because either the binary merges when it is so tight that the constituent stars touch (forming a blue straggler), or because the hardened binary receives a recoil expelling it from the cluster, or the hardening binary evolves to a cross section so small that the binary becomes essentially unresolved in further interactions.

Heggie (1975) and *Hills* (1975) studied the details of these processes and formulated the *Heggie-Hills law* of binary evolution in clusters: "soft binaries soften and hard binaries harden." An important implication of this law is that hard binaries can absorb the entire binding energy of a cluster and drive the evolution of the core of a massive cluster.

Not accessible to analytical work are *active binaries* with intermediate binding energies. Only full-scale N-body computations can treat the dynamics of the interactions accurately (e.g., *Heggie et al.*, 1996). Such binaries couple efficiently to the cluster, and efficiently exchange energy with it. The binary-binary and binary-single-star interactions form complex resonances and short-lived higher-order configurations that decay by expelling typically the least-massive member. Active binaries are thus quite efficient in exchanging partners, but more work needs to be done in order to quantify the exchange rates for typical galactic star-forming clusters.

The dynamical interactions between binaries or single stars will continue to alter the binary properties of the population as long as it remains relatively dense (i.e., until the cluster dissolves or expands significantly after residual gas removal). In a series of papers (*Kroupa*, 1995a,b; *Kroupa et al.*, 2001) it has been shown that a population initially composed entirely of binary systems with a PMS binary separation distribution can evolve into the field-like distribution through dynamical encounters, as is exemplified by the evolution of the mass-ratio distribution (Fig. 2).

Importantly, however, dynamical interactions in a cluster cannot form a significant number of binaries from an initially single star population, *or* widen an initially narrow separation distribution (*Kroupa and Burkert*, 2001). Also, the clusters within which most galactic-field stars form do not sufficiently harden an initially wide separation distribution to be consistent with the number of tight binaries. Such clusters that would lead to significant hardening would disrupt too many of the wide binaries diluting the galactic field binary-star population.

This indicates that the observed broad period distribution (Fig. 1) is already imprinted at the time of binary formation. The existing N-body simulations of young clusters tend to assume a relatively well-mixed and relaxed initial distribution of stars. However, real young clusters tend to be lumpy and unrelaxed, which alters the binary-binary interaction rate (e.g., *Goodwin and Whitworth*, 2004), but is not expected to change the general outcome [a smaller-N system with a smaller radius is *dynamically equivalent* to a larger-N system with a larger radius (*Kroupa*, 1995a)].

2.3.3. The effect of dynamics on the binary population. Small-N decay will act to modify the *birth binary population* (i.e., that produced by star formation) by reducing the overall binary fraction and by hardening the remaining binaries on a timescale of $<10^5$ yr. Small-N decay cannot occur very often as it would produce too many single PMS stars and too many hard binaries (and quite possibly too many BDs). However, it is unclear if small-N decay is rare because cores usually form only binaries and only sometimes triples, or because higher-order multiples are formed in stable, hierarchical systems.

The *initial binary population* [i.e., the population after small-N decay and internal energy redistribution processes, "eigenevolution" (see below), has modified it] is further changed by dynamical interactions within a cluster on a timescale of 1 m.y. for typical galactic star-forming clusters (*Kroupa*, 1995a,b; *Adams and Myers*, 2001). Encounters will destroy soft and active binaries, leaving mostly the hard binary population unchanged. Crucially it cannot produce more binaries in any significant numbers.

Thus both of the processes that act to modify the birth binary population into the field binary population *reduce* the binary fraction. We are led to the conclusion that the birth binary fraction must be higher than that of the field. This conclusions agrees well with observations.

Goodwin and Kroupa (2005) argue that the birth binary fraction must be almost unity for all stars, and that the low binary fraction among M dwarfs is due to the preferential destruction of low-mass binaries. *Lada* (2006), however, argues that most stars form as single stars, as most M dwarfs are single and most stars are M dwarfs. The crucial issue here is how many initial M-dwarf binaries decay? We note that in a model that assumes that stars are born with a 60% binary fraction with companion masses selected randomly from the IMF and without dynamical dissolution of the binaries leads to a population with the observed binary fraction-spectral type relationship (*Kroupa et al.*, 1993, their Fig. 11). In models that assume a 100% initial binary fraction, processing through an "average" cluster (*Lada and Lada*, 2003) also converts the binary population into the observed field population. Observations of the binarity of a very young (embedded?) M star population in an average cluster (as opposed to Taurus, which is atypical) are required to fully resolve this issue. However, for higher-mass stars, it seems almost certain that the initial binary fraction must be *significantly* higher than the field binary fraction due to their current relatively high binary fraction. It is worth noting that the observed population of high-order multiples may be formed from the final handful of stars that remain at the end of cluster dissolution, which tend to be in high-order hierarchical systems (see *Goodwin and Kroupa*, 2005).

Thus, while more work needs to be done, especially on the effects of the cluster density and on the initial distribution in star-forming regions with very different physical properties, a picture has emerged in which the observed high-multiplicity PMS population can be modified by secular dynamical evolution to produce the field population. Most (especially K dwarfs and later) stars therefore form in multiple — probably binary and triple — systems with a very wide separation distribution and a relatively flat mass-ratio distribution.

2.4. The Initial Binary Population

Given these results, a useful working hypothesis appears to be that the initial binary-star properties are invariant to star-formation conditions. The observed differences between binary populations result from different secular dynamical histories of the respective populations; i.e., due to different cluster masses and densities.

Kroupa (1995b) suggests that the field binary properties can be understood if the *birth* binary population has the following semi-empirical distribution functions:

1. Companion masses are chosen randomly from the IMF.

2. The distribution of periods is independent of primary mass, and can be described with the following functional form

$$f_{lP} = 2.5 \frac{lP - 1}{45 + (lP - 1)^2} \tag{2}$$

where $dN_{lP} = N_{tot} f_{lP} dlP$ is the number of binaries with periods in the range lP, $lP + dlP$ ($lP \equiv \log_{10}P$, P in days) and N_{tot} is the number of single-star and binary-star systems in the sample under consideration.

3. The eccentricity distribution is thermal (all binding energies are equally occupied)

$$dN = f_o f(e) de = f_o 2e de \tag{3}$$

being the number of binary systems with eccentricity in the range e, e + de.

These birth distributions need to be modified for short period binaries ($lP \lesssim 3$) through the evolution of the binding energy and angular momentum owing to dissipative processes within the young binary system termed collectively as pre-main-sequence eigenevolution. This then gives the *initial* distributions, which are evident in dynamically unevolved populations (e.g., Taurus) (*Kroupa and Bouvier*, 2003a) and can be used as the initial binary-star population in N-body modeling of stellar populations.

The distribution over binding energies and specific angular momenta can be evaluated readily given the above distribution functions. Figure 4 shows that the distribution of specific angular momenta of molecular cloud cores forms a natural extension to the distribution of specific angular momenta of the initial binary stellar population, possibly suggesting an evolutionary connection.

These semi-empirical distribution functions have been formulated for late-type stars (primary mass $m_p \lesssim 1 \ M_\odot$) as it is for these that we have the best observational constraints. It is not clear yet if they are also applicable to massive binaries. *Baines et al.* (2006) report a very high (f ≈ 0.7 ± 0.1) binary fraction among Herbig Ae/Be stars with the binary fraction increasing with increasing primary mass. Further-

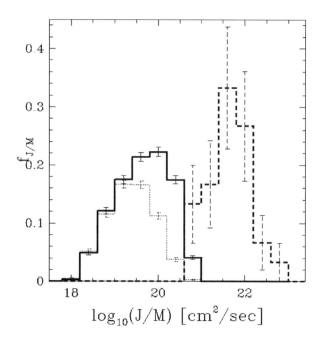

Fig. 4. The observed specific angular momentum distribution of molecular cloud cores by *Goodman et al.* (1993) is shown as the rightmost dashed histogram, while the initial binary-star population (section 2.4) is plotted as the solid histogram. It evolves to the dot-dashed histogram after passing through a typical star cluster (from *Kroupa*, 1995b).

more, they find that the circumbinary disks and the companions appear to be co-planar, thereby supporting a fragmentation origin rather than collisions or capture as the origin of massive binaries. Most O stars are believed to exist as short-period binaries with q ≈ 1 (*García and Mermilliod*, 2001), at least in rich clusters, while small q appear to be favored in less-substantial clusters such as the Orion Nebula Cluster (ONC), being consistent there with random pairing (*Preibisch et al.*, 1999). *Kouwenhoven et al.* (2005) report the A and late-type B binaries in the Scorpius OB2 association to have a mass-ratio distribution *not* consistent with random pairing. The lower limit on the binary fraction is 0.52. Perhaps the massive binaries in the ONC represent the primordial population, whereas in rich clusters and in OB associations the population has already dynamically evolved through hardening and companion exchanges to that observed there. This possibility needs to be investigated using high-precision N-body computations of young star clusters.

Given such reasonably well quantified estimates of the distribution functions of orbital elements of the primordial binary population, the problem remains as to how these distributions functions can be understood theoretically as a result of the star-formation process. *Fisher* (2004) notes that distribution functions similar to the ones derived above, and in particular a wide mass-ratio distribution, a very wide period range, and a thermal eccentricity distribution, are obtained quite naturally from a turbulent molecular cloud (see also *Burkert and Bodenheimer*, 2000).

3. THE PROPERTIES OF PRESTELLAR CORES

The gas that is just about ready to form stars arranges itself into denser structures often called prestellar "cores" (e.g., *Myers and Benson*, 1983). Often, a "typical" prestellar core is described as having a radius ~0.1 pc, density $\geq 10^4$ cm^{-3}, and velocity dispersion ~0.5 km s^{-1}. In fact, however, the idea that such cores are "typical" primarily arises from the relative ease with which nearby, isolated dense cores, which will each form fewer than a handful of stars, can be observed and modeled. It is in fact likely that accounting for the diversity in core properties is crucial to improving the match between theory and observations of the conversion of gas to (binary) stars.

3.1. What is a "Core"?

In general, observations and theory have concentrated on *isolated* and *coherent* prestellar cores such as those found in low-mass star-forming regions such as Taurus (due to the relative ease of observing such cores). It is not yet clear if legitimate analogs to these cores exist within the dense concentrations of gas that form the clusters (e.g., *Goodman et al.*, 1998). In particular, the ~0.1-pc size of isolated cores would result in them having multiple dynamical encounters in the dense environment that forms "typical" clusters such as Orion (e.g., *Lada and Lada*, 2003).

That we have not observed any 0.1-pc core analogs in dense clusters is not surprising. Even in very nearby clusters like NGC 1333 in Perseus (at ~300 pc), 0.1 pc is 1 arcmin, typical of single-dish resolution for tracers like NH$_3$ and N$_2$H$^+$, which map out gas with density $\geq 10^4$ (e.g., *Benson and Myers*, 1989; *Evans*, 1999). Thus, to find meaningful dense structures on scales significantly less than 1 arcmin, interferometry is required. Interferometers have definitively revealed substructure in the gas within clusters, but this substructure does not offer a one-to-one gas clump-to-star match the way observations of isolated cores do. Instead, regions forming many stars are associated with more dense gas than those forming fewer. This lack of one-to-one correspondence suggests that long-lived blobs associated with the formation of individual cores within clusters do not exist.

Given this, is it reasonable to extend observations and simulations of isolated cores to more typical clustered star-forming "cores"? The answer is, possibly. While large isolated cores cannot exist in clustered star-forming regions, the regions in which stars are thought to form are far smaller than the size of a whole core. Observationally, the size of binary systems is less than a few hundred AU, in agreement with theoretical expectations of the scale of fragmentation (see section 4.1). Systems of this scale would be expected to interact on timescales of >1 m.y. in a typical cluster that is significantly longer than the star-formation timescale is thought to be (see chapters by Di Francesco et al. and Ward-Thompson et al.). So, while the details of the first stages of collapse in isolated cores are probably not applicable to most star formation, the details of the final stages of frag-

mentation and star formation occurring on scales of a few hundred AU quite possibly occur in relative isolation. However, continued accretion onto cores may significantly effect the evolution of the inner protostellar system depending on the details and timescale of core and star formation in clusters.

With this in mind, we will continue to review the properties of isolated prestellar cores.

3.2. Rotation

Clearly, for cores to fragment, some angular momentum must be present, otherwise the cores will collapse onto a single, central point. The simplest source of this angular momentum is due to bulk rotation of the core. It is a relatively straightforward procedure to estimate the component of solid-body rotation present in a dense core by fitting for the gradient in observed line-of-sight velocity over the face of a core (e.g., *Goodman et al.*, 1993; *Barranco and Goodman*, 1998; *Caselli et al.*, 2002). The results of this fitting (see Fig. 4) have been used as input values of "initial angular momentum" in many calculations. While the estimates of the *component* of solid-body rotation made in this way are sound, and are thus fine to use as inputs, it is important to appreciate that cores do not really rotate as solid-bodies (*Burkert and Bodenheimer*, 2000). When the velocity measurements are put into the context of measurements of velocities on larger scales, both observations (*Schnee et al.*, 2005) and simulations (*Burkert and Bodenheimer*, 2000) show that the "rotation" is often just an artifact created by larger-scale turbulent motions.

Figure 4 gives a summary of the measured specific angular momentum for core rotation for the 29 dense cores from *Goodman et al.* (1993), which show significant rotation. The majority of cores included in the *Goodman et al.* (1993) study are isolated, low-mass cores: One should keep in mind that the rotational properties of smaller fragments that may form inside those cores as true(r) precursors to protostars remain largely unmeasured.

3.3. Nonthermal Line Widths

It has been known for many years (*Larson*, 1981; *Myers*, 1983; *Solomon et al.*, 1987; see also *Elmegreen and Scalo*, 2004a, and references therein) that the line widths inside even the most quiescent of dense cores are more than thermal. The coldest isolated dense cores have gas temperatures on the order of 10 K, and dust temperatures as low as 6 K (see chapters by Di Francesco et al. and Ward-Thompson et al.). A gas temperature of 10 K implies an H$_2$ 1σ velocity dispersion of only 0.2 km s^{-1}. Observed dispersions have a distribution from ~0.2 to 1 km s^{-1} with a peak at ~0.4 km s^{-1}, never quite reaching down to the thermal value [e.g., see the catalog of *Jijina et al.* (1999)].

The origin of the nonthermal line width in dense cores is the subject of an extensive literature, but it is fair to say that a consensus exists that "turbulence" is responsible (see, e.g., *Myers*, 1983; *Barranco and Goodman*, 1998; *Good-*

man et al., 1998; *Elmegreen and Scalo*, 2004a). Significant levels of turbulence in cores are important for fragmentation as they may provide the angular momentum to form multiple systems (see section 4.2.2) and, as mentioned above, could also be responsible for the observed rotation.

3.4. Magnetic Fields

The role of magnetic fields in supporting cores against collapse is a subject of much debate. Magnetic fields are not thought to be dynamically dominant in cores as was once thought [e.g., *Shu et al.* (1987), or indeed in the ISM as a whole (*Elmegreen and Scalo*, 2004a)]. However, they may be very important as the role of magnetic fields in the fragmentation of cores is poorly understood (see section 4.2.6).

Isolated cores are found to be statistically triaxial with a tendency toward being prolate (*Jones et al.*, 2001; *Goodwin et al.*, 2002). In contrast to this, magnetic support would tend to produce oblate cores. In addition, these magnetically supported oblate cores would tend to rotate around their short axis, which is not observed (*Goodman et al.*, 1993). This is supported by observations of the magnetic field, which show that cores are not magnetically critical (*Crutcher et al.*, 1999; see *Bourke and Goodman*, 2004, and references therein).

4. THE FRAGMENTATION OF PRESTELLAR CORES

4.1. The Physics of Collapsing Cores

During the early stages of the collapse of a prestellar core, the rate of compressional heating is low and the gas is able to cool radiatively, either by molecular line emission, or, when $\rho > 10^{-19}$ g cm^{-3}, by thermally coupling to the dust. The gas is therefore approximately isothermal (at ~10 K) with an equation of state $P \propto \rho$.

Eventually the rate of compressional heating becomes so high (due to the acceleration of the collapse), and the rate of radiative cooling so low (due to the increasing column density and dust optical depth), that the gas switches to being approximately adiabatic, with $P \propto \rho^\gamma$ (where $\gamma = 5/3$ initially for a monatomic gas, and then $\gamma = 7/5$ above ~300 K when H$_2$ becomes rotationally excited).

This behavior has been studied in detail (*Larson*, 1969; *Tohline*, 1982; *Masunaga et al.*, 1998; *Masunaga and Inutsuka*, 2000) for cores in the range 1–10 M$_\odot$ with an initial temperature of 10 K. These authors find that the switch between isothermality and adiabaticity occurs at a critical density of $\rho_{crit} \sim 10^{-13}$ g cm^{-3} [see Fig. 1 in *Bate* (1998) or Fig. 2 in *Tohline* (2002)]. Thus, as contraction proceeds and the density, ρ, increases the Jeans mass, $M_J \propto \rho^{-1/2}T^{3/2}$, decreases as long as the gas can retain the same temperature, T, while it is optically thin. Once the opacity increases such that the gas core heats up, M_J increases. The most important result of this thermal behavior is therefore that there is

a *minimum* Jeans mass that is reached at ~ρ_{crit} on the order of $M_{min} \sim 10^{-2}$ M$_\odot$. This is often referred to as the opacity limit for fragmentation.

There is an even lower minimum mass that occurs during a later isothermal phase at $\rho \sim 10^{-3}$ g cm^{-3} when molecular hydrogen dissociates at a few thousand K. It is possible that a further fragmentation episode can occur at these densities, which may account for some close binaries.

Fragmentation in cores is expected to occur at around ρ_{crit} as at lower or higher densities the Jeans mass increases (although how rapidly it rises above ρ_{crit} does depend sensitively on the γ used in the adiabatic equation of state). Thus we expect multiple systems to be formed with a typical length scale R_{form} of

$$R_{form} < \left(\frac{3M_{core}}{4\pi\rho_{crit}} \right)^{1/3} \sim 125(M_{core}/M_\odot)^{1/3} \text{ AU} \quad (4)$$

where M_{core} is the mass of the core. Interestingly, this scale matches the observed peak in the T Tauri separation distribution (see also *Sterzik et al.*, 2003).

There is a minimum separation in this picture of ~30 AU, which is the separation of two fragments of M_{min} at ρ_{crit}. It may — or may not — be significant that this is the *average* binary separation (*DM91*; see also *Sterzik et al.*, 2003). However, it would appear difficult to form binaries closer than ~20–30 AU without some hardening mechanism *or* a secondary fragmentation phase.

It should be noted that the length scales of star formation of less than a few hundred AU are several orders of magnitude smaller than the thousands of AU scales on which core properties have been observed.

4.2. Fragmentation Mechanisms

In this section we examine the main mechanisms that have been proposed to explain multiple formation. Given the complex and highly nonlinear nature of the physics in most models, numerical simulations are the main route by which the mechanisms for fragmentation have been investigated. Bulk rotation and turbulence are the two main mechanisms that have been considered to provide the angular momentum required for fragmentation to occur, and we review the theoretical work and simulations conducted on both of these mechanisms. In addition, we discuss the possible role of magnetic fields, disk fragmentation, and "secondary fragmentation."

4.2.1. Rotational fragmentation. The simplest situation in which fragmentation may well occur is in a spherical cloud with solid-body rotation and an isothermal equation of state. *Tohline* (1981), using semianalytic arguments, concluded that all such clouds should fragment. A number of simulations have shown that such clouds do fragment if $\alpha_{therm}\beta_{rot} \lesssim 0.12–0.15$, where $\alpha_{therm} = E_{therm}/|\Omega|$ is the initial thermal virial ratio (where E_{therm} is the thermal kinetic and Ω is the gravitational potential energy), and $\beta_{rot} = E_{rot}/$

$|\Omega|$ the initial rotational virial ratio (where E_{rot} is the rotational kinetic energy) (*Miyama et al.*, 1984; *Hachisu and Eriguchi*, 1984, 1985; *Miyama*, 1992; see also *Tsuribe and Inutsuka*, 1999a,b; *Tohline*, 2002).

Boss and Bodenheimer (1979) added an m = 2 azimuthal density perturbation to a standard rotating cloud (effectively creating an elongated cloud more similar to those observed than purely spherical clouds; see section 3.4). They found that with a perturbation of amplitude A = 0.5 the cloud fragments into a binary system. This simulation was repeated by *Burkert and Bodenheimer* (1993) who also found that when A = 0.1, a filament connecting the two components of the binary fragments into several smaller fragments. However, the connecting filament should not fragment, as predicted by *Inutsuka and Miyama* (1992) and demonstrated by *Truelove et al.* (1997). Indeed the "Boss and Bodenheimer test" has become a standard test for the accuracy of codes [e.g., *Truelove et al.* (1997) for adaptive mesh refinement (AMR) and *Kitsionas and Whitworth* (2002) for smoothed particle hydrodynamics (SPH)]. However, it is a rather unsatisfactory test as, while the *Truelove et al.* (1997) simulations are generally considered to have converged, no analytic solution to the problem exists. An alternative test based on the original analysis of Jeans is presented by *Hubber et al.* (2006).

The simulation of rotating clouds can be made more physical by including an adiabatic (e.g., *Tohline*, 1981; *Miyama*, 1992) or barotropic (e.g., *Bonnell*, 1994; *Bate and Burkert*, 1997; *Boss et al.*, 2000; *Cha and Whitworth*, 2003) equation of state (eos). *Bate and Burkert* (1997) showed that the Boss and Bodenheimer test *does* produce a line of fragments with a barotropic eos, but not if it remains isothermal. In addition, *Boss et al.* (2000) simulated a cloud with an m = 2, A = 0.1 perturbation using a barotropic eos *and* also with radiation transport; the second case producing a binary while the first did not, despite the similarity of the pressure-temperature relations. Both of these results suggest that fragmentation is highly sensitive to thermal inertia and radiation transport effects.

Other authors have modified the initial conditions to include effects such as different density profiles (e.g., *Myhill and Kaula*, 1992; *Burkert et al.*, 1997; *Boss*, 1996; *Boss and Myhill*, 1995; *Burkert and Bodenheimer*, 1996; *Boss et al.*, 2000; *Boss*, 1993), differential rotation (which tends to promote fragmentation) (*Myhill and Kaula*, 1992; *Boss and Myhill*, 1995; *Cha and Whitworth*, 2003), and nonspherical shapes (e.g., *Bastien*, 1983; *Bonnell and Bastien*, 1991; *Bonnell et al.*, 1991; *Nelson and Papaloizou*, 1993; *Boss*, 1993; *Sigalotti and Klapp*, 1997). The effect of increasing external pressure on the collapse of rotating cores has been investigated by *Hennebelle et al.* (2003, 2004, 2006).

4.2.2. Turbulent fragmentation. Recently, a picture of star formation as a rapid and highly dynamic process has appeared (e.g., *Elmegreen*, 2000; *Váquez-Semadeni et al.*, 2000; *Larson*, 2003; *Elmegreen and Scalo*, 2004b), as opposed to a quasistatic process (e.g., *Shu et al.*, 1987). In particular, the idea of cores evolving slowly via ambipolar diffusion (e.g., *Basu and Mouschovias*, 1994, 1995a,b; *Ciolek and Mouschovias*, 1993, 1994, 1995; *Ciolek and Basu*, 2000) has been replaced by one in which cores form in converging flows in a highly turbulent molecular cloud. This is rather good news for fragmentation, as the main effects of a quasistatic evolution are to delay fragmentation and reduce the angular momentum and turbulence in a core and organize material so that its collapse is well focused onto a central point (see also section 3.2). Simulations of core formation in a turbulent medium suggest that cores form with significant amounts of turbulence. Turbulent, rapidly formed cores also reproduce many of the observed properties of cores (e.g., *Burkert and Bodenheimer*, 2000; *Ballesteros-Paredes et al.*, 2003; *Jappsen and Klessen*, 2004) and have a mass spectrum not dissimilar to the observed core mass spectrum (e.g., *Padoan and Nordlund*, 2002, 2004; *Klessen et al.*, 2005).

Simulations of the effects of turbulence in cores focus on two different regimes: high-velocity (*Bate et al.*, 2002, 2003; *Bate and Bonnell*, 2005; *Delgado Donate et al.*, 2004a,b) and low-velocity (*Goodwin et al.*, 2004a,b) [see also *Fisher* (2004) for a semianalytic approach to multiple formation with turbulence]. The level of turbulence is usually quantified as a turbulent virial ratio $\alpha_{turb} = E_{turb}/|\Omega|$, where E_{turb} is the kinetic energy in turbulent motions and $|\Omega|$ is the gravitational potential energy (note — not any rotational property). Highly turbulent simulations focus on $\alpha_{turb} = 1$ in 50-$M_{\odot}$ (*Bate et al.*, 2002, 2003; *Bate and Bonnell*, 2005) and 5-$M_{\odot}$ (*Delgado Donate et al.*, 2004a,b) cores. Simulations of slightly turbulent cores range between $\alpha_{turb} = 0$–0.25 in 5.4-$M_{\odot}$ cores (*Goodwin et al.*, 2004a,b). In all these simulations turbulent motions are modeled using a Gaussian divergence-free random velocity field $P(k) \propto k^{-n}$ where n is usually taken to be 4 to match observations of cores for which n = 3–4 provides a good fit to the Larson relations (*Burkert and Bodenheimer*, 2000). It should be noted that the random chaotic effects introduced by variations in the initial turbulent velocity field can be very important. Therefore a statistical approach utilizing large ensembles of simulations is desirable (e.g., *Larson*, 2002).

These simulations of turbulence are different to those of turbulence in molecular clouds, which concentrate on the formation of dense cores and massive stars (e.g., *Klessen and Burkert*, 2000). This is largely due to computational limitations that do not allow the resolution of the opacity limit for fragmentation in the large-scale context of giant molecular clouds. A mass resolution of ~10^{-2} $M_{\odot}$ is required to resolve the opacity limit for fragmentation in SPH. In AMR the problem is even worse, as the Jeans length continues to fall (albeit more slowly) after the minimum Jeans mass is reached and codes must resolve few-AU scales to capture the lowest-mass fragments.

Even very low levels of turbulence ($\alpha_{turb} \sim 0.025$) are enough to allow cores to fragment, i.e., for most cores in an ensemble to form more than one star (*Goodwin et al.*,

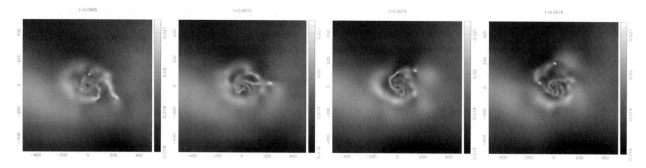

Fig. 5.   The evolution of a CAR in a turbulent molecular core from *Goodwin et al.* (2004a). The boxes are 1000 AU on a side and the timescale runs from 66.6 to 67.8 k.y. after the start of the simulation in steps of 400 yr. The gray-scale bar gives the column density in g cm^{-2}. The spiral features can become self-gravitating if their density exceeds ~10^{-12} g cm^{-3} as their Jeans length falls to ~20 AU, which can allow collapse without being shredded. This does not always occur; in the first two panels a dense knot can be seen being shredded and accreted onto the central object. Very similar evolution is also seen in AMR simulations (*Gawryszczak et al.,* 2006).

2004b). As the level of turbulence is increased, the average number of stars that form in a core increases (*Goodwin et al.,* 2004b). It has been suggested that approximately one star forms per initial Jeans mass: ~1 M$_\odot$ for these initial temperatures and densities (cf. *Bate et al.,* 2003; *Delgado Donate et al.,* 2004a). This seems to hold in highly turbulent cores; however, the number of stars forming falls with decreasing turbulence (*Goodwin et al.,* 2004b) and so this at best probably only represents a (statistical) asymptotic behavior.

In highly turbulent cores, the supersonic turbulent velocity field creates a number of condensations in shocked, converging regions that become Jeans unstable and collapse (see *Bate et al.,* 2003; *Delgado Donate et al.,* 2004a,b). However, it is unclear if this mode of fragmentation is realistic in small (certainly less than a few M$_\odot$) cores as the observed levels of nonthermal motions rule out significant highly supersonic turbulence in these cores.

In Fig. 5 we show the formation of a fragment in a mildly turbulent 5.4 M$_\odot$ core with α_{turb} = 0.05 based on observations of the isolated core L1544 (from *Goodwin et al.,* 2004a). Fragmentation occurs in a "disk-like" mode in circumstellar accretion regions (CARs) (we avoid the use of "disk" to describe these regions as they are not rotationally supported structures) that form around the first star. CARs are highly unstable structures as there is nonuniform (in space, time, and angular momentum) inflow onto them. Complex spiral instabilities form in the CAR due to this inhomogeneous infall of material. We note that these instabilities are seen in both SPH *and* AMR simulations of the same situation (*Gawryszczak et al.,* 2006) (see also section 4.4). Fragmentation occurs if the density in spiral waves becomes high enough that the Jeans length falls to the typical width of a spiral wave and the collapse time falls to a low enough fraction of the local rotation period that it may escape shredding by differential rotation. In these simulations, fragmentation occurs for some (but not all) regions that exceed ~10^{-12} g cm^{-3} in density (equating to a Jeans length of ~20 AU) beyond ~50–100 AU from the central

star. We note that the highly unstable nature of CARs makes usual applications of instability criteria such as the Toomre Q parameter impossible.

Such a mode of fragmentation is highly sensitive to the equation of state that has been adopted. It has been found that the number of fragments that form increases if γ is changed from 5/3 to 7/5 (Goodwin et al., in preparation). This is due to the sensitivity of the Jeans length with density and so to the ease with which fragmentation can occur in CARs.

The process of fragmentation in CARs is highly chaotic, relying as it does on a certain degree of "luck" in being able to reach a high enough density and avoiding shredding while collapsing. Thus it is no surprise that anywhere between 1 and 12 stars form in each core depending entirely on the details of the initial turbulent velocity field (*Goodwin et al.,* 2004a,b).

In summary, it is found that turbulent cores generally fragment into several stars: approximately one per initial Jeans mass (~1 M$_\odot$) in the core. The number of stars that form increases with increasing turbulence and is also highly sensitive to the details of the turbulent velocity field. However, only relatively high-mass cores (>5 M$_\odot$) have been investigated in turbulent simulations so far. The effect of turbulence in lower-mass cores must be investigated, as lower-mass cores appear to dominate the core mass function (*Motte et al.,* 1998, 2001; *Testi and Sargent,* 1998).

4.2.3. Disk fragmentation.   Disk fragmentation is a mechanism by which low-mass stars and BDs may be formed. In the dense environments of clusters close encounters between stars can disturb the circumstellar disks, promoting instabilities that can lead to the fragmentation of otherwise stable disks. (Note that this is rather different from the turbulent disk-like scenario described above, as these protoplanetary disks are much less massive than CARs and are also stable, rotationally supported disks as opposed to CARs.)

In a series of papers, *Boffin et al.* (1998) and *Watkins et al.* (1998a,b) found that most star-disk interactions will lead to gravitational instabilities that form new low-mass companions. These simulations generally considered massive disks where $M_{star} = M_{disk} = 0.5\,M_\odot$. *Bate et al.* (2003) find that star-disk encounters play an important role in forming binaries and also truncating disks. Star-disk encounters are also thought to play an important role in redistributing angular momentum in protoplanetary disks even if they do not cause further fragmentation (*Larson*, 2002; *Pfalzner*, 2004; *Pfalzner et al.*, 2005).

Star-disk encounters probably play a role in star formation, and may lead to the formation of BD (or even planetary) mass companions (*Thies et al.*, 2005). However, they are probably not a significant contributor to the primordial stellar binary population. This is due to the requirement that the encounters occur early in the star-formation process — during the Class 0 phase when the disk mass is still very large compared to the stellar mass — a phase that lasts for only $\sim 10^5$ yr, leaving only a small time for encounters to occur. However, the role of disk fragmentation in planet formation may well be important.

4.2.4. The role of magnetic fields. The treatments of collapse and fragmentation discussed above do not include magnetic fields. The new picture of rapid, turbulence-driven star formation combined with the lack of observational evidence for magnetically critical cores suggests that magnetic fields are not dynamically dominant. In addition, fragmentation is expected to occur at densities $\geq 10^{-13}$ g cm^{-3}, densities at which the magnetic field is expected to be decoupled from the gas due to the extremely low fractional ionization (see *Tohline*, 2002). However, possibly one of the main reasons for neglecting magnetic fields is the difficulty in including them in SPH simulations [although this is improving; see in particular *Hosking and Whitworth* (2004a,b) and *Price and Monaghan* (2004a,b)]. Magnetic fields *are* clearly present in (many) cores, even if they are not dynamically dominant, and their effects may be very important. Grid-based simulations that include magnetic fields in rotating clouds show that fragmentation can occur in these clouds, although magnetic fields appear to have a tendency to suppress fragmentation (e.g., *Hosking and Whitworth*, 2004b; *Machida et al.*, 2005b), although *Boss* (2002, 2004) claims the opposite. *Sigalotti and Klapp* (2000) find binary and higher-order multiple formation in slowly rotating ~ 1-$M_\odot$ clouds that includes a model for ambipolar diffusion.

Possibly the most extensive investigation of the effects of magnetic fields on fragmentation has been made by *Machida et al.* (2005a,b). They find that fragmentation occurs in $\sim 50\%$ of their rotating, magnetized clouds when either the rotation is relatively high or magnetic field strength relatively low. In particular, fragmentation always occurs in magnetized clouds if $\beta_{rot} > 0.05$, but it almost never occurs below this limit [see Fig. 10 of *Machida et al.* (2005b)]. Indeed, *Burkert and Balsara* (2001) conclude that once magnetic fields are strong enough to affect the dynamical evolution they will also efficiently suppress fragmentation, which means that magnetic fields cannot be important as we know that fragmentation *must* occur.

4.3. "Secondary" Fragmentation

As briefly mentioned in section 4.1, there is a second isothermal phase in the evolution of gas toward stellar densities. This occurs at a temperature of ~ 2000 K and a density of $\sim 10^{-3}$ g cm^{-3} when molecular hydrogen dissociates into atomic hydrogen. This phase occurs in the hydrostatic protostar when its radius is ~ 1 AU and — if fragmentation can occur at this stage — it may explain very close binaries.

Both *Boss* (1989) and *Bonnell and Bate* (1994) simulated the collapse of a rotating hydrostatic first object to high densities. They found that fragmentation can occur in axisymmetric instabilities or a ring formed by a centrifugal bounce. However, *Bate* (1998) found that spiral instabilities remove angular momentum and suppress further fragmentation. Recent two-dimensional simulations by *Saigo and Tomisaka* (2006) suggest that the angular momentum of the first core is a crucial factor in determining if fragmentation will occur during the second collapse.

Thus it is unclear if a secondary fragmentation phase occurs. However, we suggest that such a phase could well be responsible for the apparently high incidence of very close BD-BD binary systems (*Pinfield et al.*, 2003; *Maxted and Jeffries*, 2005), as the evolution of BD-mass hydrostatic objects occurs on a longer timescale than in stellar-mass objects. This possibility is being investigated by the lead author without any firm conclusions as yet.

4.4. Simulations vs. Observations

A summary of the simulations to date suggests that collapsing cores are easily able to fragment. However, no detailed model is currently able to correctly predict all the observed binary properties.

A successful model of star formation must produce multiple systems that generally have only two or three stars with a wide range of separations from $\ll 1$ AU to a peak at ~ 100–200 AU. At all separations, most stars must usually have quite different masses, but avoiding BDs within at least 5 AU of the primary (the BD desert).

Possibly the most significant problem at the moment is that simulations seem to form *too many* single stars (see *Bouvier et al.*, 2001; *Duchêne et al.*, 2004; *Goodwin and Kroupa*, 2005). As described in section 2.3.1, systems with $N \geq 3$ are generally unstable and decay by ejecting their lowest-mass member and hardening the remaining multiple. The ejection of members of small-N multiple systems dilutes the multiplicity of stars, as ejected stars tend to be single. Thus, many ejections will result in a far lower multiplicity fraction than is observed in young star-forming regions. *Goodwin and Kroupa* (2005) suggest that the observed multiplicity frequencies can be explained if roughly half of the cores form two stars, and half form three stars. However, these numbers are far lower than are usually found in core fragmentation simulations.

The inclusion of magnetic fields produces the opposite problem, that too few binaries are produced. *Machida et al.*

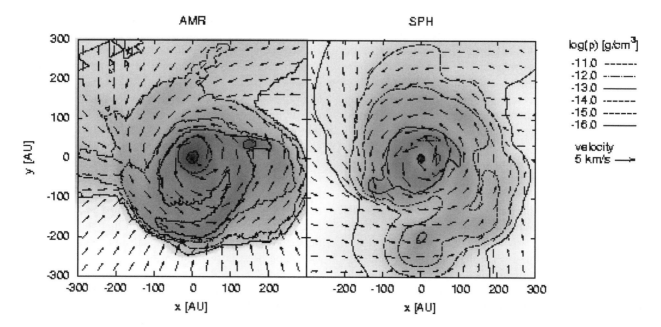

Fig. 6. Comparison of an AMR (left) and a SPH (right) simulation of a collapsing, turbulent 5.4 M$_\odot$ core showing the first "disk" fragmentation episode from *Gawryszczak et al.* (2006). The bound fragments can be seen at (0.125) AU in the AMR simulation (left) and (–200.0) AU in the SPH simulation (right). The scale and orientation of both views are identical.

(2005b) find the fragmentation does not occur in rotating, magnetized clouds when $\beta_{rot} \lesssim 0.04$ — a higher level of rotation than is observed in many cores. This problem becomes especially acute when we consider that much of the observed rotation in cores could well be due to turbulent motions rather than a bulk rotation.

A related problem is posed by the existence of a significant number (~20% of field G-dwarfs) (*DM91*) of close, unequal mass binary systems. It appears difficult to form stars much closer together than ~30 AU. In order to obtain hard binary systems, a further hardening mechanism is required. In many simulations of turbulent star formation, this hardening mechanism is provided by the ejection of low-mass components (e.g., *Bate et al.*, 2003; *Delgado Donate et al.*, 2004a,b; *Goodwin et al.*, 2004a,b; *Umbreit et al.*, 2005; *Hubber and Whitworth*, 2005; see *Ochi et al.*, 2005, for a caveat). However, ejections appear to occur rapidly, during the main accretion phase, producing many close, equal-mass binary systems (see above) in contradiction to observations of a relatively flat mass ratio distribution (*Mazeh et al.*, 1992). *White and Ghez* (2001) do find many roughly equal-mass PMS binaries <100 AU in Taurus; however, there is no trend to more equal-mass ratios at very low separations. *Fisher et al.* (2005) do find a bias toward equal-mass binaries among local field spectroscopic binaries (with separations ≲1 AU).

In general, simulations that produce a small number of stars consistent with limits on ejection do not produce hard binaries. These binaries are difficult to form in significant numbers through later dynamical interactions in a clustered environment, which tend to disrupt wide binaries but not harden them. But simulations that produce many stars tend to form too many close, equal-mass binaries and very-high-

order multiple systems (many quadruples and quintuples) and also dilute the multiplicity fraction too much through ejections.

4.5. Numerical Issues

As already discussed in the chapter by Klein et al., there is some debate about the ability of simulations to correctly resolve fragmentation. Simulations of star formation are usually conducted using SPH as opposed to AMR schemes due to the Lagrangian nature of SPH (see *Gawryszczak et al.*, 2006, for details).

No numerical scheme is perfect, and both SPH and AMR have their advantages and disadvantages. It is worth noting that a recent study by *Gawryszczak et al.* (2006) has shown that AMR and SPH converge when simulating the collapse of a slightly turbulent core. Figure 6 shows a snapshot of both the SPH and AMR simulations at roughly 76 k.y. from the start of the simulation, showing that in both numerical schemes a highly unstable CAR forms that fragments in both simulations. Gawryszczak et al. found that AMR is significantly more computationally intensive than SPH for an identical simulation. This result is not surprising, as when simulating gravitational collapse, the Lagrangian nature of SPH should prove highly efficient. The agreement of two very different methods when applied to the same physical situation increases our confidence that the results are not dominated by numerical effects.

Both SPH and AMR suffer from problems with artificial angular momentum transport. In AMR, rotation in a poorly resolved Cartesian grid is likely to transport angular momentum outward. In SPH, the use of artificial viscosity to reduce particle interpenetration in shocks produces an inward transport of angular momentum with rotation. These problems are probably responsible for the different CAR (disk) sizes between SPH and AMR seen by *Gawryszczak et al.* (2006) (see also Fig. 6).

It should be noted that *Hubber et al.* (2006) have shown that SPH supresses artificial fragmentation rather than promoting it, which suggests that the current generation of SPH simulations could *underestimate* the number of fragments that form.

We would conclude that the computational situation is not perfect and many problems with both SPH and AMR remain. However, we think that the conflict between simulation and observation is more probably due to missing and/or incorrect physics, rather than any fundamental numerical difficulties.

The computational situation is made more problematic by the need to perform ensembles of simulations to get the statistical properties of multiple systems. Even in cases where the result for any *single* set of parameters might be expected to converge, there is a large parameter space to cover, and small differences in initial conditions may make a significant difference to the result. More realistically, situations where fragmentation occurs due to nonlinear instabilities, or with turbulence (when the details of the velocity field vary), require large ensembles even for the *same* region of parameter space. No matter how detailed or correct any single simulation is, it can only be a snapshot of the outcome of a particular initial configuration. This vastly increases the computational effort required to model fragmentation and star formation.

4.6. Missing Physics

One of the greatest problems facing the simulation of core fragmentation is to correctly model the thermal physics of cores. It is not possible in the foreseeable future that we will be able to conduct hydrodynamical simulations that include a proper treatment of radiation transport as the computational expense is just too great. However, as shown by *Boss et al.* (2000) thermal effects can be very important. The use of the barotropic equation of state is at best a first approximation, and it is clear that varying the adiabatic exponent can have significant effects on fragmentation (Goodwin et al., in preparation). In particular, the barotopic equation of state is based on simulations that have used (necessarily) simplistic, spherically symmetric assumptions. It is not clear to what extent these can be applied to highly inhomogeneous cores, including disks and local density peaks. Improvements are being made including using a flux-limited diffusion approximation (*Whitehouse and Bate*, 2004) and an approximation based on the local potential as a guide to optical depth (Stamatellos et al., in preparation). However, a problem remains with any approximation in that, while it should obviously match the fully detailed radiative transfer simulations of simple situations, it is not clear if it is correct in more complex situations.

Most simulations (especially SPH simulations) do not include the effects of magnetic fields. Yet those simulations that do include magnetic fields seem to suggest that fragmentation is suppressed. This could well be a very important conclusion given that nonmagnetic models seem to overproduce stars in cores. However, the very efficient suppression of fragmentation by magnetic fields may rule out the importance of magnetic fields in the fragmentation process as we know that cores *must* fragment.

Given that fragmentation appears to occur in disk-like structures, the proper treatment of these is vital. Both AMR and SPH have problems with the artificial transport of angular momentum (see *Gawryszczak et al.*, 2006), which will affect their ability to correctly model disks. Disks are also an environment in which magnetic fields may play an important role.

Finally, very few simulations attempt to model the effects of feedback from stars as jets or through their radiation field [for some first attempts to deal with these problems see *Stamatellos et al.* (2005) and *Dale et al.* (2005)]. In particular, *Stamatellos et al.* (2005) find that the inclusion of jets may inhibit fragmentation by decreasing the inflow rate onto the disk and forcing that inflow to occur away from the poles. Many of the physical situations that may result in fragmentation are rather complex and often chaotic (turbulence being the most obvious example). Such situations will not produce any single, unique answer. Indeed, given the variety of multiple systems such a situation would not be expected. However, this does require that a statistical approach be taken when performing simulations. This vastly increases the computational effort required, as any "single" region of an already huge parameter space will require an ensemble of simulations to investigate it.

5. CONCLUSIONS

Almost all young stars are found in multiple systems with a very wide separation distribution and a fairly flat mass-ratio distribution. Thus prestellar cores must fragment into multiple stars and/or BDs with these properties. The dynamical decay of small-N systems would rapidly produce a large single-star pre-main-sequence population if large numbers of unstable systems form with N > 2 or 3. This decay would also result in large numbers of very close binary systems. Neither of these are observed, leading to the conclusion that cores must usually form only two to four stars in hierarchical systems (for N > 2).

Simulations show that most cores that contain some angular momentum — either in bulk rotation, or in turbulence — are able to fragment into multiple objects. However, these simulations have been unsuccessful in matching their results to the observed young multiple population. In particular, the distributions of separations and mass ratios from simulations tend not to fit well.

The future is somewhat rosier, however. The inclusion of more detailed physics and more realistic initial conditions may well yield better fits to observations.

Acknowledgments. S.P.G. is supported by a UK Astrophysical Fluids Facility (UKAFF) fellowship.

REFERENCES

Adams F. C. and Myers P. C. (2001) *Astrophys. J., 553,* 744–753.

Anosova Zh. P. (1986) *Astrophys. Space Sci., 124,* 217–241.

Baines D., Oudmaijer R., Porter J., and Pozzo M. (2006) *Mon. Not. R. Astron. Soc., 367,* 737–753.

Ballesteros-Paredes J., Klessen R. S., and Vázquez-Semadeni E. (2003) *Astrophys. J., 592,* 188–202.

Barranco J. A. and Goodman A. A. (1998) *Astrophys. J., 504,* 207–222.

Bastien P. (1983) *Astron. Astrophys., 119,* 109–116.

Basu S. and Mouschovias T. Ch. (1994) *Astrophys. J., 432,* 720–741.

Basu S. and Mouschovias T. Ch. (1995a) *Astrophys. J., 452,* 386–400.

Basu S. and Mouschovias T. Ch. (1995b) *Astrophys. J., 453,* 271–283.

Bate M. R. (1998) *Astrophys. J., 508,* L95–L98.

Bate M. R. and Bonnell I. A. (1997) *Mon. Not. R. Astron. Soc., 285,* 33–48.

Bate M. R. and Bonnell I. A. (2005) *Mon. Not. R. Astron. Soc., 356,* 1201–1221.

Bate M. R. and Burkert A. (1997) *Mon. Not. R. Astron. Soc., 288,* 1060–1072.

Bate M. R., Bonnell I. A., and Bromm V. (2002) *Mon. Not. R. Astron. Soc., 332,* L65–L68.

Bate M. R., Bonnell I. A., and Bromm V. (2003) *Mon. Not. R. Astron. Soc., 339,* 577–599.

Benson P. J. and Myers P. C. (1989) *Astrophys. J. Suppl., 71,* 89–108.

Boffin H. M. J., Watkins S. J., Bhattal A. S., Francis N., and Whitworth A. P. (1998) *Mon. Not. R. Astron. Soc., 300,* 1189–1204.

Bonnell I. A. (1994) *Mon. Not. R. Astron. Soc., 269,* 837–848.

Bonnell I. A. and Bastien P. (1991) *Astrophys. J., 374,* 610–622.

Bonnell I. A. and Bate M. R. (1994) *Mon. Not. R. Astron. Soc., 269,* L45–L48.

Bonnell I. A., Martel H., Bastien P., Arcoragi J-P., and Benz W. (1991) *Astrophys. J., 377,* 553–558.

Boss A. P. (1989) *Astrophys. J., 346,* 336–349.

Boss A. P. (1993) *Astrophys. J., 410,* 157–167.

Boss A. P. (1996) *Astrophys. J., 468,* 231–240.

Boss A. P. (2002) *Astrophys. J., 568,* 743–753.

Boss A. P. (2004) *Mon. Not. R. Astron. Soc., 350,* L57–L60.

Boss A. P. and Bodenheimer P. (1979) *Astrophys. J., 234,* 289–295.

Boss A. P. and Myhill E. A. (1995) *Astrophys. J., 451,* 218–224.

Boss A. P., Fisher R. T., Klein R. I., and McKee C. (2000) *Astrophys. J., 528,* 325–335.

Bourke T. L. and Goodman A. A. (2004) In *Star Formation at High Angular Resolution* (M. Burton et al., eds.), pp. 83–96. IAU Symposium 221, ASP, San Francisco.

Bouvier J., Duchêne G., Mermilliod J.-C., and Simon T. (2001) *Astron. Astrophys., 375,* 989–998.

Bouy H., Brandner W., Martín E. L., Delfosse X., Allard F., and Basri G. (2003) *Astron. J., 126,* 1526–1554.

Burkert A. and Balsara D. (2001) *Bull. Am. Astron. Soc., 33,* 885.

Burkert A. and Bodenheimer P. (1993) *Mon. Not. R. Astron. Soc., 264,* 798–806.

Burkert A. and Bodenheimer P. (1996) *Mon. Not. R. Astron. Soc., 280,* 1190–1200.

Burkert A. and Bodenheimer P. (2000) *Astrophys. J., 543,* 822–830.

Burkert A., Bate M. R., and Bodenheimer P. (1997) *Mon. Not. R. Astron. Soc., 289,* 497–504.

Caselli P., Benson P. J., Myers P. C., and Tafalla M. (2002) *Astrophys. J., 572,* 238–263.

Cha S.-H. and Whitworth A. P. (2003) *Mon. Not. R. Astron. Soc., 340,* 91–104.

Ciolek G. E. and Basu S. (2000) *Astrophys. J., 529,* 925–931.

Ciolek G. E. and Mouschovias T. Ch. (1993) *Astrophys. J., 418,* 774–793.

Ciolek G. E. and Mouschovias T. Ch. (1994) *Astrophys. J., 425,* 142–160.

Ciolek G. E. and Mouschovias T. Ch. (1995) *Astrophys. J., 454,* 195–216.

Close L. M., Siegler N., Freed M., and Biller B. (2003) *Astrophys. J., 587,* 407–422.

Crutcher R. M., Troland T. H., Lazareff B., Paubert G., and Kazés I. (1999) *Astrophys. J., 514,* L121–L124.

Dale J. E., Bonnell I. A., Clarke C. J., and Bate M. R. (2005) *Mon. Not. R. Astron. Soc., 358,* 291–304.

Delgado Donate E. J., Clarke C. J., and Bate M. R. (2003) *Mon. Not. R. Astron. Soc., 342,* 926–938.

Delgado Donate E. J., Clarke C. J., Bate M. R., and Hodgkin S. T. (2004a) *Mon. Not. R. Astron. Soc., 351,* 617–629.

Delgado Donate E. J., Clarke C. J., and Bate M. R. (2004b) *Mon. Not. R. Astron. Soc., 347,* 759–770.

Duchêne G., Bouvier J., Bontemps S., André P., and Motte F. (2004) *Astron. Astrophys., 427,* 651–665.

Duquennoy A. and Mayor M. (1991) *Astron. Astrophys., 248,* 485–524.

Durisen R. H., Sterzik M. F., and Pickett B. K. (2001) *Astron. Astrophys., 371,* 952–962.

Eggleton P. and Kiseleva L. (1995) *Astrophys. J., 455,* 640–645.

Elmegreen B. G. (2000) *Astrophys. J., 530,* 277–281.

Elmegreen B. G. and Scalo J. (2004a) *Ann. Rev. Astron. Astrophys., 42,* 211–273.

Elmegreen B. G. and Scalo J. (2004b) *Ann. Rev. Astron. Astrophys., 42,* 275–316.

Evans N. J. (1999) *Ann. Rev. Astron. Astrophys., 37,* 311–362.

Fischer D. A. and Marcy G. W. (1992) *Astrophys. J., 396,* 178–194.

Fisher J., Schröder K.-P., and Smith R. C. (2005) *Mon. Not. R. Astron. Soc., 361,* 495–503.

Fisher R. T. (2004) *Astrophys. J., 600,* 769–780.

García B. and Mermilliod J. C. (2001) *Astron. Astrophys., 368,* 122–136.

Gawryszczak A. J., Goodwin S. P., Burkert A., and Różyczka M. (2006) *Astron. Astrophys.,* submitted.

Gizis J. E., Reid I. N., Knapp G. R., Liebert J., Kirkpatrick J. D., Koerner D. W., and Burgasser A. J. (2003) *Astron. J., 125,* 3302–3310.

Gómez L., Rodriguez L. F., Loinard L., Lizano S., Poveda A., and Allen C. (2005) *Astrophys. J., 635,* 1166–1172.

Goodman A. A., Benson P. J., Fuller G. A., and Myers P. C. (1993) *Astrophys. J., 406,* 528–547.

Goodman A. A., Barranco J. A., Wilner D. J., and Heyer M. H. (1998) *Astrophys. J., 504,* 223–246.

Goodwin S. P. and Kroupa P. (2005) *Astron. Astrophys., 439,* 565–569.

Goodwin S. P. and Whitworth A. P. (2004) *Astron. Astrophys., 413,* 929–937.

Goodwin S. P., Ward-Thompson D., and Whitworth A. P. (2002) *Mon. Not. R. Astron. Soc., 330,* 769–771.

Goodwin S. P., Whitworth A. P., and Ward-Thompson D. (2004a) *Astron. Astrophys., 414,* 633–650.

Goodwin S. P., Whitworth A. P., and Ward-Thompson D. (2004b) *Astron. Astrophys., 423,* 169–182.

Goodwin S. P., Hubber D. A., Moraux E., and Whitworth A. P. (2005) *Astron. Nachricten., 326,* 1040–1043.

Hachisu I. and Eriguchi Y. (1984) *Astron. Astrophys., 140,* 259–264.

Hachisu I. and Eriguchi Y. (1985) *Astron. Astrophys., 143,* 355–364.

Haisch K. E. Jr., Greene T. P., Barsony M., and Stahler S. W. (2004) *Astron. J., 127,* 1747–1754.

Heggie D. C. (1975) *Mon. Not. R. Astron. Soc., 173,* 729–787.

Heggie D. C., Hut P., and McMillan S. L. W. (1996) *Astrophys. J., 467,* 359–369.

Hennebelle P., Whitworth A. P., Gladwin P. P., and André P. (2003) *Mon. Not. R. Astron. Soc., 340,* 870–882.

Hennebelle P., Whitworth A. P., Cha S.-H., and Goodwin S. P. (2004) *Mon. Not. R. Astron. Soc., 348,* 687–701.

Hennebelle P., Whitworth A. P., and Goodwin S. P. (2006) *Astron. Astrophys.,* in press.

Hills J. G. (1975) *Astron. J., 80,* 809–825.

Hosking J. G. and Whitworth A. P. (2004a) *Mon. Not. R. Astron. Soc., 347,* 994–1000.

Hosking J. G. and Whitworth A. P. (2004b) *Mon. Not. R. Astron. Soc., 347,* 1001–1010.

Hubber D. A. and Whitworth A. P. (2005) *Astron. Astrophys., 437,* 113–125.

Hubber D. A., Goodwin S. P., and Whitworth A. P. (2006) *Astron. Astrophys., 450,* 881–886.

Inutsuka S.-I. and Miyama S. M. (1992) *Astrophys. J., 388,* 392–399.

Jappsen A-K. and Klessen R. S. (2004) *Astron. Astrophys., 423,* 1–12.

Jijina J., Myers P. C., and Adams F. C. (1999) *Astrophys. J. Suppl., 125,* 161–236.

Jones C. E., Basu S., and Dubinski J. (2001) *Astrophys. J., 551,* 387–393.

Kitsionas S. and Whitworth A. P. (2002) *Mon. Not. R. Astron. Soc., 330,* 129–136.

Klessen R. S. and Burkert A. (2000) *Astrophys. J. Suppl., 128,* 287–319.

Klessen R. S., Ballesteros-Paredes J., Vázquez-Semadeni E., and Durán-Rojas (2005) *Astrophys. J., 620,* 786–794.

Köhler R. and Leinert C. (1998) *Astron. Astrophys., 331,* 977–988.

Kouwenhoven M. B. N., Brown A. G. A., Zinnecker H., Kaper L., and Portegies Zwart S. F. (2005) *Astron. Astrophys., 430,* 137–154.

Kroupa P. (1995a) *Mon. Not. R. Astron. Soc., 277,* 1491–1506.

Kroupa P. (1995b) *Mon. Not. R. Astron. Soc., 277,* 1507–1521.

Kroupa P. and Bouvier J. (2003a) *Mon. Not. R. Astron. Soc., 346,* 343–353.

Kroupa P. and Bouvier J. (2003b) *Mon. Not. R. Astron. Soc., 346,* 369–380.

Kroupa P. and Burkert A. (2001) *Astrophys. J., 555,* 945–949.

Kroupa P., Tout C. A., and Gilmore G. (1993) *Mon. Not. R. Astron. Soc., 262,* 545–587.

Kroupa P., Petr M. G., and McCaughrean M. J. (1999) *New Astron., 4,* 495–520.

Kroupa P., Aarseth S. J., and Hurley J. (2001) *Mon. Not. R. Astron. Soc., 321,* 699–712.

Kroupa P., Bouvier J., Duchêne G., and Moraux E. (2003) *Mon. Not. R. Astron. Soc., 346,* 354–368.

Lada C. J. (2006) *Astrophys. J., 640,* L63–L66.

Lada C. J. and Lada E. A. (2003) *Ann. Rev. Astron. Astrophys., 41,* 57–115.

Larson R. B. (1969) *Mon. Not. R. Astron. Soc., 145,* 297–308.

Larson R. B. (1981) *Mon. Not. R. Astron. Soc., 194,* 809–826.

Larson R. B. (2002) *Mon. Not. R. Astron. Soc., 332,* 155–164.

Larson R. B. (2003) *Rep. Prog. Phys., 66,* 1651–1697.

Leinert C., Zinnecker H., Weitzel N., Christou J., Ridgway S. T., Jameson R., Haas M., and Lenzen R. (1993) *Astron. Astrophys., 278,* 129–149.

Luhman K. L. (2004) *Astrophys. J., 617,* 1216–1232.

Machida M. N., Matsumoto T., Tomisaka K., and Hanawa T. (2005a) *Mon. Not. R. Astron. Soc., 362,* 369–381.

Machida M. N., Matsumoto T., Hanawa T., and Tomisaka K. (2005b) *Mon. Not. R. Astron. Soc., 362,* 382–402.

Martín E. L., Navascués D. B. y., Baraffe I., Bouy H., and Dahm S. (2003) *Astrophys. J., 594,* 525–532.

Masunaga H. and Inutsuka S. (2000) *Astrophys. J., 531,* 350–365.

Masunaga H., Miyama S. M., and Inutsuka S. (1998) *Astrophys. J., 495,* 346–369.

Mathieu R. D. (1994) *Ann. Rev. Astron. Astrophys., 32,* 465–530.

Maxted P. F. L. and Jeffries R. D. (2005) *Mon. Not. R. Astron. Soc., 362,* L45–L49.

Mayor M., Duquennoy A., Halbwachs J.-L., and Mermilliod J.-C. (1992) In *Complementary Approaches to Double and Multiple Star Research* (H. A. McAlister and W. I. Hartkopf, eds.), pp. 73–81. ASP Conf. Series 32, San Francisco.

Mazeh T., Goldberg D., Duquennoy A., and Mayor M. (1992) *Astrophys. J., 401,* 265–268.

Miyama S. M. (1992) *Publ. Astron. Soc. Japan, 44,* 193–202.

Miyama S. M., Hayashi C., and Narita S. (1984) *Astrophys. J., 279,* 621–632.

Motte F., André P., and Neri R. (1998) *Astron. Astrophys., 336,* 150–172.

Motte F., André P., Ward-Thompson D., and Bontemps S. (2001) *Astron. Astrophys., 372,* L41–L44.

Muench A. A., Lada E. A., Lada C. J., and Alves J. (2002) *Astrophys. J., 573,* 366–393.

Myers P. C. (1983) *Astrophys. J., 270,* 105–118.

Myers P. C. and Benson P. J. (1983) *Astrophys. J., 266,* 309–320.

Myhill E. A. and Kaula W. M. (1992) *Astrophys. J., 386,* 578–586.

Nelson R. P. and Papaloizou J. C. B. (1993) *Mon. Not. R. Astron. Soc., 265,* 905–920.

Ochi Y., Sugimoto K., and Hanawa T. (2005) *Astrophys. J., 623,* 922–939.

Padoan P. and Nordlund Å. (2002) *Astrophys. J., 576,* 870–879.

Padoan P. and Nordlund Å. (2004) *Astrophys. J., 617,* 559–564.

Patience J., Ghez A. M., Reid I. N., and Matthews K. (2002) *Astron. J., 123,* 1570–1602.

Pfalzner S. (2004) *Astrophys. J., 602,* 356–362.

Pfalzner S., Vogel P., Scharwächter J., and Olczak C. (2005) *Astron. Astrophys., 437,* 967–976.

Pinfield D. J., Dobbie P. D., Jameson R. F., Steele I. A., Jones H. R. A., and Katsiyannis A. C. (2003) *Mon. Not. R. Astron. Soc., 342,* 1241–1259.

Preibisch T., Balega Y., Hofmann K., Weigelt G., and Zinnecker H. (1999) *New Astron., 4,* 531–542.

Price D. J. and Monaghan J. J. (2004a) *Mon. Not. R. Astron. Soc., 348,* 123–138.

Price D. J. and Monaghan J. J. (2004b) *Mon. Not. R. Astron. Soc., 348,* 139–152.

Quist C. F. and Lindegren L. (2000) *Astron. Astrophys., 361,* 770–780.

Reid N. and Gizis J. E. (1997) *Astron. J., 113,* 2246–2269.

Reipurth B. (2000) *Astron. J., 120,* 3177–3191.

Reipurth B. and Clarke C. J. (2001) *Astron. J., 122,* 432–439.

Reipurth B. and Zinnecker H. (1993) *Astron. Astrophys., 278,* 81–108.

Saigo K. and Tomisaka K. (2006) *Astrophys. J., 645,* 381–394.

Schnee S. L., Ridge N. A., Goodman A. A., and Li J. G. (2005) *Astrophys. J., 634,* 442–450.

Shu F. H., Adams F. C., and Lizano S. (1987) *Ann. Rev. Astron. Astrophys., 25,* 23–81.

Sigalotti L. D. G. and Klapp J. (1997) *Astrophys. J., 474,* 710–718.

Sigalotti L. D. G. and Klapp J. (2000) *Astrophys. J., 531,* 1037–1052.

Söderhjelm S. (2000) *Astron. Nachricten., 321,* 165–170.

Solomon P. M., Rivolo A. R., Barrett J., and Yahil A. (1987) *Astrophys. J., 319,* 730–741.

Stamatellos D., Whitworth A. P., Boyd D. F. A., and Goodwin S. P. (2005) *Astron. Astrophys., 439,* 159–169.

Sterzik M. F. and Durisen R. H. (1998) *Astron. Astrophys., 339,* 95–112.

Sterzik M. F. and Durisen R. H. (2003) *Astron. Astrophys., 400,* 1031–1042.

Sterzik M. F., Durisen R. H., and Zinnecker H. (2003) *Astron. Astrophys., 411,* 91–97.

Testi L. and Sargent A. (1998) *Astrophys. J., 508,* L91–L94.

Thies I., Kroupa P., and Theis C. (2005) *Mon. Not. R. Astron. Soc., 364,* 961–970.

Tohline J. E. (1981) *Astrophys. J., 248,* 717–726.

Tohline J. E. (1982) *Fund. Cosmic. Phys., 8,* 1–81.

Tohline J. E. (2002) *Ann. Rev. Astron. Astrophys., 40,* 349–385.

Tokovinin A. A. and Smekhov M. G. (2002) *Astron. Astrophys., 382,* 118–123.

Truelove J. K., Klein R. I., McKee C. F. Holliman J. H. II, Howell L. H., Grrenough J. A., and Woods D. T. (1997) *Astrophys. J., 495,* 821–852.

Tsuribe T. and Inutsuka S.-I. (1999a) *Astrophys. J., 526,* 307–313.

Tsuribe T. and Inutsuka S.-I. (1999b) *Astrophys. J., 523,* L155–L158.

Umbreit S., Burkert A., Henning T., Mikkola S., and Spurzem R. (2005) *Astrophys. J., 623,* 940–951.

Vázquez-Semadeni E., Ostriker E. C., Passot T., Gammie C., and Stone J. (2000) In *Protostars and Planets IV* (V. Mannings et al., eds.), pp. 3–28. Univ. of Arizona, Tuscon.

Watkins S. J., Bhattal A. S., Boffin H. M. J., Francis N., and Whitworth A. P. (1998a) *Mon. Not. R. Astron. Soc., 300,* 1205–1213.

Watkins S. J., Bhattal A. S., Boffin H. M. J., Francis N., and Whitworth A. P. (1998b) *Mon. Not. R. Astron. Soc., 300,* 1214–1224.

White R. J. and Ghez A. M. (2001) *Astrophys. J., 556,* 265–295.

Whitehouse S. C. and Bate M. R. (2004) *Mon. Not. R. Astron. Soc., 353,* 1078–1094.

Whitworth A. P., Chapman S. J., Bhattal A. S., Disney M. J., Pongracic H., and Turner J. A. (1995) *Mon. Not. R. Astron. Soc., 277,* 727–746.

Woitas J., Leinert C., and Köhler R. (2001) *Astron. Astrophys., 376,* 982–996.

The Origin of the Initial Mass Function

Ian A. Bonnell
University of St Andrews

Richard B. Larson
Yale University

Hans Zinnecker
Astrophysikalisches Institut Potsdam

We review recent advances in our understanding of the origin of the initial mass function (IMF). We emphasize the use of numerical simulations to investigate how each physical process involved in star formation affects the resulting IMF. We stress that it is insufficient to just reproduce the IMF, but that any successful model needs to account for the many observed properties of star-forming regions, including clustering, mass segregation, and binarity. Fragmentation involving the interplay of gravity, turbulence, and thermal effects is probably responsible for setting the characteristic stellar mass. Low-mass stars and brown dwarfs can form through the fragmentation of dense filaments and disks, possibly followed by early ejection from these dense environments, which truncates their growth in mass. Higher-mass stars and the Salpeter-like slope of the IMF are most likely formed through continued accretion in a clustered environment. The effects of feedback and magnetic fields on the origin of the IMF are still largely unclear. Finally, we discuss a number of outstanding problems that need to be addressed in order to develop a complete theory for the origin of the IMF.

1. INTRODUCTION

One of the main goals for a theory of star formation is to understand the origin of the stellar initial mass function (IMF). There has been considerable observational work establishing the general form of the IMF (e.g., *Scalo*, 1986, 1998; *Kroupa*, 2001, 2002; *Reid et al.*, 2002; *Chabrier*, 2003), but as yet we do not have a clear understanding of the physics that determine the distribution of stellar masses. The aim of this chapter is to review the physical processes that are most likely involved and to discuss observational tests that can be used to distinguish between them.

Understanding the origin of the IMF is crucial as it includes the basic physics that determine our observable universe, the generation of the chemical elements, the kinematic feedback into the ISM, and overall the formation and evolution of galaxies. Once we understand the origin of the IMF, we can also contemplate how and when the IMF is likely to vary in certain environments such as the early universe and the galactic center.

There have been many theoretical ideas advanced to explain the IMF (cf. *Miller and Scalo*, 1979; *Silk and Takahashi*, 1979; *Fleck*, 1982, *Zinnecker*, 1982, 1984; *Elmegreen and Mathieu*, 1983; *Yoshii and Saio*, 1985; *Silk*, 1995, *Adams and Fatuzzo*, 1996, *Elmegreen*, 1997; *Clarke*, 1998; *Meyer et al.*, 2000; *Larson*, 2003, 2005; *Zinnecker et al.*,

1993; *Zinnecker*, 2005; *Corbelli et al.*, 2005, and references therein). Most theories are "successful" in that they are able to derive a Salpeter-slope IMF (*Salpeter*, 1955) but generally they have lacked significant predictive powers. The main problem is that it is far too easy to develop a theory, typically involving many variables, that has as its goal the explanation of a population distribution dependent on only one variable, the stellar mass. There have been a large number of analytical theories developed to explain the IMF and therefore the probability of any one of them being correct is relatively small. It is thus imperative not only for a model to "explain" the IMF, but also to develop secondary indicators that can be used to assess its likelihood of contributing to a full theory.

Recent increases of computational power are such that numerical simulations can now include many of the relevant physical processes and be used to produce a measurable IMF that can be compared with observations. This means that we no longer have to rely on analytical arguments as to what individual processes can do, but we can include these processes in numerical simulations and can test what their effect is on star formation and the generation of an IMF. Most importantly, numerical simulations provide a wealth of secondary information other than just an IMF, and these can be taken to compare directly with observed properties of young stars and star-forming regions. We thus concentrate in this review on the use of numerical simulations to assess the importance of the physical processes and guide us in our aim of developing a theory for the origin of the IMF.

The IMF is generally categorized by a segmented power-law or a log-normal type mass distribution (*Kroupa*, 2001; *Chabrier*, 2003). For the sake of simplicity, we adopt the power-law formalism of the type

$$dN \propto m^{-\alpha} dm \qquad (1)$$

but this should not be taken to mean that the IMF needs to be described in such a manner. For clarity, it should be noted that IMFs are also commonly described in terms of a distribution in log mass

$$dN \propto m^{\Gamma} d(\log m) \qquad (2)$$

where $\Gamma = -(\alpha - 1)$ (*Scalo*, 1986). The *Salpeter* (1955) slope for high-mass stars (see section 2) is then $\alpha = 2.35$ or $\Gamma = -1.35$. We also note here that the critical values of $\alpha = 2$, $\Gamma = -1$ occur when equal mass is present in each mass decade (for example, 1–10 $M_\odot$ and 10–100 $M_\odot$).

2. OBSERVED FEATURES

The most important feature of the IMF that we need to understand is the fact that there is a characteristic mass for stars at slightly less than 1 $M_\odot$. This is indicated by the occurrence of a marked flattening of the IMF below 1 $M_\odot$, such that the total mass does not diverge at either high or low stellar masses. If we can explain this one basic feature, then we will have the foundation for a complete theory of star formation. In terms of understanding the role of star formation in affecting the evolution of galaxies and their interstellar media, it is the upper-mass Salpeter-like slope that is most important. The relative numbers of massive stars determines the chemical and kinematic feedback of star formation. Other basic features of the IMF are most likely a lower, and potentially an upper, mass cutoff.

One of the most remarkable features of IMF research is that the upper-mass Salpeter slope has survived 50 years without significant revision (e.g., *Salpeter*, 1955; *Corbelli et al.*, 2005). At the same time, much work and debate has concentrated on understanding the low-mass IMF (e.g., *Reid et al.*, 2002; *Corbelli et al.*, 2005, and references therein). It appears that the form of the IMF has converged to a certain degree and is generally described as either a log-normal distribution with a power-law tail or as a series of power-laws (see Fig. 1). For ease of description, the IMF is generally given in the latter form, such as the *Kroupa* (2001) IMF

$$\begin{aligned} dN &\propto m^{-2.3} dm &&(m \geq 0.5\ M_\odot) \\ dN &\propto m^{-1.3} dm &&(0.08 \leq m \leq 0.5\ M_\odot) \qquad (3) \\ dN &\propto m^{-0.3} dm &&(m \leq 0.08\ M_\odot) \end{aligned}$$

Observational studies of the IMF in regions of star formation (e.g., *Meyer et al.*, 2000; *Zinnecker et al.*, 1993) have shown that the IMF is set early in the star-formation process. With the caveat that stellar masses (and ages) are difficult to extract during the pre-main-sequence contraction phase, most young stellar regions have mass functions that

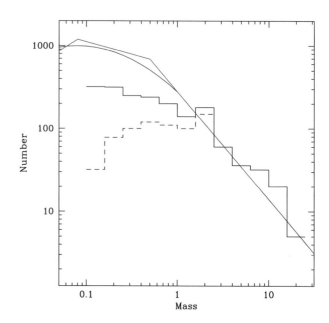

Fig. 1. The IMF for NGC3603 (Stolte et al., in preparation) is shown as a histogram in log mass [dN(log mass)] for the completeness corrected (solid) and uncorrected (dashed) populations. For comparison, the *Kroupa* (2001) segmented power-law and the *Chabrier* (2003) log-normal plus power-law IMFs are also plotted in terms of log mass.

follow a "normal" IMF. In this case, the IMF appears to be a (near) universal function of star formation in our galaxy.

One of the most pressing questions concerning the origin of the IMF is how early the mass distribution is set. Does this occur at the molecular cloud fragmentation stage or does it occur afterward due to gas accretion, feedback, etc.? The possibility that the IMF is set at the prestellar core stage has received a significant boost from observations of the clump-mass distributions in ρ Oph, Serpens, Orion, and others that appear to closely follow the stellar IMF (*Motte et al.*, 1998, 2001; *Testi and Sargent*, 1998; *Johnstone et al.*, 2000; see chapter by Lada et al.). The main assumption is that there is a direct mapping of core to stellar masses. This is uncertain for a number of reasons, including the possibility that some or most of the cores are gravitationally unbound (*Johnstone et al.*, 2000) and therefore may never form any stars. If the cores do collapse, they are likely to form binary or multiple stellar systems (e.g., *Goodwin and Kroupa*, 2005) that would affect the resulting stellar IMF, at least for core masses ≥1 $M_\odot$ (*Lada*, 2006). Finally, in order for the clump-mass spectrum to match the IMF, none of the extended mass in the system can become involved in the star-formation process. *Johnstone et al.* (2004) report that in ρ Oph only a few percent of the total mass is in the clumps. The remaining mass also explains why the clump masses vary from study to study, as the masses are likely to depend on the exact location of the clump boundaries.

Another important question concerns the universality of the IMF, especially the relative abundances of high- and low-mass stars (*Scalo*, 2005; *Elmegreen*, 2004). Although there have been occasional claims of top-heavy or truncated IMFs, they have generally relied on unresolved stellar populations and have gone away when individual stars are detected and counted. At present, there are two cases that appear to be more robust and worthy of consideration. They are both located near the galactic center, which may be an indication of the different physics there (*Larson*, 2005). First, there is the Arches cluster, which appears to have a top-heavy IMF in the resolved population (*Stolte et al.*, 2005). Caveats are that this may be influenced by mass segregation in the cluster, incompleteness, and perhaps unresolved binaries. The second case is the galactic center, where the massive stars are resolved (*Paumard et al.*, 2006), but there appears little evidence for a low-mass pre-main-sequence population based on expected X-ray fluxes and on dynamical mass estimates (*Nayashin and Sunyaev*, 2005).

3. RELEVANT OBSERVATIONAL CONSTRAINTS

It is apparent that many models have been advanced to explain the origin of the IMF. It is equally apparent that just being capable of reproducing the observed IMF is not a sufficient condition. We need observational tests and secondary indicators that can be used to distinguish between the models, whether current or in the future. In theory, most if not all observed properties of young stars (disks, velocities, clusterings) and star-forming regions (mass distributions, kinematics) should be explained by a complete model for the IMF. In practice, it is presently unclear what the implications of many of the observed properties are. Here we outline a selection of potential tests that can either be used presently or are likely to be usable in the next several years.

3.1. Young Stellar Clusters

It is becoming increasingly apparent that most stars form in groups and clusters, with the higher-mass stars forming almost exclusively in dense stellar environments. Thus, models for the IMF need to account for both the clustered nature of star formation and the fact that the environment is likely to play an important role in determining the stellar masses. For example, models for the IMF need to be able to reproduce the cluster properties in terms of stellar densities and spatial distributions of lower- and higher-mass stars.

One question is whether there is a physical correlation between the star-forming environment and the formation of massive stars. A correlation between the mass of the most massive star and the stellar density of companions is seen to exist around Herbig AeBe stars (*Testi et al.*, 1999), although this is not necessarily incompatible with random sampling from an IMF (*Bonnell and Clarke*, 1999). Recently, *Weidner and Kroupa* (2006) have suggested that observations indicate a strong correlation between the most massive star and the cluster mass, and that a random sampling model can be excluded. Estimates of the number of truly isolated massive stars are on the order of 4% or less

(*de Wit et al.*, 2005). It is therefore a necessary condition for any model for the IMF to explain how massive star formation occurs preferentially in the cores (see below) of stellar clusters where stars are most crowded.

3.2. Mass Segregation

Observations show that young stellar clusters generally have a significant degree of mass segregation with the most massive stars located in the dense core of the cluster (*Hillenbrand and Hartmann*, 1998; *Carpenter et al.*, 1997; *Garcia and Mermilliod*, 2001). For example, mass segregation is present in the Orion Nebula Cluster (ONC), where stars more massive than 5 $M_\odot$ are significantly more concentrated in the cluster core than are lower-mass stars (*Hillenbrand and Hartmann*, 1988). This suggests that either the higher-mass stars formed in the center of the clusters, or that they moved there since their formation. Massive stars are expected to sink to the center of the cluster due to two-body relaxation, but this dynamical relaxation occurs on the relaxation time, inversely proportional to the stellar mass (e.g., *Binney and Tremaine*, 1987). The young stellar clusters considered are generally less than a relaxation time old, such that dynamical mass segregation cannot have fully occurred. N-body simulations have shown that while some dynamical mass segregation does occur relatively quickly, especially for the most massive star, the degree of mass segregation present cannot be fully attributed to dynamical relaxation. Instead, the mass segregation is at least partially primordial (*Bonnell and Davies*, 1998; *Littlefair et al.*, 2003).

For example, in the ONC at 1 m.y., the massive stars need to have formed within three core radii for two-body relaxation to be able to produce the central grouping of massive stars known as the Trapezium (*Bonnell and Davies*, 1998). Putting the massive stars at radii greater than the half-mass radius of the cluster implies that the ONC would have to be at least 10 dynamical times old (3–5 m.y.) in order to have a 20% chance of creating a Trapezium-like system in the center due to dynamical mass segregation. It therefore appears to be an unavoidable consequence of star formation that higher-mass stars typically form in the center of stellar clusters. A caveat is that these conclusions depend on estimates of the stellar ages. If the systems are significantly older than is generally believed (*Palla et al.*, 2005), then dynamical relaxation is more likely to have contributed to the current mass segregation.

3.3. Binary Systems

We know that many stars form in binary systems and that the binary frequency increases with stellar mass. Thus, the formation of binary stars is an essential test for models of the IMF. While the frequency of binaries among the lower-mass stars and brown dwarfs is ≈10–30% (see chapter by Burgasser et al.), this frequency increases to ≥50% for solar-type stars (*Duquennoy and Mayor*, 1991) and up to nearly 100% for massive stars (*Mason et al.*, 1998; *Preibisch et al.*, 1999; *Garcia and Mermilliod*, 2001).

Of added importance is that many of these systems are very close, with separations less than the expected Jeans or fragmentation lengths within molecular clouds. This implies that they could not have formed at their present separations and masses but must have either evolved to smaller separations, higher masses, or both. An evolution in binary separation, combined with a continuum of massive binary systems with decreasing separation down to a few stellar radii, implies that the likelihood for binary mergers should be significant. System mass ratios also probably depend on primary mass, as high-mass stars appear to have an overabundance of similar mass companions relative to solar-type stars (*Mason et al.*, 1998; *Zinnecker*, 2003).

The fact that binary properties (frequency, separations, mass ratios) depend on the primary mass is important in terms of models for the IMF. Fragmentation is unlikely to be able to account for the increased tendency of high-mass binaries to have smaller separations and more similar masses relative to lower-mass stars, whereas subsequent accretion potentially can (*Bate and Bonnell*, 1997).

Understanding the binary properties, and how they depend on primary mass, is also crucial in determining the IMF. For example, are the two components paired randomly, or are they correlated in mass? One needs to correct for unresolved binary systems and this requires detailed knowledge of the distribution of mass ratios (*Sagar and Richtler*, 1991; *Malkov and Zinnecker*, 2001; *Kroupa*, 2001).

4. NUMERICAL SIMULATIONS

While numerical simulations provide a useful tool to test how the individual physical processes affect the star formation and resulting IMF, it should be recalled that each simulation has its particular strengths and weaknesses and that no simulation to date has included all the relevant physical processes. Therefore all conclusions based on numerical simulations should be qualified by the physics they include and their abilities to follow the processes involved.

The majority of the simulations used to study the origin of the IMF have used either grid-based methods or the particle-based smoothed particle hydrodynamics (SPH). Grid-based codes use either a fixed Eulerian grid or an adaptive grid refinement (AMR). Adaptive grids are a very important development as they provide much higher resolution in regions of high density or otherwise of interest. This allows grid-based methods to follow collapsing objects over many orders of magnitude increase in density. The resolution elements are individual cells although at least eight cells are required in order to resolve a three-dimensional self-gravitating object. Grid-based methods are well suited for including additional physics such as magnetic fields and radiation transport. They are also generally better at capturing shocks as exact solutions across neighboring grid cells are straightforward to calculate. The greatest weaknesses of grid-based methods is the necessary advection of fluid through the grid cell, especially when considering a self-gravitating fluid. Recently, *Edgar et al.* (2005) have shown how a resolved self-gravitating binary system can lose angular momentum as it rotates through an AMR grid, forcing the system to merge artificially.

In contrast, SPH uses particles to sample the fluid and a smoothing kernel with which to establish the local hydrodynamical quantities. The resolution element is the smoothing length, which generally contains ~50 individual particles, but this is also sufficient to resolve a self-gravitating object. Additionally, when following accretion flows, individual particles can be accreted. The primary asset of SPH is that it is Lagrangian and thus is ideally suited to follow the flow of self-gravitating fluids. Gravity is calculated directly from the particles such that it can easily follow a collapsing object. Following fragmentation requires resolving the Jeans mass (*Bate and Burkert*, 1997) at all points during the collapse. When the Jeans mass is not adequately resolved, fragments with masses below the resolution limit cannot be followed and are forced to disperse into the larger-scale environment. This results in the simulation only determining a lower limit to the total number of physical fragments that should form. For the same reason, SPH cannot overestimate the number of fragments that form. Tests have repeatedly shown that artificial fragmentation does not occur in SPH (*Hubber et al.*, 2006; *Bonnell and Bate*, 2006), as any clumps that contain less than the minimum number of particles cannot collapse, whether gravitationally bound or not. Young stellar objects can be represented by sink particles that accrete all gas that flows into their sink radius, and are bound to the star (*Bate et al.*, 1995). This permits simulations to follow the dynamics much longer than otherwise possible and follow the accretion of mass onto individual stars. It does exclude the possibility of resolving any disks interior to the sink-radius or their susbequent fragmentation.

Complicated fluid configurations such as occur in a turbulent medium are straightforward to follow due to the Lagrangian nature of the SPH method. Including radiative transfer and magnetic fields are more complicated due to the disorder inherent in a particle-based code. SPH also smoothes out shock fronts over at least one kernel smoothing length, but generally SPH does an adequate job of establishing the physical conditions across the shock.

The Lagrangian nature of SPH also makes it possible to trace individual fluid elements throughout a simulation in order to establish what exactly is occurring, something that is impossible with grid-based methods. Furthermore, stringent tests can be made on individual particles in order to avoid unphysical results. For example, in the classical Bondi-Hoyle accretion flows, it is necessary to resolve down below the Bondi-Hoyle radius in order to resolve the shock that allows the gas to become bound to the star. This is a grave concern in grid-based codes as otherwise the accretion can be overestimated, but is less of a worry in SPH as particles can be required to be bound before accretion occurs, even when inside the sink-radius of the accretor. This ensures that the accretion is not overestimated, but underresolved flows could result in an underestimation of the accretion rates. Particles that would shock and become bound and accreted are instead free to escape the star.

5. PHYSICAL PROCESSES

There are a number of physical processes that are likely to play an important role in the star-formation process and thus affect the resulting distribution of stellar masses. These include gravity, accretion, turbulence, magnetic fields, feedback from young stars, and other semi-random processes such as dynamical ejections.

5.1. Gravitational Fragmentation

It is clear that gravity has to play an important and potentially dominant role in determining the stellar masses. Gravity is the one force that we know plays the most important role in star formation, forcing molecular clouds with densities on the order of 10^{-20} g cm^{-3} to collapse to form stars with densities on the order of 1 g cm^{-3}. It is therefore likely that gravity likewise plays a dominant role in shaping the IMF. Gravitational fragmentation is simply the tendency for gravity to generate clumpy structure from an otherwise smooth medium. It occurs when a subpart of the medium is self-gravitating, i.e., when gravitational attraction dominates over all support mechanisms. In astrophysics, the one support that cannot be removed and is intrinsically isotropic (such that it supports an object in three dimensions) is the thermal pressure of the gas. Thus thermal support sets a minimum scale on which gravitational fragmentation can occur. The Jeans mass, based on the mass necessary for an object to be bound gravitationally against its thermal support, can be estimated by comparing the respective energies and requiring that $|E_{grav}| \geq E_{therm}$. For the simplest case of a uniform density sphere this yields

$$M_{Jeans} \approx 1.1(T_{10})^{3/2}(\rho_{19})^{-1/2}M_{\odot} \qquad (4)$$

where ρ_{19} is the gas density in units of 10^{-19} g cm^{-3} and T_{10} is the temperature in units of 10 K. If external pressure is important, then one must use the Bonnor-Ebert mass (*Ebert,* 1955; *Bonnor,* 1956), which is somewhat smaller. The corresponding Jeans length or minimum length scale for gravi-

tational fragmentation is given by

$$R_{Jeans} \approx 0.057(T_{10})^{1/2}(\rho_{19})^{-1/2} \text{ pc} \qquad (5)$$

This gives an estimate of the minimum initial separation for self-gravitating fragments.

One can see that by varying the temperature and/or the density, it is straightforward to obtain the full range of Jeans masses and thus potentially stellar masses and therefore a variation in either of these variables can produce an IMF. Generally, the temperature is low before star formation and assumed to be nearly isothermal at ≈10 K such that it is the density that primarily determines the Jeans mass. Other forms of support, such as turbulence and magnetic fields, have often been invoked to set the Jeans mass (e.g., *McKee and Tan,* 2003), but their relevance to gravitational fragmentation is doubtful due to their nonisotropic nature.

The primary requirement for gravitational fragmentation is that there exists sufficient initial structure to provide a focus for the gravity. In a smooth uniform sphere, even if subregions are gravitationally unstable, they will all collapse and merge together at the center of the cloud (*Layzer,* 1963). Some form of seeding is required such that the local free-fall time

$$t_{ff} = \left(\frac{3\pi}{32G\rho} \right)^{1/2} \qquad (6)$$

is shorter than the global free-fall time of the cloud. In principle, any small perturbations can be sufficient as long as the gas remains nearly isothermal (*Silk,* 1982), but fragmentation is effectively halted once the gas becomes optically thick and the collapse slows down (*Tohline,* 1982; *Boss,* 1986). Simple deformations in the form of sheets (*Larson,* 1985; *Burkert and Hartmann,* 2004) or filaments (*Larson,* 1985; *Bastien et al.,* 1991) are unstable to fragmentation as the local free-fall time is much shorter than that for the object as a whole. This allows any density perturbations to grow nonlinear during the collapse. More complex configurations (see Fig. 2) in the initial density

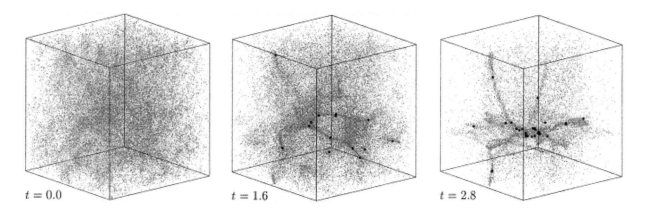

Fig. 2. The gravitational fragmentation of molecular cloud is shown from a simulation containing initial structure (*Klessen et al.,* 1998). The gravitational collapse enhances this structure, producing filaments that fragment to form individual stars. The time t is given in units of the free-fall time.

field can equally result in gravitational fragmentation (e.g., *Elmegreen and Falgarone*, 1996; *Elmegreen*, 1997, 1999; *Klessen et al.*, 1998; *Klessen and Burkert*, 2000, 2001). The origin of such density fluctuations can be due to the "turbulent" velocity field seen in molecular clouds (see below), providing the seeds for the gravitational fragmentation (*Klessen*, 2001; *Bate et al.*, 2003; *Bonnell et al.*, 2003, 2006a).

One of the general outcomes of gravitational fragmentation is that an upper limit to the number of fragments is approximately given by the number of Jeans masses present in the cloud (*Larson*, 1978, 1985; *Bastien et al.*, 1991; *Klessen et al.*, 1998, *Bate et al.*, 2003). This is easily understood as being the number of individual elements within the cloud that can be gravitationally bound. This results in an average fragment mass that is on the order of the Jeans mass at the time of fragmentation (e.g., *Klessen et al.*, 1998; *Klessen*, 2001; *Bonnell et al.*, 2004, 2006a; *Clark and Bonnell*, 2005; *Jappsen et al.*, 2005). The Jeans criterion can then be thought of as a criterion to determine the characteristic stellar mass and thus provides the foundations for the origin of the IMF. Resulting IMFs are log-normal in shape (Fig. 3) (*Klessen et al.*, 1998; *Klessen and Burkert*, 2001; *Klessen*, 2001; *Bate et al.*, 2003). The problem is then what determines the Jeans mass at the point of fragmentation. There are two possible solutions. First, the initial conditions for star formation would always have to have the same physical conditions and thus the same Jeans mass on the order of 1 $M_\odot$, which would seem unlikely. The second solution requires some additional thermal physics that set the Jeans mass at the point where fragmentation occurs (*Larson*, 2005; *Spaans and Silk*, 2000).

The coupling of gas to dust may provide the necessary physics to change from a cooling equation of state (T $\propto$ $\rho^{-0.25}$) to one including a slight heating (T $\propto$ $\rho^{-0.1}$) with increasing gas densities (*Larson*, 2005). This provides a method of setting the characteristic stellar mass, which is then *independent of the initial conditions for star formation*. Numerical simulations using a simple cooling/heating prescrip-

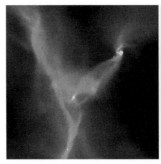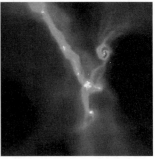

Fig. 4. The fragmentation of filamentary structure, and the formation of low-mass stars and brown dwarfs, is shown in a simulation of the formation of a small stellar cluster (*Bate et al.*, 2003).

tion to mimic the effects of this transition show that this sets the fragment mass and thus the peak of the IMF (*Jappsen et al.*, 2005; *Bonnell et al.*, 2006a). Indeed, starting from initial conditions with a Jeans mass of 5 $M_\odot$, which in an isothermal simulation provide a nearly flat (in log mass) IMF up to $\approx$5 $M_\odot$, the cooling/heating equation of state reduces this characteristic mass to below 1 $M_\odot$ (*Jappsen et al.*, 2005; *Bonnell et al.*, 2006a), allowing for an upper-mass Salpeter-like slope due to subsequent accretion (see below).

Stellar masses significantly lower than the characteristic stellar mass are also explainable through gravitational fragmentation. In a collapsing region the gas density can increase dramatically and this decreases the Jeans mass. The growth of filamentary structure in the collapse (see Fig. 4), due to the funneling of gas into local potential minima, can then provide the seeds for fragmentation to form very-low-mass objects such as brown dwarfs (*Bate et al.*, 2002a). Dense circumstellar disks also provide the necessary low Jeans mass in order to form low-mass stars and brown dwarfs. Numerical simulations of gravitational fragmentation can thus explain the characteristic stellar mass and the roughly flat (in log space) distribution of lower-mass stars and brown dwarfs (*Bate et al.*, 2003).

Gravitational fragmentation is unlikely to determine the full mass spectrum. It is difficult to see how gravitational fragmentation could account for the higher-mass stars. These stars are born in the dense cores of stellar clusters where stars are fairly closely packed. Their separations can be used to limit the sizes of any prestellar fragments via the Jeans radius, the minimum radius for an object to be gravitationally bound. This, combined with probable gas temperatures, imply a high gas density and thus a low Jeans mass (*Zinnecker et al.*, 1993). Thus, naively, it is low-mass and not high-mass stars that would be expected from a gravitational fragmentation in the cores of clusters. In general, gravitational fragmentation would be expected to instill a reverse mass segregation, the opposite of which is seen in young clusters. Similarly, although fragmentation is likely to be responsible for the formation of most binary stars, it cannot explain the closest systems nor the tendency of higher-mass stars to be in close systems with comparable mass companions.

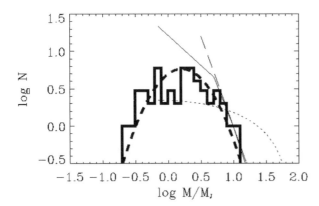

Fig. 3. The IMF that results from isothermal gravitational fragmentation (e.g., Fig. 2) is typically broad and log-normal in shape (*Klessen et al.*, 1998). The stellar masses are measured in terms of the average *initial* Jeans mass of the cloud.

5.2. Turbulence

It has long been known that supersonic motions are contained within molecular clouds (*Larson,* 1981). These motions are generally considered as being turbulent principally because of the linewidth-size relation $\sigma \propto R^{0.5}$ (*Larson,* 1981; *Heyer and Brunt,* 2004) that mimics the expectation for turbulence (*Elmegreen and Scalo,* 2004) and implies an energy cascade from large to small scales. Alternatively, the clouds could simply contain random bulk motions generated at all scales such as occurs in a clumpy shock (*Bonnell et al.,* 2006b). Nevertheless, for the purposes of this review, we define turbulence as supersonic irregular motions in the clouds that contribute to the support of these clouds (see chapter by Ballesteros-Paredes et al.). It is well known that turbulence or its equivalent can generate density structures in molecular clouds due to supersonic shocks that compress the gas (*Elmegreen,* 1993; *Vazquez-Semadeni,* 1994; *Padoan,* 1995; *Stone et al.,* 1998; *Mac Low et al.,* 1998; *Ostriker et al.,* 1999; *Mac Low and Klessen,* 2004; *Elmegreen and Scalo,* 2004) (Fig. 5). The resultant distribution of density structures, generally referred to as turbulent fragmentation, can either provide the seeds for a gravitational fragmentation (e.g., references above, especially *Mac Low and Klessen,* 2004), or alternatively could determine the IMF directly at the prestellar core phase of star formation (*Padoan et al.,* 1997, 2001; *Padoan and Nordlund,* 2002).

Turbulent fragmentation provides an attractive mechanism to explain the IMF as it involves only one physical process, which is observed to be ubiquitous in molecular clouds (*Elmegreen,* 1993; *Padoan et al.,* 1997; *Padoan and Nordlund,* 2002). Multiple compressions result in the formation of sheets and then filaments in the cloud (Fig. 5). The density ρ and widths w of these filaments are due to the

Fig. 5. The fragmentation of a turbulent medium and the formation of prestellar clumps (*Ballesteros-Paredes et al.,* 2006).

(MHD) shock conditions such that higher Mach number shocks produce higher density but thinner filaments (*Padoan and Nordlund,* 2002). Clump masses can then be derived assuming that the shock width gives the three-dimensional size of the clump ($M \propto \rho w^3$). High-velocity shocks produce high-density but small clumps, and thus the lowest-mass objects. In contrast, lower-velocity shocks produce low-density but large shocks, which account for the higher-mass clumps. Using the power-spectrum of velocities from numerical simulations of turbulence, and estimates of the density, ρ, and width, w, of MHD shocks as a function of the flow speed, *Padoan and Nordlund* (2002) derive a clump mass distribution for turbulent fragmentation. The turbulent spectrum results in a "universal" IMF slope that closely resembles the Salpeter slope. At lower masses, consideration of the likelihood that these clumps are sufficiently dense to be Jeans unstable produces a turnover and a log-normal shape into the brown dwarf regime. This is calculated from the fraction of gas that is over the critical density for a particular Jeans mass, but it does not require that this gas is in a particular core of that mass.

Numerical simulations using grid-based codes have investigated the resulting clump-mass distribution from turbulent fragmentation. While *Padoan and Nordlund* (2004) have reported results consistent with their earlier analytical models, *Ballesteros-Paredes et al.* (2006) conclude that the high-mass end of the mass distribution is not truly Salpeter but becomes steeper at higher masses. Furthermore, the shape depends on the Mach number of the turbulence, implying that turbulent fragmentation alone cannot reproduce the stellar IMF (*Ballesteros-Paredes et al.,* 2006). The difference is attributed to having multiple shocks producing the density structure, which then blurs the relation between the turbulent velocity spectrum and the resultant clump-mass distribution. Thus, the higher-mass clumps in the *Padoan and Nordlund* (2004) model have internal motions that will subfragment them into smaller clumps (see chapter by Ballesteros-Parades et al.).

The above grid-based simulations are generally not able to follow any gravitational collapse and star formation so the question remains open regarding what stellar IMF would result. SPH simulations that are capable of following the gravitational collapse and star formation introduce a further complication. These simulations find that most of the clumps are generally unbound and therefore do not collapse to form stars (*Klessen et al.,* 2005; *Clark and Bonnell,* 2005). It is only the most massive clumps that become gravitationally unstable and form stars. Gravitational collapse requires masses on the order of the unperturbed Jeans mass of the cloud, suggesting that the turbulence has played only a minor role in triggering the star-formation process (*Clark and Bonnell,* 2005). Even then, these cores often contain multiple thermal Jeans masses and thus fragment to form several stars.

In terms of observable predictions, the *Padoan and Nordlund* (2004) turbulent compression model suggests, as does gravitational fragmentation, that the minimum clump separations scale with the mass of the core. Thus, lower-mass clumps can be closely packed, whereas higher-mass cores need to be well separated. If these clumps translate directly into stars as required for turbulent compression to generate the IMF, then this appears to predict an initial configuration where the more massive stars are in the least-crowded locations. Unless they can dynamically migrate to the cores of stellar clusters fairly quickly, then their formation is difficult to attribute to turbulent fragmentation.

Turbulence has also been invoked as a support for massive cores (*McKee and Tan*, 2003) and thus as a potential source for massive stars in the center of clusters. The main idea is that the turbulence acts as a substitute for thermal support and the massive clump evolves as if it were very warm and thus had a much higher Jeans mass. The difficulty with this is that turbulence drives structures into objects and therefore any turbulently supported clump is liable to fragment, forming a small stellar cluster instead of one star. SPH simulations have shown that, in the absence of magnetic fields, a centrally condensed turbulent core fragments readily into multiple objects (*Dobbs et al.*, 2005). The fragmentation is somewhat suppressed if the gas is already optically thick and thus non-isothermal. Heating from accretion onto a stellar surface can also potentially limit any fragmentation (*Krumholz*, 2006) but is likely to arise only after the fragmentation has occurred. In fact, the difficulty really lies in how such a massive turbulent core could form in the first place. In a turbulent cloud, cores form and dissipate on dynamical timescales, suggesting that forming a long-lived core is problematic (*Ballesteros-Paredes et al.*, 1999; *Vazquez-Semadeni et al.*, 2005). As long as the region contains supersonic turbulence, it should fragment on its dynamical timescale long before it can collapse as a single entity. Even MHD turbulence does not suppress the generation of structures that will form the seeds for fragmentation (see chapter by Ballesteros-Paredes et al.).

The most probable role for turbulence is as a means for generating structure in molecular clouds. This structure then provides the finite amplitude seeds for gravitational fragmentation to occur, while the stellar masses are set by the local density and thermal properties of the shocked gas. The formation of lower-mass stars and brown dwarfs directly from turbulent compression is still an open question, as it is unclear if turbulent compression can form gravitationally bound cores at such low masses. Turbulent compression is least likely to be responsible for the high-mass slope of the IMF as numerical simulations suggest that the high-end core-mass distribution is not universal and does not follow a Salpeter-like slope (see Fig. 6).

5.3. Accretion

Gas accretion is a major process that is likely to play an important role in determining the spectrum of stellar masses. To see this, one needs to consider three facts. First, gravitational collapse is highly non-homologous (*Larson*, 1969),

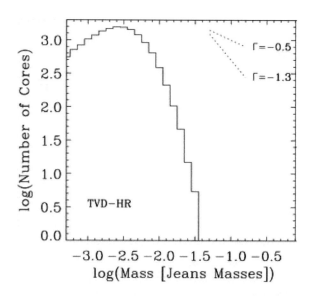

Fig. 6. The clump-mass distribution from a hydrodynamical simulation of turbulent fragmentation (*Ballesteros-Paredes et al.*, 2006). Note that the high-mass end does not follow a Salpeter slope.

with only a fraction of a stellar mass reaching stellar densities at the end of a free-fall time. The vast majority of the eventual star needs to be accreted over longer timescales. Second, fragmentation is highly inefficient, with only a small fraction of the total mass being initially incorporated into the self-gravitating fragments (*Larson*, 1978; *Bate et al.*, 2003). Third, and most important, millimeter observations of molecular clouds show that even when significant structure is present, this structure only comprises a few percent of the mass available (*Motte et al.*, 1998; *Johnstone et al.*, 2000). The great majority of the cloud mass is in a more distributed form at lower column densities, as detected by extinction mapping (*Johnstone et al.*, 2004). Young stellar clusters are also seen to have 70–90% of their total mass in the form of gas (*Lada and Lada*, 2003). Thus, a large gas reservoir exists such that if accretion of this gas does occur, it is likely to be the dominant contributor to the final stellar masses and the IMF.

Models using accretion as the basis for the IMF rely essentially on the equation

$$M_* = \dot{M}_* t_{acc} \qquad (7)$$

and by having a physical model to vary either the accretion rate $\dot{M}_*$ or the accretion timescale t_{acc} can easily generate a full distribution of stellar masses. In fact, accretion can be an extremely complex time-dependent phenomenon (e.g., *Schmeja and Klessen*, 2004) and it may occur in bursts, suggesting that we should consider the above equation in terms of a mean accretion rate and timescale. The accretion rates can be varied by being mass dependent (*Larson*, 1978; *Zinnecker*, 1982; *Bonnell et al.*, 2001b), dependent on variations of the gas density (*Bonnell et al.*, 1997, 2001a), or dependent on the relative velocity between gas and stars (*Bondi and Hoyle*, 1944; *Bate et al.*, 2003). Variations in t_{acc} (*Basu and Jones*, 2004; *Bate and Bonnell*, 2005) can be due to ejections in clusters (*Bate et al.*, 2002a) (see section 5.6 below) or feedback from forming stars (*Shu et al.*, 2004; *Dale et al.*, 2005) (see section 5.5 below).

The first models based on accretion (*Larson*, 1978, 1982; *Zinnecker*, 1982) discussed how stars compete from the available mass in a reservoir. Stars that accrete slightly more due to their initial mass or proximity to more gas (*Larson*, 1992) increase their gravitational attraction and therefore their ability to accrete. The depletion of the gas reservoir means that there is less for the remaining stars to accrete. This competitive accretion then provides a reason why there are a few high-mass stars compared to a much larger number of low-mass stars.

In a stellar cluster, the accretion is complicated by the overall potential of the system. Figure 7 shows schematically the effect of the cluster potential on the competitive accretion process. The gravitational potential is the combined potential of all the stars and gas contained in the cluster. This potential then acts to funnel gas down to the center of the cluster such that any stars located there have significantly higher accretion rates (*Bonnell et al.*, 1997, 2001a). These stars therefore have a greater ability to become higher-mass stars due to the higher gas density and due to the fact that this gas is constantly being replenished by infall from the outer part of the cluster. Stars that accrete more are also more liable to sink to the center of the potential, thereby increasing their accretion rates further. It is worthwhile noting here that this process would occur even for a static potential where the stars do not move. The gas is being drawn down to the center of the potential. It has to settle

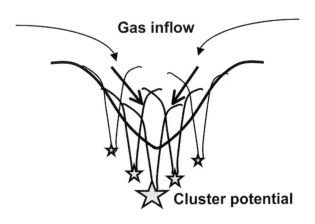

Fig. 7. A schematic diagram of the physics of accretion in a stellar cluster: The gravitational potential of the individual stars form a larger scale potential that funnels gas down to the cluster core. The stars located there are therefore able to accrete more gas and become higher-mass stars. The gas reservoir can be replenished by infall into the large-scale cluster potential.

somewhere, and unless it is already a self-gravitating fragment (i.e., a protostar), it will fall into the local potential of one of the stars.

Stars not in the center of the cluster accrete less as gas is spirited away toward the cluster center. This ensures that the mean stellar mass remains close to the characteristic mass given by the fragmentation process. Accretion rates onto individual stars depend on the local gas density, the mass of the star, and the relative velocity between the gas and the star

$$\dot{M}_* \approx \pi \rho v_{rel} R_{acc}^2 \qquad (8)$$

where R_{acc} is the accretion radius, which depends on the mass of the star (see below). The accretion radius is the radius at which gas is irrevocably bound to the star. As a cautionary note, in a stellar cluster the local gas density depends on the cluster potential and the relative gas velocity can be very different from the star's velocity in the rest frame of the cluster, as both gas and stars are experiencing the same accelerations.

Numerical simulations (*Bonnell et al.*, 2001a) show that in a stellar cluster the accretion radius depends on whether the gas or the stars dominate the potential. In the former case, the relative velocity is low and accretion is limited by the star's tidal radius. This is given by

$$R_{tidal} \approx 0.5 (M_*/M_{enc})^{\frac{1}{3}} R_* \qquad (9)$$

which measures at what distance gas is more bound to an individual star rather than being tidally sheared away by the overall cluster potential. The tidal radius depends on the star's position in the cluster, via the enclosed cluster mass M_{enc} at the radial location of the star R_*. The alternative is if the stars dominate the potential, then the relative velocity between the gas and the stars can be high. The accretion radius is then the more traditional Bondi-Hoyle radius of the form

$$R_{BH} \approx 2GM_*/(v_{rel}^2 + c_s^2) \qquad (10)$$

It is always the smaller of these two accretion radii that determines when gas is bound to the star and thus should be used to determine the accretion rates. We note again that the relative gas velocity can differ significantly from the star's velocity in the rest frame of the cluster. Using a simple model for a stellar cluster, it is straightforward to show that these two physical regimes result in two different IMF slopes because of the differing mass dependencies in the accretion rates (*Bonnell et al.*, 2001b). The tidal radius accretion has $\dot{M}_* \propto M_*^{2/3}$ and, in a $n \propto r^{-2}$ stellar density distribution, results in a relatively shallow $dN \propto M_*^{-1.5} dM_*$ (cf. *Klessen and Burkert*, 2000). Shallower stellar density distributions produce steeper IMFs. For accretion in a stellar-dominated potential, Bondi-Hoyle accretion in a uniform gas distribution results in an IMF of the form $dN \propto M_*^{-2} dM$

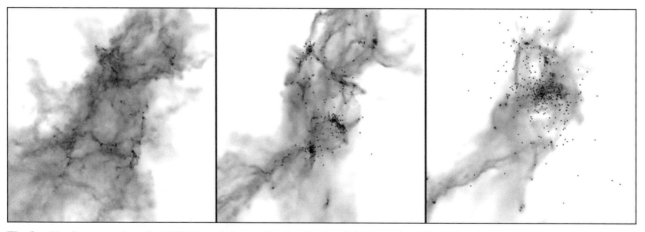

Fig. 8.   The fragmentation of a 1000 $M_\odot$ turbulent molecular cloud and the formation of a stellar cluster (*Bonnell et al.*, 2003). Note the merging of the smaller subclusters to a single big cluster.

(*Zinnecker,* 1982). To see this, consider an accretion rate based on equations (8) and (10)

$$\dot{M}_* \propto M_*^2 \tag{11}$$

with a solution

$$M_* = \frac{M_0}{1 - \beta M_0 t} \tag{12}$$

where M_0 is the initial stellar mass and β includes the dependence on gas density and velocity (assumed constant in time). From equation (12) we can derive a mass function $dN = F(M_*)dM_*$ by noting that there is a one-to-one mapping of the initial and final stellar masses (i.e., that the total number of stars is conserved and that there is a monotonic relation between initial and final masses) such that

$$F(M_*)dM_* = F(M_0)dM_0 \tag{13}$$

Using equation (12) we can easily derive

$$F(M_*) = F(M_0)(M_*/M_0)^{-2} \tag{14}$$

which, in the case where there is only a small range of initial stellar masses and for $M_* \gg M_0$, gives

$$dN \propto M_*^{-2}dM_* \tag{15}$$

whereas if the "initial" mass distribution is initially significant and decreasing with increasing masses, then the resulting IMF is steeper. In a stellar cluster with a degree of mass segregation from an earlier gas-dominated phase, this results in a steeper IMF closer to $dN \propto M_*^{-2.5}dM_*$ (*Bonnell et al.*, 2001b). This steeper IMF is therefore appropriate for the more massive stars that form in the core of a cluster because it is there that the stars first dominate the cluster potential. Although the above is a semi-analytical model and

suffers from the pitfalls described in section 1, it is comforting to note that numerical simulations do reproduce the above IMFs and additionally show that the higher-mass stars accrete the majority of their mass in the stellar-dominated regime, which should, and in this case does, produce the steeper Salpeter-like IMF (*Bonnell et al.*, 2001b).

A recent numerical simulation showing the fragmentation of a turbulent molecular cloud and the formation of a stellar cluster is shown in Fig. 8. The newly formed stars fall into local potential minima, forming small-N systems that subsequently merge to form one larger stellar cluster. The initial fragmentation produces objects with masses comparable to the mean Jeans mass of the cloud ($\approx 0.5\ M_\odot$), which implies that they are formed due to gravitational, not turbulent, fragmentation. It is the subsequent competitive accretion that forms the higher-mass stars (*Bonnell et al.*, 2004) and thus the Salpeter-like power-law part of the IMF. Overall, the simulation forms a complete stellar population that follows a realistic IMF from $0.1\ M_\odot$ to $30\ M_\odot$ (Fig. 9) (*Bonnell et al.*, 2003). Accretion forms six stars in excess of $10\ M_\odot$ with the most massive star nearly $30\ M_\odot$. Each forming subcluster contains a more massive star in its center and has a population consistent with a Salpeter IMF (*Bonnell et al.*, 2004).

One of the advantages of such a model for the IMF is that it automatically results in a mass-segregated cluster. This can be seen from the schematic in Fig. 7, which shows how the stars that are located in the core of the cluster benefit from the extended cluster potential to increase their accretion rates over what they would be in isolation. Thus stars more massive than the mean stellar mass should be relatively mass segregated from birth in the cluster. This is shown in Fig. 10, which displays the distribution of low-mass stars with the higher-mass stars located in the center of individual clusters. There is always a higher-mass star in every (sub)cluster. Even when individual subclusters merge, the massive stars quickly settle into the center of the combined potential, thereby benefitting most from any continuing accretion. One of the strong predictions of competitive

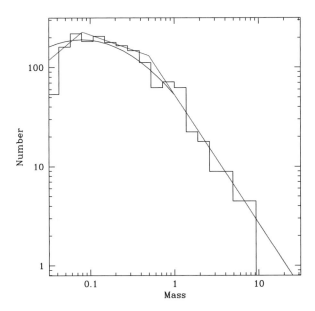

Fig. 9. The resulting IMF from a simulation of the fragmentation and competitive accretion in a forming stellar cluster (e.g., *Bonnell et al., 2003*) is shown as a function of log mass [dN(log mass)]. Overplotted is the three segment power-law IMF from *Kroupa* (2001) and *Chabrier*'s (2003) log-normal plus power-law IMF. The masses from the simulation have been rescaled to reflect an initial Jeans mass of $\approx 0.5~M_{\odot}$.

accretion is that there is a direct correlation between the formation of a stellar cluster and the most massive star it contains. Accretion and the growth of the cluster are linked such that the system always has a realistic IMF.

There have recently been some concerns raised that accretion cannot produce the high-mass IMF either due to numerical reasons (*Krumholz et al., 2006*) or due to the turbulent velocity field (*Krumholz et al., 2005a*). The numerical concern is that SPH calculations may overestimate the accretion rates if they do not resolve the Bondi-Hoyle radius. However, SPH simulations, being particle based, ensure that unphysical accretion does not occur by demanding that any gas that is accreted is bound to the star. The second concern is that accretion rates should be too low in a turbulent medium to affect the stellar masses. Unfortunately, this study assumes that gravity is negligible on large scales except as a boundary condition for the star-forming clump. This cannot be correct in a forming stellar cluster where both gas and stars undergo significant gravitational accelerations from the cluster potential. Furthermore, Krumholz et al. take a virial velocity for the clump to use as the turbulent velocity, neglecting that turbulence follows a velocity size scale $v \propto R^{1/2}$ law (*Larson*, 1981; *Heyer and Brunt*, 2004). SPH simulations show that mass accretion occurs from lower-velocity gas initially, proceeding to higher velocities when the stellar mass is larger, consistent with both the requirements of the turbulent scaling laws and Bondi-Hoyle accretion (*Bonnell and Bate*, 2006).

5.4. Magnetic Fields

Magnetic fields are commonly invoked as an important mechanism for star formation and thus need to be considered as a potential mechanism for affecting the IMF. Magnetic fields were initially believed to dominate molecular clouds with ambipolar diffusion of these fields driving the star-formation process (*Mestel and Spitzer,* 1956; *Shu et al.,* 1987). Since the realization that ambipolar diffusion takes too long, and that it would inhibit fragmentation and thus the formation of multiple stars and clusters, and crucially that supersonic motions are common in molecular clouds, the perceived role of magnetic fields has been revised to one of increasing the lifetime of turbulence (*Arons and Max,* 1975; *Lizano and Shu,* 1989). More recently, it has been shown that magnetic fields have little effect on the decay rate of turbulence as they do not fully cushion shocks (*Mac Low et al.,* 1998; *Stone et al.,* 1998). Still, magnetic fields are likely to be generally present in molecular clouds and can play an important, if still relatively unknown, role.

There have been many studies into the evolution of MHD turbulence and structure formation in molecular clouds (e.g., *Ostriker et al.,* 1999, *Vazquez-Semadeni et al.,* 2000; *Heitsch et al.,* 2001; *Tilley and Pudritz,* 2005; *Li and Nakamura,* 2004; see chapter by Balesteros-Paredes et al.). These simulations have found that both MHD and pure HD simulations result in similar clump-mass distributions. One difference is that the slightly weaker shocks in MHD turbulence shift the clump-masses to slightly higher masses.

One potential role for magnetic fields that has not been adequately explored is that they could play an important role in setting the characteristic stellar mass in terms of an

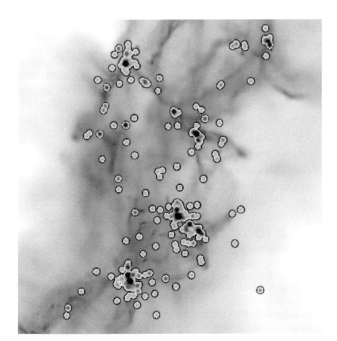

Fig. 10. The location of the massive stars (dark circles) is shown to be in the center of individual subclusters of low-mass stars (light circles) due to competitive accretion (cf. *Bonnell et al.,* 2004).

effective magnetic Jeans mass. Although in principle this is easy to derive, it is unclear how it would work in practice, as magnetic fields are intrinsically non-isotropic and therefore the analogy to an isotropic pressure support is difficult to make. Recent work on this by *Shu et al.* (2004) has investigated whether magnetic levitation, the support of the outer envelopes of collapsing cores, can set the characteristic mass. Inclusion of such models into numerical simulations is needed to verify if such processes do occur.

5.5. Feedback

Observations of star-forming regions readily display the fact that young stars have a significant effect on their environment. This feedback, including jets and outflows from low-mass stars and winds, ionization, and radiation pressure from high-mass stars, is therefore a good candidate to halt the accretion process and thereby set the stellar masses (*Silk*, 1995, *Adams and Fatuzzo*, 1996). To date, it has been difficult to construct a detailed model for the IMF from feedback as it is a rather complex process. Work is ongoing to include the effects of feedback in numerical models of star formation but have not yet been able to generate stellar mass functions (*Li and Nakamura*, 2006). In these models, feedback injects significant kinetic energy into the system that appears to quickly decay away again (*Li and Nakamura*, 2006). Overall, the system continues to evolve (collapse) in a similar way to simulations that neglect both feedback and magnetic fields (e.g., *Bonnell et al.*, 2003).

Nevertheless, we can perhaps garner some insight from recent numerical simulations including the effects of ionization from massive stars (*Dale et al.*, 2005). The inclusion of ionization from an O star into a simulation of the formation of a stellar cluster shows that the intrinsically isotropic radiation escapes in preferential directions due to the non-uniform gas distributions (see also *Krumholz et al.*, 2005b). Generally, the radiation decreases the accretion rates but does not halt accretion. In more extreme cases where the gas density is lower, the feedback halts the accretion almost completely for the full cluster. This implies that feedback can stop accretion but probably not differentially and therefore does not result in a non-uniform t_{acc}, which can be combined with a uniform $\dot{M}_*$ to form a stellar IMF.

Feedback from low-mass stars is less likely to play an important role in setting the IMF. This is simply due to the well-collimated outflows being able to deposit their energy at large distances from the star-forming environment (*Stanke et al.*, 2000). As accretion can continue in the much more hostile environment of a massive star where the feedback is intrinsically isotropic, it is difficult to see a role for well-collimated outflows in setting the IMF.

5.6. Stellar Interactions

The fact that most stars form in groups and clusters, and on smaller scales in binary and multiple systems, means that they are likely to interact with each other on timescales comparable to that for gravitational collapse and accretion. By interactions, we generally mean gravitational interactions (*Reipurth and Clarke*, 2001), although in the dense cores of stellar clusters this could involve collisions and mergers (*Bonnell et al.*, 1998, *Bally and Zinnecker*, 2005). These processes are essentially random with a probability given by the stellar density, velocity dispersion, and stellar mass. Close encounters with binary or higher-order systems generally result in an exchange of energy, which can eject the lower-mass objects of the encounter (*Reipurth and Clarke*, 2001). Such an event can quickly remove an accreting star from its gas reservoir, thereby truncating its accretion and setting the stellar mass. This process is what is seen to occur in numerical simulations of clustered star formation (*Bate et al.*, 2002a, 2003) where low-mass objects are preferentially ejected. These objects are often then limited to being brown dwarfs, whereas they could have accreted up to stellar masses had they remained in the star-forming core (see also *Price and Podsialowski*, 1995).

Numerical simulations including the dynamics of the newly-formed stars have repeatedly shown that such interactions are relatively common (*McDonald and Clarke*, 1995; *Bonnell et al.*, 1997; *Sterzik and Durisen*, 1998, 2003; *Klessen and Burkert*, 2001; *Bate et al.*, 2003; *Bate and Bonnell*, 2005), especially in small-N or subclusters where the velocity dispersion is relatively low. Thus, such a mechanism should populate the entire regime from the smallest Jeans mass formed from thermal (or turbulent) fragmentation up to the characteristic mass. This results in a relatively flat IMF (in log mass) for low-mass objects (*Klessen and Burkert*, 2001; *Bate et al.*, 2003, 2005; *Bate and Bonnell*, 2005; *Delgado-Donate et al.*, 2004).

Stellar mergers are another quasi-random event that could occur in very dense cores of stellar clusters involving mergers of intermediate- or high-mass single (*Bonnell et al.*, 1998; *Bonnell and Bate*, 2002; *Bally and Zinnecker*, 2005) or binary (*Bonnell and Bate*, 2005) stars. In either case, mergers require relatively high stellar densities on the order of 10^8 and 10^6 stars pc^{-3} respectively. These densities, although higher than generally observed, are conceivably due to a likely high-density phase in the early evolution of stellar clusters (*Bonnell and Bate*, 2002; *Bonnell et al.*, 2003). In fact, estimates of the resolved central stellar density in 30 Doradus and the Arches cluster are on the order of a few × 10^5 M$_\odot$ pc^{-3} (*Hofmann et al.*, 1995; *Stolte et al.*, 2002). Such events could play an important role in setting the IMF for the most massive stars.

5.7. Summary of Processes

From the above arguments and the expectation of the different physical processes, we can start to assess what determines the stellar masses in the various regimes (Fig. 11). A general caveat should be noted that we still do not have a thorough understanding of what magnetic fields and feedback can do, but it is worth noting that in their absence we can construct a working model for the origin of the IMF.

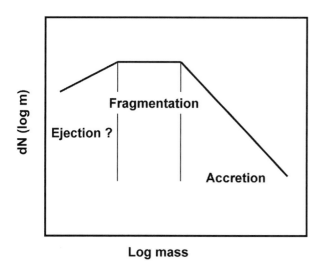

Fig. 11. A schematic IMF showing the regions that are expected to be due to the individual processes. The peak of the IMF and the characteristic stellar mass are believed to be due to gravitational fragmentation, lower-mass stars are best understood as being due to fragmentation plus ejection or truncated accretion, and higher-mass stars are understood as being due to accretion.

First of all, we conclude that the characteristic stellar mass and the broad peak of the IMF is best attributed to gravitational fragmentation and the accompanying thermal physics, which sets the mean Jeans mass for fragmentation. The broad peak can be understood as being due to the dispersion in gas densities and temperature at the point where fragmentation occurs. Turbulence is a necessary condition in that it generates the filamentary structure in the molecular clouds that facilitates the fragmentation, but does not itself set the median or characteristic stellar mass.

Lower-mass stars are most likely formed through the gravitational fragmentation of a collapsing region such that the increased gas density allows for lower-mass fragments. These fragments arise in collapsing filaments and circumstellar disks (*Bate et al.*, 2002a). A crucial aspect of this mechanism for the formation of low-mass stars and brown dwarfs is that they not be allowed to increase their mass significantly through accretion. If lower-mass stars are indeed formed in gas-dense environments to achieve the low Jeans masses, then subsequent accretion can be expected to be significant. Their continued low-mass status requires that they are ejected from their natal environment, or at least that they are accelerated by stellar interactions such that their accretion rates drop to close to zero. The turbulent compressional formation of low-mass objects (*Padoan and Nordlund*, 2004) is potentially a viable mechanism, although conflicting simulations have raised doubts as to whether low-mass gravitationally bound cores are produced that then collapse to form stars.

Finally, we conclude that the higher-mass IMF is probably due to continued accretion in a clustered environment. A turbulent compression origin for higher-mass stars is

problematic as the core-mass distribution from turbulence does not appear to be universal (*Ballesteros-Paredes et al.*, 2006). Furthermore, the large sizes of higher-mass prestellar cores generated from turbulence suggest that they should be found in low stellar density environments, not in the dense cores of stellar clusters. Nor should they then be in close, or even relatively wide, binary systems. In contrast, the ability of the cluster potential to increase accretion rates onto the stars in the cluster center is a simple explanation for more massive stars in the context of low-mass star formation. Continuing accretion is most important for the more massive stars in a forming cluster because it is these that settle, and remain, in the denser central regions. This also produces the observed mass segregation in young stellar clusters. The strong mass dependency of the accretion rates ($\propto M^2$) results in a Salpeter-like high-mass IMF.

Continued accretion and dynamical interactions can also potentially explain the existence of closer binary stars and the dependency of binary properties on stellar masses (*Bate et al.*, 2002b; *Bonnell and Bate*, 2005; *Sterzik et al.*, 2003; *Durisen et al.*, 2001). Dynamical interactions harden any existing binary (*Sterzik and Durisen*, 1998, 2003; *Kroupa*, 1995) and continued accretion increases both the stellar masses and the binding energy of the system (*Bate and Bonnell*, 1997; *Bonnell and Bate*, 2005). This can explain the higher frequency of binary systems among massive stars, and the increased likelihood that these systems are close and of near-equal masses.

6. OUTSTANDING PROBLEMS

There are outstanding issues that need to be resolved in order to fully understand the origin of the IMF. Some of these involve detailed understanding of the process (e.g., massive star formation, mass limits), whereas others include new observations that may be particularly useful in determining the origin of the IMF.

6.1. Clump-Mass Spectrum

In order for the stellar IMF to come directly from the clump-mass spectra observed in molecular clouds (e.g., *Motte et al.*, 1998; *Johnstone et al.*, 2000), a one-to-one mapping of core clump to stellar mass is required. The high frequency of multiple systems even among the youngest stars (*Duchêne et al.*, 2004) makes a one-to-one mapping unlikely for masses near solar and above. At lower masses, the reduced frequency of binary systems (e.g., *Lada*, 2006) means that a one-to-one mapping is potentially viable. Another potential difficulty is that some, especially lower-mass, clumps are likely to be transient (*Johnstone et al.*, 2000). Simulations commonly report that much of the lower-mass structure formed is gravitationally unbound (*Klessen*, 2001; *Clark and Bonnell*, 2005; *Tilley and Pudritz*, 2005). Furthermore, as such mass spectra can be understood to arise due to purely hydrodynamical effects without any self-gravity (e.g., *Clark and Bonnell*, 2006), the relevance for star formation is unclear. If the clump-mass spectrum does play an integral role in the origin of the IMF, then there should be additional evidence for this in terms of observational properties that can be directly compared. For example, the clustering and spatial mass distribution of clumps should compare directly and favorably to that of the youngest class 0 sources (e.g., *Elmegreen and Krakowski,* 2001).

6.2. Massive Stars

The formation of massive stars, with masses in excess of 10 $M_\odot$, is problematic due to the high radiation pressure on dust grains and because of the dense stellar environment in which they form. The former can actually halt the infall of gas and thus appears to limit stellar masses. Simulations to date suggest that this sets an upper-mass limit to accretion somewhere in the 10–40-$M_\odot$ range (*Wolfire and Cassinelli,* 1987; *Yorke and Sonnhalter,* 2002; *Edgar and Clarke,* 2004). Clearly, there needs to be a mechanism for circumventing this problem, as stars as massive as 80–150 $M_\odot$ exist (*Massey and Hunter,* 1998; *Weidner and Kroupa,* 2004; *Figer,* 2005). Suggested solutions include disk accretion and radiation beaming, ultrahigh accretion rates that overwhelm the radiation pressure (*McKee and Tan,* 2003), Rayleigh Taylor instabilities in the infalling gas (*Krumholz et al.,* 2005c), and stellar collisions (*Bonnell et al.,* 1998; *Bonnell and Bate,* 2002, 2005). The most complete simulations of disk accretion (*Yorke and Sonnhalter,* 2002) suggest that radiation beaming due to the star's rapid rotation, combined with disk accretion, can reach stellar masses on the order of 30–40 $M_\odot$, although with low efficiencies. What is most important for any mechanism for massive star formation is that it be put into the context of forming a full IMF (e.g., *Bonnell et al.,* 2004). The most likely scenario for massive star formation involves a combination of many of the above processes, competitive accretion in order to set the distribution of stellar masses, disk accretion, radiation beaming, and potentially Rayleigh-Taylor instabilities or even stellar mergers to overcome the radiation pressure. Any of these could result in a change in the slope of high-mass stars reflecting the change in physics.

6.3. Mass Limits

Observationally, it is unclear what limits there are on stellar masses. At low masses, the IMF appears to continue as far down as is observable. Upper-mass limits are on firmer ground observationally with strong evidence of a lack of stars higher than ≈150 $M_\odot$ even in regions where statistically they are expected (*Figer,* 2005; *Oey and Clarke,* 2005; *Weidner and Kroupa,* 2004). Physically, the only limitation on the formation of low-mass objects is likely to be the opacity limit whereby an object cannot cool faster than it contracts, setting a lower limit for a gravitationally bound object (*Low and Lynden-Bell,* 1976; *Rees,* 1976; *Boyd and Whitworth,* 2005). This sets a minimum Jeans mass on the order of 3–10 Jupiter masses. At the higher end, physical

limits could be set by radiation pressure on dust or electrons (the Eddington limit) or by physical collisions.

6.4. Clustering and the Initial Mass Function

Does the existence of a bound stellar cluster affect the high-mass end of the IMF? If accretion in a clustered environment is responsible for the high-mass IMF, then there should be a direct link between cluster properties and the presence of high-mass stars. Competitive accretion models require the presence of a stellar cluster in order for the distributed gas to be sufficient to form high-mass stars. Thus, a large-N cluster produces a more massive star than does the same number of stars divided into many small-N systems (e.g., *Weidner and Kroupa,* 2006). The combined number of stars in the small-N systems should show a significant lack of higher-mass stars. Evidence for such an environmental dependence on the IMF has recently been argued based on observations of the Vela D cloud (*Massi et al.,* 2006). The six clusters together appear to have a significant lack of higher-mass stars in relation to the expected number from a Salpeter-like IMF and the total number of stars present. A larger statistical sample of small-N systems is required to firmly establish this possibility.

7. SUMMARY

We can now construct a working model for the origin of the IMF based on the physical processes known to occur in star formation and their effects determined through numerical simulations (Fig. 11). This working model attributes the peak of the IMF and the characteristic stellar mass to gravitational fragmentation and the thermal physics at the point of fragmentation. Lower-mass stars and brown dwarfs are ascribed to fragmentation in dense regions and then ejection to truncate the accretion rates, while higher-mass stars are due to the continued competitive accretion in the dense cores of forming stellar clusters. It is worth noting that all three physical processes are primarily due to gravity and thus in combination provide the simplest mechanism to produce the IMF.

There is much work yet to be done in terms of including additional physics (magnetic fields, feedback) into the numerical simulations that produce testable IMFs. It is also important to develop additional observational predictions from the theoretical models and to use observed properties of star-forming regions to determine necessary and sufficient conditions for a full theory for the origin of the IMF. For example, competitive accretion predicts that high-mass star formation is linked to the formation of a bound stellar cluster. This can be tested by observations: The existence of significant numbers of high-mass stars in nonclustered regions or small-N clusters would argue strongly against the accretion model.

Acknowledgments. We thank the referee, B. Elmegreen, for valuable comments. H.Z. thanks the DFG for travel support to the Protostars and Planets V conference.

REFERENCES

Adams F. C. and Fatuzzo M. (1996) *Astrophys. J., 464*, 256–271.

Arons J. and Max C. E. (1975) *Astrophys. J., 196*, L77–L81.

Ballesteros-Paredes J., Vázquez-Semadeni E., and Scalo J. (1999) *Astrophys. J., 515*, 286–303.

Ballesteros-Paredes J., Gazol A., Kim J., Klessen R. S., Jappsen A.-K., and Tejero E. (2006) *Astrophys. J., 637*, 384–391.

Bally J. and Zinnecker H. (2005) *Astron. J., 129*, 2281–2293.

Bastien P., Arcoragi J.-P., Benz W., Bonnell I., and Martel H. (1991) *Astrophys. J., 378*, 255–265.

Basu S. and Jones C. E. (2004) *Mon. Not. R. Astron. Soc., 347*, L47–L51.

Bate M. R. (2005) *Mon. Not. R. Astron. Soc., 363*, 363–378.

Bate M. R. and Bonnell I. A. (1997) *Mon. Not. R. Astron. Soc., 285*, 33–48.

Bate M. R. and Bonnell I. A. (2005) *Mon. Not. R. Astron. Soc., 356*, 1201–1221.

Bate M. R. and Burkert A. (1997) *Mon. Not. R. Astron. Soc., 288*, 1060–1072.

Bate M. R., Bonnell I. A., and Price N. M. (1995) *Mon. Not. R. Astron. Soc., 277*, 362–376.

Bate M. R., Bonnell I. A., and Bromm V. (2002a) *Mon. Not. R. Astron. Soc., 332*, L65–L68.

Bate M. R., Bonnell I. A., and Bromm V. (2002b) *Mon. Not. R. Astron. Soc., 336*, 705–713.

Bate M. R., Bonnell I. A., and Bromm V. (2003) *Mon. Not. R. Astron. Soc., 339*, 577–599.

Binney J. and Tremaine S. (1987) In *Galactic Dynamics*, p. 747. Princeton Univ., Princeton.

Bondi H. and Hoyle F. (1944) *Mon. Not. R. Astron. Soc., 104*, 273–282.

Bonnell I. A. and Bate M. R. (2002) *Mon. Not. R. Astron. Soc., 336*, 659–669.

Bonnell I. A. and Bate M. R. (2005) *Mon. Not. R. Astron. Soc., 362*, 915–920.

Bonnell I. A. and Bate M. R. (2006) *Mon. Not. R. Astron. Soc., 370*, 488–494.

Bonnell I. A. and Clarke C. J. (1999) *Mon. Not. R. Astron. Soc., 309*, 461–464.

Bonnell I. A. and Davies M. B. (1998) *Mon. Not. R. Astron. Soc., 295*, 691–698.

Bonnell I. A., Bate M. R., Clarke C. J., and Pringle J. E. (1997) *Mon. Not. R. Astron. Soc., 285*, 201–208.

Bonnell I. A., Bate M. R., and Zinnecker H. (1998) *Mon. Not. R. Astron. Soc., 298*, 93–102.

Bonnell I. A., Bate M. R., Clarke C. J., and Pringle J. E. (2001a) *Mon. Not. R. Astron. Soc., 323*, 785–794.

Bonnell I. A., Clarke C. J., Bate M. R., and Pringle J. E. (2001b) *Mon. Not. R. Astron. Soc., 324*, 573–579.

Bonnell I. A., Bate M. R., and Vine S. G. (2003) *Mon. Not. R. Astron. Soc., 343*, 413–418.

Bonnell I. A., Vine S. G., and Bate M. R. (2004) *Mon. Not. R. Astron. Soc., 349*, 735–741.

Bonnell I. A., Clarke C. J., and Bate M. R. (2006a) *Mon. Not. R. Astron. Soc., 368*, 1296–1300.

Bonnell I. A., Dobbs C. L., Robitaille T. P. and Pringle J. E. (2006b) *Mon. Not. R. Astron. Soc., 365*, 37–45.

Bonnor W. B. (1956) *Mon. Not. R. Astron. Soc., 116*, 351–359.

Boss A. R. (1986) *Astrophys. J. Suppl., 62*, 519–552.

Boyd D. F. A. and Whitworth A. P. (2005) *Astron. Astrophys., 430*, 1059–1066.

Burkert A. and Hartmann L. (2004) *Astrophys. J., 616*, 288–300.

Carpenter J. M., Meyer M. R., Dougados C., Strom S. E., and Hillenbrand L. A. (1997) *Astron. J., 114*, 198–221.

Chabrier G. (2003) *Publ. Astron. Soc. Pac., 115*, 763–795.

Clark P. C. and Bonnell I. A. (2005) *Mon. Not. R. Astron. Soc., 361*, 2–16.

Clark P. C. and Bonnell I. A. (2006) *Mon. Not. R. Astron. Soc., 368*, 1787–1795.

Clarke C. J. (1998) In *The Stellar Initial Mass Function* (G. Gilmore and D. Howell, eds.), pp. 189–199. ASP Conf. Series 142, San Francisco.

Corbelli E., Palla F., and Zinnecker H. (2005) *The Initial Mass Function 50 Years Later*. ASSL Vol. 327, Springer, Dordrecht.

Dale J. E., Bonnell I. A., Clarke C. J., and Bate M. R. (2005) *Mon. Not. R. Astron. Soc., 358*, 291–304.

Delgado-Donate E. J., Clarke C. J., and Bate M. R. (2004) *Mon. Not. R. Astron. Soc., 347*, 759–770.

de Wit W. J., Testi L., Palla F., and Zinnecker H. (2005) *Astron. Astrophys., 437*, 247–255.

Dobbs C. L., Bonnell I. A., and Clark P. C. (2005) *Mon. Not. R. Astron. Soc., 360*, 2–8.

Duchêne G., Bouvier J., Bontemps S., André P., and Motte F. (2004) *Astron. Astrophys., 427*, 651–665.

Duquennoy A. and Mayor M. (1991) *Astron. Astrophys., 248*, 485–524.

Durisen R. H., Sterzik M. F., and Pickett B. K. (2001) *Astron. Astrophys., 371*, 952–962.

Ebert R. (1955) *Z. Astrophys., 37*, 217–232.

Edgar R. and Clarke C. (2004) *Mon. Not. R. Astron. Soc., 349*, 678–686.

Edgar R. G., Gawryszczak A., and Walch S. (2005) In *PPV Poster Proceedings*. Available on line at www.lpi.usra.edu/meetings/ppv2005/pdf/8005.pdf.

Elmegreen B. G. (1993) *Astrophys. J., 419*, L29–32.

Elmegreen B. G. (1997) *Astrophys. J., 486*, 944–954.

Elmegreen B. G. (1999) *Astrophys. J., 515*, 323–336.

Elmegreen B. G. (2004) *Mon. Not. R. Astron. Soc., 354*, 367–374.

Elmegreen B. G. and Falgarone E. (1996) *Astrophys. J., 471*, 816–821.

Elmegreen B. G. and Krakowski A. (2001) *Astrophys. J., 562*, 433–439.

Elmegreen B. G. and Mathieu R. D. (1983) *Mon. Not. R. Astron. Soc., 203*, 305–315.

Elmegreen B. G. and Scalo J. (2004) *Ann. Rev. Astron. Astrophys., 42*, 211–273.

Fleck R. C. (1982) *Mon. Not. R. Astron. Soc., 201*, 551–559.

Figer D. F. (2005) *Nature, 434*, 192–194.

García B. and Mermilliod J. C. (2001) *Astron. Astrophys., 368*, 122–136.

Goodwin S. P. and Kroupa P. (2005) *Astron. Astrophys., 439*, 565–569.

Heitsch F., Mac Low M.-M., and Klessen R. S. (2001) *Astrophys. J., 547*, 280–291.

Heyer M. H. and Brunt C. M. (2004) *Astrophys. J., 615*, L45–L48.

Hillenbrand L. A. and Hartmann L. W. (1998) *Astrophys. J., 492*, 540–553.

Hofmann K.-H., Seggewiss W., and Weigelt G. (1995) *Astron. Astrophys., 300*, 403–414.

Hubber D. A., Goodwin S. P., and Whitworth A. P. (2006) *Astron. Astrophys., 450*, 881–886.

Jappsen A.-K., Klessen R. S., Larson R. B., Li Y., and Mac Low M.-M. (2005) *Astron. Astrophys., 435*, 611–623.

Johnstone D., Wilson C. D., Moriarty-Schieven G., Joncas G., Smith G., Gregersen E., and Fich M. (2000) *Astrophys. J., 545*, 327–339.

Johnstone D., Di Francesco J., and Kirk H. (2004) *Astrophys. J., 611*, L45–L48.

Klessen R. S. (2001) *Astrophys. J., 556*, 837–846.

Klessen R. S. and Burkert A. (2000) *Astrophys. J. Suppl., 128*, 287–319.

Klessen R. S. and Burkert A. (2001) *Astrophys. J., 549*, 386–401.

Klessen R. S., Burkert A., and Bate M. R. (1998) *Astrophys. J., 501*, L205–L208.

Klessen R. S., Ballesteros-Paredes J., Vázquez-Semadeni E., and Durán-Rojas C. (2005) *Astrophys. J., 620*, 786–794.

Kroupa P. (1995) *Mon. Not. R. Astron. Soc., 277*, 1491–1506.

Kroupa P. (2001) *Mon. Not. R. Astron. Soc., 322*, 231–246.

Kroupa P. (2002) *Science, 295*, 82–91.

Krumholz M. R. (2006) *Astrophys. J., 641*, L45–L48.

Krumholz M. R., McKee C. F., and Klein R. I. (2005a) *Nature, 438*, 332–334.

Krumholz M. R., McKee C. F., and Klein R. I. (2005b) *Astrophys. J., 618*, L33–L36.

Krumholz M. R., Klein R. I., and McKee C. F. (2005c) In *Massive Star Birth: A Crossroads of Astrophysics* (R. Cesaroni et al., eds.), pp. 231–236. IAU Symp. 227, Cambridge Univ., Cambridge.

Krumholz M. R., McKee C. F., and Klein R. I. (2006) *Astrophys. J., 638,* 369–381.

Lada C. J. (2006) *Astrophys. J., 640,* L63–L66.

Lada C. J. and Lada E. A. (2003) *Ann. Rev. Astron. Astrophys., 41,* 57–115.

Larson R. B. (1969) *Mon. Not. R. Astron. Soc., 145,* 271–295.

Larson R. B. (1978) *Mon. Not. R. Astron. Soc., 184,* 69–85.

Larson R. B. (1981) *Mon. Not. R. Astron. Soc., 194,* 809–826.

Larson R. B. (1982) *Mon. Not. R. Astron. Soc., 200,* 159–174.

Larson R. B. (1985) *Mon. Not. R. Astron. Soc., 214,* 379–398.

Larson R. B. (1992) *Mon. Not. R. Astron. Soc., 256,* 641–646.

Larson R. B. (2003) In *Galactic Star Formation Across the Stellar Mass Spectrum* (J. M. De Buizer and N. S. van der Bliek, eds.), pp. 65–80. ASP Conf. Series 287, San Francisco.

Larson R. B. (2005) *Mon. Not. R. Astron. Soc., 359,* 211–222.

Layzer D. (1963) *Astrophys. J., 137,* 351–362.

Li Z.-Y. and Nakamura F. (2004) *Astrophys. J., 609,* L83–86.

Li Z.-Y. and Nakamura F. (2006) *Astrophys. J., 640,* L187–L190.

Littlefair S. P. Naylor T., Jeffries R. D., Devey C. R., and Vine S. (2003) *Mon. Not. R. Astron. Soc., 345,* 1205–1211.

Lizano S. and Shu F. H. (1989) *Astrophys. J., 342,* 834–854.

Low C. and Lynden-Bell D. (1976) *Mon. Not. R. Astron. Soc., 176,* 367–390.

Mac Low M.-M. and Klessen R. S. (2004) *Rev. Mod. Phys., 76,* 125–194.

Mac Low M.-M., Klessen R. S., Burkert A., and Smith M. D. (1998) *Phys. Rev. Lett., 80,* 2754–2757.

Malkov O. and Zinnecker H. (2001) *Mon. Not. R. Astron. Soc., 321,* 149–154.

Mason B. D., Gies D. R., Hartkopf W. I., Bagnuolo W. G., Brummelaar T. T., and McAlister H. A. (1998) *Astron. J., 115,* 821–847.

Massey P. and Hunter D. A. (1998) *Astrophys. J., 493,* 180–194.

Massi F., Testi L., and Vanzi L. (2006) *Astron. Astrophys., 448,* 1007–1022.

McDonald J. M. and Clarke C. J. (1995) *Mon. Not. R. Astron. Soc., 275,* 671–684.

McKee C. F. and Tan J. C. (2003) *Astrophys. J., 585,* 850–871.

Mestel L. and Spitzer L. (1956) *Mon. Not. R. Astron. Soc., 116,* 503–514.

Meyer M. R., Adams F. C., Hillenbrand L. A., Carpenter J. M. and Larson R. B. (2000) In *Protostars and Planets IV* (V. Mannings et al., eds.), pp. 121–150. Univ. of Arizona, Tucson.

Miller G. E. and Scalo J. M. (1979) *Astrophys. J. Suppl., 41,* 513–547.

Motte F., Andre P., and Neri R. (1998) *Astron. Astrophys., 336,* 150–172.

Motte F., André P., Ward-Thompson D., and Bontemps S. (2001) *Astron. Astrophys., 372,* L41–L44.

Nayakshin S. and Sunyaev R. (2005) *Mon. Not. R. Astron. Soc., 364,* L23–L27.

Oey M. S. and Clarke C. J. (2005) *Astrophys. J., 620,* L43–L47.

Ostriker E. C., Gammie C. F., and Stone J. M. (1999) *Astrophys. J., 513,* 259–274.

Padoan P. (1995) *Mon. Not. R. Astron. Soc., 277,* 377–388.

Padoan P. and Nordlund Å. (2002) *Astrophys. J., 576,* 870–879.

Padoan P. and Nordlund Å. (2004) *Astrophys. J., 617,* 559–564.

Padoan P., Nordlund A., and Jones B. J. T. (1997) *Mon. Not. R. Astron. Soc., 288,* 145–152.

Padoan P., Juvela M., Goodman A. A., and Nordlund Å. (2001) *Astrophys. J., 553,* 227–234.

Palla F., Randich S., Flaccomio E., and Pallavicini R. (2005) *Astrophys. J., 626,* L49–L52.

Paumard T., Genzel R., Martins F., Nayakshin S., Beloborodov A. M., et al. (2006) *Astrophys. J., 643,* 1011–1035.

Preibisch T., Balega Y., Hofmann K.-H., Weigelt G., and Zinnecker H. (1999) *New Astron., 4,* 531–542.

Price N. M. and Podsiadlowski P. (1995) *Mon. Not. R. Astron. Soc., 273,* 1041–1068.

Rees M. J. (1976) *Mon. Not. R. Astron. Soc., 176,* 483–486.

Reid I. N., Gizis J. E., and Hawley S. L. (2002) *Astron. J., 124,* 2721–2738.

Reipurth B. and Clarke C. (2001) *Astron. J., 122,* 432–439.

Salpeter E. E. (1955) *Astrophys. J., 121,* 161–167.

Sagar R. and Richtler T. (1991) *Astron. Astrophys., 250,* 324–339.

Scalo J. M. (1986) *Fund. Cosmic Phys., 11,* 1–278.

Scalo J. M. (1998) In *The Stellar Initial Mass Function* (G. Gilmore and D. Howell, eds.), pp. 201–236. ASP Conf. Series 142, San Francisco.

Scalo J. M. (2005) In *The Initial Mass Function 50 Years Later* (E. Corbelli et al., eds.), pp. 23–39. ASSL Vol. 327, Springer, Dordrecht.

Schmeja S. and Klessen R. S. (2004) *Astron. Astrophys., 419,* 405–417.

Shu F. H., Adams F. C., and Lizano S. (1987) *Ann. Rev. Astron. Astrophys., 25,* 23–81.

Shu F. H., Li Z.-Y., and Allen A. (2004) *Astrophys. J., 601,* 930–951.

Silk J. (1982) *Astrophys. J., 256,* 514–522.

Silk J. (1995) *Astrophys. J., 438,* L41–L44.

Silk J. and Takahashi T. (1979) *Astrophys. J., 229,* 242–256.

Spaans M. and Silk J. (2000) *Astrophys. J., 538,* 115–120.

Stanke T., McCaughrean M. J., and Zinnecker H. (2000) *Astron. Astrophys., 355,* 639–650.

Sterzik M. F. and Durisen R. H. (1998) *Astron. Astrophys., 339,* 95–112.

Sterzik M. F. and Durisen R. H. (2003) *Astron. Astrophys., 400,* 1031–1042.

Sterzik M. F., Durisen R. H., and Zinnecker H. (2003) *Astron. Astrophys., 411,* 91–97.

Stone J. M., Ostriker E. C., and Gammie C. F. (1998) *Astrophys. J., 508,* L99–L102.

Stolte A., Grebel E. K., BrandnerW., and Figer D. F. (2002) *Astron. Astrophys., 394,* 459–478.

Stolte A., Brandner W., Grebel E. K., Lenzen R., and Lagrange A.-M. (2005) *Astrophys. J., 628,* L113–L117.

Testi L. and Sargent A. I. (1998) *Astrophys. J., 508,* L91–L94.

Testi L., Palla F., and Natta A. (1999) *Astron. Astrophys., 342,* 515–523.

Tilley D. A. and Pudritz R. E. (2005) ArXiv Astrophysics e-prints, arXiv: astro-ph/0508562.

Tohline J. E. (1982) *Fund. Cosmic Phys., 8,* 1–81.

Vázquez-Semadeni E. (1994) *Astrophys. J., 423,* 681–692.

Vázquez-Semadeni E., Ostriker E. C., Passot T., Gammie C. F., and Stone J. M. (2000) In *Protostars and Planets IV* (V. Mannings et al., eds.), pp. 3–28. Univ. of Arizona, Tucson.

Vázquez-Semadeni E., Kim J., and Ballesteros-Paredes J. (2005) *Astrophys. J., 630,* L49–L52.

Weidner C. and Kroupa P. (2004) *Mon. Not. R. Astron. Soc., 348,* 187–191.

Weidner C. and Kroupa P. (2006) *Mon. Not. R. Astron. Soc., 365,* 1333–1347.

Wolfire M. G. and Cassinelli J. P. (1987) *Astrophys. J., 319,* 850–867.

Yoshii Y. and Saio H. (1985) *Astrophys. J., 295,* 521–536.

Yorke H. and Sonnhalter C. (2002) *Astrophys. J., 569,* 846–862.

Zinnecker H. (1982) *N.Y. Acad. Sci. Ann., 395,* 226–235.

Zinnecker H. (1984) *Mon. Not. R. Astron. Soc., 210,* 43–56.

Zinnecker H. (2003) In *A Massive Star Odyssey: From Main Sequence to Supernova* (K. van der Hucht et al., eds.), pp. 80–90. IAU Symposium Series 212, ASP, San Francisco.

Zinnecker H. (2005) In *The Young Local Universe, XXXIXth Rencontres de Moriond* (A. Chalabaev et al., eds.), pp. 17–22. Thé Giói, Vietnam.

Zinnecker H., McCaughrean M. J., and Wilking B. A. (1993) In *Protostars and Planets III* (E. H. Levy and J. I. Lunine, eds.), pp. 429–495. Univ. of Arizona, Tucson.

The Formation of Massive Stars

Henrik Beuther
Max-Planck-Institute for Astronomy, Heidelberg

Edward B. Churchwell
University of Wisconsin, Madison

Christopher F. McKee
University of California, Berkeley

Jonathan C. Tan
University of Florida, Gainesville

Massive stars have a profound influence on the universe, but their formation remains poorly understood. We review the current status of observational and theoretical research in this field, describing the various stages of an evolutionary sequence that begins with cold, massive gas cores and ends with the dispersal and ionization of gas by the newly formed star. The physical processes in massive star formation are described and related to their observational manifestations. Feedback processes and the relation of massive stars to star cluster formation are also discussed. We identify key observational and theoretical questions that future studies should address.

1. INTRODUCTION

Massive star formation has drawn considerable interest for several decades, but the last 10 years have witnessed a strong acceleration of theoretical and observational research in this field. One of the major conceptual problems in massive star formation arises from the radiation pressure massive stars exert on the surrounding dust and gas core (e.g., *Kahn*, 1974; *Wolfire and Cassinelli*, 1987; *Jijina and Adams*, 1996; *Yorke and Sonnhalter*, 2002; *Krumholz et al.*, 2005b). In principle, this radiation pressure could be strong enough to stop further accretion, which would imply that the standard theory of low-mass star formation had to be adapted to account for the formation of massive stars. Two primary approaches have been followed to overcome these problems: The first and more straightforward approach is to modify the standard theory quantitatively rather than qualitatively. Theories have been proposed that invoke varying dust properties (e.g., *Wolfire and Cassinelli*, 1987), increasing accretion rates in turbulent cloud cores on the order of 10^{-4}–10^{-3} M$_\odot$ yr^{-1} compared to ~10^{-6} M$_\odot$ yr^{-1} for low-mass star formation (e.g., *Norberg and Maeder*, 2000; *McKee and Tan*, 2003), accretion via disks (e.g., *Jijina and Adams*, 1996; *Yorke and Sonnhalter*, 2002), accretion through the evolving hypercompact HII region (*Keto*, 2003; *Keto and Wood*, 2006), the escape of radiation through windblown cavities (*Krumholz et al.*, 2005a), or radiatively driven Rayleigh-Taylor instabilities (*Krumholz et al.*, 2005b).

These variations to the standard picture of low-mass star formation suggest that massive stars can form within an accretion-based picture of star formation. Contrary to this, a paradigm change for the formation of massive stars has been proposed based on the observational fact that massive stars always form at the dense centers of stellar clusters: the coalescence scenario. In this scenario, the protostellar and stellar densities of a forming massive cluster are high enough (~10^8 pc^{-3}) that protostars undergo physical collisions and merge, thereby avoiding the effects of radiation pressure (*Bonnell et al.*, 1998; *Bally and Zinnecker*, 2005). Variants of the coalescence model that operate at lower stellar densities have been proposed by *Stahler at al.* (2000) and by *Bonnell and Bate* (2005). A less dramatic approach suggests that the bulk of the stellar mass is accreted via competitive accretion in a clustered environment (*Bonnell et al.*, 2004). This does not necessarily require the coalescence of protostars, but the mass accretion rates of the massive cluster members would be directly linked to the number of their stellar companions, implying a causal relationship between the cluster formation process and the formation of higher-mass stars therein.

We propose an evolutionary scenario for massive star formation, and then discuss the various stages in more detail. Following *Williams et al.* (2000), we use the term "clumps" for condensations associated with cluster formation, and the term "cores" for molecular condensations that form single or gravitationally bound multiple massive protostars. The evolutionary sequence we propose for high-mass star-forming cores is

High-mass starless cores (HMSCs)
→ high-mass cores harboring accreting low/intermediate mass protostar(s) destined to become a high-mass star(s)
→ high-mass protostellar objects (HMPOs)
→ final stars

The term HMPO is used here in a literal sense, i.e., accreting high-mass protostars. Hence, the HMPO group consists of protostars >8 $M_\odot$, which early on have not necessarily formed a detectable hot molecular core (HMC) and/or hypercompact HII region (HCHIIs, size <0.01 pc). HMCs and HCHIIs might coexist simultaneously. Ultracompact HII regions (UCHIIs, size <0.1 pc) are a transition group: Some of them may still harbor accreting protostars (hence are at the end of the HMPO stage), but many have likely already ceased accretion (hence are part of the final-star class). High-mass stars can be on the main sequence while they are deeply embedded and actively accreting as well as after they cease accreting and become final stars. The class of high-mass cores harboring accreting low/intermediate-mass protostars has not been well studied yet, but there has to be a stage between the HMSCs and the HMPOs, consisting of high-mass cores with embedded low/intermediate-mass objects. On the cluster/clump scale the proposed evolutionary sequence is

Massive starless clumps
→ protoclusters
→ stellar clusters

By definition, massive starless clumps can harbor only HMSCs (and low-mass starless cores), whereas protoclusters in principle can harbor all sorts of smaller-scale entities (low- and intermediate-mass protostars, HMPOs, HMCs, HCHIIs, UCHIIs, and even HMSCs).

This review discusses the evolutionary stages and their associated physical processes (sections 2, 3, 4, and 6), feedback processes (section 4), and cluster formation (section 5), always from an observational *and* theoretical perspective. We restrict ourselves to present-day massive star formation in a typical galactic environment. Primordial star formation, lower metallicities, or different dust properties may change this picture (e.g., *Bromm and Loeb,* 2004; *Draine,* 2003). The direct comparison of the theoretical predictions with the observational evidences and indications shows the potentials and limitations of our current understanding of high-mass star formation. We also refer to the IAU 227 Proceedings dedicated to massive star birth (*Cesaroni et al.,* 2005b).

2. INITIAL CONDITIONS

2.1. Observational Results

The largest structures within our galaxy are giant molecular clouds (GMCs) with sizes from ~20 to ~100 pc and masses between ~10^4 and ~10^6 $M_\odot$. The physical properties have been discussed in many reviews (e.g., *McKee,* 1999; *Evans,* 1999), and we summarize only the most important characteristics. A multitransition survey of GMCs in our galaxy shows that the average local density derived from an LVG analysis is n_H ~ 4×10^3–1.2×10^4 cm^{-3} and the temperature is ~10–15 K (e.g., *Sanders et al.,* 1993), giving a typical Bonnor-Ebert mass ~2 $M_\odot$. The volume-averaged densities in GMCs are n_H ~ 50 to 100 cm^{-3}; these are substantially less than the local density values, indicating that the molecular gas is highly clumped. Velocity dispersions of 2–3 km s^{-1} indicate highly supersonic internal motions given that the typical sound speed is ~0.2 km s^{-1}. These motions are largely due to turbulence (e.g., *Mac Low and Klessen,* 2004; *Elmegreen and Scalo,* 2004). Measured magnetic field strengths are of the order a few 10 μG (e.g., *Crutcher,* 1999; *Bourke et al.,* 2001). Depending on the size scales and average densities, magnetic critical masses can range from ~5×10^5 $M_\odot$ to a few solar masses, corresponding to GMCs and low-mass star-forming regions, respectively (*McKee,* 1999). Thus, although the rather low Jeans masses indicate gravitationally bound and likely unstable entities within GMCs, turbulence and magnetic stresses appear to be strong enough to support the GMCs against complete collapse on large scales.

Most important for any star formation activity, GMCs show substructures on all spatial scales. They contain dense gas clumps that are easily identifiable in the (sub)millimeter continuum and high-density molecular line tracers (e.g., *Plume et al.,* 1992; *Bronfman et al.,* 1996; *Beuther et al.,* 2002a; *Mueller et al.,* 2002; *Faundez et al.,* 2004; *Beltrán et al.,* 2006). Peak densities in such dense clumps can easily reach 10^6 cm^{-3}, and the massive dense clumps we are interested in typically have masses between a few 100 and a few 1000 $M_\odot$ (e.g., *Beuther et al.,* 2002a; *Williams et al.,* 2004; *Faundez et al.,* 2004). Massive dense clumps are the main locations where high-mass star formation is taking place. We shall concentrate on the physical properties and evolutionary stages of clumps of dense molecular gas and dust.

Most observational high-mass star formation research in the last decade has focused on HMPOs and UCHII regions. These objects have mid-infrared emission from hot dust and thus already contain an embedded massive protostellar source. Earlier evolutionary stages at the onset of massive star formation were observationally largely inaccessible because no telescope existed to identify these objects. The most basic observational characteristics of the earliest stages of massive star formation, prior to the formation of any embedded heating source, should be that they are strong cold dust and gas emitters at (sub)millimeter wavelengths, and weak or non-detections in the mid-infrared because they have not yet heated a warm dust cocoon. The advent of the mid-infrared space observatories ISO and MSX permitted for the first time identifications of large samples of potential (massive) starless clumps, the infrared dark clouds (IRDCs) (e.g., *Egan et al.,* 1998; *Bacmann et al.,* 2000; *Carey et al.,* 2000). Figure 1 shows a series of IRDCs as seen with Spitzer and the Galactic Legacy Infrared Mid-Plane Survey Extraordinaire (GLIMPSE). Various groups work currently on massive IRDCs, but so far not

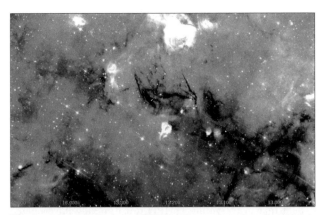

Fig. 1. Example image of IRDCs observed against the galactic background with the SPITZER GLIMPSE survey in the 8-μm band (*Benjamin et al., 2003*).

much has been published. Some initial ideas about observational quantities at the initial stages of massive cloud collapse were discussed by *Evans et al.* (2002). *Garay et al.* (2004) presented early millimeter observations of a sample of four sources; other recent statistical identifications and studies of potential HMSCs or regions at the onset of massive star formation can be found in *Hill et al.* (2005), *Klein et al.* (2005), *Sridharan et al.* (2005), and *Beltrán et al.* (2006).

These massive dense clumps have masses between a few 100 and a few 1000 $M_\odot$, sizes on the order of 0.25–0.5 pc, mean densities of 10^5 cm^{-3}, and temperatures on the order of 16 K. While the masses and densities are typical for high-mass star-forming regions, the temperatures derived, for example, from NH$_3$ observations [around 16 K (*Sridharan et al., 2005*)] are lower than those toward young HMPOs and UCHII regions [usually ≥22 K (e.g., *Churchwell et al., 1990; Sridharan et al., 2002*)]. Furthermore, the measured NH$_3$ line widths from the IRDCs are narrow as well; *Sridharan et al.* (2005) found mean values of 1.6 km s^{-1} whereas HMPOs and UCHIIs have mean values of 2.1 and 3.0 km s^{-1}, respectively (*Churchwell et al., 1990; Sridharan et al., 2002*). The narrow line widths and low temperatures support the idea that IRDCs represent an earlier evolutionary stage than HMPOs and UCHII regions, with less internal turbulence. Although subject to large uncertainties, a comparison of the virial masses calculated from NH$_3$ data with the gas masses estimated from 1.2 mm continuum emission indicates that most candidate HMSCs are virially bound and prone to potential collapse and star formation (*Sridharan et al., 2005*).

The IRDCs are not a well defined class, but are expected to harbor various evolutionary stages, from genuine HMSCs via high-mass cores harboring accreting low/intermediate-mass protostars to the youngest HMPOs. While the first stage provides good targets for studying the initial conditions of massive star formation prior to cloud collapse, the other stages are important to understand the early evolution of massive star-forming clumps. For example, the source HMSC18223-3 is probably in such an early accretion stage: Correlating high-spatial-resolution millimeter observations

from the Plateau de Bure Interferometer with Spitzer mid-infrared observations, *Beuther et al.* (2005c) studied a massive dust and gas core with no protostellar mid-infrared counterpart in the GLIMPSE data. While this could also indicate a genuine HMSC, they found relatively high temperatures [~33 K from NH$_3$(1,1) and (2,2)], an increasing N$_2$H$^+$(1–0) line width from the core edge to the core center, and so-called "green fuzzy" mid-infrared emission at the edge of the core in the IRAC data at 4.5 μm, indicative of molecular outflow emission. The outflow scenario is supported by strong non-Gaussian line-wing emission in CO(2–1) and CS(2–1). These observational features are interpreted as circumstantial evidence for early star formation activity at the onset of massive star formation. Similar results toward selected IRDCs have recently been reported by, e.g., *Rathborne et al.* (2005), *Ormel et al.* (2005), *Birkmann et al.* (2006), and *Pillai et al.* (2006). Interestingly, none of these IRDC case studies has yet revealed a true HMSC. However, with the given low number of such studies, we cannot infer whether this is solely a selection effect or whether HMSCs are genuinely rare.

Recent large-scale millimeter-continuum mapping of Cygnus-X revealed approximately the same number of infrared-quiet sources compared with infrared-bright HMPO-like regions (*Motte et al., 2005*). However, none of these infrared-quiet sources appears to be a genuine HMSC, and hence they could be part of the class of high-mass cores harboring accreting low/intermediate-mass protostars. A few studies report global infall on large spatial scales (e.g., *Rudolph et al., 1990; Williams and Garland, 2002; Peretto et al., 2006; Motte et al., 2005*), suggesting that massive clumps could form from lower-density regions collapsing during the early star formation process.

The earliest stages of massive star formation, specifically massive IRDCs, have received increased attention over the past few years, but some properties such as the magnetic field have so far not been studied at all. Since this class of objects is observationally rather new, we are expecting many exciting results in the coming years.

2.2. Theory

A central fact about GMCs is that they are turbulent (*Larson, 1981*). The level of turbulence can be characterized by the virial parameter $\alpha_{vir} = 5\sigma^2 R/GM$, where σ is the rms one-dimensional velocity dispersion, R the mean cloud radius, and M the cloud mass; α_{vir} is proportional to the ratio of kinetic to gravitational energy (*Myers and Goodman, 1988; Bertoldi and McKee, 1992*). The large-scale surveys of GMCs by *Dame et al.* (1986) and *Solomon et al.* (1987) give $\alpha_{vir} \approx 1.3$–1.4 (*McKee and Tan, 2003*), whereas regions of active low-mass star formation have $\alpha_{vir} \approx 0.9$ (e.g., *Onishi et al., 1996*). For regions of massive star formation, *Yonekura et al.* (2005) find $0.5 \leq \alpha_{vir} \leq 1.4$.

The great advance in our understanding of the dynamics of GMCs in the past decade has come from simulations of the turbulence in GMCs (e.g., *Elmegreen and Scalo, 2004; Scalo and Elmegreen, 2004; Mac Low and Klessen,*

2004). One of the primary results of these studies is that turbulence decays in less than a crossing time. A corollary of this result is that it is difficult to transmit turbulent motions for more than a wavelength. These results raise a major question: Since most of the sources of interstellar turbulence are intermittent in both space and time, how is it possible to maintain the high observed levels of turbulence in the face of such strong damping? Simulations without driven turbulence, such as those used to establish the theory of competitive accretion (e.g., *Bonnell et al.*, 2001a) (see section 3.2), reach virial parameters $\alpha_{vir} \ll 1$, far less than observed.

The results of these turbulence simulations have led to two competing approaches to the modeling of GMCs and the gravitationally bound structures (clumps) within them: as quasi-equilibrium structures or as transient objects. The first approach builds on the classical analysis of *Spitzer* (1978) and utilizes the steady-state virial theorem (*Chieze*, 1987; *Elmegreen*, 1989; *Bertoldi and McKee*, 1992; *McKee*, 1999). This model naturally explains why GMCs and the bound clumps within them have virial parameters on the order of unity (provided the magnetic field has a strength comparable to that observed) and why their mean column densities in the galaxy are ~10^{22} H cm^{-2}. In order to account for the ubiquity of the turbulence, such models must assume that (1) the turbulence actually decays more slowly than in the simulations, perhaps due to an imbalanced MHD cascade (*Cho and Lazarian*, 2003); (2) the energy cascades into the GMC or clump from larger scales; and/or (3) energy injection from star formation maintains the observed level of turbulence (*Norman and Silk*, 1980; *McKee*, 1989; *Matzner*, 2002). Recent simulations by Li and Nakamura (in preparation) are consistent with the suggestion that protostellar energy injection can indeed lead to virial motions in star-forming clumps (see section 4.2). In the alternative view, the clouds are transient and the observed turbulence is associated with their formation (*Ballesteros-Paredes et al.*, 1999; *Elmegreen*, 2000; *Hartmann et al.*, 2001; *Clark et al.*, 2005; *Heitsch et al.*, 2005). *Bonnell et al.* (2006) propose that the observed velocity dispersion in molecular clouds could be due to clumpy molecular gas passing through galactic spiral shocks. While these theories naturally account for the observed turbulence, they do not explain why GMCs have virial parameters on the order of unity, nor do they explain why clouds that by chance live longer than average do not have very low levels of turbulence. Quasi-equilibrium models predict that star formation will occur over a longer period of time than do transient cloud models. How these predictions compare with observations of high-mass star formation regions will be discussed in section 5.2.

3. HIGH-MASS PROTOSTELLAR OBJECTS

3.1. Observational Results

3.1.1. General properties. The most studied objects in massive star formation research are HMPOs and UCHII regions. This is partly because the IRAS all-sky survey per-

mitted detection and identification of a large number of such sources from which statistical studies could be undertaken (e.g., *Wood and Churchwell*, 1989a; *Plume et al.*, 1992; *Kurtz et al.* 1994; *Shepherd and Churchwell*, 1996; *Molinari et al.*, 1996; *Sridharan et al.*, 2002; *Beltrán et al.*, 2006). The main observational difference between young HMPOs/HMCs and UCHII regions is that the former are weak or non-detections in the centimeter-regime due to undetectable free-free emission [for a UCHII discussion see, e.g., *Churchwell* (2002) and the chapter by Hoare et al.]. Although in our classification typical HMCs with their high temperatures and complex chemistry are a subset of HMPOs, we expect that every young HMPO must already have heated a small central gas core to high temperatures, and it is likely that sensitive high-spatial resolution observations will reveal small HMC-type structures toward all HMPOs. This is reminiscent of the so-called hot corinos found recently in some low-mass star-forming cores (see the chapter by Ceccarelli et al.).

Many surveys have been conducted in the last decade characterizing the physical properties of massive star-forming regions containing HMPOs (e.g., *Plume et al.*, 1997; *Molinari et al.*, 1998; *Sridharan et al.*, 2002; *Beuther et al.*, 2002a; *Mueller et al.*, 2002; *Shirley et al.*, 2003; *Walsh et al.*, 2003; *Williams et al.*, 2004; *Faundez et al.*, 2004; *Zhang et al.*, 2005; *Hill et al.*, 2005; *Klein et al.*, 2005; *Beltrán et al.*, 2006). While the masses and sizes are of the same order as for the IRDCs (a few 100 to a few 1000 M$_\odot$ and on the order of 0.25–0.5 pc; see section 2), mean densities can exceed 10^6 cm^{-3}, and mean surface densities, although with a considerable spread, are reported around 1 g cm^{-2} (for a compilation see *McKee and Tan*, 2003). In contrast to earlier claims that the density distributions n $\propto$ r^{-p} of massive star-forming clumps may have power-law indices p around 1.0, several studies derived density distributions with mean power-law indices p around 1.5 (*Beuther et al.*, 2002a; *Mueller et al.*, 2002; *Hatchell and van der Tak*, 2003; *Williams et al.*, 2005), consistent with density distributions observed toward regions of low-mass star formation (e.g., *Motte and André*, 2001). However, one has to bear in mind that these high-mass studies analyzed the density distributions of the gas on cluster-scales, whereas the low-mass investigations trace scales of individual or multiple protostars. Mean temperatures (~22 K, derived from NH$_3$ observations) and NH$_3$(1,1) line widths (~2.1 km s^{-1}) are also larger for HMPOs than for IRDCs.

Furthermore, HMPOs are often associated with H$_2$O and Class II CH$_3$OH maser emission (e.g., *Walsh et al.*, 1998; *Kylafis and Pavlakis*, 1999; *Beuther et al.*, 2002c; *Codella et al.*, 2004; *Pestalozzi et al.*, 2005; *Ellingsen*, 2006). While the community agrees that both maser types are useful signposts of massive star formation (H$_2$O masers are also found in low-mass outflows), there is no general agreement what these phenomena actually trace in massive star-forming regions. Observations indicate that both species are found either in molecular outflows (e.g., *De Buizer*, 2003; *Codella et al.*, 2004) or in potential massive accretion disks (e.g., *Norris et al.*, 1998; *Torrelles et al.*, 1998). In a few cases, such as very high spatial resolution VLBI studies, it has been possible to distinguish between an origin in a disk and an outflow (e.g., *Torrelles et al.,* 2003; *Pestalozzi et al.,* 2004; *Goddi et al.,* 2005), but, in general, it is for the most part impossible to distinguish between the two possibilities.

One of the most studied properties of HMPOs are the massive molecular outflows found to be associated essentially with all stages of early massive star formation. For a discussion of this phenomenon and its implications, see section 4.

3.1.2. Massive disks. Disks are an essential property of the accretion-based formation of high-mass stars. The chapter by Cesaroni et al. provides a detailed discussion about observations and modeling of disks around massive protostellar objects. We simply summarize that massive, Keplerian disks have been observed around early-B-type stars, the best known example being IRAS20126 + 4102 (*Cesaroni et al.,* 1997, 1999, 2005a; *Zhang et al.,* 1998). Venturing further to higher-mass sources, several studies found rotating structures perpendicular to molecular outflows, indicative of an inner accretion disk (e.g., *Zhang et al.,* 2002; *Sandell et al.,* 2003; *Beltrán et al.,* 2004; *Beuther et al.,* 2005b). However, these structures are not necessarily Keplerian and could be larger-scale toroids, rotating around the central forming O-B cluster as suggested by *Cesaroni* (2005). Recently *van der Tak and Menten* (2005) conclude from 43 GHz continuum observations that massive star formation at least up to $10^5 \, L_\odot$ proceeds through accretion with associated collimated molecular outflows. A detailed theoretical and observational understanding of massive accretion disks is one of the important issues for future high-mass star formation studies.

3.1.3. Spectral energy distributions (SEDs). Spectral energy distributions (SEDs) have often been used to classify low-mass star-forming regions and to infer their physical properties (e.g., *Lada and Wilking,* 1984; *André et al.,* 1993). In massive star formation, deriving SEDs for individual high-mass protostellar sources proves to be more complicated. Problems arise because of varying spatial resolution with frequency, varying telescope sensitivity, and disk orientation to the line of sight. While we can resolve massive cluster-forming regions in the (sub)millimeter regime (e.g., *Cesaroni et al.,* 1999; *Shepherd et al.,* 2003; *Beuther and Schilke,* 2004) and at centimeter wavelengths (e.g., *Wood and Churchwell,* 1989b; *Kurtz et al.,* 1994; *Gaume et al.,* 1995, *De Pree et al.,* 2000), near-/mid- and far-infrared wavelength data for individual subsources are difficult to obtain. The earliest evolutionary stages are generally so deeply embedded that they are undetectable at near-infrared wavelengths, and until recently this was also a severe problem at mid-infrared wavelength [although a few notable exceptions exist (e.g., *De Buizer et al.,* 2002; *Linz et al.,* 2005)]. The advent of the Spitzer Space Telescope now allows deep imaging of such regions and will likely reveal many objects. However, even with Spitzer the spatial resolution in the far-infrared regime, where the SEDs at early evolutionary stages peak, is usually not good enough (16" pixels) to spatially resolve the massive star-forming clus-

ters. The only statistically relevant data at far-infrared wavelengths so far stem from the IRAS satellite, which had a spatial resolution of approximately 100". Although the IRAS data have proven very useful in identifying regions of massive star formation (e.g., *Wood and Churchwell,* 1989a; *Molinari et al.,* 1996; *Sridharan et al.,* 2002; *Beltrán et al.,* 2006), they just give the fluxes integrated over the whole star-forming cluster and thus hardly constrain the emission of individual cluster members. Nevertheless, the IRAS data were regularly employed to estimate the integrated luminosity of young massive star-forming regions by two-component gray-body fits (e.g., *Hunter et al.,* 2000; *Sridharan et al.,* 2002), and to set additional constraints on the density distributions of the regions (e.g., *Hatchell and van der Tak,* 2003). Spectral energy distribution modeling allows one to infer the characteristics of protostars in massive star-forming regions (*Osorio et al.,* 1999). *Chakrabarti and McKee* (2005) showed that the far-IR SEDs of protostars embedded in homogeneous, spherical envelopes are characterized by the density profile in the envelope and by two dimensionless parameters, the light-to-mass ratio, L/M, and the surface density of the envelope, $\Sigma = M/(\pi R^2)$. If these parameters are determined from the SED and if one knows the distance, then it is possible to infer both the mass and accretion rate of the protostar (*McKee and Chakrabarti,* 2005). *Whitney et al.* (2005) and *De Buizer et al.* (2005) have determined the effects of disks on the SEDs. Recent three-dimensional modeling by *Indebetouw et al.* (2006) shows how sensitive the SEDs, especially below 100 μm, are to the clumpy structure of the regions and to the observed line of sights. For example, they are able to fit the entire sample of UCHII regions studied by *Faison et al.* (1998) with the same clumpy model because the varying line of sights produce very different SEDs. Hence, SEDs alone do not provide sufficient information to infer the properties of clumpy sources, and it will be essential to obtain additional information by mapping these sources with powerful observatories such as the SMA, CARMA in 2006, ALMA at the end of the decade, and JWST in the next decade.

3.1.4. Chemistry. Young massive star-forming regions, and specifically HMCs, exhibit a rich chemistry from simple two-atom molecules to large organic carbon chains (e.g., *Blake et al.,* 1987; *Schilke et al.,* 1997, 2001). While these single-dish observations were not capable of resolving the chemical differences within the regions, interferometric studies toward a few sources have revealed the spatial complexity of the chemistry in HMCs (e.g., *Blake et al.,* 1996; *Wyrowski et al.,* 1999; *Beuther et al.,* 2005a). Here, we present studies toward W3(H_2O)/OH and Orion-KL as prominent chemical showcases.

The hot core region W3(H_2O) 6" east of the UCHII region W3(OH) exhibits an H_2O maser outflow and a synchrotron jet (*Alcolea et al.,* 1993; *Wilner et al.,* 1999). Follow-up observations with the Plateau de Bure Interferometer (PdBI) reveal dust emission associated with the synchrotron jet source and a large diversity of molecular line emission between the UCHII region W3(OH) and the hot core

W3(H$_2$O) (*Wyrowski et al.*, 1997, 1999). Nitrogen-bearing molecules are observed only toward W3(H$_2$O), whereas oxygen-bearing species are detected from both regions. Based on HNCO observations, *Wyrowski et al.* (1999) estimate gas temperatures toward W3(H$_2$O) of ~200 K, clearly confirming the hot core nature of the source. The differences in oxygen- and nitrogen-bearing species are manifestations of chemical evolution due to different ages of the sources.

One of the early targets of the recently completed Submillimeter Array (SMA) was the prototypical HMC Orion-KL. *Beuther et al.* (2004, 2005a, 2006) observed the region in the 865-µm and 440-µm windows and resolved the submillimeter continuum and molecular line emission at 1″ resolution. The continuum maps resolved the enigmatic source I from the hot molecular core, detected source n for the first time shortward of 7 mm and furthermore isolated a new protostellar source SMA1, emitting strong line emission. The observed 4-GHz bandpass in the 865-µm band revealed more than 145 lines from various molecular species with considerable spatial structure. Figure 2 shows an SMA example spectrum and representative line images. SiO emis-

sion is observed from the collimated northeast-southwest outflow and the more extended northwest-southeast outflow. Typical hot core molecules like CH$_3$CN and CH$_3$CH$_2$CN follow the hot core morphology known from other molecules and lower frequency observations (e.g., *Wright et al.*, 1996; *Blake et al.*, 1996; *Wilson et al.*, 2000). In contrast to this, oxygen-bearing molecules like CH$_3$OH or HCOOCH$_3$ are weaker toward the hot molecular core, but they show strong emission features a few arcseconds to the southwest, associated with the so-called compact ridge. Many molecules, in particular sulfur-bearing species like C^{34}S or SO$_2$, show additional emission further to the northeast, associated with IrC6.

Although existing chemical models predict the evolution and production paths of various molecules (e.g., *Charnley*, 1997; *van Dishoeck and Blake*, 1998; *Doty et al.*, 2002; *Nomura and Millar*, 2004; *Viti et al.*, 2004), we are certainly not at the stage where they can reliably predict the chemical structure of HMCs. Considering the complexity of the closest region of massive star formation, Orion-KL, it is essential to get a deeper understanding of the basic chemical and physical processes, because otherwise the confidence in studies of regions at larger distances is greatly diminished. On the positive side, a better knowledge of the chemical details may allow us to use molecular line observations as chemical clocks for (massive) star-forming regions.

3.2. Theory

The critical difference between low- and high-mass star formation is that low-mass stars form in a time t_{*f} short compared to the Kelvin-Helmholtz time t_{KH}, whereas high-mass stars generally have $t_{KH} \lesssim t_{*f}$ (*Kahn*, 1974). As a result, low-mass stars undergo extensive pre-main-sequence evolution after accretion has finished, whereas the highest mass stars can accrete a significant amount of mass while on the main sequence. The feedback associated with the intense radiation produced by high-mass stars will be considered in section 4; here we ask whether high-mass star formation differs significantly from low-mass star formation. At the time of the Protostars and Planets IV conference, *Stahler et al.* (2000) argued that it does.

The conventional view remains that high-mass star formation is a scaled-up version of low-mass star formation, with an accretion rate $\dot{m}_* \simeq c^3/G$, where the effective sound speed c includes the effects of thermal gas pressure, magnetic pressure, and turbulence (*Stahler et al.*, 1980). *Wolfire and Cassinelli* (1987) found that accretion rates on the order of 10^{-3} M$_\odot$ yr^{-1} are needed to overcome the effects of radiation pressure, and attributed this to the high values of c in high-mass star forming regions. By modeling the SEDs of high-mass protostars, *Osorio et al.* (1999) inferred that high-mass stars form in somewhat less than 10^5 yr. They favored a logatropic model, in which the ambient density varies as r^{-1} away from the protostar. *McKee and Tan* (2002, 2003) critiqued logatropic models and developed the turbulent core model, in which massive stars form from gravitationally

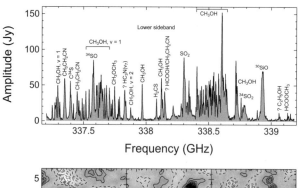

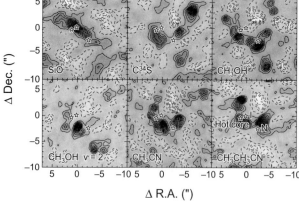

Fig. 2. Submillimeter array spectral line observations in the 850-µm band toward Orion-KL from *Beuther et al.* (2005a). The top panel shows a vector-averaged spectrum in the UV domain on a baseline of 21 m. The bottom panel presents representative images from various molecular species. Full contours show positive emission, dashed contours negative features due to missing short spacings and thus inadequate cleaning. The stars mark the locations of source I, the hot core peak position, and source n (see bottom-right panel).

bound cores supported by turbulence and magnetic fields. They argued that on scales large compared to the thermal Jeans mass, the density and pressure distributions in turbulent, gravitationally bound cores and clumps should be scale free and vary as powers of the radius (e.g., $\rho \propto r^{-k_\rho}$). As a result, the core and the star-forming clump in which it is embedded are polytropes, with $P \propto \rho^{\gamma_p}$. The gravitational collapse of a polytrope that is initially in approximate equilibrium results in an accretion rate $\dot{m}_* \propto m_*^q$, with $q = 3(1 - \gamma_p)/(4 - 3\gamma_p)$ (*McLaughlin and Pudritz*, 1997). Isothermal cores have $q = 0$, whereas logatropes ($\gamma_p \rightarrow 0$) have $q = \frac{3}{4}$. [It should be noted that the numerical simulations of *Yorke and Sonnhalter* (2002) generally have $q < 0$ due to feedback effects. This simulation differs from the turbulent core model in that the initial conditions were nonturbulent and the restriction to two dimensions overemphasizes feedback effects; see section 4.2.] Regions of high-mass star formation have surface densities $\Sigma \sim 1$ g cm^{-2} (*Plume et al.*, 1997), corresponding to visual extinctions $A_V \sim 200$ mag. *McKee and Tan* (2003) showed that the typical accretion rate and the corresponding time to form a star of mass m_{*f} in such regions are

$$\dot{m}_* \simeq 0.5 \times 10^{-3} \left(\frac{m_{*f}}{30\,M_\odot} \right)^{3/4} \Sigma_{cl}^{3/4} \left(\frac{m_*}{m_{*f}} \right)^{0.5} \frac{M_\odot}{yr}$$

$$t_{*f} \simeq 1.3 \times 10^5 \left(\frac{m_{*f}}{30\,M_\odot} \right)^{1/4} \Sigma_{cl}^{-3/4} \; yr$$

where Σ_{cl} is the surface density of the several thousand $M_\odot$ clump in which the star is forming and where they adopted $k_\rho = \frac{3}{2}$ as a typical value for the density power law in a core. The radius of the core out of which the star forms is $0.06(m_{*f}/30\,M_\odot)^{1/2}\Sigma_{cl}^{1/2}$ pc. Observed star clusters in the galaxy have surface densities comparable to those of high-mass star-forming regions, with values ranging from about 0.2 g cm^{-2} in the Orion Nebula Cluster to about 4 g cm^{-2} in the Arches Cluster (*McKee and Tan*, 2003). This work has been criticized on two grounds: First, it approximates the large-scale macroturbulence in the cores and clumps as a local pressure (microturbulence), which is equivalent to ignoring the surface terms in the virial equation (e.g., *Mac Low*, 2004). This approximation is valid provided the cores and clumps live for a number of free-fall times, so that they are in quasi-equilibrium. Evidence that clumps are quasi-equilibrium structures will be discussed in section 5.2; being smaller, cores are likely to experience greater fluctuations, so the quasi-equilibrium approximation is probably less accurate for them. Second, the turbulent core model assumes that most of the mass in the core that is not ejected by outflows will go into a single massive star (or binary). *Dobbs et al.* (2005) investigated this assumption by simulating the collapse of a high-mass core similar to that considered by *McKee and Tan* (2003). In the isothermal case, *Dobbs et al.* (2005) found that the core fragmented into

many pieces, which is inconsistent with the formation of a massive star. With a more realistic equation of state, however, only a few fragments formed, and when the heating due to the central protostar is included, even less fragmentation occurs (*Krumholz et al.*, 2005b). Furthermore, the level of turbulence in the simulations by *Dobbs et al.* (2005) is significantly less than the observed value.

Variants of the gravitational collapse model in which the accretion rate accelerates very rapidly have also been considered ($\dot{m}_* \propto m_*^q$ with $q > 1$, so that $m_* \rightarrow \infty$ in a finite time in the absence of other effects). However, such models have accretion rates that can exceed the value $\dot{m}_* \propto c^3$ that is expected on dynamical grounds. *Behrend and Maeder* (2001) assumed that the accretion rate onto a protostar is proportional to the observed mass loss rate in the protostellar outflow and found that a massive star could form in $\sim 3 \times 10^5$ yr. This phenomenological model has $q \simeq 1.5$, the value adopted in an earlier model by *Norberg and Maeder* (2000). However, it is not at all clear that the accretion rate onto the protostar is in fact proportional to the observed mass outflow rates. *Keto* (2002, 2003) modeled the growth of massive stars as being due to Bondi accretion, so that the accretion rate is $\dot{m}_* \propto m_*^2$, under his assumption that the ambient medium has a constant density and temperature. As Keto points out, the Bondi accretion model assumes that the self-gravity of the gas is negligible. The condition that the mass within the Bondi radius Gm_*/c^2 be much less than the stellar mass can be shown to be equivalent to requiring $\dot{m}_* \lesssim c^3/G$; for the value of $c \simeq 0.5$ km s^{-1} considered by Keto, this restricts the accretion rate to $\dot{m}_* \lesssim 3 \times 10^{-5}$ $M_\odot$ yr^{-1}, smaller than the values he considers. (One can show that when one generalizes the Bondi accretion model to approximately include the self gravity of the gas, the accretion rate is indeed about c^3/G if the gas is initially in virial equilibrium.) *Schmeja and Klessen* (2004) analyze mass accretion rates in the framework of gravo-turbulent fragmentation, and they find that the accretion rates are highly time-variant, with a sharp peak shortly after the formation of the protostellar core. Furthermore, in their models the peak and mean accretion rates increase with increasing mass of the final star.

Most models for (proto)stellar structure and evolution do not yet include the effects of rotation (e.g., *Meynet and Maeder*, 2005), which are expected to be relatively large given the recent accumulation of stellar material from the accretion disk. In models of gravo-turbulent fragmentation, *Jappsen and Klessen* (2004) find that the angular momentum j correlates with the core mass M like $j \propto M^{2/3}$. Furthermore, they conclude that the angular momentum evolution is approximately consistent with the contraction of initially uniform density spheres undergoing solid-body rotation. The precise amount of stellar angular momentum depends on how the accretion and outflow from the star-disk interaction region is modulated by magnetic fields and on the strength of the stellar wind (e.g., *Matt and Pudritz*, 2005). One potentially important effect is the variation in photospheric temperature from the equatorial to polar regions, which can enhance the beaming of bolometric and ionizing luminosity along the polar directions.

Alternative models for high-mass star formation have been developed by Bonnell and collaborators (e.g., *Bonnell et al.*, 1998, 2004). In the competitive accretion model, small stars ($m_* \sim 0.1$ M$_\odot$) form via gravitational collapse, but then grow by gravitational accretion of gas that was initially unbound to the star — i.e., by Bondi-Hoyle accretion, with allowance for the possibility that tidal effects can reduce the accretion radius (*Bonnell et al.*, 2001a, 2004). This model naturally results in segregating high-mass stars toward the center of the cluster, as observed. Furthermore, it gives a two-power law IMF that is qualitatively consistent with observations (*Bonnell et al.*, 2001b). Simulations by *Bonnell et al.* (2004) are consistent with this model. However, there are two significant difficulties: First, radiation pressure disrupts Bondi-Hoyle accretion once the stellar mass exceeds ~ 10 M$_\odot$ (*Edgar and Clarke*, 2004), so it is unlikely that competitive accretion can operate at masses above this. There is no evidence for a change in the IMF in this mass range, however, which suggests that competitive accretion does not determine the IMF at lower masses either. Second, competitive accretion is effective only if the virial parameter is much less than observed: Based on simulations of Bondi-Hoyle accretion in a turbulent medium, *Krumholz et al.* (2005c, 2006) show that protostars of mass $m_* \sim 0.1$ M$_\odot$ can accrete more than their initial mass in a dynamical time only if $\alpha_{vir} \lesssim 0.1(10^3$ M$_\odot/$M$_{cl})^{1/2}$. Such low values of α_{vir} do appear in the simulations, but, as discussed above, not in observed high-mass star-forming regions, which have masses M$_{cl}$ of hundreds to thousands of solar masses. Since the expected amount of mass accreted in a dynamical time is small, Krumholz et al. conclude that stars form via gravitational collapse of individual cores (Fig. 3). Bonnell et al. (this volume) argue that both difficulties can be ameliorated if the clump in which the massive stars are forming is undergoing global gravitational collapse. *Tan et al.* (2006) present arguments against such dynamical col-

lapse in the formation of star clusters. It is thus very important to observationally determine the nature of the motions in massive star-forming clumps: Are they dominated by turbulence or by collapse?

The most radical and imaginative model for the formation of high-mass stars is that they form via stellar collisions during a brief epoch in which the stellar density reaches $\sim 10^8$ stars pc^{-3} (e.g., *Bonnell et al.*, 1998; *Bonnell and Bate*, 2002), far greater than observed in any galactic star cluster [the densest region reported so far is W3 IRS5 with an approximate stellar density of 10^6 stars pc^{-3} (*Megeath et al.*, 2005)]. This model also results in an IMF that is in qualitative agreement with observation, although it must be borne in mind that the simulations to date have not included feedback. In their review, *Stahler et al.* (2000) supported the merger model, emphasizing that gas associated with protostars could increase the effective collision cross section and permit merging to occur at lower stellar densities. More recently, *Bonnell and Bate* (2005) have suggested that binaries in clusters will evolve to smaller separations due to accretion, resulting in mergers. However, a key assumption in this model is that there is no net angular momentum in the accreted gas, which makes sense in the competitive accretion model but not the gravitational collapse model. Stellar dynamical calculations by *Portegies Zwart et al.* (2004), which did not include any gas, show that at densities $\geq 10^8$ stars pc^{-3}, it is possible to have runaway stellar mergers at the center of a star cluster, which they suggest results in the formation of an intermediate-mass black hole. It should be noted that they inferred that this could have occurred based on the currently observed properties of the star cluster (although with the assumption that the tidal radius is greater than $100\times$ the core radius), not on a hypothetical ultradense state of the cluster. *Bally and Zinnecker* (2005) discuss observational approaches to testing the merger scenario, and suggest that the wide-angle outflow from OMC-1 in the Orion molecular cloud could be due to a protostellar merger that released 10^{48}–10^{49} erg. While it

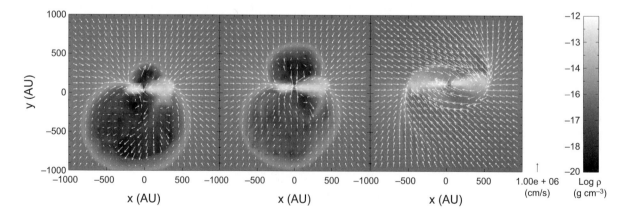

Fig. 3. Plot from *Krumholz et al.* (2005b) showing three-dimensional radiation hydrodynamic simulations of the collapse of a massive core. The slice in the XY plane at three different times shows the initial growth, instability, and collapse of a radiation bubble. The times of the three slices are 1.5×10^4, 1.65×10^4, and 2.0×10^4 yr, and the (proto)stellar masses are 21.3, 22.4, and 25.7 M$_\odot$. The density is shown in grayscale and the velocity as arrows.

is quite possible that some stellar mergers occur near the centers of some star clusters, the hypothesis that stellar mergers are responsible for a significant fraction of high-mass stars faces several major hurdles: (1) the hypothesized ultradense state would be quite luminous due to the massive stars, yet has never been observed; (2) the mass loss hypothesized to be responsible for reducing the cluster density from ~10^8 stars pc^{-3} to observed values must be finely tuned in order to decrease the magnitude of the binding energy by a large factor; and (3) it is difficult to see how this model could account for the observations of disks and collimated outflows discussed in section 3.1 and 4.1.

4. FEEDBACK PROCESSES

4.1. Observational Results

4.1.1. Hypercompact HII regions. Hypercompact HII regions (HCHIIs) are smaller, denser, and brighter than UCHII regions. Specifically, they are defined as having diameters less than 0.01 pc, consistent with being small photoionized nebulae produced by O or B stars. None have more ionizing flux than can be provided by a single O or B star. Their common properties that distinguish them from UCHII regions are:

1. They are ≥10 times smaller (≤0.01 pc) and ~100× denser than UCHII regions with emission measures ≥10^8 pc cm^{-6} (*Kurtz*, 2002, 2005).

2. They have rising radio spectral indices α (where $S_\nu \propto \nu^\alpha$) from short centimeter to millimeter wavelengths, and α ranges from ~0.3 to 1.6 with a typical value of ~1 (e.g., *Churchwell*, 2002; *Hofner et al.*, 1996). They are very faint or not detected at wavelengths longward of 1 cm. The power-law spectra span too large a range in frequency to be the transition from optically thick to thin emission in a constant density nebula.

3. In massive star-formation regions, they often appear in tight groups of two or more components (*Sewilo et al.*, 2004), reminiscent of the Trapezium in Orion.

4. Many, but not all, HCHII regions have unusually broad radio recombination lines (RRLs; FWHM ≥ 40 km s^{-1}). Some have FWHMs > 100 km s^{-1} (*Sewilo et al.*, 2004).

5. They are often (always?) coincident with strong water masers (e.g., *Hofner and Churchwell*, 1996; *Carral et al.*, 1997) and possibly other masers also, but the latter has not yet been observationally established.

What is the nature of HCHII regions? Their compactness, multiplicity, range of luminosities, and coincidence with water masers all argue for ionization by a single or possibly a binary system of late O or B star(s) at an age younger than UCHII regions. Their broad RRLs indicate highly dynamic internal structures (outflow jets, disk rotation, expansion, shocks, accretion, etc.), the nature of which is not yet understood. The fact that only about half of the HCHIIs that have observed RRLs have broad lines would argue that this phase is short-lived, perhaps only apparent in the first half of an HCHII region's lifetime, provided the accretion rate

is larger during the early stages of the HCHIIs. Their radio spectral indices have implications for the internal density structure, but here also too little observational information is available to do more than speculate at this juncture. The power-law spectra can be produced by a clumpy nebula (*Ignace and Churchwell*, 2004), but this is only one of several possibilities that needs to be investigated (e.g., *Keto*, 2003; *Tan and McKee*, 2003). It is not clear yet whether they form after the HMC stage or whether HMCs and HCHIIs coexist.

4.1.2. Outflows. Massive molecular outflows are among the most studied phenomena in massive star formation over the last decade, and the observations range from statistical studies of large samples at low spatial resolution to individual case studies at high spatial resolution. Because (massive) molecular outflows are presented in the chapter by Arce et al., here we only discuss their general properties and implications for the massive star-forming processes.

Since the early statistical work by *Shepherd and Churchwell* (1996) it is known that massive molecular outflows are an ubiquitous phenomenon in massive star formation. Early observations indicated that massive outflows appear less collimated than their low-mass counterparts, implying potentially different formation scenarios for the outflows and the massive star-forming processes (e.g., *Richer et al.*, 2000; *Shepherd et al.*, 2003). However, *Beuther et al.* (2002b) showed that outflows from HMPOs, even if observed only with single-dish instruments, are consistent with collimated outflows if one considers the large distances, projection effects, and poor angular resolution carefully (see also *Kim and Kurtz*, 2006). Interferometric follow-up studies revealed more massive star-forming regions with molecular outflows consistent with the collimated outflows known from low-mass star formation (for a compilation see *Beuther and Shepherd*, 2005). Since collimated structures are hard to maintain over a few × 10^4 yr (the typical dynamical timescales of molecular outflows) if they are associated with colliding protostars within the cluster centers; these outflow observations strongly support the accretion-based formation scenario in massive star formation.

We note that no highly collimated outflow has been observed for high-mass star-forming regions exceeding 10^5 L$_\odot$, corresponding to approximately 30 M$_\odot$ stars. Therefore, these data cannot exclude that stars more massive than that may form via different processes. However, there are other possibilities to explain the current non-observations of collimated outflows at the high-luminosity end. An easy explanation would be that these sources are so exceptionally rare that we simply have not been lucky enough to detect one. Alternatively, *Beuther and Shepherd* (2005) recently suggested an evolutionary sequence that explains qualitatively the present state of observational facts (see also the chapter by Arce et al.): To form massive early O stars via accretion, the protostellar objects have to go through lower-mass stages as well. During the early B-star stage, the accreting protostars can drive collimated outflows as observed. Growing further in mass and luminosity, they develop HCHII regions in the late O-star stage, and collimated jets and less collimated winds can coexist, producing bipolar outflows with a lower degree of collimation. In this scenario, it would be intrinsically impossible to ever observe jet-like outflows from young early O-type protostars. Alternatively, the effect may be due to greater observational confusion of the outflows from very luminous sources with those from surrounding lower-mass protostars, since the more luminous sources tend to be in richer, more distant clusters. These evolutionary models for massive molecular outflows have to be tested further against theory and observations.

4.2. Theory

Feedback processes that act against gravitational collapse and accretion of gas to protostars include radiation pressure (transmitted via dust grains, and, for sufficiently massive stars, by electron scattering), thermal pressure of photoionized gas, ram pressure from protostellar winds, and main-sequence stellar winds. These processes become increasingly important with protostellar mass and may reduce the efficiency of star formation from a given core. There is good evidence for a cutoff in the stellar IMF at around 150 $M_\odot$ (e.g., *Weidner and Kroupa*, 2004; *Figer*, 2005), but it remains to be determined whether this is due to feedback processes or to instabilities in massive stars.

For individual low-mass star formation from a core, bipolar protostellar outflows, accelerated from the inner accretion disk and star by rotating magnetic fields, appear to be the dominant feedback mechanism, probably preventing accretion from polar directions and also ejecting a fraction, up to a third, of the material accreting through the disk. This leads to star formation efficiencies from the core on the order of 30–50% (*Matzner and McKee*, 2000).

For massive protostars, forming in the same way from a core and accretion disk, one expects similar MHD-driven outflows to be present, leading to similar formation efficiencies. In addition, once the massive protostar has contracted to the main sequence (this can occur rapidly before accretion has finished), it starts to produce a large flux of ionizing photons. The resulting HCHII region is likely to be confined in all but the polar directions by the protostellar jets (*Tan and McKee*, 2003). This could provide an important potential observational diagnostic for the physics of protostellar jets, as they might be illuminated along the axis by the ionizing radiation. As the protostellar mass and ionizing flux increase, the HCHII region can eventually burn its way through the jet and begin to ionize the disk surface. If the disk is ionized out to a radius where the escape speed is about equal to the ionized gas sound speed, then a photoevaporated flow is set up, reducing accretion to the star (*Hollenbach et al.*, 1994).

Observations indicate that outflows may be less well collimated for luminosities above about $10^5 \, L_\odot$ (section 4.1). As discussed above, *Beuther and Shepherd* (2005) have suggested that this is due to a decrease in the collimation of the protostellar jet with increasing protostellar luminos-

ity. A possible mechanism for this is suggested by the work of *Fendt and Cameljic* (2002), who simulated protostellar jets with a large turbulent diffusivity and found that the collimation decreases as the diffusivity increases. Applying this picture to massive outflows, the level of turbulence in the accretion flow would have to grow as the luminosity of the protostar increases.

The importance of massive molecular outflows in driving turbulence in molecular clouds is not generally agreed upon. *Mac Low and Klessen* (2004, and references therein) argue that although molecular outflows are very energetic, they deposit most of their energy at low densities. Furthermore, since the molecular gas motions show increasing power all the way up to the largest cloud complexes, *Mac Low and Klessen* (2004) conclude that it would be hard to fathom how such large scales should be driven by embedded protostars. Contrary to this, on the relatively small scales of the clumps, if the energy of turbulent motions decays with a half-life of one dynamical time, then protostellar outflows from star formation are able to maintain turbulence if 50% of the gas mass forms stars in 20 dynamical times, and 1% of the resulting outflow energy couples to the ambient gas (*Tan*, 2006). Recently, *Quillen et al.* (2005) reported that their observations toward the low-mass starforming region NGC1333 are also consistent with outflow-driven turbulence, and, as remarked in section 2.2, *Li and Nakamura* (2006) have given theoretical support to this idea. It becomes more difficult for this mechanism to support turbulence on larger scales in the GMC involving greater gas masses; on these scales, *Matzner* (2002) has shown that energy injection by HII regions dominates that by protostellar outflows and can support the observed level of turbulence. Alternatives to protostellar driving of the turbulence in molecular clouds are discussed in section 2.2.

Radiation pressure on dust grains (well coupled to the gas at these densities) is also important for massive protostars. It has been suggested, in the context of spherical accretion models, that this leads to an upper limit to the initial mass function (*Kahn*, 1974; *Wolfire and Cassinelli*, 1987). The difficulties faced by spherical accretion models was a major motivation for the formation model via stellar collisions (*Bonnell et al.*, 1998). However, massive star formation becomes easier once a disk geometry is allowed for (e.g., *Nakano*, 1989; *Jijina and Adams*, 1996). *Yorke and Sonnhalter* (2002) used two-dimensional axially symmetric simulations to follow massive star formation from a core collapsing to a disk, including radiation pressure feedback; accretion stopped at 43 $M_\odot$ in their most massive core. They showed the accretion geometry channeled radiative flux into the polar directions and away from the disk, terming this the "flashlight effect." *Krumholz et al.* (2005a) found that cavities created by protostellar outflows increase the flashlight effect, allowing even higher final masses. The first three-dimensional simulation of massive star formation shows that instabilities facilitate the escape of radiation and allow the formation of stars significantly more massive than suggested by two-dimensional calculations (Fig. 3) (*Krumholz et al.*, 2005b).

The high accretion rates required to form massive stars tend to quench HCHII regions (*Walmsley*, 1995). For spherical accretion, the density profile in a freely infalling envelope is $n \propto r^{-3/2}$. As a result, the radius of the HCHII region is

$$R_{HCHII} = R_* \exp(S/S_{cr})$$

where S is the ionizing photon luminosity and

$$S_{cr} = \frac{\alpha^{(2)} \dot{m}_*^2}{8\pi\mu_H^2 G m_*} = 5.6 \times 10^{50} \left(\frac{\dot{m}_{*,-3}^2}{m_{*2}} \right) s^{-1}$$

(*Omukai and Inutsuka*, 2002). Here $\alpha^{(2)}$ is the recombination coefficient to excited states of hydrogen, $\dot{m}_{*,-3} = \dot{m}_*/(10^{-3} M_\odot \, yr^{-1})$ and $m_{*2} = m_*/(100 \, M_\odot)$; we have replaced m_p in their expression with $\mu_H = 2.34 \times 10^{-24}$ g, the mass per hydrogen nucleus. When the radius of the HCHII region is small enough that the infall velocity exceeds the velocity of an R-critical ionization front ($2c_i$, where $c_i \approx 10$ km s^{-1} is the isothermal sound speed of the ionized gas), the HCHII region is said to be "trapped" (*Keto*, 2002): There is no shock in the accretion flow and the HCHII region cannot undergo the classical pressure-driven expansion. The ionizing photon luminosity S increases rapidly with m_*. If the accretion rate depends on stellar mass such that the critical luminosity S_{cr} is approximately independent of mass [e.g., the standard *McKee and Tan* (2003) model has $\dot{m}_* \propto m_*^{1/2}$, so that $S_{cr} = const.$], then the radius of the HCHII region expands as $\exp(S)$ and the trapped phase is relatively brief. On the other hand, if S_{cr} increases rapidly with mass, as in the Bondi accretion model ($\dot{m}_* \propto m_*^2$), then the expansion is retarded, leading to the possibility that the trapped phase of the HCHII region could last for much of the life of the protostar (*Keto*, 2003). However, as outlined in section 3.2, the parameters adopted by *Keto* (2003) are not consistent with the neglect of the self-gravity of the gas.

The evolution of the HCHII region changes substantially due to rotation of the infalling gas. The density is significantly lower above the accretion disk (*Ulrich*, 1976), so trapped HCHII regions will generally expand out to the radius of the accretion disk; when this is larger than the gravitational radius $R_g = Gm_*/c_i^2$, then the HCHII region is no longer trapped (McKee and Tan, in preparation). *Keto and Wood* (2006) have also considered the effects of disks in massive protostars: They point out that it is possible to form an ionized accretion disk, and suggest that there is evidence for this in G10.6-04.

5. FORMATION OF STAR CLUSTERS

5.1. Observational Results

5.1.1. Initial mass function. The formation of the initial mass function (IMF) has been an important issue in star formation research since the early work by *Salpeter* (1955). For a current summary of IMF studies see *Corbelli et al.*

(2005) and references therein, and the chapter by Bonnell et al. One of the questions in the context of this review is whether the IMF is determined already at the earliest stages of cluster formation by the initial gravo-turbulent fragmentation processes of molecular clouds (e.g., *Padoan and Nordlund*, 2002; *Mac Low and Klessen*, 2004), or whether the IMF is determined by subsequent processes like competitive accretion or feedback processes from the underlying star-forming cluster (e.g., *Bonnell et al.*, 2004; *Ballesteros-Paredes et al.*, 2006). (Sub)millimeter continuum studies of young low-mass clusters have convincingly shown that the core mass function at the beginning of low-mass cluster formation already resembles the stellar IMF (*Motte et al.*, 1998; *Johnstone et al.*, 2001; *Enoch et al.*, 2006; see also the chapter by Lada et al.). Because massive star-forming regions are on average more distant, resolving these clusters is difficult. However, several single-dish studies of different high-mass star-forming regions at early evolutionary stages have shown that at high clump masses, the cumulative mass distributions are consistent with the the slope of the high-mass stellar IMF (*Shirley et al.*, 2003; *Williams et al.*, 2004; *Reid and Wilson*, 2005; *Beltrán et al.*, 2006). Furthermore, the only existing high-spatial-resolution interferometric study that resolves a massive star-forming clump into a statistically meaningful number of cores also finds the core mass distribution to be consistent with the stellar IMF (*Beuther and Schilke*, 2004). Although millimeter continuum observations alone are ambiguous as to whether the observed cores and clumps are bound or transient structures, the consistently steeper mass functions observed in millimeter continuum emission compared with the lower density tracing CO line studies (e.g., *Kramer et al.*, 1998) suggest that the millimeter-continuum sources could be bound whereas the CO sources could be transient. Furthermore, *Belloche et al.* (2001) report additional observations supporting the interpretation that the study of *Motte et al.* (1998) sampled bound sources. Combining these results from massive star-formation studies with the previous investigations in the low-mass regime, the apparent similarity between the (cumulative) clump and core mass functions and the stellar IMF supports the idea that the IMF is determined by molecular cloud structure before star formation is initiated, maybe implicating gravo-turbulent fragmentation. However, on a cautionary note, one has to keep in mind that the cumulative mass distributions from single-dish studies as reported above trace scales of cluster formation and thus probably refer to cluster-mass distributions rather than to the IMF. The only way to assess the relationship between the fragmentation of initial high-mass star-forming clumps and the resulting IMF is to carry out high-spatial-resolution interferometric (sub)millimeter line and continuum studies of a statistically significant sample of (very) young massive star-forming regions.

5.1.2. First GLIMPSE results. The GLIMPSE survey is providing an entirely new view of the inner galaxy with a higher resolution and sensitivity than ever achieved at mid-infrared wavelengths (*Benjamin et al.*, 2003). This is enabling a host of new research on massive star formation as well as many other fields of astronomy. Unfortunately, UCHII regions are generally saturated and too bright in the IRAC bands to identify the ionizing star(s) and associated clusters above the glaring diffuse PAH emission found toward all these objects. The known HCHII regions are also bright in the GLIMPSE survey. Bipolar outflows stand out in the 4.5-μm band, providing a powerful way to identify many new outflows in the inner galaxy. Numerous outflows have been identified in the survey and a catalog of them is being assembled by the GLIMPSE team. The mechanism responsible for this emission is believed to be line emission from shocked H_2 and/or CO bands; the shocks are produced by outflowing gas ramming into the ambient interstellar medium. Near-infrared spectroscopic observations of molecular H_2 or CO between 4.1 and 4.7 μm are needed to determine which interpretation is correct.

Within ~ 45° of the galactic center, hundreds of IRDCs are apparent in silhouette against the diffuse infrared background (Fig. 1). A catalog of many IRDCs is being prepared for publication by the GLIMPSE team. These clouds are optically thick at 8 μm, implying visual extinctions ≥50 mag (see section 2.1.). The GLIMPSE images are striking in part because of the large number of bubbles contained in them; there are about 1.5 bubbles per square degree on average. A catalog of 329 bubbles has been identified and a Web-accessible image archive will accompany the archive (Churchwell et al., in preparation). It is found that the bubbles are associated with HII regions and stellar clusters. Only three are associated with supernova remnants and none with planetary nebulae or Wolf Rayet stars. About one-third of the bubbles appear to be produced by the stellar winds and radiation pressure from O and B stars (i.e., massive starforming regions). About two-thirds of the bubbles have small angular diameters (typically only 3–4 arcmin) and do not coincide with a radio HII region or known cluster; these are believed to be driven by B4–B9 stars that have strong enough winds to form a resolved bubble and have enough UV radiation to excite PAH features, but not enough UV photons to ionize a detectable HII region.

One of the most exciting prospects from the GLIMPSE survey is the possibility of identifying the entire population of HMPOs and lower-mass protostars from the approximately 50 million stars in the GLIMPSE archive. This is now possible with the large archive of radiative transfer models of protostars calculated for the entire range of protostellar masses, the full range of suspected accretion rates, disk masses, and orientations of the accretion disks to the line of sight (Whitney et al., in preparation). Model photospheres for main-sequence stars and red giants are included in the model archive as well, so it is possible to distinguish reddened main-sequence stars and red giants from protostars and slightly evolved young stellar objects. What makes this archive of models powerful, however, is the model fitter that will fit the best models to observed SEDs of large numbers of sources, giving the mass, spectral types, approximate evolutionary state, and interstellar extinction for the best fit models to every source (Robitaille et al., in preparation). This will provide a powerful alternative to the classical method of estimating the global star-formation rate in the galaxy based on measured UV photon luminosities of radio HII regions and an assumed initial mass function.

5.2. Theory

In the local universe, massive star and star cluster formation are intrinsically linked: Massive stars almost always appear to form in clusters (*De Wit et al.,* 2005). We have seen that this is a natural expectation of models of star formation from cores, since the accretion rates are higher if the core is pressurized by the weight of a large clump of gas. This scenario predicts that massive stars tend to form near the center of the clump and that there can be extensive star formation (mostly of lower-mass stars) from the clump's gas while massive star formation is ongoing.

Presently there is no consensus on whether massive stars form preferentially at the centers of clusters since, although they are often observed in central locations, it is possible that they could have migrated there by dynamical interactions after their formation. *Bonnell and Davies* (1998) found that the mass segregation time of clusters with mass-independent initial velocity dispersions was similar to the relaxation time, $t_{relax} \simeq 0.1 \, N/\ln N)t_{dyn}$ for N equal mass stars, i.e., about 14 crossing timescales for N = 1000. The presence of gas should shorten these timescales (*Ostriker,* 1999). To resolve this issue, we need to measure the cluster formation time: Does it take few or many dynamical times? *Elmegreen* (2000) presented a number of arguments for rapid star formation in ~1–2 dynamical timescales, including scales relevant to star clusters. *Tan et al.* (2006) presented arguments for somewhat longer formation timescales and argued that star formation in clusters is a quasi-equilibrium process. For example, the age spread of stars in the Orion Nebula Cluster is at least 2.5 m.y. (*Palla and Stahler,* 1999), while the dynamical time is only 7×10^5 yr for the cluster as a whole, and is much shorter in the central region. A relatively long formation timescale is also consistent with the observed morphologies of protoclusters in CS molecular lines: *Shirley et al.* (2003) found approximately spherical and centrally concentrated morphologies for a large fraction of their sources, suggesting they are older than a few dynamical times. Long formation timescales mean that the observed central locations of massive stars could be due either to central formation or mass segregation (or both). A corollary of long formation timescales is that the level of turbulence in the clump must be maintained, possibly by protostellar outflows and HII regions (see sections 2.2 and 4.2). Studies of the spatial distributions of massive stars in more embedded, presumably less dynamically evolved, clusters should help to resolve this issue.

As with spatial segregation, there is also no consensus about whether there is a temporal segregation in massive star formation from the surrounding cluster: Do massive stars form early, late, or contemporaneously with the other cluster members? Late formation of massive stars was often proposed, since it was expected that once massive stars were present, they would rapidly disrupt the remaining gas with their feedback. However, *Tan and McKee* (2001) showed that the impact of feedback was much reduced in a medium composed of dense cores, virialized and orbiting supersonically in the clump potential. *Dale et al.* (2005) have carried out the first simulations of photo-ionizing feedback in clusters and have confirmed that feedback is significantly reduced in realistic, inhomogeneous clumps. The observed numbers of UCHII regions (*Kurtz et al.,* 1994) also suggest that massive star ionizing feedback can be confined inside ~0.1 pc for at least ~10^5 yr. *Hoogerwerf et al.* (2001) have proposed that four O and B stars were ejected 2.5 m.y. ago from the Orion Nebula Cluster, where massive star formation is still underway. If true, this would indicate that massive star formation occurred in both the early and late stages of cluster formation.

6. CONCLUSIONS AND FUTURE PROSPECTS

Observational evidence suggests that stars at least up to 30 M$_\odot$ form via an accretion-based formation scenario. Venturing to higher-mass objects is an important observational future task. Theoretically, stars of all masses are capable of forming via accretion processes but it remains an open question whether nature follows that path or whether other processes become more important for the highest-mass stars. Recent work suggests that the accretion-based formation scenario in turbulent molecular cloud cores is the more likely way to build most stars of all masses.

Regarding the evolutionary sequence outlined in the introduction, the HMPO and final-star stages have been studied extensively in the past, whereas the earliest stages of massive star formation, i.e., high-mass starless cores and high-mass cores harboring low/intermediate-mass protostars, are just beginning to be explored in more detail. The coming years promise important results for the initial conditions of massive star formation and the origin of the IMF.

One of the observational challenges of the coming decade is to identify and study the properties of genuine accretion disks in high-mass star formation (see also the chapter by Cesaroni et al.). Are massive accretion disks similar to their low-mass counterparts, or are they massive enough to become self-gravitating entities? Determining the nature of the broad radio recombination lines in HCHII regions requires high spatial resolution and high sensitivity, which both will be provided by the EVLA and ALMA. Ultimately, we need to understand how outflows are collimated and driven. Do they originate from the surface of the disk? What fraction of the matter that becomes unstable and begins falling toward the star/disk actually makes it into the star vs. being thrown back out via bipolar outflows? What fraction of the outflow mass is due to gas entrainment and what fraction is due to recycled infalling gas? At what evolutionary stage is the IMF actually determined? Furthermore, astrochemistry is still poorly investigated, but the advent of large correlator bandwidths now allows us to investigate the

chemical census in massive star-forming regions regularly in more detail. An important astrochemical goal is to establish chemical clocks for star-forming clumps and cores.

The observational capabilities available now and coming on line within the next few years are exciting. Just to mention a few, the Spitzer observatory, and especially the Spitzer surveys GLIMPSE and MIPSGAL, will provide an unprecedented census of star-forming regions over large parts of the galactic plane. The so-far poorly explored far-infrared spectrum will be available with the launches of SOFIA and Herschel. The SMA is currently opening the submillimeter spectral window to high spatial resolution observations, and ALMA will revolutionize (sub)millimeter interferometry and star formation research in many ways. Near- and mid-infrared interferometry is still in its infancy but early results from the VLTI are very promising. In addition, many existing observatories are upgraded to reach new levels of performance (e.g., PdBI, EVLA, CARMA). Combining the advantages of all instruments, massive star formation research is going to experience tremendous progress in the coming years.

Theorists face the same challenges as observers in understanding the formation and evolution of the molecular clouds and clumps that are the sites of massive star formation, the processes by which individual and binary massive stars form, the origin of the IMF, the strong feedback processes associated with massive star formation, and the interactions that occur in stellar clusters. Here the primary progress is likely to come from simulations on increasingly powerful computers. By the time of the next Protostars and Planets conference, it should be possible to simulate the formation of a cluster of stars in a turbulent, magnetized medium, to assess the merits of existing theoretical models, and to point the way toward a deeper understanding of massive star formation.

Acknowledgments. We wish to thank M. Krumholz and S. Lizano for helpful comments. The research of C.F.M. is supported in part by NSF grant AST-0303689. E.C. acknowledges partial support by NSF grant 978959. H.B. acknowledges financial support by the Emmy-Noether-Program of the Deutsche Forschungsgemeinschaft (DFG, grant BE2578).

REFERENCES

Alcolea J., Menten K. M., Moran J. M., and Reid M. J. (1993) *Lecture Notes in Physics, 412,* 225–228. Springer, Berlin.
André P., Ward-Thompson D., and Barsony M. (1993) *Astrophys. J., 406,* 122–141.
Bacmann A., André P., Puget J.-L., Abergel A., Bontemps S., et al. (2000) *Astron. Astrophys., 361,* 555–580.
Ballesteros-Paredes J., Hartmann L., and Vazquez-Semadeni E. (1999) *Astrophys. J., 527,* 285–297.
Ballesteros-Paredes J., Gazol A., Kim J., Klessen R. S., Jappsen A.-K., et al. (2006) *Astrophys. J., 637,* 384–391.
Bally J. and Zinnecker H. (2005) *Astron. J., 129,* 2281–2293.
Behrend R. and Maeder A. (2001) *Astron. Astrophys., 273,* 190–198.
Belloche A., André P., and Motte F. (2001) In *From Darkness to Light* (T. Montmerle and P. André, eds.), pp. 313–318. ASP Conf. Ser. 243, San Francisco.

Beltrán M. T., Cesaroni R., Neri R., Codella C., Furuya R. S., et al. (2004) *Astrophys. J., 601,* L187–L190.

Beltrán M. T., Brand J., Cesaroni C., Fontani F., Pezzuto S., et al. (2006) *Astron. Astrophys., 447,* 221–233.

Benjamin R. A., Churchwell E., Babler B. L., Bania T. M., Clemens D. P., et al. (2003) *Publ. Astron. Soc. Pac., 115,* 953–964.

Bertoldi F. and McKee C. F. (1992) *Astrophys. J., 395,* 140–157.

Beuther H. and Schilke P. (2004) *Science, 303,* 1167–1169.

Beuther H. and Shepherd D. (2005) In *Cores to Clusters* (M. S. N. Kumar et al., eds), pp. 105–119. Springer, New York.

Beuther H., Schilke P., Menten K. M., Motte F., Sridharan T. K., et al. (2002a) *Astrophys. J., 566,* 945–965.

Beuther H., Schilke P., Sridharan T. K., Menten K. M., Walmsley C. M., et al. (2002b) *Astron. Astrophys., 383,* 892–904.

Beuther H., Walsh A., Schilke P., Sridharan T. K., Menten K. M., et al. (2002c) *Astron. Astrophys., 390,* 289–298.

Beuther H., Zhang Q., Greenhill L. J., Reid M. J., Wilner D., et al. (2004) *Astrophys. J., 616,* L31–L34.

Beuther H., Zhang Q., Greenhill L. J., Reid M. J., Wilner D., et al. (2005a) *Astrophys. J., 632,* 355–370.

Beuther H., Zhang Q., Sridharan T. K., and Chen Y. (2005b) *Astrophys. J., 628,* 800–810.

Beuther H., Sridharan T. K., and Saito M. (2005c) *Astrophys. J., 634,* L185–L188.

Beuther H., Zhang Q., Reid M. J., Hunter T. R., Gurwell M., et al. (2006) *Astrophys. J., 636,* 323–331.

Birkmann S. M., Krause O., and Lemke D. (2006) *Astrophys. J., 637,* 380–383.

Blake G. A., Sutton E. C., Masson C. R., and Phillips T. G. (1987) *Astrophys. J., 315,* 621–645.

Blake G. A., Mundy L. G., Carlstrom J. E., Padin S., Scott S. L., et al. (1996) *Astrophys. J., 472,* L49–L52.

Bonnell I. A. and Bate M. R. (2002) *Mon. Not. R. Astron. Soc., 336,* 659–669.

Bonnell I. A. and Bate M. R. (2005) *Mon. Not. R. Astron. Soc., 362,* 915–920.

Bonnell I. A. and Davies M. B. (1998) *Mon. Not. R. Astron. Soc., 295,* 691–698.

Bonnell I. A., Bate M. R., and Zinnecker H. (1998) *Mon. Not. R. Astron. Soc., 298,* 93–102.

Bonnell I. A., Bate M. R., Clarke C. J., and Pringle J. E. (2001a) *Mon. Not. R. Astron. Soc., 323,* 785–794.

Bonnell I. A., Clarke C. J., Bate M. R., and Pringle J. E. (2001b) *Mon. Not. R. Astron. Soc., 324,* 573–579.

Bonnell I. A., Vine S. G., and Bate M. R. (2004) *Mon. Not. R. Astron. Soc., 349,* 735–741.

Bonnell I. A., Dobbs C. L., Robitaille T. P., and Pringle J. E. (2006) *Mon. Not. R. Astron. Soc., 365,* 37–45.

Bourke T. L., Myers P. C., Robinson G., and Hyland A. R. (2001) *Astrophys. J., 554,* 916–932.

Bromm V. and Loeb A. (2004) In *The Dense Interstellar Medium in Galaxies* (S. Pfalzner et al., eds.), pp. 3–10. Springer, Berlin.

Bronfman L., Nyman L.-A., and May J. (1996) *Astrophys. J. Suppl., 115,* 81–95.

Carey S. J., Feldman P. A., Redman R. O., Egan M. P., MacLeod J. M., et al. (2000) *Astrophys. J., 543,* L157–L161.

Carral P., Kurtz S. E., Rodríguez L. F., de Pree C., and Hofner P. (1997) *Astrophys. J., 486,* L103–L106.

Cesaroni R. (2005) *Astrophys. Space Sci., 295,* 5–17.

Cesaroni R., Felli M., Testi L., Walmsley C. M., and Olmi L. (1997) *Astron. Astrophys., 325,* 725–744.

Cesaroni R., Felli M., Jenness T., Neri R., Olmi L., et al. (1999) *Astron. Astrophys., 345,* 949–964.

Cesaroni R., Neri R., Olmi L., Testi L., Walmsley C. M., et al. (2005a) *Astron. Astrophys., 434,* 1039–1054.

Cesaroni R., Felli M., Churchwell E., and Walmsley C. M., eds. (2005b) In *Massive Star Birth: A Crossroad to Astrophysics.* IAU Symp. 227, Cambridge Univ., Cambridge.

Chakrabarti S. and McKee C. (2005) *Astrophys. J., 631,* 792–808.

Charnley S. B. (1997) *Astrophys. J., 481,* 396–405.

Chieze J.-P. (1987) *Astron. Astrophys., 171,* 225–232.

Cho J. and Lazarian A. (2003) *Mon. Not. R. Astron. Soc., 345,* 325–339.

Churchwell E. (2002) *Ann. Rev. Astron. Astrophys., 40,* 27–62.

Churchwell E., Walmsley C. M., and Cesaroni R. (1990) *Astron Astrophys. Suppl., 83,* 119–144.

Clark P. C., Bonnell I. A., Zinnecker H., and Bate M. R. (2005) *Mon. Not. R. Astron. Soc., 359,* 809–818.

Codella C., Lorenzani A., Gallego A. T., Cesaroni R., and Moscadelli L. (2004) *Astron. Astrophys., 417,* 615–624.

Corbelli B., Palla F., and Zinnecker H., eds. (2005) *The Initial Mass Function 50 Years Later.* Springer, Dordrecht.

Crutcher R. M. (1999) *Astrophys. J., 520,* 706–713.

Dale J. E., Bonnell I. A., Clarke C. J., and Bate M. R. (2005) *Mon. Not. R. Astron. Soc., 358,* 291–304.

Dame T. M., Elmegreen B. G., Cohen R. S., and Thaddeus P. (1986) *Astrophys. J., 305,* 892–908.

De Buizer J. M. (2003) *Mon. Not. R. Astron. Soc., 341,* 277–298.

De Buizer J. M., Watson A. M., Radomski J. T., Pina R. K., and Telesco C. M. (2002) *Astrophys. J., 564,* L101–L104.

De Buizer J. M., Osorio M., and Calvet N. (2005) *Astrophys. J., 635,* 452–465.

De Pree C. G., Wilner D. J., Goss W. M., Welch W. J., and McGrath E. (2000) *Astrophys. J., 540,* 308–315.

De Wit W. J., Testi L., Palla F., and Zinnecker H. (2005) *Astron. Astrophys., 437,* 247–255.

Dobbs C. L., Bonnell I. A., and Clark P. C. (2005) *Mon. Not. R. Astron. Soc., 360,* 2–8.

Doty S. D., van Dishoeck E. F., van der Tak F. F. S., and Boonman A. M. (2002) *Astron. Astrophys., 389,* 446–463.

Draine B. T. (2003) *Ann. Rev. Astron. Astrophys., 41,* 241–289.

Edgar R. and Clarke C. J. (2004) *Mon. Not. R. Astron. Soc., 349,* 678–686.

Egan M. P., Shipman R. F., Price S. D., Carey S. J., Clark F. O., et al. (1998) *Astrophys. J., 494,* L199–L202.

Ellingsen S. P. (2006) *Astrophys. J., 638,* 241–261.

Elmegreen B. G. (1989) *Astrophys. J., 338,* 178–196.

Elmegreen B. G. (2000) *Astrophys. J., 530,* 277–281.

Elmegreen B. G. and Scalo J. (2004) *Ann. Rev. Astron. Astrophys., 42,* 211–273.

Enoch M. L., Young K. E., Glenn J., Evans N. J., Golwala S., et al. (2006) *Astrophys. J., 638,* 293–313.

Evans N. J. (1999) *Ann. Rev. Astron. Astrophys., 37,* 311–362.

Evans N. J., Shirley Y. L., Mueller K. E., and Knez C. (2002) In *Hot Star Workshop III: The Earliest Stages of Massive Star Birth* (P. A. Crowther, ed.), pp. 17–31. ASP Conf. Ser. 267, San Francisco.

Faison M., Churchwell E., Hofner P., Hackwell J., Lynch D. K., et al. (1998) *Astrophys. J., 500,* 280–290.

Faundez S., Bronfman L., Garay G., Chini R., Nyman L.-A., et al. (2004) *Astron. Astrophys., 426,* 97–103.

Fendt C. and Cameljic M. (2002) *Astron. Astrophys., 395,* 1045–1060.

Figer D. F. (2005) *Nature, 34,* 192–194.

Garay G., Faundez S., Mardones D., Bronfman L., Chini R., et al. (2004) *Astrophys. J., 610,* 313–319.

Gaume R. A., Claussen M. J., de Pree C. G., Goss W. M., and Mehringer D. M. (1995) *Astrophys. J., 449,* 663–673.

Goddi C., Moscadelli L., Alef W., Tarchi A., Brand J., et al. (2005) *Astron. Astrophys., 432,* 161–173.

Hartmann L., Ballesteros-Paredes J., and Bergin E. (2001) *Astrophys. J., 562,* 852–868.

Hatchell J. and van der Tak F. F. S. (2003) *Astron. Astrophys., 409,* 589–598.

Heitsch F., Burkert A., Hartmann L. W., Slyz A. D., and Devriendt J. E. G. (2005) *Astrophys. J., 633,* L113–l116.

Hill T., Burton M., Minier V., Thompson M. A., Walsh A. J., et al. (2005) *Mon. Not. R. Astron. Soc., 363,* 405–451.

Hofner P. and Churchwell E. (1996) *Astron. Astrophys. Suppl., 120,* 283–299.

Hofner P., Kurtz S., Churchwell E., Walmsley C. M., and Cesaroni R.

(1996) *Astrophys. J., 460,* 359–371.

Hollenbach D., Johnstone D., Lizano S., and Shu F. (1994) *Astrophys. J., 428,* 654–669.

Hoogerwerf R., de Bruijne J. H. J., and de Zeeuw P. T. (2001) *Astron. Astrophys., 365,* 49–77.

Hunter T. R., Churchwell E., Watson C., Cox P., Benford D. J., et al. (2000) *Astron. J., 119,* 2711–2727.

Ignace R. and Churchwell E. (2004) *Astrophys. J., 610,* 351–360.

Indebetouw R., Whitney B. A., Johnson K. E., and Wood K. (2006) *Astrophys. J., 636,* 362.

Jappsen A.-K. and Klessen R. S. (2004) *Astron. Astrophys., 423,* 1–12.

Jijina J. and Adams F. C. (1996) *Astrophys. J., 462,* 874–884.

Johnstone D., Fich M., Mitchell G. F., and Moriarty-Schieven G. (2001) *Astrophys. J., 559,* 307–317.

Kahn F. D. (1974) *Astron. Astrophys., 37,* 149–162.

Keto E. (2002) *Astrophys. J., 580,* 980–986.

Keto E. (2003) *Astrophys. J., 599,* 1196–1206.

Keto E. and Wood K. (2006) *Astrophys. J., 637,* 850–859.

Kim K.-T. and Kurtz S. E. (2006) *Astrophys. J., 643,* 978–984.

Klein R., Posselt B., Schreyer K., Forbrich J., and Henning T. (2005) *Astrophys. J. Suppl., 161,* 361–393.

Kramer C., Stutzki J., Röhrig R., and Corneliussen U. (1998) *Astron. Astrophys., 329,* 249–264.

Krumholz M. R., McKee C. F., and Klein R. I. (2005a) *Astrophys. J., 618,* L33–L36.

Krumholz M. R., Klein R. I., McKee C. F. (2005b) In *Massive Star Birth: A Crossroads of Astro-Physics* (R. Cesaroni et al., eds.), pp. 231–236. IAU Symp. 227, Cambridge Univ., Cambridge.

Krumholz M. R., McKee C. F., and Klein R. I. (2005c) *Nature, 438,* 332–334.

Krumholz M. R., McKee C. F., and Klein R. I. (2006) *Astrophys. J., 638,* 369–381.

Kurtz S. (2002) In *Hot Star Workshop III: The Earliest Stages of Massive Star Birth* (P. A. Crowther, ed.), pp. 81–94. ASP Conf. Ser. 267, San Francisco.

Kurtz S. (2005) In *Massive Star Birth: A Crossroads of Astrophysics* (R. Cesaroni et al., eds.), pp. 111–119. IAU Symp. 227, Cambridge Univ., Cambridge.

Kurtz S., Churchwell E., and Wood D. O. S. (1994) *Astron Astrophys. Suppl., 91,* 659–712.

Kylafis N. D. and Pavlakis K. G. (1999) In *The Origins of Stars and Planetary Systems* (C. J. Lada and N. D. Kylafis, eds.), pp. 553–575. Kluwer, Dordrecht.

Lada C. L. and Wilking B. A. (1984) *Astrophys. J., 287,* 610–621.

Larson R. (1981) *Mon. Not. R. Astron. Soc., 194,* 809–826.

Li Z.-Y. and Nakamura T. (2006) *Astrophys. J.,* in press, astro-ph/0512278.

Linz H., Stecklum B., Henning T., Hofner P., and Brandl B. (2005) *Astron. Astrophys., 429,* 903–921.

Mac Low M. (2004) In *The Dense Interstellar Medium in Galaxies* (S. Pfalzner et al., eds.), pp. 379–386. Springer, Berlin.

Mac Low M. and Klessen R. S. (2004) *Rev. Mod. Phys., 76,* 125–194.

Matt S. and Pudritz R. E. (2005) *Mon. Not. R. Astron. Soc., 356,* 167–182.

Matzner C. D. (2002) *Astrophys. J., 566,* 302–314.

Matzner C. D. and McKee C. F. (2000) *Astrophys. J., 545,* 364–378.

McKee C. F. (1989) *Astrophys. J., 345,* 782–801.

McKee C. F. (1999) In *The Origins of Stars and Planetary Systems* (C. J. Lada and N. D. Kylafis, eds.), pp. 29–67. Kluwer, Dordrecht.

McKee C. F. and Chakrabarti S. (2005) In *Massive Star Birth: A Crossroads of Astrophysics* (R. Cesaroni et al., eds.), pp. 276–281. IAU Symp. 227, Cambridge Univ., Cambridge.

McKee C. F. and Tan J. C. (2002) *Nature, 416,* 59–61.

McKee C. F. and Tan J. C. (2003) *Astrophys. J., 585,* 850–871.

McLaughlin D. E. and Pudritz R. E. (1997) *Astrophys. J., 476,* 750–765.

Megeath S. T., Wilson T. L., and Corbin M. R. (2005) *Astrophys. J., 622,* 141–144.

Meynet G. and Maeder A. (2005) In *The Nature and Evolution of Disks Around Hot Stars* (R. Ignace and K. G. Gayley, eds.), pp. 15–26. ASP Conf. Ser. 337, San Francisco.

Molinari S., Brand J., Cesaroni R., and Palla F. (1996) *Astron. Astrophys., 308,* 573–587.

Molinari S., Brand J., Cesaroni R., Palla F., and Palumbo G. G. C. (1998) *Astron. Astrophys., 336,* 339–351.

Motte F. and André P. (2001) *Astron. Astrophys., 365,* 440–464.

Motte F., André P., and Neri R. (1998) *Astron. Astrophys., 336,* 150–172.

Motte F., Bontemps S., Schilke P., Lis D. C., Schneider N., et al. (2005) In *Massive Star Birth: A Crossroads of Astrophysics* (R. Cesaroni et al., eds.), pp. 151–156. IAU Symp. 227, Cambridge Univ., Cambridge.

Mueller K. E., Shirley Y. L., Evans N. J., and Jacobson H. R. (2002) *Astrophys. J. Suppl., 143,* 469–497.

Myers P. C. and Goodman A. A. (1988) *Astrophys. J., 326,* L27–L30.

Nakano T. (1989) *Astrophys. J., 345,* 464–471.

Nomura H. and Millar T. J. (2004) *Astron. Astrophys., 414,* 409–423.

Norberg P. and Maeder A. (2000) *Astron. Astrophys., 359,* 1025–1034.

Norman C. and Silk J. (1980) *Astrophys. J., 238,* 158–174.

Norris R. P., Byleveld S. E., Diamond P. J., Ellingsen S. P., Ferris R. H., et al. (1998) *Astrophys. J., 508,* 275–285.

Omukai K. and Inutsuka S. (2002) *Mon. Not. R. Astron. Soc., 332,* 59–64.

Onishi T., Mizuno A., Kawamura A., Ogawa H., and Fukui Y. (1996) *Astrophys. J., 465,* 815–824.

Ormel C. W., Shipman R. F., Ossenkopf V., and Helmich F. P. (2005) *Astron. Astrophys., 439,* 613–625.

Osorio M., Lizano S., and D'Alessio P. (1999) *Astrophys. J., 525,* 808–820.

Ostriker E. C. (1999) *Astrophys. J., 513,* 252–258.

Padoan P. and Nordlund A. (2002) *Astrophys. J., 576,* 870–879.

Palla F. and Stahler S. W. (1999) *Astrophys. J., 525,* 722–783.

Peretto N., André P., and Belloche A. (2006) *Astron. Astrophys., 445,* 979–998.

Pestalozzi M. R., Elitzur M., Conway J. E., and Booth R. S. (2004) *Astrophys. J., 603,* L113–L116.

Pestalozzi M. R., Minier V., and Booth R. S. (2005) *Astron. Astrophys., 432,* 737–742.

Pillai T., Wyrowski F., Menten K. M., and Krügel E. (2006) *Astron. Astrophys., 447,* 929–936.

Plume R., Jaffe D. T., and Evans N. J. (1992) *Astrophys. J. Suppl., 78,* 505–515.

Plume R., Jaffe D. T., Evans N. J, Martin-Pintado J., and Gomez-Gonzales J. (1997) *Astrophys. J., 476,* 730–749.

Portegies Zwart S. F., Baumgardt H., Hut P., Makino J., and McMillan S. L. W. (2004) *Nature, 428,* 724–726.

Quillen A. C., Thorndike S. L., Cunningham A., Frank A., Gutermuth R. A., et al. (2005) *Astrophys. J., 632,* 941–955.

Rathborne J. M., Jackson J. M., Chambers E. T., Simon R., Shipman R., et al. (2005) *Astrophys. J., 630,* L181–L184.

Reid M. A. and Wilson C. D. (2005) *Astrophys. J., 625,* 891–905.

Richer J. S., Shepherd D. S., Cabrit S., Bachiller R., and Churchwell E. (2000) In *Protostars and Planets IV* (V. Mannings et al., eds.), pp. 867–894. Univ. of Arizona, Tucson.

Rudolph A., Welch W. J., Palmer P., and Dubrulle B. (1990) *Astrophys. J., 363,* 528–546.

Salpeter E. E. (1955) *Astrophys. J., 121,* 161–167.

Sandell G., Wright M., and Forster J. R. (2003) *Astrophys. J., 590,* L45–L48.

Sanders D. B., Scoville N. Z., Tilanus R. P. J., Wang Z., and Zhou S. (1993) In *Back to the Galaxy* (S. S. Holt and F. Verter, eds.), pp. 311–314. AIP Conf. Proc. 278, New York.

Scalo J. and Elmegreen B. G. (2004) *Ann. Rev. Astron. Astrophys., 42,* 275–316.

Schilke P., Groesbeck T. D., Blake G. A., and Phillips T. G. (1997) *Astrophys. J., 108,* 301–337.

Schilke P., Benford D. J., Hunter T. R., Lis D. C., and Phillips T. G. (2001) *Astrophys. J. Suppl., 132,* 281–364.

Schmeja S. and Klessen R. S. (2004) *Astron. Astrophys., 419,* 405–417.

Sewilo M., Churchwell E., Kurtz S., Goss W. M., and Hofner P. (2004) *Astrophys. J., 605,* 285–299.

Shepherd D. S. and Churchwell E. (1996) *Astrophys. J., 457,* 267–276.

Shepherd D. S., Testi L., and Stark D. P. (2003) *Astrophys. J., 584,* 882–894.

Shirley Y. L., Evans N. L., Young K. E., Knez C., and Jaffe D. T. (2003) *Astrophys. J. Suppl., 149,* 375–403.

Solomon P. M., Rivolo A. R., Barrett J. W., and Yahil A. (1987) *Astrophys. J., 319,* 730–741.

Spitzer L. Jr. (1978) *Physical Processes in the Interstellar Medium.* Wiley, New York.

Sridharan T. K., Beuther H., Schilke P., Menten K. M., and Wyrowski F. (2002) *Astrophys. J., 566,* 931–944.

Sridharan T. K., Beuther H., Saito M., Wyrowski F., and Schilke P. (2005) *Astrophys. J., 634,* L57–L60.

Stahler S. W., Shu F. H., and Taam R. E. (1980) *Astrophys. J., 241,* 637–654.

Stahler S. W., Palla F., and Ho P. T. P. (2000) In *Protostars and Planets IV* (V. Mannings et al., eds.), pp. 327–351. Univ. of Arizona, Tucson.

Tan J. C. (2006) In *The Young Local Universe,* astro-ph/0407093.

Tan J. C. and McKee C. F. (2001) In *Starburst Galaxies: Near and Far* (L. Tacconi et al., eds.), pp. 188–196. Springer, Heidelberg.

Tan J. C. and McKee C. F. (2003) In *Star Formation at High Angular Resolution,* IAU Symp. 221, Poster #274, astro-ph/0309139.

Tan J. C., Krumholz M. R., and McKee C. F. (2006) *Astrophys J., 641,* L121–L124.

Torrelles J. M., Gómez J. F., Rodríguez L. F., Curiel S., Anglada G., et al. (1998) *Astrophys. J., 505,* 756–765.

Torrelles J. M., Patel N. A., Anglada G., Gómez J. F., Ho P. T. P., et al. (2003) *Astrophys. J., 598,* L115–L118.

Ulrich R. K. (1976) *Astrophys. J., 210,* 377–391.

van der Tak F. F.S. and Menten K. M. (2005) *Astron. Astrophys., 437,* 947–956.

van Dishoeck E. F. and Blake G. A. (1998) *Ann. Rev. Astron. Astrophys., 36,* 317–368.

Viti S., Collings M. P., Dever J. W., McCoustra M. R. S., and Williams D. A. (2004) *Mon. Not. R. Astron. Soc., 354,* 1141–1145.

Walmsley C. M. (1995) *Rev. Mex. Astron. Astrof., 1,* 137–148.

Walsh A. J., Burton M. G., Hyland A. R., and Robinson G. (1998) *Mon. Not. R. Astron. Soc., 301,* 640–698.

Walsh A. J., Macdonald G. H., Alvey N. D. S., Burton M. G., and Lee J.-K. (2003) *Astron. Astrophys., 410,* 597–610.

Weidner C. and Kroupa P. (2004) *Mon. Not. R. Astron. Soc., 348,* 187–191.

Whitney B. A., Robitaille T. P., Indebetouw R., Wood K., Bjorkmann J. E., et al. (2005) In *Massive Star Birth: A Crossroads of Astrophysics* (R. Cesaroni et al., eds.), pp. 206–215. IAU Symp. 227, Cambridge Univ., Cambridge.

Williams J. P. and Garland C. A. (2002) *Astrophys. J., 568,* 259–266.

Williams J. P., Blitz L., and McKee C. F. (2000) In *Protostars and Planets IV* (V. Mannings et al., eds.), pp. 97–120. Univ. of Arizona, Tucson.

Williams S. J., Fuller G. A., and Sridharan T. K. (2004) *Astron. Astrophys., 417,* 115–133.

Williams S. J., Fuller G. A., and Sridharan T. K. (2005) *Astron. Astrophys., 434,* 257–274.

Wilner D. J., Reid M. J., and Menten K. M. (1999) *Astrophys. J., 513,* 775–779.

Wilson T. L., Gaume R. A., Gensheimer P., and Johanston K. T. (2000) *Astrophys. J., 538,* 665–674.

Wolfire M. G. and Cassinelli J. P. (1987) *Astrophys. J., 319,* 850–867.

Wood D. O. S. and Churchwell E. (1989a) *Astrophys. J., 340,* 265–272.

Wood D. O. S. and Churchwell E. (1989b) *Astrophys. J. Suppl., 69,* 831–895.

Wright M. C. H., Plambeck R. L., and Wilner D. J. (1996) *Astrophys. J., 469,* 216–237.

Wyrowski F., Hofner P., Schilke P., Walmsley C. M., Wilner D. J., et al. (1997) *Astron. Astrophys., 320,* L17–L20.

Wyrowski F., Schilke P., Walmsley C. M., and Menten K. M. (1999) *Astrophys. J., 514,* L43–L46.

Yonekura Y., Asayama S., Kimura K., Ogawa H., Kanai Y., et al. (2005) *Astrophys. J., 634,* 476–494.

Yorke H. W. and Sonnhalter C. (2002) *Astrophys. J., 569,* 846–862.

Zhang Q., Hunter T. R., and Sridharan T. K. (1998) *Astrophys. J., 505,* L151–L154.

Zhang Q., Hunter T. R., Sridharan T. K., and Ho P. T. P. (2002) *Astrophys. J., 566,* 982–992.

Zhang Q., Hunter T. R., Brand J., Sridharan T. K., Cesaroni R., et al. (2005) *Astrophys. J., 625,* 864–882.

Ultracompact HII Regions and the Early Lives of Massive Stars

M. G. Hoare
University of Leeds

S. E. Kurtz and S. Lizano
Universidad Nacional Autónoma de México-Morelia

E. Keto
Harvard University

P. Hofner
New Mexico Institute of Technology and National Radio Astronomy Observatory

We review the phenomenon of ultracompact HII regions (UCHIIs) as a key phase in the early lives of massive stars. This most visible manifestation of massive star formation begins when the Lyman continuum output from the massive young stellar object becomes sufficient to ionize the surroundings from which it was born. Knowledge of this environment is gained through an understanding of the morphologies of UCHII regions, and we examine the latest developments in deep radio and mid-IR imaging. Spitzer data from the GLIMPSE survey are an important new resource in which PAH emission and the ionizing stars can be seen. These data provide good indications as to whether extended radio continuum emission around UCHII regions is part of the same structure or due to separate sources in close proximity. We review the role played by strong stellar winds from the central stars in sweeping out central cavities and causing the limb-brightened appearance. New clues to the wind properties from stellar spectroscopy and hard X-ray emission are discussed. A range of evidence from velocity structure, proper motions, the molecular environment, and recent hydrodynamical modeling indicates that cometary UCHII regions require a combination of champagne flow and bow shock motion. The frequent appearance of hot cores, maser activity, and massive young stellar objects (YSOs) ahead of cometary regions is noted. Finally, we discuss the class of hypercompact HII regions or broad recombination line objects. They are likely to mark the transition soon after the breakout of the Lyman continuum radiation from the young star. Models for these objects are presented, including photo-evaporating disks and ionized accretion flows that are gravitationally trapped. Evolutionary scenarios tracing the passage of young massive stars through these ionized phases are discussed.

1. INTRODUCTION

The time when newly formed massive stars begin to ionize their surroundings is one of the energetic events that underlines their important role in astrophysics. As they evolve, the copious amounts of UV radiation and powerful stellar winds they produce have a profound effect on the surrounding interstellar medium. Their early lives are spent deeply embedded within dense molecular cores whose high column densities absorb the optical and near-IR light from the young stars, shielding them from view. One of the first observable manifestations of a newly formed massive star is the radio free-free emission of the HII region surrounding the star. Since only the most massive stars produce significant radiation beyond the Lyman limit, embedded HII regions are a unique identifier of high-mass star formation.

The absorption of the UV and optical radiation by dust, both inside and outside the newly formed nebula, heats the grains to temperatures that range from the sublimation temperature close to the star to interstellar temperatures in the surrounding molecular cloud. Owing to the high luminosity of the massive stars, HII regions are some of the strongest infrared sources in the galaxy. Similar to the radio emission, the thermal IR radiation is little affected by extinction. Thus the combination of the radio and IR wave bands allows us to peer deep into the star-forming clouds to study the processes of star formation within.

The youngest massive stars are associated with the smallest HII regions. These are the ultracompact HII regions (UCHII) and the newly identified class of hypercompact (HCHII) regions. UCHII regions were first distinguished from merely "compact" HII regions around 25 years ago, and came to be defined observationally (*Wood and Churchwell,* 1989a) as those regions with sizes ≤ 0.1 pc, densities $\geq 10^4$ cm^{-3}, and emission measures $\geq 10^7$ pc cm^{-6}. Since then, hundreds of UCHII regions with these general properties have been found. While the division of HII regions into different degrees of compactness may be a convenient label, for the larger objects at least, it is likely to have little physical significance. Once in the expansion phase, the physics of their dynamics probably stays the same until the molecular material is cleared away and the OB star joins the field population. Of greater interest are the smallest HII regions, as they have more to tell us about the process of massive star formation. The ionized gas within the UCHII and HCHII regions not only reveals properties of the stars themselves, but also lights up the immediate surroundings to allow investigations of the density distribution and environment into which the massive stars are born. The external environment has a profound influence on the evolution of the HII regions.

The properties of UCHII regions and their immediate precursors have been reviewed previously by *Churchwell* (2002) and *Kurtz et al.* (2000). In this review we concentrate on more recent developments in the field of UCHII and HCHII regions. These include new theories of massive star-forming accretion flows and the new views of UCHII regions opened up by infrared studies on large groundbased telescopes and the Spitzer satellite. The Spitzer GLIMPSE survey has covered a large part of the inner galactic plane at unprecedented spatial resolution and depth in the 4–8-μm region where there is a local minimum in the extinction curve. High-resolution X-ray studies with Chandra are also beginning to bear on the problem. Together these promise great new insights into how OB stars are formed and interact with their environment.

2. ULTRACOMPACT HII REGIONS

2.1. Morphologies

The morphologies of UCHII regions are important since they yield clues to the state of the surrounding medium relatively soon after a massive star has formed. The common appearance of a regular morphology indicates that there are ordered physical processes occurring during massive star formation rather than just stochastic ones.

As part of their pioneering high-resolution radio surveys of massive star-forming regions, *Wood and Churchwell* (1989a) developed a morphological classification scheme for UCHII regions. Together with *Kurtz et al.* (1994), they found that 28% of UCHII regions are spherical, 26% cometary, 26% irregular, 17% core-halo, and 3% shell. The significant numbers of unresolved sources have been omitted here, since nothing can be said about their morphology. *Walsh et al.* (1998) studied a sample of southern UCHII regions and classified most of their sources as either cometary (43%) or irregular (40%) after omitting the unresolved ones. As pointed out by *Wood and Churchwell* (1989a), many of the sources classified as spherical do reveal ordered morphology when observed at higher spatial resolution. Examples are M17-UC1, which was shown to be cometary (*Felli et al.*, 1984), and G28.20-0.04, which is shell-like (*Sewilo et al.*, 2004) (see Fig. 7). If the spherical sources

of earlier studies were also discounted, then the proportion of cometaries in the well-resolved objects would be similar to that of *Walsh et al.* (1998).

As *Wood and Churchwell* (1989a) forewarned, radio interferometric observations have a limited range of spatial scales that they are sensitive to. The larger objects can quickly become over-resolved and break up into irregular sources. The snapshot nature of observations necessary to investigate significant numbers of sources also limits the dynamic range. For cometary objects this often means that at high resolution only the dense material at the head is seen and the full extent of the much weaker emission in the tail is not. Even for relatively simple objects like the archetypal cometary UCHII region, G29.96-0.02, the complete radio continuum picture is only fully revealed by time-consuming deep, multi-configuration observations (*Fey et al.*, 1995).

The recent radio studies by *De Pree et al.* (2005) also address this point. They have conducted deep, multi-configuration studies of two regions of intense massive star formation, W49A and Sgr B2, where nearly 100 UCHII regions are found. These radio data had good spatial and dynamic range, leading De Pree et al. to reassess the morphological classes. They classify about one-third of their sources as "shell-like," after omitting the unresolved fraction. At least half of these are very asymmetric and could just as easily be classified as cometary. This would take the total cometary fraction to over a third in that sample. The high dynamic range of the De Pree et al. data also led them to drop the core-halo class and instead attempt to ascertain the shape of the compact and extended emission separately. These sources often appear to be a superposition of a compact source on a different, more extended source. They also detect a significant population of UCHII regions elongated along one axis. Following *Churchwell* (2002), they classify these regions as bipolar. It is not yet clear whether the detection of these bipolar sources is due to the better-quality radio imaging or the more extreme pressure in these two environments compared to the general galactic population.

Imaging in the infrared is sensitive to all spatial scales at once, from the resolution limit up to the total size of the image. This could be several orders of magnitude of spatial dynamic range as long as the image is sensitive enough to pick up very extended, low surface brightness emission. Therefore, IR imaging can in principle overcome some of the limitations of radio interferometric snapshot data for morphological classification, although it also comes with its own disadvantages. In the near-IR, the continuum emission from UCHII regions is mostly made up of bound-free and free-free emission from the nebular gas. There are also minor contributions from scattered light, emission from very hot grains, and non-thermal-equilibrium emission from very small grains. Hence, near-IR images should show the same morphology as the radio, apart from the effect of intervening extinction. An example of this is seen in the near-IR image of G29.96-0.02 by *Fey et al.* (1995), where the nebular continuum looks just like their deep multi-configuration radio map. However, in general the extinction in the near-

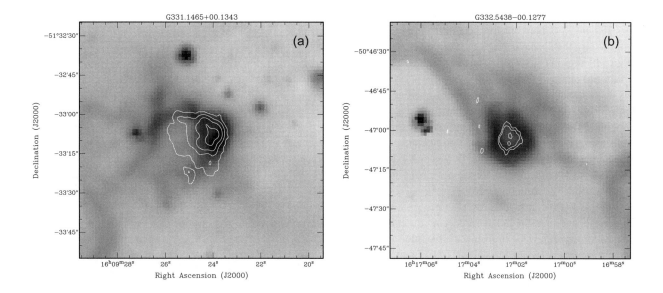

Fig. 1. **(a)** An 8.0-μm GLIMPSE image of G331.1465+00.1343 overlaid with contours of the 5-GHz radio continuum emission from ATCA with a noise level of 0.2 mJy per beam at a resolution of about 3". The object has a clear cometary morphology seen in both the radio and mid-IR image. **(b)** Same for G332.5438-00.1277. Note the "horseshoe"-shaped IR emission that is open to the NNE similar to cometary objects and not revealed by the barely resolved radio image. For both sources there is a darker region ahead of the object in the 8.0-μm image, indicating extinction due to dense molecular cloud material. From *Busfield* (2006).

IR is often too high and renders the UCHIIs invisible or cut through by dust extinction lanes.

Moving into the thermal IR reduces the total extinction, mostly eliminates scattering by the dust, and the UCHII regions become much brighter. The interstellar dust grains are heated by a combination of direct stellar radiation near the star and Lyα photons resonantly scattering in the ionized zone (e.g., *Natta and Panagia,* 1976; *Hoare et al.,* 1991). The latter process means that the dust grains never drop below temperatures of a few hundred Kelvin throughout the nebula and thus emit strongly in the mid-IR. As such, the intrinsic mid-IR morphology is similar to that seen in the radio, but again heavy extinction can intervene.

The advent of a new generation of mid-IR cameras on large telescopes has yielded high-quality mid-IR images of a number of UCHII regions. In many cases the mid-IR and radio morphology are in good agreement. Recent examples include W49A South (*Smith et al.,* 2000), G29.96-0.02 (*De Buizer et al.,* 2002a), NGC 6334F (*De Buizer et al.,* 2002b), W3(OH) (*Stecklum et al.,* 2002), and K3-50A (*Okamoto et al.,* 2003), as well as several sources observed by *Kraemer et al.* (2003). In the cometary G9.62+0.19B there are even signs of a bright spot near the expected location of the exciting star due to direct stellar heating (*De Buizer et al.,* 2003; *Linz et al.,* 2005).

There are other objects, however, where a large extinction, even at 10 and 20 μm, greatly attenuates the mid-IR emission. A good example is G5.89-0.39, where the heavy extinction in the near-IR obscures the southern half of the radio source in the N and Q bands, consistent with the large column density measured in the millimeter continuum (*Feldt*

et al., 1999). *De Buizer et al.* (2003, 2005) and *Linz et al.* (2005) show other instances where the mid-IR morphology does not follow that of the radio continuum. The most likely reason for this is extinction, which, due to the silicate features, is still high in the groundbased mid-IR windows. Indeed, the extinction at these wavelengths is only about half the value in the K band (*Draine,* 2003).

Observations by the Spitzer satellite now offer a whole new IR perspective on UCHII regions, in particular through the GLIMPSE survey of a large fraction of the inner galactic plane (*Benjamin et al.,* 2003). The 4–8-μm range of the IRAC instrument has extinction values lower than both the near- and groundbased mid-IR windows, bottoming out around the 5.8-μm channel (*Indebetouw et al.,* 2005). At these wavelengths one can expect to see some emission from the hot grains in thermal equilibrium near the exciting star. However, the IRAC filters at 3.6, 5.8, and 8.0 μm are dominated by strong PAH features from the UCHII regions. The PAH emission is strong in the photon-dominated regions (PDRs) that lie in a thin shell of neutral gas just outside the ionization front. In directions in which the HII region is ionization bounded, the PAH emission is expected to form a sheath around the nebula. If there are directions where ionizing photons can escape, the PAHs will be destroyed (*Girad et al.,* 1994).

An example for a cometary UCHII region is shown in Fig. 1a, where the 8-μm emission closely follows the radio emission around the head and sides, but is absent in the tail direction, which is likely to be density bounded. Possibly unrelated clumps and filaments can be seen in the vicinity, so caution must be exercised when interpreting the Spitzer images; information from other wavelengths, especially the radio, is also needed. Nevertheless, the uniformity and coverage of the GLIMPSE dataset makes it very useful for studies of UCHII regions. Its resolution and sensitivity can in principle yield new insights into the morphologies of UCHII regions, notwithstanding the caveats mentioned above.

Hoare et al. (in preparation) are carrying out mid-IR morphological classification utilizing the GLIMPSE dataset. They are using a sample of massive young stars color-selected from the lower-resolution mid-IR survey by the MSX satellite (*Lumsden et al., 2002*). These have been followed up with a variety of groundbased observations, including high-resolution radio continuum observations to identify the UCHII regions and massive YSOs (*Hoare et al., 2004*) (see *www.ast.leeds.ac.uk/RMS*). Preliminary findings indicate that the proportion of cometary objects is higher than in previous radio studies. Figure 1b shows an example where the radio image is barely resolved and yet the GLIMPSE 8-μm image shows a "horseshoe"-shaped nebula wrapped around the radio source as expected for PAHs in a PDR. However, the structure is open to the north-northeast, suggesting that the UCHII region is not ionization bounded in that direction. Although it does not have the classic parabolic shape of a cometary, it is suggestive of a champagne flow away from the dense molecular cloud seen in extinction ahead of the object. Other data are needed to confirm such an interpretation. A star is seen in the center of the nebula in the 3.6- and 4.5-μm GLIMPSE images and is likely to be the exciting source. This demonstrates that the IRAC wavelength range can address many aspects of UCHII regions.

Kurtz et al. (1999) found that many UCHII regions, when observed at lower angular resolution, show extended, diffuse emission in addition to the ultracompact component. The radio morphologies suggested a physical connection between the UCHII region and the extended component for a significant fraction of their sources. *Kim and Koo* (2001), using a different sample, made radio recombination line observations of UCHIIs with extended emission. The line velocities of the ultracompact and extended gas were nearly equal, supporting the idea of a physical relationship between the two components. *Ellingsen et al.* (2005) searched for extended emission in eight southern UCHIIs with associated methanol masers. They found a lesser degree of extended emission than reported for the random UCHII sample of *Kurtz et al.* (1999). Methanol masers may trace earlier stages of UCHII regions, hence the lesser degree of extended emission in the *Ellingsen et al.* (2005) sample may reflect the relative youth of those regions.

As pointed out by *Kurtz et al.* (1999), in many cases it is difficult to assess from the radio data whether the extended emission is a coherent structure excited by the same star(s). IR images can often clarify the situation; the case of G031.3948-0.2585 (IRAS 18469-0132) is shown in Fig. 2. The lower-resolution radio map of *Kurtz et al.* (1999) shows connected radio emission over 2 arcmin. In the IR image it looks more like a complex of several different sources. The

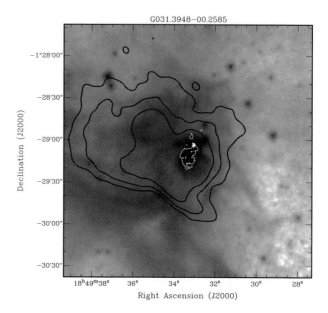

Fig. 2. GLIMPSE 8-μm image of G031.3948-00.2585 (IRAS 18469-0132). Overlaid in thick black contours is the 3.6-cm data from VLA D configuration observations and in thin white contours the VLA B configuration data also at 3.6 cm. Radio data is from *Kurtz et al.* (1999). D-array contours are at 1, 2. and 4 mJy per beam and then switch to the B-array contours at 0.5 mJy per beam up to the peak. The IR image is more suggestive of a collection of separate UCHIIs and bright-rimmed clouds than a single source responsible for the compact and extended emission.

cometary, with a size of about 15" in the high-resolution radio image, is seen along with IR counterparts to the two radio point sources. However, the diffuse radio emission, seen extending to the east and north in the low-resolution radio map, appears to be due to a combination of another larger cometary with its tail pointing northeast and an ionized bright-rimmed cloud type structure running to the south and east of the two cometary objects. Spectrophotometric identification of the individual exciting stars, together with high spatial and dynamic radio continuum and line observations, are needed to confirm the picture in detail.

2.2. Molecular Environment

In this section we examine the molecular environment within which UCHII regions exist, and in particular, that of cometary regions. Unfortunately, the typical spatial resolution of single-dish millimeter line observations is usually insufficient to resolve UCHII regions from associated star-formation activity and the molecular cores where they reside. Higher-resolution data on the molecular environment can be obtained with millimeter interferometry. However, most studies of the environs of UCHII regions have naturally tended to concentrate on those with associated hot cores. Although this benefits high-mass star-formation studies in general, it is not so useful in the search for a full understand-

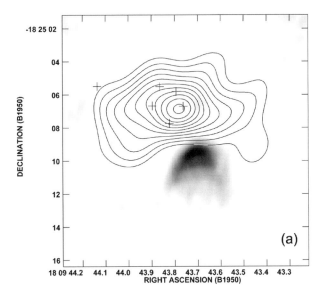
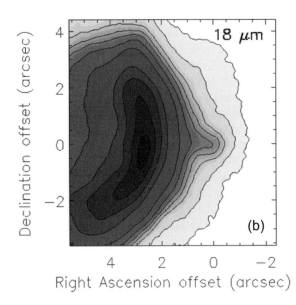

Fig. 3. (a) High-resolution map of the CS 2-1 emission (contours) tracing the dense molecular gas ahead of the cometary UCHII region G12.21-0.10 (radio continuum image in grayscale). Crosses mark water masers. From de la Fuente et al. (in preparation). (b) An 18-μm image of G29.96-0.02 showing the detection of the hot core located about 2 arcsec in front of the cometary arc of the UCHII region. From *De Buizer et al.* (2002a).

ing of UCHII region physics, a key ingredient of which is the ambient density distribution. The presence of a hot core with its own associated infall and possibly outflow in close proximity to the UCHII hampers a clear view of the molecular environment into which the older UCHII region was born.

In single-dish studies *Hofner et al.* (2000) used optically thin $C^{17}O$ transitions to determine that the molecular clumps in which UCHII regions reside are typically about 1 pc in size with densities of 10^5 cm^{-3} and temperatures of 25 K. In their survey of ^{13}CO and CS transitions, *Kim and Koo* (2003) found that the UCHII region is usually located right at the peak of the molecular line emission. When they did resolve clear density and velocity structure in the molecular gas, the pattern was consistent with a champagne flow.

Somewhat higher-resolution information on the density distribution is available from submillimeter dust continuum emission maps (*Mueller et al.*, 2002). These show very centrally condensed clouds centered on the UCHII regions themselves, but if there is an offset of the peak density then it is usually ahead of the cometary apex. Using the submillimeter dust continuum will give a more centrally peaked distribution than a pure column density tracer, as the warm dust in and around the nebula will enhance the submillimeter emission. *Hatchell and van der Tak* (2003) examined the more extended submillimeter continuum emission around UCHII regions and found that the average radial density distribution derived was consistent with the r$^{-1.5}$ expected for freefall collapse.

About half of all UCHII regions are associated with warm molecular gas as evidenced by highly excited NH$_3$ (*Cesaroni et al.*, 1992), CS (*Olmi and Cesaroni*, 1999), methanol masers (*Walsh et al.*, 1998), or other molecular tracers

of the early hot core phase (*Hatchell et al.*, 1998). Higher-resolution data showed that these hot cores were usually spatially offset from the UCHII regions by a few arcseconds (*Cesaroni et al.*, 1994). Where these are associated with cometary regions, the hot cores are always located ahead of the apex. Many cometary regions have been found to show signs of earlier stages of massive star formation such as maser sources or massive YSOs as well as hot cores in the dense regions a few arcseconds ahead of the cometary region (e.g., *Hofner et al.*, 1994; *Hofner and Churchwell*, 1996).

G29.96-0.02 is again a good case study. Interferometric observations of ^{13}CO 1-0 and $C^{18}O$ 1-0 by *Pratap et al.* (1999) and *Olmi et al.* (2003) reveal gas densities of 5×10^5 cm^{-3}. The density peaks just in front of the cometary HII region, at the location of the hot core. *Maxia et al.* (2001) found blueshifted CO and HCO$^+$ 1-0 emission in the tail region and HCO$^+$ 1-0 in absorption, redshifted with respect to the main cloud at the head. They interpreted these motions as infall onto the hot core.

Figure 3 shows a high-resolution observation of the dense molecular gas associated with the cometary region G12.21-0.10. The molecular gas lies just in front of the cometary ionization front. Observations of the NH$_3$ (2,2) line show that a hot core is present, coincident with a collection of water masers showing that star-formation activity is occurring (de la Fuente et al., in preparation). Recently these hot cores in the vicinity of UCHII regions have been detected in the mid-IR via their warm dust emission (*De Buizer et al.*, 2002a, 2003; *Linz et al.*, 2005). Figure 3 shows the first mid-IR detection of such an object ahead of the G29.96-0.02 cometary region.

In the few cases where a cometary object without obvious interference from a nearby hot core or massive young stellar object has been studied at high resolution, dense gas surrounding the head of the cometary structure has been found. Maps of W3(OH) in dense gas tracers such $C^{18}O$ and CH_3OH (*Wyrowski et al.,* 1999) reveal an arc of emission wrapped around the western end of the UCHII region, which has a champagne flow to the east (*Keto et al.,* 1995). Similarly, high-resolution observations of G34026+0.15 by *Watt and Mundy* (1999) show $C^{18}O$ 1-0 emission enveloping the head of this cometary region, while the traditional hot core tracer, methyl cyanide, peaks up right on the apex of the cometary region. Watt and Mundy interpret this emission as arising from dense molecular gas externally heated by the UCHII region rather than internally heated by a young massive star. The interpretation of molecular line emission close to the PDR around an UCHII region is complicated by the many excitation and chemical effects that are likely to be at work.

An alternative probe of dense molecular gas surrounding UCHII regions is to use molecular lines in absorption against the strong continuum background. This gives additional constraints due to the particular geometry necessary to generate them. The H_2CO lines at 2 and 6 cm have been observed at high resolution in absorption against W3(OH) and W58 C1 (*Dickel and Goss,* 1987; *Dickel et al.,* 2001). Both of these objects have a cometary structure and the H_2CO line optical depth maps of both objects show peaks consistent with the densest gas being located at the head of the cometary region. Total gas densities of 6×10^4 cm^{-3} were derived from the strengths of the absorption lines. The authors interpret other details in the line structure as arising from possible outflows from other sources in these regions.

Very high spatial resolution absorption line studies have been achieved with the NH_3 lines at 23 GHz. *Sollins et al.* (2004) found evidence of spherical infall together with some rotation in G10.6-0.4. A multi-transition study of G28.20-0.04 (*Sollins et al.,* 2005) revealed complex motions interpreted as a combination of toroidal infall and outflowing shell. These two HII regions fall into the hypercompact category (see section 3).

Lines that arise in the PDR can probe the conditions and kinematics in the neutral material. Carbon radio recombination lines are particularly useful as they can be compared to the ionized gas traced by hydrogen and helium recombination lines. *Garay et al.* (1998b) and *Gómez et al.* (1998) observed these lines toward cometary objects and found that the spatial distribution and velocity of the carbon lines are consistent with an origin in a PDR around objects undergoing a champagne flow. The carbon lines are significantly enhanced by stimulated emission and therefore arise predominantly from the nearside in front of the continuum, and hence give more specific insight into the velocity structure. *Roshi et al.* (2005) used this fact to deduce that the PDR region in the cometary object G35.20-1.74 is moving into the cloud at a few km s^{-1} by comparison with the ambient molecular cloud velocity. *Lebrón et al.* (2001) report a comprehensive study of a cometary region, G111.61+0.37, using a combination of atomic hydrogen 21-cm emission and absorption components related to ionized and molecular lines. They demonstrated that the atomic gas exhibits a champagne flow like the ionized gas, but at significantly slower velocities, consistent with entrainment.

2.3. Stellar Winds

If we are to understand the nature of UCHII regions we first need a sound knowledge of the OB stars that excite them and their stellar winds. Due to the heavy extinction, optical and UV spectroscopic diagnostics are not available. Even for the objects that are visible in the near-IR when the extinction is low enough, it is usually difficult to see them because of the very strong nebular continuum emission. One of the few stars that is clearly visible at 2 μm is the one that powers G29.96-0.02. *Martín-Hernández et al.* (2003) used the VLT to take an intermediate resolution K-band spectrum of the star building on previous work by *Watson and Hanson* (1997). By comparison with spectra of known spectral types from the field, they deduced that it was an O5–6V, and argued that this was consistent with other estimates of the spectral type. More importantly, the spectrum is consistent with field stars, which argues that the stellar wind should also have properties similar to the well-studied radiation driven OB star winds. Hence, the action of a strong wind moving at a few thousands of km s^{-1} has to be an ingredient of any model of expanding UCHII regions.

The indirect information on the radiation output from the central stars via the study of the excitation of the nebulae has been put on a much firmer basis as a result of ISO spectroscopy of a large sample. *Martín-Hernández et al.* (2002) find a consistent picture whereby the metallicity gradient in the galaxy can explain the excitation gradient due to the hardening of the stellar radiation field. However, progress is still required on the stellar atmosphere models to derive the exact stellar parameters. The use of the near-IR HeI pure recombination lines can also help constrain the stellar parameters (*Lumsden et al.,* 2003). It should now be possible to combine the new spatially resolved mid-IR imaging with multi-dimensional dusty photo-ionization modeling including PAHs to determine the dust-to-gas ratio inside UCHII regions. This is crucial to finding the fraction of Lyman continuum photons absorbed by the dust that at present bedevils the use of radio to IR luminosities in constraining the stellar effective temperature.

There is an urgent need to find and conduct detailed direct studies of the exciting stars from a range of UCHII regions. *Bik et al.* (2005) have studied the spectra of many objects found in the vicinity of UCHII regions. However, these were either OB stars in a cluster associated with, but not powering, the UCHII region, or the UCHII region itself where the spectrum is dominated by the nebular continuum and not the star. To make progress, higher spatial resolution is required to pick the star out of the nebular light. *Alvarez et al.* (2004) present a start down this road with near-IR adaptive optics imaging of several UCHII regions. Higher Strehl ratios on larger telescopes will be needed to carry out the high-spatial and high-spectral resolution observations to classify the exciting stars. The speckle imaging of K3-50A by *Hoffmann et al.* (2004) shows the promise of such techniques.

Another new line of evidence that can give insights into the stellar winds in the centers of UCHII regions comes from X-ray observations. Hard X-ray sources are now routinely found in high-mass star-forming regions. *Hofner et al.* (2002) made a study of the W3 region using the Chandra X-ray Observatory and found X-ray point sources coincident with many of the OB stars powering the more evolved HII regions, but also at the centers of some ultracompact regions. The point-like emission from the visible stars in the evolved regions, which are not likely to be accreting, probably arises in shocks within a radiatively driven stellar wind. This could also be the case for the more compact sources, but accretion shocks or wind-nebula interactions are also possible. The diffuse X-ray emission arising from the very hot post-shock region where the stellar wind interacts with the surrounding UCHII region should fill the windblown cavity. Such diffuse emission is seen on much larger scales where clusters of OB stars excite a large complex such as M17 (see chapter by Feigelson et al.). The indirect effect of X-ray emission in UCHII regions is the generation of a partially ionized layer just outside the nebula due to the penetration of hard photons into the neutral material. *Garay et al.* (1998a) observed narrow H I radio recombination line components, which they ascribed to X-ray heating.

If the wind-nebula interaction can be detected in diffuse X-rays in UCHII regions, it would in theory give an independent measure of the strength of the wind. However, the modeling of this interaction and its X-ray emission is far from straightforward. The evolution of wind-driven regions expanding into a high-pressure ambient medium was studied numerically by *García-Segura and Franco* (1996), who showed that the swept-up shells of ambient gas suffered a strong thin-shell instability, producing "elephant-trunks" and cometary globules in the shocked stellar wind. In these models, a hot bubble of shocked stellar wind drives the expansion of a swept-up shell where the ionization front is trapped.

Comeron (1997) made some initial calculations in the context of stellar winds in a champagne-flow scenario. More recently, *González-Avilés et al.* (2005) studied the evolution of an HII region driven by a strong stellar wind blowing inside a logatropic cloud in gravitational collapse toward the central star. They included thermal conductivity, which reduces the shocked stellar wind cooling time by several orders of magnitude (*Shull*, 1980). The tangential magnetic fields will suppress the thermal conduction in the direction perpendicular to the magnetic field direction. Nevertheless, the morphology and strength of magnetic fields is unknown. Thus, the magnitude of the thermal conductivity coefficient, κ, was used as a free parameter. They found that only mod-

els with the full thermal value of κ were compatible with the hard X-ray luminosities observed in the W3 sources (10^{-3} L$_\odot$) by *Hofner et al.* (2002). For this value, thermal conductivity cools the hot region of shocked stellar wind and the shells of swept-up gas are driven by the momentum of the stellar wind. The hot shocked wind is also in a very thin shell and would predict emission on 0.1-pc scales that would easily be resolved by Chandra, which was not the case for the W3 sources. Much deeper observations of this and other massive star-forming regions are being obtained and these may reveal the diffuse X-ray emission that will fully unlock this new insight into the effect of strong stellar winds in UCHII regions.

2.4. Dynamical Picture

As reviewed by *Churchwell* (1999), many models have been put forward to explain the morphologies and long lifetime of UCHIIs. *Wood and Churchwell* (1989b) were the first to argue that the sheer numbers of objects with IR colors like UCHII regions implies a lifetime on the order of 10^5 yr rather than the simple dynamical lifetime of 10^4 yr. This has since been criticized by several studies (*Codella et al.*, 1994; *Ramesh and Sridharan*, 1997; *van der Walt*, 1997; *Bourke et al.*, 2005), but nevertheless many models attempted to include ways of lengthening the lifetime of the phase. These included using thermal pressure to confine the expansion (*De Pree et al.*, 1995), although this was claimed by *Xie et al.* (1996) to predict too high an emission measure and they appealed to turbulent pressure instead. Another way is to continuously introduce neutral material to the ionized region. In mass-loading models by *Dyson et al.* (1995), *Lizano et al.* (1996), and *Redman et al.* (1998), the ablation of neutral clumps sets up a recombination front structure. These models predict mainly double-peaked line profiles that are not observed. Magnetic fields may also impede the advance of ionization fronts (*Williams et al.*, 2000) and require sensitive measurements of the field strength and geometry to test them.

For cometary objects, the bow-shock model, whereby the stellar wind from an OB star moving through a uniform medium sets up a parabolic standing shock, also continuously feeds neutral material into the nebula (*Mac Low et al.*, 1991). Several high-resolution radio recombination line studies of the dense gas at the head of cometary regions, and G29.96-0.02 in particular, observed a velocity structure that was claimed to support the bow-shock model (*Wood and Churchwell*, 1991; *Afflerbach et al.*, 1994). These pure bow-shock models required the star to be moving at velocities of about 20 km s^{-1} to impart sufficient velocity structure into the gas. This is much higher than typical members of young OB star clusters, which have velocity dispersions of only a few km s^{-1} (e.g., *Sicilia-Aguilar et al.*, 2005). Of course, there exists a class of runaway OB stars that do have much higher proper motions, probably resulting from dynamical ejections (*Hoogerwerf et al.*, 2001; see also *Gómez et al.*, 2005). However, these only make up a small frac-

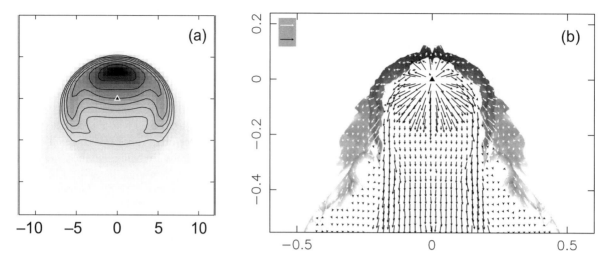

Fig. 4. Results from hydrodynamic modeling of a cometary UCHII region expanding into an exponential density gradient of scale height 0.2 pc excited by a star with stellar wind typical for an early type star and moving up the density gradient by 10 km s⁻¹. **(a)** Predicted emission measure map at age 20,000 yr with size scale in arcseconds for an adopted distance of 7 kpc. **(b)** Corresponding velocity structure of the model with white arrows showing velocities up to 30 km s⁻¹ and black arrows up to 2000 km s⁻¹. The ionization front at the head moves up the density gradient at about 3 km s⁻¹. From *Arthur and Hoare* (2006).

tion of all OB stars (*Gies,* 1987; *de Wit et al.,* 2005). The velocity dispersion of young clusters where the potential is still dominated by gas will be higher. Franco et al. (in preparation) have recently carried out numerical simulations of UCHII regions in which the exciting star is moving with speeds of up to 13 km s⁻¹ down a density gradient surrounding a hot core. They identify these stellar speeds as being consistent with viral motion due to the gravitational field of the hot core. Their motivation was to explain the occurrence of both compact and extended emission if due to a single ionizing source (see section 2.1.2).

Lower spatial resolution radio recombination line studies (e.g., *Garay et al.,* 1994; *Keto et al.,* 1995) that are sensitive to more extended emission in the tail of cometary objects indicate flowing motions consistent with the blister and champagne-flow models (*Israel,* 1978; *Yorke et al.,* 1983). Near-IR long-slit spectroscopy also showed that the velocity structure in the Brγ line in G29.96-0.02 was not consistent with a pure bow-shock picture (*Lumsden and Hoare,* 1996, 1999). The same velocity structure was recently recovered by *Zhu et al.* (2005) using long-slit echelle observations of the 12.8-μm [NeII] fine structure line. A high-quality near-IR spectrum was presented by *Martín-Hernández et al.* (2003), which allowed a more accurate comparison with the surrounding molecular gas through the presence of the H₂ S(1) line. Its velocity agreed well with millimeter molecular line measurements even though it is likely to arise in a thin layer just outside the ionized zone.

Comparison with the molecular cloud velocity is a vital part of testing dynamical models since it sets the reference frame. For instance, in G29.96-0.02, where the tail is clearly pointing toward us, the bow-shock model would predict that most of the gas is redshifted as the stellar wind shock tunnels into the molecular cloud. At the other extreme, a pure champagne-flow scenario would involve most of the gas

being blueshifted as it photo-evaporates off the molecular cloud face and flows back down the density gradient toward the observer. In G29.96-0.02, and other simple cometary objects where sufficiently good spatially resolved velocity data for the ionized and molecular gas exists (e.g., *Keto et al.,* 1995; *Garay et al.,* 1994; *Hoare et al.,* 2003; *Cyganowski et al.,* 2003), the range of velocity structure usually straddles the molecular cloud velocity.

As discussed in section 2.3, stellar winds must be included in any realistic model of expanding HII regions. The limb-brightened morphology in the ionized gas of the shell and cometary regions gives the appearance that the central regions are cleared out by a strong stellar wind. Pure champagne-flow models (*Yorke et al.,* 1983) have a centrally peaked radio structure. Numerical modeling by *Comeron* (1997) first indicated that winds can produce limb-brightened cometary structures. The action of the wind also helps to slow the expansion, as the ionization front is trapped in the dense swept-up shell, as first pointed out by *Turner and Matthews* (1984). The spherically symmetric wind-driven HII region models by *González-Aviles et al.* (2005) produced very thin shell nebular regions. They found that when the HII region is expanding into a logatropic density distribution $\rho \propto r^{-1}$, driven by stellar winds with mass-loss rates $\dot{M}_w \sim 10^{-5}$ M$_\odot$ yr⁻¹ and wind speeds $v_w \sim 1000$ km s⁻¹, the HII shells evolve from HCHII to UCHII sizes in timescales $\Delta t \sim 3$–7×10^4 yr.

Recently, new hydrodynamical models of cometary HII regions by *Arthur and Hoare* (2006) considered the action of fast (~2000 km s⁻¹), strong (Ṁ ~ 10⁻⁶ M$_\odot$) winds, typical for early type stars, within steep power-law and exponential density gradients. Figure 4a shows the emission measure map from one such model that has a limb-brightened structure very much like those seen in cometary objects. If the mass-loss rate is dropped by an order of magnitude, then the wind can no longer open up a significant cavity.

In these champagne models with a stellar wind, the ionized gas is forced to flow in a parabolic shell around the bubble blown by the stellar wind and then on down the density gradient. To achieve the right velocity structure relative to the molecular cloud a bow-shock element is also required, whereby the star is moving up the density gradient. A stellar speed in the range 5–10 km s⁻¹ is sufficient to provide enough forward motion of the gas in the head to shift those velocities relative to the molecular cloud by the typical amount observed. Figure 4b shows the velocity structure for such a combined champagne and bow-shock model. This model produces a velocity structure similar to that seen in G29.96-0.02 (Fig. 5). Such a hybrid between pure champagne and pure bow-shock models was also developed by *Cyganowski et al.* (2003) to explain the twin cometary regions in the same molecular cloud in DR21. The lifetimes predicted by such models are determined by how fast the ionization front or the star reaches the local density maximum. For G29.96-0.02 and the DR21 regions the den-

sity maximum is on the order of 0.1 pc ahead of the ionization front. For ionization fronts advancing at about 3 km s⁻¹, as in the *Arthur and Hoare* (2006) models, this gives a lifetime of a few × 10⁴ yr, similar to the timescale for the star to reach the same point. *Kawamura and Masson* (1998) measured the expansion speed of the cometary W3(OH) region to be 3–5 km s⁻¹ using proper motions.

Zhu et al. (2005) developed a model of flow around a parabolic shell for the G29.96-0.02 and Mon R2 cometary regions, including pressure gradient forces resulting from the sweeping up of material by the bow shock, but still within a constant ambient density. As with the original bow-shock models for G29.96-0.02, this required a stellar velocity of 20 km s⁻¹, resulting in a mismatch with the molecular cloud reference frame of 10 km s⁻¹. *Lumsden and Hoare* (1996, 1999) also needed to invoke a 10 km s⁻¹ shift in the opposite direction to get their semi-empirical champagne-flow model to agree with the observations, which they ascribed to expansion of the ionization front. Overall, the picture that is emerging for the few well-studied cometary regions is that the morphology, velocity structure, and lifetime of cometary UCHII regions can be explained by a combination of champagne flow down a density gradient and bow-shock motion up the density gradient. This still remains to be tested on a wider range of objects. Possible explanations of the direction of motion will be discussed by Hoare et al. (in preparation).

3. HYPERCOMPACT HII REGIONS

3.1. Observed Properties

Within the large number of UCHII regions that have been discovered, a number of objects stand out as being exceptionally small and dense. Sizes for this group are ≤0.05 pc (10,000 AU), while densities are ≥10⁶ cm⁻³ and emission measures are ≥10¹⁰ pc cm⁻⁶. In the past few years, these regions have come to be considered as a separate class, referred to as *hypercompact* HII regions (HCHII). These regions are considered a distinct class from UCHII regions — rather than merely representing the most extreme sizes and densities — primarily because they have extremely broad radio recombination line profiles, with ΔV typically 40–50 km s⁻¹, and some greater than 100 km s⁻¹ (*Gaume et al.,* 1995; *Johnson et al.,* 1998; *Sewilo et al.,* 2004). By comparison, UCHII regions typically have recombination line widths of 30–40 km s⁻¹ (*Keto et al.,* 1995; *Afflerbach et al.,* 1996). Although the precise relationship between HCHII regions and broad recombination line objects (BRLO) is not yet clear, at present the two classes have so many objects in common that we take them to be one and the same.

There are also similarities between the HCHII/BRLOs and the massive YSOs that have strong ionized stellar winds, including BN, W33A, Cep A2, GGD27, and S140 IRS 1 (*Rodríguez,* 1999; *Hoare,* 2002). These luminous (>10⁴ L⊙), embedded sources are often found without hot molecular core signatures, but are usually still driving outflows and probably accreting. We do not use the term "protostar,"

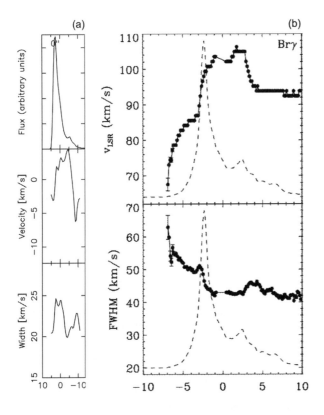

Fig. 5. **(a)** Simulated long-slit spectral data along the axis of the model cometary HII region shown in Fig. 4. The three panels show the flux, velocity centroid, and line width respectively. Velocities are with respect to the molecular cloud, which is at rest in the simulation. Horizontal axis is offset from the position of the exciting star in arcseconds. **(b)** Long-slit spectral data for a position approximately along the axis of G29.96-0.02 from *Martín-Hernández et al.* (2003). Top panel shows the velocity centroid measured for the Brγ line while the bottom shows the observed line width with the flux profile in the dotted line. The best estimate of the ambient molecular cloud velocity is v_{LSR} = 98 km s⁻¹. Horizontal axis is offset from the exciting star in arcseconds.

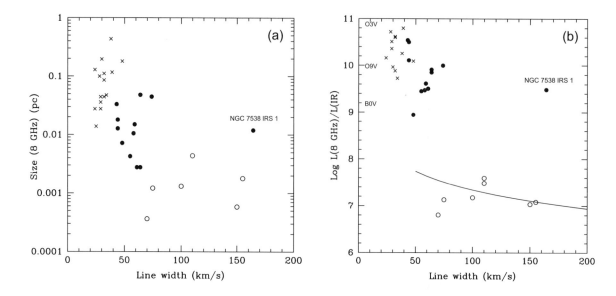

Fig. 6. (a) Size vs. line width for UCHII regions (crosses), HCHII regions (solid circles), and massive young stellar object wind sources (open circles). Line widths are FWHM. Sizes are measured or equivalent at 8 GHz and are geometric mean FWHM. Note how the HCHII regions lie between UCHII regions and the MYSOs wind sources. (b) As left, but for the ratio of the radio luminosity at 8 GHz (W Hz^{-1}) to the bolometric luminosity from the IR (L$_\odot$) (except W49A regions where radio spectral type has been used). The flux at 8 GHz has been extrapolated from higher frequencies if not measured directly, using observed or typical spectral indices. Data for UCHIIs are from *Wood and Churchwell* (1989a) and *De Pree et al.* (2004); for the HCHII regions from *Sewilo et al.* (2004), *Jaffe and Martin-Pintado* (1999), *Johnson et al.* (1998), and *De Pree et al.* (2004); and for the MYSO wind sources from *Bunn et al.* (1995) and *Nisini et al.* (1994). Sizes of MYSOs come from *Hoare* (2002) and references therein, *Rengarajan and Ho* (1996), *Snell and Bally* (1986), and *Campbell et al.* (1986). The expected ratio for optically thin HII regions for given spectral type exciting stars is indicated at top left using stellar parameters from *Smith et al.* (2002). The solid line shows the level expected for a stellar wind assuming $\dot{M} = 10^{-6}$ M$_\odot$ yr^{-1}, $v_{inf} = 2v_{FWHM}$, and L$_{(bol)} = 10^4$ L$_\odot$ (*Wright and Barlow,* 1975).

since it is likely that the luminosity of these objects is dominated by hydrogen burning and not accretion, which cannot be observationally distinguished at present.

The massive YSOs are weaker radio sources and have therefore not been the subject of radio recombination line studies. However, their IR recombination line profiles have been studied and are very broad, usually in excess of 100 km s^{-1} (e.g., *Bunn et al.,* 1995). High-resolution radio mapping has yielded both jet and equatorial wind morphologies for the nearest sources (*Hoare,* 2006; *Patel et al.,* 2005). Proper motion studies of jet sources reveal velocities of 500 km s^{-1} (*Martí et al.,* 1998; *Curiel et al.,* 2006), and the high collimation suggests a magnetohydrodynamic mechanism as in low-mass YSOs. [Note the ionized X-wind or "outflow-confined" model put forward for HCHII regions by *Tan and McKee* (2003) is more appropriate for these jet sources than the slower HCHII regions.] The equatorial wind sources can result from the pressure of stellar radiation acting on the gas on the surface of the accretion disk (*Sim et al.,* 2005).

A key difference between wind sources and UCHII regions is that in wind sources the ionized material originates from the star-disk system itself. In contrast, for expanding UCHII regions it is the surrounding molecular cloud material that is being ionized.

The observed properties of HCHII regions would appear to be intermediate between these two extremes, as illustrated in Fig. 6. The size vs. line width plot is an extension of previous plots (e.g., *Garay and Lizano,* 1999). The line widths for the MYSO wind sources are from the near-IR Brγ line, while those for the UCHII and HCHII regions are from radio recombination lines. The highest-frequency data available were used to minimize the effects of pressure broadening. For winds, the size varies with frequency and the measured or extrapolated size at 8 GHz is plotted. The UCHII region size does not vary with frequency and we assume the same for HCHII regions, although they may vary somewhat.

For the righthand plot of the radio brightness normalized to the total luminosity, the HCHII and UCHII regions are clearly much more radio-loud than the MYSO winds. The 8 GHz radio luminosity is not a straightforward property to use, because it underestimates the number of ionizing photons produced by the exciting source when the source is optically thick. At 8 GHz, HCHII regions in particular are well below their turnover frequency, and present correspondingly lower flux densities. There is also a strong dependence on the effective temperature of the ionizing source. It should be noted that the MYSO wind sources plotted (BN, W33A, S140 IRS 1, NGC 2024 IRS 2, GL 989, and GL 490) are

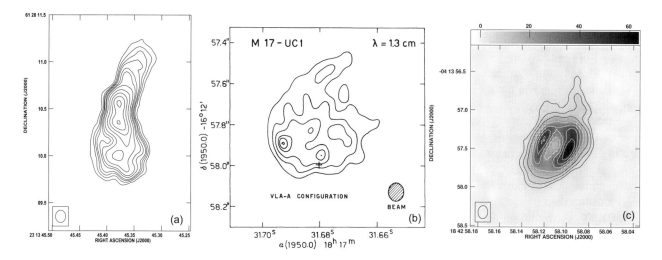

Fig. 7. The different morphologies of hypercompact HII regions. **(a)** The bipolar NGC 7538 IRS 1 (*Franco-Hernández and Rodríguez*, 2004); **(b)** the cometary M17 UC1 (*Felli et al.*, 1984); **(c)** the shell-like G28.20–0.04 (*Sewilo et al.*, 2005).

somewhat lower luminosity than most of the HII regions plotted. However, the combination of these observable parameters provides a reasonable indication as to what type of object one is dealing with. It is interesting to note that NGC 7538 IRS 1 stands out in these plots as an exceptional source due to its broad lines and strong emission.

Two different physical mechanisms may contribute to the broad linewidths of HCHII regions. Pressure broadening can be significant in such high-density regions, particularly for the high principle quantum number centimeter wave transitions observed with the VLA, owing to the n^7 dependence of the broadening [e.g., *Brocklehurst and Seaton* (1972); *Griem* (1974); see also *Keto et al.* (1995) for a discussion of the combination of dynamics and line broadening as applied to HII regions]. Bulk motion of the gas, either via accretion (e.g., *Keto*, 2002b; *Keto and Wood*, 2006), outflow (e.g., *Lugo et al.*, 2004), or rotation, could be occurring. To distinguish the relative contributions from pressure broadening and bulk motion of the gas will require high-spatial-resolution ($\leq 1"$) observations over a wide range of frequencies, including low principal quantum number transitions in the millimeter, submillimeter, and IR regions.

Apart from the common occurrence of broad recombination lines, other differences exist between the hyper- and ultracompact classes. UCHII regions typically have turnover frequencies (between the optically thick and thin regimes) of 10–15 GHz, while HCHII regions become optically thin at frequencies above 30 GHz, owing to their higher densities. Nevertheless, no HCHII regions have been found with an optically thick spectral index of +2, indicative of uniform density gas. Rather, HCHII regions show an intermediate spectral index of ~+1, which suggests nonuniform gas density. Some UCHII regions also show $\alpha \sim +1$ at centimeter wavelengths. Density gradients have been invoked to explain these intermediate spectral indices (e.g., *Franco et*

al., 2000; *Avalos et al.*, 2006). Although a density gradient and small size are suggestive of a stellar wind, the measured flux densities and the inferred electron densities are much higher than values encountered in stellar winds. A disk wind, however, may be capable of producing a density gradient, and at the high electron densities implied by the radio flux densities (*Lugo et al.*, 2004) (see section 3.2). The effect of champagne flows or influence of stellar gravity (*Keto*, 2003) lead to steep density gradients. *Ignace and Churchwell* (2004) have shown that the intermediate spectral indices can also be produced by unresolved clumps with a power-law distribution of optical depths.

Morphologies of HCHII regions are not well known; to date, only a few have been resolved by radio continuum observations. The archetypal BRLO NGC 7538 IRS 1 is bipolar, but other morphologies are seen as well. Figure 7 shows radio continuum images of NGC 7538 IRS 1, M17-UC1, and G28.20–0.04, which have, respectively, bipolar, cometary, and shell-like morphologies. The high spatial resolution observations needed to identify the nature of the broad recombination line emission will also provide much needed morphological information.

Although classifications based on observed parameters can be useful, a more fundamental definition relating to the nature of the object is preferable. One possibility is to define HCHIIs as those regions with radii less than the transonic point where the escape velocity from the star(s) within the HII region equals the ionized gas sound speed (see section 3.2). Alternatively, one might consider the place of HCHIIs in the general massive-star-formation process. In this context, the crucial element is the recognition that HCHII sizes are 10,000 AU *and smaller*. Regardless of the model of massive star formation that is adopted, it is clear that HCHII sizes are approaching a scale appropriate for individual high-mass star formation.

3.2. Theoretical Models

3.2.1. Context: Hot molecular cores. The observed properties of the HCHII regions suggest that they are very young and it is natural to associate them with the turn-on of the Lyman continuum radiation. HII regions around stars later than early-B type are not detectable by present radio telescopes. Thus, if a star is growing by accretion through the spectral type sequence from B to O (see section 3.2.2), then in the earlier stages the star and accretion flow would appear as a hot molecular core (HMC) without radio continuum emission. *Osorio et al.* (1999) modeled the spectral energy distribution of the thermal emission from dust in such HMCs accreting toward a central star. They modeled the HMCs as logatropic cores with pressure $P = P_0 \log(\rho = \rho_{ref})$, where $P_0 \sim 10^{-7}$–10^{-6} dynes cm^{-2} is the pressure constant and ρ_{ref} is a reference density (*Lizano and Shu,* 1989; *Myers and Fuller,* 1992). Logatropic cores have an equilibrium density profile $\rho \propto r^{-1}$, and a large velocity dispersion $\sigma^2 = dP/d\rho = P_0/\rho$, that increases with decreasing density, as observed in molecular clouds (e.g., *Fuller and Myers,* 1992) supporting massive envelopes of several hundred $M_\odot$. The gravitational collapse of logatropic spheres has the property that the mass accretion rate increases with time as $\dot{M} \propto t^3$ (*McLaughlin and Pudritz,* 1997). In a comparison of their model with observations of several hot molecular cores without detectable radio continuum, *Osorio et al.* (1999) found that the observed spectral energy distributions (SEDs) were consistent with central B stars with ages $\tau_{age} \lesssim 6 \times 10^4$ yr, accreting at rates of $\dot{M}_{acc} > 10^{-4}$ $M_\odot$ yr^{-1}. At these accretion rates, the main heating agent is the luminosity arising from the deceleration of the accretion flow, $L_{acc} = GM_* \dot{M}_{acc}/R_*$ (assuming purely spherical accretion). Comparison of observed and modeled SEDs suggests that some HMCs are consistent with the earlier stages of the formation of massive stars by accretion.

The logatropic core model is also attractive because it has properties that are required if massive stars are to form despite the constraints imposed by the outward force of radiation pressure and within the short timescale allowed by the main-sequence lifetime. Osorio et al. showed that their inferred accretion rates were large enough that the momentum of the accretion flow could overcome the outward force of radiation pressure on dust grains. At the risk of oversimplicity, this argument may be summarized as follows: The most naive estimate of the outward force deriving from the luminosity, assuming spherical geometry and the total absorption of the luminosity by the dust in the flow, would be L/c. A similarly simple estimate of the force deriving from the momentum of the accretion flow, again assuming spherical accretion, is $\dot{M}v$. If the momentum of the accretion flow is high enough, the flow will push the dust grains inward until they are sublimated by the higher temperatures in the center of the flow. At this point the flow is rendered essentially transparent to the stellar radiation (*Kahn,* 1974; *Wolfire and Cassinelli,* 1987).

3.2.2. Ionized accretion flow or gravitationally trapped HII regions. Once the star, possibly one of the several stars forming together, has gained sufficient mass and therefore temperature, the number of emitted Lyman continuum photons will be enough to ionize an HII region within the continuing accretion flow. The radius of ionization equilibrium, r_i, is set by the balance between ionization and recombination within the HII region. If r_i is less than the gravitational radius in the accretion flow, $r_g \equiv GM_*/a^2$, where the inward velocity equals the sound speed of the ionized gas, a (within a factor of 2), then the HII region can be trapped within the accretion flow with its boundary as a stationary R-type ionization front. This model explains why the development of an HII region does not immediately end the accretion as would be expected in the classic textbook model for the evolution of HII regions by pressure-driven expansion (*Shu,* 1992; *Spitzer,* 1978; *Dyson and Williams,* 1980). In pressure-driven expansion, the flow of the ionized gas is entirely outward, precluding accretion. The model for HII regions trapped within an accretion flow was developed to explain the observations of the inward flow of ionized gas toward the massive stars forming in the HII region G10.6–0.4 (*Keto,* 2002a).

The accretion model assumed in *Keto* (2002b) is a spherical steady-state flow (Bondi accretion) that has a density gradient everywhere less steep than $n \sim r^{-3/2}$. This is significant in that the spherical solution for ionization equilibrium does not allow for solutions in steeper density gradients (*Franco et al.,* 1990). Similar to accretion in a logatropic sphere, Bondi accretion has the property that the accretion rate depends on the stellar mass (as M^2 in Bondi accretion). As in the model of *Osorio et al.* (1999), this model relies on the momentum of a massive accretion flow to overcome the radiation pressure of the stars. Observations of G10.6–0.4 indicate that the outward force from the observed luminosity of 1.2×10^6 $L_\odot$ (*Fazio et al.,* 1978) is approximately equal to the inward force deriving from the momentum of the flow, $\dot{M}v \sim 3 \times 10^{28}$ dynes, for an estimated accretion rate of 10^{-3} $M_\odot$ yr^{-1} and velocity of 4.5 km s^{-1} at 5000 AU (*Keto,* 2002a). In this model, the accretion flow, if not reversed by radiation pressure, will end when the ionization rate, which increases with the increasing mass of the star(s), is high enough that the radius of ionization equilibrium increases beyond the gravitational radius.

While the central star is accreting mass at rates $\dot{M} \sim 10^{-3}$ $M_\odot$ yr^{-1}, the ram pressure of the accreting material will be larger than the ram pressure of the stellar wind, $\rho_{acc}v_{acc}^2 \gg \rho_w v_w^2$, and will prevent the stellar wind from blowing out. If the radiation pressure on dust grains reverses the accretion flow, or the HCHII region expands because its radius becomes larger than r_g, then a stellar wind can blow out (see also *González-Avilés et al.,* 2005).

Recently, *Keto and Wood* (2006) extended the model of *Keto* (2002b) to consider accretion flows with angular momentum as described by *Ulrich* (1976) and *Terebey et al.* (1984). In this model of accretion, the gas spirals in to the star on ballistic trajectories conserving angular momentum. An accretion disk develops at a radius r_D, where the centrifugal force balances the gravitational force, $\Gamma^2/r_D^3 = GM/r_D^2$, roughly where the infall velocity equals the rotational velocity. Here Γ is the initial specific angular momentum. Since $r_D \sim GM/v_{orbital}^2$, if $v_{orbital} < c$ at the gravitational radius, r_g, then the radius of disk formation will be within the maximum radius, r_g, of a trapped HII region. Thus the radius of disk formation, depending on the flux of ionizing photons, could be within the ionized portion of the flow, meaning that the accretion disk would form out of infalling ionized gas. Note that the central parts of the centrifugal disk should be neutral because the gas recombines at the high densities expected in these disks. With a different choice of parameters, specifically, if the ionizing photon flux or the angular momentum were greater than in the previous case, then the disk might form in the molecular portion of the flow. In this case, the accretion flow would have the same structure as that assumed in the model of photo-evaporating disks of *Hollenbach et al.* (1994) and *Johnstone et al.* (1998).

Keto and Wood (2006) proposed the following evolutionary hypothesis for the development of an HII region within an accretion flow with a disk that parallels the development of an HII region within a spherical accretion flow. In the initial stage when there is no HII region, necessarily $r_D > r_i$ and the flow is described by the massive molecular accretion disk. In the second stage, a trapped HCHII region will develop in the center of the disk. Because the gas in the disk is denser than the gas elsewhere around the star, the molecular accretion disk will not necessarily be fully ionized, but because $r_i < r_g$, the ionized surface of the disk will not be expanding off the disk. Rather there will be a limited region, contained within the HII region, with an ionized accretion flow onto the disk. However, depending on the initial angular momentum in the flow, this region may be very small with respect to the extent of the molecular disk. Outside this region, there will be a molecular accretion flow onto the disk. The third stage of evolution is defined by the condition $r_{ionized} > r_G$. In the nonspherical case, because the gas density around the star is a function of angle off the disk, the ionization radius, r_i, will also be a function of angle. If the disk is sharply defined as in the *Ulrich* (1976) and *Terebey et al.* (1984) models, then in the third stage the HII region will expand around the disk. In this third stage, because $r_i > r_g$, the surface of the disk (except for a small region in the center) will photo-evaporate with an outward flow of ionized gas off the disk, as described in the models for photo-evaporating disks.

3.2.3. Photo-evaporating disks. *Hollenbach et al.* (1994) and *Johnstone et al.* (1998) proposed that unresolved UCHII (now termed HCHII regions) arise from photo-evaporating molecular disks. The photo-evaporation of circumstellar molecular disks around massive stars occurs as the disk surface is ionized by the stellar Lyman continuum photons. *Hollenbach et al.* (1994) proposed that within the gravitational radius, r_g, the heated gas is confined in the gravitational potential well of the star. For $r > r_g$ the ionized gas can escape and an isothermal evaporative flow is established. The ionized material flows away, but is constantly replenished by the photo-evaporation of the disk. If the star has a strong stellar wind, the wind may push the confined gas within r_g out to a critical radius where the ram pressure of

the stellar wind is balanced by the thermal pressure of the photo-evaporated flow, resulting in outflow at all radii. *Yorke and Welz* (1996) and *Richling and Yorke* (1997) made hydrodynamical simulations of the evolution of photo-evaporated disks under a variety of conditions. In particular, Richling and Yorke found that scattering of ionizing photons on dust grains increases the photo-evaporation rate.

Recently, *Lugo et al.* (2004) modeled the density and velocity structure of axisymmetric isothermal winds photo-evaporated from a spatially thin Keplerian disk. They calculated the predicted free-free continuum emission of these models to match the observed spectral energy distributions of the bipolar objects MWC 349 A and NGC 7538 IRS 1. These models naturally give a bipolar morphology that is seen in a subset of the HCHII regions. The next step is to investigate whether the thermally evaporating flow can produce the high velocities seen in the broad recombination line objects.

4. FUTURE DIRECTIONS

A key test of all the proposed models for the HCHII regions and broad recombination line objects is to match the observed morphologies and velocity structures of the ionized gas. Quality datasets already exist for a few well-studied examples against which to test the models. At present there are so few objects that a wider sample is needed to determine whether well-defined morphological classes actually exist. This may be challenging since they do appear rare, presumably since the phase is short-lived, but it does mark an important transition. Further high-frequency radio recombination line observations and, where possible, IR spectroscopy are needed to separate those that are merely pressure broadened (and therefore likely to be understood in terms of the usual UCHII region dynamics) from those that need a new physical picture. We also eagerly await the results currently being obtained with the submillimeter array (SMA) and the future studies that will be made with the millimeter interferometers CARMA and ALMA. These will be able to trace the molecular gas dynamics down to the scales of the HCHII regions.

Once the ionization front expands to UCHII region scales, it is likely to leave behind the local density enhancement resulting from the gravitational collapse and begin to experience the wider environment in which the collapse took place. Significant proper motion relative to the cloud core will add to this effect. The presence of champagne flows in cometary HII regions indicates that the wider environment must have a density gradient, i.e., the cometary regions are not located at the core center. The original blister picture developed by *Israel* (1978) argued that such regions are a result of triggered star formation rather than spontaneous collapse; the latter more likely to occur within a dense core. He was considering more evolved HII regions, but they show a similar distribution of morphological types as the UCHII regions (*Fich*, 1993). Velocity studies of large, optically visible HII regions have revealed that champagne flows dominate their dynamics too (e.g., *Priestley*, 1999, *Roger et al.*,

2004). This is most easily understood as the continued evolution of cometary UCHIIs down a large-scale density gradient.

A key test of the dynamical models for cometary regions is to measure the motion of the stars. Radial velocity measurements on the star in G29.96-0.02 were attempted by *Martín-Hernández et al.* (2003). However, the resolution and signal-to-noise of the spectrum, and uncertainties in the rest wavelengths of the heavy-element spectral features uncontaminated by nebular lines, prevented the required precision. Once the EVLA and e-MERLIN telescopes come on line it may be possible to detect the radio emission directly from the free-flowing stellar wind of the exciting stars at high resolution. This would open up the possibility of proper motion measurements directly on the stars, as well as many more expansion measurements of the nebulae themselves.

The more global aspects of massive star formation will be addressed by the numerous galactic plane surveys in coming years, following the lead by the GLIMPSE survey in the mid-IR. A high-resolution radio survey of the entire northern GLIMPSE region (*www.ast.leeds.ac.uk/Cornish*) will pick out all UCHII regions across the galaxy, building on the previous blind surveys by *Giveon et al.* (2005) and references therein. These will be complemented by a submillimeter continuum survey of the plane at the JCMT, methanol maser surveys using Parkes and the Lovell telescopes (*www.jb.man.ac.uk/research/methanol*), ^{13}CO from the BU-FCRAO Galactic Ring Survey (*www.bu.edu/galacticring*), HI from the VLA survey (*www.ras.ucalgary.ca/VGPS*), and near-IR from UKIRT (*www.ukidss.org*).

Unbiased area surveys will allow the luminosity function and lifetimes to be derived and compared with simulations of the evolution of the galactic UCHII region population. Their location within the wider GMCs, the fractions that appear to be triggered or isolated, and their location relative to potential external triggers will be established on a large-scale statistical basis. The power of this sensitive, multiwavelength, high-resolution campaign on the Milky Way will yield many advances in our understanding of massive star formation.

Acknowledgments. We would like to thank J. Arthur for her input on the hydrodynamic modeling and J. Urquhart and A. Busfield for help with the figures. Useful discussions were had with T. Hartquist and J. De Buizer. We thank the referee for several useful suggestions.

REFERENCES

Afflerbach A., Churchwell E., Hofner P., and Kurtz S. (1994) *Astrophys. J., 437,* 697–704.
Afflerbach A., Churchwell E., Acord J. M., Hofner P., Kurtz S., and De Pree C. G. (1996) *Astrophys. J. Suppl., 106,* 423–446.
Alvarez C., Feldt M., Henning T., Puga E., Brandner W., and Stecklum B. (2004) *Astrophys. J. Suppl., 155,* 123–148.
Arthur J. and Hoare M. G. (2006) *Astrophys. J. Suppl., 165,* 283–306.
Avalos M., Lizano S., Rodríguez L., Franco-Hernández R., and Moran J. (2006) *Astrophys. J., 641,* 406–409.

Benjamin R. A., Churchwell E., Babler B. L., Bania T. M., Clemens D. P., et al. (2003) *Publ. Astron. Soc. Pac., 115,* 953–964.
Bik A., Kaper L., Hanson M. M., and Smits M. (2005) *Astron. Astrophys., 440,* 121–137.
Bourke T. L., Hyland A. R., and Robinson G. (2005) *Astrophys. J., 625,* 883–890.
Brocklehurst M. and Seaton M. J. (1972) *Mon. Not. R. Astron. Soc., 157,* 179–210.
Bunn J. C., Hoare M. G., and Drew J. E. (1995) *Mon. Not. R. Astron. Soc., 272,* 346–354.
Busfield A. L. (2006) Ph.D. thesis, Univ. of Leeds.
Campbell B., Persson S. E., and McGregor P. J. (1986) *Astrophys. J., 305,* 336–352.
Cesaroni R., Walmsley C. M., and Churchwell E. (1992) *Astron. Astrophys., 256,* 618–630.
Cesaroni R., Churchwell E., Hofner P., Walmsley C. M., and Kurtz S. (1994) *Astron. Astrophys., 288,* 903–920.
Churchwell E. (1999) In *The Origin of Stars and Planetary Systems* (C. Lada and N. Kylafis, eds.), pp. 515–552. NATO Science Series, Kluwer, Dordrecht.
Churchwell E. (2002) *Ann. Rev. Astron. Astrophys., 40,* 27–62.
Codella C., Felli M., and Natale V. (1994) *Astron. Astrophys., 284,* 233–240.
Comeron F. (1997) *Astron. Astrophys., 326,* 1195–1214.
Curiel S., Ho P. T. P., Patel N. A., Torrelles J. M., Rodríguez L. F., et al. (2006) *Astrophys J., 638,* 878–886.
Cyganowski C. J., Reid M. J., Fish V. L., and Ho P. T. P. (2003) *Astrophys. J., 596,* 344–349.
De Buizer J. M., Watson A. M., Radomski J. T., Pia R. K., and Telesco C. M. (2002a) *Astrophys. J., 564,* L101–L104.
De Buizer J. M., Radomski J. T., Pia R. K., and Telesco C. M. (2002b) *Astrophys. J., 580,* 305–316.
De Buizer J. M., Radomski J. T., Telesco C. M., and Pia R. K. (2003) *Astrophys. J., 598,* 1127–1139.
De Buizer J. M., Radomski J. T., Telesco C. M., and Pia R. K. (2005) *Astrophys. J. Suppl., 156,* 179–215.
De Pree C. G., Rodríguez L. F., and Goss W. M. (1995) *Rev. Mex. Astron. Astrophys., 31,* 39–44.
De Pree C. G., Wilner D. J., Mercer A. J., Davis L. E., Goss W. M., and Kurtz S. (2004) *Astrophys. J., 600,* 286–291.
De Pree C. G., Wilner D. J., Deblasio J., Mercer A. J., and Davis L. E. (2005) *Astrophys. J., 624,* L101–L104.
De Wit W. J., Testi L., Palla F., and Zinnecker H. (2005) *Astron. Astrophys., 437,* 247–255.
Dickel H. R. and Goss W. M. (1987) *Astron. Astrophys., 185,* 271–282.
Dickel H. R., Goss W. M., and De Pree C. G. (2001) *Astron. J., 121,* 391–398.
Draine B. T. (2003) *Ann. Rev. Astron. Astrophys., 41,* 241–289.
Dyson J. and Williams D. (1980) *The Physics of the Interstellar Medium,* Wiley, New York.
Dyson J., Williams R., and Redman M. (1995) *Mon. Not. R. Astron. Soc., 227,* 700–704.
Ellingsen S. P., Shabala S. S., and Kurtz S. E. (2005) *Mon. Not. R. Astron. Soc., 357,* 1003–1012.
Fazio G. G., Lada C. J., Kleinmann D. E., Wright E. L., Ho P. T. P., and Low F. J. (1978) *Astrophys. J., 221,* L77–L81.
Feldt M., Stecklum B., Henning Th., Launhardt R., and Hayward T. L. (1999) *Astron. Astrophys., 346,* 243–259.
Felli M., Churchwell E., and Massi M. (1984) *Astron. Astrophys., 136,* 53–64.
Fey A. L., Gaume R. A., Claussen M. J., and Vrba F. J. (1995) *Astrophys. J., 453,* 308–312.
Fich M. (1993) *Astrophys. J. Suppl., 86,* 475–497.
Franco J., Tenorio-Tagle G., and Bodenheimer P. (1990) *Astrophys. J., 349,* 126–140.
Franco J., Kurtz S., Hofner P., Testi L., García-Segura G., and Martos M. (2000) *Astrophys. J., 542,* L143–L146.

Franco-Hernández R. and Rodríguez L. F. (2004) *Astrophys. J., 604*, L105–L108.

Fuller G. A. and Myers P. C. (1992) *Astrophys. J., 384*, 523–527.

Garay G. and Lizano S. (1999) *Publ. Astron. Soc. Pac., 111*, 1049–1087.

Garay G., Lizano S., and Gómez Y. (1994) *Astrophys. J., 429*, 268–284.

Garay G., Lizano S., Gómez Y., and Brown R. L. (1998a) *Astrophys. J., 501*, 710–722.

Garay G., Gómez Y., Lizano S., and Brown R. L. (1998b) *Astrophys. J., 501*, 699–709.

García-Segura G. and Franco J. (1996) *Astrophys. J., 469*, 171–188.

Gaume R., Goss W., Dickel H., Wilson T., and Johnston K. (1995) *Astrophys. J., 438*, 776–783.

Giard M., Bernard J. P., Lacombe F., Normand P., and Rouan D. (1994) *Astron. Astrophys., 291*, 239–249.

Gies D. R. (1987) *Astrophys. J. Suppl., 64*, 545–563.

Giveon U., Becker R. H., Helfand D. J., and White R. L. (2005) *Astron. J., 129*, 348–354.

Gómez L., Rodríguez L. F., Loinard L., Lizano S., Poveda A., and Allen C. (2005) *Astrophys. J., 635*, 1166–1172.

Gómez Y., Lebron M., Rodriguez L. F., Garay G., Lizano S., Escalante V., and Canto J. (1998) *Astrophys. J., 503*, 297–306.

Gonzáles-Aviles M., Lizano S., and Raga A. C. (2005) *Astrophys. J., 621*, 359–371.

Griem H. R. (1974) *Spectral Line Broadening by Plasmas*, Academic, New York.

Hatchell J. and van der Tak F. F. S. (2003) *Astron. Astrophys., 409*, 589–598.

Hatchell J., Thompson M. A., Millar T. J., and MacDonald G. H. (1998) *Astron. Astrophys., 133*, 29–49.

Hoare M. G. (2002) In *Hot Star Workshop III: The Earliest Stages of Massive Star Birth* (P. A. Crowther, ed.), pp. 137–144. ASP Conf. Series 267, San Francisco.

Hoare M. G. (2006) *Astrophys J.*, in press.

Hoare M. G., Roche P. F., and Glencross W. M. (1991) *Mon. Not. R. Astron. Soc., 251*, 584–599.

Hoare M. G., Lumsden S. L., Busfield A. L., and Buckley P. (2003) In *Winds, Bubbles and Explosions* (S. J. Arthur and W. J. Henney, eds.), pp. 172–174. Rev. Mex. Astron. Astrophys. Series Conf. 15.

Hoare M. G., Lumsden S. J., Oudmaijer R. D., Busfield A. L., King T. L., and Moore T. L. J. (2004) In *Milky Way Surveys: The Structure and Evolution of Our Galaxy* (D. Clemens et al., eds.), pp. 156–158. ASP Conf. Series 317, San Francisco.

Hofmann K.-H., Balega Y. Y., Preibisch T., and Weigelt G. (2004) *Astron. Astrophys., 417*, 981–985.

Hofner P. and Churchwell E. (1996) *Astron. Astrophys., 120*, 283–299.

Hofner P., Kurtz S., Churchwell E., Walmsley C. M., and Cesaroni R. (1994) *Astrophys. J., 429*, L85–L88.

Hofner P., Wyrowski F., Walmsley C. M., and Churchwell E. (2000) *Astrophys. J., 536*, 393–405.

Hofner P., Delgado H., Whitney B., Churchwell E., and Linz H. (2002) *Astrophys. J., 579*, L95–L98.

Hollenbach D., Johnstone D., Lizano S., and Shu F. (1994) *Astrophys. J., 428*, 654–669.

Hoogerwerf R., de Bruijne J. H. J., and de Zeeuw P. T. (2001) *Astron. Astrophys., 365*, 49–77.

Ignace R. and Churchwell E. (2004) *Astrophys. J., 610*, 351–360.

Indebetouw R., Mathis J. S., Babler B. L., Meade M. R., Watson C., Whitney B. A., et al. (2005) *Astrophys. J., 619*, 931–938.

Israel F. P. (1978) *Astron. Astrophys., 90*, 769–775.

Jaffe D. and Martin-Pintado J. (1999) *Astrophys. J., 520*, 162–172.

Johnson C. O., De Pree C. G., and Goss W. M. (1998) *Astrophys. J., 500*, 302–310.

Johnstone D., Hollenbach D., and Bally J. (1998) *Astrophys. J., 499*, 758–776.

Kahn F. (1974) *Astron. Astrophys., 37*, 149–162.

Kawamura J. H. and Masson C. R. (1998) *Astrophys. J., 509*, 270–282.

Keto E. (2002a) *Astrophys. J., 568*, 754–760.

Keto E. (2002b) *Astrophys. J., 580*, 980–986.

Keto E. (2003) *Astrophys. J., 599*, 1196–1206.

Keto E. and Wood K. (2006) *Astrophys. J., 637*, 850–859.

Keto E., Welch W., Reid M., and Ho P. (1995) *Astrophys. J., 444*, 765–769.

Kim K.-T. and Koo B.-C. (2001) *Astrophys. J., 549*, 979–996.

Kim K.-T. and Koo B.-C. (2003) *Astrophys. J., 596*, 362–382.

Kraemer K. E., Jackson J. M., Kassis M., Deutsch L. K., Hora J. L., et al. (2003) *Astrophys. J., 588*, 918–930.

Kurtz S., Churchwell E., and Wood D. O. S. (1994) *Astrophys. J. Suppl., 91*, 659–712.

Kurtz S. E., Watson A. M., Hofner P., and Otte B. (1999) *Astrophys. J., 514*, 232–248.

Kurtz S., Cesaroni R., Churchwell E., Hofner P., and Walmsley C. M. (2000) In *Protostars and Planets IV* (V. Mannings et al., eds.), pp. 299–326. Univ. of Arizona, Tuscon.

Lebrón M., Rodríguez L. F., and Lizano S. (2001) *Astrophys. J., 560*, 806–820.

Linz H., Stecklum B., Henning Th., Hofner P., and Brandl B. (2005) *Astron. Astrophys., 429*, 903–921.

Lizano S. and Shu F. H. (1989) *Astrophys. J., 342*, 834–854.

Lizano S., Cantó J., Garay G., and Hollenbach D. (1996) *Astrophys. J., 468*, 739–748.

Lugo J., Lizano S., and Garay G. (2004) *Astrophys. J., 614*, 807–817.

Lumsden S. L. and Hoare M. G. (1996) *Astrophys. J., 464*, 272–285.

Lumsden S. L. and Hoare M. G. (1999) *Mon. Not. R. Astron. Soc., 305*, 701–706.

Lumsden S. L., Hoare M. G., Oudmaijer R. D., and Richards D. (2002) *Mon. Not. R. Astron. Soc., 336*, 621–636.

Lumsden S. L., Puxley P. J., Hoare M. G., Moore T. J. T., and Ridge N. A. (2003) *Mon. Not. R. Astron. Soc., 340*, 799–812.

Mac Low M.-M., Van Buren D., Wood D. O. S., and Churchwell E. (1991) *Astrophys. J., 369*, 395–409.

McLaughlin D. E. and Pudritz R. E. (1997) *Astrophys. J., 476*, 750–765.

Martí J., Rodríguez L. F., and Reipurth B. (1998) *Astrophys. J., 502*, 337–341.

Martín-Hernández N. L., Vermeij R., Tielens A. G. G. M., van der Hulst J. M., and Peeters E. (2002) *Astron. Astrophys., 389*, 286–294.

Martín-Hernández N. L., Bik A., Kaper L., Tielens A. G. G. M., and Hanson M. M. (2003) *Astron. Astrophys., 405*, 175–188.

Maxia C., Testi L., Cesaroni R., and Walmsley C. M. (2001) *Astron. Astrophys., 371*, 286–299.

Mueller K. E., Shirley Y. L., Evans N. J. II, and Jacobson H. R. (2002) *Astrophys. J. Suppl., 143*, 469–497.

Myers P. C. and Fuller G. A. (1992) *Astrophys. J., 396*, 631–642.

Natta A. and Panagia N. (1976) *Astron. Astrophys., 50*, 191–211.

Nisini B., Smith H. A., Fischer J., and Geballe T. R. (1994) *Astron. Astrophys., 290*, 463–472.

Okamoto Y., Kataza H., Yamashita T., Miyata T., Sako S., et al. (2003) *Astrophys. J., 584*, 368–384.

Olmi L. and Cesaroni R. (1999) *Astron. Astrophys., 352*, 266–276.

Olmi L., Cesaroni R., Hofner P., Kurtz S., Churchwell E., and Walmsley C. M. (2003) *Astron. Astrophys., 407*, 225–235.

Osorio M., Lizano S., and D'Alessio P. (1999) *Astrophys. J., 525*, 808–820.

Patel N. A., Curiel S., Sridharan T. K., Zhang Q., Hunter T. R., et al. (2005) *Nature, 437*, 109–111.

Pratap P., Megeath S. T., and Bergin E. A. (1999) *Astrophys. J., 517*, 799–818.

Priestley C. (1999) Ph.D. thesis, Univ. of Leeds.

Ramesh B. and Sridharan T. K. (1997) *Mon. Not. R. Astron. Soc., 284*, 1001–1006.

Redman M., Williams R., and Dyson J. (1998) *Mon. Not. R. Astron. Soc., 298*, 33–41.

Rengarajan T. N. and Ho P. T. P. (1996) *Astrophys. J., 465*, 363–370.

Richling S. and Yorke H. W. (1997) *Astron. Astrophys., 327*, 317–324.

Rodríguez L. F. (1999) In *Star Formation 1999* (T. Nakamoto, ed.), pp. 257–262. Nobeyama Radio Observatory, Nagano.

Roger R. S., McCutcheon W. H., Purton C. R., and Dewdney P. E. (2004) *Astron. Astrophys., 425,* 553–567.

Roshi A., Balser D. S., Bania T. M., Goss W. M., and De Pree C. G. (2005) *Astrophys. J., 625,* 181–193.

Sewilo M., Churchwell E., Kurtz S., Goss W., and Hofner P. (2004) *Astrophys. J., 609,* 285–299.

Sewilo M., Churchwell E., Kurtz S., Goss W. M., and Hofner P. (2005) In *Massive Star Birth: A Crossroads in Astrophysics* (R. Cesaroni et al., eds.), IAU Symposium 227 Poster Proceedings, www.arcetri.astro.it/iaus227/posters/sewilo_m.pdf.

Shu F. (1992) *The Physics of Astrophysics, Volume II, Gas Dynamics,* Univ. Science Books, Mill Valley, California.

Shull J. M. (1980) *Astrophys. J., 238,* 860–866.

Sicilia-Aguilar A., Hartmann L. W., Szentgyorgyi A. H., Fabricant D. G., Furész G., et al. (2005) *Astron. J., 129,* 363–381.

Sim S. A., Drew J. E., and Long K. S. (2005) *Mon. Not. R. Astron. Soc., 363,* 615–627.

Smith L. J., Norris R. P. F., and Crowther P. A. (2002) *Mon. Not. R. Astron. Soc., 337,* 1309–1328.

Smith N., Jackson J. M., Kraemer K. E., Deutsch L. K., Bolatto A., et al. (2000) *Astrophys. J., 540,* 316–331.

Snell R. L. and Bally J. (1986) *Astrophys. J., 303,* 683–701.

Sollins P., Zhang Q., Keto E., and Ho P. (2004) *Astrophys. J., 624,* L49–L52.

Sollins P., Zhang Q., Keto E., and Ho P. (2005) *Astrophys. J., 631,* 399–410.

Spitzer L. Jr. (1978) *Physical Processes in the Interstellar Medium,* Wiley, New York.

Stecklum B., Brandl B., Henning Th., Pascucci I., Hayward T. L., and Wilson J. C. (2002) *Astron. Astrophys., 392,* 1025–1029.

Tan J. C. and McKee C. F. (2003) In *Star Formation at High Angular Resolution,* IAU Symposium 221, Poster #276 (astro-ph/0309139).

Terebey S., Shu F., and Cassen P. (1984) *Astrophys. J., 286,* 529–551.

Turner B. E. and Matthews H. E. (1984) *Astrophys. J., 277,* 164–180.

Ulrich R. (1976) *Astrophys. J., 210,* 377–391.

van der Walt D. J. (1997) *Astron. Astrophys., 322,* 307–310.

Walsh A. J., Burton M. G., Hyland A. R., and Robinson G. (1998) *Mon. Not. R. Astron. Soc., 301,* 640–698.

Watson A. M. and Hanson M. M. (1997) *Astrophys. J., 490,* L165–L169.

Watt S. and Mundy L. G. (1999) *Astrophys. J. Suppl., 125,* 143–160.

Williams R. J. R., Dyson J. E., and Hartquist T. W. (2000) *Mon. Not. R. Astron. Soc., 314,* 315–323.

Wolfire M. and Cassinelli J. (1987) *Astrophys. J., 319,* 850–867.

Wood D. O. S. and Churchwell E. (1989a) *Astrophys. J. Suppl., 69,* 831–895.

Wood D. O. S. and Churchwell E. (1989b) *Astrophys. J., 340,* 265–272.

Wood D. O. S. and Churchwell E. (1991) *Astrophys. J., 372,* 199–207.

Wright A. E. and Barlow M. J. (1975) *Mon. Not. R. Astron. Soc., 170,* 41–51.

Wyrowski F., Schilke P., Walmsley C. M., and Menten K. M. (1999) *Astrophys. J., 514,* L43–L46.

Xie T., Mundy L. G., Vogel S. N., and Hofner P. (1996) *Astrophys. J., 473,* L131–L134.

Yorke H. W. and Welz A. (1996) *Astron. Astrophys., 315,* 555–564.

Yorke H. W., Tenorio-Tagle G., and Bodenheimer P. (1983) *Astron. Astrophys., 127,* 313–319.

Zhu Q.-F., Lacy J. H., Jaffe D. T., Greathouse T. K., and Richter M. J. (2005) *Astrophys. J., 631,* 381–398.

Disks Around Young O-B (Proto)Stars: Observations and Theory

R. Cesaroni and D. Galli
INAF–Osservatorio Astrofisico di Arcetri

G. Lodato
Institute of Astronomy

C. M. Walmsley
INAF–Osservatorio Astrofisico di Arcetri

Q. Zhang
Harvard-Smithsonian Center for Astrophysics

Disks are a natural outcome of the star-formation process, in which they play a crucial role. Luminous, massive stars of spectral type earlier than B4 are likely to be those that benefit most from the existence of accretion disks, which may significantly reduce the effect of radiation pressure on the accreting material. The scope of the present contribution is to review the current knowledge about disks in young high-mass (proto)stars and discuss their implications. The issues of disk stability and lifetime are also discussed. We conclude that for protostars of less than $\sim 20\,M_\odot$, disks with mass comparable to that of the central star are common. Above this limit the situation is unclear and there are no good examples of proto O4–O8 stars surrounded by accretion disks: In these objects only huge, massive, toroidal, nonequilibrium rotating structures are seen. It is clear, on the other hand, that the observed disks in stars of 10–$20\,M_\odot$ are likely to be unstable and with short lifetimes.

1. INTRODUCTION

Disk formation seems to be the natural consequence of star formation. For low-mass stars, this is attested to by the existence of pre-main-sequence (PMS) star disks as shown by a number of publications, including several reviews in *Protostars and Planets IV* (see, e.g., *Mathieu et al., 2000; O'Dell and Wen, 1994; O'Dell, 2001; Guilloteau et al., 1999; Simon et al., 2000; Mundy et al., 2000; Calvet et al., 2000; Hollenbach et al., 2000; Wilner and Lay, 2000*). Noticeably, all these refer only to disks in low- to intermediate-mass protostars, while the topic of disks in OB stars is reviewed for the first time in the present contribution. Millimeter observations of disks around young early-B stars (*Natta et al., 2000; Fuente et al., 2003*) give upper limits (or in one case a detection) less than $0.001\times$ the mass of the star, in contrast to PMS stars of spectral type B5 and later. Thus the evidence is that O and early-B stars lose any disks they may originally have had prior to becoming optically visible after roughly 10^6 yr. This article aims at discussing the question of what happened in those first million years. That in turn depends on understanding what happens in the embedded phase where we depend on millimeter and infrared measurements to detect disks. Before getting to this, it is worth noting that there is evidence for disks around what one presumes to be embedded young O-B stars with characteristics similar to those of the BN object in Orion (e.g., *Scoville et al., 1983; Bik and Thi, 2004*). This comes from high-reso-

lution spectra of the CO overtone bands and suggests that the lines are emitted in a disk a few AU from the star at temperatures above 1500 K. Note that disks of this size would not be detectable by the millimeter observations mentioned earlier. Nevertheless, it seems likely that also diskson AU scales disappear rapidly. *Hollenbach et al.* (2000) summarize disk-dispersal mechanisms, and many of these are more effective in the high-pressure cores where OB stars form. Stellar wind stripping and photoevaporation occur naturally in the neighborhood of hot massive stars, whereas disks on AU scales dissipate rapidly due to viscous accretion.

On the other hand, massive stars have to form, and one may need to accrete through a disk in order to form them. This is mainly because the most obvious barrier to massive star formation is the action of radiation pressure on infalling gas and dust (e.g., *Kahn, 1974; Wolfire and Cassinelli, 1987*). Current models (*Jijina and Adams, 1996; Stahler et al., 2000; Yorke and Sonnhalter, 2002; Yorke, 2004a; Krumholz et al., 2005*) suggest that the most effective way of overcoming radiation pressure is to accrete via a disk. In this way, however, forming a star requires loss of angular momentum and one way of doing this is via an outflow that may be powered by a wind (*Shu et al., 2000; Königl and Pudritz, 2000*). As discussed in section 2.1, there is evidence for outflows from young massive (proto)stars (*Richer et al., 2000*) and thus implicitly evidence for disks. Thus it seems likely that disks are an essential part of high-mass star formation but that they are short-lived and hence, by inference, high-mass star formation is a rapid process. In fact, current models (*McKee and Tan*, 2003; see also chapter by Beuther et al.) of the formation of high-mass stars in clusters suggest timescales on the order of 10^5 yr and accretion rates upward of 10^{-4} $M_\odot$ yr^{-1}.

One difficulty in understanding the observations of massive protostars and their associated disks is our ignorance of the evolution of the central star itself. One can imagine scenarios where the accreting protostar evolves rapidly to the zero-age main sequence (ZAMS) and others where the protostellar radius remains larger than the ZAMS value (and thus the effective temperature smaller) for a protracted period of time (e.g., *Nakano et al.*, 2000). In this latter case, the accretion luminosity is a considerable fraction of the bolometric luminosity. This alleviates somewhat the radiation pressure problem and postpones the photoevaporation of the disk. It also may explain why some luminous young stellar objects (YSOs) appear to have a very low Lyman continuum output as measured by the observed radio free-free emission (e.g., *Molinari et al.*, 1998; *Sridharan et al.*, 2002). In any case, the uncertainty about the central star properties gives rise to an additional uncertainty in our discussion of the surrounding disk.

Since it is useful to be guided in the first place by the observations, in section 2 of this review we summarize the observations of disks around massive protostars; by massive, we mean luminosities larger than ~10^3 $L_\odot$, corresponding on the ZAMS to a B4 spectral type (~8 $M_\odot$). This is roughly the point where the accretion (free-fall) timescale equals the Kelvin-Helmholtz timescale (e.g., *Palla and Stahler*, 1993; *Stahler et al.*, 2000). Conversely, all stars below ~8 $M_\odot$ will be called hereafter "low-mass," so we will *not* use the expression "intermediate-mass," commonly adopted in the literature for stars in the range 2–8 $M_\odot$. It is worth noting that the threshold of 8 $M_\odot$ is not to be taken at face value: This corresponds to an accretion rate of 10^{-5} $M_\odot$ yr^{-1}, while for higher or lower rates the critical mass may change significantly (*Palla and Stahler*, 1993). We will thus also consider as massive (proto)stars with slightly lower masses than 8 $M_\odot$.

In section 3, we summarize the data for the well-studied "disk" around the protostar IRAS 20126+4104, which we take to be our prototype. In sections 4 to 6, we discuss the evolution, stability, and lifetime of massive disks around massive stars. We illustrate the difficulties with classical disk theory to explain the accretion rates needed to form massive stars on short timescales and that thus "unconventional approaches" (perhaps involving companions) might be required. In section 7, we give a brief summary and draw our conclusions.

2. OBSERVATIONAL ASPECTS

In this section we illustrate the techniques adopted to search for disks in regions of high-mass star formation and the results obtained, with special attention to the physical properties of the disks. Eventually, we will draw some conclusions about the apparent scarcity of detections, in an attempt to decide whether this is due to instrumental limitations or instead related to the formation mechanism.

As already mentioned, the first step to undertake is the definition of "disk." Theoretically a disk may be defined as a long-lived, flat, rotating structure in centrifugal equilibrium. Although not easy to fulfill on an observational ground, this criterion may inspire a more "pragmatic" definition.

A morphological definition based on the disk geometry may not only be difficult but even ambiguous, given that massive YSOs (and hence their disks) are deeply embedded in the parental clumps. This makes the distinction between disk and surrounding envelope hard to establish, even at subarcsecond resolution. Disks could be identified from their spectral energy distributions (SEDs), but such an identification would be model dependent. The most reliable criterion is based on the velocity field of the rotating disk: An edge-on disk should be seen as a linear structure with a systematic velocity shift along it. However, velocity gradients may be determined also by infalling or outflowing gas, not only by rotation. To discriminate among these, one possibility is to take note of the fact that disks are probably always associated with large-scale jets/outflows, ejected along their rotation axes: This is likely true also for disks in high-mass YSOs, as the disk-outflow association is well established in a large variety of environments ranging from disks in low-mass protostars (e.g., *Burrows et al.*, 1996) to those in active galactic nuclei (e.g., *Krolik*, 1999). One may therefore compare the direction of the velocity gradient detected on a small scale (≤0.1 pc), which traces the putative disk, to that of the molecular flow (seen over ≤1 pc): If the two are parallel, the disk hypothesis can be ruled out, whereas perpendicularity supports the disk interpretation.

In conclusion, one may use as indicative of the presence of a disk the existence of a small (≤0.1 pc) molecular core, located at the geometrical center of a bipolar outflow, with a velocity gradient perpendicular to the outflow axis. This "observer's definition" is to be taken as a necessary condition: A satisfactory disk identification not only needs more robust observational evidence, but must satisfy theoretical constraints related to disk stability and lifetime. These will be discussed in sections 4 through 6.

The association between disks and outflows is very important. The fact that outflows turn out to be common in star-forming regions of all masses (*Shepherd*, 2003) suggests indirectly that disks exist also in massive YSOs. The outflow detection rate also allows one to obtain a quantitative estimate of the disk number in such regions. This is discussed in the following section.

2.1. The Outflow-Disk Connection and the Expected Number of Disks

During the last decade a number of systematic searches for molecular outflows have been carried out toward a variety of candidate high-mass YSOs. After the work by *Shepherd and Churchwell* (1996a,b), who searched for CO(1–0) line wings in a sample of ultracompact (UC) HII regions,

TABLE 1.   List of tracers used to search for disks in high-mass YSOs.

Tracer	References
CH$_3$OH masers	*Norris et al.* (1998); *Phillips et al.* (1998); *Minier et al.* (1998, 2000); *Pestalozzi et al.* (2004); *Edris et al.* (2005)
OH masers	*Hutawarakorn and Cohen* (1999); *Edris et al.* (2005)
SiO masers	*Barvainis* (1984); *Wright et al.* (1995); *Greenhill et al.* (2004)
H$_2$O masers	*Torrelles et al.* (1998); *Shepherd and Kurtz* (1999)
IR, mm, cm continuum	*Yao et al.* (2000); *Shepherd et al.* (2001); *Preibisch et al.* (2003); *Gibb et al.* (2004a); *Chini et al.* (2004); *Sridharan et al.* (2005); *Jiang et al.* (2005); *Puga et al.* (2006)
NH$_3$, C^{18}O, CS, C^{34}S, CH$_3$CN, HCOOCH$_3$	*Keto et al.* (1988); *Cesaroni et al.* (1994, 1997, 1998, 1999a, 2005); *Zhang et al.* (1998a,b, 2002); *Shepherd and Kurtz* (1999); *Olmi et al.* (2003); *Sandell et al.* (2003); *Gibb et al.* (2004b); *Beltrán et al.* (2004, 2005); *Beuther et al.* (2004b, 2005)

subsequent surveys were made by *Osterloh et al.* (1997), *Beuther et al.* (2002a), and *Zhang et al.* (2001, 2005) toward IRAS selected sources, and *Codella et al.* (2004) in UCHII regions and maser sources. The estimated outflow detection rate ranges from 39% to 90%, demonstrating that the phenomenon is ubiquitous in massive star-forming regions (SFRs). The parameters of these flows, such as mass, momentum, and mass loss rate, are at least an order of magnitude greater than those of outflows from low-mass YSOs. Nevertheless, this fact by itself does not demonstrate that the massive outflows are powered by massive YSOs, but they might arise from a cluster of low-mass stars. Based on knowledge of the properties of outflows from low-mass YSOs and assuming a Salpeter initial mass function, *Zhang et al.* (2005) have actually shown that a collection of outflows from a cluster of low-mass stars may explain the total outflow mass observed. However, the same authors conclude that this scenario is unlikely, as a randomly oriented sample of outflows should not produce a clear bipolarity such as that observed in massive flows.

In conclusion, single-dish observations indicate that outflows from massive YSOs are common. This result is confirmed by high-angular resolution observations of a limited number of objects (see, e.g., *Martí et al.*, 1993; *Cesaroni et al.*, 1997, 1999a; *Hunter et al.*, 1999; *Shepherd et al.*, 2000; *Beuther et al.*, 2002b, 2003, 2004a; *Fontani et al.*, 2004; *Su et al.*, 2004) where interferometric observations have resolved the structure of several outflows, proving that their properties are indeed different from those of flows in low-mass SFRs.

An important consequence is that disks must also be widespread in massive SFRs, if the (somewhat arbitrary) assumption that disks and outflows are strictly associated is correct. Only direct observations of the circumstellar environment on a subarcsecond scale may prove this conclusion. In the next section we will illustrate the methods used to search for circumstellar disks in massive YSOs.

2.2.   The Search for Disks in Massive Young Stellar Objects

Identifying candidate disks in massive (proto)stars requires careful selection of targets and tracers, to overcome the problems related to observations of massive YSOs. The most important of these are the large distance (a few kpc, i.e., ~10× those to low-mass objects) and richness of the environment (massive stars form in clusters), which often complicates the interpretation of the results. Consequently, sensitive and high-angular-resolution observations are needed. Albeit difficult to establish during the protostellar phase, the luminosities of massive YSOs must be significantly larger than those of low-mass stars, so that a reasonable lower limit to search for massive YSOs is ~10^3 L$_\odot$. In an attempt to bias the search toward young sources, additional criteria may be applied, such as association with maser sources (mainly water and methanol), nondetection of free-free emission from associated HII regions, and/or constraints on the IRAS colors to filter out more evolved objects. Finally, the presence of molecular outflows may be used to identify sources that are potentially more suited to host disks, for the reason discussed in the previous section. Examples of catalogs of massive protostellar candidates selected according to the previous criteria are those by *Churchwell et al.* (1990), *Plume et al.* (1992), *Tofani et al.* (1995), *Molinari et al.* (1996), and *Sridharan et al.* (2002).

As for the choice of tracers, disks are located in the densest and hottest part of the molecular clump. It is hence necessary to look for optically thin, high-temperature tracers, such as continuum emission at (sub)millimeter wavelengths and line emission from high-energy levels of low-abundance molecular species. In Table 1, we summarize the main techniques used to search for disks in massive YSOs and a number of relevant references. These may be classified in four categories depending on the tracer adopted.

2.2.1. Continuum emission. The dust in the disk is bound to emit as a gray body, characterized by temperatures from a few 10 K to several ~100 K. This means that the continuum emission must peak at far-IR wavelengths. However, at present, (sub)arcsecond imaging is possible only in the (sub)millimeter and mid-IR regimes, due to instrumental limitations and poor atmospheric transparency that make (sub)arcsecond imaging impossible in the far-IR. Multiwavelength information is nevertheless helpful to identify the possible contribution from ionized stellar winds to the continuum emission, which may be confused with dust (i.e., disk) emission, especially at 7 mm (see, e.g., *Gibb et al.,* 2004a). The major problem with continuum imaging is related to the fact that massive young (proto)stars, unlike their low-mass counterparts, are still deeply embedded in their parental clumps; this makes it very difficult to decouple the disk from the surrounding envelope, even at subarcsecond resolutions. The problem may be less critical in the near-IR, where the large extinction in the plane of the disk makes it possible to see the disk (if close to edge-on) as a dark silhouette against the bright background. Examples of this are found in M17 (*Chini et al.,* 2004), IRAS 20126+4104 (*Sridharan et al.,* 2005), and G5.89–0.39 (*Puga et al.,* 2006). Also, near-IR polarimetric imaging may be helpful in some cases (*Yao et al.,* 2000; *Jiang et al.,* 2005). Notwithstanding a few encouraging results, continuum observations alone can hardly achieve convincing evidence of the presence of disks: Line observations are needed in addition and are necessary to study the rotation velocity field.

2.2.2. Maser lines.

Maser emission is concentrated in narrow (≤1 km s⁻¹), strong (up to 10^6 Jy) lines and arises from regions ("spots") that may be as small as a few AU (e.g., *Elitzur,* 1992). These characteristics make it suitable for milliarcsecond resolution studies, which can investigate the distribution, line-of-sight velocities, and even proper motions of the maser "spots." One may thus obtain a detailed picture of the kinematics of the circumstellar environment on scales (a few 10 AU) not accessible by any other means. The shortcoming of this technique is that the strength of maser lines is very sensitive to a variety of factors (amplification path, inversion conditions, velocity gradients, etc.), which make it very hard if not impossible to derive useful physical parameters (temperature, density) from the observables. As shown in Table 1, a large number of tentative disk detections are reported in the literature using maser observations. In all cases the authors find a linear distribution of maser spots with velocity changing systematically along it. This is taken as an indication of rotation in an edge-on disk. As discussed at the beginning of section 2, such an evidence must be corroborated by the presence of a jet/outflow perpendicular to the maser distribution, which is not always the case. It hence remains unclear whether the velocity gradients observed are related to rotation, infall, or expanding motions. In fact, the interpretation of CH_3OH masers as disks rotating about massive YSOs (see, e.g., *Norris et al.,* 1998; *Pestalozzi et al.,* 2004), is questioned by the small stellar masses implied by the rotation curves (≤1 M⊙) and by the fact that in many cases the maser spots are aligned along the direction of a large-scale H_2 jet (*De Buizer,* 2003), suggesting that the spots could be part of the jet. A similar conclusion is attained from mid-IR imaging observation of the putative disk in NGC 7538 IRS 1 (*De Buizer and Minier,* 2005). Nonetheless, interesting examples of convincing disks in massive YSOs have been found. Using SiO masers, *Greenhill et al.* (2004) have imaged an expanding and rotating disk in the Orion KL region, while hydroxyl masers have been observed to trace rotation in the disk associated with the massive source IRAS 20126+4104 (*Edris et al.,* 2005). Also, the putative disk detected toward Cep A HW2 in the H_2O maser, Very Large Array (VLA) study of *Torrelles et al.* (1998) has found recent confirmation in the Submillimeter Array (SMA) observations of *Patel et al.* (2005).

2.2.3. Thermal lines.

As explained above, convincing evidence of a disk can be obtained by comparing its orientation to that of the associated jet/outflow. The latter may be revealed by mapping the wing emission of lines such as those of CO and its isotopomers, HCO^+, SiO, and other molecular species. A disk tracer to be observed simultaneously with one of these outflow tracers would hence be of great help for disk surveys. This must be searched among high-excitation transitions of rare molecular species, which are the most suitable to trace the high-density, optically thick, hot gas in the plane of the disk. Experience has shown that species such as NH_3, CH_3CN, $HCOOCH_3$, and $C^{34}S$ are very effective for this purpose. In particular, the proximity in frequency of CH_3CN to ^{13}CO makes the former an ideal tool to study the disk *and* the outflow, simultaneously. With this technique, interferometric observations at millimeter and (thanks to the advent of the SMA) submillimeter wavelengths have been carried out, leading to the detection of an ever-growing number of disk candidates in massive YSOs. A handful of sources has been observed also at centimeter wavelengths, with the VLA: The main advantage in this regime is the possibility of observing strong absorption lines against the bright continuum of background UCHII regions. Although successful, thermal-line observations must be complemented by diverse imaging and spectroscopy at various wavelengths to safely identify a disk: An illuminating demonstration of the effectiveness of this type of synergy is provided by the case of IRAS 20126+4104, which will be discussed in detail in section 3.

2.3. Evidence for Disks in High-Mass Young Stellar Objects

In recent years, the disk searches toward massive YSOs described in the previous section have detected an ever-growing number of candidates. The main discriminating feature between these disks and those in pre-main-sequence low-mass stars is the ratio between disk and stellar mass. While in the latter such a ratio is <0.1, for massive YSOs the disk mass becomes comparable to or even greater than that of the star. This poses a question about the stability and lifetime of the disks, which will be discussed in detail in section 4. Here, we note that candidate disks may be roughly divided into two types: those having mass M in excess of several 10 M⊙, for which the ratio M/M⋆ ≫ 1, and those with M ≤ M⋆. As noted by *Cesaroni* (2005a) and *Zhang* (2005), the former have in all cases luminosities typical of ZAMS O stars, whereas the latter are more likely associated with B-type stars.

Typically, the massive, rotating structures observed in YSOs with luminosities ≥10^5 L⊙ have radii of 4000–30,000 AU, masses of 60–500 M⊙, and observed rotation speeds of a few km s⁻¹. Examples are G10.62–0.38 (*Keto et al.,* 1988), G24.78+0.08 (*Beltrán et al.,* 2004, 2005), G28.20.0.05 (*Sollins et al.,* 2005b), G29.96.0.02 (*Olmi et al.,* 2003), G31.41+0.31 (*Cesaroni et al.,* 1994; *Beltrán et al.,* 2004, 2005), IRAS 18566+0408 (Zhang et al., in preparation), and NGC 7538S (*Sandell et al.,* 2003). These objects trans-

TABLE 2.   List of candidate disks in high-mass (proto)stars.

Name	L(L$_\odot$)	M(M$_\odot$)	R(AU)	M$_\star$(M$_\odot$)	Ṁ$_{out}$(M$_\odot$ yr^{-1})	t$_{out}$(10^4 yr)	References
AFGL490	2.2–4 × 10^3	3–6	≤500	8–10	6.2 × 10^{-4}	0.95	1,2
G192.16–3.82	3 × 10^3	15	500	6–10	5.6 × 10^{-4}	17	3,4
AFGL5142	4 × 10^3	4	1800	12	1.6 × 10^{-3}	2	5,6
G92.67+3.07	5 × 10^3	12	14400	4–7.5	1.7 × 10^{-4}	0.35	7
Orion BN	2.5 × 10^3–10^4	?	500	7–20*	10^{-6}	?	8,9
Orion I	4 × 10^3–10^5	?	500	≥6	10^{-3}	?	10,11
IRAS 20126+4104	~10^4	4	1600	7	8.1 × 10^{-4}	6.4	12,13,14,15,16
G35.2.–0.74N	10^4	0.15	1500	4–7	3 × 10^{-3}	2	17,18,19
Cep A HW2	~10^4	1–8	400–600	15	3 × 10^{-5}	3	20,21; but see 22
AFGL 2591	2 × 10^4	0.4–1.8	500	16	5.8 × 10^{-4}	6	23,24,25
IRAS 18089–1732	6 × 10^4	12–45	1000	<25	?	4	26,27
M17	?	4–>110	7500–20,000	<8–20	?	?	28,29

*See *Hillenbrand et al.* (2001) and references therein; values depend on adopted mass-luminosity conversion.

References: [1] *Schreyer et al.* (2002, 2006); [2] *Mitchell et al.* (1992); [3] *Shepherd and Kurtz* (1999); [4] *Shepherd et al.* (2001); [5] *Zhang et al.* (2002); [6] *Hunter et al.* (1999); [7] *Bernard et al.* (1999); [8] *Jiang et al.* (2005); [9] *Scoville et al.* (1983); [10] *Greenhill et al.* (2004); [11] *Genzel and Stutzki* (1989); [12] *Cesaroni et al.* (1997); [13] *Zhang et al.* (1998b); [14] *Cesaroni et al.* (1999a); [15] *Cesaroni et al.* (2005); [16] *Shepherd et al.* (2000); [17] *Hutawarakorn and Cohen* (1999); [18] *Fuller et al.* (2001); [19] *Dent et al.* (1985); [20] *Gómez et al.* (1999); [21] *Patel et al.* (2005); [22] Comito et al. (in preparation) challenge the disk interpretation; [23] *Hasegawa and Mitchell* (1995); [24] *van der Tak and Menten* (2005); [25] *van der Tak et al.* (2006); [26] *Beuther et al.* (2004b); [27] *Beuther et al.* (2005); [28] *Chini et al.* (2004); [29] *Sako et al.* (2005).

fer mass to the star at a rate of 2 × 10^{-3}–2 × 10^{-2} M$_\odot$ yr^{-1} (*Zhang*, 2005, and references therein); for the mass range quoted above, these correspond to timescales of ~10^4 yr. Note that this is an order of magnitude less than the typical rotation period at the outer disk radius of ~10^5 yr. This fact indicates that these massive rotating structures cannot be centrifugally supported, and should be considered as transient, nonequilibrium evolving structures. These objects might be observational examples of either the "pseudodisks" discussed by *Galli and Shu* (1993a,b) (in which, however, the flattening is induced by magnetic pinching forces) or the transient structures seen in numerical simulations of competitive accretion (*Bonnell and Bate,* 2005; see also chapter by Bonnell et al.). On the other hand, "true disks," which extend on much smaller scales, are more likely to be close to an equilibrium state where gravity from the central star (and the disk itself) is balanced by rotation, and, possibly, by radiative forces. Clearly, a stability analysis makes sense only for this class of objects. In section 4, we will therefore discuss the stability of massive disks, with the aim of providing some insight into the properties of these objects.

It is advisable to use a different terminology for non-equilibrium massive rotating structures: Following *Cesaroni* (2005a), we will call them "toroids." What is the role of toroids in the process of high-mass star formation? Given the large masses and luminosities involved, they likely host a stellar cluster rather than a single star. That is why some authors have used the term "circumcluster" toroids, as opposed to "circumstellar" disks (*Beltrán et al.,* 2005). Since their size is several times the centrifugal radius [i.e., the radius at which a centrifugal barrier occurs (see, e.g., *Terebey et al.,* 1984)], one may speculate that eventually toroids will

fragment into smaller accretion disks rotating about single stars or binary systems (*Cesaroni,* 2005b). In the following we will not consider toroids but only "true" circumstellar (or circumbinary) disks in massive YSOs.

Table 2 lists disk candidates associated with high-mass YSOs. The columns are luminosity, disk mass, disk radius, mass of the central star, outflow mass loss rate, and outflow dynamical timescale. When considering this table, several caveats are in order. Although most of the entries are to be considered bona fide disks, in some cases the interpretation is challenged [e.g., Cep A HW2 (Comito et al., in preparation)]. Also, some of the values given in the table are quite uncertain, in particular the mass of the star. While in a few cases (e.g., IRAS 20126+4104) this is obtained directly from the Keplerian rotation curve of the disk, in the majority of the objects the estimate comes from the luminosity or from the Lyman continuum of the YSO. The conversion from these quantities to stellar masses is very uncertain as it depends on the unknown evolutionary stage of the (proto)star, on the optical depth and origin of the radio emission (HII region or thermal jet), on the number of Lyman continuum photons absorbed by dust, and on the multiplicity of the system. Hence, the luminosities are affected by significant errors and the mass-luminosity relationship for this type of objects is unclear. We have thus decided to adopt the values quoted in the literature without any further analysis.

Despite the large uncertainties, one may conclude that all objects have luminosities typical of B-type ZAMS stars, i.e., below a few 10^4 L$_\odot$. Albeit very difficult to establish, accretion rates for disks seem to be ~10^{-4} M$_\odot$ yr^{-1} (*Zhang,* 2005): This implies a time of ~10^5 yr to transfer the material from the disk to the star, significantly larger than the rotation period (~10^4 yr), thus allowing the disk to reach centrifugal equilibrium. Whether this will be stable or not is a complicated issue that we will deal with in section 4. Here, we just point out that the disk timescale is comparable to the typical freefall time of molecular clumps hosting massive star formation, which in turn equals the star-formation timescale predicted by recent theoretical models [~10^5 yr (*Tan and McKee,* 2004)]. This seems to suggest that the material lost by the disk through infall onto the star and ejection in the jet/outflow is continuously replaced by fresh gas accreted from the surrounding envelope.

Naively, one may thus conclude that disks fit very well in the high-mass star-formation scenario. However, this seems to hold only for B-type stars, as the stellar masses quoted in Table 2 are never in excess of ~20 M$_\odot$ and the luminosities on the order of ~10^4 L$_\odot$, typical of stars later than B0. What about disks in more luminous/massive objects? Disk searches performed in sources with luminosities typical of O-type stars (i.e., above ~10^5 L$_\odot$) resulted in negative detections; in all cases, as already discussed, massive rotating toroids were found. One possibility is that the latter "hide" true disks in their interiors, still not detectable due to insufficient angular resolution and sensitivity. It must be taken into account that objects that luminous are mostly located at significantly larger distances than B stars. A noticeable exception is Orion, the closest high-mass star-forming region (450 pc), whose total luminosity is ~10^5 L$_\odot$. The presence of a disk rotating about source I has been clearly established by VLBI observations of the SiO masers (see Table 2). However, it remains unclear which fraction of the total luminosity is to be attributed to this object. Thanks to their recent high-frequency observations with the SMA, *Beuther et al.* (2006) have interpreted the continuum spectrum of source I as the combination of free-free emission from a uniform HII region plus thermal emission from dust. According to their fit, one expects an optically thin flux of ~45 mJy at ~200 GHz, corresponding to a B1 ZAMS star. On the other hand, proper-motion measurements of source I and BN (*Rodríguez,* 2005) have demonstrated that the two are receding from each other with speeds of 12 and 27 km s^{-1} respectively. Since the estimated mass of BN lies in the range ~7–20 M$_\odot$ (see Table 2), the ratio between the two velocities suggests a mass for source I of 16–45 M$_\odot$. While the upper limit is larger than expected for a ZAMS star of ~10^5 L$_\odot$, the lower seems roughly consistent with the estimate from the free-free emission. It is hence possible that Orion I is an early-B or late-O star. In this case, we would be still missing an example of a disk in an early-O-type star.

Is such a lack of evidence for disks in O-type stars due to an observational bias? The absence of disks in O stars could have important theoretical implications, as disks are believed to play a key role in favoring accretion onto the star through angular momentum dissipation. If no disks are present in stars more massive than ~20 M$_\odot$, alternative models of massive star formation may be needed, such as, e.g., those involving coalescence of lower-mass stars (*Bonnell and Bate,* 2005). It is hence in order to discuss all possible

technical limitations that may have hindered detection of disks in O-type stars.

2.4. Are Disks in O-type Stars Detectable?

As noted above, no clear evidence for circumstellar disks in early-O-type stars has been presented to date. But even disks in B-type stars do not seem to be easy to detect. It is difficult to draw any statistically reliable conclusion on the real number of high-mass YSOs with disks, as most studies have been conducted toward selected candidates. Only one systematic search for disks is known to date: the recent survey by Zhang et al. (personal communication). These authors used the VLA to survey 50 YSOs with luminosities between 10^3 and 10^5 L$_\odot$ in NH$_3$ inversion transitions, and detected 10 possible disks. Is a detection rate of 20% the one expected if all B-type stars form through disk accretion? And are disks in O-type stars really elusive?

We consider the possibility of an observational bias in disk searches, due to limited sensitivity, angular, and spectral resolution (in the case of line observations). Let us discuss separately the sensitivity and resolution issues.

2.4.1. Spectral and angular resolution. As explained in section 2.2, most disk searches attempt to detect the velocity gradient due to rotation about the star. To achieve such a detection, the observations must disentangle the emission of the red- and blueshifted parts of the disk, not only in space but also in velocity. One constraint is hence the angular resolution achieved by current interferometers. As for the spectral resolution, at radio wavelengths one may easily attain 0.1 km s^{-1}, sufficient to resolve lines as broad as a few km s^{-1}. The limitation is due to the line width, on the same order as the expected rotation velocity. Therefore, to detect rotation with velocity V at radius R, two conditions must be fulfilled: the separation between two diametrically opposite points must be greater than the instrumental half power beam width (HPBW), and the corresponding velocity difference must be greater than the line full width at half maximum (FWHM). These may be expressed as

$$2R > \Theta d \qquad (1)$$

$$2V(R)\sin i > W \qquad (2)$$

where i is the inclination angle with respect to the line of sight (i = 0 for a face-on disk), Θ the HPBW, d the distance to the source, and W the line FWHM. For a star with mass M$_\star$, V = $\sqrt{GM_\star/R}$, so that one obtains

$$d(\text{kpc}) < 7 \, \frac{M_\star(M_\odot)\sin^2 i}{\Theta('')W^2(\text{km s}^{-1})} \qquad (3)$$

A conservative estimate of this expression may be obtained for Θ = 1" (the HPBW of millimeter interferometers), W = 5 km s^{-1} (the typical line FWHM of a hot molecular core — to be taken as an upper limit, as the intrinsic line width is less than the observed one), and a mean value of $\langle \sin^2 i \rangle = 2/3$ assuming random orientation for the disk axes. The result is $d(kpc) < 0.19\, M_\star(M_\odot)$, which implies that the maximum distance at which a disk can be detected in, e.g., a 50-$M_\odot$ star is 9.5 kpc. This shows that disks in all massive stars should be seen up to the galactic center.

The previous analysis does not properly take into account the effect of the inclination angle, as well as other effects that may complicate the picture: infall and outflow, disk flaring, and non-Keplerian rotation. Further complication is added by the presence of multiple (proto)stars, each of these possibly associated with an outflow/disk, and by the fact that molecular species believed to be "pure" disk tracers might instead exist also in the outflow.

Naively, infall and rotation speeds should both be $\propto R^{-0.5}$ and hence comparable, so that the Keplerian pattern should be only slightly affected. However, recent observations have revealed infall with no or little rotation in the O-type (proto)stars G10.62.–0.38 (*Sollins et al.*, 2005a) and IRAS 18566+0408 (Zhang, personal communication). This occurs on spatial scales of ~1000 AU, comparable to those over which pure rotation is seen in the B-type sources of Table 2. Possible explanations must involve angular momentum dissipation, perhaps due to magnetic field braking.

Deviation from Keplerian rotation is expected to occur at large radii, where the enclosed disk mass becomes comparable to the stellar mass (see the example in section 3); in this case, however, the value of V is larger than in the Keplerian case and equation (2) is satisfied.

Outflows could "spoil" the detection of disks, because the ratio between expansion and rotation speeds may be as large as ~100. However, the molecular lines used for disk searches do not seem to trace also the outflow/jet. An illuminating example of this is given by the best-studied disk outflow/jet system to date, IRAS 20126+4104: Here the Keplerian disk [rotating at a few km s⁻¹ (*Cesaroni et al.*, 2005)] is clearly seen in the CH₃CN lines, notwithstanding the presence of a jet with velocities in excess of 100 km s⁻¹ (*Moscadelli et al.*, 2005) on the same scale as the disk. It is hence clear that the success of disk searches and the possibility of decoupling them from the associated outflows/jets depend crucially on the molecular tracers used.

Finally, disk flaring may help revealing rotation. In fact, while equation (3) depends sensitively on the inclination angle i, this result is obtained for an infinitesimally thin disk. Real disks have a finite $H(R)$ and, considering as an example IRAS 20126+4104, we can assume $H = R/2$ (see section 4.2). Under this assumption, the line of sight will lie in the plane of the disk as long as $76° < i < 104°$. For a random orientation of the disk axis, this implies that ~24% of the disks will satisfy this condition. Noticeably, such a number is very close to the 20% detection rate obtained in the previously mentioned NH₃ survey by Zhang et al. and suggests that disk inclination should not represent a serious limitation for disk searches.

2.4.2. Sensitivity. The continuum emission from disks in the (sub)millimeter regime (see also section 2.2) should

be easily detected by modern interferometers. This can be evaluated assuming a disk mass M proportional to the stellar mass $M_\star$, e.g., $M = M_\star/2$ (see Table 2) and a dust temperature of 100 K. The flux density measured at a distance d is equal to

$$S_\nu(mJy) = 85.3 \left[\frac{\nu(GHz)}{230.6} \right]^{2+\beta} M_\star(M_\odot)d^{-2}(kpc) \quad (4)$$

where a dust absorption coefficient equal to 0.005 cm² g⁻¹ $[\nu(GHz)/230.6]^\beta$ has been adopted (*Kramer et al.*, 1998). Very conservatively, one may assume a sensitivity of 30 mJy at 230.6 GHz, which sets the constraint $d(kpc) < 1.7\sqrt{M_\star(M_\odot)}$. Disks associated with stars of 10 and 50 $M_\odot$ should be detectable up to 5.4 and 12 kpc, respectively. Therefore, the main limitation is not given by the sensitivity but by confusion with the circumstellar envelope, as illustrated in section 2.2.

On the other hand, line observations are less affected by this problem, as the velocity information helps decoupling the rotating disk from the more quiescent envelope. The line intensity may be estimated under the assumption of optically thick emission. For $T \geq 100$ K, the flux density from the disk surface within a radius R must be at least

$$S_\nu = \frac{R^2 \pi B_\nu(T)}{4\pi d^2} \quad (5)$$

with B_ν black-body brightness. Here, we have assumed R^2 as a lower limit to the surface; this relies upon the fact that the disk thickness is ~R/2, so that the effective surface ranges from R^2 (edge-on) to πR^2 (face-on). In order to detect the line emission from the disk with an instrumental noise σ, one must have $S_\nu > 3\sigma$, which using equation (5) becomes $R > \sqrt{12\sigma/B_\nu(T)}d$. The radius within which the flux is measured must satisfy also equation (2). For the sake of simplicity, we assume a temperature $T = 100$ K independent of R and a noise $\sigma \simeq 0.1$ Jy at 230 GHz, thus obtaining

$$d(kpc) < 6.2 \frac{M_\star(M_\odot)\sin^2 i}{W^2(km\ s^{-1})} \quad (6)$$

For the same fiducial values adopted for equation (3), this condition takes the form $d(kpc) < 0.16\, M_\star(M_\odot)$. Once more, all stars of a few 10 $M_\odot$ should be detectable up to the galactic center.

In conclusion, albeit very rough, the previous discussion suggests that the currently available instruments should be sufficient to detect circumstellar disks around all massive stars, if the disk mass is nonnegligible with respect to the stellar mass. One must hence look for an astronomical explanation to justify the lack of detections for disks in O-type stars. As illustrated in the following, this might be found in the stability and/or lifetime of massive disks.

3. IRAS 20126+4104: THE PROTOTYPE DISK

Among the disk candidates listed in Table 2, one stands unique as the best-studied and probably most convincing example of a Keplerian disk rotating about a massive YSO. This is IRAS 20126+4104, an IRAS point source believed to be associated with the Cyg X region (*Wilking et al.,* 1989, and references therein). First recognized by *Cohen et al.* (1988) as an OH maser emitter, it was then observed in the continuum, at millimeter wavelengths (*Wilking et al.,* 1989; *Walker et al.,* 1990), and in the CO(2–1) line by *Wilking et al.* (1990), who detected a powerful, parsec-scale molecular outflow. Later on, association with H_2O and CH_3OH masers was also established (*Palla et al.,* 1991; *MacLeod and Gaylard,* 1992) and the molecular emission from high-density tracers was detected (*Estalella et al.,* 1993). These pioneering works provided evidence that the source is associated with a luminous, embedded star still in a very early stage of the evolution. The fact that no free-free continuum was detected (*Tofani et al.,* 1995) indicated that although luminous, the source had not yet developed an HII region, while the presence of a molecular outflow suggested that the YSO could be actively accreting material from the parental cloud. Eventually, observations at 3 mm with the Plateau de Bure interferometer (*Cesaroni et al.,* 1997) achieved the angular resolution needed to dissect the outflow and analyze the velocity field in the associated molecular core. Thanks to these observations, it was first recognized that the core was rotating about the outflow axis.

This finding triggered a burst of observations toward IRAS 20126+4104, thus shedding light on its properties and allowing us to draw a detailed picture that is best summarized in Fig. 1. The main features of this object may be summarized as follows:

1. The spectral energy distribution, well sampled at (sub)millimeter and mid-IR wavelengths (*Cesaroni et al.,* 1999a, and references therein), supplies us with a luminosity of $\sim 10^4$ $L_\odot$ for a distance of 1.7 kpc. Although near-IR observations of the 2.2-μm continuum reveal an embedded cluster spread over 0.5 pc (*Cesaroni et al.,* 1997), arcsecond-resolution mid-IR images (*Cesaroni et al.,* 1999a; *Shepherd et al.,* 2000; *Sridharan et al.,* 2005) prove that most of the luminosity arises from the inner 1000 AU.

2. A bipolar molecular outflow is seen in various tracers, including CO and HCO^+ (*Wilking et al.,* 1990; *Cesaroni et al.,* 1997, 1999b; *Shepherd et al.,* 2000). The kinematical outflow age is $t_{out} \simeq 6 \times 10^4$ yr, which may be taken as a lower limit to the age of the YSO powering the flow. The momentum (400 $M_\odot$ km s^{-1}) and mass-loss rate (8×10^{-4} $M_\odot$ yr^{-1}), as well as the other parameters of the flow, are typical of outflows in high-mass YSOs.

3. The orientation of the outflow axis changes by $\sim 45°$ from the large (1 pc) to the small (0.1 pc) scale. A similar behavior is observed in H_2 knots that are distributed along the outflow axis and thus describe an S-shaped pattern centered on the YSO powering the flow (*Shepherd et al.,* 2000).

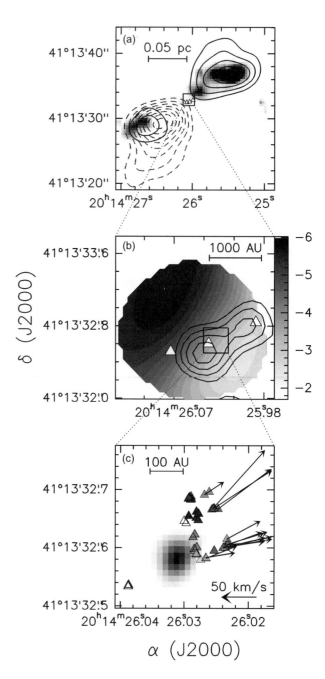

Fig. 1. Disk/outflow system in the high-mass (proto)star IRAS 20126+4104. **(a)** Overlay of the H_2 line emission at 2.2 μm (gray scale) and the bipolar outflow traced by the $HCO^+(1–0)$ line (*Cesaroni et al.,* 1997). Solid and dashed contours correspond respectively to blue- and redshifted gas. The triangles mark the positions of the H_2O maser spots detected by *Tofani et al.* (1995). **(b)** 3.6-cm continuum map [contours (*Hofner et al.,* 1999)] overlaid on a map of the velocity measured in the $C^{34}S(5–4)$ line by *Cesaroni et al.* (2005). **(c)** Distribution of the H_2O maser spots (*Moscadelli et al.,* 2000, 2005) compared to a VLA map (image) of the 7-mm continuum emission (Hofner, personal communication). The gray scale of the spots ranges from white, for the most redshifted spots, to black, for the most blueshifted. The arrows denote the absolute proper motions of the spots, measured by *Moscadelli et al.* (2005).

These features supply evidence that one is observing a precessing jet feeding the outflow, the former outlined by the shocked H_2 emission, the latter by the CO and HCO^+ line emission. *Cesaroni et al.* (2005) estimate a precession period of 2×10^4 yr, which implies that the jet/outflow has undergone at least three full precessions.

4. The collimated jet traced by the H_2 knots is also seen on the same scale in other molecular tracers such as SiO (*Cesaroni et al.*, 1999a; *Liu et al.*, 2005), NH_3 (*Zhang et al.*, 1999), and CH_3OH (*Cesaroni et al.*, 2005), although the morphology and kinematics of the gas may vary significantly from molecules tracing the preshock and those arising from the postshock material (*Cesaroni et al.*, 2005). The expansion speed of the jet over 0.5 pc is estimated ~100 km s⁻¹ (*Cesaroni et al.*, 1999a), while the component along the line of sight is relatively small, on the order of ±20 km s⁻¹. This finding and the fact that both blue- and redshifted emission overlap in space prove that the jet/outflow axis inside ≤0.5 pc from the YSO lies very close to the plane of the sky, at an angle ≤9°. Proper-motion measurements of the H_2O masers indicate a jet expansion speed in excess of ~100 km s⁻¹ at a distance of ~250 AU from the powering source (*Moscadelli et al.*, 2005), while on an intermediate scale between the one traced by H_2 (and other molecular lines) and that sampled by the H_2O masers, free-free emission from the thermal component of the jet is seen at 3.6 cm (*Hofner et al.*, 1999). All these results indicate that the jet/outflow system extends almost continuously from the neighborhoods of the YSO powering it to the outer borders of the parental molecular clump.

5. At the geometrical center of the bipolar jet/outflow, a hot molecular core of ~200 K is detected. When observed at high-angular resolution in the CH_3CN transitions, this core presents a velocity gradient roughly perpendicular to the jet. *Cesaroni et al.* (1997) interpreted this result as rotation about the YSO powering the outflow, while subsequent imaging in CH_3CN (12–11) (*Cesaroni et al.*, 1999a), NH_3 (1,1) and (2,2) (*Zhang et al.*, 1998b), and $C^{34}S(2–1)$ and (5–4) (*Cesaroni et al.*, 2005) has established that the rotation is Keplerian. However, the estimate of the stellar mass seems to differ depending on the tracer used to measure the rotation curve. In fact, going from a few 10^3 AU to 10^4 AU, such an estimate changes from 7 $M_\odot$ (*Cesaroni et al.*, 2005) to 24 $M_\odot$ (*Zhang et al.*, 1998b). This effect may be due to the disk mass enclosed inside the radius at which the velocity is measured (*Bertin and Lodato*, 1999): For small radii, the disk mass inside that radius (~4 $M_\odot$) is less than that of the star, whereas at large radii the two become comparable, thus mimicking Keplerian rotation about a bigger star. *Cesaroni et al.* (2005) have also estimated the temperature profile as a function of radius, which seems to be compatible with the "classical" law expected for geometrically thin disks heated externally by the star or internally by viscosity: $T \propto R^{-3/4}$. The mean disk temperature is on the order of ~170 K. Additional evidence of a circumstellar or possibly circumbinary disk in IRAS 20126+4104

comes also from OH and CH_3OH maser studies (*Edris et al.*, 2005), as well as from recent near- and mid-IR continuum images obtained by *Sridharan et al.* (2005). In particular, at 2.2 μm the disk is seen in absorption as a dark silhouette similar to those observed in the optical toward the proplyds in the Orion Nebula (e.g., *O'Dell and Wen*, 1994).

In conclusion, the observational results obtained for IRAS 20126+4104 provide robust evidence for the existence of a Keplerian disk associated with a precessing jet/outflow powered by a massive YSO of ~10^4 $L_\odot$. One possibility is that one is dealing with a binary system, as suggested by the jet precession, which may be caused by interaction with a companion. Indeed, the latter seems to appear as a secondary peak, close to the disk border, in the mid-IR images by *Sridharan et al.* (2005). However, it is unlikely that the mass of the companion is comparable to that of the YSO at the center of the disk, because the Keplerian pattern does not seem to be significantly perturbed by the presence of the second star. Better angular resolution and sensitivity are required to settle the binary issue.

4. FORMATION AND STABILITY OF MASSIVE DISKS

In this section, we consider some theoretical problems posed by the existence of massive (few solar masses) disks around massive protostars as in the cases listed in Table 2. One obvious question is that of their stability particularly if, as seems likely, the disk mass in some cases is comparable to the stellar mass. A second problem is posed by the high accretion rates needed both to form the massive stars on a reasonable timescale [typically 10^5 yr, according to *Tan and McKee* (2004)] and to account for the observed large outflow rates. We ask the question of whether classical "α-disk models" can account for such high accretion rates and conclude it is unlikely. Third, following the approach of *Hollenbach et al.* (2000), we consider briefly effects that might limit massive disk lifetimes in the phase just subsequent to the cessation of accretion. We commence, however, with a brief discussion of what one might naively expect to be the properties of disks associated with massive protostars.

4.1. From Clouds to Disks

From a theoretical point of view, the presence of disks around massive protostars is expected on the same physical grounds as in the case of solar-mass protostars. In a simple inside-out collapse with constant accretion rate (*Terebey et al.*, 1984), the mass of the disk increases linearly with time, $M \propto t$, whereas the disk radius increases much faster, $R \propto t^3$, or $R \propto M^3$. Thus, in principle, we expect massive, centrifugally supported disks to have much larger sizes than the low-mass disks observed around T Tauri stars.

In particular, the sizes and masses of the disks observed around young massive stars can be used to set constraints on the physical characteristics of the clouds from which they formed. It is a good approximation to express the disk angular velocity at a radius R as $\Omega(R) \approx (GM_t/R^3)^{1/2}$ (*Mestel, 1963*), where M_t is the total (star plus disk) mass of the system (the relation holds exactly only for a surface density $\Sigma \propto R^{-1}$). The specific angular momentum of the system is then $J/M_t \approx \Omega R^2 \approx (GM_t R)^{1/2} \approx 10^{-2}$ km s^{-1} pc with the values listed in Table 2 for IRAS 20126+4104. This is within the range of observed values of the specific angular momentum of molecular cloud cores in low-mass star-forming regions [$J_c/M_c \approx 10^{-3}$–10^{-1} km s^{-1} pc (*Goodman et al., 1993*)], although it is unclear if similar values pertain to massive star-forming regions as well.

In any case, the formation of disks like IRAS 20126+ 4104 from the collapse of slowly rotating clumps appears physically possible, and seems to imply that the clump's angular momentum is not reduced much during the formation of a massive disk. It is possible to obtain the radius R_i within which the mass M_t of the star plus disk system was originally contained in the parental cloud, $R_i \gtrsim (GM_t R/\Omega_c^2)^{1/4}$, where Ω_c is the cloud's angular velocity. This imposes a severe constraint on the effective sound speed in the cloud $c_{eff} \approx GM_t/2R_i \lesssim (G^3 M_t^3 \Omega_c^2/2R)^{1/8} \approx 1.2$ km s^{-1} for $\Omega_c \approx 1$ km s^{-1} pc^{-1}. Notice that the resulting accretion rate $\dot{M} \approx c_{eff}^3/G$ is $\dot{M} \approx 2 \times 10^{-4}$ M$_\odot$ yr^{-1}, which implies a timescale of ~6×10^4 yr for the formation of the IRAS 20126+4104 star plus disk system, in agreement with the inferred outflow kinematic age (see Table 2). Thus, the cloud cores where massive stars are formed must be characterized, at least in their central parts, by low or moderate levels of turbulence, similarly to their low-mass counterparts. The cold, massive ($M_c \approx 10^2$ M$_\odot$) molecular cloud cores with narrow line widths ($\Delta v \approx 1$ km s^{-1}) recently discovered by *Birkmann et al.* (2006) seem to present the required properties, and suggest that the initial conditions for the formation of massive stars should be similar to those observed in low-mass star-forming regions.

Numerical calculations of the collapse of massive molecular cores have been performed by *Yorke and Sonnhalter* (2002), who also included a detailed treatment of continuum radiation transfer. The calculations start from cold, massive, slowly rotating clouds far from equilibrium that typically contain tens of Jeans masses. The resulting mass accretion rate is strongly time-dependent, peaking at a value $\dot{M} \approx 10^{-3}$ M$_\odot$ yr^{-1} after ~10^4 yr from the onset of collapse and declining thereafter because of radiation pressure. In some cases, the infalling material flows onto a disk-like feature appearing after ~10^5 yr from the onset of collapse, which rapidly extends up to 10^4 AU in radius. The calculated density in the outer disk is ~10^{-18} g cm^{-3} and compares well with that inferred from the ratio of C^{34}S (2–1) and (5–4) lines measured in IRAS 20126+4104 at radii $R \gtrsim 2000$ AU. The calculations suggest that for disks around stars above ~20 M$_\odot$, stellar radiative forces may contribute as much as rotation to the radial support of the disk. The disk may also be short-lived, disappearing after some 10^4 yr.

4.2. Local Gravitational Stability of Massive Disks

Since *Toomre* (1964), it has been well known that a fluid, thin disk is unstable to axisymmetric gravitational disturbances if the stability parameter

$$Q = \frac{c_s \kappa}{\pi G \Sigma} < 1 \qquad (7)$$

where c_s is the sound speed, Σ the disk surface density, and κ the epicyclic frequency. The angular velocity and the epicyclic frequency are related by $\kappa = f\Omega$, where f is a numerical factor dependent on the shape of the rotation curve; for Keplerian rotation f = 1, while for a flat rotation curve f = $\sqrt{2}$.

The value of Q, the disk aspect ratio, and the total disk mass can be easily related to one another in the following way. The disk thickness H is given by

$$H = \begin{cases} c_s/\Omega & \text{if } M/M_t \ll H/R \\ c_s^2/\pi G\Sigma & \text{if } M/M_t \gtrsim H/R \end{cases} \qquad (8)$$

If we assume a power-law behavior for $\Sigma \propto R^{-1}$, so that $M(R) = 2\pi\Sigma R^2$, and we adopt the approximation $\Omega \approx (GM_t/R^3)^{1/2}$, we can rewrite Q as

$$Q = \begin{cases} \frac{2H}{R}\frac{M_t}{M} \gg 1 & \text{if } \frac{M}{M_t} \ll \frac{H}{R} \\ \left(\frac{2H}{R}\frac{M_t}{M}\right)^{1/2} \lesssim \sqrt{2} & \text{if } \frac{H}{R} \lesssim \frac{M}{M_t} \ll 1 \\ \left(\frac{4H}{R}\frac{M_t}{M}\right)^{1/2} \approx 2\sqrt{\frac{H}{R}} & \text{if } \frac{M}{M_t} \approx 1 \end{cases} \qquad (9)$$

The relationship between cumulative disk mass, aspect ratio, and stability parameter Q is also shown in Fig. 2, which shows a contour plot of Q as a function of M/M_t and of H/R based on the equation above. We wish to stress that Q is only a measure of the local stability of a massive disk; the above relationship and the figure describe the stability of the disk at a radius R, where M is the disk mass enclosed within R and the aspect ratio H/R is computed at R. We then see that, in order for the disk to be gravitationally unstable, we need $M/M_t \gtrsim 2$ H/R.

Nonaxisymmetric, spiral disturbances are generally more unstable, and thin disks can become unstable at considerably larger values of $Q \approx 3$–4. On the other hand, it is also well known that a finite thickness has a strong stabilizing effect. The general problem of finding the marginal stability of a thick disk with respect to nonaxisymmetric instabilities is, however, too complex to be treated analytically. Numerical simulations (*Lodato and Rice, 2004*, see below) seem to show that relatively thick and massive disks evolve in such a way to achieve $Q \approx 1$.

The simple estimate of Q outlined above and shown in Fig. 2 offers an easy tool to evaluate the local stability of the massive disks observed around massive stars. Let us

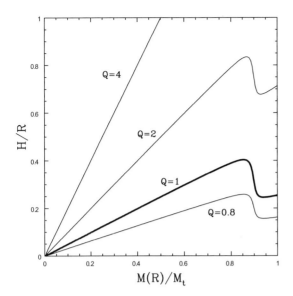

Fig. 2. Contour plot of the relationship between stability parameter Q, enclosed disk mass $M(R)/M_t$, and aspect ratio H/R. This plot describes local stability at radius R, and M has to be intended as the mass enclosed within R. The curves are computed under the assumption that $\Sigma \propto R^{-1}$. The wiggles at large disk mass indicate the transition from a Keplerian to a flat rotation curve. The thick contour corresponds to $Q = 1$, which marks the threshold between stability and instability.

take, as an example, the case of IRAS 20126+4104 and consider its properties at a radius $R \approx 1600$ AU. As we have seen before, the enclosed mass within 1600 AU is on the order of 4 $M_\odot$, while the central object mass is 7 $M_\odot$, so that $M(R)/M_t \approx 0.4$. The average temperature at R is 170 K, so that $H/R \approx 0.4$. With these estimates, the value of Q turns out to be $Q \approx 2$ (see Fig. 2). The disk is therefore expected to be stable with respect to axisymmetric disturbances but unstable with respect to spiral instabilities.

4.3. Transport Properties and Evolution of Massive Disks

From theoretical models of massive star formation (*Yorke and Sonnhalter,* 2002) we know (see above) that in order to form massive stars, the disk has to be fed at high accretion rates (on the order of 10^{-3} $M_\odot$ yr^{-1}). On the other hand, we also know from observations that this mass has to be transported at similar rates to small radii, in order to resupply the powerful outflows observed in these systems (with outflow rates in the range of 10^{-4}–10^{-3} $M_\odot$ yr^{-1}, see Table 2). The natural question that arises is therefore whether in the disk there are efficient mechanisms able to deliver the torques needed to redistribute the angular momentum within the disk at the required rates.

Torques in accretion disks are generally parameterized through the dimensionless parameter α, which measures the

strength of the viscous torques relative to the local disk pressure. The most promising mechanism of transport in disks is usually considered to be related to MHD instabilities, and in particular to the magnetorotational instability (MRI) (*Balbus and Hawley,* 1992). This kind of instability is able to provide torques with $\alpha \approx 10^{-2}$ (see *Balbus,* 2003, for a review). It is not clear, however, how efficient such processes can be in the cold outer regions of the disk, where the ionization level is expected to be small.

On the other hand, we have just shown how the massive disks observed around massive stars are likely to be gravitationally unstable. This leads to another important source of angular momentum transport, in the form of gravitational instabilities. In this context, numerical simulations play a very important role, as they are the only way to follow the dynamics of gravitationally unstable disks to the nonlinear regime. A detailed discussion of the role of gravitational instabilities is presented in the chapter by Boss et al. Here, we will only summarize the main results on this issue, in consideration of the particular properties of disks around massive stars (taking, as a prototypical example, the case of IRAS 20126+4104).

Laughlin and Bodenheimer (1994), in a pioneering work, have shown for the first time through smoothed particle hydrodynamics simulations the effectiveness of gravitational instabilities in promoting the accretion process. They found that the redistribution of matter in the disk, due to the spiral structure, occurs over a few rotational periods. However, these early simulations were limited by the particular choice of the equation of state of the gas in the disk. *Laughling and Bodenheimer* (1994), in fact, assumed that the disk is isothermal, thus inhibiting the important feedback effect on the disk stability provided by the heating of the disk due to the instability itself.

A full treatment of the disk thermodynamics, including a detailed radiation transfer through the disk, is presently impossible to achieve with current numerical techniques [although progress is being made to include a more realistic cooling (see *Johnson and Gammie,* 2003)]. The approach that has been taken more recently is to adopt some simplified prescription for the cooling of the disk and constrain the evolution under such simplified conditions (*Laughlin and Korchagin,* 1996; *Pickett et al.,* 2000; *Gammie,* 2001; *Lodato and Rice,* 2004, 2005). *Gammie* (2001) assumed that $t_{cool} = \beta\Omega^{-1}$ and found that for $\beta < 3$ the disk fragments into bound objects, whereas for $\beta > 3$ a quasisteady unstable state is reached, in which efficient redistribution of angular momentum takes place. *Lodato and Rice* (2004, 2005) have extended the analysis of *Gammie* (2001) to a full three-dimensional, global context. In this way they were able to study the effect of the development of global spiral structures on the disk evolution, which is precluded in the local simulations by Gammie. The general result of these simulations (either local or global) is that the gravitational instability saturates at an amplitude such that the dissipation provided by the instability is able to balance the imposed cooling. The typical values of α found in these simulations are in the range 0.01–0.06. More recently, *Rice et al.* (2005) have elucidated the process of fragmentation in massive disks, finding that a self-gravitating disk can provide a stress no larger than $\alpha \approx 0.06$. If the cooling time is so short that the dissipation required to balance it is larger than this maximum, the response of the disk is to fragment.

We can now estimate the maximum accretion rate expected for a centrifugally supported accretion disk. Assuming a density profile $\Sigma \propto R^{-p}$, we have

$$\dot{M} = 3\pi\nu\Sigma \approx \frac{\alpha}{2-p}\left(\frac{H}{R}\right)^2 M(R)\Omega(R) \qquad (10)$$

where $\nu = \alpha H^2 \Omega$ is the viscosity, expressed through the *Shakura-Sunyaev* (1973) prescription, and $M(R)$ is the disk mass enclosed within R. Let us consider, as an illustration, the case of IRAS 20126+4104, where the mass enclosed within R = 1600 AU is M = 4 $M_\odot$, and the aspect ratio is $H/R \approx 0.4$ (see section 4.2). With $\alpha = 0.06$ and p = 1 we find that the maximum accretion rate that can be delivered by gravitational instabilities is $\dot{M} \approx 10^{-5}$ $M_\odot$ yr^{-1}, much smaller than the values required to power the outflow (the MRI instability can only provide lower accretion rates). It is important to stress that the above numerical estimate for the mass accretion rate depends on the specific radius at which we evaluate the enclosed mass and the angular velocity at that radius. The "problem" that we are referring to here refers to the difficulty of transferring to the star ~4 $M_\odot$ from a distance of ~1600 AU through a standard accretion disk. The problem may be alleviated (or eliminated) if the density distribution in the disk is very steep (p $\approx$ 2) and most of the disk mass is concentrated at small radii where the rotation period is shorter and therefore the disk's dynamical evolution is faster.

The required high accretion rates pose in general a serious challenge for theoretical models. However, we must keep in mind that the accretion rates predicted by hydrodynamical or magnetohydrodynamical disk simulations is highly variable in time, and may be temporarily enhanced by an order of magnitude above the average value. For example, *Lodato and Rice* (2004, 2005) have compared simulations of self-gravitating disks of low (M $\ll$ M$_\star$) and high mass (M $\approx$ M$_\star$), finding that low- and high-mass disks behave somewhat differently. In the low-mass case a self-regulated state is rapidly achieved and the evolution of the disk after the development of the gravitational instability is quasi-steady. On the other hand, as M approaches M$_\star$, the temporal behavior of the disk becomes more complex and a series of recurrent episodes of spiral activity are often seen, lasting for roughly one rotation period on the order of ~10^4 yr in IRAS 20126+4104 at a radius of ~1600 AU. During such episodes, the efficiency of angular momentum transport increases by at least one order of magnitude. In this way one might expect the inner disk to be episodically resupplied to power the outflow. Such episodes are observed to last for

a few rotation periods, a timescale not inconsistent with the estimated outflow kinematic age (see Table 2). This recurrent behavior is also observed in simulations of magnetized disks, as a consequence of the interaction between gravitational and magnetic instabilities (*Fromang et al.*, 2004a,b), which can lead to a periodic variability of the accretion rate.

Finally, it should be remembered that the energy transport in self-gravitating disks might be dominated by wave transport, rather than by viscous diffusion (*Balbus and Papaloizou*, 1999). These nonlocal effects may affect significantly the rate of mass transfer through the disk and possibly lead to higher values of α. However, the evidence for these nonlocal phenomena remains controversial — see *Lodato and Rice* (2004, 2005), *Pickett et al.* (2003), and *Mejía et al.* (2005) for contrasting views.

5. DISK LIFETIME

The radiation emitted by the central star affects the structure and evolution of the surrounding disk in many ways. One of the most important effects is the possibility of photoevaporative mass flows from the surface of the disk. These processes have been studied extensively in the past, starting with *Hollenbach et al.* (1993, 1994), who proposed that the process of photoevaporation of a massive disk could lead to the formation of an ultracompact HII region.

The basic idea of the photoevaporation is that the ionizing ultraviolet radiation from the central star can produce a thin, hot layer of ionized material at the surface of the disk. Far enough from the central star, this thin layer can be hot enough to become unbound from the star and leave the disk plane to form a photoevaporative outflow. The radius outside which this outflow can be produced is obtained by equating the thermal energy of the hot gas to the gravitational binding energy — see equation (2.1) of *Hollenbach et al.* (1994). The photoevaporation outflow rate depends sensitively on the magnitude of the ionizing flux Φ

$$\dot{M}_{ph} \approx 10^{-6}\left(\frac{\Phi}{10^{47}\text{ s}^{-1}}\right)^{1/2}\left(\frac{M_\star}{10\text{ M}_\odot}\right)^{1/2} \text{M}_\odot \text{ yr}^{-1} \qquad (11)$$

where we have scaled the main parameters to values typical of a B0 star. The above estimates, based on a simple static model, neglect the effects of dust opacity. However, more realistic models (*Richling and Yorke*, 1997) confirm the estimate of an outflow rate of ~10^{-6} M$_\odot$ yr^{-1} for a typical B star, as in the case of IRAS 20126+4104.

Note that the photoevaporation outflow rate estimated above is much smaller than the typical accretion rate estimated for these systems (see sections 2.3 and 4). However, the situation might be different for O stars, where the photoevaporation lifetime becomes comparable to the accretion timescale (*Yorke*, 2004b). In order to properly assess the interplay between viscous evolution and photoevaporation (especially for very-high-mass stars, where the two processes occur on comparable timescales), we need models that incorporate both processes. Such models have been developed only in the context of low-mass young stars (*Clarke et al.*, 2001; *Matsuyama et al.*, 2003), but their extrapolation to higher masses is not straightforward.

6. THE EFFECTS OF STELLAR COMPANIONS

A process competing with photoevaporation in the dispersal of disk material at distances ~10^3 AU from the central star is the tidal stripping due to encounters with binary or cluster companions. This process may result in the truncation of the disk at smaller radii, or in the complete dispersal of the disk material depending on the impact parameter of the collision. The following two limiting cases are easy to analyze.

1. If the periastron of the companion is inside the primary's disk, the interaction is highly destructive, and results in removing all disk material external to the periastron in a single orbital transit. A smaller circumstellar disk of radius approximately one-third of the original periastron radius may survive or reform shortly after the interaction, as shown by *Clarke and Pringle* (1993) and *Hall et al.* (1996).

2. For a wide binary system, the maximum size of a circumstellar accretion disk is smaller than the Roche lobe (*Paczynski*, 1977). For stars of comparable mass M_1 and M_2 this is a fraction ~$0.4 + 0.2 \log(M_1/M_2)$ of the semimajor axis (*Paczynski*, 1971). Under these assumptions, *Hollenbach et al.* (2000) estimated a timescale of ~2×10^5 yr for dispersal of a 10^3-AU disk in a cluster with stellar density ~10^4 pc^{-3} and velocity dispersion ~1 km s^{-1}.

It is clear then that a higher frequency of wide binaries (separation ~10^3 AU) among O-type stars might help to explain the apparent scarcity of large circumstellar disks around these stars discussed in section 2. There is in fact a possible indication of a trend for increasing degree of multiplicity among stars of the earliest spectral types (*Preibisch et al.*, 2001). For example, the binary frequency in O-type stars ranges from ~40% (*Garmany et al.*, 1980) to ~60% (*Mason et al.*, 1998), whereas the binary frequency of B-type stars is generally lower, on the order of ~14% (*McAlister et al.*, 1993). Despite the large statistical uncertainties in these estimates, the possibility of efficient tidal truncation of circumstellar disks by binary companions among O-type stars is not completely negligible, especially for wide binary systems with separations comparable to the disk sizes listed in Table 2, on the order of ~10^2–10^3 AU. These systems, however, seem to be underrepresented in the Orion Trapezium cluster relative to the main-sequence field star population (*McCaughrean et al.*, 2000).

In addition to interactions with an orbiting exterior companion, a circumstellar disk around a star in a dense cluster is also subject to significant tidal forces due to the cluster's gravitational field, which may limit the disk size and affect its evolution. Tidal effects are not important if the variation of the cluster's potential $\mathcal{V}_{cl}$ over a distance on the order of the disk diameter is smaller than the gravitational

potential of the star plus disk system itself, $2 R(d\mathcal{V}_{cl}/dr) < GM_t/R$, implying $R^2 < GM_t(2d\mathcal{V}_{cl}/dr)^{-1}$.

For a cluster with central density ρ_0 and central velocity dispersion v_0, the gradient of the gravitational potential is maximum at a distance on the order of the cluster's scale radius $R_{cl} = v_0(6/4\pi G\rho_0)^{1/2}$. For a Plummer's model (see *Spitzer*, 1987), this occurs at a distance $0.71 R_{cl}$ from the cluster's center, where the enclosed mass is 19% of the total cluster mass $M_{cl} = (4/3)\pi R_{cl}^3\rho_0$. At this radius, $(d\mathcal{V}_{cl}/dr)_{max} = 0.38 GM_{cl}/R_{cl}^2$, and the condition on the disk radius becomes $R < 1.1 (M_t/M_{cl})^{1/2}R_{cl}$. Inserting appropriate numerical values, we obtain

$$R < 10^4 \left(\frac{M_t}{10\ M_\odot}\right)^{\frac{1}{2}} \left(\frac{v_0}{1\ \text{km s}^{-1}}\right)^{-\frac{1}{2}} \left(\frac{n_\star}{10^4\ \text{pc}^{-3}}\right)^{-\frac{1}{4}} \text{AU} \quad (12)$$

where we have adopted a mean stellar mass in the cluster of $1\ M_\odot$. Tidal effects thus do not seem to be significant for disks of ~10^3 AU size. For clusters like the Orion Trapezium cluster, however, with $v_0 \approx 4.5$ km s^{-1} (*Jones and Walker*, 1988) and $n_\star \approx 4 \times 10^4$ pc^{-3} (*McCaughrean and Stauffer*, 1994), tidal effects become important for $R \approx 3500$ AU, not an unrealistic value for massive circumstellar disks, and their contribution to the disk stability and lifetime must be properly taken into account.

7. SUMMARY AND CONCLUSIONS

The way high-mass stars form is matter of debate, and observations of circumstellar disks may help to settle this issue. The existence of "disks" rotating about YSOs with masses $\leq 20\ M_\odot$ and luminosities $\leq 10^4\ L_\odot$ is well established, strongly supporting a common formation scenario across the stellar mass spectrum. To date about 10 disks in massive stars have been found, among which IRAS 20126+4104 is the best studied. Here one sees a Keplerian disk about a ~$7\ M_\odot$ (proto)star associated with a precessing outflow. This suggests the presence of a binary system, with the dominant member lying at the center of the disk. Using IRAS 20126+4104 as a prototype, we have shown that the disk is likely to be gravitationally stable and is bound to develop spiral density waves. These effects may be invoked to solve the problem of transfer of material through the disk, which in a classical α disk is much less than the expected accretion rate onto the star.

What is still missing is evidence of disks in more luminous objects, namely above ~$10^5\ L_\odot$, where only huge, massive, nonequilibrium rotating structures are detected; we have named these "toroids". However, absence of evidence does not imply evidence of absence, and it is still possible that O (proto)stars are surrounded by circumstellar disks that remain undetected because of instrumental limitations. Indeed, the distance of these objects is significantly larger than for B-type stars, so that observational biases may play an important role. Also, more massive stars are expected to be associated with richer clusters: The presence of multiple outflows/disks in the same field may confuse the observations. A slightly different possibility is that disks are hidden inside the massive toroids, and one may speculate that the latter could eventually evolve into a sample of circumstellar disks. Alternatively, the formation of large disks might be inhibited in O stars: We have shown that tidal interaction with companions may be effective for this purpose, whereas disk photoevaporation by the O star occurs over too long a timescale. Consequently, disks in O stars might be truncated at small radii and hence difficult to detect with current techniques.

In conclusion, it seems plausible that massive stars form through disk accretion as well as low-mass ones. However, one cannot neglect the possibility that *no* disks are really present in early O stars; in this case alternative formation scenarios (coalescence, competitive accretion) must be invoked. The advent of new-generation instruments, such as the Atacama Large Millimeter Array (ALMA), with their high sensitivity and resolution, is bound to shed light on this important topic.

Acknowledgments.   It is a pleasure to thank A. Natta and C. Clarke for their suggestions and criticisms on the manuscript.

REFERENCES

Balbus S. A. (2003) *Ann. Rev. Astron. Astrophys., 41,* 555–597.

Balbus S. A. and Hawley J. F. (1992) *Astrophys. J., 400,* 610–621.

Balbus S. A. and Papaloizou J. C. B. (1999) *Astrophys. J., 521,* 650–658.

Barvainis R. (1984) *Astrophys. J., 279,* 358–362.

Beltrán M. T., Cesaroni R., Neri R., Codella C., Furuya R. S., Testi L., and Olmi L. (2004) *Astrophys. J., 601,* L187–L190.

Beltrán M. T., Cesaroni R., Neri R., Codella C., Furuya R. S., Testi L., and Olmi L. (2005) *Astron. Astrophys., 435,* 901–925.

Bernard J. P., Dobashi K., and Momose M. (1999) *Astron. Astrophys., 350,* 197–203.

Bertin G. and Lodato G. (1999) *Astron. Astrophys., 350,* 694–704.

Beuther H., Schilke P., Sridharan T. K., Menten K. M., Walmsley C. M., and Wyrowski F. (2002a) *Astron. Astrophys., 383,* 892–904.

Beuther H., Schilke P., Gueth F., McCaughrean M., Andersen M., Sridharan T. K., and Menten K. M. (2002b) *Astron. Astrophys., 387,* 931–943.

Beuther H., Schilke P., and Stanke T. (2003) *Astron. Astrophys., 408,* 601–610.

Beuther H., Schilke P., and Gueth F. (2004a) *Astrophys. J., 608,* 330–340.

Beuther H., Hunter T. R., Zhang Q., Sridharan T. K., Zhao J.-H., et al. (2004b) *Astrophys. J., 616,* L23–L26.

Beuther H., Zhang Q., Sridharan T. K., and Chen Y. (2005) *Astrophys. J., 628,* 800–810.

Beuther H., Zhang Q., Reid M. J., Hunter T. R., Gurwell M., et al. (2006) *Astrophys. J., 636,* 323.

Bik A. and Thi W. F. (2004) *Astron. Astrophys., 427,* L13–L16.

Birkmann S. M., Krause O., and Lemke D. (2006) *Astrophys. J., 637,* 380–383.

Bonnell I. A. and Bate M. R. (2005) *Mon. Not. R. Astron. Soc., 362,* 915–920.

Burrows C. J., Stapelfeldt K. R., Watson A. M., Krist J. E., Ballester G. E., et al. (1996) *Astrophys. J., 473,* 437–451.

Calvet N., Hartmann L., and Strom S. E. (2000) In *Protostars and Planets IV* (V. Mannings et al., eds.), pp. 377–399. Univ. of Arizona, Tucson.

Cesaroni R. (2005a) *Astrophys. Space Sci., 295,* 5–17.

Cesaroni R. (2005b) In *Massive Star Birth: A Crossroads of Astrophysics* (R. Cesaroni et al., eds.), pp. 59–69. IAU Symposium 227, Cambridge Univ., Cambridge.

Cesaroni R., Olmi L., Walmsley C. M., Churchwell E., and Hofner P. (1994) *Astrophys. J., 435,* L137–L140.

Cesaroni R., Felli M., Testi L., Walmsley C. M., and Olmi L. (1997) *Astron. Astrophys., 325,* 725–744.

Cesaroni R., Hofner P., Walmsley C. M., and Churchwell E. (1998) *Astron. Astrophys., 331,* 709–725.

Cesaroni R., Felli M., Jenness T., Neri R., Olmi L., Robberto M., Testi L., and Walmsley C. M. (1999a) *Astron. Astrophys., 345,* 949–964.

Cesaroni R., Felli M., and Walmsley C. M. (1999b) *Astron. Astrophys. Suppl., 136,* 333–361.

Cesaroni R., Neri R., Olmi L., Testi L., Walmsley C. M., and Hofner P. (2005) *Astron. Astrophys., 434,* 1039–1054.

Chini R., Hoffmeister V., Kimeswenger S., Nielbock M., Nürnberger D., Schmidtobreick L., and Sterzik M. (2004) *Nature, 429,* 155–157.

Churchwell E., Walmsley C. M., and Cesaroni R. (1990) *Astron. Astrophys. Suppl., 83,* 119–144.

Clarke C. J. and Pringle J. E. (1993) *Mon. Not. R. Astron. Soc., 261,* 190–202.

Clarke C. J., Gendrin A., and Sotomayor M. (2001) *Mon. Not. R. Astron. Soc., 328,* 485–491.

Codella C., Lorenzani A., Gallego A. T., Cesaroni R., and Moscadelli L. (2004) *Astron. Astrophys., 417,* 615–624.

Cohen R. J., Baart E. E., and Jonas J. L. (1988) *Mon. Not. R. Astron. Soc., 231,* 205–227.

De Buizer J. M. (2003) *Mon. Not. R. Astron. Soc., 341,* 277–298.

De Buizer J. M. and Minier V. (2005) *Astrophys. J., 628,* L151–L154.

Dent W. R. F., Little L. T., Kaifu N., Ohishi M., and Suzuki S. (1985) *Astron. Astrophys., 146,* 375–380.

Edris K. A., Fuller G. A., Cohen R. J., and Etoka S. (2005) *Astron. Astrophys., 434,* 213–220.

Elitzur M. (1992) *Ann. Rev. Astron. Astrophys., 30,* 75–112.

Estalella R., Mauersberger R., Torrelles J. M., Anglada G., Gómez J. F., Lopez R., and Muders D. (1993) *Astrophys. J., 419,* 698–706.

Fontani F., Cesaroni R., Testi L., Molinari S., Zhang Q., Brand J., and Walmsley C. M. (2004) *Astron. Astrophys., 424,* 179–195.

Fromang S., Balbus S. A., and De Villiers J.-P. (2004a) *Astrophys. J., 616,* 357–363.

Fromang S., Balbus S. A., Terquem C., and De Villiers J.-P. (2004b) *Astrophys. J., 616,* 364–375.

Fuente A., Rodríguez-Franco A., Testi L., Natta A., Bachiller R., and Neri R. (2003) *Astrophys. J., 598,* L39–L42.

Fuller G. A., Zijlstra A. A., and Williams S. J. (2001) *Astrophys. J., 555,* L125–L128.

Galli D. and Shu F. H. (1993a) *Astrophys. J., 417,* 243–258.

Galli D. and Shu F. H. (1993b) *Astrophys. J., 417,* 220–242.

Gammie C. F. (2001) *Astrophys. J., 553,* 174–183.

Garmany C. D., Conti P. S., and Massey P. (1980) *Astrophys. J., 242,* 1063–1076.

Genzel R. and Stutzki J. (1989) *Ann. Rev. Astron. Astrophys., 27,* 41–85.

Gibb A. G., Hoare M. G., Mundy L. G., and Wyrowski F. (2004a) In *Star Formation at High Angular Resolution* (M. Burton et al., eds.), pp. 425–430. IAU Symposium 221, Kluwer/Springer, Dordrecht.

Gibb A. G., Wyrowski F., and Mundy L. G. (2004b) *Astrophys. J., 616,* 301–318.

Gómez J. F., Sargent A. I., Torrelles J. M., Ho P. T. P., Rodríguez L. F., Cantó J., and Garay G. (1999) *Astrophys. J., 514,* 287–295.

Goodman A. A., Benson P. J., Fuller G. A., and Myers P. C. (1993) *Astrophys. J., 406,* 528–547.

Greenhill L. J., Reid M. J., Chandler C. J., Diamond P. J., and Elitzur M. (2004) In *Star Formation at High Angular Resolution* (M. Burton et al., eds.), pp. 155–160. IAU Symposium 221, Kluwer/Springer, Dordrecht.

Guilloteau S., Dutrey A., and Simon M. (1999) *Astron. Astrophys., 348,* 570–578.

Hall S. M., Clarke C. J., and Pringle J. E. (1996) *Mon. Not. R. Astron. Soc., 278,* 303–320.

Hasegawa T. I. and Mitchell G. F. (1995) *Astrophys. J., 451,* 225–237.

Hillenbrand L. A., Carpenter J. M., and Skrutskie M. F. (2001) *Astrophys. J., 547,* L53–L56.

Hofner P., Cesaroni R., Rodríguez L. F., and Martín J. (1999) *Astron. Astrophys., 345,* L43–L46.

Hollenbach D., Johnstone D., and Shu F. (1993) In *Massive Stars: Their Lives in the Interstellar Medium* (J. P. Cassinelli and E. B. Churchwell, eds.), pp. 26–34. ASP Conf. Series 35, San Francisco.

Hollenbach D., Johnstone D., Lizano S., and Shu F. (1994) *Astrophys. J., 428,* 654–669.

Hollenbach D. J., Yorke H. W., and Johnstone D. (2000) In *Protostars and Planets IV* (V. Mannings et al., eds.), pp. 401–428. Univ. of Arizona, Tucson.

Hunter T. R., Testi L., Zhang Q., and Sridharan T. K. (1999) *Astron. J., 118,* 477–487.

Hutawarakorn B. and Cohen R. J. (1999) *Mon. Not. R. Astron. Soc., 303,* 845–854.

Jiang Z., Tamura M., Fukagawa M., Hough J., Lucas P., Suto H., Ishii M., and Yang J. (2005) *Nature, 437,* 112–115.

Jijina J. and Adams F. C. (1996) *Astrophys. J., 462,* 874–887.

Jones B. F. and Walker M. F. (1988) *Astron. J., 95,* 1755–1782.

Johnson B. M. and Gammie C. F. (2003) *Astrophys. J., 597,* 131–141.

Kahn F. D. (1974) *Astron. Astrophys., 37,* 149–162.

Keto E. R., Ho P. T. P., and Haschick A. D. (1988) *Astrophys. J., 324,* 920–930.

Königl A. and Pudritz R. E. (2000) In *Protostars and Planets IV* (V. Mannings et al., eds.), pp. 759–787. Univ. of Arizona, Tucson.

Kramer C., Alves J., Lada C., Lada E., Sievers A., Ungerechts H., and Walmsley M. (1998) *Astron. Astrophys., 329,* L33–L36.

Krolik J. H. (1999) *Active Galactic Nuclei: From the Central Black Hole to the Galactic Environment,* Princeton Univ., Princeton, New Jersey.

Krumholz M. R., McKee C. F., and Klein R. I. (2005) *Astrophys. J., 618,* L33–L36.

Laughlin G. and Bodenheimer P. (1994) *Astrophys. J., 436,* 335–354.

Laughlin G. and Korchagin V. (1996) *Astrophys. J., 460,* 855–868.

Liu S.-Y. and the SMA Team (2005) In *Massive Star Birth: A Crossroads of Astrophysics* (R. Cesaroni et al., eds.), pp. 47–52. IAU Symposium 227, Cambridge Univ., Cambridge.

Lodato G. and Rice W. K. M. (2004) *Mon. Not. R. Astron. Soc., 351,* 630–642.

Lodato G. and Rice W. K. M. (2005) *Mon. Not. R. Astron. Soc., 358,* 1489–1500.

MacLeod G. C. and Gaylard M. J. (1992) *Mon. Not. R. Astron. Soc., 256,* 519–527.

Martí J., Rodríguez L. F., and Reipurth B. (1993) *Astrophys. J., 416,* 208–217.

Mason B. D., Henry T. J., Hartkopf W. I., Ten Brummelaar T., and Soderblom D. R. (1998) *Astron. J., 116,* 2975–2983.

Mathieu R. D., Ghez A. M., Jensen E. L. N., and Simon M. (2000) In *Protostars and Planets IV* (V. Mannings et al., eds.), pp. 703–730. Univ. of Arizona, Tucson.

Matsuyama I., Johnstone D., and Hartmann L. (2003) *Astrophys. J., 582,* 893–904.

McAlister H. A., Mason B. D., Hartkopf W. I., and Shara M. M. (1993) *Astron. J., 106,* 1639–1655.

McCaughrean M. J. and Stauffer J. R. (1994) *Astron. J., 108,* 1382–1397.

McCaughrean M. J., Stapelfeldt K. R., and Close L. M. (2000) In *Protostars and Planets IV* (V. Mannings et al., eds.), pp. 485–507. Univ. of Arizona, Tucson.

McKee C. F. and Tan J. C. (2003) *Astrophys. J., 585,* 850–871.

Mejía A. C., Durisen R. H., Pickett M. K., and Cai K. (2005) *Astrophys. J., 619,* 1098–1113.

Mestel L. (1963) *Mon. Not. R. Astron. Soc., 126,* 553–575.

Minier V., Booth R. S., and Conway J. E. (1998) *Astron. Astrophys., 336,* L5–L8.

Minier V., Booth R. S., and Conway J. E. (2000) *Astron. Astrophys., 362,* 1093–1108.

Mitchell G. F., Hasegawa T. I., and Schella J. (1992) *Astrophys. J., 386,* 604–617.

Molinari S., Brand J., Cesaroni R., and Palla F. (1996) *Astron. Astrophys., 308,* 573–587.

Molinari S., Brand J., Cesaroni R., Palla F., and Palumbo G. G. C. (1998) *Astron. Astrophys., 336,* 339–351.

Moscadelli L., Cesaroni R., and Rioja M. J. (2000) *Astron. Astrophys., 360,* 663–670.

Moscadelli L., Cesaroni R., and Rioja M. J. (2005) *Astron. Astrophys., 438,* 889–898.

Mundy L. G., Looney L.W., and Welch W. J. (2000) In *Protostars and Planets IV* (V. Mannings et al., eds.), pp. 355–376. Univ. of Arizona, Tucson.

Nakano T., Hasegawa T., Morino J.-I., and Yamashita T. (2000) *Astrophys. J., 534,* 976–983.

Natta A., Grinin V., and Mannings V. (2000) In *Protostars and Planets IV* (V. Mannings et al., eds.), pp. 559–587. Univ. of Arizona, Tucson.

Norris R. P., Byleveld S. E., Diamond P. J., Ellingsen S. P., Ferris R. H., et al. (1998) *Astrophys. J., 508,* 275–285.

O'Dell C. R. (2001) *Astron. J., 122,* 2662–2667.

O'Dell C. R. and Wen Z. (1994) *Astrophys. J., 436,* 194–202.

Olmi L., Cesaroni R., Hofner P., Kurtz S., Churchwell E., and Walmsley C. M. (2003) *Astron. Astrophys., 407,* 225–235.

Osterloh M., Henning Th., and Launhardt R. (1997) *Astrophys. J. Suppl., 110,* 71–114.

Paczynski B. (1971) *Ann. Rev. Astron. Astrophys., 9,* 183–208.

Paczynski B. (1977) *Astrophys. J., 216,* 822–826.

Palla F. and Stahler S. W. (1993) *Astrophys. J., 418,* 414–425.

Palla F., Brand J., Comoretto G., Felli M., and Cesaroni R. (1991) *Astron. Astrophys., 246,* 249–263.

Patel N. A., Curiel S., Sridharan T. K., Zhang Q., Hunter T. R., et al. (2005) *Nature, 437,* 109–111.

Pestalozzi M. R., Elitzur M., Conway J. E., and Booth R. S. (2004) *Astrophys. J., 603,* L113–L116.

Phillips C. J., Norris R. P., Ellingsen S. P., and McCulloch P. M. (1998) *Mon. Not. R. Astron. Soc., 300,* 1131–1157.

Pickett B. K., Cassen P., Durisen R. H., and Link R. (2000) *Astrophys. J., 529,* 1034–1053.

Pickett B. K., Mejía A. C., Durisen R. H., Cassen P. M., Berry D. K., and Link R. P. (2003) *Astrophys. J., 590,* 1060–1080.

Plume R., Jaffe D. T., and Evans N. J. II (1992) *Astrophys. J. Suppl., 78,* 505–515.

Preibisch Th., Weigelt G., and Zinnecker H. (2001) In *The Formation of Binary Stars* (H. Zinnecker and R. Mathieu, eds.), pp. 69–78. IAU Symposium 200, Kluwer/Springer, Dordrecht.

Preibisch T., Balega Y. Y., Schertl D., and Weigelt G. (2003) *Astron. Astrophys., 412,* 735–743.

Puga E., Feldt M., Alvarez C., Henning Th., Apai D., et al. (2006) *Astrophys J., 641,* 373–382.

Rice W. K. M., Lodato G., and Armitage P. J. (2005) *Mon. Not. R. Astron. Soc., 346,* L56–L60.

Richer J. S., Shepherd D. S., Cabrit S., Bachiller R., and Churchwell E. (2000) In *Protostars and Planets IV* (V. Mannings et al., eds.), pp. 867–894. Univ. of Arizona, Tucson.

Richling S. and Yorke H. W. (1997) *Astron. Astrophys., 327,* 317–324.

Rodríguez L. F. (2005) In *Massive Star Birth: A Crossroads of Astrophysics* (R. Cesaroni et al., eds.), pp. 120–127. IAU Symposium 227, Cambridge Univ., Cambridge.

Sako S., Yamashita T., Kataza H., Miyata T., Okamoto Y. K., et al. (2005) *Nature, 434,* 995–998.

Sandell G., Wright M., and Forster J. R. (2003) *Astrophys. J., 590,* L45–L48.

Schreyer K., Henning Th., van der Tak F. F. S., Boonman A. M. S., and van Dishoeck E. F. (2002) *Astron. Astrophys., 394,* 561–583.

Schreyer K., Semenov D., Henning Th., and Forbrich J. (2006) *Astrophys. J., 637,* L129–L132.

Scoville N., Kleinmann S. G., Hall D. N. B., and Ridgway S. T. (1983) *Astrophys. J., 275,* 201–224.

Shakura N. I. and Sunyaev R. A. (1973) *Astron. Astrophys., 24,* 337–355.

Shepherd D. S. (2003) In *Galactic Star Formation Across the Stellar Mass Spectrum* (J. M. De Buizer and N. S. van der Bliek, eds.), pp. 333–

344. ASP Conf. Series 287, San Francisco.

Shepherd D. S. and Churchwell E. (1996a) *Astrophys. J., 457,* 267–276.

Shepherd D. S. and Churchwell E. (1996b) *Astrophys. J., 472,* 225–239.

Shepherd D. S. and Kurtz S. E. (1999) *Astrophys. J., 523,* 690–700.

Shepherd D. S., Yu K. C., Bally J., and Testi L. (2000) *Astrophys. J., 535,* 833–846.

Shepherd D. S., Claussen M. J., and Kurtz S. E. (2001) *Science, 292,* 1513–1518.

Shu F. H., Najita J. R., Shang H., and Li Z.-Y. (2000) In *Protostars and Planets IV* (V. Mannings et al., eds.), pp. 789–813. Univ. of Arizona, Tucson.

Simon M., Dutrey A., and Guilloteau S. (2000) *Astrophys. J., 545,* 1034–1043.

Sollins P. K., Zhang Q., Keto E., and Ho P. T. P. (2005a) *Astrophys. J., 624,* L49–L52.

Sollins P. K., Zhang Q., Keto E., and Ho P. T. P. (2005b) *Astrophys. J., 631,* 399–410.

Spitzer L. (1987) *Dynamical Evolution of Globular Clusters,* Princeton Univ., Princeton, New Jersey.

Sridharan T. K., Beuther H., Schilke P., Menten K. M., and Wyrowski F. (2002) *Astrophys. J., 566,* 931–944.

Sridharan T. K., Williams S. J., and Fuller G. A. (2005) *Astrophys. J., 631,* L73–L76.

Stahler S. W., Palla F., and Ho P. T. P. (2000) In *Protostars and Planets IV* (V. Mannings et al., eds.), pp. 327–351. Univ. of Arizona, Tucson.

Su Y.-N., Zhang Q., and Lim J. (2004) *Astrophys. J., 604,* 258–271.

Tan J. C. and McKee C. F. (2004) *Astrophys. J., 603,* 383–400.

Terebey S., Shu F. H., and Cassen P. (1984) *Astrophys. J., 286,* 529–551.

Tofani G., Felli M., Taylor G. B., and Hunter T. R. (1995) *Astron. Astrophys. Suppl., 112,* 299–346.

Toomre A. (1964) *Astrophys. J., 139,* 1217–1238.

Torrelles J. M., Gómez J. F., Garay G., Rodríguez L. F., Curiel S., Cohen R. J., and Ho P. T. P. (1998) *Astrophys. J., 509,* 262–269.

van der Tak F. F. S. and Menten K. M. (2005) *Astron. Astrophys., 437,* 947–956.

van der Tak F. F. S., Walmsley C. M., Herpin F., and Ceccarelli C. (2006) *Astron. Astrophys., 447,* 1011–1025.

Walker C. K., Adams F. C., and Lada C. J. (1990) *Astrophys. J., 349,* 515–528.

Wilking B. A., Blackwell J. H., Mundy L. G., and Howe J. E. (1989) *Astrophys. J., 345,* 257–264.

Wilking B. A., Blackwell J. H., and Mundy L. G. (1990) *Astron. J., 100,* 758–770.

Wilner D. J. and Lay O. P. (2000) In *Protostars and Planets IV* (V. Mannings et al., eds.), pp. 509–532. Univ. of Arizona, Tucson.

Wolfire M. G. and Cassinelli J. P. (1987) *Astrophys. J., 319,* 850–867.

Wright M. C. H., Plambeck R. L., Mundy L. G., and Looney L. W. (1995) *Astrophys. J., 455,* L185–L188.

Yao Y., Ishii M., Nagata T., Nakaya H., and Sato S. (2000) *Astrophys. J., 542,* 392–399.

Yorke H. W. (2004a) In *Star Formation at High Angular Resolution* (M. Burton et al., eds.), pp. 141–152. IAU Symposium 221, Kluwer/Springer, Dordrecht.

Yorke H. W. (2004b) *Rev. Mex. Astron. Astrofis. Ser. Conf., 22,* 42–45.

Yorke H. W. and Sonnhalter C. (2002) *Astrophys. J., 569,* 846–862.

Zhang Q. (2005) In *Massive Star Birth: A Crossroads of Astrophysics* (R. Cesaroni et al., eds.), pp. 135–144. IAU Symposium 227, Cambridge Univ., Cambridge.

Zhang Q., Ho P. T. P., and Ohashi N. (1998a) *Astrophys. J., 494,* 636–656.

Zhang Q., Hunter T. R., and Sridharan T. K. (1998b) *Astrophys. J., 505,* L151–L154.

Zhang Q., Hunter T. R., Sridharan T. K., and Cesaroni R. (1999) *Astrophys. J., 527,* L117–L120.

Zhang Q., Hunter T. R., Brand J., Sridharan T. K., Molinari S., et al. (2001) *Astrophys. J., 552,* L167–L170.

Zhang Q., Hunter T. R., Sridharan T. K., and Ho P. T. P. (2002) *Astrophys. J., 566,* 982–992.

Zhang Q., Hunter T. R., Brand J., Sridharan T. K., Cesaroni R., et al. (2005) *Astrophys. J., 625,* 864–882.

Part III:
Outflows

Observations of Jets and Outflows from Young Stars

John Bally
University of Colorado at Boulder

Bo Reipurth
University of Hawaii

Christopher J. Davis
Joint Astronomy Centre

This review concentrates on observations of outflows from young stars during the last seven years. Recent developments include detections of an increasing number of Herbig-Haro flows at X-rays and UV wavelengths, high-resolution studies of irradiated jets with HST, wide-field imaging of parsec-scale outflows with groundbased CCDs and near-IR imagers, complete surveys of visual and near-IR emission from shocks in the vicinity of entire molecular clouds with wide-field imagers, far-infrared studies with the Infrared Space Observatory and the Spitzer Space Telescope, and high angular submillimeter, millimeter, and centimeter wavelength aperture synthesis array data cubes showing both the spatial and velocity structure of jets and outflows.

1. INTRODUCTION

Outflows are one of the manifestations of the birth of a young star that are easiest to observe. More than a half century ago, *Herbig* (1950, 1951) and *Haro* (1952, 1953) found the first examples of the peculiar nebulae that have come to be known as Herbig-Haro (HH) objects. Located in or near dark clouds in regions suspected of having undergone recent star formation, HH objects were first thought to be young stars or their associated reflection nebulae. However, by the 1970s, their spectra were interpreted as indicative of mostly low-excitation shock waves. By the 1980s some of the growing list of HH objects were found to trace highly collimated jets powered by young stars (*Dopita et al.,* 1982; *Mundt and Fried,* 1983; *Reipurth et al.,* 1986), associated with high proper-motion bipolar outflows (*Herbig and Jones,* 1981).

During the 1960s and 1970s, additional manifestations of outflow activity were discovered. These include P-Cygni profiles and other spectroscopic indicators of powerful stellar winds emerging from young stars, high-velocity OH and H_2O masers, bipolar molecular outflows (e.g., *Snell et al.,* 1980), as well as shock-excited near-IR emission lines of species such as [FeII] and H_2. By the 1980s, a number of young stellar objects (YSOs) were found to produce radio continuum jets visible at centimeter wavelengths. Although it was not at first apparent, by the early 1990s, observations and theoretical considerations made it clear that most manifestations of outflow activity were produced by shocks powered by forming stars.

The advent of large-format CCDs and the launch of the Hubble Space Telescope (HST) during the 1990s ushered in new developments. Wide-field visual wavelength surveys revealed that outflows can attain parsec-scale dimensions

with lengths exceeding 10 pc, and that they blow out of their molecular clouds (*Bally and Devine,* 1994; *Reipurth et al.,* 1997; *Eislöffel and Mundt,* 1997). Narrow-band-filter imaging revealed a new population of protostellar jets in HII regions that are rendered visible by the UV radiation fields of nearby massive stars (*Reipurth et al.,* 1998a; *Cernicharo et al.,* 1998; *Bally and Reipurth,* 2001). HST has resolved the subarcsecond scale structure and cooling layers of dozens of HH objects. HST has enabled the measurement of proper motions on images taken less than one year apart — a timescale shorter than the typical cooling time.

1.1. The Importance of Outflows

Although no theory of star formation anticipated jets and outflows, it is now clear that the production of these flows is a fundamental aspect of star formation. Collimated outflows occur in most astrophysical systems in which accretion, rotation, and magnetic fields interact. Due to their proximity, large numbers [nearly a thousand masers, HH objects, and molecular outflows are now known (*Wu et al.,* 2004)], and diversity of available tracers, protostellar outflows make ideal laboratories for the investigations of the physics, chemistry, acceleration, collimation, propagation, and impacts of these systems. The results of these studies should be relevant to many other classes of astrophysical outflow.

Outflows provide a fossil record of the mass loss and, therefore, mass-accretion histories of forming stars. Outflow symmetries provide clues about the dynamical environment of the engine; S- and Z-shaped symmetries indicate that the outflow axis has changed over time, perhaps due to precession induced by a companion, or interactions with sibling stars in a cluster. C-shaped bends indicate motion of surrounding gas (side winds), or the motion of the outflow source itself. Outflows have a profound impact on their surroundings. Jets and winds create cavities in the parent cloud and inject energy and momentum into the surrounding gas that, in the absence of massive stars, may dominate the generation of turbulence and cloud motions. The terminal shocks in outflows dissociate molecules, sputter grains, and can reset the chemical evolution of clouds to an initial state. Shocks also drive chemistry, thereby altering the chemical composition of the impacted media. Outflows may play a fundamental role in sculpting and disrupting their parent clouds. They may play a role in determining final stellar masses and the shape of the initial mass function. Outflows may also carry away some of the angular momentum of matter accreting onto the forming star.

This review will concentrate on developments since the last Protostars and Planets conference held in 1998. During the last seven years, observational and computational capabilities have increased greatly. Major new developments include the detection of X-rays from an increasing number of protostellar outflows and the measurement of the emission and absorption spectra of some HH objects in the UV (section 2). At visual and near-IR wavelengths, the formats of detectors have continued to increase with pixel counts doubling approximately every two years. This has led to the first complete imaging surveys of entire giant molecular clouds with 4-m-class telescopes in narrow-band filters sensitive to shock tracers. Eight-meter-class telescopes, adaptive optics, a new generation of high-resolution and multi-object spectrographs, and the Advanced Camera for Surveys on HST have had major impacts (section 3). The Spitzer Space Telescope is surveying the galactic plane and many of the nearby clouds, and IRAC and MIPS images are providing stunning images of the mid-IR emission from jets, outflows, and the properties of their sources. Thermal imaging with Gemini has traced entrained warm dust in jets. Upgrades at the VLA, and a new generation of millimeter-interferometers such as the SMA on Mauna Kea, are producing stunning arcsecond-resolution images of molecular jets, and have enabled the detection of many highly embedded sources invisible at shorter wavelengths (section 4).

1.2. Overview of Outflow Properties and Behavior

The velocity difference between the parent cloud and the outflow correlates with wavelength. Although weak knots of CO emission and jets can sometimes be found at velocities of several hundred km s^{-1} near the core, the bulk of millimeter-wavelength CO emission tends to have velocities of only a few to ten km s^{-1}. While the near-IR lines of H$_2$ and [FeII] tend to have velocities of several tens of km s^{-1}, the visual wavelength forbidden and recombination lines usually have radial velocities of tens to hundreds of km s^{-1}. X-ray emission is only seen from shocks with speeds higher than 300 km s^{-1}.

This trend can be understood in terms of a unified picture. The YSO accelerates a wind, sometimes collimated into a jet, with a velocity several times the escape speed from the launch region, which can range from 100 to over 500 km s^{-1}. Internal shocks form where faster ejecta overrun slower material. The very fastest shocks can sometimes be detected in X-rays. These primary flows often contain molecules such as H$_2$, CO, and SiO when launched from young Class 0/I sources; they tend to be dominated by HI and low-ionization metals when the source is more evolved. Jet speeds also tend to increase, and their densities and mass-loss rates tend to decrease with the evolutionary state of the source. In young, high-density molecular jets, these shocks can excite H$_2$O maser spots and H$_2$ emission; in mostly atomic or ionized jets from somewhat older sources, knots of Hα and/or forbidden line emission are produced.

The primary jets and winds transfer momentum and entrain their surroundings by means of shock waves propagating into the medium. These shocks tend to have much lower velocities than the jets; they can be seen in H$_2$ emission when the interaction is with a molecular cloud, or in Hα or forbidden lines when the medium is atomic or ionized. Most molecular emission observed at submillimeter, millimeter, and centimeter wavelengths is produced by gas entrained and accelerated by these secondary shocks.

There are no perfect tracers of outflows; each tracer and every transition provides information about a limited range of physical conditions. A complete picture of outflows therefore requires observations in the radio, submillimeter, IR, visual, UV, and even X-ray portions of the spectrum.

Species such as CO and other molecular transitions probe the mass and radial velocity of swept-up and entrained gas in an outflow, but only in the molecular cloud. When primary jets and winds blow out of their parent clouds, they are no longer visible in molecular transitions. Sometimes, the 21 cm line of HI can be used to trace entrained gas. These species are excited by collisions at the ambient temperature of the cloud and therefore do not require shocks to be observable. CO, other easy-to-excite molecular transitions, and HI trace the total amount of momentum injected into the cloud and the amount of mass accelerated by an outflow over its lifetime. Shocks inside a molecular cloud can sometimes be traced by the near- and mid-IR transitions of H$_2$ or, if the flow is partially ionized, by species such as [FeII]. At sufficiently low extinctions, or outside the molecular cloud once the outflow has broken out of its natal environment, atomic lines such as Hα and forbidden transitions of atoms and ions such as [OI], [OII], [OIII], [NII], and [SII] in the visual and near-IR can be used to trace shocks and post-shock cooling layers. However, these tracers disappear after a cooling time in the postshock gas. The bulk of the moving mass propagating outside a molecular environment can be traced by 21 cm HI if the medium is atomic. In practice, such observations are made very difficult by foreground and background emission from the galaxy. Illumination of the jet or entrained gas by nearby massive stars provides a better opportunity to trace the unshocked portions of outflows propagating outside molecular clouds. In recent years, the study of externally irradiated jets and outflows has shown that such flows are common in HII regions.

Many questions remain unanswered. Are the collimated outflows powered directly by the YSO, its magnetosphere, its circumstellar disk, or by a combination of these sources? Is the X-wind or the disk model more correct? Do both mechanisms coexist? How are winds collimated into jets? How do the properties of outflows change with the evolution of the driving source? How long does jet production persist? Although there are many strongly held opinions, and a few observational constraints, most of these issues remain open. The next generation of interferometers, spectographs, and ultra-high-resolution instruments will be needed to obtain definitive observations.

2. FAST SHOCKS: OBSERVATIONS OF ULTRAVIOLET AND X-RAYS

2.1. X-Ray Results

YSOs have been known to be prolific sources of X-rays and UV radiation (see the chapter by Feigelson et al.). Time-averaged X-ray luminosities range from L_x less than 10^{27} to over 10^{31} erg s^{-1} with occasional flares that can approach or exceed 1 $L_\odot$. A key feature of X-ray emission from YSOs is that it is highly variable. The launch of the XMM/Newton and Chandra (CXO) X-ray observatories has led to the detection of steady X-ray emission from shocks associated with HH objects.

Pravdo et al. (2001) detected X-ray emission from knot H in HH 2 in the HH 1/2 system in Orion. The emission region is compact (less than a few arcseconds) and arises from the fastest shocks in the entire HH 1/2 system. Proper motions (*Bally et al.,* 2002a) indicate velocities of over 400 km s^{-1}. Knot H is a high-excitation shock with strong [OIII] and radio continuum emission. This fast-moving feature is located near the axis of symmetry of HH 2 and may trace the location where a fast jet slams into much slower moving debris.

Favata et al. (2002) and *Bally et al.* (2003) detected the base of the HH 154 jet emerging from the L1551 IRS5 protobinary with XMM and CXO, respectively. The X-ray emission is offset from the location of the YSOs by about 0.5" along the jet axis. It is located close to the first 1.64 μm [FeII] knot at the base of the jet (*Pyo et al.,* 2002). The radial velocity of the [FeII] emission from the HH 154 jet extends to nearly 400 km s^{-1}. The X-ray spectrum implies a much lower column density of foreground hydrogen and dust in front of the X-ray source than the column density in front of IRS5 as determined from visual and IR extinction. Thus, the X-ray source is located near the base of the jet. *Bally et al.* (2003) proposed several models for the X-ray source including a collision front where the winds from each member of the IRS5 binary collide, a shock where the stellar wind impacts the disk surrounding the other member of the binary, or shocks formed where a wide-angle wind from one or both stars is/are collimated into a jet.

Both HH 2 and 154 are low-mass, low-luminosity protostars. *Pravdo et al.* (2004) found X-ray emission from HH 80/81, which is powered by a moderate-mass protostar with a much larger luminosity of order 10^4 $L_\odot$. As in the cases of HH 154 and HH 2, the X-ray emission is associated with the fastest components in this outflow.

Pravdo and Tsuboi (2005) found X-ray emission from the base of HH 168, the bright HH object associated with Cepheus A West powered by a cluster containing at least three early B stars in formation and a total luminosity of about 2×10^4 $L_\odot$. This feature coincides with the extended radio-continuum source HW at the base of HH 168. The plasma temperature is around $T_X \approx 6 \times 10^6$ K and the X-ray luminosity is around $L_X \approx 10^{29}$ erg s^{-1}.

Several additional HH objects or shocks may have been detected in the Orion Nebula by the COUP project (*Kastner et al.,* 2005), including HH 210, the fastest visually detected component of the shock-excited fingers of H_2 emission emerging from the 10^5 $L_\odot$ OMC1 cloud core located immediately behind the Orion Nebula. Of the more than 1000 X-ray sources found by COUP, a few appear to be associated with shocks at the bases of irradiated jets powered by young stars. *Raga et al.* (2002) have developed a simple, analytical model that can estimate the expected X-ray luminosity of an HH flow.

The detected X-ray-bright HH objects have plasma temperatures in the range $T_X = 10^6$–10^7 K. The X-ray spectrum is dominated by soft photons with energies below 1 keV. The soft-portion of the X-ray spectra are highly attenuated, implying column densities on the order of $\log N(H) = 10^{21}$–10^{22} cm^{-2}. The typical observed X-ray luminosities are around 10^{27} erg s^{-1}; this may be a severe lower limit on the intrinsic X-ray emission since most of the soft X-rays are absorbed by foreground gas and dust.

In summary, X-ray emission traces the fastest and highest excitation shocks in outflows. Most HH objects trace internal working surfaces where shock velocities are much lower than the flow velocities, indicating interactions between moving fluid elements. Thus, the majority of HH objects are dominated by emission in low-excitation tracers. X-ray emission indicates the presence of hard shocks that create high-excitation conditions with temperatures on the order of 10^6–10^7 K. Purely hydrodynamic shocks with speeds around 200–300 km s^{-1} can, in principle, produce such high temperatures. However, observations show that X-rays are only seen from HH objects with velocities larger than 300 km s^{-1} that tend to be associated with radio continuum emission. The absence of X-ray emission from shocks with speeds around 200–300 km s^{-1} indicates that postshock plasma temperatures are somewhat lower than expected from purely hydrodynamic shocks ramming stationary media consisting of a cosmic mixture of hydrogen and helium. It is possible that the hard, X-ray emitting shocks are also propagating into moving media or that postshock temperatures are lowered by other mechanisms such as the addition of coolants or very strong magnetic fields. Grains entering the X-ray emitting volume can radiate away the energy imparted by impacting electrons. Sputtering of these grains can poison the X-ray plasma by injecting metals with tightly bound inner K- and L-shell electrons, which act as efficient coolants.

2.2. Ultraviolet Results

The Space Telescope Imaging Spectrograph (STIS) was used to observe a number of young stars and their HH objects. *Grady et al.* (2005) detected outflows and disks in the immediate vicinity of several Herbig AeBe (HAeBe) stars. *Devine et al.* (2000b) discovered a bipolar jet, HH 409, from the HAeBe star HD 163296 in the Lyman α line. *Grady et al.* (2004) report a bipolar microjet, HH 669, from the bright HAeBe star, HD 104237 (DX Cha). The age of this star, and the small association of young stars around it, has been estimated to be around 5 m.y. A growing number of low mass-loss-rate outflows emerge from stars with ages of several million years. The Lyman α observations are especially sensitive to the very low mass-loss rates of these older YSOs.

Hartigan et al. (1999) obtained UV spectra of HH 46/47 with HST, finding emission from a large number of permitted resonance lines and forbidden transitions. Comparison with shock models indicates a large range of excitation conditions.

Over a dozen proplyds, young stars, and irradiated jets in the Orion Nebula have been observed by STIS at both visual and UV wavelengths. The visual-wavelength STIS observations provide much better spatial resolution than groundbased spectroscopy. With a 0.1" slit, the spectrum of the proplyd ionization front, photo-ablation flow, stellar jet, and star can be spatially resolved. Many of Orion's irradiated microjets exhibit strong kinematic asymmetries in which the brighter jet beam has a lower radial velocity than the fainter counter-jet beam (see section 3.4). The HH 514 microjet emerging from the proplyd HST2 provides an excellent example (Fig. 1) (*Bally et al.*, 2006). This behavior is similar to that of many other irradiated jets and will be discussed further below.

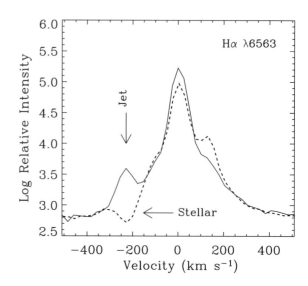

Fig. 2. The solid line shows the STIS spectrum of the Hα emission line profile about 0.2" south of the star in HST2 showing the blue-shifted emission associated with the southern lobe of the HH 514 jet. The dotted line shows the STIS spectrum of the Hα emission line from the central star in HST2 (170–337). Note the absorption feature at the velocity of the blue-shifted lobe of the HH 514 jet. From Smith et al. (in preparation).

A remarkable aspect of the HH 514 jet and source star is the presence of a dip in the stellar Hα emission line profile with exactly the same radial velocity and line width as the HH 514 counterjet at positions offset from the central star (Fig. 2). It is possible that the intense Lyman α radiation produced in the Orion Nebula, combined with self-irradiation by the central star in the proplyd, pumps the $n = 2$ level of hydrogen near the base of the jet. This process requires a very high density neutral envelope [$n(H) > 10^5$ cm^{-3}]. Thus, at the base of the HH 514 jet, the $n = 2$ level may be sufficiently populated by trapped Lyman α to enable absorption in the 6563 Å ($n = 2$–3) Hα line. The presence of an absorption dip at the velocity of the HH 514 counterjet implies that the outflow is launched as a wide-angle wind with an opening angle at least as large as the inclination angle of the jet. After producing the absorption, the wind is funneled into the highly collimated jet that propagates for tens of arcseconds to the south.

Unfortunately, the loss of STIS has brought UV and high-angular-resolution (0.1") visual wavelength studies of protostellar outflows to an end for the time being.

3. VISUAL WAVELENGTH SURVEYS

3.1. Wide-Field Surveys: Counting Shocks and Determining the Impact of Outflows

The continuing development of large format arrays in the visual and near-IR has enabled the first complete surveys of entire giant molecular clouds and star-forming complexes

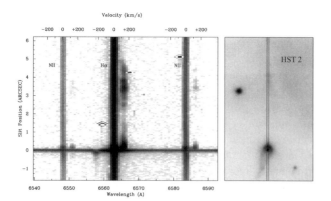

Fig. 1. (Left) A long-slit spectrum showing Hα and the [NII] doublet from the HH 514 microjet obtained with STIS on HST. (Right) An HST/WFPC2 F656N (Hα) image showing a direct image of the proplyd HST2 (170–337), the source of the HH 514 jet. The orientation and relative size of the STIS slit used to obtain the spectrum is shown. From Smith et al. (in preparation).

with narrow-band filters sensitive to selected shock tracers. Visual wavelength imaging complements spectral line surveys at IR, submillimeter, millimeter, and radio wavelengths.

Narrow-band Hα, [SII], and H_2 imaging of entire GMCs and star-forming complexes provide unbiased samples of shocks and their associated outflows. Such data are an excellent way to find new YSOs and outflows. They have been used to estimate the contribution of outflows and shocks to the generation and sustenance of chaotic, turbulent motions, and the momentum and energy budgets of molecular clouds. Wide-field Hα and [SII] imaging has been the best way to identify giant, parsec-scale outflows. In this section, we first summarize recent or ongoing surveys, highlighting the Perseus cloud as a specific example, then we review recent developments in the study of giant outflows, and finally discuss recent visual-wavelength observations of irradiated jets in HII regions.

3.2. Narrow-Band Surveys

Nearby (d < 1 kpc) clouds surveyed in one or more narrow-band filters at visual wavelengths include NGC 2264 (*Reipurth et al.*, 2004), S140 (*Bally et al.*, 2002b), Perseus (*Walawender et al.*, 2005), Orion A and B (*Bally et al.*, 2002c; Bally et al., in preparation), Taurus, Ophiuchus, Serpens, Chamaeleon I, Circinus, NGC 7000, and portions of Cepheus. Many of these datasets remain unpublished because the complete analysis has proven to be labor intensive. *Phelps and Barsony* (2004) and *Wilking et al.* (1997) presented CCD surveys of the entire ρ-Oph dark cloud, finding several dozen HH objects on the surface of this cloud. *Khanzadyan et al.* (2004a) surveyed these clouds in H_2. The Orion A molecular cloud was surveyed in its entirety in H_2 by *Stanke et al.* (1998, 2000, 2002).

Hundreds of individual shock systems in dozens of outflows and jets, many with parsec-scale dimensions, were found. A deep narrow-band CCD survey of both Orion A and B detected hundreds of HH objects including several dozens of parsec-scale HH flows, and a remarkable chain, nearly 2.5° long, terminating near HH 131 from a suspected source in or near the L1641N cluster (*Reipurth et al.*, 1998b; Bally et al., in preparation).

Walawender et al. (2005) used the Kitt Peak Mayall 4-m reflector to completely image the entire Perseus molecular cloud in Hα, [SII], and broadband SDSS I'. This work resulted in a reevaluation of energy and momentum injected into the cloud by outflows. The origin of the random motions, turbulence, and cloud structure was previously discussed by *Miesch and Bally* (1994), *Miesch et al.* (1999), and *Bally et al.* (1996). *Walawender et al.* (2005) compared the energy and momentum traced by these shocks with the total turbulent kinetic energy and momentum in the Perseus clouds as traced by CO observations. Walawender et al. also analyzed the amount of energy and momentum injected by molecular outflows. The results of these preliminary analyses can be summarized as follows: Outflows inject more than sufficient energy and momentum to drive turbulence

motions and even to disrupt individual cloud cores such as NGC 1333. For example, *Bally et al.* (1996) found that in the NGC 1333 region, there are dozens of currently active outflows. If this level of activity is typical, then in a steady state the mean time between the passage of a shock with sufficient speed to dissociate molecules in any random parcel of gas is about 10^4–10^5 yr. Thus, outflows can self-regulate star formation and even disrupt surviving portions of the cloud core in regions such as NGC 1333. However, on the scale of the entire 10^4 M$_\odot$ GMC, outflows may fail by nearly an order of magnitude to resupply the momentum being dissipated by the decay of turbulent energy. Thus, while being able to self-regulate star formation on the scale of individual cloud cores containing a few hundred solar masses, outflows may not supply the random motions observed on large scales. To maintain a steady state, other sources of energy are required. Possible sources include acceleration of the cloud by soft-UV photo-heating, ionization of the cloud by hard-UV radiation, the impacts of winds from massive stars and supernova explosions, or the injection of turbulent energy from large-scale flows such as superbubbles, or collisions with other clouds.

Noteworthy results from the Perseus imaging survey include the discovery of a giant flow, HH 280/317/492/493, emerging from the head of the cometary cloud L1451 (*Walawender et al.*, 2004). This cloud has apparently been sculpted by UV radiation from the B0 star 40 Per located over 25 pc away. The outflow propagates through a portion of the Perseus cloud sufficiently transparent to reveal background galaxies, impacts the western edge of the L1455 cloud at HH 317D, and forms an obscured H_2 emitting shock. The deflected flow powers the HH object HH 317.

Perhaps the richest source of HH jets and outflows within 400 pc of the Sun is the NGC 1333 region in Perseus (*Bally and Reipurth*, 2001; *Bally et al.*, 1996) (see Fig. 3). Bent jets, giant flows, and H_2 shocks overlap along the line of sight, and on deep images are essentially confusion limited. The HH 7/11 group traces a collimated flow emerging from a dense cloud core at roughly PA ~ 125° and originates from the vicinity of an embedded Class I source, SVS 13 (*Strom et al.*, 1976). However, *Rodríguez et al.* (1997, 1999) found several additional embedded radio sources that cluster around SVS 13. The HH 7/11 group is embedded within a high-velocity CO outflow (*Knee and Sandell*, 2000; *Rudolph et al.*, 2001). This flow is one of the few that has been detected and mapped in the 21-cm line of HI (*Lizano et al.*, 1988; *Rodríguez et al.*, 1990). The HH 7/11 region turned out even more complex with the detection of additional nearby sources; the radio source VLA 20 (*Rodríguez et al.*, 1999) and the Class 0 protostar SVS 13B, which is known to drive a highly collimated molecular outflow along a roughly north-south direction (PA ~ 170°) (*Bachiller et al.*, 1998). This flow is bright in the shock-enhanced tracer, SiO. The radio studies of *Rodríguez et al.* (1999) show that the SVS 13 region contains a 3' long chain of over a dozen radio continuum sources that are embedded in a ridge of dense molecular gas and submillimeter cores extending along PA ~ 20°

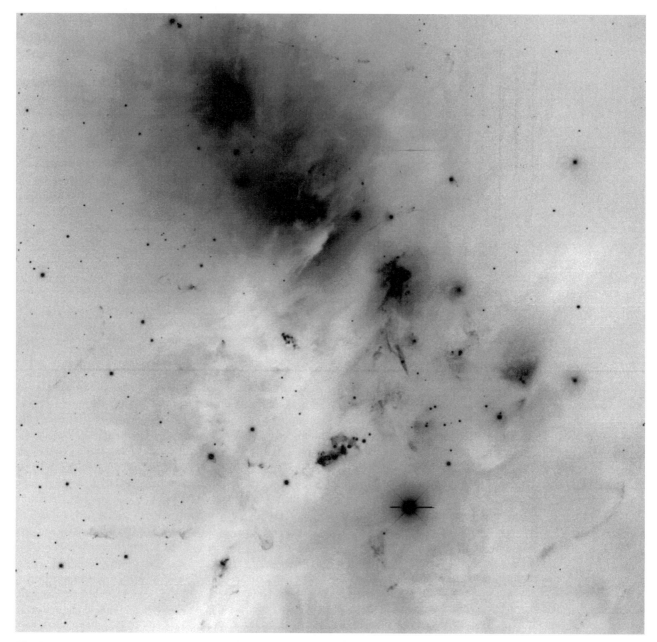

Fig. 3. The NGC 1333 region seen in the combined light of Hα and [SII]. Images obtained with the Mosaic prime-focus camera on the Mayall 4-m reflector.

(*Hatchell et al.*, 2005). The 40"-long dense core containing SVS 13 (VLA 4) includes VLA 20, VLA 3, VLA 17 (also known as SVS 13B), and VLA 2 [H$_2$O (B)]. Thus, it is perhaps not surprising that multiple outflows originate from this region. Interferometric millimeter-wave observations by *Bachiller et al.* (2000) have provided evidence that the actual driving source of the HH 7/11 outflow is indeed SVS 13. These authors produced high-resolution CO maps that show that SVS 13 is the point of origin of a high-velocity CO flow oriented toward HH 7/11.

Other noteworthy regions in Perseus include Barnard 4, which contains the HH 211 jet and is now known to be the site of a small cluster of embedded stars, and dozens of H$_2$

shocks (*Eislöffel et al.*, 2003) and HH objects (*Walawender et al.*, 2005). Barnard 1 in the central portion of Perseus (*Walawender et al.*, 2005) has also been proven to be a rich source of outflows, some of which are parsec-scale.

3.3. Giant Outflows

Bally and Devine (1994) found the first parsec-scale outflow from a low-mass star by recognizing that the HH 34 jet was merely the innermost part of a giant flow extending to HH 33/40 in the north and HH 87/88 in the south. This association was confirmed by *Devine et al.* (1997). Many more giant outflows were reported by *Ogura* (1995),

Reipurth et al. (1997), and *Eislöffel and Mundt* (1997). Most well-studied HH objects have turned out to be internal working surfaces within much larger parsec-scale giant protostellar outflows. These flows include the 10-pc-long HH 111/113/311 system in the northern portion of Orion B, the HH 1/2/401/402 system in Orion A, the HH 83/84 system on the west side of the Orion A cloud, and several nearly 10-pc-long chains emerging out of the L1641N cluster in Orion A (*Reipurth et al.*, 1998b). Since the late 1990s, dozens of giant, parsec-scale outflows have been found.

One of the nearest sites of ongoing star formation in Taurus, the L1551 cloud, has continued to reveal many surprises. In addition to the X-rays at the base of the HH 154 jet emerging from IRS5 (see above), this region contains criss-crossing jets, parsec-scale HH flows, a common CO outflow lobe inflated by several sources, and jet/wind interactions. Multiepoch images were used by *Devine et al.* (1999) to measure proper motions, which revealed that the L1551 bipolar CO outflow is powered not just by the twin jets from L1551 IRS5, but also by a highly collimated jet, HH 454, emerging from one component of the L1551NE binary located about 1' east of IRS5. Thus, the L1551 CO outflow lobes are collectively inflated by winds and jets from several individual protostars. Devine et al. noted that the bright shock HH 29 and possibly HH 28 in the blueshifted lobe of the L1551 CO outflow, together with components of HH 262 in the red-shifted CO lobe, lie exactly on the axis of HH 454 emerging from L1551NE, making this object the likely driving source. HST imaging of HH 29 by *Devine et al.* (2000a) provide supporting evidence for this hypothesis. A large but faint bow shock, HH 286, lies 10' beyond the edge of the L1551 molecular cloud and the red-shifted lobe of the main L1551 molecular outflow in a region full of background galaxies, indicating that the flows from IRS5 and NE have blown completely out of the L1551 cloud. Thus, L1551 contains a remarkable, giant outflow energized by multiple sources.

Pound and Bally (1991) found a remarkable 30' long redshifted outflow lobe that extends due west from HH 102. Although a component of L1551NE has been proposed as a driver, it is possible that there are additional highly embedded YSOs, or previously unrecognized companions to known YSOs near NE and IRS5, that are responsible for this outflow lobe.

The famous HH 30 jet in the L1551 cloud was also found to extend many arcminutes northeast of HL and XZ Tau in the L1551 cloud and to cross the lobes of the L1551 IRS5/NE common outflow lobe toward the southwest. Each of the well-known T Tauri stars in the cloud, HL and XZ Tau and LkHα 358, appears to power an outflow in its own right. Recent integral field spectroscopy and proper motion measurements by Magakian et al. (in preparation) provide evidence that the HL Tau jet is being deflected by a wide-angle wind from XZ Tau. These authors also hypothesize that a collimated outflow from LkHα 358 is deflected by the southwestern lobe of the XZ Tau wind to produce the "Hα" jet a few arcseconds southeast of HL Tau. It is remarkable

that the small, roughly 40-M$_\odot$ L1551 cloud contains about a dozen active outflows, several of which have clearly blown out of the parent cloud and are powering shocks beyond the extent of the associated CO outflow lobes.

The well-known HH 46/47 jet was discovered to be merely the inner part of a giant outflow injecting energy into the interior of the Gum Nebula by *Stanke et al.* (1999), who discovered a pair of giant bow shocks, HH 47NE and HH 47SW, over 10' from the HH 46/47 jet, making this well-studied object a member of the class of giant flows.

Bally et al. (2002c) found that the chain of HH objects HH 90 through 93 located in the Orion B cloud north of NGC 2024 form a single giant flow from the 10 L$_\odot$ Class I source IRAS 05399-0121. This is one of the first results of the complete narrow-band survey of the Orion A and B cloud discussed above.

One of the most remarkable giant bow shocks near a star-forming region is the 30' wide Hα bow shock surrounding HH 131 (*Ogura*, 1991; *Wang et al.*, 2005). The most remarkable aspect of HH 131 and the surrounding filamentary nebula is that it is located more than a full degree from the nearest edge of the Orion A molecular cloud. *Stanke* (2000) hypothesized that the chain of H$_2$ features emerging from source S3 in the southern portion of the Orion A cloud may trace the parent flow leading to HH 131. However, there are no intervening HH objects, and the orientation of the giant bow shock surrounding HH 131 suggests a source located toward the north. The narrow-band imaging survey of Orion has led to the identification of a chain of faint bow shocks and [SII]-bright filaments extending from the vicinity of V380 Ori and the L1641 cluster about 2° north of HH 131. *Wang et al.* (2005) suggest that HH 131 originated from the vicinity of the L1641N cluster. If so, then HH 131 may be the southern end of a giant flow from this cluster, which contains HH 403–406 in the north (*Reipurth et al.*, 1998b), and may contain HH 61/62/127/479/480, which lie south of L1641N between it and HH 131. If these shocks exhibit similar radial velocities and if their proper motions point toward HH 131, then this system may represent the largest giant flow found to date.

Not all shocks in star-forming regions are protostellar. The Rosette Nebula contains a set of high-excitation arcs of emission that have the morphology of bow shocks. They form a linear chain extending across the nebula from northeast to southwest. Tracing this orientation back toward the apparent source of these shocks, there are no known IRAS sources or protostars. However, this direction does contain a supernova remnant — the Monoceros N-ring — that is impinging on the outskirts of the Rosette Nebula. It is possible that outlying portions of the expanding supernova debris field energize the Rosette features, which are rendered visible by the intense UV radiation of the Rosette's massive stars.

Reipurth (2000) noted that most sources of prominent HH flows in low- to moderate-mass star-forming regions are members of multiple systems. He hypothesized that disk perturbations produced during close approaches of companion stars trigger the most energetic episodes of mass loss. As non-hierarchical multiples evolve toward hierarchical systems consisting of a tight binary and either a distant companion or an ejected star, unusually strong outbursts of outflow activity occur. Reipurth postulated that such dynamical interactions produce the giant, parsec-scale outflows, which therefore form a fossil record of the dynamical evolution of newborn binaries. In this picture, the FU Ori outbursts occur at the end of jet production, when the binary companion gradually destroys the magnetic anchoring in the inner disk that is required for jet launch and collimation (*Reipurth and Aspin, 2004*).

3.4. Irradiated Jets

The discovery of irradiated jets embedded in HII regions and in UV-rich environments (*Reipurth et al., 1998a; Cernicharo et al., 1998; Bally and Reipurth, 2001*) has flowered into a productive area of research. Irradiated outflows permit the measurement of flow properties using the standard theory of photo-ionized plasmas that can provide a more robust method of density measurement than the analysis of the highly nonlinear theory of shocks. External radiation renders visible much weaker jets and outflows than the flows seen in dark clouds, where only shock-processed gas can be seen at visual and near-IR wavelengths.

A group of irradiated jets, HH 444 to 447, were found about 20' west of the Horsehead Nebula in Orion in the σ Ori cluster (*Reipurth et al., 1998a*). These four outflows are located within about 10' of the massive O9.5 star σ Ori in a region that contains no molecular gas and very little extinction. The brightest jet, HH 444, is highly asymmetric in intensity and radial velocity. The brighter jet beam, which faces away from σ Ori, is at least four times brighter and has a radial velocity about two times slower than the faint counter jet. A pair of Hα bow shocks are visible about 1' from the source star. The other three irradiated jets in the σ Ori cluster exhibit similar asymmetries. In all cases, the brighter jet beam is slower and faces away from the illuminating star σ Ori. *Reipurth et al.* (1998a) noted an anomalous jet-like tail emerging from the source star of HH 445 but at about a 20° angle with respect to the actual jet. The Hα images show that this 10" long feature points directly toward σ Ori and that it tapers to an unresolved tip away from the source of HH 445. HH 444 and HH 447 also exhibit similar, but fainter tails whose morphology resembles the proplyds observed in the Orion Nebula. *Andrews et al.* (2004) presented high-resolution long-slit spectra of several of these jets obtained with the HIRES spectrometer at the Keck Observatory. These spectra show that, within 2" of the source star of HH 444, the main jet contains two converging radial velocity components (Fig. 4). The first branch starts near the star with a large negative radial velocity (similar to the faint counter jet, but with the opposite sign), decelerates, and merges with a second, brighter branch that starts near the star with a very low radial velocity and accelerates to join with the high-velocity branch. After the two

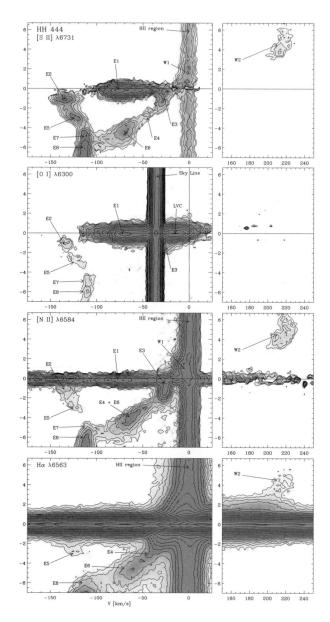

Fig. 4. High-resolution long-slit spectra showing different lines of the irradiated HH 444 jet. The jet is seen in the panels to the left, and the faint counter jet in the righthand panels. From *Andrews et al.* (2004).

branches merge, the radial velocity remains constant for many tens of arcseconds.

A likely interpretation of HH 444 is that the source star is surrounded by a dense, but compact, circumstellar envelope or disk. The jet pointing away from σ Ori interacts with an arcsecond-scale cavity in the envelope and entrains some material from it. The flow may be brightest where it interacts with and entrains material from the walls of a cavity in the envelope. The nearside of the interaction zone may be traced by the first, decelerating, branch of the HH 444 jet, while the farside may be traced by the accelerating second branch seen in the Keck HIRES spectra. The counterjet may not interact significantly with circumstellar matter, is not decelerated, and does not show merging branches of emission. UV-induced photo-ablation of the circumstellar environment by σ Ori may have removed any such medium on this side of the circumstellar disk. Thus, the observed asymmetries in the irradiated jets may be a result of interactions with a highly asymmetric circumstellar environment.

During the last five years, many more irradiated jets have been discovered in the Orion Nebula and in NGC 1333 (*Bally and Reipurth,* 2001; *Bally et al.,* 2006), in M43 (*Smith et al.,* 2005), in the NGC 1977 cluster about 0.5° north of the Orion Nebula (Bally et al., in preparation), in the Carina Nebula (Smith et al., in preparation), and in some other regions. Many of these jets exhibit asymmetries similar to the σ Ori jets.

However, a number of additional features have emerged. A significant subset of irradiated jets show extreme bends indicating deflection by a side wind, radiation pressure, or the rocket effect. The subarcsecond resolution of HST was required to identify dozens of irradiated jets in the core of the Orion Nebula (*Bally et al.,* 2000). Some, such as HH 514 emerging from the proplyd HST 2, exhibit pronounced kinematic and intensity asymmetry. HH 508, emerging from one of the four companion stars to θ¹ Ori B, the northern member of the Trapezium, is a one-sided microjet that has the highest surface brightness of any HH object in Hα, probably due to its location within 10^3 AU from an OB star. *Bally et al.* (2000) and *Bally and Reipurth* (2001) noted that about a dozen low-mass stars, located mostly in the southwestern part of the Orion Nebula, were surrounded by parabolic arcs of emission, indicating deflection of circumstellar material away from the nebular core. These were dubbed "LL Ori" objects after the prototype LL Ori (Fig. 5), which was first noted by *Gull and Sofia* (1979) to be a possible example of a wind-wind collision front. The system HH 505

contains both a jet and an LL Ori bow shock. *Masciadri and Raga* (2001) modeled the parabolic bow as a jet deflected by a side wind. They were able to reproduce the Hα morphology of the parabolic bow surrounding HH 505 and its source star, showing that the bow is produced by the weak shocks formed where the side wind interacts with the jet material that moves away from the jet axis after passing through the bow shock at the head of the jet.

The Advanced Camera for Surveys (ACS) on HST revealed dozens of additional irradiated outflows in the outskirts of the Orion Nebula (*Bally et al.,* 2006). Some of the larger outflows, such as HH 503 to 505, were previously detected on groundbased images (*Bally and Reipurth,* 2001), but many consist of subarcsecond chains of knots and compact bow shocks lost in the glare of the nebula on groundbased images. The ACS images show that the jets in many of these outflows consist of a fine series of subarcsecond knots whose proper motions can be measured on HST images taken only a year apart. The inner parts of these flows show proper motions of up to 300 km s⁻¹. The jets tend to fade to invisibility at distances ranging from less than 1 to 10" from the source to be replaced, in some flows, by a series of bow shocks. HH 502 in the southern part of the Orion Nebula is a clear example. This arcminute-scale flow has a long southern lobe containing a half-dozen bow shocks while the shorter northern lobe contains three. The flow exhibits bending away from the nebula core.

The ACS images demonstrate that many LL Ori objects, including LL Ori itself, contain jets that are frequently asymmetric. In some outflows, such as HH 505, proper motions show that jet segments are deflected away from the nebular core as they skirt along the parabolic bow that wraps around the source star and bipolar jet. *Bally et al.* (2006) show that while most LL Ori-type bows and bent jets in the southwestern quadrant of the Orion Nebula may be deflected by a large-scale outflow of plasma from the nebular core, even in the absence of such a side wind, radiation pressure acting on dust and the asymmetric photo-ablation of a neutral jet beam can also deflect irradiated jets. As the neutral jet beam emerges from the neutral circumstellar environment into the irradiated environment of the HII region interior, the photo-ionized skin of the jet expands away from the remaining neutral jet core. Because for most irradiated jets, the radiation field is highly anisotropic, the photo-ablation flow exerts a force on the neutral jet beam, deflecting it away from the illuminating star or stars.

The jet densities, mass-loss rates, and other properties can be estimated from the jet proper motions or radial velocities combined with estimates of the electron density at the jet surface determined from the [SII] line ratio, the Hα emission measure, or the apparent length of the jet (*Bally et al.,* 2006). Analysis of the irradiated jets in the Orion Nebula indicate mass-loss rates ranging from as little as 10^{-9} $M_\odot$ yr⁻¹ to more than 10^{-7} $M_\odot$ yr⁻¹. Observations of irradiated jets can detect lower mass-loss rates than the classical methods used to study outflows such as CO, radio continuum, or the various shock-excited tracers.

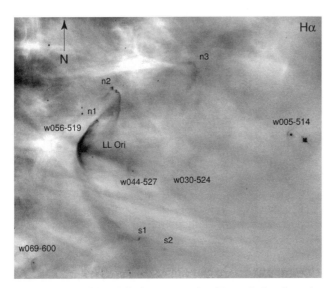

Fig. 5. An HST/WFPC2 image showing Hα emission from the parabolic bow shock and compact bow shock surrounding LL Ori in the Orion Nebula. The deflected jet is invisible in this image, but can be seen as a high-velocity feature in Fabry-Perot data cubes.

A particularly beautiful example of a jet powered by a low-mass young star embedded in a circumstellar disk and a large externally ionized proplyd was presented by *Bally et al.* (2005). The Beehive proplyd contains a several hundred AU diameter silhouette disk, a jet that can be traced for over 1' south of the source star in Fabry-Perot images, and a chain of large bow shocks, HH 540, visible on ground-based images. The proplyd ionization front contains a set of three or four concentric rings of enhanced Hα emission that appear as corrugations. These features may have been triggered by a series of mass ejections from the central star that are responsible for some of the bow shocks in the HH 540 outflow.

The Orion Nebula contains at least several giant, low radial velocity bow shocks best traced by [SII] Fabry-Perot data (*Bally et al.*, 2001). The HH 400 bow is about 2' in diameter, 8' long, and extends south from a few arcminutes south of the Bright Bar in the Orion Nebula. The bow orientation indicates that the source lies somewhere in the general vicinity of the Trapezium stars. It could be a star still embedded in one of the molecular cloud cores behind the nebula.

The Hα and [SII] survey of the entire Orion A and B clouds conducted with the NOAO 4-m telescopes and their mosaic CCD cameras (Bally et al., in preparation) reveal the presence of a 5' scale bow shock at the southwestern periphery of the Orion Nebula. This feature may be the terminus of a 30' scale outflow that emerges from the Orion Nebula core parallel and just north of Orion's Bright Bar. Many filaments and arcs of Hα emission located south and southwest of the Orion Nebula core may trace shocks or dense photo-ionized filaments in this flow. Multiepoch HST images show proper motions ranging from 20 to over 50 km s⁻¹. This large-scale flow may be several times larger than HH 400. HH 400 and this giant flow in the southwestern part of the nebula may be responsible for deflecting the bent jets and LL Ori-type bows in this quadrant of the nebula. However, it is not clear if this is a giant outflow driven by one or more embedded young stars, the leading edge of the stellar wind bubble powered by the Trapezium stars, or possibly the outflow of plasma photo-ablated from the background Orion A cloud by the Trapezium stars.

Although the nature of the giant bow shocks emerging from the Orion Nebula remains unclear, protostellar outflows have been observed to break out of their parent molecular clouds into adjacent HII regions where they become irradiated and ionized. Examples include a pair of giant bow shocks emerging into the S140 HII region from the adjacent dense cloud core (*Bally et al.*, 2002b), the quadrupolar outflow containing HH 124 and HH 571/572 from NGC 2264N (*Ogura*, 1995; *Reipurth et al.*, 2004), and the giant bow shocks emerging from embedded sources in the dense cloud cores associated with IC 1396N (*Reipurth et al.*, 2003).

Smith et al. (2004) reported the discovery of several parsec-scale irradiated outflows emerging from dusty pillars embedded within the Carina Nebula located at a distance of about 2.2 kpc (Fig. 6). The HH 666 flow emerges from a moderate luminosity IRAS source, drives a jet bright in [FeII], Hα, and [OIII], and is associated with a series of giant bow shocks. Long-slit spectra reveal a set of three distinct Hubble flows indicating a series of three ever older mass ejections that exhibit radial velocities as large as ±200 km s⁻¹.

The Carina Nebula also contains many small cometary clouds with sizes up to tens of thousands of AU in length. Many contain IR sources, presumably young stars in their heads. A few exhibit spectacular irradiated jets that are very bright in Hα, presumably because the UV and ionizing radiation field in Carina is nearly 100× stronger than the field in the Orion Nebula. The presence of cometary clouds with properties similar to Orion's proplyds that contain IR sources or visible stars in their heads and irradiated jets and outflows indicates that active star formation is still occurring within the molecular clouds embedded in and surrounding the Carina Nebula HII region.

4. INFRARED OBSERVATIONS OF MOLECULAR JETS

4.1. H₂ and [FeII] as Tracers of Jets and Bow Shocks

The 1990s saw a rapid increase in near-IR observations of outflows, particularly flows from embedded Class 0 and Class I YSOs (see the review by *Reipurth and Bally*, 2001). Observations benefited in particular from increasing array sizes and improved image quality at 4-m-class telescopes. Between 1 and 2.5 μm the most useful probes of outflows are undoubtedly H₂ and [FeII]. Both species produce a wealth of lines across the near-IR bands that allow us to measure the excitation and kinematics of shocked gas in outflows. H₂ in particular is also proving to be a powerful tracer of parsec-scale jets (*Eislöffel*, 2000; *Stanke et al.*, 2002) (see also Fig. 7) as well as outflows from massive (early-B-type) young stars (Fig. 8) (Varricatt et al., in preparation).

Rigorous efforts have been devoted to interpreting H₂ (and CO) observations of jets in terms of entrainment of ambient gas (e.g., *Downes and Cabrit*, 2003) and the physics of discrete shock fronts (e.g., *Smith et al.*, 2003a). *Rosen and Smith* (2004) claim that the H₂ 1-0 S(1)/CO J = 1–0 flux ratio may even be used as an age discriminant. Although collimated jets seem present at all evolutionary stages, from Class 0 through to optically visible TTSs, there are still cases where poorly collimated winds best explain the observations (*Lee et al.*, 2001; *Davis et al.*, 2002). Whether collimation is a factor of age or environment is not yet clear.

For individual shock features, line profiles — when combined with maps of excitation across the shock surface — can be used to constrain bow shock physics, as well as parameters pertaining to the overall outflow, such as the transverse magnetic field strength, jet density, velocity, and flow orientation (*Davis et al.*, 2001a; *Schultz et al.*, 2005). Typically, magnetically cushioned C-type shocks are required to produce the bright molecular emission in extended bow wings, while dissociative J-type shocks probably produce

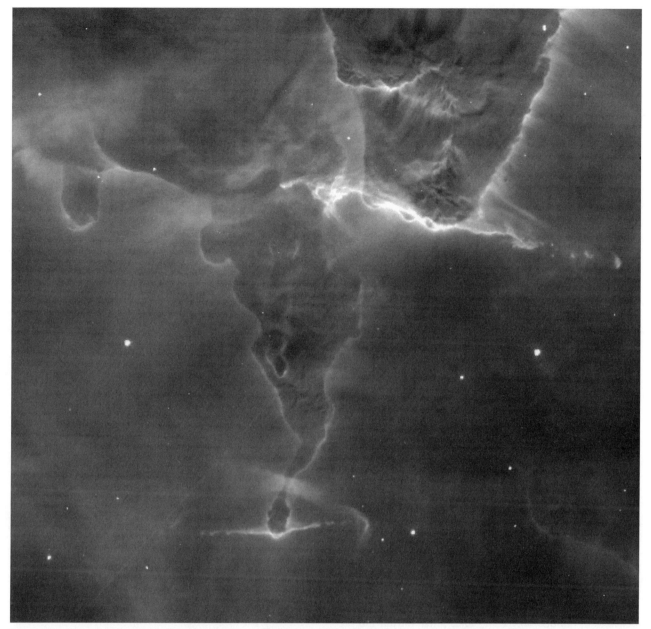

Fig. 6. A pair of bright externally irradiated jets emerging from dusty pillars north of the cluster Trumpler 14 in the Carina Nebula observed with ACS on HST with the F658N filter, which transmits Hα and [NII]. From Smith et al. (in preparation).

the [FeII] emission near the bow head (*O'Connell et al.*, 2005). Fluorescent excitation, which produces a lower rotational than vibrational temperature in H_2 excitation diagrams, can be ruled out in most cases.

In some flows, however, enhancement of molecular emission from levels at very low and very high excitation, which can be difficult to reproduce with steady-state J- and C-type shocks, respectively, have lead researchers to consider the intermediate case of a J-type shock with a magnetic precursor (e.g., *McCoey et al.*, 2004; *Giannini et al.*, 2004). Such a structure is possible only if there is a sufficiently-strong magnetic field, so that the J shock is enveloped by a C-type flow. Indeed, given the lengthy timescale for a shock wave to attain steady state (~10^5 yr), a J shock may slowly evolve

into a C shock (via a J shock plus magnetic precursor) over the lifetime of an outflow, requiring nonequilibrium shock modeling of bows in some regions (*Le Bourlot et al.*, 2002; *Flower et al.*, 2003). This also means that younger flows may drive shocks that are closer to J than C type, with correspondingly higher excitation spectra and molecular line populations closer to LTE.

Parabolic bow shapes, preshock densities of 10^4–10^5 cm^{-3}, and shock velocities (relative to the preshock medium) of 30–100 km s^{-1} generally best describe the observations (e.g., *Smith et al.*, 2003a; *Khanzadyan et al.*, 2004b; *Giannini et al.*, 2004). Inclined magnetic fields or an inhomogeneous preshock medium can produce asymmetries in observed bow shock shapes. It also seems clear that the jet itself must

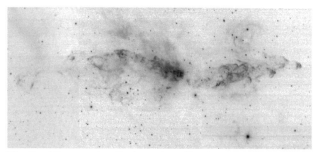

Fig. 7. The complex Cepheus A outflow in H_2. This image was obtained with the NICFPS camera on the 3.5-m Apache Point Observatory telescope in November 2004.

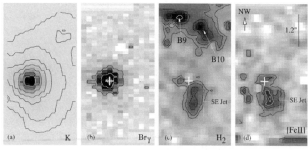

Fig. 8. Near-IR images extracted from UKIRT/UIST Integral Field spectroscopy data (*Davis et al.*, 2004). **(a)** The K-band image (2.03–2.37 μm) shows the nebulosity associated with the massive YSO IRAS 18151–1208. **(b)–(d)** Continuum-subtracted "narrow-band" images, in Brγ, H_2 1-0S(1), and [FeII] 1.644-μm emission, revealing line emission associated with accretion (Brγ) and a collimated jet (H_2 and [FeII]). The jet extends a few arcseconds to the southeast of the embedded source; the jet may in fact be driven by an unseen companion to the near-IR source, since it is offset slightly to the right. The features labeled B9 and B10 are H_2 emission knots in the extended counterflow.

be partially molecular, since not all H_2 features can be explained by the interaction of the jet with the ambient medium. The survival and collisional excitation of H_2 is only possible in oblique bow wings and, often, only if the preshock medium is already in motion. The H_2 must therefore have already passed through one or more bow shocks downwind. The H_2 must also survive these shocks, since at a density of $\sim 10^4$ cm^{-3} the timescale for H_2 reformation on grains is $\sim 2 \times 10^5$ yr, comparable to the outflow dynamical timescale (*O'Connell et al.*, 2004). On the other hand, it has recently been proposed that gas phase H_2 formation in the warm, dust-free, partially ionized regions associated with internal working surfaces could account for the H_2 emission in some jets (*Raga et al.*, 2005).

Pulsed, precessing jet scenarios can be adopted to widen flows and produce the clusters of bows often seen in images of jets, like Cep A in Fig. 7 (e.g., *Völker et al.*, 1999; *Raga et al.*, 2004; *Smith and Rosen*, 2005a). It is interesting to note that, as is the case with many HH jets from TTS, some degree of variability is usually needed to produce the range of characteristics — chains of knots, multiple bows, broad cavities, etc. — observed in molecular flows.

[FeII] is often used to complement the molecular hydrogen data, as a probe of the higher-excitation, atomic jet component (*Reipurth et al.*, 2000; *Nisini et al.*, 2002a; *Giannini et al.*, 2002, 2004). [FeII] is particularly useful in regions where extinction inhibits optical studies, for example, in outflows from low-mass Class 0 sources or high-mass YSOs, or from the base of young jets close to the embedded protostar. With combined H- and K-band spectra, basic physical parameters can be traced along the jet axis. Where extinction is less extreme, e.g., where the jet breaks out of the circumstellar envelope, combined optical and near-IR studies can likewise be used to probe differing atomic gas components; the [FeII] 1.533/1.643 μm ratio, for example, probes densities up to 10^5 cm^{-3} (Fig. 9), one order of magnitude larger than those diagnosed through the [SII] 6716/6731 ratio (*Nisini et al.*, 2005). Because the [FeII] lines in the H band derive from levels with similar excitation energies (11,000–12,000 K), this ratio is almost independent of temperature. The [FeII] 1.643 μm/[SII] 1.03 μm or the [FeII] 1.25 μm/Paβ ratio can also potentially be used to estimate the fraction of iron in the gas phase, important when considering cooling, chemistry, and grain-sputtering processes in shocks, while the [FeII] 1.643 μm/[NI] 1.04 μm ratio can be used to estimate the ionization fraction and ultimately the jet mass flux (*Nisini et al.*, 2002a; *Giannini et al.*, 2004).

It is likely that most, if not all, of the infrared outflows that are detected in H_2 and [FeII] are embedded cases of HH flows that would have been seen in the optical if the extinction had been lower. An interesting transition case is

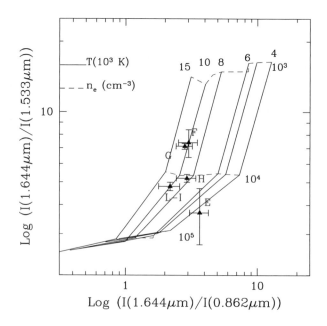

Fig. 9. Diagnostic diagram showing how [FeII] line ratios can be used to constrain the conditions in an ionized jet. The grid spans electron densities of 10^3, 10^4, and 10^5 cm^{-3} (dashed lines) and electron temperatures of 4, 5, 6, 8, 10, and 15×10^3 K (solid lines). Overplotted are line ratios extracted from near-IR spectra of the HH 1 jet. See *Nisini et al.* (2005) for details.

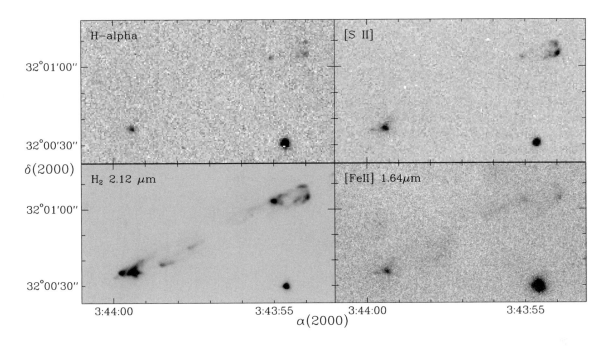

Fig. 10. The HH 211 flow emerges from a deeply embedded source in the IC 348 region in Perseus. The flow moves through a steep density gradient, and only the very outermost shock regions are visible as optical HH objects. From *Walawender et al.* (2006).

the HH 211 flow, which was originally discovered in the infrared (*McCaughrean et al.*, 1994). In the optical, only the outermost shock regions can be seen in Hα and [SII] (Fig. 10) (*Walawender et al.*, 2006).

4.2. Emission from the Jet Base: Constraining Jet Models

In the near-IR, considerable effort has also been spent on observing the central engine, particularly the region at the jet base within ~100 AU of the outflow source, where jets are collimated and accelerated. Prompted by the spectroscopic survey of *Reipurth and Aspin* (1997), and earlier optical studies of forbidden emission line (FEL) regions in TTS (e.g., *Hirth et al.*, 1997), *Davis et al.* (2001b, 2002, 2003) conducted a modest survey of Class I outflows at high spectral (although poor spatial) resolution (Fig. 11). Once again, H_2 and [FeII] proved to be important jet tracers, the H_2 delineating a slow ($v_{rad} \sim 10\text{--}30$ km s^{-1}), low-excitation, shocked molecular gas component (T ~ 2000K), with the [FeII] tracing a high-velocity ($v_{rad} \sim 50\text{--}100$ km s^{-1}), hot, dense, partially ionized region (T ~ 10,000K). While it seems likely that the [FeII] emission is closely tied to emission knots and shock fronts along the jet axis, the origin of the H_2 remains unclear; it could originate in the low-velocity wings of small bow shocks, from a turbulent boundary between the jet and ambient medium, or from a broad, cool component in a disk wind.

Spectro-astrometric techniques were employed to extract spatial information for the various kinematic components in each line. However, it was quickly realized that higher-spatial resolution was required, not least so that emission

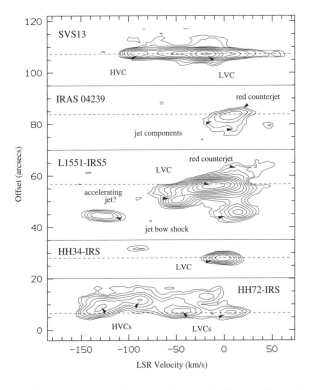

Fig. 11. Echelle spectral images showing the H_2 emission at the base of five molecular jets. The long slit was aligned with the jet axis in each region. The continuum (marked with a dashed line in each plot) has been fitted and removed from each dataset. Low-velocity H_2 is observed coincident with each outflow source, although spectro-astrometry usually indicates an offset, of the order of a few tens of AU, into the blue outflow lobe. Additional high-velocity shock features are seen along the jet axis in some flows. From *Davis et al.* (2001b).

at the "true jet base" could be distinguished from shock fronts further downwind. *Pyo et al.* (2002) and *Takami et al.* (2005) have since sought to characterize the kinematics and excitation of the atomic and molecular gas component on fine spatial scales, while *Davis et al.* (2006) have combined Subaru and Very Large Telescope (VLT) data to measure proper motions. Interestingly, in SVS 13 they measure a modest proper motion for the H_2 peak at the jet base, although they find that the [FeII] seems to be stationary. If confirmed with subsequent observations, this would suggest that the [FeII] is excited in a stationary shock, perhaps in a collimation point at the jet base.

The aim remains to constrain models of jet collimation and acceleration, distinguishing between X-wind and disk-wind models that to date have only been applied to relatively "evolved" YSOs (see the chapter by Ray et al.). Extending observational and theoretical studies to younger (Class 0/I) sources is important, particularly when one considers that a star's mass has largely been determined before the T Tauri phase, with accretion during the T Tauri phase adding only a small fraction to the final stellar mass.

Future observations, which *map* excitation and kinematics outward from the source, but also in the transverse direction across the width of the jet, will allow us to better test acceleration models, and also understand how energy and momentum are transferred from the jet to the ambient medium (e.g., *Chrysostomou et al.*, 2005). Near-IR integral field spectrometers, particularly those that offer high spectral resolution, or those with adaptive optics correction, should help further this cause.

4.3. Mid-Infrared Observations from Space

By extending infrared observations of jets to longer wavelengths, one has the opportunity to observe the very youngest flows, tracing emission from low-excitation regions (cool, molecular emission) but also from embedded, high-excitation jets (ionic emission) (reviewed by *van Dishoeck*, 2004). The Infrared Space Observatory (ISO) showed that essentially all the broadband mid-IR emission from outflows is due to lines. Again, H_2 and [FeII] are important coolants, via the pure rotational H_2 lines S(0)–S(7) between 5 and 30 μm, and the [FeII] forbidden emission lines, particularly those at 5.34, 17.94, 25.99, and 35.35 μm. The 17.03 μm H_2 v = 0–0 S(1) line, for example, from an energy level only 1015 K above the ground state, emits strongly from gas as cool as 100 K, while the [FeII] lines at 5.34 and 25.99 μm should be as strong as the 1.64 μm line, given conditions typical of protostellar jets (*Smith and Rosen*, 2005b; *Hartigan et al.*, 2004). Although the near- and mid-IR [FeII] lines trace similar flow components, the difference in extinction between 1 and 5 μm (on the order of 7) is potentially a problem and, with the launch of Spitzer, near-comparable spatial resolution is now possible.

As in the near-IR, H_2 and [FeII] line ratios can be used to test shock models and diagnose flow parameters. For example, the [FeII] 17.94 μm/5.34 μm and 25.99 μm/17.94 μm ratios can be used to constrain the density and temperature

in the ionized flow (*White et al.*, 2000), while the H_2 S(5)/S(1) vs. S(5)/S(4) ratio diagram can be used to estimate shock velocities (*Molinari et al.*, 2000). The H_2 S(3) line at 9.67 μm can also be used to measure extinction, because it lies in the middle of a silicate absorption feature. A_v can thus be evaluated by aligning (or reducing the scatter) of H_2 column densities in a Boltzmann diagram, and particularly by adjusting A_v until the S(3) line converges toward the straight line defined by the other H_2 column densities.

Observations in transitions of other ionic species (e.g., [NeII] 12.81 μm, [NeIII] 15.55 μm, [SIII] 33.5 μm, [SiII] 34.82 μm) allow one to probe faster, higher-excitation shocks. For example, the [NeII] 12.81-μm line intensity is directly related to the shock velocity, requiring J shocks in excess of ~60 km s^{-1}, while [NeIII] 15.55-μm detections imply shock velocities exceeding 100 km s^{-1} (*Hollenbach and McKee*, 1989; *Molinari and Noriega-Crespo, 2002; Lefloch et al.*, 2003). With ISO these lines were only detected in massive YSO outflows, the exception being the high-excitation knots in HH 2. Nondetections in some high-excitation HH objects from low-mass YSOs were probably due to beam-filling effects, since the emitting regions, detected in optical high-excitation tracers like [OIII] 5007 emission, are known to be very small. Spitzer is proving to be more sensitive in this respect (e.g., *Noriega-Crespo et al.*, 2004a,b; *Morris et al.*, 2004).

Some ionic transitions, like [SiII] 34.82 μm, [FeII] 25.99 μm, and [FeII] 35.35 μm, can also be excited in intense UV fields, so although line ratios can be used to distinguish shock from PDR excitation, caution needs to be taken when analyzing data in some regions (e.g., *Froebrich et al.*, 2002).

At still longer wavelengths there are many important coolants for gas at temperatures between 300 and 1500 K (*Benedettini et al.*, 2000; *Molinari et al.*, 2000; *Giannini et al.*, 2001; *Smith et al.*, 2003b). Observations out to 200 μm have the potential to bridge the gap between the cold molecular flow components traced in the (sub)millimeter and the hot atomic/ionic shock tracers seen in the optical and near-IR. Indeed, the complete IR spectral coverage provided by ISO and Spitzer leads to a direct determination of the contribution of each species to the cooling, and a better overall estimate of the energy budget in each outflow. In Class 0 and Class I YSO flows, *Nisini et al.* (2002b) find that cooling from FIR lines is equivalent to a few percent of the source bolometric luminosity, the cooling from molecular lines being larger in outflows from Class 0 sources than in flows from Class I sources. Generally, CO, H_2, and H_2O contribute more or less equally to the cooling, although emission from gaseous water was conspicuously absent in many Class I YSO outflows, perhaps because of dissociation in shocks, dissociation by the interstellar FUV field, or freeze-out onto warm dust grains. Together these molecular lines radiate perhaps 50% of the mechanical power input by shocks, as one would expect from a momentum driven flow, where the mechanical power of the flow should be comparable to the energy radiated in the dissipative shocks. C-type shocks usually best fit the observed molecular line ratios — in outflows from low *and* high-mass YSOs — with parameters similar to those inferred from near-IR observations (i.e. $v_{sh} \sim 10$–30 km s^{-1}; $n_H \sim 10^4$–10^6 cm^{-3}). However, a J-type component is also needed to explain the higher-excitation transitions (including [OI] 63-μm emission) seen in some regions, particularly in the Class I YSO flows. Note that the [CII] 158 μm/[OI] 63 μm fine-structure line ratio can be used to distinguish C from J shocks (being lower in the latter), provided the [CII] is not contaminated by emission from a PDR (*Benedettini et al., 2000; Molinari and Noriega-Crespo,* 2002).

With ISO laying the groundwork, Spitzer is now in a position to build on these early results, by *spatially distinguishing* regions of hot (1000–3000 K) molecular material from cool (200–1000 K) molecular gas in extended bow shocks and thereby fine-tuning combined C- plus J-type (or J-type plus magnetic precursor) shock models. Detecting jets from the most embedded flows, from very low-mass YSOs, and from distant, high-mass YSOs should also be possible.

Acknowledgments. C.J.D. would like to thank M. Smith, B. Nisini, D. Froebrich, and A. Noriega-Crespo for their comments and suggestions. J.B. thanks P. Hartigan for suggestions regarding the interpretation of the HH 444 jet spectra. We thank the referee, A. Raga, for helpful comments. This work was supported by the NASA Astrobiology Institute under Cooperative Agreement No. NNA04CC08A issued through the Office of Space Science.

REFERENCES

Andrews S. M., Reipurth B., Bally J., and Heathcote S. R. (2004) *Astrophys. J., 606,* 353–368.
Bachiller R., Guilloteau S., Gueth F., Tafalla M., Dutrey A., Codella C., and Castets A. (1998) *Astron. Astrophys., 339,* L49–L52.
Bachiller R., Gueth F., Guilloteau S., Tafalla M., and Dutrey A. (2000) *Astron. Astrophys., 362,* L33–L36.
Bally J. and Devine D. (1994) *Astrophys. J., 428,* L65–L68.
Bally J. and Reipurth B. (2001) *Astrophys. J., 546,* 299–323.
Bally J., Devine D., and Reipurth B. (1996) *Astrophys. J., 473,* L49–L52.
Bally J., O'Dell C. R., and McCaughrean M. J. (2000) *Astron. J., 119,* 2919–2959.
Bally J., Johnstone D., Joncas G., Reipurth B., and Mallén-Ornelas G. (2001) *Astron. J., 122,* 1508–1524.
Bally J., Heathcote S., Reipurth B., Morse J., Hartigan P., and Schwartz R. D. (2002a) *Astron. J., 123,* 2627–2657.
Bally J., Reipurth B., Walawender J., and Armond T. (2002b) *Astron. J., 124,* 2152–2163.
Bally J., Reipurth B., and Aspin C. (2002c) *Astrophys. J., 574,* L79–L82.
Bally J., Feigelson E., and Reipurth B. (2003) *Astrophys. J., 584,* 843–852.
Bally J., Licht D., Smith N., and Walawender J. (2005) *Astron. J., 129,* 355–362.
Bally J., Licht D., Smith N., and Walawender J. (2006) *Astron. J., 131,* 473–500.
Benedettini M., Giannini T., Nisini B., et al. (2000) *Astron. Astrophys., 359,* 148–158.
Cernicharo J., Lefloch B., Cox P., Cesarsky D., Esteban C., et al. (1998) *Science, 282,* 462–465.
Chrysostomou A., Bacciotti F., Nisini B., Ray T. P., Eislöffel J., Davis C. J., and Takami M. (2005) *PPV Poster Proceedings.* Available on line at http://www.lpi.usra.edu/meetings/ppv2005/pdf/8156.pdf.
Davis C. J., Hodapp K. W., and Desroches L. (2001a) *Astron. Astrophys., 377,* 285–296.

Davis C. J., Ray T. P., Desroches L., and Aspin C. (2001b) *Mon. Not. R. Astron. Soc., 326,* 524–538.
Davis C. J., Stern L., Ray T. P., and Chrysostomou A. (2002) *Astron. Astrophys., 382,* 1021–1031.
Davis C. J., Whelan E., Ray T. P., and Chrysostomou A. (2003) *Astron. Astrophys., 397,* 693–710.
Davis C. J., Varricatt W. P., Todd S. P., and Ramsay Howat S. K. (2004) *Astron. Astrophys., 425,* 981–995.
Davis C. J., Nisini B., Takami M., Pyo T. S., Smith M. D., et al. (2006) *Astrophys. J., 639,* 969–974.
Devine D., Bally J., Reipurth B., and Heathcote S. (1997) *Astron. J., 114,* 2095–2111.
Devine D., Reipurth B., and Bally J. (1999) *Astron. J., 118,* 972–982.
Devine D., Bally J., Reipurth B., Stocke J., and Morse J. (2000a) *Astrophys. J., 540,* L57–L59.
Devine D., Grady C. A., Kimble R. A., Woodgate B., Bruhweiler F. C., Boggess A., Linsky J. L., and Clampin M. (2000b) *Astrophys. J., 542,* L115–L118.
Dopita M. A., Schwartz R. D., and Evans I. (1982) *Astrophys. J., 263,* L73–L77.
Downes T. and Cabrit S. (2003) *Astron. Astrophys., 403,* 135–140.
Eislöffel J. (2000) *Astron. Astrophys., 354,* 236–246.
Eislöffel J. and Mundt R. (1997) *Astron. J., 114,* 280–287.
Eislöffel J., Froebrich D., Stanke T., and McCaughrean M. J. (2003) *Astrophys. J., 595,* 259–265.
Favata F., Fridlund C. V. M., Micela G., Sciortino S., and Kaas A. A. (2002) *Astron. Astrophys., 386,* 204–210.
Flower D. R., Le Bourlot J., Pineau des Forêts G., and Cabrit S. (2003) *Mon. Not. R. Astron. Soc., 341,* 70–80.
Froebrich D., Smith M. D., and Eislöffel J. (2002). *Astron. Astrophys., 385,* 239–256.
Giannini T., Nisini B., and Lorenzetti D. (2001) *Astrophys. J., 555,* 40–57.
Giannini T., Nisini B., Caratti o Garatti A., and Lorenzetti D. (2002) *Astrophys. J., 570,* L33–L36.
Giannini T., McCoey C., Caratti o Garatti A., Nisini B., Lorenzetti D., and Flower D. R. (2004) *Astron. Astrophys., 419,* 999–1014.
Grady C. A., Woodgate B., Torres C. A. O., Henning Th., Apai D., et al. (2004) *Astrophys. J., 608,* 809–830.
Grady C. A., Woodgate B. E., Bowers C. W., Gull T. R., Sitko M. L., et al. (2005) *Astrophys. J., 630,* 958–875.
Gull T. R. and Sofia S. (1979) *Astrophys. J., 230,* 782–785.
Haro G. (1952) *Astrophys. J., 115,* 572–572.
Haro G. (1953) *Astrophys. J., 117,* 73–82.
Hartigan P., Morse J. A., Tumlinson J., Raymond J., and Heathcote S. (1999) *Astrophys. J., 512,* 901–915.
Hartigan P., Raymond J., and Pierson R. (2004) *Astrophys. J., 614,* L69–L71.
Hatchell J., Richer J. S., Fuller G. A., Qualtrough C. J., Ladd E. F., and Chandler C. J. (2005) *Astron. Astrophys., 440,* 151–161.
Herbig G. H. (1950) *Astrophys. J., 111,* 11–14.
Herbig G. H. (1951) *Astrophys. J., 113,* 697–699.
Herbig G. H. and Jones B. F. (1981) *Astron. J., 86,* 1232–1244.
Hirth G., Mundt R., and Solf J. (1997) *Astron. Astrophys. Suppl., 126,* 437–469.
Hollenbach D. and McKee C. F. (1989) *Astrophys. J., 342,* 306–336.
Kastner J. H., Franz G., Grosso N., Bally J., McCaughrean M. J., Getman K., Feigelson E. D., and Schulz N. S. (2005) *Astrophys. J. Suppl., 160,* 511–529.
Khanzadyan T., Gredel R., Smith M. D., and Stanke T. (2004a) *Astron. Astrophys., 426,* 171–183.
Khanzadyan T., Smith M. D., Davis C. J., and Stanke T. (2004b) *Astron. Astrophys., 418,* 163–176.
Knee L. B. G. and Sandell G. (2000) *Astron. Astrophys., 361,* 671–684.
Le Bourlot J., Pineau des Forêts G., Flower D. R., and Cabrit S. (2002) *Mon. Not. R. Astron. Soc., 332,* 985–993.
Lee C. F., Stone J. M., Ostriker E. C., and Mundy L. G. (2001) *Astrophys. J., 557,* 429–442.
Lefloch B., Cernicharo J., Cabrit S., Noriega-Crespo A., Moro-Martín A., and Cesarsky D. (2003) *Astrophys. J., 590,* L41–L44.

Lizano S., Heiles C., Rodríguez L. F., Koo B. C., Shu F. H., et al. (1988) *Astrophys. J., 328,* 763–776.

Masciadri E. and Raga A. C. (2001) *Astron. J., 121,* 408–412.

McCaughrean M. J., Rayner J. T., and Zinnecker H. (1994) *Astrophys. J., 436,* L189–L192.

McCoey C., Giannini T., Flower D. R., and Caratti o Garatti A. (2004) *Mon. Not. R. Astron. Soc., 353,* 813–824.

Miesch M. S. and Bally J. (1994) *Astrophys. J., 429,* 645–671.

Miesch M. S., Scalo J., and Bally J. (1999) *Astrophys. J., 524,* 895–922.

Molinari S. and Noriega-Crespo A. (2002) *Astron. J., 123,* 2010–2018.

Molinari S., Noriega-Crespo A., Ceccarelli C., Nisini B., Giannini T., et al. (2000) *Astrophys. J., 538,* 698–709.

Morris P. W., Noriega-Crespo A., Marleau F. R., Teplitz H. I., Uchida K. I., and Armus L. (2004) *Astrophys. J. Suppl., 154,* 339–345.

Mundt R. and Fried J. W. (1983) *Astrophys. J., 274,* L83–L86.

Nisini B., Caratti o Garatti A., Giannini T., and Lorenzetti D. (2002a) *Astron. Astrophys., 393,* 1035–1051.

Nisini B., Giannini T., and Lorenzetti D. (2002b) *Astrophys. J., 574,* 246–257.

Nisini B., Bacciotti F., Giannini T., Massi F., Eislöffel J., et al. (2005) *Astron. Astrophys., 441,* 159–170.

Noriega-Crespo A., Moro-Martin A., Carey S., Morris P. W., Padgett D. L., et al. (2004a) *Astrophys. J. Suppl., 154,* 402–407.

Noriega-Crespo A., Morris P., Marleau F. R., Carey S., Boogert A., et al. (2004b) *Astrophys. J. Suppl., 154,* 352–358.

O'Connell B., Smith M. D., Davis C. J., Hodapp K. W., Khanzadyan T., and Ray T. P. (2004) *Astron. Astrophys., 419,* 975–990.

O'Connell B., Smith M. D., Froebrich D., Davis C. J., and Eislöffel J. (2005) *Astron. Astrophys., 431,* 223–234.

Ogura K. (1991) *Astron. J., 101,* 1803–1806.

Ogura K. (1995) *Astrophys. J., 450,* L23–L26.

Phelps R. L. and Barsony M. (2004) *Astron. J., 127,* 420–443.

Pound M. W. and Bally J. (1991) *Astrophys. J., 383,* 705–713.

Pravdo S. H. and Tsuboi Y. (2005) *Astrophys. J., 626,* 272–282.

Pravdo S. H., Feigelson E. D., Garmire G., Maeda Y., Tsuboi Y., and Bally J. (2001) *Nature, 413,* 708–711.

Pravdo S. H., Tsuboi Y., and Maeda Y. (2004) *Astrophys. J., 605,* 259–271.

Pyo T-S., Hayashi M., Kobayashi N., Terada H., Goto M., et al. (2002) *Astrophys. J., 570,* 724–733.

Raga A. C., Noriega-Crespo A., and Velázquez P. F. (2002) *Astrophys. J., 576,* L149–L152.

Raga A. C., Noriega-Crespo A., González R. F., and Veázquez P. F. (2004) *Astrophys. J. Suppl., 154,* 346–351.

Raga A. C., Williams D. A., and Lim A. J. (2005) *Rev. Mex. Astron. Astrofis., 41,* 137–146.

Reipurth B. (2000) *Astron. J., 120,* 3177–3191.

Reipurth B. and Aspin C. A. (1997) *Astron. J., 114,* 2700–2707.

Reipurth B. and Aspin C. A. (2004) *Astrophys. J., 608,* L65–L68.

Reipurth B. and Bally J. (2001) *Ann. Rev. Astron. Astrophys., 39,* 403–455.

Reipurth B., Bally J., Graham J. A., Lane A. P., and Zealey W. J. (1986) *Astron. Astrophys., 164,* 51–66.

Reipurth B., Bally J., and Devine D. (1997) *Astron. J., 114,* 2708–2735.

Reipurth B., Bally J., Fesen R. A., and Devine D. (1998a) *Nature, 396,* 343–345.

Reipurth B., Devine D., and Bally J. (1998b) *Astron. J., 116,* 1396–1411.

Reipurth B., Yu K. C., Heathcote S., Bally J., and Rodríguez L. F. (2000) *Astron. J., 120,* 1449–1466.

Reipurth B., Armond T., Raga A., and Bally J. (2003) *Astrophys. J., 593,* L47–L50.

Reipurth B., Yu K. C., Moriarty-Schieven G., Bally J., Aspin C., and Heathcote S. (2004) *Astron. J., 127,* 1069–1080.

Rodríguez L. F., Escalante V., Lizano S., Cantó J., and Mirabel I. F. (1990) *Astrophys. J., 365,* 261–268.

Rodríguez L. F., Anglada G., and Curiel S. (1997) *Astrophys. J., 480,* L125–L128.

Rodríguez L. F., Anglada G., and Curiel S. (1999) *Astrophys. J. Suppl., 125,* 427–438.

Rosen A. and Smith M. D. (2004) *Mon. Not. R. Astron. Soc., 347,* 1097–1112.

Rudolph A. L., Bachiller R., Rieu, N. Q., Van Trung D., Palmer P., and Welch W. J. (2001) *Astrophys. J., 558,* 204–215.

Schultz A. S. B., Burton M. G., and Brand P. W. J. L. (2005) *Mon. Not. R. Astron. Soc., 358,* 1195–1214.

Smith M. D. and Rosen A. (2005a) *Mon. Not. R. Astron. Soc., 357,* 579–589.

Smith M. D. and Rosen A. (2005b) *Mon. Not. R. Astron. Soc., 357,* 1370–1376.

Smith M. D., Khanzadyan T., and Davis C. J. (2003a) *Mon. Not. R. Astron. Soc., 339,* 524–536.

Smith M. D., Froebrich D., and Eislöffel J. (2003b) *Astrophys. J., 592,* 245–254.

Smith N., Bally J., and Brooks K. J. (2004) *Astron. J., 127,* 2793–2808.

Smith N., Bally J., Licht D., and Walawender J. (2005) *Astron. J., 129,* 382–392.

Snell R. L., Loren R. B., and Plambeck R. L. (1980) *Astrophys. J., 239,* L17–L22.

Stanke T. (2000) Ph.D. thesis, Univ. of Potsdam.

Stanke T., McCaughrean M. J., and Zinnecker H. (1998) *Astron. Astrophys., 332,* 307–313.

Stanke T., McCaughrean M. J., and Zinnecker H. (1999) *Astron. Astrophys., 350,* L43–L46.

Stanke T., McCaughrean M. J., and Zinnecker H. (2000) *Astron. Astrophys., 355,* 639–650.

Stanke T., McCaughrean M. J., and Zinnecker H. (2002) *Astron. Astrophys., 392,* 239–266.

Strom S. E., Vrba F. J., and Strom K. M. (1976) *Astron. J., 81,* 314–316.

Takami M., Chrysostomou A., Ray T. P., Davis C. J., Dent W. R. F., et al. (2005) *PPV Poster Proceedings.* Available on line at http://www.lpi.usra.edu/meetings/ppv2005/pdf/8207.pdf.

van Dishoeck E. F. (2004) *Ann. Rev. Astron. Astrophys., 42,* 119–167.

Völker R., Smith M. D., Suttner G., and Yorke H. W. (1999) *Astron. Astrophys., 343,* 953–965.

Wang M., Noumaru J., Wang H., Yang J., and Chen J. (2005) *Astron. J., 130,* 2745–2756.

Walawender J., Bally J., Reipurth B., and Aspin C. (2004) *Astron. J., 127,* 2809–2816.

Walawender J., Bally J., and Reipurth B. (2005) *Astron. J., 129,* 2308–2351.

Walawender J., Bally J., Kirk H., Johnstone D., Reipurth B., and Aspin C. (2006) *Astron. J., 132,* 467–477.

White G. W., Liseau R., Men'shchikov A. B., Justtanont K., Nisini B., et al. (2000) *Astron. Astrophys., 364,* 741–762.

Wilking B. A., Schwartz R. D., Fanetti T. M., and Friel E. D. (1997) *Publ. Astron. Soc. Pac., 109,* 549–553.

Wu Y., Wei Y., Zhao M., Shi Y., Yu W., Qin S., and Huang M. (2004) *Astron. Astrophys., 426,* 503–515.

Toward Resolving the Outflow Engine: An Observational Perspective

Tom Ray
Dublin Institute for Advanced Studies

Catherine Dougados
Laboratoire d'Astrophysique de Grenoble

Francesca Bacciotti
INAF–Osservatorio Astrofisico di Arcetri

Jochen Eislöffel
Thüringer Landessternwarte Tautenburg

Antonio Chrysostomou
University of Hertfordshire

Jets from young stars represent one of the most striking signposts of star formation. The phenomenon has been researched for over two decades and there is now general agreement that such jets are generated as a byproduct of accretion, most likely by the accretion disk itself. Thus they mimic what occurs in more exotic objects such as active galactic nuclei and microquasars. The precise mechanism for their production, however, remains a mystery. To a large degree, progress is hampered observationally by the embedded nature of many jet sources as well as a lack of spatial resolution: Crude estimates, as well as more sophisticated models, nevertheless suggest that jets are accelerated and focused on scales of a few AU at most. It is only in the past few years, however, that we have begun to probe such scales in detail using classical T Tauri stars as touchstones. Application of adaptive optics, data provided by the HST, use of specialized techniques such as spectroastrometry, and the development of spectral diagnostic tools are beginning to reveal conditions in the jet launch zone. This has helped enormously to constrain models. Further improvements in the quality of the observational data are expected when the new generation of interferometers come on line. Here we review some of the most dramatic findings in this area since Protostars and Planets IV, including indications for jet rotation, i.e., that they transport angular momentum. We will also show how measurements such as those of width and the velocity field close to the source suggest jets are initially launched as warm magnetocentrifugal disk winds. Finally, the power of the spectroastrometric technique, as a probe of the central engine in very-low-mass stars and brown dwarfs, is shown by revealing the presence of a collimated outflow from a brown dwarf for the first time, copying what occurs on a larger scale in T Tauri stars.

1. INTRODUCTION

The phenomenon of jets from young stellar objects (YSOs) has been known for over two decades. While we now have a reasonably good understanding of how they propagate and interact with their surroundings on large scales (e.g., see chapter by Bally et al.), i.e., hundreds of AU and beyond, precisely how these jets are generated remains a puzzle. The observed correlation between mass outflow and accretion through the star's disk (e.g., *Hartigan et al.,* 1995; *Cabrit et al.,* 1990) would seem to favor some sort of magnetohydrodynamic (MHD) jet launching mecha-

nism, but which one is open to question. In particular, it is not known whether jets originate from the interface between the star's magnetosphere and disk [the so-called X-wind model (see *Shang et al.,* 2002; *Shu et al.,* 2000; chapter by Shang et al.)] or from a wide range of disk radii [the disk or D-wind model (see chapter by Pudritz et al.)].

On the observational front we have begun to probe the region where the jet is generated and collimated, thus testing the various models. Moreover, since Protostars and Planets IV a number of major advances have been made thanks to the availability of high-angular-resolution imaging and spectroscopy. In particular, the use of intermediate-disper-

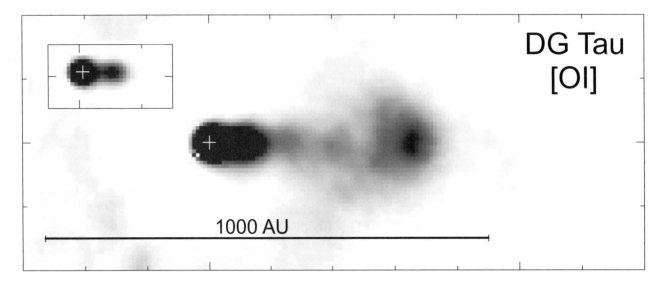

Fig. 1. Deconvolved [OI]λ6300 + continuum narrow-band image of the DG Tau jet obtained with the AO system PUEO on the Canada-France-Hawaii Telescope. The spatial resolution achieved was 0."1 = 14 AU at the distance of the Taurus Auriga Cloud. Inset (top left) is a high-contrast image near the source. Adapted from *Dougados et al.* (2002).

sion spectroscopy (long-slit, Fabry-Perot, or employing integral field units) has provided excellent contrast between the line-emitting outflow and its continuum-generating parent YSO, a prerequisite to trace outflows right back to their source (see section 2). Examples include groundbased telescopes equipped with adaptive optics (AO) and the Space Telescope Imaging Spectrograph (STIS), giving angular resolution down to 0."05 (see sections 2 and 3). In addition, as will be explained below, the technique of spectroastrometry (see section 5) is providing insights on scales of a few AU from the source. The new methodologies have brought an impressive wealth of morphological and kinematical data on the jet-launching region (≤200 AU from the YSO), providing the most stringent constraints to date for the various models. Here we will illustrate these results, e.g., measurements of the jet diameter close to the source, and, where appropriate, comparison with model predictions.

Jets may be nature's way of removing excess angular momentum from accretion disks, thereby allowing accretion to occur. Moreover, they are produced not only by young stars but by a plethora of astronomical objects from nascent brown dwarfs, with masses of around 5×10^{-2} $M_\odot$, to black holes at the center of AGN, as massive as 5×10^8 $M_\odot$. In between they are generated by X-ray binaries, symbiotic systems, planetary nebulae, and gamma-ray burst sources (e.g., *Livio*, 2004). The range of environments, and the astounding 10 orders of magnitude in mass over which the jet mechanism operates, is testimony to its robustness. Thus understanding how they are generated is of wide interest to the astrophysics community. Given the quality of data now coming on stream, and the prospect of even better angular resolution in the near future, we are hopefully close to unraveling the nature of the engine itself. Our suspicion, as supported below, is that this will be done first in the context of YSO outflows.

2. IMAGING STRUCTURES CLOSE TO THE JET BASE

YSO jets largely emit in a number of atomic and molecular lines (see, e.g., *Reipurth and Bally*, 2001; *Eislöffel et al.*, 2000). In the infrared to the ultraviolet, these lines originate in the radiative cooling zones of shocks with typical velocities from a few tens to a few hundred km s^{-1}. Thus in order to explore the morphology and kinematics of the jet-launching zone, both high-spatial-resolution narrowband imaging (on subarcsecond scales) and intermediate-resolution spectroscopy is required. Leaving aside interferometry of their meager radio continuum emission (*Girart et al.*, 2002), currently the best spatial resolution is afforded by optical/NIR instruments. Many YSOs are so deeply embedded, however, that optical/NIR observations are impossible. That said, one class of evolved YSO, namely the classical T Tauri stars (CTTS), are optically visible and have jets (see Fig. 1). CTTS therefore offer a unique opportunity to test current ejection theories. In particular, they give access to the innermost regions of the wind (≤100 AU) where models predict that most of the collimation and acceleration processes occur. Moreover, the separate stellar and accretion disk properties of these systems are well known. Here we summarize what has been learned from recent high-angular-resolution imaging studies of a number of CTTS jets. These observations have been conducted either from the ground with adaptive optics (AO) or from space with the Hubble Space Telescope (HST).

A fundamental difficulty in imaging a faint jet close to a bright CTTS is contrast with the source itself. This problem is often further exacerbated by the presence of an extended reflection nebula. Contrast with the line-emitting jet can be improved either by decreasing the width of the PSF (with AO systems from the ground or by imaging from space with, e.g., HST) and by increasing the spectral resolution. Obviously the optimum solution is a combination of both. The pioneering work of *Hirth et al.* (1997) used long-slit spectroscopy with accurate central continuum subtraction to reveal spatial extensions of a few arcseconds in the forbidden line emission of a dozen CTTS. Around the same time, narrowband imaging with the HST provided the first high-spatial-resolution images of jets from CTTS (*Ray et al.,* 1996). More recently, application of AO systems from the ground on 4-m-class telescopes, using both conventional narrow-band imaging and in combination with intermediate-spectral-resolution systems, have led to remarkable 0".2-resolution images of a number of small-scale jets from CTTS including DG Tau, CW Tau, and RW Aur (*Dougados et al., 2000; Lavalley et al., 1997; Lavalley-Fouquet et al.,* 2000). An alternative approach, pioneered by *Hartigan et al.,* (2004), is "slit-less spectroscopy" using high-spatial-resolution instruments such as the STIS onboard the HST. The advantage of this method is that, providing the dispersion of the spectrograph is large enough, one can obtain non-overlapping emission-line images covering a large range of wavelengths. This includes lines where no narrowband HST filters exist. In this way *Hartigan et al.* (2004) constructed high-spatial-resolution images of the jets from CW Tau, HN Tau, UZ Tau E, DF Tau, and the primary of DD Tau. Moreover, as will be described fully in the next section, the use of contiguous parallel long-slits, e.g., with STIS, can also provide high resolution "images" not only in individual lines but over a range of velocities [see section 3 and, e.g., *Woitas et al.* (2002)].

All these observations show that these "small-scale" T Tauri jets have complex morphologies, dominated by emission knots, that strongly resemble the striking HH jets emanating from Class 0/I sources (see chapter by Bally et al.). Turning to larger scales, deep groundbased imaging by *Mundt and Eislöffel* (1998) revealed that CTTS jets are associated with faint bow-shock-like structures, at distances of a few thousand AU. Moreover, *McGroarty and Ray* (2004) found that CTTS outflows can stretch for several parsecs, as is the case with outflows from more embedded YSOs. Thus the observations clearly suggest that the same ejection mechanism is at work at all phases of star formation.

Images taken at different epochs, usually a few years apart, can reveal how the outflow varies with time. For example, large proper motions, on the order of the jet flow velocity, have been inferred for the knots in the DG Tau and RW Aur jets (*Dougados et al., 2000; López-Martín et al.,* 2003; *Hartigan et al.,* 2004). In the DG Tau jet, a clear bow-shaped morphology is revealed for the knot located at 3" (*Lavalley et al., 1997; Dougados et al., 2000*) as well as jet wiggling, suggestive of precession. These properties, as in the younger HH flows, suggest that knots are likely to be internal working surfaces due to time variable ejection. Detailed hydrodynamical modeling, e.g., of the DG Tau jet, indicate fluctuations on timescales of 1–10 yr, although one single period does not seem to account for both the kinematics and the morphology (*Raga et al.,* 2001). Such rapid changes are also observed at the base of the younger, em-

bedded HH flows. Attempts at modeling these changes have been made. For example, numerical simulations by *Goodson et al.* (1999) of the interaction of the stellar magnetosphere with the disk predict a cyclic inflation of the former leading to eruptive ejection. The estimated timescales, however, seem too short as they are around a few stellar rotation periods. It is also possible that such short-term fluctuations in the outflow may come from disk instabilities or cyclic changes in the stellar magnetic field (as in the Sun). Further study is clearly needed to give insight into the origin of the variability process.

Imaging studies close to the YSO reveal details that are useful in discriminating between various models. For example, plots of jet width (FWHM) against distance from the source are shown in Fig. 2 for a number of CTTS outflows [HL Tau, HH 30 (*Ray et al.,* 1996); DG Tau, CW Tau, and RW Aur (*Dougados et al., 2000; Woitas et al.,* 2002)]. Similar results were found by *Hartigan et al.* (2004) for the jets from HN Tau and UZ Tau E (see Fig. 2). These studies show that at the highest spatial resolution currently achieved (0".1 = 14 AU for the Taurus Auriga star-formation region), the jet is unresolved within 15 AU from the central source. Moreover, large opening angles (20°–30°) are inferred for the HN Tau and UZ Tau E jets on scales of 15–50 AU, indicating widths at the jet source <5 AU (*Hartigan et al.,* 2004). Beyond 50 AU, HH jets seem to slowly increase in width with much smaller opening angles of a few degrees. FWHM of 20–40 AU are inferred at projected distances of around 100 AU from the central source. Such measurements

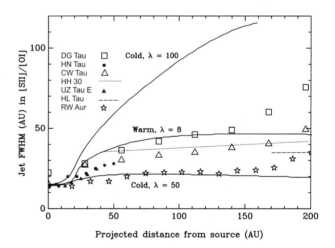

Fig. 2. Variation of jet width (FWHM) derived from [SII] and [OI] images as a function of distance from the source. Data points are from CFHT/PUEO and HST/STIS observations made by the authors as well as those of *Hartigan et al.* (2004). Overlaid (solid lines) are predicted variations based on two cold disk wind models with low efficiency (high λ) and a warm disk solution for comparison from *Dougados et al.* (2004). Note that moderate to high efficiency is favored, i.e., warm solutions. Here efficiency is measured in terms of the ratio of mass outflow to mass accretion. Models are convolved with a 14 AU (FWHM) gaussian beam. For full details see *Garcia et al.* (2001) and *Dougados et al.* (2004).

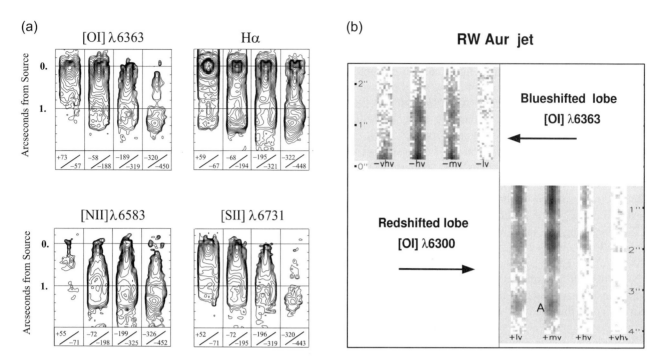

Fig. 3. Two-dimensional velocity "channel maps" of (**a**) the blueshifted jet from DG Tau and (**b**) the bipolar jet from RW Aur, reconstructed from HST/STIS multislit optical spectra. For DG Tau, the low-, medium-, high-, and very-high-velocity intervals are approximately from +60 to –70 km s⁻¹ (LV), –70 to –195 km s⁻¹ (MV), –195 to –320 km s⁻¹ (HV), and –320 to –450 km s⁻¹ (VHV) respectively. For RW Aur, velocity bins of about 80 km s⁻¹ were used starting from –5 km s⁻¹ in the approaching lobe and from +11 km s⁻¹ in the receding lobe. Note the increase in jet collimation with increasing velocity.

demonstrate that collimation is achieved on scales of a few tens of AU, i.e., close to the YSO, and appear to rule out pure hydrodynamic models for their focusing (see, e.g., *Frank and Mellema,* 1996).

Here we note that synthetic predictions of jet widths, taking into account projection and beam dilution effects, have been computed for self-similar disk wind solutions (e.g., *Ferreira,* 1997; *Casse and Ferreira,* 2000). The variation of jet diameter with distance from the source is consistent with disk wind models of moderate to high efficiency ($\dot{M}_{eject}/\dot{M}_{acc} > 0.03$) (e.g., *Garcia et al., 2001; Dougados et al., 2004*) (see also Fig. 2). Here efficiency is defined as the ratio of the total bipolar jet mass flux to the accretion rate.

3. KINEMATICS: VELOCITY PROFILES, ROTATION, ACCELERATION, AND IMPLICATIONS

Spectra of CTTS with jets frequently reveal the presence of two or more velocity components (*Hartigan et al.,* 1995). In long-slit spectra with the slit oriented along the jet direction, the so-called high-velocity component (HVC), with velocities as large as a few hundred km s⁻¹, appears more extended and of higher excitation than the low-velocity component (LVC), which has velocities in the range 10–50 km s⁻¹ (*Hirth et al., 1997; Pyo et al., 2003*).

In order to understand the nature of such components, the region of the jet base (first few hundred AU) has to be observed with sufficient spectral resolution. This can be done using integral field spectroscopy, ideally employing AO, to produce three-dimensional data cubes (two-dimensional spatial, one-dimensional radial velocity). Slices of the data-cubes then give two-dimensional images of a jet at different velocity intervals, akin to the "channel maps" of radio interferometry. For example, *Lavalley-Fouquet et al.* (2000) used OASIS to map the kinematics of the DG Tau jet with 0".5 resolution (corresponding to 70 AU at the distance of DG Tau). Even better spatial resolution, however, can be achieved with HST, although a long-slit rather than an integral field spectrograph has to be employed (*Bacciotti et al., 2000; Woitas et al., 2002*). In particular, the jets from RW Aur and DG Tau were studied with multiple exposures of a 0".1 slit, stepping the slit transversely across the outflow every 0".07. Combining the exposures together, in different velocity bins, again provided channel maps (see Fig. 3). In such images, the jets show at their base an onion-like kinematic structure, being more collimated at higher velocities and excitation. The high velocity "spine" can be identified with the HVC. The images also show, however, progressively wider, slower, and less excited layers further from the outflow axis, in a continuous transition between the HVC and the LVC. At a distance of about 50–80 AU from the source, though, the low-velocity material gradually disappears, while the axial HVC is seen to larger distances. The flow as a whole thus *appears* to accelerate on scales of 50–100 AU. This structure has recently been shown to extend to the external H₂-emitting portion of the flow (*Takami et al., 2004*). All these properties have been predicted

(a)

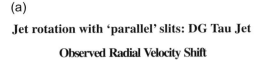

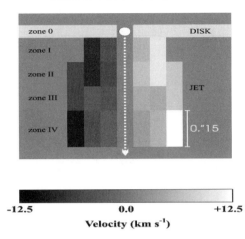

(b)

Jet rotation with `perpendicular' slits: TH 28 jet

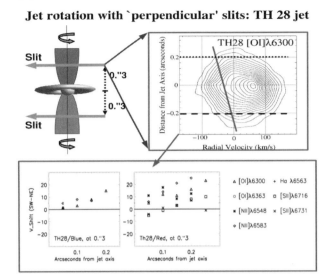

Fig. 4.   Transverse velocity shifts in the optical emission lines detected with HST/STIS across the jets from **(a)** DG Tau and **(b)** Th 28, at about 50–60 AU from the source and 20–30 AU from the outflow axis. The application of gaussian fitting and cross-correlation routines to line profiles from diametrically opposite positions centered on the jet axis revealed velocity shifts of 5–25 km s⁻¹. The values obtained suggest toroidal speeds of 10–20 km s⁻¹ at the jet boundaries.

by theoretical models of magnetocentrifugal winds (see, e.g., chapters by Shang et al. and Pudritz et al.).

The most exciting finding in recent years, however, has been the detection of radial velocity asymmetries that could be interpreted as *rotation* of YSO jets around their axes. Early hints of rotation were found in the HH 212 jet at large distance ($\approx 10^4$ AU) from the source by *Davis et al.* (2000). More recently, indications for rotation have been obtained in the first 100–200 AU of the jet channel through high-angular-resolution observations, both from space and the ground (*Bacciotti et al.*, 2002; *Coffey et al.*, 2004; *Woitas et al.*, 2005). Such velocity asymmetries have been seen in all the T Tauri jets observed with HST/STIS (DG Tau, RW Aur, CW Tau, Th 28, and HH 30), in different emission lines and using slit orientations both parallel and perpendicular to the outflow axis. For example, at optical wavelengths, systematic shifts in radial velocity, typically from 5 to 25 ± 5 km s⁻¹, were found at jet positions displaced symmetrically with respect to the outflow axis, at 50–60 AU from the source and 20–30 AU from the axis (see Fig. 4). Note that the resolving power of STIS in the optical is around 55 km s⁻¹. However, applying gaussian fitting and cross-correlation routines to the line profiles, it is possible to detect velocity shifts as small as 5 km s⁻¹. It should also be mentioned that the sense and degree of velocity asymmetry suggesting rotation were found to be consistent in different elements of various systems (e.g., both lobes of the bipolar RW Aur jet, the disk and jet lobes in HH 212 and DG Tau) and between different datasets (*Testi et al.*, 2002; *Woitas et al.*, 2005; *Coffey et al.*, 2005).

Very recently, such findings have been confirmed by the detection of systematic radial velocity shifts in the near ul-

traviolet (NUV) lines of Mg⁺ $\lambda\lambda$ 2796, 2803 (*Coffey et al.*, in preparation). Such lines are believed to arise from the fast, highly excited axial portion of the flow, and can be studied with higher angular resolution. Unfortunately, however, because of the failure of STIS in August 2003, it was only possible to study the jets from Th 28 and DG Tau in the NUV. As expected, the measurements of radial velocity in the NUV lines show slightly higher velocities in the axial region (see Fig. 5). Once again asymmetry in radial velocity across the jet is found. Assuming this is rotation, both the sense of rotation and its amplitude agree with the optical values (see Fig. 5).

Finally, velocity asymmetries compatible with jet rotation have also been detected from the ground using ISAAC on the VLT in two small-scale jets, HH 26 and HH 72, emitting in the H₂ 2.12-μm line (*Chrysostomou et al.*, 2005). The position-velocity diagrams, again based on slit positions transverse to the outflow, indicate rotation velocities of 5–10 km s⁻¹ at ≈2–3" from the source and ≈1" from the jet axis. Note that these jets are driven by Class I sources, suggesting rotation is present, as one would expect, from the earliest epochs.

The detection of rotation is interesting *per se*, as it supports the idea that jets are centrifugally launched presumably through the action of a magnetic "lever arm." Assuming that an outflow can be described by a relatively simple steady magnetohydrodynamic wind model, the application of a few basic equations, in conjunction with the observed velocity shifts, allows us to derive a number of interesting quantities that are not yet directly observable. One of these is the "foot-point radius," i.e., the location in the accretion disk from where the observed portion of the jet is launched

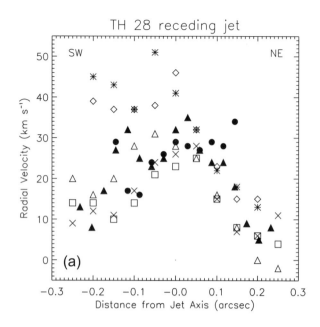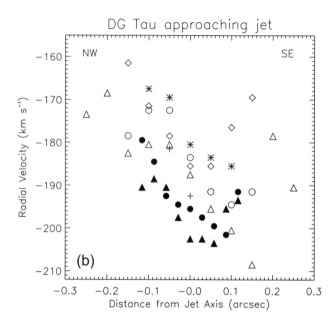

Fig. 5. Radial velocities across the jets from **(a)** Th 28 and **(b)** DG Tau, in the NUV (solid symbols) and optical (hollow symbols) lines, from HST/STIS spectra taken with the slit transverse to the outflow and at 0".3 from the star. The two datasets fit well together. The asymmetry in radial velocity from opposing sides of the jet axis, seen in both wavelength regimes, suggests rotation.

(*Anderson et al.*, 2003). The observations described here are consistent with foot-point radii between 0.5 and 5 AU from the star, with the NUV (NIR)-emitting layers coming from a location in the disk closer (farther) from the axis than the optical layers. These findings suggest that at least some of the jet derives from an extensive region of the disk [see also *Ferreira et al.* (2006), who also show that intermediate-sized magnetic lever arms are favored]. It has to be emphasized, however, that the current observations only probe the outer streamlines (*Pesenti et al.*, 2004) and accurate determination of the full transverse rotation profile is critical to constrain the models. One cannot, for example, exclude the presence of an inner X-wind at this stage, as the spatial resolution of the measurements is not yet sufficient to probe the axial region of the flow corresponding to any X-wind ejecta (but see *Ferreira et al.*, 2006). Moreover, it is not absolutely certain that flow lines can be traced back to unique foot-points. In fact, once the region beyond the acceleration zone is reached (typically a few AU above the disk), the wind is likely to undergo various kinds of MHD instabilities that complicate the geometry of the field lines. Thus a one-to-one mapping to a precise foot-point cannot be expected *a priori* (see below for alternative interpretations of the velocity asymmetries).

Assuming, however, that the current foot-point determinations are valid, these can be used to get information on the geometry of the magnetic field. In fact, from the toroidal and poloidal components of the velocity field, one can derive the ratio of the corresponding components of the magnetic field **B**. It is found that at the observed locations $B_\phi = B_p \sim 3$–4 (*Woitas et al.*, 2005). A prevalence of the toroidal field component at 50–100 AU from the star is

indeed predicted by the models that attribute the collimation of the flow to a magnetic "hoop stress" (*Königl and Pudritz*, 2000). The most important quantity derived from the observed putative rotation, however, is the amount of angular momentum carried by the jet. In the two systems for which we had sufficient information (namely, DG Tau and RW Aur), we have verified that this is between 60% and 100% of the angular momentum that the inner disk has to lose to accrete at the observed rate. Thus, the fundamental implication of the inferred rotation is that jets are likely to be the major agent for extracting excess angular momentum from the inner disk, and this could in fact be their *raison d'être*.

It should be noted that a few recent studies have proposed alternative explanations for the observed velocity asymmetries. For example, they could be produced in asymmetric shocks generated by jet precession (*Cerqueira et al.*, 2006) or by interaction with a warped disk (*Soker*, 2005). It seems unlikely, however, that these alternative models could explain why the phenomenon is so common (virtually all observed jets were found to "rotate") and why the amplitude of the velocity asymmetry is in the range predicted by disk wind theory.

Moreover, while the above developments have shed new light on how jets are generated, many puzzles remain. For example, a recent study of the disk around RW Aur suggests the disk rotates *in the opposite sense* to both of its jets (*Cabrit et al.*, 2006). Although such observations might be explained by complex interactions with a companion star, it is clear that further studies of jets and disks at high spatial/spectral resolution are needed to definitely confirm the detection of jet rotation.

Another problem is the observed asymmetry in ejection velocity between opposite lobes in a number of outflows. One "classical" example is the bipolar jet from RW Aur, in which the blueshifted material moves away from the star with a radial velocity $v_{rad} \sim 190$ km s^{-1}, while the redshifted lobe is only moving at ~ 100–110 km s^{-1}. In addition, recent measurements (*López-Martín et al.*, 2003) have shown that the proper motions of the knots in the blue- and redshifted lobes are in the same ratio as the radial velocities. This suggests that measured proper motions represent true bulk motions rather than some form of wave. Such findings beg the obvious question: If there are differences in parameters like jet velocity [see also *Hirth et al.* (1994) for other cases] in bipolar jets, are there differences in even more fundamental quantities such as mass and momentum flux?

What is the origin of the detected near-infrared H_2 emission at the base of flows? This apparently comes from layers just external to the jet seen in forbidden lines. Several such "H_2 small-scale jets" have been found recently, as in, e.g., HH 72, HH 26, HH 7-11, but H_2 winds are also associated with well-known jets from T Tauri stars, as in, e.g., DG Tau. In the latter case the emission originates from a warm (T ~ 2000 K) molecular wind with a flow length and width of 40 and 80 AU, respectively, and has a radial velocity of ~ 15 km s^{-1} (*Takami et al.*, 2004). It is not clear if such a molecular component is entrained by the axial jet, or if it is a slow external component of the same disk wind that generates the fast jet. The latter flow geometry would again agree with model predictions of magnetocentrifugal driven winds.

4. LINE DIAGNOSTICS

The numerous lines emitted by stellar jets, e.g., from transitions of O^0, S^+, N^+, Fe^+, H, and H_2, provide a wealth of useful information. By comparing various line ratios and intensities with radiative models, one is able to determine the basic physical characteristics of jets. Not only can we plot the variation of critical parameters such as temperature, ionization, and density in outflows close to their source and in different velocity channels, but also such fundamental quantities as elemental abundances and mass flux rates. Recently, spectral diagnostic methods have been extended from the optical into the near-infrared. This not only presents the prospect of probing jet conditions in very-low-velocity shock regions, but also gives the opportunity of investigating more embedded jets from less-evolved sources. These studies are complementary to the spectral analysis aimed at investigating the disk/star interaction zone and the role of accretion in determining outflow properties.

Determination of the line excitation mechanism in T Tauri stars is a long-standing issue. It is critical in particular for a detailed comparison of wind model predictions with observations. Comparison of the optical line emission properties of the DG Tau and RW Aur jets with what is expected from three different classes of excitation mechanism (mixing layers, ambipolar diffusion, planar shocks) show that line

ratios in these jets are best explained by shock excitation with moderate to large velocities (50–100 km s^{-1}) (*Lavalley-Fouquet et al.*, 2000; *Dougados et al.*, 2003), indicating that time variability plays a dominant role in the heating process.

Since Protostars and Planets IV, line diagnostics of stellar jets have developed to the point that not only can we determine the usual quantities such as the electron density (n_e) and temperature (T_e), but a number of additional ones as well, the most important of which is probably *total density*, n_H. In fact, all models for the dynamics and radiative properties of jets are highly dependent on this parameter, either directly or through derived quantities such as the jet mass or angular momentum fluxes (see sections 3 and 5).

In early studies, as now, physical quantities such as n_e and T_e were determined from line ratios sensitive to these parameters. The observations, however, were made at low spatial resolution and thus effectively integrated over the shock cooling zone (e.g., *Böhm et al.*, 1980; *Brugel et al.*, 1981). Estimates of other important quantities, such as the ionization fraction and the strength of the preshock magnetic field, had to wait for studies in which the observed line intensities and ratios were compared with predicted values based on shock models (e.g., *Hartigan et al.*, 1994). More recently, however, a simpler method, referred to as the "BE" technique (*Bacciotti and Eislöffel*, 1999) has been developed to measure $x_e = n_e/n_H$ and T_e directly from readily observed optical line ratios utilizing transitions of O^0, N^+, and S^+ (*Bacciotti and Eislöffel*, 1999; *Bacciotti*, 2002). The procedure is based on the fact that the gas producing forbidden lines is collisionally excited and no assumptions are made as to the heating mechanism. Such an approach is clearly expedient although not a substitute for a detailed model. The method assumes, for example, that the emitting gas is at a single temperature, which is not true in the cooling zone behind a shock. That said, one should consider parameters derived from the BE technique as relevant to those regions behind a shock in which the used lines peak in their emission [see the discussion and diagrams in *Bacciotti and Eislöffel* (1999)].

The basic premise of the BE method is that sulfur is singly ionized in YSO jets because of its low ionization potential. At the same time the ionization fraction of oxygen and nitrogen is assumed to be regulated by charge-exchange with hydrogen (and recombination). This provides the link between observed line ratios and x_e. The necessary conditions for these assumptions are generally fulfilled in YSO jets. Since photoionization is ignored, however, applicability of the BE technique is limited to regions far away from any strong sources of UV photons, such as at the vicinity of terminal bow-shocks. One must also assume a set of elemental abundances, appropriate for the region under study [see the discussion in Podio et al. (in preparation)].

A number of jets have been analyzed with this technique using moderate-resolution long-slit spectra and integral field spectroscopy (*Bacciotti and Eislöffel*, 1999; *Lavalley-Fouquet et al.*, 2000; *Dougados et al.*, 2000; *Nisini et al.*, 2005; Podio et al., in preparation; Medves et al., in preparation).

Typical n_e values are found to vary between 50 cm^{-3} and 3×10^3 cm^{-3}, x_e to range from 0.03 to 0.6, and T_e to decrease along a jet from a peak of $2–3 \times 10^4$ K close to the source to averages of $1.0–1.4 \times 10^4$ K on larger scales. The variation of n_e and x_e along the flow *is consistent with ionization freezing close to the source,* followed by slow nonequilibrium recombination. The mechanism that produces a high degree of ionization close to, although not coincident with, the base of the jet, however, is not known. It might, for example, derive from a series of spatially compact nonstationary shocks at the jet base (*Lavalley-Fouquet et al.,* 2000; *Massaglia et al.,* 2005). Whatever the mechanism, a comparison of various jets seems to suggest that it has similar efficiencies in all cases: Lower ionization fractions are found in the densest jets, i.e., those that recombine faster. In any event, the realization that stellar jets are only partially ionized has provided new, more accurate estimates of the total density n_H. These estimates are much bigger than previous ones and typically, on large scales, range from 10^3 to 10^5 cm^{-3}. This, in turn, implies that jets can strongly affect their environments, given their markedly increased mass, energy, and momentum fluxes. Taking into account typical emissivity filling factors, mass loss rates are found to be about $10^{-8}–10^{-7}$ $M_\odot$ yr^{-1} (Podio et al., in preparation) for CTTS jets. The associated linear momentum fluxes (calculated as $\dot{P} = v_{jet} \dot{M}$) are higher, or on the same order, as those measured in associated coaxial molecular flows, where present. This suggests that partially ionized YSO jets could drive the latter.

The BE method is well suited to analyzing large datasets, such as those provided by high-angular-resolution observations. An example of its application, to HST narrowband images of the HH 30 jet, is shown in Fig. 6 (*Bacciotti et al.,* 1999). A similar analysis of spectra taken from the ground using AO is presented in *Lavalley-Fouquet et al.* (2000). Even more interesting is the application of the technique to the two-dimensional channel maps reconstructed from parallel slit STIS data (see Fig. 3). In this way one can obtain high-angular-resolution two-dimensional maps of the various quantities of interest in the different velocity channels, as illustrated in Fig. 7 for the ionization fraction in the DG Tau jet (*Bacciotti,* 2002; Bacciotti et al., in preparation). The electron density maps, for example, confirm that n_e is highest closest to the star, nearer the axis, and at the highest velocities. At the jet base, one typically finds $0.01 < x_e < 0.4$, and total densities up to 10^6 cm^{-3}. In the same region $8 \times 10^3 < T_e < 2 \times 10^4$ K. These values can be compared with those predicted by MHD jet-launching models. Finally, from n_H, the jet diameter, and the deprojected velocity one can determine the initial mass flux in the jet $\dot{M}_{jet}$. Typical values are found to be around 10^{-7} $M_\odot$ yr^{-1}, with the colder and slower external layers of the jet contributing most to the flux. Such values can be combined with known accretion rates in these stars to produce $\dot{M}_{jet}/\dot{M}_{acc}$ ratios. Note that accretion rates are determined independently through line veiling (*Hartigan et al.,* 1995). Typical ratios in the range 0.05–0.1 are found. This seems inconsistent with cold disk wind models, although warm disk winds with moderate

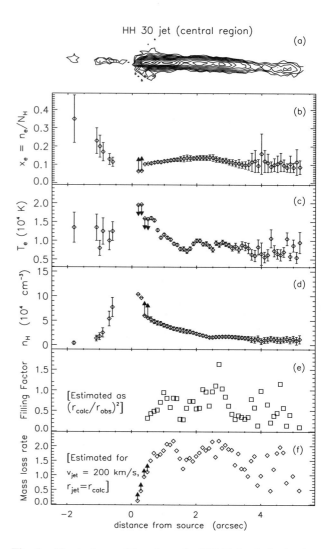

Fig. 6. Physical quantities along the HH 30 jet derived using the BE technique (see text) from HST/WFPC2 narrowband images. **(a)** [S II] emission, **(b)** ionization fraction, **(c)** electron temperature, **(d)** total density, **(e)** filling factor, and **(f)** mass flux in units of 10^{-9} $M_\odot$ yr^{-1} along the jet.

magnetic lever arms are expected to produce such ratios (*Casse and Ferreira,* 2000).

In the last few years application of line diagnostics has been extended from the optical into the near-infrared (*Nisini et al.,* 2002; *Pesenti et al.,* 2003; *Giannini et al.,* 2004; *Hartigan et al.,* 2004). As stated at the beginning of this section, this presents the prospect of not only probing jet conditions in very-low-velocity shock regions, but also the possibility of investigating more embedded, and less evolved, jets. For example, using NIR lines of Fe$^+$, one can determine not only the electron density in denser embedded regions of the jet, but also such fundamental quantities as the visual extinction A_V along the line of sight to an outflow. Note that A_V has to be known if we are to correct line ratios using lines that are far apart in wavelength. In addition, the NIR H_2 lines provide a probe of excitation conditions in low-velocity shocks near the base of the flow where some molecular species survive.

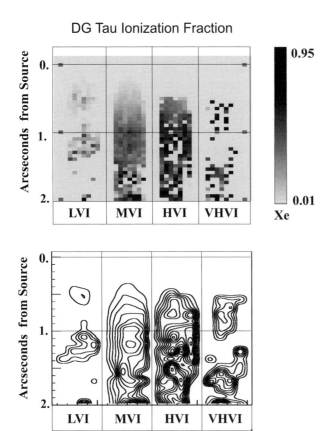

DG Tau Ionization Fraction

Fig. 7. Two-dimensional maps of the level of ionization in the first 200 AU of the DG Tau jet in the different velocity channels indicated in Fig. 3. Ionization values were derived by applying the BE technique to HST/STIS multiple spectra.

Very recently, the potential of a combined optical/NIR set of line diagnostics has been exploited (*Nisini et al.,* 2005; Podio et al., in preparation) using a variety of transitions in the 0.6–2.2-μm range. This approach has turned out to be a very useful means of determining how physical quantities vary in the different stratified layers behind a shock front as well as providing additional checks on many parameters. For example, the combination of red and NIR Fe^+ lines gives an independent estimate of T_e and n_e, which does not rely on the choice of elemental abundances. The electron density and temperature, derived from iron lines, turn out to be higher and lower, respectively, than those determined from optical lines. This is in agreement with the prediction that iron emission should, on average, come from a region within the postshock cooling zone that is farther from the shock front than those regions giving rise to the optical lines. Hence it is cooler and of higher density. Moreover, analysis of NIR Ca^+ and C^0 lines show that jets possess regions of even higher density (n_H up to 10^6 cm^{-3}). Finally, various lines have been used to estimate the depletion onto dust grains of calcium and iron with respect to solar values. The amount of depletion turns out to be quite substantial: around 30–70% for calcium and 90% for iron. This leads to the suggestion that the weak shocks present in many jets are not capable of completely destroying ambient dust grains, as expected theoretically (*Draine,* 1995).

5. PROBING THE OUTFLOW ON AU SCALES: USE OF SPECTROASTROMETRY

As this review (and others in this volume) shows, there has been a dramatic improvement over the past decade in the spatial information provided by modern instrumentation, with arguably the most detail delivered by the HST (e.g., *Bacciotti et al.,* 2000). AO also played an important role in allowing us to peek closer to the YSO itself (e.g., *Dougados et al.,* 2002). Nevertheless, it still remains true that the spatial resolution of astronomical observations are constrained by the diffraction limit of the telescope (as well as optical aberrations and atmospheric seeing in many cases). In order to probe down to the central engine, we need to achieve a resolution of ~10 AU or better, which corresponds to ≤0.05″ for the nearest star-forming regions. In other words, we need milliarcsecond resolution!

Here we briefly describe the technique of spectroastrometry and how it can be used to determine the spatiokinematic structure of sources well below the diffraction limit of the telescope, and then review some of the important results obtained using this technique.

5.1. Spectroastrometry: The Technique

The great value and appeal of spectroastrometry lies in its rather simple application and the fact that it does not require specialist equipment nor the best weather conditions. With no more than a standard CCD in a long-slit spectrograph, we have in return a tool that is capable of probing emission structures within AU scales of the YSO. An unresolved star is observed with a long-slit spectrograph and the positional centroid of the emission along the slit is determined as a function of wavelength. If the unresolved source consists of an outflow or binary with distinctive features in their spectra, the centroid position will shift relative to the continuum at the wavelength of the features. As such, the only factor that affects the accuracy of this technique is the ability to measure an accurate centroid position and this is ultimately governed by the number of photons detected and a well-sampled PSF (hence a small pixel scale, with the added requirement of a uniform CCD). Each photon detected should be within a distance, defined by the seeing disk, from the centroid and the uncertainty in this position reduces by $\sqrt{N}$ when N photons are detected. This results in a position spectrum the accuracy of which, measured in milliarcseconds, is given by $\Delta x \sim 0.5 \times$ FWHM$_{seeing}/\sqrt{N}$ (*Takami et al.,* 2003). Consequently, brighter emission lines are better suited to probing the inner regions of the YSO.

As *Bailey* (1998) explains, the method is not new and has been used in previous works, although these involved specialist techniques and equipment (*Beckers,* 1982; *Christy et al.,* 1983). Moreover, a number of authors (e.g., *Solf and Böhm,* 1993; *Hirth et al.,* 1997) have used long-slit spectroscopy to examine subarcsecond kinematic structure of the HVC and LVC in a number of outflows. *Bailey* (1998) revived interest in the technique by demonstrating that one could routinely recover information on milliarcsecond scales by taking long-slit spectra at orthogonal and antiparallel position angles, i.e., by rotating the instrument and pairing spectra taken at 0° and 180° and 90° and 270° apart. Each antiparallel pair is subtracted to remove any systematic errors introduced by the telescope and/or instrument, as such signals should be independent of rotation, whereas the true astronomical signal is reversed (Brannigan et al., in preparation).

5.2. Spectroastrometry: The Science

The first extensive use of spectroastrometry came with the study of *Bailey* (1998), who used it to prove the technique on known binaries, in the course of which two previously unknown binaries were also discovered. Since then, *Garcia et al.* (1999) and *Takami et al.* (2001, 2003) have surveyed a number of YSOs revealing jets on scales of ~10 AU from the source in a few cases. Perhaps most interesting are discoveries of *bipolar* jets at these small scales.

Forbidden lines only trace outflows in those zones with less than the critical electron density for the line. As jets tend to have higher densities closer to their source, this means that a point is reached where individual forbidden lines fade. In contrast, a permitted line traces activity all the way to the star. However, such lines, e.g., Hα, not only map outflows but magnetospheric accretion onto the YSO as well (e.g., *Muzerolle et al.,* 1998). It follows that the profile of a permitted line at the star tends to be a mixture of outflow and inflow components. As we will show, spectroastrometry is a way of unraveling these respective contributions. Now, as mentioned previously, if a spectrum is taken of a CTTS, it tends only to show blueshifted forbidden lines. This is readily understandable if the forbidden line emission comes only from an outflow and the redshifted (counterflow) at the star is obscured by an accretion disk (a view that is also endorsed by spectroastrometry). Now if we search for a spectroastrometric offset in the blueshifted wing of a permitted line we find the maximum offset at the blueshifted jet velocity and in the same direction. The offsets, however, are much smaller than those measured for the forbidden lines, implying the permitted emission tends to come from much closer to the source. Remarkably *Takami et al.* (2001, 2003) found bipolar Hα emission centered on RU Lupi and CS Cha, clearly indicating a direct line of sight to the counterflow through the accretion disk in these stars. This is interpreted as evidence of a sizable gap or dust hole in the disk at a radius of ~1–5 AU from the protostar (see also Fig. 8). The gap itself could be generated by a planetary body (see, e.g., *Varniére et al.,* 2005), although it could equally be due to the development of very large dust grains in the innermost region of the disk (see, e.g., *Watson and Stapelfeldt,* 2004). Such large dust grains would have reduced opacity. Supporting evidence for the presence of gaps is seen in the spectral energy distributions of these objects; they show mid-infrared emission dips con-

sistent with temperatures of ~200 K, coincident with the ice condensation temperature where the increase in surface density may aid planet formation (*Boss,* 1995).

Recently the technique has been used in the near-infrared. As well as allowing us to investigate younger and more embedded sources, this wavelength range also makes available other lines as probes. For example, Paβ (1.2822 μm) is found in the spectra of many T Tauri stars and was believed to exclusively trace accretion. However, *Whelan et al.* (2004) showed that this is not always the case using spectroastrometry. In particular, large spatial offsets relative to the YSO were found in the line wings (something that would not be expected in the case of accretion). Magnetospheric accretion models have always struggled to explain the detailed profiles of permitted lines (*Folha and Emerson,* 2001). Spectroastrometry seems to have resolved this problem by identifying those parts of the line attributable to an outflowing jet. It is also worth noting that both *Takami et al.* (2001) and *Whelan et al.* (2004) show evidence that suggests that offsets from the star increases with velocity, consistent with the presence of an acceleration zone.

Finally, *Whelan et al.* (2005) used spectroastrometry to report the first detection of an outflow from a brown dwarf. Their data suggest many similarities (allowing for scaling factors) between the brown dwarf outflow and those seen in CTTS (see section 6). Observations such as these suggest a universal correlation between the gravitational collapse of an object with an accretion disk and the generation of an outflow.

6. FROM BROWN DWARFS TO HERBIG Ae/Be STARS

It is conceivable that the accretion/ejection mechanism responsible for the generation and collimation of jets becomes substantially modified, or may not even operate, as one goes to sources of substantially higher or lower mass than T Tauri stars. Thus, it is an interesting question as to whether such objects also produce jets and if they are similar to those from CTTS. Certainly exploring how outflows vary with the mass of the central object (varying escape velocity, radiation field, etc.) can provide useful constraints and tests for any proposed accretion/ejection model.

Recent studies have detected signatures of accretion in a wide range of objects from brown dwarfs, with masses as low as 0.03 $M_\odot$ (*Jayawardhana et al.,* 2003; *Natta et al.,* 2004; *Mohanty et al.,* 2005), through very-low-mass (VLM) stars (*Scholz and Eislöffel,* 2004) to Herbig Ae/Be stars (*Finkenzeller,* 1985; *Böhm and Catala,* 1994; *Corcoran and Ray,* 1997) of 2–10 $M_\odot$. While the accretion rates show a large spread at any given mass, it seems that a relationship $\dot{M}_{acc} \propto \dot{M}_{obj}^2$ holds for the upper envelope of the distribution (see, e.g., *Natta et al.,* 2004). This dependence is much steeper than expected from viscous disk models with a constant viscosity parameter α, which would predict a much shallower relationship (*Natta et al.,* 2004; *Mohanty et al.,* 2005). Strongly varying disk ionization with the central object mass caused by its X-ray emission (*Muzerolle et al.,*

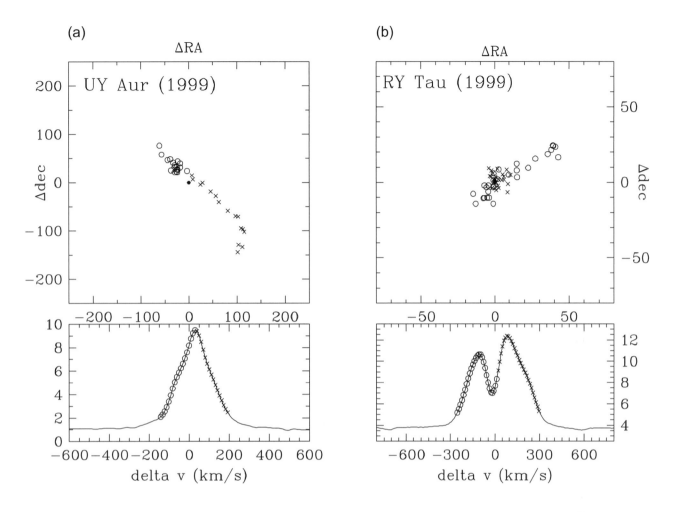

Fig. 8. Example spectroastrometry data of the Hα line. The upper panels show x-y plots of the spectroastrometric offsets at each position across the line profile (circles and crosses for blue- and redshifted components, respectively). The integrated line profile is shown below. The data were taken using the ISIS instrument on the William Herschel Telescope on La Palma. **(a)** Detection of a bipolar outflow in UY Aur, implying the presence of a disk gap in this object. **(b)** The same for RY Tau, although the evidence suggests either a binary companion or a monopolar jet with the redshifted component hidden from view.

2003) or disk accretion controlled by Bondi-Hoyle accretion from the large gas reservoir of the surrounding cloud core (*Padoan et al., 2005*) have been proposed to understand the steep relationship between accretion rate and central object mass.

In order to estimate the minimum mass of a source that we might be able to detect an outflow from, it is plausible to extrapolate the known linear correlation between ejection and accretion rates. Moreover, *Masciadri and Raga* (2004) have shown that the luminosity of a putative brown dwarf jet should scale approximately with its mass outflow rate. Thus, giving the known accretion rates of brown dwarfs, we expect their outflows to be about 100× fainter at best than those from CTTS, i.e., just within the reach of the biggest telescopes (*Whelan et al., 2005*).

The first very-low-mass object found to show forbidden lines typical of outflows from T Tauri stars, LS-RCrA-1, was discovered by *Fernández and Comerón* (2001). Its mass has been estimated as substellar (*Barrado y Navascués et al., 2004*), but the forbidden line emission could not be

resolved spatially (*Fernández and Comerón, 2005*). In the same work, however, a 4"-long jet and a 2"-long counterjet were reported on from the VLM star Par-Lup3-4. Simultaneously, *Whelan et al.* (2005) detected a spatially resolved outflow from the 60 M_J brown dwarf ρ Oph 102 using spectroastrometry. The outflow is blueshifted by about –50 km s^{-1} with respect to its source, and has characteristics similar to those from CTTS.

Moving to the other end of the mass spectrum, it is also interesting to investigate if the intermediate-mass Herbig Ae/Be stars show accretion/ejection structures like those of the lower-mass T Tauri stars. Several large-scale Herbig-Haro flows have been known for some time, which may be driven by Herbig Ae/Be stars. In many cases, however, it is not certain that the Herbig Ae/Be star, as opposed to a nearby, unrelated T Tauri star or a lower-mass companion, could be responsible. Examples include HH 39 associated with R Mon (*Herbig, 1968*), HH 218 associated with V645 Cyg (*Goodrich, 1986*), and HH 215 and HH 315 associated with PV Cep (*Neckel et al., 1987; Gomez et al., 1997; Rei-*

purth et al., 1997). The first jet that could unambiguously be traced back to a Herbig Ae/Be star was HH 398 emanating from LkHα 233 (*Corcoran and Ray, 1998*). All these outflow sources exhibit blueshifted forbidden emission lines in their optical spectra with typical velocities of a few hundred km s^{-1}, similar to CTTS.

More recently, small-scale jets from two nearby Herbig Ae stars have been found by coronographic imaging in the ultraviolet with the HST. These jets and counterjets from HD 163296 (HH 409) (*Devine et al., 2000; Grady et al., 2000*) and HD 104237 (HH 669) (*Grady et al., 2004*) are only a few arcseconds long. They are readily seen in Lyα as the contrast between the star and the shocked flow is much more favorable than in the optical: Lyα/Hα ≥ 10 according to shock models (e.g., *Hartigan et al., 1987*).

Summarizing, we find that Herbig-Haro jets both in the form of small-scale jets and parsec-scale flows are ubiquitous in CTTS. They seem to be much rarer toward the higher-mass Herbig AeBe stars, and only one clear example of an outflow from a brown dwarf is known so far. Nevertheless, this seems to indicate that the accretion/ejection structures in all objects are similar and the same physical processes are at work over the whole mass spectrum. The conditions for the generation of jets seem to be optimal in CTTS T Tauri stars, while under the more extreme environments, both in the lower-mass brown dwarfs and in the higher-mass Herbig Ae/Be stars, jet production could be less efficient. Current models of the magnetocentrifugal launching of jets from young stars will have to be tested for the conditions found in brown dwarfs and Herbig AeBe stars, in order to see if they are can reproduce the frequency and physical characteristics of the observed flows.

7. THE FUTURE: TOWARD RESOLVING THE CENTRAL AU

While it is clear from this review that high-angular-resolution observations (on scales ≈0."1) have provided important information on the launch mechanism, the true "core" of the engine lies below the so-called "Alfvén surface." This surface is located within a few AU of the disk (i.e., tens of milliarcseconds for the nearest star-forming region), so the core cannot be resolved with conventional instrumentation either from the ground or space. This, however, is about to change with the new generation of optical/NIR interferometers coming on line, opening up the exciting possibility of exploring this region for the first time. Currently being rolled out are two facilities in which Europe will play an active role: the VLT Interferometer (VLTI) at ESO-Paranal, and the Large Binocular Telescope (LBT) at Mount Graham in Arizona.

The VLTI will operate by connecting, in various combinations, up to four 8-m and four smaller (1.8-m) auxiliary telescopes. The latter can move on tracks to a number of fixed stations so as to obtain good (u,v) plane coverage. The beams, of course, from the various telescopes have to be combined in a correlator. In this regard the AMBER instrument is of particular interest to the study of YSO jet sources

as it allows for medium-resolution spectroscopy in the NIR (e.g., high-velocity Paschen β emission). AMBER can combine up three AO corrected beams, thus allowing "closure phase" to be achieved and it will provide angular resolution as small as a few milliarcseconds. In the early days of operating this instrument, incomplete (u,v) plane coverage is expected and in this case models of the expected emission are needed to interpret the observations (*Bacciotti et al., 2003*). The LBT Interferometer, in contrast to the VLTI, is formed by combining the beams of two fixed 8.4-m telescopes and hence has a fixed baseline. Effectively it will have the diffraction limited resolution of a 23-m telescope (*Herbst, 2003*) and excellent (u,v) plane coverage because of the shortness of the baseline in comparison to the telescope apertures. Moreover, it will complement the VLTI through its short projected baseline spacings, spacings that are inaccessible to the former. The LBT Interferometer will initially operate in the NIR in so-called Fizeau mode, providing high-resolution images over a relatively large field of view (unlike the VLTI). A future extension into the optical is planned, however.

Radio provides a means of obtaining high-spatial-resolution observations of YSO jets close to their source if the source is highly embedded (*Girart et al., 2002*). Their free-free emission, however, tends to be rather weak and so the number of outflows mapped so far at radio wavelengths has been quite small. This will change dramatically when the new generation of radio interferometers, in particular e-MERLIN (extended MERLIN) and EVLA (extended VLA) come on line. Although there will be modest improvements in resolution, the primary gain will be in sensitivity, allowing, in some cases, the detection of emission 20–50× weaker than current thresholds. This increase in sensitivity will be achieved through correlators and telescope links with much broader band capacities than before. Such improvements will not only lead to the detection of more jet sources, and hopefully allow meaningful statistical studies, but perhaps more importantly to the detection of nonthermal components in known outflows as already hinted at in a number of studies (see *Girart et al., 2002; Ray et al., 1997; Reid et al., 1995*). Polarization studies of such emission in turn would give us a measure of ambient magnetic field strength and direction, parameters that are poorly known at present. Finally, it is worth noting that a number of studies have shown that H_2O masers may in some instances be tracing outflows (*Claussen et al., 1998; Torrelles et al., 2005*) and not circumstellar disks as often assumed. High-resolution polarization studies of such masers can not only provide information on magnetic field strengths and direction but, when combined with multiepoch studies, information on how the magnetic fields evolve with time (*Baudry and Diamond, 1998*).

The other large interferometric facility being planned for observations at the submillimeter wavelength range is ALMA, an array of 64 12-m antennas to be built in the Atacama desert in Chile, not far from the VLTI site. With ALMA we will be able not only to routinely measure disk rotation but also conceiveably rotation in molecular outflows if present. The future for this field is therefore very bright indeed.

Acknowledgments. T.R., C.D., F.B., and J.E. wish to acknowledge support through the Marie Curie Research Training Network JETSET (Jet Simulations, Experiments and Theory) under contract MRTN-CT-2004-005592. T.R. would also like to acknowledge assistance from Science Foundation Ireland under contract 04/BRG/P02741. Finally, we wish to thank the referee for very helpful comments while preparing this manuscript.

REFERENCES

Anderson J. M., Li Z.-Y., Krasnopolsky R., and Blandford R. (2003) *Astrophys. J., 590,* L107–L110.

Bacciotti F. (2002) *Rev. Mex. Astron. Astrofis., 13,* 8–15.

Bacciotti F. and Eislöffel J. (1999) *Astron. Astrophys., 342,* 717–735.

Bacciotti F., Eislöffel J., and Ray T. P. (1999) *Astron. Astrophys., 350,* 917–927.

Bacciotti F., Mundt R., Ray T. P., Eislöffel J., Solf J., and Camenzind M. (2000) *Astrophys. J., 537,* L49–L52.

Bacciotti F., Ray T. P., Mundt R., Eislöffel J., and Solf J. (2002) *Astrophys. J., 576,* 222–231.

Bacciotti F., Testi L., Marconi A., Garcia P. J. V., Ray T. P., Eislöffel J., and Dougados C. (2003) *Astrophys. Space Sci., 286,* 157–162.

Bailey J. A. (1998) *Mon. Not. R. Astron. Soc., 301,* 161–167.

Barrado y Navascués D., Mohanty S., and Jayawardhana R. (2004) *Astrophys. J., 604,* 284–296.

Baudry A. and Diamond P. J. (1998) *Astron. Astrophys., 331,* 697–708.

Beckers J. (1982) *Opt. Acta, 29,* 361–362.

Böhm T. and Catala C. (1994) *Astron. Astrophys., 290,* 167–175.

Böhm K. H., Mannery E., and Brugel E. W. (1980) *Astrophys. J., 235,* L137–L141.

Boss A. (1995) *Science, 267,* 360–362.

Brugel E. W., Böhm K. H., and Mannery E. (1981) *Astrophys. J. Suppl., 47,* 117–138.

Cabrit S., Edwards S., Strom S. E., and Strom K. M. (1990) *Astrophys. J., 354,* 687–700.

Cabrit S., Pety J., Pesenti N., and Dougados C. (2006) *Astron. Astrophys., 452,* 897–906.

Casse F. and Ferreira J. (2000) *Astron. Astrophys., 353,* 1115–1128.

Cerqueira A. H., Velazquez P. F., Raga A. C., Vasconcelos M. J., and de Colle F. (2006) *Astron. Astrophys., 448,* 231–241.

Christy J., Wellnitz D., and Currie D. (1983) In *Current Techniques in Double and Multiple Star Research* (R. Harrington and O. Franz, eds.), IAU Colloquium 62, *Lowell Obs. Bull., 167,* 28–35.

Chrysostomou A., Bacciotti F., Nisini B., Ray T. P., Eislöffel J., Davis C. J., and Takami M. (2005) In *PPV Poster Proceedings,* www.lpi. usra.edu/meetings/ppv2005/pdf/8156.pdf.

Claussen M. J., Marvel K. B., Wootten A., and Wilking B. A. (1998) *Astrophys. J., 507,* L79–L82.

Coffey D., Bacciotti F., Ray T. P., Woitas J., and Eislöffel J. (2004) *Astrophys. J., 604,* 758–765.

Coffey D. A., Bacciotti F., Woitas J., Ray T. P., and Eislöffel J. (2005) In *PPV Poster Proceedings,* www.lpi.usra.edu/meetings/ppv2005/pdf/ 8032.pdf.

Corcoran M. and Ray T. P. (1997) *Astron. Astrophys., 321,* 189–201.

Corcoran M. and Ray T. P. (1998) *Astron. Astrophys., 336,* 535–538.

Davis C. J., Berndsen A., Smith M. D., Chrysostomou A., and Hobson J. (2000) *Mon. Not. R. Astron. Soc., 314,* 241–255.

Devine D., Grady C. A., Kimble R. A., Woodgate B., Bruhweiler F. C., Boggess A., Linsky J. L., and Clampin M. (2000) *Astrophys. J., 542,* L115–L118.

Dougados C., Cabrit S., Lavalley C., and Ménard F. (2000) *Astron. Astrophys., 357,* L61–L64.

Dougados C., Cabrit S., and Lavalley-Fouquet C. (2002) *Rev. Mex. Astron. Astrofis., 13,* 43–48.

Dougados C., Cabrit S., Lopez-Martin L., Garcia P., and O'Brien D. (2003) *Astrophys. Space Sci., 287,* 135–138

Dougados C., Cabrit S., Ferreira J., Pesenti N., Garcia P., and O'Brien D. (2004) *Astrophys. Space Sci., 293,* 45–52.

Draine B. T. (1995) *Astrophys Space Sci., 233,* 111–123.

Eislöffel J., Mundt R., Ray T. P., and Rodríguez L. F. (2000) In *Protostars and Planets IV* (V. Mannings et al., eds.), pp. 815–840. Univ. of Arizona, Tucson.

Fernández M. and Comerón F. (2001) *Astron. Astrophys., 380,* 264–276.

Fernández M. and Comerón F. (2005) *Astron. Astrophys., 440,* 1119–1126.

Ferreira J. (1997) *Astron. Astrophys., 319,* 340–359.

Ferreira J., Dougados C., and Cabrit S. (2006) *Astron. Astrophys., 453,* 785–796.

Finkenzeller U. (1985) *Astron. Astrophys., 151,* 340–348.

Folha D. F. M. and Emerson J. P. (2001) *Astron. Astrophys., 365,* 90–109.

Frank A. and Mellema G. (1996) *Astrophys. J., 472,* 684–702.

Garcia P. J. V., Thiébaut E., and Bacon R. (1999) *Astron. Astrophys., 346,* 892–896.

Garcia P. J. V., Cabrit S., Ferreira J., and Binette L. (2001) *Astron. Astrophys., 377,* 609–616.

Giannini T., McCoey C., Caratti o Garatti A., Nisini B., Lorenzetti D., and Flower D. R. (2004) *Astron. Astrophys., 419,* 999–1014.

Girart J. M., Curiel S., Rodríguez L. F., and Cantó J. (2002) *Rev. Mex. Astron. Astrofis., 38,* 169–186.

Gomez M., Kenyon S. J., and Whitney B. A. (1997) *Astron. J., 114,* 265–271.

Goodson A. P., Böhm K.-H., and Winglee R. M. (1999) *Astrophys. J., 524,* 142–158.

Goodrich R. (1986) *Astrophys. J., 311,* 882–894.

Grady C. A., Devine D., Woodgate B., Kimble R., Bruhweiler F. C., Boggess A., Linsky J. L., Plait P., Clampin M., and Kalas P. (2000) *Astrophys. J., 544,* 895–902.

Grady C. A., Woodgate B., Torres C. A. O., Henning T., Apai D., Rodmann J., Wang H., Stecklum B., Linz H., Willinger G. M., Brown A., Wilkinson E., Harper G. M., Herczeg G. J., Danks A., Vieira G. L., Malumuth E., Collins N. R., and Hill R. S. (2004) *Astrophys. J., 608,* 809–830.

Hartigan P., Raymond J., and Hartmann L. (1987) *Astrophys. J., 316,* 323–348.

Hartigan P., Morse J., and Raymond J. (1994) *Astrophys. J., 436,* 125–143.

Hartigan P., Edwards S., and Gandhour L. (1995) *Astrophys. J., 452,* 736–768.

Hartigan P., Edwards S., and Pierson R. (2004) *Astrophys. J., 609,* 261–276.

Herbig G. (1968) *Astrophys. J., 152,* 439–441.

Herbst T. (2003) *Astrophys. Space Sci., 286,* 45–53.

Hirth G. A., Mundt R., Solf J., and Ray T. P. (1994) *Astrophys. J., 427,* L99–L102.

Hirth G. A., Mundt R., and Solf J. (1997) *Astron. Astrophys. Suppl., 126,* 437–469.

Jayawardhana R., Mohanty S., and Basri G. (2003) *Astrophys. J., 592,* 282–287.

Königl A. and Pudritz R. (2000) In *Protostars and Planets IV* (V. Mannings et al., eds.), pp. 759–788. Univ. of Arizona, Tucson.

Lavalley C., Cabrit S., Dougados C., Ferruit P., and Bacon R. (1997) *Astron. Astrophys., 327,* 671–680.

Lavalley-Fouquet C., Cabrit S., and Dougados C. (2000) *Astron. Astrophys., 356,* L41–L44.

Livio M. (2004) *Baltic Astron., 13,* 273–279.

López-Martín L., Cabrit S., and Dougados C. (2003) *Astron. Astrophys., 405,* L1–L4.

Masciadri E. and Raga A. C. (2004) *Astrophys. J., 615,* 850–854.

Massaglia S., Mignone A., and Bodo G. (2005) *Astron. Astrophys., 442,* 549–554.

McGroarty F. and Ray T. P. (2004) *Astron. Astrophys., 420,* 975–986.

Mohanty S., Jayawardhana R., and Basri G. (2005) *Astrophys. J., 626,* 498–522.

Mundt R. and Eislöffel J. (1998) *Astron. J., 116,* 860–867.

Muzerolle J., Calvet N., and Hartmann L. (1998) *Astrophys. J., 492,* 743–753.

Muzerolle J., Hillenbrand L., Calvet N., Briceño C., and Hartmann L. (2003) *Astrophys. J., 592,* 266–281.

Natta A., Testi L., Muzerolle J., Randich S., Comerón F., and Persi P. (2004) *Astron. Astrophys., 424,* 603–612.

Neckel T., Staude H. J., Sarcander M., and Birkle K. (1987) *Astron. Astrophys., 175,* 231–237.

Nisini B., Caratti o Garatti A., Giannini T., and Lorenzetti D. (2002) *Astron. Astrophys., 393,* 1035–1051.

Nisini B., Bacciotti F., Giannini T., Massi F., Eislöffel J., Podio L., and Ray T. P. (2005) *Astron. Astrophys., 441,* 159–170.

Padoan P., Kritsuk A., Norman M., and Nordlund A. (2005) *Astrophys. J., 622,* L61–L64.

Pesenti N., Dougados C., Cabrit S., O'Brien D., Garcia P., and Ferreira J. (2003) *Astron. Astrophys., 410,* 155–164.

Pesenti N., Dougados C., Cabrit S., Ferreira J., O'Brien D., and Garcia P. (2004) *Astron. Astrophys., 416,* L9–L12.

Pyo T.-S., Hayashi M., Kobayashi N., Tokunaga A. T., Terada H., et al. (2003) *Astrophys. Space Sci., 287,* 21–24.

Raga A., Cabrit S., Dougados C., and Lavalley C. (2001) *Astron. Astrophys., 367,* 959–966.

Ray T. P., Mundt R., Dyson J. E., Falle S. A. E. G., and Raga A. C. (1996) *Astrophys. J., 468,* L103–L106.

Ray T. P., Muxlow T. W. B., Axon D. J., Brown A., Corcoran D., Dyson J., and Mundt R. (1997) *Nature, 385,* 415–417.

Reid M. J., Argon A. L., Masson C. R., Menten K. M., and Moran J. M. (1995) *Astrophys. J., 443,* 238–244.

Reipurth B. and Bally J. (2001) *Ann. Rev. Astron. Astrophys., 39,* 403–455.

Reipurth B., Bally J., and Devine D. (1997) *Astron. J., 114,* 2708–2735.

Scholz A. and Eislöffel J. (2004) *Astron. Astrophys., 419,* 249–267.

Shang H., Glassgold A. E., Shu F. H., and Lizano S. (2002) *Astrophys. J., 564,* 853–876.

Shu F. H., Najita J. R., Shang H., and Li Z.-Y. (2000) In *Protostars and Planets IV* (V. Mannings et al., eds.), pp. 789–814. Univ. of Arizona, Tucson.

Soker N. (2005) *Astron. Astrophys., 435,* 125–129.

Solf J. and Böhm K. H. (1993) *Astrophys. J., 410,* L31–L34.

Takami M., Bailey J. A., Gledhill T. M., Chrysostomou A., and Hough J. H. (2001) *Mon. Not. R. Astron. Soc., 323,* 177–187.

Takami M., Bailey J.A., and Chrysostomou A. (2003) *Astron. Astrophys., 397,* 675–691.

Takami M., Chrysostomou A., Ray T. P., Davis C., Dent W. R. F., Bailey J., Tamura M., and Terada H. (2004) *Astron. Astrophys., 416,* 213–219.

Testi L., Bacciotti F., Sargent A. I., Ray T. P., and Eislöffel J. (2002) *Astron. Astrophys., 394,* L31–L34.

Torrelles J. M., Patel N., Gómez J. F., Anglada G., and Uscanga L. (2005) *Astrophys. Space Sci., 295,* 53–63.

Varniére P., Blackman E. C, Frank A., and Quillen A. C. (2005) In *PPV Poster Proceedings,* www.lpi.usra.edu/meetings/ppv2005/pdf/8064.pdf.

Watson A. M. and Stapelfeldt K. R. (2004) *Astrophys. J., 602,* 860–874.

Whelan E., Ray T. P., and Davis C. J. (2004) *Astron. Astrophys., 417,* 247–261.

Whelan E .T., Ray T. P., Bacciotti F., Natta A., Testi L., and Randich S. (2005) *Nature, 435,* 652–654.

Woitas J., Ray T. P., Bacciotti F., Davis C. J., and Eislöffel J. (2002) *Astrophys. J., 580,* 336–342.

Woitas J., Bacciotti F., Ray T. P., Marconi A., Coffey D., and Eislöffel J. (2005) *Astron. Astrophys., 432,* 149–160.

Molecular Outflows in Low- and High-Mass Star-forming Regions

Héctor G. Arce
American Museum of Natural History

Debra Shepherd
National Radio Astronomy Observatory

Frédéric Gueth
Institut de Radioastronomie Millimétrique, Grenoble

Chin-Fei Lee
Harvard-Smithsonian Center for Astrophysics

Rafael Bachiller
Observatorio Astronómico Nacional

Alexander Rosen
Max-Planck-Institut für Radioastronomie

Henrik Beuther
Max-Planck-Institut für Astronomie

We review the known properties of molecular outflows from low- and high-mass young stars. General trends among outflows are identified, and the most recent studies on the morphology, kinematics, energetics, and evolution of molecular outflows are discussed, focusing on results from high-resolution millimeter observations. We review the existing four broad classes of outflow models and compare numerical simulations with the observational data. A single class of models cannot explain the range of morphological and kinematic properties that are observed, and we propose a possible solution. The impact of outflows on their cloud is examined, and we review how outflows can disrupt their surrounding environment, through the clearing of gas and the injection of momentum and energy onto the gas at distances from their powering sources from about 0.01 to a few pc. We also discuss the effects of shock-induced chemical processes on the ambient medium, and how these processes may act as a chemical clock to date outflows. Finally, future outflow research with existing and planned millimeter and submillimeter instruments is presented.

1. INTRODUCTION

As a star forms by gravitational infall, it energetically expels mass in a bipolar jet. There is strong evidence for a physical link between inflow and outflow and that magnetic stresses in the circumstellar disk-protostar system initially launch the outflowing material (see chapters by Pudritz et al., Ray et al., and Shang et al.). The ejected matter can accelerate entrained gas to velocities greater than those of the cloud, thereby creating a molecular outflow. Outflows can induce changes in the chemical composition of their host cloud and may even contribute to the decline of the infall process by clearing out dense gas surrounding the protostar. In addition, molecular outflows can be useful tools for understanding the underlying formation process of stars of all masses, as they provide a record of the mass-loss history of the system.

Protostellar outflows can be observed over a broad range of wavelengths, from the ultraviolet to the radio. In this review we will concentrate on the general characteristics and properties of molecular outflows, the entrainment process, and the chemical and physical impact of outflows on the cloud that are mainly detected through observations of molecular rotational line transitions at millimeter and submillimeter wavelengths. At these wavelengths the observations mainly trace the cloud gas that has been swept up by the underlying protostellar wind, and provide a time-integrated view of the protostar's mass-loss process and its interaction with the surrounding medium.

2. GENERAL OUTFLOW PROPERTIES

Over the last 10 years, millimeter interferometers have allowed the observation of molecular outflows at high angular resolutions (~1 to 4"), while the capability to observe mosaics of several adjacent fields has enabled mapping of complete outflows at those resolutions. Such interferometric observations give access to the internal structure of the gas surrounding protostars, and can disentangle the morphology and dynamics of the different elements that are present (i.e., protostellar condensation, infalling and outflowing gas). These high-resolution observations have been critical to the discovery of the kinematics and morphology of outflows from massive OB (proto)stars, which are typically more than a kiloparsec away.

General trends have been identified in molecular outflows from both low- and high-mass protostars, even though they display a broad diversity of sizes and shapes. These properties have been identified mostly using single-dish and interferometer observations of the CO lines. Molecular outflows exhibit a mass-velocity relation with a broken power law appearance, $dM(v)/dv \propto v^{-\gamma}$, with the slope, γ, typically ranging from 1 to 3 at low outflow velocities, and a steeper slope at higher velocities — as large as 10 in some cases (e.g., *Rodríguez et al.*, 1982; *Lada and Fich*, 1996; *Ridge and Moore*, 2001). The slope of the mass-velocity relation steepens with age and energy in the flow (*Richer et al.*, 2000). The velocity at which the slope changes is typically between 6 and 12 km s^{-1} although outflows can have CO break velocities as low as about 2 km s^{-1} and, in the youngest CO outflows, can be high as 30 km s^{-1} (see, e.g., *Richer et al.*, 2000, and references therein). The mass, force, and mechanical luminosity of molecular outflows correlate with bolometric luminosity (*Bally and Lada*, 1983; *Cabrit and Bertout*, 1992; *Wu et al.*, 2004), and many fairly collimated outflows show a linear velocity-distance relation, typically referred to as the "Hubble law," where the maximum radial velocity is proportional to position (e.g., *Lada and Fich*, 1996). Also, the degree of collimation of outflows from low- and high-mass systems appears to decrease as the powering source evolves (see below).

These observed general trends are consistent with a common outflow/infall mechanism for forming stars with a wide range of masses, from low-mass protostars up to early-B protostars. Although there is evidence that the energetics for at least some early-B stars may differ from their low-mass counterparts, the dynamics are still governed by the presence of linked accretion and outflow. A few young O stars show evidence for accretion as well, although this is not as well established as for early-B stars (e.g., *van der Tak and Menten*, 2005; see also chapter by Cesaroni et al.).

2.1. Outflows from Low-Mass Protostars

Since their discovery in the early 1980s, molecular outflows driven by young low-mass protostars (i.e., typically <1 M$_\odot$) have been extensively studied, giving rise to a detailed picture of these objects (see, e.g., reviews by *Richer et al.*, 2000; *Bachiller and Tafalla*, 1999, and references therein). The flows typically extend over 0.1–1 parsec, with outflowing velocities of 10–100 km s^{-1}. Typical momentum rates of 10^{-5} M$_\odot$ km s^{-1} yr^{-1} are observed, while the molecular outflow mass flux can be as high as 10^{-6} M$_\odot$ yr^{-1} (*Bontemps et al.*, 1996). Particular interest has been devoted to the outflows driven by the youngest, embedded protostars (ages of a few 10^3 to a few 10^4 yr, the Class 0 objects). These sources are still in their main accretion phase and are therefore at the origin of very powerful ejections of matter.

2.1.1. Molecular jets. The collimation factor (i.e., length/width, or major/minor radius) of the CO outflows, as derived from single-dish studies, ranges from ~3 to >20. There is, however, a clear trend of higher collimation at higher outflowing velocities (see, e.g., *Bachiller and Tafalla*, 1999). Interferometric maps have revealed even higher collimation factors, and, in some cases, high-velocity structures that are so collimated (opening angles less than a few degrees) that they can be described as "molecular jets."

HH 211 is the best example to date of such a molecular jet (*Gueth and Guilloteau*, 1999). At high velocity, the CO emission is tracing a highly collimated linear structure that is emanating from the central protostar. This CO jet terminates at the position of strong H$_2$ bow shocks, and shows a Hubble law velocity relation. Low-velocity CO traces a cavity that is very precisely located in the wake of the shocks. These observations strongly suggest that the propagation of one or several shocks in a protostellar jet entrain the ambient molecular gas and produces the low-velocity molecular outflow (see section 3). With an estimated dynamical timescale of ~10^3 yr, HH 211 is obviously an extremely young object. Other examples of such highly collimated, high-velocity jets include IRAS 04166+2706 (*Tafalla et al.*, 2004) and HH 212 (*Lee et al.*, 2000) — these sources are or will be in the near future the subject of more detailed investigations.

In at least SVS 13B (*Bachiller et al.*, 1998, 2000), NGC1333 IRAS 2 (*Jørgensen et al.*, 2004), NGC1333 IRAS 4 (*Choi* 2005), and HH 211 (*Chandler and Richer*, 2001; *Hirano et al.*, 2006; *Palau et al.*, 2006), the SiO emission traces the molecular jet and *not* the strong terminal shocks against the interstellar medium. This came as a surprise, as it seems to contradict the widely accepted idea that SiO is a tracer of outflow shocks, where the density is increased by several order of magnitudes (e.g., *Martín-Pintado et al.*, 1992; *Schilke et al.*, 1997; *Gibb et al.*, 2004). The lack of significant SiO emission in the terminal shocks suggests that the formation process of this molecule has a strong dependence on the shock conditions (velocity, density) and/or outflow age (see section 4.2).

The exact nature of these CO and SiO molecular jets is not yet clear. Three basic scenarios could be invoked, in which the high-velocity CO and SiO molecules (1) belong to the actual protostellar jet, (2) are entrained along the jet in a turbulent cocoon (e.g., *Stahler,* 1994; *Raga et al.,* 1995), or (3) are formed/excited in shocks that are propagating down the jet [i.e., "internal working surfaces" (*Raga and Cabrit,* 1993)]. This latter scenario would reconcile the observation of SiO in the jet and the shock-tracer nature of this molecule. The predictions of these three cases, both in terms of line properties and observed morphologies, are somewhat different but the current observations have not yet provided a clear preference for one of these scenarios.

2.1.2. More complex structures. Not all sources have structures as simple or unperturbed as the molecular jets discussed above. CO observations have also revealed a number of more complex outflow properties.

Episodic ejection events seem to be a common property of young molecular outflows. In sources such as, e.g., L 1157 (*Gueth et al.,* 1998) and IRAS 04239+2436 [HH 300 (*Arce and Goodman,* 2001b)], a limited number (2–5) of strong ejection events have taken place, each of them resulting in the propagation of a large shock. Morphologically, the flow is therefore the superposition of several shocked/outflowing gas structures, while position-velocity diagrams show multiple "Hubble wedges" [i.e., a jagged profile (*Arce and Goodman,* 2001a)]. In most of the sources, if several strong shocks are not present, a main ejection event followed by several smaller, weaker shocked areas are observed [e.g., L 1448: *Bachiller et al.* (1990); HH 111: *Cernicharo et al.* (1999); several sources: *Lee et al.* (2000, 2002)]. As noted before, even the molecular jets could include several internal shocks. Altogether, these properties suggest that the ejection phenomenon in young outflows is intrinsically episodic, or — a somewhat more attractive possibility — could be continuous but include frequent ejection bursts. This could be explained by sudden variations in the accretion rate onto the forming star that result in variations of the velocity of the ejected matter, hence the creation of a series of shocks.

Precession of the ejection direction has been established in a few sources, like Cep E (*Eislöffel et al.,* 1996), and L 1157 (*Gueth et al.,* 1996, 1998). In several other objects, the observations reveal bending or misalignment between the structures within the outflows (see, e.g., *Lee et al.,* 2000, 2002). In fact, when observed at the angular resolution provided by millimeter interferometers, many well-defined, regular bipolar outflows mapped with single-dish telescopes often reveal much more complex and irregular structures, which indicate both temporal and spatial variations of the ejection phenomenon.

Quadrupolar outflows are sources in which four lobes are observed, and seem to be driven by the same protostellar condensation. Several scenarios were proposed to explain these peculiar objects: two independent outflows (e.g.,

Anglada et al., 1991; *Walker et al.,* 1993); one single flow with strong limb-brightening, which would thus mimic four lobes (e.g., *Avery et al.,* 1990); and a single outflow but with a strong precession of the ejection direction (e.g., *Ladd and Hodapp,* 1997). The angular resolution provided by recent interferometric observations have clearly favored the first hypothesis in at least two objects [HH 288 (*Gueth et al.,* 2001) and L 723 (*Lee et al.,* 2002)]. In both cases, the two outflows are driven by two independent, nearby protostars, located in the same molecular core. It is unclear, however, whether the sources are gravitationally bound or not.

2.1.3. Time evolution. There is increasing evidence that outflow collimation and morphology changes with time (e.g., *Lee et al.,* 2002; *Arce and Sargent,* 2006). The youngest outflows are highly collimated or include a very collimated component, strongly suggesting that jet bow shock-driven models are appropriate to explain these objects. Older sources present much lower collimation factors, or — a somewhat more relevant parameter — wider opening angles, pointing toward wide-angle, wind-driven outflows (see section 3.1.1). In fact, neither the jet-driven nor the wind-driven models can explain the range of morphological and kinematic properties that are observed in all outflows (see section 3.2). This was noted by *Cabrit et al.* (1997), who compared outflow observations to morphologies and PV diagrams predicted by various hydrodynamical models. More recently, a similar conclusion was obtained by *Lee et al.* (2000, 2001, 2002) from interferometric observations of 10 outflows. One attractive scenario to reconcile all observations is to invoke the superposition of both a jet and a wind component in the underlying protostellar wind and a variation in time of the relative weight between these two components. One possible explanation for this scenario is that at very early ages only the dense collimated part of the wind can break out of the surrounding dense infalling envelope. As the envelope loses mass, through infall and outflow entrainment along the axis (see section 4.1), the less dense and wider wind component will break through, entraining the gas unaffected by the collimated component, and will eventually become the main component responsible for the observed molecular outflow.

2.2. Outflows from High-Mass Protostars

Outflows from more luminous protostars have received increasing attention in recent years, with the result that we now have a more consistent understanding of massive outflow properties and their relationship to outflows from lower luminosity objects (see, e.g., recent reviews by *Churchwell,* 1999; *Shepherd,* 2003, 2005; *Cesaroni,* 2005).

Outflows from mid- to early-B-type stars have mass outflow rates 10^{-5} to a few $\times 10^{-3}$ $M_\odot$ yr^{-1}, momentum rates 10^{-4}–10^{-2} $M_\odot$ km s^{-1} yr^{-1}, and mechanical luminosity of 10^{-1}–10^{2} $L_\odot$. O stars with bolometric luminosity (L_{bol}) of more than 10^4 $L_\odot$ generate powerful winds with wind-opening angle of about 90° within 50 AU of the star [measured from water masers in and along the flow boundaries and models derived from ionized gas emission observed with resolutions of 20–100 AU, e.g., Orion (*Greenhill et al.*, 1998) and MWC 349A (*Tafoya et al.*, 2004)]. The accompanying molecular flows can have an opening angle of more than 90° (measured from CO outflow boundaries 1000 AU to 0.1 pc from the protostar). The flow momentum rate ($>10^{-2}$ $M_\odot$ km s^{-1} yr^{-1}) is more than an order of magnitude higher than what can be produced by stellar winds, and the mechanical luminosity exceeds 10^2 $L_\odot$ (e.g., *Churchwell*, 1999; *Garay and Lizano*, 1999).

Outflows from early-B and late-O stars can be well collimated (collimation factors greater than 5) when the dynamical timescale is less than ~10^4 yr. For a few early-B (proto)stars with outflows that have a well-defined jet, the jet appears to have adequate momentum to power the larger-scale CO flow, although this relation is not as well established as it is for lower luminosity sources. For example, IRAS 20126+4104 has a momentum rate in the SiO jet of

$$2 \times 10^{-1} \left(\frac{2 \times 10^{-9}}{[\mathrm{SiO/H_2}]} \right) M_\odot \ \mathrm{km \ s^{-1} \ yr^{-1}}$$

while the CO momentum rate is 6×10^{-3} $M_\odot$ km s^{-1} yr^{-1} (*Cesaroni et al.*, 1999; *Shepherd et al.*, 2000). Although the calculated momentum rate in the SiO jet is adequate to power the CO flow, the uncertainties in the assumed SiO abundance makes this difficult to prove. Another example is IRAS 18151–1208, in which the H$_2$ jet appears to have adequate momentum to power the observed CO flow (*Beuther et al.*, 2002a; *Davis et al.*, 2004). A counterexample may be the Ceph A HW2 outflow because the momentum rate in the HCO$^+$ outflow is 20 times larger than that of the observed ionized jet. However, the jet could be largely neutral or there may be an undetected wide-angle wind component (*Gómez et al.*, 1999).

Wu et al. (2004) find that the average collimation factor for outflows from sources with $L_{bol} > 10^3$ $L_\odot$ is 2.05 compared with 2.81 for flows from lower-luminosity sources. This is true even for sources in which the angular size of the flow is at least five times the resolution. Table 1 of *Beuther and Shepherd* (2005) summarizes our current understanding of massive outflows from low-spatial-resolution single-dish studies and gives a summary of and references to 15 massive flows that have been observed at higher spatial resolution using an interferometer. Here, we discuss a few of these sources that illustrate specific characteristics of massive outflows.

2.2.1. Collimated flows. The youngest early-B protostars (~10^4 yr or less) can be jet-dominated and can have either well-collimated or poorly collimated molecular flows. In a few sources, jets tend to have opening angles, α, between 25° and 30° but they do not recollimate [e.g., IRAS 20126+4104 (*Cesaroni et al.*, 1999; *Moscadelli et al.*, 2005) or IRAS 16547–4247 (*Rodríguez et al.*, 2005a)]. Other sources appear to generate well-collimated jets (α ~ few degrees) that look like scaled-up jets from low luminosity protostars [e.g., IRAS 05358+3543 (*Beuther et al.*, 2002b)]. All

these sources are ≲10^4 yr old — they have not yet reached the main sequence. In at least one case jet activity has continued as long as 10^6 yr, although the associated molecular flow has a large opening angle and complex morphology [HH 80–81 (*Yamashita et al.*, 1989; *Martí et al.*, 1993)].

One possible collimated outflow event may have been traced to a young O5 (proto)star in the G5.89–0.39 UC HII region. The O5 star has a small excess at 3.5 μm and is along the axis of two H$_2$ knots that appear to trace a north-south molecular flow along the direction of the UC HII region expansion (*Puga et al.*, 2005). The north-south molecular flow is unresolved so it is not clear that it is collimated even if the H$_2$ knots appear to trace a collimated outflow event. Although still circumstantial, the evidence is mounting that the O5 star in G5.89 produced the north-south outflow and thus is forming via accretion (*Shepherd*, 2005, and references therein).

2.2.2. Poorly collimated flows. Poorly collimated molecular flows can be due to (1) extreme precession of the jet as in IRAS 20126+4124 (*Shepherd et al.*, 2000); (2) a wide-angle wind associated with a jet as in HH 80–81 (*Yamashita et al.*, 1989) or perhaps Ceph A HW2 (e.g., *Gómez et al.*, 1999; *Rodríguez et al.*, 2001); (3) a strong wide-angle wind that has no accompanying jet; or (4) an explosive event as seen in Orion (*McCaughrean and Mac Low*, 1997). In massive flows, collimation factors as high as 4 or 5 in the molecular gas can still be consistent with being produced by wind-blown bubbles if the cloud core is very dense and it is easier for the flow to break out of the cloud rather than widen the flow cavity. Once the flow has escaped the cloud core, the bulk of the momentum is transferred to the interclump medium.

In at least some young early-B stars, both the ionized wind near the central source and the larger-scale molecular flow are poorly collimated and there is no evidence for a well-collimated jet. Examples of sources that do not appear to have a collimated jet powering the flow include G192.16–3.82 (*Shepherd and Kurtz*, 1999, and references therein), W75N VLA2 (*Torrelles et al.*, 2003, and references therein), AFGL 490 (*Schreyer et al.*, 2006, and references therein), and the SiO flow in G5.89–0.39 [not related to the O5 star discussed above (*Sollins et al.*, 2004; *Puga et al.*, 2005)]. Sources with poorly collimated flows, no evidence for a jet, and a good determination of the dynamical age show that the ages tend to be a few × 10^5 yr old and a UC HII region exists around a new ZAMS star.

To date, extremely collimated molecular outflows have not been observed toward sources earlier than B0. It is possible that this is simply a selection effect because O stars form in dense clusters and reach the ZAMS in only a few × 10^4 yr. Thus, any collimated outflows may be confused by other flows. In a few cases, outflows appear to be due to a sudden explosive event such as that seen in Spitzer images of shocked gas in G34.26+0.15 (Churchwell, personal communication) or the H$_2$ fingers of Orion. There is now good evidence that Source I in Orion and the Becklin-Neugebauer (BN) object were within a few hundred AU from each other about 500 yr ago (*Rodríguez et al.*, 2005b). Such close encounters could disrupt the accretion process and create an explosive outflow as seen in Orion (e.g., *Bonnell et al.*, 2003).

2.2.3. Evolution. Early-B stars ($L_{bol} \sim 10^4\, L_\odot$) generate UC H II regions and reach the ZAMS in 5–9×10^4 yr while still accreting and generating strong molecular outflows (e.g., *Churchwell*, 1999; *Garay and Lizano*, 1999, and references therein). The duration of the accretion phase is about the same as in low-luminosity sources (e.g., 5–10×10^5 yr) yet the development of an HII region that expands to encompass the accretion disk midway through the formation process suggests that there is a sharp transition in the physical conditions at the base of the flow where material is lifted off the surface of the disk and collimated.

Well-collimated molecular flows from massive protostars tend to be in systems with ages less than a few $\times 10^4$ yr where the central object has not yet reached the main sequence (e.g., IRAS 05358+3543 is well-collimated over approximately 1 pc). In these young sources the effects of increased irradiation on the disk and disk-wind due to the stellar radiation field are minimal. Poorly collimated flows (opening angle greater than $50°$ that show no evidence for a more collimated component) are associated with more evolved sources that have detectable UC H II regions and the central star has reached the main sequence.

To account for the differences seen in flow morphologies from early-B to late-O stars, *Beuther and Shepherd* (2005) proposed two possible evolutionary sequences that could result in similar observable outflow signatures. In Fig. 1 we show a schematic of the proposed sequences and explain how the observed outflow morphologies can be related to O and B star evolution.

Once a massive OB star reaches the main sequence, the increased radiation from the central star generates significant Lyman continuum photons and will likely ionize the outflowing gas even at large radii. Inherently lower collimation of the ionized wind due to increased radiation pressure is suggested by the hydrodynamic simulations of *Yorke and Sonnhalter* (2002). However, the radiation pressure is still too low by a factor of 10 to 100 to produce significant changes in the collimation of the observed molecular flows (*Richer et al.*, 2000).

The larger photon flux will also increase the ionization degree in the molecular gas and produce shorter ion-neutral collisional timescales. Thus, in principle, this could improve the matter-field coupling, even aiding MHD collimation. However, other effects are likely to counteract this. In particular, if the plasma pressure exceeds the magnetic field pressure and ions are well-coupled to the field, then the outflowing, ionized gas may be able to drag the magnetic field lines into a less-collimated configuration (see, e.g., *Königl*, 1999; *Shepherd et al.*, 2003).

Turbulence could also contribute to the decollimation of molecular outflows from massive OB protostars. Increased turbulence in the disk and outflow is expected to weaken the conditions for ideal MHD and hence weaken the colli-

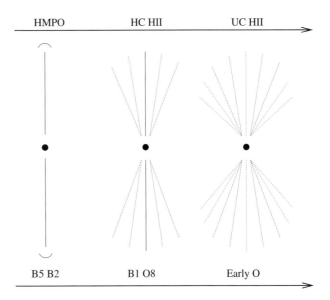

Fig. 1. Sketch of the proposed evolutionary outflow scenario put forth by *Beuther and Shepherd* (2005). The three outflow morphologies can be caused by two evolutionary sequences: (top) the evolution of a typical B1-type star from a high-mass protostellar object (HMPO) via a hypercompact H II (HCH II) region to an ultra-compact HII (UCH II) region, and (bottom) the evolution of an O5-type star that goes through B1- and O8- type stages (only approximate labels) before reaching its final mass and stellar luminosity. This evolutionary sequence appears to qualitatively fit the observations, yet it must be tested against both theory and observations.

mation effect. Turbulence could be due to higher accretion disk to stellar mass ratios ($M_{disk} > 0.3\, M_*$), making disks susceptible to local gravitational instabilities, increased radiation pressure, and high plasma temperatures. If the ions and neutrals are not well coupled in a turbulent flow, then ideal MHD begins to break down and magnetic diffusivity could significantly decollimate the molecular outflow (see, e.g., *Fendt and Cemeljic*, 2002). Furthermore, simulations by Fendt and Cemeljic find that the toroidal magnetic field component, B_ϕ, decreases with increased turbulence. Since B_ϕ is the collimating magnetic component (e.g., *Pudritz and Banerjee*, 2005), such a decrease in B_ϕ may contribute to the lower observed collimation for more-evolved massive molecular outflows.

3. MOLECULAR OUTFLOW MODELS

3.1. General Overview of Models

Several outflow models have been proposed to explain how molecular outflows from protostars are formed. Currently, outflow models can be separated into four broad classes (*Cabrit et al.*, 1997): (1) wind-driven shells, (2) jet-driven bow shocks, (3) jet-driven turbulent flows, and (4) circulation flows. In the first three, molecular outflows represent ambient material that has been entrained by a wide-angle wind or accelerated by a highly collimated jet. In the last class of models, molecular outflows are produced by deflected infalling gas. Most of the work has concentrated on simulating outflows specifically from low-mass protostars, and little work has been done on modeling outflows from high-mass stars. Many flow properties, in particular the CO spatial and velocity structure, are broadly similar across the entire luminosity range (*Richer et al.*, 2000), suggesting that similar mechanisms may be responsible for the production of molecular outflows from both low- and high-mass systems. Recent results from simulation work on the disk/outflow connection (*Pudritz and Banerjee*, 2005) as well as from observations (*Zhang et al.*, 2002; *Beuther et al.*, 2004) further indicate that molecular outflows from massive stars may be approximately modeled as scaled-up versions of their lower-mass brethren.

In the past, most studies used analytical models to try to explain the outflow morphology and kinematics. However, in the last decade, computational power has increased sufficiently to allow for multidimensional hydrodynamical (HD) simulations of protostellar outflows that include a simple molecular chemical network. Numerical modeling of the molecular cooling and chemistry, as well as the hydrodynamics, is required in these systems, which are described by a set of hyperbolic differential equations with solutions that are usually mathematically chaotic and cannot be treated analytically. Treatment of the molecular cooling and chemistry facilitates a comparison of the underlying flow with observational quantities (for example, the velocity distribution of mass vs. CO intensity, the temperature distribution of the outflowing gas, and the H_2 1–0 S(1) maps).

3.1.1. Wind-driven shell models. In the wind-driven shell model, a wide-angle radial wind blows into the stratified surrounding ambient material, forming a thin swept-up shell that can be identified as the outflow shell (*Shu et al.*, 1991; *Li and Shu*, 1996; *Matzner and McKee*, 1999). In these models, the ambient material is often assumed to be toroidal with density $\rho_a = \rho_{ao} \sin^2 \theta / r^2$, while the wind is intrinsically stratified with density $\rho_w = \rho_{wo}/(r^2 \sin^2 \theta)$, where ρ_{ao} is the ambient density at the equator and ρ_{wo} is the wind density at the pole (*Lee et al.*, 2001). This class of models is attractive as it particularly explains old outflows of large lateral extents and low collimation.

In recent years, there have been a few efforts to model wide-angle winds numerically. *Lee et al.* (2001) performed numerical HD simulations of an atomic axisymmetric wind and compared it to simulations of bow shock-driven outflows. Their wide-wind models yielded smaller values of γ (see section 2) over a narrower range (1.3–1.8), as compared to the jet models (1.5–3.5). *Raga et al.* (2004b) have included both wide-angle winds and bow-shock models in a study aimed at reproducing features of the southwest lobe of HH 46/47, with the result that a jet model is able to match enough features that they feel that it is not necessary to invoke a wide-angle wind (although it produces a reasonable

fit to the observations). In simulations by *Delamarter et al.* (2000) the wind is assumed to be spherical, even though the physical origin of such a wind is not yet clear, and it is focused toward the polar axis by the density gradients in the surrounding (infalling) torus-like environment. In these models the low-velocity γ ranges from approximately 1.3 to 1.5, similar to other studies. The MHD simulations performed by *Gardiner et al.* (2003) show that winds that have a wide opening angle at the base can produce a dense jetlike structure downstream due to MHD collimation. Very recently, axisymmetric winds have been modeled with a code that includes molecular chemistry and cooling as well as Adaptive Mesh Refinement (AMR) (*Cunningham et al.*, 2005). These last two studies produce satisfactory general outflow lobe appearance; however, no mass-velocity, position-velocity maps, or channel maps have been generated to compare with observations.

3.1.2. Turbulent jet model. In the jet-driven turbulent model, Kelvin-Helmholtz instabilities along the jet/environmental boundary lead to the formation of a turbulent viscous mixing layer, through which the cloud molecular gas is entrained (*Cantó and Raga*, 1991; *Raga et al.*, 1993; *Stahler*, 1994; *Lizano and Giovanardi*, 1995; *Cantó et al.*, 2003, and references therein). The mixing layer grows both into the environment and into the jet, and eventually the whole flow becomes turbulent. Discussion of the few existing numerical studies that investigate how molecular outflows are created by a turbulent jet is presented in a recent review by *Raga et al.* (2004a), who cite the "Torino group" as the only simulations with predictions for atomic (e.g., Hα, [SII]) emission (*Micono et al.*, 1998). The radiatively cooled jet simulations reproduce the broken power law behavior of the observationally determined mass-velocity distribution, even though molecular chemistry or cooling is not included (*Micono et al.*, 2000). However, these models produce decreasing molecular outflow velocity with distance from the powering source — opposite to that observed in most molecular outflows. An analytical model using Kelvin-Helmholtz instabilities has recently been proposed by *Watson et al.* (2004) to explain entrainment of cloud material by outflows from high-mass stars.

3.1.3. Jet bow shock model. In the jet-driven bow shock model, a highly collimated jet propagates into the surrounding ambient material, producing a thin outflow shell around the jet (*Raga and Cabrit*, 1993; *Masson and Chernin*, 1993). The physical origin of the jet is currently unclear and could even be considered as an extreme case of a highly collimated wide-angle wind without a tenuous wide-angle component. As the jet impacts the ambient material, a pair of shocks, a jet shock and a bow shock, are formed at the head of the jet. High-pressure gas between the shocks is ejected sideways out of the jet beam, which then interacts with unperturbed ambient gas through a broader bow shock surface, producing an outflow shell surrounding the jet. An episodic variation in the mass-loss rate produces a chain of knotty shocks and bow shocks along the jet axis within the outflow shell. Recent analytical models without magnetic field include *Wilkin* (1996), *Zhang and Zheng* (1997), *Smith et al.* (1997), *Ostriker et al.* (2001), and *Downes and Cabrit* (2003).

There have been two recent sets of efforts (by two different groups) to model molecular protostellar jets numerically in two or three spatial dimensions, where the mass velocity and position velocity have routinely been measured. In these simulations, a tracer associated with molecular hydrogen is followed. However, each group approaches this problem in a different way, with each approach having its own advantages and disadvantages. In an effort to resolve the postshock region, *Downes and Ray* (1999) and *Downes and Cabrit* (2003) have simulated relatively low-density, axisymmetric (two-dimensional) fast jets. Alternatively, recognizing that observed flows associated with Class 0 sources have a higher density and a complex appearance, Smith and Rosen have extended the work of *Suttner et al.* (1997) and *Völker et al.* (1999) by further investigating sets of fully three-dimensional flows (e.g., *Rosen and Smith,* 2004a). The main disadvantage of this approach is that with such high densities the postshock region will necessarily be underresolved, especially in three-dimensional flows. Both the Downes and Smith groups have included molecular hydrogen dissociation and reformation as well as ro-vibrational cooling in their hydrodynamical simulations, although the treatment of this cooling is quite different in each group. One example is that the Downes group turns off all cooling and chemistry below 1000 K, while the Smith and Rosen simulations [explained in detail in *Smith and Rosen* (2003)] include cooling and chemistry calculations at essentially all temperatures (albeit with an equilibrium assumption for some reactions). The jet flows themselves enter the grid from a limited number of zones at one side of the computational domain, with densities and temperatures that are constant radially (a top hat profile) and over time. Both groups usually model the jet as nearly completely molecular — even though there are arguments suggesting that the jet will not initially be molecular, and that H_2 might subsequently form on the internal working surfaces of the jet (*Raga et al.,* 2005). The initial jet velocities of the Downes and Smith and Rosen groups are varied with shear, pulsation, and, in the three-dimensional simulations, with precession.

These different approaches have yielded different slopes for the computed CO intensity-velocity plots. The Downes group results have tended to be steeper and closer to the nominal value of $\gamma = 2$, while the standard Rosen and Smith case has a value near 1. Much of this difference can be attributed to the difference in jet-to-ambient density ratio (see *Rosen and Smith,* 2004a), which is 1 in the Downes standard case, and 10 in the Rosen and Smith standard case. The value of γ has been shown in these simulations to evolve over time, with steeper slopes associated with older flows. Most of these simulations are quite young, but there has been a recent effort to run the simulations out to t = 2300 yr (*Keegan and Downes,* 2005). They confirm the steepening

of the mass-velocity slope up to t = 1600 yr (when $\gamma = 1.6$), and then it becomes roughly constant. The Smith and Rosen group have investigated whether fast (*Rosen and Smith,* 2004b) or slow (*Smith and Rosen,* 2005) precession has an effect on the mass-velocity slopes. While the simulations with fast precessing jets show a dependence of γ on the precession angle (generally increasing γ with the angle), some of this dependence was reduced in the slowly precessing cases. However, at this time only very young (t < 500 yr) precessing sources have been simulated.

The initially molecular jet simulations that include periodic velocity pulses exhibit position-velocity plots with a sequence of Hubble wedges, similar to that observed in molecular outflows produced by an episodic protostellar wind (see section 2.1.2). Where computed, velocity channel maps in CO from molecular jet simulations, as in *Rosen and Smith* (2004a), have a morphology similar to that of many sources [e.g., HH 211 (*Gueth and Guilloteau,* 1999)], i.e., revealing the knots within the jet at high velocities and showing the overall shape of the bow shock at low velocities.

Some recent studies show the need to expand the interpretation of molecular outflow observations beyond the simulated H_2 and CO emission from the numerical models discussed above. For example, the work of *Lesaffre et al.* (2004) includes more complex chemistry in one dimension, focusing on the unstable nature of combined C and J shocks. Also, radiation transfer with a complex chemistry has been simulated for a steady three dimensional (jet) flow, with a focus on HCO^+ emission (*Rawlings et al.,* 2004).

In addition, magnetic field effects have been included in atomic protostellar jets that are axisymmetric (*Gardiner et al.,* 2000; *Stone and Hardee,* 2000) and fully three-dimensional (*Cerqueira and de Gouveia dal Pino,* 1999, 2001) and even molecular axisymmetric protostellar jets (*O'Sullivan and Ray,* 2000). These studies show significant differences compared to simulations of jets without magnetic fields. For example, magnetic tension, either along the jet axis or as a hoop stress from a toroidal field, can help collimate and stabilize the jet — although some of the additional stability is mitigated in a pulsed jet. Some of the differences between pure HD and MHD simulations that show up in the axisymmetric cases are less prominent in three-dimensional simulations (*Cerqueira and de Gouveia dal Pino,* 2001).

3.1.4. Circulation models. In circulation models the molecular outflow is not entrained by an underlying wind or jet, it is rather formed by infalling matter that is deflected away from the protostar in a central torus of high MHD pressure through a quadrupolar circulation pattern around the protostar, and accelerated above escape speeds by local heating (*Fiege and Henriksen,* 1996a,b). The molecular outflow may still be affected by entrainment from the wind or jet, but this would be limited to the polar regions and it would not be the dominant factor for its acceleration (*Lery et al.,* 1999, 2002). Circulation models may provide a means of injecting added mass into outflows from O stars where it appears unlikely that direct entrainment can supply all the observed mass in the flow (*Churchwell*, 1999).

The most recent numerical studies of the circulation model have focused on a steady-state axisymmetric case, usually involving radiative heating, magnetic fields, and Poynting flux (*Lery*, 2003). The addition of the Poynting flux in recent versions of this model has alleviated one of its major flaws (*Lery et al.*, 2002), i.e., the inability in earlier models to generate an outflow of sufficient speed. The toroidal magnetic field in what is currently being called the "steady-state transit model" assists in the formation of a collimated fast moving flow (*Combet et al.*, 2006).

3.2.   Comparing Observations and Models

In the past ten years, molecular outflows have been mapped at high angular resolutions with millimeter interferometers, allowing us to confront the outflow models in great detail. A schematic of the predicted properties of molecular outflows produced by the different models discussed above is presented in Fig. 2. High-resolution molecular outflow observations can be used to compare the data with the outflow characteristics shown in Fig. 2 in order to establish what model best fits the observed outflow.

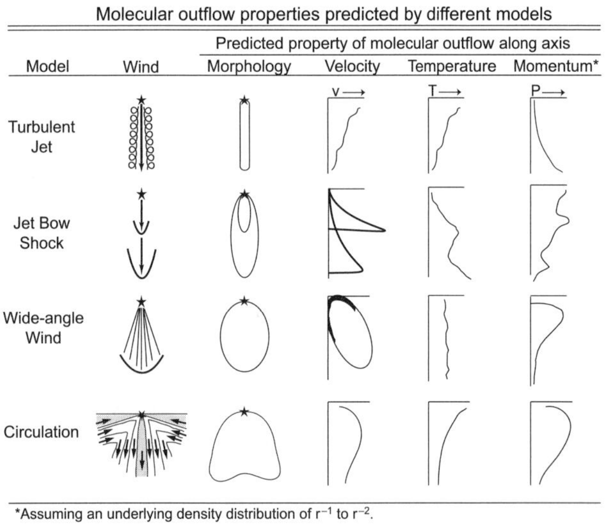

Molecular outflow properties predicted by different models

Fig. 2. Observable molecular outflow properties predicted by the four leading broad classes of models: (1) turbulent jet (*Cantó and Raga*, 1991; *Chernin and Masson*, 1995; *Bence et al.*, 1996); (2) jet bow shock (*Chernin and Masson*, 1995; *Cliffe et al.*, 1996; *Hatchell et al.*, 1999; *Lee et al.*, 2001); (3) wide-angle wind (*Li and Shu*, 1996; *Lee et al.*, 2001); and (4) circulation models (*Fiege and Henriksen*, 1996b; *Lery et al.*, 1999). In the jet-driven bow shock model, an episodic variation in jet velocity produces an internal bow shock driving an internal shell, in addition to the terminal shock. This episodic variation can also be present in the other wind models, but in this figure the effects of an episodic wind are only shown for the jet bow shock model. This figure is based on Fig. 1 of *Arce and Goodman* (2002b).

Here we focus our attention on comparing observations with the jet-driven bow shock and wide-angle wind-driven models, as most of the numerical simulations concentrate on these two models and they are the most promising models thus far. The predicted mass-velocity relationships in jet bow shock and wide-angle wind models have a slope (γ) of 1–4, in tune with observations. Each model predicts a somewhat different position-velocity (PV) relation that can be used to differentiate between these two leading molecular outflow driving mechanisms (*Cabrit et al.*, 1997; *Lee et al.*, 2000, 2001).

3.2.1. Jet-driven bow shock models vs. observations. Current jet-driven bow shock models can qualitatively account for the PV spur structure (where the outflow velocity increases rapidly toward the position of the internal and leading bow shocks; see Fig. 2), the broad range of CO velocities near H_2 shocks, and the morphological relation between the CO and H_2 emission seen in young and collimated outflows. These models are able to produce the observed outflow width for highly collimated outflows, such as L 1448, HH 211, and HH 212 (*Bachiller et al.*, 1995; *Gueth and Guilloteau*, 1999; *Lee et al.*, 2001). However, jet-driven bow shock models have difficulty producing the observed width of poorly collimated outflows, like RNO 91, VLA 05487, and L 1221 (*Lee et al.*, 2000, 2001, 2002). Jet models produce narrow molecular outflows mainly because the shocked gas in the bow shock working surfaces limits the transverse momentum (perpendicular to the jet-axis) that can be delivered to the ambient medium. In numerical simulations of jets, the width of the outflow shell is mainly determined by the effects of the leading bow shock from the jet's first impact into the ambient material (e.g., *Suttner et al.*, 1997; *Downes and Ray*, 1999; *Lee et al.*, 2001). While the jet penetration into the cloud increases roughly linearly with time, the width only grows as the one-third power of time (*Masson and Chernin*, 1993; *Wilkin*, 1996; *Ostriker et al.*, 2001).

Jets also have difficulty producing the observed outflow momenta. The transverse momentum of the outflow shell is acquired primarily near the jet head where the pressure gradient is large, and the mean transverse velocity of the shell, $\bar{v}_R$, can be approximated by $\bar{v}_R \simeq \beta c_s(R_j^2/R^2)$, where R and R_j are the outflow and jet radius, respectively, and βc_s is the velocity of the gas ejected from the working surface (*Ostriker et al.*, 2001). For example, in a 10,000-AU-wide molecular outflow driven by a 150-AU jet, and assuming $\beta c_s = 32$ km s^{-1}, the expected mean transverse velocity of the shell is only 0.03 km s^{-1}. As a result, if outflows were driven by a steady jet, the wide portions of outflow shells would exhibit extremely low velocities and very small momenta. This is inconsistent with the observations, especially in the wider flows where the well-defined cavity walls have appreciable velocities [e.g., B5-IRS1 (*Velusamy and Langer*, 1998), RNO 91 (*Lee et al.*, 2002), and L 1228 (*Arce and Sargent*, 2004)].

Systematic wandering of the jet flow axis has been argued to occur in several outflows based on outflow morphology, e.g., IRAS 20126+4104 (*Shepherd et al., 2000*) and L 1157 (*Bachiller et al., 2001*). This may mitigate the above discrepancies. The width and momentum of the outflow shell can increase because a wandering jet has a larger "effective radius" of interaction and can impact the outflow shell more directly (*Raga et al., 1993; Cliffe et al., 1996*). Some simulations show hints of widening by jet wandering (*Völker et al., 1999; Rosen and Smith, 2004a; Smith and Rosen, 2005*), but some show that a wandering jet could produce a smaller width than a steady jet (*Raga et al., 2004b*). Further calculations are needed to ascertain if motion of the jet axis at realistic levels can improve quantitative agreement with observed outflow features.

3.2.2. Wide-angle wind models vs. observations. Wide-angle winds can readily produce CO outflows with large widths but have trouble producing other commonly observed features. In this model, the outflow velocity also increases with the distance from the source, showing a lobe PV structure tilted with inclination that exhibits only a small velocity range at the tip. If the tip is not observed, the PV structure appears as a tilted parabola (see Fig. 2). As discussed in section 3.1.1, most wind-driven models assume the protostellar wind density depends on the angle from the pole (θ). If the wind velocity has a small, or no, dependence on θ, and assuming a density stratification similar to that proposed by *Li and Shu* (1996), then the outflow width, W, can be expressed in terms of the ratio of wind to ambient density at the equator, (ρ_{wo}/ρ_{ao}), the wind velocity at the pole, v_{wo}, and the outflow age, t, as $W \approx (\rho_{wo}/\rho_{ao})^{1/4}v_{wo}t$ (*Lee et al., 2001*). For (ρ_{wo}/ρ_{ao}) between 10^{-3} and 10^{-4}, a 100 km s^{-1} wind can produce an outflow width of 0.1 to 0.2 pc in 10^4 yr. Thus, the wind-driven model can produce widths consistent with observed molecular outflows in about 10^4 yr. However, these models have problems producing discrete bow-shock-type features in the entrained molecular gas, as seen in many high-resolution maps of CO outflows (e.g., *Lee et al., 2000, 2002*), and discrete position-velocity spur structures (and Hubble wedges). These features are hard to generate as the wide wind impacts all locations on the shell. Models of wide-angle pulsed winds produce a series of flat internal shocks within the outflow shell (*Lee et al., 2001*), inconsistent with the curved internal H$_2$ bow shocks typically observed in episodic outflows (see section 2.1).

One possible solution to these problems is to require the winds to have a collimated core with a strong velocity gradient with respect to θ. A disk wind driven from a large range of radii may have velocity strongly decreasing toward equatorial latitudes, because the asymptotic velocity on a given streamline in an MHD wind is characteristic of the Keplerian speed at the streamline's footpoint (see chapter by Pudritz et al.). Further work is needed to study whether this sort of modification can produce the observed outflow features.

3.2.3. A synthesis with an evolutionary scenario. A model that combines attributes of the jet and wide-angle wind models is arguably the best match to the available CO

outflow data. A two-component protostellar wind may be produced, for example, by a slow disk wind and a fast central disk-driven jet or X-wind (arising from the magnetosphere-disk boundary region). The disk wind could help collimate the X-wind into the jet component (*Ostriker, 1997*) and provide a slow wide-angle component that drives the outflow width and momentum (see chapter by Shang et al.).

Observational support for the synthesis model exist at different wavelengths. There is mounting evidence from millimeter observations that the morphology of some molecular outflows is better explained with a "dual-wind" model (e.g., *Yu et al., 1999; Arce and Goodman, 2002a; Arce and Sargent, 2004*). In the optical, the forbidden emission line profiles of T Tauri stars show two velocity components: a high-velocity component that is argued to arise in a jet and a low-velocity component that might result from a disk wind (*Kwan and Tademaru, 1995*; chapter by Ray et al.). A possible scenario is that the main driving agent producing most of the observed molecular outflow may change over time, as discussed in section 2. Numerical simulations of an evolving dual-wind model will be critical to study whether this proposed scenario can reproduce the wide range of observed features in molecular outflows from low- and high-mass protostars.

4. IMPACT OF OUTFLOWS ON SURROUNDING ENVIRONMENT

4.1. Physical Impact

Outflows from newborn stars inject momentum and energy into the surrounding molecular cloud at distances ranging from a few AU to up to tens of parsecs away from the source. Historically, most studies have concentrated on the interaction between the outflow and the surrounding core (~0.1–0.3 pc) as these scales can easily be observed with single-dish telescopes in the nearby (≤1 kpc) star-forming regions. More recently, studies using millimeter interferometer array and single telescopes with focal-plane arrays have been crucial in the understanding of the outflow's impact at smaller (<0.1 pc) and larger (≥1 pc) scales, respectively.

4.1.1. Outflow-envelope interactions. Protostellar winds originate within a few AU of the star (see chapter by Ray et al.), and so they are destined to interact with the dense circumstellar envelope — the primary mass reservoir of the forming star, with sizes in the range of 10^3 to 10^4 AU. In fact, survey studies of the circumstellar gas within 10^4 AU of low-mass YSOs show outflows contribute significantly to the observed mass loss of the surrounding dense gas (from about 10^{-8} to 10^{-4} M$_\odot$ yr^{-1}, depending on the protostar's age) and indicate there is an evolution in the outflow-envelope interaction (e.g., *Fuller and Ladd, 2002; Arce and Sargent, 2006*). As shown below, detailed studies of individual sources corroborate these results. The powerful outflows from low-mass Class 0 sources are able to modify the distribution and kinematics of the dense gas surrounding a protostar, as evidenced in L 1157 (*Gueth et al., 1997; Bel-*

trán et al., 2004b), and RNO 43 (*Arce and Sargent,* 2005), where molecular line maps show the circumstellar high-density gas has an elongated structure and a velocity gradient, at scales of 4000 AU, along the outflow axis. Similarly, in IRAM 0491 (*Lee et al.,* 2005) and HH 212 (*Wiseman et al.,* 2001) the dense gas traced by N_2H^+ and NH_3, respectively, exhibit blue- and red-shifted protrusions extending along the blue and red outflow lobes, evidence that there are strong outflow-envelope interactions in these Class 0 sources. These results clearly show that, independent of the original (i.e., pre-protostellar outflow) underlying circumstellar matter distribution, young outflows entrain dense envelope gas along the outflow axis.

Although not as powerful as those of Class 0 sources, the wide-angle outflows typically observed in Class I sources (with opening angles of ≥90°) are capable of constraining the infalling envelope to a limited volume outside the outflow lobes, as seen in the L 1228 (*Arce and Sargent,* 2004) and B5-IRS1 (*Velusamy and Langer,* 1998) outflows. The L 1228 outflow is currently eroding the surrounding envelope by accelerating high-density ambient gas along the outflow-envelope interface and has the potential to further widen the cavities, as the outflow ram pressure is about a factor of 4 higher than the infall ram pressure (*Arce and Sargent,* 2004). In RNO 91, a Class II source, the outflow exhibits an even wider opening angle of 160° that is expanding, and decreasing the volume of the infall region (*Lee and Ho,* 2005).

Widening of the outflow opening angle with age appears to be a general trend in low-mass protostars and there is ample evidence for erosion of the envelope due to outflow envelope interactions (*Velusamy and Langer,* 1998; *Arce and Sargent,* 2004; *Arce,* 2004; *Lee and Ho,* 2005; *Arce and Sargent,* 2006). Thus, it is clear that even if the pre-protostellar outflow circumstellar distribution of matter has a lower density along the polar regions (i.e., the outflow axis), as suggested by different models (i.e., *Hartmann et al.,* 1996; *Li and Shu,* 1996), outflow-envelope interactions will have an impact on the subsequent circumstellar density distribution, as they will help widen the cavity and constrain the infall region. It is tempting to extrapolate and suggest that as a young star evolves further its outflow will eventually become wide enough to end the infall process and disperse the circumstellar envelope altogether.

4.1.2. Outflow-core interactions. Strong evidence exists for the disruptive effects outflows have on their parent core — the dense gas within 0.1–0.3 pc of the young star. Direct evidence of outflow-core interaction comes from the detection of velocity shifts in the core's medium- and high-density gas in the same sense, both in position and velocity, as the high-velocity (low-density) molecular outflow traced by ^{12}CO (e.g., *Tafalla and Myers,* 1997; *Dobashi and Uehara,* 2001; *Takakuwa et al.,* 2003; *Beltrán et al.,* 2004a). The high opacity of the ^{12}CO lines hampers the ability to trace low-velocity molecular outflows in high-density regions. Therefore, other molecular species like ^{13}CO, CS, $C^{18}O$, NH_3, CH_3OH, and C_3H_2 are used to trace the high-

density gas perturbed by the underlying protostellar wind. The average velocity shifts in the dense core gas are typically lower than the average velocity of the molecular (^{12}CO) outflow, consistent with a momentum-conserving outflow entrainment process. In addition to being able to produce systematic velocity shifts in the gas, outflows have been proposed to be a major source of the turbulence in the core (e.g., *Myers et al.,* 1988; *Fuller and Ladd,* 2002; *Zhang et al.,* 2005).

Outflows can also reshape the structure of the star-forming core by sweeping and clearing the surrounding dense gas and producing density enhancements along the outflow axis. The clearing process is revealed by the presence of nebular emission resulting from the scattering of photons, from the young star, off of cavity walls created by the outflow (e.g., *Yamashita et al.,* 1989; *Shepherd et al.,* 1998; *Yu et al.,* 1999), or depressions along the outflow axis in millimeter molecular line maps of high-density tracers (e.g., *Moriarty-Schieven and Snell,* 1988; *White and Fridlund,* 1992; *Tafalla et al.,* 1997). Outflow-induced density enhancements (and shock-heated dust) in the core may be revealed by the dust continuum emission (e.g., *Gueth et al.,* 2003; *Beuther et al.,* 2004; *Sollins et al.,* 2004). A change in the outflow axis direction with time, as observed in many sources (see chapter by Bally et al.), will allow an outflow to interact with a substantial volume of the core and be more disruptive on the dense gas than outflows with a constant axis (e.g., *Shepherd et al.,* 2000; *Arce and Goodman,* 2002a). By accelerating and moving the surrounding dense gas, outflows can gravitationally unbind a significant amount of gas in the dense core, thereby limiting the star-formation efficiency of the dense gas (see *Matzner and McKee,* 2000).

The study of *Fuente et al.* (2002) shows that outflows *appear* to be the dominant mechanism able to efficiently sweep out about 90% of the parent core by the end of the pre-main-sequence phase of young intermediate-mass (Herbig Ae/Be) stars. In addition, outflows from low- and high-mass protostars have kinetic energies comparable to the gravitational binding energy of their parent core, suggesting outflows have the potential to disperse the entire core (e.g., *Tafalla and Myers,* 1997; *Tafalla et al.,* 1997). We may even be observing the last stages of the outflow-core interaction in G192.16, a massive (early B) young star, where the dense core gas is optically thin and clumpy, and the ammonia core is gravitationally unstable (*Shepherd et al.,* 2004). However, further systematic observations of a statistical sample of outflow-harboring cores at different ages are needed in order to fully understand the details of the core-dispersal mechanism and conclude whether outflows can disperse their entire parent core.

Theoretical studies indicate that shocks from a protostellar wind impacting on a dense clump of gas (i.e., a prestellar core) along the outflow's path can trigger collapse and accelerate the infall process in the impacted core (*Foster and Boss,* 1996; *Motoyama and Yoshida,* 2003). Outflow-triggered star formation has been suggested in only a handful of sources where the morphology and velocity structure of the dense gas surrounding a young protostar appears to be affected by the outflow from a nearby YSO (*Girart et al.*, 2001; *Sandell and Knee*, 2001; *Yokogawa et al.*, 2003).

4.1.3. Outflow-cloud interactions far from the source. Giant outflows from young stars of all masses are common, and they can interact with the cloud gas at distances greater than 1 pc from their source (*Reipurth et al.*, 1997; *Stanke et al.*, 2000). Outflows from low-mass protostars are able to entrain 0.1–1 $M_\odot$ of cloud material, accelerate and enhance the linewidth of the cloud gas (*Bence et al.*, 1996; *Arce and Goodman*, 2001b), and in some cases their kinetic energy is comparable to (or larger than) the turbulent energy and gravitational binding energy of their parent cloud (*Arce*, 2003). The effects of giant outflows from intermediate- and high-mass YSOs on their surroundings can be much more damaging to their surrounding environment. Studies of individual sources indicate that giant outflows are able to entrain tens to hundreds of solar masses, induce parsec-scale velocity gradients in the cloud, produce dense massive shells of swept-up gas at large (>0.5 pc) distances from the source, and even break the cloud apart (*Fuente et al.*, 1998; *Shepherd et al.*, 2000; *Arce and Goodman*, 2002a; *Benedettini et al.*, 2004). The limited number of studies in this field suggest that a single giant outflow has the *potential* to have a disruptive effect on their parent molecular cloud (e.g., *Arce*, 2003). Clearly, additional observations of giant outflows and their clouds are needed in order to quantify their disruptive potential.

Most star formation appears in a clustered mode and so multiple outflows should be more disruptive on their cloud than a single star. Outflows from a group of young stars interact with a substantial volume of their parent cloud by sweeping up the gas and dust into shells (e.g., *Davis et al.*, 1999; *Knee and Sandell*, 2000), and may be a considerable, albeit not the major, source of energy for driving the supersonic turbulent motions inside clouds (*Yu et al.*, 2000; *Williams et al.*, 2003; *Mac Low and Klessen*, 2004). It has also been suggested that past outflow events from a group of stars may leave their imprint on the cloud in the form of numerous cavities (e.g., *Bally et al.*, 1999; *Quillen et al.*, 2005). Very limited (observational and theoretical) work on this topic exists, and further observations of star-forming regions with different environments and at different evolutionary stages are essential to understand the role of outflows in the gaseous environs of young stellar clusters.

4.2. Shock Chemistry

The propagation of a supersonic protostellar wind through its surrounding medium happens primarily via shock waves. The rapid heating and compression of the region trigger different microscopic processes — such as molecular dissociation, endothermic reactions, ice sublimation, and dust grain disruption — which do not operate in the unperturbed gas. The timescales involved in the heating and in some of the "shock chemistry" processes are short (a few 10^2 to 10^4 yr), so the shocked region rapidly acquires a chemical composition distinct from that of the quiescent unperturbed medium. Given the short shock cooling times [~10^2 yr (*Kaufman and Neufeld*, 1996)], some of these high-temperature chemical processes only operate at the initial stages, as the subsequent chemical evolution is dominated by low-temperature processes. This chemical evolution, the gradual clearing of the outflow path, and the likely intrinsic weakening of the main accelerating agent, all together make the important signatures of the shock interaction (including some of the chemical anomalies) vanish as the protostellar object evolves. Chemical anomalies found in an outflow can therefore be considered as an indicator of the outflow age (e.g., *Bachiller et al.*, 2001).

The chemical impact of outflows are better studied in outflows around Class 0 sources with favorable orientation in the sky (i.e., high inclination with respect to the line of sight). With less confusion than that found around massive outflows, the shocked regions of low-mass, high-collimation outflows (which often adopt the form of well-defined bows) are well separated spatially with respect to the quiescent gas. Detail studies of these "simple" regions can help disentangle the effects of outflow shocks from other shocks in more complex regions — like in circumstellar disks, where one expects to find outflow shock effects blended with those produced by shocks triggered by the collapsing envelope (e.g., *Ceccarelli et al.*, 2000).

Shocks in molecular gas can be of C type or of J type, depending on whether the hydrodynamical variables change continuously across the shock front (e.g., *Draine and McKee*, 1993). C-type shocks are mediated by magnetic fields acting on ions that are weakly coupled with neutrals, they are slow, have maximum temperatures of about 2000–3000 K, and are nondissociative. J-type shocks are typically faster, and can reach much higher temperatures. The critical velocity at which the change between the C and J regime is produced depends on several parameters such as the preshock density (*Le Bourlot et al.*, 2002) and the presence of charged grains (*Flower and Pineau des Forts*, 2003), and it typically ranges from ~20 up to ~50 km s^{-1}. J-type shocks may also occur at relatively low velocities when the transverse component of the magnetic field is small (*Flower et al.*, 2003). Recent infrared observations of several lines of H_2, CO, H_2O, and OH, and of some crucial atomic lines, have made possible the estimate of temperature and physical conditions in a relatively large sample of outflows. It follows that the interpretation of the data from most shocked regions require a combination of C- and J-type shocks (for comprehensive reviews, see *Noriega-Crespo*, 2002; *van Dishoeck*, 2004). Such a combination of shocks can be obtained by the overlap of multiple outflow episodes as observed in several sources, and/or by the bow shock geometry, which could generate J-type shocks at the apex of the bow together with C-type shocks at the bow flanks (*Nisini et al.*, 2000; *O'Connell et al.*, 2004, 2005). C-type shocks are particularly efficient in triggering a distinct molecular chemistry in the region in which the molecules are preserved and heated to ~2000–3000 K. Moreover, molecules can also re-form in J-shocked regions when the gas rapidly cools, or in warm layers around the hottest regions. The main processes expected to dominate this shock chemistry were discussed by *Richer et al.* (2000).

Comprehensive chemical surveys have been carried out in two prototypical Class 0 sources [L 1157 (*Bachiller and Pérez-Gutiérrez,* 1997) and BHR 71 (*Garay et al.,* 1998)]. More recent observations, including high-resolution molecular maps, have been made for a sample of sources, for example, L 1157 (*Bachiller et al.,* 2001), NGC 1333 IRAS 2 (*Jørgensen et al.,* 2004), NGC 1333 IRAS 4 (*Choi et al.,* 2004), NGC 2071 (*Garay et al.,* 2000), and Cep A (*Codella et al.,* 2005). These observations have revealed that there are important differences in molecular abundances in different outflow regions. Such variations in the abundances may be linked to the time evolution of the chemistry (*Bachiller et al.,* 2001) and may also be related to variations in the abundance of the atomic carbon (*Jørgensen et al.,* 2004).

SiO exhibits the most extreme enhancement factors (up to ~10^6) with respect to the quiescent unperturbed medium. Such high enhancements are often found close to the heads (bow shocks), and along the axes, of some highly collimated outflows (e.g., *Dutrey et al.,* 1997, and references therein; *Codella et al.,* 1999; *Bachiller et al.,* 2001; *Garay et al.,* 2002; *Jørgensen et al.,* 2004; *Palau et al.,* 2006, and references therein). Sputtering of atomic Si from the dust grains is at the root of such high SiO abundances (*Schilke et al.,* 1997), a process that requires shock velocities in excess of ~25 km s^{-1}. Accordingly, the SiO lines usually present broad wings and, together with CO, the SiO emission usually reaches the highest terminal velocities among all molecular species. Moreover, recent observations of several outflows have revealed the presence of a narrow (<1 km s^{-1}) SiO line component (*Lefloch et al.,* 1998; *Codella et al.,* 1999; *Jiménez-Serra et al.,* 2004). The presence of SiO at low velocities is not well understood. Plausible explanations include that this is the signature of a shock precursor component (*Jiménez-Serra et al.,* 2004, 2005) or that SiO is indeed produced at high velocities and subsequently slowed down in timescales of ~10^4 yr (*Codella et al.,* 1999).

CH_3OH and H_2CO are also observed to be significantly overabundant in several outflows, enhanced by factors of about 100 (*Bachiller et al.,* 2001; *Garay et al.,* 2000, 2002; *Jørgensen et al.,* 2004; *Maret et al.,* 2005). These two species are likely evaporated directly from the icy dust mantles, and in many cases the terminal velocities of their line profile wings are significantly lower than that of SiO, probably because CH_3OH and H_2CO do not survive at velocities as high as those required to form SiO (*Garay et al.,* 2000). Thus, an enhancement of CH_3OH and H_2CO with no SiO may indicate the existence of a weak shock. On the other hand, after the passage of a strong shock, and once the abundances of CH_3OH, H_2CO, and SiO are enhanced in the gas phase, one would expect the SiO molecules to reincorporate to the grains while some molecules of CH_3OH and

H_2CO remain in the gas phase, as these two molecules are more volatile than SiO (their molecular depletion timescales are about a few 10^3 yr for densities of ~10^6 cm^{-3}). In this scenario enhancement of CH_3OH and H_2CO most likely marks a later stage in the shock evolution than that traced by high SiO abundances.

In several outflows HCO$^+$ high-velocity emission is only prominent in regions of the outflow that are relatively close to the driving sources (*Bachiller et al.,* 2001; *Jørgensen et al.,* 2004). In such regions, the HCO$^+$ abundance can be enhanced by a factor of ~20. This behavior can be understood if the HCO$^+$ that was originally produced through shock-induced chemistry (e.g., *Rawlings et al.,* 2004) is destroyed by dissociative recombination or by reaction with the abundant molecules of H_2O (*Bergin et al.,* 1998). Once the abundance of the gaseous H_2O decreases due to freeze-out, the abundance of HCO$^+$ may increase. A rough anti-correlation between CH_3OH and HCO$^+$ (*Jørgensen et al.,* 2004) seems to support these arguments. In other cases, HCO$^+$ emission is observed at positions close to HH objects that can be relatively distant from the driving sources. In fact, together with NH_3, HCO$^+$ is expected to be enhanced in clumps within the molecular cloud by UV irradiation from bright HH objects (*Viti and Williams,* 1999), an effect observed near HH 2 according to *Girart et al.* (2002). Nevertheless, *Girart et al.* (2005) have recently found that UV irradiation alone is insufficient to explain the measured HCO$^+$ enhancements and that strong heating (as that caused by a shock) is also needed.

The chemistry of S-bearing species is of special interest as it has been proposed to be a potential tool to construct chemical clocks to date outflows (and hence their proto-stellar driving sources). The scenario initially proposed by a number of models is that H_2S is the main reservoir of S in grain mantles, although recent observations seem to indicate that OCS is more abundant on ices than H_2S (*Palumbo et al.,* 1997; *van der Tak et al.,* 2003). Once H_2S is ejected to the gas phase by the effect of shocks, its abundance will rapidly decrease after 10^4 yr (e.g., *Charnley,* 1997) due to oxidation with O and OH, thereby producing SO (first) and SO_2 (at a later time). Models and observations indicate that the SO/H_2S and SO_2/H_2S ratios are particularly promising for obtaining the relative age of shocks in an outflow (*Charnley,* 1997; *Hatchell et al.,* 1998; *Bachiller et al.,* 2001; *Buckle and Fuller,* 2003). On the other hand, recent models by *Wakelam et al.* (2004) have shown that the chemistry of S can be more complex than previously thought since — among other reasons — the abundances of the S-bearing species critically depend on the gas excitation conditions, which in turn depend on the outflow velocity structure. *Wakelam et al.* (2005) used the SO_2/SO and the CS/SO ratios to constrain the age of the NGC 1333 IRAS 2 outflow to ≤5 × 10^3 yr. A recent study by *Codella et al.* (2005) confirms that the use of the SO/H_2S and SO_2/H_2S ratios is subject to important uncertainties in many circumstances, and that other molecular ratios (e.g., CH_3OH/H_2CS,

OCS/H$_2$CS) can be used as more effective chemical clocks to date outflows.

Recent work has revealed that chemical studies can be useful for the investigation of interstellar gas structure. For instance, *Viti et al.* (2004) have recently shown that, if the outflow chemistry is dominated by UV irradiation, clumping in the surrounding medium prior to the outflow passage is needed in order to reproduce the observed chemical abundances in some outflows. We stress, however, that this result depends on the chemical modeling and that more work is needed before it can be generalized.

5. FUTURE WORK

We discussed how the high-angular-resolution observations have revealed general properties and evolutionary trends in molecular outflows from low- and high-mass protostars. However, these results rely on a limited number of outflows maps, thus making any statistical analysis somewhat dangerous. A large sample of fully mapped outflows at different evolutionary stages, using millimeter interferometers, is needed to soundly establish an empirical model of outflow evolution, and the outflow's physical and chemical impact on its surroundings. Also, detail mapping of many outflows will enable a thorough comparison with different numerical outflow models in order to study the outflow entrainment process.

Further progress in our understanding of outflows is expected from current or planned instrument developments that aim at improving both the sensitivity and the angular resolution, while opening new frequency windows. The soon-to-be-implemented improvements to the IRAM Plateau de Bure interferometer — which include longer baselines, wider frequency coverage, and better sensitivity — as well as the soon-to-be-operational Combined Array for Research in Millimeter-wave Astronomy (CARMA), will allow multiline large-scale mosaic maps with 1" resolution (or less), required to thoroughly study the outflow physical properties (e.g., kinematics, temperature, densities), the entrainment process and the different chemical processes along the outflows' entire extent. In addition, large-scale mosaic maps of clouds with outflows will allow the study of the impact of many outflows on their parent cloud. The Atacama Large Millimeter Array (ALMA), presumably operational by 2012, will have the ability to determine high-fidelity kinematics and morphologies of even the most distant outflows in our galaxy as well as flows in nearby galaxies. The superb (subarcsecond) angular resolution will be particularly useful to study how outflows are ejected from accretion disks, how molecular gas is entrained in the outflow, and the interaction between the molecular jet/outflow and the environment very close to the protostar (i.e., the infalling envelope, and protoplanetary disk). The Expanded Very Large Array (EVLA), expected to be complete in 2012, will be critical to image the wide-opening angle and ionized outflow close to the powering source, and will allow

sensitive studies of reionization events in jets, H$_2$O masers, and SiO(1–0) in outflows.

New submillimeter facilities and telescopes under construction will soon provide sensitive observations of high excitation lines, important for the study of outflow driving and entrainment, as well as shock-induced chemical processes. The recently dedicated Submillimeter Array (SMA) is the first instrument capable of studying the warm molecular gas in the CO(6–5) line, at (sub)arcsecond resolution, allowing tracing of the outflow components closer to the driving source and closer to the jet axis than previously possible. Furthermore, the large bandwidth of the correlator allows for simultaneous multiline observations crucial for studying the various shock chemistry processes in the outflow. Also, the Herschel Space Observatory (HSO) will measure the abundances of shock tracers of great interest, in particular water, which cannot be observed from the ground. In the near future, greater computing power will make possible larger-scale numerical simulations that take advantage of adaptive grids, better and more complex cooling and chemistry functions, and the inclusion of radiative transfer and magnetic fields. Given the wealth of high-resolution data that will soon be available, numerical studies will need to compare the simulated outflows with observations in more detail, using the outflow density, kinematics, temperature, and chemical structure. In addition, simulations that run for far longer times ($\sim 10^4$–10^5 yr) than current models ($\sim 10^3$ yr) are needed to study the outflow temporal behavior and evolution. Advances in computing, perhaps including GRID technology, may even allow a version of a virtual telescope, where both numerical modelers and observers can find the best fit from a set of models for different sources.

Acknowledgments. H.G.A. is supported by an NSF Astronomy and Astrophysics Postdoctoral Fellowship under award AST-0401568. D.S. is supported by the National Radio Astronomy Observatory, a facility of the National Science Foundation operated under cooperative agreement by Associated Universities, Inc. R.B. acknowledges partial support from Spanish grant AYA2003-7584. A.R. acknowledges the support of the Visitor Theory Grant at Armagh Observatory, which hosted the author while some of the review was written. H.B. acknowledges financial support by the Emmy-Noether-Program of the Deutsche Forschungsgemeinschaft (DFG, grant BE2578).

REFERENCES

Anglada G., Estalella R., Rodríguez L. F., Torrelles J. M., Lopez R., and Cantó J. (1991) *Astrophys. J., 376*, 615–617.

Arce H. G. (2003) *Rev. Mexicana Astron. Astrofis. Conf. Series, 15*, 123–125.

Arce H. G. (2004) In *Star Formation at High Angular Resolution* (M. Burton et al., eds.), pp. 345–350. IAU Symp. 221, Kluwer, Dordrecht.

Arce H. G. and Goodman A. A. (2001a) *Astrophys. J., 551*, L171–L174.

Arce H. G. and Goodman A. A. (2001b) *Astrophys. J., 554*, 132–151.

Arce H. G. and Goodman A. A. (2002a) *Astrophys. J., 575*, 911–927.

Arce H. G. and Goodman A. A. (2002b) *Astrophys. J., 575*, 928–949.

Arce H. G. and Sargent A. I. (2004) *Astrophys. J., 612*, 342–356.

Arce H. G. and Sargent A. I. (2005) *Astrophys. J., 624*, 232–245.

Arce H. G. and Sargent A. I. (2006) *Astrophys. J.*, in press.

Avery L. W., Hayashi S. S., and White G. L. (1990) *Astrophys. J., 357,* 524–530.

Bachiller R. and Pérez-Gutiérrez M. (1997) *Astrophys. J., 487,* L93–L96.

Bachiller R. and Tafalla M. (1999) In *The Origin of Stars and Planetary System* (C. J. Lada and N. D. Kylafis, eds.), pp. 227–265. Kluwer, Dordrecht.

Bachiller R., Martín-Pintado J., Tafalla M., Cernicharo J., and Lazareff B. (1990) *Astron. Astrophys., 231,* 174–186.

Bachiller R., Guilloteau S., Dutrey A., Planesas P., and Martín-Pintado J. (1995) *Astron. Astrophys., 299,* 857–868.

Bachiller R., Guilloteau S., Gueth F., Tafalla M., Dutrey A., Codella C., and Castets A. (1998) *Astron. Astrophys., 339,* L49–L52.

Bachiller R., Gueth F., Guilloteau S., Tafalla M., and Dutrey A. (2000) *Astron. Astrophys., 362,* L33–L36.

Bachiller R., Pérez-Gutiérrez M., Kumar M. S. N., and Tafalla M. (2001) *Astron. Astrophys., 372,* 899–912.

Bally J. and Lada C. J. (1983) *Astrophys. J., 265,* 824–847.

Bally J., Reipurth B., Lada C. J., and Billawala Y. (1999) *Astron. J., 117,* 410–428.

Beltrán M. T., Girart J. M., Estalella R., and Ho P. T. P. (2004a) *Astron. Astrophys., 426,* 941–949.

Beltrán M. T., Gueth F., Guilloteau S., and Dutrey A. (2004b) *Astron. Astrophys., 416,* 631–640.

Bence S. J., Richer J. S., and Padman R. (1996) *Mon. Not. R. Astron. Soc., 279,* 866–883.

Benedettini M., Molinari S., Testi L., and Noriega-Crespo A. (2004) *Mon. Not. R. Astron. Soc., 347,* 295–306.

Bergin E. A., Neufeld D. A., and Melnick G. J. (1998) *Astrophys. J., 499,* 777–792.

Beuther H. and Shepherd D. S. (2005) In *Cores to Clusters: Star Formation with Next Generation Telescopes* (M. S. N. Kumar et al., eds.), pp. 105–119. Springer, New York.

Beuther H., Schilke P., Sridharan T. K., Menten K. M., Walmsley C. M., et al. (2002a) *Astron. Astrophys., 383,* 892–904.

Beuther H., Schilke P., Gueth F., McCaughrean M., Andersen M., et al. (2002b) *Astron. Astrophys., 387,* 931–943.

Beuther H., Schilke P., and Gueth F. (2004) *Astrophys. J., 608,* 330–340.

Bonnell I. A., Bate M. R., and Vine S. G. (2003) *Mon. Not. R. Astron. Soc., 343,* 413–418.

Bontemps S., André P., Terebey S., and Cabrit S. (1996) *Astron. Astrophys., 311,* 858–872.

Buckle J. V. and Fuller G. A. (2003) *Astron. Astrophys., 399,* 567–581.

Cabrit S. and Bertout C. (1992) *Astron. Astrophys., 311,* 858–872.

Cabrit S., Raga A., and Gueth F. (1997) In *Herbig-Haro Flows and the Birth of Stars* (B. Reipurth and C. Bertout, eds.), pp. 163–180. IAU Symp. 182, Kluwer, Dordrecht.

Cantó J. and Raga A. C. (1991) *Astrophys. J., 372,* 646–658.

Cantó J., Raga A. C., and Riera A. (2003) *Rev. Mexicana Astron. Astrofis., 39,* 207–212.

Ceccarelli C., Castets A., Caux E., Hollenbach D., Loinard L., et al. (2000) *Astron. Astrophys., 355,* 1129–1137.

Cernicharo J., Neri R., and Reipurth B. (1999) In *Herbig-Haro Flows and the Birth of Low Mass Stars* (B. Reipurth and C. Bertout, eds.), pp. 141–152. IAU Symp. 182, Kluwer, Dordrecht.

Cerqueira A. H. and de Gouveia dal Pino E. M. (1999) *Astrophys. J., 510,* 828–845.

Cerqueira A. H. and de Gouveia dal Pino E. M. (2001) *Astrophys. J., 560,* 779–791.

Cesaroni R. (2005) *Astrophys. Space Sci., 295,* 5–17.

Cesaroni R., Felli M., Jenness T., Neri R., Olmi L., et al. (1999) *Astron. Astrophys., 345,* 949–964.

Chandler C. J. and Richer J. S. (2001) *Astrophys. J., 555,* 139–145.

Charnley S. B. (1997) *Astrophys. J., 481,* 396–405.

Chernin L. M. and Masson C. R. (1995) *Astrophys. J., 455,* 182–189.

Choi M. (2005) *Astrophys. J., 630,* 976–986.

Choi M., Kamazaki T., Tatematsu K., and Panis J. F. (2004) *Astrophys. J., 617,* 1157–1166.

Churchwell E. (1999) In *The Origin of Stars and Planetary Systems* (C. J.

Lada and N. D. Kylafis, eds.), pp. 515–552. Kluwer, Dordrecht.

Cliffe J. A., Frank A., and Jones T. W. (1996) *Mon. Not. R. Astron. Soc., 282,* 1114–1128.

Codella C., Bachiller R., and Reipurth B. (1999) *Astron. Astrophys., 343,* 585–598.

Codella C., Bachiller R., Benedettini M., Caselli P., Viti S., and Wakelam V. (2005) *Mon. Not. R. Astron. Soc., 361,* 244–258.

Combet C., Lery T., and Murphy G. C. (2006) *Astrophys. J., 637,* 798–810.

Cunningham A., Frank A., and Hartmann L. (2005) *Astrophys. J., 631,* 1010–1021.

Davis C. J., Matthews H. E., Ray T. P., Dent W. R. F., and Richer J. S. (1999) *Mon. Not. R. Astron. Soc., 309,* 141–152.

Davis C. J., Varricatt W. P., Todd S. P., and Ramsay Howat S. K. (2004) *Astron. Astrophys., 425,* 981–995.

Delamarter G., Frank A., and Hartmann L. (2000) *Astrophys. J., 530,* 923–938.

Dobashi K. and Uehara H. (2001) *Publ. Astron. Soc. Japan, 53,* 799–809.

Downes T. P. and Cabrit S. (2003) *Astron. Astrophys., 403,* 135–140.

Downes T. P. and Ray T. P. (1999) *Astron. Astrophys., 345,* 977–985.

Draine B. T. and McKee C. F. (1993) *Ann. Rev. Astron. Astrophys., 31,* 373–432.

Dutrey A., Guilloteau S., and Bachiller R. (1997) *Astron. Astrophys., 325,* 758–768.

Eislöffel J., Smith M. D., Christopher J., and Ray T. P. (1996) *Astron. J., 112,* 2086–2093.

Fendt C. and Cemeljic M. (2002) *Astron. Astrophys., 395,* 1045–1060.

Fiege J. D. and Henriksen R. N. (1996a) *Mon. Not. R. Astron. Soc., 281,* 1038–1054.

Fiege J. D. and Henriksen R. N. (1996b) *Mon. Not. R. Astron. Soc., 281,* 1055–1072.

Flower D. R. and Pineau des Forêts G. (2003) *Mon. Not. R. Astron. Soc., 343,* 390–400.

Flower D. R., Le Bourlot J., Pineau des Forêts G., and Cabrit S. (2003) *Astrophys. Space Sci., 287,* 183–186.

Foster P. N. and Boss A. P. (1996) *Astrophys. J., 468,* 784–796.

Fuente A., Martín-Pintado J., Rodríguez-Franco A., and Moriarty-Schieven G. D. (1998) *Astron. Astrophys., 339,* 575–586.

Fuente A., Martín-Pintado J., Bachiller R., Rodríguez-Franco A., and Palla F. (2002) *Astron. Astrophys., 387,* 977–992.

Fuller G. A. and Ladd E. F. (2002) *Astrophys. J., 573,* 699–719.

Garay G. and Lizano S. (1999) *Publ. Astron. Soc. Pac., 111,* 1049–1087.

Garay G., Köhnenkamp I., Bourke T. L., Rodríguez L. F., and Lehtinen K. K. (1998) *Astrophys. J., 509,* 768–784.

Garay G., Mardones D., and Rodríguez L. F. (2000) *Astrophys. J., 545,* 861–873.

Garay G., Mardones D., Rodríguez L. F., Caselli P., and Bourke T. L. (2002) *Astrophys. J., 567,* 980–998.

Gardiner T. A., Frank A., Jones T. W., and Ryu D. (2000) *Astrophys. J., 530,* 834–850.

Gardiner T. A., Frank A., and Hartmann L. (2003) *Astrophys. J., 582,* 269–276.

Gibb A. G., Richer J. S., Chandler C. J., and Davis C. J. (2004) *Astrophys. J., 603,* 198–212.

Girart J. M., Estalella R., Viti S., Williams D. A., and Ho P. T. P. (2001) *Astrophys. J., 562,* L91–L94.

Girart J. M., Viti S., Williams D. A., Estalella R., and Ho P. T. P. (2002) *Astron. Astrophys., 388,* 1004–1015.

Girart J. M., Viti S., Estalella R., and Williams D. A. (2005) *Astron. Astrophys., 439,* 601–612.

Gómez J. F., Sargent A. I., Torrelles J. M., Ho P. T. P., Rodríguez L. F., et al. (1999) *Astrophys. J., 514,* 287–295.

Greenhill L. J., Gwinn C. R., Schwartz C., Moran J. M., and Diamond P. J. (1998) *Nature, 396,* 650–653.

Gueth F. and Guilloteau S. (1999) *Astron. Astrophys., 343,* 571–584.

Gueth F., Guilloteau S., and Bachiller R. (1996) *Astron. Astrophys., 307,* 891–897.

Gueth F., Guilloteau S., Dutrey A., and Bachiller R. (1997) *Astron. Astrophys., 323,* 943–952.

Gueth F., Guilloteau S., and Bachiller R. (1998) *Astron. Astrophys., 333,* 287–297.

Gueth F., Schilke P., and McCaughrean M. J. (2001) *Astron. Astrophys., 375,* 1018–1031.

Gueth F., Bachiller R., and Tafalla M. (2003) *Astron. Astrophys., 401,* L5–L8.

Hartmann L., Calvet N., and Boss A. (1996) *Astrophys. J., 464,* 387–403.

Hatchell J., Thompson M. A., Millar T. J., and MacDonald G. H. (1998) *Astron. Astrophys., 338,* 713–722.

Hatchell J., Fuller G. A., and Ladd E. F. (1999) *Astron. Astrophys., 344,* 687–695.

Hirano N., Liu S.-Y., Shang H., Ho T. P. T., Huang H.-C., et al. (2006) *Astrophys. J., 636,* L141–L144.

Jiménez-Serra I., Martín-Pintado J., Rodríguez-Franco A., and Marcelino N. (2004) *Astrophys. J., 603,* L49–L52.

Jiménez-Serra I., Martín-Pintado J., Rodríguez-Franco A., and Martín S. (2005) *Astrophys. J., 627,* L121–L124.

Jørgensen J. K., Hogerheijde M. R., Blake G. A., van Dishoeck E. F., Mundy L. G., and Schöier F. L. (2004) *Astron. Astrophys., 416,* 1021–1037.

Kaufman M. J. and Neufeld D. A. (1996) *Astrophys. J., 456,* 611–630.

Keegan R. and Downes T. P. (2005) *Astron. Astrophys., 437,* 517–524.

Knee L. B. G. and Sandell G. (2000) *Astron. Astrophys., 361,* 671–684.

Königl A. (1999) *New A Rev., 43,* 67–77.

Kwan J. and Tademaru E. (1995) *Astrophys. J., 454,* 382–393.

Lada C. J. and Fich M. (1996) *Astrophys. J., 459,* 638–652.

Ladd E. F and Hodapp K. W. (1997) *Astrophys. J., 474,* 749–759.

Le Bourlot J., Pineau des Forêts G., Flower D. R., and Cabrit S. (2002) *Mon. Not. R. Astron. Soc., 332,* 985–993.

Lee C.-F. and Ho P. T. P. (2005) *Astrophys. J., 624,* 841–852.

Lee C.-F., Mundy L. G., Reipurth B., Ostriker E. C., and Stone J. M. (2000) *Astrophys. J., 542,* 925–945.

Lee C.-F., Stone J. M., Ostriker E. C., and Mundy L. G. (2001) *Astrophys. J., 557,* 429–442.

Lee C.-F., Mundy L. G., Stone J. M., and Ostriker E. C. (2002) *Astrophys. J., 576,* 294–312.

Lee C.-F., Ho P. T. P., and White S. M. (2005) *Astrophys. J., 619,* 948–958.

Lefloch B., Castets A., Cernicharo J., and Loinard L. (1998) *Astrophys. J., 504,* L109–L112.

Lery T. (2003) *Astrophys. Space Sci., 287,* 35–38.

Lery T., Henriksen R. N., and Fiege J. D. (1999) *Astron. Astrophys., 350,* 254–274.

Lery T., Henriksen R. N., Fiege J. D., Ray T. P., Frank A., and Bacciotti F. (2002) *Astron. Astrophys., 387,* 187–200.

Lesaffre P., Chièze J.-P., Cabrit S., and Pineau des Forêts G. (2004) *Astron. Astrophys., 427,* 147–155.

Li Z.-Y. and Shu F. H. (1996) *Astrophys. J., 472,* 211–224.

Lizano S. and Giovanardi C. (1995) *Astrophys. J., 447,* 742–751.

Mac Low M.-M. and Klessen R. (2004) *Rev. Modern Phys., 76,* 125–196.

Maret S., Ceccarelli C., Tielens A. G. G. M., Caux E., Lefloch B., et al. (2005) *Astron. Astrophys., 442,* 527–538.

Martí J., Rodríguez L. F., and Reipurth B. (1993) *Astrophys. J., 416,* 208–217.

Martín-Pintado J., Bachiller R., and Fuente A. (1992) *Astron. Astrophys., 254,* 315–326.

Masson C. R. and Chernin L. M. (1993) *Astrophys. J., 414,* 230–241.

Matzner C. D. and McKee C. F. (1999) *Astrophys. J., 526,* L109–L112.

Matzner C. D. and McKee C. F. (2000) *Astrophys. J., 545,* 364–378.

McCaughrean M. J. and Mac Low M.-M. (1997) *Astron. J., 113,* 391–400.

Micono M., Massaglia S., Bodo G., Rossi P., and Ferrari A. (1998) *Astron. Astrophys., 333,* 1001–1006.

Micono M., Bodo G., Massaglia S., Rossi P., and Ferrari A. (2000) *Astron. Astrophys., 364,* 318–326.

Moriarty-Schieven G. H. and Snell R. L. (1988) *Astrophys. J., 332,* 364–378.

Moscadelli L., Cesaroni R., and Rioja M. J. (2005) *Astron. Astrophys., 438,* 889–898.

Motoyama K. and Yoshida T. (2003) *Mon. Not. R. Astron. Soc., 344,* 461–467.

Myers P. C., Heyer M., Snell R. L., and Goldsmith P. F. (1988) *Astrophys. J., 324,* 907–919.

Nisini B., Benedettini M., Giannini T., and Codella C. (2000) *Astron. Astrophys., 360,* 297–310.

Noriega-Crespo A. (2002) *Rev. Mexicana Astron. Astrofis. Conf. Ser., 13,* 71–78.

O'Connell B., Smith M. D., Davis C. J., Hodapp K. W., Khanzadyan T., and Ray T. (2004) *Astron. Astrophys., 419,* 975–990.

O'Connell B., Smith M. D., Froebrich D., Davis C. J., and Eislöffel J. (2005) *Astron. Astrophys., 431,* 223–234.

Ostriker E. C. (1997) *Astrophys. J., 486,* 291–306.

Ostriker E. C., Lee C.-F., Stone J. M., and Mundy L. G. (2001) *Astrophys. J., 557,* 443–450.

O'Sullivan S. and Ray T. P. (2000) *Astron. Astrophys., 363,* 355–372.

Palau A., Ho P. T. P., Zhang Q., Estalella R., Hirano N., et al. (2006) *Astrophys. J., 636,* L137–L140.

Palumbo M. E., Geballe T. R., and Tielens A. G. G. M. (1997) *Astrophys. J., 479,* 839–844.

Pudritz R. E. and Banerjee R. (2005) In *Massive Star Birth: A Crossroads of Astrophysics* (R. Cesaroni et al., eds.), pp. 163–173. IAU Symp. 227, Cambridge Univ., Cambridge.

Puga E., Feldt M., Alvarez C., and Henning T. (2005) Poster presented at IAU Symp. 227: *"Massive Star Birth: A Crossroads of Astrophysics."*

Quillen A. C., Thorndike S. L., Cunningham A., Frank A., Gutermuth R. A., et al. (2005) *Astrophys. J., 632,* 941–955.

Raga A. C. and Cabrit S. (1993) *Astron. Astrophys., 278,* 267–278.

Raga A. C., Cantó J., Calvet N., Rodríguez L. F., and Torrelles J. M. (1993) *Astron. Astrophys., 276,* 539–548.

Raga A. C., Cabrit S., and Cantó J. (1995) *Mon. Not. R. Astron. Soc., 273,* 422–430.

Raga A. C., Beck T., and Riera A. (2004a) *Astrophys. Space Sci., 293,* 27–36.

Raga A. C., Noriega-Crespo A., González R. F., and Velázquez P. F. (2004b) *Astrophys. J. Suppl., 154,* 346–351.

Raga A. C., Williams D. A., and Lim A. (2005) *Rev. Mexicana Astron. Astrofis., 41,* 137–146.

Rawlings J. M. C., Redman M. P., Keto E., and Williams D. A. (2004) *Mon. Not. R. Astron. Soc., 351,* 1054–1062.

Reipurth B., Bally J., and Devine D. (1997) *Astron. J., 114,* 2708–2735.

Richer R. S., Shepherd D. S., Cabrit S., Bachiller R., and Churchwell E. (2000) In *Protostars and Planets IV* (V. Mannings et al., eds.), pp. 867–894. Univ. of Arizona, Tucson.

Ridge N. A. and Moore J. T. (2001) *Astron. Astrophys., 378,* 495–508.

Rodríguez L. F., Carral P., Moran J. M., and Ho P. T. P. (1982) *Astrophys. J., 260,* 635–646.

Rodríguez L. F., Torrelles J. M., Anglada G., and Martí J. (2001) *Rev. Mexicana Astron. Astrofis., 37,* 95–99.

Rodríguez L. F., Garay G., Brooks K. J., and Mardones D. (2005a) *Astrophys. J., 626,* 953–958.

Rodríguez L. F., Poveda A., Lizano S., and Allen C. (2005b) *Astrophys. J., 627,* L65–L68.

Rosen A. and Smith M. D. (2004a) *Astron. Astrophys., 413,* 593–607.

Rosen A. and Smith M. D. (2004b) *Mon. Not. R. Astron. Soc., 347,* 1097–1112.

Sandell G. and Knee L. B. G. (2001) *Astrophys. J., 546,* L49–L52.

Schilke P., Walmsley C. M., Pineau des Forêts G., and Flower D. R. (1997) *Astron. Astrophys., 321,* 293–304.

Schreyer K., Semenov D., Henning T., and Forbrich J. (2006) *Astrophys. J., 637,* L129–L132.

Shepherd D. S. (2003) In *Galactic Star Formation Across the Stellar Mass Spectrum* (J. M. De Buizer and N. S. van der Bliek, eds.), pp. 333–344. ASP Conf. Ser. 287, San Francisco.

Shepherd D. S. (2005) Massive molecular outflows. In *Massive Star Birth: A Crossroads of Astrophysics* (R. Cesaroni et al., eds.), pp. 237–246. IAU Symp. 227, Cambridge Univ., Cambridge.

Shepherd D. S. and Kurtz S. E. (1999) *Astrophys. J., 523*, 690–700.

Shepherd D. S., Watson A. M., Sargent A. I., and Churchwell E. (1998) *Astrophys. J., 507*, 861–873.

Shepherd D. S., Yu K. C., Bally J., and Testi L. (2000) *Astrophys. J., 535*, 833–846.

Shepherd D. S., Borders T., Claussen M., Shirley Y., and Kurtz S. (2004) *Astrophys. J., 614*, 211–220.

Shu F. H., Ruden S. P., Lada C. J., and Lizano S. (1991) *Astrophys. J., 370*, L31–L34.

Smith M. D. and Rosen A. (2003) *Mon. Not. R. Astron. Soc., 339*, 133–147.

Smith M. D. and Rosen A. (2005) *Mon. Not. R. Astron. Soc., 357*, 579–589.

Smith M. D., Suttner G., and Yorke H. W. (1997) *Astron. Astrophys., 323*, 223–230.

Sollins P. K., Hunter T. R., Battat J., Beuther H., Ho P. T. P., et al. (2004) *Astrophys. J., 616*, L35–L38.

Stanke T., McCaughrean M. J., and Zinnecker H. (2000) *Astron. Astrophys., 355*, 639–650.

Stahler S. W. (1994) *Astrophys. J., 422*, 616–620.

Stone J. M. and Hardee P. E. (2000) *Astrophys. J., 540*, 192–210.

Suttner G., Smith M. D., Yorke H. W., and Zinnecker H. (1997) *Astron. Astrophys., 318*, 595–607.

Tafalla M. and Myers P. C. (1997) *Astrophys. J., 491*, 653–662.

Tafalla M., Bachiller R., Wright M. C. H., and Welch W. J. (1997) *Astrophys. J., 474*, 329–345.

Tafalla M., Santiago J., Johnstone D., and Bachiller R. (2004) *Astron. Astrophys., 423*, L21–L24.

Tafoya D., Gómez Y., and Rodríguez L. F. (2004) *Astrophys. J., 610*, 827–834.

Takakuwa S., Ohashi N., and Hirano N. (2003) *Astrophys. J., 590*, 932–943.

Torrelles J. M., Patel N. A., Anglada G., Gómez J. F., Ho P. T. P., et al. (2003) *Astrophys. J., 598*, L115–L119.

van der Tak F. F. S. and Menten K. M. (2005) *Astron. Astrophys., 437*, 947–956.

van der Tak F. F. S., Boonman A. M. S., Braakman R., and van Dishoeck E. F. (2003) *Astron. Astrophys., 412*, 133–145.

van Dishoeck E. F. (2004) *Ann. Rev. Astron. Astrophys., 42*, 119–167.

Velusamy T. and Langer W. D. (1998) *Nature, 392*, 685–687.

Viti S. and Williams D. A. (1999) *Mon. Not. R. Astron. Soc., 310*, 517–526.

Viti S., Codella C., Benedettini M., and Bachiller R. (2004) *Mon. Not. R. Astron. Soc., 350*, 1029–1037.

Völker R., Smith M. D., Suttner G., and Yorke H. W. (1999) *Astron. Astrophys., 343*, 953–965.

Wakelam V., Caselli P., Ceccarelli C., Herbst E., and Castets A. (2004) *Astron. Astrophys., 422*, 159–169.

Wakelam V., Ceccarelli C., Castets A., Lefloch B., Loinard L., et al. (2005) *Astron. Astrophys., 437*, 149–158.

Walker C. K., Carlstrom J. E., and Bieging J. H. (1993) *Astrophys. J., 402*, 655–666.

Watson C., Zweibel E. G., Heitsch F., and Churchwell E. (2004) *Astrophys. J., 608*, 274–281.

White G. J. and Fridlund C. V. M. (1992) *Astron. Astrophys., 266*, 452–456.

Wilkin F. P. (1996) *Astrophys. J., 459*, L31–L34.

Williams J. P., Plambeck R. L., and Heyer M. H. (2003) *Astrophys. J., 591*, 1025–1033.

Wiseman J., Wootten A., Zinnecker H., and McCaughrean M. (2001) *Astrophys. J., 550*, L87–L90.

Wu Y., Wei Y., Zhao M., Shi Y., Yu W., Qin S., and Huang M. (2004) *Astron. Astrophys., 426*, 503–515.

Yamashita T., Suzuki H., Kaifu N., Tamura M., Mountain C. M., and Moore T. J. T. (1989) *Astrophys. J., 347*, 894–900.

Yokogawa S., Kitamura Y., Momose M., and Kawabe R. (2003) *Astrophys. J., 595*, 266–278.

Yorke H. W. and Sonnhalter C. (2002) *Astrophys. J., 569*, 846–862.

Yu K. C., Billawala Y., and Bally J. (1999) *Astron. J., 118*, 2940–2961.

Yu K. C., Billawala Y., Smith M. D., Bally J., and Butner H. M. (2000) *Astron. J., 120*, 1974–2006.

Zhang Q. and Zheng X. (1997) *Astrophys. J., 474*, 719–723.

Zhang Q., Hunter T. R., Sridharan T. K., and Ho T. P. T. (2002) *Astrophys. J., 566*, 982–992.

Zhang Q., Hunter T. R., Brand J., Sridharan T. K., Cesaroni R., et al. (2005) *Astrophys. J., 625*, 864–882.

Jets and Bipolar Outflows from Young Stars: Theory and Observational Tests

Hsien Shang
Institute of Astronomy and Astrophysics, Academia Sinica

Zhi-Yun Li
University of Virginia, Charlottesville

Naomi Hirano
Institute of Astronomy and Astrophysics, Academia Sinica

Jets and outflows from young stars are an integral part of the star-formation process. A particular framework for explaining these phenomena is the X-wind theory. Since *Protostars and Planets IV,* we have made good progress in modeling the jet phenomena and their associated fundamental physical processes, in both deeply embedded Class I objects and more revealed classical T Tauri stars. In particular, we have improved the treatment of the atomic physics and chemistry for modeling jet emission, including reaction rates and interaction cross-sections, as well as ambipolar diffusion between ions and neutrals. We have broadened the original X-wind picture to include the winds driven magnetocentrifugally from the innermost disk regions. We have carried out numerical simulations that follow the wind evolution from the launching surface to large, observable distances. The interaction between the magnetocentrifugal wind and a realistic ambient medium was also investigated. It allows us to generalize the shell model of *Shu et al.* (1991) to unify the the jet-driven and wind-driven scenarios for molecular outflow production. In addition, we review related theoretical works on jets and outflows from young stars, and make connections between theory and recent observations, particularly those from HST/STIS, VLA, and SMA.

1. INTRODUCTION

Jets and outflows have long been recognized as an important part of star formation. They are reviewed in both *Protostars and Planets III* (PPIII) and *Protostars and Planets IV* (PPIV), and by several groups in this volume (see chapters by Bally et al., Arce et al., Pudritz et al., and Ray et al.). Our emphasis will be on the X-wind theory and related work, and connection to recent high-resolution observations. Comparisons with other efforts are made where appropriate.

Since PPIV, progress has been made in modeling both the dynamics and radiative signatures of jets and winds. There is increasing consensus that these outflows are driven by rotating magnetic fields, although many details remain unresolved. The old debate between disk winds (*Königl and Pudritz,* 2000) and X-winds (*Shu et al.,* 2000) is still with us. Ultimately, observations must be used to distinguish these and other possibilities. Predicting the radiative signatures of different dynamical models is a key step toward this goal. This effort is reviewed in the first part of the chapter (section 2 through section 5).

The X-winds and disk winds are not mutually exclusive. Both are driven magnetocentrifugally from open field lines anchored on rapidly rotating circumstellar disks. Their main distinction lies in where the field lines are anchored: near the radius of magnetospherical truncation on the disk — the X-point — for X-winds and over a wider range in disk radii for disk winds. The different outflow launching conditions envisioned in these scenarios cannot yet be probed directly by observations. It is prudent to consider both possibilities. An effort in this direction is numerical simulation of winds driven magnetocentrifugally from inner disk regions over a range in disk radii that is adjustable. The results of this effort are summarized in the second part of the chapter (section 6), which also includes a method for locating the footpoints of wind-launching magnetic field lines on the disk based on measurements of rotation speed at large distances (section 7).

Regardless of where a magnetocentrifugal wind is driven, as long as its launching region is much smaller than the region of interest, its density structure asymptotes to a characteristic distribution: nearly cylindrical stratification, first shown by *Shu et al.* (1995). An implication is that a magnetocentrifugal wind naturally has dual characteristics: a dense axial jet, surrounded by a wide-angle wind. This intrinsic structure provides a basis for unifying the jet-driven and wind-driven scenarios of molecular outflows, which are thought to be mainly the ambient material set into motion by the primary wind. We devote the last part of the review to recent theoretical and observational advances in this direction (section 8 and section 9). Concluding remarks are given in section 10.

2. THERMAL-CHEMICAL MODELING OF X-WINDS

Thermal-chemical modeling of the winds of young stellar objects is an important step toward understanding their dynamics and origins. *Ruden et al.* (1990; hereafter *RGS*) was the first to investigate in detail the thermal and ionization structure of a cold spherical wind. *Safier* (1993a) implemented similar chemistry into a self-similar disk-wind [a model generalizing the solution of *Blandford and Payne* (1982)]. *Shang et al.* (2002, hereafter *SGSL*) extended and improved upon previous chemical and physical processes of *RGS* to establish a package of diagnostic tools based on the X-wind. Concurrently, *Garcia et al.,* (2001a,b) extended the work of *Safier* (1993a,b) by computing the thermal and ionization structure of the self-similar disk-winds of *Ferreira* (1997). The advances in thermal-chemical modeling since PPIV, particularly *SGSL*, are reviewed below.

2.1. Basic Formulation

Many processes are involved in determining the thermal and ionization structure of jets. *SGSL* considered both processes that take place locally in the wind, and external contributions from mechanical disturbances and radiation on top of the background smooth flow, including new ingredients such as UV radiation and photo-ionization from the accretion hot spots where the funnel flows strike the stellar surfaces, and ionization and heating by X-rays from the secondary electrons. We begin by describing the basic formulation of the problem.

The temperature T and electron fraction x_e of a steady flow are governed by the rate equations that balance heating and cooling and ionization production and destruction. The ionization is primarily neutralized by radiative recombination. Recombination and adiabatic cooling set the two basic timescales in the flow. The former is long compared to flow time and the latter is only tens of seconds near the base. Adiabatic expansion cooling has been recognized as a "severe constraint" on potential heat sources independent of details of wind models (*RGS*). The balance of heating and cooling due to adiabatic expansion provides a rough estimate for the asymptotic temperature profile. In *RGS*, ambipolar diffusion was the dominant heat source; while in *SGSL*, other mechanical contributions play more important roles (section 2.4).

The hydrogen-based radiative and collisional processes that take place locally in a fluid element form the background network of reactions. These are ionization by Balmer continuum, H⁻ detachment, and collisional ionization; heating by ambipolar diffusion, Balmer photoionization of H, $H^+ + H^-$ neutralization, and H⁻ photodetachment; cooling by H⁻ radiative attachment, recombination of H^+, Lyα, col-

lisional ionization, and line cooling from the heavy elements. The chemistry that form and dissociate hydrogen molecules may enter the network if the wind is predominantly molecular; however, bright optical jets can be reasonably treated as atomic winds. Table 1 of *SGSL* has a summary of processes mentioned above.

The rate coefficients of most atomic processes were improved in *SGSL*. For example, the rate coefficient of the reaction $H^- + H^+ \rightarrow H(1) + H(n)$ was revised in light of newer measurements. Photo rates involving photodetachment of H⁻ were computed with better approximations. The electronic collisions and X-ray ionization affect the level population of atomic hydrogen at n = 2. The presence of X-rays also causes an indirect contribution of heating, in addition to the direct heating by elastic collisions of H atoms at the n = 2 level with X-rays. These enter when detailed processes involving the levels n = 1 and 2, and the continuum, are considered.

2.2. Effects of X-Rays

The effects of X-rays on the disk surrounding a young star have been reviewed in *Glassgold et al.* (2000, 2005b), and the chapter by Najita at al. Before reaching the outer part of the disk, the X-rays may first intercept and interact with a wind. *SGSL* considered the effects of X-rays on an X-wind using an approach similar to that developed for X-ray-irradiated disks.

In an X-wind environment, a useful estimate for the X-ray ionization rate ζ_X at a distance r = R_x (the distance of X-point to the star, effectively the base of the X-wind) from a source of X-ray luminosity L_X can be expressed as

$$\zeta_X \equiv \frac{L_X \sigma_{pe}(kT_X)}{4\pi R_x^2 kT_X} = 1.13 \times 10^{-8}\ \mathrm{s^{-1}} \left(\frac{L_X}{10^{30}\ \mathrm{erg\ s^{-1}}} \right)$$
$$\left(\frac{kT_X}{\mathrm{keV}} \right)^{-(p+1)} \left(\frac{10^{12}\ \mathrm{cm}}{R_x} \right)^2 \quad (1)$$

for a thermal spectrum of temperature T_X for the X-rays, where σ_{pe} is the energy-smoothed cosmic photoelectric absorption cross section per H nucleus. The ionization rate ζ at a distance r from the X-ray source can be further expressed with respect to the rate at the X-point as

$$\zeta \approx \zeta_X \left(\frac{R_x}{r} \right)^2 \left(\frac{kT_X}{\varepsilon_{ion}} \right) I_p(\tau_X, \xi_0) \quad (2)$$

where ε_{ion} is the energy to make an ion pair, and the function $I_p(\tau_X, \xi_0)$ describes the attenuation of the X-rays in the surrounding medium, which is a sensitive function of the X-ray optical depth τ_X calculated from the low-energy cutoff ξ_0 expressed in temperature units (see also *Glassgold et al.,* 1997; *SGSL*).

X-rays are capable of lightly ionizing the innermost jet proper and the base of the wind for typical X-ray luminosities observed in young stars through Chandra (e.g., *Feigelson et al., 2002, 2005*; chapter by Feigelson et al.) and earlier satellite missions ASCA and ROSAT (e.g., *Feigelson and Montmerle, 1999*). Together with UV photons from accretion hot spots, they can quantitatively account for the majority of the ionization at the base of the flow. The recombination timescale for an X-wind is long compared to the flow time for the bulk of the flow volume. Hence ionization created locally (in a Lagrangian frame) could be preserved and carried by the flow to large distances. Because of the $1/r^2$ drop-off in the ionization parameter ζ (equation (2)), and the peculiar $1/\varpi^2$ profile of the density (where ϖ is horizontal distance to the axis), the ionization rate is a slowly attenuated function in winds that are not very optically thick. Within the X-wind framework, X-rays can effectively ionize the inner jet throughout a hollow cone that is supported by an axially opened stellar magnetic field. At an elevated angle geometrically over the base of the X-wind, the ionizing X-rays would be less absorbed when the paths of rays go through the diverging portion of the flow.

The effective ionizing power scales approximately as $L_X/\dot{M}_w$ since the ionization rate enters the rate equation as ζ/n_H (where n_H is the volumetric density of hydrogen nuclei), when no other competitive processes are present. The ratio $L_X/\dot{M}_w$ can be interpreted as the *average* efficiency for converting accretion power into X-rays at radius R_x: $L_X/\dot{M}_w = \varepsilon_X GM_*/fR_x$ (*Shu et al., 1997*). The small factor ε_X measures the efficiency of converting accretion energy into X-rays, if the energetic events of X-ray production are ultimately powered by accretion through the twisted field lines between the star and the disk. For models where X-rays are capable of maintaining ionization fractions of a few to several percent at physical distances of interest to optical forbidden lines, the level of L_X needed may go up to 10^{31}–10^{33} erg/s. Although on the high end of the distribution of flare luminosity for typical Class I–II sources, such a level of X-ray luminosity has been seen in giant flares from protostars (chapter by Feigelson et al.; e.g., *Tsuboi et al., 2000*; *Grosso et al., 1997*). For lower-luminosity objects, additional sources of ionization must be sought (see section 2.4 below).

2.3. Ambipolar Diffusion

RGS was the first to carefully consider wind heating due to ambipolar diffusion. The authors adopted a spherically symmetric wind model with profiles of density, velocity, and Lorentz forces chosen such that the neutrals were accelerated to escape velocity by the ion-neutral drag. They concluded that the ambipolar diffusion associated with the magnetic acceleration was the dominant heating process, and adiabatic expansion the dominant cooling process. The wind plasma was lightly ionized ($<10^{-4}$) and cooled below 100 K at a distance beyond $10^4\times$ the stellar radius. Such conditions were unfavorable for optical emission in the predominantly atomic winds. Based on similar chemistry and atomic processes, *Safier (1993a)* found on the contrary that ambipolar diffusion was in fact capable of heating a self-

similar disk wind easily to 10^4 K, with an electron fraction as high as ~0.1–1 at distances of ~10^2–10^3 AU from the central star. The discrepancy is not completely resolved with an improved treatment of ambipolar diffusion (see below).

There are some uncertainties in computing the heating rate of ambipolar diffusion. The rate is given approximately by [equations (4.1) and (4.2) in *SGSL*]

$$\Gamma_{AD} = \frac{\rho_n |f_L|^2}{\gamma \rho_i (\rho_n + \rho_i)^2}, \quad f_L = \frac{1}{4\pi}(\nabla \times \mathbf{B}) \times \mathbf{B} \quad (3)$$

where ρ_n and ρ_i are the mass densities of the neutrals and the ions, respectively, and f_L the Lorentz force. When $\rho_i \ll \rho_n$, equation (3) reduces to the usual one for low-ionization situations, e.g., equation (27.19) of *Shu (1992)*. It has the important property that when ρ_n vanishes the heating rate Γ_{AD} goes to zero.

The ion-neutral momentum transfer coefficient, γ, plays the central role of determining the numerical values of ambipolar diffusion heating in different regimes. *SGSL* combined previous approximations by *Draine (1980)*, and adopted updated calculations and experiments on the collision of H^+ ions with atomic and molecular hydrogen and with helium. The exchange scattering in H^+ + H collisions from *Krstić and Schultz (1998)* was included as the sum of contributions from power-laws from both the high- and low-energy regimes. It agrees with *Draine*'s (1980) prescription only for cold clouds.

Glassgold et al. (2005a) reinvestigated the adopted forms of γ and numerical values in different energy regimes through quantum mechanical calculations of H^+ scattering by H. They fit the coefficient down to very low energy (10^{-10} eV), and did not find the traditional behavior of $1/v$ expected from a constant Langevin cross section. They showed that the early fit of *Draine (1980)* remains accurate within 15%–25%, whereas the approximation adopted by *SGSL* [their equation (4.3)] is too big by a factor of 2. The overestimate came from the assumption that the rate coefficient tends to the Langevin rate at low velocities adopted by astrophysicists (e.g., *Osterbrock, 1961*). After dropping the contribution from the assumed Langevin behavior, the revised formula that *Glassgold et al. (2005a)* suggested *SGSL* should have adopted can be used to a good approximation [equation (3.20) in *Glassgold et al. (2005a)*].

SGSL concluded that ambipolar diffusion cannot heat an X-wind in an extended volume up to the temperature of ~10^4 K and ionization fraction $x_e \sim 0.01$ or higher inferred from forbidden optical lines. Even with a better wind configuration and magnetic field geometry, their conclusion on ambipolar diffusion agrees with the earlier *RGS* findings. Adopting the revised value of γ suggested by *Glassgold et al. (2005a)* does not change this conclusion. *Garcia et al. (2001a)*, on the other hand, reached the same conclusion as *Safier (1993a)* that ambipolar diffusion is able to heat a self-similar disk wind to a temperature plateau $\simeq 10^4$ K. However, they obtained a very different profile of ionization fraction, a factor of 10 down from *Safier (1993a)*. The difference was attributed to the omission of thermal velocity in the ion-neutral momentum-exchange rates adopted by *Safier* (1993a), which overproduced the electron fraction (*Garcia et al.,* 2001a) while making no obvious change in temperature. The systematic disagreement in the heating of self-similar disk winds and nonheating of X-winds may be related to the difference in the exact configurations of the magnetic field throughout the wind, although detailed comparative studies are needed to quantify this possibility (*Garcia et al.,* 2001a; *Glassgold et al.,* 2005a).

2.4. Mechanical Heating and Ionization

To mimic the effects of time-variabilities often seen in knotty jets, we adopt here a phenomenological expression for the volumetric rate of mechanical heating (equation (5-2) of *SGSL*)

$$\Gamma_{mech} = \alpha\rho\frac{v^3}{s} \qquad (4)$$

where ρ and v are the local gas density and flow velocity in an inertial frame at rest with respect to the central star, and s is the distance that the fluid element has traveled along a streamline in the flow. The *global* coefficient $\alpha \geq 0$ phenomenologically characterizes the *average* magnitude of disturbances, possibly magnetic in origin. A choice of $\alpha \ll 1$ indicates that only a small fraction of the mechanical energy contained in the shock waves and turbulent cascades is dissipated into heat when integrated over the flow volumes of interest at the characteristic distance s. Small values of α are self-consistent for a cold flow. In the regime of weak disturbances, the variabilities can be treated as small perturbations on top of a more steady background flow.

An averaged estimate of α is adopted simply for modeling purposes. In real systems, the local efficiency of dissipation may determine an α for each fluid element. For strong disturbances, α should be very localized in nature. This approach is partly motivated by the knotty appearance of jets, which indicates time variabilities in the systems. For example, an $\alpha \approx 0.002$ adopted in *SGSL* for a slightly revealed source suggests a variation in velocity of less than 5%. Given a typical jet velocity of 300 km/s, this implies a weak shock of 15 km/s decaying over the jet length ~s. The thermal profiles generated by all the physical processes included are consistent with the underlying dynamical properties, stellar parameters, chemistry, heating, and ionizing sources in the framework of star-disk interacting systems.

The shock waves represented in equation (4) can also produce UV radiation in the Balmer and Lyman continua that ionize as well as heat the gas. In the same spirit, we can express a phenomenological ionization rate per unit volume (*Shang et al.,* 2004; hereafter *SLGS*)

$$P_e = \beta n_H v/s \qquad (5)$$

where β is the shock-ionization parameter. If the local medium is optically thin to UV radiation, the simplest way to ionize locally is to convert mechanical input into UV photons. This perhaps can help explain some local increase in ionization fraction, in addition to collisional ionization.

3. FORBIDDEN EMISSION LINES

Thermal-chemical modeling helps to bring theoretical wind models to confront observations. Predictions of the images, position-velocity diagrams, and diagnostic line ratios (section 3.1 and section 3.2) are most useful to infer emission properties from comparison with observations of bright jets of young stars. For young active Class II sources, red forbidden lines [SII], [OI], and [NII] are most commonly observed. The modeling of their properties based on physical conditions arising in the X-wind is described here.

The ionizations created by X-rays and UV photons and temperature raised by mechanical heating set the overall excitation profiles of X-winds. The general pattern of ionization along the length of jets produced in *SGSL* resembles a trend seen in optical jets from high-resolution observations (e.g., *Bacciotti and Eislöffel,* 1999, hereafter referred to as *BE*; *Bacciotti,* 2002): The electron fraction first rises, then decays along the length of jets following roughly the behavior determined by recombination. The highest values of ionization fraction inferred from the bases of jets are close to ~30–60% (e.g., *Lavalley-Fouquet et al.,* 2000; *Bacciotti,* 2002). Overall, the ranges obtained from model calculations are within the observational findings of 1–60% throughout the lengths of jets (e.g., *Ray and Bacciotti,* 2003). To excite optical forbidden lines to the observed level of fluxes and spatial extents, enough emitting volume (and area) of gas needs to be heated to temperature of $\geq$8000–10,000 K. This poses a constraint for the minimal average value of the parameter α for specific X-wind parameters. [For a review on the units of scaling adopted within X-wind models, readers can refer to *Shu et al.* (2000).]

Synthetic images best deliver the direct visual effects of emission predicted from models. The local excitation condition expressed in the electron fraction x_e and temperature T and cylindrically stratified density distribution of the wind determines the radiative properties that characterize the source. Fig. 1a shows the distributions of electron fractional abundance x_e and temperature T in a representative X-wind solution calculated in *SGSL* for an active revealed source. Fig. 1b shows the synthetic images made in [SII] and [OI] lines. Note that the emission fans out near the base of the jet, giving an impression of a conic opening near the base. The appearance differs substantially from that of isodensity contours, which is strictly cylindrical in the model [Figs. 1 and 2 in *Shang et al.* (1998)]. This example illustrates the importance of computing the excitation conditions self-consistently. They play a crucial role in determining how a jet is perceived. The main effect of heating is to change the excitation conditions of the background flow, which could be roughly modeled by steady-state solutions. The synthetic images of X-winds obtained with detailed treatment of excitation conditions strengthen the notion, put forth originally in *Shu et al.* (1995), that observed jets are merely an "optical illusion."

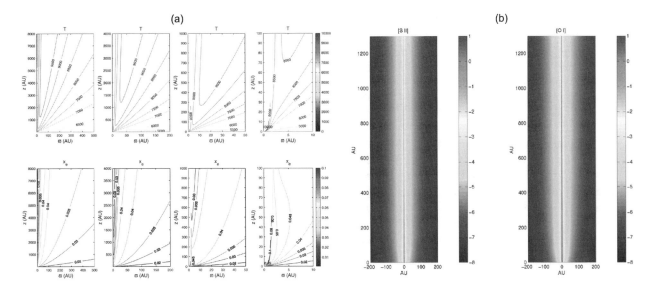

Fig. 1. (a) Temperature (upper) and ionization (lower) contours in the ϖ–z plane calculated in *SGSL* for a fiducial case characterizing an active but revealed source. The mass-loss rate adopted is $3 \times 10^{-8} M_\odot$/yr, and $L_X/\dot{M}_\varpi = 2 \times 10^{13}$ erg/g for an X-ray luminosity $L_X = 4 \times 10^{31}$ erg/s. The parameter α is 0.002 with no inclusion of β in the case shown. The units for the spatial scales are AU. (b) Synthetic images of the [SII]λ6731 (left) and [OI]λ6300 (right) brightness for the same model as in (a) adapting the methods of *Shang et al.* (1998). The $\log_{10}$ of integrated intensity is plotted in units of erg s^{-1} cm^{-2} ster^{-1}.

3.1. Optical Diagnostics

Bright optical emission lines are the best candidates to diagnose conditions arising in real jet models. Important constraints can be extracted from the relative strengths of the optical forbidden lines, with knowledge of their individual atomic structures, and physical processes of excitation and de-excitation. *BE* developed a semi-empirical approach that has been widely applied to available HST and adaptive optics data. The so-called *BE* technique is based on some simple assumptions. The jet emitting region is optically thin. The electron fraction x_e is determined solely by charge exchange with N and O. Collisional ionization and photoionization (via shocks) do not contribute until shock velocity exceeds 100 km/s. Sulfur is singly ionized because of its low first ionization potential. The ratio of [SII] doublets λ6717 and λ6731 can be used as an electron density indicator up to 2×10^4 cm^{-3}, if the two sulfur doublets are treated as a two-level system. The ratio of [OI] (λ6300 + λ6363) and [NII] (λ6548 + λ6583) and that of [SII] (λ6717 + λ6731) and [OI] (λ6300 + λ6363) are tracers of the electron fraction x_e and temperature T_e, respectively. The background abundances of each of the atomic species are assumed to be solar: N/H = 1.1×10^{-4}, O/H = 6.0×10^{-4}, and S/H = 1.6×10^{-5}. There was no implicit assumption made for the excitation mechanisms, although some uncertainties exist in the atomic and ionic physics used in *BE*. Applying *BE* to bright jet sources typically yield $0.01 < x_e < 0.6$, and $7000 < T_e < 2 \times 10^4$ (see chapter by Ray et al. for more detailed discussions). The total hydrogen nuclei n_H is derived using the upper limit of $n_e = 2 \times 10^4$ cm^{-3} for the sulfur doublets and the inferred electron fraction x_e (*Bacciotti*, 2002). Similar analyses extended to IR or semi-permitted

UV lines from the same or different species could provide independent checks on the derived parameters (n_e, x_e, T_e, n_H) and the mass-loss rates.

Cross-correlation of different lines can reveal interesting trends in the underlying physical conditions of jets. *Dougados et al.* (2000, 2002) used diagrams of relative line strengths to infer physical conditions in strong microjets DG Tau and RW Aur [e.g., Fig. 4 of *Dougados et al.* (2002)]. They found that the line ratio [NII]/[OI] increases with x_e; [SII]/[OI] decreases with increasing electron temperature and electron density for $n_e \geq n_{cr}$, the critical density. The observational data were compared with the predictions of a few models. Planar shocks (*Hartigan et al.*, 1994) trace a family of curves in the [SII] (λ6717/λ6731)-[SII]/[OI] diagram. Shock curves for different preshock densities are distinctly separated out on the [NII]/[OI]–[SII]/[OI] diagram. Curves from viscous mixing layers with neutral boundaries (*Binette et al.*, 1999) and a version of the cold disk-wind model heated by ambipolar diffusion (*Garcia et al.*, 2001a,b) follow trajectories on the diagrams that are distinct from those of shock models. On the same diagrams, Herbig-Haro objects obtained by groundbased telescopes (*Raga et al.*, 1996) follow closely shock curves of preshock density 10^2–10^3 cm^{-3}.

Line ratio diagrams have been used to constrain the X-wind model. *SGSL* constructed a line ratio diagram of [SII] λ6717/[SII]λ6731 and [SII]λ6731/[OI]λ6300 for an X-wind jet of mass-loss rate 3×10^{-8} $M_\odot$/yr, using model data points from the images shown in Fig. 1b. Observational data for the HH objects studied in *Raga et al.* (1996) and DG Tau [from Fig. 3 of *Lavalley-Fouquet et al.* (2000)] are plotted on the same diagram for comparison. Adaptive optics data for RW Aur (*Dougados et al.*, 2002) would also be located within the coverage of the model points. The model points encompass most of the observational data. With only a few sources, which are known to be strong shock excited objects (in knots or bow shocks), lying outside of the loci traced by the shape of potential curves, the distribution of physical and excitation conditions reached in an X-wind jet for a slightly revealed source indeed captures the average conditions inferred from optical jets.

From the comparison of theoretical and observational data, *SGSL* concluded that the treatment of weak shocks on top of a steady-state background flow appears to recover the excitation conditions inferred from a large set of optical jets and HH objects. The dynamical properties in fact remain close to the cold steady-state solution of the underlying X-wind model. The wide range in shock conditions inferred in the jets may not be a coincidence. Most optical emission, excited by a network of weak shocks, may be the tell-tale traces of vastly varying density structures that cannot be observationally resolved with the current instrumentation, but whose presence can be inferred from detailed modeling.

3.2. Infrared Diagnostics

Compared to optical emission, the near-IR lines have the obvious advantage of being less affected by extinction (*Reipurth et al.*, 2000). However, their radiative properties have not been theoretically explored as thoroughly. Strong near-IR lines of the abundant ion [FeII] are frequently associated with Class I sources or revealed T Tauri sources of relatively high-mass accretion rates (*Davis et al.*, 2003). Sources showing strong [FeII] and optical forbidden lines such as L1551-IRS5 (*Pyo et al.*, 2002, 2005) and DG Tau (*Pyo et al.*, 2003) have been well studied observationally at both wavelengths. RW Aur and HL Tau have recently been studied by the Subaru telescope with spectroscopy and adaptive optics (*Pyo et al.*, 2006), adding to the list of sources available for multiline modeling.

The [FeII] ion has hundreds of fine structure levels. The large number of possible transitions between the levels makes the calculation of level populations a daunting task. Most atomic data and radiative coefficients of [FeII] were not available until after the mid-1990s. *Zhang and Pradhan* (1995) included 142 fine-structure levels for transitions in IR, optical, and UV, and 10011 transitions were calculated. *Hartigan et al.* (2004) included 159 energy levels and 1488 transitions for coverage of wavelengths longer than 8000 Å. For the purpose of modeling only cooler regions of jets and [FeII] in the near-IR, a simplified non-LTE model for the lowest 16 levels under the optically thin assumption may provide a reasonable approximation, as shown in *Pesenti et al.* (2003). The level populations are computed under statistical equilibrium with electron collisional excitation and spontaneous radiative emission processes. The lowest levels of the [FeII] ion may remain collisionally dominated as suggested by *Verner et al.* (2000), as in the case of Orion Nebula. For the brightest lines whose ratios are to be taken,

results from *Pesenti et al.* (2003) and *Pradhan and Zhang* (1993) agree well for n_e ranging from 10 to 10^8 cm^{-3} and temperature from 3000 to 2×10^4 K.

Critical densities derived from the forbidden lines are often used to infer densities from which the radiation originates. The [FeII] near-IR lines have critical densities in the range of ~10^4–10^5 cm^{-3} for temperature up to 10^4 K [Table 1 in *Pesenti et al.* (2003)], sitting between the critical densities of [SII] (~10^3 cm^{-3} from 16 levels) and [OI] (~10^6 cm^{-3}). The ratio of the two brightest lines, [FeII] 1.644 μm and 1.533 μm, can be a diagnostic for n_e in the range ~10^2–10^5 cm^{-3}. This may be combined with bright transitions in the red (0.8617 and 0.8892 μm) to derive an estimate on temperature (*Nisini et al.*, 2002a). *Pesenti et al.* (2003) proposed a line-ratio diagram based on the correlation between 1.644 μm/1.533 μm and 0.8617 μm/1.257 μm [Fig. 3 in *Pesenti et al.* (2003)]. *Nisini et al.* (2005) adopted 1.644 μm/ 1.533 μm and 1.644 μm/0.862 μm as their diagnostic diagram [Fig. 6 in *Nisini et al.* (2005)]. (Note: The notation of lines in this section is changed from the optical diagnostics in section 3.1 to follow the standard practice of the infrared diagnostics for easier identification with literature.)

Multiline analysis across accessible wavelengths, including the optical forbidden lines, of [FeII], [OI], [SII], [NII], and even H_2 lines together, may sample the parameter space more completely than individual bands. Such analysis provides a check on diagnostic tools derived from individual wavebands. It also serves as a more reliable approach to infer the physical conditions from emission lines of jets. *Nisini et al.* (2005) was the first to demonstrate such an approach through spectra collected for HH1. Using line ratios from [SII], near-IR and optical [FeII], and [CaII], they found evidence for density stratification. The derived temperature also varied from 8000–11,000 K using [FeII] to 11,000–20,000K using [OI] and [NII]. This result suggests that different lines originate from distinct regions (*Nisini et al.*, 2005). The results may be a reflection of the fact that different diagnostics are sensitive to different physical conditions. The electron densities derived are higher than the values obtained by applying the *BE* technique, while the temperature follows an opposite trend. The combined approach has been applied to only a few cases to date. It may become increasingly more useful as more IR observations are become available, particularly from VLTI/AMBER.

4. RADIO CONTINUUM

Radio jets are elongated, jet-like structures seen on subarcsecond scales near stars of low and intermediate masses (e.g., *Evans et al.*, 1987; *Anglada*, 1996). They show good alignment with large-scale optical jets and outflows, usually identifiable with the youngest deeply embedded stellar objects, and have partial association with optical jets or Herbig-Haro objects. In a few YSO sources such as L1551-IRS5, DG Tau B, and HL Tau, the optical jets trace material on scales of several hundred AU and larger, while small radio jets from ionized material are only present very close to the (projected) bases of the optical jets (*Rodríguez et al.*, 1998, 2003). The emission from low-mass radio jet sources is weak, typically at mJy level.

The production of mJy radio emission has long been a problem for the theory of jets and winds in low-mass YSOs (*Anglada*, 1996). Previous workers agreed that stellar radiation produced too little ionization to account for the observations (e.g., *Rodríguez et al.*, 1989). *Rodríguez and Cantó* (1983) and *Torrelles et al.* (1985) pointed out that thermalization of a small fraction of kinetic energy of a neutral flow might provide the ionization rates inferred from early radio observations. The role of shock-produced UV radiation has also been investigated (e.g., *Hartmann and Raymond*, 1984; *Curiel et al.*, 1987; *Ghavamian and Hartigan*, 1998). *Ghavamian and Hartigan* (1998) investigated free-free emission from postshock regions that are optically thick at radio frequencies. They computed radio spectra under a variety of shock conditions ($10^3 < n < 10^9$ cm^{-3} and $30 < V < 300$ km/s). *González and Cantó* (2002) obtained thermal radio emission at the mJy level by modeling periodically driven internal shocks in a spherical wind. The radio free-free emission comes from the working surfaces produced by time-varying ejections. In this model, the emission is variable and optical thickness changes with time. At times, emission may completely disappear. *Shang et al.* (2004, hereafter *SLGS*) applied the thermal-chemical model of *SGSL* to radio jets, with wind parameters suitable for Class I sources. They concluded that UV radiation from the same shocks that heat the X-wind may play a role in the ionization structure of radio jets.

Compared with optical and IR emission, the radio emission suffers the least from dust extinction. This makes the radio free-free emission a powerful diagnostic tool for probing close to sources that are under active accretion (e.g., *Reipurth et al.*, 2002, 2004; *Torrelles et al.*, 2003). The archetype L1551-IRS5 is by far the best studied example. It shows jets in both forbidden lines and radio continua. *Rodríguez et al.* (2003) obtained VLA observations with an angular resolution of 0".1 (14 AU), and found two radio jets from the now-identified binary (e.g., *Rodríguez et al.*, 1998) at the origins of the larger-scale jets observed in both optical and NIR wavelengths (*Fridlund and Liseau*, 1998; *Itoh et al.*, 2000). For comparison with the best available observation, *SLGS* made an intensity map at 3.6 cm based on a profile like Fig. 1 for an X-wind jet of mass-loss rate 1×10^{-6} M$_\odot$/yr, similar to that inferred from HI measurements of neutral winds (*Giovanardi et al.*, 2000). As shown in Fig. 2, the elongated contours show clearly the collimation due to the cylindrically stratified density profiles of electrons to which the free-free emission is sensitive. The apparent collimation can be traced down to below the 10 AU level, beyond which radio observations become unresolved. Contour-by-contour comparisons are possible when the theoretical intensity maps are convolved with the real beams to produce synthetic radio images. Direct comparison with observed radio maps, such as the ones shown in Fig. 1 of *Rodríguez et al.* (2003), can yield important constraints on

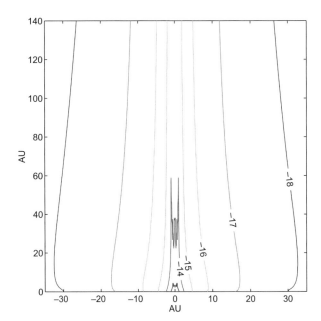

Fig. 2. Free-free intensity contours (in units erg cm^{-2} s^{-1} ster^{-1}) for the X-wind model using parameters scaled up from the *SGSL* ratio, $L_X/\dot{M}_w = 2 \times 10^{13}$ erg g^{-1}: $\dot{M}_w = 10^{-6}$ M$_\odot$ yr^{-1}, $L_X = 1.3 \times 10^{33}$ erg s^{-1}, $\alpha = 0.005$, and $\beta = 0$.

detailed physical processes such as heating and ionization. To produce radio emission at the mJy level through thermal bremsstrahlung, enough mass-loss rate is needed. A lightly ionized jet in a light wind would produce a radio emission that is well below the sensitivity of current telescopes and too small to be resolved even with interferometric arrays. The preferential detection of radio jets at mJy level and subarcsecond scales may be an observational selection effect. Radio continuum observations, particularly at the highest possible resolution, can probe the jet structure near the base in a way that complements the optical and near-IR observations.

The fluxes of radio jets are characterized by a small non-negative spectral index, $S_\nu \propto \nu^p$ and $p \geq -0.1$, consistent with a thermal origin in ionized winds. The classic example of an unresolved, constant-velocity, fully ionized, isothermal, spherical wind has $p = 0.6$. *Reynolds* (1986) showed that $p < 0.6$ occurs for an unresolved, partially opaque flow whose cross section grows more slowly than its length. The index can vary from $p = 2$ for totally opaque emission to $p = -0.1$ for totally transparent emission. Observers have usually interpreted radio data with Reynolds' model, obtaining the spectral index from total flux measurements at several wavelengths (e.g., *Rodríguez*, 1998). A more detailed analysis is now possible employing the approaches developed in section 2. A relation of S_ν-ν performed on X-wind models with self-consistent heating and ionization can best illustrate the behavior at various mass-loss rates. For mass-loss rates lower than 3×10^{-7} M$_\odot$ yr^{-1}, $S_\nu \propto \nu^{-0.1}$, indicating transparent emission. The spectra turnover around 8.3 GHz as the mass-loss rate increases to $\dot{M}_w \approx 3 \times 10^{-7}$ M$_\odot$ yr^{-1}, suggesting that at this mass-loss rate, emission from optically thick regions starts to appear. For mass-loss rates higher than the "crossover" mass rate $\dot{M}_w$, the index is approximately 0.3 — very close to that inferred for L1551. The value also matches that predicted for a collimating partially optically thin wind that is lightly ionized.

5. EMISSION FROM INNER DISK WINDS

On general energetic grounds, the fast-moving jets and winds observed in YSOs, if driven magnetocentrifugally, are expected to come from a disk region close to the central object. X-winds automatically satisfy this requirement. For disk-winds to reach a typical speed of a few hundred km/s, they would most likely be launched from the inner disk region, perhaps close to the disk truncation or corotation radii. The dynamics of inner disk-driven winds will be reviewed in section 5 (see also chapter by Pudritz et al.). Here, we concentrate on the effort in modeling emission from inner disk-winds, which parallels that described above for X-winds.

To date, emission modeling of disk winds has been carried out using self-similar solutions. *Garcia et al.* (2001a,b), for example, adopted a self-similar solution from *Ferreira* (1997). They concluded that ambipolar-diffusion heating was able to create a *warm* temperature plateau of 10^4 K as in *Safier* (1993a). However, the electron fraction was an order of magnitude below the typical ranges inferred from observations (*Garcia et al.*, 2001a; *Dougados et al.*, 2003). Even for a relatively large wind mass-loss rate of 10^{-6} M$_\odot$/yr, the average densities are lower than the inferred values of 10^5–10^6 cm^{-3} by one order of magnitude at projected distances ≤ 100 AU as in the microjets (*Dougados et al.*, 2004). Overall, these so-called *cold* disk-wind models were unable to reproduce integrated line fluxes in a large sample of classical T Tauri stars unless a large amount of additional mechanical heating is included. However, that heating may produce an ionization fraction too high to be consistent with observations (*O'Brien et al.*, 2003; *Dougados et al.*, 2004). A mechanism is needed to heat the wind efficiently without overproducing ionization. A clue for such a mechanism may come from the line ratios. In the inner regions of the DG Tau and RW Aur microjets, curves of moderate shock velocities produced by planar shock models seem to best fit the excitation condition (*Lavalley-Fouquet et al.*, 2000; *Dougados et al.*, 2002). This leads to a conclusion similar to that of *SGSL* (see section 2.4): Mild internal shocks due to time variability in the ejection process may play a role in generating the required gently varying excitation conditions.

Driven from a range of disk radii, a disk wind is expected to have a range of flow speeds. The variation in speed may account for the coexistence of a high-velocity component (HVC) and low-velocity component (LVC) observed in many sources (*Cabrit et al.*, 1999; *Garcia et al.*, 2001b). In particular, the self-similar disk-wind model appears capable of producing the two emission peaks in the position-velocity diagrams (PV) of DG Tau and L1551-IRS. The spatial extents of the LVC emission, however, exceed the model prediction (e.g., *Pesenti et al.*, 2003; *Pyo et al.*, 2002,

2003, 2005). The LVC may have a more complicated origin. The observed kinematics were best fit with self-similar solutions in which mass-loaded streamlines are coming out from disk radii of 0.07–1 AU (*Garcia et al.*, 2001b); these are inner disk winds. For the HVC, the overall observed velocity widths seem to be narrower than the model predictions. Some disk-wind models show deceleration of the HVC (due to refocusing of streamlines) that is rarely observed (*Pesenti et al.*, 2003; *Dougados et al.*, 2004). It would be interesting to see whether these discrepancies can be removed when the self-similarity assumption is relaxed.

The X-wind is an intrinsically HVC-dominated wind with a strongly density-collimated jet. The velocity profile extends smoothly to the lower-velocity range without apparently distinct peaks of emission in a steady-state model. The detailed shapes and locations of the emission peaks may be further affected by local excitation conditions inside the winds. The general features of the HVC can be modeled in a steady-state X-wind. Some well observed sources such as RW Aur and HL Tau have a predominant HVC (e.g., *Bacciotti et al.*, 1996; *Pyo et al.*, 2005b). Their observed position-velocity diagrams and images in various lines closely resemble model predictions [Figs. 2 and 3 in *Shang et al.* (1998); Fig. 4 in *SGSL*]. The much weaker LVC may come from the slight extension into the lower velocities due to the natural broadening in the x-region in real systems (*Shu et al.*, 1994), or may originate in a weaker (and slower) disk wind. The much broader, stronger, and extended LVC emission from L1551 and DG Tau may come from a separate strong disk-wind surrounding the faster X-wind jet. Possible interaction between a disk wind and X-wind is a topic that deserves future attention.

We note that the X-wind is an integral part of the disk-magnetosphere interaction, which also includes funnel flows onto the stellar surface. Possible connections between the disk winds and funnel flows, if any, remain to be elucidated.

6. INNER DISK WINDS: SIMULATIONS

Since PPIV, MHD wind launching and early propagation has risen to the main focus of a number of numerical simulations. These simulations generally fall into two categories, depending on how the disk is treated. Some workers include the disk as part of the wind simulation (e.g., *Kudoh et al.*, 2003; *von Rekowski and Brandenburg*, 2004), an approach pioneered by *Uchida and Shibata* (1985). The disk-wind system generally evolves quickly, and the long-term outcome of the simulation is uncertain, at least in the ideal MHD limit. In the presence of magnetic diffusion, steady-state solutions can be obtained numerically (*Casse and Keppens*, 2002). These solutions extended the semi-analytic self-similar disk-wind solutions (*Wardle and König*, 1993; *Li*, 1995; *Ferreira and Casse*, 2004) into the non-self-similar regime. Other workers have chosen to focus on the wind properties exclusively, treating the disk as a boundary (*Krasnopolsky et al.*, 1999, 2003; *Bogovalov and Tsinganos*, 1999; *Fendt and Čemeljić*, 2002; *Anderson et al.*, 2005a), following the original formulation of *Koldoba et al.* (1995) and

Ouyed and Pudritz (1997). This approach enables the determination of wind properties from the launching surface to large, observable distances.

The simulation setup of *Krasnopolsky et al.* (2003) is closest to that envisioned in the X-wind theory. The wind is assumed to be launched magnetocentrifugally from a Keplerian disk extending from an inner radius R_i to an outer radius R_o. As the axis is approached, the magnetic field lines are forced to become more and more vertical by symmetry and less and less capable of magnetocentrifugal wind-launching. Thus, a fast, light outflow is injected from the disk surface inside R_i, which may represent either the (coronal) stellar wind or the magnetosphere of the star envisioned in the X-wind theory. Two representative wind solutions, both with $R_i = 0.1$ AU and $R_o = 1.0$ AU, are shown in Fig. 3, after a steady state has been reached. Note that the isodensity contours become nearly parallel to the axis in the polar region at large distances. The wind solution in the lower panels of Fig. 3 has a mass loading that is more concentrated near the inner edge of the Keplerian disk. It resembles the X-wind solution shown in Fig. 1 of *Shang et al.* (1998). This cylindrical density stratification is in agreement with the asymptotic analysis of *Shu et al.* (1995) (see also *Matzner and McKee,* 1999), which predicts that the dense, axial "jet" is always surrounded by a more tenuous, wide-angle component.

Anderson et al. (2005a) carried out a parameter study of the large-scale structure of axisymmetric magnetocentrifugal winds launched from inner disks, focusing on the effects of mass loading. They found that, despite different degrees of flow collimation, the terminal speed and magnetic level arm scale with the amount of mass loading roughly as predicted analytically for a radial wind (*Spruit,* 1996). As the mass loading increases, the wind of a given magnetic field distribution changes from a "light" regime, where the field lines remain relatively untwisted, up to the Alfvén surface, to a "heavy" regime, where the field is toroidally dominated from large distances all the way to the launching surface. The existence of such heavily loaded winds has implications for mass loss from magnetized accretion disks. Whether they are stable in three dimensions is an open question.

It has been argued that magnetocentrifugal winds may be intrinsically unstable, at least outside the Alfvén surface, where the magnetic field is toroidally dominated (*Eichler,* 1993). *Lucek and Bell* (1996) studied the three-dimensional stability of a (nonrotating) jet accelerated and pinched by a purely toroidal magnetic field. They found that the m = 1 (kink) instability can grow to the point of causing the tip of the jet to fold back upon itself. Their mechanism of jet formation is, however, quite different from the magnetocentrifugal mechanism. *Ouyed et al.* (2003) carried out three-dimensional simulations of cold jets launched magnetically along initially vertical field lines from a Keplerian disk. They found that the jets become unstable beyond the Alfvén surface, but the instability is prevented from disrupting the jet by a self-regulatory process that keeps the Alfvén Mach number close to unity. *Anderson et al.* (2006) adopted as their base models steady axisymmetric winds driven magnetocentrifugally along open field lines inclined more than 30° away from the rotation axis. They increased the mass loading on one half of the launching surface by a factor of $10^{1/2}$ and decreased that on the other half by the same factor. The strongly perturbed winds settle into a new, nonaxisymmetric steady state. There is no evidence for the growth of any instability, even in cases where the magnetic field is toroidally dominated all the way to the launching surface. One possibility is that their magnetocentrifugal winds are stabilized by the strong axial magnetic field enclosed by the wind, as envisioned in *Shu et al.* (1995), although this possibility remains to be firmly established. More discussion of outflow simulations is given in the chapter by Pudritz et al.

7. SIGNATURE OF WIND ROTATION

In the ballistic wind region well outside the fast magnetosonic surface, an approximate relation exists between the poloidal velocity component in the meridian plane $v_{p,\infty}$ and toroidal component $v_{\phi,\infty}$ at a given location (of distance ϖ_∞ from the axis) and the angular speed Ω_0 at the footpoint of the magnetic field line passing through that location by (*Anderson et al.,* 2003)

$$\Omega_0 = \frac{v_{p,\infty}^2/2}{v_{\phi,\infty}\varpi_\infty} \qquad (6)$$

This relation follows from the fact that both the energy and

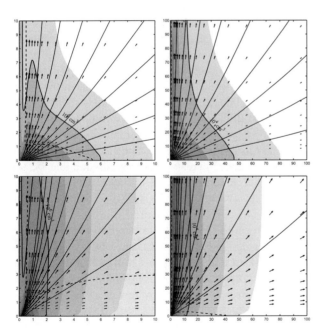

Fig. 3. Streamlines (light solid) and isodensity contours (heavy solid lines and shades) of two representative steady-wind solutions on the 10-AU (left panels) and 10^2-AU (right panels) scale. The dashed line is the fast magnetosonic surface, and the arrows are poloidal velocity vectors. The wind solution in the lower panels appears better collimated in density. It has a mass loading that is more concentrated near the inner edge of the Keplerian disk.

angular momentum in the wind are extracted by the same agent, the magnetic field, from the underlying disk. The angular momentum is extracted by a magnetic torque, which brakes the disk rotation. The energy extracted is simply the work done by the rotating disk against the braking torque (*Spruit,* 1996). To extract more energy out by a given torque, the field lines must rotate faster, which in turn means that they must be anchored closer to the star. At large distances well outside the fast surface, most of the magnetic energy is converted into the kinetic energy of the wind, and most of the angular momentum extracted magnetically will also be carried by fluid rotation. Therefore, the fluid energy and angular momentum at large distances are related through the angular speed at the footpoint. Note that all quantities on the righthand side of equation (6) are in principle measurable.

From these measurements, one can deduce the rotation rate at the footpoint, and thus the wind-launching radius, approximately from

$$\varpi_0 = 0.7 \left(\frac{\varpi_\infty}{10 \text{ AU}} \right)^{2/3} \left(\frac{v_{\phi,\infty}}{10 \text{ km s}^{-1}} \right)^{2/3}$$
$$\left(\frac{v_{\phi,\infty}}{100 \text{ km s}^{-1}} \right)^{-4/3} \left(\frac{M_*}{1 \text{ M}_\odot} \right)^{1/3} \text{ AU} \qquad (7)$$

provided that the stellar mass M_* is known independently.

The above technique for locating the wind-launching region was applied to the low-velocity component of the DG Tau wind, for which detailed velocity field is available from HST/STIS observations (*Bacciotti et al.,* 2000, 2002). These observations allow one to derive not only the line-of-sight (radial) velocity component but also the rotational velocity. Since the inclination of the flow axis is known for this source, one can make corrections for the projection effects to obtain the true poloidal and toroidal velocities. The result is shown in Fig. 4. The straight lines connect an observing location where data are available and the location on the disk where we infer the flow in that region originates. One can think of these lines loosely as "streamlines."

The LVC of the DG Tau wind appears to be launched from a region on the disk extending from ~0.3 to 4 AU from the central star [the exact range depends somewhat on the distribution of emissivity inside the jet (see *Pesenti et al.,* 2004)]. That is, the spatially extended, relatively low-velocity flow appears to be a disk wind, as has been suspected for some time (*Kwan and Tademaru,* 1988). *Anderson et al.* (2003) have also estimated the so-called "Alfvén radius" along each streamline, which is simply the square root of the specific angular momentum divided by the rotation rate $\varpi_A = (v_{\phi,\infty} \varpi_\infty / \Omega_0)^{1/2}$. It turns out that the Alfvén radius is a factor of 2–3 times the footpoint radius (the Alfvén points are indicated by the filled triangles in the figure). This implies that the mass-loss rate in the wind is about 10–25% of the mass accretion rate through the disk, if the disk angular momentum removal is dominated by the magnetocentrifugal wind.

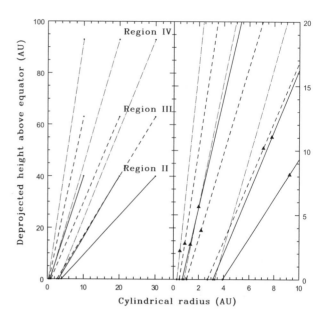

Fig. 4. Calculated "streamlines" for DG Tau. The left panel shows the observation points from Region II (solid lines), III (dashed), and IV (dash-dotted) of *Bacciotti et al.* (2002) connected to the calculated footpoints of the flow. The right is a blowup of the inner region, showing where the flow originates from the disk. Also shown is the location of the Alfvén surface along each "streamline" as filled triangles.

There is a high-velocity component of more than 200 km s^{-1} in the DG Tau system. It is not spatially resolved in the lateral direction, and is likely originated within 0.3 AU of the star. It could be an X-wind confined by the disk wind. One can in principle use the same technique to infer where the high-velocity component originates if its emission can be spatially resolved. This may be achieved in the future through optical interferometers. A potential difficulty is that the highly collimated HVC may be surrounded (and even confined) by an outer outflow (perhaps the LVC), which could mask its rotation signature along the line of sight. The projection effect can create a false impression that the toroidal velocity in a rotating wind increases with the distance from the axis (*Pesenti et al.,* 2004). Also, both the LVC and HVC could be intrinsically asymmetric with respect to the axis, which could create velocity gradients that mimic rotation. This possibility is strengthened by the observation that the disk in RW Aur appears to rotate in the opposite sense to the purported rotation measured in the wind (*Cabrit et al.,* 2006; see also chapter by Ray et al.). The apparent rotation signatures should be interpreted with caution (for more discussion, see chapter by Ray et al.).

8. DENSE CORE ENVIRONMENT

As a wind driven from close to the central stellar object propagates outward, it interacts with the dense core material that is yet to be accreted. To model the interaction properly, one needs to determine the core structure, which depends on how the cores are formed. One school of thought is that the cores are produced by shocks in supersonically turbulent clouds (e.g., *Mac Low and Klessen,* 2004). Although a small fraction of such cores can have subsonic infall motions, the majority expand transonically or even supersonically. The rapid contraction may be difficult to reconcile with the observational results that only a fraction of dense cores show clear evidence for infall and the contraction speeds inferred for the best infall candidates are typically half the sound speed (*Myers,* 1999). Subsonic contraction, on the other hand, is the hallmark of the standard scenario of core formation in magnetically subcritical clouds through ambipolar diffusion (*Nakano,* 1984; *Shu et al.,* 1987; *Mouschovias and Ciolek,* 1999). Predominantly quiescent cores are formed in subcritical clouds even in the presence of strong turbulence (*Nakamura and Li,* 2005). The magnetically regulated core formation is consistent with Zeeman measurements and molecular line observations of L1544 (*Ciolek and Basu,* 2000), arguably the best observed starless core (*Tafalla et al.,* 1998).

Dynamically important magnetic fields introduce anisotropy to the mass distribution of the core. This anisotropy is illustrated in *Li and Shu* (1996b), who considered the self-similar equilibrium configurations supported partly by thermal pressure and partly by the static magnetic field. These configurations are described by

$$\rho(r,\theta) = \frac{a^2}{2\pi Gr^2} R(\theta); \quad \Phi(r,\theta) = \frac{4\pi a^2 r}{G^{1/2}} \phi(\theta) \quad (8)$$

where $R(\theta)$ and $\phi(\theta)$ are dimensionless angular functions of mass density and magnetic flux, respectively. They are solved from the equations of force balance along and across the field direction. The solutions turn out to be a linear sequence of singular isothermal toroids, characterized by a parameter H_0, the fractional overdensity supported by the magnetic field above that supported by thermal pressure. As H_0 increases, the toroid becomes more flattened. These toroids provide plausible initial conditions for protostellar collapse calculations.

Allen et al. (2003a,b) carried out protostellar collapse calculations starting from magnetized singular isothermal toroids, with or without rotation. Examples of nonrotating collapse are shown in Fig. 5 in self-similar coordinates. The dynamical collapse of magnetized toroids proceeds in a self-similar fashion as in the classical nonmagnetized (*Shu,* 1977) or strongly magnetized (*Li and Shu,* 1997) limit. Prominent in all collapse solutions of nonzero H_0 is the dense flattened structure in the equatorial region. It is the pseudodisk first discussed in *Galli and Shu* (1993a,b).

For rotating toroids, *Allen et al.* (2003b) concluded that for magnetic fields of reasonable strength, the rotation is braked so efficiently during the protostellar accretion phase that the formation of rotationally supported disks is suppressed in the ideal MHD limit. Nonideal effects, such as ambipolar diffusion or magnetic reconnection, must be considered for the all-important protostellar disks to appear in

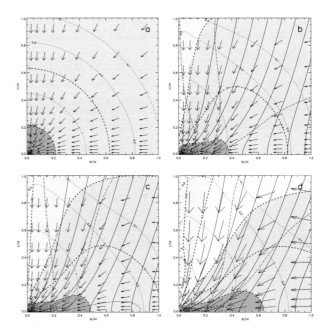

Fig. 5. Collapse solutions for different degrees of magnetization characterized by $H_0 = 0$, 0.125, 0.25, and 0.5. The contours of constant self-similar density $4\pi Gt^2\rho$ are plotted as dashed lines, with the shades highlighting the high-density regions. The magnetic field lines are plotted as solid lines, with contours of constant β (ratio of thermal and magnetic pressures; dash-dot-dashed) superposed. Representative velocity vectors are shown as arrows, and dotted lines are contours of constant speed.

the problem of star formation. Most of the angular momentum of the collapsed material is removed by a low-speed wind. Magnetic braking-driven winds have been obtained in the two-dimensional simulations of *Tomisaka* (1998, 2002) and three-dimensional simulations of *Machida et al.* (2004) and *Banerjee and Pudritz* (2005). It is tempting to identify these winds with bipolar molecular outflows; however, they move too slowly (with a typical speed of a few times the sound speed) to be identified as such. Nevertheless, the slow wind can modify the ambient environment for the fast jet/wind driven from close to the central stellar object, and should be included in a complete theory of molecular outflows. It would be interesting to look for observational signatures of magnetic braking-driven (slow) outflows, perhaps around the youngest protostars.

9. MOLECULAR OUTFLOWS

9.1. Unified Wind-driven Shell Model

Models for molecular outflows generally fall into two categories: jet-driven and wind-blown (see chapter by Arce et al.). The properties of outflows driven by winds were first quantified in *Shu et al.* (1991, hereafter *SRLL*). It is often stated that the shell model of *SRLL* is applicable to the class of broad, "classical" CO outflows, but not to the newer class of highly collimated sources (*Bachiller and Tafalla,* 1999). Recent calculations have shown that this need not be the case (*Shang et al., 2006*). Here, we focus on the advances on the wind-driven shell model since PPIV, starting with a quick overview of the *SRLL* model.

The *SRLL* model is characterized by two dimensionless functions, P and Q, that specify the angular distributions of wind momentum per steradian and ambient density

$$\rho(r,\theta) = \frac{a^2}{2\pi Gr^2} Q(\theta), \quad \frac{\dot{M}_w v_w}{4\pi} P(\theta) \quad (9)$$

where a is the isothermal sound speed, G the gravitational constant, and $\dot{M}_w$ and v_w represent the mass-loss rate and velocity of the wind. If the ambient medium is swept into a thin shell by the wind in a momentum conserving fashion, the angular distribution of the speed, and thus the shape, of the shell is determined by a bipolarity function $\mathcal{B} \propto (P/Q)^{1/2}$.

Current observations are unable to provide detailed information on the functions $P(\theta)$ and $Q(\theta)$. They can be determined theoretically in idealized situations. For example, one can use as the angular distribution $Q(\theta)$ the function $R(\theta)$ in equation (8), which describes the density distribution of a magnetized singular isothermal toroid. This was first done in *Li and Shu (1996a)*, who also adopted an idealized angular distribution for wind momentum $P(\theta) \propto 1/\sin^2\theta$ that is motivated by an asymptotic theory of magnetocentrifugal winds (*Shu et al., 1995*). For a toroid of overdensity parameter $H_0 = 0.5$ [which gives an aspect ratio of column density contours of about 2 : 1, as typically observed in low-mass cores (*Myers et al., 1991*)], the bipolarity function $\mathcal{B}$ corresponds to a shell of hour-glass shape, with a relatively wide opening near the base. This simple analytic model can explain several features commonly observed in molecular outflows, including mass-velocity relation and Hubble-type expansion. The fact that the shell in the model does not close on itself raises the question of its applicability to relatively young sources whose shells of outflowing material are often observed to be closed.

Shang et al. (2006) reconsidered the above model of molecular outflow formation through numerical simulations. The simulations, carried out using Zeus2D (*Stone and Norman, 1992*), allowed them to relax some of the simplifying assumptions made in the analytic models. Examples of the numerically computed shells are shown in Fig. 6. These are obtained by running a cylindrically stratified, toroidally magnetized wind of Alfvén Mach number 6 into toroids of different degrees of flattening, characterized by the overdensity parameter $H_0 = 0.25$, 0.5, 1.0, and 1.5. Common to all cases is the apparent two-component density structure: a dense axial jet, which is part of the primary wind, and a dense, closed, shell, which encases the jet. Between the jet and shell lies the wide-angle wind.

Note that the same density-stratified wind produces shells of varying degrees of collimation, some of which are jet-like. The jet-like shells are produced in relatively weakly

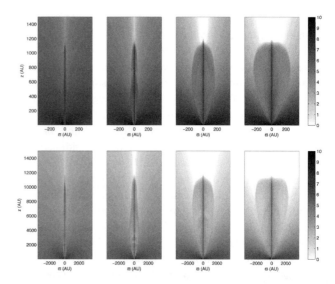

Fig. 6. Snapshots of density structures at 100 (top) and 1000 (bottom) years for a wind of mass-loss rate ($1 \times 10^{-6} \, M_\odot/\text{yr}$) and Alfvén Mach number 6. The toroids are shown in the order of $H_0 = 0.25$, 0.5, 1, and 1.5 (from left to right). The color bar shows density variation in logarithmic ($\log_{10}$) scale.

magnetized toroids (with values of H_0 less than unity), which have a relatively narrow low-density polar funnel through which the primary wind escapes. Even in more strongly magnetized toroids, the shells are still significantly elongated, despite the fact that the density distribution in the ambient toroids is strongly flattened. The elongation is due, to a large extent, to the anisotropy in the momentum distribution intrinsic to the shell-driving primary wind. In this picture, both the dense, axial "jet" part of the wind and the more tenuous "wide-angle" component participate in shaping the shell structure: The jet controls the length of the shell and the wide-angle component (together with the lateral ambient density distribution) controls the width. The jet-driven and wind-blown scenarios are thus unified in the single framework of *SRLL*, which appears capable of producing both classical and jet-like molecular outflows. In particular, the fast SiO jets observed in the prototype of jet-like outflows, HH211, find a natural explanation in this unified model (see section 9). The flattened NH_3 core, elongated CO shell, and highly collimated H_2 jet observed in HH212 (*Lee et al., 2006*) also strongly resemble the prominent features shown in Fig. 5.

Dynamical collapse is expected to modify the ambient environment through which the primary wind propagates, particularly near the central star (see Fig. 5). A future refinement of the shell model of *SRLL* would be to include the collapse-induced modification to the ambient density distribution, which is expected to broaden the base of the wind-driven cavity, especially at late times, when the collapsing region has expanded to large distances. The evolution of the collapsing envelope may be a key factor in controlling the outflow evolution from Class 0 to Class I to Class II described in the chapter by Arce et al. Indeed, one may heuristically view the sequence of models with increasing degrees of toroid flattening in Fig. 6 as an evolutionary sequence in time: As the collapse empties out an increasingly larger region near the star, the outflow cavity becomes wider near the base in time.

Another improvement would be in treating the wind-ambient interface, where mixing is expected due to Kelvin-Helmholtz instability. *Cunningham et al.* (2005) treated the interface using adaptive mesh refinement (AMR) techniques, extending the work of *Gardiner et al.* (2003) by including molecular, ionic, and atomic species, cooling functions, and molecule recombination and dissociation. They were able to partially resolve the strongly cooling, shocked layers around the outflow lobes. They found that the shocked wind and ambient medium are not completely mixed along the walls of the wind-blown cavity, but noted that this result might be affected by their limited resolution, despite the fact that AMR was used.

9.2. HH211 and SiO Emission

As mentioned earlier, the wind-driven shell model was thought to be unable to explain the class of highly collimated molecular outflows. It turns out, however, that the prototype of the class, HH211, may have structures that most closely correspond to the predictions of the unified shell model.

The HH211 outflow in the IC 348 molecular cloud complex (D ~ 315 pc) was discovered in the NIR H_2 emission (*McCaughrean et al.*, 1994). It is driven by a low-luminosity (3.6 $L_\odot$) Class 0 protostar and is considered to be extremely young, with a dynamical age of only ~750 yr. The high-resolution (1.5") CO J = 2–1 observations by *Gueth and Guilloteau* (1999) revealed a remarkable structure: The low-velocity CO delineates a pair of shells whose tips are associated with NIR H_2 emission, while the high-velocity CO traces a narrower feature whose velocity increases linearly with distance from the star. PV diagrams of the CO emission show parabolic shapes characteristic of wind-driven shells (*Lee et al.*, 2000). It is plausible that the CO emission traces a shell of ambient material swept up by a (wide-angle) primary wind.

Bisecting the CO lobe is a narrow jet of thermal SiO emission. The thermal emission from SiO is considered to trace the dense shocked gas because its extremely low gas-phase abundance (<10^{-12}) in quiescent regions (*Ziurys et al.*, 1989; *Martín-Pintado et al.*, 1992) can be enhanced by several orders of magnitude by means of shocks. The SiO emission is detected in J = 1–0 (*Chandler and Richer*, 2001), J = 5–4 (*Gibb et al.*, 2004), and 8–7 and 11–10 (*Nisini et al.*, 2002b). Since some of these lines have critical densities higher than ~10^6 cm^{-3} and the energy level of J = 11 is higher than 100 K, their detection means that the SiO jet is much denser and warmer than the lower-velocity shell com-

ponent. High-resolution (1–2") observations of SiO J = 8–7 (*Palau et al.*, 2006), 5–4 (*Hirano et al.*, 2006), and 3–2 (Hirano et al., in preparation) have been done using SMA and NMA. As shown in Fig. 7, the SiO jet is resolved into a chain of knots separated by 4" (~1000 AU). The SiO knots have their H_2 counterparts except the ones in close vicinity of the central source where the extinction is large. The knotty structure is more prominent in the higher SiO transition. The innermost knots located at ±2" from the source are prominent in the J = 8–7 and 5–4 maps, while barely seen in the J = 1–0 map. *Hirano et al.* (2006) estimated that these knots have a temperature >300–500 K and density $(0.5–1) \times 10^7$ cm^{-3}. These densities and temperatures are much higher than those inferred for the CO shell component, indicating that the SiO jet has a different origin. The most likely possibility is that the SiO jet traces the densest part of the primary wind. This identification is strengthened by the PV diagrams of SiO J = 8–7 and 5–4 emission (Fig. 8), which show a large (projected) velocity dispersion (~30 km s^{-1}) close to the star. This velocity feature, absent from the CO emission, is reminiscent of that predicted in the X-wind the-

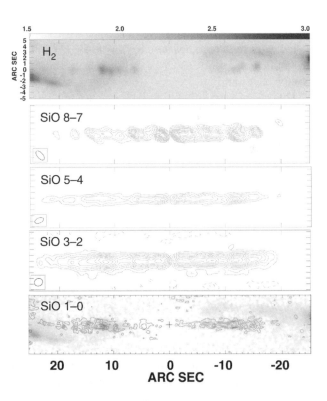

Fig. 7. HH211 outflow observed with the near-infrared H_2 υ = 1–0 S(1) emission [gray scale in the top panel (*Hirano et al.*, 2006a)] and the SiO J = 8–7 (*Palau et al.*, 2006), J = 5–4 (*Hirano et al.*, 2006), J = 3–2 (Hirano et al., in preparation.), and J = 1–0 (*Chandler and Richer*, 2001). All the images are rotated by –26° from the equatorial coordinates. The contours represent the integrated SiO emission in the blueshifted velocity range (positive offset from the position of the driving source, HH211-mm, along the major axis) and in the redshifted velocity range (negative offset).

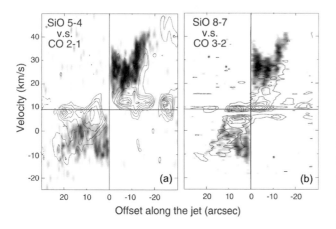

Fig. 8. Position-velocity (p-v) plots of the SiO and CO along the major axis of the HH211 outflow. **(a)** The SiO J = 5–4 (gray scale) superposed on the CO J = 2–1 [contours from *Gueth and Guilloteau* (1999)]. **(b)** The SiO J = 8–7 (gray scale) and the CO J = 3–2 (contours) from *Palau et al.* (2006). The vertical and horizontal straight lines indicate the position of the continuum source and the systemic velocity, respectively.

ory for optical forbidden lines (*Shang et al.*, 1998). It may be indicative of a (perhaps unsteady) wide-angle wind that is stratified in density, as envisioned in our unified model of molecular outflows.

The spectacular SiO jet of HH211 is not unique. It is seen in at least one other young Class 0 source, L1448-mm (*Bachiller et al.*, 1991; *Guilloteau et al.*, 1992; *Dutrey et al.*, 1997). The SiO abundance is estimated to be ~10^{-6} (*Martín-Pintado et al.*, 1992; *Bachiller et al.*, 1991; *Hirano et al.*, 2006) in both L1448-mm and HH211, implying an enhancement by a factor of $\geq 10^5$ above the value in the ambient quiescent clouds. The bright SiO jet appears to fade away quickly as the source evolves. SiO emission is also observed at the tips of outflow cavities in some sources, including L1157-mm (*Mikami et al.*, 1992; *Zhang et al.*, 1995, 2000; *Gueth et al.*, 1998; *Bachiller et al.*, 2001), IRAS 03282 (*Bachiller et al.*, 1994), BHR71 (*Garay et al.*, 1998), and NGC 1333 IRAS 2A (*Jørgensen et al.*, 2004). For these sources, the typical SiO abundance is ~10^{-9}–10^{-8} (e.g., *Bachiller*, 1996). Narrow SiO lines have been detected in relatively quiescent regions around SVS 13 (*Lefloch et al.*, 1998; *Codella et al.*, 1999), B1 (*Martín-Pintado et al.*, 1992; *Yamamoto et al.*, 1992), IRAS 00338+6312, and Cep A (*Codella et al.*, 1999). The estimated SiO abundance is about 10^{-11}–10^{-10} (*Lefloch et al.*, 1998; *Codella et al.*, 1999), the lowest among the three classes described here. These three classes of SiO sources can plausibly be put into an evolutionary sequence (*Bachiller and Tafalla*, 1999). Detailed calculations that take into account the evolution of both the primary wind and the ambient medium, as well as SiO chemistry, are needed to quantify the sequence. The unified wind-driven shell model provides a framework for quantifying the outflow evolution.

10. CONCLUDING REMARKS

We have reviewed the progress on the X-wind model and related work since PPIV. In particular, we showed that when detailed excitation conditions and radiative processes are modeled, X-winds are capable of explaining the gross characteristics of many observations, ranging from optical forbidden lines to thermal radio emission. To make model predictions match observational data, a mechanical heating is required to keep the wind warm enough to emit in optical and radio. The heating is prescribed phenomenologically, and may be related to wind variabilities, which produce shocks of varying strengths. How the variabilities are generated remains little explored, and should be a focus of future investigation.

Dynamical models of jets and winds in YSOs have shown some convergence. The magnetocentrifugal mechanism remains the leading candidate for outflow production. The modeling effort is hampered, however, by our limited knowledge of the physical conditions near the wind-disk interface. These include both the thermal and ionization structure of the region, as well as the magnetic field distribution. Detailed calculations that treat the coupling between the wind, the disk, and perhaps the stellar magnetosphere self-consistently may unify the disk-wind and X-wind picture. An improved model of the jets and winds will also deepen our understanding of the molecular outflows that they drive.

Acknowledgments. The authors would like to thank the referee for his careful reading and thorough review of the manuscript. This work is supported by the Theoretical Institute for Advanced Research in Astrophysics (TIARA) operated under Academia Sinica and the National Science Council Excellence Projects program in Taiwan administered through grant numbers NSC94-2752-M-007-001 and NSC94-2752-M-001-001, and NASA grant NAG5-12102 and NSF AST-0307368 in the U.S.

REFERENCES

Allen A., Li Z.-Y., and Shu F. H. (2003a) *Astrophys. J., 599*, 351–362.
Allen A., Li Z.-Y., and Shu F. H. (2003b) *Astrophys. J., 599*, 363–379.
Anderson J. M., Li Z.-Y., Krasnopolsky R., and Blandford R. D. (2003) *Astrophys. J., 590*, L107–L110.
Anderson J. M., Li Z.-Y., Krasnopolsky R., and Blandford R. D. (2005a) *Astrophys. J., 630*, 945–957.
Anderson J. M., Li Z.-Y., Krasnopolsky R., and Blandford R. D. (2006) *Astrophys. J.*, in press.
Anglada G. (1996) In *Radio Emission from the Stars and the Sun* (A. R. Taylor and J. M. Paredes, eds.), pp. 3–14. ASP Conf. Series 93, San Francisco.
Bacciotti F. (2002) *Rev. Mex. Astron. Astrofis. Ser. Conf., 13*, 8–15.
Bacciotti F. and Eislöffel J. (1999) *Astron. Astrophys., 342*, 717–735.
Bacciotti F., Hirth G. A., and Natta A. (1996) *Astron. Astrophys., 310*, 309–314.
Bacciotti F., Mundt R., Ray T. P., Eislöffel J., Solf J., and Camezind M. (2000) *Astrophys. J., 537*, L49–L52.
Bacciotti F., Ray T. P., Mundt R., Eislöffel J., and Solf J. (2002) *Astrophys. J., 576*, 222–231.
Bachiller R. (1996) *Ann. Rev. Astron. Astrophys., 34*, 111–154.
Bachiller R. and Tafalla M. (1999) In *The Origin of Stars and Planetary*

Systems (C. J. Lada and N. D. Kylafis eds.), pp. 227–266. Kluwer, Dordrecht.

Bachiller R., Martín-Pintado J., and Fuente A. (1991) *Astron. Astrophys., 243*, L21–L24.

Bachiller R., Terebey S., Jarrett T., Martín-Pintado J., Beichman C. A., and van Buren D. (1994) *Astrophys. J., 437*, 296–304.

Bachiller R., Pérez Gutiérrez M., Kumar M. S. N., and Tafalla M. (2001) *Astron. Astrophys., 372*, 899–912.

Banerjee R. and Pudritz R. E. (2006) *Astrophys. J., 641*, 949–960.

Binette L., Cabrit S., Raga A., and Cantó J. (1999) *Astron. Astrophys., 346*, 260–266.

Blandford R. D. and Payne D. G. (1982) *Mon. Not. R. Astron. Soc., 199*, 883–903.

Bogovalov S. and Tsinganos K. (1999) *Mon. Not. R. Astron. Soc., 305*, 211–224.

Cabrit S., Ferreira J., and Raga A. C. (1999) *Astron. Astrophys., 343*, L61–L64.

Cabrit S., Pety J., Pesenti N., and Dougados C. (2006) *Astron. Astrophys., 452*, 897–906.

Casse F. and Keppens R. (2002) *Astrophys. J., 581*, 988–1001.

Chandler C. J. and Richer J. S. (2001) *Astrophys. J., 555*, 139–145.

Ciolek G. E. and Basu S. (2000) *Astrophys. J., 529*, 925–931.

Codella C., Bachiller R., and Reipurth B. (1999) *Astron. Astrophys., 343*, 585–598.

Cunningham A., Frank A., and Hartmann L. (2005) *Astrophys. J., 631*, 1010–1021.

Curiel S., Cantó J., and Rodríguez L. F. (1987) *Rev. Mex. Astron. Astrofis., 14*, 595–602.

Davis C. J., Whelan E., Ray T. P., and Chrysostomou A. (2003) *Astron. Astrophys., 397*, 693–710.

Dougados C., Cabrit S., Lavalley C., and Ménard F. (2000) *Astron. Astrophys., 357*, L61–L64.

Dougados C., Cabrit S., and Lavalley-Fouquet C. (2002) *Rev. Mex. Astron. Astrofis. Ser. Conf., 13*, 43–48.

Dougados C., Cabrit S., López-Martín L., Garcia P., and O'Brien D. (2003) *Astrophys. Space Sci., 287*, 135–138.

Dougados C., Cabrit S., Ferreira J., Pesenti N., Garcia P., and O'Brien D. (2004) *Astrophys. Space Sci., 293*, 45–52.

Draine B. T. (1980) *Astrophys. J., 241*, 1021–1038.

Dutrey A., Guilloteau S., and Bachiller R. (1997) *Astron. Astrophys., 325*, 758–768.

Eichler D. (1993) *Astrophys. J., 419*, 111–116.

Evans N. J. II, Levreault R. M., Beckwith S., and Skrutskie M. (1987) *Astrophys. J., 320*, 364–375.

Feigelson E. D. and Montmerle T. (1999) *Ann. Rev. Astron. Astrophys., 37*, 363–408.

Feigelson E. D., Broos P., Gaffney J. A. III, Garmire G., Hillenbrand L. A., Pravdo S. H., Townsley L., and Tsuboi Y. (2002) *Astrophys. J., 574*, 258–292.

Feigelson E. D., Getman K., Townsley L., Garmire G., Preibisch T., Grosso N., Montmerle T., Muench A., and McCaughrean M. (2005) *Astrophys. J. Suppl., 160*, 379–389.

Fendt C. and Čemeljić M. (2002) *Astron. Astrophys., 395*, 1045–1060.

Ferreira J. (1997) *Astron. Astrophys., 319*, 340–359.

Ferreira J. and Casse F. (2004) *Astrophys. Space Sci., 292*, 479–492.

Fridlund C. V. M. and Liseau R. (1998) *Astrophys. J., 499*, L75–L77.

Galli D. and Shu F. (1993a) *Astrophys. J., 417*, 220–242.

Galli D. and Shu F. (1993b) *Astrophys. J., 417*, 243–258.

Garay G., Köhnenkamp I., Bourke T. L., Rodríguez L. F., and Lehtinen K. K. (1998) *Astrophys. J., 509*, 768–784.

Garcia P. J. V., Ferreira J., Cabrit S., and Binette L. (2001a) *Astron. Astrophys., 377*, 589–608.

Garcia P. J. V., Cabrit S., Ferreira J., and Binette L. (2001b) *Astron. Astrophys., 377*, 609–616.

Gardiner T. A., Frank A., and Hartmann L. (2003) *Astrophys. J., 582*, 269–276.

Ghavamian P. and Hartigan P. (1998) *Astrophys. J., 501*, 687–698.

Gibb A. G., Richer J. S., Chandler C. J., and Davis C. J. (2004) *Astrophys. J., 603*, 198–212.

Giovanardi C., Rodríguez L. F., Lizano S., and Cantó J. (2000) *Astrophys. J., 538*, 728–737.

González R. F. and Cantó J. (2002) *Astrophys. J., 580*, 459–467.

Glassgold A. E., Najita J., and Igea J. (1997) *Astrophys. J., 480*, 344–350.

Glassgold A. E., Feigelson E. D., and Montmerle T. (2000) In *Protostars and Planets IV* (V. Mannings et al., eds.), pp. 429–456. Univ. of Arizona, Tucson.

Glassgold A. E., Krsti P. S., and Schultz D. R. (2005a) *Astrophys. J., 621*, 808–816.

Glassgold A. E., Feigelson E. D., Montmerle T., and Wolk S. (2005b) In *Chondrites and the Protoplanetary Disk* (A. N. Krot et al., eds.), pp. 165–180. ASP Conf. Series 341, San Francisco.

Grosso N., Montmerle T., Feigelson E. D., Andre P., Casanova S., and Gregorio-Hetem J. (1997) *Nature, 387*, 56–58.

Gueth F. and Guilloteau S. (1999) *Astron. Astrophys., 343*, 571–584.

Gueth F., Guilloteau S., and Bachiller R. (1998) *Astron. Astrophys., 333*, 287–297.

Guilloteau S., Bachiller R., Fuente A., and Lucas R. (1992) *Astron. Astrophys., 265*, L49–L52.

Hartigan P., Morse J. A., and Raymond J. (1994) *Astrophys. J., 436*, 125–143.

Hartigan P., Raymond J., and Pierson R. (2004) *Astrophys. J., 614*, L69–L71.

Hartmann L. and Raymond J. C. (1984) *Astrophys. J., 276*, 560–571.

Hirano N., Liu S.-Y., Shang H., Ho P. T. P., Huang H.-C., Kuan Y.-J., McCaughrean M. J., and Zhang Q. (2006) *Astrophys. J., 636*, L141–L144.

Itoh Y., Kaifu N., Hayashi M., Hayashi S. S., Yamashita T., Usuda T., Noumaru J., Maihara T., Iwamuro F., Motohara K., Taguchi T., and Hata R. (2000) *Publ. Astron. Soc. Japan, 52*, 81–86.

Jørgensen J. K., Hogerheijde M. R., Blake G. A., van Dishoeck E. F., Mundy L. G., and Schöier F. L. (2004) *Astron. Astrophys., 415*, 1021–1037.

Koldoba A. V., Ustyugova G. V., Romanova M. M., Chechetkin V. M., and Lovelace R. V. E. (1995) *Astrophys. Space Sci., 232*, 241–261.

Königl A. and Pudritz R. E. (2000) In *Protostars and Planets IV* (V. Mannings et al., eds.), pp. 759–788. Univ. of Arizona, Tucson.

Krasnopolsky R., Li Z.-Y., and Blandford R. (1999) *Astrophys. J., 526*, 631–642.

Krasnopolsky R., Li Z.-Y., and Blandford R. (2003) *Astrophys. J., 595*, 631–642.

Krstić P. S. and Schultz D. R. (1998) *Atomic and Plasma-Material Data for Fusion, 8*.

Kudoh T., Matsumoto R., and Shibata K. (2003) *Astrophys. Space Sci., 287*, 99–102.

Kwan J. and Tademaru E. (1988) *Astrophys. J., 332*, L41–L44.

Lavalley-Fouquet C., Cabrit S., and Dougados C. (2000) *Astron. Astrophys., 356*, L41–L44.

Lee C.-F., Mundy L. G., Reipurth B., Ostriker E. C., and Stone J. M. (2000) *Astrophys. J., 542*, 925–945.

Lee C.-F., Ho P. T. P., Beuther H., Bourke T. L., Zhang Q., Hirano N., and Shang H. (2006) *Astrophys. J., 639*, 292–302.

Lefloch B., Castets A., Cernicharo J., and Loinard L. (1998) *Astrophys. J., 504*, L109–L112.

Li Z.-Y. (1995) *Astrophys. J., 444*, 848–860.

Li Z.-Y. and Shu F. H. (1996a) *Astrophys. J., 468*, 261–268.

Li Z.-Y. and Shu F. H. (1996b) *Astrophys. J., 472*, 211–224.

Li Z.-Y. and Shu F. H. (1997) *Astrophys. J., 475*, 237–250.

Lucek S. G. and Bell A. R. (1996) *Mon. Not. R. Astron. Soc., 281*, 245–256.

Mac Low M.-M. and Klessen R. S. (2004) *Rev. Modern Phys., 76*, 125–194.

Machida M. N., Tomisaka K., and Matsumoto T. (2004) *Mon. Not. R. Astron. Soc., 348*, L1–L5.

Martín-Pintado J., Bachiller R., and Fuente A. (1992) *Astron. Astrophys., 254*, 315–326.

Matzner C. D. and McKee C. F. (1999) *Astrophys. J., 526,* L109–L112.

McCaughrean M. J., Rayner J. T., and Zinnecker H. (1994) *Astrophys. J., 436,* L189–L192.

Mikami H., Umemoto T., Yamamoto S., and Saito S. (1992) *Astrophys. J., 392,* L87–L90.

Mouschovias T. Ch. and Ciolek G. E. (1999) In *The Origin of Stars and Planetary Systems* (C. J. Lada and N. D. Kylafis eds.), pp. 305–340. Kluwer, Dordrecht.

Myers P. C. (1999) In *The Origin of Stars and Planetary Systems* (C. J. Lada and N. D. Kylafis eds.), pp. 67–96. Kluwer, Dordrecht.

Myers P. C., Fuller G. A., Goodman A. A., and Benson P. J. (1991) *Astrophys. J., 376,* 561–572.

Nakamura F. and Li Z.-Y. (2005) *Astrophys. J., 631,* 411–428.

Nakano T. (1984) *Fund. Cosmic Phys., 9,* 139–231.

Nisini B., Caratti o Garatti A., Giannini T., and Lorenzetti D. (2002a) *Astron. Astrophys., 393,* 1035–1051.

Nisini B., Codella C., Giannini T., and Richer J. S. (2002b) *Astron. Astrophys., 395,* L25–L28.

Nisini B., Bacciotti F., Giannini T., Massi F., Eislöffel J., Podio L., and Ray T. P. (2005) *Astron. Astrophys., 441,* 159–170.

O'Brien D., Garcia P., Ferreira J., Cabrit S., and Binette L. (2003) *Astrophys. Space Sci., 287,* 129–134.

Osterbrock D. E. (1961) *Astrophys. J., 134,* 270–272.

Ouyed R. and Pudritz R. E. (1997) *Astrophys. J., 482,* 712–732.

Ouyed R., Clarke D. A., and Pudritz R. E. (2003) *Astrophys. J., 582,* 292–319.

Palau A., Ho P. T. P., Zhang Q., Estalella R., Hirano N., Shang H., Lee C.-F., Bourke T. L., Beuther H., and Kuan Y.-J. (2006) *Astrophys. J., 636,* L137–L140.

Pesenti N., Dougados C., Cabrit S., O'Brien D., Garcia P., and Ferreira J. (2003) *Astron. Astrophys., 410,* 155–164.

Pesenti N., Dougados C., Cabrit S., Ferreira J., Casse F., Garcia P., and O'Brien D. (2004) *Astron. Astrophys., 416,* L9–L12.

Pradhan A. K. and Zhang H. L. (1993) *Astrophys. J., 409,* L77–L79.

Pyo T.-S., Hayashi M., Kobayashi N., Terada H., Goto M., Yamashita T., Tokunaga A. T., and Itoh Y. (2002) *Astrophys. J., 570,* 724–733.

Pyo T.-S., Kobayashi N., Hayashi M., Terada H., Goto M., Takami H., Takato N., Gaessler W., Usuda T., Yamashita T., Tokunaga A. T., Hayano Y., Kamata Y., Iye M., and Minowa Y. (2003) *Astrophys. J., 590,* 340–347.

Pyo T.-S., Hayashi M., Kobayashi N., Tokunaga A. T., Terada H., Tsujimoto M., Hayashi S. S., Usuda T., Yamashita T., Takami H., Takato N., and Nedachi K. (2005) *Astrophys. J., 618,* 817–821.

Pyo T.-S., Hayashi M., Kobayashi N., Tokunaga A. T., Terada H., et al. (2006) *Astrophys. J., 649,* in press.

Raga A. C., Böhm K.-H., and Cantó J. (1996) *Rev. Mex. Astron. Astrofis., 32,* 161–174.

Ray T. P. and Bacciotti F. (2003) *Rev. Mex. Astron. Astrofis. Ser. Conf., 15,* 106–111.

Reipurth B., Yu K. C., Heathcote S., Bally J., and Rodríguez L. F. (2000) *Astron. J., 120,* 1449–1466.

Reipurth B., Rodríguez L. F., Anglada G., and Bally J. (2002) *Astrophys. J., 125,* 1045–1053.

Reipurth B., Rodríguez L. F., Anglada G., and Bally J. (2004) *Astrophys. J., 127,* 1736–1746.

Reynolds S. P. (1986) *Astrophys. J., 304,* 713–720.

Rodríguez L. F. (1998) *Rev. Mex. Astron. Astrofis., 7,* 14–20.

Rodríguez L. F. and Cantó J. (1983) *Rev. Mex. Astron. Astrofis., 8,* 163–173.

Rodríguez L. F., Myers P. C., Cruz-González I., and Terebey S. (1989) *Astrophys. J., 347,* 461–467.

Rodríguez L. F., D'Alessio P., Wilner D. J., Ho P. T. P., Torrelles J. M., Curiel S., Gómez Y., Lizano S., Pedlar A., Cantó J., and Raga A. C. (1998) *Nature, 395,* 355–357.

Rodríguez L. F., Porras A., Claussen M. J., Curiel S., Wilner D. J., and Ho P. T. P. (2003) *Astrophys. J., 586,* L137–L139.

Ruden S. P., Glassgold A. E., and Shu F. H. (1990) *Astrophys. J., 361,* 546–569.

Safier P. N. (1993a) *Astrophys. J., 408,* 115–147.

Safier P. N. (1993b) *Astrophys. J., 408,* 148–159.

Shang H., Shu F. H., and Glassgold A. E. (1998) *Astrophys. J., 493,* L91–L94.

Shang H., Glassgold A. E., Shu F. H., and Lizano S. (2002) *Astrophys. J., 564,* 853–876.

Shang H., Lizano S., Glassgold A. E., and Shu F. (2004) *Astrophys. J., 612,* L69–L72.

Shang H., Allen A., Li Z.-Y., Liu C.-F., Chou M.-Y., and Anderson J. (2006) *Astrophys. J., 649,* in press.

Shu F. H. (1977) *Astrophys. J., 214,* 488–497.

Shu F. H. (1992) In *The Physics of Astrophysics: Gas Dynamics, Volume II.* Univ. Science Books, Herndon.

Shu F. H., Adams F. C., and Lizano S. (1987) *Ann. Rev. Astron. Astrophys., 25,* 23–81.

Shu F. H., Ruden S. P., Lada C. J., and Lizano S. (1991) *Astrophys. J., 370,* L31–L34.

Shu F., Najita J., Ostriker E., Wilkin F., Ruden S., and Lizano S. (1994) *Astrophys. J., 429,* 781–796.

Shu F. H., Najita J., Ostriker E. C., and Shang H. (1995) *Astrophys. J., 455,* L155–L158.

Shu F. H., Shang H., Glassgold A. E., and Lee T. (1997) *Science, 277,* 1475–1479.

Shu F. H., Najita J. R., Shang H., and Li Z.-Y. (2000) In *Protostars and Planets IV* (V. Mannings et al., eds.), pp. 789–814. Univ. of Arizona, Tucson.

Spruit H. C. (1996) In *Evolutionary Processes in Binary Stars* (R. A. M. J. Wijers et al., eds.), pp. 249–286. Kluwer, Dordrecht.

Stone J. M. and Norman M. L. (1992) *Astrophys. J. Suppl., 80,* 791–818.

Tafalla M., Mardones D., Myers P. C., Caselli P., Bachiller R., and Benson P. J. (1998) *Astrophys. J., 504,* 900–914.

Tomisaka K. (1998) *Astrophys. J., 502,* L163–L167.

Tomisaka K. (2002) *Astrophys. J., 575,* 306–326.

Torrelles J. M., Ho P. T. P., Rodríguez L. F., and Cantó J. (1985) *Astrophys. J., 288,* 595–603.

Torrelles J. M., Patel N. A., Anglada G., Gómez J. F., Ho P. T. P., Lara L., Alberdi A., Cantó J., Curiel S., Garay G., and Rodríguez L. F. (2003) *Astrophys. J., 598,* L115–L119.

Tsuboi Y., Imanishi K., Koyama K., Grosso N., and Montmerle T. (2000) *Astrophys. J., 532,* 1089–1096.

Uchida Y. and Shibata K. (1985) *Publ. Astron. Soc. Japan, 37,* 515–535.

Verner E. M., Verner D. A., Baldwin J. A., Ferland G. J., and Martin P. G. (2000) *Astrophys. J., 543,* 831–839.

von Rekowski B. and Brandenburg A. (2004) *Astron. Astrophys., 420,* 17–32.

Wardle M. and Königl A. (1993) *Astrophys. J., 410,* 218–238.

Yamamoto S., Mikami H., Saito S., Kaifu N., Ohishi M., and Kawaguchi K. (1992) *Publ. Astron. Soc. Japan, 44,* 459–467.

Zhang H. L. and Pradhan A. K. (1995) *Astron. Astrophys., 293,* 953–966.

Zhang Q., Ho P. T. P., Wright M. C. H., and Wilner D. J. (1995) *Astrophys. J., 451,* L71–L74.

Zhang Q., Ho P. T. P., and Wright M. C. H. (2000) *Astron. J., 119,* 1345–1351.

Ziurys L. M., Friberg P., and Irvine W. M. (1989) *Astrophys. J., 343,* 201–207.

Disk Winds, Jets, and Outflows:
Theoretical and Computational Foundations

Ralph E. Pudritz
McMaster University

Rachid Ouyed
University of Calgary

Christian Fendt
Max-Planck Institute for Astronomy, Heidelberg

Axel Brandenburg
Nordic Institute for Theoretical Physics

We review advances in the theoretical and computational studies of disk winds, jets, and outflows, including the connection between accretion and jets, the launch of jets from magnetized disks, the coupled evolution of jets and disks, the interaction of magnetized young stellar objects with their surrounding disks and the relevance to outflows, and finally, the link between jet formation and gravitational collapse. We also address the predictions the theory makes about jet kinematics, collimation, and rotation that have recently been confirmed by high-spatial- and high-spectral-resolution observations. Disk winds have a universal character that may account for jets and outflows during the formation of massive stars as well as brown dwarfs.

1. INTRODUCTION

The close association of jets and outflows with protostellar accretion disks is one of the hallmarks of the accretion picture of low-mass star formation. The most energetic outflow phase occurs during gravitational collapse of a molecular cloud core — the so-called Class 0 phase — when much of its envelope is still raining down onto the forming protostellar disk and the disk accretion rate is high. Later, in the T Tauri star (TTS) stage, when most of the original core has been accreted and the young stellar object (YSO) is being fed by lower accretion rates through the surrounding Keplerian accretion disk, the high-speed jet becomes optically visible. When the disk disappears in the weak-lined TTS (WTTS) phase, the jet goes with it.

The most comprehensive theoretical picture that we have for these phenomena is that jets are highly collimated, hydromagnetic disk winds whose torques efficiently extract disk angular momentum and gravitational potential energy. Jets also sweep up ambient molecular gas and drive large-scale molecular outflows. A disk wind was first suggested as the origin of jets from accretion disks around black holes in the seminal paper by *Blandford and Payne* (1982) (*BP82*), and was soon proposed as the mechanism for protostellar jets (*Pudritz and Norman, 1983, 1986*).

Several major observational breakthroughs have taken place in the study of jets and outflows since the Protostars and Planets IV (PPIV) conference held in 1998. Direct, high-resolution spectro-imaging and adaptive optics methods discovered the rotation of protostellar jets (*Bacciotti et al., 2003*). These observations also revealed that jets have an onion-like, velocity structure (with the highest speeds being closest to the outflow axis). This work provides strong support for the idea that jets originate as centrifugally driven MHD winds from extended regions of their surrounding disks (see chapter by Ray et al.). Recently, outflows and disks have also been discovered around massive stars (see chapter by Arce et al.) as well as brown dwarfs (e.g., *Bourke et al., 2005*), implying that the mechanism is important across the entire stellar mass spectrum.

Major advances in the theoretical modeling of these systems have also occurred, due primarily to a variety of MHD computational studies. Simulations now resolve the global evolution of disks and outflows, track the interaction of disks with central magnetized stars, and even follow the generation of outflows during gravitational collapse. These studies show that jets and disks are closely coupled and that jet dynamics scales to YSOs of all masses.

Our review examines the theory of the central engine of jets and its exploration through the use of computer simulations. We focus mainly on developments since PPIV and refer to the review by *Königl and Pudritz* (2000) (*KP00*) for a discussion of the earlier literature, as well as *Pudritz* (2003) and *Heyvaerts* (2003) for more technical background. We first discuss the basic theory of disk winds and their kinematics. We then switch to computational studies of jets from accretion disks treated as boundary conditions, as well as global simulations including the disk. We then examine the innermost regions of the disk where the stellar magnetosphere interacts with the disk, as well as the surface of the star that may drive an accretion-powered outflow. Finally, we discuss how outflows are generated during the early stages of the gravitational collapse.

2. THEORY OF DISK WINDS

An important insight into the nature of the engine for jets can be gleaned from the observed ratio of the momentum transport rate (or thrust) carried by the CO molecular outflow to the thrust that can be provided by the bolometric luminosity of the central star (e.g., *Cabrit and Bertout*, 1992)

$$F_{outflow}/F_{rad} = 250(L_{bol}/10^3 \ L_\odot)^{-0.3} \qquad (1)$$

This relation has been confirmed and extended by the analysis of data from over 390 outflows, ranging over six decades up to $10^6 \ L_\odot$ in stellar luminosity (*Wu et al.*, 2004). It suggests that jets from both low- and high-mass systems are probably driven by a single, nonradiative mechanism.

Jets are observed to have a variety of structures and time-dependent behavior: from internal shocks and moving knots to systems of bow shocks that suggest longtime episodic outbursts. They show wiggles and often have cork-screw-like structure, suggesting the presence either of jet precession, the operation of nonaxisymmetric kink modes, or both. Given the highly nonlinear behavior of the force balance equation for jets (the so-called Grad-Shafranov equation), theoretical work has focused on tractable and idealized time-independent, and axisymmetric or self-similar models (e.g., *BP82*) of various kinds.

2.1. Conservation Laws and Jet Kinematics

Conservation laws are the gold standard in physics, and play a significant role in understanding astrophysical jets. This is because whatever the details (e.g., the asymptotics, the crossing of critical points, the way that matter is loaded onto field lines within the disks, etc.), conservation laws strongly constrain the flux of mass, angular momentum, and energy. What cannot be constrained by these laws will depend on the general physics of the disks such as on how matter is loaded onto field lines.

Jet dynamics can be described by the time-dependent equations of ideal MHD. The evolution of a magnetized, rotating system that is threaded by a large-scale field **B** involves (1) the continuity equation for a conducting gas of density ρ moving at velocity **v** (which includes turbulence); (2) the equation of motion for the gas that undergoes pressure (p), gravitational (with potential Φ), and Lorentz forces; (3) the induction equation for the evolution of the magnetic field in the moving gas where the current density is $j = (c/4\pi)\nabla \times \mathbf{B}$; (4) the energy equation, where e is the internal energy per unit mass; and, (5) the absence of magnetic

monopoles. These are written as

$$\frac{\partial \rho}{\partial t} + \nabla \cdot (\rho \mathbf{v}) = 0 \qquad (2)$$

$$\rho \left(\frac{\partial \mathbf{v}}{\partial t} + (\mathbf{v} \cdot \nabla) \mathbf{v} \right) + \nabla p + \rho \nabla \Phi - \frac{\mathbf{j} \times \mathbf{B}}{c} = 0 \qquad (3)$$

$$\frac{\partial \mathbf{B}}{\partial t} - \nabla \times (\mathbf{v} \times \mathbf{B}) = 0 \qquad (4)$$

$$\rho \left(\frac{\partial e}{\partial t} + (\mathbf{v} \cdot \nabla)e \right) + p(\nabla \cdot \mathbf{v}) = 0 \qquad (5)$$

$$\nabla \cdot \mathbf{B} = 0 \qquad (6)$$

We specialize to the restricted case of stationary, as well as two-dimensional (axisymmetric) flows, from which the conservation laws follow. It is useful to decompose vector quantities into poloidal and toroidal components (e.g., magnetic field $\mathbf{B} = \mathbf{B_p} + B_\phi \hat{\mathbf{e}}_\phi$). In axisymmetric conditions, the poloidal field Bp can be derived from a single scalar potential a(r, z) whose individual values, a = const, define the surfaces of constant magnetic flux in the outflow and can be specified at the surface of the disk (e.g., *Pelletier and Pudritz*, 1992) (*PP92*).

Conservation of mass and magnetic flux along a field line can be combined into a single function k that is called the "mass load" of the wind, which is a constant along a magnetic field line

$$\rho \mathbf{v_p} = k \mathbf{B_p} \qquad (7)$$

This function represents the mass load per unit time, per unit magnetic flux of the wind. For axisymmetric flows, its value is preserved on each ring of field lines emanating from the accretion disk. Its value on each field line is determined by physical conditions — including dissipative processes — near the disk surface. It may be more revealingly written as

$$k(a) = \frac{\rho v_p}{B_p} = \frac{d\dot{M}_w}{d\psi} \qquad (8)$$

where $d\dot{M}_w$ is the mass flow rate through an annulus of cross-sectional area dA through the wind and dΨ is the amount of poloidal magnetic flux threading through this same annulus. The mass load profile is a function of the footpoint radius r_0 of the wind on the disk.

The toroidal field in rotating flows derives from the induction equation

$$B_\phi = \frac{\rho}{k}(v_\phi - \Omega_0 r) \qquad (9)$$

where Ω_0 is the angular velocity of the disk at the midplane. This result shows that the strength of the toroidal field in the jet depends on the mass loading as well as the jet density. Denser winds should have stronger toroidal fields. We note however, that the density does itself depend on the value of k. Equation (9) also suggests that at higher-mass loads, one has lower toroidal field strengths. This can be reconciled however, since it can be shown from the conservation laws (see below) that the value of k is related to the density of the outflow at the Alfvén point on a field line; $k = (\rho_A/4\pi)^{1/2}$ (e.g., *PP92*). Thus, higher-mass loads correspond to denser winds and when this is substituted into equation (9), we see that this also implies stronger toroidal fields. We show later that jet collimation depends on hoop stress through the toroidal field and thus the mass load must have a very important effect on jet collimation (section 3.2).

Conservation of angular momentum along each field line leads to the conserved angular momentum per unit mass

$$l(a) = rv_\phi - \frac{rB_\phi}{4\pi k} = \text{const} \qquad (10)$$

The form for l reveals that the total angular momentum is carried by both the rotating gas (first term) as well by the twisted field (second term), the relative proportion being determined by the mass load.

The value of l(a) that is transported along each field line is fixed by the position of the Alfvén point in the flow, where the poloidal flow speed reaches the Alfvén speed for the first time ($m_A = 1$). It is easy to show that the value of the specific angular momentum is

$$l(a) = \Omega_0 r_A^2 = (r_A/r_0)^2 l_0 \qquad (11)$$

where $l_0 = v_{K,0}r_0 = \Omega_0 r_0^2$ is the specific angular momentum of a Keplerian disk. For a field line starting at a point r_0 on the rotor (disk in our case), the Alfvén radius is $r_A(r_0)$ and constitutes a lever arm for the flow. The result shows that the angular momentum per unit mass that is being extracted from the disk by the outflow is a factor of $(r_A/r_0)^2$ greater than it is for gas in the disk. For typical lever arms, one particle in the outflow can carry the angular momentum of 10 of its fellows left behind in the disk.

Conservation of energy along a field line is expressed as a generalized version of Bernoulli's theorem (this may be derived by taking the dot product of the equation of motion with $\mathbf{B_p}$). Thus, there is a specific energy e(a) that is a constant along field lines, which may be found in many papers (e.g., *BP82*; *PP92*). Since the terminal speed $v_p = v_\infty$ of the disk wind is much greater than its rotational speed, and for cold flows, the pressure may also be ignored, one finds the result

$$v_\infty \simeq 2^{1/2}\Omega_0 r_A = (r_A/r_0)v_{esc,0} \qquad (12)$$

There are three important consequences for jet kinematics here: (1) the terminal speed exceeds the local escape speed from its launch point on the disk by the lever arm ratio; (2) the terminal speed scales with the Kepler speed

as a function of radius, so that the flow will have an onion-like layering of velocities, the largest inside, and the smallest on the larger scales, as seen in the observations; and (3) the terminal speed depends on the depth of the local gravitational well at the footpoint of the flow — implying that it is essentially scalable to flows from disks around YSOs of any mass and therefore universal.

Another useful form of the conservation laws is the combination of energy and angular momentum conservation to produce a new constant along a field line (e.g., *PP92*); $j(a) \equiv e(a) - \Omega_0 l(a)$. This expression has been used (*Anderson et al.*, 2003) to deduce the rotation rate of the launch region on the Kepler disk, where the observed jet rotation speed is $v_{\phi,\infty}$ at a radius r_∞ and which is moving in the poloidal direction with a jet speed of $v_{p,\infty}$. Evaluating j for a cold jet at infinity and noting that its value (calculated at the foot point) is $j(a_0) = -(3/2)v_{K,0}^2$, one solves for the Kepler rotation at the point on the disk where this flow was launched

$$\Omega_0 = v_{p,\infty}^2/(2v_{\phi,\infty}r_\infty) \qquad (13)$$

When applied to the observed rotation of the large velocity component (LVC) of the jet DG Tau (*Bacciotti et al.*, 2002); this yields a range of disk radii for the observed rotating material in the range of disk radii, 0.3–4 AU, and the magnetic lever arm is $r_A/r_0 \simeq 1.8$–2.6.

2.2. Angular Momentum Extraction

How much angular momentum can such a wind extract from the disk? The angular momentum equation for the accretion disk undergoing an external magnetic torque may be written

$$\dot{M}_a \frac{d(r_0v_0)}{dr_0} = -r_0^2 B_\phi B_z|_{r_0,H} \qquad (14)$$

where we have ignored transport by MRI turbulence or spiral waves. By using the relation between poloidal field and outflow on the one hand, as well as the link between the toroidal field and rotation of the disk on the other, the angular momentum equation for the disk yields one of the most profound scaling relations in disk wind theory — namely, the link between disk accretion and mass outflow rate (for details, see *KP00*; *PP92*)

$$\dot{M}_a \simeq (r_A/r_0)^2 \dot{M}_w \qquad (15)$$

The observationally well known result that in many systems, $\dot{M}_w/\dot{M}_a \simeq 0.1$ is a consequence of the fact that lever arms are often found in numerical and theoretical work to be $r_A/r_0 \simeq 3$ — the observations of DG Tau being a perfect example. Finally, we note that the angular momentum that is observed to be carried by these rotating flows (e.g., DG Tau) is a consistent fraction of the excess disk angular momentum — from 60 to 100% (e.g., *Bacciotti*, 2004), which is consistent with the high extraction efficiency discussed here.

2.3. Jet Power and Universality

These results can be directly connected to the observations of momentum and energy transport in the molecular outflows. Consider the total mechanical power that is carried by the jet, which may be written as (e.g., *Pudritz,* 2003)

$$L_{jet} = \frac{1}{2} \int_{r_i}^{r_j} dM_w v_\infty^2 \simeq \frac{GM_* \dot{M}_a}{2r_i} \left(1 - \frac{r_i^2}{r_j^2} \right) \simeq \frac{1}{2} L_{acc} \quad (16)$$

This explains the observations of Class 0 outflows wherein $L_w/L_{bol} \simeq 1/2$, since the main luminosity of the central source at this time is due to accretion and not nuclear reactions. (The factor of 1/2 arises from the dissipation of some accretion energy as heat at the inner boundary.) The ratio of wind to stellar luminosity decreases at later stages because the accretion luminosity becomes relatively small compared to the bolometric luminosity of the star as it nears the ZAMS.

This result states that the wind luminosity taps the gravitational energy release through accretion in the gravitational potential of the central object — and is a direct consequence of Bernoulli's theorem. This, and the previous results, imply that jets may be produced in any accreting system. The lowest mass outflow that has yet been observed corresponds to a proto-brown dwarf of luminosity $\simeq 0.09\ L_\odot$, a stellar mass of only 20–45 M_{Jup}, and a very-low-mass disk $<10^{-4}\ M_\odot$ (*Bourke et al.,* 2005).

It should be possible, therefore, for lower-luminosity jets to be launched from the disks around jovian-mass planets. Recent hydrodynamical simulations of circumstellar accretion disks containing and building up an orbiting protoplanetary core have numerically proven the existence of a circumplanetary subdisk in almost Keplerian rotation close to the planet (*Kley et al.,* 2001). The accretion rate of these subdisks is about $\dot{M}_{cp} = 6 \times 10^{-5}\ M_{Jup}\ yr^{-1}$ and is confirmed by many independent simulations. With that, the circumplanetary disk temperature may reach values up to 2000 K, indicating a sufficient degree of ionization for matter-field coupling and would also allow for strong equipartition field strength (*Fendt,* 2003).

The possibility of a planetary-scale MHD outflow, similar to the larger-scale YSO disk winds, is indeed quite likely because of (1) the fact that the numerically established existence of circumplanetary disks is a natural feature of the formation of massive planets; and (2) the feasibility of a large-scale magnetic field in the protoplanetary environment (*Quillen and Trilling,* 1998; *Fendt,* 2003). One may show, moreover, that the outflow velocity is on the order of the escape speed for the protoplanet, at about 60 km s⁻¹ (*Fendt,* 2003).

On very general grounds, disk winds are also likely to be active during massive star formation (e.g., *Königl,* 1999). Such outflows may already start during the early collapse phase when the central YSO still has only a fraction of a solar mass (e.g., *Pudritz and Banerjee,* 2005). Such early outflows may actually enhance the formation of massive stars via disk accretion by punching a hole in the infalling envelope and releasing the building radiation pressure (e.g., *Krumholz et al.,* 2005).

2.4. Jet Collimation

In the standard picture of hydromagnetic winds, collimation of an outflow occurs because of the increasing toroidal magnetic field in the flow resulting from the inertia of the gas. Beyond the Alfvén surface, equation (8) shows that the ratio of the toroidal field to the poloidal field in the jet is on the order of $B_\phi/B_p \simeq r/r_A \gg 1$, so that the field becomes highly toroidal. In this situation, collimation is achieved by the tension force associated with the toroidal field, which leads to a radially inward directed component of the Lorentz force (or "z-pinch"); $F_{Lorentz,r} \simeq j_z B_\phi$. The stability of such systems is examined in the next section.

In *Heyvaerts and Norman* (1989) it was shown that two types of solution are possible depending upon the asymptotic behavior of the total current intensity in the jet

$$I = 2\pi \int_0^r j_z(r', z') dr' = (c/2) r B_\phi \quad (17)$$

In the limit that $I \to 0$ as $r \to \infty$, the field lines are paraboloids that fill space. On the other hand, if the current is finite in this limit, then the flow is collimated to cylinders. The collimation of a jet therefore depends upon its current distribution — and hence on the radial distribution of its toroidal field — which, as we saw earlier, depends on the mass load. Mass loading therefore must play a very important role in controlling jet collimation.

It can be shown (*Pudritz et al.,* 2006) (*PRO*) that for a power-law distribution of the magnetic field in the disk, $B_z(r_0, 0) \propto r_0^{\mu - 1}$, and an injection speed at the base of a (polytropic) corona that scales as the Kepler speed, that the mass load takes the form $k \propto r_0^{1 - \mu}$. In this regime, the current takes the form $I(r, z) \propto r_0^{\mu - (1/2)}$. Thus, the current goes to zero for models with $\mu < -1/2$, and that these therefore must be wide-angle flows. For models with $\mu > -1/2$, however, the current diverges, and the flow should collimate to cylinders.

These results predict that jets should show different degrees of collimation depending on how they are mass loaded (*PRO*). As an example, neither the highly centrally concentrated, magnetic field lines associated with the initial split-monopole magnetic configuration used in simulations by *Romanova et al.* (1997), nor the similar field structure invoked in the X-wind (see review by *Shu et al.,* 2000) should become collimated in this picture. On the other hand, less centrally (radially) concentrated magnetic configurations such as the potential configuration of *Ouyed and Pudritz* (1997a) (*OPI*) and *BP82* should collimate to cylinders.

This result also explains the range of collimation that is observed for molecular outflows. Models for observed outflows fall into two general categories: the jet-driven bow-shock picture, and a wind-driven shell picture in which the molecular gas is driven by an underlying wide-angle wind component such as given by the X-wind (see review by *Cabrit et al., 1997*). A survey of molecular outflows by *Lee et al.* (2000) found that both mechanisms are needed in order to explain the full set of systems observed.

Finally, we note that apart from these general theorems on collimation, *Spruit et al.* (1997) has proposed that a sufficiently strong poloidal field that is external to the flow could also force its collimation. According to this criterion, such a poloidal field could collimate jets provided that the field strength decreases no slower than $B_p \sim r^{-\mu}$ with $\mu \leq 1.3$.

3. SIMULATIONS: DISKS AS JET ENGINES

Computational approaches are necessary if we are to open up and explore the vast spaces of solutions to the highly nonlinear jet problem. The first simulations of non-steady MHD jets from accretion disks were performed by *Uchida and Shibata* (1985) and *Shibata and Uchida* (1986). These early simulations were based on initial states that were violently out of equilibrium and demonstrated the role of magnetic fields in launching and accelerating jets to velocities on the order of the Keplerian velocity of the disk.

The published simulations of nonradiative ideal MHD YSO jets differ in their assumed initial conditions, such as the magnetic field distribution on the disk, the conditions in the plasma above the disk surfaces, the state of the initial disk corona, and the handling of the gravity of the central star. Nevertheless, they share common goals: to establish and verify the four important stages in jet evolution, namely, (1) ejection, (2) acceleration, (3) collimation, and (4) stability. In the following, we describe a basic physical approach for setting up clean numerical simulations and how this leads to advances in our understanding of disk winds and outflows.

3.1. Two-Dimensional Simulations

The simulations we discuss here have been studied in greater detail in *Ouyed and Pudritz* (1997a) (*OPI*); *Ouyed and Pudritz* (1997b) (*OPII*); *Ouyed et al.* (1997) (*OPS*); *Ouyed and Pudritz* (1999) (*OPIII*); and *PRO*. We use these for pedagogical purposes in this review. To see animations of the simulations presented here as well as those performed by other authors, the interested reader is directed to the "animations" link at *www.capca.ucalgary.ca* (*CAPCA*). These simulations were run using the ZEUS (two-dimensional, three-dimensional, and MP) code, which is arguably the best-documented and utilized MHD code in the literature (*Stone and Norman*, 1992, 1994). It is an explicit, finite difference code that runs on a staggered grid. The equations generally solved are those of ideal MHD listed in section 1, equations (2)–(6).

The evolution of the magnetic field (induction equation (4) above) is followed by the method of constrained transport. In this approach, if $\nabla \cdot \mathbf{B} = 0$ holds for the initial magnetic configuration, then it remains so for all later times to machine accuracy. The obvious way of securing this con-

dition is to use an initial vector potential $\mathbf{A}(\mathbf{r}, \mathbf{z}, t = 0)$ that describes the desired initial magnetic field at every point in the computational domain.

To ensure a stable initial state that allows for a tractable simulation, three simple rules are useful (OPS and OPI): (1) use a corona that is in hydrostatic balance with the central object (to ensure a perfectly stable hydrostatic equilibrium of the corona, the point-mass gravitational potential should be relocated to zone centers; see section 4.2 in *OPI* for details); (2) put the disk in pressure balance with the corona above it, and (3) use a force-free magnetic field configuration to thread the disk and corona. The initial magnetic configurations should be chosen so that no Lorentz force is exerted on the initial (nonrotating) hydrostatic corona described above.

The *BP82* self-similar solution for jets uses a simple polytropic equation of state, with an index $\gamma = 5/3$. Their solution is possible because this choice eliminates the many complications that arise from the energy equation, while preserving the essence of the outflow problem. This Ansatz corresponds to situations where the combined effects of heating and cooling simulates a tendency toward a locally constant entropy. This choice also greatly simplifies the numerical setup and allows one to test the code against analytic solutions.

Another key simplification in this approach is to examine the physics of the outflow for fixed physical conditions in the accretion disk. Thus, the accretion disk at the base of the corona — and in pressure balance with the overlying atmosphere — is given a density profile that it maintains at all times in the simulation, since the disk boundary conditions are applied to the "ghost zones" and are not part of the computational domain. In part, this simplification may be justified by the fact that typically, accretion disks will evolve on longer timescales than their associated jets.

The hydrostatic state one arrives at has a simple analytic solution that was used as the initial state for all of our simulations and was adopted and further developed by several groups. These studies include, e.g., low plasma-β monopole-like field distributions (*Romanova et al.,* 1997); a magneto-gravitational switch mechanism (*Meier et al.,* 1997); the disk accretion-ejection process (*Kudoh et al.,* 1998); a self-adjusting disk magnetic field inclination (*Krasnopolsky et al.,* 1999, 2003); the interrelation between the jet's turbulent magnetic diffusivity and collimation (*Fendt and Čemeljić,* 2002); a variation of the disk rotation profile (*Vitorino et al.,* 2002); or dynamo-maintained disk magnetic fields (*von Rekowski and Brandenburg,* 2004). More elaborate setups wherein the initial magnetic field configuration originates on the surface of a star and connects with the disk have also been examined (section 5) (e.g., *Hayashi et al.,* 1996; *Goodson et al.,* 1997; *Fendt and Elstner,* 1999; *Keppens and Goedbloed,* 2000).

The lesson from these varied simulations seems to be that all roads lead to a disk field. Thus, the twisting of a closed magnetic field that initially threads the disk beyond the co-rotation radius rapidly inflates the field and then disconnects it from the star, thereby producing an open disk-field line.

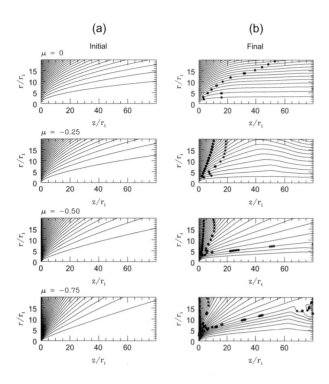

Fig. 1. (a) Initial magnetic field configurations for winds with $\mu = 0$ (OPI), $\mu = -0.25$ (BP), $\mu = -0.5$ (PP), and $\mu = -0.75$ (steep). (b) Final magnetic field configurations (at t = 400) for each case, with Alfvén points (filled circles) and fast magnetosonic points (stars) marked. Note the more open magnetic — and streamline — structures as μ goes down. Adapted from *PRO*.

Likewise, simulations including dynamo-generated fields in the disk lead to a state that resembles our initial setup (see section 4). Thus, from the numerical point of view, a "fixed-disk" simulation is general and useful.

The setup requires five physical quantities to be specified at all points of the disk surface at all times (see *OPI* for all details). These are the disk density $\rho(r_0)$; components of the vertical and toroidal magnetic field, $B_z(r_0)$ and $B_\phi(r_0)$; and velocity components in the disk, $v_z(r_0)$ and $v_\phi(r_0)$, where r_0 is the radius [cylindrical coordinates are adopted with the disk located at z = 0; $r_0 = r(z = 0)$]. The remaining field component $B_r(r_0)$ is determined by the solenoidal condition, while the radial inflow speed through the disk is neglected [$v_r(r_0) \simeq 0$] since it is far smaller than the sound speed in a real disk. The model is described by five parameters defined at the inner disk radius r_i, three of which describe the initial corona. The two additional parameters describe the disk physics and are v_i, which scales the toroidal field in the disk $B_\phi = v_i \times (r_i/r_0)$, and the (subsonic) injection speed of material from the disk into the base of the corona, $v_{inj} = v_z(r_0)/v_\phi(r_0) \simeq 0.001$. Simulations were typically run with (500 × 200) spatial zones to resolve a physical region of (80 r_i × 20 r_i) in the z and r directions, respectively. A resolution of 10 zones per r_i provides enough dynamical range to accurately follow the smooth acceleration above the disk surface. The simulations were run up to 400 t_i (where t_i is the Kepler time for an orbit at the inner edge of the disk).

A series of magnetic configurations is shown in Fig. 1 where the disk field is modeled as $B_z(r_0, 0) \propto r_0^{\mu - 1}$. The initial configurations range from the rather well collimated (such as the potential configuration of *OP97*, $\mu = 0.0$; and the Blandford-Payne configuration, $\mu = -0.25$), to the initially more open configurations of Pelletier-Pudritz (*PP92*; $\mu = -0.5$), and steeper ($\mu = -0.75$).

A good example of a configuration that evolves into a stationary jet was studied by *OPI* and is shown in the upper panels of Fig. 1). The simulation shows the existence of an acceleration region very close to the disk surface. The acceleration from the disk occurs by a centrifugal effect whereby, at some point along sufficiently inclined field lines, centrifugal force dominates gravity and gas is flung away like beads on a wire. Thus, a toroidal field component is created because the field lines corotate with the underlying disk. The inertia of the matter in the flow region ultimately forces the field to fall behind the rotation of the disk, which produces the toroidal field component. More precisely, beyond the Alfvén surface (shown as dots in Fig. 1b), the hoop stress induced by the self-generated B_ϕ eventually dominates, which provides the collimation. The ratio of the toroidal to poloidal magnetic field along illustrative field lines (e.g., Fig. 3 in *OPS*) clearly shows that the predominant magnetic field in jets beyond their fast magnetosonic (FM) surfaces is the toroidal field component. The gas is eventually collimated into cylinders parallel to the disk's axis starting at the Alfvén surface to distances much beyond the FM surface (see Fig. 1).

The final velocities achieved by such winds are of the order of 2× the Kepler velocity, along a given field line (e.g., Fig. 3 in *OPS*), which translates to roughly 100–300 km s^{-1} for a standard YSO. The general trend is that the jet solutions, dominated mainly by the poloidal kinetic energy, are very efficient in magnetically extracting angular momentum and energy from the disk, as confirmed by simulations performed by other groups. In 1000 years, for example, and for a standard YSO, the simulations imply that the disk winds can carry a total energy of 3×10^{43} ergs, sufficient to produce the observed molecular outflows

In *OPII* an initial vertical field configuration was used. The simulations show that the strong toroidal magnetic field generated recollimates the flow toward the disk's axis and, through MHD shocks, produces knots. The knot generator is located at a distance of about $z \simeq 8r_i$ from the surface of the disk (*OPS*). Knots propagate down the length of the jet at speeds less than the diffuse component of the outflow. The knots are episodic, and are intrinsic to the jet and not the accretion disk, in this calculation.

A different initial state was used by *Krasnopolsky et al.* (1999), who introduce and maintain throughout the simulation a strong outflow on the outflow axis. Otherwise, they choose an initial disk field distribution that is the same as *PP92*, a mass flux density $\rho v_z \propto r_0^{3/2}$, so that k = const. The values of B_r and B_ϕ at the disk surface were not fixed in their simulations. (This does not guarantee that the disk or the boundary at the base will remain Keplerian over time.) The fixed $B_\phi \propto r_0^{-1}$ profile adopted in the *OP* setup ensures exactly that — see the discussion around equation (3.46) in *OPI*. To evade the problems with the initial setup mentioned above, these authors continuously launch a cylindrical wind on the axis. This imposed jet introduces currents that could significantly affect the stability and collimation of the disk wind. Their disk mass loading would predict that the disk wind should not be well collimated (see section 2.4), whereas their simulation does appear to collimate. This suggests that their on-axis jet may be playing a significant role in the simulation results.

3.2. The Role of Mass Loading

The mass-load k, defined in equation (8), can be established by varying the coronal density profile while keeping ρv_p constant (*Anderson et al.*, 2005), the disk rotational profile since $v_p = v_{inj} v_\phi$ (e.g., *Vitorino et al.*, 2002), the distribution of the poloidal magnetic field on the disk (*PRO*), or the distribution of both the disk magnetic field and the mass flow profile (*Fendt*, 2006).

The prediction that the mass load determines the collimation of the jet was tested in simulations by *PRO*. Figure 1 shows that simulations with $\mu = 0$, –0.25 collimate into cylinders, while the $\mu = -0.5$ case (*PP92*) transitions toward the wide-angle flow seen in the $\mu = -0.75$ simulation. These results confirm the theory laid out in section 2.4. We also note that each of the simulations mentioned above results in a unique rotation profile of the jet that might in principle be observable. Going from the potential case to the Pelletier-Pudritz configuration, the radial profiles of the rotational velocity of the jets scale as power laws

$$v_\phi(r, \infty) \propto r^a \qquad (18)$$

where a = –0.76; –0.66; –0.46 respectively. The observation of a rotational profile would help pick out a unique mass loading in this model (*PRO*).

The mass loading also affects the time-dependent behavior of jets. *As the mass load is varied, a transition from stationary, to periodic, and sometimes to a discontinuous dying jet occurs.* For example, in some of the simulations it was found that low-mass loads for jets lead to rapid, episodic behavior while more heavily mass-loaded systems tend to achieve stationary outflow configurations (see *OPIII*).

It is interesting that the simulations of *Anderson et al.* (2005) did not find any nonsteady behavior for *low*-mass loading (differently defined than equation (8)), but instead find an instability for *high*-mass loading. They suggest that the origin of the nonsteady behavior for high-mass loading is that the initially dominant toroidal field they impose could be subject to the kink instability (e.g., *Cao and Spruit*, 1994). The very large mass loads and large injection speeds [$v_{inj} = (0.01, 0.1)$] they use drive an instability probably related to excessive magnetic braking. Some of the main differences with *OP* are (1) the use of the nonequilibrium set up of *Krasnopolsky et al.* (1999), which introduces a strong current along the axis that could strongly affect the stability properties of these outflows; and (2) the large injection

speeds make the sonic surface too close to the disk (the condition that $c_s \ll v_K$ is not satisfied) and does not provide enough dynamical range for the gas launched from the disk to evolve smoothly. Numerical instabilities reminiscent of what was found by *Anderson et al.* (2005) were observed by *OP* and these often disappear with sufficiently high resolution (i.e., dynamical range). We suspect that this type of instability will disappear when a proper disk is included in the simulations (e.g., *Casse and Keppens*, 2002).

3.3. Three-Dimensional Simulations

The stability of jets is one of the principal remaining challenges in the theory. It is well known that the purely toroidal field configurations that are used to help confine static, three-dimensional, tokamak plasmas are unstable (e.g., *Roberts*, 1967; *Bateman*, 1980). The resulting kink, or helical (m = 1) mode instability, derived from a three-dimensional linear stability analysis is powered by the free energy in the toroidal field, namely $B_\phi^2/8\pi$ (*Eichler*, 1993). Why are real three-dimensional jets so stable over great distances in spite of the fact that they are probably threaded by strong toroidal fields?

The three-dimensional simulations needed to investigate the importance of kink modes are still rare. Early attempts include *Lucek and Bell* (1996) and *Hardee and Rosen* (1999 and references therein) who performed three-dimensional simulations of "equilibrium" jets, and found that these uniform, magnetized jet models remain Kelvin-Helmholtz (K-H) stable to low-order, surface helical and elliptical modes (m = 1, 2), provided that jets are on average sub-Alfvénic. This is in accord with the prediction of linear stability analysis. However, most configurations for jet simulations use rather *ad hoc* prescriptions for the initial toroidal field configuration so that it is difficult to assess how pertinent the results are to the case of a jet that establishes its own toroidal field as the jet is accelerated from the accretion disk. In general, the available analytic and numerical results for the stability of simple jets show that the fastest-growing modes are of K-H type. These K-H instabilities are increasingly stabilized for super-Alfvénic jets, as M_{FM} is increased much beyond unity. It is also generally known that sub-Alfvénic jets are stable. Taken together, these results suggest that three-dimensional jets are the most prone to K-H instabilities a bit beyond their Alfvén surface, a region wherein their destabilizing super-Alfvénic character cannot yet be offset by the stabilizing effects engendered at large super-FM numbers.

There have, as yet, only been a few attempts to simulate three-dimensional disk winds. Contrary to what may be intuitive, it is inadvisable to perform these three-dimensional simulations in cylindrical coordinates. For one thing, special treatment must be given to the "wedge zones" that abut the z-axis (no longer a symmetry axis in three dimension), and velocities that pass through the z-axis pose a very difficult numerical problem. Second, even with such technical details in hand, plane waves are badly disrupted on passing through the z-axis, and this provides an undesirable bias to what should be an unbiased (no axis should be preferred over another) three-dimensional calculation. Fortunately, using Cartesian coordinates to simulate cylinders is feasible with careful setups, as has been demonstrated in *Ouyed et al.* (2003) (*OCP*); and *Ouyed* (2003). These authors used a three-dimensional version of the *OPI* setup.

The central result of this study is that jets survive the threatening nonaxisymmetric kink (m = 1) mode. The simulated jets maintain their long-term stability through a self-limiting process wherein the average Alfvénic Mach number within the jet is maintained to order unity. This is accomplished in at least two ways: (1) Poloidal magnetic field is concentrated along the central axis of the jet, forming a "backbone" in which the Alfvén speed is sufficiently high to reduce the average jet Alfvénic Mach number to unity; and (2) the onset of higher-order Kelvin-Helmholtz "flute" modes (m ≥ 2) reduce the efficiency with which the jet material is accelerated, and transfer kinetic energy of the outflow into the stretched, poloidal field lines of the distorted jet. This too has the effect of increasing the Alfvén speed, and thus reducing the Alfvénic Mach number. The jet is able to survive the onset of the more destructive m = 1 mode in this way. It was further discovered that jets go into alternating periods of low and high activity as the disappearance of unstable modes in the sub-Alfvénic regime enables another cycle of acceleration to super-Alfvénic speeds. The period of the episodic (low vs. high) behavior uncovered in the case of the wobbling jet (Fig. 2a) is on the order of 200 t_i. For a typical young stellar object of mass 0.5 M$_\odot$ and for $r_i \sim 0.05$ AU (i.e., $t_i \sim 0.9$ d), this would correspond to a minimum period of roughly 180 d. This is consistent with the temporal variations of the observed YSOs outflow velocities, which appears to occur on timescales of a few years (*Woitas et al.*, 2002).

Jets ultimately settle into relatively stable endstates involving either corkscrew (Fig. 2b) or wobbling types of structure (Fig. 2a). The difference in the two type of jets can be traced to the difference in the v_ϕ profiles imposed at the accretion disk (see *OCP* for more details). This trend has been uncovered in three-dimensional simulations of large-scale jets where wiggled jets have also been observed after the jets survived the destructive era (*Uchida et al.*, 1992; *Todo et al.*, 1993; *Nakamura et al.*, 2001; *Nakamura and Meier*, 2004). For completeness, we should mention the recent similar simulations performed by *Kigure and Shibata* (2005), which included the disk self-consistently. These studies also show that nonaxisymmetric perturbations dominate the dynamics. However, these simulations were only run for very short times, never exceeding two inner-disk orbital periods due to numerical obstacles reported by the authors. In *OCP* simulations, it was shown that the Cartesian grid (also used by Kigure and Shibata) induces artificial modes in the early stages of the simulations and only after many orbital periods do these become washed out before the real modes enter the dynamics. Another suggested endstate for three-dimensional jets is a minimum-energy "Taylor state," wherein the jet maintains comparable poloidal and toroidal field components (*Königl and Choudhuri*, 1985). This alternative does not pertain to our simplified simula-

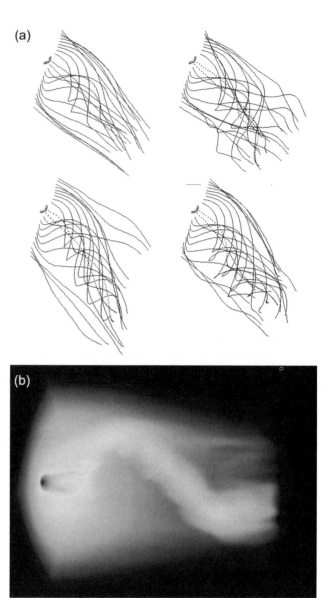

Fig. 2. (a) Cylindrical jets: Snapshots of 20 magnetic field lines of the three-dimensional simulations performed in *OCP*, at t = 50, 130, 210, and 240. The two central magnetic field lines (dotted lines) originate on the central compact object (illustrated by the semisphere to the left). The disk axis is along the diagonal of the frame (on a 45° angle). Notice the violent jet behavior in the in-between frames when the kink mode appears. The jet eventually aligns itself with the original disk's axis, acquiring a cylindrically collimated shape. (b) Corkscrew jets: Map of the jet column density at t = 320 for a simulation with a v_f profile, imposed at the accretion disk, that is different from the simulation shown in the upper four panels (see *OCP*). The jet has settled into a quasi-steady-state structure in the shape of a "corkscrew" (grayscale arranged from white to black to represent low and high values of the density). The disk (not visible) is on the lefthand side of the image, and outflow is from left to right. Adapted from *OCP*.

tions since we do not allow for internal magnetic energy dissipation.

In summary, a large body of different simulations have converged to show that jets (1) are centrifugally disk winds; (2) are collimated by the "hoop stress" engendered by their toroidal fields; (3) achieve two types of configuration depending on the mass loading that takes place in the underlying accretion disk — those that achieve collimation toward a cylinder, and those that have a wide-angle structure; (4) achieve two types of regimes depending on the mass loading that takes place in the underlying accretion disk — those that achieve a stationary state, and those that are episodic; (5) achieve stability by a combination of MHD mode coupling and "backbone" effect leading to a self-regulating mechanism that saturates the instabilities; and (6) achieve different morphologies ranging from cylindrical wobbling to corkscrew structure, depending on the profile of v_ϕ imposed on the underlying accretion disk.

4. COUPLED DISK-JET EVOLUTION

While it is physically useful to regard the accretion disk as a boundary condition for the jet, this ignores critical issues such as the self-consistent radial distribution of the mass loading, or of the threading magnetic field across the disk. These and other degrees of freedom can be found by solving the combined disk-outflow problem, to which we now turn.

Significant progress on the launch of disk winds has been made by theoretical studies of a restricted class of self-similar two-dimensional models (e.g., *Wardle and Königl,* 1993; *Li,* 1995, 1996; *Ferreira,* 1997; *Casse and Ferreira,* 2000). Since centrifugally driven outflows can occur for completely cold winds (*BP82*), the models have examined both "cold" and "warm" (i.e., with some kind of corona) conditions. In the former class, material from some height above the disk midplane must move upward and eventually be accelerated outward in a jet. Therefore, there must be a direct link between the physics of magnetized disks and the origin of outflows.

Calculations show that in order to match the properties of jets measured by *Bacciotti et al.* (2000), warm wind solutions are preferred, wherein a disk corona plays a central role. In this situation, gas pressure imbalance will assist with feeding the jet. A warm disk corona is expected on general grounds because it is a consequence of the magnetorotational instability (MRI), as the vertically resolved disk simulations of *Miller and Stone* (2000) have shown.

A semianalytic model of the radial and vertical disk structure, which includes self-consistently the outflow, has also been presented by *Campbell* (2003). These solutions demonstrate explicitly that the outflow contributes to the loss of angular momentum in the disk by channeling it along field lines into the outflow. For self-consistent solutions to be possible, the turbulent Mach number has to be between 0.01 and 0.1.

Models of outflows and jets generally assume ideal (nonresistive) MHD. However, inside the disk nonideal effects must become important, because the accreting matter would otherwise never be able to get onto the field lines that thread the disk and connect it with the outflow. In recent two-dimensional models, *Casse and Keppens* (2002, 2004) assumed a resistivity profile, analogous to the Shakura and Sunyaev prescription. This assumes the presence of an underlying turbulence within the disk. Only fairly large values of the corresponding α_{SS} parameter of around 0.1 have been used. The subscript SS refers to *Shakura and Sunyaev* (1973), who were the first to introduce this viscosity parameter. In all cases the system evolves to a steady equilibrium. Using resistive simulations in a different context, *Kuwabara et al.* (2005) demonstrated that a substantial amount of energy can be transported by Poynting flux if the poloidal field falls off with distance no faster than r^{-2}. Otherwise, the fast magnetosonic point is located closer to the Alfvén point and the jet will be dominated by kinetic energy, which is the case in the simulations of *Casse and Keppens* (2004).

The general stability of disk-outflow solutions is still being debated and the result may depend on the detailed assumptions about the model. In the solutions discussed here the accretion stress comes entirely from the large-scale magnetic field rather than some small-scale turbulence. As emphasized by *Ferreira and Casse* (2004), real disks have a turbulent viscosity just as they have turbulent magnetic diffusivity or resistivity. However, if the accretion stress does come entirely from the large-scale magnetic field, the wind-driven accretion flows may be unstable (*Lubow et al.,* 1994). *Königl* (2004) has shown recently that there are in fact two distinct solution branches: a stable and an unstable one. He argues that real disk-wind systems would correspond to the stable branch.

Finally, the idea of a gently flared accretion disk is likely to be merely the result of the modeler's simplification rather than observational reality. Indeed, accretion disks can be warped due to various instabilities that can be driven by radiation from the central object (e.g., *Pringle,* 1996) or, more likely, by the outflow itself (*Schandl and Meyer,* 1994). If a system is observed nearly edge-on, a warp in the accretion disk produces periodic modulations of the light curve. As *Pinte and Ménard* (2004) have demonstrated, this may be the case in AA Tau. Observations frequently reveal major asymmetries in bipolar outflows, which may be traced back to the corresponding asymmetries in the disk itself. The possible causes for these asymmetries may be either an externally imposed asymmetry such as one-sided heating by a nearby OB association, or an internal symmetry breaking of the disk-wind solution as a result increased rotation. Examples of the latter are familiar from the study of mean field dynamo solutions of accretion disks (*Torkelsson and Brandenburg,* 1994).

4.1. Dead Zones

So far it has been assumed that the magnetic field is well coupled to the disk. This is certainly valid for most of the envelope of an accretion disk, which is ionized by a combination of cosmic rays, as well as X-rays from the central YSO. However, the deeper layers of the disk are strongly shielded from these ionizing agents and the degree of ionization plumets.

This dense layer, which encompasses the bulk of the disk's column density, cannot maintain a sufficiently high electron fraction, and is referred to as the dead zone (*Gam-*

mie, 1996). It is the poorly ionized region within which the MRI fails to grow as a consequence of the diffusivity of the field. Recent work of *Fleming and Stone* (2003) shows that, although the local Maxwell stress drops to negligible values in the dead zones, the Reynolds stress remains approximately independent of height and never drops below approximately 10% of the maximum Maxwell stress, provided the column density in that zone is less that 10× the column of the active layers. The nondimensional ratio of stress to gas pressure is just the viscosity parameter, α_{SS}. *Fleming and Stone* (2003) find typical values of a few times 10^{-4} in the dead zones and a few times 10^{-3} in the MRI-active layers. *Inutsuka and Sano* (2005) have questioned the very existence of dead zones, and proposed that the turbulent dissipation in the disk provides sufficient energy for the ionization. On the other hand, their calculation assumes a magnetic Reynolds number that is much smaller than expected from the simulations (see *Matsumura and Pudritz,* 2006) (*MP06*).

The radial extent of the dead zone depends primarily upon the disk column density as well as effects of grains and radiation chemistry [e.g., *Sano et al.,* 2000; *Matsumura and Pudritz,* 2005 (*MP05*); *MP06*]. Inside the dead zone the magnetic Reynolds number tends to be below a certain critical value that is somewhere between 1 and 100 (*Sano and Stone,* 2002), making MRI-driven turbulence impossible. Estimates for the radial extent of the dead zone range from 0.7 to 100 AU (*Fromang et al.,* 2002) to 2–20 AU in calculations by *Semenov et al.* (2004). For *Chiang et al.* (2001) models of disks that are well constrained by the observations, and whose surface density declines as $\Sigma \propto r_0^{-3/2}$, *MP05* find that the extent of the dead zone is robust — typically extending out to 15 AU, and is fairly independent of the ionizing environment of the disk.

For smaller radii, thermal and UV ionization are mainly responsible for sustaining some degree of ionization. The significance of the reduced value of α_{SS} in the dead zones is that it provides a mechanism for stopping the inward migration of Jupiter-sized planets (*MP05*; *MP06*). Jets can still be launched from the well-coupled surface layer above the dead zone (e.g., *Li,* 1996; *Campbell,* 2000).

When the MRI is inactive in the body of the disk, alternative mechanisms of angular momentum transport are still possible. In protostellar disks there are probably at least two other mechanisms that might contribute to the accretion torque: density waves (*Różyczka and Spruit,* 1993) and the interaction with other planets in the disk (*Goodman and Rafikov,* 2001).

A more controversial alternative is to drive turbulence by a nonlinear finite amplitude instability (*Chagelishvili et al.,* 2003). While *Hawley et al.* (1999) have presented general arguments against this possibility, *Afshordi et al.* (2005) and other groups have continued investigating the so-called bypass to turbulence. The basic idea is that successive strong transients can maintain a turbulent state in a continuously excited manner. *Lesur and Longaretti* (2005) have recently

been able to quantify more precisely the critical Reynolds number required for instability. They have also highlighted the importance of pressure fluctuations that demonstrate that the general argument by *Balbus et al.* (1996) is insufficient.

Finally, the role of vertical (convectively stable) density stratification in disks and the possibility of the so-called stratorotational instability has been proposed as a possible mechanism for disk turbulence (*Dubrulle et al.,* 2005). This instability was recently discovered by *Molemaker et al.* (2001) and the linear stability regime was analyzed by *Shalybkov and Rüdiger* (2005). However, the presence of noslip radial boundary conditions that are relevant to experiments and used in simulations are vital. Indeed, the instability vanishes for an unbounded regime, making it irrelevant for accretion disks (*Umurhan,* 2006).

4.2. Advected Versus Dynamo-Generated Magnetic Fields

One of the central unresolved issues in disk wind theory is the origin of the threading magnetic field. Is it dragged in by the gravitational collapse of an original magnetized core, or is it generated *in situ* by a disk-dynamo of some sort? The recent direct detection of a rather strong, true disk field of strength 1 kG at 0.05 AU in FU Ori provides new and strong support for the disk-wind mechanism (*Donati et al.,* 2005). The observation technique uses high-efficiency, high-resolution spectropolarimetry and holds much promise for further measurements. This work provides excellent evidence that distinct fields exist in disks in spite of processes such as ambipolar diffusion that might be expected to reduce them.

4.2.1. Turbulence effects. Accretion disks are turbulent in the well-coupled lower corona and beyond. They are also intrinsically three-dimensional and unsteady. Questions regarding the stability of disks concern therefore only the mean (azimuthally averaged) state, where all turbulent eddies are averaged out. The averaged equations used to obtain such solutions incorporate a turbulent viscosity. In this connection we must explain that the transition from a macroscopic viscosity to a turbulent one is more than just a change in coefficients, because one is really entering a new level of description that is uncertain in many respects. For example, in the related problem of magnetic diffusion it has been suggested that in MRI-driven turbulence, the functional form of turbulent magnetic diffusion may be different, such that it operates mainly on the system scale and less efficiently on smaller scales (*Brandenburg and Sokoloff,* 2002). It is therefore important to use direct simulations to investigate the stability in systems that might be unstable according to a mean field description. There is no good example relevant to protostellar disks, but related experience with radiatively dominated accretion disks, where it was found that the so-called Lightman-Eardly instability might not lead to the breakup of the disk (*Turner,* 2004), should provide enough reason to treat the mean-field stability problem of protostellar disks with caution. Also, one cannot exclude that circulation patterns found in simulations of the two-dimensional mean-field equations (e.g., *Kley and Lin*, 1992) may take a different form if the fully turbulent three-dimensional problem was solved.

The turbulent regions of disks will generally be capable of dynamo action, wherein an ordered field is generated by feeding on the energy that sustains the turbulence. These two aspects are in principle interlinked, as has been demonstrated some time ago using local shearing box simulations. These simulations show not only that the magnetic field is generated from the turbulence by dynamo action, but also that the turbulence itself is a consequence of the magnetic field via the MRI (*Brandenburg et al.*, 1995; *Stone et al.*, 1996).

In comparing numerical dynamos with observational reality, it is important to distinguish between two different types of dynamos: small-scale and large-scale dynamos (referring primarily to the typical physical scale of the field). Both types of dynamos have in general a turbulent component, but large-scale dynamos have an additional component on a scale that is larger than the typical scale of the turbulence. Physically, this can be caused by the effects of anisotropies, helicity, and/or shear. These large-scale dynamos are amenable to mean field modeling (see below). Small-scale dynamos, on the other hand, tend to be quite prominent in simulations, but they are not described by mean field models. However, small-scale dynamos could be an artifact of using unrealistically large magnetic Prandtl numbers in the simulations. Protostellar disks have magnetic Prandtl numbers, ν/η, around 10^{-8}, so the viscosity of the gas, ν, is much smaller than the diffusivity of the field, η; see Table 1 of *Brandenburg and Subramanian* (2005). High magnetic Prandtl numbers imply that the field is advected with the flow without slipping too much, whereas for low numbers the field readily slips significantly with respect to the flow. At the resistive scale, where the small-scale dynamo would operate fastest, the velocity is still in its inertial range where the spatial variation of the velocity is much rougher than for unit magnetic Prandtl number (*Boldyrev and Cattaneo*, 2004). This tends to inhibit small-scale dynamo action (*Schekochihin et al.*, 2005). In many simulations, especially when a subgrid scale prescription is used, the magnetic Prandtl number is effectively close to unity. As a consequence, the production of a small-scale field may be exaggerated. It is therefore possible that in real disks large-scale dynamo action is much more prominent than what is currently seen in simulations.

In mean field models, only the large-scale field is modeled. In addition to a turbulent magnetic diffusivity that emerges analogously with the turbulent viscosity mentioned above, there are also nondiffusive contributions such as the so-called α effect, wherein the mean electromotive force can acquire a component parallel to the mean field, $\alpha_{dyn}\overline{\mathbf{B}}$; see *Brandenburg and Subramanian* (2005) for a recent review. For symmetry reasons, α_{dyn} has on average opposite signs

above and below the midplane. Simulations of *Ziegler and Rüdiger* (2000) confirm the unconventional negative sign of α_{dyn} in the upper disk plane, which was originally found in *Brandenburg et al.* (1995). This has implications for the expected field parity, which is expected to be dipolar (antisymmetric about the midplane); see *Campbell et al.* (1998) and *Bardou et al.* (2001). Although a tendency toward dipolar fields is now seen in some global accretion disk simulations (*De Villiers et al.*, 2005), this issue cannot as yet be regarded as conclusive.

4.2.2. Outflows from dynamo-active disks. Given that the disk is capable of dynamo action, what are the relative roles played by ambient and dynamo-generated fields? This question has so far only been studied in limiting cases. The first global simulations were axisymmetric, in which case dynamo action is impossible by the *Cowling* (1934) theorem. Nevertheless, such simulations have demonstrated quite convincingly that a jet can be launched by winding up the ambient field and thereby driving a torsional Alfvén wave (*Matsumoto et al.*, 1996). These simulations are scale invariant and, although they were originally discussed in the context of active galaxies, after appropriate rescaling, the same physics applies also to stellar disk outflows.

When the first three-dimensional simulations became available an immediate issue was the demonstration that MRI and dynamo action are still possible (*Hawley*, 2000). Although the simulations were applied to model black hole accretion disks, many aspects of such models are sufficiently generic that they carry over to protostellar disks as well. These simulations showed that the dynamo works efficiently and produces normalized accretion stresses ($\alpha_{SS} \approx 0.1$) that exceed those from local simulations ($\alpha_{SS} \approx 0.01$). The global simulations did not at first seem to show any signs of outflows, but this was mainly a matter of looking at suitable diagnostics, which emphasizes the tenuous outflows rather than the much denser disk dynamics (*De Villiers et al.*, 2005). Although these simulations have no ambient field, a large-scale field develops in the outflow region away from the midplane. However, it is difficult to run such global simulations long enough to exclude a dependence of the large-scale field on the initial conditions. The outflows are found to be uncollimated such that most of the mass flux occurs in a conical shell with half-opening angle of 25°–30° (*De Villiers et al.*, 2005).

These very same properties are also shared by mean field simulations, where the three-dimensional turbulent dynamics is modeled using axisymmetric models (*von Rekowski et al.*, 2003). In these simulations (Fig. 3), where the field is entirely dynamo-generated, there is an uncollimated outflow with most of the mass flux occurring in a conical shell with a half-opening angle of about 25°–30°, just like in the three-dimensional black hole simulations. The field strength generated by dynamo action in the disk is found to scale as $B_p \propto r^{-2}$. This is very similar to the scaling for poloidal field strength found in collapse simulations (see *Banerjee and Pudritz*, 2006).

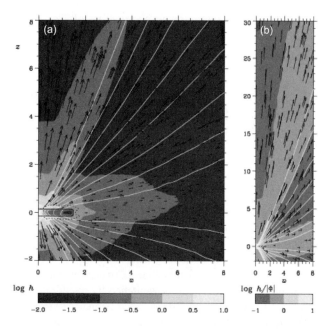

Fig. 3. Outflow from a dynamo active accretion disk driven by a combination of pressure driving and magnetocentrifugal acceleration. The extent of the domain is [0, 8] × [−2, 30] in nondimensional units (corresponding to about [0, 0.8] AU × [−0.2, 3] AU in dimensional units). **(a)** Velocity vectors, poloidal magnetic field lines, and grayscale representation of the specific enthalpy h in the inner part of the domain. (b) Velocity vectors, poloidal magnetic field lines, and normalized specific enthalpy h/|Φ| in the full domain. Adapted from *von Rekowski et al.* (2003).

The conical outflows discussed above have tentatively been compared with the observed conical outflows inferred for the BN/KL region in the Orion nebula out to a distance of 25–60 AU from its origin (*Greenhill et al.*, 1998). One may wonder whether collimated outflows are only possible when there is an ambient field (see section 3). This conclusion might be consistent with observations of the Taurus-Auriga molecular cloud. *Ménard and Duchêne* (2004) found that, although T Tauri stars are oriented randomly with respect to the ambient field, there are no bright and extended outflows when the axis is perpendicular to the ambient field. Note that *Spruit and Uzdensky* (2005) have argued that the efficiency of dragging an ambient field toward the star have been underestimated in the past, provided the field segregates into many isolated flux bundles.

In summing up this section we note that all magnetic disk models produce outflows of some form. However, the degree of collimation may depend on the possibility of an ambient field. An important ingredient that seems to have been ignored in all combined disk-jet models is the effect of radiation. There are also many similarities between the outflows in models with and without explicitly including the disk structure. Although these simulations are nonideal (finite viscosity and resistivity), the various Lagrangian invariants (mass loading parameter, angular velocity of mag-

netic field lines, as well as the angular momentum and Bernoulli constants) are still very nearly constant along field lines outside the disk.

5. JETS AND STAR-DISK INTERACTION

In this section we review the recent development concerning the interaction of young stellar magnetospheres with the surrounding accretion disk and the contribution that these processes may have to the formation of jets.

Perhaps the most important clue that star-disk interaction is an important physical process comes from the observation that many YSOs are observed to spin at only a small fraction ($\approx10\%$) of their breakup speed (e.g., *Herbst et al.*, 2002). They must therefore undergo a significant spindown torque that almost cancels the strong spin-up that arising from gas accretion from the inner edge of the Keplerian disk.

Three possible spin-down torques have been proposed: (1) disk-locking, a magnetospheric connection between the disk and the star that dumps accreted angular momentum back out into the disk beyond the co-rotation radius (*Königl*, 1991); (2) an X-wind, a centrifugally driven outflow launched from the very inner edge of the accretion disk that intercepts the angular momentum destined for the star (see chapter by Shang et al.); and (3) an accretion-powered stellar wind, in which the slowly rotating, central star drives a massive ($\dot{M}_{w,*} \simeq 0.1 \, \dot{M}_a$) stellar wind that carries off the bulk of the angular momentum as well as a fraction of the energy that is deposited into the photosphere by magnetospheric accretion (*Matt and Pudritz*, 2005b). The last possibility implies that the central object in an accretion disk also drives an outflow, which is in addition to the disk wind. Separating the disk wind from the outflow from the central object would be difficult, but the shear layer that separates them might generate instabilities and shocks that might be diagnostics.

The strength of the star-disk coupling depends on the strength of the stellar magnetic field. Precise stellar magnetic field measurements exist for nine TTSs, while for seven other stars statistically significant fields have been found (see *Symington et al.*, 2005b, and references therein). Zeeman circular-polarization measurements in HeI of several classical TTs give direct evidence of kG magnetic fields with strong indication of considerable field variation on timescales of years (*Symington et al.*, 2005b). For a sample of stars a magnetic field strength up to 4 kG fields could be derived (DF Tau, BP Tau); as for others, notably the jet source DG Tau, no significant field could be detected at the time of observation.

5.1. Basic Processes

5.1.1. Variable accretion from a tipped dipole. A central dipolar field inclined to the rotation axis of star and disk may strongly disturb the axisymmetry of the system. Photometric and spectroscopic variability studies of AA Tau give evidence for time-dependent magnetospheric accretion on timescales on the order of a month. Monte Carlo modeling shows that the observed photo-polarimetric variability may arise by warping of the disk that is induced by a tipped magnetic dipole (*O'Sullivan et al., 2005*). More general investigations of the warping process by numerical simulations show that the warp could evolve into a steady state, precessing rigidly (*Pfeiffer and Lai, 2004*). Disks can be warped by the magnetic torque that arises from the a slight misalignment between the disk and star's rotational axis (*Lai, 1999*). Disk warping may also operate in the absence of a stellar magnetosphere since it can be induced by the interaction between a large-scale magnetic field that is anchored in the disk and the disk electric current. This leads to a warping instability and to the retrograde precession of magnetic jets/outflows (*Lai, 2003*).

Three-dimensional radiative transfer models of the magnetospheric emission line profile based on the warped disk density and velocity distribution obtained by numerical MHD simulations give gross agreement with observations with a variability somewhat larger than observed (*Symington et al., 2005a*).

5.1.2. Magnetic flux. Compared to the situation of a pure disk magnetic field, the magnetic field of the star may add substantial magnetic flux to the system. For a polar field strength B_0 and a stellar radius $R_\star$, the large-scale stellar dipolar field (ignoring angular variations)

$$B_{p,\star}(r) \simeq 40 \, G \left(\frac{B_0}{1 \, kG} \right) \left(\frac{r}{3 \, R_\star} \right)^{-3} \tag{19}$$

has to be compared to the accretion disk poloidal magnetic field, which is provided either by a disk dynamo or by advection of ambient interstellar field

$$B_{p,disk} < B_{eq}(r) = 20 \, G \alpha^{-1/2} \left(\frac{\dot{M}_a}{10^{-6} \, M_\odot \, yr^{-1}} \right)^{1/2} \cdot$$
$$\left(\frac{M_\star}{M_\odot} \right)^{1/4} \left(\frac{H/r}{0.1} \right)^{-1/2} \left(\frac{r}{10 \, R_\odot} \right)^{-5/4} \tag{20}$$

where B_{eq} is the equipartition field strength in the disk. This flux will not remain closed, but will inflate and open up as magnetic field lines are sheared and extract gravitational potential energy from the accreting flow (e.g., *Uzdensky et al., 2002; Matt and Pudritz, 2005a*). These field lines therefore effectively become a disk field, and therefore follow the processes of disk-wind production we have already discussed.

The additional Poynting flux that threads the disk may assist the jet launching by MHD forces and may serve as an additional energy reservoir for the kinetic energy of the jet, implying greater asymptotic jet speed (Michel scaling) (*Michel, 1969; Fendt and Camenzind, 1996*).

5.1.3. Disk locking versus stellar winds. The spin of the star will depend on how angular momentum arriving from the disk is dealt with. In the magnetic "disk-locking" picture, the threading field of the disk will rearrange the global angular momentum budget. The torque on the star by the accretion of disk matter is

$$\tau_a = \dot{M}_a (GM_\star r_{in})^{1/2} \tag{21}$$

(e.g., *Matt and Pudritz, 2005a*), with the disk accretion rate $\dot{M}_{acc}$, the stellar mass $M_\star$, and the disk inner radius r_{in} inside the corotation radius $r_{co} = (GM_\star)^{1/3}\Omega_\star^{-2/3}$. On the other hand, if "disk locking" is present, the stellar rotation may be decelerated by the magnetic torque due to stellar field lines connecting the star with the accretion disk outside R_{co}. The differential magnetic torque acting on a disk annulus of dr width is

$$d\tau_{mag} = r^2 B_\phi B_z dr \tag{22}$$

While B_z in principle follows from assuming a central dipolar field, the induction of toroidal magnetic fields is model dependent (disk resistivity, poloidal field structure). This is why recent numerical simulations of dipole-disk interaction that simultaneously evaluate the poloidal and toroidal field components have become extremely valuable.

If indeed the star loses angular momentum to the disk (this is not yet decided by the simulations; see below), both disk accretion and jet formation are affected. In order to continue accretion, excess angular momentum has to be removed from the accreting matter. A disk jet can be an efficient way to do this, as has been supposed in the X-wind picture.

The central stellar magnetic field may launch a strong stellar wind to rid itself of the accreted angular momentum. Such an outflow will interact with the surrounding disk wind. If true, observed YSO jets may consist of two components — the stellar wind and the disk wind — with strength depending on intrinsic (yet unknown) parameters. The stellar wind (open field lines of stellar magnetosphere) exerts a spin-down torque upon the star of magnitude

$$\dot{M}_{wind,\star} \Omega_\star r_A^2 = 3 \times 10^{36} \frac{g \, cm^2}{s \, yr} \left(\frac{\Omega_\star}{10^{-5} \, s^{-1}} \right)^{1/2} \cdot$$
$$\left(\frac{r_A}{30 \, R_\odot} \right)^2 \left(\frac{\dot{M}_{wind,\star}}{10^{-9} \, M_\odot \, yr^{-1}} \right) \tag{23}$$

As a historical remark we note that the model topology of dipole-plus-disk field were introduced for protostellar jet formation more then 20 years ago in the MHD simulations of *Uchida and Shibata* (1984). Further investigations considering the detailed physical processes involved in disk truncation and channeling the matter along the dipolar field lines by *Camenzind* (1990), *Königl* (1991), and *Shu et al.* (1994) resulted in a breakthrough of these ideas to the protostellar jet community.

5.2. Numerical Simulations of Star-Disk Interaction

The numerical simulation of the magnetospheric star-disk interaction is technically most demanding since one must treat a complex model geometry in combination with strong gradients in magnetic field strength, density, and resistivity. In general, this may imply a large variation in physical timescales for the three components of disk, jet, and magnetosphere, which all have to be resolved numerically. Essentially, numerical modeling of the star-disk interaction requires a *diffusive and viscous MHD code including radiative transfer*. In addition, realistic models for reconnection processes and disk opacity are needed. Then, such a code has to run with *high spatial and temporal resolution* on a global scale.

Compared to the situation about a decade ago when there was still only *Uchida and Shibata*'s (1984) initial (although 10 years old) simulation available, today huge progress has been made with several groups (and also codes) competing in the field. Early simulations were able to follow the evolution only for a few rotations of the inner disk (note that 100 rotations at the co-rotation radius correspond to 0.3 Keplerian rotations at 10 co-rotation radii) (*Hayashi et al.*, 1996; *Miller and Stone*, 1997; *Goodson et al.*, 1997). Among the problems involved is the initial condition of the simulation, in particular the nature of the disk model, which could be numerically treated. A steady initial corona will strongly shear with the Keplerian disk, leading to current sheet (thus pressure gradients) along the disk surface. If the simulations run long enough, this could be an intermittent feature. However, the danger exists that the artificial current sheet will fatally destroy the result of the simulation. Applying a Shakura-Sunyaev disk for the initial disk structure, the code should also consider α_{SS} viscosity. Otherwise the initial disk evolution is not self-consistent (see, e.g., *Goodson et al.*, 1997).

The next step is to increase the grid resolution and to redo the axisymmetric simulations. *Goodson et al.* (1999) and *Goodson and Winglee* (1999) were able to treat several hundreds of (inner) disk rotations on a global grid of 50 AU extension. The main result of these simulations is a two-component flow consisting of a fast and narrow axial jet and a slow disk wind, both launched in the inner part of the disk (r < 1 AU). An interesting feature is that the narrow axial jet is actually a collimation in density and not in velocity. Close to the inner disk radius repetitive reconnection processes are seen on timescales of a couple of rotation periods. The dipolar field inflates and an expanding current sheet builds up. After field reconnection along the current sheet, the process starts all over again. The oscillatory behavior leads to the ejection of axial knots. On average the central dipolar magnetosphere remains present and loaded with plasma. Forbidden emission line intensity maps of these simulations have been calculated and allow for direct comparison with the observations. However, the numerically derived time and length scales of axial knots were still

too different from what is observed. The magnetospheric origin of jets (stellar dynamo) in favor of a pure disk origin ("primordial field") has also been stressed by *Matt et al.* (2002).

In order to be able to perform long-term simulations of dipolar magnetospheres interacting with the accretion disk, *Fendt and Elstner* (1999, 2000) neglected the evolution of the disk structure and instead assume that the disk is a fixed boundary condition for the outflow (as in *OPI*). After 2000 rotations a quasi-steady state was obtained with a two-component outflow from disk and star. The outflow expands almost radially without signature of collimation on the spatial scale investigated (20 × 20 inner disk radii). One consequence of this very long simulation is that the axial narrow jet observed in other simulations might be an intermittent feature launched in the early phase of the simulation as the initial field is reconfigured to a new equilibrium. The axial outflow in this simulations is massive but slow, but tends to develop axial instabilities for a lower mass loading. Clearly, since the disk structure was not been taken into account, nothing could be said about the launching process of the outflow out of the disk.

In a series of ideal MHD simulations Romanova and collaborators succeeded in working out a detailed and sufficiently stable numerical model of magnetospheric disk interaction. They were the first to simulate the axisymmetric funnel flow from the disk inner radius onto the stellar surface (*Romanova et al.*, 2002) on a global scale ($R_{max} = 50\ R_{in}$) and for a sufficiently long period of time (300 rotations) in order to reach a steady state in the accretion funnel. The funnel flow with free-falling material builds up as a gap opens up between disk and star. The authors further investigated the angular momentum exchange due to the disk "locking" by the funnel flow. Slowly rotating stars seem to break the inner disk region and spin up, while rapid rotators would accelerate the inner disk region to super-Keplerian velocity and slow down themselves. Only for certain stellar rotational periods could a "torque-less" accretion be observed. Strong outflows have not been observed for the parameter space investigated, probably due to the matter-dominated corona, which does not allow the dipolar field to open up.

Further progress has been achieved, extending these simulations to three dimensions (*Romanova et al.*, 2003, 2004). For the first time it has been possible to investigate the interaction of an inclined stellar dipolar magnetosphere with the surrounding disk. For zero inclination, axisymmetric funnels are found as in the axisymmetric simulations. For nonzero inclination, the accretion splits in two main streams following the closest path from the disk to the stellar surface, as is seen in Fig. 4. Magnetic braking changes the disk structure in several ways: its density structure (gaps, rings), its velocity structure (sub-Keplerian region), and its geometry (warping). The slowly rotating star is spun up by accreting matter but this acceleration only weakly depends on inclination. For completeness we should note that for these simu-

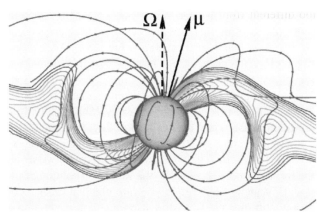

Fig. 4. Slice through a funnel stream. Density contours from ρ = 0.2 to ρ = 2.0. The density of the disk corona is ρ = 0.01–0.02 selected magnetic field lines. The rotational axis and the magnetic moment are indicated as **Ω** and μ. Adapted from *Romanova et al.* (2004).

lations a clever approach for the numerical grid has been applied, the "cubed grid," which does not obey the singular axes as for spherical coordinates.

The star-disk coupling by the stellar magnetosphere was also investigated by *Küker et al.* (2003). These simulations have been performed in axisymmetry, but an advanced disk model has been applied. Taking into account α-viscosity, a corresponding eddy magnetic diffusivity, and radiative energy transport, the code enables the authors to treat a realistic *accretion* disk in their MHD simulations. As a result the simulations could be advanced to very long physical time steps into a quasi-stationary state of the disk evolution. The authors show that the commonly assumed 1000 G magnetosphere is not sufficient to open up a gap in the disk. Unsteady outflows may be launched outside the co-rotation radius with mass loss rates of about 10% of the accretion rate. The authors note that they were unable to detect the narrow axial jet seen previously in other publications (in agreement with *Fendt and Elstner,* 2000). Magnetic braking of the star may happen in some cases, but is overwhelmed by the accretion torque still spinning up the star (however, these simulations treat the stellar surface as an inner boundary condition).

Using a different approach, the boundary condition of a stellar *dipolar* magnetosphere or a large-scale disk field were relaxed. This allows for the interaction of a stellar dipole with a *dynamo*-generated disk magnetic field (*von Rekowski and Brandenburg,* 2004), or the evolution of a dynamo-generated stellar field affected by a surrounding disk with its own disk dynamo (*von Rekowski and Brandenburg,* 2006) to be studied. The results of this work are in agreement with previous studies, and directly prove that a disk surrounding a stellar magnetosphere may actually develop its own magnetic field strong enough to launch MHD outflows. In agreement with results by *Goodson and Winglee* (1999), accretion tends to be unsteady and alternating between con-

nected and disconnected states. Even for realistic stellar fields of several hundred gauss the magnetic spin-down torque is insufficient to overcome the spin-up torque from the accretion flow itself. As shown by the authors, angular momentum exchange is complex and may vary in sign along the stellar surface, braking some parts of the star and accelerating others. The dynamo-generated stellar field may reach up to 750 G and switches in time between dipolar and quadrupolar symmetry, while the dynamo-generated 50 G disk field is of dipolar symmetry. In general, the dynamo-generated stellar field is better suited to drive a stellar wind. The simulations show that for these cases stellar wind braking is more efficient than braking by the star-disk coupling, confirming the results of *Matt and Pudritz* (2005a).

Recent studies of torque-less accretion (*Long et al.,* 2005) compare cases of (1) a weak stellar magnetic field within a dense corona and (2) a strong field in a lower-density corona. They investigate the role of quasi-periodic field line reconnection in coupling the disk and stellar fields. Unlike previous works, these authors conclude that magnetic interaction effectively locks the stellar rotation to about 10% of the breakup velocity, a value that actually depends on the disk accretion rate and the stellar magnetic moment. While they correctly stress the importance of dealing with the exact balance between open and closed field lines and their corresponding angular momentum flux, the exact magnetospheric structure in the innermost region will certainly depend on the resistivity (diffusivity) of the disk and corona material. This, however, is not included in their treatment as an ideal MHD code has been applied.

We summarize this section noting that tremendous progress has occurred in the numerical simulation of the star-disk interaction. The role of numerical simulations is pivotal in this area because the mechanisms involved are complex, strongly interrelated, and often highly time-dependent. It is fair to say that numerical simulations of the star-disk interaction have not yet shown the launching of a jet flow comparable to the observations.

6. JETS AND GRAVITATIONAL COLLAPSE

Jets are expected to be associated with gravitational collapse because disks are the result of the collapse of rotating molecular cloud cores. One of the first simulations to show how jets arise during gravitational collapse is the work of *Tomisaka* (1998, 2002). Here, the collapse of a magnetized core within a rotating cylinder of gas gave rise to the formation of a disk from which a centrifugally driven disk wind was produced. Analytical work by *Krasnopolsky and Königl* (2002) provides an interesting class of self-similar solutions for the collapse of rotating, magnetized, isothermal cores including ambipolar diffusion that could be matched onto *BP82* outflows.

Given the importance of Bonnor-Ebert (B-E) spheres in the physics of star formation and gravitational collapse, recent efforts have focused on understanding the evolution

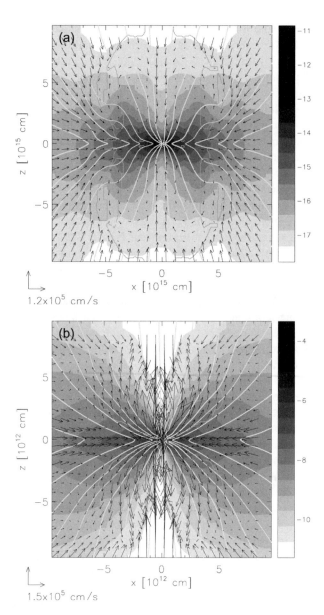

Fig. 5. (a) Large-scale outflow (scale of hundreds of AU) and (b) small-scale disk wind (scale of a fraction of an AU) and jet formed during the gravitational collapse of a magnetized B-E, rotating cloud core. Cross-sections through the disk and outflows are shown. The blue contour (see colorized electronic version of journal paper) marks the Alfvén surface. Snapshots taken of an adaptive mesh calculation at about 70,000 years into the collapse. Adapted from *Banerjee and Pudritz* (2006).

of magnetized B-E spheres. Whereas purely hydrodynamic collapses of such objects never show outflows (e.g., *Foster and Chevalier,* 1993; *Banerjee et al.,* 2004), the addition of a magnetic field produces them. *Matsumoto and Tomisaka* (2004) have studied the collapse of rotating B-E spheres wherein the magnetic axis is inclined with the initial rotation axis. They observe that after the formation of an adiabatic core, outflow is ejected along the local magnetic field lines. Eventually, strong torques result in the rapid alignment of the rotation and magnetic field vectors.

The collapse of a magnetized, rotating, B-E sphere with molecular cooling included (but not dust) was carried out by *Banerjee and Pudritz* (2006) using the FLASH adaptive mesh refinement code. The results of this simulation are shown in Fig. 5, which shows the end state (at about 70,000 yr) of the collapse of a B-E sphere that is chosen to be precisely the Bok globule observed by *Alves et al.* (2001), whose mass is 2.1 $M_\odot$ and radius R = 1.25 × 10⁴ AU at an initial temperature of 16 K. Two types of outflow can be seen: (1) an outflow that originates at scale of ≈130 AU on the forming disk that consists of a wound up column of toroidal magnetic field whose pressure gradient pushes out a slow outflow; and (2) a disk wind that collimates into a jet on scale of 0.07 AU. A tight protobinary system has formed in this simulation, whose masses are still very small, ≤10⁻² $M_\odot$, which is much less than the mass of the disk at this time, ≈10⁻¹ $M_\odot$. The outer flow bears the hallmark of a magnetic tower, first observed by *Uchida and Shibata* (1985), and studied by *Lynden-Bell* (2003). Both flow components are unbound, with the disk wind reaching 3 km s⁻¹ at 0.4 AU, which is quite super-Alfvénic and above the local escape speed. The outflow and jet speeds will increase as the central mass grows.

We conclude that the theory and computation of jets and outflows is in excellent agreement with new observations of many kinds. Disk winds are triggered during magnetized collapse and persist throughout the evolution of the disk. They efficiently tap accretion power, transport a significant part portion of the disk's angular momentum, and can achieve different degrees of collimation depending on their mass loading. Accretion-powered stellar winds may also solve the stellar angular momentum problem. We are optimistic that observations will soon be able to test the universality of outflows, all the way from circumplanetary disks to those around O stars.

Acknowledgments. We are indebted to Nordita for hosting an authors' meeting, T. Ray for stimulating discussions, R. Banerjee and B. Reipurth for careful reads of the manuscript, and an anonymous referee for a very useful report. The research of R.E.P. and R.O. is supported by grants from NSERC of Canada. C.F. acknowledges travel support by the German science foundation to participate in the PPV conference.

REFERENCES

Afshordi N., Mukhopadhyay B., and Narayan R. (2005) *Astrophys. J., 629,* 373–382.
Alves J. F., Lada C. J., and Lada E. A. (2001) *Nature, 409,* 159–161.
Anderson J. M., Li Z.-Y., Krasnopolsky R., and Blandford R. D. (2003) *Astrophys. J., 590,* L107–L110.
Anderson J. M., Li Z.-Y., Krasnopolsky R., and Blandford R. D. (2005) *Astrophys. J., 630,* 945–957.
Bacciotti F. (2004) *Astrophys. Space Sci., 293,* 37–44.
Bacciotti F., Mundt R., Ray T. P., Eislöffel J., Solf J., and Camenzind M. (2000) *Astrophys. J., 537,* L49–L52.
Bacciotti F., Ray T. P., Mundt R., Eislöffel J., and Solf J. (2002) *Astrophys. J., 576,* 222–231.
Bacciotti F., Ray T. P., Eislöffel J., Woitas J., Solf J., Mundt R., and Davis C. J. (2003) *Astrophys. Space Sci., 287,* 3–13.

Balbus S. A., Hawley J. F., and Stone J. M. (1996) *Astrophys. J., 467,* 76–86.

Banerjee R. and Pudritz R. E. (2006) *Astrophys. J., 641,* 949–960.

Banerjee R., Pudritz R. E., and Holmes L. (2004) *Mon. Not. R. Astron. Soc., 355,* 248–272.

Bardou A., Rekowski B. v., Dobler W., Brandenburg A., and Shukurov A. (2001) *Astron. Astrophys., 370,* 635–648.

Bateman G. (1980) *Magneto-Hydrodynamical Instabilities.* MIT Press, Cambridge.

Blandford R. D. and Payne D. G. (BP82) (1982) *Mon. Not. R. Astron. Soc., 199,* 883–903.

Boldyrev S. and Cattaneo F. (2004) *Phys. Rev. Lett., 92,* 144501.

Bourke T. L., Crapsi A., Myers P. C., Evans N. J., et al. (2005) *Astrophys. J., 633,* L129–L132.

Brandenburg A. and Sokoloff D. (2002) *Geophys. Astrophys. Fluid Dyn., 96,* 319–344.

Brandenburg A. and Subramanian K. (2005) *Phys. Rep., 417,* 1–209.

Brandenburg A., Nordlund A., Stein R. F., and Torkelsson U. (1995) *Astrophys. J., 446,* 741–754.

Cabrit S. and Bertout C. (1992) *Astron. Astrophys., 261,* 274–284.

Cabrit S., Raga A. C., and Gueth F. (1997) In *Herbig-Haro Flows and the Birth of Low Mass Stars* (B. Reipurth and C. Bertout, eds.), pp. 163–180. Kluwer, Dordrecht.

Camenzind M. (1990) *Rev. Modern Astron., 3,* 234–265.

Campbell C. G. (2000) *Mon. Not. R. Astron. Soc., 317,* 501–527.

Campbell C. G. (2003) *Mon. Not. R. Astron. Soc., 345,* 123–143.

Campbell C. G., Papaloizou J. C. B., and Agapitou V. (1998) *Mon. Not. R. Astron. Soc., 300,* 315–320.

Cao X. and Spruit H. C. (1994) *Astron. Astrophys., 287,* 80–86.

Casse F. and Ferreira J. (2000) *Astron. Astrophys., 353,* 1115–1128.

Casse F. and Keppens R. (2002) *Astrophys. J., 581,* 988–1001.

Casse F. and Keppens R. (2004) *Astrophys. J., 601,* 90–103.

Chagelishvili G. D., Zahn J.-P., Tevzadze A. G., and Lominadze J. G. (2003) *Astron. Astrophys., 402,* 401–407.

Chiang E. I., Joung M. K., Creech-Eakman M. J., Qi C., Kessler J. E., Blake G. A., and van Dishoeck E. F. (2001) *Astrophys. J., 547,* 1077–1089.

Cowling T. G. (1934) *Mon. Not. R. Astron. Soc., 94,* 39–48.

De Villiers J.-P., Hawley J. F., Krolik J. H., and Hirose S. (2005) *Astrophys. J., 620,* 878–888.

Donati J.-F., Paletou F, Bouvier J., and Ferreira J. (2005) *Nature, 438,* 466–469.

Dubrulle B., Marie L., Normand C., Richard D., Hersant F., and Zahn J.-P. (2005) *Astron. Astrophys., 429,* 1–13.

Eichler D. (1993) *Astrophys. J., 419,* 111–116.

Fendt C. (2003) *Astron. Astrophys., 411,* 623–635.

Fendt C. (2006) *Astrophys. J.,* in press (astro-ph/0511611).

Fendt C. and Camenzind M. (1996) *Astron. Astrophys., 313,* 591–604.

Fendt C. and Čemeljić M. (2002) *Astron. Astrophys., 395,* 1045–1060.

Fendt C. and Elstner D. (1999) *Astron. Astrophys., 349,* L61–L64.

Fendt C. and Elstner D. (2000) *Astron. Astrophys., 363,* 208–222.

Ferriera J. (1997) *Astron. Astrophys., 319,* 340–359.

Ferreira J. and Casse F. (2004) *Astrophys. J., 601,* L139–L142.

Fleming T. and Stone J. M. (2003) *Astrophys. J., 585,* 908–920.

Foster P. N. and Chevalier R. A. (1993) *Astrophys. J., 416,* 303–311.

Fromang S., Terquem C., and Balbus S. A. (2002) *Mon. Not. R. Astron. Soc., 329,* 18–28.

Gammie C. F. (1996) *Astrophys. J., 457,* 355–362.

Goodman J. and Rafikov R. R. (2001) *Astrophys. J., 552,* 793–802.

Goodson A. P. and Winglee R. M. (1999) *Astrophys. J., 524,* 159–168.

Goodson A. P., Winglee R. M., and Böhm K.-H. (1997) *Astrophys. J., 489,* 199–209.

Goodson A. P., Böhm K.-H., and Winglee R. M. (1999) *Astrophys. J., 524,* 142–158.

Greenhill L. J., Gwinn C. R., Schwartz C., Moran J. M., and Diamond P. J. (1998) *Nature, 396,* 650–653.

Hardee P. and Rosen A. (1999) *Astrophys. J., 524,* 650–666.

Hawley J. F. (2000) *Astrophys. J., 528,* 462–479.

Hawley J. F., Balbus S. A., and Winters W. F. (1999) *Astrophys. J., 518,* 394–404.

Hayashi M. R., Shibata K., and Matsumoto R. (1996) *Astrophys. J., 468,* L37–L40.

Herbst W., Bailer-Jones C. A. L., Mundt R., Meisenheimer K., and Wackermann R. (2002) *Astron. Astrophys., 398,* 513–532.

Heyvaerts J. (2003) In *Accretion Discs, Jets, and High Energy Phenomena in Astrophysics* (V. Beskin et al., eds.), p. 3. Springer-Verlag, Berlin.

Heyvaerts J. and Norman C. (1989) *Astrophys. J., 347,* 1055–1081.

Inutsuka S.-I. and Sano T. (2005) *Astrophys. J., 628,* L155–L158.

Keppens R. and Goedbloed J. P. (2000) *Astrophys. J., 530,* 1036–1048.

Kigure H. and Shibata K. (2005) *Astrophys. J., 634,* 879–900.

Kley W. and Lin D. N. C. (1992) *Astrophys. J., 397,* 600–612.

Kley W., D'Angelo G., and Henning T. (2001) *Astrophys. J., 547,* 457–464.

Königl A. (1991) *Astrophys. J., 370,* L39–L43.

Königl A. (1999) *New Astron. Rev., 43,* 67–77.

Königl A. (2004) *Astrophys. J., 617,* 1267–1271.

Königl A. and Choudhuri A. R. (1985) *Astrophys. J., 289,* 173–187.

Königl A. and Pudritz R. E. (KP00) (2000) In *Protostars and Planets IV* (V. Mannings et al., eds.), pp. 759–787. Univ. of Arizona, Tucson.

Krasnopolsky R. and Königl A. (2002) *Astrophys. J., 580,* 987–1012.

Krasnopolsky R., Li Z.-Y., and Blandford R. D. (1999) *Astrophys. J., 526,* 631–642.

Krasnopolsky R., Li Z.-Y., and Blandford R. D. (2003) *Astrophys. J., 595,* 631–642.

Krumholz M. R., McKee C. F., and Klein R. I. (2005) *Astrophys. J., 618,* L33–L36.

Kudoh T., Matsumoto R., and Shibata K. (1998) *Astrophys. J., 508,* 186–199.

Küker M., Henning T., and Rüdiger G. (2003) *Astrophys. J., 589,* 397–409; erratum: *Astrophys. J., 614,* 526.

Kuwabara T., Shibata K., Kudoh T., and Matsumoto R. (2005) *Astrophys. J., 621,* 921–931.

Lai D. (1999) *Astrophys. J., 524,* 1030–1047.

Lai D. (2003) *Astrophys. J., 591,* L119–L122.

Lee C.-F., Mundy L. G., Reipurth B., Ostriker E. C., and Stone J. M. (2000) *Astrophys. J., 542,* 925–945.

Lesur G. and Longaretti P.-Y. (2005) *Astron. Astrophys., 444,* 25–44.

Li Z.-Y. (1995) *Astrophys. J., 444,* 848–860.

Li Z.-Y. (1996) *Astrophys. J., 465,* 855–868.

Long M., Romanova M. M., and Lovelace R. V. E. (2005) *Astrophys. J., 634,* 1214–1222.

Lubow S. H., Papaloizou J. C. B., and Pringle J. E. (1994) *Mon. Not. R. Astron. Soc., 268,* 1010–1014.

Lucek S. G. and Bell A. R. (1996) *Mon. Not. R. Astron. Soc., 281,* 245–256.

Lynden-Bell D. (2003) *Mon. Not. R. Astron. Soc., 341,* 1360–1372.

Matsumoto R., Uchida Y., Hirose S., Shibata K., Hayashi M. R., Ferrari A., Bodo G., and Norman C. (1996) *Astrophys. J., 461,* 115–126.

Matsumoto T. and Tomisaka K. (2004) *Astrophys. J., 616,* 266–282.

Matsumura S. and Pudritz R. E. (2005) *Astrophys. J., 618,* L137–L140.

Matsumura S. and Pudritz R. E. (2006) *Mon. Not. R. Astron. Soc., 365,* 572–584.

Matt S. and Pudritz R. E. (2005a) *Mon. Not. R. Astron. Soc., 356,* 167–182.

Matt S. and Pudritz R. E. (2005b) *Astrophys. J., 632,* L135–L138.

Matt S., Goodson A. P., Winglee R. M., and Böhm K.-H. (2002) *Astrophys. J., 574,* 232–245.

Ménard F. and Duchêne G. (2004) *Astron. Astrophys., 425,* 973–980.

Meier D. L., Edgington S., Godon P., Payne D. G., and Lind, K. R. (1997) *Nature, 388,* 350–352.

Michel F. C. (1969) *Astrophys. J., 158,* 727–738.

Miller K. A. and Stone J. M. (1997) *Astrophys. J., 489,* 890–902.

Miller K. A. and Stone J. M. (2000) *Astrophys. J., 376,* 214–419.

Molemaker M. J., McWilliams J. C., and Yavneh I. (2001) *Phys. Rev. Lett., 86,* 5270–5273.

Nakamura M. and Meier D. (2004) *Astrophys. J., 617,* 123–154.

Nakamura M., Uchida Y., and Hirose S. (2001) *New Astron., 6,* 61–78.

Ouyed R. (2003) *Astrophys. Space Sci., 287,* 87–97.

Ouyed R. and Pudritz R. E. (OPI) (1997a) *Astrophys. J., 482,* 712–732.

Ouyed R. and Pudritz R. E. (OPII) (1997b) *Astrophys. J., 484,* 794–809.

Ouyed R. and Pudritz R. E. (OPIII) (1999) *Mon. Not. R. Astron. Soc., 309,* 233–244.

Ouyed R., Pudritz R. E., and Stone J. M. (OPS) (1997) *Nature, 385,* 409–414.

Ouyed R., Clarke D. A., and Pudritz R. E. (OCP) (2003) *Astrophys. J., 582,* 292–319.

O'Sullivan M., Truss M., Walker C., Wood K., Matthews O., et al. (2005) *Mon. Not. R. Astron. Soc., 358,* 632–640.

Pelletier G. and Pudritz R. E. (PP92) (1992) *Astrophys. J., 394,* 117–138.

Pfeiffer H. P. and Lai D. (2004) *Astrophys. J., 604,* 766–774.

Pinte C. and Ménard F. (2004) In *The Search for Other Worlds* (S. S. Holt and D. Demings, eds.), pp. 123–126. American Institute of Physics, New York.

Pringle J. E. (1996) *Mon. Not. R. Astron. Soc., 281,* 357–361.

Pudritz R. E. (2003) In *Accretion discs, Jets, and High Energy Phenomena in Astrophysics* (V. Beskin et al., eds.), pp. 187–230. Springer-Verlag, Berlin.

Pudritz R. E. and Banerjee B. (2005) In *Massive Star Birth: A Crossroads of Astrophysics* (R. Cesaroni et al., eds.), pp. 163–173. Cambridge Univ., Cambridge.

Pudritz R. E. and Norman C. A. (1983) *Astrophys. J., 274,* 677–697.

Pudritz R. E. and Norman C. A. (1986) *Astrophys. J., 301,* 571–586.

Pudritz R. E., Rogers C., and Ouyed R. (PRO) (2006) *Mon. Not. R. Astron. Soc., 365,* 1131–1148.

Quillen A. C. and Trilling D. E. (1998) *Astrophys. J., 508,* 707–713.

Roberts P. H. (1967) *Introduction to Magneto-Hydrodynamics.* Longmans, London.

Romanova M. M., Ustyugova G. V., Koldoba A. V., Chechetkin V. M., and Lovelace R. V. (1997) *Astrophys. J., 482,* 708–711.

Romanova M. M., Ustyugova G. V., Koldoba A. V., and Lovelace R. V. E. (2002) *Astrophys. J., 578,* 420–438.

Romanova M. M., Toropina O. D., Toropin Y. M., and Lovelace R. V. E. (2003) *Astrophys. J., 588,* 400–407.

Romanova M. M., Ustyugova G. V., Koldoba A. V., and Lovelace R. V. E. (2004) *Astrophys. J., 610,* 920–932.

Różyczka M. and Spruit H. C. (1993) *Astrophys. J., 417,* 677–686.

Sano T. and Stone J. M. (2002) *Astrophys. J., 570,* 314–328.

Sano T., Miyama S. M., Umebayashi T., and Nakano T. (2000) *Astrophys. J., 543,* 486–501.

Schandl S. and Meyer F. (1994) *Astron. Astrophys., 289,* 149–161.

Schekochihin A. A., Haugen N. E. L., Brandenburg A., Cowley S. C., Maron J. L., and McWilliams J. C. (2005) *Astrophys. J., 625,* L115–L118.

Semenov D., Wiebe D., and Henning T. (2004) *Astron. Astrophys., 417,* 93–106.

Shakura N. I. and Sunyaev R. A. (1973) *Astron. Astrophys., 24,* 337–355.

Shalybkov D. and Rüdiger G. (2005) *Astron. Astrophys., 438,* 411–417.

Shibata K. and Uchida Y. (1986) *Publ. Astron. Soc. Japan, 38,* 631–660.

Shu F., Najita J., Ostriker E., Wilkin F., Ruden S., and Lizano S. (1994) *Astrophys. J., 429,* 781–796.

Shu F. H., Najita J. R., Shang H., and Li Z.-Y. (2000) In *Protostars and Planets IV* (V. Mannings et al., eds.), pp. 789–813. Univ. of Arizona, Tucson.

Spruit H. C. and Uzdensky D. A. (2005) *Astrophys. J., 629,* 960–968.

Spruit H. C., Foglizzo T., and Stehle R. (1997) *Mon. Not. R. Astron. Soc., 288,* 333–342.

Stone J. M. and Norman M. L. (1992) *Astrophys. J. Suppl., 80,* 753–790.

Stone J. M. and Norman M. L. (1994) *Astrophys. J., 433,* 746–756.

Stone J. M., Hawley J. F., Gammie C. F., and Balbus S. A. (1996) *Astrophys. J., 463,* 656–673.

Symington N. H., Harries T. J., and Kurosawa R. (2005a) *Mon. Not. R. Astron. Soc., 356,* 1489–1500.

Symington N. H., Harries T. J., Kurosawa R., and Naylor T. (2005b) *Mon. Not. R. Astron. Soc., 358,* 977–984.

Todo Y., Uchida Y., Sato T., and Rosner R. (1993) *Astrophys. J., 403,* 164–174.

Tomisaka K. (1998) *Astrophys. J., 502,* L163–L167.

Tomisaka K. (2002) *Astrophys. J., 575,* 306–326.

Torkelsson U. and Brandenburg A. (1994) *Astron. Astrophys., 283,* 677–692.

Turner N. J. (2004) *Astrophys. J., 605,* L45–L48.

Uchida Y. and Shibata K. (1984) *Publ. Astron. Soc. Japan, 36,* 105–118.

Uchida Y. and Shibata K. (1985) *Proc. Astron. Soc. Japan, 37,* 515–535.

Uchida Y., McAllister A., Strong K. T., Ogawara Y., Shimizu T., Matsumoto R., and Hudson H. S. (1992) *Proc. Astron. Soc. Japan, 44,* L155–L160.

Umurhan O. M. (2006) *Mon. Not. R. Astron. Soc., 365,* 85–100.

Uzdensky D. A., König A., and Litwin C. (2002) *Astrophys. J., 565,* 1191–1204.

Vitorino B. F., Jatenco-Pereira V., and Opher R. (2002) *Astron. Astrophys., 384,* 329.

von Rekowski B., Brandenburg A., Dobler W., and Shukurov A. (2003) *Astron. Astrophys., 398,* 825–844.

von Rekowski B. and Brandenburg A. (2004) *Astron. Astrophys., 420,* 17–32.

von Rekowski B. and Brandenburg A. (2006) *Astron. Nachr., 327,* 53–71.

Wardle M. and Königl A. (1993) *Astrophys. J., 410,* 218–238.

Woitas J., Ray T. P., Bacciotti F., Davis C. J., and Eislöffel J. (2002) *Astrophys. J., 580,* 336–342.

Wu Y., Wei Y., Zhao M., Shi Y., Yu W., Qin S., and Huang M. (2004) *Astron. Astrophys., 426,* 503–515.

Ziegler U. and Rüdiger G. (2000) *Astron. Astrophys., 356,* 1141–1148.

Part IV:

Young Stars and Clusters

The Rotation of Young Low-Mass Stars and Brown Dwarfs

William Herbst
Wesleyan University

Jochen Eislöffel
Thüringer Landessternwarte, Tautenburg

Reinhard Mundt
Max-Planck-Institute for Astronomy, Heidelberg

Alexander Scholz
University of Toronto

We review the current state of our knowledge concerning the rotation and angular momentum evolution of young stellar objects and brown dwarfs from a primarily observational viewpoint. There has been a tremendous growth in the number of young, low-mass objects with measured rotation periods over the last five years, due to the application of wide field imagers on 1–2-m-class telescopes. Periods are typically accurate to 1% and available for about 1700 stars and 30 brown dwarfs in young clusters. Discussion of angular momentum evolution also requires knowledge of stellar radii, which are poorly known for pre-main-sequence stars. It is clear that rotation rates at a given age depend strongly on mass; higher-mass stars (0.4–1.2 $M_\odot$) have longer periods than lower-mass stars and brown dwarfs. On the other hand, specific angular momentum is approximately independent of mass for low-mass pre-main-sequence stars and young brown dwarfs. A spread of about a factor of 30 is seen at any given mass and age. The evolution of rotation of solar-like stars during the first 100 m.y. is discussed. A broad, bimodal distribution exists at the earliest observable phases (~1 m.y.) for stars more massive than 0.4 $M_\odot$. The rapid rotators (50–60% of the sample) evolve to the ZAMS with little or no angular momentum loss. The slow rotators continue to lose substantial amounts of angular momentum for up to 5 m.y., creating the even broader bimodal distribution characteristic of 30–120-m.y.-old clusters. Accretion disk signatures are more prevalent among slowly rotating PMS stars, indicating a connection between accretion and rotation. Disks appear to influence rotation for, at most, ~5 m.y., and considerably less than that for the majority of stars. This time interval is comparable to the *maximum* lifetime of accretion disks derived from near-infrared studies, and may be a useful upper limit to the time available for forming giant planets. If the dense clusters studied so far are an accurate guide, then the typical solar-like star may have only ~1 m.y. for this task. There is less data available for very-low-mass stars and brown dwarfs but the indication is that the same mechanisms are influencing their rotation as for the solar-like stars. However, it appears that both disk interactions and stellar winds are less efficient at braking these objects. We also review our knowledge of the various types of variability of these objects over as broad a mass range as possible with particular attention to magnetically induced cool spots and magnetically channeled variable mass accretion.

1. INTRODUCTION

The study of stellar rotation during the pre-main-sequence (PMS) and zero-age main-sequence (ZAMS) phase provides important clues to the solution of the angular momentum problem of star formation. It is also intimately connected to the evolution of the circumstellar disk out of which planets are believed to form. The angular momentum problem, simply stated, is that the specific angular momentum (j = J/M where J is angular momentum and M is mass) of dense molecular cloud cores, the birth places of low-mass stars, is 5–6 orders of magnitude higher than that of solar-

type stars on the ZAMS (e.g., *Bodenheimer*, 1995). We note that by the time they reach the age of the Sun (~5 G.y.), solar-type stars have lost an additional 1–2 orders of magnitude of j due to the cumulative effect of torques from their magnetized coronal winds.

Undoubtedly, the full angular momentum problem is solved by a combination of factors occurring throughout the star-formation process, not just a single event at a specific time (*Bodenheimer*, 1995). Important processes at early stages include magnetic torques between the collapsing molecular cloud core and the surrounding interstellar medium as well as the deposition of large amounts of angular momentum in the orbital motions of a circumstellar disk, planetary system, and/or binary star. At later stages the redistribution of angular momentum within the disk and J-loss by magnetically driven outflows and jets become important. Recent HST/STIS studies of optical jets from T Tauri stars (TTS) (e.g., *Bacciotti et al.*, 2002; *Woitas et al.*, 2005) show that these jets are rotating. In the framework of magnetically driven outflows, the derived rotation velocities of these jets imply large angular momentum loss rates. The magnetohydrodynamic (MHD) models suggested for the acceleration of bipolar outflows from young stars also involve a strong magnetic coupling between the stellar magnetic field and the inner parts of the circumstellar accretion disk. Hence, it seems likely that the rotation of PMS stars will be influenced by these processes.

On the observational side, the first studies of rotation to include a substantial number of TTS employed high-resolution spectra to measure line broadening (e.g., *Vogel and Kuhi*, 1981; *Hartmann et al.*, 1986; *Bouvier et al.*, 1986). The resulting measurements of rotational velocity (V_{rot}) are inherently uncertain due to the unknown inclination of the stellar rotation axis to the line of sight. It is also difficult to measure V_{rot} for slow rotators since there is very little line broadening in that case. Nonetheless, it was clear from these studies that, in general, TTS rotated much slower (about a factor of 10, on average) than their critical velocities, although details of the distribution such as its true breadth and bimodal nature were not apparent.

Starting in the mid-1980s the technique of directly measuring the rotation period from photometric monitoring of a spinning, spotted surface was applied to a growing number of TTS (*Rydgren and Vrba*, 1983; *Herbst et al.*, 1986, 1987; *Bouvier and Bertout*, 1989). Through a proper analysis of these brightness modulations, the rotation period (P), which is independent of inclination angle, can be measured to an accuracy of about 1%, even for the slowest rotators. We note that P, or its equivalent, angular velocity ($\omega = 2\pi/P$) is currently the most accurately known stellar parameter for most PMS stars and it is known for a large number of them.

The amount of available rotation period data has skyrocketed during the last six years. This advance occurred as a result of the application of widefield optical imaging devices to the problem. For the first time, it became possible to simultaneously monitor hundreds of PMS stars in young open clusters such as the Orion Nebula Cluster (ONC), NGC 2264, and IC 348 (see, e.g., *Mandel and Herbst*, 1991; *Attridge and Herbst*, 1992; *Choi and Herbst*, 1996; *Stassun et al.*, 1999; *Herbst et al.*, 2000a,b, 2001, 2002; *Rebull*, 2001; *Carpenter et al.*, 2001; *Lamm et al.*, 2004, 2005; *Makidon et al.*, 2004; *Littlefair et al.*, 2005; *Kiziloglu et al.*, 2005; Nordhagen et al., in preparation). These imaging devices also provided the first extensive measurements of rotation periods for very-low-mass stars (VLMSs) and brown dwarfs (BDs) in PMS and ZAMS clusters (*Scholz*, 2004; *Scholz and Eislöffel*, 2004a,b, 2005). Altogether ~1700 periods of low-mass PMS stars and young VLMSs and BDs are currently available in the literature. In addi-

tion, for several ZAMS clusters there are extensive rotation period data available on both solar-type stars (see references in *Herbst and Mundt*, 2005) and VLMSs and BDs (see *Scholz*, 2004; *Scholz and Eislöffel*, 2004b, and references therein).

In this article we provide an overview of the rotation period data currently available on PMS objects over the broadest possible mass range that can be investigated with the photometric monitoring technique, i.e., from about 1.5 $M_\odot$ down into the substellar mass regime. Furthermore, we discuss how our present knowledge of the rotation properties of these objects in ZAMS clusters constrains our current understanding of their angular momentum evolution during the first few million years. Our chapter is structured as follows: In section 2 we will describe the observational method employed in measuring rotation periods. The next two sections deal with the rotation properties and the empirical results on angular momentum evolution for solar-type stars, VLMSs, and BDs. In section 5 we review our knowledge of the variability of these objects, the importance of disk-rotation interactions, and magnetically channeled mass accretion over as broad a mass regime as possible. Finally, in section 6 we discuss the confrontation between observation and theory, in particular the model popularly known as "disk locking."

2. OBSERVATIONAL METHODS

The method of determining stellar rotation periods by monitoring the motion of a surface spot of very different temperature from the surrounding photosphere can be traced back to Christopher Scheiner and Galileo Galilei, who employed it to determine the rotation rate of the Sun at low latitude (see, e.g., *Tassoul*, 2000). Modern application of this method exploits the photometric variability induced by the spot or spot group as it is carried around by the star's rotation. Sufficiently dense photometric monitoring over at least a couple of cycles will reveal periodicity in many PMS stars' brightness variations that can be linked with certainty to rotation (*Rydgren and Vrba*, 1983; *Bouvier et al.*, 1986; *Stassun et al.*, 1999; *Rhode et al.*, 2001). This method only works for stars of spectral class G, K, or M (~1.5 $M_\odot$ and less) and for BDs. It is most effective for the mid-K to early-M stars (~0.5 $M_\odot$), where the spot amplitudes are largest (see section 5).

The value of j, the specific angular momentum at the surface, of a spherically symmetric, uniformly rotating star depends on only two variables: rotation period (P) and radius (R). In general, we may expect a star's surface rotation rate to be a function of latitude, as is well known to be the case for the Sun. Remarkably, this effect has never been convincingly demonstrated for any T Tauri star (see, however, *Herbst et al.*, 2005). On the contrary, one finds that measured rotation periods are stable to within the errors of their determination, typically 1% (*Cohen et al.*, 2004; *Kiziloglu et al.*, 2005; Nordhagen et al., in preparation). This presumably means that spots on PMS stars are generally confined

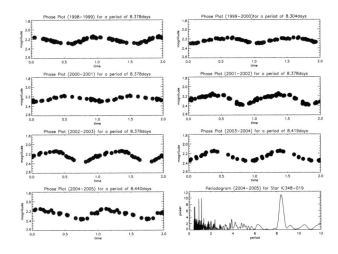

Fig. 1. Light curve of HMW 19, a WTTS in IC 348, based on seven years of photometric monitoring in the Cousins I band at Wesleyan University (Nordhagen et al., in preparation). The bottom right panel shows the periodogram function for one season. Note how the light curve shape and amplitude changes slightly from year to year while the period remains essentially constant to within the error of measurement.

to a small range of latitudes or that their surfaces are rotating in much more rigid fashion than the Sun, or both. It is normally assumed by astronomers working in this area that the measured periodicity in brightness is the rotation rate applicable to all latitudes on the surface of the star.

An example of the data is shown in Fig. 1 from the work of Nordhagen et al. (in preparation). Typically, one searches for periodicity using the Lomb-Scargle periodogram technique, which is effective for unevenly spaced datasets. Evaluation of the *false alarm probability* (FAP) associated with a peak of any given power can be tricky and is best done with a Monte Carlo simulation of the data. Most authors in this field have adopted relatively conservative criteria generally equivalent to a FAP of about 0.01. Data obtained from a single observatory (longitude) have an unavoidable 1 d^{-1} natural frequency embedded in them, imposed by the rotation of Earth. Truly periodic objects therefore normally have more than one significant peak in their periodograms due to the beat phenomenon. Separating true periods from beat periods can be difficult and often requires qualitative judgments about which period does the best job of phasing the data. Most disagreements about periods in the literature arise from this complication. Continued monitoring, or a v sin i measurement, will permit resolution of the question. Occasionally a star will show periodicity at one-half its true rotation period due to the existence of spots in opposite hemispheres of longitude. Examples are V410 Tau (*Vrba et al.*, 1988) and CB 34V (*Tackett et al.*, 2003).

It is expected on general physical grounds (e.g., *Tassoul*, 2000) that rotation rate will vary with depth into a star and helioseismology has shown this to be the case for the Sun, especially within its convective zone (e.g., *Thompson et al.*, 2003). Since we are very far from having any way of determining, either observationally or theoretically, what the dependence of rotation with depth in a PMS star might be like, there is realistically at present no way of measuring how j might vary in a radial sense. From a purely empirical point of view one can, therefore, discuss only the *surface value* of j. Models, of course, can be and have been constructed with assumed rotation laws (e.g., uniform with depth) and with interior mass distributions satisfying the usual constraints of stellar structure. A common assumption is that fully convective stars are rigid rotators, although rigid rotation is not what is observed in the solar convection zone. These assumptions, of course, do allow one to estimate a value of j applicable to the whole star at the expense of increased uncertainty due to the inability to test critical assumptions.

Finally, we note that the surface value of j for a spherical star [see *Herbst and Mundt* (2005) for a discussion of nonspherical stars] depends on stellar radius, a notoriously difficult quantity to measure for PMS stars. Debates in the literature on rotational evolution during the PMS stage often center on how to evaluate the somewhat bewildering data on luminosity and effective temperatures of such stars, from which their stellar radii are inferred. Rotation periods are relatively easy to determine, are highly accurate, and are generally not the source of disagreement about interpretation in this field. Radii, on the other hand, are hard to determine for any individual star, show a large scatter among stars of apparently the same mass, within a single cluster, and are at the root of some recent debates in the literature over how to interpret data on stellar rotation (*Rebull et al.*, 2002, 2004; *Lamm et al.*, 2004, 2005; *Makidon et al.*, 2004; *Herbst and Mundt*, 2005). The problem is exacerbated by the fact that the relative ages of PMS stars of the same mass are set by their relative radii. If radii are in error then ages are in error and evolutionary trends become difficult to discern. We return to this difficulty in what follows but first give an an overview of recent empirical results on rotation.

3. ROTATION OF YOUNG STELLAR OBJECTS

With ~1700 rotation periods measured, there is now a fairly good empirical understanding of spin rates of PMS stars and their dependence on spectral type (or mass). A couple of surprising results have emerged from this, the first of which is the breadth of the rotation period distribution, which extends at least over a factor of 30 for all well-observed mass ranges. The second surprising fact is that the measured rotation period distributions are highly mass-dependent. For stars with 0.4 < M < 1.5 M$_\odot$, the period distribution ranges from ~0.6 d to ~20 d and is clearly bimodal with peaks near 2 and 8 d in the ONC and near 1 and 4 d in NGC 2264 (see Fig. 2). For stars with masses below 0.4 M$_\odot$ (or 0.25 M$_\odot$ depending on which PMS models one adopts) the median of the distribution is about a factor of 2 shorter in both clusters (*Herbst et al.*, 2001; *Lamm et al.*, 2004, 2005).

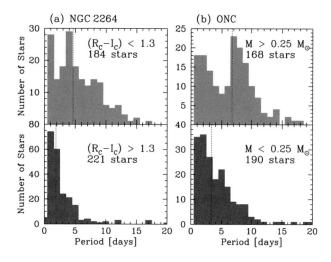

Fig. 2. The rotation period distribution in NGC 2264 and the ONC divided by mass range. Vertical lines indicate the median value of each sample. It is clear that the higher-mass stars in both clusters have longer rotation periods and exhibit bimodal distributions. It is also clear that, within both mass ranges, stars in the ONC tend to have longer rotation periods. The division for the ONC is actually by effective temperature and translates to 0.4 $M_\odot$ for some PMS models.

A recent comparison of the j-distributions of solar-type PMS stars in the ONC and NGC 2264 with solar-type ZAMS stars (0.4–1.2 $M_\odot$) shows that the j-distributions of the PMS and ZAMS sample match very well for high j-values, but *not* for low j-values (*Herbst and Mundt, 2005*). For the ZAMS stars the j-distribution extends by about a factor of 3 toward lower values. This means that the rapid rotators among the optically observable solar-like PMS stars do not lose much, if any, angular momentum on their way to the ZAMS, while the slow rotators do continue to experience some braking. This again indicates that most of the angular momentum problem is solved by the time these stars become observable in the optical. There is some indication that much younger and much deeper embedded PMS stars (i.e., protostars) have about a factor of 2 higher V_{rot} values than their optically visible counterparts (see *Covey et al., 2005*, and references therein), suggesting significant angular momentum loss during this phase. Such an increased angular momentum loss seems to be in accordance with the substantial bipolar outflow activity during this phase.

As mentioned in the introduction, the measured periods vary by about a factor of 30 from 0.6 to about 20 d for solar-like stars. This wide range is present at all spectral types and masses and seen in all samples (clusters and associations) where enough data exist. A difficulty in interpreting these data is that PMS stars contract rather rapidly and if they conserve angular momentum will spin up rapidly. For example, a star with an 8-d rotation period at 1 m.y. will have a 5-d rotation period at 2 m.y. if it conserves angular momentum at its surface. It is, therefore, critical to compare

samples at the same age, if one wants to discern a mass dependence.

The dependence of rotation on mass among PMS stars of a common age was first demonstrated for the ONC by *Herbst et al.* (2001). They found that lower-mass stars, in general, rotate faster than their higher-mass counterparts. In terms of (surface) j, however, there is little or no dependence on mass. It appears that lower-mass stars spin faster primarily because they have smaller radii. When separated at a spectral class of about M2, corresponding to a mass of between 0.25 and 0.4 $M_\odot$ (depending on the model adopted), one finds that the higher-mass stars have a distinctly bimodal period distribution with peaks near 2 and 8 d, while the lower-mass stars, in addition to having a shorter-period median, may have a somewhat smoother distribution, perhaps characterized by a single mode, although this is uncertain.

Lamm (2003) and collaborators (*Lamm et al.,* 2004, 2005) found similar results for another young cluster in which the stars can reasonably be regarded as mostly coeval. His results are shown in Fig. 2, where the bimodal nature of the higher-mass stars in both the ONC and NGC 2264 is clearly seen, as is the more rapid rotation and, perhaps, single-mode distribution of the lower-mass stars. The figure also demonstrates that, for PMS stars of the same mass range, those in NGC 2264 rotate about twice as fast as those in the ONC. Statistical analyses of the distributions confirms this claim; NGC 2264 stars cannot have been drawn from the same parent population as ONC stars.

The interpretation of the difference in rotation rates between the clusters favored by Lamm and collaborators is that NGC 2264 is a factor of 2 older than the ONC and that a significant fraction of the stars have spun up as they contracted, roughly conserving angular momentum. At the same time, not all the stars could have spun up since there remains a fairly well-populated tail of slow rotators at all masses among the NGC 2264 stars. There is no evidence that these slower rotators are, as a group, larger (i.e., younger) than the rapid rotators in the cluster, so an age spread within the cluster does not seem to account for the breadth of the rotation distribution. We discuss in more detail below the interpretation of the broad, bimodal rotation distribution. First, however, we turn to a discussion of the evolution of the rotation of higher-mass stars with time over a broader time frame (~100 m.y.).

4. ANGULAR MOMENTUM EVOLUTION: EMPIRICAL RESULTS

Since the rotation rates of PMS stars are mass dependent it is only appropriate to discuss the evolution of j with time within restricted mass regimes. One such important regime, in part because it includes the Sun, is 0.4–1.2 $M_\odot$. Two other regimes are the very-low-mass PMS stars, corresponding to spectral class M2.5 and later, and the brown dwarfs. At present, the only mass range that is reasonably well constrained by observations is the solar-like range because ZAMS rotation periods are known for a significant number of such stars from studies of young clusters such as the Pleiades, IC 2602, and the α Per cluster (see references in *Herbst and Mundt,* 2005). Here we will discuss the time evolution of j for the 0.4–1.2 M$_\odot$ stars in some detail and then briefly say how this may depend on mass.

4.1. Solar-like Stars (0.4–1.2 M$_\odot$)

It has been known for more than a decade that ZAMS stars of around 1 M$_\odot$ display an enormous range of rotation rates, larger even than is seen among PMS stars. Photometric monitoring of spotted stars in three clusters with ages around 30–120 m.y. has provided periods for about 150 ZAMS (or close to ZAMS) stars (see references in *Herbst and Mundt,* 2005). Radii are well known for these stars, so it is possible to determine the j-distribution with some certitude. This provides a "goal" for the evolution of the PMS distributions.

Early studies of the evolution from PMS to MS assumed that stars started from a somewhat narrow j-distribution (e.g., *Bouvier,* 1994) and that the breadth observed on the ZAMS developed during the PMS phase. A common assumption was that all PMS stars had 8-d periods to start with, based on the larger peak in the distribution of the ONC stars and the corresponding peak for classical T Tauri stars (CTTS) (*Bouvier et al.,* 1993). With increased data samples it is now clear that most of the breadth of the ZAMS population is already built in at the earliest observable PMS phases, represented by the ONC. The current problem divides into understanding how the broad PMS distribution at 1 m.y. came into existence and how rotation evolves between the PMS and ZAMS.

The first question is difficult to constrain with data because Class 0 and I protostars are rare and it is hard to ascertain essential data on them, in particular their masses, radii, and rotation rates. No rotation periods have yet been discovered for protostars. Line-broadening measurements, however, have been made for 38 Class I/flat spectrum objects by *Covey et al.* (2005). Although the sample is necessarily small and heterogeneous, the authors do find an average v sin i for the sample of 38 km/s, which they argue is significantly larger than for CTTS. Since protostars should be, if anything, larger than CTTS, this implies that, as a group, they have larger values of surface angular momenta than CTTS. These data, therefore, reinforce the view that the protostellar or very early PMS stage is a time during which large amounts of angular momentum are lost. This is when the jets and winds that are needed to carry off angular momentum are most prominent and active. This is also the relatively brief period of time (~1 m.y.) when a majority of stars must lose substantial amounts of angular momentum to reduce their j-values to the levels observed in the ONC.

The second aspect of understanding the origin of the ZAMS j-distribution is more amenable to an observational approach, but is not without its own set of complications.

The goal is to understand how a 1-m.y.-old j-distribution, as represented by the ONC, evolves to a 100-m.y.-old distribution, as represented by the ZAMS clusters, and the problem is that there are very few signposts along the way. That is, one requires a substantial number of stars of homogeneous origin (or at least common age) and data on young clusters are scarce. At present, there are really only two PMS clusters with enough periods known to reasonably define the rotation distributions: the ONC and NGC 2264. IC 348 may soon be included among these (see Nordhagen et al., in preparation).

Figure 3 shows the observed situation for these clusters and for the set of ZAMS (or nearly ZAMS) stars drawn from the Pleiades, IC 2602, and the α Per clusters (*Herbst and Mundt,* 2005). There are many *caveats* and considerations in comparing these samples and the interested reader is referred to the original work for the details. Here we touch only on the main points, the first of which is that there are clear, statistically significant differences between the three j-distributions represented by the ONC, NGC 2264, and the ZAMS clusters. If one assumes that each sample had similar initial rotation parameters, then these differences can be interpreted as an evolutionary sequence.

The age of the ONC is about 1 m.y. and the average radius of its 0.4–1.2 M$_\odot$ stars is 2.1 R$_\odot$. NGC 2264 stars of the same mass (spectral class) range are significantly smaller, with a mean radius of 1.7 R$_\odot$, implying an age of about 2 m.y. ZAMS cluster stars are, of course, even smaller, and

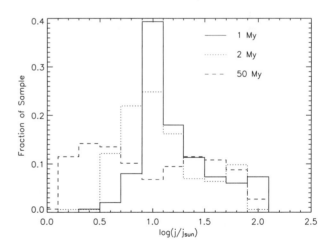

Fig. 3. The observed j-distribution of the ONC, NGC 2264, and combined three other clusters corrected for wind losses. It is clear that there is little change on the high-j side, implying that rapidly rotating stars nearly conserve angular momentum as they evolve from the PMS to the MS. However, there is a broadening of the distribution on the low-j side, which is already noticeable in the comparison of the 1-m.y.-old ONC with the 2-m.y.-old NGC 2264 cluster and becomes quite dramatic when comparing the PMS and ZAMS clusters. This indicates that slowly rotating PMS stars must lose substantial additional amounts (factor of 3 or more) of their surface angular momentum during contraction to the ZAMS.

individual radii can be employed in calculating j for them since their effective temperatures and luminosities are much more secure than for the PMS stars. Note that radius evolution is very nonlinear over the course of the first 30 m.y.; most of the contraction occurs within the initial few million years.

It is evident, to a first approximation, from Fig. 3 that the evolution of j occurs in a bimodal fashion. The high-j stars (i.e., the rapid rotators) evolve with essentially no loss of angular momentum from ONC age to the ZAMS clusters. In this interpretation, no physical mechanism beyond conservation of angular momentum is required to explain the evolution of about half the sample. Empirically, there is no need for angular momentum loss (by disk locking, stellar winds, or any other process) for the 50–60% of the sample that is already among the rapid rotators at ONC age. Conversely, there is no need to invoke any special mechanism to explain the fast rotation of these stars. They were already spinning rapidly at 1 m.y. and they have spun even faster as they contracted to the ZAMS.

On the other hand, the initially slowly rotating stars in the ONC follow a different evolution. It is clearly seen that they continue to lose angular momentum as they evolve to NGC 2264 age and beyond. The ZAMS j-distribution is about a factor of 3–5 broader than the ONC j-distribution entirely because of the low-j stars. To summarize, we can understand to a first approximation the evolution from PMS to ZAMS in terms of only two processes: angular momentum conservation, which applies for the initially rapidly rotating half of the sample, and a braking mechanism that applies for the initially slowly rotating half. The distribution broadens on one side only — the slowly rotating side. This is an important clue to the braking mechanism, which we discuss in section 6.

4.2. Very-Low-Mass Stars (<0.4 M$_\odot$)

The two largest homogeneous samples of rotation periods, in the ONC and NGC 2264, include about 200 periods per cluster for very-low-mass stars, an object class here defined as stars with spectral class later than M2, corresponding to M < 0.4 M$_\odot$ or so, depending on the stellar models chosen. Together with smaller VLM period samples for slightly older objects, most notably in the σ Ori cluster (*Scholz and Eislöffel*, 2004a), the ε Ori cluster (*Scholz and Eislöffel*, 2005), and the much older Pleiades (*Scholz and Eislöffel*, 2004b), they allow us to make meaningful statistical comparisons between periods for VLM stars and their higher-mass siblings.

The initial period distribution of VLM stars is well-defined by the large samples in the ONC and NGC 2264. Periods usually range from a few hours up to 10 d in both clusters, a dynamic range similar to solar-like stars. As evident in the bottom panel of Fig. 2, however, slow rotators are clearly much rarer in the VLM regime. For example, in NGC 2264 only 4% of the VLM stars have P > 10 d. This

change is reflected in the median period, which is 3.33 d for VLM stars in the ONC and 1.88 d in NGC 2264, about a factor of 2 lower than for more massive stars (see Fig. 2). The period distribution in the slightly older clusters σ Ori and ε Ori is, as shown by *Scholz* (2004) and *Scholz and Eislöffel* (2005), roughly comparable to the NGC 2264 sample, both in median period and limiting values. These similar distributions are not surprising in consideration of the similar ages estimated for the clusters.

First attempts to model the angular momentum evolution in the VLM regime have been made by *Terndrup et al.* (1999) and *Sills et al.* (2000), both based on rotational velocity data of clusters with ages between 30 and 700 m.y. Both papers arrive at the conclusion that the rotational braking by stellar winds changes in the VLM regime: The so-called "saturation limit" (ω_{crit}) probably drops quickly at very low masses, with the result that basically all VLM stars rotate with $\omega > \omega_{crit}$, and are considered to be in the "saturated" regime of the rotation-activity relation. As a consequence, the rotational braking by stellar winds follows an exponential law [$\omega \propto \exp(t)$] rather than the established Skumanich law ($\omega \propto \sqrt{t}$) for stars of solar-like mass.

With the available VLM period data, it is now possible to have a more detailed look at their angular momentum evolution. Based on the VLM periods in σ Ori (age ~3 m.y.) and the Pleiades (age ~120 m.y.), *Scholz and Eislöffel* (2004b) investigated the rotational evolution on timescales of ~100 m.y. by using the σ Ori periods as starting points and evolving these periods forward in time, taking into account basic angular momentum regulation mechanisms. That way, evolutionary tracks in the period-age diagram were produced that can then be compared with period distributions in older clusters.

Figure 4 shows the results from *Scholz and Eislöffel* (2004b). Plotted are the VLM periods for σ Ori and the Pleiades plus evolutionary tracks. The dotted lines show the evolution assuming angular momentum conservation, i.e., the tracks are completely determined by the contraction process. It is clear that this is not a reasonable description of the available period data, because the models predict an upper period limit <20 h in the Pleiades, whereas the observed periods range up to ~40 h. Furthermore, it is evident from Fig. 4 that the most rapid rotators in σ Ori would rotate at periods well below 1 h at the age of the Pleiades (i.e., faster than the breakup velocity). We would like to point out that this upper period limit in the Pleiades is nicely confirmed by v sin i data from *Terndrup et al.* (2000). Thus, some kind of rotational braking must play a role for VLM objects on these timescales. The second model (dash-dotted lines) includes a Skumanich-type wind-braking law as observed for solar-type stars. In this case, the model predicts an upper period limit >100 h in the Pleiades, significantly higher than the observed value. Thus, the Skumanich braking is too strong and cannot be applied in the VLM regime. In a third model (dashed lines), the model uses an exponential wind-braking approach. Except for the two

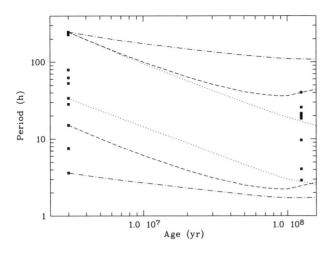

Fig. 4. Rotation period of VLM stars in the σ Ori cluster and the Pleiades as a function of age (from *Scholz and Eislöffel,* 2004b). The lines show model calculations, which use the periods in σ Ori as starting values. Evolutionary tracks are shown for three assumptions: angular momentum conservation (dotted lines), Skumanich-type wind losses (dashed-dotted lines), and exponential wind braking (dashed lines).

fastest rotators in σ Ori, model and period agree fairly well. Thus, in agreement with *Terndrup et al.* (2000), it was found that the period evolution for VLM objects on timescales of ~100 m.y. is mainly determined by contraction and exponential angular momentum loss by stellar winds.

How can we explain the failure of the best-fitting model for the fastest rotators in σ Ori? First of all, the existence of these fast rotators in σ Ori has been confirmed independently by *Zapatero-Osorio et al.* (2003), who found a brown dwarf with a period of only 3.1 h in this cluster (see section 4.3). It has to be mentioned that these objects rotate fast in terms of their breakup period, which is around 3–5 h for VLM stars at the age of σ Ori. Their fast rotation might change the physics of these objects, e.g., it can be expected that they are strongly oblate. It is, perhaps, not surprising that the simple models by *Scholz and Eislöffel* (2004b) cannot provide a correct description of the ultrafast rotators. Clearly more sophisticated modeling has to be done for these objects.

The fact that the rotation of VLM objects follows a weak exponential braking law rather than the Skumanich law has been interpreted in terms of the magnetic field structure of these objects (*Barnes,* 2003; *Scholz and Eislöffel,* 2004b). VLM objects are fully convective throughout their lifetime, and will never develop a radiative core. As a consequence, they are probably not able to host a solar-type, large-scale dynamo, which is believed to operate in the transition layer between convective and radiative zone. Alternative mechanisms to explain the magnetic activity in the VLM regime are the so-called turbulent dynamo (*Durney et al.,* 1993)

and the α² (*Chabrier and Kueker,* 2006). Both types of dynamos predict reduced Alfven radii in comparison with the solar-type αω dynamo and thus only weak braking by stellar winds. Therefore, the results from the rotational evolution analysis are consistent with the possibility of a change of dynamo mechanism in the VLM regime.

It should be noted that the cited models are not able to constrain the influence of more rapid angular-loss mechanisms (e.g., from a disk interaction) on the rotational evolution, because they operate on timescales of 100 m.y. Disk-interaction timescales, for example, are probably only a few million years (see section 6). To assess the validity of efficient loss mechanisms such as the disk-interaction hypothesis in the VLM regime, it is clear that younger object samples have to be considered. This has been done by *Lamm et al.* (2005), in their comparison of the period samples in the ONC and NGC 2264. They found that the period evolution for VLM stars from ~1 m.y. (ONC) to ~2 m.y. (NGC 2264) can be described with a scenario of "imperfect" disk locking, in the sense that the rotation of the stars is not actually "locked" with constant period. Instead, it spins up, but not as fast as it would with the assumption of angular momentum conservation. Thus, the disk brakes the rotation somewhat, but the interaction between star and disk is less efficient than for solar-mass stars (see section 6 for more discussion of this).

We conclude that the mechanisms of angular momentum regulation in the VLM regime are similar to solar-mass stars, but the efficiency of these mechanisms is a function of mass. They appear to be less efficient in the VLM regime, resulting in more rapidly rotating objects on the ZAMS. We also note that, since the radii of the VLM stars are smaller, at all ages, than their solar-like siblings, for a given amount of specific angular momentum they spin faster. As *Herbst et al.* (2001) have shown, j does not vary much with mass in the ONC, so the faster rotation of the VLM stars at 1 m.y. may be entirely a result of their having contracted to smaller sizes than the higher-mass stars.

4.3. Brown Dwarfs

In this section we discuss the available rotational data for brown dwarfs (BDs), i.e., substellar objects intermediate in mass between stars and planets. BDs are defined as objects with masses below the hydrogen-burning mass limit [~0.08 M$_\odot$ (*Chabrier and Baraffe,* 1997)]. Since it is only rarely possible to determine the object mass directly, the effective temperature and luminosity in combination with stellar evolutionary tracks is usually used to identify BDs. Therefore, in the following we refer to "brown dwarfs" as objects whose spectral types and luminosities classify them as substellar, although some of them might have masses slightly higher than the substellar limit, because of uncertainties in spectral typing and atmosphere modeling.

Following the discovery of the first BDs (*Nakajima et al.,* 1995; *Rebolo et al.,* 1995), hundreds of them have been identified in star-forming regions (e.g., *Comerón et al.,* 2000; *López Martín et al.,* 2004), open clusters (e.g., *Zapatero-Osorio et al.,* 1997; *Barrado y Navascués et al.,* 2001), and in the field (*Kirkpatrick et al.,* 1999; *Phan-Bao et al.,* 2001). A large number of follow-up studies led to rapid progress in our understanding of the physical properties of these objects. For example, about 30 rotation periods have been measured for BDs at different ages, complemented by rotational velocity (v sin i) data. Here, we review the available periods and their implications for our understanding of rotation and the angular momentum evolution of brown dwarfs.

The first rotation period for a likely BD was published for an object in the open cluster α Per, which has an age of ~50 m.y. (*Martín and Zapatero-Osorio,* 1997). In the following years, periods have been measured for three BDs in the ~1-m.y.-old Cha I star-forming region (*Joergens et al.,* 2003), three objects in the ~3-m.y.-old σ Ori cluster (*Bailer-Jones and Mundt,* 2001; *Zapatero-Osorio et al.,* 2003), one object in the ~120-m.y.-old Pleiades cluster (*Terndrup et al.,* 1999), and three so-called "ultracool dwarfs" in the field (*Bailer-Jones and Mundt,* 2001; *Clarke et al.,* 2002), with late-M or early-L spectral types. We would like to caution that it is not, in all cases, unambiguously clear that the photometrically derived period corresponds to the rotation period. This is particularly important for the ultracool field dwarfs, which, in many cases, have relatively sparse sampling and, perhaps, less-stable surface features, factors that hamper a reliable period detection. In this review, we include discussion only of periods that authors themselves consider to be likely rotation periods.

It is interesting to note that until 2003 all known periods for objects with ages >1 m.y. were shorter than 1 d, providing the first evidence that evolved BDs are, in general, very rapid rotators. This is supported by spectroscopic rotational velocity data, indicating that the majority of evolved BDs have v sin i > 10 km s⁻¹ (*Mohanty and Basri,* 2003; *Bailer-Jones,* 2004). By contrast, the rotation periods for the three youngest objects in the sample, i.e., those in Cha I, are in the range of 2 to 4 d.

In the last two years, deep widefield monitoring campaigns have more than doubled the number of known periods for BDs. In the young open clusters σ Ori and ε Ori, *Scholz and Eislöffel* (2004a, 2005) measured photometric periods for 18 probably substellar objects, 9 for each cluster. Both clusters belong to the young population of the Ori OB1b association, which has an age of about 3 m.y., although the σ Ori objects are on average probably slightly younger than those in ε Ori (*Sherry,* 2003). In σ Ori, the periods cover a range from 5.8 to 74 h with a median of 14.7 h, whereas in ε Ori the total range is 4.1 to 88 h with a median of 15.5 h. Thus, the BD periods in both clusters are comparable. Additionally, periods for two likely substellar members of the Pleiades (*Scholz and Eislöffel,* 2004b) and IC 4665 [age ~40 m.y. (*Scholz,* 2004)] have been published. In total, the rotation sample for BDs (or objects very close to the substellar limit, see above) comprises 31 periods. In

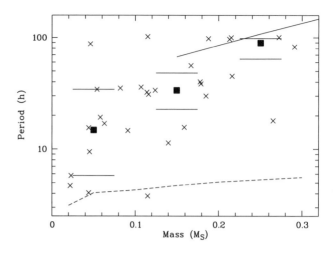

Fig. 5. Period vs. mass for the ε Ori cluster (from *Scholz and Eislöffel,* 2005). Filled squares mark the period median, horizontal lines the quartiles. The solid line is the period median for the ONC, the dashed line the breakup limit for ε Ori.

the following, we will discuss the period-mass relationship and rotational evolution in the substellar regime based on this dataset.

To separate age and mass effects, the period-mass relationship has to be studied for each age separately. In Fig. 5 (from *Scholz and Eislöffel,* 2005) period vs. mass is plotted for the ε Ori objects (crosses), where one of the largest BD period samples is available. The figure additionally shows the median period for certain mass bins (filled squares), together with the period-mass relationship in the ONC (solid line). We have already demonstrated in section 4.2 that the average period decreases with decreasing object mass in the VLM star regime. As can be seen in Fig. 5, this trend continues well down into the substellar regime. The median in the BD regime is clearly lower than for VLM stars. The same result is obtained for the period sample in the σ Ori cluster. It is particularly interesting that the BD period range in young open clusters extends down to the breakup period, which is the physical limit of the rotational velocity. In Fig. 5 the breakup limit is overplotted as a dashed line [calculated using the evolutionary tracks of *Baraffe et al,* (1998)]. For masses below 0.1 M⊙ a subsample of objects rotates very close to breakup. Since extremely fast rotation affects the evolution of the objects, as we know from massive stars, this has to be taken into account in future evolutionary models for rapidly rotation BDs.

All available BD periods are plotted in Fig. 6 as a function of age. For the matter of simplicity, we assigned an arbitrary age of 1 G.y. to all evolved field objects. Whereas very young BDs have periods ranging from a few hours up to 4 d, all periods for older objects are shorter than 15 h. Please note that the upper period limit in the youngest clusters may be set, in part, by the upper period detection limit.

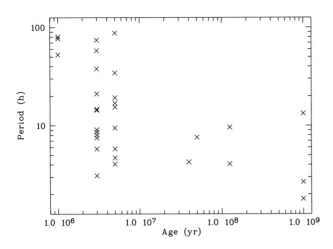

Fig. 6. Rotation period of brown dwarfs (and objects very close to the substellar limit) as a function of age. For simplicity, periods for evolved field dwarfs were all plotted at an age of 1 G.y.

In the σ Ori cluster, however, the detection limit is ~10 d (*Scholz and Eislöffel*, 2004a), leading us to the tentative conclusion that 100 h might be a realistic value for the upper period limit at very young ages. Two clear trends can be seen from Fig. 6, related to the upper and the lower period limit, which we will discuss separately in the following paragraphs.

First, the upper period limit is apparently more or less constant for ages <5 m.y. and is decreased by a factor of about 6–10 in the more evolved clusters with ages >40 m.y. The second aspect of Fig. 6 that we wish to discuss is the lower period limit, which appears to be in the range of a few hours at all evolutionary stages. The only exception is the (very sparse) period sample for Cha I at 1 m.y., but this might be related to small number statistics or a time series sampling unable to detect short periods. Within the statistical uncertainties, the lower period limit is constant with age. Thus, for a fraction of ultrafast rotating BDs the period changes by less than a factor of 2 on timescales of ~1 G.y. This is surprising, because on the same timescales BDs contract and we should therefore expect a rotational acceleration at least by a factor of about 10. One possible explanation is that we have not yet found the fastest rotators among the evolved BDs. Assuming angular momentum conservation, we should expect objects with periods down to ~0.5 h at ages >200 m.y., i.e., when the contraction process is finished. Whether these objects exist or not has to be probed by future observations.

If the lower period limit for evolved BDs, however, is really in the range of a few hours, as indicated by the available period data, the fastest rotators among the BDs have to experience strong angular momentum loss on timescales of 1 G.y. It is unclear what mechanism could be responsible for this rotational braking. To summarize this section, while the rotation data on BDs are still very scanty com-

pared to the low-mass stars, they appear to be a natural extension of the phenomena observed for stars. There is nothing yet to suggest that their rotation properties and evolution are discontinuous in any way from stars.

5. OVERVIEW OF VARIABILITY, SPOTS, ACCRETION, AND MAGNETIC STAR-DISK INTERACTIONS IN YOUNG STELLAR OBJECTS AND BROWN DWARFS

Most of our knowledge of the rotational properties of the objects discussed here is based on variability studies. Therefore we regard it as important to give an overview on this subject, with particular attention to magnetically induced cool stellar spots and magnetically channeled variable mass accretion, the principle variability mechanisms in weak T Tauri stars (WTTS) and CTTS, respectively.

A detailed study of the various sources of TTS variability has been carried out by *Herbst et al.* (1994), based on a large electronic UBVRI catalog with about 10,000 entries for several hundred stars. A further variability study, with particular attention to periodic variations, has been carried out by *Bouvier et al.* (1995). On the basis of these two and related studies one can distinguish at least five types of common PMS variability, at least the first two of which are also seen in BDs. These are:

1. Periodic variability caused by rotational modulation of the stellar flux by an asymmetric distribution of cool spots or spot groups on the stellar surface. This type of variability is more frequently seen in WTTS but can also be observed in the CTTS. An example is shown in Fig. 1. The typical amplitudes for these variations range from about 0.03–0.3 mag in the V band, with the most extreme values reaching 0.8 mag in V and 0.5 mag in I. Spot sizes and temperatures have been derived from the observed amplitudes and the derived spot coverage factors range from a few percent up to 30% [for ΔV = 0.5 mag, see *Bouvier et al.* (1995)].

2. Irregular variations, probably caused by highly variable, magnetically channeled accretion from the circumstellar disk onto the star. The accretion rate onto the star is not only variable in time but the accretion zones are certainly not uniformly distributed over the stellar surface. The complex interaction between the stellar magnetosphere and the inner disk is evidently highly dynamic and time dependent. The typical amplitudes of the resulting (largely) irregular variations are a factor of 2–5 larger in V than those of the periodic variations observed in many WTTS. Variations by 1.5 mag in V within a few days are not unusual and some stars can vary that much within hours. This type of variability is designated as Type II by *Herbst et al.* (1994).

3. Periodic variations due to hot spots. This type of variation (also known as Type IIp) is only seen in CTTS and the hot spots are presumably at the base of the magnetic channels. The periodicity typically persists for only a couple of rotation cycles. Since the magnetic field configuration

is highly unstable, the size and location of these spots is correspondingly changing within a few rotation periods. This is quite in contrast to the cool spots, which may last for hundreds to thousands of rotations. The amplitudes of the rotational modulation by hot spots is typically a factor of 2–3 larger in V than those seen in WTTS due to cool spots, but more extreme cases have been observed in some CTTS.

4. Flare-like variations, mainly seen in the U and B band in the WTTS. This type of variability is probably also present in CTTS, but difficult to distinguish from the strong irregular variability.

5. UX Ori-type variability (also referred to as UXors) is mostly seen in early type TTS (earlier than K0) and in Herbig Ae/Be stars. It is designated as Type III variability by *Herbst et al.* (1994). The variation amplitudes can be very large (up to 2.8 mag in V), but the timescales are about a factor of 2–10 longer compared to the irregular variations of CTTS. Also, the stars often get bluer when fainter and the Hα flux does not correlate with the continuum flux as it does in the CTTS. *Herbst and Shevchenko* (1999) discuss this type of variability in detail. Its cause is still uncertain, although many authors believe it derives from variations in circumstellar extinction.

The photometric behavior of numerous periodically variable WTTS on timescales of several years has been studied in the ONC by *Choi and Herbst* (1996) and *Steinhauer et al.* (1996) and in IC 348 by *Cohen et al.* (2004) and Nordhagen et al. (in preparation). In the latter case the study now extends over seven years. These investigators found clear variations in the amplitude and lightcurve shape of these periodic variables on timescales of less than one year, probably due to changes in the spot sizes and spot distributions (see Fig. 1). In no case, however, did they find definitive evidence for a change in period by more than the measurement limit of about 1%, indicating that differential rotation in WTTS is much less than in the Sun or that spots are confined to certain latitude zones. This conclusion is also supported by the very similar rotation period values found in the ONC by *Herbst et al.* (2002) in comparison to previous studies by *Stassun et al.* (1999) and *Herbst et al.* (2000a). In NGC 1333 there may be an "exception that proves the rule": A T Tauri star in that region has been found with a period of 5.6 d over three seasons and then 4.6 d over the next two seasons, a change exceeding 20% (*Herbst et al., 2005*). This proves that such stars can be found, but they are clearly exceedingly rare.

It is apparent in the data that the amplitudes due to rotational modulations by cool spots strongly decrease with age. While WTTS commonly have spot amplitudes of 0.1–0.3 mag in V, this drops to about 0.02 mag for stars in the Hyades (600 m.y.) and at least another factor of 10 in the case of our Sun (*Strassmeier, 1992*). The same is true for VLM stars. While the amplitudes for the VLM stars in NGC 2264 (see Fig. 7) are similar to those observed in the Pleiades (*Scholz and Eislöffel, 2004b*) (which could be due to

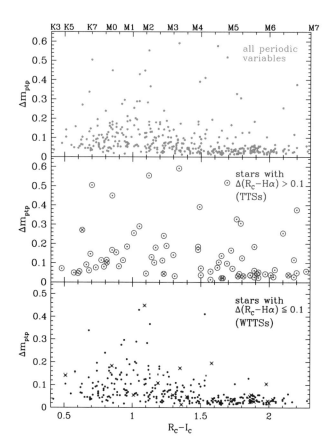

Fig. 7. The peak-to-peak variation of the 405 periodic variables in NGC 2264 found by *Lamm et al.* (2004) as a function of their R-I color. In the top panel all 405 periodic variables are shown. The diagram in the middle panel contains only 89 stars with strong Hα emission. In the bottom panel only the 316 stars with weak Hα emission (WTTS) are shown. Note the strong decrease in amplitude, by about a factor of 3, for the WTTS in the bottom panel with R–I ≥ 1.5 (from Mundt et al., in preparation).

biasing of detections toward large amplitudes) they drop by about a factor of 3 for late-field M stars (Rockenfeller et al., in preparation). It is obvious that the decreasing amplitudes with increasing age makes it more difficult to study photometrically the rotational properties of ZAMS and older clusters.

The dependence of amplitudes of the periodic light variations in PMS stars on mass (or effective temperature) has been investigated by Mundt et al. (in preparation) for the case of NGC 2264 over a broad range of spectral types from ~K3 to ~M6.5. This investigation is based on the data of *Lamm et al.* (2004, 2005). The main results are illustrated in Fig. 7, which shows the peak-to-peak variation as a function of the $(R_C–I_C)$ color for three different subsamples of the periodic variables. The peak-to-peak amplitudes were derived by fitting a sine wave to the phased light curves. In the top panel of Fig. 7, all 405 periodic variables are shown. In the middle panel only the peak to peak variations of the 89 stars with large Hα indices [Δ(R_C–Hα) > 0.1 mag, i.e., strong Hα emitters] are shown. It is evident that these stars show a large scatter in their peak-to-peak variations, i.e., we probably deal with a mixture of stars in which either cool or hot spots are responsible for the periodic variability. Most important for our discussion is the bottom panel of Fig. 7, which shows the peak-to-peak variations for the remaining 316 stars, which are mostly WTTS due to their weak Hα emission [i.e., Δ(R_C–Hα) ≤ 0.1 mag]. From this panel it is clearly evident that the peak-to-peak variations of the cooler WTTS with (R_C–I_C) ≥ 1.5 mag show on average a factor of ~3 smaller peak-to-peak variations than stars with (R_C–I_C) ≤ 1.3 mag. This impressive decrease in the peak-to-peak variations for stars with (R_C–I_C) ≥ 1.5 mag is practically independent of the period of the investigated stars. Only for the slowest rotators might there be a tendency for somewhat higher peak-to-peak variations. We note that the objects with (R_C–I_C) ≥ 2.0 mag are all VLM stars (≤0.1–0.15 $M_\odot$) with some of them probably falling below the substellar limit. A similar decrease in the peak-to-peak variations of the periodic variables with decreasing mass was found by *Scholz and Eislöffel* (2004) in the Pleiades.

This tremendous decrease in the peak-to-peak variations of the coolest periodically variable WTTS either implies that the spot coverage of the stellar surface, the asymmetry of the spot distribution, and/or the contrast between the spot and the photospheric environment has decreased. It has been argued by Mundt et al. (in preparation) that the spot coverage (spot size) has probably decreased as a result of the much poorer coupling between the magnetic fields and the atmospheric plasma caused by the low ionization fraction in the atmosphere of these very cool objects. The change in peak-to-peak variations of WTTS due to cool spots was also investigated by *Bouvier et al.* (1995), but only for spectral types between about G0 and K7 and only for a sample of 23 stars. In this spectral range the peak-to-peak variations apparently *increase* with decreasing temperature. Such a behavior is not inconsistent with the data displayed in Fig. 7, since there the median value for the peak-to-peak variations, in fact, increases until (R_C–I_C) ~ 1 mag (~M1).

It is quite obvious from Fig. 7 and the work summarized above that periodic light modulations due to cool spots are observable over a very large mass range; i.e., from about 1.5 $M_\odot$ well down into the substellar mass regime. We know that the same is true for variable mass accretion and accretion related activity phenomena. Such CTTS-like phenomena are observable in young VLM stars and young BDs down to masses near the deuterium-burning limit (see, e.g., *Mohanty et al.*, 2005a, for a review). Examples of such accretion-related activity are numerous and include strong line emission (e.g., *Mohanty et al.*, 2005b), variable line emission (e.g., *Barrado y Navascués et al.*, 2003), irregular variations in the continuum flux (e.g., *Zapatero-Osorio et al.*, 2003; *Scholz and Eislöffel*, 2004a, 2005), and bipolar emission line jets (*Whelan et al.*, 2005). It remains for us to discuss the confrontation between theory and observation that involves the connection between rotation, magnetic fields, and accretion disks in all the mass ranges discussed here, from stellar to substellar.

6. INTERPRETATIONS: COMPARISON OF EMPIRICAL RESULTS WITH THEORY

The leading theory to account for the slower than critical rotation rates observed for T Tauri stars is commonly known as "disk locking" and was first proposed by *Camenzind* (1990) and *Königl* (1991). It has been worked out in more detail by *Shu et al.* (1994) and others, most recently by *Long et al.* (2005). Its principal success in the rotation arena is in predicting an equilibrium rotation period in the 2–8-d range, characteristic of the observations, when other parameters, such as surface magnetic field strength and accretion rate, are set to nominally representative values. Observational credence is given to this picture by the fact that a statistically significant anticorrelation does exist between angular velocity and various disk indicators such as near-infrared excess and Hα equivalent width (*Edwards et al.*, 1993; *Herbst et al.*, 2002; *Lamm et al.*, 2004, 2005; *Dahm and Simon*, 2005; *Rebull et al.*, 2005). Figure 8 shows the anticorrelation for one cluster, the ONC.

Note that these anticorrelations, including the one in Fig. 8, generally indicate that rapid rotators are very unlikely to have disks, while slow rotators may or may not have them. This is actually what one would expect from a disk-locking (or, more generally, disk regulation) scenario. In particular, a star does not respond instantaneously to the loss of angular velocity regulation. It takes time for a star

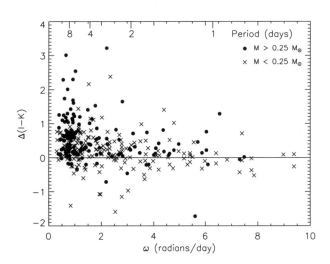

Fig. 8. Infrared excess emission depends on rotation for PMS stars in the ONC (from *Herbst et al.*, 2002). The indicated mass range is based on a spectral class range and would translate into masses greater than 0.4 $M_\odot$ by some models. Since infrared excess is an indication of an accretion disk, this figure clearly demonstrates the rotation disk connection.

to contract and spin up. Some stars that are no longer regulated, i.e., have lost their accretion disks recently, would be expected not to have had time yet to spin up. They would appear as slow rotators that lacked accretion disks. It would be harder, in this scenario, to explain rapid rotators with disks and, indeed, such stars are rare in the sample. It would be interesting to inquire more deeply if there are, in fact, any such cases that cannot be ascribed to errors of observation.

In spite of its successes, the disk-locking theory has been controversial over the years. Many authors have pointed to various shortcomings of either the theory or its confrontation with observation (e.g., *Stassun et al.,* 1999, 2001; *Rebull,* 2001; *Bouvier et al.,* 2004; *Uzdensky,* 2004; *Littlefair et al.,* 2005; *Matt and Pudritz,* 2004, 2005a,b). The interested reader is referred to the meeting report by *Stassun and Terndrup* (2003) and references therein, as well as to the recent review by *Mathieu* (2004), for further discussion. Here we note that the difficulties appear to us to be rooted in the natural complexity of the phenomena and our limited ability to either model them or to obtain a sufficient amount of accurate data to empirically constrain them. As with all MHD processes, theoretical progress can only be made with simplifying assumptions such as axisymmetric magnetic fields and steady accretion, which we know in the case of T Tauri stars are not realistic. For example, if the geometry of the problem were truly axisymmetric for a typical star then we would not observe cyclic photometric variations with the rotation period. If accretion were truly steady, we would not see the large-amplitude irregular photometric variations characteristic of CTTS.

Time-dependent magnetic accretion models may result in a decreased braking of the stellar rotation rates in comparison to the simple disk-locking scenario (*Agapitou and Papaloizou,* 2000). It has been argued by *Matt and Pudritz* (2005a,b) that the magnetic braking of PMS stars may not be due to a magnetic star-disk interaction, but may result from a magnetically driven wind emanating directly from the star. If the high rotation rates of Class I protostars compared to Class II protostars (CTTS) observed by *Covey et al.* (2005) are confirmed by further studies, they may imply angular momentum loss rates higher than predicted by the disk-locking scenario. Note that in the disk-locking scenario the Class II sources should rotate as fast as the Class I sources. If they in fact rotate a factor of 2 slower, it may demand a breaking mechanism more efficient than disk-locking to account for the observations. A strong magnetically driven wind, as proposed by *Matt and Pudritz* (2005a,b), is certainly an idea that deserves further study. An observational argument in favor of a wind emanating directly from the stellar surface is the broad and deep P Cygni profiles observed in some CTTS (see, e.g., *Mundt,* 1984). The deep blueshifted absorptions of these line profiles would be much harder to understand if the wind acceleration region is far from the stellar surface, as would be the case for a disk wind.

From the theoretical side, part of the difficulty in testing the disk-locking theory is that it is hard to pin down specific, testable predictions that are based on realistic models, especially when parameters such as the magnetic field strength are involved, which are hard to observe. Some attempts to test the theory in detail have met with mixed success (*Johns-Krull and Gafford,* 2002). Observationally, we are further faced with the problem of a very broad distribution of rotation rates at any given mass and age as well as the difficulties of even establishing mass and age for PMS stars. Magnetic phenomena are notoriously complex and this apparition of their importance is no exception. Considerations such as these have led *Lamm et al.* (2004, 2005), for example, to employ the terminology "imperfect" disk locking to account for the data. The notion that a star's rotation is fully controlled by its interaction with a disk during all of its PMS phase in a way that can accurately be described by a current disk-locking theory is probably oversimplified. Nonetheless, the concept of disk regulation of, or at least effect on, rotation seems undeniable in the light of the observed correlations, such as seen in Fig. 8.

Indeed, there is no way to understand the evolution of rotation of stars from PMS to ZAMS without invoking significant braking for a significant amount of time (cf. Fig. 3). All attempts to model this have employed such "disk-locking" (e.g., *Armitage and Clarke,* 1996; *Bouvier et al.,* 1997; *Krishnamurthi et al.,* 1997; *Sills et al.,* 2000; *Barnes et al.,* 2001; *Tinker et al.,* 2002; *Herbst et al.,* 2002; *Barnes,* 2003; *Rebull et al.,* 2004; *Herbst and Mundt,* 2005). Usually, the approach is to include the "disk-locking time" as a parameter in the models. In its simplest form this is the time that the period remains constant, after which it is allowed to change in response to the contraction of the star.

Obviously a real star would not be expected to maintain a constant rotation period for times on the order of 1 m.y., even if the disk-locking theory were strictly true, since other parameters of the problem, such as magnetic field strength and accretion rate, are likely to vary on this timescale (or shorter — perhaps much shorter). Hence, the parameterized models are only approximations of reality. What is interesting and significant is that it is simply not possible to model the rotational evolution of solar-like (0.4–1.2 $M_\odot$) stars from PMS to ZAMS without significant braking for about half of them. And, the observations show that the half that must be braked is the half already slowly rotating (Fig. 3), precisely the same stars that are most likely to show evidence of circumstellar disks (Fig. 8). We note that the maximum braking times found by various authors are on the same order, normally around 5–10 m.y., comparable to the maximum lifetime of accretion disks derived from near-infrared studies (e.g., *Haisch et al.,* 2001). Of course, it is possible that the influence of a disk on rotation may wane before its detectability in the infrared, so if disk-locking times are somewhat shorter than disk-detectability times, this may not be surprising. Nonetheless, rotation studies and infrared excess emission studies appear to concur, indicating that substantial gaseous accretion disks have disappeared by about 5 m.y. for almost all stars and by about 1 m.y. for half of them. This, in turn, implies that the era for gas-giant-planet formation has ended. Terrestrial planets may, of course, continue to form around such stars for much longer periods of time.

Finally, we would like to make some comments on the disk-locking scenario and on magnetic star-disk coupling in general as it applies to VLMs and BDs. As discussed above, the disk-locking scenario is quite successful, in some ways, in explaining important aspects of the rotational evolution of low-mass stars with masses larger than about 0.4 M$_\odot$. On the other hand, for VLM stars below this mass limit, the evidence is less convincing (see, e.g., *Lamm et al.,* 2005). Nevertheless, it appears from the available data that those VLM objects showing evidence for active accretion (e.g., strong Hα emission) do rotate on average much slower than those stars without any accretion indicators. This means that these low-mass objects do indeed lose angular momentum, but probably at a lower rate than in the case of precise disk locking. Therefore the term "moderate angular momentum loss" was proposed by *Lamm et al.* (2005).

The efficiency of disk locking is likely to decrease in the VLM star and BD regimes. Unfortunately, it is very difficult to test these ideas observationally because of the difficulties of determining rotation periods, masses, and ages for large samples of VLMs and BDs. In particular, we lack a sufficient amount of data for ZAMS VLMs that could provide a goal for the PMS evolution analogous to what is available for the solar-type stars. As argued in section 4.2, rotational braking by stellar winds is also probably less efficient for VLM objects, either because they have no solar-like dynamo or because of little coupling between gas and magnetic field due to low gas temperatures in the atmosphere. As already noted above, the period distribution at very young ages extends down to the breakup limit (see Fig. 5). This might have an effect on the fastest rotators, in the sense that strong centrifugal forces remove angular momentum and thus brake the rotation, as argued by *Scholz and Eislöffel* (2005). If and how this can explain the observed period evolution has to be investigated with detailed future modeling.

Even in the substellar regime there is some evidence for disk braking or some other form of angular momentum loss related to disks. Accreting, very young objects are nearly exclusively slow rotators, whereas nonaccretors cover a broad range of rotation rates (*Scholz and Eislöffel,* 2004a; *Mohanty et al.,* 2005a). The magnetic braking of the disk in the PMS phase is probably less strong than for solar-like stars, but it still might be able to prevent the objects from conserving angular momentum during the first few million years of their evolution. After that, there seems to be very little chance for significantly slowing the rotation of BDs and their rotational evolution is probably dominated by conservation of angular momentum as they continue to contract. Since the radii of substellar objects are expected to decrease by a factor of ~3, due to hydrostatic contraction, on timescales of ~200 m.y. (*Baraffe et al.,* 1998), the periods should decrease by a factor of ~9 on the same timescale, assuming angular momentum conservation. Thus, the decline of the upper period limit between 5 m.y. and the older clusters seen on Fig. 6 probably just reflects conservation of angu-

lar momentum and is roughly consistent with it quantitatively. Finally, while at young ages, substellar objects are clearly able to maintain magnetic activity that should lead to mild rotational braking by stellar winds on timescales of about 100 m.y., they seem to lose their magnetic signatures as they age and cool. Objects with spectral type L, into which these objects evolve, have been found to be too cool to maintain significant chromospheric and coronal activity (*Mohanty and Basri,* 2003). Hence, there is probably little or no rotational braking by stellar winds on very long timescales, explaining why the upper period limit on Fig. 6 stays at a very low levels for a very long time. Again, it will clearly be necessary to expand the data sample in order to test these ideas more rigorously.

To summarize, in the time since Protostars and Planets IV we have seen a tremendous growth, by a factor of 10 or more, in the number of PMS stars for which we know the rotation period. Coupled with similar data for young clusters we now have a good, statistical picture of the evolution of surface angular momentum for solar-like stars from 1 m.y. to the ZAMS (Fig. 3). To a first approximation, at 1 m.y. the stars already divide into a slower rotating half and a more rapidly rotating half. The division is exaggerated over the next few million years as the slower-rotating stars continue to suffer substantial rotational braking while the faster rotators spin up in rough agreement with angular momentum conservation as they contract toward the ZAMS. The braking is disk related (see Fig. 8) and persists for around 5–6 m.y.

At the same time we have begun to probe into the low-mass and substellar-mass regimes with the same photometric technique, although the amplitudes of the variations (Fig. 7) and faintness of the objects make it more difficult to find rotation periods. Nonetheless, the data show similar kinds of behavior for these stars (Fig. 5), although braking by both disks and winds appears to become increasingly less efficient as one progresses to smaller-mass objects (Fig. 6). It is in this mass regime that we expect the next five years to bring particular progress, since there is so much to be done. It is also likely that the improved data on disks coming from the Spitzer Infrared Telescope and elsewhere will help sharpen the observational tests relevant to disk locking and other theories. Finally, the difficult problem of what happens during the first 1 m.y. (i.e., the protostellar and early PMS phases) to produce such a broad rotational distribution already in the ONC will hopefully become clearer as data on the highly embedded objects continue to accumulate.

Acknowledgments. W.H. gratefully acknowledges the continued support of NASA through its Origins of Solar System Program and the support and hospitality of the staff of the Max-Planck-Institute for Astronomy in Heidelberg during extended visits. J.E. and A.S. were partially supported by the Deutsche Forschungsgemeinschaft grants Ei 409/11–1 and 11–2. We thank the referee, K. Stassun, for his helpful comments on the first draft of this chapter.

REFERENCES

Agapitou V. and Papaloizou J. C. B. (2000) *Mon. Not. R. Astron. Soc.*, *317*, 273–288.

Armitage P. J. and Clarke C. (1996) *Mon. Not. R. Astron. Soc.*, *280*, 458–468.

Attridge J. M. and Herbst W. (1992) *Astrophys. J.*, *398*, L61–L64.

Bacciotti F., Ray T. P., Mundt R., Eislöffel J., and Solf J. (2002) *Astrophys. J.*, *576*, 222–231.

Bailer-Jones C. A. L. (2004) *Astron. Astrophys.*, *419*, 703–712.

Bailer-Jones C. A. L. and Mundt R. (2001) *Astron. Astrophys.*, *367*, 218–235.

Baraffe I., Chabrier G., Allard F., and Hauschildt P. H. (1998) *Astron. Astrophys.*, *337*, 403–412.

Barnes S. A. (2003) *Astrophys. J.*, *586*, 464–479.

Barnes S. A., Sofia S., and Pinsonneault M. (2001) *Astrophys. J.*, *548*, 1071–1080.

Barrado y Navascués D., Stauffer J. R., Briceno C., Patten B., Hambly N. C., and Adams J. D. (2001) *Astrophys. J. Suppl.*, *134*, 103–114.

Barrado y Navascués D., Béjar V. J. S., Mundt R., Martin E. L., Rebolo R., Zapatero-Osorio M. R., and Bailer-Jones C. A. L. (2003) *Astron. Astrophys.*, *404*, 171–185.

Bodenheimer P. (1995) *Ann. Rev. Astron. Astrophys.*, *33*, 199–238.

Bouvier J. (1994) In *Cool Stars; Stellar Systems; and the Sun; Eighth Cambridge Workshop* (J.-P. Caillault, ed.), pp. 151–158. ASP Conf. Series 64, San Francisco.

Bouvier J. and Bertout C. (1989) *Astron. Astrophys.*, *211*, 99–114.

Bouvier J., Bertout C., Benz W., and Mayor M. (1986) *Astron. Astrophys.*, *165*, 110–119.

Bouvier J., Cabrit S., Fernandez M., Martin E. L., and Matthews J. M. (1993) *Astron. Astrophys. Suppl.*, *101*, 485–505.

Bouvier J., Covino E., Kovo O., Martin E. L., Matthews J. M., Terranegra L., and Beck S. C. (1995) *Astron. Astrophys.*, *299*, 89–107.

Bouvier J., Forestini M., and Allain S. (1997) *Astron. Astrophys.*, *326*, 1023–1043.

Bouvier J., Dougados C., and Alencar S. H. P. (2004) *Astrophys. Space Sci.*, *292*, 659–664.

Camenzind M. (1990) *Rev. Modern Astron.*, *3*, 234–265.

Carpenter J. M., Hillenbrand L. A., and Skrutskie M. F. (2001) *Astron. J.*, *121*, 3160–3190.

Chabrier G. and Baraffe I. (1997) *Astron. Astrophys.*, *327*, 1039–1053.

Chabrier G. and Kueker L. (2006) *Astron. Astrophys.*, *446*, 1027–1038.

Choi P. I. and Herbst W. (1996) *Astron. J.*, *111*, 283.

Clarke F. J., Tinney C. G., and Covey K. R. (2002) *Mon. Not. R. Astron. Soc.*, *332*, 361–366.

Comerón F., Neuhäuser R., and Kaas A. A. (2000) *Astron. Astrophys.*, *359*, 269–288.

Cohen R. E., Herbst W., and Williams E. C. (2004) *Astron. J.*, *127*, 1594–1601.

Covey K. R., Greene T. P., Doppmann G. W., and Lada C. J. (2005) *Astron. J.*, *129*, 2765–2776.

Dahm S. and Simon T. (2005) *Astron. J.*, *129*, 829–855.

Durney B. R., De Young D. S., and Roxburgh I. W. (1993) *Solar Phys.*, *145*, 207–225.

Edwards S., Strom S. E., Hartigan P., Strom K. M., Hillenbrand L. A., Herbst W., Attridge J., Merrill K. M., Probst R., and Gatley I. (1993) *Astron. J.*, *106*, 372–382.

Haisch K. E., Lada E. A., and Lada C. J. (2001) *Astrophys. J.*, *553*, L153–156.

Hartmann L., Hewett R., Stahler S., and Mathieu R. D. (1986) *Astrophys. J.*, *309*, 275–293.

Herbst W. and Mundt R. (2005) *Astrophys. J.*, *633*, 967–985.

Herbst W. and Shevchenko V. S. (1999) *Astron. J.*, *118*, 1043.

Herbst W., Booth J. F., Chugainov P. F., Zajtseva G. V., Barksdale W., Covino E., Terranegra L., Vittone A., and Vrba F. (1986) *Astrophys. J.*, *310*, L71–L75.

Herbst W., Booth J. F., Koret D. L., Zajtseva G. V., et al. (1987) *Astron. J.*, *94*, 137–149.

Herbst W., Herbst D. K., Grossman E. J., and Weinstein D. (1994) *Astron. J.*, *108*, 1906–1923.

Herbst W., Rhode K. L., Hillenbrand L. A., and Curran G. (2000a) *Astron. J.*, *119*, 261–280.

Herbst W., Maley J. A., and Williams E. C. (2000b) *Astron. J.*, *120*, 349–366.

Herbst W., Bailer-Jones C. A. L., and Mundt R. (2001) *Astrophys. J.*, *554*, L197–L200.

Herbst W., Bailer-Jones C. A. L., Mundt R., Meisenheimer K., and Wackermann R. (2002) *Astron. Astrophys.*, *396*, 513–532.

Herbst W., Francis A., Dhital S., Tresser N., Lin L., and Williams E. C. (2005) *Bull. Am. Astron. Soc.*, *37*, 74–10.

Joergens V., Fernández M., Carpenter J. M., and Neuhäuser R. (2003) *Astrophys. J.*, *594*, 971–981.

Johns-Krull C. M. and Gafford A. (2002) *Astrophys. J.*, *573*, 685–698.

Kirkpatrick J. D., Reid I. N., Liebert J., Cutri R. M., Nelson B., Beichman C. A., Dann C. C., Monet D. G., Gizis J. E., and Skrutskie M. F. (1999) *Astrophys. J.*, *519*, 802–833.

Kiziloglu Ü., Kiziloglu N., and Baykal A. (2005) *Astron. J.*, *130*, 2766–2777.

Köngil A. (1991) *Astrophys. J.*, *370*, L39–L43.

Krishnamurthi A., Pinsonneault M. H., Barnes S., and Sofa S. (1997) *Astrophys. J.*, *480*, 303–323.

Lamm M. H. (2003) Ph.D. Thesis, Univ. of Heidelberg.

Lamm M. H., Bailer-Jones C. A. L., Mundt R., Herbst W., and Scholz A. (2004) *Astron. Astrophys.*, *417*, 557–581.

Lamm M. H., Mundt R., Bailer-Jones C. A. L., and Herbst W. (2005) *Astron. Astrophys.*, *430*, 1005–1026.

Littlefair S. P., Naylor T., Burningham B., and Jeffries R. D. (2005) *Mon. Not. R. Astron. Soc.*, *358*, 341–352.

Long M., Romanova M. M., and Lovelace R. V. E. (2005) *Astrophys. J.*, *634*, 1214–1222.

López Martí B., Eislöffel J., Scholz A., and Mundt R. (2004) *Astron. Astrophys.*, *416*, 555–576.

Makidon R. B., Rebull L. M., Strom S. E., Adams M. T., and Patten B. M. (2004) *Astron. J.*, *127*, 2228–2245.

Mandel G. N. and Herbst W. (1991) *Astrophys. J.*, *383*, L75–L78.

Martín E. L. and Zapatero-Osorio M. R. (1997) *Mon. Not. R. Astron. Soc.*, *286*, 17–20.

Mathieu R. D. (2004) In *Stellar Rotation* (A. Maeder and P. Eenens, eds.), pp. 113–122. IAU Symposium 215, ASP, San Francisco.

Matt S. and Pudritz R. E. (2004) *Astrophys. J.*, *632*, L135–L138.

Matt S. and Pudritz R. E. (2005a) *Mon. Not. R. Astron. Soc.*, *356*, 167–182.

Matt S. and Pudritz R. E. (2005b) *Astrophys. J.*, *632*, L135–L138.

Mohanty S. and Basri G. (2003) *Astrophys. J.*, *583*, 451–472.

Mohanty S., Jayawardhana R., and Basri G. (2005a) *Mem. Soc. Astron. Ital.*, *76*, 303–308.

Mohanty S., Jayawardhana R., and Basri G. (2005b) *Astrophys. J.*, *626*, 498–522.

Mundt R. (1984) *Astrophys. J.*, *280*, 749–770.

Nakajima T., Oppenheimer B. R., Kulkarni S. R., Golimowsk D. A., Matthews K., and Durrance S. T. (1995) *Nature*, *378*, 463.

Phan-Bao N., Guibert J., Crifo F., Delfosse X., Forveille T., Borsenberger J., Epchtein N., Fouqué P., and Simon G. (2001) *Astron. Astrophys.*, *380*, 590–598.

Rebolo R., Zapatero-Osorio M. R., and Martín E. L. (1995) *Nature*, *377*, 129.

Rebull L. M. (2001) *Astron. J.*, *121*, 1676–1709.

Rebull L. M., Wolff S. C., Strom S. E., and Makidon R. B. (2002) *Astron. J.*, *124*, 546–559.

Rebull L. M., Wolff S. C., and Strom S. E. (2004) *Astron. J.*, *127*, 1029–1051.

Rebull L. M., Stauffer J. R., Megeath T., Hora J., and Hartmann L. (2005) *Bull. Am. Astron. Soc.*, *207*, 185.08.

Rhode K. L., Herbst W., and Mathieu R. D. (2001) *Astron. J.*, *122*, 3258–3279.

Rydgren A. E. and Vrba F. J. (1983) *Astrophys. J.*, *267*, 191–198.

Scholz A. (2004) Ph.D. thesis, Univ. of Jena.

Scholz A. and Eislöffel J. (2004a) *Astron. Astrophys., 419,* 249–267.

Scholz A. and Eislöffel J. (2004b) *Astron. Astrophys., 421,* 259–271.

Scholz A. and Eislöffel J. (2005) *Astron. Astrophys., 429,* 1007–1023.

Sherry W. H. (2003) Ph.D. thesis, State Univ. of New York, Stony Brook.

Shu F., Najita J., Ostriker E., Wilkin F., Ruden S., and Lizano S. (1994) *Astrophys. J., 429,* 781–796.

Sills A., Pinsonneault M. H., and Terndrup D. M. (2000) *Astrophys. J., 543,* 335–347.

Stassun K. G. and Terndrup D. (2003) *Publ. Astron. Soc. Pac., 115,* 505–512.

Stassun K. G., Mathieu R. D., Mazeh T., and Vrba F. J. (1999) *Astron. J., 117,* 2941–2979.

Stassun K. G., Mathieu R. D., Vrba F. J., Mazeh T., and Henden A. (2001) *Astron. J., 121,* 1003–1012.

Steinhauer A. J. B., Herbst W., and Henden A. (1996) *Bull. Am. Astron. Soc., 28,* 884.

Strassmeier K. G. (1992) In *Robotic Telescopes in the 1990s* (A. V. Filippenko, ed.), pp. 39–52. ASP Conf. Series 34, San Francisco.

Tackett S., Herbst W., and Williams E. C. (2003) *Astron. J., 126,* 348–352.

Tassoul J.-L. (2000) *Stellar Rotation,* Cambridge Univ., New York. 256 pp.

Terndrup D. M., Krishnamurthi A., Pinsonneault M. H., and Stauffer J. R. (1999) *Ann. Rev. Astron. Astrophys., 41,* 599–643.

Terndrup D. M., Stauffer J. R., Pinsonneault M. H., Sills A., Yuan Y., Jones B. F., Fischer D., and Krishnamurthi A. (2000) *Astron. J., 119,* 1303–1316.

Thompson M. J., Christensen-Dalsgaard J., Miesch M. S., and Toomre J. (2003) *Astron. J., 118,* 1814–1818.

Tinker J., Pinsonneault M., and Terndrup D. M. (2002) *Astrophys. J., 564,* 877–886.

Uzdensky D. A. (2004) *Astrophys. Space Sci., 292,* 573–585.

Vogel S. N. and Kuhi L. V. (1981) *Astrophys. J., 245,* 960–976.

Vrba F. J., Herbst W., and Booth J. F. (1988) *Astron. J., 96,* 1032–1039.

Whelan E. T., Ray T. P., Bacciotti F., Natta A., Testi L., and Randich S. (2005) *Nature, 435,* 652–654.

Woitas J., Bacciotti F., Ray T. P., Marconi A., Coffey D., and Eislöffel J. (2005) *Astron. Astrophys., 432,* 149–160.

Zapatero-Osorio M. R., Rebolo R., and Martín E. L. (1997) *Astron. Astrophys., 317,* 164–170.

Zapatero-Osorio M. R., Caballero J. A., Béjar V. J. S., and Rebolo R. (2003) *Astron. Astrophys., 408,* 663–673.

X-Ray Properties of Young Stars and Stellar Clusters

Eric Feigelson and Leisa Townsley
Pennsylvania State University

Manuel Güdel
Paul Scherrer Institute

Keivan Stassun
Vanderbilt University

Although the environments of star and planet formation are thermodynamically cold, substantial X-ray emission from 10–100 MK plasmas is present. In low-mass pre-main-sequence stars, X-rays are produced by violent magnetic reconnection flares. In high-mass O stars, they are produced by wind shocks on both stellar and parsec scales. The recent Chandra Orion Ultradeep Project, XMM-Newton Extended Survey of Taurus, and Chandra studies of more distant high-mass star-forming regions reveal a wealth of X-ray phenomenology and astrophysics. X-ray flares mostly resemble solar-like magnetic activity from multipolar surface fields, although extreme flares may arise in field lines extending to the protoplanetary disk. Accretion plays a secondary role. Fluorescent iron line emission and absorption in inclined disks demonstrate that X-rays can efficiently illuminate disk material. The consequent ionization of disk gas and irradiation of disk solids addresses a variety of important astrophysical issues of disk dynamics, planet formation, and meteoritics. New observations of massive star-forming environments such as M 17, the Carina Nebula, and 30 Doradus show remarkably complex X-ray morphologies including the low-mass stellar population, diffuse X-ray flows from blister HII regions, and inhomogeneous superbubbles. X-ray astronomy is thus providing qualitatively new insights into star and planet formation.

1. INTRODUCTION

Star and planet formation is generally viewed as a hydrodynamic process involving gravitational collapse of interstellar material at low temperatures, 10–100 K in molecular cloud cores and 100–1500 K in protoplanetary disks. If thermodynamical equilibrium holds, this material should be neutral except in localized HII regions where the bolometric ultraviolet emission from massive O-star photoionization is present. However, stars have turned out to be sources of intense X-rays at almost every stage of early formation and evolution, from low-mass brown dwarfs to massive O stars, to an extent that the stellar environment is ionized and heated (beyond effects due to ultraviolet radiation) out to considerable distances and thus made accessible to magnetic fields.

X-ray observations reveal the presence of highly ionized plasma with temperatures of 10^7–10^8 K. In lower-mass stars, the X-ray emission is reminiscent of X-rays observed on the Sun, particularly the plasma explosively heated and confined in magnetic loops following magnetic reconnection events. X-ray flares with luminosities orders of magnitude more powerful than seen in the contemporary Sun are frequently seen in young stars. Evidence for an impulsive phase is seen in radio bursts and in U-band enhancements preceding X-ray flares, thought to be due to the bombard-

ment of the stellar surface by electron beams. Thus, young stars prolifically accelerate particles to relativistic energies. In rich young stellar clusters, X-rays are also produced by shocks in O-star winds, on both small ($<10^2$ $R_\star$) and large (parsec) scales. If the region has been producing rich clusters for a sufficiently long time, the resulting supernova remnants will dominate the X-ray properties.

X-ray studies with the Chandra and XMM-Newton space observatories are propelling advances of our knowledge and understanding of high-energy processes during the earliest phases of stellar evolution. In the nearest young stars and clusters (d < 500 pc), they provide detailed information about magnetic reconnection processes. In the more distant and richer regions, the X-ray images are amazingly complex with diffuse plasma surrounding hundreds of stars exhibiting a wide range of absorptions. We concentrate here on results from three recent large surveys: the Chandra Orion Ultradeep Project (COUP), based on a nearly continuous 13-day observation of the Orion Nebula region in 2003; the XMM-Newton Extended Survey of Taurus (XEST) that maps ~5 deg^2 of the Taurus Molecular Cloud (TMC); and an ongoing Chandra survey of high-mass star-formation regions across the galactic disk. Because the XEST study is discussed in specific detail in a closely related chapter (see chapter by Güdel et al.) together with optical and infrared surveys, we present only selected XEST results. This volume has another closely related chapter: Bally et al. discuss X-ray emission from high-velocity protostellar Herbig-Haro outflows. The reader interested in earlier X-ray studies is referred to reviews by *Feigelson and Montmerle* (1999), *Glassgold et al.* (2000), *Favata and Micela* (2003), *Paerels and Kahn* (2003), and *Güdel* (2004).

The COUP is particularly valuable in establishing a comprehensive observational basis for describing the physical characteristics of flaring phenomena and elucidating the mechanisms of X-ray production. The central portion of the COUP image, showing the PMS population around the bright Trapezium stars and the embedded OMC-1 populations, is shown in Plate 1 (*Getman et al.*, 2005a). X-rays are detected from nearly all known optical members except for many of the bolometrically fainter M stars and brown dwarfs. Conversely, 1315 of 1616 COUP sources (81%) have clear cluster member counterparts and ≈75 (5%) are new obscured cloud members; most of the remaining X-ray sources are extragalactic background sources seen through the cloud (*Getman et al.*, 2005b).

X-ray emission and flaring is thus ubiquitous in PMS stars across the initial mass function (IMF). The X-ray luminosity function (XLF) is broad, spanning $28 < \log L_x[\mathrm{erg/s}] < 32$ (0.5–8 keV), with a peak around $\log L_x[\mathrm{erg/s}] \sim 29$ (*Feigelson et al.*, 2005). For comparison, the contemporary Sun emits $26 < \log L_x[\mathrm{erg/s}] < 27$, with flares up to 10^{28} erg/s, in the same spectral band. Results from the more distributed star formation clouds surveyed by XEST reveal a very similar X-ray population as in the rich cluster of the Orion Nebula, although confined to stars with masses mostly below 2 $M_\odot$ (see chapter by Güdel et al.), although there is some evidence the XLF is not identical in all regions (section 4.1). There is no evidence for an X-ray-quiet, nonflaring PMS population.

The empirical findings generate discussion on a variety of astrophysical implications, including the nature of magnetic fields in young stellar systems, the role of accretion in X-ray emission, the effects of X-ray irradiation of protoplanetary disks and molecular clouds, and the discovery of X-ray flows from HII regions. A number of important related issues are not discussed here, including discovery of heavily obscured X-ray populations, X-ray identification of older PMS stars, the X-ray emission of intermediate mass Herbig Ae/Be stars, the enigmatic X-ray spectra of some O stars, the generation of superbubbles by OB clusters and their multiple supernova remnants, and the large scale starburst conditions in the galactic center region and other galactic nuclei.

2. FLARING IN PRE-MAIN-SEQUENCE STARS

2.1. The Solar Model

Many lines of evidence link the PMS X-ray properties to magnetic activity on the Sun and other late-type magnetically active stars such as dMe flare stars, spotted BY Dra variables, and tidally spun-up RS CVn post-main-sequence

binaries. These systems have geometrically complex multipolar magnetic fields in arcades of loops rooted in the stellar photospheres and extending into the coronae. The field lines become twisted and tangled by gas convection and undergo explosive magnetic reconnection. The reconnection immediately accelerates a population of particles with energies tens of keV to several MeV; this is the "impulsive phase" manifested by gyrosynchrotron radio continuum emission, blue optical/UV continuum, and, in the Sun, high γ-ray and energetic particle fluences. These particles impact the stellar surface at the magnetic footprints, immediately heating gas that flows upward to fill coronal loops with X-ray emitting plasma. It is this "gradual phase" of the flare that is seen with X-ray telescopes. *Schrijver and Zwaan* (2000) and *Priest and Forbes* (2002) review the observations and physics of solar and stellar flares.

Extensive multiwavelength properties of PMS stars indicate they are highly magnetically active stars (*Feigelson and Montmerle*, 1999). Hundreds of Orion stars, and many in other young stellar populations, have periodic photometric variations from rotationally modulated starspots covering 10–50% of the surface (*Herbst et al.*, 2002). A few of these have been subject to detailed Doppler mapping showing spot structure. Radio gyrosynchrotron emission from flare electrons spiraling in magnetic loops has been detected in several dozen PMS stars (*Güdel*, 2002). A few nearby PMS stars have Zeeman measurements indicating that kilo-Gauss fields cover much of the surface (*Johns-Krull et al.*, 2004). In the COUP and XEST studies, temperatures inferred from time-averaged spectra extends the T_{cool}–T_{hot} and T–L_x trends found in the Sun and older stars to higher levels (*Preibisch et al.*, 2005; Telleschi et al., in preparation). X-ray spectra also show plasma abundance anomalies that are virtually identical to those seen in older magnetically active stars (*Scelsi et al.*, 2005; Maggio et al., in preparation).

Taking advantage of the unprecedented length of the COUP observation, *Flaccomio et al.* (2005) find rotational modulation of X-ray emission in at least 10% of Orion PMS stars with previously determined rotation periods from optical monitoring. An example is shown in Fig. 1a. Amplitudes of variability range from 20% to 70% and X-ray periods generally agree with the optical periods. In a few cases, it is half the optical value, implying X-ray-emitting regions on opposite hemispheres. This result indicates that in at least some PMS stars, the X-rays are emitting from relatively long-lived structures lying low in the corona that are inhomogeneously distributed around the star. Similar X-ray rotational modulations are seen in the Sun and a few older stars.

Wolk et al. (2005) examined the flaring behavior of a complete sample of solar analogs (0.9 < M/M$_\odot$ < 1.2) in the Orion Nebula Cluster. The lightcurve in Fig. 1b shows one of the more spectacular flares in this subsample reaching $\log L_x(\mathrm{peak})[\mathrm{erg/s}] = 32.2$. Most flares show solar-type rapid rise and slower decay; the decay phase can last from <1 h to >3 d. The brightness and spectral variations during the decay phases of this and similarly powerful Orion flares have been analyzed by *Favata et al.* (2005a) using a loop

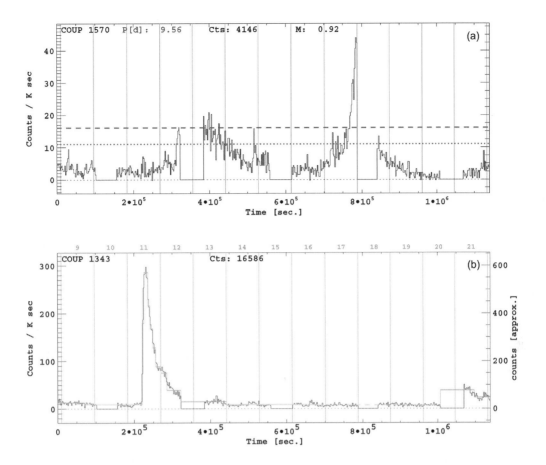

Fig. 1. Two of 1400+ X-ray lightcurves from the Chandra Orion Ultradeep Project. The abscissa is time spanning 13.2 d, and the ordinate gives X-ray count rate in the 0.5–8 keV band. **(a)** An Orion star showing typical PMS flaring behavior superposed on a rotational modulation of the "characteristic" emission. From *Flaccomio et al.* (2005). **(b)** COUP #1343 = JW 793, a poorly characterized PMS star in the Orion Nebula, showing a spectacular solar-type flare. From *Wolk et al.* (2005) and *Favata et al.* (2005a).

model previously applied to "gradual" (i.e., powerful events spanning several hours) solar and stellar flares. The result for COUP #1343 and other morphologically simple cases is clear: The drop in X-ray emission and plasma temperature seen in PMS stellar flares is completely compatible with that of older stars. In some COUP flares, the decay shows evidence of continued or episodic reheating after the flare peak, a phenomenon also seen in solar flares and in older stars.

The intensity of PMS flaring is remarkably high. In the solar analog sample, flares brighter than L_x(peak) $\geq 2 \times 10^{30}$ erg/s occur roughly once a week (*Wolk et al.*, 2005). The most powerful flares have peak luminosities up to several times 10^{32} erg/s (*Favata et al.*, 2005a). The peak plasma temperature are typically T (peak) $\simeq 100$ MK but sometimes appear much higher. The time-integrated energy emitted in the X-ray band during flares in solar-mass COUP stars is $\log E_x[\text{erg}] \simeq 34$–36. An even more remarkable flare with $\log E_x[\text{erg}] \simeq 37$ was seen by ROSAT from the nonaccreting Orion star Parenago 1724 in 1992 (*Preibisch et al.*, 1995). These values are far above solar flaring levels: The COUP flares are $10^2\times$ stronger and $10^2\times$ more frequent than the

most powerful flares seen in the contemporary Sun; the implied fluence of energetic particles may be $10^5\times$ above solar levels (*Feigelson et al.*, 2002).

The Orion solar analogs emit a relatively constant "characteristic" X-ray level about three-fourths of the time (see Fig. 1). The X-ray spectrum of this characteristic state can be modeled as a two-temperature plasma with one component $T_{cool} \simeq 10$ MK and the other component $T_{hot} \simeq 30$ MK. These temperatures are much higher than the quiescent solar corona. The concept of "microflaring" or "nanoflaring" for the Sun has been widely discussed (*Parker*, 1988) and has gained favor in studies of older magnetically active stars based on lightcurve and spectral analysis (*Kashyap et al.*, 2002; *Güdel et al.*, 2003, *Arzner and Güdel*, 2004). These latter studies of dMe flare stars indicate that a power-law distribution of flare energies, dN/dE $\propto$ E$^{-\alpha}$, is present with $\alpha \simeq 2.0$–2.7. The energetics is clearly dominated by smaller flares. The COUP lightcurves vary widely in appearance, but collectively can also be roughly simulated by a power law with $\alpha \simeq 2.0$–2.5 without a truly quiescent component (*Flaccomio et al.*, 2005). Thus, when reference is made to the more easily studied superflares, one must always re-

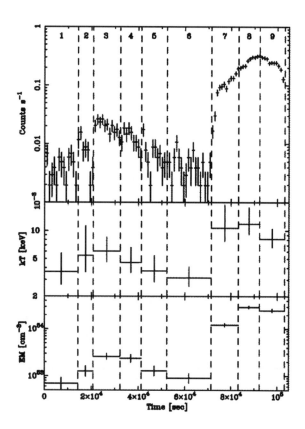

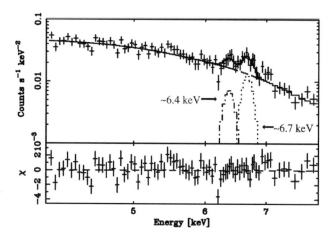

Fig. 2. X-ray flare from Class I protostar YLW 16A in the ρ Ophiuchi cloud, observed with Chandra. The flare has an unusual morphology and the spectrum shows very hot plasma temperatures with strong emission from the fluorescent 6.4 keV line of neutral iron attributable to reflection off the protoplanetary disk. From *Imanishi et al.* (2001).

member that many more weaker flares are present and may have comparable or greater astrophysical effects. Not infrequently, secondary flares and reheating events are seen superposed on the decay phase of powerful flares (e.g., *Gagné et al.*, 2004; *Favata et al.*, 2005a).

One puzzle with a solar model for PMS flares is that some show unusually slow rises. The nonaccreting star LkHα 312 exhibited a 2-h fast rise with peak temperature $T \simeq 88$ MK, followed by a 6-h slower rise to $\log L_x$(peak) [erg/s] = 32.0 (*Grosso et al.*, 2004). The flare from the Class I protostar YLW 16A in the dense core of the ρ Ophiuchi cloud showed a remarkable morphology with two rise phases and similar temperature structure (Fig. 2) (*Imanishi et al.*, 2001). Other flares seen with COUP show roughly symmetrical rise and fall morphologies, sometimes extending over 1–2 days (*Wolk et al.*, 2005). It is possible that some of these variations are due to the stellar rotation where X-ray structures are emerging from eclipse, but they are currently poorly understood.

By monitoring young stars with optical telescopes simultaneous with X-ray observations, the early impulsive phase of PMS flares can be revealed. This has been achieved with distributed groundbased telescopes and in space: The XMM-Newton satellite has an optical-band telescope coaligned with the X-ray telescope. During the impulsive phase, electron beams accelerated after the reconnection event

bombard the stellar chromosphere, which produces a burst of short-wavelength optical and UV radiation. XMM-Newton observation of the Taurus PMS star GK Tau shows both uncorrelated modulations as well as a strong U-band burst preceding an X-ray flare in good analogy with solar events (*Audard et al.*, in preparation) (Fig. 3a). Groundbased optical and Hα monitoring of the Orion Nebula during the COUP campaign revealed one case of an I-band spike simultaneous with a very short X-ray flare of intermediate brightness (*Stassun et al.*, 2006) (Fig. 3b).

2.2. The Role of Accretion

It was established in the 1980s and 1990s that elevated levels of X-ray emission in PMS stars appears in both "classical" T Tauri stars, with optical/infrared signatures of accretion from a protoplanetary disk, and "weak-lined" T Tauri stars, without these signatures. This basic result is confirmed but with some important refinements — and controversy — from recent studies.

While the presence or absence of a K-band emitting inner disk does not appear to influence X-ray emission, the presence of accretion has a *negative* impact on X-ray production (*Flaccomio et al.*, 2003; *Stassun et al.*, 2004; *Preibisch et al.*, 2005; Telleschi et al., in preparation). This is manifested as a statistical decrease in X-rays by a factor of

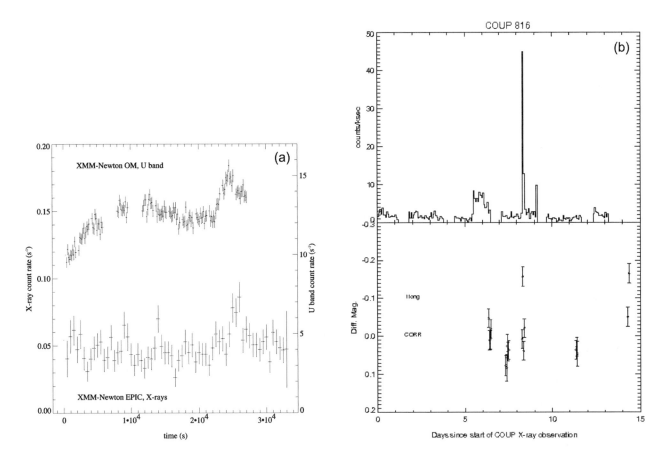

Fig. 3. Detection of the "white light" component during the impulsive phases of PMS X-ray flares. **(a)** Short-term behavior of the classical T Tau binary GK Tau in U-band light (upper curve) and X-rays (lower curve) from the XEST survey. The lightcurve covers approximately 9 h. From Audard et al. (in preparation). **(b)** Rapid X-ray flare (top panel) from COUP #816 = JW 522, an obscured PMS star in the Orion Nebula Cluster, apparently accompanied by impulsive I-band emission. This COUP X-ray lightcurve spans 13.2 d. From *Stassun et al.* (2006).

2–3 in accreting vs. nonaccreting PMS stars, even after dependencies on mass and age are taken into account. The effect does not appear to arise from absorption by accreting gas; e.g., the offset appears in the hard 2–8-keV band where absorption is negligible. The offset is relatively small compared to the 10^4 range in the PMS X-ray luminosity function, and flaring behavior is not affected in any obvious way. One possible explanation is that mass-loaded accreting field lines cannot emit X-rays (*Preibisch et al.,* 2005). If a magnetic reconnection event liberates a certain amount of energy, this energy would heat the low-density plasma of non-accretors to X-ray emitting temperatures, while the denser plasma in the mass-loaded magnetic field lines would be heated to much lower temperatures. The remaining field lines that are not linked to the disk would have low coronal densities and continue to produce solar-like flares. Note that the very young accreting star XZ Tau shows unusual temporal variations in X-ray absorption that can be attributed to eclipses by the accretion stream (*Favata et al.,* 2003).

The optical observations conducted simultaneously with the COUP X-ray observations give conclusive evidence that accretion does not produce or suppress flaring in the great majority of PMS stars (*Stassun et al.,* 2006). Of the 278 Orion stars exhibiting variations in both optical and X-ray bands, not a single case is found where optical variations (attributable to either rotationally modulated starspots or to changes in accretion) have an evident effect on the X-ray flaring or characteristic emission.

An example from the XEST survey is shown in Fig. 3a where the slow modulation seen in the first half is too rapid for effects due to rotation, but on the other hand shows no equivalent signatures in X-rays. The optical fluctuations are therefore unrelated to flare processes and, in this case, are likely due to variable accretion (Audard et al., in preparation). Similarly, a Taurus brown dwarf with no X-ray emission detected in XEST showed a slow rise by a factor of 6 over eight hours in the U-band flux (*Grosso et al.,* 2006). Such behavior is uncommon for a flare, and because this brown dwarf is accreting, mass streams may again be responsible for producing the excess ultraviolet flux.

The simplest interpretation of the absence of statistical links between accretion and X-ray luminosities and the absence of simultaneous optical/X-ray variability is that different magnetic field lines are involved with funneling gas

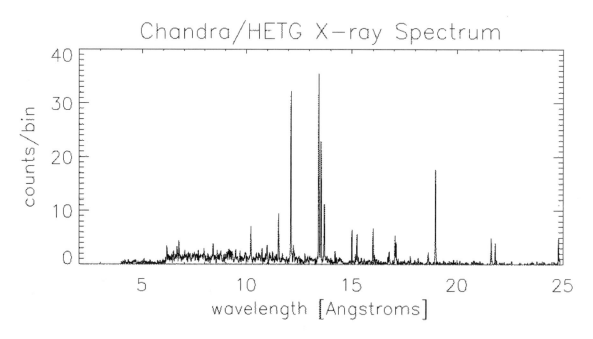

Fig. 4. High-resolution transmission grating spectrum of the nearest classical T Tauri star, TW Hya. The spectrum is softer than other PMS stars, and the triplet line ratios imply either X-ray production in a high-density accretion shock or irradiation by ultraviolet radiation. From *Kastner et al.* (2002).

from the disk and with reconnection events producing X-ray plasma. There is no evidence that the expected shock at the base of the accretion column produces the X-rays seen in COUP and XEST.

There are some counterindications to these conclusions. A huge increase in X-ray emission was seen on long time-scales from the star illuminating McNeil's Nebula, which exhibited an optical/infrared outburst attributed to the onset of rapid accretion (*Kastner et al.,* 2004). In contrast, however, X-rays are seen before, during, and after outburst of the EXor star V1118 Ori, with a lower temperature seen when accretion was strongest (*Audard et al.,* 2005). These findings suggest that accretion, and perhaps the inner disk structure, might sometimes affect magnetic field configurations and flaring in complicated ways.

The biggest challenge comes from TW Hya, the nearest and brightest accreting PMS star. It has an X-ray spectrum much softer than most COUP or other PMS sources (Fig. 4). Due to its proximity to the Sun, TW Hya is sufficiently bright in X-rays to be subject to detailed high-resolution spectroscopy using transmission gratings on the Chandra and XMM-Newton telescopes (*Kastner et al.,* 2002; *Stelzer and Schmitt,* 2004; *Ness and Schmitt,* 2005). According to the magnetospheric accretion scenario, accreted material crashes onto the stellar surface with velocities of up to several hundred km/s, which should cause $\sim 10^6$ K shocks in which strong optical and UV excess emission and perhaps also soft X-ray emission is produced (*Lamzin,* 1999). Density-sensitive triplet line ratios of Ne IX and O VII are saturated, indicating either that the emitting plasma has densities $\log n_e[\mathrm{cm}^{-3}] \sim 13$, considerably higher than the $\log n_e[\mathrm{cm}^{-3}] \sim$

10 characteristic of low-level coronal emission although reminiscent of densities in flares. However, these densities were measured during an observation dominated by relative quiescence with no hot plasma present. Alternatively, the high triplet ratios might be induced by plasma subject to strong ultraviolet irradiation. A similar weak forbidden line in the O VII triplet is seen in the accreting PMS star BP Tau (*Schmitt et al.,* 2005), and similar soft X-ray emission is seen from the Herbig Ae star HD 163296 (*Swartz et al.,* 2005).

If the plasma material in TW Hya is drawn from the disk rather than the stellar surface, one must explain the observed high Ne/Fe abundance ratio that is similar to that seen in flare plasmas. One possibility is that the abundance anomalies do not arise from the coronal first ionization potential effect, but rather from the depletion of refractory elements into disk solids (*Brinkman et al.,* 2001; *Drake et al.,* 2005). This model, however, must confront models of the infrared disk indicating that grains have sublimated in the disk around 4 AU, returning refractory elements back into the gas phase (*Calvet et al.,* 2002).

Finally, we note that current X-ray instrumentation used for PMS imaging studies is not very sensitive to the cooler plasma expected from accretion shocks, and that much of this emission may be attenuated by line-of-sight interstellar material. The possibility that some soft accretion X-ray emission is present in addition to the hard flare emission is difficult to firmly exclude. But there is little doubt that most of the X-rays seen with Chandra and XMM-Newton are generated by magnetic reconnection flaring rather than the accretion process.

2.3. The Role of Disks

There are strong reasons from theoretical models to believe that PMS stars are magnetically coupled to their disks at the corotation radii typically 5–10 R_* from the stellar surface (e.g., *Hartmann*, 1998; *Shu et al.*, 2000). This hypothesis unifies such diverse phenomena as the self-absorbed optical emission lines, the slow rotation of accreting PMS stars, and the magnetocentrifugal acceleration of Herbig-Haro jets. However, there is little *direct* evidence for magnetic field lines connecting the star and the disk. Direct imaging of large-scale magnetic fields in PMS stars is only possible today using very long baseline interferometry at radio wavelengths where an angular resolution of 1 mas corresponds to 0.14 AU at the distance of the Taurus or Ophiuchus clouds. But only a few PMS stars are sufficiently bright in radio continuum for such study. One of the components of T Tau S has consistently shown evidence of magnetic field extensions to several stellar radii, perhaps connecting to the inner border of the accretion disk (*Loinard et al.*, 2005).

But X-ray flares can provide supporting evidence for star-disk magnetic coupling. An early report of star-disk fields arose from a sequence of three powerful flares with separations of ~20 h from the Class I protostar YLW 15 in the ρ Oph cloud (*Tsuboi et al.*, 2000). Standard flare plasma models indicated loop lengths around 14 $R_\odot$, and periodicity might arise from incomplete rotational star-disk coupling (*Montmerle et al.*, 2000). However, it is also possible that the YLA 15 flaring is not truly periodic; many cases of multiple flares without periodicities are seen in the COUP lightcurves.

Analysis of the most luminous X-ray flares in the COUP study also indicates that huge magnetic structures can be present. *Favata et al.* (2005a) reports analysis of the flare decay phases in sources such as COUP #1343 (Fig. 1) using models that account for reheating processes, which otherwise can lead to overestimation of loop lengths. The combination of very high luminosities (log L_x(peak)[erg/s] ≈ 31–32), peak temperatures in excess of 100 MK, and very slow decays appear to require loops much larger than the star, up to several 10^{12} cm or 5–20 R_*. Recall that these flares represent only the strongest ~1% of all flares observed by COUP; most flares from PMS stars are much weaker and likely arise from smaller loops. This is clearly shown in some stars by the rotational modulation of the nonflaring component in the COUP study (*Flaccomio et al.*, 2005).

Given the typical 2–10-d rotation periods of PMS stars, is seems very doubtful such long flaring loops would be stable if both footpoints were anchored to the photosphere. Even if MHD instabilities are not important, gas pressure and centrifugal forces from the embedded plasma may be sufficient to destroy such enormous coronal loops (*Jardine and Unruh*, 1999). *Jardine et al.* (2006) develop a model of magnetically confined multipolar coronae of PMS stars where accretion follows some field lines while others contain X-ray emitting plasma; the model also accounts for ob-

served statistical relations between X-ray properties and stellar mass.

The magnetospheres of PMS stars are thus likely to be quite complex. Unlike the Sun where only a tiny fraction of the photosphere has active regions, intense multipolar fields cover much of the surface in extremely young stars. Continuous microflaring is likely responsible for the ubiquitous strong 10–30-MK plasma emission. Other field lines extend several stellar radii: Some are mass-loaded with gas accreting from the circumstellar disk, while others may undergo reconnection producing the most X-ray-luminous flares.

3. THE EVOLUTION OF MAGNETIC ACTIVITY

The COUP observation provides the most sensitive, uniform, and complete study of X-ray properties for a PMS stellar population available to date. When combined with studies of older stellar clusters, such as the Pleiades and Hyades, and of volume-limited samples in the solar neighborhood, evolutionary trends in X-ray emission can be traced. Since chromospheric indicators of magnetic activity (such as Ca II line emission) are confused by accretion, and photospheric variations from rotationally modulated starspots are too faint to be generally measured in most older stars, X-ray emission is the only magnetic indicator that can be traced in stellar populations from 10^5 to 10^{10} yr. The result from the PMS to the giga-year-old disk population is shown in Fig. 5a (*Preibisch and Feigelson*, 2005). The two critical advantages here over other measures of magnetic activity evolution are the complete samples (and correct treatment of nondetections) and stratification by mass. The latter is important because the mass-dependence of X-ray luminosities (for unknown reasons) differs in PMS and main-sequence stars (*Preibisch et al.*, 2005).

If one approximates the decay of magnetic activity as a power law, then evolution in the 0.5 < M < 1.2 $M_\odot$ mass range is approximately power-law with $L_x \propto \tau^{-3/4}$ over the wide range of ages 5 < log τ [yr] < 9.5. Other X-ray studies of older disk stars suggest that the decay rate steepens: $L_x \propto \tau^{-3/2}$ or τ^{-2} over 8 < log τ [yr] < 10 (*Güdel et al.*, 1997; *Feigelson et al.*, 2004). Note, however, that *Pace and Pasquini* (2004) find no decay in chromospheric activity in a sample of solar mass stars after 3 G.y. These results are similar to, but show more rapid decay than, the classical *Skumanich* (1972) $\tau^{-1/2}$ relation that had been measured for main-sequence stars only over the limited age range 7.5 < log τ [yr] < 9.5. The COUP sample also exhibits a mild decay in magnetic activity for ages 5 < log τ [yr] < 7 within the PMS phase, although the trend is dominated by star-to-star scatter (*Preibisch and Feigelson*, 2005).

While these results would appear to confirm and elaborate the long-standing rotation-age-activity relationship of solar-type stars, the data paint a more complex picture. The Chandra Orion studies show that the rotation-activity relation is largely absent at 1 m.y. (*Stassun et al.*, 2004; *Preibisch et al.*, 2005). This finding suggests the somewhat

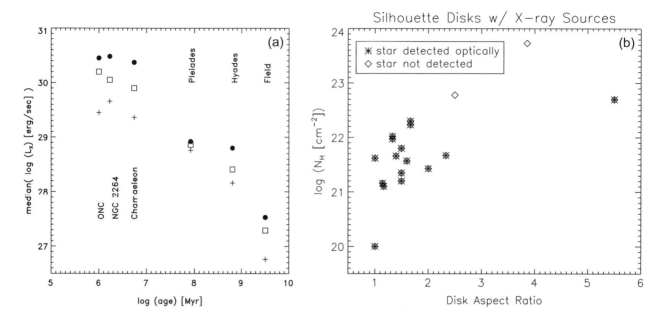

Fig. 5. **(a)** Evolution of the median X-ray luminosities for stars in different mass ranges: 0.9–1.2 $M_\odot$ (solid circles), 0.5–0.9 $M_\odot$ (open squares), and 0.1–0.5 $M_\odot$ (plusses). From *Preibisch and Feigelson* (2005). **(b)** The link between soft X-ray absorption and proplyd inclination is the first measurement of gas column densities in irradiated disks. From *Kastner et al.* (2005).

surprising result that the activity-age decay is strong across the entire history of solar-type stars but is not entirely attributable to rotational deceleration. The PMS magnetic fields may either be generated by a solar-type dynamo that is completely saturated, or by a qualitatively different dynamo powered by turbulence distributed throughout the convective interior rather than by rotational shear at the tachocline. At the same time, the Orion studies show a small *positive* correlation between rotation period and X-ray activity, similar to that seen in the "supersaturated" regime of main-sequence stars. It is also possible that this effect is due to a sample bias against slowly rotating, X-ray-weak Orion stars (*Stassun et al.*, 2004).

The XEST findings in the Taurus PMS population give a different result, suggesting that an unsaturated solar-type dynamo may in fact be present in PMS stars when rotation periods are longer than a few days (Briggs et al., in preparation). It is possible that the somewhat more evolved Taurus sample, compared to the Orion Nebula Cluster stars, has produced sufficiently prominent radiative zones in some of these late-type PMS stars to put a solar-type dynamo into operation.

The origins of magnetic fields in PMS stars are thus still not well established. It is possible that both tachoclinal and convective dynamos are involved, as discussed by *Barnes* (2003a,b). There is a hint of a transition between convective and rotational dynamos in the plot of L_x/L_{bol} against mass in Orion stars. The X-ray emissivity for many stars drops precipitously for masses above 2–3 $M_\odot$, which is also the boundary between lower-mass convective and higher-mass radiative interiors (*Feigelson et al.*, 2003). Another

possible influence is that accretion in younger PMS stars alters convection and thereby influences the magnetic field generation process (*Siess et al.*, 1999; *Stassun et al.*, 2004).

The magnetic activity history for M stars with masses 0.1–0.4 $M_\odot$ appears to be different than more massive PMS stars (Fig. 5a). Only a mild decrease in X-ray luminosity, and even a mild increase in L_x/L_{bol}, is seen over the $6 < \log \tau$ [yr] < 8 range, though the X-ray emission does decay over giga-year timescales. This result may be related to the well-established fact that the low-mass M stars have much longer rotational slow-down times than solar-type stars. But the difference in behavior compared to higher-mass stars could support the idea that the dynamos in PMS and dM stars both arise from a convective turbulent dynamo. These issues are further discussed in *Mullan and MacDonald* (2001), *Feigelson et al.* (2003), *Barnes* (2003a), and *Preibisch et al.* (2005).

An unresolved debate has emerged concerning the onset of X-ray emission in PMS stars. There is no question that magnetic activity with violent flaring is common among Class I protostars with ages ~10^5 yr (*Imanishi et al.*, 2001; *Preibisch*, 2004) (Fig. 2). The question is whether X-ray emission is present in Class 0 protostars with ages ~10^4 yr. There is one report of hard X-rays from two Class 0 protostars in the OMC 2/3 region (*Tsuboi et al.*, 2001); however, other researchers classify the systems as Class I (*Nielbock et al.*, 2003). An X-ray emitting protostar deeply embedded in the R Corona Australis cloud core has a similarly uncertain Class 0 or I status (*Hamaguchi et al.*, 2005). In contrast, a considerable number of well-established Class 0 protostars appear in Chandra images and are not detected;

e.g., in the NGC 1333 and Serpens embedded clusters (*Getman et al.,* 2002; *Preibisch,* 2004). However, because Class 0 stars are typically surrounded by very dense gaseous envelopes, it is possible that the X-ray nondetections arise from absorption rather than an absence of emission. An interesting new case is an intermediate-mass Class 0 system in the IC 1396N region that exhibits extremely strong soft X-ray absorption (*Getman et al.,* 2006b).

4. X-RAY STARS AND HOT GAS IN MASSIVE STAR-FORMING REGIONS

Most stars are born in massive star-forming regions (MSFRs) where rich clusters containing thousands of stars are produced in molecular cloud cores. Yet, surprisingly little is known about the lower mass populations of these rich clusters. Beyond the Orion Molecular Clouds, near-infrared surveys like 2MASS are dominated by foreground or background stars, and the initial mass functions are typically measured statistically rather than by identification of individual cluster members. X-ray surveys of MSFRs are important in this respect because they readily discriminate young stars from unrelated objects that often contaminate JHK images of such fields, especially for those young stars no longer surrounded by a dusty circumstellar disk. Furthermore, modern X-ray telescopes penetrate heavy obscuration (routinely $A_V \sim 10$–100 mag, occasionally up to 1000 mag) with little source confusion or contamination from unrelated objects to reveal the young stellar populations in MSFRs.

The O and Wolf-Rayet (WR) members of MSFRs have been catalogued, and the extent of their UV ionization is known through HII region studies. But often little is known about the fate of their powerful winds. The kinetic power of a massive O star's winds injected into its stellar neighborhood over its lifetime is comparable to the input of its supernova explosion. Theorists calculate that wind-blown bubbles of coronal-temperature gas should be present, but no clear measurement of this diffuse plasma had been made in HII regions prior to Chandra. X-ray studies also detect the presence of earlier generations of OB stars through the shocks of their supernova remnants (SNRs). In very rich and long-lived star-forming cloud complexes, SNRs combine with massive stellar winds to form superbubbles and chimneys extending over hundreds of parsecs and often into the galactic halo. O stars are thus the principal drivers of the interstellar medium.

The MSFR X-ray investigations discussed here represent only a fraction of this rapidly growing field. A dozen early observations of MSFRs by Chandra and XMM-Newton are summarized by *Townsley et al.* (2003). Since then, Chandra has performed observations of many other regions, typically revealing hundreds of low-mass PMS stars, known and new high-mass OB stars, and occasionally diffuse X-ray emission from stellar winds or SNRs. In addition to those discussed below, these include NGC 2024 in the Orion B mo-

lecular cloud (*Skinner et al.,* 2003), NGC 6193 in Ara OB1 (*Skinner et al.,* 2005), NGC 6334 (*Ezoe et al.,* 2006), NGC 6530 ionizing Messier 8 (*Damiani et al.,* 2004), the Arches and Quintuplet galactic center clusters (*Law and Yusef-Zadeh,* 2004; *Rockefeller et al.,* 2005), and Westerlund 1, which has an X-ray pulsar (*Muno et al.,* 2006; *Skinner et al.,* 2006). Chandra studies of NGC 6357, M 16, RCW 49, W 51A, W 3, and other regions are also underway. Both XMM-Newton and Chandra have examined rich clusters in the Carina Nebula (*Evans et al.,* 2003, 2004; *Albacete Colombo et al.,* 2003), NGC 6231 at the core of the Sco OB1 association, and portions of Cyg OB2.

4.1. Cepheus B, RCW 38, and Stellar Populations

Each Chandra image of a MSFR shows hundreds, sometimes over a thousand, unresolved sources. For regions at distances around $d \simeq 1$–3 kpc, only a small fraction (typically 3–10%) of these sources are background quasars or field galactic stars. The stellar contamination is low because PMS stars are typically 100-fold more X-ray luminous than 1–10-G.y.-old main-sequence stars (Fig. 5a). Since Chandra source positions are accurate to 0.2"–0.4", identifications have little ambiguity except for components of multiple systems. The XLF of a stellar population spans 4 orders of magnitude; 2 orders of magnitude of this range arises from a correlation with stellar mass and bolometric magnitude (*Preibisch et al.,* 2005). This means that the X-ray flux limit of a MSFR observation roughly has a corresponding limit in K-band magnitude and mass. Day-long exposures of regions $d \simeq 2$ kpc away are typically complete to $\log L_x[\text{erg/s}] \sim 29.5$, which gives nearly complete samples down to $M \simeq 1 M_\odot$ with little contamination. We outline two recent studies of this type.

A 27-h Chandra exposure of the stellar cluster illuminating the HII region RCW 38 ($d \simeq 1.7$ kpc) reveals 461 X-ray sources, of which 360 are confirmed cluster members (*Wolk et al.,* 2006). Half have near-infrared counterparts, of which 20% have K-band excesses associated with optically thick disks. The cluster is centrally concentrated with a half-width of 0.2 pc and a central density of 100 X-ray stars/pc². Obscuration of the cluster members, seen both in the soft X-ray absorption column and near-infrared photometry, is typically $10 < A_V < 20$ mag. The X-ray stars are mostly unstudied; particular noteworthy are 31 X-ray stars that may be new obscured OB stars. Assuming a standard IMF, the total cluster membership is estimated to exceed 2000 stars. About 15% of the X-ray sources are variable, and several show plasma temperatures exceeding 100 MK.

A recent Chandra study was made of the Sharpless 155 HII region on the interface where stars from the Cep OB3b association ($d = 725$ pc) illuminates the Cepheus B molecular cloud core (*Getman et al.,* 2006a). Earlier, a few ultra-compact HII regions inside the cloud indicated an embedded cluster is present, but little was known about the embedded population. The 8-h exposure shows 431 X-ray sources, of which 89% are identified with K-band stars. Sixty-four highly absorbed X-ray stars inside the cloud provide the best census of the embedded cluster, while the 321 X-ray stars outside the cloud provide the best census of this portion of the Cep OB3b cluster. Surprisingly, the XLF of the unobscured sample has a different shape from that seen in the Orion Nebula Cluster, with an excess of stars around log L_x[erg/s] $\simeq 29.7$ or M $\simeq 0.3$ $M_\odot$. It is not clear whether this arises from a deviation in the IMF or some other cause, such as sequential star formation generating a noncoeval population. The diffuse X-rays in this region, which has only one known O star, are entirely attributable to the integrated contribution of fainter PMS stars.

4.2. M 17 and X-Ray Flows in HII Regions

For OB stars excavating an HII region within their nascent molecular cloud, diffuse X-rays may be generated as fast winds shock the surrounding media (*Weaver et al., 1977*). Chandra has clearly discriminated this diffuse emission from the hundreds of X-ray-emitting young stars in M 17 and the Rosette Nebula (*Townsley et al., 2003*).

Perhaps the clearest example of diffuse X-ray emission in MSFRs is the Chandra observation of M 17, a bright blownout blister HII region on the edge of a massive molecular cloud (d $\simeq 1.6$ kpc). The expansion of the blister HII region is triggering star formation in its associated giant molecular cloud, which contains an ultracompact HII region, water masers, and the dense core M 17SW. M 17 has 100 stars earlier than B9 (for comparison, the Orion Nebula Cluster has 8), with 14 O stars. The Chandra image is shown in Plate 3, along with an earlier, wider-field image from the ROSAT satellite. Over 900 point sources in the ~172 × 172 field are found (Broos et al., in preparation).

The diffuse emission of M 17 is spatially concentrated eastward of the stellar cluster and fills the region delineated by the photodissociation region and the molecular cloud. The X-ray spectrum can be modeled as a two-temperature plasma with T = 1.5 MK and 7 MK, and a total intrinsic X-ray luminosity (corrected for absorption) of $L_{x,diffuse} = 3 \times 10^{33}$ erg/s (*Townsley et al., 2003*). The X-ray plasma has mass M ~ 0.1 $M_\odot$ and density 0.1–0.3 cm^{-3} spread over several cubic parsecs. It represents only ~10^4 yr of recent O-wind production; past wind material has already flowed eastward into the galactic interstellar medium.

The diffuse emission produced by the M 17 cluster, and similar but less dramatic emission by the Rosette Nebula cluster, gives new insight into HII region physics. The traditional HII region model developed decades ago by Strömgren and others omitted the role of OB winds, which were not discovered until the 1960s. The winds play a small role in the overall energetics of HII regions, but they dominate the momentum and dynamics of the nebula with $\frac{1}{2}\dot{M}v_w^2$ ~ 10^{36-37} erg/s for a typical early-O star. If completely surrounded by a cold cloud medium, an O star should create a "wind-swept bubble" with concentric zones: a freely expanding wind, a wind termination shock followed by an X-ray emitting zone, the standard T = 10^4 K HII region, the ionization front, and the interface with the cold interstellar environment (*Weaver et al., 1977; Capriotti and Kozminski, 2001*).

These early models predicted L_x ~ 10^{35} erg/s from a single embedded O star, 2 orders of magnitude brighter than the emission produced by M 17 (*Dunne et al., 2003*). Several explanations for this discrepancy can be envisioned: Perhaps the wind energy is dissipated in a turbulent mixing layer (*Kahn and Breitschwerdt, 1990*), or the wind terminal shock may be weakened by mass-loading of interstellar material (e.g., *Pittard et al., 2001*). Winds from several OB stars may collide and shock before they hit the ambient medium (*Cantó et al., 2000*). Finally, a simple explanation may be that most of the kinetic energy of the O star winds remains in a bulk kinetic flow into the galactic interstellar medium (*Townsley et al., 2003*).

4.3. Trumpler 14 in the Carina Nebula

The Carina complex at d $\simeq 2.8$ kpc is a remarkably rich star-forming region containing 8 open clusters with at least 64 O stars, several WR stars, and the luminous blue variable η Car. The presence of WR stars may indicate past supernovae, although no well-defined remnant has been identified. One of these clusters is Tr 14, an extremely rich, young (~1 m.y.), compact OB association with ~30 previously identified OB stars. Together with the nearby Trumpler 16 cluster, it has the highest concentration of O3 stars known in the galaxy. Over 20 years ago, an Einstein Observatory X-ray study of the Carina star-forming complex detected a few dozen high-mass stars and diffuse emission attributed to O-star winds (*Seward and Chlebowski, 1982*). Chandra studies show that thousands of the lower-mass stars in these young clusters were likely to be contributing to this diffuse flux; a major goal is to determine the relative contributions of stars, winds, and SNRs to the extended emission in the Carina Nebula.

Plate 4a shows a 16-h Chandra exposure centered on HD 93129AB, the O2I/O3.5V binary at the center of Tr 14 (Townsley et al., in preparation). Over 1600 members of the Tr 14 and Tr 16 clusters can be identified from the X-ray point sources and extensive diffuse emission is clearly present. The diffuse emission surrounding Tr 14 is quite soft with subsolar elemental abundances, similar to the M 17 OB-wind shocks. But the much brighter diffuse emission seen far from the massive stellar clusters is less absorbed and requires enhanced abundances of Ne and Fe. This supports models involving old "cavity" supernova remnants that exploded inside the Carina superbubble (e.g., *Chu et al., 1993*).

Chandra resolves the two components of HD 93129AB separated by ~3": HD 93129B shows a typical O-star soft X-ray spectrum (T ~ 6 MK), while HD 93129A shows a similar soft component plus a T ~ 35 MK component that dominates the total X-ray luminosity. This hard spectrum and high X-ray luminosity are indicative of a colliding-wind binary (*Pittard et al.,* 2001), in agreement with the recent finding that HD 93129A is itself a binary (*Nelan et al.,* 2004). Other colliding wind binaries are similarly identified in the cluster.

4.4. The Starburst of 30 Doradus

Plate 4b shows a 6-h Chandra exposure of 30 Dor in the Large Magellanic Cloud, the most luminous giant extragalactic HII region and "starburst cluster" in the Local Group. 30 Dor is the product of multiple epochs of star formation, which have produced multiple SNRs seen with ROSAT as elongated plasma-filled superbubbles on ~100-pc scales (*Wang and Helfand,* 1991).

The new Chandra image shows a bright concentration of X-rays associated with the R136 star cluster, the bright SNR N157B to the southwest, a number of new widely distributed compact X-ray sources, and diffuse structures that fill the superbubbles produced by the collective effects of massive stellar winds and their past supernova events (*Townsley et al.,* 2006a). Some of these are center-filled while others are edge-brightened, indicating a complicated mix of viewing angles and perhaps filling factors. Comparison of the morphologies of the diffuse X-ray emission with the photodissociation region revealed by Hα imaging and cool dust revealed by infrared imaging with the Spitzer Space Telescope shows a remarkable association: The hot plasma clearly fills the cavities outlined by ionized gas and warm dust. Spectral analysis of the superbubbles reveals a range of absorptions ($A_V = 1$–3 mag), plasma temperatures (T = 3–9 MK), and abundances. About 100 X-ray sources are associated with the central massive cluster R136 (*Townsley et al.,* 2006b). Some are bright, hard X-ray point sources in the field likely to be colliding-wind binaries, while others are probably from ordinary O and WR stellar winds.

5. X-RAY EFFECTS ON STAR AND PLANET FORMATION

5.1. X-Ray Ionization of Molecular Cloud Cores

One of the mysteries of galactic astrophysics is why most interstellar molecular material is not engaged in star formation. Large volumes of most molecular clouds are inactive, and some clouds appear to be completely quiescent. A possible explanation is that star formation is suppressed by ionization: Stellar ultraviolet will ionize the outer edges of clouds, and galactic cosmic rays may penetrate into their cores (*Stahler and Palla,* 2005). Even very low levels of ionization will couple the mostly neutral gas to magnetic fields, inhibiting gravitational collapse until sufficient ambipolar diffusion occurs.

The X-ray observations of star-forming regions demonstrate that a third source of ionization must be considered: X-rays from the winds and flares of deeply embedded X-ray sources. The X-ray ionization zones, sometimes called X-ray dissociation regions (XDRs) or Röntgen spheres, do

not have sharp edges like ultraviolet Strömgren spheres, but rather extend to large distances with decreasing effect (*Hollenbach and Tielens,* 1997).

The COUP observation provides a unique opportunity to calculate realistic XDRs in two molecular cloud cores: OMC-1 or the Becklin-Neugebauer region, and OMC-1 South. Several dozen embedded X-ray stars are seen in these clouds (Plate 1), and each can be characterized by X-ray luminosity, spectrum, and line-of-sight absorption (*Grosso et al.,* 2005). Figure 6 illustrates the X-ray properties of deeply embedded objects. COUP #554 is a young star with a strong infrared-excess in the OMC-1 South core. The Chandra spectrum shows soft X-ray absorption of log $N_H[cm^{-2}]$ = 22.7, equivalent to A_V ~ 30 mag, and the light-curve exhibits many powerful flares at the top of the XLF with peak X-ray luminosities reaching ~10^{32} erg/s. COUP #632 has no optical or K-band counterpart and its X-ray spectrum shows the strongest absorption of all COUP sources: log $N_H[cm^{-2}] \simeq$ 23.9 or A_V ~ 500 mag.

Using the COUP source positions and absorptions, we can roughly place each star into a simplified geometrical model of the molecular cloud gas, and calculate the region around each where the X-ray ionization exceeds the expected uniform cosmic-ray ionization. Plate 2 shows the resulting XDRs in OMC-1 from the embedded Becklin-Neugebauer cluster (Lorenzani et al., in preparation). Here a significant fraction of the volume is dominated by X-ray ionization. In general, the ionization of cloud cores ≈0.1 pc in size will be significantly altered if they contain clusters with more than ~50 members.

5.2. X-Ray Irradiation of Protoplanetary Disks

The circumstellar disks around PMS stars where planetary systems form were generally considered to consist of cool, neutral molecular material in thermodynamic equilibrium with ~100–1000 K temperatures. But there is a growing understanding that they are not closed and isolated structures. A few years ago, discussion concentrated on ultraviolet radiation from O stars that can photoevaporate nearby disks (*Hollenbach et al.,* 2000). More recently, considerable theoretical discussion has focused on X-ray irradiation from the host star (Fig. 7). This is a rapidly evolving field and only a fraction of the studies can be mentioned here. Readers are referred to reviews by *Feigelson* (2005) and *Glassgold et al.* (2005) for more detail.

X-ray studies provide two lines of empirical evidence that the X-rays seen with our telescopes actually do efficiently irradiate protoplanetary disks. First, the 6.4-keV fluorescent line of neutral iron is seen in several embedded COUP stars with massive disks, as shown in Fig. 2 (*Imanishi et al.,* 2001). This line is only produced when hard X-rays illuminate >1 g/cm² of material; this is too great to be intervening material and must be reflection off of a flattened structure surrounding the X-ray source (*Tsujimoto et al.,* 2005; *Favata et al.,* 2005b). Second, X-ray spectra of PMS stars with inclined disks show more absorption than

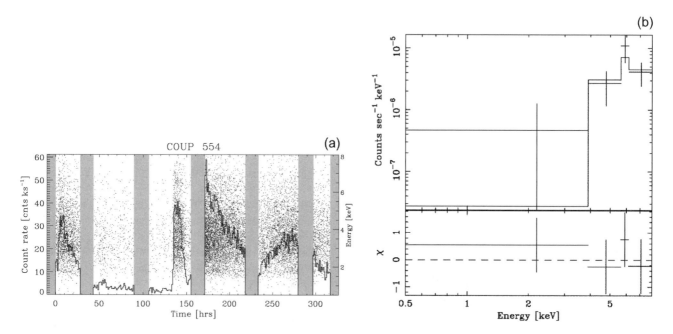

Fig. 6. X-ray properties of two stars deeply embedded in the OMC-1 South molecular cloud core from the Chandra Orion Ultradeep Project. **(a)** Lightcurve of COUP #554 over 13.2 d where the histogram shows the integrated brightness (lefthand vertical axis) and the dots show the energies of individual photons (righthand vertical axis). **(b)** Spectrum of COUP #632 showing very strong absorption at energies below 4 keV. From *Grosso et al.* (2005).

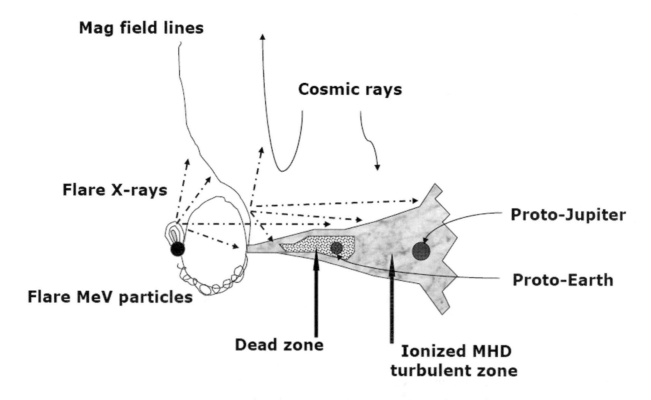

Fig. 7. Cartoon illustrating sources of energetic irradiation (galactic cosmic rays, flare X-rays, flare particles) and their possible effects on protoplanetary disks (ionization of gas and induction of MHD turbulence, layered accretion structure, spallation of solids). From *Feigelson* (2005).

spectra from stars with face-on disks. This is most clearly seen in the COUP survey, where column densities log $N_H[cm^{-2}] \sim 23$ are seen in edge-on proplyds imaged with the Hubble Space Telescope (*Kastner et al.,* 2005). This demonstrates the deposition of ionizing radiation in the disk and gives a rare measurement of the gas (rather than dust) content of protoplanetary disks.

Having established that X-ray emission, particularly X-ray flaring, is ubiquitous in PMS stars, and that at least some disks are efficiently irradiated by these X-rays, one can now estimate the X-ray ionization rate throughout a disk. The result is that X-rays penetrate deeply toward the midplane in the jovian planet region, but leave a neutral "dead zone" in the terrestrial planet region (e.g., *Igea and Glassgold,* 1999; *Fromang et al.,* 2002). The ionization effect of X-rays is many orders of magnitude more important than that of cosmic rays. However, differing treatments of metal ions and dust leads to considerable differences in the inferred steady-state ionization level of the disk (*Ilgner and Nelson,* 2006). The theory of the X-ray ionization *rate* thus appears satisfactory, but calculations of the X-ray ionization *fraction* depend on poorly established recombination rates.

X-ray ionization effects become important contributors to the complex and nonlinear interplay between the thermodynamics, dynamics, gas-phase chemistry, and gas-grain interactions in protoplanetary disks. One important consequence may be the induction of the magnetorotational instability, which quickly develops into a full spectrum of MHD turbulence including both vertical and radial mixing. The radial viscosity associated with the active turbulent zone may cause the flow of material from the outer disk into the inner disk, and thereby into the bipolar outflows and onto the protostar. This may solve a long-standing problem in young stellar studies: A completely neutral disk should have negligible viscosity and thus cannot efficiently be an accretion disk. Ionization-induced turbulence should affect planet formation and early evolution in complex ways: suppressing gravitational instabilities, concentrating solids, producing density inhomogeneities that can inhibit Type I migration of protoplanets, diminishing disk gaps involved in Type II migration, and so forth. It is thus possible that X-ray emission plays an important role in regulating the structure and dynamics of planetary systems, and the wide range in X-ray luminosities may be relevant to the diversity of extrasolar planetary systems.

PMS X-rays are also a major source of ionization at the base of outflows from protostellar disks that produce the emission line Herbig-Haro objects and molecular bipolar outflows (*Shang et al.,* 2002). This is a profound result: If low-mass PMS stars were not magnetically active and profusely emitting penetrating photoionizing radiation, then the coupling between the Keplerian orbits in the disk and the magnetocentrifugal orbits spiralling outward perpendicular to the disks might be much less efficient than we see.

X-ray ionization of a molecular environment will induce a complex series of molecular-ion and radical chemical reactions (e.g., *Aikawa and Herbst,* 1999; *Semenov et al.,*

2004). CN, HCO$^+$, and C$_2$H abundances may be good tracers of photoionization effects, although it is often difficult to distinguish X-ray and ultraviolet irradiation from global disk observations. X-ray heating may also lead to ice evaporation and enhanced gaseous abundances of molecules such as methanol. X-ray absorption also contributes to the warming of outer molecular layers of the disk. In the outermost layer, the gas is heated to 5000 K, far above the equilibrium dust temperature (*Glassgold et al.,* 2004). This may be responsible for the strong rovibrational CO and H$_2$ infrared bands seen from several young disks.

Finally, PMS flaring may address several long-standing characteristics of ancient meteorites that are difficult to explain within the context of a quiescent solar nebula in thermodynamic equilibrium:

1. Meteorites reveal an enormous quantity of flash-melted chondrules. While many explanations have been proposed, often with little empirical support, it is possible that they were melted by the $>10^8$ X-ray flares experienced by a protoplanetary disk during the era of chondrule melting. Melting might either be produced directly by the absorption of X-rays by dustballs (*Shu et al.,* 2001) or by the passage of a shock along the outer disk (*Nakamoto et al.,* 2005).

2. Certain meteoritic components, particularly the calcium-aluminum-rich inclusions (CAIs), exhibit high abundances of daughter nuclides of short-lived radioisotopic anomalies that must have been produced immediately before or during disk formation. Some of these may arise from the injection of recently synthesized radionuclides from supernovae, but other may be produced by spallation from MeV particles associated with the X-ray flares (*Feigelson et al.,* 2002). Radio gyrosynchrotron studies already demonstrate that relativistic particles are frequently present in PMS systems.

3. Some meteoritic grains that were free-floating in the solar nebula show high abundances of spallogenic ^{21}Ne excesses correlated with energetic particle track densities (*Woolum and Hohenberg,* 1993). The only reasonable explanation is irradiation by high fluences of MeV particles from early solar flares.

We thus find that X-ray astronomical studies of PMS stars have a wide variety of potentially powerful effects on the physics, chemistry, and mineralogy of protoplanetary disks and the environment of planet formation. These investigations are still in early stages, and it is quite possible that some of these proposed effects may prove to be uninteresting while others prove to be important.

6. SUMMARY

The fundamental result of X-ray studies of young stars and star-formation regions is that material with characteristic energies of keV (or even MeV) per particle is present in environments where the equilibrium energies of the bulk material are meV. Magnetic reconnection flares in lower-mass PMS stars, and wind shocks on different scales in

O stars, produce these hot gases. Although the X-ray luminosities are relatively small, the radiation effectively penetrates and ionizes otherwise neutral molecular gases and may even melt solids. X-rays from PMS stars may thus have profound effects on the astrophysics of star and planet formation.

The recent investigations outlined here from the Chandra and XMM-Newton observatories paint a rich picture of X-ray emission in young stars. Both the ensemble statistics and the characteristics of individual X-ray flares strongly resemble the flaring seen in the Sun and other magnetically active stars. Astrophysical models of flare cooling developed for solar flares fit many PMS flares well. PMS spectra show the same abundance anomalies seen in older stars. Rotationally modulated X-ray variability of the nonflaring characteristic emission show that the X-ray emitting structures lie close to the stellar surface and are inhomogeneously distributed in longitude. This is a solid indication that the X-ray emitting structures responsible for the observed modulation are in most cases multipolar magnetic fields, as on the Sun.

At the same time, the analysis of the most powerful flares indicates that very long magnetic structures are likely present in some of the most active PMS stars, quite possibly connecting the star with its surrounding accretion disk. The evidence suggests that both coronal-type and star-disk magnetic field lines are present in PMS systems, in agreement with current theoretical models of magnetically funnelled accretion.

There is a controversy over the X-ray spectra of a few of the brightest accreting PMS stars. TW Hya shows low plasma temperatures and emission lines, suggesting an origin in accretion shocks rather than coronal loops. However, it is a challenge to explain the elemental abundances and to exclude the role of ultraviolet irradiation. Simultaneous optical observations during the COUP X-ray observation clearly shows that the bulk of X-ray emission does not arise from accretion processes. Perhaps counterintuitively, various studies clearly show that accreting PMS stars are statistically weaker X-ray emitters than nonaccretors. A fraction of the magnetic field lines in accreting PMS stars are likely to be mass loaded and cannot reach X-ray temperatures.

X-ray images of high-mass star-forming regions are incredibly rich and complex. Each image shows hundreds or thousands of magnetically active PMS stars with ages ranging from Class I (and controversially, Class 0) protostars to zero-age main-sequence stars. Hard X-rays are often emitted so that Chandra can penetrate up to $A_V \simeq 500$ mag into molecular cloud material. Chandra images of MSFRs also clearly reveal for the first time the fate of O-star winds: The interiors of some HII regions are suffused with a diffuse 10 MK plasma, restricting the 10^4 K gas to a thin shell. The concept of a Strömgren sphere must be revised in these cases. Only a small portion of the wind energy and mass appears in the diffuse X-ray plasma; most likely flows unimpeded into the galactic interstellar medium. The full population of stars down to ~1 $M_\odot$ is readily seen in X-ray im-

ages of MSFRs, with little contamination from extraneous populations. This may lead, for example, to X-ray-based discrimination of close OB binaries with colliding winds and identification of intermediate-mass PMS stars that are not accreting. In the most active and long-lived MSFRs, cavity SNRs and superbubbles coexist with, and may dominate, the stellar and wind X-ray components. X-ray studies thus chronicle the life cycle of massive stars from proto-O stars to colliding O winds, to supernova remnants and superbubbles. These star-forming regions represent the building blocks of galactic-scale star formation and starburst galaxies.

Some of the issues discussed here are now well developed while others are still in early stages of investigation. It is unlikely that foreseeable studies will give qualitatively new information on the X-ray properties of low-mass PMS stars than obtained from the many studies emerging from the COUP and XEST projects. In-depth analysis of individual objects, especially high-resolution spectroscopic study, represents an important area ripe for follow-up exploration. The many X-ray studies of MSFRs now emerging should give large new samples of intermediate-mass stars, and new insights into the complex physics of OB stellar winds on both small and large scales. Although Chandra and XMM-Newton have relatively small fields, a commitment to wide-field mosaics of MSFR complexes like W3-W4-W5 and Carina could give unique views into the interactions of high-mass stars and the galactic interstellar medium. Deep X-ray exposures are needed to penetrate deeply to study the youngest embedded systems. Finally, the next generation of high-throughput X-ray telescopes should bring new capabilities to perform high-resolution spectroscopy of the X-ray emitting plasmas. Today, theoretical work is urgently needed on a host of issues raised by the X-ray findings: magnetic dynamos in convective stars, accretion and reconnection in disk-star magnetic fields, flare physics at levels far above those seen in the Sun, and possible effects of X-ray ionization of protoplanetary disks.

Acknowledgments.   E.D.F. recognizes the excellent work by K. Getman and the other 36 scientists in the COUP collaboration. E.D.F. and L.K.T. benefit from discussions with their Penn State colleagues P. Broos, G. Garmire, K. Getman, M. Tsujimoto, and J. Wang. Penn State work is supported by the National Aeronautics and Space Administration (NASA) through contract NAS8-38252 and Chandra Awards G04-5006X, G05-6143X, and SV4-74018 issued by the Chandra X-ray Observatory Center, operated by the Smithsonian Astrophysical Observatory for and on behalf of NASA under contract NAS8-03060. M.G. warmly acknowledges the extensive work performed by XEST team members. The XEST team has been financially supported by the Space Science Institute (ISSI) in Bern, Switzerland. XMM-Newton is an ESA science mission with instruments and contributions directly funded by ESA Member States and the U.S. (NASA). K.G.S. is grateful for funding support from NSF CAREER grant AST-0349075.

REFERENCES

Aikawa Y. and Herbst E. (1999) *Astron. Astrophys., 351,* 233–246.
Albacete Colombo J. F., Méndez M., and Morrell N. I. (2003) *Mon. Not. R. Astron. Soc., 346,* 704–718.

Arzner K. and Güdel M. (2004) *Astrophys. J., 602,* 363–376.

Audard M., Güdel M., Skinner S. L., Briggs K. R., Walter F. M., et al. (2005) *Astrophys. J., 635,* L81–L84.

Barnes S. A. (2003a) *Astrophys. J., 586,* 464–479.

Barnes S. A. (2003b) *Astrophys. J., 586,* L145–L147.

Brinkman A. C., Behar E., Güdel M., Audard M., den Boggende A. J. F., et al. (2001) *Astron. Astrophys., 365,* L324–L328.

Calvet N., D'Alessio P., Hartmann L., Wilner D., Walsh A., and Sitko M. (2002) *Astrophys. J., 568,* 1008–1016.

Cantó J., Raga A. C., and Rodríguez L. F. (2000) *Astrophys. J., 536,* 896–901.

Capriotti E. R. and Kozminski J. F. (2001) *Publ. Astron. Soc. Pac., 113,* 677–691.

Chu Y.-H., Mac Low M.-M., Garcia-Segura G., Wakker B., and Kennicutt R. C. (1993) *Astrophys. J., 414,* 213–218.

Damiani F., Flaccomio E., Micela G., Sciortino S., Harnden F. R. Jr., and Murray S. S. (2004) *Astrophys. J., 608,* 781–796.

Drake J. J., Testa P., and Hartmann L. (2005) *Astrophys. J., 627,* L149–L152.

Dunne B. C., Chu Y.-H., Chen C.-H. R., Lowry J. D., Townsley L., et al. (2003) *Astrophys. J., 590,* 306–313.

Evans N. R., Seward F. D., Krauss M. I., Isobe T., Nichols J., et al. (2003) *Astrophys. J., 589,* 509–525.

Evans N. R., Schlegel E. M., Waldron W. L., Seward F. D., Krauss M. I., et al. (2004) *Astrophys. J., 612,* 1065–1080.

Ezoe Y., Kokubun M., Makishima K., Sekimoto Y., and Matsuzaki K. (2006) *Astrophys. J., 638,* 860–877.

Favata F. and Micela G. (2003) *Space Sci. Rev., 108,* 577–708.

Favata F., Giardino G., Micela G., Sciortino S., and Damiani F. (2003) *Astron. Astrophys., 403,* 187–203.

Favata F., Flaccomio E., Reale F., Micela G., Sciortino S., et al. (2005a) *Astrophys. J. Suppl., 160,* 469–502.

Favata F., Micela G., Silva B., Sciortino S., and Tsujimoto M. (2005b) *Astron. Astrophys. 433,* 1047–1054.

Feigelson E. D. (2005) In *Proc. 13th Cool Stars Workshop* (F. Favata et al., eds.), pp. 175–183. ESA SP-560, Noordwijk.

Feigelson E. D. and Montmerle T. (1999) *Ann. Rev. Astron. Astrophys., 37,* 363–408.

Feigelson E. D., Garmire G. P., and Pravdo S. H. (2002) *Astrophys. J., 572,* 335–349.

Feigelson E. D., Gaffney J. A., Garmire G., Hillenbrand L. A., and Townsley L. (2003) *Astrophys. J., 584,* 911–930.

Feigelson E. D., Hornschemeier A. E., Micela G., Bauer F. E., Alexander D. M., et al. (2004) *Astrophys. J., 611,* 1107–1120.

Feigelson E. D., Getman K., Townsley L., Garmire G., Preibisch T., et al. (2005) *Astrophys. J. Suppl., 160,* 379–389.

Flaccomio E., Micela G., and Sciortino S. (2003) *Astron. Astrophys., 402,* 277–292.

Flaccomio E., Micela G., Sciortino S., Feigelson E. D., Herbst W., et al. (2005) *Astrophys. J. Suppl., 160,* 450–468.

Fromang S., Terquem C., and Balbus S. A. (2002) *Mon. Not. R. Astron. Soc., 329,* 18–28.

Gagné M., Skinner S. L., and Daniel K. J. (2004) *Astrophys. J., 613,* 393–415.

Getman K. V., Feigelson E. D., Townsley L., Bally J., Lada C. J., and Reipurth B. (2002) *Astrophys. J., 575,* 354–377.

Getman K. V., Flaccomio E., Broos P. S., Grosso N., Tsujimoto M., et al. (2005a) *Astrophys. J. Suppl., 160,* 319–352.

Getman K. V., Feigelson E. D., Grosso N., McCaughrean M. J., Micela G., et al. (2005b) *Astrophys. J. Suppl., 160,* 353–378.

Getman K. V., Feigelson E. D., Townsley L., Broos P., Garmire G., Tsujimoto M. (2006a) *Astrophys. J. Suppl., 163,* 306–334.

Getman K. V., Feigelson E. D., Garmire G., Broos R., and Wang J. (2006b) *Astrophys. J.,* in press.

Glassgold A. E., Feigelson E. D., and Montmerle T. (2000) In *Protostars and Planets IV* (V. Mannings et al., eds.), pp. 429–456. Univ. of Arizona, Tucson.

Glassgold A. E., Najita J., and Igea J. (2004) *Astrophys. J., 615,* 972–990.

Glassgold A. E., Feigelson E. D., Montmerle T., and Wolk S. (2005) In *Chondrites and the Protoplanetary Disk* (A. N. Krot et al., eds.), pp. 161–180. ASP Conf. Series 341, San Francisco.

Grosso N., Montmerle T., Feigelson E. D., and Forbes T. G. (2004) *Astron. Astrophys., 419,* 653–665.

Grosso N., Feigelson E. D., Getman K. V., Townsley L., Broos P., et al. (2005) *Astrophys. J. Suppl., 160,* 530–556.

Grosso N., et al. (2006) *Astron. Astrophys.,* in press.

Güdel M. (2002) *Ann. Rev. Astron. Astrophys., 40,* 217–261.

Güdel M. (2004) *Astron. Astrophys. Rev., 12,* 71–237.

Güdel M., Guinan E. F., and Skinner S. L. (1997) *Astrophys. J., 483,* 947–960.

Güdel M., Audard M., Kashyap V. L., Drake J. J., and Guinan E. F. (2003) *Astrophys. J., 582,* 423–442.

Hamaguchi K., Corcoran M. F., Petre R., White N. E., Stelzer B., et al. (2005) *Astrophys. J., 623,* 291–301.

Hartmann L. (1998) *Accretion Processes in Star Formation,* Cambridge Univ., New York.

Herbst W., Bailer-Jones C. A. L., Mundt R., Meisenheimer K., and Wackermann R. (2002) *Astron. Astrophys., 396,* 513–532.

Hollenbach D. J. and Tielens A. G. G. M. (1997) *Ann. Rev. Astron. Astrophys., 35,* 179–216.

Hollenbach D. J., Yorke H. W., and Johnstone D. (2000) In *Protostars and Planets IV* (V. Mannings et al., eds.), pp. 401–416. Univ. of Arizona, Tucson.

Igea J. and Glassgold A. E. (1999) *Astrophys. J., 518,* 848–858.

Imanishi K., Koyama K., and Tsuboi Y. (2001) *Astrophys. J., 557,* 747–760.

Ilgner M. and Nelson R. P. (2006) *Astron. Astrophys., 445,* 223–232.

Jardine M. and Unruh Y. C. (1999) *Astron. Astrophys., 346,* 883–891.

Jardine M., Collier Cameron A., Donati J.-F., Gregory S. G., and Wood K. (2006) *Mon. Not. R. Astron. Soc., 367,* 917–927.

Johns-Krull C. M., Valenti J. A., and Saar S. H. (2004) *Astrophys. J., 617,* 1204–1215.

Kahn F. D. and Breitschwerdt D. (1990) *Mon. Not. R. Astron. Soc., 242,* 209–214.

Kashyap V. L., Drake J. J., Güdel M., and Audard M. (2002) *Astrophys. J., 580,* 1118–1132.

Kastner J. H., Huenemoerder D. P., Schulz N. S., Canizares C. R., and Weintraub D. A. (2002) *Astrophys. J., 567,* 434–440.

Kastner J. H., Richmond M., Grosso N., Weintraub D. A., Simon T., et al. (2004) *Nature, 430,* 429–431.

Kastner J. H., Franz G., Grosso N., Bally J., McCaughrean M. J., et al. (2005) *Astrophys. J. Suppl., 160,* 511–529.

Lamzin S. A. (1999) *Astron. Lett., 25,* 430–436.

Law C. and Yusef-Zadeh F. (2004) *Astrophys. J., 611,* 858–870.

Loinard L., Mioduszewski A. J., Rodríguez L. F., González R. A., Rodríguez M. I., and Torres R. M. (2005) *Astrophys. J., 619,* L179–L182.

Montmerle T., Grosso N., Tsuboi Y., and Koyama K. (2000) *Astrophys. J., 532,* 1097–1110.

Mullan D. J. and MacDonald J. (2001) *Astrophys. J., 559,* 353–371.

Muno M. P., Clark J. S., Crowther P. A., Dougherty S. M., de Grijs R., et al. (2006) *Astrophys. J., 636,* L41–L44.

Nakamoto T. and Miura H. (2005) In *PPV Poster Proceedings,* www.lpi.usra.edu/meetings/ppv2005/pdf/8530.pdf.

Nelan E. P., Walborn N. R., Wallace D. J., Moffat A. F. J., Makidon R. B., et al. (2004) *Astron. J., 128,* 323–329.

Ness J.-U. and Schmitt J. H. M. M. (2005) *Astron. Astrophys., 444,* L41–L44.

Ness J.-U. and Schmitt J. H. M. M. (2006) *Astron. Astrophys.,* in press.

Nielbock M., Chini R., and Müller S. A. H. (2003) *Astron. Astrophys., 408,* 245–256.

Pace G. and Pasquini L. (2004) *Astron. Astrophys., 426,* 1021–1034.

Paerels F. B. S. and Kahn S. M. (2003) *Ann. Rev. Astron. Astrophys., 41,* 291–342.

Parker E. N. (1998) *Astrophys. J., 330,* 474–479.

Pittard J. M., Hartquist T. W., and Dyson J. E. (2001) *Astron. Astrophys., 373,* 1043–1055.

Preibisch T. (2004) *Astron. Astrophys., 428,* 569–577.

Preibisch T. and Feigelson E. D. (2005) *Astrophys. J. Suppl., 160,* 390–400.

Preibisch T., Neuhäuser R., and Alcalá J. M. (1995) *Astron. Astrophys., 304,* L13–L16.

Preibisch T., Kim Y.-C., Favata F., Feigelson E. D., Flaccomio E., et al. (2005) *Astrophys. J. Suppl., 160,* 401–422.

Priest E. R. and Forbes T. G. (2002) *Astron. Astrophys. Rev., 10,* 313–377.

Rockefeller G., Fryer C. L., Melia F., and Wang Q. D. (2005) *Astrophys. J., 623,* 171–180.

Scelsi L., Maggio A., Peres G., and Pallavicini R. (2005) *Astron. Astrophys., 432,* 671–685.

Schmitt J. H. M. M., Robrade J., Ness J.-U., Favata F., and Stelzer B. (2005) *Astron. Astrophys., 432,* L35–L38.

Schrijver C. J. and Zwaan C. (2000) *Solar and Stellar Magnetic Activity,* Cambridge Univ., New York.

Semenov D., Weibe D., and Henning Th. (2004) *Astron. Astrophys., 417,* 93–106.

Seward F. D. and Chlebowski T. (1982) *Astrophys. J., 256,* 530–542.

Shang H., Glassgold A. E., Shu F. H., and Lizano S. (2002) *Astrophys. J., 564,* 853–876.

Shu F. H., Najita J. R., Shang H., and Li Z.-Y. (2000) In *Protostars and Planets IV* (V. Mannings et al., eds.), pp. 789–814. Univ. of Arizona, Tucson.

Shu F. H., Shang H., Gounelle M., Glassgold A. E., and Lee T. (2001) *Astrophys. J., 548,* 1029–1050.

Siess L., Forestini M., and Bertout C. (1999) *Astron. Astrophys., 342,* 480–491.

Skinner S., Gagné M., and Belzer E. (2003) *Astrophys. J., 598,* 375–391.

Skinner S. L., Zhekov S. A., Palla F., and Barbosa C. L. D. R. (2005) *Mon. Not. R. Astron. Soc., 361,* 191–205.

Skinner S. L., et al. (2006) *Astrophys. J. Lett., 639,* L35–L38.

Skumanich A. (1972) *Astrophys. J., 171,* 565–567.

Stahler S. W. and Palla F. (2005) *The Formation of Stars,* Wiley-VCH, New York.

Stassun K. G., Ardila D. R., Barsony M., Basri G., and Mathieu R. D. (2004) *Astron. J., 127,* 3537–3552.

Stassun K. G., van den Berg M., Feigelson E., and Flaccomio E. (2006) *Astrophys. J.,* in press.

Stelzer B. and Schmitt J. H. M. M. (2004) *Astron. Astrophys., 418,* 687–697.

Swartz D. A., Drake J. J., Elsner R. F., Ghosh K. K., Grady C. A., et al. (2005) *Astrophys. J., 628,* 811–816.

Townsley L. K., Feigelson E. D., Montmerle T., Broos P. S., Chu Y.-H., and Garmire G. P. (2003) *Astrophys. J., 593,* 874–905.

Townsley L. K., Broos P. S., Feigelson E. D., Brandl B. R., Chu Y.-H., Garmire G. P., and Pavlov G. G. (2006a) *Astron. J., 131,* 2140–2163.

Townsley L. K., Broos P. S., Feigelson E. D., Garmire G. P., and Getman K. V. (2006b) *Astron. J., 131,* 2164–2184.

Tsuboi Y., Imanishi K., Koyama K., Grosso N., and Montmerle T. (2000) *Astrophys. J., 532,* 1089–1096.

Tsuboi Y., Koyama K., Hamaguchi K., Tatematsu K., Sekimoto Y., et al. (2001) *Astrophys. J., 554,* 734–741.

Tsujimoto M., Feigelson E. D., Grosso N., Micela G., Tsuboi Y., et al. (2005) *Astrophys. J. Suppl., 160,* 503–510.

Wang Q. and Helfand D. J. (1991) *Astrophys. J., 373,* 497–508.

Weaver R., McCray R., Castor J., Shapiro P., and Moore R. (1977) *Astrophys. J., 218,* 377–395.

Wolk S. J., Harnden F. R. Jr., Flaccomio E., Micela G., Favata F., et al. (2005) *Astrophys. J. Suppl., 160,* 423–449.

Wolk S. J., et al. (2006) *Astrophys. J.,* in press.

Woolum D. S. and Hohenberg C. (1993) In *Protostars and Planets III* (E. H. Levy and J. I. Lunine, eds.), pp. 903–919. Univ. of Arizona, Tucson.

The Taurus Molecular Cloud: Multiwavelength Surveys with XMM-Newton, the Spitzer Space Telescope, and the Canada-France-Hawaii Telescope*

Manuel Güdel
Paul Scherrer Institut

Deborah L. Padgett
California Institute of Technology

Catherine Dougados
Laboratoire d'Astrophysique de Grenoble

The Taurus Molecular Cloud (TMC) ranks among the nearest and best-studied low-mass star-formation regions. It contains numerous prototypical examples of deeply embedded protostars with massive disks and outflows, classical and weak-lined T Tauri stars, jets and Herbig-Haro objects, and a growing number of confirmed brown dwarfs. Star formation is ongoing, and the cloud covers all stages of pre-main-sequence stellar evolution. We have initiated comprehensive surveys of the TMC, including in particular (1) a deep X-ray survey of about 5 deg^2 with XMM-Newton; (2) a near- to mid-infrared photometric survey of ≈30 deg^2 with the Spitzer Space Telescope, mapping the entire cloud in all available photometric bands; and (3) a deep optical survey using the Canada-France-Hawaii Telescope. Each wavelength regime contributes to the understanding of different aspects of young stellar systems. XMM-Newton and Spitzer mapping of the central TMC is a real breakthrough in disk characterization, offering the most detailed studies of correlations between disk properties and high-energy magnetic processes in any low-mass star-forming region, extending also to brown dwarfs in which disk physics is largely unexplored. The optical data critically complements the other two surveys by allowing clear source identification with 0.8" resolution, identifying substellar candidates, and, when combined with NIR data, providing the wavelength baseline to probe NIR excess emission. We report results and correlation studies from these surveys. In particular, we address the physical interpretation of our new X-ray data, discuss the entire young stellar population from embedded protostars to weak-lined T Tau stars and their environment, and present new results on the low-mass population of the TMC, including young brown dwarfs.

1. INTRODUCTION

In a modern picture of star formation, complex feedback loops regulate mass accretion processes, the ejection of jets and outflows, and the chemical and physical evolution of disk material destined to form planets. Observations in X-rays with Chandra and XMM-Newton penetrate dense molecular envelopes, revealing an environment exposed to high levels of X-ray radiation. In a complementary manner, observations in the infrared with the Spitzer Space Telescope (Spitzer) now obtain detailed infrared photometry and spectroscopy with diagnostics for disk structure and chemical composition of the gas and dust in the circumstellar environment. Furthermore, optical surveys have reached a sensitivity and area coverage with which a detailed census of the substellar population has become possible.

Near- to far-infrared (IR) emission originates predominantly in the dusty environment of the forming stars, either in contracting gaseous envelopes or in circumstellar disks. IR excess (relative to the photospheric contributions) has been successfully used to model disk geometry, the structure of the envelope, and also the composition and structure of dust grains (*d'Alessio et al.*, 1999).

X-rays play a crucial role in studies of star formation, both physically and diagnostically. They may be generated at various locations in young stellar systems, such as in a "solar-like" coronal/magnetospheric environment, in shocks forming in accretion funnel flows (e.g., *Kastner et al.*, 2002), or in jets and Herbig-Haro flows (e.g., *Pravdo et al.*, 2001;

---

*With significant contributions from Lori Allen (CfA), Marc Audard (Columbia University), Jerôme Bouvier (Grenoble), Kevin Briggs (PSI), Sean Carey (Caltech), Elena Franciosini (Palermo), Misato Fukagawa (Caltech), Nicolas Grosso (Grenoble), Sylvain Guieu (Grenoble), Dean Hines (Space Science Institute), Tracy Huard (CfA), Eugene Magnier (IfA Honolulu), Eduardo Martín (IAC, Spain), François Ménard (Grenoble), Jean-Louis Monin (Grenoble), Alberto Noriega-Crespo (Caltech), Francesco Palla (Florence), Luisa Rebull (Caltech), Luigi Scelsi (Palermo), Alessandra Telleschi (PSI), and Susan Terebey (CalStateLA).

Bally et al., 2003; *Güdel et al.,* 2005). By ionizing circumstellar material, X-rays also determine some of the prevalent chemistry (*Glassgold et al.,* 2004) while making the gas accessible to magnetic fields. Ionization of the circumstellar-disk surface may further drive accretion instabilities (*Balbus and Hawley,* 1991). Many of these X-ray related issues are summarized in the chapter by Feigelson et al., based on recent X-ray observations of star-forming regions.

1.1.   The Taurus Molecular Cloud Complex

The Taurus Molecular Cloud (TMC) has played a fundamental role in our understanding of low-mass star formation. At a distance around 140 pc (e.g., *Loinard et al.,* 2005; *Kenyon et al.,* 1994), it is one of the nearest star-formation regions (SFR) and reveals characteristics that make it ideal for detailed physical studies. One of the most notable properties of the TMC in this regard is its structure in which several loosely associated but otherwise rather isolated molecular cores each produce one or only a few low-mass stars, different from the much denser cores in ρ Oph or in Orion. TMC features a low stellar density of only 1–10 stars pc^{-2} (e.g., *Gómez et al.,* 1993). Strong mutual influence due to outflows, jets, or gravitational effects are therefore minimized. Furthermore, most stars in TMC are subject to relatively modest extinction, providing access to a broad spectrum of stars at all evolutionary stages from Class 0 sources to near-zero-age main-sequence T Tau stars. TMC has also become of central interest for the study of substellar objects, in particular brown dwarfs (BD), with regard to their evolutionary history and their spatial distribution and dispersal (*Reipurth and Clarke,* 2001; *Briceño et al.,* 2002).

TMC has figured prominently in star-formation studies at all wavelengths. It has provided the best-characterized sample of classical and weak-lined T Tau stars [CTTS and WTTS, respectively, or "Class II" and "Class III" objects (*Kenyon and Hartmann,* 1995)]; most of our current picture of low-density star formation is indeed based on IRAS studies of TMC (*Beichman et al.,* 1986; *Myers et al.,* 1987; *Strom et al.,* 1989; *Kenyon et al.,* 1990; *Weaver and Jones,* 1992; *Kenyon and Hartmann,* 1995). Among the key results from TMC studies as listed in *Kenyon and Hartmann* (1995) figure the following: (1) More than 50% of the TMC objects have IR excess beyond the photospheric contribution, correlating with other activity indicators (Hα, UV excess, etc.) and indicating the presence of warm circumstellar material predominantly in the form of envelopes for Class I protostars and circumstellar disks for Class II stars. (2) Class III sources (mostly to be identified with WTTS) are distinctly different from lower classes by not revealing optically thick disks or signatures of accretion. (3) Star formation has been ongoing at a similar level during the past 1–2 m.y., with the Class-I protostars having ages of typically 0.1–0.2 m.y. (4) There is clear support for an evolutionary sequence Class I → II → III, although there is little luminosity evolution along with this sequence, indicating different evolu-

tionary speeds for different objects. The infall timescale is a few times 10^5 yr, while the disk phase amounts to a few times 10^6 yr.

TMC has also been well-studied at millimeter wavelengths, having better high-resolution molecular line maps than any other star-forming region (*Onishi et al.* 2002). This region has been a major site for searches of complex molecules (*Kaifu et al.,* 2004); many molecular transitions have been mapped across a large area, and the results will be interesting to compare with large-scale IR dust maps of the TMC (see Fig. 3 below).

In X-rays, TMC has again played a key role in our understanding of high-energy processes and circumstellar magnetic fields around pre-main-sequence stars. Among the key surveys are those by *Feigelson et al.* (1987), *Walter et al.* (1988), *Bouvier* (1990), *Strom et al.* (1990), *Damiani et al.* (1995), and *Damiani and Micela* (1995) based on Einstein Observatory observations, and the work by *Strom and Strom* (1994), *Neuhäuser et al.* (1995), and *Stelzer and Neuhäuser* (2001) based on ROSAT. These surveys have characterized the overall luminosity behavior of TTS, indicated a dependence of X-ray activity on rotation, and partly suggested X-ray differences between CTTS and WTTS.

1.2.   Is the Taurus Molecular Cloud Anomalous?

Although TMC has been regarded, together with the ρ Oph dark cloud, as the prototypical low-mass star-forming region, a few apparent peculiarities should be mentioned. TMC contains an anomalously large fraction of binaries (*Ghez et al.,* 1993) compared with other SFRs (e.g., Orion) or with field stars. In TMC, about two-thirds of all members are bound in multiple systems, with an average separation of about 0.3" (e.g., *Leinert et al.,* 1993; *Mathieu,* 1994; *Simon et al.,* 1995; *Duchêne et al.,* 1999; *White and Ghez,* 2001; *Hartigan and Kenyon,* 2003). Also, TMC cloud cores are comparatively small and low-mass when compared with cores in Orion or Perseus (*Kun,* 1998).

TMC has also been found to be deficient of lowest-mass stars and BDs, with a mass distribution significantly enriched in 0.5–1 M$_\odot$ stars, compared to Orion samples (*Briceño et al.,* 2002). The formation of BDs may simply be different in the low-density environment of TMC compared to the dense packing of stars in Orion.

Furthermore, for reasons very poorly understood, TMC differs from other SFRs significantly with regard to X-ray properties. Whereas no X-ray activity-rotation correlation (analogous to that in main-sequence stars) is found for samples in the Orion star-forming regions, perhaps suggesting that all stars are in a saturated state (*Flaccomio et al.,* 2003a; *Preibisch et al.,* 2005a), the X-ray activity in TMC stars has been reported to decrease for increasing rotation period (e.g., *Neuhäuser et al.,* 1995; *Damiani and Micela,* 1995; *Stelzer and Neuhäuser,* 2001). Also, claims have been made that the X-ray behavior of TMC CTTS and WTTS is significantly different, CTTS being less luminous than WTTS (*Strom and Strom,* 1994; *Damiani et al.,* 1995; *Neuhäuser et al.,* 1995; *Stelzer and Neuhäuser,* 2001). This contrasts with other star-forming regions (*Flaccomio et al.,* 2000; *Preibisch and Zinnecker,* 2001), although recent reports reveal a similar segregation also for Orion and some other SFRs (*Flaccomio et al.,* 2003b; *Preibisch et al.,* 2005a). Some of these discrepancies may be due to selection and detection bias (e.g., WTTS are predominantly identified in X-ray studies, in contrast to CTTS).

2. NEW SURVEYS OF THE TAURUS CLOUDS

2.1. The Need for New Surveys

Some of the anomalies mentioned above may be strongly influenced by survey bias from the large size of TMC (25+ deg²). For example, with the exception of IRAS and 2MASS, all the IR surveys of TMC focus on limited regions around dense molecular cloud cores of TMC (or of other star-forming regions) since these are easy to map using inefficient point-and-shoot strategies with small IR arrays. Systematic and deep investigations of more distributed low-mass star formation such as that in TMC are lacking. The exception in X-rays is the ROSAT All-Sky Survey obtained in the early 1990s. However, this survey was of comparatively low sensitivity and recorded only rather soft X-ray photons, thus leaving all embedded sources and most of the lowest-mass stellar and BD population undetected. Angular resolution has also provided serious ambiguity.

A suite of new telescopes and satellite observatories now permits reconsideration of the issues mentioned in section 1. We have started a large multiwavelength project to map significant portions of TMC in X-rays, the optical, near-infrared, and mid-infrared. The instruments used, XMM-Newton in X-rays, Spitzer in the near-/mid-infrared range, and the Canada-France-Hawaii Telescope (CFHT) in the optical range, provide an observational leap forward. Table 1 compares characteristic limitations of our surveys with limitations defined by previous instruments, represented here by the ROSAT Position-Sensitive Proportional Counter [PSPC All-Sky Survey for X-rays (*Neuhäuser et al.,* 1995); also some pointing observations (*Stelzer and Neuhäuser,* 2001)], IRAS [25 μm for mid-infrared data (*Kenyon and Hartmann,* 1995)], and the KPNO 0.9-m telescope CCD surveys [for the optical (*Briceño et al.,* 1998, 2002; *Luhman,* 2000, 2004; *Luhman et al.,* 2003)].

2.2. Scientific Goals of the New Surveys

2.2.1. The XMM-Newton X-ray survey. Our XMM-Newton X-ray survey maps approximately 5 deg² (Fig. 1) of the denser molecular cloud areas with limiting sensitivities around $L_X \approx 10^{28}$ erg s⁻¹, sufficient, for the first time, to detect nearly every lightly absorbed, normal CTTS and WTTS and a significant fraction of BDs and protostars. The goals of this survey are (1) to collect X-ray spectra and light curves from a statistically meaningful sample of TMC objects, and to characterize them in terms of X-ray emission

TABLE 1. Characteristics of previous and new TMC surveys.

Parameter	Previous Surveys	New Surveys
Luminosity detection limit		
X-rays [erg s⁻¹]	$\approx 10^{29}$	$\approx 10^{28}$
mid-IR [$L_\odot$]	> 0.1	0.001
optical* [$L_\odot$]	$\approx 10^{-3}$	$\approx 10^{-4}$
Flux detection limit		
X-rays [erg cm⁻² s⁻¹]	$\approx 4 \times 10^{-14}$	$\approx 4 \times 10^{-15}$
mid-IR [24 μm, mJy]	500	1.2
optical [I, mag]	≈ 21	≈ 24
Mass detection limit		
X-rays	$\approx 0.1\ M_\odot$	$\approx 0.05\ M_\odot$
mid-IR, 24 μm	$\approx 0.5\ M_\odot$	$\approx 1\ M_{Jup}$
optical*	$\approx 15{-}20\ M_{Jup}$	$\approx 10\ M_{Jup}$
Accessible object types†		
X-rays	II, III	I–III, BD, (HH)
mid-IR	0–III	0–III, BD, Ph, DD
optical	I–III, BD	I–III, BD
Angular resolution (FWHM, respective pixel size)		
X-rays	$\approx 25''$	$\approx 4.5''$
mid-IR, 24 μm	$1' \times 5'$	$\approx 6''$
optical	$\approx 0.67''/$pix	$\approx 0.2''/$pix
Energy or wavelength range		
X-rays [keV]	0.1–2.4	0.2–15
mid-IR [μm]	1.2–100	3.6–160
optical [nm]	800–1000	780–1170
Spectral resolution or filter width		
X-rays [E/ΔE]	≈ 1	$\approx 10{-}50‡$
mid-IR 24 μm [Δλ]	5.8 μm	4.7 μm
Spatial extent		
X-rays	all-sky	5 deg²
mid-IR	all-sky	30 deg²
optical	12.4 deg²	34 deg²

*Derived from the DUSTY models of *Chabrier et al.* (2000) for $A_V < 5$ mag and age < 5 m.y.

†IR classes; HH = Herbig-Haro, DD = debris disks, Ph = photospheres (Class III).

‡X-ray gratings: ≈ 300 (selected targets).

measure distributions, temperatures, X-ray luminosities, and variability; (2) to interpret X-ray emission in the context of other stellar properties such as rotation, mass accretion, and outflows; (3) to investigate changes in the X-ray behavior as a young stellar object evolves; (4) to obtain a census of X-ray emitting objects at the stellar mass limit and in the substellar regime (BDs); (5) to study in what sense the stellar environment influences the X-ray production, and vice versa; and (6) to search for new, hitherto unrecognized TMC members.

The outstanding characteristics of the survey are its sensitivity, its energy resolution, and its energy coverage. The TMC population is detected nearly completely in the surveyed fields, thus suppressing potential bias that previous surveys may have been subject to. In particular, a significant

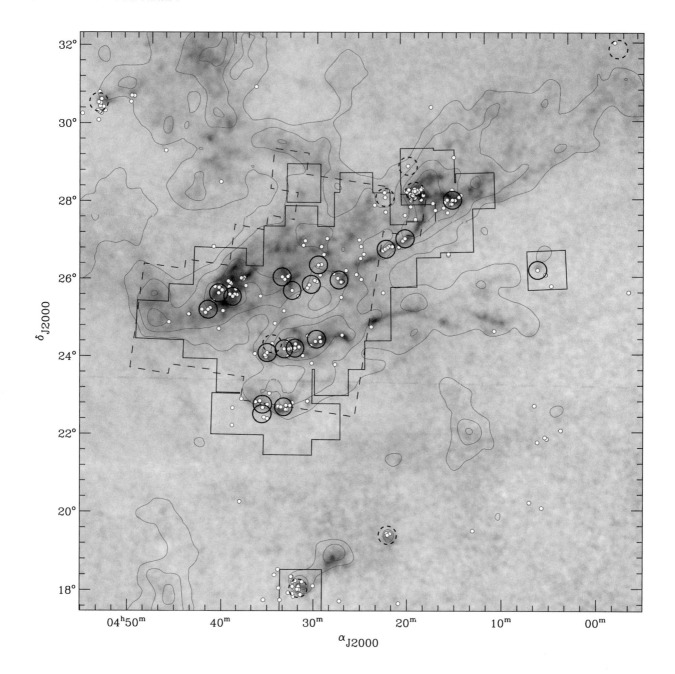

Fig. 1. Map of the TMC region. The grayscale background map is an extinction (A_V) map from *Dobashi et al.* (2005). The outlines of the CFHT and the Spitzer surveys are indicated by the solid and dashed polygons, respectively. The large (0.5° diameter) circles show the fields of view of the XMM-Newton X-ray survey (the dashed circles marking fields from separate projects also used for the survey). Small white dots mark the positions of individual young stars. Note the outlying XMM-Newton fields around SU Aur (NE corner), ζ Per (NW corner), and L1551 and T Tau (south). Figure courtesy of N. Grosso.

fraction of the TMC BDs can now be studied and their high energy properties put into context with T Tauri stars. Moderate energy resolution permits a detailed description of the thermal plasma properties together with the measurement of the absorbing gas columns that are located predominantly in the Taurus clouds themselves, and even in the immediate circumstellar environment in the case of strongly absorbed objects. Photoelectric absorption acts more strongly on the softer photons. ROSAT detected no TMC protostars because photons below 2 keV are almost completely absorbed. In

contrast, harder photons to which XMM-Newton is sensitive penetrate large absorbing gas columns around protostars.

X-rays can be used to detect new candidate members of the TMC. We note in particular that WTTS are best identified through X-ray surveys (*Neuhäuser et al.*, 1995). Placement on the HR diagram, X-ray spectral and temporal properties, and infrared photometry together can characterize new sources as being likely members of the association.

2.2.2. The Spitzer infrared survey. We have mapped the majority of the main TMC (≈30 deg²) (Fig. 1) using the

Spitzer IRAC and MIPS instruments in order to take a deep census of young stars and disks to below the deuterium-burning limit. These data play a crucial role in characterizing circumstellar material and BDs in our multiwavelength surveys of TMC. The goals are (1) to survey the nearest example of a low-density star-forming region completely and objectively for stars and BDs; (2) to learn whether there is a "distributed" component of isolated star formation far removed from the multiple "centers" within the TMC complex; (3) to determine (via combination with groundbased data) what the distribution of stellar ages, masses, and disk lifetimes are; (4) to carry out a definitive search for disks in transition between optically thick accretion disks and post-planet-building disks and to use these data to constrain the timescales for planet building; (5) to discover which optical/2MASS/X-ray sources are the counterparts of known IRAS sources, as well as new Spitzer far-IR sources.

2.2.3. The Canada-France-Hawaii Telescope optical survey. Our CFHT TMC survey was conducted as part of a larger-scale program targeted toward young galactic clusters (P.I. J. Bouvier). It is a comprehensive, deep optical survey of the TMC down to 24 mag in I (23 in z) of 34 deg², covering most of the XMM-Newton and Spitzer fields (Fig. 1). The specific goals of the CFHT survey are (1) to obtain a complete census of the low-mass star population down to I ≈ 22, well into the substellar regime; (2) to obtain accurate optical I-band photometry critical to derive fundamental parameters (reddening, luminosity estimates) for low-luminosity TMC members and investigate the occurrence of IR excesses; and (3) to obtain deep images at subarcsecond resolution for (a) a proper source identification and (b) the detection and mapping of new large-scale HH flows and embedded envelopes. These characteristics make the TMC CFHT survey the largest optical survey of the TMC sensitive well into the substellar regime. These data are critical for a proper identification of the sources and the determination of their fundamental parameters (effective temperature, reddening, luminosities). In particular, they play a central role in the detection and characterization of the TMC substellar population.

3. SURVEY STRATEGIES AND STATUS

3.1. The XMM-Newton Survey

Our XMM-Newton survey comprises 19 exposures obtained coherently as part of a large project (*Güdel et al.,* 2006a), complemented by an additional 9 exposures obtained in separate projects, bringing the total exposure time to 1.3 Ms. The former set of exposures used all three EPIC cameras in full-frame mode with the medium filter, collecting photons during approximately 30 ks each. Most of the additional fields obtained longer exposures (up to 120 ks) but the instrument setup was otherwise similar. The spatial coverage of our survey is illustrated in Fig. 1. The survey includes approximately 5 deg² in total, covering a useful energy range of ≈0.3–10 keV with a time resolution down to a few seconds. The circular fields of view with a diameter

of 0.5° each were selected such that they cover the densest concentrations of CO gas, which also show the strongest accumulations of TMC stellar and substellar members. We also obtained near-ultraviolet observations with the Optical Monitor (OM), usually through the U-band filter, and in exceptional cases through filters at somewhat shorter wavelengths. The OM field of view is rectangular, with a side length of 17'. An on-axis object could be observed at high time resolution (0.5 s).

Data were processed using standard software and analysis tools. Source identification was based on procedures involving maximum-likelihood and wavelet algorithms. In order to optimize detection of faint sources, periods of high background local particle radiation levels were cut out.

A typical exposure with an average background contamination level reached a detection threshold of $\approx 9 \times 10^{27}$ erg s^{-1} on-axis and $\approx 1.3 \times 10^{28}$ erg s^{-1} at 10' off-axis for a lightly absorbed X-ray source with a thermal spectrum characteristic of T Tau stars (see below) subject to a hydrogen absorption column density of $N_H = 3 \times 10^{21}$ cm^{-2}. This threshold turns out to be appropriate to detect nearly every T Tau star in the surveyed fields.

We interpreted all spectra based on thermal components combined in such a way that they describe emission measure distributions similar to those previously found for nearby pre-main-sequence stars or active zero-age main-sequence stars (*Telleschi et al.,* 2005; *Argiroffi et al.,* 2004; *Garcia et al.,* 2005; *Scelsi et al.,* 2005). An absorbing hydrogen column density was also fitted to the spectrum.

3.2. The Spitzer Survey

In order to study material around very-low-mass young stars and BDs, we require the use of both the Spitzer imaging instruments (IRAC and MIPS). The short IRAC bands detect the photospheres of BDs and low-mass stars. Using the models of *Burrows et al.* (2003), we find that 2–12 s exposures are required to detect BDs down to 1 M_{Jup} in the 4.5-μm IRAC band (see Table 2). Young BDs without disks will require spectroscopic confirmation since they will resemble M stars. The exquisite sensitivity of IRAC also makes it susceptible to saturation issues for the solar-type population of the Taurus clouds. In order to mitigate these effects, we have observed with the "high dynamic range"

TABLE 2. Spitzer TMC survey estimated sensitivity.

Instrument	Band (μm)	5σ Point Source Sensitivity (mJy)	Redundancy
MIPS	24	1.2	10
MIPS	70	~44	5
MIPS	160	~500; 2.64 MJy/sr	~1
—	—	(extended source)	(bonus band)
IRAC	3.6	0.017	2
IRAC	4.5	0.025	2
IRAC	5.8	0.155	2
IRAC	8.0	0.184	2

mode, which obtains a short (0.4 s) frame together with a 10.4 s one. Total IRAC and MIPS sensitivities for our maps are listed in Table 2, together with the number of independent exposures performed per point in the map (data redundancy). Disks are indicated by measured flux densities well above photospheric levels and/or stellar colors inconsistent with a Rayleigh-Jeans spectrum between any pair of bands. The presence of disks is revealed by comparison of the 8- and 24-μm bands to each other and the shorter IRAC bands. For MIPS, we chose fast scan for efficient mapping at 24 μm. The resulting 1.2-mJy 5σ sensitivity for MIPS 24 μm enables detection of disks 20× fainter than the low-luminosity edge-on disk HH 30 IRS (*Stapelfeldt and Moneti,* 1999). This is sufficient to detect ≥few M_{Jup} disks around ≥0.007 $M_\odot$ 3-m.y.-old BDs (*Evans et al.,* 2003).

Due to the proximity of TMC, it subtends an area of more than 25 deg² on the sky. Spitzer is the first space observatory with modern sensitivity that possesses the observing efficiency to map the region as a single unit within a feasible observing time. Our observing program attempted to survey the TMC nearly completely and objectively within a limited time (134 h). In particular, we chose our coverage of TMC (illustrated in Fig. 1) to overlap as many CFHT and XMM-Newton pointings as possible while mapping all the known contiguous ¹³CO clouds. In addition, in order to obtain imaging of the Heiles 2 and Lynds 1536 clouds most efficiently by mapping in long strips, we obtained a considerable portion of adjacent off-cloud area south of the TMC. This region is invaluable for assessing the distribution of sources with excess off the cloud, as well as galactic and extragalactic contamination. The MIPS data were obtained using the scan mapping mode (*Rieke et al.,* 2004), which uses a cryogenic mirror to perform image motion compensation for short exposures while the telescope is slewing. The "fast scan" mode is by far the most efficient way for Spitzer to map large areas quickly, covering more than 1 deg²/h. Due to the lack of an internal scan mirror, IRAC maps at the slow speed of 0.33 deg²/h at the minimal depth of our survey. Because each point in the IRAC map has a total redundancy of only two exposures, cosmic-ray removal is difficult.

Since TMC is located virtually on the ecliptic plane, we mapped twice at 8 and 24 μm to identify asteroids. A waiting period of 3–6 h between mapping epochs allowed asteroids with a minimum motion of 3.5"/h to move enough so that they can be identified by comparing images from the two epochs. Initial results suggest there are several thousand asteroids detected in our maps. These will be eliminated from the catalog of young stellar objects. However, our strategy enables a statistical asteroid study with the largest low ecliptic latitude survey yet envisioned for Spitzer. Spitzer observations of TMC were performed in February and March of 2005. Although the bulk of the data were released to the team in May, the field that includes Lynds 1495 remained embargoed until March 2006. The team spent several months in 2005 developing algorithms to mitigate remaining instrumental signatures in the data, many of which are aggravated by the presence of bright point sources and extended nebu-

losity. Removal of radiation artifacts from the IRAC data has been a considerable challenge due to the minimal repeat coverage of our mapping strategy. However, as of the time of the Protostars and Planets V conference, a prototype reprocessing code has been implemented for IRAC. Using this software and a robust mosaicing algorithm that uses the short HDR frame to eliminate the brightest cosmic rays, a 1.5 deg² region centered around 04ʰ 26ᵐ 10.9ˢ + 27° 18'10" (J2000) has been fully reprocessed in the IRAC and MIPS bands, source extracted, and bandmerged. The MIPS 24-μm data are most affected by the presence of thousands of asteroids, which outnumber the stars at this wavelength. It is therefore necessary to separately extract sources from each of the MIPS 24-μm observing epochs, then bandmerge these lists to achieve a reliable source list minus asteroids. One of the most interesting preliminary data products is a complete 160-μm map of the entire available TMC region (29 deg²). This map is presented in Fig. 2. The areas of brightest 160-μm emission correspond well to the ¹³CO clouds.

3.3. The Canada-France-Hawaii Telescope Survey

The CFHT TMC survey has been performed in four successive periods on the Canada-France-Hawaii Telescope with the CFH12k and MEGACAM large-scale optical cameras (see Table 3 for a detailed journal). Taking into account overlapping fields, a total effective surface of 34 deg² has been surveyed down to an I-band sensitivity limit of I_C =

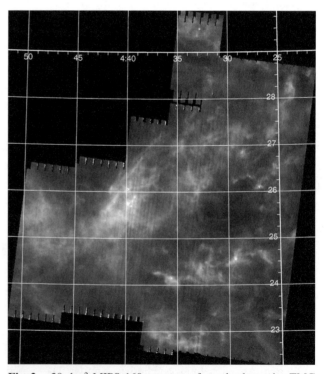

Fig. 2.   29 deg² MIPS 160-μm map of nearly the entire TMC region observed in the Spitzer TMC survey. The brightest regions reach an intensity of over 200 MJy/sr. The image has been slightly smoothed to fill in some narrow gaps in coverage. Effective exposure time is about 3 s.

TABLE 3.   Overview of the CFHT optical survey of the Taurus cloud.

Instrument	FOV (deg^2)	Pixel ("/pixel)	Date	Area (deg^2)	Band	Completeness Limit	Mass Limit ($A_V = 5$, 5 m.y.)
CFHT12k	0.33	0.21	1999–2001	3.6	R, I, z'	23, 22, 21	15 M_{Jup}
CFHT12k	0.33	0.21	2002	8.8	I, z'	22, 21	15 M_{Jup}
Megacam	1	0.19	2003–2004	34	i', z'	24, 23	10 M_{Jup}

24, i.e., well into the substellar regime (see below). It encompasses more than 80% of the known stellar population. This is to date the largest optical survey of the TMC sensitive well into the substellar regime. The outline of the survey, overlaid on the CO gas-density map, is presented in Fig. 1 together with the known TMC population.

The technical characteristics of the CFH12k and MEGA-CAM cameras are presented in *Cuillandre et al.* (2000) and *Boulade et al.* (2003), respectively. The first half of the CFH12k survey (centered on the densest part of TMC, from 1999 to 2001), has been obtained as part of the CFHT director's discretionary time, while the remaining larger set of data has been obtained, in service mode, as part of a larger key program devoted to the study of young clusters. Currently, 28 of the total 34 deg^2 have been reduced and analyzed. We refer below to this as the primary CFHT survey. Data reduction, performed at CFHT using elements of the Elixir system (*Magnier and Cuillandre,* 2004), included bias and dark subtraction, flat-fielding, fringing correction, bad pixel removal, and individual frame combination. Point source detection was performed on the combined I + z' images. For the CFH12k data, PSF fitting photometry was extracted with the PSFex routine from the SExtractor program (*Bertin and Arnouts,* 1996), while aperture photometry was obtained for the MEGACAM data with the same program. Photometric catalogs were combined, using the transformation between CFH12k (I, Z') and MEGACAM (i', z') photometric systems, computed with overlapping fields. The survey yielded more than 10^6 sources detected down to I = 24 and z' = 23. From the turnover at the faint end of the magnitude distribution, we estimate the completeness limits of our optical photometric survey to be I = 21.8 and z' = 20.9, which corresponds to a mass completeness limit of 15 M_{Jup} for $A_V < 5$ and age < 5 m.y., according to the pre-main-sequence DUSTY models of *Chabrier et al.* (2000). On the bright side, the saturation limits are i' = 12.5 and z' = 12. Follow-up studies at NIR wavelengths (1–2 μm) are planned. As part of the UKIRT Infrared Deep Sky Survey most of the MEGACAM fields will be observed down to K = 18.

4.   OVERVIEW OF THE POPULATION

4.1.   Fundamental Parameters

We have compiled a comprehensive catalog of all sources thought to be members of the TMC, collecting photometry, effective surface temperature, bolometric luminosities of the stars, L_*, derived mostly from near-IR photometry (see, e.g., *Briceño et al.,* 2002), extinctions A_V and A_J, masses, radii, Hα equivalent widths, rotation periods and v sin i values, mass accretion and outflow rates, and some further parameters from the published literature.

There is considerable spread in some of the photometry and A_V (or A_J) estimates for a subsample of stars, resulting in notable differences in derived L_* and masses. We have coherently re-derived ages and masses from the original L_* and T_{eff} using *Siess et al.* (2000) evolutionary tracks. For the final list of parameters, we have typically adopted the values in the recent compilation by *Briceño et al.* (2002) or, if not available, the catalogs of *Kenyon and Hartmann* (1995) and *Briceño et al.* (1998). Binary component information is mostly from *White and Ghez* (2001) and *Hartigan and Kenyon* (2003). Further parameters were complemented from the studies by *Luhman et al.* (2003), *Luhman* (2004), *White and Hillenbrand* (2004), and *Andrews and Williams* (2005).

4.2.   Known Protostellar and Stellar Population

The bright protostellar and stellar population of TMC has long been studied in the infrared. *Strom et al.* (1989) concluded that about half the young stellar population of TMC above 1 $L_\odot$ were surrounded by optically thick disks as demonstrated by their IRAS detections. This work was extended and complemented by the work of *Kenyon and Hartmann* (1995), who added fainter association members and ground-based photometry out to 5 μm to the SEDs. ISOCAM observed the L1551 field in the southern TMC, detecting an additional 15 YSO candidates (*Galfalk et al.,* 2004). Spitzer IRAC photometry of 82 known Taurus association members is reported in *Hartmann et al.* (2005). This study finds that the CTTS are cleanly separated from the WTTS in the [3.6]–[4.5] vs. [5.8]–[8.0] color-color diagram. The WTTS are tightly clustered around 0 in both colors, and the CTTS form a locus around [3.6]–[4.5] ≈ 0.5 and [5.8]–[8.0] ≈ 0.8. A similar conclusion is reached by *Padgett et al.* (2006) who obtained pointed photometry of 83 WTTS and 7 CTTS in Taurus, Lupus, Chamaeleon, and Ophiuchus at distances of about 140–180 pc, with ages most likely around 0.5–3 m.y. They find that only 6% of WTTS show excess at 24 μm, with a smaller percentage showing IRAC excesses. Unfortunately, it is currently not possible in every case to distinguish a true WTTS (pre-main-sequence star without strong Hα emission) from X-ray bright zero-age main-sequence stars projected onto the cloud. Thus, current samples of

WTTS and possibly "weak" BDs may be contaminated with older objects, skewing the disk frequency for these sources.

Class I "protostars" are perhaps more easily studied in TMC than elsewhere due to the lack of confusion in the large long-wavelength IRAS beams. One troubling aspect of the placement of Class Is in the standard picture of star formation (*Adams et al.,* 1987) is that the TMC Class Is typically show luminosities no higher than, and in many cases lower than, the Class II T Tauri stars (*Kenyon and Hartmann,* 1995). This issue has led to controversy regarding whether Class I sources are at an earlier evolutionary state than Class II T Tauri stars (*Eisner et al.,* 2005; *White and Hillenbrand,* 2004). It is hoped that Spitzer can boost the number of known Class I sources and elucidate their spectral properties, helping to determine the true nature of these objects.

4.3. X-Ray Sources

The detection statistics of our X-ray survey is summarized in Table 4 (we have added one BD detection from a complementary Chandra field). An important point for further statistical studies is that the X-ray sample of detected CTTS and WTTS is nearly complete for the surveyed fields (as far as the population is known). The few remaining, undetected objects are either heavily absorbed, have unclear YSO classification, are objects that have been very poorly studied before, or are very-low-mass objects. Some may not be genuine TMC members. In contrast, previous X-ray surveys did not detect the intrinsically fainter TTS population, potentially introducing bias into statistical correlations and population studies. It is little surprising that some of the protostars remained undetected given their strong photoelectric absorption. The detection rate of BDs (53%) is also very favorable; the remaining objects of this class are likely to be intrinsically fainter than our detection limit rather than being excessively absorbed by gas (A_V of those objects typically being no more than few magnitudes).

X-rays can efficiently be used to find new candidate TMC association members if X-ray information (luminosity, temporal, and spectral characteristics) is combined with information from the optical/near-infrared (placement on the HR diagram, L_*, age) and from the mid-infrared (presence of disks or envelopes). Scelsi et al. (in preparation) have thus identified several dozen of potential candidates of the TMC population. Follow-up studies will be needed to confirm these candidates.

TABLE 4.   X-ray detection statistics.

Object Type	Members Surveyed	Detections	Detection Fraction
Protostars	20	9	45%
CTTS	62	54	87%
WTTS	49	48	98%
BDs	19	10	53%

4.4.   Bright 24-μm Sources

The scientific goal of the TMC Spitzer survey is to obtain a complete census of the stellar content of these clouds down to the hydrogen-burning limit. Our Spitzer maps have sufficient sensitivity to detect 1-M_{Jup} young BDs and optically thin disks around solar-type stars at the distance of TMC. However, a complication of our survey is that the small size of Spitzer limits its spatial resolution, making the task of distinguishing faint stars from galaxies difficult, especially in the presence of optical extinction. Unfortunately, the IR spectral energy distributions of extincted stars with infrared excesses strongly resemble the SEDs of IR bright galaxies (*Evans et al.,* 2003). Experience with the galactic First Look Survey (*Padgett et al.,* 2004) and the c2d Legacy program (cf. *Young et al.,* 2005) have shown that strong 24-μm emission is an excellent signpost of young stellar objects. The extragalactic Spitzer surveys performed by the GTOs and the Extragalactic First Look Survey have established that extragalactic sources dominate the sky at a flux level of 1 mJy at 24 μm, but are fewer than 1 per square degree at 10 mJy (*Papovich et al.,* 2004). Thus, in a region of known star formation, strong 24-μm sources are more likely to be galactic than extragalactic sources. Although our analysis of the TMC Spitzer maps is incomplete, we have assembled SEDs for the bright ($\geq$10 mJy) 24-μm sources over more than 15 deg². About 100 sources were found in this preliminary list, of which 56 have no SIMBAD identifier. By analogy with SEDs of the known young stellar objects of the cloud that were also recovered by this technique, we believe that some of the previously unknown 24-μm sources may represent the brightest stars with disks among the YSOs that were too faint for IRAS to detect. SEDs for four of the new bright 24-μm sources are presented in Fig. 3.

5.   THE SUBSTELLAR SAMPLE

The TMC is a particularly interesting target for searches of young substellar objects. It has a large extension, so it can be studied for ejection effects; there are no bright stars there to irradiate and disturb the stellar surroundings; the census of stellar members is relatively complete down to M2V spectral types (*Kenyon and Hartmann,* 1995), and its spatial distribution is known (*Gómez et al.,* 1993). The average low extinction ($A_V = 1$) associated with this cloud as well as its young age combine to provide a high sensitivity to very-low-mass objects in the optical domain. However, its large spatial extent ($\approx$100 deg²) requires mapping an extensive area. Significant breakthrough in this domain has been made possible with the recent availability of large-scale optical cameras.

Searches for substellar objects by *Briceño et al.* (1998), *Luhman* (2000, 2004 and references therein), *Briceño et al.* (2002), and *Luhman et al.* (2003) have revealed a factor of 1.4 to 1.8 deficit of BDs with respect to stars in TMC compared to the Trapezium cluster. This result has been interpreted as an indication that substellar object formation

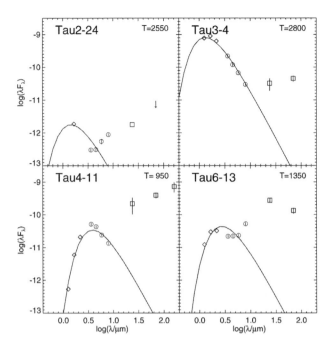

Fig. 3. 2MASS NIR + Spitzer IRAC and MIPS spectral energy distributions of four bright 24-μm sources discovered in the course of the survey. Temperatures indicated in the plots are effective temperatures of the plotted photosphere.

depends on the environment. However, all these previous studies were concentrated on the immediate vicinity of the high-stellar-density regions. If BDs are stellar embryos ejected from their birth sites early in their evolution as proposed by *Reipurth and Clarke* (2001), a significant fraction of the substellar content of the central parts of the cloud could have scattered away and may have been missed. This being the main scientific driver of the CFHT survey, we describe recent CFHT results from the search for substellar objects in TMC based on *Martín et al.* (2001) and *Guieu et al.* (2006) below, together with aspects from Spitzer and XMM-Newton.

5.1. New Taurus Molecular Cloud Very-Low-Mass Members

Substellar photometric candidates are identified from their location in combined optical/NIR color-magnitude and color-color diagrams (see Fig. 4). The full details of the selection process are given in *Guieu et al.* (2006). Complementary near-infrared photometry, taken from the 2MASS catalog, is critical to reduce the strong expected galactic contamination, primarily from background giants. Residual contamination is still expected to be at the 50% level. In order to properly assess TMC membership, spectroscopic follow-up of the photometrically selected candidates is therefore mandatory. The criteria used to assess TMC membership are detailed in *Guieu et al.* (2006). They rely on estimates of the surface gravity, obtained both from spectral fitting and measurements of the NaI equivalent widths. The level of Hα emission is used as an additional indicator of youth. At the median age of the TMC population of 3 m.y., the pre-main-sequence models of *Chabrier et al.* (2000) predict the stellar/substellar boundary to lie at a spectral type between M6 and M6.5V.

The photometric selection procedure yielded, over the primary 28 deg² CFHT survey, 37 TMC mid- to late-M spectral type new candidate members with i' < 20 (magnitude limit set to allow a proper spectral type determination). TMC membership has been confirmed spectroscopically for 21 of these sources (*Martín et al.*, 2001; *Guieu et al.*, 2006), 16 of which have spectral types later than M6.5V, i.e., are likely substellar. These new findings bring to 33 the current published census of TMC BDs, thus allowing for a preliminary statistical study of their properties.

5.2. A Peculiar Substellar Initial Mass Function in the Taurus Molecular Cloud?

When all published optical surveys are combined, the census for very-low-mass TMC objects is now complete down to 30 M_{Jup} and $A_V \leq 4$ over an effective surface area of 35 deg² (≈30% of the total cloud surface). This mass completeness, derived from the DUSTY model of *Chabrier et al.* (2000), is set both by the 2MASS completeness lim-

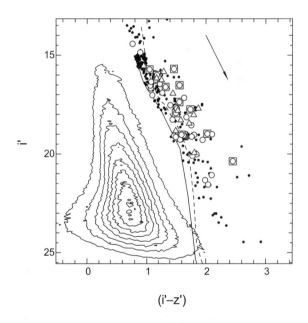

Fig. 4. Observed i'/(i'–z') color-magnitude diagram used to select low-mass TMC candidates. Small black dots are candidate TMC members. The photometric mid- to late-M candidates observed spectroscopically by *Guieu et al.* (2006) are displayed by black open circles. Triangles are previously known TMC members. Squares identify the spectroscopically confirmed 21 new TMC members from *Guieu et al.* (2006) and *Martín et al.* (2001). The two steep solid curves show the locations of the 1-m.y. and 10-m.y. isochrones from the DUSTY model of *Chabrier et al.* (2000) at the TMC distance. The arrow indicates a reddening vector of $A_V = 4$ mag. Figure adapted from *Guieu et al.* (2006).

its (J = 15.25, H = 14.4, K = 13.75) and the typical sensitivity limit of optical spectroscopic observations (i' ≤ 20).

Briceño et al. (2002) have introduced the substellar-to-stellar ratio

$$R_{ss} = \frac{N(0.03 < M/M_\odot < 0.08)}{N(0.08 < M/M_\odot < 10)}$$

as a measure of the relative abundance of BDs. In their pioneering study, targeted toward the high-density aggregates, *Briceño et al.* (2002) determined a value of R_{ss} (0.13 ± 0.05) lower by a factor of 2 than the one found in the Trapezium cluster. This study thus suggested that the relative abundance of BDs in star-forming regions may depend on initial conditions, such as the stellar density. However, in a recent, more extensive study covering 12 deg², *Luhman* (2004) found that the BD deficit in TMC (with respect to Orion) could be less pronounced than previously thought. Combining the recent new members from the CFHT survey with previously published results, *Guieu et al.* (2006) derive an updated substellar-to-stellar ratio in TMC of R_{ss} = 0.23 ± 0.05. This value is now in close agreement with the Trapezium value of R_{ss} (Trapezium) = 0.26 ± 0.04 estimated by *Briceño et al.* (2002), using the same evolutionary models and treating binary systems in the same manner. It also appears consistent with the more recent values derived for IC 348 by *Slesnick et al.* (2004), the Pleiades by *Moraux et al.* (2003), and computed from the galactic disk system IMF of *Chabrier* (2003) [see *Monin et al.* (2005) for a compilation of these values]. These new findings seem to suggest a universal 20–25% value for the relative abundance of BDs in young clusters. The fact that the estimate of the substellar-to-stellar ratio in TMC has kept increasing as larger areas were surveyed suggests that this ratio may depend on the local stellar density. Indeed, *Guieu et al.* (2006) find evidence for a deficit of the abundance of BDs of a factor of ≈2 in the central regions of the TMC aggregates (on scales of ≈0.5 pc) with respect to the more distributed population. As discussed in *Guieu et al.* (2006), this result may be an indication for spatial segregation resulting from ejection of the lowest-mass members and seem to favor dynamical evolution of small N-body systems as the formation process of substellar objects [see *Guieu et al.* (2006) for a full discussion].

5.3. X-Ray Properties of Taurus Molecular Cloud Brown Dwarfs

The area surveyed by the 28 pointings of XMM-Newton, combined with one Chandra archival observation, allow us to study the X-ray emission of 19/33 TMC BDs (*Grosso et al.*, 2006a). Among these, 10 BDs are detected, yielding a detection rate of ≈53%. One BD displayed an impulsive flare, demonstrating variability in X-rays over periods of a few hours. The detection rate of TMC BDs thus appears similar to the one in Orion where it reaches ≈50% for $A_V < 5$ (*Preibisch et al.*, 2005b). As in Orion, there is a tendency

to detect earlier (hotter) BD, with spectral types earlier than M7–M8.

There is appreciable scatter in the X-ray luminosities. The most luminous examples show L_X on the order of ≈10^{29} erg s⁻¹. No trend is seen for L_X/L_* with spectral type, i.e., the efficiency of magnetic field production and coronal heating appears to be constant in low-mass stellar and substellar objects (Fig. 5).

5.4. Disk and Accretion Properties of Taurus Molecular Cloud Brown Dwarfs

There is now ample evidence that TMC BDs experience accretion processes similar to the more massive TTS. Near-infrared L-band excesses have been detected in TMC substellar sources, indicating a disk frequency of ≈50% (*Liu et al.*, 2003; *Jayawardhana et al.*, 2003). Broad asymmetric Hα emission profiles characteristic of accretion are reported in a few TMC BDs (*Jayawardhana et al.*, 2002, 2003; *White and Basri*, 2003; *Muzerolle et al.*, 2005, and references therein; *Mohanty et al.*, 2005). Extending the study of *Barrado y Navascues and Martín* (2003), *Guieu et al.* (2006) find that the fraction of BDs in TMC with levels of Hα emission in excess of chromospheric activity to be 42%, similar to the low-mass TTS.

Of the 12 BD candidates optically selected from the CFHT survey, half have strong excesses in the infrared as measured from Spitzer data (*Guieu et al.*, 2005). Most of these diverge from the predicted photospheric fluxes at 5.8 μm, and all six are strongly detected at 24 μm. The other substellar candidates have IRAC fluxes indistinguishable from a late-M photosphere, and only one is detected at 24 μm. These results are similar to those found for TMC BD in the literature by *Hartmann et al.* (2005). Both studies find that the BD with excess ("classical" BD or CBD) have disk properties indistinguishable from classical T Tauri

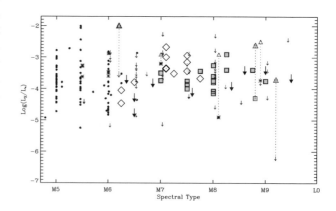

Fig. 5. L_X/L_* of low-mass stars and BDs as a function of spectral class, including samples of late-type main-sequence field stars (asterisks) (*Fleming et al.*, 1993), the Orion Nebula Cluster BD sample (filled squares, triangles, and small arrows) and T Tauri stars later than M5 (filled dots) from *Preibisch et al.* (2005b), and our sample of TMC detections (diamonds) and upper limits (arrows). Figure courtesy of N. Grosso.

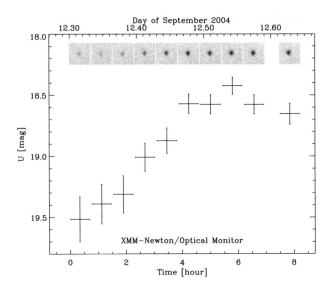

Fig. 6. U-band light curve of 2MASS J04552333+3027366 during an XMM-Newton observation. The slow increase may be ascribed to an accretion event. The insets show U-band images from which the fluxes have been extracted. Background limiting magnitudes are indicated by thick horizontal lines (*Grosso et al.,* 2006b).

stars. Similarly, the "weak" BDs have purely photospheric colors similar to the WTTS. Further analysis and modeling is required to determine whether the "classical" BDs have unusually flat disks, as suggested by *Apai et al.* (2002).

There is support for BD variability in the U-band observations obtained simultaneously by the Optical Monitor (OM) onboard XMM-Newton. The OM observed 13 of the 19 X-ray-surveyed BDs in the U band. Only one BD was detected, 2MASS J04552333+3027366, for which the U-band flux increased by a factor of about 2–3 in ≈6 hours (Fig. 6) (*Grosso et al.,* 2006b). The origin of this behavior can be explained by several different mechanisms: either rotational modulation of a hot or dark spot, or a coronal magnetic flare, or variable accretion. This BD was not detected in X-rays at any time. It is known to accrete at a rate of 10^{-10} $M_\odot$ yr^{-1} (*Muzerolle et al.,* 2005). Assuming that the relation $\log L_{acc} \propto \log L_U$ that applies to TTS (*Gullbring et al.,* 1998) is also valid for BDs, the accretion rate must have increased by a factor of about 2–3 during the observing time to explain the increase observed in the U band.

All these results argue for a continuous $M/\dot{M}$ relation through the stellar/substellar boundary, as illustrated, e.g., in Fig. 5 of *Muzerolle et al.* (2005).

5.5. Implication for Substellar Formation Model

The fact that the abundance of BDs (down to 30 M_{Jup}) relative to stars is found to be the same (≈25%) in the diffuse TMC *and* in the high-density Orion Nebula Cluster seems to suggest that there is no strong dependency of the substellar IMF on initial molecular cloud conditions, in particular gas density and level of turbulence. This fact and the

increase of the BD abundance with decreasing stellar density found by *Guieu et al.* (2006) could be best explained if a fraction of the distributed population in TMC is formed of low-mass stars and substellar objects ejected from the aggregates through rapid dynamical decay in unstable small N-body systems (the ejected-embryo model). Indeed, such a result is predicted by the dynamical evolution studies of *Kroupa and Bouvier* (2003) and the recent subsonic turbulent fragmentation simulations of *Goodwin et al.* (2004). For a detailed discussion, see *Guieu et al.* (2006) and *Monin et al.* (2006).

It has often been argued that the presence of accretion/outflow activity in BDs would be incompatible with an ejection-formation scenario. However, even truncated disks in ejected objects can survive for a few million years, a period consistent with the TMC age. The viscous timescale of a disk around a central mass M varies as $M^{-1/2}$, so for a disk truncated at $R_{out} = 10$ AU, $\tau_{visc} \approx 2$ m.y. around a 50 M_{Jup} BD (for $\alpha \approx 10^{-3}$ at 10 AU). Furthermore, an accretion rate of $\dot{M} = 10^{-11}$ $M_\odot$ yr^{-1} and a disk mass of 10^{-4} $M_\odot$ results in a similar lifetime of a few million years. So, there is no contradiction between BD ejection and the presence of (possibly small) accretion disks at an age of a few million years.

6. X-RAYS AND MAGNETIC ACTIVITY

6.1. X-Ray Luminosity

Figure 7 shows the distribution of the ratio between X-ray luminosity L_X and (stellar, photospheric) bolometric luminosity L_* as a function of L_* (the latter derived from optical or near-IR data, see section 4.1 for references) for all spectrally modeled TTS and protostars, and also including the *detected* BDs (we exclude the peculiar sources dis-

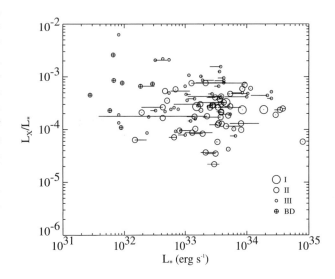

Fig. 7. Plot of L_X/L_* as a function of L_* for all X-ray detected (and spectrally modeled) stars and BDs (but excluding Herbig stars). Symbol size, from largest to smallest: protostars (IR Class I) — CTTS (or IR Class II) — WTTS (or IR Class III). Circles with crosses: BDs. The error bars indicate ranges of L_* given in the literature.

cussed in section 6 below; some objects were observed twice with different L_X — we use the logarithmic averages of L_X in these cases). We do not give errors for L_X because most objects are variable on short and long timescales (hours to days), typically within a factor of 2 outside obvious, outstanding flares. Most stars cluster between $L_X/L_* = 10^{-4}$–10^{-3} as is often found in star-forming regions (*Güdel*, 2004, and references therein). The value $L_X/L_* = 10^{-3}$ corresponds to the saturation value for rapidly rotating main-sequence stars (see below). We note a trend for somewhat lower levels of L_X/L_* for higher L_* (typically, more massive stars). What controls the X-ray luminosity level? Given the trend toward saturation in Fig. 7, one key parameter is obviously L_*. Al-though for pre-main-sequence stars there is no strict correlation between L_* and stellar mass, it is interesting that we find a rather well-developed correlation between L_X and mass M [Fig. 8; masses derived from T_{eff} and L_* based on *Siess et al.* (2000) isochrones] that has been similarly noted in Orion (*Preibisch et al.*, 2005a). Part of this correlation might be explained by higher-mass stars being larger, i.e., providing more surface area for coronal active regions. The correlation between surface area and L_X is, however, considerably weaker than the trend shown in Fig. 8.

We plot in Fig. 9 the L_X/L_* distribution separately for CTTS and WTTS (Telleschi et al., in preparation; the average L_X is used for objects observed twice). Because our samples are nearly complete, there is little bias by detection limits. The distributions are close to log-normal, and corresponding Gaussian fits reveal that WTTS are on average more X-ray luminous (mean of distribution: $\log L_X/L_* = -3.39 \pm 0.06$) than CTTS (mean: $\log L_X/L_* = -3.73 \pm 0.05$), although the widths of the distributions are similar. This finding parallels earlier reports on less complete samples (*Stelzer and Neuhäuser*, 2001), ruling out detection bias as a

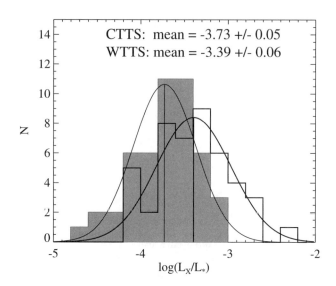

Fig. 9. Comparison of the L_X/L_* distributions for CTTS (shaded histogram) and WTTS (solid), together with log-normal fits. The CTTS sample is on average less luminous (normalized to L_*) than the WTTS sample. The errors in the plot indicate the error of the means of the distributions (Telleschi et al., in preparation).

cause for this difference. A similar segregation into two X-ray populations has not been identified in most other SFRs [e.g., *Preibisch and Zinnecker* (2001); but see recent results on the Orion Nebula Cluster in *Preibisch et al.* (2005a)]. The cause of the difference seen in TMC may be evolutionary (stellar size, convection zone depth), or related to the presence of accretion disks or the accretion process itself. We will return to this point in section 6.3 below.

6.2. Rotation and Activity

Rotation plays a pivotal role for the production of magnetic fields in main-sequence stars, and thus for the production of ionizing (ultraviolet and X-ray) radiation. The rotation period P is controlled by the angular momentum of the young star inherited from the contracting molecular cloud, by the further contraction of the star, but possibly also by magnetic fields that connect the star to the inner border of the circumstellar disk and thus apply torques to the star. Strictly speaking, in the standard (solar) α–ω dynamo theory, it is *differential* rotation that, together with convection, produces magnetic flux near the boundary between the convection zone and the radiative core. Because the younger T Tau stars are fully convective, a dynamo of this kind is not expected, but alternative dynamo theories based entirely on convective motion have been proposed (e.g., *Durney et al.*, 1993). It is therefore of prime interest to understand the behavior of a well-defined sample of T Tau stars.

In cool main-sequence stars, a rotation-activity relation is found for P exceeding a few days (the limit being somewhat dependent on the stellar mass or spectral type), approximately following $L_X \propto P^{-2.6}$ (*Güdel et al.*, 1997; *Flaccomio*

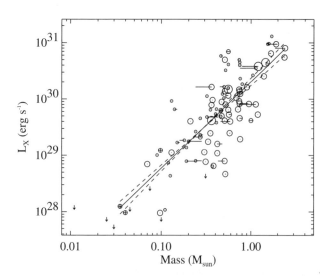

Fig. 8. X-ray luminosity L_X vs. stellar mass (excluding Herbig stars). A clear correlation is visible. Key to the symbols is as in Fig. 7 (after Telleschi et al., in preparation).

et al., 2003a). Given the role of the convective motion, a better independent variable may be the Rossby number R = P/τ where τ is the convective turnover time (*Noyes et al.*, 1984). If the rotation period is smaller than a few days, the X-ray luminosity saturates at a value of $L_X/L_* \approx 10^{-3}$ and stays at this level down to very short periods.

Corresponding studies of TTS have produced conflicting results. Although a relation has been indicated in TMC (*Stelzer and Neuhäuser*, 2001), samples in other star-forming regions show stars at saturated X-ray luminosities all the way to periods of ≈20 d (e.g., *Preibisch et al.*, 2005a). There is speculation that these stars are still within the saturation regime because their Rossby number remains small enough for the entire range of P, given the long convective turnover times in fully convective stars.

Our nearly complete sample of TTS (for the surveyed area) permits an unbiased investigation of this question with the restriction that we know P for only 25 TTS (13 CTTS and 12 WTTS) in our sample. Another 23 stars (15 CTTS and 8WTTS) have measured projected rotational velocities v sin i that imply upper limits to P once the stellar radius is known. In a statistical sample with random orientation of the rotation axes, the average of sin i is π/4, which we used for estimates of P if only v sin i was known. The stellar radii were calculated from T_{eff} and the (stellar) L_*.

The resulting trend is shown in Fig. 10 [the average L_X is used for stars with multiple observations (Briggs et al., in preparation)]. First, it is evident that the sample of CTTS with measured P rotates, on average, less rapidly than WTTS (characteristically, P ≈ 8 d and 4 d, respectively). Figure 10 shows that the rotation-activity behavior is clearly different from that of main-sequence solar-mass stars in that L_X/L_* remains at a saturation level up to longer periods. This is not entirely surprising given that the same is true for less-massive main-sequence K- and M-type stars that are more representative of the TTS sample (*Pizzolato et al.*, 2003). The trend is even clearer when plotting the average X-ray surface flux, in particular for periods exceeding ≈5 d (Briggs et al., in preparation), supporting previous ROSAT studies (*Stelzer and Neuhäuser*, 2001).

Why this finding is at variance from findings in Orion (*Preibisch et al.*, 2005a) is unclear. One possibility are unknown biases (Briggs et al., in preparation). A more likely reason is the (on average) larger age of TMC in which a larger fraction of stars may have developed a radiative core (Briggs et al., in preparation).

6.3. Accretion and Disks

In the standard dynamo interpretation, the (on average) slower rotation of the CTTS compared to WTTS and their (on average) slightly lower L_X are well explained by a decreasing dynamo efficiency with decreasing rotation rate. This is the conventional explanation for the activity-rotation relation in aging main-sequence stars. The relation suggested above could, however, be mimicked by the CTTS sample, rotating less rapidly, being subject to suppressed X-ray production for another reason than the decreasing efficiency of the rotation-induced dynamo. We already found that the average L_X/L_* is smaller by a factor of 2 for CTTS compared to WTTS (section 6.1). The most obvious distinction between CTTS and WTTS is active accretion from the disk to the star for the former class.

There are two arguments against this explanation. First, a rotation-activity relation holds *within* the CTTS sample, and there is no obvious correlation between P and the mass accretion rate, Ṁ, for that sample. And second, when investigating the coronal properties L_X, L_X/L_* (and also average coronal temperature T_{av}) as a function of the mass accretion rate (as given by *White and Ghez*, 2001; *Hartigan and Kenyon*, 2003; *White and Hillenbrand*, 2004), we see no trend over 3 orders of magnitude in Ṁ [Fig. 11 for L_X (Telleschi et al., in preparation)]. Mass accretion rate therefore does not seem to be a sensitive parameter that determines overall X-ray coronal properties. It therefore rather seems that CTTS produce, on average, lower L_X because they are typically rotating more slowly, which may be related to disk-locked rotation enforced by star-disk magnetic fields (e.g., *Montmerle et al.*, 2000).

7. JETS AND OUTFLOWS

Shock speeds in the high-velocity component of protostellar jets may be sufficient to shock-heat plasma to X-ray temperatures. The shock temperature is $T \approx 1.5 \times 10^5 v_{100}^2$ K where v_{100} is the shock front speed relative to a target, in units of 100 km s⁻¹ (*Raga et al.*, 2002). Jet speeds in TMC are typically of order v = 300–400 km s⁻¹ (*Eislöffel and Mundt*, 1998; *Anglada*, 1995; *Bally et al.*, 2003), allowing

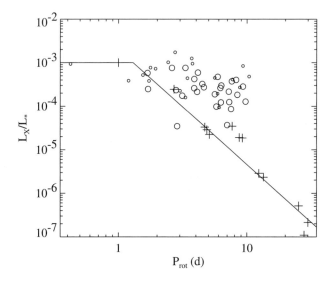

Fig. 10. The ratio L_X/L_* as a function of rotation period for the TMC sample (Briggs et al., in preparation). Symbols are as in Fig. 7. The crosses and the schematic power-law fit respective to horizontal saturation law apply to a sample of solar analogs on the main sequence (*Güdel et al.*, 1997).

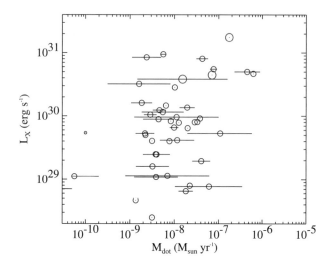

Fig. 11. Scatter plot of L_X vs. the (range of) mass accretion rates reported in the literature. No trend is evident. Symbols are as in Fig. 7. From Telleschi et al. (in preparation).

for shock speeds of similar magnitude. If a flow shocks a standing medium at 400 km s⁻¹, then T ≈ 2.4 MK. X-rays have been detected from the L1551 IRS-5 protostellar jet about 0.5–1″ away from the protostar, while the central star is entirely absorbed by molecular gas (*Bally et al.*, 2003).

X-rays cannot be traced down to the acceleration region or the collimation region of most protostellar jets because of the considerable photoelectric absorption, in particular of the very soft X-ray photons expected from shocks (energy ≲0.5 keV). An interesting alternative is provided by the study of strong jets and microjets driven by optically revealed T Tau stars. *Hirth et al.* (1997) surveyed TMC CTTS for evidence of outflows and microjets on the 1″ scale, identifying low-velocity (tens of km s⁻¹) and high-velocity (up to hundreds of km s⁻¹) flow components in several of them.

X-ray observations of these jet-driving CTTS have revealed new X-ray spectral phenomenology in at least three, and probably four, of these objects in TMC [DG Tau A, GV Tau A, DP Tau, and tentatively CW Tau — see Fig. 12 (*Güdel et al.*, 2005; *Güdel et al.*, 2006b)]. They share X-ray spectra that are composed of two different emission components subject to entirely different photoelectric absorption. The soft component, subject to very low absorption ($N_H \approx 10^{21}$ cm⁻²), peaks at 0.7–0.8 keV where Fe XVII produces strong emission lines, suggestive of low temperatures. This is borne out by spectral modeling, indicating temperatures of 2–5 MK. Such temperatures are not common to TTS. A much harder but strongly absorbed component (N_H several times 10^{22} cm⁻²) indicates extremely hot (several tens of MK) plasma.

These objects show flares in their X-ray light curves (Fig. 13), but such variability is so far seen only in the hard component while the soft component is steady. Evidently, these "two-absorber" spectra require that *two physically unrelated X-ray sources are present around these objects.*

All these stars are strong accretors ($\dot{M}$ on the order of 10^{-7}–10^{-6} M$_\odot$ yr⁻¹). However, as shown above, the TMC sample reveals no significant relation between mass accretion rate and coronal properties. The distinguishing property of these objects is, in contrast, the presence of well-developed, protostar-like jets and outflows with appreciable mass-loss rates (10^{-7}–10^{-6} M$_\odot$ yr⁻¹).

A tentative interpretation is the following (*Güdel et al.*, 2005; *Güdel et al.*, 2006b): The flaring in the hard component occurs on timescales of hours, suggesting ordinary coronal active regions. The preceding U band bursts signal the

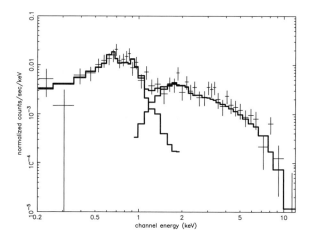

Fig. 12. Average spectrum of DG Tau A. Also shown is the fit to the spectrum (black histogram) and its two constituents, the soft and the hard components. From *Güdel et al.* (2006b).

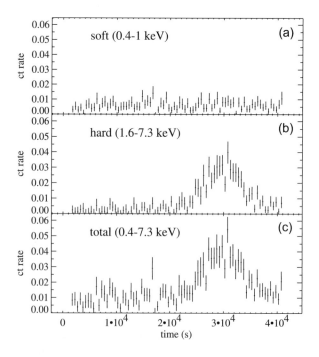

Fig. 13. X-ray light curves of DG Tau A. **(a)** Soft component, **(b)** hard component, and **(c)** total X-ray light curve. From *Güdel et al.* (2006b).

initial chromospheric heating before plasma is evaporated into the coronal magnetic loops. The flaring active regions are therefore likely to be of modest size, well connected to the surface active regions. The excess absorption is probably due to cool gas that streams in from the disk along the magnetic field lines, enshrouding the magnetosphere with absorbing material. This increases the photoelectric absorption of X-rays but does not increase the optical extinction because the gas streams are very likely to be depleted of dust (the latter being evaporated farther away from the star). As for the cool X-ray component, although its temperature is also compatible with shock heating of material in accretion columns close to the star (e.g., *Kastner et al., 2002*), the low photoabsorption makes this interpretation problematic and prefers a location outside the magnetosphere. An obvious location of the cool, soft X-ray sources are shocks forming near the base or the collimation region of the jet (e.g., *Bally et al., 2003*). Jet speeds of several hundred km s^{-1} support this model, as do estimated X-ray luminosities [see *Güdel et al.* (2005), based on the theory of *Raga et al.* (2002)].

If this model is correct, then the consequences are far reaching: Distributed, large-scale X-ray sources may efficiently ionize larger parts of the circumstellar environment than the central star alone, and in particular the disk surface, thus inducing disk accretion instabilities (*Balbus and Hawley*, 1991) and altering the disk chemistry (*Feigelson and Montmerle*, 1999; *Glassgold et al.*, 2004).

8. SUMMARY

The Taurus Molecular Cloud provides unequaled insight into the detailed physical processes of low-mass star formation, environmental issues, astrochemistry aspects, and evolutionary scenarios down to the substellar level. New observatories now available help us tackle outstanding problems with unprecedented sensitivity, spectral and spatial resolution. Of particular interest to star-formation studies are the new X-ray observatories (XMM-Newton and Chandra), the Spitzer Space Telescope in the infrared, and deep, large-scale optical surveys such as the CFHT survey summarized here.

Combining the X-ray, infrared, and optical population studies, there is considerable potential for detection of new Taurus members, some of which may be strongly embedded or extinced by their disks. Joint multiwavelength studies have been particularly fruitful for the characterization of brown dwarfs, which have been amply detected by all three studies and are now supporting a model in which a fraction of these objects are ejected from denser stellar aggregates.

The surveys also deepen previous studies of properties of T Tau stars and protostars (e.g., rotation-activity relations, disk properties, etc.), while at the same time opening the window to new types of phenomenology such as accretion events on brown dwarfs or X-ray emission perhaps forming at the base of accelerating jets. Further insight is expected to be obtained from high-resolution (optical, IR, and X-ray) spectroscopy that should probe composition and structure of accretion disks and heated X-ray sources.

Acknowledgments. We thank our referee for constructive and helpful comments on our paper. We acknowledge extensive contributions to this work by the three TMC teams (XMM-Newton, Spitzer, and CFHT). M.G. specifically acknowledges help from K. Arzner, M. Audard, K. Briggs, E. Franciosini, N. Grosso (particularly for the preparation of Fig. 1), G. Micela, F. Palla, I. Pillitteri, L. Rebull, L. Scelsi, and A. Telleschi. D.P. thanks J. Bouvier, T. Brooke, N. Evans, P. Harvey, J.-L. Monin, K. Stapelfeldt, and S. Strom for contributions. C.D. wishes to thank the whole CFHT TMC survey team, in particular F. Ménard, J.-L. Monin, S. Guieu, E. Magnier, E. Martín, and the CFHT astronomers (J. C. Cuillandre, T. Forveille, P. Martin, N. Manset, J. Shapiro) and director (G. Fahlman) as well as the Terapix MEGACAM data reduction center staff (Y. Mellier, G. Missonier). Research at PSI has been financially supported by the Swiss National Science Foundation (grant 20-66875.01). We thank the International Space Science Institute (ISSI) in Bern for further significant financial support to the XMM-Newton team. XMM-Newton is an ESA science mission with instruments and contributions directly funded by ESA Member States and the U.S. (NASA).

REFERENCES

Adams F. C., Lada C. J., and Shu F. H. (1987) *Astrophys. J., 312,* 788–806.

Andrews S. M. and Williams J. P. (2005) *Astrophys. J., 631,* 1134–1160.

Anglada G. (1995) *Rev. Mex. Astron. Astrophys., 1,* 67–76.

Apai D., Pascucci I., Henning Th., Sterzik M. F., Klein R., Semenov D., Günther E., and Stecklum B. (2002) *Astrophys. J., 573,* L115–L117.

Argiroffi C., Drake J. J., Maggio A., Peres G., Sciortino S., and Harnden F. R. (2004) *Astrophys. J., 609,* 925–934.

Balbus S. A. and Hawley J. F. (1991) *Astrophys. J., 376,* 214–233.

Bally J., Feigelson E., and Reipurth B. (2003) *Astrophys. J., 584,* 843–852.

Barrado y Navascués D. and Martín E. (2003) *Astron. J., 126,* 2997–3006.

Beichman C. A., Myers P. C., Emerson J. P., Harris S., Mathieu R., Benson P. J., and Jennings R. E. (1986) *Astrophys. J., 307,* 337–349.

Bertin E. and Arnouts S. (1996) *Astron. Astrophys. Suppl., 117,* 393–404.

Boulade O., et al. (2003) In *Instrument Design and Performance for Optical/Infrared Ground-based Telescopes* (M. Iye and A. F. M. Moorwood, eds.), pp. 72–81. International Society for Optical Engineering.

Bouvier J. (1990) *Astron. J., 99,* 946–964.

Briceño C., Hartmann L., Stauffer J., and Martín E. (1998) *Astron. J., 115,* 2074–2091.

Briceño C., Luhman K. L., Hartmann L., Stauffer J. R., and Kirkpatrick J. D. (2002) *Astrophys. J., 580,* 317–335.

Burrows A., Sudarsky D., and Lunine J. I. (2003) *Astrophys. J., 596,* 587–596.

Chabrier G. (2003) *Publ. Astron. Soc. Pac., 115,* 763–795.

Chabrier G., Baraffe I., Allard F., and Hauschildt P. (2000) *Astrophys. J., 542,* 464–472.

Cuillandre J., Luppino G. A., Starr B. M., and Isani S. (2000) *SPIE, 4008,* 1010–1021.

d'Alessio P., Calvet N., Hartmann L., Lizano S., and Cantó J. (1999) *Astrophys. J., 527,* 893–909.

Damiani F. and Micela G. (1995) *Astrophys. J., 446,* 341–349.

Damiani F., Micela G., Sciortino S., and Harnden F. R. Jr. (1995) *Astrophys. J., 446,* 331–340.

Dobashi K., Uehara H., Kandori R., Sakurai T., Kaiden M., Umemoto T., and Sato F. (2005) *Publ. Astron. Soc. Japan, 57,* S1–S368.

Duchêne G., Monin J.-L., Bouvier J., and Ménard F. (1999) *Astron. Astrophys., 351,* 954–962.

Durney B. R., De Young D. S., and Roxburgh I. W. (1993) *Solar Phys., 145,* 207–225.

Eislöffel J. and Mundt R. (1998) *Astron. J., 115,* 1554–1575.

Eisner J. A., Hillenbrand L. A., Carpenter J. M., and Wolf S. (2005) *Astrophys. J., 635*, 396–421.

Evans N. J., et al. (2003) *Publ. Astron. Soc. Pac., 115*, 965–980.

Feigelson E. D. and Montmerle T. (1999) *Ann. Rev. Astron. Astrophys., 37*, 363–408.

Feigelson E. D., Jackson J. M., Mathieu R. D., Myers P. C., and Walter F. M. (1987) *Astron. J., 94*, 1251–1259.

Flaccomio E., Micela G., Sciortino S., Damiani F., Favata F., Harnden F. R. Jr., and Schachter J. (2000) *Astron. Astrophys., 355*, 651–667.

Flaccomio E., Micela G., and Sciortino S. (2003a) *Astron. Astrophys., 402*, 277–292.

Flaccomio E., Damiani F., Micela G., Sciortino S., Harnden F. R. Jr., Murray S. S., and Wolk S. J. (2003b) *Astrophys. J., 582*, 398–409.

Fleming T. A., Giampapa M. S., Schmitt J. H. M. M., and Bookbinder J. A. (1993) *Astrophys. J., 410*, 387–392.

Galfalk M., et al. (2004) *Astron. Astrophys., 420*, 945–955.

Garcia-Alvarez D., Drake J. J., Lin L., Kashyap V. L., and Ball B. (2005) *Astrophys. J., 621*, 1009–1022.

Ghez A. M., Neugebauer G., and Matthews K. (1993) *Astron. J., 106*, 2005–2023.

Glassgold A. E., Najita J., and Igea J. (2004) *Astrophys. J., 615*, 972–990.

Gómez M., Hartmann L., Kenyon S. J., and Hewett R. (1993) *Astron. J., 105*, 1927–1937.

Grosso N., et al. (2006a) *Astron Astrophys.*, in press.

Grosso N., Audard M., Bouvier J., Briggs K., and Güdel M. (2006b) *Astron. Astrophys.*, in press.

Goodwin S. P., Whitworth A. P., and Ward-Thompson D. (2004) *Astron. Astrophys., 419*, 543–547.

Güdel M. (2004) *Astron. Astrophys. Rev., 12*, 71–237.

Güdel M., Guinan E. D., and Skinner S. L. (1997) *Astrophys. J., 483*, 947–960.

Güdel M., Skinner S. L., Briggs K. R., Audard M., Arzner K., and Telleschi A. (2005) *Astrophys. J., 626*, L53–L56.

Güdel M., et al. (2006a) *Astron Astrophys.*, in press.

Güdel M., et al. (2006b) *Astron Astrophys.*, in press.

Guieu S., Pinte C., Monin J.-L., Ménard F., Fukagawa M., Padgett D., Carey S., Noriega-Crespo A., and Rebull L. (2005) In *PPV Poster Proceedings*, www.lpi.usra.edu/meetings/ppv2005/pdf/8096.pdf.

Guieu S., Dougados C., Monin J.-L., Magnier E., and Martín E. L. (2006) *Astron. Astrophys., 446*, 485–500.

Gullbring E., Hartmann L., Briceño C., and Calvet N. (1998) *Astrophys. J., 492*, 323–341.

Hartmann L., Megeath S. T., Allen L., Luhman K., Calvet N., D'Alessio P., Franco-Hernandez R., and Fazio G. (2005) *Astrophys. J., 629*, 881–896.

Hartigan P. and Kenyon S. J. (2003) *Astrophys. J., 583*, 334–357.

Hirth G. A., Mundt R., and Solf J. (1997) *Astron. Astrophys. Suppl., 126*, 437–469.

Jayawardhana R., Mohanty S., and Basri G. (2002) *Astrophys. J., 578*, L141–L144.

Jayawardhana R., Mohanty S., and Basri G. (2003) *Astrophys. J., 592*, 282–287.

Kaifu N., et al. (2004) *Publ. Astron. Soc. Japan, 56*, 69–173.

Kastner J. H., Huenemoerder D. P., Schulz N. S., Canizares C. R., and Weintraub D. A. (2002) *Astrophys. J., 567*, 434–440.

Kenyon S. J. and Hartmann L. (1995) *Astrophys. J. Suppl., 101*, 117–171.

Kenyon S. J., Hartmann L. W., Strom K. M., and Strom S. E. (1990) *Astrophys. J., 99*, 869–887.

Kenyon S. J., Dobrzycka D., and Hartmann L. (1994) *Astron. J., 108*, 1872–1880.

Kroupa P. and Bouvier J. (2003) *Mon. Not. R. Astron. Soc., 346*, 343–353.

Kun M. (1998) *Astrophys. J. Suppl., 115*, 59–89.

Leinert Ch., Zinnecker H., Weitzel N., Christou J., Ridgway S. T., Jameson R., Haas M., and Lenzen R. (1993) *Astron. Astrophys., 278*, 129–149.

Liu M. C., Najita J., and Tokunaga A. T. (2003) *Astrophys. J., 585*, 372–391.

Loinard L., Mioduszewski A. J., Rodríguez L. F., González R. A., Rodríguez M. I., and Torres R. M. (2005) *Astrophys. J., 619*, L179–L182.

Luhman K. L. (2000) *Astrophys. J., 544*, 1044–1055.

Luhman K. L. (2004) *Astrophys. J., 617*, 1216–1232.

Luhman K. L., Briceño C., Stauffer J. R., Hartmann L., Barrado y Navascués D., and Caldwell N. (2003) *Astrophys. J., 590*, 348–356.

Magnier E. A. and Cuillandre J.-C. (2004) *Publ. Astron. Soc. Pac., 116*, 449–464.

Martín E. L., Dougados C., Magnier E., Ménard F., Magazzù A., Cuillandre J.-C., and Delfosse X. (2001) *Astrophys. J., 561*, L195–L198.

Mathieu R. D. (1994) *Ann. Rev. Astron. Astrophys., 32*, 465–530.

Mohanty S., Jayawardhana R., and Basri G. (2005) *Astrophys. J., 626*, 498–522.

Monin J.-L., Dougados C., and Guieu S. (2005) *Astron. Nachr., 326*, 996–1000.

Montmerle T., Grosso N., Tsuboi Y., and Koyama K. (2000) *Astrophys. J., 532*, 1097–1110.

Moraux E., Bouvier J., Stauffer J. R., and Cuillandre J.-C. (2003) *Astron. Astrophys., 400*, 891–902.

Muzerolle J., Luhman K. L., Briceño C., Hartmann L., and Calvet N. (2005) *Astrophys. J., 625*, 906–912.

Myers P. C., Fuller G. A., Mathieu R. D., Beichman C. A., Benson P. J., Schild R. E., and Emerson J. P. (1987) *Astrophys. J., 319*, 340–357.

Neuhäuser R., Sterzik M. F., Schmitt J. H. M. M., Wichmann R., and Krautter J. (1995) *Astron. Astrophys., 297*, 391–417.

Noyes R. W., Hartmann L. W., Baliunas S. L., Duncan D. K., and Vaughan A H. (1984) *Astrophys. J., 279*, 763–777.

Onishi T., Mizuno A., Kawamura A., Tachihara K., and Fukui Y. (2002) *Astrophys. J., 575*, 950–973.

Padgett D. L., et al. (2004) *Astrophys. J. Suppl., 154*, 433–438.

Padgett D. L., et al. (2006) *Astrophys. J., 645*, 1283–1296.

Papovich C., et al. (2004) *Astrophys. J. Suppl., 154*, 70–74.

Pizzolato N., Maggio A., Micela G., Sciortino S., and Ventura P. (2003) *Astrophys. J., 397*, 147–157.

Pravdo S. H., Feigelson E. D., Garmire G., Maeda Y., Tsuboi Y., and Bally J. (2001) *Nature, 413*, 708–711.

Preibisch T. and Zinnecker H. (2001) *Astron. J., 122*, 866–875.

Preibisch T., et al. (2005a) *Astrophys. J. Suppl., 160*, 401–422.

Preibisch T., et al. (2005b) *Astrophys. J. Suppl., 160*, 582–593.

Raga A. C., Noriega-Crespo A., and Velázquez P. F. (2002) *Astrophys. J., 576*, L149–L152.

Reipurth B. and Clarke C. (2001) *Astron. J., 122*, 432–439.

Rieke G. H., et al. (2004) *Astrophys. J. Suppl., 154*, 25–29.

Scelsi L., Maggio, A., Peres G., and Pallavicini R. (2005) *Astron. Astrophys., 432*, 671–685.

Siess L., Dufour E., and Forestini M. (2000) *Astron. Astrophys., 358*, 593–599.

Simon M., et al. (1995) *Astrophys. J., 443*, 625–637.

Slesnick C. L., Hillenbrand L. A., and Carpenter J. M. (2004) *Astrophys. J., 610*, 1045–1063.

Stapelfeldt K. R. and Moneti A. (1999) In *The Universe as Seen by ISO* (P. Cox and M. F. Kessler, eds.), pp. 521–524. ESA, Noordwijk.

Stelzer B. and Neuhäuser R. (2001) *Astron. Astrophys., 377*, 538–556.

Strom K. M. and Strom S. E. (1994) *Astrophys. J., 424*, 237–256.

Strom K. M., Strom S. E., Edwards S., Cabrit S., and Skrutskie M. (1989) *Astron. J., 97*, 1451–1470.

Strom K. M., Strom S. E., Wilkin F. P., Carrasco L., Cruz-Gonzalez I., Recillas E., Serrano A., Seaman R. L., Stauffer J. R., Dai D., and Sottile J. (1990) *Astrophys. J., 362*, 168–190.

Telleschi A., Güdel M., Briggs K., Audard M., Ness J.-U., and Skinner S. L. (2005) *Astrophys. J., 622*, 653–679.

Walter F. M., Brown A., Mathieu R. D., Myers P. C., and Vrba F. J. (1988) *Astron. J., 96*, 297–325.

Weaver W. B. and Jones G. (1992) *Astrophys. J. Suppl., 78*, 239–266.

White R. J. and Basri G. (2003) *Astrophys. J., 582*, 1109–1122.

White R. J. and Ghez A. M. (2001) *Astrophys. J., 556*, 265–295.

White R. J. and Hillenbrand L. A. (2004) *Astrophys. J., 616*, 998–1032.

Young K. E., et al. (2005) *Astrophys. J., 628*, 283–297.

The Low-Mass Populations in OB Associations

César Briceño
Centro de Investigaciones de Astronomía

Thomas Preibisch
Max-Planck-Institut für Radioastronomie

William H. Sherry
National Optical Astronomy Observatory

Eric E. Mamajek
Harvard-Smithsonian Center for Astrophysics

Robert D. Mathieu
University of Wisconsin–Madison

Frederick M. Walter
Stony Brook University

Hans Zinnecker
Astrophysikalisches Institut Potsdam

Low-mass stars ($0.1 \leq M \leq 1\ M_\odot$) in OB associations are key to addressing some of the most fundamental problems in star formation. The low-mass stellar populations of OB associations provide a snapshot of the fossil star-formation record of giant molecular cloud complexes. Large-scale surveys have identified hundreds of members of nearby OB associations, and revealed that low-mass stars exist wherever high-mass stars have recently formed. The spatial distribution of low-mass members of OB associations demonstrate the existence of significant substructure ("subgroups"). This "discretized" sequence of stellar groups is consistent with an origin in short-lived parent molecular clouds within a giant molecular cloud complex. The low-mass population in each subgroup within an OB association exhibits little evidence for significant age spreads on timescales of ~10 m.y. or greater, in agreement with a scenario of rapid star formation and cloud dissipation. The initial mass function (IMF) of the stellar populations in OB associations in the mass range $0.1 \leq M \leq 1\ M_\odot$ is largely consistent with the field IMF, and most low-mass pre-main-sequence stars in the solar vicinity are in OB associations. These findings agree with early suggestions that *the majority of stars in the galaxy were born in OB associations*. The most recent work further suggests that a significant fraction of the stellar population may have their origin in the more spread out regions of OB associations, instead of all being born in dense clusters. Groundbased and spacebased (Spitzer Space Telescope) infrared studies have provided robust evidence that primordial accretion disks around low-mass stars dissipate on timescales of a few million years. However, on close inspection there appears to be great variance in the disk dissipation timescales for stars of a given mass in OB associations. While some stars appear to lack disks at ~1 m.y., a few appear to retain accretion disks up to ages of ~10–20 m.y.

1. INTRODUCTION

Most star formation in normal galaxies occurs in the cores of the largest dark clouds in spiral arms, known as giant molecular clouds (GMCs). A GMC may give rise to one or more star complexes known as OB associations, first defined and recognized by *Ambartsumian* (1947) as young expanding stellar systems of blue luminous stars. These generally include groups of T Tauri stars or T associations (*Kholopov*, 1959; *Herbig*, 1962; *Strom et al.*, 1975) as well as clusters, some containing massive ($M \gtrsim 10\ M_\odot$) stars, but all teeming with solar-like and lower-mass stars.

Although we now recognize OB associations as the prime sites for star formation in our galaxy, much of our knowl-

345

edge of star formation is based on studies of low-mass (M ≤ 1 M$_\odot$) pre-main-sequence (PMS) stars located in nearby T associations, like the ~1–2 m.y. old Taurus, Lupus, and Chamaeleon star-forming regions. The view of star formation conveyed by these observations is probably biased to the particular physical conditions found in these young, quiescent regions. In contrast, the various OB associations in the solar vicinity are in a variety of evolutionary stages and environments, some containing very young objects (ages ≤1 m.y.) still embedded in their natal gas (e.g., Orion A and B clouds, Cep OB2), others in the process of dispersing their parent clouds, like λ Ori and Carina, while others harbor more evolved populations, several million years old, that have long since dissipated their progenitor clouds (like Scorpius-Centaurus and Orion OB1a). The low-mass populations in these differing regions are key to investigating fundamental issues in the formation and early evolution of stars and planetary systems:

1. *Slow vs. rapid protostellar cloud collapse and molecular cloud lifetimes.* In the old model of star formation (see *Shu et al.,* 1987) protostellar clouds contract slowly until ambipolar diffusion removes enough magnetic flux for dynamical (inside-out) collapse to set in. It was expected that the diffusion timescale of ~10 m.y. should produce a similar age spread in the resulting populations of stars, consistent with the ≤40 m.y. early estimates of molecular cloud lifetimes (see discussion in *Elmegreen,* 1990). Such age spreads should be readily apparent in color-magnitude or HR diagrams for masses ≤1 M$_\odot$. However, the lack of even ~10-m.y.-old, low-mass stars in and near molecular clouds challenged this paradigm, suggesting that star formation proceeds much more rapidly than previously thought, even over regions as large as 10 pc in size (*Ballesteros-Paredes et al.,* 1999), and therefore that cloud lifetimes over the same scales could be much shorter than 40 m.y. (*Hartmann et al.,* 1991).

2. *The shape of the IMF.* Whether OB associations have low-mass populations according to the field IMF, or if their IMF is truncated, is still a debated issue. There have been many claims for IMF cutoffs in high-mass star-forming regions (see, e.g., *Slawson and Landstreet,* 1992; *Leitherer,* 1998; *Smith et al.,* 2001; *Stolte et al.,* 2005). However, several well-investigated massive star-forming regions show *no* evidence for an IMF cutoff [see *Brandl et al.* (1999) and *Brandner et al.* (2001) for the cases of NGC 3603 and 30 Dor, respectively], and notorious difficulties in IMF determinations of distant regions may easily lead to wrong conclusions about IMF variations (e.g., *Zinnecker et al.,* 1993; *Selman and Melnick,* 2005). An empirical proof of a field-like IMF, rather than a truncated IMF, has important consequences not only for star-formation models but also for scenarios of distant starburst regions; e.g., since most of the stellar mass is then in low-mass stars, this limits the amount of material that is enriched in metals via nucleosynthesis in massive stars and that is then injected back into the interstellar medium by the winds and supernovae of the massive stars.

3. *Bound vs. unbound clusters.* While many young stars are born in groups and clusters, most disperse rapidly; few clusters remain bound over timescales >10 m.y. The conditions under which bound clusters are produced are not clear. Studies of older, widely spread low-mass stars around young clusters might show a time sequence of cluster formation, and observations of older, spreading groups would yield insight into how and why clusters disperse.

4. *Slow vs. rapid disk evolution.* Early studies of near-infrared dust emission from low-mass young stars suggested that most stars lose their optically thick disks over periods of ~10 m.y. (e.g., *Strom et al.,* 1993), similar to the timescale suggested for planet formation (*Podosek and Cassen,* 1994). However, there is also evidence for faster evolution in some cases; for example, half of all ~1-m.y.-old stars in Taurus have strongly reduced or absent disk emission (*Beckwith et al.,* 1990). The most recent observations of IR emission from low-mass PMS stars in nearby OB associations like Orion suggest that the timescales for the dissipation of the inner disks can vary even in coeval populations at young ages (*Muzerolle et al.,* 2005).

5. *Triggered vs. independent star formation.* Although it is likely that star formation in one region can "trigger" more star formation later in neighboring areas, and there is evidence for this from studies of the massive stars in OB populations (e.g., *Brown,* 1996), proof of causality and precise time sequences are difficult to obtain without studying the associated lower-mass populations. In the past, studies of the massive O and B stars have been used to investigate sequential star formation and triggering on large scales (e.g., *Blaauw,* 1964, 1991, and references therein). However, OB stars are formed essentially on the main sequence (e.g., *Palla and Stahler,* 1992, 1993) and evolve off the main sequence on a timescale on the order of 10 m.y. (depending upon mass and amount of convective overshoot), thus they are not useful tracers of star-forming histories on timescales of several million years, while young low-mass stars are. Moreover, we cannot investigate cluster structure and dispersal or disk evolution without studying low-mass stars. Many young individual clusters have been studied at both optical and infrared wavelengths (cf. *Lada and Lada,* 2003), but these only represent the highest-density regions, and do not address older and/or more widely dispersed populations. In contrast to their high-mass counterparts, low-mass stars offer distinct advantages for addressing the aforementioned issues. They are simply vastly more numerous than O, B, and A stars, allowing statistical studies not possible with the few massive stars in each region. Their spatial distribution is a fossil imprint of recently completed star formation, providing much needed constraints for models of molecular cloud and cluster formation and dissipation; with velocity dispersions of ~1 km s^{-1} (e.g., *de Bruijne,* 1999) the stars simply have not traveled far from their birth sites (~10 pc in 10 m.y.). Low-mass stars also provide better kinematics, because it is easier to obtain accurate radial velocities from the many metallic lines in G-, K-, and M-type stars than it is from O- and B-type stars.

2. SEARCHES FOR LOW-MASS PRE-MAIN-SEQUENCE STARS IN OB ASSOCIATIONS

Except for the youngest, mostly embedded populations in the molecular clouds, or dense, optically visible clusters like the Orion Nebula Cluster (ONC), most of the low-mass stellar population in nearby OB associations is widely spread over tens or even hundreds of square degrees on the sky. Moreover, it is likely that after ~4 m.y. the stars are no longer associated with their parent molecular clouds, making it difficult to sort them out from the field population. Therefore, a particular combination of various instruments and techniques is required to reliably single out the low-mass PMS stars. The main strategies that have been used to identify these populations are objective prism surveys, X-ray emission, proper motions, and, more recently, variability surveys.

2.1. Objective Prism Surveys

The TTS originally were identified as stars of late spectral types (G-M), with strong emission lines (especially Hα) and erratic light variations, spatially associated with regions of dark nebulosity (*Joy*, 1945). Stars resembling the original variables first identified as TTS are currently called "strong emission" or Classical TTS (CTTS). Subsequent spectroscopic studies of the Ca II H and K lines and the first X-ray observations with the Einstein X-ray observatory (*Feigelson and De Campli*, 1981; *Walter and Kuhi*, 1981) revealed surprisingly strong X-ray activity in TTS, exceeding the solar levels by several orders of magnitude, and also revealed a population of X-ray strong objects lacking the optical signposts of CTTS, like strong Hα emission. These stars, initially called "naked-T Tauri stars" (*Walter and Myers*, 1986), are now widely known as "weak-line" TTS after *Herbig and Bell* (1988). The CTTS/WTTS dividing line was set at W(Hα) = 10 Å. In general, the excess Hα emission in WTTS seems to originate in enhanced solar-type magnetic activity (*Walter et al.*, 1988), while the extreme levels observed in CTTS can be explained by a combination of enhanced chromospheric activity and emission coming from accretion shocks in which material from a circumstellar disk is funneled along magnetic field lines onto the stellar photosphere (section 5). Recently, *White and Basri* (2003) revisited the WTTS/CTTS classification and suggested a modified criterion that takes into account the contrast effect in Hα emission as a function of spectral type in stars cooler than late K.

The strong Hα emission characteristic of low-mass young stars, and in particular of CTTS, encouraged early large-scale searches using photographic plates and objective prisms on widefield instruments like Schmidt telescopes [e.g., *Sanduleak* (1971), in Orion]. These very-low-resolution spectroscopic surveys [typical dispersions of ~500 Å/mm to ~1700 Å/mm at Hα (cf. *Wilking et al.*, 1987; *Briceño et al.*, 1993)] provided large area coverage, allowed estimates of

spectral types and a qualitative assessment of the strength of prominent emission lines, like the hydrogen Balmer lines or the Ca II H and K lines. *Liu et al.* (1981) explored a 5° × 5° region in Per OB2 and detected 25 candidate TTS. *Ogura* (1984) used the 1-m Kiso Schmidt to find 135 Hα-emitting stars in Mon OB1. *Wilking et al.* (1987) detected 86 emission line objects over 40 deg² in the ρ Ophiuchi complex. *Mikami and Ogura* (2001) searched an area of 36 deg² in Cep OB3 and identified 68 new emission line sources. In the Orion OB1 association, the most systematic search was that done with the 1-m Kiso Schmidt (e.g., *Wiramihardja et al.*, 1989, 1993), covering roughly 150 deg² and detecting ~1200 emission line stars, many of which were argued to be likely TTS. *Weaver and Babcock* (2004) recently identified 63 Hα-emitting objects in a deep objective prism survey of the σ Orionis region.

The main limitation of this technique is the strong bias toward Hα-strong PMS stars; few WTTS can be detected at the resolution of objective prisms (cf. *Briceño et al.*, 1999). *Briceño et al.* (2001) find that only 38% of the 151 Kiso Hα sources falling within their ~34 deg² survey area in the Orion OB1a and 1b subassociations are located above the ZAMS in color-magnitude diagrams, and argue that the Kiso survey is strongly contaminated by foreground main-sequence stars (largely dMe stars). The spatial distribution of the Kiso sources has been useful to outline the youngest regions in Orion, where the highest concentrations of CTTS are located (*Gómez and Lada*, 1998), but these samples can be dominated by field stars in regions far from the molecular clouds, in which the CTTS/WTTS fraction is small. Therefore, as with other survey techniques, objective prism studies require follow-up spectroscopy to confirm membership.

2.2. X-Ray Surveys

Young stars in all evolutionary stages, from class I protostars to ZAMS stars, show strong X-ray activity [for recent reviews on the X-ray properties of YSOs see *Feigelson and Montmerle* (1999) and *Favata and Micela* (2003)]. After the initial Einstein studies, the ROSAT and ASCA X-ray observatories increased considerably the number of observed star-forming regions, and thereby the number of known X-ray-emitting TTS. Today, XMM-Newton and Chandra allow X-ray studies of star-forming regions at unprecedented sensitivity and spatial resolution.

X-ray observations are a well-established tool to find young stars. For nearby OB associations, which typically cover areas in the sky much larger than the field-of-view of X-ray observatories, deep and spatially complete observations are usually not feasible. However, large-scale shallow surveys have been conducted with great success. The ROSAT All Sky Survey (RASS) provided coverage of the whole sky in the 0.1–2.4-keV soft X-ray band. With a mean limiting flux of about 2×10^{-13} erg s^{-1} cm^2 this survey provided a spatially complete, flux-limited sample of X-ray sources that led to the detection of hundreds of candidate

PMS stars in star-forming regions all over the sky (see *Neu-häuser*, 1997).

The X-ray luminosities of young stars for a given age, mass, and bolometric luminosity can differ by several orders of magnitude. Until recently it was not even clear whether all young stars are highly X-ray active, or whether an "X-ray quiet" population of stars with suppressed magnetic activity may exist, which would have introduced a serious bias in any X-ray selected sample. The Chandra Orion Ultradeep Project (*Getman et al.*, 2005), a 10-day-long observation of the Orion Nebular Cluster, has provided the most comprehensive dataset ever acquired on the X-ray emission of PMS, and solved this question by providing definitive information on the distribution of X-ray luminosities in young stars. It found no indications for "X-ray quiet" TTS, and established that 50% of the TTS have log $(L_X/L_{bol}) \geq -3.5$, while 90% have log $(L_X/L_{bol}) \geq -4.5$ (*Preibisch et al.*, 2005; see also chapter by *Feigelson et al.*). Since the RASS flux limit corresponds to X-ray luminosities of about 5×10^{29} erg s^{-1} at the distance of the nearest OB associations (~140 pc), this implies that the RASS data are essentially complete only for M = 1 M$_\odot$ PMS stars in those regions, while only a fraction of the X-ray brightest subsolar-mass PMS stars are detected. A caveat of the RASS surveys for PMS stars is that these samples can be significantly contaminated by foreground, X-ray active zero-age main-sequence stars (*Briceño et al.*, 1997). These limitations have to be kept in mind when working with X-ray selected samples; at any rate, follow-up observations are necessary to determine the nature of the objects.

2.3. Proper Motion Surveys

The recent availability of ever-deeper, all-sky catalogs of proper motions (like *Hipparcos* and the Tycho family of catalogs) has aided the effort in identifying the low-mass members of the nearest OB associations. The proper motions of members of a few of the nearest OB associations are on the order of tens of mas yr^{-1} (*de Zeeuw et al.*, 1999). With proper motions whose errors are less than a few mas yr^{-1}, one can attempt to kinematically select low-mass members of nearby associations. The nearest OB association, Sco-Cen, has been the most fruitful hunting ground for identifying low-mass members by virtue of their proper motions. Current proper motion catalogs (e.g., Tycho-2, UCAC) are probably adequate to consider kinematic selection of low-mass stars in at least a few other nearby groups (e.g., Vel OB2, Tr 10, α Per, Cas-Tau, Cep OB6). The very small (<10 mas yr^{-1}) proper motions for some of the other nearby OB associations (e.g., Ori OB1, Lac OB1, Col 121) will preclude any attempts at efficient selection of low-mass members via proper motions, at least with contemporary astrometric catalogs.

The *Hipparcos* survey of the nearest OB associations by *de Zeeuw et al.* (1999) was able to identify dozens of FGK-type stars as candidate members. *de Zeeuw et al.* (1999) predicted that ~37/52 (71%) of their GK-type *Hipparcos*

candidates would be bona fide association members, and indeed *Mamajek et al.* (2002) found that 22/30 (73%) of a subsample of candidates, located in Sco-Cen, could be spectroscopically confirmed as PMS stars. *Hoogerwerf* (2000) used the ACT and TRC proper motion catalogs (p.m. errors ≃ 3 mas yr^{-1}) to identify thousands of candidate Sco-Cen members down to V ~ 12. Unfortunately, the vast majority of stars in the ACT and TRC catalogs, and their descendant (Tycho-2), do not have known spectral types or parallaxes (in contrast to the *Hipparcos* catalog), and hence the contamination level is large. *Mamajek et al.* (2002) conducted a spectroscopic survey of an X-ray and color-magnitude-selected subsample of the Hoogerwerf proper-motion-selected sample and found that 93% of the candidates were bona fide PMS association members. The high-quality proper motions also enabled the estimate of individual parallaxes to the Upper Centaurus Lupus (UCL) and Lower Centaurus Crux (LCC) members, reducing the scatter in the HR diagram (*Mamajek et al.*, 2002). In a survey of 115 candidate Upper Sco members selected solely via STARNET proper motions (p.m. errors of ~5 mas yr^{-1}), *Preibisch et al.* (1998) found that *none* were PMS stars. The lesson learned appears to be that proper motions *alone* are insufficient for efficiently identifying low-mass members of nearby OB associations. However, when proper motions are used in conjunction with color-magnitude, X-ray, spectral type, or parallax data (or some combination thereof), finding low-mass associations can be a very efficient task.

2.4. Photometric Surveys: Single-Epoch Observations

Single-epoch photometric surveys are frequently used to select candidate low-mass members of young clusters or associations. Most studies use broadband, optical filters that are sensitive to the temperatures of G-, K-, and M-type stars. Near-IR color-magnitude diagrams (CMDs) are not as useful for selecting low-mass PMS stars because NIR colors are similar for all late-type stars.

Candidate low-mass association members are usually selected by their location in the CMD above the zero-age main sequence (ZAMS). This locus is usually defined by either a known (spectroscopically confirmed) population of PMS stars or because the PMS population of the association is clearly visible as a concentration on the CMD (e.g., Fig. 1). Single-epoch photometry is most effective in regions such as σ Ori or the ONC where the proximity and youth of the cluster make the low-mass PMS members brighter than the bulk of the field stars at the colors of K- and M-type stars.

The main advantage of photometric selection is that for a specified amount of time on any given telescope a region of the sky can be surveyed to a fainter limit than can be done by a variability survey or a spectroscopic survey. Also, photometric selection can identify low-mass association members with very low amplitude variability. The disadvantage of single-epoch photometric selection is that there is inevitably some contamination by foreground field stars and

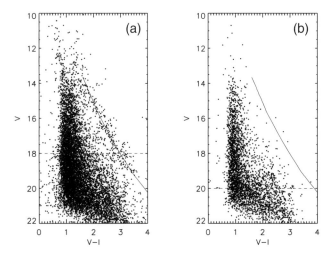

Fig. 1. (a) V vs. V-IC color-magnitude diagram of 9556 stars in 0.89 deg² around σ Ori (from *Sherry et al., 2004*). The solid line is a 2.5-m.y. isochrone (*Baraffe et al.*, 1998, 2001) at a distance of 440 pc (*de Zeeuw et al.*, 1999), extending from 1.2 M$_\odot$ at V ~ 13.5 down to ~0.2 M$_\odot$ at V ~ 18 (the completeness limit, indicated by the dashed line). This isochrone marks the expected position of the PMS locus for Orion OB1b. There is a clear increase in the density of stars around the expected position of the PMS locus. (b) The same color-magnitude diagram (CMD) for the 0.27 deg² control fields from *Sherry et al.* (2004). The isochrone (solid line) is the same as in the left panel. The dashed line marks the fainter completeness limit of the control fields.

background giants. As with the other techniques, it is impossible to securely identify any individual star as a low-mass member of the association without spectroscopic follow-up. In small areas with a high density of low-mass association members such as the σ Ori cluster or the ONC, single-epoch photometry can effectively select the low-mass population because field star contamination is fairly small (*Sherry et al.*, 2004; *Kenyon et al.*, 2005). But in large areas with a lower density of low-mass association members, such as Orion OB1b (Orion's belt) or Orion OB1a (northwest of the belt), the field star contamination can be large enough to make it difficult to even see the PMS locus.

2.5. Photometric Surveys: Variability

Variability in T Tauri stars has been intensively studied over the years (e.g., *Herbst et al.*, 1994), but mostly as follow-up observations of individual young stars that had been identified by some other means. Building on the availability of large-format CCD cameras installed on widefield telescopes it has now become feasible to conduct multi-epoch, photometric surveys that use variability to pick out candidate TTS over the extended areas spanned by nearby OB associations. In Orion, two major studies have been conducted over the past few years. *Briceño et al.* (2001, 2005a) have done a VRI variability survey using the Quest I CCD Mosaic Camera installed on the Venezuela 1-m Schmidt,

over an area of ≥150 deg² in the Orion OB1 association. In their first release, based on some 25 epochs and spanning an area of 34 deg², they identified ~200 new, spectroscopically confirmed, low-mass members, and a new 10-m.y.-old clustering of stars around the star 25 Ori (*Briceño et al.*, in preparation). *McGehee et al.* (2005) analyzed nine repeated observations over 25 deg² in Orion, obtained with the Sloan Digital Sky Survey (SDSS). They selected 507 stars that met their variability criterion in the SDSS g-band. They did not obtain follow-up spectra of their candidates, rather, they apply their observations in a statistical sense to search for photometric accretion-related signatures in their lower-mass candidate members. *Slesnick et al.* (2005) are using the Quest II CCD Mosaic Camera on the Palomar Schmidt to conduct a BRI, multi-epoch survey of ~200 deg² in Upper Sco, and *Carpenter et al.* (2001) has used repeated observations made with 2 MASS in a 0.86° × 6° strip centered on the ONC to study the near-IR variability of a large sample of young, low-mass stars.

With more wide-angle detectors on small- and medium-sized telescopes, and projects like LSST coming on line within less than a decade, variability promises to grow as an efficient means of selecting large samples of candidate low-mass PMS stars down to much lower masses than available to all-sky X-ray surveys like ROSAT, and without the bias toward CTTS of objective prism studies. However, as with every technique there are limitations involved, e.g., temporal sampling and a bias toward variables with larger amplitudes, especially at the faint end, are issues that need to be explored.

2.6. Spectroscopy

As already emphasized in the previous paragraphs, low- to moderate-resolution follow-up spectroscopy is essential to confirm membership of PMS stars. However, the observational effort to identify the widespread population of PMS stars among the many thousands of field stars in the large areas spanned by nearby OB associations is huge, and up to recently has largely precluded further investigations. With the advent of extremely powerful multiple object spectrographs, such as 2dF at the Anglo-Australian Telescope, Hydra on the WIYN 3.5-m and the CTIO 4-m telescopes, and now Hectospec on the 6.5-m MMT, large-scale spectroscopic surveys have now become feasible.

One of the most powerful approaches to unbiased surveys of young low-mass stars is to use the presence of strong 6708 Å Li I absorption lines as a diagnostic of the PMS nature of a candidate object (e.g., *Dolan and Mathieu*, 1999, 2001). Because Li I is strongly diminished in very early phases of stellar evolution, a high Li content is a reliable indication for the youth of a star (e.g., *Herbig*, 1962; *D'Antona and Mazzitelli*, 1994). However, Li depletion is not only a function of stellar age, but also of stellar mass and presumably even depends on additional factors like stellar rotation (cf. *Soderblom*, 1996). Not only PMS stars, but also somewhat older, although still relatively young zero-age main-sequence stars, e.g., the G- and K-type stars in the ~10^8-yr-old Pleiades (cf. *Soderblom et al.*, 1993), can display Li absorption lines. In order to classify stars as PMS, one thus has to consider a spectral type dependent threshold for the Li line width. Such a threshold can be defined by the upper envelope of Li measurements in young clusters of main-sequence stars with ages between ~30 m.y. and a few 100 m.y. such as IC 2602, IC 4665, IC 2391, α Per, and the Pleiades. Any star with a Li line width considerably above this threshold should be younger than ~30 m.y. and can therefore be classified as a PMS object.

In addition to Li, other spectroscopic signatures can be used as youth indicators, such as the K I and Na I absorption lines that are typically weaker in low-mass PMS objects compared to field M-type dwarfs (*Martín et al.*, 1996; *Luhman*, 1999), and radial velocities (if high-resolution spectra are available).

3. AGES AND AGE SPREADS OF LOW-MASS STARS

Stellar ages are usually inferred from the positions of the stars in the HR diagram by comparison with theoretical PMS evolutionary models. Photometry of young clusters and associations has shown that low-mass members occupy a wide swath on the CMD or the H-R diagram (NGC 2264, ONC, σ Ori). A common interpretation has been that this spread is evidence of an age spread among cluster members (e.g., *Palla and Stahler*, 2000). This interpretation assumes that the single-epoch observed colors and magnitudes of low-mass PMS stars accurately correspond to the temperatures and luminosities of each star. However, it has to be strongly emphasized that the masses and especially the ages of the individual stars read off from their position in these diagrams are generally *not* identical to their true masses and ages. Several factors can cause considerable deviations of an individual star's position in the HR-diagram from the locations predicted by theoretical models for a given age and mass. Low-mass PMS stars exhibit variability ranging from a few tenths in the WTTS up to several magnitudes in the CTTS (*Herbst et al.*, 1994). Furthermore, binarity can add a factor of 2 (~0.75 mag) spread to the CMD. For regions spanning large areas on the sky there will be an additional spread caused by the distribution of distances along the line of sight among individual stars. Another source of uncertainty are the calibrations used to derive bolometric luminosities and effective temperatures, and finally, the choice of evolutionary tracks used, that in some cases yield different ages for intermediate (~2–5 M$_\odot$) and low-mass stars located in the same region [e.g., the ONC (*Hillenbrand*, 1997; see also *Hartmann*, 2001)]. Therefore, any age estimates in young star-forming regions must account for the significant spread that a coeval population will necessarily have on the CMD, which translates into a spread on the HR diagram.

As an example, Fig. 2 shows the HR diagram containing all Upper Sco association members from *Preibisch et al.* (2002); the diagram also shows the main sequence and a

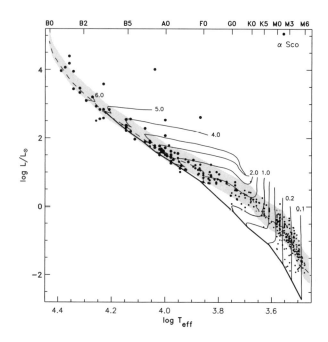

Fig. 2. HR diagram for the Upper Sco association members from the study of *Preibisch et al.* (2002). The lines show the evolutionary tracks from the *Palla and Stahler* (1999) PMS models, some labeled by their masses in solar units. The thick solid line shows the main sequence. The 5-m.y. isochrone is shown as the dashed line; it was composed from the high-mass isochrone from *Bertelli et al.* (1994) for masses 6–30 M$_\odot$, the *Palla and Stahler* (1999) PMS models for masses 1–6 M$_\odot$, and the *Baraffe et al.* (1998) PMS models for masses 0.02–1 M$_\odot$. The gray shaded band shows the region in which one expects 90% of the member stars to lie, based on the assumption of a common age of 5 m.y. for all stars and taking proper account of the uncertainties and the effects of unresolved binaries (for details see text).

5-m.y. isochrone. Not only the majority of the low-mass stars, but also most of the intermediate- and high-mass stars, lie close to or on the 5-m.y. isochrone. There clearly is a considerable scatter that may seem to suggest a spread in stellar ages. In the particular case of Upper Sco shown here, in addition to the other effects mentioned above, the most important factor for the apparent scatter is the relatively large spread of individual stellar distances [~±20 pc around the mean value of 145 pc (*de Bruijne*, 1999, and personal communication)] in this very nearby and extended region, which causes the luminosities to be either over- or underestimated when a single distance is adopted for all sources. A detailed discussion and statistical modeling of these effects is given in *Preibisch and Zinnecker* (1999) and *Preibisch et al.* (2002). In the later work these authors showed that the observed HR diagram for the low-mass stars in Upper Sco is consistent with the assumption of a *common stellar age of about 5 m.y.*; there is no evidence for an age dispersion, although small ages spreads of ~1–2 m.y. cannot be excluded by the data. *Preibisch et al.* (2002) showed that the derived age is also robust when taking into account the uncertainties of the theoretical PMS models. It is remarkable that the isochronal age derived for the low-mass stars is consistent with previous and independent age determinations based on the nuclear and kinematic ages of the massive stars (*de Zeeuw and Brand*, 1985; *de Geus et al.*, 1989), which also yielded 5 m.y. This very good agreement of the *independent* age determinations for the high-mass and the low-mass stellar population shows that *low- and high-mass stars are coeval* and thus have formed together. Furthermore, the absence of a significant age dispersion implies that all stars in the association have formed more or less simultaneously. Therefore, the star-formation process must have started rather suddenly and everywhere at the same time in the association, and also must have ended after at most a few million years. The star-formation process in Upper Sco can thus be considered as a *burst of star formation*.

Sherry (2003) compared the observed spread across the V vs. V-I_C CMD of low-mass members of the σ Ori cluster to the spread that would be expected based upon the known variability of WTTS, the field binary fraction, the photometric errors of his survey, and a range of simple star-formation histories. The observed spread was consistent with the predicted spread of an isochronal population. Sherry concluded that the bulk of the low-mass stars must have formed over a period of less than ~1 m.y. A population with a larger age spread would have been distributed over a larger region of the CMD than was observed.

Burningham et al. (2005) also examined the possibility of an age spread among members of the σ Ori cluster. They used two epoch R and i' observations of cluster members taken in 1999 and 2003 to estimate the variability of each cluster member. They then constructed a series of simple models with a varying fraction of equal mass binaries. They found that the observed spread on the CMD was too large to be fully accounted for by the combined effects of observational errors, variability (over 1–4 years), and binaries. They conclude that the larger spread on the CMD could be accounted for by either a longer period of accretion driven variability, an age spread of ≤2 m.y. (using a distance of 440 pc or 4 m.y. using a distance of 350 pc), or a combination of long-term variability and a smaller age spread. However, their result is not necessarily in contradiction with the findings of *Sherry* (2003), especially if the actual variability is larger than they estimate based on their two epoch observations.

In the Orion OB1 association, the V and I_C-band CMDs for spectroscopically confirmed TTS (*Briceño et al.*, 2005a; Briceño et al., in preparation), which mitigate the spread caused by the variability of individual sources by plotting their mean magnitudes and colors, provide evidence that the Ori OB1b subassociation is older (~4 m.y.) than the often quoted age of about 1.7 m.y. (*Brown et al.*, 1994), with a rather narrow age distribution regardless of the tracks used (Fig. 3). The same dataset for the older OB1a subassociation also shows a narrow range of ages, with a mean value of ~8 m.y.

The λ Ori region also shows that the age distribution and star-formation history is spatially dependent. *Dolan and Mathieu* (2001) (see also *Lee et al.*, 2005) found that age

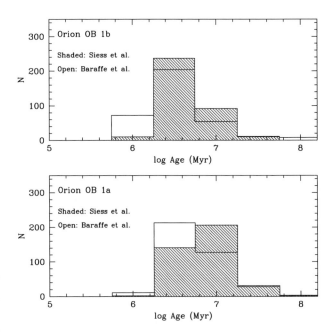

Fig. 3. Distribution of ages of ~1000 newly identified TTS in a ~60 deg² area spanning the Orion OB1a and 1b subassociations (Briceño et al., in preparation). Ages were derived using the *Baraffe et al.* (1998) and *Siess et al.* (2000) evolutionary tracks and isochrones. The mean ages are ~4 m.y. for Ori OB1b and ~8 m.y. for Ori OB1a.

distributions of high-mass and solar-type stars in the region show several critical features: (1) both high- and low-mass star formation began concurrently in the center of the SFR roughly 6–8 m.y. ago; (2) low-mass star formation ended in the vicinity of λ Ori roughly 1 m.y. ago; (3) low-mass star formation rates near the B30 and B35 clouds reached their maxima later than did low-mass star formation in the vicinity of λ Ori; (4) low-mass star formation continues today near the B30 and B35 clouds. As with the Ori OB1 associations, this varied star formation history reflects the rich interplay of the massive stars and the gas.

The accumulated evidence for little, if any, age spreads in various star-forming regions provides a natural explanation for the "post-T Tauri problem" [the absence of "older" TTS in star-forming regions like Taurus, assumed to have been forming stars for up to tens of millions of years (*Herbig*, 1978)], and at the same time implies relatively short lifetimes for molecular clouds. These observational arguments support the evolving picture of star formation as a fast and remarkably synchronized process in molecular clouds (section 6).

4. THE INITIAL MASS FUNCTION IN OB ASSOCIATIONS

The IMF is the utmost challenge for any theory of star formation. Some theories suggest that the IMF should vary systematically with the star-formation environment (*Larson*, 1985), and for many years star formation was supposed to be a bimodal process (e.g., *Shu and Lizano,* 1988) according to which high- and low-mass stars should form in totally different sites. For example, it was suggested that increased heating due to the strong radiation from massive stars raises the Jeans mass, so that the bottom of the IMF would be truncated in regions of high-mass star formation. In contrast to the rather quiescent environment in small low-mass clusters and T associations (like the Taurus molecular clouds), forming stars in OB association are exposed to the strong winds and intense UV radiation of the massive stars, and, after a few million years, also affected by supernova explosions. In such an environment, it may be harder to form low-mass stars, because, e.g., the lower-mass cloud cores may be completely dispersed before protostars can even begin to form.

Although it has been long established that low-mass stars can form alongside their high-mass siblings (e.g., *Herbig,* 1962) in nearby OB associations, until recently it was not well known what quantities of low-mass stars are produced in OB environments. If the IMF in OB associations is not truncated and similar to the field IMF, it would follow that most of their total stellar mass ($\geq 60\%$) is found in low-mass (<2 M$_\odot$) stars. This would then imply that most of the current galactic star formation is taking place in OB associations. Therefore, the typical environment for forming stars (and planets) would be close to massive stars and not in isolated regions like Taurus.

In Fig. 4 we show the empirical mass function for Upper Sco as derived in *Preibisch et al.* (2002), for a total sample of 364 stars covering the mass range from 0.1 M$_\odot$ up to 20 M$_\odot$. The best-fit multipart power law function for the probability density distribution is given by

$$\frac{dN}{dM} \propto \begin{cases} M^{-0.9 \pm 0.2} & \text{for } 0.1 \leq M/M_\odot < 0.6 \\ M^{-2.8 \pm 0.5} & \text{for } 0.6 \leq M/M_\odot < 2 \\ M^{-2.6 \pm 0.5} & \text{for } 2 \leq M/M_\odot < 20 \end{cases} \quad (1)$$

or, in shorter notation, α (0.1–0.6) = -0.9 ± 0.2, α (0.6–2.0) = -2.8 ± 0.5, α (2.0–20) = -2.6 ± 0.3. For comparison, the plot also shows two different field IMF models, the *Scalo* (1998) model, which is given by α (0.1–1) = -1.2 ± 0.3, α (1–10) = -2.7 ± 0.5, α (10–100) = -2.3 ± 0.5, and the *Kroupa* (2002) model with α (0.02–0.08) = -0.3 ± 0.7, α (0.08–0.5) = -1.3 ± 0.5, α (0.5–100) = -2.3 ± 0.3.

While the slopes of the fit to the empirical mass function of Upper Sco are not identical to those of these models, they are well within the ranges of slopes derived for similar mass ranges in other young clusters or associations, as compiled in *Kroupa* (2002). Therefore, it can be concluded that, within the uncertainties, the general shape of the Upper Sco mass function is consistent with recent field star and cluster IMF determinations.

In the Orion OB1 association *Sherry et al.* (2004) find that the mass function for the σ Ori cluster is consistent with the *Kroupa* (2002) IMF. In the immediate vicinity of λ Ori,

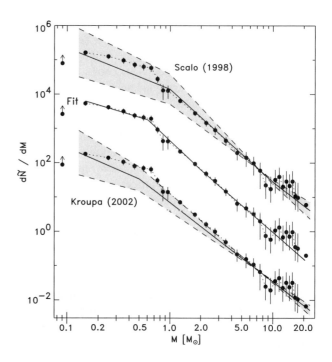

Fig. 4. Comparison of the mass function derived for the Upper Sco association with different mass function measurements for the field (from *Preibisch et al.,* 2002). The Upper Sco mass function is shown three times by the solid dots connected with the dotted lines, multiplied by arbitrary factors. The middle curve shows the original mass function; the solid line is our multipart power law fit. The upper curve shows our mass function multiplied by a factor of 30 and compared to the *Scalo* (1998) IMF (solid line); the gray shaded area delimited by the dashed lines represents the range allowed by the errors of the model. The lower curve shows our mass function multiplied by a factor of 1/30 and compared to the *Kroupa* (2002) IMF (solid line).

Barrado y Navascues et al. (2004) combined their deep imaging data with the surveys of *Dolan and Mathieu* (1999, 2001) (limited to the same area) to obtain an initial mass function from 0.02 to 1.2 M$_\odot$. They find that the data indicate a power law index of $\alpha = -0.60 \pm 0.06$ across the stellar-substellar limit and a slightly steeper index of $\alpha = 0.86 \pm 0.05$ over the larger mass range of 0.024 M$_\odot$ to 0.86 M$_\odot$, much as is found in other young regions.

Over the entire λ Ori star-forming region, *Dolan and Mathieu* (2001) were able to clearly show that the IMF has a spatial dependence. *Dolan and Mathieu* (1999) had found that within the central ~3.5 pc around λ Ori the low-mass stars were deficient by a factor of 2 compared to the field IMF. Outside this central field, *Dolan and Mathieu* (2001) showed the low-mass stars to be overrepresented compared to the *Miller and Scalo* (1978) IMF by a factor of 3. A similar overrepresentation of low-mass stars is also found at significant confidence levels when considering only stars associated with B30 and B35.

Thus the global IMF of the λ Ori SFR resembles the field, while the local IMF appears to vary substantially across the region. No one place in the λ Ori SFR creates the field IMF by itself. Only the integration of the star-formation process over the entire region produces the field IMF.

5. DISK EVOLUTION

The presence of circumstellar disks around low-mass pre-main-sequence stars appears to be a natural consequence of the star-formation process; these disks play an important role both in determining the final mass of the star and as the potential sites for planet formation. Although we have a good general understanding of the overall processes involved, many important gaps still remain. For instance, the timescales for mass accretion and disk dissipation are still matters of debate. An example is the discovery of a seemingly long-lived accreting disk around the ~25-m.y.-old, late-type (M3) star St 34 located in the general area of the Taurus dark clouds (*Hartmann et al.*, 2005a; *White and Hillenbrand*, 2005).

How circumstellar disks evolve and whether their evolution is affected by environmental conditions are questions that at present can only be investigated by looking to low-mass PMS stars, and in particular the best samples can now be drawn from the various nearby OB associations. First, low-mass young stars like T Tauri stars, in particular those with masses ~1 $M_\odot$, constitute good analogs of what the conditions may have been in the early solar system. Second, as we have discussed in section 3, OB associations can harbor many stellar aggregates, with distinct ages, such as in Orion OB1, possibly the result of star-forming events (triggered or not) occurring at various times throughout the original GMC. Some events will have produced dense clusters while others are responsible for the more spread out population. The most recent events are easily recognizable by the very young (≤1 m.y.) stars still embedded in their natal gas, while older ones may be traced by the ~10-m.y. stars that have long dissipated their parent clouds. This is why these regions provide large numbers of PMS stars in different environments, but presumably sharing the same "genetic pool," that can allow us to build a differential picture of how disks evolve from one stage to the next.

Disks are related to many of the photometric and spectroscopic features observed in T Tauri stars. The IR emission originates by the contribution from warm dust in the disk, heated at a range of temperatures by irradiation from the star and viscous dissipation (e.g., *Meyer et al.*, 1997). The UV excesses, excess continuum emission (veiling), irregular photometric variability, and broadened spectral line profiles (particularly in the hydrogen lines and others like Ca II) are explained as different manifestations of gas accretion from a circumstellar disk. In the standard magnetospheric model (*Königl*, 1991), the accretion disk is truncated at a few stellar radii by the magnetic field of the star; the disk material falls onto the photosphere along magnetic field lines at supersonic velocities, creating an accretion shock that is thought to be largely responsible for the excess UV and continuum emission (*Calvet and Gullbring*, 1998). The in-

falling material also produces the observed broadened and P Cygni profiles observed in hydrogen lines (*Muzerolle et al.*, 1998a,b, 2001). Disk accretion rates for most CTTS are on the order of 10^{-8} $M_\odot$ yr^{-1} at ages of 1–2 m.y. (e.g., *Gullbring et al.*, 1998; *Hartmann*, 1998; *Johns-Krull and Valenti*, 2001).

Comparative studies of near-IR emission and accretion-related indicators (Hα and Ca II emission, UV excess emission) at ages ~1–10 m.y. offer insight into how the innermost part of the disk evolves. One way to derive the fraction of stellar systems with inner disks is counting the number of objects showing excess emission in the JHKL near-IR bands. The availability of 2MASS has made JHK studies of young populations over wide spatial scales feasible. More recently, the emission of T Tauri stars in the Spitzer IRAC and MIPS bands has been characterized by *Allen et al.* (2004) and *Hartmann et al.* (2005b). Another approach is determining the number of objects that exhibit strong Hα and Ca II emission (CTTS), or UV excesses; these figures provide an indication of how many systems are actively accreting from their disks.

So far, the most extensive studies of how disk fractions change with time have been conducted in the Orion OB1 association. *Hillenbrand et al.* (1998) used the I_c-K color to derive a disk fraction of 61–88% in the ONC, and the Ca II lines in emission, or "filled-in," as a proxy for determining an accretion disk frequency of ~70%. *Rebull et al.* (2000) studied a region on both sides of the ONC and determined a disk accretion fraction in excess of 40%. *Lada et al.* (2000) used the JHKL bands to derive a ONC disk fraction of 80–85% in the low-mass PMS population. *Lada et al.* (2004) extended this study to the substellar candidate members and found a disk fraction of ~50%. In their ongoing large-scale study of the Orion OB1 association, *Calvet et al.* (2005a) combined UV, optical, JHKL, and 10-μm measurements in a sample of confirmed members of the 1a and 1b subassociations to study dust emission and disk accretion. They showed evidence for an overall decrease in IR emission with age, interpreted as a sign of dust evolution between the disks in Ori OB1b (age ~4 m.y.), Ori OB1a (age ~8 m.y.), and those of younger populations like Taurus (age ~2 m.y.). *Briceño et al.* (2005b) used IRAC and MIPS on Spitzer to look for dusty disks in Ori OB1a and 1b. They confirm a decline in IR emission by the age of Orion OB1b, and find a number of "transition" disk systems (~14% in 1b and ~6% in 1a), objects with essentially photospheric fluxes at wavelengths ≤4.5 μm and excess emission at longer wavelengths. These systems are interpreted as showing signatures of inner disk clearing, with optically thin inner regions stretching out to one or a few AU (*Calvet et al.*, 2002, 2005b; *Uchida et al.*, 2004; *D'Alessio et al.*, 2005); the fraction of these transition disks that are still accreting is low (~5–10%), hinting at a rapid shutoff of the accretion phase in these systems [similar results have been obtained by *Sicilia-Aguilar et al.* (2006) in Cep OB2; see below]. *Haisch et al.* (2000) derived an IR-excess fraction of ~86% in the ≤1-m.y.-old NGC 2024 embedded cluster.

Their findings indicate that the majority of the sources that formed in NGC 2024 are presently surrounded by, and were likely formed with, circumstellar disks.

One of the more surprising findings in the λ Ori region was that, despite the discovery of 72 low-mass PMS stars within 0.5° (~3.5 pc) of λ Ori, only two of them showed strong Hα emission indicative of accretion disks (*Dolan and Mathieu*, 1999). *Dolan and Mathieu* (2001) expanded on this result by examining the distribution of Hα emission along an axis from B35 through λ Ori to B30. The paucity of Hα emission-line stars continues from λ Ori out to the two dark clouds, at which point the surface density of Hα emission-line stars increases dramatically. Even so, many of the Hα stars associated with B30 and B35 have ages similar to PMS stars found in the cluster near λ Ori. Yet almost none of the latter show Hα emission. This strongly suggests that the absence of Hα emission from the central PMS stars is the result of an environmental influence linked to the luminous OB stars.

The nearby Scorpius-Centaurus OB association [d ~ 130 pc (*de Zeeuw et al.*, 1999)] has been a natural place to search for circumstellar disks, especially at somewhat older ages (up to ~10 m.y.). *Moneti et al.* (1999) detected excess emission at the ISOCAM 6.7- and 15-μm bands in 10 X-ray selected WTTS belonging to the more widely spread population of Sco-Cen, albeit at levels significantly lower than in the much younger (age ~1 m.y.) Chamaeleon I T association. In a sample of X-ray and proper motion-selected late-type stars in the Lower Centaurus Crux (LCC, age ~17 m.y.) and Upper Centaurus Lupus (UCL, age ~15 m.y.) in Sco-Cen, *Mamajek et al.* (2002) find that only 1 out of 110 PMS solar-type stars shows both enhanced Hα emission and a K-band excess indicative of active accretion from a truncated circumstellar disk, suggesting timescales of ~10 m.y. for halting most of the disk accretion. *Chen et al.* (2005) obtained Spitzer Space Telescope MIPS observations of 40 F- and G-type common proper motion members of the Sco-Cen OB association with ages between 5 and 20 m.y. They detected 24 μm excess emission in 14 objects, corresponding to a disk fraction of ≥35%.

In Perseus, *Haisch et al.* (2001) used JHKL observations to estimate a disk fraction of ~65% in the 2–3-m.y.-old (*Luhman et al.*, 1998) IC 348 cluster, a value lower than in the younger NGC 2024 and ONC clusters in Orion, suggestive of a timescale of 2–3 m.y. for the disappearance of approximately one-third of the inner disks in IC 348. The ~4-m.y.-old Tr 37 and the ~10-m.y.-old NGC 7160 clusters in Cep OB2 have been studied by *Sicilia-Aguilar et al.* (2005), who derive an accreting fraction of ~40% for Tr 37, and 2–5% (one object) for NGC 7160. *Sicilia-Aguilar et al.* (2006) used Spitzer IRAC and MIPS observations to further investigate disk properties in Cep OB2. About 48% of the members exhibit IR excesses in the IRAC bands, consistent with their inferred accreting disk fraction. They also find a number of "transition" objects (10%) in Tr 37. They interpret their results as evidence for differential evolution

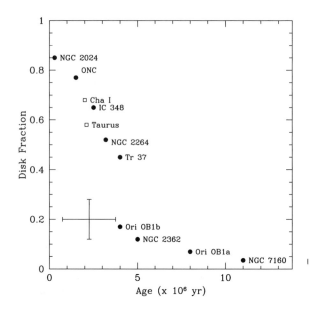

Fig. 5. Inner disk fraction around low-mass young stars (0.2 ≤ M ≤ 1 M☉) as a function of age in various nearby OB associations (solid dots). The ONC point is an average from estimates by *Hillenbrand et al.* (1998), *Lada et al.* (2000), and *Haisch et al.* (2001). Data in the JHKL bands from *Haisch et al.* (2001) were used for IC 348, NGC 2264, and NGC 2362. Note that for the more distant regions like NGC 2362 and NGC 2264, the mass completeness limit is ~1 M☉. Results for Orion OB1a and 1b are from JHK data in *Briceño et al.* (2005a). In Tr 37 and NGC 7160 we used values by *Sicilia-Aguilar et al.* (2005, 2006) from accretion indicators and JHK (3.6 μm) measurements. As a comparison, we also plot disk fractions (open squares) derived in Taurus (from data in *Kenyon and Hartmann*, 1995) and for Chamaeleon from *Gómez and Kenyon* (2001). A conservative error bar is indicated at the lower left corner.

in optically thick disks as a function of age, with faster decreases in the IR emission at shorter wavelengths, suggestive of a more rapid evolution in the inner disks.

All these studies agree on a general trend toward rapid inner disk evolution (Fig. 5); it looks like the disappearance of the dust in the innermost parts of the disk, either due to grain growth and settling or to photoevaporation (*Clarke et al.*, 2001), is followed by a rapid shutoff of accretion. However, both these investigations and newer findings also reveal exceptions that lead to a more complex picture. *Muzerolle et al.* (2005) find that 5–10% of all disk sources in the ~1-m.y.-old NGC 2068 and 2071 clusters in Orion show evidence of significant grain growth, suggesting a wide variation in timescales for the onset of primordial disk evolution and dissipation. This could also be related to the existence of CTTS and WTTS in many regions at very young ages, the WTTS showing no signatures of inner, optically thick disks. Did some particular initial conditions favor very fast disk evolution in these WTTS (binarity)? Finally, the detection of long-lived disks implies that dust dissipation and the halting of accretion do not necessarily follow a universal trend. This may have implications for the formation of planetary systems; if slow accretion processes are the dominant formation mechanism for jovian planets then long-lived disks may be ideal sites to search for evidence for protoplanets.

6. THE ORIGIN OF OB ASSOCIATIONS: BOUND VS. UNBOUND STELLAR GROUPS

Blauuw (1964) in his masterly review first provided some clues as to the origin of OB associations, alluding to two ideas: "originally small, compact bodies with dimensions of several parsec or less" or "regional star formation, more or less simultaneous, but scattered over different parts of a large cloud complex." Similarly, there are at present two (competing) models that attempt to explain the origin of OB associations:

Model A: origin as expanding dense embedded clusters
Model B: origin in unbound turbulent giant molecular clouds

We will label Model A as the KLL model (*Kroupa et al.,* 2001; *Lada and Lada,* 2003), while Model B will be named the BBC model (for Bonnell, Bate, and Clark) (see also *Clark et al.,* 2005).

The jury is still out on which model fits the observations better. It may be that both models contain elements of truth and thus are not mutually exclusive. Future astrometric surveys, Gaia in particular, will provide constraints to perform sensitive tests on these models, but that will not happen for about another decade. We refer the reader to the review of *Brown et al.* (1999), who also points out the problems involved in the definition of an OB association and the division of these vast stellar aggregates into subgroups (see *Brown et al.,* 1997).

We now discuss both models in turn. The KLL model proposes that ultimately the star-formation efficiency in most embedded, incipient star clusters is too low (≤5%) for these clusters to remain bound (see references in *Elmegreen et al.,* 2000; *Lada and Lada,* 2003) after the massive stars have expelled the bulk of the lower-density, leftover cluster gas that was not turned into stars, or did not get accreted. Therefore the cluster will find itself globally unbound, although the cluster *core* may remain bound (*Kroupa et al.,* 2001) and survive. Such an originally compact but expanding cluster could evolve into an extended subgroup of an OB association after a few million years. Essentially, the KLL model describes how the dense gas in a bound and virialized GMC is partly converted into stars, most of which are born in dense embedded clusters, the majority of which disperse quickly (so-called cluster "infant mortality").

The BBC model supposes that GMCs need not be regarded as objects in virial equilibrium, or even bound, for them to be sites of star formation. Globally unbound or marginally bound GMC can form stellar groups or clusters very quickly, over roughly their crossing time (cf. *Elmegreen et al.,* 2000) [recent simulations also show that GMC them-

selves can form very quickly from atomic gas, in 1–2 m.y. (see chapter by Ballesteros-Paredes et al.)]. The unbound state of the GMC ensures that the whole region is dispersing while it is forming stars or star clusters in locally compressed sheets and filaments, due to the compressive nature of supersonic turbulence ("converging flows"). The mass fraction of compressed cloud gas is low, affecting only ~10% of the GMC. In this model, the spacing of the OB stars that define an OB association would be initially large (larger than a few parsec), rather than compact (≤1 pc) as in the KLL model. Furthermore, no sequential star formation triggered by the thermal pressure of expanding HII regions "a la *Elmegreen and Lada* (1977)" may be needed to create OB subgroups, unlike in the KLL model. Delayed SN triggered star formation in adjacent transient clouds at some distance may still occur and add to the geometric complexity of the spatial and temporal distribution of stars in giant OB associations [e.g., NGC 206 in M31 and NGC 604 in M33 (*Maíz-Apellániz et al.,* 2004)].

A problem common to both models is that neither one explains very well the spatial and temporal structure of OB associations, i.e., the fact that the subgroups seem to form an age sequence (well known in Orion OB1 and Sco OB2). One way out would be to argue that the subgroups are not always causally connected and did not originate by sequential star formation as in the *Elmegreen and Lada* (1977) paradigm. Star formation in clouds with supersonic turbulence occurring in convergent flows may be of a more random nature and only mimic a causal sequence of triggering events [e.g., in Sco OB2, the UCL and LCC subgroups, with ages of ~15 and ~17 m.y. from the *D'Antona and Mazzitelli* (1994) tracks, have hardly an age difference at all]. This is a problem for the KLL model, as sequential star formation can hardly generate two adjacent clusters (subgroups) within such a short timespan. It is also a problem for the BBC model, but for a different reason. The fact that star formation must be rapid in unbound transient molecular clouds is in conflict with the ages of the Sco OB2 subgroups (1, 5, 15, and 17 m.y.), if all subgroups formed from a single coherent (long-lived) GMC. Still, the observational evidence suggesting that Upper Sco can be understood in the context of triggered star formation (see section 7.2) can be reconciled with these models if we consider a scenario with multiple star-formation sites in a turbulent large GMC, in which triggering may easily take place. It remains to be seen if subgroups of OB associations had some elongated minimum size configuration, as *Blaauw* (1991) surmised, on the order of 20 pc × 40 pc. If so, OB associations are then something fundamentally different from embedded clusters, which would have many ramifications for the origin of OB stars [However, a caveat with tracing back minimum size configurations (*Brown et al.,* 1997) is that, even using modern *Hipparcos* proper motions, there is a tendency to obtain overestimated dimensions because present proper motions cannot resolve the small velocity dispersion. This situation should improve with Gaia, and with the newer census of low-mass stars, that can potentially provide statistically robust samples to trace the past kinematics of these regions.]

7. CONSTRAINTS ON RAPID AND TRIGGERED/SEQUENTIAL STAR FORMATION

7.1. The Duration of Star Formation

One of the problems directly related to the properties of the stellar populations in OB associations are the lifetimes of molecular clouds. Age estimates for GMCs can be very discordant, ranging from $\sim 10^8$ yr (*Scoville et al.,* 1979; *Scoville and Hersh,* 1979) to just a few 10^6 yr (e.g., *Elmegreen et al.,* 2000; *Hartmann et al.,* 2001; *Clark et al.,* 2005). Molecular cloud lifetimes bear importantly on the picture we have of the process of star formation. Two main views have been contending among the scientific community during the past few years. In the standard picture of star formation, magnetic fields are a major support mechanism for clouds (*Shu et al.,* 1987). Because of this, the cloud must somehow reduce its magnetic flux per unit mass if it is to attain the critical value for collapse. One way to do this is through ambipolar diffusion, in which the gravitational force pulls mass through the resisting magnetic field, effectively concentrating the cloud and slowly "leaving the magnetic field behind." From these arguments it follows that timescale for star formation should be on the order of the diffusion time of the magnetic field, $t_D \sim 5 \times 10^{13}$ (n_i/n_{H_2}) yr (*Hartmann,* 1998), which will be important only if the ionization inside the cloud is low $(n_i/n_{H_2} \lesssim 10^{-7})$, in which case $t_D \sim 10^7$ yr; therefore, the so called "standard" picture depicts star formation as a "slow" process. This leads to relevant observational consequences. If molecular clouds live for long periods before the onset of star formation, then we should expect to find a majority of starless dark clouds; however, the observational evidence points to quite the contrary. Almost all cloud complexes within ~ 500 pc exhibit active star formation, harboring stellar populations with ages ~ 1–10 m.y. Another implication of "slow" star formation is that of *age spreads* in star-forming regions. If clouds such as Taurus last for tens of millions of years there should exist a population of PMS sequence stars with comparable ages (the "post-T Tauri problem"). Many searches for such "missing population" were conducted in the optical and in X-rays, in Taurus, and in other regions (see *Neuhäuser,* 1997). The early claims by these studies of the detection of large numbers of older T Tauri stars widely spread across several nearby star-forming regions were countered by *Briceño et al.* (1997), who showed that these samples were composed of an admixture of young, X-ray active ZAMS field stars and some true PMS sequence members of these regions. Subsequent high-resolution spectroscopy confirmed this idea. As discussed in section 4, presently there is little evidence for the presence of substantial numbers of older PMS stars in and around molecular clouds.

Recent widefield optical studies in OB associations like Sco-Cen (*Preibisch et al.,* 2002), Orion (*Dolan and Mathieu,*

2001; *Briceño et al.,* 2001, 2005a; Briceño et al., in preparation), and Cepheus (*Sicilia-Aguilar et al.,* 2005), show that the groupings of stars with ages ≥ 4–5 m.y. have mostly lost their natal gas. The growing notion is that not only do molecular clouds form stars rapidly, but that they are transient structures, dissipating quickly after the onset of star formation. This dispersal seems to be effective in both low-mass regions as well as in GMC complexes that give birth to OB associations. The problem of accumulating and then dissipating the gas quickly in molecular clouds has been addressed by *Hartmann et al.* (2001). The energy input from stellar winds of massive stars, or more easily from SN shocks, seems to be able to account for the dispersal of the gas on short timescales in the high-density regions typical of GMC complexes, as well as in low-density regions, like Taurus or Lupus; if enough stellar energy is input into the gas such that the column density is reduced by factors of only 2–3, the shielding could be reduced enough to allow dissociation of much of the gas into atomic phase, effectively "dissipating" the molecular cloud.

7.2. Sequential and Triggered Star Formation

Preibisch et al. (2002) investigated the star-formation history in Upper Sco. A very important aspect in this context is the spatial extent of the association and the corresponding crossing time. The bulk (70%) of the *Hipparcos* members (and thus also the low-mass stars) lie within an area of 11° diameter on the sky, which implies a characteristic size of the association of 28 pc. They estimated that the original size of the association was probably about 25 pc. *de Bruijne* (1999) showed that the internal velocity dispersion of the *Hipparcos* members of Upper Sco is only 1.3 km s^{-1}. This implies a lateral crossing time of 25 pc/1.3 km s^{-1} ~ 20 m.y. It is obvious that the lateral crossing time is much (about an order of magnitude) larger than the age spread of the association members [which is <2 m.y. as derived by *Preibisch and Zinnecker* (1999)]. This finding clearly shows that some external agent is required to have coordinated the onset of the star-formation process over the full spatial extent of the association. In order to account for the small spread of stellar ages, the triggering agent must have crossed the initial cloud with a velocity of at least ~ 15–25 km s^{-1}. Finally, some mechanism must have terminated the star-formation process at most about 1 m.y. after it started. Both effects can be attributed to the influence of massive stars.

In their immediate surroundings, massive stars generally have a destructive effect on their environment; they can disrupt molecular clouds very quickly and therefore prevent further star formation. At somewhat larger distances, however, the wind- and supernova-driven shock waves originating from massive stars can have a constructive rather than destructive effect by driving molecular cloud cores into collapse. Several numerical studies (e.g., *Boss,* 1995; *Foster and Boss,* 1996; *Vanhala and Cameron,* 1998; *Fukuda and Hanawa,* 2000) have found that the outcome of the impact of a shock wave on a cloud core mainly depends on the type of the shock and its velocity: In its initial, adiabatic phase, the shock wave is likely to destroy ambient clouds; the later, isothermal phase, however, is capable of triggering cloud collapse if the velocity is in the right range. Shocks traveling faster than about 50 km s^{-1} shred cloud cores to pieces, while shocks with velocities slower than about 15 km s^{-1} usually cause only a slight temporary compression of cloud cores. Shock waves with velocities in the range of ~15–45 km s^{-1}, however, are able to induce collapse of molecular cloud cores. A good source of shock waves with velocities in that range are supernova explosions at a distance between ~10 pc and ~100 pc. Other potential sources of such shock waves include wind-blown bubbles and expanding H II regions. Observational evidence for star-forming events triggered by shock waves from massive stars has been discussed in, e.g., *Carpenter et al.* (2000), *Walborn et al.* (1999), *Yamaguchi et al.* (2001), *Efremov and Elmegreen* (1998), *Oey and Massey* (1995), and *Oey et al.* (2005).

For the star burst in Upper Sco, a very suitable trigger is a supernova explosion in the Upper Centaurus-Lupus association that happened about 12 m.y. ago. The structure and kinematics of the large H I loops surrounding the Scorpius-Centaurus association suggest that this shock wave passed through the former Upper Sco molecular cloud just about 5–6 m.y. ago (*de Geus,* 1992). This point in time agrees very well with the ages found for the low-mass stars as well as the high-mass stars in Upper Sco, which have been determined above in an absolutely independent way. Furthermore, since the distance from Upper Centaurus-Lupus to Upper Sco is about 60 pc, this shock wave probably had precisely the properties (v ~ 20–25 km s^{-1}) that are required to induce star formation according to the modeling results mentioned above. Thus, the assumption that this supernova shock wave triggered the star-formation process in Upper Sco provides a self-consistent explanation of all observational data.

The shock-wave crossing Upper Sco initiated the formation of some 2500 stars, including 10 massive stars upward of 10 M$_\odot$. When the newborn massive stars "turned on," they immediately started to destroy the cloud from inside by their ionizing radiation and their strong winds. This affected the cloud so strongly that after a period of ≤1 m.y. the star-formation process was terminated, probably simply because all the remaining dense cloud material was disrupted. This explains the narrow age distribution and why only about 2% of the original cloud mass was transformed into stars. About 1.5 m.y. ago the most massive star in Upper Sco, probably the progenitor of the pulsar PSR J1932+1059, exploded as a supernova. This explosion created a strong shock wave, which fully dispersed the Upper Sco molecular cloud and removed basically all the remaining diffuse material.

It is interesting to note that this shock wave must have crossed the ρ Oph cloud within the last 1 m.y. (*de Geus,* 1992). The strong star-formation activity we witness right now in the ρ Oph cloud might therefore be triggered by this shock wave (see *Motte et al.,* 1998) and would represent the third generation of sequential triggered star formation in the Scorpius-Centaurus-Ophiuchus complex.

Other relatively nearby regions have also been suggested as scenarios for triggered star formation. In Cepheus, a large-scale ring-like feature with a diameter of 120 pc has been known since the time of the Hα photographic atlases of H II regions (*Sivan,* 1974). *Kun et al.* (1987) first identified the infrared emission of this structure in IRAS 60- and 100-μm sky flux maps. The Cepheus bubble includes the Cepheus OB2 association (Cep OB2), which is partly made up of the Tr 37 and NGC 7160 open clusters, and includes the H II region IC 1396. *Patel et al.* (1995, 1998) mapped ~100 deg^2 in Cepheus in the J = 1–0 transition of CO and ^{13}CO. Their observations reveal that the molecular clouds are undergoing an asymmetrical expansion away from the galactic plane. They propose a scenario in which the large-scale bubble was blown away by stellar winds and photoionization from the first generation of OB stars, which are no longer present (having exploded as supernovae). The ~10-m.y.-old (*Sicilia-Aguilar et al.,* 2004) NGC 7160 cluster and evolved stars such as μ Cephei, VV Cephei, and ν Cephei are the present-day companions of those first OB stars. *Patel et al.* (1998) show that the expanding shell becomes unstable at ~7 m.y. after the birth of the first OB stars. The estimated radius of the shell at that time (~30 pc) is consistent with the present radius of the ring of O- and B-type stars that constitute Cep OB2. Within a factor of ≤2, this age is also consistent with the estimated age for the Tr 37 cluster (~4 m.y.) (*Sicilia-Aguilar et al.,* 2004). Once the second generation of massive stars formed, they started affecting the dense gas in the remaining parent shell. The gas around these O stars expanded in rings like the one seen in IC 1396. The dynamical timescale for this expansion is on the order of 1–3 m.y., consistent with the very young ages (~1–2 m.y.) of the low-mass stars in the vicinity of IC 1396 (*Sicilia-Aguilar et al.,* 2004). This H II region is interpreted as the most recent generation of stars in Cep OB2.

In the Orion OB1 association *Blaauw* (1964) proposed that the ONC is the most recent event in a series of star-forming episodes within this association. The increasing ages between the ONC, Ori OB1b, and Ori OB1a have been suggested to be a case for sequential star formation (*Blaauw,* 1991). However, until now it has been difficult to investigate triggered star formation in Ori OB1 because of the lack of an unbiased census of the low-mass stars over the entire region. This situation is changing with the new large-scale surveys (e.g., *Briceño et al.,* 2005a) that are mapping the low-mass population of Ori OB1 over tens of square degrees; we may soon be able to test if Orion can also be interpreted as a case of induced, sequential star formation.

8. CONCLUDING REMARKS

Low-mass stars (0.1 ≤ M ≤ 1 M$_\odot$) in OB associations are essential for understanding many of the most fundamental problems in star formation, and important progress has been made during the past years by mapping and characterizing these objects. The newer large-scale surveys reveal that low-mass stars exist wherever high-mass stars are found, not only in the dense clusters, but also in a much more widely distributed population. As ever-increasing numbers of low-mass stars are identified over large areas in older regions like Orion OB1a, their spatial distribution shows substructure suggestive of a far more complex history than would be inferred from the massive stars.

The low-mass stellar populations in OB associations provide a snapshot of the IMF just after the completion of star formation, and before stars diffuse into the field population. The IMF derived from OB associations is consistent with the field IMF. The large majority of the low-mass PMS stars in the solar vicinity are in OB associations, therefore this agrees with early suggestions (*Miller and Scalo, 1978*) that the majority of stars in the galaxy were born in OB associations.

Since Protostars and Planets IV, the recent large surveys for low-mass members in several OB associations have allowed important progress on studies of early circumstellar disk evolution. With large scale surveys like 2MASS, large IR imagers, and now the Spitzer Space Telescope, we have unprecedented amounts of data sensitive to dusty disks in many regions. Overall disks largely dissipate over timescales of a few million years, either by dust evaporation, or settling and growth into larger bodies like planetesimals and planets; exactly which mechanisms participate in this evolution may depend on initial conditions and even on the environment. The actual picture seems more complex; current evidence supports a wide range of disk properties even at ages of ~1 m.y., and in some regions disks somehow manage to extend their lifetimes, surviving for up to ~10–20 m.y.

As the census of low-mass stars in nearby OB associations are extended in the coming years, our overall picture of star formation promises to grow even more complex and challenging.

Acknowledgments. We thank the referee, K. Luhman, for his thorough review and useful suggestions that helped us improve this manuscript. We also are grateful to A. G. A. Brown for helpful comments. C.B. acknowledges support from NASA Origins grant NGC-5 10545. E.M. is supported through a Clay Postdoctoral Fellowship from the Smithsonian Astrophysical Observatory. R.M. appreciates the support of the National Science Foundation. F.W. acknowledges support from NSF grant AST-030745 to Stony Brook University. T.P. and H.Z. are grateful to the Deutsche Forschungsgemeinschaft for travel support to attend the Protostars and Planets V conference.

REFERENCES

Allen L. E., Calvet N., D'Alessio P., Merin B., Hartmann L., et al. (2004) *Astrophys. J. Suppl., 154*, 363–366.

Ambartsumian V. A. (1947) In *Stellar Evolution and Astrophysics, Armenian Acad. of Sci.* (German translation, 1951, Abhandl, *Sowjetischen Astron., 1*, 33).

Ballesteros-Paredes J., Hartmann L., and Vázquez-Semadeni E. (1999) *Astrophys. J., 527*, 285–297.

Baraffe I., Chabrier G., Allard F., and Hauschildt P. H. (1998) *Astron. Astrophys., 337*, 403–412.

Baraffe I., Chabrier G., Allard F., and Hauschildt P. H. (2001) In *The Formation of Binary Stars* (H. Zinnecker and R. D. Mathieu, eds.), pp. 483–491. IAU Symposium 200, ASP, San Francisco.

Barrado y Navascués D., Stauffer J. R., Bouvier J., Jayawardhana R., and Cuillandre J-C. (2004) *Astrophys. J., 610*, 1064–1078.

Beckwith S. V. W., Sargent A. I., Chini R S., and Guesten R. (1990) *Astron. J., 99*, 924–945.

Bertelli G., Bressan A., Chiosi C., Fagotto F., and Nasi E. (1994) *Astron. Astrophys. Suppl., 106*, 275–302.

Blaauw A. (1964) *Ann. Rev. Astron. Astrophys., 2*, 213–246.

Blaauw A. (1991) In *The Physics of Star Formation and Early Stellar Evolution* (C. J. Lada and N. D. Kylafis, eds.), pp. 125–154. NATO Advanced Science Institutes Series C, Vol. 342, Kluwer, Dordrecht.

Boss A. P. (1995) *Astrophys. J., 439*, 224–236.

Brandl B., Brandner W., Eisenhauer F., Moffat A. F. J., Palla F., and Zinnecker H. (1999) *Astron. Astrophys., 352*, L69–L72.

Brandner W., Grebel E. K., Barbá R. H., Walborn N. R., and Moncti A. (2001) *Astron. J., 122*, 858–865.

Briceño C., Calvet N., Gómez M., Hartmann L., Kenyon S., and Whitney B. (1993) *Publ. Astron. Soc. Pac., 105*, 686–692.

Briceño C., Hartmann L. W., Stauffer J., Gagné M., Stern R., and Caillault J. (1997) *Astron. J., 113*, 740–752.

Briceño C., Hartmann L., Calvet N., and Kenyon S. (1999) *Astron. J., 118*, 1354–1368.

Briceño C., Vivas A. K., Calvet N., Hartmann L., et al. (2001) *Science, 291*, 93–96.

Briceño C., Calvet N., Hernández J., Vivas A. K., Hartmann L., Downes J. J., and Berlind P. (2005a) *Astron. J., 129*, 907–926.

Briceño C., Calvet N., Hernández J., Hartmann L., Muzerolle J., D'Alessio P., and Vivas A. K. (2005b) In *Star Formation in the Era of Three Great Observatories*, available on line at http://cxc.harvard.edu/stars05/agenda/pdfs/bricenoc.pdf.

Brown A. G. A. (1996) *Publ. Astron. Soc. Pac., 108*, 459–459.

Brown A. G. A., de Geus E. J., and de Zeeuw P. T. (1994) *Astron. Astrophys., 289*, 101–120.

Brown A. G. A., Dekker G., and de Zeeuw P. T. (1997) *Mon. Not. R. Astron. Soc., 285*, 479–492.

Brown A. G. A., Blaauw A., Hoogerwerf R., de Bruijne J. H. J., and de Zeeuw P. T. (1999) In *The Origin of Stars and Planetary Systems* (C. J. Lada and N. D. Kylafis, eds.), pp. 411–440. Kluwer, Dordrecht.

Burningham B., Naylor T., Littlefair S. P., and Jeffries R. D. (2005) *Mon. Not. R. Astron. Soc., 363*, 1389–1397.

Calvet N. and Gullbring E. (1998) *Astrophys. J., 509*, 802–818.

Calvet N., D'Alessio P., Hartmann L., Wilner D., Walsh A., and Sitko M. (2002) *Astrophys. J., 568*, 1008–1016.

Calvet N., Briceño C., Hernández J., Hoyer S., Hartmann L., et al. (2005a) *Astron. J., 129*, 935–946.

Calvet N., D'Alessio P., Watson D. M., Franco-Hernández R., Furlan E., et al. (2005b) *Astrophys. J., 630*, L185–L188.

Carpenter J. M., Heyer M. H., and Snell R. L. (2000) *Astrophys. J. Suppl., 130*, 381–402.

Carpenter J. M., Hillenbrand L. A., and Strutskie M. F. R. (2001) *Astron. J., 121*, 3160–3190.

Chen C. H., Jura M., Gordon K. D., and Blaylock M. (2005) *Astrophys. J., 623*, 493–501.

Clark P. C., Bonnell I. A., Zinnecker H., and Bate M. R. (2005) *Mon. Not. R. Astron. Soc., 359*, 809–818.

Clarke C. J., Gendrin A., and Sotomayor M. (2001) *Mon. Not. R. Astron. Soc., 328*, 485–491.

D'Alessio P., Hartmann L., Calvet N., Franco-Hernández R., Forrest W., et al. (2005) *Astrophys. J., 621*, 461–472.

D'Antona F. and Mazzitelli I. (1994) *Astrophys. J. Suppl., 90*, 467–500.

de Bruijne J. H. J. (1999) *Mon. Not. R. Astron. Soc., 310*, 585–617.

de Geus E. J. (1992) *Astron. Astrophys., 262*, 258–270.

de Geus E. J., de Zeeuw P. T., and Lub J. (1989) *Astron. Astrophys., 216*, 44–61.

de Zeeuw P. T. and Brand J. (1985) In *Birth and Evolution of Massive Stars and Stellar Groups* (H. van denWoerden and W. Boland, eds.), pp. 95–101. Reidel, Dordrecht.

de Zeeuw P. T., Hoogerwerf R., de Bruijne J. H. J., Brown A. G. A., and Blaauw A. (1999) *Astron. J., 117,* 354–399.

Dolan Ch. J. and Mathieu R. D. (1999) *Astron. J., 118,* 2409–2423.

Dolan Ch. J. and Mathieu R. D. (2001) *Astron. J., 121,* 2124–2147.

Efremov Y. and Elmegreen B. G. (1998) *Mon. Not. R. Astron. Soc., 299,* 643–652.

Elmegreen B. G. (1990) In *The Evolution of the Interstellar Medium* (L. Blitz, ed.), pp. 247–271. ASP Conf. Series 12, San Francisco.

Elmegreen B. G. and Lada C. J. (1977) *Astrophys. J., 214,* 725–741.

Elmegreen B. G., Efremov Y., Pudritz R. E., and Zinnecker H. (2000) In *Protostars and Planets IV* (V. Mannings et al., eds.), pp. 179–202. Univ. of Arizona, Tucson.

Favata F. and Micela G. (2003) *Space Sci. Rev., 108,* 577–708.

Feigelson E. D. and DeCampli W. M. (1981) *Astrophys. J., 243,* L89–L93.

Feigelson E. D. and Montmerle T. (1999) *Ann. Rev. Astron. Astrophys., 37,* 363–408.

Foster P. N. and Boss A. P. (1996) *Astrophys. J., 468,* 784–796.

Fukuda N. and Hanawa T. (2000) *Astrophys. J., 533,* 911–923.

Getman K. V., Flaccomio E., Broos P. S., Grosso N., Tsujimoto M., et al. (2005) *Astrophys. J. Suppl., 160,* 319–352.

Gómez M. and Kenyon S. J. (2001) *Astron. J., 121,* 974–983.

Gómez M. and Lada C. (1998) *Astron. J., 115,* 1524–1535.

Gullbring E., Hartmann L., Briceño C., and Calvet N. (1998) *Astrophys. J., 492,* 323–341.

Haisch K. E. Jr., Lada E. A., and Lada C. J. (2000) *Astron. J., 120,* 1396–1409.

Haisch K. E. Jr., Lada E. A., and Lada C. J. (2001) *Astrophys. J., 553,* L153–L156.

Hartmann L. (1998) *Accretion Processes in Star Formation,* Cambridge Univ., Cambridge.

Hartmann L. (2001) *Astron. J., 121,* 1030–1039.

Hartmann L., Stauffer J. R., Kenyon S. J., and Jones B. F. (1991) *Astron. J., 101,* 1050–1062.

Hartmann L., Ballesteros-Paredes J., and Bergin E. A. (2001) *Astrophys. J., 562,* 852–868.

Hartmann L., Calvet N., Watson D. M., D'Alessio P., Furlan E., et al. (2005a) *Astrophys. J., 628,* L147–L150.

Hartmann L., Megeath S. T., Allen L., Luhman K., Calvet N., et al. (2005b) *Astrophys. J., 629,* 881–896.

Herbig G. H. (1962) *Adv. Astron. Astrophys., 1,* 47–103.

Herbig G. H. (1978) In *Problems of Physics and Evolution of the Universe* (L. V. Mirzoyan, ed.), pp. 171–188. Publ. Armenian Acad. Sci., Yerevan.

Herbig G. H. and Bell K. R. (1988) *Lick Observatory Bulletin,* Lick Observatory, Santa Cruz.

Herbst W., Herbst D. K., Grossman E. J., and Weinstein D. (1994) *Astron. J., 108,* 1906–1923.

Hillenbrand L. A. (1997) *Astron. J., 113,* 1733–1768.

Hillenbrand L. A., Strom S. E., Calvet N., Merrill K. M., Gatley I., et al. (1998) *Astron. J., 116,* 1816–1841.

Hoogerwerf R. (2000) *Mon. Not. R. Astron. Soc., 313,* 43–65.

Johns-Krull C. M. and Valenti J. A. (2001) *Astrophys. J., 561,* 1060–1073.

Joy A. H. (1945) *Astrophys. J., 102,* 168–200.

Kenyon S. J. and Hartmann L. W. (1995) *Astrophys. J. Suppl., 101,* 117–171.

Kenyon M. J., Jeffries R. D., Naylor T., Oliveira J. M., and Maxted P. F. L. (2005) *Mon. Not. R. Astron. Soc., 356,* 89–106.

Kholopov P. N. (1959) *Sov. Astron., 3,* 425–433.

Königl A. (1991) *Astrophys. J., 370,* L39–L43.

Kroupa P. (2002) *Science, 295,* 82–91.

Kroupa P., Aarseth S., and Hurley J. (2001) *Mon. Not. R. Astron. Soc., 321,* 699–712.

Kun M., Balazs L. G., and Toth I. (1987) *Astrophys. Space Sci., 134,* 211–217.

Lada C. J. and Lada E. A. (2003) *Ann. Rev. Astron. Astrophys., 41,* 57–115.

Lada C. J., Muench A. A., Haisch K. E. Jr., Lada E. A., Alves J. F., et al. (2000) *Astron. J., 120,* 3162–3176.

Lada C. J., Muench A. A., Lada E. A., and Alves J. F. (2004) *Astron. J., 128,* 1254–1264.

Larson R. (1985) *Mon. Not. R. Astron. Soc., 214,* 379–398.

Lee H-T., Chen W. P., Zhang Z-W., and Hu J-Y. (2005) *Astrophys. J., 624,* 808–820.

Leitherer C. (1998) In *The Stellar Initial Mass Function (38th Herstmonceux Conference)* (G. Gilmore and D. Howell, eds.), pp. 61–88. ASP Conf. Series 142, San Francisco.

Liu C. P., Zhang C. S., and Kimura H. (1981) *Chinese Astron. Astrophys., 5,* 276–281.

Luhman K. L. (1999) *Astrophys. J., 525,* 466–481.

Luhman K. L., Rieke G. H., Lada C. J., and Lada E. A. (1998) *Astrophys. J., 508,* 347–369.

Maíz-Apellániz J., Pérez E., and Mas-Hesse J. M. (2004) *Astron. J., 128,* 1196–1218.

Mamajek E. E., Meyer M. R., and Liebert J. W. (2002) *Bull. Am. Astron. Soc., 34,* 762–762.

Martín E. L., Rebolo R., and Zapatero-Osorio M. R. (1996) *Astrophys. J., 469,* 706–714.

McGehee P. M., West A. A., Smith J. A., Anderson K. S. J., and Brinkmann J. (2005) *Astron. J., 130,* 1752–1762.

Meyer M., Calvet N., and Hillenbrand L. A. (1997) *Astron. J., 114,* 288–300.

Mikami T. and Ogura K. (2001) *Astrophys. Space Sci., 275,* 441–462.

Miller G. E. and Scalo J. M. (1978) *Publ. Astron. Soc. Pac., 90,* 506–513.

Moneti A., Zinnecker H., Brandner W., and Wilking B. (1999) In *Astrophysics with Infrared Surveys: A Prelude to SIRTF* (M. D. Bicay et al., eds.), pp. 355–358. ASP Conf. Series 177, San Francisco.

Motte F., André P., and Neri R. (1998) *Astron. Astrophys., 336,* 150–172.

Muzerolle J., Calvet N., and Hartmann L. (1998a) *Astrophys. J., 492,* 743–753.

Muzerolle J., Hartmann L., and Calvet N. (1998b) *Astron. J., 116,* 455–468.

Muzerolle J., Calvet N., and Hartmann L. (2001) *Astrophys. J., 550,* 944–961.

Muzerolle J., Young E., Megeath S. T., and Allen L. (2005) In *Star Formation in the Era of Three Great Observatories,* available on line at http://cxc.harvard.edu/stars05/agenda/pdfs/muzerolle_disks_splinter.pdf.

Neuhäuser R. (1997) *Science, 267,* 1363–1370.

Oey M. S. and Massey P. (1995) *Astrophys. J., 542,* 210–225.

Oey M. S., Watson A. M., Kern K., and Walth G. L. (2005) *Astron. J., 129,* 393–401.

Ogura K. (1984) *Publ. Astron. Soc. Japan, 36,* 139–148.

Palla F. and Stahler S. W. (1992) *Astrophys. J., 392,* 667–677.

Palla F. and Stahler S. W. (1993) *Astrophys. J., 418,* 414–425.

Palla F. and Stahler S. W. (1999) *Astrophys. J., 525,* 772–783.

Palla F. and Stahler S. W. (2000) *Astrophys. J., 540,* 255–270.

Patel N. A., Goldsmith P. F., Snell R. L., Hezel T., and Xie T. (1995) *Astrophys. J., 447,* 721–741.

Patel N. A., Goldsmith P. F., Heyer M. H., Snell R. L., and Pratap P. (1998) *Astrophys. J., 507,* 241–253.

Preibisch Th. and Zinnecker H. (1999) *Astron. J., 117,* 2381–2397.

Preibisch Th., Guenther E., Zinnecker H., Sterzik M., Frink S., and Röser S. (1998) *Astron. Astrophys., 333,* 619–628.

Preibisch Th., Brown A. G. A., Bridges T., Guenther E., and Zinnecker H. (2002) *Astron. J., 124,* 404–416.

Preibisch Th., Kim Y-C., Favata F., Feigelson E. D., Flaccomio E., et al. (2005) *Astrophys. J. Suppl., 160,* 401–422.

Podosek F. A. and Cassen P. (1994) *Meteoritics, 29,* 6–25.

Rebull L. M., Hillenbrand L. A., Strom S. E., Duncan D. K., Patten B. M., et al. (2000) *Astron. J., 119,* 3026–3043.

Sanduleak N. (1971) *Publ. Astron. Soc. Pac., 83,* 95–97.

Scalo J. M. (1998) In *The Stellar Initial Mass Function* (G. Gilmore and D. Howel, eds.), pp. 201–236. ASP Conf. Series 142, San Francisco.

Scoville N. Z. and Hersh K. (1979) *Astrophys. J., 229,* 578–582.

Scoville N. Z., Solomon P. D., and Sanders D. B. (1979) In *The Large-Scale Characteristics of the Galaxy* (W. B. Burton, ed.), pp. 277–283. Reidel, Dordrecht.

Selman F. J. and Melnick J. (2005) *Astron. Astrophys., 443,* 851–861.

Sherry W. H. (2003) Ph.D. thesis, State Univ. of New York, Stony Brook.

Sherry W. H., Walter F. M., and Wolk S. J. (2004) *Astron. J., 128,* 2316–2330.

Shu F. H. and Lizano S. (1988) *Astrophys. Lett. Comm., 26,* 217–226.

Shu F. H., Adams F. C., and Lizano S. (1987) *Ann. Rev. Astron. Astrophys., 25,* 23–81.

Sicilia-Aguilar A., Hartmann L. W., Briceño C., Muzerolle J., and Calvet N. (2004) *Astron. J., 128,* 805–821.

Sicilia-Aguilar A., Hartmann L. W., Hernández J., Briceño C., and Calvet N. (2005) *Astrophys. J., 130,* 188–209.

Sicilia-Aguilar A., Hartmann L. W., Calvet N., Megeath S. T., Muzerolle J., et al. (2006) *Astrophys. J., 638,* 897–919.

Siess L., Dufour E., and Forestini M. (2000) *Astron. Astrophys., 358,* 593–599.

Sivan J. P. (1974) *Astron. Astrophys. Suppl. Ser., 16,* 163–172.

Slawson R. W. and Landstreet J. D. (1992) *Bull. Am. Astron. Soc., 24,* 824–824.

Slesnick C. L., Carpenter J. M., and Hillenbrand L. A. (2005) In *PPV Poster Proceedings,* available on line at http://www.lpi.usra.edu/meetings/ppv2005/pdf/8365.pdf.

Smith L. J. and Gallagher J. S. (2001) *Mon. Not. R. Astron. Soc., 326,* 1027–1040.

Soderblom D. R. (1996) In *Cool Stars, Stellar Systems and the Sun IX* (R. Pallavicini and A. K. Dupree, eds.), pp. 315–324. ASP Conf. Series 109, San Francisco.

Soderblom D. R., Jones B. F., Balachandran S., Stauffer J. R., Duncan D. K., et al. (1993) *Astron. J., 106,* 1059–1079.

Stolte A., Brandner W., Grebel E. K., Lenzen R., and Lagrange A. M. (2005) *Astrophys. J., 628,* L113–L117.

Strom S. E., Strom K. M., and Grasdalen G. L. (1975) *Ann. Rev. Astron. Astrophys., 13,* 187–216.

Strom S. E., Edwards S., and Skrutskie M. F. (1993) In *Protostars and Planets III* (E. H. Levy and J. I. Lunine, eds.), pp. 837–866. Univ. of Arizona, Tucson.

Uchida K. I., Calvet N., Hartmann L., Kemper F., Forrest W. J., et al. (2004) *Astrophys. J. Suppl., 154,* 439–442.

Vanhala H. A. T. and Cameron A. G. W. (1998) *Astrophys. J., 508,* 291–307.

Walborn N. R., Barbá R. H., Brandner W., Rubio M., Grebel E. K., and Probst R. G. (1999) *Astron. J., 117,* 225–237.

Walter F. M. and Kuhi L. V. (1981) *Astrophys. J., 250,* 254–261.

Walter F. M. and Myers P. C. (1986) In *Fourth Cambridge Workshop on Cool Stars, Stellar Systems, and the Sun* (M. Zeilik and D. M. Gibson, eds.), pp. 55–57. Lecture Notes in Physics Vol. 254, Springer-Verlag, Berlin.

Walter F. M., Brown A., Mathieu R. D., Myers P. C., and Vrba F. J. (1988) *Astron. J., 96,* 297–325.

Weaver W. B. and Babcock A. (2004) *Publ. Astron. Soc. Pac., 116,* 1035–1038.

White R. J. and Basri G. (2003) *Astrophys. J., 582,* 1109–1122.

White R. J. and Hillenbrand L. A. (2005) *Astrophys. J., 621,* L65–L68.

Wilking B. A., Schwartz R. D., and Blackwell J. H. (1987) *Astron. J., 94,* 106–110.

Wiramihardja S. D., Kogure T., Yoshida S., Ogura K., and Nakano M. (1989) *Publ. Astron. Soc. Japan, 41,* 155–174.

Wiramihardja S. D., Kogure T., Yoshida S., Ogura K., and Nakano M. (1993) *Publ. Astron. Soc. Japan, 45,* 643–653.

Yamaguchi R., Mizuno N., Onishi T., Mizuno A., and Fukui Y. (2001) *Publ. Astron. Soc. Japan, 53,* 959–969.

Zinnecker H., McCaughrean M. J., and Wilking B. A. (1993) In *Protostars and Planets III* (E. H. Levy and J. I. Lunine, eds.), pp. 429–495. Univ. of Arizona, Tucson.

The Structure and Evolution of Young Stellar Clusters

Lori Allen, S. Thomas Megeath, Robert Gutermuth,
Philip C. Myers, and Scott Wolk
Harvard-Smithsonian Center for Astrophysics

Fred C. Adams
University of Michigan

James Muzerolle and Erick Young
University of Arizona

Judith L. Pipher
University of Rochester

We examine the properties of embedded clusters within 1 kpc using new data from the Spitzer Space Telescope, as well as recent results from 2MASS and other groundbased near-infrared surveys. We use surveys of entire molecular clouds to understand the range and distribution of cluster membership, size, and surface density. The Spitzer data demonstrate clearly that there is a continuum of star-forming environments, from relative isolation to dense clusters. The number of members of a cluster is correlated with the cluster radius, such that the average surface density of clusters having a few to a thousand members varies by a factor of only a few. The spatial distributions of Spitzer-identified young stellar objects frequently show elongation, low density halos, and subclustering. The spatial distributions of protostars resemble the distribution of dense molecular gas, suggesting that their morphologies result directly from the fragmentation of the natal gas. We also examine the effects of the cluster environments on star and planet formation. Although far-UV (FUV) and extreme-UV (EUV) radiation from massive stars can truncate disks in a few million years, fewer than half the young stars in our sample (embedded clusters within 1 kpc) are found in regions of strong FUV and EUV fields. Typical volume densities and lifetimes of the observed clusters suggest that dynamical interactions are not an important mechanism for truncating disks on solar system size scales.

1. INTRODUCTION

Since the publication of *Protostars and Planets IV* (PPIV), there have been significant advances in observations of young stellar clusters from X-ray to millimeter wavelengths. But while much of the recent work has concentrated on the stellar initial mass function (IMF) or protoplanetary disk evolution (e.g., *Lada and Lada,* 2003), less attention has been directed to discerning the structure of young embedded clusters, and the evolution of that structure during the first few million years. Physical properties of young embedded clusters, such as their shapes, sizes, and densities, should inform theories of cluster formation. In this contribution, we describe recent results in which these properties are obtained for a representative sample of young (1–3 m.y.), nearby (d ≤ 1 kpc), embedded clusters.

This contribution is motivated by three recent surveys made with the Spitzer Space Telescope: the Spitzer Young Stellar Cluster Survey, which includes Spitzer, near-IR, and millimeter-wave images of 30 clusters; the Spitzer Orion Molecular Cloud Survey, which covers 6.8 square degrees

in Orion; and the Cores to Disks (c2d) Legacy program, which surveyed several nearby molecular clouds (*Evans et al.,* 2003). These surveys provide a comprehensive census of nearly all the known embedded clusters in the nearest kiloparsec, ranging from small groups of several stars to rich clusters with several hundred stars. A new archival survey from Chandra (ANCHORS) is providing X-ray data for many of the nearby clusters. Since PPIV, the Two Micron All Sky Survey (2MASS) has become widely used as an effective tool for mapping large regions of star formation, particularly in the nearby molecular clouds. This combination of X-ray, near-IR, and mid-IR data is a powerful means for studying embedded populations of pre-main-sequence stars and protostars.

Any study of embedded clusters requires some method of identifying cluster members, and we begin by briefly reviewing methods which have progressed rapidly since PPIV, including work from X-ray to submillimeter wavelengths, but with an emphasis on the mid-infrared spectrum covered by Spitzer. Beyond section 2 we focus almost entirely on recent results from Spitzer, rather than a review of the literature. In section 3, we discuss the cluster properties derived from large-scale surveys of young embedded clusters in nearby molecular clouds, including their sizes, spatial distributions, surface densities, and morphologies. In section 4 we consider the evolution of young embedded clusters as the surrounding molecular gas begins to disperse. In section 5 we discuss theories of embedded cluster evolution, and in section 6 consider the impact of the cluster environment on star and planet formation. Our conclusions are presented in section 7.

2. METHODS OF IDENTIFYING YOUNG STARS IN CLUSTERS

2.1. Near- and Mid-Infrared

Young stellar objects (YSOs) can be identified and classified on the basis of their mid-infrared properties (*Adams et al.*, 1987; *Wilking et al.*, 1989; *Myers and Ladd*, 1993). Here we review recent work on cluster identification and characterization based primarily on data from the Spitzer Space Telescope.

Megeath et al. (2004) and *Allen et al.* (2004) developed YSO classification schemes based on color-color diagrams from observations taken with the Infrared Array Camera (IRAC) on Spitzer. Examining models of protostellar envelopes and circumstellar disks with a wide range of plausible parameters, they found that the two types of objects should occupy relatively distinct regions of the diagram. Almost all the Class I (star + disk + envelope) models exhibited the reddest colors, not surprisingly, with the envelope density and central source luminosity having the most significant effect on the range of colors. The Class II (star + disk) models included a treatment of the inner disk wall at the dust sublimation radius, which is a significant contributor to the flux in the IRAC bands. Models of the two classes generally occupy distinct regions in color space, indicating that they can be identified fairly accurately from IRAC data even in the absence of other information such as spectra.

Comparison of these loci with YSOs of known types in the Taurus star-forming region shows reasonably good agreement (*Hartmann et al.*, 2005). Some degeneracy in the IRAC color space does exist; Class I sources with low envelope column densities, low mass infall rates, or certain orientations may have the colors of Class II objects. The most significant source of degeneracy is from extreme reddening due to high extinction, which can cause Class II objects to appear as low-luminosity Class I objects when considering wavelengths $\lambda \leq 10$ μm.

The addition of data from the 24-μm channel of the Multiband Imaging Photometer for Spitzer (MIPS) provides a longer wavelength baseline for classification, particularly useful for resolving reddening degeneracy between Class I and II. It is also crucial for robust identification of evolved disks, both "transition" and "debris," which lack excess emission at shorter wavelengths due to the absence of dust close to the star. Such 24-μm observations are limited, how-

ever, by lower sensitivity and spatial resolution compared to IRAC, as well as the generally higher background emission seen in most embedded regions. *Muzerolle et al.* (2004) delineated Class I and II loci in an IRAC/MIPS color-color diagram of one young cluster based on the 3.6–24-μm spectral slope.

The choice of classification method depends partly on the available data; not all sources are detected (or observed) in the 2MASS, IRAC, and MIPS bands. IRAC itself is significantly more sensitive at 3.5 and 4.5 μm than at 5.8 and 8 μm, so many sources may have IRAC detections in only the two shorter wavelengths, and require a detection in one or more near-IR band to classify young stars (*Gutermuth et al.*, 2004; *Megeath et al.*, 2005; *Allen et al.*, 2005). Gutermuth et al. (in preparation) refined the IRAC + near-IR approach by correcting for the effects of extinction, estimated from the H-K color, and developed new classification criteria based on the extinction-corrected colors.

It is useful to compare some of the different classification schemes. In Fig. 1 we plot first a comparison of the IRAC model colors from *Allen et al.* (2004), *Hartmann et al.* (2005), and *Whitney et al.* (2003). In general, the models predict a similar range of IRAC colors for both Class I and Class II sources. Also in Fig. 1 we plot the same sample of IRAC data (NGC 2068/71) from Muzerolle et al. (in preparation) in three color-color planes that correspond to the classification methods discussed above. In all diagrams, only those sources with detections in the three 2MASS bands, the four IRAC bands, and the MIPS 24-μm band were included. For the sake of comparison with pre-Spitzer work, the points are coded according to their K–24-μm SED slope. Prior to Spitzer, a commonly used four-class system was determined by the 2–10-μm (or 2–20-μm) slope (α), in which $\alpha > 0.3$ = Class I, $-0.3 \leq \alpha < 0.3$ = "flat" spectrum, $-1.6 \leq \alpha < -0.3$ = Class II, and $\alpha < -1.6$ = Class III (photosphere) (*Greene et al.*, 1994). A few of the sources in Fig. 1 have been observed spectroscopically and determined to be T Tauri stars, background giants, or dwarfs unassociated with the cluster. These are indicated. The diagrams also show the adopted regions of color space used to roughly distinguish between Class I and Class II objects.

Classifications made with these methods are in general agreement with each other, although some differences are apparent. For example, roughly 30% of Class I objects identified with the *Allen et al.* (2004) method and detected at 24 μm appear as Class II objects in the IRAC/MIPS-24 color space; however, many of these are borderline "flat spectrum" sources where the separation between Class I and II is somewhat arbitrary and may not be physically meaningful.

These classification methods implicitly assume that all objects that exhibit infrared excess are YSOs. However, there can be contamination from other sources, including evolved stars, AGN, quasars, and high-redshift dusty galaxies. Since most of these unrelated objects are faint high-redshift AGN (*Stern et al.*, 2005), we have found that a magnitude cut of $m_{3.6} < 14$ will remove all but approximately 10 non-YSOs per square degree within each of the IRAC-

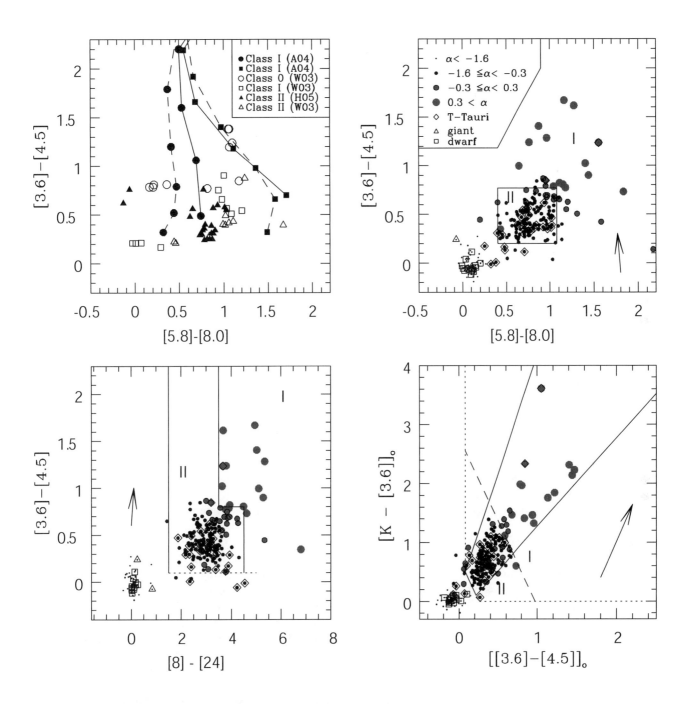

Fig. 1. Identifying and classifying young stars using near- and mid-infrared measurements. In the panel at top left, a comparison of predicted IRAC colors from *Allen et al.* (2004) (A04), *Hartmann et al.* (2005) (H05), and *Whitney et al.* (2003) (W03). Triangles represent Class II models with T_{eff} = 4000 K and a range of accretion rates, grain size distributions, and inclinations. Squares and circles are Class I/0 models for a range of envelope density, centrifugal radius, and central source luminosity. In the remaining panels, we plot the data for the embedded cluster NGC 2068/71 (Muzerolle et al., in preparation). Point types are coded according to the measured SED slope between 2 and 24 μm. Spectroscopically confirmed T Tauri, giant, and dwarf stars are indicated. In the top right panel, the large rectangle marks the adopted domain of Class II sources; the Class I domain is above and to the right (adapted from *Allen et al.*, 2004). In the bottom right panel (Gutermuth et al., in preparation), dereddened colors are separated into Class I and II domains by the dashed line. Diagonal lines outline the region where most of the classifiable sources are found. In the bottom left panel, the approximate domains of Class I and II sources are indicated by the solid lines. The dotted line represents the adopted threshold for excess emission at 3.6 and 4.5 μm; sources below this that exhibit large [8]–[24] excess are probably disks with large optically thin or evacuated holes (adapted from *Muzerolle et al.*, 2004). Arrows show extinction vectors for A_V = 30 (Flaherty et al., in preparation). These figures show that the various color planes considered here yield similar results when used to classify Spitzer sources.

only Class I and Class II loci, and all but a few non-YSOs per square degree from the IRAC/MIPS-24 loci, while retaining most if not all of the cluster population.

2.2. Submillimeter and Millimeter

The youngest sources in star-forming regions are characterized by strong emission in the submillimeter and far-infrared, but ususally weak emission shortward of 24 μm. These "Class 0" objects were first discovered in submillimeter surveys of molecular clouds (*André et al.,* 1993). They are defined as protostars with half or more of their mass still in their envelopes, and emitting at least 0.5% of their luminosity at submillimeter wavelengths. Motivated in part by the discovery of Class 0 objects, observers have imaged many embedded clusters in their dust continuum emission at millimeter and submillimeter wavelengths, revealing complex filamentary structure and many previously unknown sources (e.g., *Motte et al.,* 1998, 2001; *Sandell and Knee,* 2001; *Nutter et al.,* 2005).

These submillimeter and millimeter wavelength images generally have tens to hundreds of local maxima, but only a small fraction of these are "protostars" having an internal heating source; the rest are "starless cores" having a maximum of column density but no internal heating source. The standard way to determine whether a submillimeter source is a protostar or a starless core is to search for coincidence with an infrared point source, such as a Spitzer source at 24 or 70 μm, or a radio continuum point source, such as a VLA source at 6 cm wavelength. For example, the protostars NGC 1333-IRAS 4A, 4B, and 4C in Fig. 2 are each detected at 850 μm (*Sandell and Knee,* 2001), and each has a counterpart in VLA observations (*Rodriguez et al.,* 1999) and in 24-μm Spitzer observations, but not in the IRAC bands. In a few cases, Class 0 protostars such as VLA 1623 have been identified from their submillimeter emission and their radio continuum, but not from their mid-infrared emission, because their mid-infrared emission is too heavily extinguished (*André et al.,* 2000).

2.3. X-Ray

Elevated X-ray emission is another signature of youth: Young stellar objects have typical X-ray luminosity 1000× that of the Sun. The e-folding decay time for this X-ray luminosity is a few hundred million years (see, e.g., *Micela et al.,* 1985; *Walter and Barry,* 1991; *Dorren et al.,* 1995; *Feigelson and Montmerle,* 1999). Although the X-ray data of young stellar clusters will be contaminated by AGN and other sources, this contamination can be reduced by identifying optical/infrared counterparts to the X-ray sources. X-ray sources where the ratio of the X-ray luminosity to the bolometric luminosity (L_X/L_{bol}) ranges from 0.1% to 0.01% are likely pre-main-sequence stars. In contrast to the infrared techniques described in section 2.1, which can only identify Class 0/I and II sources; X-ray observations can readily detect Class II and Class III objects, with perhaps some bias toward Class III objects (*Flacomio et al.,* 2003).

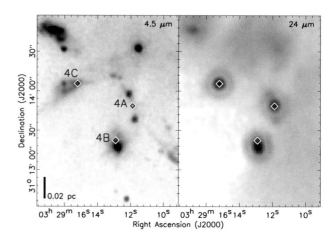

Fig. 2. IRAC 4.5-μm and MIPS 24-μm images of IRAS-4 in NGC 1333. MIPS detects each of the three VLA sources, while IRAC detects their outflows but not the driving sources.

The main limitation of X-ray observations is the lack of sensitivity toward lower mass stars. A complete sample of stars requires a sensitivity toward sources with luminosities as low as 1027 erg cm^{-2} s^{-1} (*Feigelson et al.,* 2005); the sensitivity of most existing observations are an order of magnitude higher. The observed X-ray luminosity is also affected by extinction. Depending on the energy of the source, the sensitivity can be reduced by a factor of ten for sources at $A_V \sim 10$ (*Wolk et al.,* 2006).

2.4. Emission Lines and Variability

Among other techniques for identifying young cluster members, spectroscopic surveys for emission lines and photometric surveys for variability have been used successfully at visible and near-IR wavelengths. The most common means of identifying young stars spectroscopically is through detection of optical emission lines, in particular Hα at 6563 Å (*Herbig and Bell,* 1988). Large-scale objective prism (*Wilking et al.,* 1987; *Wiramihardja et al.,* 1989) and later, wide-field multiobject spectroscopy (e.g., *Hillenbrand et al.,* 1993) have been effective in identifying young stars in clusters and throughout molecular clouds; however, they miss the deeply embedded members that are optically faint or invisible. This problem is partly alleviated by large-scale surveys for photometric variability in the optical and near-IR. Recent near-IR surveys by *Kaas* (1999) and *Carpenter et al.* (2001, 2002) have been successful at identifying young cluster members in Serpens, Orion, and Chamaeleon, respectively.

2.5. Star Counts

Much of the work on the density, size, and structure of embedded clusters has relied on using star counts; indeed, the distribution of 2.2-μm sources were used to identify clusters in the Orion B cloud in the seminal work of *Lada et al.* (1991). Instead of identifying individual stars as mem-

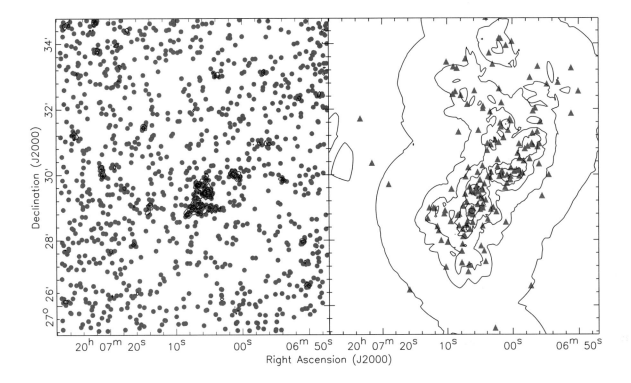

Fig. 3. IRAS 20050 surface densities derived from the statistical technique applied to all stars (left), and from identifying the infrared-excess sources (right). In the left panel, all sources having K < 16 are plotted as a function of their position. Contours show the surface density of K-band sources, starting at 1450 pc^{-2} (5σ above median field star density) and increasing at intervals of 750 pc^{-2}. In the right panel, sources with infrared excess emission are plotted, and contours of their surface density are shown for 10, 60, 160, 360, 760, and 1560 pc^{-2}. The statistical technique (left) yields a higher peak surface density (~6000 pc^{-2} at the center) than the IR-excess technique (~3000 pc^{-2}), but the latter is more sensitive to the spatially extended population of young stars.

bers, methods based on star counts include all detected sources and employ a statistical approach toward membership, in which an average density of background stars is typically estimated and subtracted out. In this analysis, the star counts are typically smoothed to produce surface density maps; a variety of smoothing algorithms are in the literature (*Gomez et al.*, 1993; *Gladwin et al.*, 1999; *Carpenter*, 2000; *Gutermuth et al.*, 2005; *Cambresy et al.*, 2006).

The degree of contamination by foreground and background stars is the most significant limitation for star-count methods, and the efficacy of using star counts depends strongly on the surface density of contaminating stars. In many cases, the contamination can be minimized by setting a K-band brightness limit (*Lada et al.*, 1991; *Gutermuth*, 2005). To estimate the position-dependent contamination by field stars, models or measurements of the field star density can be combined with extinction maps of the molecular cloud (*Carpenter*, 2000; *Gutermuth et al.*, 2005; *Cambresy*, 2006). These maps are subtracted from the surface density of observed sources to produce maps of the distribution of embedded stars; however, these maps are still limited by the remaining Poisson noise from the subtracted stars.

Star-count methods have the advantage that they do not discriminate against sources without infrared excess, bright X-ray emission, variability, or some other indication of youth. On the other hand, they only work in regions where the surface density of member stars is higher than the statistical noise from contaminating field stars. In Fig. 3 we show maps of the IRAS 20050 cluster derived from the K-band star counts and from the distribution of infrared excess sources. In the case of IRAS 20050, we find that the star-count method provides a better map of the densest regions (due in part to confusion with bright nebulosity and sources in the Spitzer data), while the lower-density regions surrounding these peaks are seen only in the distribution of Spitzer-identified infrared excess sources (due to the high density of background stars).

3. THE STRUCTURE AND EVOLUTION OF CLUSTERS: OBSERVATIONS

3.1. Identifying Clusters in Large-Scale Surveys of Molecular Clouds

Unlike gravitationally bound open clusters or globular clusters, embedded clusters are not isolated objects. In most cases, molecular cloud complexes contain multiple embedded clusters as well as distributed populations of relatively isolated stars. Recent large-scale surveys and all-sky catalogs are now providing new opportunities to study the properties of embedded clusters through surveys of entire molecular clouds. The advantage of studying clusters by surveying entire molecular clouds is twofold. First, the surveys provide an unbiased sample of both the distributed and clustered populations within a molecular cloud. Second, the surveys result in an unbiased measurement of the distribution of cluster properties within a single cloud or ensemble of clouds. For the remainder of this discussion, we will use the word "cluster" to denote embedded clusters of young stars. Most of these clusters will not form bound open clusters (*Lada and Lada*, 1995).

We now concentrate on two recent surveys for young stars in relatively nearby (<1 kpc) molecular clouds. *Carpenter* (2000) used the 2MASS 2nd incremental point source catalog to study the distribution of young stars in the Orion A, Orion B, Perseus, and Monoceros R2 clouds. Since the 2nd incremental release did not cover the entire sky, only parts of the Orion B and Perseus clouds were studied. More recently, Spitzer has surveyed a number of molecular clouds. We discuss here new results from the Spitzer Orion Molecular Cloud Survey (Megeath et al., in preparation) and the Cores to Disks (c2d) Legacy program survey of the Ophiuchus Cloud (Allen et al., in preparation). We use these data to study the distribution of the number of cluster members, the cluster radius, and the stellar density in this small sample of clouds.

The advantage of using these two surveys is that they draw from different techniques to identify populations of young stellar objects. The analysis of the 2MASS data relies on star-counting methods (section 2.5), while the Spitzer analysis relies on identifying young stars with infrared-excesses from combined Spitzer and 2MASS photometry (section 2.1) (Megeath et al., in preparation). The 2MASS analysis is limited by the systematic and random noise from the background star subtraction, making the identification of small groups and distributed stars subject to large uncertainties. The Spitzer analysis is limited to young stars with disks or envelopes. A significant number of young stars in embedded clusters do not show excesses; this fraction may range from 20% to as much as 50% for 1–3-m.y. clusters (*Haisch et al.*, 2001).

Carpenter (2000) identified stellar density peaks more than six times the RMS background noise, and defined a cluster as all stars in a closed 2σ contour surrounding these peaks. Megeath et al. (in preparation) defined clusters as groups of 10 or more IR-excess sources in which each member is within a projected distance of 0.32 pc of another member (corresponding to a density of 10 stars pc^{-2}). Only groups of ten or more neighbors are considered clusters. The clusters identified in the Spitzer survey are shown in Fig. 4.

3.2. The Fraction of Stars in Large Clusters

It is now generally accepted that most stars form in clusters (*Lada and Lada*, 1995), but quantitative estimates of the fraction of stars that form in large clusters, small clusters, groups, and relative isolation are still uncertain. *Porras*

et al. (2003) compiled a list of all known groups and clusters with more than 5 members within 1 kpc of the Sun, while *Lada and Lada* (2003) compiled the properties of a sample of 76 clusters with more than 36 members within 2 kpc. Although these compilations are not complete, they probably give a representative sample of clusters in the nearest 1–2 kpc. In the sample of *Porras et al.* (2003), 80% of the stars are found in clusters with $N_{star} \geq 100$, and the more numerous groups and small clusters contain only a small fraction of the stars (see also *Lada and Lada*, 2003).

In Fig. 5, we plot the fraction of members from the 2MASS and Spitzer surveys as a function of the number of cluster size. Following the work of *Porras et al.* (2003), we divide the distribution into four sizes: $N_{star} \geq 100$, $100 > N_{star} \geq 30$, $30 > N_{star} \geq 10$, and $N_{star} < 10$. The main difference from the previous work is that we include a bin for $N_{star} < 10$; these we refer to as the distributed population. All the observed molecular clouds appear to contain a distributed population. *Carpenter* (2000) estimated that the fraction of stars in the distributed population were 0%, 20%, 27%, and 44% for the Orion B, Perseus, Orion A, and Mon R2 cloud, respectively, although the estimated fraction ranged from 0–66%, 13–41%, 0–61%, and 26–59%, depending on the assumptions made in the background star subtraction. In the combined Spitzer survey sample, the fraction of distributed stars is 32–11%, 26–24%, and 25–21% for the Ophiuchus, Orion A, and Orion B clouds respectively. The uncertainty is due to contamination from AGN: The higher fraction assumes no contamination, and the lower number assumes that the distributed population contains 10 AGN for every square degree of map size. The actual value will be in between those; extinction from the cloud will lower the density of AGN, and some of the contaminating AGN will be found toward clusters. In total, these measurements suggest that typically 20–25% of the stars are in the distributed population.

There are several caveats with this analysis. The first is the lack of completeness in the existing surveys. *Carpenter* (2000) considered the values of N_{star} as lower limits due to incompleteness and due to the masking of parts of the clusters to avoid artifacts from bright sources. Completeness is also an issue in the center of the Orion Nebula Cluster (ONC) for the Spitzer measurements. Also, we have not corrected the Spitzer data for the fraction of stars that do not show infrared excesses; the actual number of stars may be as much as a factor of two higher (*Gutermuth et al.*, 2004).

Another uncertainty is in the definition of the clusters. The clusters identified by these two methods are not entirely consistent. For example, in Orion A there is an uncertainty in the boundaries of the ONC. There is a large halo of stars surrounding this cluster, and the fraction of young stars in large clusters is dependent on whether stars are grouped in the ONC, in nearby smaller groups, or the distributed population. Both the 2MASS and the Spitzer data lead to an expansive definition of this cluster, extending beyond the Orion nebula and incorporating the OMC 2/3 and NGC 1977 regions, as well the L 1641 North group for

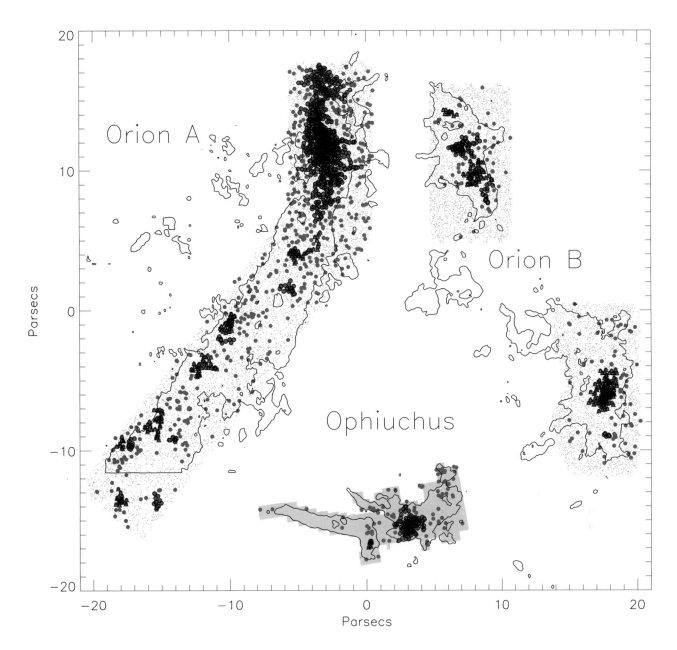

Fig. 4. The spatial distribution of all Spitzer identified infrared excess sources from the combined IRAC and 2MASS photometry of Orion A (left), Orion B (right), and Ophiuchus (bottom center). The contours outline the Bell Labs ^{13}CO maps for the Orion A and B clouds (*Bally et al.,* 1987; *Miesch and Bally,* 1994), and an A_V map of Ophiuchus (Huard, personal communication, 2006). The small gray dots show all the detections in the Spitzer 3.6- and 4.5-μm bands with magnitudes brighter than 15th and uncertainties less than 0.15. The large gray dots are the sources with infrared excesses. The black circles and triangles are sources found in clusters using the method described in section 3.1; the two symbols are alternated so that neighboring clusters can be differentiated. Note that there are two clusters in the Orion A cloud that are below the lower boundary of the Bell Labs map. Each of the clouds has a significant distributed population of IR-excess sources.

the 2MASS analysis. The resulting cluster contains a significant number of stars in a relatively low stellar density environment far from the O stars exciting in the nebula, which differs significantly from the environment of the dense core of the cluster embedded in the Orion nebula. The treatment of the ONC is critical to this analysis: 50% (for the 2MASS sample) to 76% (for the Spitzer sample) of the stars in large clusters ($N_{star} \geq 100$) are found in the ONC.

A final caveat is that these results apply to the current epoch of star formation in the nearest kiloparsec. While the largest cluster within 1 kpc is the ONC with 1000–2000 members, a growing number of young super star clusters, which contain many thousands of stars, have been detected in our galaxy. Super star clusters may bridge the gap between embedded clusters in the nearest kiloparsec, and the progenitors of the globular clusters that formed earlier in

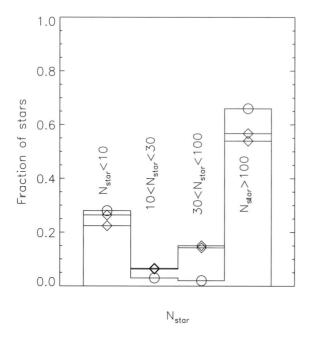

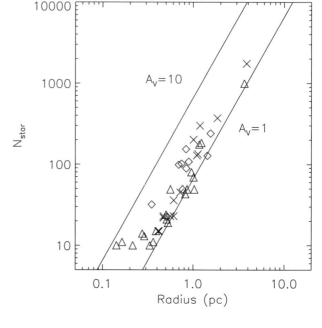

Fig. 5. The distribution of the fraction of stars in clusters taken from *Carpenter* (2000) (circles) and the Spitzer surveys of Orion and Ophiuchus (diamonds). The Spitzer surveys show a range, depending on whether corrections are made for background AGN. In both the 2MASS and Spitzer surveys, the distributed population ($N_{star} < 10$) accounts for more than 20% of the total number of stars.

Fig. 6. N_{star} vs. cluster radius for the 2MASS survey (crosses) of *Carpenter* (2000), the Spitzer Ophiuchus and Orion surveys (triangles) of Megeath et al. (in preparation) and Allen et al. (in preparation), and the Spitzer young stellar cluster survey (diamonds). Lines of constant column density are shown for a column density for $A_V = 1$ and $A_V = 10$. The average surface density of cluster members varies by less than an order of magnitude.

our galaxy's history. Thus, the distribution of cluster sizes we have derived may not be representative for other regions of the galaxy, or early epochs in our galaxy's evolution.

3.3. The Surface Density of Stars in Embedded Clusters

In a recent paper, *Adams et al.* (2006) found a correlation between the number of stars in a cluster and the radius of the cluster, using the tabulated cluster properties in *Lada and Lada* (2003). They found that the correlation is even stronger if only the 2MASS identified clusters from *Carpenter* (2000) were used, in which case the parameters were derived in a uniform manner. The same correlation is seen in a sample of clusters defined by Spitzer-identified IR-excess sources. This correlation is shown for the 2MASS and Spitzer samples in Fig. 6. This relationship shows that while N_{star} varies over 2 orders of magnitude and the cluster radius ($R_{cluster}$) varies by almost 2 orders of magnitude, the average surface density of cluster members ($N_{star}/\pi R_{cluster}^2$) varies by less than 1 order of magnitude. The lower surface ($A_V = 1$) envelope of this correlation may result in part from the methods used to identify clusters. In particular, for the many clusters surrounded by large, low surface density halos of stars, the measured radius and density of these clusters depends on the threshold surface density or spatial sepa-

ration used to distinguish the cluster stars from those in the halos. We can convert the surface densities of members into column densities of mass by assuming an average stellar mass of 0.5 $M_\odot$. Assuming a standard abundance of H, and the typical conversion from hydrogen column density to A_V, we plot lines of constant A_V in Fig. 6. In this figure the clusters are bracketed by lines equivalent to $A_V \sim 1$ and $A_V \sim 10$. Interestingly, this result is similar to one of Larson's laws for molecular clouds, that the average column density of gas in molecular clouds is independent of cloud size and mass (*Larson*, 1985; see also the chapter by Blitz et al.).

3.4. The Spatial Structure of Embedded Clusters

One of the major goals of the Spitzer young stellar cluster and Orion surveys is to systematically survey the range of cluster morphologies by identifying the young stellar objects with disks and envelopes in these clusters. An initial result of this effort is displayed for 10 clusters in Fig. 7, which shows the surface density of IR-excess sources. In this section, we give a brief overview of the common structures found in embedded clusters, both in the literature and in the sample of clusters imaged with Spitzer. We also discuss Infrared Space Observatory (ISO) and Spitzer observations of the youngest objects in these regions, the Class I and 0 sources.

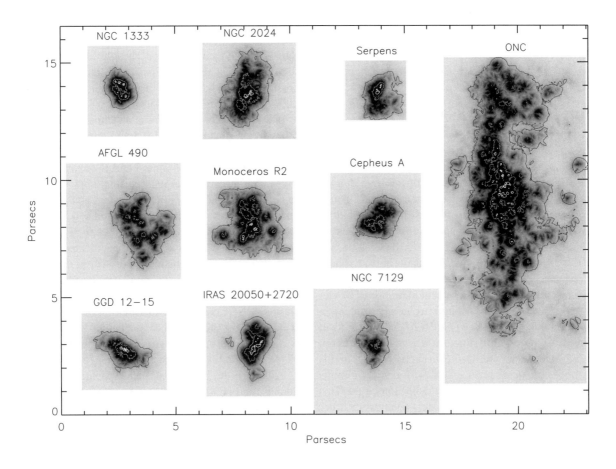

Fig. 7. The distribution of infrared excess sources in ten clusters surveyed with Spitzer. The contours are at 1, 10, and 100 IR-excess sources pc^{-2}. These data clearly show that clusters are not circularly symmetric, but are often elongated. Some of the clusters, such as IRAS 20050, show distinct clumpy structure, although much of the small-scale structure seen in the highest contours is due to statistical fluctuations in the smoothing scale. The three most circularly symmetric clusters are Cepheus A, AFGL 490, and NGC 7129; the irregular structure in these clusters is due in part to statistical fluctuations in regions of lower surface density.

Many of the clusters shown in Fig. 7 appear elongated; this had also been evident in some of the earlier studies of clusters (*Carpenter et al.*, 1997; *Hillenbrand and Hartmann*, 1998). To quantify this asymmetry, *Gutermuth et al.* (2005; Gutermuth et al., in preparation) compared the distribution of stars as a function of position angle to Monte Carlo simulations of circularly symmetric clusters, and demonstrated that the elongation is statistically significant in three of the six clusters in their sample. The elongation appears to be a result of the primordial structure in the cloud; for the two elongated clusters that have 850-µm dust continuum maps, the elongation of the cluster is aligned with filamentary structure seen in the parental molecular cloud. This suggests that the elongation results from the formation of the clusters in highly elongated, or filamentary, molecular clouds.

Not all clusters are elongated. *Gutermuth et al.* (2005) found no significant elongation of the NGC 7129 cluster, a region that also showed a significantly lower mean and peak stellar surface density than the more elongated clusters in his sample. Since the cluster was also centered in a cavity

in the molecular cloud (see section 5), they proposed that the lack of elongation was due to the expansion of the cluster following the dissipation of the molecular gas. However, not all circularly symmetric clusters are easily explained by expansion; Gutermuth et al. (in preparation) find two deeply embedded clusters with no significant elongation or clumps, but no sign of the gas dispersal evident in NGC 7129. These two clusters, Cepheus A and AFGL 490, show azimuthal symmetry, which may reflect the primordial structure of the cluster.

Examination of Fig. 7 reveals another common structure: low-density halos surrounding the dense centers, or cores, of the clusters. With the exception of AFGL 490 and perhaps Cepheus A, all the clusters in Fig. 7 show cores and halos. The core-halo structure of clusters has been studied quantitatively through azimuthally smoothed radial density profiles (*Muench et al.*, 2003). Although these density profiles can be fit by power laws, King models, or exponential functions (*Lada and Lada*, 1995; *Horner et al.*, 1997; *Hillenbrand and Hartmann*, 1998; *Gutermuth*, 2005), the resulting fits and their physical implications can be misleading. As pointed out by *Hartmann* (2004), azimuthally averaged density profiles can be significantly steepened by elongation [*Hartmann* (2004) argues this for molecular cores, but the same argument applies to clusters]. A more sophisticated treatment is required to study the density profiles of elongated clusters.

It has long been noted that young stellar clusters are sometimes composed of multiple subclusters (*Lada et al.,* 1996; *Chen et al.,* 1997; *Megeath and Wilson,* 1997; *Allen et al.,* 2002; *Testi,* 2002). Clusters with multiple density peaks or subclusters were classified as heirarchical clusters by *Lada and Lada* (2003). In some cases it is difficult to distinguish between two individual clusters and subclusters within a single cluster. Examples are the NGC 2068 and NGC 2071 clusters in the Orion B cloud (Fig. 4). These appear as two peaks in a more extended distribution of stars, although the cluster identification method described in section 3.1 separated the two peaks into two neighboring clusters. In the sample of *Gutermuth et al.* (2005; Gutermuth et al., in preparation), clumpy structure was most apparent in the IRAS 20050 cluster (see also *Chen et al.,* 1997). In this cluster, the subclusters are asociated with distinct clumps in the 850-μm map of the associated molecular cloud. This suggests that like elongation, subclusters result from structures in the parental molecular cloud.

3.5.   The Distribution of Protostars

If the observed morphologies of embedded clusters result from the filamentary and clumpy nature of the parental molecular clouds, then the younger Class 0/I objects, which have had the least time to move away from their star formation sites, should show more pronounced structures than the older, pre-main-sequence Class II and Class III stars. *Lada et al.* (2000) found a deeply embedded population of young stellar objects with large K-L colors toward the ONC; these protostar candidates showed a much more elongated and clumpy structure than the young pre-main-sequence stars in the Orion nebula. Using the methods described in section 2.1, we have identified Class 0/I and II objects in clusters using combined Spitzer and groundbased near-IR photometry. In Fig. 8, we plot the distribution of Class 0/I and II sources for four clusters in our sample. In the L 1688 and IRAS 20050 clusters, the protostars fall preferentially in small subclusters and are less widely distributed than the Class II objects. In the Serpens and GGD 12–15 clusters, the protostars are organized into highly elongated distributions. An interesting example containing multiple elongated distributions of protostars is the "spokes" cluster of NGC 2264, which shows several linear chains of protostars extending from a bright infrared source (*Teixeira et al.,* 2006). These chains, which give the impression of spokes on a wheel, follow filamentary structures in the molecular cloud. These data support the view that the elongation and subclustering are indeed the result of the primordial distribution of the parental dense gas. It is less clear whether the

observed halos result from dynamical evolution or originate *in situ* in less-active regions of star formation surrounding the more active cluster cores. The current data suggest that the halos are at least in part primordial; Class 0/I objects are observed in the halos of many clusters (*Gutermuth et al.,* 2004; *Megeath et al.,* 2004).

The spacing of protostars is an important constraint on the physical mechanisms for fragmentation and possible subsequent interactions by protostars. *Kaas et al.* (2004) analyzed the spacing of Class I and II objects identified in ISO imaging of the Serpens cluster. They calculated the separations of pairs of Class I objects, and found that the distribution of these separations peaked at 0.12 pc. In comparison, the distribution of separations for Class II objects show only a broad peak at 0.2–0.6 pc; this reflects the more spatially confined distributions of protostars discussed in the previous section. *Teixeira et al.* (2006) performed a similar analysis for the sample of protostars identified in the spokes cluster of NGC 2264. The distribution of nearest neighbor separations for this sample peaked at 0.085 pc; this spacing is similar to the Jeans length calculated from observations of the surrounding molecular gas.

Although the observed typical spacing of protostars in Serpens and NGC 2264 apears to be ~0.1 pc, as shown in Fig. 8, dense groups of protostars are observed in both these regions (and others) in which several Class I/0 sources are found within a region 0.1 pc in diameter. This is the distance a protostar could move in 100,000 yr (the nominal protostellar lifetime) at a velocity of 1 km s^{-1}. This suggests that if the velocity dispersion of protostars is comparable to the turbulent velocity dispersion observed in molecular clouds, interactions between protostars may occur, particularly in dense groupings. On the other hand, observations of some dense star-forming clumps show motions through their envelopes much less than 1 km s^{-1} (*Walsh et al.,* 2004). The densest grouping of protostars so far identified in the Spitzer survey is found in the spokes cluster. One of the protostars in the spokes has been resolved into a small system of 10 protostars by groundbased near-IR imaging and by Spitzer IRAC imaging. These protostars are found in a region 10,000 AU in diameter. It is not clear whether these objects are in a bound system, facilitating interactions as the sources orbit within the system, or whether the stars are drifting apart as the molecular gas binding the region is dispersed by the evident outflows (*Young et al.,* 2006). It should be noted that this group of 10 protostars appears to be the only such system in the spokes cluster. Thus, although dense groups of protostars are present in star-forming regions, they may not be common.

4.   GAS DISRUPTION AND THE LIFETIME OF EMBEDDED CLUSTERS

In the current picture of cluster evolution, star formation is terminated when the parental gas has dispersed. An understanding of the mechanisms and timescales for the disruption of the gas is necessary for understanding the dura-

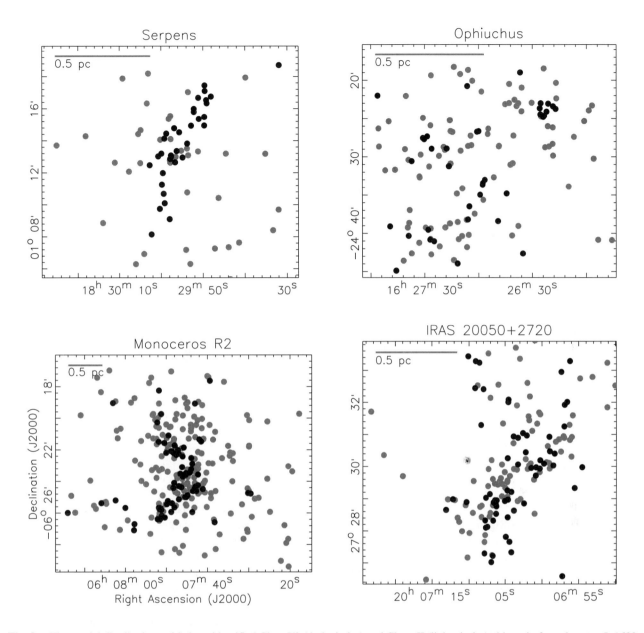

Fig. 8. The spatial distributions of Spitzer identified Class I/0 (dark circles) and Class II (light circles) objects in four clusters: L 1688 in Ophiuchus, Serpens, Mon R2, and IRAS 20050. The Class I/0 sources are often distributed along filamentary structures, while the Class II sources are more widely distributed. Many small groups of protostars are dense enough that interactions between individual objects may occur.

tion of star formation in clusters, the lifetime and eventual fate of the clusters, and the ultimate star formation efficiency achieved in a molecular cloud.

The most massive stars have a disproportionate effect on cluster evolution. Massive O stars can rapidly disrupt the parental molecular cloud through their ionizating radiation. The effect of the disruption is not immediate; once massive stars form in a molecular core, star formation may continue in the cluster while the massive star remains embedded in an ultracompact HII region. Examples of clusters in this state within 1 kpc of the Sun are the GGD 12–15 and Mon R2 clusters. The timescale for the disruption of the core is equivalent to the lifetime of the ultracompact HII

region (*Megeath et al.,* 2002); this lifetime is thought to be ~10,000 yr for the solar neighborhood (*Comeron and Torra,* 1996; *Casussus et al.,* 2000).

In our sample of nearby embedded clusters, most systems do not contain O stars. However, a number of partially embedded clusters in the nearest 1 kpc show evidence for significant disruption by B-type stars. Due to the partial disruption of the clouds, the clusters in these regions are found in cavities filled with emission from UV heated polycyclic aromatic hydrocarbons (*Gutermuth et al.,* 2004). The timescale for the disruption by B stars can be estimated using measurements of the ages of the clusters. In our survey of nearby regions, we have three examples of regions with such cavities: NGC 7129 (earliest member B2), IC 348 (earliest member B5), and IC 5146 (earliest member B0-1), with ages of 2 m.y., 3 m.y. and 1 m.y., respectively (*Hillenbrand et al.,* 1992; *Hillenbrand,* 1995; *Herbig and Dahm,* 2002; *Luhman et al.,* 2003). The presence of large, UV illuminated cavities in these regions suggest that the nonionizing FUV from B stars may be effective at heating and evaporating molecular cloud surfaces in cases where intense FUV radiation from O stars is not present. For example, in the case of NGC 7129, *Morris et al.* (2004) find that the temperature at the molecular cloud surface has been heated to 700 K by the FUV radiation. Future work is needed to determine if the high temperatures created by the FUV radiation can lead to substantial evaporative flows.

In regions without OB stars, however, some other mechanism must operate. An example is IRAS 20050. Based on SCUBA maps, as well as the reddening of the members, *Gutermuth et al.* (2005) found that the cluster is partially offset from the associated molecular gas, suggesting that the gas had been partially dispersed by the young stars Although this region contains no OB stars, it displays multiple outflows (*Chen et al.,* 1997). Another example may be the NGC 1333 cloud, where *Quillen et al.* (2005) found evidence of wind-blown cavities in the molecular gas. In these regions, outflows may be primarily responsible for dissipating the dense molecular gas (e.g., *Matzner and McKee,* 2000).

It is important to note that star formation continues during the gas dissipation process. Even when the gas around the main cluster has been largely disrupted (such is the case in the ONC, IC 348, and NGC 7129), star formation continues on the outskirts of the cluster in regions where the gas which has not been removed. Thus, the duration of star formation in these regions appears similar to the gas dispersal time of ~1–3 m.y. Older clusters have not been found partially embedded in their molecular gas (*Leisawitz et al.,* 1989).

5. EARLY CLUSTER EVOLUTION

Theories of cluster formation are reviewed elsewhere in this volume (see chapters by Ballesteros-Paredes et al. and Bonnell et al.). Here we will discuss the dynamical evolution of young clusters during the first few million years.

Although most stars seem to form within clusters of some type (see section 4), only about 10% of stars are born within star-forming units that are destined to become open clusters. As a result, for perhaps 90% of forming stars, the destruction of their birth aggregates is an important issue. Star formation in these systems is not 100% efficient, so a great deal of cluster gas remains in the system. This gaseous component leaves the system in a relatively short time (a few million years; see above) and its departure acts to unbind the cluster. At the zeroth level of understanding, if the star-formation efficiency (SFE) is less than 50%, then a substantial amount of unbinding occurs when gas is removed. However, this description is overly simple. The stars

in the system will always have a distribution of velocities. When gas is removed, stars on the high-velocity tail of the distribution will always leave the system (even for very high SFE) and those on the extreme low-velocity tail will tend to stay (even for low SFE). The fraction of stars that remain bound after gas removal is thus a smooth function of star-formation efficiency [several authors have tried to calculate the function (see *Lada et al.,* 1984; *Adams,* 2000; *Boily and Kroupa,* 2003a,b)]. The exact form of the bound fraction, $f_b(\varepsilon)$, which is a function of SFE, depends on many other cluster properties: gas removal rates, concentration of the cluster, total depth of the cluster potential well, the distribution functions for the stellar velocities (radial vs. isotropic), and the spatial profiles of the gaseous and stellar components (essentially, the SFE as a function of radial position). At the crudest level, the bound fraction function has the form $f_b \approx \sqrt{\varepsilon}$, but the aforementioned complications allow for a range of forms.

The manner in which a cluster spreads out and dissolves after its gas is removed is another important problem. After gas removal, clusters are expected to retain some stars as described above, but such systems are relatively short-lived. For example, consider a cluster with N = 100 in its early embedded phase, before gas removal. After the gas leaves, typically one-half to two-thirds of the stars will become unbound along with the gas. The part of the cluster that remains bound will thus contain only N = 30–50 stars. Small groups with N < 36 have relaxation times that are shorter than their crossing times (*Adams,* 2000), and such small units will exhibit different dynamical behavior than their larger counterparts. In particular, such systems will relax quickly and will not remain visible as clusters for very long.

As more data are taken, another mismatch between theory and observations seems to be emerging: The theoretical calculations described above start with an established cluster with a well-defined velocity distribution function, and then remove the gaseous component and follow the evolution. Given the constant column density relationship for clusters (section 3.4), that the velocity of the stars are virialized, and assuming that 30% of the cluster mass is in stars (see *Lada and Lada,* 2003), then the crossing time for the typical cluster in our sample is ~1 m.y. (although it can be shorter in the dense centers of clusters). As a result, in rough terms, the gas removal time, the duration of star formation, and the crossing time are comparable. This implies that partially embedded clusters may not have enough time to form relaxed, virial clusters; this in turn may explain in part the range of morphologies discussed in section 3.

6. EFFECTS OF CLUSTERS ON STAR AND PLANET FORMATION

The radiation fields produced by the cluster environment can have an important impact on stars and planets formed within. Both the EUV and FUV radiation can drive disk evaporation (*Shu et al.,* 1993; *Johnstone et al.,* 1998; *Störzer and Hollenbach,* 1999; *Armitage,* 2000). In the modest-sized clusters of interest here (100–1000 stars), the mass loss driven by FUV radiation generally dominates (e.g., *Adams et al.,* 2004), although EUV photoevaporation can also be important (*Shu et al.,* 1993; *Johnstone et al.,* 1998; *Störzer and Hollenbach,* 1999; *Armitage,* 2000). For clusters with typical cluster membership, e.g., with N_{star} = 300 (section 3.1), the average solar system is exposed to a FUV flux of G ≈ 1000–3000 (*Adams et al.,* 2006), where G = 1 corresponds to a flux of 1.6×10^{-3} erg cm^{-2} s^{-1}. FUV fluxes of this magnitude will evaporate a disk orbiting a solar-type star down to a truncation radius of about 50 AU over a timescale of 4 m.y. As a result, planet-forming disks are relatively immune in the regions thought to be relevant for making giant gaseous planets. Forming solar systems around smaller stars are more easily evaporated for two reasons. First, the central potential well is less deep, so the stellar gravity holds less tightly onto the disk gas, which is more easily evaporated. Second, we expect the disk mass to scale linearly with stellar mass so that disks around smaller stars have a smaller supply and can be evaporated more quickly. With these disadvantages, M stars with 0.25 $M_\odot$ can be evaporated down to 10 AU in 4 m.y. with an FUV radiation field of G = 3000. In larger clusters with more massive stars, *Adams et al.* (2004) find that regions with strong FUV and EUV can affect disks around solar mass stars on solar system size scales, truncating an initially 100 AU disk to a radius of 30 AU in 4 m.y.

A full assessment of the importance of UV radiation on disks needs to be informed by the observed properties of clusters. What fraction of stars in the Spitzer and 2MASS samples are found in clusters with significant EUV radiation fields? We use the presence of an HII region as an indicator of a EUV field. In the Spitzer sample (the Orion A, Orion B, and Ophiuchus clouds), the two clusters with HII regions contain 45% of the IR-excess sources. In the 2MASS sample (Orion A, Orion B, Perseus, and Mon R2), 55% of the young stars are found in the four clusters with HII regions. Thus, a significant fraction of stars is found in clusters with HII regions. However, in both the 2MASS and Spitzer samples most of the stars found in clusters with HII regions are found in the ONC. The ONC has a radius of 4 pc and many of the low-mass stars in this cluster are more than a parsec away from the massive stars, which are concentrated in the center of the cluster. Thus, the fraction of stars exposed to a significant EUV field appears to be less than 50%. However, a more systematic determination of this fraction should be made as data become available.

In addition to driving photoevaporation, EUV radiation (and X-rays) can help ionize the disk gas. This effect is potentially important. One of the most important mechanisms for producing disk viscosity is through magneto-rotational instability (MRI), and this instability depends on having a substantial ionization fraction in the disk. One problem with this idea is that the disk can become too cold and the ionization fraction can become too low to sustain the turbulence. If the background environment of the cluster provides enough EUV radiation, then the cluster environment

can be important for helping drive disk accretion. Clusters can also have an affect on the processes of star and planet formation through dynamical interactions. This raises a variant of the classic question of nature vs. nurture: Are the properties of the protostars and the emergent stars influenced by interactions, or are they primarily the result of initial conditions in a relatively isolated collapse? The numerical simulations of cloud collapse and cluster formation (*Bate et al.,* 2003; *Bonnell et al.,* 2003, 2004) predict that interactions are important, with the individual protostars competively accreting gas from a common reservoir as they move through the cloud, and dynamical interactions between protostars resulting in ejections from the cloud.

We assess the importance of interactions given our current understanding of cluster structure. The density of clusters, and of protostars in clusters, suggest that if stars move with velocities similar to the turbulent gas velocity (~1 km s^{-1}), interactions can occur in the lifetime of a protostar (100,000 yr). *Gutermuth et al.* (2005) found typical stellar densities of 104 stars pc^{-3} in the cores of two young clusters. If the velocity dispersion is 1 km s^{-1}, most protostars will pass within 1000 AU — the size of a protostellar envelope — of another star or protostars within a protostellar lifetime. The observed spacing of Class I/0 sources discussed in section 3.2 also suggests that interactions can occur in some cases. At these distances protostars could compete for gas or interact through collisions of their envelopes. Interestingly, recent data suggest that, at least in some clusters, the observed prestellar clumps that make up the initial states for star formation are not moving dynamically, but rather have subvirial velocities (*Walsh et al.,* 2004; *Peretto et al.,* 2006). If these clusters are typical, then interactions between protostars in clusters would be minimal.

Given the observed surface densities of clusters, is it possible that a cluster could result from the collapse of individual, noninteracting prestellar cores (i.e., nature over nurture?). If the starting density profile of an individual star formation event can be modeled as an isothermal sphere, then its radial size would be given by $r = GM_*/2a^2 \approx 0.03$ pc (where we use a typical stellar mass of $M_* = 0.5\ M_\odot$ and sound speed a = 0.2 km s^{-1}). A spherical volume of radius R = 1 pc can thus hold about 37,000 of these smaller spheres (in a close-packed configuration). Thus, we can conclude that there is no *a priori* geometrical requirement for the individual star-forming units to interact.

Once a star sheds or accretes its protostellar envelope, direct collisions are relatively rare because their cross sections are small. Other interactions are much more likely to occur because they have larger cross sections. For example, the disks around newly formed stars can interact with each other or with passing binaries and be truncated (*Ostriker,* 1994; *Kobayashi and Ida,* 2001). In rough terms, these studies indicate that a passing star can truncate a circumstellar disk down to a radius r_d that is one-third of the impact parameter. In addition, newly formed planetary systems can interact with each other, and with passing binary star systems, and change the planetary orbits (*Adams and Laughlin,*

2001). In a similar vein, binaries and single stars can interact with each other, exchange partners, form new binaries, and/or ionize existing binaries (*Rasio et al.,* 1995; *McMillan and Hut,* 1996).

To affect a disk on a solar system (40 AU) scale requires a close approach at a distance of 100 AU or less. *Gutermuth et al.* (2005) estimated the rate of such approaches for the dense cores of clusters. They estimate that for the typical density of 104 stars per pc^{-3}, the interaction time is 10^7 yr, longer than the lifetime of the cluster. For N-body models of the modest-sized clusters of interest here (100–1000 members), the typical star/disk system is expected to experience about one close encounter within 1000 AU over the next ~5 m.y. while the cluster remains intact; close encounters within 100 AU are rare (e.g., *Smith and Bonnell,* 2001; *Adams et al.,* 2006). Given that lifetime of the cluster is less than 5 m.y., these models again indicate a minimal effect on nascent solar systems.

7. CONCLUSIONS

1. *The distribution of cluster properties:* Systematic surveys of giant molecular clouds from 2MASS and Spitzer, as well as targeted surveys of individual clusters, are providing the first measurements of the range and distribution of cluster properties in the nearest kiloparsec. Although most stars appear in groups or clusters, in many star-forming regions there is a significant distributed component. These results suggest that there is a continuum of star-forming environments from relative isolation to dense clusters. Theories of star formation must take into account (and eventually explain) this observed distribution. The 2MASS and Spitzer surveys also show a correlation between number of member stars and the radii of clusters, such that the average surface density of stars varies by a factor of only ~5.

2. *The structure of young stellar clusters:* Common cluster morphologies include elongation, low-density halos, and subclustering. The observed cluster and molecular gas morphologies are similar, especially when only the youngest Class I/0 sources are considered. This similarity suggests that these morphologies (except possibly halos) result from the distribution of fragmentation sites in the parental cloud, and not the subsequent dynamical evolution of the cluster. Consequently, the surface densities and morphologies of clusters are important constraints on models of the birth of clusters.

3. *The evolution of clusters:* The evolution of clusters is driven initially by the formation of stars, and then later on by the dissipation of gas. Gas dissipation appears to be driven by different processes in different regions, including photoevaporation by EUV radiation from O stars, photoevaporation by FUV radiation from B stars, and outflows from lower-mass stars. Much of the gas appears to be dissipated in 3 m.y., which is a few times the crossing time and the duration of star formation in these clusters. With these short timescales, clusters probably never reach dynamical equilibrium in the embedded phase. The survival of clusters as the gas is dispersed is primarily a function of the size of the cluster, the efficiency of star formation, and the rate at which the gas is dispersed.

4. *The impact of clustering on star and planet formation:* FUV and EUV radiation from massive stars can effectively truncate disks in a few million years. EUV radiation is needed to affect disks around solar type stars on solar system scales (<40 AU) in the lifetime of the cluster. Within our sample of molecular clouds, fewer than 50% of the stars are found in regions with strong EUV fields. The observed spacing of protostars suggest that dynamical interactions and competitive accretion may occur in the denser regions of the observed clusters. However, evidence of subvirial velocities of prestellar condensations in at least one cluster hints that these interactions may not be important. Given the densities and lifetimes of the observed clusters, dynamical interactions do not appear to be an important mechanism for truncating disks on solar system size scales.

Acknowledgments. This work is based in part on observations made with the Spitzer Space Telescope, which is operated by the Jet Propulsion Laboratory, California Institute of Technology, under NASA Contract 1407. Support for this work was provided by NASA through Contract Numbers 1256790 and 960785, issued by JPL/Caltech. P.C.M. acknowledges a grant from the Spitzer Legacy Science Program to the "Cores to Disks" team and a grant from the NASA Origins of Solar Systems Program. S.J.W. received support from Chandra X-ray Center Contract NAS8-39073. F.C.A. is supported by NASA through the Terrestrial Planet Finder Mission (NNG04G190G) and the Astrophysics Theory Program (NNG04GK56G0).

REFERENCES

Adams F. C. (2000) *Astrophys. J., 542,* 964–973.

Adams F. C. and Laughlin G. (2001) *Icarus, 150,* 151–162.

Adams F. C., Lada C. J., and Shu F. H. (1987) *Astrophys. J., 312,* 788–806.

Adams F. C., Hollenbach D., Laughlin G., and Gorti U. (2004) *Astrophys. J., 611,* 360–379.

Adams F. C., Proszkow E. M., Fatuzzo M., and Myers P. C. (2006) *Astrophys. J., 641,* 504–525.

Allen L. E., Myers P. C., Di Francesco J., Mathieu R., Chen H., and Young E. (2002) *Astrophys. J., 566,* 993–1004.

Allen L. E., Calvet N., D'Alessio P., Merin B., Hartmann L., Megeath S. T., et al. (2004) *Astrophys. J. Suppl., 154,* 363–366.

Allen L. E., Hora J. L., Megeath S. T., Deutsch L. K., Fazio G. G., Chavarria L., and Dell R. D. (2005) In *Massive Star Birth: A Crossroads of Astrophysics* (R. Cesaroni et al., eds.), pp. 352–357. Cambridge Univ., Cambridge.

André P., Ward-Thompson D., and Barsony M. (1993) *Astrophys. J., 406,* 122–141.

André P., Ward-Thompson D., and Barsony M. (2000) In *Protostars and Protoplanets IV* (V. Mannings et al., eds.), pp. 59–96. Univ. of Arizona, Tucson.

Armitage P. J. (2000) *Astron. Astrophys., 362,* 968–972.

Bally J., Stark A. A., Wilson R. W., and Langer W. D. (1987) *Astrophys. J., 312,* L45–L49.

Bate M. R., Bonnell I. A., and Bromm V. (2003) *Mon. Not. R. Astron. Soc., 339,* 577–599.

Boily C. M. and Kroupa P. (2003a) *Mon. Not. R. Astron. Soc., 338,* 665–672.

Boily C. M. and Kroupa P. (2003b) *Mon. Not. R. Astron. Soc., 338,* 673–686.

Bonnell I. A., Bate M., and Vine S. G. (2003) *Mon. Not. R. Astron. Soc., 343,* 413–418.

Bonnell I. A., Vine S. G., and Bate M. (2004) *Mon. Not. R. Astron. Soc., 349,* 735–741.

Cambrésy L., Petropoulou V., Kontizas M., and Kontizas E. (2006) *Astron. Astrophys., 445,* 999–1003.

Carpenter J. M. (2000) *Astron. J., 120,* 3139–3161.

Carpenter J. M., Meyer M. R., Dougados C., Strom S. E., and Hillenbrand L. A. (1997) *Astron. J., 114,* 198–221.

Carpenter J. M., Hillenbrand L. A., and Skrutskie M. F. (2001) *Astron. J., 121,* 3160–3190.

Carpenter J. M., Hillenbrand L. A., SkrutskieM. F., and Meyer M. R. (2002) *Astron. J., 124,* 1001–1025.

Casassus S., Bronfman L., May J., and Nyman L. A. (2000) *Astron. Astrophys., 358,* 514–520.

Chen H., Tafalla M., Greene T. P., Myers P. C., and Wilner D. J. (1997) *Astrophys. J., 475,* 163–172.

Comeron F. and Torra J. (1996) *Astron. Astrophys., 314,* 776–784.

Dorren J. D., Guedel M., and Guinan E. F. (1995) *Astrophys. J., 448,* 431–436.

Evans N. J. II, Allen L. E., Blake G. A., Boogert A. C. A., Bourke T., et al. (2003) *Publ. Astron. Soc. Pac., 115,* 965–980.

Feigelson E. D. and Montmerle T. (1999) *Ann. Rev. Astron. Astrophys., 37,* 363–408.

Feigelson E. D., Getman K., Townsley L., Garmire G., Preibisch T., Grosso N., Montmerle T., Muench A., and McCaughrean M. (2005) *Astrophys. J. Suppl., 160,* 379–389.

Flaccomio E., Damiani F., Micela G., Sciortino S., Harnden F. R., Murray S. S., and Wolk S. J. (2003) *Astrophys. J., 582,* 398–409.

Gladwin P. P., Kitsionas S., Boffin H. M. J., and Whitworth A. P. (1999) *Mon. Not. R. Astron. Soc., 302,* 305–313.

Gomez M., Hartmann L., Kenyon S. J., and Hewett R. (1993) *Astron. J., 105,* 1927–1937.

Greene T. P., Wilking B. A., André, P., Young, E. T., and Lada C. J. (1994) *Astrophys. J., 434,* 614–626.

Gutermuth R. A. (2005) Ph.D. thesis, Univ. of Rochester.

Gutermuth R. A., Megeath S. T., Muzerolle J., Allen L. E., Pipher J. L., Myers P. C., and Fazio G. G. (2004) *Astrophys. J. Suppl., 154,* 374–378.

Gutermuth R. A., Megeath S. T., Pipher J. L.,Williams J. P., Allen L. E., Myers P. C., and Raines S. N. (2005) *Astrophys. J., 632,* 397–420.

Haisch K. E., Lada E. A., and Lada C. J. (2001) *Astrophys. J., 553,* L153–L156.

Hartmann L. (2004) In *Star Formation at High Angular Resolution* (M. Burton et al., eds.), pp. 201–211. ASP Conf. Series 221, San Francisco.

Hartmann L., Megeath S. T., Allen L., Luhman K., Calvet N., et al. (2005) *Astrophys. J., 629,* 881–896.

Herbig G. H. and Bell K. R. (1998) *Lick Observatory Bulletin, Santa Cruz: Lick Observatory, 1995,* VizieR Online Data Catalog, 5073.

Herbig G. H. and Dahm S. E. (2002) *Astron. J., 123,* 304–327.

Hillenbrand L. A. (1995) Ph.D. thesis, Univ. of Massachusetts.

Hillenbrand L. A. and Hartmann L. (1998) *Astrophys. J., 492,* 540.

Hillenbrand L. A., Strom S. E., Vrba F. J., and Keene J. (1992) *Astrophys. J., 397,* 613–643.

Hillenbrand L. A., Massey P., Strom S. E., and Merrill K. M. (1993) *Astron. J., 106,* 1906–1946.

Horner D. J., Lada E. A., and Lada C. J. (1997) *Astron. J., 113,* 1788–1798.

Johnstone D., Hollenbach D. J., and Bally J. (1998) *Astrophys. J., 499,* 758–776.

Kaas A. A. (1999) *Astron. J., 118,* 558–571.

Kaas A. A., Olofsson G., Bontemps S., André P., Nordh L., et al. (2004) *Astron. Astrophys., 421,* 623–642.

Kobayashi H. and Ida S. (2001) *Icarus, 153,* 416–429.

Lada C. J. and Lada E. A. (1995) *Astron. J., 109,* 1682–1696.

Lada C. J. and Lada E. A. (2003) *Ann. Rev. Astron. Astrophys., 41,* 57–115.

Lada C. J., Margulis M., and Dearborn D. (1984) *Astrophys. J., 285,* 141–152.

Lada C. J., Alves J., and Lada E. A. (1996) *Astron. J., 1111,* 1964–1976.

Lada C. J., Muench A. A., Haisch K. E., Lada E. A., Alves J. F., Tollestrup E. V., and Willner S. P. (2000) *Astron. J., 120,* 3162–3176.

Lada E. A., Evans N. J. II, Depoy D. L., and Gatley I. (1991) *Astrophys. J., 371,* 171–182.

Larson R. B. (1985) *Mon. Not. R. Astron. Soc., 214,* 379–398.

Leisawitz D., Bash F. N., and Thaddeus P. (1989) *Astrophys. J. Suppl., 70,* 731–812.

Luhman K. L., Stauffer J. R., Muench A. A., Rieke G. H., Lada E. A., Bouvier J., and Lada C. J. (2003) *Astrophys. J., 593,* 1093–1115.

Matzner C. D. and McKee C. F. (2000) *Astrophys. J., 545,* 364–378.

McMillan S. L. W. and Hut P. (1996) *Astrophys. J., 467,* 348–358.

Megeath S. T. and Wilson T. L. (1997) *Astron. J., 114,* 1106–1120.

Megeath S. T., Biller B., Dame, T. M., Leass E., Whitaker R. S., and Wilson T. L. (2002) *Hot Star Workshop III: The Earliest Stages of Massive Star Formation* (P. A. Crowther, ed.), pp. 257–265. ASP Conf. Series 267, San Francisco.

Megeath S. T., Allen L. E., Gutermuth R. A., Pipher J. L., Myers P. C., et al. (2004) *Astrophys. J. Suppl., 154,* 367–373.

Megeath S. T., Flaherty K. M., Hora J., Allen L. E., Fazio G. G., et al. (2005) *Massive Star Birth: A Crossroads of Astrophysics* (R. Cesaroni et al., eds.), pp. 383–388. Cambridge Univ., Cambridge.

Micela G., Sciortino S., Serio S., Vaiana G. S., Bookbinder J., Golub L., Harnden F. R., and Rosner R. (1985) *Astrophys. J., 292,* 172–180.

Miesch M. S. and Bally J. (1994) *Astrophys. J., 429,* 645–671.

Morris P. W., Noriega-Crespo A., Marleau F. R., Teplitz H. I., Uchida K. I., and Armus L. (2004) *Astrophys. J. Suppl., 154,* 339–343.

Motte F., André P., and Neri R. (1998) *Astron. Astrophys., 336,* 150–172.

Motte F., André P., Ward-Thompson D., and Bontemps S. (2001) *Astron. Astrophys., 372,* L41–L44.

Muench A. A., Lada E. A., Lada C. J., Elston R. J., Alves J. F., Horrobin M., Huard T. H., et al. (2003) *Astron. J., 125,* 2029–2049.

Muzerolle J., Megeath S. T., Gutermuth R. A., Allen L. E., Pipher J. L., et al. (2004) *Astrophys. J. Suppl., 154,* 379–384.

Myers P. C. and Ladd E. F. (1993) *Astrophys. J., 413,* L47–L50.

Nutter D. J., Ward-Thompson D., and André P. (2005) *Mon. Not. R. Astron. Soc., 357,* 975–982.

Ostriker E. C. (1994) *Astrophys. J., 424,* 292–318.

Peretto N., André P., and Belloche A. (2006) *Astron. Astrophys., 445,* 879–998.

Porras A., Christopher M., Allen L., Di Francesco J., Megeath S. T., and Myers P. C. (2003) *Astron. J., 126,* 1916–1924.

Quillen A. C., Thorndike S. L., Cunninghman A., Frank A., Gutermuth R. A., Blackmann E. G., Pipher J. L, and Ridge N. (2005) *Astrophys. J., 632,* 941–955.

Rasio F. A., McMillan S., and Hut P. (1995) *Astrophys. J., 438,* L33–L36.

Rodríguez L. F., Anglada G., and Curiel S. (1999) *Astrophys. J. Suppl., 125,* 427–438.

Sandell G. and Knee L. B. G. (2001) *Astrophys. J., 546,* L49–L52.

Shu F. H., Johnstone D., and Hollenbach D. J. (1993) *Icarus, 106,* 92–101.

Smith K. W. and Bonnell I. A. (2001) *Mon. Not. R. Astron. Soc., 322,* L1–L4.

Stern D., Eisenhardt P., Gorjian V., Kochanek C. S., and Caldwell N. (2005) *Astrophys. J., 631,* 163–168.

Störzer H. and Hollenbach D. (1999) *Astrophys. J., 515,* 669–684.

Teixeira P. S., Lada C. J., Young E. T., Marengo M., and Muench A., et al. (2006) *Astrophys. J., 636,* L45–L48.

Testi L. (2002) In *Modes of Star Formation and the Origin of Field Populations* (E. K. Grebel and W. Brandner, eds.), pp. 60–70. ASP Conf. Series 285, San Francisco.

Walsh A. J., Myers P. C., and Burton M. G. (2004) *Astrophys. J., 614,* 194–202.

Walter F. M. and Barry D. C. (1991) *The Sun in Time* (C. P. Sonett et al., eds.), pp. 633–657. Univ. of Arizona, Tucson.

Whitney B. A., Wood K., Bjorkman J. E., and Cohen M. (2003) *Astrophys. J., 598,* 1079–1099.

Wilking B. A., Schwartz R. D., and Blackwell J.-H. (1987) *Astron. J., 94,* 106–110.

Wilking B. A., Lada C. J., and Young E. T. (1989) *Astrophys. J., 340,* 823–852.

Wiramihardja S. D., Kogure T., Yoshida S., Ogura K., and Nakano M. (1989) *Publ. Astron. Soc. Pac., 41,* 155–174.

Wolk S., Spitzbart B. D., Bourke T. L., and Alves J. (2006) *Astron. J., 132,* 1100–1125.

Young E. T., Teixeira P., Lada C. J., Muzerolle J., Persson S. E., et al. (2006) *Astrophys. J., 642,* 972–978.

Part V:
Young Binaries — Brown Dwarfs

New Observational Frontiers in the Multiplicity of Young Stars

Gaspard Duchêne
Laboratoire d'Astrophysique de Grenoble

Eduardo Delgado-Donate
Stockholm Observatory

Karl E. Haisch Jr.
Utah Valley State College

Laurent Loinard and Luis F. Rodríguez
Universidad Nacional Autónoma de México

It has now been known for over a decade that low-mass stars located in star-forming regions are very frequently members of binary and multiple systems, even more so than main-sequence stars in the solar neighborhood. This high multiplicity rate has been interpreted as the consequence of the fragmentation of small molecular cores into a few seed objects that accrete to their final mass from the remaining material and dynamically evolve into stable multiple systems, possibly producing a few ejecta in the process. Analyzing the statistical properties of young multiple systems in a variety of environments therefore represents a powerful approach to place stringent constraints on star-formation theories. In this contribution, we first review a number of recent results related to the multiplicity of T Tauri stars. We then present a series of studies focusing on the multiplicity and properties of optically undetected, heavily embedded protostars. These objects are much younger than the previously studied pre-main-sequence stars, and they therefore offer a closer look at the primordial population of multiple systems. In addition to these observational avenues, we present new results of a series of numerical simulations that attempt to reproduce the fragmentation of small molecular cores into multiple systems, and compare these results to the observations.

1. INTRODUCTION

The prevalence of binary and higher-order multiple systems is a long-established observational fact for field low-mass stars (*Duquennoy and Mayor,* 1991; *Fischer and Marcy,* 1992). For over a decade, it has been known that young pre-main-sequence stars are also often found in multiple systems. The chapter by *Mathieu et al.* (2000) in the previous volume in this series has summarized various statistical surveys for visual multiple systems among T Tauri stars in star-forming regions as well as of zero-age main-sequence stars in open clusters. These early multiplicity surveys have shown that multiple systems are ubiquitous among young stellar objects (YSOs) and further revealed an environment-dependent trend. The multiplicity rate in all stellar clusters, even those with the youngest ages such as the Orion Trapezium cluster, is in excellent agreement with that observed in main-sequence field stars. On the other hand, the least dense T Tauri populations, like the Taurus-Auriga and Ophiuchus clouds, show a factor of ~2 multiplicity excess, relative to field low-mass stars. However, it remained impossible to decide whether this behavior was

the consequence of an intrinsic difference in the fragmentation process, or the result of dynamical disruptive interactions acting on timescales shorter than 1 m.y. in stellar clusters (see *Patience and Duchêne,* 2001, for a review).

The purpose of this chapter is to review a variety of observational results concerning the multiplicity of young low-mass stars in order to update the view presented by *Mathieu et al.* (2000). In addition, we present some numerical results related to the fragmentation and subsequent evolution of low-mass prestellar cores. These models make predictions that can be readily tested with the observational results discussed here. Throughout this chapter, we focus on low-mass stellar objects with masses roughly ranging from 0.1 to 2 $M_\odot$. The multiplicity of young substellar objects is discussed in detail in the chapters by Burgasser et al. and Luhman et al., whereas the multiplicity of higher-mass objects is addressed in the chapter by Beuther et al. Other chapters in this volume present complementary insights on the subject: Goodwin et al. present more numerical results on the collapse and fragmentation of molecular cores, as well as on the dynamical evolution of small stellar aggregates; Whitworth et al., Ballesteros-Paredes et al., and Klein et al. discuss the collapse of larger-scale molecular cores; and Mathieu et al. and Monin et al. focus on various properties (dynamical masses and disks properties, respectively) of known T Tauri binary stars.

2. AN UPDATE ON THE MULTIPLICITY OF YOUNG LOW-MASS STARS

As mentioned above, the first efforts to study young multiple systems were focused on determining the average number of wide companions per object in well-known pre-main-sequence populations. Attempting to account for these multiplicity rates has led to various theories that involve the fragmentation of molecular cores and the subsequent dynamical evolution of aggregates of stars embedded in gaseous clouds. So far, the observed multiplicity rate of T Tauri populations alone has not proved entirely conclusive, and since the review by *Mathieu et al.* (2000), the focus of statistical studies of young multiple systems has shifted to other areas. Before probing much younger, still embedded multiple systems (section 3), we review here a number of studies on T Tauri multiple systems that go beyond the surveys that were conducted in the 1990s.

2.1. Multiplicity in Young Nearby Associations

The clear dichotomy between high- and low-multiplicity star-forming regions has usually been considered evidence of an environment-dependent star-formation scenario. However, since most stars form in stellar clusters, one could also consider that the rare molecular clouds that host too many companions are exceptions for yet undetermined reasons. Over the last decade, several groups of a few tens of stars with ages typically between ~10 and 50 m.y. have been identified in the Sun's vicinity based on their common three-dimensional motion and youth indicators (*Zuckerman and Song*, 2004). Most members of these associations are low-mass pre-main-sequence stars. Therefore, these co-moving groups represent additional, nearby populations of young stars whose multiplicity could be expected to resemble that of the Taurus-Auriga population given their low stellar densities.

Soon after their discovery, systematic searches for visual companions were conducted in some of these groups in order to complement the previous surveys. For instance, *Chauvin et al.* (2002, 2003) and *Brandeker et al.* (2003) conducted surveys for visual companions in the TW Hya, Tucana-Horologium, and MBM 12 groups; we include MBM 12 in this discussion despite the continuing debate regarding its distance [~65 pc according to *Hearty et al.* (2000), revised upward to ~275 pc by *Luhman* (2001)]. In their review of all known nearby associations, *Zuckerman and Song* (2004) marked those systems that were discovered to be multiple in these surveys or during pointed observations of individual objects. The average multiplicity rate in these associations range from 20% to over 60% but the small sample sizes preclude clear conclusions on any

individual association. Averaging all associations listed in *Zuckerman and Song*'s review, and adding the MBM 12 surveys from *Chauvin et al.* (2002) and *Brandeker et al.* (2003), the average number of visual companions per member is 38.2 ± 3.6%. Because of the range of distances to the stars involved in this survey, it is difficult to compare this to previous surveys of T Tauri stars, which were surveyed over a more homogeneous separation range. Nonetheless, the observed multiplicity rate in nearby young associations appears to be high, possibly as high as in Taurus-like populations. Future dedicated studies sampling a uniform separation range will help reinforce this conclusion.

2.2. Multiplicity of the Lowest-Mass T Tauri Stars

Most of the T Tauri multiplicity surveys summarized in *Mathieu et al.* (2000) were focusing on objects with masses in the range $0.5–2 M_\odot$, essentially because of the limited sensitivity of high-angular resolution devices at that time. Multiplicity surveys conducted in recent years have, therefore, focused primarily on the multiplicity rate of the lowest-mass T Tauri stars in known star-forming regions in order to determine the mass-dependency of the properties of multiple systems.

We first focus on systematic surveys for multiplicity among low-mass ($0.1–0.5 M_\odot$) T Tauri stars. White et al. (in preparation) have obtained high-angular-resolution datasets on ~50 such objects in Taurus-Auriga that represent a nice comparison sample to the early surveys of *Ghez et al.* (1993) and *Leinert et al.* (1993), for instance. They find that even low-mass T Tauri stars have a high multiplicity rate, although with a decreasing trend toward the lowest stellar masses. In addition, they find that multiple systems in which the primary has a mass $M_A \lesssim 0.4 M_\odot$ are confined to mass ratios higher than $M_B/M_A \sim 0.6$, and very rarely have projected separations larger than ~200 AU despite sensitive searches. Apart from the overall multiplicity excess among T Tauri stars, these trends are in line with the results of multiplicity surveys among lower-mass main-sequence field stars (*Marchal et al.*, 2003; *Halbwachs et al.*, 2003). These various mass-dependencies must be explained by models of fragmentation and early evolution of multiple systems.

Extending this approach to and beyond the substellar limit, *Bouy et al.* (2003) in the Pleiades and *Kraus et al.* (2005) in Upper Scorpius found that the trends for lower metalicity, tighter, and preferentially equal-mass systems are amplified in the brown dwarf regime, although the exact multiplicity frequency of young brown dwarfs is still under debate. Pre-main-sequence brown dwarfs are discussed in more detail elsewhere in this volume (see chapters by Luhman et al. and Burgasser et al.).

2.3. Dynamical Masses of Binary T Tauri Systems

The masses of T Tauri stars are usually determined through their location in the HR diagram in comparison to pre-main-sequence evolutionary models. However, there is a long-standing debate on the validity domain of these models, which all include some, but not all, of the key physical ingredients (e.g., *Baraffe et al., 2002*). Empirical mass determinations for T Tauri stars have been attempted for many years, largely through the study of binary and multiple systems. A thorough analysis of the confrontation of current evolutionary models with empirical mass determinations of T Tauri stars has recently been presented by *Hillenbrand and White* (2004).

There are only a handful of known pre-main-sequence eclipsing binaries (*Mathieu et al., 2000; Covino et al., 2000, 2004; Stassun et al., 2004*), but fortunately optical/near-infrared follow-up studies of known, tight T Tauri binary systems can be used to estimate dynamical masses for non-embedded YSOs. This was first done through statistical means by observing small transverse motion of a number of binary systems, and assuming random orientation of the orbits (*Ghez et al., 1995; Woitas et al., 2001*). This led to the conclusion that the average total system mass for T Tauri multiple systems is about 1.7 $M_\odot$. This will be compared to dynamical masses of embedded multiple systems in section 3.4.

In recent years, dynamical masses were determined for individual systems for which a substantial time coverage of their short orbital period could be achieved, and therefore a Keplerian orbit could be adjusted to the data (*Steffen et al., 2001; Tamazian et al., 2002; Duchêne et al., 2003;* Konopacky et al., in preparation). Mass estimates range from 0.7 to 3.7 $M_\odot$ and are generally in agreement with model predictions. Significant uncertainties due to the limited orbital coverage and poor distance estimates are still left, but within a few years, substantial progress could be achieved. When this is done, it will be possible to discriminate between evolutionary models. For more details, we refer the reader to the chapter by Mathieu et al.

2.4. Spectroscopic T Tauri Binaries

Due to the distance to most star-forming regions, visual binaries are usually detected if their separation exceeds 5–10 AU. However, for solar-type main-sequence field stars, roughly a third of all companions have tighter separations (*Duquennoy and Mayor, 1991*), and cannot be spatially resolved even with current high-angular-resolution devices. These tight systems are particularly interesting, as their formation mechanism may differ dramatically from that of wider systems: Indeed, fragmentation of prestellar cores occurs on much larger linear scales than the sub-AU separation of spectroscopic binaries. On the other hand, it appears that neither fission (*Tohline, 2002*) nor the orbital decay of wider pairs induced by accretion or dynamical interaction within unstable multiple systems (*Bate et al., 2002*) is able to produce the observed large population of systems with orbital periods of a few days to a few months. One must therefore try to detect spectroscopic binaries among T Tauri stars, which is no easy task when considering their strong activity, including the frequent veiling and additional

line emission induced by accretion onto, and magnetic activity in the vicinity of, the central object.

First estimates of the frequency of T Tauri spectroscopic binaries were available since *Mathieu* (1994), but larger samples of objects have been spectroscopically monitored since then. Most noticeably, *Melo* (2003) has surveyed 59 T Tauri stars in 4 nearby star-forming regions during 3 campaigns over 2 years. He found four new double-lined spectroscopic binaries but could not determine their orbits due to limited time coverage. Within the interval $1^d \leq P_{orb} \leq 100^d$, the proportion of companions is on the order of a few percent. Once incompleteness corrections are taken into account, Melo finds almost twice as many companions as among field stars; however, statistical uncertainties are large enough that this is not significant (see Fig. 1). While there could be as strong a multiplicity excess among short-period spectroscopic systems as there is for visual companions, it will not be possible to demonstrate it before much larger samples are monitored.

2.5. High-Order T Tauri Multiple Systems

As soon as the first multiplicity surveys were conducted among T Tauri stars, a few high-order multiple systems were identified, all triples and quadruples. T Tau, the eponymous low-mass pre-main-sequence star, is itself a triple system (*Koresko, 2000*). The number of high-order multiple systems was too limited to pursue any statistical analysis of their frequency and properties at the time of the review

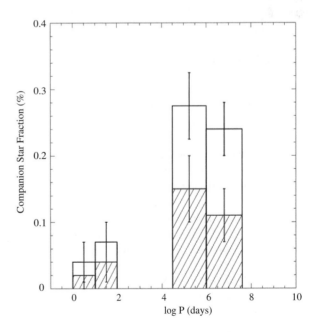

Fig. 1. Orbital period distribution for T Tauri stars in several southern hemisphere star-forming regions (open histogram) compared to that of solar-type field stars [hatched histogram, from *Duquennoy and Mayor* (1991)]; figure from *Melo* (2003). Note the nonsignificant excess of spectroscopic binaries.

by *Mathieu et al.* (2000). Among field solar-type stars, the frequency of high-order multiples is currently being revised from 10× to 4× lower than that of binary systems as high-angular-resolution techniques and large surveys begin to expose the closer and wider companions (*Duquennoy and Mayor,* 1991; *Tokovinin and Smekhov,* 2002). Overall, high-order multiple systems may appear to be of limited numerical strength, but their dynamical importance may be much higher. Estimating the frequency of these high-order systems may therefore turn out to be a more stringent constraint on fragmentation models than the average number of companions per star, irrespective of the system's number of components.

A first survey dedicated to the search for triple systems was conducted by *Koresko* (2002) in Ophiuchus, where he focused on already known binaries. Among 14 targets, he found 2 clear cases of triple systems and 5 more may also be triples, suggesting a high frequency of high-order multiples. More recently, Correia et al. (in preparation) targeted 55 known binaries (from the list of *Reipurth and Zinnecker,* 1993) located in various star-forming regions. They identified 15 triple and quadruple systems, i.e., a ratio of high-order multiples to binaries on the order of 4, similar to recent findings of *Tokovinin and Smekhov* (2002) for main-sequence systems. However, it must be emphasized that the imaging surveys of *Koresko* (2002) and Correia et al. (in preparation) are only sensitive to companions wider than ~10 AU, so that the actual number of triple and quadruple systems may be much higher. Interestingly, *Melo* (2003) suggested that short-period, spectroscopic T Tauri systems have a tendency to host more visual companions that single stars. This result was recently confirmed by *Sterzik et al.* (2005), suggesting that the formation of sub-AU spectroscopic systems may be related to the presence of a third component on a stable outer orbit. For instance, *Kiseleva et al.* (1998) suggested that the combination of Kozai cycles and tidal friction within triple systems with high relative inclination could result in the shrinkage of the inner orbit down to periods of only a few days. While the actual frequency of high-order multiples among T Tauri stars is not yet firmly established, it is likely to place stringent constraints on star-formation models.

Among the important properties of triple and higher-order multiple systems, the relative orientation of the inner and outer orbits can play an important role in the dynamical evolution of the systems, and may also provide insight on their formation mechanism. Among field triple systems, *Sterzik and Tokovinin* (2002) have found a "moderate" alignment of the inner and outer orbits' angular momentum vectors. Unfortunately, there are almost no T Tauri multiple systems for which a similar study can be performed at this point, mostly because of the long orbital periods associated with visual binaries at the distance of the closest star-forming regions. In the unique case of the young hierarchical system V 773 Tau, however, *Duchêne et al.* (2003) have argued that the inner 51-day orbital period is almost coplanar with the outer 46-yr orbital period, although the existing dataset is insufficient to solve all ambiguities in the orbital

solution. In coming years, the increasing number of astrometric orbital solutions for binary young stellar objects, along with the capacity of long-baseline interferometers to spatially resolve known spectroscopic systems, will enable the study of the relative inclination of orbits within pre-main-sequence triple systems.

3. EMBEDDED MULTIPLE PROTOSTARS

As summarized in the previous section, a high multiplicity rate is already established 1 m.y. into the evolution of low-mass stars, as demonstrated by the many observations of populations of T Tauri stars. However, with these observations only, the observed dichotomy between young clusters and loose associations cannot be unambiguously explained with a single mechanism: Different pre-collapse conditions and/or a differential dynamical evolution could be involved. Numerical analysis has shown that close encounters within dense clusters can substantially decrease the frequency of wide companions in less than 1 m.y. (e.g., *Kroupa,* 1995), and that nonhierarchical systems decay to stable configurations through few-body interactions in less than a hundred crossing times, i.e., in 0.1 m.y. or even less (e.g., *Anosova,* 1986; *Sterzik and Durisen,* 1998). It is therefore critical to conduct multiplicity studies in the youngest possible stellar populations in order to determine the "initial conditions" of the evolution of multiple systems. This is why the observational effort in this field has shifted in recent years toward the study of the multiplicity of even younger systems, namely embedded protostellar objects. The existence of extremely young (Class 0) multiple systems (e.g., *Wootten,* 1989; *Loinard,* 2002; *Chandler et al.,* 2005) shows that the formation of multiples most probably occurs very shortly after the apparition of the initial protostellar seeds. Class 0 and Class I sources represent objects whose age is believed to be on the order of a few × 10^4 yr and a few × 10^5 yr, respectively. While they may already be too old to be considered pristine from the point of view of dynamical evolution, these objects provide an opportunity to probe an intermediate stage of the star-formation process, where some evolution has already taken place but, hopefully, is not yet over. Furthermore, it is possible that the youngest (Class 0) protostars have suffered only very little evolution. In this section, we focus on such embedded multiple systems in order to assert some of their basic properties, and how they compare to more evolved T Tauri multiple systems.

3.1. A Combination of Observational Approaches

A few embedded multiple systems were already known at the time of *Protostars and Planets IV* [see, e.g., the discussion of L1551 IRS5 in *Mathieu et al.* (2000)]. However, the first statistical surveys of such objects have only been conducted in the last few years with the advent of a newer generation of instruments. Because they are still enshrouded in their dusty cocoons, the youngest protostars are not detectable at visible wavelengths. They are often dim even in the near- and mid-infrared, and emit most of their luminosity at far-infrared and submillimeter wavelengths, where the angular resolution currently available remains limited. Fortunately, high-angular-resolution techniques are now available in the near-infrared, mid-infrared, and radio domains.

The most embedded, Class 0 protostars are often associated with relatively bright and compact radio emission. Indeed, they frequently power supersonic jets that generate free-free emission detectable at radio wavelengths (*Rodríguez,* 1997), they are surrounded by accretion disks whose thermal dust emission is sometimes still detectable in the centimeter regime (*Loinard et al.,* 2002; *Rodríguez et al.,* 1998, 2005a), and they often have active magnetospheres (*Dulk,* 1985; *Feigelson and Montmerle,* 1999; *Berger et al.,* 2001; *Güdel,* 2002). The former two mechanisms produce emission on linear scales of tens to hundreds of astronomical units, whereas the nonthermal emission related to active magnetospheres is usually thought to result from the interaction of mildly relativistic electrons with the strong magnetic fields (a few kGauss) that are often present at the surface of young, low-mass stars (e.g., *Valenti and Johns-Krull,* 2004; *Symington et al.,* 2005). This process therefore produces emission on very small scales, typically a few stellar radii. Interferometric radio observations (7 mm $\leq \lambda \leq$ 6 cm) can supply images of YSOs with such high angular resolution and excellent astrometric quality: NRAO's Very Large Array (VLA) and Very Long Baseline Array (VLBA) connected interferometers provide typical angular resolution of $0\rlap{.}''1$ and $0\rlap{.}''001$ in combination with astrometric accuracies of $0\rlap{.}''01$ and $0\rlap{.}''0001$, respectively. The combination of these two assets makes it possible with radio interferometry to identify tight binaries among those protostars that emit centimeter radio waves, and study their orbital motions.

Class I protostars, which are more evolved and less deeply embedded, are strong mid-infrared emitters. The 10 μm radiation from each component does not originate from the star itself, but from a "photosphere" of surrounding dust heated to several hundred degrees. According to the radiative transfer model of *Chick and Cassen* (1997), the 10-μm photosphere is located about 1 AU from a low-mass protostar. In the mid-infrared regime, direct imaging on the newest generation of instruments on 6–10-m telescopes provides deep, diffraction-limited ($\leq 0\rlap{.}''3$) images that are extremely sensitive to protostars, and whose spatial resolution largely surpasses current space capabilities. At such a spatial resolution, the individual components of Class I sources should be point sources in the mid-infrared, and easy to disentangle from one another.

As far as groundbased observations are concerned, the near-infrared regime currently offers the best combination of spatial resolution (diffraction limit on the order of $0\rlap{.}''05$ for 8–10-m telescopes), sensitivity, and field of view. The spectral energy distribution (SED) of Class I protostars extends into the near-infrared and, as in the mid-infrared, the light emitted by these objects comes from a very small photosphere that remains unresolved. Near-infrared observations have therefore become one of the most powerful approaches to study the multiplicity of Class I protostars.

Observations in this regime not only allow the discovery of very tight companions, but also provide a probe of the evolutionary state of individual components: the latter is directly related to the spectral index of YSOs in the near- to mid-infrared regime (*Lada,* 1987).

Taking advantage of these complementary techniques, we will now discuss several recent results concerning young embedded multiple systems: their average multiplicity rate, their evolutionary status, and high-precision astrometric follow-up studies (orbital motions, possible dynamical decays).

3.2. The Multiplicity of Low-Mass Embedded Protostars

3.2.1. Radio imaging surveys. One of the first systematic surveys for multiplicity of embedded YSOs was conducted by *Looney et al.* (2000) with the BIMA interferometer at 2.7 mm. Although companions as wide as 15,000 AU could be identified in this survey, one must be cautious that such wide systems may not be physically bound. Considering a 2000 AU upper limit for projected separation, a value frequently used for other multiplicity surveys, and focusing on the Class 0 and Class I sources in their sample, 3 out of the 16 independent targets actually are binaries and none is of higher multiplicity. Still, this is quite a high multiplicity rate considering the relatively large lower limit on projected separation (60–140 AU depending on the molecular cloud): It is somewhat higher than the rate observed for solar-like field stars. Furthermore, it must be emphasized that millimeter observations have a limited sensitivity to YSOs in multiple systems with separations of a few hundred AU because their disks are much reduced as a result of internal dynamics (*Osterloh and Beckwith,* 1995; *Jensen et al.,* 1994). Overall, the limited size of the sample studied by *Looney et al.* (2000) warrants statistically robust conclusions, and the main outcome of this pioneering work was to confirm the prevalence of multiple systems on a wide range of spatial scales among the youngest embedded protostars.

In parallel to this study, *Reipurth* (2000) also emphasized that low-mass embedded objects are frequently binary or multiple. In his study of 14 sources driving giant Herbig-Haro flows (mostly Class 0 and Class I sources), based on a variety of high-angular-resolution datasets, he concluded that they had an observed multiplicity rate between 80% and 90%, and that more than 50% of them were higher-order multiples. He further suggested that strong outflows could be the consequence of accretion outbursts occurring during the dynamical decay of unstable high-order multiple systems. If this statement is correct, the extremely high multiplicity rate may be overestimated due to a selection bias. In posterior VLA continuum studies on a larger sample of mostly Class I sources and with a uniform observing strategy, *Reipurth et al.* (2002, 2004) found a binary frequency of 33% in the separation range from 0.5" to 12" for a sample of 21 embedded objects located between 140 and 800 pc from the Sun. Within the uncertainties, this binary frequency is comparable to the observed binary frequency among

T Tauri stars in a similar separation range. Among this new sample, four out of seven objects that drive giant molecular flows were found to have companions, a marginally higher rate that is not sufficient for any conclusions to be reached on the multiplicity-outflow connection.

3.2.2. Near-infrared imaging surveys. Independent systematic surveys of the multiplicity of embedded protostars were later conducted at near-infrared wavelengths (1–4 μm) on samples of several tens of Class I and "flat spectrum" embedded sources. A first series of surveys (*Haisch et al., 2002, 2004; Duchêne et al., 2004*) were conducted using widefield near-infrared cameras, which permitted seeing-limited observations. Sources in the Perseus, Taurus-Auriga, Chamaeleon, Serpens, and ρ Ophiuchi molecular clouds were targeted. To derive a robust multiplicity rate, we define here a "restricted" companion star fraction by focusing on the 300–1400 AU projected separation range, for which all targets have been observed, and by retaining only those companions that satisfy ΔK ≤ 4 mag, following *Haisch et al.* (2004). Merging all surveys into one large sample, there are 119 targets, for which 19 companions are identified within these limits. This corresponds to a multiplicity rate of 16.0 ± 3.4%, with all clouds presenting entirely consistent rates. The multiplicity rate found for Class I sources in Taurus and ρ Ophiuchi is in excellent agreement with those obtained for T Tauri stars in the same star-forming regions, and is about twice as high as that observed for late-type field dwarfs (see Fig. 2). A few systems present very large near-infrared flux ratios (up to ΔK ~ 6 mag); if the physical nature of these pairs is confirmed, these companions could be candidate proto-brown dwarfs, whose frequency should then be compared to the (very rare) occurrence of wide star-brown dwarf systems among field stars. Alternatively, these systems could be understood as the association of two objects whose evolutionary states are different, with the fainter star still being much more deeply embedded. Only subsequent mid-infrared imaging and/or high-resolution spectroscopy will help disentangle these two options.

To study in more detail the frequency and properties of multiple systems, higher-spatial-resolution observations are required. Indeed, such observations enable the discovery of many more companions and, in particular, of a number of stable hierarchical triple and higher-order multiple systems. Duchêne et al. (in preparation) have recently conducted an adaptive optics imaging survey of 44 Class I protostars located in the Taurus-Auriga, ρ Ophiuchi, Serpens, and Orion (L 1641) molecular clouds. The diffraction-limited images they obtained on the 8-m VLT allowed companions as close as 0".1 (<20 AU in the closest clouds) to be resolved, providing an order of magnitude improvement in spatial resolution from previous surveys, and identifying a dozen subarcsecond companions that were not known previously. Combining these observations with direct images of the same sources from previous surveys, and concentrating on the 36–1400 AU separation range, there are a total of 23 com-

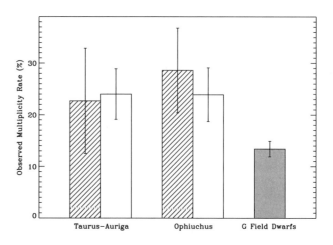

Fig. 2. Observed multiplicity rates in the projected separation range 110–1400 AU for Class I protostars (hatched histograms) and T Tauri stars (open histogram) in the Taurus-Auriga and Ophiuchus molecular clouds; adapted from *Duchêne et al.* (2004). The multiplicity rate for field solar-type stars from *Duquennoy and Mayor* (1991) is shown for reference.

panions fulfilling the ΔK ≤ 4 mag criteria, representing a total multiplicity rate of 52.2 ± 7.5%. Within current statistical uncertainties, observations are consistent with the hypothesis that Class I multiple systems have essentially the same properties in all clouds. Among this high-angular resolution sample, six systems are triple, including five in which the ratio of projected separations is high enough to consider that they are hierarchically stable. No higher-order system was identified in this survey. The observed proportion of triple systems (~14%) is comparable to that observed among T Tauri stars (see section 2.5), and may be higher than the proportion of such systems among field stars.

3.2.3. Spectroscopic surveys. The radio and near-infrared surveys conducted so far have been able to identify companions with separations of several tens to hundreds of AUs. However, as mentioned already, many companions to low-mass field stars are on much tighter orbits. While VLBA observations already provide an opportunity to resolve some tight spectroscopic binary systems if they are strong radio emitters (see section 3.4), it remains difficult to conduct a systematic analysis of the frequency of spectroscopic binaries among embedded protostars with imaging techniques. An alternative approach is to obtain high-resolution spectra of protostars to identify spectroscopic binaries. Despite the difficulty induced by the partially opaque envelope that surrounds these objects, this spectroscopic effort is now ongoing, both at visible and near-infrared wavelengths (see chapter by Greene et al.). Using single-epoch radial velocity measurements for a sample of 31 Class I and flat-spectrum protostars in several star-forming regions, *Covey et al.* (2006) found 4 objects whose radial velocities significantly depart from that of the surrounding local gas velocity. None of the objects were found to be double-line spectroscopic binaries, and their discrepant radial velocities either indicate that they have been ejected after an unstable multibody interaction, or that they are single-line spectroscopic binaries. Long-term monitoring of these objects will reveal their true nature. At any rate, this preliminary study shows that systematic searches for spectroscopic binaries among embedded protostars are now feasible. We can therefore hope to rapidly complement existing imaging surveys to almost completely cover the entire range of orbital periods from a few days up to separations of several thousand AU. Comparing such surveys to the known properties of field stars would be important to determine whether the tightest systems actually form through a different mechanism than wide pairs.

3.2.4. Implications for star formation scenarios. Overall, these surveys for multiplicity among embedded protostars consistently support the general scenario in which all star-forming regions produce a very high fraction of binary and higher-order multiple systems, at least as high as that observed among T Tauri stars in the most binary-rich regions like Taurus-Auriga. While there is marginal evidence for a decrease in multiplicity rate between the most and least embedded protostellar sources (*Duchêne et al.,* 2004), no significant evolution of the multiple system population has been found in regions like Taurus or Ophiuchus within the ≤1-m.y. timescale during which protostars evolve into optically bright T Tauri stars. The absence of mini-clusters of five or more sources within ~2000 AU implies that if cores frequently fragment into many independent seeds, they must decay into unbound stable configurations within a very short timescale, on the order of ~10^5 yr at most. The relatively low number of protostars found to be single in the surveys presented here (≤50%), however, suggests that cores can rarely result in the formation of unstable quadruple or higher-order multiples, supporting the point of view recently presented by *Goodwin and Kroupa* (2005). This and other star-formation models are further discussed in section 4.2.

One of the most intriguing results that arises from these surveys is the finding that there seems to be no influence of environmental conditions on the multiplicity of embedded protostars, as opposed to what is observed for T Tauri multiple systems. Namely, Class I protostars in Orion (in the L 1641 cloud) show as high a multiplicity excess over field stars as all categories of YSOs in the Taurus and ρ Ophiuchi clouds. This seems to favor a scenario in which the end result of core fragmentation is independent of large-scale physical conditions, but rather is sensitive only to small-scale physics, which may well be very similar in all molecular clouds. As already discussed by *Kroupa et al.* (1999), dynamical disruptions among an initial population of binary systems can account for the deficit of wide binaries observed for optically bright YSOs in dense clusters, even though the population of "primordial multiple systems" has universal properties. Based on the present observations, we may conclude that such disruptive encounters, which rep-resent a distinct process from the internal decay of unstable multiple systems, is likely to occur in the densest clusters on a timescale of ~few × ~10^5 yr.

3.3. Evolutionary Status within Multiple Protostars

Haisch et al. (in preparation) have recently obtained new mid-infrared observations of 64 Class I and flat-spectrum objects in the Perseus, Taurus, Chamaeleon I and II, ρ Ophiuchi, and Serpens dark clouds. They detected 45/48 (94%) of the single sources, 16/16 (100%) of the primary components, and 12/16 (75%) of the secondary/triple components of the binary/multiple objects surveyed. The 10-μm fluxes, in conjunction with JHKL photometry from *Haisch et al.* (2002, 2004), were used to construct SEDs for the individual binary/multiple components. Each source was classified using the least-squares fit to the slope of its SED between 2.2 and 10 μm in order to quantify their nature. The classification scheme of *Greene et al.* (1994) has been adopted in our analysis as it is believed to correspond well to the physical stages of evolution of YSOs (e.g., *André and Montmerle,* 1994). A Class I object is one in which the central YSO has attained essentially its entire initial main-sequence mass, but is still surrounded by a remnant infalling envelope and an accretion disk. Flat-spectrum YSOs are characterized by spectra that are strongly veiled by continuum emission from hot, circumstellar dust. Class II sources are surrounded by accretion disks, while Class III YSOs have remnant, or absent, accretion disks. Thus, the progression from the very red Class I YSO → flat spectrum → Class II → Class III has been frequently interpreted as representing an evolutionary sequence, even though *Reipurth* (2000) has suggested that more violent transitions from the embedded to the optically bright stages could occur when components are ejected from unstable multiple systems.

3.3.1. Nature of "mixed" systems. While the composite SEDs for all YSOs in the Haisch et al. (in preparation) study are either Class I or flat-spectrum, the individual source components sometimes display Class II, or in one case Class III, spectral indices. The SED classes of the primary and secondary components are frequently different. For example, a Class I object may be found to be paired with a flat spectrum source, or a flat-spectrum source paired with a Class II YSO. Such behavior is not consistent with what one typically finds for T Tauri stars, where the companion of a classical T Tauri star also tends to be a classical T Tauri star (*Prato and Simon,* 1997; *Duchêne et al.,* 1999), although mixed pairings have been previously observed among Class II YSOs (e.g., *Ressler and Barsony,* 2001). Indeed, there appears to be a higher proportion of mixed Class I/flat-spectrum systems (67%) (Haisch et al., in preparation) than of mixed CTTS/WTTS systems (25%) (*Hartigan and Kenyon,* 2003; *Prato et al.,* 2003; *McCabe et al.,* 2006; see chapter by Monin et al.).

While several of the Class I/flat-spectrum binary components lie in regions of a JHKL color-color diagram that are not consistent with their SED classes, they all lie in their expected locations in a KLN color-color diagram. In fact, in this diagram, one can see a clear progression from the very red Class I YSO → flat spectrum → Class II (Haisch et al., in preparation). Taken together with the above discussion, this demonstrates the fact that while in most cases the SED class reflects the evolutionary state of the YSO, there may be instances in which the SED class does not yield the correct evolutionary state. The rigorously correct way to determine an objects' evolutionary state is to obtain multiwavelength imaging data for each source and quantitatively compare these data to models produced using three-dimensional radiative transfer codes (*Whitney et al., 2003*).

Visual extinctions, A_v, have been determined for all binary/multiple components, except the Class I sources, for which accurate dereddened colors cannot be derived using infrared color-color diagrams. In general, the individual binary/multiple components suffer very similar extinctions, A_v, suggesting that most of the line-of-sight material is either foreground to the molecular cloud or circumbinary in nature.

3.3.2. Notes on selected objects. Among the various mid-infrared surveys to date, several sources deserve specific mention. A detailed study of WL 20, the only nonhierarchical triple system among embedded protostars, by *Ressler and Barsony (2001)* and *Barsony et al. (2002)* has suggested that disk interaction has resulted in enhanced accretion onto one component of this system, WL 20S. This tidally induced disk disturbance could explain the Class I SED of this object, although it is probably coeval, at an age of several million years, with its Class II companions. On the other hand, the recent high-angular-resolution images obtained in the near-infrared by Duchêne et al. (in preparation) as part of their multiplicity survey revealed a completely unexpected morphology that does not seem to be consistent with a ~1 AU opaque envelope photosphere. Rather, it seems like WL 20S is a normal T Tauri system whose circumstellar material has such a geometry that only scattered light reaches the observer shortward of ~5 μm, emphasizing the need for high-angular-resolution images to determine the actual nature of an embedded YSO.

IRS 43 (also known as YLW 15) in Ophiuchus was found to be a binary VLA source with 0.″6 separation (*Girart et al., 2000*). IRS 43 is also part of a wide-binary system with GY 263 (*Allen et al., 2002*). *Haisch et al. (2002)* find IRS 43 to be multiple at 10 μm but single in the near-infrared. The brighter mid-infrared source in IRS 43 corresponds to VLA 2 and the heavily veiled, Class I near-infrared source. VLA 1 is an embedded protostar, undetected in the near-infrared, and possibly in the Class 0 to Class I transition and powering a Herbig-Haro outflow. Its mid-infrared emission appears slightly resolved with a diameter of ~16 AU, possibly tracing circumstellar material from both the envelope and the disk (*Girart et al., 2004*). Both VLA/mid-infrared sources associated with IRS 43 are embedded in extended, faint near-infrared nebulosity imaged with HST/NICMOS (*Allen et al., 2002*). Strikingly, the near- to mid-infrared

properties of YLW 15 suggest that VLA 1 is a more embedded YSO, or alternatively, less luminous than VLA 2, whereas orbital proper motions of this binary system by *Curiel et al. (2003)* indicate that VLA 1 is more massive than VLA 2. This is apparently against the expected evolutionary scenario, in which one expects that the more massive YSO in a binary system is the more evolved and more luminous YSO.

Another source, ISO-Cha I 97 in Chamaeleon I, was detected as a single star in the near-infrared; however, mid-infrared observations have revealed that this source is in fact binary (Haisch et al., in preparation). The K-band sensitivity limit from *Haisch et al. (2004)*, combined with its 10-μm flux, yields an extremely steep lower limit to the spectral index that places ISO-Cha I 97 in a class of YSO that has heretofore been rarely known. Three such objects have been recently reported, the Class 0 object Cep E mm (*Noriega-Crespo et al., 2004*), source X_E in R CrA (*Hamaguchi et al., 2005*), and source L1448 IRS 3 A (*Ciardi et al., 2003*; *Tsujimoto et al., 2005*). Further very steep spectrum YSOs are expected to be discovered with the Spitzer Space Telescope.

Spatially resolved mid-infrared spectroscopy of the Class I/flat-spectrum protostellar binary system, SVS 20 in the Serpens cloud core, has been recently obtained by *Ciardi et al. (2005)*. SVS 20S, the more luminous of the two sources, exhibits a mid-infrared emission spectrum peaking near 11.3 μm, while SVS 20N exhibits a shallow amorphous silicate absorption spectrum with a peak optical depth of τ ~ 0.3. After removal of the line-of-sight extinction by the molecular common envelope, the "protostar-only" spectra are found to be dominated by strong amorphous olivine emission peaking near 10 μm. There is also evidence for emission from crystalline forsterite and enstatite associated with both SVS 20S and SVS 20N. The presence of crystalline silicate in such a young binary system indicates that the grain processing found in more evolved Herbig Ae/Be and T Tauri pre-main-sequence stars likely begins at a relatively young evolutionary stage, while mass accretion is still ongoing. A third component to the system was found by Duchêne et al. (in preparation), making the analysis of the system even more complex.

Finally, *Meeus et al. (2003)* have presented mid-infrared spectroscopy of three T Tauri stars in the young Chamaeleon I dark cloud, CR Cha, Glass I, and VW Cha, in which the silicate emission band at 9.7 μm is prominent. This emission was modeled with a mixture of amorphous olivine grains of different size, crystalline silicates, and silica. The fractional mass of these various components was found to change widely from star to star. While the spectrum of CR Cha is dominated by small amorphous silicates, in VW Cha (and in a lesser degree in Glass I), there is clear evidence of a large amount of processed dust in the form of crystalline silicates and large amorphous grains. Interestingly, the two objects with an "evolved" dust population are associated with a tight companion, leading to the intriguing speculation that multiplicity may accelerate dust processing in circumstellar disks.

3.4. High-Accuracy Astrometry of Embedded Multiples

As pointed out in section 3.1, VLA and VLBA observations can provide a view of embedded protostars with an unsurpassed resolution and astrometric accuracy. This allows a number of studies that are impossible to conduct with shorter wavelength observations. We summarize here some recent results that take full advantage of VLA and VLBA capabilities to study embedded multiple systems.

3.4.1. Orbital motion within embedded multiple systems. For a very limited subset of tight radio binaries associated with low-luminosity Class 0 and Class I deeply embedded sources, orbital motions have now been detected by comparing images taken at several epochs. It is very important to determine the system masses in order to constrain the time evolution of the central sources, and, for instance, to determine the fate of the material located in the circumstellar envelopes of Class I sources. Comparable estimates are already available for T Tauri stars (see section 2.3).

Due to slightly poorer spatial resolution of VLA observations compared to the highest angular resolution near-infrared datasets, the embedded multiple systems for which orbital motion was detected typically have orbital periods of hundreds of years. Therefore, the multi-epoch observations to date only cover a small fraction of the orbit, and the exact orbital parameters cannot be derived. However, a reasonable estimate of the mass of the system can still be obtained assuming circular orbits, if the inclination can be guessed from independent means. The four sources where this could be achieved are IRAS 16293–2422 (*Loinard*, 2002; *Chandler et al.*, 2005) and YLW 15 (*Curiel et al.*, 2003) in ρ Ophiuchi, and L 1527 (*Loinard et al.*, 2002) and L 1551 (*Rodríguez et al.*, 2003) in Taurus-Auriga. The mass estimates, respectively, are $2.8 \pm 0.7 \, M_\odot$; $1.7 \pm 0.8 \, M_\odot$; $1.0 \pm 0.5 \, M_\odot$, and $1.2 \pm 0.5 \, M_\odot$.

The mean value is therefore $1.7 \pm 0.7 \, M_\odot$, confirming that these low-luminosity, deeply embedded protostars are most likely the precursors of solar-type stars. Noticeably, this average mass is not significantly lower than that derived for T Tauri binary systems. Although small number statistics and selection biases preclude definitive conclusions at this stage, this seems to imply that the mass of the remnant envelopes in these systems are already significantly lower than the stellar seeds themselves; this is expected for Class I sources but may be more surprising for Class 0 sources. Future astrometric follow-up studies on these and other embedded tight binary systems will help clarify this issue.

3.4.2. Spectroscopic systems resolved with interferometry. Most radio observations of embedded protostars have been conducted with the VLA. As mentioned above, the VLBA offers an angular resolution and an astrometric precision two orders of magnitude better than the VLA, but can detect YSOs that emit bright and compact nonthermal radio emission. Although all protostars probably do generate nonthermal radiation at some level, those that are cur-

rently easily detectable represent only a limited sample. Recently, Loinard and coworkers have started to monitor 10 protostars in the Taurus and ρ Ophiuchi star-forming regions that are known to be nonthermal radio sources. Interestingly, at least two of these systems were found to be multiple with separations of only a few times 0″.001, or a few tenths of an AU at the distance of these molecular clouds. The clearest case is that of V773 Tau, which was previously known to be a spectroscopic binary (*Welty*, 1995), and had been found to be a double in previous VLBI observations (*Phillips et al.*, 1996). While V773 Tau is a T Tauri system, more embedded systems are among the sample studied in this survey and, if they are found to have a tight companion, they would represent the earliest stage at which spectroscopic binaries can actually be spatially resolved. In parallel to this VLBA approach, near- and/or mid-infrared long baseline interferometers (such as Keck or VLTI) could resolve these systems in the near future, allowing for powerful infrared-to-radio analysis. In any case, these very tight systems clearly have a much shorter period (≤1 yr) than the embedded multiple systems resolved so far, so one should be able to accurately measure their masses in a comparatively shorter time.

3.4.3. Hints of disruption? With high-angular resolution and high astrometric accuracy, following the relative motion within multiple systems may reveal departures from bound Keplerian orbits. In particular, the possibility of studying the internal dynamics of unstable multiple systems, if some can be found, is extremely appealing. A special case of this decay is the violent disruption of a triple system into an ejected single star and a stable binary system. Only radio observations with the VLA combine high resolution and astrometric accuracy with a relatively long time baseline. It is therefore not a surprise that the first two claims of possible disruption of young multiple systems have come from such observations.

Using multi-epoch archival VLA observations of the triple system T Tauri over almost 20 years, *Loinard et al.* (2003) have found that the trajectory of one of its components could not be easily explained in terms of a stable Keplerian orbit. Follow-up near-infrared observations have cast doubt upon this conclusion as the orbit of the putative ejected component has been seen to slow down and curve as if it were on a Keplerian orbit (*Furlan et al.*, 2003; *Tamazian*, 2004; *Beck et al.*, 2004). However, no satisfying orbit has been fit simultaneously to all infrared and radio observations, and the exact correspondence of the detected sources at both wavelengths is still under debate (*Johnston et al.*, 2004, 2005; *Loinard et al.*, 2005). Following the system in both wavelength regimes for a few more years should yield a final conclusion to this issue.

More recently, *Rodríguez et al.* (2005b) and *Gómez et al.* (2006) have shown that three of the four compact radio sources around the Orion Becklin-Neugebauer/Kleinmann-Low region are moving away from a common point of origin, where they must have been located about 500 years ago (Fig. 3). These three radio sources are apparently associated

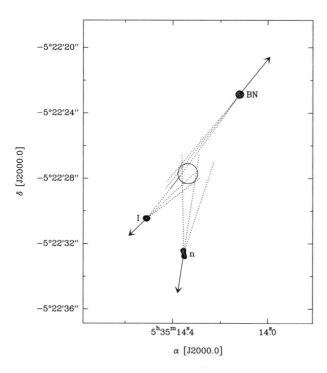

Fig. 3. Proper motion of three radio sources in the Becklin-Neugebauer/Kleinman-Low region, from *Gómez et al.* (2006). They all trace back to the same point in space and time, 500 years ago.

with relatively massive young stars (M > 8 M$_\odot$), suggesting that a massive multiple system disintegrated around that time. Although a different point of origin for the Becklin-Neugebauer object and an eightfold longer timescale has been advocated by *Tan* (2004), there is little doubt that this system was formed as an unstable multiple system, and has very recently experienced a dynamical ejection event.

These two cases remain open to discussion, and are insufficient to assess the exact relevance of few-body encounters to the final rates of multiplicity in stellar systems. However, they demonstrate the potential of radio interferometry to tackle this important issue, and should be pursued in upcoming years.

4. CORE FRAGMENTATION AND EARLY DYNAMICAL EVOLUTION OF MULTIPLE SYSTEMS

In this section, we review some of the most recent numerical simulations that aim at following the processes of collapse and fragmentation of a prestellar core as well as the subsequent dynamical interactions within the multiple systems resulting from fragmentation. More specifically, we concentrate on three different models of star formation, namely those by Goodwin and collaborators (*Goodwin et al.*, 2004a,b), Sterzik and collaborators (*Sterzik and Durisen*, 1995, 1998; *Durisen et al.*, 2001; *Sterzik et al.*, 2003),

and Delgado-Donate and collaborators (*Delgado-Donate et al.*, 2004a,b), with special emphasis on the latter two. A more detailed description of other simulations can be found in the chapters by Whitworth et al. and Goodwin et al.

4.1. A Brief Overview of Current Simulations

For some time, numerical models with predictive power on the statistical properties of young stars (e.g. multiplicity fractions, mass ratio, semimajor axis distributions) had to rely on pure N-body integration of the breakup of small clusters of point masses (*Sterzik and Durisen*, 1995, 1998; *Durisen et al.*, 2001). The masses, location, and velocities of the stars had to be selected at the outset, and subsequently the orbital evolution was calculated. This approach to multiple star "formation" has the advantage of being easy and fast to calculate, so that many realizations of the same initial conditions could be run. However, it completely neglects the modeling of gas fragmentation, collapse, and accretion, a highly demanding task from a computational point of view. Yet gas is a fundamental ingredient of the star-formation process, not only during the fragmentation and collapse stage, but also during the embedded phase of the life of a star. Gaseous material accumulates in the form of accretion disks around the protostars, and these disks can modify substantially both the orbital parameters of a protobinary (*Artymowicz and Lubow*, 1996; *Bate and Bonnell*, 1997; *Ochi et al.*, 2005) and the outcome of dynamical encounters with other cluster members (*McDonald and Clarke*, 1995). Furthermore, under adequate physical conditions, disks can fragment, and in doing so, produce a second generation of objects (*Gammie*, 2001; *Lodato and Rice*, 2005). Gas also acts on the large scale throughout the embedded phase of star formation by providing a substantial contribution to the gravitational potential of the system. In this manner, gas can affect the mass evolution and motion of both single and multiple stars, hence the binary pairing outcomes, through the action of gravitational drag (*Bonnell et al.*, 1997, 2003; *Delgado-Donate et al.*, 2003). Although they prioritize the N-body dynamics over the gas dynamical processes, the simulations by Sterzik and collaborators provide useful constraints on what effects dynamical interactions alone can have on the star-formation process. Interestingly, these calculations provide the best match to date to the mass dependence of the multiplicity fraction of stars, as is constrained by our present observational knowledge. They do so by means of a two-step procedure (*Durisen et al.*, 2001), whereby the stellar masses are picked randomly from a stellar mass function, subject to the additional constraint that the total cluster mass equals a value also picked randomly from a cluster mass function. This way, they alleviate the usually steepening effect of the process known as dynamical biasing — i.e., the strong trend of the two most massive stars in a cluster to pair together — on the multiplicity-vs.-primary mass curve, by having a significant number of clusters where the two most massive stars both have low masses. This major success of N-body models is unmatched by current gas dynamical calculations, which so far are just able to give a positive but too steep dependence of the multiplicity fraction on primary mass (see section 4.2). This success should not mask an obvious caveat, however: It remains unclear at present how a cluster may break into subunits that fragment into stars with the appropriate mass spectrum set by the two-step process.

Early star-formation models that included the effect of gas did so to study the formation of binary stars from clouds subject to some kind of specific initial instability (see reviews by *Sigalotti and Klapp,* 2001; *Tohline,* 2002). These models have been of great importance, but a caveat remained: They produce a low number of objects in a more or less predictable fashion. Other models tried to take into account large numbers of stars embedded in a big gas cloud (*Bonnell et al.,* 2001), e.g., by utilizing point masses with the ability to accrete and interact with the gas and other stars ("sink particles") (*Bate et al.,* 1995), but once more, with positions and velocities selected at the outset. These models focused mostly on the study of the resulting initial mass function. The purely gas dynamical models had to be refined and taken to a larger scale, while the aim of point masses-in-gas models had to shift to the study of the properties of multiple stars, if star-formation models were to match the predictive power of N-body models. The earliest model to take such step was that by *Bate et al.* (2003), who applied more general "turbulent" initial conditions to a relatively large (for theory standards) 50 $M_\odot$ cloud and followed its fragmentation and collapse down to the opacity limit for fragmentation. For higher densities, pressure-supported objects were replaced by "sink particles," and thus the simulation could be followed well beyond the formation of the first objects. This calculation showed the power of the combination of more realistic initial conditions and a refined numerical scheme blending gas with N-body dynamics and, beyond any doubt, it meant a great leap forward in star-formation studies; but, obviously, it had some shortcomings as well. Among them was the high computational expense involved, and the fact that the evolution of the cloud could not be followed for as long as it would be desirable in order to ensure the stability of most multiples. Thus, complementary calculations were necessary, and these were performed mainly by Delgado-Donate and collaborators and Goodwin and collaborators. The *Goodwin et al.* (2004a,b) and *Delgado-Donate et al.* (2004a,b) simulations model the fragmentation of small (≈ 5 $M_\odot$) molecular clouds subject to different degrees of internal turbulent motion. Their models basically differ in the resolution employed, and the different subsets of parameter space studied: Most importantly for this review, *Goodwin et al.* (2004a,b) study subsonic turbulence, whereas *Delgado-Donate et al.* (2004a,b) impose supersonic to hypersonic random velocity fields. Both sets of simulations address the solution of the fluid equations (smoothed particle hydrodynamics, or SPH) and the dynamic creation of point masses to replace collapsed gas fragments in a similar manner. In the Delgado-Donate et al. calculations, SPH simulations follow the gas-domi-

nated stage during ≈ 0.5 m.y. before switching to an N-body integration followed until the stability of most of the multiples could be guaranteed. Each set of initial conditions is run 10 to 20 times, varying only the spectrum of the turbulent velocity field imposed initially, in order to obtain statistically significant average properties.

Finally, some numericists focused on the largest scales, and tried to study the collapse and fragmentation of large clouds (100 to a few 1000 $M_\odot$) (*Mac Low and Klessen,* 2004; *Padoan and Nordlund,* 2002). These models reproduce the filamentary structure observed in molecular clouds, and find that cores are not quasistatic structures, but rather grow in mass by accretion and merge hierarchically until a specific core mass function (resembling the initial mass function at the high mass end) is built. While these simulations provide an important step for our understanding of the collapse of entire molecular clouds, they have little, if any, predictive power regarding the properties of individual or multiple stars. We do not consider these simulations in this review, which focusses on testing models against some of the most constraining observational data available, the properties of multiple stars at the earliest stages. These simulations are discussed in the chapters by Ballesteros-Paredes et al. and Klein et al..

4.2. Predictions and Comparisons with Observations

These numerical simulations make a number of predictions regarding the statistical properties of young multiple systems. Ideally, these predictions should be tested against the observational results summarized in *Mathieu et al.* (2000) and in previous sections of this chapter, for instance. However, because of the limited parameter space that has been explored to date, and the daunting task of including all relevant physical processes, such predictions may still be premature. In the following, we consider a few such predictions, focusing primarily on general trends rather than detailed quantitative predictions, and briefly mention some other models that could be relevant for the formation of multiple systems.

4.2.1. Triples and higher-order multiples. The *Delgado-Donate et al.* (2004a,b) simulations produce a wealth of multiple systems. The companion frequency (average number of companions per primary) at 0.5 m.y. after the initiation of star formation is close to 100%, whereas the frequency of multiple systems (ratio of all binary and higher-order systems to all primaries) is $\approx 20\%$; in other words, for each binary/multiple system, there are four isolated single objects. Clearly, multiple star formation is a major channel for star formation in turbulent flows, as found also by *Bate et al.* (2003), *Bate* (2005), and *Goodwin et al.* (2004a,b), but in these simulations, high-order multiples are more frequent than binaries. The systems can adopt a variety of configurations, like binaries orbiting binaries or triples. Such exotic systems have been observed, and currently, the occurrence of high-order multiples among main-sequence field stars is on the order of ~15–25% (e.g., *Tokovinin,* 2004). A similar proportion of multiple systems was found in both T Tauri and embedded protostars populations (see section 2.5 and 3.2). The *Goodwin et al.* (2004b) calculations, characterized by a very low ratio of initial turbulent energy to thermal energy, produce a significantly lower number of stars per core, and match better the observed multiplicity fractions. This has led *Goodwin and Kroupa* (2005) to propose that the main mode for star formation involves the breakup of a core into two to three stars, a larger number being a rare outcome. This is a revealing constraint on star-formation theories although it does not settle the question of how this main mode of two to three fragments comes to be (low turbulence is a possibility but there may be others) and which multiplicity properties we should expect from it.

4.2.2. Multiplicity as a function of age. *Delgado-Donate et al.* (2004a,b) found that the companion frequency decreases during the first few million years of N-body evolution, as many of the initial multiple systems are unstable. It must be noted that, although the canonical timescale for dynamical breakup of an unstable multiple is at most 10^5 yr, this timescale is significantly increased in these gas-dynamical simulations because of two effects. First, numerous companions form on orbits with very large separations, up to orbital periods approaching the 10^5-yr timescale. Second, it is found in gas-dynamical calculations that star formation always occurs in bursts, which repeat themselves with decreasing intensity for several cluster freefall times, i.e., several × 10^5 yr, adding new stars to already formed, and maybe already stable, multiples. While the former effect may be a somewhat artificial consequence of the selected initial conditions, the latter is very robust, and likely applies to most star-formation scenarios.

The total companion frequency is seen to rapidly decay from ≈100% to ≈30%. This internal decay affects mostly low-mass outliers, which are released in vast amounts to the field. It might be expected that in a real cluster the companionship would drop even further — or sooner — as star-forming cores do not form in isolation but close to one another. Weakly bound outliers might have been stripped sooner by torques from other cluster members [see simulations by *Kroupa et al.* (1999), for instance]. The total frequency of binary and high-order multiples, on the other hand, varies little after the first few × 10^5 yr. Thus, although a multiple system is still likely to evolve further toward its hierarchical stable configuration in longer timescales, the relative frequency of singles, binaries, triples, and so on, seems to be essentially established after a few × 10^5 yr.

Based on these models, one may expect that the frequency of binary/high-order multiples among T Tauri stars, which are already a few million years old, should be essentially the same as that of main-sequence field stars. While this is true for young clusters, loose associations clearly show a higher companion frequency, in disagreement with this prediction. It must be emphasized that the simulations discussed here consider prestellar cores as isolated entities, whereas core-core interactions may play an important role in shaping the outcome of the star-formation process. None-

theless, to reconcile these simulations with the observations, one can argue that a large number of the single objects produced by the simulations are not included in the observed samples of T Tauri stars, either because they are of too low mass or because they have already been expelled from the molecular cloud owing to their high ejection velocities (see below) and the shallow potential well of the cloud. Both explanations have their own weaknesses: On the one hand, the ratio of stellar-mass objects to brown dwarfs is on the order of 4 in the Taurus molecular cloud (*Guieu et al.,* 2006) and low-mass stars are frequently members of binary and multiple systems (see section 2.2), and on the other hand, the much younger embedded protostars — although they may already be too old to be compared with the simulations — do not appear to have a dramatically lower proportion of single stars (see section 3.2).

4.2.3. Multiplicity as a function of primary mass. The *Delgado-Donate et al.* (2004a,b) models find a positive dependence of the multiplicity fraction on primary mass (see Fig. 4), in qualitative agreement with observations. The low- and high-mass end of the distribution, however, do not match satisfactorily the observations; the dependence on mass is too steep. The models fail to produce as many low-mass binaries as observed because of their extremely low binding energy, making them prone to disruption in an environ-

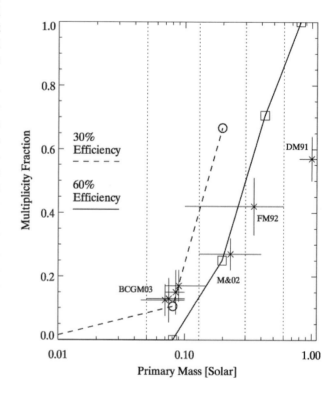

Fig. 4. Predicted mass-dependence of the multiplicity rate (solid and dashed lines), from *Delgado-Donate et al.* (2004b). Observational data points are for field stars [*Duquennoy and Mayor* (1991), *Fischer and Marcy* (1992), *Marchal et al.* (2003), *Bouy et al.* (2003), in order of decreasing primary mass].

ment dominated by dynamical interactions. Observationally, a binary fraction of at least 15% is seen among field brown dwarfs (e.g., *Bouy et al.,* 2003, *Martín et al.,* 2003), and values as high as 30–40% have been suggested (*Maxted and Jeffries,* 2005; but see *Joergens,* 2006). Also, the high-mass end shows a paucity of singles. This is a common outcome of all "turbulent" star-formation simulations to date, and stems from the fact that the most massive members of the cluster are always in binaries and thus the binary fraction at the high-mass end is close to 100%, whereas even in the Taurus-Auriga cloud, there are a number of ≥ 1 $M_\odot$ T Tauri stars that have no known companion. Calculations by *Goodwin et al.* (2004a,b) also find similar problems to fit the observed multiplicity dependence of primary mass. It is likely that by simulating the evolution of an ensemble of clouds with different initial masses, thus following the successful prescription of a two-step procedure pioneered by *Durisen et al.* (2001), the problem would be lessened, but it is unclear at the moment — until parameter space is more widely investigated — whether it would solve it completely or not. The problem of the formation of a significant population of low-mass/brown dwarf binaries would remain, and it has been suggested that initial conditions less prone to fragmentation or resolution effects may provide a possible solution to this riddle (Clarke and Delgado-Donate, in preparation). As mentioned before, *Durisen et al.* (2001) and *Sterzik and Durisen* (2003) extensively discuss how a steep multiplicity vs. mass correlation can be smoothed, and manage to do so by means of their two-step mass selection.

4.2.4. Velocity dispersion. The simulations show that single and binary stars attain comparable velocities in the range 1–10 km s⁻¹, whereas higher-order multiples display lower-velocity dispersions. This kinematic segregation as a function of N is the expected outcome of the breakup of unstable multiples, whereby the ejected objects (typically singles, or less often binaries) acquire large velocities, whereas the remaining more massive multiple recoils with a lower speed. Among the singles and binaries, the peak of the velocity distribution is on the order of a few km s⁻¹, in the range of the cloud random velocities. A similar velocity distribution is produced by N-body models (*Sterzik and Durisen,* 2003) and models with lower levels of turbulence (*Goodwin et al.,* 2004a,b). Therefore, we would expect low-mass star-forming regions like Taurus, where a local kinematic segregation may survive against the influence of large-scale dynamics, to display an overabundance of multiple systems in the densest regions, from where the high-speed low-mass singles would have escaped. This prediction was made by *Delgado-Donate et al.* (2004b), and has been recently supported by the simulations of *Bate* (2005).

On the observational side, we note that the radial velocity outliers found by *Covey et al.* (2006) and discussed in section 3.2 could be ejected protostars; only a long-term monitoring of their radial velocities will help determine their status. From another perspective, the most recent survey for low-mass Taurus members by *Guieu et al.* (2006), covering several times the area of previous surveys, has found

that the fraction of brown dwarfs increases as one moves away from the densest cores, known before to be overabundant in binaries. This could be an indication for an average larger speed for the lower-mass single objects, as predicted. A complete analysis of the spatial distribution of multiple systems in the Taurus cloud has not been performed yet, but would provide a crucial test of this prediction. Furthermore, there are some caveats on whether the census of low-mass stars in the extended area covered by *Guieu et al.* (2006) is complete. More observational work is still required to test this critical prediction of numerical simulations.

4.2.5. Other models. So far, we have reviewed models based on the pure N-body breakup of small clusters and models where an "initial" turbulent velocity field is imposed to the cloud, so that its decay triggers the formation of structure until the Jeans instability takes over and produce multiple fragmentation. However, there are researchers that advocate a less dynamic view of star formation. This more quiescent star-formation mode may be thought of as an extension of the *Shu et al.* (1987) paradigm of single-star formation to multiple systems, whereby a core in quasistatic equilibrium collapses from the inside out in such a way that only a few independent fragments are formed. The statistical properties of multiple systems formed in this way are almost impossible to predict in the absence of a detailed physical framework for this "quiescent fragmentation," but can be constrained *a posteriori.* To simultaneously match the high frequency of companions to low-mass stars and the paucity of quadruple and higher-order multiples among populations of T Tauri stars and embedded protostars, *Goodwin and Kroupa* (2005) have concluded that this star-formation mode must produce primarily binary and triple systems. *Goodwin et al.*'s (2004a,b) low turbulence simulations are the closest we have at the moment to a paradigm of not-so-dynamic star formation. Alternatively, *Sterzik et al.* (2003) also find that clusters with low N are a better match for current observations.

In addition, there exists the possibility that the numerical scheme used in most models reviewed in this section, i.e., SPH, may not perform entirely satisfactorily in some of the regimes modeled, especially when shear flows or voids are involved. There are alternatives to SPH, most of them based on adaptive mesh refinement techniques, that could offer a different view on the problem. Efforts by Padoan and collaborators go in that direction, as well as those by Klein and coworkers, but the complexity of the codes and the implementation of sink particles or their equivalent for grid codes, essential to follow star-formation calculations beyond the formation of the first star, have proved a serious obstacle so far to produce simulations comparable in predictive power to the SPH ones. In addition, the role of feedback in the star-formation process, e.g., through outflows, has never been included in such simulations, although it may be important. The effects of photoionizing feedback by massive stars have been preliminarily studied by *Dale et al.* (2005), who find that photoionization fronts may have both a positive and negative effect in star formation, by triggering fragmentation and collapse at the HII fronts, or disrupting incoming accretion flows respectively. These new developments are likely to shed light on some of the issues over which theory stays the furthest apart from observations.

5. CONCLUSIONS AND PERSPECTIVES

Binary and higher-order multiple systems represent the preferred outcome of the star-formation process; studying the statistical properties of these systems at various evolutionary stages, therefore, offers indirect constraints on the core fragmentation and on the subsequent dynamical interaction between gas and stars. While the (high) frequency of million-year-old T Tauri stars has now been long established, there has been tremendous progress in recent years: New populations of young stars have been surveyed, the frequency of high-order multiple and spectroscopic binaries among T Tauri stars is being accurately estimated, and embedded protostars have for the first time been surveyed for multiplicity. In the meantime, numerical simulations describing the fragmentation and dynamical evolution of prestellar cores toward fully formed stars and multiple systems have made tremendous progress, and while they may not yet allow for a fine comparison of predictions with observations, they already predict significant trends that can be tested.

All these studies have provided important clues toward the star-formation process, but a number of open questions remain to be solved. For instance, the apparent uniformity of the multiplicity rate of embedded protostars independent of environment is quite puzzling given the strong dependence to initial conditions of all numerical simulations of core fragmentation. Another surprising observational result is the existence of a fairly large proportion of low-binding energy multiple systems, which rarely survive the violent early star-and-gas dynamical evolution in numerical simulations of collapse and fragmentation. The absence of aggregates of more than four to five stars on scales of a couple thousand AU is also surprising, as they seem to be ubiquitous in numerical simulations of the fragmentation and collapse of gas clouds. Could we be missing a number of young stars in molecular clouds, completely biasing our multiplicity surveys? If so, then we may wonder how useful the traditional Class 0–I–II–III evolutionary sequence really is: If stars usually form as part of unstable multiple systems, then many stars probably "jump" from one category to another over very short timescales, which could have dramatic consequences for their circumstellar environments. The existence of systems pairing stars of different evolutionary categories (including the "infrared companion" systems among T Tauri stars, which are not discussed in this chapter) could be footprints of this violent evolution, and would deserve increased attention in upcoming years.

To investigate these and many other issues described above, continuing both survey and follow-up efforts related to the multiplicity of young stars appears as a crucial endeavor for the future. While high-angular-resolution ground-based infrared methods are bound to provide important new results, one must also remember that radio interferometric observations have the potential to complement infrared surveys in two major ways: first, by allowing the study of the youngest and most embedded protostars so far inaccessible to other techniques, and second, by providing images of extremely tight systems, which are so far only known to be binaries because of spectroscopy. They may also offer opportunities to examine the results of few-body disruption of initially nonhierarchical systems. In parallel to these observational efforts, more numerical simulations must be run to sample a wider parameter space than has been currently explored, and a general effort to include as many physical effects as possible must be undertaken in upcoming years to allow for direct comparisons of simulations with observations.

REFERENCES

Allen L. E., Myers P. C., Di Francesco J., Mathieu R., Chen H., and Young E. (2002) *Astrophys. J., 566,* 993–1004.
André P. and Montmerle T. (1994) *Astrophys. J., 420,* 837–862.
Anosova J. P. (1986) *Astrophys. Space Sci., 124,* 217–241.
Artymowicz P. and Lubow S. H. (1996) *Astrophys. J., 467,* L77–L80.
Baraffe I., Chabrier G., Allard F., and Hauschildt P. H. (2002) *Astron. Astrophys., 382,* 563–572.
Barsony M., Greene T. P., and Blake G. A. (2002) *Astrophys. J., 572,* L75–L78.
Bate M. R. (2005) *Mon. Not. R. Astron. Soc., 363,* 363–378 .
Bate M. R. and Bonnell I. A. (1997) *Mon. Not. R. Astron. Soc., 285,* 33–48.
Bate M. R., Bonnell I. A., and Price N. M. (1995) *Mon. Not. R. Astron. Soc., 277,* 362–376.
Bate M. R., Bonnell I. A., and Bromm V. (2002) *Mon. Not. R. Astron. Soc., 336,* 705–713.
Bate M. R., Bonnell I. A., and Bromm V. (2003) *Mon. Not. R. Astron. Soc., 339,* 577–599.
Beck T. L., Schaefer G. H., Simon M., Prato L., Stoesz J. A., and Howell R. R. (2004) *Astrophys. J., 614,* 235–251.
Berger E., Ball S., Becker K. M., Clarke M., Frail D. A., Fukuda T. A., Hoffman I. M., Mellon R., Momjian E., Murphy N. W., Teng S. H., Woodruff T., Zauderer B. A., and Zavala R. T. (2001) *Nature, 410,* 338–340.
Bonnell I. A., Bate M. R., Clarke C. J., and Pringle J. E. (1997) *Mon. Not. R. Astron. Soc., 285,* 201–208.
Bonnell I. A., Bate M. R., Clarke C. J., and Pringle J. E. (2001) *Mon. Not. R. Astron. Soc., 323,* 785–794.
Bonnell I. A., Bate M. R., and Vine S. G. (2003) *Mon. Not. R. Astron. Soc., 343,* 413–418.
Bouy H., Brandner W., Martín E. L., Delfosse X., Allard F., and Basri G. (2003) *Astron. J., 126,* 1526–1554.
Brandeker A., Jayawardhana R., and Najita J. (2003) *Astron. J., 126,* 2009–2014.
Chandler C. J., Brogan C. L., Shirley Y. L., and Loinard L. (2005) *Astrophys. J., 632,* 371–396.
Chauvin G., Ménard F., Fusco T., Lagrange A.-M., Beuzit J.-L., Mouillet D., and Augereau J.-C. (2002) *Astron. Astrophys., 394,* 949–956.
Chauvin G., Thomson M., Dumas C., Beuzit J.-L., Lowrance P., Fusco T., Lagrange A.-M., Zuckerman B., and Mouillet D. (2003) *Astron. Astrophys., 404,* 157–162.
Chick K. M. and Cassen P. (1997) in *Herbig-Haro Flows and the Birth of Stars* (B. Reipurth and C. Bertout, eds.), p. 207. IAU Symposium 182, Kluwer, Dordrecht.
Ciardi D. R., Telesco C. M., Williams J. P., Fisher R. S., Packham C.,

Piña R., and Radomski J. (2003) *Astrophys. J., 585*, 392–397.

Ciardi D. R., Telesco C. M., Packham C., Gómez Martin C., Radomski J. T., De Buizer J. M., Phillips C. J., and Harker D. E. (2005) *Astrophys. J., 629*, 897–902.

Covey K. R., Greene T. P., Doppmann G. W., and Lada C. J. (2006) *Astron. J., 131*, 512–519.

Covino E., Catalano S., Frasca A., Marilli E., Fernández M., Alcalá J. M., Melo C., Paladino R., Sterzik M. F., and Stelzer B. (2000) *Astron. Astrophys., 361*, L49–L52.

Covino E., Frasca A., Alcalá J. M., Paladino R., and Sterzik M. F. (2004) *Astron. Astrophys., 427*, 637–649.

Curiel S., Girart J. M., Rodríguez L. F., and Cantó J. (2003) *Astrophys. J., 582*, L109–L113.

Dale J. E., Bonnell I. A., Clarke C. J., and Bate M. R. (2005) *Mon. Not. R. Astron. Soc., 358*, 291–304.

Delgado-Donate E. J., Clarke C. J., and Bate M. R. (2003) *Mon. Not. R. Astron. Soc., 342*, 926–938.

Delgado-Donate E. J., Clarke C. J., and Bate M. R. (2004a) *Mon. Not. R. Astron. Soc., 347*, 759–770.

Delgado-Donate E. J., Clarke C. J., Bate M. R., and Hodgkin S. T. (2004b) *Mon. Not. R. Astron. Soc., 351*, 617–629.

Duchêne G., Monin J.-L., Bouvier J., and Ménard F. (1999) *Astron. Astrophys., 351*, 954–962.

Duchêne G., Ghez A. M., McCabe C., and Weinberger A. J. (2003) *Astrophys. J., 592*, 288–298.

Duchêne G., Bouvier J., Bontemps S., André P., and Motte F. (2004) *Astron. Astrophys., 427*, 651–665.

Dulk G. A. (1985) *Ann. Rev. Astron. Astrophys., 23*, 169–224.

Duquennoy A. and Mayor M. (1991) *Astron. Astrophys., 248*, 485–524.

Durisen R. H., Sterzik M. F., and Pickett B. K. (2001) *Astron. Astrophys., 371*, 952–962.

Feigelson E. D. and Montmerle T. (1999) *Ann. Rev. Astron. Astrophys., 37*, 363–408.

Fischer D. A. and Marcy G. W. (1992) *Astrophys. J., 396*, 178–194.

Furlan E., Forrest W. J., Watson D. M., Uchida K. I., Brandl B. R., Keller L. D., and Herter T. L. (2003) *Astrophys. J., 596*, L87–L90.

Gammie C. F. (2001) *Astrophys. J., 553*, 174–183.

Ghez A. M., Neugebauer G., and Matthews K. (1993) *Astron. J., 106*, 2005–2023.

Ghez A. M., Weinberger A. J., Neugebauer G., Matthews K., and McCarthy D. W. Jr. (1995) *Astron. J., 110*, 753–765.

Girart J. M., Rodríguez L. F., and Curiel S. (2000) *Astrophys. J., 544*, L153–L156.

Girart J. M., Curiel S., Rodríguez L. F., Honda M., Cantó J., Okamoto Y. K., and Sako S. (2004) *Astron. J., 127*, 2969–2977.

Gómez L., Rodríguez L. F., Loinard L., Lizano S., Poveda A., and Allen C. (2006) *Astrophys. J., 635*, 1166–1172.

Goodwin S. P. and Kroupa P. (2005) *Astron. Astrophys., 439*, 565–569.

Goodwin S. P.,Whitworth A. P., and Ward-Thompson D. (2004a) *Astron. Astrophys., 414*, 633–650.

Goodwin S. P., Whitworth A. P., and Ward-Thompson D. (2004b) *Astron. Astrophys., 423*, 169–182.

Greene T. P., Wilking B. A., André P., Young E. T., and Lada C. J. (1994) *Astrophys. J., 434*, 614–626.

Güdel M. (2002) *Ann. Rev. Astron. Astrophys., 40*, 217–261.

Guieu S., Dougados C., Monin J.-L., Magnier E., and Martín E. L. (2006) *Astron. Astrophys., 446*, 485–500.

Haisch K. E. Jr., Barsony M., Greene T. P., and Ressler M. E. (2002) *Astron. J., 124*, 2841–2852.

Haisch K. E. Jr., Greene T. P., Barsony M., and Stahler S. W. (2004) *Astron. J., 127*, 1747–1754.

Halbwachs J. L., Mayor M., Udry S., and Arenou F. (2003) *Astron. Astrophys., 397*, 159–175.

Hamaguchi K., Corcoran M. F., Petre R., White N. E., Stelzer B., Nedachi K., Kobayashi N., and Tokunaga A. T. (2005) *Astrophys. J., 623*, 291–301.

Hartigan P. and Kenyon S. J. (2003) *Astrophys. J., 583*, 334–357.

Hearty T., Fernández M., Alcalá J. M., Covino E., and Neuhäuser R. (2000) *Astron. Astrophys., 357*, 681–685.

Hillenbrand L. A. and White R. J. (2004) *Astrophys. J., 604*, 741–757.

Jensen E. L. N., Mathieu R. D., and Fuller G. A. (1994) *Astrophys. J., 429*, L29–L32.

Joergens V. (2006) *Astron. Astrophys., 446*, 1165–1176.

Johnston K. J., Fey A. L., Gaume R. A., Hummel C. A., Garrington S., Muxlow T., and Thomasson P. (2004) *Astrophys. J., 604*, L65–L69.

Johnston K. J., Fey A. L., Gaume R. A., Claussen M. J., and Hummel C. A. (2005) *Astron. J., 128*, 822–828.

Kiseleva L. G., Eggleton P. P., and Mikkola S. (1998) *Mon. Not. R. Astron. Soc., 300*, 292–302.

Koresko C. D. (2000) *Astrophys. J., 531*, L147–L149.

Koresko C. D. (2002) *Astron. J., 124*, 1082–1088.

Kraus A. L., White R. J., and Hillenbrand L. A. (2005) *Astrophys. J., 633*, 452–459.

Kroupa P. (1995) *Mon. Not. R. Astron. Soc., 277*, 1491–1506.

Kroupa P., Petr M. G., and McCaughrean M. J. (1999) *New Astron., 4*, 495–520.

Lada C. J. (1987) in *Star Forming Regions* (M. Peimbert and J. Jugaku, eds.), pp. 1–17. Reidel, Dordrecht.

Leinert Ch., Zinnecker H., Weitzel N., Christou J., Ridgway S. T., Jameson R., Hass M., and Lenzen R. (1993) *Astron. Astrophys., 278*, 129–149.

Lodato G. and Rice W. K. M. (2005) *Mon. Not. R. Astron. Soc., 358*, 1489–1500.

Loinard L. (2002) *Rev. Mex. Astron. Astrofis., 38*, 61–69.

Loinard L., Rodríguez L. F., D'Alessio P., Wilner D. J., and Ho P. T. P. (2002) *Astrophys. J., 581*, L109–L113.

Loinard L., Rodríguez L. F., and Rodríguez M. I. (2003) *Astrophys. J., 587*, L47–L50.

Loinard L., Mioduszewski A. J., Rodríguez L. F., González R. A., Rodríguez M. I., and Torres R. M. (2005) *Astrophys. J., 619*, L179–L182.

Looney L. W., Mundy L. G., and Welch W. J. (2000) *Astrophys. J., 529*, 477–498.

Luhman K. L. (2001) *Astrophys. J., 560*, 287–306.

Mac Low M.-M. and Klessen R. S. (2004) *Rev. Modern Phys., 76*, 125–194.

Marchal L., Delfosse X., Forveille T., Ségransan D., Beuzit J.-L., Udry S., Perrier C., and Mayor M. (2003) in *Brown Dwarfs* (E. Martín, ed.), p. 311. IAU Symposium 211, ASP, San Francisco.

Martín E. L., Barrado y Navascués D., Baraffe I., Bouy H., and Dahm S. (2003) *Astrophys. J., 594*, 525–532.

Mathieu R. D. (1994) *Ann. Rev. Astron. Astrophys., 32*, 465–530.

Mathieu R. D., Ghez A. M., Jensen E. L. N., and Simon M. (2000) in *Protostars and Planets IV* (V. Mannings et al., eds.), pp. 703–730. Univ. of Arizona, Tucson.

Maxted P. F. L. and Jeffries R. D. (2005) *Mon. Not. R. Astron. Soc., 362*, L45–L49.

McCabe C., Ghez A. M., Prato L., Duchêne G., Fisher R. S., and Telesco C. (2006) *Astrophys. J., 636*, 932–951.

McDonald J. M. and Clarke C. J. (1995) *Mon. Not. R. Astron. Soc., 275*, 671–684.

Meeus G., Sterzik M., Bouwman J., and Natta A. (2003) *Astron. Astrophys., 409*, L25–L29.

Melo C. H. F. (2003) *Astron. Astrophys., 410*, 269–282.

Noriega-Crespo A., Moro-Martin A., Carey S., Morris P. W., Padgett D. L., Latter W. B., and Muzerolle J. (2004) *Astrophys. J. Suppl., 154*, 402–407.

Ochi Y., Sugimoto K., and Hanawa T. (2005) *Astrophys. J., 623*, 922–939.

Osterloh M. and Beckwith S. V. W. (1995) *Astrophys. J., 439*, 288–302.

Padoan P. and Nordlund A. (2002) *Astrophys. J., 576*, 870–879.

Patience J. and Duchêne G. (2001) in *The Formation of Binary Stars* (H. Zinnecker and R. D. Mathieu, eds.), pp. 181–190. IAU Symposium 200, ASP, San Francisco.

Phillips R. B., Lonsdale C. J., Feigelson E. D., and Deeney B. D. (1996) *Astron. J., 111*, 918–929.

Prato L. and Simon M. (1997) *Astrophys. J., 474*, 455–463.

Prato L., Greene T. P., and Simon M. (2003) *Astrophys. J., 584*, 853–874.

Reipurth B. (2000) *Astron. J., 120,* 3177–3191.

Reipurth B. and Zinnecker H. (1993) *Astron. Astrophys., 278,* 81–108.

Reipurth B., Rodríguez L. F., Anglada G., and Bally J. (2002) *Astron. J., 124,* 1045–1053.

Reipurth B., Rodríguez L. F., Anglada G., and Bally J. (2004) *Astron. J., 127,* 1736–1746.

Ressler M. E. and Barsony M. (2001) *Astron. J., 121,* 1098–1110.

Rodríguez L. F. (1997) in *Herbig-Haro Flows and the Birth of Stars* (B. Reipurth and C. Bertout, eds.), p. 83. IAU Symposium 182, Kluwer, Dordrecht.

Rodríguez L. F., D'Alessio P., Wilner D. J., Ho P. T. P., Torrelles J. M., Curiel S., Gómez Y., Lizano S., Pedlar A., Cantó J., and Raga A. C. (1998) *Nature, 395,* 355–357.

Rodríguez L. F., Porras A., Claussen M. J., Curiel S., Wilner D. J., and Ho P. T. P. (2003) *Astrophys. J., 586,* L137–L139.

Rodríguez L. F., Loinard L., D'Alessio P., Wilner D. J., and Ho P. T. P. (2005a) *Astrophys. J., 621,* L133–L136.

Rodríguez L. F., Poveda A., Lizano S., and Allen C. (2005b) *Astrophys. J., 627,* L65–L68.

Shu F. H., Adams F. C., and Lizano S. (1987) *Ann. Rev. Astron. Astrophys., 25,* 23–81.

Sigalotti L., Di G., and Klapp J. (2001) *Astron. Astrophys., 378,* 165–179.

Stassun K. G., Mathieu R. D., Vaz L. P. R., Stroud N., and Vrba F. J. (2004) *Astrophys. J. Suppl., 151,* 357–385.

Steffen A. T., Mathieu R. D., Lattanzi M. G., Latham D. W., Mazeh T., Prato L., Simon M., Zinnecker H., and Loreggia D. (2001) *Astron. J., 122,* 997–1006.

Sterzik M. F. and Durisen R. H. (1995) *Astron. Astrophys., 304,* L9–L12.

Sterzik M. F. and Durisen R. H. (1998) *Astron. Astrophys., 339,* 95–112.

Sterzik M. F. and Durisen R. H. (2003) *Astron. Astrophys., 400,* 1031–1042.

Sterzik M. F. and Tokovinin A. A. (2002) *Astron. Astrophys., 384,* 1030–1037.

Sterzik M. F., Durisen R. H., and Zinnecker H. (2003) *Astron. Astrophys., 411,* 91–97.

Sterzik M. F., Melo C. H. F., Tokovinin A. A., and van der Bliek N. (2005) *Astron. Astrophys., 434,* 671–676.

Symington N. H., Harries T. J., Kurosawa R., and Naylor T. (2005) *Mon. Not. R. Astron. Soc., 358,* 977–984.

Tamazian V. S. (2004) *Astron. J., 127,* 2378–2381.

Tamazian V. S., Docobo J. A., White R. J., and Woitas J. (2002) *Astrophys. J., 578,* 925–934.

Tan J. C. (2004) *Astrophys. J., 607,* L47–L50.

Tohline J. E. (2002) *Ann. Rev. Astron. Astrophys., 40,* 349–385.

Tokovinin A. A. (2004) *Rev. Mex. Astron. Astrofis., 21,* 7–14.

Tokovinin A. A. and Smekhov M. G. (2002) *Astron. Astrophys., 382,* 118–123.

Tsujimoto M., Kobayashi N., and Tsuboi Y. (2005) *Astron. J., 130,* 2212–2219.

Valenti J. A. and Johns-Krull C. M. (2004) *Astrophys. Space Sci., 292,* 619–629.

Welty A.D. (1995) *Astron. J., 110,* 776–781.

Whitney B. A., Wood K., Bjorkman J. E., and Wolff M. J. (2003) *Astrophys. J., 591,* 1049–1063.

Woitas J., Köhler R., and Leinert Ch. (2001) *Astron. Astrophys., 369,* 249–262.

Wooten A. (1989) *Astrophys. J., 337,* 858–864.

Zuckerman B. and Song I. (2004) *Ann. Rev. Astron. Astrophys., 42,* 685–721.

Disk Evolution in Young Binaries: From Observations to Theory

J.-L. Monin
Laboratoire d'Astrophysique de Grenoble

C. J. Clarke
Institute of Astronomy, Cambridge

L. Prato
Lowell Observatory

C. McCabe
Jet Propulsion Laboratory

The formation of a binary system surrounded by disks is the most common outcome of stellar formation. Hence studying and understanding the formation and the evolution of binary systems and associated disks is a cornerstone of star-formation science. Moreover, since the components within binary systems are coeval and the sizes of their disks are fixed by the tidal truncation of their companion, binary systems provide an ideal "laboratory" in which to study disk evolution under well-defined boundary conditions. Since the previous edition of Protostars and Planets, large diameter (8–10 m) telescopes have been optimized and equipped with adaptive optics systems, providing diffraction-limited observations in the near-infrared where most of the emission of the disks can be traced. These cutting-edge facilities provide observations of the inner parts of circumstellar and circumbinary disks in binary systems with unprecedented detail. It is therefore a timely exercise to review the observational results of the last five years and to attempt to interpret them in a theoretical framework. In this paper, we review observations of several inner disk diagnostics in multiple systems, including hydrogen emission lines (indicative of ongoing accretion), K–L and K–N color excesses (evidence of warm inner disks), and polarization (indicative of the relative orientations of the disks around each component). We examine to what degree these properties are correlated within binary systems and how this degree of correlation depends on parameters such as separation and binary mass ratio. These findings will be interpreted both in terms of models that treat each disk as an isolated reservoir and those in which the disks are subject to resupply from some form of circumbinary reservoir, the observational evidence for which we will also critically review. The planet-forming potential of multiple star systems is discussed in terms of the relative lifetimes of disks around single stars, binary primaries, and binary secondaries. Finally, we summarize several potentially revealing observational problems and future projects that could provide further insight into disk evolution in the coming decade.

1. INTRODUCTION

It is now a matter of common knowledge that the majority of stars in star-forming regions are in binary or higher-order multiple systems (*Ghez et al.,* 1993; *Leinert et al.,* 1993; *Reipurth and Zinnecker,* 1993; *Simon et al.,* 1995). Likewise, it is undisputed that many of the younger stars in these regions exhibit evidence for circumstellar disks and/or accretion. Putting these two facts together, an inescapable conclusion is that disks typically form and evolve in the environment of a binary/multiple-star system.

This prompts a number of obvious questions. Can the distribution of dust and gas in young binaries provide a "smoking gun" for the binary-formation process? Is disk evolution, and perhaps the possible formation of planets, radically affected by the binary environment and, if so, how does this depend on binary separation and mass ratio? Alternatively, if the influence of binarity on disk evolution is rather mild, we can at least use binary systems as well-controlled laboratories, constituting coeval stars with disk outer radii set by tidal truncation criteria, to study disk evolution as a function of stellar mass.

However, it is not possible to address any of these issues unless we can disentangle the disk/accretion signatures produced by each component in the binary. Given that the separation distribution for binaries in the nearest populous star forming regions, such as Taurus-Aurigae, peaks at ~0.3" [≡40 AU (e.g., *Mathieu,* 1994)], this necessitates the use of high-resolution photometry and spectroscopy. Such an enterprise has only become possible in the past decade.

We review what has been learned in recent years about the distribution of dust and gas within young binary systems. We mainly highlight observational developments since Protostars and Planets IV, for example, the discovery of a population of so-called passive disks (*McCabe et al.,* 2006) in low-mass secondaries and the use of polarimetry to constrain the orientations of disks in young binaries (e.g., *Jensen et al.,* 2004; *Monin et al.,* 2006). We also discuss circumbinary disks and profile in detail a few systems that have been the subject of intense observational scrutiny. In addition, it is timely to examine the statistical properties of resolved binaries that have been accumulating in the literature over the past decade. We have therefore combined the results from a number of relatively small-scale studies in order to assemble around 60 resolved pairs and use this dataset to examine the relationship between binarity and disk evolution.

In this chapter we progress through a description of disk/accretion *diagnostics* (and their application to resolved binary star studies; section 2) to highlighting some recent results on disk *structure* in binaries (section 3) to a statistical analysis of the relationship between binarity and disk *evolution* (section 4). In section 4.5 we briefly consider how the insights of the preceding section can be applied to the question of planet formation in binaries. Section 5 examines future prospects and potential projects to advance our understanding of disk structure and evolution in young binaries.

2.   INNER DISK DIAGNOSTICS IN YOUNG BINARIES

How do we know when either circumstellar or circumbinary disks are present in a young multiple-star system? It took over a decade of observations to confirm the existence of simple circumstellar disk structures after the original observational and theoretical introduction of the concept in the early 1970s (e.g., *Strom et al.,* 1971, 1972; *Lynden-Bell and Pringle,* 1974). The paradigm is yet more complicated for a binary systems with multiple disks. Over the last two decades, direct means of imaging circumstellar disks have become available to astronomers, beginning with millimeter observations in the mid-1980s and ending with the recent development of high-angular-resolution laser guide star adaptive optics. Most critically to this chapter, the last decade has witnessed unprecedented improvements in our ability not only to directly image disks and to indirectly infer their presence, but also to detect disks around both stellar components of extremely close binary systems, as well as larger, circumbinary structures.

In this section we summarize the methods used to determine the presence of circumstellar and circumbinary disks in multiples. The section is divided into two parts: disk diagnostics and accretion diagnostics. In this manner we distinguish between observations that detect the disks themselves, directly or indirectly, and the observations that are sensitive to the presence of accretion processes, indicating that material is flowing from a disk onto the central star, and thus betraying the existence of the disk indirectly. The end of this section summarizes the database that we compiled in the process of writing this chapter.

An excellent inner disk diagnostic that we do not explore is the emission of molecular lines. This topic is reviewed in the chapter by Najita et al. as well as in *Najita et al.* (2003) and references therein. Outer disk diagnostics, such as submillimeter and millimeter observations, and more narrowly applied accretion diagnostics, such as forbidden line emission and ultraviolet excesses, are also neglected here because they do not appear in our analysis. These tools are either not as relevant to our component resolved studies or have not yet been widely applied to many binary observations. Relevant references may be found in *Dutrey et al.* (1996) and *Jensen et al.* (1996) (submillimeter and millimeter), *Hartigan and Kenyon* (2003) ([OI] emission lines), and *Gullbring et al.* (1998).

2.1.   Background

Lynden-Bell and Pringle's (1974) prescient disk model for classical T Tauri stars (CTTS) accounted for a number of characteristics of these objects, including atomic emission lines and the relatively flat λF_λ distribution of light at infrared wavelengths. Pioneering infrared observations of young stars indicated the presence of a strong excess (*Mendoza,* 1966, 1968) above expected photospheric value (*Johnson,* 1966) for T Tauri stars. These were largely interpreted as indicative of either a spherical dust shell around the young stars studied (*Strom et al.,* 1971, 1972; *Strom,* 1972) or freefree emission from circumstellar gaseous envelopes (*Breger and Dyke,* 1972; *Strom et al.,* 1975), although the suggestion of a circumstellar disk structure was raised as early as 1971 (*Strom et al.,* 1971, 1972). Early analysis of the IRAS satellite data (e.g., *Rucinski,* 1985) and direct imaging of disklike structures around HL Tau and L1551 IRS5 (*Grasdalen et al.,* 1984; *Beckwith et al.,* 1984; *Strom et al.,* 1985) provided ultimately convincing evidence in support of disks.

2.2.   Disk Diagnostics

2.2.1. Near- and mid-infrared excesses.   Optically thick but physically thin dusty circumstellar disks around T Tauri stars reprocess stellar flux and give rise to excess thermal radiation at wavelengths greater than ~1 μm. At larger disk radii, the equilibrium dust temperature is lower; thus, different circumstellar disk regions are studied in different wavelength regimes. For low-mass stars, the JHK (1–2 μm) colors sample the inner few tenths of an AU, the L band (3.5 μm) about twice that distance, and the N band (10 μm) the inner ~1–2 AU. IRAS and Spitzer data sample radii of several to tens of AU. The exact correspondences depend on the luminosity of the central star and the disk properties and geometry [e.g., scale height of the dust, degree of flaring, particle size distribution (see, e.g., *D'Alessio,* 2003; *Chiang et al.,* 2001; *Malbet et al.,* 2001)]. This results in a spectral energy distribution where the disk and star contributions can be disentangled (Fig. 1). For the nearby star-forming regions, binary separation distributions typically peak at ~0.3" (*Simon et al.,* 1995; *Patience et al.,* 2002); for a distance

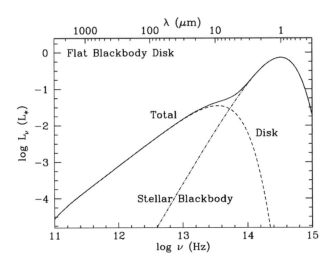

Fig. 1. SED of a flat reprocessing disk from *Chiang and Goldreich* (1997). The dashed line corresponds to the disk emission, the dot-dash line to the stellar photosphere, and the solid line shows the total flux.

of 150 pc, this corresponds to 45 AU. Therefore, most stars in binaries should have a direct impact on the circumstellar environments of their companions, at least at radii of several to a few dozen AU from the individual stars.

The shortest wavelength JHK colors, although useful (e.g., *Hillenbrand et al.*, 1998), are susceptible to contamination from reflected light and are highly sensitive to the disk inner gap size (e.g., *Haisch et al.*, 2000). L-band data, where the contribution from the T Tauri stellar photosphere decreases, offers a far more reliable indicator of the innermost circumstellar dust (e.g., *Haisch et al.*, 2001) and reveals a much larger proportion of stars with disks. In the proceedings from the Protostars and Planets III conference, *Edwards et al.* (1993) summarized the relationships between the K–L disk colors and the winds off of, and accretion flows on to, T Tauri stars. This establishment of the usefulness of K–L colors as a disk diagnostic coincided with early studies of binary colors (*Leinert and Haas*, 1989; *Ghez et al.*, 1991). Since the mid-1990s this diagnostic has been widely used as a convenient indicator of circumstellar disks in small separation binaries (e.g., *Tessier et al.*, 1994; *Chelli et al.*, 1995; *Geoffray and Monin*, 2001; *White and Ghez*, 2001; *Prato et al.*, 2003; *McCabe et al.*, 2006). At 8–10-m-class telescopes, an angular resolution of ~0.1" is achievable in the L band.

N-band observations are sensitive to dusty disk material that may surround a young star even in the absence of an innermost disk and a corresponding near-infrared excess (e.g., *Koerner et al.*, 2000; *Prato et al.*, 2001). For more than 30 years, observations have been made of young stars at 10 and 20 μm (e.g., *Strom et al.*, 1972, 1975; *Knacke et al.*, 1973; *Rydgren et al.*, 1976; *Skrutskie et al.*, 1990; *Stassun et al.*, 2001), but, with few exceptions (e.g., *Ghez et al.*, 1994), high-angular-resolution mid-infrared observations

required the development of a new generation of cameras in the late 1990s for the largest existing (8–10-m-class) telescopes. The Keck telescopes, for example, provide a 0.25" diffraction limit at 10 μm. Over 80% of the known, angularly resolved, young binary N-band measurements, and most of the angularly resolved Q-band (~20 μm) measurements, have only recently been published in *McCabe et al.* (2006). Although far-infrared spacebased observations do not provide the requisite angular resolution to distinguish between close binary components (Spitzer's diffraction limit at 160 μm is about half a minute of arc), ALMA will provide unprecedented resolution in the far infrared/submillimeter regime (see section 5).

2.2.2. Polarization. Linear polarization maps of young stars typically show an axisymmetric, or "centrosymmetric," pattern. By the late 1980s, these observations were interpreted by *Bastien and Ménard* (1988, 1990) as the result, in part, of light scattering from optically thick circumstellar disks. A prescient remark from the *Protostars and Planets III* paper of *Basri and Bertout* (1993) notes that "High resolution near-infrared polarization maps are, however, becoming possible with the advent of 256 × 256 detectors and AO . . ." Indeed, by the late 1990s, stunning detail in the polarization maps of *Close et al.* (1998), *Potter et al.* (2000), and *Kuhn et al.* (2001) illustrated the power of polarization observations for the study of circumstellar and circumbinary disks. *Monin et al.* (1998) applied the tool of polarization to a sample of wide (8–40") binaries in Taurus. Most recently, *Jensen et al.* (2004) and *Monin et al.* (2006) mapped polarization around more than three dozen small separation (~1–10") binaries (section 3.1). Given that polarization observations can identify the orientation of a circumstellar disk, this provides a unique way in which to test the alignment of disks in binary systems.

2.3. Accretion Diagnostics: Permitted Atomic Line Emission

The prolific work in the 1940s of Joy (e.g., *Joy and van Biesbroeck*, 1944) and later of Herbig (e.g., *Herbig*, 1948) on T Tauri-type emission line stars laid the foundations for the study of emission lines in young binaries. Although the source of hydrogen emission lines in young stars was variously attributed as the result of free-free emission, chromospheric activity, and stellar winds (e.g., *Strom et al.*, 1975; *Herbig*, 1989; *Edwards et al.*, 1987), by the late 1980s *Bertout et al.* (1988) and others had established a model for magnetospheric accretion. *Strom et al.* (1989) canonized the nominal 10 Å distinction in Hα ($\lambda = 6563$ Å) line emission between weak-lined (WTT) and classical (CTT) T Tauri stars, which is still — somewhat indiscriminately — used today, albeit with slight modifications (e.g., *Martín*, 1998).

As the high frequency of young star binaries became established in the mid-1990s, hydrogen emission lines were recognized as a useful approach for the study of circumstellar material around each star in the system (e.g., *Hartigan et al.*, 1994; *Brandner and Zinnecker*, 1997; *Prato and*

Simon, 1997; *Duchêne et al.*, 1999; *White and Ghez*, 2001; *Prato et al.*, 2003). Infrared Brγ ($\lambda = 2.16$ μm) observations (*Prato and Simon*, 1997; *Muzerolle et al.*, 1998; *Prato et al.*, 2003) provide a means of measuring emission lines with the best possible seeing at longer wavelengths, as well as a method of detecting infrared components not readily seen in visible light.

2.4. The Young Star Binary Database

Observations of young binaries that resolve the circumstellar disk and accretion diagnostics of each component involve access to large telescopes with adaptive optics capabilities or spacebased observatories. Limited access to such facilities has meant that the results of such studies have often been published in papers describing a relatively small number of objects, from which it has proved impossible to derive statistically secure results. We have therefore combined many studies into a single database. In order to qualify for inclusion in the database, it is only necessary for the binary components to be angularly resolved and located in a region with a distance estimate such that the separation of the pair in the plane of the sky is known. Because we restrict the database to resolved systems, we exclude systems with semimajor axis a < 14 AU. In order to avoid contamination with chance projections we also exclude systems with a > 1400 AU. These limits correspond to binaries in the angular separation range of 0.1–10.0" at the 140-pc distance of Taurus.

Although the information available is incomplete for a number of objects, we have ~60 systems where the spectral type of each component is known, as well as the presence or absence of disks and/or accretion for each component. We shall return to the statistical properties of these systems, and the implications for disk evolution in binaries, in section 4. The database is available at *www.lowell.edu/ users/lprato/compil_binaires_cmc5.html*. We welcome additions, revisions, and comments.

3. DISK STRUCTURE IN YOUNG BINARIES

3.1. Disk Orientations in Binary Systems

A binary system with disks possesses many more degrees of freedom than an isolated star. Both stars can have a disk, they orbit around each other, and the entire system can be surrounded by a circumbinary disk. This defines four planes: two circumstellar disks, the stellar binary orbit, and the circumbinary disk. In a single young star system, only one plane is potentially present, that of one disk. In this section we examine some recent observational and theoretical results that shed light on the respective orientations of the multiple planes present in a binary.

If a binary forms through the fragmentation of a disk, then the disks that form around each star are expected to be mutually aligned and also to be aligned with the binary orbit. The same is true for binaries that form through core fragmentation, provided the angular momentum vectors of

the parent core material are well aligned and provided that the initial core geometry (or the result of any perturbation inducing fragmentation) does not introduce any other symmetry planes into the problem. Although *Papaloizou and Terquem* (1995) showed that tidal effects may sometimes induce subsequent misalignment of the disks and orbital planes, *Lubow and Ogilvie* (2000) show that the required conditions are unlikely to be met in practice. We shall therefore assume that a binary that is created with all its planes aligned will remain in this state throughout its Class 0, Class I, and Class II phases.

If any of the above conditions are not met, however, the binary will be created with some planes misaligned. For example, *Bonnell et al.* (1992) showed that if the initial cloud is elongated and if the rotation axis is oriented arbitrarily with respect to the cloud axis, then the initial disk and binary orbital planes are misaligned: In this case, the disk planes (which reflect the angular momentum of the core) are parallel, and misaligned with the binary orbit (which reflects the symmetry of the initial core). On the other hand, all planes may be misaligned either in the case that the angular momentum distribution of the initial core is complex or that the fragmentation involves more than two bodies. There are therefore a number of routes by which misaligned systems can be created and may be manifest among Class 0 systems. This does not, however, imply that these systems will remain misaligned during subsequent evolutionary phases, owing to the fact that both tidal effects and accretion onto the protobinary can bring the system into alignment at a later stage. Therefore the detection of misaligned systems is an unambiguous sign of misaligned formation, whereas aligned systems may either have been created that way or else have subsequently evolved into this state.

At the earliest evolutionary stages, it now seems inescapable that at least some systems contain misaligned disks. In these systems, jet orientation provides an observable proxy for disk orientation since jets are always launched perpendicular to the inner disk: The detection of multiple jets emanating with different position angles from a small region is thus an unambiguous sign of misalignment (*Reipurth et al.*, 1993; *Gredel and Reipurth*, 1993; *Davis et al.*, 1994; *Böhm and Solf*, 1994; *Eislöffel et al.*, 1996). In all cases, the parent multiple systems are either unresolved or are known to be wide binaries (i.e., with a > 100 AU). Less directly, the observation of changes in jet position angle have been interpreted as the result of jet precession [or "wobble" (*Bate et al.*, 2000)], induced by misalignment between the disk and the orbital plane of a putative companion (*Chandler et al.*, 2005; *Hodapp et al.*, 2005). The observed rates of change of jet position angle are thought to be consistent with the presence of unresolved binary companions with separations in the range several to ~100 AU. However, not all observed instances of changes in jet direction can necessarily be explained in these terms (*Eislöffel and Mundt*, 1997).

The expected timescale on which strongly misaligned disks should be brought into rough alignment by tidal torques is about 20 binary orbital periods (*Bate et al.*, 2000);

it is thus only in rather wide binaries (i.e., with a > 100 AU) that we should expect misalignment throughout their Class 0 and Class I stages. However, as the misalignment angle (δ) evolves toward zero, the rate of alignment becomes proportional to δ and hence the system may be expected to remain in a mildly misaligned state over considerably longer periods.

How are these expectations borne out by observations of Class II sources (with typical ages of a few $\times$ 10^6 yr)? The most obvious approach is through direct imaging of disks in pre-main-sequence binaries. Unfortunately, despite the recent deployment of a range of instruments offering high angular resolution on very large telescopes, circumstellar disks in TTS multiple systems have only been imaged in few cases [HK Tau (*Stapelfeldt et al.,* 1998), HV Tau (*Monin and Bouvier,* 2000; *Stapelfeldt et al.,* 2003), LkHα 263 (*Jayawardhana et al.,* 2002)]. In each of these systems only one disk is detectable via imaging and is seen edge on, a favorable orientation for detection. In all three systems, the observed edge-on disk is oriented in a direction quite different from the projection of the binary orbit on the sky: Therefore we see immediately that *at least some disks in binaries remain misaligned with the binary orbital plane during the Class II phase.*

Several properties of these imaged systems are noteworthy: First, they are all wide binaries (a > several hundred AU) and are thus consistent with the estimate given above that disks in binaries closer than ~100 AU should be brought into alignment during the Class 0 or Class I phase. Second, for HK Tau and LkHα 263, the companion to the star with the edge-on disk is itself a close binary system. Third, in each of these systems only one disk is detectable through imaging, although there is some spectroscopic evidence that the other component does possess a disk. The fact that these other disks are not detected through direct imaging implies that they are not themselves viewed close to edge on and we can thus infer that the disks in these systems are not parallel with each other. However, since only a slight tilt of the other disk away from edge on can abruptly reduce its detectability as the central star becomes visible directly, this observation only excludes an alignment between the disks to within ≈15°.

Since the publication of *Protostars and Planets IV,* various studies have been performed to determine the orientation of binary disks relative to each other in the plane of the sky. Following the theoretical computations by *Bastien and Ménard* (1990) and the previous measurements of *Monin et al.* (1998), *Wolf et al.* (2001), *Jensen et al.* (2004), and *Monin et al.* (2006) have used polarimetric observations to determine the relative orientation of disks in the plane of the sky. The position angle of the integrated linear polarization of the scattered starlight is parallel to the equatorial plane of the disk, provided that its inclination is sufficiently large to mask the direct light from the star (*Monin et al.,* 2006). One caveat of this method is that it does not reveal the actual three-dimensional orientations of disks; two disks with parallel polarization could be differently inclined along the line of sight. In principle, this other orientation angle

can be obtained from v sin i and rotation period measurements, but these are quite rare and difficult to obtain in close binaries and are in any case subject to errors attributable to uncertainties in the stellar radius. However, *Wolf et al.* (2001) have shown from statistical arguments that if the relative polarization position angle difference distribution peaks at zero, then the disks tend to be parallel.

The net result of these studies is twofold: Disk polarizations tend to be close to (but not exactly) parallel in binary systems, but there exist systems with misaligned polarization, with a few objects having polarization position angle differences of ~90°. *Jensen et al.* (2004) argue that it is unlikely that this result is compromised by dilution of the polarization signal from each disk by interstellar polarization since, among other evidence, they note that disk polarization tends *not* to be parallel in the case that one or other of the two components is itself a close binary system [see Fig. 2, where we have merged the results from *Jensen et al.* (2004) and *Monin et al.* (2006)]. This supports the notion that the polarization differences are intrinsic, since there should be no correlation between the degree of contamination by interstellar polarization and the multiplicity of the system studied. On the other hand, there is a plausible physical reason for this result: Owing to the much larger angular momentum contained in a close binary pair than a simple star-disk system, the timescale for torquing a binary into alignment with the orbital plane of the wider pair is evidently much longer than that for the alignment of a disk.

All the above discussion relates to binaries that are wide enough to be imaged (typically wider than 100 AU). However, in the case of spectroscopic binaries, there is the possi-

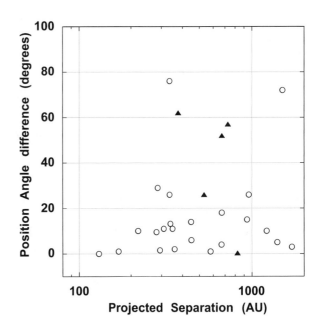

Fig. 2. Adapted from *Jensen et al.* (2004) and *Monin et al.* (2006); binaries are plotted as empty circles and higher-order multiples as filled triangles. Note the suggestion in the data that disks in triples and quadruples are proportionally less aligned than in pure binary systems.

bility of determining the system inclination from the orbital solution and then comparing this with the inclination determined from direct imaging of the circumbinary disk (albeit on a much larger scale). In the small number of systems where this has proved possible, the evidence is for alignment between the plane of the spectroscopic binary and its circumbinary disk (see *Mathieu et al., 1997; Prato et al., 2001, 2002; Simon et al., 2000*).

Finally, among *main-sequence* solar-type binaries, it is found that the stellar orbital planes are aligned with the binary orbit for binaries closer than 40 AU (*Hale, 1994*), as one would expect, given the short predicted alignment timescales for closer binaries.

In summary, we have plenty of examples (through imaging and polarization studies) of misaligned systems among wider binaries (i.e., with a > 100 AU or so), implying that at least some of these systems must be formed in a misaligned state. By the Class II phase, it would appear that binaries in this separation range constitute a mixture of aligned and misaligned systems (*Monin et al., 2006; Jensen et al., 2004*). This may imply that wider binaries are formed in both aligned and misaligned states, or, alternatively, that all such binaries are born in the misaligned state and are brought into alignment through tidal torques (which should operate on a roughly 10^6-yr timescale for binaries of this separation). In the case of closer binaries, where direct imaging is not possible, observational evidence for disk alignment can be derived only in the case of spectroscopic binaries with imaged circumbinary disks and also through the fossil evidence contained in stellar spin vectors within main-sequence binaries. Both these lines of evidence point to close binaries being aligned during the main disk-accretion stage. This is expected, given the short predicted alignment timescales for close binaries, and therefore gives us no information about the *initial* state of alignment of these systems.

3.2. A Sampling of Circumbinary Disks

Only a few circumbinary disks have been imaged directly. *Ménard et al.* (1993) proposed a circumbinary disk to explain NIR images of Haro 6-10, and in 1994, the circumbinary disk that still today remains the most impressive to date was found by *Dutrey et al.* (1994) around GG Tau. The majority of the currently inferred circumbinary disks are proposed to explain SED emission from warm dust in disks with a central hole where the binary resides. With the ever-growing number of discoveries of pre-main-sequence spectroscopic binaries, the number of putative circumbinary disks in these closer systems has increased. On the other hand, in the case of wide binaries, very few circumbinary disks have been directly imaged and, moreover, the low upper limits for circumbinary disk masses from millimeter measurements (*Jensen et al., 1996*) (see also section 4.4) suggest that circumbinary disks are weak or absent in the majority of these systems. However, this conclusion remains provisional on two grounds. First, the very small number of circumbinary disks that have been imaged might not be as surprising as originally thought, when one considers also the relatively low rate of detection of circumstellar disks by direct imaging; only when the system geometry is very favorable can the disk be imaged easily (see section 3.1 above). Second, there is at least one case in which a circumbinary disk that has been imaged in CO lines is not detectable in dust as probed by the millimeter continuum (see discussion of SR 24 N below). We therefore cannot rule out that wide binaries either possess low-mass circumbinary disks that escape detection in the dust continuum (corresponding to disk masses ≤ 1 M_J) or else that some process, such as grain growth, is reducing the dust emission in these systems. Such a process may be at work in the GG Tau circumbinary disk (see section 3.2.1 below).

In the case of wider binaries (a > 20 AU), the argument in favor of circumbinary disks as a necessary reservoir for the resupply of circumstellar disks has weakened since its orginal proposal by *Prato and Simon* (1997): Our analysis described in section 4 below shows that mixed systems (i.e., pairs containing both a CTTS and a WTTS) are in fact common. It is likely that, in wider binaries, circumstellar disks evolve in relative isolation, and resupply might not be a necessity. In closer binaries, resupply remains a necessity on the grounds that the circumstellar disk lifetimes in these close systems would otherwise be too short to explain the incidence of component CTT stars. In these closer systems, circumbinary disks, as evidenced by their contribution to the spectral energy distribution, remain a good candidate for the resupply reservoir. Indeed, in various objects, signatures of accretion episodes from the circumbinary environment onto the central objects, presumably via their associated circumstellar disks, have been detected. In this section we examine in more detail several circumbinary disk systems and discuss their properties in terms of disk evolution, circumbinary accretion, and potential for planet formation.

3.2.1. GG Tau. Discovered by *Simon and Guilloteau* (1992), this circumbinary disk orbits the 0.25" separation pair GG Tau A and has been spatially resolved in the optical (*Krist et al., 2002, 2005*), near-infrared (*Roddier et al., 1996; McCabe et al., 2002; Duchêne et al., 2004*) and in the millimeter (continuum and ^{12}CO) (e.g., *Guilloteau et al., 1999*). *Beust and Dutrey* (2005) investigated the GG Tau A orbit and the inner ring gap and find that a binary orbital solution with a = 62 AU and e = 0.35 could be consistent with the data; in this study, the presence of the circumbinary disk is used to add dynamical constraints to the central binary system. Using a collection of images at various wavelengths, *Duchêne et al.* (2004) have shown that grain growth is at work in the midplane of the GG Tau circumbinary ring within a stratified structure. This shows that the processes leading to planet formation might be at work in circumbinary disks as well as in circumstellar disks.

3.2.2. SR 24 N. The binary separation in this system is of the same order as GG Tau's, 32 AU. *Andrews and Williams* (2005) have observed a 250-AU structure in this system, probably a circumbinary disk. An interesting feature of their observations is that this disk shows no emission in the continuum, possibly as the result of a central gap inside the disk, and is seen only in CO line emission. This suggests that other wide circumbinary disks could have been missed by continuum observations, and thus could be more frequent than previously thought. K–L measurements by *McCabe et al.* (2006) indicate that both components of SR 24 N are themselves CTTS.

3.2.3. GW Ori. GW Ori is a spectroscopic binary with an orbital period of 242 d (*Mathieu et al.,* 1991) and a separation slightly more than 1 AU. These authors used a circumbinary disk model to reproduce the mid-infrared excess at 20 μm: GW Ori is one of those spectroscopic binaries in which a large emitting region is needed to explain the submillimeter flux. With an estimated stellar separation of ~1 AU, this requires an extended circumbinary structure. The presence of circumbinary material was even confirmed by *Mathieu et al.* (1995), who found that independently of any specific disk model, the extended (≈500 AU) submillimeter emission of GW Ori was circumbinary in origin.

3.2.4. DQ Tau. Like GW Ori, this 0.1-AU-separation spectroscopic binary possesses excess emission at longer wavelengths, indicating the presence of circumbinary material around the central stars. Further observations have revealed evidence for accretion bursts near the binary periastron in the form of photometric variability (*Mathieu et al.,* 1997) and increased veiling (*Basri et al.,* 1997). These results are consistent with the prediction by *Artymowicz and Lubow* (1996), who showed that accretion streams are likely to link the inner edge of the circumbinary disk to the stars. Thus DQ Tau is an example of a binary where replenishment from a circumbinary structure is at work.

3.2.5. V4046 Sgr. This pair has an orbital period of 2.4 d and an eccentricity close to zero. *Artymovicz and Lubow*'s (1996) models of accretion from the circumbinary environment predict that mass ratio, $q(M_2/M_1)$, ~1, low-eccentricity binaries should not experience accretion bursts. However, *Stempels and Gahm* (2004) have recently observed spectroscopic features that can be explained by the presence of gas concentrations in corotation with the central binary. These gas accumulations might provide further evidence for accretion from the circumbinary environment.

3.2.6. AK Sco. AK Sco is an eccentric spectroscopic binary with q ~ 1 and a separation of 0.14 AU. The circumbinary disk needed to explain the spectral energy distribution possesses an inner hole of radius ~0.4 AU within which the binary resides. This is consistent with the prediction of *Artymowicz and Lubow* (1996) for the inner rim of a circumbinary disk in such a system. Like DQ Tau, it also shows evidence of accretion bursts related to the orbital motion, but not near periastron (*Alencar et al.,* 2003). Indeed, the Hα equivalent width peaks at the orbital phase when the stars are farthest apart.

These puzzling results show that the search for clear signs of circumbinary accretion onto the central system of young binaries is ongoing. However, if circumbinary environment replenishment occurs only when the binaries are sufficiently close, imaging such systems will be very difficult. Future interferometric measurements might allow us to disentangle the various possible modes of accretion.

4. DISK EVOLUTION IN YOUNG BINARIES

4.1. The Need for Resolved Observations of Young Binaries

A problem with using ensembles of T Tauri stars for discerning evolutionary trends is that one has to make judgements about the ages of the stars concerned. Some studies have used pre-main-sequence evolutionary tracks to ascribe ages to individual systems (e.g., *Hartmann et al.,* 1998; *Armitage et al.,* 2003), whereas others simply assumed that all stars in a given star-forming region have a similar age (e.g., *Haisch et al.,* 2000). In each case, the assignment of age is subject to uncertainties as a result of both the uncertainties in the pre-main-sequence tracks and the additional errors introduced by placing unresolved systems, as opposed to individual stars, in the HR diagram.

In binaries, however, we know *a priori* that the components are coeval, at least to within ~10^5 yr (i.e., to within a small fraction of the average ages of T Tauri stars). This statement is based on theoretical models for binary formation: The only possibility for binaries forming in a significantly noncoeval fashion is via star-disk capture. A number of studies, however, have shown that this is likely to be a very minor source of binary systems, even in dense environments like the Orion Nebula Cluster (*Clarke and Pringle,* 1991; *Scally and Clarke,* 2001). Therefore, without any need to rely on the accuracy of pre-main-sequence tracks, we can use binary stars as stellar pairs that are guaranteed to be coeval.

In recent years, each of the diagnostics described in section 2 has been used extensively to study the timescale and nature of evolutionary processes in protostellar disks. Typically these studies have not separated the individual components in binaries closer than an arcsecond or so. Because closer binaries constitute more than half of the systems in the best-studied region, Taurus Aurigae, this means that conclusions on disk evolution based on these studies are subject to considerable uncertainties.

For example, the designation of spectral types, and hence masses, to unresolved systems is unreliable; likewise, the detection of a disk diagnostic in an unresolved system does not in itself indicate whether it is the primary or the secondary or both components that possess a disk. These two factors introduce considerable uncertainties when using such data to investigate how disk evolutionary processes depend on stellar mass.

Another potential problem resulting from using unresolved data relates to the case in which the distribution of some observed property in T Tauri systems is used to infer the *rate* at which systems pass through various evolutionary stages. Evidently, this analysis is compromised in the case that the observed property is the sum of quantities arising from the individual binary components, whose evolution may not be synchronized. For example, the distribution of T Tauri stars in the K–L, K–N two-color plane has been used to deduce the relative amounts of time that stars spend with disks that are respectively optically thick or optically thin ("transition disks") or undetectable (*Kenyon and Hartmann*, 1995). This study revealed the striking result that very few systems were located in the transition region of the two-color plane, and has motivated the quest for disk-clearing models that can effect a rapid dispersal of the inner disk (*Armitage et al.*, 1999; *Clarke et al.*, 2001; *Alexander et al.*, 2005). *Prato and Simon* (1997) recognized that interpretation of this diagram is complicated by the existence of binaries and argued that the small numbers of systems with colors characteristic of transition objects implies that mixed binary pairs (i.e., one star with a disk and one without a disk) must be relatively rare. Our analysis in section 4.3 below shows that mixed pairs do in fact occur quite frequently in systems whose components have very disparate masses; in this case, however, the infrared colors of the unresolved system are then dominated by that of the primary and so such systems do not frequently end up in the transition region.

In summary, although studies of disk evolution based on unresolved systems are indeed valuable, they represent a rather blunt instrument compared with that provided by studies that resolve the individual components of binary systems. The value of this latter data can only be exploited if we first use it to answer a fundamental question: To what extent is disk evolution affected if the disk in question is located in a binary system? Depending on the answer to this question, we can *either* use the data to explore the influence of binarity on disk evolution *or* use the binary environment as just representing samples of coeval stars of various masses. We will return to this issue in section 4.3 below.

4.2. Overview of the Database: Separation Distribution of Binaries and Associated Selection Effects

We have classified the binaries in the database for which we have been able to assess the presence of a disk in each component as CC, CW, WC, and WW. Here C denotes a CTTS (accreting, disk-possessing) star and W a WTTS (nonaccreting, generally diskless) star. The first and second letter refer to the primary and secondary, respectively. The designation of C or W for each component is based primarily on the criterion of *Martín* (1998) for the equivalent width of Hα as a function of spectral type. In the minority of systems for which this is not available, the presence of Brγ is used instead. In the absence of information on either of these diagnostics a cut-off in near-infrared color of K–L = 0.3 or mid-infrared color of K–N = 2.0 is employed in-

TABLE 1.   Numbers of binaries in the database according to classification; see text for details.

CC	29	38	7
CW	11	14	1
CP	2	2	0
WC	4	6	1
WP	1	1	0
WW	12	21	1

stead. We also consider two additional categories, CP and WP, in which the primary is a CTTS or a WTTS and the secondary is a "passive disk" object: a nonaccreting star that while generally lacking any near-infrared excess also possesses a significant mid-infrared excess, indicating the presence of an inner dust disk hole (*McCabe et al.*, 2006).

Table 1 lists the numbers of objects of each type in the database that satisfy certain criteria. The lefthand column lists the number of objects of each type that have the most complete information (i.e., binary separation and spectral type for each component). Objects in the lefthand column have not been reported as possessing additional unresolved companions (at <0.1" separation) to one of the components, a feature that would disrupt the accretion flow in that region. The second column (which includes those in the first column) covers the larger sample of systems with known separations but not necessarily spectral types for both components. Objects in this column also have no reported additional close companions. The third column lists the number of systems with additional close companions.

To some extent, the numbers in Table 1 reflect observational selection effects. For example, it is possible that binaries with W primaries are underrepresented in this sample: Comparison of in Table 1 with the total numbers of stars in the Taurus aggregates that are classified as CTTS and WTTS, 100 and 70, respectively (Guieu, personal communication), suggests a mild deficit of binaries with WTT primaries. Any underrepresentation is likely to result from the relative disincentive to make high-angular-resolution observations of objects that show no obvious accretion signatures in their combined spectra. We would expect this underrepresentation to be more acute at small separations (where resolved observations require more effort) and, in the case of WCs, in low-mass-ratio objects, where the accretion signatures of the secondary are not obvious in the combined spectrum. In addition, relatively few objects have been scrutinized at N band, so that further systems may subsequently be transferred from the CW/WW to the CP/WP category; we have been rather conservative in our assignment of passive systems in Table 1, and so have not included several systems judged to be marginal passive candidates according to *McCabe et al.* (2006).

In Fig. 3 we plot the cumulative separation distributions of the binaries in the central column of Table 1, with the histograms (in descending order at a = 500 AU) representing CCs, WWs, CWs, and WCs. There is no statistically significant difference between any of these distributions: In the case of the two categories of binary with the largest subsample numbers, the CCs and the WWs, a KS test indicates that in the case that the two subsamples were drawn from the same parent distribution, the probability that the samples would be at least as different from each other as observed is 25%. There is some theoretical expectation that disk evolution should be accelerated in closer systems (see below), which might in principle lead to an excess of WWs at small separations. Although the fraction of close binaries is somewhat higher for WWs (i.e., 57% of WWs have separation less than 100 AU compared with only 38% of CCs), this difference is not statistically significant, possibly implying that accelerated disk evolution at small separations is not occurring in the binaries in our sample, which are rarely closer than ~20 AU. On the other hand, as we mentioned above, there is an observational selection effect against the discovery of closer systems with a W primary, so that this might mask any evidence for accelerated disk evolution in closer binaries.

Figure 3 also demonstrates that mixed pairs (WCs, and, to a lesser extent, CWs) are more concentrated at larger separations, although again the relatively small numbers of these systems yields a statistically insignificant result. The KS probability of either the mixed binary samples having a different separation distribution from the CC or WW samples is never less than 25%. We are less inclined to ascribe this tendency to an observational selection effect, since there is no reason why WCs should be underrepresented at small radii compared with WWs, or why CWs should be underrepresented compared with CCs at small separations.

The numbers of mixed systems (CWs or WCs) compared with CCs is a measure of the difference in lifetimes of the disks around each component. Synchronized evolution would imply mixed systems should be very rare, whereas a large difference in lifetimes would imply that mixed systems should be abundant. Including also the 4 passive systems

as mixed systems, the total numbers of CCs compared with mixed systems is 37 compared with 24; we have avoided the complicating factor of close companions by using the systems in the middle column of Table 1. This implies that the average lifetime of the shorter-lived disk is ~60% of the longer-lived disk. A further point to make about the mixed systems is that the number of mixed systems with a CTTS primary compared with a WTTS primary is 17 compared with 7. Evidently, there is a tendency for the primary's disk to be longer lived, although this is not universally the case.

We therefore conclude that when one combines all the available data from the literature, mixed systems are much less rare than was previously thought. It would appear that the reason that we need to revise our conclusions is that the incidence of mixed systems varies between different star-forming regions (see also *Prato and Monin,* 2001). Thus among the CCs and WWs in the middle column of Table 1, around half are located in Taurus. However, only 20% of the mixed systems are located in Taurus. Thus early studies (e.g., *Prato and Simon,* 1997) whose targets were mainly in Taurus contained relatively few mixed systems. We can only speculate as to why the fraction of mixed systems should vary from region to region. One obvious possibility is if the mixed phase corresponds to a particular range of ages and if different star-forming regions have different fractions of stars in the relevant age range.

4.3. The Distribution of Binaries in the a–q Plane

To make further progress, we must examine how various categories of binaries are distributed in the plane of mass ratio vs. separation. This necessitates using the more restricted subsample listed in the lefthand column of Table 1, for which we have spectral-type information for each component. We have checked that the separation distribution of the subsample is consistent with that of the full sample; although the difference is not statistically significant, we note that there happens to be a deficit of wide (>500 AU) CC binaries in the subsample compared with the full sample, which is manifest as the lack of solid dots in the righthand portion of Fig. 4.

In Fig. 4, filled circles represent the CCs, open circles the CWs, filled triangles the CPs, and open triangles the WCs. We do not include the WWs in this plot since they contain no information about *differential* disk evolution. We note that we expect the selection effects to be similar for all the binaries with CTTS primaries and that we expect the selection bias against low-q and low-a systems to be more severe for the systems with WTTS primaries.

We have placed binaries in Fig. 4 using the correlation between spectral type and mass for 1-m.y.-old stars given in *Hillenbrand and White* (2004). The necessity of having an optical spectral type for each star means our sample of 29 CCs and 11 CWs has excluded any binary containing an infrared companion or Class I source. For each binary we then calculate q_{DM} (i.e., the mass ratio M_2/M_1) using the pre-main-sequence tracks of *D'Antona and Mazzitelli*

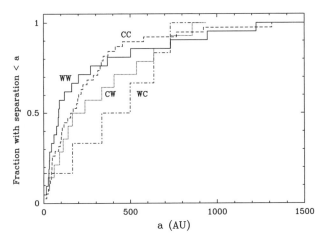

Fig. 3. Cumulative separation distribution of the four different binary categories.

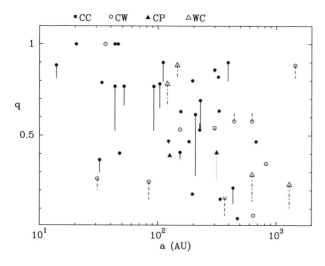

Fig. 4. Binaries from the lefthand column of Table 1 plotted in the q, a plane.

(1994). For a subset of systems for which both spectral types are later than K3, we also compute q_{BCAH}, using the pre-main-sequence tracks of *Baraffe et al.* (1998), also listed in *Hillenbrand and White* (2004). In Fig. 4, we plot q_{DM} in each case but link q_{DM} to the corresponding value of q_{BCAH} in the systems where both components lie in the range where q_{BCAH} can be computed. We use different dashes for different type of pairs. The length of the vertical lines gives some indication of the uncertainties inherent in pre-main-sequence tracks, although cannot in any sense be regarded as an error bar on q. Despite the strong disagreement between the tracks in certain ranges of spectral type, we nevertheless find that both set of tracks are in broad agreement as to whether binary systems are high or low q. In the quantitative analysis of the q distributions described below, we use q_{DM} as this is the only quantity that is available for all systems in our sample.

There are several striking features in this figure. As we have already noted, it first demonstrates that mixed systems are not rare and that many of the mixed systems are binaries with low q. On theoretical grounds (see below), one might expect that systems where the secondary's disk is exhausted before the primary's (i.e., the CWs and the CPs) would be low-q binaries. This is borne out with marginal statistical significance when one compares the q distribution of the CCs with the combined population of CWs and CPs. If we restrict our sample to binaries closer than 1000 AU in order to reduce the risk of picking up chance projections in our sample, we find that a KS test reveals that the two q distributions are different at the 2σ level. A KS test assesses the statistical significance of the *maximum* difference between the two datasets, which in this case refers to the fact that 11/28 CCs have q < 0.6 whereas for CWs and CPs the combined figure is 11/13. We also note that systems in which the primary's disk is exhausted first are relatively rare, i.e.,

for a < 1000 AU the total number of WCs and WPs is 4, compared with the 13 mixed systems with a CTTS primary in this separation range. From Fig. 4, we see that these four mixed systems with WTTS primaries are not found preferentially at low q, in contrast to what appears to be the case for the mixed systems with CTTS primaries. However, we caution that there may be a selection effect against the detection of low-q mixed systems with WTTS primaries at small separations.

Further analysis of this figure [i.e., division of the (a, q) domain into different regimes] is rendered difficult by the small total number of objects, so any trends that might appear to be qualitatively significant do not correspond to an impressively significant KS statistic. For example, we draw attention to the fact that for binaries closer than 100 AU, this being the canonical scale of disks around young stars (*Vicente and Alves,* 2005), there are *no* examples of pairs in which the primary's disk is exhausted first (i.e., WCs or WPs) and that two-thirds of the mixed systems have q < 0.5 compared with only two-tenths of the CCs having such low values of q.

This behavior is qualitatively consistent with what is expected theoretically in the case in which the disks around each star evolve in isolation, with their outer radii set by tidal truncation in the binary potential. Tidal truncation of disks occurs at a radius equal to a factor R_{tidal} times the binary separation, where R_{tidal} is plotted in Fig. 5 (*Armitage et al.,* 1999; *Papaloizou and Pringle,* 1987).

Evidently, for binaries at fixed separation, the secondary's disk is always tidally truncated to a smaller radius, but the difference only becomes significant for q < ~0.5. In the case of disks that are not continually replenished from an external reservoir, the tidal limitation of the disks around secondaries at a smaller radius leads to a more rapid accretion of the secondary's disk (*Armitage et al.,* 1999). This can be readily understood as disk accretion depends on viscous redistribution of angular momentum, which, in a freely expanding disk, occurs on a longer and longer timescale the disk spreads outward. If a disk is tidally truncated, however, angular momentum is tidally transferred to the binary orbit

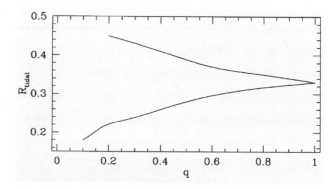

Fig. 5. Truncation radius as a fraction of the semimajor axis of the binary orbit vs. q: upper line for primary, lower line for secondary (from *Armitage et al.,* 1999).

at the point that the disk grows to the tidal truncation radius. Hence the disk dispersal timescale is roughly given by the disk's viscous timescale at the tidal truncation radius. For a disk with surface density profile of the form R^{-a}, the viscous timescale at radius R scales roughly as R^{2-a}. Hence, for a in the range 1–1.5 (*Beckwith and Sargent, 1991; Hartmann et al., 1998*), we have that the viscous timescale at the tidal radius R_T scales as $R_T^{0.5-1}$. Putting this scaling together with Fig. 4, we can therefore see that for binaries with q > 0.5, the viscous timescales at R_T are sufficiently similar that the disks should evolve more or less synchronously. The phase during which the secondary has exhausted its disk, but the primary has not, is relatively brief. On the other hand, for lower q in the range observed, we expect the viscous timescales at R_T for the two components to have a difference on the order of unity. This means that the time spent by a system as a CW is comparable with the time spent as a CC, and hence, as observed, the two sorts of system should occur in roughly equal numbers.

At larger separations, a > 100 AU, the picture is apparently rather different since mixed systems with WTTS primaries now start to appear. This suggests that we are entering a regime where the tidal truncation condition exerted by the binary is no longer the critical factor in determining which disk is exhausted first, a result that is perfectly comprehensible in the limit that the binary separation is much larger than typical disk sizes. We also note that the data for the wider binaries (where the disks evolve without obvious reference to their location in a binary) provides good evidence that disk lifetime is not a strong function of stellar mass. As an example, Sz 30 and Sz 108 are mixed systems with identical separations (630 AU) and similar spectral types for each component (M0.5–M2 and M0–M4.5 respectively). Nevertheless, in the former system it is the secondary that has lost its disk and in the latter it is the primary. Because we cannot appeal to noncoevality to explain this difference, we must assume that the lifetime of isolated disks is not a strong function of stellar mass in the range 0.1–1 $M_\odot$, and, hence, that presumably the initial conditions in the disk (such as initial mass or radius) instead dictate disk lifetime.

4.4. Implications for Disk Resupply

Early studies of binaries in which accretion diagnostics were separated for each component concluded that mixed systems are rare (see discussion in *Prato and Monin, 2001*), leading *Prato and Simon* (1997) to argue that the disks around each component must be sustained and then dissipated in a synchronized manner. It is hard to understand synchronized dispersal unless it is effected by some external agent. On the other hand, a low fraction of mixed systems can be explained if both components are fed from a common reservoir over most of the disk lifetime and if, once the reservoir is exhausted, the dispersal of both disks is relatively rapid. This explanation was favoured by Prato and Simon on the grounds that continued replenishment is the only way to explain the presence of accretion diagnostics

in the closest binaries (a < a few AU), for which the viscous timescale of their (highly truncated) disks is much less than the system age. In these closest binaries, there is good evidence for circumbinary disks (*Jensen and Mathieu,* 1997), which can plausibly continue to feed the central binary (*Mathieu et al.,* 1997). In wider binaries, however, i.e., a in the range a few to ~100 AU, upper limits on circumbinary disk masses are ~5 M_J (*Jensen et al., 1996*) and therefore inadequate to provide substantial replenishment of circumstellar disks. In these wider systems, it is instead necessary to invoke replenishment through infall from an extended envelope. Possible evidence for such an envelope is provided by the millimeter study of young binaries by *Jensen and Akeson* (2003), who found that their interferometric measurements contained 46–85% of the flux found in previous, single-dish measurements (*Beckwith et al., 1990*). *Jensen and Akeson* (2003) therefore speculated that the additional flux originated in an envelope on scales of >700 AU, with the caution that the flux difference could be due to a flux calibration issue. However, as it is possible to conceal large quantities of cold dust at large distances from the binary without contributing significantly to the millimetre flux (*Lay et al., 1994*), it is impossible to use this observation to constrain whether the extended emission contains a viable mass for resupplying the binaries' circumstellar disks.

Our analysis here, however, indicates that mixed systems are, in fact, common, and thus does not require continued replenishment of disks for the binaries in our sample (which mostly have separations >20 AU). Our results do not require there to be no replenishment, but imply that such replenishment must occur over a minor fraction of the disks' lifetimes or else be concentrated on to the primary's disk at late times. This latter possibility is in conflict with numerical simulations of infall onto protobinaries (*Artymowicz,* 1983; *Bate,* 1997) [but see *Ochi et al.* (2005) for a recent contrary view on this issue]. The simplest interpretation of our results, however, is that the disks evolve in isolation and that disk tidal truncation in the binary potential results in the secondary disk being dissipated somewhat prior to the primary's disk.

4.5. Implications for Planet Formation in Binaries

How do these findings bear on the probability that planets are located in binary systems? The presence of a binary companion may render the existence of planets less likely in two ways. First, binarity restricts the regions of orbital parameter space in which planets can exist in stable, circumstellar orbits, ruling out orbital radii that are within a factor of the binary separation, modulo the mass ratio q. For example, *Holman and Weigert* (1999) have conducted a study of the long-term orbital stability of planets in binary systems and find that a companion star orbiting beyond more than 5× the planetary orbital radius does not strongly threaten the planet's orbital stability. Second, if binarity reduces disk lifetimes (in the primary or secondary or both), then it may reduce the probability of planet formation, since there may be insufficient time for slow processes (such as those involved in the core-accretion model) to operate before the disk is dispersed. For example, *Thebault et al.* (2004) find that the formation of the observed planet at 2 AU in the 18-AU binary γ Cephei requires the presence of a long-lived and massive gas disk. In the absence of such gas, secular perturbations by the binary companion generate too high a velocity dispersion among the planetesimals for runaway accretion to proceed.

The present study, however, finds that the influence of binarity on circumstellar disk lifetime is rather mild in the systems with separations >20 AU. The fact that the separation distribution of diskless binaries is indistinguishable from that of binaries with disks suggests that disk dispersal is not strongly accelerated for the closer binaries in this sample. Concerning differential evolution between the disks around primaries and secondaries, we found that the overall statistics of mixed systems vs. CC systems implied that the shorter-lived disk (usually the secondary's) had a mean lifetime of ~60% that of the longer-lived disk. Unless there are processes in planet formation for which a factor of 2 difference in disk lifetime is critical, we conclude that *circumstellar planet formation is not likely to be strongly suppressed in the case of binary secondaries*. We therefore expect planets to be formed around both components in binary systems wider than ~20 AU. The recent numerical simulations of *Lissauer et al.* (2004) and *Quintana et al.* (2005) (see also *Barbieri et al.,* 2002) are in good agreement with this result.

The observational situation regarding the detection of planets in binary systems is strongly skewed by the selection criteria used in Doppler reflex motion surveys, as these tend to exclude known binaries on the grounds that binary orbital motion makes it harder to detect a planetary companion. Among the >150 G to M stars hosting planetary companions, only 25 are binary or multiple systems, hosting a total of 31 planets (*www.exoplanets.org*) (*Eggenberger et al.,* 2004, 2006; *Mugrauer et al.,* 2005). Therefore, only around 15% of known planets are in binary or multiple systems.

Figures 6 and 7 show the orbital properties of the binary systems known to host Doppler reflex motion planets. As expected, the sample is strongly biased toward larger separations: Planet search programs do not typically monitor binaries with separations less than ~2″, corresponding to separations in the range >2 AU at the distances of the target stars (*Valenti and Fischer,* 2005). Because the median binary separation for G stars is 30 AU (*Duquennoy and Mayor,* 1991), it is evident that a large fraction of binaries have been excluded from such surveys. There is also the possibility of an observational bias toward low q on the grounds that low-mass companions are more likely to have been overlooked when initially selecting the radial velocity targets.

It is immediately obvious from Fig. 7 that the ratio of binary semimajor axis to planet semimajor axis (a_b/a_p) is extremely large, generally in the range 100–1000 and in all cases >10. It is therefore unsurprising, on the grounds of orbital stability, that planets are found in these systems. Moreover, the binaries in Fig. 7 are in the same separation range that we have studied in section 4.2, where we found

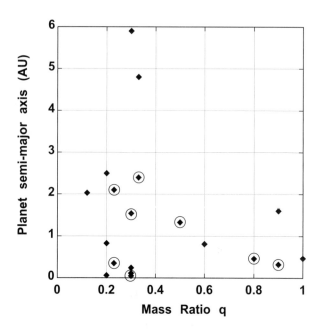

Fig. 6. Distance from the planet to its central star (component of a binary) vs. mass ratio. The circled points are the ones for binaries with separation less than 500 AU.

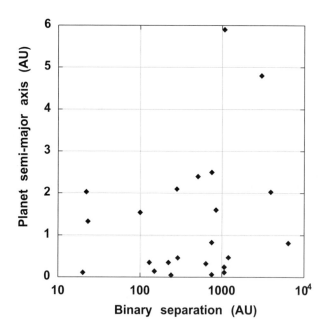

Fig. 7. Distance from the planet to its central star (component of a binary) vs. binary separation.

little apparent dependence of disk lifetime on binary separation. We would therefore not expect planet formation to be suppressed in these systems on the basis of reduced disk lifetime.

We stress that the current data cannot be used to determine whether planets are preferentially found around binary primaries or secondaries, since in almost all cases it is only the primary that has been a radial velocity target.

In only two systems is the planet detected around the secondary component (16 Cyg and HD 178911). Likewise, it would be premature to derive the statistics of *circumbinary planets*. To date, there is one system, HD 202206, that might be described as containing a circumbinary planet, although the mass ratio of the central binary is extremely low: The central companion is itself in the brown dwarf/planetary regime (*Correia et al.*, 2005). From a theoretical point of view, *Moriwaki and Nakagawa* (2004) have claimed that in the case of a binary of separation 1 AU, planetesimal accretion should be able to proceed undisturbed at radii greater than ~13 AU from the barycenter. This relatively large region in which planet formation might be expected to be suppressed in the circumbinary disk means that it may be problematic to detect planets through radial velocity measurements around all but the closest binaries. *Quintana et al.* (2005) calculate, however, that for binary separations of <0.2 AU, the growth of planetesimals into a system of terrestrial planets is statistically indistinguishable from similar simulations for single stars. Surveys for planets around single-lined, spectroscopic binaries (e.g., Eggenberger, personal communication, 2005) have only recently begun. When data are available, they should provide interesting constraints.

5. FUTURE DIRECTIONS

The most formidable obstacle to furthering our understanding of disk evolution in young binaries is the relatively small size of our database. Although our compilation of around 60 binaries with complete spectral type and disk diagnostic information for each component represents tremendous progress in the last decade, it is nevertheless too small a sample for us to be able to divide it into subcategories according to, e.g., separation and subsequently derive statistically significant results. There are, however, good prospects for increasing the sample size. In our database of ~170 total systems, we estimate that we can derive complete properties for approximately another half-dozen systems based on extant data. An additional 28 systems with separations of >1″ can be characterized with a 2–3-m-class telescope in a site with good seeing, such as Mauna Kea or Cerro Tololo. A further three dozen systems have separations between 0.1″ and 1.0″. For these pairs it would be straightforward to characterize each component with low-resolution spectroscopy behind an adaptive optics system, or an integral field spectrometer unit, at a 6–10-m-class facility. The results of such observations would more than double the young binary sample. Furthermore, our database was compiled from a limited number of references and is certainly far from complete. We anticipate the ongoing compilation of additional objects and improvement in the quantity and quality of data for objects already listed.

Larger samples of binaries with known properties in a variety of star-forming regions with a range of estimated ages will allow us to test the extremely intriguing notion of the regional dependence of the fraction of mixed systems. The data in this paper, as well as data obtained in the earlier studies of *Prato and Simon* (1997), *Prato and Monin*

(2001), and *Hartigan and Kenyon* (2003), suggest a low fraction of mixed pairs in the Taurus region. Could this be the result of a younger age for Taurus than the other regions from which our sample is culled? Is it simply a selection effect, or a result of small number statistics? If a real and age-dependent effect, the mixed system fraction may yield a unique and sensitive approach to estimating the ages of star-forming regions.

With high-resolution spectroscopy of both components in young binaries more detailed properties may be examined. For example, with multiple-epoch observations, hierarchical spectroscopic binaries might be identified in binary component stars. The individual rotation properties of the stars in close pairs could also be examined and compared with the circumstellar disk properties to better understand the evolution of angular momentum in young binaries (*Armitage et al.*, 1999). High-resolution observations of accretion line diagnostics, such as hydrogen emission lines, could provide a unique approach to the measurement of how accretion is apportioned between the two stars in spectroscopic binaries. Such observations at infrared wavelengths would provide a better opportunity to observe emission lines from both stars, even for systems with large continuum flux ratios (e.g., *Prato et al.*, 2002).

An interesting problem raised in *McCabe et al.* (2006) is the origin of the passive disk phenomenon. By combining resolved near- and mid-infrared observations with longer-wavelength Spitzer data and astrophysical information for the binary stars themselves, i.e., masses, it will be possible to test the premise set up in *Clarke et al.* (2001) and *Takeuchi et al.* (2005), namely that a population of young systems with large inner disk holes exists around higher-mass stars that have previously been identified as WTTS.

The advent of very-high-resolution interferometry, in both the optical-infrared as well as in the millimeter regimes, will provide an unprecedented view of the orientations of disks in binaries even at circumstellar scales. Progress has already been made using the Keck Interferometer (*Patience et al.*, 2005) and the VLTI (Malbet et al., in preparation). The ALMA interferometer, anticipated for first light in the next three to four years at partial capacity, will provide unprecedented images of the cool, dusty disk structures.

These new generations of facilities will enable entirely new studies, which will go far beyond the issue of simple existence of disks in binary systems. Instead, it will be possible to measure how *disk properties* vary as a function of binary properties such as separation, mass ratio, angular momentum, magnetic field strength, etc. For example, an instrument such as ALMA will enable us to study disk particle-size distributions as a function of binary separation. Optical-infrared interferometers could provide data on inner disk structure as a function of magnetic field strength. Numerous such exciting possibilities for future study exist.

Acknowledgments. We thank the anonymous referee for useful comments that helped improve the quality of the paper. C.C. thanks the LAOG and J.-L.M. thanks the IoA for their generous hospitality while writing this paper. L.P. acknowledges the contribution of helpful information from K. Strom, S. Strom, and G.

Marcy. We are grateful to A. Eggenberger for discussions of her research in advance of publication. This work has made use of the ADS database.

REFERENCES

Alencar S. H. P., Melo C. H. F., Dullemond C. P., Andersen J., Batalha C., Vaz L. P. R., and Mathieu R. D. (2003) *Astron. Astrophys., 409,* 1037–1053.

Alexander R. D., Clarke C. J., and Pringle J. E. (2005) *Mon. Not. R. Astron. Soc., 358,* 283–290.

Andrews S. and Williams J. (2005) *Astrophys. J., 619,* L175–L178.

Armitage P. J., Clarke C. J., and Tout C. A. (1999) *Mon. Not. R. Astron. Soc., 304,* 425–433.

Armitage P. J., Clarke C. J., and Palla F. (2003) *Mon. Not. R. Astron. Soc., 342,* 1139–1146.

Artymowicz P. (1983) *Acta. Astron., 33,* 223–230.

Artymowicz P. and Lubow S. H. (1996) *Astrophys. J., 467,* L77–L80.

Baraffe I., Chabrier G., Allard F., and Hauschildt P. (1998) *Astron. Astrophys., 337,* 403–412.

Barbieri M., Marzar F., and Scholl H. (2002) *Astron. Astrophys., 396,* 219–224.

Basri G. and Bertout C. (1993) In *Protostars and Planets III* (E. H. Levy and J. I. Lunine, eds), pp. 543–566. Univ. of Arizona, Tucson.

Basri G., Johns-Krull C. M., and Mathieu R. D. (1997) *Astron. J., 114,* 781–792.

Bastien P. and Ménard F. (1988) *Astrophys. J., 326,* 334–338.

Bastien P. and Ménard F. (1990) *Astrophys. J., 364,* 232–241.

Bate M. R. (1997) *Mon. Not. R. Astron. Soc., 285,* 16–32.

Bate M. R., Bonnell I. A., Clarke C. J., Lubow S. H., Ogilvie G. I., Pringle J. E., and Tout C. A. (2000) *Mon. Not. R. Astron. Soc., 317,* 773–781.

Beckwith S. V. W. and Sargent A. I. (1991) *Astrophys. J., 381,* 250–258.

Beckwith S. V. W., Zuckerman B., Skrutskie M. F., and Dyck H. M. (1984) *Astrophys. J., 287,* 793–800.

Beckwith S. V. W., Sargent A. I., Chini R. S., and Güstem R. (1990) *Astron. J., 99,* 924–945.

Bertout C., Basri G., and Bouvier J. (1988) *Astrophys. J., 330,* 350–373.

Beust H. and Dutrey A. (2005) *Astron. Astrophys., 439,* 585–594.

Bonnell I. A., Arcoragi J.-P., Martel H., and Bastien P. (1992) *Astrophys. J., 400,* 579–594.

Böhm K.-H. and Solf J. (1994) *Astrophys. J., 430,* 277–290.

Brandner W. and Zinnecker H. (1997) *Astron. Astrophys., 321,* 220–228.

Breger M. and Dyck H. M. (1972) *Astrophys. J., 175,* 127–134.

Chandler C. J., Brogan C. L., Shirley Y. L., and Loinard L. (2005) *Astrophys. J., 632,* 371–396.

Chelli A., Cruz-Gonzalez I., and Reipurth B. (1995) *Astron. Astrophys. Suppl., 114,* 135–142.

Chiang E. I. and Goldreich P. (1997) *Astrophys. J., 490,* 368–376.

Chiang E. I., Joung M. K., Creech-Eakman M. J., Qi C., Kessler J. E., Blake G. A., and van Dishoeck E. F. (2001) *Astrophys. J., 547,* 1077–1089.

Clarke C. J. and Pringle J. E. (1991) *Mon. Not. R. Astron. Soc., 249,* 584–587.

Clarke C. J., Gendrin A., and Sotomayor M. (2001) *Mon. Not. R. Astron. Soc., 328,* 485–491.

Close L. M., Dutrey A., Roddier F., Guilloteau S., Roddier C., Northcott M., Ménard F., Duvert G., Graves J. E., and Potter D. (1998) *Astrophys. J., 499,* 883–888.

D'Alessio P. (2003) *Rev. Mex. Astron. Astrofís., 18,* 14–28.

D'Antona D. and Mazzitelli I. (1994) *Astrophys. J. Suppl., 90,* 467–500.

Davis C. J., Mundt R., and Eislöffel J. (1994) *Astrophys. J., 437,* L55–L58.

Duchêne G., Monin J.-L., Bouvier J., and Ménard F. (1999) *Astron. Astrophys., 351,* 954–962.

Duchêne G., McCabe C., Ghez A. M., and Macintosh B. A. (2004) *Astrophys. J., 606,* 969–982.

Duquennoy A. and Mayor M. (1991) *Astron. Astrophys., 248,* 485–524.

Dutrey A., Guilloteau S., and Simon M. (1994) *Astron. Astrophys., 286,* 149–159.

Dutrey A., Guilloteau S., Duvert G., Prato L., Simon M., Schuster K., and Ménard F. (1996) *Astron. Astrophys., 409,* 493–504.

Correia A. C. M., Udry S., Mayor M., Laskar J., Naef D., Pepe F., Queloz D., and Santos N. C. (2005) *Astron. Astrophys., 440,* 751–758.

Edwards S., Cabrit S., Strom S. E., Heyer I., Strom K. M., and Anderson E. (1987) *Astrophys. J., 321,* 473–495.

Edwards S., Ray T., and Mundt R. (1993) In *Protostars and Planets III* (E. H. Levy and J. I. Lunine, eds.), pp. 567–603. Univ. of Arizona, Tucson.

Eggenberger A., Udry S., and Mayor M. (2004) *Astron. Astrophys., 417,* 353–360.

Eggenberger A., Mayor M., Naef D., Pepe F., Queloz D., Santos N. C., Udry S., and Lovis C. (2006) *Astron. Astrophys., 447,* 1159.

Eislöffel J. and Mundt R. (1997) *Astron. J., 114,* 280–287.

Eislöffel J., Smith M. D., Davis C. J., and Ray T. P. (1996) *Astron. J., 112,* 2086–2095.

Geoffray H. and Monin J.-L. (2001) *Astron. Astrophys., 369,* 239–248.

Ghez A. M., Neugebauer G., Gorham P. W., Haniff C. A., Kulkarni S. R., Matthews K., Koresko C., and Beckwith S. V. W. (1991) *Astron. J., 102,* 2066–2072.

Ghez A. M., Neugebauer G., and Matthews K. (1993) *Astron. J., 106,* 2005–2023.

Ghez A. M., Emerson J. P., Graham J. R., Meixner M., and Skinner C. J. (1994) *Astrophys. J., 434,* 707–712.

Grasdalen G. L., Strom S. E., Strom K. M., Capps R. W., Thompson D., and Castelaz M. (1984) *Astrophys. J., 283,* L57–L60.

Gredel R. and Reipurth B. (1993) *Astrophys. J., 407,* L29–L32.

Guilloteau S., Dutrey A., and Simon M. (1999) *Astron. Astrophys., 348,* 570–578.

Gullbring E., Hartmann L., Briceno C., and Calvet N. (1998) *Astrophys. J., 492,* 323–341.

Haisch K. E., Lada E. A., and Lada C. J. (2000) *Astron. J., 120,* 1396–1409.

Haisch K. E., Lada E. A., and Lada C. J. (2001) *Astrophys. J., 553,* L153–L156.

Hale A. (1994) *Astron. J., 107,* 306–332.

Hartigan P. and Kenyon S. J. (2003) *Astrophys. J., 583,* 334–357.

Hartigan P., Strom K. M., and Strom S. E. (1994) *Astrophys. J., 427,* 961–977.

Hartmann L., Calvet N., Gullbring E., and D'Alessio P. (1998) *Astrophys. J., 495,* 385–400.

Herbig G. H. (1948) *Publ. Astron. Soc. Pac., 60,* 256–267.

Herbig G. H. (1989) In *The Formation and Evolution of Planetary Systems* (H. Weaver and L. Danly, eds.), pp. 296–312. Cambridge Univ., Cambridge.

Hillenbrand L. A. and White R. J. (2004) *Astrophys. J., 604,* 741–757.

Hillenbrand L. A., Strom S. E., Calvet N., Merrill K. M., Gatley I., Makidon R. B., Meyer M. R., and Skrutskie M. F. (1998) *Astron. J., 116,* 1816–1841.

Hodapp K. W., Bally J., Eislöffel J., and Davis C. J. (2005) *Astron. J., 129,* 1580–1588.

Holman M. J. and Wiegert P. A. (1999) *Astron. J., 117,* 621–628.

Jayawardhana R., Luhman K. L., D'Alessio P., and Stauffer J. R. (2002) *Astrophys. J., 571,* L51–L54.

Jensen E. L. N. and Akeson R. L. (2003) *Astrophys. J., 584,* 875–881.

Jensen E. L. N. and Mathieu R. D. (1997) *Astron. J., 114,* 301–316.

Jensen E. L. N., Mathieu R. D., and Fuller G. A. (1996) *Astrophys. J., 458,* 312–326.

Jensen E. L. N., Mathieu R. D., Donar A. X., and Dullighan A. (2004) *Astrophys. J., 600,* 789–803.

Joy A. H. and van Biesbroeck G. (1944) *Publ. Astron. Soc. Pac., 56,* 123–124.

Johnson H. L. (1966) *Ann. Rev. Astron. Astrophys., 4,* 193–206.

Kenyon S. J. and Hartmann L. (1995) *Astrophys. J. Suppl., 101,* 117–171.

Knacke R. F., Strom K. M., Strom S. E., Young E., and Kunkel W. (1973) *Astrophys. J., 179,* 847–854.

Koerner D. W., Jensen E. L. N., Cruz K. L., Guild T. B., and Gultekin K. (2000) *Astrophys. J., 533,* L37–L40.

Krist J. E., Stapelfeldt K. R., and Watson A. M. (2002) *Astrophys. J., 570,* 785–792.

Krist J. E., Stapelfeldt K. R., Golimowski D. A., Ardila D. R., Clampin M., et al. (2005) *Astron. J., 130,* 2778–2787.

Kuhn J. R., Potter D., and Parise B. (2001) *Astrophys. J., 553,* L189–L191.

Lay O. P., Carlstrom J. E., Hills R. E., and Phillips T. G. (1994) *Astrophys. J., 434,* L75–L78.

Leinert Ch. and Haas M. (1989) *Astrophys. J., 342,* L39–L42.

Leinert Ch., Zinnecker H., Weitzel N., Christou J., Ridgway S. T., Jameson R., Haas M., and Lenzen R. (1993) *Astron. Astrophys., 278,* 129–149.

Lissauer J. J., Quintana E. V., Chambers J. E., Duncan M. J., and Adams F. C. (2004) *Rev. Mex. Astron. Astrofis., 22,* 99–103.

Lynden-Bell D. and Pringle J. E. (1974) *Mon. Not. R. Astron. Soc., 168,* 603–637.

Lubow S. H. and Ogilvie G. I. (2000) *Astrophys. J., 538,* L326–L340.

Martín E. L. (1998) *Astron. J., 115,* 351–357.

Malbet F., Lachaume R., and Monin J.-L. (2001) *Astron. Astrophys., 379,* 515–528.

Mathieu R. D. (1994) *Ann. Rev. Astron. Astrophys., 32,* 465–530.

Mathieu R. D., Adams F. C., and Latham D. W. (1991) *Astron. J., 101,* 2184–2198.

Mathieu R. D., Adams F. C., Fuller G. A., Jensen E. L. N., Koerner D. W., and Sargent A. I. (1995) *Astron. J., 109,* 2655–2669.

Mathieu R. D., Stassun K., Basri G., Jensen E. L. N., Johns-Krull C. M., Valenti J. A., and Hartmann L. W. (1997) *Astron. J., 113,* 1841–1854.

McCabe C., Duchêne G., and Ghez A. M. (2002) *Astrophys. J., 575,* 974–988.

McCabe C., Ghez A. M., Prato L., Duchêne G., Fisher R. S., and Telesco C. (2006) *Astrophys. J., 636,* 932–951.

Ménard F., Monin J.-L., Angelucci F., and Rouan D. (1993) *Astrophys. J., 414,* L117–L120.

Mendoza V. and Eugenio E. (1966) *Astrophys. J., 143,* 1010.

Mendoza V. and Eugenio E. (1968) *Astrophys. J., 151,* 977.

Monin J.-L. and Bouvier J. (2000) *Astron. Astrophys., 356,* L75–L78.

Monin J.-L., Ménard F., and Duchêne G. (1998) *Astron. Astrophys., 339,* 113–122.

Monin J.-L., Ménard F., and Peretto N. (2006) *Astron. Astrophys., 446,* 201–210.

Moriwaki K. and Nakagawa Y. (2004) *Astrophys. J., 609,* 1065–1070.

Mugrauer M., Neuhauser R., Seifahrt A., Mazeh T., and Guenther E. (2005) *Astron. Astrophys., 440,* 1051–1060.

Muzerolle J., Hartmann L., and Calvet N. (1998) *Astron. J., 116,* 2965–2974.

Najita J., Carr J. S., and Mathieu R. D. (2003) *Astrophys. J., 589,* 931–952.

Ochi Y., Sugimoto K., and Hanawa T. (2005) *Astrophys. J., 623,* 922–939.

Papaloizou J. C. B. and Pringle J. E. (1987) *Mon. Not. R. Astron. Soc., 225,* 267–283.

Papaloizou J. C. B. and Terquem C. E. J. M. L. J. (1995) *Mon. Not. R. Astron. Soc., 274,* 987–1001.

Patience J., Ghez A. M., Reid I. N., and Matthews K. (2002) *Astron. J., 123,* 1570–1602.

Patience J., Akeson R. L., Jensen E. L. N., and Sargent A. I. (2005) In *Protostars and Planets V,* www.lpi.usra.edu/meetings/ppv2005/pdf/8603.pdf.

Potter D. E., Close L. M., Roddier F., Roddier C., Graves J. E., and Northcott M. (2000) *Astrophys. J., 540,* 422–428.

Prato L. and Monin J.-L. (2001) In *The Formation of Binary Stars* (H. Zinnecker and R. D. Mathieu, eds.), pp. 313–322. IAU Symposium 200, ASP, San Francisco.

Prato L. and Simon M. (1997) *Astrophys. J., 474,* 455–463.

Prato L., Ghez A. M., Piña R. K., Telesco C. M., Fisher R. S., Wizinowich P., Lai O., Acton D. S., and Stomski P. (2001) *Astrophys. J., 549,* 590–598.

Prato L., Simon M., Mazeh T., Zucker S., and McLean I. S. (2002) *Astrophys. J., 579,* L99–L102.

Prato L., Greene T. P., and Simon M. (2003) *Astrophys. J., 584,* 853–874.

Quintana E. V., Lissauer J. J., Adams F. C., Chambers J. E., and Duncan M. J. (2005) In *Protostars and Planets V,* www.lpi.usra.edu/meetings/ppv/8621.pdf.

Reipurth B. and Zinnecker H. (1993) *Astron. Astrophys., 278,* 81.

Reipurth B., Heathcote S., Roth M., Noriega-Crespo A., and Raga A. C. (1993) *Astrophys. J., 408,* L49–L52.

Roddier C., Roddier F., Northcott M. J., Graves J. E., Jim K., et al. (1996) *Astrophys. J., 463,* 326–335.

Rucinski S. M. (1985) *Astron. J., 90,* 2321–2330.

Rydgren A. E., Strom S. E., and Strom K. M. (1976) *Astrophys. J. Suppl., 30,* 307–336.

Scally A. and Clarke C. J. (2001) *Mon. Not. R. Astron. Soc., 325,* 449–456.

Simon M. and Guilloteau S. (1992) *Astrophys. J., 397,* L47–L49.

Simon M., Ghez A. M., Leinert Ch., Cassar L., Chen W. P., Howell R. R., Jameson R. F., Matthews K., Neugebauer G., and Richichi A. (1995) *Astrophys. J., 443,* 625–637.

Simon M., Dutrey A., and Guilloteau S. (2000) *Astrophys. J., 545,* 1034–1043.

Skrutskie M. F., Dutkevitch D., Strom S. E., Edwards S., Strom K. M., and Shure M. A. (1990) *Astron. J., 99,* 1187–1195.

Stapelfeldt K. R., Krist J. E., Ménard F., Bouvier J., Padgett D. L., and Burrows C. J. (1998) *Astrophys. J., 502,* L65–L69.

Stapelfeldt K. R., Ménard F., Watson A. M., Krist J. E., Dougados C., Padgett D. L., and Brandner W. (2003) *Astrophys. J., 589,* 410–418.

Stassun K. G., Mathieu R. D., Vrba F. J., Mazeh T., and Henden A. (2001) *Astron. J., 121,* 1003–1012.

Stempels H. C. and Gahm G. F. (2004) *Astron. Astrophys., 421,* 1159–1168.

Strom S. E. (1972) *Publ. Astron. Soc. Pac., 84,* 745–756.

Strom K. M., Strom S. E., and Yost J. (1971) *Astrophys. J., 165,* 479–488.

Strom K. M., Strom S. E., Breger M., Brooke A. L., Yost J., Grasdalen G., and Carrasco L. (1972) *Astrophys. J., 173,* L65–L70.

Strom S. E., Strom K. M., and Grasdalen G. L. (1975) *Ann. Rev. Astron. Astrophys., 13,* 187–216.

Strom S. E., Strom K. M., Grasdalen G. L., Capps R. W., and Thompson D. (1985) *Astron. J., 90,* 2575–2580.

Strom K. M., Strom S. E., Edwards S., Cabrit S., and Skrutskie M. F. (1989) *Astron. J., 97,* 1451–1470.

Takeuchi T., Clarke C. J., and Lin D. N. C. (2005) *Astrophys. J., 627,* 286–292.

Tessier E., Bouvier J., and Lacombe F. (1994) *Astron. Astrophys., 283,* 827–834.

Thebault P., Marzari F., Scholl, H., Turrini, D., and Barbieri M. (2004) *Astron. Astrophys., 427,* 1097–1104.

Valenti J. A. and Fischer D. A. (2005) *Astrophys. J. Suppl., 159,* 141–166.

Vicente S. M. and Alves J. (2005) *Astron. Astrophys., 441,* 195–205.

White R. J. and Ghez A. M. (2001) *Astrophys. J., 556,* 265–295.

Wolf S., Stecklum B., and Henning T. (2001) In *The Formation of Binary Stars* (H. Zinnecker and R. D. Mathieu eds.), pp. 295–304. IAU Symposium 200, ASP, San Francisco.

Dynamical Mass Measurements of Pre-Main-Sequence Stars: Fundamental Tests of the Physics of Young Stars

R. D. Mathieu
University of Wisconsin–Madison

I. Baraffe
École Normale Supérieure, Lyon

M. Simon
State University of New York–Stony Brook

K. G. Stassun
Vanderbilt University

R. White
University of Alabama in Huntsville

There are now 23 dynamical mass measurements for pre-main-sequence (PMS) stars of less than 2 $M_\odot$, with most of the measured stars having masses greater than 0.5 $M_\odot$. The masses of two PMS brown dwarfs have also been precisely measured. The most important application of these dynamical mass measurements has been to provide tests of theoretical masses derived from PMS stellar evolution models. On average, most models in use today predict stellar masses to within 20%; however, the predictions for individual stars can be in error by 50% or more. Now that dynamical mass measurements are relatively abundant, and will become more so with the application of groundbased optical/infrared interferometers, the primary limitations to such tests have become systematic errors in the determination of the stellar properties necessary for the comparison with evolutionary models, such as effective temperature, luminosity, and radii. *Additional dynamical mass determinations between 0.5 $M_\odot$ and 2 $M_\odot$ will not likely improve the constraints on evolutionary models until these systematic uncertainties in measurements of stellar properties are reduced.* The nature and origin of these uncertainties, as well as the dominant physical issues in theoretical PMS stellar evolution models, are discussed. There are immediately realizable possibilities for improving the characterizations of those stars with dynamical mass measurements. Additional dynamical mass measurements for stars below 0.5 $M_\odot$ are also very much needed.

1. INTRODUCTION

Prior to the measurement by *Popper* (1987) of the mass of the pre-main-sequence (PMS) secondary of the eclipsing binary EK Cep, every mass cited for a PMS star had been derived by comparing the star's location in the Hertzsprung-Russell (HR) diagram to the predictions of theoretical evolutionary models, which track how a star of given mass evolves in luminosity and temperature with age. These theoretical assignments of stellar masses were entirely unconstrained by direct mass measurements of PMS stars. Furthermore, different models can predict masses that differ by as much as a factor of 2 or more.

The masses so assigned to PMS stars are at the very foundation of our understanding of star and planet formation. These masses define the initial mass function that delimits the outcome of the star-formation process. They set the energy scale available to explain processes ranging from accretion to outflow. They allow us to link young stars with older generations. And these masses permit us to identify stars that can serve as proxies for the Sun-Earth system at an early age.

Equally importantly, stellar masses represent a key observational interface for theoretical stellar evolution models. Such models provide our chronology of early stellar evolution and thereby touch upon the most basic of questions, including the timescale for circumstellar disk evolution and planet formation.

Thus the importance of the measurement of accurate PMS stellar masses in order to test theoretical masses can hardly be overstated. As the authors of this paper, we have the good fortune to be writing one of the first reviews of PMS masses in which much of the text discusses actual mass measurements for PMS stars, rather than what needs to be known and how that knowledge will be gained in the very near future!

TABLE 1. Dynamical masses and stellar properties of 23 pre-main-sequence stars.

Name	Mass (M$_\odot$)	Technique*	Reference	Radius (R$_\odot$)	SpT	Log (T$_{eff}$)	Log (L/(L$_\odot$)	Reference
RS Cha A	1.858 ± 0.016	EB	A91	2.137 ± 0.055	A8	3.883 ± 0.010	1.144 ± 0.044	M00
RS Cha B	1.821 ± 0.018	EB	A91	2.338 ± 0.055	A8	3.859 ± 0.010	1.126 ± 0.043	M00
MWC 480	1.65 ± 0.07	DK	Si00	—	A2-3	3.948 ± 0.015	1.243 ± 0.10	HW
TY CrA B	1.64 ± 0.01	EB	C98	2.080 ± 0.140	—	3.690 ± 0.035	0.380 ± 0.145	C98
045251+3016 A	1.45 ± 0.19	AS	St01	—	K5	3.643 ± 0.015	−0.167 ± 0.053	St01
BP Tau	1.32 +0.20/−0.12†	DK	D03	—	K7	3.608 ± 0.012	−0.78 ± 0.10	J99
0529.4+0041 A	1.27 ± 0.01	EB	C04	1.44 ± 0.05	K1-2	3.716 ± 0.012	0.137 ± 0.073	C04
EK Cep B	1.124 ± 0.012	EB	P87	1.320 ± 0.015	—	3.755 ± 0.015	0.190 ± 0.070	P87
UZ Tau Ea	1.016 ± 0.065†	DKS	P02	—	M1	3.557 ± 0.015	−0.201 ± 0.124	P02
V1174 Ori A	1.009 ± 0.015	EB	S04	1.339 ± 0.015	K4.5	3.650 ± 0.011	−0.193 ± 0.048	S04
LkCa 15	0.97 ± 0.03†	DK	Si00	—	K5	3.643 ± 0.015	−0.165 ± 0.10	HW
0529.4+0041 B	0.93 ± 0.01	EB	C04	1.35 ± 0.05	—	3.625 ± 0.015	−0.283 ± 0.11	C04
GM Aur	0.84 ± 0.05†	DK	S01	—	K7	3.602 ± 0.015	0.598 ± 0.10	HW
045251+3016 B	0.8 ± 0.09	AS	St01	—	M2	3.535 ± 0.015	−0.830 ± 0.086	St01
V1174 Ori B	0.731 ± 0.008	EB	S04	1.065 ± 0.011	—	3.558 ± 0.011	−0.761 ± 0.058	S04
DL Tau	0.72 ± 0.11†	DK	Si00	—	K7-M0	3.591 ± 0.015	0.005 ± 0.10	HW
HD 98800 Ba	0.699 ± 0.064	AS	B05	—	—	3.623 ± 0.016	0.330 ± 0.075	B05
HD 98800 Bb	0.582 ± 0.051	AS	B05	—	—	3.602 ± 0.016	0.167 ± 0.038	B05
DM Tau	0.55 ± 0.03†	DK	Si00	—	M1	3.557 ± 0.015	−0.532 ± 0.10	HW
CY Tau	0.55 ± 0.33†	DK	Si00	—	M2	3.535 ± 0.015	−0.491 ± 0.10	HW
UZ Tau Eb	0.294 ± 0.027†	DKS	P02	—	M4	3.491 ± 0.015	−0.553 ± 0.124	HW
2M0535−05 A	0.0541 ± 0.0046	EB	S06	0.669 ± 0.034	M6.5	3.423 ± 0.016	−1.699 ± 0.078	S06
2M0535−05 B	0.0340 ± 0.0027	EB	S06	0.511 ± 0.026	—	3.446 ± 0.016	−1.848 ± 0.076	S06

*EB = eclipsing binary, DK = disk kinematics, AS = astrometric and spectroscopic, DKS = disk kinematics and spectroscopic (to divide total mass in double-lined system).
†Uncertainty in the distance is not included in the mass uncertainty.

References: A91 = Andersen (1991); Si00 = Simon et al. (2000); C98 = Casey et al. (1998); St01 = Steffen et al. (2001); D03 = Dutrey et al. (2003); C04 = Covino et al. (2004); P87 = Popper (1987); P02 = Prato et al. (2002); S04 = Stassun et al. (2004); B05 = Boden et al. (2005); S06 = Stassun et al. (2006); M00 = Mamajek et al. (2000); J99 = Johns-Krull et al. (1999); HW = Hillenbrand and White (2004).

This paper first reviews in section 2 the present capabilities for dynamical mass measurements of PMS stars, with an emphasis on random and systematic uncertainties, and then summarizes the present dynamical mass measurements in Table 1. In section 3 we compare these mass measurements with theoretical mass values, and note that theory tends to underpredict masses at a marginally significant level. A key point of this paper is that the limitation in these comparisons has become the determination of accurate stellar properties (effective temperature, luminosity, and radius) by which to derive theoretical mass values for those stars with measured masses. The uncertainties in these stellar properties are also discussed in section 3. Section 4 presents a review of present stellar evolution theory for very-low-mass stars and brown dwarfs, and identifies the primary physical issues for the theory. Finally, in section 5 we briefly discuss the very bright future in this field.

2. DYNAMICAL MASS MEASUREMENTS

Dynamical analyses applied to Keplerian motions of companion stars or circumstellar material provide the most reliable measurements of stellar masses. In the past two decades astronomers have applied several well-established techniques in order to measure dynamical masses of young

stars, including analyzing eclipsing binaries, angularly resolved spectroscopic binaries, and young stars with circumstellar disks.

Table 1 presents dynamical mass measurements for 23 young (t < 10 m.y.) stars with M < 2 M$_\odot$ (all the results of which we are aware at this writing). Column 2 presents the measured mass and uncertainty (random errors only; systematic uncertainties not included). Columns 3 and 4 identify the dynamical technique used and the reference. Column 5 gives the stellar radius for the EBs. Columns 6, 7, 8, and 9 list the spectral type when directly determined, effective temperature T$_{eff}$, luminosity, L, and the references for these quantities.

Mass measurements with accuracies better than 10% are necessary to begin to distinguish between the present suite of PMS evolutionary tracks. Table 1 shows that all the techniques can provide mass measurements of young stars with such a *precision* or better. However, each technique has unique limitations on its *accuracy*, which we briefly discuss in turn.

2.1. Techniques and Uncertainties

2.1.1. Eclipsing binaries (EB). Eclipsing binaries are particularly valuable, for in addition to two mass measurements they can also provide direct measurements of the stellar radii and the ratio of the effective temperatures. As with most things valuable, PMS EBs are also rare, and to date only six have been discovered. Nonetheless, they provide 10 of the dynamical mass measurements in Table 1.

The eclipsing geometry by itself constrains the orbital inclination to be ≈90°, and with a standard lightcurve analysis the inclination can be determined with a precision of ~0.1%. In the case that the EB is also a double-lined spectroscopic binary (and indeed all the PMS EBs discovered to date have been), the precision of the masses of both stars is then limited only by the precision of the double-lined orbit solution (i.e., by the number and precision of the radial velocity measurements). Importantly, at no point in the determination of the masses is the distance to the system required.

Subtle effects in the radial-velocity and lightcurve analyses introduce systematic uncertainties. Radial velocities of double-lined systems are typically obtained via cross-correlation techniques that can produce systematic offsets in radial velocities of the primary stars relative to their secondaries (e.g., *Rucinski,* 1999). These effects can be minimized by selecting templates that are well matched to the target stars in spectral type, by avoiding orbital phases subject to strong line blending, and by employing techniques designed either to accurately measure primary and secondary radial velocities when blended [e.g., TODCOR (*Zucker and Mazeh,* 1994)] or fully disentangle the spectra of multiple components (e.g., *Simon et al.,* 1994; *Hadrava,* 1997). For the EBs listed in Table 1, where the radial-velocity amplitudes are large and blending of cross-correlation peaks is negligible over most of the orbit, these systematic errors in practice represent minor effects (i.e., ~1%). For PMS stars a more significant effect is that of starspots, which introduce asymmetries in the line profiles and hence cause errors in the radial-velocity measurements that vary with the stellar rotation; such radial-velocity errors can be a few percent of the orbital semi-amplitudes. However, sophisticated lightcurve analyses provide detailed information about spots that permit these effects to be well modeled in the orbit solution (e.g., *Stassun et al.,* 2004). Systematics in the lightcurve analysis directly affect the masses only insofar as the determination of the inclination is concerned, the main uncertainty usually being the (wavelength-dependent) limb-darkening law. Again, in practice these systematic errors are minor (i.e., <1%).

Taking these systematic uncertainties into account, masses of PMS EBs have now been measured with accuracies of 1–2% (e.g., *Popper,* 1987; *Andersen,* 1991; *Stassun et al.,* 2004).

2.1.2. Astrometric/spectroscopic binaries (AS). Combination of relative astrometric and double-lined spectroscopic orbit solutions (AS) provides mass measurements for both stars, as well as a geometrical distance to the binary. (Both masses can also be derived from the combination of astrometric and single-lined spectroscopic orbit solutions if

the distance is well known.) Until recently, studies of angularly resolved binaries ("visual binaries," or VBs) and spectroscopic binaries (SBs) have been entirely disconnected since the binaries detected as SBs had orbital separations that at the distance of star-forming regions fell far below the angular resolution limits of the available techniques. This situation has changed dramatically, first with application of the high-angular resolution capabilities of the Fine Guidance Sensors (FGS) of the Hubble Space Telescope, and now with adaptive optics (AO) at large telescopes and groundbased optical and infrared interferometers. Dynamical mass measurements of four stars from two AS systems are presented in Table 1. We anticipate that the summary table in the Protostars and Planets VI review will show a large increase in the number of AS binaries (section 5).

The application of the AS technique to the two PMS binaries in Table 1 employed wholly or in part FGS data near instrumental angular resolution limits. As such, systematic errors in the calibration are an important challenge that may propagate into the masses via the inclination angle i.

Looking to the very near future, new applications of the AS technique will rely in large part on interferometric data. The angular resolution limits of these observations are more than sufficient to achieve excellent mass-measurement precisions for PMS SBs in nearby star-forming regions. The limitations are that both components must "fit" within the beam of a single telescope, and that at least one star is bright enough for AO guiding and fringe tracking. These are not necessarily the same thing because often the first is done in the visible and the second at the wavelength of observation.

If these limitations do not preclude an AS approach to a binary, then the primary concern is adequate orbital phase coverage, especially given the longer periods of the resolvable SBs in more distant star-forming regions. As these interferometric observations become more standard as a technique, both the time baselines and the number of measurements will increase and be less of a limitation. The photometric variability of PMS stars can also be a concern, particularly when the orbit solution is derived directly from visibilities.

Given present uncertainties in distance determinations, AS solutions are best achieved when the secondary star is detected spectroscopically; the resultant mass ratio makes a distance measurement unnecessary. Recently, detections of secondary spectra in PMS SBs have been greatly enhanced through near-infrared, high-resolution spectroscopy (*Steffen et al.,* 2001; *Prato et al.,* 2002).

2.1.3. Disk rotation curves (DK). The circumstellar disks around some young stars provide an additional method for measuring dynamical masses. When the disk mass is small compared to the star's mass (or, if a circumbinary disk, the binary mass), the disk rotation is Keplerian, and the amplitude of the rotation curve is determined by the mass of the star. Of course, the observed rotation curve must be corrected for inclination, which is typically derived from the morphology of the observed disk emission.

Typically disk rotation curves have been observed via millimeter interferometry in CO lines (*Guilloteau and Dutrey*, 1998; *Simon et al.*, 2000).

Measuring a mass with this technique requires knowing the linear radius in the disk at which a velocity is measured. Consequently, deriving a mass measurement requires a measured distance to the star. The uncertainty on the distance propagates linearly into the uncertainty on the mass. At present, the accuracies of distance measurements to individual young stars are generally poorer (e.g., on the order of 15% in Taurus-Auriga) than the internal precisions of the mass measurements. (Only the latter are given in Table 1.) *Simon et al.* (2000) reduce the effect of this systematic uncertainty in their tests of evolutionary models by doing the comparison in the quantity L/M^2. Since the luminosity L depends on the distance squared, this parameter is distance independent.

Additionally, the DK technique can be frustrated by line-of-sight molecular-line emission of the host molecular cloud at or near the systemic velocity of the disk. Of course, both the DK and AS techniques are useless if observations reveal that the targets are essentially face-on, or are unresolvable as disks or VBs.

2.2. Exemplars at the Forefront

Until ALMA comes on line in the next decade, we anticipate that most new mass measurements for young stars will derive from eclipsing binaries and angularly resolved spectroscopic binaries. With this in mind, we highlight here two recently studied systems that exemplify the capabilities and exciting scientific potential of applying these techniques to young stars.

The recently discovered Orion Nebula EB 2M0535-05 is a particularly important case study, for it comprises two young brown dwarfs (BDs) (*Stassun et al.*, 2006). With masses of 0.034 $M_\odot$ and 0.054 $M_\odot$, presently these stars alone constrain PMS stellar evolutionary theory below 0.3 $M_\odot$.

Both the colors and spectra indicate that the primary BD has a spectral class of M6.5 ± 0.5, or a surface temperature of T_1 = 2650 ± 100 K (*Slesnick et al.*, 2004). The measured temperature ratio of T_2/T_1 = 1.054 ± 0.006 implies a surface temperature for the secondary of T_2 = 2790 ± 105 K. The luminosities of the BDs are calculated from the measured radii and surface temperatures, which when combined with the observed apparent magnitude and an appropriate bolometric correction yields a distance of 435 ± 55 parsec, assuming no extinction. Extinction by as much as 0.75 visual magnitudes may be present, in which case the distance would be slightly smaller, 420 ± 55 parsec. In either case the derived distance is in agreement with typical distances adopted for the Orion Nebula region [e.g., 480 ± 80 parsec from *Genzel and Stutzki* (1989)]. Additional evidence for the association of 2M0535-05 with Orion is provided by its center-of-mass velocity, which is within 1 km s⁻¹ of the systemic radial velocity of the Orion Nebula cluster.

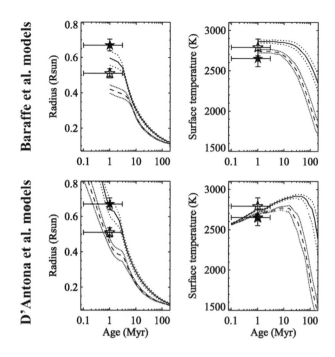

Fig. 1. Comparison of the brown dwarf components of the eclipsing binary 2M0535-05 with evolution models of *Baraffe et al.* (1998) and *D'Antona and Mazzitelli* (1997).

The Orion Nebula star cluster is very young, with an age that has been estimated to be 1^{+2}_{-1} m.y.; thus *Stassun et al.* (2006) suggest a similar age for 2M0535-05. If any remaining disk material is co-planar with the binary (i.e., in an edge-on disk), the near-infrared colors of 2M0535-05 limit the amount of such material available for further accretion onto the BDs. The currently observed masses are therefore unlikely to change significantly over time; these BDs will forever remain BDs.

Encouragingly, the physical properties determined for the BDs in 2M0535-05 are broadly consistent with the most basic theoretical expectations (see Fig. 1). The fact that 2M0535-05 comprises BDs that are both large and luminous — and even simply that they are of M spectral type — is a testament to the generally good predictive power of current theoretical models of young BDs. At the same time, there is a highly unexpected result in the ratio of their surface temperatures, T_2/T_1 = 1.054 ± 0.006; the less-massive BD is hotter than its higher-mass companion. Such a reversal of temperatures with mass is not predicted by standard theoretical models for coeval brown dwarfs, in which temperature increases monotonically with mass.

This result may be a clue to the formation history of brown dwarfs, if interpreted as evidence for noncoevality of the two BDs. Alternatively, the influence of magnetic fields and surface activity on convection may be affecting the energy flow in one or both stars. Or perhaps this result serves as yet another cautionary lesson on the difficulty of determining effective temperatures of very-low-mass PMS objects. Improvement in atmosphere models and the subsequent calculation of spectral energy distributions may be required (see also section 3.1.1). 2M0535-05 promises to serve as an important benchmark in our understanding of young brown dwarfs.

Many PMS mass measurements in the near future will derive from the application of optical/infrared interferometers coming on line to young spectroscopic binaries with existing and new orbit solutions. The first application of the AS technique to PMS stars was done by *Steffen et al.* (2001), who angularly resolved 045251+3016 with the FGS (ρ < 0.05"). The resultant astrometric orbital solution provided the inclination angle of the system, after which the masses were measured with precisions of 12%.

More recently, *Boden et al.* (2005) combined FGS data with groundbased K-band Keck interferometric visibility data to derive an astrometric orbital solution (Fig. 2) for HD 98800B, one binary in this PMS quadruple system. Notably, the filled circles in Fig. 2 do not represent separation measurements, but phases where Keck visibility data were obtained. These visibility functions were compared directly to model predictions in order to constrain the astrometric orbit, without an intermediate determination of separation. *Boden et al.* (2005) find that the component masses, luminosities, and effective temperatures of HD 98800B are inconsistent with solar-metallicity evolutionary tracks; they note that a lower metal abundance by a factor of 2–3 would resolve the discrepancy. Their study highlights a significant complication with the AS method, for even if a system is adequately resolved for orbital solutions it can be difficult

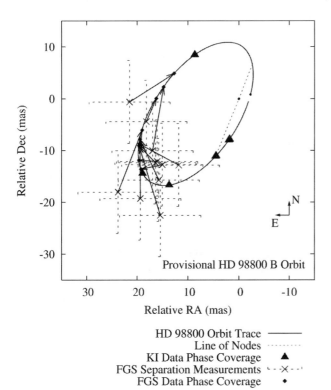

Fig. 2. Astrometric data and orbit solution for HD98800 B (*Boden et al.*, 2005). Triangles are FGS measurements, circles are groundbased interferometry.

to derive accurate stellar parameters, such as angularly resolved spectral energy distributions, by which to compare with theory. Careful analyses of the double-lined spectra are thus important.

This forefront observational study bodes well for a significant number of masses being measured via interferometry and spectroscopy in the very near future. Indeed, Haro 1–14c is under interferometric study and on the verge of yielding masses; the secondary is likely to have a mass ≈0.4 M⊙ (*Simon and Prato*, 2004; *Schaefer*, 2004). Ultimately, the lower-mass secondary stars in these binaries may prove the most valuable, as they provide tests of low-mass PMS evolutionary tracks.

3. COMPARISONS OF DYNAMICAL MASSES WITH PRE-MAIN-SEQUENCE STELLAR EVOLUTION MODELS

3.1. Physical Properties of Pre-Main-Sequence Stars

The dynamically determined masses of PMS stars offer powerful tests of PMS stellar evolution models. However, conducting these tests requires not only accurate stellar mass measurements, but also accurately determined stellar properties, such as luminosities, effective temperatures, or radii, for comparisons with model predictions. For the 23 stars listed in Table 1, we have assembled the current best determinations of their stellar effective temperatures and luminosities; for the 10 stars in eclipsing systems, we also list their stellar radii. The methods by which these stellar properties have been determined merit some discussion.

3.1.1. Effective temperature.   About half of the effective temperatures presented in Table 1 have been derived via assigned spectral types. These spectral types are typically accurate to within one spectral subclass, which corresponds to approximately ±150 K in effective temperature.

However, the uncertainty in the appropriate spectral type-temperature conversion scale at least doubles this uncertainty. Typically a dwarf-like temperature scale is assumed (e.g., *Legget et al.*, 1996), but slightly hotter temperature scales have been proposed for M-type PMS stars to account for their less-than-dwarf surface gravities (*White et al.*, 1999; *Luhman et al.*, 2003). Typically the proposed increase in temperatures is on the order of 100 K, but can become larger for the coolest stars. In Table 1, we use a temperature scale appropriate for dwarf stars (*Hillenbrand and White*, 2004).

Flux-calibrated spectral energy distributions (SEDs) or comparisons with synthetically generated spectra may provide the best determinations of effective temperature. This latter technique has been used for the case of BP Tau (*Johns-Krull et al.*, 1999), and bears real promise for future reanalyses of the binaries in Table 1.

For stars in EBs, analysis of the relative eclipse depths provides a very precise measure of the ratio of the stellar surface fluxes in the bandpass of the light curve, which in turn provides the *ratio* of effective temperatures via bolometric corrections from stellar atmospheres. EB effective temperature ratios are extremely precise, <1%. However, the accuracy of these effective temperature ratios is ultimately limited by the accuracy of the model atmospheres used in the lightcurve synthesis. Finally, determining the effective temperatures for each star separately requires an external determination of the effective temperature of one of the stars, usually the primary. This determination is subject to the same uncertainties described above.

3.1.2. Radii. Direct measurements of PMS stellar radii are provided only by EBs. With well-sampled lightcurves, covering completely both eclipses (ingress and egress included), and additional information concerning the luminosity ratio (e.g., from spectroscopy in double-lined systems), the stellar radii (relative to the orbital semimajor axis) can be as precise as 1–2%. The effects of limb darkening complicate the analysis somewhat, but lightcurves at multiple wavelengths usually insure that these effects do not limit the accuracy of the measurements. Conversion to absolute radii requires accurate radial-velocity curves for both components. The median precision of the stellar radii measurements in Table 1 is 4%.

3.1.3. Luminosities. Luminosities are typically derived from a broadband photometric measurement combined with an extinction correction (uncertainty on the order of 20% in the optical), a bolometric correction (uncertainty of less than 10%, except for the latest-type PMS stars), and a distance measurement. While a direct measure of the distance is provided for double-lined binaries with an AS solution, distances for the remaining stars are typically taken as those of their associated star-forming regions. Uncertainties in such distances, and in the location along the line of sight within the association, yield luminosity uncertainties as large as ~40%.

For the stars in EBs, the individual luminosities are determined directly from the temperatures and radii via Stefan's law ($L = 4\pi R^2 \sigma T_{\text{eff}}^4$). For the 10 stars in eclipsing systems, the median precision of the luminosities is 15%, driven almost entirely by uncertainties in the effective temperatures.

Finally, we note that with knowledge of bolometric and extinction corrections, EBs yield an independent distance determination. Such distance measurements can be used to control measures derived for other stellar parameters, or to provide additional information on the distances to star-forming regions.

3.1.4. Other uncertainties. In addition to these known uncertainties, other characteristics of young stars likely bias the determined stellar properties in ways that are difficult to account for. Stars that are still actively accreting show excess optical and ultraviolet emission (e.g., *Basri and Batalha*, 1990; *Hartigan et al.*, 1991) and excess infrared emission from warm circumstellar dust (e.g., *Strom et al.*, 1989). These additional sources of radiation can confuse determinations of both effective temperatures and luminosities. Additionally, young stars can generate magnetic fields and associated cool-temperature star spots covering a large fraction of the stellar surface. Such spots confuse spectral analyses, bolometric and reddening corrections, etc., in as yet poorly quantified ways. Finally, many EBs have been found to be members of triple systems (e.g., V1174 Ori and TY CrA). Disentangling the light from the third star adds additional complexity and uncertainty.

These issues are particularly well illustrated by considering the three stars UZ Tau Ea, V1174 Ori B, and DM Tau. While these three stars are reported in the literature to have identical effective temperatures and luminosities that are the same to within 0.5 dex, their dynamically determined masses differ widely: 1.016 ± 0.065 M$_\odot$, 0.731 ± 0.008 M$_\odot$, and 0.55 ± 0.03 M$_\odot$, respectively. If the random uncertainties in the dynamical masses are reasonably assessed, then the inferred stellar properties must be in error. (Note that the uncertainty in distance has not been incorporated in the uncertainty for DM Tau.)

Cognizant of these uncertainties, we proceed to compare theoretical masses to dynamical mass measurements in order to test the success of theoretical PMS stellar evolution models. We follow the procedure of *Hillenbrand and White* (2004), who previously performed such a comparison for 17 of the 23 stars listed in Table 1, and add comparisons for the 6 stars in binaries identified more recently (V1174 Ori, HD 98800B, and 2M0535-05).

3.2. Comparison of Dynamical and Theoretical Masses

We compare the dynamically determined masses with the predictions of six evolutionary models that are widely used by the community to describe the physical properties of young objects. These models include *Swenson et al.* (1994, hereafter *S94*) (approximately the series F models; 0.15–5.00 M$_\odot$); *D'Antona and Mazzitelli* (1997) (hereafter *DM97*) (0.017–3.00 M$_\odot$); *Baraffe et al.* (1998, hereafter *BCAH98*) (B1.0; 0.035–1.20 M$_\odot$); *BCAH98* (B1.9; 0.035–1.20 M$_\odot$); *Palla and Stahler* (1999, hereafter *PS99*) (0.1–6.0 M$_\odot$); and *Siess et al.* (2000, hereafter *SDF00*) (0.1–7.0 M$_\odot$). The physics in these models are given in section 4. The *BCAH98* models B1.0 and B1.9 differ only in the values of the mixing length α, which are 1.0 and 1.9 respectively.

Figure 3 shows the comparisons of the theoretical masses predicted by these evolutionary models with the dynamically determined masses. The uncertainties indicated by the error bars include random uncertainties in the dynamical masses, and the random uncertainties in effective temperatures and luminosities (Table 1) as propagated into the theoretical masses.

The comparison in Fig. 3 shows a consistent tendency for the measured dynamical masses to be higher than the theoretical masses. Above 1.02 M$_\odot$ the *mean* differences are only on the order of 10% for each of the theoretical models. Indeed, all models predict PMS masses that are consistent in the mean with dynamically measured masses to better than 1.6σ.

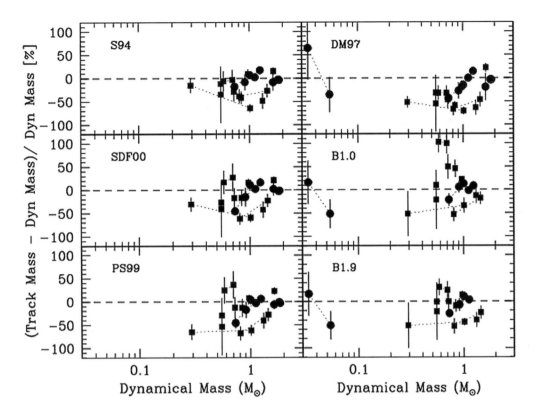

Fig. 3. Percentage differences between theoretically and dynamically determined stellar masses vs. dynamical stellar mass. Circles are components of eclipsing systems and squares are not. Error bars indicate only the random uncertainties in both the dynamical and theoretical masses.

Below 1.02 M$_\odot$ the mean differences increase to ≈20% for the *S94*, *PS99*, and *SDF00* models, and to 43% for the *DM97* models. These differences in the mean are significant at the 2.5σ or higher level and begin to suggest meaningful discrepancies between the observations and the models.

The notable exceptions are the two sets of *BCAH98* models, which in the mean continue to predict masses consistent with dynamical values at the 10% level (<1.4σ) for these lower-mass stars.

Curiously, the B1.0 models produce a standard deviation about the mean of 51%, substantially larger than any other model (in the range of 20–30%). These standard deviations represent reasonable estimates for the uncertainties in mass determinations for any given PMS star.

Currently there is only one star, UZ Tau Eb, with a dynamical mass between the BD mass limit and 0.5 M$_\odot$. The trend of low theoretical masses continues for this star, although the difference for each theoretical model is not statistically significant when UZ Tau Eb is considered as a single case.

Other results generally support these findings. In a study of double-lined spectroscopic binaries having precisely determined mass ratios, *Palla and Stahler* (2001) showed that their evolutionary models predict consistent mass ratios for a sample of PMS stars above 1 M$_\odot$. Using high-dispersion IR spectroscopy, *Prato et al.* (2002) studied a sample of four

low-mass-ratio binaries, with companion masses as low as ~0.2 M$_\odot$ (based on theoretical primary masses and a dynamical mass ratio). They found that models predicted mass ratios that, while statistically consistent, were systematically less than the dynamically determined ratio. This can be interpreted again as the models underpredicting masses in the subsolar regime of the secondary stars.

Finally, *Stassun et al.* (2006) (section 2.2) present the first dynamical mass estimates for two young BDs. These masses are an order of magnitude lower than previous dynamical measurements and constrain evolutionary models at the very uncertain low-mass end. The theoretical masses of *BCAH98* and *DM97* (the only sets of theoretical models that extend this cool) agree with the observed masses to within a factor of 2.

As pointed out by *Hillenbrand and White* (2004), a hotter temperature scale for young M dwarfs would systematically shift the PMS theoretical masses to larger values, and could reconcile these discrepancies at subsolar masses. A uniform shift in temperature scale by 150 K for stars below 1 M$_\odot$ would bring the *S94*, *PS99*, and *SDF00* models into good agreement; the *DM* models require a shift closer to 500 K.

Since PMS stars have surface gravities intermediate between those of dwarfs and giants, they may have intermediate temperatures as well. For comparison, while M0 dwarfs and giants have similar temperatures, M4 giants are systematically warmer by ~500 K (e.g., *Perrin et al., 1998*). *Luhman et al.* (2003) propose a specific intermediate temperature scale for stars cooler than spectral type M0; the values were chosen to produce coeval ages for the T Tauri quadruple GG Tauri and for members of the IC 348 cluster using the *BCAH98* $\alpha = 1.9$ evolutionary models.

Application of the *Luhman et al.* (2003) temperature scale improves the agreement of dynamical and theoretical masses for most models, but makes the theoretical masses of the *BCAH98* $\alpha = 1.0$ models 22% higher than the dynamical masses. However, all these adjustments are within the 3σ range for the uncertainties on the means. As such, the case for a particular temperature scale is not yet compelling.

Alternatively, these comparisons may provide guidance for resolving questions about the physics of stellar evolution theory. Such issues are presented in detail in section 4.

These comparisons only hint at the potential value of a large sample of dynamical mass measurements for testing both observational techniques and stellar evolution theory. Another key finding from Fig. 3 is that the scatter of the differences between theoretical and dynamical masses for individual stars is larger than can be explained by the assigned random uncertainties. This scatter is not reduced when considering only those stars with the most precise dynamical mass measurements.

As such, we suggest that this additional scatter derives primarily from errors in the determination of effective temperatures and luminosities. *Additional dynamical mass determinations in the mass range of 0.5–2.0 M_⊙ will not greatly improve the constraints on evolutionary models until these uncertainties in determining stellar properties are resolved.* Thus the marginal levels of significance of the discrepancies discussed in this section will only be reduced with improved determination of the stellar properties that link each star to theoretical models of stellar evolution.

3.3.   Comparison of Observation and Theory in the Mass-Radius Plane

The PMS EBs provide constraints on evolutionary models that are independent of many of the uncertainties in determining effective temperatures and luminosities. In addition to having very accurately determined dynamical masses, the eclipsing pairs offer a direct measure of the stellar radii. Thus comparison with theoretical models can be done in the mass-radius (M-R) plane, thereby preserving the accuracy of the measurements and avoiding issues related to uncertain temperature scales. Specifically, EBs allow tests of the mass-radius relationships of theoretical models, given an assumption of coevality for the pair of stars within each EB.

Three of the PMS EBs — 0529.4+0041 AB, V1174 Ori AB, and 2M0535 AB — consist of two PMS stars that are significantly different in mass and temperature. Such pairs offer the most interesting constraints on theoretical models since they span a broad range of predicted proper-

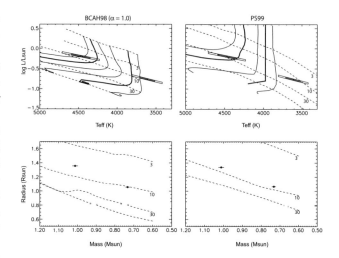

Fig. 4. L-T$_{eff}$ and mass-radius (M-R) plots comparing the measurements of the eclipsing binary V1174 Ori with the PMS models of BCAH98 ($\alpha = 1.0$) (left) and PS99 (right). Dashed lines represent isochrones with ages in million years as labeled. Solid lines represent mass tracks, with those appropriate to the dynamical masses (1.01 M$_⊙$ and 0.73 M$_⊙$) emphasized. Nested parallelograms represent the 1σ and 2σ confidence intervals in T$_{eff}$ and log L. Note that the positions of the primary and secondary in these uncertainty domains are highly correlated, in the sense that a hotter primary is associated with a hotter secondary.

ties. Of these three systems, V1174 Ori has masses and radii determined with sufficient accuracy (<2%) to permit strong conclusions, and so we present this EB in some detail here.

In Fig. 4 V1174 Ori is compared in the M-R plane with the models of *BCAH98* $\alpha = 1.9$ and *PS99* $\alpha = 1.5$. These models were selected to show the improved performance of models with lower convective efficiency, a general trend found by *Stassun et al.* (2004). For such models the current generation of PMS tracks is performing well in predicting the PMS mass-radius relationship — arguably two of the most fundamental physical parameters of young stars. This agreement may indicate that the equations of state in the theoretical models are accurate (see also the discussion in section 4.4.6). Importantly, though, different theoretical models still yield different ages.

Interestingly, success of models in the M-R plane does not necessarily translate to success in the T$_{eff}$-L plane. The *PS99* models yield isochrones parallel to the observed M-R relationship, yet fail to simultaneously yield coevality in the T$_{eff}$-L plane. Even more strikingly, while both sets of tracks can match the T$_{eff}$-L position of the 1.01 M$_⊙$ primary at the 2σ level or better, they fail miserably at simultaneously matching the T$_{eff}$-L position of the 0.73 M$_⊙$ secondary. For example, the *PS99* mass tracks corresponding to the secondary mass are 500 K too warm.

Similar analyses (and results) are done for a suite of current models by *Stassun et al.* (2004). Importantly, the accurate temperature ratio for V1174 Ori clarifies that a simple shift in the temperature scale cannot resolve the discrepancies for all models. The mass tracks are too compressed in effective temperature to ever simultaneously fit the locations of these stars.

3.4. Lithium as a Test of Pre-Main-Sequence Evolution Models

While these tests of PMS models focus either on global physical properties (e.g., mass, radius) or on surface properties (e.g., effective temperature, luminosity), the predicted evolution of abundances (e.g., lithium and deuterium) may provide a powerful probe of PMS stellar interiors, particularly with respect to convection. *Stassun et al.* (2004) examined the Li abundances of all PMS stars with dynamical mass determinations (Fig. 5) and were able to draw several conclusions. First, the observed pattern of increased Li depletion with decreasing mass is, qualitatively, as predicted; the deeper convective zones of cooler stars, over this mass range, lead to more efficient depletion. More quantitatively, the absolute level of depletion observed again favors models with inefficient convective mixing. For example, the observed Li depletion for the components of V1174 Ori agrees well with the *BCAH98* $\alpha = 1.0$ models, while these models with $\alpha = 1.9$ predict at least two orders of magnitude greater depletion than observed in the secondary.

The Li data again reveal likely problems with the determination of stellar parameters. For example, the observed Li depletion of DL Tau is an order of magnitude too low for its mass. This same star is discordant in terms of its placement in the H-R diagram relative to other stars of similar empirical mass; the $0.73 \, M_\odot$ secondary of V1174 Ori is 450 K cooler than the inferred temperature for DL Tau, even though their masses are nearly identical. These discrepancies may possibly be tied to DL Tau being one of the most actively accreting stars in the sample.

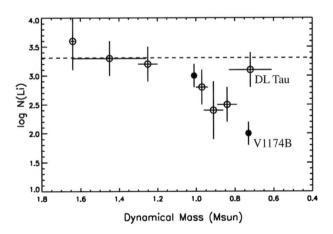

Fig. 5. Lithium abundances for all PMS stars with dynamical mass determinations and Li abundance measurements in the literature. The cosmic abundance is represented by the horizontal dashed line.

4. THEORETICAL MODELS OF PRE-MAIN-SEQUENCE STELLAR EVOLUTION

The comparisons discussed in the previous section suggest interesting discrepancies between observations and theoretical models at young ages, and specifically possible systematic underestimates of theoretically predicted masses. Unfortunately, the uncertainties in these theoretical masses resulting from uncertainties in the stellar properties do not yet permit unambiguous tests of PMS evolutionary models. With the recent increase in the number of dynamical mass measurements, the reduction of these uncertainties must be a critical objective for progress in the field. In parallel, it is important to identify the key physical uncertainties affecting the theory, and to assess whether the mass comparisons might provide guidance for improvement of the theory.

Important progress has been made within the past few years regarding the theory of very-low-mass stars (VLMS) ($M < 1 \, M_\odot$) and BDs ($M < 0.075 \, M_\odot$). The main improvements concern the equation of state (EOS) of dense plasmas and the modeling of cool, dense atmospheres. These theoretical efforts have yielded both a better understanding of these objects and good agreement with observations of older objects (age $\gg 10$ m.y.).

Although several shortcomings remain (as discussed in section 4.4), the improved reliability of the current theory for VLMS and BDs allows us to return to a thorough analysis of theoretical models of young objects. From the theoretical viewpoint, young objects represent a formidable challenge given the extra level of complexity from processes such as accretion, rapid rotation, and magnetic activity that are characteristic of the early phases of stellar evolution.

4.1. Physics of Low-Mass Stars and Brown Dwarfs

One of the main theoretical achievements of the past years in the modeling of VLMS concerns the description of their thermodynamic properties. VLMS and BDs are dense, cool objects, with typical central densities on the order of $100–1000 \, \mathrm{g \, cm^{-3}}$ and central temperatures lower than 10^7 K. Under such conditions, a correct EOS for the description of their inner structure must take into account strong interactions between particles, resulting in important departures from a perfect gas EOS (cf. *Chabrier and Baraffe*, 1997). Important progress has been made by *Saumon et al.* (1995), who developed an EOS specifically designed for VLMS, BDs, and giant planets. Since the EOS determines the mechanical structure of these objects, and thus the mass-radius relationship, it can be tested against direct observations of stellar radii obtained via EBs, planetary transits, or interferometric measurement. Also, several high-pressure shockwave experiments have been conducted in order to probe the EOS of deuterium under conditions characteristic of the interior of these objects. The Saumon-Chabrier-VanHorn EOS was found to adequately reproduce the experimental pressure-density profiles of gas gun shock compression experiments at pressures below 1 Mbar, probing the domain

TABLE 2.   Input physics in current theoretical PMS stellar evolution models.

Authors	EOS	Atmosphere	Convection	l_{mix}/H_P
B97	SCVH	Gray atmosphere: solve RT equation	MLT	1.5
BCAH98	SCVH	Nongray atmosphere: NextGen	MLT	1.0; 1.9
DM94	*Magni and Mazzitelli* (1979)	Gray approx: $T(\tau)$ relationship	MLT; CM	1.2
DM97	*Magni and Mazzitelli* (1979)	Grey approx: $T(\tau)$ relationship	CM	
PS99	*Pols et al.* (1995)	Gray approx: $T(\tau)$ relationship	MLT	1.5
SDF00	*Pols et al.* (1995)	Nongray atmosphere: Uppsala models (Plez)	MLT	1.6
S94	*Eggleton et al.* (1973)	Gray approx: $T(\tau)$ relationship	MLT	1.957

of molecular hydrogen dissociation. However, discrepancies were found with the experimental temperatures. Recent experiments at higher pressure, testing the domain of pressure-ionization (P ~ 1–3 Mbar), unfortunately give results different from the Saumon-Chabrier-VanHorn EOS. Robust comparisons between experiments and theory in this critical pressure regime cannot be done before this discrepancy is resolved.

Another essential physical ingredient for theoretical models of VLMS and BDs concerns atmosphere models. VLMS are characterized by effective temperatures from 5000 K down to 2000 K and surface gravities of log g ~ 3–5.5, while BDs cover a much cooler temperature regime extending down to 100 K. (Young BDs, however, remain relatively hot, with temperatures in excess of 2000 K.)

Such effective temperatures allow the presence of stable molecules, in particular metal oxides and hydrides (TiO, VO, FeH, CaH, MgH), which are the major absorbers in the optical, and CO and H_2O, which dominate in the infrared. These molecules cause strong nongray effects and significant departures of the spectral energy distribution from blackbody emission. Another difficulty inherent in cool dwarf atmospheres is the presence of convection in the optically thin layers. This is due to the molecular hydrogen recombination (H + H → H_2), which lowers the adiabatic gradient and favors the onset of convective instability.

Since radiative equilibrium is no longer satisfied in such atmospheres, the usual procedure of imposing outer boundary conditions based on standard $T(\tau)$ relationships from gray atmosphere models is incorrect. *Chabrier and Baraffe* (1997, 2000) show that as soon as molecules form in the atmosphere (i.e, for $T_{eff} \le 4000$ K), the use of standard $T(\tau)$ relationships and gray outer boundary conditions, like the well-known Eddington approximation, overestimates the effective temperature for a given mass and yields a higher hydrogen-burning minimum mass. An accurate surface boundary condition based on nongray atmosphere models is therefore required for evolutionary models.

4.2. Pre-Main-Sequence Evolutionary Models

Various sets of tracks can be found in the literature, and detailed comparisons are given by *Siess et al.* (2000) and *Baraffe et al.* (2002). Here we briefly comment on the main input physics of the models used in the analyses of section 3. Table 2 summarizes the key physical inputs in each

model. Only the *BCAH98* and *SDF00* models use nongray atmosphere models for outer boundary conditions. *D'Antona and Mazzitelli* (1994, hereafter *DM94*), *DM97*, and *PS99* use approximate boundary conditions based on $T(\tau)$ relationships assuming a gray approximation and radiative equilibrium. Consequently, for a similar treatment of convection, the *DM94* models for VLMS are usually hotter than the *BCAH98* models. The *Burrows et al.* (1997, hereafter *B97*) models for VLMS and hot BDs ($T_{eff} > 2000$ K) are based on gray atmosphere models obtained by solving the radiative transfer equation. Such an approximation, although it represents an improvement over the previous $T(\tau)$ relationships, still overestimates effective temperature at a given mass compared to evolution models based on full nongray atmospheres.

Above 4000 K, the choice of the outer boundary condition (BC) has less influence and the treatment of convection becomes more crucial. Convection in most of the models is treated within the framework of the mixing length theory (MLT), with a mixing length $1 < l_{mix} < 2$. Only *DM94* and *DM97* differ by using the Canuto-Mazzitelli (CM) formalism.

The treatment of convection affects the temperature gradient in convective zones. In solar-type stars, above T_{eff} ~ 4000 K, it affects primarily the radius of an object, and to a lesser extent its luminosity. For such objects, an increase in the mixing length yields a decrease of the radius (for a given mass at a given age) and thus an increase of the effective temperature, shifting Hayashi lines toward hotter effective temperature (see *Baraffe et al.*, 2002, for a discussion). This is why the *DM97* Hayashi lines behave differently with respect to models of other groups, as noted by *SDF00*. The Canuto-Mazzitelli treatment of convection yields results at odds with three-dimensional hydrodynamic simulations for the outer thermal profile of the Sun (see e.g., *Nordlund and Stein,* 1999), and does not provide an accurate treatment of convection in optically thin media, at least for solar-type stars and low-mass stars (see, e.g., *Chabrier and Baraffe,* 2000).

As a last remark, we note that the comparison of observations with evolutionary models based on BCs provided by nongray atmosphere models are much more instructive, since it confronts *both* the inner atmosphere profile (which determines the BC and fixes the M-T_{eff} and M-L relationship) and the outer atmosphere profile and resultant spectral synthesis (through synthetic spectra and colors).

4.3.   Successes of the Theory

Before applying current theory to the complex case of young objects, it is essential to confront first the more accessible and accurate observations of old VLMS and BDs. Several successes of the theory, including color-magnitude diagrams of globular clusters, mass-magnitude and mass-radius relationships, and near-IR color-magnitude diagrams for open clusters (see *Chabrier et al.,* 2005, and references therein), tell us that uncertainties due to the input physics have been considerably reduced. Figure 6 displays a comparison between observed and predicted mass-radius relationships and shows the good agreement with observations. This agreement is achieved down to the very bottom of the main sequence with the recent observation of the smallest H-burning object known (*Pont et al.,* 2005). The same theory applied to giant planets, where the description of pressure ionization and H_2 dissociation is even more crucial, also provides excellent agreement with measured radii of exoplanet transits (*Baraffe et al.,* 2005). This gives confidence in the underlying physics describing the mechanical structure of low-mass objects. The general description of spectral properties of M dwarfs is also satisfactory, although with some problems to be discussed below.

4.4.   Failures of the Theory

Although current models for VLMS and BDs provide generally good agreement with observations, several shortcomings remain. In this subsection we describe some of the known failings of the models. In the next subsection we discuss possible explanations and improvements that are still required in various domains.

1. A shortcoming pointed out in *BCAH98* concerns the optical colors (V-I) and (R-I) for solar metallicity models. These are significantly too blue for objects fainter than $M_V \sim 10$.

2. In the near-IR, current atmosphere models, based on the most updated molecular line lists, do not provide satisfactory agreement with observed color-magnitude diagrams (see *Allard et al.,* 2000, for discussion).

3. A discrepancy between observed and theoretical radii of EBs with low-mass components (M < 1 $M_\odot$) has been pointed out for several systems (see *Torres et al.,* 2006, and references therein). The predicted values are usually 10–15% smaller than observed.

4. Recent observations of young binary systems suggest a problem in the mass-luminosity relationship of young low-mass objects (*Close et al.,* 2005; *Reiners et al.,* 2005), in the same sense as found by *Hillenbrand and White* (2004) and discussed in section 3. These observations and their interpretation are still controversial, especially with respect to AB Dor (*Luhman and Potter,* 2006; *Luhman et al.,* 2005).

5. *Mohanty et al.* (2004a,b) pointed out a problem in the spectroscopic analysis of young, low-gravity objects. Comparing observed and synthetic spectra, they use molecular bands of TiO and lines of neutral atomic alkalis to determine the effective temperatures and surface gravities of M-type PMS objects. They find gravities that are not in agreement with predictions from evolutionary models for the coolest objects and argue that uncertainties in the models may be responsible for this discrepancy.

6. A discrepancy between the age derived from Li depletion and the age derived from isochrones has been recently pointed out for young systems (*Song et al.,* 2002; *White and Hillenbrand,* 2005).

7. Finally, we note that there remain as yet unquantified theoretical uncertainties in the accuracies of theoretical spectra. Synthetic spectra are typically only tested through reference to the solar spectrum (e.g., *Johns-Krull et al.,* 1999). These uncertainties also propagate into theoretical spectral energy distributions, from which bolometric corrections are derived.

4.5.   Physical Issues

Among this list of discrepancies between observations and models, problems 1 through 3 are robust and clearly indicate problems with our current modeling of low-mass objects. Points 4 through 6 are to be taken with caution and must await confirmation. The conclusions reached by the studies mentioned in points 4 through 6 depend strongly on details of the observational analyses that are not yet secure. Finally, point 7 requires improvement of synthetic spectra in order to quantify current uncertainties.

Here we consider the key physical issues that might lead to resolution of these discrepancies between observation and theory.

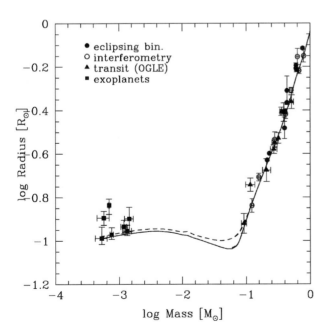

Fig. 6.   Observed and theoretical mass–radius relationships. Observations are compared to the models of BCAH98 for different ages [0.5 G.y. (dashed curve) and 1 G.y. (solid curve)].

4.5.1. Molecular opacities. The problems mentioned in points 1 and 2 concerning shortcomings in optical colors and near-IR colors/spectra point to still inaccurate molecular linelists. The linelists and/or oscillator strenghts of TiO and H_2O, which are important absorbers in the optical and the near-IR respectively, are still imperfect. This may not only affect the colors and spectra, but may also bear consequences for the atmospheric structure and thus on the evolutionary properties (e.g., luminosity and effective temperature vs. mass and age).

The introduction in current atmosphere models of the AMES-TiO linelist reduces the mismatch in optical colors found in *BCAH98*, but still not to a satisfactory level (*Allard et al.*, 2000; *Chabrier et al.*, 2000). This seems to point to remaining uncertainties in TiO opacities. Also, the treatment of the opacities for MgH, CaH, CrH, FeH, and VO is still uncertain in current atmosphere models, affecting spectroscopic analysis in the optical (F. Allard, personal communication). These uncertainties must be remembered especially when performing spectroscopic analyses based on TiO lines. Because of these uncertainties, the gravities and effective temperatures derived by *Mohanty et al.* (2004a,b) are to be taken with caution.

With respect to water, an important source of opacity affecting both SEDs and thermal atmosphere profiles, the most recent linelists available still provide an unsatisfactory agreement with observed spectra of M dwarfs (for details, see *Allard et al.*, 2000; *Jones et al.*, 2002) and with color-magnitude diagrams in the near-IR (see, e.g., Fig. 6 of *Chabrier et al.*, 2000).

4.5.2. Line broadening. Under density and pressure conditions characteristic of cool atmospheres, the treatment of spectral line broadening provides another source of uncertainties. We stress that these uncertainties may affect spectroscopic analyses devoted to the determination of fundamental parameters (gravity, effective temperature) from line profile fitting. This may be another explanation for the discrepancy pointed out in the *Mohanty et al.* (2004a,b) analysis. Theoretical efforts are now being devoted to the modeling of absorption profiles perturbed by He and molecular H_2, a complex fundamental problem in physics (*Allard et al.*, 2005). Such theoretical improvement will hopefully reduce the uncertainties due to the treatment of collisional line broadening in the next generation of atmosphere models.

4.5.3. Convection. The treatment of convection is known to be an important source of uncertainty in the evolution of stars with masses $M > 0.6 M_\odot$ at any age (see *Chabrier and Baraffe*, 2000; *Baraffe et al.*, 2002). The effect of a variation of the mixing length l_{mix}, used in the mixing length theory (MLT), on evolutionary tracks for solar-type stars is well known and is illustrated, for example, in Fig. 2b of *Baraffe et al.* (2001).

For masses $M < 0.6 M_\odot$, the superadiabatic layers retract appreciably and the transition from convective to radiative outer layers is characterized by an abrupt transition from a fully adiabatic to a radiative structure with a very small entropy jump. This means that during most of the evolution, *except at early ages* (see below), the sensitivity of the evolutionary models to l_{mix} is small for this mass range. Multidimensional hydrodynamical simulations for conditions characteristic of M-dwarf atmospheres, $T_{eff} \leq 3000$ K, log g = 5, have been conducted by *Ludwig et al.* (2002). These simulations confirm the aforementioned small entropy jump found in the one-dimensional models described by MLT, illustrating the large efficiency of atmospheric convection for these objects due to the formation of molecules. Under these circumstances, the three-dimensional simulations show that MLT does indeed provide a correct thermal profile, providing a value of $l_{mix} > H_P$ (with H_P a pressure scale height), at least for high gravities (log g > 4) and older objects (t $\gg$ 10 m.y.).

As emphasized in *Baraffe et al.* (2002) and shown in Fig. 7, the evolution of very young objects with $M < 0.6 M_\odot$ and gravities log g < 4 can be affected by the treatment of convection. *Ludwig et al.* (2006) extended their three-dimensional numerical simulations of convection to atmosphere models with gravities of log g < 4, i.e., appropriate for PMS stars and young BDs. They find values of $l_{mix} \sim 2 H_P$ to match the entropy of the deep regions of the convective envelope, whereas a larger value of l_{mix}, between 2.5 H_P and 3 H_P, is required to match the thermal structure of the deep photosphere (*Ludwig et al.*, 2006). This means that current spectral analysis of PMS objects and young BDs based on MLT atmosphere models calculated with $l_{mix} = 2 H_P$ could significantly overestimate the effective temperature. Work is currently underway to determine a better calibration of l_{mix} in such low-gravity atmosphere models in order to reduce the uncertainties on the thermal profile in the deep interior, which may affect the evolutionary proper-

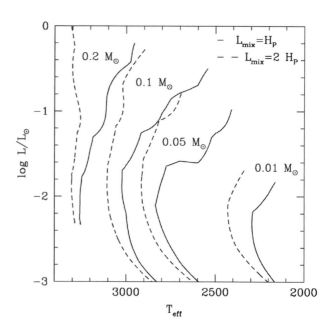

Fig. 7. Effect of the mixing length parameter on evolutionary tracks of VLMS and BDs (*Baraffe et al.*, 2002).

ties at young ages, and in the outer layers, where the spectrum emerges.

4.5.4. Accretion. Accretion is an important process that may affect the early phase of evolution of stars and brown dwarfs. Signatures of accretion onto young objects are now observed over a wide range of masses down to the substellar regime.

In the VLMS and BD regime observed accretion rates for ages greater than 1 m.y. are rather low, ranging from 5×10^{-9} $M_\odot$ yr^{-1} to $\sim 10^{-12}$ $M_\odot$ yr^{-1}, with a sharp decrease of the rate with mass (roughly proportional to M^2) (*Mohanty et al.*, 2005, and references therein). Theoretical and observational arguments suggest that accretion rates increase with younger age so that rates are significantly larger at ages earlier than 1 m.y. (*Henriksen et al.*, 1997; *Mohanty et al.*, 2005).

The effects of accretion on the structure and evolution of young stars have been widely investigated since the seminal work by *Stahler* (1988), and a summary can be found in the chapter by Chabrier et al. Here we only briefly discuss the main effects.

As shown by *Hartmann et al.* (1997), assuming that most of the thermal energy released by accretion is radiated away instead of being added to the stellar interior, accreting low-mass stars are expected to be more compact than their non-accreting counterparts with the same mass and same age. Consequently, an accreting object looks older in a L-T$_{eff}$ diagram than a nonaccreting object of the same mass and age. Thus, ages assigned from nonaccreting tracks can be overestimates. Effects can be even more drastic if a non-negligible fraction of the thermal energy from the accretion shock is transferred to the interior. Several authors have investigated such possibilities (see the chapter by Chabrier et al. and references therein) and found that convection could be inhibited, with profound modifications on the stellar structure. The amount of thermal energy released from accretion and added to the stellar interior is poorly known, since it depends highly on the details of the accretion mechanisms and the properties of the accretion shock.

4.5.5. Initial conditions. Most low-mass PMS models available in the literature, including those considered here, start from arbitrary initial conditions that are totally independent of the outcome of the prior protostellar collapse and accretion phases. The initial configuration is a fully convective object starting its contraction along the Hayashi line from arbitrarily large radii. Evolution starts prior to or at central deuterium ignition, with initial central temperatures $\sim 5 \times 10^5$ K.

According to studies of the protostellar collapse and accretion phases, such initial conditions are oversimplified and low-mass objects could form with significantly smaller radii (*Hartmann et al.*, 1997, and references therein). *Baraffe et al.* (2002) demonstrated the arbitrariness of current initial conditions and starting times for evolutionary tracks, and emphasized the large uncertainty in assigning ages to objects younger than a few million years based on current PMS tracks.

To resolve these substantial uncertainties requires self-consistent evolution from the three-dimensional protostellar collapse phase to the subsequent PMS evolution, a significant theoretical challenge in the field of star formation and evolution.

4.5.6. Magnetic activity. Understanding the effects of magnetic activity, usually linked to rapid rotation, on the inner structure and atmosphere of low-mass objects is still far from reach (*Chabrier and Kueker*, 2006). However, there are suggestions about the nature of these effects and their possible importance. As an example, the discrepancy of 10–15% found between observed and predicted radii of some main-sequence EBs may be related to the magnetic activity of the components. This idea arises from the fact that inactive stars agree well with model predictions, whereas the most active ones appear systematically too large (see, e.g., *Torres et al.*, 2006). The inhibition of convective heat transport due to strong magnetic fields and/or the presence of numerous spots on the stellar surface could be responsible for such structural changes.

If such effects are confirmed, the constraints provided by the EB mass-radius relationship on current PMS evolutionary models that do not account for these effects may be limited. We even speculate that the agreement mentioned in section 3.3 between the observed M-R relationship and current generations of models may be fortuitous. The discrepancy between data and models in the T$_{eff}$-L plane shown in Fig. 5, opposite of the good agreement found in the M-R plane, reveals the existence of remaining problems.

A quantitative estimate of such effects is a difficult task. Modelers have begun to explore them (e.g., *Ventura et al.*, 1998; *Mullan and MacDonald*, 2001), but their treatments remain very simplified. The huge progress of multidimensional magnetohydrodynamic simulations expected in the near future will definitely improve our understanding of these effects.

Another effect related to magnetic activity concerns the formation of lithium lines. Lithium abundance analysis in the presence of strong chromospheres should be taken with caution. As suggested by *Pavlenko et al.* (1995), Li I lines may be significantly affected by the presence of a chromosphere, which could reduce their strength. Such an effect would thus yield an incorrect determination of the level of Li depletion in young objects. This could explain the discrepancies mentioned in section 3 between V1174 Ori and DL Tau. A systematic search for possible correlations between the level of lithium depletion and the level of H$_\alpha$ emission (as one measure of chromospheric strength) may provide some clues to address this problem.

4.6. Final Remarks

In the previous sections, we have tried to highlight the challenges and new problems that theorists and observers are facing when analyzing young stars. Theorists are working hard to resolve the physical issues listed above. But at the time this review is being written, we are not able to estimate their quantitative effects on observable quantities.

Looking to the future, rapid progress is expected with respect to opacities and convection. But improving our knowledge of initial conditions and of magnetic activity is a major challenge for the future.

5. VISIONS OF THE FUTURE

Of the 23 masses listed in Table 1, 20 have been contributed since the Protostars and Planets IV meeting in 1998! Both the greater availability of classical instrumentation and powerful innovative facilities now being built suggest that future progress will be at least as rapid. The capability of measuring masses precisely over more than a decade of stellar mass will continue to be driven by technological developments. The number of PMS EBs is being increased by extensive photometric surveys enabled by large-format detectors at modest-aperture telescopes. It has proven difficult to apply the DK technique to young stars estimated to have stellar masses less than 0.5 $M_\odot$ because the CO emission of their disks is weak (*Schaefer*, 2004). ALMA may be expected to advance this technique through its sensitivity and its access to a new and very large sample of young star targets in the southern sky. The development of multibaseline IR interferometers such as VLTI, KI, and CHARA will enable the measurement of astrometric orbits of a large sample of SBs. Similarly, the advent of laser-guided AO removes the limitation that the young star targets, often in dark clouds, be located near suitable "natural" guide stars. The integral-field high-resolution IR spectrographs now planned will speed the identification of SB2s with low-mass secondary stars in compact clusters. And in the more distant future, astrometric space missions will provide distance estimates of remarkable accuracy for young stars, converting relative astrometric orbits and rotation curves into accurate masses. The most important products of this variety of techniques will be dynamical mass measurements for stars below 0.5 $M_\odot$.

While mass is the fundamental stellar parameter, it is not the only parameter necessary for a comparison with theoretical stellar evolution models. For each star we also need a combination of luminosity, effective temperature, and radius. A promising route to effective temperatures is detailed comparison of high-dispersion spectra with synthetic spectra. As an example, the temperature of the young T Tauri stars Hubble 4 and TW Hydrae have been determined with precisions of 56 K and 24 K, respectively (*Johns-Krull et al.*, 2004; *Yang et al.*, 2005). In addition, this analysis permits precise determination of any excess continuum emission caused by accretion and metallicity [(M/H) = −0.08 ± 0.05, in the case of Hubble 4], both of which can bias luminosity and temperature estimates.

Recent advances in atmosphere models and more complete molecular opacity tables now allow codes to produce synthetic spectra that agree remarkably well with those of young stars, even at very low masses. In principle, such analyses can allow for a direct determination of both stellar temperature and surface gravity. Our description of the remaining uncertainties on opacities, line broadening, and convection suggests that present atmosphere models may not yet provide very accurate measures. Still, the early successes auger well for this approach.

Finally, for the young stars that are surprisingly nearby — as close as 20 pc (*Zuckerman and Song*, 2004) — interferometric measurements of their diameters may enable astronomers to directly measure the radii of stars not in EBs (*Simon*, 2006).

The essential message of this review is that with the acquisition of larger numbers of dynamical mass measurements for PMS stars, the limitation in testing stellar evolution theory via stellar masses has become the determination of comparably accurate theoretical mass predictions for these same stars. We suggest that presently the limitation is primarily in the measurement of effective temperatures and luminosities. We emphasize that there are immediately realizable prospects for improving effective temperature and luminosity measurements for young stars. Thus we urge the community to apply itself to the determination of much improved theoretical masses for the 23 stars with measured dynamical masses.

Acknowledgments.   The authors would like to thank referee L. P. Vaz for a very constructive review. The U.S. authors would like to gratefully acknowledge funding support from the U.S. National Science Foundation.

REFERENCES

Allard F., Hauschildt P. H., and Schwenke D. (2000) *Astrophys. J., 540,* 1005–1015.

Allard N., Allard F., and Kielkopf J. F. (2005) *Astron. Astrophys., 440,* 1195–1201.

Andersen J. (A91) (1991) *Astron. Astrophys. Rev., 3,* 91–126.

Baraffe I., Chabrier G., Allard F., and Hauschildt P. H. (BCAH98) (1998) *Astron. Astrophys., 337,* 403–412.

Baraffe I., Chabrier G., Allard F., and Hauschildt P. H. (2001) In *From Darkness to Light: Origin and Evolution of Young Stellar Clusters* (T. Montmerle and P. André, eds.), pp. 571–580. ASP Conf. Series 243, San Francisco.

Baraffe I., Chabrier G., Allard F., and Hauschildt P. H. (BCAH02) (2002) *Astron. Astrophys., 381,* 563–572.

Baraffe I., Chabrier G., Barman T. S., Selsis F., Allard F., and Hauschildt P. H. (2005) *Astron. Astrophys., 436,* L47–L51.

Basri G. and Batalha C. (1990) *Astrophys. J., 363,* 654–669.

Boden A. F., Sargent A. I., Akeson R. L., Carpenter J. M., Torres G., et al. (B05) (2005) *Astrophys. J., 635,* 442–451.

Burrows A., Marley M., Hubbard W. B., Lunine J. I., Guillot T., et al. (1997) *Astrophys. J., 491,* 856–875.

Casey B. W., Mathieu R.D., Vaz L. P. R., Andersen J., and Suntzeff N. B. (C98) (1998) *Astron. J., 115,* 1617–1633.

Chabrier G. and Baraffe I. (1997) *Astron. Astrophys., 327,* 1039–1053.

Chabrier G. and Baraffe I. (2000) *Ann. Rev. Astron. Astrophys., 38,* 337–377.

Chabrier G. and Kueker M. (2006) *Astron. Astrophys., 446,* 1027–1038.

Chabrier G., Baraffe, I., Allard F., and Hauschildt P. H. (2000) *Astrophys. J., 542,* 464–472.

Chabrier G., Baraffe I., Allard F., and Hauschildt P. H. (2005) In *Resolved Stellar Populations,* invited review talk, Cancun, Mexico, April 2005 (astro-ph/0509798).

Close L., Lenzen R., Guirado J. C., Nielsen E. L., Mamajek E. E., et al. (2005) *Nature, 433,* 286–289.

Covino E., Frasca A., Alcala J. M., Paladino R., and Sterzik M. F. (C04) (2004) *Astron. Astrophys., 427,* 637–649.

D'Antona F. and Mazzitelli I. (DM94) (1994) *Astrophys. J. Suppl., 90,* 467–500.

D'Antona F. and Mazzitelli I. (DM97) (1997) *Mem. Soc. Astron. Ital., 68,* 807–822.

Dutrey A., Guilloteau S., and Simon M. (D03) (2003) *Astron. Astrophys., 402,* 1003–1011.

Eggleton P. P., Faulkner J., and Flannery B. P. (1973) *Astron. Astrophys., 23,* 325–330.

Genzel R. and Stutzki J. (1989) *Ann. Rev. Astron. Astrophys., 27,* 41–85.

Guilloteau S. and Dutrey A. (1998) *Astron. Astrophys., 339,* 467–476.

Hadrava P. (1997) *Astron. Astrophys. Suppl., 122,* 581–584.

Hartigan P., Kenyon S. J., Hartmann L., Strom S. E., Edwards S., et al. (1991) *Astrophys. J., 382,* 617–635.

Hartmann L., Cassen P., and Kenyon S. J. (1997) *Astrophys. J., 475,* 770–785.

Henriksen R., Andre P., and Bontemps S. (1997) *Astron. Astrophys., 323,* 549–565.

Hillenbrand L. A. and White R .J. (HW) (2004) *Astrophys. J., 604,* 741–757.

Johns-Krull C. M., Valenti J. A., and Koresko C. (J99) (1999) *Astrophys. J., 516,* 900–915.

Johns-Krull C. M., Valenti J. A., and Saar S. H. (2004) *Astrophys. J., 617,* 1204–1215.

Jones H., Pavlenko Y., Viti S., and Tennyson J. (2002) *Mon. Not. R. Astron. Soc., 330,* 675–684.

Leggett S. K., Allard F., Berriman G., Dahn C. C., and Hauschildt P. H. (1996) *Astrophys. J. Suppl., 104,* 117–143.

Ludwig H., Allard F., and Hauschildt P. H. (2002) *Astron. Astrophys., 395,* 99–115.

Ludwig H., Allard F., and Hauschildt P. H. (2006) *Astron. Astrophys.,* in press.

Luhman K. and Potter D. (2006) *Astrophys. J., 638,* 887.

Luhman K. L., Stauffer J. R., Muench A. A., Rieke G. H., Lada E. A., et al. (2003) *Astrophys. J., 593,* 1093–1115.

Luhman K., Stauffer J., and Mamajek E. (2005) *Astrophys. J., 628,* L69–L72.

Magni G. and Mazzitelli I. (1979) *Astron. Astrophys., 72,* 134–147.

Mamajek E. E., Lawson W. A., and Feigelson E. D. (M00) (2000) *Astrophys. J., 544,* 356–374.

Mohanty S., Basri G., Jayawardhana R., Allard F., Hauschildt P., and Ardila D. (2004a) *Astrophys. J., 609,* 854–884.

Mohanty S., Jayawardhana R., and Basri G. (2004b) *Astrophys. J., 609,* 885–905.

Mohanty S., Jayawardhana R., and Basri G. (2005) *Astrophys. J., 626,* 498–522.

Mullan D. J. and MacDonald J. (2001) *Astrophys. J., 559,* 353–371.

Nordlund A. and Stein R. F. (1999) In *Theory and Tests of Convection in Stellar Structure* (A. Gimenez et al., eds.), pp. 91–102. ASP Conf. Series 173, San Francisco.

Palla F. and Stahler S. W. (PS99) (1999) *Astrophys. J., 525,* 772–783.

Palla F. and Stahler S. W. (2001) *Astrophys. J., 553,* 299–306.

Pavlenko Y. V., Rebolo R., Martin E. L., and Garcia Lopez R. J. (1995) *Astron. Astrophys., 303,* 807–818.

Perrin G., Coude Du Foresto V., Ridgway S. T., Mariotti J.-M., et al. (1998) *Astron. Astrophys., 331,* 619–626.

Pols O. R., Tout C. A., Eggleton P. P., and Han Z. (1995) *Mon. Not. R. Astron. Soc., 274,* 964–974.

Pont F., Melo C. H. F., Bouchy F., Udry S., Queloz D., et al. (2005) *Astron. Astrophys., 433,* L21–L24.

Popper D. (P87) (1987) *Astrophys. J., 313,* L81–L83.

Prato L., Simon M., Mazeh T., McLean I. S., Norman D., and Zucker S. (P02) (2002) *Astrophys. J., 569,* 863–871.

Reiners A., Basri G., and Mohanty S. (2005) *Astrophys. J., 634,* 1346–1352.

Rucinski S. M. (1999) In *Precise Stellar Radial Velocities* (J. B. Hearnshaw and C. D. Scarfe, eds.), pp. 82–90. IAU Colloquium 170, ASP Conf. Series 185, San Francisco.

Saumon D., Chabrier G., and Van Horn H. M. (1995) *Astrophys. J. Suppl., 99,* 713–741.

Schaefer G. (2004) Ph.D. dissertation, State University of New York at Stony Brook.

Siess L., Dufour E., and Forestini M. (SDF00) (2000) *Astron. Astrophys., 358,* 593–599.

Simon K. P., Sturm E., and Fiedler A. (1994) *Astron. Astrophys., 292,* 507–518.

Simon M. (2006) In *ESO Workshop on the Power of Optical/IR Interferometry* (F. Paresce and A. Richichi, eds.), in press.

Simon M. and Prato L. (2004) *Astrophys. J., 613,* L69–L71.

Simon M., Dutrey A., and Guilloteau S. (Si00) (2000) *Astrophys. J., 545,* 1034–1043.

Slesnick C., Hillenbrand L., and Carpenter J. M. (2004) *Astrophys. J., 610,* 1045–1063.

Song I., Bessell M., and Zuckerman B. (2002) *Astrophys. J., 581,* L43–L46.

Stahler S. (1988) *Astrophys. J., 332,* 804–825.

Stassun K. G., Mathieu R. D., Vaz L. P. R., Stroud N., and Vrba F. J. (S04) (2004) *Astrophys. J. Suppl., 151,* 357–385.

Stassun K. G., Mathieu R. D., and Valenti J. (S06) (2006) *Nature, 440,* 311.

Steffen A. T., Mathieu R. D., Lattanzi M. G., Latham D. W., Mazeh T., et al. (St01) (2001) *Astron. J., 122,* 997–1006.

Strom K. M., Strom S. E., Edwards S., Cabrit S., and Skrutskie M. F. (1989) *Astron. J., 97,* 1451–1470.

Swenson F. J., Faulkner J., Rogers F. J., and Iglesias C. A. (S94) (1994) *Astrophys. J., 425,* 286–302.

Torres G., Sandberg C. H., Marschall L. A., Sheets H. A., and Mader J. A. (2006) *Astrophys. J., 640,* 1018.

Ventura P., Zeppieri A., Mazzitelli I., and D'Antona F. (1998) *Astron. Astrophys., 331,* 1011–1021.

White R. and Hillenbrand L. (2005) *Astrophys. J., 621,* L65–L68.

White R. J., Ghez A. M., Reid I. N., and Schultz G. (1999) *Astrophys. J., 520,* 811–821.

Yang H., Johns-Krull C. M., and Valenti J. A. (2005) *Astrophys. J., 635,* 466–475.

Zucker S. and Mazeh T. (1994) *Astrophys. J., 420,* 806–810.

Zuckerman B. and Song I. (2004) *Ann. Rev. Astron. Astrophys., 42,* 685–721.

Not Alone: Tracing the Origins of Very-Low-Mass Stars and Brown Dwarfs Through Multiplicity Studies

Adam J. Burgasser
Massachusetts Institute of Technology

I. Neill Reid
Space Telescope Science Institute

Nick Siegler and Laird Close
University of Arizona

Peter Allen
Pennsylvania State University

Patrick Lowrance
Spitzer Science Center

John Gizis
University of Delaware

The properties of multiple stellar systems have long provided important empirical constraints for star-formation theories, enabling (along with several other lines of evidence) a concrete, qualitative picture of the birth and early evolution of normal stars. At very low masses (VLM; $M \leq 0.1 M_\odot$), down to and below the hydrogen-burning minimum mass, our understanding of formation processes is not as clear, with several competing theories now under consideration. One means of testing these theories is through the empirical characterization of VLM multiple systems. Here, we review the results of various VLM multiplicity studies to date. These systems can be generally characterized as closely separated (93% have projected separations $\Delta < 20$ AU), near equal-mass (77% have $M_2/M_1 \geq 0.8$) and occurring infrequently (perhaps 10–30% of systems are binary). Both the frequency and maximum separation of stellar and brown dwarf binaries steadily decrease for lower system masses, suggesting that VLM binary formation and/or evolution may be a mass-dependent process. There is evidence for a fairly rapid decline in the number of loosely bound systems below ~0.3 $M_\odot$, corresponding to a factor of 10–20 increase in the minimum binding energy of VLM binaries as compared to more massive stellar binaries. This wide-separation "desert" is present among both field (~1–5 G.y.) and older (>100 m.y.) cluster systems, while the youngest (≤10 m.y.) VLM binaries, particularly those in nearby, low-density star-forming regions, appear to have somewhat different systemic properties. We compare these empirical trends to predictions laid out by current formation theories, and outline future observational studies needed to probe the full parameter space of the lowest-mass multiple systems.

1. INTRODUCTION

The frequency of multiple systems and their properties are key constraints for studies of stellar formation and evolution. Binary and multiple stars are common in the galaxy, and the physical properties of the components in these systems can be significantly influenced by dynamical and co-evolutionary processes. Furthermore, successful theories of star formation must take into account the creation of multiples and empirical multiplicity trends as functions of mass, age, and metallicity.

The main focus of this review is multiplicity in very-low-mass (VLM; $M \leq 0.1 M_\odot$) stars and brown dwarfs. However, to put these results in the proper context, we start with a brief review of our current understanding of multiplicity among higher-mass stars (also see chapter by Duchêne et al.). The standard references for binary frequency are *Duquennoy and Mayor* (1991, hereafter *DM91*; also *Abt and Levy*, 1976; *Abt*, 1978; *Mayor et al.*, 1992) for solar-type stars and *Fischer and Marcy* (1992, hereafter *FM92*; also *Henry and McCarthy*, 1990; *Reid and Gizis*, 1997a; *Halbwachs et al.*, 2003; *Delfosse et al.*, 2004) for early-type M

427

dwarfs. The *DM91* survey combined spectroscopic, astrometric, and direct imaging of 164 G dwarfs; 44% of those stars were identified as binaries, with incompleteness corrections increasing the binary fraction to $f_{bin} \sim 65\%$. These corrections include 8% attributed to VLM companions; as discussed further below, more recent observations show that the actual correction is much lower. The *FM92* survey covered 72 M2–M5 dwarfs within 20 pc, and derived $f_{bin} = 42 \pm 9\%$, significantly lower than the *DM91* G-dwarf survey. While both surveys include nearby stars, neither comprises a *volume-complete* sample.

Recent surveys of solar-type stars have concentrated on VLM companions. Radial velocity (RV) surveys (e.g., *Marcy and Butler,* 2000; *Udry et al.,* 2003) have shown that less than 0.5% of solar-type stars have brown dwarf companions within ~5 AU. *Guenther et al.* (2005) find $f_{bin}^{BD} < 2\%$ for projected separations $\Delta < 8$ AU among Hyades stars; this is in contrast with $f_{bin} \sim 13\%$ for stellar-mass companions at those separations (*DM91*). At larger separations, imaging surveys of young solar neighborhood stars (members of the TW Hydrae, Tucanae, Horologium, and β Pic associations) (*Zuckerman and Song,* 2004) find $f_{bin}^{BD} \sim 6 \pm 4\%$ for $\Delta > 50$ AU (*Neuhäuser et al.,* 2003), similar to the brown dwarf companion fraction measured for field stars for separations of 30–1600 AU (*Metchev,* 2005). These fractions are ~3× lower than the hydrogen-burning companion rate over the same separation range. At the widest separations ($\Delta > 1000$ AU), *Gizis et al.* (2001) find that solar-type stars have comparable numbers of brown-dwarf and M-dwarf companions, although this result is based on a very small number of VLM companions.

Besides the overall binary fraction, the mass distribution of companions sets constraints on formation models. Figure 1 shows the results for late-F to K stars [$0.5 < (B–V) < 1.0$] within 25 pc of the Sun, breaking down the sample by projected separation/orbital semimajor axis. The left panels compare the mass distribution of companions against a schematic representation of the initial mass function (*Reid et al.,* 1999, 2002a); the right panels compare the mass ratio ($q \equiv M_2/M_1$) distributions against the VLM dwarf data assembled in this review (cf. Fig. 3). Clearly, low q binary systems are more common at all separations among solar-type stars than in VLM dwarfs. We return to this issue in section 2.2.3. At small separations ($\Delta < 10$ AU), there is an obvious deficit of low-mass companions (with the exception of planetary companions) as compared to the distribution expected for random selection from the field-star mass function. The notorious brown-dwarf desert (e.g., *Marcy and Butler,* 2000) extends well into the M-dwarf regime. This result is consistent with the original analysis of *Mazeh and Goldberg* (1992) of the mass ratio distribution of spectroscopic binaries, although their more recent study of proper motion stars (*Goldberg et al.,* 2003) finds a bimodal distribution, with peaks at $q \sim 0.8$ and ~ 0.2 (see also *Halbwachs et al.,* 2003). The deficit in low-mass companions is less pronounced at intermediate separations, while it is possible that observational selection effects (e.g., sensitivity limitations) might account for the small discrepancy for $q < 0.2$ in the wide-binary sample.

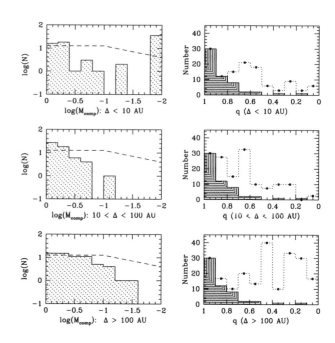

Fig. 1. Mass and mass ratio distributions of companions to late-F to K-type dwarfs within 25 pc of the Sun (*Reid et al.,* 2002a), segregated by projected separation/orbital semimajor axis. The left panels plot the mass distribution of companions, with the dashed lines providing a schematic representation of the initial mass function. The right panels plot the mass-ratio distributions (dotted histograms), with the solid histogram showing the mass ratio distribution for VLM dwarfs (no segregation of separations; see Fig. 3). These distributions are normalized at q = 1 bin.

In the case of M dwarfs, attention has focused on the nearest stars. *Delfosse et al.* (2004) recently completed a spectroscopic and adaptive optics (AO) imaging survey of M dwarfs within 9 pc that is effectively complete for stellar mass companions. Combining their results with the imaging surveys by *Oppenheimer et al.* (2001) and *Hinz et al.* (2002), they derive an overall binary fraction of 26% for M dwarfs. For a more detailed breakdown with spectral type, we can turn to the northern 8-pc sample (*Reid and Gizis,* 1997a; *Reid et al.,* 2003). Those data indicate binary fractions of $24_{-7}^{+13}\%$ for spectral types M0–M2.5 (4/17 systems), $27_{-7}^{+5}\%$ for M3–M4.5 (12/45 systems), and $31_{-9}^{+13}\%$ for M5–M9 (5/16 systems; uncertainties assume a binomial distribution), where the spectral type refers to the primary star in the system; the overall binary frequency is $f_{bin} = 27_{-4}^{+5}\%$. These results, based on volume-limited samples, confirm that M dwarfs have significantly lower multiplicity than more massive solar-type stars. This is consistent with an overall trend of decreasing multiplicity with decreasing mass (cf. A and B stars have overall multiplicity fractions as high as 80%) (*Shatsky and Tokovinin,* 2002; *Kouwenhoven et al.,* 2005). These changes in multiplicity properties with mass among hydrogen-burning stars emphasize that we must consider VLM dwarfs as part of a continuum, not as a distinct species unto themselves. (We note that, even with 30% binarity for M dwarfs, most stars still reside in multiple systems. As a numerical example, consider a volume-limited sample of 100 stellar systems: 20 are type G or earlier, 10 are type K, and 70 are type M. Assuming binary fractions of 70%, 50%, and 30%, respectively, these 100 systems include 140 stars, 80 in binaries and 60 in isolated systems. Higher-order multiples only serve to increase the companion fraction.)

2. OBSERVATIONS OF VERY-LOW-MASS BINARIES

2.1. Very-Low-Mass Binary Systems

With the discovery of hundreds of VLM dwarf stars and brown dwarfs over the past decade (see reviews by *Basri*, 2000; *Oppenheimer et al.*, 2000; *Kirkpatrick*, 2005), it is now possible to examine systems with primaries down to 100× less massive than the Sun. In this regime, formation mechanisms are under considerable debate (see chapters by Bonnell et al., Goodwin et al., Klein et al., Luhman et al., and Whitworth et al.). Hence, accurate assessment of the multiplicity and systemic properties of VLM stars and brown dwarfs are essential for constraining current theoretical work.

Searches for VLM binaries — defined here as having a total system mass $M_{tot} < 0.2$ $M_\odot$ and primary mass $M_1 < 0.1$ $M_\odot$ (cf. *Siegler et al.*, 2005) — have been conducted predominantly through high-resolution imaging surveys, using both groundbased [including natural and, quite recently, laser guide star adaptive optics (AO)] and spacebased facilities. Major surveys have targeted both nearby field sources (*Koerner et al.*, 1999; *Reid et al.*, 2001; *Bouy et al.*, 2003; *Burgasser et al.*, 2003; *Close et al.*, 2002, 2003; *Gizis et al.*, 2003; *Siegler et al.*, 2003, 2005; *Law et al.*, 2006; Allen et al., in preparation; Billères et al., in preparation; Burgasser et al., in preparation; Reid et al., in preparation) and young clusters and associations (*Martín et al.*, 1998, 2000a, 2003; *Neuhäuser et al.*, 2002; *Kraus et al.*, 2005; *Luhman et al.*, 2005; *Bouy et al.*, 2006). A smaller number of high-resolution spectroscopic surveys for closely separated binaries have also taken place (*Basri and Martín*, 1999; *Joergens and Guenther*, 2001; *Reid et al.*, 2002b; *Guenther and Wuchterl*, 2003; *Kenyon et al.*, 2005; *Joergens*, 2006). Only one eclipsing system has been discovered so far via photometric monitoring (*Stassun et al.*, 2006). Observations leading to the identification of low-mass multiple systems has been accompanied by resolved photometry and spectroscopy, allowing characterization of the colors, luminosities, and spectral characteristics of several binary components. Astrometric and radial velocity monitoring has led to mass measurements or constraints for five VLM systems to date (*Basri and Martín*, 1999; *Lane et al.*, 2001b; *Bouy et al.*, 2004a; *Brandner et al.*, 2004; *Zapatero Osorio et al.*, 2004; *Stassun et al.*, 2006).

In Table 1 we list 75 VLM binary systems published in the literature or reported to us as of 2005. The mass criteria correspond to field dwarf binary components later than spectral type ~M6; younger systems may include earlier spectral types. Table 1 provides a subset of the compiled data for these sources, given in more complete detail through an online database maintained by N. Siegler (see http://paperclip.as.arizona.edu/~nsiegler/VLM_binaries).

2.2. General Properties of Very-Low-Mass Binaries

Large-scale, high-resolution imaging surveys in the field have converged to similar conclusions on the general properties of VLM field binaries. Compared to their higher-mass stellar counterparts, VLM binaries are (1) rarer ($f_{bin} \approx 10$–30%; however, see discussion below); (2) more closely separated (93% have $\Delta < 20$ AU); and (3) more frequently in near-equal mass configurations (77% have $q \geq 0.8$).

Analogous imaging surveys in young open clusters (e.g., Pleiades, α Persei) find similar trends, although the youngest ($\leq$10 m.y.) associations (e.g., Chamaeleon I, Upper Scorpius, Orion) appear to exhibit somewhat different properties. We discuss these broad characterizations in detail below.

2.2.1. The binary fraction. Magnitude-limited imaging surveys for VLM stars and brown dwarfs in the field with spectral types M6 and later have generally yielded *observed* binary fractions of ~20%; taking into consideration selection effects (e.g., *Burgasser et al.*, 2003) lowers this fraction to 7–15% for $\Delta \gtrsim 2$–3 AU and $q \gtrsim 0.4$–0.5 (*Bouy et al.*, 2003; *Burgasser et al.*, 2003; *Close et al.*, 2003; *Siegler et al.*, 2005). *Burgasser et al.* (2003) deduced $f_{bin} = 9^{+11}_{-4}$% for a small sample of L and T dwarfs using the $1/V_{max}$ technique (*Schmidt et al.*, 1968); *Bouy et al.* (2003) deduced a volume-limited fraction of $f_{bin} \sim 15$%. Over the same separation ($\Delta > 2$ AU) and mass ratio ($q > 0.5$) ranges, these multiplicity rates are less than half of those of M dwarfs (*FM92*; *Close et al.*, 2003) and G dwarfs (*DM91*; *Bouy et al.*, 2003). Similarly, HST imaging surveys of the 125-m.y. Pleiades open cluster (*Martín et al.*, 2000a, 2003; *Bouy et al.*, 2006) found a resolved binary fraction of 13–15% for $\Delta > 7$ AU for components at and below the hydrogen-burning limit. On the other hand, *Kraus et al.* (2005) found $f_{bin} = 25^{+16}_{-8}$% for a small sample of 0.04–0.1 $M_\odot$ members of Upper Scorpius over the range $\Delta = 5$–18 AU, somewhat higher than, but still consistent with, other field and open cluster results.

One problem with resolved imaging surveys is their inherent selection against tightly bound systems ($\Delta \lesssim 2$–3 AU for the field dwarfs and nearby associations, $\Delta \lesssim 10$–15 AU for more distant star-forming regions). Here, one must generally turn to high-resolution spectroscopic surveys of VLM stars, currently few in number and with as yet limited follow-up. *Reid et al.* (2002b) deduced a double-lined spectroscopic binary (SB2) fraction of 6^{+7}_{-2}% for a sample of M7–M9.5 field dwarfs. *Guenther and Wuchterl* (2003) identified two SB2s and marginally significant RV variations in the active M9 LP 944-20 (which they attribute to either the

TABLE 1.   Known very-low-mass binaries.

Source Name (1)	Separation (mas) (2)	(AU) (3)	Spectral Types (4)	Estimated Masses (M☉) (5)	(M☉) (6)	q (7)	Estimated Period (yr) (8)	Age (m.y.) (9)	Association or Note (10)	Ref. (11)
Cha Hα8	—	—	M6.5 + [M6.5:]	0.070	—	—	—	2	Cha I; RV	37
2MASS J0253202+271333AB	—	—	M8 + [M8:]	0.092	0.092:	1:	—	—	SB2	8; 42
2MASS J0952219-192431AB	—	—	M7 + [M7:]	0.098	0.098:	1:	—	—	SB2	8; 43
LHS 292AB	—	—	M7 + [M7:]	0.098	0.098:	1:	—	—	SB2	8,28; 76
2MASS J2113029-100941AB	—	—	M6 + [M6:]	0.085	0.085:	1:	—	—	SB2	28; 42
PPl 15AB	—	0.03*	M7 + [M8:]	0.070*	0.060*	0.86	0.0159*	120	Pleiades; SB2	1; 47,60
2MASS J0535218-054608AB	—	0.04†	M6.5 + [M6.5]	0.054†	0.034†	0.63	0.0268†	1	Orion; SB2, EB	79
2MASS J15344984-2952274AB	65	0.9	T5.5 + [T5.5]	0.035	0.035	1.00	4	—		5; 38
GJ 569BC	103	0.90*	M8.5 + M9.0	0.071*	0.054*	0.76	2.4*	300	Ursa Major; triple	2; 3,33,40,75
GJ 1001BC	87	1.0	L4.5 + [L4.5]	0.068	0.068	1.00	4	—	triple	25; 35,36,52
LP 349-25AB	125	1.3	M8 + [M9]	0.090	0.085	0.94	4	—		31; 42
SDSS J092615.38+584720.9AB	70	1.4	T4.5 + [T4.5]	0.050	0.050	1.00	/	—		69, 71
GJ 417BC	70	1.5	L4.5 + [L6]	0.073	0.070	0.96	7	—	triple	4; 39,40
2MASS J0920122+351742AB	70	1.5	L6.5 + [T:]	0.068	0.068	1.00	6	—		7; 39,69,78
2MASS J2252107-173014AB	140	1.9	L6 + [T2]	0.070	0.060	0.86	10	—		32; 58,59
2MASS J1847034+552243AB	82	1.9	M7 + [M7.5]	0.098	0.094	0.96	8	—		23; 43
2MASS J0700366+315727	170	2.0	L3.5 + [L6.5]	0.075	0.071	0.95	10	—		78; 43
DENIS PJ035726.9-441730AB	98	2.2	M9.5 + [L1.5]	0.085	0.080	0.91	11	—		4,13
HD 130948BC	134	2.4	L4 + [L4]	0.070	0.060	0.86	14	—	triple	6; 26,40
SDSS J042348.57-041403.5AB	164	2.5	L7 + T2	0.060	0.050	0.83	16	—		68; 43,70,71
2MASS J0746425+200032AB	220	2.5a	L0 + L1.5	0.085*	0.066*	0.78	11*	300		4,7,17; 20,39,41,61,71
ε IndiBC	732	2.6	T1 + T6	0.045	0.027	0.60	22	1300	triple	16; 40
2MASS J1430436+291541AB	88	2.6	L2 + [L2:]	0.076	0.075	0.99	15	—		4; 43
2MASS J1728114+394859AB	131	2.7	L7 + [L8]	0.069	0.066	0.96	16	—		4,13; 39
LP 213-68AB	122	2.8	M8 + [L0]	0.092	0.084	0.91	15	—	triple	17,53
LHS 2397Aab	207	3.0	M8 + [L7.5]	0.090	0.068	0.76	18	—		10; 36,42,82
LSPM 1735+2634AB	290	3.2	[M9:] + [M9:]	0.082	0.074	0.90	14	—		51; 83
LHS 1070BC	446	3.4*	M8.5 + [M9]	0.070*	0.068*	0.97	16*	—	quadruple	18; 74
2MASS J0856479+223518AB‡	98	3.4	L3: + [L:]	0.071	0.064	0.90	24	—		4; 43
2MASS J1017075+130839AB	104	3.4	L2 + [L2]	0.076	0.076	1.00	23	—		4; 43
SDSS 2335583-001304AB	57	3.5	L1: + [L4:]	0.079	0.074	0.94	24	—		4; 81
2MASS J1600054+170832AB	57	3.5	L1 + [L3]	0.078	0.075	0.96	23	—		4,13; 39
LP 415-20AB	119	3.6	M7 + [M9.5]	0.095	0.079	0.83	22	625	Hyades	9; 42
2MASS J2255432-2739466AB	282	3.8	T6 + [T8]	0.033	0.024	0.73	43	—		5; 38,77
SDSS J153417.05+161546.1AB	106	3.8	T1.5 + [T5.5]	0.050	0.040	0.80	35	—		15
SDSS J102109.69-030420.1AB	160	3.9	T1 + T5	0.060	0.050	0.83	33	—		69; 70,72
2MASS J1426316+155701AB	152	4.0	M8.5 + [L1]	0.088	0.076	0.86	27	—		17; 42
2MASS J2140293+162518AB	155	4.0	M8.5 + [L2]	0.092	0.078	0.85	27	—		17; 42
2MASS J15530228+1532369AB	340	4.4	T7 + [T7]	0.040	0.030	0.75	49	—		69; 39
2MASS J1239272+551537AB	211	4.5	L5 + [L5]	0.071	0.071	1.00	35	—		4,13; 39
2MASS J2206228-204705AB	168	4.5	M8 + [M8]	0.092	0.091	0.99	31	—		17; 42
2MASS J0850359+105716AB	160	4.7	L6 + [L8]	0.050	0.040	0.80	39	—		7; 41,52,70
2MASS J1750129+442404AB	158	4.9	M7.5 + [L0]	0.095	0.084	0.88	36	—		9; 42
USco-109AB‡	34	4.9	M6 + [M7.5]	0.070	0.040	0.57	46	5	Up Sco	29; 45,65
2MASS J2101154+175658AB	234	5.4	L7 + [L8]	0.068	0.065	0.96	49	—		4,13; 39
Kelu-1AB	291	5.4	L2 + [L4]	0.060	0.055	0.92	52	—		24; 41,52
2MASS J0429184-312356AB	531	5.8	M7.5 + [L0]	0.094	0.079	0.84	48	—		23,78; 43
2MASS J0147328-495448AB	190	5.8	M8 + [M9]	0.086	0.080	0.93	47	—		78
2MASS J2152260+093757AB	250	6.0	L6: + [L6:]	0.069	0.069	1.00	55	—		78
MHO Tau 8AB	44	6.2	M6 + [M6.5]	0.100	0.070	0.70	53	2	Taurus	55; 56
DENIS J122815.2-154733AB	275	6.4a	L6 + [L6]	0.065*	0.065*	1.00	44*	—		11; 41,64,71
DENIS J100428.3-114648AB	146	6.8	L0: + [L2:]	0.080	0.076	0.95	63	—		4
2MASS J2147436+143131AB	322	7.0	M8 + [L0]	0.084	0.078	0.93	65	—		4,13; 42
DENIS J185950.9-370632AB	60	7.7	L0 + [L3]	0.084	0.076	0.90	76	5	R-CrA	20; 57
2MASS J1311391+803222AB	267	7.7	M8.5 + [M9]	0.089	0.087	0.98	72	—		17; 42
IPMBD 29AB	58	7.8	L1 + [L4]	0.045	0.038	0.84	106	120	Pleiades	14; 47
2MASS J1146345+223053AB	290	7.9	L3 + [L4]	0.055	0.055	1.00	94	—		7,12; 41,52
CFHT-Pl-12AB	62	8.3	M8 + [L4]	0.054	0.038	0.70	111	120	Pleiades	14; 47,62
2MASS J1127534+741107AB	246	8.4	M8 + [M9]	0.092	0.087	0.95	80	—		17; 42
2MASS J1449378+235537AB	134	8.5	L0 + [L4]	0.084	0.075	0.89	88	—		4,13; 39
LP 475-855AB	294	8.5	M7.5 + [M9.5]	0.091	0.080	0.88	85	625	Hyades	9; 42
DENIS J020529.0-115925AB	510	9.2	L7 + [L7]	0.070	0.070	1.00	105	—	poss. Triple	12; 52,41
USco-66AB	70	10.2	M6 + [M6]	0.070	0.070	1.00	120	5	Up Sco	29; 45,65
2MASS J17072343-0558249AB	950	10.4	M9 + L3	0.090	0.060	0.67	125	—		67
GJ 337CD	530	10.9	L8 + [T:]	0.055	0.055	1.00	150	—	quadruple	30; 50,67
2MASS J0915341+042204AB	730	11.0	L7 + [L7]	0.070	0.070	1.00	138	—		78
IPMBD 25AB	94	12.6	M7 + [L4]	0.063	0.039	0.62	200	120	Pleiades	14; 47
DENIS J144137.3-094559AB	420	14.3	L1 + [L1]	.072	0.072	1.00	200	—	triple	4,48; 39,80
2MASS J2331016-040618AB	573	15.0	M8 + [L7]	0.093	0.067	0.72	200	—	triple	4,13,17; 40,42
USco-55AB	122	17.7	M5.5 + [M6]	0.100	0.070	0.70	250	5	Up Sco	29; 45,65
CFHT-Pl-18AB	330	34.6	M8 + M8	0.090	0.090	1.00	680			4; 19

TABLE 1.   (continued).

Source Name (1)	Separation (mas) (2)	Separation (AU) (3)	Spectral Types (4)	Estimated Masses $(M_\odot)$ (5)	Estimated Masses $(M_\odot)$ (6)	q (7)	Estimated Period (yr) (8)	Age (m.y.) (9)	Association or Note (10)	Ref. (11)
DENIS J220002.0-303832.9AB	1090	38.2	M8 + L0	0.085	0.083	0.98	800	—		66
2MASS J1207334-393254AB	776	41.1	M8.5 + L:	0.024	0.004	0.17	2250	8	TW Hyd	22,63; 34
DENIS J055146.0-443412.2AB	2200	220.0	M8.5 + L0	0.085	0.079	0.93	11500	—		27
2MASS J11011926-7732383AB	1440	241.9	M7 + M8	0.050	0.025	0.50	20000	2	Cha I	21; 44

*Parameters derived or estimated from orbital motion measurements.
†Parameters for 2MASS J0535218-054608AB based on both spectroscopic orbit and eclipsing light curve; see *Stassun et al.* (2006).
‡Candidate binary.

Uncertain values are indicated by colons. [1] Name of binary; [2] angular separation in mas; [3] projected separation (Δ) in AU, or semimajor axis of orbit as noted; [4] spectral types of binary components — for sources without resolved spectroscopy, primary spectral type is for combined light data, secondary spectral type is estimated from photometric flux ratios (as indicated by brackets); [5] estimated primary mass in $M_\odot$, taken as the average of the reported mass ranges; masses determined from orbital dynamics are indicated; [6] estimated secondary mass in $M_\odot$, taken as the average of the reported mass ranges; masses determined from orbital dynamics are indicated; [7] $q \equiv M_2/M_1$, as reported or calculated from columns [5]–[6]; [8] estimated orbital period in yr, assuming circular orbit with semimajor axis $a = 1.26\Delta$ (*FM92*); sources with period measurements from orbital measurements are indicated; [9] estimated age in million years of binary if member of a moving group or association, or companion to a age-dated star; [10] additional notes, including cluster association; [11] references as given below; discovery references are listed first, followed by references for additional data (spectral types, distance measurements/estimates, orbital measurements) separated by a semicolon.

References: (1) *Basri and Martín* (1999); (2) *Martín et al.* (2000b); (3) *Lane et al.* (2001b); (4) *Bouy et al.* (2003); (5) *Burgasser et al.* (2003); (6) *Potter et al.* (2002); (7) *Reid et al.* (2001); (8) *Reid et al.* (2002b); (9) *Siegler et al.* (2003); (10) *Freed et al.* (2003); (11) *Martín et al.* (1999); (12) *Koerner et al.* (1999); (13) *Gizis et al.* (2003); (14) *Martín et al.* (2003); (15) *Liu et al.* (in preparation); (16) *McCaughrean et al.* (2004); (17) *Close et al.* (2003); (18) *Leinert et al.* (2001); (19) *Martín et al.* (2000a); (20) *Bouy et al.* (2004b); (21) *Luhman* (2004); (22) *Chauvin et al.* (2004); (23) *Siegler et al.* (2005); (24) *Liu and Leggett* (2005); (25) *Golimowski et al.* (2004); (26) *Goto et al.* (2002); (27) *Billères et al.* (2005); (28) *Guenther and Wuchterl* (2003); (29) *Kraus et al.* (2005); (30) *Burgasser et al.* (2005a); (31) *Forveille et al.* (2005); (32) *Reid et al.* (2006); (33) *Kenworthy et al.* (2001); (34) *Mamajek* (2005); (35) *Leggett et al.* (2002); (36) *van Altena et al.* (1995); (37) *Joergens* (2006); (38) *Tinney et al.* (2003); (39) *Kirkpatrick et al.* (2000); (40) *Perryman* (1997); (41) *Dahn et al.* (2002); (42) *Gizis et al.* (2000b); (43) *Cruz et al.* (2003); (44) *Whittet et al.* (1997); (45) *de Zeeuw et al.* (1999); (46) *Kenyon et al.* (1994); (47) *Percival et al.* (2005); (48) *Stephens et al.* (2001); (50) *Wilson et al.* (2001); (51) *Law et al.* (2006); (52) *Kirkpatrick et al.* (1999); (53) *Gizis et al.* (2000a); (54) *Delfosse et al.* (1997); (55) *White et al.* (in preparation); (56) *Briceño et al.* (1998); (57) *Casey et al.* (1998); (58) *Kendall et al.* (2004); (59) *McGovern* (2005); (60) *Martín et al.* (1996); (61) *Bouy et al.* (2004a); (62) *Stauffer et al.* (1998); (63) *Chauvin et al.* (2005a); (64) *Brandner et al.* (2004); (65) *Ardila et al.* (2000); (66) *Burgasser and McElwain* (2006); (67) McElwain and Burgasser (in preparation); (68) *Burgasser et al.* (2005b), (69) Burgasser et al. (in preparation); (70) *Vrba et al.* (2004); (71) *Geballe et al.* (2002); (72) *McLean et al.* (2003); (73) *Burgasser et al.* (2002); (74) *Leinert et al.* (2000); (75) *Zapatero-Osorio et al.* (2004); (76) *Dahn et al.* (1986); (77) *Burgasser et al.* (1999); (78) Reid et al. (in preparation); (79) *Stassun et al.* (2006); (80) *Seifahrt et al.* (2005); (81) H. Bouy, personal communication (2005); (82) *Tinney* (1996); (83) *Lépine and Shara* (2005).

presence of a low-mass companion or magnetic-induced activity) in a sample of 25 M5.5–L1.5 field and cluster dwarfs. Including all three objects implies an observed binary fraction of $12^{+10}_{-4}\%$, although this value does not take into consideration selection biases. *Joergens* (2006) detected one RV variable, the M6.5 Cha Hα8, among a sample of nine VLM stars and brown dwarfs in the 2-m.y. Cha I association, implying an observed fraction of $11^{+18}_{-4}\%$, again subject to sampling and selection biases. *Kenyon et al.* (2005) identified four possible spectroscopic binaries (SBs) among VLM stars and brown dwarfs in the 3–7 m.y. σ Orionis cluster on the basis of RV variations over two nights. They derive $f_{bin} > 7$–17% for $\Delta < 1$ AU (after correcting for selection effects) and a best-fit fraction of 7–19% (for their sample A) depending on the assumed underlying separation distribution. However, none of the sources from this particular study have had sufficient follow-up to verify RV variability, and cluster membership for some of the targets have been called into question. A more thorough analysis of sensitivity and sampling biases in these SB studies has been done by *Maxted and Jeffries* (2005), who find $f_{bin} = 17$–30% for $\Delta < 2.6$ AU, and an overall binary fraction of 32–45% (assuming $f_{bin} = 15$% for $\Delta > 2.6$ AU). This result suggests that imaging studies may be missing a significant fraction of VLM systems hiding in tightly separated pairs. However, as orbital properties have only been determined for two SB systems so far [PPl 15 (*Basri and Martín*, 1999) and 2MASS 0535-0546 (*Stassun et al.*, 2006)], individual separations and mass ratios for most VLM SB binaries remain largely unconstrained.

Two recent studies (*Pinfield et al.*, 2003; *Chapelle et al.*, 2005) have examined the fraction of unresolved (overluminous) binary candidates among VLM stars and brown dwarfs in young associations. Contrary to other studies, these groups find much larger binary fractions, as high as 50% in the *Pinfield et al.* (2003) study of the Pleiades and Praesepe for q > 0.65. This study also finds a binary fraction that increases with decreasing mass, in disagreement with results in the field (see below); the *Chapelle et al.* (2005) study finds evidence for the opposite effect in the 0.9-G.y. Praesepe cluster. Both studies have been controversial due to the lack of membership confirmation, and hence likelihood of contamination, and the possible influence of variability on the identification of overluminous sources. Nevertheless, both of these studies and the SB results suggest that a higher VLM binary fraction than that inferred from imaging studies, perhaps 30% or more, is possible.

2.2.2. The separation distribution. Figure 2 plots the histogram of projected separations/orbital semimajor axes for 70 binaries in Table 1 (SB systems without orbital measurements are not included). This distribution exhibits a clear peak around 3–10 AU, with 53 ± 6% of known VLM binaries encompassing this range. Again, because imaging surveys (from which most of the objects in Table 1 are drawn) can only resolve systems down to a minimum angular scale (typically 0''.05–0''.1 for AO and HST programs), the decline in this distribution at small separations is likely a selection effect. Results from SB studies remain as yet unclear in this regime. *Basri and Martín* (1999) have sug-

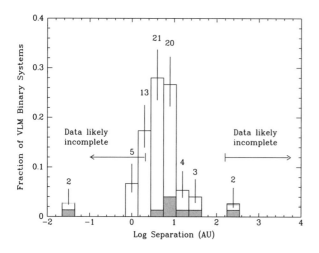

Fig. 2. Distribution of separations/orbital semimajor axes for known VLM binary systems (Table 1). The number of VLM binary systems in each 0.3 dex bin is labeled, and uncertainties (vertical lines) are derived from a binomial distribution. Note that SBs with unknown separations are not plotted but included in the total number of binaries for scaling the distribution. The distribution peaks at $\Delta \sim 3$–10 AU, with steep declines at shorter and longer separations. While there is likely observational incompleteness for $\Delta \lesssim 3$ AU, the sharp drop in binary systems with $\Delta \gtrsim 20$ AU is a real, statistically robust feature. The shaded bins represent the eight systems with ages <10 m.y. While the statistics are still small, the separation distribution of these young binaries is flatter, and suggests a peak at wider separations than that of the field and older cluster binaries.

gested that very close binaries are common based on the detection of one (PPl 15) in a small spectral sample. The analysis of *Maxted and Jeffries* (2005) suggest that there may be as many or more binaries with $\Delta \lesssim 3$ AU as those with $\Delta \gtrsim 3$ AU. At the extreme, *Pinfield et al.* (2003) estimate that 70–80% of VLM binaries in the Pleiades have $\Delta < 1$ AU, although this result has not been corroborated by similar studies in the Pleiades (*Bouy et al.*, 2006) and Praesepe (*Chapelle et al.*, 2005). In any case, as the peak of the observed separation distribution lies adjacent to the incompleteness limit, closely separated systems likely comprise a nonnegligible fraction of VLM binaries.

The steep decline in the separation distribution at larger separations is, on the other hand, a statistically robust feature. While high-resolution imaging surveys are limited in this domain by field of view (typically 10–20″ for HST and AO studies), this only excludes systems with $\Delta \gtrsim 150$ AU for a typical VLM field source (distances ~30 pc) or $\Delta \gtrsim$ 200–1000 AU for young cluster systems. Even wider separations for hundreds of VLM field dwarfs should be detectable — and are not found — in the original surveys from which they were identified [e.g., 2MASS, DENIS, and SDSS; however, see *Billères et al.* (2005)]. In open clusters, deep imaging has demonstrated a consistent lack of wide companions to VLM dwarfs. An upper limit of $f_{bin} <$ 8% for $\Delta > 11$ AU is derived for the 90-m.y. α Per open

cluster (*Martín et al.*, 2003), similar to the 5% upper limit for $\Delta > 15$ AU measured for 32 VLM members of the 2-m.y. IC-348 cluster (*Luhman et al.*, 2005). *Lucas et al.* (2005) measure an upper limit of 2% for wide VLM binaries ($\Delta >$ 150 AU) in the 1-m.y. Trapezium cluster based on a two-point correlation function. In contrast, 93% of the known VLM binaries have $\Delta \leq 20$ AU. Hence, a "wide brown dwarf binary desert" is evidenced for VLM stars and brown dwarfs (*Martín et al.*, 2000a), a potential clue to their formation.

While survey results have generally been negative for wide VLM binaries, two — 2MASS J11011926-7732383AB (hereafter 2MASS 1101-7732AB) (*Luhman*, 2004) and DENIS J055146.0-443412.2AB (hereafter DENIS 0551-4434AB) (*Billères et al.*, 2005) — have been identified serendipitously. These systems have projected separations $\gtrsim 200$ AU, over 10× wider than the vast majority of VLM binary systems. A third low-mass binary not included in Table 1, GG Tau BaBb (a.k.a. GG Tau/c) (*Leinert et al.*, 1991; *White et al.*, 1999), with estimated primary and total system masses of 0.12 and 0.16 $M_\odot$, respectively, also has a projected separation greater than 200 AU. Interestingly, two of these three systems are members of very young, loose associations. We discuss these sources further in section 2.4.2.

The separation distribution of VLM stars therefore peaks at or below ~3–10 AU, corresponding to orbital periods of ≤ 40 yr. This is quite different from the separation distribution of G dwarfs, which shows a broad peak around 30 AU (periods of ~170 yr) (*DM91*); and the M dwarf distribution, which peaks between 4–30 AU (periods of 9–270 yr) (*FM92*). There is a suggestion in this trend of decreasing separations as a function of mass, as discussed further below.

2.2.3. The mass ratio distribution. Figure 3 shows the distribution of mass ratios for 70 of the binaries in Table 1 (not including SBs without mass estimates). These ratios were derived by a variety of methods, including comparison of component fluxes to evolutionary models (e.g., *Chabrier et al.*, 2000), analytic relations (e.g., *Burrows et al.*, 2001) and direct estimates from orbital motion measurements. Despite these different techniques, a comparison of all the data shows congruence with individual studies. The mass ratio distribution for VLM systems is strongly peaked at near unity ratios; over half of the known VLM binaries have q > 0.9 and 77^{+4}_{-5}% have $q \geq 0.8$.

As with the separation distribution, it is important to consider selection effects in the observed mass ratios. Most pertinent is the detectability of secondaries in low q binaries, which may be too faint for direct imaging or of insufficient mass to induce a measureable RV variation in the primary's spectrum. The former case is an important issue for field binaries, as low-mass substellar companions fade to obscurity over time. However, most imaging and spectroscopic surveys to date are sensitive down to $q \geq 0.5$, while a sharp dropoff is clearly evident at the highest-mass ratios. Hence, while the number of low-mass ratio systems may be underestimated, the q ~ 1 peak is not the result of this bias.

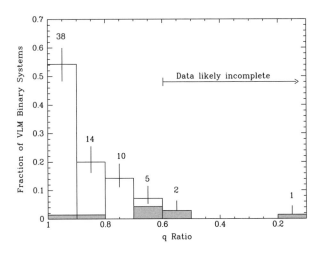

Fig. 3. Mass ratio distribution of known VLM binary systems (Table 1). The number of VLM binary systems in each 0.1 fractional bin is labeled, and uncertainties are derived from a binomial distribution. Note that SBs with unknown mass ratios are not plotted and not included in the total number of binaries when scaling the distribution. The distribution peaks near unity for binary systems with $\Delta \geq 3$–4 AU, and matches a power law. Note that incompleteness is likely for $q \lesssim 0.6$. The shaded bins represent the eight systems with ages <10 m.y. While the statistics are still small, the mass ratio distribution of these young systems suggests a flatter distribution than that of field and older cluster binaries.

A second effect is the preferential discovery of unresolved equal-mass systems in widefield surveys. As such systems are twice as bright as their single counterparts, they are ~3× more likely to be found than single sources in a magnitude-limited survey. Systems with lower-mass ratios are not as overluminous and are less affected by this bias. *Burgasser et al.* (2003) examined this impact of this bias on a small sample of L- and T-dwarf binaries and found it to be significant only for $q \lesssim 0.6$. Hence, this bias cannot be responsible for the $q \sim 1$ peak.

VLM (field and open cluster) binaries therefore show a clear preference for equal-mass systems, in contrast to the majority of F–K stellar systems (Fig. 1). It is worth noting that M dwarfs in the 8-pc sample show a similar, although less pronounced, $q \sim 1$ peak (*Reid and Gizis,* 1997a), again suggesting a mass-dependent trend.

2.2.4. Higher-order multiples. Thus far we have focused on VLM binaries, but higher-order multiples (triples, quadruples, etc.) are also abundant among more massive stars, comprising perhaps 15–25% of all multiple stellar systems (*Tokovinin,* 2004; see chapter by Duchêne et al.). Several VLM binaries are components of higher-order multiple systems with more massive stars. *Burgasser et al.* (2005a) have even suggested a higher binary fraction for brown dwarfs that are widely separated companions to massive stars. Higher-order multiples are currently rare among purely VLM systems, however. The LP 213-67/LP 213-68AB system is one exception, with the three components

(spectral types M6.5, M8, and L0) forming a wide hierarchical triple with separations of 340 AU and 2.8 AU (*Gizis et al.,* 2000a; *Close et al.,* 2003). DENIS 0205-1159AB may also have a third component, marginally resolved through high-resolution imaging (*Bouy et al.,* 2005). Considering both systems as VLM triples, the ratio of high-order multiples to binaries is only $3^{+4}_{-1}\%$, quite low in comparison to higher-mass stars. This may be due to selection effects, however, as the already tight separations of VLM binaries implies that the third component of a (stable) hierarchical triple must be squeezed into an extremely small orbit. Indeed, this could argue against a large fraction of higher-order VLM systems. On the other hand, undiscovered wide tertiaries (as in LP 213-67/LP 213-68AB) may be present around some of these systems. Additional observational work is needed to determine whether higher-order VLM multiples are truly less common than their stellar counterparts.

2.3. Statistical Analysis: Bayesian Modeling

To examine the observed binary properties of resolved VLM stars in more detail, we performed a Bayesian statistical analysis of imaging surveys to date. The Bayesian approach allows the incorporation of many disparate datasets, and the easy assimilation of nondetections, into a unified analysis of a single problem (*Sivia,* 1996). We focused our analysis on the surveys of *Koerner et al.* (1999), *Reid et al.* (2001), *Bouy et al.* (2003), *Close et al.* (2003), *Gizis et al.* (2003), *Siegler et al.* (2005), and Allen et al. (in preparation). The Bayesian statistical method employed is similar to that described in *Allen et al.* (2005).

We first constructed a set of parameterized companion distribution models in terms of orbital semimajor axis (a) and companion mass ratio. For the semimajor axis distribution we use a Gaussian in log AU given by

$$P(a_0, \sigma_a) = \frac{1}{\sqrt{2\sigma_a^2}} e^{-(\log(a) - \log(a_0))^2/2\sigma_a^2} \quad (1)$$

where a_0 is the peak of the Gaussian and σ_a is the logarithmic half-width, both variable parameters. This formulation is prompted by the results of *DM91* and *FM92* (however, see *Maxted and Jeffries,* 2005). For the mass-ratio model, we assume a power law of the form

$$P(N, \gamma) = N \frac{q^\gamma}{\int_0^1 q^\gamma} \quad (2)$$

where the normalization factor is defined to be the overall binary fraction (i.e., f_{bin}), and γ is a variable parameter.

In order to compare the model distributions to the data, we transform them to observables, namely the log of the projected separation (log Δ) and the difference in magnitude between the secondary and the primary (ΔM). The former is computed by transforming the semimajor axis distribution as

$$\Delta = a\sqrt{\cos^2(\phi)\sin^2(i) + \sin^2(\phi)} \qquad (3)$$

where we assume a uniform distribution of circular orbits over all possible inclinations (i) and phases (ϕ). The transformation of the q distribution to a ΔM distribution is done by assigning each mass ratio a range of possible luminosities for ages between 10 m.y. and 10 G.y. using evolutionary models from *Burrows et al.* (2001).

The transformed model distributions are then compared to the observed distributions via a Bayesian statistical method, as described in *Allen et al.* (2005). The models are directly compared to the data after being convolved with a window function, which describes how many times a bin in observational space (Δ, ΔM) was observed in a particular survey. In this way we do not analyze the models where there is no data, and the relative frequency of observations is taken into account.

The output posterior distribution is four-dimensional [log(a_0), σ_a, N, γ] and is impossible to display in its entirety. Instead, we show marginalized distributions (Fig. 4), collapsing the posterior distribution along different parameter axes. These distributions have a nonnegligible dispersion, as parameter spaces outside the observational window function (e.g., very tight binaries) add considerable uncertainty to the statistical model. Nevertheless, the distributions are well-behaved and enable us to derive best-fit values and uncertainties for the various parameters. The overall binary fraction is reasonably well constrained, $N = 22^{+8}_{-4}\%$, with a long tail in its probability distribution to higher rates. The remaining parameters are log(a_0) = $0.86^{+0.06}_{-0.18}$ log(AU), $\sigma_a =$

$0.24^{+0.08}_{-0.06}$ log(AU), and $\gamma = 4.8^{+1.4}_{-1.6}$ (all listed uncertainties are 68% confidence level).

The mass ratio and projected separation distributions inferred from the best-fit parameters are shown in Fig. 5. The best-fit binary fraction is 22%, but after applying our window function the expected resolved fraction is ~17%, slightly higher than but consistent with the observed f_{bin} from imaging surveys (section 2.2.1). The best-fit mass ratio distribution is highly peaked near q = 1, similar to the data but somewhat flatter than observed due to selection effects in the empirical samples. This nevertheless confirms that

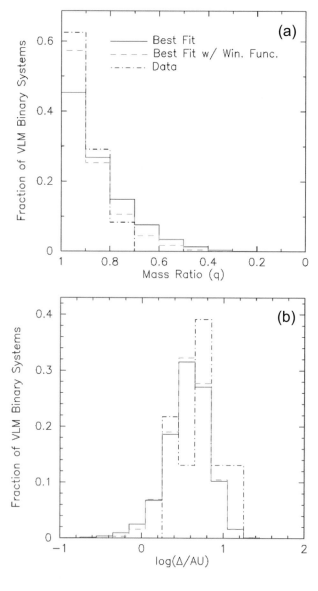

Fig. 5. **(a)** The fraction of VLM binaries with a given q for the best fit model (solid line), the best fit model view through the window function (dashed line), and the data used in the Bayesian analysis (dot-dashed line). Note how the window function overemphasizes the high mass ratio systems. **(b)** The projected separation distribution for the best fit model (lines are the same as the top panel).

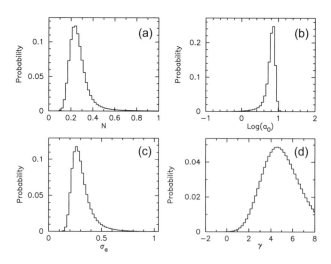

Fig. 4. Posterior probability distributions of the four companion model parameters: **(a)** overall binary fraction (N); **(b)** center of the semimajor axis distribution [log(a_0)]; **(c)** width of the semimajor axis distribution (σ_a); **(d)** mass ratio distribution power law index (γ).

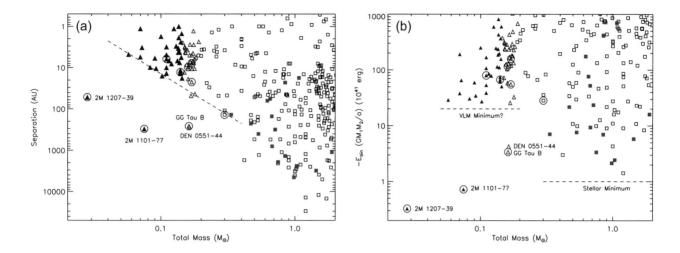

Fig. 6. **(a)** Separation (in AU) vs. total system mass (in $M_\odot$) for known binary systems. Stellar binaries from *Close et al.* (1990); *DM91*; *FM92*; *Reid and Gizis* (1997b); and *Tokovinin* (1997) are shown as open squares; stellar-brown dwarf systems compiled by *Reid et al.* (2001) are shown as filled squares. The 68 binaries from Table 1 with measured projected separations and estimated masses are plotted as triangles; filled triangles indicate substellar primaries. Systems younger than 10 m.y. are encircled. The dotted line indicates the maximum separation/system mass relation for VLM stellar and substellar binaries proposed by *Burgasser et al.* (2003), indicating that lower-mass systems are more tightly bound (see also *Close et al.*, 2003). However, three young systems (GG Tau BaBb, 2MASS 1101-7732AB, and 2MASS 1207-3932AB), and the field binary DENIS 0551-4434AB, all appear to contradict these trends. **(b)** Same systems but this time comparing binding energy ($-E_{bind} = GM_1M_2/a$) to total system mass. As first pointed out in *Close et al.* (2003), the widest VLM field binaries are 10–20× more tightly bound than the widest stellar binaries, with the singular exception of DENIS 0551-4434AB. On the other hand, the three young VLM systems GG Tau BaBb, 2MASS 1101-7732AB, and 2MASS 1207-3932AB are much more weakly bound.

the mass ratio distribution is fundamentally peaked toward high q values.

The best-fit value for the peak of the semimajor axis distribution is ~7 AU, implying a peak in the projected separation distribution of about 3.5 AU, matching well with the data (Fig. 5b). The best-fit width of this distribution is quite narrow, implying very few wide systems (>20 AU ~ 1%) and very few close systems (<1 AU ~ 2–3%). It is important to stress that the imaging data provide weak constraints on closely separated binaries, and the latter fraction may be somewhat higher (cf. *Maxted and Jeffries*, 2005). On the other hand, the constraint on the wide binary fraction (1% or less) is the most robust result of this analysis. Between all the surveys considered here there are over 250 unique fields that have been probed for companions out to hundreds of AU with no detections. Hence, such pairings are exceptionally rare.

2.4. Discussion

2.4.1. On the preference of tight binaries. The sharp decline in the VLM binary fraction for $\Delta > 20$ AU is not a feature shared with more massive stellar systems, which can extend from 0.1 AU to 0.1 pc. However, the decline is consistent with the observed trend of smaller mean separations, and smaller maximum separations, from A to M field binaries. This is demonstrated in Fig. 6, which plots projected separations/semimajor axes vs. total system mass for stellar

and substellar field and cluster binaries. The maximum separations (Δ_{max}) of these systems show a striking dependence on total system mass. Prior to the discovery of the wide pairs 2MASS 1101-7732AB and DENIS 0551-4434AB, *Burgasser et al.* (2003) found a power-law relation between Δ_{max} and total system mass, $\Delta_{max} = 1400(M_{tot}/M_\odot)^2$ AU, that appeared to fit all VLM systems known at that time. Similarly, *Close et al.* (2003) found a linear relation of $\Delta_{max} = 23.2(M_{tot}/0.185\ M_\odot)$ AU for VLM binaries, corresponding to a minimum escape velocity $V_{esc} = 3.8$ km s^{-1}. This was greater than a minimum value of $V_{esc} = 0.6$ km s^{-1} inferred for more massive stellar systems, and both results indicate that lower-mass binaries are progressively more tightly bound. *Close et al.* (2003) further pointed out a possible "break" in the minimum binding energies of stellar and VLM binaries, also shown in Fig. 6. Around $M_{tot} \approx 0.3\ M_\odot$, the majority of wide VLM systems appear to be 10–20× more strongly bound than the widest stellar systems.

More recently, exceptions to the empirical trends shown in Fig. 6 have been identified, including the widely separated VLM systems 2MASS 1101-7732AB, DENIS 0551-4434AB, and GG Tau BaBb. In addition, the extremely low mass ($M_{tot} \approx 0.03\ M_\odot$) brown dwarf pair 2MASS J12073346-3932549AB (hereafter 2MASS 1207-3932AB) (*Chauvin et al.*, 2004, 2005a), identified in the 8-m.y. TW Hydrae moving group (*Gizis*, 2002), falls well outside the mass/Δ_{max} limits outlined above. Such "exceptions" have called into question whether current empirical separation limits are representative of VLM systems in general, and can be considered robust constraints for formation models, or if wide binaries are a normal (if rare) mode of VLM binary formation. These questions remain under debate.

2.4.2. Do evolution or environment play a role in VLM binary properties? That three of the four weakly bound VLM systems are in young (≤10 m.y.), low-density associations may be an important clue to their formation and existence, and encourages a closer examination of the multiplicity properties of such objects in general. The shaded histograms in Figs. 2 and 3 delineate the separation and mass ratio distributions, respectively, of the eight binaries in Table 1 that are members of clusters or associations younger than 10 m.y. These distributions, although based on small number statistics, are nevertheless compelling. Young systems show a much broader range of separations, spanning $0.04 \leq \Delta \leq 240$ AU, with $25^{+19}_{-9}\%$ (2/8) having $\Delta > 20$ AU (as compared to $5^{+4}_{-1}\%$ of older VLM systems). The mass ratio distribution is also quite flat, with a *statistically significant* shortfall in the relative number of $q \geq 0.8$ binaries ($25^{+19}_{-9}\%$ vs. $81^{+4}_{-6}\%$). Assuming that the older field sources predominately originate from young clusters (*Lada and Lada,* 2003), these differences suggest an *evolution* of VLM binary properties over a timescale of 5–10 m.y.

However, care must be taken when interpreting these data, as selection effects can distort the underlying distributions. Because the youngest brown dwarfs are still quite warm and luminous, imaging surveys in young clusters can generally probe much smaller masses — and hence smaller mass ratios — than equivalent surveys of older clusters or in the field. In addition, with the exception of some nearby moving groups (e.g., TW Hydra, Ursa Major), most of the youngest clusters lie at larger distances, so closely separated systems ($\Delta \leq 10$ AU) cannot be generally resolved through direct imaging. This biases young samples against the close separations typical of field binaries. So in fact there may be many more closely separated young VLM pairs, or many more widely separated, small-mass-ratio older VLM pairs, than currently known.

What about older VLM members of the galactic thick disk and halo? Unfortunately, current imaging searches for companions to low-mass subdwarfs are not yet capable of detecting substellar companions directly, and radial velocity surveys of the necessary frequency are not yet complete. *Gizis and Reid* (2000) imaged nine VLM metal-poor (M subdwarf) primaries with HST, and found that none had companions down to the hydrogen-burning limit. This sample has been extended to a total of 28 M subdwarfs within 60 pc, but all appear single (Riaz and Gizis, in preparation). Taken at face value, this result ($f_{bin} < 6\%$) suggests that halo VLM doubles with separations in the range 5–100 AU are rarer than those in the disk population. However, given the danger of unknown selection biases, the possibility of metallicity effects, and the still small numbers of the empirical sample, this result should be taken with caution.

The current data also support the possibility that environment may play a role in the multiple properties of VLM

systems. The three young, widely separated binaries discussed above all reside in loose associations that have average stellar densities of 0.01–1 pc⁻³ (e.g., *Luhman,* 2004; *Mamajek,* 2005), too low for stellar encounters to have a significant disruptive effect (*Weinberg et al.,* 1987). This is in contrast to high-density star-formation regions such as Orion, where average densities of 10^4 pc⁻³ (*Hillenbrand,* 1997) are sufficient for stellar encounters to disrupt ~10 AU VLM binaries over a ~10-m.y. timescale (*Weinberg et al.,* 1987; *Burgasser et al.,* 2003). The influence of stellar density has been cited for observed differences in multiplicity among solar-mass stars in various clusters (e.g., *Ghez et al.,* 1993; *Scally et al.,* 1999; *Patience and Duchêne,* 2001; *Lada and Lada,* 2003; see also chapter by Duchêne et al.), so differences among VLM binaries should not be surprising. This scenario can also explain the paucity of wide binaries in the field. Dense embedded clusters, in which wide binaries can be easily disrupted (cf. *Kroupa,* 1995a,b,c) contribute perhaps 70–80% of the stars in the galaxy (*Lada and Lada,* 2003). The few wide systems created in less-dense clusters or associations would therefore comprise a negligible fraction of all VLM binaries in the field (cf. *Kroupa and Bouvier,* 2003). This scenario is compelling, but requires better statistics to be tested sufficiently.

3. CONFRONTING THE MODELS

With a full analysis of the empirical properties of VLM multiple systems in hand, we now examine how the predictions of current star and brown dwarf formation theories compare. Detailed discussion on the current modeling efforts are provided in the chapters of Ballesteros-Paredes et al., Goodwin et al., Klein et al., and Whitworth et al. Comparison of formation theories with the general properties (mass function, disk fraction, etc.) of low-mass stars and brown dwarfs are provided in the chapters of Duchêne et al. and Luhman et al. Here we focus primarily on the predictions for VLM multiplicity.

3.1. Fragmentation and Dynamical Evolution

Undoubtedly, gravitational contraction of dense cores in molecular clouds provides the fundamental building blocks for stellar and substellar objects. However, the details of the contraction and subsequent evolution of the cores remain critical details under considerable debate. This is particularly the case for VLM star and brown dwarfs whose masses are significantly below the Jeans mass (~1 $M_\odot$), and as such cannot be formed efficiently through basic contraction scenarios (e.g., *Shu et al.,* 1987). The inclusion of additional physics, such as magnetic field effects (*Boss,* 2001, 2002, 2004) and turbulent fragmentation, has brought some resolution to this problem, and has enabled a new generation of VLM formation models.

Turbulent fragmentation (*Henriksen,* 1986, 1991; *Larson,* 1992; *Elmegreen,* 1997, 1999, 2000), in which gas flows collide, are compressed, and form gravitationally unstable clumps, has pushed the fragmentation mass limit down to the effective opacity limit, on the order of 0.01 M$_\odot$ (*Bate et al.,* 2002). *Boyd and Whitworth* (2005) have modeled the turbulent fragmentation of two-dimensional sheets and found a protostellar mass distribution that extends to ~0.003 M$_\odot$. *Padoan and Nordlund* (2004) and *Padoan et al.* (2005) have studied three-dimensional turbulent fragmentation of a molecular cloud using an adaptive mesh refinement code, and are also capable of producing cores as small as ~0.003 M$_\odot$. In these studies, no predictions are made on the overall multiplicity of the protostars. However, fragmented cores naturally lead to the creation of gravitationally bound, high-order multiple systems, as confirmed in multiplicity studies of Class 0 and I protostars (*Haisch et al.,* 2002, 2004; *Reipurth et al.,* 2002, 2004; *Duchêne et al.,* 2004; see chapter by Duchêne et al.), and therefore provide a natural framework for the creation of VLM multiple systems.

However, it is well known that N-body groups are generally dynamically unstable, and dynamical scattering will dissolve such systems in a few crossing times (~10^5 yr), preferentially removing the lowest-mass members (e.g., *Kroupa et al.,* 1999). The scattering of low-mass bodies will also limit the accretion of gas and dust onto initially substellar cores, which would otherwise build up to stellar masses. These ideas have led to the so-called "ejection" model for brown dwarf formation (*Reipurth and Clarke,* 2001), in which brown dwarfs (and presumably VLM stars) are simply stellar embryos ejected from their nascent cloud. This model has received a great deal of attention recently, as its qualitative multiplicity predictions — a small fraction of multiples and a preference for strongly bound binaries (close separations and near-unity mass ratios) — appear to fall in line with observational results.

The most comprehensive simulations of this scenario, incorporating both smoothed particle hydrodynamics (SPH) modeling for fragmentation and accretion and N-body simulations for dynamic interactions, have been produced by M. Bate and collaborators (*Bate et al.,* 2002, 2003; *Bate and Bonnell,* 2005), and are described in detail in the chapter by Whitworth et al. Their original simulation of a 50-M$_\odot$ cloud produced only one brown dwarf-brown dwarf binary system, still accreting and dynamically unstable at the end of the simulation, implying a VLM binary fraction of ≲5%. It was immediately recognized that this fraction may be too low when compared to observations (*Close et al.,* 2003). Later simulations (*Bate and Bonnell,* 2005) found that higher VLM binary fractions (up to 8%) were possible in denser clouds. The highest-density simulation also produced stable wide (>60 AU) VLM binary systems when low-mass cores were ejected in the same direction and became bound. It is important to note that the two wide young VLM systems currently known are, on the contrary, associated with *low-density* associations.

While the *Bate et al.* (2002, 2003) simulations have provided a great leap forward in the modeling of low-mass star formation, their relevance to the observed properties of VLM binaries are hindered by necessary computational approximations. First, sink particles encompassing all bound gas within 5 AU are used when densities exceed 10^{-11} g cm^{-3}. This approximation rules out any binaries more closely separated than this limit, encompassing a majority of VLM systems (see section 2.2.2). Second, a softened Newtonian potential is employed below separations of 5 AU (down to 1 AU), which enhances the disruption of binary pairs with smaller separations (*Delgado-Donate et al.,* 2004). Again, as the peak of the observed VLM binary separation distribution falls within this range, it is possible that the Bate et al. simulations underpredict the number of VLM binary systems. Because the simulations are computationally expensive, only one simulation is undertaken for a given set of initial conditions, resulting in poor statistics. In addition, the simulations are allowed to run for a limited time (~0.3 m.y.), so long-term evolution of unstable multiples is left unresolved.

More recent SPH + N-body simulations have attempted to tackle these issues by reducing the scale of the simulation. Studies by *Delgado-Donate et al.* (2004) and *Goodwin et al.* (2004a,b) have focused on smaller clouds (~5 M$_\odot$) and have performed multiple simulations to improve statistical results. The Delgado-Donate et al. simulations were based on the same format as the *Bate et al.* (2002, 2003) work and proceeded in two steps: An SPH + N-body simulation of the gas and sink particles was first conducted for ~0.5 m.y., followed by an N-body simulation of the resulting protostellar cores for a subsequent 10 m.y. This allows an examination of both early fragmentation and accretion on the formation and disruption of bound systems, and the dynamical relaxation of high-order multiples over time. While brown dwarfs were frequently found in multiple systems containing more massive stellar components, particularly at early times (~1 m.y.), none of the simulations produced purely VLM binaries, again indicating a disagreement between theory (or at least the modeling of the theory) and observations. A strong trend of binary fraction with primary mass is found, although this trend is perhaps too strong (underestimating VLM multiplicity and overestimating stellar multiplicity). The SPH simulations of *Goodwin et al.* (2004a,b), which tested variations of the cloud's initial turbulent energy spectrum, also failed to produce any VLM binaries within 0.3 m.y. In retrospect, both sets of simulations may be hindered by their use of 5-AU sink particles and softened Newtonian potentials, and both groups intend to address these limitations (M. Bate, personal communication).

Pure N-body simulations have focused on the dynamical evolution of small-N clusters of protostars, and (because they are less computationally intensive) have generally produced more robust statistical predictions for VLM multiples than SPH simulations. *Sterzik and Durisen* (2003) simulated the dynamical interactions of closely separated, small-N clusters and were able to broadly reproduce the empirical trends, including an increasing binary fraction and median separation with increasing primary mass (cf. Fig. 6), a brown dwarf binary fraction of ~10%, and a median VLM binary separation of 3 AU. *Umbreit et al.* (2005) studied the decay of widely separated accreting triple systems (incorporating momentum transfer with N-body dynamics) and found that VLM systems hundreds of AU apart were efficiently hardened to a distribution of that peaks at 3 AU, with a long tail to wider separations. These simulations predict few very tight brown dwarf binary systems, although this may be because dissipative forces were not included. One drawback to both of these studies is that they do not take into account interactions with the larger star-forming environment, which appear to be important in the SPH simulations of, e.g., *Bate et al.* (2002, 2003). There are plans to study these effects in detail (S. Umbreit, personal communication).

In short, dynamical simulations appear to reproduce many of the observed properties of VLM binaries, both in terms of quantitative results (binary fraction and separation distribution) and overall trends (mass dependence on binary fraction and mean separations). SPH + N-body simulations, on the other hand, generally underpredict the number of VLM binaries, and the lack of statistics makes the assessment of other multiple properties difficult to verify. The shortcomings of SPH simulations are likely related to the use of large sink particles. Decreasing the size of these sink particles, and the imposed smooth potential for close interactions, should be a priority.

3.2. Other Formation Mechanisms

For completeness, we briefly touch upon two other modes of star formation that may be relevant to the creation of VLM multiples. Disk fragmentation can occur when gravitational instabilities in massive circumstellar disks form, either through dynamical interactions with a passing bare star or another disk, or spontaneously through tidal or spiral instabilities. Most disk fragmentation simulations use SPH codes to test the outcomes of different encounter geometries, with results depending largely on the alignment of the angular momentum axes of the interacting pair. The simulations of *Lin et al.* (1998), *Watkins et al.* (1998a,b), and *Boss* (2000) have all successfully produced substellar mass objects through this process [the simulations of *Bate et al.* (2002, 2003) have also produced protostellar cores through disk interactions]. However, the disk mass necessary to produce such objects is nearly 0.1 $M_\odot$, and is hence unlikely around a VLM primary. Therefore, while the disk fragmentation scenario appears quite capable of producing single brown dwarfs from disks around massive stars, it does little to explain the production of VLM binary systems.

Another VLM formation mechanism recently explored by *Whitworth and Zinnecker* (2004) is photoevaporation. This process occurs when a substantial prestellar core (a few 0.1 $M_\odot$) is compressed and stripped by the ionizing radiation front of a nearby massive O or B star. *Whitworth and Zinnecker* (2004) do not discuss binary formation explicitly, but is possible in principle if the initial core was frag-

mented. This scenario also requires the presence of massive young stars, making it appropriate for high-mass star-formation environments such as Orion, but not for low-mass environments such as Taurus or Cha I. Therefore, photoevaporation cannot be a universal mechanism for VLM multiple formation.

4. FUTURE OBSERVATIONAL DIRECTIONS

Despite the large assemblage of VLM binaries now in place (Table 1), it should be clear that the search for VLM binaries should continue, particularly by broadening the multiplicity parameter space sampled. As such, search efforts should focus on low-mass ratio systems (q ≤ 0.5), particularly in the field; very tight systems (Δ ≤ 3 AU); and very wide systems (Δ ≥ 150 AU) with moderate to low mass ratios (q ≤ 0.8).

High-resolution imaging will remain an important tool in the discovery and characterization of VLM binaries, particularly with the implementation of laser guide star (LGS) AO systems on 5–10-m-class telescopes (e.g., Palomar, Keck, VLT). LGS AO greatly increases the number of VLM systems that are accessible from the ground. Groundbased AO enables the examination of larger samples in the field and in nearby moving groups and young star-forming regions, as well as the ability to astrometrically monitor systems on decadal time periods, long after HST is decommissioned. Future studies combining AO imaging with spectroscopy will permit refined characterization of VLM binary components; note that most of the systems listed in Table 1 lack resolved spectroscopy. AO plus coronagraphy, the latter used successfully to identify several VLM companions to nearby, more massive stars (e.g., *Oppenheimer et al.*, 2001; *Lowrance et al.*, 2005) will facilitate the detection of low-mass-ratio systems around VLM primaries, probing well into the so-called "planetary mass" regime.

For the tightest binaries, high-resolution spectroscopy remains an important tool for search and characterization. Efforts thus far have been largely conducted at optical wavelengths. While suitable for young brown dwarfs with M spectral types, optical spectroscopy becomes increasingly limited for L dwarfs, T dwarfs, and cooler objects that are extremely faint at these wavelengths. Hence, future studies should focus their efforts using high-throughput, high-resolution infrared spectrographs (e.g., *Simon and Prato*, 2004). Short- and long-term spectroscopic monitoring campaigns of VLM samples should be pursued to identify sufficiently complete samples and to determine systemic properties. Of the few RV variable VLM binary candidates identified to date (*Guenther and Wuchterl*, 2003; *Kenyon et al.*, 2005; *Joergens*, 2006), most have only two to four epochs of observation, and parameters such as separation, mass ratio, etc., remain largely unknown. Combining astrometric monitoring with spectroscopic monitoring for closely separated resolved systems (e.g., Gliese 569Bab) (*Zapatero Osorio et al.*, 2004) will permit precise orbital solutions, leading to component mass and semimajor axis measurements, and enabling the examination of other multiplicity properties such as eccentricity distributions and spin/orbit angular momentum alignment.

Tight binaries can also be probed by searches for eclipsing systems. For substellar objects, this is a particularly powerful technique, as the near-constancy of evolved (i.e., field) brown dwarf radii over a broad range of masses (*Burrows and Liebert*, 1993) implies that eclipse depths for edge-on geometries depend only on the relative fluxes of the components, while grazing transits can span a larger range of inclinations for a given separation. To date, only one eclipsing substellar system has been identified in the ~1-m.y. ONC, 2MASS J0535218-054608 (*Stassun et al.*, 2006). To the best of our knowledge no large surveys for eclipsing field VLM binaries have been undertaken. While eclipsing systems will likely be rare, the success and scientific yield of transiting extrasolar planet searches (e.g., *Charbonnaeu et al.*, 2000) should inspire dedicated programs in this direction.

Interferometric observations can also probe tighter binaries than direct imaging, encouraging studies in this direction. Current facilities (e.g., Palomar, Keck, VLT) are limited in sensitivity, however; only the closest mid-type M dwarfs have been observed thus far (*Lane et al.*, 2001a; *Segransan et al.*, 2003). Increasing the throughput of these systems, or making use of future spacebased facilities (e.g., SIM, TPF-I), may eventually make interferometry a viable observational method in the VLM regíme.

For widely separated VLM companions, the most extensive limits to date arise from the shallow, widefield surveys from which most of these objects were identified (e.g., 2MASS, DENIS, and SDSS). Only a few dedicated widefield programs are now underway (*Billères et al.*, 2005; Allen et al., in preparation). Deep, but not necessarily high-resolution, imaging surveys around large samples of VLM primaries would provide better constraints on the frequency and properties of such systems. Such surveys will benefit from proper motion analysis and component spectroscopy, allowing bona fide systems to be extracted from the vast number of unrelated projected doubles. Searches for wide companions to young nearby stars have identified a few VLM objects (e.g., *Chauvin et al.*, 2005b; *Neuhäuser et al.*, 2005), and the case of 2MASS 1207-3934AB proves that widely separated low-mass companions can exist around VLM primaries. Future searches for equivalent systems, particularly in the field, will test the veracity of the apparent wide-separation desert.

Finally, careful selection of binary search samples should be of high priority. Current imaging and spectroscopic field samples are largely based on compilations from magnitude-limited surveys, and are therefore inherently biased. The examination of *volume-limited* VLM samples (e.g., *Cruz et al.*, 2003) is necessary to eliminate these biases. Similarly, many cluster binary surveys fail to concurrently verify cluster membership, leading to contamination issues (e.g., CFHT-Pl-18) (*Martín et al.*, 1998, 2000a). Studies have begun to address this (e.g., *Luhman*, 2004), but more work

is needed. Finally, given the suggestion of age and/or environmental effects in binary properties, comparison of large, complete samples for several clusters of different ages will probe the origins of multiplicity properties and over what timescales VLM binaries evolve.

Acknowledgments. This review would not have been possible without helpful input from several researchers that have contributed to the study of VLM stars and brown dwarf binaries. We acknowledge specific discussions with G. Basri, I. Bonnell, A. Boss, G. Duchêne, T. Forveille, R. Jeffries, V. Joergens, C. Lada, M. Liu, K. Luhman, E. Martín, E. Mamajek, S. Metchev, K. Noll, P. Padoan, D. Pinfield, B. Reipurth, K. Stassun, M. Sterzik, S. Umbreit, R. White, and A. Whitworth. We also thank our anonymous referee for her/his insightful criticisms. Support for this work has been provided by NASA through funding for Hubble Space Telescope program GO-10559.

REFERENCES

Abt H. A. (1978) In *Protostars and Planets* (T. Gehrels, ed.), pp. 323–355. Univ. of Arizona, Tucson.
Abt H. A. and Levy S. G. (1976) *Astrophys. J. Suppl., 30,* 273–306.
Allen P. R., Koerner D. W., Reid I. N., and Trilling D. E. (2005) *Astrophys. J., 625,* 385–397.
Ardila D., Martín E., and Basri G. (2000) *Astron. J., 120,* 479–487.
Basri G. (2000) *Ann. Rev. Astron. Astrophys., 38,* 485–519.
Basri G. and Martín E. (1999) *Astron. J., 118,* 2460–2465.
Bate M. R. and Bonnell I. A. (2005) *Mon. Not. R. Astron. Soc., 356,* 1201–1221.
Bate M. R., Bonnell I. A., and Bromm V. (2002) *Mon. Not. R. Astron. Soc., 332,* L65–L68.
Bate M. R., Bonnell I. A., and Bromm V. (2003) *Mon. Not. R. Astron. Soc., 339,* 577–599.
Billères M., Delfosse X., Beuzit J.-L., Forveille T., Marchal L., et al. (2005) *Astron. Astrophys., 440,* L55–L58.
Boss A. P. (2000) *Astrophys. J., 536,* L101–L104.
Boss A. P. (2001) *Astrophys. J., 551,* L167–L170.
Boss A. P. (2002) *Astrophys. J., 568,* 743–753.
Boss A. P. (2004) *Mon. Not. R. Astron. Soc., 350,* L57–L60.
Bouy H., Brandner W., Martín E. L., Delfosse X., Allard F., et al. (2003) *Astron. J., 126,* 1526–1554.
Bouy H., Brandner W., Martín E. L., Delfosse X., Allard F., et al. (2004a) *Astron. Astrophys., 423,* 341–352.
Bouy H., Brandner W., Martín E. L., Delfosse X., Allard F., et al. (2004b) *Astron. Astrophys., 424,* 213–226.
Bouy H., Martín E. L., Brandner W., and Bouvier J. (2005) *Astron. J., 129,* 511–517.
Bouy H., Moraux E., Bouvier J., Brandner W., Martín E. L., et al. (2006) *Astrophys. J., 637,* 1056–1066.
Boyd D. F. A. and Whitworth A. P. (2005) *Astron. Astrophys., 430,* 1059–1066.
Brandner W., Martín E. L., Bouy H., Köhler R., Delfosse X., et al. (2004) *Astron. Astrophys., 428,* 205–208.
Briceño C., Hartmann L., Stauffer J., and Martín E. (1998) *Astron. J., 115,* 2074–2091.
Burgasser A. J. and McElwain M. W. (2006) *Astron. J., 131,* 1007–1014.
Burgasser A. J., Kirkpatrick J. D., Brown M. E., Reid I. N., Gizis J. E., et al. (1999) *Astrophys. J., 522,* L65–L68.
Burgasser A. J., Kirkpatrick J. D., Brown M. E., Reid I. N., Burrows A., et al. (2002) *Astrophys. J., 564,* 421–451.
Burgasser A. J., Kirkpatrick J. D., Reid I. N., Brown M. E., Miskey C. L., et al. (2003) *Astrophys. J., 586,* 512–526.
Burgasser A. J., Kirkpatrick J. D., and Lowrance P. J. (2005a) *Astron. J., 129,* 2849–2855.
Burgasser A. J., Reid I. N., Leggett S. K., Kirkpatrick J. D., Liebert J., et al. (2005b) *Astrophys. J., 634,* L177–L180.

Burrows A. and Liebert J. (1993) *Rev. Mod. Phys., 65,* 301–336.

Burrows A., Hubbard W. B., Lunine J. I., and Liebert J. (2001) *Rev. Mod. Phys., 73,* 719–765.

Casey B. W., Mathieu R. D., Vaz L. P. R., Andersen J., and Suntzeff N. B. (1998) *Astron. J., 115,* 1617–1633.

Chabrier G., Baraffe I., Allard F., and Hauschildt P. (2000) *Astrophys. J., 542,* 464–472.

Chappelle R. J., Pinfield D. J., Steele I. A., Dobbie P. D., and Magazzú A. (2005) *Mon. Not. R. Astron. Soc., 361,* 1323–1336.

Charbonneau D., Brown T. M., Latham D. W., and Mayor M. (2000) *Astrophys. J., 529,* L45–L48.

Chauvin G., Lagrange A.-M., Dumas C., Zuckerman B., Mouillet D., et al. (2004) *Astron. Astrophys., 425,* L29–L32.

Chauvin G., Lagrange A.-M., Dumas C., Zuckerman B., Mouillet D., et al. (2005a) *Astron. Astrophys., 438,* L25–L28.

Chauvin G., Lagrange A.-M., Zuckerman B., Dumas C., Mouillet D., et al. (2005b) *Astron. Astrophys., 438,* L29–L32.

Close L. M., Richer H. B., and Crabtree D. R. (1990) *Astron. J., 100,* 1968–1980.

Close L. M., Siegler N., Freed M., and Biller B. (2003) *Astrophys. J., 587,* 407–422.

Close L. M., Siegler N., Potter D., Brandner W., and Liebert J. (2002) *Astrophys. J., 567,* L53–L57.

Cruz K. L., Reid I. N., Liebert J., Kirkpatrick J. D., and Lowrance P. J. (2003) *Astron. J., 126,* 2421–2448.

Dahn C. C., Liebert J., and Harrington R. S. (1986) *Astron. J., 91,* 621–625.

Dahn C. C., Harris H. C., Vrba F. J., Guetter H. H., Canzian B., et al. (2002) *Astron. J., 124,* 1170–1189.

de Zeeuw P. T., Hoogerwerf R., de Bruijne J. H. J., Brown A. G. A., and Blaauw A. (1999) *Astron. J., 117,* 354–399.

Delfosse X., Tinney C. G., Forveille T., Epchtein N., Bertin E., et al. (1997) *Astron. Astrophys., 327,* L25–L28.

Delfosse X., Beuzit J.-L., Marchal L., Bonfils X. C., Perrier C., et al. (2004) In *Spectroscopically and Spatially Resolving the Components of the Close Binary Stars* (R. W. Hilditch et al.), pp. 166–174. ASP Conf. Series 318, San Francisco.

Delgado-Donate D. J., Clarke C. J., Bate M. R., and Hodgkin S. T. (2004) *Mon. Not. R. Astron. Soc., 351,* 617–629.

Duchêne G., Bouvier J., Bontemp S., André P., and Motte F. (2004) *Astron. Astrophys., 427,* 651–665.

Duquennoy A. and Mayor M. (1991) *Astron. Astrophys., 248,* 485–524.

Elmegreen B. G. (1997) *Astrophys. J., 486,* 944–954.

Elmegreen B. G. (1999) *Astrophys. J., 527,* 266–284.

Elmegreen B. G. (2000) *Astrophys. J., 530,* 277–281.

Fischer D. A. and Marcy G. W. (1992) *Astrophys. J., 396,* 178–194.

Forveille T., Beuzit J.-L., Delorme P., Ségransan D., Delfosse X., et al. (2005) *Astron. Astrophys., 435,* L5–L9.

Freed M., Close L., and Siegler N. (2003) *Astrophys. J., 584,* 453–458.

Geballe T. R., Knapp G. R., Leggett S. K., Fan X., Golimowski D. A., et al. (2002) *Astrophys. J., 564,* 466–481.

Ghez A. M., Neugebauer G., and Matthews K. (1993) *Astron. J., 106,* 2005–2023.

Gizis J. E. (2002) *Astrophys. J., 575,* 484–492.

Gizis J. E. and Reid I. N. (2000) *Publ. Astron. Soc. Pac., 112,* 610–613.

Gizis J. E., Monet D. G., Reid I. N., Kirkpatrick J. D., and Burgasser A. J. (2000a) *Mon. Not. R. Astron. Soc., 311,* 385–388.

Gizis J. E., Monet D. G., Reid I. N., Kirkpatrick J. D., Liebert J., et al. (2000b) *Astron. J., 120,* 1085–1099.

Gizis J. E., Kirkpatrick J. D., Burgasser A. J., Reid I. N., Monet D. G., et al. (2001) *Astrophys. J., 551,* L163–L166.

Gizis J. E., Reid I. N., Knapp G. R., Liebert J., Kirkpatrick J. D., et al. (2003) *Astron. J., 125,* 3302–3310.

Goldberg D., Mazeh T., and Latham D. W. (2003) *Astrophys. J., 591,* 397–405.

Golimowski D. A., Henry T. J., Krist J. E., Dieterich S., Ford H. C., et al. (2004) *Astron. J., 128,* 1733–1747.

Goodwin S. P., Whitworth A. P., and Ward-Thompson D. (2004a) *Astron. Astrophys., 414,* 633–650.

Goodwin S. P., Whitworth A. P., and Ward-Thompson D. (2004b) *Astron. Astrophys., 423,* 169–182.

Goto M., Kobayashi N., Terada H., Gaessler W., Kanzawa T., et al. (2002) *Astrophys. J., 567,* L59–L62.

Guenther E. W. and Wuchterl G. (2003) *Astron. Astrophys., 401,* 677–683

Guenther E. W., Paulson D. B., Cochran W. D., Patience J., Hatzes A. P., et al. (2005) *Astron. Astrophys., 442,* 1031–1039.

Haisch K. E. Jr., Barsony M., Greene T. P., and Ressler M. E. (2002) *Astron. J., 124,* 2841–2852.

Haisch K. E. Jr., Greene T. P., Barsony M., and Stahler S. W. (2004) *Astron. J., 127,* 1747–1754.

Halbwachs J. L., Mayor M., Udry S., and Arenou F. (2003) *Astron. Astrophys., 397,* 159–175.

Henriksen R. N. (1986) *Astrophys. J., 310,* 189–206.

Henriksen R. N. (1991) *Astrophys. J., 377,* 500–509.

Henry T. J. and McCarthy D. W. Jr. (1990) *Astrophys. J., 350,* 334–347.

Hillenbrand L. A. (1997) *Astron. J., 113,* 1733–1768.

Hinz J. L., McCarthy D. W., Simons D. A., Henry T. J., Kirkpatrick J. D., et al. (2002) *Astron. J., 123,* 2027–2032.

Joergens V. (2006) *Astron. Astrophys., 448,* 655–663

Joergens V. and Guenther E. W. (2001) *Astron. Astrophys., 379,* L9–L12.

Kendall T. R., Delfosse X., Martín E. L., and Forveille T. (2004) *Astron. Astrophys., 416,* L17–L20.

Kenworthy M., Hofmann K.-H., Close L., Hinz P., Mamajek E., et al. (2001) *Astrophys. J., 554,* L67–L70.

Kenyon M. J., Jeffries R. D., Naylor T., Oliveira J. M., and Maxted P. F. L. (2005) *Mon. Not. R. Astron. Soc., 356,* 89–106.

Kenyon S. J., Dobrzycka D., and Hartmann L. (1994) *Astron. J., 108,* 1872–1880.

Kirkpatrick J. D. (2005) *Ann. Rev. Astron. Astrophys., 43,* 195–245.

Kirkpatrick J. D., Reid I. N., Liebert J., Cutri R. M., Nelson B., et al. (1999) *Astrophys. J., 519,* 802–833.

Kirkpatrick J. D., Reid I. N., Liebert J., Gizis J. E., Burgasser A. J., et al. (2000) *Astron. J., 120,* 447–472.

Koerner D. W., Kirkpatrick J. D., McElwain M. W., and Bonaventura N. R. (1999) *Astrophys. J., 526,* L25–L28.

Kouwenhoven M. B. N., Brown A. G. A., Zinnecker H., Kaper L., and Portegies Zwart S. F. (2005) *Astron. Astrophys., 430,* 137–154.

Kraus A. L., White R. J., and Hillenbrand L. A. (2005) *Astrophys. J., 633,* 452–459.

Kroupa P. (1995a) *Mon. Not. R. Astron. Soc., 277,* 1491–1506.

Kroupa P. (1995b) *Mon. Not. R. Astron. Soc., 277,* 1507–1521.

Kroupa P. (1995c) *Mon. Not. R. Astron. Soc., 277,* 1522–1540.

Kroupa P. and Bouvier J. (2003) *Mon. Not. R. Astron. Soc., 346,* 343–353.

Kroupa P., Petr M. G., and McCaughrean M. J. (1999) *New Astron., 4,* 495–520.

Lada C. J. and Lada E. A. (2003) *Ann. Rev. Astron. Astrophys., 41,* 57–115.

Lane B. F., Boden A. F., and Kulkarni S. R. (2001a) *Astrophys. J., 551,* L81–L83.

Lane B. F., Zapatero Osorio M. R., Britton M. C., Martín E. L., and Kulkarni S. R. (2001b) *Astrophys. J., 560,* 390–399.

Larson R. B. (1992) *Mon. Not. R. Astron. Soc., 256,* 641–646.

Law N. M., Hodgkin S. T., and Mackay C. D. (2006) *Mon. Not. R. Astron. Soc., 368,* 1917–1924.

Leggett S. K., Golimowski D. A., Fan X., Geballe T. R., Knapp G. R., et al. (2002) *Astrophys. J., 564,* 452–465.

Leinert Ch., Haas M., Mundt R., Richichi A., and Zinnecker H. (1991) *Astron. Astrophys., 250,* 407–419.

Leinert Ch., Allard F., Richichi A., and Hauschildt P. H. (2000) *Astron. Astrophys., 353,* 691–706.

Leinert Ch., Jahreiss H., Woitas J., Zucker S., Mazeh T., et al. (2001) *Astron. Astrophys., 367,* 183–188.

Lépine S. and Shara M. M. (2005) *Astron. J., 129,* 1483–1522.

Lin D. N. C., Laughlin G., Bodenheimer P., and Rozyczka M. (1998) *Science, 281,* 2025–2027.

Liu M. C. and Leggett S. K. (2005) *Astrophys. J., 634,* 616–624.

Lowrance P. J., Becklin E. E., Schneider G., Kirkpatrick J. D., Weinberger A. J., et al. (2005) *Astron. J., 130*, 1845–1861.

Lucas P. W., Roche P. F., and Tamura M. (2005) *Mon. Not. R. Astron. Soc., 361*, 211–232.

Luhman K. L. (2004) *Astrophys. J., 614*, 398–403.

Luhman K. L., McLeod K. K., and Goldenson N. (2005) *Astrophys. J., 623*, 1141–1156.

Mamajek E. E. (2005) *Astrophys. J., 634*, 1385–1394.

Marcy G. W. and Butler R. P. (2000) *Publ. Astron. Soc. Pac., 112*, 137–140.

Martín E. L., Rebolo R., and Zapatero Osorio M. R. (1996) *Astrophys. J., 469*, 706.

Martín E. L., Basri G., Brandner W., Bouvier J., Zapatero Osorio M. R., et al. (1998) *Astrophys. J., 509*, L113–L116.

Martín E. L., Brandner W., and Basri G. (1999) *Science, 283*, 1718–1720.

Martín E. L., Brandner W., Bouvier J., Luhman K. L., Stauffer J., et al. (2000a) *Astrophys. J., 543*, 299–312.

Martín E. L., Koresko C. D., Kulkarni S. R., Lane B. F., and Wizinowich P. L. (2000b) *Astrophys. J., 529*, L37–L40.

Martín E. L., Barrado y Navascués D., Baraffe I., Bouy H., and Dahm S. (2003) *Astrophys. J., 594*, 525–532.

Maxted P. F. L. and Jeffries R. D. (2005) *Mon. Not. R. Astron. Soc., 362*, L45–L49.

Mayor M., Duquennoy A., Halbwachs J. L., and Mermilliod J. C. (1992) In *Complementary Approaches to Double and Multiple Star Research* (H. MacAlister and W. I. Hartkopf, eds.), pp. 73-81. ASP Conf. Series 32, San Francisco.

Mazeh T. and Goldberg D. (1992) *Astrophys. J., 394*, 592–598.

McCaughrean M., Close L. M., Scholz R.-D., Lenzen R., Biller B., et al. (2004) *Astron. Astrophys., 413*, 1029–1036.

McGovern M. R. (2005) Ph.D. thesis, Univ. of California, Los Angeles.

McLean I. S., McGovern M. R., Burgasser A. J., Kirkpatrick J. D., Prato L., et al. (2003) *Astrophys. J., 596*, 561–586.

Metchev S. A. (2005) Ph.D. thesis, California Institute of Technology, Pasadena.

Neuhäuser R., Brandner W., Alves J., Joergens V., and Comerón F. (2002) *Astron. Astrophys., 384*, 999–1011.

Neuhäuser R., Guenther E. W., Alves J., Huélamo N., Ott T., et al. (2003) *Astron. Nachr., 324*, 535–542.

Neuhäuser R., Guenther E. W., Wuchterl G., Mugrauer M., Bedalov A., et al. (2005) *Astron. Astrophys., 435*, L13–L16.

Oppenheimer B. R., Golimowski D. A., Kulkarni S. R., Matthews K., Nakajima T., et al. (2001) *Astron. J., 121*, 2189–2211.

Oppenheimer B. R., Kulkarni S. R. and Stauffer J. R. (2000) In *Protostars and Planets IV* (V. Mannings et al., eds.), pp. 1313–1338. Univ. of Arizona, Tucson.

Padoan P. and Nordlund A. (2004) *Astrophys. J., 617*, 559–564.

Padoan P., Kritsuk A., Norman M. L., and Nordlund A. (2005) *Mem. Soc. Astron. Ital., 76*, 187–192.

Patience J. and Duchêne G. (2001) In *The Formation of Binary Stars* (H. Zinnecker and R. D. Mathieu, eds.), p. 181. IAU Symposium 200, Kluwer, Dordrecht.

Percival S. M., Salaris M., and Groenewegen M. A. T. (2005) *Astron. Astrophys., 429*, 887–894.

Perryman M. A. C., Lindegren L., Kovalevsky J., Hoeg E., Bastian U., et al. (1997) *Astron. Astrophys., 323*, L49–L52.

Pinfield D. J., Dobbie P. D., Jameson R. F., Steele I. A., Jones H. R. A., et al. (2003) *Mon. Not. R. Astron. Soc., 342*, 1241–1259.

Potter D., Martín E. L., Cushing M. C., Baudoz P., Brandner W., et al. (2002) *Astrophys. J., 567*, L133–L136.

Reid I. N. and Gizis J. E. (1997a) *Astron. J., 113*, 2246–2269.

Reid I. N. and Gizis J. E. (1997b) *Astron. J., 114*, 1992–1998.

Reid I. N., Kirkpatrick J. D., Liebert J., Burrows A., Gizis J. E., et al. (1999) *Astrophys. J., 521*, 613–629.

Reid I. N., Gizis J. E., Kirkpatrick J. D., and Koerner D. W. (2001) *Astron. J., 121*, 489–502.

Reid I. N., Gizis J. E., and Hawley S. L. (2002a) *Astron. J., 124*, 2721–2738.

Reid I. N., Kirkpatrick J. D., Liebert J., Gizis J. E., Dahn C. C., et al. (2002b) *Astron. J., 124*, 519–540.

Reid I. N., Cruz K. L., Laurie S. P., Liebert J., Dahn C. C., et al. (2003) *Astron. J., 125*, 354–358.

Reid I. N., Lewitus E., Burgasser A. J., and Cruz K. L. (2006) *Astrophys. J., 639*, 1114–1119.

Reipurth B. and Clarke C. (2001) *Astron. J., 122*, 432–439.

Reipurth B., Rodríguez L. F., Anglada G., and Bally J. (2002) *Astron. J., 124*, 1045–1053.

Reipurth B., Rodríguez L. F., Anglada G., and Bally J. (2004) *Astron. J., 127*, 1736–1746.

Scally A., Clarke C., and McCaughrean M. J. (1999) *Mon. Not. R. Astron. Soc., 306*, 253–256.

Schmidt M. (1968) *Astrophys. J., 151*, 393–409.

Ségransan D., Kervella P., Forveille T., and Queloz D. (2003) *Astron. Astrophys., 397*, L5–L8.

Seifahrt A., Guenther E., and Neuhäuser R. (2005) *Astron. Astrophys., 440*, 967–972.

Shatsky N. and Tokovinin A. (2002) *Astron. Astrophys., 382*, 92–103.

Shu F. H., Adams F. C., and Lizano S. (1987) *Ann. Rev. Astron. Astrophys., 25*, 23–81.

Siegler N., Close L. M., Mamajek E. E., and Freed M. (2003) *Astrophys. J., 598*, 1265–1276.

Siegler N., Close L. M., Cruz K. L., Martín E. L., and Reid I. N. (2005) *Astrophys. J., 621*, 1023–1032.

Simon M. and Prato L. (2004) *Astrophys. J., 613*, L69–L71.

Sivia D. (1996) *Data Analysis.* Claredon, Oxford.

Stassun K. G., Mathieu R. D., and Valenti J. A. (2006) *Nature, 440*, 311–314.

Stauffer J. R., Schultz G., and Kirkpatrick J. D. (1998) *Astrophys. J., 499*, L199–L203.

Stephens D. C., Marley M. S., Noll K. S., and Chanover N. (2001) *Astrophys. J., 556*, L97–L101.

Sterzik M. F. and Durisen R. H. (2003) *Astron. Astrophys., 400*, 1031–1042.

Tinney C. G. (1996) *Mon. Not. R. Astron. Soc., 281*, 644–658.

Tinney C. G., Burgasser A. J., and Kirkpatrick J. D. (2003) *Astron. J., 126*, 975–992.

Tokovinin A. A. (1997) *Astron. Astrophys. Suppl., 124*, 71–76.

Tokovinin A. A. (2004) *Rev. Mex. Astron. Astrofis., 21*, 7–14.

Udry S., Mayor M., and Queloz D. (2003) In *Scientific Frontiers in Research on Extrasolar Planets* (D. Deming and S. Seager, eds.), pp. 17–26. ASP Conf. Series 294, San Francisco.

Umbreit S., Burkert A., Henning T., Mikkola S., and Spurzem R. (2005) *Astrophys. J., 623*, 940–951.

van Altena W. F., Lee J. T., and Hoffleit E. D. (1995) *The General Catalogue of Trigonometric Stellar Parallaxes, 4th edition.* Yale Univ. Observatory, New Haven.

Vrba F. J., Henden A. A., Luginbuhl C. B., Guetter H. H., Munn J. A., et al. (2004) *Astron. J., 127*, 2948–2968.

Watkins S. J., Bhattal A. S., Boffin H. M. J., Francis N., and Whitworth A. P. (1998a) *Mon. Not. R. Astron. Soc., 300*, 1205–1213.

Watkins S. J., Bhattal A. S., Boffin H. M. J., Francis N., and Whitworth A. P. (1998b) *Mon. Not. R. Astron. Soc., 300*, 1214–1224.

Weinberg M. D., Shapiro S. L., and Wasserman I. (1987) *Astrophys. J., 312*, 367–389.

White R. D., Ghez A. M., Reid I. N., and Schultz G. (1999) *Astrophys. J., 520*, 811–821.

Whittet D. C. B., Prusti T., Franco G. A. P., Gerakines P. A., Kilkenny D., et al. (1997) *Astron. Astrophys., 327*, 1194–1205.

Whitworth A. P. and Zinnecker H. (2004) *Astron. Astrophys., 427*, 299–306.

Wilson J. C., Kirkpatrick J. D., Gizis J. E., Skrutskie M. F., Monet D. G., et al. (2001) *Astron. J., 122*, 1989–2000.

Zapatero Osorio M. R., Lane B. F., Pavlenko Ya., Martín E. L., Britton M., et al. (2004) *Astrophys. J., 615*, 958–971.

Zuckerman B. and Song I. (2004) *Ann. Rev. Astron. Astrophys., 42*, 685–721.

The Formation of Brown Dwarfs: Observations

Kevin L. Luhman
The Pennsylvania State University

Viki Joergens
The University of Leiden

Charles Lada
Smithsonian Astrophysical Observatory

James Muzerolle and Ilaria Pascucci
The University of Arizona

Russel White
The University of Alabama

We review the current state of observational work on the formation of brown dwarfs, focusing on their initial mass function, velocity and spatial distributions at birth, multiplicity, accretion, and circumstellar disks. The available measurements of these various properties are consistent with a common formation mechanism for brown dwarfs and stars. In particular, the existence of widely separated binary brown dwarfs and a probable isolated proto-brown dwarf indicate that some substellar objects are able to form in the same manner as stars through unperturbed cloud fragmentation. Additional mechanisms such as ejection and photoevaporation may play a role in the birth of some brown dwarfs, but there is no observational evidence to date to suggest that they are the key elements that make it possible for substellar bodies to form.

1. INTRODUCTION

Although many of the details are not perfectly understood, stars and giant planets are generally believed to form through the collapse of molecular cloud cores and the accretion of gas by rocky cores in circumstellar disks, respectively. In comparison, the formation of objects intermediate between stars and planets — free-floating and companion brown dwarfs — has no widely accepted explanation. *A priori,* one might expect that brown dwarfs form in the same manner as stars, just on a much smaller scale. However, although self-gravitating objects can form with initial masses of only ~1 M_{Jup} in simulations of the fragmentation of molecular cloud cores, these fragments continue to accrete matter from their surrounding cores, usually to the point of eventually reaching stellar masses (*Boss,* 2001; *Bate et al.,* 2003). Thus, standard cloud fragmentation in these models seems to have difficulty in making brown dwarfs. One possible explanation is that the simulations lack an important piece of physics (e.g., turbulence), and brown dwarfs are able to form through cloud fragmentation despite their predictions (e.g., *Padoan and Nordlund,* 2004). Another possibility is that a brown dwarf is born when cloud fragmentation is modified by an additional process that prematurely halts accretion during the protostellar stage, such as

dynamical ejection (*Reipurth and Clarke,* 2001; *Boss,* 2001; *Bate et al.,* 2002) or photoevaporation by ionizing radiation from massive stars (*Kroupa and Bouvier,* 2003; *Whitworth and Zinnecker,* 2004). This uncertainty surrounding the formation of brown dwarfs has motivated a great deal of theoretical and observational work over the last decade.

In this paper, we review the current observational constraints on the formation process of brown dwarfs (BDs), which complements the theoretical review of this topic provided in the chapter by Whitworth et al. By the nature of the topic of this review, we focus on observations of BDs at young ages ($\tau < 10$ m.y.), although we also consider properties of evolved BDs that provide insight into BD formation (e.g., multiplicity). A convenient characteristic of young BDs is their relatively bright luminosities and warm temperatures compared to their older counterparts in the solar neighborhood, making them easier to observe. However, because the luminosities and temperatures of young BDs are continuous extensions of those of stars, positively identifying a young object as either a low-mass star or a BD is often not possible. The mass estimates for a given object vary greatly with the adopted evolutionary models and the manner in which observations are compared to the model predictions. Using the models of *Baraffe et al.* (1998) and *Chabrier et al.* (2000) and the temperature scale of *Luhman*

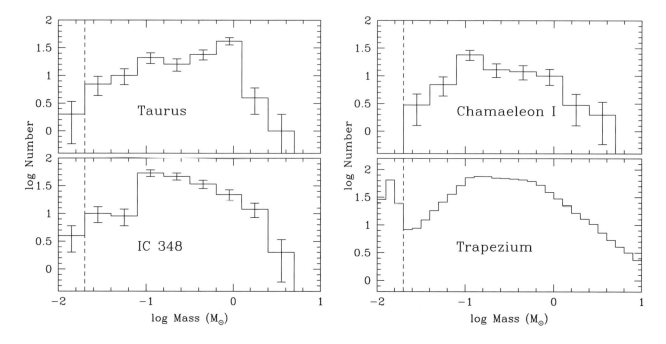

Fig. 1.   IMFs for Taurus (*Luhman,* 2004c), IC 348 (*Luhman et al.,* 2003b), Chamaeleon I (Luhman, in preparation), and the Trapezium Cluster (*Muench et al.,* 2002). The completeness limits for these measurements are near 0.02 M$_\odot$ (dashed lines). In the units of this diagram, the Salpeter slope is 1.35.

et al. (2003b), the hydrogen-burning mass limit at ages of 0.5–3 m.y. corresponds to a spectral type of ~M6.25, which is consistent with the dynamical mass and spectral type of the first known eclipsing binary BD (*Stassun et al.,* 2006) and other observational tests (*Luhman and Potter,* 2006). Therefore, we will treat young objects later than M6 as BDs for the purposes of this review.

2.   INITIAL MASS FUNCTION

One of the most fundamental properties of BDs is their initial mass function (IMF). Because BDs are brightest when they are young, star-forming regions and young clusters are the best sites for finding them in large numbers and at low masses, which is necessary for measuring statistically significant IMFs. Spectroscopic surveys for BDs have been performed toward many young populations (τ < 10 m.y.) during the last decade, including IC 348 (*Luhman et al.,* 1998, 2003b, 2005a; *Luhman,* 1999), Taurus (*Briceño et al.,* 1998, 2002; *Martín et al.,* 2001b; *Luhman,* 2000, 2004c, 2006; *Luhman et al.,* 2003a; *Guieu et al.,* 2006), Chamaeleon I (*Coméron et al.,* 1999, 2000, 2004; *Neuhäuser and Comerón,* 1999; *Luhman,* 2004a,b; *Luhman et al.,* 2004), Ophiuchus (*Luhman et al.,* 1997; *Wilking et al.,* 1999; *Cushing et al.,* 2000), Upper Scorpius (*Ardila et al.,* 2000; *Martín et al.,* 2004), Orion (*Hillenbrand,* 1997; *Lucas et al.,* 2001; *Slesnick et al.,* 2004), NGC 2024 (*Levine et al.,* 2006), NGC 1333 (*Wilking et al.,* 2004), TW Hya (*Gizis,* 2002; *R. Scholz et al.,* 2005), λ Ori (*Barrado y Navascués et al.,* 2004b), and σ Ori (*Barrado y Navascués et al.,* 2001, 2002; *Béjar et al.,* 1999, 2001; *Martín et al.,* 2001a; *Zapatero Osorio*

et al., 1999, 2000, 2002a,b,c; *Martín and Zapatero Osorio,* 2003).

We now examine the IMF measurements for IC 348, Chamaeleon I, Taurus, and the Trapezium, which exhibit the best combination of number statistics, completeness, and dynamic range in mass among the young populations studied to date. These IMFs are shown in Fig. 1. Because the same techniques and models were employed in converting from data to masses for each population, one can be confident in the validity of any differences in these IMFs. For the Trapezium, we use the IMF derived through infrared (IR) luminosity function modeling by *Muench et al.* (2002) (see also *Luhman et al.,* 2000; *Hillenbrand and Carpenter,* 2000; *Lucas et al.,* 2005). The spectroscopically determined IMF for IC 348 agrees well with the IMF derived by *Muench et al.* (2003) through the same kind of luminosity function analysis, which suggests that the Trapezium IMF from *Muench et al.* (2002) can be reliably compared to the spectroscopic IMFs for IC 348, Chamaeleon I, and Taurus. We quantify the relative numbers of BDs and stars with the ratio

$$\mathcal{R} = N(0.02 \leq M/M_\odot \leq 0.08)/N(0.08 < M/M_\odot \leq 10)$$

The IMFs for Taurus, IC 348, and Orion exhibit $\mathcal{R}$ = 0.18 ± 0.04, 0.12 ± 0.03, and 0.26 ± 0.04, respectively. Because the IMF measurement for Chamaeleon I is preliminary, a reliable BD fraction is not yet available. These BD fractions for Taurus and IC 348 are a factor of 2 lower than the value for Orion. However, upon spectroscopy of a large sample of BD candidates in the Trapezium, *Slesnick et al.* (2004) found a population of faint objects with stellar masses, possibly seen in scattered light, which had contaminated previous photometric IMF samples and resulted in overestimates of the BD fraction in this cluster. After they corrected for this contamination, the BD fraction in the Trapezium was a factor of only ~1.4 higher than the value in Taurus from *Luhman* (2004c). Through a survey of additional areas of Taurus, *Guieu et al.* (2006) have recently discovered 17 new low-mass stars and BDs. By combining these data with the previous surveys, they measured a BD fraction that is still higher, and thus closer to the value for Orion. However, *Luhman* (2006) finds that their higher BD fraction is due to a systematic offset between the spectral types of *Guieu et al.* (2006) and the classification system used for the previously known late-type members of Taurus (*Luhman,* 1999; *Briceño et al.,* 2002). In summary, according to the best available data, the BD fractions in Taurus and IC 348 are lower than in the Trapezium, but by a factor that is smaller than that reported in earlier studies.

Because the mass-luminosity relation is a function of age for BDs at any age, and the ages of individual field BDs are unknown, a unique, well-sampled IMF of field BDs cannot be constructed. When substellar mass functions are instead compared in terms of power-law slopes (Salpeter is 1.35), the latest constraints in the field from *Chabrier* (2002) ($\alpha \leq 0$) and *Allen et al.* (2005) ($-1.5 \leq \alpha \leq 0$) are consistent with the mildly negative slopes exhibited by the data for star-forming regions in Fig. 1. Thus, data for both star-forming regions and the solar neighborhood are consistent with stars outnumbering BDs by a factor of ~5–8. If BDs form through ejection and have higher-velocity dispersions than stars as predicted by *Kroupa and Bouvier* (2003) (but not *Bate et al.,* 2003), then the BD fraction would be higher in the field than in star-forming regions since BD members would be quickly ejected from the latter. However, current data show no evidence of such a difference.

In addition to the abundance of BDs relative to stars, the minimum mass at which BDs can form also represents a fundamental constraint for theories of BD formation. For several of the star-forming regions cited in this section, BDs with conclusive evidence of membership and accurate spectral classifications have been discovered down to optical spectral types of M9.5, corresponding to masses of ~10–20 M_{Jup}. Additional BDs have been reported at cooler and fainter levels, most notably in σ Ori. However, some of these objects lack clear evidence of membership and instead could be field dwarfs (*Burgasser et al.,* 2004, and references therein). Finally, *Kirkpatrick et al.* (2006) recently discovered a young L dwarf in the field that is probably comparable in mass (6–25 M_{Jup}) to the least-massive BDs found in young clusters.

3. KINEMATICS AND POSITIONS AT BIRTH

Some models for the formation of BDs via embryo ejection predict that BDs are born with higher-velocity dispersions than stars and thus are more widely distributed in star-forming regions than their stellar counterparts (*Reipurth and Clarke,* 2001; *Kroupa and Bouvier,* 2003). Meanwhile, other

models of ejection (*Bate et al.,* 2003) and models in which BDs form in a star-like manner predict that stars and BDs should have similar spatial and velocity distributions. Because normal dynamical evolution of a cluster can produce mass-dependent distributions like those of the first set of ejection models (*Bonnell and Davies,* 1998), clusters that are old or dense are not suitable for testing their predictions (*Moraux and Clarke,* 2005). Therefore, based on their youth and low stellar densities, the Taurus and Chamaeleon star-forming region are ideal sites for comparing the positions and kinematics of stars and BDs.

Precise radial velocities of low-mass stars and BDs in Chamaeleon I measured from high-resolution spectra are slightly less dispersed (0.9 ± 0.3 km/s) but still consistent with those of stars (1.3 ± 0.3 km/s) (*Joergens and Guenther,* 2001; *Joergens,* 2006b). The BDs do not show a high-velocity tail as predicted by some models of the ejection scenario (*Sterzik and Durisen,* 2003; *Umbreit et al.,* 2005). Similar results have been found for Taurus (*White and Basri,* 2003; *Joergens,* 2006b). While the absence of a significant mass dependence of the velocities is consistent with some models of the ejection scenario (*Bate et al.,* 2003; *Delgado-Donate et al.,* 2004), the observed global radial velocity dispersion (BDs and stars) for Chamaeleon I members is smaller than predicted by any model of the ejection scenario.

Over time, the surveys for BDs in Taurus have encompassed steadily larger areas surrounding the stellar aggregates (see references in previous section). These data have exhibited no statistically significant differences in the spatial distribution of the high- and low-mass members of Taurus (*Briceño et al.,* 2002; *Luhman,* 2004c; *Guieu et al.,* 2006). This result has been established definitively by the completion of a BD survey of 225 deg² encompassing all of Taurus (*Luhman,* 2006). As shown in Fig. 2, the BDs follow the spatial distribution of the stellar members, and there is no evidence of a large, distributed population of BDs. As with the kinematic properties, these spatial data are consistent with a common formation mechanism for stars and BDs and some models for ejection (*Bate et al.,* 2003), but not others (*Kroupa and Bouvier,* 2003).

4. MULTIPLICITY

As with stars, the multiplicity properties of BDs (frequency, separation, and mass ratio distributions) are intimately tied to their formation. As discussed in the chapter by Whitworth et al., embryo-ejection scenarios predict few binaries and only close orbits, while isolated fragmentation models allow for higher binary frequencies and larger maximum separations. Therefore, accurately characterizing the multiplicity of BDs can help distinguish between these scenarios (and others). Moreover, the identification of very-low-mass companions to BDs will delimit better the types of environments in which planets can form. The chapter by Burgasser et al. provides a comprehensive review of the observational and theoretical work on the binary properties of BDs. In this section, we discuss highlights of the latest observational work and their implications for the origin of BDs.

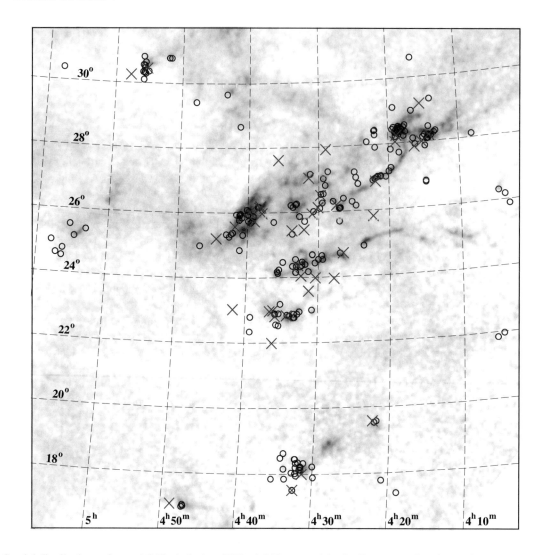

Fig. 2.   Spatial distributions of stars (≤M6, circles) and BDs (>M6, crosses) in the Taurus star-forming region shown with a map of extinction (grayscale) (*Dobashi et al.,* 2005).

4.1. Brown Dwarf Companions to Stars: The Brown Dwarf Desert

Among companions at separations less than a few AU from solar-type stars, radial velocity surveys have revealed a paucity of BDs (20–80 M_{Jup}) relative to giant planets and stellar companions (*Marcy and Butler,* 2000). Deficiencies in substellar companions have been observed at wider separations as well, as illustrated in Fig. 8 from *McCarthy and Zuckerman* (2004), which compared published frequencies of stellar and substellar companions as a function of separation. At separations less than 3 AU, the frequency of BD companions is ~0.1% (<0.5%) (*Marcy and Butler,* 2000) and the frequency of stellar companions is 13 ± 3% (*Duquennoy and Mayor,* 1991; *Mazeh et al.,* 1992), indicating that BDs are outnumbered by stars among close companions by a factor of ~100 (>20). In comparison, the ratio of the frequencies of stellar and substellar companions is between ~3 and 10 at wider separations (*McCarthy and Zuckerman,* 2004), which is comparable to the ratio of the numbers of stars and BDs in

isolation (~5–8, section 2). Thus, for solar-type primaries, only the close companions exhibit a true desert of BDs. The similarity in the abundances of BDs among wider companions and free-floating objects suggests that they arise from a common formation mechanism (e.g., core fragmentation.)

4.2. Binary Brown Dwarfs

Binary surveys of members of the solar neighborhood have found progressively smaller binary fractions, smaller average and maximum separations, and larger mass ratios ($q \equiv M_2/M_1$) with decreasing primary mass from stars to BDs (*Duquennoy and Mayor,* 1991; *Fischer and Marcy,* 1992; *Reid et al.,* 2001; *Bouy et al.,* 2003; *Burgasser et al.,* 2003; *Close et al.,* 2003; *Gizis et al.,* 2003; *Siegler et al.,* 2005). To help identify the sources of these trends (e.g., formation mechanism, environment), it is useful to compare the field data to measurements in young clusters and star-forming regions. Because young clusters have greater distances than the nearest stars and BDs in the field, the range of separations probed in binary surveys of young clusters is usually smaller than that for field objects. As a result, accurate measurements of the multiplicity as a function of mass are difficult for young populations. However, for one of the best studied star-forming regions, Taurus-Auriga, White et al. (in preparation) and *Kraus et al.* (2006) have measured the binary fraction (for a = 9–460 AU, q ≥ 0.09, defined for completeness) as a function of mass from 1.5 to 0.015 $M_\odot$. As shown in Fig. 3, the binary fraction in Taurus declines steadily with primary mass, which resembles the trend observed for the solar neighborhood. This behavior can be explained by simple random pairing from the same mass function without the presence of different formation mechanisms at high and low masses. Out of the 17 Taurus members with spectral types cooler than M6 (≤0.1 $M_\odot$), none have spatially resolved companions, which again is consistent with the small separations of a < 20 AU that have been observed for most of the binary low-mass stars and BDs in the field. A similar paucity of wide low-mass pairs has been found in other young regions, including IC 348 (*Duchêne et al.*, 1999; *Luhman et al.*, 2005c), Chamaeleon I (*Neuhäuser et al.*, 2002), Corona Australis (*Bouy et al.*, 2004), Upper Scorpius (*Kraus et al.*, 2005), and the Trapezium Cluster in Orion (*Lucas et al.*, 2005). However, in both the field and in young clusters, a few wide binary low-mass stars and BDs have been found at projected separations that range from 33 to 41 AU (*Harrington et al.*, 1974; *Martín et al.*, 2000; *Chauvin et al.*, 2004, 2005; *Phan-Bao et al.*, 2005) to beyond 100 AU (*White et al.*, 1999; *Gizis et al.*, 2001; *Luhman*, 2004b, 2005; *Billères et al.*, 2005; *Bouy et al.*, 2006). Because these wide binaries are weakly bound and extremely fragile, it would seem unlikely that they have been subjected to violent dynamical interactions, suggesting that some low-mass stars and BDs are able to form without the involvement of ejection, apparently through standard, unperturbed cloud fragmentation. Indeed, in embryo-ejection simulations, *Bate et al.* (2002) found that "because close dynamical interactions are involved in their formation . . . binary brown dwarf systems that do exist must be close, ≤10 AU." However, in more recent calculations by *Bate and*

Bonnell (2005), a wide binary BD was able to form when two BDs were simultaneously ejected in similar directions.

Because the surveys for binary BDs cited above employed direct imaging, they were not sensitive to very close binaries (a ≤ 1 and a ≤ 10 AU for the field and clusters, respectively), making the resulting binary fractions only lower limits. Spectroscopic monitoring for radial velocity variations provides a means of identifying the closest companions, which is essential for assessing whether the formation mechanism of companions in substellar multiple systems changes with separation. The first free-floating BD to be discovered, PPL 15 (*Stauffer et al.*, 1994), turned out to be a spectroscopic binary with companions of nearly equal mass in a six-day orbit (*Basri and Martín*, 1999). More recently, *Guenther and Wuchterl* (2003) started a systematic survey for close companions to 25 low-mass stars and BDs in the field, finding two candidate double-lined spectroscopic binaries. An additional object in their sample was found to be a binary by *Reid et al.* (2002). *Joergens and Guenther* (2001) started a similar survey for close low-mass binaries in the Chamaeleon I star-forming region. Among a subsample of 10 low-mass objects (M ≤ 0.12 $M_\odot$, M5–M8), none show signs of companions down to the masses of giant planets for orbital periods P < 40 d, corresponding to separations of a < 0.1 AU (*Joergens*, 2006a). For Cha Hα 8 (M6.5), data recorded across a longer period of time does indicate the existence of a spectroscopic companion of planetary or BD mass with an orbital period of several months to a few years, as shown in Fig. 4. In a combination of the above old and young samples, three (~9%) and four (~11%) ob-

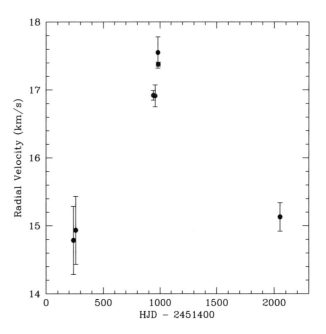

Fig. 4. Radial velocity data for the young low-mass object Cha Hα 8 (M6.5) recorded with UVES/VLT; significant variability occurring on timescales of months to years hint at a companion at a > 0.2 AU and M sin i ≥ 6 M_J (*Joergens*, 2006a).

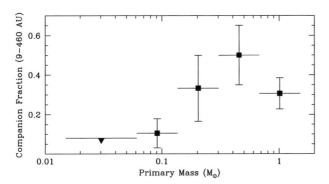

Fig. 3. Binary fraction over the separation range 9–460 AU vs. primary mass for young stars and BDs in the Taurus star-forming region (White et al., in preparation; *Kraus et al.*, 2006).

jects have possible companions at P < 100 d and P < 1000 d, respectively. For comparison, the frequencies of binaries among solar-type field stars are 7% and 13% in these same period ranges (*Duquennoy and Mayor,* 1991).

5. ACCRETION

The past 10–15 years has seen the establishment of a disk-accretion paradigm in low-mass T Tauri stars that explains many of their observed characteristics. The picture centers on the concept of magnetospheric accretion, whereby the stellar magnetic field truncates the circumstellar disk and channels accreting material out of the disk plane and onto the star (see chapter by Bouvier et al. and references therein). Models of magnetospheric accretion successfully describe many features of classical T Tauri stars (CTTSs), including the broad, asymmetric permitted line emission (e.g., *Muzerolle et al.,* 2001) and blue/UV continuum excess (e.g., *Calvet and Gullbring,* 1998). The investigation of accretion and disk signatures in lower-mass objects extending below the substellar limit is a natural extension of the CTTS studies, helping to address the origin of BDs.

Accretion in young BDs is fundamental to our understanding of formation mechanisms, and hence has undergone considerable scrutiny following the discovery of the first substellar objects in nearby star-forming regions. Spectroscopy of the first known young BDs (e.g., *Luhman et al.,* 1997; *Briceño et al.,* 1998; *Coméron et al.,* 1999) revealed that many were superficially similar to CTTSs in terms of emission line activity. In particular, equivalent widths of Hα emission in many cases exceeded levels typical of chromospheric activity in low-mass CTTSs and main sequence dMe stars, suggesting the presence of accretion. With the advent of 8–10-m-class telescopes, high-resolution optical spectroscopy of young BDs became possible. As a result, research by many groups over the last five years has provided conclusive evidence of ongoing accretion in many young substellar systems. This evidence includes the presence of broad, asymmetric Balmer line profiles, continuum veiling of photospheric absorption features, and in a few cases forbidden line emission, all similar to features seen in CTTSs.

5.1. Diagnostics

The first demonstration of accretion infall in a very-low-mass object was presented by *Muzerolle et al.* (2000) for the Taurus member V410 Anon 13 [~0.1 M$_\odot$ (*Briceño et al.,* 2002)]. The Hα profile for this object shows a clear infall asymmetry similar to that commonly seen in CTTSs, albeit with a narrower line width and a lack of opacity-broadened wings. Such features indicated ballistic infall at velocities consistent with the object's mass and radius, and a much lower mass accretion rate (Ṁ) than typical of higher-mass CTTSs with similar ages. Modeling of the profile in fact yielded an extremely small value of Ṁ ~ 5 × 10^{-12} M$_\odot$ yr^{-1}, a mere trickle in comparison with the average rate of ~10^{-8} M$_\odot$ yr^{-1} for solar-mass CTTSs (*Gullbring et al.,* 1998).

Evidence for accretion in many other very-low-mass stars and BDs has since accumulated by various techniques. *White and Basri* (2003) were the first to publish measurements of continuum veiling from accretion shock emission from objects near and below the substellar limit, providing more direct measures of mass accretion rates that were again lower than the typical of CTTSs. However, measurable veiling has turned out to be very rare in substellar accretors because of their small accretion rates. Models of substellar accretion shock emission (*Muzerolle et al.,* 2000) show that measurable veiling is produced only when Ṁ > 10^{-10} M$_\odot$ yr^{-1}. Since Hα emission from the accretion flow is detectable at much lower Ṁ, Hα emission line profiles remain the most sensitive accretion diagnostics available for young BDs. Chromospheric emission associated with magnetic activity appears to be a common feature of both young and older field dwarfs (e.g., *Mohanty and Basri,* 2003), producing generally larger Hα equivalent widths at later spectral types as a result of decreasing contrast with the photosphere; Hα equivalent widths of 20 Å are not uncommon in young objects with spectral types M5 or later. Spectral type-dependent equivalent width criteria have been proposed by several authors (*Barrado y Navascués and Martín,* 2003; *White and Basri,* 2003). However, line profiles offer the most unambiguous discriminant, as chromospheric emission produces much narrower and symmetric profiles compared to the broader and often asymmetric accretion profiles, as illustrated in Fig. 5.

A popular high-resolution accretion criterion has been the Hα 10% line velocity width. Accretion profile widths are related to the maximum ballistic infall velocity

$$V_{inf} \sim \sqrt{\frac{2GM_*}{R_*}\left(1 - \frac{1}{R_m}\right)} \sim 160 \text{ km s}^{-1}$$

for M$_*$ = 0.05 M$_\odot$, R$_*$ = 0.5 R$_\odot$, and a magnetospheric truncation radius R$_m$ = 3 R$_*$ (see *Muzerolle et al.,* 2003). Most chromospheric profiles generally exhibit velocity half-widths ≤70 km s^{-1}, much lower than the characteristic infall velocity. Broadening from rapid rotation can create larger line widths, as has been observed, but this is rare [(typical rotation velocities are v sin i ≤ 20 km s^{-1} (*Muzerolle et al.,* 2003, 2005)] and in any case can be checked with v sin i measurements. Also, objects with large Hα velocity widths consistent with infall tend to have larger Hα *equivalent* widths than those with chromospheric profiles and also tend to correlate with the presence of other accretion signatures such as other permitted and forbidden emission lines. Adopting a 10% line width threshold of V$_{10}$ ≥ 180–200 km s^{-1} gives reasonably accurate accretor identifications in BDs (*Jayawardhana et al.,* 2003b; *Muzerolle et al.,* 2005), although occasional misidentifications can occur for rapidly rotating nonaccretors or pole-on accretors.

Dozens of substellar accretors have now been identified down to masses approaching the deuterium-burning limit and with ages from 1 to 10 m.y. (e.g., *Jayawardhana et al.,* 2003b; *Mohanty et al.,* 2005; *Muzerolle et al.,* 2003, 2005).

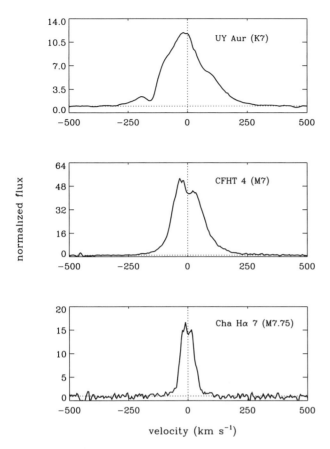

Fig. 5. Comparison of Hα profiles of (from top to bottom) a typical CTTS (*Muzerolle et al., 1998*), a typical substellar accretor (note the smaller line width indicative of the much smaller gravitational potential), and a substellar nonaccretor exhibiting the narrow and symmetric profile produced by chromospheric emission (*Muzerolle et al., 2005*).

Such a statistically robust sample has allowed systematic studies of accretion properties across nearly the entire range of substellar masses yet identified. For instance, magnetospheric accretion requires disk material to be present at or within the corotation radius, which should be detectable at near- or mid-IR wavelengths. Comparing known substellar accretors in Chamaeleon I and Ophiuchus with IR excesses detected by ISO at 6.7 and 14.3 μm (*Natta et al., 2004*) and Spitzer at 3.6–8 μm (*Luhman et al., 2005d*), 3/10 and 7/10 in each region, respectively, exhibit both accretion and disks, while 3/10 in each region show disks but no accretion. There are no cases of accretion without disk signatures. The objects with disks and lacking accretion signatures may simply be accreting at rates below the observable threshold. The larger fraction of these in Chamaeleon I compared to Ophiuchus may be a reflection of the slightly older age of the former, so that the disks have evolved to lower accretion rates on average (see below). Indeed, many studies have now found strong indications of a decreasing fraction of accreting objects with age, both above and below the substellar limit. Typical values range from 30 to 60% in 1–

3-m.y.-old regions such as Taurus and Chamaeleon I, but drop significantly to 0–5% in 3–5-m.y.-old regions such as σ Ori and Upper Scorpius (*Muzerolle et al., 2005; Mohanty et al., 2005*). These numbers are consistent with similar declines observed in the accretor fraction at stellar masses, indicating similar evolutionary timescales for accretion between stars and BDs. The same result is found for the disk fractions of stars and BDs, as we discuss in section 6.2.

5.2. Substellar Accretion Rates

The results summarized above show that the overall accretion characteristics are essentially continuous across the substellar boundary, which is consistent with stars and BDs forming via the same accretion processes. A more quantitative assessment can be made from measurements of mass accretion rates. Most of the estimates of $\dot{M}$ based on line profile modeling (*Muzerolle et al., 2003, 2005*) and secondary IR calibrators such as Paschen β and Brackett γ (*Natta et al., 2004*) and the CaII triplet (*Mohanty et al., 2005*) have shown that very small accretion rates are in fact typical of very-low-mass young objects. The average value for substellar accretors is roughly 2–3 orders of magnitude lower than that of 1-m.y.-old CTTSs. A clear trend of decreasing accretion rate with decreasing mass was found by *Muzerolle et al.* (2003) and subsequently extended down to $M \sim 0.02\ M_\odot$ by *Mohanty et al.* (2005) and *Muzerolle et al.* (2005), with a functional form of $\dot{M} \propto M^2$ (Fig. 6). The surprising correlation between mass and accretion rate has

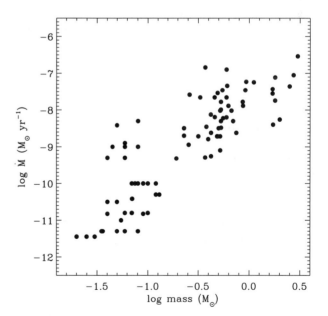

Fig. 6. Mass accretion rate as a function of substellar and stellar mass for objects in Taurus (1 m.y.), Cha I (2 m.y.), IC 348 (2 m.y.), and Ophiuchus (0.5 m.y.) (*Gullbring et al., 1998; White and Ghez, 2001; Muzerolle et al., 2000, 2003, 2005; Natta et al., 2004; Mohanty et al., 2005*). These regions exhibit similar accretion rates at a given mass, except for slightly higher rates in Ophiuchus.

profound implications for BD origins. The lack of any obvious shift in the correlation about the substellar limit implies a continuity in the formation processes of stars and BDs. This lends support to BD formation via fragmentation and collapse of low-mass cloud cores. However, the physical origins of accretion in young stellar objects need to be better understood before definitive conclusions can be made.

What is the source of the mass-accretion correlation? The general theory of viscously accreting disks does not predict a strong relation between these two quantities. The answer may lie in the essentially unknown source of viscosity needed to drive accretion. A commonly invoked mechanism is the Balbus-Hawley instability, which requires sufficient ionization of disk gas to effectively couple with magnetic fields. X-ray activity from the stellar magnetic field is a potential ionization source (*Glassgold et al.*, 2004); *Muzerolle et al.* (2003) suggested that the observed correlation $L_X \propto M_*^2$ (e.g., *Feigelson et al.*, 2003) may then be related to the similar dependence of accretion rate on mass. However, a comparison between L_X and $\dot{M}$ for the small number of objects for which both quantities have been measured reveals no statistically significant correlation, although the mass range covered is not very large. More observations of both quantities are needed; a particularly interesting analysis would be to compare accretion variability at the onset of and subsequent to an X-ray flare event.

More recently, *Padoan et al.* (2005) have proposed a modified Bondi-Hoyle accretion model in which young stars and BDs accrete primarily from the large-scale medium in which they are moving rather than from their disks alone. In this case, the accretion rate can be determined by the density and sound speed of the surrounding gas and the relative velocity between the object and that material. The resultant relation produces the correct mass dependence. However, it is not clear how a Bondi-Hoyle flow would interact with the disk. For instance, it may be incorporated into the disk prior to reaching the star; note that the observed accretion diagnostics are inconsistent with spherical infall onto the stellar surface. If so, the Bondi-Hoyle relation may in fact determine the rate of residual infall onto the disk, but not necessarily the rate of accretion onto the star, which is what is measured. In addition, the model cannot explain accretors located in low-density regions far from molecular clouds, such as the well-known CTTS TW Hydrae. Comparisons of mass accretion rates vs. surrounding cloud temperatures and densities need to be made to further assess the applicability of this model.

5.3. Jets and Outflows

Many other similarities in accretion activity between stars and BDs have been found, including photometric and line profile variability (*Caballero et al.*, 2004; *Scholz and Eislöffel*, 2004, 2005; *A. Scholz et al.*, 2005), detections of H_2 emission in the UV (*Gizis et al.*, 2005), possible detections of UV continuum excesses (*McGehee et al.*, 2005), and evidence of accretion-generated outflows such as blueshifted absorption and forbidden emission (*Fernández and*

Comerón, 2001; *Muzerolle et al.*, 2003; *Barrado y Navascués et al.*, 2004a; *Luhman*, 2004c; *Mohanty et al.*, 2005). Among the four accretors at M6 or later from *Muzerolle et al.* (2003), one shows forbidden line emission, while the two Class I objects at M6 from *White and Hillenbrand* (2004) show forbidden line emission. Based on small number statistics, jet signatures appear to be less often associated with accretion signatures for low-mass stars and BDs than for stars. However, this may stem from lower mass loss rates in substellar jets producing emission that is more difficult to detect (*Masciadri and Raga*, 2004). *White and Hillenbrand* (2004) found that the ratio of mass loss to mass accretion rate is the same for objects with both high and low mass accretion rates, though with considerable dispersion. Thus, the low accretion rates inferred for BDs likely correspond to diminished mass loss rates and less luminous forbidden line emission, possibly below typical detection levels. Overall, the sparse data on jets from young accreting BDs are similar to those of higher-mass CTTSs, but on a smaller and less energetic scale.

In addition to the above indirect evidence for outflows provided by forbidden line emission, *Whelan et al.* (2005) and *Bourke et al.* (2005) have spatially resolved outflows toward ISO 102 in Ophiuchus (also known as GY 202) and L1014-IRS through optical forbidden lines and millimeter CO emission, respectively. Although *Whelan et al.* (2005) referred to ISO 102 as a BD, the combination of its M6 spectral type from *Natta et al.* (2002), the evolutionary models of *Chabrier et al.* (2000), and the temperature scale of *Luhman et al.* (2003b) suggest that it could have a stellar mass of ~0.1 $M_\odot$. It appears likely that L1014-IRS has a substellar mass (*Young et al.*, 2004; *Huard et al.*, 2006), but this is difficult to confirm because of its highly embedded nature. The molecular outflow detected toward this object by *Bourke et al.* (2005) is one of the smallest known outflows in terms of its size, mass, and energetics.

6. CIRCUMSTELLAR DISKS

The collapse of a cloud core naturally produces a circumstellar disk via angular momentum conservation. Thus, understanding the formation of BDs requires close scrutiny of their circumstellar disks. In addition, as with stars, studying disks around BDs should provide insight into if and how planet formation occurs around these small bodies. In this section, we summarize our knowledge of disks around BDs and discuss the resulting implications for the origin of BDs, the evolution of their disks, and the formation of planets.

6.1. Detections of Disks

Although resolved images of disks around BDs are not yet available, there is mounting evidence for their existence through detections of IR emission above that expected from stellar photospheres alone. Modeling of the IR spectral energy distributions (SEDs) of young BDs showing excess emission strongly suggests that the emitting dust resides in disk configurations. For instance, *Pascucci et al.* (2003) considered different shell and disk geometries for a BD system in Taurus and demonstrated that spherically distributed dust with a mass estimated from the millimeter measurements (*Klein et al., 2003*) would produce much more extinction than observed toward the BD. In comparison, when the same material is modeled as a disk, the SED can be well reproduced without conflicting with the observed low extinction.

Excess emission in the K and L bands has been observed for several young objects at M6–M8 and for a few as late as M8.5 (*Luhman, 1999, 2004c; Lada et al., 2000, 2004; Muench et al., 2001; Liu et al., 2003; Jayawardhana et al., 2003a*). Excesses at longer, mid-IR wavelengths have been detected for CFHT 4 (M7) (*Pascucci et al., 2003; Apai et al., 2004*), Cha Hα 1 (M7.75) (*Persi et al., 2000; Comerón et al., 2000; Natta and Testi, 2001; Sterzik et al., 2004*), 2MASS 1207-3932 (M8) (*Sterzik et al., 2004*), GY 141 (M8.5) (*Comerón et al., 1998*), and several late-type objects in Ophiuchus (*Testi et al., 2002; Natta et al., 2002; Mohanty et al., 2004*). *Klein et al.* (2003) has extended these detections of circumstellar material to millimeter wavelengths for CFHT 4 and IC 348-613 (M8.25).

Because the Spitzer Space Telescope (*Werner et al., 2004*) is far more sensitive beyond 3 μm than any other existing facility, it is capable of detecting disks for BDs at very low masses. To search for circumstellar disks around BDs at the lowest possible masses, *Luhman et al.* (2005b) obtained mid-IR images (3.6–8 μm) of the Chamaeleon I star-forming region with the Infrared Array Camera (IRAC) (*Fazio et al., 2004*) on Spitzer. In these data, they detected mid-IR excess emission from the coolest and least-massive known BD in the cluster, OTS 44 (*Oasa et al., 1999*), which has a spectral type of ≥M9.5 and a mass of M ~ 15 M_{Jup} (*Luhman et al., 2004*). By obtaining even deeper IRAC images of Chamaeleon I and combining them with optical and near-IR images from the Hubble Space Telescope and the CTIO 4-m telescope, *Luhman et al.* (2005e) discovered a BD that is twice as faint as OTS 44 and exhibits mid-IR excess emission. By comparing the bolometric luminosity of this object, Cha 1109-7734, to the luminosities predicted by the evolutionary models of *Chabrier et al.* (2000) and *Burrows et al.* (1997), *Luhman et al.* (2005e) estimated a mass of 8^{+7}_{-3} M_{Jup}, placing it within the mass range observed for extrasolar planetary companions [M ≤ 15 M_{Jup} (*Marcy et al., 2005*)]. *Luhman et al.* (2005e) successfully modeled the mid-IR excess emission for Cha 1109-7734 in terms of an irradiated viscous accretion disk with $\dot{M} \lesssim 10^{-12}$ $M_{\odot}$ yr^{-1}, as shown in Fig. 7, making it the least-massive BD observed to have a circumstellar disk and demonstrating that the basic ingredients for making planets are present around free-floating planetary-mass bodies.

6.2. Disk Fractions and Lifetimes

Extensive work has been done in measuring disk fractions for stars (e.g., *Kenyon and Hartmann, 1995; Hillenbrand et al., 1998; Haisch et al., 2001*), which typically consists of IR photometry of a significant fraction of a young

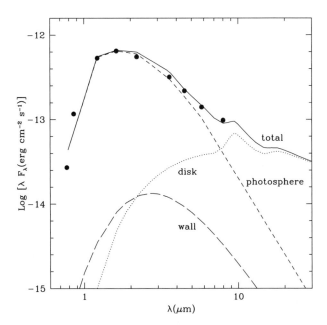

Fig. 7. SED of the least-massive BD known to harbor a disk, Cha 1109-7734 (points) (*Luhman et al., 2005e*). Relative to the distribution expected for its photosphere (~M9.5, short dashed line), this BD exhibits significant excess emission at wavelengths longer than 5 μm. The excess flux is modeled in terms of emission from a circumstellar accretion disk (dotted line) and a vertical wall at the inner disk edge (long dashed line). The sum of this disk model and the photosphere (solid line) is a reasonable match to the data for Cha 1109-7734.

stellar population and identification of the objects with excess emission. Attempts have been made to extend measurements of this kind to low-mass stars and BDs. Using JHKL' photometry, *Jayawardhana et al.* (2003a) searched for excess emission among 53 objects in IC 348, Taurus, σ Ori, Chamaeleon I, the TW Hya association, Upper Scorpius, and Ophiuchus, 27 of which are later than M6 and thus likely to be substellar. For the individual populations, the disk fractions for the stars and BDs exhibited large statistical errors of ~25%. For a sample combining Chamaeleon I, IC 348, Taurus, and U Sco, the number statistics were better, and *Jayawardhana et al.* (2003a) found a disk fraction of 40–60%. Their disk/no-disk classifications agreed well with those based on the Spitzer data from *Luhman et al.* (2005d) for types of ≤M6. However, the two objects later than M6 in IC 348 and Chamaeleon I that were reported to have disks by *Jayawardhana et al.* (2003a) show no excess emission in the Spitzer colors. *Liu et al.* (2003) also performed an L'-band survey of low-mass objects. They considered a sample of 7 and 32 late-type members of Taurus and IC 348, respectively, 12 of which have optical spectral types later than M6. For their entire sample of low-mass stars and BDs, *Liu et al.* (2003) found a disk fraction of 77 ± 15%, which is a factor of 2 larger than measurements for IC 348 from Spitzer (*Luhman et al., 2005d*). Nine of 10 objects with E(K–L') > 0.2 in the data from *Liu et al.* (2003) did exhibit significant excesses in the Spitzer colors, but the putative detections of disks with smaller L' excesses were not confirmed by Spitzer. Any bona fide detection of a disk at L' would be easily verified with Spitzer given that the contrast of a disk relative to the central object increases with longer wavelengths.

Because disks around BDs produce little L'-band emission compared to stellar systems, the L'-band surveys were not able to reliably detect BD disks. Meanwhile, BD disk excesses are larger at longer wavelengths, but measurements of this kind are feasible for only a small number of the brighter, more-massive objects with most telescopes. In comparison, because Spitzer is highly sensitive and can survey large areas of sky, it can reliably and efficiently detect disks for BDs at very low masses and for large numbers of BDs in young clusters. *Luhman et al.* (2005d) used IRAC on Spitzer to obtain mid-IR images of IC 348 and Chamaeleon I, which encompassed 25 and 18 spectroscopically confirmed low-mass members of the clusters, respectively (>M6, M ≤ 0.08 M$_\odot$). They found that 42 ± 13% and 50 ± 17% of the two samples exhibit excess emission indicative of circumstellar disks. In comparison, the disk fractions for stellar members of these clusters are 33 ± 4% and 45 ± 7% (M0–M6, 0.7 M$_\odot$ ≥ M ≥ 0.1 M$_\odot$). The similarity of the disk fractions of stars and BDs indicates that the raw materials for planet formation are available around BDs as often as around stars and supports the notion that stars and BDs share a common formation history. However, as with the continuity of accretion rates from stars to BDs from section 5.2, these results do not completely exclude some scenarios in which BDs form through a distinct mechanism. For instance, during formation through embryo ejection, the inner regions of disks that emit at mid-IR wavelengths could survive, although one might expect these truncated disks to have shorter lifetimes than those around stars.

When disk fractions for stellar populations across a range of ages (0.5–30 m.y.) are compared, they indicate that the inner disks around stars have lifetimes of ~6 m.y. (*Haisch et al.,* 2001). Accurate measurements of disk fractions for BDs are available only for IC 348 and Chamaeleon I, both of which have ages near 2 m.y., and so a comparable estimate of the disk lifetime for BDs is not currently possible. However, the presence of a disk around a BD in the TW Hya association (*Mohanty et al.,* 2003; *Sterzik et al.,* 2004), which has an age of 10 m.y., does suggest that the lifetime of BD disks might be similar to that of stars.

6.3. Disk Mass

A few estimates of dust masses of disks around BDs have been obtained through deep single-dish millimeter observations. In a survey of 9 young BDs and 10 field BDs, *Klein et al.* (2003) detected disks around two of the young objects. Because the millimeter emission is optically thin, fluxes could be converted to total disk masses assuming dust emission coefficients typical to disks for low-mass stars and the standard gas-to-dust mass ratio of 100. They derived disk masses of 0.4–6 M$_{Jup}$, which are a few percent of the BD masses, thus suggesting that disk masses scale with the mass of the central object down to the substellar regime. Similar measurements for larger samples of low-mass stars and BDs are needed to confirm such a trend.

6.4. Disk Geometry

Various theoretical studies have shown that the disk geometry strongly impacts the SED and have investigated the link between disk geometry and dust evolution (e.g., *Dullemond and Dominik,* 2005). Flared disks are those with opening angles increasing with the disk radius as a consequence of vertical hydrostatic equilibrium (*Kenyon and Hartmann,* 1987). Such geometry characterizes the early phases of the disk evolution prior to dust processing (grain growth and dust settling; see also the chapter by Natta et al.). Flatter disk geometries supposedly represent the evolutionary stage after flared disks (see also the chapter by Dullemond et al.). Because flared disks intercept more stellar radiation than flat ones, especially at large distances from the star, flared disks produce larger mid- and far-IR fluxes and a more prominent silicate emission feature than flat ones (e.g., *Chiang and Goldreich,* 1997). *Walker et al.* (2004) calculated that, under the assumption of vertical hydrostatic equilibrium, BD disks should be highly flared with disk scale heights three times larger than those derived for disks around CTTSs.

The work by *Natta and Testi* (2001) represents the first attempt to investigate the geometry of disks around low-mass objects. The authors used scaled-down T Tauri disks to reproduce ISO mid-IR measurements of two low-mass stars (Cha Hα 2 and 9, M5.25 and M5.5) and one BD (Cha Hα 1, M7.75) in the Chamaeleon I star-forming region. They considered passive flared and flat disks and made a number of simplifying assumptions following the method of *Chiang and Goldreich* (1997, 1999). Passive disks are appropriate for BDs because of their very low accretion rates (section 5.2). They concluded that models of flared disks were required to fit the SEDs of these three objects. A similar approach was used by the authors to investigate a larger sample of nine low-mass stars and BDs in the Ophiuchus star-forming region (*Testi et al.,* 2002; *Natta et al.,* 2002). A more careful inspection of the flared and flat disk predictions revealed that two ISOCAM broadband measurements were not always sufficient to determine the disk geometry. *Apai et al.* (2002) used groundbased mid-IR narrowband photometry to probe the silicate emission feature in the disk of Cha Hα 2 and thus add an important new constraint to the disk models from *Natta and Testi* (2001). Their measurements ruled out the presence of strong silicate emission that was predicted by *Natta and Testi* (2001) and found that a flatter disk structure was required to fit the observed SED.

Recent ground- and spacebased measurements have provided more comprehensive SEDs for about a dozen low-mass stars and BDs (*Pascucci et al.,* 2003; *Mohanty et al.,* 2004; *Apai et al.,* 2004; *Sterzik et al.,* 2004; *Furlan et al.,*

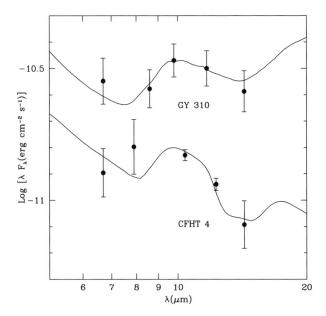

Fig. 8. Comparison of geometries for two BD disks. The best model fit for GY 310 consists of a flared disk with dust dominated by small submicrometer grains (*Mohanty et al., 2004*). A good match to the SED of CFHT 4 can be achieved with a disk model with a little flaring and micrometer-sized grains (*Apai et al., 2004*). The SED of GY 310 has been shifted up by 0.5 dex.

2005a; *Hartmann et al., 2005; Muzerolle et al., 2006*; Pascucci et al., in preparation). The modeling of these disks shows that flared, flat and intermediate flaring geometries all occur in BD disks (see Fig. 8). A similar trend is found for disks around more massive stars (e.g., *Furlan et al., 2005b*). As with studies of disks at stellar masses, samples of BD disks from a greater variety of ages and star-forming conditions are needed to distinguish between the effects of evolution and environment on disk structure.

6.5. Dust Processing

Grain growth and dust settling are thought to represent the first steps of planet formation (e.g., *Henning et al., 2006*). Studies of disks around intermediate-mass stars also indicate a possible link between grain growth and crystallinity, with high crystallinity measured in disks having grains larger than the dominant submicrometer interstellar grains (e.g., *van Boekel et al., 2005*). Determining whether BD disks evolve into planetary systems requires first identifying the presence of such dust processing. Because dust settling is related to the disk geometry, some evidence of dust processing can be gained by the kind of SED modeling described in the previous section. For instance, *Mohanty et al.* (2004) concluded that the SED of a young BD in Ophiuchus, GY 310, was consistent with a flared disk geometry and small interstellar grains, while *Apai et al.* (2004) found that the SED of CFHT 4 was indicative of a flat disk structure (Fig. 8). The latter authors also found that the peak

position of the silicate emission feature and the line-to-continuum flux ratio demonstrated that the emission was dominated by grains about 10× larger than the dominant 0.1-μm interstellar grains. Fitting the emission and the overall continuum required a disk with intermediate flaring. This work indicated that young BD disks process dust in a similar fashion as disks around stars (e.g., *Przygodda et al., 2003*).

The recent Spitzer spectroscopy of disks around low-mass stars and BDs has confirmed the results from the SED modeling. Most of the spectra show a silicate emission feature broader than that from the interstellar medium and peaks that are indicative of crystalline grains (see Fig. 9). *Furlan et al.* (2005a) found that the disk model of V410 Anon 13 also required a reduced gas-to-dust ratio, which was suggestive of some settling. A quantitative analysis of the dust composition of disks in Chamaeleon I reveals large grains and high crystallinity mass fractions (~40%) for the majority of the sources (*Apai et al., 2005*). In addition, most of the SEDs are consistent with flatter disk structures than those predicted by vertical hydrostatic equilibrium. These results demonstrate that dust processing is largely independent of stellar properties and mainly determined by local processes in the disk.

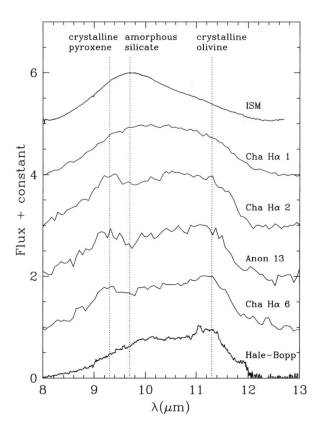

Fig. 9. Spitzer spectra of disks around low-mass stars and BDs (*Furlan et al., 2005a; Apai et al., 2005*). The spectra have been continuum-subtracted and normalized to the peak emission in the range between 7.6 and 13.5 μm. For comparison we show the spectra of the amorphous silicate-dominated interstellar medium and the crystalline-rich comet Hale-Bopp.

6.6. Planet Formation

The identification of grain growth, crystallization, and dust settling in BD disks indicate that the first steps leading to planet formation occur in disks around BDs. In fact, there now appears to be tantalizing evidence for planet formation at a more advanced stage around a low-mass object in IC 348, source 316 from *Luhman et al.* (2003b). This object has a spectral type of M6.5, indicating a mass near the hydrogen-burning mass limit, and no strong signature of accretion based on its small Hα equivalent width. Spitzer photometry has revealed strong excess emission at 24 μm, indicating a substantial disk (Fig. 10). Interestingly, no excess emission is seen at wavelengths shortward of 8 μm, strongly suggesting the presence of an inner hole in the disk that is cleared of at least small dust grains. Disk models of the SED require an inner hole size of 0.5–1 AU to fit the observations (*Muzerolle et al.,* 2006). IC 348-316 is thus the first low-mass object known to possess significant inner disk clearing akin to that seen in higher-mass CTTSs such as CoKu Tau/4 (*D'Alessio et al.,* 2005). The origin of this clearing is a matter of considerable debate. *Muzerolle et al.* (2006) ruled out the photoevaporation model, in which a photoevaporative wind generated by UV radiation from the central object can remove material from the inner disk (e.g., *Clarke et al.,* 2001), because the mass loss timescale is much longer than the age of IC 348-316 (1–3 m.y.) given plausible UV flux from an accretion shock or chromo-sphere. Other possibilities include inside-out dust coagulation into meter- or kilometer-sized planetesimals, or the rapid formation of a single planet that is preventing further accretion from the outer disk. For the latter scenario, *Muzerolle et al.* (2006) estimated a plausible mass range of $M_p \sim 2.5–25\ M_\oplus$. This type of SED analysis is not conclusive proof of the presence of a planetary companion, but nevertheless it strongly suggest that the same steps to planet formation interpreted from observations of disks around stars are also possible in disks around BDs. Higher-resolution data for IC 348-316 through Spitzer spectroscopy should better constrain the nature of its inner disk hole.

7. SUMMARY

We summarize the current observations of BDs that are relevant to their formation as follows:

1. The least-massive known free-floating BDs have masses of $\sim 10\ M_{Jup}$. No conclusive measurement of the minimum mass of BDs is yet available.

2. Stars outnumber BDs at 20–80 M_{Jup} by a factor of ~5–8 in star-forming regions. This ratio is consistent with data for BDs in the solar neighborhood, although the larger uncertainties in the field data allow for modest differences from star-forming regions (factor of a few).

3. Stars and BDs share similar velocity and spatial distributions in the available data for star-forming regions.

4. In the original BD desert observed at separations less than 3 AU from solar-type primaries, BD companions are less common than stellar companions by a factor of ~100. BDs are outnumbered by stars at larger separations as well, but the size of the deficiency (~3–10) is smaller than at close separations, and is consistent with the deficiency of BDs among isolated objects (~5–8). These data suggest that wider stellar and substellar companions form in the same manner as their free-floating counterparts, and that a true BD desert for solar-type stars is restricted to small separations.

5. For both star-forming regions and the solar neighborhood, binary fractions decrease continuously with mass from stars to BDs and most binary BDs have small separations, although a few wide systems do exist.

6. Accretion rates decrease continuously with mass from stars to BDs (as $\dot{M} \propto M^2$).

7. Circumstellar disks have been found around BDs with masses as low as $\sim 10\ M_{Jup}$.

8. The disk fraction of BDs is similar to that of stars at ages of a few million years. BDs also appear to have similar disk lifetimes, although a definitive statement is not possible with available data.

9. Disks around BDs exhibit a range of geometries from flat to flared, and some of these disks experience grain growth and settling and may develop inner holes, which are possible signatures of planet formation. All these characteristics are also found among disks around stars.

All these data are consistent with a common formation mechanism for BDs and stars. In particular, the existence of widely separated binary BDs and a likely isolated proto-

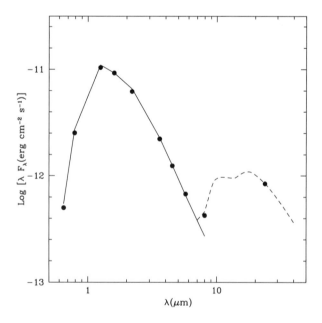

Fig. 10. SED of an object near the hydrogen-burning limit in IC 348 (points) (*Muzerolle et al.,* 2005) compared to the median SED of late-type members of IC 348 that lack IR excess emission (solid line) (*Lada et al.,* 2006). This object exhibits excess emission only at λ ≥ 8 μm, which has been fit with a model of a disk with an inner hole (dashed line).

BD (*Young et al.*, 2004; *Bourke et al.*, 2005; *Huard et al.*, 2006) indicate that some BDs are able to form in the same manner as stars through unperturbed cloud fragmentation. It remains possible that additional mechanisms such as ejection and photoevaporation influence the birth of some BDs, just as they likely do with stars. However, it appears that they are not essential ingredients in making it possible for these small bodies to form.

REFERENCES

Allen P. R., Koerner D. W., Reid I. N., and Trilling D. E. (2005) *Astrophys. J., 625*, 385–397.

Apai D., Pascucci I., Henning Th., Sterzik M. F., Klein R., et al. (2002) *Astrophys. J., 573*, L115–L117.

Apai D., Pascucci I., Sterzik M. F., van der Bliek N., Bouwman J., Dullemond C. P., and Henning Th. (2004) *Astron. Astrophys., 426*, L53–L57.

Apai D., Pascucci I., Bouwman J., Natta A., Henning Th., and Dullemond C. P. (2005) *Science, 310*, 834–836.

Ardila D., Martín E., and Basri G. (2000) *Astron. J., 120*, 479–487.

Barrado y Navascués D. and Martín E. L. (2003) *Astron. J., 126*, 2997–3006.

Barrado y Navascués D., Zapatero Osorio M. R., Béjar V. J. S., Rebolo R., Martín E. L., Mundt R., and Bailer-Jones C. A. L. (2001) *Astron. Astrophys., 377*, L9–L13.

Barrado y Navascués D., Zapatero Osorio M. R., Martín E. L., Béjar V. J. S., Rebolo R., and Mundt R. (2002) *Astron. Astrophys., 393*, L85–L88.

Barrado y Navascués D., Mohanty S., and Jayawardhana R. (2004a) *Astrophys. J., 604*, 284–296.

Barrado y Navascués D., Stauffer J. R., Bouvier J., Jayawardhana R., and Cuillandre J.-C. (2004b) *Astrophys. J., 610*, 1064–1078.

Baraffe I., Chabrier G., Allard F., and Hauschildt P. H. (1998) *Astron. Astrophys., 337*, 403–412.

Basri G. and Martín E. L. (1999) *Astrophys. J., 118*, 2460–2465.

Bate M. R. and Bonnell I. A. (2005) *Mon. Not. R. Astron. Soc., 356*, 1201–1221.

Bate M. R., Bonnell I. A., and Bromm V. (2002) *Mon. Not. R. Astron. Soc., 332*, L65–L68.

Bate M. R., Bonnell I. A., and Bromm V. (2003) *Mon. Not. R. Astron. Soc., 339*, 577–599.

Béjar V. J. S., Zapatero Osorio M. R., and Rebolo R. (1999) *Astrophys. J., 521*, 671–681.

Béjar V. J. S., Martín E. L., Zapatero Osorio M. R., Rebolo R., Barrado y Navascués D., et al. (2001) *Astrophys. J., 556*, 830–836.

Billères M., Delfosse X., Beuzit J.-L., Forveille T., Marchal L., and Martín E. L. (2005) *Astron. Astrophys., 440*, L55–L58.

Bonnell I. A. and Davies M. B. (1998) *Mon. Not. R. Astron. Soc., 295*, 691–698.

Boss A. (2001) *Astrophys. J., 551*, L167–L170.

Bourke T. L., Crapsi A., Myers P. C., Evans N. J., Wilner D. J., et al. (2005) *Astrophys. J., 633*, L129–L132.

Bouy H., Brandner W., Martín E. L., Delfosse X., Allard F., and Basri G. (2003) *Astron. J., 126*, 1526–1554.

Bouy H., Brandner W., Martín E. L., Delfosse X., Allard F., et al. (2004) *Astron. Astrophys., 424*, 213–226.

Bouy H., Martín E. L., Brandner W., Zapatero Osorio M. R., Béjar V. J. S., et al. (2006) *Astron. Astrophys., 451*, 177–186.

Briceño C., Hartmann L., Stauffer J., and Martín E. (1998) *Astron. J., 115*, 2074–2091.

Briceño C., Luhman K. L., Hartmann L., Stauffer J. R., and Kirkpatrick J. D. (2002) *Astrophys. J., 580*, 317–335.

Burgasser A. J., Kirkpatrick J. D., Reid I. N., Brown M. E., Miskey C. L., and Gizis J. E. (2003) *Astrophys. J., 586*, 512–526.

Burgasser A. J., Kirkpatrick J. D., McGovern M. R., McLean I. S., Prato L., and Reid I. N. (2004) *Astrophys. J., 604*, 827–831.

Burrows A., Marley M., Hubbard W. B., Lunine J. I., Guillot T., et al. (1997) *Astrophys. J., 491*, 856–875.

Caballero J. A., Béjar V. J. S., Rebolo R., and Zapatero Osorio M. R. (2004) *Astron. Astrophys., 424*, 857–872.

Calvet N. and Gullbring E. (1998) *Astrophys. J., 509*, 802–818.

Chabrier G. (2002) *Astrophys. J., 567*, 304–313.

Chabrier G., Baraffe I., Allard F., and Hauschildt P. H. (2000) *Astrophys. J., 542*, 464–472.

Chauvin G., Lagrange A.-M., Dumas C., Zuckerman B., Mouillet D., et al. (2004) *Astron. Astrophys., 425*, L29–L32.

Chauvin G., Lagrange A.-M., Dumas C., Zuckerman B., Mouillet D., et al. (2005) *Astron. Astrophys., 438*, L25–L28.

Chiang E. I. and Goldreich P. (1997) *Astrophys. J., 490*, 368–376.

Chiang E. I. and Goldreich P. (1999) *Astrophys. J., 519*, 279–284.

Clarke C. J., Gendrin A., and Sotomayor M. (2001) *Mon. Not. R. Astron. Soc., 328*, 485–491.

Close L. M., Siegler N., Freed M., and Biller B. (2003) *Astrophys. J., 587*, 407–422.

Comerón F., Rieke G. H., Claes P., Torra J., and Laureijs R. J. (1998) *Astron. Astrophys., 335*, 522–532.

Comerón F., Rieke G. H., and Neuhäuser R. (1999) *Astron. Astrophys., 343*, 477–495.

Comerón F., Neuhäuser R., and Kaas A. A. (2000) *Astron. Astrophys., 359*, 269–288.

Comerón F., Reipurth B., Henry A., and Fernández M. (2004) *Astron. Astrophys., 417*, 583–596.

Cushing M. C., Tokunaga A. T., and Kobayashi N. (2000) *Astron. J., 119*, 3019–3025.

D'Alessio P., Hartmann L., Calvet N., Franco-Hernández R., Forrest W. J., et al. (2005) *Astrophys. J., 621*, 461–472.

Delgado-Donate E. J., Clarke C. J., and Bate M. R. (2004) *Mon. Not. R. Astron. Soc., 347*, 759–770.

Dobashi K., Uehara H., Kandori R., Sakurai T., Kaiden M., et al. (2005) *Publ. Astron. Soc. Japan, 57*, S1–S386.

Duchêne G., Bouvier J., and Simon T. (1999) *Astron. Astrophys., 343*, 831–840.

Dullemond C. P. and Dominik C. (2005) *Astron. Astrophys., 434*, 971–986.

Duquennoy A. and Mayor M. (1991) *Astron. Astrophys., 248*, 485–524.

Fazio G. G., Hora J. L., Allen L. E., Ashby M. L. N., Barmby P., et al. (2004) *Astrophys. J. Suppl., 154*, 10–17.

Feigelson E. D., Gaffney J. A., Garmire G., Hillenbrand L. A., and Townsley L. (2003) *Astrophys. J., 584*, 911–930.

Fernández M. and Comerón F. (2001) *Astron. Astrophys., 380*, 264–276.

Fischer D. A. and Marcy G. W. (1992) *Astrophys. J., 396*, 178–194.

Furlan E., Calvet N., D'Alessio P., Hartmann L., Forrest W. J., et al. (2005a) *Astrophys. J., 621*, L129–L132.

Furlan E., Calvet N., D'Alessio P., Hartmann L., Forrest W. J., et al. (2005b) *Astrophys. J., 628*, L65–L68.

Gizis J. E. (2002) *Astrophys. J., 575*, 484–492.

Gizis J. E., Kirkpatrick J. D., Burgasser A., Reid I. N., Monet D. G., et al. (2001) *Astrophys. J., 551*, L163–L166.

Gizis J. E., Reid I. N., Knapp G. R., Liebert J., Kirkpatrick J. D., et al. (2003) *Astron. J., 125*, 3302–3310.

Gizis J. E., Shipman H. L., and Harvin J. A. (2005) *Astrophys. J., 630*, L89–L91.

Glassgold A. E., Najita J., and Igea J. (2004) *Astrophys. J., 615*, 972–990.

Guenther E. W. and Wuchterl G. (2003) *Astron. Astrophys., 401*, 677–683.

Guieu S., Dougados C., Monin J.-L., Magnier E., and Martin E. L. (2006) *Astron. Astrophys., 446*, 485–500.

Gullbring E., Hartmann L., Briceño C., and Calvet N. (1998) *Astrophys. J., 492*, 323–341.

Haisch K. E., Lada E. A., and Lada C. J. (2001) *Astrophys. J., 553*, L153–L156.

Harrington R. S., Dahn C. C., and Guetter H. H. (1974) *Astrophys. J., 194*, L87–L87.

Hartmann L., Megeath S. T., Allen L. E., Luhman K. L., Calvet N., et al. (2005) *Astrophys. J., 629,* 881–896.

Henning Th., Dullemond C. P., Wolf S., and Dominik C. (2006) In *Planet Formation: Theory, Observation and Experiments* (H. Klahr and W. Brandner, eds.), pp. 112–128. Cambridge Univ., Cambridge.

Hillenbrand L. A. (1997) *Astron. J., 113,* 1733–1768.

Hillenbrand L. A. and Carpenter J. M. (2000) *Astrophys. J., 540,* 236–254.

Hillenbrand L. A., Strom S. E., Calvet N., Merrill K. M., Gatley I., et al. (1998) *Astron. J., 116,* 1816–1841.

Huard T. L., Myers P. C., Murphy D. C., Crews L. J., Lada C. J., et al. (2006) *Astrophys. J., 640,* 391–401.

Jayawardhana R., Ardila D. R., Stelzer B., and Haisch K. E. (2003a) *Astron. J., 126,* 1515–1521.

Jayawardhana R., Mohanty S., and Basri G. (2003b) *Astrophys. J., 592,* 282–287.

Joergens V. (2006a) *Astron. Astrophys., 446,* 1165–1176.

Joergens V. (2006b) *Astron. Astrophys., 448,* 655–663.

Joergens V. and Guenther E. (2001) *Astron. Astrophys., 379,* L9–L12.

Kenyon S. J. and Hartmann L. (1987) *Astrophys. J., 323,* 714–733.

Kenyon S. J. and Hartmann L. (1995) *Astrophys. J. Suppl., 101,* 117–171.

Kirkpatrick J. D., Barman T. S., Burgasser A. J., McGovern M. R., McLean I. S., et al. (2006) *Astrophys. J., 639,* 1120–1128.

Klein R., Apai D., Pascucci I., Henning Th., and Waters L. B. F. M. (2003) *Astrophys. J., 593,* L57–L60.

Kraus A. L., White R. J., and Hillenbrand L. A. (2005) *Astrophys. J., 633,* 452–459.

Kraus A. L., White R. J., and Hillenbrand L. A. (2006) *Astrophys. J.,* in press.

Kroupa P. and Bouvier J. (2003) *Mon. Not. R. Astron. Soc., 346,* 369–380.

Lada C. J., Muench A. A., Haisch K. E., Lada E. A., Alves J. F., et al. (2000) *Astron. J., 120,* 3162–3176.

Lada C. J., Muench A. A., Lada E. A., and Alves J. F. (2004) *Astron. J., 128,* 1254–1264.

Lada C. J., Muench A. A., Luhman K. L., Allen L., Hartmann L., et al. (2006) *Astron. J., 131,* 1574–1607.

Levine J. L., Steinhauer A., Elston R. J., and Lada E. A. (2006) *Astrophys. J.,* in press.

Liu M. C., Najita J., and Tokunaga A. T. (2003) *Astrophys. J., 585,* 372–391.

Lucas P. W., Roche P. F., Allard F., and Hauschildt P. H. (2001) *Mon. Not. R. Astron. Soc., 326,* 695–721.

Lucas P. W., Roche P. F., and Tamura M. (2005) *Mon. Not. R. Astron. Soc., 361,* 211–232.

Luhman K. L. (1999) *Astrophys. J., 525,* 466–481.

Luhman K. L. (2000) *Astrophys. J., 544,* 1044–1055.

Luhman K. L. (2004a) *Astrophys. J., 602,* 816–842.

Luhman K. L. (2004b) *Astrophys. J., 614,* 398–403.

Luhman K. L. (2004c) *Astrophys. J., 617,* 1216–1232.

Luhman K. L. (2005) *Astrophys. J., 633,* L41–L44.

Luhman K. L. (2006) *Astrophys. J., 645,* 676–687.

Luhman K. L. and Potter D. (2006) *Astrophys. J., 638,* 887–896.

Luhman K. L., Liebert J., and Rieke G. H. (1997) *Astrophys. J., 489,* L165–L168.

Luhman K. L., Rieke G. H., Lada C. J., and Lada E. A. (1998) *Astrophys. J., 508,* 347–369.

Luhman K. L., Rieke G. H., Young E. T., Cotera A. S., Chen H., et al. (2000) *Astrophys. J., 540,* 1016–1040.

Luhman K. L., Briceño C., Stauffer J. R., Hartmann L., Barrado y Navascués D., and Caldwell N. (2003a) *Astrophys. J., 590,* 348–356.

Luhman K. L., Stauffer J. R., Muench A. A., Rieke G. H., Lada E. A., et al. (2003b) *Astrophys. J., 593,* 1093–1115.

Luhman K. L., Peterson D. E., and Megeath S. T. (2004) *Astrophys. J., 617,* 565–568.

Luhman K. L., Lada E. A., Muench A. A., and Elston R. J. (2005a) *Astrophys. J., 618,* 810–816.

Luhman K. L., D'Alessio P., Calvet N., Allen L. E., Hartmann L., et al. (2005b) *Astrophys. J., 620,* L51–L54.

Luhman K. L., McLeod K. K., and Goldenson N. (2005c) *Astrophys. J., 623,* 1141–1156.

Luhman K. L., Lada C. J., Hartmann L., Muench A. A., Megeath S. T., et al. (2005d) *Astrophys. J., 631,* L69–L72.

Luhman K. L., Adame L., D'Alessio P., Calvet N., Hartmann L., et al. (2005e) *Astrophys. J., 635,* L93–L96.

Marcy G. W. and Butler R. P. (2000) *Publ. Astron. Soc. Pac., 112,* 137–140.

Marcy G., Butler R. P., Fischer D., Wright J. T., Tinney C. G., and Jones H. R. A. (2005) *Progr. Theor. Phys. Suppl., 158,* 24–42.

Martín E. L. and Zapatero Osorio M. R. (2003) *Astrophys. J., 593,* L113–L116.

Martín E. L., Brander W., Bouvier J., Luhman K. L., Stauffer J., et al. (2000) *Astrophys. J., 543,* 299–312.

Martín E. L., Zapatero Osorio M. R., Barrado y Navascués D., Béjar V. J. S., Rebolo R., et al. (2001a) *Astrophys. J., 558,* L117–L121.

Martín E. L., Dougados C., Magnier E., Ménard F., Magazzù A., et al. (2001b) *Astrophys. J., 561,* L195–L198.

Martín E. L., Delfosse X., and Guieu S. (2004) *Astron. J., 127,* 449–454.

Masciadri E. and Raga A. C. (2004) *Astrophys. J., 615,* 850–854.

Mazeh T., Goldberg D., Duquennoy A., and Mayor M. (1992) *Astrophys. J., 265,* 265–268.

McCarthy C. and Zuckerman B. (2004) *Astron. J., 127,* 2871–2884.

McGehee P. M., West A. A., Smith J. A., Anderson K. S. J., and Brinkmann J. (2005) *Astron. J., 130,* 1752–1762.

Mohanty S. and Basri G. (2003) *Astrophys. J., 583,* 451–472.

Mohanty S., Jayawardhana R., and Barrado y Navascués D. (2003) *Astrophys. J., 593,* L109–L112.

Mohanty S., Jayawardhana R., Natta A., Fujiyoshi T., Tamura M., and Barrado y Navascués D. (2004) *Astrophys. J., 609,* L33–L36.

Mohanty S., Jayawardhana R., and Basri G. (2005) *Astrophys. J., 626,* 498–522.

Moraux E. and Clarke C. (2005) *Astron. Astrophys., 429,* 895–901.

Muench A. A., Alves J., Lada C. J., and Lada E. A. (2001) *Astrophys. J., 558,* L51–L54.

Muench A. A., Lada E. A., Lada C. J., and Alves J. (2002) *Astrophys. J., 573,* 366–393.

Muench A. A., Lada E. A., Lada C. J., Elston R. J., Alves J. F., et al. (2003) *Astron. J., 125,* 2029–2049.

Muzerolle J., Hartmann L., and Calvet N. (1998) *Astron. J., 116,* 455–468.

Muzerolle J., Briceño C., Calvet N., Hartmann L., Hillenbrand L., and Gullbring E. (2000) *Astrophys. J., 545,* L141–L144.

Muzerolle J., Calvet N., and Hartmann L. (2001) *Astrophys. J., 550,* 944–961.

Muzerolle J., Hillenbrand L., Calvet N., Briceño C., and Hartmann L. (2003) *Astrophys. J., 592,* 266–281.

Muzerolle J., Luhman K. L., Briceño C., Hartmann L., and Calvet N. (2005) *Astrophys. J., 625,* 906–912.

Muzerolle J., Adame L., D'Alessio P., Calvet N., Luhman K. L., et al. (2006) *Astrophys. J., 643,* 1003–1010.

Natta A. and Testi L. (2001) *Astron. Astrophys., 376,* L22–L25

Natta A., Testi L., Comerón F., Oliva E., D'Antona F., et al. (2002) *Astron. Astrophys., 393,* 597–609.

Natta A., Testi L., Muzerolle J., Randich S., Comerón F., and Persi P. (2004) *Astron. Astrophys., 424,* 603–612.

Neuhäuser R. and Comerón F. (1999) *Astron. Astrophys., 350,* 612–616.

Neuhäuser R., Brandner W., Alves J., Joergens V., and Comerón F. (2002) *Astron. Astrophys., 384,* 999–1011.

Oasa Y., Tamura M., and Sugitani K. (1999) *Astrophys. J., 526,* 336–343.

Padoan P. and Nordlund A. (2004) *Astrophys. J., 617,* 559–564.

Padoan P., Kritsuk A., Norman M. L., and Nordlund A. (2005) *Astrophys. J., 622,* L61–L64.

Pascucci I., Apai D., Henning Th., and Dullemond C. P. (2003) *Astrophys. J., 590,* L111–L114.

Persi P., Marenzi A. R., Olofsson G., Kaas A. A., Nordh L., et al. (2000) *Astron. Astrophys., 357,* 219–224.

Phan-Bao N., Martín E. L., Reylé C., Forveille T., and Lim J. (2005) *Astron. Astrophys., 439,* L19–L22.

Przygodda F., van Boekel R., Àbrahàm P., Melnikov S. Y., Waters L. B. F. M., and Leinert Ch. (2003) *Astron. Astrophys., 412,* L43–L46.

Reid I. N., Gizis J. E., Kirkpatrick J. D., and Koerner D. W. (2001) *Astron. J., 121,* 489–502.

Reid I. N., Kirkpatrick J. D., Liebert J., Gizis J. E., Dahn C. C., and Monet D. G. (2002) *Astron. J., 124,* 519–540.

Reipurth B. and Clarke C. (2001) *Astron. J., 122,* 432–439.

Scholz A. and Eislöffel J. (2004) *Astron. Astrophys., 419,* 249–267.

Scholz A. and Eislöffel J. (2005) *Astron. Astrophys., 429,* 1007–1023.

Scholz A., Jayawardhana R., and Brandeker A. (2005) *Astrophys. J., 629,* L41–L44.

Scholz R.-D., McCaughrean M. J., Zinnecker H., and Lodieu N. (2005) *Astron. Astrophys., 430,* L49–L52.

Siegler N., Close L. M., Cruz K. L., Martín E. L., and Reid I. N. (2005) *Astrophys. J., 621,* 1023–1032.

Slesnick C. L., Hillenbrand L. A., and Carpenter J. M. (2004) *Astrophys. J., 610,* 1045–1063.

Stassun K., Mathieu R. D., and Valenti J. A. (2006) *Nature, 440,* 311–314.

Stauffer J. R., Hamilton D., and Probst R. (1994) *Astron. J., 108,* 155–159.

Sterzik M. F. and Durisen R. H. (2003) *Astron. Astrophys., 400,* 1031–1042.

Sterzik M. F., Pascucci I., Apai D., van der Bliek N., and Dullemond C. P. (2004) *Astron. Astrophys., 427,* 245–250.

Testi L., Natta A., Oliva E., D'Antona F., Comeron F., et al. (2002) *Astrophys. J., 571,* L155–L159.

Umbreit S., Burkert A., Henning Th., Mikkola S., and Spurzem R. (2005) *Astrophys. J., 623,* 940–951.

van Boekel R., Min M., Waters L. B. F. M., de Koter A., Dominik C., et al. (2005) *Astron. Astrophys., 437,* 189–208.

Walker C.,Wood K., Lada C. J., Robitaille T., Bjorkman J. E., and Whitney B. (2004) *Mon. Not. R. Astron. Soc., 351,* 607–616.

Werner M. W., Roellig T. L., Low F. J., Rieke G. H., Rieke M., et al. (2004) *Astrophys. J. Suppl., 154,* 1–9.

Whelan E. T., Ray T. P., Bacciotti F., Natta A., Testi L., and Randich S. (2005) *Nature, 435,* 652–654.

White R. J. and Basri G. (2003) *Astrophys. J., 582,* 1109–1122.

White R. J. and Ghez A. M. (2001) *Astrophys. J., 556,* 265–295.

White R. J. and Hillenbrand L. A. (2004) *Astrophys. J., 616,* 998–1032.

White R. J., Ghez A. M., Reid I. N., and Schultz G. (1999) *Astrophys. J., 520,* 811–821.

Whitworth A. P. and Zinnecker H. (2004) *Astron. Astrophys., 427,* 299–306.

Wilking B. A., Greene T. P., and Meyer M. R. (1999) *Astron. J., 117,* 469–482.

Wilking B. A., Meyer M. R., Greene T. P., Mikhail A., and Carlson G. (2004) *Astron. J., 127,* 1131–1146.

Young C. H., Jørgensen J. K., Shirley Y. L., Kauffmann J., Huard T., et al. (2004) *Astrophys. J. Suppl., 154,* 396–401.

Zapatero Osorio M. R., Béjar V. J. S., Rebolo R., Martín E. L., and Basri G. (1999) *Astrophys. J., 524,* L115–L118.

Zapatero Osorio M. R., Béjar V. J. S., Martín E. L., Rebolo R., Barrado y Navascués D., et al. (2000) *Science, 290,* 103–107.

Zapatero Osorio M. R., Béjar V. J. S., Martín E. L., Barrado y Navascués D., and Rebolo R. (2002a) *Astrophys. J., 569,* L99–L102.

Zapatero Osorio M. R., Béjar V. J. S., Martín E. L., Rebolo R., Barrado y Navascués D., et al. (2002b) *Astrophys. J., 578,* 536–542.

Zapatero Osorio M. R., Béjar V. J. S., Pavlenko Y., Rebolo R., Allende P., et al. (2002c) *Astron. Astrophys., 384,* 937–953.

The Formation of Brown Dwarfs: Theory

Anthony Whitworth
Cardiff University

Matthew R. Bate
University of Exeter

Åke Nordlund
University of Copenhagen

Bo Reipurth
University of Hawaii

Hans Zinnecker
Astrophysikalisches Institut, Potsdam

We review five mechanisms for forming brown dwarfs: (1) turbulent fragmentation of molecular clouds, producing very-low-mass prestellar cores by shock compression; (2) collapse and fragmentation of more massive prestellar cores; (3) disk fragmentation; (4) premature ejection of protostellar embryos from their natal cores; and (5) photoerosion of pre-existing cores overrun by HII regions. These mechanisms are not mutually exclusive. Their relative importance probably depends on environment, and should be judged by their ability to reproduce the brown dwarf IMF, the distribution and kinematics of newly formed brown dwarfs, the binary statistics of brown dwarfs, the ability of brown dwarfs to retain disks, and hence their ability to sustain accretion and outflows. This will require more sophisticated numerical modeling than is presently possible, in particular more realistic initial conditions and more realistic treatments of radiation transport, angular momentum transport, and magnetic fields. We discuss the minimum mass for brown dwarfs, and how brown dwarfs should be distinguished from planets.

1. INTRODUCTION

The existence of brown dwarfs was first proposed on theoretical grounds by *Kumar* (1963) and *Hayashi and Nakano* (1963). However, more than three decades then passed before brown dwarfs were observed unambiguously (*Rebolo et al.,* 1995; *Nakajima et al.,* 1995; *Oppenheimer et al.,* 1995). Brown dwarfs are now observed routinely, and it is therefore appropriate to ask how brown dwarfs form, and in particular to ascertain (1) whether brown dwarfs form in the same way as H-burning stars, and (2) whether there is a clear distinction between the mechanisms that produce brown dwarfs and those that produce planets.

In section 2 we argue that the mechanisms forming brown dwarfs are no different from those forming low-mass H-burning stars, on the grounds that the statistical properties of brown dwarfs (mass function, clustering properties, kinematics, binary statistics, accretion rates, etc.) appear to form a smooth continuum with those of low-mass H-burning stars. Understanding how brown dwarfs form is therefore the key to understanding what determines the minimum mass for star formation. In section 3 we review the basic physics of star formation, as it applies to brown dwarfs, and derive key analytic results, in particular the minimum mass for opacity-limited fragmentation in various different formation scenarios. We have deliberately assembled most of the mathematical analysis in this one section. In sections 4 to 8 we consider five different mechanisms that may be involved in the formation of brown dwarfs. Section 4 explores the possibility that turbulent fragmentation of molecular clouds produces prestellar cores of such low mass that they inevitably collapse to form brown dwarfs. Section 5 considers more massive prestellar cores, and the possibility that they fragment dynamically as they collapse, thereby spawning protostars with a range of masses. The collapse of such cores ceases when the gas starts to heat up (due to adiabatic compression) and/or when rotation becomes important. Unfortunately, the interplay of dynamics and adiabatic compression, while likely to play a critical role (e.g., *Boss et al., 2000*), is hard to analyze. On the other hand, rotational effects can be analyzed systematically, and therefore section 6 explores the formation of brown dwarfs by gravitational instabilities in disks, considering first isolated relaxed disks, then unrelaxed disks, and finally interacting disks in clusters. Section 7 reviews the process of competitive accretion, which determines how an ensemble of protostellar embryos evolves to populate higher masses; and then considers the N-body processes that (1) may eject brown dwarfs and low-mass stars from their natal cores, thereby terminating accretion and effectively capping their masses, and (2) may influence the binary statistics and clustering properties of brown dwarfs. Section 8 considers the formation of brown dwarfs by photoerosion of pre-existing cores that are overrun by HII regions. Section 9 presents numerical simulations of the birth of a whole star cluster in a large protocluster core; in these simulations, the mechanisms of section 5 (collapse and fragmentation), section 6 (disk fragmentation), and section 7 (premature ejection) all occur simultaneously and in tandem, and the collective properties of brown dwarfs formed can be extracted for comparison with observation. In section 10 we summarize the principal conclusions of this review.

Comprehensive discussions of the observational properties of brown dwarfs, and how they may constrain formation mechanisms, are contained in the chapters by Luhman et al. (which deals with the entire observational picture) and Burgasser et al. (which deals specifically with the issue of multiplicity). Therefore we offer only a brief summary of the observations in section 2. Likewise, a critique of simulations of core fragmentation is given in the chapter by Goodwin et al.; simulations of disk fragmentation are discussed and compared in detail in the chapter by Durisen et al.; and the origin of the IMF is treated in the chapter by Bonnell et al., so we have limited our consideration to those aspects that pertain specifically to the origin of brown dwarfs.

2. EVIDENCE THAT BROWN DWARFS FORM LIKE LOW-MASS HYDROGEN-BURNING STARS

We shall assume that brown dwarfs form in the same way as H-burning stars, i.e., on a dynamical timescale, by gravitational instability, and with initially uniform elemental composition (reflecting the composition of the interstellar matter out of which they form). (By implication, we distinguish *brown dwarfs* from *planets*, a term that we will here reserve for objects that form on a much longer timescale, by the amalgamation of a rocky core and — if circumstances allow — the subsequent accretion of a gaseous envelope. This results in planets having an initially fractionated elemental composition with an overall deficit of light elements.) If this is the correct way to view the formation of brown

dwarfs, then brown dwarfs should not be distinguished from stars. Many stars fail to burn helium, and most fail to burn carbon, without forfeiting the right to be called stars. The reason for categorizing brown dwarfs as stars is that the statistical properties of brown dwarfs appear to form a continuum with those of low-mass H-burning stars (and not with those of high-mass planets).

2.1. The Initial Mass Function, Clustering Statistics, and Velocity Dispersion

The initial mass function (IMF) is apparently continuous across the H-burning limit at ~0.075 $M_\odot$, in the Trapezium cluster (*Slesnick et al., 2004*), in the σ Orionis cluster (*Bejar et al., 2001*), in Taurus (*Luhman, 2004*), in IC 348 (*Luhman et al., 2003*), in the Pleiades (*Moraux et al., 2003*), in the field stars of the disk (*Chabrier, 2003*), and even possibly in the halo (*Burgasser et al., 2003b*). If the IMF is fitted by a power law across the H-burning limit, $dN/dM \propto M^{-\alpha}$, estimates of α fall in the range 0.4 to 0.8 (e.g., *Moraux et al., 2003*). The IMF appears to extend down to a few Jupiter masses (e.g., *Zapatero Osorio et al., 2002*; *McCaughrean et al., 2002*; *Lucas et al., 2005*). The continuity of the IMF across the H-burning limit is not surprising, since the processes that determine the mass of a low-mass star are presumed to occur at relatively low densities (ρ ≲ 10^{-8} g cm^{-3}) and temperatures (T ≲ 2000 K), long before the material involved knows whether it will reach sufficiently high density (ρ ≳ 1 g cm^{-3}) to be supported in perpetuity by electron degeneracy pressure before or after it reaches sufficiently high temperature (T ≳ 10^7 K) to burn hydrogen.

In the Trapezium cluster, and in Taurus, brown dwarfs appear to be homogeneously mixed with H-burning stars (*Lucas and Roche, 2000*; *Briceño et al., 2002*; *Luhman, 2004*), and in Chamaeleon I their kinematics are also essentially indistinguishable (*Joergens, 2006b*). Although they have been searched for — as possible signatures of formation by ejection — neither a greater velocity dispersion of brown dwarfs in very young clusters, nor a diaspora of brown dwarfs around older clusters, has been found.

2.2. Binary Statistics

Multiple systems involving brown dwarf secondaries can be categorized based on whether the primary is a Sun-like star or another brown dwarf.

1. Sun-like primaries. Sun-like stars seldom have brown dwarf companions. At close separations (≲5 AU), the frequency of companions with masses in the range 0.01 to 0.1 $M_\odot$ is ~0.5% (*Marcy and Butler, 2000*); this figure is known rather accurately due to the numerous Doppler surveys aimed at detecting extrasolar planets. Since the frequency of companions outside this mass range (both exoplanets below, and H-burning stars above) is much higher, the paucity of brown dwarf companions is termed the "brown dwarf desert." However, the chapter by Luhman et al. points out that the paucity of close brown dwarf companions to Sun-like stars may simply reflect the overall paucity of brown dwarfs. At larger separations (≥100 AU), only about 20 brown dwarf companions to Sun-like stars have been found to date, indicating that these systems are also rare (frequency ~1 to 2%), although this estimate is based on more limited statistics (e.g., *Gizis et al.*, 2001; *McCarthy and Zuckerman*, 2004).

2. *Brown-dwarf primaries.* BD-BD binary systems (and binary systems with very-low-mass H-burning primaries) have an observed multiplicity of ~10–20% for separations greater than ~2 AU (*Bouy et al.*, 2003; *Burgasser et al.*, 2003a; *Close et al.*, 2003; *Gizis et al.*, 2003), and all but 5 of the ~75 known BD-BD binaries have separations less than ~20 AU (see Table 1 in the chapter by Burgasser et al.). Below ~2 AU, the multiplicity is unknown because most surveys to date have been imaging surveys that cannot resolve close systems. Only a few spectroscopic BD-BD binaries have been discovered (*Basri and Martin*, 1999; *Stassun et al.*, 2006; *Kenyon et al.*, 2005). Some authors speculate that the overall multiplicity might be ~30–50%, based on the positions of brown dwarfs in color-magnitude diagrams (*Pinfield et al.*, 2003) and the statistics of radial velocity variations (*Maxted and Jeffries*, 2005). However, *Joergens* (2006a) has examined the radial velocities of 10 BDs and very-low-mass H-burning stars, and finds no binaries with separations less than ~0.1 AU, and only two objects with variability on timescales greater than 100 days that might indicate companions at greater than ~0.2 AU. Thus, the peak in the separation distribution for BD-BD binaries is likely to be at ~1 to 4 AU. In contrast, the distribution of separations in binaries with Sun-like primaries peaks at ~30 AU (*Duquennoy and Mayor*, 1991). The mass ratio distribution also seems to be dependent on primary mass, with BD-BD binaries having a distribution that peaks toward equal masses ($q_{PEAK} \sim 1$), while binaries with Sun-like primaries have $q_{PEAK} \sim 0.3$. The implication is that, as the primary mass decreases, (1) the multiplicity decreases, (2) the distribution of semimajor axes shifts to smaller separations and becomes narrower (logarithmically), and (3) the distribution of mass ratios shifts toward unity — with these trends all continuing across the divide between H-burning stars and brown dwarfs.

3. *Exotica.* Finally, we note that some brown dwarfs are components in more complex systems. There are at least six systems currently known in which an H-burning star is orbited at large distance (≥50 AU) by a BD-BD binary. Indeed, preliminary results suggest that wide brown dwarf companions to H-burning stars are 2–3× more likely to be in a close BD-BD binary than are field brown dwarfs (*Burgasser et al.*, 2005). Cases also exist in which a close binary with H-burning components is orbited by a wide brown dwarf, while *Bouy et al.* (2005) have recently reported the discovery of what is likely to be a triple brown dwarf system. However, the statistics of these exotic systems are currently too small to interpret quantitatively.

2.3. Disks, Accretion, and Outflows

Young brown dwarfs are observed to have infrared excesses indicative of circumstellar disks, like young H-burning stars (*Muench et al.*, 2001; *Natta and Testi*, 2001; *Jayawardhana et al.*, 2003; *Mohanty et al.*, 2004). Brown-dwarf disk lifetimes are estimated to be 3–10 m.y., again like H-burning stars. From their Hα emission-line profiles, there is evidence for ongoing magnetospheric accretion onto brown dwarfs (*Scholz and Eislöffel*, 2004), and the inferred accretion rates form a continuous distribution with those for H-burning stars, fitted approximately by $\dot{M} \sim 10^{-8} \, M_\odot \, \mathrm{yr}^{-1} \, (M/M_\odot)^2$ (*Muzerolle et al.*, 2003, 2005). To sustain these estimated accretion rates, brown dwarfs only require rather low-mass disks ~$10^{-4} \, M_\odot$. Finally, the spectra of brown dwarfs also show forbidden emission lines suggestive of outflows like those from H-burning stars (*Fernández and Comerón*, 2001; *Natta et al.*, 2004), and recently an outflow from a brown dwarf has been resolved spatially (*Whelan et al.*, 2005). Thus, in the details of their circumstellar disks, accretion rates, and outflows, young brown dwarfs appear to mimic H-burning stars very closely, and to differ significantly only in scale.

2.4. Rotation and X-Rays

The rotational properties of brown dwarfs also appear to connect smoothly with those of very-low-mass H-burning stars (e.g., *Joergens et al.*, 2003). There is a decrease in the amplitude of periodic photometric variations with decreasing mass, presumably because the decreasing surface temperature leads to weaker coupling between the gas and the magnetic field and hence smaller spots. As with H-burning stars, brown dwarfs show evidence for braking by their accretion disks; those with strong accretion signatures (broad Hα emission) are exclusively slow rotators. X-ray emission is detected from M-type brown dwarfs, with properties very similar to the X-ray emission from very-low-mass H-burning stars (*Preibisch et al.*, 2005), but none is detected from L-type brown dwarfs, which are too cool to support surface magnetic activity.

2.5. Synopsis

Given the continuity of statistical properties between brown dwarfs and H-burning stars, it is probably unhelpful to distinguish the formation of brown dwarfs from the formation of stars, and in the rest of this review we will only use the H-burning limit at ~0.075 $M_\odot$ as a reference point in the range of stellar masses. The D-burning limit at ~0.012 $M_\odot$ falls in the same category. We will then define a star as any object forming on a dynamical timescale by gravitational instability. With this definition, there may well be a small overlap between the mass range of stars and that of planets. Given that in the immediate future we are unlikely to know too much more than the masses and radii of the lowest-mass objects, and certainly not their internal composition, we will simply have to accept that there is a gray area in the range 0.001–0.01 $M_\odot$ that may harbor both stars and planets, and possibly even hybrids.

It follows that understanding how brown dwarfs form is important, not just for its own sake, but because it is a key element in understanding why most stars have masses in the range 0.01 $M_\odot$–100 $M_\odot$ — and hence why there are lots of hospitable stars like the Sun with long-lived habitable zones, and enough heavy elements to produce rocky planets and life. The high-mass cutoff is probably due to the fact that radiation pressure makes it hard to form the highest-mass stars. The low-mass cutoff is probably due to the opacity limit. By studying brown dwarf formation we seek to confirm and quantify the low-mass cutoff.

3.   STAR-FORMATION THERMODYNAMICS

In this section we review the basic thermodynamics of gravitational collapse and fragmentation. The first three subsections deal — respectively — with three-dimensional collapse and hierarchical fragmentation (section 3.1); two-dimensional one-shot fragmentation of a shock compressed layer (section 3.2); and fragmentation of a disk (section 3.3). We describe and contrast these different environments, and in each case estimate the minimum mass for star formation. In section 3.3 we conclude that brown dwarf formation by fragmentation of disks around Sun-like stars is more likely in the cooler outer parts (R ≳ 100 AU), and this may explain why brown dwarf companions to Sun-like stars almost always have large separations. In section 3.4 we explain why impulsive compression does not promote cooling of an optically thick fragment. In section 3.5 we suggest that close BD-BD binaries may be formed by secondary fragmentation promoted by the softening of the equation of state when H_2 dissociates, and enhanced cooling due to the opacity gap. In section 3.6 we speculate that close BD-BD binaries may form in the outer parts of disks around Sun-like stars.

We caution that analytic estimates cannot capture all the nonlinear effects that are likely to occur, and are probably important, in a process as chaotic as star formation; they are therefore only indicative. A full understanding of any mode of star formation requires detailed simulations with all potentially influential physical effects included. However, as long as converged, robust simulations with all this physics properly included remain beyond the compass of current supercomputers, analytic estimates provide useful insights into the trends to be expected.

In sections 3.1 to 3.3, we will be mainly concerned with contemporary star formation in the disk of the Milky Way, and therefore with molecular hydrogen at temperatures T ≲ 100 K where the rotational levels are not strongly excited. In this regime the adiabatic exponent is $\gamma \simeq 5/3$ and the isothermal sound speed is a ≃ 0.06 km s^{-1} (T/K)$^{1/2}$. With these assumptions, our estimates will not apply to the hot gas (T > 10^3 K) where the equation of state is softened by effects like H_2 dissociation (e.g., the inner regions of disks). We will

also assume that the metallicity is approximately solar, and that the Rosseland- and Planck-mean opacities due to dust are (to order of magnitude) the same, $\bar{\kappa}_R(T) \simeq \bar{\kappa}_P(T) \simeq \kappa_1(T/K)^\beta$ with $\kappa_1 = 10^{-3}$ cm^2 g^{-1} and emissivity index $\beta = 2$ in the far-infrared and submillimeter. With this assumption, our estimates will again not apply to the hot gas (T > 10^3 K) where dust sublimates and the opacity falls abruptly with increasing temperature (before picking up at even higher temperatures, due to the H$^-$ ion). In sections 3.4 to 3.6, we relax these assumptions.

3.1.   Three-Dimensional Collapse and Hierarchical Fragmentation

In a uniform three-dimensional medium, an approximately spherical fragment of mass M_{F3} will only condense out if it is sufficiently massive

$$M_{F3} > M_{J3} \simeq \left[\frac{\pi^5 a^6}{36G^3\rho} \right]^{1/2} \qquad (1)$$

or equivalently, if it is sufficiently small and dense

$$\rho_{F3} > \rho_{J3} \simeq \frac{\pi^5 a^6}{36G^3 M_{F3}^2} \qquad (2)$$

(Subscript F is for fragment, J is for Jeans, and 3 is for three-dimensional.) The timescale on which the fragment condenses out is

$$t_{F3} \simeq \left[\frac{3\pi}{32G\rho} \right]^{1/2} \left[1 - \left(\frac{M_{J3}}{M_{F3}} \right)^{2/3} \right]^{-1/2} \qquad (3)$$

The molecular-cloud gas from which stars are forming today in the Milky Way is expected to be approximately isothermal, with T ~ 10 K, as long as it can radiate efficiently via molecular lines and dust continuum. Therefore it has been argued, following *Hoyle* (1953), that star formation proceeds in molecular clouds by a process of hierarchical fragmentation in which an initially massive low-density cloud (destined to form a protocluster of stars) satisfies condition (1) and starts to contract. Once its density has increased by a factor of f^2, M_{J3} is reduced by a factor of f^{-1}, and hence parts of the cloud can condense out independently, thereby breaking the cloud up into ≤ f subclouds. Moreover, as long as the gas remains approximately isothermal, the process can repeat itself recursively, breaking the cloud up into ever smaller "sub-sub...subclouds."

The process ends when the smallest sub-sub...subclouds are so optically thick, and/or collapsing so fast, that the PdV work being done on them cannot be radiated away fast enough and they heat up. This process is presumed to determine the minimum mass for star formation (e.g., *Rees,* 1976; *Low and Lynden-Bell,* 1976), and is usually referred to as the opacity limit, but see *Masunaga and Inutsuka* (1999) for a more accurate discussion. To estimate M_{MIN3} we first formulate the PdV heating rate for a spherical fragment, neglecting the background radiation field,

$$\mathcal{H} \equiv -P\frac{dV}{dt} = -\frac{3M_{F3}a^2}{R_{F3}}\frac{dR_{F3}}{dt} \sim \frac{3M_{F3}a^2}{R_{F3}}\left[\frac{GM_{F3}}{R_{F3}}\right]^{1/2} \quad (4)$$

where in putting $dR_{F3}/dt \sim -(GM_{F3}/R_{F3})^{1/2}$ we are assuming the collapse is dynamical. By comparison, the maximum radiative luminosity of a spherical fragment is

$$\mathcal{L} \simeq \frac{4\pi R_{F3}^2 \sigma_{SB} T^4}{(\bar{\tau}_R(T) + \bar{\tau}_P^{-1}(T))} \quad (5)$$

where the optical depths are given by $\bar{\tau}_R(T) \simeq \bar{\tau}_P(T) \simeq 3M_{F3}\kappa_1(T/K)^2/4\pi R_{F3}^2$.

If we follow *Rees* (1976) and assume that the fragment is marginally optically thick, we can put $(\bar{\tau}_R(T) + \bar{\tau}_P^{-1}(T)) \simeq 2$, and the requirement that $\mathcal{L} \geq \mathcal{H}$ then reduces to

$$\rho_{F3} \lesssim \rho_{C3} \simeq \left[\frac{\pi^{29}}{2^2 3^5 5^6}\frac{\bar{m}^{24}a^{36}}{G^3 M_{FS}^2 c^{12} h^{18}}\right]^{1/7} \quad (6)$$

Conditions (2) and (6) require $\rho_{J3} < \rho_{C3}$ and hence

$$M_{F3} \geq M_{MIN3} \simeq \left[\frac{5^2\pi^2}{2^4 3^3}\right]^{1/4}\frac{m_{PL}^3}{\bar{m}^2}\left[\frac{a}{c}\right]^{1/2} \quad (7)$$

Here $m_{PL} = (hc/G)^{1/2} = 5.5 \times 10^{-5}$ g is the Planck mass, and so M_{MIN3} is essentially the Chandrasekhar mass times a factor of $(a/c)^{1/2} \sim 10^{-3}$. We also note the relatively weak dependence of M_{MIN3} on T ($\propto T^{1/4}$) and the relatively strong dependence on $\bar{m}$ ($\propto \bar{m}^{-9/4}$). For contemporary local star formation we substitute $\bar{m} \simeq 4.0 \times 10^{-24}$ g and $a \simeq 1.8 \times 10^4$ cm s^{-1} to obtain $M_{MIN3} \sim 0.004$ M$_\odot$.

In general, the limiting fragment will not necessarily be marginally optically thick, but it is trivial to substitute for $\bar{\tau}_R(T)$ and $\bar{\tau}_P(T)$, and it turns out that, *for contemporary star formation*, the value of M_{MIN3} is unchanged. This is because — coincidentally — the limiting fragment in contemporary star formation *is* marginally optically thick.

There are, however, some serious problems with three-dimensional hierarchical fragmentation. There is no conclusive evidence that it operates in nature, and nor does it seem to occur in numerical simulations of star formation. This is because a protofragment inevitably condenses out more slowly than the larger structure of which it is part, by virtue of the fact that it is, at every stage, less Jeans unstable than that larger structure (see equation (3)). Therefore the protofragment is very likely to be merged with other nearby fragments before its condensation becomes nonlinear. In addition, even if a protofragment starts off with mass $\sim M_{J3}$, it

will subsequently increase its mass by a large factor through accretion before its condensation becomes nonlinear. Finally, individual fragments will be backwarmed by the ambient radiation field from other cooling fragments, which in principle fill a significant fraction of the celestial sphere, and again this will tend to increase M_{MIN3}.

3.2. Two-Dimensional One-Shot Fragmentation of a Shocked Layer

In fact, three-dimensional hierarchical fragmentation may be an inappropriate paradigm for star formation in molecular clouds. There is growing evidence that, once a molecular cloud is assembled, star formation proceeds very rapidly, essentially "in a crossing time" (*Elmegreen*, 2000). In this scenario, star formation occurs in molecular clouds only where two or more turbulent flows of sufficient density collide with sufficient ram pressure to produce a shock-compressed layer or filament that then fragments to produce prestellar cores; in cases where flows converge simultaneously from several directions, isolated cores may even form (see *Padoan and Nordlund*, 2002, 2004) (see also section 4).

A basic model of this scenario can be constructed with relatively few parameters by considering two flows having preshock density ρ, colliding at relative speed v, to produce a shock-compressed layer. If the effective isothermal sound speed in the resulting layer is a, the density is $\sim\rho(v/a)^2$. The layer is initially contained by the ram pressure of the inflowing gas, and until it fragments it has a rather flat density profile. It fragments at time t_{F2}, while it is still accumulating, and the fastest-growing fragments have mass M_{F2}, radius R_{F2} (in the plane of the layer), and half-thickness Z_{F2} (perpendicular to the plane of the layer) given by

$$t_{F2} \sim (2a/\pi G\rho v)^{1/2} \quad (8)$$

$$M_{F2} \sim (8a^7/\pi G^3\rho v)^{1/2} \quad (9)$$

$$R_{F2} \sim (2a^3/\pi G\rho v)^{1/2} \quad (10)$$

$$Z_{F2} \sim (8a^5/\pi G\rho v^3)^{1/2} \quad (11)$$

(see *Whitworth et al.*, 1994a,b). We note that (1) this mode of fragmentation is "two-dimensional" because the motions that assemble a fragment out of the shock-compressed layer are initially largely in the plane of the layer; (2) it is "one-shot" in the sense of not being hierarchical; (3) the fragments are initially flattened objects ($R_{F2}/Z_{F2} \sim v/2a \gg 1$); (4) M_{F2} is not simply the standard three-dimensional Jeans mass (M_{J3}, equation (1)) evaluated at the postshock density and velocity dispersion — it is larger by a factor of $\sim(v/a)^{1/2}$; and (5) in reality the colliding flows will contain density substructure that acts as seeds for fragmentation and gives rise to a range of M_{F2} values.

From equation (9), we see that the fragment mass decreases monotonically with the mass-flux in the colliding flows, ρv. As in hierarchical fragmentation, a fragment in a shock-compressed layer will only condense out if it is able to remain approximately isothermal by radiating efficiently. The PdV heating rate for a flattened fragment in a layer is

$$\mathcal{H} \equiv -P \frac{dV_{F2}}{dt} \simeq \frac{\rho v^2}{4} \frac{2\pi R_{F2}^2 Z_{F2}}{t_{F2}} \simeq \frac{2a^5}{G} \quad (12)$$

The radiative cooling rate of the fragment is

$$\mathcal{L} \simeq \frac{2\pi R_{F2}^2 \sigma_{SB} T^4}{(\bar{\tau}_R(T) + \bar{\tau}_P^{-1}(T))} \simeq \frac{8\pi^5 \bar{m}^4 a^{11}/15c^2 h^3 G\rho v}{(\tau_R(T) + \bar{\tau}_P^{-1}(T))} \quad (13)$$

where the optical depths are now given by $\bar{\tau}_R(T) = \bar{\tau}_P(T) = (2a\rho v/\pi G)^{1/2} \kappa_1 (T/K)^2$. The requirement that $\mathcal{L} \geq \mathcal{H}$ then reduces to a limit on the mass flux in the colliding flows

$$\rho v \leq \frac{4\pi^5 \bar{m}^4 a^6/15c^2 h^3}{(\bar{\tau}_R(T) + \bar{\tau}_P^{-1}(T))} \quad (14)$$

If we assume the fragment is marginally optically thick, and set $(\bar{\tau}_R(T) + \bar{\tau}_P^{-1}(T)) \simeq 2$, we obtain

$$M_{F2} \geq M_{MIN2} \simeq \frac{(30)^{1/2}}{\pi^3} \frac{m_{PL}^3}{\bar{m}^2} \left[\frac{a}{c}\right]^{1/2} \quad (15)$$

and for contemporary local star formation, this gives $M_{MIN2} \sim 0.001 \, M_\odot$. Once again, if we treat the completely general case by including the optical-depth terms, we obtain essentially the same value for M_{MIN2}, because — purely coincidentally — the limiting mass for contemporary local star formation is again marginally optically thick.

Although one-shot two-dimensional fragmentation of a shock-compressed layer and three-dimensional hierarchical fragmentation give essentially the same expression for the minimum mass ($M_{MIN2} \sim M_{MIN3}$), they give very different expressions for the Jeans mass ($M_{J2} \neq M_{J3}$), and there are other important differences. In particular, two-dimensional one-shot fragmentation bypasses all the problems associated with three-dimensional hierarchical fragmentation. There is no backwarming because there are no other local fragments filling the part of the celestial sphere into which a fragment radiates (i.e., perpendicular to the layer). More importantly, there is little likelihood of fragments merging, since fragments with $M_{F2} \sim M_{J2}$ condense out faster than any other structures in a layer [whereas fragments in a three-dimensional medium with $M_{F3} \sim M_{J3}$ condense out slower than all larger structures in that medium (e.g., *Larson*, 1985)]. Finally, because condensation in a layer is so fast, growth by accretion is limited.

Boyd and Whitworth (2004) have analyzed the radiative cooling of a fragmenting layer, taking into account *not only* the PdV heating of a condensing fragment, *but also* the ongoing accretion (as matter continues to flow into the layer)

and the energy dissipated in the accretion shock. They find that for T $\sim$ 10 K, the minimum mass is 0.0027 $M_\odot$. This fragment starts with mass $\sim$0.0011 $M_\odot$, but continues to grow by accretion as it condenses out.

3.3. Fragmentation of a Circumstellar Disk

Another scenario that may be more relevant to contemporary star formation than three-dimensional hierarchical fragmentation is fragmentation of a circumstellar disk. There are three critical issues here: (1) Under what circumstances does a disk fragment gravitationally? (2) Can the resulting fragments cool fast enough to condense out? (3) Can the resulting fragments lose angular momentum fast enough to condense out? The last two issues are critical, because if a fragment cannot cool and lose angular momentum fast enough, it is likely to bounce and be sheared apart. For simplicity we consider an equilibrium disk.

1. The condition for an isolated disk to fragment gravitationally is that the surface density, Σ, be sufficiently large

$$\Sigma \geq \Sigma_T \simeq \frac{\alpha \varepsilon}{\pi G} \quad (16)$$

where a is the local sound speed and ε is the local epicyclic frequency (*Toomre*, 1964). Condition (16) is also the condition for spiral modes to develop in the disk, and these will have the effect of redistributing angular momentum. As a result, the inner parts of the disk may simply accrete onto the central primary star and the outer parts may disperse *without fragmenting* (*Laughlin and Bodenheimer*, 1994; *Nelson et al.*, 1998). Thus, if fragmentation is to occur, it must occur on a dynamical timescale.

2. The condition for a fragment to cool fast enough to condense out is therefore that the fragment can radiate away, on a dynamical timescale, the thermal energy being delivered by compression. *Gammie* (2001) has shown that for a Keplerian disk, this condition can be written as a constraint on the cooling time

$$t_{COOL} \leq \frac{3}{\Omega} \quad (17)$$

where Ω is the local orbital angular speed.

We assume that, as a disk forms, Σ increases sufficiently slowly that it does not greatly exceed Σ_T when the disk becomes unstable. It then follows that the radius, growth time, and mass of the fastest growing fragment are

$$R_{FD} \simeq \frac{a}{\varepsilon} \quad (18)$$

$$t_{FD} \simeq \frac{1}{\varepsilon} \quad (19)$$

$$M_{FD} \simeq \frac{a^3}{G\varepsilon} \qquad (20)$$

The compressional heating rate for a fragment is thus

$$\mathcal{H} = P\frac{dV}{dt} \simeq \frac{3M_{FD}a^2}{2t_{FD}} \simeq \frac{3a^5}{2G} \qquad (21)$$

and the radiative cooling rate is

$$\mathcal{L} \simeq \frac{2\pi R_{FD}^2 \sigma_{SB} T^4}{(\bar{\tau}_R(T) + \bar{\tau}_P^{-1}(T))} \simeq \frac{4\pi^6 \bar{m}^4 a^{10}/15c^2 h^3 \varepsilon^2}{(\bar{\tau}_R(T) + \bar{\tau}_P^{-1}(T))} \qquad (22)$$

where now $\bar{\tau}_R(T) \simeq \bar{\tau}_P(T) \simeq a\varepsilon\bar{\kappa}(T)/\pi G$. Consequently the requirement that $\mathcal{L} \geq \mathcal{H}$ reduces to

$$\frac{\varepsilon^2}{a^5} \lesssim \frac{4\pi^6 G\bar{m}^4/45c^2 h^3}{(\bar{\tau}_R(T) + \bar{\tau}_P^{-1}(T))} \qquad (23)$$

To illustrate the discussion we consider the specific case of a Keplerian disk around a Sun-like star, with

$$\varepsilon(D) \simeq 2 \times 10^{-7} \text{ s}^{-1} \left(\frac{M_\star}{M_\odot}\right)^{1/2} \left(\frac{D}{AU}\right)^{-3/2} \qquad (24)$$

$$T(D) \simeq 300 \text{ K} \left(\frac{L_\star}{L_\odot}\right)^{1/4} \left(\frac{D}{AU}\right)^{-1/2} \qquad (25)$$

and hence $a(D) \simeq 1 \text{ km s}^{-1} (L_\star/L_\odot)^{1/8} (D/AU)^{-1/4}$, where D is distance from the Sun-like star. The fastest-growing fragment then has mass

$$M_{FD} \simeq 3 \times 10^{-5} \text{ M}_\odot \left(\frac{M_\star}{M_\odot}\right)^{-1/2} \left(\frac{L_\star}{L_\odot}\right)^{3/8} \left(\frac{D}{AU}\right)^{3/4} \qquad (26)$$

and condition (23) is only satisfied for

$$D \gtrsim 90 \text{ AU} \left(\frac{M_\star}{M_\odot}\right)^{3/7} \left(\frac{L_\star}{L_\odot}\right)^{-1/7} \qquad (27)$$

$$M_{FD} \gtrsim 0.003 \text{ M}_\odot \left(\frac{M_\star}{M_\odot}\right)^{-5/28} \left(\frac{L_\star}{L_\odot}\right)^{15/56} \qquad (28)$$

3. Angular momentum is removed from a condensing fragment by gravitational torques, in the same way that angular momentum is redistributed in the disk as a whole when $\Sigma \gtrsim \Sigma_T$. This is a nonlinear and stochastic process, and there is no analytic estimate of the rate at which it occurs.

Therefore the condition for a fragment to lose angular momentum fast enough to condense out has to be evaluated by numerical simulation. The chapter by Durisen et al. deals with this problem.

3.4. Nonlinear Thermodynamics

Impulsive triggers that produce rapid compression will always help to amplify self-gravity, because the freefall time varies as $t_{FF} \propto \rho^{-1/2}$. However, there is only a very restricted temperature range within which rapid compression will help an optically thick protofragment to cool more rapidly (and thereby avoid the likelihood of its bouncing and being sheared apart). Suppose that the Rosseland-mean opacity is given by $\bar{\kappa}_R \propto \rho^\alpha T^\beta$, and that the gas has a ratio of specific heats γ.

Then if a fragment condenses out quasistatically, its cooling time varies as $t_{COOL} \propto \rho^{\alpha + (1 + \beta)/3}$, so t_{COOL} only decreases as fast as t_{FF}, with increasing density, if $\beta < -(6\alpha + 5)/2$. This condition is only likely to be satisfied in the temperature range where refractory dust sublimates (1500–3000 K).

If instead the fragment is compressed impulsively — and therefore adiabatically — its cooling time varies as $t_{COOL} \propto \rho^{\alpha + 4/3 + (\beta - 3)(\gamma - 1)}$, so now t_{COOL} only decreases as fast as t_{FF}, with increasing density, if $\beta < 3 - (6\alpha + 11)/6(\gamma - 1)$. For $T \lesssim 100$ K, we have $\gamma \simeq 5/3$ and so enhanced cooling requires $\beta \lesssim -1/4$, which is unlikely. For $100 \text{ K} \lesssim T \lesssim 1000$ K, $\gamma \simeq 7/5$ and enhanced cooling requires $\beta \lesssim -19/12$, which is even less likely. For $1000 \text{ K} \lesssim T \lesssim 3000$ K, H_2 dissociation gives $\gamma \sim 1.1$ and enhanced cooling requires $\beta \lesssim -15$, which may occur during sublimation, but then necessarily only over a small temperature range. For $3000 \text{ K} \lesssim T \lesssim 10,000$ K, $\gamma \sim 5/3$ and H^- opacity gives $\alpha \sim 1/2$ and $\beta \sim 4$, so $t_{COOL} \propto \rho^{7/2}$. At even higher temperatures where Kramers opacity dominates $t_{COOL} \propto \rho^{-2}$, and where electron scattering opacity dominates $t_{COOL} \propto \rho^{-2/3}$, but by this stage a fragment is very opaque and very strongly bound.

3.5. Forming Close Brown Dwarf–Brown Dwarf Binaries

It is possible that there is a secondary fragmentation regime at $T \sim 2000$ K, due to the softening of the equation of state caused by H_2 dissociation and the enhanced cooling that occurs in the opacity gap between dust sublimation and H^- opacity. A spherical cloud of mass M in hydrostatic equilibrium at this temperature has radius

$$R \sim 4 \text{ AU} \left(\frac{M}{0.1 \text{ M}_\odot}\right) \qquad (29)$$

If the cloud fragments into a binary and virializes, the binary should have a separation of the same order. Equation (29) is actually a good mean fit to the values plotted on Fig. 6 of the chapter by Burgasser et al., suggesting that BD-BD binaries may be produced by secondary fragmentation facilitated by H_2 dissociation and/or the opacity gap.

3.6. Forming Close Brown Dwarf–Brown Dwarf Binaries in Disks

Burgasser et al. (2005) have noted that — modulo the small-number statistics involved — a brown dwarf in a wide orbit ($\gtrsim$200 AU) about a Sun-like star is apparently more likely to be in a close BD-BD binary system ($\lesssim$20 AU) than a brown dwarf in the field. If this trend is confirmed, it suggests that BD-BD binaries are formed in disks, and then may be ejected. For a close BD-BD binary in a wide orbit around a Sun-like star, the internal binding energy of the BD-BD binary is typically comparable to or larger than the binding energy of the BD-BD binary to the Sun-like star, and therefore some BD-BD binaries should be able to survive ejection.

Moreover, we know from the Toomre criterion that for fragments condensing out of disks, losing angular momentum is a critical issue. The smaller a condensation becomes, the slower the rate at which angular momentum can be lost by gravitational torques. Therefore such a condensation may be strongly disposed to binary fragmentation when its thermal support is weakened at T ~ 2000 K. Since only a subset of brown dwarfs is in close BD-BD binaries, this suggestion does not require that all brown dwarfs are formed in disks.

4. FORMING VERY-LOW-MASS PRESTELLAR CORES IN TURBULENT CLOUDS

A number of surveys (*Testi and Sargent,* 1998; *Peng et al.,* 1998; *Motte et al.,* 1998, 2001; *Motte and André,* 2001; *Johnstone et al.,* 2000; *Sandell and Knee,* 2001; *Onishi et al.,* 2002; *Reid and Wilson,* 2005) have noted the close similarity between the core mass function (CMF) and the stellar initial mass function (IMF). Cores with brown dwarf masses have been observed in several of these surveys, but since these low-mass cores are usually below the completeness limit, their statistics are unreliable. However, the similarity between the CMF and the IMF suggests that each core gives rise to a similar low number of stars, whose final masses are heavily influenced by the mass of the core. Indeed, protostars are generally observed inside relatively well-defined envelopes, as implied by the definition of Class 0 and Class I objects (*André et al.,* 1993, 2000; *Tachihara et al.,* 2002). This is not to say that the final masses of stars are the same as the masses of the prestellar cores in which they form. Protostars are associated with outflows, which can remove a significant fraction of the mass of the protostellar envelope, and cores may split into more than one star (see also *Hosking and Whitworth,* 2004; *Stamatellos et al.,* 2005). However, low-mass cores appear typically to harbor only one or two protostars (e.g., *Tachihara et al.,* 2002), and the overall binary statistics are inconsistent with a larger number of stars being formed from each core (*Goodwin and Kroupa,* 2005).

4.1. Core Structure and Low-Mass Core Formation

Observed prestellar cores generally have Bonnor-Ebert (BE)-like density profiles, with relatively sharp outer edges (*Bacmann et al.,* 2000; *Motte and André,* 2001; *Kirk et al.,* 2005). Starless cores have flat BE-like profiles, while cores with detected Class 0 or Class I protostars have more centrally peaked density profiles, presumably as a result of their deeper potential wells. Since well-defined cores with sharp edges define finite mass reservoirs for the stars that form inside them, the core observations reinforce the notion that the masses of stars are, at least statistically, strongly influenced by the masses of the cores within which they form (*Padoan and Nordlund,* 2002, 2004).

It is therefore important to consider whether it is possible to form brown dwarfs directly, from correspondingly low-mass cores. The standard argument against brown dwarfs being formed directly is that the density needs to be very high for a fragment with brown dwarf mass to collapse gravitationally. If one considers low-amplitude perturbations on top of a typical mean density of 10^4 to 10^5 cm^{-3}, he will conclude that only fluctuations on the order of the Jeans mass (1–3 $M_\odot$; equation (1)) are able to collapse in a typical molecular cloud. However, many cores actually have very high density contrast relative to their surroundings, and so it is inappropriate to use the Jeans mass at the mean density as an estimate of the resulting protostellar mass.

It is also important to realize that cores having sufficiently high density to form brown dwarfs directly need not — indeed *must* not — be very common, and therefore we can appeal to exceptional circumstances to generate them. As a measure of how exceptional the circumstances must be, consider the IMF (in the form $d\mathcal{N}/d \ln M$). At high masses, the IMF falls off with the *Salpeter* (1955) exponent –1.35, and if this power-law IMF continued unbroken down to the brown dwarf regime, $10^{4 \times 1.35} \simeq 250{,}000$ brown dwarfs with M ~ 0.01 $M_\odot$ would be formed for every massive star with M ~ 100 $M_\odot$. Instead, the IMF peaks near ~0.3 $M_\odot$ and falls off at low masses; adopting the *Chabrier* (2003) IMF for the Milky Way, we find that the number of brown dwarfs formed with M ~ 0.01 $M_\odot$ is actually about equal to the number of massive stars formed with M ~ 100 $M_\odot$, as predicted theoretically by *Zinnecker* (1984).

In order for a high-density core to form, its mass must be concentrated into a small volume. If **u** is the fluid velocity, accumulation of mass is measured by the quantity $-\nabla \cdot \mathbf{u} = $ D ln ρ/Dt (the co-moving time derivative of log density), and so individual cores form at local maxima of $-\nabla \cdot$ **u**.

The flow toward a convergence point is generally supersonic, and so standoff shocks develop, separating the unshocked upstream gas from the shocked and nearly stagnant downstream gas. The standoff shocks correspond to density jumps ~$\mathcal{M}^2$ if the shocked gas is dominated by gas pressure; and ~$\mathcal{M}_A$ (the Alfvénic Mach number) if the shocked gas is dominated by magnetic pressure. As matter accumulates around the convergence point, a deepening gravitational potential well develops, and the accumulated and stagnant gas forms a growing Bonnor-Ebert like core, stabilized by the external ram pressure, $\rho|\mathbf{u}|^2$.

If the density at the center of the core increases to more than about 14× the boundary density (just inside the shock surface), the core becomes gravitationally unstable and collapses. However, even before it becomes unstable, such a core has — by virtue of the shock jump at its boundary — a total density contrast (between its center and the surrounding inflowing unshocked gas) that can greatly exceed a factor of 14. Moreover, if the inflowing gas runs out before the core has become sufficiently massive and dense to collapse, then the core will expand and disperse due to the decrease in ram pressure at its boundary.

In the last column of Table 3 in *Bacmann et al.* (2000), at least 8 of the 9 cores listed have density contrasts exceeding 14, but no strong indication of collapse. Bacmann et al. conclude that the structures are best fitted with Bonnor-Ebert spheres, but since they cannot explain how the cores can be stable at these high-density contrasts, they go on to discuss other density profiles and models, which do not fit the data as well as BE-spheres. Similar core profiles have been observed in many other surveys. There is thus direct observational evidence for the creation of prestellar cores with large density contrasts between their centers and the surrounding medium, as expected when cores are produced as stagnant structures by supersonic flows. Indeed, cores formed in numerical simulations of molecular cloud turbulence have internal velocity dispersions and rotation velocities that are entirely consistent with those of observed cores (cf. Figs. 8 and 9 in *Nordlund and Padoan*, 2003).

With this scenario in mind, we now investigate semiquantitatively the range of densities that is possible, and the circumstances under which the densities attained would be high enough to lead to the collapse of brown dwarf mass cores. Any particular converging flow may be broadly characterized by three parameters: the upstream density ρ_0, the Mach number $\mathcal{M}$ of the upstream flow, and the degree of focusing of the upstream flow toward a three-dimensional convergence point. As a measure of the latter we take the ratio f of the surface area that encloses the upstream flow to the surface area of the standoff shock that surrounds the central BE-like structure. Since the upstream flow is supersonic, its motion is essentially inertial (i.e., constant velocity), until it encounters the standoff shock. Thus, by mass conservation, its density increases from ρ_0 to $f\rho_0$. At the standoff shock the density increases by an additional factor $\mathcal{M}^2$ for hydroshocks (or $\mathcal{M}_A$ for MHD shocks). Finally, from just inside the shock to the center of the BE-like core the density increases by a further factor of ≤ 14 (~14, in the marginally stable case). The total density increase from the upstream source to the core center is ~14 $f\mathcal{M}^2$. This factor has the right dependence to account for the formation of brown dwarf mass cores, in that it can be arbitrarily large, albeit with decreasing probability.

As an example, consider the formation of a core with temperature ~10 K and central density ~10^7 cm^{-3}, sufficiently high to form a star with M ~ 0.1 M$_\odot$. If the average cloud density is ~10^4 cm^{-3} and the Mach number is 5, f only needs to be ~3; i.e., the upstream flow needs to be focused onto a 3× smaller standoff shock surface. This must be rather common, and indeed at M ~ 0.1 M$_\odot$ the IMF has hardly dropped below its maximum.

Repeating this estimate for a core with M ~ 0.01 M$_\odot$ one finds that the flow needs to be much better focused. The standoff shock surface now needs to be ~300× smaller than the upstream source area, and so the linear size of the core needs to be ~17× smaller than the scale of the upstream flow. This does require a rather exceptional focusing of the upstream flow, but then brown dwarfs with M ~ 0.01 M$_\odot$ actually are very rare. The fraction of brown dwarfs that form in this manner can only be ascertained quantitatively by performing very-high-resolution numerical simulations.

Figure 1 illustrates an example intermediate between these two extremes, i.e., a core with final mass ~0.03 M$_\odot$.

4.2. Envelope Breakout and Continued Accretion

The sharp density transitions surrounding accreting cores produce more well-defined cores than would exist in a subsonic medium. Nevertheless, there will be some continued accretion after a collapsed object is formed, until the upstream supply of gas is exhausted. Except toward the end of the process, such cores correspond to Class 0 objects, which by definition have not yet accreted half their final mass (*André et al.*, 2000).

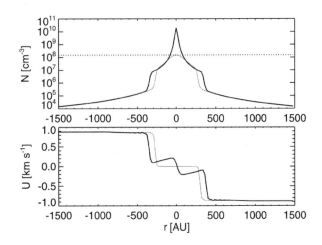

Fig. 1. Density and velocity profiles, through a core forming from a spherically symmetric convergent flow. The thin line represents an early stage when the ~0.015 M$_\odot$ core approximates to a stable and stagnant BE-like configuration; the boundary shock is at ~300 AU. The thick line represents a later stage after the core has become unstable and started to collapse; the boundary shock is now at ~350 AU, because the core is more massive (~0.025 M$_\odot$). Outside the shock is the convergent inertial flow. Note the very large density contrast, even between the gas at the edge of the core and the ambient medium at ~1500 AU.

Class I and II objects are observed to accrete from low-mass envelopes (*Motte and André*, 2001) and have much-reduced accretion rates, $\leq 10^{-6}$ M$_\odot$ yr^{-1}. Once the natal envelope is consumed there is no longer a strong coupling between the collapsed object (moving under the N-body influence of neighboring objects and gas) and the surrounding medium, and one expects the collapsed object to pick up speed relative to the surrounding gas, and to then accrete in a manner similar to Bondi-Hoyle accretion (*Padoan et al.*, 2005; *Krumholz et al.*, 2005b). In a turbulent medium the problem is more complicated (*Krumholz et al.*, 2005a), but the scaling is more or less as for Bondi-Hoyle accretion, with a prefactor $\Phi \sim 1$ to 5 that accounts for the complications. Quantitative estimates indicate that the accretion after breakout from the natal core is insignificant.

4.3. The Viability of Turbulent Fragmentation

Although turbulent fragmentation generates a mass function for prestellar cores, matching broadly the observed stellar IMF, there are two caveats that should be born in mind when considering the formation of brown dwarfs.

First, in turbulent fragmentation (*Padoan and Nordlund*, 2002), brown dwarfs are produced by a small subset of low-mass cores, i.e., those that are sufficiently dense to collapse gravitationally. At its low-mass end, the overall (i.e., fully sampled) CMF should be dominated by a much larger number of transient (i.e., nonprestellar) cores. Moreover, many of these transient cores will be only slightly less dense than the ones that spawn brown dwarfs. They will therefore be detectable in submillimeter continuum observations, and they should also be somewhat longer lived. Thus, even allowing for selection effects, the observed CMF should fall much less steeply with decreasing mass than the stellar IMF. Recent estimates of the CMF (Nutter and Ward-Thompson, in preparation), which probe to lower masses by using longer exposures, actually suggest the opposite, but completeness remains a concern (e.g., *Kirk et al.*, 2006).

Second, turbulent fragmentation predicts a ratio of brown dwarfs to H-burning stars, $\mathcal{R}$, which is exponentially sensitive to the Alfvénic Mach number on the largest scales ($\mathcal{M}_A$) and to the mean cloud density (n). Regions with smaller values of $\mathcal{M}_A$ and/or n should generate stellar populations with significantly fewer brown dwarfs, and regions with larger values should generate stellar populations with significantly more brown dwarfs. However, in nature $\mathcal{R}$ appears to vary little over a wide range of local star-forming environments (see chapter by Luhman et al.).

This problem can be overcome if nature selects a narrow range of $\mathcal{M}_A$ and n. For example, *Whitworth* (2005) has noted that contemporary star formation may only proceed rapidly if the gas couples thermally to the dust (so that it can avail itself of broadband — as distinct from molecular-line — cooling). This requires that the ram pressure in shocks producing prestellar cores exceeds a critical value, which converts into the constraint nT$\mathcal{M}_A^2 \geq$ P$_{CRIT} \sim 10^5$ cm^{-3} K. This constraint may help to select the combinations of $\mathcal{M}_A$ and

n that reproduce the observed ratio of brown dwarfs to H-burning stars.

5. COLLAPSE AND FRAGMENTATION OF LARGE PRESTELLAR CORES

While a very-low-mass prestellar core (≤ 0.1 M$_\odot$) must collapse to form either a single brown dwarf or a multiple brown dwarf system, larger prestellar cores (≥ 1 M$_\odot$) are expected to form clusters of stars having a range of masses. We can identify five mechanisms that may play a role in determining the final stellar masses: (1) During the approximately isothermal initial collapse phase, as the pressure becomes increasingly unimportant and the collapse approaches freefall, self-gravity will amplify any existing density substructure. (2) Then, when the density reaches n$_{H_2} \sim 10^{11}$ cm^{-3}, the gas becomes optically thick and switches rather suddenly from approximate isothermality to approximate adiabaticity. At this juncture, a network of shock waves develops to slow the collapse down, and nonlinear interactions between these shock waves produce and amplify further substructure (sheets, filaments, and isolated prestellar cores). (3) Some of these structures will have sufficient angular momentum to form disks, and these may then fragment due to rotational instability (see section 6). Finally, once a protostellar embryo (i.e., an object that is sufficiently well bound to be treated dynamically as a single entity) has condensed out of a fragment, its subsequent evolution and final mass will be determined by (4) competitive accretion and (5) dynamical interaction (see section 7). Here we concentrate on mechanisms (1) and (2), since these are the ones that distinguish the evolution of a high-mass core from the low-mass cores considered in section 4.

High-mass prestellar cores (≥ 10 M$_\odot$) invariably display nonlinear internal density structure, and at the typical density (n$_{H_2} \sim 10^5$ cm^{-3}) and temperature (T ~ 10 K) in a prestellar core, the Jeans mass is M$_{J3} \sim 0.8$ M$_\odot$ (equation (1)). Therefore, in the absence of a significant magnetic field, they are very likely to fragment during collapse.

Even the smallest cores are usually far from spherical, and have been modeled as being either prolate or oblate (*Myers et al.*, 1991; *Ryden*, 1996; *Jones et al.*, 2001; *Goodwin et al.*, 2002; *Curry*, 2002; *Myers*, 2005), with prolate models being favored statistically, although in reality cores are probably triaxial — or even more complicated — in their full three-dimensional structure (*Jones et al.*, 2001; *Goodwin et al.*, 2002). Gravity works to enhance anisotropies in collapsing objects (*Lin et al.*, 1965) with collapse occurring fastest along the shortest axis to form sheets and filaments, which then subsequently fragment (*Bastien*, 1983; *Inutsuka and Miyama*, 1992). Thus, it is unsurprising that hydrodynamical simulations of both oblate and prolate cores are prone to fragmentation (e.g., *Bastien et al.*, 1991; *Boss*, 1996).

Prestellar cores also tend to have complex internal velocity fields (*Larson*, 1981; *Myers and Benson*, 1983; *Arquilla and Goldsmith*, 1985), but since prestellar cores can only be observed from a single direction, the interpretation of these velocities is difficult. If cores are assumed to be in solid-body rotation, the ratio of rotational to gravitational energy is typically $\beta \equiv \mathcal{R}/|\Omega| \sim 0.03$, with some cases as high as $\beta \sim 0.1$ (*Goodman et al.*, 1993; *Barranco and Goodman*, 1998). However, the observed velocities are more likely to be turbulent in nature, i.e., less well ordered than solid-body rotation (*Myers and Gammie*, 1999; *Burkert and Bodenheimer*, 2000). Quite low levels of turbulence (e.g., *Goodwin et al.*, 2004a), and/or global rotation (e.g., *Boss*, 1986; *Bonnell and Bate*, 1994a; *Hennebelle et al.*, 2004; *Cha and Whitworth*, 2003) are sufficient to make a collapsing core fragment into a small ensemble of protostellar embryos.

Unfortunately, collapse and fragmentation can only be explored by means of numerical simulations, and there is a huge and poorly constrained range of admissible initial conditions, which makes the extraction of robust theorems very hard (see, e.g., *Hennebelle et al.*, 2004). Moreover, almost all simulations to date use barotropic equations of state [i.e., $P = P(\rho)$]. These barotropic equations of state are designed to mimic the expected thermal behavior of protostellar gas, but they do not capture the thermal inertia effects that become important at the juncture when the gas starts to heat up due to adiabatic compression, and it appears (*Boss et al.*, 2000) that these thermal inertia effects play a critical, deterministic role in gravitational fragmentation. Proper treatment of the energy equation and the associated radiative transport (e.g., *Whitehouse and Bate*, 2006) is needed to make these simulations more realistic.

6. DISK FRAGMENTATION

We organize our discussion of disk fragmentation under the headings (1) isolated, relaxed disks; (2) unrelaxed disks; and (3) interacting disks.

6.1. Isolated Relaxed Disks

The dynamical fragmentation of isolated relaxed disks is discussed in detail in the chapter by Durisen et al. Although the emphasis there is on the genesis of planets, the same issues pertain to the formation of brown dwarfs, viz., under what circumstances do disks become unstable against fragmentation? Is equation (16) a sufficient condition for gravitational instability (in which case, what is the precise value of Σ_T) or is it also necessary for Σ to increase rapidly (in order to avoid the disk simply being accreted and dispersed by torques due to spiral density waves)? Does equation (17) determine whether fragments can cool fast enough to condense out? What role is played by thermodynamic effects like H_2 dissociation, dust sublimation, and convection? What are the properties of dust in disks, and how well mixed are the gas and dust? Does the survival of a fragment depend on its ability to lose angular momentum rapidly? Some of these issues are only beginning to be investigated.

Rice et al. (2003) have corroborated equation (17) numerically by performing SPH simulations of disk fragmentation with a parameterized cooling law of the form $du/dt = -u\Omega/\beta$ (where u is the specific internal energy and Ω the orbital angular speed). Endemic fragmentation occurs with $\beta = 3$ but not with $\beta = 5$. However, because this cooling law results in indefinite cooling, whereas there is a limit to the supply of rotational energy that can be tapped through shock heating, by the time fragmentation occurs the temperatures have dropped to rather low values, which may be hard to realize in nature.

Rafikov (2005) argues that the opacity of gravitationally unstable disks is so high, and the cooling times are therefore so long, that fragments can only condense out on a dynamical timescale in the outer, cooler regions of massive disks. (Although the treatment is somewhat different, his conclusion resonates with the simple analysis we have presented in section 3.3 and carries the same caveats.)

Johnson and Gammie (2003) point out that the effects of opacity may be mitigated in temperature regimes where dust sublimates and hence the opacity decreases abruptly with increasing temperature. However, dust sublimation effects are confined to the temperature range $T \geq 200$ K, and the most critical effects occur at $T \geq 2000$ K, so they are probably only relevant to fragmentation close to the central star ($D \leq 10$ AU).

Cai et al. (2006) report simulations of disk evolution taking radiation transport into account, and conclude that — even at low metallicities — the opacity in the hot inner parts of a disk is too high to allow fragmentation on a dynamical timescale, in agreement with *Rafikov* (2005).

In contrast, *Boss* (2004) presents simulations of disk evolution taking radiation transport into account and concludes that fragments of planetary mass do condense out — or at least that gravitationally bound fragments form that would subsequently condense out. He argues that the protofragments in his simulations bypass the effects of high opacity by transporting energy convectively. However, it is not clear how convection can cool a fragment that is condensing out on a dynamical timescale. The coherent small-scale motions he attributes to convection may actually be manifestations of a fragment that is unable to cool and is bouncing — or will soon bounce — prior to being sheared apart. This needs to be investigated further, but the numerical complexities are considerable.

6.2. Unrelaxed Disks

In numerical simulations of the collapse and fragmentation of intermediate- and high-mass prestellar cores, the first single protostars to form usually quickly acquire massive circumstellar disks, and secondary protostars then condense out of these disks. This pattern is common for rotating cores (e.g., *Bonnell*, 1994; *Bonnell and Bate*, 1994a,b; *Turner et al.*, 1995; *Whitworth et al.*, 1995; *Burkert and Bodenheimer*, 1996; *Burkert et al.*, 1997), for turbulent cores (e.g., *Bate et al.*, 2002b, 2003; *Goodwin et al.*, 2004b), and for cores that are subjected to a sudden increase in external pressure (e.g., *Hennebelle et al., 2004*). The disks thus formed fragment before they have time to relax to an equilibrium state. Indeed, the material accreting onto the disk is often quite lumpy, and this helps to seed fragmentation. Under this circumstance, gravitational fragmentation is more likely simply because protofragments are launched directly into the nonlinear regime of gravitational instability, rather than having first to grow through the linear phase.

Even so, a protofragment still has to be able to cool and lose angular momentum on a dynamical timescale, if it is to condense out, rather than bouncing and then being sheared apart. Fragmentation of unrelaxed disks can only be studied by means of numerical simulations, and to date no simulations of the process have been performed that treat properly the energy equation and the associated transport of radiation. Since young protostars have relatively high luminosities, an important consideration will be irradiation of the disk by the primary protostar at its center and any other nearby protostars.

6.3. Interacting Disks

Another way in which disk fragmentation can be triggered is by an impulsive interaction with another disk, or with a naked star. In the dense protocluster environment where most stars are presumed to form, such interactions must be quite frequent, since many very young protostellar disks have diameters ≥300 AU and the mean separation between neighboring protostars in a typical cluster is ≤3000 AU. Indeed, ~50% of solar-type stars end up in binary systems with semimajor axes a ≤ 1000 AU, so the notion of a disk evolving in the gravitational field of a single, isolated protostar is probably rather artificial.

Boffin et al. (1998) and *Watkins et al.* (1998a,b) have simulated parabolic interactions between two protostellar disks, and between a single protostellar disk and a naked protostar. All possible mutual orientations of spin and orbit are sampled, and the gas is assumed to behave isothermally, which is probably a reasonable assumption, since the disks are large (initial radius 1000 AU) and most of the secondary protostars form at large distances (periastra ≥ 100 AU). The critical parameter turns out to be the effective shear viscosity in the disk. If the Shakura-Sunyaev parameter is low, $\alpha_{SS} \sim 10^{-3}$, most of the secondary protostars have masses in the range 0.001 $M_\odot$ to 0.01 $M_\odot$. Conversely, if α_{SS} is larger, $\alpha_{SS} \sim 10^{-2}$, most of the secondary protostars have masses in the range 0.01 $M_\odot$ to 0.1 $M_\odot$. The formation of low-mass companions is most efficient for interactions in which the orbital and spin angular momenta are aligned; on average 2.4 low-mass companions are formed per interaction in this case. If the orbital and spin angular momenta are randomly oriented, then on average 1.2 companions are formed per interaction. It is important that such simulations be repeated, with a proper treatment of the energy equation and the associated energy transport, to check their fidelity, and to establish whether low-mass companions can form at closer periastra.

7. PREMATURE EJECTION OF PROTOSTELLAR EMBRYOS

The scenario where brown dwarfs form by premature ejection is closely linked to — but ultimately independent of — the notion of competitive accretion. In the ejection hypothesis, brown dwarfs are simply protostellar embryos that get separated from their reservoir of accretable material at an early stage, and so in the context of brown dwarf formation it is irrelevant whether there are other stars competing for this same material. The importance of competitive accretion for intermediate- and high-mass stars is argued in the chapter by Bonnell et al. The contentious issue is whether protostellar embryos — i.e., the first, very-low-mass (~0.003 $M_\odot$) high-density star-like objects — exist for long enough to do much competing. At one extreme, it is argued that a protostellar embryo forms following the non-homologous collapse of a much larger gravitationally unstable core, and therefore it accretes mainly from its own co-moving placenta. At the other extreme, it is argued that a protostellar embryo quickly becomes sufficiently decoupled from the ambient gas that it can roam around competing with other embryos for the same reservoir of accretable gas. For the purpose of this section, we assume that nature cleaves to the second extreme.

7.1. Competitive Accretion in Gas-rich Protoclusters

Once an ensemble of protostellar embryos has formed, still deeply embedded in its parental prestellar core and/or parental disk, the individual embryos evolve by accreting gas, and by interacting dynamically with one another. The accretion histories of individual embryos differ due to their varying circumstances, leading to a spectrum of protostellar masses, extending from high masses down to below the H-burning limit. Those that spend a long time moving slowly through the dense gas near the center of the core can grow to high mass. Conversely, those that spend most of their time moving rapidly through the diffuse gas in the outer reaches of the core do not grow much. This process of "competitive accretion" may be a major element in the origin of the IMF, as first pointed out by *Zinnecker* (1982).

Over the past decade, many numerical simulations of the formation of star clusters by fragmentation and competitive accretion have been performed (e.g., *Chapman et al.*, 1992; *Turner et al.*, 1995; *Whitworth et al.*, 1995; *Bonnell et al.*, 1997, 2001a,b; *Klessen et al.*, 1998; *Klessen and Burkert*, 2000, 2001; *Bate et al.*, 2002a,b, 2003; *Goodwin et al.*, 2004b, 2004c). *Bonnell et al.* (1997, 2001a,b) have shown through numerical simulations and analytical arguments that competitive accretion in large-N protoclusters can reproduce the general form of the IMF.

7.2. Unstable Multiple Systems

Multiplicity studies of main-sequence and evolved stars have revealed that 15–25% of all stars, when studied in sufficient detail, are triple or higher-order multiples (e.g., *Toko-*

vinin and Smekhov, 2002; Tokovinin, 2004). It follows that the formation of multiple stars is an important element of the star-formation process.

The multiplicity fraction among PMS stars is poorly known, partly because of the difficulty in studying stars in the embedded phase, but it appears to be at least as high as — and probably much higher than — that for more evolved stars (e.g., *Reipurth,* 2000; *Looney et al.,* 2000; *Koresko,* 2002; *Reipurth et al.,* 2002, 2004; *Haisch et al.,* 2004) (see also chapters by Duchêne et al., Goodwin et al., and Burgasser et al.). Most stars are formed in embedded clusters, and there is increasing evidence that the primary building blocks of clusters are small subclusters of ≤20 stars that quickly dissolve and merge to form the more extended cluster (e.g., *Teixeira et al.,* 2006).

Nonhierarchical multiple systems, in which the time-averaged distances between components are comparable, are inherently unstable (e.g., *van Albada,* 1968). Within about 100 crossing times a triple system is likely to have ejected one member, most likely the least-massive component, since the ejection probability scales approximately as the inverse third power of the mass (e.g., *Anosova,* 1986; *Mikkola and Valtonen,* 1986). Although most nonhierarchical systems disintegrate in this way, the existence of numerous stable hierarchical triple systems shows that this is not always the case. Ejected members leave with a velocity that, to first order, is comparable to the velocity attained at pericenter in the close triple encounter, and depends on the geometry of the encounter, the energy and angular momentum of the system, and the masses of the components (e.g., *Standish,* 1972; *Monaghan,* 1976; *Sterzik and Durisen,* 1995, 1998; *Armitage and Clarke,* 1997). Ejection of brown dwarfs from systems that are marginally hierarchical is also possible (*Jiang et al.,* 2004).

The disintegration of a small multiple system is most likely to occur during the deeply embedded Class 0 phase, while massive accretion from a surrounding envelope is still taking place (*Reipurth,* 2000). If the ejection leads to an escape, then the accretion halts and the final mass of the object is capped (*Klessen et al.,* 1998).

7.3. Dynamical Ejection as a Source of Brown Dwarfs

If a protostellar embryo is ejected from its natal core with a mass below the H-burning limit, then it becomes a brown dwarf (*Reipurth and Clarke,* 2001). All that is needed for this to happen is for a prestellar core to spawn more than two protostellar embryos; for at least one of them to be less massive than 0.075 $M_\odot$ at the outset; and for one of them to stay less than 0.075 $M_\odot$ long enough to be ejected by dynamical interaction with the other embryos.

Simulations suggest that forming more than two protostellar embryos in a collapsing core is routine, as is the ejection of brown dwarfs and very-low-mass stars (e.g., *Bate et al.,* 2002a; *Delgado-Donate et al.,* 2003, 2004; *Goodwin et al.,* 2004a,b). However, these simulations may be misleading: (1) They use sink particles (*Bate et al.,* 1995), and

thereby create protostellar embryos that at the outset are inevitably low-mass (~0.005 $M_\odot$), very prone to dynamical interaction (being effectively point masses, albeit with gravity softening), and unable to merge. Thus, although formation by ejection seems inevitable, it may be less efficient than these simulations suggest. As discussed by *Goodwin and Kroupa* (2005) and *Hubber and Whitworth* (2005), the observed binary statistics are incompatible with too many ejections, and the number of protostellar embryos undergoing dynamical interactions within a single prestellar core should be relatively small ($\mathcal{N} \leq 4$). (2) They do not include magnetic fields. *Boss* (2002) argues that a magnetic field may promote fragmentation of a collapsing core by inhibiting the formation of a central density peak; however, his code does not capture all the possible MHD effects, in particular the anisotropy of magnetic pressure and the torques exerted by a twisted field. In contrast, *Hosking and Whitworth* (2004) have simulated the collapse of a rotating core with imperfect MHD, and find that fragmentation is inhibited. More work is needed.

In their original paper, *Reipurth and Clarke* (2001) conjectured that brown dwarfs formed by ejection could have a higher-velocity dispersion than more massive H-burning stars, and that this might be detected either in observations of the velocity dispersions of young star clusters or as a diaspora of brown dwarfs around more evolved clusters. They also pointed out that violent ejections would result in smaller accretion disks, and therefore a shortened quasi-T Tauri phase. Numerical simulations of star-cluster formation in highly turbulent massive cores (e.g., *Bate et al.,* 2003; *Bate and Bonnell,* 2005) find that the brown dwarfs do indeed have quite a high-velocity dispersion (~2 to 4 km s^{-1}), but they are difficult to distinguish from the low-mass H-burning stars, because both are frequently ejected from small dynamically unstable groups. Simulations of low-density star-forming regions by *Goodwin et al.* (2005) find that the velocity dispersions are typically somewhat lower (~0.5 to 1 km s^{-1}), and again the velocity dispersions of brown dwarfs and H-burning stars are hard to distinguish, because they are partially masked by the velocity dispersion between the different cores in which small subgroups (≤5) of stars are born. For the same reasons, segregation of brown dwarfs from H-burning stars may be difficult to detect. Moreover, given their low ejection speeds, some brown dwarfs formed by ejection retain significant disks (see Fig. 2), and are therefore presumably well able to sustain accretion and outflows, as observed (see section 2.3 and chapter by Luhman et al.). Thus kinematical and spatial information on brown dwarfs, and the signatures of accretion and outflow, provide less-powerful constraints on brown dwarf formation than were initially surmised by *Reipurth and Clarke* (2001).

7.4. Dynamical Interactions in Early Stellar Evolution

The binary statistics of brown dwarfs provide another potential constraint on formation mechanisms (see section 2.2 and the chapters by Burgasser et al., Duchêne et al., Good-

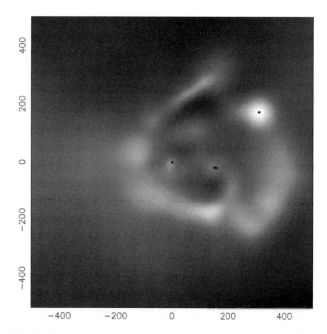

Fig. 2. Four stars formed in a low-mass core (~5 M$_\odot$) with a low initial level of turbulence (E$_{TURB}$ ~ 0.05 |E$_{GRAV}$|) (*Goodwin et al.,* 2004b). Near the center is a triple system containing a close binary (barely resolved pair of black dots). The object toward the upper righthand corner is a brown dwarf that has been ejected and is unbound from the triple. It has a significant disk (M$_{DISK}$ ~ 0.001 M$_\odot$). The frame is 10^3 AU across, and the time is 0.073 m.y. since the start of collapse.

win et al., and Luhman et al.). Formation by ejection may be incompatible with the relatively high frequency of close BD-BD binaries, if this is confirmed (e.g., *Maxted and Jeffries,* 2005; *Joergens,* 2006a). As well as resulting in the ejection of some brown dwarfs, dynamical interactions must also influence the binary statistics (distributions of primary mass, mass ratio, semimajor axis, and eccentricity) of the brown dwarfs and low-mass H-burning stars that do not get ejected (e.g., *Sterzik and Durisen,* 2003; *Kroupa and Bouvier,* 2003). *Hubber and Whitworth* (2005) have shown that if they take the observed distributions of core mass, radius, and rotation rate, convert each core into a ring of four to five stars with masses drawn from a log-normal distribution having dispersion σ$_{\log_{10}[M]}$ = 0.6, and then follow the dissolution of the ring by pure N-body dynamics, they reproduce rather well the observed distribution of multiplicity and binary statistics in young clusters, as a function of primary mass. *Umbreit et al.* (2005) have investigated the disintegration of nonhierarchical accreting triple systems, and find that they are able to produce the observed separation distribution of close binary brown dwarfs. Thus dynamical interactions may make important contributions, both to the formation of brown dwarfs, and to their binary statistics.

In addition to producing brown dwarfs and very-low-mass stars, and helping to shape the lower-mass end of the IMF and the statistics of binary and higher multiple systems, dynamical interactions may be the key to understanding a variety of other phenomena in early stellar evolution.

Reipurth (2000) notes that the different sizes and separations of shocks in Herbig-Haro flows can be understood as a fossil record of the dynamical evolution of a newly formed binary. He also notes that dynamical interactions may on occasion lead to a departure from the standard evolutionary picture of a star passing smoothly from the Class 0 stage through the Class III stage. Instead, stochastic dynamical interactions may lead to the sudden ejection of an object from the Class 0 or I stage, resulting in its abrupt appearance as a Class II or even a Class III object. The infrequent young binaries with infrared companions may be related to such events. Finally, the FU Orionis eruptions may be related to the formation of a close binary, and could result from the viscous interactions of circumstellar material as the two components in a newly formed binary spiral together (*Bonnell and Bastien,* 1992; *Reipurth and Aspin,* 2004).

8. PHOTOEROSION OF PRE-EXISTING CORES

A fifth — and somewhat separate — mechanism for forming brown dwarfs is to start with a pre-existing core of standard mass (i.e., ≤M$_\odot$) and have it overrun by an HII region (*Hester et al.,* 1996). As a result, an ionization front (IF) starts to eat into the core, "photoeroding" it. The IF is preceded by a compression wave (CW), and when the CW reaches the center, a protostar is created, which then grows by accretion. At the same time, an expansion wave (EW) is reflected and propagates outward, setting up the inflow that feeds accretion onto the central protostar. The outward-propagating EW soon meets the inward-propagating IF, and shortly thereafter the IF finds itself ionizing gas that is so tightly bound to the protostar that it cannot be unbound by the act of ionization. All the material interior to the IF at this juncture ends up in the protostar. On the basis of a simple semianalytic treatment, *Whitworth and Zinnecker* (2004) show that the final mass is given by

$$\sim 0.01 \, M_\odot \left(\frac{a_1}{0.3 \, km \, s^{-1}} \right)^6 \left(\frac{\dot{N}_{LyC}}{10^{50} \, s^{-1}} \right)^{-\frac{1}{3}} \left(\frac{n_O}{10^3 \, cm^{-3}} \right)^{-\frac{1}{3}} \quad (30)$$

where a$_1$ is the sound speed in the neutral gas of the core, $\dot{N}_{LyC}$ is the rate at which the exciting star(s) emit ionizing photons, and n$_O$ is the density in the HII region.

This mechanism is rather robust, in the sense that it produces very-low-mass stars for a wide range of initial conditions, and these conditions are likely to be realized in nature. Indeed, the evaporating gaseous globules (EGGs) identified in M16 by *Hester et al.* (1996) — and subsequently in other HII regions — would appear to be pre-existing cores being photoeroded in the manner we have described (e.g., *McCaughrean and Andersen,* 2002). However, the mechanism is also very inefficient, in the sense that it usually takes a rather massive pre-existing prestellar core to form a single brown dwarf or very-low-mass H-burning star. Moreover, the mechanism can only work in the immediate vicinity of an OB star, so it cannot explain the formation of all brown dwarfs, and another mechanism is required to explain those seen in star-formation regions like Taurus. Nonetheless, if the majority of stars are born in Trapezium-like clusters, rather than Taurus-like regions, then photoerosion should remain a contender for producing some brown dwarfs. Brown dwarfs formed by photoerosion should include close BD-BD binaries. It is unclear whether they can retain significant accretion disks.

9. SIMULATIONS OF CLUSTER FORMATION BY TURBULENT CLOUD COLLAPSE

With several different likely mechanisms for the production of brown dwarfs, the question arises as to which, if any, is the dominant formation mechanism. This is only likely to be answered through numerical simulations that are able to model the full star-formation process, including all the relevant physical ingredients (gravity, hydrodynamics, magnetic fields, radiative transfer, and chemistry). There is a huge effort underway to perform such simulations, but there are formidable numerical challenges to overcome.

The most comprehensive simulations to date are those of *Bate et al.* (2002a, 2003), *Bate and Bonnell* (2005), and *Bate* (2005). These model the collapse and fragmentation of turbulent molecular clouds with mass $M \sim 50 \, M_\odot$, initial diameter 0.18 to 0.38 pc, and initial temperature $T \sim 10$ K, to form small stellar clusters containing ~50 stars, including numerous brown dwarfs. The clouds are seeded with a power spectrum of supersonic velocity structure that matches the scaling of velocity dispersion with length scale observed in molecular clouds (*Larson,* 1981) and is allowed to decay during the simulations. The key difference between these simulations and earlier ones is that they are able to produce large numbers of objects from which statistical

quantities can be derived (e.g., the form of the IMF), but simultaneously they also resolve down to the opacity limit for fragmentation (section 3) and so they are able to follow the formation of all the stars that the clouds produce. They also resolve gaseous disks with radii down to ~10 AU and binaries with separations greater than ~1 AU. On the other hand, they do not include magnetic fields, radiative transfer, chemistry, or feedback. Therefore, for instance, they are unable to investigate the fraction of brown dwarfs that might form via photoerosion.

With these caveats in mind, the simulations generate clusters that are quite realistic. Starting with initial conditions typical of the molecular clouds in our galaxy, the simulations produce roughly equal numbers of stars and brown dwarfs, with an IMF that is roughly compatible with that observed, at low masses; the high-mass end of the IMF ($\geq M_\odot$) is not usefully constrained by these simulations. All stars, including brown dwarfs, originate from the fragmentation of dense filaments of molecular gas and from disk fragmentation, as illustrated in Fig. 3. Crucially, however, those that end up as brown dwarfs are those that avoid accreting large amounts of gas from the surrounding cloud. They are able to avoid becoming stars because they are ejected dynamically from unstable multiple systems, thereby terminating their accretion. In general it is easier to form brown dwarfs by disk fragmentation, because the resulting protostellar embryos are then born in a dense multiple system, and they can therefore be ejected quickly, before acquiring too much mass. In contrast, protostellar embryos formed by filament fragmentation are born in relative isolation, and they first have to fall down the filament into a dense cluster before they are ejected; their final mass is therefore less likely to stay below the H-burning limit. Thus, these simulations

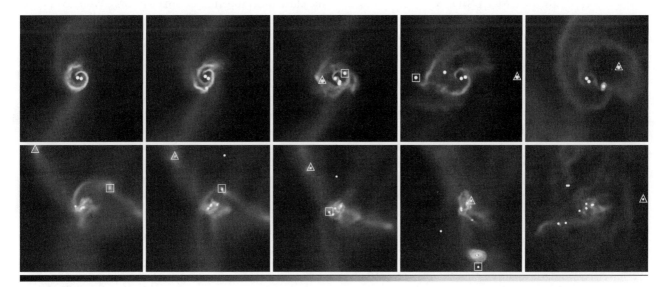

Fig. 3. Two sequences illustrating the brown dwarf formation mechanisms that occur in the simulations of *Bate et al.* (2002a,b, 2003). The upper sequence shows two brown dwarfs (square and triangle) forming in a circumbinary disk, while the lower sequence shows two brown dwarfs (square and triangle) forming in a filament. In both cases, these objects remain as brown dwarfs because they are dynamically ejected before they can accrete sufficient mass to ignite H-burning. In the upper case, they are ejected from a multiple system formed by disk fragmentation. In the lower case, the two brown dwarfs form in separate filaments, but then fall into, and are ejected from, the multiple system that exists at the intersection of the two filaments. Each panel is 600 AU across.

support the ejection hypothesis for the origin of brown dwarfs, with disk fragmentation also helping to generate dynamically unstable systems.

The brown dwarfs formed in these simulations do not frequently have companions. With three individual calculations now published, the overall frequency of BD-BD and very-low-mass binaries is ~5%. Most of these systems are closer than ~20 AU. Sun-like stars with brown dwarf companions at separations less than 10 AU are also very rare, at ~2%. Wider brown dwarf companions at hundreds or thousands of AU are much more common, although none of these systems have reached dynamic stability when the simulations are terminated. Most of these results are consistent with current observations, with the exception of the BD-BD binary frequency, where the value from the simulations is about three times lower than the observed value (~15%). There are at least two possible reasons for this. First, most observed and calculated BD-BD binaries have separations ≤20 AU, but the simulations are unable to resolve disks ≤10 AU, so the discrepancy might be solved with better resolution. If this is not the case, missing physics may be the problem (e.g., the effects of radiative transfer on disk fragmentation). Finally, only one brown dwarf that was ejected during any of the simulations had a resolved disk (radius $R \geq 10$ AU). This implies that brown dwarfs formed from highly turbulent cores should only have small disks. In contrast, the simulations of cores with low levels of turbulence performed by *Goodwin et al.* (2004a,b) produce brown dwarfs with somewhat larger disks (see Fig. 2). This indicates that disk size may be a function of birth environment. Although many brown dwarfs are observed to have disks from their spectral energy distributions, the distribution of their sizes is not currently known.

10. CONCLUSIONS

Since the statistical properties of brown dwarfs appear to form a continuum with those of low-mass H-burning stars, we have argued that brown dwarfs form like stars, i.e., on a dynamical timescale and by gravitational instability, with an homogeneous initial elemental composition, the same as the interstellar medium from which they form. In this regard brown dwarfs are distinct from planets, which we define to be objects that form on a longer timescale, by the accumulation of a rocky core and — if circumstances allow — the subsequent acquisition of a gaseous envelope, leading to an initially fractionated elemental composition and a deficit of light elements.

We have evaluated the minimum mass for a brown dwarf by considering the cooling required for a very-low-mass prestellar core to condense out on a dynamical timescale. We have treated several different scenarios: hierarchical three-dimensional fragmentation; one-shot two-dimensional fragmentation of a shock-compressed layer; and fragmentation of a Toomre-unstable disk. All three cases yield values of M_{MIN} in the range 0.001–0.004 $M_\odot$ for contemporary star formation in the solar vicinity. This suggests that there may be some overlap between the range of masses occupied by stars and planets. In hotter environments and at earlier epochs, M_{MIN} was probably larger, and brown dwarfs were therefore less common. We also suggest that the thermodynamics of disks make it easier for proto brown dwarfs to condense out at large radii ≥100 AU, and that these proto brown dwarfs may fragment to produce close BD-BD binaries, due to the dissociation of H_2 and the opacity gap at $T \sim 2000$ K.

We have discussed five possible mechanisms for forming brown dwarfs. Turbulent fragmentation of molecular clouds may deliver prestellar cores of such low mass that their subsequent collapse (and possible fragmentation) can only yield brown dwarfs. The collapse and fragmentation of more massive cores is likely to deliver protostellar embryos with a wide range of masses, many of which are too low to support hydrogen burning. Similarly, disk fragmentation is likely to deliver low-mass protostellar embryos. These protostellar embryos may undergo competitive accretion (as a result of which some of them evolve to higher mass) and dynamical interactions (as a result of which some of them are ejected before they reach H-burning masses). Lastly, cores that find themselves overrun by an HII region may be photoeroded by the resulting ionization front and end up spawning brown dwarfs. None of these mechanisms is mutually exclusive, and in the most advanced simulations of cluster formation, collapse and fragmentation, disk fragmentation, competitive accretion, and dynamical ejection all occur concurrently. However, these simulations do not capture all the deterministic physics. It requires a fully radiative, effectively inviscid, three-dimensional magnetohydrodynamical simulation to evaluate properly the thermal effects that influence the minimum mass for star formation, the angular momentum transport processes that influence the binary statistics, and the N-body dynamics that influence the clustering properties of brown dwarfs. Work to develop and validate such codes is ongoing.

We believe that all the proposed mechanisms operate in nature, and that once they have been properly modeled, *the ultimate task will be to determine their relative contributions to the overall brown dwarf population.* These relative contributions may depend on environment, metallicity, and epoch, and may therefore lead to local and/or cosmological variations in the ratio of brown dwarfs to H-burning stars, and in the binary and accretion statistics of brown dwarfs.

REFERENCES

André Ph., Ward-Thompson D., and Barsony M. (1993) *Astrophys. J., 406,* 122–141.

André Ph., Ward-Thompson D., and Barsony M. (2000) In *Protostars and Planets IV* (V. Mannings et al., eds.), pp. 59–96. Univ. of Arizona, Tucson.

Anosova J. P. (1986) *Astrophys. Space Sci., 124,* 217–241.

Armitage P. J. and Clarke C. J. (1997) *Mon. Not. R. Astron. Soc., 285,* 540–546.

Arquilla R. and Goldsmith P. F. (1985) *Astrophys. J., 297,* 436–454.

Bacmann A., André Ph., Puget J.-L., Abergel A., Bontemps S., and Ward-Thompson D. (2000) *Astron. Astrophys., 361,* 555–580.

Barranco J. A. and Goodman A. A. (1998) *Astrophys. J., 504,* 207–222.

Basri G. and Martín E. L. (1999) *Astron. J., 118,* 2460–2465.

Bastien P. (1983) *Astron. Astrophys., 119,* 109–116.

Bastien P., Arcoragi J.-P., Benz W., Bonnell I. A., and Martel H. (1991) *Astrophys. J., 378,* 255–265.

Bate M. R. (2005) *Mon. Not. R. Astron. Soc., 363,* 363–378.

Bate M. R. and Bonnell I. A. (2005) *Mon. Not. R. Astron. Soc., 356,* 1201–1221.

Bate M. R., Bonnell I. A., and Price N. M. (1995) *Mon. Not. R. Astron. Soc., 277,* 362–376.

Bate M. R., Bonnell I. A., and Bromm V. (2002a) *Mon. Not. R. Astron. Soc., 332,* L65–L68.

Bate M. R., Bonnell I. A., and Bromm V. (2002b) *Mon. Not. R. Astron. Soc., 336,* 705–713.

Bate M. R., Bonnell I. A., and Bromm V. (2003) *Mon. Not. R. Astron. Soc., 339,* 577–599.

Béjar V. J. S., Martín E. L., Zapatero Osorio M. R., Rebolo R., Barrado y Navacués D., et al. (2001) *Astrophys. J., 556,* 830–836.

Boffin H. M. J.,Watkins S. J., Bhattal A. S., Francis N., and Whitworth A. P. (1998) *Mon. Not. R. Astron. Soc., 300,* 1189–1204.

Bonnell I. A. (1994) *Mon. Not. R. Astron. Soc., 269,* 837–848.

Bonnell I. A. and Bastien P. (1992) *Astrophys. J., 401,* L31–L34.

Bonnell I. A. and Bate M. R. (1994a) *Mon. Not. R. Astron. Soc., 269,* L45–L48.

Bonnell I. A. and Bate M. R. (1994b) *Mon. Not. R. Astron. Soc., 271,* 999–1004.

Bonnell I. A., Bate M. R., Clarke C. J., and Pringle J. E. (1997) *Mon. Not. R. Astron. Soc., 285,* 201–208.

Bonnell I. A., Bate M. R., Clarke C. J., and Pringle J. E. (2001a) *Mon. Not. R. Astron. Soc., 323,* 785–794.

Bonnell I. A., Clarke C. J., Bate M. R., and Pringle J. E. (2001b) *Mon. Not. R. Astron. Soc., 324,* 573–579.

Boss A. P. (1986) *Astrophys. J. Suppl., 62,* 519–552.

Boss A. P. (1996) *Astrophys. J., 468,* 231–240.

Boss A. P. (2002) *Astrophys. J., 568,* 743–753.

Boss A. P. (2004) *Astrophys. J., 610,* 456–463.

Boss A. P., Fisher R. T., Klein R. I., and McKee C. F. (2000) *Astrophys. J., 528,* 325–335.

Bouy H., Brandner W., Martín E. L., Delfosse X., Allard F., and Basri G. (2003) *Astron. J., 126,* 1526–1554.

Bouy H., Martín E. L., Brandner W., and Bouvier J. (2005) *Astron. J., 129,* 511–517.

Boyd D. F. A. and Whitworth A. P. (2004) *Astron. Astrophys., 430,* 1059–1066.

Briceño C., Luhman K. L., Hartmann L., Stauffer J. R., and Kirkpatrick J. D. (2002) *Astrophys. J., 580,* 317–335.

Burgasser A. J., Kirkpatrick J. D., Reid I. N., Brown M. E., Miskey C. L., and Gizis J. E. (2003a) *Astrophys. J., 586,* 512–526.

Burgasser A. J., Kirkpatrick J. D., Burrows A., Liebert J., Reid I. N., et al. (2003b) *Astrophys. J., 592,* 1186–1192.

Burgasser A. J., Kirkpatrick J. D., and Lowrance P. J. (2005) *Astron. J., 129,* 2849–2855.

Burkert A. and Bodenheimer P. (1996) *Mon. Not. R. Astron. Soc., 280,* 1190–1200.

Burkert A. and Bodenheimer P. (2000) *Astrophys. J., 543,* 822–830.

Burkert A., Bate M. R., and Bodenheimer P. (1997) *Mon. Not. R. Astron. Soc., 289,* 497–504.

Cai K., Durisen R. H.,Michael S., Boley A. C., Mejía A. C., Pickett M. K., and D'Alessio P. (2006) *Astrophys. J., 636,* L149–L152.

Cha S.-H. and Whitworth A. P. (2003) *Mon. Not. R. Astron. Soc., 340,* 91–104.

Chabrier G. (2003) *Publ. Astron. Soc. Pac., 115,* 763–795.

Chapman S. J., Pongracic H., Disney M. J., Nelson A. H., Turner J. A., and Whitworth A. P. (1992) *Nature, 359,* 207–210.

Close L. M., Siegler N., Freed M., and Biller B. (2003) *Astrophys. J., 587,* 407–422.

Curry C. L. (2002) *Astrophys. J., 576,* 849–859.

Delgado-Donate E. J., Clarke C. J., and Bate M. R. (2003) *Mon. Not. R. Astron. Soc., 342,* 926–938.

Delgado-Donate E. J., Clarke C. J., and Bate M. R. (2004) *Mon. Not. R. Astron. Soc., 347,* 759–770.

Duquennoy A. and Mayor M. (1991) *Astron. Astrophys., 248,* 485–524.

Elmegreen B. G. (2000) *Astrophys. J., 530,* 277–281.

Fernández M. and Comerón F. (2001) *Astron. Astrophys., 380,* 264–276.

Gammie C. F. (2001) *Astrophys. J., 553,* 174–183.

Gizis J. E., Kirkpatrick J. D., Burgasser A., Reid I. N., Monet D. G., et al. (2001) *Astrophys. J., 551,* L163–L166.

Gizis J. E., Reid I. N., Knapp G. R., Liebert J., Kirkpatrick J. D., et al. (2003) *Astron. J., 125,* 3302–3310.

Goodman A. A., Benson P. J., Fuller G. A., and Myers P. C. (1993) *Astrophys. J., 406,* 528–547.

Goodwin S. P. and Kroupa P. (2005) *Astron. Astrophys., 439,* 565–569.

Goodwin S. P., Ward-Thompson D., and Whitworth A. P. (2002) *Mon. Not. R. Astron. Soc., 330,* 769–771.

Goodwin S. P., Whitworth A. P., and Ward-Thompson D. (2004a) *Astron. Astrophys., 414,* 633–650.

Goodwin S. P., Whitworth A. P., and Ward-Thompson D. (2004b) *Astron. Astrophys., 419,* 543–547.

Goodwin S. P., Whitworth A. P., and Ward-Thompson D. (2004c) *Astron. Astrophys., 423,* 169–182.

Goodwin S. P., Hubber D. A., Moraux E., and Whitworth A. P. (2005) *Astron. Nachr., 326,* 1040–1043.

Haisch K. E., Greene T. P., Barsony M., and Stahler S.W. (2004) *Astron. J., 127,* 1747–1754.

Hayashi C. and Nakano T. (1963) *Prog. Theor. Phys., 30,* 460–474.

Hennebelle P., Whitworth A. P., Cha S.-H., and Goodwin S. P. (2004) *Mon. Not. R. Astron. Soc., 348,* 687–701.

Hester J. J., Scowen P. A., Sankrit R., Lauer T. R., Ajhar E. A., et al. (1996) *Astron. J., 111,* 2349–2360.

Hosking J. G. and Whitworth A. P. (2004) *Mon. Not. R. Astron. Soc., 347,* 1001–1010.

Hoyle F. (1953) *Astrophys. J., 118,* 513–528.

Hubber D. A. and Whitworth A. P. (2005) *Astron. Astrophys., 437,* 113–125.

Inutsuka S.-I. and Miyama S. M. (1992) *Astrophys. J., 388,* 392–399.

Jayawardhana R., Ardila D. R., Stelzer B., and Haisch K. E. Jr. (2003) *Astron. J., 125,* 1515–1521.

Jiang I.-G., Laughlin G., and Lin D. N. C. (2004) *Astron. J., 127,* 455–459.

Joergens V. (2006a) *Astron. Astrophys., 446,* 1165–1176.

Joergens V. (2006b) *Astron. Astrophys., 448,* 655–663.

Joergens V., Fernández M., Carpenter J. M., and Neuhäuser R. (2003) *Astrophys. J., 594,* 971–981.

Johnson B. M. and Gammie C. F. (2003) *Astrophys. J., 597,* 131–141.

Johnstone D., Wilson C. D., Moriarty-Schieven G., Giannakopoulou-Creighton J., and Gregersen E. (2000) *Astrophys. J. Suppl., 131,* 505–518.

Jones C. E., Basu S., and Dubinski J. (2001) *Astrophys. J., 551,* 387–393.

Kenyon M. J., Jeffries R. D., Naylor T., Oliveira J. M., and Maxted P. F. L. (2005) *Mon. Not. R. Astron. Soc., 356,* 89–106.

Kirk H., Johnstone D., and Di Francesco J. (2006) *Astrophys. J., 646,* 1009–1023.

Kirk J. M., Ward-Thompson D., and André Ph. (2005) *Mon. Not. R. Astron. Soc., 360,* 1506–1526.

Klessen R. S. and Burkert A. (2000) *Astrophys. J. Suppl., 128,* 287–319.

Klessen R. S. and Burkert A. (2001) *Astrophys. J., 549,* 386–401.

Klessen R. S., Burkert A., and Bate M. R. (1998) *Astrophys. J., 501,* L205–L208.

Koresko C. D. (2002) *Astron. J., 124,* 1082–1088.

Kroupa P. and Bouvier J. (2003) *Mon. Not. R. Astron. Soc., 346,* 369–380.

Krumholz M. R., McKee C. F., and Klein R. I. (2005a) *Nature, 438,* 332–334.

Krumholz M. R., McKee C. F., and Klein R. I. (2005b) *Astrophys. J., 618,* 757–768.

Kumar S. S. (1963) *Astrophys. J., 137,* 1121–1125.

Larson R. B. (1981) *Mon. Not. R. Astron. Soc., 194,* 809–826.

Larson R. B. (1985) *Mon. Not. R. Astron. Soc., 214,* 379–398.

Laughlin G. and Bodenheimer P. (1994) *Astrophys. J., 436,* 335–354.

Lin C. C., Mestel L., and Shu F. H. (1965) *Astrophys. J., 142*, 1431–1446.

Looney L. W., Mundy L. G., and Welch W. J. (2000) *Astrophys. J., 529*, 477–498.

Low C. and Lynden-Bell D. (1976) *Mon. Not. R. Astron. Soc., 176*, 367–390.

Lucas P. W. and Roche P. F. (2000) *Mon. Not. R. Astron. Soc., 314*, 858–864.

Lucas P. W., Roche P. F., and Tamura M. (2005) *Mon. Not. R. Astron. Soc., 361*, 211–232.

Luhman K. L. (2004) *Astrophys. J., 617*, 1216–1232.

Luhman K. L., Stauffer J. R., Muench A. A., Rieke G. H., Lada E. A., Bouvier J., and Lada C. J. (2003) *Astrophys. J., 593*, 1093–1115.

Marcy G. W. and Butler R. P. (2000) *Publ. Astron. Soc. Pac., 112*, 137–140.

Masunaga H. and Inutsuka S. (1999) *Astrophys. J., 510*, 822–827.

Maxted P. and Jeffries R. (2005) *Mon. Not. R. Astron. Soc., 362*, L45–L49.

McCarthy C. and Zuckerman B. (2004) *Astron. J., 127*, 2871–2884.

McCaughrean M. J. and Andersen M. (2002) *Astron. Astrophys., 389*, 513–518.

McCaughrean M. J., Zinnecker H., Anderson M., Meeus G., and Lodieu N. (2002) *The Messenger, 109*, 28–36.

Mikkola S. and Valtonen M. J. (1986) *Mon. Not. R. Astron. Soc., 223*, 269–278.

Mohanty S., Jayawardhana R., Natta A., Fujiyoshi T., Tamura M., and Barrado y Navascués D. (2004) *Astrophys. J., 609*, L33–L36.

Monaghan J. J. (1976) *Mon. Not. R. Astron. Soc., 176*, 63–72.

Moraux E., Bouvier J., Stauffer J. R., and Cuillandre J.-C. (2003) *Astron. Astrophys., 400*, 891–902.

Motte F. and André Ph. (2001) *Astron. Astrophys., 365*, 440–464.

Motte F., André Ph., and Neri R. (1998) *Astron. Astrophys., 336*, 150–172.

Motte F., André Ph., Ward-Thompson D., and Bontemps S. (2001) *Astron. Astrophys., 372*, L41–L44.

Muench A. A., Alves J., Lada C. J., and Lada E. A. (2001) *Astrophys. J., 558*, L51–L54.

Muzerolle J., Hillenbrand L., Calvet N., Briceño C., and Hartmann L. (2003) *Astrophys. J., 592*, 266–281.

Muzerolle J., Luhman K. L., Briceño C., Hartman L., and Calvet N. (2005) *Astrophys. J., 625*, 906–912.

Myers P. C. (2005) *Astrophys. J., 623*, 280–290.

Myers P. C. and Benson P. J. (1983) *Astrophys. J., 266*, 309–320.

Myers P. C. and Gammie C. F. (1999) *Astrophys. J., 522*, L141–L144.

Myers P. C., Fuller G. A., Goodman A. A., and Benson P. J. (1991) *Astrophys. J., 376*, 561–572.

Nakajima T., Oppenheimer B. R., Kulkarni S. R., Golimowski D. A., Matthew K., and Durrance S. T. (1995) *Nature, 378*, 463–465.

Natta A. and Testi L. (2001) *Astron. Astrophys., 376*, L22–L25.

Natta A., Testi L., Muzerolle J., Randich S., Comerón F., and Persi P. (2004) *Astron. Astrophys., 424*, 603–612.

Nelson A. F., Benz W., Adams F. C., and Arnett D. (1998) *Astrophys. J., 502*, 342–371.

Nordlund Å. and Padoan P. (2003) In *Turbulence and Magnetic Fields in Astrophysics* (E. Falgarone and T. Passot, eds.), pp. 271–298. LNP Vol. 614, Springer, Berlin.

Onishi T., Mizuno A., Kawamura A., Tachihara K., and Fukui Y. (2002) *Astrophys. J., 575*, 950–973.

Oppenheimer B. R., Kulkarni S. R., Matthews K., and Nakajima T. (1995) *Science, 270*, 1478–1479.

Padoan P. and Nordlund Å. (2002) *Astrophys. J., 576*, 870–879.

Padoan P. and Nordlund Å. (2004) *Astrophys. J., 617*, 559–564.

Padoan P., Kritsuk A., Norman M. L., and Nordlund Å. (2005) *Astrophys. J., 622*, L61–L64.

Peng R., Langer W. D., Velusamy W. T., Kuiper T. B. H., and Levin S. (1998) *Astrophys. J., 497*, 842–849.

Pinfield D. J., Dobbie P. D., Jameson R. F., Steele I. A., Jones H. R. A., and Katsiyannis A. C. (2003) *Mon. Not. R. Astron. Soc., 342*, 1241–1259.

Preibisch T., McCaughrean M. J., Grosso N., Feigelson E. D., Flaccomio E., et al. (2005) *Astrophys. J. Suppl., 160*, 582–593.

Rafikov R. R. (2005) *Astrophys. J., 621*, L69–L72.

Rebolo R., Zapatero Osorio M. R., and Martín E. L. (1995) *Nature, 377*, 129–131.

Rees M. J. (1976) *Mon. Not. R. Astron. Soc., 176*, 483–486.

Reid M. A. and Wilson C. D. (2005) *Astrophys. J., 625*, 891–905.

Reipurth B. (2000) *Astron. J., 120*, 3177–3191.

Reipurth B. and Aspin C. (2004) *Astrophys. J., 608*, L65–L68.

Reipurth B. and Clarke C. J. (2001) *Astron. J., 122*, 432–439.

Reipurth B., Rodríguez L. F., Anglada G., and Bally J. (2002) *Astron. J., 124*, 1045–1053.

Reipurth B., Rodríguez L. F., Anglada G., and Bally J. (2004) *Astron. J., 127*, 1736–1746.

Rice W. K. M., Armitage P. J., Bate M. R., and Bonnell I. A. (2003) *Mon. Not. R. Astron. Soc., 339*, 1025–1030.

Ryden B. S. (1996) *Astrophys. J., 471*, 822–831.

Salpeter E. E. (1955) *Astrophys. J., 121*, 161–167.

Sandell G. and Knee L. B. G. (2001) *Astrophys. J., 546*, L49–L52.

Scholz A. and Eislöffel J. (2004) *Astron. Astrophys., 419*, 249–267.

Slesnick C. L., Hillenbrand L. A., and Carpenter J. M. (2004) *Astrophys. J., 610*, 1045–1063.

Stamatellos D., Whitworth A. P., Boyd D. F. A., and Goodwin S. P. (2005) *Astron. Astrophys., 439*, 159–169.

Standish E. M. (1972) *Astron. Astrophys., 21*, 185–191.

Stassun K. G., Mathieu R. D., and Valenti J. A. (2006) *Nature, 440*, 311.

Sterzik M. F. and Durisen R. H. (1995) *Astron. Astrophys., 304*, L9–L12.

Sterzik M. F. and Durisen R. H. (1998) *Astron. Astrophys., 339*, 95–112.

Sterzik M. F. and Durisen R. H. (2003) *Astron. Astrophys., 400*, 1031–1042.

Tachihara K., Onishi T., Mizuno A., and Fukui Y. (2002) *Astron. Astrophys., 385*, 909–920.

Teixeira P. S., Lada C. J., Young E. T., Marengo M., Muench A., et al. (2006) *Astrophys. J., 636*, L45–L48.

Testi L. and Sargent A. I. (1998) *Astrophys. J., 508*, L91–L94.

Tokovinin A. A. (2004) In *The Environment and Evolution of Double and Multiple Stars* (C. Allen and C. Scarfe, eds.), pp. 7–14. IAU Colloquium 191, Rev. Mex. Astron. Astrofis. Ser. Conf. 21.

Tokovinin A. A. and Smekhov M. G. (2002) *Astron. Astrophys., 382*, 118–123.

Toomre A. (1964) *Astrophys. J., 139*, 1217–1238.

Turner J. A., Chapman S. J., Bhattal A. S., Disney M. J., Pongracic H., and Whitworth A. P. (1995) *Mon. Not. R. Astron. Soc., 277*, 705–726.

Umbreit S., Burkert A., Henning T., Mikkola S., and Spurzem R. (2005) *Astrophys. J., 623*, 940–951.

van Albada T. S. (1968) *Bull. Astron. Inst. Netherlands, 19*, 479–499.

Watkins S. J., Bhattal A. S., Boffin H. M. J., Francis N., and Whitworth A. P. (1998a) *Mon. Not. R. Astron. Soc., 300*, 1205–1213.

Watkins S. J., Bhattal A. S., Boffin H. M. J., Francis N., and Whitworth A. P. (1998b) *Mon. Not. R. Astron. Soc., 300*, 1214–1224.

Whelan E. T., Ray T. P., Bacciotti F., Natta A., Testi L., and Randich S. (2005) *Nature, 435*, 652–654.

Whitehouse S. C. and Bate M. R. (2006) *Mon. Not. R. Astron. Soc., 367*, 32.

Whitworth A. P. (2005) In *Cores to Clusters* (M. S. N. Kumar et al., eds.), pp. 15–29. Astrophys. Space Sci. Library Ser. 324, Kluwer, Dordrecht.

Whitworth A. P. and Zinnecker H. (2004) *Astron. Astrophys., 427*, 299–306.

Whitworth A. P., Bhattal A. S., Chapman S. J., Disney M. J., and Turner J. A. (1994a) *Mon. Not. R. Astron. Soc., 268*, 291–298.

Whitworth A. P., Bhattal A. S., Chapman S. J., Disney M. J., and Turner J. A. (1994b) *Astron. Astrophys., 290*, 421–427.

Whitworth A. P., Chapman S. J., Bhattal A. S., Disney M. J., Pongracic H., and Turner J. A. (1995) *Mon. Not. R. Astron. Soc., 277*, 727–746.

Zapatero Osorio M. R., Béjar V. J. S., Martín E. L., Rebolo R., Barrado y Navasqués D., et al. (2002) *Astrophys. J., 578*, 536–542.

Zinnecker H. (1982) *N.Y. Acad. Sci. Ann., 395*, 226–235.

Zinnecker H. (1984) *Mon. Not. R. Astron. Soc., 210*, 43–56.

Part VI:
Circumstellar Disks

Magnetospheric Accretion in Classical T Tauri Stars

J. Bouvier and S. H. P. Alencar
Laboratoire d'Astrophysique Grenoble

T. J. Harries
University of Exeter

C. M. Johns-Krull
Rice University

M. M. Romanova
Cornell University

The inner 0.1 AU around accreting T Tauri stars hold clues to many physical processes that characterize the early evolution of solar-type stars. The accretion-ejection connection takes place at least in part in this compact magnetized region around the central star, with the inner disk edge interacting with the star's magnetosphere, thus leading simultaneously to magnetically channeled accretion flows and to high-velocity winds and outflows. The magnetic star-disk interaction is thought to have strong implications for the angular momentum evolution of the central system, the inner structure of the disk, and possibly halting the migration of young planets close to the stellar surface. We review here the current status of magnetic field measurements in T Tauri stars, the recent modeling efforts of the magnetospheric accretion process, including both radiative transfer and multi-dimensional numerical simulations, and summarize current evidence supporting the concept of magnetically channeled accretion in young stars. We also discuss the limits of the models and highlight observational results that suggest the star-disk interaction is a highly dynamical and time-variable process in young stars.

1. THE MAGNETIC ACCRETION PARADIGM

T Tauri stars are low-mass stars with an age of a few million years, still contracting down their Hayashi tracks toward the main sequence. Many of them, the so-called classical T Tauri stars (CTTSs), show signs of accretion from a circumstellar disk (see, e.g., *Ménard and Bertout*, 1999, for a review). Understanding the accretion process in T Tauri stars is one of the major challenges in the study of pre-main-sequence evolution. Indeed, accretion has a significant and long-lasting impact on the evolution of low-mass stars by providing both mass and angular momentum. The evolution and ultimate fate of circumstellar accretion disks have also become increasingly important issues since the discovery of extrasolar planets and planetary systems with unexpected properties. Deriving the properties of young stellar systems, and of their associated disks and outflows, is therefore an important step toward the establishment of plausible scenarios for star and planet formation.

The general paradigm of magnetically controlled accretion onto a compact object is used to explain many of the most fascinating objects in the universe. This model is a seminal feature of low-mass star formation, but it is also encountered in theories explaining accretion onto white dwarf stars [the AM Her stars (e.g., *Warner,* 2004)], accretion onto pulsars [the pulsating X-ray sources (e.g., *Ghosh*

and Lamb, 1979a)], and accretion onto black holes at the center of AGNs and microquasars (*Koide et al.,* 1999). Strong surface magnetic fields have long been suspected to exist in TTSs based on their powerful X-ray and centrimetric radio emissions (*Montmerle et al.,* 1983; *André,* 1987). Surface fields on the order of 1–3 kG have recently been derived from Zeeman broadening measurements of CTTS photospheric lines (*Johns-Krull et al.,* 1999a, 2001; *Guenther et al.,* 1999) and from the detection of electron cyclotron maser emission (*Smith et al.,* 2003). These strong stellar magnetic fields are believed to significantly alter the accretion flow in the circumstellar disk close to the central star.

Based on models originally developed for magnetized compact objects in X-ray pulsars (*Ghosh and Lamb,* 1979a) and *assuming* that T Tauri magnetospheres are predominantly dipolar on the large scale, *Camenzind* (1990) and *Königl* (1991) showed that the inner accretion disk is expected to be truncated by the magnetosphere at a distance of a few stellar radii above the stellar surface for typical mass accretion rates of 10^{-9}–10^{-7} $M_\odot$ yr^{-1} in the disk (*Basri and Bertout,* 1989; *Hartigan et al.,* 1995; *Gullbring et al.,* 1998). Disk material is then channeled from the disk inner edge onto the star along the magnetic field lines, thus giving rise to magnetospheric accretion columns. As the freefalling material in the funnel flow eventually hits the stellar surface, accretion shocks develop near the magnetic poles. The basic

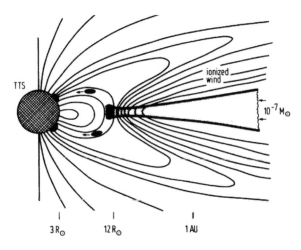

Fig. 1. A sketch of the basic concept of magnetospheric accretion in T Tauri stars (from *Camenzind*, 1990).

concept of magnetospheric accretion in T Tauri stars is illustrated in Fig. 1.

The successes and limits of current magnetospheric accretion models in accounting for the observed properties of classical T Tauri systems are reviewed in the next sections. Section 2 summarizes the current status of magnetic field measurements in young stars, section 3 provides an account of current radiative transfer models developed to reproduce the observed line profiles thought to form at least in part in accretion funnel flows, section 4 reviews current observational evidence for a highly dynamical magnetospheric accretion process in CTTSs, and section 5 describes the most recent two-dimensional and three-dimensional numerical simulations of time-dependent star-disk magnetic interaction.

2. MAGNETIC FIELD MEASUREMENTS

2.1. Theoretical Expectations for T Tauri Magnetic Fields

While the interaction of a stellar magnetic field with an accretion disk is potentially very complicated (e.g., *Ghosh and Lamb,* 1979a,b), we present here some results from the leading treatments applied to young stars.

The theoretical idea behind magnetospheric accretion is that the ram pressure of the accreting material ($P_{ram} = 0.5 \rho v^2$) will at some point be offset by the magnetic pressure ($P_B = B^2/8\pi$) for a sufficiently strong stellar field. Where these two pressures are equal, if the accreting material is sufficiently ionized, its motion will start to be controlled by the stellar field. This point is usually referred to as the truncation radius (R_T). If we consider the case of spherical accretion, the ram pressure becomes

$$B^2 = \frac{\dot{M}v}{r^2} \qquad (1)$$

If we then assume a dipolar stellar magnetic field where $B = B_*(R_*/r)^3$ and set the velocity of the accreting material equal to the freefall speed, the radius at which the magnetic

field pressure balances the ram pressure of the accreting material is

$$\frac{R_T}{R_*} = \frac{B_*^{4/7}R_*^{5/7}}{\dot{M}^{2/7}(2GM_*)^{1/7}} = 7.1B_3^{4/7}\dot{M}_{-8}^{-2/7}M_{0.5}^{-1/7}R_2^{5/7} \qquad (2)$$

where B_3 is the stellar field strength in kG, $\dot{M}_{-8}$ is the mass accretion rate in units of $10^{-8}\,M_\odot\,yr^{-1}$, $M_{0.5}$ is the stellar mass in units of $0.5\,M_\odot$, and R_2 is the stellar radius in units of $2\,R_\odot$. Then, for $B_* = 1$ kG and typical CTTS properties ($M_* = 0.5\,M_\odot$, $R_* = 2\,R_\odot$, and $\dot{M} = 10^{-8}\,M_\odot\,yr^{-1}$), the truncation radius is about 7 stellar radii.

In the case of disk accretion, the coefficient above is changed, but the scaling with the stellar and accretion parameters remains the same. In accretion disks around young stars, the radial motion due to accretion is relatively low while the Keplerian velocity due to the orbital motion is only a factor of 2.5 lower than the freefall velocity. The low radial velocity of the disk means that the disk densities are much higher than in the spherical case, so that the disk ram pressure is higher than the ram pressure due to spherical freefall accretion. As a result, the truncation radius will move closer to the star. In this regard, equation (2) gives an upper limit for the truncation radius. As we will discuss below, this may be problematic when we consider the current observations of stellar magnetic fields. In the case of disk accretion, another important point in the disk is the corotation radius, R_{CO}, where the Keplerian angular velocity is equal to the stellar angular velocity. Stellar field lines that couple to the disk outside R_{CO} will act to slow the rotation of the star down, while field lines that couple to the disk inside R_{CO} will act to spin the star up. Thus, the value of R_T relative to R_{CO} is an important quantity in determining whether the star speeds up or slows down its rotation. For accretion onto the star to proceed, we have the relation $R_T < R_{CO}$. This follows from the idea that at the truncation radius and interior to that, the disk material will be locked to the stellar field lines and will move at the same angular velocity as the star. Outside R_{CO} the stellar angular velocity is greater than the Keplerian velocity, so that any material there that becomes locked to the stellar field will experience a centrifugal force that tries to fling the material away from the star. Only inside R_{CO} will the net force allow the material to accrete onto the star.

Traditional magnetospheric accretion theories as applied to stars (young stellar objects, white dwarfs, and pulsars) suggest that the rotation rate of the central star will be set by the Keplerian rotation rate in the disk near the point where the disk is truncated by the stellar magnetic field when the system is in equilibrium. Hence these theories are often referred to as disk-locking theories. For CTTSs, we have a unique opportunity to test these theories since all the variables of the problem (stellar mass, radius, rotation rate, magnetic field, and disk accretion rate) are measureable in principle (see *Johns-Krull and Gafford,* 2002). Under the assumption that an equilibrium situation exists, *Königl* (1991), *Collier Cameron and Campbell* (1993), and *Shu et al.* (1994) have all analytically examined the interaction be-

TABLE 1.   Predicted magnetic field strengths.

Star	M_* $(M_\odot)$	R_* $(R_\odot)$	$\dot{M} \times 10^8$ $(M_\odot\,yr^{-1})$	P_{rot} (days)	$B_*^{[1]}$ (G)	$B_*^{[2]}$ (G)	$B_*^{[3]}$ (G)	R_{CO} (R_*)	$\bar{B}_{obs}$ (kG)
AA Tau	0.53	1.74	0.33	8.20	810	240	960	8.0	2.57
BP Tau	0.49	1.99	2.88	7.60	1370	490	1620	6.4	2.17
CY Tau	0.42	1.63	0.75	7.90	1170	390	1380	7.7	
DE Tau	0.26	2.45	2.64	7.60	420	164	490	4.2	1.35
DF Tau	0.27	3.37	17.7	8.50	490	220	570	3.4	2.98
DK Tau	0.43	2.49	3.79	8.40	810	300	950	5.3	2.58
DN Tau	0.38	2.09	0.35	6.00	250	80	300	4.8	2.14
GG Tau A	0.44	2.31	1.75	10.30	890	320	1050	6.6	1.57
GI Tau	0.67	1.74	0.96	7.20	1450	450	1700	7.9	2.69
GK Tau	0.46	2.15	0.64	4.65	270	90	320	4.2	2.13
GM Aur	0.52	1.78	0.96	12.00	1990	660	2340	10.0	
IP Tau	0.52	1.44	0.08	3.25	240	60	280	5.2	
TW Hya	0.70	1.00	0.20	2.20	900	240	1060	6.3	2.61
T Tau	2.11	3.31	4.40	2.80	390	110	460	3.2	2.39

Magnetic field values come from applying the theory of [1] *Königl* (1991), [2] *Collier Cameron and Campbell* (1993), or [3] *Shu et al.* (1994). These are the equatorial field strengths assuming a dipole magnetic field.

tween a dipolar stellar magnetic field (aligned with the stellar rotation axis) and the surrounding accretion disk. As detailed in *Johns-Krull et al.* (1999b), one can solve for the surface magnetic field strength on a CTTS implied by each of these theories given the stellar mass, radius, rotation period, and accretion rate. For the work of *Königl* (1991), the resulting equation is

$$
\begin{aligned}
B_* = 3.43 \left(\frac{\varepsilon}{0.35} \right)^{7/6} \left(\frac{\beta}{0.5} \right)^{-7/4} \left(\frac{M_*}{M_\odot} \right)^{5/6} \times \\
\left(\frac{\dot{M}}{10^{-7}\,M_\odot\,yr^{-1}} \right)^{1/2} \left(\frac{R_*}{R_\odot} \right)^{-3} \left(\frac{P_*}{1\,dy} \right)^{7/6} \, kG
\end{aligned}
\tag{3}
$$

In the work of *Collier Cameron and Campbell* (1993) the equation for the stellar field is

$$
\begin{aligned}
B_* = 1.10\gamma^{-1/3} \left(\frac{M_*}{M_\odot} \right)^{2/3} \left(\frac{\dot{M}}{10^{-7}\,M_\odot\,yr^{-1}} \right)^{23/40} \times \\
\left(\frac{R_*}{R_\odot} \right)^{-3} \left(\frac{P_*}{1\,dy} \right)^{29/24} \, kG
\end{aligned}
\tag{4}
$$

Finally, from *Shu et al.* (1994), the resulting equation is

$$
\begin{aligned}
B_* = 3.38 \left(\frac{\alpha_x}{0.923} \right)^{-7/4} \left(\frac{M_*}{M_\odot} \right)^{5/6} \left(\frac{\dot{M}}{10^{-7}\,M_\odot\,yr^{-1}} \right)^{1/2} \times \\
\left(\frac{R_*}{R_\odot} \right)^{-3} \left(\frac{P_*}{1\,dy} \right)^{7/6} \, kG
\end{aligned}
\tag{5}
$$

All these equations contain uncertain scaling parameters (ε, β, γ, α_x) that characterize the efficiency with which the stellar field couples to the disk or the level of vertical shear in the disk. Each study presents a best estimate for these parameters allowing the stellar field to be estimated (Table 1). Observations of magnetic fields on CTTSs can then serve as a test of these models.

To predict magnetic field strengths for specific CTTSs, we need observational estimates for certain system parameters. We adopt rotation periods from *Bouvier et al.* (1993, 1995) and stellar masses, radii, and mass accretion rates from *Gullbring et al.* (1998). Predictions for each analytic study are presented in Table 1. Note that these field strengths are the equatorial values. The field at the pole will be twice these values and the average over the star will depend on the exact inclination of the dipole to the observer, but for i = 45% the average field strength on the star is ~1.4× the values given in the Table 1. Because of differences in underlying assumptions, these predictions are not identical, but they do have the same general dependence on system characteristics. Consequently, field strengths predicted by the three theories, while different in scale, nonetheless have the same pattern from star to star. Relatively weak fields are predicted for some stars (DN Tau, IP Tau), but detectably strong fields are expected on stars such as BP Tau.

2.2. Measurement Techniques

Virtually all measurements of stellar magnetic fields make use of the Zeeman effect. Typically, one of two general aspects of the Zeeman effect is utilized: (1) Zeeman broadening of magnetically sensitive lines observed in intensity spectra, or (2) circular polarization of magnetically sensitive lines. Due to the nature of the Zeeman effect, the splitting due to a magnetic field is proportional to λ^2 of the transition. Compared with the λ^1 dependence of Doppler line broadening mechanisms, this means that observations in the infrared (IR) are generally more sensitive to the presence of magnetic fields than optical observations.

The simplest model of the spectrum from a magnetic star assumes that the observed line profile can be expressed as $F(\lambda) = F_B(\lambda) \times f + F_Q(\lambda) \times (1-f)$, where F_B is the spectrum formed in magnetic regions, F_Q is the spectrum formed in nonmagnetic (quiet) regions, and f is the flux weighted surface filling factor of magnetic regions. The magnetic spectrum, F_B, differs from the spectrum in the quiet region not only due to Zeeman broadening of the line, but also because magnetic fields affect atmospheric structure, causing changes in both line strength and continuum intensity at the surface. Most studies *assume* that the magnetic atmosphere is in fact the same as the quiet atmosphere because there is no theory to predict the structure of the magnetic atmosphere. If the stellar magnetic field is very strong, the splitting of the σ components is a substantial fraction of the line width, and it is easy to see the σ components sticking out on either side of a magnetically sensitive line. In this case, it is relatively straightforward to measure the magnetic field strength, B. Differences in the atmospheres of the magnetic and quiet regions primarily affect the value of f. If the splitting is a small fraction of the intrinsic line width, then the resulting observed profile is only subtly different from the profile produced by a star with no magnetic field and more complicated modeling is required to be sure all possible nonmagnetic sources (e.g., rotation and pressure broadening) have been properly constrained.

In cases where the Zeeman broadening is too subtle to detect directly, it is still possible to diagnose the presence of magnetic fields through their effect on the equivalent width of magnetically sensitive lines. For strong lines, the Zeeman effect moves the σ components out of the partially saturated core into the line wings where they can effectively add opacity to the line and increase the equivalent width. The exact amount of equivalent width increase is a complicated function of the line strength and Zeeman splitting pattern (*Basri et al.*, 1992). This method is primarily sensitive to the product of B multiplied by the filling factor f (*Basri et al.*, 1992; *Guenther et al.*, 1999). Since this method relies on relatively small changes in the line equivalent width, it is very important to be sure other atmospheric parameters that affect equivalent width (particularly temperature) are accurately measured.

Measuring circular polarization in magnetically sensitive lines is perhaps the most direct means of detecting magnetic fields on stellar surfaces, but is also subject to several limitations. When viewed along the axis of a magnetic field, the Zeeman σ components are circularly polarized, but with opposite helicity; and the σ component is absent. The helicity of the σ components reverses as the polarity of the field reverses. Thus, on a star like the Sun that typically displays equal amounts of + and – polarity fields on its surface, the net polarization is very small. If one magnetic polarity does dominate the visible surface of the star, net circular polarization is present in Zeeman sensitive lines, resulting in a wavelength shift between the line observed through right-

and left-circular polarizers. The magnitude of the shift represents the surface averaged line of sight component of the magnetic field (which on the Sun is typically less than 4 G even though individual magnetic elements on the solar surface range from ~1.5 kG in plages to ~3.0 kG in spots). Several polarimetric studies of cool stars have generally failed to detect circular polarization, placing limits on the disk-averaged magnetic field strength present of 10–100 G (e.g., *Vogt*, 1980; *Brown and Landstreet*, 1981; *Borra et al.*, 1984). One notable exception is the detection of circular polarization in segments of the line profile observed on rapidly rotating dwarfs and RS CVn stars where Doppler broadening of the line "resolves" several independent strips on the stellar surface (e.g., *Donati et al.*, 1997; *Petit et al.*, 2004; *Jardine et al.*, 2002).

2.3. Mean Magnetic Field Strength

TTSs typically have v sin i values of 10 km s^{-1}, which means that observations in the optical typically cannot detect the actual Zeeman broadening of magnetically sensitive lines because the rotational broadening is too strong. Nevertheless, optical observations can be used with the equivalent width technique to detect stellar fields. *Basri et al.* (1992) were the first to detect a magnetic field on the surface of a TTS, inferring a value of $B_f = 1.0$ kG on the NTTS Tap 35. For the NTTS Tap 10, *Basri et al.* (1992) find only an upper limit of $B_f < 0.7$ kG. *Guenther et al.* (1999) apply the same technique to spectra of five TTSs, claiming significant field detections on two stars; however, these authors analyze their data using models off by several hundred K from the expected effective temperature of their target stars, always a concern when relying on equivalent widths. As we saw above, observations in the IR will help solve the difficulty in detecting direct Zeeman broadening. For this reason and given the temperature of most TTSs (K7–M2), Zeeman broadening measurements for these stars are best done using several Ti I lines found in the K band. Robust Zeeman broadening measurements require Zeeman insensitive lines to constrain nonmagnetic broadening mechanisms. Numerous CO lines at 2.31 μm have negligible Landé-g factors, making them an ideal null reference.

It has now been shown that the Zeeman insensitive CO lines are well fitted by models with the same level of rotational broadening as that determined from optical line profiles (*Johns-Krull and Valenti*, 2001; *Johns-Krull et al.*, 2004; *Yang et al.*, 2005). In contrast, the 2.2-μm Ti I lines cannot be fitted by models without a magnetic field. Instead, the observed spectrum is best fit by a model with a superposition of synthetic spectra representing different regions on the star with different magnetic field strengths. Typically, the field strengths in these regions are assumed to have values of 0, 2, 4, and 6 kG and only the filling factor of each region is solved for. The resulting magnetic field distribution is unique because the Zeeman splitting produced by a 2-kG field is comparable to the nonmagnetic width of the Ti I spectral lines. In other words, the Zeeman resolution of the Ti I lines is about 2 kG (see *Johns-Krull et al.*, 1999b, 2004).

The intensity-weighted mean magnetic field strength, $\bar{B}$, over the entire surface of most TTSs analyzed to date is ~2.5 kG, with field strengths reaching at least 4 kG and probably even 6 kG in some regions. Thus, magnetic fields on TTSs are stronger than on the Sun, even though the surface gravity on these stars is lower by a factor of 10. On the Sun and other main-sequence stars, magnetic field strength seems to be set by an equipartition of gas and magnetic pressure. In contrast, the photospheres of TTSs are apparently dominated by magnetic pressure, rather than gas pressure (see also *Johns-Krull et al., 2004*). Strong magnetic fields are ubiquitous on TTSs. By fitting IR spectra, magnetic field distributions for several TTSs have now been measured (*Johns-Krull et al., 1999b, 2001, 2004; Yang et al., 2005*). Many of these field strengths are reported in Table 1.

2.4. Magnetic Field Topology

Zeeman broadening measurements are sensitive to the distribution of magnetic field strengths, but they have limited sensitivity to magnetic geometry. In contrast, circular polarization measurements for individual spectral lines are sensitive to magnetic geometry, but they provide limited information about field strength. The two techniques complement each other well, as we demonstrate below.

Most existing magnetospheric accretion models assume that intrinsic TTS magnetic fields are dipolar, but this would be unprecedented for cool stars. The higher order components of a realistic multipolar field will fall off more rapidly with distance than the dipole component, so at the inner edge of the disk a few stellar radii from the surface, it is likely that the dipolar component of the stellar field will dominate. However, at the stellar surface the magnetic field is likely to be more complicated. In support of this picture is the fact that spectropolarimetric observations do not detect polarization in photospheric absorption lines: *Brown and Landstreet* (1981) failed to detect polarization in T Tau and two FU Ori objects; *Johnstone and Penston* (1986) observed three CTTSs and reported a marginal field detection for RU Lup, but they were not able to confirm the signal in a subsequent observation, perhaps because of rotational modulation (*Johnstone and Penston, 1987*); *Donati et al.* (1997) find no evidence for a strong dipolar field component in the three TTSs they observed; *Johns-Krull et al.* (1999a) failed to detect polarization in the photospheric lines of BP Tau; and *Valenti and Johns-Krull* (2004) do not detect significant polarization in the photospheric lines of four CTTSs each observed over a rotation period. *Smirnov et al.* (2003) report a marginal detection of circular polarization in the lines of T Tau corresponding to a field of ~150 ± 50 G; however, *Smirnov et al.* (2004) and *Daou et al.* (2006) failed to confirm this detection, placing an upper limit on the field of ≤120 G for T Tau.

However, *Johns-Krull et al.* (1999a) did discover circular polarization in CTTS emission line diagnostics that form predominantly in the accretion shock at the surface of the star. This circular polarization signal is strongest in the narrow component of the He I 5876-Å emission line, but it is

also present in the Ca I infrared triplet lines. The peak value of B_z is 2.5 kG, which is comparable to our measured values of $\bar{B}$. Circular polarization in the He I 5876-Å emission line has now been observed in a number of CTTSs (*Valenti et al., 2004; Symington et al., 2005b*). Note that since this polarization is detected in a line associated with the accretion shock on CTTSs, it forms over an area covering typically <5% of the stellar surface (*Valenti et al., 1993; Calvet and Gullbring, 1998*). While the field in this 5% of the star appears to be highly organized (and as discussed below may trace the dipole component of the field at the surface), the lack of polarization detected in photospheric lines forming over the entire surface of the star strongly rules out a global dipole geometry for the entire field.

Figure 2 shows measurements of B_z on six consecutive nights. These measurements were obtained at McDonald Observatory, using the Zeeman analyzer described by *Johns-Krull et al.* (1999a). The measured values of B_z vary smoothly on rotational timescales, suggesting that uniformly oriented magnetic field lines in accretion regions sweep out a cone in the sky, as the star rotates. Rotational modulation implies a lack of symmetry about the rotation axis in the accretion or the magnetic field or both. For example, the inner edge of the disk could have a concentration of gas that co-rotates with the star, preferentially illuminating one sector of a symmetric magnetosphere. Alternatively, a single large-scale magnetic loop could draw material from just one sector of a symmetric disk.

Figure 2 shows one interpretation of the He I polarization data. Predicted values of B_z are shown for a simple model consisting of a single magnetic spot at latitude ϕ that rotates with the star. The magnetic field is assumed to be

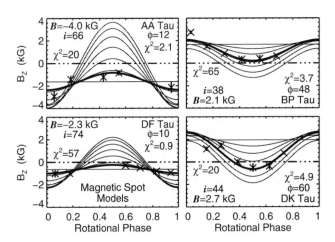

Fig. 2. Variations in the circular polarization of the He I emission line as a function of rotation phase for four CTTSs. Polarization levels are translated into B_z values in the line formation region. Vertical bars centered on each measurement (×) give the 1σ uncertainty in the field measurement. Solid lines show predicted rotational modulation in B_z for a single magnetic spot at latitudes (ϕ) ranging from 0° to 90° in 15° increments. The best-fit latitude is shown in the thick solid curve.

radial with a strength equal to our measured values of $\bar{B}$. Inclination of the rotation axis is constrained by measured v sin i and rotation period, except that inclination (1) is allowed to float when it exceeds 60° because v sin i measurements cannot distinguish between these possibilities. Predicted variations in B_z are plotted for spot latitudes ranging from 0° to 90° in 15° increments. The best fitting model is shown by the thick curve. The corresponding spot latitude and reduced χ^2 are given on the right side of each panel. The null hypothesis (that no polarization signal is present) produces very large values of χ^2, which are given on the left side of each panel. In all four cases, this simple magnetic spot model reproduces the observed B_z time series. The He I rotationally modulated polarization combined with the lack of detectable polarization in photospheric absorption lines as described above paints a picture in which the magnetic field on TTSs displays a complicated geometry at the surface that gives way to a more ordered, dipole-like geometry a few stellar radii from the surface where the field intersects the disk. The complicated surface topology results in no net polarization in photospheric absorption lines, but the dipole-like geometry of the field at the inner disk edge means that accreting material follows these field lines down to the surface so that emission lines formed in the accretion shock preferentially illuminate the dipole component of the field, producing substantial circular polarization in these emission lines.

2.5. Confronting Theory with Observations

At first glance, it might appear that magnetic field measurements on TTSs are generally in good agreement with theoretical expectations. Indeed, the IR Zeeman broadening measurements indicate mean fields on several TTSs of ~2 kG, similar in value to those predicted in Table 1 (recall the field values in the table are the equatorial values for a dipolar field, and that the mean field is about 1.5× these equatorial values). However, in detail the field observations do not agree with the theory. This can be seen in Fig. 3, where we plot the measured magnetic field strengths vs. the predicted field strengths from *Shu et al.* (1994) (see Table 1). Clearly, the measured field strengths show no correlation with the predicted field strengths. The field topology measurements give some indication to why there may be a lack of correlation: The magnetic field on TTSs are not dipolar, and the dipole component to the field is likely to be a factor of ~10 or more lower than the values predicted in Table 1. As discussed in *Johns-Krull et al.* (1999b), the three studies that produce the field predictions in Table 1 involve uncertain constants that describe the efficiency with which stellar field lines couple to the accretion disk. If these factors are much different than estimated, it may be that the required dipole components to the field are substantially less than the values given in the table. On the other hand, equation (2) was derived assuming perfect coupling of the field and the matter, so it serves as a firm upper limit to RT as discussed in section 2.1. Spectropolarimetry of TTSs indicates that the

dipole component of the magnetic field is ≤0.1 kG (*Valenti and Johns-Krull*, 2004; *Smirnov et al.*, 2004; *Daou et al.*, 2006). Putting this value into equation (2), we find $R_T \leq$ 1.9 R_* for typical CTTS parameters. Such a low value for the truncation radius is incompatible with rotation periods of 7–10 days as found for many CTTSs [Table 1 and, e.g., *Herbst et al.* (2002)].

Does this then mean that magnetospheric accretion does not work? Independent of the coupling efficiency between the stellar field and the disk, magnetospheric accretion models predict correlations between stellar and accretion parameters. As shown in section 2.3, the fields on TTSs are found to all be rather uniform in strength. Eliminating the stellar field, *Johns-Krull and Gafford* (2002) then looked for correlation among the stellar and accretion parameters, finding little evidence for the predicted correlations. This absence of the expected correlations had been noted earlier by *Muzerolle et al.* (2001). On the other hand, *Johns-Krull and Gafford* (2002) showed how the models of *Ostriker and Shu* (1995) could be extended to take into account nondipole field geometries. Once this is done, the current data do reveal the predicted correlations, suggesting magnetospheric accretion theory is basically correct as currently formulated. So then how do we reconcile the current field measurements with this picture? While the dipole component of the field is small on TTSs, it is clear the stars possess strong fields over most, if not all, of their surface but with a complicated surface topology. Perhaps this can lead to a strong enough field so that $R_T \sim 6 R_*$ as generally suggested by observations of CTTS phenomena. More complicated numerical modeling of the interaction of a complex geometry field with an accretion disk will be required to see if this is feasible.

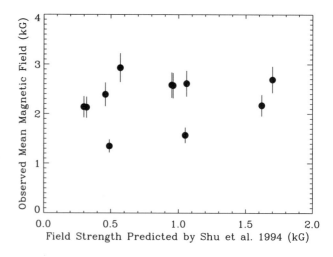

Fig. 3. Observed mean magnetic field strength determined from IR Zeeman broadening measurements as a function of the predicted field strength from Table 1 for the theory of *Shu et al.* (1994). No statistically significant correlation is found between the observed and predicted field strengths.

3. SPECTRAL DIAGNOSTICS OF MAGNETOSPHERIC ACCRETION

Permitted emission line profiles from CTTSs, in particular the Balmer series, show a wide variety of morphologies including symmetric, double-peaked, P Cygni, and inverse P Cygni (IPC) type (*Edwards et al., 1994*): Common to all shapes is a characteristic line width indicative of bulk motion within the circumstellar material of hundreds of km s⁻¹. The lines themselves encode both geometrical and physical information on the accretion process and its rate, and the challenge is to use the profiles to test and refine the magnetospheric accretion model.

Interpretation of the profiles requires a translational step between the physical model and the observable spectra; this is the process of radiative-transfer (RT) modeling. The magnetospheric accretion paradigm presents a formidable problem in RT, since the geometry is two- or three-dimensional, the material is moving, and the radiation-field and the accreting gas are decoupled (i.e., the problem is non-LTE). However, the past decade has seen the development of increasingly sophisticated RT models that have been used to model line profiles (both equivalent width and shape) in order to determine accretion rates. In this section we describe the development of these models, and characterize their successes and failures.

Current models are based on idealized axisymmetric geometry, in which the circumstellar density structure is calculated assuming freefall along dipolar field lines that emerge from a geometrically thin disk at a range of radii encompassing the co-rotation radius. It is assumed that the kinetic energy of the accreting material is completely thermalized, and that the accretion luminosity, combined with the area of the accretion footprints (rings) on the stellar surface, provide the temperature of the hot spots. The circumstellar density and velocity structure is then fully described by the mass accretion rate and the outer and inner radii of the magnetosphere in terms of the photospheric radius of the star (*Hartmann et al., 1994*).

A significant, but poorly constrained, input parameter for the models is the temperature structure of the accretion flow. This is a potential pitfall, as the form of the temperature structure may have a significant impact on the line source functions, and therefore the line profiles themselves. Self-consistent radiative equilibrium models (*Martin, 1996*) indicate that adiabatic heating and cooling via bremsstrahlung dominate the thermal budget, whereas Hartmann and co-workers (*Hartmann et al., 1994*) adopt a simple volumetric heating rate combined with a schematic radiative cooling rate that leads to a temperature structure that goes as the reciprocal of the density. Thus the temperature is low near the disk, and passes through a maximum (as the velocity increases and density decreases) before the stream cools again as it approaches the stellar surface (and the density increases once more).

With the density, temperature, and velocity structure of the accreting material in place, the level populations of the particular atom under consideration must be calculated under the constraint of statistical equilibrium. This calculation is usually performed using the Sobolev approximation, in which it is assumed that the conditions in the gas do not vary significantly over a length scale given by

$$l_S = v_{therm}/(dv/dr) \qquad (6)$$

where v_{therm} is the thermal velocity of the gas and dv/dr is the velocity gradient. Such an approximation is only strictly valid in the fastest parts of the accretion flow. Once the level populations have converged, the line opacities and emissivities are then computed, allowing the line profile of any particular transition to be calculated.

The first models computed using the method outlined above were presented by *Hartmann et al.* (1994), who adopted a two-level atom approximation. It was demonstrated that the magnetospheric accretion model could reproduce the main characteristics of the profiles, including IPC profiles and blue-shifted central emission peaks. The original Hartmann et al. model was further improved by *Muzerolle et al.* (2001). Instead of using a two-level approximation, they solved statistical equilibrium (still under Sobolev) for a 20-level hydrogen atom. Their line profiles were computed using a direct integration method, which, unlike the Sobolev approach, allows the inclusion of Stark broadening effects. It was found that the broadening was most significant for Hα, with the line reaching widths of ~500 km s⁻¹, a width that significantly exceeds the doppler broadening due to infall alone, and in much better agreement with observation. The Hβ model profiles were found to be in broad agreement with the observations, in terms of the velocity of the emission peak (*Alencar and Basri, 2000*), and in the asymmetry of the profiles (*Edwards et al., 1994*). Figure 4 shows model Hα profiles as a function of mass accretion rate and accretion flow temperature; one can see that for typical CTTS accretion rates the line profiles are broadly symmetric although slightly blue-shifted — the reduced optical depth for the lower accretion rate models yields the IPC morphology.

Axisymmetric models are obviously incapable of reproducing the wide range of variability that is observed in the emission lines of CTTSs (section 4). Although the addition of further free parameters to models naturally renders them more arbitrary, the observational evidence for introducing such parameters is compelling. Perhaps the simplest extension is to break the axisymmetry of the dipole, leaving curtains of accretion in azimuth; models such as these have been proposed by a number of observers attempting to explain variability in CTTSs and are observed in MHD simulations (*Romanova et al., 2003*). Synthetic time-series for a CTTS magnetosphere structured along these lines were presented by *Symington et al.* (2005a). It was found that some gross characteristics of the observed line profiles were produced using a "curtains" model, although the general level of variability predicted is larger than that observed, suggesting that the magnetosphere may be characterized by

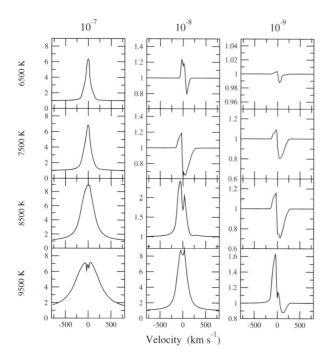

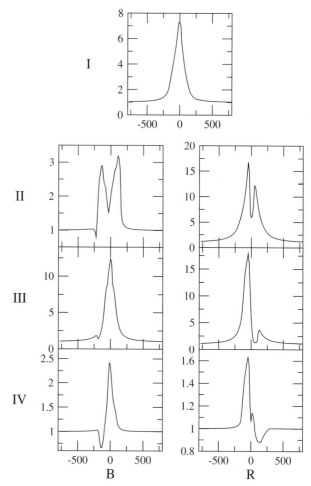

Fig. 4. Hα model profiles for a wide range of mass accretion rate and accretion flow maximum temperature (from *Kurosawa et al., 2006*). The profiles are based on canonical CTTS parameters (R = 2 R$_\odot$, M = 0.5 M$_\odot$, T = 4000 K) viewed at an inclination of 55°. The maximum temperature of the accretion flow is indicated along the left of the figure, while the accretion rate (in M$_\odot$ yr^{-1}) is shown along the top.

Fig. 5. Sample Hα model profiles (*Kurosawa et al., 2006*) that characterize the morphological classification (Types I–IV B/R) by *Reipurth et al.* (1996). The combination of magnetospheric accretion, the accretion disk, and the collimated disk wind can reproduce the wide range of Hα profiles seen in observations. The horizontal axes are velocities in km s^{-1} and the vertical axes are continuum normalized intensities.

a high degree of axisymmetry, broken by higher-density streams that produce the variability.

The emission line profiles of CTTSs often display the signatures of outflow as well as infall, and recent attempts have been made to account for this in RT modeling. *Alencar et al.* (2005) investigated a dipolar accretion geometry combined with a disk wind in order to model the line profile variability of RW Aur. They discovered that magnetospheric accretion alone could not simultaneously model Hα, Hβ, and NaD profiles, and found that the wind contribution to the lines profiles is quite important in that case.

The broad range of observed Hα profiles (Fig. 5) can be reproduced by hybrid models (*Kurosawa et al., 2006*) combining a standard dipolar accretion flow with an outflow (e.g., Fig. 6). Obviously, spectroscopy alone is insufficient to uniquely identify a set of model parameters for an individual object, although by combining spectroscopy with other probes of the circumstellar material, one should be able to reduce the allowable parameter space considerably. For example, linear spectropolarimetry provides a unique insight into the accretion process; scattering of the line emission by circumstellar dust imprints a polarization signature on the line that is geometry dependent. An Hα spectropolarimetric survey by *Vink et al.* (2005a) revealed that 9 out of 11 CTTSs showed a measurable change in polarization through the line, while simple numerical models by

Vink et al. (2005b) demonstrate that this polarization may be used to gauge the size of the disk inner hole.

The RT models described above are now routinely used to determine mass accretion rates across the mass spectrum from Herbig AeBe (*Muzerolle et al., 2004*) stars to brown dwarfs (*Lawson et al.,* 2004; *Muzerolle et al.,* 2005), and in the CTTS mass regime at least the accretion rates derived from RT modeling have been roughly calibrated against other accretion-rate measures, such as the UV continuum (e.g., *Muzerolle et al.,* 2001). However, one must be aware of the simplifying assumptions that underlie the models and must necessarily impact on the validity of any quantity derived from them, particularly the mass accretion rate. Magnetic field measurements (section 2) and time-series spectroscopy (section 4) clearly show us that the geometry of the magnetosphere is far from a pristine axisymmetric dipole, but instead probably consists of many azimuthally dis-

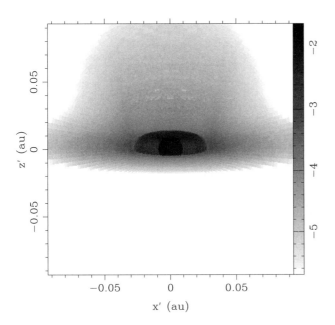

Fig. 6. A simulated Hα image of an accreting CTTS with an outflow (log $\dot{M}_{acc}$ = –8, log $\dot{M}_{wind}$ = –9) viewed at an inclination of 80°. The wind emission is negligible compared to the emission from the magnetosphere, and the lower half of the wind is obscured by the circumstellar disk (*Kurosawa et al., 2006*).

tributed funnels of accretion, curved by rotation and varying in position relative to the stellar surface on the timescale of a few stellar rotation periods. Furthermore, the temperature of the magnetosphere and the mass accretion rate are degenerate quantities in the models, with a higher-temperature magnetosphere producing more line flux for the same accretion rate. This means that brown dwarf models require a much higher accretion stream temperature than those of CTTSs in order to produce the observed line flux, and although the temperature is grossly constrained by the line broadening (which may preclude lower temperature streams), the thermal structure of the accretion streams is still a problem. Despite these uncertainties, and in defense of the BD models, it should be noted that the low accretion rates derived are consistent with both the lack of optical veiling (*Muzerolle et al., 2003a*) and the strength of the Ca II λ8662 line (*Mohanty et al., 2005*).

Current models do not match the line core particularly well, which is often attributed to a breakdown of the Sobolev approximation; co-moving frame calculations (which are many orders of magnitude more expensive computationally) may be required. An additional problem with current RT modeling is the reliance on fitting a single profile (current studies have almost always been limited to Hα); one that rarely shows an IPC profile (*Edwards et al., 1994; Reipurth et al., 1996*) is vulnerable to contamination by outflows (e.g., *Alencar et al., 2005*) and may be significantly spatially extended (*Takami et al., 2003*). Even in modeling a single line, it is fair to say that the state-of-the art is some way short of line profile fitting; the best fits reported in the literature may match the observation in terms of peak intensity, equivalent width, or in the line wings, but are rarely convincing reproductions of the observations in detail. Only by simultaneously fitting several lines may one have confidence in the models, particularly if those lines share a common upper/lower level (Hα and Paβ, for example). Although such observations are in the literature (e.g., *Edwards et al., 1994; Folha and Emerson, 2001*), their usefulness is marginalized by the likely presence of significant variability between the epochs of the observations at the different wavelengths: Simultaneous observations of a wide range of spectral diagnostics are required. Despite the caveats described above, line profile modeling remains a useful (and in the BD case the only) route to the mass accretion rate, and there is real hope that the current factor of ~5 uncertainties in mass accretion rates derived from RT modeling may be significantly reduced in the future.

4. OBSERVATIONAL EVIDENCE FOR MAGNETOSPHERIC ACCRETION

Observations seem to globally support the magnetospheric accretion concept in CTTSs, which includes the presence of strong stellar magnetic fields, the existence of an inner magnetospheric cavity of a few stellar radii, magnetic accretion columns filled with freefalling plasma, and accretion shocks at the surface of the stars. While this section summarizes the observational signatures of magnetospheric accretion in T Tauri stars, there is some evidence that the general picture applies to a much wider range of mass, from young brown dwarfs (*Muzerolle et al., 2005; Mohanty et al., 2005*) to Herbig AeBe stars (*Muzerolle et al., 2004; Calvet et al., 2004; Sorelli et al., 1996*).

In recent years, the rapidly growing number of detections of strong stellar magnetic fields at the surface of young stars seem to put the magnetospheric accretion scenario on a robust ground (see section 2). As expected from the models, given the typical mass accretion rates (10^{-9}–10^{-7} $M_\odot$ yr^{-1}) (*Gullbring et al., 1998*) and magnetic field strengths (2–3 kG) (*Valenti and Johns-Krull, 2004*) obtained from the observations, circumstellar disk inner holes of about 3–9 R_* are required to explain the observed line widths of the CO fundamental emission, which likely come from gas in Keplerian rotation in the circumstellar disk of CTTSs (*Najita et al., 2003*). There has also been evidence for accretion columns through the common occurrence of inverse P Cygni profiles with red-shifted absorptions reaching several hundred km s^{-1}, which indicates that gas is accreted onto the star from a distance of a few stellar radii (*Edwards et al., 1994*).

Accretion shocks are inferred from the rotational modulation of light curves by bright surface spots (*Bouvier et al., 1995*), and modeling of the light curves suggests hot spots covering about 1% of the stellar surface. The theoretical prediction of accretion shocks and its associated hot excess emission are also supported by accretion shock models that successfully reproduce the observed spectral energy distributions of optical and UV excesses (*Calvet and Gullbring,* 1998; *Ardila and Basri,* 2000; *Gullbring et al.,* 2000). In these models, the spectral energy distribution of the excess emission is explained as a combination of optically thick emission from the heated photosphere below the shock and optically thin emission from the preshock and postshock regions. *Gullbring et al.* (2000) also showed that the high mass accretion rate CTTSs have accretion columns with similar values of energy flux as the moderate to low mass accretion rate CTTSs, but their accretion columns cover a larger fraction of the stellar surface (filling factors ranging from less than 1% for low accretors to more than 10% for the high one). A similar trend was observed by *Ardila and Basri* (2000), who found from the study of the variability of IUE spectra of BP Tau that the higher the mass accretion rate, the bigger the hot spot size.

Statistical correlations between line fluxes and mass accretion rates predicted by magnetospheric accretion models have also been reported for emission lines in a broad spectral range, from the UV to the near-IR (*Johns-Krull et al.,* 2000; *Beristain et al.,* 2001; *Alencar and Basri,* 2000; *Muzerolle et al.,* 2001; *Folha and Emerson,* 2001). However, in recent years, a number of observational results indicate that the idealized steady-state axisymmetric dipolar magnetospheric accretion models cannot account for many observed characteristics of CTTSs.

Recent studies showed that accreting systems present strikingly large veiling variability in the near-IR (*Eiroa et al.,* 2002; *Barsony et al.,* 2005), pointing to observational evidence for time-variable accretion in the inner disk. Moreover, the near-IR veiling measured in CTTSs is often larger than predicted by standard disk models (*Folha and Emerson,* 1999; *Johns-Krull and Valenti,* 2001). This suggests that the inner disk structure is significantly modified by its interaction with an inclined stellar magnetosphere and thus departs from a flat disk geometry. Alternatively, a "puffed" inner disk rim could result from the irradiation of the inner disk by the central star and accretion shock (*Natta et al.,* 2001; *Muzerolle et al.,* 2003b). In mildly accreting T Tauri stars, the dust sublimation radius computed from irradiation models is predicted to lie close to the co-rotation radius (3–9 R_* ≈ 0.03–0.08 AU), although direct interferometric measurements tend to indicate larger values (0.08–0.2 AU) (*Akeson et al.,* 2005).

Observational evidence for an inner disk warp has been reported by *Bouvier et al.* (1999, 2003) for AA Tau, as expected from the interaction between the disk and an *inclined* stellar magnetosphere (see section 5). Inclined magnetospheres are also necessary to explain the observed periodic variations over a rotational timescale in the emission line and veiling fluxes of a few CTTSs (*Johns and Basri,* 1995; *Petrov et al.,* 1996, 2001; *Bouvier et al.,* 1999; *Batalha et al.,* 2002). These are expected to arise from the variations of the projected funnel and shock geometry as the star rotates. An example can be seen in Fig. 7, which shows the periodic modulation of the Hα line profile of the CTTS AA Tau as the system rotates, with the development of a high-velocity red-shifted absorption component when

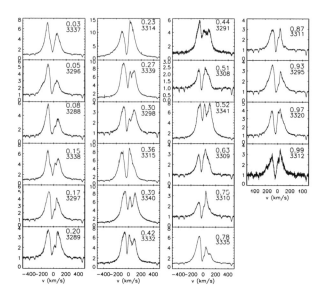

Fig. 7. The rotational modulation of the Hα line profile of the CTTS AA Tau (8.2-d period). Line profiles are ordered by increasing rotational phase (top panel number) at different Julian dates (bottom panel number). Note the development of a high-velocity red-shifted absorption component in the profile from phase 0.39 to 0.52, when the funnel flow is seen against the hot accretion shock (from Bouvier et al., in preparation).

the funnel flow is seen against the hot accretion shock. Sometimes, however, multiple periods are observed in the line flux variability and their relationship to stellar rotation is not always clear (e.g., *Alencar and Batalha,* 2002; *Oliveira et al.,* 2000). The expected correlation between the line flux from the accretion columns, and the continuum excess flux from the accretion shock, is not always present either (*Ardila and Basri,* 2000; *Batalha et al.,* 2002), and the correlations predicted by static *dipolar* magnetospheric accretion models are generally not seen (*Johns-Krull and Gafford,* 2002).

Winds are generally expected to be seen as forbidden emission lines or the blue-shifted absorption components of permitted emission lines. Some permitted emission line profiles of high-mass accretion rate CTTSs, however, do not always look like the ones calculated with magnetospheric accretion models, and this could be in part due to a strong wind contribution to the emission profiles, given the high optical depth of the wind in these cases (*Muzerolle et al.,* 2001; *Alencar et al.,* 2005). Accretion-powered hot winds originating at or close to the stellar surface have recently been proposed to exist in CTTSs with high mass accretion rates (*Edwards et al.,* 2003). These winds are inferred from the observations of P Cygni profiles of the He I line (10780 Å) that present blue-shifted absorptions extending up to –400 km/s. *Matt and Pudritz* (2005) have argued that such stellar winds can extract a significant amount of the star's angular momentum, thus helping regulate the spin of CTTSs. Turbulence could also be important and help explain the very wide (±500 km s⁻¹) emission line profiles commonly observed in Balmer and Mg II UV lines (*Ardila et al.,* 2002).

Synoptic studies of different CTTSs highlighted the dynamical aspect of the accretion/ejection processes, which only recently has begun to be studied theoretically by numerical simulations (see section 5). The accretion process appears to be time dependent on several timescales, from hours for nonsteady accretion (*Gullbring et al.,* 1996; *Alencar and Batalha,* 2002; *Stempels and Piskunov,* 2002; *Bouvier et al.,* 2003) to weeks for rotational modulation (*Smith et al.,* 1999; *Johns and Basri,* 1995; *Petrov et al.,* 2001), and from months for global instabilities of the magnetospheric structure (*Bouvier et al.,* 2003) to years for EX Ori and FU Ori eruptions (e.g., *Reipurth and Aspin,* 2004; *Herbig,* 1989).

One reason for such a variability could come from the interaction between the stellar magnetosphere and the inner accretion disk. In general, magnetospheric accretion models assume that the circumstellar disk is truncated close to the co-rotation radius and that field lines threading the disk co-rotate with the star. However, many field lines should interact with the disk in regions where the star and the disk rotate differentially. Possible evidence has been reported for differential rotation between the star and the inner disk (*Oliveira et al.,* 2000) through the presence of an observed time delay of a few hours between the appearance of high-velocity red-shifted absorption components in line profiles formed in different regions of the accretion columns. This was interpreted as resulting from the crossing of an azimuthally twisted accretion column on the line of sight. Another possible evidence for twisted magnetic field lines by differential rotation leading to reconnection events has been proposed by *Montmerle et al.* (2000) for the embedded protostellar source YLW 15, based on the observations of quasi-periodic X-ray flaring. A third possible evidence was reported by *Bouvier et al.* (2003) for the CTTS AA Tau. On timescales on the order of a month, they observed significant variations in the line and continuum excess flux, indicative of a smoothly varying mass accretion rate onto the star. At the same time, they found a tight correlation between the radial velocity of the blue-shifted (outflow) and red-shifted (inflow) absorption components in the Hα emission line profile. This correlation provides support for a physical connection between time-dependent inflow and outflow in CTTSs. *Bouvier et al.* (2003) interpreted the flux and radial velocity variations in the framework of magnetospheric inflation cycles due to differential rotation between the star and the inner disk, as observed in recent numerical simulations (see section 5). The periodicity of such instabilities, as predicted by numerical models, is yet to be tested observationally and will require monitoring campaigns of chosen CTTSs lasting for several months.

5. NUMERICAL SIMULATIONS OF MAGNETOSPHERIC ACCRETION

Significant progress has been made in recent years in the numerical modeling of magnetospheric accretion onto a rotating star with a dipolar magnetic field. One of the main problems is to find adequate initial conditions that do not destroy the disk in the first few rotations of the star and do not influence the simulations thereafter. In particular, one must deal with the initial discontinuity of the magnetic field between the disk and the corona, which usually leads to significant magnetic braking of the disk matter and artificially fast accretion onto the star on a dynamical timescale. Specific quasi-equilibrium initial conditions were developed, which helped to overcome this difficulty (*Romanova et al.,* 2002). In axisymmetric (two-dimensional) simulations, the matter of the disk accretes inward slowly, on a viscous timescale as expected in actual stellar disks. The rate of accretion is regulated by a viscous torque incorporated into the numerical code through the α prescription, with typically $\alpha_v = 0.01–0.03$.

Simulations have shown that the accretion disk is disrupted by the stellar magnetosphere at the magnetospheric or truncation radius R_T, where the gas pressure in the disk is comparable to the magnetic pressure, $P_{ram} = B^2/8\pi$ (see section 2). In this region matter is lifted above the disk plane due to the pressure force and falls onto the stellar surface supersonically along the field lines, forming funnel flows (*Romanova et al.,* 2002). The location of the inner disk radius oscillates as a result of accumulation and reconnection of the magnetic flux at this boundary, which blocks or "permits" accretion (see discussion of this issue below), thus leading to nonsteady accretion through the funnel flows. Nevertheless, simulations have shown that the funnel flow is a quasi-stationary feature during at least 50–80 rotation periods of the disk at the truncation radius, P_0, and recent simulations with improved numerical schemes indicate that this structure survives for more than 1000 P_0 (*Long et al.,* 2005). Axisymmetric simulations thus confirmed the theoretical ideas regarding the structure of the accretion flow around magnetized CTTSs. As a next step, similar initial conditions were applied to full three-dimensional simulations of disk accretion onto a star with an *inclined* dipole, a challenging problem that required the development of new numerical methods [e.g., the "inflated cube" grid (cf. *Koldoba et al.,* 2002; *Romanova et al.,* 2003, 2004a)]. Simulations have shown that the disk is disrupted at the truncation radius R_T, as in the axisymmetric case, but the magnetospheric flow to the star is more complex. Matter flows around the magnetosphere and falls onto the stellar surface supersonically. The magnetospheric structure varies depending on the misalignment angle of the dipole, but settles into a quasi-stationary state after a few P_0, as demonstrated by recent simulations run up to 40 P_0 (*Kulkarni and Romanova,* 2005). In both two-dimensional and three-dimensional simulations the fluxes of matter and angular momentum to or from the star vary in time; however, they are smooth on average. This average value is determined by the properties of the accretion disk.

Numerical simulation studies have shown that a star may either spin up, spin down, or be in rotational equilibrium when the net torque on the star vanishes. Detailed investigation of the rotational equilibrium state has shown that the rotation of the star is then *locked* at an angular velocity Ω_{eq} that is smaller by a factor of ~0.67–0.83 than the angular

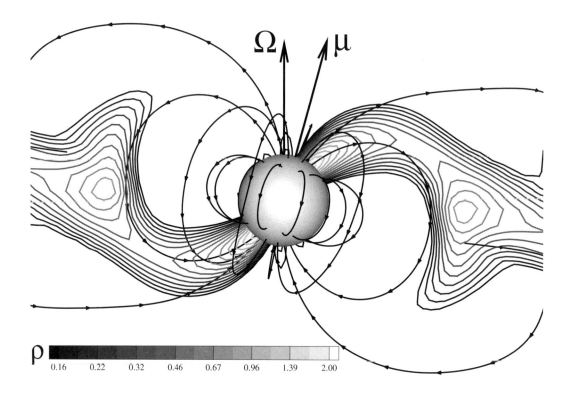

Fig. 8. A slice of the funnel stream obtained in three-dimensional simulations for an inclined dipole ($\Theta = 15°$). The contour lines show density levels, from the minimum (dark) to the maximum (light). The corona above the disk has a low density but is not shown. The thick lines depict magnetic field lines (from *Romanova et al., 2004a*).

velocity at the truncation radius (*Long et al.*, 2005). The corresponding "equilibrium" co-rotation radius $R_{CO} \approx (1.3-1.5) R_T$ is close to that predicted theoretically (e.g., *Ghosh and Lamb*, 1978, 1979b; *Königl*, 1991). Recently, the disk-locking paradigm was challenged by a number of authors (e.g., *Agapitou and Papaloizou*, 2000; *Matt and Pudritz*, 2004, 2005). The skepticism was based on the fact that the magnetic field lines connecting the star to the disk may inflate and open (e.g., *Aly and Kuijpers*, 1990; *Lovelace et al.*, 1995; *Bardou*, 1999; *Uzdensky et al.*, 2002), resulting in a significant decrease of angular momentum transport between the star and the disk. Such an opening of field lines was observed in a number of simulations (e.g., *Miller and Stone*, 1997; *Romanova et al.*, 1998; *Fendt and Elstner*, 2000). Several factors, however, tend to restore an efficient disk-star connection. One of them is that the inflated field lines have a tendency to reconnect and close again (*Uzdensky et al.*, 2002). Furthermore, there is always a region of closed field lines connecting the inner regions of the disk with the magnetosphere, which provides angular momentum transport between the disk and the star (e.g., *Pringle and Rees*, 1972; *Ghosh and Lamb*, 1979b). This is the region where matter accretes through funnel flows and efficiently transports angular momentum to or from the star. This torque tends to bring a star in co-rotation with the inner regions of the disk. There is always, however, a smaller but noticeable negative torque either connected with the region $r > R_{CO}$ (*Ghosh and Lamb*, 1978,1979b), if the field lines

are closed in this region, or associated with a wind that carries angular momentum out along the open field lines connecting the star to a low-density corona. Simulations have shown that the wind is magnetically dominated (*Long et al.*, 2005; *Romanova et al.*, 2005), although the possibility of an accretion-driven *stellar* wind has also been discussed (*Matt and Pudritz*, 2005). The spin-down through magnetic winds was proposed earlier by *Tout and Pringle* (1992). Both torques are negative so that in rotational equilibrium a star rotates *slower* than the inner disk. Thus, the result is similar to the one predicted earlier theoretically, although the physics of the spin-down contribution may be different. Axisymmetric simulations of the *fast rotating* CTTSs have shown that they efficiently spin down through both disk-magnetosphere interaction and magnetic winds (*Romanova et al.*, 2005; *Ustyugova et al.*, 2006). For instance, it was shown that a CTTS with an initial period P = 1 d spins down to the typically observed periods of about a week in less than 10^6 yr.

Three-dimensional simulations of disk accretion onto a star with a misaligned dipolar magnetic field have shown that at the nonzero misalignment angle Θ, where Θ is an angle between the magnetic moment μ_* and the rotational axis Ω_* of the star (with the disk axis aligned with Ω_*); matter typically accretes in two and, in some cases, several streams (*Koldoba et al.*, 2002; *Romanova et al.*, 2003, 2004a). Figure 8 shows a slice of the magnetospheric stream at $\Theta = 15°$. The density and pressure of the flow increase toward the star as a result of the convergence of the flow. They are also larger in the central regions of the funnel streams. Thus, the structure of the magnetospheric flow depends on the density. The high-density part is channeled in narrow funnel streams, while the low-density part is wider, with accreting matter blanketing the magnetosphere nearly completely (*Romanova et al.,* 2003) (see Fig. 9). The spectral lines that form in the funnel streams are red-shifted or blue-shifted depending on the angle Θ and viewing angle i and their strength is modulated by the rotation of the star.

Matter in the funnel flows falls onto the star's surface and forms *hot spots*. The shape of the spots and the distribution of different parameters (density, velocity, pressure) in the spots reflect those in the cross-section of the funnel streams (*Romanova et al.,* 2004a). Figure 10 shows an example of magnetospheric flows and hot spots at different Θ. At relatively small angles, Θ ≤ 30°, the spots have the shape of a bow, while at very large angles, Θ ≥ 60°, they have a shape of a bar crossing the surface of the star near the magnetic pole. The density, velocity, and pressure are the largest in the central regions of the spots and decrease outward (see Fig. 10). The temperature also increases toward the center of the spots because the kinetic energy flux is the largest there. The rotation of the star with surface hot spots leads to variability with one or two peaks per period

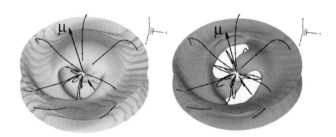

Fig. 9. Three-dimensional simulations show that matter accretes onto the star through narrow, high-density streams (right panel) surrounded by lower-density funnel flows that blanket nearly the whole magnetosphere (left panel).

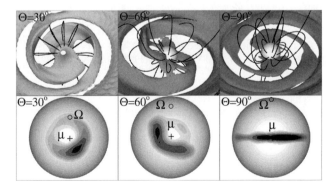

Fig. 10. *Top panels:* Matter flow close to the star at different misalignment angles Θ. *Bottom panels:* The shape of the corresponding hot spots. Darker regions correspond to larger density. From *Romanova et al.* (2004a).

depending on Θ and i. The two peaks are typical for larger Θ and i. The position of the funnel streams on the star is determined by both the angular velocity of the star and that of the inner radius of the disk. In the rotational equilibrium state, the funnel flows usually settle in a particular "favorite" position. However, if the accretion rate changes slightly, say, increases, then the truncation radius decreases accordingly and the angular velocity at the foot-point of the funnel stream on the disk is larger. As a result, the other end of the stream at the surface of the star changes its position by a small amount. Thus, the location of the spots "wobbles" around an equilibrium position depending on the accretion rate (*Romanova et al.,* 2004a). The variation of the accretion rate also changes the size and the brightness of the spots.

The disk-magnetosphere interaction leads to the thickening of the inner regions of the disk, which eases the lifting of matter to the funnel flow. Matter typically accumulates near the closed magnetosphere, forming a denser ring (*Romanova et al.,* 2002) that brakes into a spiral structure in the case of misaligned dipole (*Romanova et al.,* 2003, 2004a). Typically, two trailing spiral arms are obtained (see Fig. 10). Three-dimensional simulations have also shown that when accretion occurs onto a tilted dipole, the inner regions of the disk are slightly warped. This results from the tendency of disk material to flow along the magnetic equator of the misaligned dipole (*Romanova et al.,* 2003). Such a warping is observed for medium misalignment angles, 30 < Θ < 60°. Disk warping in the opposite direction (toward the magnetic axis of the dipole) was predicted theoretically when the disk is strongly diamagnetic (*Aly,* 1980; *Lipunov and Shakura,* 1980; *Lai,* 1999). The warping of the inner disk and the formation of a spiral structure in the accretion flow may possibly be at the origin of the observed variability of some CTTSs (*Terquem and Papaloizou,* 2000; *Bouvier et al.,* 2003).

Progress has also been made in the modeling of outflows from the vicinity of the magnetized stars. Such outflows may occur from the disk-magnetosphere boundary (*Shu et al.,* 1994), from the disk (*Blandford and Payne,* 1982; *Pudritz and Norman,* 1986; *Lovelace et al.,* 1991, 1995; *Casse and Ferreira,* 2000; *Pudritz et al.,* 2006), or from the star (*Matt and Pudritz,* 2005). Magneto-centrifugally driven outflows were first investigated in pioneering short-term simulations by *Hayashi et al.* (1996) and *Miller and Stone* (1997) and later in longer-term simulations with a fixed disk (*Ouyed and Pudritz,* 1997; *Romanova et al.,* 1997; *Ustyugova et al.,* 1999; *Krasnopolsky et al.,* 1999; *Fendt and Elstner,* 2000). Simulations including feedback on the inner disk have shown that the process of the disk-magnetosphere interaction is nonstationary: The inner radius of the disk oscillates, and matter accretes to the star and outflows quasiperiodically (*Goodson et al.,* 1997, 1999; *Hirose et al.,* 1997; *Matt et al.,* 2002; *Kato et al.,* 2004; *Romanova et al.,* 2004b; *Von Rekowski and Brandenburg,* 2004; *Romanova et al.,* 2005), as predicted by *Aly and Kuijpers* (1990). The characteristic timescale of variability is determined by a number of factors, including the timescale of diffusive penetration of the inner disk matter through the external regions of the magnetosphere (*Goodson and Winglee,* 1999). It was earlier suggested that reconnection of the magnetic flux at the disk-magnetosphere boundary may lead to X-ray flares in CTTSs (*Hayashi et al.,* 1996; *Feigelson and Montmerle,* 1999), and evidence for very large flaring structures has been recently reported by *Favata et al.* (2005).

So far, simulations were done for a dipolar magnetic field. Observations suggest a nondipolar magnetic field near the stellar surface (see section 2) (see also, e.g., *Safier,* 1998; *Kravtsova and Lamzin,* 2003; *Lamzin,* 2003; *Smirnov et al.,* 2005). If the dipole component dominates on the large scale, many properties of magnetospheric accretion will be similar to those described above, including the structure of the funnel streams and their physical properties. However, the multipolar component will probably control the flow near the stellar surface, possibly affecting the shape and the number of hot spots. Simulations of accretion to a star with a multipolar magnetic field are more complicated, and should be done in the future.

6. CONCLUSIONS

Recent magnetic field measurements in T Tauri stars support the view that the accretion flow from the inner disk onto the star is magnetically controlled. While typical values of 2.5 kG are obtained for photospheric fields, it also appears that the field topology is likely complex on the small scales ($R \leq R_*$), while on the larger scale ($R \gg R_*$) a globally more organized but weaker (~0.1 kG) magnetic component dominates. This structure is thought to interact with the inner disk to yield magnetically channeled accretion onto the star. Observational evidence for magnetospheric accretion in classical T Tauri star is robust (inner disk truncation, hot spots, line profiles) and the rotational modulation of accretion/ejection diagnostics observed in some systems suggests that the stellar magnetosphere is moderately inclined relative to the star's rotational axis. Realistic three-dimensional numerical models have capitalized on the observational evidence to demonstrate that many properties of accreting T Tauri stars could be interpreted in the framework of magnetically controlled accretion. One of the most conspicuous properties of young stars is their extreme variability on timescales ranging from hours to months, which can sometimes be traced to instabilities or quasi-periodic phenomena associated to the magnetic star-disk interaction.

Much work remains to be done, however, before reaching a complete understanding of this highly dynamical and time-variable process. Numerical simulations still have to incorporate field geometries more complex than a tilted dipole, e.g., the superposition of a large-scale dipolar or quadrupolar field with multipolar fields at smaller scales. The modeling of emission line profiles now starts to combine radiative transfer computations in both accretion funnel flows and associated mass loss flows (disk winds, stellar winds), which indeed appears necessary to account for the large variety of line profiles exhibited by CTTSs. These models also have to address the strong line profile variability, which occurs on a timescale ranging from hours to weeks in accreting T Tauri stars. These foreseen developments must be driven by intense monitoring of typical CTTSs on all timescales from hours to years, which combines photometry, spectroscopy, and polarimetry in various wavelength domains. This will provide strong constraints on the origin of the variability of the various components of the star-disk interaction process (e.g., inner disk in the near-IR, funnel flows in emission lines, hot spots in the optical or UV, magnetic reconnections in X-rays, etc.).

The implications of the dynamical nature of magnetospheric accretion in CTTSs are plentiful and remain to be fully explored. They range from the evolution of stellar angular momentum during the pre-main-sequence phase (e.g., *Agapitou and Papaloizou,* 2000), the origin of inflow/outflow short term variability (e.g., *Woitas et al.,* 2002; *Lopez-Martin et al.,* 2003), the modeling of the near infrared veiling of CTTSs and of its variations, both of which will be affected by a nonplanar and time-variable inner disk structure (e.g., *Carpenter et al.,* 2001; *Eiroa et al.,* 2002), and possibly the halting of planet migration close to the star (*Lin et al.,* 1996).

Acknowledgments.   S.A. acknowledges financial support from CNPq through grant 201228/2004-1. Work of M.M.R. was supported by NASA grants NAG5-13060 and NAG5-13220, and by NSF grants AST-0307817 and AST-0507760.

REFERENCES

Agapitou V. and Papaloizou J. C. B. (2000) *Mon. Not. R. Astron. Soc., 317,* 273–288.
Akeson R. L., Boden A. F., Monnier J. D., Millan-Gabet R., Beichman C., et al. (2005) *Astrophys. J., 635,* 1173–1181.
Alencar S. H. P. and Basri G. (2000) *Astron. J., 119,* 1881–1900.
Alencar S. H. P. and Batalha C. (2002) *Astrophys. J., 571,* 378–393.
Alencar S. H. P., Basri G., Hartmann L., and Calvet N. (2005) *Astron. Astrophys., 440,* 595–608.
Aly J. J. (1980) *Astron. Astrophys., 86,* 192–197.
Aly J. J. and Kuijpers J. (1990) *Astron. Astrophys., 227,* 473–482.
André P. (1987) In *Protostars and Molecular Clouds* (T. Montmerle and C. Bertout, eds.), pp. 143–187. CEA, Saclay.
Ardila D. R. and Basri G. (2000) *Astrophys. J., 539,* 834–846.
Ardila D. R., Basri G., Walter F. M., Valenti J. A., and Johns-Krull C. M. (2002) *Astrophys. J., 567,* 1013–1027.
Bardou A. (1999) *Mon. Not. R. Astron. Soc., 306,* 669–674.
Barsony M., Ressler M. E., and Marsh K. A. (2005) *Astrophys. J., 630,* 381–399.
Basri G. and Bertout C. (1989) *Astrophys. J., 341,* 340–358.
Basri G., Marcy G. W., and Valenti J. A. (1992) *Astrophys. J., 390,* 622–633.
Batalha C., Batalha N. M., Alencar S. H. P., Lopes D. F., and Duarte E. S. (2002) *Astrophys. J., 580,* 343–357.
Beristain G., Edwards S., and Kwan J. (2001) *Astrophys. J., 551,* 1037–1064.
Blandford R. D. and Payne D. G. (1982) *Mon. Not. R. Astron. Soc., 199,* 883–903.
Borra E. F., Edwards G., and Mayor M. (1984) *Astrophys. J., 284,* 211–222.
Bouvier J., Cabrit S., Fenandez M., Martin E. L., and Matthews J. M. (1993) *Astron. Astrophys., 272,* 176–206.
Bouvier J., Covino E., Kovo O., Martín E. L., Matthews J. M., et al. (1995) *Astron. Astrophys., 299,* 89–107.

Bouvier J., Chelli A., Allain S., Carrasco L., Costero R., et al. (1999) *Astron. Astrophys., 349,* 619–635.

Bouvier J., Grankin K. N., Alencar S. H. P., Dougados C., Fernàndez M., et al. (2003) *Astron. Astrophys., 409,* 169–192.

Brown D. N. and Landstreet J. D. (1981) *Astrophys. J., 246,* 899–904.

Calvet N. and Gullbring E. (1998) *Astrophys. J., 509,* 802–818.

Calvet N., Muzerolle J., Briceño C., Fernandez J., Hartmann L., et al. (2004) *Astron. J., 128,* 1294–1318.

Camenzind M. (1990) *Rev. Mex. Astron. Astrofis., 3,* 234–265.

Carpenter J. M., Hillenbrand L. A., and Skrutskie M. F. (2001) *Astron. J., 121,* 3160–3190.

Casse F. and Ferreira J. (2000) *Astron. Astrophys., 361,* 1178–1190.

Collier Cameron A. C. and Campbell C. G. (1993) *Astron. Astrophys., 274,* 309–318.

Daou A. G., Johns-Krull C. M., and Valenti J. A. (2006) *Astron. J., 131,* 520–526.

Donati J.-F., Semel M., Carter B. D., Rees D. E., and Collier Cameron A. (1997) *Mon. Not. R. Astron. Soc., 291,* 658–682.

Edwards S., Hartigan P., Ghandour L., and Andrulis C. (1994) *Astron. J., 108,* 1056–1070.

Edwards S., Fischer W., Kwan J., Hillenbrand L., and Dupree A. K. (2003) *Astrophys. J., 599,* L41–L44.

Eiroa C., Oudmaijer R. D., Davies J. K., de Winter D., Garzn F., et al. (2002) *Astron. Astrophys., 384,* 1038–1049.

Favata F., Flaccomio E., Reale F., Micela G., Sciortino S., et al. (2005) *Astrophys. J. Suppl., 160,* 469–502.

Feigelson E. D. and Montmerle T. (1999) *Ann. Rev. Astron. Astrophys., 37,* 363–408.

Fendt C. and Elstner D. (2000) *Astron. Astrophys., 363,* 208–222.

Folha D. F. M. and Emerson J. P. (1999) *Astron. Astrophys., 352,* 517–531.

Folha D. F. M. and Emerson J. P. (2001) *Astron. Astrophys., 365,* 90–109.

Ghosh P. and Lamb F. K. (1978) *Astrophys. J., 223,* L83–L87.

Ghosh P. and Lamb F. K. (1979a) *Astrophys. J., 232,* 259–276.

Ghosh P. and Lamb F. K. (1979b) *Astrophys. J., 234,* 296–316.

Goodson A. P. and Winglee R. M. (1999) *Astrophys. J., 524,* 159–168.

Goodson A. P., Winglee R. M., and Böhm K.-H. (1997) *Astrophys. J., 489,* 199–209.

Goodson A. P., Böhm K.-H., and Winglee R. M. (1999) *Astrophys. J., 524,* 142–158.

Guenther E. W., Lehmann H., Emerson J. P., and Staude J. (1999) *Astron. Astrophys., 341,* 768–783.

Gullbring E., Barwig H., Chen P. S., Gahm G. F., and Bao M. X. (1996) *Astron. Astrophys., 307,* 791–802.

Gullbring E., Hartmann L., Briceño C., and Calvet N. (1998) *Astrophys. J., 492,* 323–341.

Gullbring E., Calvet N., Muzerolle J., and Hartmann L. (2000) *Astrophys. J., 544,* 927–932.

Hartigan P., Edwards S., and Ghandour L. (1995) *Astrophys. J., 452,* 736–768.

Hartmann L., Hewett R., and Calvet N. (1994) *Astrophys. J., 426,* 669–687.

Hayashi M. R., Shibata K., and Matsumoto R. (1996) *Astrophys. J., 468,* L37–L40.

Herbig G. H. (1989) In *Low Mass Star Formation and Pre-Main Sequence Objects* (B. Reipurth, ed.), pp. 233–246. ESO, Garching.

Herbst W., Bailer-Jones C. A. L., Mundt R., Meisenheimer K., and Wackermann R. (2002) *Astron. Astrophys., 396,* 513–532.

Hirose S., Uchida Y., Shibata K., and Matsumoto R. (1997) *Publ. Astron. Soc. Jap., 49,* 193–205.

Jardine M., Collier Cameron A., and Donati J.-F. (2002) *Mon. Not. R. Astron. Soc., 333,* 339–346.

Johns C. M. and Basri G. (1995) *Astrophys. J., 449,* 341–364.

Johns-Krull C. M. and Gafford A. D. (2002) *Astrophys. J., 573,* 685–698.

Johns-Krull C. M. and Valenti J. A. (2001) *Astrophys. J., 561,* 1060–1073.

Johns-Krull C. M., Valenti J. A., Hatzes A. P., and Kanaan A. (1999a) *Astrophys. J., 510,* L41–L44.

Johns-Krull C. M., Valenti J. A., and Koresko C. (1999b) *Astrophys. J., 516,* 900–915.

Johns-Krull C. M., Valenti J. A., and Linsky J. L. (2000) *Astrophys. J., 539,* 815–833.

Johns-Krull C. M., Valenti J. A., Saar S. H., and Hatzes A. P. (2001) In *Magnetic Fields Across the Hertzsprung-Russell Diagram* (G.Mathys et al., eds.), pp. 527–532. ASP Conf. Series 248, San Francisco.

Johns-Krull C. M., Valenti J. A., and Saar S. H. (2004) *Astrophys. J., 617,* 1204–1215.

Johnstone R. M. and Penston M. V. (1986) *Mon. Not. R. Astron. Soc., 219,* 927–941.

Johnstone R. M. and Penston M. V. (1987) *Mon. Not. R. Astron. Soc., 227,* 797–800.

Kato Y., Hayashi M. R., and Matsumoto R. (2004) *Astrophys. J., 600,* 338–342.

Koide S., Shibata K., and Kudoh T. (1999) *Astrophys. J., 522,* 727–752.

Koldoba A. V., Romanova M. M., Ustyugova G. V., and Lovelace R. V. E. (2002) *Astrophys. J., 576,* L53–L56.

Königl A. (1991) *Astrophys. J., 370,* L39–L43.

Krasnopolsky R., Li Z.-Y., and Blandford R. (1999) *Astrophys. J., 526,* 631–642.

Kravtsova A. S. and Lamzin S. A. (2003) *Astron. Lett., 29,* 612–620.

Kulkarni A. K. and Romanova M. M. (2005) *Astrophys. J., 633,* 349–357.

Kurosawa R., Harries T. J., and Symington N. H. (2006) *Mon. Not. R. Astron. Soc.,* in press.

Lai D. (1999) *Astrophys. J., 524,* 1030–1047.

Lamzin S. A. (2003) *Astron. Reports, 47,* 498–510.

Lawson W. A., Lyo A.-R., and Muzerolle J. (2004) *Mon. Not. R. Astron. Soc., 351,* L39–L43.

Lin D. N. C., Bodenheimer P., and Richardson D. C. (1996) *Nature, 380,* 606–607.

Lipunov V. M. and Shakura N. I. (1980) *Soviet Astron. Lett., 6,* 14–17.

Long M., Romanova M. M., and Lovelace R. V. E. (2005) *Astrophys. J., 634,* 1214–1222.

López-Martín L., Cabrit S., and Dougados C. (2003) *Astron. Astrophys., 405,* L1–L4.

Lovelace R. V. E., Berk H. L., and Contopoulos J. (1991) *Astrophys. J., 379,* 696–705.

Lovelace R. V. E., Romanova M. M., and Bisnovatyi-Kogan G. S. (1995) *Mon. Not. R. Astron. Soc., 275,* 244–254.

Martin S. C. (1996) *Astrophys. J., 470,* 537–550.

Matt S. and Pudritz R. E. (2004) *Astrophys. J., 607,* L43–L46.

Matt S. and Pudritz R. E. (2005) *Astrophys. J., 632,* L135–L138.

Matt S., Goodson A. P., Winglee R. M., and Böhm K.-H. (2002) *Astrophys. J., 574,* 232–245.

Ménard F. and Bertout C. (1999) In *The Origin of Stars and Planetary Systems* (C. J. Lada and N. D. Kylafis, eds), p. 341. Kluwer, Dordrecht.

Miller K. A. and Stone J. M. (1997) *Astrophys. J., 489,* 890–902.

Mohanty S. M., Jayawardhana R., and Basri G. (2005) *Astrophys. J., 626,* 498–522.

Montmerle T., Koch-Miramond L., Falgarone E., and Grindlay J. E. (1983) *Astrophys. J., 269,* 182–201.

Montmerle T., Grosso N., Tsuboi Y., and Koyama K. (2000) *Astrophys. J., 532,* 1097–1110.

Muzerolle J., Calvet N., and Hartmann L. (2001) *Astrophys. J., 550,* 944–961.

Muzerolle J., Hillenbrand L., Calvet N., Briceño C., and Hartmann L. (2003a) *Astrophys. J., 592,* 266–281.

Muzerolle J., Calvet N., Hartmann L., and D'Alessio P. (2003b) *Astrophys. J., 597,* L149–L152.

Muzerolle J., D'Alessio P., Calvet N., and Hartmann L. (2004) *Astrophys. J., 617,* 406–417.

Muzerolle J., Luhman K. L., Briceño C., Hartmann L., and Calvet N. (2005) *Astrophys. J., 625,* 906–912.

Najita J., Carr J. S., and Mathieu R. D. (2003) *Astrophys. J., 589,* 931–952.

Natta A., Prusti T., Neri R., Wooden D., Grinin V. P., and Mannings V. (2001) *Astron. Astrophys., 371,* 186–197.

Oliveira J. M., Foing B. H., van Loon J. T., and Unruh Y. C. (2000) *Astron. Astrophys., 362,* 615–627.

Ostriker E. C. and Shu F. H. (1995) *Astrophys. J., 447,* 813–828.

Ouyed R. and Pudritz R. E. (1997) *Astrophys. J., 482,* 712–732.

Petit P., Donati J.-F.,Wade G. A., Landstreet J. D., Bagnulo S., et al. (2004) *Mon. Not. R. Astron. Soc., 348,* 1175–1190.

Petrov P. P., Gullbring E., Ilyin I., Gahm G. F., Tuominen I., et al. (1996) *Astron. Astrophys., 314,* 821–834.

Petrov P. P., Gahm G. F., Gameiro J. F., Duemmler R., Ilyin I. V., et al. (2001) *Astron. Astrophys., 369,* 993–1008.

Pringle J. E. and Rees M. J. (1972) *Astron. Astrophys., 21,* 1P–9P.

Pudritz R. E. and Norman C. A. (1986) *Astrophys. J., 301,* 571–586.

Pudritz R. E., Rogers C. S., and Ouyed R. (2006) *Mon. Not. R. Astron. Soc., 365,* 1131–1148.

Reipurth B. and Aspin C. (2004) *Astrophys. J., 608,* L65–L68.

Reipurth B., Pedrosa A., and Lago M. T. V. T. (1996) *Astron. Astrophys. Suppl., 120,* 229–256.

Romanova M. M., Ustyugova G. V., Koldoba A. V., Chechetkin V. M., and Lovelace R. V. E. (1997) *Astrophys. J., 482,* 708–711.

Romanova M. M., Ustyugova G. V., Koldoba A. V., Chechetkin V. M., and Lovelace R. V. E. (1998) *Astrophys. J., 500,* 703–713.

Romanova M. M., Ustyugova G. V., Koldoba A. V., and Lovelace R. V. E. (2002) *Astrophys. J., 578,* 420–438.

Romanova M. M., Ustyugova G. V., Koldoba A. V., Wick J. V., and Lovelace R. V. E. (2003) *Astrophys. J., 595,* 1009–1031.

Romanova M. M., Ustyugova G. V., Koldoba A. V., and Lovelace R. V. E. (2004a) *Astrophys. J., 610,* 920–932.

Romanova M. M., Ustyugova G. V., Koldoba A. V., and Lovelace R. V. E. (2004b) *Astrophys. J., 616,* L151–L154.

Romanova M. M., Ustyugova G. V., Koldoba A. V., and Lovelace R. V. E. (2005) *Astrophys. J., 635,* L165–L168.

Safier P. N. (1998) *Astrophys. J., 494,* 336–341.

Shu F., Najita J., Ostriker E., Wilkin F., Ruden S., and Lizano S. (1994) *Astrophys. J., 429,* 781–796.

Smirnov D. A., Lamzin S. A., Fabrika S. N., and Valyavin G. G. (2003) *Astron. Astrophys., 401,* 1057–1061.

Smirnov D. A., Lamzin S. A., Fabrika S. N., and Chuntonov G. A. (2004) *Astron. Lett., 30,* 456–460.

Smirnov D. A., Romanova M. M., and Lamzin S. A. (2005) *Astron. Lett., 31,* 335–339.

Smith K.W., Lewis G. F., Bonnell I. A., Bunclark P. S., and Emerson J. P. (1999) *Mon. Not. R. Astron. Soc., 304,* 367–388.

Smith K., Pestalozzi M., Güdel M., Conway J., and Benz A. O. (2003) *Astron. Astrophys., 406,* 957–967.

Sorelli C., Grinin V. P., and Natta A. (1996) *Astron. Astrophys., 309,* 155–162.

Stempels H. C. and Piskunov N. (2002) *Astron. Astrophys., 391,* 595–608.

Symington N. H., Harries T. J., and Kurosawa R. (2005a) *Mon. Not. R. Astron. Soc., 356,* 1489–1500.

Symington N. H., Harries T. J., Kurosawa R., and Naylor T. (2005b) *Mon. Not. R. Astron. Soc., 358,* 977–984.

Takami M., Bailey J., and Chrysostomou A. (2003) *Astron. Astrophys., 397,* 675–984.

Terquem C. and Papaloizou J. C. B. (2000) *Astron. Astrophys., 360,* 1031–1042.

Tout C. A. and Pringle J. E. (1992) *Mon. Not. R. Astron. Soc., 256,* 269–276.

Ustyugova G. V., Koldoba A. V., Romanova M. M., and Lovelace R. V. E. (1999) *Astrophys. J., 516,* 221–235.

Ustyugova G. V., Koldoba A. V., Romanova M. M., and Lovelace R. V. E. (2006) *Astrophys. J., 646,* 304–318.

Uzdensky D. A., Königl A., and Litwin C. (2002) *Astrophys. J., 565,* 1191–1204.

Valenti J. A. and Johns-Krull C. M. (2004) *Astrophys. Space Sci. Ser., 292,* 619–629.

Valenti J. A., Basri G., and Johns C. M. (1993) *Astron. J., 106,* 2024–2050.

Vink J. S., Drew J. E., Harries T. J., Oudmaijer R. D., and Unruh Y. (2005a) *Mon. Not. R. Astron. Soc., 359,* 1049–1064.

Vink J. S., Harries T. J., and Drew J. E. (2005b) *Astron. Astrophys., 430,* 213–222.

Vogt S. S. (1980) *Astrophys. J., 240,* 567–584.

von Rekowski B. and Brandenburg A. (2004) *Astron. Astrophys., 420,* 17–32.

Warner B. (2004) *Publ. Astron. Soc. Pac., 116,* 115–132

Woitas J., Ray T. P., Bacciotti F., Davis C. J., and Eislöffel J. (2002) *Astrophys. J., 580,* 336–342.

Yang H., Johns-Krull C. M., and Valenti J. A. (2005) *Astrophys. J., 635,* 466–475.

Interferometric Spectroimaging of Molecular Gas in Protoplanetary Disks

Anne Dutrey and Stéphane Guilloteau
Observatoire de Bordeaux

Paul Ho
Academia Sinica Institute of Astronomy and Astrophysics and
Smithsonian Astrophysical Observatory

Protoplanetary disks are found to orbit around low- and intermediate-mass stars. Current theories predict that these disks are the likely sites for planet formation. In this review, we summarize the improvement in our knowledge of their observed molecular properties since Protostars and Planets IV (PPIV). This is timely since a new facility, the Submillimeter Array (SMA), has recently begun operation and has opened the submillimeter atmospheric windows to interferometry, allowing studies of warmer gas and dust in disks at subarcsecond resolution. Using results from the IRAM array and the SMA, we focus on two complementary main topics: (1) the determination of the physical structure of the disks from multitransition CO isotopic analysis of high-angular-resolution millimeter interferometric data and (2) the observations of molecules other than CO (and isotopes), which enable investigations of the chemistry in protoplanetary disks. In particular, we emphasize how to handle the available data to provide relevant constraints on the thermal, physical, and chemical structure of the disks as a function of radius, within the current limitations in sensitivity and angular resolution of the existing arrays. These results suggest the importance of photodissociation effects and X-ray heating. They also reveal unexpected results, such as the discovery of non-Keplerian rotation in the AB Aur disk. We also discuss how to extrapolate these results in the context of the tremendous capabilities of the ALMA project currently under construction in Chile.

1. INTRODUCTION

Over the last 15 years, observations from the optical up to the millimeter-wavelength domains have revealed that low- and intermediate-mass stars of ages around a million years, such as T Tauri and Herbig Ae stars, are surrounded by circumstellar disks. These disks are often called "protoplanetary" because they still contain enough original material (from the parent cloud) to form giant planets. Indeed, although dust is the easiest tracer of protoplanetary disks, molecules represent more than 70% of the total mass in disks. H_2, which does not deplete on dust grains, is by far the most abundant molecule in disks but it remains difficult to observe because of the lack of a dipole moment. Even if a few direct H_2 detections are now possible, the observed H_2 lines mainly trace the warm gas located in the inner disks (R < 10–20 AU), while, in many cases, disks are known to extend out to several 100 AU. These outer regions, which may even contain most of the mass, are cold and can only be characterized by molecules having rotational lines at low energy levels and detectable with large millimeter and submillimeter interferometers. After H_2, CO is the most abundant molecule (even if it can deplete on dust grains), and its lowest rotational lines are easily excited by collisions with H_2 in disks. Only heterodyne arrays give the sensitivity, resolving power, and spectral resolution needed to map cold molecular disks. In the last 10 years, millimeter spectroscopic studies of disks have shown that they are in Keplerian rotation. More recently, the SMA (*Ho et al.*, 2004) has opened a new era of submillimeter interferometry, while the IRAM Plateau de Bure interferometer (PdBI) routinely provides images of 0.6" resolution at 1.3 mm. Since PPIV, observations of molecular disks have considerably improved, and millimeter/submillimeter arrays have provided many new direct constraints on the physical and chemical structure of disks that could not be addressed by disk continuum observations.

In this review we focus on the recent molecular results obtained with the SMA and the IRAM array. We summarize in section 2 the sensitivity that can be achieved by observing CO transitions in disks with the PdBI and the SMA. We also calculate the brightness temperature for transitions J = 1–0 up to J = 3–2 and describe new methods of analysis for interferometric data. In section 3, we present the CO disk properties as inferred from SMA and IRAM array observations. Section 4 is dedicated to new results for some specific disks because they provide new quantitative information on the disk physics since PPIV. Particular attention is given to the observation of molecular chemistry in section 5. Then we conclude by presenting the sensitivity of ALMA. This paper focuses on line data and outer disks only. In these proceedings, more information on continuum emission can be found in the chapter by Testi et al., and more information on chemistry can be found in the chapter by Natta et al.

2. MOLECULAR LINE FORMATION AND DISK MODELS

The line-formation process in protoplanetary disks is complicated because of the combination of strong velocity, temperature, and density gradients. The regular pattern of rotational motions provides a direct link between the projected velocity and position in the disk, a property that provides some effective superresolution, under the guidance of an applicable model. In addition, protoplanetary disks have two specific properties that make the derivation of physical parameters from observations particularly robust and relatively simple: (1) Power laws are good approximations for the radial dependence of many physical quantities; (2) as a result, molecular column densities can be derived from the observation of a single (partially optically thin) transition.

However, radiative transfer models are required to estimate the line brightness distribution as a function of projected velocity. This implies the use of dedicated codes, whose precision should be matched to the sensitivity of the observations to be analyzed. Such detailed models have become necessary only because the observations now provide sufficient sensitivity. In this section, we describe the available tools and their use.

2.1. Disk Description

Disks models (such as those described in *Dutrey et al.,* 1994) are usually based on the description of *Pringle* (1981): a geometrically thin disk in hydrostatic equilibrium with sharp inner and outer edges. Temperature, velocity, and surface density assume power-law radial dependencies:

For the kinetic temperature: $T(r) = T_o(r/r_o)^{-q}$.

For the surface density: $\Sigma(r) = \Sigma_o(r/r_o)^{-p}$.

For the velocity: $V(r) = V_o(r/r_o)^{-v}$, with $v = 0.5$ for Keplerian disks.

We start by the simplest approach, which assumes power-law dependencies vs. radius. This is of course appropriate for the velocity, and power laws have been shown to be a good approximation for the temperature distribution (e.g., *Chiang and Goldreich,* 1997). Using power laws for the surface density is more debatable. It is often justified for the mass distribution based on the α prescription of the viscosity with a constant accretion rate, but may not be applicable to any arbitrary molecule. Given the limited spatial dynamic range provided by current (sub-)millimeter arrays, more sophisticated prescriptions are not warranted.

Since the disks are assumed to be in hydrostatic equilibrium, the volume density is given by

$$n(r,z) = n(r,0)e^{-\left(\frac{z}{H(r)}\right)^2}$$

where the midplane density $n(r,0)$ and the hydrostatic scale height $H(r)$ are given by

$$n(r,0) = \frac{\Sigma(r)}{\sqrt{\pi}H(r)} = n_o(r/r_o)^{-s}$$

$$H(r) = \sqrt{\frac{2kr^3T(r)}{GM_*\overline{m}}} = h_o(r/r_o)^h$$

and also follows power laws with $h = 1 + v - q/2$ and $s = p + h = p + 1 + v - q/2$. G, M_*, and $\overline{m}$ are the gravitational constant, the stellar mass (the disk is not self-gravitating), and the mean molecular weight, respectively.

The definition given above implies that $H(r) = \sqrt{2}c_s/\Omega$ (where c_s is the sound speed and Ω the angular velocity). Other groups take $H(r) = c_s/\Omega$ (*D'Alessio et al.,* 1998; *Chiang and Goldreich,* 1997). As a consequence the scale height given in this description is $\sqrt{2}$ larger, and this must be properly taken into account in comparisons.

Figure 1 presents the density structure of a representative disk, the DM Tau disk, with characteristic curves overlaid on it. For any given molecular line, two curves are especially important: the height at which the density for thermalization is reached, and the height at which an opacity on the order of 1 is obtained. By comparing the two, one can readily derive whether non-LTE effects have to be considered or not. These curves are given for the CO J = 2–1 tran-

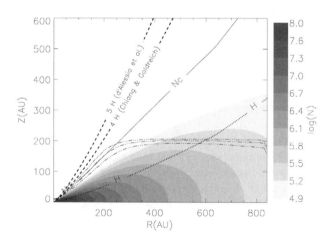

Fig. 1. Density structure of the DM Tau disk, with the critical density for thermalization of the J = 2–1 transition, NC, indicated, as well as the τ = 0.5, 1, 2 opacity curves (dash-dotted lines). The superheated layers of the *Chiang and Goldreich* (1997) model and of those of *D'Alessio et al.* (1998, 1999) are also indicated. From *Dartois et al.* (2003).

sition in Fig. 1, using a depletion factor of 10, showing that this transition remains thermalized throughout the disk, as the $\tau = 1$ curve is below the critical density line. Note that higher CO lines, J = 3–2 and above, are more optically thick and have higher critical densities, and therefore may be sensitive to non-LTE effects.

2.2. Analysis of Thermalized Lines

For the analysis of thermalized lines a ray-tracing code integrating the radiative transfer equation step by step along the line of sight is enough, although care must be taken to use sufficient resolution for the extreme velocities, which originate from the inner parts of the disk.

The brightness temperature T_b of a protoplanetary disk can be easily expressed as a function of radius r for a face-on disk. In the optically thick case

$$T_b(r) = T(r) = T_o \times (r/r_o)^{-q} \qquad (1)$$

while in the optically thin case

$$T_b(r) = \tau T(r) \qquad (2)$$

which, for the $J + 1 \rightarrow J$ transition of a linear molecule, is

$$T_b(r) = \frac{8\pi^3}{3h}\mu^2 \cdot \frac{\left(e^{-\frac{E_J}{kT(r)}} - e^{-\frac{E_{J+1}}{kT(r)}}\right)}{Z\Delta V}$$
$$(2J+1)\frac{X(r)}{m_{H_2}\bar{m}}T(r)\Sigma(r)$$

$$T_b(r) = \frac{8\pi^3}{3h}\mu^2 \cdot e^{-\frac{E_J}{kT(r)}}\frac{\left(1 - e^{-\frac{h\nu}{kT(r)}}\right)}{Z\Delta V} \qquad (3)$$
$$(2J+1)\frac{X(r)}{m_{H_2}\bar{m}}T_o\Sigma_o(r/r_o)^{-(p+q)}$$

where X is the molecular abundance relative to H_2 (Σ being a mass surface density in this formula) and Z is the partition function. The local linewidth ΔV is given by $\Delta V = \sqrt{v_{th}^2 + v_{turb}^2}$, where v_{th} is the thermal linewidth and v_{turb} a turbulent term usually on the order of $\simeq 0.1$ km s^{-1} (*Dartois et al., 2003*).

At high enough temperature, the expression can be simplified. The partition function is

$$Z \simeq \frac{kT}{hB}$$

and the Rayleigh Jeans approximation, $h\nu \ll kT$, can be used. For the J = 1–0 transition, equation (3) shows that

$$T_b(r) \propto X(r)\Sigma(r)/T(r)$$

if the linewidth is constant or

$$T_b(r) \propto X(r)\Sigma(r)/T(r)^{3/2}$$

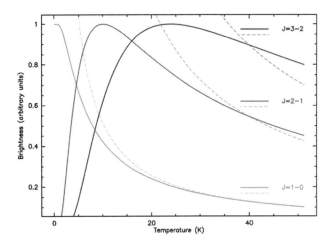

Fig. 2. Brightness, given in arbitrary units, vs. kinetic temperature for CO lines J = 1–0, J = 2–1, and J = 3–2 in a Keplerian disk. Both a high-temperature approximation and exact formula are drawn. Only the J = 1–0 transition appears well represented by the high-temperature approximation in the whole disk. From *Dartois et al.* (2003).

if the linewidth is thermal. Figure 2 compares the approximation with the exact formula.

For higher transitions, a similar asymptotic behavior is obtained if $E_{J+1} \ll kT$, but at lower temperature the exponential term in equation (3) plays a significant role. For typical temperatures of protoplanetary disks, the J = 2–1 transition brightness is $T_b(r) \propto X(r)\Sigma(r)/T(r)$ to first order.

For inclined disks, the above treatment is a simplification indicating the first-order behavior as a function of radial distance from the star. Because of the velocity shear due to the disk rotation, the line opacity is a complex function of the position and projected velocity, but the general trend remains.

For the CO abundances found in disks (*Dutrey et al.,* 1997), the first rotational transitions of ^{12}CO are optically thick throughout the disk. It is convenient to call the disk layer where ^{12}CO J = 2–1 reaches an opacity of ~1 along the line of sight as the "CO surface" of the disk. Rarer CO isotopologues appear partially optically thin in the outer regions, but remain optically thick in the inner parts. In this case, a single line can provide measurement of both the temperature and the (molecular) surface density, provided the power law distribution holds everywhere. Optically thick lines are direct tracers of the kinetic temperature at $\tau \simeq 1$ along the line of sight. This property can thus be used to sample not only the radial but also the vertical temperature gradient in the disk. This method is illustrated by Fig. 3 for a pole-on disk and was first developed by *Dartois et al.* (2003) in the case of the DM Tau disk (see also section 2.5).

2.3. Sensitivity of Millimeter/Submillimeter Arrays

Many gas disks are large enough (outer radius $R_{out} \simeq$ 200–800 AU) to allow millimeter/submillimeter arrays to measure the molecular brightness distribution vs. radius. To

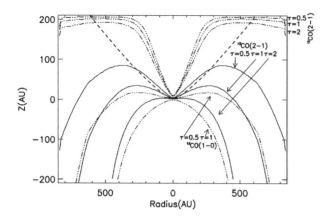

Fig. 3. Surface at which the opacity along the line of sight reaches one for the various CO isotopes overlaid on the density structure (gray scale). This figure assumes the canonical density of DM Tau disk (taking into account a depletion factor of 10 for the CO abundance). Assuming this disk to be representative of its category, this shows that transitions above the J = 2–1 line are not thermalized, at least in the outer layer. From *Dartois et al.* (2003).

understand the limitation of the measurements, it is important to compare the expected line brightness with the array sensitivity.

Figure 4 presents the expected brightness temperature as a function of radius for the J = 2–1 and J = 3–2 transitions of the CO isotopologues, and of the dust emission at the corresponding frequencies, and compare them with the sen-

sitivity of the SMA and PdBIs, estimated using the canonical values in system temperature, efficiency, etc. Note that although the continuum sensitivities could be improved by increasing the detector bandwidth, this is not the case for the spectral lines where the bandwidth is limited by the intrinsic line width. Only better receivers or more collecting area can improve the situation in this case.

The brightness temperature clearly shows the two expected regimes for the partially optically thin transitions: $T_b(r) = T_k(r)$ at small radii, and $T_b(r) \propto X(r)\Sigma(r)$ at large radii. Depending on the relative importance of the two regimes, the analysis of the observations will suffer from different limitations. The temperature determination will be difficult if insufficient resolution is available, while the surface density determination is limited by the low S/N in the outer regions. The apparent sizes will be largely dependent on the optical depth, rather than on the true outer radius. This is the reason why the (apparent) observed size of a dust disk always appears smaller than the size of its ^{12}CO emission. Fortunately, for both the SMA and the PdBI, many CO disks can be observed with a similar level of sensitivity.

2.4. Fourier Plane Analysis and χ^2 Minimization Techniques

Figure 4 clearly shows that sensitivity is a limitation even with the best arrays available so far. This problem is further amplified by the limited uv coverage due to the small number of antennas in all existing millimeter/submillimeter arrays. The IRAM array has six 15-m-diameter antennas, while the SMA has eight 6-m-diameter antennas. This lim-

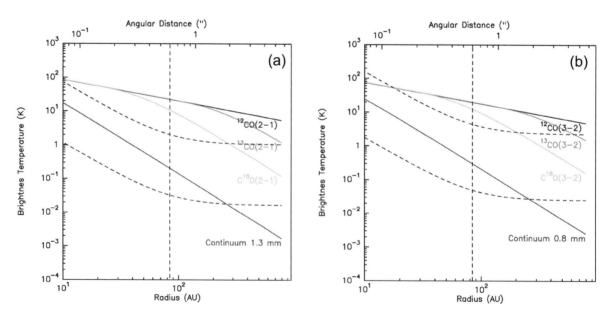

Fig. 4. (a) Brightness temperature for CO (and isotopologues) in J = 2–1 transition for a Keplerian disk orbiting a central object of ~1 $M_\odot$ (thick lines) compared to the sensitivity of the IRAM Plateau de Bure array (dashed lines). (b) Same for the SMA in the transition J = 3–2. We assume LTE conditions, which are valid for J = 2–1, but the J = 3–2 line is not necessarily thermalized everywhere. This curve is only an upper limit to the expected signal. The angular resolution is here 0.6″. This corresponds to the best resolution obtained so far. The diagram has been calculated for a T Tauri disk located at the Taurus distance (140 pc).

ited uv coverage implies significant side-lobe levels in the dirty images, and deconvolution is required to remove them. However, deconvolution is a nonlinear process that increases the noise level, especially in the case of weak extended structures. As a consequence, comparing disk models to CLEANed images is currently a method that remains inaccurate. With nine 6-m-diameter antennas and six 10.4-m diameter antennas, CARMA will improve the synthesized beam issue, but will remain sensitivity limited. With 50 12-m-diameter antennas, these problems will become much less significant with ALMA since the sampling of the uv plane will be significantly higher (1225 instantaneous baselines). The addition of the ACA to ALMA will also improve the sampling of the short spacings, which can be important for the submillimeter data.

Accordingly, the optimal way to analyze the data is to compare directly the "raw" data, i.e., the observed visibilities, with the predicted visibilities from the model images. Weighting each visibility data point by its (known) thermal noise allows the definition of a least-square distance between the model and the data, in the usual χ^2 sense

$$\chi^2 = \Sigma_n \Sigma_i (Re(mod_{i,n}) - Re(obs_{i,n}))^2 \times W_i \\ + \Sigma_n \Sigma_i (Im(mod_{i,n}) - Im(obs_{i,n}))^2 \times W_i \quad (4)$$

where $Re(a)$ and $Im(a)$ are the real and imaginary parts of the visibility a, and $a_{i,n}$ is the visibility i for velocity channel n. The weight W_i is derived from the system temperature T_{sys}, the spectral resolution Δv, the integration time τ, the effective collecting area of one antenna A_{eff}, and the loss of efficiency introduced by the correlator η

$$W_i = \frac{1}{\sigma_i^2} \text{ with } \sigma_i = \frac{\sqrt{2}kT_{sys}}{A_{eff}\eta\sqrt{\tau\Delta v}}$$

Since, for the LTE case, the model images are characterized by a reasonably small number of parameters (x_0, y_0, PA, V_{LSR}, ΔV, R_{out}, V_0, v, T_0, q, Σ_0, p, and i) (the inner radius being in general not significantly constrained because of the lack of sensitivity; see Fig. 4), χ^2 minimization techniques can be used to determine the best model parameters and, ideally, their error bars. This was first used by *Guilloteau and Dutrey* (1998) using a simple grid-mapping to find the best model for CO in DM Tau. This is, however, slow, and does not properly consider the coupling between the parameters in the error determination. These two limitations have been recently overcome by *Piétu et al.* (2005), who used a more sophisticated and faster minimization routine that takes into account the coupling between the parameters (but not the skewness of the distribution, i.e., asymmetric error bars). Analysis of the data inside the uv plane with χ^2 minimization is becoming a standard procedure (although the derivation of error bars is not always performed), and a similar method has been recently applied by *Qi et al.* (2004) to the SMA CO J = 3–2 data of TW Hya (see also section 4.2).

However, neither method includes the systematic bias, which may be introduced by using an inappropriate disk model.

2.5. An Example: DM Tau

Because it is isolated from the nearby molecular clouds, DM Tau was among the first single stars around which a bona fide protoplanetary Keplerian disk was detected (*Guilloteau and Dutrey,* 1994). This T Tauri star, located at 140 pc, is surrounded by a large CO disk of ~800 AU. It was the first object for which the method of uv analysis described above was developed, allowing the first measurement of its stellar mass $M_* = 0.5\ M_\odot$ (*Guilloteau and Dutrey,* 1998). Because of its large CO disk, it was also the first object on which a multi-isotope study of CO was performed (*Dartois et al.,* 2003). Since the ^{13}CO J = 1–0 and J = 2–1 sample different disk layers, a global analysis of these lines permit the derivation of the vertical kinetic temperature gradient. In the case of DM Tau, the measured gradient is in agreement with predictions from models of disks externally heated by their central star (e.g., *D'Alessio et al.,* 1999). The midplane is cooler (~13 K) than the CO disk surface (~30 K at 100 AU). This appears in the region of the disk where the dust is still optically thick to the stellar radiation while it is already optically thin to its own emission, around r ~ 50–200 AU in the DM Tau case. Beyond r ≥ 200 AU, where the dust becomes optically thin to both processes, the temperature profile appears vertically isothermal.

Figure 5 also shows that a significant fraction of the DM Tau disk has a temperature that is below the CO freezeout point (17 K), but there remains enough CO in the gas phase to allow the J = 2–1 line of the main isotope to be optically thick. This is also the case for most of the sources discussed here, and there is no satisfactory explanation of this chemical puzzle so far.

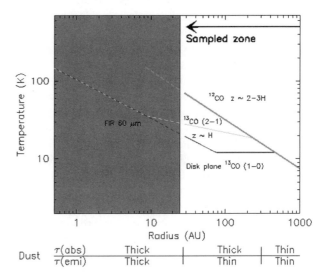

Fig. 5. Vertical temperature gradient in the DM Tau disk inferred from multi-transition, multi-isotope CO analysis. From *Dartois et al.* (2003).

2.6. Subthermal Excitation

The previous sections show how thermalized transitions can be analyzed using a simple power-law model for the temperature, and show that the temperatures derived from isotopologues can provide insights into the radial and vertical temperature structure of the disk. If the same analysis is used for a transition that remains subthermally excited (in the regions where its opacity is >1), such an analysis will yield the excitation temperature of the line instead of the kinetic temperature. If all lines of the molecule shared the same excitation temperature (i.e., if $T_{ex} = T_{rot}$, the rotation temperature), the molecular column density would still be correctly determined. In general, though, T_{ex} will decrease with the energy levels, and the above assumption will underestimate the partition function, leading to a slight underestimation of the column density.

A more accurate determination of the column density requires solving the coupled equations of statistical equilibrium and radiative transfer. This introduces four additional complexities: (1) the disk temperature structure must be known; (2) the disk density structure must be known; (3) the location of the molecules within the disk must be known; and (4) the coupled equations must be solved to sufficient accuracy.

The last complexity is actually the least difficult to solve. For this purpose, two kinds of models are currently being used: (1) approximate solutions based on escape probability and one-dimensional radiative transfer, such as those developed by Piétu et al. (in preparation); (2) full two-dimensional Monte-Carlo radiative transfer codes (*Hogerheijde and van der Tak,* 2000; *Pavlyuchenkov and Shustov,* 2004).

The first method remains fast enough for use in χ^2 minimization routines, the escape probability technique being only marginally slower than a simple LTE approximation. The second method is in theory more accurate from the point of view of the radiative transfer, provided the Monte-Carlo noise is adequately controlled, but current implementations are too slow for use in χ^2 minimization. Furthermore, it should be stressed that the gain in accuracy is not necessarily significant: the main limitation in accuracy remains currently in the interferometric observations, because all molecular lines (except those of ^{12}CO) are relatively weak. Moreover, all techniques suffer from the required a priori knowledge of the temperature and density structure of the disk, and of the localization of the molecules within this structure. The related uncertainties are often much larger than those introduced by the approximation made to solve the coupled statistical and radiative transfer equations.

For the disk structure, two methods are presently in use: (1) The gas kinetic temperature is determined from the CO line analysis, and the radial density distribution from the fit of the millimeter continuum images (method used by Dutrey, Guilloteau, and collaborators). (2) The kinetic temperature and density structures are determined by modeling the spectral energy distribution (and in some cases, interferometric dust map) of the dust disk with a model similar to those of *Chiang and Goldreich* (1997) or *d'Alessio et al.* (1998). This method supposes (a) a dust grain distribution and composition, and (b) that dust and gas have the same kinetic temperature, a good approximation over the whole disk (with the exception of the disk atmosphere where the density is low). This method was recently used to determine the best physical parameters of the gas disk orbiting TW Hya by *Qi et al.* (2004). In both methods, hydrostatic equilibrium is assumed. The largest source of uncertainty remains the radial density distribution, which is coupled to the dust properties.

Finally, the localization of the molecules within the disk remains an unsolved problem. One can either assume full vertical mixing, or use vertical distributions provided by chemical models. Given all the combined uncertainties, it remains unclear whether a full non-LTE analysis yields more sensible results than those provided by the simple T_{rot} approximation.

3. PROTOPLANETARY DISK PROPERTIES

The first extensive detailed analysis of CO emission from disks around T Tauri and HAeBe stars using the above methods was performed by *Simon et al.* (2000). They used the ^{12}CO J = 2 → 1 emission to characterize the disks around 12 stars, showing the existence of large (often >300 AU) disks and demonstrating unambiguously the Keplerian nature of the rotation. More recently, Piétu et al. (in preparation) have extended the CO isotopologue line analysis performed for DM Tau to a small sample of T Tauri (LkCa 15) and Herbig Ae stars (MWC 480, AB Aur, and HD 34282).

Together with other studies of HAeBe stars by *Mannings and Sargent* (1997, 2000), this allows some robust conclusions to be reached:

1. Large disks (CO outer radius of ~200–1000 AU) exist.

2. The intrinsic linewidths are small, with a nonthermal component on the order of ~0.1–0.2 km s⁻¹.

3. All disks studied so far are in Keplerian rotation (with the exception of AB Aur, which is described in section 6). Since the mass of the disks is negligible, χ^2 minimization as those described in the section above allow direct measurements of the stellar mass (see also Fig. 6).

4. For both Herbig Ae and (classical) T Tauri disks, the kinetic temperature at the CO disk surface (defined as the surface where $\tau \simeq 1$ is reached for the ^{12}CO J = 2–1 transition) is in agreement with the model of flared disks heated by the central star, $T(r) \propto r^{-0.6}$.

5. Multiline, multi-isotope analysis of CO reveal the existence of a vertical temperature gradient in outer disks that seems compatible with disk models, with disk midplanes cooler than CO surfaces.

6. Disks orbiting around Herbig Ae stars are hotter than those found around T Tauri stars. ^{13}CO observations also reveal that the temperature close to the disk midplane is larger and above the temperature of freezeout of CO.

7. The degree of CO depletion clearly decreases with the effective temperature of the star, with CO being essen-

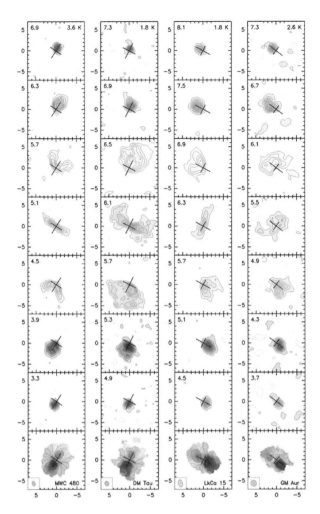

Fig. 6. CO J = 2–1 channel maps of disks around T Tauri and Herbig Ae stars in Taurus-Auriga, observed with the IRAM array. These objects are isolated from their parent clouds. The channel patterns appear very similar from one disk to another. The analysis of the kinematics show that all these disks are in Keplerian rotation. Because of the velocity gradient and the limited local line-width, only a small fraction of the disk contributes to the line emission in each velocity channel. From *Simon et al.* (2000).

tially undepleted in AB Aur, demonstrating that sticking onto dust grains plays a significant role in the depletion process.

8. Beyond a radius of ~150 AU, outer disks orbiting T Tauri stars have a temperature that is below 17 K, the CO freezeout temperature. However, a significant amount of CO is still present (optically thick ^{12}CO J = 2 → 1 line) even at these low temperatures. This remains to be explained by chemical models.

9. Finally, when both ^{13}CO and ^{12}CO data exist, the outer radius derived from ^{13}CO is smaller than the one derived from ^{12}CO. The differences are consistent with the behavior expected from selective photodissociation due to self-shielding.

Table 1 summarizes the known properties of outer disks inferred by millimeter and submillimeter arrays so far. The properties above apply to classical T Tauri stars (CTTS) or

HAe stars. In addition, *Duvert et al.* (2000) searched for disks around weak-lined T Tauri stars (WTTS). No bona fide WTTS showed significant ^{12}CO emission, and the only disk found was around V836 Tau, a star that displays characteristics intermediate between Class II and Class III objects. The disk was unresolved, the observations indicating a radius ≃120 AU. These observations, although statistically not significant, suggest that most of the outer disk disappears on the same timescale as the inner disk traced by the near-IR excess.

Finally, it should be emphasized that interferometric observations of spectral lines do provide with high accuracy two parameters of the disk geometry: the position angle and, more importantly, the inclination, which is essential for the modeling of the SED.

4. SOME INTERESTING CASE STUDIES

The properties derived above for the disks of T Tauri and HAe stars are not necessarily representative of all stars. In particular, the disk size is likely biased toward large values because the large disks provide the stronger emissions. Furthermore, the initial studies have focused on objects well isolated from molecular clouds, in order to avoid confusion with the emission from the molecular cloud.

Indeed, recent results have revealed stars with atypical disk properties. We focus in this section on three well-known stars. AB Aur is considered as the prototype of the Herbig Ae star. TW Hya is the closest (and among the oldest) T Tauri star surrounded by a disk and BP Tau is considered, from optical observations, as the reference for a classical T Tauri star surrounded by an active accretion disk. Thanks to the ability to trace directly the gas component and the disk kinematics, the millimeter/submillimeter interferometric observations provide information that are in two cases, at least, in apparent contradiction with the simple picture obtained from optical observations.

TABLE 1. Typical values for protoplanetary
disks around T Tauri stars.

Physical Parameter	Ref. Value	Exponent
Outer radius	(AU)	—
R_{out}	2000–1000	—
Turbulent linewidth	(km s^{-1})	
Δv	~0.1–0.2	—
Column density*	(g cm^{-2})	
$\Sigma(r) = \Sigma_{100} \times (r/100\ AU)^{-p}$	0.8	~1.5
Kinetic temperature†	(K)	
$T(r) = T_{100} \times (r/100\ AU)^{-q}$	~30	~0.6
Velocity law‡		0.53 ± 0.01
$V(r) = V100(r/100\ AU)^v$	—	

*Of gas + dust, assuming a gas to dust ratio of 100; values from *Dartois et al.* (2003).

†T_{100} corresponds to the CO surface (J = 2–1 line); values from *Dartois et al.* (2003).

‡From *Simon et al.* (2000).

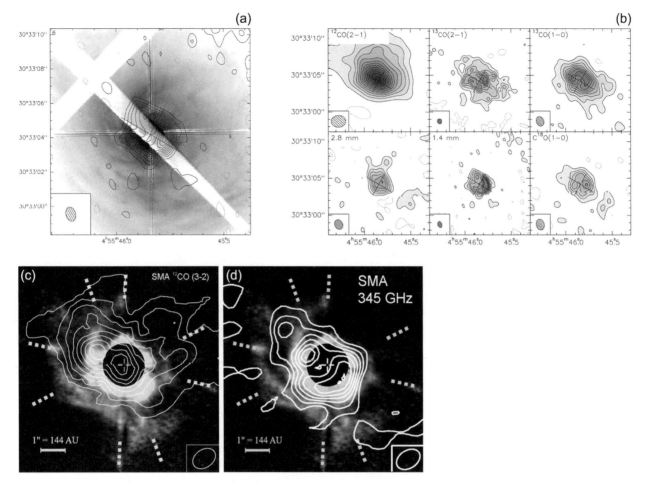

Fig. 7. **(a)** 1.4-mm continuum data on AB Aur (in contours) superimposed on the HST image from *Grady et al.* (1999), in false color. The angular resolution is 0.85 × 0.59" at PA 18°. **(b)** A montage displaying high-resolution images of the continuum emission at 2.8 mm and 1.4 mm, and of the integrated line emission of ^{12}CO J = 2 → 1 ^{13}CO J = 2 → 1 and ^{13}CO J = 1 → 0 transitions. The (lower-resolution) emission at 110 GHz is also presented. From *Piétu et al.* (2005). **(c, d)** Disk of AB Aur observed with the SMA and superimposed to the Subaru Near-Infrared image. **(c)** CO J = 3–2; angular resolution is 1.04" × 0.72" (natural weighing). **(d)** Dust emission; angular resolution is 0.95" × 0.66" (obtained by removing the data inside the first 30 kλ). The SMA dust images clearly reveal a depression in the center. The submillimeter dust peaks at the location of the inner NIR arms. From *Lin et al.* (2006).

4.1. AB Aur

In many aspects, AB Aur is taken as the prototype of the Herbig Ae star. The star is an A0 star of ~10^6 yr. Modeling of the SED shows that the star is a Group I source having a flared disk *(Meeus et al., 2001)*. It is also surrounded by a large reflection nebula *(Grady et al., 1999)* extending up to ~10,000 AU (Fig. 7a). More recently, using the Subaru telescope, *Fukagawa et al.* (2004) found that at medium scale (~100–500 AU) the material around the star presents a spiral pattern. *Semenov et al.* (2005), using the IRAM 30-m and PdBI, have shown that AB Aur is surrounded by a large envelope and a circumstellar disk traced by HCO⁺.

4.1.1. Plateau de Bure interferometer data. Recent subarcsecond images of AB Aur obtained by *Piétu et al.* (2005) with the IRAM PdBI in the isotopologues of CO, and in continuum at 3 and 1.3 mm reveal that the environment of AB Aur is very different from the protoplanetary disks observed so far with millimeter arrays. These observations also allow the authors to trace the structure of the circumstellar material in regions where optical and IR mapping is impossible because of the emission from the star itself.

In Fig. 7a, the HST image from *Grady et al.* (1999) is given in gray scale while contours correspond to the thermal dust emission observed at 1.3 mm by *Piétu et al.* (2005) (the cross shows the star location). The millimeter continuum emission is not centrally peaked but is dominated by a bright, asymmetric ("spiral-like") feature at about 140 AU from the central star. Little emission is associated with the star itself.

The molecular emission, shown in Figs. 7b and 7c, reveals that AB Aur is surrounded by a very extended flattened low-mass gaseous structure ("disk") that is also not centrally peaked. Bright molecular emission is also found toward the continuum asymmetry. The large-scale molecular structure suggests the AB Aur "disk" is inclined between ~25° and 35°. The significant asymmetry of the continuum and molecular emission prevents an accurate determination of the inclination of the inner disk part. Surprisingly, the

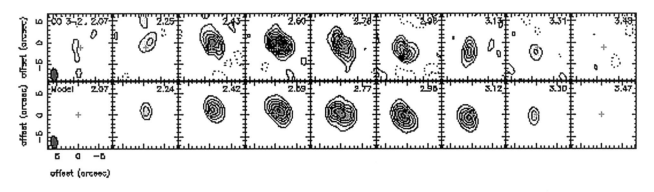

Fig. 8. Analysis of TW Hya. Channel map for the CO J = 3–2 SMA observations (top) and for the best model (bottom). From *Qi et al.* (2004).

analysis of the CO line kinematics reveal that the disk rotation is non-Keplerian, the exponent of the velocity law differing from 0.5 at the 10σ level (0.40 ± 0.01).

4.1.2. Submillimeter Array analysis. AB Aur has also been observed in the ^{12}CO J = 3–2 line and continuum at 0.8 mm with the SMA by *Lin et al.* (2006). The dust emission is not centrally peaked but exhibits a lower surface density in the center. This is clearly seen in Figs. 7c and 7d. The Subaru near-infrared image has been superimposed to the SMA continuum data (from *Lin et al.*, 2006). The submillimeter continuum emission is dominated by bright asymmetric spots that follow the spiral pattern visible in the near-infrared images.

Contrary to the continuum, the ^{12}CO J = 3–2 emission peaks at the stellar position. Moreover, in several velocity channels, the CO J = 3–2 traces the innermost spiral arm seen in the NIR. The kinematics of the ^{12}CO data cannot be fitted by Keplerian rotation and the CO emission that follows the innermost spiral arm exhibits outward radial motions.

4.1.3. Comparison. It is clear that both SMA and PdBI data reveal that the circumstellar material presents many departures from a symmetric structure. In particular, both sets of data indicate that (1) spiral-like features are observed as in the near-infrared; (2) there is a clear (and large) departure from Keplerian rotation; and (3) at millimeter/submillimeter wavelengths, the inner disk has been cleared or, at least, exhibits a lower column density than the outer disk.

The interpretation, however, remains unclear. Using "standard" dust properties to derive the disk mass, the disk is not sufficiently massive to be unstable against its own gravity. Although dust absorption coefficients are in general rather uncertain in disks, making mass estimates quite unreliable, in this case the disk mass is corroborated by the CO abundances. This disk mass appears consistent with a normal CO abundance for the Taurus region. As the disk of AB Aur is warm enough (about 30 K in the spiral regions), it is indeed expected that CO does not deplete onto grains.

One possibility would be the formation of a planet or low-mass companion in the inner disk. This is likely a possible explanation for the inner hole, but it remains unclear whether the spiral structure and strong departure from a

Keplerian rotation can be sustained at such large distances from the perturbing object.

The other possibility is that it is still in an early phase of star formation in which the Keplerian regime is not yet fully established. The latter interpretation is supported by the existence of a large envelope around AB Aur, and is reinforced by the fact that the dust observed at millimeter wavelengths appears less evolved than in most protoplanetary disks.

These observations alone do not allow a definite answer and more modeling of the first phase of disk and planet formation are required to properly interpret these data.

4.2. TW Hya

Located at a distance of 56 pc, TW Hya is the closest known T Tauri star. With an age of ~5 × 10⁶ yr, it is also one of the oldest stars that still exhibits a CO and gas disk (*Kastner et al.*, 1997). Recent observations performed by *Qi et al.* (2004) with the SMA in ^{12}CO J = 3–2 confirm that the size of the disk is small, with a radius of ~170 AU, and show that the disk is not perfectly pole-on but is tilted by ~6° from face-on.

A detailed modeling was performed by *Qi et al.* (2004) on these data. They used two-dimensional Monte Carlo codes to constrain the disk properties. Channel maps of the observations and of the best model are shown in Fig. 8. In the modeling, the disk thermal structure was mainly derived from the spectral energy distribution assuming reasonable dust properties, and the gas was assumed to be fully coupled to the dust. While this is a reasonable assumption for the lower CO transitions, the CO J = 3–2 line becomes optically thick in a high layer above the disk plane, where the density is not necessarily sufficient to provide full coupling between the gas and dust temperature. The height at which CO J = 3–2 becomes optically thick depends significantly on the disk mass and CO abundance, and will vary from one disk to another. Indeed, *Qi et al.* (2004) reported that the predicted CO J = 3–2 intensities were always lower than the observed ones, and mentioned that this was most likely due to such an effect.

To check this, *Qi et al.* (2006) used the SMA to map the CO J = 6–5 emission in the TW Hya disk. These new data reveal, as with the existing CO J = 2–1 and J = 3–2, a rotating disk. A detailed simultaneous modeling of the J = 3–2 and J = 6–5 data show that additional heating is needed to explain the line intensities and that, at least in the outer disk layers, the vertical temperature gradient of the gas is steeper than that of the dust.

A natural heating source is provided by the X-ray emission coming from the star: TW Hya is a strong X-ray emitter with a typical luminosity of 2×10^{30} ergs s^{-1} (*Kastner et al.*, 1999). Inclusion of an idealized X-ray heating in the disk model provides a better simultaneous fit of the CO J = 2–1 and J = 3–2 (performed with χ^2 minimization) and also of the J = 6–5 data. The agreement with J = 6–5 transition is not perfect, but these observations can be considered as the first observed evidence of X-ray heating of molecular gas in T Tauri disks.

4.3. BP Tau

In many aspects, BP Tau can be considered as the prototype of the CTTS. The object exhibits a high accretion rate of ~3×10^{-8} M$_\odot$/yr from its circumstellar disk, which also produces a strong excess emission in the ultraviolet, visible, and NIR (*Gullbring et al.*, 1998). Such a high accretion rate is not surprising since the star is likely very young (6×10^5 yr) (*Gullbring et al.*, 1998).

Recent CO J = 2–1 and 1.3-mm continuum images obtained with the IRAM array have shown a weak and small CO and dust disk (*Simon et al.*, 2000). The disk is small since its radius is on the order of ≈120 AU and in Keplerian rotation around a (1.3 ± 0.2)(D/140 pc) M$_\odot$ mass star. Moreover, contrary to what is observed in other T Tauri disks, the detailed analysis of the CO data shows that the J = 2–1 transition is marginally optically thin (*Dutrey et al.*, 2003). The disk mass can be estimated from the millimeter continuum emission by assuming a gas-to-dust ratio of 100; it is very small (~1.2×10^{-3} M$_\odot$). In reference to this mass, the CO depletion factor is estimated to be ~150 with respect to H$_2$. This CO depletion remains unique compared to other T Tauri CO disks, even if one takes into account possible uncertainties such as a lower gas-to-dust ratio or a higher value for the dust absorption coefficient. With a kinetic temperature of about ~50 K at 100 AU (derived from the CO data), BP Tau has also the hottest outer disk found so far around a T Tauri star.

One possibility would be that a significant fraction of the disk might be superheated (above the blackbody temperature) similarly to a disk atmosphere (see model from *Chiang and Goldreich*, 1997). This could explain both the relatively high temperature and the low disk mass. *Dutrey et al.* (2003) have estimated the fraction of small grains ($\alpha \leq 0.1$ μm) still present in the disk to reach in the visible $\tau_V = 1$ at the disk midplane. The mass of small grains is about 10% of the total mass of dust (1.2×10^{-5} M$_\odot$) derived from the millimeter continuum data.

In view of the TW Hya results, another alternative would be that X-ray (or UV radiation, BP Tau being a very ac-

tively accreting object) provides an additional heating of the gas. A more sophisticated modeling, using the appropriate dust and gas densities, would be required to test this idea.

Whatever the cause of the heating, BP Tau remains unique so far for its low CO abundance. It suggests that BP Tau may be a transient object in the phase of clearing its outer disk. This also shows that optical observations alone do not allow comprehensive studies of the disk physics.

5. MOLECULAR CHEMISTRY IN DISKS

Chemistry is an essential agent in the shaping of the gas disks. Molecular (and atomic) abundances play a key role in determining the cooling rate of the gas, and thereby in the thermal balance of the disk. To measure this structure, it is essential to understand where in the disk the molecules are located.

Many examples are given above that illustrate this requirement. Among them, one can cite the process of CO selective photodissociation at the disk outer edge in DM Tau (*Dartois et al.*, 2003) or the role of the X-ray heating in the gas disk surface (*Qi et al.*, 2006). Both processes require a deep understanding of the disk physical structure (at least the geometry, density, and gas and dust temperature) and of the dust properties (e.g., size of the dust particles). So far, only the disk geometry can be easily constrained.

Because molecules are in general much less abundant than ^{13}CO, and the line emission much weaker (with the exception of HCO$^+$ due to its large dipole moment), the sensitivity of the arrays becomes a serious problem for the study of chemistry (see section 2.3). This explains why so far no comprehensive observational study of the disk chemistry has been published, and that chemical models still refer to the single-dish work of *Dutrey et al.* (1997) or *van Zadelhoff et al.* (2001). Observational studies of the chemistry in disks are currently limited to detections of the more abundant species: HCO$^+$, CN, C$_2$H, CS, HCN, H$_2$CO, and DCO$^+$. The high CN/HCN abundance ratio observed in DM Tau by *Dutrey et al.* (1997), as well as the detection of C$_2$H, are strongly suggestive of photon-dominated processes.

Another complexity is related to the possible non-LTE effects in the line excitation as discussed in section 2.6. As a consequence, chemical studies in disks are even more difficult to handle from the point of view of the observations than from the point of view of the theory. A first systematic attempt has been performed by Piétu and collaborators, who recently completed a molecular survey of DM Tau and a few sources including LkCa15 with the IRAM array. Figure 9 shows the results of the χ^2 minimization, assuming LTE conditions, for the observed molecular transitions in DM Tau. Such an analysis provides a first-order indication on the molecular abundance gradients in the disk. Chemical models will have to explain the general trends found, e.g., for the HCO$^+$/CO or CN/HCN ratios, as well as the magnitude of the abundances. However, a detailed comparison with chemical models will require proper incorporation of

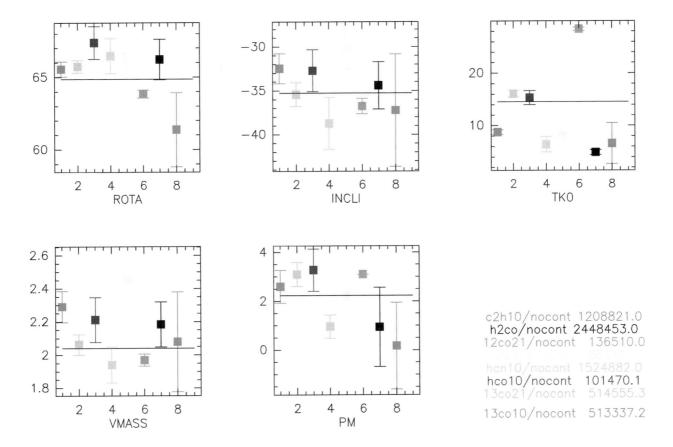

Fig. 9. Fits of the molecular data obtained with the IRAM PdBI array for DM Tau. From left to right and top to bottom, the parameters are: position angle (ROTA, in degrees), inclination (INCLI, in degrees), temperature at 100 AU (TKO, in K), radial velocity at 100 AU (VMASS, in km/s), and radial index of the surface density law (PM). The consistency of the error bars on all fits for the geometry and the kinematics clearly demonstrate the robustness of the method and the correct derivation of the error bars. From Piétu et al. (in preparation).

the disk structure and non-LTE analysis, since some transitions appear subthermal.

Note that proper analysis of any line detection in disks requires adequate handling of the line-formation process, and in particular of the disk kinematics. Failure to do so can result in misinterpretation of spectral lines, as shown by *Guilloteau et al.* (2006) for HDO in DM Tau. *Guilloteau et al.* (2006) demonstrate that the line-to-continuum ratio of an absorption line from a circumstellar disk is limited to the ratio of intrinsic linewidth (thermal + turbulent) to the projected rotation velocity at the disk edge. Except for edge-on disks, this ratio is less than about 0.1–0.2, showing that the detection of HDO in absorption claimed by *Ceccarelli et al.* (2005) cannot be real.

6. SUMMARY AND PERSPECTIVES

Millimeter and submillimeter interferometric studies of spectral lines have revealed a number of essential features of the protoplanetary disks. CO images have shown the existence of large disks, and have been used to constrain stellar masses and disk inclinations, two key parameters for SED modeling. The observations of the CO isotopologues have

provided the first direct evidence of the thermal structure of disks, constraining radial and vertical gradients. Although currently biased toward the largest and most isolated disks, observations have also revealed the diversity of the protoplanetary disks. The recent discovery of non-Keplerian rotation around AB Aur may prove important for the understanding of the disk-formation process. At the other extreme, objects like BP Tau also indicate that we do not yet understand the causes of the diversity in disk properties.

Because of the intrinsic faintness of the line emission, studies of other molecules than CO isotopologues have not been performed extensively. Large observational projects are now close to completion, and should lead to significant progress in this area over the next few years. However, the complexity of the analysis due to the non-LTE line formation process will have to be considered when comparing with chemical models. Significant progress in this area is nevertheless expected, thanks to a combination of several independent improvements:

1. The advent of submillimeter observations with the SMA, which will allow some multiline studies of molecules, although this will be restricted to very few objects given the sensitivity limitation.

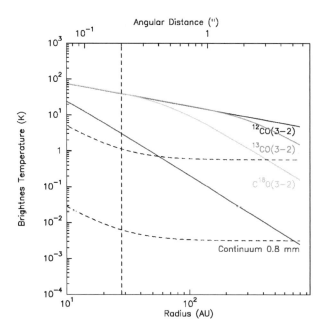

Fig. 10. Same as Fig. 4 but for ALMA. Brightness temperature for CO (and isotopomers) in J = 3–2 transition for a Keplerian disk orbiting a central object of ~1 $M_\odot$ (thick lines) compared to the sensitivity of ALMA with the whole array. Note that the angular resolution here is 0.2″. This has been done for a T Tauri disk located at the Taurus distance (140 pc).

2. The progress in modeling, especially radiative transfer codes, which are now accurate and fast enough to be used systematically.

3. The improvement in sensitivity of the millimeter arrays that is expected in the next two years, thanks to new dual-polarization receivers of the IRAM PdBI and the advent of the CARMA array.

4. The improvement in continuum sensitivity, due to increased continuum bandwidth, which will allow better comparisons of dust and gas distributions.

However, sensitivity will remain a concern, and good coordination between the major arrays, as well as use of the best analysis techniques, will be essential to get the best constraints on the chemistry of disks from the available instruments.

Despite these improvements, the studies of molecular disks will largely remain confined to the nearest star-formation regions, Taurus-Auriga, the TW Hya association, and ρ Oph, which are not representative of the bulk of the star-formation process. Going beyond this limitation and accessing the nearest star-forming cluster, Orion, will require a further step in sensitivity that can only be provided by ALMA.

ALMA will also allow a major breakthrough in the study of less-abundant molecules, such as H_2CO or isotopologues of the other molecules: At the Taurus distance, it will allow meaningful comparison of the distributions of dust and molecules on more than a decade in radius (see Fig. 10.)

Acknowledgments. The SMA is a joint project between the Smithsonian Astrophysical Observatory and the Academia Sinica Institute of Astronomy and Astrophysics, and is funded by the Smithsonian Institution and Academia Sinica. IRAM is supported by INSU/CNRS (France), MPG (Germany), and IGN (Spain). V. Piétu and E. Dartois are acknowledged for many fruitful discussions on protoplanetary disks. S.-Y. Lin and C. Qi are thanked for providing material for this chapter. A.D. thanks the CNFA for providing her travel funds to attend the Protostars and Planets V conference. This work was also partially supported by the French "Programme National de Physico-Chimie du Milieu Interstellaire."

REFERENCES

Ceccarelli C., Dominik C., Caux E., Lefloch B., and Caselli P. (2005) *Astrophys. J., 631,* L81–L84.
Chiang E. I. and Goldreich P. (1997) *Astrophys. J., 490,* 368–376.
D'Alessio P., Canto J., Calvet N., and Lizano S. (1998) *Astrophys. J., 500,* 411–427.
D'Alessio P., Calvet N., Hartmann L., Lizano S., and Canto J. (1999) *Astrophys. J., 527,* 893–909.
Dartois E., Dutrey A., and Guilloteau S. (2003) *Astron. Astrophys., 399,* 773–787.
Dutrey A., Guilloteau S., and Simon M. (1994) *Astron. Astrophys., 286,* 149–159.
Dutrey A., Guilloteau S., and Guélin M. (1997) *Astron. Astrophys., 317,* L55–L58.
Dutrey A., Guilloteau S., and Simon M. (2003) *Astron. Astrophys., 402,* 1003–1011.
Duvert G., Guilloteau S., Ménard F., Simon M., and Dutrey A. (2000) *Astron. Astrophys., 355,* 165–170.
Fukagawa M., Hayashi M., Tamura M., Itoh Y., Hayashi S. S., et al. (2004) *Astrophys. J., 605,* L53–L56.
Grady C. A., Woodgate B., Bruhweiler F. C., Boggess A., Plait P., et al. (1999) *Astrophys. J., 523,* L151–L154.
Guilloteau S. and Dutrey A. (1994) *Astron. Astrophys., 291,* L23–L26.
Guilloteau S. and Dutrey A. (1998) *Astron. Astrophys., 339,* 467–476.
Guilloteau S., Pietu V., Dutrey A., and Guélin M. (2006) *Astron. Astrophys., 448,* L5–L8.
Gullbring E., Hartmann L., Briceno C., and Calvet N. (1998) *Astrophys. J., 492,* 323–341.
Ho P., Moran J., and Lo K. Y. (2004) *Astrophys. J., 616,* L1–L6.
Hogerheijde M. R. and van der Tak F. F. S. (2000) *Astrophys. J., 362,* 697–710.
Kastner J. H., Zuckermann B., Weintraub D. A., and Forveille T. (1997) *Science, 277,* 67–71.
Kastner J. H., Huenemoerder D. P., Schulz N. S., and Weintraub D. A. (1999) *Astrophys. J., 525,* 837–844.
Lin S. Y., Ohashi N., Lim J., Ho P. T. P., Fukagawa M., and Tamura M. (2006) *Astrophys. J., 645,* 1297–1304.
Mannings V. and Sargent A. I. (1997) *Astrophys. J., 490,* 792–802.
Mannings V. and Sargent A. I. (2000) *Astrophys. J., 529,* 391–401.
Meeus G., Waters L. B. F. M., Bouwman J., van den Ancker M. E., Waelkens C., and Malfait K. (2001) *Astron. Astrophys., 365,* 476–490.
Pavlyuchenkov Y. N. and Shustov B. M. (2004) *Astron. Rep., 48,* 315–326.
Piétu V., Guilloteau S., and Dutrey A. (2005) *Astron. Astrophys., 443,* 945–954.
Pringle J. E. (1981) *Ann. Rev. Astron. Astrophys., 19,* 137–162.
Semenov D., Pavlyuchenkov Y. N., Schreyer K., Henning T., Dullemond C., and Bacmann A. (2005) *Astrophys. J., 621,* 853–874.
Simon M., Dutrey A., and Guilloteau S. (2000) *Astrophys. J., 545,* 1034–1043.
Qi C., Ho P. T. P., Wilner D. J., Takakuwa S., Hirano N., et al. (2004) *Astrophys. J., 616,* L11–L14.
Qi C., Wilner D. J., Calvet N., Bourke T. L., Blake G. A., Hogerheijde M. R., Ho P. T. P., and Bergin E. (2006) *Astrophys. J., 636,* L157–L160.
van Zadelhoff G.-J., van Dishoeck E. F., Thi W.-F., and Blake G. A. (2001) *Astron. Astrophys., 377,* 566–580.

Gaseous Inner Disks

Joan R. Najita
National Optical Astronomy Observatory

John S. Carr
Naval Research Laboratory

Alfred E. Glassgold
University of California, Berkeley

Jeff A. Valenti
Space Telescope Science Institute

As the likely birthplaces of planets and an essential conduit for the buildup of stellar masses, inner disks are of fundamental interest in star and planet formation. Studies of the gaseous component of inner disks are of interest because of their ability to probe the dynamics, physical and chemical structure, and gas content of this region. We review the observational and theoretical developments in this field, highlighting the potential of such studies to, e.g., measure inner disk truncation radii, probe the nature of the disk-accretion process, and chart the evolution in the gas content of disks. Measurements of this kind have the potential to provide unique insights on the physical processes governing star and planet formation.

1. INTRODUCTION

Circumstellar disks play a fundamental role in the formation of stars and planets. A significant fraction of the mass of a star is thought to be built up by accretion through the disk. The gas and dust in the inner disk (r < 10 AU) also constitute the likely material from which planets form. As a result, observations of the gaseous component of inner disks have the potential to provide critical clues to the physical processes governing star and planet formation.

From the planet-formation perspective, probing the structure, gas content, and dynamics of inner disks is of interest, since they all play important roles in establishing the architectures of planetary systems (i.e., planetary masses, orbital radii, and eccentricities). For example, the lifetime of gas in the inner disk (limited by accretion onto the star, photoevaporation, and other processes) places an upper limit on the timescale for giant planet formation (e.g., *Zuckerman et al.,* 1995).

The evolution of gaseous inner disks may also bear on the efficiency of orbital migration and the eccentricity evolution of giant and terrestrial planets. Significant inward orbital migration, induced by the interaction of planets with a gaseous disk, is implied by the small orbital radii of extrasolar giant planets compared to their likely formation distances (e.g., *Ida and Lin,* 2004). The spread in the orbital radii of the planets (0.05–5 AU) has been further taken to indicate that the timing of the dissipation of the inner disk sets the final orbital radius of the planet (*Trilling et al.,*

2002). Thus, understanding how inner disks dissipate may impact our understanding of the origin of planetary orbital radii. Similarly, residual gas in the terrestrial planet region may play a role in defining the final masses and eccentricities of terrestrial planets. Such issues have a strong connection to the question of the likelihood of solar systems like our own.

An important issue from the perspective of both star and planet formation is the nature of the physical mechanism that is responsible for disk accretion. Among the proposed mechanisms, perhaps the foremost is the magnetorotational instability (*Balbus and Hawley,* 1991), although other possibilities exist. Despite the significant theoretical progress that has been made in identifying plausible accretion mechanisms (e.g., *Stone et al.,* 2000), there is little observational evidence that any of these processes are active in disks. Studies of the gas in inner disks offer opportunities to probe the nature of the accretion process.

For these reasons, it is of interest to probe the dynamical state, physical and chemical structure, and the evolution of the gas content of inner disks. We begin this chapter with a brief review of the development of this field and an overview of how high-resolution spectroscopy can be used to study the properties of inner disks (section 1). Previous reviews provide additional background on these topics (e.g., *Najita et al.,* 2000). In sections 2 and 3, we review recent observational and theoretical developments in this field, first describing observational work to date on the gas in inner disks, and then describing theoretical models for the surface and interior regions of disks. In section 4, we look to the future, highlighting several topics that can be explored using the tools discussed in sections 2 and 3.

1.1. Historical Perspective

One of the earliest studies of gaseous inner disks was the work by Kenyon and Hartmann on FU Orionis objects. They showed that many of the peculiarities of these systems could be explained in terms of an accretion outburst in a disk surrounding a low-mass young stellar object (YSO) (cf. *Hartmann and Kenyon*, 1996). In particular, the varying spectral type of FU Ori objects in optical to near-infrared spectra, evidence for double-peaked absorption line profiles, and the decreasing widths of absorption lines from the optical to the near-infrared argued for an origin in an optically thick gaseous atmosphere in the inner region of a rotating disk. Around the same time, observations of CO vibrational overtone emission, first in the BN object (*Scoville et al.*, 1983) and later in other high- and low-mass objects (*Thompson*, 1985; *Geballe and Persson*, 1987; *Carr*, 1989), revealed the existence of hot, dense molecular gas plausibly located in a disk. One of the first models for the CO overtone emission (*Carr*, 1989) placed the emitting gas in an optically thin inner region of an accretion disk. However, only the observations of the BN object had sufficient spectral resolution to constrain the kinematics of the emitting gas.

The circumstances under which a disk would produce emission or absorption lines of this kind were explored in early models of the atmospheres of gaseous accretion disks under the influence of external irradiation (e.g., *Calvet et al.*, 1991). The models interpreted the FU Ori absorption features as a consequence of midplane accretion rates high enough to overwhelm external irradiation in establishing a temperature profile that decreases with disk height. At lower accretion rates, the external irradiation of the disk was expected to induce a temperature inversion in the disk atmosphere, producing emission rather than absorption features from the disk atmosphere. Thus the models potentially provided an explanation for the FU Ori absorption features and CO emission lines that had been detected.

By the time of Protostars and Planets IV (PPIV) (*Najita et al.*, 2000), high-resolution spectroscopy had demonstrated that CO overtone emission shows the dynamical signature of a rotating disk (*Carr et al.*, 1993; *Chandler et al.*, 1993), thus confirming theoretical expectations and opening the door to the detailed study of gaseous inner disks in a larger number of YSOs. The detection of CO fundamental emission (section 2.3) and emission lines of hot H_2O (section 2.2) had also added new probes of the inner disk gas.

Seven years later, at Protostars and Planets V, we find both a growing number of diagnostics available to probe gaseous inner disks as well as increasingly detailed information that can be gleaned from these diagnostics. Disk diagnostics familiar from PPIV have been used to infer the intrinsic line broadening of disk gas, possibly indicating evidence for turbulence in disks (section 2.1). They also dem-

onstrate the differential rotation of disks, provide evidence for nonequilibrium molecular abundances (section 2.2), probe the inner radii of gaseous disks (section 2.3), and are being used to probe the gas dissipation timescale in the terrestrial planet region (section 4.1). Along with these developments, new spectral line diagnostics have been used as probes of the gas in inner disks. These include transitions of molecular hydrogen at UV, near-infrared, and mid-infrared wavelengths (sections 2.4 and 2.5) and the fundamental rovibrational transitions of the OH molecule (section 2.2). Additional potential diagnostics are discussed in section 2.6.

1.2. High-Resolution Spectroscopy of Inner Disks

The growing suite of diagnostics can be used to probe inner disks using standard high-resolution spectroscopic techniques. Although inner disks are typically too small to resolve spatially at the distance of the nearest star-forming regions, we can utilize the likely differential rotation of the disk along with high spectral resolution to separate disk radii in velocity. At the warm temperatures (~100–5000 K) and high densities of inner disks, molecules are expected to be abundant in the gas phase and sufficiently excited to produce rovibrational features in the infrared. Complementary atomic transitions are likely to be good probes of the hot inner disk and the photodissociated surface layers at larger radii. By measuring multiple transitions of different species, we should therefore be able to probe the temperatures, column densities, and abundances of gaseous disks as a function of radius.

With high spectral resolution we can resolve individual lines, which facilitates the detection of weak spectral features. We can also work around telluric absorption features, using the radial velocity of the source to shift its spectral features out of telluric absorption cores. This approach makes it possible to study a variety of atomic and molecular species, including those present in the Earth's atmosphere.

Gaseous spectral features are expected in a variety of situations. As already mentioned, significant vertical variation in the temperature of the disk atmosphere will produce emission (absorption) features if the temperature increases (decreases) with height (*Calvet et al.*, 1991; *Malbet and Bertout*, 1991). In the general case, when the disk is optically thick, observed spectral features measure only the atmosphere of the disk and are unable to probe directly the entire disk column density, a situation familiar from the study of stellar atmospheres.

Gaseous emission features are also expected from regions of the disk that are optically thin in the continuum. Such regions might arise as a result of dust sublimation (e.g., *Carr*, 1989) or as a consequence of grain growth and planetesimal formation. In these scenarios, the disk would have a low continuum opacity despite a potentially large gas column density. Optically thin regions can also be produced by a significant reduction in the total column density of the disk. This situation might occur as a consequence of giant planet formation, in which the orbiting giant planet carves out a "gap" in the disk. Low column densities would also be characteristic of a dissipating disk. Thus, we should be able to use gaseous emission lines to probe the properties of inner disks in a variety of interesting evolutionary phases.

2. OBSERVATIONS OF GASEOUS INNER DISKS

2.1. CO Overtone Emission

The CO molecule is expected to be abundant in the gas phase over a wide range of temperatures, from the temperature at which it condenses on grains (~20 K) up to its thermal dissociation temperature (~4000 K at the densities of inner disks). As a result, CO transitions are expected to probe disks from their cool outer reaches (>100 AU) to their innermost radii. Among these, the overtone transitions of CO ($\Delta v = 2$, $\lambda = 2.3$ μm) were the emission line diagnostics first recognized to probe the gaseous inner disk.

CO overtone emission is detected in both low- and high-mass young stellar objects, but only in a small fraction of the objects observed. It appears more commonly among higher-luminosity objects. Among the lower-luminosity stars, it is detected from embedded protostars or sources with energetic outflows (*Geballe and Persson*, 1987; *Carr*, 1989; *Greene and Lada*, 1996; *Hanson et al.*, 1997; *Luhman et al.*, 1998; *Ishii et al.*, 2001; *Figueredo et al.*, 2002; *Doppmann et al.*, 2005). The conditions required to excite the overtone emission, warm temperatures (≥2000 K) and high densities (>10^{10} cm^{-3}), may be met in disks (*Scoville et al.*, 1983; *Carr*, 1989; *Calvet et al.*, 1991), inner winds (*Carr*, 1989), or funnel flows (*Martin*, 1997).

High-resolution spectroscopy can be used to distinguish among these possibilities. The observations typically find strong evidence for the disk interpretation. The emission line profiles of the v = 2–0 bandhead in most cases show the characteristic signature of bandhead emission from symmetric, double-peaked line profiles originating in a rotating disk (e.g., *Carr et al.*, 1993; *Chandler et al.*, 1993; *Najita et al.*, 1996; *Blum et al.*, 2004). The symmetry of the observed line profiles argues against the likelihood that the emission arises in a wind or funnel flow, since inflowing or outflowing gas is expected to produce line profiles with red- or blueshifted absorption components (alternatively line asymmetries) of the kind that are seen in the hydrogen Balmer lines of T Tauri stars (TTS). Thus high-resolution spectra provide strong evidence for rotating inner disks.

The velocity profiles of the CO overtone emission are normally very broad (>100 km s^{-1}). In lower-mass stars (~1 M$_\odot$), the emission profiles show that the emission extends from very close to the star, ~0.05 AU, out to ~0.3 AU (e.g., *Chandler et al.*, 1993; *Najita et al.*, 2000). The small radii are consistent with the high excitation temperatures measured for the emission (~1500–4000 K). Velocity resolved spectra have also been modeled in a number of high-mass stars (*Blum et al.*, 2004; *Bik and Thi*, 2004), where the CO emission is found to arise at radii ~3 AU.

The large near-infrared excesses of the sources in which CO overtone emission is detected imply that the warm emitting gas is located in a vertical temperature inversion region in the disk atmosphere. Possible heating sources for the temperature inversion include external irradiation by the star at optical through UV wavelengths (e.g., *Calvet et al.*, 1991; *D'Alessio et al.*, 1998) or by stellar X-rays (*Glassgold et al.*, 2004; henceforth *GNI04*), turbulent heating in the disk atmosphere generated by a stellar wind flowing over the disk surface (*Carr et al.*, 1993), or the dissipation of turbulence generated by disk accretion (*GNI04*). Detailed predictions of how these mechanisms heat the gaseous atmosphere are needed in order to use the observed bandhead emission strengths and profiles to investigate the origin of the temperature inversion.

The overtone emission provides an additional clue that suggests a role for turbulent dissipation in heating disk atmospheres. Since the CO overtone bandhead is made up of closely spaced lines with varying interline spacing and optical depth, the emission is sensitive to the intrinsic line broadening of the emitting gas (as long as the gas is not optically thin). It is therefore possible to distinguish intrinsic line broadening from macroscopic motions such as rotation. In this way, one can deduce from spectral synthesis modeling that the lines are suprathermally broadened, with line widths approximately Mach 2 (*Carr et al.*, 2004; *Najita et al.*, 1996). *Hartmann et al.* (2004) find further evidence for turbulent motions in disks based on high-resolution spectroscopy of CO overtone absorption in FU Ori objects.

Thus disk atmospheres appear to be turbulent. The turbulence may arise as a consequence of turbulent angular momentum transport in disks, as in the magnetorotational instability (MRI) (*Balbus and Hawley*, 1991) or the global baroclinic instability (*Klahr and Bodenheimer*, 2003). Turbulence in the upper disk atmosphere may also be generated by a wind blowing over the disk surface.

2.2. Hot Water and OH Fundamental Emission

Water molecules are also expected to be abundant in disks over a range of disk radii, from the temperature at which water condenses on grains (~150 K) up to its thermal dissociation temperature (~2500 K). Like the CO overtone transitions, the rovibrational transitions of water are also expected to probe the high-density conditions in disks. While the strong telluric absorption produced by water vapor in Earth's atmosphere will restrict the study of cool water to space or airborne platforms, it is possible to observe from the groundwater that is much hotter than the Earth's atmosphere. Very strong emission from hot water can be detected in the near-infrared even at low spectral resolution [e.g., SVS-13 (*Carr et al.*, 2004)]. More typically, high-resolution spectroscopy of individual lines is required to detect much weaker emission lines.

For example, emission from individual lines of water in the K and L bands have been detected in a few stars (both low and high mass) that also show CO overtone emission

(*Carr et al.,* 2004; *Najita et al.,* 2000; *Thi and Bik,* 2005). Velocity resolved spectra show that the widths of the water lines are consistently narrower than those of the CO emission lines. Spectral synthesis modeling further shows that the excitation temperature of the water emission (typically ~1500 K) is less than that of the CO emission. These results are consistent with both the water and CO originating in a differentially rotating disk with an outwardly decreasing temperature profile. That is, given the lower dissociation temperature of water (~2500 K) compared to CO (~4000 K), CO is expected to extend inward to smaller radii than water, i.e., to higher velocities and temperatures.

The Δv=1 OH fundamental transitions at 3.6 μm have also been detected in the spectra of two actively accreting sources, SVS-13 and V1331 Cyg, that also show CO overtone and hot water emission (Carr et al., in preparation). As shown in Fig. 1, these features arise in a region that is crowded with spectral lines of water and perhaps other species. Determining the strengths of the OH lines will therefore require making corrections for spectral features that overlap closely in wavelength.

Spectral synthesis modeling of the detected CO, H_2O, and OH features reveals relative abundances that depart significantly from chemical equilibrium (cf. *Prinn,* 1993), with the relative abundances of H_2O and OH a factor of 2–10 below that of CO in the region of the disk probed by both diagnostics (*Carr et al.,* 2004; Carr et al., in preparation; see also *Thi and Bik,* 2005). These abundance ratios may arise from strong vertical abundance gradients produced by the external irradiation of the disk (see section 3.4).

2.3. CO Fundamental Emission

The fundamental (Δv=1) transitions of CO at 4.6 μm are an important probe of inner disk gas in part because of their broader applicability compared, e.g., to the CO overtone

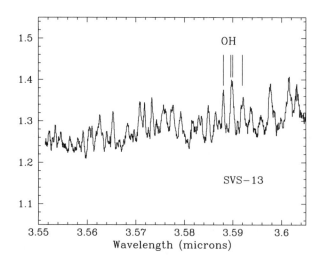

Fig. 1. OH fundamental rovibrational emission from SVS-13 on a relative flux scale.

lines. As a result of their comparatively small A-values, the CO overtone transitions require large column densities of warm gas (typically in a disk temperature inversion region) in order to produce detectable emission. Such large column densities of warm gas may be rare except in sources with the largest accretion rates, i.e., those best able to tap a large accretion energy budget and heat a large column density of the disk atmosphere. In contrast, the CO fundamental transitions, with their much larger A-values, should be detectable in systems with more modest column densities of warm gas, i.e., in a broader range of sources. This is borne out in high-resolution spectroscopic surveys for CO fundamental emission from TTS (*Najita et al.,* 2003) and Herbig AeBe stars (*Blake and Boogert,* 2004) that detect emission from essentially all sources with accretion rates typical of these classes of objects.

In addition, the lower temperatures required to excite the CO v=1–0 transitions make these transitions sensitive to cooler gas at larger disk radii, beyond the region probed by the CO overtone lines. Indeed, the measured line profiles for the CO fundamental emission are broad (typically 50–100 km s^{-1} FWHM) and centrally peaked, in contrast to the CO overtone lines, which are typically double-peaked. These velocity profiles suggest that the CO fundamental emission arises from a wide range of radii, from ≤0.1 AU out to 1–2 AU in disks around low-mass stars, i.e., the terrestrial planet region of the disk (*Najita et al.,* 2003).

CO fundamental emission spectra typically show symmetric emission lines from multiple vibrational states (e.g., v = 1–0, 2–1, 3–2); lines of ^{13}CO can also be detected when the emission is strong and optically thick. The ability to study multiple vibrational states as well as isotopic species within a limited spectral range makes the CO fundamental lines an appealing choice to probe gas in the inner disk over a range of temperatures and column densities. The relative strengths of the lines also provide insight into the excitation mechanism for the emission.

In one source, the Herbig AeBe star HD141569, the excitation temperature of the rotational levels (~200 K) is much lower than the excitation temperature of the vibrational levels (v = 6 is populated), which is suggestive of UV pumping of cold gas (*Brittain et al.,* 2003). The emission lines from the source are narrow, indicating an origin at ≥17 AU. The lack of fluorescent emission from smaller radii strongly suggests that the region within 17 AU is depleted of gaseous CO. Thus detailed models of the fluorescence process can be used to constrain the gas content in the inner disk region (S. Brittain, personal communication).

Thus far HD141569 appears to be an unusual case. For the majority of sources from which CO fundamental is detected, the relative line strengths are consistent with emission from thermally excited gas. They indicate typical excitation temperatures of 1000–1500 K and CO column densities of ~10^{18} cm^{-2} for low-mass stars. These temperatures are much warmer than the dust temperatures at the same radii implied by spectral energy distributions (SEDs) and the expectations of some disk atmosphere models (e.g., *D'Alessio et al.,*

1998). The temperature difference can be accounted for by disk atmosphere models that allow for the thermal decoupling of the gas and dust (section 3.2).

For CTTS systems in which the inclination is known, we can convert a measured HWZI velocity for the emission to an inner radius. The CO inner radii, thus derived, are typically ~0.04 AU for TTS (*Najita et al., 2003; Carr et al.*, in preparation), smaller than the inner radii that are measured for the dust component either through interferometry (e.g., *Eisner et al., 2005; Akeson et al., 2005a; Colavita et al.,* 2003; see chapter by Millan-Gabet et al.) or through the interpretation of SEDs (e.g., *Muzerolle et al., 2003*). This shows that gaseous disks extend inward to smaller radii than dust disks, a result that is not surprising given the relatively low sublimation temperature of dust grains (~1500–2000 K) compared to the CO dissociation temperature (~4000 K). These results are consistent with the suggestion that the inner radius of the dust disk is defined by dust sublimation rather than by physical truncation (*Muzerolle et al., 2003; Eisner et al., 2005*).

Perhaps more interestingly, the inner radius of the CO emission appears to extend up to and usually within the corotation radius (i.e., the radius at which the disk rotates at the same angular velocity as the star; Fig. 2). In the current paradigm for TTS, a strong stellar magnetic field truncates the disk near the corotation radius. The coupling between the stellar magnetic field and the gaseous inner disk regulates the rotation of the star, bringing the star into corotation with the disk at the coupling radius. From this region emerge both energetic (X-)winds and magnetospheric accretion flows (funnel flows) (*Shu et al., 1994*). The velocity extent of the CO fundamental emission shows that gaseous circumstellar disks indeed extend inward beyond the dust destruction radius to the corotation radius (and beyond), providing the material that feeds both X-winds and funnel flows. Such small coupling radii are consistent with the rotational rates of young stars.

It is also interesting to compare the distribution of inner radii for the CO emission with the orbital radii of the "close-in" extrasolar giant planets (Fig. 3). Extrasolar planets discovered by radial velocity surveys are known to pile up near a minimum radius of 0.04 AU. The similarity between these distributions is roughly consistent with the idea that the truncation of the inner disk can halt the inward orbital migration of a giant planet (*Lin et al., 1996*). In detail, however, the planet is expected to migrate slightly inward of the truncation radius, to the 2:1 resonance, an effect that is not seen in the present data. A possible caveat is that the wings of the CO lines may not trace Keplerian motion or that the innermost gas is not dynamically significant. It would be interesting to explore this issue further since the results impact our understanding of planet formation and the origin of planetary architectures. In particular, the existence of a stopping mechanism implies a lower efficiency for giant planet formation, e.g., compared to a scenario in which multiple generations of planets form and only the last generation survives (e.g., *Trilling et al., 2002*).

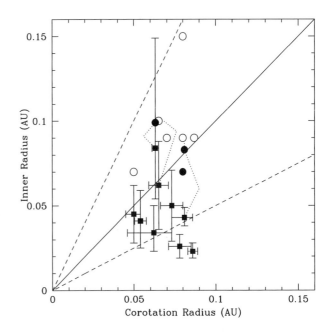

Fig. 2. Gaseous inner disk radii for TTS from CO fundamental emission (filled squares) compared with corotation radii for the same sources. Also shown are dust inner radii from near-infrared interferometry (filled circles) (*Akeson et al.,* 2005a,b) or spectral energy distributions (open circles) (*Muzerolle et al., 2003*). The solid and dashed lines indicate an inner radius equal to, twice, and one-half the corotation radius. The points for the three stars with measured inner radii for both the gas and dust are connected by dotted lines. Gas is observed to extend inward of the dust inner radius and typically inward of the corotation radius.

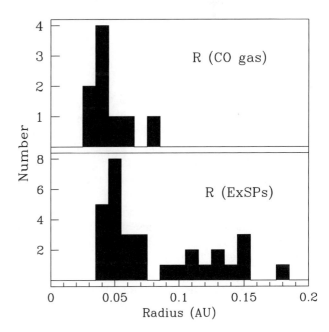

Fig. 3. The distribution of gaseous inner radii, measured with the CO fundamental transitions, compared to the distribution of orbital radii of short-period extrasolar planets. A minimum planetary orbital radius of ~0.04 AU is similar to the minimum gaseous inner radius inferred from the CO emission-line profiles.

2.4. Ultraviolet Transitions of Molecular Hydrogen

Among the diagnostics of inner disk gas developed since PPIV, perhaps the most interesting are those of H_2. H_2 is presumably the dominant gaseous species in disks, due to high elemental abundance, low depletion onto grains, and robustness against dissociation. Despite its expected ubiquity, H_2 is difficult to detect because permitted electronic transitions are in the far ultraviolet (FUV) and accessible only from space. Optical and rovibrational infrared transitions have radiative rates that are 14 orders of magnitude smaller.

Considering only radiative transitions with spontaneous rates above 10^7 s^{-1}, H_2 has about 9000 possible Lyman-band (B-X) transitions from 850 to 1650 Å and about 5000 possible Werner-band (C-X) transitions from 850 to 1300 Å (*Abgrall et al.,* 1993a,b). However, only about 200 FUV transitions have actually been detected in spectra of accreting TTS. Detected H_2 emission lines in the FUV all originate from about two dozen radiatively pumped states, each more than 11 eV above ground. These pumped states of H_2 are the only ones connected to the ground electronic configuration by strong radiative transitions that overlap the broad Lyα emission that is characteristic of accreting TTS (see Fig. 4). Evidently, absorption of broad Lyα emission pumps the H_2 fluorescence. The two dozen strong H_2 transitions that happen to overlap the broad Lyα emission are all pumped out of high-v and/or high-J states at least 1 eV above ground (see inset in Fig. 4). This means some mechanism must excite H_2 in the ground electronic configuration, before Lyα pumping can be effective. If the excitation mechanism is thermal, then the gas must be roughly 10^3 K to obtain a significant H_2 population in excited states.

H_2 emission is a ubiquitous feature of accreting TTS. Fluoresced H_2 is detected in the spectra of 22 out of 24 accreting TTS observed in the FUV by HST/STIS (*Herczeg et al.,* 2002, 2005; *Walter et al.,* 2003; *Calvet et al.,* 2004; *Bergin et al.,* 2004; *Gizis et al.,* 2005; unpublished archival data). Similarly, H_2 is detected in all eight accreting TTS observed by HST/GHRS (*Ardila et al.,* 2002) and all four published FUSE spectra (*Wilkinson et al.,* 2002; *Herczeg et al.,* 2002, 2004, 2005). Fluoresced H_2 was even detected in 13 out of 39 accreting TTS observed by IUE, despite poor sensitivity (*Brown et al.,* 1981; *Valenti et al.,* 2000). Fluoresced H_2 has not been detected in FUV spectra of nonaccreting TTS, despite observations of 14 stars with STIS (*Calvet et al.,* 2004; unpublished archival data), 1 star with GHRS (*Ardila et al.,* 2002), and 19 stars with IUE (*Valenti et al.,* 2000). However, the existing observations are not sensitive enough to prove that the circumstellar H_2 column decreases contemporaneously with the dust continuum of the inner disk. When accretion onto the stellar surface stops, fluorescent pumping becomes less efficient because the strength and breadth of Lyα decreases significantly and the H_2 excitation needed to prime the pumping mechanism may become less efficient. COS, if installed on HST, will have

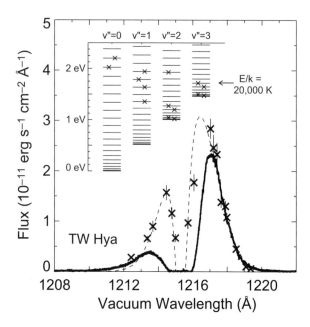

Fig. 4. Lyα emission from TW Hya, an accreting T Tauri star, and a reconstruction of the Lyα profile seen by the circumstellar H_2. Each observed H_2 progression (with a common excited state) yields a single point in the reconstructed Lyα profile. The wavelength of each point in the reconstructed Lyα profile corresponds to the wavelength of the upward transition that pumps the progression. The required excitation energies for the H_2 *before* the pumping is indicated in the inset energy-level diagram. There are no low excitation states of H_2 with strong transitions that overlap Lyα. Thus, the H_2 must be very warm to be preconditioned for pumping and subsequent fluorescence.

the sensitivity to set interesting limits on H_2 around nonaccreting TTS in the TW Hya association.

The intrinsic Lyα profile of a TTS is not observable at Earth, except possibly in the far wings, due to absorption by neutral hydrogen along the line of sight. However, observations of H_2 line fluxes constrain the Lyα profile seen by the fluoresced H_2. The rate at which a particular H_2 upward transition absorbs Lyα photons is equal to the total rate of observed downward transitions out of the pumped state, corrected for missing lines, dissociation losses, and propagation losses. If the total number of excited H_2 molecules before pumping is known (e.g., by assuming a temperature), then the inferred pumping rate yields a Lyα flux point at the wavelength of each pumping transition (Fig. 4).

Herczeg et al. (2004) applied this type of analysis to TW Hya, treating the circumstellar H_2 as an isothermal, self-absorbing slab. Figure 4 shows reconstructed Lyα flux points for the upward pumping transitions, assuming the fluoresced H_2 is at 2500 K. The smoothness of the reconstructed Lyα flux points implies that the H_2 level populations are consistent with thermal excitation. Assuming an H_2 temperature warmer or cooler by a few hundred degrees leads to unrealistic discontinuities in the reconstructed Lyα flux points. The reconstructed Lyα profile has a narrow absorption component that is blueshifted by –90 km s⁻¹, presumably due to an intervening flow.

The spatial morphology of fluoresced H$_2$ around TTS is diverse. *Herczeg et al.* (2002) used STIS to observe TW Hya with 50 mas angular resolution, corresponding to a spatial resolution of 2.8 AU at a distance of 56 pc, finding no evidence that the fluoresced H$_2$ is extended. At the other extreme, *Walter et al.* (2003) detected fluoresced H$_2$ up to 9 arcsec from T Tau N, but only in progressions pumped by H$_2$ transitions near the core of Lyα. Fluoresced H$_2$ lines have a main velocity component at or near the stellar radial velocity and perhaps a weaker component that is blueshifted by tens of km s⁻¹ (Herczeg et al., in preparation). These two components are attributed to the disk and the outflow, respectively. TW Hya has H$_2$ lines with no net velocity shift, consistent with formation in the face-on disk (*Herczeg et al.,* 2002). On the other hand, RU Lup has H$_2$ lines that are blueshifted by 12 km s⁻¹, suggesting formation in an outflow. In both of these stars, absorption in the blue wing of the CII 1335 Å wind feature strongly attenuates H$_2$ lines that happen to overlap in wavelength, so in either case H$_2$ forms inside the warm atomic wind (*Herczeg et al.,* 2002, 2005).

The velocity widths of fluoresced H$_2$ lines (after removing instrumental broadening) range from 18 km s⁻¹ to 28 km s⁻¹ for the seven accreting TTS observed at high spectral resolution with STIS (*Herczeg et al.,* 2006). Line width does not correlate well with inclination. For example, TW Hya (nearly face-on disk) and DF Tau (nearly edge-on disk) both have line widths of 18 km s⁻¹. Thermal broadening is negligible, even at 2000 K. Keplerian motion, enforced corotation, and outflow may all contribute to H$_2$ line width in different systems. More data are needed to understand how velocity widths (and shifts) depend on disk inclination, accretion rate, and other factors.

2.5. Infrared Transitions of Molecular Hydrogen

Transitions of molecular hydrogen have also been studied at longer wavelengths, in the near- and mid-infrared. The v=1–0 S(1) transition of H$_2$ (at 2 μm) has been detected in emission in a small sample of classical T Tauri stars (CTTS) and one weak T Tauri star (WTTS) (*Bary et al.,* 2003, and references therein). The narrow emission lines (≤10 km s⁻¹), if arising in a disk, indicate an origin at large radii, probably beyond 10 AU. The high temperatures required to excite these transitions thermally (1000s K), in contrast to the low temperatures expected for the outer disk, suggest that the emission is nonthermally excited, possibly by X-rays (*Bary et al.,* 2003). The measurement of other rovibrational transitions of H$_2$ is needed to confirm this.

The gas mass detectable by this technique depends on the depth to which the exciting radiation can penetrate the disk. Thus, the emission strength may be limited either by the strength of the radiation field, if the gas column density is high, or by the mass of gas present, if the gas column density is low. While it is therefore difficult to measure total gas masses with this approach, clearly nonthermal processes can light up cold gas, making it easier to detect.

Emission from a WTTS is surprising since WTTS are thought to be largely devoid of circumstellar dust and gas, given the lack of infrared excesses and the low accretion rates for these systems. The Bary et al. results call this assumption into question and suggest that longer-lived gaseous reservoirs may be present in systems with low accretion rates. We return to this issue in section 4.1.

At longer wavelengths, the pure rotational transitions of H$_2$ are of considerable interest because molecular hydrogen carries most of the mass of the disk, and these mid-infrared transitions are capable of probing the ~100 K temperatures that are expected for the giant planet region of the disk. These transitions present both advantages and challenges as probes of gaseous disks. On the one hand, their small A-values make them sensitive, in principle, to very large gas masses (i.e., the transitions do not become optically thick until large gas column densities $N_H = 10^{23}$–10^{24} cm⁻² are reached). On the other hand, the small A-values also imply small critical densities, which allows the possibility of contaminating emission from gas at lower densities not associated with the disk, including shocks in outflows and UV excitation of ambient gas.

In considering the detectability of H$_2$ emission from gaseous disks mixed with dust, one issue is that the dust continuum can become optically thick over column densities $N_H \ll 10^{23}$–10^{24} cm⁻². Therefore, in a disk that is optically thick in the continuum (i.e., in CTTS), H$_2$ emission may probe smaller column densities. In this case, the line-to-continuum contrast may be low unless there is a strong temperature inversion in the disk atmosphere, and high signal-to-noise observations may be required to detect the emission. In comparison, in disk systems that are optically thin in the continuum (e.g., WTTS), H$_2$ could be a powerful probe as long as there are sufficient heating mechanisms (e.g., beyond gas-grain coupling) to heat the H$_2$.

A thought-provoking result from ISO was the report of approximately Jupiter masses of warm gas residing in ~20-m.y.-old debris disk systems (*Thi et al.,* 2001) based on the detection of the 28 μm and 17 μm lines of H$_2$. This result was surprising because of the advanced age of the sources in which the emission was detected; gaseous reservoirs are expected to dissipate on much shorter timescales (section 4.1). This intriguing result is thus far unconfirmed by either groundbased studies (*Richter et al.,* 2002; *Sheret et al.,* 2003; *Sako et al.,* 2005) or studies with Spitzer (e.g., *Chen et al.,* 2004).

Nevertheless, groundbased studies have detected pure rotational H$_2$ emission from some sources. Detections to date include AB Aur (*Richter et al.,* 2002). The narrow width of the emission in AB Aur (~10 km s⁻¹ FWHM), if arising in a disk, locates the emission beyond the giant planet region. Thus, an important future direction for these studies is to search for H$_2$ emission in a larger number of sources and at higher velocities, in the giant planet region of the disk. High-resolution mid-infrared spectrographs on >3-m telescopes will provide the greater sensitivity needed for such studies.

2.6. Potential Disk Diagnostics

In a recent development, *Acke et al.* (2005) have reported high-resolution spectroscopy of the [OI] $\lambda6300$ Å line in Herbig AeBe stars. The majority of the sources show a narrow (<50 km s^{-1} FWHM), fairly symmetric emission component centered at the radial velocity of the star. In some cases, double-peaked lines are detected. These features are interpreted as arising in a surface layer of the disk that is irradiated by the star. UV photons incident on the disk surface are thought to photodissociate OH and H$_2$O, producing a nonthermal population of excited neutral oxygen that decays radiatively, producing the observed emission lines. Fractional OH abundances of $\sim10^{-7}$–10^{-6} are needed to account for the observed line luminosities.

Another recent development is the report of strong absorption in the rovibrational bands of C$_2$H$_2$, HCN, and CO$_2$ in the 13–15-μm spectrum of a low-mass Class I source in Ophiuchus, IRS 46 (*Lahuis et al.*, 2006). The high excitation temperature of the absorbing gas (400–900 K) suggests an origin close to the star, an interpretation that is consistent with millimeter observations of HCN that indicate a source size ≪100 AU. Surprisingly, high dispersion observations of rovibrational CO (4.7 μm) and HCN (3.0 μm) show that the molecular absorption is *blueshifted* relative to the molecular cloud. If IRS 46 is similarly blueshifted relative to the cloud, the absorption may arise in the atmosphere of a nearly edge-on disk. A disk origin for the absorption is consistent with the observed relative abundances of C$_2$H$_2$, HCN, and CO$_2$ (10^{-6}–10^{-5}), which are close to those predicted by *Markwick et al.* (2002) for the inner region of gaseous disks (≤2 AU; see section 3). Alternatively, if IRS 46 has approximately the same velocity as the cloud, then the absorbing gas is blueshifted with respect to the star and the absorption may arise in an outflowing wind. Winds launched from the disk, at AU distances, may have molecular abundances similar to those observed if the chemical properties of the wind change slowly as the wind is launched. Detailed calculations of the chemistry of disk winds are needed to explore this possibility. The molecular abundances in the inner disk midplane (section 3.3) provide the initial conditions for such studies.

3. THERMAL-CHEMICAL MODELING

3.1. General Discussion

The results discussed in the previous section illustrate the growing potential for observations to probe gaseous inner disks. While, as already indicated, some conclusions can be drawn directly from the data coupled with simple spectral synthesis modeling, harnessing the full diagnostic potential of the observations will likely rely on detailed models of the thermal-chemical structure (and dynamics) of disks. Fortu-

nately, the development of such models has been an active area of recent research. Although much of the effort has been devoted to understanding the outer regions of disks (~100 AU) (e.g., *Langer et al.,* 2000; chapters by Bergin et al. and Dullemond et al.), recent work has begun to focus on the region within 10 AU.

Because disks are intrinsically complex structures, the models include a wide array of processes. These encompass heating sources such as stellar irradiation (including far UV and X-rays) and viscous accretion; chemical processes such as photochemistry and grain surface reactions; and mass transport via magnetocentrifugal winds, surface evaporative flows, turbulent mixing, and accretion onto the star. The basic goal of the models is to calculate the density, temperature, and chemical abundance structures that result from these processes. Ideally, the calculation would be fully self-consistent, although approximations are made to simplify the problem.

A common simplification is to adopt a specified density distribution and then solve the rate equations that define the chemical model. This is justifiable where the thermal and chemical timescales are short compared to the dynamical timescale. A popular choice is the α-disk model (*Shakura and Sunyaev,* 1973; *Lynden-Bell and Pringle,* 1974), in which a phenomenological parameter α characterizes the efficiency of angular momentum transport; its vertically averaged value is estimated to be $\sim10^{-2}$ for T Tauri disks on the basis of measured accretion rates (*Hartmann et al.,* 1998). Both vertically isothermal α-disk models and the Hayashi minimum mass solar nebula (e.g., *Aikawa et al.,* 1999) were adopted in early studies.

An improved method removes the assumption of vertical isothermality and calculates the vertical thermal structure of the disk including viscous accretion heating at the midplane (specified by α) and stellar radiative heating under the assumption that the gas and dust temperatures are the same (*Calvet et al.,* 1991; *D'Alessio et al.,* 1999). Several chemical models have been built using the D'Alessio density distribution (e.g., *Aikawa and Herbst,* 1999; *GNI04*; *Jonkheid et al.,* 2004).

Starting about 2001, theoretical models showed that the gas temperature can become much larger than the dust temperature in the atmospheres of outer (*Kamp and van Zadelhoff,* 2001) and inner (*Glassgold and Najita,* 2001) disks. This suggested the need to treat the gas and dust as two independent but coupled thermodynamic systems. As an example of this approach, *Gorti and Hollenbach* (2004) have iteratively solved a system of chemical rate equations along with the equations of hydrostatic equilibrium and thermal balance for both the gas and the dust.

The chemical models developed so far are characterized by diversity as well as uncertainty. There is diversity in the adopted density distribution and external radiation field (UV, X-rays, and cosmic rays; the relative importance of these depends on the evolutionary stage) and in the thermal and chemical processes considered. The relevant heating processes are less well understood than line cooling. One issue is how UV, X-ray, and cosmic rays heat the gas. Another is the role of mechanical heating associated with various flows in the disk, especially accretion (*GNI04*). The chemical processes are also less certain. Our understanding of astrochemistry is based mainly on the interstellar medium, where densities and temperatures are low compared to those of inner disks, except perhaps in shocks and photon-dominated regions. New reaction pathways or processes may be important at the higher densities ($>10^7$ cm^{-3}) and higher radiation fields of inner disks. A basic challenge is to understand the thermal-chemical role of dust grains and PAHs. Indeed, perhaps the most significant difference between models is the treatment of grain chemistry. The more sophisticated models include adsorption of gas onto grains in cold regions and desorption in warm regions. Yet another level of complexity is introduced by transport processes that can affect the chemistry through vertical or radial mixing.

An important practical issue in thermal-chemical modeling is that self-consistent calculations become increasingly difficult as the density, temperature, and number of species increase. Almost all models employ truncated chemistries with with somewhere from 25 to 215 species, compared with 396 in the UMIST database (*Le Teuff et al.,* 2000). The truncation process is arbitrary, determined largely by the goals of the calculations. *Wiebe et al.* (2003) have an objective method for selecting the most important reactions from large databases. Additional insights into disk chemistry are offered in the chapter by Bergin et al.

3.2. The Disk Atmosphere

As noted above, *Kamp and van Zadelhoff* (2001) concluded in their model of debris disks that the gas and dust temperature can differ, as did *Glassgold and Najita* (2001) for T Tauri disks. The former authors developed a comprehensive thermal-chemical model where the heating is primarily from the dissipation of the drift velocity of the dust through the gas. For T Tauri disks, stellar X-rays, known to be a universal property of low-mass YSOs, heat the gas to temperatures thousands of degrees hotter than the dust temperature.

Figure 5 shows the vertical temperature profile obtained by *Glassgold et al.* (2004) with a thermal-chemical model based on the dust model of *D'Alessio et al.* (1999) for a generic T Tauri disk. Near the midplane, the densities are high enough to strongly couple the dust and gas. At higher altitudes, the disk becomes optically thin to the stellar optical and infrared radiation, and the temperature of the (small) grains rises, as does the still closely coupled gas temperature. However, at still higher altitudes, the gas responds strongly to the less-attenuated X-ray flux, and its temperature rises much above the dust temperature. The presence of a hot X-ray heated layer above a cold midplane layer was obtained independently by *Alexander et al.* (2004).

GNI04 also considered the possibility that the surface layers of protoplanetary disks are heated by the dissipation of mechanical energy. This might arise through the inter-

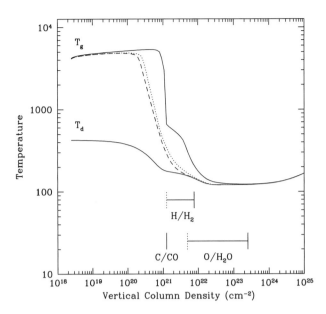

Fig. 5. Temperature profiles from *GNI04* for a protoplanetary disk atmosphere. The lower solid line shows the dust temperature of *D'Alessio et al.* (1999) at a radius of 1 AU and a mass accretion rate of 10^{-8} M$_\odot$ yr^{-1}. The upper curves show the corresponding gas temperature as a function of the phenomenological mechanical heating parameter defined by equation (1), $\alpha_h = 1$ (solid line), 0.1 (dotted line), and 0.01 (dashed line). The $\alpha_h = 0.01$ curve closely follows the limiting case of pure X-ray heating. The lower vertical lines indicate the major chemical transitions, specifically CO forming at $\sim 10^{21}$ cm^{-2}, H$_2$ forming at $\sim 6 \times 10^{21}$ cm^{-2}, and water forming at higher column densities.

action of a wind with the upper layers of the disk or through disk angular momentum transport. Since the theoretical understanding of such processes is incomplete, a phenomenological treatment is required. In the case of angular momentum transport, the most widely accepted mechanism is the MRI (*Balbus and Hawley,* 1991; *Stone et al.,* 2000), which leads to the local heating formula

$$\Gamma_{acc} = \frac{9}{4} \alpha_h \rho c^2 \Omega \qquad (1)$$

where ρ is the mass density, c is the isothermal sound speed, Ω is the angular rotation speed, and α_h is a phenomenological parameter that depends on how the turbulence dissipates. One can argue, on the basis of simulations by *Miller and Stone* (2000), that midplane turbulence generates Alfvén waves, which, upon reaching the diffuse surface regions, produce shocks and heating. Wind-disk heating can be represented by a similar expression on the basis of dimensional arguments. Equation (1) is essentially an adaptation of the expression for volumetric heating in an α-disk model, where α can in general depend on position. *GNI04* used the notation α_h to distinguish its value in the disk atmosphere from the usual midplane value.

In the top layers fully exposed to X-rays, the gas temperature at 1 AU is ~5000 K. Further down, there is a warm transition region (500–2000 K) composed mainly of atomic hydrogen but with carbon fully associated into CO. The conversion from atomic H to H_2 is reached at a column density of $\sim 6 \times 10^{21}$ cm^{-2}, with more complex molecules such as water forming deeper in the disk. The location and thickness of the warm molecular region depends on the strength of the surface heating. The curves in Fig. 5 illustrate this dependence for a T Tauri disk at r = 1 AU. With $\alpha_h = 0.01$, X-ray heating dominates this region, whereas with $\alpha_h > 0.1$, mechanical heating dominates.

Gas-temperature inversions can also be produced by UV radiation operating on small dust grains and PAHs, as demonstrated by the thermal-chemical models of *Jonkheid et al.* (2004) and *Kamp and Dullemond* (2004). *Jonkheid et al.*(2004) use the *D'Alessio et al.* (1999) model and focus on the disk beyond 50 AU. At this radius, the gas temperature can rise to 800 K or 200 K, depending on whether small grains are well mixed or settled. For a thin disk and a high stellar UV flux, *Kamp and Dullemond* (2004) obtain temperatures that asymptote to several 1000 K inside 50 AU. Of course these results are subject to various assumptions that have been made about the stellar UV, the abundance of PAHs, and the growth and settling of dust grains.

Many of the earlier chemical models, oriented toward outer disks (e.g., *Willacy and Langer*, 2000; *Aikawa and Herbst*, 1999, 2001; *Markwick et al.*, 2002), adopt a value for the *stellar* UV radiation field that is 10^4 times larger than galactic at a distance of 100 AU. This choice can be traced back to early IUE measurements of the stellar UV beyond 1400 Å for several TTS (*Herbig and Goodrich*, 1986). Although the UV flux from TTS covers a range of values and is undoubtedly time-variable, detailed studies with IUE (e.g., *Valenti et al.*, 2000; *Johns-Krull et al.*, 2000) and FUSE (e.g., *Wilkinson et al.*, 2002; *Bergin et al.*, 2003) indicate that it decreases into the FUV domain with a typical value $\sim 10^{-15}$ erg cm^{-2} s^{-1} Å^{-1}, much smaller than earlier estimates. A flux of $\sim 10^{-15}$ erg cm^{-2} s^{-1} Å^{-1} at Earth translates into a value at 100 AU of ~100 times the traditional Habing value for the interstellar medium. The data in the FUV range are sparse, unfortunately, as a function of age or the evolutionary state of the system. More measurements of this kind are needed since it is obviously important to use realistic fluxes in the crucial FUV band between 912 and 1100 Å where atomic C can be photoionized and H_2 and CO photodissociated (*Bergin et al.*, 2003; see also chapter by Bergin et al.).

Whether stellar FUV or X-ray radiation dominates the ionization, chemistry, and heating of protoplanetary disks is important because of the vast difference in photon energy. The most direct physical consequence is that FUV photons cannot ionize H, and thus the abundance of C provides an upper limit to the ionization level produced by the photoionization of heavy atoms, $x_e \sim 10^{-4}$–10^{-3}. Next, FUV photons are absorbed much more readily than X-rays, although this depends on the size and spatial distribution of the dust

grains, i.e, on grain growth and sedimentation. Using realistic numbers for the FUV and X-ray luminosities of TTS, we estimate that $L_{FUV} \sim L_X$. The rates used in many early chemical models correspond to $L_X \ll L_{FUV}$. This suggests that future chemical modeling of protoplanetary disks should consider both X-rays and FUV in their treatment of ionization, heating, and chemistry.

3.3. The Midplane Region

Unlike the warm upper atmosphere of the disk, which is accessible to observation, the optically thick midplane is much more difficult to study. Nonetheless, it is extremely important for understanding the dynamics of the basic flows in star formation such as accretion and outflow. The important role of the ionization level for disk accretion via the MRI was pointed out by *Gammie* (1996). The physical reason is that collisional coupling between electrons and neutrals is required to transfer the turbulence in the magnetic field to the neutral material of the disk. Gammie found that galactic cosmic rays cannot penetrate beyond a surface layer of the disk. He suggested that accretion only occurs in the surface of the inner disk (the "active region") and not in the much thicker midplane region (the "dead zone") where the ionization level is too small to mediate the MRI.

Glassgold et al. (1997) argued that the galactic cosmic rays never reach the inner disk because they are blown away by the stellar wind, much as the solar wind excludes galactic cosmic rays. They showed that YSO X-rays do almost as good a job as cosmic rays in ionizing surface regions, thus preserving the layered accretion model of the MRI for YSOs. *Igea and Glassgold* (1999) supported this conclusion with a Monte Carlo calculation of X-ray transport through disks, demonstrating that scattering plays an important role in the MRI by extending the active surface layer to column densities greater than 10^{25} cm^{-2}, approaching the galactic cosmic ray range used by *Gammie* (1996). This early work showed that the theory of disk ionization and chemistry is crucial for understanding the role of the MRI for YSO disk accretion and possibly for planet formation. Indeed, *Glassgold et al.* (1997) suggested that Gammie's dead zone might provide a good environment for the formation of planets.

These challenges have been taken up by several groups (e.g., *Sano et al.*, 2000; *Fromang et al.*, 2002; *Semenov et al.*, 2004; *Kunz and Balbus*, 2004; *Desch*, 2004; *Matsumura and Pudritz*, 2003, 2005; *Ilgner and Nelson*, 2006a,b). *Fromang et al.* (2002) discussed many of the issues that affect the size of the dead zone: differences in the disk model, such as a Hayashi disk or a standard α-disk; temporal evolution of the disk; the role of a small abundance of heavy atoms that recombine mainly radiatively; and the value of the magnetic Reynolds number. *Sano et al.* (2000) explored the role played by small dust grains in reducing the electron fraction when it becomes as small as the abundance of dust grains. They showed that the dead zone decreases and eventually vanishes as the grain size increases or as sedimentation toward the midplane proceeds. More recently,

Inutsuka and Sano (2005) have suggested that a small fraction of the energy dissipated by the MRI leads to the production of fast electrons with energies sufficient to ionize H_2. When coupled with vertical mixing of highly ionized surface regions, Inutsuka and Sano argue that the MRI can self generate the ionization it needs to be operative throughout the entire disk.

Recent chemical modeling (*Semenov et al.*, 2004; *Ilgner and Nelson*, 2006a,b) confirms that the level of ionization in the midplane is affected by many microphysical processes. These include the abundances of radiatively recombining atomic ions, molecular ions, small grains, and PAHs. The proper treatment of the ions represents a great challenge for disk chemistry, one made particularly difficult by the lack of observations of the dense gas at the midplane of the inner disk. Thus the uncertainties in inner disk chemistry preclude definitive quantitative conclusions about the midplane ionization of protoplanetary disks. Perhaps the biggest wild card is the issue of grain growth, emphasized anew by *Semenov et al.* (2004). If the disk grain size distribution were close to interstellar, then the small grains would be effective in reducing the electron fraction and producing dead zones. But significant grain growth is expected *and* observed in the disks of YSOs, limiting the extent of dead zones (e.g., *Sano et al.*, 2000).

The broader chemical properties of the *inner* midplane region are also of great interest since most of the gas in the disk is within one or two scale heights. The chemical composition of the inner midplane gas is important because it provides the initial conditions for outflows and for the formation of planets and other small bodies; it also determines whether the MRI operates. Relatively little work has been done on the mid-plane chemistry of the inner disk. For example, *GNI04* excluded N and S species and restricted the carbon chemistry to species closely related to CO. However, *Willacy et al.* (1998), *Markwick et al.* (2002), and *Ilgner et al.* (2004) have carried out interesting calculations that shed light on a possible rich organic chemistry in the inner disk.

Using essentially the same chemical model, these authors follow mass elements in time as they travel in a steady accretion flow toward the star. At large distances, the gas is subject to adsorption, and at small distances to thermal desorption. In between it reacts on the surface of the dust grains; on being liberated from the dust, it is processed by gas-phase chemical reactions. The gas and dust are assumed to have the same temperature, and all effects of stellar radiation are ignored. The ionizing sources are cosmic rays and ^{26}Al. Since the collisional ionization of low ionization potential atoms is ignored, a very low ionization level results. *Markwick et al.* (2002) improve on *Willacy et al.* (1998) by calculating the vertical variation of the temperature, and *Ilgner et al.* (2004) consider the effects of mixing. Near 1 AU, H_2O and CO are very abundant, as predicted by simpler models, but *Markwick et al.* (2002) find that CH_4 and CO have roughly equal abundances. Nitrogen-bearing molecules, such as NH_3, HCN, and HNC are also predicted to be abundant, as are a variety of hydrocarbons such as CH_4,

C_2H_2, C_2H_3, C_2H_4, etc. Markwick et al. also simulate the presence of penetrating X-rays and find increased column densities of CN and HCO^+. Despite many uncertainties, these predictions are of interest for our future understanding of the midplane region.

Infrared spectroscopic searches for hydrocarbons in disks may be able to test these predictions. For example, *Gibb et al.* (2004) searched for CH_4 in absorption toward HL Tau. The upper limit on the abundance of CH_4 relative to CO (<1%) in the absorbing gas may contradict the predictions of *Markwick et al.* (2002) if the absorption arises in the disk atmosphere. However, some support for the Markwick et al. model comes from a recent report by *Lahuis et al.* (2006) of a rare detection by Spitzer of C_2H_2 and HCN in absorption toward a YSO, with ratios close to those predicted for the inner disk (section 2.6).

3.4. Modeling Implications

An interesting implication of the irradiated disk atmosphere models discussed above is that the region of the atmosphere over which the gas and dust temperatures differ includes the region that is accessible to observational study. Indeed, the models have interesting implications for some of the observations presented in section 2. They can account roughly for the unexpectedly warm gas temperatures that have been found for the inner disk region based on the CO fundamental (section 2.3) and UV fluorescent H_2 transitions (section 2.4). In essence, the warm gas temperatures arise from the direct heating of the gaseous component and the poor thermal coupling between the gas and dust components at the low densities characteristic of upper disk atmospheres. The role of X-rays in heating disk atmospheres has some support from the results of *Bergin et al.* (2004); they suggested that some of the UV H_2 emission from TTS arises from excitation by fast electrons produced by X-rays.

In the models, CO is found to form at a column density $N_H \simeq 10^{21}$ cm^{-2} and temperature ~1000 K in the radial range 0.5–2 AU [Fig. 5 (*GNI04*)], conditions similar to those deduced for the emitting gas from the CO fundamental lines (*Najita et al.*, 2003). Moreover, CO is abundant in a region of the disk that is predominantly atomic hydrogen, a situation that is favorable for exciting the rovibrational transitions because of the large collisional excitation cross section for H + CO inelastic scattering. Interestingly, X-ray irradiation alone is probably insufficient to explain the strength of the CO emission observed in actively accreting TTS. This suggests that other processes may be important in heating disk atmospheres. *GNI04* have explored the role of mechanical heating. Other possible heating processes are FUV irradiation of grains and or PAHs.

Molecular hydrogen column densities comparable to the UV fluorescent column of ~5 × 10^{18} cm^{-2} observed from TW Hya are reached at 1 AU at a total vertical hydrogen column density of ~5 × 10^{21} cm^{-2}, where the fractional abundance of H_2 is ~10^{-3} [Fig. 5 (*GNI04*)]. Since Lyα photons must traverse the entire $\sim 5 \times 10^{21}$ cm^{-2} in order to excite the emission, the line-of-sight dust opacity through this column must be relatively low. Observations of this kind, when combined with atmosphere models, may be able to constrain the gas-to-dust ratio in disk atmospheres, with consequent implications for grain growth and settling.

Work in this direction has been carried out by *Nomura and Millar* (2005). They have made a detailed thermal model of a disk that includes the formation of H$_2$ on grains, destruction via FUV lines, and excitation by Lyα photons. The gas at the surface is heated primarily by the photoelectric effect on dust grains and PAHs, with a dust model appropriate for interstellar clouds, i.e., one that reflects little grain growth. Both interstellar and stellar UV radiation are included, the latter based on observations of TW Hya. The gas temperature at the surface of their flaring disk model reaches 1500 K at 1 AU. They are partially successful in accounting for the measurements of *Herczeg et al.* (2002), but their model fluxes fall short by a factor of 5 or so. A likely defect in their model is that the calculated temperature of the disk surface is too low, a problem that might be remedied by reducing the UV attenuation by dust and by including X-ray or other surface-heating processes.

The relative molecular abundances that are predicted by these nonturbulent, layered model atmospheres are also of interest. At a distance of 1 AU, the calculations of *GNI04* indicate that the relative abundance of H$_2$O to CO is $\sim 10^{-2}$ in the disk atmosphere for column densities $<10^{22}$ cm^{-2}; only at column densities $>10^{23}$ cm^{-2} are H$_2$O and CO comparably abundant. The abundance ratio in the atmosphere is significantly lower than the few relative abundances measurements to date (0.1–0.5) at <0.3 AU (*Carr et al.,* 2004; section 2.2). Perhaps layered model atmospheres, when extended to these small radii, will be able to account for the abundant water that is detected. If not, the large water abundance may be evidence of strong vertical (turbulent) mixing that carries abundant water from deeper in the disk up to the surface. Thus, it would be of great interest to develop the modeling for the sources and regions where water is observed in the context of both layered models and those with vertical mixing. Work in this direction has the potential to place unique constraints on the dynamical state of the disk.

4. CURRENT AND FUTURE DIRECTIONS

As described in the previous sections, significant progress has been made in developing both observational probes of gaseous inner disks as well as the theoretical models that are needed to interpret the observations. In this section, we describe some areas of current interest as well as future directions for studies of gaseous inner disks.

4.1. Gas Dissipation Timescale

The lifetime of gas in the inner disk is of interest in the context of both giant and terrestrial planet formation. Since significant gas must be present in the disk in order for a gas giant to form, the gas dissipation timescale in the giant planet region of the disk can help to identify dominant pathways for the formation of giant planets. A short dissipation timescale favors processes such as gravitational instabilities that can form giant planets on short timescales (<1000 yr) (*Boss,* 1997; *Mayer et al.,* 2002). A longer dissipation timescale accommodates the more leisurely formation of planets in the core accretion scenario (few–10 m.y.) (*Bodenheimer and Lin,* 2002).

Similarly, the outcome of terrestrial planet formation (the masses and eccentricities of the planets and their consequent habitability) may depend sensitively on the residual gas in the terrestrial planet region of the disk at ages of a few million years. For example, in the picture of terrestrial planet formation described by *Kominami and Ida* (2002), if the gas column density in this region is $\gg 1$ g cm^{-2} at the epoch when protoplanets assemble to form terrestrial planets, gravitational gas drag is strong enough to circularize the orbits of the protoplanets, making it difficult for them to collide and build Earth-mass planets. In contrast, if the gas column density is $\ll 1$ g cm^{-2}, Earth-mass planets can be produced, but gravitational gas drag is too weak to recircularize their orbits. As a result, only a narrow range of gas column densities around ~ 1 g cm^{-2} is expected to lead to planets with the Earth-like masses and low eccentricities that we associate with habitability on Earth.

From an observational perspective, relatively little is known about the evolution of the gaseous component. Disk lifetimes are typically inferred from infrared excesses that probe the dust component of the disk, although processes such as grain growth, planetesimal formation, and rapid grain inspiraling produced by gas drag (*Takeuchi and Lin,* 2005) can compromise dust as a tracer of the gas. Our understanding of disk lifetimes can be improved by directly probing the gas content of disks and using indirect probes of disk gas content such as stellar accretion rates [see *Najita* (2006) for a review of this topic].

Several of the diagnostics decribed in section 2 may be suitable as direct probes of disk gas content. For example, transitions of H$_2$ and other molecules and atoms at mid- through far-infrared wavelengths are thought to be promising probes of the giant planet region of the disk (*Gorti and Hollenbach,* 2004). This is a important area of investigation currently for the Spitzer Space Telescope and, in the future, for Herschel and 8- to 30-m groundbased telescopes.

Studies of the lifetime of gas in the terrestrial planet region are also in progress. The CO transitions are well suited for this purpose because the transitions of CO and its isotopes probe gas column densities in the range of interest (10^{-4}–1 g cm^{-2}). A current study by Najita, Carr, and Mathieu, which explores the residual gas content of optically thin disks (*Najita,* 2004), illustrates some of the challenges in probing the residual gas content of disks. First, given the well-known correlation between infrared excess and accretion rate in young stars (e.g., *Kenyon and Hartmann,* 1995), CO emission from sources with optically thin inner disks may be intrinsically weak if accretion contributes significantly to heating disk atmospheres. Thus, high signal-to-noise spectra may be needed to detect this emission. Second, since the line emission may be intrinsically weak, structure in the stellar photosphere may complicate the identification of emission features. Figure 6 shows an example in which CO absorption in the stellar photosphere of TW Hya likely veils weak emission from the disk. Correcting for the stellar photosphere would not only amplify the strong v=1–0 emission that is clearly present (cf. *Rettig et al.*, 2004), but would also uncover weak emission in the higher vibrational lines, confirming the presence of the warmer gas probed by the UV fluorescent lines of H_2 (*Herczeg et al.*, 2002).

Stellar accretion rates provide a complementary probe of the gas content of inner disks. In a steady accretion disk, the column density Σ is related to the disk accretion rate $\dot{M}$ by a relation of the form $\Sigma \propto \dot{M}/\alpha T$, where T is the disk temperature. A relation of this form allows us to infer Σ from $\dot{M}$ given a value for the viscosity parameter α. Alternatively, the relation could be calibrated empirically using measured disk column densities.

Accretion rates are available for many sources in the age range 0.5–10 m.y. (e.g., *Gullbring et al.*, 1998; *Hartmann*

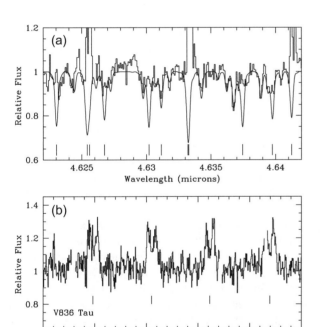

Fig. 6. **(a)** Spectrum of the transitional disk system TW Hya at 4.6 μm (histogram). The strong emission in the v = 1–0 CO fundamental lines extend above the plotted region. Although the model stellar photospheric spectrum (light solid line) fits the weaker features in the TW Hya spectrum, it predicts stronger absorption in the low vibrational CO transitions (indicated by the lower vertical lines) than is observed. This suggests that the stellar photosphere is veiled by CO emission from warm disk gas. **(b)** CO fundamental emission from the transitional disk system V836 Tau. Vertical lines mark the approximate line centers at the velocity of the star. The velocity widths of the lines locate the emission within a few AU of the star, and the relative strengths of the lines suggest optically thick emission. Thus, a large reservoir of gas may be present in the inner disk despite the weak infrared excess from this portion of the disk.

et al., 1998; *Muzerolle et al.*, 1998, 2000). A typical value of $10^{-8}\ M_\odot\ yr^{-1}$ for TTS corresponds to a(n active) disk column density of ~100 g cm^{-2} at 1 AU for $\alpha = 0.01$ (*D'Alessio et al.*, 1998). The accretion rates show an overall decline with time with a large dispersion at any given age. The existence of 10-m.y.-old sources with accretion rates as large as $10^{-8}\ M_\odot\ yr^{-1}$ (*Sicilia-Aguilar et al.*, 2005) suggests that gaseous disks may be long lived in some systems.

Even the lowest measured accretion rates may be dynamically significant. For a system like V836 Tau (Fig. 6), a ~3-m.y.-old (*Siess et al.*, 1999) system with an optically thin inner disk, the stellar accretion rate of $4 \times 10^{-10}\ M_\odot\ yr^{-1}$ (*Hartigan et al.*, 1995; *Gullbring et al.*, 1998) would correspond to ~4 g cm^{-2} at 1 AU. Although the accretion rate is irrelevant for the buildup of the stellar mass, it corresponds to a column density that would favorably impact terrestrial planet formation. More interesting perhaps is St34, a TTS with a Li depletion age of 25 m.y.; its stellar accretion rate of $2 \times 10^{-10}\ M_\odot\ yr^{-1}$ (*White and Hillenbrand*, 2005) suggests a dynamically significant reservoir of gas in the inner disk region. These examples suggest that dynamically significant reservoirs of gas may persist even after inner disks become optically thin and over the timescales needed to influence the outcome of terrestrial planet formation.

The possibility of long-lived gaseous reservoirs can be confirmed by using the diagnostics described in section 2 to measure total disk column densities. Equally important, a measured disk column density, combined with the stellar accretion rate, would allow us to infer a value for viscosity parameter α for the system. This would be another way of constraining the disk accretion mechanism.

4.2. Nature of Transitional Disk Systems

Measurements of the gas content and distribution in inner disks can help us to identify systems in various states of planet formation. Among the most interesting objects to study in this context are the transitional disk systems, which possess optically thin inner and optically thick outer disks. Examples of this class of objects include TW Hya, GM Aur, DM Tau, and CoKu Tau/4 (*Calvet et al.*, 2002, 2005; *Rice et al.*, 2003; *Bergin et al.*, 2004; *D'Alessio et al.*, 2005). It was suggested early on that optically thin inner disks might be produced by the dynamical sculpting of the disk by orbiting giant planets (*Skrutskie et al.*, 1990; see also *Marsh and Mahoney*, 1992).

Indeed, optically thin disks may arise in multiple phases of disk evolution. For example, as a first step in planet formation (via core accretion), grains are expected to grow into planetesimals and eventually rocky planetary cores, producing a region of the disk that has reduced continuum opacity but is gas-rich. These regions of the disk may therefore show strong line emission. Determining the fraction of sources in this phase of evolution may help to establish the relative timescales for planetary core formation and the accretion of gaseous envelope.

If a planetary core accretes enough gas to produce a low-mass giant planet (~1 M_J), it is expected to carve out a gap in its vicinity (e.g., *Takeuchi et al.,* 1996). Gap-crossing streams can replenish an inner disk and allow further accretion onto both the star and planet (*Lubow et al.,* 1999). The small solid angle subtended by the accretion streams would produce a deficit in the emission from both gas and dust in the vicinity of the planet's orbit. We would also expect to detect the presence of an inner disk. Possible examples of systems in this phase of evolution include GM Aur and TW Hya in which hot gas is detected close to the star as is accretion onto the star (*Bergin et al.,* 2004; *Herczeg et al.,* 2002; *Muzerolle et al.,* 2000). The absence of gas in the vicinity of the planet's orbit would help to confirm this interpretation.

Once the planet accretes enough mass via the accretion streams to reach a mass ~5–10 M_J, it is expected to cut off further accretion (e.g., *Lubow et al.,* 1999). The inner disk will accrete onto the star, leaving a large inner hole and no trace of stellar accretion. CoKu Tau/4 is a possible example of a system in this phase of evolution (cf. *Quillen et al.,* 2004) since it appears to have a large inner hole and a low to negligible accretion rate (<few × 10^{-10} $M_\odot$ yr^{-1}). This interpretation predicts little gas anywhere within the orbit of the planet.

At late times, when the disk column density around 10 AU has decreased sufficiently that the outer disk is being photoevaporated away faster than it can resupply material to the inner disk via accretion, the outer disk will decouple from the inner disk, which will accrete onto the star, leaving an inner hole that is devoid of gas and dust [the "UV Switch" model (*Clarke et al.,* 2001)]. Measurements of the disk gas column density and the stellar accretion rate can be used to test this possibility. As an example, TW Hya is in the age range (~10 m.y.) where photoevaporation is likely to be significant. However, the accretion rate onto the star, gas content of the inner disk (sections 2 and 4), as well as the column density inferred for the outer disk [32 g cm^{-2} at 20 AU based on the dust SED (*Calvet et al.,* 2002)] are all much larger than is expected in the UV switch model. Although this mechanism is, therefore, unlikely to explain the SED for TW Hya, it may explain the presence of inner holes in less massive disk systems of comparable age.

4.3. Turbulence in Disks

Future studies of gaseous inner disks may also help to clarify the nature of the disk accretion process. As indicated in section 2.1, evidence for suprathermal line broadening in disks supports the idea of a turbulent accretion process. A turbulent inner disk may have important consequences for the survival of terrestrial planets and the cores of giant planets. An intriguing puzzle is how these objects avoid Type-I migration, which is expected to cause the object to lose angular momentum and spiral into the star on short timescales (e.g., *Ward,* 1997). A recent suggestion is that if disk accretion is turbulent, terrestrial planets will scatter off turbulent fluctuations, executing a "random walk" that greatly in-

creases the migration time as well as the chances of survival (*Nelson et al.,* 2000; see chapter by Papaloizou et al.).

It would be interesting to explore this possible connection further by extending the approach used for the CO overtone lines to a wider range of diagnostics to probe the intrinsic line width as a function of radius and disk height. By comparing the results to the detailed predictions of theoretical models, it may be possible to distinguish between the turbulent signature, produced, e.g., by the MRI instability, from the turbulence that might be produced by, e.g., a wind blowing over the disk.

A complementary probe of turbulence may come from exploring the relative molecular abundances in disks. As noted in section 3.4, if relative abundances cannot be explained by model predictions for nonturbulent, layered accretion flows, a significant role for strong vertical mixing produced by turbulence may be implied. Although model dependent, this approach toward diagnosing turbulent accretion appears to be less sensitive to confusion from wind-induced turbulence, especially if one can identify diagnostics that require vertical mixing from deep down in the disk. Another complementary approach toward probing the accretion process, discussed in section 4.1, is to measure total gas column densities in low column density, dissipating disks in order to infer values for the viscosity parameter α.

5. SUMMARY AND CONCLUSIONS

Recent work has lent new insights on the structure, dynamics, and gas content of inner disks surrounding young stars. Gaseous atmospheres appear to be hotter than the dust in inner disks. This is a consequence of irradiative (and possibly mechanical) heating of the gas as well as the poor thermal coupling between the gas and dust at the low densities of disk atmospheres. In accreting systems, the gaseous disk appears to be turbulent and extends inward beyond the dust sublimation radius to the vicinity of the corotation radius. There is also evidence that dynamically significant reservoirs of gas can persist even after the inner disk becomes optically thin in the continuum. These results bear on important star- and planet-formation issues such as the origin of winds, funnel flows, and the rotation rates of young stars; the mechanism(s) responsible for disk accretion; and the role of gas in the determining the architectures of terrestrial and giant planets. Although significant future work is needed to reach any conclusions on these issues, the future for such studies is bright. Increasingly detailed studies of the inner disk region should be possible with the advent of powerful spectrographs and interferometers (infrared and submillimeter) as well as sophisticated models that describe the coupled thermal, chemical, and dynamical state of the disk.

Acknowledgments. We thank S. Strom, who contributed significantly to the discussion on the nature of transitional disk systems. We also thank F. Lahuis and M. Richter for sharing manuscripts of their work in advance of publication. A.E.G. acknowledges support from the NASA Origins and NSF Astronomy programs. J.S.C. and J.R.N. also thank the NASA Origins program for its support.

REFERENCES

Abgrall H., Roueff E., Launay F., Roncin J. Y., and Subtil J. L. (1993a) *Astron. Astrophys. Suppl., 101,* 273–321.

Abgrall H., Roueff E., Launay F., Roncin J. Y., and Subtil J. L. (1993b) *Astron. Astrophys. Suppl., 101,* 323–362.

Acke B., van den Ancker M. E., and Dullemond C. P. (2005) *Astron. Astrophys., 436,* 209–230.

Aikawa Y. and Herbst E. (1999) *Astrophys. J., 526,* 314–326.

Aikawa Y. and Herbst E. (2001) *Astron. Astrophys., 371,* 1107–1117.

Aikawa Y., Umebayashi T., Nakano T., and Miyama S. M. (1999) *Astrophys. J., 519,* 705–725.

Akeson R. L., Walker C. H., Wood K., Eisner J. A., Scire E., et al. (2005a) *Astrophys. J., 622,* 440–450.

Akeson R. L., Boden A. F., Monnier J. D., Millan-Gabet R., Beichman C., et al. (2005b) *Astrophys. J., 635,* 1173–1181.

Alexander R. D., Clarke C. J., and Pringle J. E. (2004) *Mon. Not. R. Astron. Soc., 354,* 71–80.

Ardila D. R., Basri G., Walter F. M., Valenti J. A., and Johns-Krull C. M. (2002) *Astrophys. J., 566,* 1100–1123.

Balbus S. A. and Hawley J. F. (1991) *Astrophys. J., 376,* 214–222.

Bary J. S., Weintraub D. A., and Kastner J. H. (2003) *Astrophys. J., 586,* 1138–1147.

Bergin E., Calvet N., D'Alessio P., and Herczeg G. J. (2003) *Astrophys. J., 591,* L159–L162.

Bergin E., Calvet N., Sitko M. L., Abgrall H., D'Alessio P., et al. (2004) *Astrophys. J., 614,* L133–L137.

Bik A. and Thi W. F. (2004) *Astron. Astrophys., 427,* L13–L16.

Blake G. A. and Boogert A. C. A. (2004) *Astrophys. J., 606,* L73–L76.

Blum R. D., Barbosa C. L., Damineli A., Conti P. S., and Ridgway S. (2004) *Astrophys. J., 617,* 1167–1176.

Bodenheimer P. and Lin D. N. C. (2002) *Ann. Rev. Earth Planet. Sci., 30,* 113–148.

Boss A. P. (1997) *Science, 276,* 1836–1839.

Brittain S. D., Rettig T. W., Simon T., Kulesa C., DiSanti M. A., and Dello Russo N. (2003) *Astrophys. J., 588,* 535–544.

Brown A., Jordan C., Millar T. J., Gondhalekar P., and Wilson R. (1981) *Nature, 290,* 34–36.

Calvet N., Patino A., Magris G., and D'Alessio P. (1991) *Astrophys. J., 380,* 617–630.

Calvet N., D'Alessio P., Hartmann L, Wilner D., Walsh A., and Sitko M. (2002) *Astrophys. J., 568,* 1008–1016.

Calvet N., Muzerolle J., Briceño C., Hernández J., Hartmann L., Saucedo J. L., and Gordon K. D. (2004) *Astron. J., 128,* 1294–1318.

Calvet N., D'Alessio P., Watson D. M., Franco-Hernández R., Furlan E., et al. (2005) *Astrophys. J., 630,* L185–L188.

Carr J. S. (1989) *Astrophys. J., 345,* 522–535.

Carr J. S., Tokunaga A. T., Najita J., Shu F. H., and Glassgold A. E. (1993) *Astrophys. J., 411,* L37–L40.

Carr J. S., Tokunaga A. T., and Najita J. (2004) *Astrophys. J., 603,* 213–220.

Chandler C. J., Carlstrom J. E., Scoville N. Z., Dent W. R. F., and Geballe T. R. (1993) *Astrophys. J., 412,* L71–L74.

Chen C. H., Van Cleve J. E., Watson D. M., Houck J. R., Werner M. W., Stapelfeldt K. R., Fazio G. G., and Rieke G. H. (2004) *AAS Meeting Abstracts, 204,* 4106.

Clarke C. J., Gendrin A., and Sotomayor M. (2001) *Mon. Not. R. Astron. Soc., 328,* 485–491.

Colavita M., Akeson R., Wizinowich P., Shao M., Acton S., et al. (2003) *Astrophys. J., 592,* L83–L86.

D'Alessio P., Canto J., Calvet N., and Lizano S. (1998) *Astrophys. J., 500,* 411.

D'Alessio P., Calvet N., Hartmann L., Lizano S., and Cantoó J. (1999) *Astrophys. J., 527,* 893–909.

D'Alessio P., Hartmann L., Calvet N., Franco-Hernández R., Forrest W. J., et al. (2005) *Astrophys. J., 621,* 461–472.

Desch S. (2004) *Astrophys. J., 608,* 509.

Doppmann G. W., Greene T. P., Covey K. R., and Lada C. J. (2005) *Astrophys. J., 130,* 1145–1170.

Eisner J. A., Hillenbrand L. A., White R. J., Akeson R. L., and Sargent A. E. (2005) *Astrophys. J., 623,* 952–966.

Figuerêdo E., Blum R. D., Damineli A., and Conti P. S. (2002) *Astron. J., 124,* 2739–2748.

Fromang S., Terquem C., and Balbus S. A. (2002) *Mon. Not. R. Astron. Soc., 339,* 19.

Gammie C. F. (1996) *Astrophys. J., 457,* 355–362.

Geballe T. R. and Persson S. E. (1987) *Astrophys. J., 312,* 297–302.

Gibb E. L., Rettig T., Brittain S., and Haywood R. (2004) *Astrophys. J., 610,* L113–L116.

Gizis J. E., Shipman H. L., and Harvin J. A. (2005) *Astrophys. J., 630,* L89–L91.

Glassgold A. E. and Najita J. (2001) in *Young Stars Near Earth,* ASP Conf. Ser. vol. 244 (R. Jayawardhana and T. Greene, eds.) pp. 251–255. ASP, San Francisco.

Glassgold A. E., Najita J., and Igea J. (1997) *Astrophys. J., 480,* 344–350.

Glassgold A. E., Najita J., and Igea J. (2004) *Astrophys. J., 615,* 972–990.

Gorti U. and Hollenbach D. H. (2004) *Astrophys. J., 613,* 424–447.

Greene T. P. and Lada C. J. (1996) *Astron. J., 112,* 2184–2221.

Gullbring E., Hartmann L., Briceño C., and Calvet N. (1998) *Astrophys. J., 492,* 323–341.

Hanson M. M., Howarth I. D., and Conti P. S. (1997) *Astrophys. J., 489,* 698–718.

Hartigan P., Edwards S., and Ghandour L. (1995) *Astrophys. J., 452,* 736–768.

Hartmann L. and Kenyon S. (1996) *Ann. Rev. Astron. Astrophys., 34,* 207–240.

Hartmann L., Calvet N., Gullbring E., and D'Alessio P. (1998) *Astrophys. J., 495,* 385–400.

Hartmann L., Hinkle K., and Calvet N. (2004) *Astrophys. J., 609,* 906–916.

Herbig G. H. and Goodrich R. W. (1986) *Astrophys. J., 309,* 294–305.

Herczeg G. J., Linsky J. L., Valenti J. A., Johns-Krull C. M., and Wood B. E. (2002) *Astrophys. J., 572,* 310–325.

Herczeg G. J., Wood B. E., Linsky J. L., Valenti J. A., and Johns-Krull C. M. (2004) *Astrophys. J., 607,* 369–383.

Herczeg G. J., Walter F. M., Linsky J. L., Gahm G. F., Ardila D. R., et al. (2005) *Astron. J., 129,* 2777–2791.

Herczeg G. J., Linsky J. L., Walter F. M., Gahm G. F., and Johns-Krull C. M. (2006) *Astrophys. J. Suppl., 165,* 256–282.

Ida S. and Lin D. N. C. (2004) *Astrophys. J., 604,* 388–413.

Igea J. and Glassgold A. E. (1999) *Astrophys. J., 518,* 848–858.

Ilgner M. and Nelson R. P. (2006a) *Astron. Astrophys., 445,* 205–222.

Ilgner M. and Nelson R. P. (2006b) *Astron. Astrophys., 445,* 223–232.

Ilgner M., Henning Th., Markwick A. J., and Millar T. J. (2004) *Astron. Astrophys., 415,* 643–659.

Inutsuka S. and Sano T. (2005) *Astrophys. J., 628,* L155–L158.

Ishii M., Nagata T., Sato S., Yao Y., Jiang Z., and Nakaya H. (2001) *Astron. Astrophys., 121,* 3191–3206.

Jonkheid B., Faas F. G. A., van Zadelhoff G.-J., and van Dishoeck E. F. (2004) *Astron. Astrophys., 428,* 511–521.

Johns-Krull C. M., Valenti J. A., and Linsky J. L. (2000) *Astrophys. J., 539,* 815–833.

Kamp I. and Dullemond C. P. (2004) *Astrophys. J., 615,* 991–999.

Kamp I. and van Zadelhoff G.-J. (2001) *Astron. Astrophys., 373,* 641–656.

Kenyon S. J. and Hartmann L. (1995) *Astrophys. J. Suppl., 101,* 117–171.

Klahr H. H. and Bodenheimer P. (2003) *Astrophys. J., 582,* 869–892.

Kominami J. and Ida S. (2002) *Icarus, 157,* 43–56.

Kunz M. W. and Balbus S. A. (2004) *Mon. Not. R. Astron. Soc., 348,* 355–360.

Lahuis F., van Dishoeck E. F., Boogert A. C. A., Pontoppidan K. M., Blake G. A., et al. (2006) *Astrophys. J., 636,* L145–L148.

Langer W. D., van Dishoeck E. F., Bergin E. A., Blake G. A., Tielens A. G. G. M., et al. (2000) In *Protostars and Planets IV* (V. Mannings et al., eds.), pp. 29–57. Univ. of Arizona, Tucson.

Le Teuff Y., Markwick A., and Millar T. (2000) *Astron. Astrophys., 146,* 157–168.

Ida S. and Lin D. N. C. (2004) *Astrophys. J., 604,* 388–413.

Lin D. N. C., Bodenheimer P., and Richardson D. C. (1996) *Nature, 380,* 606–607.

Lubow S. H., Seibert M., and Artymowicz P. (1999) *Astrophys. J., 526,* 1001–1012.

Luhman K. L., Rieke G. H., Lada C. J., and Lada E. A. (1998) *Astrophys. J., 508,* 347–369.

Lynden-Bell D. and Pringle J. E. (1974) *Mon. Not. R. Astron. Soc., 168,* 603–637.

Malbet F. and Bertout C. (1991) *Astrophys. J., 383,* 814–819.

Markwick A. J., Ilgner M., Millar T. J., and Henning Th. (2002) *Astron. Astrophys., 385,* 632–646.

Marsh K. A. and Mahoney M. J. (1992) *Astrophys. J., 395,* L115–L118.

Martin S. C. (1997) *Astrophys. J., 478,* L33–L36.

Matsumura S. and Pudritz R. E. (2003) *Astrophys. J., 598,* 645–656.

Matsumura S. and Pudritz R. E. (2005) *Astrophys. J., 618,* L137–L140.

Mayer L., Quinn T., Wadsley J., and Stadel J. (2002) *Science, 298,* 1756–1759.

Miller K. A. and Stone J. M. (2000) *Astrophys. J., 534,* 398–419.

Muzerolle J., Hartmann L., and Calvet N. (1998) *Astron. J., 116,* 2965–2974.

Muzerolle J., Calvet N., Briceño C., Hartmann L., and Hillenbrand L. (2000) *Astrophys. J., 535,* L47–L50.

Muzerolle J., Calvet N., Hartmann L., and D'Alessio P. (2003) *Astrophys. J., 597,* L149–152.

Najita J. (2004) In *Star Formation in the Interstellar Medium* (D. Johnstone et al., eds.), pp. 271–277. ASP Conf. Series 323, San Francisco.

Najita J. (2006) In *A Decade of Extrasolar Planets Around Normal Stars* (M. Livio, ed.), STScI Symp. Series 19, Cambridge Univ., Cambridge, in press.

Najita J., Carr J. S., Glassgold A. E., Shu F. H., and Tokunaga A. T. (1996) *Astrophys. J., 462,* 919–936.

Najita J., Edwards S., Basri G., and Carr J. (2000) In *Protostars and Planets IV* (V. Mannings et al., eds.), pp. 457–483. Univ. of Arizona, Tucson.

Najita J., Carr J. S., and Mathieu R. D. (2003) *Astrophys. J., 589,* 931–952.

Nelson R. P., Papaloizou J. C. B., Masset F., and Kley W. (2000) *Mon. Not. R. Astron. Soc., 318,* 18–36.

Nomura H. and Millar T. J. (2005) *Astron. Astrophys., 438,* 923–938.

Prinn R. (1993) In *Protostars and Planets III* (E. Levy and J. Lunine, eds.), pp. 1005–1028. Univ. of Arizona, Tucson.

Quillen A. C., Blackman E. G., Frank A., and Varnière P. (2004) *Astrophys. J., 612,* L137–L140.

Rettig T. W., Haywood J., Simon T., Brittain S. D., and Gibb E. (2004) *Astrophys. J., 616,* L163–L166.

Rice W. K. M., Wood K., Armitage P. J., Whitney B. A., and Bjorkman J. E. (2003) *Mon. Not. R. Astron. Soc., 342,* 79–85.

Richter M. J., Jaffe D. T., Blake G. A., and Lacy J. H. (2002) *Astrophys. J., 572,* L161–L164.

Sako S., Yamashita T., Kataza H., Miyata T., Okamoto Y. K., et al. (2005) *Astrophys. J., 620,* 347–354.

Sano T., Miyama S., Umebayashi T., and Nakano T. (2000) *Astrophys. J., 543,* 486–501.

Scoville N., Kleinmann S. G., Hall D. N. B., and Ridgway S. T. (1983) *Astrophys. J., 275,* 201–224.

Semenov D., Widebe D., and Henning Th. (2004) *Astron. Astrophys., 417,* 93–106.

Shakura N. I. and Sunyaev R. A. (1973) *Astron. Astrophys., 24,* 337–355.

Sheret I., Ramsay Howat S. K., and Dent W. R. F. (2003) *Mon. Not. R. Astron. Soc., 343,* L65–L68.

Shu F., Najita J., Ostriker E., Wilkin F., Ruden S., and Lizano S. (1994) *Astrophys. J., 429,* 781–796.

Sicilia-Aguilar A., Hartmann L. W., Hernández J., Briceño C., and Calvet N. (2005) *Astron. J., 130,* 188–209.

Siess L., Forestini M., and Bertout C. (1999) *Astron. Astrophys., 342,* 480–491.

Skrutskie M. F., Dutkevitch D., Strom S. E., Edwards S., Strom K. M., and Shure M. A. (1990) *Astron. J., 99,* 1187–1195.

Stone J. M., Gammie C. F., Balbus S. A., and Hawley J. F. (2000) In *Protostars and Planets IV* (V. Mannings et al., eds.), pp. 589–611. Univ. of Arizona, Tucson.

Takeuchi T. and Lin D. N. C. (2005) *Astrophys. J., 623,* 482–492.

Takeuchi T., Miyama S. M., and Lin D. N. C. (1996) *Astrophys. J., 460,* 832–847.

Thi W. F. and Bik A. (2005) *Astron. Astrophys., 438,* 557–570.

Thi W. F., van Dishoeck E. F., Blake G. A., van Zadelhoff G. J., Horn J., et al. (2001) *Astrophys. J., 561,* 1074–1094.

Thompson R. (1985) *Astrophys. J., 299,* L41–L44.

Trilling D. D., Lunine J. I., and Benz W. (2002) *Astron. Astrophys., 394,* 241–251.

Valenti J. A., Johns-Krull C. M., and Linsky J. L. (2000) *Astrophys. J. Suppl., 129,* 399–420.

Walter F. M., Herczeg G., Brown A., Ardila D. R., Gahm G. F., Johns-Krull C. M., Lissauer J. J., Simon M., and Valenti J. A. (2003) *Astron. J., 126,* 3076–3089.

Ward W. R. (1997) *Icarus, 126,* 261–281.

White R. J. and Hillenbrand L. A. (2005) *Astrophys. J., 621,* L65–L68.

Wiebe D., Semenov D., and Henning Th. (2003) *Astron. Astrophys., 399,* 197–210.

Wilkinson E., Harper G. M., Brown A., and Herczeg G. J. (2002) *Astron. J., 124,* 1077–1081.

Willacy K. and Langer W. D. (2000) *Astrophys. J., 544,* 903–920.

Willacy K., Klahr H. H., Millar T. J., and Henning Th. (1998) *Astron. Astrophys., 338,* 995–1005.

Zuckerman B., Forveille T., and Kastner J. H. (1995) *Nature, 373,* 494–496.

Multiwavelength Imaging of Young Stellar Object Disks: Toward an Understanding of Disk Structure and Dust Evolution

Alan M. Watson
Universidad Nacional Autónoma de México

Karl R. Stapelfeldt
Jet Propulsion Laboratory, California Institute of Technology

Kenneth Wood
University of St. Andrews

François Ménard
Laboratoire d'Astrophysique de Grenoble

We review recent progress in high-resolution imaging of scattered light from disks around young stellar objects. Many new disks have been discovered or imaged in scattered light, and improved instrumentation and observing techniques have led to better disk images at optical, near-infrared, and thermal-infrared wavelengths. Multiwavelength datasets are particularly valuable, as dust particle properties have wavelength dependencies. Modeling the changes in scattered-light images with wavelength gives direct information on the dust properties. This has now been done for several different disks. The results indicate that modest grain growth has taken place in some of these systems. Scattered-light images also provide useful constraints on the disk structure, especially when combined with long-wavelength SEDs. There are tentative suggestions in some disks that the dust may have begun to settle. The next few years should see this work extended to many more disks; this will clarify our understanding of the evolution of protoplanetary dust and disks.

1. INTRODUCTION

This chapter reviews the progress since the Protostars and Planets IV meeting (*McCaughrean et al.,* 2000) in observations and modeling of scattered-light images of disks around young stellar objects (YSOs). For many years these disks were studied in the infrared and at millimeter wavelengths, but without the spatial resolution necessary to reveal their detailed structure. With the advent of the Hubble Space Telescope (HST) and groundbased adaptive optics (AO) systems, these disks were seen in scattered optical and near-infrared starlight. The first images of disks in scattered light showed them as never before. These images currently provide the highest-spatial-resolution images of disks and provide unique information on disk structure and dust properties.

Most disks around young low- and intermediate-mass stars fall into one of two categories on the basis of their gross observational properties. YSO disks are optically thick at visible and near-infrared wavelengths, are rich in molecular gas, and are found around Class I and Class II systems. Debris disks are optically thin at optical and near-infrared wavelengths, have only trace quantities of gas, and are found around Class III and older systems. Our current understand-ing of the evolution of dust in disks around young low- and intermediate-mass stars is that these observational categories correspond to very distinct phases. In YSO disks our understanding is that gas and dust from molecular cores is processed through the disk and provides the raw material both for accretion and outflows. The dust is processed first in the dense molecular core and then in the disk itself. The dust grows and suffers chemical processing. Dust growth is hypothesized to result in the production of planetesimals, which form rocky planets and the cores of gas giant planets. In debris disks, our understanding is that dust is present largely as a result of collisions between planetesimals. Thus, YSO disks are thought to be characterized by dust growth whereas debris disks are thought to characterized by planetesimal destruction.

In this chapter we focus on observations and modeling of scattered-light images of YSO disks around low- and intermediate-mass stars. That is, we focus on the phase in which dust is expected to grow. Of course, scattered-light observations and models are not the only means to explore these disks. The chapter by Dutrey et al. discusses millimeter and submillimeter observations of the gas and dust components and the chapters by Bouvier et al., Najita et al., and Millan-Gabet et al. cover observations of the inner disk region, no-

   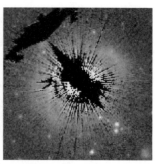

Fig. 1.   Scattered-light images of four face-on or intermediate inclination YSO disks newly resolved since 1998. Left to right: TW Hya face-on disk (*Krist et al., 2000*), HD 100546 (*Grady et al., 2001*), HD 141569A (*Clampin et al., 2003*), and HD 163296 (*Grady et al., 2000*).

tably interferometry, and their interpretations. The chapter by Monin et al. discusses aspects of disks that are peculiar to binaries. Further from the subject of our chapter, the chapter by Cesaroni et al. discusses disks around young high-mass stars and the chapter by Meyer et al. reviews debris disks around solar-type stars.

In the following sections we summarize the data available on scattered-light disks around YSOs, imaging techniques, and what we can learn about disk structure and dust properties from modeling multiwavelength images and SEDs. We also speculate about future advances.

2.   OBSERVATIONAL PROGRESS

At the time of the Protostars and Planets IV meeting, held in 1998, only a dozen YSO disks had been resolved in scattered light and another half dozen in silhouette against the Orion Nebula. Since then, the number of resolved disks in these two categories has doubled and quadrupled, respectively. The new discoveries stem largely from targeted high-contrast imaging of greater numbers of YSOs and widefield imaging surveys of star-forming clouds. In the optical and near-infrared, the broader application of established observing techniques, rather than any instrumental improvements, has driven the recent expansion in the number of disks discovered. At longer mid-infrared wavelengths, larger telescopes equipped with more potent mid-infrared imagers became available. These results are described in this section and images of some of the new disks are shown in Figs. 1 and 2. A complete list of published YSO disks resolved in scattered light appears in Table 1. Most of these are around T Tau stars, although some are around Herbig AeBe stars. An online catalog of circumstellar disks resolved in scattered light, thermal emission, or molecular lines is now available at *www.circumstellardisks.org*.

2.1.   Optical and Near-Infrared Imaging

2.1.1. Coronagraphic and direct imaging.   The most general case, and also the most observationally challenging, is the detection of disk scattered light in the presence of direct starlight. Subtraction of a fiducial reference star

is almost always necessary to reveal the disk, and a coronagraph is usually employed to suppress stellar diffraction. Various HST instruments have produced excellent images of disks around bright stars, while groundbased AO imaging of these disks is often hampered by the instability of the point-spread function.

The brightness of Herbig AeBe stars makes them excellent disk-imaging targets, and three have been found to show interesting circumstellar nebulosity. HD 100546 has a disk viewed from high latitudes that contains what appear to be two spiral arms (*Pantin et al., 2000; Grady et al., 2001; Augereau et al., 2001*). HD 163296 shows an inclined disk with a hint of a cleared central zone and bifurcation into upper and lower reflection nebulosities (*Grady et al., 2000*). The nebulosity around AB Aur shows a wealth of structure (*Grady et al., 1999; Fukagawa et al., 2004*), but it is unclear how much is associated with the disk and how much with a more extended envelope. *Fukagawa et al.* (2003) report a possible disk around HD 150193A.

The transitional disk HD 141569A has been the subject of a great deal of observational and theoretical work. The first images in the near-infrared by *Weinberger et al.* (1999) and *Augereau et al.* (1999) appeared to show a large radial clearing in the disk at a radius of 250 AU. Optical imaging by *Mouillet et al.* (2001) and *Clampin et al.* (2003) revealed an asymmetric spiral-like feature and showed that the cleared region was not completely empty. HD 141569A is in a multiple system. Dynamical studies suggest that stellar flybys or a recent periastron passage by the companions could be the origin of the observed spiral feature (*Augereau and Papaloizou, 2004; Ardila et al., 2005; Quillen et al., 2005*).

Perhaps the most significant new disk imaged around a directly visible star since Protostars and Planets IV is that of TW Hya (*Krist et al., 2000; Trilling et al., 2001; Weinberger et al., 2002*). This face-on disk is also the closest T Tauri star disk, at a distance of 56 pc, and has been well studied at millimeter wavelengths. The disk radial profile shows a sharp break in slope at a radius of 140 AU that is seen in multiple independent datasets. The physical origin of this sudden fading in the outer disk is unclear. Long-slit spectra of the TW Hya disk with HST STIS (*Roberge et al., 2005*) find that the disk has a roughly neutral color from 5000 Å to 8000 Å (see their Fig. 5), consistent with previous broadband imaging in the same spectral region (*Krist et al.,* 2000) and near-infrared (*Weinberger et al.,* 2002).

Despite the interesting results highlighted above, most YSO disks (those whose presence is inferred from infrared and submillimeter excess emission) remain undetectable in scattered light. There are probably multiple reasons for this fact. Disks with outer radii smaller than 0.5", the typical inner working angle cutoff for current coronagraphs and PSF subtraction techniques, would not extend far enough from their host star to be readily detected. This cannot be the whole explanation, however; the disk of MWC 480 has a large outer radius in millimeter wave CO emission (*Simon et al.,* 2000), but repeated attempts to detect it in scattered

light have failed (*Augereau et al.,* 2001; *Grady et al.,* 2005). There are other examples. Many disks may simply be intrinsically fainter than the ones that have been imaged in scattered light to date; they may be geometrically flatter (and thus tending toward self-shadowing), depleted of small particles, or have undergone chemical processing that has reduced their albedo. With the typical YSO disk currently undetectable against the direct light of its parent star, improved coronagraphic instrumentation will be needed to understand the diversity in disk scattering properties and their underlying causes.

2.1.2. Edge-on disks. Edge-on, optically thick disks naturally occult their central stars and in the process present their vertical structure to direct view. Observations of edge-

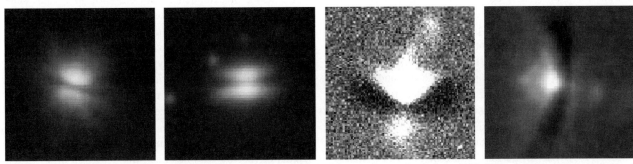

Fig. 2. Scattered-light images of four edge-on or silhouette YSO disks newly resolved since 1998. Left to right: Edge-on disks of HV Tau C (*Stapelfeldt et al.,* 2003), 2MASSI1628137-243139 (*Grosso et al.,* 2003), IRAS 04158+2805 (Ménard et al., in preparation), and Orion 216-0939 (*Smith et al.,* 2005).

TABLE 1.   YSO disks imaged in scattered light.

Object Name	Type	Outer Radius	Recent Reference
TW Hya	face-on	220 AU	*Roberge et al.* (2005)
HD 141569A	transition	370 AU	*Clampin et al.* (2003)
HD 100546	spiral	360 AU	*Grady et al.* (2005)
HD 163296		450 AU	*Grady et al.* (2005)
HD 150193A		190 AU	*Fukagawa et al.* (2003)
AB Aur	disk + envelope	>300 AU	*Fukagawa et al.* (2004)
CB 26	edge-on	380 AU	*Stecklum et al.* (2004)
CRBR 2422.8-3423	edge-on	105 AU	*Pontoppidan et al.* (2005)
DG Tau B	edge-on	270 AU	*Padgett et al.* (1999)
GG Tau	CB ring	260 AU	*Krist et al.* (2005)
GM Aur		500 AU	*Schneider et al.* (2003)
HH 30	edge-on	225 AU	*Watson and Stapelfeldt* (2004)
HK Tau B	edge-on	105 AU	*McCabe et al.* (2003)
HV Tau C	edge-on	85 AU	*Stapelfeldt et al.* (2003)
Haro 6-5B	edge-on	280 AU	*Padgett et al.* (1999)
IRAS 04302+2247	edge-on	420 AU	*Wolf et al.* (2003)
UY Aurigae	CB ring	2100 AU	*Potter et al.* (2000)
2MASSI J1628137-243139	edge-on	300 AU	*Grosso et al.* (2003)
IRAS 04325+2402	edge-on	30 AU	*Hartmann et al.* (1999)
OphE-MM3	edge-on	105 AU	*Brandner et al.* (2000a)
ASR 41	shadow	<3100 AU	*Hodapp et al.* (2004)
LkHα 263 C	edge-on	150 AU	*Jayawardhana et al.* (2002)
Orion 114-426	edge-on	620 AU	*Shuping et al.* (2003)
Orion 216-0939	edge-on	600 AU	*Smith et al.* (2005)
PDS 144N	edge-on	400 AU	*Perrin et al.* (2006)

on disks require high spatial resolution but not high contrast, so AO systems are competitive. Furthermore, the absence of stellar PSF artifacts make edge-on systems particularly amenable to scattered light modeling. Two of the first known examples, HH 30 and HK Tau B, have been extensively studied over the past few years, and are discussed at greater length in sections 2.4, 5.1, and 5.2 below. Additional examples are very valuable for comparative studies of disk scale heights, flaring, and dust properties. Finding new edge-on disks and imaging them across a wide range of wavelengths are high priorities for future research.

The edge-on disk CRBR 2422.8 was discovered in the ρ Oph cloud core by *Brandner et al.* (2000a). A model for the source combining scattered light images and mid-infrared spectra is presented by *Pontoppidan et al.* (2005). Near ρ Oph lies another new edge-on disk, 2MASSI1628137-243139. *Grosso et al.* (2003) discovered and present models for this object and note a peculiar near-IR color difference between the two lobes of its reflection nebula. Three new edge-on disks have been found via AO as companions to brighter stars. *Jayawardhana et al.* (2002) and *Chauvin et al.* (2002) discovered LkHα 263C in a quadruple system, and *Jayawardhana et al.* (2002) present initial models suggesting a disk mass of 0.002 $M_\odot$. PDS 144N is a compact edge-on disk around an Ae star in a binary system (*Perrin et al.*, 2006). The mysterious nature of HV Tau C was finally resolved by *Monin and Bouvier* (2000) to be an edge-on disk around the tertiary star. Modeling of HST images of HV Tau C by *Stapelfeldt et al.* (2003) indicates that this disk also possesses a small circumstellar envelope. Finally, a new kind of edge-on disk has been identified around the Perseus source ASR 41: An extended shadow from a (presumably) compact disk is cast across foreground cloud material (*Hodapp et al.*, 2004).

2.1.3. Silhouette disks. The Orion Nebula continues to be a unique and fertile ground for finding new resolved disks. Several hundred compact ionized globules ("proplyds") likely contain circumstellar disks, but have a morphology dominated by photoevaporative processing in their HII region environment. Some of these contain internal silhouettes clearly reminiscent of disks; there is also a distinct category of pure silhouette disks lacking any external ionization. In both cases, the disk silhouettes are visible in Hα images as foreground absorption to the HII region. New silhouette sources are reported in a comprehensive paper by *Bally et al.* (2000) and by *Smith et al.* (2005). A particularly interesting new source is Orion 216-0939; like Orion 114-426 (*McCaughrean et al.*, 1998; *Shuping et al.*, 2003), this is a giant edge-on silhoutte more than 1000 AU in diameter, and with bipolar reflection nebulae.

Despite the discovery of proplyds in M8 (*Stecklum et al.*, 1998) and NGC 3603 (*Brandner et al.*, 2000b) and despite HST imaging of NGC 2024, NGC 2264, M16, M17, and Carinae, no silhouette disks have been identified in HII regions other than Orion. Ménard et al. (in preparation) found the first example of a silhouette disk in nearby Taurus clouds: IRAS 04158+2805 shows a cone of scattered light, a jet, and a silhouette 3000 AU in diameter projected in front of diffuse Hα emission near V892 Tau.

2.2. Mid-Infrared Imaging

The usual objective when imaging YSOs in the mid-infrared is to resolve extended thermal emission from their inner disks. At the distances of the nearest star-forming clouds, extended 10–20-μm thermal emission has only been detected around the luminous Ae stars HD 100546, AB Aur, and V892 Tau (*Liu et al.*, 2003, 2005; *Chen and Jura*, 2003; *Pantin et al.*, 2005). In lower-luminosity T Tauri stars, the 10–20 μm emission is spatially unresolved — even in the relatively nearby case of TW Hya (*Weinberger et al.*, 2002). The absence of extended 20-μm emission in T Tauri stars may pose a problem for models of flared-disk spectral energy distributions (*Chiang and Goldreich*, 1997). These models postulate a superheated disk upper layer where stellar radiation is predominantly absorbed by small dust particles, which then radiate inefficiently at longer wavelengths. Large grains or flatter disks may be needed to explain the fact that most of these sources are unresolved in the mid-infrared.

One of the more significant and surprising disk imaging results of recent years has been the detection of *scattered light* from three YSO disks in the mid-infrared. *McCabe et al.* (2003) found that the edge-on disk of HK Tau B appears as an extended source in sensitive 10-μm Keck images. The good alignment of this 10-μm nebulosity with the optical scattered light (*Stapelfeldt et al.*, 1998), its extent well beyond a reasonable diameter for disk thermal emission, and its monotonically declining flux density from 2 to 10 μm argue that this is scattered light. It is not clear if the original source of emission is the star or the inner part of the disk. The fact that some edge-on disks might be seen entirely via scattered light in the mid-infrared was already indicated by Infrared Space Observatory photometry of HH 30 (*Wood et al.*, 2002); HK Tau B is the first resolved example. A second is in the case of GG Tau, where *Duchêne et al.* (2004) clearly detect the circumbinary ring in deep 3.8-μm images. By the same arguments as above, this must also be scattered light. A third source, PDS 144N (*Perrin et al.*, 2006), is an Ae star that appears as a spectacular bipolar nebula in 10-μm images; the relative contributions of scattered light and PAH emission in this image are still being assessed. As discussed in section 5.3 below, scattered light at these long wavelengths is a powerful diagnostic of large grains in circumstellar disks. Additional examples of resolved mid-IR scattered light from YSO disks can be expected in the future.

2.3. Polarimetric Imaging

Imaging polarimetry can confirm the presence of scattered light nebulosity and offer clues to the location of embedded illuminating sources. The strength of the observed polarization depends on the dust grain properties, the scattering geometry, the degree of multiple scattering, and the polarization induced by any foreground cloud material. Spatially resolved polarimetry has been reported for only a few disks. The best example is GG Tau, where the geometry of the circumbinary ring is very well understood, and thus the polarimetric results can be readily interpreted. *Silber et al.* (2000) found that backscattered light from the far side of the ring was very highly polarized at 1 µm, requiring that the scattered light originate from submicrometer-sized grains.

A new application for polarimetry has emerged in high-contrast AO. Groundbased AO systems provide a diffraction-limited image core, but also an extended, uncontrolled seeing halo. The light from this halo can overwhelm the faint nebulosity of a circumstellar disk. Differential polarimetry exploits the fact that scattered light from YSO disks is highly polarized while the seeing halo has virtually zero polarization. By simultaneously imaging in two polarizations, the unpolarized halo can be removed, and the polarized disk light more clearly seen. *Potter* (2003) and *Perrin et al.* (2004) report detections of several YSO nebulosities with this technique; most are circumstellar envelopes. A few disks have also been studied with this technique, notably TW Hya (*Apai et al.,* 2004) and LkCa 15 (*Potter,* 2003). The latter has not been detected in several conventional unpolarized imaging searches.

2.4. Variability

T Tauri stars commonly show photometric variability on timescales of a few days to months. In the youngest stars, these variations are now thought to be the result of hot spots, variation in accretion rate, and occultation by warps in the inner disk, all of which are natural consequences of the magnetospheric accretion mechanism discussed in the chapter by Bouvier et al. Disks also show variability, and in this section we describe the best-studied cases, HH 30 and AA Tau.

2.4.1. Photometric variability in HH 30. Similarly to their central stars, disks show photometric variability. The scattered light from the disk should follow as the star brightens and fades. The integrated magnitude of HH 30 varies on timescales of a few days over a range of more than one magnitude in both V and I (*Wood et al.,* 2000). The range is slightly greater in V than in I. Early photometry suggested that the variability was periodic, but subsequent studies have not confirmed this.

The range of variability of the disk is likely to be a lower limit on the range of variability of the star, as the multiple optical paths taken by scattered light will likely act to smooth the stellar variations to some degree, both because of the finite speed of light (173 AU per day) and because the disk is illuminated by light from a range of stellar azimuths. Cool spots produce stellar variability with ranges of no more than one magnitude (*Herbst et al.,* 1994), and so cannot explain the variability of the disk. Some other mechanism must be at work in HH 30, perhaps hot spots, varia-

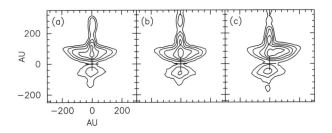

Fig. 3. HST/WFPC2 images showing the morphological variability of HH 30 (Watson and Stapelfeldt, in preparation). The images are from **(a)** 2001 February, **(b)** 1995 January, and **(c)** 1998 March. All images were taken in a broad filter centered at 675 nm. The images show the left side brighter, both sides having similar brightnesses, and the right side brighter.

tions in accretion rate, or occultations. This is consistent with the strong veiling component observed by *White and Hillenbrand* (2004).

2.4.2. Morphological variability in HH 30. More interesting and unexpected are the quantitative changes in the morphology of the scattered light (*Burrows et al.,* 1996; *Stapelfeldt et al.,* 1999; *Cotera et al.,* 2001; Watson and Stapelfeldt, in preparation). These include changes in the contrast between the brighter and fainter nebulae over a range of more than one magnitude and changes in the lateral contrast between the two sides of the brighter nebula over a range of more than one magnitude (see Fig. 3).

The timescales for the variability are uncertain, but are less than one year. Thus, the variability is not due to changes in the disk structure at radii of 100 AU but rather changes in the pattern of illumination at radii of 1 AU or less. Thus, this *morphological* variability is fascinating as it allows us to peek into the central part of the disk and possibly constrain the geometry of the accretion region.

However, before we use the variability to constrain anything, we need to understand its origin. Two mechanisms have been suggested; *Wood and Whitney* (1998) have suggested that inclined hot spots could illuminate the outer parts of disk like lighthouses, and *Stapelfeldt et al.* (1999) have suggested that warps in the inner disk could cast shadows over the outer disk.

In the case of HH 30, one might hope to distinguish between the mechanisms on the basis of the color of the morphological variability. HST observations show no significant differences between V and I. Naively, one would expect illumination by hot spots to produce stronger variability in the blue, whereas optically thick shadowing should be neutral in color. However, spectroscopic studies show that the veiling component is often relatively flat in the red (*Basri and Batalha,* 1990; *White and Hillenbrand,* 2004), so the lack of color is inconclusive.

Determining the timescale might help to distinguish between the mechanisms, although if the putative warp in the inner disk is locked to the star as it appears to be in AA Tau

TABLE 2.   Publicly available scattering and thermal equilibrium codes.

Code	URL
HO-CHUNK	gemelli.colorado.edu/~bwhitney/codes/codes.html
MC3D	www.mpia-hd.mpg.de/FRINGE/SOFTWARE/mc3d/mc3d.html
Pinball	www.astrosmo.unam.mx/~a.watson/
RADMC and RADICAL	http://www.mpia-hd.mpg.de/homes/dullemon/radtrans/

(discussed below), this will also be inconclusive. HST observations show that the asymmetry can change from one side to the other in no more than about six months but do not significantly constrain a lower limit on the period or, indeed, demonstrate that the variability is periodic rather than stochastic. Since the net polarization vectors of the two sides of the bright nebula are not parallel (see Fig. 6 of *Whitney and Hartmann*, 1992), the asymmetry should produce a polarimetric signature. Ongoing polarimetric monitoring of HH 30 may help to determine the timescale of the asymmetry.

2.4.3. AA Tau. AA Tau seems to be a prototype for both mechanisms suggested to explain the morphological variability of HH 30, apparently possessing both inclined hot spots and occulting inner-disk warps (*Bouvier et al.*, 1999; *Ménard et al.*, 2003; *O'Sullivan et al.*, 2005), both the result of an inclined stellar magnetic dipole. The disk around AA Tau has recently been imaged in scattered light by Grady et al. (in preparation), although previous observations with similar sensitivities did not detect the disk. It seems that the disk was finally detected because it was observed at an epoch in which the inner-disk warp was at least partially occulting the star.

2.4.4. Other disks. It would be useful to be able to compare the variability in HH 30 to other disks, to understand what is unusual and what is common. In this respect, recent second epoch observations of several objects by *Cotera et al.* (in preparation) are very useful. Nevertheless, since most disks have been observed at only one or two epochs, there is little that we can say about their variability.

3. MODELING TECHNIQUES

3.1. Radiation Transfer

There are several techniques for simulating scattered light images of disks and envelopes, including the single-scattering approximation (e.g., *Dent*, 1988; *Burrows et al.*, 1996; *D'Alessio et al.*, 1999) and direct integration of the equation of radiation transfer under the assumption of isotropic scattering (e.g., *Dullemond and Dominik*, 2004). However, by far the most common techniques are Monte Carlo simulations or integrations (e.g., *Lefèvre et al.*, 1982, 1983; *Bastien and Ménard*, 1988, 1990; *Whitney and Hartmann*, 1992, 1993; *Lopez et al.*, 1995; *Burrows et al.*, 1996; *Whitney et al.*, 1997, 2003a,b, 2004; *Lucas and Roche*,

1997, 1998; *Stapelfeldt et al.*, 1999; *Wood et al.*, 1998, 1999, 2001; *Lucy*, 1999; *Wolf et al.*, 1999, 2002, 2003; *Bjorkman and Wood*, 2001, *Cotera et al.*, 2001; *Watson and Henney*, 2001; *Schneider et al.*, 2003; *Stamatellos and Whitworth*, 2003, 2005; *Watson and Stapelfeldt*, 2004). Faster computers and improved algorithms allow these simulations to be fast and incorporate anisotropic scattering, polarization, and fully three-dimensional circumstellar geometries and illuminations.

For optical and near-infrared simulations one can normally assume that scattered *starlight* dominates the images and there is no contribution from dust reprocessing. With this assumption, an image at a specific wavelength and orientation can be calculated in a few minutes on current computers. Simulations at longer wavelengths must include thermal reprocessing and calculate thermal equilibrium (e.g., *D'Alessio et al.*, 1998, 1999; *Whitney et al.*, 2003a,b).

Some groups have made their scattering and thermal equilibrium codes publicly available (see Table 2), and these tools are now being used by the community.

3.2. Density Distributions

Modelers take different approaches to the density distribution in disks. One extreme is to assume disks are vertically isothermal, the dust is well mixed with the gas, and the surface density and scale heights are power laws in the radius. Another extreme is to solve self-consistently for the temperature using thermal equilibrium, to solve for the vertical density distribution using pressure equilibrium, to solve for the surface density assuming an accretion mechanism with a constant mass-transfer rate, and to include dust settling. In between are many intermediate approaches, for example, solving for thermal equilibrium and vertical pressure equilibrium but imposing a surface-density law.

The approaches are complementary. The simple power-law disks have very little physics, but have many "knobs" that can be arbitrarily adjusted to represent a wide range of disk-density distributions and thereby cover the very real uncertainties in our understanding of these objects. On the other hand, the approaches that incorporate more and more physics have fewer and fewer "knobs." In one sense, they are more realistic, but only to the extent that our understanding of disk physics is correct. Unfortunately, there are real gaps in our knowledge. For example, thermal equilibrium depends on the dust opacity, which is not well known; the

TABLE 3. Parameters constrained by scattered-light images.

Inclination	Parameters
Edge-on	inclination, mass-opacity, forward scattering, and scale height
Intermediate	inclination, mass-opacity, forward scattering, and outer radius
Face-on	inclination, radial dependence of scale height

details of accretion are still actively being researched, with disk viscosities uncertain by orders of magnitude; and disks may not have a constant inward mass-transfer rate.

Both approaches are useful, but is important to understand the strengths and weaknesses of each. Simple parameterized models allow one to investigate the dependence of other properties on the density distribution (e.g., *Chauvin et al.*, 2002; *Watson and Stapelfeldt*, 2004), but are limited in what they can tell us about disk physics. On the other hand, models with more physics are necessary to test and advance our understanding of disks (e.g., *D'Alessio et al.*, 1998, 1999; *Schneider et al.*, 2003; *Calvet et al.*, 2005), but our knowledge of the input physics in these models is still incomplete. For these reasons, future modeling efforts will continue to tailor their approaches to modeling the density distribution according to the specific problem being addressed.

3.3. Fitting

Determining how well scattered-light images can be modeled is important for testing physical models of flared disks and collapsing envelopes. Many studies have been successful at reproducing the overall morphology and intensity pattern of scattered-light images (e.g., *Lucas and Roche*, 1997; *Whitney et al.*, 1997; *Wood et al.*, 2001). However, quantitative model fitting is more convincing and provides a better test of physical models. *Burrows et al.* (1996) and *Stapelfeldt et al.* (1998) applied least-squares fitting techniques to single-scattering models of the disks of HH 30 and HK Tau B and determined their physical structure and scattering properties. More recently, *Watson and Stapelfeldt* (2004) applied least-squares fitting techniques to multiple scattering models of the disk of HH 30 and allowed the density structure and dust-scattering properties to be free parameters. Their results provide strong constraints on the circumstellar density, structure, and dust properties (see section 5.2). Glauser et al. (in preparation) have recently fitted models to scattered-light images, polarization images, and the SED of IRAS 04158+2805.

4. DISK STRUCTURE FROM SCATTERED-LIGHT IMAGES

Interpreting scattered-light images of optically thick disks is an inverse problem. The physics that takes us from athree-dimensional distribution of emissivity and opacity to a two-dimensional distribution of surface brightness removes a great deal of information. In this section we discuss what information is lost and what can be recovered. The orientation from which a disk is viewed is a critical factor in this. For this reason, we classify disks as edge-on, intermediate-inclination, and face-on. The parameters that can be most directly constrained in each case are summarized in Table 3.

4.1. Edge-On Disks

Opaque material close to the equatorial plane of edge on optically thick disks occults the star; all that is seen at optical and near-infrared wavelengths are two nebulae formed by light scattered by material away from the equatorial plane. These nebulae tend to be dominated by material in the segment of the disk closest to the observer. The information present in the images can be summarized as follows:

1. *The brightness ratio of the nebulae.* This is largely sensitive to the inclination of the disk with respect to the observer. When a symmetrical star-disk system is observed in its equatorial plane, the nebulae should have equal brightness. As the disk is tilted with respect to the observer, the nebula that is closer to the line of sight to the star dominates (see the top row of Fig. 4).

2. *The minimum separation of the nebulae.* The minimum separation occurs on the projected axis of an axisymmetric disk. It is largely sensitive to the total mass-opacity product of the disk. As the mass opacity product of a disk is increased, the dark lane separating the two nebulae widens (see the second row of Fig. 4). A good example of this is the decreasing separation of the nebulae of HH 30 with increasing wavelength (e.g., Figs. 6 and 8 of *Cotera et al.*, 2001; Fig. 2 of *Watson and Stapelfeldt*, 2004). One cannot use the width of the dark lane to constrain the mass or the absolute opacity individually without additional information. This is because the appearance of the disk at optical and near-infrared wavelengths is a scattering problem, and in the equations governing such problems the mass density and opacity per unit mass always appear as a product. On the other hand, one can use changes in the width of the lane to quantify relative changes in the opacity as a function of wavelength (see section 5.2).

3. *The increase in separation of the nebulae with increasing projected distance from the star.* By this we refer to the degree of curvature of the boundary of the dark lane separating the two nebulae or the degree of apparent flaring of the nebulae. This is largely sensitive to the effective scale

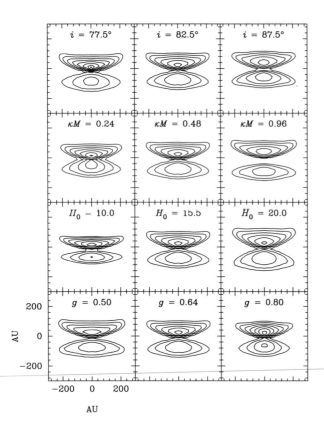

Fig. 4. Scattered-light models of an edge-on disk. The center column shows model A1 of *Burrows et al.* (1996). The left and right columns show models with, from top to bottom, different inclinations from face-on, different mass-opacity products κM (in cm^2 g^{-1} M$_\odot$), scale-height normalizations H_0 (in AU), and phase function asymmetry parameters g. The contours have the same level in each panel and are spaced by factors of 2. See Burrows et al. for precise definitions of the density distribution and parameters.

height of dust in the outer part of the disk (say, at the radii that dominate the observed scattered light). Larger scale heights lead to greater curvature or flaring (see the third row of Fig. 4). Good examples of this are the relatively flared nebulae of HH 30, which has a scale height of 16–18 AU at 100 AU, and the very flat nebulae of HK Tau B, which has a scale height of about 4 AU at 50 AU (see section 5.1).

4. *The vertical extent of each individual nebula.* By this we mean the apparent height of each nebula in comparison to its apparent diameter. Again, this is largely sensitive to the effective scale height of dust in the outer part of the disk. Larger scale heights lead to more extended nebulae (see the third row of Fig. 4). Good examples of this are again the vertically extended nebulae of HH 30 and the narrow nebulae of HK Tau B.

5. *The degree to which the surface brightness of the nebulae drops with increasing projected distance from the star.* This is largely sensitive to the degree to which forward scattering dominates the phase function. Enhanced forward scattering produces a more centrally concentrated distribution and a more pronounced decrease in surface brightness with

increasing projected distance (see the fourth row of Fig. 4). The light that emerges close to the projected disk axis has scattered through relatively small angles and the light that emerges far from the projected disk axis has scattered through relatively large angles. Increased forward scattering enhances the former and diminishes the latter. A good example of this is the better fits for HH 30 obtained by *Watson and Stapelfeldt* (2004) with high values of the phase function asymmetry parameter g.

However, much information is lost or ambiguous:

1. *The luminosity of the star.* All the light we see is processed through the disk, either by scattering or by absorption and thermal emission.

2. *The inner region of the disk.* The observed disks appear to be sufficiently flared that the light scattered from the outer parts comes from a relatively small range of angles and passes either completely above the inner disk or through regions of the inner disk with similar extinctions. In these cases, one could remove the inner 30 AU of the disk and the observed pattern of scattered light would not change, although the total brightness of scattered light might. An exception to this general rule may be HH 30, in which is it is possible that parts of the inner disk shadow parts of the outer disk (see section 2.4). Another exception may be disks in which the inner part of the disk shadows the outer part completely.

3. *The outer radius.* Scattered-light images of edge-on disks sometimes show a relatively sharp outer cutoff and this is often interpreted as the outer radius of the disk. However, it is possible that the disk continues beyond the bright nebulae, but that it is shadowed. Such an extension would likely not be seen in scattered light. There is evidence that this is the case in the sihouette disk Orion 114-426, where the scattered light nebulae do not extend to the edge of the sihouette (see Fig. 2d of *McCaughrean et al.,* 1998). An extended disk like this might also be detectable in CO.

4. *The radial dependence of the surface density and effective scale height.* Neither of these have strong effects on the models, at least as long as they are constrained to lie within plausible ranges. Worse, even their subtle effects are degenerate; putting more mass at larger radii produces similar effects in the images to increasing the scale height at larger radii (*Burrows et al., 1996; Watson and Stapelfeldt, 2004*).

It is clear from this that scattered-light images of edge-on optically-thick disks provide limited but important information on disk properties. Those that can be most cleanly separated are the inclination, the relative opacities at different wavelengths, the scale height in the outer part of the disk, and the degree of forward scattering in the phase function.

4.2. Intermediate-Inclination Disks

Intermediate-inclination disks are those in which the star is directly visible but both nebulae are still present (although one may be lost in the noise). In contrast to edge-on disks,

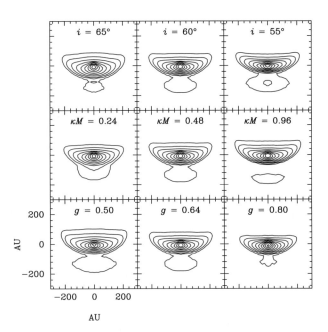

Fig. 5. Scattered-light models of an intermediate-inclination disk. The center column shows model A1 of *Burrows et al.* (1996) inclined to 60° from face-on. The left and right columns show models with, from top to bottom, different inclinations from face-on, different mass-opacity products κM (in cm² g⁻¹ M_⊙), and phase function asymmetry parameters g. The contours have the same level in each panel and are spaced by factors of 2. See Burrows et al. for precise definitions of the density distribution and parameters.

in principle one nebulae extends over all azimuths from the star, although some azimuths may be lost in the glare of direct light from the star. Scattered-light images of intermediate-angle optically thick disks have not been as extensively studied as edge-on disks, but preliminary studies (*Schneider et al.*, 2003; *Quijano*, 2005) suggest that the following information is present:

1. *The brightness ratio of the nebulae.* This depends principally on the mass-opacity product and the inclination (see the top two rows of Fig. 5).

2. *The separation of the nebulae.* Like the brightness ratio, this depends principally on the mass-opacity product and the inclination (see the top two rows of Fig. 5).

3. *The degree to which the surface brightness drops with increasing projected distance from the star.* This depends principally on the degree to which forward scattering dominates the phase function (see the third row of Fig. 5).

4. *The ratio of scattered light to unscattered light.* That is, the relative brightness of the nebulae and the star. This depends on the scale height in the outer disk, the albedo, the degree to which forward scattering dominates the phase function, the inclination, and the mass-opacity product. (A complication here is that light scattered very close to the star is difficult to distinguish from direct light from the star.)

5. *The outer radius.* The presence of two nebulae in many systems suggests that there is not a cold, collapsed,

optically thick disk that extends significantly beyond the bright nebulae. This allows one to constrain the outer radius of the optically thin disk with some degree of confidence. For example, GM Aur shows two nebulae that suggest that the true outer radius of the optically thick disk is around 300 AU (*Schneider et al.*, 2003).

Again, much information is lost:

1. *The inner region of the disk.* In current images, information on the inner disk is lost under the PSF of the star, even when observing in polarized light (e.g., *Apai et al.*, 2004). Infrared interferometry, covered in the chapter by Millan-Gabet et al., can provide a great deal of information on the very innermost part of the disk at radii of less than 1 AU. Coronagraphic images from space or, perhaps, from extreme AO systems will be required to recover information on scales larger than those available to interferometers but smaller than those lost under current PSFs.

2. *The radial dependence of the surface density and effective scale height.* As with edge-on disks, neither of these have strong effects on the observed morphology.

Scattered-light images of intermediate-angle optically thick disks are thus more difficult to interpret than those of edge-on disks. The only parameter that seems to be cleanly separated is the outer radius of the disk. The mass-opacity product and the inclination are to a large degree degenerate (compare, for example, the first two rows of Fig. 5, which show similar changes in the brighter nebula), but if millimeter data or an SED are available to constrain the inclination, the mass-opacity product can be obtained from scattered-light images. The dust asymmetry parameter can be constrained if disk structure is known with adequate certainty.

4.3. Face-On Disks

Face-on disks are those in which departures from radial symmetry are dominated by inclination (combined with the phase function) rather than vertical structure (at least for axisymmetric star-disk systems). The only example we have of such a disk around a YSO is TW Hya (*Krist et al.*, 2000; *Trilling et al.*, 2001; *Weinberger et al.*, 2002; *Apai et al.*, 2004; *Roberge et al.*, 2005).

The information present in scattered-light images of face-on optically thick disks is:

1. *The ellipticity.* This is by definition determined by the inclination and phase function in axisymmetric star-disk systems. Departures from constant ellipticity can provide a fascinating insight into nonaxisymmetric illumination or disk structure. For example, *Roberge et al.* (2005) show evidence that the disk around TW Hya is significantly elliptical between 65 and 140 AU and essentially circular beyond this region; they suggest that this suggests that the disk is warped. (See the first row of Fig. 6.)

2. *The dependence of the surface brightness on radius.* This depends essentially on the radial dependence of the scale height. If the disk is well-mixed, this provides infor-

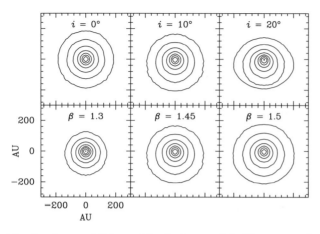

Fig. 6. Scattered-light models of a face-on disk. The center column shows model A1 of *Burrows et al.* (1996) inclined to 10° from face-on. The left and right columns show models with, from top to bottom, different inclinations from face-on and different indices β in the scale-height power law. The contours have the same level in each panel and are spaced by factors of 2. See Burrows et al. for precise definitions of the density distribution and parameters.

mation on the radial dependence of the temperature. (See the second row of Fig. 6.)

3. *The ratio of scattered light to unscattered light.* That is, the relative brightness of the nebula and the star. This depends principally on the scale height in the outer disk, the albedo, the phase function, and the mass-opacity product.

Information on the inner disk is not present for the same reasons as in the case of intermediate-inclination disks. Information on the outer disk radius is not present for the same reasons as in the case of edge-on disks.

5. HIGHLIGHTS OF APPLIED DISK MODELING

In this section, we present four examples of scientific results from the modeling of scattered-light disks around YSOs. Our first three are related to the evolution of dust, which is also discussed in the chapters by *Natta et al. and Dominik et al.* The fourth concerns evidence for an inner hole in the disk around GM Aur, which may have been cleared by a planet (see the chapter by Papaloizou et al.). While many disks have been imaged, the cases discussed below are perhaps the best examples of how disk structure and dust properties have been derived from modeling of multiwavelength image datasets.

5.1. Evidence for Dust Settling

The gas scale height in disks is determined by the balance between the stellar gravitational force, which compresses the disk toward its midplane, and gas pressure, which acts to puff the disk up. In vertical hydrostatic equilibrium in a vertically isothermal disk, the gas scale height

is related to the local *gas temperature* according to

$$H(r) = \sqrt{kT(r)r^3/GM_*m}$$

where m is the mean molecular weight of the disk gas. Direct determination of the disk temperature is difficult, but reasonable results can be obtained by modeling the infrared and millimeter spectrum. A complication is that disks likely have vertical temperature inversions (*Calvet et al.,* 1991, 1992; *Chiang and Goldreich,* 1997; *D'Alessio et al.,* 1998).

If the dust and gas are well mixed, then the dust follows the gas density distribution and will have the same scale height. As we have seen in section 4.1, the dust scale height can be determined from high-resolution scattered-light images. It is important to note that the height of the scattering surface above the disk plane is not equal to the dust scale height; rather, the scattering surface defines the locus of points with optical depth unity between the star and individual disk volume elements, and can be any number of dust scale heights above the midplane. The primary observational indicators of the scale height are the vertical extent of the nebula, the degree of curvature of the nebulae, and the sharpness of the dark lane. Disks with large scale heights appear more vertically extended ("fluffy") and have curved nebulae, whereas those with small scale heights are vertically narrow and have almost parallel nebulae. The exact value for the scale height must be determined by fitting models to the images.

The comparison of the scale heights for the gas and dust offers a unique opportunity to test the assumption that disks are vertically well mixed. *Burrows et al.* (1996) derived an equivalent temperature from the dust scale height for the HH 30 disk at 100 AU radius under the assumption that the disk was vertically well mixed. The result was broadly consistent with expectations from a simple thermal model, and suggested that the assumption that the disk was vertically well mixed was correct.

Since Protostars and Planets IV, dust scale heights have been derived for several other YSO disks. In two cases, HV Tau C and IRAS 04302+2247, initial scale height derivations implied unreasonably high equivalent temperatures and it is difficult to imagine why the dust would be more extended than the gas (*Stapelfeldt et al.,* 2003; *Wolf et al.,* 2003). These appear to be caused by the presence of circumstellar envelopes in addition to the disks, with the envelopes producing more diffuse nebulosity than expected for a pure disk. By adding an envelope to the density distribution, it was possible to remove this effect, and derive dust scale heights more consistent with simple thermal models. Scale height values for these and three other disks are shown in Table 4. The scale heights in the original references have been extrapolated to a reference radius of 50 AU to facilitate comparison.

The dust scale heights measured to date fall into two groups with the values differing by almost a factor of 2. The three objects with larger scale heights all have outflows

TABLE 4. Scale heights.

Object	H at 50 AU	$T_{equivalent}$	Reference
HH 30	6.3 AU	51 K	*Watson and Stapelfeldt* (2004)
IRAS 04302+2247	6.1 AU	48 K	*Wolf et al.* (2003)
HV Tau C	6.5 AU	35 K	*Stapelfeldt et al.* (2003)
GM Aur	3.4 AU	18 K	*Schneider et al.* (2003)
HK Tau B	3.8 AU	8 K	*Stapelfeldt et al.* (1998)

indicating ongoing accretion, whereas the two objects with smaller scale heights have little or no outflow activity. This could be observational evidence that accreting disks are systematically more puffed-up (warmer) than nonaccreting disks. The small equivalent temperature implied for the HK Tau B disk merits particular attention. Unfortunately, the infrared spectral energy distribution for this disk is incomplete, so this value cannot be compared to a well-constrained disk thermal model. However, it is highly unlikely that the gas in the disk could actually be as cold as 8 K. Instead, this could be a case where the assumption that the dust and gas are well mixed is not correct. Instead, it appears more likely that the dust has decoupled from the gas and partially settled toward the disk midplane, an expected stage of disk evolution (*Dubrulle et al.* 1995; *Dullemond and Dominik*, 2004). In this case, the equivalent temperature derived from the dust scale height would be lower than the true gas temperature.

The vertical scale height of dust in a YSO circumstellar disk could be a key indicator of its structure and evolutionary state. It would be very valuable to accumulate scale height measurements for larger numbers of edge-on disks, to see how unique the HK Tau B results are and uncover the full diversity of scale heights in the YSO disk population.

5.2. Evidence for Dust Growth from the Opacity Law

In optically thin media, the color of scattered light directly depends on the wavelength dependence of the grain opacity. Small grains like those in the ISM have a much higher scattering cross section at shorter wavelengths, and thus optically thin nebulae will have strongly blue colors relative to their illuminating star. However, YSO disks are optically thick in the optical and near-infrared. In this situation the color of the scattered light no longer depends on the grain opacity, and the disk will appear spectrally neutral relative to the star — even if small grains are the dominant scatterers. Neutral colors have been observed for the disks of TW Hya (*Krist et al.*, 2000; *Weinberger et al.*, 2002; *Roberge et al.*, 2005) and GM Aur (*Schneider et al.*, 2003), consistent with this expectation. A mechanism that can produce nonstellar colors in reflection from an optically thick disk is a wavelength-dependent dust albedo or phase function, but this is thought to be a small effect. Given these considerations, how can scattered light from an optically thick disk be used to constrain the grain properties?

The answer is to study changes in nebular spatial structure as a function of wavelength. For small ISM-like particles, the $\tau = 1$ scattering surface is located in lower-density regions above the disk midplane at optical wavelengths, and shifts into higher-density regions nearer the midplane at near-infrared wavelengths. Conversely, large grains acting as gray scatterers would produce a reflection nebula whose spatial structure would not vary with wavelength.

For edge-on disks, the key observable is a narrowing of the central dust lane as the object is imaged at progressively longer wavelengths (see the second row of Fig. 4). This behavior is seen in Orion 114-426 (*McCaughrean et al.*, 1998), IRAS 04302+2247 (*Padgett et al.*, 1999), HH 30 (*Cotera et al.*, 2001), and HV Tau C (*Stapelfeldt et al.*, 2003). It provides clear evidence that the scattering dust grains are dominated by small particles. But how small? A quantitative answer can be derived by fitting scattered light models to multiwavelength image datasets. This was first done in the case of HH 30 by *Cotera et al.* (2001). From modeling of HST near-infrared images, they found that the dust lane thickness changed less quickly with wavelength than expected for standard interstellar grains, and interpreted this as evidence for grain growth in the disk. *Watson and Stapelfeldt* (2004) extended this analysis by including optical images and by considering a wider range of possible density structures, but found essentially the same result: The most likely ratio of grain opacities between 0.45 and 2.0 μm is 2.0 for the HH 30 disk, vs. a value of 10 expected for ISM grain models.

While the disk of HH 30 appears to show some grain evolution, modeling of multiwavelength images of the edge-on disks of HV Tau C (*Stapelfeldt et al.*, 2003) and IRAS 04302+2247 (*Wolf et al.*, 2003) finds grain opacity ratios consistent with standard ISM grains. Both of these sources possess circumstellar envelopes in addition to disks, so the presence of primitive grains could reflect this ongoing infall from the ISM onto the disks. The disk of HK Tau B shows only subtle changes in its dust lane thickness between optical and near-IR images, and may represent a more evolved system; a firm conclusion on its dust properties can be expected from future model fitting. Dust properties in the giant edge-on silhouette disk Orion 114-426 are uncertain; *Throop et al.* (2001) found the radial extent of the silhouette to be achromatic between 0.66 and 1.87 μm; *Shuping et al.* (2003) found that it was chromatic between 1.87 and

4.05 μm; while *McCaughrean et al.* (1998) showed that the dust lane thickness between the lobes of reflected light was clearly chromatic between 1.1 and 2.0 μm. Additional modeling of this source is needed. Observations and modeling of a broader sample of edge-on disks offer an opportunity to probe the diversity of dust properties across the variables of disk age, disk enviroment, and accretion signatures, and should be vigorously pursued.

5.3. Evidence for Dust Stratification from the Phase Function

GG Tau is a binary T Tauri star with 0.3″ (42 AU) projected separation. It hosts the most prominent example of a circumbinary ring of dust and gas. The ring has been studied in the millimeter lines and continuum (*Guilloteau et al.,* 1999; *Wood et al.,* 1999), near-infrared scattered light (*Roddier et al.,* 1996; *Wood et al.,* 1999; *Silber et al.,* 2000; *McCabe et al.,* 2002), and optical scattered light (*Krist et al.,* 2002, 2005). Through these studies, the density structure of the GG Tau ring is perhaps now the best understood of all YSO disks. A key feature of the ring is its intermediate inclination of 37° from face-on. This spatially separates both the foreground and background parts of the ring from each other and from the central binary. This "clean" configuration allows the relative strength of forward scattering and backscattering to be directly measured. This quantity can be a powerful diagnostic of dust properties in the ring.

The scattering phase function strongly favors forward scattering when the grain size is comparable to the wavelength. At wavelengths much larger than the grain size, scattering becomes more isotropic. A comprehensive study of phase function effects in the GG Tau ring was recently carried out by *Duchêne et al.* (2004). Using new images taken at 3.8 μm, the longest wavelength to date at which scattered light from the ring has been detected, and existing images at shorter wavelengths, Duchêne et al. modeled the wavelength dependence of the phase function from 0.8 to 3.8 μm. Highlights from their results are shown in Fig. 7. The results show that the 3.8-μm scattered light must arise from dust grains larger than those in the ISM, whereas the 0.8-μm scattered light must simultaneously originate from much smaller particles. Additional evidence for small particles is provided by the 1.0-μm polarimetry of *Silber et al.* (2000), who found the backscattered light was highly polarized and thus dominated by submicrometer grains.

The GG Tau phase function results indicate that no single power-law distribution of grain sizes can simultaneously account for the observations. *Duchêne et al.* (2004) suggest that these results can be explained by a vertically stratified disk in which the grain size increases toward the midplane. In this view, large grains responsible for the 3.8-μm scattered light would be located in a denser region closer to the disk midplane, a region that shorter-wavelength photons cannot reach. The stratification could be due to dust settling to the ring midplane. Alternatively, it might be the

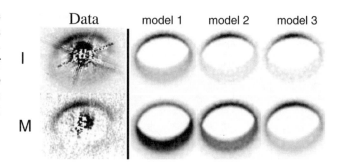

Fig. 7. A comparison of data and models for the GG Tau circumbinary ring, in the I band (0.8 μm; top row) and M band (3.8 μm; bottom row). The ring is inclined 37° from face-on, with the forward-scattering edge projected above the binary, and backscattering side projected below the binary. The key observable is the brightness contrast between the front and back sides. The first model has a maximum grain size of 0.3 μm. It matches the I band image well, but in the M band it predicts too much backscattering and not enough forward scattering. The second model has a maximum grain size of 0.9 μm. In this model, the backscattered flux becomes too small in the I band, and still too large in the M band. A reasonable match to the M band image is obtained in the third model (maximum grain size of 1.5 μm), but the I band backscattering is underpredicted. No single grain size distribution accounts for the phase function effects at both wavelengths. Figure and results from *Duchêne et al.* (2004).

case that the preference for scattering at grain sizes comparable to the wavelength is so strong that the large grains dominate the 3.8-μm scattered light, even though they are less numerous than the small grains. In that case, vertical dust settling would not necessarily have taken place.

This picture is very attractive. However, *Krist et al.* (2005) point out that current models of GG Tau do not fully explain the observations. For example, no current model simultaneously reproduces the total brightness of the disk, its color, and its azimuthal variation. Furthermore, the ratio of brightness of the near and far parts of the disk has been observed to vary with time. Additional observations and models will be required to confirm the suggestion of a stratified disk. It would be very valuable to perform a similar study in other disk systems; unfortunately, the GG Tau ring is currently unique.

5.4. Combining Scattered Light Images and Spectral Energy Distributions

GM Aur has a disk viewed at an intermediate inclination. The disk is clearly seen in WFPC2 and NICMOS images after careful subtraction of a reference PSF. Scattered light is detected between 0.4″ and 2.1″ (55 and 300 AU) from the star. Modeling of the scattered light images by *Schneider et al.* (2003) derived a dust vertical scale height of 8 AU at a radius of 100 AU and a disk inclination 56° from face-on, and demonstrated that there was also a remnant circumstellar envelope. However, the coronagraphic occulting spot used for the observations blocked any information on the properties of the inner disk. To access that region, Schneider et al. turned to the infrared spectral energy distribution (SED). Emission between 2 and 70 μm probes the disk temperature structure inside 40 AU, and thus is complementary to the results of the scattered-light modeling. Using the disk model derived from the scattered light images, and appropriate assumptions about grain properties, *Schneider et al. w*ere able to reproduce GM Aur's SED, including the millimeter continuum points (see Fig. 8). The very small near-IR excess emission requires that the inner part of the disk be optically thin. *Rice et al.* (2003) showed that the inner region may have been cleared by a jovian-mass planet.

More recently, *Calvet et al.* (2005) have presented Spitzer IRS spectroscopy of GM Aur. This data requires an inner optically thin region that extends to 24 AU from the star. This is much larger than previous estimates based on modeling of broadband mid-infrared photometry. The Spitzer IRS data show that the inner disk is not empty, but contains a small amount of small dust grains that produce the silicate

emission feature. To fit the long-wavelength SED, the average grain size in the outer disk must be larger than in the inner, optically thin region.

Scattered-light images do not generally provide information on the inner part of the disk. In addition to IRAS data, high-quality mid-infrared spectra from Spitzer are becoming available for many YSO disks. Scattered-light images reduce many of the degrees of freedom in modeling disk spectra and SEDs, leading directly to more robust SED models. Future combined studies of multiwavelength disk images, continuum spectra, and SEDs will provide critical tests for models of disk evolution.

6. PROSPECTS FOR FUTURE ADVANCES

6.1. Science Goals

The science goals for scattered-light imaging of disks around YSOs center on gaining a better understanding of disk structure, disk evolution, and dust evolution in the context of planet formation.

A key goal is to determine an evolutionary sequence for dust, especially dust growth and changes in shape and composition. One manifestation of dust evolution is to change the dust-opacity law. Work in HV Tau C and HH 30 has shown that the opacity law can be determined from HST images in the optical and near-infrared using current modeling codes. Extending this work is simply a matter of obtaining the relevant observations and "cranking the handle." Observations of thermal emission at longer wavelengths will help to restrict the degeneracies in the scattered-light modeling. We would expect to find disks with different opacity laws, some with near-ISM opacity laws like HV Tau C and others with less chromatic opacity laws like HH 30. The science will come from comparing the opacity laws with models for dust evolution and from correlating the opacity laws with system properties such as age, accretion activity, disk mass, and stellar multiplicity.

The evidence for dust settling discussed in section 5.1 is as yet tentative. To advance, we need to be able to confidently compare the equivalent temperatures derived from the dust scale heights to the real temperatures in the disk or, more generally, the vertical distributions of the dust and the gas. This will require detailed calculations of the structure of the disk that solve for the vertical structure under the assumption of thermal equilibrium but allow for uncertainties in the radial structure. Again, including mid-infrared and millimeter data will be very useful.

Of course, we are ultimately interested in detecting young planets in disks, and especially those planets that are still accreting. Direct detection will be difficult (see the chapters by Beuzit et al. and Beichman et al.), but we can hope to observe the gap cleared by a planet as it grows. At the moment, these are most clearly seen in infrared spectra, and the contribution of scattered-light images is mainly to reduce the number of degrees of freedom in the disk geometry and

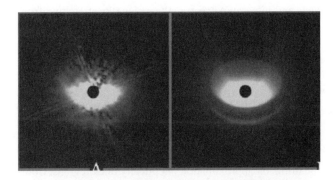

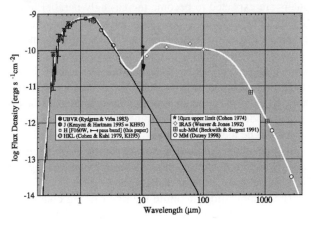

Fig. 8. Scattered light images and SEDs of the classical T Tauri star GM Aur (*Schneider et al.,* 2003). The top left panel shows the scattered-light image and the top right a model. The lower panel shows the SED. The dark line is a model for the stellar photosphere emission. The white line shows a model combining the stellar emission and the disk excess emission, with disk model parameters taken from scattered light results. For the photometry references, see *Schneider et al.* (2003).

inclination and thereby improve the reliability of the models for the infrared emission.

We would like to understand the transition between optically thick protoplanetary disks and optically thin debris disks in order to determine the timescale over which planetesimals and gas — the building blocks of planet formation — are present. Advances are being made in this field with Spitzer imaging and spectroscopy. Scattered-light imaging can contribute because small amounts of dust can potentially be detected even at radii at which infrared detections are difficult, although the realization of this goal will require advances in high-contrast imaging.

6.2.   Instrument Advances

The biggest instrumental advance for the study of YSO disks will be the advent of the Atacama Large Millimeter Array (ALMA). With spatial resolution surpassing that of HST, very-high-resolution spectroscopy for chemical and kinematic studies, and the sensitivity to study the disks of low-mass stars as far away as Orion, ALMA will have a major impact on the field. However, even in the era of ALMA, scattered-light imaging will still make significant contributions to disk studies. First, scattered light traces the surface where stellar photons deliver energy to the disk. It will still be necessary to characterize this interaction region if the disk temperature structure, and thus its chemical nature, is to be understood. Second, while millimeter wavelengths probe large particle sizes in the dust population, scattered-light imaging provides information on the small particles. Both are needed to provide a full picture of disk grain properties and their time evolution. The combination of millimeter maps and scattered-light images at comparable resolutions will be a powerful synthesis for disk science efforts.

The imaging performance of large groundbased telescopes can be expected to continue its evolution. Particularly important will be new extreme AO instruments at the Gemini and VLT observatories. Their improved contrast performance should enable additional detections of YSO disks in the near- and mid-infrared. The maturation and increasing application of differential imaging polarimetry should also yield exciting results. Several concepts for extremely large (D ~ 30 m) telescopes are now being studied. When realized (maybe not until after Protostars and Planets VI), these facilities will provide a threefold advance in spatial resolution. Higher-resolution images will improve our knowledge of all aspects of disk structure. The inner holes in systems such as TW Hya and perhaps GM Aur should be resolvable. An exciting possibility is the detection of hot young planets near and perhaps even within YSO disks and characterization of their dynamical interactions.

Among future space missions in design and development, two will provide important capabilities for disk scattered-light imaging. The NASA/ESA James Webb Space Telescope will have a 6.5-m primary mirror and operate from 1 to 28 μm. It will be a superb telescope for imaging disks in mid-infrared scattered light and will provide roughly 0.2" resolution. In the near-infrared, it will offer almost three times better resolution than HST, but is unlikely to provide improved contrast. From the point of view of high-contrast imaging, the Terrestrial Planet Finder Coronagraph mission would be extremely exciting, as it would be able to detect scattered light from disks as tenuous as our own solar system's zodiacal light. However, as this mission may be more than a decade away, several groups are proposing smaller coronagraphic space telescopes that might be realized sooner.

While waiting for these future developments, the expanded application of existing scattered-light imaging capabilities (AO and mid-IR imaging from large groundbased telescopes and high-contrast imaging with HST) should continue unabated.

6.3.   Modeling Advances

The new observations described in this review and the observational advances outlined above suggest that a wealth of detailed data on circumstellar disks will become available. What will this data demand from the codes and models?

Future codes will need to produce high-resolution images and integrated spectra at wavelengths stretching from the optical to the millimeter. They should be able to model three-dimensional distributions of sources and opacity and should incorporate accurate dust scattering phase functions, polarization, and the effects of aligned grains. Evidence for dust growth and sedimentation requires that codes no longer restrict themselves to homogeneous dust properties, but must be able to treat multiple dust species with different spatial distributions. They will probably solve for radiation transfer using Monte Carlo techniques in optically thin regions and the diffusion approximation in optically thick regions. Computers are expected to become increasingly parallel in the future. Many codes can already run in parallel, but those that cannot will need to be modified to do so. In this context, Monte Carlo algorithms have the advantage over classical algorithms as they often have natural parallelism. Many radiation transfer codes now have the capability to determine the density structure from given disk physics or incorporate density structures from dynamical simulations. These will be increasingly useful in providing detailed tests of disk structure. On the other hand, parameterized density models will continue to be a useful tool for mitigating our incomplete knowledge of disk physics.

The combination of future data and future codes will allow us to study dust properties, dust settling, disk structure, disk-planet interactions, accretion, and disk evolution. In the longer term, advances in techniques for numerical simulations coupled with increases in computing power and parallel processing make accurate radiation hydrodynamic simulations of disk formation and evolution a distant but realistic goal.

Perhaps the most important near-term work is to apply scattered-light modeling techniques across already extant disk image datasets. Observers have provided a significant number of new, high-quality images, but the corresponding modeling efforts have not kept pace.

Acknowledgments. We thank an anonymous referee for comments that helped to improve this chapter. This work was partially supported by the Centro de Radioastronomía y Astrofísica of the Universidad Nacional Autónoma de México, HST GO program 9424 funding to the Jet Propulsion Laboratory of the California Institute of Technology, and the Programme National de Physique Stellaire (PNPS) of CNRS/INSU, France.

REFERENCES

Apai D., Pascucci I., Brandner W., Henning Th., Lenzen R., et al. (2004) *Astron. Astrophys., 415,* 671–676.

Ardila D. R., Lubow S. H., Golimowski D. A., Krist J. E., Clampin M., et al. (2005) *Astrophys. J., 627,* 986–1000.

Augereau J.-C. and Papaloizou J. C. B. (2004) *Astron. Astrophys., 414,* 1153–1164.

Augereau J.-C., Lagrange A.-M., Mouillet D., and Ménard F. (1999) *Astron. Astrophys., 350,* L51–L54.

Augereau J.-C., Lagrange A.-M., Mouillet D., and Ménard F. (2001) *Astron. Astrophys., 365,* 78–89.

Bally J., O'Dell C. R., and McCaughrean M. J. (2000) *Astron. J., 119,* 2919–2959.

Basri G. and Batalha C. (1990) *Astrophys. J., 363,* 654–669.

Bastien P. and Ménard F. (1988) *Astrophys. J., 326,* 334–338.

Bastien P. and Ménard F. (1990) *Astrophys. J., 364,* 232–241.

Bjorkman J. E. and Wood K. (2001) *Astrophys. J., 554,* 615–623.

Bouvier J., Chelli A., Allain S., Carrasco L., Costero R., et al. (1999) *Astron. Astrophys., 349,* 619–635.

Brandner W., Sheppard S., Zinnecker H., Close L., Iwamuro F., et al. (2000a) *Astron. Astrophys., 364,* L13–L18.

Brandner W., Grebel E. K., Chu Y.-H., Dottori H., Brandl B., et al. (2000b) *Astron. J., 119,* 292–301.

Burrows C. J., Stapelfeldt K. R., Watson A. M., Krist J. E., et al. (1996) *Astrophys. J., 473,* 437–451.

Calvet N., Patino A., Magris G. C., and D'Alessio P. (1991) *Astrophys. J., 380,* 617–630.

Calvet N., Magris G. C., Patino A., and D'Alessio P. (1992) *Rev. Mex. Astron. Astrofís., 24,* 27–42.

Calvet N., D'Alessio P., Watson D. M., Franco-Hernández R., Furlan E., et al. (2005) *Astrophys. J., 630,* L185–L189.

Chauvin G., Ménard F., Fusco T., Lagrange A.-M., Beuzit J.-L., et al. (2002) *Astron. Astrophys., 394,* 949–956.

Chen C. H. and Jura M. (2003) *Astrophys. J., 591,* 267–274.

Chiang E. I. and Goldreich P. (1997) *Astrophys. J., 490,* 368–376.

Clampin M., Krist J. E., Ardila D. R., Golimowski D. A., et al. (2003) *Astron. J., 126,* 385–392

Cotera A. S., Whitney B. A., Young E., Wolff M. J., Wood K., et al. (2001) *Astrophys. J., 556,* 958–969.

D'Alessio P., Cantó J., Calvet N., and Lizano S. (1998) *Astrophys. J., 500,* 411–427.

D'Alessio P., Calvet N., Hartmann L., Lizano S., and Cantó J. (1999) *Astrophys. J., 527,* 893–909.

Dent W. R. F. (1988) *Astrophys. J., 325,* 252–265.

Dubrulle B., Morfill G., and Sterzik M. (1995) *Icarus, 114,* 237–246.

Duchêne G., McCabe C., Ghez A. M., and Macintosh B. A. (2004) *Astrophys. J., 606,* 969–982.

Dullemond C. P. and Dominik C. (2004) *Astron. Astrophys., 421,* 1075–1086.

Fukugawa M., Tamura M., Itoh Y., Hayashi S. S., and Oasa Y. (2003) *Astrophys. J., 590,* L49–L52.

Fukagawa M., Hayashi M., Tamura M., Itoh Y., Hayashi S. S., et al. (2004) *Astrophys. J., 605,* L53–L56.

Grady C. A., Woodgate B., Bruhweiler F. C., Boggess A., Plait P., et al. (1999) *Astrophys. J., 523,* L151–L154.

Grady C. A., Devine D., Woodgate B., Kimble R., Bruhweiler F. C., et al. (2000) *Astrophys. J., 544,* 895–902.

Grady C. A., Polomski E. F., Henning Th., Stecklum B., Woodgate B. E., et al. (2001) *Astron. J., 122,* 3396–3406.

Grady C. A., Woodgate B. E., Bowers C. W., Gull T. R., Sitko M. L., et al. (2005) *Astrophys. J., 630,* 958–975.

Grosso N., Alves J., Wood K., Neuhäuser R., Montmerle T., and Bjorkman J. E. (2003) *Astrophys. J., 586,* 296–305.

Guilloteau S., Dutrey A., and Simon M. (1999) *Astron. Astrophys., 348,* 570–578.

Hartmann L., Calvet N., Allen L., Chen H., and Jayawardhana R. (1999) *Astron. J., 118,* 1748.

Herbst W., Herbst D. K., Grossman E. J., and Weinstein D. (1994) *Astron. J., 108,* 1906–1923.

Hodapp K. W., Walker C. H., Reipurth B., Wood K., Bally J., et al. (2004) *Astrophys. J., 601,* L79–L82.

Jayawardhana R., Luhman K. L., D'Alessio P., and Stauffer J. R. (2002) *Astrophys. J., 571,* L51–L54.

Krist J. E., Stapelfeldt K. R., Ménard F., Padgett D. L., and Burrows C. J. (2000) *Astrophys. J., 538,* 793–800.

Krist J. E., Stapelfeldt K. R., and Watson A. M. (2002) *Astrophys. J., 570,* 785–792.

Krist J. E., Stapelfeldt K. R., Golimowski D. A., Ardila D. R., Clampin M., et al. (2005) *Astron. J., 130,* 2778–2787.

Lefèvre J., Bergeat J., and Daniel J.-Y. (1982) *Astron. Astrophys., 114,* 341–346.

Lefèvre J., Bergeat J., and Daniel J.-Y. (1983) *Astron. Astrophys., 121,* 51–58.

Liu W. M., Hinz P. M., Meyer M. R., Mamajek E. E., Hoffmann W. F., and Hora J. L. (2003) *Astrophys. J., 598,* L111–L114.

Liu W. M., Hinz P. M., Hoffmann W. F., Brusa G., Miller D., and Kenworthy M. A. (2005) *Astrophys. J., 618,* L133–L136.

Lopez B., Mékarnia D., and Lefèvre J. (1995) *Astron. Astrophys., 296,* 752–760.

Lucas P. W. and Roche P. F. (1997) *Mon. Not. R. Astron. Soc., 286,* 895–919.

Lucas P. W. and Roche P. F. (1998) *Mon. Not. R. Astron. Soc., 299,* 699–722.

Lucy L. B. (1999) *Astron. Astrophys., 344,* 282–288.

McCabe C., Duchene G., and Ghez A. M. (2002) *Astrophys. J., 575,* 974–988.

McCabe C., Duchêne G., and Ghez A. M. (2003) *Astrophys. J., 588,* L113–L116.

McCaughrean M. J., Chen H., Bally J., Erickson E., Thompson R., et al. (1998) *Astrophys. J., 492,* L157–L161.

McCaughrean M. J., Stapelfeldt K. R., and Close L. M. (2000) In *Protostars and Planets IV* (V. Mannings et al., eds.), pp. 485–507. Univ. of Arizona, Tucson.

Ménard F., Bouvier J., Dougados C., Mel'nikov S. Y., and Grankin K. N. (2003) *Astron. Astrophys., 409,* 163–167.

Monin J.-L. and Bouvier J. (2000) *Astron. Astrophys., 356,* L75–L78.

Mouillet D., Lagrange A. M., Augereau J.-C., and Ménard F. (2001) *Astron. Astrophys., 372,* L61–L64.

O'Sullivan M., Truss M., Walker C., Wood K., Matthews O., et al. (2005) *Mon. Not. R. Astron. Soc., 358,* 632–640.

Padgett D. L., Brandner W., Stapelfeldt K. R., Strom S. E., Terebey S., and Koerner D. (1999) *Astron. J., 117,* 1490–1504.

Pantin E., Waelkens C., and Lagage P. O. (2000) *Astron. Astrophys., 361,* L9–L12.

Pantin E., Bouwman J., and Lagage P. O. (2005) *Astron. Astrophys., 437,* 525–530.

Perrin M. D., Graham J. R., Kalas P., Lloyd J. P., Max C. E., et al. (2004) *Science, 303,* 1345–1348.

Perrin M. D., Duchêne G., Kalas P., and Graham J. R. (2006) *Astrophys. J., 645,* 1272.

Pontoppidan K. M., Dullemond C. P., van Dishoeck E. F., Blake G. A., Boogert A. C. A., et al. (2005) *Astrophys. J., 622,* 463–481.

Potter D. E. (2003) Ph.D. thesis, Univ. of Hawaii.

Potter D. E., Close L. M., Roddier F., Rodder C., Graves J. E., and Northcott M. (2000) *Astrophys. J., 540,* 422–428.

Quijano B. (2005) Undergraduate thesis, Univ. Michoacana de San Nicolas de Hidalgo, Morelia, Mexico.

Quillen A. C., Varnière P., Minchev I., and Frank A. (2005) *Astron. J., 129,* 2481–2495.

Rice W. K. M., Wood K., Armitage P. J., Whitney B. A., and Bjorkman J. E. (2003) *Mon. Not. R. Astron. Soc., 342,* 79–85.

Roberge A., Weinberger A. J., and Malmuth E. M. (2005) *Astrophys. J., 622,* 1171–1181.

Roddier C., Roddier F., Northcott M. J., Graves J. E., and Jim K. (1996) *Astrophys. J., 463,* 326–335.

Schneider G., Wood K., Silverstone M. D., Hines D. C., Koerner D. W., et al. (2003) *Astron. J., 125,* 1467–1479.

Shuping R. Y., Bally J., Morris M., and Throop H. (2003) *Astrophys. J., 587,* L109–L112.

Silber J., Gledhill T., Duchêne G., and Ménard F. (2000) *Astrophys. J., 536,* L89–L92.

Simon M., Dutrey A., and Guilloteau S. (2000) *Astrophys. J., 545,* 1034–1043.

Stecklum B., Henning T., Feldt M., Hayward T. L., Hoare M. G., et al. (1998) *Astron. J., 115,* 767–776.

Stecklum B., Launhardt R., Fischer O., Henden A., Leinert Ch., and Meusinger H. (2004) *Astrophys. J., 617,* 418–424.

Smith N., Bally J., Licht D., and Walawender J. (2005) *Astron. J., 129,* 382–392.

Stamatellos D. and Whitworth A. P. (2003) *Astron. Astrophys., 407,* 941–955

Stamatellos D. and Whitworth A. P. (2005) *Astron. Astrophys., 439,* 153–158.

Stapelfeldt K. R., Krist J. E., Ménard F., Bouvier J., Padgett D. L., and Burrows C. J. (1998) *Astrophys. J., 502,* L65–L69.

Stapelfeldt K. R., Watson A. M., Krist J. E., Burrows C. J., Crisp D., et al. (1999) *Astrophys. J., 516,* L95–L98.

Stapelfeldt K. R., Ménard F.,Watson A. M., Krist J. E., Dougados C., et al. (2003) *Astrophys. J., 589,* 410–418.

Throop H. B., Bally J., Esposito L. W., and McCaughrean M. (2001) *Science, 292,* 1686–1689.

Trilling D. E., Koerner D. W., Barnes J. W., Ftaclas C., and Brown R. H. (2001) *Astrophys. J., 552,* L151–L154.

Watson A. M. and Henney W. J. (2001) *Rev. Mex. Astron. Astrofis., 37,* 221–236.

Watson A. M. and Stapelfeldt K. R. (2004) *Astrophys. J., 602,* 860–874.

Weinberger A. J., Becklin E. E., Schneider G., Smith B. A., Lowrance P. J., et al. (1999) *Astrophys. J., 525,* L53–L56.

Weinberger A. J., Becklin E. E., Schneider G., Chiang E. I., Lowrance P. J., et al. (2002) *Astrophys. J., 566,* 409–418.

White R. J. and Hillenbrand L. A. (2004) *Astrophys. J., 616,* 998–1032.

Whitney B. A. and Hartmann L. (1992) *Astrophys. J., 395,* 529–539.

Whitney B. A. and Hartmann L. (1993) *Astrophys. J., 402,* 605–622.

Whitney B. A., Kenyon S. J., and Gomez M. (1997) *Astrophys. J., 485,* 703–734.

Whitney B. A., Wood K., Bjorkman J. E., and Wolff M. J. (2003a) *Astrophys. J., 591,* 1049–1063.

Whitney B. A., Wood K., Bjorkman J. E., and Cohen M. (2003b) *Astrophys. J., 598,* 1079–1099.

Whitney B. A., Indebetouw R., Bjorkman J. E., and Wood K. (2004) *Astrophys. J., 617,* 1177–1190.

Wood K. and Whitney B. (1998) *Astrophys. J., 506,* L43–L45.

Wood K., Kenyon S. J., Whitney B. A., and Turnbull M. (1998) *Astrophys. J., 497,* 404–418.

Wood K., Crosas M., and Ghez A. M. (1999) *Astrophys. J., 516,* 335–341.

Wood K., Wolk S. J., Stanek K. Z., Leussis G., Stassun K., et al. (2000) *Astrophys. J., 542,* L21–L24.

Wood K., Smith D., Whitney B. A., Stassun K., Kenyon S. J., et al. (2001) *Astrophys. J., 561,* 299–307.

Wood K., Wolff M. J., Bjorkman J. E., and Whitney B. (2002) *Astrophys. J., 564,* 887–895.

Wolf S., Henning Th., and Stecklum B. (1999) *Astron. Astrophys., 349,* 839–850.

Wolf S., Voshchinnikov N. V., and Henning Th. (2002) *Astron. Astrophys., 385,* 365–376.

Wolf S., Padgett D. L., and Stapelfeldt K. R. (2003) *Astrophys. J., 588,* 373–386.

The Circumstellar Environments of Young Stars at AU Scales

Rafael Millan-Gabet
California Institute of Technology

Fabien Malbet
Laboratoire d'Astrophysique de Grenoble

Rachel Akeson
California Institute of Technology

Christoph Leinert
Max-Planck-Institut für Astronomie

John Monnier
University of Michigan

Rens Waters
University of Amsterdam

We review recent advances in our understanding of the innermost regions of the circumstellar environment around young stars, made possible by the technique of long-baseline interferometry at infrared wavelengths. Near-infrared observations directly probe the location of the hottest dust. The characteristic sizes found are much larger than previously thought, and strongly correlate with the luminosity of the central young stars. This relation has motivated in part a new class of models of the inner disk structure. The first mid-infrared observations have probed disk emission over a larger range of scales, and spectrally resolved interferometry has for the first time revealed mineralogy gradients in the disk. These new measurements provide crucial information on the structure and physical properties of young circumstellar disks, as initial conditions for planet formation.

1. INTRODUCTION

Stars form from collapsing clouds of gas and dust and in their earliest infancies are surrounded by complex environments that obscure our view at optical wavelengths. As evolution proceeds, a stage is revealed with three main components: the young star, a circumstellar disk, and an infalling envelope. Eventually, the envelope dissipates, and the emission is dominated by the young star-disk system. Later on, the disk also dissipates to very tenuous levels. It is out of the young circumstellar disks that planets are expected to form, and therefore understanding their physical condition is necessary before we can understand the formation process. Of particular interest are the inner few AU (astronomical unit), corresponding to formation sites of terrestrial-type planets, and to migration sites for gas giants presumably formed further out in the disk (see chapter by Udry et al.).

A great deal of direct observational support exists for the scenario outlined above, and in particular for the existence of circumstellar disks around young stars (see chapters by Dutrey et al. and by Watson et al.). In addition, the spectroscopic and spectrophotometric characteristics of these systems (i.e., spectral energy distributions, emission lines) are also well described by the disk hypothesis. However, current optical and millimeter-wave imaging typically probe scales of hundreds to thousands of AU with resolutions of tens of AU, and models of unresolved observations are degenerate with respect to the spatial distribution of material. As a result, our understanding of even the most general properties of the circumstellar environment at few AU or smaller spatial scales is in its infancy.

Currently, the only way to achieve sufficient angular resolution to directly reveal emission within the inner AU is through interferometry at visible and infrared wavelengths, here referred to as *optical* interferometry. An interferometer with a baseline length of $B = 100$ m (typical of current facilities) operating at near- to mid-infrared wavelengths (NIR, MIR, typically H to N bands, $\lambda_0 = 1.65–10$ μm) probes 1800 300 K material and achieves an angular resolution $\sim\lambda_0/2B \sim 2–10$ milliarcseconds (mas), or 0.25–1.5 AU at a distance typical of the nearest well-known star-forming regions (150 pc). This observational discovery space is illustrated in Fig. 1, along with the domains corresponding to various young stellar object (YSO) phenomena and complementary techniques and instruments. Optical interferometers are ideally suited to directly probe the innermost regions of the circumstellar environment around young stars, and indeed using this technique surprising and rapid progress has been made, as the results reviewed in this chapter will show.

Optical, as well as radio, interferometers operate by coherently combining the electromagnetic waves collected by

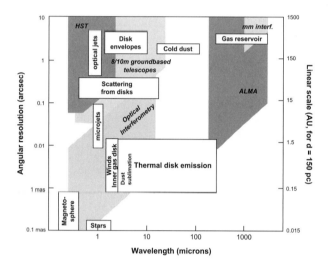

Fig. 1. Observational phase-space (spectral domain and angular resolution) for optical interferometers and for complementary techniques (shaded polygons). Also outlined over the most relevant phase-space regions (rectangular boxes) are the main physical phenomena associated with young stellar objects.

two or more telescopes. Under conditions that apply to most astrophysical observations, the amplitude and phase of the resulting interference patterns (components of the complex *visibility*) are related via a two-dimensional Fourier transform to the brightness distribution of the object being observed. For a detailed description of the fundamental principles, variations in their practical implementation, and science highlights in a variety of astrophysics areas we refer to the reviews by *Monnier* (2003) and *Quirrenbach* (2001). The reader may also be interested in consulting the proceedings of topical summer schools and workshops such as *Lawson* (2000), *Perrin and Malbet* (2003), *Garcia et al.* (2004), and *Paresce et al.* (2006) (online proceedings of the Michelson Summer Workshops may also be found at *msc.caltech. edu/michelson/workshop.html*).

While interferometers are capable of model-independent imaging by combining the light from many telescopes, most results to date have been obtained using two telescopes (a single-baseline). The main characteristics of current facilities involved in these studies are summarized in Table 1. Interferometer data, even with sparse spatial frequency coverage, provides direct constraints about source geometry and thus contains some of the power of direct imaging. However, with such data alone only the simplest objects can be constrained. Therefore, more typically, a small number of fringe visibility amplitude data points are combined with a spectral energy distribution (SED) for fitting simple physically motivated geometrical models, such as a point source representing the central star surrounded by a Gaussian or ring-like brightness (depending on the context) representing the source of infrared excess flux. This allows us to determine "characteristic sizes" at the wavelength of observa-

TABLE 1. Long-baseline optical interferometers involved in YSO research.

Facility	Instrument*	Wavelength Coverage†	Number of Telescopes‡	Telescope Diameter (m)	Baseline Range (m)	Best Resolution (mas)§
PTI	V^2	1.6–2.2 µm [44]	2 [3]	0.4	80–110	1.5
IOTA	V^2, IONICS	1.6–2.2 µm	3	0.4	5–38	4.5
ISI	Heterodyne	11 µm	3	1.65	4–70	16.2
KI	V_2	1.6–2.2 µm [22]	2	10	85	2.0
KI	Nuller	8–13 µm [34]	2	10	85	9.7
VLTI	MIDI	8–13 µm [250]	2 [8]	8.2/1.8	8–200	4.1
VLTI	AMBER	1–2.5 µm [10^4]	3 [8]	8.2/1.8	8–200	0.6
CHARA	V^2	1.6–2.2 µm	2 [6]	1	50–350	0.4

*V^2 refers to a mode in which only the visibility amplitude is measured, often implemented in practice as a measurement of the (unbiased estimator) square of the visibility amplitude.

†The maximum spectral resolution available is given in square brackets.

‡The number of telescopes that can be simultaneously combined is given, along with the total number of telescopes available in the array in square brackets.

§In each case, the best resolution is given as half the fringe spacing, $\lambda_{min}/(2B_{max})$, for the longest physical baseline length and shortest wavelength available.

References: [1] PTI (Palomar Testbed Interferometer), *Colavita et al.* (1999); [2] IOTA (Infrared-Optical Telescope Array), *Traub et al.* (2004); [3] ISI (Infrared Spatial Interferometer), *Hale et al.* (2000); [4] KI (Keck Interferometer), *Colavita et al.* (2004); [5] KI Nuller, *Serabyn et al.* (2004); [6] MIDI (MID-infrared Interferometric instrument), *Leinert et al.* (2003); [7] AMBER (Astronomical Multiple BEam Recombiner), *Malbet et al.* (2004); [8] CHARA (Center for High Angular Resolution Astronomy interferometer), *ten Brummelaar et al.* (2005).

tion for many types of YSOs (e.g., T Tauri, Herbig Ae/Be, FU Orionis). The interferometer data can also can be compared to predictions of specific physical models and this exercise has allowed to rule out certain classes of models, and provided crucial constraints to models that can be made to reproduce the spectral and spatial observables.

A note on nomenclature: Before the innermost disk regions could be spatially resolved, as described in this review, disk models were tailored to fit primarily the unresolved spectrophotometry. The models used can be generally characterized as being geometrically thin and optically thick, with simple radial temperature power laws with exponents q = –0.75 [flat disk heated by accretion and/or stellar irradiation (*Lynden-Bell and Pringle*, 1974)] or shallower [flared disks (*Kenyon and Hartmann*, 1987; *Chiang and Goldreich*, 1997)]. In this review, we will refer to these classes of models as "classical," to be distinguished from models with a modified inner disk structure that have recently emerged in part to explain the NIR interferometer measurements.

The outline of this review is as follows. Sections 2 and 3 summarize the observational state of the art and emerging interpretations, drawing a physically motivated distinction between inner and outer disk regions that also naturally addresses distinct wavelength regimes and experimental techniques. We note that these developments are indeed recent, with the first preliminary analyses of interferometric observations of YSOs being presented (as posters) at the Protostars and Planets IV Conference in 1998. In section 4 we highlight additional YSO phenomena that current facilities are addressing, such as the disk-wind connection and young star mass determination, using combination of interferometry with high-resolution line spectroscopy. In section 5 we present a top-level summary of the main well-established results, and highlight remaining important open questions likely to be experimentally and theoretically addressed in the coming years. In section 6 we describe additional phenomena yet to be explored and new instrumental capabilities that promise to enable this future progress.

2. THE INNER DISK

At near-infrared (NIR) wavelengths (~1–2.4 μm) optical interferometers probe the hottest circumstellar material located at stellocentric distances ≤1 AU. In this review these regions are referred to as the *inner disk*. For technical reasons, it is at NIR wavelengths that interferometric measurements of YSOs first became possible, and provided the first direct probes of inner disk properties. Until only a few years ago, relatively simple models of accretion disks were adequate to reproduce most observables. However, the new observations with higher spatial resolution have revealed a richer set of phenomena.

2.1. Herbig Ae/Be Objects

In Herbig Ae/Be (HAeBe) objects, circumstellar material (protoplanetary disk, remnant envelope, or both) sur-

rounds a young star of intermediate mass [~2–10 M$_\odot$ (see, e.g., the review by *Natta et al.*, 2000)]. A significant number of these objects fall within the sensitivity limits of small interferometers, hence the first YSO studies containing relatively large samples focused on objects in this class.

AB Aur was the first "normal" (i.e., non-outburst, see section 2.3) YSO to be resolved at NIR wavelengths. In observations at the Infrared Optical Telescope Array (IOTA) AB Aur appeared clearly resolved in the H and K spectral bands on 38-m baselines, by an amount corresponding to a characteristic diameter of 0.6 AU for the circumstellar NIR emission (*Millan-Gabet et al.*, 1999). The measured size was unexpectedly large in the context of then-current disk models of HAeBe objects (e.g., *Hillenbrand et al.*, 1992), which predicted NIR diameter of 0.2 AU based on optically thick, geometrically thin circumstellar disks with a small dust-free inner hole. This conclusion was reinforced by the completed IOTA survey of *Millan-Gabet et al.* (2001), finding characteristic NIR sizes of 0.6–6 AU for objects with a range of stellar properties (spectral types A2–O9).

Due to the limited position angle coverage of these first single-baseline observations, the precise geometry of the NIR emission remained ambiguous. *Millan-Gabet et al.* (2001) found that the NIR sizes were similar at 1.65 and 2.2 μm, suggesting a steep temperature gradient at the inner disk edge. These early observations revealed no direct evidence for "disk-like" morphologies; indeed, for the few objects with size measurements at multiple position angles, the data indicated circular symmetry (as if from spherical halos or face-on disks, but not inclined disks).

The first unambiguous indication that the NIR emission arises in disk-like structures came from single-telescope imaging of the highly luminous YSOs LkHα 101 and MWC 349-A, using a single-telescope interferometric technique (aperture masking at the Keck I telescope). MWC 349-A appeared clearly elongated (*Danchi et al.*, 2001) and LkHα 101 presented an asymmetric ring-like morphology, interpreted as a bright inner disk edge occulted on one side by foreground cooler material in the outer regions of the flared disk (*Tuthill et al.*, 2001). Moreover, the location of the LkHα 101 ring was also found to be inconsistent with predictions of classical disk models and these authors suggested that an optically thin inner cavity (instead of optically thick disk midplane) would result in larger dust sublimation radii, consistent with the new size measurements.

Could simple dust sublimation of directly heated dust be setting the NIR sizes of other HAeBe objects as well? The idea was put to the test by *Monnier and Millan-Gabet* (2002). Inspired by the LkHα 101 morphology and interpretation, these authors fit a simple model consisting of a central star surrounded by a thin ring for all objects that were measured interferometrically at the time [IOTA and aperture masking, plus first YSO results from the Palomar Testbed Interferometer (PTI) (*Akeson et al.*, 2000)]. Indeed, the fitted ring radii are clearly correlated with the luminosity of the central star, and follow the expected relation R ∝ L$_\star^{1/2}$. Furthermore, by using realistic dust properties it was

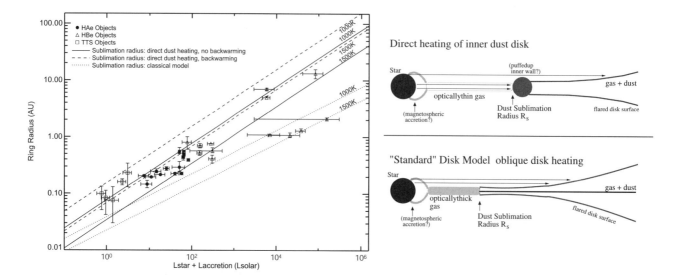

Fig. 2. Measured sizes of HAeBe and T Tauri objects vs. central luminosity (stellar + accretion shock); and comparison with sublimation radii for dust directly heated by the central luminosity (solid and dashed lines) and for the oblique heating implied by the classical models (dotted line). A schematic representation of the key features of inner disk structure in these two classes of models is also shown. Data for HAeBe objects are from IOTA (*Millan-Gabet et al.*, 2001), Keck aperture masking (*Danchi et al.*, 2001; *Tuthill et al.*, 2001), PTI (*Eisner et al.*, 2004), and KI (*Monnier et al.*, 2005a). Data for T Tauri objects are from PTI (*Akeson et al.*, 2005b) and KI (*Akeson et al.*, 2005a; *Eisner et al.*, 2005). For clarity, for objects observed at more than one facility, we include only the most recent measurement.

also found that the measured NIR sizes are consistent with the dust sublimation radii of relatively large grains (≥ 1 µm) with sublimation temperatures in the range 1000–2000 K.

Following *Monnier and Millan-Gabet* (2002), in Fig. 2 we have constructed an updated diagram of NIR size vs. central luminosity based on all existing data in the literature. We include data for both HAeBe and T Tauri objects, the latter being discussed in detail in the next section, and we also illustrate schematically the essential ingredients (i.e., location of the inner dust disk edge) of the models to which the data are being compared. For HAeBe objects, disk irradiation is dominated by the stellar luminosity, and therefore we neglect heating by accretion shock luminosity (which may play a significant role for the most active lower stellar luminosity T Tauri objects, as discussed in the next section).

Indeed, it can be seen that for HAe and late HBe objects, over more than two decades in stellar luminosity, the measured NIR sizes are tightly contained within the sublimation radii of directly heated gray dust with sublimation temperatures of 1000–1500 K under the assumption that dust grains radiate over the full solid angle (e.g., no backwarming, solid lines in Fig. 2). If instead we assume that the dust grains emit only over 2π steradian (e.g., full backwarming), then the corresponding sublimation temperature range is 1500–2000 K.

The most luminous objects (the early spectral type HBe objects), on the other hand, have NIR sizes in good agreement with the classical model, indicating that for these objects the disk extends closer in to the central star. *Monnier and Millan-Gabet* (2002) hypothesized that even optically

thin low-density gas inside the dust destruction radius can scatter UV stellar photons, partly shielding the dust at the inner disk edge and reducing the sublimation radius. *Eisner et al.* (2003, 2004) argue instead that the distinction may mark a transition from magnetospheric accretion (late types) to a regime for the early types where optically thick disks in fact do extend to the stellar surface, either as a result of higher accretion rates or weak stellar magnetic fields. We note that using polarimetry across H_α lines originating in the inner gas disk at ~few $R_\star$, *Vink et al.* (2005) (and references therein) also find (in addition to flattened structures consistent with disk-like morphology) a marked difference in properties between HBe objects and HAe and T Tauri objects, qualitatively consistent with the NIR size properties described above. However, not *all* high-luminosity HBe objects are in better agreement with the classical model; there are notable exceptions (e.g., LkHα 101, MWC 349-A, see Fig. 2), and *Monnier et al.* (2005a) have noted that these may be more evolved systems, where other physical processes, such as dispersal by stellar winds or erosion by photoevaporation, have a dominant effect in shaping the inner disk.

In parallel with these developments, and in the context of detailed modeling of the NIR SED, *Natta et al.* (2001) similarly formulated the hypothesis that the inner regions of circumstellar disks may be largely optically thin to the stellar photons such that a directly illuminated "wall" forms at the inner dust-disk edge. The relatively large surface area of this inner dust wall is able to produce the required levels of NIR flux (the so-called NIR SED "bump"). The simple blackbody surface of the initial modeling implementation has since been replaced by more realistic models that take additional physical effects into account: self-consistent treatment of radiative transfer and hydrostatic vertical structure of the inner wall and corresponding outer disk shadowing (*Dullemond et al.*, 2001), optical depth and scale height of gas inside the dust destruction radius (*Muzerolle et al.*, 2004), and inner rim curvature due to density-dependent dust sublimation temperature (*Isella and Natta*, 2005).

Independent of model details, however, it is clear that the optically thin cavity disk model with a "puffed-up" inner dust wall can explain a variety of observables: the ring-like morphology, the characteristic NIR sizes, and the NIR SED. We also emphasize that although the discussion of the interferometer data presented above was centered around the size-luminosity diagram, the same conclusions have been obtained by modeling SEDs and visibility measurements for individual sources using physical models of both classical and puffed-up inner rim models.

Although the characteristic NIR sizes have now been well established for a relatively large sample, few constraints still exist as to the actual geometry of the resolved NIR emission. In this respect, most notable are the PTI observations of *Eisner et al.* (2003, 2004), which provided the first evidence for elongated emission, consistent with inclined disks. Further, the inclinations found can explain the residual scatter (50%) in the size-luminosity relation for HAe and late HBe systems (*Monnier et al.*, 2005a).

It is important to note that three different interferometers have now reached the same conclusions regarding the characteristic NIR sizes of HAeBe objects, in spite of having vastly different field-of-views (FOV) (3, 1, and 0.05 arcsec FWHM for IOTA, PTI, and KI respectively). This indicates that the interpretation is not significantly effected by non-modeled flux entering the FOV, from the large-scale (scattered) emission often found around these objects (e.g., *Leinert et al.*, 2001). Indeed, if not taken into account in a composite model, such incoherent flux would lead to overestimated characteristic sizes, and still needs to be considered as a (small) possible effect for some objects.

2.2. T Tauri Objects

Only the brightest T Tauri objects can be observed with the small aperture interferometers, so early observations were limited in number. The first observations were taken at the PTI and were of the luminous sources T Tau N and SU Aur (*Akeson et al.*, 2000, 2002). Later observations (*Akeson et al.*, 2005a) added DR Tau and RY Tau to the sample, and employed all three PTI baselines to constrain the disk inclination. Using simple geometrical models (or simple power-law disk models) the characteristic NIR sizes (or disk inner radii) were found to be larger than predicted by SED-based models (*e.g., Malbet and Bertout*, 1995), similar to the result for the HAe objects. For example, inclined ring models for SU Aur and RY Tau yield radii of $10 R_\star$ and $11 R_\star$ respectively, substantially larger than magnetic truncation radii of $3-5 R_\star$ expected from magneto-

spheric accretion models (*Shu et al.*, 1994). If the inner dust disk does not terminate at the magnetic radius, what then sets its radial location? Interestingly, these initial T Tauri observations revealed NIR sizes that were also consistent with sublimation radii of dust directly heated by the central star, suggesting perhaps that similar physical processes, decoupled from magnetospheric accretion, set the NIR sizes of both T Tauri and HAe objects. *Lachaume et al.* (2003) were the first to use physical models to fit the new visibility data, as well as the SEDs. Using a two-layer accretion disk model, these authors find satisfactory fits for SU Aur (and FU Ori, see section 2.3), in solutions that are characterized by the midplane temperature being dominated by accretion, while the emerging flux is dominated by reprocessed stellar photons.

Analysis of T Tauri systems of more typical luminosities became possible with the advent of large-aperture infrared interferometers. Observations at the KI (*Colavita et al.*, 2003; *Eisner et al.*, 2005; *Akeson et al.*, 2005a) continued to find large NIR sizes for lower-luminosity stars, in many cases even larger than would be expected from extrapolation of the HAe relation.

Current T Tauri measurements and comparison with dust sublimation radii are also summarized in Fig. 2. *Eisner et al.* (2005) found better agreement with puffed-up inner wall models, and in Fig. 2 we use their derived inner disk radii. Following *Muzerolle et al.* (2003) and *D'Alessio et al.* (2004), the central luminosity for T Tauri objects includes the accretion luminosity, released when the magnetic accretion flow encounters the stellar surface, which can contribute to the location of the dust sublimation radius for objects experiencing the highest accretion rates (in Fig. 2, accretion luminosity only affects the location of three objects significantly: AS205-A, PX Vul, and DR Tau, having $L_{accretion}/L_\star \sim 10.0$, 1.8, and 1.2 respectively.)

It can be seen that many T Tauri objects follow the HAe relation, although several are even larger given the central luminosity (the T Tauri objects located above the 1000 K solid line in Fig. 2 are BP Tau, GM Aur, LkCa 15, and RW Aur).

It must be noted, however, that current measurements of the lowest-luminosity T Tauri objects have relatively large errors. Furthermore, the ring radii plotted in Fig. 2 depend on the relative star/disk NIR flux contributions, most often derived from an SED decomposition that for T Tauri objects is considerably more uncertain than for the HAeBe objects, due to their higher photometric variability, and the fact that the stellar SEDs peak not far (~1 μm) from the NIR wavelengths of observation. We note that an improved approach was taken by *Eisner et al.* (2005), who derived stellar fluxes at the KI wavelength of observation from near-contemporaneous optical veiling measurements and extrapolation to NIR wavelengths based on model atmospheres. These caveats illustrate the importance of obtaining more direct estimates of the excess NIR fluxes, via veiling measurements (e.g., *Johns-Krull and Valenti*, 2001, and references therein).

Under the interpretation that the T Tauri NIR sizes trace the dust sublimation radii, different properties for the dust located at the disk inner edge appear to be required to explain all objects. In particular, if the inner dust-disk edge becomes very abruptly optically thick (e.g., strong back-warming), the sublimation radii become larger by a factor of 2, in better agreement with some of the low-luminosity T Tauri objects (dashed lines in Fig. 2). As noted by *Monnier and Millan-Gabet* (2002), smaller grains also lead to larger sublimation radii, by a factor of $\sqrt{Q_R}$, where Q_R is the ratio of dust absorption efficiencies at the temperatures of the incident and reemitted field ($Q_R \simeq 1$–10 for grain sizes 1.0–0.03 μm, the incident field of a stellar temperature of 5000 K typical of T Tauri stars, and reemission at 1500 K).

Alternatively, an evolutionary effect could be contributing to the size-luminosity scatter for the T Tauri sample. *Akeson et al.* (2005a) note that the size discrepancy (measured vs. sublimation radii) is greatest for objects with the lowest ratio of accretion to stellar luminosity. If objects with lower accretion rates are older (e.g., *Hartmann et al.*, 1998) then other physical processes such as disk dispersal (*Clarke et al.*, 2001; see also review by *Hollenbach et al.*, 2000) may be at play for the most "discrepant" systems mentioned above [the size derived for GM Aur contains additional uncertainties, given the evidence in this system for a dust gap at several AU (*Rice et al.*, 2003) and the possibility that it contributes significant scattered NIR light].

In detailed modeling work of their PTI data, *Akeson et al.* (2005b) used Monte Carlo radiative transfer calculations (*Bjorkman and Wood*, 2001) to model the SEDs and infrared visibilities of SU Aur, RY Tau, and DR Tau. These models addressed two important questions concerning the validity of the size interpretation based on fits to simple star + ring models: (1) Is there significant NIR thermal emission from *gas* inside the dust destruction radius, and (2) is there significant flux arising in large-scale (scattered) emission? The modeling code includes accretion and shock/boundary luminosity as well as multiple scattering and this technique naturally accounts for the radiative transfer effects and the heating and hydrostatic structure of the inner wall of the disk. In these models, the gas disk extends to within a few stellar radii and the dust disk radius is set at the dust sublimation radius. For SU Aur and RY Tau, the gas within the inner dust disk was found to contribute substantially to the NIR emission. Modeling of HAeBe circumstellar disks by *Muzerolle et al.* (2004) found that emission from the inner gas exceeded the stellar emission for accretion rates >10^{-7} M$_{\odot}$ yr^{-1}. Thus, even as a simple geometrical representation, a ring model at the dust sublimation temperature may be too simplistic for sources where the accretion luminosity is a substantial fraction of the total system luminosity. Moreover, even if the NIR gas emission were located close to the central star and were essentially unresolved, its presence would affect the SED decomposition (amount of NIR excess attributed to the dust "ring"), leading to incorrect (underestimated) ring radii. The second finding from the radiative models is that the extended emission (here meaning emission from scales larger than 10 mas) was less than 6%

of the total emission for all sources, implying that scattered light is not a dominant component for these systems.

Clearly, the new spatially resolved observations just described, although basic, constitute our only direct view into the physical conditions in young disks on scales ≤1 AU, and therefore deliver crucial ingredients as preconditions for planet formation. As pointed out by *Eisner et al.* (2005) the interpretation that the measured NIR sizes correspond to the location of the innermost disk dust implies that dust is in fact present at terrestrial planet locations. Conversely, the measured radii imply that no dust exists, and therefore planet formation is unlikely, inside these relatively large ~0.1–1 AU inner dust cavities. Finally, that the dust disk stops at these radii may also have implications for planet migration, depending on whether migration is halted at the inner radius of the gas (*Lin et al.*, 1996) or dust (*Kuchner and Lecar*, 2002) disk (see also chapter by Papaloizou et al.).

2.3. Detailed Tests of the Disk Hypothesis: FU Orionis

FU Orionis objects are a rare type of YSO believed to be T Tauri stars surrounded by a disk that has recently undergone an episode of accretion rate outburst (see, e.g., review by *Hartmann and Kenyon*, 1996). During the outburst, the system brightens by several visual magnitudes, followed by decade-long fading. The disk luminosity dominates the emission at all wavelengths, and is expected to be well represented by thermal emission by disk annulii that follow the canonical temperature profile T ∝ r$^{-3/4}$. With this prescription, the model has few parameters that are well constrained by the SEDs alone (except for inclination effects). In principle then, these systems are ideal laboratories for testing the validity of disk models. The total number of known objects in this class is small [e.g., 20 objects in the recent compilation by *Ábrahám et al.* (2004)], and the subset that is observable by current optical interferometers is even smaller (about 6 objects, typically limited by instrumental sensitivity at short wavelengths 0.55–1.25 μm).

The very first YSO to be observed with an optical interferometer was in fact the prototype for this class, FU Orionis itself (*Malbet et al.*, 1998). Indeed, the first K-band visibility amplitudes measured by the PTI agreed well with the predictions of the canonical disk model. This basic conclusion has been reinforced in a more recent study using multi-baseline, multi-interferometer observations (*Malbet et al.*, 2005). The detailed work on FU Ori will ultimately close the debate on its nature: Current evidence favors that it is a T Tauri star surrounded by a disk undergoing massive accretion (*Hartmann and Kenyon*, 1985, 1996) rather than a rotating supergiant with extreme chromosphere activity (*Herbig et al.*, 2003); and under that interpretation the physical (temperature law) and geometrical (inclination and orientation) parameters describing the disk have been established with unprecedented detail.

Three more FU Orionis objects have been resolved in the NIR: V1057 Cyg [PTI, *Wilkin and Akeson* (2003); KI, *Millan-Gabet et al.* (2006a)], V1515 Cyg, and ZCMa-SE

(*Millan-Gabet et al.*, 2006a). In contrast to the FU Ori case, these three objects appear more resolved in the NIR than expected from a T ∝ r$^{-3/4}$ model, and simple exponent adjustments of such a single power-law model does not allow simultaneous fitting of the visibilities and SEDs. On the other hand, these objects are also known to possess large mid-infrared fluxes in excess of emission by flat disks, usually attributed to a large degree of outer disk flaring or, more likely, the presence of a dense dust envelope (*Kenyon and Hartmann*, 1991). The low visibilities measured, particularly for V1057 Cyg and V1515 Cyg, can be explained by K-band flux (~10% of total) in addition to the thermal disk emission, resulting from scattering through envelope material over scales corresponding to the interferometer FOV (50 mas FWHM). Contrary to initial expectations, the likely complexity of the circumstellar environment of most FU Orionis objects will require observations at multiple baselines and wavelengths in order to disentangle the putative multiple components, discriminate between the competing models, and perform detailed tests of accretion disk structure.

3. THE OUTER DISK

At mid-infrared (MIR) wavelengths, and for 50–100-m baselines, optical interferometers probe the spatial distribution and composition of ~few 100 K circumstellar gas and dust, with 8–27-mas resolution, or 1–4 AU at a distance of 150 pc. The MIR radiation of circumstellar disks in the N band (8–13 μm) comes from a comparatively wide range of distances from the star (see Fig. 3), and we now focus the discussion on circumstellar dust in this several-AU transition region between the hottest dust in the sublimation region close to the star (section 2), and the much cooler disk regions probed by millimeter interferometry (see chapter by Dutrey et al.) and by scattered light imaging (see chapter by Watson et al.). As a matter of interpretational convenience, at these longer MIR wavelengths the flux contribution from scattered light originating from these relatively small spatial scales is negligible, allowing the consideration of fewer model components and physical processes.

3.1. Disk Sizes and Structure

The geometry of protoplanetary disks is of great importance for a better understanding of the processes of disk evolution, dissipation, and planet formation. The composition of the dust in the upper disk layers plays a crucial role in this respect, since the optical and UV photons of the central star, responsible for determining the local disk temperature and thus the disk vertical scale-height, are absorbed by these upper disk layer dust particles. The disk scale-height plays a pivotal role in the process of dust aggregation and settling, and therefore impacts the global disk evolution. In addition, the formation of larger bodies in the disk can lead to structure, such as density waves and gaps.

Spatially resolved observations of disks in the MIR using single telescopes have been limited to probing relatively

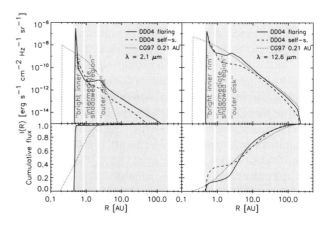

Fig. 3. Predicted radial distribution of NIR (left) and MIR (right) light for typical disk models. DD04 (*Dullemond and Dominik*, 2004) refers to a two-dimensional radiative transfer and hydrostatic equilibrium Monte Carlo code featuring a puffed up inner rim; CG97 to the gradually flaring model of *Chiang and Goldreich* (1997). We note that the CG97 disk model can not self-consistently treat the inner disk once the temperature of the "superheated" dust layer increases beyond the dust sublimation temperature. This only effects the NIR light profiles and here the disk was artificially truncated at 0.21 AU. The central star is assumed to be of type A0 with M = 2.5 M$_\odot$ and R = 2.0 R$_\odot$. The upper panels show the intensity profiles, and the lower panels the cumulative brightness contributions, normalized to 1. Three qualitatively different disk regions are indicated by shadowing. This figure illustrates that NIR observations almost exclusively probe the hot inner rim of the circumstellar disks, while MIR observations are sensitive to disk structure over the first dozen or so AU. Courtesy of R. van Boekel.

large-scale (thousands of AU) thermal emission in evolved disks around main-sequence stars (e.g., *Koerner et al.*, 1998; *Jayawardhana*, 1998) or scattering by younger systems (e.g., *McCabe et al.*, 2003). Single-aperture interferometric techniques [e.g., nulling (*Hinz et al.*, 2001; *Liu et al.*, 2003, 2005) or segment-tilting (*Monnier et al.*, 2004b)] as well as observations at the Infrared Spatial Interferometer (ISI) (*Tuthill et al.*, 2002) have succeeded in resolving the brightest young systems on smaller scales for the first time, establishing MIR sizes of approximately tens of AU. Although the observed samples are still small, and thus far concern mostly HAe objects, these initial measurements have already provided a number of interesting, albeit puzzling results: MIR characteristic sizes may not always be well predicted by simple (flared) disk models (some objects are "undersized" compared to these predictions, opposite to the NIR result), and may not always connect in a straightforward way with measurements at shorter (NIR) and longer (millimeter) wavelengths.

These pioneering efforts were, however, limited in resolution (for the single-aperture techniques) or sensitivity (for the heterodyne ISI), and probing the MIR emission on small scales has only recently become possible with the advent of the new-generation long-baseline MIR instruments [VLTI/MIDI, *Leinert et al.* (2003); KI/Nuller, *Serabyn et al.* (2004)].

Two main factors determine the interferometric signature of a dusty protoplanetary disk in the MIR: the overall geometry of the disk, and the composition of the dust in the disk "atmosphere." The most powerful way to distentangle these two effects is via spectrally resolved interferometry, and this is what the new generation of instruments such as MIDI at the VLTI are capable of providing.

First results establishing the MIR sizes of young circumstellar disks were obtained by VLTI/MIDI for a sample of seven HAeBe objects (*Leinert et al.*, 2004). The characteristic sizes measured, based on visibilities at the wavelength of 12.5 μm, are in the range 1–10 AU. Moreover, as shown in Fig. 4, they are found to correlate with the IRAS [12]–[25] color, with redder objects having larger MIR sizes. These observations lend additional support to the SED-based classification of *Meeus et al.* (2001) and *Dullemond and Dominik* (2004), whereby redder sources correspond to disks with a larger degree of flaring, which therefore also emit thermally in the MIR from larger radii, compared to sources with flatter disks.

A more powerful analysis is possible by exploiting the spectral resolution capabilities of the MIDI instrument. Based on the reddest object in the *Leinert et al.* (2004) sample (HD 100546), the spectrally resolved visibilities also support the *Meeus et al.* (2001) classification: This object displays a markedly steeper visibility drop between 8 and 9 μm, and lower visibilities at longer wavelengths than the

rest of the objects in the sample, implying the expected larger radius for the MIR emission. These first observations have also been used to test detailed disk models that were originally synthesized to fit the SEDs of individual objects (*Dominik et al.*, 2003); without any feedback to the models *Leinert et al.* (2004) find encouraging qualitative agreement between the predicted spectral visibility shapes and the interferometer data.

It remains to be seen whether direct fitting to the interferometer data will solve the remaining discrepancies [see, e.g., the work of *Gil et al.* (2006) on the somewhat unusual object 51 Oph]. In general, changes in both disk geometry and dust composition strongly affect the MIR spectral visibilities. However, simulations by *van Boekel et al.* (2005) have shown that the problem is tractable and even just a few well-chosen baselines (and wavelengths) permit a test of some of the key features of recently proposed disk models, such as the presence of the putative puffed-up inner rim and the degree of outer disk flaring (which in these models is influenced by shadowing by the inner rim).

Clearly, the spectrally and spatially resolved disk observations made possible by instruments such as VLTI/MIDI will prove crucial toward further exploring exciting prospects for linking the overall disk structure, the properties of midplane and surface layer dust, and their time evolution toward the planet-building phase (see also the chapter by Dominik et al.).

3.2. Dust Mineralogy and Mixing

The MIR spectral region contains strong resonances of abundant dust species, both O-rich (amorphous or crystalline silicates) and C-rich (polycyclic aromatic hydrocarbons, or PAHs). Therefore, MIR spectroscopy of optically thick protoplanetary disks offers a rich diagnostic of the chemical composition and grain size of dust in the upper disk layers or "disk atmosphere." At higher spectral resolution, $R \sim$ few 100, gas emission such as [NeII] lines and lines of the Hund and higher hydrogen series can be observed. It should be noted that these observations do not constrain the properties of large (>2–4 μm) grains, since these have little or no spectral structure at MIR wavelengths. In the submicrometer and micrometer size range, however, different silicate particles can be well distinguished on the basis of the shape of their emission features. Clearly, valuable information can be obtained from spatially unresolved observations — the overall SED constrains the distribution of material with temperature (or radius), and MIR spectroscopy provides the properties of dust. However, strong radial gradients in the nature of the dust are expected, both in terms of size and chemistry, that can only be observed with spatially resolved observations.

Van Boekel et al. (2004) have demonstrated the power of spectrally resolved MIR interferometry, by spatially resolving three protoplanetary disks surrounding HAeBe stars across the N band. The correlated spectra measured by VLTI/MIDI correspond to disk regions at 1–2 AU. By combining

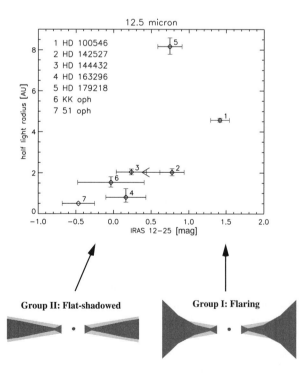

Fig. 4. Relation between 12.5-μm sizes (measured as half-light radius) and infrared slope (measured by IRAS colors) for the first HAeBe stars observed with MIDI on the VLTI. This size-color relation is consistent with expectations from the SED classification of *Meeus et al.* (2001) into objects with flared vs. flat outer disks.

these with unresolved spectra, the spectrum corresponding to outer disk regions at 2–20 AU can also be deduced. These observations have revealed radial gradients in dust crystallinity, particle size (grain growth), and, at least in one case (HD 142527), chemical composition. These early results have revived the discussion of radial and out-of-the plane mixing.

Interstellar particles observed in the direction of the galactic center are mainly small amorphous olivine grains ($Fe_{2-2x}Mg_{2x}SiO_4$) with some admixture of pyroxene grains ($Fe_{1-x}Mg_xSiO_3$) (*Kemper et al.,* 2004). In our planetary system, comets like Halley, Levi, or Hale-Bopp (*Hanner et al.,* 1994; *Crovisier et al.,* 1997) show the features of crystalline forsterite (an Mg-rich olivine, x = 1), in particular the conspicuous emission at 11.3 µm.

The VLTI/MIDI observations have revealed that crystallinization in young circumstellar disks can be a very efficient process, by suggesting a very high fraction of crystalline dust in the central 1–2 AU, and that the outer 2–20 AU disk possesses a markedly lower crystalline fraction, but still much higher than in the interstellar medium. Combined with the very young evolutionary status of these systems (as young as 1 m.y.), these observations imply that protoplanetary disks are highly crystalline well before the onset of planet formation. The presence of crystalline dust in relatively cold disk regions (or in solar system comets) is surprising, given that temperatures in these regions are well below the glass temperature of ~1000 K, but can be explained by chemical models of protoplanetary disks that include the effects of radial mixing and local processes in the outer disk (*Gail,* 2004; *Bockelée-Morvan et al.,* 2002). The radial gradient found in dust chemistry is also in qualitative agreement with the predictions of these models. Finally, the width of the silicate feature in the MIDI spectra corresponding to the inner and outer disk have revealed a radial gradient in (amorphous) grain size, with small grains (broader feature) being less abundant in the inner disk regions. This is perhaps expected, given that grain aggregation is a strong function of density (see chapter by Natta et al.). Figure 5 includes an example illustrating radial dust mineralogy gradients for the HAe object HD 144432.

These radial gradients in dust mineralogy are of course not restricted to HAeBe stars, but are expected in all YSO disks. In Fig. 5 we show preliminary VLTI/MIDI results for two low-mass objects of different evolutionary status, TW Hya and RY Tau (Leinert and Schegerer, in preparation). For TW Hya, the spatially unresolved N-band spectrum (shown on the right) shows no strong evidence for crystalline silicates, while the correlated flux spectrum (shown on the left) shows a weak but highly processed silicate band. Preliminary modeling indicates that the inner disk region of TW Hya contains a high fraction of crystalline silicates (30%), and that the amorphous grains have aggregated to larger units. As with the HAeBe stars, not all T Tauri objects show the same features; e.g., the effect is qualitatively similar but quantitatively less pronounced in RY Tau, as can be seen in Fig. 5.

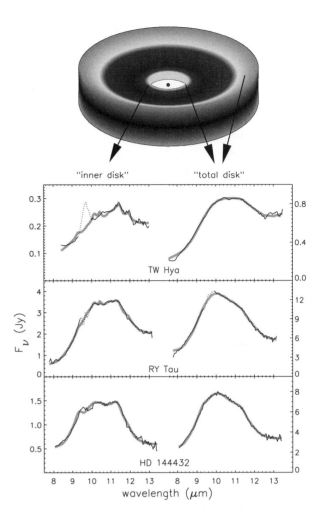

Fig. 5. Evidence for radial changes of dust composition in the circumstellar disks of young stars. Infrared spectra of the inner ≈1–2 AU are shown on the left, and of the total disk (≈1–20 AU) on the right. The different shapes of the silicate emission lines testify that in the inner disk — as probed by the interferometric measurement — the fraction of crystalline particles is high while the fraction of small amorphous particles is significantly reduced. TW Hya (0.3 $L_\odot$) and RY Tau (18 $L_\odot$) approximately bracket the luminosity range of T Tauri stars; the 10 $L_\odot$ HAe star HD 144432 (from *van Boekel et al.,* 2004) is included to emphasize the similarity between these two classes of young objects. Courtesy of R. van Boekel.

4. OTHER PHENOMENA

4.1. Outflows and Winds

The power of spectrally resolved interferometric measurements has recently become available in the NIR spectral range as well [VLTI/AMBER, *Malbet et al.* (2004); CHARA/MIRC, *Monnier et al.* (2004a)]. In addition to providing the detailed wavelength dependence of inner disk continuum (dust) emission, these capabilities enable detailed studies of the physical conditions and kinematics of the gaseous components in which emission and absorption lines arise (e.g., Br_γ, CO, and H_2 lines as probes of hot winds, disk rotation, and outflows, respectively).

As a rather spectacular example of this potential, during its first commissioning observations the VLTI/AMBER instrument spatially resolved the luminous HBe object MWC 297, providing visibility amplitudes as a function of wavelength at intermediate spectral resolution R = 1500 across a 2.0–2.2-μm band, and in particular a finely sampled Br_γ emission line (*Malbet et al., 2006*). The interferometer visibilities in the Br_γ line are ~30% lower than those of the nearby continuum, showing that the Br_γ emitting region is significantly larger than the NIR continuum region.

Known to be an outflow source (*Drew et al., 1997*), a preliminary model has been constructed by *Malbet et al.* (2006) in which a gas envelope, responsible for the Br_γ emission, surrounds an optically thick circumstellar disk (the characteristic size of the line-emitting region being 40% larger than that of the NIR disk). This model is successful at reproducing the new VLTI/AMBER measurements as well as previous continuum interferometric measurements at shorter and longer baselines (*Millan-Gabet et al., 2001*; *Eisner et al., 2004*), the SED, and the shapes of the H_α, H_β, and Br_γ emission lines.

The precise nature of the MWC 297 wind, however, remains unclear; the limited amount of data obtained in these first observations cannot, for example, discriminate between a stellar or disk origin for the wind, or between competing models of disk winds (e.g., *Casse and Ferreira, 2000*; *Shu et al., 1994*). These key questions may, however, be addressed in follow-up observations of this and similar objects, perhaps exploiting enhanced spectral resolution modes (up to R = 10,000 possible with VLTI/AMBER) and closure phase capabilities.

Quirrenbach et al. (2006) have also presented preliminary VLTI/MIDI results of rich spectral content for the wind environment in another high-luminosity YSO, MWC 349-A. The general shape of the N-band spectrally resolved visibilities is consistent with disk emission, as in the *Leinert et al.* (2004) results (section 3.1). In addition, the visibility spectrum contains unprecedented richness, displaying over a dozen lines corresponding to wind emission in forbidden ([NeII], [ArIII], [SIV]) and H recombination lines. The relative amplitudes of the visibilities measured in the continuum and in the lines indicate that the forbidden line region is larger than the dust disk, and that the recombination region is smaller than the dust disk (at least along some directions). Moreover, differential phases measured with respect to the continuum also display structure as a function of wavelength, showing clear phase shifts that encode information about the relative locations of the emitting regions for the strongest emission lines.

4.2. Multiplicity and Stellar Masses

Most young (and main-sequence) stars are members of multiple systems, likely as a result of the star-formation process itself (see chapters by Goodwin et al. and Duchene et al.). The measurement of the physical orbits of stars in multiple systems provide the only direct method for measuring stellar masses, a fundamental stellar parameter. In turn, by placing stars with measured masses in an HR diagram, models of stellar structure and evolution can be critically tested and refined, provided the mass measurements have sufficient accuracy (see chapter by Mathieu et al.). Generally speaking, dynamical and predicted masses agree well for 1–10-M$_\odot$ stars in the main sequence. However, fundamental stellar properties are much less well known for pre-main-sequence (PMS) stars, particularly of low mass (*Hillenbrand and White, 2004*), and call for mass measurements with better than 10% accuracy.

Optical interferometers can spatially resolve close binaries (having separations of order ~1 mas, or 0.10 AU at 150 pc), and establish the apparent astrometric orbits, most notably its inclination (for eclipsing binaries, the inclination is naturally well constrained). In combination with radial velocity measurements using Doppler spectroscopy, a full solution for the physical orbit and the properties of the system can then be obtained, in particular the individual stellar masses and luminosities. This method has proved very fruitful for critical tests of stellar models for non-PMS stars (e.g., *Boden et al., 2005a*); however, few simultaneous radial velocity and astrometric measurements exist for PMS stars.

Boden et al. (2005b) performed the first direct measurement of PMS stellar masses using optical interferometry, for the double-lined system HD 98800-B (a pair in a quadruple system located in the TW Hya association). Using observations made at the KI and by the fine guidance sensors onboard the Hubble Space Telescope, and in combination with radial velocity measurements, these authors establish a preliminary orbit that allowed determination of the (subsolar) masses of the individual components with 8% accuracy. Comparison with stellar models indicate the need for subsolar abundances for both components, although stringent tests of competing models will only become possible when more observations improve the orbital phase coverage and thus the accuracy of the stellar masses derived.

Naturally, disk phenomena (circumstellar and/or circumbinary) are also often observed in young multiple systems (see chapter by Monin et al.), and a number of recent observations also address this interesting and more complicated situation. Indeed, successful modeling of the individual HD 98800-B components by *Boden et al.* (2005b) requires a small amount of extinction ($A_V = 0.3$) toward the B components, not required toward the A components, and possibly due to obscuration by circumbinary disk material around the B system [originally hypothesized by *Tokovinin* (1999)]. As another example, based on a low-level oscillation in the visibility amplitude signature in the PTI data for FU Ori, *Malbet et al.* (2005) claim the detection of an off-centered spot embedded in the disk that could be physically interpreted as a young stellar or protoplanetary companion (located at ~10 AU), and could possibly be at the origin of the FU Ori outburst itself.

As noted by *Mathieu* (1994), spectroscopic detection of radial velocity changes is challenging in the presence of strong emission lines, extreme veiling, or rapid rotation, all of which are common in PMS stars. Therefore, imaging techniques are unique discovery tools for PMS multiples. For the closest systems in particular, interferometric techniques can reveal companions with separations ~10× smaller than single-aperture techniques (e.g., speckle or adaptive optics). Therefore, the increasing number of new detections will also add an important new sample to statistical studies aimed at determining the multiplicity fraction for young stars; its dependence on separation, stellar properties, and evolutionary status; and implications for circumstellar disk survival in close binary environments.

5. SUMMARY AND OPEN QUESTIONS

5.1. Summary

Long-baseline interferometers operating at near- and mid-infrared wavelengths have spatially resolved the circumstellar emission of a large number of YSOs of various types (60 objects published to date), most within the last few years. While the new observables are relatively basic in most cases (broadband visibility amplitudes), they have provided fundamentally new direct information (characteristic sizes and crude geometry), placing powerful new constraints on the nature of these circumstellar environments. In addition, more powerful observational capabilities combining high spatial resolution and spectral resolution have just begun to deliver their potential as unique tools to study YSO environments in exquisite detail and to provide breakthrough new molecular and kinematic data.

The principal well-established observational facts may be summarized as follows:

1. For HAe, late HBe, and T Tauri objects the measured NIR characteristic sizes are much larger (by factors of ~3–7) than predicted by previous-generation models that had been reasonably successful at reproducing the unresolved spectrophotometry. In particular, the measured sizes are larger than initial inner dust hole radii estimates and magnetospheric truncation radii. The measured NIR sizes strongly correlate with the central luminosity, and the empirical NIR size vs. luminosity relation (Fig. 2) suggests that the NIR emission is located at radii corresponding to sublimation radii for dust directly heated by the central star. These measurements have motivated in part the development of a new class of models for the origin of the NIR disk emission (the "puffed-up" inner wall models), which not only provide qualitative agreement with the interferometer data, but also solves the SED "NIR bump" problem for HAe objects.

2. Some (but not all) of the earliest-type HBe objects are different than the later types in terms of their NIR size scales. Many are inferred to have relatively small inner dust holes in better agreement with the classical disk picture, while others have larger NIR sizes consistent with the inner rim models (or perhaps even larger than predictions of either competing disk model, e.g., LkHα 101).

3. FU Ori, the prototype for the subclass of YSOs expected to have disks that dominate the total luminosity, is

well described by the canonical model temperature law (based on modeling the NIR interferometry data and SED). For other systems in this class, however, the observations reveal more complexity and the need for data at multiple baselines and wavelengths in order to discriminate between competing scenarios.

4. Young disks have also been resolved at MIR wavelengths, revealing characteristic sizes that correlate with red color (IRAS 12–25 m). This link appears to support an SED-based classification based on the degree of outer disk flaring, although the current sample of measured MIR sizes is too small to extract a firm conclusion.

5. Spectrally resolved MIR visibilities have emerged as a powerful tool to probe the dust mineralogy in young disks. Silicate dust appears to be deficient in small (0.1 μm) amorphous grains and more crystalline (less amorphous) close to the central star when compared to the disk as a whole.

6. The power of spectrally resolved visibilities, both NIR and MIR, has also been dramatically demonstrated in observations of lines corresponding to the gas component in wind phenomena associated with high-mass young stars.

7. Masses have been measured for young stars for the first time using interferometric techniques, in combination with spectroscopic radial velocity techniques, and a number of systems are being pursued with the primary goal of providing useful constraints to state-of-the-art models of stellar structure and evolution.

5.2. Open Questions

The new interferometric observations have allowed a direct view of the inner regions of YSO accretion disks for the first time. While this work has motivated significant additions to our standard picture of these regions (such as the puffed-up inner rim and spatially varying dust properties), continued theoretical development will require more sophisticated and complete interferometry data. In this section, we outline the most pressing unresolved issues exposed by the progress describe above:

1. A disk origin for the resolved NIR emission has not been unambiguously established. Most notably, *Vincovik et al.* (2006, and references therein) consider that the remaining scatter in the size-luminosity relation for HAeBe objects indicates that the bright-ring model does not adequately describe all the objects with a uniform set of parameters, and propose instead a disk plus compact halo model, where the resolved NIR emission in fact corresponds to the compact halo component. Because the halo need not be spherically symmetric, the elongations detected at the PTI do not rule it out. Effectively reigniting the disk vs. envelope debate of the previous decade (see, e.g., *Natta et al.*, 2000), *Vincovik et al.* (2006) also show that this model equivalently reproduces the SED, including the NIR bump.

2. The detailed properties of the putative puffed-up inner rim remain to be established. Assumptions about the grain size and density directly impact the expected rim location (dust sublimation radius), and based on size comparisons (Fig. 2) current observations appear to indicate that a

range of properties may be needed to explain all disks, particularly among the T Tauri objects. HAe objects have NIR sizes consistent with optically thin dust and relatively large grains (~1 µm). The lowest-luminosity T Tauri objects, however, favor either optically thick dust or smaller grains, or a combination of both. Currently, however, the T Tauri sample is still small, several objects are barely resolved, and part of the scatter could be due to unreliable SED decomposition due to photometric variability and uncertain stellar properties. Detailed modeling of individual objects and crucial supporting observations such as infrared veiling (to directly establish the excess flux) are needed to help resolve these uncertainties.

3. Using IR spectroscopy, it has been established that the *gas* in young disk systems extends inward of the interferometrically deduced inner dust radii (see chapter by Najita et al.). Whether or not magnetospheric accretion theories can accommodate these observations remains to be explored. We note that current interpretations of the CO spectroscopy do not take into account the inner rim, relying instead on previous-generation models that have been all but ruled out for the inner few AU. Finally, effects of NIR emission by gas in the inner disk, both on the measured visibilities and on the SED decomposition (both of which affect the inferred NIR sizes), need to be quantified in detail on a case-by-case basis.

4. Large-scale (~50–1000 mas) "halos" contributing small but nonnegligible NIR flux (~10%) are often invoked as an additional component needed to satisfactorily model some YSO disks [e.g., the FU Orionis objects, or most recently in the IOTA-3T HAeBe sample of *Monnier et al.* (2006)]. The origin of this halo material is, however, unclear, as is its possible connection with the compact halos proposed by *Vincovik et al.* (2006). The observational evidence should motivate further work on such multicomponent models.

5. Late (HAe, late HBe) and early HBe systems appear to have different NIR scale properties (see section 2.1), although there are notable exceptions, and the sample is small. Differences between these types have also been noted in terms of their millimeter emission (absent in most early HBe, in contrast to the late Herbig and T Tauri objects), perhaps due to outer-disk photoevaporation (*Hollenbach et al.*, 2000; see also chapter by Dullemond et al.). Whether the observed differences in inner disk structure are due to different accretion mechanisms (disk accretion vs. magnetospheric accretion) or to gas opacity effects remains to be investigated.

6. Do all FU Orionis objects conform to the canonical accretion disk model? If so, the additional question arises as to why their temperature structure would be that simple. Their light curves are fading, indicating a departure from a steady-state disk. Indeed, the high outburst accretion rates are expected to decay, and to do so in a radially dependent manner, leading to nonstandard temperature laws. First observations using VLTI/MIDI do not indicate a simple connection between the NIR and MIR radial disk structure. In-

deed, fitting the MIR visibilities and SED of V1647 Ori requires a very flat $q = -0.5$ radial temperature exponent (*Ábrahám et al.*, 2006), and preliminary analysis of FU Ori itself (*Quanz*, 2006) indicates that the model constructed by *Malbet et al.* (2005) to fit the NIR AMBER data and the SED does not reproduce (overestimates) the measured N-band visibilities.

7. The overall disk structure needs to be secured. While the NIR data probe primarily the hottest dust, the MIR observations are sensitive to a wider range of temperatures. Given the complexity of the disk structure under consideration, correctly interpreting the MIR and NIR data for the same object is proving challenging. Assumptions made for modeling the NIR data, the MIR dust features, and CO line profiles are generally not consistent with each other. Excitingly, the joint modeling of near- and mid-IR high-resolution data can yield the temperature profile of the disk surface layers.

8. Chemical models of protoplanetary disks, including the effects of radial transport and vertical mixing, must explain the observed dust mineralogy and the differences seen among different types of objects. Also, we note that the effect of the inner rim (which contributes ~20% of the MIR flux — see Fig. 3) on the interpretation of the observed silicate features and inferred mineralogy gradients must be assessed in detail.

6. FUTURE PROSPECTS

This review marks the maturation of two-telescope (single-baseline) observations using infrared interferometry. However, much more information is needed to seriously constrain the next generation of models (and the proliferating parameters) and to provide reliable density and temperature profiles as initial conditions for planet-formation theories. Fortunately, progress with modern interferometers continues to accelerate and will provide the unique new observations that are needed. In this section, we briefly discuss the expected scientific impact on yet-unexplored areas that can be expected from longer baselines, multiwavelength high-resolution data, spectral line and polarization capabilities, closure phase data, and aperture synthesis imaging. We will end with suggestions for disk modelers who wish to prepare for the new kinds of interferometric data in the pipeline.

6.1. Detailed Disk Structure

All interferometric observations of YSO disks to date have used baselines ≤100 m, corresponding to angular resolution of ~2 and 11 mas at 2.2 and 10 µm respectively. While sufficient to resolve the overall extent of the disks, longer baselines are needed to probe the internal structure of this hot dust emission. For instance, current data cannot simultaneously constrain the inner radius and thickness of the in-ner rim dust emission, the latter often assumed to be "thin" (10–25% of radius). Nor can current data independently determine the fraction of light from the disk compared to star (critical input to the models), relying instead on SED decomposition.

Longer baselines from the CHARA interferometer (330 m) and from the VLTI interferometer (200 m) can significantly increase the angular resolution and allow the rim emission to be probed in detail. For instance, if the NIR emission is contained in a thin ring, as expected for "hot inner wall" models, there will be a dramatic signal in the second lobe of the interferometer visibility response. Competing "halo" models predict smoother fall-offs in brightness (and visibility) and thus new long-baseline data will provide further definitive and completely unique constraints on the inner disk structure of YSOs [the first long-baseline CHARA observations of YSOs were presented by *Monnier et al.* (2005b)].

While even two-telescope visibility data provide critical constraints, ultimately we strive for aperture synthesis imaging, as routinely done at radio wavelengths. The first steps have recently been taken with closure phase results on HAeBe objects using IOTA/IONIC3 (*Millan-Gabet et al.*, 2006b; *Monnier et al.*, 2006). Closure phases are produced by combining light from three or more telescopes, and allow "phase" information to be measured despite atmospheric turbulence (see, e.g., *Monnier*, 2000). While this phase information is essential for image reconstruction, the closure phase gives unambiguous signal of "skewed" emission, such as would arise from a flared disk viewed away from the pole. "Skew" in this context indicates a deviation from a centro-symmetric brightness.

Closure phases are interesting for measuring the vertical structure of disks. Early hot inner wall models (e.g., *Dullemond et al.*, 2001) posited vertical inner walls. Recently, the view was made more realistic by *Isella and Natta* (2005) by incorporating pressure-dependent dust sublimation temperatures to curve the inner rim away from the midplane. Closure phases can easily distinguish between these scenarios because vertical walls impose strong skewed emission when a disk is viewed at intermediate inclination angles, while curved inner walls appear more symmetric on the sky unless viewed nearly edge-on. NIR closure phase data that can be used to measure the curvature and inner rim height through model fitting of specific sources will soon be available from IOTA/IONIC3, VLTI/AMBER, and CHARA/MIRC.

Aperture synthesis imaging of selected YSOs using the VLTI and CHARA arrays should also be possible within the next five years. By collecting dozens of closure phase triangles and hundreds of visibilities, simple images can be created, as has been demonstrated for LkHα 101 using the aperture-masking technique. The first steps in this direction have been presented for the triple T Tauri system GW Ori (also believed to contain circumstellar and circumbinary disks) by *Berger et al.* (2005), who present preliminary separation vectors and the first reconstructed image of a young multiple system. Ultimately, YSO imaging will allow the first model-independent tests of disk theories; all current interpretations are wed to specific (albeit general) models and the first images may be eye-opening.

As described earlier (section 3), it is essential to know the properties of circumstellar dust particles in YSO disks in order to interpret a broad range of observations, most especially in standard SED modeling. Interferometers can measure dust properties, such as size distribution and composition, using the combination of NIR, MIR, and polarization measurements. Efforts are being explored at the PTI, IOTA, VLTI, and CHARA interferometers to measure the disk emission in linearly polarized light [see *Ireland et al.* (2005) for similar first results for dust around AGB stars]. Because scattering is heavily dependent on grain size, the polarization-dependent size estimates at different wavelengths will pinpoint the sizes of the grains present in the inner disk.

With its nulling capability at MIR wavelengths, the KI (*Serabyn et al.*, 2004) will soon address a wide range of YSO phenomena, from spectrally resolved MIR disk structure in young disks to exozodiacal emission (*Serabyn et al.*, 2000), the latter being a crucial element in the selection of favorable targets for future spacebased planet-finding missions [e.g., Terrestrial Planet Finder/Darwin (*Fridlund*, 2003)].

Protoplanets forming in circumstellar disks are expected to carve fine structure, such as gaps, opening the possibility of directly detecting this process via high dynamic range interferometric techniques (e.g., highly accurate visibility amplitude measurements). While some initial simulations are pessimistic that new interferometers can resolve such disk structures (*Wolf et al.*, 2002), it is clear that theorists have only begun to investigate the many possibilities and interferometers are gearing up to meet the necessary observational challenges ahead.

Finally, although outside the scope of this review, we point out the tremendous potential of the upcoming Atacama Large Millimeter Array (ALMA) (*Wootten*, 2003). In particular, the combination of infrared, millimeter, and submillimeter interferometry will probe the entire YSO disk at all scales.

6.2. Dynamics

New spectral-line (e.g., VLTI/AMBER) capabilities have exciting applications. For instance, spectrointerferometry can measure the Keplerian rotation curves (in CO) for disks to derive dynamical masses of young stars as has been done using millimeter-wave interferometry (*Simon et al.*, 2000). Clearly, the potentially very powerful combination of spectroscopy and interferometry has only begun to be explored.

The dynamic timescale for inner disk material can be ≤1 yr, thus we might expect to see changes with time. The higher the angular resolution, the greater the possibility that we can identify temporal changes in the circumstellar structures. If disk inhomogeneities are present, new measurements will discover them and track their orbital motions. Evidence of inner disk dynamics has been previously inferred from reflection nebulosity [e.g., HH 30 (*Stapelfeldt et al.,* 1999)] and photometric variability (e.g., *Hamilton et al.,* 2001), but only recently directly imaged at AU scales [LkHα 101 (*Tuthill et al.,* 2002)].

6.3. The Star-Disk-Outflow Connection

The interface between the star, or its magnetosphere, and the circumstellar disk holds the observational key into how disks mediate angular momentum transfer as stars accrete material. The relevant spatial scales (1 to few $R_\star$) may be resolvable by the longest baselines available, providing unique tests of magnetospheric accretion theories (see the chapter by Bouvier et al.).

In addition to being surrounded by pre-planetary disks, YSOs are often associated with outflow phenomena, including optical jets, almost certainly associated with the disk accretion process itself (see chapter by Bally et al.). However, the precise origin of the jet phenomena is unknown, as is the physical relation between the young star, the disk, and jets (see chapter by Ray et al.). Discriminating among competing models, e.g., jets driven by external large-scale magnetic fields (the "disk wind" models or stellar fields models) or by transient local disk fields, can be helped by determining whether the jets launch near the star, or near the disk. Observationally, progress depends in part on attaining sufficient spatial resolution to probe the jet-launching region directly, a natural task for optical interferometers (e.g., *Thiebaut et al.,* 2003).

6.4. New Objects

Infrared interferometry can directly address important questions concerning the early evolution of high mass (O-B) protostars (see the chapter by Cesaroni et al. for a review of the pressing issues), and observational work in this area using infrared interferometry has just begun [with the VLTI/MIDI observations of *Quirrenbach et al.* (2006); *Feldt et al.* (2006)]. Most notably, it is not presently known what role, if any, accretion via circumstellar disks plays in the formation of high-mass stars. If present, interferometers can resolve circumstellar emission, and determine whether or not it corresponds to a circumstellar disk. Further, by studying disk properties as a function of stellar spectral type, formation timescales and disk lifetimes can be constrained.

Current studies have mainly targeted young disk systems, i.e., Class II (*Adams et al.,* 1987). More evolved, Class III, disks have remained thus far unexplored [the only exception being V380 Ori in *Akeson et al.* (2005a)], due to the sensitivity limitations that affected the first observations. Moreover, directly establishing the detailed disk structure of transition objects in the planet-building phase and of debris disks on sub-AU to tens of AU scales will likely be an energetic area of investigation as the field progresses toward higher dynamic range and spatial frequency coverage capabilities. First steps have been taken in this direction, with NIR observations that resolve the transition object TW Hya (*Eisner et al.,* 2006) and detect dust in the debris disk around Vega (*Ciardi et al.,* 2001; *Absil et al.,* 2006).

Finally, we note that prospects also exist for the direct detection of exoplanets from the ground, including interferometric techniques, and we refer the reader to the chapter by Beuzit et al.

6.5. Comments on Modeling

A number of new physical ingredients must be introduced into current disk models in order to take advantage of interferometer data. Disk models should take into account both vertical variations (settling) and radial dust differences (processing and transport). Also, the structure of the inner wall itself depends on the physics of dust destruction, and may depend on gas cooling and other physical effects not usually explicitly included in current codes. Perhaps codes will need to calculate separate temperatures for each grain size and type to accurately determine the inner rim structure. Last, we stress that the many free parameters in these models can only be meaningfully constrained through a data-driven approach, fitting not only SEDs (normal situation) but NIR and MIR interferometry simultaneously for many individual sources. With these new models, disk-mass profiles and midplane physical conditions will be known, providing crucial information bridging the late stages of star formation to the initial conditions of planet formation.

Acknowledgments. We thank R. van Boekel for providing Figs. 3 and 5 and for helpful discussions. We also thank many of our colleagues for illuminating discussions while preparing this review, in particular M. Benisty, J.-P. Berger, A. Boden, C. P. Dullemond, J. Eisner, L. Hillenbrand, and A. Quirrenbach.

REFERENCES

Ábrahám P., Kóspál Á., Csizmadia Sz., Kun M., Moór A., et al. (2004) *Astron. Astrophys., 428,* 89–97.
Ábrahám P., Mosoni, L., Henning Th., Kóspál Á., Leinert Ch., et al. (2006) *Astron. Astrophys., 449,* L13–L16.
Absil O., di Folco E., Mérand A., Augereau J.-C., Coudé Du Foresto V., et al. (2006) *Astron. Astrophys., 452,* 237–244.
Adams F. C., Lada C. J., and Shu F. H. (1987) *Astrophys. J., 312,* 788–806.
Akeson R. L., Ciardi D. R., van Belle G. T., Creech-Eakman M. J., and Lada E. A. (2000) *Astrophys. J., 543,* 313–317.
Akeson R. L., Ciardi D. R., van Belle G. T., and Creech-Eakman M. J. (2002) *Astrophys. J., 566,* 1124–1131.
Akeson R. L., Boden A. F., Monnier J. D., Millan-Gabet R., Beichman C., et al. (2005a) *Astrophys. J., 635,* 1173–1181.
Akeson R. L., Walker C. H., Wood K., Eisner J. A., and Scire E. (2005b) *Astrophys. J., 622,* 440–450.
Berger J.-P., Monnier J. D., Pedretti E., Millan-Gabet R., Malbet F., et al. (2005) In *PPV Poster Proceedings,* www.lpi.usra.edu/meetings/ppv2005/pdf/8398.pdf.
Bjorkman J. E. and Wood K. (2001) *Astrophys. J., 554,* 615–623.
Boden A. F., Torres G., and Hummel C. A. (2005a) *Astrophys. J., 627,* 464–476.
Boden A. F., Torres G., and Hummel C. A. (2005b) *Astrophys. J., 627,* 464–476.
Bockelée-Morvan D., Gautier D., Hersant F., Huré J.-M., and Robert F. (2002) *Astron. Astrophys., 384,* 1107–1118.
Casse F. and Ferreira J. (2000) *Astron. Astrophys., 353,* 1115–1128.
Chiang E. and Goldreich P. (1997) *Astrophys. J., 490,* 368.
Ciardi D. R., van Belle G. T., Akeson R. L., Thompson R. R., Lada E. A., et al. (2001) *Astrophys. J., 559,* 1147–1154.

Colavita M. M., Wallace J. K., Hines B. E., Gursel Y., Malbet F., et al. (1999) *Astrophys. J., 510,* 505–521.

Colavita M. M., Akeson R., Wizinowich P., Shao M., Acton S., et al. (2003) *Astrophys. J., 529,* L83–L86.

Colavita M. M., Wizinowich P. L., and Akeson R. L. (2004) *Proc. SPIE, 5491,* 454.

Clarke C. J., Gendrin A., and Sotomayor M. (2001) *Mon. Not. R. Astron. Soc., 328,* 485–491.

Crovisier J., Akeson R., Wizinowich P., Shao M., Acton S., et al. (1997) *Science, 275,* 1904–1907.

D'Alessio P., Calvet N., Hartmann L., Muzerolle J., and Sitko M. (2004) In *Star Formation at High Angular Resolution* (M. G. Burton et al., eds.), pp. 403–410. IAU Symposium 221, ASP, San Francisco.

Danchi W. C., Tuthill P. G., and Monnier J. D. (2001) *Astrophys. J., 562,* 440–445.

Dominik C., Dullemond C. P., Waters L. B. F. M., and Walch S. (2003) *Astron. Astrophys., 398,* 607–619.

Dullemond C. P. and Dominik C. (2004) *Astron. Astrophys., 417,* 159–168.

Dullemond C. P., Dominik C., and Natta A. (2001) *Astrophys. J., 560,* 957–969.

Drew J., Busfield G., Hoare M. G., Murdoch K. A., Nixon C. A., et al. (1997) *Mon. Not. R. Astron. Soc., 286,* 538–548.

Eisner J. A., Busfield G., Hoare M. G., Murdoch K. A., and Nixon C. A. (2003) *Astrophys. J., 588,* 360–372.

Eisner J. A., Lane B. F., Hillenbrand L. A., Akeson R. L., and Sargent A. I. (2004) *Astrophys. J., 613,* 1049–1071.

Eisner J. A., Hillenbrand L. A., White R. J., Akeson R. L., and Sargent A. I. (2005) *Astrophys. J., 623,* 952–966.

Eisner J. A., Chiang E. I., and Hillenbrand L. A. (2006) *Astrophys. J., 637,* L133–L136.

Feldt M., Pascucci I., Chesnau O., Apai D., Henning Th., et al. (2006) In *The Power of Optical/IR Interferometry: Recent Scientific Results and 2nd Generation VLTI Instrumentation* (F. Paresce et al., eds.), in press. ESO Astrophysics Symposia, Springer-Verlag, Garching.

Fridlund M. C. (2003) *Proc. SPIE, 5491,* 227.

Gail H.-P. (2004) *Astron. Astrophys., 413,* 571–591.

Garcia P. J. V., Glindeman A., Henning T., and Malbet F., eds. (2004) *The Very Large Telescope Interferometer — Challenges for the Future,* Astrophysics and Space Science Volume 286, Nos. 1–2, Kluwer, Dordrecht.

Gil C., Malbet F., Schoeller M., Chesnau O., Leinert Ch., et al. (2006) In *The Power of Optical/IR Interferometry: Recent Scientific Results and 2nd Generation VLTI Instrumentation* (F. Paresce et al., eds.), in press. ESO Astrophysics Symposia, Springer-Verlag, Garching.

Hamilton C., Herbst W., Shih C., and Ferro A. J. (2001) *Astrophys. J., 554,* L201–L204.

Hanner M. S., Lynch D. K., and Russell R. W. (1994) *Astrophys. J., 425,* 274–285.

Hale D., Bester M., Danchi W. C., Fitelson W., Hoss S., et al. (2000) *Astrophys. J., 537,* 998–1012.

Hartmann L. and Kenyon S. J. (1985) *Astrophys. J., 299,* 462–478.

Hartmann L. and Kenyon S. J. (1996) *Ann. Rev. Astron. Astrophys., 34,* 207–240.

Hartmann L., Calvet N., Gullbring E., and D'Alessio P. (1998) *Astrophys. J., 495,* 385.

Herbig G. H., Petrov P. P., and Duemmler R. (2003) *Astrophys. J., 595,* 384–411.

Hillenbrand L. A. and White R. J. (2004) *Astrophys. J., 604,* 741–757.

Hillenbrand L. A., Strom S. E., Vrba F. J., and Keene J. (1992) *Astrophys. J., 397,* 613–643.

Hinz P. M., Hoffmann W. F., and Hora J. L. (2001) *Astrophys. J., 561,* L131–L134.

Hollenbach D., Yorke H. W., and Johnstone D. (2000) In *Protostars and Planets IV* (V. Mannings et al., eds.), pp. 401–428. Univ. of Arizona, Tucson.

Ireland M., Tuthill P. G., Davis J., and Tango W. (2005) *Mon. Not. R. Astron. Soc., 361,* 337–344.

Isella A. and Natta A. (2005) *Astron. Astrophys., 438,* 899–907.

Jayawardhana R., Fisher S., Hartmann L., Telesco C., Pina R., et al. (1998) *Astrophys. J., 503,* L79.

Johns-Krull C. M. and Valenti V. A. (2001) *Astrophys. J., 561,* 1060–1073.

Kemper F., Vriend W. J., and Tielens A. G. G. M. (2004) *Astrophys. J., 609,* 826–837.

Kenyon S. J. and Hartmann L. W. (1987) *Astrophys. J., 323,* 714–733.

Kenyon S. J. and Hartmann L. W. (1991) *Astrophys. J., 383,* 664–673.

Koerner D. W., Ressler M. E., Werner M. W., and Backman D. E. (1998) *Astrophys. J., 503,* L83.

Kuchner M. and Lecar M. (2002) *Astrophys. J., 574,* L87–L89.

Lachaume R., Malbet F., and Monin J.-L. (2003) *Astron. Astrophys., 400,* 185–202.

Lawson P., ed. (2000) *Principles of Long Baseline Interferometry,* Jet Propulsion Laboratory, Pasadena.

Leinert Ch., Haas M., Abraham P., and Richichi A. (2001) *Astron. Astrophys., 375,* 927.

Leinert Ch., Graser U., Przygodda F., Waters L. B. F. M., Perrin G., et al. (2003) *Astrophys. Space Sci., 286,* 73–83.

Leinert Ch., van Boekel R., Waters L. B. F. M., Chesneau O., Malbet F., et al. (2004) *Astron. Astrophys., 423,* 537–548.

Lin D. N. C., Bodenheimer P., and Richardson D. C. (1996) *Nature, 380,* 606–607.

Liu W. M., Hinz P. M., Meyer M. R., Mamajek E. E., et al. (2003) *Astrophys. J., 598,* L111–L114.

Liu W. M., Hinz P. M., Hoffmann W. F., Brusa G., Miller D., et al. (2005) *Astrophys. J., 618,* L133–L136.

Lynden-Bell D. and Pringle J. E. (1974) *Mon. Not. R. Astron. Soc., 168,* 603–637.

Malbet F. and Bertout C. (1995) *Astron. Astrophys. Suppl., 113,* 369.

Malbet F., Berger J.-P., Colavita M. M., Koresko C. D., Beichman C., et al. (1998) *Astrophys. J., 507,* L149–L152.

Malbet F., Driebe T. M., Foy R., Fraix-Burnet D., Mathias P., et al. (2004) *Proc. SPIE, 5491,* 1722.

Malbet F., Lachaume R., Berger J.-P., Colavita M. M., di Folco E., et al. (2005) *Astron. Astrophys., 437,* 627–636.

Malbet F., Benisty M., de Wit W. J., Kraus S., Meilland A., et al. (2006) *Astron. Astrophys.,* in press.

Mathieu R. D. (1994) *Ann. Rev. Astron. Astrophys., 32,* 465–530.

McCabe C., Duchene G., and Ghez A. (2003) *Astrophys. J., 588,* L113–L116.

Meeus G., Waters L. B. F. M., Bouwman J., van den Ancker M. E., Waelkens C., et al. (2001) *Astron. Astrophys., 365,* 476–490.

Millan-Gabet R., Schloerb F. P., Traub W. A., Malbet F., Berger J.-P., and Bregman J. D. (1999) *Astrophys. J., 513,* L131–L143.

Millan-Gabet R., Schloerb F. P., and Traub W. A. (2001) *Astrophys. J., 546,* 358–381.

Millan-Gabet R., Monnier J. D., Akeson R. L., Hartmann L., Berger J.-P. et al. (2006a) *Astrophys. J., 641,* 547–555.

Millan-Gabet R., Monnier J. D., Berger J.-P., Traub W. A., Schloerb F. P., et al. (2006b) *Astrophys J., 645,* L77–L80.

Monnier J. D. (2000) In *Principles of Long Baseline Interferometry* (P. Lawson, ed.), pp. 203–226. Jet Propulsion Labratory, Pasadena.

Monnier J. D. (2003) *Rep. Progr. Phys., 66,* 789–897.

Monnier J. D. and Millan-Gabet R. (2002) *Astrophys. J., 579,* 694–698.

Monnier J. D., Berger J.-P., Millan-Gabet R., and Ten Brummelaar T. A. (2004a) *Proc. SPIE, 5491,* 1370.

Monnier J. D., Tuthill P. G., Ireland M. J., Cohen R., and Tannirkulam A. (2004b) In *AAS Meeting 2005,* 17.15.

Monnier J. D., Millan-Gabet R., Billmeier R., Akeson R. L., Wallace D., Berger J.-P., et al. (2005a) *Astrophys. J., 624,* 832–840.

Monnier J. D., Pedretti E., Millan-Gabet R., Berger J.-P., Traub W., et al. (2005b) In *PPV Poster Proceedings,* www.lpi.usra.edu/meetings/ppv2005/pdf/8238.pdf.

Monnier J. D., Berger J.-P., Millan-Gabet R., Traub W., Schloerb F. P., et al. (2006) *Astrophys J.,* in press.

Muzerolle J., Calvet N., Hartmann L., and D'Alessio P. (2003) *Astrophys. J., 597,* L149–L152.

Muzerolle J., D'Alessio P., Calvet N., and Hartmann L. (2004) *Astrophys. J., 617,* 406–417.

Natta A., Grinin V., and Mannings V. (2000) In *Protostars and Planets IV* (V. Mannings et al., eds.), pp. 559–588. Univ. of Arizona, Tucson.

Natta A., Prusti T., Neri R., Wooden D., Grinin V. P., et al. (2001) *Astron. Astrophys., 371,* 186–197.

Paresce F., Richichi A., Chelli A., and Delplancke F., eds. (2006) *The Power of Optical/IR Interferometry: Recent Scientific Results and 2nd Generation VLTI Instrumentation.* ESO Astrophysics Symposia, Springer-Verlag, Garching.

Perrin G. and Malbet F., eds. (2003) *Observing with the VLTI.* EAS Publications Series, Vol. 6.

Quanz S. P. (2006) In *The Power of Optical/IR Interferometry: Recent Scientific Results and 2nd Generation VLTI Instrumentation* (F. Paresce et al., eds.), in press. ESO Astrophysics Symposia, Springer-Verlag, Garching.

Quirrenbach A. (2001) *Ann. Rev. Astron. Astrophys., 39,* 353–401.

Quirrenbach A., Albrecht S., and Tubbs R. N. (2006) In *Stars with the B[e] Phenomenon* (M. Kraus and A. S. Miroshnichenko, eds.), in press. ASP Conf. Series, San Francisco.

Rice W. K. M., Wood K., Armitage P. J., Whitney B. A., and Bjorkman J. E. (2003) *Mon. Not. R. Astron. Soc., 342,* 79–85.

Serabyn E., Colavita M. M., and Beichman C. A. (2000) In *Thermal Emission Spectroscopy and Analysis of Dust, Disks, and Regoliths* (M. L. Sitko et al., eds.), pp. 357–365. ASP Conf. Series 196, San Francisco.

Serabyn E., Booth A. J., Colavita M. M., Creech-Eakman M. J., Crawford S. L., et al. (2004) *Proc. SPIE, 5491,* 806.

Shu F., Najita J., Ostriker E., Wilkin F., Ruden S., et al. (1994) *Astrophys. J., 429,* 781–807.

Simon M., Dutrey A., and Guilloteau S. (2000) *Astrophys. J., 545,* 1034–1043.

Stapelfeldt K., Watson A. M., Krist J. E., Burrows C. J., Crisp D., et al. (1999) *Astrophys. J., 516,* L95–L98.

ten Brummelaar T. A., McAlister H. A., Ridgway S. T., Bagnuolo W. G. Jr., Turner N. H., et al. (2005) *Astrophys. J., 628,* 453–465.

Thiébaut E., Garcia P. J. V., and Foy R. (2003) *Astrophys. J. Suppl., 286,* 171.

Tokovinin A. (1999) *Astron. Lett., 25,* 669–671.

Traub W. A., Berger. J.-P., Brewer M. K., Carleton N. P., Kern P. Y., et al. (2004) *Proc. SPIE, 5491,* 482.

Tuthill P., Monnier J. D., and Danchi W. C. (2001) *Nature, 409,* 1012–1014.

Tuthill P., Monnier J. D., Danchi W. C., Hale D. D. S., and Townes C. H. (2002) *Astrophys. J., 577,* 826–838.

van Boekel R., Min M., Leinert Ch., Waters L. B. F. M., Richichi A., et al. (2004) *Nature, 432,* 479–482.

van Boekel R., Dullemond C. P., and Dominik C. (2005) *Astron. Astrophys., 441,* 563–571.

Vink J. S., Drew J. E., Harries T. J., Oudmaijer R. D., and Unruh Y. (2005) *Mon. Not. R. Astron. Soc., 359,* 1049.

Vinković D., Ivezić Ž., Jurkić T., and Elitzur M. (2006) *Astrophys. J., 636,* 348–361.

Wilkin F. P. and Akeson R. L. (2003) *Astrophys. Space Sci., 286,* 145–150.

Wolf S., Gueth F., Henning T., and Kley W. (2002) *Astrophys. J., 566,* L97.

Wootten A. (2003) *Proc. SPIE, 4837,* 110.

Models of the Structure and Evolution of Protoplanetary Disks

C. P. Dullemond
Max-Planck-Institute for Astronomy, Heidelberg

D. Hollenbach
NASA Ames Research Center

I. Kamp
Space Telescope Division of the European Space Agency

P. D'Alessio
Centro de Radioastronomía y Astrofísica, México

We review advances in the modeling of protoplanetary disks. This review will focus on the regions of the disk beyond the dust sublimation radius, i.e., beyond 0.1–1 AU, depending on the stellar luminosity. We will be mostly concerned with models that aim to fit spectra of the dust continuum or gas lines, and derive physical parameters from these fits. For optically thick disks, these parameters include the accretion rate through the disk onto the star, the geometry of the disk, the dust properties, the surface chemistry, and the thermal balance of the gas. For the latter we are mostly concerned with the upper layers of the disk, where the gas and dust temperature decouple and a photoevaporative flow may originate. We also briefly discuss optically thin disks, focusing mainly on the gas, not the dust. The evolution of these disks is dominated by accretion, viscous spreading, photoevaporation, and dust settling and coagulation. The density and temperature structure arising from the surface layer models provide input to models of photoevaporation, which occurs largely in the outer disk. We discuss the consequences of photoevaporation on disk evolution and planet formation.

1. INTRODUCTION

Dusty circumstellar disks have been the focus of intense observational interest in recent years, largely because they are thought to be the birthplaces of planetary systems. These observational efforts have yielded many new insights on the structure and evolution of these disks. In spite of major developments in spatially resolved observations of these disks, much of our knowledge of their structure is still derived from spatially *un*resolved spectroscopy and spectral energy distributions (SEDs). The interpretation of this information (as well as spatially resolved data) requires the use of theoretical models, preferentially with as much realism and self-consistency as possible. Such disk models have been developed and improved over many years. When they are in reasonable agreement with observations they can also serve as a background onto which other processes are modeled, such as chemistry, grain growth, and ultimately the formation of planets.

This chapter reviews the development of such self-consistent disk structure models, and discusses the current status of the field. We focus on the regions of the disk beyond the dust sublimation radius, since the very inner regions are discussed in the chapter by Najita et al. To limit our scope further, we restrict our review to models primarily aimed at a comparison with observations. We will start with a concise resumé of the formation and viscous evolution of disks (section 2). This sets the radial disk structure as a function of time. We then turn our attention to the vertical structure, under the simplifying assumption that the gas temperature equals the dust temperature everywhere (section 3). While this assumption is valid in the main body of the disk, it breaks down in the disk surface layers. Section 4 treats the gas physics and chemistry of these surface layers, where much of the spectra originate. Photoevaporation flows also originate from the warm surface layers, and affect the disk evolution, which is the topic of section 5.

2. FORMATION AND VISCOUS EVOLUTION OF DISKS

The formation of stars and planetary systems starts with the gravitational collapse of a dense molecular cloud core. Since such a core will always have some angular momentum at the onset of collapse, most of the infalling matter will not fall directly onto the protostar, but form a disk around

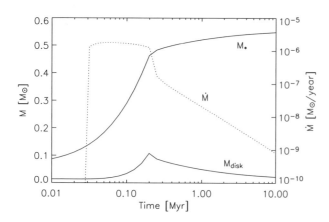

Fig. 1. Evolution of various disk and star quantities as a function of time after the onset of collapse of the cloud core (after *Hueso and Guillot,* 2005). Solid lines: stellar mass (upper) and disk mass (lower). Dotted line: accretion rate *in the disk*. In this model the disk is formed at t ≃ 0.03 m.y., causing the jump in the dotted line at this point. The collapse phase is finished by 2×10^5 yr.

it (e.g., *Terebey et al.,* 1984; *Yorke et al.,* 1993) or fragment into a multiple stellar system (e.g., *Matsumoto and Hanawa,* 2003). Because of the complexity of the latter, we focus on the single-star-formation scenario here. While matter falls onto the disk, viscous stresses and gravitational torques within the disk will transport angular momentum to its outer regions. As a consequence of this, most of the disk matter moves inward, adding matter to the protostar, while some disk matter moves outward, absorbing all the angular momentum (*Lynden-Bell and Pringle,* 1974). During its formation and evolution a disk will spread out to several 100 AU or more (*Nakamoto and Nakagawa,* 1994, henceforth *NN94*; *Hueso and Guillot,* 2005, henceforth *HG05*). This spreading is only stopped when processes such as photoevaporation (this chapter), stellar encounters (*Scally and Clarke,* 2001; *Pfalzner et al.,* 2005), or a binary companion (*Artymowicz and Lubow,* 1994) truncate the disk from the outside. During the collapse phase, which lasts a few × 10⁵ yr, the accretion rate within the disk is very high ($\dot{M} \sim 10^{-5}$–10^{-6} $M_\odot$/yr), but quickly drops to $\dot{M} \sim 10^{-7}$–10^{-9} $M_\odot$/yr once the infall phase is over (*NN94*; *HG05*). The optical and ultraviolet excess observed from classical T Tauri stars (CTTSs) and Herbig Ae/Be stars (HAeBes) confirms that this ongoing accretion indeed takes place (*Calvet et al.,* 2000, and references therein). In Fig. 1 we show the evolution of various disk and star parameters.

2.1. Anomalous Viscosity

An issue that is still a matter of debate is what constitutes the viscosity required for disk accretion, particularly after infall has ceased. Molecular viscosity is too small to account for the observed mass accretion rates. Turbulent and mag-

netic stresses, however, can constitute some kind of anomalous viscosity. The magnetorotational instability (MRI), present in weakly magnetized disks, is the most accepted mechanism to drive turbulence in disks and transport angular momentum outward (*Balbus and Hawley,* 1991; *Stone and Pringle,* 2001; *Wardle,* 2004, and references therein).

There is a disk region [0.2 < r < 4 AU, for typical CTTS disk parameters according to *D'Alessio et al.* (1998)] in which the ionization fraction is smaller than the minimum value required for the MRI. In this region neither thermal ionization (requiring a temperature higher than 1000 K), cosmic-ray ionization (requiring a mass surface density smaller than ~100 g/cm²) (*Jin,* 1996; *Gammie,* 1996), nor X-rays (*Glassgold et al.,* 1997a,b) are able to provide a sufficient number of free electrons to have MRI operating near the midplane. *Gammie* (1996) proposed a layered accretion disk model, in which a "dead zone" is encased between two actively accreting layers. The precise extent of this dead zone is difficult to assess, because the number density of free electrons depends on detailed chemistry as well as the dust grain size distribution, since dust grains tend to capture free electrons (*Sano et al.,* 2000, and references therein). If the disk dust is like the interstellar dust, the MRI should be inhibited in large parts of the disk (*Ilgner and Nelson,* 2006), although this is still under debate (e.g., *Semenov et al.,* 2004; see chapter by Bergin et al.).

There are also other (nonmagnetic) mechanisms for anomalous viscosity, like the baroclinic instability (*Klahr and Bodenheimer,* 2003) or the shear instability (*Dubrulle et al.,* 2005), which are still subject to some controversy (see recent review by *Gammie and Johnson,* 2005). Angular momentum can also be transferred by global torques, such as through gravitational spiral waves (*Tohline and Hachisu,* 1990; *Laughlin and Bodenheimer,* 1994; *Pickett et al.,* 2003, and references therein) or via global magnetic fields threading the disk (*Stehle and Spruit,* 2001), possibly with hydromagneticwinds launched along them (*Blandford and Payne,* 1982; *Reyes-Ruiz and Stepinski,* 1996).

2.2.   α-Disk Models for Protoplanetary Disks

To avoid having to solve the problem of viscosity in detail, but still be able to produce sensible disk models, *Shakura and Sunyaev* (1973) introduced the "α-prescription," based on dimensional arguments. In this recipe the vertically averaged viscosity ν at radius r is written as $\nu = \alpha H_p c_s$, where H_p is the pressure scale height of the disk and c_s is the isothermal sound speed, both evaluated at the disk midplane where most of the mass is concentrated. The parameter α summarizes the uncertainties related to the sources of anomalous viscosity, and is often taken to be on the order of $\alpha \simeq 10^{-2}$ for sufficiently ionized disks.

From conservation of angular momentum, the mass surface density Σ of a *steady* disk (i.e., with a constant mass accretion rate $\dot{M}$), for radii much larger than the disk inner radius r_{in}, can be written as $\Sigma(r) \approx \dot{M}/3\pi\nu$. With $H_p = c_s/\Omega_K$, where Ω_K is the Keplerian angular frequency, we see that for $r \gg r_{in}$

$$\Sigma(r) = K \frac{\dot{M}}{r^{3/2} \alpha T_c(r)} \qquad (1)$$

where $T_c(r)$ is the midplane temperature of the disk at radius r and K is a constant with the value $K \equiv \sqrt{GM_\star} \mu m_p / 3\pi k$. Here μ is the mean molecular weight in units of the proton mass m_p, G is the gravitational constant, k is Boltzmann's constant, and $M_\star$ is the stellar mass. As we will show in section 3, most of the disk is "irradiation dominated," and consequently has temperature given approximately by $T_c \sim r^{-1/2}$. This results in the surface density going as $\Sigma \sim r^{-1}$. This surface density distribution is less steep than the so-called "minimum mass solar nebula" (MMSN), given by $\Sigma \sim r^{-3/2}$ (*Weidenschilling*, 1977; *Hayashi*, 1981). Strictly speaking, the MMSN does not necessarily represent the mass distribution at any instant, but the minimum mass that has passed through the disk during its lifetime (*Lissauer*, 1993, and references therein).

In reality, protoplanetary disks are not quite steady. After the main infall phase is over, the disk is not supplied anymore with new matter, and the continuing accretion onto the star will drain matter from the disk (see Fig. 1). In addition, the disk viscously expands and is subject to photoevaporation (see section 5). The timescale for "viscous evolution" depends on radius and is given by $t_{vis} \simeq r^2/\nu$, which for typical CTTS parameters is 1 m.y. at $r \simeq 100$ AU. Since for irradiated disks $t_{vis} \propto r$, the outer regions evolve the slowest, yet they contain most of the mass. These regions ($\gtrsim 50$–100 AU) therefore form a reservoir of mass constantly resupplying the inner regions. The latter can thus be approximately described by steady accretion disk models.

There might also be dramatic variability taking place on shorter timescales, as shown by FU Ori- and EX Lupi-type outbursts (*Gammie and Johnson*, 2005, and references therein). These outbursts can have various triggering mechanisms, such as thermal instability (*Kawazoe and Mineshige*, 1993; *Bell and Lin*, 1994), close passage of a companion star (*Bonnell and Bastien*, 1992; *Clarke and Syer*, 1996), or mass accumulation in the dead zone followed by gravitational instability (*Gammie*, 1996; *Armitage et al.*, 2001). Disks are therefore quite time-varying, and constant α steady disk models should be taken as zeroth-order estimates of the disk structure.

Given the challenges of understanding the disk viscosity from first principles, attempts have been made to find observational constraints on disk evolution (*Ruden and Pollack*, 1991; *Cassen*, 1996; *Hartmann et al.*, 1998; *Stepinski*, 1998). For example, *Hartmann et al.* (1998) study a large sample of CTTSs and find a decline in mass accretion rate with time, roughly described as $\dot{M} \sim t^{-1.5}$, which they compare to the analytic similarity solutions of *Lynden-Bell and Pringle* (1974) for the expanding disk. A similar type of observational constraint is the recently found rough corre-

lation $\dot{M} \propto M_\star^2$ (*Muzerolle et al.*, 2003a, 2005; *Natta et al.*, 2004). High-angular-resolution millimeter-wave continuum imaging can also help to constrain the mass surface density distribution. With this technique *Wilner et al.* (2000) concluded that $\Sigma \propto r^{-1}$ for TW Hydra. *Kitamura et al.* (2002) find $\Sigma \sim r^{-p}$, with $p = 0$–1 for a sample of T Tauri stars.

3. VERTICAL STRUCTURE OF DUSTY DISKS

With the radial structure following from accretion physics, as described above, the next issue is the vertical structure of these disks. Many authors have modeled this with full time-dependent two- and three-dimensional (magneto/radiation) hydrodynamics (e.g., *Boss*, 1996, 1997; *Yorke and Bodenheimer*, 1999; *Fromang et al.*, 2004). While this approach is obviously very powerful, it suffers from large computational costs, and often requires strong simplifying assumptions in the radiative transfer to keep the problem tractable. For comparison to observed spectra and images these models are therefore less practical. Most observation-oriented disk structure models split the disk into a series of (nearly independent) annuli, each constituting a one-dimensional or two-layer local vertical structure problem. In this section we review this kind of "one + one-dimensional" models, and their two- and three-dimensional generalizations.

3.1. Basic Principles

The main objective of the models described in this section is the determination of the density and temperature structure of the disk. For a given surface density $\Sigma(r)$, and a given gas temperature structure $T_g(r, z)$ (where z is the vertical coordinate measured upward from the midplane), the vertical density distribution $\rho(r, z)$ can be readily obtained by integrating the vertical equation of hydrostatics

$$\frac{dP}{dz} = -\rho \Omega_K^2 z \qquad (2)$$

where $P = \rho c_s^2$ with $c_s^2 \equiv kT_g/\mu m_p$. The main complexity of a disk model lies in the computation of the *temperature* structure. Since the main source of opacity is the dust, most models so far make the assumption that the gas temperature is equal to the dust temperature, so that the gas temperature determination reduces to solving a dust continuum radiative transfer problem. In section 4 we will relax this assumption, but until then we will keep it.

The temperature of the disk is set by a balance between heating and cooling. The disk cools by thermal emission from the dust grains at infrared wavelengths. This radiation is what is observed as infrared dust continuum radiation from such disks. Line cooling is only a minor coolant, and only plays a role for T_g when gas and dust are thermally

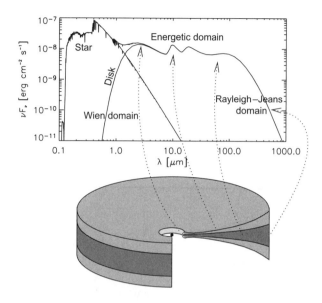

Fig. 2. Buildup of the SED of a flaring protoplanetary disk and the origin of various components: The near-infrared bump comes from the inner rim, the infrared dust features from the warm surface layer, and the underlying continuum from the deeper (cooler) disk regions. Typically the near- and mid-infrared emission comes from small radii, while the far-infrared comes from the outer disk regions. The (sub-)millimeter emission mostly comes from the midplane of the outer disk. Scattering is not included here.

decoupled. Dust grains can be heated in part by radiation from other grains in the disk. The iterative absorption and reemission of infrared radiation by dust grains in the disk causes the radiation to propagate through the disk in a diffusive way. Net energy input comes from absorption of direct stellar light in the disk's surface layers, and from viscous dissipation of gravitational energy in the disk due to accretion. For most disks around CTTSs and Herbig Ae/Be stars the heating by stellar radiation is dominant over the viscous heating (except in the very inner regions). Only for strongly accreting disks does the latter dominate.

Once the temperature structure is determined, the SED can be computed. The observable thermal emission of a dusty disk model consists of three wavelength regions (see Fig. 2). The main portion of the energy is emitted in a wavelength range depending on the minimum and maximum temperature of the dust in the disk. We call this the "energetic domain" of the SED, which typically ranges from 1.5 μm to about 100 μm. At shorter wavelength the SED turns over into the "Wien domain." At longer wavelengths the SED turns over into the "Rayleigh-Jeans domain," a steep, nearly power law profile with a slope depending on grain properties and disk optical depth (see chapter by Natta et al.). Differences in disk geometry are mainly reflected in the energetic domain of the SED, while the submillimeter and millimeter fluxes probe the disk mass.

3.2. A First Confrontation with Observations

It is quite challenging to solve the entire disk structure according to the above principles. Early disk models were therefore often based on strong simplifications. An example of such a model is a perfectly flat disk being irradiated by the star due to the star's nonnegligible size (*Adams and Shu*, 1986; *Friedjung*, 1985). The stellar radiation impinges onto the flat disk under an irradiation angle $\varphi \simeq 0.4 r_\star / r$ (with $r_\star$ the stellar radius). Neglecting viscous dissipation, the effective temperature of the disk is set by a balance between the irradiated flux $(1/2)\varphi L_\star / 4\pi r^2$ (with $L_\star$ the stellar luminosity) and blackbody cooling σT_{eff}^4, which yields $T_{eff} \propto r^{-3/4}$. The energetic domain of its SED therefore has a slope of $\nu F_\nu \propto \nu^s$ with $s = 4/3 = 1.33$, which follows from the fact that any disk with $T_{eff} \propto r^{-q}$ has an SED slope of $s = (4q - 2)/q$. This steep slope arises because most of the stellar radiation is absorbed and reemitted at small radii where the disk is hot. This produces strong emission at short wavelength. The long wavelength flux is weak because only little stellar radiation is absorbed at large radii. Observations of CTTSs, however, show SED slopes typically in the range $s = 0.6$ to 1 (*Kenyon and Hartmann*, 1995), i.e., much less steep. The SEDs of Herbig Ae/Be stars show a similar picture, but with a somewhat larger spread in s, although it must be kept in mind that the determination of the slope of a bumpy SED like in Fig. 2 is somewhat subjective. *Meeus et al.* (2001, henceforth *M01*) divide the SEDs of Herbig Ae/Be stars into two groups: those with strong far-infrared flux (called "Group I," having slope $s \simeq -1 \ldots 0.2$) and those with weak far-infrared flux (called "Group II," having slope $s \simeq 0.2 \ldots 1$). All but the most extreme Group II sources have a slope that is clearly inconsistent with that of a flat disk. Note, at this point, that the Meeus "Group I" and "Group II" are unrelated to the Lada "Class I" and "Class II" classification (both Meeus Group I and II are members of Lada Class II).

A number of authors have employed another model to interpret their observations of protoplanetary disks: that of a steady accretion disk heated by viscous dissipation (*Rucinski*, 1985; *Bertout et al.*, 1988; *Hillenbrand et al.*, 1992). These models are based on the model by *Shakura and Sunyaev* (1973). A detailed vertical structure model of such a disk was presented by *Bell et al.* (1997). The luminosity of such disks, including the magnetospheric accretion column, is $L_{accr} = GM_\star \dot{M}/r_\star$. For $r \gg r_{in}$ the effective temperature of such disks is given by $\sigma T_{eff}^4 = 3\dot{M}\Omega_K^2/8\pi$ (with σ the Stefan-Boltzmann constant), yielding an SED slope of $s = 4/3$, like for passive flat disks (*Lynden-Bell*, 1969; see solid lines of Fig. 6). Therefore these models are not very successful either, except for modeling very active disks like FU Orionis (FUOr) outbursts (see *Bell and Lin*, 1994).

3.3. Flaring Disk Geometry

It was recognized by *Kenyon and Hartmann* (1987) that a natural explanation for the strong far-infrared flux (i.e., shallow SED slope) of most sources is a flaring ("bowl-shaped") geometry of the disk's surface, as depicted in Fig. 2. The flaring geometry allows the disk to capture a significant portion of the stellar radiation at large radii where the disk is cool, thereby boosting the mid- to far-infrared emission.

The flaring geometry adds an extra term to the irradiation angle: $\varphi \simeq 0.4\, r_\star/r + r\,d(H_s/r)/dr$ (*Chiang and Goldreich*, 1997, henceforth *CG97*), where H_s is the height above the midplane where the disk becomes optically thick to the impinging stellar radiation. In the same way as for the flat disks the thermal balance determines the T_{eff} of the disk, but this now depends strongly on the shape of the disk: $H_s(r)$. The pressure scale height H_p, on the other hand, depends on the midplane temperature T_c by $H_p = \sqrt{kT_c r^3/\mu m_p GM_\star}$ (with $M_\star$ the stellar mass). If we set $T_c = T_{eff}$ and if the ratio $\chi \equiv H_s/H_p$ is known, then the system of equations is closed and can be solved (see appendix in *Chiang et al.*, 2001). For the special case that χ is constant we obtain $H_s \propto r^{9/7}$, *a posteriori* confirming that the disk indeed has a "bowl" shape. In general, though, χ must be computed numerically, and depends on the dust opacity of the disk upper layers. The resulting temperature profile is typically about $T_c \propto r^{-0.5}$.

The total (thermal + scattering) luminosity of such a nonaccreting flaring disk is $L_{disk} = C\, L_\star$, where C is the *covering fraction* of the disk. The covering fraction is the fraction of the starlight that is captured by the material in the disk. With a large enough disk inner radius ($r_{in} \gg r_\star$) one can write $C \simeq \max[H_s(r)/r]$. For an infinitely thin disk extending from $r_\star$ to $r \to \infty$ one has $C = 0.25$. The *observed* flux ratio $F_{disk}/F_\star$ may deviate from C by about a factor of 2 due to the anisotropy of disk emission.

In addition to irradiation by the star, the outer regions of a strongly accreting flaring disk (like an FUor object) can also be irradiated by the accretion luminosity from the inner disk (*Kenyon and Hartmann*, 1991; *Bell*, 1999; *Lachaume*, 2004) and by the emission from the magnetospheric accretion column or boundary layer (*Muzerolle et al.*, 2003b).

3.4. Warm Dust Surface Layer

A closer look at the physics of an irradiation-dominated disk (be it flat or flared) reveals that its surface temperature is generally higher than its interior temperature (*Calvet et al.*, 1991; *Malbet and Bertout*, 1991; *CG97*). Dust grains in the surface layers are directly exposed to the stellar radiation, and are therefore hotter than dust grains residing deep in the disk, which only "see" the infrared emission by other dust grains. The temperature difference is typically a factor of 2–4 for non- or weakly accreting disks (see curve labeled "–9" in Fig. 4). For nonnegligible accretion, on the other hand, the disk is heated from inside as well, producing a temperature minimum somewhere between the equatorial plane and the surface layer (see other curves in Fig. 4). Because of the shallow incidence angle of the stellar radiation $\varphi \ll 1$, the *vertical* optical depth of this warm surface layer is very low. The layer produces optically thin emis-

sion at a temperature higher than the effective temperature of the disk. The frequency-integrated flux of this emission is the same as that from the disk interior. As a consequence, the thermal radiation from these surface layers produces dust features in *emission*. This is exactly what is seen in nearly all non-edge-on T Tauri and Herbig Ae/Be star spectra (e.g., *M01*; *Kessler-Silacci et al.*, 2006), indicating that these disks are nearly always dominated by irradiation.

3.5. Detailed Models for Flaring Disks

3.5.1. Disk structure. Armed with the concepts of disk flaring and hot surface layers, a number of authors published detailed one + one-dimensional disk models and two-layer (surface + interior) models with direct applicability to observations. The aforementioned *CG97* model [with refinements described in *Chiang et al.* (2001)] is a handy two-layer model for the interpretation of SEDs and dust emission features from *nonaccreting* ("passive") disks. *Lachaume et al.* (2003) extended it to include viscous dissipation.

The models by *D'Alessio et al.* (1998) solve the complete one + one-dimensional disk structure problem with diffusive radiative transfer, including stellar irradiation and viscous dissipation (using the α prescription). The main input parameters are a global (constant) mass accretion rate $\dot{M}$ and α. The surface density profile $\Sigma(r)$ is calculated self-consistently. This model shows that the disk can be divided into three zones: an outer zone in which the disk is dominated by irradiation, an inner zone where viscous dissipation dominates the energy balance, and an intermediate zone where the midplane temperature is dominated by the viscous dissipation but the surface temperature by irradiation (see Fig. 3). In the intermediate zone the vertical disk thickness is set by $\dot{M}$ and α but the infrared spectrum is still powered by irradiation. In Fig. 4 the vertical structure of the disk is shown, for fixed Σ but varying $\dot{M}$ for r = 1 AU.

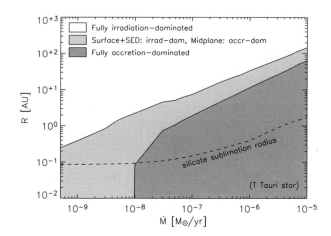

Fig. 3. Characteristic radii of a disk around a 0.9 $M_\odot$ star with $T_\star = 4000$ K and $R_\star = 1.9\, R_\odot$ at different accretion rates. The silicate sublimation radius is the dust "inner rim." Figure based on models by *D'Alessio et al.* (1998).

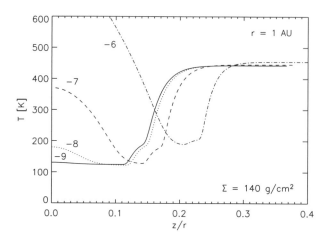

Fig. 4. Vertical temperature distribution of an irradiated α disk at 1 AU, for a fixed Σ (chosen to be that of a disk model with $\dot{M} = 10^{-8} M_\odot$/yr for α = 0.01), but varying $\dot{M}$, computed using the models of *D'Alessio et al.* (1998). The labels of the curves denote the 10-log of the accretion rate in $M_\odot$/yr.

The models described by *Dullemond et al.* (2002) apply exact one-dimensional wavelength-dependent radiative transfer for the vertical structure, but these models do not include viscous dissipation.

3.5.2. Dust growth and sedimentation. Models of the kind discussed above describe the SEDs of CTTSs reasonably well. However, *D'Alessio et al.* (1999) argue that they tend to slightly overproduce far-infrared flux and have too thick dark lanes in images of edge-on disks. They also show that the percentage of expected edge-on disks appears to be overpredicted. They suggest that dust sedimentation could help to solve this problem. *Chiang et al.* (2001) find similar results for a subset of their Herbig Ae/Be star sample: the Meeus Group II sources (see also *CG97*). They fit these sources by mimicking dust settling through a reduction of the disk surface height. Self-consistent computations of dust sedimentation produce similar SEDs and confirm the dust settling idea (*Miyake and Nakagawa*, 1995; *Dullemond and Dominik*, 2004b, henceforth *DD04b*; *D'Alessio et al.*, 2006). The disk thickness and far-infrared flux can also be reduced by grain growth (*D'Alessio et al.*, 2001; *Dullemond and Dominik*, 2004a). The chapter by Dominik et al. discusses such models of grain growth and sedimentation in detail.

From comparing infrared and (sub-)millimeter spectra of the same sources (*Acke et al.*, 2004), it is clear that small and big grains coexist in these disks. The (sub-)millimeter spectral slopes usually require millimeter-sized grains near the midplane in the outer regions of the disk, while infrared dust emission features clearly prove that the disk surface layers are dominated by grains no larger than a few micrometers (see chapter by Natta et al.). It appears that a bimodal grain size distribution can fit the observed spectra: a portion of submicrometer grains in the surface layers responsible for the infrared dust emission features and a por-

tion of millimeter-sized grains in the disk interior accounting for the (sub-)millimeter emission (*Natta et al.*, 2001).

3.6. The Dust "Inner Rim"

The very inner part of the disk is dust-free due to dust sublimation (see chapter by Najita et al. for a discussion of this region). The dusty part of the disk can therefore be expected to have a relatively abrupt inner edge at about 0.5 AU for a 50 $L_\odot$ star (scaling roughly with $\sqrt{L_\star}$). If the gas inward of this dust inner rim is optically thin, which appears to be mostly the case (*Muzerolle et al.*, 2004), then this dust inner rim is illuminated by the star at a ~90° angle, and is hence expected to be much hotter than the rest of the disk behind it, which is irradiated under a shallow angle φ ≪ 1 (*Natta et al.*, 2001). Consequently it must be hydrostatically "puffed-up," although this is still under debate. *Natta et al.* (2001) showed that the emission from such a hot inner rim can explain the near-infrared bump seen in almost all Herbig Ae/Be star SEDs (see, e.g., *M01*). This is a natural explanation, since dust sublimation occurs typically around 1500 K, and a 1500 K blackbody bump fits reasonably well to the near-infrared bumps in those sources. *Tuthill et al.* (2001) independently discovered a bright half-moon ring around the Herbig Be star LkHa 101, which they attribute to a bright inner disk rim due to dust sublimation. *Dullemond et al.* (2001, henceforth *DDN01*) extended the *CG97* model to include such a puffed-up rim, and *Dominik et al.* (2003) showed that the Meeus sample of Herbig Ae/Be stars can be reasonably well fitted by this model. However, for Meeus Group II sources these fits required relatively small disks (see, however, section 3.7).

The initial rim models were rather simplified, treating it as a vertical blackbody "wall" (*DDN01*). *Isella and Natta* (2005) improved this by noting that the weak dependence of the sublimation temperature on gas density is enough to strongly round off the rim. Rounded-off rims appear to be more consistent with observations than the vertical ones: Their flux is less inclination dependent, and their images on the sky are not so much one-sided. There is still a worry, though, whether the rims can be high enough to fit sources with a strong near-infrared bump.

With near-infrared interferometry the rim can be spatially resolved, and thus the models can be tested. The measurements so far do not yet give images, but the measured "visibilities" can be compared to models. In this way one can measure the radius of the rim (e.g., *Monnier et al.*, 2005; *Akeson et al.*, 2005) and its inclination (e.g., *Eisner et al.*, 2003). Moreover, it can test whether indeed the near-infrared emission comes from the inner rim of the dust disk in the first place [some doubts have been voiced by *Vinković et al.* (2003)]. We refer to the chapter by Millan-Gabet et al. for a more in-depth discussion of interferometric measurements of disks.

The inner rim model has so far been mainly applied to Herbig Ae/Be stars because the rim appears so apparent in the near-infrared (NIR). But *Muzerolle et al.* (2003b) showed that it also applies to T Tauri stars. In that case, however, the luminosity from the magnetospheric accretion shock is required in addition to the stellar luminosity to power the inner rim emission.

In addition to being a strong source of NIR flux, the "puffed-up" inner dust rim might also be responsible for the irregular few-day-long extinction events observed toward UX Orionis stars (*Natta et al., 2001; Dullemond et al., 2003*). The latter authors argued that this only works for self-shadowed (or only weakly flaring) disks (see section 3.7).

In Fig. 5 we summarize in a qualitative way how the disk geometry (inner rim, flaring) affects the SED shape of an irradiated passive disk. In Fig. 6 the SEDs of actively accreting disks are shown, in which the irradiation by the central star is ignored. In reality, both the accretional heating and the irradiation by the central star must be included in the models simultaneously.

3.7. Two-Dimensional Radiative Transfer in Disk Models

The models described so far are all based on an approximate one + one-dimensional (or two-layer) irradiation-angle description. In reality the structure of these disks is two-dimensional, if axisymmetry can be assumed, and three-dimensional if it cannot. Over the last 10 years many multidimensional dust continuum radiative transfer programs and algorithms were developed for this purpose (e.g., *Whitney et al., 1992; Lucy, 1999; Wolf et al., 1999; Bjorkman and Wood, 2001; Nicolinni et al., 2003; Steinacker et al., 2003*). Most applications of these codes assume a given density distribution and compute spectra and images. There

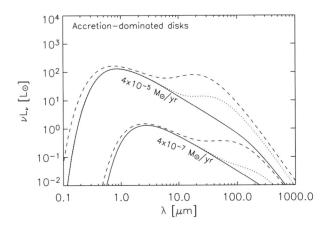

Fig. 6. Overall SED shape for accreting disks *without* stellar irradiation for two accretion rates. A simple Shakura-Sunyaev model is used here with gray opacities. Solid line: pure Shakura-Sunyaev model (star not included); dotted line: model with disk-self-irradiation included; dashed line: model with disk-self-irradiation *and* irradiation by the magnetospheric accretion column on the star included.

is a vast literature on such applications that we will not review here (see chapter by Watson et al.). But there is a trend to include the self-consistent vertical density structure into the models by iterating between radiative transfer and the vertical pressure balance equation (*Nomura, 2002; Dullemond, 2002, henceforth D02; Dullemond and Dominik, 2004a, henceforth DD04a; Walker et al., 2004*). The main improvements of two-dimensional/three-dimensional models over one + one-dimensional models is their ability to account for radial radiative energy diffusion in the disk, for cooling of the outer disk in radial direction, for the complex three-dimensional structure of the dust inner rim, and in general for more realistic model images.

In addition to this, two-dimensional/three-dimensional models allow for a "new" class of disk geometries to be investigated. The one + one-dimensional models can, because of their reliance on an irradiation angle φ, only model disk geometries that are either flat or flared. In principle, however, there might be circumstances under which, roughly speaking, the surface of the outer disk regions lies within the shadow of the inner disk regions (although the concept of "shadow" must be used with care here). These shadowed regions are cooler than they would be if the disk was flaring, but the two-dimensional/three-dimensional nature of radiative transfer prevents them from becoming entirely cold and flat. For Herbig Ae/Be stars the origin of the shadow might be the puffed-up inner rim (*D02; DD04a*), while for T Tauri stars it might be the entire inner flaring disk region out to some radius (*DD04b*).

Although the concept of "self-shadowing" is still under debate, it might be linked to various observable features of protoplanetary disks. For instance, *DD04a* showed

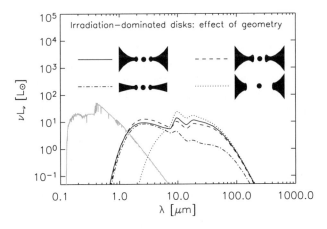

Fig. 5. Overall SED shape for *nonaccreting* disks with stellar irradiation, computed using the two-dimensional radiative transfer tools from *Dullemond and Dominik* (2004a). The stellar spectrum is added in grayscale. Scattered light is not included in these SEDs. Solid line is normal flaring disk with inner dust rim; dashed line is when the rim is made higher; dot-dashed line is when the flaring is reduced (or when the disk becomes "self-shadowed"); dotted line is when the inner rim is at 10× larger radius.

that self-shadowed disks produce SEDs consistent with Meeus Group II sources, while flaring disks generally produce Group I-type SEDs, unless the disk outer radius is very small. It might also underly the observed correlation between SED shape and submillimeter slope (*Acke et al.,* 2004). Moreover, self-shadowed disks, when spatially resolved in scattered light, would be much dimmer than flaring disks.

4. GAS TEMPERATURE AND LINE SPECTRA

Although the dust in disks is generally more easily observed, there is an obvious interest in direct observations of the gas. It dominates the mass, sets the structure, and impacts dust dynamics and settling in these disks. Moreover, it is important to estimate how long disks remain gas-rich, and whether this is consistent with the formation timescale of gas giant planets (*Hubickyj et al.,* 2004). Unfortunately, gas lines such as CO rotational, H_2 rotational, and atomic fine structure lines often probe those surface regions of disks in which the gas temperature is difficult to compute. The disk models we described above assume that the gas temperature in the disk is always equal to the local dust temperature. While this is presumably true for most of the matter deep within optically thick disks, in the tenuous surface layers of these disks (or throughout optically thin disks) the densities become so low that the gas will thermally decouple from the dust. The gas will acquire its own temperature, which is set by a balance between various heating and cooling processes. These processes depend strongly on the abundance of various atomic and molecular species, which, for their part, depend strongly on the temperature. The gas temperature, density, chemistry, radiative transfer, and radiation environment are therefore intimately intertwined and have to be studied as a single entity. This greatly complicates the modeling effort, and the first models that study this in detail have only recently been published.

This chapter focuses on stationary models, i.e., models that are in chemical, thermal, and hydrostatic equilibrium. For the tenuous regions of disks the chemical timescales are short enough that this is valid, in contrast to the longer chemical timescales deeper in the disk (e.g., *Aikawa and Herbst,* 1999; *Willacy and Langer,* 2000). The models constructed so far either solve the gas temperature/chemistry for a *fixed* gas density structure (*Jonkheid et al.,* 2004; *Kamp and Dullemond,* 2004), or include the gas density in the computation to obtain a self-consistent thermochemical hydrostatic structure (*Gorti and Hollenbach,* 2004; *Nomura and Millar,* 2005).

4.1. Basic Gas Physics

The physics and chemistry of the surface layers of protoplanetary disks strongly resembles that of photon-dominated regions (PDRs) (*Tielens and Hollenbach,* 1985; *Yamashita et al.,* 1993). In those surface layers the gas temperature generally greatly exceeds the dust temperature. But the dust-gas coupling gradually takes over the gas temperature balance as one gets deeper into the disk, typically beyond a vertical column depth of $A_V \simeq 1$, and forces the gas temperature to the dust temperature.

The uppermost surface layer contains mostly atomic and ionized species, since the high UV irradiation effectively dissociates all molecules (*Aikawa et al.,* 2002). The photochemistry is driven by the stellar UV irradiation and/or in case of nearby O/B stars, by external illumination. In flaring disk models, the stellar UV radiation penetrates the disk under an irradiation angle φ like the one described in the previous section. This radiation gets diluted with increasing distance from the central star and attenuated by dust and gas along an *inclined* path into the disk. The stellar UV radiation therefore penetrates less deep into the disk than external UV radiation. As one goes deeper into the surface layer, the gas becomes molecular (see chapter by Bergin et al.).

The thermal balance of the gas in disks is solved by equating all relevant heating and cooling processes. For this gas thermal balance equation, a limited set of key atomic and molecular species is sufficient: e.g., H_2, CO, OH, H_2O, C^+, O, Si^+, and various other heavy elements. For most atoms and molecules, the statistical equilibrium equation has to include the pumping of the fine structure and rotational levels by the cosmic background radiation, which become important deep in the disk, where stellar radiation cannot penetrate. The full radiative transfer in chemical models is very challenging, and therefore generally approximated by a simple escape probability approach, where the pumping and escape probability are derived from the optical depth of the line [similar to the approach of *Tielens and Hollenbach* (1985) for PDRs]. Even though the emitted photons travel in all directions, the optical depth used for this escape probability is the line optical depth in the *vertical* direction where the photons most readily escape.

One of the most critical ingredients of these models is the UV and X-ray radiation field (stellar and external), which can be split into the far-ultraviolet (FUV, 6–13.6 eV), the extreme-ultraviolet (EUV, 13.6–100 eV), and X-ray ($\geq$100 eV) regime. In the literature the far ultraviolet radiation field (FUV) is often represented by a single parameter G0 describing the integrated intensity between 912 and 2000 Å normalized to that of the typical interstellar radiation field (*Habing,* 1968). However, several papers have shown the importance of a more detailed description of the radiation field for calculations of the chemistry and the gas heating/cooling balance (*Spaans et al.,* 1994; *Kamp and Bertoldi,* 2000; *Bergin et al.,* 2003; *Kamp et al.,* 2006; *Nomura and Millar,* 2005). For instance, in T Tauri stars the radiation field is dominated by strong Lyα emission, which has consequences for the photodissociation rate of molecules that can be dissociated by Lyα photons. The photoelectric heating process, on the other hand, depends strongly on the overall shape of the radiation field, which is much steeper in the case of cool stars. A similar problem appears in the X-ray spectra of cool M stars, which are dominated by line emission.

FUV-induced grain photoelectric heating is often a dominant heating process for the gas in the irradiated surface layers. The FUV photon is absorbed by a dust grain or a polycyclic aromatic hydrocarbon (PAH) molecule, which ejects an energetic electron to the gas, and heats the gas via the thermalization of the energetic electron. Its efficiency and thus the final gas temperature depends strongly on the grain charge, dust grain size, and composition (PAHs, silicates, graphites, ices, etc.). X-rays from the central star also heat only the uppermost surface layers, as their heating drops off monotonically with column, and gets quite small by columns on the order of 10^{21} cm^{-2}.

4.2. Surfaces of Optically Thick Disks

This subsection focuses on the warm surface layers of the optically thick disk at $A_V < 1$, measured vertically downward, where gas and dust temperatures decouple. Modeling of these surface layers is not affected by the optically thick disk interior and depends mainly on the local UV/X-ray flux and the gas density. The typical hydrogen gas number density in these layers is roughly $n_H(A_V = 1) \simeq 10^7$ (100 AU/r) δ^{-1} cm^{-3}. The location of the $A_V = 1$ surface depends on the ratio δ of the dust surface area per hydrogen nucleus to the interstellar value, which is roughly 10^{-21} cm^2/H. Assuming a surface density that drops linearly with radius and $\Sigma(1$ AU$) = 1000$ g cm^{-2}, the fractional column density Σ_{surf}/Σ contained in the surface layer ($T_{gas} \neq T_{dust}$) is usually small, $\Sigma_{surf}/\Sigma \simeq 1.5 \times 10^{-6} \delta^{-1}$ (r/AU), but increases linearly with radius.

4.2.1. Gas temperatures. The detailed temperature structure of the surface layers of optically thick young disks was studied for the first time by *Jonkheid et al.* (2004), *Kamp and Dullemond* (2004), and *Nomura and Millar* (2005). These models neglect EUV irradiation and start with neutral hydrogen in the top layers. Figure 7 shows the vertical structure in a disk model with 0.01 M$_\odot$ at 100 AU around a 0.5 M$_\odot$ T Tauri star [from models of *Kamp and Dullemond* (2004); note that the density structure in these models is not iterated with the gas temperature]. Very high in the atmosphere at particle densities as low as n < 10^5 cm^{-3} ($A_V \leq 10^{-3}$), the gas temperature is set by a balance between photoelectric heating and fine structure line cooling of neutral oxygen (*Kamp and Dullemond*, 2004; *Jonkheid et al.*, 2004). This leads to gas temperatures of several hundred K. Deeper in the disk, for $A_V > 0.01$, molecules can shield themselves from the dissociating FUV radiation. As soon as the fraction of molecular hydrogen becomes larger than 1%, H$_2$ line cooling becomes important. Molecular line emission — mainly CO and H$_2$ — cools the gas down to below 100 K before the densities are high enough for gas and dust to thermally couple. As gas temperature drops below ~100 K, H$_2$ no longer contributes to the cooling. Instead, CO, which has a rich rotational spectrum at low temperatures, becomes an important coolant. At larger radii the FUV flux from the central star drops as well as the density of the surface layer, leading to lower gas temperatures. At

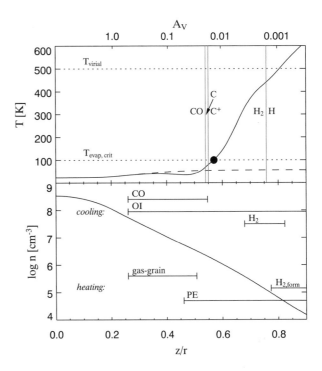

Fig. 7. Upper panel: gas (solid line) and dust (dashed line) temperatures in a vertical cut through the T Tauri star model at 100 AU. Overplotted are the most important chemical transitions from atomic to molecular species. The black filled circle shows the point where a photoevaporative flow can be initiated (see section 5). Lower panel: same vertical cut for the hydrogen number density (fixed input density distribution). Overplotted are the ranges over which the respective heating/cooling processes contribute more than 10% to the total heating/cooling rate at that depth ("PE" means photoelectric heating; "H$_2$, form" means heating through molecular hydrogen formation; see section 5 for the definition of $T_{evap,crit}$ and T_{virial}).

distances r ≳ 100 AU the gas temperature is too low for the endothermic destruction of H$_2$ by O atoms and hence the surface layer at those distances contains substantial fractions of molecular hydrogen.

4.2.2. Implications for the disk structure. Detailed models of the gas temperature have shown that gas and dust are collisionally coupled at optical depth $A_V > 1$. Thus the basic assumption $T_{gas} = T_{dust}$ of the disk structure models presented in the previous section is justified for the disk interior. The main effect of the higher gas temperatures in the warm surface layer is an enhanced flaring of the disk surface (*Nomura and Millar*, 2005).

4.2.3. Observations and comparison with models. The pure rotational lines of H$_2$ such as J = 2–0 S(0) [28 μm], J = 3–1 S(1) [17 μm], J = 4–2 S(2) [12 μm], and J = 6–4 S(4) [8 μm] trace the warm gas (100–200 K) in the disks. Even though there is some controversy about detection of those lines with different instruments (*Thi et al.*, 2001; *Richter et al.*, 2002), there is a tentative detection of H$_2$ in AB Aurigae using the Texas Echelon Cross Echelle Spectrograph (TEXES) at the Infrared Telescope Facility (*Richter et al.*, 2002). *Bary et al.* (2003) report v = 1–0 S(1) [2.12 µm] emission in high-resolution spectra (R ~ 60,000) of the T Tauri stars GG Tau A, LkCa 15, TW Hya, and DoAr 21. This emission most likely arises in the low-density, high-temperature upper surfaces beyond 10 AU. According to the disk models, warm H_2 exists indeed in the optically thin surface layers, where $T_{gas} \gg T_{dust}$ and the observed fluxes can be reproduced (*Nomura and Millar*, 2005). Figure 8 reveals the effect of UV fluorescence on the line strength. The UV fluorescent lines, which are an excellent probe of the inner disk (r < few AU), are discussed in the chapter by Najita et al. The detection of the mid-IR H_2 lines at low spectral resolution (e.g., with Spitzer) is hindered by the low line-to-continuum ratio.

The impact of detailed gas modeling differs for the various emission lines. CO, which forms deeper in the disk, is generally less affected than fine structure lines such as [O I] and [C II] that form in the uppermost surface layers, where $T_{gas} \gg T_{dust}$ (Fig. 9). Since the gas temperature in those layers is set by photoelectric heating, dust settling leads to lower temperatures and thus to weaker line emission.

The [O I] 6300-Å line is another tracer of the physics in the tenuous surface layers (see also chapter by Bergin et al.). It has been detected in a number of externally illuminated proplyds in the Orion nebula (*Johnstone et al.*, 1998) as well as in T Tauri and Herbig Ae/Be stars (*Acke et al.*, 2005; *Acke and van den Ancker*, 2006). *Störzer and Hollenbach* (1999) explain the emission in the Orion proplyds by the photodissociation of the OH molecule, which leaves about 50% of the atomic oxygen formed in the upper $^1D_2$ level of the 6300-Å line. *Acke et al.* (2005) find indication of Keplerian rotation from the [O I] line profiles. However, they need OH abundances higher than those predicted from

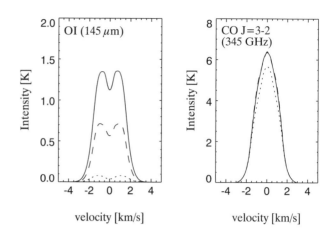

Fig. 9. Impact of detailed gas temperature modeling and dust settling on the emission lines of oxygen and CO (*Jonkheid et al.*, 2004): $T_{gas} = T_{dust}$ (dotted line), detailed gas energy balance (solid line), dust settling (dashed line).

disk models to fit the emission from the disks around Herbig Ae/Be stars. Gas models of those disks reveal the presence of a high-temperature reservoir (few 1000 K); hence the [O I] line might arise partly from thermal excitation at radii smaller than 100 AU (*Kamp et al.*, 2006). Resolved [O I] 6300-Å line emission from the disk around the Herbig Ae star HD100546 (*Acke and van den Ancker*, 2006) shows that the emission is spread between ~1 and 100 AU and supports the presence of a gap at ~10 AU as reported initially by *Bouwman et al.* (2003).

4.3. Optically Thin Disks

As protoplanetary disks evolve, the dust grains grow to at least centimeter sizes and the disks become optically thin. In addition, as we shall discuss in section 5, the gas in the disk ultimately disappears, turning the disk into a debris disk. It is therefore theoretically conceivable that there exists a transition period in which the disk has become optically thin in dust continuum, but still contains a detectable amount of gas. The source HD141569 (5 m.y.) might be an example of this, as *Brittain et al.* (2003) observed UV-excited warm CO gas from the inner rim at ~17 AU, and *Dent et al.* (2005) observed cold gas further out (J = 3–2). Measuring the gas mass in such transition disks sets a timescale for the planet-formation process. The Spitzer Legacy Science Program "Formation and Evolution of Planetary Systems" (FEPS) has set upper limits on gas masses of ~0.1 M_J around solar-type stars with ages greater than 10 m.y. (*Meyer et al.*, 2004; *Hollenbach et al.*, 2005; *Pascucci et al.*, 2005).

4.3.1. Disk models. Several groups have so far studied these transition phases of protoplanetary disks: *Gorti and Hollenbach* (2004) modeled the disk structure and gas/dust emission from intermediate-aged disks around low-mass stars; *Kamp and Bertoldi* (2000), *Kamp and van Zadelhoff* (2001), and *Kamp et al.* (2003) modeled the gas chemistry

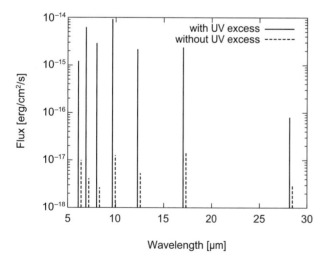

Fig. 8. The mid-infrared line spectra of molecular hydrogen from a T Tauri disk model ($M_\star = 0.5\ M_\odot$, $R_\star = 2\ R_\odot$, $T_{eff} = 4000$ K, $\dot{M} = 10^{-8}\ M_\odot/yr$) with (solid line) and without (dotted line) UV excess (*Nomura and Millar*, 2005).

and line emission from A-type stars such as β Pictoris and Vega. *Jonkheid et al.* (2006) studied the gas chemical structure and molecular emission in the transition phase disk around HD141569A. These models are all based on the same physics as outlined above for the optically thick protoplanetary disks. The disks are still in hydrostatic equilibrium, so that the disk structure in these low-mass disks is similar to that in the more massive disks with the midplane simply removed. However, some fundamental differences remain: The minimum grain size in these disks is typically a few micrometers, much larger than in the young protoplanetary disks; in addition, the dust may have settled toward the midplane, and much of the solid mass may reside in larger particles (a > 1 cm) than can be currently observed. This reduces the grain opacity and the dust-to-gas mass ratio compared to the younger optically thick disks. At radial midplane gas column densities smaller than 10^{23} cm^{-2}, these disks are optically thin to stellar UV and ~1-KeV X-ray photons and the gas is mostly atomic. At radial columns greater than that, the gas opacity becomes large enough to shield H_2 and CO, allowing significant molecular abundances. For disks extended to 100 AU, very little mass (very roughly $\geq 10^{-3}$ M$_J$) is needed to provide this shielding.

4.3.2. Comparison with observations. Since the continuum remains optically thin, the mid-IR spectrum is dominated by fine-structure emission lines from ions such as [Fe II] and [Si II]; large columns of neutral sulfur are common, leading to strong [S I] emission (Fig. 10). However, the strength and thus detectability of these lines depends on the abundances of heavy metals in late phases of disk evolution, which is uncertain, especially the more refractory Fe. In the opaque molecular regions somewhat closer to the star, the gas temperature exceeds 100 K and molecular hydrogen emission is produced. The S(0) and S(1) H_2 lines stay optically thin over a large range of disk masses and *if detected* are more diagnostic of disk mass than other fine structure lines (*Gorti and Hollenbach*, 2004). While the strongest molecular bands of H_2O (important coolant in disk midplane) and OH are similar in strength to the fine structure lines, the H_2 lines are weak and detection can be significantly hampered by the low line-to-continuum ratio (weak narrow line against the bright dust thermal background). These mid-IR lines generally originate from 1 to 10 AU.

Kamp et al. (2003) have shown that beyond 40 AU the dominant coolant for the latest tenuous stages of disk evolution is the [C II] 158-μm line. The fine structure lines of C, O, and C$^+$ trace only the surface of these tenuous disks: [O I] becomes rapidly optically thick and C$^+$ and C turn into CO as soon as UV CO and H_2 bands become optically thick, and stellar UV cannot penetrate any further. Since typical gas temperatures are higher than in molecular clouds, CO lines from the upper rotational levels (J = 4–3) are predicted to be stronger than the lower J lines. *Dent et al.* (2005) have recently detected the CO J = 3–2 line in HD141569 and disk modeling by *Jonkheid et al.* (2006) shows that the profile excludes a significant contribution from gas inward of ~80 AU and estimate the total gas mass to be 80 M$_\oplus$.

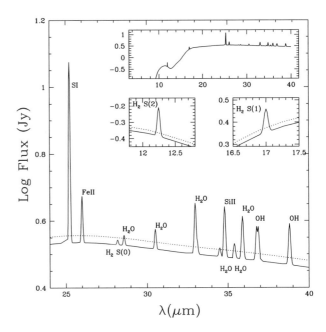

Fig. 10. Mid-infrared spectrum in the 24–40-μm wavelength region showing the dust continuum and dominant gas emission lines for a disk model with M$_{gas}$ = 10^{-2} M$_J$ and M$_{dust}$ = 10^{-5} M$_J$ (with dust defined as particles smaller than 1 mm). A distance to the disk of 30 pc and a spectral resolving power R = 600 is assumed (*Gorti and Hollenbach*, 2004).

5. PHOTOEVAPORATION OF A DISK BY ITS CENTRAL STAR

5.1. Introduction

The above section has shown that in the surface layers of the disk the gas temperature can become very high, greatly exceeding the dust temperature. The warm surface gas can flow off the disk and escape the gravity of the star. Since the heating process responsible for these high temperatures is the *radiation* from the central star or a nearby O star, this process is called "photoevaporation." The viscous evolution (i.e., accretion and spreading) of the disk, discussed in section 2, can be strongly affected by this photoevaporation process. Typically, it significantly shortens the "lifetime" of a disk compared to pure viscous evolution. Photoevaporation can also create inner holes or truncate the outer disk. This has relevance to observations of such disks, such as the percentage of young stars with infrared excess vs. their age (*Haisch et al.*, 2001; *Carpenter et al.*, 2005), or the inferred "large inner holes" of some disks (e.g., *Calvet et al.*, 2002; *Bouwman et al.*, 2003; *Forrest et al.*, 2004; *D'Alessio et al.*, 2005). It has also far-reaching consequences for the formation of planets, as we will discuss below.

Photoevaporation has already been discussed in earlier reviews (*Hollenbach et al.*, 2000; *Hollenbach and Adams*, 2004; *Richling et al.*, 2006). However, these reviews mainly focused on the heating by a nearby massive star (such as the famous case of the proplyds in Orion). In contrast, in this section we will exclusively review recent results on photoevaporation by the central star, consistent with the previous sections that focus on heating and photodissociation by the central star. Progress in this field since Protostars and Planets IV (PPIV) has been mostly theoretical, since observations of diagnostic gas spectral lines for the case of photoevaporation by the central, low-mass star requires greater sensitivity, spectral resolution, and spatial resolution than currently available. We will, however, discuss the implications for the observed "inner holes" and disk lifetimes.

5.2. The Physics of Photoevaporation

5.2.1. Basic concepts.   Photoevaporation results when stellar radiation heats the disk surface and resulting thermal pressure gradients drive an expanding hydrodynamical flow to space. As shown in section 4 the main heating photons lie in the FUV, EUV, and X-ray energy regimes. X-rays, however, were shown to be of lesser importance for photoevaporation (*Alexander et al., 2004b*), and we will not consider them further.

There are two main sources of the strong EUV and FUV excesses observed in young low-mass stars: accretion luminosity and prolonged excess chromospheric activity. Recent work (*Alexander et al., 2004a*) has shown that EUV photons do not penetrate accretion columns, so that accretion cannot provide escaping EUV photons to power photoevaporation. *Alexander et al.* (2005) present indirect observational evidence that an active chromosphere may persist in T Tauri stars even without strong accretion, and that EUV luminosities of $\Phi_{EUV} > 10^{41}$ photons/s may persist in low-mass stars for extended ($\geq 10^{6.5}$–10^7 yr) periods to illuminate their outer disks. FUV photons may penetrate accretion columns and also are produced in active chromospheres. They are measured in nearby, young, solar-mass stars with little accretion and typically (with great scatter) have luminosity ratios $L_{FUV}/L_{bol} \sim 10^{-3}$ or $\Phi_{FUV} \sim 10^{42}$ photons/s.

EUV photons ionize the hydrogen in the very upper layers of the disk and heat it to a temperature of $\sim 10^4$ K, independent of radius. FUV photons penetrate deeper into the disk and heat the gas to T $\sim$ 100–5000 K, depending on the intensity of the FUV flux, the gas density, and the chemistry (as was discussed in section 4). Whether the EUV or FUV heating is enough to drive an evaporative flow depends on how the resulting thermal speed (or sound speed) compares to the local escape speed from the gravitationally bound system. A characteristic radius for thermal evaporation is the "gravitational radius" r_g, where the sound speed equals the escape speed

$$r_g = \frac{GM_\star \mu m_p}{kT} \sim 100 \text{ AU} \left(\frac{T}{1000 \text{ K}}\right)^{-1} \left(\frac{M_\star}{M_\odot}\right) \quad (3)$$

Early analytic models made the simple assumption that photoevaporation occurred for r > r_g, and that the warm surface was gravitationally bound for r < r_g. However, a closer

look at the gas dynamics shows that this division happens not at r_g but at about 0.1–0.2 r_g (*Liffman*, 2003; *Adams et al.*, 2004; *Font et al.*, 2004), and that this division is not entirely sharp. In other words, photoevaporation happens *mostly* outside the "critical radius" $r_{cr} \sim 0.15 \, r_g$, although a weak evaporation occurs inside of r_{cr}. Since these are important new insights since PPIV, we devote a subsubsection to them below.

With T $\sim 10^4$ K the critical radius for EUV-induced photoevaporation is r_{cr}(EUV) $\sim$ 1–2($M_\star/M_\odot$) AU. However, there is no fixed r_{cr}(FUV) because the FUV-heated gas has temperatures that depend on FUV flux and gas density, i.e., on r and z. Therefore, r_{cr}(FUV) depends on r and z, and may range from 3 to 150 AU for solar-mass stars.

The evaporative mass flux $\dot{\Sigma}$ depends not only on the temperature of the photon-heated gas, but also on the vertical penetration depth of the FUV/EUV photons. For EUV photons this is roughly set for r < $r_{cr} \sim$ 1 AU by the Strömgren condition that recombinations in the ionized layer equal the incident ionizing flux. Neglecting dust attenuation, this penetration column can be expressed as A_V(EUV) $\sim$ $0.05 \delta \Phi_{41}^{1/2}(r/AU)^{-1/2}$, where $\Phi_{41} \equiv \Phi_{EUV}/10^{41}$ photons/s and δ is the ratio of the dust surface area per hydrogen to the interstellar dust value (see section 4 and note that δ can be much smaller than unity if dust has settled or coagulated). Outside 1 AU, the penetration depth falls even faster with r, roughly as $r^{-3/2}$ (see *Hollenbach et al.*, 1994). On the other hand, the FUV penetration depth is set by dust attenuation, or A_V(FUV) $\sim$ φ, where we recall that φ is the irradiation angle and depends on disk flaring (see section 3). In general A_V(EUV) $\ll A_V$(FUV), so the EUV-ionized skin of the disk lies on top of the FUV-heated gas surface layer.

The penetration depth is an important quantity because it sets the density at the base of the photoevaporative flow: The deeper the penetration depth, the higher the density. The flux of outflowing matter is proportional to the product of local density and sound speed within this heated layer. This is why the complex surface structure models of section 4 are so important for FUV-driven photoevaporation. For EUV-driven photoevaporation, on the other hand, the situation is less complicated, since the temperature in the ionized skin of the disk is independent of r and z, as long as z > z_b, where z_b is the bottom of the ionized layer, i.e., the base of the flow. For this simple case, the evaporative mass flux originates at z_b, which is where the highest-density gas at temperature $T_{EUV} \simeq 10^4$ K resides.

Although FUV-heated layers have lower temperatures than the EUV-heated skin they are at higher densities and may equally well initiate the flow and determine the mass flux as EUV photons [see *Johnstone et al.* (1998) for a similar situation for externally illuminated disks]. Gorti and Hollenbach (in preparation, henceforth *GH06*) find that the FUV-photoevaporative flow typically originates at vertical heights where T $\sim$ 100–200 K, yielding $r_{cr} \sim$ 50–100 AU. For r > 50 AU, the FUV photoevaporation dominates. On the other hand, EUV photons (with $r_{cr} \sim$ 1 AU) affect the planet-forming regions at r $\ll$ 50 AU more than the FUV photons.

5.2.2. Photoevaporation as a Bernoulli flow.

One way to understand why the disk can evaporate at radii as small as 0.2 r_g is to consider the evaporative flow as a Bernoulli flow along streamlines (*Liffman, 2003; Adams et al., 2004*). These streamlines initially rise nearly vertically out of the disk and then bend over to become asymptotically radially outward streamlines. If a streamline starts at $r > r_g$, then the flow rapidly goes through a sonic point and achieves the sound speed c_s near the base of the flow. The one-sided mass flux rate in the flow is then $\dot{\Sigma} \simeq 2\rho_b c_s$, where ρ_b is the mass density of the gas at the base, and a factor of 2 arises due to the two sides of the disk.

On the other hand, if a streamline starts at $r \ll r_g$, the gas at its base lies deep in the gravitational potential. As a simplification let us now treat these streamlines as if they are entirely radial streamlines (ignoring their vertical rise out of a disk). Then the standard atmospheric solution has a density that falls off from r to roughly r_g as $\exp(-r_g/2r)$. The gas flows subsonically and accelerates, as it slowly expands outward, until it passes through a sonic point at $r_s \lesssim 0.5\ r_g$ (0.5 r_g is the classic Parker wind solution for zero rotation). For $r \ll r_g$, the mass flux is reduced considerably by the rapid fall-off of the density from r to r_s. For $r < r_g$, the mass flux is roughly given by the density at r_s × the sound speed × the dilution factor $(r_s/r)^2$ that accounts for mass conservation between r and r_s: $\dot{\Sigma} \simeq 2\rho_b e^{-r_g/2r} c_s (r_s/r)^2$. Assuming the same ρ_b and c_s at all r, we see that $\dot{\Sigma}(0.2\ r_g) \simeq 0.5\ \dot{\Sigma}(r_g)$ and that $\dot{\Sigma}(0.1\ r_g) \simeq 0.17\ \dot{\Sigma}(r_g)$. This demonstrates that $r_{cr} \sim 0.15\ r_g$ for this simplified case, and that even for $r \lesssim r_{cr}$ evaporation is weak, but not zero. In Fig. 7 the base of the flow is marked with the large dot (although that figure shows a static, non-evaporating model with only FUV heating). In that figure, T_{virial} is the temperature such that the sound speed equals the escape speed; $T_{evap,crit} \equiv 0.2\ T_{virial}$ is roughly where the photoevaporation flow originates (i.e., where $r = r_{cr}$).

5.2.3. Mass loss rates for EUV-induced flows.

Although central-star FUV models are not yet published, several central-star EUV models have appeared in the literature. *Hollenbach et al.* (1994) first outlined the essential physics of EUV-induced flows by the central star and presented an approximate analytic solution to the mass loss rate for a disk larger than r_g. The basic physics is the Strömgren relation, $\Phi_{EUV} \simeq \alpha_r n_e^2 r^3$, where α_r is the hydrogen recombination coefficient and n_e is the electron density in the ionized surface gas. This sets the hydrogen nucleus (proton) number density at the base of the flow $n_b \propto \Phi_{EUV}^{1/2}$, and therefore an identical proportionality for the mass loss rate

$$\dot{M}_{EUV} \sim 4 \times 10^{-10} \left(\frac{\Phi_{EUV}}{10^{41}\ s^{-1}} \right)^{0.5} \left(\frac{M_\star}{M_\odot} \right)^{0.5} \qquad (4)$$

in units of $M_\odot$/yr. Radiation hydrodynamical simulations (*Yorke and Welz, 1996; Richling and Yorke, 1997*) find a similar power-law index for the dependence of the mass-loss rate on the EUV photon rate of the central star. This result applies for both high- and low-mass central stars, and is valid for a weak stellar wind. The effect of a strong stel-

lar wind is such that the ram pressure reduces the scale height of the atmosphere above the disk and the EUV photons are allowed to penetrate more easily to larger radii. This increases the mass-loss rate from the outer parts of the disk. It is noteworthy that the diffuse EUV field, caused by recombining electrons and protons in the disk's ionized atmosphere inside r_{cr}, controls the EUV-induced mass-loss rates (*Hollenbach et al., 1994*) for disks with no or small inner holes ($<r_{cr}$). This effect negates any potential for self-shadowing of the EUV by the disk.

5.3. Evolution of Photoevaporating Disks

5.3.1. Case without viscous evolution.

Let us first assume a disk that does not viscously evolve; it simply passively undergoes photoevaporation. For disks with size $r_d < r_{cr}$, the photoevaporation proceeds from outside in. The mass flux rate at r_d is much higher than inside of r_d, because the gas at r_d is least bound. In addition, disk surface densities generally fall with r (see section 2). Therefore, the disk shrinks as it photoevaporates, and most of the mass flux comes from the outer disk radius. However, for disks with $r_d > r_{cr}$, two types of disk evolution may occur. For EUV photons, *Hollenbach et al.* (1994) showed that the mass flux $\dot{\Sigma}$ beyond r_{cr} goes roughly as $r^{-2.5}$ if there is no inner hole extending to r_{cr}. The timescale for complete evaporation at r goes as $\Sigma(r)/\dot{\Sigma}(r)$. As long as Σ does not drop faster than $r^{-2.5}$, the disk will evaporate first at $r \sim r_{cr}$ and, once a gap forms there, will then steadily erode the disk from this gap outward.

If, on the other hand, $\Sigma(r)/\dot{\Sigma}(r)$ decreases with r, then the disk shrinks from outside in as in the $r_d < r_{cr}$ case. The photoevaporation by the FUV from the central star has not yet been fully explored, but preliminary work by *GH06* suggests that the mass flux $\dot{\Sigma}$ in the outer disks around solar mass stars *increases* with r. In this case, the disk evaporates from the outside in for most generally assumed surface density laws, which decrease with r. The combined effect of EUV and FUV photoevaporation then is likely to erode the disk outward from r_{cr}(EUV) ~1 AU by the EUV flow and inward from the outer disk radius by the FUV flow, sandwiching the intermediate radii.

5.3.2. Case with viscous evolution.

Now let us consider a disk that is actively accreting onto the star (see section 2). In general, if the photoevaporation drills a hole somewhere in the disk or "eats" its way from the outside in, the forces of viscous spreading tend to move matter toward these photoevaporation regions, which can accelerate the dissipation of the disk. If the disk has a steady accretion rate $\dot{M}$, then a gap forms once $\dot{M}_{evap} \propto r^2\dot{\Sigma}$ exceeds $\dot{M}$. Since $r^2\dot{\Sigma} \propto r^{-0.5}$ for EUV photoevaporation beyond r_{cr}, the gap first forms at the minimum radius ($\sim r_{cr}$) and then works its way outward. *Clarke et al.* (2001) presented time-dependent computations of the evolution of disks around low-mass stars with $\Phi_{EUV} \simeq 10^{41-43}$ photons/s. Their model combines EUV photoevaporation with a viscous evolution code. However, they used the current hypothesis at that time that evaporation only occurs outside r_g. After ~10^6–10^7 yr of viscous evolution relatively unperturbed by photoevaporation, the viscous accretion inflow rates fall below the photoevaporation rates at r_g. At this point, a gap opens up at r_g and the inner disk rapidly (on an inner disk viscous timescale of ~10^5 yr) drains onto the central star or spreads to r_g where it evaporates. In this fashion, an inner hole is rapidly produced extending to r_g.

Alexander et al. (2006a,b) extended the work of *Clarke et al.* (2001) to include the effect of $r_{cr} < r_g$, and to treat the outward EUV evaporation of the disk beyond r_{cr} ~ 1 AU. They show that once the inner hole is produced, the diffuse flux from the atmosphere of the inner disk is removed and the attenuation of the direct flux by this same atmosphere is also removed. This enhances the EUV photoevaporation rate by the direct EUV flux from the star, and the effect magnifies as the inner hole grows as $\dot{M}_{EUV} \propto r_{inner}^{1/2}$ again derivable from a simple Strömgren criterion. The conclusion is that the outer disk is very rapidly cleared once the inner hole forms (see Fig. 11).

The rapid formation of a cleared out inner hole almost instantly changes the nature and appearance of the disk. The above authors compare their model favorably with a number of observations: (1) the rapid transition from classical T Tauri stars to weak line T Tauri stars, (2) the almost simultaneous loss of the outer disk (as detected by submillimeter measurements of the dust continuum) with the inner disk (as detected by near IR observations of very hot dust near the star), and (3) the SED observations of large (3–10 AU) inner holes in those sources (see dotted line of Fig. 5) with evidence for low accretion rates and intermediate-mass outer disks such as the source CoKu Tau/4. Figure 12 shows the evolutionary tracks of their models with $\Phi_{EUV} = 10^{42}$ photons/s compared to the observations of weak-line T Tauri stars (WTTSs) and CTTSs.

In a similar vein, *Armitage et al.* (2003) used the combination of EUV photoevaporation and viscous dispersal, together with an assumed dispersion of a factor of 3 in the initial disk mass, to explain the dispersion in the lifetime and accretion rates of T Tauri disks. They found that the models predict a low fraction of binaries that pair a classical T Tauri star with a weak-lined T Tauri star. Their models are in better agreement with observations of disk lifetimes in binaries than models without photoevaporation.

Going one step further, *Takeuchi et al.* (2005) constructed models combining viscous evolution, EUV photoevaporation, and the differential radial motion of grains and gas. Their models predicted that for low-mass stars with a low photoevaporation rate, dust-poor gas disks with an inner hole would form (WTTs), whereas for high-mass stars (evolved Herbig Ae/Be) with a high photoevaporation rate, gas-poor dust rings would form.

Matsuyama et al. (2003) pointed out that if the EUV luminosity is created by accretion onto the star, then, as the accretion rate diminishes, the EUV luminosity drops and the timescale to create a gap greatly increases. Even worse, as discussed above, the EUV photons are unlikely to escape the accretion column. Only if the EUV luminosity remains

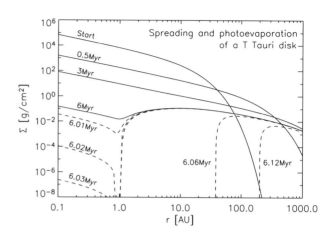

Fig. 11. Evolution of the surface density of a EUV photoevaporating disk. This simulation starts from a given disk structure of about 0.05 $M_\odot$ (marked with "Start" in the figure). Initially the disk accretes and viscously spreads (solid lines). At t = 6 × 10^6 yr the photoevaporation starts affecting the disk. Once the EUV photoevaporation has drilled a gap in the disk at ~1 AU, the destruction of the disk goes very rapidly (dashed lines). The inner disk quickly accretes onto the star, followed by a rapid erosion of the outer disk from inside out. In this model the disk viscosity spreads to >1000 AU; however, FUV photoevaporation (not included) will likely truncate the outer disk. Adapted from *Alexander et al.* (2006b).

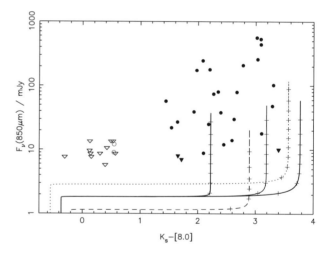

Fig. 12. Near- to mid-infrared color (in magnitudes) vs. 850 μm flux for photoevaporation/viscous evolution models. The data are taken from *Hartmann et al.* (2005) and *Andrews and Williams* (2005): 850-μm detections (circles) and upper limits (triangles) are plotted for both CTTSs (filled symbols) and WTTSs (open symbols). Evolutionary tracks are shown for models with stellar masses 0.5 (dashed), 1.0 (solid), and 2.0 $M_\odot$ (dotted), at a disk inclination of i = 60° to the line of sight. The thick tracks to the right and left show the 1-$M_\odot$ model at i = 0 and i = 80°, respectively. Crosses are added every 1 m.y. to show the temporal evolution. Initially the (optically thin) 850-μm flux declines slowly at constant (optically thick) infrared color. However, once the viscous accretion rate falls below the photoevaporation rate, the disk is rapidly cleared from the inside out. Adapted from *Alexander et al.* (2006b).

high due to chromospheric activity does EUV photoevaporation play an important role in the evolution of disks around isolated low-mass stars. *Alexander et al.* (2005) argue this is the case. *Ruden* (2004) provides a detailed analytic analysis that describes the evolution of disks in the presence of viscous accretion and photoevaporation and compares his results favorably with these two groups.

5.4. Effect on Planet Formation

The processes that disperse the gas influence the formation of planets. Small dust particles follow the gas flow. If the gas is dispersed before the dust can grow, all the dust will be lost in the gas dispersal and planetesimals and planets will not form. Even if there is time for particles to coagulate and build sufficiently large rocky cores that can accrete gas (*Pollack et al.*, 1996; *Hubickyj et al.*, 2004), the formation of gas giant planets like Jupiter and Saturn will be suppressed if the gas is dispersed before the accretion can occur. Furthermore, gas dispersal helps trigger gravitational instabilities that may lead to planetesimal formation (*Goldreich and Ward*, 1973; *Youdin and Shu*, 2002; *Throop et al.*, 2005), affects planet migration (e.g., *Ward*, 1997), and influences the orbital parameters of planetesimals and planets (*Kominami and Ida*, 2002).

5.4.1. Gas-rich versus gas-poor giant planets in the solar system. *Shu et al.* (1993) showed that with $\Phi_{EUV} \sim 10^{41}$ photons/s, the early Sun could have photoevaporated the gas beyond Saturn before the cores of Neptune and Uranus formed, leaving them gas poor. However, this model ignored photoevaporation inside r_g. The current work by *Adams et al.* (2004) would suggest rather rapid photoevaporation inside 10 AU, and make the timing of this scenario less plausible. FUV photoevaporation (either from external sources or from the central star) may provide a better explanation. Preliminary results from *GH06* suggest that the early Sun did not produce enough FUV generally to rapidly remove the gas in the outer giant-planet regions. Adams et al. and *Hollenbach and Adams* (2005) discuss the external illumination case, which looks more plausible.

5.4.2. Truncation of the Kuiper belt. A number of observations point to the truncation of Kuiper belt objects (KBOs) beyond about 50 AU (e.g., *Allen et al.*, 2002; *Trujillo and Brown*, 2001). *Adams et al.* (2004) and *Hollenbach and Adams* (2004, 2005) show that photoevaporation by a nearby massive star could cause truncation of KBOs at about 100 AU, but probably not 50 AU. The truncation is caused by the gas dispersal before the dust can coagulate to sizes that survive the gas dispersal, and that can then later form KBOs. Models of FUV photoevaporation by the early Sun are needed.

5.4.3. Formation of planetesimals. In young disks, dust settles toward the midplane under the influence of the stellar gravity and coagulates. Once coagulated dust has concentrated in the midplane, the roughly centimeter-sized particles can grow further by collisions or by local gravitational instability (*Goldreich and Ward*, 1973; *Youdin and Shu*, 2002). A numerical model by *Throop and Bally* (2005) follows the evolution of gas and dust independently and considers the effects of vertical sedimentation and external photoevaporation. The surface layer of the disk becomes dust depleted, which leads to dust-depleted evaporating flows. Because of the combined effects of the dust settling and the gas evaporating, the dust-to-gas ratio in the disk midplane is so high that it meets the gravitational instability criteria of *Youdin and Shu* (2002), indicating that kilometer-sized planetesimals could spontaneously form. These results imply that photoevaporation may even trigger the formation of planetesimals. Presumably, photoevaporation by the central star may also produce this effect.

6. SUMMARY AND OUTLOOK

In this chapter we have given a brief outline of how disks form and viscously evolve, what their structure is, what their spectra look like in dust continuum and in gas lines, and how they might come to their end by photoevaporation and viscous dispersion. The disk structure in dust and gas is summarized in Fig. 13. Obviously, due to the broadness of the topic we had to omit many important issues. For in-

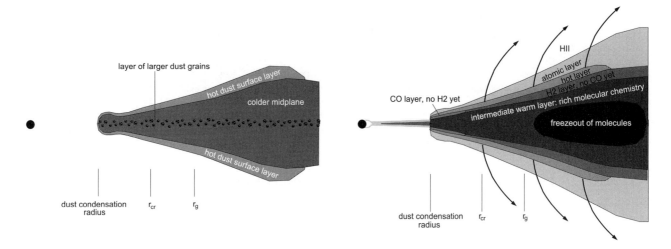

Fig. 13. Pictograms of the structure of a flaring protoplanetary disk, in dust (left) and gas (right).

stance, the formation of disks is presumably much more chaotic than the simple picture we have discussed. In recent years there is a trend to outfit even the observation-oriented workhorse models with ever-more-detailed physics. This is not a luxury, since the volume of observational data (both spectral and spatial) is increasing dramatically, as shown by various other chapters in this book. For instance, with observational information about dust growth and sedimentation in disks, it will be necessary to include realistic dust evolution models into the disk models. Additionally, with clear evidence for nonaxial symmetry in many disks (e.g., *Fukagawa et al., 2004*) modelers may be forced to abandon the assumption of axial symmetry. The thermal-chemical study of the gas in the disk surface layers is a rather new field, and more major developments are expected in the next few years, both in theory and in the comparison to observations. These new insights will also influence the models of FUV photoevaporation, and thereby the expected disk lifetime.

A crucial step to aim for in the future is the unification of the various aspects discussed here. They are all intimately connected and mutually influence each other. Such a unification opens up the perspective of connecting seemingly unrelated observations and thereby improving our understanding of the bigger picture.

Acknowledgments. The authors would like to thank A. Glassgold, M. Walmsley, K. Pontoppidan, and J. Joergensen for proofreading and providing very useful comments.

REFERENCES

Acke B. and van den Ancker M. E. (2006) *Astron. Astrophys., 449,* 267.
Acke B., van den Ancker M. E., Dullemond C. P., van Boekel R., and Waters L. B. F. M. (2004) *Astron. Astrophys., 422,* 621–626.
Acke B., van den Ancker M. E., and Dullemond C. P. (2005) *Astron. Astrophys., 436,* 209–230.
Adams F. C. and Shu F. H. (1986) *Astrophys. J., 308,* 836–853.
Adams F. C., Hollenbach D., Laughlin G., and Gorti U. (2004) *Astrophys. J., 611,* 360–379.
Aikawa Y. and Herbst E. (1999) *Astron. Astrophys., 351,* 233–246.
Aikawa Y., van Zadelhoff G. J., van Dishoeck E. F., and Herbst E. (2002) *Astron. Astrophys., 386,* 622–632.
Akeson R. L., Walker C. H., Wood K., Eisner J. A., Scire E., Penprase B., Ciardi D. R., van Belle G. T., Whitney B., and Bjorkman J. E. (2005) *Astrophys. J., 622,* 440–450.
Alexander R. D., Clarke C. J., and Pringle J. E. (2004a) *Mon. Not. R. Astron. Soc., 348,* 879–884.
Alexander R. D., Clarke C. J., and Pringle J. E. (2004b) *Mon. Not. R. Astron. Soc., 354,* 71–80.
Alexander R. D., Clarke C. J., and Pringle J. E. (2005) *Mon. Not. R. Astron. Soc., 358,* 283–290.
Alexander R. D., Clarke C. J., and Pringle J. E. (2006a) *Mon. Not. R. Astron. Soc., 369,* 216.
Alexander R. D., Clarke C. J., and Pringle J. E. (2006b) *Mon. Not. R. Astron. Soc., 369,* 229.
Allen R. L., Bernstein G. M., and Malhotra R. (2002) *Astron. J., 124,* 2949–2954.
Andrews S. M. and Williams J. P. (2005) *Astrophys. J., 631,* 1134–1160.
Armitage P. J., Livio M., and Pringle J. E. (2001) *Mon. Not. R. Astron. Soc., 324,* 705–711.

Armitage P. J., Clarke C. J., and Palla F. (2003) *Mon. Not. R. Astron. Soc., 342,* 1139–1146.
Artymowicz P. and Lubow S. H. (1994) *Astrophys. J., 421,* 651–667.
Balbus S. and Hawley J. (1991) *Astrophys. J., 376,* 214–233.
Bary J. S., Weintraub D. A., and Kastner J. H. (2003) *Astrophys. J., 586,* 1136–1147.
Bell K. R. (1999) *Astrophys. J., 526,* 411–434.
Bell K. R. and Lin D. N. C. (1994) *Astrophys. J., 427,* 987–1004.
Bell K. R., Cassen P. M., Klahr H. H., and Henning T. (1997) *Astrophys. J., 486,* 372–387.
Bergin E., Calvet N., D'Alessio P., and Herczeg G. J. (2003) *Astrophys. J., 591,* L159–L162.
Bertout C., Basri G., and Bouvier J. (1988) *Astrophys. J., 330,* 350–373.
Bjorkman J. E. and Wood K. (2001) *Astrophys. J., 554,* 615–623.
Blandford R. D. and Payne D. G. (1982) *Mon. Not. R. Astron. Soc., 199,* 883–903.
Bonnell I. and Bastien P. (1992) *Astrophys. J., 401,* L31–L34.
Boss A. P. (1996) *Astrophys. J., 469,* 906–920.
Boss A. P. (1997) *Astrophys. J., 478,* 828–828.
Bouwman J., de Koter A., Dominik C., and Waters L. B. F. M. (2003) *Astron. Astrophys., 401,* 577–592.
Brittain S. D., Rettig T. W., Simon T., Kulesa C., DiSanti M. A., and Dello Russo N. (2003) *Astrophys. J., 588,* 535–544.
Calvet N., Patino A., Magris G. C., and D'Alessio P. (1991) *Astrophys. J., 380,* 617–630.
Calvet N., Hartmann L., and Strom S. E. (2000) In *Protostars and Planets IV* (V. Mannings et al., eds.), pp. 377–399. Univ. of Arizona, Tucson.
Calvet N., D'Alessio P., Hartmann L., Wilner D., Walsh A., and Sitko M. (2002) *Astrophys. J., 568,* 1008–1016.
Carpenter J. M., Wolf S., Schreyer K., Launhardt R., and Henning T. (2005) *Astron. J., 129,* 1049–1062.
Cassen P. (1996) *Meteoritics & Planet. Sci., 31,* 793–806.
Chiang E. I. and Goldreich P. (1997) *Astrophys. J., 490,* 368–376.
Chiang E. I., Joung M. K., Creech-Eakman M. J., Qi C., Kessler J. E., Blake G. A., and van Dishoeck E. F. (2001) *Astrophys. J., 547,* 1077–1089.
Clarke C. J. and Syer D. (1996) *Mon. Not. R. Astron. Soc., 278,* L23–L27.
Clarke C. J., Gendrin A., and Sotomayor M. (2001) *Mon. Not. R. Astron. Soc., 328,* 485–491.
D'Alessio P., Canto J., Calvet N., and Lizano S. (1998) *Astrophys. J., 500,* 411–427.
D'Alessio P., Calvet N., Hartmann L., Lizano S., and Cantó J. (1999) *Astrophys. J., 527,* 893–909.
D'Alessio P., Calvet N., and Hartmann L. (2001) *Astrophys. J., 553,* 321–334.
D'Alessio P., Hartmann L., Calvet N., Franco-Hernández R., Forrest W. J., Sargent B., et al. (2005) *Astrophys. J., 621,* 461–472.
D'Alessio P., Calvet N., Hartmann L., Franco-Hernández R., and Servín H. (2006) *Astrophys. J., 638,* 314–335.
Dent W. R. F., Greaves J. S., and Coulson I. M. (2005) *Mon. Not. R. Astron. Soc., 359,* 663–676.
Dominik C., Dullemond C. P., Waters L. B. F. M., and Walch S. (2003) *Astron. Astrophys., 398,* 607–619.
Dubrulle B., Marié L., Normand C., Richard D., Hersant F., and Zahn J.-P. (2005) *Astron. Astrophys., 429,* 1–13.
Dullemond C. P. (2002) *Astron. Astrophys., 395,* 853–862.
Dullemond C. P. and Dominik C. (2004a) *Astron. Astrophys., 417,* 159–168.
Dullemond C. P. and Dominik C. (2004b) *Astron. Astrophys., 421,* 1075–1086.
Dullemond C. P., Dominik C., and Natta A. (2001) *Astrophys. J., 560,* 957–969.
Dullemond C. P., van Zadelhoff G. J., and Natta A. (2002) *Astron. Astrophys., 389,* 464–474.
Dullemond C. P., van den Ancker M. E., Acke B., and van Boekel R. (2003) *Astrophys. J., 594,* L47–L50.

Eisner J. A., Lane B. F., Akeson R. L., Hillenbrand L. A., and Sargent A. I. (2003) *Astrophys. J., 588*, 360–372.

Font A. S., McCarthy I. G., Johnstone D., and Ballantyne D. R. (2004) *Astrophys. J., 607*, 890–903.

Forrest W. J., Sargent B., Furlan E., D'Alessio P., Calvet N., Hartmann L., et al. (2004) *Astrophys. J. Suppl., 154*, 443–447.

Friedjung M. (1985) *Astron. Astrophys., 146*, 366–368.

Fromang S., Balbus S. A., and De Villiers J.-P. (2004) *Astrophys. J., 616*, 357–363.

Fukagawa M., Hayashi M., Tamura M., Itoh Y., Hayashi S. S., and Oasa Y. e. a. (2004) *Astrophys. J., 605*, L53–L56.

Gammie C. F. (1996) *Astrophys. J., 457*, 355–362.

Gammie C. F. and Johnson B. M. (2005) In *Chondrites and the Protoplanetary Disk* (A. N. Krot et al., eds.), pp. 145–164. ASP Conf. Series 341, San Francisco.

Glassgold A. E., Najita J., and Igea J. (1997a) *Astrophys. J., 480*, 344–350.

Glassgold A. E., Najita J., and Igea J. (1997b) *Astrophys. J., 485*, 920–920.

Goldreich P. and Ward W. R. (1973) *Astrophys. J., 183*, 1051–1062.

Gorti U. and Hollenbach D. (2004) *Astrophys. J., 613*, 424–447.

Habing H. J. (1968) *Bull. Astron. Inst. Netherlands, 19*, 421–431.

Haisch K. E., Lada E. A., and Lada C. J. (2001) *Astrophys. J., 553*, L153–L156.

Hartmann L., Calvet N., Gullbring E., and D'Alessio P. (1998) *Astrophys. J., 495*, 385–400.

Hartmann L., Megeath S. T., Allen L., Luhman K., Calvet N., D'Alessio P., Franco-Hernandez R., and Fazio G. (2005) *Astrophys. J., 629*, 881–896.

Hayashi C. (1981) In *Fundamental Problems in the Theory of Stellar Evolution* (D. Sugimoto et al., eds.), pp. 113–126. IAU Symposium 93, Reidel, Dordrecht.

Hillenbrand L. A., Strom S. E., Vrba F. J., and Keene J. (1992) *Astrophys. J., 397*, 613–643.

Hollenbach D. and Adams F. C. (2004) In *Debris Disks and the Formation of Planets* (L. Caroff et al., eds.), pp. 168–183. ASP Conf. Series 324, San Francisco.

Hollenbach D. and Adams F. (2005) In *Star Formation in the Interstellar Medium* (D. Lin et al., eds.), p. 3. ASP Conf. Series 323, San Francisco.

Hollenbach D., Johnstone D., Lizano S., and Shu F. (1994) *Astrophys. J., 428*, 654–669.

Hollenbach D. J., Yorke H. W., and Johnstone D. (2000) In *Protostars and Planets IV* (V. Mannings et al., eds.), pp. 401–428. Univ. of Arizona, Tucson.

Hollenbach D., Gorti U., Meyer M., Kim J. S., Morris P., Najita J., et al. (2005) *Astrophys. J., 631*, 1180–1190.

Hubickyj O., Bodenheimer P., and Lissauer J. J. (2004) *Rev. Mex. Astron. Astrofis. Conf. Ser., 22*, 83–86.

Hueso R. and Guillot T. (2005) *Astron. Astrophys., 442*, 703–725.

Ilgner M. and Nelson R. P. (2006) *Astron. Astrophys., 445*, 205–222.

Isella A. and Natta A. (2005) *Astron. Astrophys., 438*, 899–907.

Jin L. (1996) *Astrophys. J., 457*, 798–804.

Johnstone D., Hollenbach D., and Bally J. (1998) *Astrophys. J., 499*, 758–776.

Jonkheid B., Faas F. G. A., van Zadelhoff G.-J., and van Dishoeck E. F. (2004) *Astron. Astrophys., 428*, 511–521.

Jonkheid B., Kamp I., Augereau J.-C., and van Dishoeck E. F. (2006) *Astron. Astrophys., 453*, 163–171.

Kamp I. and Bertoldi F. (2000) *Astron. Astrophys., 353*, 276–286.

Kamp I. and Dullemond C. P. (2004) *Astrophys. J., 615*, 991–999.

Kamp I. and van Zadelhoff G.-J. (2001) *Astron. Astrophys., 373*, 641–656.

Kamp I., van Zadelhoff G.-J., van Dishoeck E. F., and Stark R. (2003) *Astron. Astrophys., 397*, 1129–1141.

Kamp I., Dullemond C. P., Hogerheijde M., and Emilio Enriquez J. (2006) In *Astrochemistry — Recent Successes and Current Challenges* (D. C. Lis et al., eds.), in press. IAU Symposium 231, Cambridge Univ., Cambridge.

Kawazoe E. and Mineshige S. (1993) *Publ. Astron. Soc. Japan, 45*, 715–725.

Kenyon S. J. and Hartmann L. (1987) *Astrophys. J., 323*, 714–733.

Kenyon S. J. and Hartmann L. (1991) *Astrophys. J., 383*, 664–673.

Kenyon S. J. and Hartmann L. (1995) *Astrophys. J. Suppl., 101*, 117–171.

Kessler-Silacci J., Augereau J.-C., Dullemond C., Geers V., Lahuis F., et al. (2006) *Astrophys. J., 639*, 275–291.

Kitamura Y., Momose M., Yokogawa S., Kawabe R., Tamura M., and Ida S. (2002) *Astrophys. J., 581*, 357–380.

Klahr H. H. and Bodenheimer P. (2003) *Astrophys. J., 582*, 869–892.

Kominami J. and Ida S. (2002) *Icarus, 157*, 43–56.

Lachaume R. (2004) *Astron. Astrophys., 422*, 171–176.

Lachaume R., Malbet F., and Monin J.-L. (2003) *Astron. Astrophys., 400*, 185–202.

Laughlin G. and Bodenheimer P. (1994) *Astrophys. J., 436*, 335–354.

Liffman K. (2003) *Publ. Astron. Soc. Australia, 20*, 337–339.

Lissauer J. J. (1993) *Ann. Rev. Astron. Astrophys., 31*, 129–174.

Lucy L. B. (1999) *Astron. Astrophys., 344*, 282–288.

Lynden-Bell D. (1969) *Nature, 223*, 690–694.

Lynden-Bell D. and Pringle J. E. (1974) *Mon. Not. R. Astron. Soc., 168*, 603–637.

Malbet F. and Bertout C. (1991) *Astrophys. J., 383*, 814–819.

Matsumoto T. and Hanawa T. (2003) *Astrophys. J., 595*, 913–934.

Matsuyama I., Johnstone D., and Hartmann L. (2003) *Astrophys. J., 582*, 893–904.

Meeus G., Waters L. B. F.M., Bouwman J., van den Ancker M. E., Waelkens C., and Malfait K. (2001) *Astron. Astrophys., 365*, 476–490.

Meyer M. R., Hillenbrand L. A., Backman D. E., Beckwith S. V. W., Bouwman J., et. al. (2004) *Astrophys. J. Suppl., 154*, 422–427.

Miyake K. and Nakagawa Y. (1995) *Astrophys. J., 441*, 361–384.

Monnier J. D., Millan-Gabet R., Billmeier R., Akeson R. L., Wallace D., et. al. (2005) *Astrophys. J., 624*, 832–840.

Muzerolle J., Hillenbrand L., Calvet N., Briceño C., and Hartmann L. (2003a) *Astrophys. J., 592*, 266–281.

Muzerolle J., Calvet N., Hartmann L., and D'Alessio P. (2003b) *Astrophys. J., 597*, L149–L152.

Muzerolle J., D'Alessio P., Calvet N., and Hartmann L. (2004) *Astrophys. J., 617*, 406–417.

Muzerolle J., Luhman K. L., Briceño C., Hartmann L., and Calvet N. (2005) *Astrophys. J., 625*, 906–912.

Nakamoto T. and Nakagawa Y. (1994) *Astrophys. J., 421*, 640–650.

Natta A., Prusti T., Neri R., Wooden D., Grinin V. P., and Mannings V. (2001) *Astron. Astrophys., 371*, 186–197.

Natta A., Testi L., Muzerolle J., Randich S., Comerón F., and Persi P. (2004) *Astron. Astrophys., 424*, 603–612.

Niccolini G., Woitke P., and Lopez B. (2003) *Astron. Astrophys., 399*, 703–716.

Nomura H. (2002) *Astrophys. J., 567*, 587–595.

Nomura H. and Millar T. J. (2005) *Astron. Astrophys., 438*, 923–938.

Pascucci I., Meyer M., Gorti U., Hollenbach D., Hillenbrand L., Carpenter J., et al. (2005) In *PPV Poster Proceedings*, www.lpi.usra.edu/meetings/ppv2005/pdf/8468.pdf.

Pfalzner S., Umbreit S., and Henning T. (2005) *Astrophys. J., 629*, 526.

Pickett B. K., Mejía A. C., Durisen R. H., Cassen P. M., Berry D. K., and Link R. P. (2003) *Astrophys. J., 590*, 1060–1080.

Pollack J. B., Hubickyj O., Bodenheimer P., Lissauer J. J., Podolak M., and Greenzweig Y. (1996) *Icarus, 124*, 62–85.

Reyes-Ruiz M. and Stepinski T. F. (1996) *Astrophys. J., 459*, 653–665.

Richling S. and Yorke H. W. (1997) *Astron. Astrophys., 327*, 317–324.

Richling S., Hollenbach D., and Yorke H. (2006) In *Planet Formation* (H. Klahr and W. Brandner, eds.), p. 38. Cambridge Univ., Cambridge

Richter M. J., Jaffe D. T., Blake G. A., and Lacy J. H. (2002) *Astrophys. J., 572*, L161–L164.

Rucinski S. M. (1985) *Astron. J., 90*, 2321–2330.

Ruden S. P. (2004) *Astrophys. J., 605*, 880–891.

Ruden S. P. and Pollack J. B. (1991) *Astrophys. J., 375*, 740–760.

Sano T., Miyama S. M., Umebayashi T., and Nakano T. (2000) *Astrophys. J., 543*, 486–501.

Scally A. and Clarke C. (2001) *Mon. Not. R. Astron. Soc., 325,* 449–456.

Semenov D., Wiebe D., and Henning T. (2004) *Astron. Astrophys., 417,* 93–106.

Shakura N. I. and Sunyaev R. A. (1973) *Astron. Astrophys., 24,* 337–355.

Shu F. H., Johnstone D., and Hollenbach D. (1993) *Icarus, 106,* 92–101.

Spaans M., Tielens A. G. G. M., van Dishoeck E. F., and Bakes E. L. O. (1994) *Astrophys. J., 437,* 270–280.

Stehle R. and Spruit H. C. (2001) *Mon. Not. R. Astron. Soc., 323,* 587–600.

Steinacker J., Henning T., Bacmann A., and Semenov D. (2003) *Astron. Astrophys., 401,* 405–418.

Stepinski T. F. (1998) *Astrophys. J., 507,* 361–370.

Stone J. M. and Pringle J. E. (2001) *Mon. Not. R. Astron. Soc., 322,* 461–472.

Störzer H. and Hollenbach D. (1999) *Astrophys. J., 515,* 669–684.

Takeuchi T., Clarke C. J., and Lin D. N. C. (2005) *Astrophys. J., 627,* 286–292.

Terebey S., Shu F. H., and Cassen P. (1984) *Astrophys. J., 286,* 529–551.

Thi W. F., van Dishoeck E. F., Blake G. A., van Zadelhoff G. J., Horn J., Becklin E. E., Mannings V., Sargent A. I., van den Ancker M. E., Natta A., and Kessler J. (2001) *Astrophys. J., 561,* 1074–1094.

Throop H. B. and Bally J. (2005) *Astrophys. J., 623,* L149–L152.

Tielens A. G. G. M. and Hollenbach D. (1985) *Astrophys. J., 291,* 722–754.

Tohline J. E. and Hachisu I. (1990) *Astrophys. J., 361,* 394–407.

Trujillo C. A. and Brown M. E. (2001) *Astrophys. J., 554,* L95–L98.

Tuthill P. G., Monnier J. D., and Danchi W. C. (2001) *Nature, 409,* 1012–1014.

Vinković D., Ivezić Ž., Miroshnichenko A. S., and Elitzur M. (2003) *Mon. Not. R. Astron. Soc., 346,* 1151–1161.

Walker C., Wood K., Lada C. J., Robitaille T., Bjorkman J. E., and Whitney B. (2004) *Mon. Not. R. Astron. Soc., 351,* 607–616.

Ward W. R. (1997) *Icarus, 126,* 261–281.

Wardle M. (2004) *Astrophys. Space Sci., 292,* 317–323.

Weidenschilling S. J. (1977) *Astrophys. Space Sci., 51,* 153–158.

Whitney B. A. and Hartmann L. (1992) *Astrophys. J., 395,* 529–539.

Willacy K. and Langer W. D. (2000) *Astrophys. J., 544,* 903–920.

Wilner D. J., Ho P. T. P., Kastner J. H., and Rodríguez L. F. (2000) *Astrophys. J., 534,* L101–L104.

Wolf S., Henning T., and Stecklum B. (1999) *Astron. Astrophys., 349,* 839–850.

Yamashita T., Handa T., Omodaka T., Kitamura Y., Kawazoe E., Hayashi S. S., and Kaifu N. (1993) *Astrophys. J., 402,* L65–L67.

Yorke H. W. and Bodenheimer P. (1999) *Astrophys. J., 525,* 330–342.

Yorke H. W., Bodenheimer P., and Laughlin G. (1993) *Astrophys. J., 411,* 274–284.

Youdin A. N. and Shu F. H. (2002) *Astrophys. J., 580,* 494–505.

Yorke H. W. and Welz A. (1996) *Astron. Astrophys., 315,* 555–564.

Evolution of Circumstellar Disks Around Normal Stars: Placing Our Solar System in Context

Michael R. Meyer
The University of Arizona

Dana E. Backman
SOFIA/SETI Institute

Alycia J. Weinberger
Carnegie Institution of Washington

Mark C. Wyatt
University of Cambridge

Over the past 10 years abundant evidence has emerged that many (if not all) stars are born with circumstellar disks. Understanding the evolution of post-accretion disks can provide strong constraints on theories of planet formation and evolution. In this review, we focus on developments in understanding (1) the evolution of the gas and dust content of circumstellar disks based on observational surveys, highlighting new results from the Spitzer Space Telescope; (2) the physical properties of specific systems as a means to interpret the survey results; (3) theoretical models used to explain the observations; (4) an evolutionary model of our own solar system for comparison to the observations of debris disks around other stars; and (5) how these new results impact our assessment of whether systems like our own are common or rare compared to the ensemble of normal stars in the disk of the Milky Way.

1. INTRODUCTION

At the first Protostars and Planets conference in 1978, the existence of circumstellar disks around Sun-like stars was in doubt, with most researchers preferring the hypothesis that young stellar objects were surrounded by spherical shells of material unlike the solar nebula thought to give rise to the solar system (*Rydgren,* 1978). By the time of Protostars and Planets II, experts in the field had accepted that young stars were surrounded by circumstellar disks although the evidence was largely circumstantial (*Harvey,* 1985). At that meeting, Fred Gillett and members of the IRAS team announced details of newly discovered debris disks, initially observed as part of the calibration program (*Aumann et al.,* 1984). At Protostars and Planets III (PPIII), it was well established that many stars are born with circumstellar accretion disks (*Strom et al.,* 1993), and at Protostars and Planets IV (PPIV), it was recognized that many of these disks must give rise to planetary systems (*Marcy et al.,* 2000). Over the last 15 years, debris disks have been recognized as playing an important role in helping us understand the formation and evolution of planetary systems (*Backman and Paresce,* 1993; *Lagrange et al.,* 2000; see also *Zuckerman,* 2001). After PPIV, several questions remained. How do debris disks evolve around Sun-like stars? When do gas-rich disks transition to debris disks? Can we infer the presence of extrasolar planets from spectral energy distributions (SEDs) and/or resolved disk morphology? Is there any connection between debris disks and the radial velocity planets? Is there evidence for differences in disk evolution as a function of stellar mass?

In answering these questions, our objective is no less than to understand the formation and evolution of planetary systems through observations of the gas and dust content of circumstellar material surrounding stars as a function of stellar age. By observing how disks dissipate from the post-accretion phase through the planet-building phase, we can hope to constrain theories of planet formation (cf. chapters by Durisen et al. and Lissauer and Stevenson). By observing how debris disks generate dust at late times and comparing those observations with physical models of planetary system dynamics, we can infer the diversity of solar system architectures as well as attempt to understand how they evolve with time.

Today, we marvel at the wealth of results from the Spitzer Space Telescope and high-contrast images of spectacular individual systems. Detection statistics that were very uncertain with the sensitivity of the Infrared Astronomical Satellite (IRAS) and the Infrared Space Observatory (ISO) now can be compared with models of planetary system evolution, placing our solar system in context. Advances in planetary system dynamical theory and the discovery and characterization of the Kuiper belt (see chapter by Chiang et al.) have proceeded in parallel and further contribute to our understanding of extrasolar planetary systems. We attempt to compare observations of disks surrounding other stars to our current understanding of solar system evolution. Our ultimate goal is to learn whether or not solar systems like our own are common or rare among stars in the disk of the Milky Way and what implications this might have on the frequency of terrestrial planets that might give rise to life.

Our plan for this contribution is as follows. In section 2, we describe recent results from observational surveys for gas and dust surrounding normal stars. Next we describe detailed studies of individual objects in section 3. In section 4, we review modeling approaches used in constraining physical properties of disks from the observations. Section 5 describes a toy model for the evolution of our solar system that we use to compare with the ensemble of observations. Finally, in section 6 we attempt to address whether or not planetary systems like our own are common or rare in the Milky Way galaxy and summarize our conclusions.

2. EVOLUTION OF CIRCUMSTELLAR DISKS

In order to study the evolution of circumstellar disks astronomers are forced to observe Sun-like stars at a variety of ages in an attempt to create a history, hoping that on average, a younger population of similar mass stars can be assumed to be the evolutionary precursors of the older. Although deriving ages of stars across the Hertzsprung-Russell (H-R) diagram is fraught with uncertainty (e.g., *Stauffer,* 2004), it is a necessary step in studies of disk evolution. Such studies, combined with knowledge of our own solar system, are the only observational tools at our disposal for constraining theories of planet formation.

2.1. Statistics from Dust Surveys

2.1.1. Circumstellar dust within 10 AU.   Nearly all stars are thought to be born with circumstellar disks (*Beckwith and Sargent,* 1996; *Hillenbrand et al.,* 1998) and recent work has shown that these disks dissipate on timescales on the order of 3 m.y. (*Haisch et al.,* 2001). However, these results are based largely on the presence of near-infrared excess emission, which only traces optically thick hot dust within 0.1 AU of the central star. Indeed, the presence of an inner disk appears to correlate with the presence of spectroscopic signatures indicating active accretion onto the star (*Hartigan et al.,* 1995; see also chapter by Bouvier et al.). As active disk accretion diminishes (*Hartmann et al.,* 1998), the fraction of young stars in clusters that show evidence for optically thick inner disks diminishes. Yet what is often overlooked is that the very data that suggest a typical inner disk lifetime of <3 m.y. also *suggest a dispersion of inner disk lifetimes from 1 to 10 m.y.*

What has remained unclear until recently is how these primordial disks left over from the formation of the young star dissipate at larger radii and whether the termination of accretion represents an end of the gas-rich phase of the cir-

cumstellar disk. Even at the time of PPIII, it was recognized that young stars (with ages <3 m.y.) lacking optically thick near-infrared excess emission but possessing optically thick mid-infrared emission were rare (*Skrutskie et al.,* 1990). This suggested that the transition time between optically thick and thin from <0.1 AU to >3 AU was rapid, ≪1 m.y. (*Wolk and Walter,* 1996; *Kenyon and Hartmann,* 1995; *Simon and Prato,* 1995).

It is important to distinguish between surveys for primordial disks, gas- and dust-rich disks left over from the star-formation process, and debris disks, where the opacity we see is dominated by grains released through collisions of larger parent bodies. Often this distinction is made based on whether remnant gas is left in the system. With a gas-to-dust ratio >1.0, dust dynamics are influenced by their interaction with the gas (*Takeuchi and Artymowicz,* 2001). In the absence of gas, one can argue based on the short dust lifetimes that observed dust is likely recently generated through collisions in a planetesimal belt (*Backman and Paresce,* 1993; *Jura et al.,* 1998). Observations that constrain the evolution of the gas content in disks are described below.

Recent work has shown that even optically thin mid-infrared emission (tracing material between 0.3 and 3 AU) is rare around Sun-like stars with ages 10–30 m.y. *Mamajek et al.* (2004) performed a survey for excess emission around Sun-like stars in the 30-m.y.-old Tucana-Horologium association and found no evidence for excess within a sample of 20 stars down to dust levels $<2 \times 10^{-6}$ $M_\oplus$ for warm dust in micrometer-sized grains. Similar studies by *Weinberger et al.* (2004) of stars in the β Pic moving group as well as TW Hya association (both ~10 m.y. old) uncovered only a handful of stars with mid-infrared excess emission. These results are being confirmed with cluster studies undertaken with Spitzer. As part of the Formation and Evolution of Planetary Systems (FEPS) Legacy Science Program a survey has been conducted searching for warm dust at wavelengths from 3.6 to 8.0 μm around 74 Sun-like stars with ages of 3–30 m.y. *Silverstone et al.* (2006) reported only five detections from this survey and all of those were rare examples of long-lived optically thick disks. *It appears that circumstellar disk material between 0.1 and 1 AU typically drops below detectable levels on timescales comparable to the cessation of accretion.* These levels are probably below what our solar system might have looked like at comparable ages (3–30 m.y.).

However, Spitzer is uncovering a new population of transitional disks at mid-infrared wavelengths in the course of several young cluster surveys (*Forrest et al.,* 2004; *Calvet et al.,* 2005). *Chen et al.* (2005) find that ~30% of Sun-like stars in the subgroups of the 5–20-m.y. Sco Cen OB association exhibit 24-μm excess emission, higher than that found by *Silverstone et al.* (2006) at shorter wavelengths. *Low et al.* (2005) find examples of mid-IR excess at 24 μm in the 10-m.y. TW Hya association. The 24-μm emission is thought to trace material >1 AU, larger radii than the material traced by emission from 3 to 10 μm. Preliminary results from the FEPS program suggests that there is some evolution in the fraction of Sun-like stars with 24-μm excess (but no excess in the IRAC bands) from 3 to 300 m.y. This brackets major events in our own solar system evolution with the terrestrial planets thought to have mostly formed in <30 m.y. and the late heavy bombardment at >300 m.y. (see section 5).

It is interesting to note that there is now a small (five-member) class of debris disks with only strong mid-infrared excess and weak or absent far-IR/submillimeter excess emission: BD +20°307 at an age of >300 m.y. (*Song et al.,* 2005), HD 69830 at ~2 G.y. (*Beichman et al.,* 2005b), HD 12039 at 30 m.y. (*Hines et al.,* 2006), HD 113766 at 15 m.y. (*Chen et al.,* 2005), and HD 98800 at 10 m.y. (*Low et al.,* 1999; *Koerner et al.,* 2000). In the two older systems, BD +20°307 and HD 69830, this excess is almost entirely silicate emission from small grains. These objects are rare; only 1–3% of all systems surveyed. Whether they represent a short-lived transient phase that all stars go through, or a rare class of massive warm debris disks, is not yet clear (section 4.4).

2.1.2. Circumstellar disks at radii >10 AU. Surveys at far-infrared (>30 μm) and submillimeter wavelengths trace the coolest dust at large radii. Often, this emission is optically thin and is therefore a good tracer of total dust mass. Early surveys utilizing the IRAS satellite focused on large optically thick disks and envelopes surrounding young stellar objects within 200 pc, the distance of most star-forming regions (*Strom et al.,* 1993), and main-sequence stars within 15 pc because of limitations in sensitivity (*Backman and Paresce,* 1993). Submillimeter work suggested that massive circumstellar disks dissipate within 10 m.y. (*Beckwith et al.,* 1990; *Andrews and Williams,* 2005). Submillimeter surveys of field stars indicated that "typical" submillimeter emission from dust surrounding main-sequence stars diminished as t^{-2} (*Zuckerman and Becklin,* 1993).

Several new far-infrared studies were initiated with the launch of ISO by the European Space Agency (ESA) and the advent of the submillimeter detector SCUBA on the James Clerk Maxwell Telescope (JCMT). *Meyer and Beckwith* (2000) describe surveys of young clusters with the ISOPHOT instrument on ISO, which indicated that far-infrared emission became optically thin on timescales comparable to the cessation of accretion (about 10 m.y.). *Habing et al.* (1999, 2001) suggested that there was another discontinuity in the evolutionary properties of debris disks surrounding isolated A stars at an age of approximately 400 m.y. *Spangler et al.* (2001) conducted a large survey including both clusters and field stars, finding that dust mass diminished at $t^{-1.8}$ as if the dust removal mechanism was P-R drag (see section 4 below). Based on the data available at the time, and limitations in sensitivity from ISO, it was unclear how to reconcile these disparate conclusions based on comparable datasets. For a small sample of Sun-like stars, *Decin et al.* (2003) found that 10–20% (5/33) of Milky Way G stars, regardless of their age, have debris disks, compara-

ble to results obtained previously for A stars (*Backman and Paresce,* 1993).

Recent work with Spitzer offers a new perspective. From the FEPS program, surveys for cold debris disks surrounding G stars have led to several new discoveries (*Meyer et al.,* 2004; *Kim et al.,* 2005). Over 40 debris disk candidates have been identified from a survey of 328 stars and no strong correlation of cold dust mass with stellar age has been found. *Bryden et al.* (2006) (see also *Beichman et al.,* 2005a) have completed a volume-limited survey of nearby Sun-like stars with probable ages between 1 and 3 G.y. Overall the Spitzer statistics suggest a cold debris disk frequency of 10–20% surrounding Sun-like stars with a weak dependence on stellar age (Fig. 1). It should be noted that our own solar system cold dust mass would be undetectable in these surveys and it is still difficult to assess the mean and dispersion in cold disk properties based on the distribution of upper limits.

Submillimeter surveys of dust mass probe the coldest dust presumably at the larger radii. *Wyatt et al.* (2003) report observations of low-mass companions to young early-type stars (see also *Jewitt et al.,* 1994) indicating a lifetime of 10–60 m.y. for the massive primordial disk phase. *Carpenter et al.* (2005) (see also *Liu et al.,* 2004) combined these data with a new survey from the FEPS sample and found that the distribution of dust masses (and upper limits) from 1 to 3 m.y. is distinguished (with higher masses) from that found in the 10–30-m.y.-old sample at the 5σ level (Fig. 1). The data do not permit such a strong statement concerning the intermediate-age (3–10 m.y.) sample. *Najita and Williams* (2005) conducted a detailed study of

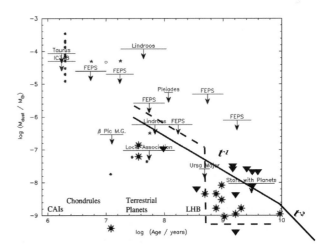

Fig. 1. Evolution of circumstellar dust mass based on submillimeter observations from *Carpenter et al.* (2005). Overplotted are Spitzer 70-μm detections from the FEPS program (stars) and upper limits (triangles). Slopes of t^{-1} and t^{-2} are shown as solid lines, along with a toy model for the evolution of our solar system (denoted with a dashed line) indicating an abrupt transition in dust mass associated with the late-heavy bombardment (LHB). Timescales associated with the formation of calcium-aluminum-rich inclusions (CAIs), chondrules, and terrestrial planets are also shown.

~15 individual objects and find that debris disks do not become colder (indicating larger radii for the debris) as they get older, surrounding Sun-like stars in contrast to the predictions of *Kenyon and Bromley* (2004). Again we note that these surveys would not detect the submillimeter emission from our own Kuiper debris belt (see section 5 below). In contrast, *Greaves et al.* (2004) point out that the familiar τ Ceti is 30× more massive than our solar system debris disk, even at comparable ages. *Greaves et al.* (2006) studied the metallicities of debris disk host stars showing that their distribution is indistinguishable from that of field stars, in contrast to the exoplanet host stars, which are metal-rich (*Fischer and Valenti,* 2005). Implications of the detected debris disk dust masses and their expected evolution are discussed in section 4 and compared to the evolution of our solar system in section 5.

The picture that emerges is complex, as illustrated in Fig. 1. In general, we observe diminished cold dust mass with time as expected from models of the collisional evolution of debris belts (see section 4). However, at any one age there is a wide dispersion of disk masses. Whether this dispersion represents a range of initial conditions in disk mass, a range of possible evolutionary paths, or is evidence that many disks pass through short-lived phases of enhanced dust production is unclear. One model for the evolution of our solar system suggests a rapid decrease in observed dust mass associated with the dynamic rearrangement of the solar system at 700 m.y. (and a tenfold decrease in the mass of colliding parent bodies). If that model is correct, we would infer that our solar system was an uncommonly bright debris disk at early times, and uncommonly faint at late times (see section 6).

2.2. Statistics from Gas Surveys

While most energy is focused on interpreting dust observations in disks, it is the gas that dominates the mass of primordial disks and is the material responsible for the formation of giant planets. Observational evidence for the dissipation of gas in primordial disks surrounding young Sun-like stars is scant. Millimeter-wave surveys (see chapter by Dutrey et al.) are ongoing and confirm the basic results: (1) classical T Tauri stars with excess emission from the near-IR through the submillimeter are gas-rich disks with some evidence for Keplerian support; and (2) complex chemistry and gas-grain interactions affect the observed molecular abundances. In a pioneering paper, *Zuckerman et al.* (1995) suggested that gas-rich disks dissipate within 10 m.y. Recent work on disk accretion rates of material falling ballistically from the inner disk onto the star by *Lawson et al.* (2004) could be interpreted as indicating that gas-rich primordial disks typically dissipate on timescales of 3–10 m.y. Other approaches include observations of warm molecular gas through near-infrared spectroscopy (see chapter by Najita et al.), UV absorption line spectroscopy of cold gas for favorably oriented objects (see next section), and millimeter-wave surveys for cold gas in remnant disks. One debris disk that showed evidence for gas in the early work of *Zuckerman et al.* (1995), the A star 49 Ceti, was recently confirmed to have CO emission by *Dent et al.* (2005). Transient absorption lines of atomic gas with abundances enhanced in refractory species would suggest the recent accretion of comet-like material (*Lecavelier des Etangs et al.,* 2001).

Since most of the mass in molecular clouds, and presumably in circumstellar disks from which giant planets form, is molecular hydrogen, it would be particularly valuable to constrain the mass in H_2 directly from observations. ISO provided tantalizing detections of warm H_2 at 12.3, 17.0, and 28.2 μm tracing gas from 50 to 200 K in both primordial and debris disks (*Thi et al.,* 2001a,b). However, followup observations with high-resolution spectroscopic observations (with a much smaller beam size) have failed to confirm some of these observations (*Richter et al.,* 2002; *Sheret et al.,* 2003). Several surveys for warm molecular gas are underway with Spitzer. *Gorti and Hollenbach* (2004) present a series of models for gas-rich disks with various gas-to-dust ratios. The initial stages of grain growth in planet-forming disks, the subsequent dissipation of the primordial gas disk, and the onset of dust production in a debris disk suggest a wide range of observable gas to dust ratios (see the chapter by Dullemond et al.). *Hollenbach et al.* (2005) placed upper limits of 0.1 M_{Jup} to the gas content of the debris disk associated with HD 105, a 30-m.y.-old Sun-like star observed as part of the FEPS project. *Pascucci et al.* (2006) have presented results for a survey finding no gas surrounding 15 stars with ages from 5 to 400 m.y. (nine of which are younger than 30 m.y.) at levels comparable to HD 105. Either these systems have already formed extrasolar giant planets, or they never will. Future work will concentrate on a larger sample of younger systems with ages of 1–10 m.y. in order to place stronger constraints on the timescale available to form gas giant planets.

3. PHYSICAL PROPERTIES OF INDIVIDUAL SYSTEMS

In order to interpret results from the surveys described above, we need to understand in detail the composition and structure of debris disks. Presumably, the dust (see section 4.1) reflects the composition of the parent planetesimal populations, so measuring the elemental composition, organic fraction, ice fraction, and ratio of amorphous to crystalline silicates provides information on the thermal and coagulation history of the small bodies. These small bodies are not only the building blocks of any larger planets, they could be an important reservoir for delivering volatiles to terrestrial planets (e.g., *Raymond et al.,* 2004). Additionally, the grain size distribution reflects the collisional state of the disk (see section 4.1). The structure of the disk may reflect the current distribution of planetesimals and therefore the system's planetary architecture (see section 5).

The literature on resolved images of circumstellar disks begins with the pioneering observations of β Pic by *Smith and Terrile* (1984). Since PPIV, there has been a significant increase in spatially resolved information on debris disks in two regimes: scattering and emission. Resolving scattered visual to near-infrared light requires high-contrast imaging such as that delivered by HST, because the amount of scattered light is at most 0.5% of the light from the star. Resolving thermal emission requires a large-aperture telescope because dust is warm closer to the star and so disks appear quite small in the infrared.

Compositional information is obtained from scattered-light albedos and colors, from mid-infrared spectroscopy that reveals solid-state features, and from fitting the slopes observed in spectral energy distributions. Resolved imaging breaks degeneracies in disk model fits and can be used to investigate changes in composition with location. Structural information is best at the highest spatial resolution and includes observations of warps, rings, nonaxisymmetric structures, and offset centers of symmetry.

Sensitivity to grain size depends on wavelength and each regime provides information on grains within approximately a range of 0.1–10× the wavelength (Fig. 2). For example, scattered visible and near-infrared light mostly probes grains smaller than 2 µm and submillimeter emission mostly probes grains >100 µm in size.

3.1. Debris Disks Resolved in Scattered Light

The number of debris disks resolved in scattered light has increased from one at the time of PPIII (β Pic) to two at the time of PPIV (HR 4796) to about 10 today (see Table 1). The detection of 55 Cnc reported in PPIV seems to be spurious (*Schneider et al.,* 2001; *Jayawardhana et al.,* 2002), and HD 141569 is not gas-free (*Jonkheid et al.,* 2005) and therefore is not counted here as a debris disk.

The scattered-light colors are now known for six debris disks (see Table 1). In many of these an asymmetry factor (g) has also been measured; larger grains are generally more

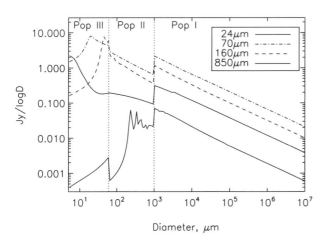

Fig. 2. Contribution of different grain sizes to the fluxes observed in different wavebands in the Vega disk (*Wyatt,* 2006). The units of the y-axis are flux per log particle diameter so that the area under the curve indicates the contribution of different-sized particles to the total flux for a given wavelength. The different wavebands probe different ranges in the size distribution and so are predicted to see very different structures.

forward-scattering. For disks in which the mid-infrared emission has also been resolved, the amount of scattered light compared to the mid-infrared emission from the same physical areas enables a calculation of the albedo [albedo = Qsca/(Qsca + Qabs)]. The albedo of canonical *Draine and Lee* (1984) astronomical silicates is such that (for 0.5–1.6-µm observations) grains smaller than 0.1 µm Rayleigh scatter and are blue, grains larger than 2 µm scatter neutrally, and grains in between appear slightly red. In the case of a power-law distribution of grain sizes, such as that of a collisional cascade (equation (2)), the scattering is dominated by the smallest grains. Thus the colors in Table 1 have been explained by tuning the smallest grain size to give the appropriate color. Rarely has the scattered color been modeled simultaneously with other constraints on similar-sized grains such as 8–13-µm spectra. If observations of scattered light at longer wavelengths continue to show red colors, the fine tuning of the minimum grain size of astronomical silicates will fail to work. More realistic grains may be porous aggregates where the voids may contain ice. Few optical constants for these are currently available in the literature.

3.2. Debris Disks Resolved in Submillimeter Emission

Resolved observations from JCMT/SCUBA in the submillimeter at 850 µm by *Holland et al.* (1998) and *Greaves et al.* (1998) led the way in placing constraints on cold dust morphologies for four disks (Fomalhaut, Vega, β Pic, and ε Eri), including rings of dust at Kuiper-belt-like distances from stars and resolving clumps and inner holes. Since PPIV, higher-spatial-resolution images at 350–450 µm revealed additional asymmetries interpreted as indications for planets (*Holland et al.,* 2003; *Greaves et al.,* 2005; *Marsh et al.,* 2005). Perhaps most excitingly, the structure of the disk surrounding ε Eri appears to be rotating about the star. A longer time baseline for the motion of disk clumps will reveal the mass and eccentricity of the planet responsible for their generation (*Greaves et al.,* 2005). Finally, three additional disks — τ Ceti (*Greaves et al.,* 2004), HD 107146 (*Williams et al.,* 2004), and η Corvi (*Wyatt et al.,* 2005) — were resolved by JCMT/SCUBA. Interferometric imaging of one debris disk, Vega, allowed the first measurement of structure at a wavelength of 1 mm (*Koerner et al.,* 2001; *Wilner et al.,* 2002). Again, the presence of clumps could be explained by the influence of a planet (*Wyatt,* 2003). It is interesting to note that three A-type stars, with masses up to twice that of the Sun and luminosities up to tens of times higher, show dynamical evidence for planets.

3.3. Debris Disks Resolved in Infrared Emission

Groundbased 8-m-class telescopes provide the best spatial resolution for imaging disks, but are hampered by low sensitivity; only two debris disks (β Pic and HR 4796) are definitively resolved at 12–25 µm from the ground.

Spitzer, with its ten-times-smaller aperture, is able to resolve only nearby disks. With MIPS, Spitzer has surprised

TABLE 1.   Resolved debris disk properties.

Star	Spectral Type	Age (m.y.)	Size (AU)	Scattered Light			Resolved in Emission?	References
				Color	g	albedo		
HR 4796A	A0	8	70	red (V-J)	0.15	0.1–0.3	yes	[1,2,3,4]
HD 32297	A0	10?	400	blue (R-J)	not avail.	0.5	no	[5,6]
β Pic	A5	12	10–1000	neutral-red (V-I)	0.3–0.5	0.7	yes	[7,8,9,10]
AU Mic	M1	12	12–200	neutral-blue (V-H)	0.4	0.3	no	[11,12,13,14]
HD 181327	F5	12	60–86	red (V-J)	0.3	0.5	no	[15]
HD 92945	K1	20–150	120–146	Red (V-I)	not avail.	not avail.	no	[16]
HD 107146	G2	30–250	130	red (V-I)	0.3	0.1	yes	[17,18]
Fomalhaut	A3	200	140	not avail.	0.2	0.05	yes	[19,20]
HD 139664	F5	300	110	not avail.	not avail.	0.1	no	[21]
HD 53143	K1	1000	110	not avail.	not avail.	0.06	no	[21]
Saturn's Rings	—	—	—	red (B-I)	–0.3	0.2–0.6	—	[22]
				Emission (Additional)				
Vega	A0	200	>90					[23,24,25]
ε Eridani	K2	<1000	60					[26,27]
η Corvi	F2	~1000	100					[28]
τ Ceti	G8	~5000	55					[29]

Note: The size given is the approximate radius or range of radii. It remains to be seen if the younger systems, particularly HD 32297, really are gas-free debris disks. The size for HD 32297 is the inner disk; it has a large circumstellar nebulosity as well (*Kalas,* 2005).

References: [1] *Schneider et al.* (1999); [2] G. Schneider and J. Debes (personal communication); [3] *Jayawardhana et al.* (1998); [4] *Koerner et al.* (1998); [5] *Schneider et al.* (2005); [6] *Kalas* (2005); [7] *Artymowicz et al.* (1989); [8] *Kalas and Jewitt* (1995); [9] *Telesco et al.* (2005); [10] *Golimowski et al.* (2005); [11] *Kalas et al.* (2004); [12] *Liu* (2004); [13] *Metchev et al.* (2005); [14] *Krist et al.* (2005); [15] *Schneider et al.* (2006); [16] Clampin et al. (in preparation); [17] *Ardila et al.* (2004); [18] *Williams et al.* (2004); [19] *Wyatt and Dent* (2002); [20] *Kalas et al.* (2005); [21] *Kalas et al.* (2006); [22] *Cuzzi et al.* (1984); [23] *Holland et al.* (1998); [24] *Wilner et al.* (2002); [25] *Su et al.* (2005); [26] *Greaves et al.* (2005); [27] *Marengo et al.* (2005); [28] *Wyatt et al.* (2005); [29] *Greaves et al.* (2004).

observers with images of β Pic, ε Eri, Fomalhaut, and Vega that look quite different from their submillimeter morphologies. If Spitzer's sensitivity picked up the Wien tail of the submillimeter grain emission or if the smaller mid-infrared emitting grains were co-located with their larger progenitor bodies, then the morphologies would be the same. In the case of Fomalhaut, the MIPS 24-μm flux originates in a zodiacal-like region closer to the star *and* the planetesimal ring while the 70-μm flux does indeed trace the ring (*Stapelfeldt et al.,* 2004) (Fig. 3). As for the solar system, there may be separate populations of planetesimals (analogous to the asteroid and Kuiper belts) generating dust.

Surprisingly, however, the 24- and 70-μm images of Vega actually have larger radii than the submillimeter ring or millimeter clumps (*Su et al.,* 2005). This emission seems to trace small grains ejected by radiation pressure. Vega is only slightly more luminous than Fomalhaut, so the minimum grain size generated in collisions within the disk would have to finely tuned to below the blowout size for Vega and above the blowout size for Fomalhaut for a unified disk model (see equation (3)). ε Eri looks about as expected with the 70-μm emission from the region of the submillimeter ring (*Marengo et al.,* 2005). An inner dust population might be expected if Poynting-Robertson (P-R) drag is important for the

dust dynamics of this system (see section 4). The absence of close-in dust may indicate that it is ejected by the postulated planet. It is also interesting that Spitzer did not resolve any of the other nearby disks imaged, including ones resolved in the submillimeter such as β Leo [however, see new results on η Corvi by Bryden et al. (in preparation)]. It is possible in these cases that the grain sizes are so large that Spitzer cannot see the Wien-side of such cold emission and/or that their viewing geometries (nearly face-on) were unfavorable.

Spatially resolved spectroscopy has been obtained for only one debris disk, β Pic. These spectra provided information on collision rates, with small silicate grains only observed within 20 AU of the star and thermal processing, with crystalline silicate fractions higher closer to the star (*Weinberger et al.,* 2003; *Okamoto et al.,* 2004). Of the stars in Table 1 with measured scattered light, only β Pic, HR 4796, and Fomalhaut have been resolved in the infrared.

Only silicates with D < 4 μm show silicate emission. In the zodiacal dust, this is only ~10% and the "typical" grain size is 100 μm (*Love and Brownlee,* 1993). Without resolving disks, the line-to-continuum ratio of the mid-infrared silicate bands at 10–20 μm, which in principle reflects the proportion of small grains, can be diluted by flux from cold

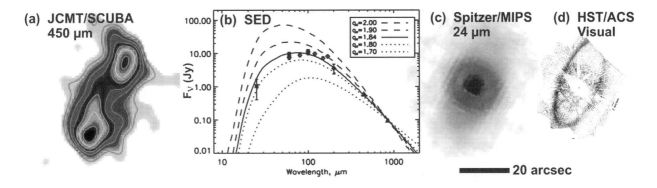

Fig. 3. The Fomalhaut disk is one of the few to have been resolved in **(a)** the submillimeter (*Holland et al.,* 2003), **(c)** the thermal infrared (*Stapelfeldt et al.,* 2004), and **(d)** scattered visual light (*Kalas et al.,* 2005). **(b)** When only mid-infrared total fluxes and the submillimeter images were available, *Wyatt and Dent* (2002) made models using compact silicate grains. The addition of the mid-infrared images allows a separation between warm (T ~ 150 K) dust in an inner portion of the ring not seen in the submillimeter and the outer colder ring. The addition of the scattered light image allows a more accurate determination of the ring geometry including a direct detection of the offset center of symmetry, similar to that observed in HR 4796 (*Wyatt et al.,* 1999). In future work, the silicate model must be tuned to fit the dust scattered light (albedo) as well as emissivity.

grains. Many debris disks with 12-μm excess show no silicate emission (*Jura et al.,* 2004), with the implication that their grains are larger than 10 μm. The unfortunate consequence is that direct compositional information is hard to acquire.

3.4. Detections of Remnant Gas

A useful definition of a debris disk is that it is gas free, because then the dust dynamics are dominated by the processed described in section 4 unmodified by gas drag (*Takeuchi and Artymowicz,* 2001). However, debris disks can have small amounts of gas released in the evaporation of comets or destructive grain-grain collisions. The most sensitive gas measurements are made with ultraviolet absorption spectroscopy of electronic transitions. These transitions are strong and trace atomic and molecular gas at a wide range of temperatures. Yet since absorption spectroscopy probes only a single line of sight, it is only very useful for edge-on disks and it remains uncertain how to go from measured column densities to total disk masses.

The edge-on disks around the coeval β Pic and AU Mic provide strong constraints on the persistence of gas into the debris disk phase. The total measured gas mass in β Pic is 7×10^{-4} M$_\oplus$, while the upper limit (set by limits on HI) is 0.03 M$_\oplus$ (*Roberge et al.,* 2006). Because the CO/H$_2$ ratio is more like that of comets than of the ISM (CO is actually more abundant than H$_2$), the gas is presumably second-generation just as the dust is (*Lecavelier des Etangs et al.,* 2001). In AU Mic, the upper limit to the gas mass from the nondetection of molecular hydrogen is 0.07 M$_\oplus$ (*Roberge et al.,* 2005). β Pic and AU Mic differ in luminosity by a factor of 90, but both were able to clear their primordial gas in under 12 m.y. Similar upper limits on the gas mass are also observed for the slightly younger, slightly less edge-on disk around HR 4796A (*Chen,* 2002).

Beyond total mass, a detailed look at the β Pic disk reveals a wide range of atomic species in absorption with an up-to-date inventory given in *Roberge et al.* (2006). In addition, the spatial distribution of gas in β Pic is also imaged by long-slit high-spectral-resolution spectroscopy (*Brandeker et al.,* 2004). Atomic gas species such as sodium, iron, and calcium are all distributed throughout the disk with Keplerian line-of-sight velocities. The observation that iron, which should experience strong radiation pressure and be ejected on orbital timescales, has such low velocities remains a puzzle (*Lagrange et al.,* 1998). At this time, the best explanation for why the gas is not ejected by radiation pressure is that the ions strongly couple via Coulomb forces enhanced by an overabundance of carbon gas (*Fernandez et al.,* 2006; *Roberge et al.,* 2006). Most of the gas in the disk is ionized by a combination of stellar and interstellar UV. Remaining puzzles are the vertical distribution of calcium gas, which is actually located predominantly away from the mid-plane (*Brandeker et al.,* 2004), and why there exists such a large overabundance of carbon in the stable gas (*Roberge et al.,* 2006).

4. OVERVIEW OF DEBRIS DISK MODELS

4.1. Basic Dust Disk Physics

As described above, knowledge concerning general trends in the evolution of dust as a function of radius (see section 2), as well as detailed information concerning particle composition and size distribution (see section 3), abounds. However, *understanding* these trends and placing specific systems in context requires that we interpret these data in the context of robust physical theory. Models of debris disks have to explain two main observations: the radial location of the dust and its size distribution. There are two competing physical processes that determine how these distributions differ from that of the parent planetesimals that are feeding the dust disk.

4.1.1. Collisions. All material in the disk is subject to collisions with other objects, both large and small. If the collision is energetic enough, the target particle is destroyed (a catastrophic collision) and its mass redistributed into smaller particles. Lower-energy collisions result in cratering of the target particle or accretion of the target and impactor. It is catastrophic collisions that replenish the dust we see in debris disks, and collisional processes are responsible for shaping a disk's size distribution.

Both experimental (*Fujiwara et al.,* 1989) and numerical (*Benz and Asphaug,* 1999) work has been used to determine the specific incident energy required to catastrophically destroy a particle, Q_D^*. This energy depends on particle composition, as well as the relative velocity of the collision (v_{rel}), but to a greater extent is dependent on the size of the target. It is found to lie in the range $Q_D^* = 10^0$–10^6 J kg^{-1}, which means that for collision velocities of ~1 km s^{-1} particles are destroyed in collisions with other particles that are at least x = 0.01–1 times their own size, D. The collision velocity depends on the eccentricities and inclinations of the particles' orbits, the mean values of which may vary with particle size after formation (e.g., *Weidenschilling et al.,* 1997). For planetesimal growth to occur, both have to be relatively low, ~10^{-3}, to prevent net destruction of particles. However, to initiate a collisional cascade something must have excited the velocity dispersion in the disk to allow collisions to be catastrophic. Models that follow the collisional evolution of planetesimal belts from their growth phase through to their cascade phase show that this switch may occur after the formation of a planet-sized object (*Kenyon and Bromley,* 2002b, 2004) or from excitation by a passing star (*Kenyon and Bromley,* 2002a).

A particle's collisional lifetime is the mean time between catastrophic collisions. This can be worked out from the catastrophic collision rate, which is the product of the relative velocity of collisions and the volume density of the cross-sectional area of impactors larger than xD. For the smallest particles in the distribution, for which collisions with any other member of the distribution is catastrophic, their collisional lifetime is given by

$$t_{coll} = t_{per}/4\pi\tau_{eff} \qquad (1)$$

where t_{per} is the orbital period and τ_{eff} is the surface density of cross-sectional area in the disk, which when multiplied by the absorption efficiency of the grains, gives the disk's face-on optical depth (*Wyatt and Dent,* 2002). Larger particles have longer collisional lifetimes. In an infinite collisional cascade in which the outcome of collisions is self-similar (in that the size distribution of collision fragments is independent of the target size), collisions are expected to result in a size distribution with

$$n(D) \propto D^{-3.5} \qquad (2)$$

(*Dohnanyi,* 1969; *Tanaka et al.,* 1996). Such a distribution has most of its mass in the largest planetesimals, but most of its cross-sectional area in the smallest particles.

4.1.2. Radiation pressure and Poynting-Robertson drag. Small grains are affected by their interaction with stellar radiation, which causes a force on the grains that is parameterized by β, the ratio of the radiation force to stellar gravity. This parameter depends on the size of the grain, and to a lesser extent on its composition. For large particles, β can be approximated by

$$\beta = (0.4 \ \mu m/D)(2.7 \ g \ cm^{-3}/\rho)(L_\star/M_\star) \qquad (3)$$

where ρ is the grain density and $L_\star$ and $M_\star$ are in units of $L_\odot$ and $M_\odot$ (*Burns et al.,* 1979). However, this relation breaks down for particles comparable in size to the wavelength of stellar radiation, for which a value of β is reached that is independent of particle size (*Gustafson,* 1994).

The radial component of the radiation force is known as radiation pressure. For grains with β > 0.5 (or D < D_{bl}), which corresponds to submicrometer-sized grains near a Sun-like star, radiation pressure causes the grains to be blown out of the system on hyperbolic trajectories as soon as they are created. Since grains with β = 1 have no force acting on them, the blowout timescale can be estimated from the orbital period of the parent planetesimal

$$t_{bl} = \sqrt{a^3/M_\star} \qquad (4)$$

where a is the semimajor axis of the parent in AU, and t_{bl} is the time to go from a radial distance of a to 6.4a. In the absence of any further interaction, such grains have a surface density distribution that falls off $\propto r^{-1}$.

The tangential component of the radiation force in known as Poynting-Robertson (P-R) drag. This acts on all grains and causes their orbits to decay into the star (where the grains evaporate) at a rate $\dot{a} = -2\alpha/a$, where $\alpha = 6.24 \times 10^{-4} \ M_\star/\beta$. Thus the evolution from a to the star takes

$$t_{pr} = 400(a^2/M_\star)/\beta \qquad (5)$$

in years. In the absence of any further interaction, such grains have a surface density distribution that is constant with the distance from the star.

4.1.3. Other processes. Other physical processes acting on dust in debris disks range from gas drag to stellar wind drag, Lorentz forces on charged grains, and sublimation. Many of these have been determined to be unimportant in the physical regimes of debris disks. However, it is becoming clear that for dust around M stars the force of the stellar wind is important both for its drag component (*Plavchan et al.,* 2005) and its pressure component (*Strubbe and Chiang,* 2006; *Augereau and Beust,* 2006). Gas drag may also be important in young debris disks. While the quantity of gas present is still poorly known, if the gas disk is sufficiently dense then gas drag can significantly alter the orbital evolution of dust grains. This can result in grains migrating to different radial locations from where they were created, with different sizes ending up at different locations (e.g., *Takeuchi and Artymowicz,* 2001; *Klahr and Lin,* 2001; *Ardila et al.,* 2005; *Takeuchi et al.,* 2005). For β Pic it has been estimated that gas drag becomes important when the gas to dust ratio exceeds 1 (*Thébault and Augereau,* 2005).

4.2. Model Regimes

A debris disk that is not subjected to the stochastic mass-loss processes discussed in sections 4.4 and 5 will evolve in steady-state losing mass through radiation processes acting on small grains: through P-R drag and consequently evaporation close to the star, or through collisional grinding down and consequently blowout by radiation pressure. The competition between collisions and P-R drag was explored in *Wyatt* (2005), who modeled the dust distribution expected if a planetesimal belt at r_0 is creating dust of just one size (see Fig. 4). The resulting distribution depends only on the parameter $\eta_0 = t_{pr}/t_{coll}$. If the disk is dense ($\eta_0 \gg 1$), then collisions occur much faster than P-R drag and the dust remains confined to the planetesimal belt, whereas if the disk is tenuous ($\eta_0 \ll 1$), then the dust suffers no collisions before reaching the star and the dust distribution is flat as expected by P-R drag. While this is a simplification of the processes going on in debris disks, which are creating dust of a range of sizes, it serves to illustrate the fact that disks operate in one of two regimes: collisional or P-R drag-dominated. These regimes are discussed in more detail below.

4.2.1. Collisionally dominated disks. In a collisionally dominated disk ($\eta_0 \gg 1$) it is possible to ignore P-R drag, since the cumulative migration of particles over all generations from planetesimal to micrometer-sized grain is negligible (e.g., *Wyatt et al.,* 1999). This is because P-R drag lifetimes increase $\propto D$, whereas collisional lifetimes increase $\propto D^{0.5}$ (assuming the distribution of equation (2)), meaning that the migration undergone before a collision becomes vanishingly small for large particles.

There are two components to a collisionally dominated debris disk dynamically bound grains at the same radial location as the planetesimals, and unbound grains with an r^{-1} distribution beyond that. The short lifetime of the unbound grains (equation (4)) suggests that their number density should be extremely tenuous, and should fall below that expected from an extrapolation of the collisional cascade distribution. However, recent observations indicate that in some imaged debris disks they are being replenished at a rate sufficient for these grains to dominate certain observations (e.g., *Telesco et al.,* 2000; *Augereau et al.,* 2001; *Su et al.,* 2005), implying a comparable cross-sectional area in these particles to that in bound grains as currently observed.

The size distribution in a collisionally dominated disk varies somewhat from that given in equation (2), since that assumes an infinite collisional cascade. If the number of blowout grains falls below that of the collisional cascade

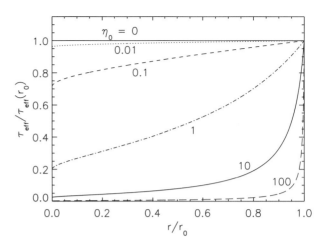

Fig. 4. Surface density distribution of dust created in a planetesimal belt at r_0 that evolves due to collisions (which remove dust) and P-R drag (which brings it closer to the star) (*Wyatt,* 2005). Assuming the dust is all of the same size, the resulting distribution depends only on η_0, the ratio of the collisional lifetime to that of P-R drag.

distribution, then since these particles would be expected to destroy particles just larger than themselves, their low number causes an increase in the equilibrium number of particles just above the blowout limit. This in turn reduces the equilibrium number of slightly larger particles, and so on; i.e., this causes a wave in the size distribution that continues up to larger sizes (*Thébault et al.,* 2003). If, on the other hand, larger quantities of blowout grains are present (e.g., because their number is enhanced by those driven out from closer to the star), then this can actually reduce the equilibrium number of particles just above the blowout limit (*Krivov et al.,* 2000).

The long-term evolution of a collisionally dominated disk was considered by *Dominik and Decin* (2003). They considered the case where the dust disk is fed by planetesimals of a given size, D_c, and showed how collisions cause the number of those planetesimals, N_c, to follow

$$N_c(t) = N_c(0)/[1 + 2t/t_c(0)] \qquad (6)$$

where t_c is the collisional lifetime of the colliding planetesimals at t = 0. In other words, the evolution is flat until the disk is old enough for the majority of the planetesimals to have collided with each other (i.e., when $t > t_c$), thus eroding their population, at which point their number falls off $\propto t^{-1}$. Since the size distribution connecting the dust to the number of planetesimals is given by equation (2), it follows that the cross-sectional area of emitting dust has the same flat or t^{-1} evolution as does the total mass of material in the disk that is dominated by planetesimals of size D_c. *Dominik and Decin* (2003) also noted ways of changing the evolution, e.g., by introducing stirring.

The quantity of blowout grains in the disk does not follow the same evolution, since their number is determined by the equilibrium between the rate at which the grains are created and that at which they are lost (eqn. 4). The rate at which they are created depends on details of the physics of collisions, but since the rate at which dust is produced by planetesimals is $\propto N_c^2$, it follows that their population should fall off $\propto t^0$ or t^{-2} depending on whether $t < t_c$ or $t > t_c$.

4.2.2. Poynting-Robertson drag-dominated disks. A conclusion shared by *Dominik and Decin* (2003) and *Wyatt* (2005) is that none of the debris disks detected with current instrumentation is in the P-R drag-dominated regime. *Wyatt* (2005) explained this as a consequence of the fact that such disks are of too low mass for their emission to be comparable to that of the stellar photosphere. Thus the detection of such disks requires calibration to a few % in the mid- to far-IR, or discovery in the submillimeter. However, the zodiacal cloud (and presumably dust from the Kuiper belt) in the solar system is a good example of a P-R drag-dominated disk.

It is not possible to completely ignore collisions in a P-R drag-dominated disk, since, while the smallest dust makes it to the star without suffering a collision, the largest grains are in a collisionally dominated regime, with intermediate sizes having distributions closer to that of $\eta_0 = 1$ in Fig. 4. Matters are complicated by the way P-R drag affects the size distribution. If collisional processes in a planetesimal belt are assumed to create dust at a rate that results in the size distribution of equation (2) in the planetesimal belt, then since small dust migrates faster than small dust (equations (3) and (4)) then the size distribution of the dust affected by P-R drag should follow

$$n(D) \propto D^{-2.5} \qquad (7)$$

(*Wyatt et al.*, 1999), a distribution in which most of the cross-sectional area is in the largest particles in that distribution. In other words, the cross-sectional area should be dominated by grains for which P-R drag and collisional lifetimes are roughly equal, with that size varying with distance from the planetesimal belt. This reasoning is in agreement with observations of the size distribution of interplanetary dust in the vicinity of Earth (*Love and Brownlee*, 1993; *Wyatt et al.*, 1999; *Grogan et al.*, 2001). *Dominik and Decin* (2003) also looked at the evolution of P-R drag-dominated disks within the model described above. They concluded that the quantity of visible grains should fall off $\propto t^{-2}$.

4.3. Formation of Inner Hole

Perhaps the most important discovery about debris disks is the fact that there are inner holes in their dust distribution. It is often suggested that planet-sized bodies are required interior to the inner edge of the debris disk to maintain the inner holes, because otherwise the dust would migrate inward due to P-R drag, thus filling in the central cavity (*Roques et al.*, 1994). It is certainly true that a planet could maintain an inner hole by a combination of trapping

the dust in its resonances (*Liou and Zook,* 1999), scattering the dust outward (*Moro-Martin and Malhotra*, 2002), and accreting the dust (*Wyatt et al.*, 1999). However, a planet is not required to prevent P-R drag filling in the holes in the detected debris disks, since in sufficiently dense disks, collisional grinding down already renders P-R drag insignificant (*Wyatt*, 2005).

What the inner holes do require, however, is a lack of colliding planetesimals in this region. One possible reason for the lack of planetesimals close to the star is that they have already formed into planet-sized objects, since planet-formation processes proceed much faster closer to the star (*Kenyon and Bromley*, 2002b). Any remaining planetesimals would then be scattered out of the system by chaos induced by perturbations from these larger bodies (e.g., *Wisdom*, 1980).

4.4. Steady-State Versus Stochastic Evolution

Much of our understanding of debris disks stems from our understanding of the evolution of the zodiacal cloud. This was originally assumed to be in a quasisteady state. However, models of the collisional evolution of the asteroid belt, and the dust produced therein, showed significant peaks in dust density occur when large asteroids collide, releasing quantities of dust sufficient to affect to total dust content in the inner solar system (*Dermott et al.*, 2002). Further evidence for the stochastic evolution of the asteroid belt came from the identification of asteroid families created in the recent (last few million years) breakup of large asteroids (*Nesvorný et al.*, 2003). The link of those young families to the dust band features in the zodiacal cloud structure (*Dermott et al.*, 2002) and to peaks in the accretion rate of ^{3}He by Earth (*Farley et al.*, 2005) confirmed the stochastic nature of the inner solar system dust content, at least on timescales of several million years. More recently, the stochastic nature of the evolution of debris disks around A stars has been proposed by *Rieke et al.* (2005) based on the dispersion of observed disk luminosities. Several debris disks are observed to have small grains (with very short lifetimes) at radii inconsistent with steady-state configurations over the lifetime of the star (e.g., *Telesco et al.*, 2005; *Song et al.*, 2005; *Su et al.*, 2005).

The arguments described previously considered the steady-state evolution of dust created in a planetesimal belt at single radius. The same ideas are still more generally applicable to stochastic models, since a situation of quasisteady state is reached relatively quickly, at least for small dust for which radiation and collision processes balance on timescales on the order of 1 m.y. (depending on disk mass and radius).

Stochastic evolution of the type seen in the zodiacal cloud arises from the random input of dust from the destruction of large planetesimals. Whether it is possible to witness the outcome of such events in extrasolar debris disks is still debated for individual objects. This is unlikely to be the case for dust seen in the submillimeter, since the large dust mass observed requires a collision between two large planetesimals (>1400 km for dust seen in Fomalhaut), and while such events may occur, the expected number of such objects makes witnessing such an event improbable (*Wyatt and Dent*, 2002). Observations at shorter wavelengths (and closer to the star) probe lower dust masses, however, and these observations may be sensitive to detecting such events (*Telesco et al.*, 2005; *Kenyon and Bromley*, 2005).

Debris disk evolution may also be affected by external influences. One such influence could be stirring of the disk by stars that pass by close to the disk (*Larwood and Kalas*, 2001; *Kenyon and Bromley*, 2002b). However, the low frequency of close encounters with field stars means this cannot account for the enhanced dust flux of all debris disk candidates, although such events may be common in the early evolution of a disk when it is still in a dense cluster environment. Another external influence could be the passage of the disk through a dense patch of interstellar material, which either replenishes the circumstellar environment with dust or erodes an extant but low-density debris disk (*Lissauer and Griffith*, 1989; *Whitmire et al.*, 1992).

Other explanations that have been proposed to explain sudden increases in dust flux include the sublimation of supercomets scattered close to the star (*Beichman et al.*, 2005b).

It is also becoming evident that the orbits of the giant planets have not remained stationary over the age of the solar system (*Malhotra*, 1993; *Gomes et al.*, 2005). The recently investigated stochastic component of giant planet orbital evolution can explain many of the features of the solar system, including the period of late heavy bombardment (LHB), which rapidly depleted the asteroid and Kuiper belts, leading to enhanced collision rates in the inner solar system. Such an event in an extrasolar system would dramatically increase its dust flux for a short period, but would likely do so only once in the system's lifetime. A similar scenario was also proposed by *Thommes et al.* (1999) to explain the LHB wherein the giant cores that formed between Jupiter and Saturn were thrown outward into the Kuiper belt by chaos at a late time. Again, this would result in a spike in the dust content of an extrasolar system. These ideas are applied to our own solar system in the next section.

5. COMPARISON TO OUR SOLAR SYSTEM

Our asteroid belt (AB) and Kuiper-Edgeworth belt (KB) contain planetesimals that accreted during the earliest epochs of the solar system's formation, plus fragments from subsequent collisions (e.g., *Bottke et al.*, 2005; *Stern and Colwell*, 1997). Collisions in both belts should generate populations of dust grains analogous to extrasolar debris disks. The dust population extending from the AB is directly observed as the zodiacal cloud, whereas dust associated with the KB is as yet only inferred (*Landgraf et al.*, 2002). An observer located 30 pc from the present solar system would receive approximately 70 μJy at 24 μm and 20 μJy at 70 μm

from the AB plus zodi cloud, in contrast to 40 mJy (24 μm) and 5 mJy (70 μm) from the Sun. The luminosity of the KB dust component is less certain but flux densities from 30 pc of about 0.4 mJy at 24 μm and 4 mJy at 70 μm correspond to an estimated KB planetesimal mass of $0.1 M_\oplus$ (see below for details of these calculations).

The solar system's original disk contained much more solid mass in the AB and KB zones than at present. A minimum-mass solar nebula would have had $3.6 M_\oplus$ of refractory material in the primordial AB between r = 2.0 and 4.0 AU, whereas now the AB contains only $5 \times 10^{-4} M_\oplus$, and only $2 \times 10^{-4} M_\oplus$ if the largest object Ceres is excluded. In contrast, the masses of Earth and Venus are close to the minimum-mass nebular values for their respective accretion zones. Likewise, the primordial KB must have had 10–$30 M_\oplus$ so that the observed population of large objects could have formed in less than 10^8 years before gravitational influence of the planets made further accretion impossible (*Stern*, 1996). The present KB contains no more than a few $\times 0.1 M_\oplus$ based on discovery statistics of massive objects (discussed by *Levison and Morbidelli*, 2003) and upper limits to IR surface brightness of collisionally evolved dust (*Backman et al.*, 1995; *Teplitz et al.*, 1999).

How did the missing AB and KB masses disappear? It is unlikely that purely "internal" collisional processes followed by radiation pressure-driven removal of small fragments is responsible for depletion of either belt. Persistence of basaltic lava flows on Vesta's crust is evidence that the AB contained no more than $0.1 M_\oplus$ 6 m.y. after the first chondrules formed (*Davis et al.*, 1985; *Bottke et al.*, 2005). This implies a factor of at least 40× depletion of the AB zone's mass by that time, impossible for purely collisional evolution of the original amount of material (reviewed by *Bottke et al.*, 2005) (cf. section 4). Also, the present AB collisional "pseudo-age," i.e., the model timescale for a purely self-colliding AB to reach its present density, is on the order of twice the current age of the solar system. This is further indication that the AB's history includes significant depletion processes other than comminution. Proposed depletion mechanisms include sweeping of secular resonances through the AB as the protoplanetary disk's gas dispersed (*Nagasawa et al.*, 2005) and as Jupiter formed and perhaps migrated during the solar system's first 10 m.y. or so (*Bottke et al.*, 2005).

Similarly, several investigators have concluded that the primordial KB was depleted by outward migration of Neptune that swept secular resonances through the planetesimal population, tossing most of the small objects either inward to encounter the other planets or outward into the KB's "scattered disk." That scenario neatly explains several features of the present KB in addition to the mass depletion (*Levison and Morbidelli*, 2003; *Gomes et al.*, 2005; see the chapter by Levison et al.). This substantial reorganization of the solar system could have waited a surprisingly long time, as much as 1.0 G.y., driven by slow evolution of the giant planets' orbits before becoming chaotic (*Gomes et al.*, 2005). The timing is consistent with the epoch of the

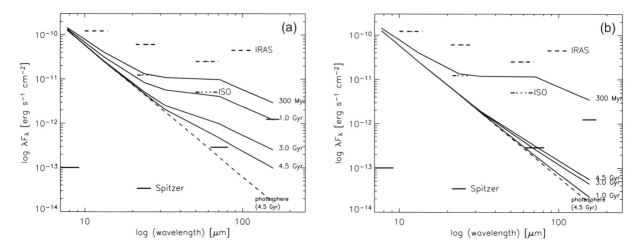

Fig. 5. Toy models for the evolution of the solar system spectral energy distribution from an age of 300 m.y. to 4.5 G.y.: **(a)** with no late heavy bombardment shown and **(b)** including the LHB as discussed in the text. The long wavelength excess in **(b)** grows with time after the LHB because the Kuiper belt has not yet reached equilibrium between dust production and dust removal. Also shown are the 3σ sensitivity limits of IRAS, ISO, and Spitzer for a Sun-like star at a distance of 30 pc for comparison.

LHB discerned in the lunar cratering record. Furthermore, *Strom et al.* (2005) point out that, because Jupiter should have migrated inward as part of the same process driving Neptune outward, the AB could have been decimated (perhaps for the second time) at the same late era as the KB.

The simple model employed herein to track the history of the solar system's IR SED involves calculating the collisional evolution of the AB and KB. Each belt is divided into 10 radial annuli that evolve independently. At each time step (generally set to 10^6 yr) for each annulus is calculated (1) the number of parent-body collisions, (2) the fragment mass produced and subtracted from the parent-body reservoir, and (3) mass lost via "blowout" of the smallest particles plus P-R drift from the belt inward toward the Sun. Parent-body collisions are considered only statistically so the model has no capacity to represent "spikes" from occasional large collisions as discussed in section 4.4. Mass in grains that would be rapidly ejected via radiation pressure "blowout" is removed from the model instantaneously when created. If the collision timescale for bound grains of a given size and location is shorter than the P-R removal time, those grains are not allowed to drift interior to the belt and contribute to the inner zodiacal cloud. Thus, based on the theory explained in the previous section, if the belt fragment density is above a certain threshold, the net mass loss is only outward via direct ejection, not inward. The terrestrial planets are not considered as barriers to P-R drift, but Neptune is assumed able to consume or deflect all grains, so the model dust surface density is set to zero between 4 and 30 AU. The system SED is calculated using generic grain emissivity that depends only on particle size. An indication that the model works well is that it naturally predicts the observed zodiacal dust density as the output from the observed AB large-body mass and spatial distributions without fine-tuning.

Our simple results agree with *Bottke et al.* (2005) and others' conclusion that the AB and KB must both have been subject to depletions by factors of 10–100 sometime during their histories because simple collisional evolution would not produce the low-mass belts we see today. The present AB and KB masses cannot be produced from the likely starting masses without either (1) an arbitrary continuous removal of parent bodies with an exponential timescale for both belts on the order of 2 G.y., shown in Fig. 5a, or (2) one sudden depletion event shown in Fig. 5b, which involves collisionally evolving the starting mass for 0.5 G.y., then reducing each belt mass by amounts necessary to allow collisional evolution to resume and continue over the next 4.0 G.y. to reach the observed low masses of the two belts.

The toy model scenario shown in Fig. 5a simply predicts that a planetary system would have significant 10–30-μm flux up to an age of 1 G.y. The general lack of observed mid-IR excesses in Spitzer targets older than 30 m.y. could mean that (1) most systems do not have belts at temperatures like our asteroid belt, or (2) most have LHB-like events earlier in their histories. A corollary of the present depleted AB having a large-body collision timescale of 10 G.y. is that the AB/zodi system is nearly constant in time (e.g., equation (6)). Thus, extrapolating backward by *Dominik and Decin* (2003) t^{-1} or t^{-2} scaling laws is inappropriate (cf. Fig. 1). Our model predicts that the AB/zodi and KB IR luminosities both would only decrease by about 30% in 4.0 G.y. after the LHB or equivalent major clearing. This agrees with the lunar cratering record indicating that the AB, the Earth-crossing asteroid population, and the zodiacal cloud have had nearly constant density for at least the past 3.0 G.y.

The chapter by Levison et al. discusses the idea that our planetary system had a traumatic reorganization about 700 m.y. after its formation. An intriguing extension of this idea is that extrasolar debris disk systems that seem brighter than their age cohorts may represent LHB-like events, i.e., collisions of small bodies excited by planet migration that can occur late in a system's development at a time determined by details of the original planetary system architecture. Our "toy" model results compared with Spitzer observations (e.g., *Kim et al., 2005*) support a picture in which many systems evolve according to the principles articulated by *Dominik and Decin* (2003) unless interrupted by an LHB event that might occur almost any time in the system's history. After the LHB event the system is nearly clear of planetesimals and dust and evolves very slowly. It remains to be seen whether observations with Spitzer can distinguish between hypotheses such as single supercollisions or late episodes of debris belt clearing.

6. SUMMARY AND IMPLICATIONS FOR FUTURE WORK

Based on the discussions presented above, it is clear that the question of how common solar systems like our own might be depends in part on what radius in the disk one looks and at what age the comparison is made. We summarize our main results as follows: (1) Warm circumstellar material inside of 1 AU dissipates rapidly on timescales comparable to the cessation of accretion; (2) the gas content of disks much older than 10 m.y. is incapable of forming giant planets; (3) while massive analogs to our asteroid belt lacking outer disks appear to be rare overall (1–3%), warm disks (lacking inner hot dust) seem to enjoy a preferred epoch around stars with ages between 10 and 300 m.y.; (4) cold outer disks (analogous to our own Kuiper belt, but much more massive) are found around 10–20% of Sun-like stars; (5) resolved images of disks are crucial in order to remove degeneracies in debris disk modeling from SEDs alone; (6) most debris disks observed to date are collisionally dominated dust systems and do not require the dynamical action of planets to maintain the observed inner holes; (7) at least some disks are observed in a short-lived phase of evolution and are not examples of the most massive debris disks; and (8) comparing the ensemble of observations of disks surrounding other stars as a function of age to the evolution of our solar system requires detailed understanding of its dynamical evolution including the late-heavy bombardment era. Yet in affecting these comparisons, we must remember that we do not yet have the sensitivity to observe tenuous debris disks comparable to our own asteroid belt or Kuiper belt.

It is unclear whether debris systems significantly more massive (and therefore more easily detectable) than our own represent a more-favorable or less-favorable condition for planet formation. It may be that systems with planets might arise from disks with higher mass surface density and thus stronger debris signatures at early times than disks lacking planets. However, if events comparable to the dynamical rearrangement of our solar system (perhaps related to the lu-

nar late-heavy bombardment) are common in planetary systems within the first few hundred million years, we might expect that debris disks lacking planets might be brighter than those with planets at late times. *Beichman et al.* (2005a) present preliminary evidence that there may be some connection between the presence of a massive debris disk and a radial velocity planet within 5 AU. It is interesting to note that extrapolations of the detection frequency of extrasolar planets as a function of radius beyond current survey limits (see chapter by Udry et al.) suggest a frequency of extrasolar giant planets >1 M_{Jup} ~ 10–20% out to 20 AU, consistent with our debris disk statistics for G stars.

How do results on debris disks compare as a function of stellar mass? On theoretical grounds, one can argue that the mass of a circumstellar disk should not exceed ~10–25% the mass of the central star (*Shu et al., 1990*). Indeed, *Natta et al.* (2000) presents evidence that the disk masses around early type pre-main-sequence stars are more massive than their lower-mass T Tauri counterparts. *Muzerolle et al.* (2003) also show that disk accretion rates appear to correlate with stellar mass. Historically, debris disks have been more commonly associated with A stars rather than G or M stars, but that has been largely attributable to a selection effect: It is easier to see smaller amounts of dust surrounding higher luminosity objects in flux-limited surveys. *Rieke et al.* (2005) present evidence for a diminution in the frequency of mid-IR excess emission surrounding A stars over 100–300 m.y. Their data indicate that over and above an envelope of decay consistent with a t^{-1} falloff, several objects show evidence for greater dust generation rates consistent with their interpretation of stochastic processes in planetesimal disks (see sections 3 and 4 above). In general, the overall picture of A-star debris disk evolution is remarkably consistent with that presented for Sun-like stars, suggesting that stellar mass does not play a defining role in debris disk evolution. In contrast, primordial disks around higher-mass stars are more massive, and have shorter lifetimes (*Hillenbrand et al.*, 1998; *Lada et al.*, 2006), than disks around lower-mass stars.

Greaves and Wyatt (2003) also present evidence from ISO observations concerning the frequency of debris disks as a function of mass. They find that debris surrounding A stars is more common than around G stars, even for stars of the same age (although the observations were sensitive to different amounts of debris as a function of stellar luminosity). They suggest that the difference is due to characteristic lifetimes of debris becoming an increasing fraction of the main-sequence lifetime for higher-mass (shorter-lived) stars, possibly because disk mass correlates with star mass. *Plavchan et al.* (2005) present a survey for warm inner debris surrounding young M dwarfs. They explain their lack of detections, which is contrary to expectations from the timescale for P-R drag as a function of stellar luminosity, due to the effects of an enhanced particulate wind from late-type stars compared to early-type stars. Yet, it is clear from recent work on low-mass stars and brown dwarfs that they too possess primordial circumstellar disks when they are young (see chapter by Luhman et al.; *Apai et al., 2005*); however, their evolutionary properties are as yet unclear. Spitzer studies of debris disks surrounding low-mass stars and brown dwarfs at longer wavelengths are now underway. Combining data on A stars, G dwarfs, and M dwarfs, there is to date no evidence for wildly divergent evolutionary histories for debris disks as a function of stellar mass averaged over main-sequence lifetimes. Observed differences to date can be explained in part by differences in dust-mass upper limits as a function of stellar luminosity and assuming that the typical star to initial disk mass is roughly constant. It is important to remember that most *Sun-like* stars in the disk of the Milky Way are binary (*Duquennoy and Mayor,* 1991), while the binary fraction of low-mass stars and brown dwarfs may be lower (see chapter by Burgasser et al.). It is clear that the evolution of disks in the pre-main-sequence phase can be influenced by the presence or absence of a companion (see chapter by Monin et al.; *Jensen et al.,* 1996). Preliminary results from Spitzer suggest that debris disk evolution is not a strong function of multiplicity, and may even be enhanced in close binaries (Trilling et al., in preparation).

What are the implications for the formation of terrestrial planets in disks surrounding stars of all masses in the disk of the Milky Way? We know that primordial accretion disks commonly surround very young stars (approaching 100%), and that gas-rich disks around more (less) massive stars are bigger (smaller), but last shorter (longer) amounts of time. Because of the surface density of solids in the disk, more massive disks surrounding higher-mass stars will probably form planetesimals faster. What is unclear is whether disks surrounding intermediate-mass stars (with shorter gas disk lifetimes) retain remnant gas needed to damp the eccentricities of forming planetesimals to create planetary systems like our own (*Kominami and Ida,* 2002). Yet the planetesimal growth time in disks surrounding low-mass stars and brown dwarfs might be prohibitive given the low surface densities of solids (see, however, *Beaulieu et al.,* 2006). Perhaps, just like Goldilocks, we will find that terrestrial planets in stable circular orbits are found in abundance around Sun-like stars from 0.3 to 3 AU. Whether these planets have liquid water and the potential for life as we know it to develop will depend on many factors (see chapter by Gaidos and Selsis). As results from Spitzer and other facilities continue to guide our understanding in the coming years, we can look forward to steady progress. Hopefully, new observational capabilities and theoretical insights will provide answers to some of these questions by the time of Protostars and Planets VI.

Acknowledgments. We would like to thank the referee for helpful comments that improved the manuscript, as well as the conference organizers and manuscript editors for their efforts. M.R.M. and D.B. would like to acknowledge members of the FEPS project for their continued collaboration (in particular J. S. Kim and F. Fan for assistance with Fig. 5) supported through a grant from JPL. M.R.M. is supported in part through the LAPLACE node of NASA's Astrobiology Institute.

REFERENCES

Andrews S. and Williams J. (2005) *Astrophys. J., 631,* 1134–1106.
Apai D., Pascucci I., Bouwman J., Natta A., Henning T., and Dullemond C. P. (2005) *Science, 310,* 834–836.
Ardila D., Golimowski D., Krist J., Clampin M., Williams J., et al. (2004) *Astrophys. J., 617,* L147–L150.
Ardila D., Lubow S., Golimowski D., Krist J., Clampin M., et al. (2005) *Astrophys. J., 627,* 986–1000.
Artymowicz P., Burrows C., and Paresce F. (1989) *Astrophys. J., 337,* 494–513.
Augereau J. C. and Beust H. (2006) *Astron. Astrophys., 455,* 987–999.
Augereau J. C., Nelson R. P., Lagrange A. M., Papaloizou J. C. B., and Mouillet D. (2001) *Astron. Astrophys., 370,* 447–455.
Aumann H., Beichman C., Gillet F., de Jong T., Houck J., et al. (1984) *Astrophys. J., 278,* L23–L27.
Backman D. and Paresce F. (1993) In *Protostars and Planets III* (E. H. Levy and J. I. Lunine, eds.), pp. 1253–1304. Univ. of Arizona, Tucson.
Backman D. E., Dasgupta A., and Stencel R. E. (1995) *Astrophys. J., 450,* L35–L38.
Beaulieu J.-P., Bennett D., Fouqu P., Williams A., Dominik M., et al. (2006) *Nature, 439,* 437–440.
Beckwith S. V. W. and Sargent A. I. (1996) *Nature, 383,* 139–144.
Beckwith S. V. W., Sargent A. I., Chini R. S., and Guesten R. (1990) *Astron. J., 99,* 924–945.
Beichman C., Bryden G., Rieke G., Stansberry J., Trilling D., et al. (2005a) *Astrophys. J., 622,* 1160–1170.
Beichman C., Bryden G., Gautier T., Stapelfeldt K., Werner M., et al. (2005b) *Astrophys. J., 626,* 1061–1069.
Benz W. and Asphaug E. (1999) *Icarus, 142,* 5–20.
Bottke W. F., Durda D. D., Nesvorný D., Jedicke R., Morbidelli A., et al. (2005) *Icarus, 179,* 63–94.
Brandeker A., Liseau R., Olofsson G., and Fridlund M. (2004) *Astron. Astrophys., 413,* 681–691.
Bryden G., Beichman C., Trilling D., Rieke G., Holmes E., et al. (2006) *Astrophys. J., 636,* 1098–1113.
Burns J. A., Lamy P. L., and Soter S. (1979) *Icarus, 40,* 1–48.
Calvet N., D'Alessio P., Watson D., Franco-Hernandez R., Furlan E., et al. (2005) *Astrophys. J., 630,* L185–L188.
Carpenter J. M., Wolf S., Schreyer K., Launhardt R., and Henning T. (2005) *Astron. J., 129,* 1049–1062.
Chen C. H. (2002) *Bull. Am. Astron. Soc., 34,* 1145.
Chen C. H., Jura M., Gordon K. D., and Blaylock M. (2005) *Astrophys. J., 623,* 493–501.
Cuzzi J. N., Lissauer J. J., Esposito L. W., Holberg J. B., Marouf E. A., Tyler G. L., and Boishchot A. (1984) In *Planetary Rings* (R. Greenberg and A. Brahic, eds.), pp. 73–199. IAU Colloquium 75, Kluwer, Dordrecht.
Davis D. R., Chapman C. R., Weidenschilling S. J., and Greenberg R. (1985) *Icarus, 63,* 30–53.
Decin G., Dominik C., Waters L. B. F. M., and Waelkens C. (2003) *Astrophys. J., 598,* 636–644.
Dent W. R. F., Greaves J. S., and Coulson I. M. (2005) *Mon. Not. R. Astron. Soc., 359,* 663–676.
Dermott S. F., Durda D. D., Grogan K., and Kehoe T. J. J. (2002) In *Asteroids III* (W. F. Bottke Jr. et al., eds.), pp. 423–442. Univ. of Arizona, Tucson.
Dohnanyi J. W. (1969) *J. Geophys. Res., 74,* 2531–2554.
Dominik C. and Decin G. (2003) *Astrophys. J., 598,* 626–635.
Draine B. and Lee H. (1984) *Astrophys. J., 285,* 89–108.
Duquennoy A. and Mayor M. (1991) *Astron. Astrophys., 248,* 485–524.
Farley K. A., Ward P., Garrison G., and Mukhopadhyay S. (2005) *Earth Planet. Sci. Lett., 240,* 265–275.

Fernandez R., Brandeker A., and Wu Y. (2006) *Astrophys. J., 643,* 509–522.

Fischer D. and Valenti J. (2005) *Astrophys. J., 622,* 1102–1117.

Forrest W., Sargent B., Furlan E., D'Alessio P., Calvet N., et al. (2004) *Astrophys. J. Suppl., 154,* 443–447.

Fujiwara A., Cerroni P., Davis D., Ryan E., and di Martino M. (1989) In *Asteroids II* (R. P. Binzel et al., eds.), pp. 240–265. Univ. of Arizona, Tucson.

Golimowski D. A., Ardila D. R., Clampin M., Krist J. E., Ford H. C., Illingworth G. D., et al. (2005) In *PPV Poster Proceedings*, www.lpi.usra.edu/meetings/ppv2005/pdf/8488.pdf.

Gomes R., Levison H. F., Tsiganis K., and Morbidelli A. (2005) *Nature, 435,* 466–469.

Gorti U. and Hollenbach D. (2004) *Astrophys. J., 613,* 424–447.

Greaves J. and Wyatt M. (2003) *Mon. Not. R. Astron. Soc., 345,* 1212–1222.

Greaves J., Holland W., Moriarty-Schieven G., Jenness T., Dent W., et al. (1998) *Astrophys. J., 506,* L133–L137.

Greaves J. S., Wyatt M. C., Holland W. S., and Dent W. R. F. (2004) *Mon. Not. R. Astron. Soc., 351,* L54–L58.

Greaves J., Holland W., Wyatt M., Dent W., Robson E., et al. (2005) *Astrophys. J., 619,* L187–L190.

Greaves J. S., Fischer D. A., and Wyatt M. C. (2006) *Mon. Not. R. Astron. Soc., 366,* 283–286.

Grogan K., Dermott S. F., and Durda D. D. (2001) *Icarus, 152,* 251–267.

Gustafson B. (1994) *Ann. Rev. Earth Planet. Sci., 22,* 553–595.

Habing H., Dominik C., Jourdain de Muizon M., Kessler M., Laureijs R., et al. (1999) *Nature, 401,* 456–458.

Habing H., Dominik C., Jourdain de Muizon M., Laureijs R., Kessler M., et al. (2001) *Astron. Astrophys., 365,* 545–561.

Haisch K., Lada E., and Lada C. (2001) *Astrophys. J., 553,* L153–L156.

Hartigan P., Edwards S., and Ghandour L. (1995) *Astrophys. J., 452,* 736–768.

Hartmann L., Calvet N., Gullbring E., and D'Alessio P. (1998) *Astrophys. J., 495,* 385–400.

Harvey P. M. (1985) In *Protostars and Planets II* (D. C. Black and M. S. Matthews, eds.), pp. 484–492. Univ. of Arizona, Tucson.

Hillenbrand L., Strom S., Calvet N., Merrill K., Gatley I., et al. (1998) *Astron. J., 116,* 1816–1841.

Hines D., Backman D., Bouwman J., Hillenbrand L., Carpenter J., et al. (2006) *Astrophys. J., 638,* 1070–1079.

Holland W., Greaves J., Zuckerman B., Webb R., McCarthy C., et al. (1998) *Nature, 392,* 788–790.

Holland W., Greaves J., Dent W., Wyatt M., Zuckerman B., et al. (2003) *Astrophys. J., 582,* 1141–1146.

Hollenbach D., Gorti U., Meyer M., Kim J., Morris P., et al. (2005) *Astrophys. J., 631,* 1180–1190.

Jayawardhana R., Fisher S., Hartmann L., Telesco C., Pina R., and Fazio G. (1998) *Astrophys. J., 503,* L79–L82.

Jayawardhana R., Holland W., Kalas P., Greaves J., Dent W., et al. (2002) *Astrophys. J., 570,* L93–L96.

Jensen E. L. N., Mathieu R. D., and Fuller G. A. (1996) *Astrophys. J., 458,* 312–326.

Jewitt D. C. (1994) *Astron. J., 108,* 661–665.

Jonkheid B., Kamp I., Augereau J.-C., and van Dishoeck E. F. (2005) In *Astrochemistry Throughout the Universe* (D. Lis et al., eds.), p. 49. Cambridge Univ., Cambridge.

Jura M., Malkan M., White R., Telesco C., Pina R., and Fisher R. S. (1998) *Astrophys. J. 505,* 897–902.

Jura M., Chen C., Furlan E., Green J., Sargent B., et al. (2004) *Astrophys. J. Suppl., 154,* 453–457.

Kalas P. (2005) *Astrophys. J., 635,* L169–L172.

Kalas P. and Jewitt D. (1995) *Astron. J., 110,* 794–804.

Kalas P., Liu M., and Matthews B. (2004) *Science, 303,* 1990–1992.

Kalas P., Graham J., and Clampin M. (2005) *Nature, 435,* 1067–1070.

Kalas P., Graham J., Clampin M. C., and Fitzgerald M. P. (2006) *Astrophys. J., 637,* L57–L60.

Kenyon S. and Bromley B. (2002a) *Astron. J., 123,* 1757–1775.

Kenyon S. and Bromley B. (2002b) *Astrophys. J., 577,* L35–L38.

Kenyon S. and Bromley B. (2004) *Astron. J., 127,* 513–530.

Kenyon S. and Bromley B. (2005) *Astron. J., 130,* 269–279.

Kenyon S. and Hartmann L. (1995) *Astrophys. J. Suppl., 101,* 117–171.

Kim J., Hines D., Backman D., Hillenbrand L., Meyer M., et al. (2005) *Astrophys. J., 632,* 659–669.

Klahr H. and Lin D. (2001) *Astrophys. J., 554,* 1095–1109.

Koerner D., Ressler M., Werner M., and Backman D. (1998) *Astrophys. J., 503,* L83–L87.

Koerner D., Jensen E., Cruz K., Guild T., and Gultekin K. (2000) *Astrophys. J., 533,* L37–L40.

Koerner D., Sargent A., and Ostroff N. (2001) *Astrophys. J., 560,* L181–L184.

Kominami J. and Ida S. (2002) *Icarus, 157,* 43–56.

Krist J., Ardila D., Golimowski D., Clampin M., Ford H., et al. (2005) *Astron. J., 129,* 1008–1017.

Krivov A., Mann I., and Krivova N. (2000) *Astron. Astrophys., 362,* 1127–1137.

Lada C., Meunch A., Luhman K., Allen L., Hartmann L., et al. (2006) *Astron. J., 131,* 1574.

Lagrange A.-M., Beust H., Mouillet D., Deleuil M., Feldman P., et al. (1998) *Astron. Astrophys., 330,* 1091–1108.

Lagrange A., Backman D., and Artymowicz P. (2000) In *Protostars and Planets IV* (V. Mannings et al., eds.), pp. 639–672. Univ. of Arizona, Tucson.

Landgraf M., Liou J.-C., Zook H., and Grün E. (2002) *Astron. J., 123,* 2857–2861.

Larwood J. and Kalas P. (2001) *Mon. Not. R. Astron. Soc., 323,* 402–416.

Lawson W., Lyo A., and Muzerolle J. (2004) *Mon. Not. R. Astron. Soc., 351,* L39–L43.

Lecavelier des Etangs A., Vidal-Madjar A., Roberge A., Feldman P., Deleuil M., et al. (2001) *Nature, 412,* 706–708.

Levison H. F. and Morbidelli A. (2003) *Nature, 426,* 419–421.

Liou J.-C. and Zook H. (1999) *Astron. J., 118,* 580–590.

Lissauer J. and Griffith C. (1989) *Astrophys. J., 340,* 468–471.

Liu M. C. (2004) *Science, 305,* 1442–1444.

Liu M. C., Matthews B. C., Williams J. P., and Kalas P. G. (2004) *Astrophys. J., 608,* 526–532.

Love S. G. and Brownlee D. E. (1993) *Science, 262,* 550–552.

Low F., Hines D., and Schneider G. (1999) *Astrophys. J., 520,* L45–L48.

Low F., Smith P., Werner M., Chen C., Krause V., et al. (2005) *Astrophys. J., 631,* 1170–1179.

Malhotra R. (1993) *Nature, 365,* 819–821.

Mamajek E., Meyer M., Hinz P., Hoffmann W., Cohen M., and Hora J. (2004) *Astrophys. J., 612,* 496–510.

Marcy G., Cochran W., and Mayor M. (2000) In *Protostars and Planets IV* (V. Mannings et al., eds.), pp. 1285–1311. Univ. of Arizona, Tucson.

Marengo M., Backman D., Megeath T., Fazio G., Wilner D., et al. (2005) In *PPV Poster Proceedings*, www.lpi.usra.edu/meetings/ppv2005/pdf/8566.pdf.

Marsh K., Velusamy T., Dowell C., Grogan K., and Beichman C. (2005) *Astrophys. J., 620,* L47–L50.

Metchev S., Eisner J., Hillenbrand L., and Wolf S. (2005) *Astrophys. J., 622,* 451–462.

Meyer M. R. and Beckwith S. (2000) In *ISO Survey of a Dusty Universe* (D. Lemke et al., eds.), pp. 341–352. LNP Vol. 548, Springer-Verlag, Heidelberg.

Meyer M., Hillenbrand L., Backman D., Beckwith S., Bouwman J., et al. (2004) *Astrophys. J. Suppl., 154,* 422–427.

Moro-Martín A. and Malhotra R. (2002) *Astron. J., 124,* 2305–2321.

Muzerolle J., Hillenbrand L., Calvet N., Briceño C., and Hartmann L. (2003) *Astrophys. J., 592,* 266–281.

Nagasawa M., Lin D., and Thommes E. (2005) *Astrophys. J., 635,* 578–598.

Najita J. and Williams J. (2005) *Astrophys. J., 635,* 625–635.

Natta A., Grinin V., and Mannings V. (2000) In *Protostars and Planets IV* (V. Mannings et al., eds.), pp. 559–587. Univ. of Arizona, Tucson.

Nesvorný D., Bottke W., Levison H., and Dones L. (2003) *Astrophys. J., 591*, 486–497.

Okamoto Y., Kataza H., Honda M., Yamashita T., Onaka T., et al. (2004) *Nature, 431*, 660–663.

Pascucci I., Gorti U., Hollenbach D., Najita J., Meyer M., et al. (2006) *Astrophys J.*, in press (astro-ph/0606669).

Plavchan P., Jura M., and Lipscy S. (2005) *Astrophys. J., 631*, 1161–1169.

Raymond S., Quinn T., and Lunine J. (2004) *Icarus, 168*, 1–17.

Richter M., Jaffe D., Blake G., and Lacy J. (2002) *Astrophys. J., 572*, L161–L164.

Rieke G., Su K., Stansberry J., Trilling D., Bryden G., et al. (2005) *Astrophys. J., 620*, 1010–1026.

Roberge A., Weinberger A., Redfield S., and Feldman P. (2005) *Astrophys. J., 626*, L105–L108.

Roberge A., Feldman P., Weinberger A., Deleuil M., and Bouret J. (2006) *Nature, 441*, 724–726.

Roques F., Scholl H., Sicardy B., and Smith B. (1994) *Icarus, 108*, 37–58.

Rydgren A. (1978) In *Protostars and Planets* (T. Gehrels, ed.), pp. 690–698. Univ. of Arizona, Tucson.

Schneider G., Smith B., Becklin E., Koerner D., Meier R., et al. (1999) *Astrophys. J., 513*, L127–L130.

Schneider G., Becklin E., Smith B., Weinberger A., Silverstone M., and Hines D. (2001) *Astron. J., 121*, 525–537.

Schneider G., Silverstone M. D., and Hines D. C. (2005) *Astrophys. J., 629*, L117–L120.

Schneider G., Silverstone M. D., Hines D. C., Augereau J.-C., Pinte C., et al. (2006) *Astrophys. J., 650*, 414.

Sheret I., Ramsay Howat S., and Dent W. (2003) *Mon. Not. R. Astron. Soc., 343*, L65–L68.

Shu F., Tremaine S., Adams F., and Ruden S. (1990) *Astrophys. J., 358*, 495–514.

Silverstone M., Meyer M., Mamajek E., Hines D., Hillenbrand L., et al. (2006) *Astrophys. J., 639*, 1138–1146.

Simon M. and Prato L. (1995) *Astrophys. J., 450*, 824–829.

Skrutskie M., Dutkevitch D., Strom S., Edwards S., Strom K., and Shure M. (1990) *Astron. J., 99*, 1187–1195.

Smith B. and Terrile R. (1984) *Science, 226*, 1421–1424.

Song I., Zuckerman B., Weinberger A., and Becklin E. (2005) *Nature, 436*, 363–365.

Spangler C., Sargent A., Silverstone M., Becklin E., and Zuckerman B. (2001) *Astrophys. J., 555*, 932–944.

Stapelfeldt K., Holmes E., Chen C., Rieke G., Su K., et al. (2004) *Astrophys. J. Suppl., 154*, 458–462.

Stauffer J. (2004) In *Debris Disks and the Formation of Planets* (L. Caroff et al., eds.), pp. 100–111. ASP Conf. Series 324, San Francisco.

Stern S. (1996) *Astron. J., 112*, 1203–1211.

Stern S. and Colwell J. (1997) *Astrophys. J., 490*, 879–882.

Strom R., Malhotra R., Ito T., Yoshida F., and Kring D. (2005) *Science, 309*, 1847–1850.

Strom S., Edwards S., and Skrutskie M. (1993) In *Protostars and Planets III* (E. H. Levy and J. I. Lunine, eds.), pp. 837–866. Univ. of Arizona, Tucson.

Strubbe L. and Chiang E. (2006) *Astrophys. J., 648*, 652–665.

Su K., Rieke G., Misselt K., Stansberry J., Moro-Martin A., et al. (2005) *Astrophys. J., 628*, 487–500.

Takeuchi T. and Artymowicz P. (2001) *Astrophys. J., 557*, 990–1006.

Takeuchi T., Clarke C., and Lin D. (2005) *Astrophys. J., 627*, 286–292.

Tanaka H., Inaba S., and Nakazawa K. (1996) *Icarus, 123*, 450–455.

Telesco C., Fisher R., Pia R., Knacke R., Dermott S., et al. (2000) *Astrophys. J., 530*, 329–341.

Telesco C., Fisher R., Wyatt M., Dermott S., Kehoe T., et al. (2005) *Nature, 433*, 133–136.

Teplitz V., Stern S., Anderson J., Rosenbaum D., Scalise R., and Wentzler P. (1999) *Astrophys. J., 516*, 425–435.

Thébault P., Augereau J., and Beust H. (2003) *Astron. Astrophys., 408*, 775–788.

Thébault P. and Augereau J.-C. (2005) *Astron. Astrophys., 437*, 141–148.

Thi W., Blake G., van Dishoeck E., van Zadelhoff G., Horn J., et al. (2001a) *Nature, 409*, 60–63.

Thi W., van Dishoeck E., Blake G., van Zadelhoff G., Horn J., et al. (2001b) *Astrophys. J., 561*, 1074–1094.

Thommes E., Duncan M., and Levison H. (1999) *Nature, 402*, 635–638.

Weidenschilling S., Spaute D., Davis D., Marzari F., and Ohtsuki K. (1997) *Icarus, 128*, 429–455.

Weinberger A., Becklin E., and Zuckerman B. (2003) *Astrophys. J., 584*, L33–L37.

Weinberger A., Becklin E., Zuckerman B., and Song I. (2004) *Astron. J., 127*, 2246–2251.

Whitmire D., Matese J., and Whitman P. (1992) *Astrophys. J., 388*, 190–195.

Williams J., Najita J., Liu M., Bottinelli S., Carpenter J., et al. (2004) *Astrophys. J., 604*, 414–419.

Wilner D., Holman M., Kuchner M., and Ho P. (2002) *Astrophys. J., 569*, L115–L119.

Wisdom J. (1980) *Astron. J., 85*, 1122–1133.

Wolk S. and Walter F. (1996) *Astron. J., 111*, 2066–2076.

Wyatt M. (2003) *Astrophys. J., 598*, 1321–1340.

Wyatt M. (2005) *Astron. Astrophys., 433*, 1007–1012.

Wyatt M. (2006) *Astrophys. J., 639*, 1153–1165.

Wyatt M. and Dent W. (2002) *Mon. Not. R. Astron. Soc., 334*, 589–607.

Wyatt M., Dermott S., Telesco C., Fisher R., Grogan K., et al. (1999) *Astrophys. J., 527*, 918–944.

Wyatt M., Dent W., and Greaves J. (2003) *Mon. Not. R. Astron. Soc., 342*, 876–888.

Wyatt M., Greaves J., Dent W., and Coulson I. (2005) *Astrophys. J., 620*, 492–500.

Zuckerman B. (2001) *Ann. Rev. Astron. Astrophys., 39*, 549–580.

Zuckerman B. and Becklin E. (1993) *Astrophys. J., 414*, 793–802.

Zuckerman B., Forveille T., and Kastner J. (1995) *Nature, 373*, 494–496.

Part VII:

Planet Formation and Extrasolar Planets

Formation of Giant Planets

Jack J. Lissauer
NASA Ames Research Center

David J. Stevenson
California Institute of Technology

The observed properties of giant planets, models of their evolution, and observations of protoplanetary disks provide constraints on the formation of gas giant planets. The four largest planets in our solar system contain considerable quantities of hydrogen and helium; these gases could not have condensed into solid planetesimals within the protoplanetary disk. Jupiter and Saturn are mostly hydrogen and helium, but have larger abundances of heavier elements than does the Sun. Neptune and Uranus are primarily composed of heavier elements. The transiting extrasolar planet HD 149026 b, which is slightly more massive than Saturn, appears to have comparable amounts of light gases and heavy elements. The other observed transiting exoplanets are primarily hydrogen and helium, but may contain supersolar abundances of heavy elements. Spacecraft flybys and observations of satellite orbits provide estimates of the gravitational moments of the giant planets in our solar system, which in turn provide information on the internal distribution of matter within Jupiter, Saturn, Uranus, and Neptune. Atmospheric thermal structure and heat flow measurements constrain the interior temperatures of these planets. Internal processes may cause giant planets to become more compositionally differentiated or alternatively more homogeneous; high-pressure laboratory experiments provide data useful for modeling these processes. The preponderance of evidence supports the core nucleated gas accretion model. According to this model, giant planets begin their growth by the accumulation of small solid bodies, as do terrestrial planets. However, unlike terrestrial planets, the giant planet cores grow massive enough to accumulate substantial amounts of gas before the protoplanetary disk dissipates. The primary question regarding the core nucleated growth model is under what conditions can planets develop cores sufficiently massive to accrete gas envelopes within the lifetimes of gaseous protoplanetary disks.

1. INTRODUCTION

The two largest planets in our solar system, Jupiter and Saturn, are composed predominantly of hydrogen and helium; these two lightest elements also comprise more than 10% of the masses of Uranus and Neptune. Moreover, most extrasolar planets thus far detected are believed (or known) to be gas giants. Helium and molecular hydrogen do not condense under conditions found in star-forming regions and protoplanetary disks, so giant planets must have accumulated them as gases. Therefore, giant planets must form prior to the dissipation of protoplanetary disks. Optically thick dust disks typically survive for only a few million years (chapters by Briceño et al. and Wadhwa et al.), and protoplanetary disks have lost essentially all their gases by the age of $<10^7$ yr (chapter by Meyer et al.), implying that giant planets formed on this timescale or less.

Jupiter and Saturn are generally referred to as *gas giants*, even though their constituents are not gases at the high pressures that most of the material in Jupiter and Saturn is subjected to. Analogously, Uranus and Neptune are frequently referred to as *ice giants*, even though the astrophysical ices such as H_2O, CH_4, H_2S, and NH_3 that models suggest make up the majority of their mass (*Hubbard et al.,* 1995) are in fluid rather than solid form. Note that whereas H and He *must* make up the bulk of Jupiter and Saturn because no other elements can have such low densities at plausible temperatures, it is possible that Uranus and Neptune are primarily composed of a mixture of "rock" and H/He.

Giant planets dominate our planetary system in mass, and our entire solar system in angular momentum (contained in their orbits). Thus, understanding giant planet formation is essential for theories of the origins of terrestrial planets, and important within the understanding of the general process of star formation.

The giant planets within our solar system also supported *in situ* formation of satellite systems. The Galilean satellite system is particularly impressive and may contain important clues to the last stages of giant planet formation (*Pollack and Reynolds,* 1974; *Canup and Ward,* 2002; *Mosqueira and Estrada,* 2003a,b). Ganymede and Callisto are roughly half water ice, and Callisto has most of this ice mixed with rock. It follows that conditions must be appropriate for the condensation of water ice at the location where Ganymede formed, and conditions at Callisto must have allowed formation of that body on a timescale exceeding about 10^5 yr, so that water ice would not melt and lead to a fully differentiated structure. The more distant irregular satellite systems of the giant planets may provide constraints on gas in the outer reaches of the atmospheres of giant planets (*Pollack et al.*, 1979; cf. *Tsui*, 2002).

The extrasolar planet discoveries of the past decade have vastly expanded our database by increasing the number of planets known by more than an order of magnitude. The distribution of known extrasolar planets is highly biased toward those planets that are most easily detectable using the Doppler radial velocity technique, which has been by far the most effective method of discovering exoplanets. These extrasolar planetary systems are quite different from our solar system; however, it is not yet known whether our planetary system is the norm, quite atypical, or somewhere in between.

Nonetheless, some unbiased statistical information can be distilled from available exoplanet data (*Marcy et al.*, 2004, 2005; chapter by Udry et al.): Roughly 1% of Sun-like stars (late F, G, and early K spectral class main-sequence stars that are chromospherically quiet, i.e., have inactive photospheres) have planets more massive than Saturn within 0.1 AU. Approximately 7% of Sun-like stars have planets more massive than Jupiter within 3 AU. Planets orbiting interior to ~0.1 AU, a region where tidal circularization timescales are less than stellar ages, have small orbital eccentricities. The median eccentricity observed for planets on more distant orbits is 0.25, and some of these planets travel on very eccentric orbits. Within 5 AU of Sun-like stars, Jupiter-mass planets are more common than planets of several Jupiter masses, and substellar companions that are more than 10× as massive as Jupiter are rare. Stars with higher metallicity are much more likely to host detectable planets than are metal-poor stars (*Gonzalez*, 2003; *Santos et al.*, 2003), with the probability of hosting an observable planet varying as the square of stellar metallicity (*Fischer and Valenti*, 2005). Low-mass main-sequence stars (M dwarfs) are significantly less likely to host one or more giant planets with orbital period(s) of less than a decade than are Sun-like stars. Multiple-planet systems are more common than if detectable planets were randomly assigned to stars (i.e., than if the presence of a planet around a given star was not correlated with the presence of other planets around that same star). Most transiting extrasolar giant planets are predominantly hydrogen (*Charbonneau et al.*, 2000; *Burrows et al.*, 2003; *Alonso et al.*, 2004), as are Jupiter and Saturn. However, HD 149026 b, which is slightly more massive than Saturn, appears to have comparable amounts of hydrogen + helium vs. heavy elements (*Sato et al.*, 2005), making its bulk composition intermediate between Saturn and Uranus, but more richly endowed in terms of total amount of "metals" than any planet in our solar system.

Transit observations have also yielded an important negative result: Hubble Space Telescope photometry of a large number of stars in the globular cluster 47 Tucanae failed to detect any transiting inner giant planets, even though ~17 such transiting objects would be expected were the frequency of such planets in this low metallicity cluster the same as that for Sun-like stars in the solar neighborhood (*Gilliland et al.*, 2000).

Various classes of models have been proposed to explain the formation of giant planets and brown dwarfs. Following *Lissauer* (2004) and consistent with current IAU nomenclature, these definitions are used in this chapter:

Star: self-sustaining fusion is sufficient for thermal pressure to balance gravity.

Stellar remnant: dead star — no more fusion, i.e., thermal pressure sustained against radiative losses by energy produced from fusion is no longer sufficient to balance gravitational contraction.

Brown dwarf: substellar object with substantial deuterium fusion — more than half the object's original inventory of deuterium is ultimately destroyed by fusion.

Planet: negligible fusion (<13 Jupiter masses, or M_{Jup}), plus it orbits one or more stars and/or stellar remnants.

The mass function of young compact objects in star-forming regions extends down through the brown dwarf mass range to below the deuterium-burning limit (*Zapatero Osorio et al.*, 2000; chapter by Luhman et al.; cf. *Marley et al.*, 2006). This observation, together with the lack of any convincing theoretical reason to believe that the collapse process that leads to stars cannot also produce substellar objects (*Wuchterl and Tscharnuter*, 2003; chapter by Whitworth et al.), strongly implies that most isolated (or distant companion) brown dwarfs and isolated high-planetary-mass objects form via the same collapse process as do stars.

By similar reasoning, the "brown dwarf desert," a profound dip over the range ~10–50 M_{Jup} in the mass function of companions orbiting within several AU of Sun-like stars (*Marcy et al.*, 2004; chapter by Udry et al.), strongly suggests that the vast majority of extrasolar giant planets formed via a mechanism different from that of stars. Within our solar system, bodies up to the mass of Earth consist almost entirely of condensable material, and even bodies of mass ~15 $M_\oplus$ (Earth masses) consist mostly of condensable material. (The definition of "condensable" is best thought of as the value of the specific entropy of the constituent relative to that for which the material can form a liquid or solid. Hydrogen and helium within protoplanetary disks have entropies far in excess of that required for condensation, even if they are compressed isothermally to pressures on the order of 1 bar, even for a temperature of only a few tens of degrees. Thus, H_2 and He remain in a gaseous state.) The fraction of highly volatile gases increases with planet mass through Uranus/Neptune, to Saturn, and finally to Jupiter, which is still enriched in condensables at least threefold compared to the Sun (*Young*, 2003). This gradual, nearly monotonic relationship between mass and composition argues for a unified formation scenario for all the planets and smaller bodies. Moreover, the continuum of observed extrasolar planetary properties, which stretches to systems not very dissimilar to our own, suggests that extrasolar planets formed in a similar way to the planets within our solar system.

Models for the formation of gas giant planets were reviewed by *Wuchterl et al.* (2000). Star-like direct quasi-spherical collapse is not considered viable, both because of the observed brown dwarf desert mentioned above and theoretical arguments against the formation of Jupiter-mass objects via fragmentation (*Bodenheimer et al.,* 2000a). The theory of giant planet formation that is favored by most researchers is the *core nucleated accretion model*, in which the planet's initial phase of growth resembles that of a terrestrial planet, but the planet becomes sufficiently massive (several $M_\oplus$) that it is able to accumulate substantial amounts of gas from the surrounding protoplanetary disk.

According to the variant of the core nucleated accretion model analyzed by *Pollack et al.* (1996), *Bodenheimer et al.* (2000b), and *Hubickyj et al.* (2005), the formation and evolution of a giant planet is viewed to occur in the following sequence: (1) Dust particles in the solar nebula form planetesimals that accrete one another, resulting in a solid core surrounded by a low-mass gaseous envelope. Initially, runaway accretion of solids occurs, and the accretion rate of gas is very slow. As the solid material in the planet's feeding zone is depleted, the rate of solids accretion tapers off. The gas accretion rate steadily increases and eventually exceeds the accretion rate of solids. (2) The protoplanet continues to grow as the gas accretes at a relatively constant rate. The mass of the solid core also increases, but at a slower rate. [The term "solids" is conventionally used to refer to the entire condensed (solid + liquid) portion of the planet. Accretion energy (and radioactive decay) heats a growing planet, and can cause material that was accreted in solid form to melt and vaporize. Vaporization of ices and other heavy compounds can significantly affect the properties of the planet's atmosphere, and its ability to radiate energy and to accrete more gas. In contrast, melting *per se* has little effect on the overall growth of the planet, apart from the capacity of the melt to release or trap gases.] Eventually, the core and envelope masses become equal. (3) Near this point, the rate of gas accretion increases in runaway fashion, and the protoplanet grows at a rapidly accelerating rate. The first three parts of the evolutionary sequence are referred to as the nebular stage, because the outer boundary of the protoplanetary envelope is in contact with the solar nebula, and the density and temperature at this interface are those of the nebula. (4) The gas accretion rate reaches a limiting value defined by the rate at which the nebula can transport gas to the vicinity of the planet. After this point, the equilibrium region of the protoplanet contracts, and gas accretes hydrodynamically into this equilibrium region. This part of the evolution is considered to be the transition stage. (5) Accretion is stopped by either the opening of a gap in the disk as a consequence of the tidal effect of the planet, accumulation of all nearby gas, or dissipation of the nebula. Once accretion stops, the planet enters the isolation stage. The planet then contracts and cools to the present state at constant mass.

Aside from core nucleated accretion, the only giant planet formation scenario receiving significant attention is the *gas instability model*, in which a giant planet forms directly from the contraction of a clump that was produced via a gravitational instability in the protoplanetary disk. Numerical calculations show that 1 M_{Jup} clumps can form in sufficiently gravitationally unstable disks (e.g., *Boss,* 2000; *Mayer et*

al., 2002). However, weak gravitational instabilities excite spiral density waves; density waves transport angular momentum that leads to spreading of a disk, lowering its surface density and making it more gravitationally stable. Rapid cooling and/or mass accretion is required to make a disk highly unstable. Thus, long-lived clumps can only be produced in protoplanetary disks with highly atypical physical properties (*Rafikov,* 2005). Additionally, gas instabilities would yield massive stellar-composition planets, requiring a separate process to explain the smaller bodies in our solar system and the heavy element enhancements in Jupiter and Saturn. The existence of intermediate objects like Uranus and Neptune is particularly difficult to account for in such a scenario. Furthermore, metal-rich stars are more likely to host observable extrasolar planets than are metal-poor stars (*Fischer and Valenti,* 2005; chapter by Udry et al.); this trend is consistent with the requirement of having sufficient condensables to form a massive core, but runs contrary to the requirement of rapid disk cooling needed to form long-lived clumps via gravitational instabilities (*Cai et al.,* 2006). See the chapter by Durisen et al. for a more extensive discussion of the gas instability model.

We review the constraints on formation provided by the internal structure of giant planets in section 2. In section 3, we summarize recent models of giant planet growth via core nucleated accretion. These models have some important shortcomings, and the issues remaining to be resolved are highlighted in section 4. We conclude this chapter with a brief summary.

2. MODELS OF GIANT PLANETS

The central issues for giant planet models are these: Do they have cores (of heavy elements) and, if so, what do those cores tell us about how the planet formed? The existence of heavy element enrichments in the solar system's four giant planets is not in doubt, because the mean densities of these planets are higher than the expected value for adiabatic bodies of solar composition. However, the existence of a core is less easily established, especially if the core is a small fraction of the total mass, as is likely in the case of Jupiter (Fig. 1).

Moreover, the presence or absence of a core does not automatically tell us whether or not a core existed at the time of planet formation. It is possible that the current core is an eroded remnant (less massive than the primordial core) or even enhanced because of rainout of heavy elements from the planet's envelope.

Seismology is by far the best method for establishing the existence and nature of a core, but we lack this approach for the giant planets since (unlike the Sun) the normal mode excitation is expected to be too small to be detectable at present. Dynamical approaches exist (e.g., measurement of the precession constant, as used to determine Earth's moment of inertia to high precision), but have not yet been implemented, since they require close-in, long-lived orbiters. We rely mostly on the old and very nonunique approach of

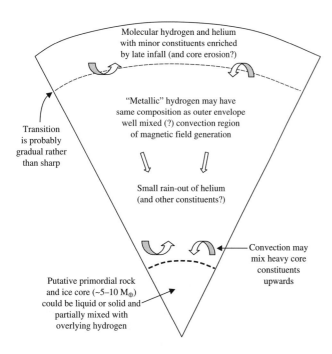

Fig. 1. Schematic cross-sectional view of the interior of Jupiter. A similar structure applies for Saturn, except that the molecular region is thicker and the presence of a core is more certain.

interpreting the gravitational response of the planet to its own rotation (see, e.g., *Podolak et al.*, 1993). In the tradition of Radau-Darwin, the change in gravity field arising from rotation of a hydrostatic body can be related to the moment of inertia of the body, and this in turn can be related to the degree of central concentration of matter within the planet. In the more rigorous approach used for giant planets, there is no possibility of deriving a moment of inertia, but the gravitational moments are nonetheless constraints on moments of the density structure derived from models of planetary internal structure, provided the body is hydrostatic and uniformly rotating. Hydrostaticity is confirmed to a high degree of accuracy by comparing the actual shape of the planet to that predicted by potential theory, and the expected level of differential rotation is unlikely to be sufficiently large to affect the determination of the presence or absence of a core.

A major uncertainty of this approach lies in the high-pressure behavior of hydrogen. This uncertainty has persisted for decades and may even have become worse in the sense that there was unfounded optimism in our understanding a few decades ago. In both Jupiter and Saturn, most of the mass resides in the region of greatest uncertainty, roughly in the pressure range between 0.5 and 10 mbar (5×10^{10} to 10^{12} Pa). At lower pressures, hydrogen is a simple molecular fluid with no significant dissociation or ionization. Above 10 mbar, hydrogen approaches the behavior of a nearly ideal Coulomb plasma (protons and degenerate electrons). At intermediate pressures, hydrogen is highly nonideal and relevant experiments are difficult. We still do not

know whether hydrogen undergoes a first-order phase transition (the so-called molecular to metallic transition, although if it exists it cannot be described in such simple language). However, the shape of the pressure-density relationship remains uncertain even if one accepts (as most experts do) that there is no first-order phase transition at the temperatures relevant to the giant planets. One way to appreciate this difficulty is to ask what error in the equation of state for hydrogen corresponds to a 1 $M_\oplus$ error in heavy elements? Roughly speaking, this is in proportion to the corresponding fraction of the planet's total mass, which is only 0.3% in the case of Jupiter and about 1% in the case of Saturn. Since the uncertainty in the equation of state is as much as several percent in the least well understood pressure range, the corresponding error in the estimated abundance of heavy elements may be as large as 5–10 $M_\oplus$.

Detailed reviews of giant planet structure include *Hubbard et al.* (2002) and *Guillot* (2005). The most complete modeling effort is the work of Guillot and collaborators. Simple coreless models of Jupiter are marginally capable of satisfying all the data. These models have a primordial solar hydrogen/helium ratio, but are enriched in heavy elements to the extent of about 10 $M_\oplus$. The most likely value for the mass of Jupiter's core is in the range of 5–10 $M_\oplus$. To a first approximation, it does not matter (in terms of explaining the mean density) whether the heavy elements are in a core or distributed internally. This heavy element enrichment is possibly greater than the observed threefold enrichment in some heavy elements in the atmosphere. However, there is no reliable determination of oxygen (as water) and no direct way of detecting the rocky component remotely, since those elements condense and form a cloud deck far below the observable atmosphere. It is common practice in models to assume a uniform mixing of the heavy elements outside a core. In some models, a jump in composition is assumed at the hypothesized molecular-metallic hydrogen phase transition. The heat flow and convection-generated magnetic field suggest a well-mixed region that extends down to at least about the megabar region. However, there is no observational or theoretical requirement that the planet be homogeneous outside the hypothetical high-density (rock and ice) core. It is also important to realize that the common practice of placing a separate core of heavy elements at the centers of these planets is governed by simplicity, rather than by observation. To varying degrees, the "core" could have a fuzzy boundary with the overlying hydrogen-rich envelope.

The heavy element fraction of Saturn is larger than that of Jupiter and as a consequence we have a more confident conclusion despite somewhat less accurate data (see Table 1). The models indicate that there is indeed a core, several to 20 $M_\oplus$, with a preferred value of ~10 $M_\oplus$. The latest gravity data results from Cassini are consistent with this. There is an uncertainty for Saturn that does not arise for Jupiter: We do not know the rotation rate with high accuracy. Saturn kilometric radiation (SKR) emissions have changed their period by six minutes between Voyager (1980) and

TABLE 1. Mass fractions.

	Primordial Solar	Jupiter	Saturn
Hydrogen	0.712	~0.68	~0.65
Helium	0.27	0.235 in atmosphere; ~0.26 assumed for the planet average	Not known for the atmosphere; 0.25 or less assumed for planet average
Volatile heavy elements ("ices")	0.013	~0.04 (assuming oxygen is enriched enriched in a similar ratio to carbon, etc.)	Not known but could be ~0.1 (assuming oxygen is enriched along with carbon, etc.); may be higher deep down
Involatile heavy elements ("rock")	0.005	Not known but presumed enriched in deep envelope and possible core	Not known but presumed enriched in deep envelope and core

now, and there is currently no generally accepted understanding of the connection between this period and the rotation period of the deep interior (*Stevenson*, 2006).

Uranus and Neptune are far less well understood than are Jupiter and Saturn. However, there is no doubt that they are mostly ice and rock, yet also possess ~2 $M_\oplus$ of gas each. Their atmospheres are estimated to have solar hydrogen to helium ratios, but the uncertainty is large because this determination is based on the pressure-induced absorption features of hydrogen, a method that has been unreliable for Jupiter and Saturn. The amount of hydrogen extractable from the ices is in principle about 20% of the total mass (assuming the hydrogen was delivered as water, methane, and ammonia), and this is marginally close to the hydrogen mass required by interior models. Moreover, there is the possibility that methane would decompose into carbon and hydrogen at extreme pressures. However, the atmospheres of Uranus and Neptune are highly enriched in methane (thus limiting the possibility of massive decomposition of this compound to very deep regions), and there is no experimental or theoretical evidence for extensive decomposition of water or ammonia under the conditions encountered inside these bodies. Consequently, it is not plausible to derive even 1 $M_\oplus$ of predominantly hydrogen gas from the breakdown of hydrogen-bearing ice or rock, even leaving aside the dubious proposition that such decomposed hydrogen would rise to the outer regions of the planet. This gas appears to have come from H_2 and He within the solar nebula. Uranus and Neptune (or precursor components massive enough to capture adequate amounts of gas) must have formed largely in the presence of the solar nebula, a very stringent constraint on the formation of solid bodies. While ice-rich embryos as small as ~0.1 $M_\oplus$ could conceivably have captured such gas mixed with steam within high mean molecular weight atmospheres (*Stevenson*, 1982, 1984; *Lissauer et al.*, 1995), there remain many open questions about this process, and most models suggest that Uranus and Neptune reached a substantial fraction of their current masses prior to the dispersal of the solar nebula.

It is often supposed that the presence or absence of a core in Jupiter (for example) can be placed in one-to-one correspondence with the presence or absence of a nucleating body that caused the inflow of gas to form the much more massive envelope. However, there is no neat correspondence between mode of giant planet formation and current presence of a core. One could imagine core nucleated accretion even if there is no core remaining, because the core might become mixed into the overlying envelope by convective processes. One could also imagine making a core in the low-density protoplanet phase by rainout. We now discuss each of these processes in more detail.

Core mix-up (or erosion) can be thought of as analogous to the following simple fluid dynamical experiment. Suppose one took a pot that has a layer of salt at the bottom, and then gently (or not so gently) added water. One then heats the pot from below to stimulate convection. Under what circumstances will the salt end up fully mixed with the water (assuming saturation is not reached)? In analogy with giant planets, one must ignore diffusion across the depth of the pot, since the diffusion time within a planet is longer than the age of the universe. Under these circumstances, the relevant consideration is the work done by convection (or initial stirring) compared to the work that must be done to mix the material. The work done by convection is determined by the buoyancy flux integrated through time. In giant planets, this is dominated by cooling. (In terrestrial planets it is dominated by radiogenic heating.) In accordance with the virial theorem, the contraction of the planet changes the gravitational energy by about the same amount as the change in internal energy (work done against gravity); see *Hubbard* (1984). However, the buoyancy production is directly related to the amount by which the planet has cooled, since that cooling is expressed in luminosity and the luminosity comes mainly from convective transport of buoyant fluid elements.

If the relevant part of the planet cools by an amount ΔT and the coefficient of thermal expansion is α, then the total available work is on the order of $Mg\alpha\Delta TH$, where g is gravity and H is the height through which the buoyant elements rise. Since the salt (or ice and rock, in the case of the planets) has a very different density from the background fluid (hydrogen for the planets), the work done lifting a high-density mass ΔM is $\sim \Delta MgH$. Accordingly, $\Delta M < M\alpha\Delta T$. The temperature drop over the age of the solar system is roughly comparable to the actual temperature now (e.g., the deep interior of Jupiter may have cooled from 40,000 K to 20,000 K). Deep within giant planets, $\alpha T \sim 0.05$. Consequently it is possible in principle to mix up on the order of 5% of the mass of the planet (15 $M_\oplus$ for Jupiter, 5 $M_\oplus$ for Saturn). In practice, the real amount is likely to be less than this by as much as an order of magnitude, for three reasons. First, the heat flow (or equivalently, the buoyancy production) at the top of the core is far less than that associated with the cooling of the overlying hydrogen (the main source of luminosity in these planets). For example, the heat content of a rock and ice mixture is about an order of magnitude less than the heat content of the same mass of hydrogen at the same temperature, because the latter has much lower molecular weight and hence much higher heat capacity. If one relied instead on radioactive heat, then the available heat flow would be only a few times the terrestrial value (per unit area), whereas the intrinsic heat flux of Jupiter (per unit area) is 30× greater. The second reason for lowering the expected erosion comes from consideration of the actual fluid dynamics of mixing. It is well established from laboratory experiments that this is typically an order of magnitude less efficient than the highest efficiency permitted from purely energetic arguments (*Turner*, 1973). Third, convective upward mixing of heavy molecules might be strongly inhibited by a compositional gradient. In those circumstances, the mixing will certainly be slower because of the lower diffusivity of the heavy atoms relative to the diffusivity of heat. This is the regime of "double diffusive convection" (*Turner*, 1973).

This discussion omits consideration of the difficult question of what happens when there are impacts of large embryos during the formation of the planet. The mixing that occurs during that early phase is not so readily analyzed by the arguments presented above and has been inadequately explored. The material is less degenerate, so that thermal differences have a potentially greater ability to cause compositional mixing. But the complexity of the fluid dynamics and shock processes make quantitative analysis difficult. In the corresponding problem of giant impacts during formation of terrestrial planets (and formation of the Moon), it has become apparent that one cannot rely on low-resolution smoothed particle hydrodynamics (SPH) simulations for quantitative assessment of the mixing. The reason is that the mixing involves Rayleigh-Taylor instabilities that grow most rapidly at length scales smaller than the resolution scale of these simulations. A similar problem will arise in the consideration of large ice and rock bodies hitting partially assembled giant planets.

It is possible that Uranus and Neptune provide the greatest insight into these issues of core formation and structure. This might seem surprising given our relatively poor un-

derstanding of their internal structure (*Guillot*, 2005). However, the ice and rock components of these planets are dominant relative to the hydrogen component, and plausible models that fit the gravity field clearly require some mixing of the constituents. End-member models consisting of a discrete rock core, ice mantle, and hydrogen/helium envelope are not permitted. To the extent that the formation of these planets is similar to that for Jupiter and Saturn (except of course for the lack of a late-stage large addition of gas), this would seem to suggest that there was considerable mixing even during the early stages.

Turning to the opposite problem of core rainout, making a core this way once the material is dense and degenerate is unlikely because the high temperatures and dilution make it thermodynamically implausible. Suppose we have a constituent of atomic (or molar) abundance x relative to the overwhelmingly predominant hydrogen. Let the Gibbs energy cost of mixing this constituent to the atomic level in hydrogen be ΔG_x. The physical origin of this energy is primarily quantum mechanical and arises from the mismatch of the electronic environments of the host and the inserted atom or molecule (including the work done in creating the cavity within the host). In the plausible situation of approximately ideal mixing, the solubility limit of this constituent is then $\sim e^{-\Delta G_x/kT}$, where k is Boltzmann's constant (*Stevenson*, 1998). In the case of water (or oxygen), the expected average concentration relative to hydrogen is about 0.001. At 10,000 K, $kT \sim 1$ eV, and rainout could begin (starting from a higher temperature, undersaturated state) for ΔG_x (in eV) $\sim -\ln 0.001 \sim 7$. This is a large energy, especially when one considers that an electronically unfavorable choice (helium) has smaller ΔG_x of perhaps only a few eV and even neon (depleted by a factor of 10 in the atmosphere of Jupiter) apparently has a lower insertion energy than 7 eV. The basic physical point is that the least soluble constituents in the deep interior are expected to be atoms with very tightly bound electrons (the noble gases), and the insolubility of helium is aided by its higher abundance. In fact, the observed inferred rainout of helium is small for Jupiter, corresponding to ~10% of the helium mass, which is in turn about one-fourth of the total mass of Jupiter. The total helium rainout is accordingly about 5–10 $M_\oplus$. (This is unlikely to form a discrete core and is in any event not what is meant when one talks of a high-density core in these planets, since the density of helium is only roughly twice that of hydrogen.) Rainout of material suitable for a rock and ice core is less favorable both because it is much less abundant by atomic number and because it is electronically more compatible with the metallic state of hydrogen deep within the planet. Indeed, water is expected to be a metal at conditions not far removed from the metallization of hydrogen and is certainly ionic at lower pressures within Jupiter.

As with erosion, this discussion does not cover the potentially important case of very early nondegenerate conditions. In the early work on giant gaseous protoplanets, it was proposed that cores could rain out, much in the same way as rock or ice can condense and settle to the midplane of the solar nebula, i.e., under very low density conditions

(*Decampli and Cameron*, 1979). This is the process still advocated in some more recent work (e.g., *Boss, 2002*). The difficulty of this picture lies in the fact that as the material settles deeper and the protoplanet contracts, the combination of adiabatic heating and gravitational energy release is likely to cause the solid and liquid iron/silicates to undergo evaporation. This can be avoided only for rather low mass protoplanets. This nonetheless remains the most plausible way of forming a core in a planet that formed via gaseous instability. Note, however, that this will not produce a core that is more massive than that predicted by solar abundances, unless the appropriate amount of gas is lost at a later stage. Models of this kind have the appearance of special pleading if they are to explain the entire set of giant planets (including Uranus and Neptune).

It seems likely that whatever model one favors for giant planet formation, it should allow for the formation of a core, since Saturn probably has a core and one must in any event explain Uranus and Neptune. It would be contrived to attribute a different origin for Jupiter than for the other giant planets. It seems likely, therefore, that the formation of giant planets is closest to a "bottom-up" scenario that proceeded through formation of a solid embryo followed by the accumulation of gas.

3. GIANT PLANET FORMATION MODELS

The core nucleated accretion model relies on a combination of planetesimal accretion and gravitational accumulation of gas. According to this scenario, the initial stages of growth of a gas giant planet are identical to those of a terrestrial planet. Dust settles toward the midplane of the protoplanetary disk and agglomerates into (at least) kilometer-sized planetesimals, which continue to grow into larger solid bodies via pairwise inelastic collisions. As the planet grows, its gravitational potential well gets deeper, and when its escape speed exceeds the thermal velocity of gas in the surrounding disk, it begins to accumulate a gaseous envelope. The gaseous envelope is initially optically thin and isothermal with the surrounding protoplanetary disk, but as it gains mass it becomes optically thick and hotter with increasing depth. While the planet's gravity pulls gas from the surrounding disk toward it, thermal pressure from the existing envelope limits accretion. For much of the planet's growth epoch, the primary limit on its accumulation of gas is its ability to radiate away the gravitational energy provided by accretion of planetesimals and envelope contraction; this energy loss is necessary for the envelope to further contract and allow more gas to reach the region in which the planet's gravity dominates. The size of the planet's gravitational domain is typically a significant fraction of the planet's Hill sphere, whose radius, R_H, is given by

$$R_H = \left(\frac{M}{3 M_\star} \right)^{1/3} r \qquad (1)$$

where M and $M_\star$ are the masses of the planet and star, respectively, and r is the distance between these two bodies. Eventually, increases in the planet's mass and radiation of energy allow the envelope to shrink rapidly. At this point, the factor limiting the planet's growth rate is the flow of gas from the surrounding protoplanetary disk.

The rate and manner in which a forming giant planet accretes solids substantially affect the planet's ability to attract gas. Initially accreted solids form the planet's core, around which gas is able to accumulate. Calculated gas accretion rates are very strongly increasing functions of the total mass of the planet, implying that rapid growth of the core is a key factor in enabling a planet to accumulate substantial quantities of gas prior to dissipation of the protoplanetary disk. Continued accretion of solids acts to reduce the planet's growth time by increasing the depth of its gravitational potential well, but has counteracting affects by providing additional thermal energy to the envelope (from solids that sink to or near the core) and increased atmospheric opacity from grains that are released in the upper parts of the envelope. Major questions remain to be answered regarding solid-body accretion in the giant planet region of a protoplanetary disk, with state-of-the-art models providing a diverse set of predictions.

Because of the complexity of the physics and chemistry involved in giant planet formation, the large range of distance scales, the long time (compared to orbital and local thermal times) required for accumulation, and the uncertainties in initial conditions provided by the protoplanetary disks, detailed planet-growth models have focused on specific aspects of the problem, and ignored or provided greatly simplified treatments of other processes. The solid-accretion scenarios incorporated into envelope models to date have been quite simplified, and in some cases completely *ad hoc*. These issues are discussed in section 3.1.

A planet on the order of one to several $M_\oplus$ is able to capture an atmosphere from the protoplanetary disk because the escape speed from its surface is large compared to the thermal velocity of gas in the disk. However, such an atmosphere is very tenuous and distended, with thermal pressure pushing outward to the limits of the planet's gravitational reach and thereby limiting further accretion of gas. The key factor governing the planet's evolution at this stage is its ability to radiate energy so that its envelope can shrink and allow more gas to enter the planet's gravitational domain. Evolution occurs slowly, and hydrostatic structure is generally a good approximation. However, the stability of the planet's atmosphere against hydrodynamically induced ejection must be calculated. The basic physical mechanisms operating during this stage of growth appear to be qualitatively understood, but serious questions remain regarding the ability of planets to pass through this stage sufficiently rapidly to complete their growth while adequate gas remains in the protoplanetary disk. This timescale issue is being addressed by numerical simulations. Models of this phase of a giant planet's growth are reviewed in section 3.2.

Once a planet has enough mass for its self-gravity to compress the envelope substantially, its ability to accrete additional gas is limited only by the amount of gas available. Hydrodynamic limits allow quite rapid gas flow to the planet in an unperturbed disk. But the planet alters the disk by accreting material from it and by exerting gravitational torques on it. Both of these processes can lead to gap formation and isolation of the planet from the surrounding gas. Hydrodynamic simulations lend insight into these processes, and are discussed briefly in section 3.3.

Radial motion of the planet and disk material can affect both the planet's growth and its ultimate orbit. Much of a protoplanetary disk is ultimately accreted by the central star (chapter by Bouvier et al.). Small dust grains are carried along with the gas, but millimeter and larger particles can suffer a secular drag if they orbit within a gaseous disk that rotates slower than the Keplerian velocity because the gas is partially supported against stellar gravity by a radial pressure gradient (*Adachi et al.,* 1976). Such gas drag can cause substantial orbital decay for bodies up to kilometer sizes (*Weidenschilling,* 1977). Once growing planets reach lunar to Mars size, their *gravitational* interactions with the surrounding disk can lead to substantial radial migration. Radial migration of a planet can have major consequences for its growth, ultimate orbit, and even survival. This process is reviewed in depth in the chapter by Papaloizou et al., but its relationship with planetary growth is briefly commented upon in section 3.4.

3.1. Growth of the Core

Models of solid planet growth do a fairly good job of explaining the origin of terrestrial planets in our solar system (e.g., *Agnor et al.,* 1999; *Chambers,* 2001), and can be applied with modification to the growth of planetary bodies at greater distances from the Sun and other stars (*Quintana et al.,* 2002; *Barbieri et al.,* 2002; *Quintana and Lissauer,* 2006). Most models of terrestrial planet growth start with a "minimum mass" disk, containing the observed heavy element components in the planets spread out smoothly into a disk, plus enough gas to make the disk's composition the same as that of the proto-Sun. The disk is assumed to be relatively quiescent, with the Sun already largely formed and close to its current mass (*Safronov,* 1969). Micrometer-sized dust, composed of surviving interstellar grains and condensates formed within the protoplanetary disk, moves mostly with the dominant gaseous component of the disk. But it gradually agglomerates and settles toward the midplane of the disk. If the disk is laminar, then the solids can collapse into a layer that is thin enough for collective gravitational instabilities to occur (*Edgeworth,* 1949; *Safronov,* 1960; *Goldreich and Ward,* 1973); such instabilities would have produced planetesimals of ~1 km radius at 1 AU from the Sun. If the disk is turbulent, then gravitational instabilities are suppressed because the dusty layer remains too thick. Under such circumstances, continued growth via pairwise agglomeration depends upon (currently unknown) sticking and disruption probabilities for collisions among larger grains (*Weidenschilling and Cuzzi,* 1993). The mechanism for growth from centimeter to kilometer sizes remains

one of the major controversies in terrestrial planet growth (*Youdin and Shu,* 2002; chapter by Dominik et al.). Nonetheless, theoretical models suggest that gravitational instabilities are more likely to occur farther from the star and that ices are stickier than rock. Moreover, many small to moderate-sized bodies are observed in the Kuiper belt beyond the orbit of Neptune (chapter by Cruikshank et al.) and probably smaller but still macroscopic bodies are inferred as parents to the observed dust seen in second-generation debris disks around Vega, Beta Pictoris, and many other stars (chapter by Meyer et al.). Thus, growth of solid bodies to multikilometer sizes in at least the inner portions of the ice-condensation region of most protoplanetary disks seems quite likely.

Once solid bodies reach kilometer size (using parameters that are appropriate for the terrestrial region of the protosolar disk), gravitational interactions between pairs of solid planetesimals provide the dominant perturbation of their basic Keplerian orbits. Electromagnetic forces, collective gravitational effects, and in most circumstances, gas drag, play minor roles. These planetesimals continue to agglomerate via pairwise mergers. The rate of solid-body accretion by a planetesimal or planetary embryo (basically a large planetesimal) is determined by the size and mass of the planetesimal/planetary embryo, the surface density of planetesimals, and the distribution of planetesimal velocities relative to the accreting body. Assuming perfect accretion, i.e., that all physical collisions are completely inelastic, this stage of growth is initially quite rapid, especially in the inner regions of a protoplanetary disk, and large bodies form quickly. The planetesimal accretion rate, $\dot{M}_Z$, is given by

$$\dot{M}_Z = \pi R^2 \sigma_Z \Omega F_g \qquad (2)$$

where R is the radius of the accreting body, σ_Z is the surface density of solid planetesimals in the solar nebula, Ω is the orbital frequency, and F_g is the gravitational enhancement factor, which is the ratio of the total effective accretion cross section to the geometric cross section. If the velocity dispersion of the bodies is large compared to the Keplerian shear of the disk across the body's accretion zone, the two-body approximation yields

$$F_g = 1 + \left(\frac{v_e}{v} \right)^2 \qquad (3)$$

where v is the velocity dispersion and v_e is the escape velocity from the body's surface. The evolution of the planetesimal size distribution is determined by the gravitationally enhanced collision cross section, which favors collisions between bodies having larger masses and smaller relative velocities.

Planetesimal growth regimes are sometimes characterized as either orderly or runaway. In orderly growth, particles containing most of the mass double their masses in about the same amount of time as the largest particle. When the relative velocity between planetesimals is comparable to or larger than the escape velocity, $v \geq v_e$, the growth rate is approximately proportional to R^2, and there is an orderly growth of the entire size distribution. When the relative velocity is small, $v \ll v_e$, the growth rate is proportional to R^4. In this situation, the planetary embryo rapidly grows much larger than any other planetesimal in its accretion zone. By virtue of its large, gravitationally enhanced cross section, this runaway particle doubles its mass faster than the smaller bodies do, and detaches itself from the mass distribution (*Wetherill and Stewart, 1989; Ohtsuki et al., 2002*).

Eventually a runaway body can grow so large that it transitions from dispersion-dominated growth to shear-dominated growth (*Lissauer, 1987*). Dynamical friction, which drives the distribution of planetesimal velocities toward a state of equipartition of kinetic energy of random motion (e.g., *Stewart and Wetherill, 1988*), reduces the random motions of the more massive bodies, so proximate embryos collide and merge. At this stage, larger embryos take longer to double in mass than do smaller ones, although embryos of all masses continue their runaway growth relative to surrounding planetesimals. This phase of rapid accretion of planetary embryos is known as oligarchic growth (*Kokubo and Ida, 1998*).

The self-limiting nature of runaway/oligarchic growth implies that massive planetary embryos form at regular intervals in semimajor axis. The agglomeration of these embryos into a small number of widely spaced terrestrial planets necessarily requires a stage characterized by large orbital eccentricities. The large velocities imply small collision cross sections (equation (3)) and hence long accretion times. Growth via binary collisions proceeds until the spacing of planetary orbits become dynamically isolated from one another, i.e., sufficient for the configuration to be stable to gravitational interactions among the planets for the lifetime of the system (*Safronov, 1969; Wetherill, 1990; Lissauer, 1993, 1995; Agnor et al., 1999; Chambers, 2001; Laskar, 2000*).

The early phases of growth from planetesimals are likely to be similar in the more distant regions of protoplanetary disks. However, the rate at which accretion of solids takes place depends upon the surface density of condensates and the orbital frequency (equation (2)), both of which decrease with heliocentric distance. Thus, the high-velocity final growth stage that takes $O(10^8)$ yr in the terrestrial planet zone (*Safronov, 1969; Wetherill, 1980; Agnor et al., 1999; Chambers, 2001*) would require $O(10^9)$ yr in the giant planet zone (*Safronov, 1969*). This is far longer than any modern estimates of the lifetimes of gas within protoplanetary disks, implying that giant planet cores must form via rapid runaway/oligarchic growth (chapter by Meyer et al.). Moreover, particles far from their stars are physically small compared to the size of their gravitational domains (Hill spheres), and giant planets eventually grow large enough that escape speeds from accreting planets exceed the escape velocity from stellar orbit at their locations.

For shear-dominated accretion, the mass at which an embryo becomes isolated from the surrounding disk is given by

$$M_{iso} = \frac{(8\pi\sqrt{3}r^2\sigma_Z)^{3/2}}{(3 M_\star)^{1/2}} \tag{4}$$

where r is the distance from the star (*Lissauer, 1993*). In the inner part of protoplanetary disks, Kepler shear is too great to allow the accretion of solid planets larger than a few $M_\oplus$ on any timescale unless surface densities are considerably above that of the minimum mass solar nebula or a large amount of radial migration occurs. Larger solid planets are permitted farther from stars, but the duration of the final, high-velocity, stages of growth (*Safronov, 1969*) are far longer than the observed lifetimes of protoplanetary disks. The epoch of runaway/rapid oligarchic growth lasts only millions of years or less near 5 AU, and can produce ~10 $M_\oplus$ cores in disks only a few times the minimum mass solar nebula (*Lissauer, 1987*). The masses at which planets become isolated from the disk, thereby terminating the runaway/rapid oligarchic growth epoch, are likely to be comparably large at greater distances from the star. However, at these large distances, random velocities of planetesimals must remain quite small for accretion rates to be sufficiently rapid for embryos to approach isolation mass within the lifetimes of gaseous disks. Indeed, if planetesimal velocities become too large, material is more likely to be ejected to interstellar space than accreted by the planetary embryos.

The fact that Uranus and Neptune contain much less H_2 and He than Jupiter and Saturn suggests that Uranus and Neptune never quite reached runaway gas accretion conditions, possibly due to a slower accretion of planetesimals (*Pollack et al., 1996*). Theoretical difficulties with forming planets at Uranus/Neptune distances have been discussed in greater detail by *Lissauer et al.* (1995) and *Thommes et al.* (2003). New models are being proposed to address these problems by allowing rapid runaway accretion of a very small number of planetary embryos (cores) beyond 10 AU. In the model presented by *Weidenschilling* (2005), an embryo is scattered from the Jupiter-Saturn region into a massive disk of small planetesimals. The embryo is several orders of magnitude more massive than are the individual planetesimals surrounding it, but still far less massive than the aggregate of the surrounding disk of planetesimals. Dynamical friction is thus able to circularize the orbit of the embryo without substantially exciting planetesimal eccentricities. *Goldreich et al.* (2004a,b) propose that (at least in the Uranus/Neptune region) planetesimals between growing embryos are ground down to very small sizes and are forced into low-inclination, nearly circular orbits by frequent mutual collisions. Planetary embryos can accrete rapidly because of their large, gravitationally enhanced collision cross sections in dynamically cold disks such as those in the models of *Weidenschilling* (2005) and of *Goldreich et al.* (2004a,b). Alternatively, *Thommes et al.* (2003) suggest that the cores and possibly also the gaseous envelopes of Uranus and Neptune accreted between or just exterior to the orbits of Jupiter and Saturn, and were subsequently scattered out to their current locations by gravitational perturbations of these two giant planets (see also *Tsiganis et al.,* 2005). Alternatively/additionally, Uranus and Neptune may have avoided gas runaway as a result of the removal of gas from the outer regions of the disk via photoevaporation (*Hollenbach et al.,* 2000).

Published simulations of the accumulation of giant planet *atmospheres* use simplified prescriptions for the planet's accretion of *solids.* In some cases, the accretion rate of solids is assumed to be constant (*Bodenheimer and Pollack,* 1986; *Ikoma et al.,* 2000). In others, an isolated planetary embryo grows by runaway accretion in a disk of much smaller planetesimals, as discussed in the following paragraph. The actual accretion of solids by a planet is more complex, variable in time and highly stochastic, and most likely including the occasional impact of a large body. But as discussed above, there are many open questions regarding the growth of solid cores at the locations of the giant planets within our solar system. Thus, more sophisticated models do not necessarily provide better approximations of actual core growth rates. Moreover, these simplified models illuminate several key aspects of how accretion of solids controls the rate of envelope (gas) accumulation.

The most sophisticated thermal models of the accumulation of massive gaseous envelopes by planets (*Pollack et al.,* 1996; *Bodenheimer et al.,* 2000b; *Alibert et al.,* 2004, 2005; *Hubickyj et al.,* 2005) assume runaway growth of an isolated (or nearly isolated) planet. An updated version (*Greenzweig and Lissauer,* 1992) of the classical theory of planetary growth (*Safronov,* 1969) is used, employing equations (2) and (3) with R replaced by R_{capt}, the effective (geometric) capture radius of the protoplanet for a planetesimal of a given size (including regions of the envelope sufficiently dense to capture planetesimals). These models begin with the growing protoplanet embedded in a disk of monodisperse planetesimal size and uniform surface density. The protoplanet's feeding zone is assumed to be an annulus extending to a radial distance of about 4 R_H on either side of its orbit (*Kary and Lissauer,* 1994). The feeding zone grows as the planet gains mass, and random scattering spreads the unaccreted planetesimals uniformly over the feeding zone. Radial migration of planetesimals into and out of the feeding zone is not considered in the models of *Pollack et al.* (1996), *Bodenheimer et al.* (2000b), and *Hubickyj et al.* (2005). However, some of the simulations by these authors terminate solids accretion at a predetermined core mass, thereby mimicking the effects of planetesimal accretion by competing embryos.

Alibert et al. (2004, 2005) incorporated planetary migration, thereby allowing the planet to move into regions of the disk with undepleted reservoirs of planetesimals. In some cases, they follow the simultaneous accumulation of multiple planets, and in these simulations one planet can migrate into a region already depleted of planetesimals as a consequence of accretion by another core. However, planetary orbits rapidly decay into the Sun in those simulations that include migration at rates predicted by theoretical mod-

els of interactions of planets with a minimum mass solar nebula. Thus, Alibert et al. arbitrarily reduce planetary migration rates by a factor of ~30; it isn't clear that this is a better approximation than that of completely ignoring migration, as done by *Hubickyj et al.* (2005) and others.

In order for cores to reach the required masses prior to isolation from their planetesimal supplies (equation (4)), models that do not incorporate migration (e.g., *Hubickyj et al.,* 2005) need to assume that the surface mass density of solids in Jupiter's region was at least 2–3× as large as the value predicted by "classical" minimum mass models of the protoplanetary disk (*Weidenschilling,* 1977; *Hayashi,* 1981). This is fully consistent with disk observations, and with models suggesting both that the giant planets in our solar system formed closer to one another than they are at present (*Fernandez and Ip,* 1984; *Hahn and Malhotra,* 1999; *Thommes et al.,* 1999; chapter by Levison et al.) and that a large number of icy planetesimals were ejected from the giant planet region to the Oort cloud as well as to interstellar space (e.g., *Dones et al.,* 2004). Models in which cores migrate relative to the planetesimal disk (e.g., *Alibert et al.,* 2005), or in which solids can be concentrated by diffusive redistribution of water vapor (*Stevenson and Lunine,* 1988), baroclinic instabilities (*Klahr and Bodenheimer,* 2006), or gravitational instabilities (*Durisen et al.,* 2005), can form planets in lower-mass disks. But all models are subject to the stronger constraints of heavy element abundances in giant planets and disk lifetime.

Inaba et al. (2003) have performed simulations of giant planet growth that incorporate a more sophisticated treatment of solid-body accretion. In their model, multiple planetary embryos stir smaller planetesimals to high enough velocities that planetesimal collisions are highly disruptive. Inaba et al. include envelope thermal evolution (albeit using a more simplified treatment than that employed by the above-mentioned groups) and planetesimal accretion cross sections that are enhanced by the presence of the envelope (*Inaba and Ikoma,* 2003). As a result of the competition between nearby growing cores, they require an initial surface mass density at 5 AU of about twice that of *Hubickyj et al.* (2005) for core growth to occur on timescales consistent with observational constraints on disk lifetimes. Specifically, with a solid surface density 25 g cm^{-2} at 5 AU and assuming full interstellar grain opacity within the protoplanet's atmosphere, they can form Jupiter possessing a ~20 $M_\oplus$ core in <4 m.y. If they reduce the grain opacity by a factor of 100, they get a Jupiter with a 7 $M_\oplus$ core in 5 × 10^6 yr in a disk with surface density 12.5 g cm^{-2}. They are not able to form Saturn in either of these cases.

3.2. Gas Accretion: Tenuous Extended Envelope Phase

The escape velocity from a planetary embryo with M > 0.1 $M_\oplus$ is larger than the sound speed in the surrounding gaseous protoplanetary disk at temperatures where ice can condense, so such an embryo can begin to accumulate a quasistatic atmosphere. As the atmosphere/envelope grows, it becomes optically thick to outgoing thermal radiation, and its lower reaches can get much warmer and denser than the gas in the surrounding protoplanetary disk. It undergoes Kelvin-Helmholtz contraction as the energy released by the accretion of planetesimals and gas is radiated away at the photosphere. A thick atmosphere expands the accretion cross section of the planet, especially for small solid bodies. At this stage, the key processes are the accretion of solids and the radiation of thermal energy. Most detailed models of this phase are spherically symmetric (one-dimensional). The energy released by accretion of planetesimals and envelope contraction heats the envelope and regulates the rate of contraction. This in turn controls how rapidly additional gas can enter the domain of the planet's gravitational reach and be accreted. Because the opacity is sufficiently high, much of the growing planet's envelope transports energy via convection. However, the distended very-low-density outer region of the envelope has thermal gradients that are too small for convection, but is so large that it acts as an efficient thermal blanket if it is sufficiently dusty to be moderately opaque to outgoing radiation.

During the runaway planetesimal accretion epoch, the protoplanet's mass increases rapidly (Fig. 2). The internal temperature and thermal pressure increase as well, preventing substantial amounts of nebular gas from falling onto the protoplanet. When the rate of planetesimal accretion decreases, gas falls onto the protoplanet more rapidly.

As a planet grows, its envelope mass is a sensitive function of the total mass, with the gaseous fraction increasing rapidly as the planet accretes (*Pollack et al.*, 1996). Accretion initially proceeds slowly, governed by the growth of the mass of the solid core and release of thermal energy from the envelope. When the envelope reaches a mass comparable to that of the core, the self-gravity of the gas becomes substantial, and the envelope contracts when more gas is added, so further accretion is governed by the availability of gas rather than thermal considerations. The time required to reach this epoch of rapid gas accretion is governed primarily by three factors: the mass of the solid core (larger core mass implies more rapid accretion); the rate of energy input from continued accretion of solids (such energy keeps the envelope large and slows further accretion of gas); and the opacity of the envelope (low opacity allows the radiation of energy that enables the envelope to cool and shrink, making room for more gas to be accreted). These three factors appear to be key in determining whether giant planets are able to form within the lifetimes of protoplanetary disks. For example, in a disk with initial $\sigma_Z = 10$ g/cm^2 at 5.2 AU from a 1 $M_\odot$ star, a planet whose atmosphere has 2% interstellar opacity forms with a 16 $M_\oplus$ core in 2.3×10^6 yr; in the same disk, a planet whose atmosphere has full interstellar opacity forms with a 17 $M_\oplus$ core in 6.3×10^6 yr; a planet whose atmosphere has 2% interstellar opacity but stops accreting solids at 10 $M_\oplus$ forms in 0.9 m.y., whereas

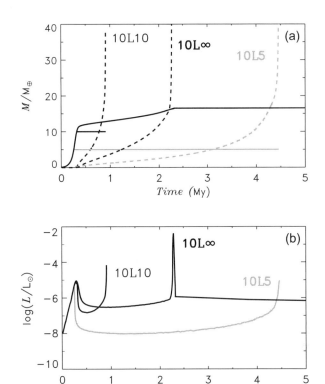

Fig. 2. Evolution of a giant protoplanet with $\sigma_{init,Z} = 10$ g/cm^2 and grain opacity at 2% interstellar value. Details of the calculation are presented in *Hubickyj et al.* (2005). **(a)** The mass is plotted as a function of time, with the solid lines referring to the solids component of the planet, the dotted lines to the gaseous component, and the dot-dashed lines the total mass. Thick black curves: no solid accretion cutoff; thin black curves: solid accretion cutoff at 10 $M_\oplus$; gray curves: solid accretion cutoff at 5 $M_\oplus$. **(b)** The luminosity is plotted on a logarithmic scale as a function of time. Note that the cutoff runs are halted when the gas accretion rate reaches a limiting value defined by the rate at which the solar nebula can transport gas to the vicinity of the planet, whereas the planet in the run with no cutoff stops growing when $M_p = 1$ M_{Jup}. The existence of a sharp peak in planetary luminosity during the phase of rapid gas accretion is physically plausible, but is likely to be somewhat lower and broader than shown in the plot because gas accretion almost certainly tapers off less abruptly than assumed for this calculation. Figure courtesy of O. Hubickyj.

if accretion of solids is halted at 3 $M_\oplus$ accretion of a massive envelope requires 12 m.y. (*Hubickyj et al.*, 2005). Thus, if Jupiter's core mass is significantly less than 10 $M_\oplus$, then it presents a problem for formation models.

As estimates of the lifetimes of protoplanetary disks have decreased, a major concern has been whether or not giant planets can form faster than typical disks are dispersed, ~ 2–5×10^6 yr. Planets can indeed form rapidly if they have sufficiently massive cores that accrete early and then stop growing and/or if the outer regions of their envelopes are transparent to outgoing radiation (have low opacities). But what are realistic values for these parameters? Observational constraints are quite weak. Limits upon the masses and locations of the heavy element components of the giant planets within our solar system were discussed in section 2. Atmospheric opacities and how the accretion rate of solids depends on time are quantities derived from planet formation models, and at present their values are quite ill constrained.

The ability of a planetary core to accrete gas does not depend strongly on the outer boundary conditions (temperature and pressure) of the surrounding disk, as long as there is adequate gas to be accreted (*Mizuno, 1980; Stevenson, 1982; Pollack et al., 1996*). The primary reason why giant planet formation is believed not to occur within a few AU of a star is the difficulty of forming a sufficiently massive core in the high Kepler shear environment of this region (*Lissauer, 1987; Bodenheimer et al., 2000b*).

The composition of the atmosphere of a giant planet is largely determined by how much heavy material was mixed with the lightweight material in the planet's envelope. Accretion energy can lead to evaporation of planetary ices, and their mixing into the atmosphere can increase its mean molecular weight, allowing it to shrink and more gas to be trapped (*Stevenson, 1982*). As the envelope becomes more massive, late-accreting planetesimals sublimate before they can reach the core, thereby enhancing the heavy element content of the envelope considerably.

In the detailed thermal calculations of giant planet envelope accumulation performed to date, the accumulation of solids governs the accretion of gas. Yet apart from increasing the planet's total mass, the effect of the extended gaseous envelope on the accretion rate of solids is minimal. But this would not be the case if most of the mass of condensed material in the disk near the planet's orbit resided in very small solid bodies, nor if the planet migrated relative to solids in the disk (*Kary et al., 1993*).

3.3.   Gas Accretion: Hydrodynamic Phase

As discussed in section 3.2, a protoplanet accumulates gas at a gradually increasing rate until its gas component is comparable to its heavy element mass. The rate of gas accretion then accelerates rapidly, and a gas runaway occurs (*Pollack et al., 1996; Hubickyj et al., 2005*). The gas runaway continues as long as there is gas in the vicinity of the protoplanet's orbit.

The protoplanet may cut off its own supply of gas by gravitationally clearing a gap within the disk (*Lin and Papaloizou, 1979*). Such gaps have been observed around small moons within Saturn's rings (*Showalter, 1991; Porco et al., 2005*). *D'Angelo et al.* (2003) used a three-dimensional adaptive mesh refinement code to follow the flow of gas onto accreting giant planets of various masses embedded within a gaseous protoplanetary disk. *Bate et al.* (2003) performed three-dimensional simulations of this problem using the ZEUS hydrodynamics code. In unperturbed disks, flows would increase with planet mass indefinitely. Using parameters appropriate for a moderately viscous minimum mass solar nebula protoplanetary disk at 5 AU, both groups

found that <10 $M_\oplus$ planets do not perturb the protoplanetary disk enough to significantly affect the amount of gas that flows toward them. Gravitational torques on the disk by larger planets drive away gas. Hydrodynamic limits on gas accretion reach to a few × 10^{-2} $M_\oplus$ per year for planets in the ~50–100 $M_\oplus$ range, and then decline as the planet continues to grow. An example of gas flow around/to a 1 M_{Jup} planet is shown in Fig. 3. These calculations do not include the thermal pressure on the nebula from the hot planet, which is found to be the major accretion-limiting factor for planets up to a few tens of $M_\oplus$ by the simulations discussed in section 3.2.

Calculations incorporating both hydrodynamic flows of gas in the disk and thermal physics of the planet are needed to fully understand the gas accretion rate by a growing planet. But it appears that the primary factor limiting growth of a planet smaller than a few dozen Earth masses is its ability to radiate energy, thereby allowing its envelope to shrink so that more gas can flow into the planet's gravitational domain. For planets larger than ~100 $M_\oplus$, thermal pressure from the envelope does not limit growth, but gravitational torques limit the flow of gas from the disk. *Bate et al.* (2003) find that gas accretion rates decline precipitously for planets more than a few times the mass of Jupiter, but that planets up to ~5 M_{Jup} can double in mass within a million years for nominal disk parameters. Thus, disks must be largely dispersed within ~10^4–10^6 yr after the onset of rapid accretion of gaseous envelopes by giant planets in order to explain the observed distribution (chapter by Udry et al.) of planetary masses.

3.4.   Migration

A major uncertainty associated with the emergence of planets is their predicted orbital migration as a consequence of the gravitational torque between the disk and the planet (*Goldreich and Tremaine, 1980; Ward, 1986; Bate et al., 2003*). Planets that are more massive than Mars may be able to migrate substantial distances prior to the dispersal of the gaseous disk. Planets that are not massive enough to clear gaps around their orbits undergo Type 1 migration as a consequence of the difference between the repulsive torques that they exert on material interior and exterior to their orbits; in the linear regime these torques vary quadratically with planet mass, so migration rates are proportional to planetary mass. Planets that clear gaps around their orbits are subjected to Type 2 migration, by which they are dragged along with the evolving disk. Thus, it is quite possible that giant planets form several AU from their star and then migrate inward to the locations at which most extrasolar planets have been observed. Disk-induced migration is considered to be the most likely explanation for the "giant vulcan" planets with orbital periods of less than a week, because the Keplerian shear close to a star makes *in situ* formation of such objects quite unlikely (*Bodenheimer et al., 2000b*). *Livio and Pringle* (2003) find no basis to suggest that planetary migration is sensitive to disk metallicity, and conclude

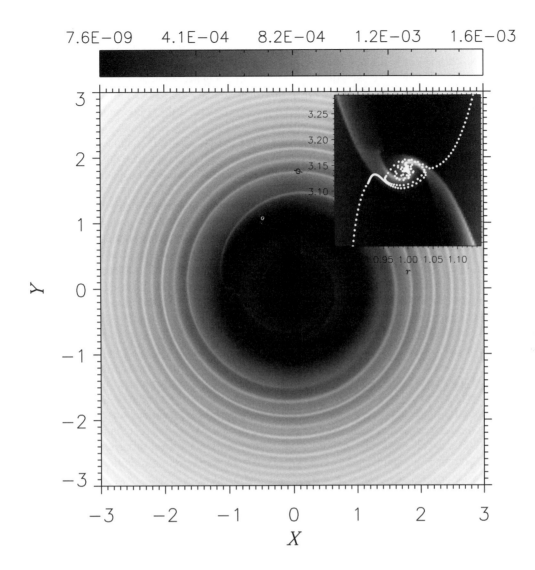

Fig. 3.   The surface mass density of a gaseous disk containing a Jupiter-mass planet on a circular orbit located 5.2 AU from a 1 M$_\odot$ star. The ratio of the scale height of the disk to the distance from the star is 1/20, and the dimensionless viscosity at the location of the planet is $\alpha = 4 \times 10^{-3}$. The distance scale is in units of the planet's orbital distance, and surface density of 10^{-4} corresponds to 33 g/cm^2. The inset shows a closeup of the disk region around the planet, plotted in cylindrical coordinates. The two series of white dots indicate actual trajectories (real particle paths, not streamlines) of material that is captured in the gravitational well of the planet and eventually accreted by the planet. Figure courtesy of G. D'Angelo. See *D'Angelo et al.* (2005) for a description of the code used.

that the correlations between the presence of observable planets and stellar metallicity probably results from a higher likelihood of giant planet formation in metal-rich disks.

The difficulty with the migration models is that they predict that planets migrate *too rapidly*, especially Type 1 migration in the Earth- to Neptune-mass range that planetary cores grow through in the core nucleated accretion scenario. (Planets formed directly via gravitational instabilities would avoid the danger of Type 1 migration, but would be subject to a greater amount of Type 2 migration as a consequence of their early formation within a massive disk.) Moreover, because predicted migration rates increase as a planet moves inward, most migrating planets should be consumed by their star. However, a planet may end up in very close 51 Peg-like orbits if stellar tides can counteract the migration or if the

disk has a large inner hole (*Lin et al.*, 2000). Resolution of this rapid migration dilemma may require the complete and nonlinear analysis of the disk response to the protoplanet. Alternatively/additionally, planets may stop migrating if they approach a density enhancement interior to their orbits, which equalizes the positive and negative torques upon them in either a quasiequilibrium or stochastic manner that allows some "lucky" planets to survive (*Laughlin et al.*, 2004), corotation torques might be able to slow down the migration of ~10 M$_\oplus$ objects (*D'Angelo et al.*, 2003), and the small amounts of gas that leak into almost clear gaps may slow the migration of more massive planets. See *Ward and Hahn* (2000), *Masset and Papaloizou* (2003), *Thommes and Lissauer* (2005), and the chapter by Papaloizou et al. for more extensive discussions of planetary migration.

Many of the known extrasolar giant planets move on quite eccentric ($0.2 < e < 0.7$) orbits. These orbital eccentricities may be the result of stochastic gravitational scatterings among massive planets, some of which have subsequently merged or been ejected to interstellar space (*Weidenschilling and Marzari, 1996; Levison et al., 1998; Ford et al., 2001*), by perturbations of a binary companion (*Holman et al., 1997*), or by past stellar companions if the now-single stars were once members of unstable multiple-star systems (*Laughlin and Adams, 1998*). However, as neither scattering nor migration offer a simple explanation for those planets with nearly circular orbits and periods from a few weeks to a few years, the possibility of giant planet formation quite close to stars should not be dismissed (*Bodenheimer et al., 2000b*).

Most of the observed extrasolar giant planets orbit between a few tenths of an AU and a few AU from their star, i.e., they are located much closer to their stars than Jupiter is from our Sun. These planets may have formed farther from their star and migrated inward, but without a stopping mechanism, which isn't known at these distances, they would have fallen into the star. *Lissauer* (2001) suggested that the orbits could be explained if disks cleared from the inside outward, leaving the planets stranded once they were too far interior to the disk for strong gravitational coupling to persist. Observations of the 2:1 resonant planets orbiting GJ 876 by *Marcy et al.* (2001; see also *Rivera et al., 2005*) support such a model, as do data that imply that the star CoKu Tau/4 has a disk with an inner hole (*Forrest et al., 2004*).

4. OUTSTANDING QUESTIONS

What are the data on giant planet composition and structure telling us? Are the enhancements in heavy elements in the atmospheres of the giant planets within our solar system the result of mixing of material throughout the planet, and thereby reflective of the planets' bulk compositions, or were they produced by a late veneer of planetesimal accretion or accretion of gas from a nebula depleted in H_2 and He (Guillot and Hueso, in preparation)? And why do they appear to be dominated by very low condensation temperature planetesimals, which seem required to produce the comparable enrichments of gasses of different volatilities (*Owen et al., 1999*)? What are the masses of the planetary cores, and are these reflective of core masses during the accretionary epoch, or have they been increased by settling or reduced by convective mixing? Progress on answering this question depends mainly on improvements in our understanding of the high pressure behavior of hydrogen.

The core nucleated accretion model provides a sound general framework for understanding the formation of giant planets. According to this scenario, giant planets begin their growth as do terrestrial planets and smaller bodies, but they become massive enough to gravitationally accrete substantial amounts of the abundant light gases prior to the dispersal of the protoplanetary disk. However, many first-order questions remain:

How rapid do solid cores accrete in the giant planet formation region? The solid core provides a gravitational potential well for the gas to fall into. Counteracting this tendency, ongoing accretion of solids provides additional heating that expands the planet's envelope, limiting accretion of gas, especially if the solids sink deeply into the gravitational potential well, down to or near the core (*Pollack et al., 1996*). Additionally, if continued accretion of solids provide a substantial amount of small grains that persist in the planet's radiative atmosphere, the resulting thermal blanket reduces planetary luminosity.

Are the atmospheres of growing giant planets good thermal blankets (high optical depth to outgoing radiation caused by the presence of abundant small grains) or nearly transparent? Models suggest that the ability to radiate energy is a key factor in determining how rapidly an atmosphere contracts, thereby allowing the planet to continue to grow. Low-opacity atmospheres allow giant planets to form much more rapidly and/or with significantly smaller cores than do high-opacity atmospheres. Small grains are provided to the planet both from the disruption and ablation of accreted planetesimals and entrained in the accreted gas, but the amounts and residence times are quite uncertain. Such grains are not present in large quantities in the atmospheres of giant planets in our solar system, nor are they detected in cool brown dwarfs (chapter by Marley et al.), but do they settle downward fast enough to allow the atmosphere to be transparent during the formation epoch (*Podolak, 2003*)?

How does a (growing or fully formed) giant planet interact with the surrounding protoplanetary disk? Models of a planet gravitationally clearing a gap around itself and accretion of material through a partially formed gap give "reasonable" results. But predicted migration rates are simply too rapid for the survival of as many giant planets as are observed within our solar system and around nearby Sun-like stars. Either giant planets form much more readily than predicted by models (perhaps because disks are significantly more massive) and the survivors that we see are a tiny fraction of the bodies formed, or migration rates have been substantially overestimated.

5. SUMMARY

The smoothness of the distribution of masses of young M stars, free-floating brown dwarfs, and even free-floating objects somewhat below the deuterium-burning limit argues strongly that these bodies formed in the same manner, i.e., via collapse, in some cases augmented by fragmentation. In contrast, the mass gap in nearby companions to Sun-like stars (the brown dwarf desert) is convincing evidence that most if not all of the known giant planets formed in a different manner.

Various models for giant planet formation have been proposed. According to the prevailing core nucleated accretion model, giant planets begin their growth by the accumulation of small solid bodies, as do terrestrial planets. However, unlike terrestrial planets, the growing giant planet cores become massive enough that they are able to accumulate substantial amounts of gas before the protoplanetary disk dissipates. The primary question regarding the core accretion model is whether planets can accrete very massive gaseous envelopes within the lifetimes of typical gaseous protoplanetary disks. Another important question is whether or not proto-Jupiter's core was sufficiently massive to capture large quantities of hydrogen and helium.

The main alternative giant planet formation scenario is the disk-instability model, in which gaseous planets form directly via gravitational instabilities within protoplanetary disks. The formation of giant planets via gas instability has never been demonstrated for realistic disk conditions. Moreover, this model has difficulty explaining the supersolar abundances of heavy elements in Jupiter and Saturn, and it does not explain the origin of planets like Uranus and Neptune. Nonetheless, it is possible that some giant planets form via disk instability, most likely in the regions of protoplanetary disks distant from the central star, where Keplerian shear is small and orbital timescales are long. Additionally, a few planets probably form via fragmentation of molecular cloud cores during collapse or are captured via exchange reactions involving (usually young) binary/multiple stars.

Most models for extrasolar giant planets suggest that they formed as Jupiter and Saturn are believed to have (in nearly circular orbits, far enough from the star that ice could condense), and subsequently migrated to their current positions, although some models allow for *in situ* formation. Gas giant planet formation may or may not be common, because the gas within most of protoplanetary disks could be depleted before solid planetary cores grow large enough to gravitationally trap substantial quantities of gas. Additionally, an unknown fraction of giant planets migrate into their star and are consumed, or are ejected into interstellar space via perturbations of neighboring giant planets, so even if giant planet formation is common, these planets may be scarce.

While considerable progress toward understanding the internal structure and formation of giant planets has been made recently, major questions remain. As we continue to place new data and simulation results into the jigsaw puzzle, some present pieces will surely need to be repositioned or discarded. With the wealth of new information being provided, we expect the picture to become clearer in the near future.

Acknowledgments. We thank O. Hubickyj and an anonymous referee for providing us with constructive comments on the manuscript. This work was partially supported by the NASA Outer Planets Research Program under grant 344-30-99-02 (J.J.L.), and by NASA's Planetary Geology and Geophysics Program (D.J.S.).

REFERENCES

Adachi I., Hayashi C., and Nakazawa K. (1976) *Prog. Theor. Phys., 56,* 1756–1771.
Agnor C. B., Canup R. M., and Levison H. F. (1999) *Icarus, 142,* 219–237.
Alibert Y., Mordasini C., and Benz W. (2004) *Astron. Astrophys., 417,* L25–L28.
Alibert Y., Mousis O., Mordasini C., and Benz W. (2005) *Astrophys. J., 626,* L57–L60.
Alonso R., Brown T. M., Torres G., Latham D. W., Sozzetti A., et al. (2004) *Astrophys. J., 613,* L153–L156.
Barbieri M., Marzari F., and Scholl H. (2002) *Astron. Astrophys., 396,* 219–224.
Bate M. R., Lubow S. H., Ogilvie G. I., and Miller K. A. (2003) *Mon. Not. R. Astron. Soc., 341,* 213–229.
Bodenheimer P. and Pollack J. B. (1986) *Icarus, 67,* 391–408.
Bodenheimer P., Burket A., Klein R., and Boss A. P. (2000a) In *Protostars and Planets IV* (V. Mannings et al., eds.), pp. 675–701. Univ. of Arizona, Tucson.
Bodenheimer P., Hubickyj O., and Lissauer J. J. (2000b) *Icarus, 143,* 2–14.
Boss A. P. (2000) *Astrophys. J., 536,* L101–L104.
Boss A. P. (2002) *Earth Planet. Sci. Lett., 202,* 513–523.
Burrows A., Sudarsky D., and Hubbard W. B. (2003) *Astrophys. J., 594,* 545–551.
Cai K., Durisen R. H., Michael S., Boley A. C., Mejía A. C., et al. (2006) *Astrophys. J., 636,* L149–L152.
Canup R. M. and Ward W. R. (2002) *Astron. J., 124,* 3404–3423.
Chambers J. E. (2001) *Icarus, 152,* 205–224.
Charbonneau B., Brown T. M., Latham D. W., and Mayor M. (2000) *Astrophys. J., 529,* L45–L48.
D'Angelo G., Kley W., and Henning T. (2003) *Astrophys. J., 586,* 540–561.
D'Angelo G., Bate M. R., and Lubow S. H. (2005) *Mon. Not. R. Astron. Soc., 358,* 316–332.
Decampli W. M. and Cameron A. G. W. (1979) *Icarus, 38,* 367–391.
Dones L., Weissman P. R., Levison H. F., and Duncan M. J. (2004) In *Comets II* (M. C. Festou et al., eds.), pp. 153–174. Univ. of Arizona, Tucson.
Durisen R. H., Cai K., Meijia A. C., and Pickett M. K. (2005) *Icarus, 173,* 417–424.
Edgeworth K. E. (1949) *Mon. Not. R. Astron. Soc., 109,* 600–609.
Fernandez J. A. and Ip W. H. (1984) *Icarus, 58,* 109–120.
Fischer D. A. and Valenti J. (2005) *Astrophys. J., 622,* 1102–1117.
Ford E. B., Havlickova M., and Rasio F. A. (2001) *Icarus, 150,* 303–313.
Forrest W. J., Sargent B., Furlan E., D'Alessio P., Calvet N., et al. (2004) *Astrophys. J., 154,* 443–447.
Gilliland R. L., Brown T. M., Guhathakurta P., Sarajedini A., Milone E. F., et al. (2000) *Astrophys. J., 545,* L47–L51.
Goldreich P. and Tremaine S. (1980) *Astrophys. J., 241,* 425–441.
Goldreich P. and Ward W. R. (1973) *Astrophys. J., 183,* 1051–1062.
Goldreich P., Lithwick Y., and Sari R. (2004a) *Ann. Rev. Astron. Astrophys., 42,* 549–601.
Goldreich P., Lithwick Y., and Sari R. (2004b) *Astrophys. J., 614,* 497–507.
Gonzalez G. (2003) *Revs. Mod. Phys., 75,* 101–120.
Greenzweig Y. and Lissauer J. J. (1992) *Icarus, 100,* 440–463.
Guillot T. (2005) *Ann. Rev. Earth Planet Sci., 33,* 493–530.
Hahn J. M. and Malhotra R. (1999) *Astrophys. J., 117,* 3041–3053.
Hayashi C. (1981) *Prog. Theor. Phys. Suppl., 70,* 35–53.
Hollenbach D., Yorke H. W., and Johnstone D. (2000). In *Protostars and Planets IV* (V. Mannings et al., eds.), pp. 401–428. Univ. of Arizona, Tucson.
Holman M. J., Touma J., and Tremaine S. (1997) *Nature, 386,* 254–256.
Hubbard W. B. (1984) *Planetary Interiors.* Van Nostrand Reinhold, New York.
Hubbard W. B., Podolak M., and Stevenson D. J. (1995) In *Neptune and Triton* (D. P. Cruikshank, ed.), pp. 109–138. Univ. of Arizona, Tucson.
Hubbard W. B., Burrows A., and Lunine J. I. (2002) *Ann. Rev. Astron. Astrophys., 40,* 103–136.
Hubickyj O., Bodenheimer P., and Lissauer J. J. (2005) *Icarus, 179,* 415–431.
Ikoma M., Nakazawa K., and Emori H. (2000) *Astron. J., 537,* 1013–1025.
Inaba S. and Ikoma M. (2003) *Astron. Astrophys., 410,* 711–723.
Inaba S., Wetherill G. W., and Ikoma M. (2003) *Icarus, 166,* 46–62.
Kary D. M. and Lissauer J. J. (1994) In *Numerical Simulations in Astro-*

physics, Modeling the Dynamics of the Universe (J. Franco et al., eds.), pp. 364–373. Cambridge Univ., Cambridge.

Kary D. M., Lissauer J. J., and Greenzweig Y. (1993) *Icarus, 106,* 288–307.

Klahr H. and Bodenheimer P. (2006) *Astrophys. J., 639,* 432–440.

Kokubo E. and Ida S. (1998) *Icarus, 131,* 171–178.

Laskar J. (2000) *Phys. Rev. Lett., 84,* 3240–3243.

Laughlin G. and Adams F. C. (1998) *Astrophys. J., 508,* L171–L174.

Laughlin G., Steinacker A., and Adams F. (2004) *Astrophys. J., 608,* 489–496.

Levison H. F., Lissauer J. J., and Duncan M. J. (1998) *Astron. J., 116,* 1998–2014.

Lin D. N. C. and Papaloizou J. (1979) *Mon. Not. R. Astron. Soc., 186,* 799–812.

Lin D. N. C., Papaloizou J. C. B., Terquem C., Bryden G., and Ida S. (2000). In *Protostars and Planets IV* (V. Mannings et al., eds.), pp. 1111–1134. Univ. of Arizona, Tucson.

Lissauer J. J. (1987) *Icarus, 69,* 249–265.

Lissauer J. J. (1993) *Ann. Rev. Astron. Astrophys., 31,* 129–174.

Lissauer J. J. (1995) *Icarus, 114,* 217–236.

Lissauer J. J. (2001) *Nature, 409,* 23–24.

Lissauer J. J. (2004) In *Extrasolar Planets: Today and Tomorrow* (J. P. Beaulieu et al., eds.), pp. 271–271. ASP Conf. Series 321, San Francisco.

Lissauer J. J., Pollack J. B., Wetherill G. W., and Stevenson D. J. (1995) In *Neptune and Triton* (D. P. Cruikshank, ed.), pp. 37–108. Univ. of Arizona, Tucson.

Livio M. and Pringle J. E. (2003) *Mon. Not. Roy. Astron. Soc., 346,* L42–L44.

Marcy G. W., Butler R. P., Fischer D., Vogt S. S., Lissauer J. J., and Rivera E. J. (2001) *Astrophys. J., 556,* 296–301.

Marcy G. W., Butler R. P., Fischer D. A., and Vogt S. S. (2004) In *Extrasolar Planets: Today and Tomorrow* (J. P. Beaulieu et al., eds.), pp. 3–14. ASP Conf. Series 321, San Francisco.

Marcy G. W., Butler R. P., Fischer D. A., Vogt S., Wright J. T., Tinney C. G., and Jones H. R. A. (2005) *Prog. Theor. Phys. Suppl., 158,* 24–42.

Marley M. S., Fortney J. J., Hubickyj O., Bodenheimer P., and Lissauer J. J. (2006) *Astrophys. J.,* in press.

Masset F. S. and Papaloizou J. C. B. (2003) *Astrophys. J., 588,* 494–508.

Mayer L., Quinn T., Wadsley J., and Standel J. (2002) *Science, 298,* 1756–1759.

Mizuno H. (1980) *Prog. Theor. Phys., 64,* 544–557.

Mosqueira I. and Estrada P. R. (2003a) *Icarus, 163,* 198–231.

Mosqueira I. and Estrada P. R. (2003b) *Icarus, 163,* 232–255.

Ohtsuki K., Stewart G. R., and Ida S. (2002) *Icarus, 155,* 436–453.

Owen T., Mahaffy P., Niemann H. B., Atreya S., Donahue T., Bar-Nun A., and de Pater I. (1999) *Nature, 402,* 269–270.

Podolak M. (2003) *Icarus, 165,* 428–437.

Podolak M., Hubbard W. B., and Pollack J. B. (1993) In *Protostars and Planets III* (E. H. Levy and J. I. Lunine, eds.), pp. 1109–1147. Univ. of Arizona, Tucson.

Pollack J. B. and Reynolds R. T. (1974) *Icarus, 21,* 248–253.

Pollack J. B., Burns J. A., and Tauber M. E. (1979) *Icarus, 37,* 587–611.

Pollack J. B., Hubickyj O., Bodenheimer P., Lissauer J. J., Podolak M., and Greenzweig Y. (1996) *Icarus, 124,* 62–85.

Porco C. C., Baker E., Barbara J., Beurle K., Brahic A., et al. (2005) *Science, 307,* 1226–1236.

Quintana E. V. and Lissauer J. J. (2006) *Icarus, 185,* in press.

Quintana E. V., Lissauer J. J., Chambers J. E., and Duncan M. J. (2002) *Astrophys. J., 576,* 982–996.

Rafikov R. R. (2005) *Astrophys. J., 621,* L69.

Rivera E. J., Lissauer J. J., Butler R. P., Marcy G. W., Vogt S. S., et al. (2005) *Astrophys. J., 634,* 625–640.

Safronov V. S. (1960) *Ann. Astrophys., 23,* 979–979.

Safronov V. S. (1969) In *Evolution of the Protoplanetary Cloud and Formation of the Earth and Planets.* Nauka, Moscow. (Translated in 1972 as NASA TTF-677.)

Santos N. C., Israelian G., Mayor M., Rebolo R., and Udry S. (2003) *Astron. Astrophys., 398,* 363–376.

Sato B., Fischer D. A., Henry G. W., Laughlin G., Butler R. P., et al. (2005) *Astrophys. J., 633,* 465–473.

Showalter M. R. (1991) *Nature, 351,* 709–713.

Stevenson D. J. (1982) *Planet. Space Sci., 30,* 755–764.

Stevenson D. J. (1984) In *Lunar Planet. Sci. XV,* pp. 822–823.

Stevenson D. J. (1998) *J. Phys. Condensed Matter, 10,* 11227–11234.

Stevenson D. J. (2006) *Nature, 441,* 34–35.

Stevenson D. J. and Lunine J. I. (1988) *Icarus, 75,* 146–155.

Stewart G. R. and Wetherill G. W. (1988) *Icarus, 74,* 542–553.

Thommes E. W. and Lissauer J. J. (2005) In *Astrophysics of Life* (M. Livio et al., eds.), pp. 41–53. Cambridge Univ., Cambridge.

Thommes E. W., Duncan M. J., and Levison H. F. (1999) *Nature, 402,* 635–638.

Thommes E. W., Duncan M. J., and Levison H. F. (2003) *Icarus, 161,* 431–455.

Tsiganis K., Gomes R., Morbidelli A., and Levison H. F. (2005) *Nature, 435,* 459–461.

Tsui K. H. (2002) *Planet. Space Sci., 50,* 269–276.

Turner J. S. (1973) *Earth Planet. Sci. Lett., 17,* 369–374.

Ward W. R. (1986) *Icarus, 67,* 164–180.

Ward W. R. and Hahn J. (2000) In *Protostars and Planets IV* (V. Mannings et al., eds.), pp. 1135–1155. Univ. of Arizona, Tucson.

Weidenschilling S. J. (1977) *Mon. Not. R. Astron. Soc., 180,* 57–70.

Weidenschilling S. J. (2005) *Space Sci. Rev., 116,* 53–66.

Weidenschilling S. J. and Cuzzi J. N. (1993) In *Protostars and Planets III* (E. H. Levy and J. I. Lunine, eds.), pp. 1031–1060. Univ. of Arizona, Tucson.

Weidenschilling S. J. and Marzari F. (1996) *Nature, 384,* 619–621.

Wetherill G. W. (1980) *Ann. Rev. Astron. Astrophys., 18,* 77–113.

Wetherill G. W. (1990) *Ann. Rev. Earth Planet. Sci., 18,* 205–256.

Wetherill G. W. and Stewart G. R. (1989) *Icarus, 77,* 330–357.

Wuchterl G. and Tscharnuter W. M. (2003) *Astron. Astrophys., 398,* 1081–1090.

Wuchterl G., Guillot T., and Lissauer J. J. (2000) In *Protostars and Planets IV* (V. Mannings et al., eds.), pp. 1081–1109. Univ. of Arizona, Tucson.

Youdin A. N. and Shu F. H. (2002) *Astrophys. J., 580,* 494–505.

Young R. E. (2003) *New Astron. Rev., 47,* 1–51.

Zapatero Osorio M. R., Béjar V. J. S., Martín E. L., Rebolo R., Barrado Y., Navascués D., Bailer-Jones C. A., and Mundt R. (2000) *Science, 290,* 103–107.

Gravitational Instabilities in Gaseous Protoplanetary Disks and Implications for Giant Planet Formation

Richard H. Durisen
Indiana University

Alan P. Boss
Carnegie Institution of Washington

Lucio Mayer
Eidgenössische Technische Hochschule Zürich

Andrew F. Nelson
Los Alamos National Laboratory

Thomas Quinn
University of Washington

W. K. M. Rice
University of Edinburgh

Protoplanetary gas disks are likely to experience gravitational instabilities (GIs) during some phase of their evolution. Density perturbations in an unstable disk grow on a dynamic timescale into spiral arms that produce efficient outward transfer of angular momentum and inward transfer of mass through gravitational torques. In a cool disk with sufficiently rapid cooling, the spiral arms in an unstable disk form self-gravitating clumps. Whether gas giant protoplanets can form by such a disk instability process is the primary question addressed by this review. We discuss the wide range of calculations undertaken by ourselves and others using various numerical techniques, and we report preliminary results from a large multicode collaboration. Additional topics include triggering mechanisms for GIs, disk heating and cooling, orbital survival of dense clumps, interactions of solids with GI-driven waves and shocks, and hybrid scenarios where GIs facilitate core accretion. The review ends with a discussion of how well disk instability and core accretion fare in meeting observational constraints.

1. INTRODUCTION

Gravitational instabilities (GIs) can occur in any region of a gas disk that becomes sufficiently cool or develops a sufficiently high surface density. In the nonlinear regime, GIs can produce local and global spiral waves, self-gravitating turbulence, mass and angular momentum transport, and disk fragmentation into dense clumps and substructure. The particular emphasis of this review article is the possibility (*Kuiper,* 1951; *Cameron,* 1978), recently revived by *Boss* (1997, 1998a), that the dense clumps in a disk fragmented by GIs may become self-gravitating precursors to gas giant planets. This particular idea for gas giant planet formation has come to be known as the disk instability theory. We provide here a thorough review of the physics of GIs as currently understood through a wide variety of techniques and offer tutorials on key issues of physics and methodology. The authors assembled for this paper were deliberately chosen to represent the full range of views on the subject. Although we disagree about some aspects of GIs and about some interpretations of available results, we have labored hard to present a fair and balanced picture. Other recent reviews of this subject include *Boss* (2002c), *Durisen et al.* (2003), and *Durisen* (2006).

2. PHYSICS OF GRAVITATIONAL INSTABILITIES

2.1. Linear Regime

The parameter that determines whether GIs occur in thin gas disks is

$$Q = c_s \kappa / \pi G \Sigma \tag{1}$$

where c_s is the sound speed, κ is the epicyclic frequency at which a fluid element oscillates when perturbed from circular motion, G is the gravitational constant, and Σ is the surface density. In a nearly Keplerian disk, $\kappa \approx$ the rotational angular speed Ω. For axisymmetric (ring-like) disturbances, disks are stable when $Q > 1$ (*Toomre*, 1964). At high Q values, pressure, represented by c_s in equation (1), stabilizes short wavelengths, and rotation, represented by κ, stabilizes long wavelengths. The most unstable wavelength when $Q < 1$ is given by $\lambda_m \approx 2\pi^2 G\Sigma/\kappa^2$.

Modern numerical simulations, beginning with *Papaloizou and Savonije* (1991), show that nonaxisymmetric disturbances, which grow as multiarmed spirals, become unstable for $Q \lesssim 1.5$. Because the instability is both linear and dynamic, small perturbations grow exponentially on the timescale of a rotation period $P_{rot} = 2\pi/\Omega$. The multiarm spiral waves that grow have a predominantly trailing pattern, and several modes can appear simultaneously (*Boss*, 1998a; *Laughlin et al.*, 1998; *Nelson et al.*, 1998; *Pickett et al.*, 1998). Although the star does become displaced from the system center of mass (*Rice et al.*, 2003a) and one-armed structures can occur (see Fig. 1 of *Cai et al.*, 2006), one-armed modes do not play the dominant role predicted by *Adams et al.* (1989) and *Shu et al.* (1990).

2.2. Nonlinear Regime

Numerical simulations (see also sections 3 and 4) show that, as GIs emerge from the linear regime, they may either saturate at nonlinear amplitude or fragment the disk. Two major effects control or limit the outcome: disk thermodynamics and nonlinear mode coupling. At this point, the disks also develop large surface distortions.

2.2.1. Disk thermodynamics. As the spiral waves grow, they can steepen into shocks that produce strong localized heating (*Pickett et al.*, 1998, 2000a; *A. Nelson et al.*, 2000). Gas is also heated by compression and through net mass transport due to gravitational torques. The ultimate source of GI heating is work done by gravity. What happens next depends on whether a balance can be reached between heating and the loss of disk thermal energy by radiative or convective cooling. The notion of a balance of heating and cooling in the nonlinear regime was described as early as the mid-1960s by *Goldreich and Lynden-Bell* (1965) and has been used as a basis for proposing α-treatments for GI-active disks (*Paczyński*, 1978; *Lin and Pringle*, 1987). For slow to moderate cooling rates, numerical experiments, such as in Fig. 1, verify that thermal self-regulation of GIs can be achieved (*Tomley et al.*, 1991; *Pickett et al.*, 1998, 2000a, 2003; *A. Nelson et al.*, 2000; *Gammie*, 2001; *Boss*, 2003; *Rice et al.*, 2003b; *Lodato and Rice*, 2004, 2005; *Mejía et al.* 2005; *Cai et al.*, 2006). Q then hovers near the instability limit, and the nonlinear amplitude is controlled by the cooling rate.

2.2.2. Nonlinear mode coupling. Using second- and third-order governing equations for spiral modes and comparing their results with a full nonlinear hydrodynamics treatment, *Laughlin et al.* (1997, 1998) studied nonlinear mode coupling in the most detail. Even if only a single mode initially emerges from the linear regime, power is quickly

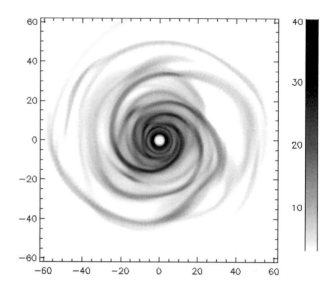

Fig. 1. Grayscale of effective temperature T_{eff} in degrees Kelvin for a face-on GI-active disk in an asymptotic state of thermal self-regulation. This figure is for the *Mejía et al.* (2005) evolution of a 0.07-$M_\odot$ disk around a 0.5-$M_\odot$ star with $t_{cool} = 1$ outer rotation period at 4500 yr. The frame is 120 AU on a side.

distributed over modes with a wide variety of wavelengths and number of arms, resulting in a self-gravitating turbulence that permeates the disk. In this gravitoturbulence, gravitational torques and even Reynolds stresses may be important over a wide range of scales (*Nelson et al.*, 1998; *Gammie*, 2001; *Lodato and Rice*, 2004; *Mejía et al.*, 2005).

2.2.3. Surface distortions. As emphasized by *Pickett et al.* (1998, 2000a, 2003), the vertical structure of the disk plays a crucial role, both for cooling and for essential aspects of the dynamics. There appears to be a relationship between GI spiral modes and the surface or f-modes of stratified disks (*Pickett et al.*, 1996; *Lubow and Ogilvie*, 1998). As a result, except for isothermal disks, GIs tend to have large amplitudes at the surface of the disk. Shock heating in the GI spirals can also disrupt vertical hydrostatic equilibrium, leading to rapid vertical expansions that resemble hydraulic jumps (*Boley et al.*, 2005; *Boley and Durisen*, 2006). The resulting spiral corrugations can produce observable effects (e.g., masers) (*Durisen et al.*, 2001).

2.3. Heating and Cooling

Protoplanetary disks are expected to be moderately thin, with H/r ~ 0.05–0.1, where H is the vertical scale height and r is the distance from the star. For hydrostatic equilibrium in the vertical direction, $H \approx c_s/\Omega$. The ratio of disk internal energy to disk binding energy $\sim c_s^2/(r\Omega)^2 \sim (H/r)^2$ is then $\lesssim 1\%$. As growing modes become nonlinear, they tap the enormous store of gravitational energy in the disk. Simulation of the disk energy budget must be done accurately and include all relevant effects, because it is the disk temperature, through c_s in equation (1), that determines whether the disk becomes or remains unstable, once the central mass, which governs most of κ, and the disk mass distribution Σ have been specified.

2.3.1. Cooling. There have been three approaches to cooling: make simple assumptions about the equation of state (EOS), include idealized cooling characterized by a cooling time, or treat radiative cooling using realistic opacities.

2.3.1.1. Equation of state. This approach has been used to study mode coupling (e.g., *Laughlin et al.*, 1998) and to examine disk fragmentation in the limits of isentropic and isothermal behavior (e.g., *Boss,* 1998a, 2000; *Nelson et al.,* 1998; *Pickett et al.,* 1998, 2003; *Mayer et al.,* 2004a). Isothermal evolution of a disk, where the disk temperature distribution is held fixed in space or when following fluid elements, effectively assumes rapid loss of energy produced by shocks and PdV work. Isentropic evolution, where specific entropy is held fixed instead of temperature, is a more moderate assumption but is still lossy because it ignores entropy generation in shocks (*Pickett et al.,* 1998, 2000a). Due to the energy loss, we do not refer to such calculations as "adiabatic." Here, we restrict adiabatic evolution to mean cases where the fluid is treated as an ideal gas with shock heating included via an artificial viscosity term in the internal energy equation but no radiative cooling. Such calculations are adiabatic in the sense that there is no energy loss by the system. Examples include a simulation in *Pickett et al.* (1998) and simulations in *Mayer et al.* (2002, 2004a). Mayer et al. use adiabatic evolution throughout some simulations, but in others that are started with a locally isothermal EOS, they switch to adiabatic evolution as the disk approaches fragmentation.

2.3.1.2. Simple cooling laws. Better experimental control over energy loss is obtained by adopting simple cooling rates per unit volume $\Lambda = \varepsilon/t_{cool}$, where ε is the internal energy per unit volume. The t_{cool} is specified either as a fixed fraction of the local disk rotation period P_{rot}, usually by setting $t_{cool}\Omega$ = constant (*Gammie,* 2001; *Rice et al.,* 2003b; *Mayer et al.,* 2004b, 2005) or t_{cool} = constant everywhere (*Pickett et al.,* 2003; *Mejía et al.,* 2005). In the *Mayer et al.* (2004b, 2005) work, the cooling is turned off in dense regions to simulate high optical depth. Regardless of t_{cool} prescription, the amplitude of the GIs in the asymptotic state (see Fig. 1), achieved when heating and cooling are balanced, increases as t_{cool} decreases. In addition to elucidating the general physics of GIs, such studies address whether GIs are intrinsically a local or global phenomenon (*Laughlin and Różyczka,* 1996; *Balbus and Papaloizou,* 1999) and whether they can be properly modeled by a simple α prescription. When t_{cool} is globally constant, the transport induced by GIs is global with high mass inflow rates (*Mejía et al.,* 2005; Michael et al., in preparation); when $t_{cool}\Omega$ is constant, transport is local, except for thick or very massive disks, and the inflow rates are well characterized by a constant α (*Gammie,* 2001; *Lodato and Rice,* 2004, 2005).

2.3.1.3. Radiative cooling. The published literature on this so far comes from only three research groups (*A. Nel-*

son et al., 2000; *Boss,* 2001, 2002b, 2004a; *Mejía,* 2004; *Cai et al.,* 2006), but work by others is in progress. Because solar-system-sized disks encompass significant volumes with small and large optical depth, this becomes a difficult three-dimensional radiative hydrodynamics problem. Techniques will be discussed in section 3.2. For a disk spanning the conventional planet-forming region, the opacity is due primarily to dust. Complications that have to be considered include the dust size distribution; its composition, grain growth, and settling; and the occurrence of fast cooling due to convection.

Let us consider how the radiative cooling time depends on temperature T and metallicity Z. Let $\kappa_r \sim ZT^{\beta_r}$ and $\kappa_p \sim ZT^{\beta_p}$ be the Rosseland and Planck mean opacities, respectively, and let $\tau \sim \kappa_r$ be the vertical optical depth to the midplane. For large τ

$$t_{cool} \sim T/T_{eff}^4 \sim T^{-3}\tau \sim T^{-3+\beta_r}Z \tag{2}$$

for small τ

$$t_{cool} \sim T/\kappa_p T^4 \sim T^{-3-\beta_p}/Z \tag{3}$$

For most temperature regimes, where no major dust constituent is condensing or vaporizing, we expect $-3 < \beta < 3$, so t_{cool} increases as T decreases. As Z increases, t_{cool} increases in optically thick regions, but decreases in optically thin ones.

2.3.2. Heating. In addition to the internal heating caused by GIs through shocks, compression, and mass transport, there can be heating due to turbulent dissipation (*A. Nelson et al.,* 2000) and other sources of shocks. In addition, a disk may be exposed to one or more external radiation fields due to a nearby OB star (e.g., *Johnstone et al.,* 1998), an infalling envelope (e.g., *D'Alessio et al.,* 1997), or the central star (e.g., *Chiang and Goldreich,* 1997). These forms of heat input can be comparable to or larger than internal sources of heating and can influence Q and the surface boundary conditions. Only crude treatments have been done so far for envelope irradiation (*Boss,* 2001, 2002b; *Cai et al.,* 2006) and for stellar irradiation (*Mejía,* 2004).

2.4. Fragmentation

As shown first by *Gammie* (2001) for local thin-disk calculations and later confirmed by *Rice et al.* (2003b) and *Mejía et al.* (2005) in full three-dimensional hydro simulations, disks with a fixed t_{cool} fragment for sufficiently fast cooling, specifically when $t_{cool}\Omega \lesssim 3$, or, equivalently, $t_{cool} \lesssim P_{rot}/2$. Finite thickness has a slight stabilizing influence (*Rice et al.,* 2003b; *Mayer et al.,* 2004a). When dealing with realistic radiative cooling, one cannot apply this simple fragmentation criterion to arbitrary initial disk models. One has to apply it to the asymptotic phase after nonlinear behavior is well developed (*Johnson and Gammie,* 2003). Cooling times can be much longer in the asymptotic state than they are initially (*Cai et al.,* 2006, Boley et al., in preparation).

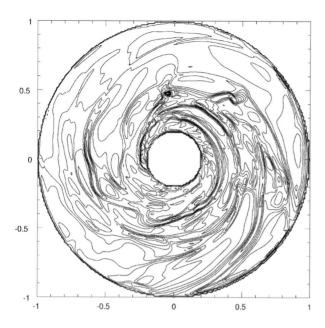

Fig. 2. Midplane density contours for the isothermal evolution of a 0.09-$M_\odot$ disk around a 1-$M_\odot$ star. A multi-Jupiter mass clump forms near 12 o'clock by 374 yr. The frame in the figure is 40 AU on a side. Adapted from *Boss* (2000).

For disks evolved under isothermal conditions, where a simple cooling time cannot be defined, local thin-disk calculations show fragmentation when $Q \lesssim 1.4$ (*Johnson and Gammie*, 2003). This is roughly consistent with results from global simulations (e.g., *Boss*, 2000; *Nelson et al.*, 1998; *Pickett et al.*, 2000a, 2003; *Mayer et al.*, 2002, 2004a). Figure 2 shows a classic example of a fragmenting disk.

Although there is agreement on conditions for fragmentation, two important questions remain. Do real disks ever cool fast enough for fragmentation to occur, and do the fragments last long enough to contract into permanent protoplanets before being disrupted by tidal stresses, shear stresses, physical collisions, and shocks?

3. NUMERICAL METHODS

A full understanding of disk evolution and the planet formation process cannot easily be obtained using a purely analytic approach. Although numerical methods are powerful, they have flaws and limitations that must be taken into account when interpreting results. Here we describe some commonly used numerical techniques and their limitations.

3.1. Hydrodynamics

Numerical models have been implemented using one or the other of two broad classes of techniques to solve the hydrodynamic equations. Each class discretizes the system in fundamentally different ways. On one hand, there are particle-based simulations using smoothed particle hydrodynamics (SPH) (*Benz*, 1990; *Monaghan*, 1992), and, on

the other, grid-based techniques (e.g., *Tohline*, 1980; *Fryxell et al.*, 1991; *Stone and Norman*, 1992; *Boss and Myhill*, 1992; *Pickett*, 1995).

SPH uses a collection of particles distributed in space to represent the fluid. Each particle is free to move in response to forces acting on it, so that the particle distribution changes with the system as it evolves. The particles are collisionless, meaning that they do not represent actual physical entities, but rather points at which the underlying distributions of mass, momentum, and energy are sampled. In order to calculate hydrodynamic quantities such as mass density or pressure forces, contributions from other particles within a specified distance, the smoothing length, are weighted according to a smoothing kernel and summed in pairwise fashion. Mutual gravitational forces are calculated by organizing particles into a tree, where close particles are treated more accurately than aggregates on distant branches.

Grid-based methods use a grid of points, usually fixed in space, on which fluid quantities are defined. In the class of finite difference schemes, fluxes of mass, momentum, and energy between adjacent cells are calculated by taking finite differences of the fluid quantities in space. Although not commonly used in simulations of GIs, the piecewise parabolic method (PPM) of *Collela and Woodward* (1984) represents an example of the class of finite volume schemes. For our purposes, an important distinguishing factor is that while finite difference and SPH methods may require artificial viscosity terms to be added to the equations to ensure numerical stability and produce correct dissipation in shocks, PPM does not.

3.2. Radiative Physics

In section 2.3, we describe a number of processes by which disks may heat and cool. In this section, we discuss various code implementations and their limitations.

Fixed EOS evolution is computationally efficient because it removes the need to solve an equation for the energy balance. On the other hand, the gas instantly radiates away all heating due to shocks and, for the isothermal case, due to compressional heating as well. As a consequence, the gas may compress to much higher densities than are realistic, biasing a simulation toward GI growth and fragmentation even when a physically appropriate temperature or entropy scale is used. Although fixed t_{cool} represents a clear advance over fixed EOS, equations (2) and (3) show that increasing the temperature, which makes the disk more stable, also decreases t_{cool}. So it is incorrect to view short global cooling times as necessarily equivalent to more rapid GI growth and fragmentation. In order for fragmentation to occur, one needs both a short t_{cool} and a disk that is cool enough to be unstable (e.g., *Rafikov*, 2005).

The most physically inclusive simulations to date employ radiative transport schemes that allow t_{cool} to be determined by disk opacity. Current implementations (section 2.3) employ variants of a radiative diffusion approximation in regions of medium to high optical depth τ, integrated from infinity toward the disk midplane. On the other hand, radiative losses actually occur from regions where $\tau \lesssim 1$, and so the treatment of the interface between optically thick and thin regions strongly influences cooling. Three groups have implemented different approaches.

A. Nelson et al. (2000) assume that the vertical structure of the disk can be defined at each point as an atmosphere in thermal equilibrium. In this limit, the interface can be defined by the location of the disk photosphere, where $\tau = 2/3$ (see, e.g., *Mihalas*, 1977). Cooling at each point is then defined as that due to a blackbody with the temperature of the photosphere. *Boss* (2001, 2002b, 2004a, 2005) performs a three-dimensional radiative diffusion treatment for the optically thick disk interior (*Bodenheimer et al.*, 1990), coupled to an outer boundary condition where the temperature is set to a constant for $\tau < 10$, τ being measured along the radial direction. *Mejía* (2004) and *Cai et al.* (2006) use a similar radiative diffusion treatment in their disk interior, but they define the interface using $\tau = 2/3$, measured vertically, above which an optically thin atmosphere model is self-consistently grafted onto the outward flux from the interior. As discussed in section 4.2, results for the three groups differ markedly, indicating that better understanding of radiative cooling at the disk surface will be required to determine the fate of GIs.

3.3. Numerical Issues

The most important limitations facing numerical simulations are finite computational resources. Simulations have a limited duration with a finite number of particles or cells, and they must have boundary conditions to describe behavior outside the region being computed. A simulation must distribute grid cells or particles over the interesting parts of the system to resolve the relevant physics and avoid errors associated with incorrect treatment of the boundaries. Here we describe a number of requirements for valid simulations and pitfalls to be avoided.

For growth of GIs, simulations must be able to resolve the wavelengths of the instabilities underlying the fragmentation. *Bate and Burkert* (1997) and *Truelove et al.* (1997) each define criteria based on the collapse of a Jeans unstable cloud that links a minimum number of grid zones or particles to either the physical wavelength or mass associated with Jeans collapse. *Nelson* (2006) notes that a Jeans analysis may be less relevant for disk systems because they are flattened and rotating rather than homogeneous and instead proposes a criterion based on the Toomre wavelength in disks. Generally, grid-based simulations must resolve the appropriate local instability wavelength with a minimum of four to five grid zones in each direction, while SPH simulations must resolve the local Jeans or Toomre mass with a minimum of a few hundred particles.

Resolution of instability wavelengths will be insufficient to ensure validity if either the hydrodynamics or gravitational forces are in error. For example, errors in the hydrodynamics may develop in SPH and finite-difference methods

because a viscous heating term must be added artificially to model shock dissipation and, in some cases, to ensure numerical stability. In practice, the magnitude of dissipation depends in part on cell dimensions rather than just on physical properties. Discontinuities may be smeared over as many as ~10 or more cells, depending on the method. Further, *Mayer et al.* (2004a) have argued that, because it takes the form of an additional pressure, artificial viscosity may by itself reduce or eliminate fragmentation. On the other hand, artificial viscosity can promote the longevity of clumps (see Fig. 3 of *Durisen*, 2006).

Gravitational force errors develop in grid simulations from at least two sources. First, when *Pickett et al.* (2003) place a small blob within their grid, errors occur in the self-gravitation force of the blob that depend on whether the cells containing it have the same spacing in each coordinate dimension. Ideally, grid zones would have comparable spacing in all directions, but disks are both thin and radially extended. Use of spherical and cylindrical grids tends to introduce disparity in grid spacing. Second, *Boss* (2000) shows that maximum densities inside clumps are enhanced by orders of magnitude as additional terms in his Poisson solver, based on a Y_{lm} decomposition, are included. SPH simulations encounter a different source of error because gravitational forces must be softened in order to preserve the collisionless nature of the particles. *Bate and Burkert* (1997) and *Nelson* (2006) each show that large imbalances between the gravitational and pressure forces can develop if the length scales for each are not identical, possibly inducing fragmentation in simulations. On the other hand, spatially and temporally variable softening implies a violation of energy conservation. Quantifying errors from sources such as insufficiently resolved shock dissipation or gravitational forces cannot be reliably addressed except by experimentation. Results of otherwise identical simulations performed at several resolutions must be compared, and identical models must be realized with more than one numerical method (as in section 4.4), so that deficiencies in one method can be checked against strengths in another.

The disks relevant for GI growth extend over several orders of magnitude in radial range, while GIs may develop large amplitudes only over some fraction of that range. Computationally affordable simulations therefore require both inner and outer radial boundaries, even though the disk may spread radially and spiral waves propagate up to or beyond those boundaries. In grid-based simulations, *Pickett et al.* (2000b) demonstrate that numerically induced fragmentation can occur with incorrect treatment of the boundary. Studies of disk evolution must ensure that treatment of the boundaries does not produce artificial effects.

In particle simulations, where there is no requirement that a grid be fixed at the beginning of the simulation, boundaries are no less a problem. The smoothing in SPH requires that the distribution of neighbors over which the smoothing occurs be relatively evenly distributed in a sphere around each particle for the hydrodynamic quantities to be well defined. At currently affordable resolutions (~10^5–10^6 particles), the smoothing kernel extends over a large fraction of a disk scale height, so meeting this requirement is especially challenging. Impact on the outcomes of simulations has not yet been quantified.

4. KEY ISSUES

4.1. Triggers for Gravitational Instabilities

When disks become unstable, they may either fragment or enter a self-regulated phase depending on the cooling time. It is therefore important to know how and when GIs may arise in real disks and the physical state of the disk at that time. Various mechanisms for triggering GIs are conceivable, but only a few have yet been studied in any detail. Possibilities include the formation of a massive disk from the collapse of a protostellar cloud (e.g., *Laughlin and Bodenheimer,* 1994; *Yorke and Bodenheimer,* 1999), clumpy infall onto a disk (*Boss,* 1997, 1998a), cooling of a disk from a stable to an unstable state, slow accretion of mass, accumulation of mass in a magnetically dead zone, perturbations by a binary companion, and close encounters with other star/disk systems (*Boffin et al.,* 1998; *Lin et al.,* 1998). A few of these will be discussed further, with an emphasis on some new results on effects of binarity.

Several authors start their disks with stable or marginally stable Q values and evolve them to instability either by slow idealized cooling (e.g., *Gammie,* 2001; *Pickett et al.,* 2003; *Mejía et al.,* 2005) or by more realistic radiative cooling (e.g., *Johnson and Gammie,* 2003; *Boss,* 2005, 2006; *Cai et al.,* 2006). To the extent tested, fragmentation in idealized cooling cases are consistent with the Gammie criterion (section 2.4). With radiative cooling, as first pointed out by *Johnson and Gammie* (2003), it is difficult to judge whether a disk will fragment when it reaches instability based on its initial t_{cool}. When *Mayer et al.* (2004a) grow the mass of a disk while keeping its temperature constant, dense clumps form in a manner similar to clump formation starting from an unstable disk. A similar treatment of accretion needs to be done using realistic radiative cooling. Simulations like these suggest that, in the absence of a strong additional source of heating, GIs are unavoidable in protoplanetary disks with sufficient mass (~0.1 $M_\odot$ for a ~1-$M_\odot$ star).

A disk evolving primarily due to magnetorotational instabilites (MRIs) may produce rings of cool gas in the disk midplane where the ionization fraction drops sufficiently to quell MRIs (*Gammie,* 1996; *Fleming and Stone,* 2003). Dense rings associated with these magnetically dead zones should become gravitationally unstable and may well trigger a localized onset of GIs. This process might lead to disk outbursts related to FU Orionis events (*Armitage et al.,* 2001) and induce chondrule-forming episodes (*Boley et al.,* 2005).

A phase of GIs robust enough to lead to gas giant protoplanet formation might be achieved through external trig-

gers, like a binary star companion or a close encounter with another protostar and its disk. A few studies have explored the effects of binary companions on GIs. *Nelson* (2000) follows the evolution of disks in an equal-mass binary system with a semimajor axis of 50 AU and an eccentricity of 0.3 and finds that the disks are heated by internal shocks and viscous processes to such an extent as to become too hot for gas giant planet formation either by disk GIs or by core accretion, because volatile ices and organics are vaporized. In a comparison of the radiated emission calculated from his simulation to those from the L1551 IRS5 system, *Nelson* (2000) finds that the simulation is well below the observed system and therefore that the temperatures in the simulation are underestimates. He therefore concludes that "planet formation is unlikely in equal-mass binary systems with a ~ 50 AU." Currently, over two dozen binary or triple star systems have known extrasolar planets, with binary separations ranging from ~10 AU to ~10^3 AU, so some means must be found for giant planet formation in binary star systems with relatively small semimajor axes.

Using idealized cooling, *Mayer et al.* (2005) find that the effect of binary companions depends on the mass of the disks involved and on the disk cooling rate. For a pair of massive disks (M ~ 0.1 $M_\odot$), formation of permanent clumps can be suppressed as a result of intense heating from spiral shocks excited by the tidal perturbation (left panel of Fig. 3). Clumps do not form in such disks for binary orbits having a semimajor axis of ~60 AU even when $t_{cool} < P_{rot}$. The temperatures reached in these disks are >200 K and would vaporize water ice, hampering core accretion, as argued by *Nelson* (2000). On the other hand, pairs of less-massive disks (M ~ 0.05 $M_\odot$) that would not fragment in isolation since they start with Q ~ 2 can produce permanent clumps provided that $t_{cool} \lesssim P_{rot}$. This is because the tidal perturbation is weaker in this case (each perturber is less massive) and the resulting shock heating is thus diminished. Finally, the behavior of such binary systems approaches that seen in simulations of isolated disks once the semimajor axis grows beyond 100 AU (right panel of Fig. 3).

Calculations by *Boss* (2006) of the evolution of initially marginally gravitationally stable disks show that the presence of a binary star companion could help to trigger the formation of dense clumps. The most likely explanation for the difference in outcomes between the models of *Nelson* (2000) and *Boss* (2006) is the relatively short cooling times in the latter models (~1 to 2 P_{rot}) (see *Boss,* 2004a) compared to the effective cooling time in *Nelson* (2000) of ~15 P_{rot} at 5 AU, dropping to ~P_{rot} at 15 AU. Similarly, some differences in outcomes between the results of *Boss* (2006) and *Mayer et al.* (2005) can be expected based on different choices of the binary semimajor axes and eccentricities and differences in the thermodynamics. For example, *Mayer et al.* (2005) turn off cooling in regions with densities higher than 10^{-10} g cm^{-3} to account for high optical depths.

Overall, the three different calculations agree that excitation or suppression of fragmentation by a binary companion

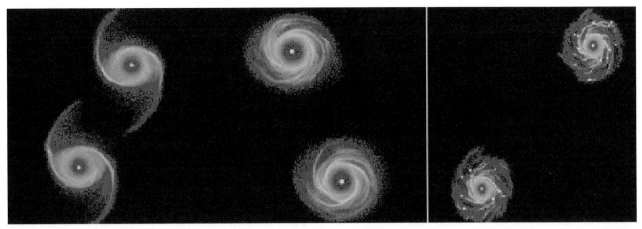

Fig. 3. Face-on density maps for two simulations of interacting $M = 0.1\ M_\odot$ protoplanetary disks in binaries with $t_{cool} = 0.5\ P_{rot}$ viewed face-on. The binary in the left panel has a nearly circular binary orbit with an initial separation of 60 AU and is shown after first pericentric passage at 150 yr (left) and then at 450 yr (right). Large tidally induced spiral arms are visible at 150 yr. The right panel shows a snapshot at 160 yr from a simulation starting from an initial orbital separation that is twice as large. In this case, fragmentation into permanent clumps occurs after a few disk orbital times. Adapted from *Mayer et al.* (2005).

depends sensitively on the balance between compressional/shock heating and cooling. This balance appears to depend on the mass of the disks involved. Interestingly, lighter disks are more likely to fragment in binary systems according to both *Mayer et al.* (2005) and *Boss* (2006).

4.2. Disk Thermodynamics

As discussed in sections 2.2 and 4.1, heating and cooling are perhaps the most important processes affecting the growth and fate of GIs. Thermal regulation in the nonlinear regime leads naturally to systems near their stability limit where temporary imbalances in one heating or cooling term lead to a proportionate increase in a balancing term. For fragmentation to occur, a disk must cool quickly enough, or fail to be heated for long enough, to upset this self-regulation. A complete model of the energy balance that includes all relevant processes in a time-dependent manner is beyond the capabilities of the current generation of models. It requires knowledge of all the following: external radiation sources and their influence on the disk at each location, the energy loss rate of the disk due to radiative cooling, dynamical processes that generate thermal energy through viscosity or shocks, and a detailed equation of state to determine how much heating any of those dynamical processes generate. Recent progress toward understanding disk evolution has focused on the more limited goals of quantifying the sensitivity of results to various processes in isolation.

In a thin, steady state α-disk, the heating and cooling times are the same and take a value (*Pringle,* 1981; *Gammie,* 2001)

$$t_{cool} = \frac{4}{9}[\gamma(\gamma-1)\alpha\Omega]^{-1} \qquad (4)$$

For $\alpha \sim 10^{-2}$ and $\gamma = 1.4$, equation (4) gives $\sim 12\ P_{rot}$. This

is a crude upper limit on the actual timescale required to change the disk thermodynamic state. External radiative heating from the star and any remaining circumstellar material can contribute a large fraction of the total heating (*D'Alessio et al.,* 1998; *A. Nelson et al.,* 2000), as will any internal heating due to globally generated dynamical instabilities that produce shocks. Each of these processes actually makes the disk more stable by heating it, but, as a consequence, dynamical evolution slows until the disk gains enough mass to become unstable again. The marginally stable state will then be precariously held because the higher temperatures mean that all the heating and cooling timescales, i.e., the times required to remove or replace all the disk thermal energy, are short (equations (2) and (3)). When the times are short, any disruption of the contribution from a single source may be able to change the thermodynamic state drastically within only a few orbits, perhaps beyond the point where balance can be restored.

A number of models (section 2.3) have used fixed EOS evolution instead of a full solution of an energy equation to explore disk evolution. A fixed EOS is equivalant to specifying the outcomes of all heating and cooling events that may occur during the evolution, short-circuiting thermal feedback. If, for example, the temperature or entropy is set much too high or too low, a simulation may predict either that no GIs develop in a system, or that they inevitably develop and produce fragmentation, respectively. Despite this limitation, fixed EOS have been useful to delineate approximate boundaries for regions of marginal stability. Since the thermal state is fixed, disk stability (as quantified by equation (1)) is essentially determined by the disk's mass and spatial dimensions, through its surface density. Marginal stability occurs generally at $Q \approx 1.2$ to 1.5 for locally isentropic evolutions, with a tendency for higher Q being required to ensure stability with softer EOS (i.e., with lower γ values) (*Boss*, 1998a; *Nelson et al.*, 1998; *Pickett et al.*, 1998, 2000a; *Mayer et al.*, 2004a). At temperatures appropriate for observed systems (e.g., *Beckwith et al.*, 1990), these Q values correspond to disks more massive than ~0.1 $M_\star$ or surface densities $\Sigma \gtrsim 10^3$ g/cm^2.

As with their fixed EOS cousins, models with fixed t_{cool} can quantify boundaries at which fragmentation may set in. They represent a clear advance over fixed EOS evolution by allowing thermal energy generated by shocks or compression to be retained temporarily, and thereby enabling the disk's natural thermal regulation mechanisms to determine the evolution. Models that employ fixed cooling times can address the question of how violently the disk's thermal regulation mechanisms must be disrupted before they can no longer return the system to balance. An example of the value of fixed t_{cool} calculations is the fragmentation criterion $t_{cool} \lesssim 3\Omega^{-1}$ (see section 2.4).

The angular momentum transport associated with disk self-gravity is a consequence of the gravitational torques induced by GI spirals (e.g., *Larson*, 1984). The viscous α parameter is actually a measure of that stress normalized by the local disk pressure. As shown in equation (4) and reversing the positions of t_{cool} and α, the stress in a self-gravitating disk depends on the cooling time and on the EOS through the specific heat ratio. As long as the dimensionless scale height is $H \lesssim 0.1$, global simulations by *Lodato and Rice* (2004) with $t_{cool}\Omega$ = constant confirm Gammie's assumption that transport due to disk self-gravity can be modeled as a local phenomenon and that equation (4) is accurate. *Gammie* (2001) and *Rice et al.* (2005) show that there is a maximum stress that can be supplied by such a quasisteady, self-gravitating disk. Fragmentation occurs if the stress required to keep the disk in a quasisteady state exceeds this maximum value. The relationship between the stress and the specific heat ratio γ results in the cooling time required for fragmentation increasing as γ decreases. For γ = 7/5, the cooling time below which fragmentation occurs may be more like 2 P_{rot}, not the $3/\Omega \approx P_{rot}/2$ obtained for γ = 2 (*Gammie*, 2001; *Mayer et al.*, 2004b; *Rice et al.*, 2005).

Important sources of stress and heating in the disk that lie outside the framework of Gammie's local analysis are global gravitational torques due to low-order GI spiral modes. There are two ways this can happen: a geometrically thick massive disk (*Lodato and Rice*, 2005) and a fixed global t_{cool} = constant (*Mejía et al.*, 2005). Disks then initially produce large-amplitude spirals, resulting in a transient burst of global mass and angular momentum redistribution. For t_{cool} = constant and moderate masses, disks eventually settle into a self-regulated asymptotic state, but with gravitational stresses dominated by low-order global modes (Michael et al., in preparation). For the very massive $t_{cool}\Omega$ = constant disks, recurrent episodic redistributions occur. In all these cases, the heating in spiral shocks is spatially and temporally very inhomogeneous, as are fluctuations in all thermodynamic variables and the velocity field.

The most accurate method to determine the internal thermodynamics of the disk is to couple the equations of radiative transport to the hydrodynamics directly. All heating or cooling due to radiation will then be properly defined by the disk opacity, which depends on local conditions. This is important because some fraction of the internal heating will be highly inhomogeneous, occurring predominantly in compressions and shocks as gas enters a high-density spiral structure, or at high altitudes where waves from the interior are refracted and steepen into shocks (*Pickett et al.*, 2000a) and where disks may be irradiated (*Mejía*, 2004; *Cai et al.*, 2006). Temperatures and the t_{cool} that depend on them will then be neither simple functions of radius, nor a single globally defined value. Depending on whether the local cooling time of the gas inside the high-density spiral structure is short enough, fragmentation will be more or less likely, and additional hydrodynamic processes such as convection may become active if large enough gradients can be generated.

Indeed, recent simulations of *Boss* (2002a, 2004a) suggest that vertical convection is active in disks when radiative transfer is included, as expected for high τ according to *Ruden and Pollack* (1991). This is important because convection will keep the upper layers of the disk hot, at the expense of the dense interior, so that radiative cooling is more efficient and fragmentation is enhanced. The results have not yet been confirmed by other work and therefore remain somewhat controversial. Simulations by *Mejía* (2004) and *Cai et al.* (2006) are most similar to those of *Boss* (2002a, 2004a) and could have developed convection sufficient to induce fragmentation, but none seems to occur. No fragmentation occurs in *A. Nelson et al.* (2000) either, where convection is implicitly assumed to be efficient through their assumption that the entropy of each vertical column is constant. Recent reanalysis of their results reveals $t_{cool} \sim 3$ to 10 P_{rot}, depending on radius, which is too long to allow fragmentation. These t_{cool} are in agreement with those seen by *Cai et al.* (2006) and by Boley et al. (in preparation) for solar metallicity. The *A. Nelson et al.* (2000) results are also interesting because their comparison of the radiated output to SEDs observed for real systems demonstrates that substantial additional heating beyond that supplied by GIs is required to reproduce the observations, perhaps further inhibiting fragmentation in their models. However, using the same temperature distribution between 1 and 10 AU now used in Boss's GI models, combined with temperatures outside this region taken from models by *Adams et al.* (1988), *Boss and Yorke* (1996) are able to reproduce the SED of the T Tauri system. It is unclear at present why their results differ from those of *A. Nelson et al.* (2000).

The origins of the differences between the three studies are uncertain, but possibilities include differences of both numerical and physical origin. The boundary treatment at the optically thick/thin interface is different in each case (see section 3.2), influencing the efficiency of cooling, as are the numerical methods and resolutions. Boss and the Cai/Mejía group each use three-dimensional grid codes but with spherical and cylindrical grids respectively, and each with a different distribution of grid zones, while Nelson et al. use a two-dimensional SPH code. Perhaps significantly, Cai/Mejía assume their ideal gas has $\gamma = 5/3$ while Boss adopts an EOS that includes rotational and vibrational states of hydrogen, so that $\gamma \approx 7/5$ for typical disk conditions. It is possible that differences in the current results may be explained if the same sensitivities to γ seen in fixed EOS and fixed cooling simulations also hold when radiative transfer is included. Boss and Cai (in preparation) are now conducting direct comparison calculations to isolate the cause of their differences. The preliminary indication is that the radiative boundary conditions may be the critical factor.

Discrepant results for radiatively cooled models should not overshadow the qualitative agreement reached about the relationship between disk thermodynamics and fragmentation. If the marginally unstable state of a self-regulated disk is upset quickly enough by an increase in cooling or decrease in heating, the disk may fragment. What is still very unclear is whether such conditions can develop in real planet-forming disks. It is key to develop a full three-dimensional portrait of the disk surface, so that radiative heating and cooling sources may be included self-consistently in numerical models. Important heating sources will include the envelope, the central star, neighboring stars, and self-heating from other parts of the disk, all of which will be sensitive to shadowing caused by corrugations in the disk surface that develop and change with time due to the GIs themselves. Preliminary studies of three-dimensional disk structure (*Boley and Durisen,* 2006) demonstrate that vertical distortions, analogous to hydraulic jumps, will in fact develop (see also *Pickett et al.,* 2003). If these corrugations are sufficient to cause portions of the disk to be shadowed, locally rapid cooling may occur in the shadowed region, perhaps inducing fragmentation.

An implicit assumption of the discussion above is that the opacity is well known. In fact, it is not. The dominant source of opacity is dust, whose size distribution, composition, and spatial distribution will vary with time (*Cuzzi et al.,* 2001; *Klahr,* 2003) (see also section 5 below), causing the opacity to vary as a result. So far, no models of GI evolution have included effects from any of these processes, except that Nelson et al. model dust destruction while Cai and Mejía consider opacity due to large grains. Possible consequences are a misidentification of the disk photospheric surface if dust grains settle toward the midplane, or incorrect radiative transfer rates in optically thick regions if the opacities themselves are in error.

4.3. Orbital Survival of Clumps

Once dense clumps form in a gravitationally unstable disk, the question becomes one of survival: Are they transient structures or permanent precursors of giant planets? Long-term evolution of simulations that develop clumps is difficult because it requires careful consideration of not only the large-scale dynamical processes that dominate formation but also physical processes that exert small influences over long timescales (e.g., migration and transport due to

viscosity). It also requires that boundary conditions be handled gracefully in cases where a clump or the disk itself tries to move outside the original computational volume.

On a more practical level, the extreme computational cost of performing such calculations limits the time over which systems may be simulated. As a dense clump forms, the temperatures, densities, and fluid velocities within it all increase. As a result, time steps, limited by the Courant condition, can decrease to as little as minutes or hours as the simulation attempts to resolve the clump's internal structure. So far only relatively short integration times of up to a few $\times$ 10^3 yr have been possible. Here, we will focus on the results of simulations and refer the reader to the chapters by Papaloizou et al. and Levison et al. for discussions of longer-term interactions.

In the simplest picture of protoplanet formation via GIs, structures are assumed to evolve along a continuum of states that are progressively more susceptible to fragmentation, presumably ending in one or more bound objects that eventually become protoplanets. *Pickett et al.* (1998, 2000a, 2003) and *Mejía et al.* (2005) simulate initially smooth disks subject to growth of instabilities and, indeed, find growth of large-amplitude spiral structures that later fragment into arclets or clumps. Instead of growing more and more bound, however, these dense structures are sheared apart by the background flow within an orbit or less, especially when shock heating is included via an artificial viscosity. This suggests that a detailed understanding of the thermodynamics inside and outside the fragments is critical for understanding whether fragmentation results in permanently bound objects.

Assuming that permanently bound objects do form, two additional questions emerge. First, how do they accrete mass and how much do they accrete? Second, how are they influenced by the remaining disk material? Recently, *Mayer et al.* (2002, 2004a) and *Lufkin et al.* (2004) have used SPH calculations to follow the formation and evolution of clumps in simulations covering up to 50 orbits (roughly 600 yr), and Mayer et al. (in preparation) are extending these calculations to several thousand years. They find that, when a locally isothermal EOS is used well past initial fragmentation, clumps grow to ~10 M_J within a few hundred years. On the other hand, in simulations using an ideal gas EOS plus bulk viscosity, accretion rates are much lower ($<10^{-6}$ $M_\odot$/ yr), and clumps do not grow to more than a few M_J or ~1% of the disk mass. The assumed thermodynamic treatment has important effects not only on the survival of clumps, but also on their growth.

Nelson and Benz (2003), using a grid-based code and starting from a 0.3-M_J seed planet, show that accretion rates this fast are unphysically high because the newly accreted gas cannot cool fast enough, even with the help of convection, unless some localized dynamical instability is present in the clump's envelope. So, the growth rate of an initially small protoplanet may be limited by its ability to accept additional matter rather than the disk's ability to supply it. They note (see also *Lin and Papaloizou,* 1993; *Bryden et al.,* 1999, 2000; *Kley,* 1999; *Lubow et al.,* 1999; *R. Nelson et al.,*

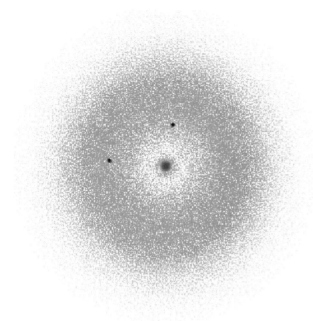
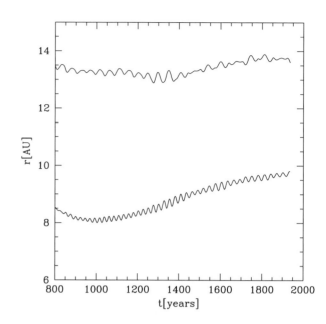

Fig. 4. The orbital evolution of two clumps (right) formed in a massive, growing protoplanetary disk simulation described in *Mayer et al.* (2004). A face-on view of the system after 2264 yr of evolution is shown on the left, using a grayscale density map (the box is 38 AU on a side). In the right panel, the orbital evolution of the two clumps is shown. Overall, both clumps migrate outward.

2000) that the accretion process after formation is self-limiting at a mass comparable to the largest planet masses yet discovered (see chapter by Udry et al.).

Figure 4 shows one of the extended Mayer et al. simulations, containing two clumps in one disk realized with 2×10^5 particles, and run for about 5000 yr (almost 200 orbits at 10 AU). There is little hint of inward orbital migration over a timescale of a few thousand years. Instead, both clumps appear to migrate slowly outward. *Boss* (2005) uses sink particles ("virtual planets") to follow a clumpy disk for about 1000 yr. He also finds that the clumps do not migrate rapidly. In both works, the total simulation times are quite short compared to the disk lifetime and so are only suggestive of the longer-term fate of the objects. Nevertheless, the results are important because they illustrate shortcomings in current analytic models of migration.

Although migration theory is now extremely well developed (see chapter by Papaloizou et al.), predictions for migration at the earliest phases of protoplanet formation by GIs are difficult to make, because many of the assumptions on which the theory is based are not well satisfied. More than one protoplanet may form in the same disk, they may form with masses larger than linear theory can accommodate, and they may be significantly extended rather than the point masses assumed by theory. If the disk remains massive, it may also undergo gravitoturbulence that changes the disk's mass distribution on a short enough timescale to call into question the resonance approximations in the theory. If applicable in the context of these limitations, recent investigations into the character of corotation resonances (see chapter by Papaloizou et al.) and vortex excitation (*Koller et al.*, 2003) in the corotation region may be of particular

interest, because a natural consequence of these processes is significant mass transport across the clump's orbit and reduced inward migration, which is in fact seen in the above simulations.

4.4. Comparison Test Cases

Disk instability has been studied so far with various types of grid codes and SPH codes that have different relative strengths and weaknesses (section 3). Whether different numerical techniques find comparable results with nearly identical assumptions is not yet known, although some comparative studies have been attempted (*Nelson et al.*, 1998). Several aspects of GI behavior can be highly dependent on code type. For example, SPH codes require artificial viscosity to handle shocks such as those occurring along spiral arms. Numerical viscosity can smooth out the velocity field in overdense regions, possibly inhibiting collapse (*Mayer et al.*, 2004a) but, at the same time, possibly increasing clump longevity if clumps form (see Fig. 3 of *Durisen*, 2006). Gravity solvers that are both accurate and fast are a robust feature of SPH codes, while gravity solvers in grid codes can under-resolve the local self-gravity of the gas (*Pickett et al.*, 2003). Both types of codes can lead to spurious fragmentation or suppress it when a force imbalance between pressure and gravity results at scales comparable to the local Jeans or Toomre length due to lack of resolution (*Truelove et al.*, 1997; *Bate and Burkert*, 1997; *Nelson*, 2006).

Another major code difference is in the setup of initial conditions. Although both Eulerian grid-based and Lagrangian particle-based techniques represent an approximation to the continuum fluid limit, noise levels due to discreteness are typically higher in SPH simulations. Initial perturbations are often applied in grid-based simulations to seed GIs (either random or specific modes or both) (e.g., *Boss,* 1998a), but are not required in SPH simulations, because they already have built-in Poissonian noise at the level of $\sqrt{N}/N$ or more, where N is the number of particles. In addition, the SPH calculation of hydrodynamic variables introduces small-scale noise at the level of $1/N_{neigh}$, where N_{neigh} is the number of neighboring particles contained in one smoothing kernel. Grid-based simulations require boundary conditions that restrict the dynamic range of the simulations. For example, clumps may reach the edge of a computational volume after only a limited number of orbits (*Boss,* 1998a, 2000; *Pickett et al.,* 2000a). Cartesian grids can lead to artificial diffusion of angular momentum in a disk, a problem that can be avoided using a cylindrical grid (*Pickett et al.,* 2000a) or spherical grid (*Boss and Myhill,* 1992). *Myhill and Boss* (1993) find good agreement between spherical and Cartesian grid results for a nonisothermal rotating protostellar collapse problem, but evolution of a nearly equilibrium disk over many orbits in a Cartesian grid is probably still a challenge.

In order to understand how well different numerical techniques can agree on the outcome of GIs, different codes need to run the same initial conditions. This is being done in a large, ongoing code-comparison project that involves eight different codes, both grid-based and SPH. Among the grid codes, there are several adaptive mesh refinement (AMR) schemes. The comparison is part of a larger effort involving several areas of computational astrophysics (*krone.physik. unizh.ch/_moore/wengen/tests.html*). The system chosen for the comparison is a uniform temperature, massive, and initially very unstable disk with a diameter of about 20 AU. The disk is evolved isothermally and has a Q profile that decreases outward, reaching a minimum value ~1 at the disk edge. The disk model is created using a particle representation by letting its mass grow slowly, as described in *Mayer et al.* (2004a). This distribution is then interpolated onto the various grids.

Here we present the preliminary results of the code comparisons from four codes: two SPH codes called GASOLINE (*Wadsley et al.,* 2004) and GADGET2 (*Springel et al.,* 2001; *Springel,* 2005), the Indiana University code with a fixed cylindrical grid (*Pickett,* 1995; *Mejía,* 2004), and the Cartesian AMR code called FLASH (*Fryxell et al.,* 2000).

Readers should consult the published literature for detailed descriptions, but we briefly enumerate some basic features. FLASH uses a PPM-based Riemann solver on a Cartesian grid with directional splitting to solve the Euler equations, and it uses an iterative multigrid Poisson solver for gravity. Both GASOLINE and GADGET2 solve the Euler equations using SPH and solve gravity using a treecode, a binary tree in the case of GASOLINE and an oct-tree in the case of GADGET2. Gravitational forces from individual particles are smoothed using a spline kernel softening, and they both adopt the *Balsara* (1995) artificial viscosity that minimizes shear forces on large scales. The Indiana code is a finite difference grid-based code that solves the equations of hydrodynamics using the Van Leer method. Poisson's equation is solved at the end of each hydrodynamic step by a Fourier transform of the density in the azimuthal direction, direct solution by cyclic reduction of the transform in (r, z), and a transform back to real space (*Tohline,* 1980). The code's Von Neumann-Richtmeyer artificial bulk viscosity is not used for isothermal evolutions.

The two SPH codes are run with fixed gravitational softening, and the local Jeans length (see *Bate and Burkert,* 1997) before and after clump formation is well resolved. Runs with adaptive gravitational softening will soon be included in the comparison. Here we show the results of the runs whose initial conditions were generated from the 8×10^5 particles setup, which was mapped onto a $512 \times 512 \times 52$ Cartesian grid for FLASH and onto a $512 \times 1024 \times 64$ (r, ϕ, z) cylindrical grid for the Indiana code. Comparable resolution (cells for grids or gravity softening for SPH runs) is available initially in the outer part of the disk, where the Q parameter reaches its minimum. In the GASOLINE and GADGET2 runs, the maximum spatial resolution is set by the gravitational softening at 0.12 AU. Below this scale, gravity is essentially suppressed. The FLASH run has a initial resolution of 0.12 AU at 10 AU, comparable with the SPH runs. The Indiana code has the same resolution as FLASH in the radial direction but has a higher azimuthal resolution of 0.06 AU at 10 AU.

As can be seen from Fig. 5, the level of agreement between the runs is satisfactory, although significant differences are noticeable. More clumps are seen in the Indiana code simulation. On the other end, clumps have similar densities in FLASH and GASOLINE, while they appear more

Fig. 5. Equatorial slice density maps of the disk in the test runs after about 100 yr of evolution. The initial disk is 20 AU in diameter. From left to right are the results from GASOLINE and GADGET2 (both SPH codes), from the Indiana cylindrical-grid code, and from the AMR Cartesian-grid code FLASH. The SPH codes adopt the shear-reduced artificial viscosity of *Balsara* (1995).

fluffy in the Indiana code than in the other three. The causes are probably different gravity solvers and the nonadaptive nature of the Indiana code. Even within a single category of code, SPH or grid-based, different types of viscosity, both artificial and numerical, might be more or less diffusive and affect the formation and survival of clumps. In fact, tests show that more fragments are present in SPH runs with shear-reduced artificial viscosity than with full shear viscosity.

Although still in an early stage, the code comparison has already produced one important result, namely that, once favorable conditions exist, widespread fragmentation is obtained in high-resolution simulations using any of the standard numerical techniques. On the other hand, the differences already noticed require further understanding and will be addressed in a forthcoming paper (Mayer et al., in preparation). Although researchers now agree on conditions for disk fragmentation, no consensus yet exists about whether or where real disks fragment or how long fragments really persist. Answers to these questions require advances over current techniques for treating radiative physics and compact structures in global simulations.

5. INTERACTIONS WITH SOLIDS

The standard model for the formation of giant gaseous planets involves the initial growth of a rocky core that, when sufficiently massive, accretes a gaseous envelope (*Bodenheimer and Pollack,* 1986; *Pollack et al.,* 1996). In this scenario, the solid particles in the disk must first grow from micrometer-sized dust grains to kilometer-sized planetesimals that then coagulate to form the rocky core.

In a standard protoplanetary disk, the gas pressure near the disk midplane will generally decrease with increasing radius, resulting in an outward pressure gradient that causes the gas to orbit with sub-Keplerian velocities. The solid particles, on the other hand, do not feel the gas pressure and orbit with Keplerian velocities. This velocity difference results in a drag force that generally causes the solid particles to lose angular momentum and to spiral inward toward the central star with a radial drift velocity that depends on the particle size (*Weidenschilling,* 1977).

While this differential radial drift can mix together particles of different size and allow large grains to grow by sweeping up smaller grains (*Weidenschilling and Cuzzi,* 1993), it also introduces a potential problem. Depending on the actual disk properties, the inward radial velocity for particles with sizes between 1 cm and 1 m can be as high as 10^4 cm s^{-1} (*Weidenschilling,* 1977), so that these particles could easily migrate into the central star before becoming large enough to decouple from the disk gas. If these particles do indeed have short residence times in the disk, it is difficult to envisage how they can grow to form the larger kilometer-sized planetesimals that are required for the subsequent formation of the planetary cores.

The above situation is only strictly valid in smooth, laminar disks with gas pressures that decrease monotonically with increasing radius. If there are any regions in the disk that have local pressure enhancements, the situation can be very different. In the vicinity of a pressure enhancement, the gas velocity can be either super- or sub-Keplerian depending on the local gas pressure gradient. The drag force can then cause solid particles to drift outward or inward, respectively (*Haghighipour and Boss,* 2003a,b). The net effect is that the solid particles should drift toward pressure maxima. A related idea is that a baroclinic instability could lead to the production of long-lived, coherent vortices (*Klahr and Bodenheimer,* 2003) and that solid particles would drift towards the center of the vortex where the enhanced concentration could lead to accelerated grain growth (*Klahr and Henning,* 1997). The existence of such vortices is, however, uncertain (*Johnson and Gammie,* 2006).

An analogous process could occur in a self-gravitating disk, where structures formed by GI activity, such as the centers of the spiral arms, are pressure and density maxima. In such a case, drag force results in solid particles drifting toward the centers of these structures, with the most significant effect occurring for those particles that would, in a smooth, laminar disk, have the largest inward radial velocities.

If disks around very young protostars do indeed undergo a self-gravitating phase, then we would expect the resulting spiral structures to influence the evolution of the solid particles in the disk (*Haghighipour and Boss,* 2003a,b). A GI-active disk will also transport dust grains small enough to remain tied to the gas across distances of many AU in only 1000 yr or so (*Boss,* 2004b), a potentially important process for explaining the components of primitive meteorites (see chapter by Alexander et al.). *Boley and Durisen* (2006) show that, in only one pass of a spiral shock, hydraulic jumps induced by shock heating can mix gas and entrained dust radially and vertically over length scales ~H through the generation of huge breaking waves. The presence of chondrules in primitive chondritic meteorites is circumstantial evidence that the solar nebula experienced a self-gravitating phase in which spiral shock waves provided the flash heating required to explain their existence (*Boss and Durisen,* 2005a,b; *Boley et al.,* 2005).

To test how a self-gravitating phase in a protostellar disk influences the evolution of embedded particles, *Rice et al.* (2004) perform three-dimensional self-gravitating disk simulations that include particles evolved under the influence of both disk self-gravity and gas drag. In their simulations, they consider both 10-m particles, which, for the chosen disk parameters, are only weakly coupled to the gas, and 50-cm particles that are significantly influenced by the gas drag. Figure 6a shows the surface density structure of the 10-m particles one outer rotation period after they were introduced into the gas disk. The structure in the particle disk matches closely that of the gas disk (not shown), showing that these particles are influenced by the gravitational force of the gas disk, but not so strongly influenced by gas drag. Figure 6b shows the surface density structure of 50-cm particles at the same epoch. Particles of this size are influenced by gas drag and Fig. 6b shows that, compared to the 10-m particles, these particles become strongly concentrated into the GI-induced spiral structures.

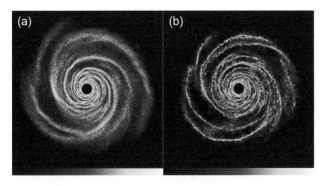

Fig. 6. Surface density structure of particles embedded in a self-gravitating gas disk. **(a)** The distribution of 10-m-radius particles is similar to that of the gas disk, because these particles are not influenced strongly by gas drag. **(b)** 50-cm particles are strongly influenced by gas drag and become concentrated into the GI spirals with density enhancements of an order of magnitude or more. Adapted from *Rice et al.* (2004).

The ability of solid particles to become concentrated in the center of GI-induced structures suggests that, even if giant planets do not form directly via GIs, a self-gravitating phase may still play an important role in giant planet formation. The solid particles may achieve densities that could accelerate grain growth either through an enhanced collision rate or through direct gravitational collapse of the particle subdisk (*Youdin and Shu,* 2002). *Durisen et al.* (2005) also note that dense rings can be formed near the boundaries between GI-active and inactive regions of a disk (e.g., the central disk in Fig. 1). Such rings are ideal sites for the concentration of solid particles by gas drag, possibly leading to accelerated growth of planetary embryos. Even if processes like these do not contribute directly to planetesimal growth, GIs may act to prevent the loss of solids by migration toward the proto-Sun. The complex and time-variable structure of GI activity should increase the residence time of solids in the disk and potentially give them enough time to become sufficiently massive to decouple from the disk gas.

6. PLANET FORMATION

The relatively high frequency (~10%) of solar-type stars with giant planets that have orbital periods less than a few years suggests that longer-period planets may be quite frequent. Perhaps ~12–25% of G dwarfs may have gas giants orbiting within ~10 AU. If so, gas giant planet formation must be a fairly efficient process. Because roughly half of protoplanetary disks disappear within 3 m.y. or less (*Bally et al.,* 1998; *Haisch et al.,* 2001; *Briceño et al.,* 2001; *Eisner and Carpenter,* 2003), core accretion may not be able to produce a high frequency of gas giants. There is also now strong theoretical (*Yorke and Bodenheimer,* 1999) and observational (*Osorio et al.,* 2003; *Rodríguez et al.,* 2005; *Eisner et al.,* 2005) evidence that disks around very young protostars should indeed be sufficiently massive to experience GIs. *Rodríguez et al.* (2005) show a 7-mm VLA image of

a disk around a Class 0 protostar that may have a mass half that of the central star.

Hybrid scenarios may help remove the bottleneck by concentrating meter-sized solids, but it is not clear that they can shorten the overall timescale for core accretion, which is limited by the time needed for the growth of 10-$M_\oplus$ cores and for accretion of a large gaseous envelope. *Durisen et al.* (2005) suggest that the latter might be possible in dense rings, but detailed calculations of core growth or envelope accretion in the environment of a dense ring do not now exist. Disk instability, on the other hand, has no problem forming gas giants rapidly in even the shortest-lived protoplanetary disk. Most stars form in regions of high-mass star formation (*Lada and Lada,* 2003) where disk lifetimes should be the shortest due to loss of outer disk gas by UV irradiation.

There is currently disagreement about whether GIs are stronger in low-metallicity systems (*Cai et al.,* 2006) or whether their strength is relatively insensitive to the opacity of the disk (*Boss,* 2002a). In either case, if disk instability is correct, we would expect that even low-metallicity stars could host gas giant planets. The growth of cores in the core accretion mechanism is hastened by higher metallicity through the increase in surface density of solids (*Pollack et al.,* 1996), although the increased envelope opacity, which slows the collapse of the atmosphere, works in the other direction (*Podolak,* 2003). The recent observation of a Saturn-mass object, orbiting the metal-rich star HD 149026, with a core mass equal to approximately half the planet's mass (*Sato et al.,* 2005), has been suggested as a strong confirmation of the core accretion model. It has, however, yet to be shown that the core accretion model can produce a core with such a relatively large mass. If this core was produced by core accretion, it seems that it never achieved a runaway growth of its envelope; yet, in the case of Jupiter, the core accretion scenario requires efficient accumulation of a massive envelope around a relatively low-mass core.

The correlation of short-period gas giants with high metallicity stars is often interpreted as strong evidence in favor of core accretion (*Laws et al.,* 2003; *Fischer et al.,* 2004; *Santos et al.,* 2004). The *Santos et al.* (2004) analysis, however, shows that even the stars with the lowest metallicities have detectable planets with a frequency comparable to or higher than that of the stars with intermediate metallicities. *Rice et al.* (2003c) have shown that the metallicity distribution of systems with at least one massive planet ($M_{pl} > 5 M_J$) on an eccentric orbit of moderate semimajor axis does not have the same metal-rich nature as the full sample of extrasolar planetary systems. Some of the metallicity correlation can be explained by the observational bias of the spectroscopic method in favor of detecting planets orbiting stars with strong metallic absorption lines. The residual velocity jitter typically increases from a few meters per second for solar metallicity to 5–16 m/s for stars with 1/4 the solar metallicity or less. In terms of extrasolar planet search space, this could account for as much as a factor of 2 difference in the total number of planets detected by spectroscopy. A spectroscopic search of 98 stars in the Hyades cluster, with a metallicity 35% greater than solar, found nothing, whereas about 10 hot Jupiters should have been found, assuming the same frequency as in the solar neighborhood (*Paulson et al., 2004*).

Jones (2004) found that the average metallicity of planet host stars increased from ~0.07 to ~0.24 dex for planets with semimajor axes of ~2 AU to ~0.03 AU, suggesting a trend toward shortest-period planets orbiting the most metal-rich stars. Similarly, *Sozzetti* (2004) showed that both metal-poor and metal-rich stars have increasing numbers of planets as the orbital period increases, but only the metal-rich stars have an excess of the shortest-period planets. This could imply that the metallicity correlation is caused by inward orbital migration, if low-metallicity stars have long-period giant planets that seldom migrate inward.

Lower disk metallicity results in slower Type II inward migration (*Livio and Pringle, 2003*), the likely dominant mechanism for planet migration (see chapter by Papaloizou et al.). This is because with increased metallicity, the disk viscosity ν increases. In standard viscous accretion disk theory (e.g., *Ruden and Pollack, 1991*) $\nu = \alpha c_s H$. Lower disk metallicity leads to lower disk opacity, lower disk temperatures, lower sound speeds, and a thinner disk. As ν decreases with lowered metallicity, the timescale for Type II migration increases. *Ruden and Pollack* (1991) found that viscous disk evolution times increased by a factor of about 20 when ν decreased by a factor of 10. It remains to be seen if this effect is large enough to explain the rest of the correlation. If disk instability is operative and if orbital migration is the major source of the metallicity correlation, then metal-poor stars should have planets on long-period orbits.

Disk instability may be necessary to account for the long-period giant planet in the M4 globular cluster (*Sigurdsson et al., 2003*), where the metallicity is 1/20 to 1/30 solar metallicity. The absence of short-period Jupiters in the 47 Tuc globular cluster (*Gilliland et al., 2000*) with 1/5 solar metallicity could be explained by the slow rate of inward migration due to the low metallicity. Furthermore, if 47 Tuc initially contained OB stars, photoevaporation of the outer disks may have occurred prior to inward orbital migration of any giant planets, preventing their evolution into short-period planets, although other factors (i.e., crowding) can also be important in these clusters.

The M dwarf GJ 876 is orbited by a pair of gas giants (as well as a much smaller mass planet) and other M dwarfs have giant planets as well (*Butler et al., 2004*), although apparently not as frequently as the G dwarfs. *Laughlin et al.* (2004) found that core accretion was too slow to form gas giants around M dwarfs because of the longer orbital periods. Disk instability does not a have similar problem for M dwarfs, and disk instability predicts that M, L, and T dwarfs should have giant planets.

With disk instability, one Jupiter mass of disk gas has at most ~6 $M_\oplus$ of elements suitable to form a rock/ice core. The preferred models of the jovian interior imply that Jupiter's core mass is less than ~3 $M_\oplus$ (*Saumon and Guillot, 2004*); Jupiter may even have no core at all. These models seem to be consistent with formation by disk instability and

inconsistent with formation by core accretion, which requires a more massive core. As a result, the possibility of core erosion has been raised (*Saumon and Guillot, 2004*). If core erosion can occur, core masses may lose much of their usefulness as formation constraints. Saturn's core mass appears to be larger than that of Jupiter (*Saumon and Guillot, 2004*), perhaps ~15 $M_\oplus$, in spite of it being the smaller gas giant. Core erosion would only make Saturn's initial core even larger. Disk instability can explain the larger saturnian core mass (*Boss et al., 2002*). Proto-Saturn may have started out with a mass larger than that of proto-Jupiter, but its excess gas may have been lost by UV photoevaporation, a process that could also form Uranus and Neptune. Disk instability predicts that inner gas giants should be accompanied by outer ice giant planets in systems that formed in OB associations due to strong UV photoevaporation. In low-mass star-forming regions, disk instability should produce only gas giants, without outer ice giants.

Disk instability predicts that even the youngest stars should show evidence of gas giant planets (*Boss, 1998b*), whereas core accretion requires several million years or more to form gas giants (*Inaba et al., 2003*). A gas giant planet seems to be orbiting at ~10 AU around the 1-m.y.-old star CoKu Tau/4 (*Forrest et al., 2004*), based on a spectral energy distribution showing an absence of disk dust inside 10 AU (for an alternative perspective, see *Tanaka et al., 2005*). Several other 1-m.y.-old stars show similar evidence for rapid formation of gas giant planets. The direct detection of a possible gas giant planet around the 1-m.y.-old star GQ Lup (*Neuhäuser et al., 2005*) similarly requires a rapid planet formation process.

We conclude that there are significant observational arguments to support the idea that disk instability, or perhaps a hybrid theory where core accretion is accelerated by GIs, might be required to form some if not all gas giant planets. Given the major uncertainties in the theories, observational tests will be crucial for determining the relative proportions of giant planets produced by the competing mechanisms.

Acknowledgments. R.H.D. was supported by NASA grants NAG5-11964 and NNG05GN11G and A.P.B. by NASA grants NNG05GH30G, NNG05GL10G, and NCC2-1056. Support for A.F.N. was provided by the U.S. Department of Energy under contract W-7405-ENG-36, for which this is publication LA-UR–05-7851. We would like to thank S. Michael for invaluable assistance in manuscript preparation, R. Rafikov and an anonymous referee for substantive improvements, and A. C. Mejía, A. Gawryszczak, and V. Springel for allowing us to premier their comparison calculations in section 4.4. FLASH was in part developed by the DOE-supported ASC/Alliance Center for Astrophysical Thermonuclear Flashes at the University of Chicago and was run on computers at Warsaw's Interdisciplinary Center for Mathematical and Computational Modeling.

REFERENCES

Adams F. C., Shu F. H., and Lada C. J. (1988) *Astrophys. J., 326,* 865–883.

Adams F. C., Ruden S. P., and Shu F. H. (1989) *Astrophys. J., 347,* 959–976.

Armitage P. J., Livio M., and Pringle J. E. (2001) *Mon. Not. R. Astron. Soc., 324,* 705–711.

Balbus S. A. and Papaloizou J. C. B. (1999) *Astrophys. J., 521,* 650–658.

Bally J., Testi L., Sargent A., and Carlstrom J. (1998) *Astron. J., 116,* 854–859.

Balsara D. S. (1995) *J. Comput. Phys., 121,* 357–372.

Bate M. R. and Burkert A. (1997) *Mon. Not. R. Astron. Soc., 228,* 1060–1072.

Beckwith S. V. W., Sargent A. I., Chini R. S., and Güsten R. (1990) *Astron. J., 99,* 924–945.

Benz W. (1990) In *The Numerical Modeling of Nonlinear Stellar Pulsations* (J. R. Buchler, ed.), pp. 269–288. Kluwer, Boston.

Boffin H. M. J., Watkins S. J., Bhattal A. S., Francis N., and Whitworth A. P. (1998) *Mon. Not. R. Astron. Soc., 300,* 1189–1204.

Bodenheimer P. and Pollack J. B. (1986) *Icarus, 67,* 391–408.

Bodenheimer P., Yorke H. W., Różyczka M., and Tohline J. E. (1990) *Astrophys. J., 355,* 651–660.

Boley A. C. and Durisen R. H. (2006) *Astrophys. J., 641,* 534–546.

Boley A. C., Durisen R. H., and Pickett M. K. (2005) In *Chondrites and the Protoplanetary Disk* (A. N. Krot et al., eds.), pp. 839–848. ASP Conf. Series 341, San Francisco.

Boss A. P. (1997) *Science, 276,* 1836–1839.

Boss A. P. (1998a) *Astrophys. J., 503,* 923–937.

Boss A. P. (1998b) *Nature, 395,* 141–143.

Boss A. P. (2000) *Astrophys. J., 536,* L101–L104.

Boss A. P. (2001) *Astrophys. J., 563,* 367–373.

Boss A. P. (2002a) *Astrophys. J., 567,* L149–L153.

Boss A. P. (2002b) *Astrophys. J., 576,* 462–472.

Boss A. P. (2002c) *Earth Planet. Sci. Lett., 202,* 513–523.

Boss A. P. (2003) In *Lunar Planet. Sci. XXXIV,* Abstract #1075.

Boss A. P. (2004a) *Astrophys. J., 610,* 456–463.

Boss A. P. (2004b) *Astrophys. J., 616,* 1265–1277.

Boss A. P. (2005) *Astrophys. J., 629,* 535–548.

Boss A. P. (2006) *Astrophys. J., 641,* 1148–1161.

Boss A. P. and Durisen R. H. (2005a) *Astrophys. J., 621,* L137–L140.

Boss A. P. and Durisen R. H. (2005b) In *Chondrites and the Protoplanetary Disk* (A. N. Krot et al., eds.), pp. 821–838. ASP Conf. Series 341, San Francisco.

Boss A. P. and Myhill E. A. (1992) *Astrophys. J. Suppl., 83,* 311–327.

Boss A. P. and Yorke H. W. (1996) *Astrophys. J., 496,* 366–372.

Boss A. P., Wetherill G. W., and Haghighipour N. (2002) *Icarus, 156,* 291–295.

Briceño C., Vivas A. K., Calvet N., Hartmann L., Pachecci R., et al. (2001) *Science, 291,* 93–96.

Bryden G., Chen X., Lin D. N. C., Nelson R. P., and Papaloizou J. C. B. (1999) *Astrophys. J., 514,* 344–367.

Bryden G., Lin D. N. C., and Ida S. (2000) *Astrophys. J., 544,* 481–495.

Butler R. P., Vogt S. S., Marcy G. W., Fischer D. A., Wright J. T., et al. (2004) *Astrophys. J., 617,* 580–588.

Cai K., Durisen R. H., Michael S., Boley A. C., Mejía A. C., Pickett M. K., and D'Alessio P. (2006) *Astrophys. J., 636,* L149–L152.

Cameron A. G. W. (1978) *Moon Planets, 18,* 5–40.

Chiang E. I. and Goldreich P. (1997) *Astrophys. J., 490,* 368–376.

Colella P. and Woodward P. R. (1984) *J. Comp. Phys., 54,* 174–201.

Cuzzi J. N., Hogan R. C., Paque J. M., and Dobrovolskis A. R. (2001) *Astrophys. J., 546,* 496–508.

D'Alessio P., Calvet N., and Hartmann L. (1997) *Astrophys. J., 474,* 397–406.

D'Alessio P., Cantó J., Calvet N., and Lizano S. (1998) *Astrophys. J., 500,* 411–427.

Durisen R. H. (2006) In *A Decade of Extrasolar Planets Around Normal Stars* (M. Livio, ed.), in press. Cambridge Univ., Cambridge.

Durisen R. H., Mejía A. C., Pickett B. K., and Hartquist T. W. (2001) *Astrophys. J., 563,* L157–L160.

Durisen R. H., Mejía A. C., and Pickett B. K. (2003) *Rec. Devel. Astrophys., 1,* 173–201.

Durisen R. H., Cai K., Mejía A. C., and Pickett M. K. (2005) *Icarus, 173,* 417–424.

Eisner J. A. and Carpenter J. M. (2003) *Astrophys. J., 598,* 1341–1349.

Eisner J. A., Hillenbrand L. A., Carpenter J. M., and Wolf S. (2005) *Astrophys. J., 635,* 396–421.

Fischer D., Valenti J. A. and Marcy G. (2004) In *Stars as Suns: Activity, Evolution, and Planets* (A. K. Dupree and A. O. Benz, eds.), pp. 29–38. IAU Symposium 219, ASP, San Francisco.

Forrest W. J., Sargent B., Furlan E., D'Alessio P., Calvet N., et al. (2004) *Astrophys. J. Suppl., 154,* 443–447.

Fleming T. and Stone J. M. (2003) *Astrophys. J., 585,* 908–920.

Fryxell B., Arnett D., and Müller E. (1991) *Astrophys. J., 367,* 619–634.

Fryxell B., Olson K., Ricker P., Timmes F. X., Zingale M., et al. (2000) *Astrophys. J. Suppl., 131,* 273–334.

Gammie C. F. (1996) *Astrophys. J., 457,* 355–362.

Gammie C. F. (2001) *Astrophys. J., 553,* 174–183.

Gilliland R. L., Brown T. M., Guhathakurta P., Sarajedini A., Milone E. F., et al. (2000) *Astrophys. J., 545,* L47–L51.

Goldreich P. and Lynden-Bell D. (1965) *Mon. Not. R. Astron. Soc., 130,* 125–158.

Haghighipour N. and Boss A. P. (2003a) *Astrophys. J., 583,* 996–1003.

Haghighipour N. and Boss A. P. (2003b) *Astrophys. J., 598,* 1301–1311.

Haisch K. E., Lada E. A., and Lada C. J. (2001) *Astrophys. J., 553,* L153–L156.

Inaba S., Wetherill G. W., and Ikoma M. (2003) *Icarus, 166,* 46–62.

Johnson B. M. and Gammie C. F. (2003) *Astrophys. J., 597,* 131–141.

Johnson B. M. and Gammie C. F. (2006) *Astrophys. J., 636,* 63–74.

Johnstone D., Hollenbach D., and Bally J. (1998) *Astrophys. J., 499,* 758–776.

Jones H. R. A. (2004) In *The Search for Other Worlds: Fourteenth Astrophysics Conference,* pp. 17–26. AIP Conf. Proc. 713, New York.

Klahr H. H. (2003) In *Scientific Frontiers in Research on Extrasolar Planets* (D. Deming and S. Seager, eds.), pp. 277–280. ASP Conf. Series 294, San Francisco.

Klahr H. H. and Bodenheimer P. (2003) *Astrophys. J., 582,* 869–892.

Klahr H. H. and Henning T. (1997) *Icarus, 128,* 213–229.

Kley W. (1999) *Mon. Not. R. Astron. Soc., 303,* 696–710.

Koller J., Li H., and Lin D. N. C. (2003) *Astrophys. J., 596,* L91–94.

Kuiper G. P. (1951) In *Proceedings of a Topical Symposium* (J. A. Hynek, ed.), pp. 357–424. McGraw-Hill, New York.

Lada C. J. and Lada E. A. (2003) *Ann. Rev. Astron. Astrophys., 41,* 57–115.

Larson R. B. (1984) *Mon. Not. R. Astron. Soc., 206,* 197–207.

Laughlin G. and Bodenheimer P. (1994) *Astrophys. J., 436,* 335–354.

Laughlin G. and Różyczka M. (1996) *Astrophys. J., 456,* 279–291.

Laughlin G., Korchagin V., and Adams F. C. (1997) *Astrophys. J., 477,* 410–423.

Laughlin G., Korchagin V., and Adams F. C. (1998) *Astrophys. J., 504,* 945–966.

Laughlin G., Bodenheimer P., and Adams F. C. (2004) *Astrophys. J., 612,* L73–L76.

Laws C., Gonzalez G., Walker K. M., Tyagi S., Dodsworth J., et al. (2003) *Astron. J., 125,* 2664–2677.

Lin D. N. C. and Papaloizou J. C. B. (1993) In *Protostars and Planets III* (E. H. Levy and J. I. Lunine, eds.), pp. 749–835. Univ. of Arizona, Tucson.

Lin D. N. C. and Pringle J. E. (1987) *Mon. Not. R. Astron. Soc., 225,* 607–613.

Lin D. N. C., Laughlin G., Bodenheimer P., and Różyczka M. (1998) *Science, 281,* 2025–2027.

Livio M. and Pringle J. E. (2003) *Mon. Not. R. Astron. Soc., 346,* L42–L44.

Lodato G. and Rice W. K. M. (2004) *Mon. Not. R. Astron. Soc., 351,* 630–642.

Lodato G. and Rice W. K. M. (2005) *Mon. Not. R. Astron. Soc., 358,* 1489–1500.

Lubow S. H. and Ogilvie G. I. (1998) *Astrophys. J., 504,* 983–995.

Lubow S. H., Siebert M., and Artymowicz P. (1999) *Astrophys. J., 526,* 1001–1012.

Lufkin G., Quinn T., Wadsley J., Stadel J., and Governato F. (2004) *Mon. Not. R. Astron. Soc., 347,* 421–429.

Mayer L., Quinn T., Wadsley J., and Stadel J. (2002) *Science, 298,* 1756–1759.

Mayer L., Quinn T., Wadsley J., and Stadel J. (2004a) *Astrophys. J., 609,* 1045–1064.

Mayer L., Wadsley J., Quinn T., and Stadel J. (2004b) In *Extrasolar Planets: Today and Tomorrow* (J.-P. Beaulieu et al., eds.), pp. 290–297. ASP Conf. Series 321, San Francisco.

Mayer L., Wadsley J., Quinn T., and Stadel J. (2005) *Mon. Not. R. Astron. Soc., 363,* 641–648.

Mejía A. C. (2004) Ph.D. dissertation, Indiana University.

Mejía A. C., Durisen R. H., Pickett M. K., and Cai K. (2005) *Astrophys. J., 619,* 1098–1113.

Mihalas D. (1977) *Stellar Atmospheres.* Univ. of Chicago, Chicago.

Monaghan J. J. (1992) *Ann. Rev. Astron. Astrophys., 30,* 543–574.

Myhill E. A. and Boss A. P. (1993) *Astrophys. J. Suppl., 89,* 345–359.

Nelson A. F. (2000) *Astrophys. J., 537,* L65–L69.

Nelson A. F. (2006) *Mon. Not. R. Astron. Soc.,* in press.

Nelson A. F. and Benz W. (2003) *Astrophys. J., 589,* 578–604.

Nelson A. F., Benz W., Adams F. C., and Arnett D. (1998) *Astrophys. J., 502,* 342–371.

Nelson A. F., Benz W., and Ruzmaikina T. V. (2000) *Astrophys. J., 529,* 357–390.

Nelson R. P., Papaloizou J. C. B., Masset F., and Kley W. (2000) *Mon. Not. R. Astron. Soc., 318,* 18–36.

Neuhäuser R., Guenther E. W., Wuchterl G., Mugrauer M., Bedalov A., and Hauschildt P. H. (2005) *Astron. Astrophys., 435,* L13–L16.

Osorio M., D'Alessio P., Muzerolle J., Calvet N., and Hartmann L. (2003) *Astrophys. J., 586,* 1148–1161.

Paczyński B. (1978) *Acta Astron., 28,* 91–109.

Papaloizou J. C. B. and Savonije G. (1991) *Mon. Not. R. Astron. Soc., 248,* 353–369.

Paulson D. B., Saar S. H., Cochran W. D., and Henry G. W. (2004) *Astron. J., 127,* 1644–1652.

Pickett B. K. (1995) Ph.D. dissertation, Indiana University.

Pickett B. K., Durisen R. H., and Davis G. A. (1996) *Astrophys. J., 458,* 714–738.

Pickett B. K., Cassen P., Durisen R. H., and Link R. P. (1998) *Astrophys. J., 504,* 468–491.

Pickett B. K., Cassen P., Durisen R. H., and Link R. P. (2000a) *Astrophys. J., 529,* 1034–1053.

Pickett B. K., Durisen R. H., Cassen P., and Mejía A. C. (2000b) *Astrophys. J., 540,* L95–L98.

Pickett B. K., Mejía A. C., Durisen R. H., Cassen P. M., Berry D. K., and Link R. P. (2003) *Astrophys. J., 590,* 1060–1080.

Podolak M. (2003) *Icarus, 165,* 428–437.

Pollack J. B., Hubickyj O., Bodenheimer P., Lissauer J. J., Podolak M., and Greenzweig Y. (1996) *Icarus, 124,* 62–85.

Pringle J. E. (1981) *Ann. Rev. Astron. Astrophys., 19,* 137–162.

Rafikov R. R. (2005) *Astrophys. J., 621,* L69–L72.

Rice W. K. M., Armitage P. J., Bate M. R., and Bonnel I. A. (2003a) *Mon. Not. R. Astron. Soc., 338,* 227–232.

Rice W. K. M., Armitage P. J., Bate M. R., and Bonnell I. A. (2003b) *Mon. Not. R. Astron. Soc., 339,* 1025–1030.

Rice W. K. M., Armitage P. J., Bate M. R. and Bonnell I. A. (2003c) *Mon. Not. R. Astron. Soc., 346,* L36–L40.

Rice W. K. M., Lodato G., Pringle J. E., Armitage P. J., and Bonnell I. A. (2004) *Mon. Not. R. Astron. Soc., 355,* 543–552.

Rice W. K. M., Lodato G., and Armitage P. J. (2005) *Mon. Not. R. Astron. Soc., 364,* L56–L60.

Rodríguez L. F., Loinard L., D'Alessio P., Wilner D. J., and Ho P. T. P. (2005) *Astrophys. J., 621,* 133–136.

Ruden S. P. and Pollack J. B. (1991) *Astrophys. J., 375,* 740–760.

Santos N. C., Israelian G., and Mayor M. (2004) *Astron. Astrophys., 415,* 1153–1166.

Sato B., Fischer D. A., Henry G. W., Laughlin G., Butler R. P., et al. (2005) *Astrophys. J., 633,* 465–473.

Saumon D. and Guillot T. (2004) *Astrophys. J., 609,* 1170–1180.

Shu F. H., Tremaine S., Adams F. C., and Ruden S. P. (1990) *Astrophys. J., 358,* 495–514.

Sigurdsson S., Richer H. B., Hansen B. M., Stairs I. H., and Thorsett S. E. (2003) *Science, 301,* 193–196.

Springel V. (2005) *Mon. Not. R. Astron. Soc., 364,* 1105–1134.

Springel V., Yoshida N., and White S. D. M. (2001) *New Astron., 6,* 79–117.

Sozzetti A. (2004) *Mon. Not. R. Astron. Soc., 354,* 1194–1200.

Stone J. M. and Norman M. L. (1992) *Astrophys. J. Suppl., 80,* 753–790.

Tanaka H., Himeno Y., and Ida S. (2005) *Astrophys. J., 625,* 414–426.

Tohline J. E. (1980) *Astrophys. J., 235,* 866–881.

Toomre A. (1964) *Astrophys. J., 139,* 1217–1238.

Tomley L., Cassen P., and Steiman-Cameron T. Y. (1991) *Astrophys. J., 382,* 530–543.

Truelove J. K., Klein R. I., McKee C. F., Holliman J. H. II, Howell L. H., and Greenough J. A. (1997) *Astrophys. J., 489,* L179–L183.

Wadsley J., Stadel J., and Quinn T. (2004) *New Astron., 9,* 137–158.

Weidenschilling S. J. (1977) *Mon. Not. R. Astron. Soc., 180,* 57–70.

Weidenschilling S. J. and Cuzzi J. N. (1993) In *Protostars and Planets III* (E. H. Levy and J. I. Lunine, eds.), pp. 1031–1060. Univ. of Arizona, Tucson.

Yorke H. W. and Bodenheimer P. (1999) *Astrophys. J., 525,* 330–342.

Youdin A. N. and Shu F. H. (2002) *Astrophys. J., 580,* 494–505.

Gaseous Planets, Protostars, and Young Brown Dwarfs: Birth and Fate

G. Chabrier, I. Baraffe, and F. Selsis
Ecole Normale Supérieure de Lyon

T. S. Barman
University of California, Los Angeles

P. Hennebelle
Ecole Normale Supérieure, Paris

Y. Alibert
University of Bern

We review recent theoretical progress aimed at understanding the formation and the early stages of evolution of giant planets, low-mass stars, and brown dwarfs. Calculations coupling giant planet formation, within a modern version of the core accretion model that includes planet migration and disk evolution, and subsequent evolution yield consistent determinations of the planet structure and evolution. Uncertainties in the initial conditions, however, translate into large uncertainties in the luminosity at early stages. It is thus not possible to say whether young planets are faint or bright compared with low-mass young brown dwarfs. We review the effects of irradiation and evaporation on the evolution of short-period planets and argue that substantial mass loss may have occurred for these objects. Concerning star formation, geometrical effects in protostar core collapse are examined by comparing one-dimensional and three-dimensional calculations. Spherical collapse is shown to significantly overestimate the core inner density and temperature and thus to yield incorrect initial conditions for pre-main-sequence or young brown dwarf evolution. Accretion is also shown to occur nonspherically over a very limited fraction of the protostar surface. Accretion affects the evolution of young brown dwarfs and yields more compact structures for a given mass and age, thus fainter luminosities, confirming previous studies for pre-main-sequence stars. This can lead to severe misinterpretations of the mass and/or age of young accreting objects from their location in the Hertzsprung-Russell (HR) diagram. Since accretion covers only a limited fraction of the protostar surface, we argue that newborn stars and brown dwarfs should appear rapidly over an extended area in the HR diagram, depending on their accretion history, rather than on a well-defined birth line. Finally, we suggest that the distinction between planets and brown dwarfs be based on an observational diagnostic, reflecting the different formation mechanisms between these two distinct populations, rather than on an arbitrary, confusing definition.

1. INTRODUCTION

One of the fundamental questions of astrophysics remains the characterization of the formation of planets and stars. The mass ranges of the most massive planets and of the least-massive brown dwarfs certainly overlap in the ~1–10 M_{Jup} range; it is thus interesting to explore our understanding of the planet- and star-formation mechanisms in a common review.

The growing number of discovered extrasolar giant planets, ranging now from Neptune-mass to few Jupiter-mass objects, has questioned our understanding of planet formation and evolution. The significant fraction of exoplanets in close orbit to their parent star, in particular, implies a revision of our standard scenario of planet formation. Indeed, these objects are located well within the so-called ice line and could not have formed *in situ*. This strongly favors planet migration as a common process in planet formation. This issue is explored in section 2 where we present consistent calculations between a revised version of the core accretion model, which does take planet migration into account, and subsequent evolution. In this section, we also review our current understanding of the effects of irradiation and evaporation on the evolution of short-period planets, hot Neptunes, and hot Jupiters, and review present uncertainties in the determination of the evaporation rates. In section 3, we briefly review our current understanding of protostellar core collapse and we show that nonspherical calculations are required to obtain proper accretion histories, densities, and thermal profiles for the prestellar core. The effect of accretion on the early contracting phase of pre-main-sequence stars and young brown dwarfs, and a review of observational determinations of accretion rates, are considered in section 4. Finally, throughout this review, we have adopted as the definition of *planet* an object formed by the three-step process described in section 2.1, characterized by a central rocky/icy core built by accretion of planetesimals in a protostellar nebula. In contrast, genuine *brown dwarfs* are defined in this review as gaseous objects of similar composition as the parent cloud from which they formed by collapse. This issue is discussed in section 5 and observational diagnostics to differentiate brown dwarfs from planets, based on their different formation mechanisms, are suggested. Section 6 is devoted to the conclusion.

2. GASEOUS PLANETS: BIRTH AND EVOLUTION

2.1. Planet Formation

The conventional planet formation model is the core accretion model as developed by *Pollack et al.* (1996, hereafter *P96*). One of the major difficulties faced by this model is the long timescale necessary to form a gaseous planet like Jupiter, a timescale significantly larger than typical disk lifetimes, ≤ 10 m.y. Reasonable timescales can be achieved only at the expense of arbitrary assumptions such as, e.g., nebula mean opacities reduced to 2% of the ISM value in some temperature range or solid surface density significantly larger than the minimum mass solar nebula value (*Hubickyj et al.*, 2005). This leaves the standard core accretion model in an uncomfortable situation. This model has been extended recently by *Alibert et al.* (2004, 2005, hereafter *A05*) by including the effects of migration and disk evolution during the planet-formation process. The occurence of migration during planet formation is supported by the discovery of numerous extrasolar giant planets at very short distance to their parent stars, well within the so-called ice line, about 5 AU for the solar nebula conditions. Below this limit, above ice melting temperature, the insufficient surface density of solids that will eventually form the planet core, and the lack of a large reservoir of gas, prevent *in situ* formation of large gaseous planets.

Moreover, inward migration of the planet should arise from angular momentum transfer due to gravitational interactions between the gaseous disk and the growing planet (*Lin and Papaloizou*, 1986; *Ward*, 1997; *Tanaka et al.*, 2002). Taking into account the migration of a growing planet solves the long-lasting timescale problem of the core-accretion scenario. Indeed, when migration is included, the planet-feeding zone never becomes depleted in planetesimals. As a result, the so-called phase 1 (see *P96*), dominated by accretion of solid material, is lengthened, whereas phase 2, dominated by gas accretion, is shortened appreciably. During the last so-called phase 3, runaway gas accretion occurs and the predominantly H/He envelope is attracted onto the core. Phase 3 is very short compared to

phases 1 and 2, and phase 2 essentially determines the formation timescale of the planet. The planet can thus form now on a timescale consistent with disk lifetimes, i.e., a few million years for a Jupiter (see *A05*).

In the models of *Bodenheimer et al.* (2000a) and *Hubyckij et al.* (2005), which are based on the *P96* formalism, the calculations proceed in three steps: (1) the planet is bounded by its Roche lobe ($R_p = R_L$) [or more precisely by $Min(R_L, R_{acc})$ where $R_{acc} = GM/c_s^2$ is the accretion radius and c_s the local sound velocity in the disk] so that the temperature and pressure at the planet surface are the ones of the surrounding nebula. Note that in *P96* calculations, opacity of the nebula is a key ingredient. (2) The planet external radius is the one obtained when the maximum gas accretion rate is reached. In *P96*, this value is fixed to 1×10^{-2} $M_\oplus$ yr^{-1}. At this stage, the external conditions have changed ($R_p < R_L$). Matter falls in free fall from the Roche lobe to the planet radius, producing a shock luminosity. (3) Once the planet reaches its *predefined* final mass, the accretion rate is set to 0 and the boundary conditions become the ones of a cooling isolated object, $L = 4\pi\sigma R^2 T_{eff}^4$ and $\kappa_R P_{ph} = \frac{2}{3}g$, where κ_R denotes the mean Rosseland opacity. The planet surface radius is essentially fixed by the accretion shock conditions (see, e.g., Fig. 1d of *Hubickyj et al.*, 2005). This value, however, remains highly uncertain, as its correct determination would imply a proper treatment of the radiative shock. In *A05*, phase 1 is similar to step 1 described above, except that the planet migration from an initial arbitrary location and the disk evolution are taken into account, so that the thermodynamic conditions of the surrounding nebula, as well as the distance to the star and thus the planet Roche lobe radius, change with time. The planet's final mass is set by the accretion rate limit, and is thus not defined *a priori*. Note that, because of the disk evolution and/or the creation of a gap around the planet, the accretion rate limit is 1 to 2 orders of magnitude smaller than the one in *P96* at the end of phase 1 and reaches essentially 0 with time, a fact supported by three-dimensional hydrodynamical simulations (*D'Angelo et al.*, 2003; *Kley and Dirksen*, 2006). Eventually the planet opens a gap when its Hill radius becomes equal to the disk density scale height and migration stops or declines until the disk is dissipated (see *A05* for details). The planet radius cannot be defined precisely in this model as it results from the competing effects of gas accretion and planet contraction with changing boundary conditions as the planet migrates inward and the disk evolves. In any event, the final stages of accretion are likely to occur within streams (see, e.g., *Lubow et al.*, 1999), i.e., non-spherically and, as mentioned above, the planet final radius remains highly uncertain, at least in any one-dimensional calculation.

The migration rate, in particular type I migration for low-mass planet seeds, remains an ill-defined parameter in these calculations. The observed frequency of extrasolar planets implies a rate significantly smaller than estimates done for laminar disks (*Tanaka et al.*, 2002). Numerical modeling of turbulent disks yields significantly reduced migration rates

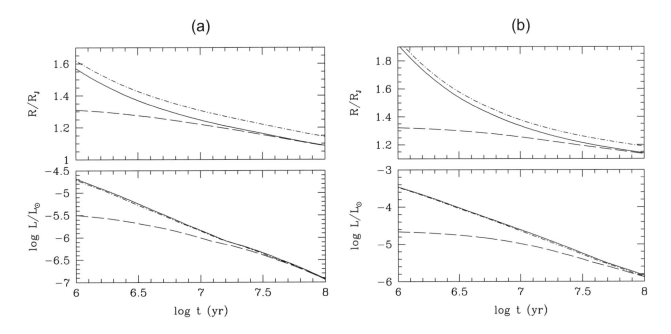

Fig. 1. Evolution of the radius and the luminosity for **(a)** 1-M_{Jup} and **(b)** 4-M_{Jup} planet with a 6 $M_\oplus$ solid core and $M_{Z,env}/M_{env}$ = 10%, for two different initial radii (solid vs. dashed lines, see text). The dot-dashed lines portray the cooling of coreless, pure gaseous brown dwarfs of solar composition with similar initial radii as for the solid lines; the differences reflect the influence of the presence of a central core on the evolution.

(*Nelson and Papaloizou*, 2004; see also *D'Angelo et al.*, 2003). It has been suggested recently that stochastic migration, i.e., protoplanets following a random walk through the disk due to gravitational interaction with turbulent density fluctuations in the disk, may provide a means of preventing at least some planetary cores from migrating into the central star due to type I migration (*Nelson*, 2005). Based on these arguments, and for lack of better determinations, *A05* divide the aforementioned rate of *Tanaka et al.* (2002) by a factor of 10 to 100. As noted by these authors, numerical tests show that, provided the rate is small enough to preserve planet survival, its exact value affects the extent of migration but *not* the formation timescale, nor the planet final structure and internal composition.

2.2. Planet Evolution

2.2.1. Nonirradiated planets. We first examine the evolution of young planets far enough from their parent star for irradiation effects to be neglected. In order for the evolution to be consistent with the formation model, the planet structure includes now a central core surrounded by an envelope enriched in heavy elements. These conditions are given by the formation model described in section 2.1, performed for different initial parameters (initial orbital distance, dust-to-gas ratio in the disk, photoevaporation rate, disk initial surface mass). The planets are found to form with essentially the same core mass ($M_{core} \simeq 6\ M_\oplus$) independent of the planet final mass, whereas the heavy element mass fraction in the envelope deposited by the accreted planetesi-

mals is found to increase substantially with decreasing total mass (*Baraffe et al.*, 2006). The hydrogen/helium equation of state (EOS) is the Saumon, Chabrier, and VanHorn EOS (*Saumon et al.*, 1995), whereas the thermodynamic properties of the heavy material relevant to the planet structure [ice, dunite ($\equiv Mg_2SiO_4$), iron] are calculated with the ANEOS EOS (*Thompson and Lauson*, 1972). In the present calculations, we assume that the core is made of dunite, as representative of rock, yielding typical mean densities in the core ~6–7 g cm⁻³. Comparative calculations with water ice cores, corresponding to a lower mean density ~3 g cm⁻³, change only slightly the mass-radius relationship for planets of identical core and total mass. As mentioned above, the specific heat of the core is calculated with the ANEOS EOS so that the core contributes to the planet thermal evolution. Figure 1 displays the evolution of the radius and luminosity for 1 and 4 M_{Jup} planets, respectively. The solid and long-dashed lines correspond to different initial radii for the newborn planet, namely 3 and 1.3 R_J for the 1 M_{Jup} planet and 4 and 1.3 R_J for the 4 M_{Jup} planet, respectively. The 1.3 R_J case is similar to the calculations of *Fortney et al.* (2005), based on the aforementioned formation model of *Hubickyj et al.* (2005). Note that these values are comfortably smaller than the Roche lobe limits at 5.2 AU from a Sun [≈530 R_J and ≈830 R_J for a 1 M_{Jup} and a 4 M_{Jup} planet, respectively (*Eggleton*, 1983)]. The t = 0 age for the planet evolution corresponds to the end of its formation process, just after the runaway gas accretion (phase 3) has terminated. This planet-formation timescale, namely ~2–3 m.y., should thus be added to the ages displayed in Fig. 1 for the planet evolution. As seen in the figure, the difference between these initial conditions, namely a factor of ~2–3 in radius, affects the evolution of the planet for 10^7–10^8 yr, depending on its mass. This reflects the significantly different thermal timescales at the begining of the evolution (t = 0) for the different initial radii, namely $t_{KH} = GM^2/R_L = 3 \times 10^5$ and ~5×10^7 yr, respectively, for 1 M_{Jup}. The smaller the initial radius, the larger the consequences. Unfortunately, as mentioned above, uncertainties in the models of planet formation prevent an accurate determination of the initial radius of the newborn planet. Changing the maximum accretion rate or the opacity in *P96*, for example, or resolving the radiation transfer in the accretion shock, will very likely affect the planet radius within a large factor. Therefore, at least within the present uncertainties of the planet-formation models, young gaseous planets with cores and heavy elements in their envelopes can easily be 10× brighter than suggested by the calculations of *Fortney et al.* (2005) and thus are not necessarily "faint" in the sense that they can be as bright as pure gaseous, solar composition H/He objects of the same mass, i.e., low-mass brown dwarfs. In the same vein, the initial gravity of the planet cannot be determined precisely and can certainly vary within at least an order of magnitude between log g ~ 2 and log g ~ 3 for a Jupiter mass. Detections of young exoplanet luminosities with reasonable age determinations, i.e., within ≤10 m.y. uncertainty, for instance in young clusters, would provide crucial information to help narrowing these uncertainties.

2.2.2. Effect of irradiation. We now examine the effects of irradiation on the evolution of close-in exoplanets, the so-called "hot-Jupiter" and "hot-Neptune" objects. Inclusion of the effect of irradiation of the parent star on the structure and evolution of short-period exoplanets has been considered by several authors. However, only a few of these calculations are based on consistent boundary conditions between the internal structure and the *irradiated* atmosphere profiles. Such a proper boundary condition, implying consistent opacities in the atmosphere and interior structure calculations, is determinant for correct evolutionary calculations of irradiated planets because of the growing external radiative zone, which pushes the internal adiabat to deeper levels (*Guillot et al.*, 1996; *Seager and Sasselov*, 1998; *Barman et al.*, 2001, 2005). The outgoing flux at the surface of the planet now includes the contribution from the incoming stellar flux $\mathcal{F}_\star$

$$\mathcal{F}_{out} = \sigma T_4^{eff} + \mathcal{F}_{inc} = \sigma T_4^{eff} + f\left(\frac{R_\star}{a}\right)^2 \mathcal{F}_\star$$
$$= \sigma T_4^{eff} + (1 - A)\mathcal{F}_{inc} + A\mathcal{F}_{inc} \qquad (1)$$

In equation (1), σT_4^{eff} denotes the intrinsic internal flux of the planet, A the Bond albedo, and the last term on the righthand side of the equation is the reflected part of the spectrum. The factor f is a geometrical factor characterizing the stellar flux redistribution over the planet surface (f = 1 implies the flux is redistributed over π steradians; f = 1/2

implies that it is redistributed over the dayside only, as intuitively expected for tidally locked planets; and f = 1/4 implies that it is distributed over the entire planet surface). *Burkert et al.* (2005) have performed hydrodynamic calculations related to the heating of the nightside of synchronously locked planets. With reasonable assumptions for the opacity in the atmosphere, these authors find that the temperature difference between the dayside and the nightside could be in the ~200–300 K range, not enough to make an appreciable difference in the radius. Previous estimates (*Showman and Guillot*, 2002; *Curtis and Showman*, 2005; *Iro et al.*, 2005), however, predict day/night temperature differences about twice this value, and this issue needs to be further explored. From equation (1), the evolution of the irradiated planet now reads

$$L = -\int_M T \frac{dS}{dt} + 4\pi R_p^2 \sigma T_{eq}^4 + L_{reflected} \qquad (2)$$

where $T_{eq}^4 = \frac{1-A}{\sigma}\mathcal{F}_{inc} = (1 - A)f\left(\frac{R_\star}{a}\right)^2 T_\star^4$ denotes the planet equilibrium temperature, i.e., the temperature it would reach after exhausting all its internal heat content and contraction work ($T_{eff} \to 0$).

As shown in *Chabrier et al.* (2004) and *Baraffe et al.* (2005), consistent calculations between the irradiated atmospheric structure and the internal structure, which fix the boundary condition for the planet photospheric radius, reproduce the radii of all observed transit planets so far, without additional sources of internal heating, except for HD 209458b, which remains a puzzle (see Fig. 1 of *Baraffe et al.*, 2005). These calculations were based on planet interior models composed entirely of hydrogen and helium and do not include either a central core or heavy element enrichment in the envelope. The effect of a central rocky core on irradiated planet evolution has been examined by *Bodenheimer et al.* (2003) but with simplified (Eddington) boundary conditions between the atmosphere and the interior. These authors found that for planets more massive than about 1 M_{Jup} the decrease in radius induced by the presence of a core is about 5%, in agreement with previous estimates for nonirradiated planets (*Saumon et al.*, 1996). The effect, however, will be larger for less-massive planets, including the recently discovered hot Neptunes. This issue has been addressed recently by *Baraffe et al.* (2006), with proper, frequency-dependent atmosphere models. These authors find that, for a Saturn-mass planet (~100 $M_\oplus$), the difference in radius between a pure H/He planet and a planet with a 6 $M_\oplus$ core and a mass fraction of heavy element in the envelope Z = $M_{Z,env}/M_{env}$ = 10%, as predicted by the formation model, is $R_Z/R_{H/He} \simeq 0.92$, i.e., a ~ 9% effect, possibly within present limits of detection.

A point of concern in the present calculations is that the boundary condition between the irradiated atmospheric profile and the interior profile is based on atmosphere models of solar composition. Most of the transiting planets, however, orbit stars that are enriched in metals and the planet atmosphere is supposed to have the same enrichment. Calculations including such an enrichment are underway (see section 5). The effect, however, is likely to be small for two reasons. First, the enrichment of the parent stars remain modest, with a mean value (M/H) ≈ 0.2–0.3 (*Santos et al.,* 2005). Second, irradiated atmospheric profiles display an extensive radiative zone (see above) so that gravitational settling may occur even though, admittedly, various mixing mechanisms (e.g., decay of gravitational waves, convective overshooting, winds) could keep gaseous heavy elements suspended in radiative regions. Planets at large enough orbital distances for the effect of irradiation on the atmospheric thermal profile to be negligible, however, should display significant heavy element enrichment in their atmosphere, as observed for the giant planets of the solar system.

2.2.3. Evaporation. The question of the long-term stability of gaseous close-in extrasolar giant planets has been raised since the discovery of 51 Peg b. In the framework of Jeans approximation, the evaporation rate Φ (hydogen atoms cm^{-2} s^{-1}) is given by *Chamberlain and Hunten* (1987)

$$\Phi = \frac{n_{exo}}{2\sqrt{\pi}} \sqrt{\frac{2kT_{exo}}{m}} \exp(-X)(1 + X) \qquad (3)$$

where n_{exo} and T_{exo} are the number density and the temperature at the exobase (the level at which the mean free path of hydrogen atoms equals the scale height) and $X = v_\infty^2/v_0^2$ is the escape parameter, $v_\infty = (2GM_p/R_p)^{1/2}$ the planet escape velocity, and $v_0 = (2kT/m)^{1/2}$ the mean thermal velocity at T_{exo}. The first estimates of the evaporation rate of hot Jupiters (e.g., *Guillot et al.,* 1996) were obtained by using the equilibrium temperature T_{eq} instead of the unknown value of T_{exo}. For a typical 51 Peg b-like hot Jupiter (1 M_{Jup}, $T_{eq} \simeq 1300$ K), the escape parameter $X = v_\infty^2/v_0^2$ is then found to be larger than 150, whereas escape rates become significant for values below 20. On this basis, hot Jupiters were claimed to be stable over the lifetime of their star. However, T_{eq} is not the relevant temperature for thermal escape, which occurs in the exosphere, where heating is due to XUV irradiation. With simple assumptions, several authors estimated that the exospheric temperature could be of the order of 10,000 K (X < 20) and thus attempted the observation of the escaping H (*Moutou et al.,* 2001). *Lammer et al.* (2003, hereafter *L03*) showed that the conditions allowing the use of Jeans approximation (hydrostatic equilibrium and negligible cooling by the escape itself) are not met in hot Jupiters because of the considerable heating by stellar XUV. The application of Jeans escape yield unrealistically high exospheric temperatures (X < 1) in contradiction with the required hydrostatic hypothesis. They concluded that hot Jupiters should experience hydrodynamic escape, without a defined exobase, where the upper atmosphere is continuously flowing to space and maintained at low temperature (≪10,000 K) by its expansion. In this *blow-off* model, the

escape rate of the main atmospheric component, H, is only limited by the stellar XUV energy absorbed by the planet and is given by

$$\dot{M} = 3\left(\frac{R_{XUV}}{R_p}\right)^3 \varepsilon \mathcal{F}_\star/(G\rho) \qquad (4)$$

where ρ is the mean planetary density and $\mathcal{F}_\star$ is the stellar flux, averaged over the whole planet surface, including both the contribution in the 1–1000 Å wavelength interval and the 1215 Å Lyman-α line. R_{XUV} is the altitude of the (infinitely thin) layer where all the incoming XUV energy is absorbed, while R_p is the radius observed in the visible during a transit. Here, ε would represent the heating efficiency, or the fraction of the incoming XUV flux that is effectively used for the escape. *L03* applied a hydrodynamic model (*Watson et al.,* 1981) and estimated $R_{XUV}/R_p \approx 3$ for orbital distances closer than 0.1 AU. By assuming $\varepsilon = 1$ (or, in other words, that escape and expansion are the only cooling processes) they inferred the physical upper limit for the XUV-induced thermal escape rate to be 10^{12} g/s for HD 209458b at the present time. Considering the evolution of XUV emission of main-sequence G stars (*Ribas et al.,* 2005) and the significantly lower density of young gaseous planets implies rates 10–100× higher in the early history of the hot Jupiters. Using these simple arguments, *L03* suggested that hot Jupiters could have been initially much more massive, although more detailed models are needed to better estimate the effective hydrodynamic escape rate.

Independent of this theoretical approach, *Vidal-Madjar et al.* (2003, hereafter *VM03*) measured the absorption in the Lyman-α line of HD 209458, using STIS onboard HST, during the transit of its planet. The decrease of luminosity they found is equivalent to the transit of a $R_{Ly\alpha} = 3 R_p$ opaque disk. Although this observation seems to be consistent with *L03*, a larger but optically thin hydrogen cloud can also account for the observation. In fact, by noticing that the Roche lobe radius of the planet was 3–4 R_p, *VM03* concludes that part of the observed hydrogen must consist in an escaping cometary-like tail. They estimated that the absorption implies an escape rate not lower than 10^{10} g/s.

The truncation of the expanded atmosphere by the Roche lobe, which was not considered by *L03*, has obviously to be taken into account in the mass loss process. *Lecavelier et al.* (2004) proposed a *geometrical blow-off* model in which a hot exobase (~10,000 K), defined according to Jeans approximation, reaches the Roche lobe radius. This yields enhanced loss rates compared to a classical Jeans calculation that would not take into account the gravity field and the tidal distorsion of the atmosphere. *Jaritz et al.* (2005) argued that, although geometrical blow-off should occur for *some* of the known hot Jupiters, HD 209458b expands hydrodynamically up to 3 R_p without reaching the L1 Lagrange point at which the Roche lobe overflow occurs. If confirmed, the debated observation of O and C in the expanded atmosphere of HD 209548b (*Vidal-Madjar et al.,* 2004) would favor the hydrodynamic regime, which is required to drag heavy species up to the escaping layers. However, the STIS instrument is no longer operational and similar observations will have to wait new EUV space observatories. Another indirect confirmation of the hydrodynamical regime is the absence of an H_α signature beyond R_p (*Winn et al.,* 2004). This can be explained by the low temperature (<5000 K) expected in the hydrodynamically expanding atmosphere. *Yelle* (2004) published a detailed model of the photochemistry, radiative budget, and physical structure of the expanding upper atmosphere of hot Jupiters and derived a loss rate of 10^8 g/s, about a factor of 100 lower than the value inferred by *VM03* from the observation. Recently, *Tian et al.* (2005) published an improved, multilayer hydrodynamical model (compared to Watson), in which the energy deposition depth and the radiative cooling are taken into account. Rates on the order of 5×10^{10} g/s are found, although they also depend on an arbitrary heating efficiency μ. It is important to note that the composition of the expanding atmosphere in heavy elements can dramatically affect its behavior, mainly by modifying the radiative transfer (absorption and cooling).

Nonthermal escape is much more difficult to estimate as it depends on the unknown magnetic field of the planet and stellar wind. Thermal escape is usually considered as the dominant mass loss process (*Griessmeier et al.,* 2004), but considering the complexity of the magnetic coupling between the star and the planet at orbital distances closer than 0.045 AU, unexpected nonthermal processes may still dominate the evaporation of some short-period exoplanets.

VM03 and *L03* both suggested that the evaporation could lead to the loss of a significant fraction of the initial planetary mass and even to the evaporation of the whole planet, possibly leaving behind a dense core. In order to investigate the possible effects on the mass-radius evolution of close-in exoplanets, *Baraffe et al.* (2004, 2005) included the maximum XUV-limited loss from *L03* in the simulated evolution of a coreless gaseous giant planet, taking also into account the time dependency of the stellar XUV luminosity, calibrated on observations (*Ribas et al.,* 2005). These studies showed that, even at the maximum loss rate, evaporation affects the long-term evolution of the radius only *below an initial critical mass*. For initial masses below this critical mass, the planet eventually vanishes in a very short but dramatic runaway expansion. This critical mass depends of course on the escape rate considered and drops to values much below 1 M_{Jup} when using lower rates like the ones predicted by Yelle, by Tian et al., and by Lecavelier et al. (*Baraffe et al.,* 2006). One interesting result of the Baraffe et al. work needing further attention is that evaporation does not seem to explain the surprisingly large visible radius (R_p) of HD 209458b, except if this planet is presently seen in its last and brief agony, which seems extremely unlikely. The explanation for the large observed radius of HD 209458b thus remains an open question.

One may wonder whether this runaway evaporation phase can be studied with hydrostatic atmosphere models

and quasistatic evolution models. Atmospheric hydrostatic equilibriumis valid for values of the escape parameter X > 30. For a hot Jupiter at 0.045 AU, values of X below 30 are found in the thermosphere, where the temperature increases above 7000 K, at R > 1.1 R_p (see, e.g., *Yelle,* 2004). Such levels, with number densities n < 10^9 cm^{-3}, lie well above the levels where the boundary con-dition applies, i.e., near the photosphere with gas pressures P ~ 10^{-5}–10 bar. The quasistatic evolution assumption is justified by the fact that, even though the characteristic timescale of evaporation, $M/\dot{M}$, can become comparable to or even shorter than the Kelvin-Helmholtz timescale, t_{KH} ~ 2 Gm^2/(RL), it remains much larger than any hydrodynami-cal timescale. The present runaway phase, indeed, refers to a *thermal* runaway like, e.g., thermal pulses in AGB stars, characterized by a thermal timescale. Quasistatic evolution thus remains appropriate to study this mass loss process, at least until truly hydrodynamic processes affect the planet photosphere.

More recently, *Baraffe et al.* (2006) examined the possibility for lower-mass hot-Neptune planets (1 M_{Nep} = 18 $M_\oplus$ ≃ 0.06 M_{Jup}) to be formed originally as larger gaseous giants that experienced significant mass loss during their evolution. Depending on the value of the evaporation rates, these authors showed that presently observed (few-gigayear-old) Neptune-mass irradiated planets may originate from objects of over 100 $M_\oplus$ if the evaporation rate reaches the maximum *L03* value. For ~10–20× lower rates, as suggested, e.g., by the hydrodynamical calculations of *Tian et al.* (2005), the hot Neptunes would originate from objects of ~50 $M_\oplus$, meaning that the planet has lost more than two-thirds of its original mass. For rates a factor of 100 smaller than *L03*, the effect of evaporation is found to become more modest, but a planet could still lose about one-fourth of its original mass due to stellar-induced evaporation. These calculations, even though hampered by the large uncertainty in the evaporation rates, show that low-mass irradiated planets that lie below the aforementioned critical initial mass may have originally formed as objects with larger gaseous envelopes. This provides an alternative path to their formation besides other scenarios such as the core-collision model (*Brunini and Cionco,* 2005).

3. GRAVITATIONAL COLLAPSE OF PRESTELLAR CORES

After having examined the status of planet formation and evolution, we now turn to the formation and the early stages of evolution of stars and brown dwarfs. In this section, we first review our current knowledge of the gravitational collapse of a protostar. We then will focus on the importance of nonspherical effects in the collapse.

3.1. One-Dimensional Models

Numerous authors have extensively considered the one-dimensional collapse of a spherical cloud. One of the most difficult aspects of the problem is the treatment of the cooling of the gas due to collisional excitation of gas molecules, particularly during the late phase of the collapse when the gas becomes optically thick. Radiative transfer calculations coupled to hydrodynamics are then required. However, as noted originally by *Hayashi and Nakano* (1965) and confirmed by various calculations (*Larson*, 1969; *Masunaga et al.*, 1998; *Lesaffre et al.*, 2005), the gas remains nearly isothermal for densities up to 10^8–10^9 cm^{-3}, making the isothermal assumption a fair and attractive simplification.

3.1.1. The isothermal phase. The isothermal phase has been extensively investigated both numerically and analytically. In particular, a family of self-similar solutions of the gravitational contraction has been studied in detail by *Penston* (1969), *Larson* (1969), *Hunter* (1977), *Shu* (1977), and *Whitworth and Summers* (1985). As shown by these authors, there is a two-dimensional continuous set of solutions (taking into account the solutions that present weak discontinuities at the sonic point) determined, for example, by the value of the central density with bands of allowed and forbidden values. Two peculiar cases have been carefully studied, the so-called Larson-Penston and Shu solutions. The first case presents supersonic velocities (up to 3.3 c_s for large radius, where c_s is the isothermal sound velocity) and is representative of very dynamical collapses. The second case assumes a quasistatic prestellar phase so that, at t = 0, the density profile corresponds to the singular isothermal sphere (SIS) and is given by $\rho_{SIS} \simeq c_s^2/2\pi Gr^2$. A rarefaction wave that propagates outward is launched and the collapse is inside-out. For both solutions the outer density profile is $\propto r^{-2}$, whereas in the neighborhood of the central singularity, the density is $\propto r^{-1.5}$.

Although the self-similar solutions depart significantly from the numerical calculations, they undoubtedly provide a physical hint on the collapse and the broad features described above appear to be generic and are observed in the simulations. Following the work of *Foster and Chevalier* (1993), various studies have focused on the collapse of a nearly critical Bonnor-Ebert sphere (*Ogino et al.*, 1999; *Hennebelle et al.*, 2003). This scenario presents a number of interesting features that agree well with observations of dense cores like those observed in the Taurus molecular cloud (*Tafalla et al.*, 1998; *Bacmann et al.*, 2000; *Belloche et al.*, 2002). Namely, (1) the density profile is approximately flat in the center during the prestellar phase; (2) during the prestellar phase there are (subsonic) inward velocities in the outer layers of the core, while the inner parts are still approximately at rest; and (3) there is an initial short phase of rapid accretion (notionally the Class 0 phase), followed by a longer phase of slower accretion (the Class I phase). This last feature is an important difference with the self-similar solutions, which have a constant accretion rate. The typical accretion rates obtained numerically are between the value of the Shu solution ($\dot{M}_{SIS} \simeq c_s^3/G$) and the Larson-Penston solution (about $50 \times c_s^3/G$).

Motivated by the observations of much faster infall (see, e.g., *Di Francesco et al.*, 2001), triggered collapses have been considered (*Boss*, 1995; *Hennebelle et al.*, 2003, 2004; *Motoyama and Yoshida*, 2003). Much larger accretion rates, higher cloud densities, and supersonic infall can be obtained

in this context. A close comparison between a strongly triggered collapse model and the Class 0 protostar IRAS 4A has been performed with success by *André et al.* (2004).

3.1.2. Second collapse and formation of a young stellar object. When the density becomes larger than $\simeq 10^{10}$ cm^{-3} the gas becomes optically thick. The isothermal phase ends and the thermal structure of the collapsing cloud is nearly adiabatic. A thermally supported core forms (*Larson*, 1969; *Masunaga et al.*, 1998). When matter piles up by accretion onto this hydrostatic core, its temperature and density increase because of the stronger self-gravitating field. When the density of the first Larson core reaches about 10^{-7} g cm^{-3}, temperature is about 2000 K and the H_2 molecules start to dissociate (*Saumon et al.*, 1995). Most of the gravitational energy goes into molecular dissociation energy so that the effective adiabatic exponent $\gamma = 1 + \frac{dLnT}{dLn\rho}$ drops to about 1.1, significantly below the critical value $\gamma = 4/3$ (*Larson*, 1969; *Masunaga and Inutsuka*, 2000). Thermal pressure is therefore unable to support the hydrostatic core and the collapse restarts.

During the second collapse the temperature is roughly constant and close to 2000 K. When all the H_2 molecules have been dissociated into atomic hydrogen, the effective adiabatic exponent increases again above $\gamma = 4/3$ and the star forms. The timescale of the second collapse is about the freefall time of the first Larson core, ~1 yr, very small compared with the timescale of the first collapse, which is about 1 m.y.

Both the first and second Larson cores are bounded during all the collapse of the cloud by an accretion shock in which the kinetic energy of the infalling material is converted into heat. The effect of the accretion shock onto the protostar has been first considered by *Stahler et al.* (1980) and *Stahler* (1988). The influence of accretion on the evolution of the protostar will be examined in section 4.

3.2. Influence of Rotation and Magnetic Field

Here we examine the main influence of rotation and magnetic field on the cloud collapse, leaving aside three-dimensional effects, which are considered in section 3.3.

3.2.1. Effects of rotation. Rotation induces a strong anisotropy in the cloud, slowing down and finally stopping the equatorial material. *Ulrich* (1976) studied exact solutions for a rotating and collapsing cold gas and showed that the equatorial density of the collapsing *envelope* is larger than in the absence of rotation. This has been further confirmed by *Terebey et al.* (1984) using an analytical solution that generalizes the collapse of the SIS (*Shu*, 1977) in the case of a slowly rotating cloud. In the case of a 1 M$_\odot$ initially slowly rotating core ($\beta = E_{rot}/E_{grav} \simeq 2\%$), *Hennebelle et al.* (2004) estimate that the equatorial density of the collapsing envelope in the inner part of the cloud ($\lesssim 2000$ AU) can be 2–3× higher than the axial one for a slow collapse and up to 10× higher in case of strongly compressed clouds.

The formation, growth, and evolution of the rotationally supported disk has been modeled analytically by *Cassen and Moosman* (1981) and *Stahler et al.* (1994). The growth of the disk drastically depends on the angular momentum distribution, j. The centrifugal radius is approximately $r_d \simeq j^2/GM_{int}$, where M_{int} is the mass inside the sphere of radius r_d. Therefore, for initial conditions corresponding to a SIS in solid-body rotation, $M_{int} \propto r$ and $j \propto r^2$, implying $r_d \propto M_{int}^3$. On the contrary, starting with a uniform density sphere in solid-body rotation, $M_{int} \propto r^3$ and $r_d \propto M_{int}^{1/3}$, which implies much bigger disks. Such disks are indeed found in hydrodynamical simulations of collapsing dense core initially in slow rotation. For 1 $M_\odot$ dense cores with $\beta \simeq 2\%$ the size of the disk during the Class 0 phase is about 200 AU.

The effect of the rotation on the forming protostar itself has been weakly explored. Two-dimensional equilibrium sequences of rotating protostars have been calculated by *Durisen et al.* (1989).

3.2.2. Effects of the magnetic field. The magnetic field has been proposed to be the main support of the dense cores against the gravitational collapse (*Shu et al.*, 1987) and the explanation for the low star-formation efficiency in the galaxy. Although this theory is now challenged by the origin of the support being mainly due to turbulence (for a recent review, see *Mac Low and Klessen*, 2004), the magnetic field certainly plays an important role in the formation of the protostar.

The magnetically controlled collapse has been carefully investigated with one-dimensional numerical simulations (e.g., *Mouschovias et al.*, 1985). It has been found that the collapse proceeds in two main phases; first, a quasistatic contraction of the flattened cloud occurs through ambipolar diffusion, and second, once a supercritical core has developed, it collapses dynamically. Quantitative estimates of the prestellar cloud lifetime are given in *Basu and Mouschovias* (1995). In strongly subcritical clouds (initial mass-to-flux ratio over critical mass-to-flux ratio smaller than 1/10) the formation of the protostar requires about 15 freefall times, whereas in a transcritical cloud (initial mass-to-flux ratio equal to critical mass-to-flux ratio), it requires about 3 freefall times. *Ciolek and Basu* (2000) showed that the collapse of the well-studied prestellar cloud L1544 is compatible with this core being transcritical. Note that, although the ambipolar diffusion timescale is much larger than the admitted star-formation timescale, namely a few dynamical timescales, recent two-dimensional simulations of compressible turbulence by *Li and Nakamura* (2004) suggest that enhanced ambipolar diffusion occurs through shock compression.

The transfer of angular momentum is another important effect of magnetic fields. It occurs through the emission of torsional Alfvén waves, which carry away the angular momentum (*Mouschovias and Paleologou*, 1980; *Basu and Mouschovias*, 1995). Since this process is more efficient if the rotation axis is perpendicular to the field lines (instead of parallel), alignment between the magnetic field and the rotation axis is rapidly achieved. During the supercritical core formation epoch the angular velocity achieves a limiting profile proportional to 1/r (*Basu*, 1998). Such a profile leads to centrifugal disks growing as $r_d \propto M_{int}$ and thus in-

termediate between the very massive disks found in hydrodynamical simulations and the low-mass disks predicted by the SIS in solid-body rotation model.

A very important difference between hydro and MHD cases is the presence of outflows in the latter ones, which have been found only recently in numerical simulations of collapsing protostellar core. They are described in the next section.

Finally, magnetic fields may induce a different mode of accretion. Motivated by the observations of T Tauri stars, which are surrounded by a disk from which they accrete material while having rotation velocities too small to be compatible with the conservation of angular momentum, *Königl* (1991) proposed that most of the accreted matter may be channeled along the magnetic field lines from the disk to the poles of the star. The angular momentum is then extracted from the infalling gas by the magnetic field. The accretion onto the star occurring over a small fraction of its surface, significant differences with the case of spherical accretion are expected (*Hartmann et al.*, 1997), an issue addressed in section 4.

3.3. Three-Dimensional Models

3.3.1. Axisymmetry breaking, transport of angular momentum, and fragmentation. One of the main new effects that appear in three-dimensional calculations of a collapsing cloud is the axisymmetry breaking of the centrifugal disk. This occurs when its rotational energy reaches about 40% of its gravitational energy. Strong spiral modes develop that exert a gravitational torque leading to a very efficient outward transport of angular momentum, allowing accretion onto the central object to continue. This effect has been modeled analytically (*Laughlin and Różyczka*, 1996) and found by many authors in the numerical simulations (e.g., *Matsumoto and Hanawa*, 2003).

The fragmentation of the dense cores and the formation of multiple systems are major challenges in the field, and entire chapters of this book are dedicated to this subject. We refer to those as well as to the review of *Bodenheimer et al.* (2000b) for a comprehensive discussion of this topic.

3.3.2. Multidimensional treatment of the second collapse. The second collapse leading to the formation of the protostar has been modeled in two or three dimensions by various authors with two main motivations, namely modeling outflows and jets and explaining the formation of close binaries. Due to the large range of dynamical scales involved in the problem, the first calculations started from the first Larson core (*Boss*, 1989; *Bonnell and Bate*, 1994). With the increase of computational power, calculations starting from prestellar core densities (e.g., 10^4 cm^{-3}) have been performed (*Bate*, 1998; *Tomisaka*, 1998; *Banerjee and Pudritz*, 2006). For computational reasons, the radiative transfer has not been calculated self-consistently yet. Instead, piecewise polytropic equations of state that mimic the thermodynamics of the cloud are often used (*Bate*, 1998; *Jappsen et al.*, 2005). More recently, *Banerjee and Pudritz* (2006) used a tabulated cooling function that takes into account the microphysics of the gas with an approximated opacity.

Bonnell and Bate (1994) conclude that fragmentation is possible during the second collapse. However since the mass of the stars is on the order of the Jeans mass, it is very small (0.01 $M_\odot$) and therefore they have to accrete most of their final mass. *Banerjee and Pudritz* (2006) form a close binary (with a separation of about 3 $M_\odot$) as well in their MHD adaptive mesh refinement calculations. Like *Bate* (1998), they find that inside the large outer disk (60–200 AU) an inner disk of about 1 AU forms.

Tomisaka (1998) and *Banerjee and Pudritz* (2006) report outflows and jets during the collapse that contribute to carry away a large amount of angular momentum. The physical mechanisms responsible for the launching of these outflows can be understood in terms of magnetic tower (*Lynden-Bell, 2003*). An annulus of highly wound magnetic field lines is created by the rotational motions and pushes the surrounding infalling material outward. The physics involved in the jet are somehow different and are based on the magnetocentrifugal mechanism proposed by *Blandford and Payne* (1982).

In the three-dimensional simulations below, we investigate inner core formation resulting from the collapse of a 10^{-3} $M_\odot$ Bonnor-Ebert sphere with densities and temperatures characteristic of the second core, namely $\rho \simeq 10^{-9}$ g cm^{-3} and $T \simeq 2000$ K. We focus on the influence of tridimensional effects on the accretion geometry and on the inner profile of the core. Figure 2 displays the evolution of the accretion rate $\dot{M}$ during the second collapse as well as the average angle of accretion $\langle \cos \theta \rangle$, i.e., the average angle between the vertical axis in spherical coordinates and the infalling gas. As seen in the figure, the accretion rate decreases immediately from a large value close to the Larson-Penston prediction to a smaller Bondi-Hoyle or Shu like value, c_s^3/G, and accretion occurs over a very limited fraction of the protostar surface, $\langle \cos \theta \rangle < 0.3$ (spherical accretion would imply $\langle \cos \theta \rangle = 0.5$), so that most of the surface can freely radiate its energy. This is important for the subsequent evolution of the object, as examined in the next section. The consequences of three-dimensional effects on the density and temperature profiles of the protostar are illustrated in Fig. 3, which displays the equatorial density and temperature profiles of the second Larson core at four time steps. Rotation leads to lower central densities and temperatures and to a more extended central core, as noted already by *Boss* (1989). These features are relevant for the internal energy transport — radiation vs. convection — and the initial deuterium burning. They also confirm that spherical collapse, although providing interesting qualitative information, cannot provide accurate initial conditions for PMS tracks as it will overestimate (1) the internal temperature of the protostar and (2) the surface fraction covered by accretion, thus preventing the object to contract at a proper rate.

4. EFFECT OF ACCRETION ON THE EARLY EVOLUTION OF LOW-MASS OBJECTS

4.1. Observed Accretion Rates

Intensive investigations of accretion in young clusters and star-formation regions show signatures of this process over a wide range of masses, down to the substellar regime (see recent work by *Kenyon et al.,* 2005; *Mohanty et al.,* 2005; *Muzerolle et al.,* 2005, and references therein). In the youngest observed star-forming regions, such as ρ Ophiuchus with an age $\lesssim$1 m.y., the fraction of accretors is greater than 50%, independent of the mass (*Mohanty et al.,* 2005). This fraction decreases significantly with age, a fact interpreted as a decrease of the accretion rates below the observational limits, $\lesssim$10–12 $M_\odot$/yr. The timescale for accretion rates to drop below such a measurable limit is ~5 m.y. In some cases, however, accretion continues up to ~10 m.y. Note, however, that these age estimates for young clusters remain very uncertain, since they are usually based on evolutionary tracks that are not reliable at such ages (*Baraffe et al.,* 2002). Indeed, as demonstrated in *Baraffe et al.* (2002), unknown initial conditions and unknown convection efficiency (mimicked in stellar evolution calculations by the mixing length parameters) during the early PMS contracting phase, characterized by short Kelvin-Helmholtz timescales ($\lesssim$10^6 yr), can drastically affect the contraction track of a young object in the Herzsprung-Russell (HR) diagram. Therefore the age and/or mass of young objects cannot be determined accurately from observations, leading to very uncertain inferred disk lifetimes. However, even though the absolute timescales are uncertain, the trend of accretion rates decreasing with time is less questionable.

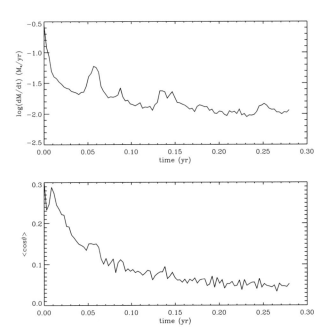

Fig. 2. Accretion rate (in $M_\odot$/yr) and average angle of accretion during the three-dimensional simulation of the second collapse of a 10^{-3}-$M_\odot$ core.

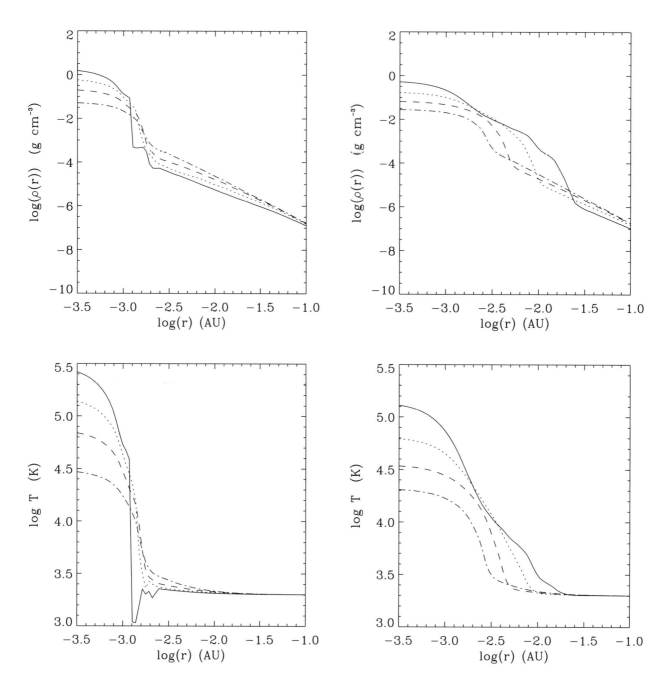

Fig. 3. Radial density and temperature profiles along the equatorial direction during the collapse of a 10^{-3}-$M_\odot$ core, for four different time steps. Left column: one-dimensional (spherical) collapse (dash-dot = 2.860 yr, dash = 2.866 yr, dot = 2.879 yr, solid = 2.906 yr); right column: three-dimensional collapse of a rotating core (dash-dot = 3.594 yr, dash = 3.606 yr, dot = 3.634 yr, solid = 3.704 yr). Note the different behavior of the accretion shock in the two cases.

A sharp decrease of accretion rates with mass is also observed, with a correlation $\dot{M} \propto M^2$, all the way from solar-mass stars to the smallest observed accreting brown dwarfs, i.e., ~0.015 $M_\odot$(*Muzerolle et al., 2005*). Typically, in the low-mass star regime (M ~ 0.2–1 $M_\odot$), the accretion rates vary between 10^{-10} $M_\odot$/yr and 10^{-7} $M_\odot$/yr, whereas below ~0.2 $M_\odot$ and down to the brown dwarf (BD) regime, accretion rates range from ~5 × 10^{-9} $M_\odot$/yr to 10^{-12} $M_\odot$/yr (*Muzerolle et al.,* 2003; *Natta et al.,* 2004; *Mohanty et al.,* 2005). Last but not least, observations now show similari-

ties of accretion properties between higher-mass stars and low-mass objects, including brown dwarfs, suggesting that stars and brown dwarfs share similar formation histories.

4.2. Modeling the Effect of Accretion in Young Objects

On the theoretical front, *Stahler* (1983, 1988) has investigated the effect of *spherical accretion* onto protostars, defining the concept of a birth line, a locus in the HR diagram where young stars first become optically visible when accretion ends. Stahler suggested that when the infall of material onto the protostar, responsible for its obscuration, ceases abruptly, the central object becomes an optically bright T Tauri star.

Since this benchmark work, progress in the observations of young objects have now shown that accretion occurs rapidly through a *disk*, as discussed in sections 3.2 and 3.3 and illustrated in Fig. 2. The timescale for disk accretion is much longer than the strongly embedded protostellar phase, as illustrated by the short lifetime of the Class 0 objects compared with Class I. Several studies have investigated the effect of accretion geometry on evolutionary tracks for low-mass and high-mass stars. These calculations generally assume that (1) accretion takes place over a small fraction δ of the stellar surface and (2) a dominant fraction of the accretion luminosity is radiated away and thus does not modify the protostar internal energy content (*Mercer-Smith et al.*, 1984; *Palla and Stahler*, 1992; *Hartmann et al.*, 1997; *Siess et al.*, 1997), in contrast to the assumptions of *Stahler* (1988). Under these conditions, the luminosity of the accreting object is given by

$$L = \delta \cdot L_{acc} + L_D$$
$$- (1 - \delta) \int_M \left\{ T \left(\frac{dS}{dt} \right)_m - T \left(\frac{\partial S}{\partial m} \right)_t \dot{m} \right\} dm' \quad (5)$$

On the righthand side of equation (5), the first term is the accreted luminosity, supposed to be entirely radiated away, L_D is the D-burning luminosity, including freshly accreted deuterium, while the last term stems from the extra entropy at constant time due to the accreted mass [where $\dot{m} \equiv \dot{m}(m')$ is the accreted rate per mass shell]. Assumption (1) is indeed relevant for thin disk accretion from a boundary layer or for magnetospheric accretion where the gas falls onto the star following magnetic accretion columns. It implies that most of the stellar photosphere can radiate freely and is unaffected by a boundary layer or accretion shocks. Assumption (2) depends on the details of the accretion process, which remain very uncertain. In a attempt to study the impact of such an assumption on evolutionary models, one can assume that some fraction of the accreted matter internal energy is transferred to the protostar outer layers, the other fraction being radiated away. This extra supply of internal energy, per unit mass of accreted matter, is proportional to the gravitational energy, $\varepsilon GM/R$, with $\varepsilon < 1$ a free parameter. As pointed out by *Hartmann et al.* (1997), the structure of an accreting object before or after ignition of deuterium, and the fact that it will be fully convective or will develop radiative layers, strongly depends on ε, and to a lesser extent on assumption (1). For large values of ε, convection can indeed be inhibited, even after deuterium ignition (see, e.g., *Mercer-Smith et al.*, 1984). Deuterium burning in the protostellar phase is also a central issue. The key role played by deuterium burning on the properties of an accreting object and its location in the HR diagram was

highlighted by *Stahler* (1983, 1988). Whether the deuterium fusion occurs in a fully convective object or in radiative layers is thus an important issue that affects significantly the structure of an accreting object.

Assuming that only a very small fraction of the thermal energy released by accretion is added to the stellar interior, most of it being radiated away, *Hartmann et al.* (1997) (see also *Siess et al.*, 1999) showed that, depending on its evolutionary stage, an accreting low-mass star expands less or contracts more than a nonaccreting similar object. Consequently, an accreting object looks older in a HR diagram, because of its smaller radiating surface for the same internal flux, compared to a nonaccreting object at the same mass and age. This stems essentially from the accretion timescale becoming on the order of the Kelvin-Helmholtz timescale, for a given accretion rate, $M/\dot{M} \approx t_{KH}$, so that the contracting object does not have time to expand to the radius it would have in the absence of accretion. An extension of these studies to the brown dwarf regime confirms these results, in the case of significant accretion rate and no thermal energy addition due to accretion ($\varepsilon = 0$) (Gallardo et al., in preparation). Figure 4 shows the effect of accretion on the radius of an object with initial mass 0.05 $M_\odot$, with accretion rate $\dot{M} = 10^{-8}$ $M_\odot$/yr and $\delta = 0$, $\varepsilon = 0$. At any time, its structure is more compact than that of a nonaccreting object of the same mass (dashed curve in Fig. 4), as mentioned above and as expected for accretion onto a fully convective object (*Prialnik and Livio*, 1985). The smaller radius, and thus the smaller luminosity, affect the location of

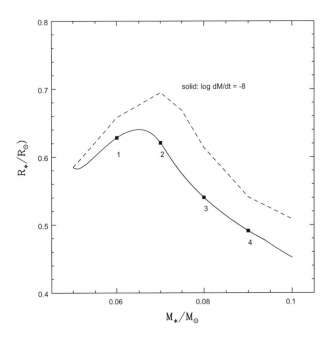

Fig. 4. Effect of accretion on the mass-radius relationship of an accreting brown dwarf with initial mass 0.05 $M_\odot$ and accretion rate $\dot{M} = 10^{-8}$ $M_\odot$/yr (solid line). The dashed line indicates the radius of a nonaccreting object with the same mass and age as its accreting counterpart. Ages for the accreting object, in million years, are indicated by the numbers.

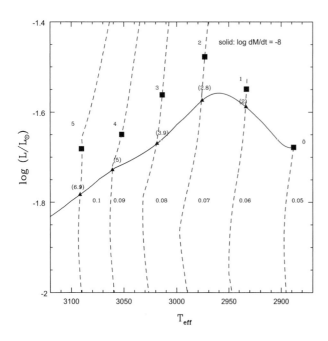

Fig. 5. Evolution in the HR diagram of an accreting brown dwarf, with initial mass 0.05 M$_\odot$ and accretion rate $\dot{M}$ = 10^{-8} M$_\odot$/yr (solid line). The vertical dashed lines are cooling tracks of nonaccreting low-mass objects, with masses indicated near the curves (from 0.05 M$_\odot$ to 0.1 M$_\odot$). The square symbols indicate the position of nonaccreting objects with the same age (indicated by the numbers near the squares, in million years) and same mass as the accreting counterpart (indicated by a triangle just below the corresponding square). The numbers in brackets (close to the triangles) give the age (in million years) of a nonaccreting object at the position indicated by the triangle.

the accreting brown dwarf in a HR diagram, as illustrated in Fig. 5. As seen on this figure, assigning an age or a mass to an observed young object of a given luminosity using nonaccreting tracks can significantly overestimate its age, at least with the present accretion parameters. The effect of various accretion rates (see below) and of finite values of ε is under study. This again illustrates the uncertainty in age determination based on evolutionary tracks at young ages.

4.3. Perspectives

The calculations presented above for an accreting brown dwarf have been done with no or small transfer of internal energy from the accretion shock to the brown dwarf interior. But our understanding of accretion mechanism is still too poor to exclude the release of a large amount of energy due to accretion at deep levels. As mentioned previously, although current observations indicate low accretion rates ($\dot{M} \ll 10^{-8}$ M$_\odot$/yr) for brown dwarfs at ages ≥1 m.y., they also point to rates decreasing with increasing time, suggesting significantly larger accretion rates at early times (≪1 m.y.). If large amounts of matter are accreted, even through a disk, one expects a significant amount of thermal

energy to be added to the object internal energy (*Hartmann et al.*, 1997; *Siess et al.*, 1999). If this were the case, we would expect important modifications of the structure of the surface layers, with possible inhibition of convection as predicted for more massive objects (*Mercer-Smith et al.*, 1984; *Palla and Stahler*, 1992), and thus a larger impact on ages and locations in the HR diagram than displayed in Fig. 5. Such effects of accretion need to be explored in detail in order to get a better characterization of their impact on the early evolution of low-mass stars and brown dwarfs and thus of the uncertainties in mass and/or age determinations for young low-mass objects.

5. BROWN DWARF VERSUS PLANET: OBSERVABLE SIGNATURES

The "planetary status" of objects below the deuterium-burning limit, ~13 M$_{Jup}$ (*Saumon et al.*, 1996; *Chabrier et al.*, 2000), remains the subject of heated debate. The debate was recently intensified by the direct image of an object below this mass limit, 2M1207b, orbiting a young brown dwarf at a *projected* orbital distance ≥55 AU (*Chauvin et al.*, 2005). The present IAU working definition of a "planet" relies primarily on mass — not on the formation mechanism. However, to understand the formation mechanisms of very-low-mass objects, it is critical that we be able to single out those that formed in a disk by a three-step process as described in section 2.1 (core accretion followed by gas capture) from low-mass, *no-deuterium-burning* objects that potentially formed by gravitational collapse of a molecular cloud fragment. According to the definition adopted in the present review, the former would be identified as genuine *planets* while the latter would be *brown dwarfs*. It is interesting, by the way, to note that D-burning is advocated to distinguish BDs from planets, whereas stars with masses below and above the limit for ignition of the CNO cycle share the same "star" denomination. A common "brown dwarf" denomination should thus be used for D-burning or not D-burning BDs. Indeed, D-burning is essentially inconsequential for the long-term evolution of these objects, in contrast to steady hydrogen burning, which yields nuclear equilibrium and determines completely the fate of the object, star, or brown dwarf (see, e.g., *Chabrier and Baraffe*, 2000, their Figs. 2 and 6).

In the coming decades, direct imaging surveys are certain to yield a sizeable number of objects below 13 M$_{Jup}$ orbiting stars and brown dwarfs beyond a few AU — a region unlikely to be well sampled by radial velocity surveys. Without a disk signpost, it will be difficult to distinguish long-period planets from very-low-mass brown dwarfs, based on their different formation history. A very-low-mass brown dwarf (that never burned deuterium) could well be mistaken for a massive planet (see section 2.2.1). Observable features that can distinguish between these two types of objects are greatly needed.

Possible formation signatures could be contained in the atmospheric abundance patterns of planets and their mass-

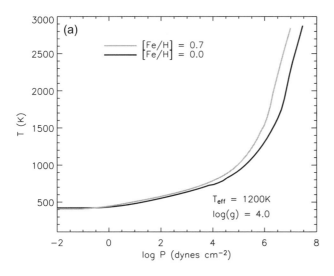
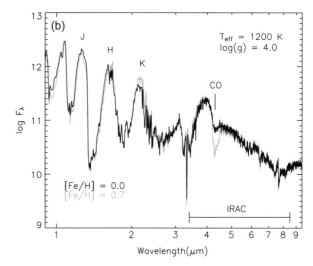

Fig. 6. (a) Temperature vs. pressure for a young Jupiter-mass planet atmosphere model with solar and 5× solar metal abundances [i.e., (Fe/H) = 0.7]. (b) Model spectra for the same conditions. The Spitzer IRAC filter is indicated.

luminosity relationships. As mentioned in section 2.1, a planet recently forged in a disk by the three-step process will experience a brief period of bombardment that enriches its atmosphere and interior in metals compared to its parent star abundances, as observed for our jovian planets (*Barshay and Lewis,* 1978; *Fegley and Lodders,* 1994; *Bézard et al.,* 2002; see also chapter by Marley et al.). Brown dwarfs, on the other hand, should retain the abundance pattern of the cloud from which they formed and, in the case of BDs in binaries, should have abundances similar to their primary star. The metallicity distribution of planet-hosting stars found by radial velocity surveys already suggests that planet formation is favored in metal-rich environments, thus making an *abundance test* even more attractive. Recent interpretations of Spitzer observations for two extrasolar planets are suggestive of nonsolar C and O abundances (see chapter by Marley et al. and references therein).

Enhanced metallicity leaves its mark on the interior, atmospheric structure, and emergent spectrum in a variety of ways. As mentioned in section 2.2, the presence of a large heavy element content in the planet interior will affect its mechanical structure, i.e., its mass-radius relationship. It will also modify its atmospheric structure. Figure 6 compares model atmospheric structures for a young [T_{eff} = 1200 K, log(g) = 4.0], cloud-free, nonirradiated planet-mass object with solar and 5× solar abundances. As the atmospheric opacities increase with increasing metallicity, a natural warming occurs in the deeper layers of the atmosphere. This warming of the atmospheric structure will have a direct impact on the evolution and predicted mass-luminosity relationship.

Figure 6b illustrates the spectral differences between these two models. Clearly the most prominent effect is seen around 4.5 μm, where the increased absorption is due to an increase in CO. Since this CO band falls in the Spitzer IRAC

(3–8 μm) coverage, significant metallicity enhancements in planets could set them apart from typical brown dwarfs on an IRAC color-color diagram. There is also a noticeable increase in the K-band (~2.2 μm) flux.

The main purpose of this section is simply to point out one avenue to explore; however, clearly a great deal of work must be done before a concise picture of the expected abundance patterns in planets is developed. Nonequilibrium CO chemistry, for example, is predicted to occur in cool so-called T-dwarf BDs (*Fegley and Lodders,* 1996; *Saumon et al.,* 2003). Moreover, brown dwarfs forming by gravitational collapse will certainly have abundance patterns as varied as their stellar associations, some being relatively metal rich, e.g., the Hyades (*Taylor and Joner,* 2005). Additionally, metallicity effects in broadband photometry could well be obscured by other competing factors like gravity. Also, our own solar system planets show a range of C-O abundance ratios and varying levels of CO atmospheric enhancement due to vertical mixing. Careful examinations of all these effects are necessary before any reliable spectral diagnostic can be used to distinguish low-mass brown dwarfs from planets. Such a diagnostic, however, has the virtue of relying on a physical distinction between two distinct populations in order to stop propagating confusion with improperly used "planet" denominations.

6. CONCLUSION

In this review, we have (nonexhaustively) explored our present understanding of the formation and the early evolution of gaseous planets and protostars and brown dwarfs. We now have consistent calculations between the planet formation, and thus its core mass and global heavy element enrichment, and the subsequent evolution after disk dissipation. These calculations are based on a revised version of the core-accretion model for planet formation, which includes planet migration and disk evolution, providing an appealing scenario to solve the long-standing timescale problem in the standard core-accretion scenario. Uncertainties in the initial conditions of planet formation, unfortunately, lead to large uncertainties in the initial radius of the newborn planet. Given the dependence of the thermal Kelvin-Helmholtz timescale on radius, this translates into large uncertainties on the characteristic luminosity of young planets, over about 10^7 yr for a 1 M_{Jup} planet. Thus, it is impossible to say whether young planets are bright or faint, and what their initial gravity is for a given mass, and therefore whether their evolution will differ from the one of young low-mass brown dwarfs. Conversely, future observations of young planets in disks of reasonably well-determined ages will enable us to constrain these initial conditions.

We have explored the effect of multidimensional collapse on the accretion properties and mechanical and thermal structures of protostellar cores. These calculations demonstrate that, within less than a freefall time, accretion occurs nonspherically, covering only a very limited fraction of the surface, so that most of the protostar surface can radiate freely into space. Spherical collapse is shown to overestimate the inner density and temperature of the prestellar core, yielding inaccurate initial conditions for PMS contracting tracks. This is important for initial deuterium burning and for energy transport, as three-dimensional inner structures have cooler temperatures and more extended cores. This issue, however, cannot be explored correctly with numerical tools available today as it requires multidimensional implicit codes. The effect of accretion on the contraction of young brown dwarfs was also explored. Even though preliminary, these calculations confirm previous results for pre-main-sequence stars, namely that, for accretion timescales comparable to the Kelvin-Helmholtz timescale, the accreting object has a smaller radius than its nonaccreting counterpart, for the same mass and age, and therefore has a fainter luminosity. This smaller radius, along with the possible contribution from the accretion disk luminosity, can lead to inaccurate determinations of young object ages and masses from their location in an HR diagram, stressing further the questionable validity of mass-age calibrations and disk-lifetime estimates from effective temperature and luminosity determinations in young clusters. These calculations also suggest that, because of the highly nonspherical accretion, young stars or brown dwarfs will be visible shortly after the second collapse and, depending on their various accretion histories, will appear over an extended region of the HR diagram, even though being coeval. This seems to be supported by the dispersion of low-mass objects observed in young stellar clusters or star-forming regions when placed in an HR diagram (see, e.g., Fig. 11 of *Chabrier*, 2003). This suggests that, in spite of all its merits, the concept of a well-defined birth line is not a correct representation, as star formation rather leads to scatter over an extended area in the HR diagram.

Finally, we suggest the deuterium-burning official distinction between brown dwarfs and planets to be abandoned as it relies on a stellar (in a generic sense, i.e., including brown dwarfs) quasistatic formation scenario that now seems to be superseded by a dynamical gravoturbulent picture. Star formation and planet formation very likely overlap in the ~few M_{Jup} range and a physically motivated distinction between these two different populations should reflect their different formation mechanisms. Within the general paradigm that brown dwarfs and stars form predominantly from the gravoturbulent collapse of a molecular cloud and should retain the composition of the parent cloud, and that planets form dominantly from planetesimal and gas accretion in a disk and thus should be significantly enriched in heavy elements compared to their parent star, we propose that these distinctions should be revealed by different mechanical (mass-radius) and spectroscopic signatures. Further exploration of this diagnostic is necessary and will hopefully be tested by *direct* obervations of genuine exoplanets.

Acknowledgments. The authors are very grateful to W. Benz for useful discussions on planet formation and to the anonymous referee for insightful comments.

REFERENCES

Alibert Y., Mordasini C., and Benz W. (2004) *Astron. Astrophys., 417,* L25–L28.

Alibert Y., Mordasini C., Benz W., and Winisdoerffer C. (A05) (2005) *Astron. Astrophys., 434,* 343–353.

André P., Bouwman J., Belloche A., and Hennebelle P. (2004) *Astron. Space Sci., 292,* 325–337.

Bacmann A., Andrè P., Puget J.-L., Abergel A., Bontemps S., and Ward-Thompson D. (2000) *Astron. Astrophys., 361,* 555–580.

Banerjee R. and Pudritz R. (2006) *Astrophys. J., 641,* 949–960.

Baraffe I., Chabrier G., Allard F., and Hauschildt P. H. (2002) *Astron. Astrophys., 382,* 563–572.

Baraffe I., Selsis F., Chabrier G., Barman T., Allard F., Hauschildt P., and Lammer H. (2004) *Astron. Astrophys., 419,* L13–L16.

Baraffe I., Chabrier G., Barman T., Selsis F., Allard F., and Hauschildt P. H. (2005) *Astron. Astrophys., 436,* 47–51.

Baraffe I., Alibert Y., Chabrier G., and Benz W. (2006) *Astron. Astrophys., 450,* 1221.

Barman T., Hauschildt P., and Allard F. (2001) *Astrophys. J., 556,* 885–895.

Barman T., Hauschildt P., and Allard F. (2005) *Astrophys. J., 632,* 1132–1139.

Barshay S. S. and Lewis J. S. (1978) *Icarus, 33,* 593–611.

Basu S. (1998). *Astrophys. J., 509,* 229–237.

Basu S. and Mouschovias T. C. (1995) *Astrophys. J., 453,* 271–283.

Bate M. (1998) *Astrophys. J., 508,* L95–L98.

Belloche A., André P., Despois D., and Blinder S. (2002) *Astron. Astrophys., 393,* 927–947.

Bézard B., Lellouch E., Strobel D., Maillard J.-P., and Drossart P. (2002) *Icarus, 159,* 95–111.

Blandford R. D. and Payne D. G. (1982) *Mon. Not. R. Astron. Soc., 199,* 883–903.

Bodenheimer P., Hubickyj O., and Lissauer J. (2000a) *Icarus, 143,* 2–14.

Bodenheimer P., Burkert A., Klein R. I., and Boss A. P. (2000b) In *Protostars and Planets IV* (V. Mannings et al., eds.), pp. 675–685. Univ. of Arizona, Tucson.

Bodenheimer P., Laughlin G., and Lin D. (2003) *Astrophys. J., 592,* 555–563.

Bonnell I. A. and Bate M. R. (1994) *Mon. Not. R. Astron. Soc., 271,* 999–1004.

Boss A. P. (1989) *Astrophys. J., 346,* 336–349.

Boss A. P. (1995) *Astrophys. J., 439,* 224–236.

Brunini A. and Cionco R. G. (2005) *Icarus, 177,* 264–268.

Burkert A., Lin D. N. C., Bodenheimer P. H., Jones C. A., and Yorke H. W. (2005) *Astrophys. J., 618,* 512–523.

Cassen P. and Moosman A. (1981) *Icarus, 48,* 353–376.

Chabrier G. (2003) *Publ. Astron. Soc. Pac., 115,* 763–795.

Chabrier G. and Baraffe I. (2000) *Ann. Rev. Astron. Astrophys., 38,* 337–377.

Chabrier G., Baraffe I., Allard F., and Hauschildt P. H. (2000) *Astrophys. J., 542,* L119–L122.

Chabrier G., Barman T., Baraffe I., Allard F., and Hauschildt P. H. (2004) *Astrophys. J., 603,* L53–L56.

Chamberlain J. W. and Hunten D. M. (1987) In *Theory of Planetary Atmospheres,* pp. 493–501. International Geophysics Series, Vol. 36, Academic, Orlando.

Chauvin G., Lagrange A.-M., Dumas C., Zuckerman B., Mouillet D., et al. (2005) *Astron. Astrophys., 438,* L25–L28.

Ciolek G. E. and Basu S. (2000) *Astrophys. J., 529,* 925–931.

Curtis S. C. and Showman A. P. (2005) *Astrophys. J., 629,* L45–L48.

D'Angelo G., Kley W., and Henning T. (2003) *Astrophys. J., 586,* 540–561.

Di Francesco J., Myers P. C., Wilner D. J., Ohashi N., and Mardones D. (2001) *Astrophys. J., 562,* 770–789.

Durisen R. H., Yang S., Cassen P., and Stahler S. W. (1989) *Astrophys. J., 345,* 959–971.

Eggleton P. P. (1983) *Astrophys. J., 268,* 368–369.

Fegley B. J. and Lodders K. (1994) *Icarus, 110,* 117–154.

Fegley B. J. and Lodders K. (1996) *Astrophys. J., 472,* L37–L40.

Fortney J. J., Marley M. S., Hubickyj O., Bodenheimer P., and Lissauer J. J. (2005) *Astron. Nach., 326,* 925–929.

Foster P. and Chevalier R. (1993) *Astrophys. J., 416,* 303–311.

Griessmeier J. M., Stadelmann A., Penz T., et al. (2004) *Astron. Astrophys., 425,* 753–762.

Guillot T., Burrows A., Hubbard W. B., Lunine J. I., and Saumon D. (1996) *Astrophys. J., 459,* L35–L38.

Hartmann L., Cassen P., and Kenyon S. J. (1997) *Astrophys. J., 475,* 770–785.

Hayashi C. and Nakano T. (1965) *Progr. Theor. Phys., 34,* 754.

Hennebelle P., Whitworth A., Gladwin P., and André P. (2003) *Mon. Not. R. Astron. Soc., 340,* 870–882.

Hennebelle P., Whitworth A., Cha S.-H., and Goodwin S. (2004) *Mon. Not. R. Astron. Soc., 348,* 687–701.

Hubickyj O., Bodenheimer P., and Lissauer J. (2005) *Icarus, 179,* 415–431.

Hunter C. (1977) *Astrophys. J., 218,* 834–845.

Iro N., Bézard B., and Guillot T. (2005) *Astron. Astrophys., 436,* 719–727.

Jaritz G. F., Endler S., Langmayr D., et al. (2005) *Astron. Astrophys., 439,* 771–775.

Jappsen A.-K., Klessen R. S., Larson R. B., Li Y., and Mac Low M.-M. (2005) *Astron. Astrophys., 435,* 611–623.

Kenyon M. J., Jeffries R. D., Naylor T., Oliveira J. M., and Maxted P. F. L. (2005) *Mon. Not. R. Astron. Soc., 356,* 89–106.

Kley W. and Dirksen G. (2006) *Astron. Astrophys., 447,* 369–377.

Königl A. (1991). *Astrophys. J., 370,* 39–43.

Lammer H., Selsis F., Ribas I., et al. (L03) (2003) *Astrophys. J., 598,* L121–L124.

Larson R. (1969) *Astrophys. J., 145,* 271–295.

Laughlin G. and Różyczka M. (1996) *Astrophys. J., 456,* 279–291.

Lecavelier des Etangs A., Vidal-Madjar A., McConnell J. C., and Hébrard G. (2004) *Astron. Astrophys., 418,* L1–L4.

Lesaffre P., Belloche A., Chièze J. P., and André P. (2005) *Astron. Astrophys., 443,* 961–971.

Li Z.-Y. and Nakamura F. (2004) *Astrophys. J., 609,* L83–L86.

Lin D. and Papaloizou J. (1986) *Astrophys. J., 309,* 846–857.

Lubow S. H., Seibert M., and Artymowicz P. (1999) *Astrophys. J., 526,* 1001–1012.

Lynden-Bell D. (2003) *Mon. Not. R. Astron. Soc., 341,* 1360–1372.

Mac Low M.-M. and Klessen R. S. (2004) *Rev. Mod. Phys., 76,* 125–194.

Masunaga H. and Inutsuka S.-I. (2000) *Astrophys. J., 531,* 350–365.

Masunaga H., Miyama S., and Inutsuka S.-I. (1998) *Astrophys. J., 495,* 346–369.

Matsumoto T. and Hanawa T. (2003) *Astrophys. J., 595,* 913–934.

Mercer-Smith J. A., Cameron A. G., and Epstein R. I. (1984) *Astrophys. J., 279,* 363–366.

Mohanty S., Jayawardhana R., and Basri G. (2005) *Astrophys. J., 626,* 498–522.

Motoyama K. and Yoshida T. (2003) *Mon. Not. R. Astron. Soc., 344,* 461–467.

Mouschovias T. Ch. and Paleologou E. V. (1980) *Astrophys. J., 237,* 877–899.

Mouschovias T. Ch., Paleologou E. V., and Fiedler R. A. (1985) *Astrophys. J., 291,* 772–797.

Moutou C., Coustenis A., Schneider J., et al. (2001) *Astron. Astrophys., 371,* 260–266.

Muzerolle J., Hillenbrand L., Calvet N., Briceno C., and Hartmann L. (2003) *Astrophys. J., 592,* 266–281.

Muzerolle J., Luhman K., Briceno C., Hartmann L., and Calvet N. (2005) *Astrophys. J., 625,* 906–912.

Natta A., Testi L., Randich S., and Muzerolle J. (2004) *Mem. Soc. Astron. Ital., 76,* 343–347.

Nelson R. P. and Papaloizou J. (2004) *Mon. Not. R. Astron. Soc., 350,* 849–864.

Nelson R. P. (2005) *Astron. Astrophys., 443,* 1067–1085.

Ogino S., Tomisaka K., and Nakamura F. (1999) *Publ. Astron. Soc. Japan, 51,* 637–651.

Palla F. and Stahler S. (1992) *Astrophys. J., 392,* 667–677.

Penston M. (1969) *Mon. Not. R. Astron. Soc., 145,* 457–485.

Pollack J. B., Hubickyj O., Bodenheimer P., Lissauer J. J., Podolak M., and Greenzweig Y. (P96) (1996) *Icarus, 124,* 62–85.

Prialnik D. and Livio M. (1985) *Mon. Not. R. Astron. Soc., 216,* 37–52.

Ribas I., Guinan E. F., Güdel M., and Audard M. (2005) *Astrophys. J., 622,* 680–694.

Santos N. C., Israelian G., Mayor M., Bento J. P., Almeida P. C., et al. (2005) *Astron. Astrophys., 437,* 1127–1133.

Saumon D., Chabrier G., and Van Horn H. M. (1995) *Astrophys. J. Suppl., 99,* 713–741.

Saumon D., Hubbard W. B., Burrows A., Guillot T., Lunine J. I., and Chabrier G. (1996) *Astrophys. J., 460,* 993–1018.

Saumon D., Marley M. S., Lodders K., and Freedman R. S. (2003) In *Brown Dwarfs* (E. Martín, ed.), pp. 345–349. IAU Symposium 211, ASP, San Francisco.

Seager S. and Sasselov D. D. (1998) *Astrophys. J., 502,* L157–L160.

Showman A. and Guillot T. (2002) *Astron. Astrophys., 385,* 166–180.

Shu F. H. (1977) *Astrophys. J., 214,* 488–497.

Shu F. H., Adams F. C., and Lizano S. (1987) *Ann. Rev. Astron. Astrophys., 25,* 23–72.

Siess L., Forestini M., and Bertout C. (1997) *Astron. Astrophys., 326,* 1001–1012.

Siess L., Forestini M., and Bertout C. (1999) *Astron. Astrophys., 342,* 480–491.

Stahler S. W. (1983) *Astrophys. J., 274,* 822–829.

Stahler S. W. (1988) *Astrophys. J., 332,* 804–825.

Stahler S.W., Shu F. H., and Taam R. E. (1980) *Astrophys. J., 241,* 637–654.

Stahler S. W., Korycansky D. G., Brothers M. J., and Touma J. (1994) *Astrophys. J., 431,* 341–358.

Tafalla M., Mardones D., Myers P. C., Caselli P., Bachiller R., and Benson P. J. (1998) *Astrophys. J., 504,* 900–914.

Tanaka H., Takeuchi T., and Ward W. R. (2002) *Astrophys. J., 565,* 1257–1274.

Taylor B. J. and Joner M. D. (2005) *Astrophys. J. Suppl., 159,* 100–117.

Terebey S., Shu F. H., and Cassen P. (1984) *Astrophys. J., 286,* 529–551.

Thompson S. L. and Lauson H. S. (1972) In *Technical Report SCRR-61 0714,* Sandia National Laboratories.

Tian F., Toon O. B., Pavlov A. A., and De Sterck H. (2005) *Astrophys. J., 621,* 1049–1060.

Tomisaka K. (1998) *Mon. Not. R. Astron. Soc., 502,* L163–L167.

Ulrich R. K. (1976) *Astrophys. J., 210,* 377–391.

Vidal-Madjar A., Lecavelier des Etangs A., Désert J. M., et al. (VM03) (2003) *Nature, 422,* 143–146.

Vidal-Madjar A., Désert J. M., Lecavelier des Etangs A., et al. (2004) *Astrophys. J., 604,* L69–L72.

Ward W. R. (1997) *Astrophys. J., 482,* L211–L214.

Watson A. J., Donahue T. M., and Walker J. C. G. (1981) *Icarus, 48,* 150–166.

Whitworth A. and Summers D. (1985) *Mon. Not. R. Astron. Soc., 214,* 1–25.

Winn J. N., Suto Y., Turner E. L., et al. (2004) *Publ. Astron. Soc. Japan, 56,* 655–662.

Yelle R. V. (2004) *Icarus, 170,* 167–179.

The Diverse Origins of Terrestrial-Planet Systems

Makiko Nagasawa
National Astronomical Observatory of Japan

Edward W. Thommes
Canadian Institute for Theoretical Astrophysics

Scott J. Kenyon
Smithsonian Astrophysical Observatory

Benjamin C. Bromley
University of Utah

Douglas N. C. Lin
University of California, Santa Cruz

We review the theory of terrestrial planet formation as it currently stands. In anticipation of forthcoming observational capabilities, the central theoretical issues to be addressed are (1) what is the frequency of terrestrial planets around nearby stars; (2) what mechanisms determine the mass distribution, dynamical structure, and stability of terrestrial-planet systems; and (3) what processes regulated the chronological sequence of gas and terrestrial planet formation in the solar system? In the context of solar system formation, the last stage of terrestrial planet formation will be discussed along with cosmochemical constraints and different dynamical architectures together with important processes such as runaway and oligarchic growth. Observations of dust around other stars, combined with models of dust production during accretion, give us a window on exoterrestrial planet formation. We discuss the latest results from such models, including predictions that will be tested by next-generation instruments such as GMT and ALMA.

1. INTRODUCTION

Our home in the solar system: Third planet, 150,000,000 km from the Sun, 70% ocean, possessing a large moon and a moderate climate. How was it born, and how did it grow? How ubiquitous are planets like it in the universe? Such questions have been asked throughout human history. Recent advances in both theory and observation have brought us closer to the answers.

Following the first discovery of an extrasolar planet around 51 Peg, more than 150 planets have been announced, including such fascinating specimens as the multiplanet system of Upsilon Andromeda, and the Neptune-mass planet in 55 Cnc (e.g., *Mayor and Queloz,* 1995; *Butler et al.,* 1999; *McArthur et al.,* 2004). In just 10 years, observational techniques have come to within an order of magnitude of being able to find an Earth-mass planet. The detected extrasolar planets turned on their head the standard models of planetary formation (e.g., *Safronov,* 1969; *Hayashi et al.,* 1985) that were proposed in the 1970s. The seemingly sensible architecture of our solar system, with four terrestrial planets orbiting inside 2 AU, plus gas and icy planets orbiting outside the "snow line," all on reassuringly circular orbits, turned out to be far from the only possible configuration. On the contrary, short orbital periods, high eccentricities, and large planetary masses have made for a diversity of planetary systems unimagined a decade ago (*Marcy et al.,* 2004; *Mayor et al.,* 2004). The theory of planet formation has likewise made rapid progress, helped by these observations and by increases in computing power. We can now numerically simulate many aspects of the formation process in great detail. Thus we have moved well beyond the dawn of planetary science, into an era of fast-paced development. It is thus an exciting time to review the current state of the theory of terrestrial planet formation.

2. BACKGROUND

In the standard scenario, terrestrial planets form through (1) dust aggregation and settling in the protoplanetary disk, (2) planetesimal formation from grains in a thin midplane, (3) protoplanet accretion from planetesimals, and (4) final accumulation by giant impacts. Here, we begin with an overview of the first stages and then focus on chaotic growth, the last stage of terrestrial planetary formation. Although analytic studies are the foundation for this picture, numerical calculations are essential to derive the physical properties of planetary systems. There are two broad classes of simulations. The statistical approach (*Wetherill and Stewart, 1989, 1993*) can follow collisional growth — including collisional disruption and evolution of dust grains — over long times with modest computational effort (*Inaba et al., 2001*, and references therein). This approach works best for the first three stages, where statistical approximations are accurate and large-scale dynamical interactions are small. Direct N-body calculations (*Kokubo and Ida, 1998; Chambers, 2001*) are computationally expensive but can follow key dynamical phenomena such as resonant interactions important during the final stages of planet growth. All calculation benefit from advances in computing power. In particular, N-body simulations no longer require an artificial scaling parameter to speed up the evolution and are thus more reliable (*Kokubo and Ida, 2000*).

2.1. From Dust to Planetesimals

The growth of kilometer-sized planetesimals involves complex interactions between dust and gas within the protoplanetary disk (e.g., *Weidenschilling and Cuzzi, 1993; Ward, 2000;* chapter by Dominik et al.). The vertical component of the star's gravity pulls millimeter-sized and larger grains toward the midplane, where they settle into a thin disk. Smaller grains are more strongly coupled to turbulence in the gas and remain suspended above the midplane. The gas disk is partly pressure-supported and rotates at slightly less than Keplerian velocity. The dust grains thus feel a headwind and undergo orbital decay by aerodrag. The drag is strongest for meter-sized particles (e.g., *Adachi et al., 1976; Tanaka and Ida, 1999*), which fall into the star from 1 AU in ~100 yr. It is not yet clear whether meter-sized objects have enough time to grow directly to kilometer-sized planetesimals that are safe from gas drag (see chapter by Dominik et al.).

Dynamical instability mechanisms sidestep the difficulties of direct accumulation of planetesimals (e.g., *Goldreich and Ward, 1973; Youdin and Shu, 2002*). If the turbulence in the gas is small, the dusty midplane becomes thinner and thinner until groups of particles overcome the local Jeans criterion — where their self-gravity overcomes tidal forces from the central star — and "collapse" into larger objects. Although this gravitational instability is promising, it is still uncertain whether the turbulence in the disk is large enough to prevent the instability (*Weidenschilling and Cuzzi, 1993; Weidenschiling, 1995*). The range of planetesimal sizes produced by the instability is also uncertain.

2.2. From Planetesimals to Protoplanets

Once planetesimals reach kilometer sizes, they begin to interact gravitationally. Collisions produce mergers and tend to circularize the orbits. Long-range gravitational interactions exchange kinetic energy (dynamical friction) and angular momentum (viscous stirring), redistributing orbital energy among planetesimals. Gas drag damps the orbits of planetesimals. For planetesimals without an external perturber, the collisional cross-section is

$$\sigma = \pi d^2 f_g = \pi d^2 \left(1 + \frac{v_{esc}^2}{v^2} \right) \qquad (1)$$

where d is the radius of a particle with mass m, v is its velocity relative to a Keplerian orbit, and $v_{esc}^2 = 2\,Gm/d$ is the escape velocity (*Wetherill, 1990; Ohtsuki et al., 1993; Kortenkamp et al., 2000*). With no outside perturbations, $v \approx v_{esc}$, and gravitational focusing factors, $f_g \sim 1$, are small. Thus, growth is slow and orderly (*Safronov, 1969*).

During orderly growth, long range gravitational interactions between growing planetesimals become important. Small particles damp the orbits of larger particles; larger particles stir up the orbits of smaller particles (*Wetherill and Stewart, 1993; Kokubo and Ida, 1995*). In cases where gas drag is negligible compared to viscous stirring, the relative orbital velocities of the largest ("l") and smallest ("s") bodies reach an approximate steady-state with

$$\frac{v_l}{v_s} = \left(\frac{\Sigma_l}{\Sigma_s} \right)^n \qquad (2)$$

where $n \approx 1/4$ to $1/2$ and Σ is the surface density (*Kokubo and Ida, 1996; Goldreich et al., 2004*). When planetesimals are small, $\Sigma_l/\Sigma_s \ll 1$ and $v_l/v_s < 1$. As planetesimals grow, their escape velocity also grows, which leads to $f_g \gg 1$ and the onset of runaway growth.

Runaway growth depends on a positive feedback between dynamical friction and gravitational focusing. Dynamical friction produces a velocity distribution that declines roughly monotonically with increasing mass (equation (2)). Although gas drag damps the orbits of these objects and keeps their velocities less than v_{esc}, dynamical friction and viscous stirring maintain the small velocities of the largest objects. Because they have the largest v_{esc} and the smallest v, the largest objects have the largest gravitational cross sections and the largest growth rates (equation (1)). Thus, a few large objects grow fastest and "run away" from the ensemble of planetesimals (*Wetherill and Stewart, 1989, 1993; Kokubo and Ida, 1996*).

Runaway growth also depends on a broad size distribution. For planetesimals with masses, m_1 and m_2, and relative velocities, v_1 and v_2, stirring from dynamical friction is $\sim (m_1 v_1^2 - m_2 v_2^2)$. With $v_1 \approx v_2$ initially, dynamical friction becomes important when $m_2 \gtrsim 10\, m_1$. For planetesimals at 1 AU, dynamical friction dominates the evolution when $m_2 \sim (10^4 - 10^6)\, m_1 \sim 10^{20} - 10^{24}$ g for planetesimals with radii of 1–10 km ($m_1 \sim 10^{16} - 10^{19}$ g) (*Ohtsuki et al., 1993*). The typical timescale for planetesimals to reach $m_2 \sim 10^{24}$ g is $\sim 10^4$–10^5 yr at 1 AU (*Wetherill and Stewart*, 1989, 1993; *Weidenschilling et al.*, 1997).

As a few protoplanets contain an ever-increasing fraction of the total mass, two processes halt runaway growth (*Ida and Makino*, 1993; *Kokubo and Ida*, 1998). Protoplanets stir up leftover planetesimals, reducing their gravitational cross-sections. With less mass in planetesimals, dynamical friction between the smallest planetesimals and the largest protoplanets cannot maintain the low velocities of the largest objects, reducing gravitational cross-sections further (equations (1) and (2)). Because the largest protoplanet stirs up its surroundings the most, its growth slows down the most, allowing smaller protoplanets to catch up before their growth rates also slow down. This process results in an "oligarchy" of (locally) similar-mass protoplanets (*Kokubo and Ida*, 1998). Protoplanets reach oligarchy faster closest to the central star, where the surface density of dust is largest and the orbital period is shortest. Thus, the onset of oligarchy sweeps from the inside to the outside of the disk. Although smaller "oligarchs" grow faster than larger ones, oligarchs continue to sweep up leftover planetesimals and thus contain a larger and larger fraction of the total mass in the disk.

During oligarchic growth, protoplanets isolate themselves from their neighbors. Mutual gravitational interactions push them apart to maintain relative separations of ~ 10 Hill radii (r_H) (*Kokubo and Ida*, 1995, 1998). Dynamical friction circularizes their orbits. The large orbital separations and circular orbits yield a maximum "isolation mass," derived from the mass of planetesimals within the $\sim 10 \, r_H$ annulus, ~ 0.1–$0.2 \, M_\oplus$, roughly 10–20% of an Earth mass (see below). The typical timescale to reach this isolation mass is $\sim 10^5$–10^6 yr (*Kokubo and Ida*, 1998).

2.3. From Protoplanets to Planets

With isolation masses a small fraction of the mass of the Earth or Venus, dynamical interactions among the oligarchs must become more energetic to complete terrestrial planet formation. When oligarchs have accumulated most of the planetesimals in their isolation zones, dynamical friction between planetesimals and oligarchs cannot balance dynamical interactions among the oligarchs. Oligarchy ends. Chaotic growth, where planets grow by giant impacts and continued accretion of small planetesimals, begins.

The transition from oligarchy to chaos depends on the balance between damping and dynamical interactions. With no external perturber — a massive planet or a binary companion — the transition occurs when the surface density in oligarchs roughly equals the surface density in planetesimals (*Goldreich et al.*, 2004; *Kenyon and Bromley*, 2006). The transition occurs in ~ 10 m.y. (e.g., *Chambers et al.*, 1996). Dynamical perturbations by a jovian planet or a binary companion shorten this timescale (*Ito and Tanikawa*, 1999); damping processes lengthen it (*Iwasaki et al.*, 2001, 2002). Once isolation is overcome, protoplanets grow to planets in

~ 10–300 m.y. (e.g., *Chambers and Wetherill*, 1998; *Chambers*, 2001; *Kenyon and Bromley*, 2006).

2.4. Outcomes of Numerical Simulations

Attempts to simulate the assembly of terrestrial planets using statistical and N-body approaches began in the late 1970s (*Greenberg et al.*, 1978). Although statistical methods have evolved from single annulus (*Wetherill and Stewart*, 1993) to multiannulus techniques (*Weidenschilling et al.*, 1997), the main result is robust: All calculations yield 10–40 isolated Mars-sized protoplanets in roughly circular orbits and many leftover planetesimals on highly eccentric orbits. During oligarchic growth, fragmentation of the leftovers introduces uncertainties in the timescales and outcomes, as described in sections 3 and 4.

With tens to a few hundred protoplanets remaining at the end of oligarchic growth, direct N-body simulations provide the only way to follow the orbital interactions over hundreds of millions of years. Indeed, the first direct N-body (*Cox and Lewis*, 1980; *Lecar and Aarseth*, 1986) and Monte-Carlo Öpik-Arnold scheme simulations (e.g., *Wetherill*, 1985, 1996) demonstrate that collisions of Moon-sized objects yield "solar systems" containing at least one planet with $m \geq M_\oplus/3$ at ~ 1 AU around a solar-mass star. However, these simulations also show that the outcomes are stochastic and sensitive to the initial number and orbit distributions of the oligarchs. Thus, each set of initial conditions requires many realizations to derive statistically meaningful results.

Several groups have performed full N-body simulations of chaotic growth with and without dynamical influence from jovian planets (e.g., *Chambers and Wetherill*, 1998; *Agnor et al.*, 1999; *Chambers*, 2001; *Kominami and Ida*, 2002, 2004; *Kokubo et al.*, 2006). The simulations start with ~ 20–200 oligarchs in circular or modestly elliptical orbits, with a radial surface density that declines with semimajor axis. Multiple runs (~ 10–200) having identical initial conditions except for, typically, random variations in orbital phase provide a first measure of the repeatability of the results. These calculations yield similar results for the masses, spin angular momenta, and orbital properties of planets and leftover oligarchs. On timescales of 100–300 m.y., most simulations yield planets with (1) masses and semimajor axes similar to the terrestrial planets in the solar system, (2) orbital eccentricities and inclinations somewhat larger than those in the solar system, and (3) spins dominated by the last few giant impacts. Thus, planets similar to Earth and Venus are an inevitable outcome of chaotic growth. Mars appears to be a leftover oligarch.

In the context of these models, the circular orbits of the Earth and Venus in the solar system require an additional damping mechanism. Dynamical friction (*Chambers*, 2001) or gravitational drag by the remnant gas disk (e.g., *Ward*, 1989, 1993; *Artymowicz*, 1993; *Agnor and Ward*, 2002) are good candidates. However, damping tends to prevent the orbital instability needed to initiate chaotic growth (*Iwasaki et al.*, 2001, 2002). Other interactions with the gas, such as type I migration, tend to induce orbital decay in Mars-sized or larger planets (*Goldreich and Tremaine*, 1980; *Ward*, 1986; *McNeil et al.*, 2005; chapter by Papaloizou et al.). These results suggest that the timescales for gas depletion and chaotic growth must be roughly comparable (section 4.3) (*Kominami and Ida*, 2002, 2004).

With this background set, we describe some useful analytic approximations for the runaway and oligarchic growth phases (section 3), review recent simulations of chaotic growth (section 4), and then introduce the dynamical shake-up model (section 4.3). We conclude with a final section describing the implications for future observing capabilities in section 5.

3. EVOLUTION TO CHAOTIC GROWTH

Although numerical calculations are required to predict the time evolution of planetesimals and protoplanets, analytic derivations clarify basic physical processes and yield important estimates for the evolution of solid objects (*Lissauer*, 1987; *Lissauer and Stewart*, 1993; *Goldreich et al.*, 2004). These results also provide basic input for numerical calculations (*Ohtsuki et al.*, 1993, 2002). Here, we outline several results to introduce recent numerical simulations.

Most treatments of terrestrial planet formation begin with a prescription for the surface density Σ of gas and dust in the disk. The number, mass, and orbital separations of oligarchs depend on this prescription (*Kokubo and Ida*, 2002). Here, we adopt a disk with surface density

$$\Sigma_d = \Sigma_{d1}\left(\frac{a}{1\ \text{AU}}\right)^{-\alpha} \quad \text{dust} \qquad (3)$$

$$\Sigma_g = \Sigma_{g1}\left(\frac{a}{1\ \text{AU}}\right)^{-\alpha} \quad \text{gas} \qquad (4)$$

where a is the semimajor axis. In the minimum mass solar nebula (MMSN), $\alpha = 3/2$, $\Sigma_{d1} \approx 7$ g cm^{-2}, and $\Sigma_{g1} \approx 1700$ g cm^{-2} (*Hayashi et al.*, 1985).

3.1. Some Analytic Estimates

To derive the growth rate, we start with equation (1) and adopt subscripts "f" for the field planetesimals, "olig" for the oligarchs, and an asterisk (*) for the central star. The disk scale height is $h = v_f/\Omega_{\text{kep}}$, where $\Omega_{\text{kep}} = (Gm_*/a^3)^{1/2}$ is the orbital frequency. For a mass density $\rho_d = \Sigma_d\Omega_{\text{kep}}/v_f$, the growth rate of particles is (e.g., *Wetherill*, 1980)

$$\frac{dm}{dt} \sim \pi d^2\Sigma_d(1 + (v_{\text{esc}}/v_f)^2)\Omega_{\text{kep}} \qquad (5)$$

The timescale for planetesimals to grow to an Earth mass is

$$t_{\text{grow}} = \left(\frac{1}{m}\frac{dm}{dt}\right)^{-1}$$

$$= 50\left(\frac{\Sigma_{d1}}{10\ \text{g cm}^{-2}}\right)^{-1}\left(\frac{\rho_m}{3\ \text{g cm}^{-3}}\right)^{2/3}\left(\frac{m_*}{M_\odot}\right)^{-1/2} \times \qquad (6)$$

$$\left(\frac{m}{M_\oplus}\right)^{1/3}\left(\frac{a}{1\ \text{AU}}\right)^{3/2-\alpha}\left(1 + \frac{v_{\text{esc}}^2}{v_f^2}\right)^{-1}\ \text{m.y.}$$

where ρ_m is the bulk density of a planetesimal and $M_\odot$ is the mass of the Sun.

Initially, v_f is set by the balance of excitation from viscous stirring between field particles and damping due to gas drag. Adopting the drag force a spherical particle feels in the gas (e.g., *Adachi et al.*, 1976) and a gas density $\rho_g \sim \Sigma_g/2h$, the random velocity of a field particle is

$$v_f \propto \Sigma_d^{1/5}\left(\frac{\Sigma_g}{h}\right)^{-1/5}m^{-1/15}v_{\text{esc,f}} \equiv Cv_{\text{esc,f}} \qquad (7)$$

where $C < 1$ (*Ida and Makino*, 1993).

When an oligarch grows among field planetesimals, $v_f \equiv C'v_{\text{esc,olig}}$ with $C' < 1$ (v_f is now set by the balance between viscous stirring due to the oligarchs and gas drag). In the runaway and oligarchic phases, respectively, the growth rates of particles are roughly

$$t_{\text{grow}} \propto m^{1/3}\left[1 + \frac{1}{C^2}\left(\frac{m}{m_f}\right)^{2/3}\right]^{-1} \qquad (8)$$

$$t_{\text{grow}} \propto m^{1/3}\left[1 + \frac{1}{C'^2}\left(\frac{m}{m_{\text{olig}}}\right)^{2/3}\right]^{-1} \qquad (9)$$

For particles with $m \gg m_f$, $t_{\text{grow}} \propto m^{-1/3}$ in the runaway stage. Larger bodies grow faster. Field plantesimals ($m = m_f$ in equation (8)) have $t_{\text{grow,f}} \propto m^{1/3}$ and undergo orderly growth (*Safronov*, 1969). When runaway growth ends, oligarchs ($m = m_{\text{olig}}$ in equation (9)) and field planetesimals ($m/m_{\text{olig}} \ll 1$) both formally undergo orderly growth, $t_{\text{grow}} \propto m^{1/3}$. However, collisions among planetesimals cease to be accretional when $v_{\text{esc,olig}} \gg v_{\text{esc,f}}$. Thus, planetesimal growth is eventually inhibited during oligarchic growth (see section 3.2 below).

Oligarchs have orbital separations $\Delta a = br_H$, with $b \sim 10$ (*Lissauer*, 1987; *Kokubo and Ida*, 1998) and

$$r_H \equiv \left(\frac{m_1 + m_2}{3\ m_*}\right)^{1/3}\frac{a_1 + a_2}{2} \sim \left(\frac{2\ m}{3\ m_*}\right)^{1/3}a \qquad (10)$$

If an oligarch consumes all planetesimals with semimajor axes between $a - 0.5\ br_H$ and $a + 0.5\ br_H$, it reaches the isolation mass, $M_{iso} = 2\pi a \Sigma_d b r_H$ (*Lissauer*, 1987)

$$\frac{M_{iso}}{M_\oplus} = \frac{4\pi^{3/2} b^{3/2}}{3^{1/2}} \left(\frac{\Sigma_d a^2}{m_*} \right)$$
$$\sim 0.16 \left(\frac{b}{10} \right)^{3/2} \left(\frac{a}{1 \text{ AU}} \right)^{3-3\alpha/2} \left(\frac{\Sigma_{d1}}{10 \text{ g cm}^{-2}} \right)^{3/2} \quad (11)$$

From equations (6) and (11), *Kokubo and Ida* (2002) derived the growth time of an isolated body

$$\frac{t_{grow}}{1 \text{ m.y.}} \approx 0.3 \left(\frac{\Sigma_{d1}}{10 \text{ g cm}^{-2}} \right)^{-9/10} \left(\frac{b}{10} \right)^{1/10} \left(\frac{a}{1 \text{ AU}} \right)^{(9\alpha+16)/10} \quad (12)$$

3.2. Observational Tests: Debris Disks

Deriving robust tests of planet formation theory requires a bridge from analytic estimates to observational predictions. Numerical simulations provide this bridge. Idealized calculations test the analytic results for the important timescales and physical processes during runaway and oligarchic growth (e.g., *Ohtsuki et al.*, 1993, 2002; *Kokubo and Ida*, 1996, 1998, 2002). More complete calculations yield starting points for observational tests (*Wetherill and Stewart*, 1989, 1993; *Weidenschilling et al.*, 1997). Recent advances in computing power enable more complete simulations and promise robust tests of planet formation theories.

There are few constraints on planetesimal formation. Observations of disks in T Tauri stars provide some evidence for grain growth (see chapters by Dutrey et al. and Ménard et al.), but there is little information on the timescale for planetesimal formation (see chapters by Dominik et al. and Natta et al.). Once planetesimals form, numerical simulations suggest a rapid transition from runaway growth to oligarchic growth, $\lesssim 10^5$ yr (e.g., *Wetherill and Stewart*, 1993). Near-infrared (near-IR) colors of T Tauri stars provide some observational support for a similarly rapid transition from a dusty disk (of planetesimals) to a relatively dust-free disk (of oligarchs) (*Kenyon and Hartmann*, 1995).

Once oligarchs form, observations can provide clean tests of planet formation theory. As protoplanets stir their surroundings, collisions between planetesimals produce debris instead of mergers (*Wetherill and Stewart*, 1993; *Kenyon and Bromley*, 2002; see also *Agnor and Asphaug*, 2004; *Leinhardt and Richardson*, 2005). Debris production leads to a collisional cascade, where leftover planetesimals are ground to dust. In the terrestrial zone, dust has an equilibrium temperature of ~200–400 K and emits radiation at wavelengths of 5–30 μm, where Spitzer operates.

The onset of the collisional cascade is tied to the evolution of the largest objects and the material properties of planetesimals. During oligarchic growth, $v_f \lesssim v_{esc,olig}$. Substantial debris production begins when the center of mass collision energy is comparable to the typical binding energy

of a leftover planetesimal (*Wetherill and Stewart*, 1993). For equal mass leftovers, this limit yields $v_f \approx 10 \, Q_d^{1/2}$ (*Kenyon and Bromley*, 2005), where Q_d is the "disruption" energy needed to eject roughly half the mass of the colliding objects (*Benz and Asphaug*, 1999)

$$Q_d = Q_b d^{\alpha_b} + \rho_m Q_g d^{\alpha_g} \quad (13)$$

Here, $Q_b d^{\alpha_b}$ is the bulk strength and $\rho_m Q_g d^{\alpha_g}$ is approximately the gravitational binding energy. For rocky planetesimals with $\rho_m = 3$ g cm^{-3}, laboratory measurements and numerical simulations suggest $Q_b \approx 6 \times 10^7$ erg g^{-1}, $\alpha_b \approx -0.4$, $Q_g \approx 0.4$ erg cm^{-3}, and $\alpha_g \approx 1.25–1.5$, which yield $Q_d \sim 10^8$–10^9 erg g^{-1} for 10-km objects (*Housen and Holsapple*, 1990, 1999; *Holsapple*, 1994; *Durda et al.*, 1998, 2004; *Benz and Asphaug*, 1999; *Michel et al.*, 2001). Thus, the collisional cascade begins when $v_{esc,olig} \approx 1$ km s^{-1}, corresponding to the formation of 1000-km objects.

The collisional cascade produces copious amounts of dust, which absorb and scatter radiation from the central star. Following the growth of protoplanets, the cascade begins at the inner edge of the disk and moves outward. For calculations with a solar-type central star, it takes ~0.1 m.y. for dust to form throughout the terrestrial zone (0.4–2 AU). The timescale is ~1 m.y. for the terrestrial zone of an A-type star (3–20 AU). As the collisional cascade proceeds, protoplanets impose structure on the disk (Fig. 1a). Bright rings form along the orbits of growing protoplanets; dark bands indicate where a large protoplanet has swept up dust along its orbit. In some calculations, the dark bands are shadows, where optically thick dust in the inner disk prevents starlight from shining on the outer disk (*Grogan et al.*, 2001; *Kenyon and Bromley*, 2002, 2004a, 2005; *Durda et al.*, 2004).

In the terrestrial zones of A-type and G-type stars, the dust emits mostly at mid-IR wavelengths. In calculations with G-type central stars, formation of a few lunar mass objects at 0.4–0.5 AU leads to copious dust production in a few thousand years (Fig. 1b). As protoplanets form farther out in the disk, the disk becomes optically thick and the mid-IR excess saturates. Once the orbits of oligarchs start to overlap (~1 m.y.), the largest objects sweep the disk clear of small planetesimals. The mid-IR excess fades. During this decline, occasional large collisions generate large clouds of debris that produce remarkable spikes in the mid-IR excess (*Kenyon and Bromley*, 2002, 2005).

In A-type stars, the terrestrial zone lies at greater distances than in G-type stars. Thus, debris formation in calculations with A-type stars begins later and lasts longer than in models with G-type stars (Fig. 1b). Because the disks in A-type stars contain more mass, they produce larger mid-IR excesses. At later times, individual collisions play a smaller role, which leads to a smoother evolution in the mid-IR excess with time (see *Kenyon and Bromley*, 2005). Although the statistics for G-type stars are incomplete, current observations suggest that mid-IR excesses are larger and last longer for A-type stars than for G-type stars (see *Rieke et al.*, 2005; chapter by Meyer et al.).

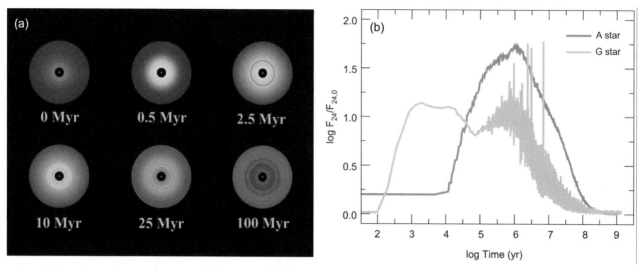

Fig. 1. Evolution of debris disks in the terrestrial zone. For an A-type star with a luminosity of ~50 L$_\odot$, the range in blackbody temperatures of planetesimals at 3–20 AU (425–165 K) is similar to the range in the solar system at 0.4–2 AU (440–200 K). **(a)** Images of a disk extending from 3–20 AU around an A-type star (*Kenyon and Bromley,* 2005). The intensity scale indicates the surface brightness of dust, with black the lowest intensity and white the highest intensity. **(b)** Mid-IR excess for two debris disk models (*Kenyon and Bromley,* 2004b, 2005). The light gray line plots the ratio of the 24-μm flux from a debris disk at 0.4–2 AU disk relative to the mid-IR flux from a G-type star. The dark gray line shows the evolution for the A-star disk shown in **(a)**.

Collisional cascades and debris disk formation may impact the final masses of terrestrial planets. Throughout oligarchic growth, ~25–50% of the initial mass in planetesimals is converted into debris. For solar-type stars, the disk is optically thick, so oligarchs probably accrete the debris before some combination of gas drag, Poynting-Robertson drag, and radiation pressure remove it. In the disks of A-type stars, the debris is more optically thin. Thus, these systems may form lower-mass planets per unit surface density than disks surrounding less massive stars. Both of these assertions require tests with detailed numerical calculations.

3.3. Observational Tests: Cosmochemistry

Radioactive dating provides local tests of the timescales for oligarchic and chaotic growth (see chapter by Wadhwa et al.). The condensation of Ca-Al-rich inclusions and formation of chondrules at a solar age $t_\odot$ ~ a few million years indicate short timescales for planetesimal formation (and perhaps runaway growth). At later times, lunar samples suggest the Moon was fully formed at $t_\odot$ ~ 50 m.y. (e.g., *Halliday et al.,* 2000; *Wood and Halliday,* 2005). The difference between the abundance of radioactive Hf in primitive meteorites and in Earth's mantle suggests that Earth's core differentiated ~30 m.y. after the first CAI condensed out of the solar nebula (*Yin et al.,* 2002). The Hf data suggest the martian core probably formed in ~15 m.y. (*Jacobsen,* 2005). With U-Pb data implying a somewhat later time, ~65–85 m.y. (*Halliday,* 2004), the Hf-W timescale probably measures the time needed to form most of the core, while the U-Pb timescale refers to the last stages of core segregation (*Sasaki and Abe,* 2005; *Wood and Halliday,* 2005).

For planet formation theory, including the giant impact model for the formation of the Moon, these data suggest

(1) the first oligarchs formed at $t_\odot \lesssim 10$ m.y., (2) the era of chaotic growth occured at $t_\odot$ ~ 15–80 m.y., and (3) massive oligarchs sufficient to produce the Moon from a giant impact existed at $t_\odot$ ~ 40–50 m.y. To confront the models with these results, we now consider the most recent numerical models of chaotic growth.

4. NEW SIMULATIONS OF CHAOTIC GROWTH

Chaotic growth is the most delicate phase of terrestrial planet formation. In the current picture, oligarchs in roughly circular orbits evolve into a chaotic system and begin colliding. Once these giant collisions produce several Earth-mass planets, the oligarchs evolve back into an orderly system with roughly circular orbits. Recent research efforts focus on (1) N-body simulations of oligarchs, to derive the frequency of Earth-like planets (including the abundance of water) as a function of initial conditions (section 4.1); (2) hybrid simulations of planet growth, to understand how the evolution depends on the mass in leftover planetesimals and fragmentation processes (section 4.2); and (3) N-body simulations where secular resonances dominate the evolution of oligarchs, to investigate how the masses and orbital properties of planets depend on the relative mass in the disk and in gas giant planets (section 4.3).

4.1. N-Body Calculations

N-body simulations of chaotic growth begin with an ensemble of N oligarchs distributed throughout the terrestrial zone. The oligarchs have an initial range of masses m_i, an initial surface density distribution (equation (3)), initial spacing b, initial spin distribution, and, in some cases, gas gi-

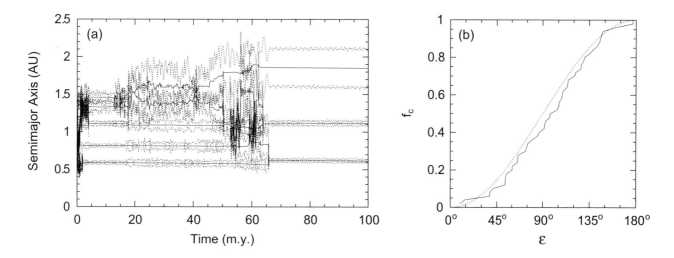

Fig. 2. N-body results for chaotic growth. **(a)** A single simulation from *Kokubo et al.* (2006) (their Fig. 1). The simulation starts with 16 protoplanets. The curves show the semimajor axes (solid lines) and peri- and aphelion distances (dashed lines) as a function of time. **(b)** Cumulative histogram of the obliquity for an ensemble of calculations as in the left panel (from Kokubo et al., in preparation). The solid line shows results (ε in deg) for the largest planet from 50 calculations. The dotted line shows a random distribution, where $2f_c = \int_0^\varepsilon \sin \varepsilon d\varepsilon$.

TABLE 1. Comparison of models with the solar system.

Planet	Terrestrial Planets a (AU)	Mass ($M_\oplus$)	$\langle e \rangle$	$\langle i_{inv} \rangle$ (rad)	Model Results Region (AU)	N	Mass ($M_\oplus$)	e	i_{inv} (rad)
Mercury	0.387	0.0553	0.16	0.12	0–0.5	14	0.68	0.042	0.038
Venus	0.723	0.815	0.031	0.026	0.5–0.85	36	0.80	0.037	0.045
Earth	1.00	1.00	0.028	0.021	0.85–1.25	39	0.86	0.036	0.031
Mars	1.52	0.107	0.19	0.073	>1.25	28	0.33	0.048	0.056

ants with initial masses and orbits or a gas disk with an initial surface density (equation (4)). To understand the range of possible outcomes, an ensemble of calculations with the same set of initial conditions yields statistical estimates for the physical properties of planets and leftover oligarchs.

Most simulations adopt a standard model with $\Sigma_{d1} \sim 8$–12 g cm^{-2}, $\alpha = 3/2$, $b \simeq 10$, and $\rho_m = 3$ g cm^{-3} (e.g., *Chambers and Wetherill*, 1998; *Agnor et al.*, 1999; *Chambers*, 2001; *Kokubo et al.*, 2006). Here, we summarize aspects of *Kokubo et al.* (2006), who consider a baseline model with $\Sigma_{d1} = 10$ g cm^{-2}, and note similarities and differences between this calculation and other published results. Figure 2a shows one run of out 200 with a variety of Σ_d from *Kokubo et al.* (2006).

In this standard case, N-body simulations yield two (2 ± 0.6) major planets with 80% of the initial mass. The masses of the largest and second-largest planets are 1.27 ± 0.25 $M_\oplus$ and 0.66 ± 0.23 $M_\oplus$. The planets have orbital elements (a, e, i) = (0.75 ± 0.20 AU, 0.11 ± 0.07, 0.06 ± 0.04, largest) and (1.12 ± 0.53, 0.12 ± 0.05, 0.10 ± 0.08, second largest). For comparison, Table 1 (left column) lists the orbital data for the solar system, with a and e averaged over 10 m.y. from J2000 and i measured in radians relative to the invari-

able plane. Aside from e and i, these calculations account for the general appearance of the inner solar system.

The obliquity distribution derived from the N-body calculations provides another probe of terrestrial planet formation. In general, the results are similar to the distribution expected for planets formed from an ensemble of giant impacts with random spins (*Agnor et al.*, 1999; *Chambers*, 2001). Figure 2b shows a cumulative fraction of the obliquities of the largest planets from Kokubo et al. (in preparation). The derived distribution is roughly random. Any comparison with the solar system requires much care. The obliquities of terrestrial planets today (~perpendicular to the orbital planes) have evolved considerably due to spin-orbit coupling, tidal interactions, and perhaps other physical processes (e.g., *Ward*, 1973; *Lissauer and Kary*, 1991).

4.1.1. Dependence on conditions for the disk.   Because every set of initial conditions requires many simulations to derive a statistical measure of possible outcomes, measuring the sensitivity of outcomes to the initial conditions is a major task. Recent advances in computing speed make this problem tractable. *Kokubo et al.* (2006) derive statistical uncertainties from 200 simulations. Within a few years, larger sets of calculations will improve these estimates. For

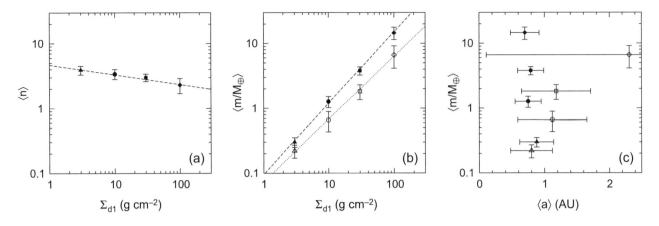

Fig. 3. Results of chaotic growth for $\Sigma_{d1} = 3$ (triangles), 10 (circles), 30 (squares), and 100 (diamonds) g cm^{-2}. **(a)** Average number of planets as a function of surface density (*Kokubo et al.*, 2006, their Fig. 5). **(b)** Masses of the largest (filled circles) and second largest (open circles) planets as functions of surface density (*Kokubo et al.*, 2006, their Fig. 7). **(c)** Masses of the largest (filled circles) and second largest (open circles) planets as a function of semimajor axis (*Kokubo et al.*, 2006, their Fig. 6). The error bars show standard deviations derived from 20 calculations for 100–300 m.y. The lines correspond to the least-squares fits.

calculations without an external perturber (a giant planet or binary companion star), there is fairly good agreement in how outcomes depend on the total mass in oligarchs M_{tot}, the shape of the surface density distribution, the initial spacing of oligarchs, and the bulk properties (mass density, water content, rotational spin, etc.) of the oligarchs. Better sampling of available phase space will provide better constraints on the outcomes.

The growth of terrestrial planets is most sensitive to the initial mass in solid material and the radial distribution of this material in the disk. For simulations with $\Sigma_{d1} = 1$–100 g cm^{-2}, $\alpha = 0$–2.5, b = 6–12, and $\rho_m = 3$–6 g cm^{-3}, larger planets form in more massive disks. Low-mass disks produce a larger number of low-mass planets (e.g., *Wetherill*, 1996; *Chambers and Cassen*, 2002; *Raymond et al.*, 2005a; *Kenyon and Bromley*, 2006; *Kokubo et al.*, 2006). *Kokubo et al.* (2006) derive (see Fig. 3)

$$\langle n \rangle \sim 3.5 \left(\frac{M_{tot}}{2\,M_\oplus} \right)^{-0.15} \quad (14)$$

for the average number of planets and

$$\langle M_1 \rangle \sim 1.0 \left(\frac{M_{tot}}{2\,M_\oplus} \right)^{1.1} M_\oplus \sim 0.5\,M_{tot} \quad (15)$$

$$\langle M_2 \rangle \sim 0.60 \left(\frac{M_{tot}}{2\,M_\oplus} \right)^{0.98} M_\oplus \sim 0.3\,M_{tot} \quad (16)$$

for the masses of the largest (M_1) and the second largest (M_2) planets. The dependence of $\langle n \rangle$ and $\langle M_{1,2} \rangle$ on the surface density is weaker than for oligarchic growth (–1/2 and 3/2, respectively). This may be because the growth mode

is more global, being no longer just scaled by the local feeding zone. The location of the largest planet (a_1) is independent of Σ_{d1} in their simulation range. They obtained a numerical fit of $\langle a_1 \rangle \sim 0.90(\bar{a}/1\ \mathrm{AU})^{1.7}$ AU, where, $\bar{a}$ is the mean semimajor axis of protoplanets. The second planet is further separated in response to the stronger repulsion as Σ_{d1} increases. For the same reason, the eccentricities and inclinations of the largest planets increase with Σ_{d1}. Qualitatively, these results can be understood as follows: In more massive disks, oligarchs have larger isolation masses (equation (11)) and thus merge to produce more massive planets. More massive planets have larger Hill radii and thus clear out larger volumes, yielding fewer planets in stable orbits. Conversely, lower-mass disks produce lower-mass oligarchs (see also *Kenyon and Bromley*, 2006).

The radial surface density gradient, α, sets the final radial mass distribution of planets and their composition. Calculations with small α produce larger, more well-mixed planets at larger semimajor axes than calculations with larger α (*Chambers and Cassen*, 2002; *Raymond et al.*, 2005a; *Kokubo et al.*, 2006). For fixed M_{tot}, the number and masses of terrestrial planets are fairly insensitive to α. When the surface density gradient is shallow (small α), there is relatively more mass at larger heliocentric distances, which leads to the production of more massive oligarchs. These oligarchs are less isolated, are easily scattered throughout the computational grid, and are thus radially well mixed. In calculations with steep surface density gradients (large α), most of the mass is concentrated in a small range of heliocentric distances, so oligarchs are more isolated and harder to scatter throughout the grid. Thus, shallow surface density gradients allow for easier water delivery to the innermost terrestrial planets from oligarchs at large semimajor axes (e.g., *Raymond et al.*, 2005a).

The orbital e and i are sensitive to the total mass in oligarchs (e.g., *Kokubo et al.*, 2006) and to the initial mass range in oligarchs (e.g., *Chambers*, 2001). For a fixed mass range and number of oligarchs, larger M_{tot} can produce larger e and i. A mass range allows dynamical friction to reduce e and i of the most massive bodies; at the same time, such simulations tend to produce more planets in the end. Because N-body calculations are still restricted in the maximum number of bodies that can be evolved for long times, detailed simulations of the transition between oligarchic and chaotic growth (i.e., starting with a significant fraction of the mass still in planetesimals) remain challenging to perform, and it is unclear what role leftover planetesimals play in the final e and i distributions of planets. We revisit this issue in section 4.2.

Despite the uncertainty in the final e and i, most Earth-mass planets form close to 1 AU (*Chambers*, 2001; *Raymond et al.*, 2005a; *Kokubo et al.*, 2006). Massive planets are less likely to form at a ≤ 0.5 AU because typical surface density profiles do not have enough mass to produce a planet as large as Earth. At larger radii, perturbations from Jupiter tend to inhibit formation, as discussed next.

4.1.2. Perturbations from gas giant planets. Long-range gravitational interactions with gas giant planets beyond 3–5 AU impact the formation of terrestrial planets in two ways. If gas giants are fully formed before the runaway growth phase, stirring by gas giants can slow runaway growth and delay oligarchic growth. Once oligarchs form, stirring promotes orbit crossing and the growth of oligarchs from giant impacts. The final configuration and composition of terrestrial planets then depends on the masses and orbital parameters of gas giants.

For gas giants in circular orbits, the importance of long-range stirring depends on the Hill radii of the planets. Because Jupiter has $r_H \sim 0.1$ a and large dynamical interactions require separations less than 5–6 r_H, a Jupiter at 5–6 AU (as in the solar system) has little impact on planet formation at 1 AU. However, Jupiter is effective at removing material outside 2 AU and effectively ends planet formation in the asteroid belt once it reaches its final mass (*Wetherill*, 1996; *Chambers and Wetherill*, 1998; *Levison and Agnor*, 2003). A Jupiter at 3.5 AU limits the formation of terrestrial planets outside 1 AU; a Jupiter at 10 AU allows formation of terrestrial planets in the asteroid belt. Because r_H scales with mass, more massive gas giants prevent planet formation throughout the terrestrial zone.

For elliptical orbits, the dynamical reach of the gas giant for gravitational scattering is ~5–6 r_H + ae. Elliptical orbits also lead to perturbations from mean-motion resonances and secular interactions. Thus gas giants on elliptical orbits affect planet formation over larger volumes (*Chambers and Cassen*, 2002; *Levison and Agnor*, 2003; *Raymond et al.*, 2004). For calculations with $e_{J,S} = 0.1$, Jupiter and Saturn rapidly clear the asteroid belt and prevent "wetter" oligarchs from colliding with drier oligarchs inside 1.5 AU. Thus, a massive, wet, and stable Earth depends on the fairly circular orbits of Jupiter and Saturn.

4.1.3. Extrasolar systems. The discovery of extrasolar planets with masses ranging from Neptune up to 10–20 M_J

opens up amazing vistas in calculations of terrestrial and gas giant planet formation. For terrestrial planets, the main issue is whether an extrasolar planetary system has enough phase space to allow the formation of a stable planet. There are two broad cases: (1) wide systems where the gas giants (or companion stars) have semimajor axes a ≳ 1–2 AU, and (2) compact systems where close-in gas giants have a ≲ 0.1 AU.

Terrestrial planets can form in wide systems that parallel the structure of the solar system (*Heppenheimer*, 1978; *Marzari and Scholl*, 2000; *Kortenkamp and Wetherill*, 2000; *Kortenkamp et al.*, 2001; *Thébault et al.*, 2002). In ε Eri and 47 UMa, terrestrial planets can form at ≤0.8 AU, well inside the orbits of the gas giant planets, but these planets may not be stable (*Jones et al.*, 2001; *Thébault et al.*, 2002; *Laughlin et al.*, 2002). Stable terrestrial planets can form around both components of the wide binary α Cen (*Quintana et al.*, 2002; *Barbieri et al.*, 2002). Unlike the solar system, uncertainties in the orbital parameters of extrasolar gas giants complicates identifying stable regions for lower-mass planets. As new observations reduce these uncertainties, we will have a better picture of possible outcomes for terrestrial planet formation in specific systems.

For systems with close-in gas giants, terrestrial planet formation depends on the availability of material and the orbital parameters of the gas giants. Current models suggest that close-in gas giants form at 5–10 AU and then migrate inward (see chapter by Papaloizou et al.). Gas giant migration through the terrestrial zone removes protoplanets, scattering them into the outer solar system or pushing them into orbits closer to the central star. The amount of material left behind depends on the physics of the migration episode (e.g., *Armitage*, 2003; *Fogg and Nelson*, 2005). However, most calculations suggest that planets can form during or after migration. The masses and orbits of these planets then depend on the mass and orbit of the close-in giant planet (e.g., *Raymond et al.*, 2005b; *Zhou et al.*, 2005).

4.1.4. Water delivery. For conditions in most protosolar nebula models, the disk temperature at 1 AU is too large for volatile molecules to condense out of the gas. Thus, current theory requires a system to deliver water to Earth. The relatively high D/H ratio of Earth's oceans, ~7× the ratio expected in the protosolar nebula, also suggests a delivery system rather than direct absorption (see, however, *Abe et al.*, 2000; *Drake*, 2005). There are two possible sources of water outside Earth's orbit: the asteroid belt, where Jupiter effectively crushes and scatters large objects into the inner solar system (e.g., *Morbidelli et al.*, 2000), and the Kuiper belt, where interactions with Neptune and then Jupiter may allow icy objects from the outer solar system to collide with Earth (e.g., *Levison et al.*, 2001; *Gomes et al.*, 2005). Current evidence may favor the asteroid belt, where the D/H ratio derived from carbonaceous chondrites is closer to the ratio in Earth's oceans than the ratio derived from comets (see *Balsiger et al.*, 1995; *Meier et al.*, 1998; *Bockelée-Morvan et al.*, 1998; *Dauphas et al.*, 2000; *Drake and Righter*, 2002).

Raymond et al. (2004, 2005b) calculate how planetesimals or oligarchs from the asteroid belt deliver water to planets near 1 AU (see also *Lunine et al., 2003*). Due to the stochastic nature of collisions with massive objects, these results suggest that the water content of terrestrial planets is highly variable. The water abundance on Earth, Venus, and Mars is also sensitive to Jupiter's efficiency at cleaning material out of the asteroid belt.

4.1.5. Eccentricity damping. In addition to water delivery, the final e and i of terrestrial planets are important targets for N-body models. Aside from leftover small oligarchs, two processes can circularize the orbits of terrestrial planets. If a large fraction of the mass in leftover planetesimals cycles through the collisional cascade (section 3.2), small dust grains might also damp the eccentricities of planets (*Goldreich et al., 2004*). Because the mass of a single grain is much less than the mass of a single planetesimal, a small total mass in grains can effectively damp massive oligarchs. *Goldreich et al.* (2004) note that oligarchs might also accrete grains more rapidly than leftover planetesimals, shortening the chaotic growth phase and promoting circular orbits. *Kenyon and Bromley* (2006) (also Kenyon and Bromley, in preparation) are testing this possibility.

Interactions with the residual gas disk also damp the eccentricities of terrestrial planets (*Artymowicz, 1993; Agnor and Ward, 2002; Kominami and Ida, 2002; Tanaka et al., 2002; Tanaka and Ward, 2004*). Although gas drag on growing oligarchs is small, coupling between a planet and the Lindblad and corotation resonances of the gas disk (e.g., *Goldreich and Tremaine, 1980*) can be significant. For $\alpha = 3/2$, the tidal torques damp e on a timescale

$$
\begin{aligned}
\tau_{damp} &\simeq \left(\frac{m}{M_\odot}\right)^{-1}\left(\frac{\Sigma_g a^2}{M_\odot}\right)^{-1}\left(\frac{c_s}{a\Omega_K}\right)^4 \Omega_K^{-1} \\
&\simeq 5 \times 10^2 \left(\frac{m}{M_\oplus}\right)^{-1}\left(\frac{\Sigma_{g1}}{\Sigma_{MMSN}}\right)^{-1}\left(\frac{a}{1\ AU}\right)^2 yr
\end{aligned}
\tag{17}
$$

where c_s is the sound velocity. Because interactions with the disk also drag the planet toward the Sun, damping the eccentricity without losing the planet requires a "remnant" gas disk with a small fraction of the mass in a MMSN. In a large set of numerical simulations, *Kominami and Ida* (2002, 2004) included this gravitational drag using

$$
\mathbf{f}_{i,grav} = -\frac{(\mathbf{v} - \mathbf{v}'_K)}{\tau_{damp}}\exp(-t/\tau_{deple})
\tag{18}
$$

where $\mathbf{v}'_K$ is the Keplerian velocity and τ_{deple} is the gas disk depletion timescale. This form is motivated by observations of young stars, where the gas disk disappears on a timescale of 1–10 m.y. Although damping with the residual gas disk is effective, these simulations tend to produce many low-mass

planets. As the gas disk disappears, these planets go through another chaotic growth phase that produces large e as in the N-body simulations of section 4.1.

4.2. Hybrid Calculations

Standard N-body calculations of chaotic growth have several successes and failures. The simulations produce two planets with (m, a) similar to Earth and Venus and several more with (m, a) similar to Mars and Mercury. While the models can explain the inverse relation between m and e and provide some understanding of the solid-body rotation rates, they fail to account for the nearly circular, low-inclination orbits of Earth and Venus. In calculations with many low-mass oligarchs, *Chambers* (2001) noted that dynamical friction helped to circularize orbits of Earth-mass planets. This result suggests that calculations including leftover planetesimals, which contain roughly half the mass at the onset of chaotic growth, might yield more circular orbits for Earth-mass planets. To test this and other ideas, *Bromley and Kenyon* (2006) developed a hybrid, multiannulus coagulation + N-body code that follows the joint evolution of planetesimals and oligarchs.

In the hybrid code, a coagulation algorithm treats the evolution of planetesimals into oligarchs using particle-in-a-box techniques (*Kenyon and Bromley, 2004a*). A direct N-body calculation follows the evolution of oligarchs into planets. The "promotion mass" m_{pro} sets the transition from the coagulation grid to the N-body grid; for most applications $m_{pro} \approx 10^{25}$ g ($\Sigma_{d1}/8$ g cm^{-2}) (*Kenyon and Bromley, 2006*). This code uses the particle-in-a-box and Fokker-Planck formalisms to treat the interactions between oligarchs in the N-body grid and planetesimals in the coagulation grid. *Bromley and Kenyon* (2006) describe tests of the hybrid code and show that it reproduces several previous calculations of terrestrial planet formation (*Weidenschilling et al., 1997; Chambers, 2001*). *Kenyon and Bromley* (2006) describe how the results depend on m_{pro}.

Figure 4 shows the evolution of oligarchs in one evolutionary sequence using the hybrid code. Following a short runaway growth phase, large objects with $m \geq m_{pro}$ appear in a wave that propagates out through the planetesimal grid. As these oligarchs continue to accrete planetesimals, dynamical friction maintains their circular orbits and they evolve into "isolated" objects. Eventually, large oligarchs start to interact dynamically at the inner edge of the grid; this wave of chaotic interactions moves out through the disk until all oligarchs interact dynamically. Once a few large oligarchs contain most of the mass in the system, dynamical friction starts to circularize their orbits. This process excites the lower-mass oligarchs and leftover planetesimals, which are slowly accreted by the largest oligarchs.

Comparisons between the results of hybrid and N-body calculations show the importance of including planetesimals in the evolution. Both approaches produce a few terrestrial-mass planets in roughly circular orbits. Because dynamical

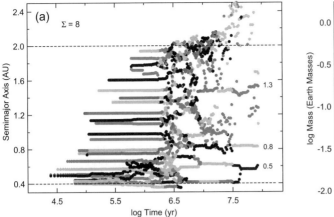
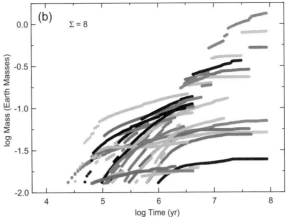

Fig. 4. Evolution of oligarchs in the terrestrial zone. The hybrid calculation starts with 1-km planetesimals ($\rho_m = 3$ g cm^{-3}) in a disk with $\Sigma_{d1} = 8$ g cm^{-2} and $\alpha = 1$. The planetesimal disk contains 40 annuli extending from 0.4 AU to 2 AU. **(a)** The time evolution of semimajor axis shows three phases that start at the inner edge of the grid and propagate outward: (1) after runaway growth, isolated oligarchs with $m \geq 4 \times 10^{25}$ g enter the grid; (2) oligarchs develop eccentric orbits, collide, and merge; and (3) a few massive oligarchs eventually contain most of the mass and develop roughly circular orbits. The legend indicates masses (in $M_\oplus$) for the largest oligarchs. **(b)** The mass evolution of oligarchs shows an early phase of runaway growth (steep tracks) and a longer phase of oligarchic growth (relatively flat tracks), which culminate in a chaotic phase where oligarchs grow by captures of other oligarchs (steps in tracks).

friction between leftover planetesimals and the largest oligarchs is significant, hybrid calculations produce planets with more circular orbits than traditional N-body calculations. In most hybrid calculations, lower-mass planets have more eccentric orbits than the most massive planets, as observed in the solar system. In both approaches, the final masses of the planets grow with the initial surface density; the number of planets is inversely proportional to surface density. However, the overall evolution is faster in hybrid calculations: Oligarchs start to interact earlier and produce massive planets faster.

In hybrid calculations, the isolation mass and the number of oligarchs are more important as local quantities than as global quantities. As waves of runaway, oligarchic, and chaotic growth propagate from the inner disk to the outer disk, protoplanets growing in the inner disk become isolated at different times compared to protoplanets growing in the outer disk. Thus, the isolation mass in hybrid models is a function of heliocentric distance, initial surface density, and *time*, which differs from the classical definition (equation (11)). It is not yet clear how this change in the evolution affects the final masses and orbital properties of terrestrial planets.

4.3. The Role of Gas Depletion in the Final Assembly of Terrestrial Planets

The solar system is full of resonances that produce delicate structure in planetary orbits. In the asteroid belt, for example, there are few objects in the n:m orbital resonances, where an asteroid makes n revolutions of the Sun for every

m revolutions of Jupiter. Because asteroids orbit within 5–6 $r_{H,J}$ ~ 2.5–3 AU of Jupiter's orbit, Jupiter constantly stirs up the asteroids and modifies their orbits. For asteroids in orbital resonance, Jupiter stirring always peaks at the same orbital phase, which leads to a secular (instead of random) change in orbital parameters and eventually ejection from the resonance. In contrast, the Trojan satellites of Jupiter occupy the 1:1 resonance, roughly the stable L4 and L5 points in the restricted three-body problem. Many KBOs occupy the n:m orbital resonances with Neptune, where they are also relatively safe from Neptune's gravitational perturbations (see chapter by Chiang et al.).

Resonances are also an important part of planet formation. When large planets start to form, the massive gaseous disk limits their gravitational reach. As the disk dissipates, planets begin to shape their surroundings. For terrestrial planets, the ν_5 resonance — effectively the semimajor axis where the precession of the line of apsides of an orbit is in phase with the precession of Jupiter's orbit — can play an important role in the transition from oligarchic growth to chaotic growth. As the disk dissipates, the ν_5 resonance sweeps from Jupiter's orbit through the terrestrial zone to its current location at ~0.6 AU (*Ward et al.,* 1976), transforming a system of isolated oligarchs into a chaotic system of merging oligarchs with overlapping orbits (*Nagasawa et al.,* 2000). As chaotic growth ends, the last remnants of the disk circularize the orbits of the remaining planets. To illustrate how this process might produce the architecture of the inner solar system, we now describe this dynamical shake-up model (*Nagasawa et al.,* 2005; Lin et al., in preparation; Thommes et al., in preparation).

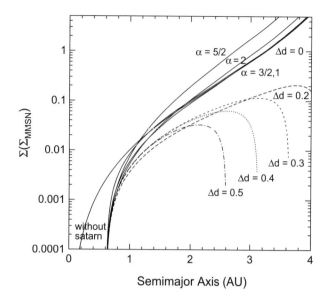

Fig. 5. Evolution with disk depletion of the ν_5 resonances for different power laws ($\alpha = 1$, 3/2, 2, and 5/2) of the surface density of the gas disk (solid lines). The edges of the gaps are located at $a_J(1 \pm \Delta d)$ and $a_S(1 \pm \Delta d)$. We examine $\Delta d = 0$, 0.2, 0.3, 0.4, and 0.5 in the case of $\alpha = 3/2$.

Several features of the sweeping ν_5 resonance make it an attractive feature of terrestrial planet formation. The resonance passes through the terrestrial region almost independently of the details of disk depletion (*Nagasawa et al.,* 2000). It is also largely independent of the disk radial surface density profile. For cases when Jupiter and Saturn have their current orbits, Fig. 5 shows that the position of the resonance is not sensitive to α or to the presence of gaps in the disk. When the jovian planets produce gaps in the disk, the resonance occurs once the planetary perturbation surpasses the disk's perturbation and the rotation of periastra of the protoplanets changes to prograde from retrograde (*Ward,* 1981). Although the timing of resonance passage is slightly delayed, gap formation does not significantly modify the overall evolution of the resonance inside 2 AU.

In addition to exciting e, the sweeping secular resonance can trap protoplanets and push them inward. As the resonance sweeps across a planet, it reduces the angular momentum without changing energy. The resonance rapidly excites e without changing a. As gravitational drag damps e, it extracts energy from the orbit and the orbits shrinks. Under the right conditions, the planet keeps pace with the sweeping resonance (*Nagasawa et al.,* 2005) (see below).

4.3.1. Basic calculation. To perform numerical simulations with sweeping secular resonances, Thommes et al. (in preparation) begin with an algorithm based on the SyMBA symplectic integrator (*Duncan et al.,* 1998). They add a dissipational force to include resonant planet-disk interactions and modify the central force to include the gravitational potential of the disk on the precession rates of the embedded protoplanets. The dissipation is an extra radial acceleration (equation (18)), which changes the energy, but not the angular momentum, of an orbit. Simulations including the much smaller azimuthal component of equation (18) differ negligibly.

These calculations begin with an analog of the inner solar system at a stage when all objects are isolated from their neighbors. The terrestrial region, 0.5–3 AU, is seeded with an ensemble of oligarchs in nearly circular orbits with $\Sigma_{d1} = 7$ g cm^{-2} and $\alpha = 1$ in equation (3). The oligarchs have separations of 10 r_H (*Kokubo and Ida,* 2000), which yields objects ranging in mass from several times 10^{-2} M$_\oplus$ at 0.5 AU to several times 10^{-1} M$_\oplus$ at 3 AU. A "proto-Jupiter" and "proto-Saturn" start with 30 M$_\oplus$ and e = 0.075 at their current semimajor axes (5.2 and 9.5 AU), which roughly corresponds to their states just prior to accretion of a gaseous envelope. The model assumes gas accretion at a linear rate sufficient for Jupiter (Saturn) to reach its current mass at 1.5×10^5 yr (5×10^5 yr) (see *Pollack et al.,* 1996). Although this prescription assumes a strong coincidence between the formation times of Jupiter and Saturn, Fig. 5 shows that Saturn has little effect on the evolution.

For the gas, the calculations assume an exponentially decaying power-law surface density

$$\Sigma_g = 2000 \left(\frac{r}{1 \text{ AU}} \right)^{-1} \exp\left(\frac{-t}{5 \text{ m.y.}} \right) \text{ g cm}^{-2} \quad (19)$$

with a flared disk scale height (*Hayashi,* 1981)

$$h = \frac{\sqrt{2} c_s}{\Omega_K} \sim H_1 \left(\frac{r}{1 \text{ AU}} \right)^{5/4} \quad (20)$$

where H_1 is the scale height at 1 AU. The models adopt $H_1 = 0.05$ AU as the standard case and also examine $H_1 = 0.04$ and 0.06 AU. Although the location of the secular resonance is independent of the mechanism for gas dissipation, the depletion timescale is important. When the gas disk inside Jupiter's orbit dissipates rapidly, the gas cannot damp e and i as the resonance moves inward. Unless there is another source of damping, such as dynamical friction due to leftover planetesimals or the collisional cascade, the final e and i of the terrestrial planets remain large.

Figure 6 shows a standard simulation. The resonance shakes up the orbits of the protoplanets and pushes them inward, leading to orbit-crossing and many mergers. As the resonance sweeps inward, it leaves behind a Mars-mass object (t ~ 20 m.y.) with a semimajor axis close to that of Mars. A little later, an Earth-mass object leaves the resonance at a ~ 1 AU; a second object of similar mass follows the resonance almost to 0.6 AU. Both massive objects have small e, comparable to Earth and Venus. Thus, the model yields a good approximation of the present-day inner solar system in ~25 m.y., when "Earth" and "Venus" suffer their last impact with another protoplanet.

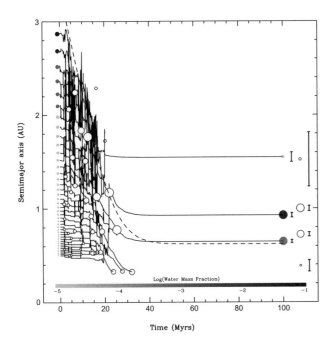

Fig. 6. A simulation of secular resonance sweeping with collisional evolution. Initial protoplanet semimajor axes and masses are shown at left (solid circles, area ∝ mass). Each body's semimajor axis is then plotted as a function of time, together with any mergers that occur (open circles, area ∝ merger product). The path of the ν_5 resonance is also shown (dashed line). Final semimajor axes and masses (solid circles, area ∝ mass) are shown on the right side, with the vertical bars indicating the final peri- and apocenter locations of each. For comparison, the present-day solar system planets are shown at far right (open circles) together with their eccentricities. Also, three different initial water abundances are adopted for the protoplanets: 10^{-5} inside 2 AU, 5×10^{-2} outside 2.5 AU, and 10^{-3} in between. The shading of the final planets indicates their resultant water content.

The timing of the formation of the Mars, Earth, and Venus analogs in this model is consistent with recent cosmochemical evidence suggesting that Mars' core is complete before Earth's core (section 3.3). Although this simulation has no analog to a Moon-forming giant impact at 40–50 m.y., other simulations have at least one collision after the secular resonance passes through the system. Thus, the timing of the Moon's formation is not a strong constraint on the dissipation timescale. Simulations also indicate that the disk depletion time must exceed $\tau_{depl} \approx 3$ m.y.; otherwise, there is not enough gas left to damp eccentricities once planets reach $m \sim M_\oplus$.

In these models, ν_5 drives material from the asteroid belt into the inner solar system. Thus, the dynamical shake-up model is more effective at delivering water to Earth than gravitational scattering alone. Figure 6 shows that most of the material incorporated into the Earth analog originated beyond 2 AU, where water is more abundant. The Mars analog also forms out of wet material, while the Venus analog contains drier material within 1.5 AU.

4.3.2. Variations in outcomes. Figure 7 shows results for multiple 100-m.y. calculations with a variety of initial conditions. The size of the circle is proportional to $m^{1/3}$. Roughly 75–80% of the calculations yield planets with $e \lesssim 0.1$. A slightly smaller fraction, ~70%, form three to four planets. In other simulations, a smaller merger efficiency during the gaseous stage leaves behind more than three to four planets. Some systems become unstable after disk depletion and produce planets with $0.1 \lesssim e \lesssim 0.2$. Many configurations with more than five planets at 100 m.y. will probably develop large e within 1–5 G.y.

Secular resonant trapping tends to migrate protoplanets inward with the ν_5 resonance (Fig. 6). To maintain resonant trapping *Nagasawa et al.* (2005) derived

$$\left(\frac{a}{1\ \text{AU}}\right)^4 \geq 26 \left(\frac{e_J}{0.05}\right)^{-2} \left(\frac{m}{M_\oplus}\right)^{-1} \left(\frac{t_{depl}}{1\ \text{m.y.}}\right)^{-1} \quad (21)$$

The condition for the isolated protoplanet is a > 2 AU with the parameters we used in simulations and a ≥ 1.5 AU with τ_{deple} = 1–10 m.y. and e_J < 0.1. The protoplanets beyond 2 AU are delivered to the terrestrial region. Inside of 2 AU, the secular resonance promotes orbital crossings of oligarchs, but does not trap them. There, semimajor axis decay only happens as a result of (brief) eccentricity damping after collisions or scatterings. Therefore, the planetary accumulation inside 2 AU is similar to that of the standard model, other than the final small eccentricities and inclinations.

Figure 8 shows histograms of the semimajor axes of the three largest and smaller planets in multiple realizations of the dynamical shake-up model, using the same standard parameters as in Fig. 6. The largest planet typically ends up

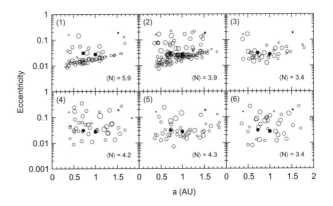

Fig. 7. The final eccentricity vs. semimajor axis of planets with different parameters. The resultant jovian eccentricity is ~0.07 (top panels) and ~0.035 (bottom panels). The damping timescale is $\hat{\tau}_e$ = 2 (left panels), 5 (middle panels), 10 (right panels), where $1000\ \hat{\tau}$ = $\tau_{damp}(m/M_\oplus)(a/1\ \text{AU})^{-3/2}$ yr. Panel 2 consists of 30 runs. Other panels consist of 10 runs. The average number of formed planets in one run is also shown in the panels. Filled circles are current terrestrial planets.

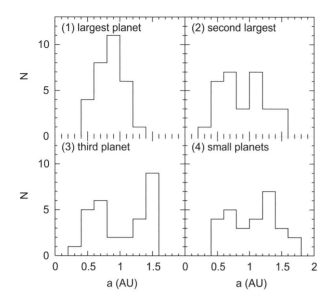

Fig. 8. The semimajor axis histogram of (1) the largest planet in a simulation, (2) the second largest planet, (3) the third planet, and (4) smaller planets. The averaged masses of them are 1.26, 0.85, 0.39, 0.23 $M_\oplus$, respectively. The largest planet has a peak at about 1 AU.

at 0.8–0.9 AU; the second largest planet lies close to this planet at either 0.5–0.7 AU or 1–1.2 AU. In their direct N-body calculations, *Kokubo et al.* (2006) derived similar results. Smaller planets are distributed throughout 0.5–1.5 AU, with a preference of 1.3 AU, slightly inside the orbit of Mars. Although the model does not simulate the formation of the asteroid belt, migration of protoplanets might account for e and i of asteroids and the loss of primordial material from the asteroid belt.

Table 1 compares average orbital elements of these calculations with the orbits of the terrestrial planets. The model produces good analogs for Earth and Venus, but the Mars and Mercury analogs are more massive than their real counterparts despite the loss of ~1 $M_\oplus$ swept through the inner edge (0.3 AU) of a typical simulation. Because the gas damps e for all planets, e and i are independent of mass. Thus, while the results for the Earth and Venus analogs are reasonably close to those in the solar system, the derived e and i for Mercury and Mars are low. Investigating how these results depend on the initial conditions is a top priority for the next set of simulations.

Calculations with multiple damping efficiencies and e_J illustrate the sensitivity of the results on some of the initial conditions (Fig. 7). All models produce some planets with small e. However, strong damping yields more lower-mass planets, which start to interact dynamically once the gas is depleted (see also *Iwasaki et al.*, 2001, 2002). Weak damping cases resemble calculations of chaotic growth. Because planets leftover after the gas disappears in the strong damping case will eventually form massive planets with eccentric

orbits, the strong and weak damping limits tend to produce planets on eccentric orbits. The moderate damping cases have the best chance of forming a few massive planets with fairly circular orbits.

5. CONCLUSION AND IMPLICATIONS

The numerical calculations summarized here have several interesting consequences for the evolution of terrestrial planets. The transition from oligarchic to chaotic growth and the final accretion phase finish on timescales, approximately a few million years to ~80 m.y., before radiometric evidence suggests the formation of Earth was fairly complete (*Yin et al.*, 2002). Planets are also fully formed well before the estimated time of the late heavy bombardment, ~100–600 m.y. after the formation of the Sun (*Tera et al.*, 1974; *Ryder*, 2002; *Koeberl*, 2003).

Throughout the chaotic growth and cleanup phases, numerical calculations produce many lunar- to Mars-sized objects on highly eccentric orbits. These objects are good candidates for the "giant impactor" that collided with Earth to produce the Moon (*Hartmann and Davis*, 1975; *Cameron and Ward*, 1976; *Benz et al.*, 1986; *Canup*, 2004). As the numerical models become more complete, predicted mass and eccentricity distributions will yield better estimates for the probability of these events.

Historically, the solar system provided the only test of models for planet formation. In the last decade, however, radial velocity surveys (see chapter by Udry et al.) and transit observations (see chapter by Charbonneau et al.) have detected over 150 extrasolar gas giants around nearby stars. Aside from Kepler (*www.kepler.arc.nasa.gov*), detections of terrestrial planets require advanced technology on new facilities. Here, we describe a few tests of debris disk simulations and some prospects for direct detections of terrestrial planets.

Direct detections of terrestrial planets require telescopes with large aperture and high spatial resolution. When terrestrial planets first form, they are probably molten, with effective temperatures ~1500 K and relative luminosities $L_p/L_* \sim 10^{-7}$ to 10^{-6}. As terrestrial planets cool, their temperatures and luminosities fall to $T \sim 200$–400 K and $L_p/L_* \sim 10^{-10}$. Thus, mid-IR observations can reveal planets around very young stars. Optical and near-IR images are necessary to detect older terrestrial planets.

Within the next 10–20 yr, new facilities will enable direct detections of terrestrial planets. The Terrestrial Planet Finder (*planetquest.jpl.nasa.gov/TPF/*) aims to detect habitable worlds around nearby F, G, and K stars, with a projected launch date ca. 2016. On the same timescale, large groundbased telescopes such as the Giant Magellan Telescope (GMT) (*www.gmto.org/*) may yield similar results. Current techniques in adaptive optics allow detections of point sources with $L_p/L_* \leq 10^{-6}$ at 1.65 μm (*Codona and Angel*, 2004; *Codona*, 2004). Modest improvements in this technology allow direct detections of molten "Earths"; more dramatic improvements may enable direct detections of cooler terrestrial planets.

Detecting debris from terrestrial planets is much easier. With maximum L_{debris}/L_* ~ 10–100 at 24 μm, IR excess emission from the dusty leftovers of terrestrial planet formation is easily detected around stars with ages ~1–10 m.y. (Fig. 1). As planets disperse this debris, scattered light and thermal emission from dust is detectable with current technology for ~100–300 m.y. For example, *Chen and Jura* (2001) and *Song et al.* (2005) report dust from massive asteroid belts around ζ Lep and BD+20°307, two stars with ages of 100–300 m.y.

Recent comparisons between theory and large samples of A-type and G-type stars are encouraging: the observations show the slow decline in 24-μm excess predicted by theory (*Dominik and Decin, 2003; Rieke et al., 2005; Najita and Williams,* 2005). At each stellar age, however, the data appear to show a larger range in dust temperatures and luminosities than predicted by theory (*Najita and Williams,* 2005). These comparisons suggest that external perturbations and stochastic events modify the relatively smooth evolution of IR excess in Fig. 1.

Groundbased and satellite projects can test other results of debris disk simulations. The four-year Kepler mission should detect two to five eclipses from the debris clouds produced by large binary collisions (*Kenyon and Bromley,* 2005). The groundbased projects OGLE, PASS, and TRES may also have the sensitivity to detect debris clouds (*Konacki et al., 2004; Alonso et al., 2004; Deeg et al., 2004; Pont et al.,* 2004). As part of a much deeper all-sky survey, Pan-STARRS (*panstarrs.ifa.hawaii.edu/public/index.html*) might also detect eclipses from debris clouds. To detect the unique signal from a large two-body collision, these projects need to distinguish real events with diminishing depth and lengthening duration from spurious signals and from repetitive eclipses with constant depth and duration.

New observations and better numerical calculations promise a better understanding of debris disks and the late stages of terrestrial planet formation. Larger samples from Spitzer and groundbased telescopes will enable better statistical comparisons. New facilities such as ALMA and GMT will provide resolved observations of debris disks in the terrestrial zones of nearby stars, enabling probes of the masses and orbits of planets in addition to the structure of the debris. These data will challenge efforts to build more complete numerical simulations of debris disk formation and the late stages of planet formation.

Acknowledgments. The authors acknowledge J. Chambers for his helpful review. We also thank E. Kokubo for his kind offer of data and M. Holman for useful discussions. M.N. is supported by MEXT (KAKENHI 16077202, 18740281). E.W.T. is supported by the Natural Sciences and Engineering Research Council of Canada, and by CITA. B.C.B. and S.J.K. acknowledge support from the NASA Astrophysics Theory Program (grant NAG5-13278) and supercomputing support from the JPL Institutional Computing and Information Services and the NASA Directorates of Aeronautics Research, Science, Exploration Systems, and Space Operations. This work was partially supported by NASA under grants NAGS5-11779 and NNG04G-191G, by JPL under grant 1228184, and by NSF under grant AST-9987417 through DNCL.

REFERENCES

Abe Y., Ohtani E., Okuchi T., Righter K., and Drake M. (2000) In *Origin of the Earth and Moon* (R. M. Canup and K. Righter, eds.), pp. 413–433. Univ. of Arizona, Tucson.

Adachi I., Hayashi C., and Nakazawa K. (1976) *Prog. Theor. Phys., 56,* 1756–1771.

Agnor C. and Asphaug E. (2004) *Astrophys. J., 613,* L157–L160.

Agnor C. B. and Ward W. R. (2002) *Astrophys. J., 567,* 579–586.

Agnor C. B., Canup R. M., and Levison H. F. (1999) *Icarus, 142,* 219–237.

Alonso R., Brown T., Torres G., Latham D. W., Sozzetti A., et al. (2004) *Astrophys. J., 613,* L153–L156.

Armitage P. J. (2003) *Astrophys. J., 582,* L47–L50.

Artymowicz P. (1993) *Astrophys. J., 419,* 166–180.

Balsiger H., Altwegg K., and Geiss J. (1995) *J. Geophys. Res., 100,* 5827–5834.

Barbieri M., Marzari F., and Scholl H. (2002) *Astron. Astrophys., 396,* 219–224.

Butler R. P., Marcy G. W., Fischer D. A., Brown T. M., Contos A. R., et al. (1999) *Astrophys. J., 526,* 916–927.

Benz W. and Asphaug E. (1999) *Icarus, 142,* 5–20.

Benz W., Slattery W. L., and Cameron A. G. W. (1986) *Icarus, 66,* 515–535.

Bockelee-Morvan D., Gautier D., Lis D. C., Young K., Keene J., et al. (1998) *Icarus, 133,* 147–162.

Bromley B. C. and Kenyon S. J. (2006) *Astron. J., 131,* 2737–2748.

Cameron A. G. W. and Ward W. R. (1976) In *Lunar Planet. Sci. VII,* p. 120.

Canup R. (2004) *Ann. Rev. Astron. Astrophys., 42,* 441–475.

Chambers J. E. (2001) *Icarus, 152,* 205–224.

Chambers J. E. and Cassen P. (2002) *Meteoritics & Planet. Sci., 37,* 1523–1540.

Chambers J. E. and Wetherill G. W. (1998) *Icarus, 136,* 304–327.

Chambers J. E., Wetherill G. W., and Boss A. P. (1996) *Icarus, 119,* 261–268.

Chen C. H. and Jura M. (2001) *Astrophys. J., 560,* L171–L174.

Codona J. L. (2004) *Proc. SPIE, 5490,* 379–388.

Codona J. L. and Angel R. (2004) *Astrophys. J., 604,* L117–L120.

Cox L. P. and Lewis J. S. (1980) *Icarus, 44,* 706–721.

Drake M. J. (2005) *Meteoritics & Planet. Sci., 40,* 519–527.

Drake M. J. and Righter K. (2002) *Nature, 416,* 39–44.

Dauphas N., Robert F., and Marty B. (2000) *Icarus, 148,* 508–512.

Deeg H. J., Alonso R., Belmonte J. A., Alsubai K., Horne K., et al. (2004) *Publ. Astron. Soc. Pac., 116,* 985–995.

Dominik C. and Decin G. (2003) *Astrophys. J., 598,* 626–635.

Duncan M. J., Levison H. F., and Lee M. H. (1998) *Astron. J., 116,* 2067–2077.

Durda D. D., Greenberg R., and Jedicke R. (1998) *Icarus, 135,* 431–440.

Durda D. D., Bottke W. F., Enke B. L., Merline W. J., Asphaug E., et al. (2004) *Icarus, 170,* 243–257.

Fogg M. J. and Nelson R. P. (2005) *Astron. Astrophys., 441,* 791–806.

Goldreich P. and Tremaine S. (1980) *Astrophys. J., 241,* 425–441.

Goldreich P. and Ward W. R. (1973) *Astrophys. J., 183,* 1051–1062.

Goldreich P., Lithwick Y., and Sari R. (2004) *Ann. Rev. Astron. Astrophys., 42,* 549–601.

Gomes R., Levison H. F., Tsiganis K., and Morbidelli A. (2005) *Nature, 435,* 466–469.

Greenberg R., Hartmann W. K., Chapman C. R., and Wacker J. F. (1978) *Icarus, 35,* 1–26.

Grogan K., Dermott S. F., and Durda D. D. (2001) *Icarus, 152,* 251–267.

Halliday A. N. (2004) *Nature, 427,* 505–590.

Halliday A. N., Lee D-C., and Jacobsen S. B. (2000) In *Origin of the Earth and Moon* (R. M. Canup and K. Righter, eds.), pp. 45–62. Univ. of Arizona, Tucson.

Hayashi C. (1981) *Prog. Theor. Phys. Suppl., 70,* 35–53.

Hayashi C., Nakazawa K., and Nakagawa Y. (1985) In *Protostars and Planets II* (D. C. Black and M. S. Matthews, eds.), pp. 1100–1153. Univ. of Arizona, Tucson.

Hartmann W. K. and Davis D. R. (1975), *Icarus, 24,* 504–514.

Heppenheimer T. A. (1978) *Astron. Astrophys., 65,* 421–426.

Holsapple K. L. (1994) *Planet. Space Sci., 42,* 1067–1078.

Housen K. and Holsapple K. (1990) *Icarus, 84,* 226–253.

Housen K. and Holsapple K. (1999) *Icarus, 142,* 21–33.

Ida S. and Makino J. (1993) *Icarus, 106,* 210-227, 875–889.

Inaba S., Tanaka H., Nakazawa K., Wetherill G. W., and Kokubo E. (2001) *Icarus, 149,* 235–250.

Ito T. and Tanikawa K. (1999) *Icarus, 139,* 336–349.

Iwasaki K., Tanaka H., Nakazawa K., and Emori H. (2001) *Publ. Astron. Soc. Japan, 53,* 321–329.

Iwasaki K., Emori H., Nakazawa K., and Tanaka H. (2002) *Publ. Astron. Soc. Japan, 54,* 471–479.

Jacobsen S. B. (2005) *Ann. Rev. Earth Planet. Sci., 33,* 531–570.

Jones B. W., Sleep P. N., and Chambers J. E. (2001) *Astron. Astrophys., 366,* 254–262.

Kenyon S. J. and Bromley B. C. (2002) *Astrophys. J., 577,* L35–L38.

Kenyon S. J. and Bromley B. C. (2004a) *Astron. J., 127,* 513–530.

Kenyon S. J. and Bromley B. C. (2004b) *Astrophys. J., 602,* L133–L136.

Kenyon S. J. and Bromley B. C. (2005) *Astron. J., 130,* 269–279.

Kenyon S. J. and Bromley B. C. (2006) *Astron. J., 131,* 1837–1850.

Kenyon S. J. and Hartmann L. W. (1995) *Astrophys. J. Suppl., 101,* 117–171.

Koeberl C. (2003) *Earth Moon Planets, 92,* 79–87.

Kokubo E. and Ida S. (1995) *Icarus, 114,* 247–257.

Kokubo E. and Ida S. (1996) *Icarus, 123,* 180–191.

Kokubo E. and Ida S. (1998) *Icarus, 131,* 171–178.

Kokubo E. and Ida S. (2000) *Icarus, 143,* 15–27.

Kokubo E. and Ida S. (2002) *Astrophys. J., 581,* 666–680.

Kokubo E., Kominami J., and Ida S. (2006) *Astrophys. J., 642,* 1131–1139.

Kominami J. and Ida S. (2002) *Icarus, 157,* 43–56.

Kominami J. and Ida S. (2004) *Icarus, 167,* 231–243.

Konacki M., Torres G., Sasselov D. D., Pietrzyski G., Udalski A., et al. (2004) *Astrophys. J., 609,* L37–L40.

Kortenkamp S. J. and Wetherill G. W. (2000) *Icarus, 143,* 60–73.

Kortenkamp S. J., Kokubo E., and Weidenschilling S. J. (2000) In *Origin of the Earth and Moon* (R. M. Canup and K. Righter, eds.), pp. 75–84. Univ. of Arizona, Tucson.

Kortenkamp S. J., Wetherill G. W., and Inaba S. (2001) *Science, 293,* 1127–1129.

Laughlin G., Chambers J., and Fischer D. (2002) *Astrophys. J., 579,* 455–467.

Lecar M. and Aarseth S. J. (1986) *Astrophys. J., 305,* 564–579.

Levison H. F. and Agnor C. (2003) *Astron. J., 125,* 2692–2713.

Levison H. F., Dones L., Chapman C. R., Stern S. A., Duncan M. J., and Zahnle K. (2001) *Icarus, 151,* 286–306.

Leinhardt Z. M. and Richardson D. C. (2005) *Astrophys. J., 625,* 427–440.

Lissauer J. J. (1987) *Icarus, 69,* 249–265.

Lissauer J. J. and Kary D. M. (1991) *Icarus, 94,* 126–159.

Lissauer J. J. and Stewart G. R. (1993) In *Protostars and Planets III* (E. H. Levy and J. I. Lunine, eds.), pp. 1061–1088. Univ. of Arizona, Tucson.

Lunine J. I., Chambers J., Morbidelli A., and Leshin L. A. (2003) *Icarus, 165,* 1–8.

Marcy G. W., Butler R. P., Fischer D. A., and Vogt S. S. (2004) In *Extra-solar Planets: Today and Tomorrow* (J. P. Beaulieu et al., eds.), pp. 3–14. ASP Conf. Series 321, San Francisco.

Marzari F. and Scholl H. (2000) *Astrophys. J., 543,* 328–339.

Mayor M. and Queloz D. (1995) *Nature, 378,* 355–359.

Mayor M., Udry S., Naef D., Pepe F., Queloz D., et al. (2004) *Astron. Astrophys., 415,* 391–402.

McArthur B. E., Endl M., Cochran W. D., Benedict G. F., Fischer D. A., et al. (2004) *Astrophys. J., 614,* L81–L84.

McNeil D., Duncan M. J., and Levison H. F. (2005) *Astron J., 130,* 2884–2899.

Meier R., Owen T. C., Jewitt D. C., Matthews H. E., Senay M., et al. (1998) *Science, 279,* 1707–1710.

Michel P., Benz W., Tanga P., and Richardson D. C. (2001) *Science, 294,* 1696–1700.

Morbidelli A., Chambers J., Lunine J. I., Petit J. M., Robert F., et al. (2000) *Meteoritics & Planet. Sci., 35,* 1309–1320.

Nagasawa M., Tanaka H., and Ida S. (2000) *Astron. J., 119,* 1480–1497.

Nagasawa M., Lin D. N. C., and Thommes E. (2005) *Astrophys. J., 635,* 578–598.

Najita J. and Williams J. P. (2005) *Astrophys. J., 635,* 625–635.

Ohtsuki K., Ida S., Nakagawa Y., and Nakazawa K. (1993) In *Protostars and Planets III* (E. H. Levy and J. I. Lunine, eds.), pp. 1089–1107. Univ. of Arizona, Tucson.

Ohtsuki K., Stewart G. R., and Ida S. (2002) *Icarus, 155,* 436.

Pollack J. B., Hubickyj O., Bodenheimer P., Lissauer J. J., Podolak M., et al. (1996) *Icarus, 124,* 62–85.

Pont F., Bouchy F., Queloz D., Santos N. C., Melo C., et al. (2004) *Astron. Astrophys., 426,* L15–L18.

Quintana E. V., Lissauer J. J., Chambers J. E., and Duncan M. (2002) *Astrophys. J., 576,* 982–996.

Raymond S. N., Quinn T., and Lunine J. I. (2004) *Icarus, 168,* 1–17.

Raymond S. N., Quinn, T., and Lunine J. I. (2005a) *Astrophys. J., 632,* 670–676.

Raymond S. N., Quinn T, and Lunine J. I. (2005b) *Icarus, 177,* 256–263.

Rieke G. H., Su K. Y. L., Stansberry J. A., Trilling D., Bryden G., et al. (2005) *Astrophys. J., 620,* 1010–1026.

Ryder G. (2002) *J. Geophys. Res., 107,* 6–26.

Safronov V. (1969) *Evolution of the Protoplanetary Cloud and Formation of the Earth and Planets.* Nauka, Moscow.

Sasaki T. and Abe Y. (2005) In *PPV Poster Proceedings,* www.lpi.usra.edu/meetings/ppv2005/pdf/8221.pdf.

Song I., Zuckerman B., Weinberger A. J., and Becklin E. E. (2005) *Nature, 436,* 363–365.

Tanaka H. and Ida S. (1999) *Icarus, 139,* 350–366.

Tanaka H. and Ward W. R. (2004) *Astrophys. J., 602,* 388–395.

Tanaka H., Takeuchi T., and Ward W. R. (2002) *Astrophys. J., 565,* 1257–1274.

Tera F., Papanastassiou D. A., and Wasserburg G. J. (1974) *Earth Planet. Sci., 22,* L1–L21.

Thébault P., Marzari F., and Scholl H. (2002) *Astron. Astrophys., 384,* 594–602.

Ward W. R. (1973) *Science, 181,* 260–262.

Ward W. R. (1981) *Icarus, 47,* 234–264.

Ward W. R. (1986) *Icarus, 67,* 164–180.

Ward W. R. (1989) *Astrophys. J., 345,* L99–L102.

Ward W. R. (1993) *Icarus, 106,* 274–287.

Ward W. R. (2000) In *Origin of the Earth and Moon* (R. M. Canup and K. Righter, eds.), pp. 75–84. Univ. of Arizona, Tucson.

Ward W. R., Colombo G., and Franklin F. A. (1976) *Icarus, 28,* 441–452.

Weidenschiling S. J. (1995) *Icarus, 116,* 433–435.

Weidenschilling S. J. and Cuzzi J. N. (1993) In *Protostars and Planets III* (E. H. Levy and J. I. Lunine, eds.), pp. 1031–1060. Univ. of Arizona, Tucson.

Weidenschilling S. J., Spaute D., Davis D. R., Marzari F., and Ohtsuki K. (1997) *Icarus, 128,* 429–455.

Wetherill G. W. (1980) *Ann. Rev. Astron. Astrophys., 18,* 77–113.

Wetherill G. W. (1985) *Science, 228,* 877–879.

Wetherill G. W. (1990) *Icarus, 88,* 336–354.

Wetherill G. W. (1996) *Icarus, 119,* 219–238.

Wetherill G. W. and Stewart G. R. (1989) *Icarus, 77,* 330–357.

Wetherill G. W. and Stewart G. R. (1993) *Icarus, 106,* 190–209.

Wood B. J. and Halliday A. N. (2005) *Nature, 437,* 1345–1348.

Yin Q., Jacobsen S. B., Yamashita K., Blichert-Toft J., Telouk P., et al. (2002) *Nature, 418,* 949–952.

Youdin A. N. and Shu F. H (2002) *Astrophys. J., 580,* 494–505.

Zhou J.-L., Asrseth S. J., Lin D. N. C., and Nagasawa M. (2005) *Astrophys. J., 631,* L85–L88.

Disk-Planet Interactions During Planet Formation

J. C. B. Papaloizou
University of Cambridge

R. P. Nelson
Queen Mary, University of London

W. Kley
Universität Tübingen

F. S. Masset
Saclay, France, and Universidad Nacional Autónoma de México

P. Artymowicz
University of Toronto at Scarborough and University of Stockholm

The discovery of close orbiting extrasolar giant planets led to extensive studies of disk-planet interactions and the forms of migration that can result as a means of accounting for their location. Early work established the type I and type II migration regimes for low-mass embedded planets and high-mass gap-forming planets respectively. While providing an attractive means of accounting for close orbiting planets initially formed at several AU, inward migration times for objects in the Earth-mass range were found to be disturbingly short, making the survival of giant planet cores an issue. Recent progress in this area has come from the application of modern numerical techniques that make use of up-to-date supercomputer resources. These have enabled higher-resolution studies of the regions close to the planet and the initiation of studies of planets interacting with disks undergoing magnetohydrodynamic turbulence. This work has led to indications of how the inward migration of low- to intermediate-mass planets could be slowed down or reversed. In addition, the possibility of a new very fast type III migration regime, which can be directed inward or outward, that is relevant to partial gap-forming planets in massive disks has been investigated.

1. INTRODUCTION

The discovery of extrasolar planets around Sun-like stars (51 Pegasi b) (*Mayor and Queloz,* 1995; *Marcy and Butler,* 1995, 1998) has revealed a population of close orbiting giant planets with periods of typically a few days, the so-called "hot Jupiters." The difficulties associated with forming such planets *in situ*, either in the critical core mass accumulation followed by gas accretion scenario, or the gravitational instability scenario for giant planet formation, has led to the realization of the potential importance of large-scale migration in forming or young planetary systems.

This in turn led to more intensive theoretical development of disk protoplanet interaction theory that had already led to predictions of orbital migration (see *Lin and Papaloizou,* 1993; *Lin et al.,* 2000, and references therein). At the time of Protostars and Planets IV (PPIV), the type I and type II migration regimes, the former applying to small-mass embedded protoplanets and the latter to gap-forming massive protoplanets, had become apparent. Both these regimes predicted disturbingly short radial infall times that in

the type I case threatened the survival of embryo cores in the 1–15 $M_\oplus$ regime before they could accrete gas to become giant planets. The main questions to be addressed were how to resolve the type I migration issue and to confirm that type II migration applicable to giant planets could indeed account for the observed radial distribution and the hot Jupiters.

Here, we review recent progress in the field of disk-planet interactions in the context of orbital migration. For reasons of space constraint we shall not consider the problem of excitation or damping of orbital eccentricity. The most recent progress in this area has come from carrying out large-scale two- and three-dimensional simulations that require the most up-to-date supercomputer resources. This has enabled the study of disk-planet interactions in disks undergoing magnetohydrodynamic (MHD) turbulence; the study of the regions close to the planet using high-resolution multigrid techniques led to suggestions for the possible resolution of the type I issue and revealed another possible type III migration regime. However, the complex nature of these problems makes them challenging numerically and as a consequence numerical convergence has not been attained in some cases.

In sections 2, 3, and 4 we review type I migration, type II migration, and type III migration respectively. In section 5 we review recent work on disk-planet interactions in disks with MHD turbulence and in section 6 we give a summary.

2. TYPE I MIGRATION

When the mass of the protoplanet is small the response it induces in the disk can be calculated using linear theory. When the disk flow is nonmagnetic and laminar, density waves propagate both outward and inward away from the protoplanet. These waves carry positive and negative angular momentum respectively and a compensating tidal torque is applied accordingly to the orbit, resulting in type I migration.

2.1. The Tidal Torque

The problem of determining the evolution of the planet orbit amounts to an evaluation of tidal torques. For a sufficiently small planet mass (an upper limit of which will be specified below) one supposes that the gravitational potential of the protoplanet forces small perturbations. The hydrodynamic equations are then linearized about a basic state consisting of an unperturbed axisymmetric accretion disk and the response calculated. The gravitational potential ψ of a proplanet in circular orbit is expressed as a Fourier series in the form

$$\psi(r, \varphi, t) = \sum_{m=0}^{\infty} \psi_m(r) \cos\{m[\varphi - \omega_p t]\} \qquad (1)$$

where φ is the azimuthal angle and $2\pi/(\omega_p)$ is the orbital period of the planet of mass M_p at orbital semimajor axis a. The total torque acting on the disk is given by $\Gamma = -\int_{Disk} \Sigma \vec{r} \times \nabla\psi d^2 r$ where Σ is the surface density of the disk.

An external forcing potential $\psi_m(r, \varphi)$ with azimuthal mode number m that rotates with a pattern frequency ω_p in a disk with angular velocity $\Omega(r)$ triggers a response that exchanges angular momentum with the orbit whenever, neglecting effects due to pressure, $m(\Omega - \omega_p)$ is equal to either 0 or $\pm\kappa$, with, for a Keplerian disk to adequate accuracy, $\kappa \equiv \Omega$ being the epicyclic frequency. The first possibility occurs when $\Omega = \omega_p$ and thus corresponds to a corotation resonance. The second possibility corresponds to an inner Lindblad resonance located inside the orbit for $\Omega = \omega_p + \kappa/m$ and an outer Lindblad resonance outside the orbit for $\Omega = \omega_p - \kappa/m$.

2.1.1. Torques at Lindblad resonances. Density waves are launched at Lindblad resonances and as a consequence, a torque acts on the planet. It is possible to solve the wave excitation problem using the WKB method. In that approximation an analytic expression for the torque can be found. The torque arising from the component of the potential with azimuthal mode number m is found for a Keplerian disk to

be given by

$$\Gamma_m^{LR} = \frac{sign(\omega_p - \Omega)\pi^2\Sigma}{3\Omega\omega_p}\Psi^2 \qquad (2)$$

with

$$\Psi = r\frac{d\psi_m}{dr} + \frac{2m^2(\Omega - \omega_p)}{\Omega}\psi_m \qquad (3)$$

where the expression has to be evaluated at the location of the resonance.

The derivation of this torque formula in the context of satellite (planet) interaction with a gaseous disk can be found in *Goldreich and Tremaine* (1979) and *Lin and Papaloizou* (1979, 1993). In a Keplerian disk, the torque exerted on the planet from an outer Lindblad resonance is negative corresponding to a drag, and the torque due to an inner Lindblad resonance is positive corresponding to an acceleration.

The total torque may be obtained by summing contributions over m. However, when doing so it must be borne in mind that the above analysis, appropriate to a cold disk, is only valid when $\xi = mc_s/(r\Omega) \ll 1$.

For finite ξ the positions of the Lindblad resonances are modified, being now given by

$$m^2(\Omega - \omega_p)^2 = \Omega^2(1 + \xi^2) \qquad (4)$$

where c_s is the sound speed.

The effective positions of the resonances are shifted with respect to the cold disk case. In particular, noting that $c_s = H\Omega$, with $H \ll r$ being the disk semithickness, one sees that when $m \rightarrow \infty$, Lindblad resonances pile up at

$$r = a \pm \frac{2H}{3} \qquad (5)$$

Physically these locations correspond to where the disk flow relative to the planet becomes sonic so that they are naturally the points from where the density waves are first launched (*Goodman and Rafikov*, 2001).

In addition, a correction factor has to be applied to the expression for the torque acting on the disk, which now reads (see *Artymowicz*, 1993; *Ward*, 1997; *Papaloizou and Larwood*, 2000)

$$\Gamma_m^{LR} = \frac{sign(\omega_p - \Omega)\pi^2\Sigma}{3\Omega\omega_p\sqrt{1 + \xi^2}(1 + 4\xi^2)}\Psi^2 \qquad (6)$$

This together with the shift in Lindblad resonance locations ensures that when contributions are summed over m, they decrease rapidly for $\xi \gg 1$, a phenomenon known as the torque cutoff.

2.1.2. Differential Lindblad torque. The total outer (resp. inner) Lindblad torque are obtained by summing over all individual components

$$\Gamma_{\text{OLR(ILR)}} = \sum_{m=1(2)}^{+\infty} \Gamma_m^{\text{OLR(ILR)}} \tag{7}$$

These are referred to as one-sided Lindblad torques. They scale with h^{-3}, where $h = H/r$ is the disk aspect ratio (*Ward, 1997*).

In Fig. 1 one can see that the torque cutoff occurs at larger m values in the thinner disk (the outer torque value peaks around m ~ 8–9 for h = 0.07, while it peaks around m ~21–22 for h = 0.03). Also, there is for both disk aspect ratios a very apparent mismatch between the inner and the outer torques, the former being systematically smaller than the latter. If we consider the torque of the disk acting on the planet, then the outer torques are negative and the inner ones positive, and the total torque is therefore negative. As a consequence, *migration is directed inward* and leads to a decay of the orbit onto the central object (*Ward, 1997*). It can be shown that the relative mismatch of inner and outer torques scales with the disk thickness (*Ward, 1997*). Since one-sided torques scale as h^{-3}, the migration rate scales with h^{-2}.

There are several reasons for the torque asymmetry that conspire to make the differential Lindblad torque a sizable fraction of the one-sided torque in a h = $O(10^{-1})$ disk (*Ward, 1997*). Most importantly, for a given m value, the inner Lindblad resonances lie further from the orbit than the corresponding outer Lindblad resonances. From this it follows that the relative disk motion becomes sonic further away inside the orbit, making the launching of density waves less efficient in the inner regions. Note that the net torque on the disk is positive, making that on the planet negative and therefore producing inward migration. This is found to be the case for disks with reasonable density profiles (see equation (9) below).

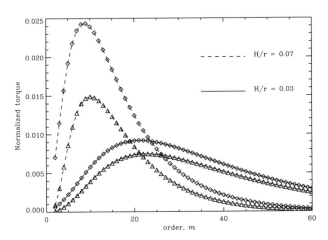

Fig. 1. Individual inner and outer torques (in absolute value) in a h = 0.07 and h = 0.03 disk, as a function of m. For each disk thickness, the upper curve (diamonds) shows the outer torque and the lower one (triangles) the inner torque. These torques are normalized to $\Gamma_0 = \pi q^2 \Sigma a^4 \omega_p^2 h^{-3}$.

2.1.3. Linear corotation torque. The angular momentum exchange at a corotation resonance corresponds to different physical processes than at a Lindblad resonance. At the latter the perturbing potential tends to excite epicyclic motion, and, in a protoplanetary disk, the angular momentum deposited is evacuated through pressure-supported waves. On the contrary, theses waves are evanescent in the corotation region, and are unable to remove the angular momentum brought by the perturber (*Goldreich and Tremaine,* 1979).

The corotation torque exerted on a disk by an external perturbing potential with m fold symmetry is given by

$$\Gamma_m^{\text{CR}} = \frac{m\pi^2\psi^2}{2(d\Omega/dr)} \frac{d}{dr}\left(\frac{\Sigma}{B}\right) \tag{8}$$

to be evaluated at the corotation radius. ψ is the amplitude of the forcing potential, and $B = \kappa^2/(4\Omega)$ the second Oort's constant. Since B is half the flow vorticity, the corotation torque scales with the gradient of (the inverse of) the specific vorticity, sometimes also called the vortensity. The corotation torque therefore cancels out in a $\Sigma \propto r^{-3/2}$ disk, such as the standard minimum mass solar nebula (MMSN) with h = 0.05.

In most cases, a disk sharp edge being a possible exception, the corotation torque can be safely neglected when estimating the migration timescale in the linear regime. Indeed, even the fully unsaturated corotation torque amounts at most to a few tens of percent of the differential Lindblad torque (*Ward, 1997; Tanaka et al.,* 2002), while *Korycansky and Pollack* (1993) find through numerical integrations that the corotation torque is an even smaller fraction of the differential Lindblad torque than given by analytical estimates.

2.1.4. Nonlinear effects. In a frame that corotates with the perturbation pattern, if inviscid, the flow in the neighborhood of corotation consists of libration islands, in which fluid elements librate on closed streamlines, between regions in which fluid elements circulate in opposite senses. This is true for general perturbations and not just those with m fold symmetry (see, e.g., Fig. 2, which applies to the corotation or coorbital region associated with a planet in circular orbit). In linear theory, the period of libration, which tends to infinity as the perturbation amplitude (or planet mass) tends to zero, is such that complete libration cycles do not occur and angular momentum exchange rates are appropriate only to sections at definite radial locations.

However, in actual fact, fluid elements in the libration region exchange zero net angular momentum with the perturbation during one complete libration cycle. Accordingly, if a steady state of this type can be set up, in full nonlinear theory the net corotation torque is zero. When this is the case the corotation resonance is said to be saturated.

But note that, when present, viscosity can cause an exchange of angular momentum between librating and circulating fluid elements that results in a net corotation torque. Then saturation is prevented. This is possible if the viscous timescale across the libration islands is smaller than the li-

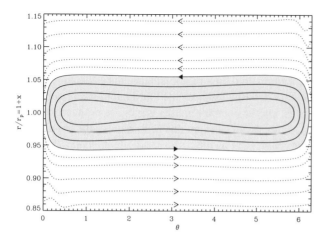

Fig. 2. Horseshoe region with closed streamlines as viewed in a frame corotating with a planet in circular orbit (shaded area). The planet is located at r = 1 and θ = 0 or 2π.

bration time (see *Goldreich and Sari*, 2003; *Ogilvie and Lubow*, 2003). It is found that for small viscosity, the corotation torque is proportional to ν (*Balmforth and Korycansky*, 2001), while at large viscosity one obtains the torque induced as material flowing through the orbit passes by the perturbing planet (*Masset*, 2001).

Note that these saturation properties are not captured by a linear analysis, since saturation requires a finite libration time, hence a finite resonance width. In the linear limit, the corotation torque appears as a discontinuity at corotation of the advected angular momentum flux, which corresponds to infinitely narrow, fully unsaturated libration islands.

2.2. Type I Migration Drift Rate Estimates

There are a number of estimates of the type I migration rate in the literature that are based on summing resonant torque contributions (see *Ward*, 1997; *Papaloizou and Larwood*, 2000, and references therein). Calculations using two-dimensional models have to soften the planet potential to obtain agreement with results obtained from three-dimensional calculations, but such agreement may be obtained for reasonable choices of the softening parameter (see *Papaloizou and Terquem*, 2006).

The most recent linear calculations by *Tanaka et al.* (2002) that take into account three-dimensional effects, and are based upon the value of the total tidal torque, including the corotation torque (fully unsaturated since it is a linear estimate), given

$$\tau \equiv a/\dot{a} = (2.7 + 1.1\alpha)^{-1} \frac{M_*^2}{M_p \Sigma a^2} h^2 \omega_p^{-1} \qquad (9)$$

for a surface density profile $\Sigma \propto r^{-\alpha}$. For an Earth-mass planet around a solar-mass star at r = 1 AU, in a disk with $\Sigma = 1700$ g cm^{-2} and h = 0.05, this translates into $\tau = 1.6 \times 10^5$ yr.

This semi-analytic estimate has been verified by means of three-dimensional numerical simulations (*Bate et al.*, 2003; *D'Angelo et al.*, 2003a). Both find an excellent agreement in the limit of low mass, thus they essentially validate the linear analytical estimate. However, while *Bate et al.* (2003) find agreement with the linear results for all planet masses, *D'Angelo et al.* (2003a) find very long migration rates for intermediate masses, i.e., for Neptune-sized objects (see Fig. 3). Additional two-dimensional and three-dimensional high-resolution numerical simulations by Masset et al. (in preparation) show that this *migration offset* from the linear results is a robust phenomenon whose strength varies (1) with departure from the $\Sigma \propto r^{-3/2}$ relation, (2) with the value of the viscosity, and (3) with the disk thickness. The transition from linear to the offset regime is apparently caused by the onset of nonlinear effects that could be related to corotation torques whose strength also increase with departure from $\Sigma \propto r^{-3/2}$ (see section 2.1.3).

The type I migration timescale is very short, much shorter than the buildup time of the $M_p \sim 5$–$15\ M_\oplus$ solid core of a giant planet (see, e.g., *Papaloizou and Nelson*, 2005, for a discussion). Hence, the existence of type I migration causes potential difficulties for the accumulation scenario for these massive cores. This remains a problem within the framework of planet formation theory (see the discussion below for possible resolutions).

The influence of the disk's self-gravity on type I migration has been analyzed through numerical and semi-analyti-

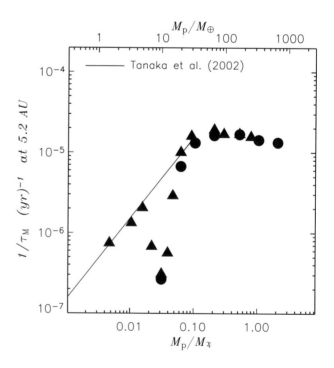

Fig. 3. Migration rate for different planet masses for three-dimensional fully nonlinear nested grid simulations. The symbols denote different approximations (smoothening) for the potential of the planet. The solid line refers to linear results for type I migration by *Tanaka et al.* (2002) (see equation (9). Adapted from *D'Angelo et al.* (2003a).

cal methods (*Nelson and Benz,* 2003; *Pierens and Huré,* 2005). It increases the migration rate but the effect is small for typical disk parameters.

Papaloizou and Larwood (2000) incorporate a nonzero planet eccentricity in their torque calculations and find that in general the torques weaken with increasing eccentricity and can even reverse once the eccentricity exceeds h by some factor on the order of unity. Thus a process that maintains eccentricity could potentially help to stall the migration process. A similar effect occurs if the disk has a global nonaxisymmetric distortion such as occurs if it has a finite eccentricity. This can also result in a weakening of the tidal interaction and a stalling of the torques under appropriate conditions (see *Papaloizou,* 2002).

Recent attempts to include more detailed physics of the protoplanetary disk, such as opacity transitions and their impact on the disk profile (*Menou and Goodman,* 2004), or radiative transfer and the importance of shadowing in the planet vicinity (*Jang-Condell and Sasselov,* 2005), have led to lower estimates of the type I migration rates that might help resolve the accretion/migration timescale discrepancy. The aforementioned offset, as well as effects due to magnetic fields (see, e.g., *Terquem,* 2003) and their associated turbulence (see below), may help to extend the type I migration timescale and allow proto giant planet cores to form.

3. TYPE II MIGRATION

When the planet grows in mass the disk response cannot be treated any longer as a linear perturbation. The flow perturbation becomes nonlinear and the planetary wake turns into a shock in its vicinity. Dissipation by these shocks as well as the action of viscosity leads to the deposition of angular momentum, pushes material away from the planet, and a gap opens. The equilibrium width of the gap is determined by the balance of gap-closing viscous and pressure forces and gap-opening gravitational torques.

To obtain rough estimates, the condition that the planet's gravity be strong enough to overwhelm pressure in its neighborhood is that the radius of the Hill sphere exceed the disk semi-thickness or

$$a \left(\frac{M_p}{3 M_*} \right)^{1/3} > H \qquad (10)$$

The condition that angular momentum transport by viscous stresses be interrupted by the planetary tide is approximately

$$\left(\frac{M_p}{M_*} \right) > (40 a^2 \Omega)/\nu \qquad (11)$$

For more discussion of these aspects of gap opening, see *Lin and Papaloizou* (1993), *Bryden et al.* (1999), and *Crida et al.* (2006).

Interestingly, for the standard parameters h = 0.05 and $\nu/(a^2\Omega) = 10^{-5}$, equation (10) gives $M_p/M_* > 3.75 \times 10^{-4}$,

while equation (11) gives $M_p/M_* > 4 \times 10^{-4}$. Note that this result obtained from simple estimates is in good agreement with that obtained from Fig. 3.

Accordingly, for typical protoplanetary disk parameters, we can expect that a planet with a mass exceeding that of Saturn will begin to open a visible gap. Using equation (9) for the type I migration rate together with equation (10), we can estimate the mimimum drift time at the marginal gap opening mass to be given by

$$(a/\dot{a}) = 3^{-2/3}(5.4 + 2.2\alpha)^{-1} \left(\frac{M_p}{M_*} \right)^{-1/3} \frac{M_*}{\pi \Sigma a^2} P_{orb} \qquad (12)$$

with P_{orb} being the orbital period. For a MMSN with $4\pi\Sigma a^2 = 2 \times 10^{-3} M_*$, and $M_p/M_* = 4 \times 10^{-4}$ at 5.2 AU, this gives a minimum drift time of only $\sim 3 \times 10^4$ yr. This simply obtained estimate is in good agreement with the results presented in Fig. 3.

3.1. Numerical Modeling

Currently, heavy reliance is placed on numerical methods to analyze the dynamics of the planet-disk interaction, the density structure of the disk, and the resulting gravitational torques acting on the planet. The corresponding migration regime is called type II migration (e.g., *Lin and Papaloizou,* 1986; *Ward,* 1997).

The first modern hydrodynamical calculations of planet disk interaction in the type II regime were performed by *Bryden et al.* (1999), *Kley* (1999), and *Lubow et al.* (1999). Since protoplanetary accretion disks are assumed to be vertically thin, these first simulations used a two-dimensional (r – φ) model of the accretion disk. The vertical thickness H of the disk is incorporated by assuming a given radial temperature profile T(r) ∝ r⁻¹, which makes the ratio H/r constant. Typically the simulations assume H/r = 0.05 so that at each radius, r, the Keplerian speed is 20× faster than the local sound speed. Initial density profiles typically have power laws for the surface density $\Sigma \propto r^{-s}$ with s between 0.5 and 1.5. More recently, fully three-dimensional models have been calculated. These have used the same kind of isothermal equation of state (*Bate et al.,* 2003; *D'Angelo et al.,* 2003a).

The viscosity is dealt with by solving the Navier Stokes equations with the kinematic viscosity ν taken as constant or given by an α-prescription $\nu = \alpha c_s H$, where α is a constant. From observations of protostellar disks, values lying between 10^{-4} and 10^{-2} are inferred for the α parameter but there is great uncertainty. Full MHD calculations have shown that the viscous stress-tensor ansatz may give (for sufficiently long time averages) a reasonable approximation to the *mean* flow in a turbulent disk (*Papaloizou and Nelson,* 2003). The embedded planets are assumed to be point masses (using a smoothed or softened potential). The disk influences their orbits through gravitational torques, which cause orbital evolution. The planets may also accrete mass from the surrounding disk (*Kley,* 1999).

3.2. Viscous Laminar Disks

The type of modeling outlined in the previous section yields in general smooth density and velocity profiles, and we refer to those models as *viscous laminar disk* models, in contrast to models that do not assume an *a priori* given viscosity and rather model the turbulent flow directly (see below).

A typical result of such a viscous computation obtained with a 128×280 grid (in $r - \varphi$) is displayed in Fig. 4. Here, the planet with mass $M_p = 1\ M_{Jup}$ and semimajor axis $a_p = 5.2$ AU is *not* allowed to move and remains on a fixed circular orbit, an approximation that is made in many simulations. Clearly seen are the major effects an embedded planet has on the structure of the protoplanetary accretion disk. The gravitational force of the planet leads to spiral wave patterns in the disk. In the present calculation (Fig. 4) there are two spirals in the outer disk and in the inner disk. The tightness of the spiral arms depends on the temperature (i.e., h) of the disk; the smaller the temperature, the tighter the spirals. The density gap at the location of the planet discussed above is also visible.

To obtain more insight into the flow near the planet and to calculate accurately the torques of the disk acting on the planet, the nested-grid approach described above has been used together with a variable grid size (*D'Angelo et al., 2002, 2003a; Bate et al., 2003*). Such a grid system is fixed and therefore not adaptive. The planet is located at the center of the finest grid.

The result for a two-dimensional computation using six grids is displayed in Fig. 5; for more details see also *D'Angelo et al.* (2002). The top lefthand base grid has a resolution of 128×440 and each subgrid has a size of 64×64

with a refinement factor of 2 from level to level. It is noticeable that the spiral arms inside the Roche lobe of a high-mass planet are detached from the global outer spirals. The top righthand panel of Fig. 5 indicates that the outer spirals fade away exterior to the one around the planet. The two-armed spiral around the planet extends deep inside the Roche lobe and enables the accretion of material onto the planet. The nested-grid calculations have recently been extended to three dimensions, and a whole range of planetary masses has been investigated (*D'Angelo et al., 2003a*). In the three-dimensional case the spiral arms are weaker and accretion occurs primarily from regions above and below the midplane of the disk.

3.3. The Migration Rate

Such high-resolution numerical computations allow for a detailed computation of the torque exerted by the disk material onto the planet, as well as its mass accretion rate (for migration rates, see Fig. 3).

The consequences of accretion and migration have been studied by numerical computations that do not hold the planet fixed at some radius but rather follow the orbital evolution of the planet (*Nelson et al., 2000*), allowing planetary growth. The typical migration and accretion timescales are on the order of 10^5 yr, while the accretion timescale may be slightly smaller. This is in very good agreement with the estimates obtained from the models using a fixed planet. These simulations show that during their inward migration they grow up to about $4\ M_{Jup}$.

The inward migration time of 10^5 yr can be understood as the natural viscous evolution time of the local accretion disk. When the planet mass is not too large, and it makes a gap in the disk, it tends to move as a disk gas particle would and thus move inward on the viscous timescale $\tau \sim a^2/\nu$ (*Lin and Papaloizou, 1986*). But note that when the mass of the planet exceeds the disk mass in its neighborhood on a scale a, the migration rate decreases because of the relatively large inertia of the planet (see, e.g., *Syer and Clarke, 1995; Ivanov et al., 1999*).

The consequence of the inclusion of thermodynamic effects (viscous heating and radiative cooling) on the gap-formation process and type II migration has been studied by *D'Angelo et al.* (2003b). In two-dimensional calculations an increased temperature in the circumplanetary disk has been found. This has interesting consequences for the possible detection of an embedded protoplanet. The effect that self-gravity of the disk has on migration has been analyzed through numerical simulations (*Nelson and Benz, 2003*). For typically expected protostellar disk masses the influence is rather small.

3.4. Consequences for Evolution in Young Planetary Systems

A number of studies with the object of explaining the existence and distribution of giant planets interior to the "snow line" at 2 AU that make use of type II migration have

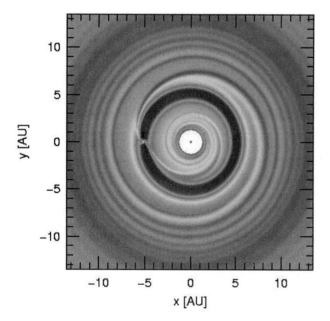

Fig. 4. Surface density profile for an initially axisymmetric disk model 200 orbital orbits after the introduction of a planet that subsequently remains on a fixed circular. The mass ratio is 10^{-3}.

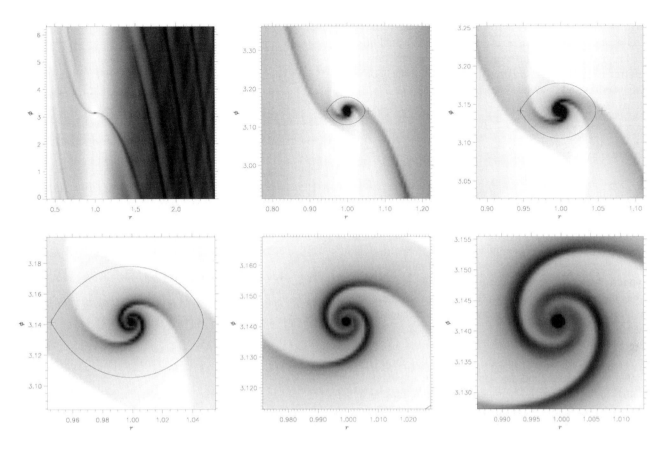

Fig. 5. Density structure of a 1 M_{Jup} planet on each level of the nested grid system, consisting of six grid levels in total. The top left panel displays the total computational domain. The line indicates the size of the Roche lobe.

been performed (e.g., *Trilling et al.,* 1998, 2002; *Armitage et al.,* 2002; *Alibert et al.,* 2004; *Ida and Lin,* 2004). These assume formation beyond the snow line followed by inward type II migration that is stopped byeither disk dispersal, Roche lobe overflow, stellar tides, or entering a stellar magnetospheric cavity. Reasonable agreement with observations is attained. But type I migration has to be suppressed, possibly by one of the mechanisms discussed in this review.

4. TYPE III MIGRATION

The terminology type III migration refers to migration for which an important driver is material flowing through the coorbital region. Consider an inwardly (resp. outwardly) migrating planet. Material of the inner disk (resp. outer disk) has to flow across the coorbital region and it executes one U-turn in the horseshoe region (see Fig. 2) to do so. By doing this, it exerts a corotation torque on the planet that scales with the migration rate [for further analysis and discussion, see *Masset and Papaloizou* (2003), *Artymowicz* (2004), and *Papaloizou* (2005)].

The specific angular momentum that a fluid element near the separatrix takes from the planet when it switches from an orbit with radius $a - x_s$ to $a + x_s$ is $\Omega a x_s$ where x_s is the radial half-width of the horseshoe region estimated to be 2.5 Hill sphere radii (Artymowicz, in preparation).

The torque exerted on a planet migrating at a rate $\dot{a}$ by the inner or outer disk elements as they cross the planet orbit on a horseshoe U-turn is accordingly to lowest order in x_s/a

$$\Gamma_2 = (2\pi a \Sigma_s \dot{a}) \cdot (\Omega a x_s) \qquad (13)$$

where Σ_s is the surface density at the upstream separatrix. The system of interest for the evaluation of the sum of external torques is composed of the planet, all fluid elements trapped in libration in the horseshoe region (with mass MHS), and the Roche lobe content (with mass MR), because these components migrate together.

The drift rate of this system is then given by

$$\tfrac{1}{2}(M_p + M_{HS} + M_R) \cdot (\Omega \dot{a}) = (4\pi a x_s \Sigma_s) \cdot \tfrac{1}{2}(\Omega a \dot{a}) + \Gamma_{LR} \qquad (14)$$

which can be rewritten as

$$M_p' \cdot \tfrac{1}{2}(\Omega \dot{a}) = (4\pi a \Sigma_s x_s - M_{HS}) \cdot \tfrac{1}{2}(\Omega a \dot{a}) + \Gamma_{LR} \qquad (15)$$

where $M_p' = M_p + M_R$ is all the mass content within the Roche lobe, which from now on for convenience we refer to as the planet mass. The first term in the first bracket of the r.h.s. of equation (15) corresponds to the horseshoe region surface multiplied by the upstream separatrix surface density, hence it is the mass that the horseshoe region would have if it had a uniform surface density equal to the up-stream surface density. The second term is the actual horseshoe region mass. The difference between these two terms is referred to in *Masset and Papaloizou* (2003) as the coorbital mass deficit and denoted δm. Thus we have

$$\tfrac{1}{2}\dot{a}\Omega a(M'_p - \delta m) = \Gamma_{LR} \qquad (16)$$

Equation (16) gives a drift rate

$$\dot{a} = \frac{2\Gamma_{LR}}{\Omega a(M'_p - \delta m)} \qquad (17)$$

This drift rate is faster than the standard estimate in which one neglects δm. This comes from the fact that the coorbital dynamics alleviate the differential Lindblad torque task by advecting fluid elements from the upstream to the downstream separatrix. The angular momentum exchanged with the planet by doing so produces a positive feedback on its migration.

As δm tends to M'_p, most of the angular momentum change of the planet and its coorbital region is balanced by that of the orbit-crossing circulating material, making migration increasingly cost effective.

When $\delta m \geq M'_p$, the above analysis, assuming a steady migration ($\dot{a}$ constant), is no longer valid. Migration may undergo a runaway, leading to a strongly time-varying migration rate. Runaway (also denoted type III or fast) migration is therefore a mode of migration of planets that deplete their coorbital region and are embedded in sufficiently massive disks, so that the above criterion be satisfied. The critical disk mass above which a planet of given mass undergoes a runaway depends on the disk parameters (aspect ratio and effective viscosity). The limit has been considered by *Masset and Papaloizou* (2003) for different disk aspect ratios and a kinematic viscosity $\nu = 10^{-5}$. The type III migration domain that was found for a disk with $H/r = 0.04$ is indicated in Fig. 6.

Once the migration rate becomes fast enough that the planet migrates through the coorbital region in less than a libration period, the analysis leading to equation (16) becomes invalid. A recent analysis by Artymowicz (in preparation) indicates how this equation should be modified when this occurs. Consider the torque term

$$\dot{a}\Omega a\delta m/2$$

in equation (16) above. This can be regarded as the corotation torque. It can be considered to be generated as follows: Material enters the horseshoe region close to and behind the planet from the region into which the planet migrates at a maximum radial separation corresponding to the full half-width x_s. However, the turn at the opposite side of the horseshoe region 2π round in azimuth occurs at a reduced maximum radial separation $x_s - \Delta$ due to the radial migration of the planet. Consideration of Keplerian circular orbits

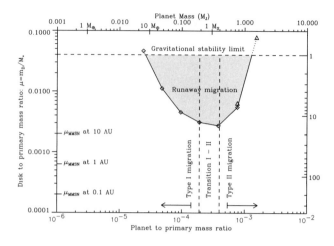

Fig. 6. Runaway limit domain for a $H/R = 0.04$ and $\nu = 10^{-5}$ disk, with a surface density profile $\Sigma \propto r^{-3/2}$. The variable $m_D = \pi\Sigma r^2$ is featured on the y axis. It is meant to represent the local disk mass, and therefore depends on the radius.

gives (Artymowicz, in preparation)

$$\Delta = x_s\left(1 - \sqrt{1 - |\dot{a}|\dot{a}_f^{-1}}\right) \qquad (18)$$

where

$$\frac{\dot{a}_f}{a} = \frac{3x_s^2}{8\pi a^2}\Omega \qquad (19)$$

gives the critical or fast migration rate for which the horseshoe region can just extend the full 2π in azimuth. For larger drift rates it contracts into a tadpole-like region and the dynamics are no longer described by the above analysis. Instead, we should replace the square root in equation (18) by zero. For smaller drift rates, the torques exerted at the horseshoe turns on opposite sides of the planet are proportional to x_s^3 and $(x_s - \Delta)^3$ respectively. This is because these torques are proportional to the product of the flow rate, specific angular momentum transferred, and the radial width, each also being proportional to the radial width. As these torques act in opposite senses, the corotation torque should be proportional to $x_s^3 - (x_s - \Delta)^3$, or $x_s^3[1 - (1 - |\dot{a}|\dot{a}_f^{-1})^{3/2}]$. Note that this factor, which applies to both librating and nonlibrating material, being $(3/2)x_s^3|\dot{a}|\dot{a}_f^{-1}$ for small drift rates, provides a match to equation (16) provided the a multiplying δm is replaced by $(2/3)\dot{a}_f\,\text{sign}(\dot{a})[1 - |\dot{a}|\dot{a}_f^{-1})^{3/2}]$.

Equation (16) then becomes

$$\tfrac{1}{2}\Omega a(M'_p\dot{a} - \tfrac{2}{3}\delta m\dot{a}_f\,\text{sign}(\dot{a})(1 - (1 - |\dot{a}|\dot{a}_f^{-1})^{3/2})) = \Gamma_{LR} \quad (20)$$

Interestingly, when Lindblad torques are small, equation (20) now allows for the existence of steady fast migration rates that can be found by setting the lefthand side to zero. Assuming without loss of generality that $a > 0$, and setting $Z = \dot{a}/\dot{a}_f$, these states satisfy

$$Z = (2/3)M_\Delta(1 - (1 - Z)^{3/2}) \quad (21)$$

with $M_\Delta = (\delta m/M'_p)$. Equation (21) gives rise to a bifurcation from the solution $Z = 0$ to fast migration solutions when $M_\Delta > 1$ or when the coorbital mass deficit exceeds the planet mass (see Fig. 7). The "fast" rate, $Z = 1$, occurs when $M_\Delta = 3/2$. For larger M_Δ, $Z = 2 M_\Delta/3$.

Fast migration, for the same disk profile and planet mass, can be directed either outward or inward, depending on the initial conditions. This type of planetary migration is found to depend on its migration history, the "memory" of this history being stored in the way the horseshoe region is populated, i.e., in the preparation of the coorbital mass deficit. Note that owing to the strong variation of the drift rate, the horseshoe streamlines are not exactly closed, so that the coorbital mass deficit can be lost and the runaway can stall. This has been observed in some numerical simulations, whereas others show sustained fast migration episodes for Saturn- or Jupiter-mass planets that can vary the semimajor axis by large factors in less than 100 orbits (e.g., see Fig. 8). To date, it is still unclear how long such episodes can last for, and what the conditions are, if any, for them to stall or to be sustained for a long period.

Because of the need to take into account complex coorbital flows in a partially gap-forming regime close to the planet, the problem of type III migration is very numerically challenging and therefore, not unexpectedly, issues of adequate numerical resolution and convergence remain outstanding. *D'Angelo et al.* (2005) have undertaken numerical simulations of runaway migration using a nested-grid system that can give high resolution within the Roche lobe, but not elsewhere in the simulation, and found that the outcome, stated to be the suppression of type III migration, was highly dependent on the torque calculation prescription

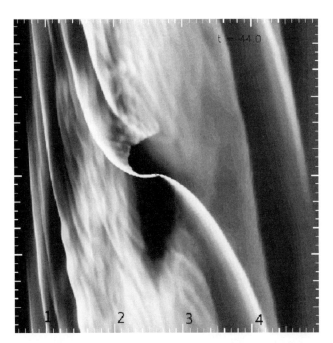

Fig. 8. Density contour plot taken from a PPM simulation with variable adaptive mesh of a Jupiter-mass planet in a disk 2.5× more massive than the MMSN (Artymowicz, in preparation). After being placed on a positive density gradient, the planet increased its semimajor axis by a factor of 2.6 in only 44 orbits. A trapped coorbital low-density region is clearly visible and the migration speed corresponds to $M_\Delta \approx 3$.

(more precisely on whether or not the Roche lobe material was taken into account in this calculation) and on the mesh resolution. One of the main subtleties of coorbital dynamics is to properly take into account the inertia of all the material trapped (even approximately) in libration with the planet, be it the horseshoe or circumplanetary material. Including the Roche lobe material in the torque calculation, while for other purposes is is assumed to be non-self-gravitating, introduces a discrepancy between the inertial mass of the migrating object (the point-like object plus the Roche lobe content) and its active gravitational mass (the mass of the point-like object). This unphysical discrepancy can be large and can severely alter the migration properties, especially at high resolution where the Roche lobe is flooded by disk material in a manner than strongly depends on the equation of state. On the other hand, *Masset and Papaloizou* (2003) consider the Roche lobe content as a whole, referred to as the planet for simplicity, assuming that for a given mass M'_p of this object, there always exists a point-like object with mass $M_p < M'_p$ such that the whole Roche lobe content has mass M'_p. In this case, one needs to exclude the Roche lobe content from the torque calculation (since the forces originating from within the Roche lobe are not external forces), and this way one naturally gets inertial and active gravitational masses of the Roche lobe content (migrating object) that both amount to M'_p.

Another way of looking at this issue is to realize that from considerations of angular momentum conservation, the

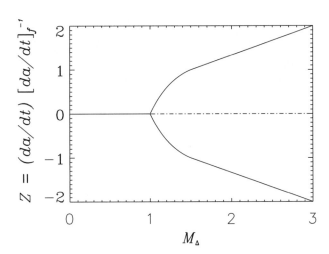

Fig. 7. Bifurcation diagram derived from equation (21). This is extended to negative $\frac{da}{dt}$ by making the curve symmetric about the M_Δ axis. The bifurcation to fast migration occurs for $M_\Delta = 1$.

angular momentum changes producing the migration can be evaluated at large distances from the Roche lobe. Suppose that the material inside the Roche lobe had some asymmetric structure that produced a torque on the point-like mass M_p. Then, in order to sustain this structure in any kind of steady state, external interaction would have to provide an exactly counterbalancing torque that could be measured in material exterior to the Roche lobe (for more details, see review by *Papaloizou and Terquem, 2006*). In addition to the above issues, the effect of viscosity on the libration region through its action on the specific vorticity profile or the consequences of specific vorticity generation at gap edges (e.g., *Koller et al., 2003*) has yet to be considered.

What are the potential consequences for forming planets? First, the MMSN is not massive enough to allow giant planets to experience runaway migration. This is most likely for planets with masses comparable to that of Saturn in disks that do not need to be more massive than a few times the MMSN.

Thus runaway migration, should it occur, makes the tendency for the migration rate to be a maximum, in the mass range associated with the onset of gap formation and type II migration, more pronounced. This may be related to the fact that most of the extrasolar planets known as "hot Jupiters," with a semimajor axis a < 0.06 AU, happen to have subjovian masses. Assuming that type I migration is suppressed by, e.g., MHD turbulence (see below), a forming protoplanet, as it passes through the domain of fast migration, would migrate very fast toward the central object and at the same time it would accrete gas from the nebula. If the protoplanet is able to form a deep gap before it reaches the central regions of the disk, or stellar magnetosphere, it enters the slow, type II migration regime, having at least about a Jupiter mass. Otherwise, it reaches the central regions still as a subjovian object.

5. TURBULENT PROTOPLANETARY DISKS

The majority of calculations examining the interaction between protoplanets and protoplanetary disks have assumed that the disk is laminar. The mass accretion rates inferred from observations of young stars, however, require an anomalous source of viscosity to operate in these disks. The most likely source of angular momentum transport is MHD turbulence generated by the magnetorotational instability (MRI) (*Balbus and Hawley, 1991*). Numerical simulations performed using both the local shearing box approximation (see *Balbus and Hawley, 1998*, for a review) and global cylindrical disk models (e.g., *Papaloizou and Nelson, 2003*, and references therein) indicate that the nonlinear outcome of the MRI is vigorous turbulence and dynamo action, whose associated stresses can account for the observed accretion rates inferred for T Tauri stars.

These studies assumed that the approximation of ideal MHD was appropriate. The ionization fraction in cool, dense protoplanetary disks, however, is probably small in the planet-forming region between 1 and 10 AU. Only the surface layers of the disk, which are exposed to external sources

of ionization such as X-rays from the central star or cosmic rays, are likely to be sufficiently ionized to sustain MHD turbulence (e.g., *Gammie, 1996*; *Fromang et al., 2002*). However, this involves complex chemical reaction networks and the degree of depletion of dust grains, which itself may vary while there is ongoing planet formation.

Recent work has examined the interaction between planets of various masses and turbulent protoplanetary disks. These studies have usually simulated explicitly MHD turbulence arising from theMRI. We now review the results of these studies.

5.1. Low-Mass Protoplanets in Turbulent Disks

The interaction between low-mass, non-gap-forming protoplanets and turbulent disks has been examined by *Papaloizou et al.* (2004), *Nelson and Papaloizou* (2004), *Nelson* (2005), and *Laughlin et al.* (2004). These calculations show that interaction between embedded planets and density fluctuations generated by MHD turbulence can significantly modify type I migration, at least over timescales equal to the duration of simulations that are currently feasible (t ~ 150 planet orbits), leading to a process of "stochastic migration" rather than the monotonic inward drift expected for planets in laminar disks. Figure 9 shows snapshots of the midplane density for six 10-$M_\oplus$ planets (noninteracting) embedded in a turbulent disk with H/R = 0.07, and show that the turbulent density fluctuations are of higher amplitude than the spiral wakes generated by the planet (*Nelson and Papaloizou, 2004*; *Nelson, 2005*). Indeed, typical surface density fluctuations generated by turbulence in simulations are typically $\delta\Sigma/\Sigma \approx 0.15$–$0.3$, with peak fluctuations being $O(1)$. Thus, on the scale of the disk thickness H, den-

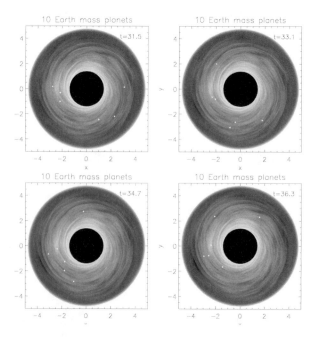

Fig. 9. Orbital evolution of six 10 $M_\oplus$ protoplanets embedded in a turbulent protoplanetary disk.

sity fluctuations can contain more than an Earth mass in a disk model a few times more massive than a minimum mass nebula.

Figure 10 shows the variation in the semimajor axes of the planets shown in Fig. 9, and Fig. 11 shows the running mean of the torque for one of the planets. It is clear that the usual inward type I migration is disrupted by the turbulence, and the mean torque does not converge toward the value obtained in a laminar disk for the duration of the simula-

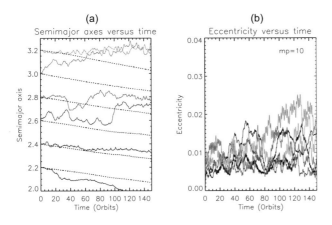

Fig. 10. **(a)** Variation of semimajor axis with time (measured in planet orbits at r = 2.3). The dotted lines represent the trajectories of 10-$M_\oplus$ planets in equivalent laminar disks. **(b)** Variation of eccentricity. For 10-$M_\oplus$ bodies, eccentricity damping due to coorbital Lindblad torques maintains low eccentricities.

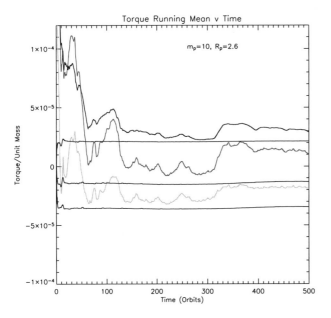

Fig. 11. Running time average of the torque per unit mass for the planet in Figs. 9 and 10 whose initial orbit is at r = 2.6. The ultimately lowest but not straight line corresponds to the torque exerted by the outer disk, the ultimately uppermost line corresponds to the inner disk torque, and the line ultimately between them corresponds to the total torque. The three horizontal lines correspond to the inner, outer, and total torque exerted on a planet in laminar disk.

tion. A key question is whether the stochastic torques can continue to overcome type I migration over timescales up to disk lifetimes. A definitive answer will require very long global simulations.

One can test the possibility that the response of the turbulent disk is that of a laminar disk whose underlying density is given by the time-averaged value, superposed on which are Gaussian-distributed fluctuations with a characteristic recurrence time ~ the orbital period. If such a picture applies then the time-averaged torque experienced by the protoplanet, $\overline{T}$, can be expressed as

$$\overline{T} = \langle T \rangle + \frac{\sigma_T}{\sqrt{t_{tot}}} \tag{22}$$

where σ_T is the standard deviation of the torque amplitude, $\langle T \rangle$ is the underlying type I torque, and t_{tot} is the total time elapsed, measured in units of the characteristic time for the torque amplitude to vary. Convergence toward the underlying type I value is expected to begin once the two terms on the righthand side become equal. For 10-$M_\oplus$ protoplanets, simulations indicate that $\sigma_T \simeq 10 \langle T \rangle$, with a simple estimate of the time for fluctuations to recur being approximately equal to one-half the planet orbital period (*Nelson and Papaloizou*, 2004; *Nelson*, 2005). The torque convergence time is then ≃50 planet orbits. Interestingly, the simulations presented in Fig. 10 were run for ≃150 planet orbits, and do not show a tendency for inward migration.

Analysis of the stochastic torques suggests that they vary on a range of characteristic times from the orbital period to the run times of the simulations themselves (*Nelson*, 2005). This feature can in principle allow a planet to overcome type I migration for extended time periods. It appears to at least partially explain why the simulations do not show inward migration on the timescale predicted by equation (22). The origin of these long timescale fluctuations is currently unknown.

In addition, the picture described above of linear superposition of stochastic fluctuations on an underlying type I torque may be incorrect due to nonlinear effects. Density fluctuations occurring within the disk in the planet vicinity are substantial, such that the usual bias between inner and outer type I torques may not be recovered easily, invalidating the assumptions leading to equation (22). To see this, consider a density fluctuation on the order of unity with length scale on the order of H a distance on the order of H from the planet. The characteristic specific torque acting on it is $G\Sigma R$. Given their stochastic nature, one might expect the specific torque acting on the planet to oscillate between $\pm G\Sigma R$. Note that this exceeds the net specific torque implied by equation (9) by a factor of $T_f \sim [(M_*/M_p)(h)^3]h^{-1}$. From the discussion in section 3, the first factor should exceed unity for an embedded planet. Thus, such an object is inevitably subject to large torque fluctuations. The strength of the perturbation of the planet on the disk is measured by the dimensionless quantity $(M_p/M_*)(h)^{-3}$. This perturbation might be expected to produce a bias in the underlying stochastic torques. If it produces a nonzero mean, corresponding to the typical fluctuation reduced by a factor T_f, this becomes comparable to the laminar type I value. However, there is no reason to suppose the exact type I result should be recovered. But note the additional complication that the concept of such a mean may not have much significance in practical cases, if large fluctuations can occur on the disk evolutionary timescale, such that it is effectively not established.

We comment that, as indicated in the plots in Fig. 11 at small times, the one-sided torque fluctuations can be more than an order of magnitude larger than expected type I values. However, such fluctuations occur on an orbital timescale and if averages over periods of 50 orbits are considered, values more like type I values are obtained. Accordingly, the large fluctuations are not associated with large orbital changes.

Further work is currently underway to clarify the role of turbulence on modifying type I migration, including type I migration in vertically stratified turbulent disks.

5.2. High-Mass Protoplanets in Turbulent Disks

The interaction of high-mass, gap-forming planets with turbulent protoplanetary disks has been considered in a number of papers (*Nelson and Papaloizou, 2003; Winters et al., 2003; Papaloizou et al., 2004; Nelson and Papaloizou, 2004*). Figure 12 shows the midplane density for a turbulent disk with an embedded 5 M_{Jup} protoplanet. As expected from the discussion presented in section 3, gap formation occurs because the Roche lobe radius exceeds the disk scale height, and tidal torques locally overwhelm viscous torques in the disk.

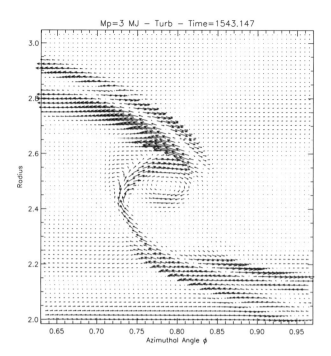

Fig. 13.   Magnetic field lines in the vicinity of the protoplanet. Field lines link the protoplanetary disk with the circumplanetary disk within the planet Hill sphere.

Papaloizou et al. (2004) considered the transition from fully embedded to gap-forming planets using local shearing box and global simulations of turbulent disks. These simulations showed that gap formation begins when $(M_p/M_*)(R/H)^3 \simeq 1$, which is the condition for the disk reponse to the planet gravity being nonlinear. The viscous stress in simulations with zero net magnetic flux (as normally considered here) typically gives rise to an effective $\alpha \approx 5 \times 10^{-3}$, such that the viscous criterion for gap formation is satisfied when the criterion for nonlinear disk response is satisfied.

Global simulations allow the net torque on the planet due to the disk to be calculated and hence the migration time to be estimated. Simulations presented in *Nelson and Papaloizou* (2003, 2004) for massive planets indicate migration times of ~10^5 yr, in line with expections for type II migration.

A number of interesting features arise in simulations of gap-forming planets embedded in turbulent disks. The magnetic field topology is significantly modified in the vicinity of the protoplanet, as illustrated by Fig. 13. The field is compressed and ordered in the postshock region associated with the spiral wakes, increasing the magnetic stresses there. Accretion of gas into the protoplanet Hill sphere causes advection of field into the circumplanetary disk that forms there, and this field then links between the protoplanetary disk and the circumplanetary disk, apparently contributing to magnetic braking of the circumplanetary material. Indeed, comparison between a simulation of a 3 M_{Jup} protoplanet in a viscous laminar disk and an equivalent turbulent disk simulation suggests that mass accretion onto the planet may be enhanced by this effect. Figure 14 shows the

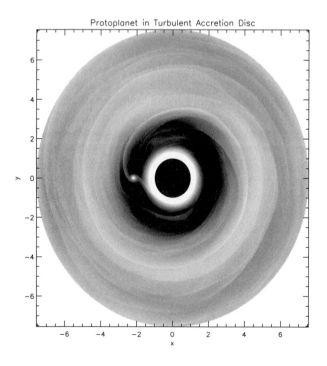

Fig. 12.   Snapshot of the disk midplane density for a 5 M_{Jup} protoplanet embedded in a turbulent disk.

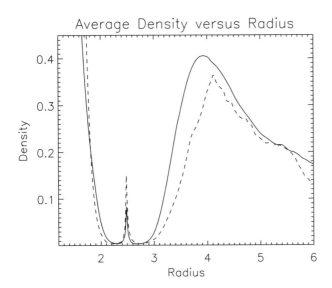

Fig. 14. Average density in the vicinity of the planet as a function of radius for a laminar viscous disk (solid line) and for a disk with MHD turbulence (dashed line). More mass has settled onto the planet in the latter case, possibly due to magnetic braking.

average density in the vicinity of the planet for a laminar disk run and a turbulent disk simulation, indicating that the softened point mass used to model the planet has accreted more gas in the turbulent disk run. In addition, the gap generated tends to be deeper and wider in turbulent disks than in equivalent viscous, laminar disks (*Nelson and Papaloizou,* 2003; *Papaloizou et al.,* 2004). It is worth noting, however, that the global simulations are of modest resolution, and more high-resolution work needs to be done to examine these issues in greater detail.

6. SUMMARY AND DISCUSSION

We have reviewed recent progress in the field of disk-planet interactions in the context of orbital migration. This has mainly come from large-scale two- and three-dimensional simulations that have utilized the most up-to-date supercomputer resources. These have allowed the application of high-resolution multigrid methods and the study of disks with MHD turbulence interacting with planets. Simulations of both laminar and turbulent disks have been carried out and while the structure of the protoplanetary disk is uncertain, both these types are valuable.

In the case of type I migration it has become clear that the three-dimensional simulations discussed above agree with results obtained from linear analysis and summing contributions from resonant torques (*Tanaka et al.,* 2002) in predicting that planets in the $M_\oplus$ range in axisymmetric smoothly varying laminar disks, modeling the MMSN, undergo a robust inward migration on a 10^6-yr timescale. This is a threat to the viability of the core accumulation scenario and a resolution has to be sought.

Several potential resolutions involving departures from the simple disk model have been suggested. These include invoking sharp radial opacity variations (*Menou and Goodman,* 2004), large-scale nonaxisymmetric distortions (e.g., an eccentric disk) (*Papaloizou,* 2002), or a large-scale toroidal magnetic field (*Terquem,* 2003).

If the disk is magnetically active, type I drift may be disrupted by stochastic migration even for very extended periods of time (*Nelson and Papaloizou,* 2004). Even if this is unable to keep type I migration at bay for the entire disk lifetime, the increased mobility of the cores may have important consequences for their buildup through accretion (*Rice and Armitage,* 2003). But note that even within the context of laminar disks, simulations have revealed that type I migration has to be suppressed in the 10-$M_\oplus$ range by weak nonlinear effects (*D'Angelo et al.,* 2003a).

Gap formation in a standard MMSN disk is found to occur for planets with mass exceeding that of ~Saturn, at which mass the migration rate is a maximum leading to an infall time of 3×10^4 yr. Beyond that mass the rate decreases with the onset of type II migration on the viscous timescale. In more massive disks the maximum migration rate at a Saturn mass is potentially enhanced by the positive feedback from coorbital torques leading to a fast type III migration regime (*Masset and Papaloizou,* 2003; *Artymowicz,* 2004). However, the operation of coorbital torques needs to be clarified with further high-resolution studies. But it is nonetheless interesting to speculate that the maximum migration rate at a Saturn mass is connected to the "hot Jupiters," which tend to have subjovian masses (see section 4 above). Once gap formation has occurred, type II migration ensues, which appears to allow models that assume formation beyond the snow line to produce giant planet distributions in accord with observations (e.g., *Trilling et al.,* 1998, 2002; *Alibert et al.,* 2004).

REFERENCES

Alibert Y., Mordasini C., and Benz W. (2004) *Astron. Astrophys., 417,* L25–L28.
Armitage P. J., Livio M., Lubow S. H., and Pringle J. E. (2002) *Mon. Not. R. Astron. Soc., 334,* 248–256.
Artymowicz P. (1993) *Astrophys. J., 419,* 155–165.
Artymowicz P. (2004) *Publ. Astron. Soc. Pac., 324,* 39–49.
Balbus S. A. and Hawley J. F. (1991) *Astrophys. J., 376,* 214–233.
Balbus S. A. and Hawley J. F. (1998) *Rev. Mod. Phys., 70,* 1–53.
Balmforth N. J. and Korycansky D. G. (2001) *Mon. Not. R. Astron. Soc., 316,* 833–851.
Bate M. R., Lubow S. H., Ogilvie G. I., and Miller K. A. (2003) *Mon. Not. R. Astron. Soc., 341,* 213–229.
Bryden G., Chen X., Lin D. N. C., Nelson R. P., and Papaloizou J. C. B. (1999) *Astrophys. J., 514,* 344–367.
Crida A., Morbidelli A., and Masset F. S. (2006) *Icarus, 181,* 587–604.
D'Angelo G., Henning T., and Kley W. (2002) *Astron. Astrophys., 385,* 647–670.
D'Angelo G., Henning T., and Kley W. (2003a) *Astrophys. J., 599,* 548–576.
D'Angelo G., Kley W., and Henning T. (2003b) *Astrophys. J., 586,* 540–561.
D'Angelo G., Bate M., and Lubow S. (2005) *Mon. Not. R. Astron. Soc., 358,* 316–332.
Fromang S., Terquem C., and Balbus S. A. (2002) *Mon. Not. R. Astron. Soc., 329,* 18–28.
Gammie C. F. (1996) *Astrophys. J., 457,* 355–362.

Goldreich P. and Sari R. (2003) *Astrophys. J., 585*, 1024–1037.

Goldreich P. and Tremaine S. (1979) *Astrophys. J., 233*, 857–871.

Goodman J. and Rafikov R. R. (2001) *Astrophys. J., 552*, 793–802.

Ida S. and Lin D. N. C. (2004) *Astrophys. J., 616*, 567–572.

Ivanov P. B., Papaloizou J. C. B., and Polnarev A. G. (1999) *Mon. Not. R. Astron. Soc., 307*, 79–90.

Jang-Condell H. and Sasselov D. D. (2005) *Astrophys. J., 619*, 1123–1131.

Kley W. (1999) *Mon. Not. R. Astron. Soc., 303*, 696–710.

Koller J., Li H., and Lin D. N. C. (2003) *Astrophys. J., 596*, L91–L94.

Korycansky D. G. and Pollack J. B. (1993) *Icarus, 102*, 150–165.

Laughlin G., Steinacker A., and Adams F. C. (2004) *Astrophys. J., 608*, 489–496.

Lin D. N. C. and Papaloizou J. C. B. (1979) *Mon. Not. R. Astron. Soc., 186*, 799–830.

Lin D. N. C. and Papaloizou J. C. B. (1986) *Astrophys. J., 309*, 846–857.

Lin D. N. C. and Papaloizou J. C. B. (1993) In *Protostars and Planets III* (E. H. Levy and J. I. Lunine eds.), pp. 749–835. Univ. of Arizona, Tucson.

Lin D. N. C., Papaloizou J. C. B., Terquem C., Bryden G., and Ida S. (2000) In *Protostars and Planets IV* (V. Mannings et al., eds.), pp. 1111–1134. Univ. of Arizona, Tucson.

Lubow S. H., Seibert M., and Artymowicz P. (1999) *Astrophys. J., 526*, 1001–1012.

Marcy G. W. and Butler R. P. (1995) *Bull. Am. Astron. Soc., 27*, 1379–1384.

Marcy G. W. and Butler R. P. (1998) *Ann. Rev. Astron. Astrophys., 36*, 57–97.

Mayor M. and Queloz D. (1995) *Nature, 378*, 355–359.

Masset F. (2001) *Astrophys. J., 558*, 453–462.

Masset F. (2002) *Astron. Astrophys., 387*, 605–623.

Masset F. and Papaloizou J. C. B. (2003) *Astrophys. J., 588*, 494–508.

Mayor M. and Queloz D. (1995) *Nature, 378*, 355–359.

Menou K. and Goodman J. (2004) *Astrophys. J., 606*, 520–531.

Nelson A. F. and Benz W. (2003) *Astrophys. J., 589*, 578–604.

Nelson R. P. (2005) *Astron. Astrophys., 443*, 1067–1085.

Nelson R. P. and Papaloizou J. C. B. (2003) *Mon. Not. R. Astron. Soc., 339*, 993–1005.

Nelson R. P. and Papaloizou J. C. B. (2004) *Mon. Not. R. Astron. Soc., 350*, 849–864.

Nelson R. P., Papaloizou J. C. B., Masset F., and Kley W. (2000) *Mon. Not. R. Astron. Soc., 318*, 18–36.

Ogilvie G. I. and Lubow S. H. (2003) *Astrophys. J., 587*, 398–406.

Papaloizou J. C. B. (2002) *Astron. Astrophys., 388*, 615 631.

Papaloizou J. C. B. (2005) *Cel. Mech. Dyn. Astron., 91*, 33–57.

Papaloizou J. C. B. and Larwood J. D. (2000) *Mon. Not. R. Astron. Soc., 315*, 823–833.

Papaloizou J. C. B. and Nelson R. P. (2003) *Mon. Not. R. Astron. Soc., 350*, 983–992.

Papaloizou J. C. B. and Nelson R. P. (2005) *Astron. Astrophys., 433*, 247–265.

Papaloizou J. C. B. and Terquem C. (2006) *Rep. Prog. Phys., 69*, 119–180.

Papaloizou J. C. B., Nelson R. P., and Snellgrove M. D. (2004) *Mon. Not. R. Astron. Soc., 350*, 829–848.

Pierens A. and Huré J. M. (2005) *Astron. Astrophys., 433*, L37–L40.

Rice W. K. M. and Armitage P. J. (2003) *Astrophys. J., 598*, L55–L58.

Syer D. and Clarke C. J. (1995) *Mon. Not. R. Astron. Soc., 277*, 758–766.

Tanaka H., Takeuchi T., and Ward W. R. (2002) *Astrophys. J., 565*, 1257–1274.

Terquem C. E. J. M. L. J. (2003) *Mon. Not. R. Astron. Soc., 341*, 1157–1173.

Trilling D. E., Benz W., Guillot T., Lunine J. I., Hubbard W. B., and Burrows A. (1998) *Astrophys. J., 500*, 428–439.

Trilling D. E., Lunine J. I., and Benz W. (2002) *Astron. Astrophys., 394*, 241–251.

Ward W. R. (1997) *Icarus, 126*, 261–281.

Winters W. F., Balbus S. A., and Hawley J. F. (2003) *Astrophys. J., 589*, 543–555.

Planet Migration in Planetesimal Disks

Harold F. Levison
Southwest Research Institute, Boulder

Alessandro Morbidelli
Observatoire de la Côte d'Azur

Rodney Gomes
Observatório Nacional, Rio de Janeiro

Dana Backman
USRA and SETI Institute, Moffett Field

Planets embedded in a planetesimal disk will migrate as a result of angular momentum and energy conservation as the planets scatter the planetesimals that they encounter. A surprising variety of interesting and complex dynamics can arise from this apparently simple process. In this chapter, we review the basic characteristics of planetesimal-driven migration. We discuss how the structure of a planetary system controls migration. We describe how this type of migration can cause planetary systems to become dynamically unstable and how a massive planetesimal disk can save planets from being ejected from the planetary system during this instability. We examine how the solar system's small-body reservoirs, particularly the Kuiper belt and Jupiter's Trojan asteroids, constrain what happened here. We also review a new model for the early dynamical evolution of the outer solar system that quantitatively reproduces much of what we see. And finally, we briefly discuss how planetesimal-driven migration could have affected some of the extrasolar systems that have recently been discovered.

1. INTRODUCTION

Our understanding of the origin and evolution of planets has drastically transformed in the last decade. Perhaps the most fundamental change was the realization that planets, in general, may not have formed in the orbits in which we see them. Indeed, planets may have migrated huge distances after they were born, as many of the extrasolar planetary systems show (*Moorhead and Adams,* 2005; *Papaloizou and Terquem,* 2006).

There are three main dynamical mechanisms that can cause such a wholesale evolution in planetary orbits. First, at early times when the natal protoplanetary gas disk is still in existence, gravitational interactions between the disk and a planet could cause a significant amount of orbital evolution (see chapter by Papaloizou et al. for a discussion). Second, after the gas disk is gone, if there still is a significant number of planetesimals in the system, the planets can migrate as a result of gravitational encounters with these objects. In particular, if a planet is immersed in a sea of small bodies, it will recoil every time it gravitationally scatters one of these objects. In general, since the small objects can come in from any direction, this will force the planets to undergo a small random walk in semimajor axis. However, since the sinks for these objects, for example, ejection from the system or encountering a neighboring planet, tend to lie on one side of the planet in question or another, there will be a net flux of material inward or outward. The planet must move in the opposite direction in order to conserve energy and angular momentum. In the absence of strong gravitational perturbations from other planets, the semimajor axis of planet in question will smoothly change with time. In the remainder of this chapter, we refer to this type of migration as *simple* migration. Third, planetary systems can suffer a dynamical instability (*Levison et al.,* 1998) that can lead to a short but violent period of orbital evolution during which the planetary eccentricities are increased to large values (*Rasio and Ford,* 1996). If the instability can be damped by some process (like dynamical friction with a disk), the planets can once again evolve onto nearly circular orbits, but in very different locations from where they formed. It has been suggested that this kind of instability occurred in the outer parts of our planetary system (*Thommes et al.,* 1999; *Levison et al.,* 2001; *Tsiganis et al.,* 2005).

The primary foci of this chapter are the second and the third mechanisms described above; namely migration in planetesimal disks. Although we claimed at the opening of this chapter that the role that planetesimal-driven migration played in the evolution of planetary systems was only acknowledged within the last decade, the idea itself is more than 20 years old. The first discussion of this process was presented by *Fernández and Ip* (1984). Their paper describes the response of Jupiter, Saturn, Uranus, and Neptune to a remnant disk of planetesimals. Although many of the details have changed as our ability to perform orbital integrations has improved, these authors found the basic result that is still held true today.

The importance of the work by Fernández and Ip was not really appreciated until the discovery of the Kuiper belt nearly a decade later, with its numerous objects on eccentric orbits in mean-motion resonances with Neptune. *Malhotra* (1993, 1995) first showed that these orbits could be the natural result of Neptune's outward migration and concluded that Neptune must have migrated about 7 AU in order to explain the eccentricities that we see (cf. section 3).

Much work as been done on this topic since Fernández and Ip's and Malhotra's groundbreaking papers. This literature is the topic of the remainder of the chapter, although it should be noted that this chapter is intended more as discussion of the current state-of-the-art than a review paper. In section 2 and section 3 we describe some of the basic physics that govern planet migration and resonant capture, respectively. In section 4, we examine simple planetesimal-driven migration in the solar system. In section 5 we look at how instabilities in the orbits of the planets, coupled with gravitational interactions with a massive planetesimal disk, could lead to significant changes in the orbits of the planets. We review a new model for the early dynamical evolution of the giant planets of the solar system in section 6, and planetesimal-driven migration in extrasolar systems in section 7. In section 8 we discuss some caveats and limitations of the N-body simulations on which most of the content of this chapter is based. We present our concluding remarks in section 9.

2. BASIC PRINCIPLES OF SIMPLE MIGRATION

The migration history for each individual planet is complicated because it is dependent on the details of how angular momentum flows through the system, namely on the distribution and the evolution of the mass and the angular momentum of the planet-crossing objects. These quantities, in turn, are determined by the sources and sinks for these particles. As a result, attempts to develop analytic theories for migration have studied very simple systems consisting of a single planet in a dynamically cold disk (*Murray et al.,* 1998; *Ida et al.,* 2000, hereafter *IBLT00*). So, the goal of this section is not to develop a comprehensive analytical model, but to develop a toy model to help cultivate a quali-

tative understanding of some of the important physical processes involved. Much of what we present is based on the work in *Gomes et al.* (2004, hereafter *GML04*). We start with a very simple model.

2.1. A Simple Model

IBLT00 shows that the rate of change of a planet's semimajor axis, a, is

$$\frac{da}{dt} = -\frac{2}{M_p}\sqrt{a}\,\dot{H}_\times \qquad (1)$$

where M_p is the mass of the planet, and $\dot{H}_\times$ is the rate of transfer of angular momentum from the planetesimals to the planet. This equation assumes that the eccentricity of the planet is small and $G \equiv 1$. Using the particle-in-a-box approximation, it is possible to show that $\dot{H}_\times = \varepsilon\bar{k}M(t)a^{-1}$, where ε is a quantity that contains fundamental constants and information about the geometry of the planet-encountering region, $M(t)$ is the total mass in planet-encountering orbits, and $\bar{k}$ is the average change of angular momentum per encounter, per unit mass planetesimal. Thus

$$\frac{da}{dt} = -2\varepsilon\bar{k}\frac{M(t)}{M_p}\frac{1}{\sqrt{a}} \qquad (2)$$

The evolution of M(t) can be approximated by the equation

$$\dot{M}(t) = -M(t)/\tau + 2\pi a|\dot{a}|\Sigma(a) \qquad (3)$$

where the first term in the r.h.s. represents the decay of the planetesimal population due to the planetesimal's finite dynamical lifetime, and the second term stands for the planetesimals that, because of the change in the planet's position, enter the region where they can be scattered by the planet for the first time. In equation (3) $\Sigma(a)$ is the surface density of the "virgin" (i.e., not yet scattered) planetesimal disk at heliocentric distance a. Substituting equation (2) into equation (3) we get

$$\dot{M}(t) = (-\tau^{-1} + 4\pi\varepsilon|\bar{k}|\sqrt{a}\,\Sigma(a)/M_p)M(t) \qquad (4)$$

Let's assume for simplicity that the term $\alpha \equiv -\tau^{-1} + 4\pi\varepsilon|\bar{k}|\sqrt{a}\,\Sigma(a)/M_p$ does not significantly change with time — an approximation that is clearly not true, but it allows us to get the essence of the planet's behavior. Under this assumption, equation (4) becomes an exponential equation, which is solvable and thus will allow us to get some insight into how the planet migrates.

If α is negative, then M(t) decays exponentially to 0 and the planet (from equation (1)) stops migrating. In this case, the loss of planetesimals due to their finite dynamical lifetime is not compensated by the acquisition of new planetesimals into the scattering region. Therefore, the planet runs "out of fuel." *GML04* called this migration mode *damped migration*. Conversely, if α is positive, M(t) grows exponentially and the planet's migration accelerates. In this case the acquisition of new planetesimals due to the migration exceeds the losses, and the migration is sustained. Thus, we will call this migration mode *sustained migration*.

2.2. The Direction of Migration

One of the limitations of the above description is that it does not contain any direct information about the direction of migration. This information is hidden in $\bar{k}$. Recall that the migration process is driven by gravitational encounters between planets and disk particles. To zeroth order, during an encounter the two objects are in a Kepler orbit about one another. Since the energy of this orbit must be conserved, all the encounter can do is rotate the relative velocity vector between the pair. Thus, the consequences of such an encounter can be effectively computed in most of the cases using an impulse approximation (*Öpik*, 1976; also see *IBLT00*). Using this approach, it is easy to compute that on average (i.e., averaged on all impact parameters and relative orientations) the planetesimals that cause a planet to move outward are those whose z-component of the angular momentum $H = \sqrt{a(1 - e^2)} \cos i$ is larger than that of the planet (H_p). The opposite is true for the planetesimals with $H < H_p$ (*Valsecchi and Manara*, 1997). In these formulae, e and i are the semimajor axis, eccentricity, and inclination of the planetesimal. Thus, $\bar{k}$ is a function of the angular momentum distribution of objects on planet-encountering orbits, and it is positive if more material has $H > H_p$, zero if the average H is the same as H_p, and negative otherwise.

The main physical effect that was not included in the derivation of equation (4) was the influence that the particles both entering and leaving the planet-encountering region can have on $\bar{k}$. For a single planet in a disk, there are two main sinks for the particles. Particles can hit the planet (this is the only sink that *IBLT00* considered). Since the chance of hitting the planet is roughly independent of the sign of $H–H_p$, in general this sink will not effect $\bar{k}$ and thus it is migration neutral. In addition, a planet can eject particles. These particles remove energy and thus the planet must move inward in response. The ability of a planet to eject particles depends on the dynamical excitation of the disk. We characterize the latter by the parameter $v' \equiv v_{enc}/v_c$, where v_{enc} is the typical encounter velocity of disk particles with the planet, and v_c is the planets's circular velocity. No matter how massive a planet is, it cannot eject a particle if the particle's $v' < v^\star$, where $v^\star = \sqrt{2} - 1 \approx 0.4$. However, v' is only conserved if the planet is on a circular orbit. In cases where that is not true, ejection can occur even if the disk is initially cold.

Multiple planet systems allow for another sink — particles can be transferred from one planet to another. The

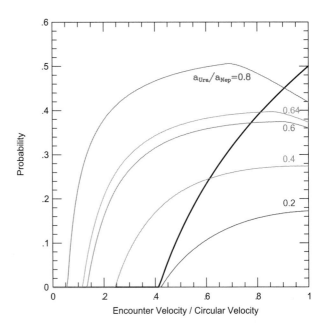

Fig. 1. The probability of a particular dynamical outcome resulting from an encounter between a small body and Neptune, as a function of the encounter velocity. The thick black curve shows the probability of ejection as determined by equation (5). The gray curves show the probability that the encounter will lead to the object being scattered onto a bound "Uranus-crossing" orbit for different semimajor axes of Uranus. These curves were determined from simple analytic arguments or simple numerical experiments (see text).

best example is provided by the solar system's four giant planets. Numerical experiments show that in the interaction with a disk of planetesimals Jupiter moves inward, but the other three giants move outward (*Fernández and Ip*, 1984; *Hahn and Malhotra*, 1999; *Gomes et al.*, 2004). We can illustrate why Neptune moves outward with the following hand-waving argument. An object's velocity in an inertial frame is $\vec{v} = \vec{v}_c + \vec{v}_{enc}$. Assuming that after the encounter $\vec{v}_{enc}$ is pointing in a random direction, the probability of ejection is

$$P_{eject} = \left(\frac{v'^2 + 2v' - 1}{4v'} \right) \qquad (5)$$

(the thick black curve in Fig. 1).

The probability that an object is transferred to Uranus (i.e., acquires an orbit with perihelion distance $q < a_{Ura}$) depends on v' and on the ratio of Uranus' semimajor axis to that of Neptune (a_{Ura}/a_{Nep}). It can be evaluated using a Monte Carlo method, if the assumption used to derive equation (5) is again made. The gray curves in Fig. 1 show the results for five different values of a_{Ura}/a_{Nep}. Currently $a_{Ura}/a_{Nep} = 0.64$. Figure 1 shows that if $a_{Ura}/a_{Nep} \geq 0.3$ and $v' \leq$

0.5, Neptune is more likely to transfer objects to Uranus then to eject them. This explains why this planet moves outward.

So far we have considered only the planetesimal sinks in the planet-migration process. In addition to these, there are two ways in which new particles can be added to the planet-crossing population. The first of these is actually a source/sink pair caused by the migration process itself. As the planet moves, so does the planet-encountering region and thus some particles leave this region while new particles enter it. If the planet is moving outward, the particles that leave have $H < H_p$, while those that enter have $H > H_p$. The opposite is true if the planet is moving inward. Thus, this process tends to support any migration that was started by another mechanism.

Resonances with the planet are another source for particles. Objects originally in the chaotic regions of these resonances can have their eccentricities pumped until their orbits start to encounter the planet (cf. *Duncan et al., 1995*). The effect that this source has on $\bar{k}$ depends on the planetesimal surface density profile and on the strength of the resonances.

To summarize, the actual migration behavior of planets depends on a competition between the various sources and sinks. If the material in the planet-encountering region is removed faster than it is replenished, the planet's migration rate will decay to zero. We call this behavior *damped* migration. If the planet-encountering region is replenished faster than it is depleted, the migration is said to be *sustained*. Sustained migration is divided into two types. If the migration is nourished by particles being fed into the planet-encountering region by the migration itself and the sinks do not require the presence of other planets, we refer to it as *runaway* migration. If other planets are required, we name the migration mode *forced* migration.

3. RESONANT CAPTURE DURING PLANET MIGRATION

One consequence of a planet's orbital migration is that the mean-motion resonances (MMRs) with the planet also move. During this process, disk planetesimals that are "swept" by a MMR can be captured in it. Resonance capture is a complicated process and an active subject of research in the field of nonlinear dynamics (see, e.g., *Wiggins,* 1991; *Neishtadt,* 1975, 1987, 1997; *Henrard,* 1982; *Malhotra,* 1993, 1995). The evolution of a particle interacting with a moving resonance depends sensitively on initial conditions, the nature of the resonance, the rate of evolution due to dissipative effects, etc. A model that has been studied in detail is that of a single resonance in the so-called *adiabatic approximation*. In the framework considered in this chapter, this model would correspond to a single planet on a circular orbit, migrating slowly and monotonically. The adiabatic condition is met if the time required for a resonance to move by a heliocentric distance range comparable to the resonance width is much longer than the libration

timescale inside the resonance (which itself is much longer than the orbital timescale). In this case, the probability of resonance capture has been calculated semi-analytically (*Henrard and Lemaitre,* 1983; *Borderies and Goldreich,* 1984).

In essence, resonance capture can occur only in exterior MMRs (which for the resonance called the "$j : j + k$ MMR" implies $k > 0$) if the planet is moving outward, and in interior MMRs ($k < 0$) if the planet is moving inward (*Henrard and Lemaitre,* 1983; *Neishtadt,* 1987; *Tsiganis et al.,* 2005). However, if the particle is swept in the correct direction, capture into the resonance is not guaranteed. For instance, in the adiabatic approximation, capture into the 2:3 MMR with Neptune (where many Kuiper belt objects are seen, including Pluto) is certain only if Neptune is migrating outward and the initial eccentricity of the particle is less than ~0.03. The capture probability decreases monotonically (but not linearly!) for higher initial eccentricities: It is less than 10% for e > 0.15.

If the object is captured into the resonance, it then moves with the resonance as the planets continues in its migration. During this evolution, the eccentricity of the object increases monotonically at a rate determined by the migration rate of the planet, which gives the relationship (*Malhotra,* 1995)

$$e_{final}^2 = e_{initial}^2 + \frac{j}{j + k} \ln \frac{a_{p,final}}{a_{p,initial}} \qquad (6)$$

where $a_{p,initial}$ is the semimajor axis of the planet when the body enters into resonance, $e_{initial}$ is the eccentricity of the body at that time, $a_{p,final}$ is the semimajor axis of the planet at the time of consideration, and e_{final} is the eccentricity of the object at the same instant.

Resonances, however, are not stable at all eccentricities. If the eccentricity is large enough, the resonance cannot protect the objects from close encounters with the planet. Thus, in this picture (see *Malhotra,* 1995), as the planet migrates, planetesimals are captured into MMRs, move together with the resonances while growing their orbital eccentricities until they reach the instability limit, and start to be scattered by the planet. The resonant population remains in roughly a steady state as long as the resonance remains in the disk, because new objects enter into the resonance while large eccentricity objects leave it. If the resonance passes beyond the edge of the disk, it is no longer refilled with new objects. The resonant population decays as the resonance moves away from the edge, while the minimum eccentricity of the resonant population grows, so that the low-eccentricity portion of the resonance becomes empty.

The formula in equation (6) is useful to deduce some properties of the migration. For instance, if a resonance is populated with objects up to an eccentricity equal to e_{max} (and the latter is smaller than the threshold value for instability), it means that the planet migrated a distance $\delta a_p = \exp\{[(j + k)/j]e_{max}^2\}$. *Malhotra* (1995), observing that the eccentricity of Kuiper belt objects in the 2:3 MMR is smaller than 0.32, deduced that Neptune migrated at least 7 AU (i.e., it formed at a ≤ 23 AU).

Unfortunately, reality is not as simple as the adiabatic model predicts. If the resonance is surrounded by a chaotic layer, as it is the case if the planet's eccentricity is not zero or the inclination of the particle is large, the computation of the capture probability with semi-analytic techniques is essentially impossible, because it depends also on the diffusion speed inside the chaotic layer (*Henrard and Morbidelli*, 1993). Numerical simulations of migration of Neptune in more realistic models of the planetary system show that the capture probability is much less sensitive on the particles' orbital eccentricity than the adiabatic theory predicts, and resonance capture is quite likely also at large eccentricity (*Chiang et al.*, 2003; *Gomes*, 2003; *Hahn and Malhotra*, 2005).

Another difference between reality and the adiabatic model concerns the eccentricity growth of resonant objects during the migration. While in the adiabatic model the eccentricity grows monotonically, in reality there can be secular terms forcing large-amplitude oscillations of the eccentricity in resonant objects. For example, *Levison and Morbidelli* (2003) showed that if a sufficiently large amount of planetesimal mass has been accumulated in a MMR, the planet feels perturbations from this material that cause new frequencies to appear in the planet's precession spectrum. These frequencies are near those of the resonant objects. Thus, the particles can resonate with frequencies in the planetary motion that they themselves induced, which produces large oscillations in their eccentricity. In this situation, *Levison and Morbidelli* (2003) showed that the resonant population can extend down to e ~ 0 even when the resonance has moved 10 AU past the edge of the disk.

Another complication to this story occurs if there are relatively large objects in the planetesimal disk that the planets are migrating through. Resonant capture requires that the migration of the planet is smooth. If the planet has jumps in semimajor axis due to the encounters with other planets or massive planetesimals, the locations of its MMRs jump as well. If the amplitude of these jumps is of order of the resonance width or larger, the particles trapped in the resonances will be released. A model of stochastic migration in planetesimal disks has been recently developed by *Murray-Clay and Chiang* (2005).

4. SIMPLE MIGRATION IN THE SOLAR SYSTEM

As described above, the magnitude and direction of planetesimal-driven migration is determined by a complex interaction of various dynamical sources and sinks of disk particles. Thus, this migration process is best studied through numerical experiments. In this section, we review what we know about the migration of the four giant planets of the solar system. Many researchers have studied this issue (*Hahn and Malhotra*, 1999; *Gomes*, 2003; *GML04*); however, we base our discussion on that in *GML04*.

4.1. Migration in Extended Disks

The simulations that follow all start with the following initial conditions. Jupiter, Saturn, Uranus, and Neptune are 5.45, 8.7, 15.5, and 17.8 AU, respectively. The planets are surrounded by a massive disk that extends from 18 AU to 50 AU and has a surface density variation as r⁻¹, which is the typically assumed value for protoplanetary disks (cf. *Hayashi*, 1981). The disk's outer edge was chosen to correspond to the current edge of the classical Kuiper belt (*Allen et al.*, 2001a,b; *Trujillo and Brown*, 2001). The initial disk mass was varied between 40 $M_\oplus$ and 200 $M_\oplus$. The systems were evolved using the techniques in *GML04* (see section 8).

Figure 2a shows a snapshot of the semimajor axis — eccentricity distribution of the planets and the planetesimals for the 50 $M_\oplus$ simulation. Figure 2b shows the semimajor axes of the four giant planets with time. As discussed in section 2, Neptune, Uranus, and Saturn move outward, while Jupiter moves inward.

The black curves in Fig. 3 show the temporal evolution of Neptune's semimajor axis for runs of different disk masses. The 40-$M_\oplus$ and 45-$M_\oplus$ runs are example of damped migration (section 2). After a fast start, Neptune's outward motion slows down, and well before 10⁹ yr, the planet reaches a quasi-asymptotic semimajor axis that is well within the outer edge of the disk. The part of the disk outside the orbit of Neptune preserves its original surface density, while the part interior to this distance is completely depleted.

A major change in Neptune's behavior occurs when the disk mass is increased ≥50 $M_\oplus$. As before, Neptune experiences a fast initial migration, after which it slows down. Then it undergoes an approximately linear migration be-

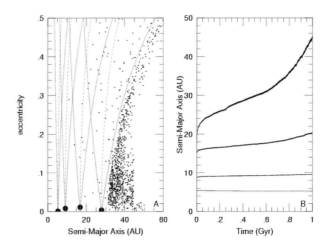

Fig. 2. Migration simulation from *GML04* of the four giant planets in a 50-$M_\oplus$ disk that extends to 50 AU. **(a)** A snapshot of the system at 330 m.y. Semimajor axis and eccentricity of the planets (filled dots) and of the planetesimals (points). The solid lines define the limits for planetary crossing orbits, while the dotted lines show where H = H_p for zero-inclination orbits. **(b)** Evolution of the semimajor axes of the four giant planets.

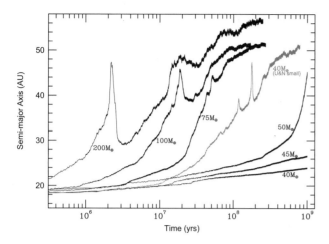

Fig. 3. The temporal evolution of Neptune's semimajor axis for seven different simulations based on *GML04*. In all cases the disk was truncated at 50 AU, but the disk mass and the mass of the ice giants varied from run to run. In particular, the black curves show runs where the ice giants are their normal mass, while the one gray shows a run with Uranus and Neptune one-third of their current masses.

tween 100 and 600 m.y. Finally, Neptune's migration accelerates toward the disk's edge, where it eventually stops. This evolution suggests that the surface density of this disk is near a critical value that separates damped migration from sustained migration (see section 2). In all cases with more massive disks, Neptune final location was near the edge of the disk.

The transition from the linear to the accelerating phase in the 50-$M_\oplus$ run is due to the variations in the number of particles trapped in Neptune's MMRs. Recall from section 3 as Neptune migrates, disk particles become trapped in its MMRs. The resonant particles effect migration because they effectively increase Neptune's inertial mass. During adiabatic migration, the number of particles in the resonances is roughly constant as long as the resonance is still in the disk. In this run, Neptune accelerates as its 1:2 MMR moves out of the disk probably because the number of objects in the resonance drops, new particles not being captured.

There is another important transition in Neptune's behavior when the disk mass is increased to values larger than ~100 $M_\oplus$: The migration passes from a "forced" to a "runaway" mode. This change leads to a very interesting new phenomenon: Neptune's migration is no longer monotonic. Figure 3 shows that in the case of the highest-mass disk, Neptune reaches ~50 AU in less than 3 m.y., and then comes back to within 30 AU almost equally as fast. Similar episodes of acceleration and return are also visible in other high-mass runs. This type of bounce is possible because, in runaway migration, objects are left behind in an excited disk as Neptune moves outward, instead of being transferred to the inner planets or ejected. When Neptune reaches the edge of the disk, the number of objects with H > H_{Nep} drops, and so the remaining objects interior to Neptune can pull the

planet inward. Thus, Neptune reverses direction and starts a runaway *inward* migration. The same argument described above applies, so that this migration ends only when the region of the disk partially depleted by Uranus is encountered again. Notice however, that Neptune always ends up near the original edge of the disk at 50 AU.

Finally, the gray curve in the Fig. 3 shows a run in a 40-$M_\oplus$ disk, but where the masses of the ice giants are one-third of their current values. Note that while migration is damped in a 40-$M_\oplus$ disk when the ice giants are at their current mass, it is runaway in this case. This result shows that the transition between these two forms of migration occurs at a smaller disk mass for smaller planetary masses. This result gives an important constraint on the formation time of Uranus and Neptune. *Levison and Stewart* (2001) showed that standard planet-formation scenarios cannot form Uranus and Neptune in their current orbits, so these planets probably formed much closer to the Sun. However, if the ice giants would have migrated to their current orbits when they were much smaller then they are now, as the gray curve in the Fig. 3 suggests, how did they reach their current masses? This conundrum has a solution if the planets were fully formed when there was still enough gas in the system that the gravitational interactions with the gas prevented the planets from moving outward. This implies that these planets formed very early (≤10 m.y.) (*Podosek and Cassen*, 1994; *Hollenbach et al.*, 2000; *Thi et al.*, 2001), and is consistent with the capture of a primordial atmosphere of several Earth masses of H and He (*Pollack et al.*, 1996).

4.2. Constraints from the Kuiper Belt

The question naturally arises whether we can determine what kind of migration actually occurred in the solar system, and from this the mass and structure of the original protoplanetary disk. In addition to the current orbits of the giant planets, the structure of the Kuiper belt supplies crucial clues to the solar system's ancient history because this structure still carries the signatures of early evolution of the planetary system.

Three characteristics of the Kuiper belt are important: (1) The Kuiper belt only contains about 0.1 $M_\oplus$ of material (*Jewitt et al.*, 1996; *Chiang and Brown*, 1999; *Trujillo et al.*, 2001; *Gladman et al.*, 2001; *Bernstein et al.*, 2004). This is surprising given that accretion models predict that ≥10 $M_\oplus$ must have existed in this region in order for the objects that we see to grow (*Stern*, 1996; *Stern and Colwell*, 1997a; *Kenyon and Luu*, 1998, 1999). (2) The Kuiper belt is dynamically excited. Again, this is unexpected since accretion models predict that relative velocities between objects must have originally been small in order for the objects that we see to grow. (3) The Kuiper belt apparently ends near 50 AU (*Trujillo and Brown*, 2001; *Allen et al.*, 2001a,b).

If we assume that the primordial Kuiper belt must have contained at least ~10 $M_\oplus$ between 40–50 AU in order to grow the observed objects, the above simulations suggest a scenario first proposed by *Hahn and Malhotra* (1999). In this model the disk contained ~45 M⊕ of material between 20 and 50 AU. Neptune started at ~22 AU and migrated to ~30 AU, where it stopped because its migration was damped. This left enough mass in the Kuiper belt to account for the growth of the known objects there. This scenario has a problem, however. If Neptune stopped at 30 AU because its migration was damped, how did the Kuiper belt lose >99% of its mass? Two general ideas have been proposed for the mass depletion of the Kuiper belt: (1) the dynamical excitation of the vast majority of Kuiper belt objects to the Neptune-crossing orbits after which they were removed, and (2) the collisional comminution of most of the mass of the Kuiper belt into dust.

GML04 studied Scenario (1), including the dynamical effects of the escaping Kuiper belt objects on Neptune. They concluded that any reasonable dynamical depletion mechanism would have forced Neptune to migrate into the Kuiper belt. Thus, Scenario (1) can be ruled out. Scenario (2) is also faced with some significant problems: (a) The orbital excitation of the cold classical Kuiper belt does not seem to be large enough to remove as much mass as is required (*Stern and Colwell*, 1997b); (b) substantial collisional grinding does not occur unless the physical strength of small Kuiper belt objects are extremely small (*Kenyon and Bromley*, 2004), much smaller than predicted by SPH collision calculations; and (c) most of the wide binaries in the cold population would not have survived the collisional grinding phase (*Petit and Mousis*, 2004). These problems led *GML04* to conclude that the collisional grinding scenario is also probably not viable.

4.3. Migration in Disks Truncated at 30 AU

GML04 argued that the current location of Neptune and the mass deficiency of the Kuiper belt imply that the protoplanetary disk possessed an edge at about 30 AU (see also *Levison and Morbidelli*, 2003). In their study of migration in such a disk, they found that a planet does not necessarily stop at the exact location of the edge. Indeed, since angular momentum must be conserved during the migration process, the final location of the planets depends more on the total angular momentum in the disk than on the location of the edge. To illustrate this, Fig. 4 shows Neptune's migration in six disks that are initially spread between 10 and 30 AU, but with masses varying from 20 to 100 M⊕.

The disk with 20 M⊕ has a subcritical surface density. Neptune exhibits a damped migration and stalls well within the disk. Therefore a massive annulus is preserved between a few AU beyond the planet's location and the original outer edge of the disk. The disks with 30 M⊕ and 35 M⊕ have a surface density close to the critical value. In both cases, when the planet reaches ~26 AU, the unstable region of the disk [which extends up to a distance of about one-sixth of the planet's semimajor axis (*Duncan et al.*, 1995)] reaches the edge of the disk. The planet starts to feel the disk truncation and its migration is rapidly damped. The final location is 2 AU inside the original disk edge, but the entire

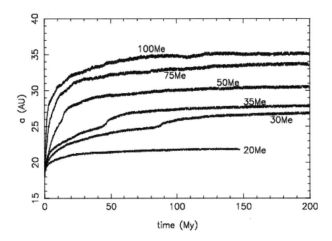

Fig. 4. Examples of Neptune's migration in disks with an outer edge at 30 AU and masses equal to between 20 and 100 M⊕. Reproduced from *GML04* (their Fig. 10). Note that a direct comparison cannot be made between these total masses and those in the runs shown in Figs. 2 and 3 because the disk was larger in the earlier runs.

region beyond the planet has been depleted. More massive disks have supercritical densities. In the case of 50 M⊕ the planet stops almost exactly at the disk's edge, while in the other cases it goes several AU beyond it.

Thus, *GML04* concluded that a disk with an outer edge close to 30 AU, the exact value depending on the disk's mass, can explain Neptune's current semimajor axis. There are at least five mechanisms that could have truncated the disk at such a small heliocentric distance, prior to planetary accretion: (1) A passing star tidally strips the Kuiper belt after the observed Kuiper belt objects formed (*Ida et al.*, 2000; *Kobayashi and Ida*, 2001; *Levison et al.*, 2004). (2) An edge formed prior to planetesimal formation due to aerodynamic drag (*Youdin and Shu*, 2002). (3) An edge formed during planet accretion due to size-dependent radial migration caused by gas drag (*Weidenschilling*, 2003). (4) Nearby early-type stars photoevaporated the outer regions of the solar nebula before planetesimals could form (*Hollenbach and Adams*, 2004). (5) Magnetohydrodynamic instabilities in the outer regions of the disk prevented the formation of planetesimals in these regions (*Stone et al.*, 1998). We stress that a small truncation radius is *not* in contradiction with the existence of the Kuiper belt beyond 40 AU. In fact, the entire Kuiper belt could have been pushed out from within the disk's edge during Neptune's migration. We return to the issue of the Kuiper belt in section 6.

5. DYNAMICAL INSTABILITIES AS A MIGRATION PROCESS

Up to this point we have been discussing "simple" migration. However, there is another way in which the interaction between planets and small bodies can result in a large change in the planetary radial distribution. First, a global instability in the planetary system increases the planets' eccentricities and semimajor axis separations; then, the interaction between the disk particles and the planets circularize the planetary orbits. Eventually, a final phase of "simple" migration can follow.

The above idea was first suggested by *Thommes et al.* (1999). They postulated that the four giant planets formed in such a compact configuration that their orbits were dynamically unstable. Plate 5 shows four snapshots from one of *Thommes et al.*'s (1999) simulations, where the ice giants were hypothesized to have formed between Jupiter and Saturn. Almost immediately (~10^4 yr), the ice giants are scattered out from between Jupiter and Saturn into a preexisting planetesimal disk, where gravitational interactions with the disk particles eventually circularize their orbits. The gravitational interaction between the planetesimals and the scattered cores comes in two flavors. First, there is a secular response by the disk to the eccentricities of the ice giants. This can clearly be seen in the lower left panel of the figure (t = 180,000 yr), where the objects between 20 and 30 AU have their eccentricities systematically pumped. Since this region of the disk is more massive than the ice giants, this secular response lifts the perihelion distances of the ice giants away from Saturn's orbit, thus saving them from ejection into interstellar space. Second, *dynamical friction*, which occurs as a large object is moving through a sea of background particles (*Chandrasekhar*, 1943; although see *Binney and Tremaine*, 1987, for a discussion), further circularizes the ice giants' orbits. The problem with the evolution illustrated in Plate 5 is that disks massive enough to circularize the ice giants typically force them to migrate too far. We address this issue in the next section.

6. THE NICE MODEL OF THE EARLY DYNAMICAL EVOLUTION OF THE GIANT PLANETS

A new model of the dynamical evolution of the outer solar system has been presented in a recent series of papers (*Tsiganis et al.*, 2005, hereafter *TGML05*; *Morbidelli et al.*, 2005a, hereafter *MLTG05*; *Gomes et al.*, 2005, hereafter *GLMT05*). This is the most comprehensive model to date and it reproduces most of the characteristics of the outer planetary system at an unprecedented level. We refer to this model as the "Nice model" because it was developed at the Nice Observatory. Since it makes use of many of the ideas explored in this chapter, we describe it in detail below.

6.1. The Dynamical Evolution of Giant Planet Orbits

The initial motivation for the Nice model was the desire to understand the orbital eccentricities and inclinations of Jupiter and Saturn, which can reach values of ~10% and ~2°, respectively. Planetary formation theories suggest that these planets should have formed on circular and coplanar orbits. In addition, the final stage of planetesimal-driven

migration, which is the topic of this chapter, quickly damps any preexisting eccentricities and inclinations. Thus, the initial work that led to the Nice model was a set of simulations intended to solve the mystery of the origins of Jupiter and Saturn's eccentricities and inclinations.

In particular, the hypothesis studied in *TGML05* was that orbital excitation could take place if, during migration, two planets crossed a low-order MMR. Saturn is currently located interior to the 2:5 MMR and exterior to the 1:2 MMR with Jupiter. If the initial planetary configuration was initially sufficiently compact, then — given that these two planets had to migrate on opposite directions — they would have crossed their 1:2 MMR, which is the strongest of the MMRs.

TGML05 performed a series of numerical integrations of the above idea. In all these simulations the four giant planets were started on nearly circular and coplanar orbits, with Saturn placed a few tenths of an AU interior to the 1:2 MMR. Starting Saturn interior to the resonance is a basic assumption of the Nice model. However, recent hydrodynamical simulations of two planets embedded in a gaseous disk (*Morbidelli et al.*, 2005b) not only suggest that this assumption is reasonable, but may even be required if one wants to avoid migrating Jupiter into the Sun via planet-gas disk interactions. However, whether reasonable or not, the Nice model can only valid if this assumption is true. *TGML05* initially placed the ice giants just outside Saturn's orbit. The planets were surrounded by massive disks containing between 30 and 50 M$_\oplus$ of planetesimals, truncated at ~30 AU.

The typical evolution from *TGML05* is shown in Fig. 5. After a phase of slow migration, Jupiter and Saturn encounter the 1:2 MMR, at which point their eccentricities jumped to values comparable to the ones currently observed, as pre-

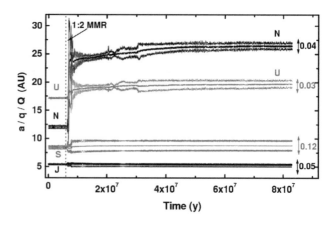

Fig. 5. Orbital evolution of the giant planets from one of *TGML05*'s N-body simulations. The values of a, q, and Q are plotted for each planet. The separation between the upper and lower curves is a measure of the eccentricity of the orbit. The maximum eccentricity of each orbit, computed over the last 2 m.y. of evolution, is noted on the plot. The vertical dotted line marks the epoch of 1:2 MMR crossing. The planetesimal disk contained initially 35 M$_\oplus$. Reproduced from *TGML05* (their Fig. 1).

dicted by adiabatic theory. The sudden jump in the eccentricities of Jupiter and Saturn has a drastic effect on the planetary system as a whole. The perturbations that Jupiter and Saturn exert on Uranus and Neptune force the ice giants' orbits to become unstable. Both ice giants are scattered outward and penetrate the disk (although less violently than in the simulation presented in section 5). Then the eccentricities and inclinations of the ice giants are damped by the disk as described in section 5, and the planetary system is stabilized. The planets stop migrating when the disk is almost completely depleted. As shown in Fig. 5, not only their final semimajor axes, but also their final eccentricities are close to the observed values. In this run, the two ice giants exchange orbits. This occurred in ~50% of *TGML05*'s simulations.

TGML05 found that in roughly 70% of their simulations both ice giants were saved from ejection by the planetesimal disk and evolved onto nearly-circular orbits. They divided these so-called "successful" runs into two groups: (1) those in which there were no encounters between an ice giant and a gas giant, and (2) those in which Saturn suffered encounters with one or both ice giants. Figure 6 shows the mean and standard deviation of the orbital elements for group (1), in gray, and group (2), in black. The orbital elements for the real planets are also shown as filled black dots. When no encounters with Saturn occur, the final eccentricities and inclinations of the planets, as well as the semimajor axis of Uranus, tend to be systematically smaller than the observed values. On the other hand, when distant encounters between Saturn and one of the ice giants also occurs, the final planetary system looks very similar to the actual outer solar system. This is the first time that a numerical model has quantitatively reproduced the orbits of the giant planets of the solar system.

Although there are many free parameters in the initial conditions of the Nice model (for example, the initial orbits of the planets, the mass of the disk, and the inner and outer edges of the disk), there are only two that effect the final location of the planets. The first is the assumption that the disk was truncated near 30 AU. As described above, *TGML05* made this assumption in order to circumvent the Kuiper belt mass-depletion problem. The only other parameter important in determining the orbits of the planets is the initial mass of the disk. All other parameters mainly affect the timing of the resonance crossing, a fact used in section 6.3, but not the final orbits of the planets themselves.

TGML05 also found a systematic relationship between final orbital separation of Jupiter and Saturn at the end of migration and the initial mass of the disk. For disk masses larger than ~35–40 $M_\oplus$, the final orbital separation of Jupiter and Saturn tends to be larger than is actually observed. Indeed, for disks of 50 $M_\oplus$, Saturn was found to cross the 2:5 MMR with Jupiter. In addition, the final eccentricities of the two gas giants were too small, because they had experienced too much dynamical friction. Thus, an initial disk mass of ~35 $M_\oplus$ was favored. The fact both the semimajor axes and the eccentricities/inclinations of Jupiter and

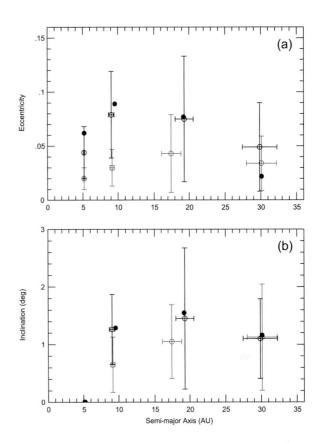

Fig. 6. Comparison of *TGML05*'s synthetic final planetary systems with the outer solar system. **(a)** Proper eccentricity vs. semimajor axis. **(b)** Proper inclination vs. semimajor axis. Proper eccentricities and inclinations are defined as the maximum values acquired over a 2-m.y. timespan and were computed from numerical integrations. The inclinations are measured relative to Jupiter's orbital plane. These values for the real planets are presented with filled black dots. The open gray dots mark the mean of the proper values for the runs of group 1 (no encounters for Saturn), while the open black dots mark the same quantities for the runs of group 2. The error bars represent 1σ of the measurements. Reproduced from *TGML05* (their Fig. 2).

Saturn are reproduced in the same integrations strongly supports this model: There is no *a priori* reason that a, e, and i should all be matched in the same runs.

6.2. Small-Body Reservoirs

Further support for the Nice model comes from the small-body reservoirs. Indeed, jovian Trojans, which are small objects in the 1:1 MMR with Jupiter, supply an important test for *TGML05*'s scenario. *Gomes* (1998) and *Michtchenko et al.* (2001) studied the effects of planetesimal-driven migration on the Trojan asteroids. They found that the Trojans were violently unstable if Jupiter and Saturn crossed the 1:2 MMR with one another. Thus, these authors concluded the Jupiter and Saturn could *not* have crossed this resonance. So, is *TGML05* wrong?

The issue was addressed in the second of the three Nice model papers, *MLTG05*. The authors of this paper pointed out that the dynamical evolution of any gravitating system is time reversible. So, if the planetary system evolves into a configuration so that trapped objects can leave the Trojan points, it must be possible that other bodies can enter the same region and be temporally trapped. Consequently, a transient Trojan population can be created if there is a source for these objects. In this case, the source is the very bodies that are forcing the planets to migrate. When Jupiter and Saturn get far enough from the 2:1 MMR so that the coorbital region becomes stable again, the population that happens to be there at that time remains trapped, becoming the population of permanent jovian Trojans.

MLTG05 performed a series of N-body simulations to study this idea and found that bodies can indeed be permanently captured in Trojan orbits. Assuming that the original planetesimal disk contained 35 $M_\oplus$ (as in *TGML05*), they predict that there should be between ~4 × 10^{-6} $M_\oplus$ and ~3 × 10^{-5} $M_\oplus$ of Trojans with libration amplitude D < 30°. This can be favorably compared to the total mass of the Trojans, which, using data from the Sloan Digital Sky Survey and updated numbers on density and albedo, was reevaluated by *MLTG05* to be 1.1 × 10^{-5} (also see discussions by *Jewitt et al.*, 2000; *Yoshida and Nakamura*, 2005). One of the surprising aspects of the Trojan population is its broad inclination distribution, which extends up to ~40° and cannot be explained by traditional capture scenarios. *MLTG05* finds that they can reproduce this as well. Since this model is the only one available that can explain these features, the Trojans represent observational evidence for the 1:2 resonance crossing proposed by *TGML05*.

Jupiter is not the only planet in the outer solar system that has Trojan asteroids. Neptune currently is known to have four such objects. These objects are also explained by the Nice model. *TGML05* found that objects can be trapped in Neptune's 1:1 MMR during the time when Neptune's eccentricity is being damped by the planetesimal disk. Indeed, the Nice model supplies the only known Neptune Trojan formation mechanism that can explain the high inclination (25°) of 2005 TN$_{53}$. [For example, the most widely quoted model, *Chiang and Lithwick* (2005), does not accomplish this.]

The Kuiper belt also presents an important test of the Nice model. Any model of the outer solar system evolution must explain the main orbital properties of the Kuiper belt objects (see *Morbidelli et al.*, 2003, for a review): (1) the presence of objects trapped in Neptune's MMRs (some shown as gray vertical lines in Fig. 7a); (2) the abrupt end of the *classical* Kuiper belt at or near the location of the 1:2 MMR [we define as the classical Kuiper belt the collection of objects that are nonresonant and fall below the stability limit determined by *Duncan et al.* (1995), shown by a gray curve in Fig. 7a]; (3) the dearth of objects with 45 ≤ a ≤ 48 AU and e < 0.1; (4) the apparent coexistence in the classical Kuiper belt of two populations: a dynamically *cold* population — made of objects on orbits with inclinations i < 4 — and a *hot* population — whose inclinations can be as large as 30°, and possibly larger (*Brown*, 2001). These populations have different size distribution (*Levison and Stern*, 2001; *Bernstein et al.*, 2004) and different colors (*Tegler and Romanishin*, 2003; *Trujillo and Brown*, 2002).

The investigation of the formation of the Kuiper belt in the framework of the Nice model is in progress, and we briefly discuss here the preliminary, unpublished results.

For reasons discussed above, the Nice model assumes that the protoplanetary disk was truncated near ~30 AU. This implies that the Kuiper belt that we see today had to be pushed outward from the initial disk during Neptune's orbital evolution. The envisioned mechanism is the following. When Neptune becomes eccentric (e_N ~ 0.3) many of its MMRs are very wide. Numerical simulations show that for e_N > 0.2, the entire region inside the 1:2 MMR is covered with overlapping resonances and thus is entirely chaotic. Consequently, it fills with disk particles scattered by Neptune. As Neptune's eccentricity damps, many of these particles become permanently trapped in the Kuiper belt. Resonances do not overlap beyond the 1:2 MMR and thus this resonance forms a natural outer boundary for the trapped population.

Figure 7b shows the result of this process according to a set of new numerical simulations we performed to study the above idea. First, note that most of the nonresonant particles above the stability curve would be lost if the simulation were carried until 4 G.y. Given this, there is remarkably good agreement between the two populations. There is an edge to the main Kuiper belt at the location of the 1:2 MMR. In addition, the resonance populations are clearly seen. The latter are not acquired by the resonances via the standard migration capture scenario of *Malhotra* (1993, 1995), but by the mechanism proposed by *Gomes* (2003).

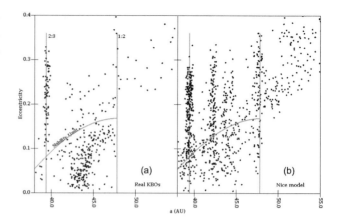

Fig. 7. The eccentricity-semimajor axis distribution of objects in the Kuiper belt. The dots show the objects. The two vertical gray lines show Neptune's 2:3 MMR and 1:2 MMR. The gray curve shows a stability limit determined by *Duncan et al.* (1995). **(a)** The real multi-opposition Kuiper belt objects, as released by the MPC in November 2005. **(b)** Resulting distribution 1 G.y. after the 1:2 MMR planetary instability, from our new dynamical model based on the Nice scenario.

The model in Fig. 7b also reproduces the dearth of low-eccentricity objects beyond ~45 AU. The inclination distribution of the main Kuiper belt is also a reasonable match to the data. This model predicts that roughly 0.1% of the original disk material should currently be in the Kuiper belt. Assuming an initial disk mass of 35 $M_\oplus$ (as in *TGML05*) leads to a Kuiper belt mass of 0.04 $M_\oplus$, which is consistent with observations. And finally, there is a crude correlation between the formation location in the planetesimal disk and the final Kuiper belt inclination for a particle. This relationship might be able to explain the observed relationship between physical characteristics and Kuiper belt inclination. This work is not yet complete. However, it seems to explain many of the characteristics of the Kuiper belt with unprecedented quality, thus also supporting the framework of the Nice model.

6.3. The Lunar Late Heavy Bombardment

The "late heavy bombardment" (LHB) was a phase in the impact history of the Moon that occurred roughly 3.9 G.y. ago, during which the lunar basins with known dates formed. There is an ongoing debate about whether the LHB was either the tail-end of planetary accretion or a spike in the impact rate lasting ≤100 m.y. ["terminal cataclysm" (*Tera et al.*, 1974)]. Although the debate continues, we believe that there is growing evidence that the LHB was indeed a spike. The LHB was recently reviewed in *Hartmann et al.* (2000) and *Levison et al.* (2001).

A spike in the impact flux in the inner solar system most likely required a major upheaval in the orbits of the planets that destabilized one or more small-body reservoirs, which, until that time, had been stable (cf. *Levison et al.*, 2001). The Nice model naturally provides such an upheaval when Jupiter and Saturn cross the 1:2 MMR. The problem investigated in *GLMT05* (the third paper in the Nice trilogy) was how to delay the resonance crossing for ~700 m.y. In the simulations in *TGML05*, planet migration started immediately because planetesimals were placed close enough to the planets to be violently unstable. While this type of initial condition was reasonable for the goals of that work, it is unlikely. Planetesimal-driven migration is probably not important for planetary dynamics as long as the gaseous massive solar nebula exists. The initial conditions for the migration simulations should represent the system that existed at the time the nebula dissipated, i.e., one in which the dynamical lifetime of the disk particles is longer than the lifetime of the nebula. *GLMT05* found that for planetary systems like those used as initial conditions in the Nice model, this inner edge of the planetesimal disk is roughly 1 AU beyond the orbit of the outermost ice giant.

In this configuration, *GLMT05* found that the 1:2 MMR crossing was delayed by between 350 m.y. and 1.1 G.y., depending on various parameters. They concluded that the global instability caused by the 1:2 MMR crossing of Jupiter and Saturn could be responsible for the LHB, since the estimated date of the LHB falls in the range of the crossing times that they found. *GLMT05* also found that, for an initial disk mass of 35 $M_\oplus$ (as in *TGML05*), after the resonance crossing roughly 8×10^{21} g of cometary material impacted the Moon. This value is consistent with the estimate of ~6×10^{21} g from the observed basins (*Levison et al.*, 2001). Moreover, the asteroid belt would have been perturbed during this event, thereby supplying additional impactors. Overall, this material arrived over a relatively short period of time (~50 m.y.). Thus, this model produces a spike in the impact rate of about the correct duration.

One of the requirements of *GLMT05*'s model is that the mass of external planetesimal disk not significantly evolve between the time when the solar nebula disperses and the time of the LHB. In principle, the disk mass can change as a result of either the action of dynamics or the action of collisions, although the simulations in *GLMT05* already show that dynamics are not a factor. A preliminary study of the collisional evolution of such a disk by *O'Brien et al.* (2005) suggests that there is a reasonable range of parameters over which the primordial transneptunian disk is able to remain massive (~35 $M_\oplus$) for 700 m.y. In addition, the final size distribution of transneptunian bodies in these calculations is consistent with that inferred for the Kuiper belt by *Bernstein et al.* (2004). However, it should be noted that *O'Brien et al.* (2005) studied only a small fraction of the available parameter space and thus more such studies are necessary in order to settle this issue.

At this juncture, let us take stock of the Nice model. It quantitatively reproduces the semimajor axes, eccentricities, and inclinations of the giant planets. In addition, it reproduces the Trojan asteroids of both Jupiter and Neptune. Indeed, it reproduces, for the first time, the orbits of Jupiter's Trojans and quantitatively predicts the amount of material that should be in this population. It also is the most successful model to date at reproducing the total mass and orbital distribution of the Kuiper belt. It accomplishes all this with very few important free parameters. As described in section 6.1, if one accepts the need to truncate the initial planetesimal disk at 30 AU, the initial disk mass is the only parameter that significantly affects these results. It is a strength of the Nice model that a single value of disk mass (35 $M_\oplus$) can produce all the characteristics we just listed. And finally, as *GLMT05* has shown, this model quantitatively reproduces the so-called late heavy bombardment of the Moon. These accomplishments are unparalleled by any other model. Not only is there no other model that can explain all these characteristics, but the level of agreement between the model and observations is unprecedented.

7. MIGRATION IN EXTRASOLAR PLANETESIMAL DISKS

In this section we discuss some behaviors that planets might have followed in extrasolar planetesimal disks. It is very difficult, if not impossible, to be exhaustive because, as we saw in section 2, planet migration in planetesimal disks depends crucially on the specific features of the system: the number of planets, their separations, their masses and mass ratios, the disk's mass and radial extent, its radial surface density profile, etc. Thus, we concentrate only on four main aspects: the origin of warm Jupiters, the evolution of two-planet systems, the runaway of medium-mass planets out to very large distances from the parent star, and the triggering of late instabilities.

7.1. Planetesimal-driven Migration and the Origin of Hot and Warm Jupiters

We have seen in section 2 that a single giant planet (of about 1 M_{Jup} or more) embedded in a planetesimal disk migrates inward, because it ejects most of the planetesimals that it interacts with. Assuming a minimal mass planetesimal disk [the solid component of the *Hayashi* (1981) minimum mass nebula], *Murray et al.* (1998) found that Jupiter would have had a very damped migration, so that it would not have moved significantly. However, if the density of the disk is enhanced by a large factor (15–200), *Murray et al.* showed that the migration can be in a runaway mode. This can push the planet inward to distances from the central star that are comparable to those observed in extrasolar systems.

Although extremely massive disks are required to produce the so-called "hot Jupiters" (a ≲ 0.1 AU), more moderate (but still massive) disks can produce the "warm Jupiters" (a ~ 1 AU). In particular, the advantage of planetesimal-driven migration over gas-driven migration for the origin of warm Jupiters is that the stopping mechanism is much more natural. If the radial surface density profile of the planetesimal disk is shallower than $1/r^2$, the runaway migration turns eventually into a damped migration, as the planet moves inward. Thus, the planet suddenly stops.

7.2. Migration in Two-Planet Systems

Two-planet systems are interesting because they have many of the characteristics of systems with a larger number of planets. Thus, insight into the behavior of systems, in general, can be gained by studying systems with two planets. Unfortunately, as explained above, it is not possible to completely explore this issue because of the large number of parameters involved. In this subsection, we investigate with new, unpublished simulations the case of two giant planets in the current orbits of Jupiter and Saturn with a disk stretching from 6 to 20 AU containing a total mass of 1.2× that of the sum of the masses of the planets. Thus, in all cases the system is in the *forced* migration mode. When Jupiter and Saturn have their current masses, Jupiter migrates inward and Saturn outward (the black curves in Plate 6a), just like in the four-planet case.

In the first series of runs, we looked at the effect that the total mass of the planets has on migration. It might be expected that in a system with two massive planets, the outer one ejects more particles than it passes to the inner one and thus both planets migrate inward. However, our simulations show that the above expectation is not correct. By scaling the mass of the disk and of the planets by factors of 3 and 10, we obtain the evolutions shown by the blue

and red curves in Plate 6a, respectively. These behaviors are similar to one another, the migration timescale being the only significant difference. We believe that the above argument is wrong because, although encounters are more common for larger planets due to their larger gravitational cross-sections, the relative distribution of velocity changes is roughly independent of mass. The larger gravitational cross-sections for the more massive planets lead to the faster migration times, not to a different evolutionary pattern.

In the second series of runs, we kept the total mass of the planets constant (3× the summed masses of Jupiter and Saturn), but varied their mass ratio. The results of these simulations are shown in Plate 6b for systems where the ratio of the mass of the inner planet to that of the outer ranged from 0.5 to 3.3. We find that for mass ratios ≥2, the outer planet always migrates outward. However, for mass ratios less than roughly 2, the inner planet becomes less effective at removing particles crossing the orbit of the outer planet, and thus the outer planet is more likely to eject them. Thus, after a short period of outward migration, the outward planet migrates inward. Also note that, for at least the conditions studied here, the outer planet inward migration is faster than that of the inner planet. These results suggest that planetesimal migration can lead to resonant trapping between giant planets, as is observed in many extrasolar planetary systems (*Schneider*, 2004). It could also drive a planetary system into an unstable configuration.

7.3. Driving Planets Far from Their Star

Most of the observed debris disks show features like gaps, warps, asymmetric clumps, and even spiral waves, usually attributed to the presence of embedded planets. For instance, *Wyatt* (2003) showed that the features of the Vega disk could be due to a Neptune-mass planet migrating from 40 to 65 AU in 56 m.y. Similarly, *Wyatt et al.* (2005) modeled the observations of the η Corvi disk with a Neptune-mass planet moving from 80 to 105 AU in 25 m.y. Other features in the β Pictoris (*Wahhaj et al.*, 2003) and ε Eridani disks (*Ozernoy et al.*, 2000; *Greaves et al.*, 2005) have been modeled with planets at several tens of AU. In the most extreme model, the spiral features of the HD141569 disk have been modeled with one planet of 0.2–2 M_{Jup} at ~250 AU and a Saturn-mass planet at 150 AU (*Wyatt*, 2005). These models call for an exploration of the possibility that planets migrate very far from the central star.

We have seen in section 4 that if our planetary system had been embedded in a massive disk truncated at 50 AU, Neptune would have had migrated very quickly in the runaway mode to the edge of the disk. Figure 8 shows the evolution of Neptune if the same disk were extended to 200 AU, with a radial surface density profile ∝ 1/r. Neptune reaches a heliocentric distance larger than 110 AU. Then, without having reached the edge of the disk, it bounces back, reversing its migration direction. This abrupt change in the migration behavior happens because, when the planet migrates sufficiently fast, the timescale for planetesimals encountering the planet becomes comparable to, or longer

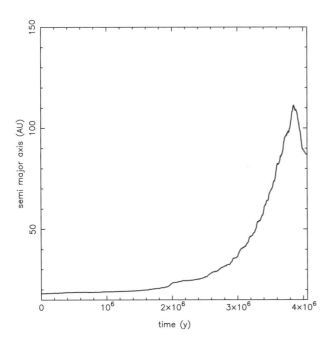

Fig. 8. The migration of Neptune in a very massive planetesimal disk, extended from 20 to 200 AU and with 6 M$_\oplus$ of material in each AU-wide annulus. From *GML04.*

than, that for passing through the planet-crossing region due to the migration of the planet itself. So, in coordinates that move with the planet, most particles simply drift inward while keeping their eccentricities roughly constant. The net result is that $\bar{k}$ in equation (2) gradually decreases with time.

The planet does not respond to the reduction of $\bar{k}$ by gradually slowing down its migration because, contemporaneously, the amount of mass in the planet-crossing region (M in equation (2)) increases at a faster rate than $\bar{k}$ decreases. Consequently, the magnitude of $\dot{a}_p$ does not decrease with time. However, when $\bar{k} = 0$, $\dot{a}_p$ becomes zero and the migration abruptly stops. When this happens, the planet finds itself in an unstable situation. If the excited disk interior to planet slightly overpowers the particles from the outer disk, the planet initiates a runaway migration inward.

It is worth stressing, however, that the kind of migration illustrated in Fig. 8 most likely only works for medium-mass planets, like Neptune. Moving a planet as large as Jupiter with a planetesimal disk probably requires a disk that is too massive to be believable.

7.4. Late Instabilities

The striking success of the Nice model for the early evolution of the solar system (see section 6) suggests that gravitational instabilities can be very important in the history of a planetary system. Another piece of evidence comes from outside the solar system. Many of the known extrasolar planets are on very eccentric orbits. It has been argued the most natural explanation for this astonishing result is that these systems also suffered from a violent, global rearrangement of their planets' orbits (*Rasio and Ford,* 1996;

Weidenschilling and Marzari, 1996; *Levison et al.,* 1998; *Papaloizou and Terquem,* 2001; *Moorhead and Adams,* 2005).

Why do planetary systems become unstable? There is nothing in the physics of the planet-formation process that guarantees that a planetary system will be stable on a timescale longer than that characterizing planet formation itself. Thus, a planetary system could remain quiescent for hundreds of millions of years and then become completely unstable (*Levison et al.,* 1998).

As the Nice model and the runs in section 7.2. illustrate, it is also possible that the gravitational interaction between the planets and a massive small-body reservoir could drive a planetary system into an unstable configuration. In fact, it could be quite generic that, at the end of the gas-disk phase, planetesimals are only on orbits with a dynamical lifetime longer than the nebula dissipation time, hence driving a *slow* migration that leads, at some point, to instability. Thus, late heavy bombardments may not be the rule, but we can expect them to occur in a fairly good fraction of the multiplanet systems.

Indeed, the recent Spitzer observation of the SEDs of nearby main-sequence A stars (*Rieke et al.,* 2005) and solar-type stars (*Kim et al.,* 2005) revealed some main-sequence systems with ages between 100 m.y. and 3 G.y. that have unexpectedly bright infrared excesses indicating large amounts of circumstellar dust. The A-star sample included systems with ages in the few × 100 m.y. range that have 24-μm flux densities 1.5–2× brighter than predicted for the stellar photospheres alone. Those excesses correspond to bolometric fractional luminosities L_{dust}/L_{star} of a few × 10^{-4} for estimated dust temperatures of 75 to 175 K, i.e., the dust intercepts that fraction of the central star's total output and reradiates the energy into the thermal IR. For A stars those dust temperatures correspond roughly to radii of 10 to 60 AU. The minimum mass of dust required to produce those excesses is on the order of only 10^{20} kg, equivalent to a single object a few hundred kilometers in diameter.

Similarly, the solar-type star sample showed that more than 15% of systems with ages into the gigayear range have IR excesses prominent at 70 μm, corresponding to temperatures of roughly 40 to 75 K and "Kuiper belt-like" radii of 20 to 50 AU. Note that the minimum dust masses required to produce prominent excesses at lower temperatures and longer wavelengths around the solar-type stars are typically 10^{-3}–10^{-2} M$_\oplus$, 2 orders of magnitude larger than for the A stars. A total planetesimal mass of 3–10 M$_\oplus$ is required to produce that much dust in collisional equilibrium for a belt with dimensions like our Kuiper belt. That mass range is deduced by either scaling limits on Kuiper belt mass from IR flux limits for a regime in which P-R drag dominates dust dynamics (*Backman et al.,* 1995), or by scaling dust mass to parent-body mass for a collision-dominated regime.

An inferred belt mass of 3–10 M$_\oplus$ is significant because evolution solely by collisional grinding of a Kuiper belt-sized system cannot produce both the amount of dust that we see and as massive a remnant after 1 G.y.: To produce this much dust, the planetesimals have a collisional evolution timescale an order of magnitude shorter than the system age. Thus, the exceptional systems observed in these samples are certainly not the late stages of ordinary evolution of originally supermassive planetesimal belts. Instead they must represent either recent single large collisions of lunar-mass bodies, or possibly more common late-epoch instabilities like our solar system's LHB (see also chapter by Meyer et al.).

8. CAVEATS AND LIMITATIONS OF NUMERICAL SIMULATIONS

Much of the information presented in this chapter is based on examples taken from numerical simulations. Thus, a natural question arises about the reliability of such calculations. Modern simulations of planet migration are performed using symplectic N-body integrators (*Duncan et al.,* 1998; *Chambers,* 1999), which have very good conservation properties for energy and angular momentum, even during close encounters. However, even with modern computers some simplifications are required in order to make these calculations feasible. In particular, the disk is usually modeled with only 1000–10,000 equal mass particles, which interact gravitationally with the planets but do not interact with each other. This method of handling the disk has three main limitations.

First, the typical mass of an individual disk particle in the simulation is on the order of 0.005–0.1 $M_\oplus$, depending on the total mass of the disk and the number of particles used. Even if we cannot exclude the presence of bodies with the mass of the Moon (0.01 $M_\oplus$) or even of Mars (0.1 $M_\oplus$) in the disk, they probably carry only a small fraction of the total disk mass. Thus, the individual particle mass used in these simulations is definitely unrealistic. As we have seen in section 3, large disk particles lead to a stochastic component of planet migration, and thus if the disk particles are unreasonably large, there could be a spurious component to the migration. This stochasticity affects the capture of particles in resonances and their subsequent release. In turn, this has an impact not only on the disk's structure, but also on the mean planet migration rate. In addition, the stochastic oscillation of the planet's semimajor axis tends to sustain the migration even in cases where, in reality, the migration should be damped. *GML04* showed examples where Neptune's migration is damped in 40-$M_\oplus$ or 45-$M_\oplus$ disks modeled with 10,000 particles (see Fig. 3), but is in a forced mode when the same disks are modeled using only 1000 particles. *GML04* argued, however, that a disk of a few thousand particles is most likely adequate for most simulations.

In any case, whenever an interesting dynamical phenomenon is observed in the disk, it is good practice to check if the phenomenon persists in a simulation where the disk's particles are massless, and the planets are forced to migrate by including a suitable drag term in the planet's equations of motion (e.g., *Malhotra,* 1993). If the phenomenon persists, then it is most likely real. However, if it disappears,

then it should be viewed with some skepticism and further tests should be done.

A second limitation with our techniques is the lack of gravitational interactions among the disk particles. We see two possible implications of this simplification. First, the precession rates of the orbits of the particles are incorrect, and thus the locations of secular resonances are wrong. It is very difficult to prevent this error, unless the mutual interactions among the particles are taken into account. But this option is often prohibitively time consuming. Thus, our advice is to be very suspicious of results that heavily rely on specific secular resonances. Another implication is that, if the disk is dynamically excited at a specific location (for instance, by a resonance), this excitation does not propagate as a wave through the disk, as it might if the particles' collective effects were correctly modeled (*Ward and Hahn,* 1998, 2003). In turn, this lack of propagation does not damp the excitation at the location where it is triggered. However, wave propagation is possible only if the excitation of the disk is very small [e < 0.01, i < 0.3° (*Hahn,* 2003)]. Thus, the relevance of this phenomenon remains to be proven in a realistic model of the solar system primordial disk, which is stirred by a number of processes, including the gravitational scattering from a number of Pluto-mass bodies embedded in the disk. In any case, neither of these processes are likely to have a major effect on the planet migration situations discussed in this chapter because the main dynamical driver of migration is relatively close encounters between planets and disk particles. They could have secondary effects, however. For example, if they are active, collective effects could modify the state of the disk particles being fed into the planet-crossing region.

A third limitation is the lack of collisional interaction between the disk particles. At some level, inelastic collisions necessarily damp the dynamical excitation of the disk. However, the importance of this process is a function of many parameters including the surface density of the disk, its dynamical state, and the sizes of individual disk particles. The potential importance of this process was recently pointed out by *Goldreich et al.* (2004), who considered a case in which the collisional damping is so efficient that any dynamical excitation of the disk due to the planets is almost instantaneously dissipated. Thus, the disk of planetesimals acts as an infinite sink of orbital excitation. Goldreich et al. argue that this mechanism can have a huge impact on the process of planet formation and presumably also of planet migration. The disk particle size distribution required for this to happen, however, is rather extreme — essentially all planetesimals need to be ~1 cm in diameter.

One of the fundamental issues with the Goldreich et al. scenario is whether such an extreme size distribution could actually arise in nature and, if it did, whether it could last long enough to be dynamically important before evolving, either by coagulation (e.g., formation of larger objects) or collisional grinding of itself to dust. However, the existence of both the asteroid and the Kuiper belts suggest that, at least by the time the planets formed, the planetesimals had a full size distribution ranging up to the sizes of Ceres and Pluto, which, at first glance, seems to be inconsistent with the basic Goldreich et al. scenario. So, although there are many intriguing aspects to this scenario, we believe that it needs to be studied in more detail before its viability can be determined. (The chapter by Chiang et al. presents a new scenario for the formation of the Kuiper belt based on Goldreich et al.'s ideas. However, our ongoing numerical simulations of this scenario have shown that it is not viable, because it leads to a solar system structure that is inconsistent with observations. In particular, we are finding that it leads to planetary systems with too many ice giants, at least one of which is on a nearly circular orbit beyond 30 AU.)

There are published works that will allow us to evaluate whether collisions are important in planetary migration situations when more moderate size distributions are considered. For instance, using a self-consistent planetesimal collision model that includes fragmentation and accretion of debris, *Leinhardt and Richardson* (2005) showed that the runaway growth of planetary embryos is virtually indistinguishable from that obtained in simulations that do not take collisions into account. Considering a realistic size distribution of the disk's planetesimals, *Charnoz and Morbidelli* (2003) showed that the collisional evolution of particles ejected from the Jupiter-Saturn region is moderate, and would affect only a minor portion of the mass. And finally, *Charnoz and Brahic* (2001) studied the process of scattering of planetesimals by a jovian-mass planet taking into account the effect of collisions on the dynamics. They found that Jupiter can still eject particles from its neighborhood and produce a scattered disk of bodies. These results seem to indicate that collisions are probably not important in the overall dynamical evolution of the disk, and consequently on the migration process.

In conclusion, despite the fact that collective effects and collisions are definitely important and must effect, at some level, the evolution of an evolving planetary system, we believe that their role is not significant in the type of planet migration studied in this chapter. Thus, the simulations presented in this chapter should capture the essence of real evolution. We stress that the fact that the simulations presented to support the Nice model reproduce the observed structure of the solar system so well argues that the neglected phenomena only play a secondary role.

9. CLOSING COMMENTS

In this chapter we reviewed the many ways in which massive planetesimal disks can drive changes in the orbits of planets. A planet's orbit can be modified as a result of gravitational encounters with objects from this disk. In particular, if a planet is immersed in a sea of small bodies, it will recoil every time it gravitationally scatters one of these objects. This process allows for the exchange of an-

gular momentum and orbital energy between the planets and the disk. If the disk mass is on the order of the mass of the planets, the planetary orbits can be profoundly modified.

Although there are many ideas in the literature concerning how planetesimal migration could have occurred in the solar system, there is general agreement that it did indeed happen. At the very minimum, the orbital element distribution of objects in the Kuiper belt says that Neptune migrated outward by at least ~7 AU (*Malhotra*, 1995). Our understanding of this process in other systems is very limited, however. This is primarily due to the very large number of unknown parameters. However, some of the strange configurations of the extrasolar planetary systems so far discovered could be the result of planetesimal-driven migration. Thus, this field is ripe for further study.

Acknowledgments. H.F.L. thanks NASA's Origins and Planetary Geology and Geophysics programs for supporting his involvement in the research related to this chapter. A.M. is also grate-ful to CNFA for supporting travel to the Protostars and Planets V conference.

REFERENCES

Allen R. L., Bernstein G. M., and Malhotra R. (2001a) *Astrophys. J., 549,* L241–L244.
Allen R. L., Bernstein G. M., and Malhotra R. (2001b) *Astron. J., 124,* 2949–2954.
Backman D. E., Dasgupta A., and Stencel R. E. (1995) *Astrophys. J., 405,* L35.
Bernstein G. M., Trilling D. E., Allen R. L., Brown M. E., Holman M., and Malhotra R. (2004) *Astron. J., 128,* 1364–1390.
Binney J. and Tremaine S. (1987) *Galactic Dynamics.* Princeton Univ., Princeton.
Borderies N. and Goldreich P. (1984) *Cel. Mech., 32,* 127–136.
Brown M. E. (2001) *Astron. J., 121,* 2804–2814.
Chambers J. E. (1999) *Mon. Not. R. Astron. Soc., 304,* 793–799.
Chandrasekhar S. (1943) *Astrophys. J., 97,* 255–262.
Charnoz S. and Brahic A. (2001) *Astron. Astrophys., 375,* L31–L34.
Charnoz S. and Morbidelli A. (2003) *Icarus, 166,* 141–156.
Chiang E. I. and Brown M. E. (1999) *Astron. J., 118,* 1411–1422.
Chiang E. I. and Lithwick Y. (2005) *Astrophys. J., 628,* 520–532.
Chiang E. I., Jordan A. B., Millis R. L., Buie M. W., Wasserman L. H., Elliot J. L. Kern S. D., Trilling D. E., Meech K. J., and Wagner R. M. (2003) *Astron. J., 126,* 430–443.
Duncan M. J., Levison H. F., and Budd S. M. (1995) *Astron. J., 110,* 3073–3081.
Duncan M. J., Levison H. F., and Lee M. H. (1998) *Astron. J., 116,* 2067–2077.
Fernandez J. A. and Ip W.-H. (1984) *Icarus, 58,* 109–120.
Gladman B., Kavelaars J. J., Petit J.-M., Morbidelli A., Holman M. J., and Loredo T. (2001) *Astron. J., 122,* 1051–1066.
Goldreich P., Lithwick Y., and Sari R. (2004) *Astrophys. J., 614,* 497–507.
Gomes R. S. (1998) *Astron. J., 116,* 2590–2597.
Gomes R. S. (2003) *Icarus, 161,* 404–418.
Gomes R. S., Morbidelli A., and Levison H. F. (GML04) (2004) *Icarus, 170,* 492–507.
Gomes R., Levison H. F., Tsiganis K., and Morbidelli A. (GLMT05) (2005) *Nature, 435,* 466–469.
Greaves J. S., Holland W. S., Wyatt M. C., Dent W. R. F., Robson E. I., et al. (2005) *Astrophys. J., 619,* L187–L190.

Hahn J. M. (2003) *Astrophys. J., 595,* 531–549.

Hahn J. M. and Malhotra R. (1999) *Astron. J., 117,* 3041–3053.

Hahn J. M. and Malhotra R. (2005) *Astron. J., 130,* 2392–2414.

Hartmann W. K., Ryder G., Dones L., and Grinspoon D. (2000) in *Origin of the Earth and Moon* (R. M. Canup and K. Righter, eds.), pp. 493–512. Univ. of Arizona, Tucson.

Hayashi C. (1981) In *Fundamental Problems in the Theory of Stellar Evolution* (D. Sugimoto et al., eds.), pp. 113–126. Reidel, Dordrecht.

Henrard J. (1982) *Cel. Mech., 27,* 3–22.

Henrard J. and Lemaitre A. (1983) *Icarus, 55,* 482–494.

Henrard J. and Morbidelli A. (1993) *Phys. D, 68,* 187–200.

Hollenbach D. and Adams F. C. (2004) In *Debris Disks and the Formation of Planets* (L. Caroff et al, eds.), p. 168. ASP Conf. Series 324, San Francisco.

Hollenbach D. J., Yorke H. W., and Johnstone D. (2000) In *Protostars and Planets IV* (V. Mannings et al., eds.), p. 401. Univ. of Arizona, Tucson.

Ida S., Bryden G., Lin D. N. C., and Tanaka H. (IBLT00) (2000) *Astrophys. J., 534,* 428–445.

Jewitt D., Luu J., and Chen J. (1996) *Astron. J., 112,* 1225–1232.

Jewitt D. C., Trujillo C. A., and Luu J. X. (2000) *Astron. J., 120,* 1140–1147.

Kenyon S. J. and Bromley B. C. (2004) *Astron. J., 128,* 1916–1926.

Kenyon S. J. and Luu J. X. (1998) *Astron. J., 115,* 2136–2160.

Kenyon S. J. and Luu J. X. (1999) *Astron. J., 118,* 1101–1119.

Kim J. S., Hines D. C., Backman D. E., Hillenbrand L. A., Meyer M. R., et al. (2005) *Astrophys. J., 632,* 659.

Kobayashi H. and Ida S. (2001) *Icarus, 153,* 416–429.

Leinhardt Z. M. and Richardson D. C. (2005) *Astrophys. J., 625,* 427–440.

Levison H. F. and Morbidelli A. (2003) *Nature, 426,* 419–421.

Levison H. F. and Stern S. A. (2001) *Astron. J., 121,* 1730–1735.

Levison H. F. and Stewart G. R. (2001) *Icarus, 153,* 224–228.

Levison H. F., Lissauer J. J., and Duncan M. J. (1998) *Astron. J., 116,* 1998–2014.

Levison H. F., Dones L., Chapman C. R., Stern S. A., Duncan M. J., and Zahnle K. (2001) *Icarus, 151,* 286–306.

Levison H. F., Morbidelli A., and Dones L. (2004) *Astron. J., 128,* 2553–2563.

Malhotra R. (1993) *Nature, 365,* 819–821.

Malhotra R. (1995) *Astron. J., 110,* 420–429.

Michtchenko T. A., Beaugé C., and Roig F. (2001) *Astron. J., 122,* 3485–3491.

Morbidelli A., Brown M. E., and Levison H. F. (2003) *Earth Moon Planets, 92,* 1–27.

Morbidelli A., Levison H. F., Tsiganis K., and Gomes R. (MLTG05) (2005a) *Nature, 435,* 462–465.

Morbidelli A., Crida A., and Masset F. (2005b) *AAS/DPS Meeting #37,* Abstract #25.07.

Moorhead A. V. and Adams F. C. (2005) *Icarus, 178,* 517–539.

Murray N., Hansen B., Holman M., and Tremaine S. (1998) *Science, 279,* 69.

Murray-Clay R. A. and Chiang E. I. (2005) *Astrophys. J., 619,* 623–638.

Neishtadt A. I. (1975) *J. Appl. Math. Mech., 39,* 594–605.

Neishtadt A. I. (1987) *Prikl. Mat. Mek., 51,* 750–757.

Neishtadt A. I. (1997) *Cel. Mech., 65,* 1–20.

O'Brien D. P., Morbidelli A., and Bottke W. F. (2005) *AAS/DPS Meeting #37,* Abstract #29.14.

Opik E. J. (1976) *Interplanetary Encounters: Close-Range Gravitational Interactions.* Elsevier, Amsterdam.

Ozernoy L. M., Gorkavyi N. N., Mather J. C., and Taidakova T. A. (2000) *Astrophys. J., 537,* L147–L151.

Papaloizou J. C. B. and Terquem C. (2001) *Mon. Not. R. Astron. Soc., 325,* 221–230.

Papaloizou J. C. B. and Terquem C. (2006) *Rep. Prog. Phys., 69,* 119–180.

Petit J.-M. and Mousis O. (2004) *Icarus, 168,* 409–419.

Podosek F. A. and Cassen P. (1994) *Meteoritics, 29,* 6–55.

Pollack J. B., Hubickyj O., Bodenheimer P., Lissauer J. J., Podolak M., and Greenzweig Y. (1996) *Icarus, 124,* 62–85.

Rasio F. A. and Ford A. B. (1996) *Science, 274,* 954–956.

Rieke G. H., Su K. Y. L., Stansberry J. A., Trilling D., Bryden G., Muzerolle J., White B., Gorlova N., Young E. T., Beichman C. A., Stapelfeldt K. R., and Hines D. C. (2005) *Astrophys. J., 620,* 1010–1026.

Schneider J. (2004) *Extra Solar Planets Encyclopaedia,* vo.obspm.fr/exoplanetes/encyclo/encycl.html.

Stern S. A. (1996) *Astron. J., 112,* 1203–1211.

Stern S. A. and Colwell J. E. (1997a) *Astron. J., 112,* 841–849.

Stern S. A. and Colwell J. E. (1997b) *Astrophys. J., 490,* 879–885.

Stone J. M., Ostriker E. C., and Gammie C. F. (1998) *Astrophys. J., 508,* L99–L102.

Tegler S. C. and Romanishin W. (2003) *Icarus, 161,* 181–191.

Tera F., Papanastassiou D. A., and Wasserburg G. J. (1974) *Earth Planet. Sci., 22,* L1–21.

Thi W. F., Blake G. A., van Dishoeck E. F., van Zadelhoff G. J., Horn J. M. M., Becklin E. E., Mannings V., Sargent A. I., Van den Ancker M. E., and Natta A. (2001) *Nature, 409,* 60–63.

Thommes E. W., Duncan M. J., and Levison H. F. (1999) *Nature, 402,* 635–638.

Trujillo C. A. and Brown M. E. (2001) *Astrophys. J., 554,* L95–L98.

Trujillo C. A. and Brown M. E. (2002) *Astrophys. J., 566,* 125–128.

Trujillo C. A., Jewitt D. C., and Luu J. X. (2001) *Astron. J., 122,* 457–473.

Tsiganis K., Gomes R., Morbidelli A., and Levison H. F. (TGML05) (2005) *Nature, 435,* 459–461.

Valsecchi A. and Manara G. B. (1997) *Astron. Astrophys., 323,* 986–998.

Wahhaj Z., Koerner D. W., Ressler M. E., Werner M. W., Backman D. E., and Sargent A. I. (2003) *Astrophys. J., 584,* L27–L31.

Ward W. R. and Hahn J. M. (1998) *Astron. J., 116,* 489–498.

Ward W. R. and Hahn J. M. (2003) *Astron. J., 125,* 3389–3397.

Weidenschilling S. J. (2003) *Icarus, 165,* 438–442.

Weidenschilling S. J. and Marzari F. (1996) *Nature, 384,* 619–621.

Wiggins S. (1991) *Chaotic Transport in Dynamical Systems.* Springer-Verlag, New York.

Wyatt M. C. (2003) *Astrophys. J., 598,* 1321–1340.

Wyatt M. C. (2005) *Astron. Astrophys., 440,* 937–948.

Wyatt M. C., Greaves J. S., Dent W. R. F., and Coulson I. M. (2005) *Astrophys. J., 620,* 492–500.

Yoshida F. and Nakamura T. (2005) *Astron. J., 130,* 2900–2911.

Youdin A. N. and Shu F. H. (2002) *Astrophys. J., 580,* 494–505.

A Decade of Radial-Velocity Discoveries in the Exoplanet Domain

Stéphane Udry
Geneva University

Debra Fischer
San Francisco State University

Didier Queloz
Geneva University

Since the detection of a planetary companion orbiting 51 Peg one decade ago, close to 200 extrasolar planets have been unveiled by radial-velocity measurements. They exhibit a wide variety of characteristics, including large masses with small orbital separations, high eccentricities, multiplanet architectures, and orbital period resonances. Here, we discuss the statistical distributions of orbital parameters and host star properties in the context of constraints they provide for planet-formation models. We expect that radial-velocity surveys will continue to provide important discoveries. Thanks to ongoing instrumental developments and improved observing strategies, Neptune-mass planets in short-period orbits have recently been detected. We foresee continued improvement in radial-velocity precision that will reveal Neptune-mass planets in longer-period orbits and planets down to a few Earth masses in short-period orbits. The next decade of Doppler observations should expand the mass distribution function of exoplanets to lower masses. Finally, the role of radial-velocity followup measurements of transit candidates is emphasized.

1. INTRODUCTION

Before 1995, the solar system was the only known example of a planetary system in orbit around a Sun-like star, and the question of its uniqueness was more a philosophical than a scientific matter. The discovery of an exoplanet orbiting the Sun-like star, 51 Peg (*Mayor and Queloz,* 1995), changed this fact and led to a steadily increasing number of exoplanet detections. During the ensuing years, we learned first that gas giant planets are common and that the planetary-formation process may produce a surprising variety of configurations: masses considerably larger than Jupiter, planets moving on highly eccentric orbits, planets orbiting closer than 10 stellar radii, planets in resonant multiplanet systems, and planets orbiting components of stellar binaries. Understanding the physical reasons for such wide variations in outcome remains a central issue in planet-formation theory. The role of observations is to provide constraints that will help theoreticians to model the large variety of properties observed for extrasolar planets.

From the mere 7 or 8 exoplanets known at the time of the Protostars and Planets IV conference [and the 17 candidates published in the proceedings (*Marcy et al.,* 2000)], the number of known exoplanets has now surpassed 200. With this larger sample, statistically significant trends now appear in the distribution of orbital elements and host-star properties. The features of these distributions are fossil traces of the processes of formation or evolution of exoplanet systems and help to constrain the planet-formation models.

Here we present a census of the main statistical results obtained from spectroscopic observations over the past decade. In addition to the orbital properties described in sections 2 and 4, and the primary-star characteristics discussed in section 5, we will discuss the evolution of radial velocity measurements over the past two years, namely (1) the role played by followup radial-velocity measurements in confirming and characterizing planetary objects among the many candidates detected by photometric-transit programs (section 6) and (2) the development of specially designed high-resolution spectrographs achieving precisions for radial velocities below the 1 m s^{-1} limit (section 3). This extreme precision opens the possibility for detection of Earth-type planets with radial-velocity measurements (section 7).

2. ORBITAL PROPERTIES OF EXOPLANETS

As a result of the increase in the temporal baseline of the large radial-velocity planet searches (Lick, Keck, AAT, ELODIE, CORALIE programs) and the initiation of new large surveys [e.g., HARPS planet search (*Mayor et al.,* 2003)] and metallicity-biased searches for hot Jupiters (*Fischer et al.,* 2005; *Da Silva et al.,* 2006), there is a large sample of known extrasolar planets. This lends some confidence to observed trends in statistical distributions of the

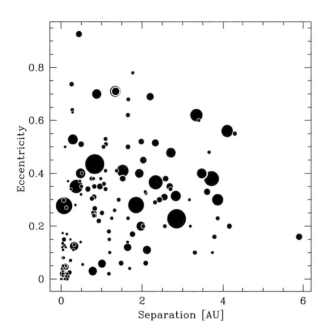

Fig. 1. Separation-eccentricity diagram for the complete sample of presently known extrasolar planets. The size of the dots is proportional to the minimum mass of the planet candidates ($m_2 \sin i \leq 18\ M_{Jup}$).

planet properties. The most remarkable overarching feature of the sample is the variety of orbital characteristics. This variety challenges the conventional views of planetary formation. A global visual illustration of these properties is given in Fig. 1, displaying orbital eccentricities as a function of planet-star separations for the complete sample of known extrasolar planets. Several of the planet properties (close proximity to the star, large eccentricity, high mass) are clearly apparent in the figure. The goal now is to interpret the observed orbital distributions in terms of constraints for the planet-formation models.

The determination of statistical properties of giant planets should be derived from surveys that are themselves statistically well defined (e.g., volume limited) and that have well-understood detection thresholds in the various planet, primary-star, and orbital parameters. There are several programs that meet these requirements, including the volume-limited CORALIE planet-search program (*Udry et al.,* 2000) and the magnitude-limited FGKM Keck survey (*Marcy et al.,* 2005). In the diagrams, we present detected planet candidates from all radial-velocity surveys and note that the discussed properties agree with those presented from single well-defined programs as well.

2.1. Giant Extrasolar Planets in Numbers

The most fundamental property that can be obtained from a planet-search program is the fraction of surveyed stars that host detected planets. Given a typical Doppler precision of a few meters per second and duration of observations, this planet occurrence rate is only defined for a

particular parameter space: planets with masses larger than m_{lim} and orbital periods shorter than P_{lim}. The minimum rate is obtained just by counting the fraction of stars hosting planets in this particular slice of parameter space. For planets more massive than $0.5\ M_{Jup}$, *Marcy et al.* (2005) find in the Lick+Keck+AAT sample that 16/1330 = 1.2% of the stars host hot Jupiters ($P \leq 10$ d, i.e., $a \leq 0.1$ AU for a solar-mass star) and 6.6% of stars have planets within 5 AU. In the volume-limited CORALIE sample (including stellar binaries), for the same m_{lim}, we count 9/1650 = 0.5% occurrence of hot Jupiters and, overall, that 63/1650 = 3.8% of stars have planets within 4 AU. As binaries with separations closer than 2″ to 6″ are usually eliminated from planet-search programs (along with rapidly rotating stars), if we restrict ourselves to stars *suitable* for planet search (i.e., not binary and with v sin i ≤ 6 km s^{-1}), then we find for CORALIE that 9/1120 = 0.8% of stars have giant planets with separations less than 0.1 AU and 63/1120 = 5.6% of stars have planets at separations out to 4 AU. Within Poisson error bars, and including a correction to account for the smaller separation range considered with CORALIE, these two large samples are in good agreement.

The true occurrence rate of gas giant planets can be better approximated by estimating the detection efficiency (as a function of planet mass and orbital period) using Monte Carlo simulations. This has not yet been done for the largest surveys. However, for the ELODIE program (magnitude-limited sample of stars cleaned from known binaries), although dominated by small number statistics errors, *Naef et al.* (2005) estimate for planets more massive than $0.5\ M_{Jup}$ a corrected fraction $0.7 \pm 0.5\%$ of hot Jupiters with $P \leq 5$ d and $7.3 \pm 1.5\%$ of planets with periods smaller than 3900 d. A similar analysis has been carried out by *Cumming et al.* (1999) for the Lick survey and by *Endl et al.* (2002) for the planet-search program with the ESO Coudé-echelle spectrometer. In the overlapping parameter space, all these analyses show good agreement.

With the continuously increasing time span of the surveys and the improvement in our ability to detect smaller-mass planets, we expect the fraction of stars hosting planets to increase substantially from these estimated minimum values, perhaps to values higher than 50%, taking into account that the number of detected planets is a rising function of decreasing planet masses and the rise in planet detections at wide separations (see sections 2.2 and 2.3).

2.2. Planetary Mass Distribution

Even after the detection of just a few extrasolar planets it became clear that these objects could not be considered as the low-mass tail of stellar companions in binary systems (with low $m_2 \sin i$ because of nearly face-on orbital inclinations). The strong bimodal aspect of the secondary mass distribution to solar-type primaries (Fig. 2) has generally been considered as the most obvious evidence of different formation mechanisms for stellar binaries and planetary systems. The interval between the two populations (the

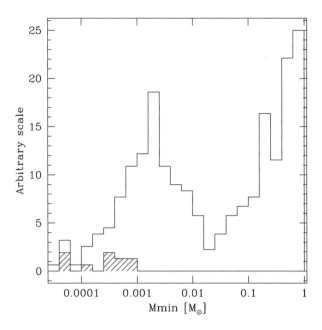

Fig. 2. Minimum mass distribution of secondaries to solar-type primaries. The stellar binaries are from *Halbwachs et al.* (2003). The hatched histogram represents HARPS planets (section 3).

brown-dwarf desert) corresponding to masses between ~20 and ~60 M_{Jup} is almost empty, at least for orbital periods shorter than a decade. However, there is probably an overlap of these two distributions; at this point, it is not easy to differentiate *low-mass brown dwarfs* from *massive planets* just from their $m_2 \sin i$ measurements, without additional information on the formation and evolution of these systems. (A dedicated working group of the IAU has proposed a *working definition* of a "planet" based on the limit in mass at 13 M_{Jup} for the ignition of deuterium burning.)

Toward the low-mass side of the planetary mass distribution, a clear power-law type rise is observed (Fig. 2). *Marcy et al.* (2005) proposed $dN/dM \propto M^{-1.05}$ for their FGKM sample. This fit is not affected by the unknown $\sin i$ distribution (*Jorissen et al.*, 2001), which simply scales in the vertical direction. The low-mass edge of this distribution is poorly defined because of observational incompleteness; the lowest-mass planets are difficult to detect because the radial-velocity variations are smaller. It is then likely that there is a large population of sub-Saturn-mass planets. This trend is further supported by accretion-based planet-formation models. In particular, large numbers of "solid" planets are expected (*Ida and Lin,* 2004a, 2005; *Alibert et al.,* 2004, 2005) (see also section 3).

2.3. Period Distribution of Giant Extrasolar Planets

Figure 3 displays the orbital period distribution for the known exoplanet sample. The numerous giant planets orbiting very close to their parent stars (P < 10 d) were completely unexpected before the first exoplanet discoveries. The standard model (e.g., *Pollack et al.,* 1996) suggests that

giant planets form first from ice grains in the outer region of the system where the temperature of the stellar nebula is cool enough. Such grain growth provides the supposed requisite solid core around which gas could rapidly accrete (*Safronov,* 1969) over the lifetime of the protoplanetary disk (~10^7 yr). The detection of planets well inside the ice line requires that the planets undergo a subsequent migration process moving them close to the central star (see, e.g., *Lin et al.,* 1996; *Ward,* 1997; see also the chapter by Papaloizou et al. for an updated review). Alternative points of view invoke *in situ* formation (*Bodenheimer et al.,* 2000; *Wuchterl et al.,* 2000), possibly triggered through disk instabilities (see the chapter by Durisen et al.). Note, however, that even in such cases, subsequent disk-planet interactions leading to migration is expected to take place as soon as the planet has formed. The observed pileup of planets with periods around 3 d is believed to be the result of migration and requires a stopping mechanism to prevent the planets from falling onto the stars (see, e.g., *Udry et al.,* 2003, and references therein for a more detailed discussion).

Another interesting feature of the period distribution is the rise of the number of planets with increasing distance from the parent star. This is not an observational bias as equivalent mass candidates are more easily detected at shorter periods with the radial-velocity technique. The decrease of the distribution beyond 10 yr coincides with, and is almost certainly a result of, the limited duration of most of the radial-velocity surveys. The overall distribution can

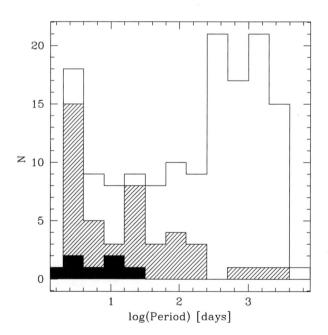

Fig. 3. Period distribution of known gaseous giant planets detected by radial-velocity measurements and orbiting dwarf primary stars. The hatched part of the histogram represents "light" planets with $m_2 \sin i \leq 0.75$ M_{Jup}. For comparison, the period distribution of known Neptune-mass planets (section 3) is given by the filled histogram. (Note, however, that there is still very high observational incompleteness for these low-mass planets.)

then be understood as being comprised of two parts: a main distribution rising with increasing periods [as for binary stars (*Halbwachs et al.,* 2003)], the maximum of which is still undetermined; and a second distribution of planets that have migrated inward. The visible lack of planets with orbital periods between 10 and 100 d is real, and appears to be the intersection between the other two distributions.

A minimum flat extrapolation of the distribution to larger distances would approximately double the occurrence rate of planets (*Marcy et al.,* 2005). This conservative extrapolation hints that a large population of yet-undetected Jupiter-mass planets may exist between 5 and 20 AU. This is of prime importance for the direct-imaging projects under development on large telescopes such as the VLT or Gemini Planet Finder (see the chapter by Beuzit et al.) and space-based imaging missions such as NASA's Terrestrial Planet Finder or ESA's Darwin.

2.4. Period Mass Distribution

The orbital-period distribution highlights the role of migration processes underlying the observed configuration of exoplanet systems. An additional correlation is seen between orbital period and planet mass. This correlation is illustrated in Fig. 4, showing the mass-period diagram for the known exoplanets orbiting dwarf primaries.

The most obvious characteristic in Fig. 4 is the paucity of massive planets on short-period orbits (*Zucker and Mazeh,*

2002; *Udry et al.,* 2002; *Pätzold and Rauer,* 2002). This is not an observational bias as these candidates are the easiest ones to detect. Even more striking, when we neglect the multiple-star systems (section 2.5), a complete void of candidates remains in the diagram for masses larger than ~2 M_{Jup} and periods smaller than ~100 d. The only candidate left is HD 168443 b, a member of a possible multi-brown-dwarf system (*Marcy et al.,* 2001; *Udry et al.,* 2002).

Migration scenarios may naturally result in a paucity of close-in massive planets. For example, type II migration (where the planet clears a gap in the disk) has been shown to be less effective for massive planets; i.e., massive planets are stranded at wider separations than low-mass planets. Alternatively, when a migrating planet reaches small separations from the star, some process related to planet-star interactions could promote mass transfer from the planet to the star, decreasing the mass of the migrating planet (e.g., *Trilling et al.,* 1998), or could cause massive planets to fall into the central star (*Pätzold and Rauer,* 2002).

Another interesting feature of the period distribution is the rise in the maximum planet mass with increasing distance from the host star (Fig. 5) (*Udry et al.,* 2003). While it is true that Doppler detectability for lower-mass planets declines with increasing distance from the star, the massive planets are easily detected at small separations, yet they preferentially reside in more distant orbits. This can be understood in the context of the migration scenario as well. More massive planets are expected to form further out in

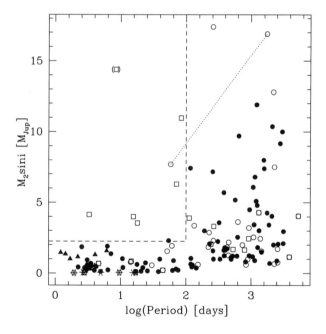

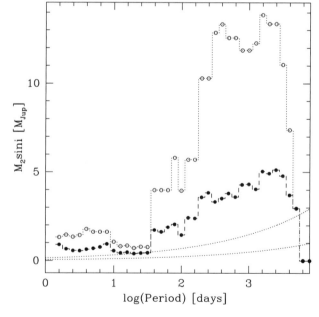

Fig. 4. Period-mass distribution of known extrasolar planets orbiting dwarf stars. Open squares represent planets orbiting one component of a binary system whereas dots are for "single" stars. Open dots represent planets in multiplanet systems. Asterisks represent Neptune-mass planets. Dashed lines are limits at 2.25 M_{Jup} and 100 days. The dotted line connects the two "massive" components orbiting HD 168443.

Fig. 5. Mean (filled circle) or highest (average on the three highest values; open circles) mass of planets in period smoothing windows of width log P (days) = 0.2. Although massive planets are easy to detect in shorter-period orbits, an increase in the maximum planet mass with increasing distance from the star is observed. Detection limits for velocity semiamplitudes K of 10 and 30 m s^{-1} (M_1 = 1 $M_\odot$, e = 0) are represented by the dotted lines.

the protoplanetary disk, where raw materials for accretion are abundant and the longer orbital path provides a larger feeding zone. Then, migration may be more difficult to initiate as a larger portion of the disk has to be disturbed to overcome the inertia of the planet. This notion is further supported by the observation that hot Jupiters have statistically lower masses ($m_2 \sin i \leq 0.75\ M_{Jup}$) that may migrate more easily (Fig. 3).

It has also been suggested that multiplanet chaotic interactions preferentially move low-mass (low-inertia) planets either inward or outward in the system, whereas massive (high-inertia) planets are harder to dislodge from their formation site (*Rasio and Ford*, 1996; *Weidenschilling and Marzari*, 1996; *Marzari and Weidenschilling*, 2002; see also the chapter by Levison et al.). One weakness of this hypothesis is that the frequency of short-period planets is difficult to reproduce with reasonable assumptions for these models (*Ford et al.*, 2001, 2003).

As discussed above, the observations empirically point to a decrease in the efficiency of migration with increasing planet mass. Simulations of migrating planets in viscous disks are consistent with this observation (*Trilling et al.*, 1998, 2002; *Nelson et al.*, 2000). Therefore, it seems reasonable to expect that a large number of massive planets may reside on long-period orbits, and yet may still be undetected because of the time duration of the present surveys. Younger primary stars among them, less amenable to radial-velocity searches because of the intrinsic astrophysical noise of the star, will be suitable targets for direct imaging searches (see the chapter by Beuzit et al.). Lower-mass planets could exist on long-period orbits as well; however, these planets are difficult to detect with precisions $\geq 3\ m\ s^{-1}$. Low-mass, distant planets orbiting chromospherically quiet stars may be detected with extreme precision radial velocities with demonstrated stability over a decade or more (see section 3).

2.5. Giant Planets in Multiple Stellar Systems

Among the ~200 extrasolar planets discovered to date, at least 30 are known to orbit one of the members of a double- or multiple-star system (*Patience et al.*, 2002; *Eggenberger et al.*, 2004; *Mugrauer et al.*, 2004, 2005). These systems cover a large range of binary projected separations: from ~20 AU for two spectroscopic binaries to more than 1000 AU for wide visual systems. Although the sample is not large, some differences between planets orbiting binary components and those orbiting single stars can be seen in the mass-period (Fig. 4) and eccentricity-period (Fig. 6) diagrams. As pointed out by *Zucker and Mazeh* (2002), the most massive short-period planets are all found in binary- or multiple-star systems. The planets orbiting a component star of a multiple-star system also tend to have a very low eccentricity when their orbital period is shorter than about 40 d (*Eggenberger et al.*, 2004). The only exception is the "massive" companion of HD 162020, which is probably a low-mass brown dwarf (*Udry et al.*, 2002). These observations suggest that some kind of migration process has been

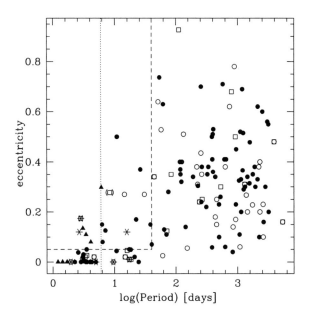

Fig. 6. Period-eccentricity diagram of the known extrasolar planets. Open squares represent planets orbiting one of the components of a binary system whereas dots are for "single" stars. Open dots represent planets in multiplanet systems. Planets detected in metallicity-biased or photometric-transit surveys are indicated by filled triangles. Stared symbols are for Neptune-mass planets. The parentheses "( )" locate HD 162020. The dotted line is indicative of an observed tidal circularization period around 6 d (*Halbwachs et al.*, 2005) and the dashed lines limit the e > 0.05 and P < 40 d domain (see section 2.5).

at work in the history of these systems. The properties of the five short-period planets orbiting in multiple-star systems seem, however, difficult to reconcile with the current models of planet formation and evolution, at least if we want to invoke a single mechanism to account for all the characteristics of these planets.

Even if the stellar orbital parameters for planet-bearing binary stars are not exactly known, we have some information like the projected separations of the systems or stellar properties. No obvious correlation between the properties of these planets and the known orbital characteristics of the binaries or of the star masses are found yet, however. Due to the limitations of the available observational techniques, most detected objects are giant (Jupiter-like) planets; the existence of smaller-mass planets in multiple-star systems is still an open question.

Searches for extrasolar planets using the radial-velocity technique have shown that giant planets exist in certain types of multiple-star systems. The number of such planets is still low, perhaps in part because close binaries are difficult targets for radial-velocity surveys and are excluded from Doppler samples. However, even if the detection and characterization of planets in binaries are more difficult to carry out than the study of planets around single stars, it is still worth doing because of the new constraints and information it may provide on planet formation and evolution. In particular, circumbinary planets offer a complete unexplored new field of investigations.

2.6. Giant Planet Eccentricities

Extrasolar planets with orbital periods longer than about 6 d have eccentricities significantly larger than those of giant planets in the solar system (Fig. 6). Their median eccentricity is e = 0.29. The eccentricity distribution for these exoplanets resembles that for binary stars, spanning almost the full range between 0 and 1. Planets with periods smaller than 6 d are probably tidally circularized (see below).

The origin of the eccentricity of extrasolar giant planets has been suggested to arise from several different mechanisms: the gravitational interaction between multiple giant planets (*Rasio and Ford,* 1996; *Weidenschilling and Marzari,* 1996; *Lin and Ida,* 1997); interactions between the giant planets and planetesimals in the early stages of the system formation (*Levison et al.,* 1998); or the secular influence of an additional, passing-by (*Zakamska and Tremaine,* 2004), or bounded companion in the system (see *Tremaine and Zakamska,* 2004, for a comprehensive review of the question).

The latter effect seems particularly interesting in some cases. The mean velocity of several planets with eccentric orbits shows a drift, consistent with the presence of a long-period companion. The gravitational perturbation arising from the more distant companion could be responsible for the observed high orbital eccentricity. This effect has been suggested as an eccentricity pumping mechanism for the planet orbiting 16 Cyg B (*Mazeh et al.,* 1997). However, *Takeda and Rasio* (2005) have shown that such a process would produce an excessive number of both very high (e ≥ 0.6) and very low (e ≤ 0.1) eccentricities, requiring at least one additional mechanism to reproduce the observed eccentricity distribution. In fact, none of the proposed eccentricity-inducing mechanisms alone is able to reproduce the observed eccentricity distribution.

For small periastron distance, giant planets are likely to undergo tidal circularization. For periods smaller than ~6 d, nearly all gaseous giant planets are in quasicircular orbits (e ≤ 0.05; Fig. 6) (*Halbwachs et al.,* 2005). The few border cases, with eccentricities around 0.1, have been recently detected with few observations in surveys biased for short-period orbits (metallicity-biased or photometric-transit searches) and have very uncertain eccentricity estimates (even compatible with zero). With more radial-velocity data spanning several orbits, the measured orbital eccentricities may decline. Alternatively, an additional companion may ultimately be found in some of these systems. In multiple-planet systems, a single Keplerian model can absorb some of the longer-period trend in mean velocities, artificially inflating the orbital eccentricity. Additional companions could also tidally pump up eccentricity in short-period systems.

Correlations can also be seen between eccentricity and period, and between eccentricity and mass. The more massive planets (i.e., more massive than 5 M_{Jup}) exhibit systematically higher eccentricities than do the lower-mass planets (*Marcy et al.,* 2005). This cannot be a selection effect (larger induced radial-velocity variation). If planets form initially in circular orbits, the high eccentricities of the most massive planets are puzzling. Such massive planets have the largest inertial resistance to perturbations that are necessary to drive them out of their initial circular orbits. Note that the more massive planets are also found at wider separations (section 2.3), and therefore eccentricity and orbital period are coupled. The long-period planets have usually only been observed for one period and are rarely well covered in phase. This could lead to an overestimate of the derived eccentricity in some Keplerian fits (*Butler et al.,* 2000), but overall it seems unlikely that improper modeling is entirely responsible for the observed correlation.

Finally, as seen in Fig. 6, a few long-period, low-eccentricity candidates are emerging from the surveys. They form a small subsample of so-called solar system *analogs*.

3. THE QUEST FOR VERY HIGH PRECISION

3.1. Down Below the Mass of Neptune

After a decade of discoveries in the field of extrasolar giant planets, mainly coming from large high-precision radial-velocity surveys of solar-type stars, the quest for other worlds has now passed a new threshold. Most of the detected planets are gaseous giants similar to our own Jupiter, with typical masses of a few hundreds of Earth masses. However, in the past year, seven planets with masses in the Uranus to Neptune range (6–21 $M_\oplus$) have been detected (Table 1). Because of their small mass and location in the system, close to their parent stars, they may well be composed mainly of a large rocky/icy core, and it is possible that they either lost most of their gaseous atmosphere or simply formed without accumulating a substantial one.

TABLE 1.   Summary table for the recently discovered Neptune-mass planets.

Planet	Reference	P (d)	$m_2 \sin i$ ($M_\oplus$)	(o-c) (m s^{-1})	q (10^{-5})
μ Ara c	[1]	9.6	14	0.9	4.2
55Cnc e	[2]	2.81	14	5.4	4.7
HD 4308 b	[3]	15.6	14	1.3	5.4
HD 190360 c	[4]	17.1	18	3.5	6.0
Gl 876 d	[5]	1.94	6	4.6	6.0
Gl 436 b	[6]	2.6	21	5.3	16.0
Gl 581 b	[7]	4.96	17	2.5	17.1

The parameter q = $m_2 \sin i/m_1$ and (o-c) is the residuals (RMS) around the Keplerian solution. The lowest $m_2 \sin i$ of 6 $M_\oplus$ is obtained for Gl 876 d while the lowest q of 4.2×10^{-5} is achieved on μ Ara c.

References: [1] *Santos et al.* (2004a); [2] *McArthur et al.* (2004); [3] *Udry et al.* (2006); [4] *Vogt et al.* (2005); [5] *Rivera et al.* (2005); [6] *Butler et al.* (2004); [7] *Bonfils et al.* (2005a).

These planetary companions, together with recently detected sub-Saturn-mass planets on intermediate-period orbits, populate the lower end of the planetary mass distribution, a region still strongly affected by detection incompleteness (Fig. 2). The discovery of very-low-mass planets so close to the detection threshold of radial-velocity surveys suggests that this kind of object may be rather common. The very existence of such planets is yet another unexpected observation for theorists. Indeed, a prediction had already been made that planets with masses between 1 and 0.1 M_{Sat} and semimajor axes of 0.1 to 1 AU would be rare [the so-called planet desert (*Ida and Lin,* 2004a)]. At least for the moment, observations seem to be at odds with the predictions (although very little is known about the actual populating of this hypothetic desert). In any case, the search and eventual detection of planets with even lower mass will set firmer constraints to planetary system formation and evolution models.

This detection of very-low-mass planets follows from the development of a new generation of instruments capable of radial-velocity measurements of unprecedented quality. One workhorse for high-precision work is the ESO high-resolution HARPS fiber-fed echelle spectrograph especially designed for planet-search programs and astroseismology. HARPS has already proven to be the most precise spectrovelocimeter to date, reaching an instrumental radial-velocity accuracy at the level of 1 m s^{-1} over months to years (*Mayor et al.,* 2003; *Lovis et al.,* 2005), and even better on a short-term basis (*Bouchy et al.,* 2005a). The Keck telescope with an upgraded detector for the HIRES spectrometer is also approaching 1 m s^{-1} precision, with demonstrated stability since August 2004.

Another fundamental change that allowed progress in planet detection toward the very low masses is the application of a careful observing strategy to reduce the perturbing effect of stellar oscillations that can obscure the tiny reflex velocity signal induced by Neptune-mass planets.

As recently as only a couple of years ago, the behavior of the stars below 3 m s^{-1} was completely unknown. However, astroseismology observations carried out with HARPS have made it clear that the achieved precision is no longer set by instrumental characteristics but rather by the stars themselves (*Mayor et al.,* 2003; *Bouchy et al.,* 2005a). Indeed, stellar p-mode oscillations on short timescales (minutes) and stellar jitter (activity-induced noise) on longer timescales (days) can and do induce significant radial-velocity changes at the level of accuracy of the HARPS measurements. Even chromospherically quiet G and K dwarfs show oscillation modes of several tens of centimeters per second each, which might add up to radial-velocity amplitudes as large as several meters per second. As a consequence, any exposure with a shorter integration time than the oscillation period of the star, or even shorter than mode-interference variation timescales, might fall arbitrarily on a peak or on a valley of these mode interferences and thus introduce additional radial-velocity "noise." This phenomenon could, therefore, seriously compromise the ability to detect very-low-mass planets around solar-type stars by means of the radial-velocity technique.

To minimize these effects as much as possible, stars for very-high-precision radial-velocity measurements have first to be chosen as slowly rotating, nonevolved, and low-activity stars. Then, in order to average out stellar oscillations, the observations have to be designed to last at least 15 to 30 min on target. This strategy is now applied to stars in the "high-precision" part of the HARPS and Keck planet-search programs. An illustration of the obtained results is given by the histogram of the radial-velocity dispersion of the HARPS high-precision survey (Fig. 7). The distribution mode is just below 2 m s^{-1}, and the peak decreases rapidly toward higher values. More than 80% of the stars show dispersion smaller than 5 m s^{-1}, and more than 35% have dispersions below 2 m s^{-1}. It must be noted that the computed dispersion includes photon-noise error, wavelength-calibration error, stellar oscillations and jitter, and, in particular, "pollution" by known extrasolar planets (hatched portion of Fig. 7) and still undetected planetary companions. The recently announced 14 $M_\oplus$ planets orbiting μ Ara (Fig. 8) and HD 4308 (Table 1) are part of this HARPS "high-precision" subsample.

3.2. Gaseous Versus Solid-Planet Properties at Short Periods

Although the number of known Neptune-mass planets is small, it is interesting to see how their orbital parameters compare with properties of giant extrasolar planets. Because of the tiny radial-velocity amplitude they induce on the

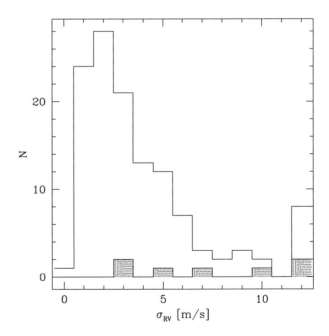

Fig. 7. Histogram of the observed radial-velocity dispersion (σ_{RV}) of the stars in the HARPS "high-precision" subprogram (124 stars with more than 3 measurements). The position of the planets detected with HARPS is indicated by the hatched area.

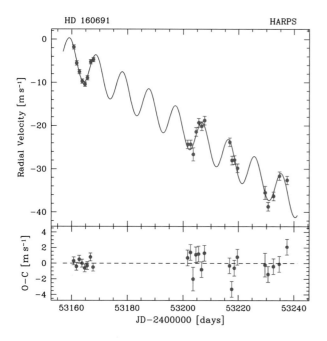

Fig. 8. HARPS measurements of μ Ara that unveiled the 14 M$_\oplus$ planet on a 9.55-d orbit. The overall r.m.s. of the residuals around the planet Keplerian solution, corrected from a long-term drift due to additional planets in the system, amounts to only 0.9 m s^{-1}, and is even as low as 0.43 m s^{-1} for the first eight points obtained by nightly averaging radial velocities measured during a one-week astroseismology campaign (*Santos et al.,* 2004a).

primary stars, limiting possible detections to short periods, a "meaningful" comparison can only be done for giant planets with periods smaller than ~20 d.

The distribution of short-period giant planets strongly peaks at periods around 3 d (Fig. 3). On the contrary, despite the mentioned detectability bias, the period distribution of Neptune-mass planets is rather flat up to 15 d. We also observe that orbits of Neptune-mass planets have small eccentricities (Fig. 6). In particular, for periods between 9 and 15 d (three out of the seven candidates), the mean eccentricity value is much smaller than the one of giant planets. At periods smaller than 6 d, orbits are supposed to be tidally circularized, especially if these planets are "solid." However, among them, the largest observed eccentricities are for 55 Cnc e (P = 2.8 d and e = 0.17) and Gl 436 (P = 2.6 d and e = 0.12). The former is a member of a multiplanet system, which might explain the nonzero eccentricity of the inner small-mass planet (section 2.6); however, the problem is more difficult for the latter case. Another difference between giant and Neptune-mass planets can also be found in the parent-star metallicity distribution (see section 5.2).

Although the number of objects does not constitute a statistically significant sample, these small differences may hint that giant gaseous and "solid" planets form two distinct populations, with different properties. More detections are, however, needed to consider this question in a more convincing way.

4. MULTIPLE PLANET SYSTEMS

There are more than 170 planet-hosting stars for the more than 200 known extrasolar planets. Seventeen of these stars have multiple-planet systems rather than single planets. One further system, HD 217107, shows an additional curved drift of the residuals of the single-planet Keplerian solution that is compatible with a second planetary companion. The orbital characteristics of these systems are summarized in Table 2. The most prolific of them is 55 Cnc, with four detected planets. ν And, HD 37124, Gl 876, and μ Ara (HD 160691) each have three planets. Finally, there are a total of 11 known double-planet systems.

Among planet-bearing stars, ~12% are known multiple-planet systems. Thus, the probability of finding a second planet is enhanced by a factor of 2 over the ~6% probability of finding the first planet. The fraction of known multiplanet systems is certainly a lower limit. One challenge is that low-amplitude trends from more distant, longer-period sibling planets are easily absorbed into single-planet Keplerian models. Detection of additional planets is easier in systems where the more distant planet is greater than a few times the mass of Jupiter since such systems will produce larger-velocity amplitudes. However, the mass histogram (Fig. 2) shows that high-mass planets are uncommon. A second challenge exists for systems with small orbital period ratios like Gl 876. There, dynamical interactions between planets can complicate Keplerian fitting of the observations and delay characterization and announcement of a second planet. As a result, while one orbital period is sufficient for a single-planet system with velocity amplitudes greater than 10 m s^{-1} (~3σ detection), longer phase coverage is generally required to disentangle additional components. The longest-running, high-precision survey is the 15-year planet search at Lick Observatory. This sample of 100 stars includes the multiplanet systems 55 Cnc, Ups And, Gl 876, and 47 UMa. Half the planet-hosting stars from that sample now have more than one detected planet. For the somewhat younger ELODIE planet-search program in Haute-Provence, begun in 1994 and expanded in 1996, 25% of the stars with detected planets host more than one planet.

In light of the challenges that preclude detection of multiplanet systems and given the high fraction of multiplanet systems in the older long-running search programs, it seems likely that most stars form *systems of planets* rather than isolated, single planets. New techniques, complementary to radial velocities, to discover exoplanets with imaging, interferometry, or astrometry will very probably exploit the sizable fraction of multiple-planet systems when designing their programs.

4.1. Mean-Motion-Resonance Systems

It is tempting to categorize multiplanet systems as either hierarchical or resonance systems. Among the known multiplanet systems, at least eight (nearly half) are in mean-motion resonances (MMR) and four of these are in the low-order 2:1 resonance. Figure 9 shows the ratio of orbital

TABLE 2. Orbital parameters of multiplanet systems.

Star ID	P (d)	e	$m_2 \sin i$ (M_{Jup})	a (AU)	Rem
HD 75732 b	14.67	0.02	0.78	0.115	55 Cnc
HD 75732 c	43.9	0.44	0.22	0.24	3:1 (c:b)
HD 75732 d	4517	0.33	3.92	5.26	
HD 75732 e	2.81	0.17	0.045	0.038	
HD 9826 b	4.617	0.012	0.69	0.06	ν And
HD 9826 c	241.5	0.28	1.89	0.83	
HD 9826 d	1284	0.27	3.75	2.53	~16:3 (d:c)
HD 37124 b	154.5	0.06	0.61	0.53	
HD 37124 c	843.6	0.14	0.60	1.64	
HD 37124 d	2295.0*	0.2	0.66	3.19	~8:3 (d:c)
Gl 876 b	60.94	0.025	1.93	0.21	2:1 ± 0.02 (b:c)
Gl 876 c	30.10	0.27	0.56	0.13	
Gl 876 d	1.938	0.0	0.023	0.021	
HD 160691 b	629.6	0.26	1.67	1.5	μ Ara
HD 160691 c	9.55	0.0	0.044	0.09	
HD 160691 d	2530	0.43	1.22	4.17	4:1 ±0.25 (d:b)
HD 12661 b	262.5	0.35	2.37	0.83	
HD 12661 c	1684	0.02	1.86	2.60	~13:2 ±0.8 (c:b)
HD 217107 b	7.12	0.13	1.35	0.10	
HD 217107 c	>10,000†	—	>10	>20	
HD 168443 b	58.11	0.53	7.64	0.29	
HD 168443 c	1764	0.22	17.0	2.85	
HD 169830 b	225.6	0.31	2.88	0.81	
HD 169830 c	2102	0.33	4.04	3.60	
HD 190360 b	2891	0.36	1.56	3.92	
HD 190360 c	17.1	0.01	0.057	0.13	
HD 202206 b	256.2	0.43	17.5	0.83	
HD 202206 c	1297	0.28	2.41	2.44	~5:1 ± 0.07 (c:b)
HD 38529 b	14.3	0.25	0.84	0.13	
HD 38529 c	2182	0.35	13.2	3.68	
HD 73526 b	187.5	0.39	2.07	0.66	
HD 73526 c	376.9	0.40	2.30	1.05	2:1 ± 0.01 (c:b)
HD 74156 b	51.6	0.64	1.86	0.29	
HD 74156 c	2025	0.58	6.19	3.40	
HD 82943 b	219.5	0.39	1.82	0.75	
HD 82943 c	439.2	0.02	1.75	1.20	2:1 ± 0.01 (c:b)
HD 95128 b	1089	0.06	2.54	2.09	47 UMa
HD 95128 c	2594	0.00	0.76	3.73	
HD 108874 b	395.4	0.07	1.36	1.05	
HD 108874 c	1606	0.25	1.02	2.68	4:1 ± 0.1 (c:b)
HD 128311 b	458.6	0.25	2.18	1.10	
HD 128311 c	928	0.17	3.20	1.77	2:1 ± 0.03 (c:b)

*See *Vogt et al.* (2005) for alternate orbital solution.
†Period not covered.

Values are from the literature or updated from *Butler et al.* (2006). Period resonances are indicated in the "Rem" column.

periods (defined as the longer period divided by the shorter period) for multiplanet systems listed in Table 2. Uncertainties in the derived orbital periods (*Butler et al.,* 2006) are shown as error bars. Except for HD 37124, which has an uncertain Keplerian model, orbital ratios less than or equal to 4:1 are all very close to integral period ratios with low orders (MMR) of 2:1, 3:1, or 4:1. The outer two planets orbiting ν And are close to a 16:3 MMR and HD 12661

may be in a 13:2 MMR. No mean-motion resonances are observed close to the exact ratio of 5:1 or 6:1. However, uncertainties in the orbital solution for HD 12661 allow for the possibility of a 6:1 MMR and the stability study of HD 202206 (*Correia et al.,* 2005) suggests that the system is trapped in the 5:1 resonance. In this latter case the 5:1 resonance could indicate that the planet formed in a circumbinary disk, as the inner "planet" has a minimum mass of

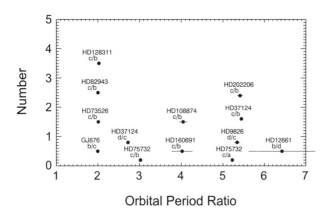

Fig. 9. The ratio of longer to shorter orbital periods is shown for multiplanet systems in Table 2. Uncertainties in the orbital periods are propagated as error bars in the period ratio. The low-order MMR at 2:1 appears to be quite narrow with 2 ± 0.01:1. Four of the 18 systems (including uncovered periods) reside in a 2:1 resonance.

17 M_{Jup}. Beyond the 4:1 MMR, the orbital period ratios quickly stray from integral ratios. This suggests that if planets are close enough, it is likely that resonance capture will occur. Conversely, resonance capture seems less effective if the orbital period ratio is greater (i.e., the planets do not make a close approach), although longer orbital periods are not as precisely determined.

Kley et al. (2004) model the resonant capture of planets and find that for the 2:1 MMR, their models predict (1) a larger mass for the outer planet, and (2) higher eccentricity for the inner planet. We find that the orbital eccentricity is higher for the inner planet in three of the four 2:1 resonance systems. In the fourth system, HD 73526, the eccentricities for both components are comparable. We find that the outer planet is more massive (assuming coplanar orbits) in Gl 876 and HD 128311. The outer planet is only slightly more massive in HD 73526, and it is slightly less massive in the Keplerian model for HD 82943 (*Mayor et al.,* 2004).

The orbital parameters of multiplanet systems seem indistinguishable from those of single-planet systems. For example, Figs. 4 and 6 compare the period mass and eccentricity distributions of multiple- and single-planet systems.

4.2. Dynamics: Planet-Planet Interactions

The presence of two or more interacting planets in a system dramatically increases our potential ability to constrain and understand the processes of planetary formation and evolution. Short-term dynamical interactions are of particular interest because of the directly observable consequences. Among them, the observed $P_i/P_j = 2/1$ resonant systems are very important because, when the planet orbital separations are not too large, planet-planet gravitational interactions become nonnegligible during planetary "close" encounters, and will noticeably influence the system evolution on a timescale on the order of a few times the long period. The radial-velocity variations of the central star will

then differ substantially from velocity variations derived assuming the planets are executing independent Keplerian motions (Fig. 10). In the most favorable cases, the orbital-plane inclinations, not otherwise known from the radial-velocity technique, can be constrained since the amplitude of the planet-planet interaction directly scales with their true masses. Several studies have been conducted in this direction for the Gl 876 system (*Laughlin et al.,* 2005; *Rivera et al.,* 2005) hosting two planets at fairly small separations (2/1 resonance). The results of the Newtonian modeling of the Gl 876 system have validated the method, improving notably the determination of the planetary orbital elements and also unveiling the small-mass planet embedded in the very inner region of the system (Tables 1 and 2).

Another useful application of the dynamical analysis of a multiplanet system is the localization of the resonances in the system that shape its overall structure. Stability studies are also mandatory to ensure the long-term viability of the systems observed now.

5. PRIMARY STAR PROPERTIES

Additional information to constrain planet-formation models comes from the study of the planet hosts themselves. In particular, the mass and metallicity of the parent stars seem to be of prime importance for models of planet formation (*Ida and Lin,* 2004b, 2005; *Benz et al.,* 2005).

5.1. Metallicity Correlation of Stars with Giant Planets

A correlation between the presence of Doppler-detected gas giant planets and high metallicity in the host stars was noted in the early years of extrasolar planet detection (*Gonzalez,* 1997, 1998; *Gonzalez et al.,* 1999; *Gonzalez and Laws,* 2000; *Fuhrmann et al.,* 1997, 1998; *Santos et al.,*

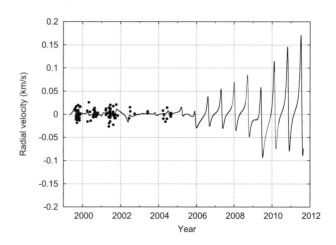

Fig. 10. Temporal differences between the radial velocities predicted by the two-Keplerian models and the numerical integration of the system HD 202206 (*Correia et al.,* 2005). Residuals of the CORALIE measurements around the Keplerian solution are displayed as well.

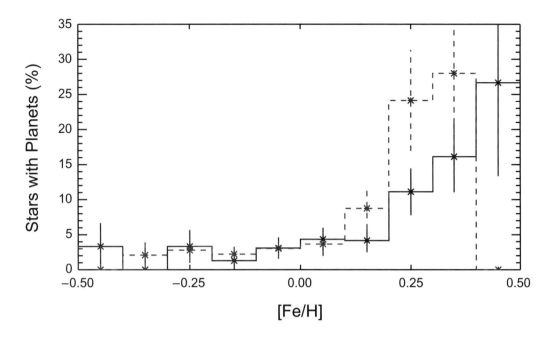

Fig. 11. The percentage of stars with exoplanets is shown as a function of stellar metallicity. Here, the dashed line shows the results of *Santos et al.* (2004b) for 875 CORALIE nonbinary stars and the solid line shows the analysis of 1040 Lick, Keck, and AAT stars (*Fischer and Valenti*, 2005). Although based on different metallicity estimates and on different star samples, the two distributions agree within the error bars.

2000, 2003). This observation led to debate over the origin of the planet-metallicity correlation. One explanation posited that high metallicity enhanced planet formation because of the increased availability of small particle condensates, the building blocks of planetesimals. Another argument suggested that enhanced stellar metallicity could be pollution of the stellar convective zone resulting from late-stage accretion of gas-depleted material. A third explanation invoking the possibility that planet migration is somewhat controlled by the dust content of the disk — thus leading to an observed bias in favor of close-in planets around metal-rich stars — seems to be reasonably ruled out by current models (*Livio and Pringle*, 2003). The two main mechanisms result in different stellar structures; in the first case, the star is metal-rich throughout, while in the latter case, the convective zone has significantly higher metallicity than the stellar interior.

At the time of the early observation of the planet metallicity correlation only a handful of planet-bearing stars were known, and the comparison metallicity distributions came from volume-limited studies, carried out by different researchers at a time when systematic offsets of 0.1 dex in metallicity results were common. Eventually, systematic, homogeneous studies of all stars on planet-search surveys were completed (*Santos et al.*, 2001) with the further requirement that the stars have enough observations to have found a Jupiter-like planet with an orbital period out to 4 yr (*Fischer et al.*, 2004; *Santos et al.*, 2004b, 2005; *Fischer and Valenti*, 2005). Rather than checking the metallicity of planet-bearing stars, the presence of gas giant planets orbiting stars with known metallicity was assessed for well over 1500 stars on ongoing Doppler planet surveys. Figure 11 shows the percentage of stars with planets as a function of metallicity from 1040 stars on the Lick, Keck, and AAT planet surveys [solid line (*Fischer and Valenti*, 2005)] and the percentage of stars with planets from 875 stars on the CORALIE survey [nonbinary and with more than five observations; dashed line (*Santos et al.*, 2004b)]. The occurrence of planets as a function of metallicity was fit by *Fischer and Valenti* (2005) with a power law

$$P(\text{planet}) = 0.03 \times \left(\frac{(N_{Fe}/N_H)}{(N_{Fe}/N_H)_\odot} \right)^2$$

Thus, the probability of forming a gas giant planet is roughly proportional to the square of the number of metal atoms, and increases by a factor of 5 when iron abundance is increased by a factor of 2, from [Fe/H] = 0 to [Fe/H] = 0.3.

The self consistent analysis of high-resolution spectra for more than 1500 stars on planet-search surveys also distinguished between the two enrichment hypotheses. Metallicity was not observed to increase with decreasing convective zone depth for main-sequence stars, suggesting that pollution through accretion was not responsible for the observed metallicity enhancement of planet-bearing stars. This argument is, however, questioned by *Vauclair* (2004), invoking thermohaline convection (metallic fingers) that might dilute the accreted matter inside the star and thus reconcile the overabundances expected in the case of accretion of planetary material with the observations of stars of different masses. Even more important in terms of discarding the pollution hypothesis, the analysis of subgiants in the sample showed that subgiants with planets had high metallicity, while subgiants without detected planets had a metallicity distribution similar to main-sequence stars without detected planets. Since significant mixing of the convective zone takes place along the subgiant branch, subgiants would have diluted accreted metals in the convective zone. The fact that high metallicity persisted in subgiants with planets demonstrated that these stars were metal-rich throughout. The existence of a planet metallicity correlation supports core accretion over gravitational instability as the formation mechanism for gas giant planets with orbital periods as long as 4 yr.

The observed relation between stellar metal content and planet occurrence has motivated metallicity-biased planet-search programs targeting short-period planets to look for hot Jupiters, which are ideal candidates for a photometric transit-search followup. These surveys are successful (*Fischer et al.*, 2005; *Sato et al.*, 2005; *Bouchy et al.*, 2005b; *Da Silva et al.*, 2006) (see section 6). However, the built-in bias of the sample has to be kept in mind when examining possible statistical relations between the star metallicity and other orbital or stellar parameters. Up to now, no clear correlation between metallicity and orbital parameters has been observed.

5.2. Metallicity of Stars Hosting Neptune-Mass Planets

It is well established that the detected giant planets preferentially orbit metal-rich stars. What is the situation for the newly found Neptune-mass planets? If, as proposed by several authors (see, e.g., *Lecavelier et al.*, 2004; *Baraffe et al.*, 2004, 2005, and references therein), the new *hot-Neptune* planets are the remains of evaporated ancient giant planets, their host stars should also follow the metallicity trend observed for their giant progenitor hosts. This does not seem to be the case, considering that the seven known planets with $m_2 \sin i \leq 21\ M_\oplus$ (Table 1) have metallicities of 0.33, 0.35, 0.02, 0.14, –0.03, –0.25, and –0.31, respectively [the metallicity of the three M dwarfs comes from the photometric calibration derived by *Bonfils et al.* (2005b)]. Although the statistics are still poor, the spread of these values over the nearly full range of planet-host metallicities (Fig. 12) suggests a different relation between metal content and planet frequency for the icy/rocky planets in regard to the giant ones.

It is worth remarking that three of the Neptune-mass candidates orbit M-dwarf primaries. Recent Monte Carlo simulations by *Ida and Lin* (2005) show that planet formation around small-mass primaries tends to form planets with lower masses in the Uranus/Neptune domain. A similar result that favors lower-mass planets is also observed for solar-type stars in the case of the low metallicity of the protostellar nebula (*Ida and Lin*, 2004b; *Benz et al.*, 2005). Future developments in the planet-formation models and new detections of very-low-mass planets will help to better understand these two converging effects.

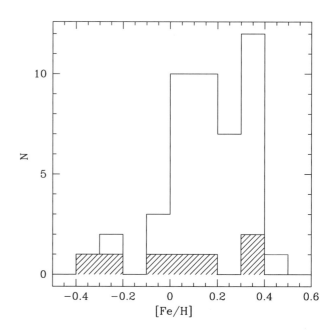

Fig. 12. Metallicity distribution of the sample of extrasolar planet hosts for planets with shorter periods than 20 d. Stars with Neptune-mass planets are indicated by the shaded portion of the histogram.

5.3. Primary Mass Effect

The mass of the primary star also appears to be an important parameter for planet-formation processes. In the case of low-mass stars, results from ongoing surveys indicate that giant gaseous planets are rare around M dwarfs in comparison to FGK primaries. The only known system with two giant planets is Gl 876 (Table 2). In particular, no hot Jupiter has been detected close to an M dwarf. This result, however, still suffers from small number statistics. On the other hand, as seen above, three of the five planets found to orbit an M dwarf have masses below $21\ M_\oplus$ and are probably "solid" planets. Thus, the occurrence rate for planets around M dwarfs appears to be directly dependent on the domain of planet masses considered.

For more massive primaries, new surveys targeting earlier, rotating A–F dwarfs (*Galland et al.*, 2005a,b) and programs surveying G–K giant stars (*Setiawan et al.*, 2005; *Sato et al.*, 2004; *Hatzes et al.*, 2005) are starting to provide interesting candidates. The detected planets are generally massive (>5 M_{Jup}), but it is still too early to reach a conclusion about a "primary mass" effect as those programs are still strongly observationally biased (larger-mass primaries and short time baseline for the surveys).

6. FOLLOWUP OF TRANSITING PLANETS

In recent years, groundbased transit searches have produced a number of planetary transiting candidates (see the chapter by Charbonneau et al.). The most successful of these searches to date has been the OGLE survey, which announced close to 180 possible transiting planets (*Udalski et al.*, 2002a,b). These new detections stimulated intensive followup observations to detect the radial-velocity signatures induced by the orbiting body. Surprisingly, these studies revealed that most of the systems were rather eclipsing binaries of small stars (M dwarfs) in front of F–G dwarfs, eclipsing binaries in blended multiple stellar systems (triple, quadruple), or grazing stellar eclipses, all mimicking photometric planetary transits (*Bouchy et al.*, 2005c; *Pont et al.*, 2005). The spectroscopic followup demonstrated the difficulty of the interpretation of shallow transit lightcurves without complementary radial-velocity measurements. The magnitude of the OGLE planetary candidates ranges from V ~ 16 to 17.5; close to the faint capability of an accurate fiber-fed spectrograph like FLAMES on the VLT and probably beyond the capability of slit spectroscopy with iodine self-calibration. This implies that a deeper photometric transit survey would face serious difficulties in confirming the planetary nature of the transiting object by Doppler followup.

To date, six planets have been detected from transit surveys and confirmed by radial velocities. Five of these have been found by the OGLE survey (*Udalski et al.*, 2002a,b) and one by the TrES network (*Alonso et al.*, 2004). Three of the OGLE planets have periods smaller than 2 d (very hot Jupiters). Such short periods, although easy to detect, are not found in the radial-velocity surveys, suggesting that those objects are about 10× less numerous than hot Jupiters [2.5 ≤ P ≤ 10 d (*Gaudi et al.*, 2005)]. In addition to the photometrically detected planets, three planets identified by radial-velocity measurements have been found transiting in front of their parent stars.

When transit photometry is combined with high-precision radial-velocity measurements, it is possible to derive an accurate mass and radius (Table 3) as well as the mean planet density. These important values constrain planetary interior models as well as the planet-evolution history. It is interesting to note here that the majority of planets for which we know both mass and radius have been found by transit survey despite the fact that more than 200 planets have been identified by radial-velocity searches. This is a consequence

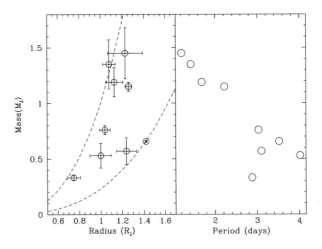

Fig. 13. Mass-radius and mass-period diagrams of transiting planets with radius and accurate mass estimates. On the left panel, the dashed lines indicate isodensity contours of 0.3 and 1.3 g cm^{-3}.

of the low probability of finding a transiting configuration among the planets found by radial-velocity surveys, while most of all the transiting candidates found so far can be followed up with radial-velocity measurements. On the other hand, the three planets transiting the brightest stars have been found first by radial velocities as transit surveys are mainly targeting crowded fields with fainter stars.

The derived density of transiting extrasolar planets covers a surprisingly wide range of values from 0.3 to 1.3 g cm^{-3} (Fig. 13). The "problem" of the anomalously large radius and low density of HD 209458 b is clearly not shared by all very close planets since planets with a similar mass are found to have a different density. This demonstrates a surprising diversity and reveals our lack of detailed understanding of the physics of irradiated giant planets.

The distribution of planets in a period vs. mass diagram shows an intriguing correlation (Fig. 13). Transiting planets seem to lie on a well-defined line of mass decreasing with increasing orbital period. This puzzling observation, pointed out by *Mazeh et al.* (2005), could be the consequence of mechanisms such as thermal evaporation (*Lecavelier et al.*, 2004; *Baraffe et al.*, 2004, 2005) or Roche limit mass transfer (*Ford and Rasio*, 2005). It is worth noting the location of HD 149026 b, below the relation, that could be a result of its different structure with a large core (*Sato et al.*, 2005; *Charbonneau et al.*, 2006). Even more surprising in the diagram is the complete lack of candidates above the relation. Why are we missing more massive transiting planets at P = 3–4 d? No convincing explanation has been proposed yet for this puzzling observation.

7. THE FUTURE OF RADIAL VELOCITIES

An important lesson from the past few years is that the radial-velocity technique has not reached its "limits" yet in the domain of exoplanets. In fact, the future of radial velocity studies is still bright.

TABLE 3.   List of planets with both radius (from transit) and mass estimate (from accurate radial velocities).

Object	Period (d)	Mass (M_{Jup})	Radius (R_{Jup})
OGLE-TR-10 b	3.101	0.63 ± 0.14	1.31 ± 0.09
OGLE-TR-56 b	1.212	1.24 ± 0.13	1.25 ± 0.08
OGLE-TR-111 b	4.016	0.52 ± 0.13	0.97 ± 0.06
OGLE-TR-113 b	1.432	1.35 ± 0.22	1.08 ± 0.06
OGLE-TR-132 b	1.690	1.19 ± 0.13	1.13 ± 0.08
TrES-1	3.030	0.73 ± 0.04	1.08 ± 0.05
HD 209458 b	3.525	0.66 ± 0.01	1.355 ± 0.005
HD 189733 b	2.219	1.15 ± 0.04	1.26 ± 0.08
HD 149026 b	2.876	0.33 ± 0.02	0.73 ± 0.06

Data from *Alonso et al.* (2004); *Moutou et al.* (2004); *Pont et al.* (2004, 2005); *Bouchy et al.* (2005b,c); *Winn et al.* (2005).

1. Recent discoveries indicate that a population of Neptune- and Saturn-mass planets remains to be discovered below 1 AU. The improved precision of the radial velocity surveys will address this issue in the near future, thereby providing us with useful new constraints on planet-formation theories. With the precision level now achieved for radial-velocity measurements, a new field in the search for extrasolar planets is at hand, allowing the detection of companions of a few Earth masses around solar-type stars. Very-low-mass planets (<10 $M_\oplus$) might be more frequent than the previously found giant worlds.

2. As described above, radial-velocity followup measurements are mandatory to determine the mass of transiting companions and then to calculate their mean densities. These observations establish the planetary nature of the companions and provide important parameters to constrain planetary atmosphere and interior models. This is important in view of the expected results of the space missions COROT and Kepler, which should provide hundreds of transiting planets of various sizes and masses. When a transit signal is detected, the orbital period is then known. As a result, radial-velocity followup is less demanding, both in terms of the number and precision of the acquired Doppler measurements. For example, a 2 $M_\oplus$ planet on a 4-d orbit induces on a Sun-mass star a radial-velocity amplitude of about 80 cm s^{-1} that will be possible to detect with only a "few" high-precision radial-velocity measurements, provided that the period of the system is known in advance. In this context, the most exciting aspect is the opportunity to explore the mass-radius relation down to the Earth-mass domain.

3. The threshold of the lowest-mass planet detectable by the Doppler technique keeps decreasing. The domain below the 1 m s^{-1} level has not yet been explored. Results obtained with the HARPS spectrograph show that, even if stars are intrinsically variable in radial velocity (at modest levels) due to acoustic modes, it is nevertheless possible on the short term to reach precisions well below 1 m s^{-1} by applying an adequate observational strategy. One open issue, however, remains unsolved: the behavior of the stars on a longer timescale, where stellar jitter and spots may impact the final achievable accuracy. In this case, an accurate preselection of the stars is needed to select good candidates and optimize the use of telescope time. In addition, line bisector analysis and followup of activity indicators such as log(R'_{HK}), as well as photometric measurements, may flag suspect results.

The discovery of an extrasolar planet by means of the Doppler technique requires either that the radial-velocity signal induced by the planet is significantly higher than the dispersion, or that very-high-cadence observations are obtained. A large number of observations with excellent phase coverage is critical for ruling out false positives, particularly given the relatively high number of free parameters in the orbital solution for multiplanet systems. A large number of measurements will help to mitigate the challenges of low-amplitude detections, but will demand an enormous investment of observing time. Thus, as long as we are willing to devote sufficient resources in terms of telescope time

and advance designed spectrographs (high-level temperature and pressure control), it should in principle be possible to detect Earth-like planets (*Pepe et al.*, 2005).

Note added in proof: During the production of this volume, a system with three Neptune-mass planets has been discovered around HD 69830 (*Lovis et al.*, 2006), as well as four transiting planets (*McCullough et al.*, 2006; *Bakos et al.*, 2006; *Cameron et al.*, 2006), significantly enlarging two small-statistic subsamples of meaningful exoplanets.

REFERENCES

Alibert Y., Mordasini C., and Benz W. (2004) *Astron. Astrophys., 417,* L25–L28.

Alibert Y., Mordasini C., Benz W., and Winisdoerffer C. (2005) *Astron. Astrophys., 434,* 343–353.

Alonso R., Brown T., Torres G., Latham D., Sozzetti A., et al. (2004) *Astrophys. J., 613,* L153–L156.

Bakos G. A., et al. (2006) *Astrophys. J.,* in press.

Baraffe I., Selsis F., Chabrier G., Barman T., Allard F., Hauschildt P. H., and Lammer H. (2004) *Astron. Astrophys., 419,* L13–L16.

Baraffe I., Chabrier G., Barman T., Selsis F., Allard F., and Hauschildt P. H. (2005) *Astron. Astrophys., 436,* L47–L51.

Benz W., Mordasini C., Alibert Y., and Naef D. (2005) In *Tenth Anniversary of 51 Peg b: Status of and Prospects for Hot Jupiter Studies* (L. Arnold et al., eds.), pp. 24–34. Frontier Group, Paris.

Bodenheimer P., Hubickyj O., and Lissauer J. (2000) *Icarus, 143,* 2–14.

Bonfils X., Forveille T., Delfosse X., Udry S., Mayor M., et al. (2005a) *Astron. Astrophys., 443,* L15–L18.

Bonfils X., Delfosse X., Udry S., Santos N. C., Forveille T., and Ségransan D. (2005b) *Astron. Astrophys., 442,* 635–642.

Bouchy F., Bazot M., Santos N. C., Vauclair S., and Sosnowska D. (2005a) *Astron. Astrophys., 440,* 609–614.

Bouchy F., Udry S., Mayor M., Moutou C., Pont F., Iribarne N., et al. (2005b) *Astron. Astrophys., 444,* L15–L19.

Bouchy F., Pont F., Melo C., Santos N. C., Mayor M., Queloz D., and Udry S. (2005c) *Astron. Astrophys., 431,* 1105–1121.

Butler P., Marcy G., Vogt S., and Fischer D. (2000) In *Planetary Systems in the Universe* (A. Penny et al., eds), p. 1. IAU Symposium 202, ASP, San Francisco.

Butler P., Vogt S., Marcy G., Fischer D., Wright J., et al. (2004) *Astrophys. J., 617,* 580–588.

Butler P., Wright J., Marcy G., Fischer D., Vogt S., et al. (2006) *Astrophys. J., 646,* 505–522.

Cameron A. C., et al. (2006) *Mon. Not. R. Astron. Soc.,* in press.

Charbonneau D., Winn J., Latham D., Bakos G., Falco E., et al. (2006) *Astrophys. J., 636,* 445–452.

Correia A., Udry S., Mayor M., Laskar J., Naef D., et al. (2005) *Astron. Astrophys., 440,* 751–758.

Cumming A., Marcy G., and Butler P. (1999) *Astrophys. J., 526,* 890–915.

Da Silva R., Udry S., Bouchy F., Mayor M., Moutou C., et al. (2006) *Astron. Astrophys. 446,* 717–722.

Eggenberger A., Udry S., and Mayor M. (2004) *Astron. Astrophys., 417,* 353–360.

Endl M., Kürster M., Els S., Hatzes A., Cochran W., et al. (2002) *Astron. Astrophys., 392,* 671–690.

Fischer D. and Valenti J. (2005) *Astrophys. J., 622,* 1102–1117.

Fischer D., Valenti J., and Marcy G. (2004) In *Stars as Suns: Activity, Evolution and Planets* (A. Dupree and A. Benz, eds.), p. 29. IAU Symposium 219, ASP, San Francisco.

Fischer D., Laughlin G., Butler P., Marcy G., Johnson J., et al. (2005) *Astrophys. J., 620,* 481–486.

Ford E. and Rasio F. (2005) In *PPV Poster Proceedings,* www.lpi.usra.edu/meetings/ppv2005/pdf/8360.pdf.

Ford E., Havlickova M., and Rasio F. (2001) *Icarus, 150,* 303–313.

Ford E., Rasio F., and Yu K. (2003) In *Scientific Frontiers in Research on Extrasolar Planets* (D. Deming and S. Seager, eds.), pp. 181–188. ASP Conf. Series 294, San Francisco.

Fuhrmann K., Pfeiffer M., and Bernkopf J. (1997) *Astron. Astrophys., 326,* 1081–1089.

Fuhrmann K., Pfeiffer M., and Bernkopf J. (1998) *Astron. Astrophys., 336,* 942–952.

Galland F., Lagrange A.-M., Udry S., Chelli A., Pepe F., et al. (2005a) *Astron. Astrophys., 443,* 337–345.

Galland F., Lagrange A.-M., Udry S., Chelli A., Pepe F., et al. (2005b) *Astron. Astrophys., 444,* L21–L24.

Gaudi B. S., Seager S., and Mallen-Ornelas G. (2005) *Astrophys. J., 623,* 472–481.

Gonzalez G. (1997) *Mon. Not. R. Astron. Soc., 285,* 403–412.

Gonzalez G. (1998) *Astron. Astrophys., 334,* 221–238.

Gonzalez G. and Laws C. (2000) *Astron. J., 119,* 390–396.

Gonzalez G., Wallerstein G., and Saar S. (1999) *Astrophys. J., 511,* L111–L114.

Halbwachs J.-L., Mayor M., Udry S., and Arenou F. (2003) *Astron. Astrophys., 397,* 159–175.

Halbwachs J.-L., Mayor M., and Udry S. (2005) *Astron. Astrophys., 431,* 1129–1137.

Hatzes A. P., Guenther E., Endl M., Cochran W., Döllinger M., and Bedalov A. (2005) *Astron. Astrophys., 437,* 743–751.

Ida S. and Lin D. (2004a) *Astrophys. J., 604,* 388–413.

Ida S. and Lin D. (2004b) *Astrophys. J., 616,* 567–672.

Ida S. and Lin D. (2005) *Astrophys. J., 626,* 1045–1060.

Jorissen A., Mayor M., and Udry S. (2001) *Astron. Astrophys., 379,* 992–998.

Kley W., Peitz J., and Bryden G. (2004) *Astron. Astrophys., 414,* 735–747.

Laughlin G., Butler P., Fischer D., Marcy G., Vogt S., and Wolf A. (2005) *Astrophys. J., 612,* 1072–1078.

Lecavelier des Etangs A., Vidal-Madjar A., McConnell J. C., and Hebrard G. (2004) *Astron. Astrophys., 418,* L1–L4.

Levison H. F., Lissauer J., and Duncan M. J. (1998) *Astron. J., 116,* 1998–2014.

Lin D. and Ida S. (1997) *Astrophys. J., 477,* 781–791.

Lin D., Bodenheimer P., and Richardson D. C. (1996) *Nature, 380,* 606–607.

Livio M. and Pringle J. E. (2003) *Mon. Not. R. Astron. Soc., 346,* L42–L44.

Lovis C., Mayor M., Bouchy F., Pepe F., Queloz D., et al. (2005) *Astron. Astrophys., 437,* 1121–1126.

Lovis C., Mayor M., Pepe F., Alibert Y., Benz W., et al. (2006) *Nature, 441,* 305–309.

Marcy G., Cochran W., and Mayor M. (2000) In *Protostars and Planets IV* (V. Mannings et al., eds.), pp. 1285–1311. Univ. of Arizona, Tucson.

Marcy G., Butler P., Vogt S., Liu M., Laughlin G., et al. (2001) *Astrophys. J., 555,* 418–425.

Marcy G., Butler P., Fischer D., Vogt S., Wright J., et al. (2005) *Prog. Theor. Phys. Suppl., 158,* 24–42.

Marzari F. and Weidenschilling S. J. (2002) *Icarus, 156,* 570–579.

Mayor M. and Queloz D. (1995) *Nature, 378,* 355–358.

Mayor M., Pepe F., Queloz D., Bouchy F., Rupprecht G., et al. (2003) *The Messenger, 114,* 20–24.

Mayor M., Udry S., Naef D., Pepe F., Queloz D., et al. (2004) *Astron. Astrophys., 415,* 391–402.

Mazeh T., Krymolowski Y., and Rosenfeld G. (1997) *Astrophys. J., 477,* L103–L106.

Mazeh T., Zucker S., and Pont F. (2005) *Mon. Not. R. Astron. Soc., 356,* 955–957.

McArthur B., Endl M., Cochran W., Benedict G. F., Fischer D., et al. (2004) *Astrophys. J., 614,* L81–L84.

McCullough P. R., Stys J. E., Valenti J. A., Johns-Krull C. M., Janes K. A., et al. (2006) *Astrophys. J., 648,* 1228–1238.

Moutou C., Pont F., Bouchy F., and Mayor M. (2004) *Astron. Astrophys., 424,* L31–L34.

Mugrauer M., Neuhäuser R., Mazeh T., Alves J., and Guenther E. (2004) *Astron. Astrophys., 425,* 249–253.

Mugrauer M., Neuhäuser R., Seifahrt A., Mazeh T., and Guenther E. (2005) *Astron. Astrophys., 440,* 1051–1060.

Naef D., Mayor D., Beuzit J.-L., Perrier C., Queloz D., Sivan J.-P., and Udry S. (2005) In *13th Cool Stars, Stellar Systems and the Sun* (F.

Favata et al., eds.), pp. 833–836. ESA SP-560, Noordwijk, The Netherlands.

Nelson R., Papaloizou J., Masset F., and Kley W. (2000) *Mon. Not. R. Astron. Soc., 318,* 18–36.

Patience J., White R. J., Ghez A., McCabe C., McLean I., et al. (2002) *Astrophys. J., 581,* 654–665.

Pätzold M. and Rauer H. (2002) *Astrophys. J., 568,* L117–L120.

Pepe F., Mayor M., Queloz D., Benz W., Bertaux J.-L., et al. (2005) *The Messenger, 120,* 22–25.

Pollack J. B., Hubickyj O., Bodenheimer P., Lissauer J., Podolak M., and Greenzweig Y. (1996) *Icarus, 124,* 62–85.

Pont F., Bouchy F., Queloz D., Santos N. C., Melo C., Mayor M., and Udry S. (2004) *Astron. Astrophys., 426,* L15–L18.

Pont F., Bouchy F., Melo C., Santos N. C., Mayor M., Queloz D., and Udry S. (2005) *Astron. Astrophys., 438,* 1123–1140.

Rasio F. and Ford E. (1996) *Science, 274,* 954–956.

Rivera E. J., Lissauer J., Butler P., Marcy G., Vogt S., et al. (2005) *Astrophys. J., 634,* 625–640.

Safronov V. S. (1969) *Evoliutsiia Doplanetnogo Oblaka,* Izdatel'stvo Nauka, Moscow.

Santos N. C., Israelian G., and Mayor M. (2000) *Astron. Astrophys., 363,* 228–238.

Santos N. C., Israelian G., and Mayor M. (2001) *Astron. Astrophys., 373,* 1019–1031.

Santos N. C., Israelian G., Mayor M., Rebolo R., and Udry S. (2003) *Astron. Astrophys., 398,* 363–376.

Santos N. C., Bouchy F., Mayor M., Pepe F., Queloz D., et al. (2004a) *Astron. Astrophys., 426,* L19–L23.

Santos N. C., Israelian G., and Mayor M. (2004b) *Astron. Astrophys., 415,* 1153–1166.

Santos N. C., Israelian G., Mayor M., Bento J., Almeida P., et al. (2005) *Astron. Astrophys., 437,* 1127–1133.

Sato B., Ando H., Kambe E., Takeda Y., Izumiura H., et al. (2004) *Astrophys. J., 597,* L157–L160.

Sato B., Fischer D., Henry G., Laughlin G., Butler P., et al. (2005) *Astrophys. J., 633,* 465–473.

Setiawan J., Rodmann J., da Silva L., Hatzes A., Pasquini L., et al. (2005) *Astron. Astrophys., 437,* L31–L34.

Takeda G. and Rasio A. (2005) *Astrophys. J., 627,* 1001–1010.

Tremaine S. and Zakamska N. (2004) In *The Search for Other Worlds* (S. S. Holt and D. Deming, eds.), pp. 257–260. AIP Conf. Proc. 713, New York.

Trilling D. E., Benz W., Guillot T., Lunine J., Hubbard W., Burrows A., et al. (1998) *Astrophys. J., 500,* 428–439.

Trilling D. E., Lunine J., and Benz W. (2002) *Astron. Astrophys., 394,* 241–251.

Udalski A., Paczynski B., Zebrun K., Szymaski M., Kubiak M., et al. (2002a) *Acta Astron., 52,* 1–37.

Udalski A., Zebrun K., Szymanski M., Kubiak M., Soszynski I., et al. (2002b) *Acta Astron., 52,* 115–128.

Udry S., Mayor M., Naef D., Pepe F., Queloz D., et al. (2000) *Astron. Astrophys., 356,* 590–598.

Udry S., Mayor M., Naef D., Pepe F., Queloz D., Santos N. C., and Burnet M. (2002) *Astron. Astrophys., 390,* 267–279.

Udry S., Mayor M., and Santos N. C. (2003) *Astron. Astrophys., 407,* 369–376.

Udry S., Mayor M., Benz W., Bertaux J.-L., Bouchy F., et al. (2006) *Astron. Astrophys., 447,* 361–367.

Vauclair S. (2004) *Astrophys. J., 605,* 874–879.

Vogt S., Butler P., Marcy G., Fischer D., Henry G., et al. (2005) *Astrophys. J., 632,* 638–658.

Ward W. (1997) *Icarus, 126,* 261–281.

Weidenschilling S. J. and Marzari F. (1996) *Nature, 384,* 619–621.

Winn J., Noyes R. W., Holman M., Charbonneau D., Ohta Y., et al. (2005) *Astrophys. J., 631,* 1215–1226.

Wuchterl G., Guillot T., and Lissauer J. (2000) In *Protostars and Planets IV* (V. Mannings et al., eds.), pp. 1081–1109. Univ. of Arizona, Tucson.

Zakamska N. and Tremaine S. (2004) *Astron. J., 128,* 869–877.

Zucker S. and Mazeh T. (2002) *Astrophys. J., 568,* L113–L116.

When Extrasolar Planets Transit Their Parent Stars

David Charbonneau
Harvard-Smithsonian Center for Astrophysics

Timothy M. Brown
High Altitude Observatory

Adam Burrows
University of Arizona

Greg Laughlin
University of California, Santa Cruz

When extrasolar planets are observed to transit their parent stars, we are granted unprecedented access to their physical properties. It is only for transiting planets that we are permitted direct estimates of the planetary masses and radii, which provide the fundamental constraints on models of their physical structure. In particular, precise determination of the radius may indicate the presence (or absence) of a core of solid material, which in turn would speak to the canonical formation model of gas accretion onto a core of ice and rock embedded in a protoplanetary disk. Furthermore, the radii of planets in close proximity to their stars are affected by tidal effects and the intense stellar radiation. As a result, some of these "hot Jupiters" are significantly larger than Jupiter in radius. Precision follow-up studies of such objects (notably with the spacebased platforms of the Hubble and Spitzer Space Telescopes) have enabled direct observation of their transmission spectra and emitted radiation. These data provide the first observational constraints on atmospheric models of these extrasolar gas giants, and permit a direct comparison with the gas giants of the solar system. Despite significant observational challenges, numerous transit surveys and quick-look radial velocity surveys are active, and promise to deliver an ever-increasing number of these precious objects. The detection of transits of short-period Neptune-sized objects, whose existence was recently uncovered by the radial-velocity surveys, is eagerly anticipated. Ultraprecise photometry enabled by upcoming space missions offers the prospect of the first detection of an extrasolar Earth-like planet in the habitable zone of its parent star, just in time for Protostars and Planets VI.

1. OVERVIEW

The month of October 2005, in which the Protostars and Planets V conference was held, marked two important events in the brief history of the observational study of planets orbiting nearby, Sun-like stars. First, it was the 10-year anniversary of the discovery of 51 Pegb (*Mayor and Queloz,* 1995), whose small orbital separation implied that similar hot Jupiters could be found in orbits nearly co-planar to our line of sight, resulting in mutual eclipses of the planet and star. Second, October 2005 heralded the discovery of the ninth such transiting planet (*Bouchy et al.,* 2005a). This select group of extrasolar planets has enormous influence on our overall understanding of these objects: The nine transiting planets are the only ones for which we have accurate estimates of key physical parameters such as mass, radius, and, by inference, composition. Furthermore, precise

monitoring of these systems during primary and secondary eclipse has permitted the direct study of their atmospheres. As a result, transiting planets are the only ones whose physical structure and atmospheres may be compared in detail to the planets of the solar system, and indeed October 2005 was notable for being the month in which the number of objects in the former category surpassed the latter.

Our review of this rapidly evolving field of study proceeds as follows. In section 2, we consider the physical structure of these objects, beginning with a summary of the observations (section 2.1) before turning to their impact on our theoretical understanding (section 2.2). In section 3, we consider the atmospheres of these planets, by first summarizing the challenges to modeling such systems (section 3.1), and subsequently reviewing the detections and upper limits, and the inferences they permit (section 3.2). We end by considering the future prospects (section 4) for learning

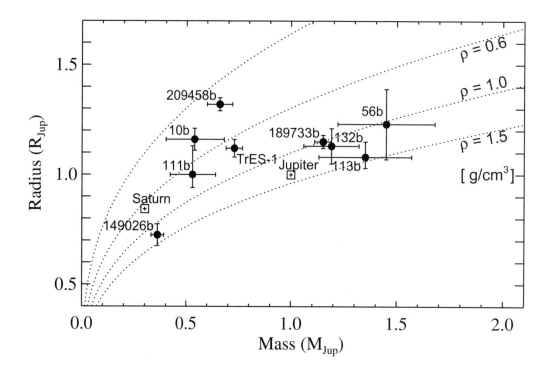

Fig. 1. Masses and radii for the nine transiting planets, as well as Jupiter and Saturn. The data are tabulated in Table 1, and are gathered from *Bakos et al.* (2006), *Bouchy et al.* (2004, 2005b), Brown et al. (in preparation), *Charbonneau et al.* (2006), Holman et al. (in preparation), Knutson et al. (in preparation), *Laughlin et al.* (2005a), *Moutou et al.* (2004), *Pont et al.* (2004), *Sato et al.* (2005), *Sozzetti et al.* (2004), *Torres et al.* (2004a), and *Winn et al.* (2005).

about rocky planets beyond the solar system through the detection and characterization of such objects in transiting configurations.

2. PHYSICAL STRUCTURE

2.1. Observations

2.1.1. Introduction. When a planet transits, we can accurately measure the orbital inclination, i, allowing us to evaluate the planetary mass M_{pl} directly from the minimum mass value $M_{pl} \sin i$ determined from radial-velocity observations and an estimate of the stellar mass, $M_\star$. The planetary radius, R_{pl}, can be obtained by measuring the fraction of the parent star's light that is occulted, provided a reasonable estimate of the stellar radius, $R_\star$, is available. With the mass and radius in hand, we can estimate such critically interesting quantities as the average density and surface gravity. Hence, the information gleaned from the transiting planets allows us to attempt to unravel the structure and composition of the larger class of extrasolar planets, to understand formation and evolution processes (including orbital evolution), and to elucidate physical processes that may be important in planetary systems generically. Figure 1 shows the mass-radius relation for the nine known transiting planets, with Jupiter and Saturn added for comparison. It is fortunate that the present small sample of objects spans a moderate range in mass and radius, and appears to contain

both a preponderance of planets whose structure is fairly well described by theory, as well as a few oddities that challenge our present knowledge.

We begin by describing how the objects shown in Fig. 1 were identified and characterized, and, along the way, we illuminate the limitations that these methods imply for our efforts to understand extrasolar planets as a class. By definition, transiting planets have their orbits oriented so that Earth lies nearly in their orbital plane. This is an uncommon occurrence; assuming random orientation of planetary orbits, the probability that a planet with orbital eccentricity, e, and longitude of periastron, ϖ, produces transits visible from Earth is given by

$$F_{tr} = 0.0045 \left(\frac{1 \text{ AU}}{a} \right) \left(\frac{R_\star + R_{pl}}{R_\odot} \right) \left[\frac{1 + e \cos\left(\frac{\pi}{2} - \varpi\right)}{1 - e^2} \right]$$

which is inversely proportion to a, the orbital semimajor axis. All known transiting planets have orbital eccentricities consistent with zero, for which the last factor in the above equation reduces to unity.

The radii of jovian planets are typically only about 10% of the stellar radii. The transits known to date result in a 0.3–3% diminution of the stellar flux reaching Earth. These transits last for 1.5–3.5 h, and accurate groundbased characterizations of these events are challenging. The paucity and subtlety of the transits make it necessary to use great care to reduce the random errors and systematic biases that plague the estimation of the planets' fundamental properties (section 2.1.4).

2.1.2. Methods of detection. The presently known transiting planets have all been detected by one of the two following means, both foreseen by *Struve* (1952): (1) photometric detection of transit-like events, with subsequent confirmation of planetary status via radial-velocity measurements; and (2) radial-velocity detection of a planet with subsequent measurement of photometric transits. Radial-velocity detection has the advantage that the planetary nature of the target object is generally unambiguous. Its disadvantage is that it requires substantial observing time on large telescopes to identify each planetary system, and only then can the relatively cheap process of searching for photometric transits begin. Direct photometric transit searches simultaneously monitor large numbers of stars in a given field of view, but suffer from a very high rate of astrophysical false positives (section 2.1.3).

Successful photometric transit searches have so far adopted one of two basic strategies, using either moderate-sized or very small telescopes to search either fainter or brighter stars. Five transiting planets (OGLE-TR-10b, 56b, 111b, 113b, and 132b) have been detected by the Optical Gravitational Lensing Experiment (OGLE) survey (*Udalski et al.*, 2002a,b,c, 2003, 2004), which uses a 1.3-m telescope. The parent stars of these planets are faint (typically V = 16.5). The large-telescope follow-up observations needed to verify their planetary status, to measure the stellar reflex velocities, and to estimate the planetary masses and radii have been conducted by several groups (*Bouchy et al.*, 2004, 2005b; *Dreizler et al.*, 2002; *Konacki et al.*, 2003a,b, 2004, 2005; *Moutou et al.*, 2004; *Pont et al.*, 2004; *Torres et al.*, 2004a,b, 2005).

The Trans-Atlantic Exoplanet Survey (TrES) employed a network of three automated small-aperture (10-cm), wide-field (6° × 6°) telescopes (*Brown and Charbonneau*, 2000; *Dunham et al.*, 2004; *Alonso*, 2005) to detect the planet TrES-1 (*Alonso et al.*, 2004; *Sozzetti et al.*, 2004). Its parent star (V = 11.8) is significantly brighter than the OGLE systems, but fainter than the transiting-planet systems detected by radial-velocity surveys (below). Because of this relative accessibility, TrES-1 has also been the subject of intensive follow-up observations, as detailed later in this review.

Numerous other photometric transit surveys are active at the current time. The BEST (*Rauer et al.*, 2004), HAT (*Bakos et al.*, 2004), KELT (*Pepper et al.*, 2004), Super-WASP (*Christian et al.*, 2004), Vulcan (*Borucki et al.*, 2001), and XO (*McCullough et al.*, 2005) surveys, and the proposed PASS (*Deeg et al.*, 2004) survey, all adopt the small-aperture, wide-field approach, whereas the EXPLORE (*Mallen-Ornelas et al.*, 2003) project employs larger telescopes to examine fainter stars. The benefits of surveying stellar clusters (*Janes*, 1996; *Pepper and Gaudi*, 2005) have motivated several surveys of such systems, including EXPLORE/OC (*von Braun et al.*, 2005), PISCES (*Mochejska et al.*, 2005,

2006), and STEPSS (*Burke et al.*, 2004; *Marshall et al.*, 2005). An early, stunning null result was the Hubble Space Telescope (HST) survey of 34,000 stars in the globular cluster 47 Tuc, which points to the interdependence of the formation and migration of hot Jupiters on the local conditions, namely crowding, metallicity, and initial proximity to O and B stars (*Gilliland et al.*, 2000).

Finally, three transiting planets were first discovered by radial-velocity surveys. These include HD 209458b, the first transiting planet discovered (*Charbonneau et al.*, 2000; *Henry et al.*, 2000; *Mazeh et al.*, 2000), and the two most recently discovered transiting planets, HD 149026b (*Sato et al.*, 2005) and HD 189733b (*Bouchy et al.*, 2005a). The latter two objects were uncovered by quick-look radial velocity surveys targeted at identifying short-period planets of metal-rich stars [respectively, the N2K Survey (*Fischer et al.*, 2005) and the Elodie Metallicity-Biased Search (*da Silva et al.*, 2006)]. Given the preference of radial-velocity surveys for bright stars, it is not surprising that all three systems are bright (7.6 < V < 8.2), making them natural targets for detailed follow-up observations. As we shall see below, HD 209458b has been extensively studied in this fashion. Similar attention has not yet been lavished on the other two, but only because of their very recent discovery.

2.1.3. Biases and false alarms. Photometric transit surveys increase their odds of success by simultaneously observing as many stars as possible. Hence, their target star fields are moderately to extremely crowded, and the surveys must therefore work near the boundary of technical feasibility. The constraints imposed by the search method influence which kinds of planets are detected.

Photometric transit searches are strongly biased in favor of planets in small orbits, since such objects have a greater probability of presenting an eclipsing configuration (section 2.1.1). Moreover, most transit searches require a minimum of two (and usually three) distinct eclipses to be observed, both to confirm the reality of the signal, and to permit an evaluation of the orbital period, P. Since larger orbits imply longer orbital periods and fewer chances for transits to occur, small orbits are preferred for transit surveys with only a limited baseline. This is frequently the regime in which single-site surveys operate. However, multisite surveys that monitor a given field for several months (e.g., HAT, TrES) frequently achieve a visibility (the fraction of systems of a given period for which the desired number of eclipse events would be observed) nearing 100% for periods up to 6 d. As a result, such surveys do not suffer this particular bias, although admittedly only over a limited range of periods. Similarly, a stroboscopic effect can afflict single-site surveys, favoring orbital periods near integer numbers of days and may account for the tendency of the longer-period transiting planet periods to clump near 3 and 3.5 d (*Pont et al.*, 2004, *Gaudi et al.*, 2005). This situation occurs if the campaign is significantly shorter in duration than that required to achieve complete visibility across the desired range or orbital periods. However, for observing campaigns for which more than adequate phase coverage has been obtained, the opposite is true, and periods near integer and half-integer values are disfavored. The limiting example of this situation would be a single-site campaign consisting of thousands of hours of observations, which nonetheless would be insensitive to systems with integer periods, if their eclipses always occur when the field is below the horizon.

Most field surveys operate in a regime limited by the signal-to-noise of their time series [which are typically searched by an algorithm than looks for statistically significant, transit-like events (e.g., *Kovács et al.,* 2002)], and for which the number of stars increases with decreasing flux (a volume effect). An important detection bias for surveys operating under such conditions has been discussed by *Pepper et al.* (2003) and described in detail by *Gaudi et al.* (2005), *Gaudi* (2005), and *Pont et al.* (2005). These surveys can more readily detect planets with shorter periods and larger radii orbiting fainter stars, and since such stars correspond to a large distance (hence volume), they are much more numerous. As a result, any such survey will reflect this bias, which cannot be corrected merely by improving the cadence, baseline, or precision of the time series (although improving the latter will reduce the threshold of the smallest planets that may be detected).

Most ongoing transit surveys are plagued by a high rate of candidate systems displaying light curves that precisely mimic the desired signal, yet are not due to planetary transits. We can divide such false positives into three broad categories: Some are true *statistical* false positives, resulting from selecting an overly permissive detection threshold whereby the light-curve search algorithm flags events that result purely from photometric noise outliers (*Jenkins et al.,* 2002). The second source is *instrumental,* due to erroneous photometry, often resulting from leakage of signal between the photometric apertures of nearby stars in a crowded field. However, the dominant form, which we shall term *astrophysical* false positives, result from eclipses among members of double- or multiple-star systems. Grazing eclipses in binary systems can result in transit-like signals with depths and durations that resemble planetary ones (*Brown,* 2003), and this effect is especially pronounced for candidate transits having depths greater than 1%. (For equal-sized components, roughly 20% of eclipsing systems have eclipse depths that are less than 2% of the total light.) In these cases the eclipse shapes are dissimilar (grazing eclipses produce V-shapes, while planetary transits have flat bottoms), but in noisy data, this difference can be difficult to detect. A false alarm may also occur when a small star transits a large one (e.g., an M dwarf eclipsing a main-sequence F star). Since the lowest-mass stars have Jupiter radii, it is not surprising that such systems mimic the desired signal closely: They produce flat-bottomed transits with the correct depths and durations. Larger stars eclipsing even larger primaries can also mimic the desired signal, but a careful analysis of the transit shape can often reveal the true nature of the system (*Seager and Mallen-Ornelas,* 2003). Other useful diagnostics emerge from careful analysis of the light curve outside of eclipses. These can reveal weak secondary eclipses, periodic variations due to tidal dis-

tortion or gravity darkening of the brighter component, or significant color effects. Any of these variations provides evidence that the eclipsing object has a stellar mass as opposed to a planetary mass (*Drake,* 2003; *Sirko and Paczyński,* 2003; *Tingley,* 2004). In the absence of these diagnostics, the stellar nature of most companions is easily revealed by low-precision (1 km s^{-1}) radial velocity measurements, since even the lowest-mass stellar companions cause reflex orbital motions of tens of km s^{-1} (for examples, see *Latham,* 2003; *Charbonneau et al.,* 2004; *Bouchy et al.,* 2005b; *Pont et al.,* 2005).

The most troublesome systems are hierarchical triple stars in which the brightest star produces the bulk of the system's light, and the two fainter ones form an eclipsing binary. In such cases, the depths of the eclipses are diluted by light from the brightest member, and often radial-velocity observations detect only the bright component as well. Given neither radial-velocity nor photometric evidence for a binary star, such cases can easily be mistaken for transiting planets. Correct identification then hinges on more subtle characteristics of the spectrum or light curve, such as line profile shapes that vary with the orbital period (*Mandushev et al.,* 2005; *Torres et al.,* 2004b, 2005) or color dependence of the eclipse depth (*O'Donovan et al.,* 2006). Because of the large preponderance of false alarms over true planets, it is only after all the above tests have been passed that it makes sense to carry out the resource intensive high-precision radial-velocity observations that establish beyond question that the transiting object has a planetary mass.

2.1.4. Determining the radii and masses. After transiting planets are identified, an arsenal of observing tools is available (and necessary) for their characterization. An accurate estimate of M_{pl} requires precise radial-velocity measurements (from which the orbital elements P, e, and ϖ are also determined), as well as an estimate of $M_\star$. The former are gathered with high-dispersion echelle spectrographs fed by large telescopes. For bright parent stars, precision of a few m s^{-1} (compared to reflex orbital speeds of 50–200 m s^{-1}) can be obtained with convenient exposure times, so that uncertainties in the velocity measurements do not dominate the estimate of M_{pl}. In this regime, the greatest source of uncertainty is the value of $M_\star$ itself. Given the difficulty of estimating the ages of field stars, comparison with grids of stellar models (e.g., *Girardi et al.,* 2002) suggests that mass estimates are likely to be in error by as much as 5%. This uncertainty could be removed by measuring the orbital speed of the planet directly. Several efforts have sought to recover the reflected-light spectrum of the planet in a series of high-resolution stellar spectra spanning key phases of the orbital period, but have achieved only upper limits (*Charbonneau et al.,* 1999; *Collier Cameron et al.,* 2002; *Leigh et al.,* 2003a,b). (These results also serve to constrain the wavelength-dependent planetary albedo, a topic to which we shall return in section 3.2.2.) For faint parent stars, the radial-velocity estimates become more expensive and problematic, and contribute significantly to the final error budget for M_{pl}. Interestingly, the most intractable uncertainty concerning masses of nontransiting planets,

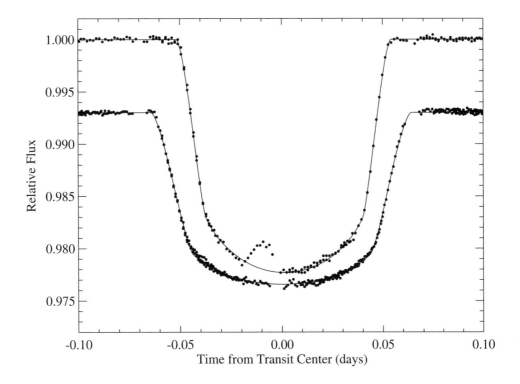

Fig. 2. HST photometric light curves of transits of TrES-1 (top) (Brown et al., in preparation) and HD 209458 (bottom) (*Brown et al.* 2001) (offset by −0.007 for clarity). The shorter orbital period and the smaller size of the TrES-1 star result in a transit that is shorter in duration than that of HD 209458. Similarly, the smaller star creates a deeper transit for TrES-1, despite the fact that HD 209458b is the larger planet; the planetary sizes also affect the duration of ingress and egress. The TrES-1 data reveal a "hump" centered at a time of −0.01 d. This is likely the result of the planet occulting a starspot (or complex of starspots) on the stellar surface.

namely the value of sin i, is exquisitely well determined by fits to the transit light curve.

Analysis of moderate-precision light curves (obtained with groundbased telescopes) nonetheless yield a tight constraint on the ratio $R_{pl}/R_\star$. However, fits to such data exhibit a fundamental degeneracy among the parameters R_{pl}, $R_\star$, and i, whereby the planet and stellar radii may be reduced in proportion so as to preserve the transit depth, and the orbital inclination may be correspondingly increased so as to conserve the chord length across the star. The uncertainty in R_{pl} is typically dominated by such degeneracies. Determining the value of R_{pl} requires fitting eclipse curves [facilitated by the analytic formulae of *Mandel and Agol* (2002)] subject to independent estimates of $M_\star$, $R_\star$, and the stellar limb-darkening coefficients. If sufficient photometric precision can be achieved, the value of $R_\star$ may be derived from the light curve itself. This results in a reduced uncertainty on the value of R_{pl}, due to its weaker dependence on $M_\star$, $(\Delta R_{pl}/R_{pl}) \simeq 0.3(\Delta M_\star/M_\star)$; see *Charbonneau* (2003). For illustrative examples of the degeneracies that result from such fits, see *Winn et al.* (2005), Holman et al. (in preparation), and *Charbonneau et al.* (2006).

HST has yielded spectacular transit light curves for two bright systems, HD 209458 (*Brown et al.,* 2001) and TrES-1 (Brown et al., in preparation), which are shown in Fig. 2. The typical precision of these light curves is 10^{-4} per 1-min integration, sufficient to extract new information from rela-

tively subtle properties of the light curve, such as the duration of the ingress and egress phases, and the curvature of the light curve near the transit center. In practice, such data have permitted a simultaneous fit that yields estimates of R_{pl}, $R_\star$, i, and the stellar limb-darkening coefficients, thus reducing the number of assumed parameters to one: $M_\star$. *Cody and Sasselov* (2002) point out that the combined constraint on $(M_\star, R_\star)$ is nearly orthogonal to that resulting from light-curve fitting, serving to reduce the uncertainty in R_{pl}. Further improvements can result from the simultaneously fitting of multicolor photometry under assumed values for the stellar-limb darkening, which serves to isolate the impact parameter (hence i) of the planet's path across the star and break the shared degeneracy among R_{pl}, $R_\star$, and i (*Jha et al.,* 2000; *Deeg et al.,* 2001). Recently, Knutson et al. (in preparation) have analyzed a spectrophotometric HST dataset spanning 290–1060 nm, and the combined effect of the constraints described above has been to permit the most precise determination of an exoplanet radius to date (HD 209458b; $R_{pl} = 1.320 \pm 0.025\ R_{Jup}$).

2.1.5. Further characterization measurements. High-resolution stellar spectra obtained during transits can be used to determine the degree of alignment of the planet's orbital angular momentum vector with the stellar spin axis. As the planet passes in front of the star, it produces a characteristic time-dependent shift of the photospheric line profiles that stems from occultation of part of the rotating stellar surface.

This phenomenon is known as the Rossiter-McLaughlin effect (*Rossiter,* 1924; *McLaughlin,* 1924), and has long been observed in the spectra of eclipsing binary stars. *Queloz et al.* (2000) and *Bundy and Marcy* (2000) detected this effect during transits of HD 209458. A full analytic treatment of the phenomenon in the context of transiting extrasolar planets has been given by *Ohta et al.* (2005). *Winn et al.* (2005) analyzed the extensive radial velocity dataset of HD 209458, including 19 measurements taken during transit. They found that the measurements of the radial velocity of HD 209458 during eclipse exhibit an effective half-amplitude of $\Delta v \simeq$ 55 m s^{-1}, indicating a line-of-sight rotation speed of the star of v sin i$_\star$ = 4.70 ± 0.16 km s^{-1}. They also detected a small asymmetry in the Rossiter-McLaughlin anomaly, which they modeled as arising from an inclination, λ, of the planetary orbit relative to the apparent stellar equator of λ = –4.4° ± 1.4°. Interestingly, this value is smaller than the λ = 7° tilt of the solar rotation axis relative to the net angular momentum vector defined by the orbits of the solar system planets (see *Beck and Giles,* 2005). *Wolf et al.* (2006) carried out a similar analysis for HD 149026, and found λ = 12° ± 14°. For these planets, the timescales for tidal coplanarization of the planetary orbits and stellar equators are expected to be on the order of 10^{12} yr (*Winn et al.,* 2005; *Greenberg,* 1974; *Hut,* 1980), indicating that the observed value of λ likely reflects that at the end of the planet-formation process.

Pertubations in the timing of planetary transits may be used to infer the presence of satellites or additional planetary companions (*Brown et al.,* 2001; *Miralda-Escudé,* 2002). *Agol et al.* (2005) and *Holman and Murray* (2005) have shown how nontransiting terrestrial-mass planets could be detected through timing anomalies. Although HST observations have yielded the most precise timing measurements to date [with a typical precision of 10 s; see tabulation for HD 209458 in *Wittenmyer et al.* (2005)], the constraints from groundbased observations can nonetheless be used to place interesting limits on additional planets in the system, as was recently done for TrES-1 (*Steffen and Agol,* 2005).

Precise photometry can also yield surprises, as in the "hump" seen in Fig. 2. This feature likely results from the planet crossing a large sunspot (or a complex of smaller ones), and thus is evidence for magnetic activity on the surface of the star. Such activity may prove to be an important noise source for timing measurements of the sort just described, but it is also an interesting object of study in its own right, allowing periodic monitoring of the stellar activity along an isolated strip of stellar latitude (*Silva,* 2003).

2.2. Theory and Interpretation

2.2.1. Overview and uncertainties. Transiting planets give us the opportunity to test our understanding of the physical structure of giant planets. In particular, structural models of the known transiting planets must be able to account for the wide range of radiation fluxes to which these planets are subjected, and they must recover the observed range of radii. In general, as the planetary mass decreases, a given external energy input has an increasingly larger influence on the size and interior structure of the planet. For hot Jupiters, the absorbed stellar flux creates a radiative zone in the subsurface regions that controls the planetary contraction, and ultimately dictates the radius. Models of transiting giant planets straddle the physical characteristics of brown dwarfs and low-mass stars, as well as the solar system giants (for an overall review, see *Burrows et al.,* 2001).

The construction of structural models for giant planets is difficult because a number of key physical inputs are poorly constrained. This situation holds equally for extrasolar planets and for the exquisitely observed outer planets of the solar system. A benefit of robust determinations of the parameters for a growing range of planets is that uncertain aspects of the theory can become increasingly constrained. Indeed, transit observations have the potential to clarify some of the core questions regarding giant planets.

The dominant uncertainty regarding the overall structure of gas giants is in the equation of state (see review of *Guillot,* 2005). The interiors of solar system and extrasolar giant planets consist of partially degenerate, partially ionized atomic-molecular fluids (*Hubbard,* 1968). The pressure in the interiors of most giant planets exceeds 10 Mbar, and central temperatures range from T$_c$ ≃ 10^4 for Uranus and Neptune to T$_c$ ≃ 3 × 10^4 for objects such as HD 209458b. This material regime lies beyond the point where hydrogen ionizes and becomes metallic, although the details of the phase transition are still uncertain (*Saumon et al.,* 2000; *Saumon and Guillot,* 2004). The equation of state of giant planet interiors is partially accessible to laboratory experiments, including gas-gun (*Holmes et al.,* 1995), laser-induced shock compression (*Collins et al.,* 1998), pulsed-power shock compression (*Knudson et al.,* 2004), and convergent shock wave (*Boriskov et al.,* 2003) techniques. These experiments can achieve momentary pressures in excess of 1 Mbar, and they appear to be approaching the molecular to metallic hydrogen transition. Unfortunately, these experiments report diverging results. In particular, they yield a range of hydrogen compression factors relevant to planetary cores that differ by ~50%. Furthermore, the laboratory experiments are in only partial agreement with first-principles quantum mechanical calculations of the hydrogen equation of state (*Militzer and Ceperley,* 2001; *Desjarlais,* 2003; *Bonev et al.,* 2004), and uncertainties associated with the equations of state of helium and heavier elements are even more severe (*Guillot,* 2005). At present, therefore, structural models must adopt the pragmatic option of choosing a thermodynamically consistent equation of state that reproduces either the high- or low-compression results (*Saumon and Guillot,* 2004).

Another uncertainty affecting the interior models is the existence and size of a radial region where helium separates from hydrogen and forms downward-raining droplets. The possibility that giant planet interiors are helium-stratified has nontrivial consequences for their structures, and

TABLE 1. Properties of the transiting planets.

Planet	$M_\star$ $M_\odot$	P days	$T_{eff,\star}$ K	$R_\star$ $R_\odot$	R_{pl} R_{Jup}	M_{pl} M_{Jup}	$T_{eq,pl}$ K	R_{pl} 20-$M_\oplus$ core	R_{pl} no core
OGLE-TR-56b	1.04±0.05	1.21	5970±150	1.10±0.10	1.23±0.16	1.45±0.23	1800±130	1.12±0.02	1.17±0.02
OGLE-TR-113b	0.77±0.06	1.43	4752±130	0.76±0.03	$1.08^{+0.07}_{-0.05}$	1.35±0.22	1186± 78	1.07±0.01	1.12±0.01
OGLE-TR-132b	1.35±0.06	1.69	6411±179	1.43±0.10	1.13±0.08	1.19±0.13	1870±170	1.13±0.02	1.18±0.02
HD 189733b	0.82±0.03	2.22	5050± 50	0.76±0.01	1.15±0.03	1.15±0.04	1074± 58	1.07±0.01	1.11±0.01
HD 149026b	1.30±0.10	2.88	6147± 50	1.45±0.10	0.73±0.05	0.36±0.03	1533± 99	0.98±0.02	1.15±0.02
TrES-1	0.87±0.03	3.00	5214± 23	0.83±0.03	1.12±0.04	0.73±0.04	1038± 61	1.02±0.01	1.10±0.00
OGLE-TR-10b	1.00±0.05	3.10	6220±140	1.18±0.04	1.16±0.05	0.54±0.14	1427± 88	1.01±0.02	1.13±0.01
HD 209458b	1.06±0.13	3.52	6099± 23	1.15±0.05	1.32±0.03	0.66±0.06	1314± 74	1.02±0.01	1.12±0.00
OGLE-TR-111b	$0.82^{+0.15}_{-0.02}$	4.02	5070±400	$0.85^{+0.10}_{-0.03}$	$1.00^{+0.13}_{-0.06}$	0.53±0.11	930±100	0.97±0.02	1.09±0.01

ultimately, their sizes. In the case of Saturn, the zone of helium rain-out may extend all the way to the center, possibly resulting in a distinct helium shell lying on top of a heavier-element core (*Fortney and Hubbard*, 2003).

Uncertainties in the equation of state, the bulk composition, and the degree of inhomogeneity allow for a depressingly wide range of models for the solar system giants that are consistent with the observed radii, surface temperatures, and gravitational moments. In particular (*Saumon and Guillot*, 2004), one can construct observationally consistent models for Jupiter with core masses ranging from 0 to 12 $M_\oplus$, and an overall envelope heavy-element content ranging from 6 to 37 $M_\oplus$. This degeneracy must be broken in order to distinguish between the core accretion (*Mizuno*, 1980; *Pollack et al.*, 1996; *Hubickyj et al.*, 2004) and gravitational instability (*Boss*, 1997, 2000, 2004) hypotheses for planet formation. Fortunately, the growing dataset of observed masses and radii from the transiting extrasolar planets suggests a possible strategy for resolving the tangle of uncertainties. The extreme range of temperature conditions under which hot Jupiters exist, along with the variety of masses that are probed, can potentially provide definitive constraints on the interior structure of these objects.

2.2.2. Comparison to observations. Following the discovery of 51 Pegb (*Mayor and Queloz*, 1995), models of jovian-mass planets subject to strong irradiation were computed (*Lin et al.*, 1996; *Guillot et al.*, 1996). These models predicted that short-period jovian-mass planets with effective temperatures of roughly 1200 K would be significantly larger than Jupiter, and the discovery that HD 209458b has a large radius initially seemed to confirm these calculations. In general, R_{pl} is a weak function of planet mass, reflecting the overall n = 1 polytropic character of giant planets (*Burrows et al.*, 1997, 2001).

In order to evaluate the present situation, we have collected the relevant quantities for the nine transiting planets in Table 1. In particular, we list the most up-to-date estimates of P, $R_\star$, $M_\star$, R_{pl}, M_{pl}, as well as the stellar effective temperature, $T_{eff,\star}$. We also list the value of the planetary equilibrium temperature, $T_{eq,pl}$, which is calculated by assuming the value for the Bond albedo, A, recently estimated for TrES-1 (A = 0.31 ± 0.14) (*Charbonneau et al.*, 2005) (section 3.2.3). The precision of the estimates of the physi-

cal properties varies considerably from star to star. By drawing from the Gaussian distributions corresponding to the uncertainties in Table 1 and the quoted value for A, we can estimate the uncertainty for M_{pl} and $T_{eq,pl}$ for each planet. Thereafter, for a particular choice of M_{pl} and $T_{eq,pl}$ and fixing the planetary age at 4.5 G.y., we can compute theoretical radii. For this task, we use the results of *Bodenheimer et al.* (2003), who computed models for insolated planets ranging in mass from 0.11 to 3.0 M_{Jup}. To evaluate the radii differences that arise from different heavy-element fractions, separate sequences were computed for models that contain and do not contain 20-$M_\oplus$ solid cores, and both predictions are listed in Table 1. The models have been calibrated so that, for the evolution of Jupiter up to the age of 4.5 G.y., a model with a core gives the correct Jupiter radius to within 1%. Planetary age can also have a significant effect on R_{pl}. For example, the evolutionary models of *Burrows et al.* (2004a) for OGLE-TR-56b (1.45 M_{Jup}) yield transit radii of $R_{pl} \simeq 1.5$ R_{Jup} at 100 m.y., and ~1.25 R_{Jup} after 2 G.y. In general, however, R_{pl} evolves only modestly beyond the first 500 m.y., and hence the uncertainties in the ages of the parent stars (for which such young ages may generally be excluded) introduce errors of only a few percent into the values of R_{pl}.

The models use a standard Rosseland mean photospheric boundary condition, and as such, are primarily intended for cross-comparison of radii. The obtained planetary radii are, however, in excellent agreement with baseline models obtained by groups employing detailed frequency-dependent atmospheres (e.g., *Burrows et al.*, 2004b; *Chabrier et al.*, 2004; *Fortney et al.*, 2006). The models assume that the surface temperature is uniform all the way around the planet, even though the rotation of the planet is likely tidally locked. Hydrodynamic simulations of the atmosphere that aim, in part, to evaluate the efficiency with which the planet redistributes heat from the dayside to the nightside have been performed by *Cho et al.* (2003), *Showman and Guillot* (2002), *Cooper and Showman* (2005), and *Burkert et al.* (2005) under various simplifying assumptions. There is no agreement on what the temperature difference between the dayside and the nightside should be (section 3.2.4), and it depends on the assumed opacity in the atmosphere. *Burkert et al.* (2005) suggest that with a reasonable opacity, the difference could be 200 K, not enough to make an appreciable difference in the radius.

A number of interesting conclusions regarding the bulk structural properties of the transiting planets can be drawn from Table 1. First, the baseline radius predictions display (1σ) agreement for seven of the nine known transiting planets. Second, the planets whose radii are in good agreement with the models span the full range of masses and effective temperatures. The models do not appear to be systematically wrong in some particular portion of parameter space. Although the reported accuracies of the basic physical parameters are noticeably worse for the OGLE systems than for the brighter targets, the constraints are nonetheless useful to address models of their physical structure and, in particular, the presence or absence of a solid core. Specifically, the baseline models in Table 1 indicate that the presence of a solid core in a 0.5 M_{Jup} planet with $T_{eff,pl}$ = 1500 K leads to a radius reduction of roughly 0.1 R_{Jup}. This difference generally exceeds the uncertainty in the estimate of R_{pl}.

In the standard core-accretion paradigm for giant planet formation, as reviewed by *Lissauer* (1993), a jovian planet arises from the collisional agglomeration of a solid 10-$M_\oplus$ core over a period of several million years, followed by a rapid accretion of hundreds of Earth masses of nebular gas, lasting roughly 10^5 yr. The competing gravitational instability hypothesis (e.g., *Boss,* 1997, 2004) posits that gas-giant planets condense directly from spiral instabilities in protostellar disks on a dynamical timescale of less than 10^3 yr. *Boss* (1998) points out that solid particles in the newly formed planet can precipitate to form a core during the initial contraction phase. Only 1% of the matter in the planet is condensible, however, so a jovian-mass planet that formed by this process will have a core that is much less massive than one that formed by the core-accretion scenario.

Among the seven planets that show agreement with the baseline models, it is presently difficult to discern the presence of a core. However, the "transit radius" effect (*Burrows et al.,* 2003) (section 2.2.3) will tend to systematically increase the observed radii above the model radii listed in Table 1 (which correspond to a 1-bar pressure level). Similarly, signal-to-noise-limited field transit surveys bias the mean radius of planets so detected to a value larger than that of the intrinsic population (*Gaudi,* 2005). Taking both effects into account lends favor to the models with cores. Clearly, more transiting planets and more precise determinations of their properties are necessary, as are more physically detailed models. We note that the identification of lower-mass transiting planets (for which the effect of a solid core is prominent) would be particularly helpful to progress in these questions. Several groups (e.g., *Gould et al.,* 2003; *Hartman et al.,* 2005; Pepper and Gaudi, in preparation) have considered the prospects for groundbased searches for planets with radii of that of Neptune or less.

2.2.3. The transit radius effect. When a planet occults its parent star, the wavelength-dependent value of R_{pl} so inferred is not necessarily the canonical planetary radius at a pressure level of 1 bar (*Lindal et al.,* 1981; *Hubbard et al.,* 2001), which we have used for the baseline predictions in Table 1. As such, the measured radius is approximately the impact parameter of the transiting planet at which the optical depth to the stellar light along a chord parallel to the star-planet line of centers is unity. This is not the optical depth in the radial direction, nor is it associated with the radius at the radiative-convective boundary. Hence, since the pressure level to which the transit beam is probing near the planet's terminator is close to one *milli*bar (*Fortney et al.,* 2003), there are typically 5–10 pressure scale heights difference between the measured value of R_{pl} and either the radiative-convective boundary (≥1000 bar) and the 1-bar radius. [If, as discussed in *Barman et al.* (2002), the transit radius is at pressures well below the 1-mbar level, then the effect would be even larger.] Furthermore, exterior to the radiative-convective boundary, the entropy is an increasing function of radius. One consequence of this fact is significant radial inflation vis à vis a constant entropy atmosphere. Both of these effects result in an apparent increase of perhaps 0.1 R_{Jup} (~7%) in the theoretical radius for HD 209458b and 0.05 R_{Jup} (~4%) for OGLE-TR-56b.

2.2.4. Explaining the oddballs. Two of the planets, HD 209458b and HD 149026b, have radii that do not agree at all with the predictions. HD 209458b is considerably larger than predicted, and HD 149026b is too small (Fig. 3). These discrepancies indicate that the physical structures of the transiting planets can depend significantly on factors other than M_{pl} and $T_{eff,pl}$. It would appear that hot Jupiters are imbued with individual personalities.

While the radius of HD 209458b is certainly broadly consistent with a gas-giant composed primarily of hydrogen, studies by *Bodenheimer et al.* (2001) and *Guillot and Showman* (2002) were the first to make it clear that a standard model of a contracting, irradiated planet can recover $R_{pl} \simeq 1.35\ R_{Jup}$ for HD 209458b only if the deep atmosphere is unrealistically hot. A number of resolutions to this conundrum have been suggested. *Bodenheimer et al.* (2001) argue that HD 209458b might be receiving interior tidal heating through ongoing orbital circularization. This hypothesis was refined by *Bodenheimer et al.* (2003), who computed grids of predicted planetary sizes under a variety of conditions, and showed that the then-current radial velocity dataset for HD 209458b was consistent with the presence of an undetected planet capable of providing the requisite eccentricity forcing. The tidal-heating hypothesis predicts that HD 209458b is caught up in an anomalous situation, and that the majority of hot Jupiter-type planets will have considerably smaller radii than that observed for HD 209458b. Recent analyses by *Laughlin et al.* (2005b) and *Winn et al.* (2005) indicate that the orbital eccentricity of HD 209458b is close to zero. This conclusion is further buttressed by the timing of the secondary eclipse by *Deming et al.* (2005a) (discussed in greater detail in section 3.2.3), which places stringent upper limits on the eccentricity, except in the unlikely event that the orbit is precisely aligned

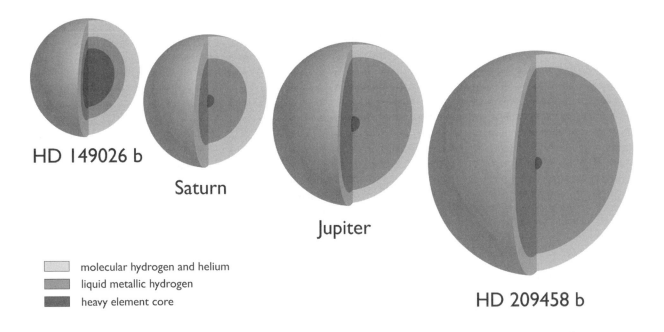

Fig. 3. Cut-away diagrams of Jupiter, Saturn, and the two oddball extrasolar planets, drawn to scale. The observed radius of HD 149026b implies a massive core of heavy elements that makes up perhaps 70% of the planetary mass. In contrast, the radius of HD 209458b intimates a coreless structural model, as well as an additional energy source to explain its large value.

to our line of sight. Thus the eccentricity appears to be below the value required to generate sufficient tidal heating to explain the inflated radius.

Guillot and Showman (2002) proposed an alternate hypothesis in which strong insolation-driven weather patterns on the planet drive the conversion of kinetic energy into thermal energy at pressures of tens of bars. They explored this idea by modifying their planet-evolution code to include a radially adjustable internal energy source term. They found that if kinetic wind energy is being deposited at adiabatic depths with an efficiency of 1%, then the large observed radius of the planet can be explained. Their hypothesis predicts that other transiting planets with similar masses and at similar irradiation levels should be similar in size to HD 209458b. The subsequent discovery that TrES-1 has a considerably smaller radius despite its similar temperature, mass, and parent-star metallicity is evidence against the kinetic heating hypothesis, since it is not clear why this mechanism should act upon only HD 209458b.

Recently, an attractive mechanism for explaining the planet's large size has been advanced by *Winn and Holman* (2005), who suggest that the anomalous source of heat arises from obliquity tides that occur as a result of the planet being trapped in a Cassini state (e.g., *Peale,* 1969). In a Cassini state, a planet that is formed with a nonzero obliquity is driven during the course of spin synchronization to a final state in which spin precession resonates with orbital precession. When caught in a Cassini state, the planet is forced to maintain a nonzero obliquity, and thus experiences continued tidal dissipation as a result of orbital libration. Order-

of-magnitude estimates indicate that the amount of expected tidal dissipation could generate enough heat to inflate the planet to the observed size.

HD 149026b presents a problem that is essentially the opposite to that of HD 209458b. Both the mass (0.36 M_{Jup}) and the radius (0.73 R_{Jup}) are considerably smaller than those of the other known transiting extrasolar planets. Curiously, HD 149026 is the only star of a transiting planet to have a metallicity that is significantly supersolar, [Fe/H] = 0.36. The observed radius is 30% smaller than the value predicted by the baseline model with a core of 20 $M_\oplus$. Clearly, a substantial enrichment in heavy elements above solar composition is required. The mean density of the planet, 1.17 g cm^{-3}, is 1.7× that of Saturn, which itself has roughly 25% heavy elements by mass. On the other hand, the planet is not composed entirely of water or silicates, or else the radius would be on the order of 0.4 or 0.28 R_{Jup}, respectively (*Guillot et al.,* 1996; *Guillot,* 2005). Models by *Sato et al.* (2005) and by *Fortney et al.* (2006) agree that the observed radius can be recovered if the planet contains approximately 70 $M_\oplus$ of heavy elements, either distributed throughout the interior or sequestered in a core (Fig. 3).

The presence of a major fraction of heavy elements in HD 149026b has a number of potentially interesting ramifications for the theory of planet formation. *Sato et al.* (2005) argue that it would be difficult to form this giant planet by the gravitational instability mechanism (*Boss,* 2004). The large core also presents difficulties for conventional models of core accretion. In the core-accretion theory, which was developed in the context of the minimum-mass solar nebula, it is difficult to prevent runaway gas accretion from occurring onto cores more massive than 30 $M_\oplus$, even if abundant infalling planetesimals heat the envelope and delay the Kelvin-Helmholtz contraction that is required to let more gas into the planet's Hill sphere. The current structure of HD 149026b suggests that it was formed in a gas-starved environment, yet presumably enough gas was present in the protoplanetary disk to drive migration from its probable formation region beyond 1–2 AU from the star inward to the current orbital separation of 0.043 AU. Alternately, a metal-rich disk would likely be abundant in planetesimals, which may in turn have promoted the inward migration of the planet via planetesimal scattering (*Murray et al.*, 1998).

3. ATMOSPHERES

By the standards of the solar system, the atmospheres of the close-in planets listed in Table 1 are quite exotic. Located only 0.05 AU from their parent stars, these gas giants receive a stellar flux that is typically 10^4 that which strikes Jupiter. As a result, a flurry of theoretical activity over the past decade has sought to predict (and, more recently, interpret) the emitted and reflected spectra of these objects (e.g., *Seager and Sasselov*, 1998; *Seager et al.*, 2000; *Barman et al.*, 2001; *Sudarsky et al.*, 2003; *Allard et al.*, 2003; *Burrows et al.*, 2004b; *Burrows*, 2005). Observations promise to grant answers to central questions regarding the atmospheres of the planets, including the identity of their chemical constituents, the presence (or absence) of clouds, the fraction of incident radiation that is absorbed (and hence the energy budget of the atmosphere), and the ability of winds and weather patterns to redistribute heat from the dayside to the nightside. For a detailed review of the theory of extrasolar planet atmospheres, see the chapter by Marley et al. We summarize the salient issues below (section 3.1), and then proceed to discuss the successful observational techniques, and resulting constraints to date (section 3.2).

3.1. Theory

3.1.1. Overview.   In order to model the atmospheres and spectra of extrasolar giant planets in general, and hot Jupiters in particular, one must assemble extensive databases of molecular and atomic opacities. The species of most relevance, and which provide diagnostic signatures, are H_2O, CO, CH_4, H_2, Na, K, Fe, NH_3, N_2, and silicates. The chemical abundances of these and minority species are derived using thermochemical data and minimizing the global free energy. Non-equilibrium effects in the upper atmospheres require chemical networks and kinetic coefficients. With the abundances and opacities, as well as models for the stellar spectrum, one can embark upon calculations of the atmospheric temperature, pressure, and composition profiles and of the emergent spectrum of an irradiated planet. With atmospheric temperatures in the 1000–2000 K range, CO, not CH_4, takes up much of the carbon in the low-pressure outer atmosphere, and N_2, not NH_3, sequesters most of the nitro-

gen. However, H_2O predominates in the atmospheres for both hot and cooler giants. Perhaps most striking in the spectrum of a close in giant planet is the strong absorption due to the sodium and potassium resonance doublets. These lines are strongly pressure-broadened and likely dominate the visible spectral region. The major infrared spectral features are due to H_2O, CO, CH_4, and NH_3. H_2 collision-induced absorption contributes very broad features in the infrared.

A self-consistent, physically realistic evolutionary calculation of the radius, $T_{eff,pl}$, and spectrum of a giant planet in isolation requires an outer boundary condition that connects radiative losses, gravity (g), and core entropy (S). When there is no irradiation, the effective temperature determines both the flux from the core and the entire object. A grid of $T_{eff,pl}$, g, and S, derived from detailed atmosphere calculations, can then be used to evolve the planet (e.g., *Burrows et al.*, 1997; *Allard et al.*, 1997). However, when a giant planet is being irradiated by its star, this procedure must be modified to include the outer stellar flux in the calculation that yields the corresponding S-$T_{eff,pl}$-g relationship. This must be done for a given external stellar flux and spectrum, which in turn depends upon the stellar luminosity spectrum and the orbital distance of the giant planet. Therefore, one needs to calculate a new S-$T_{eff,pl}$-g grid under the irradiation regime of the hot Jupiter that is tailormade for the luminosity and spectrum of its primary and orbital distance. With such a grid, the radius evolution of a hot Jupiter can be calculated, with its spectrum as a byproduct.

3.1.2. The day-night effect and weather.   A major issue is the day-night cooling difference. The gravity and interior entropy are the same for the dayside and nightside. For a synchronously rotating hot Jupiter, the higher core entropies needed to explain a large measured radius imply higher internal fluxes on a nightside if the day and the night atmospheres are not coupled (e.g., *Guillot and Showman*, 2002). For strongly irradiated giant planets, there is a pronounced inflection and flattening in the temperature-pressure profile that is predominantly a result of the near balance at some depth between countervailing incident and internal fluxes. The dayside core flux is suppressed by this flattening of the temperature gradient and the thickening of the radiative zone due to irradiation. However, *Showman and Guillot* (2002), *Menou et al.* (2003), *Cho et al.* (2003), *Burkert et al.* (2005), and *Cooper and Showman* (2005) have recently demonstrated that strong atmospheric circulation currents that advect heat from the dayside to nightside at a wide range of pressure levels are expected for close-in giant planets. *Showman and Guillot* (2002) estimate that below pressures of 1 bar the nightside cooling of the air can be quicker than the time it takes the winds to traverse the nightside, but that at higher pressures the cooling timescale is far longer. Importantly, the radiative-convective boundary in a planet such as HD 209458b is very deep, at pressures of perhaps 1000 bar. This may mean that due to the coupling of the dayside and nightside via strong winds at depth, the temperature-pressure profiles at the convective boundary on both sides are similar. This would imply that the core cooling rate is roughly the same in both hemispheres. Since the planet brightness inferred during secondary eclipse (section 3.2.3) depends upon the advection of stellar heat to the nightside, such data can provide constraints on the meteorology and general circulation models.

Almost complete redistribution of heat occurs in the case of Jupiter, where the interior flux is independent of latitude and longitude. However, the similarity in Jupiter of the day and night temperature-pressure profiles and effective temperatures is a consequence not of the redistribution of heat by rotation or zonal winds, but of the penetration into the convective zone on the dayside of the stellar irradiation (*Hubbard*, 1977). Core convection then redistributes the heat globally and accounts for the uniformity of the temperature over the entire surface. Therefore, whether direct heating of the convective zone by the stellar light is responsible, as it is in our own jovian planets, for the day-night smoothing can depend on the ability of the stellar insolation to penetrate below the radiative-convective boundary. This does not happen for a hot Jupiter. Clearly, a full three-dimensional study will be required to definitively resolve this thorny issue.

Clouds high in the atmospheres of hot Jupiters with $T_{eff,pl} > 1500$ K would result in wavelength-dependent flux variations. Cloud opacity tends to block the flux windows between the molecular absorption features, thereby reducing the flux peaks. Additionally, clouds reflect some of the stellar radiation, increasing the incident flux where the scattering opacity is high. This phenomenon tends to be more noticeable in the vicinity of the gaseous absorption troughs.

3.2. Observations

The direct study of extrasolar planets orbiting mature (G.y.) Sun-like stars may proceed without the need to image the planet (i.e., to spatially separate the light of the planet from that of the star). Indeed, this technical feat has not yet been accomplished. Rather, the eclipsing geometry of transiting systems permits the spectrum of the planet and star to be disentangled through monitoring of the variation in the combined system light as a function of the known orbital phase. Detections and meaningful upper limits have been achieved using the following three techniques.

3.2.1. Transmission spectroscopy. The technique of transmission spectroscopy seeks to ratio stellar spectra gathered during transit with those taken just before or after this time, the latter providing a measurement of the spectrum of the isolated star. Wavelength-dependent sources of opacity in the upper portions of the planetary atmosphere, or in its exosphere, will impose absorption features that could be revealed in this ratio. This technique can be viewed as probing the wavelength-dependent variations in the inferred value of R_{pl}.

The first composition signature (*Charbonneau et al.*, 2002) was detected with the HST STIS spectrograph. The team measured an increase in the transit depth of $(2.32 \pm$

$0.57) \times 10^{-4}$ for HD 209458 in a narrow bandpass centered on the sodium resonance lines near 589 nm. Ruling out alternate explanations of this diminution, they conclude that the effect results from absorption due to atomic sodium in the planetary atmosphere, which indeed had been unanimously predicted to be a very prominent feature at visible wavelengths (*Seager and Sasselov*, 2000; *Hubbard et al.*, 2001; *Brown*, 2001). Interestingly, the detected amplitude was roughly one-third that predicted by baseline models that incorporated a cloudless atmosphere and a solar abundance of sodium in atomic form. *Deming et al.* (2005b) follow the earlier work of *Brown et al.* (2002) to achieve strong upper limits on the CO bandhead at 2.3 µm. Taken together, the reduced amplitude of the sodium detection and the upper limits on CO suggest the presence of clouds high in the planetary atmosphere (e.g., *Fortney et al.*, 2003), which serve to truncate the effective size of the atmosphere viewed in transmission. *Fortney* (2005) considers the slant optical depth and shows that even a modest abundance of condensates or hazes can greatly reduce the size of absorption features measured by this technique. Alternately, non-LTE effects may explain the weaker-than-expected sodium feature (*Barman et al.*, 2005).

Planetary exospheres are amenable to study by this method, as the increased cross-sectional area (compared to the atmospheres) implies a large potential signal. *Vidal-Madjar et al.* (2003) observed a $15 \pm 4\%$ transit depth of HD 209458 when measured at Lyα. The implied physical radius exceeds the Roche limit, leading them to conclude that material is escaping the planet (*Lecavelier des Etangs et al.*, 2004; *Baraffe et al.*, 2004). However, the minimum escape rate required by the data is low enough to reduce the planetary mass by only 0.1% over the age of the system. More recently, *Vidal-Madjar et al.* (2004) have claimed detection of other elements, with a lower statistical significance. Significant upper limits on various species at visible wavelengths have been presented by *Bundy and Marcy* (2000), *Moutou et al.* (2001, 2003), *Winn et al.* (2004), and *Narita et al.* (2005).

3.2.2. Reflected light. Planets shine in reflected light with a visible-light flux f_{pl} (relative to that of their stars, $f_\star$) of

$$\left(\frac{f_{pl}}{f_\star}\right)_\lambda (\alpha) = \left(\frac{R_{pl}}{a}\right)^2 p_\lambda \Phi_\lambda(\alpha)$$

where a is the orbital separation, p_λ is the geometric albedo, and $\Phi_\lambda(\alpha)$ is the phase function, which describes the relative flux at a phase angle α to that at opposition. Even assuming an optimistic values for p_λ, hot Jupiters present a flux ratio of less than 10^{-4} that of their stars. See *Marley et al.* (1999), *Seager et al.* (2000), and *Sudarsky et al.* (2000) for theoretical predictions of the reflection spectra and phase functions of hot Jupiters.

The first attempts to detect this modulation adopted a spectroscopic approach, whereby a series of spectra span-

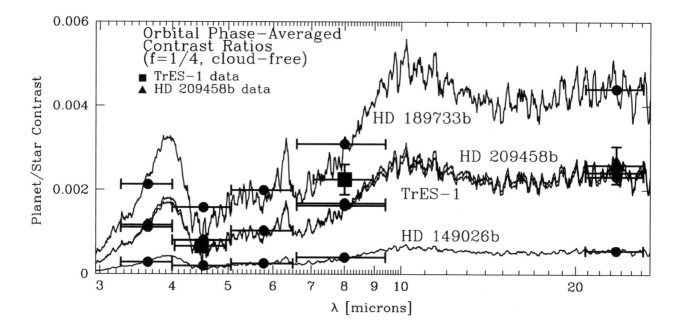

Fig. 4. The orbital phase-averaged planet-to-star flux-density ratio as a function of wavelength (λ, in μm) for the models of the four known transiting extrasolar planets for which such observations might be feasible (*Burrows et al.,* 2005; Burrows et al., in preparation). The bandpass-integrated predicted values are shown as filled circles, with the bandwidths indicated by horizontal bars. The measured values for TrES-1 at 4.5 μm and 8.0 μm (*Charbonneau et al.,* 2005) are shown as filled squares, and the observed value at 24 μm for HD 209458b (*Deming et al.,* 2005a) is shown as a filled triangle. The extremely favorable contrast for HD 189733, and the extremely challenging contrast ratio for HD 149026, both result primarily from the respective planet-to-star surface area ratios.

ning key portions of the orbital phase are searched for the presence of a copy of the stellar spectrum. For nontransiting systems, this method is complicated by the need to search over possible values of the unknown orbital inclination. The secondary spectrum should be very well separated spectroscopically, as the orbital velocities for these hot Jupiters are typically 100 km s^{-1}, much greater than the typical stellar line widths of <15 km s^{-1}. Since the method requires multiple high signal-to-noise ratio, high-dispersion spectra, only the brightest systems have been examined. A host of upper limits have resulted for several systems (e.g., *Charbonneau et al.,* 1999; *Collier Cameron et al.,* 2002; *Leigh et al.,* 2003a,b), typically excluding values of $p_\lambda > 0.25$ averaged across visible wavelengths. These upper limits assume a functional dependence for $\Phi_\lambda(\alpha)$ as well as a gray albedo, i.e., that the planetary spectrum is a reflected copy of the stellar spectrum.

Spacebased platforms afford the opportunity to study the albedo and phase function in a straightforward fashion by seeking the photometric modulation of the system light. The MOST satellite (*Walker et al.,* 2003) should be able to detect the reflected light from several hot Jupiters (*Green et al.,* 2003) or yield upper limits that will severely constrain the atmospheric models, and campaigns on several systems are completed or planned (*Rowe et al.,* 2006). The upcoming Kepler mission (*Borucki et al.,* 2003) will search for this effect, and should identify 100–760 nontransiting hot Jupiters with orbital periods of P < 7 d (*Jenkins and Doyle,* 2003).

3.2.3. Infrared emission. At infrared wavelengths, the secondary eclipse (i.e., the decrement in the system flux due to the passage of the planet behind the star) permits a determination of the planet-to-star brightness ratio. Since the underlying stellar spectrum may be reliably assumed from stellar models (e.g., *Kurucz,* 1992), such estimates afford the first direct constraints on the emitted spectra of planets orbiting other Sun-like stars. In the Rayleigh-Jeans limit, the ratio of the planetary flux to that of the star is

$$\left(\frac{f_{pl}}{f_\star}\right) \simeq \frac{T_{eq,pl}}{T_{eff,\star}}\left(\frac{R_{pl}}{R_\star}\right)^2$$

The last factor is simply the transit depth. From Table 1, we can see that the typical ratio of stellar to planetary temperatures is 3.5–5.5, leading to predicted secondary eclipse amplitudes of several millimagnitudes.

Charbonneau et al. (2005) and *Deming et al.* (2005a) have recently employed the remarkable sensitivity and stability of the Spitzer Space Telescope to detect the thermal emission from TrES-1 (4.5 μm and 8.0 μm) and HD 209458b (24 μm) (Fig. 4). These measurements provide estimates of the planetary brightness temperatures in these three bands, which in turn can be used to estimate (under several assumptions) the value of $T_{eq,pl}$ and A of the planets. Observations of these two objects in the other Spitzer bands shown in Fig. 4 (as well as the 16-μm photometric band of the IRS peak-up array) are feasible. Indeed, at the time of writing, partial datasets have been gathered for all four planets shown in Fig. 4. The results should permit a detailed search for the presence of spectroscopically dominant molecules, notably, CH_4, CO, and H_2O. Using the related technique of occultation spectroscopy, *Richardson et al.* (2003a,b) have analyzed a series of infrared spectra spanning a time before, during, and after secondary eclipse, and present useful upper limits on the presence of planetary features due to these molecules.

Williams et al. (2006) have outlined a technique by which the spatial dependence of the planetary emission could be resolved in longitude through a careful monitoring of the secondary eclipse. Such observations, as well as attempts to measure the phase variation as the planet orbits the star (and hence presents a different face to Earth) are eagerly anticipated to address numerous models of the dynamics and weather of these atmospheres (section 3.1.2). The elapsed time between the primary and secondary eclipse affords a stringent upper limit on the quantity e cos ϖ, and the relative durations of the two events constrains e sin ϖ (*Kallrath and Milone*, 1999; *Charbonneau*, 2003). The resulting limits on e are of great interest in gauging whether tidal circularization is a significant source of energy for the planet (section 2.2.4).

3.2.4. Inferences from the infrared detections. Varying planet mass, planet radius, and stellar mass within their error bars alters the resulting predicted average planet-star flux ratios only slightly (*Burrows et al.*, 2005). Similarly, and perhaps surprisingly, adding Fe and forsterite clouds does not shift the predictions in the Spitzer bands by an appreciable amount. Moreover, despite the more than a factor of 2 difference in the stellar flux at the planet, the predictions for the planet-star ratios for TrES-1 and HD 209458b are not very different. *Fortney et al.* (2005) explore the effect of increasing the metallicity of the planets, and find a better agreement to the red 4.5/8.0-μm color of TrES-1 with a enrichment factor of 3–5. *Seager et al.* (2005) show that models with an increased carbon-to-oxygen abundance produce good fits to the HD 209458b data, but conclude that a wide range of models produces plausible fits. *Barman et al.* (2005) examine the effect of varying the efficiency for the redistribution of heat from the dayside to the nightside, and find evidence that models with significant redistribution (and hence more isotropic temperatures) are favored.

In Fig. 4, there is a hint of the presence of H_2O, since it is expected to suppress flux between 4 and 10 μm. This is shortward of the predicted 10-μm peak in planet-star flux ratio, which is due to water's relative abundance and the strength of its absorption bands in that wavelength range. Without H_2O, the fluxes in the 3.6–8.0-μm bands would be much greater. Hence, a comparison of the TrES-1 and HD 209458b data suggests, but does not prove, the presence of water. Seeing (or excluding) the expected slope between the 5.8-μm and 8.0-μm bands and the rise from 4.5 μm to 3.6 μm would be more revealing in this regard. Furthermore, the relative strength of the 24-μm flux ratio in comparison with

the 3.6-μm, 4.5-μm, and 5.8-μm channel ratios is another constraint on the models, as is the closeness of the 8.0-μm and 24-μm ratios. If CH_4 is present in abundance, then the 3.6-μm band will test this. However, the preliminary conclusion for these close-in Jupiters is that CH_4 should not be in evidence. Models have difficulty fitting the precise depth of the 4.5-μm feature for TrES-1. It coincides with the strong CO absorption predicted to be a signature of hot Jupiter atmospheres. However, the depth of this feature is only a weak function of the CO abundance. A CO abundance 100× larger than expected in chemical equilibrium lowers this flux ratio at 4.5 μm by only ~25%. Therefore, while the 4.5-μm data point for TrES-1 implies that CO has been detected, the exact fit is problematic.

In sum, Spitzer observations of the secondary eclipses of the close-in transiting giant planets will provide information on the presence of CO and H_2O in their atmospheres, as well as on the role of clouds in modifying the planet-to-star flux ratios over the 3–25-μm spectral range. Furthermore, there is good reason to believe that the surface elemental abundances of extrasolar giant planets are not the same as the corresponding stellar elemental abundances, and Spitzer data across the available band passes will soon better constrain the atmospheric metallicities and C/O ratios of these planets. Moreover, and most importantly, the degree to which the heat deposited by the star on the dayside is advected by winds and jet streams to the nightside is unknown. If this transport is efficient, the dayside emissions probed during secondary eclipse will be lower than the case for inefficient transport. There is already indication in the data for HD 209458b and TrES-1 that such transport may be efficient (e.g., *Barman et al.*, 2005), but much more data are needed to disentangle the effects of the day-night heat redistribution, metallicity, and clouds and to identify the diagnostic signatures of the climate of these extrasolar giant planets. The recently detected hot Jupiter, HD 189733b (*Bouchy et al.*, 2005a), is a veritable goldmine for such observations (Fig. 4), owing to the much greater planet-to-star contrast ratio (*Deming et al.*, 2006).

4. FUTURE PROSPECTS

With the recent radial-velocity discoveries of planets with masses of 7–20 $M_\oplus$ (e.g., *Bonfils et al.*, 2005; *Butler et al.*, 2004; *McArthur et al.*, 2004; *Rivera et al.*, 2005; *Santos et al.*, 2004), the identification of the first such object in a transiting configuration is eagerly awaited. The majority of these objects have been found in orbit around low-mass stars, likely reflecting the increased facility of their detection for a fixed Doppler precision. Despite the smaller expected planetary size, the technical challenge of measuring the transits will be alleviated by the smaller stellar radius, which will serve to make the transits deep (but less likely to occur). Due to the low planetary mass, the influence of a central core (section 2.2.2) will be much more prominent. Furthermore, the reduced stellar size and brightness implies that atmospheric observations (section 3.2) will be feasible. The radial-velocity surveys monitor few stars later than M4V, but transiting planets of even later spectral types could be identified by a dedicated photometric monitoring campaign of several thousand of the nearest targets. An Earth-sized planet orbiting a late M dwarf with a week-long period would lie within the habitable zone and, moreover, would present the same infrared planet-to-star brightness ratio as that detected (section 3.2.3). We note the urgency of locating such objects (should they exist), due to the limited cryogenic lifetime of Spitzer.

The excitement with which we anticipate the results from the Kepler (*Borucki et al.,* 2003) and COROT (*Baglin,* 2003) missions cannot be overstated. These projects aim to detect scores of rocky planets transiting Sun-like primaries, and the Kepler mission in particular will be sensitive to year-long periods and hence true analogs of Earth. Although direct follow-up of such systems (section 3.2) with extant facilities appears precluded by signal-to-noise considerations, future facilities (notably the James Webb Space Telescope) may permit some initial successes. We conclude that the near-future prospects for studies of transiting planets are quite bright (although they may dim, periodically), and we anticipate that the current rapid pace of results will soon eclipse this review — just in time for Protostars and Planets VI.

Acknowledgments. A.B. acknowledges support from NASA through grant NNG04GL22G. G.L. acknowledges support from the NASA OSS and NSF CAREER programs. We thank S. Gaudi for illuminating suggestions.

REFERENCES

Agol E., Steffen J., Sari R., and Clarkson W. (2005) *Mon. Not. R. Astron. Soc., 359,* 567–579.

Allard F., Hauschildt P. H., Alexander D. R., and Starrfield S. (1997) *Ann. Rev. Astron. Astrophys., 35,* 137–177.

Allard F., Baraffe I., Chabrier G., Barman T. S., and Hauschildt P. H. (2003) In *Scientific Frontiers in Research on Extrasolar Planets* (D. Deming and S. Seager, eds.), pp. 483–490. ASP Conf. Series 294, San Francisco.

Alonso R. (2005) Ph.D. thesis, Univ. of La Laguna.

Alonso R., Brown T. M., Torres G., Latham D. W., Sozzetti A., et al. (2004) *Astrophys. J., 613,* L153–L156.

Baglin A. (2003) *Adv. Space Res., 31,* 345–349.

Bakos G., Noyes R. W., Kovács G., Stanek K. Z., Sasselov D. D., et al. (2004) *Publ. Astron. Soc. Pac., 116,* 266–277.

Bakos G. Á., Knutson H., Pont F., Moutou C., Charbonneau D., et al. (2006) *Astrophys J.,* in press.

Baraffe I., Selsis F., Chabrier G., Barman T. S., Allard F., et al. (2004) *Astron. Astrophys., 419,* L13–L16.

Barman T. S., Hauschildt P. H., and Allard F. (2001) *Astrophys. J., 556,* 885–895.

Barman T. S., Hauschildt P. H., Schweitzer A., Stancil P. C., Baron E., et al. (2002) *Astrophys. J., 569,* L51–L54.

Barman T. S., Hauschildt P. H., and Allard F. (2005) *Astrophys. J., 632,* 1132–1139.

Beck J. G. and Giles P. (2005) *Astrophys. J., 621,* L153–L156.

Bodenheimer P., Lin D. N. C., and Mardling R. A. (2001) *Astrophys. J., 548,* 466–472.

Bodenheimer P., Laughlin G., and Lin D. N. C. (2003) *Astrophys. J., 592,* 555–563.

Bonev G. V., Militzer B., and Galli G. (2004) *Phys. Rev. B, 69,* 014101.

Bonfils X., Forveille T., Delfosse X., Udry S., Mayor M., et al. (2005) *Astron. Astrophys., 443,* L15–L18.

Boriskov G. V., et al. (2003) *Dokl. Phys., 48,* 553–555.

Borucki W. J., Caldwell D., Koch D. G., Webster L. D., Jenkins J. M., et al. (2001) *Publ. Astron. Soc. Pac., 113,* 439–451.

Borucki W. J., Koch D. G., Lisssauer J. J., Basri G. B., Caldwell J. F., et al. (2003) *Proc. SPIE, 4854,* 129–140.

Boss A. P. (1997) *Science, 276,* 1836–1839.

Boss A. P. (1998) *Astrophys. J., 503,* 923–937.

Boss A. P. (2000) *Astrophys. J., 536,* L101–L104.

Boss A. P. (2004) *Astrophys. J., 610,* 456–463.

Bouchy F., Pont F., Santos N. C., Melo C., Mayor M., et al. (2004) *Astron. Astrophys., 421,* L13–L16.

Bouchy F., Udry S., Mayor M., Pont F., Iribane N., et al. (2005a) *Astron. Astrophys., 444,* L15–L19.

Bouchy F., Pont F., Melo C., Santos N. C., Mayor M., et al. (2005b) *Astron. Astrophys., 431,* 1105–1121.

Brown T. M. (2001) *Astrophys. J., 553,* 1006–1026.

Brown T. M. (2003) *Astrophys. J., 593,* L125–L128.

Brown T. M. and Charbonneau D. (2000) In *Disks, Planetesimals, and Planets* (F. Garzón et al., eds.), pp. 584–589. ASP Conf. Series 219, San Francisco.

Brown T. M., Charbonneau D., Gilliland R. L., Noyes R. W., and Burrows A. (2001) *Astrophys. J., 551,* 699–709.

Brown T. M., Libbrecht K. G., and Charbonneau D. (2002) *Publ. Astron. Soc. Pac., 114,* 826–832.

Bundy K. A. and Marcy G. W. (2000) *Publ. Astron. Soc. Pac., 112,* 1421–1425.

Burke C. J., Gaudi B. S., DePoy D. L., Pogge R. W., and Pinsonneault M. H. (2004) *Astron. J., 127,* 2382–2397.

Burkert A., Lin D. N. C., Bodenheimer P. H., Jones C. A., and Yorke H. W. (2005) *Astrophys. J., 618,* 512–523.

Burrows A. (2005) *Nature, 433,* 261–268.

Burrows A., Marley M., Hubbard W. B., Lunine J. I., Guillot T., et al. (1997) *Astrophys. J., 491,* 856–875.

Burrows A., Hubbard W. B., Lunine J. I., and Liebert J. (2001) *Rev. Mod. Phys., 73,* 719–765.

Burrows A., Sudarsky D., and Hubbard W. B. (2003) *Astrophys. J., 594,* 545–551.

Burrows A., Hubeny I., Hubbard W. B., Sudarsky D., and Fortney J. J. (2004a) *Astrophys. J., 610,* L53–L56.

Burrows A., Sudarsky D., and Hubeny I. (2004b) *Astrophys. J., 609,* 407–416.

Burrows A., Hubeny I., and Sudarsky D. (2005) *Astrophys. J., 625,* L135–L138.

Butler R. P., Vogt S. S., Marcy G. W., Fischer D. A., Wright J. T., et al. (2004) *Astrophys. J., 617,* 580–588.

Chabrier G., Barman T., Baraffe I., Allard F., and Hauschildt P. H. (2004) *Astrophys. J., 603,* L53–L56.

Charbonneau D. (2003) In *Scientific Frontiers in Research on Extrasolar Planets* (D. Deming and S. Seager, eds.), pp. 449–456. ASP Conf. Series 294, San Franciso.

Charbonneau D., Noyes R. W., Korzennik S. G., Nisenson P., Jha S., et al. (1999) *Astrophys. J., 522,* L145–L148.

Charbonneau D., Brown T. M., Latham D. W., and Mayor M. (2000) *Astrophys. J., 529,* L45–L48.

Charbonneau D., Brown T. M., Noyes R. W., and Gilliland R. L. (2002) *Astrophys. J., 568,* 377–384.

Charbonneau D., Brown T. M., Dunham E. W., Latham D. W., Looper D. L., et al. (2004) In *The Search for Other Worlds* (S. Holt and D. Deming, eds.), pp. 151–160. AIP Conf. Proc. 713, New York.

Charbonneau D., Allen L. E., Megeath S. T., Torres G., Alonso R., et al. (2005) *Astrophys. J., 626,* 523–529.

Charbonneau D., Winn J. N., Latham D. W., Bakos G., Falco E., et al. (2006) *Astrophys. J., 636,* 445–452.

Cho J. Y.-K., Menou K., Hansen B. M. S., and Seager S. (2003) *Astrophys. J., 587,* L117–L120.

Christian D. J., Pollacco D. L., Clarkson W. I., Collier Cameron A., Evans N., et al. (2004) In *The 13th Cool Stars Workshop* (F. Favata, ed.), pp. 475–477. ESA Spec. Publ. Series, Noordwijk, The Netherlands.

Cody A. M. and Sasselov D. D. (2002) *Astrophys. J., 569,* 451–458.

Collier Cameron A., Horne K., Penny A., and Leigh C. (2002) *Mon. Not. R. Astron. Soc., 330,* 187–204.

Collins G.W., da Silva L. B., Celliers P., et al. (1998) *Science, 281,* 1178–1181.

Cooper C. S. and Showman A. P. (2005) *Astrophys. J., 629,* L45–L48.

da Silva R., Udry S., Bouchy F., Mayor M., Moutou C., et al. (2006) *Astron. Astrophys., 446,* 717–722.

Deeg H. J., Garrido R., and Claret A. (2001) *New Astron., 6,* 51–60.

Deeg H. J., Alonso R., Belmonte J. A., Alsubai K., Horne K., et al. (2004) *Publ. Astron. Soc. Pac., 116,* 985–995.

Deming D., Seager S., Richardson L. J., and Harrington J. (2005a) *Nature, 434,* 740–743.

Deming D., Brown T. M., Charbonneau D., Harrington J., and Richardson L. J. (2005b) *Astrophys. J., 622,* 1149–1159.

Deming D., Harrington J., Seager S., and Richardson L. J. (2006) *Astrophys. J., 644,* 560–564.

Desjarlais M. P. (2003) *Phys Rev. B, 68,* 064204.

Drake A. J. (2003) *Astrophys. J., 589,* 1020–1026.

Dreizler S., Rauch T., Hauschildt P., Schuh S. L., Kley W., et al. (2002) *Astron. Astrophys., 391,* L17–L20.

Dunham E. W., Mandushev G. I., Taylor B. W., and Oetiker B. (2004) *Publ. Astron. Soc. Pac., 116,* 1072–1080.

Fischer D., Laughlin G., Butler R. P., Marcy G., Johnson J., et al. (2005) *Astrophys. J., 620,* 481–486.

Fortney J. J. (2005) *Mon. Not. R. Astron. Soc., 364,* 649–653.

Fortney J. J. and Hubbard W. B. (2003) *Icarus, 164,* 228–243.

Fortney J. J., Sudarsky D., Hubeny I., Cooper C. S., Hubbard W. B., et al. (2003) *Astrophys. J., 589,* 615–622.

Fortney J. J., Marley M. S., Lodders K., Saumon D., and Freedman R. S. (2005) *Astrophys. J., 627,* L69–L72.

Fortney J. J., Saumon D., Marley M. S., Lodders K., and Freedman R. (2006) *Astrophys. J., 642,* 495–504.

Gaudi B. S. (2005) *Astrophys. J., 628,* L73–L76.

Gaudi B. S., Seager S., and Mallen-Ornelas G. (2005) *Astrophys. J., 623,* 472–481.

Gilliland R. L., Brown T. M., Guhathakurta P., Sarajedini A., Milone E. F., et al. (2000) *Astrophys. J., 545,* L47–L51.

Girardi L., Bertelli G., Bressan A., Chiosi C., Groenewegen M. A. T., et al. (2002) *Astron. Astrophys., 391,* 195–212.

Gould A., Pepper J., and DePoy D. L. (2003) *Astrophys. J., 594,* 533–537.

Green D., Matthews J., Seager S., and Kuschnig R. (2003) *Astrophys. J., 597,* 590–601.

Greenberg R. (1974) *Icarus, 23,* 51–58.

Guillot T. (2005) *Ann. Rev. Earth Planet. Sci., 33,* 493–530.

Guillot T. and Showman A. P. (2002) *Astron. Astrophys., 385,* 156–165.

Guillot T., Burrows A., Hubbard W. B., Lunine J. I., and Saumon D. (1996) *Astrophys. J., 459,* L35–L38.

Hartman J. D., Stanek K. Z., Gaudi B. S., Holman M. J., and McLeod B. A. (2005) *Astron. J., 130,* 2241–2251.

Henry G. W., Marcy G. W., Butler R. P., and Vogt S. S. (2000) *Astrophys. J., 529,* L41–L44.

Holman M. J. and Murray N. W. (2005) *Science, 307,* 1288–1291.

Holmes N. C., Ross M, and Nellis W. J. (1995) *Phys. Rev. B., 52,* 15835–15845.

Hubbard W. B. (1968) *Astrophys. J., 152,* 745–754.

Hubbard W. B. (1977) *Icarus, 30,* 305–310.

Hubbard W. B., Fortney J. J., Lunine J. I., Burrows A., Sudarsky D., et al. (2001) *Astrophys. J., 560,* 413–419.

Hubickyj O., Bodenheimer P., and Lissauer J. J. (2004) In *Gravitational Collapse: From Massive Stars to Planets* (G. García- Segura et al., eds.), pp. 83–86. Rev. Mex. Astron. Astrophys. Conf. Series, UNAM.

Hut P. (1980) *Astron. Astrophys., 92,* 167–170.

Janes K. (1996) *J. Geophys. Res., 101,* 14853–14860.

Jenkins J. M. and Doyle L. R. (2003) *Astrophys. J., 595,* 429–445.

Jenkins J. M., Caldwell D. A., and Borucki W. J. (2002) *Astrophys. J., 564,* 495–507.

Jha S., Charbonneau D., Garnavich P. M., Sullivan D. J., Sullivan T., et al. (2000) *Astrophys. J., 540,* L45–L48.

Kallrath J. and Milone E. F. (1999) In *Eclipsing Binary Stars: Modeling and Analysis,* pp. 60–64. Springer, New York.

Knudson M. D., Hanson D. L., Bailey J. E., Hall C. A., Asay J. R., et al. (2004) *Phys. Rev. B, 69,* 144–209.

Konacki M., Torres G., Jha S., and Sasselov D. D. (2003a) *Nature, 421,* 507–509.

Konacki M., Torres G., Sasselov D. D., and Jha S. (2003b) *Astrophys. J., 597,* 1076–1091.

Konacki M., Torres G., Sasselov D. D., Pietrzynski G., Udalski A., et al. (2004) *Astrophys. J., 609,* L37–L40.

Konacki M., Torres G., Sasselov D. D., and Jha S. (2005) *Astrophys. J., 624,* 372–377.

Kovács G., Zucker S., and Mazeh T. (2002) *Astron. Astrophys., 391,* 369–377.

Kurucz R. (1992) In *The Stellar Populations of Galaxies* (B. Barbuy and A. Renzini, eds.), pp. 225–232. Kluwer, Dordrecht.

Latham D. W. (2003) In *Scientific Frontiers in Research on Extrasolar Planets* (D. Deming and S. Seager, eds.), pp. 409–412. ASP Conf. Series 294, San Francisco.

Laughlin G., Wolf A., Vanmunster T., Bodenheimer P., Fischer D., et al. (2005a) *Astrophys. J., 621,* 1072–1078.

Laughlin G., Marcy G. W., Vogt S. S., Fischer D. A., and Butler R. P. (2005b) *Astrophys. J., 629,* L121–L124.

Lecavelier des Etangs A., Vidal-Madjar A., McConnell J. C., and Hébrard G. (2004) *Astron. Astrophys., 418,* L1–L4.

Leigh C., Collier Cameron A., Udry S., Donati J.-F., Horne K., et al. (2003a) *Mon. Not. R. Astron. Soc., 346,* L16–L20.

Leigh C., Collier Cameron A., Horne K., Penny A., and James D. (2003b) *Mon. Not. R. Astron. Soc., 344,* 1271–1282.

Lin D. N. C., Bodenheimer P., and Richardson D. C. (1996) *Nature, 380,* 606–607.

Lindal G. F., Wood G. E., Levy G. S., Anderson J. D., Sweetnam D. N., et al. (1981) *J. Geophys. Res., 86,* 8721–8727.

Lissauer J. J. (1993) *Ann. Rev. Astron. Astrophys., 31,* 129–174.

Mallen-Ornelas G., Seager S., Yee H. K. C., Minniti D., Gladders M. D., et al. (2003) *Astrophys. J., 582,* 1123–1140.

Mandel K. and Agol E. (2002) *Astrophys. J., 580,* L171–L174.

Mandushev G., Torres G., Latham D. W., Charbonneau D., Alonso R., et al. (2005) *Astrophys. J., 621,* 1061–1071.

Marley M. S., Gelino C., Stephens D., Lunine J. I., and Freedman R. (1999) *Astrophys. J., 513,* 879–893.

Marshall J. L., Burke C. J., DePoy D. L., Gould A., and Kollmeier J. A. (2005) *Astron. J., 130,* 1916–1928.

Mayor M. and Queloz D. (1995) *Nature, 378,* 355–359.

Mazeh T., Naef D., Torres G., Latham D. W., Mayor M., et al. (2000) *Astrophys. J., 532,* L55–L58.

McArthur B. E., Endl M., Cochran W. D., Benedict G. F., Fischer D. A., et al. (2004) *Astrophys. J., 614,* L81–L84.

McCullough P. R., Stys J. E., Valenti J. A., Fleming S. W., Janes K. A., et al. (2005) *Publ. Astron. Soc. Pac., 117,* 783–795.

McLaughlin D. B. (1924) *Astrophys. J., 60,* 22–31.

Menou K, Cho J. Y-K., Hansen B. M. S., and Seager S. (2003) *Astrophys. J., 587,* L113–L116.

Militzer B. and Ceperley D. M. (2001) *Phys. Rev. E, 63,* 066404.

Miralda-Escudé J. (2002) *Astrophys. J., 564,* 1019–1023.

Mizuno H. (1980) *Prog. Theor. Phys., 64,* 544–557.

Mochejska B. J., Stanek K. Z., Sasselov D. D., Szentgyorgyi A. H., Bakos G. Á., et al. (2005) *Astron. J., 129,* 2856–2868.

Mochejska B. J., Stanek K. Z., Sasselov D. D., Szentgyorgyi A. H., Adams E., et al. (2006) *Astron. J., 131,* 1090–1105.

Moutou C., Coustenis A., Schneider J., St. Gilles R., Mayor M., et al. (2001) *Astron. Astrophys., 371,* 260–266.

Moutou C., Coustenis A., Schneider J., Queloz D., and Mayor M. (2003)

Astron. Astrophys., 405, 341–348.

Moutou C., Pont F., Bouchy F., and Mayor M. (2004) *Astron. Astrophys., 424*, L31–L34.

Murray N., Hansen B., Holman M, and Tremaine S. (1998) *Science, 279*, 69–72.

Narita N., Suto Y., Winn J. N., Turner E. L., Aoki W., et al. (2005) *Publ. Astron. Soc. Japan, 57*, 471–480.

O'Donovan F. T., Charbonneau D., Torres G., Mandushev G., Dunham E. W., et al. (2006) *Astrophys. J.,* in press.

Ohta Y., Taruya A., and Suto Y. (2005) *Astrophys. J., 622*, 1118–1135.

Peale S. J. (1969) *Astron. J., 74*, 483–489.

Pepper J. and Gaudi B. S. (2005) *Astrophys. J., 631*, 581–596.

Pepper J., Gould A., and Depoy D. L. (2003) *Acta Astron., 53*, 213–228.

Pepper J., Gould A., and Depoy D. L. (2004) In *The Search for Other Worlds* (S. Holt and D. Deming, eds.), pp. 185–188. AIP Conf. Series 713, New York.

Pollack J. B., Hubickyj O., Bodenheimer P., Lissauer J. J., Podolak M., et al. (1996) *Icarus, 124*, 62–85.

Pont F., Bouchy F., Queloz D., Santos N. C., Melo C., et al. (2004) *Astron. Astrophys., 426*, L15–L18.

Pont F., Bouchy F., Melo C., Santos N. C., Mayor M., et al. (2005) *Astron. Astrophys., 438*, 1123–1140.

Queloz D., Eggenberger A., Mayor M., Perrier C., Beuzit J. L., et al. (2000) *Astron. Astrophys., 359*, L13–L17.

Rauer H., Eislöffel J., Erikson A., Guenther E., Hatzes A. P., et al. (2004) *Publ. Astron. Soc. Pac., 116*, 38–45.

Richardson L. J., Deming D., and Seager S. (2003a) *Astrophys. J., 597*, 581–589.

Richardson L. J., Deming D., Wiedemann G., Goukenleuque C., Steyert D., et al. (2003b) *Astrophys. J., 584*, 1053–1062.

Rivera E. J., Lissauer J. J., Butler R. P., Marcy G. W., Vogt S. S., et al. (2005) *Astrophys. J., 634*, 625–640.

Rossiter R. A. (1924) *Astrophys. J., 60*, 15–21.

Rowe J. F., Matthews J. M., Seager S., Kuschnig R., Guenther D. B., et al. (2006) *Astrophys. J.,* in press.

Santos N. C., Bouchy F., Mayor M., Pepe F., Queloz D., et al. (2004) *Astron. Astrophys., 426*, L19–L23.

Sato B., Fischer D. A., Henry G. W., Laughlin G., Butler R. P., et al. (2005) *Astrophys. J., 633*, 465–473.

Saumon D. and Guillot T. (2004) *Astrophys. J., 460*, 993–1018.

Saumon D., Chabrier G., Wagner D. J., and Xie X. (2000) *High Pressure Res., 16*, 331–343.

Seager S. and Mallen-Ornelas G. (2003) *Astrophys. J., 585*, 1038–1055.

Seager S. and Sasselov D. D. (1998) *Astrophys. J., 502*, L157–L161.

Seager S. and Sasselov D. D. (2000) *Astrophys. J., 537*, 916–921.

Seager S., Whitney B. A., and Sasselov D. D. (2000) *Astrophys. J., 540*, 504–520.

Seager S., Richardson L. J., Hansen B. M. S., Menou K., Cho J. Y-K., et al. (2005) *Astrophys. J., 632*, 1122–1131.

Showman A. P. and Guillot T. (2002) *Astron. Astrophys., 385*, 166–180.

Silva A. V. R. (2003) *Astrophys. J., 585*, L147–L150.

Sirko E. and Paczyński B. (2003) *Astrophys. J., 592*, 1217–1224.

Sozzetti A., Yong D., Torres G., Charbonneau D., Latham D. W., et al. (2004) *Astrophys. J., 616*, L167–L170.

Steffen J. H. and Agol E. (2005) *Mon. Not. R. Astron. Soc., 364*, L96–L100.

Struve O. (1952) *Observatory, 72*, 199–200.

Sudarsky D., Burrows A., and Pinto P. (2000) *Astrophys. J., 538*, 885–903.

Sudarsky D., Burrows A., and Hubeny I. (2003) *Astrophys. J., 588*, 1121–1148.

Tingley B. (2004) *Astron. Astrophys., 425*, 1125–1131.

Torres G., Konacki M., Sasselov D. D., and Jha S. (2004a) *Astrophys. J., 609*, 1071–1075.

Torres G., Konacki M., Sasselov D. D., and Jha S. (2004b) *Astrophys. J., 614*, 979–989.

Torres G., Konacki M., Sasselov D. D., and Jha S. (2005) *Astrophys. J., 619*, 558–569.

Udalski A., Paczynski B., Zebrun K., Szymaski M., Kubiak M., et al. (2002a) *Acta Astron., 52*, 1–37.

Udalski A., Zebrun K., Szymanski M., Kubiak M., Soszynski I., et al. (2002b) *Acta Astron., 52*, 115–128.

Udalski A., Szewczyk O., Zebrun K., Pietrzynski G., Szymanski M., et al. (2002c) *Acta Astron., 52*, 317–359.

Udalski A., Pietrzynski G., Szymanski M., Kubiak M., Zebrun K., et al. (2003) *Acta Astron., 53*, 133–149.

Udalski A. Szymanski M. K., Kubiak M., Pietrzynski G., Soszynski I., et al. (2004) *Acta Astron., 54*, 313–345.

Vidal-Madjar A., Lecavelier des Etangs A., Désert J.-M., Ballester G. E., Ferlet R., et al. (2003) *Nature, 422*, 143–146.

Vidal-Madjar A., Désert J.-M., Lecavelier des Etangs A., Hébrard G., Ballester G. E., et al. (2004) *Astrophys. J., 604*, L69–L72.

von Braun K., Lee B. L., Seager S., Yee H. K. C., Mallén-Ornelas G., et al. (2005) *Publ. Astron. Soc. Pac., 117*, 141–159.

Walker G., Matthews J., Kuschnig R., Johnson R., Rucinski S., et al. (2003) *Publ. Astron. Soc. Pac., 115*, 1023–1035.

Williams P. K. G., Charbonneau D., Cooper C. S., Showman A. P., and Fortney J. J. (2006) *Astrophys. J.,* in press.

Winn J. and Holman M. J. (2005) *Astrophys. J., 628*, L159–L162.

Winn J. N., Suto Y., Turner E. L., Narita N., Frye B. L., et al. (2004) *Publ. Astron. Soc. Japan, 56*, 655–662.

Winn J. N., Noyes R. W., Holman M. J., Charbonneau D., Ohta Y., et al. (2005) *Astrophys. J., 631*, 1215–1226.

Wittenmyer R. A., Welsh W. F., Orosz J. A., Schultz A. B., Kinzel W., et al. (2005) *Astrophys. J., 632*, 1157–1167.

Wolf A., Laughlin G., Henry G. W., Fischer D. A., Marcy G., et al. (2006) *Astrophys. J.,* in press.

Direct Detection of Exoplanets

J.-L. Beuzit
Laboratoire d'Astrophysique de Grenoble

D. Mouillet
Observatoire Midi-Pyrénées

B. R. Oppenheimer
American Museum of Natural History

J. D. Monnier
University of Michigan, Ann Arbor

Direct detection of exoplanets from the ground is now within reach of existing astronomical instruments. Indeed, a few planet candidates have already been imaged and analyzed and the capability to detect (through imaging or interferometry) young, hot, Jupiter-mass planets exists. We present here an overview of what such detection methods can be expected to do in the near and far term. These methods will provide qualitatively new information about exoplanets, including spectroscopic data that will mature the study of exoplanets into a new field of comparative exoplanetary science. Spectroscopic study of exoplanet atmospheres promises to reveal aspects of atmospheric physics and chemistry as well as internal structure. Astrometric measurements will complete orbital element determinations partially known from the radial velocity surveys. We discuss the impact of these techniques, on three different timescales, corresponding to the currently available instruments, the new "Planet Finder" systems under development for 8- to 10-m telescopes, foreseen to be in operation in 5–10 years, and the more ambitious but more distant projects at the horizon of 2020.

1. INTRODUCTION

Since the discovery of a planet around a solar-type star 10 years ago by *Mayor and Queloz* (1995), the study of exoplanets has developed into one of the primary research areas in astronomy today. More than 170 exoplanets have been found orbiting stars of spectral types F to M, with a significant fraction in multiplanet systems (see the chapter by Udry et al. for a review). These exoplanets have been discovered using indirect detection methods, in which only the planet's influence on the host star is observed.

Indirect detection techniques include radial velocity measurements, which detect the movement of a star due to a planet's gravitational influence (see *Marcy et al., 2003*; chapter by Udry et al.); photometric transit observations, which detect the variation in the integrated stellar flux due to a planetary companion passing through the line of sight to its host star (see chapter by Charbonneau et al.); as well as astrometry, which also detects stellar motion (*Sozzetti, 2005*); and gravitational microlensing, which involves unrepeatable observations biased toward planets with short orbital periods (*Mao and Paczynski, 1991*; *Gould and Loeb, 1992*).

Almost all the currently known exoplanets have been detected by radial velocity measurements. Due to the unknown inclination angle of the orbit, only a lower limit of the mass can actually be derived for each individual planet candidate. The statistical distribution of exoplanets can still be obtained thanks to the large number of detections. These exoplanets typically have masses similar to those of the giant, gaseous planets in our own solar system. They are therefore generally referred to as extrasolar giant planets, or EGPs. Since 2004, a few planets with minimum masses ranging between 6 and 25 $M_\oplus$ have been detected by radial velocity measurements (*Butler et al., 2004*; *Rivera et al., 2005*; *Bonfils et al., 2005*). Very recently, *Beaulieu et al.* (2006) have discovered a 5.5 $M_\oplus$ planet by gravitational lensing, assuming a mass of 0.2 $M_\odot$ for the host star, based on the peak of the IMF at that mass.

These indirect methods have proven to be very successful in detecting exoplanets, but they only provide limited information about the planets themselves. For example, radial velocity detections allow derivation of a planet's orbital period and eccentricity as well as a lower limit to its mass due to the unknown inclination angle of the orbit. Photometric transit observations provide information about a planet's radius and, with great effort, limited measurements of the composition of its upper atmosphere (*Jha et al., 2000*). In addition, large radial velocity surveys with sufficient precision only started about a decade ago. Thus, they are sensitive only to exoplanets with relatively small orbital periods, typically corresponding to objects at distance smaller than a few AU from their parent stars. Currently, these surveys are completely insensitive to planets at separations comparable to those of Jupiter and Saturn in our solar system. Finally, the accuracy of radial velocity measurements strongly biases the detections based on the type of the host stars toward old, quiet, and solar mass (G-K) stars. Future extensive search for transiting planets will suffer similar biases regarding orbital periods and stellar types.

Direct detection and spectroscopy of the radiation from these exoplanets is necessary to determine their physical parameters, such as temperature, pressure, chemical composition, and atmospheric structure. These parameters are critically needed to constrain theories of planet formation and evolution. Furthermore, direct detection enables the study of planets in systems like our own. In these respects, direct detection is complementary to the indirect methods, especially to the radial velocity technique.

However, direct observation of exoplanets is still at the edge of the current capabilities, able to reveal only the most favorable cases of very young and massive planets at large distances from the central star (see section 2.1). The major challenge for direct study of the vast and seemingly diverse population of exoplanets resides in the fact that most of the planets are believed to be 10^6–$10^{12}\times$ fainter than their host stars, at separations in the subarcsecond regime. This requires both high-contrast and spatial resolution.

Various instrumental approaches have been proposed for the direct detection of exoplanets, either from the ground or from space. In particular, several concepts of high-order adaptive optics (AO) systems dedicated to groundbased large (and small) telescopes have been published during the past 10 years (*Angel and Burrows, 1995; Dekany et al., 2000; Mouillet et al., 2002; Macintosh et al., 2002; Oppenheimer et al., 2003*). Interferometric systems, using either a nulling or a differential phase approach, have also been described (see section 4). Space missions have been proposed, relying either on coronagraphic imaging [TPF-C observatory (see chapter by Stapelfeldt et al.)] or on interferometry [TPF-I and Darwin projects (e.g., *Mennesson et al., 2005; Kaltenegger and Fridlund, 2005*, and references therein)].

In this chapter, we present the scientific objectives of the direct detection of exoplanets, concentrating on groundbased approaches (see chapter by Stapelfeldt et al. for space-based approaches). We discuss the principle, performance, and impact of the high-contrast imaging and interferometric techniques on three different timescales, corresponding to the currently available instruments, the new "Planet Finder" systems under development and foreseen to be in operation in about five years, and the more ambitious projects on the 2020 horizon.

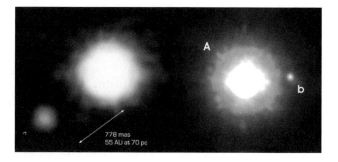

Fig. 1. Examples of detection of planetary mass objects using current AO instruments. (Left) 2M1207 found by *Chauvin et al.* (2005a). (Right) GQ Lupi found by *Neuhäuser et al.* (2005). Both objects are estimated by some researchers to have planetary mass. See section 2.1 for more detail.

2. SCIENCE

2.1. Direct Detections

Observations obtained in the past few years have reached contrasts compatible with the detection of Jupiter-mass planets in the most favorable case of very young ages (10^6 to a few 10^7 yr) when EGPs are still warm. Detection is typically possible outside a ~0.5" radius from the host star, corresponding to a few tens of AU at the typical distances of young nearby stellar associations, i.e., closer than 100 pc (*Zuckerman and Song, 2004*).

A few AO surveys of some of these associations conducted in recent years (e.g., *Chauvin et al., 2002; Neuhäuser et al., 2003; Brandeker et al., 2003; Beuzit et al., 2004; Masciadri et al., 2005*) indeed resulted in the first three direct detections of exoplanet candidates. These discoveries include a companion around the brown dwarf 2MASSWJ 1207334-393254 (hereafter 2M1207), a member of the young TW Hydrae Association, by *Chauvin et al. (2005a)*; around the classical T Tauri star GQ Lup by *Neuhäuser et al. (2005)*; and around AB Pic, a member of the large Tucana-Horologium association, by *Chauvin et al. (2005b)*. Figure 1 illustrates the detection of two of these planet candidates, 2M1207 B and GQ Lup.

The analyses of the proper motions, colors, and low-resolution spectra of these objects definitely confirm their status as co-moving, very-low-mass, companions. An estimation of the actual mass of these companions requires the determination of their luminosity and temperature through theoretical models, such as the Tucson (*Burrows et al., 1997*) and Lyon (*Chabrier et al., 2000; Baraffe et al., 2002*) models. These models are known to be uncertain at early ages, typically <100 m.y., and the derived mass estimates, and therefore whether these companions are brown dwarfs or planets, should be considered with caution (see *Neuhäuser, 2006*).

In any case, these observational results can now be directly compared to theoretical models. Such detections at very large orbital separations (55 to 250 AU) also possibly constrain the formation mechanisms of such planetary mass objects (see, for example, chapters by Lissauer et al., Durisen et al., Papaloizou et al., and Chabrier et al. for a discussion of these mechanisms).

These first direct detections are very encouraging and they clearly open a new window for exoplanet characterization. As with the radial velocity techniques, meaningful astrophysical results can only be derived when a larger sample of observations will be available. The rest of this paper presents a wider perspective of the primary goals of such an approach, the expected science impact, and the corresponding observational requirements.

2.2. Primary Science Goals

In order to test these theories, it is crucial to obtain both statistical and direct information on these planets. A variety of observational techniques, direct and indirect, must ideally be used in parallel to reach this goal. As mentioned above, indirect detection methods suffer from ambiguities in interpretation: For instance, the observation of an occultation (or transit) provides the radius of the companion, but without knowing the object's mass, it is impossible to know, in many cases, whether the object is in fact a planet, brown dwarf (whose radii are similar to EGPs), or white dwarf (whose radii are similar to Earth-sized planets). Furthermore, radial velocity techniques are limited to only certain types of stars. Direct detection, in contrast, can survey, in theory, any type of star, including searches around degenerates such as white dwarfs and brown dwarfs for planetary mass companions. Direct detection is also far more efficient, even for long-period planets, since it requires only two or three observations to confirm a detection, and only one to achieve spectra, rather than the multitude of observations over long periods of time necessary for radial velocity searches.

Direct detection of exoplanets in various planetary systems represents a great leap in our knowledge of these objects, as well as a rigorous maturation of the field. Some of the expected scientific returns from these observations are detailed below.

1. A removal of the sin i ambiguity that plagues the radial velocity measurements will not only remove statistical biases inherent in this technique (e.g., *Han et al.,* 2001), but also tests whether planetary systems generally have a single orbital plane for all planets (as seen in our solar system). This is a crucial aspect of formation theories (e.g., *Butler et al.,* 1999).

2. Detection of planets around active stars is only possible through direct detection, and perhaps marginally by astrometry and transit monitoring. In particular, young stars, which are choice targets for the study of planet formation, are particularly active. These kinds of studies can lead to an understanding of planet evolution.

3. Complete orbit determination leads to improved statistical understanding of planets as a function of mass, parent star type, and age (i.e., orbital evolution). A significant fraction of the orbit must be covered for such studies, requiring long-term monitoring campaigns commensurate with that of indirect methods (10–20 years).

4. Measurement of planet colors will constitute the first direct tests of atmospheric models, and will provide information on the presence of clouds in the planetary atmosphere, for instance.

5. Determination of the intrinsic luminosity of planets at various ages will yield additional constraints on models of planet evolution.

6. Spectroscopy of exoplanets. Surprisingly, the heavy-element composition of solar gas giants is still poorly understood (e.g., *Guillot,* 1999). Spectral features of CH_4, NH_3, and H_2O are observed at resolutions R ~ 20–30. For brown dwarfs and luminous planets, narrow features of KI, NaI, and FeH are observed in the J band and require R ~ 1000 to be characterized. Major changes in spectral appearance occur as the amount and height of dust in the atmosphere changes, significantly modifying the heavy elements available for formation of molecules (*Burrows and Sharp,* 1999). Of particular interest is the comparison of these features in different planets along with such information as the metallicity of the parent star.

7. Measurement of spectral variability in planets due to activity or weather. Planets with effective temperatures between 200 and 300 K are expected to have atmospheres dominated by water clouds (e.g., Earth) (*Guillot,* 1999). Such clouds yield activity detectable by high dynamic range observations through three interrelated processes: (a) inhomogeneities and rotation, (b) meteorological activity, and (c) the presence of dynamical waves. The third process may be the most promising, as it is thought to be responsible for the presence of Jupiter's 5-μm hot spots. In the case of planets with water clouds and a higher effective temperature, the effect should be more pronounced and displaced to shorter wavelengths.

8. Detection of exoplanetary ring systems. From our current limited statistics (i.e., the solar system), a substantial fraction of the discovered gas giants may have rings. These can be detected with polarimetry. In addition, changes in the projection of the rings with orbital phase will produce changes in a planet's spectrum. The effect of rings on planets where reflected light is an insignificant fraction of the emergent radiation must still be understood theoretically (i.e., how does luminosity modulate or degree of polarization change), but for systems dominated by reflected star light, the impact of rings has been evaluated (Figs. 3–16 in *Arnold and Schneider,* 2004). Luminosity modulation and polarization depend on ring geometry and opacity. In some configurations these effects may augment detectability of the planet.

9. Polarimetry of planets in general. Although the fractional polarization from self-luminous planets is expected to be very small (typically lower than 1%), significant asymmetries in cloud distributions can induce higher polarization. In reflection, more than 20% polarization is seen at some orbital phases in the solar system. For a planet at the very short orbital distance of 0.3 AU (0.1" at 3 pc), with 20%

polarization and an albedo similar to that of Jupiter (0.3), the star-planet contrast will be on the order of 17.6 mag, with a strong dependence on the orbital phase and system inclination. Such systems might be the first exoplanets imaged in reflected light.

10. Detection of moons. If our solar system is any indication of planets in general, most exoplanets will harbor moons. These can be detected in the future through searches for transits of the systems found, particularly in edge-on configurations. Moons the size of Earth and even smaller may be detectable with 1% photometry of EGPs.

11. Evidence for biological activity. It would be irresponsible for us not to mention the possibility that some of the first signs of life outside the solar system may come in the distant future from spectroscopic observations of planets. A cottage industry of theorists modeling the chemical disequilibrium induced by biological activity on planets has emerged in anticipation of such results, and long-term monitoring of the Earth-shine spectrum is designed to build up a library of spectra (at least of a 4.6-G.y. Earth-like planet) with which to compare future exoplanet spectra (e.g., *Kasting and Catling,* 2003; *Woolf et al.,* 2002).

These science objectives indicate an incredibly wealthy field, one that routinely draws unprecedented interest from the public and is responsible for many new students entering astronomy. The record attendance at the Protostars and Planets V conference is the best testimonial to this statement. The bottom line is that this field of research is broad, interdisciplinary, and exciting in that it is attacking some of the most fundamental questions science can ask. A galaxy of worlds awaits our inspection. The future of astrophysics and planetary science lies in comparative exoplanetary science.

2.3.   Survey Design

The frequency of planets in the mass and period ranges accessible to the proposed techniques is essentially unprobed (Fig. 2). This means that understanding planets in general requires collection of a large sample, through which salient properties and generalizations can be made and tied to theory. Direct observations require extensive surveys of hundreds of target stars. Such dedicated surveys must be optimized to ensure observing efficiency and data consistency compatible with statistically relevant analyses. Indeed, advanced survey design and optimization is already being explored, in particular by *Brown* (2005) in the context of the Terrestrial Planet Finder. However, some simpler conceptions of surveys can be imagined, depending on the type of observation, expected system performance and telescope design. The following is an expurgated, but representative, list of such surveys.

1. Young and nearby stars: Young nearby stellar associations are ideal targets for the direct detection of exoplanets, since (a) they are typically closer than 100 pc and therefore allow the exploration of their circumstellar environment at very small distances from the host star, and (b) substellar objects are hotter and brighter when young (~10 m.y.) and

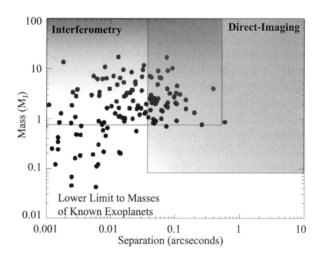

Fig. 2. Mass vs. separation diagram illustrating the complementarity of the different techniques used for extrasolar planet detection. The black circles indicate known planets discovered through indirect methods. The shaded regions show the approximate parameter space accessible, by 2020, to the direct techniques employing interferometry and high-resolution imaging. Both direct and indirect techniques are complementary. See text for a more complete discussion.

can be more easily detected than evolved companions, because the contrast between star and planet is smaller. The first direct detection of planetary mass objects used such associations to induce an age bias in the sample (see section 2.1 above).

2. Nearest active stars: Moderately young stars (younger than ~1 G.y.) in the solar neighborhood (within 50 pc of the Sun) offer major advantages for direct imaging due to their proximity and favorable contrast (again due to youth). For the most active of these stars, radial velocity detection of planets is very difficult if not impossible since the stars exhibit large intrinsic rotational velocities. Direct detection is thus the only possibility for exploring the planetary systems of these stars. For stars with distances from 10 to 50 pc, translating into separations of 1–100 AU or periods of 1–1000 years for a solar mass primary, direct observation probes a separation range much larger than, and complementary to, that of radial velocities. For stars closer than 10 pc, radial velocity and direct imaging techniques overlap, allowing precise mass and luminosity estimates. These feed directly into an observed mass-luminosity relation. Finally, the closest of these stars will also offer the only opportunities for detecting planets by reflected light at short orbital distances in the near future.

3. Late-type stars (M dwarfs): These stars represent even more favorable targets for direct detection since the contrast between the star and the planet is much lower.

4. Stars with planets known from radial velocity surveys, especially stars that exhibit long-term residuals in their radial velocity curves: Such residuals indicate the presence of a more distant companion, perhaps more easily found through direct methods. When direct imaging instruments

come into operation, we expect that the sample of planets discovered by radial velocity techniques will have increased significantly and include planets with longer periods. Such planets are at larger separations from their stars, which is favorable for direct detection.

2.4. Key Scientific Requirements

The scientific requirements driving the design of instruments for the direct exoplanet detection are based on the size of our own solar system (which determines the angular resolution and field of view needed to survey the nearest 1000 or so stars), the spectra of planets in our solar system, and theoretical models of EGPs (e.g., *Burrows*, 2005, and references therein, which set contrast requirements).

1. Sensitivity to wavelengths from 0.6 to 2.5 μm: Giant gaseous planets are dominated by CH_4 features in the J and H bands, while for terrestrial planets, the wavelength range 700–900 nm is particularly interesting, especially with astrophysically unique features like the O_2 band. Wavelengths shorter than 700 nm have enhanced polarization in solar system objects, feeding into some of the science objectives mentioned above. Observations longward of 2.5 μm from the ground are heavily impeded by the terrestrial atmosphere, and are restricted to spacebased observations.

2. Extremely high contrast: Young EGPs are believed to be typically 15 mag ($10^6×$) fainter than their host stars, while an analog of Earth is between 25 and 35 mag (10^{10}–$10^{14}×$) fainter, depending on the wavelength of observation (e.g., *Segura et al.*, 2005).

3. Very high angular resolution: Access to angular separations as small as 20 mas is needed to resolve the radial scale of Earth's orbit at a distance of 25 pc, within which there are thousands of stars to survey. This angle is also called the "inner working angle."

4. Total field-of-view extending to 2″ to 4″ in diameter: A field-of-view of 2″ corresponds to an exploration region of 1 AU in radius at 1 pc, roughly the distance to the nearest star. A field-of-view of 4″ covers a region similar to the scale of our own solar system at distances of around 10 pc.

5. Relative astrometry at better than 10 mas precision: This permits discrimination of background objects from bona fide companions and accurate measurement of orbital motion on timescales of less than a few years around stars within 10 pc.

Early instruments in this endeavor will be limited in many respects. Achieving these goals requires an acute awareness of the technology available. For example, the sensitivity of high-resolution AO approaches (which will be discussed in more detail below) is limited to stars with visual magnitudes of about 9 or 10, and interferometers struggle to operate effectively on stars four or five times as bright. Clearly, extending this performance to fainter stars is necessary to enable surveys of thousands of stars. In addition, due to the faintness of the planets and the various noise sources impeding their detection, spectral resolution over the desired wavelength range will be limited to R ~ 30 typically in the near term. Differential polarization from the ground is already making progress with the detection of protoplanetary disks and is planned to be included in the upcoming groundbased exoplanet imaging systems. Ultimately, much of this work will have to be done in space. However, spacebased missions have other limitations, such as duration, which means that the groundbased observations will always play a complementary role. The spacebased missions are described in the chapter by Stapelfeldt et al.

3. HIGH-CONTRAST IMAGING

3.1. System Overview

The primary challenge in imaging exoplanets is the presence of a star at the center of the field of view and up to $10^{12}×$ brighter than the planet. Eliminating the light of this star without damaging that of the planet represents a major technical challenge and, in the case of imaging (interferometry is discussed in section 4), requires new technology. In particular we identify the following challenges:

1. Correct the point-spread function (PSF): (a) confine the disturbing stellar light into a coherent, diffraction-limited pattern that is understood and controlled and can then be canceled (see next item); (b) concentrate the planetary flux, making the planet's peak brightness higher relative to the background due to the star light; and (c) reduce the incoherent stellar halo. If S is the image Strehl ratio (which is essentially the fraction of coherent energy in this regime), then the ratio of planet brightness to stellar halo background scales as S/(1–S). Note that with this asymptotic function, very large gains are made as S → 1.

2. Cancel the coherent part of the stellar PSF by coronagraphy: This reduces the fundamental Poisson noise limit and dramatically relaxes constraints on PSF calibration, such as flat-field accuracy and stability.

3. Calibrate the residual (usually incoherent) stellar halo including residuals from atmospheric effects (evolving on millisecond timescales) but also much slower instrumental effects. Any *a priori* information able to distinguish an artifact from a planetary signal is to be used in this case for calibration purposes, such as the point-like shape of the planet, spectral and/or polarization signatures, or perhaps the difference in coherence properties between the stellar halo and planet wavefronts.

Actually, these three steps are closely entangled in any practical system design: Adaptive optics correction for PSF improvement contributes to measurements and calibration of the PSF halo. Such information can be used to improve stellar halo suppression in real time, for example. In any case, these three steps are the fundamental guiding principles in instrument design and data analysis.

3.2. Current State of the Art: Necessary Subsystems

3.2.1. Correction: Adaptive optics. The primary source of disturbances to PSF quality is due to wavefront perturbations imposed during propagation through the inhomogeneous atmosphere. Adaptive optics is a technique that corrects these disturbances. The perturbations are measured with a dedicated wavefront sensor and corrections are applied by an optically conjugated deformable mirror (DM) in real-time. The maximum performance for a given system, usually quantified as S (above), is limited by technological constraints such as the number of actuators of the DM, the servo loop bandpass, and the overall system stability and calibration accuracy. The whole system acts as a filter, correcting most low-temporal (>10 ms) and spatial frequencies (>20 cm) of the wavefront perturbations. Remaining high spatial frequency defects translate in the image plane as a residual seeing-like halo extending outward from the "control radius" $\lambda/2d$ [where d is the interactuator spacing as projected on the telescope primary (*Oppenheimer et al.*, 2003)]. Within this AO "control radius," the departure from a pure diffraction pattern is due to remaining very fast wavefront variations and high-frequency aliasing.

The typical performance achieved by current systems is on the order of S ~ 50–60% on 8-m telescopes (*Clénet et al.*, 2004; *van Dam et al.*, 2004; *Stoesz et al.*, 2004) in the near-infrared and up to 90% on 4-m telescopes (*Oppenheimer et al.*, 2004). S ~ 93% in the near-infrared on 8-m telescopes is the operating goal of the next generation of instruments, requiring technological improvements. Such improvements include wavefront sensor detectors, microdeformable mirrors, and a new generation of real-time calculators, as a few examples.

With a reasonable interactuator spacing corresponding to d ~ 10–20 cm (matched to the approximate Fried parameter for the telescope and site), the corrected region of the image extends to 0.8" radius from the central star, which is well suited for planet searches (section 2.4). Progress on wavefront sensor detectors allows achieving correction on stars as faint as 11th mag, depending on the wavelength used for guiding, even with the required spatial and temporal sampling. This is compatible with investigations around a large target sample as discussed in section 2.

When trying to reach even higher levels of correction (S > 90%), new limiting factors appear and must be taken into account. This includes atmospheric scintillation effects and wavefront perturbation chromatism, among many others [see for example *Dekany* (2004) or *Fusco et al.* (2005) for in-depth discussions of these effects].

3.2.2. Cancellation: Coronagraphy. The basic purpose of coronagraphy (*Lyot*, 1939; *Sivaramakrishnan et al.*, 2001) is to cancel the coherent part of the stellar energy over the observed field of view. In addition to the intrinsic merit of reducing the Poisson noise in the PSF wings (by decreasing the number of photons in the image), coronagraphy also reduces the interference between the residual incoherent and coherent wavefronts that form pinned speckles (*Bloemhof et al.*, 2001; *Aime and Soummer*, 2004). This strongly relaxes the constraints on the quality of all optical components located *after* the coronagraphic devices.

Considering a well-corrected incoming wavefront, when the phase defects remain small, the corresponding image can be approximated by the sum of a pure diffraction pattern (Airy pattern in the case of a circular aperture) and

of an incoherent halo (shaped as the phase defect power spectrum density). In a perfect coronagraph the diffraction pattern is completely removed. In practice, however, no perfect coronagraph can be built and trade-offs must be made. The important parameters for a coronagraph are (1) the level of star light rejection, (2) the inner working angle, and (3) the transmission of the off-axis planet light.

Various coronagraphic concepts have been proposed in recent years, supported by impressive efforts in simulation and prototyping. These concepts include amplitude focal masks coupled with pupil stops or amplitude pupil masks, phase masks, pupil amplitude separation, and recombination (e.g., *Soummer et al.*, 2003; *Soummer*, 2005; *Guyon et al.*, 2005; *Baudoz et al.*, 2005). These concepts present various levels of efficiency and sensitivity with respect to chromatism, pupil shape, residual image defects, and residual tilt.

Laboratory experiments have convincingly demonstrated that, for future systems, the limitation should essentially come from the residual halo associated with phase defects rather than from the efficiency of the coronagraph itself (*Trauger et al.*, 2004). This preliminary conclusion should be revisited for even more ambitious projects (whether based on the ground or in space).

3.2.3. Calibration: Differential techniques. Several differential techniques have been proposed combining AO systems and coronagraphs, including differential imaging instruments [*Lenzen et al.* (2004); *Close et al.* (2005) for the Simultaneous Differential Imager, or SDI, on NACO], Integral Field Spectrographs (IFS) (*Berton et al.*, 2006), and differential polarimeters (*Gisler et al.*, 2004).

In the case of differential detection, the final performance is not limited by remaining wavefront phase defects resulting in speckles that could mimic a faint planet, because these techniques are designed to remove such speckles. Instead the limitations are due to the much smaller noise induced by the differences in the various measurement channels (see also differential phase measurement in interferometry, which uses a similar approach, in section 4).

In spectral differential techniques, these limitations come in particular from (1) overall chromatic effects (speckle structure chromatism, residuals of the atmospheric dispersion correction, potential coronagraph chromatism, impact of the optical beam-shift due to atmospheric refraction, and defects in optical components located before the dispersion corrector), and (2) distinct optical aberrations if distinct optical paths are involved (such as in SDI). When looking for very high contrasts, the remaining level of differential aberrations is a critical issue. In particular, it should be noted that these differential aberrations add to common aberrations: The impact on the image is then a combination of both types of aberrations and scales as $\sigma^2_{common} \times \sigma_{differential}$. As already underlined in the previous paragraphs, this strongly reinforces the requirement for very high image quality on the coronagraph.

The information obtained with spectral differential techniques is particularly well suited to the search and study of cold giant planets exhibiting in particular deep and low-resolution molecular absorption features in the near-infrared [e.g., CH_4 (*Oppenheimer et al.*, 1995, 1998)]. But, on the other hand, these techniques are likely to become much less efficient when searching for terrestrial-type exoplanets in the future (especially when observing from the ground, where the atmospheric features of the exoplanet are likely to significantly overlap with our own atmosphere absorption).

Differential polarimetry permits a similar approach, when comparing simultaneous images in two distinct polarization states. Difficulties associated with the atmospheric dispersion disappear here, but similar limitations related to differential instrumental polarization are encountered in this case if the images in the two polarization states are separated and propagated through distinct optical components. Such limitations were already encountered in existing instruments for very high polarization accuracy (such as in solar physics). A solution has been proposed in the ZIMPOL concept (*Gisler et al.*, 2004) with the use of an high-frequency (kHz regime) polarization switch on a single optical path. In that case, a polarization accuracy of 10^{-6} could be reached, which can translate into star-to-planet contrasts higher than 10^{-8}. For such a performance level, the fundamental photon noise limit dominates with the consequence that only the brightest nearby stars (until the advent of the 30-m class Extremely Large Telescopes) can be surveyed. Also, the most favorable potential planetary targets would be close-in (<1 AU) planets with a significant reflected (and polarized) light component.

In principle, any defect measured and identified as being an instrumental artifact by differential measurements could be corrected by a high-accuracy corrective device in a closed loop. In that sense, such a technique is very similar to the two-stage adaptive correction systems being studied now. For example, such dual-stage systems are under consideration in the Planet Finder projects just beginning [Gemini system (*Macintosh et al.*, 2004) and VLT project (*Beuzit et al.*, 2006)].

3.3. Achieved and Expected Performance

3.3.1. Existing systems. Today, the 4- and 8-m-class telescopes are equipped with instruments providing diffraction-limited imaging in the near-infrared. Even if they remain "general use" instruments, deep investigations of the outer (typically >0.5") stellar environments can be carried out with classical coronagraphy and/or saturated images, opening the observational window to direct detection of orbiting planets. Accordingly, these searches must primarily focus on distant massive planets around young and low-mass stars, with extensive surveys of the newly identified young nearby stellar associations (see section 2.1 for a discussion of the first direct detections obtained with the NACO instrument on the VLT).

Figure 3 shows an example of an image from the Lyot Project, a dedicated system comprising an optimized Lyot-style coronagraph behind a high-order AO system on a 3.6-m telescope (*Oppenheimer et al.*, 2004). This project is

Fig. 3. An H-band image from the Lyot Project Coronagraph, an optimized, diffraction-limited, classical coronagraph fitted behind an AO system with d = 10 cm on a 3.67-m telescope, currently the highest-order AO correction available (*Oppenheimer et al.*, 2004). The companion (confirmed through common proper motion astrometry) to this nearby star is 11.3 mag ($10^{-4.5}$) times fainter than the star at a separation of 1.8". Greater contrast with this system is possible with differential polarimetry.

currently surveying nearby stars with contrast ratios in the H band up to 10^{-6} within 2" of the central star and has implemented a differential polarimeter with a sensitivity to polarized objects up to 15× fainter. Within two years an IFS will be installed behind this system to provide early results on much fainter unpolarized companions.

The definite scientific impact of the current search programs, on a very short timescale, provides only a glimpse into the existing EGPs in the outer environments (greater than a few tens of AUs) of certain stellar classes. Even if the expected number of detections of planetary mass objects will not allow refined statistics, first indications on the existence and separation of such objects will be extremely valuable. For positive detections, comparison with models (including both internal structure and atmospheres of cool objects) will address the questions of temperature, composition (molecules and dust), and evolution through near-infrared colors, low-resolution spectroscopy, and estimations of bolometric luminosities. However, to gain access to a larger sample of targets and to smaller separations, in the 1–10-AU domain, future instruments designed for much higher contrast and smaller inner working angles are required.

3.3.2. Instruments under development. As mentioned above, optimized high-contrast instruments require a consistent and well-focused global design. Far superior performance is expected on existing telescopes using dedicated instruments instead of the "general use" AO systems. Such dedicated planet hunting instruments are currently under

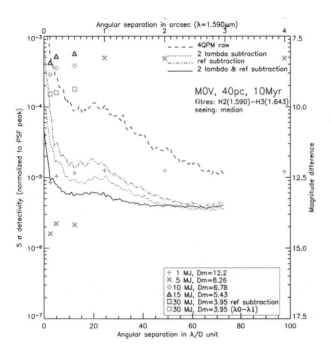

Fig. 4. Expected performance of the VLT Planet Finder instrument for the detection of young planetary systems using the simultaneous differential imaging technique. The best performance is achieved in the H band with the use of a reference star (double subtraction, i.e., two-wavelength subtraction and residual pattern subtraction from the reference target observation). In this case the host star has an M0 type, is aged 10 m.y., and is located at 40 pc from the Sun. Planets with 1 M_J are detected in the H2–H3 filter pair (respectively 1.590 and 1.643 μm) beyond 0.15". Planets with 5 M_J are easily detected, and spectrophotometry is feasible beyond 0.1" (detection in all NIR filters) without a reference star.

development, targeted for the study of the inner environment of bright objects: NICI (*Ftaclas et al.,* 2003), Subaru's Hi-CIAO, Gemini's Planet Imager (*Macintosh et al.,* 2004), and VLT's Planet Finder (*Beuzit et al.,* 2006) (Fig. 4). These instruments will gradually become operational with increasing performance between 2006 and 2010.

With performance of 10^{-6}–10^{-7} in contrast at 0.5" from the star or closer, Jupiter-mass companions will be accessible around a large number of young (<100 m.y.) stars. Also, slightly more massive planets (5–10 M_J) could be detected around older and closer stars at shorter physical separations, in the 1–10-AU range. The increased sensitivity of the new generation of wavefront sensor detectors permit AO correction for host stars with magnitudes up to 12 with far better precision. This allows surveys to cover several hundred stars.

This generation of instruments, mostly operated in survey mode, rather than general standard small allocations of time for pointed observations, will definitely improve the statistical understanding of the population of EGPs, covering a far broader range of orbital separations range and stellar types. These studies are in direct complement to previous RV and photometric transit searches. Because of the improved statistics, it will be possible to test the impact of parameters such as the stellar age, mass, and metallicity, as well as the presence of other (inner or outer) planets on the planet population and to place the brown dwarf and planet in relative scientific context.

Long integration times on the detected planets will provide a finer characterization of their spectra at low spectral resolution (R ~ 30). For the closest separations (~1–5 AU), monitoring the orbital motion permits determination of refined orbital parameters, variability, and derivation of dynamical masses.

Finally, direct detections of the reflected light of close-in (<1 AU) planets around very nearby and bright stars can in principal be achieved with long observation times at short wavelengths (~0.6–0.9 μm) with a photon-noise-limited (i.e., artifact-free) instrument such as the proposed ZIMPOL concept on for the VLT Planet Finder (*Gisler et al.,* 2004). Such measurements combine all the benefits of improved AO and advanced coronagraphy with far more accurate differential measurements of the planetary polarization signal. Indeed, polarized, reflected light from a planet is not a function of age, unlike intrinsic radiation.

3.3.3. Future systems (2015/2020). The potential of groundbased observations after 2015 will essentially depend on a number of technological developments that yet remain to be demonstrated. Reaching even higher contrasts than those foreseen for the coming generation of systems, in the 10^8 to several 10^{10} range, i.e., Δm > 25, at small separations (~0.1") is motivated by the search for lower-mass exoplanets, ultimately down to the direct detection of Earth analogs. Similar performance will be required also for fine characterization of previously detected EGPs, with higher SNR and higher spectral resolution, R ~ 1000, and very small signal variations along the planet orbit.

Reaching such extreme performances in contrast will definitively require very detailed system engineering, based on the experience obtained with the previous generations of instruments, and accounting for other effects such as atmospheric scintillation (*Masciadri et al.,* 2004). These observations will also require much larger collecting areas (ELT class) for the detection of terrestrial planets.

Both the development of larger telescopes and of new instruments of unprecedented precision appears very challenging but can be seen as realistic, with appropriate development plans and corresponding resources, on the ~2015–2020 timescale. Efforts are being made both in the U.S. and in Europe to start the design of such telescopes (Thirty Meter Telescope, OWL) and corresponding instruments (PFI, EPICS). However, one should note that this new generation of facilities will certainly not extend the sample of accessible stellar targets. On the contrary, both the planet faintness and the even higher-precision AO correction required will restrict the potential targets to the brightest stars.

Results on the lowest planetary masses (1–10 $M_\oplus$) will be detections at relatively low SNR and low spectral resolution. Because a larger variety of atmospheric composition and physical parameters can be expected for terrestrial planets, further spectral diagnosis and higher SNR will be required with longer integration times for positive detections. Ultimately, the refined characterization of the atmosphere and the search for biomarkers will be even an order of magnitude more difficult. Indeed, specific known features (like O_2, NO_2, the vegetation red-edge signature, and others) often have only small absorption depths (requiring corresponding high SNR spectra) or are strongly blended with other absorption features, including our own atmospheric absorption. At this stage, a complete analysis in terms of complexity, cost, schedule, and risk is required to compare the relative interests of observations from the ground or from space. For the ultimate performance goals, the fundamental limitations associated with atmosphere (for both image perturbation and absorption) will increasingly defavor the groundbased approach even if larger collecting areas can be achieved.

4. INTERFEROMETRY

While AO methods have achieved incredible dynamic range for imaging beyond the first or second Airy ring of a telescope PSF, only long-baseline interferometers can directly search for planets orbiting parent stars within ~100 mas (1 AU at 10 pc) — at least until the next generation of giant telescopes are built. While interferometers have a resolution advantage, they suffer from signal-to-noise disadvantages since collecting area is usually smaller, net optical throughput is low, and Poisson noise from unattenuated stellar light usually dominates the noise budget. Fortunately, planets in sub-AU orbits are considerably warmer than their more distant siblings, relaxing the dynamic-range requirement for direct detection. In this section, we will outline the current observational approaches being attempted for exoplanet detection.

From basic signal-to-noise considerations, Earth-mass planets are not realistically detectable with conventional interferometers based around 8-m-class telescopes (or smaller ones). However, RV surveys have uncovered a diversity of planets with a range of masses and semimajor axes in nearby systems. Indeed, massive planets at ≤1 AU distances are exciting to study from the perspective of comparative planetology since we lack solar system analogs. This includes the fascinating subset of planets known as "hot Jupiters," including the first exoplanet 51 Peg.

As a mechanism to focus discussion (and admit humility), we concentrate specifically on detectability of hot Jupiters since they will prove easiest to detect. Hot Jupiters are typically 0.05–0.10 AU from their parent star with expected surface temperatures between 1000 and 1500 K. Calculations (e.g., *Guillot et al.,* 1996; *Burrows et al.,* 1997; *Goukenleuque et al.,* 2000; *Sudarsky et al.,* 2000; *Barman et al.,* 2001) indicate the planet-to-star flux ratio to be a > 10^{-4}, a number that will only sound large to readers finishing the last section of this chapter. Figure 5 shows an example of an early calculation by *Seager et al.* (2000) that

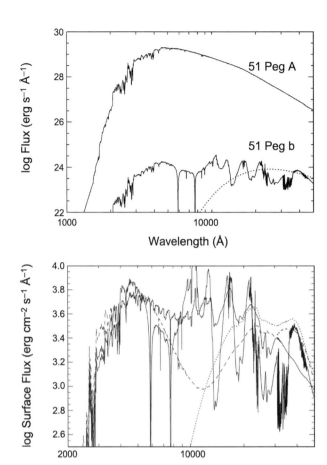

Fig. 5. Calculations of the emergent spectrum of 51 Peg b published by *Seager et al.* (2000) under various assumptions (their Figs. 3 and 2 respectively). The top panel explores different dust size distributions, while the bottom panel compares a fiducial model spectrum to that of the host star. The flux ratio is expected to be ~10^4 in the near-IR observing bands.

suggests the relative ease of near-infrared detection. Note that these planets should exhibit a rich molecular spectrum dependent on atmospheric density and temperature structure, making even a low-resolution spectrum measurement incredibly exciting.

4.1. Narrow-Angle Astrometry

The interferometric techniques discussed here are related to, but distinct from, "astrometric" search strategies that attempt to detect physical motion of the host star in the plane of the sky (e.g., *Gatewood,* 1976). This technique is analogous to the radial velocity methods and does not directly detect the planet itself (only the host star's "wobble"). The modern equivalent is called "narrow-angle astrometry," which uses special-purpose long-baseline optical interferometers built with a "dual-star" module to simultaneously observe a target and reference star. This technique has been well-studied, and is being pursued by Keck and VLT interferometers and the Space Interferometry Mission. Recent exciting results have been reported for the Palomar Testbed Interferometer (*Lane and Muterspaugh, 2004; Muterspaugh et al., 2005*), achieving 10–100 μas relative astrometry between components of close (subarcsecond) binaries. Demonstrated performance is adequate to detect short-period Jupiter-mass planets if present in their (binary) sample.

4.2. Differential Phase

The first "direct" interferometric detection technique we will discuss is called the "differential phase" method. It is a multiwavelength approach to find massive exoplanets by detecting a very slight photocenter shift (of the star + planet system) between different infrared filters. This occurs because of wavelength-dependent flux ratios caused by molecular opacity in the planet's atmosphere and the large temperature difference between star and planet [i.e., the *differential phase* method (e.g., *Akeson and Swain,* 1999; *Lopez and Petrov,* 2000)]. Figure 6 illustrates how an interferometer can be used to detect a faint companion (planet) to a star. In the left panel of this figure, an optical interferometer is represented as a Young's two-slit experiment (*Born and Wolf,* 1965). Flat wavefronts from a distant source impinge on the slits and produce an interference pattern on an illuminated screen. Multiple sources would produce multiple such fringe patterns, incoherently adding together in power. The right panel illustrates how the presence of a planet would cause a small reduction in fringe contrast (visi-

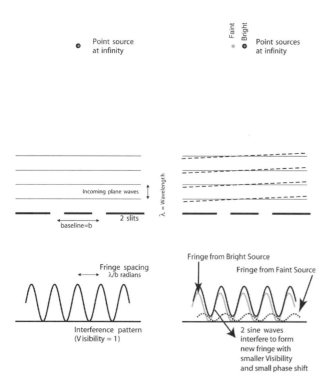

Fig. 6. The figure illustrates how the presence of a planet causes a phase shift in the stellar fringe observed by a long-baseline optical interferometer.

bility) and would shift the fringe phase a correspondingly tiny amount.

In the limit of high brightness ratio (very appropriate here!), we can be more quantitative. Assuming a planet-to-star brightness ratio of $\alpha = \frac{F_{planet}}{F_*}$, projected baseline $\vec{b}$, observing wavelength λ, and planet-star separation $\vec{\delta}$, then the normalized *complex visibility* $\tilde{V}$ observed by an interferometer is simply (assuming both components are not resolved by the interferometer)

$$\tilde{V} = \frac{1 + \alpha e^{-2\pi i \frac{\vec{b}}{\lambda} \cdot \vec{\delta}}}{1 + \alpha} \qquad (1)$$

In the limit of $\alpha \ll 1$ appropriate for exoplanet detection experiments, the last equation takes a simpler form when we consider the visibility amplitude |V| and phase Φ_V separately, $\tilde{V} = |V| e^{-i\Phi_V}$

$$|V| \approx 1 - 2\alpha \sin^2\left(\pi \frac{\vec{b}}{\lambda} \cdot \vec{\delta}\right) \qquad (2)$$

$$\Phi_V \approx \alpha \sin\left(2\pi \frac{\vec{b}}{\lambda} \cdot \vec{\delta}\right) \qquad (3)$$

Hence, by measuring Φ_V (or alternatively |V|) at various baselines $\vec{b}$ and wavelengths λ, both the brightness ratio spectrum $\alpha(\lambda)$ and the separation vector $\vec{\delta}$ between the star and planet can be determined. If the stellar photosphere is partially resolved (as will often be the case), the phase signal is boosted, easing detection. Unfortunately, the atmosphere at visible and infrared wavelengths defeats this simplistic approach, and this is detailed below.

We can see how telescope-specific phase delays caused by atmospheric turbulence (or anything else) affect the measured visibility amplitude and phase by considering the same idealized interferometer sketched again in Fig. 7. If the pathlength above one slit is changed (due to a pocket of warm air moving across the aperture, for example), the interference pattern will be shifted by an amount depending on the difference in pathlength of the two legs in this simple interferometer. If the extra pathlength is half the wavelength, the fringe pattern will shift by half a fringe, or π radians. The phase shift is completely independent of the slit separation, and only depends on slit-specific phase delays. The corruption of this phase information has serious consequences. For the search for exoplanets, the tiny signal of the planet, as encoded by a baseline-dependent phaseshift, is lost.

But all is not lost: By comparing fringes at one wavelength to that at another, we can track the fluctuating atmosphere term and recover an estimate of the "differential phase." This method should be relatively insensitive to the atmospheric turbulence. This method also benefits from the bright flux of the target star, a beacon through the changing atmosphere acting much like a natural guide star in AO.

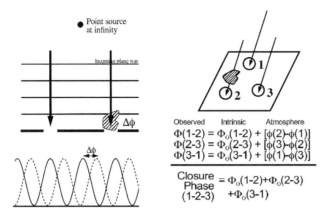

Fig. 7. (Left) In an interferometer, a phase delay above an aperture causes a phase shift in the detected fringe pattern. (Right) Phase errors introduced at any telescope causes equal but opposite phase shifts, canceling out in the *closure phase* (after *Readhead et al., 1988*).

However, recent studies of line-of-sight variability of atmospheric water vapor (*Colavita et al., 2004*) indicate that differential chromatic dispersion might be more difficult to calibrate for *differential phase* methods than originally expected.

We note in passing that atmospheric turbulence affects the calibration of visibility amplitudes as well, since phase irregularities over each telescope aperture degrade spatial coherence. Calibrating visibility amplitudes to the precision

needed for detecting planets is considered even more difficult than phase methods (due to poorer statistical properties), although it has been discussed (*Coudé du Foresto, 2000*). Table 1 summarizes the current active efforts to detect exoplanets using differential phases.

4.3. Precision Closure-Phase Method

There is another interferometry technique that is robust to both atmospheric phase shifts and also differential chromatic dispersion. Consider the right panel of Fig. 7, in which a phase delay is introduced above telescope 2 in a three-telescope interferometer. This causes a phase shift in the fringe detected between telescopes 1–2, as discussed in the last section. Note that a phase shift is also induced for fringes between telescopes 2–3; however, this phase shift is equal but *opposite* to the one for telescopes 1–2. Hence, the sum of three fringe phases, between 1–2, 2–3, and 3–1, is insensitive to the phase delay above telescope 2. This argument holds for arbitrary phase delays above any of the three telescopes. In general, the sum of three phases around a closed triangle of baselines, the *closure phase*, is a good interferometric observable; that is, it is independent of telescope-specific phase shifts induced by the atmosphere or optics.

The idea of closure phase was first introduced to compensate for poor phase stability in early radio VLBI work (*Jennison, 1958*). Application at higher frequencies was first mentioned by *Rogstad* (1968), but only much later carried out in the visible/infrared through aperture masking experi-

---

TABLE 1.   Exoplanet searches using interferometry.

Interferometer	Method	Goals
CHARA Interferometer	Precision/differential closure phases	Measure low-res spectrum of hot Jupiter in H/K bands
	Precision visibilities	Hot Jupiter detection using FLUOR combiner
Keck Interferometer	Differential phase*	Hot Jupiter detection using 1–5-µm wavelength range
	Nulling	Mid-infrared search for zodiacal dust might uncover close-in planet
	Narrow-angle astrometry (I)	Planet detection using Keck-Keck and outrigger array[†]
Navy Prototype Interferometer	Imaging	Detect hot Jupiter transit by imaging stellar photosphere in visible
Palomar Testbed Interferometer	Narrow-angle astrometry	Detect massive planets around subarcsecond binaries
Very Large Telescope Interferometer	Differential phase	Low-res spectrum of hot/warm exoplanets (H/K bands)
	Differential closure phase	Hot Jupiters and perhaps more difficult planets
	Narrow-angle astrometry	PRIMA[‡] instrument will allow long-term astrometry
	Nulling[§]	GENIE instrument meant to detect low-mass companions (K,L,M bands)
	Future Plans	
Antarctic Plateau Interferometer	Differential phase/closure phase	Hot Jupiter planet characterization (3–5 µm)
Darwin (ESA)	Nulling Space Interferometer	Terrestrial planet finding (mid-infrared)
Large Binocular Telescope Interferometer	Nulling	Detect (warm) massive exoplanets in thermal IR
Space Interferometry Mission (NASA)	Astrometry	Detect low-mass exoplanets via induced planet wobble
Terrestrial Planet Finder Interferometer (NASA)	Nulling Space Interferometer	Terrestrial planet finding (mid-infrared)

*Development suspended due to NASA budget cuts.
[†] Decision to begin construction expected end of 2006.
[‡] Expected commissioning 2007.
[§] Implementation under review at ESO/VLTI.

ments (*Baldwin et al., 1986; Haniff et al., 1987; Readhead et al., 1988*). Currently, six separate-element interferometers have reported obtaining closure-phase measurements in the optical (visible/infrared), first at the Cambridge Optical Aperture Synthesis Telescope (COAST) (*Baldwin et al., 1996*), followed by the Navy Prototype Optical Interferometer (NPOI) (*Benson et al., 1997*), the Infrared Optical Telescope Array (IONIC3 combiner) (*Monnier et al., 2004b*), the Infrared Spatial Interferometer (ISI) (*Weiner et al., 2006*), VLT Interferometer (AMBER combiner), and most recently at the CHARA Interferometer (MIRC combiner).

While differential-phase methods have been the focus of much current activity, precision measurements of only *closure phases* can also be used to detect faint companions. As discussed, the closure phase is formed by summing the interferometer phases on three baselines around a triangle of telescopes, and this quantity is immune to atmospheric phase delays. *Ségransan et al. (2000)* and *Monnier (2002)* recently discussed how closure phases are immune to dominant calibration problems of differential phase, and that they can also be used to solve for all the parameters of a binary system without needing to measure any visibility amplitudes. Figure 8 shows the expected closure-phase signature of the hot Jupiter ν And b, one of the best candidates for direct detection, as it would appear to the CHARA Interferometer.

The derivation of the planet-induced fringe phase (last section) can be used to prove that the closure phase yields an observable with the same magnitude as differential phase methods, *when the planet-star separation is clearly resolved.* Unfortunately, the closure-phase amplitude scales like baseline to the third power for short baselines, making detections of partially resolved systems difficult or impossible. Note that differential phase scales linearly with baseline in this case and thus remains a viable planet detection even

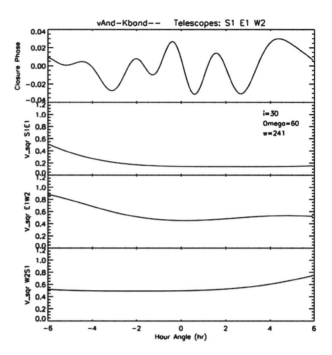

Fig. 8. The top panel shows the predicted closure-phase signal for CHARA interferometer observations of ν And b. The closure phase is significantly boosted because the star itself is partially-resolved by the 300-m baselines of this interferometer (see bottom three panels showing the predicted visibility squared for this observation).

for relatively short baselines (e.g., the 85-m Keck-Keck baseline).

Table 2 contains the list of hot Jupiter systems, planets with semimajor axes within 0.1 AU of their parent stars. Included in this table are important observing parameters and one can see a few favorable targets with very bright

TABLE 2.   Target list of (nearby) exoplanets with semimajor axes ≤0.1 AU.

Star Name	RA (J2000)	Dec (J2000)	Spectral Type	Distance (pc)	Semimajor Axis AU (mas)	M sin i (M_J)	V mag	K mag	Minimum Baseline*
HD 83443 b	09 37 11.8281	–43 16 19.939	K0V	43.5	0.038 (0.87)	0.35	8.24	~6.3	195
HD 46375 b	06 33 12.6237	+05 27 46.532	K1IV	33.4	0.041 (1.23)	0.249	7.84	~5.8	139
HD 179949 b	19 15 33.2278	–24 10 45.668	F8V	27.0	0.045 (1.67)	0.84	6.25	~4.9	102
HD 187123 b	19 46 58.1130	+34 25 10.288	G5	47.9	0.042 (0.88)	0.52	7.86	~6.3	194
τ Boo b[†]	13 47 15.7429	+17 27 24.862	F6IV	15.6	0.0462 (2.96)	3.87	4.50	~3.3	57
HD 75289 b	08 47 40.3894	–41 44 12.452	G0Ia	28.9	0.046 (1.59)	0.42	6.36	~5.0	107
HD 209458 b	22 03 10.8	+18 53 04	G0V	47.1	0.045 (0.96)	0.69	7.65	~6.2	178
HD 76700 b	08 53 55.5153	–66 48 03.571	G6V	59.7	0.049 (0.82)	0.197	8.13	~6.5	207
51 Peg b[†]	22 57 27.9805	+20 46 07.796	G2.5IV	15.4	0.0512 (3.33)	0.46	5.49	~4.0	51
ν And b[†]	01 36 47.8428	+41 24 19.652	F8V	13.5	0.059 (4.37)	0.69	4.09	~2.7	39
HD 49674 b	6 51 30.5164	+40 52 03.923	G0	40.7	0.057 (1.40)	0.12	8.10	~6.7	122
HD 68988 b	08 18 22.1731	+61 27 38.599	G0	58.9	0.071 (1.21)	1.90	8.21	~6.8	141
HD 168746 b	18 21 49.7832	–11 55 21.66	G5	43.1	0.065 (1.51)	0.23	7.95	~6.4	113
HD 217107 b[†]	22 58 15.5413	–02 23 43.386	G8IV	19.7	0.07 (3.55)	1.28	6.18	~4.4	48
HD 162020 b	7 50 38.3575	–40 19 06.056	K2V	31.3	0.088 (2.81)	1.08	9.18	~7.0	61
HD 108147 b	12 25 46.2686	–64 01 19.516	F8V	38.6	0.104 (2.69)	0.41	7.00	~5.7	63

*Minimum baseline needed to resolve planet and star pair at H band (1.65 μm): $\Delta\Theta = \frac{\lambda}{2B}$

[†] Denotes a favorable target.

infrared magnitudes (e.g., K ~ 4). Also, the minimum baseline needed to resolve known hot Jupiters at 1.65 μm is shown in the last column of the table. Essentially, all these sources can be resolved by VLTI and CHARA interferometers (longest baseline 220 m and 330 m respectively), thus permitting precision closure-phase work as well as differential phase studies.

Current published measurement precision of closure phases is only 0.3°–5° (*Tuthill et al.,* 2000; *Benson et al.,* 1997; *Young et al.,* 2000; *Millour et al.,* 2006); for reference, a typical closure phase for a binary with brightness ratio of 10^4 is ~0.01°. Recently, IOTA interferometer has demonstrated a calibration stability of 0.1° in closure phase using the IONIC combiner, an integrated optics device that minimizes drifts by miniaturizing the waveguides (*Berger et al.,* 2003). Improving the 3 orders of magnitudes needed to detect even the brightest possible exoplanet is a daunting challenge. Figure 9 shows the results from a recent study of closure-phase stability at the CHARA Interferometer using the new MIRC combiner (*Monnier et al.,* 2004a). These data were taken under very poor seeing on the very first night of CHARA closure phases, thus significantly better results should be straightforward to achieve. The closure phase here appears stable over 2 h with no detectable drifts. The formal closure-phase error for the average of the whole dataset is 0.03°, nearly sufficient to detect an extrasolar planet (see Fig. 8). While there are surely unconsidered systematic effects (perhaps due to birefringence or drifts in optical alignment) that will degrade the sensitivity of the precision closure-phase technique, the lack of any "show-stopper" effects, like differential atmospheric dispersion for

the differential phase methods, strongly motivates current efforts.

4.4. Nulling Interferometry

Another approach to detecting exoplanets is *nulling* interferometry. First demonstrated by *Hinz et al.* (1998) on the MMT telescope, a nulling interferometer introduced an extra (achromatic) phase delay in one arm of the interferometer so that light from the central star *destructively* interferes. Since the null depends on the incident angle of the starlight, one can tune the interferometer to selectively null out the star while allowing light from the planet to still be measured by the interferometer [more details can be found in *Serabyn et al.* (2000)]. A nuller is the interferometry equivalent of a coronagraph on single-aperture telescope, minimizing noise from Poisson fluctuations of stellar light and relaxing the calibration requirements on the dynamic range.

Nulling interferometry is very difficult to carry out from the ground because of atmospheric turbulence. Not only do the path lengths have to match, requiring active tracking of piston fluctuations at nanometer-level precision, but the wavefront across each telescope must also be corrected using AO in order to match the amplitude of the fringes a well. For these reasons, significant gains using nulling from the ground is only expected at longer, mid-infrared wavelengths. Because nulling interferometry is the foundation of all spacebased interferometers aiming to find Earth-like planets, groundbased projects are focused on technology development and measuring zodiacal dust properties of future observing targets (exozodiacal dust emission may be the dominant background noise source for Earth-mass planet detection using nulling).

The Keck Interferometer project (NASA) is beginning a survey of nearby stars using nulling interferometry (in the thermal infrared), and first results of 100-to-1 nulls on the sky were recently reported (Colavita, personal communication). The explicit aim of this survey is not to discover exoplanets, but rather to measure the exozodiacal light in the thermal infrared. However, it has been suggested that the survey strategy will also be sensitive to close-in Jupiter-mass planets (Kuchner, personal communication) if present.

The VLTI originally was to have used the GENIE instrument, a nuller testbed optimized for 2–5 μm, for a zodiacal light and faint companion survey at somewhat shorter wavelengths. However, the ultimate fate for GENIE is still being considered and it is currently not known if this instrument will see skytime at the VLTI or anywhere. Lastly, the Large Binocular Telescope Interferometer (LBTI) has ambitious plans for mid-infrared nulling in order to map zodiacal dust and search for faint companions. First light for the LBTI is a number of years away, but the project is rapidly progressing.

4.5. Immediate Outlook

Excitement in the interferometry field is motivated by the convergence of many developments: newly commissioned

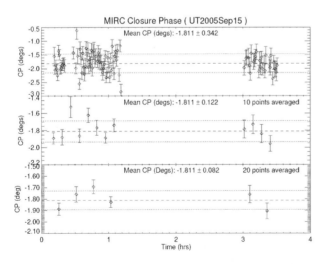

Fig. 9. The results of a closure-phase stability test at the CHARA interferometer under very poor seeing conditions using the MIRC Combiner. This dataset currently represents the highest-precision closure phase ever achieved on a star — corresponding to ±0.03 in ~2 h. In the top panel, each point is a 45-s average for observations of test source β Tau. The middle panel shows averages of 10 points, while the bottom panel shows the results for 20 data points. The final precision is within about an order of magnitude of that required for planet detection.

large-aperture and long-baseline interferometers (VLTI, Keck, CHARA), new combiner architectures optimized for precision calibration of (closure) phases (AMBER, IONIC3, MIRC), and the recent discovery of *bright* hot Jupiter systems that should be "easy" to detect (51 Peg, τ Boo, ν And). A back-of-the-envelope calculation for CHARA will further motivate and justify our optimism and excitement. We know from Table 2 that the CHARA interferometer has sufficiently long baselines to resolve all the planet-star separations listed. However, it is not obvious if CHARA has the light-collecting capability to detect such a small phase shift in the closure phase. Assuming a net CHARA optical efficiency of 5%, H = 4 mag (e.g., 51 Peg), bandwidth 0.05 μm, 1-m telescopes, and coherent integration time of 0.1 s, we will detect $\sim 5 \times 10^4$ photons per telescope per coherence time. Depending on the exact beam combination strategy, each 0.1-s fringe measurement should have a SNR~ 200, yielding a closure-phase measurement with about ~0.50° ($\sqrt{3}\frac{1}{200}$ rad). Improving this to 0.0011° (e.g., 5σ detection of 10^{-4} companion) would require an additional 10,000 measurements (total onsource integration time of ~30 min). Thus, it is theoretically possible to measure the low-resolution spectrum of the 51 Peg companion with less than a night of integration time with 1-m telescopes, *if precision calibration schemes can be developed*.

Currently, the biggest limitation of today's systems is low optical throughput. The above calculation assumed 5% throughput, while measured values are 3–10× worse depending on seeing. For most current interferometers, optical throughput is ~1% and efforts are being concentrated on improving this performance. While large-aperture telescope arrays like VLTI and Keck have plenty of aperture to overcome these losses, fairly limited observing time is allocated to interferometry since these telescopes are used by the general astronomy community.

Table 1 summarizes all the major efforts to detect exoplanets using current and planned interferometers, with special emphasis on current groundbased projects. Most likely, the first detection will come from the innovative AMBER combiner on the Very Large Telescope Interferometer within the next two years. However, there is intense competition at many of the world's interferometers, so keep your eyes open for new and unexpected discoveries.

Acknowledgments. J.D.M. thanks M. Zhao for his help in creating some of the figures in this chapter. J.L.B. and D.M. are very grateful to the VLT "Planet Finder" Phase A study team, and especially A.-M. Lagrange and C. Moutou, who greatly contributed to the discussion and results presented in this paper. B.R.O. thanks his collaborators on the Lyot Project for their dedication and hard work.

REFERENCES

Aime C. and Soummer R. (2004) *Astrophys. J., 612,* L85–L88.
Akeson R. L. and Swain M. R. (1999) In *Working on the Fringe: Optical and IR Interferometry from Ground and Space* (S. C. Unwin and R. V. Stachnik, eds.), pp. 89–94. ASP Conf. Series 194, San Francisco.
Angel R. and Burrows A. (1995) *Nature, 374,* 678–679.

Arnold L. and Schneider J. (2004) *Astron. Astrophys., 420,* 1153–1162.
Baldwin J. E., Haniff C. A., Mackay C. D., and Warner P. J. (1986) *Nature, 320,* 595–597.
Baldwin J. E., Beckett M. G., Boysen R. C., Burns D., Buscher D. F., Cox G. C., Haniff C. A., Mackay C. D., Nightingale N. S., Rogers J., Scheuer P. A. G., Scott T. R., Tuthill P. G., Warner P. J., Wilson D. M. A., and Wilson R. W. (1996) *Astron. Astrophys., 306,* L13–L16.
Barman T. S., Hauschildt P. H., and Allard F. (2001) *Astrophys. J., 556,* 885–895.
Baraffe I., Chabrier G., Allard F., and Hauschildt P. (2002) *Astron. Astrophys., 382,* 563–572.
Baudoz P., Boccaletti A., Rabbia Y., and Gay J. (2005) *Publ. Astron. Soc. Pac., 117,* 1104–1111.
Beaulieu J.-P., Bennett D. P., Fouqué P., Williams A., Dominik M., Jorgensen U.-G., Kubas D., et al. (2006) *Nature, 439,* 437–440.
Benson J., Hutter D., Elias N. M. I., Bowers P. F., Johnston K. J., Hajian A. R., Armstrong J. T., Mozurkewich D., Pauls T. A., Rickard L. J., Hummel C. A., White N. M., Black D., and Denison C. (1997) *Astron. J., 114,* 1221–1226.
Berger J.-P., Haguenauer P., Kern P. Y., Rousselet-Perraut K., Malbet F., Gluck S., Lagny L., Schanen-Duport I., Laurent E., Delboulbe A., Tatulli E., Traub W. A., Carleton N., Millan-Gabet R., Monnier J. D., and Pedretti E. (2003) In *Interferometry for Optical Astronomy II* (W. A. Traub, ed.), pp. 1099–1106. SPIE Proc. 4838, International Society for Optical Engineering, Bellingham.
Berton A., Feldt M., Gratton R., Hippler S., and Henning T. (2006) *New Astron. Rev., 49,* 661–669.
Beuzit J.-L., Chauvin G., Delfosse X., Forveille T., Lagrange A.-M., Marchal L., Mayor M., Ménard F., Mouillet D., Perrier C., Ségransan D., and Udry S. (2004) In *Astronomy with High Contrast Imaging* (C. Aime and R. Soummer, eds.), pp. 319–336. EAS Publications Series, Paris.
Beuzit J.-L., Feldt M., Mouillet D., Moutou C., Dohlen K., Puget P., Fusco T., et al. (2006) In *Direct Imaging of Exoplanets: Science and Techniques* (C. Aime and F. Vakili, eds.), pp. 317–322. IAU Colloquium 200, Cambridge Univ., Cambridge.
Bloemhof E. E., Dekany R. G., Troy M., and Oppenheimer B. R. (2001) *Astrophys. J., 558,* L71–L76.
Bonfils X., Forveille T., Delfosse X., Udry S., Mayor M., Perrier C., Bouchy F., Pepe F., Queloz D., and Bertaux J.-L. (2005) *Astron. Astrophys., 443,* L15–L18.
Born M. and Wolf E. (1965) *Principles of Optics. Electromagnetic Theory of Propagation, Interference and Diffraction of Light,* 3rd edition.
Brandeker A., Jayawardhana R., and Najita J. (2003) *Astron. J., 126,* 2009–2014.
Brown R. A. (2005) *Astrophys. J., 624,* 1010–1024.
Burrows A. (2005) *Nature, 433,* 261–268.
Burrows A. and Sharp C. M. (1999) *Astrophys. J., 512,* 843–863.
Burrows A., Marley M., Hubbard W. B., Lunine J. I., Guillot T., Saumon D., Freedman R., Sudarsky D., and Sharp C. (1997) *Astrophys. J., 491,* 856–875.
Butler R. P., Marcy G. W., Fischer D. A., Brown T. M., Contos A. R., Korzennik S. G., Nisenson P., and Noyes R. W. (1999) *Astrophys. J., 526,* 916–927.
Butler R. P., Vogt S. S., Marcy G. W., Fischer D. A., Wright J. T., Henry G. W., Laughlin G., and Lissauer J. J. (2004) *Astrophys. J., 617,* 580–588.
Chabrier G., Baraffe I., Allard F., and Hauschildt P. (2000) *Astrophys. J., 542,* 464–472.
Chauvin G., Ménard F., Fusco T., Lagrange A.-M., Beuzit J.-L., Mouillet D., and Augereau J.-C. (2002) *Astron. Astrophys., 394,* 949–956.
Chauvin G., Lagrange A.-M., Dumas C., Zuckerman B., Mouillet D., Song I., Beuzit J.-L., and Lowrance P. (2005a) *Astron. Astrophys., 438,* L25–L28.
Chauvin G., Lagrange A.-M., Zuckerman B., Dumas C., Mouillet D., Song I., Beuzit J.-L., Lowrance P., and Bessell M. S. (2005b) *Astron. Astrophys., 438,* L29–L32.
Clénet Y., Kasper M., Ageorges N., Lidman C., Fusco T., Marco O., Hartung M., Mouillet D., Koehler B., Rousset G., and Hubin N.

(2004) In *Advancements in Adaptive Optics* (D. B. Calia et al., eds.), pp. 107–117. SPIE Proc. 5490, International Society for Optical Engineering, Bellingham.

Close L. M., Lenzen R., Guirado J. C., Nielsen E. L., Mamajek E. E., Brandner W., Hartung M., Lidman C., and Biller B. (2005) *Nature, 433,* 286–289.

Colavita M. M., Swain M. R., Akeson R. L., Koresko C. D., and Hill R. J. (2004) *Publ. Astron. Soc. Pac., 116,* 876–885.

Coudé du Foresto V. (2000) In *From Exoplanets to Cosmology: The VLT Opening Symposium* (J. Bergeron and A. Renzini, eds.), p. 560. Proc. ESO Symp., Springer-Verlag, Berlin.

Dekany R. G. (2004) In *Second Backaskog Workshop on Extremely Large Telescopes* (G. E. Jabbour and J. T. Rantala, eds.), pp. 12–20. SPIE Proc. 5382, International Society for Optical Engineering, Bellingham.

Dekany R. G., Nelson J. E., and Bauman B. J. (2000) In *Optical Design, Materials, Fabrication, and Maintenance* (P. Dierickx, ed.), pp. 212–225. SPIE Proc. 4003, International Society for Optical Engineering, Bellingham.

Ftaclas C., Martin E. L., and Toomey D. (2003) In *Brown Dwarfs* (E. Martin, ed.), p. 521. IAU Symp. 211, Astronomical Society of the Pacific, San Francisco.

Fusco T., Rousset G., Beuzit J.-L., Mouillet D., Dohlen K., Conan R., Petit C., and Montagnier G. (2005) In *Astronomical Adaptive Optics Systems and Applications II* (R. K. Tyson and M. Lloyd-Hart, eds.), pp. 178–189. SPIE Proc. 5903, International Society for Optical Engineering, Bellingham.

Gatewood G. (1976) *Icarus, 27,* 1–12.

Gisler D., Schmid H. M., Thalmann C. P., Hans P., Stenflo J. O., Joos F., Feldt M., et al. (2004) In *Ground-based Instrumentation for Astronomy* (A. F. M. Moorwood and I. Masanori, eds.), pp. 463–474. SPIE Proc. 5492, International Society for Optical Engineering, Bellingham.

Goukenleuque C., Bézard B., Joguet B., Lellouch E., and Freedman R. (2000) *Icarus, 143,* 308–323.

Gould A. and Loeb A. (1992) *Astrophys. J., 396,* 104–114.

Guillot T. (1999) *Science, 286,* 72–77.

Guillot T., Burrows A., Hubbard W. B., Lunine J. I., and Saumon D. (1996) *Astrophys. J., 459,* L35–L38.

Guyon O., Pluzhnik E. A., Galicher R., Martinache F., Ridgway S. T., and Woodruff R. A. (2005) *Astrophys. J., 622,* 744–758.

Han I., Black D. C., and Gatewood G. (2001) *Astrophys. J., 548,* L57–L60.

Haniff C. A., Mackay C. D., Titterington D. J., Sivia D., and Baldwin J. E. (1987) *Nature, 328,* 694–696.

Hinz P. M., Angel J. R. P., Hoffmann W. F., McCarthy D. W., McGuire P. C., Cheselka M., Hora J. L., and Woolf N. J. (1998) *Nature, 395,* 251–253.

Jennison R. C. (1958) *Mon. Not. R. Astron. Soc., 118,* 276–284.

Jha S., Charbonneau D., Garnavich P. M., Sullivan D. J., Sullivan T., Brown T. M., and Tonry J. L. (2000) *Astrophys. J., 540,* L45–L48.

Kaltenegger L. and Fridlund M. (2005) *Adv. Space Res., 36,* 1114–1122.

Kasting J. F. and Catling D. (2003) *Ann. Rev. Astron. Astrophys., 41,* 429–463.

Lane B. F. and Muterspaugh M. W. (2004) *Astrophys. J., 601,* 1129–1135.

Lenzen R., Close L., Brandner W., Biller B., and Hartung M. (2004) In *Ground-based Instrumentation for Astronomy* (A. F. M. Moorwood and I. Masanori, eds), pp. 970–977. SPIE Proc. 5492, International Society for Optical Engineering, Bellingham.

Lopez B. and Petrov R. G. (2000) In *From Exoplanets to Cosmology: The VLT Opening Symposium* (J. Bergeron and A. Renzini, eds.), p. 565. Proc. ESO Symp., Springer-Verlag, Berlin.

Lyot B. (1939) *Mon. Not. R. Astron. Soc., 99,* 538–580.

Macintosh B. A., Olivier S. S., Bauman B. J., Brase J. M., Carr E., Carrano C. J., Gavel D. T., Max C. E., and Patience J. (2002) In *Adaptive Optics Systems and Technology II* (R. K. Tyson et al., eds.), pp. 60–68. SPIE Proc. 4494, International Society for Optical Engineering, Bellingham.

Macintosh B. A., Bauman B., Wilhelmsen E. J., Graham J. R., Lockwood

C., Poyneer L., Dillon D., Gavel D. T. et al. (2004) In *Advancements in Adaptive Optics* (D. Bonaccini et al., eds.), pp. 359–369. SPIE Proc. 5490, International Society for Optical Engineering, Bellingham.

Mao S. and Paczynski B. (1991) *Astrophys. J., 374,* L37–L40.

Marcy G. W., Butler R. P., Fischer D. A., and Vogt S. S. (2003) In *Scientific Frontiers in Research on Exoplanets* (D. Deming and S. Seager, eds), pp. 1–16. ASP Conf. Series 294, San Francisco.

Masciadri E., Feldt M., and Hippler S. (2004) *Astrophys. J., 613,* 572–579.

Masciadri E., Mundt R., Henning T., Alvarez C., and Barrado y Navascués D. (2005) *Astrophys. J., 625,* 1004–1018.

Mayor M. and Queloz D. (1995) *Nature, 378,* 355–359.

Mennesson B., Léger A., and Ollivier M. (2005) *Icarus, 178,* 570–588.

Millour F., Vannier M., Petrov R. G., Lopez B., and Rantakyrö F. (2006) In *Direct Imaging of Exoplanets: Science and Techniques* (C. Aime and F. Vakili, eds.), pp. 291–296. IAU Colloquium 200, Cambridge Univ., Cambridge.

Monnier J. D. (2002) In *Eurowinter School: Observing with the Very Large Telescope Interferometer* (G. Perrin and F. Malbet, eds.), pp. 213–226.

Monnier J. D., Berger J.-P., Millan-Gabet R., and Ten Brummelaar T. A. (2004a) In *New Frontiers in Stellar Interferometry* (W. A. Traub, ed.), p. 1370. SPIE Proc. 5491, International Society for Optical Engineering, Bellingham.

Monnier J. D., Traub W. A., Schloerb F. P., Millan-Gabet R., Berger J.-P., Pedretti E., Carleton N. P., Kraus S., Lacasse M. G., Brewer M., Ragland S., Ahearn A., Coldwell C., Haguenauer P., Kern P., Labeye P., Lagny L., Malbet F., et al. (2004b) *Astrophys. J., 602,* L57–L60.

Mouillet D., Beuzit J.-L., Chauvin G., and Lagrange A.-M. (2002) In *Scientific Drivers for ESO Future VLT/VLTI Instrumentation* (J. Bergeron and G. Monnet, eds.), pp. 258–263. Proc. ESO Workshop, Springer-Verlag, Berlin.

Muterspaugh M. W., Lane B. F., Konacki M., Burke B. F., Colavita M. M., Kulkarni S. R., and Shao M. (2005) *Astron. J., 130,* 2866–2875.

Neuhäuser R., Guenther E., Alves J., Húelamo N., Ott T., and Eckart A. (2003) *Astron. Nachricht., 324,* 535–542.

Neuhäuser R., Guenther E. W., Wuchterl G., Mugrauer M., Bedalov A., and Hauschildt P. H. (2005) *Astron. Astrophys., 435,* L13–L16.

Neuhäuser R. (2006) In *ESO Workshop Proceedings on Multiple Stars,* in press (astro-ph/0509906).

Oppenheimer B. R., Kulkarni S. R., Matthews K., and Nakajima T. (1995) *Science, 270,* 1478–1481.

Oppenheimer B. R., Kulkarni S. R., Matthews K., and van Kerkwijk M. H. (1998) *Astrophys. J., 502,* 932–941.

Oppenheimer B. R., Sivaramakrishnan A., and Makidon R. B. (2003) In *The Future of Small Telescopes In The New Millennium. Volume III — Science in the Shadows of Giants* (T. D. Oswalt, ed.) pp. 155–169. Astrophysics and Space Science Library, Kluwer, Dordrecht.

Oppenheimer B. R., Digby A. P., Newburgh L., Brenner D., Shara M., Mey M., Mandeville C., Makidon R. B., Sivaramakrishnan A., Soummer R., Graham J. R., Kalas P., Perrin M. D., Roberts L. C. Jr., Kuhn J. R., Whitman K., and Lloyd J. P. (2004) In *Advancements in Adaptive Optics* (D. B. Calia et al., eds.), pp. 433–442. SPIE Proc. 5490, International Society for Optical Engineering, Bellingham.

Readhead A., Nakajima T., Pearson T., Neugebauer G., Oke J., and Sargent W. (1988) *Astron. J., 95,* 1278–1296.

Rivera E. J., Lissauer J. J., Butler R. P., Marcy G. W., Vogt S. S., Fischer D. A., Brown T. M., Laughlin G., and Henry G. W. (2005) *Astrophys. J., 634,* 625–640.

Rogstad D. H. (1968) *Applied Optics, 7,* 585–588.

Seager S., Whitney B. A., and Sasselov D. D. (2000) *Astrophys. J., 540,* 504–520.

Ségransan D., Beuzit J.-L., Delfosse X., Forveille T., Mayor M., Perrier-Bellet C., and Allard F. (2000) In *Interferometry in Optical Astronomy* (P. Léna and A. Quirrenbach, eds.), pp. 269–276. SPIE Proc. 4006, International Society for Optical Engineering, Bellingham.

Segura A., Kastin J. F., Meadows V., Cohen M., Scalo J., Crisp D., Butler R. A. H., and Tinetti G. (2005) *Astrobiology, 5,* 706–725.

Serabyn E., Colavita M. M., and Beichman C. A. (2000) In *Thermal*

Emission Spectroscopy and Analysis of Dust, Disks, and Regoliths (M. L. Sitko et al., eds.), pp. 357–365. ASP Conf. Series 196, San Francisco.

Sivaramakrishnan A., Koresko C. D., Makidon R. B., Berkefeld T., and Kuchner M. J. (2001) *Astrophys. J., 552,* 397–415.

Soummer R. (2005) *Astrophys. J., 618,* L161–L164.

Soummer R., Dohlen K., and Aime C. (2003) *Astron. Astrophys., 403,* 369–381.

Sozzetti A. (2005) *Publ. Astron. Soc. Pac., 117,* 1021–1048.

Stoesz J. A., Véran J.-P., Rigaut F. J., Herriot G., Jolissaint L., Frenette D., Dunn J., and Smith M. (2004) In *Advancements in Adaptive Optics* (D. B. Calia et al., eds.), pp. 67–78. SPIE Proc. 5490, International Society for Optical Engineering, Bellingham.

Sudarsky D., Burrows A., and Pinto P. (2000) *Astrophys. J., 538,* 885–903.

Trauger J. T., Burrows C., Gordon B., Green J. J., Lowman A. E., Moody D., Niessner A. F., Shi F., and Wilson D. (2004) In *Optical, Infrared, and Millimeter Space Telescopes* (J. C. Mather, ed.), pp. 1330–1336. SPIE Proc. 5487, International Society for Optical Engineering, Bellingham.

Tuthill P. G., Monnier J. D., Danchi W. C., Wishnow E. H., and Haniff C. A. (2000) *Publ. Astron. Soc. Pac., 112,* 555–565.

van Dam M. A., Le Mignant D., and Macintosh B. A. (2004) *Appl. Optics, 43,* 5458–5467.

Weiner J., Tatebe K., Hale D. D. S., Townes C. H., Monnier J. D., Ireland M., Tuthill P. G., Cohen R., Barry R. K., Rajagopal J., and Danchi W. C. (2006) *Astrophys. J., 636,* 1067–1077.

Woolf N. J., Smith P. S., Traub W. A., and Jucks K. W. (2002) *Astrophys. J., 574,* 430–433.

Young J. S., Baldwin J. E., Boysen R. C., Haniff C. A., Lawson P. R., Mackay C. D., Pearson D., Rogers J., Saint.-Jacques D., Warner P. J., Wilson D. M. A., and Wilson R. W. (2000) *Mon. Not. R. Astron. Soc., 315,* 635–645.

Zuckerman B. and Song I. (2004) *Ann. Rev. Astron. Astrophys., 42,* 685–721.

Atmospheres of Extrasolar Giant Planets

Mark S. Marley and Jonathan Fortney
NASA Ames Research Center

Sara Seager
Carnegie Institute of Washington

Travis Barman
University of California at Los Angeles

The key to understanding an extrasolar giant planet's spectrum — and hence its detectability and evolution — lies with its atmosphere. Now that direct observations of thermal emission from extrasolar giant planets (EGPs) are in hand, atmosphere models can be used to constrain atmospheric composition, thermal structure, and ultimately the formation and evolution of detected planets. We review the important physical processes that influence the atmospheric structure and evolution of EGPs and consider what has already been learned from the first generation of observations and modeling. We pay particular attention to the roles of cloud structure, metallicity, and atmospheric chemistry in affecting detectable properties through Spitzer Space Telescope observations of the transiting giant planets. Our review stresses the uncertainties that ultimately limit our ability to interpret EGP observations. Finally we will conclude with a look to the future as characterization of multiple individual planets in a single stellar system leads to the study of comparative planetary architectures.

1. INTRODUCTION

Atmospheres of planets serve as gatekeepers, controlling the fate of incident radiation and regulating the loss of thermal energy. Atmospheres are also archives, preserving gases that reflect the formation and the evolution of a planet. Thus a complete characterization of an extrasolar giant planet (EGP) entails understanding its thermal evolution through time, bulk and atmospheric composition, and atmospheric structure. To date transit spectroscopy has probed the chemistry of the upper atmosphere of one EGP, and broadband measurements of the flux emitted by two EGPs were reported in 2005. Many more such observations will follow as we await the direct imaging and resultant characterization of many EGPs around nearby stars.

This review focuses on the physics of giant planet atmospheres and the models that describe them. We first approach these planets from a theoretical perspective, paying particular attention to those aspects of planetary models that directly relate to understanding detectability, characterization, and evolution. We stress the modeling uncertainties that will ultimately limit our ability to interpret observations. We will review the observations of the transiting giant planets and explore the constraints these observations place on their atmospheric structure, composition, and evolution. Unlike purely radial velocity detections, direct imaging will allow characterization of the atmosphere and bulk composition of extrasolar planets, and provide data that will shed light on their formation and evolution through time. We will explore what plausibly can be learned from the first generation of EGP observations and discuss likely degeneracies in interpretation that may plague early efforts at characterization.

2. OVERVIEW OF GIANT PLANET ATMOSPHERES

The core accretion theory describing the formation of giant planets (*Wetherill and Steward*, 1989; *Lissauer*, 1993) suggests that any planet more massive than about 10 $M_\oplus$ should have accreted a gaseous envelope from the surrounding planetary nebula. This leads to the expectation that any massive planet will have a thick envelope of roughly nebular composition surrounding a denser core of rock and ice. For this review we implicitly adhere to this viewpoint. Because subsequent processes, such as bombardment by planetesimals, can lead to enhancements of the heavier elements, we do not expect the composition of the planetary atmosphere to precisely mirror that of the nebula or the parent star. Observed enhancements of carbon in solar system giant planets (Fig. 1), for example, range from a factor of about 3 at Jupiter to about 30× solar abundance at Uranus and Neptune.

Departures from nebular abundance provide a window to the formation and evolution history of a planet. The near-uniform enrichment of heavy elements in the atmosphere of Jupiter (*Owen et al.*, 1999) has been interpreted as evidence that planetesimals bombarded the atmosphere over time (e.g., *Atreya et al.*, 2003). Direct collapse of Jupiter from nebular gas would not result in such a pattern of en-

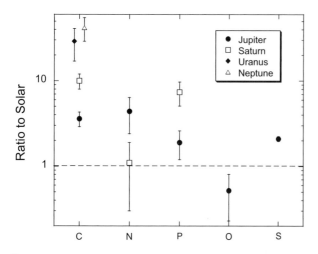

Fig. 1. Measured atmospheric composition of solar system giant planets (neglecting the noble gases) expressed as a ratio to solar abundance (*Lodders, 2003*). Jupiter and Saturn abundances are as discussed in *Lodders* (2004), *Visscher and Fegley* (2005), and *Flasar et al.* (2005). Uranus and Neptune abundances are reviewed in *Fegley et al.* (1991) and *Gautier et al.* (1995).

richment. A major goal of future observations should be to determine if most EGPs are similarly enriched in atmospheric heavy elements above that in the atmosphere of their primary stars.

Other outstanding questions relate to the thermal structure, evolution, cloud and haze properties, and photochemistry of EGPs. For discussion it is useful to distinguish between cooler, Jupiter-like planets and those giant planets that orbit very close to their primary stars, the "hot Jupiters." While the ultimate goals for characterizing both types of planets are similar, the unique atmospheres of the two classes raise different types of questions. For the hot Jupiters, most research has focused on the horizontal and vertical distribution of incident stellar radiation in their atmospheres and the uncertain role of photochemical processes in altering their equilibrium atmospheric composition. The available data from the transiting hot Jupiters also challenges conventional atmospheric models as their emergent flux seems to be grayer than expected.

Cooler, more Jupiter-like planets have yet to be directly detected. Consequently most research focuses on predicting the albedos and phase curves (variation of brightness as a planet orbits its star caused by the angular dependence of atmospheric scattering) of these objects to aid their eventual detection and characterization. As with the solar system giants, most of the scattered light reflecting from extrasolar EGPs will emerge from their cloud decks. Thus developing an understanding of which species will be condensed at which orbital distances and — critically — the vertical distribution of those condensates is required to facilitate their characterization. Finally, for both types of planets, second-order effects, including photochemistry and nonequilibrium chemical abundances, can play surprisingly large roles in controlling the observed planetary spectra.

2.1. Atmospheric Temperature and Evolution

A key diagnostic of the thermal state of a giant planet atmosphere is the effective temperature, T_{eff}. The total luminosity, L, of a planet with radius R arises from the combination of emission of absorbed incident stellar energy $\pi \mathcal{F}_*$ and the intrinsic internal heat flux L_{int}

$$L = 4\pi R^2 \sigma T_{eff}^4 = (1 - \Lambda)\pi R^2(\pi \mathcal{F}_*) + L_{int} \qquad (1)$$

The Bond albedo, Λ, measures the fraction of incident energy scattered back to space from the atmosphere. The reradiation of thermalized solar photons (the first term on the righthand side) makes up about 60% of the total luminosity of Jupiter and Saturn. For a hot Jupiter, 99.99% of the planet's luminosity is due to reradiation of absorbed stellar photons. As a giant planet ages, the contribution to the total luminosity from cooling of the interior, L_{int}, falls as this energy is radiated away.

For a remotely detected planet, equation (1) can be used to constrain the planetary radius. Given independent measurements of the total emitted infrared flux and the reflected visible flux — and assuming the internal flux is either negligible or precisely known from models — then the planetary radius can be inferred with an accuracy limited only by that of the optical and thermal infrared photometry. However, for EGPs the internal flux will not be well constrained since this quantity varies with the age and mass of the planet, both of which will not be known perfectly well. Furthermore, for solar system planets, models only predict well the internal luminosity of Jupiter. Standard cooling models underpredict the internal luminosity of Saturn and overpredict that of Uranus by a sizeable margin. Realistic errors in remotely sensed EGP radii will be dominated by the uncertainty in L_{int} and could easily exceed 25%, particularly for objects that have high internal luminosities compared to absorbed incident radiation (e.g., planets somewhat younger or more massive than Jupiter). Thus bulk composition inferred from the measured mass and radius will be highly uncertain.

To aid in the interpretation of observations of a given object, modelers frequently compute a one-dimensional, globally averaged temperature profile (connecting temperature to pressure or depth vertically through the atmosphere). Reflected and emitted spectra can be computed from such a profile (section 2.2). However, since the fraction of the incident stellar radiation varies over a globe, one must first choose what fraction, f, of the normal incidence stellar constant should strike the upper layers of a one-dimensional atmosphere model. Setting f = 1 results in a model atmosphere that is only correct for the planet's substellar point. Combined with an appropriate choice for the mean solar incidence angle, setting f = 1/2 gives a dayside average while f = 1/4 gives a planet-wide average. The latter is the usual choice for models of solar system atmospheres, since radiative time constants are typically long compared to rotation periods, allowing the atmosphere to come to equilibrium with the mean incident flux. Such an "average" profile may be less meaningful for tidally locked planets, depending on

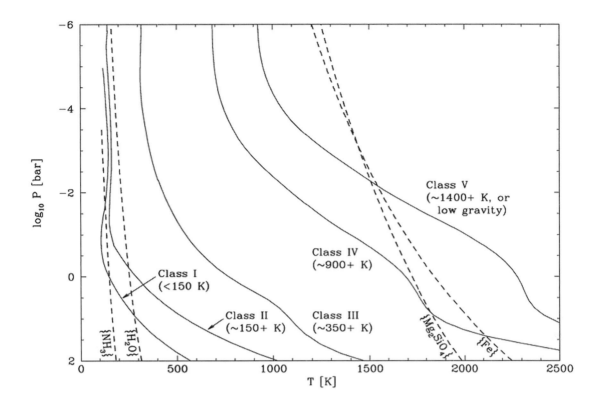

Fig. 2. Model temperature-pressure (T-P) profiles for cloudless atmospheres illustrating the giant planet classification scheme proposed by *Sudarsky et al.* (2003). Vertical dashed lines identify condensation curves for iron (Fe), forsterite (Mg$_2$SiO$_4$), water (H$_2$O), and ammonia (NH$_3$) in a solar-composition atmosphere. The base of cloud for a given condensate is predicted to occur where the atmosphere T-P profile crosses the species' condensation curve. The Class V planets have high iron and silicate clouds, Class III planets are relatively cloudless, and Class I planets have high ammonia clouds, like Jupiter.

the atmospheric temperature, radiative time constant, and circulation.

The internal energy of a giant planet (L$_{int}$), a remnant of its formation, is transported through the bulk of the planet's fluid interior by efficient convection, as first discussed by *Hubbard* (1968). Whether a giant planet is at 0.05, 0.5, or 5 AU from its parent star, the rate at which this internal energy is lost is controlled by the planet's atmosphere. In general, the closer a planet is to its parent star, or the smaller its flux from the interior, the deeper the boundary between the atmospheric radiative zone and the convective deep interior will be (Fig. 2). Models indicate that the radiative/convective boundary is at ~0.5 bar in Jupiter and can range from 10 to ~1 kbar in hot Jupiters (*Guillot et al.*, 1996; *Barman et al.*, 2001; *Sudarsky et al.*, 2003). Cooling and contraction is slow for planets with deeper radiative zones because the flux carried by the atmosphere is proportional to the atmosphere's temperature gradient (see *Guillot and Showman*, 2002).

Connecting planetary age to total luminosity or effective temperature presents a number of challenges. First, evolution models depend upon average planetary atmospheric conditions since the rate of cooling of the interior is governed by the mean energy loss of the entire planet. For hot Jupiters, "mean" conditions may involve subtleties of radia-

tive transport, dynamics, and convection. *Guillot and Showman* (2002) have shown that cooling and contraction are hastened for models that include temperature inhomogeneities at deep levels, rather than a uniform atmosphere, given the same incident flux. Recently, *Iro et al.* (2005) have computed time-dependent radiative models for HD 209458b, including energy transport due to constant zonal winds of up to 2 km/s. They find that at altitudes deeper than the 5-bar pressure level, as the timescale for the atmosphere to come into radiative equilibrium becomes very long, pressure-temperature profiles around the planet become uniform with longitude and time, and match a single "mean" profile computed using f = 1/4. This clearly suggests that model atmosphere grids computed with f = 1/4 are most nearly correct for use as boundary conditions for evolution models. Models that use f = 1/2 (such as *Burrows et al.*, 2003; *Baraffe et al.*, 2003; *Chabrier et al.*, 2004) overestimate the effect of stellar irradiation on the evolution of giant planets, as they assume *all* regions of the atmosphere receive the flux of the dayside.

A second difficulty with evolution models relates to the depth at which incident stellar energy is deposited. For Jupiter-like atmospheres (Fig. 2) the deep interior of the planet is connected to the visible atmosphere by a continuous adiabat. Thus absorbed stellar energy simply adds to the internal en-

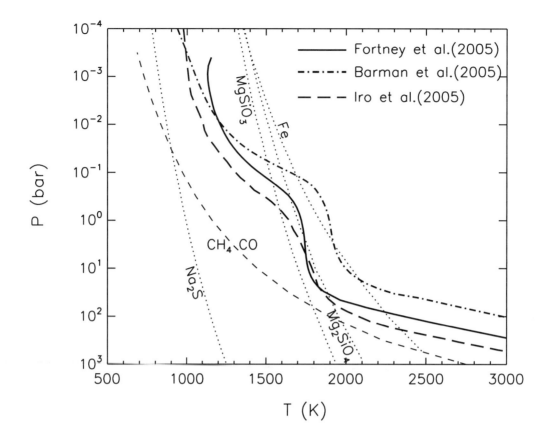

Fig. 3. Comparison of model atmosphere profiles for hot Jupiter HD 209458 computed by three groups (labeled). Each profile assumes somewhat different intrinsic luminosity, L_{int}, so differences at depth are not significant. All three profiles assume global redistribution of incident energy, or f = 4. Differences at lower pressure presumably arise from different methods for computing chemical equilibria, opacities, and radiative transfer assumptions. The spread in models provides an estimate of the current uncertainty in modeling these objects. In addition to the internal luminosity the size of the deep isothermal layer depends upon the behaviors at high pressure of the opacities of the major atmospheric constituents, which are poorly known.

ergy budget of the planet and — with appropriate book-keeping — atmosphere models appropriate for isolated, nonirradiated objects can be used to compute the evolution. Evolution models computed in this limit indeed work well for Jupiter (*Hubbard,* 1977). For hot Jupiters, however, the deep adiabat is separated by a radiative, isothermal region from that part of the atmosphere that is in equilibrium with the incident radiation (Figs. 2 and 3). As discussed by *Guillot and Showman* (2002), using atmosphere models suitable for isolated objects as the boundary conditions for hot-Jupiter evolution calculations (e.g., *Burrows et al.,* 2000) significantly overestimates the temperature of the atmosphere at pressures of ~1 bar, leading one to overestimate the effect of irradiation, and predict contraction that is too slow. In contrast, *Bodenheimer et al.* (2001, 2003) and *Laughlin et al.* (2005) have computed hot-Jupiter evolution models where contraction is likely too fast. For their atmospheric boundary calculation, the atmospheric pressure at which the Rosseland mean optical depth reaches 2/3 (~1 mbar in their models) is assigned the planetary effective temperature, which itself is calculated after assuming a Bond albedo. This method assumes a very inefficient penetration of stellar flux into the planet's atmosphere, compared to detailed atmosphere

models. Temperatures at higher atmospheric pressures are underestimated, leading to an underestimation of the effect of irradiation. The ideal solution is to compute individualized atmosphere models to use as boundary conditions for many timesteps in the evolutionary calculation.

2.2. Spectra of Giant Planets

The reflected and emitted spectra of giant planet atmospheres are controlled by Rayleigh and Mie scattering from atmospheric gases, aerosols, and cloud particles, and by absorption and emission of gaseous absorbers. Scattering of incident light usually dominates in the blue, giving way to absorption by the major molecular components at wavelengths greater than about 0.6 μm. The major absorbers in the optical are methane and, for warmer planets, water. Sodium and potassium are major optical absorbers in the atmospheres of the hot Jupiters. Generally speaking, in strong molecular bands photons are absorbed before they can scatter back to space. In the continua between bands photons scatter before they are absorbed. The continuum flux from a given object is thus controlled by Mie scattering from its clouds and hazes and Rayleigh scattering from the column of clear gas above the clouds. Longward of about 3–5 μm for the cooler planets and at shorter wavelengths for the warmest, scattering gives way to thermal emission.

The likelihood of absorption, and hence the depth of a given band, depends upon the molecular opacity at a given wavelength and the column abundance of the principal gaseous absorbers (e.g., methane and water) above a scattering layer. The column abundance of an absorber, in turn, depends upon the gravity (known for solar system planets), the height of a cloud layer, and its mixing ratio. Thus the spectra, even at low resolution, of EGPs are sensitive to their atmospheric temperature, metallicity, cloud structure, and mass. In principle, by comparing observed spectra to models, one can infer these properties from the data. Experience with the giant planets of our solar system, however, has shown that degeneracies in cloud properties and molecular abundances can be difficult to disentangle without broad, high-resolution spectral coverage.

As with the optical and near-infrared wavelengths, the thermal emission of EGPs is sculpted by molecular opacities. In regions of low absorption, planets brightly emit from deep, warm layers of the atmosphere. Within strongly absorbing regions flux arises from higher, cooler layers (unless there is a stratospheric temperature inversion). Bright emission in the window around 5 μm was flagged by *Marley et al.* (1996) as a diagnostic of the entire class of EGPs (but see the caveat in section 3.4 below). This opacity window is responsible for the well-known "5-μm hot spots" of Jupiter (*Westphal et al.*, 1974). Solar system giant planets also exhibit true emission features arising from a temperature inversion above the tropopause, notably in the 7.8-μm methane band that plays an important role in the stratospheric energy budget. Photochemically produced ethane and acetylene also exhibit emission in some giant planet atmospheres.

The *Galileo* atmosphere entry probe provided a test of the ability of remote observers to accurately measure the abundance of gases in a giant planet atmosphere. Prior to *Galileo*'s arrival at Jupiter the methane abundance was estimated to lie in the range of 2.0–3.6× solar at Jupiter and 2–6× solar at Saturn. *Galileo* measured Jupiter's methane abundance to be 2.9 ± 0.5× solar (see review by *Young*, 2003) and recent observations by Cassini have pinned Saturn's methane abundance at 10 ± 2× solar (*Flasar et al.*, 2005; *Lodders*, 2004). In both cases (at least some) remotely sensed values were accurate (e.g., *Buriez and de Bergh*, 1981). Remotely determining the abundance of condensed gases, such as ammonia or water, is more problematic and pre-*Galileo* measurements were not as accurate. Fortunately, ammonia will not condense in planets just slightly warmer than Jupiter. In young or more massive planets, water will be in the vapor phase as well, which should allow for more accurate abundance retrievals.

3. MODEL ATMOSPHERES

Model atmospheres predict the appearance of EGPs. They thus facilitate the design of optimal detection strategies and play a role in interpreting observations. A typical model recipe begins with assumptions about the atmospheric elemental composition, the atmospheric chemistry, the internal heat flow of the planet, the incident stellar flux, and various radiative transfer assumptions (e.g., is the atmosphere in local thermodynamic equilibrium?). A cloud model for the treatment of atmospheric condensates and the relevant gaseous opacities are also required ingredients. When combined with a suitable method for handling atmospheric radiative and convective energy transport, the modeling process yields the thermal structure of the atmosphere and the reflected and emitted spectrum. Of course, as in any recipe, models of giant planet atmospheres are only as good as the quality of the ingredients and the assumptions. Neglected physical processes, including some that might initially seem to be of only second-order importance, can in fact have first-order effects and spoil the predictions, at least in certain spectral regions. In this section we summarize typical inputs into atmosphere models and discuss their relative contributions to the accuracy of the final product. Figure 3 provides a comparison of hot-Jupiter profiles computed by three different groups. The differences between the profiles give an indication of the uncertainty in our understanding of these atmospheres just due to varying modeling techniques.

3.1. Chemistry

Perhaps the most elemental input to an atmosphere model is the assumed composition of the atmosphere. To date most models of EGP atmospheres have assumed solar composition. However, the best estimate of "solar" has changed over time (see *Lodders*, 2003, for a review) and of course the composition of the primary star will vary between each planetary system. Notably the carbon and oxygen abundances of the solar atmosphere remain somewhat uncertain (*Asplund*, 2005) and vary widely between stars (*Allende-Prieto et al.*, 2002).

Given a set of elemental abundances, a chemical equilibrium calculation provides the abundances of individual species at any given temperature and pressure. Several subtleties enter such a calculation, particularly the treatment of condensates. In an atmosphere subject to a gravitational field, condensates are removed by settling from the atmosphere above the condensation level. Thus "equilibrium" reactions that might take place were the gas to be kept in a sealed container will not proceed in a realistic atmosphere. A canonical example (e.g., *Fegley and Lodders*, 1994) is that under pure equilibrium, sulfur-bearing gases would not appear in Jupiter's atmospheres since sulfur reacts with iron grains at low temperature to form FeS. In fact, when iron condenses near 1600 K in Jupiter's deep atmosphere, the grains fall out of the atmosphere, allowing sulfur to remain as H_2S at low temperature. The removal of condensates from the atmosphere by sedimentation is sometimes termed "rainout," but this term can be confusing since, rigorously, "rain" refers only to the sedimentation of liquid water. Some early brown dwarf models did not properly account for sedimentation, but most recent modeling efforts do include this effect (see *Marley et al., 2002*, for a more complete discussion). For a recent, detailed review of the atmospheric chemistry of EGPs and brown dwarfs, see *Lodders and Fegley* (2006).

Finally, photochemistry, discussed further below, can alter atmospheric composition. Trace gases produced by the photolysis of methane in Jupiter's atmosphere, for example, are both important UV absorbers and emitters in the thermal infrared. As such, they play important roles in the stratospheric energy balance. Photochemical products may include important absorbers or hazes that may substantially alter the spectra of EGPs and cloud their interpretation. *Yelle* (2004) thoroughly discusses the photochemistry of the upper atmosphere of EGPs and predicts thermospheres heated to over 10,000 K by the extreme ultraviolet flux impinging on the top of these planets' atmospheres. The high temperatures drive vigorous atmospheric escape by hydrogen, producing an extended cloud surrounding the planet that has been observed in transit by *Vidal-Madjar et al.* (2003) for HD 209458b. Despite the high escape flux, a negligible fraction of the total mass of the planet escapes over time (*Yelle, 2004*).

3.2. Opacities

For the temperature-pressure regimes found in the atmospheres of all but the hottest EGPs, the most important gaseous absorbers are H_2O, CH_4, NH_3, Na, and K. In addition, the pressure-induced continuum opacity arising from collisions of H_2 with H_2 and He is particularly important in the thermal infrared. Other species found in brown dwarf atmospheres play a role in the hottest planets orbiting close to their primary stars. Freedman and Lodders (in preparation) review the current state of the various opacity databases used in atmospheric modeling. For the cool atmospheres most likely to be directly imaged, the opacities are fairly well known. The greatest shortcomings of the current opacities are the lack of a hot molecular line list for CH_4 and the highly uncertain treatment of the far wings of collisionally broadened lines. Neither is a major limitation for most EGP modeling applications.

3.3. Clouds and Hazes

Clouds and hazes play a crucial role in controlling giant planet spectra. In the absence of such scattering layers, red photons would penetrate to deep layers of an EGP atmosphere and generally be absorbed before they could be scattered (*Marley et al., 1999*), leading to very low reflectivity in the red and near-infrared. Planets with bright high water clouds, for example, tend to exhibit a bright continuum from scattered starlight punctuated by a few absorption bands. Likewise silicate and iron clouds in the atmospheres of the close-in planets play major roles in controlling their spectra (*Seager and Sasselov, 1998*). Furthermore, for a given cloudy planet, reflected and emitted spectra are sensitive to the vertical distribution, fractional global coverage, size distribution, and column number density of cloud particles.

Unfortunately, clouds are notoriously difficult to model, even in Earth's atmosphere where the representation of clouds is a leading source of uncertainty in terrestrial global atmospheric circulation models. Real clouds are a product of upward, downward, and horizontal transport of condensable vapor and solid or liquid condensate. Their detailed structure depends on a number of highly local factors, including the availability of condensation nuclei and the degree of supersaturation as well as a host of microphysical properties of the condensate. Approaches applying a one-dimensional atmosphere model to what is intrinsically a three-dimensional problem are certainly overly simplistic. Nevertheless, given the paucity of information, simple one-dimensional models currently provide the most workable approach.

A number of cloud models have been developed for solar system studies. Perhaps the most widely used has been an approach focusing on microphysical time constants developed by *Rossow* (1978). An important shortcoming of such an approach is that the time constants sensitively depend upon a variety of highly uncertain factors, particularly the degree of supersaturation. *Ackerman and Marley* (2001) and *Marley et al.* (2003) review the physics employed by the most popular cloud models. *Ackerman and Marley* (2001) proposed a simple one-dimensional cloud model that accounts for vertical transport of condensate and condensable gas, including a variable describing the efficiency of particle sedimentation. This model has had success fitting the cloudy atmospheres of L-type brown dwarfs, but is not able to predict such quantities as fractional global cloudiness or account for the rapidity of the L- to T-type brown dwarf transition. Other modeling approaches are discussed by *Tsuji* (2005), *Helling et al.* (2004), and *Cooper et al.* (2003), which range from purely phenomenological efforts to detailed numerical microphysical models of dust nucleation. Given the strong influence of clouds on EGP spectra, *Sudarsky et al.* (2000, 2003) have suggested classifying EGPs on the basis of which cloud layers are present or absent from the visible atmosphere (see section 3.7). The suggestion is appealing but might be difficult to apply in practice for transitional cases, hazey planets, or for objects with only limited spectral data.

Perhaps an even more challenging problem is atmospheric photochemical hazes. All the solar system giant planets are strongly influenced by hazes produced by the ultraviolet photolysis of methane. Figure 4 compares the incident stellar fluxes at two transiting planets with that received by Jupiter. The maximum wavelengths at which ultraviolet photons can photolyze various molecules are shown. At Jupiter, solar Lyman-α is an important contribution of the far-UV flux. Although the primary stars of the hot Jupiters may lack substantial Lyman-α flux, given the proximity of the planets the integrated continuum radiation capable of photolyzing major molecules is comparable to or greater than that received by Jupiter. At Jupiter methane photolysis is the main driver of photochemistry since water and H_2S are trapped in clouds far below the upper atmosphere. In the atmospheres of hot Jupiters, this will not be the case and

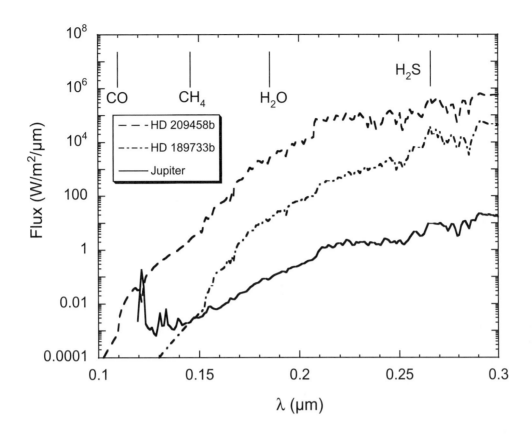

Fig. 4. Incident flux at the top of the atmospheres of several transiting planets compared to that received by Jupiter. Vertical lines denote the approximate maximum wavelengths at which various molecules can be dissociated. Incident spectra at HD 149026b and TrES-1 are similar to HD 209458b and HD 189733b (G4.5V), respectively, and are not shown for clarity. Model stellar spectra from *Kurucz* (1993).

these molecules will be rapidly photolyzed, perhaps providing important sources for photochemical haze production. Ultimately, haze optical depths depend upon production rates, condensation temperatures, microphysical processes, and mixing rates in the nominally stable stratosphere that in turn depend upon the atmospheric structure and poorly understood dynamical processes. To date only *Liang et al.* (2004) have considered this issue and then only in the context of hot-Jupiter atmospheres, which they found to be too warm for condensation of photochemical hydrocarbon compounds. They did not consider O- or S-derived photochemical products. Since optically thick hazes can substantially alter the idealized EGP spectra and phase functions computed to date, much more work on their production is needed. In any case, disentangling the effects of clouds, hazes, and uncertain atmospheric abundances in the atmospheres of EGPs will likely require high-quality spectra obtained over a large spectral range.

3.4. Dynamics and Mixing

An important limitation to conventional one-dimensional models of mean atmospheric structure is the neglect of vertical mixing. Vertical transport plays an important role when the dynamical timescale is short compared to a particular chemical equilibrium timescale, as is the case for CO in the atmosphere of Jupiter (*Prinn and Barshay,* 1977; *Fegley and Prinn,* 1988; *Yung et al.,* 1988) and cool T-type dwarfs (*Fegley and Lodders,* 1996; *Griffith and Yelle,* 1999; *Saumon et al.,* 2003; *Golimowski et al.,* 2004). While methane is most abundant in the visible atmospheres, in the deep atmosphere, where temperatures are higher, the abundance of CO is substantially larger. Since the C-O bond is very strong, the conversion time from CO to CH_4 in a parcel of rising gas is correspondingly long. This allows vertical mixing through the atmosphere to transport CO from the deep atmosphere to the visible atmosphere. CO absorbs strongly in the M photometric band and excess CO in T-dwarf atmospheres depresses the flux in this window region by up to a magnitude below that predicted by pure chemical equilibrium models (*Saumon et al.,* 2003). If this mechanism also depresses the flux of cool EGPs, the utility of this spectral region for planet detection may not be as great as predicted (e.g., *Burrows et al.,* 2001).

Another interesting effect is that of atmospheric dynamics. At lower pressures the radiative timescales are shorter than at higher pressures. For a tidally locked hot Jupiter, this will likely mean that that the upper atmosphere quickly adjusts to the flux it receives from the parent star, but deeper layers (P > 1 bar) will adjust much more sluggishly, and the dynamic transport of energy will be important. This is only beginning to be studied in detail (*Showman and Guillot,* 2002; *Cho et al.,* 2003; *Burkert et al.,* 2005; *Cooper and Showman,* 2005). Infrared observations as a function of planetary phase, with the Spitzer Space Telescope and perhaps other platforms, will enable constraints to be placed on atmospheric dynamics of HD 209458b and other planets.

3.5.   Albedos and Phase Curves

Albedos are often of interest as they allow for a simple parameterization of the expected brightness of a planet. Spectra of outer solar system planets with atmospheres are commonly reported as geometric albedo spectra, which is simply the reflectivity of a planet measured at opposition. Other albedo definitions include the wavelength-averaged geometric albedo and the Bond albedo, Λ in equation (1), which measures the ratio of scattered to incident light. Unfortunately, the extrasolar planet literature on albedos has become somewhat muddled and terms are not always carefully defined. Generic "albedos" are often cited with no definition. Yet different albedo varieties can differ from each other by several tenths or more. For example, the commonly referenced Lambert sphere has a geometric albedo of 2/3 while an infinitely deep Rayleigh scattering atmosphere would have a wavelength-independent geometric albedo of 3/4, yet both have $\Lambda = 1$. The two differ in the angular dependence of their scattered radiation. For absorbing atmospheres the Bond albedo depends upon the incident spectrum. Since proportionately more red photons are absorbed than blue photons (which tend to scatter before absorption), an identical planet will have a different Bond albedo under the light of a red star than a blue one even though the geometric albedos are identical (*Marley et al.,* 1999).

In any case, more information is needed to fully predict or interpret the flux observed by a distant observer of an extrasolar planet. Geometry dictates that extrasolar planets are most detectable near quadrature and not detectable at true opposition since they would be hidden by their star, thus a general description of the phase dependence of the scattered and emitted radiation is required. Phase information on solar system giant planets has long been used to constrain cloud particle sizes and atmospheric structure. For example, *Voyager 1,* which did not visit Uranus but instead imaged it from afar, observed the planet at high phase angles not reachable from Earth to help constrain the scattering phase function of its clouds and Bond albedo (*Pollack et al.,* 1986). *Dyudina et al.* (2005) recently relied upon *Voyager* observations to derive phase curves of Jupiter and Saturn. *Marley et al.* (1999) computed phase curves for model EGPs by relying upon scattering tables computed by *Dlugach and Yanovitskij* (1974). More recently, *Sudarsky et al.* (2005) presented a suite of model phase curves for EGPs. The differences between their model calculations and the observed phase curve of Jupiter (compare their Figs. 4 and 6) demonstrates that interpretation of specific planets will always be challenging since the specifics of particle size and composition, hazes, and overall atmospheric structure will likely make each giant planet discovered unique.

3.6.   Atmosphere Models

Although pioneered by *Kuiper* (1952), giant planetary atmosphere modeling entered the modern era with the work of *Trafton* (1967) and *Hogan et al.* (1969). Following the *Voyager 1* and *2* traverses of the outer solar system, substantially more complex models were developed to explore the atmospheric energy budgets, the thermal structure, and the reflected and emitted spectra (*Appleby and Hogan,* 1984; *Appleby,* 1986; *Marley and McKay,* 1999) of each giant planet. (Note that these are *forward* models that combine first principle information about planetary atmospheres to reproduce the observed atmospheric thermal structure. There is also a very rich literature of inverse models that aid in the interpretation of specific datasets.) These authors modeled the equilibrium one-dimensional radiative convective thermal profiles of these atmospheres, including deposition of incident radiation, by modeling the atmospheric radiative transfer given observed atmospheric abundances and cloud properties. The models generally well reproduced observed spectra and thermal structure. This success provides an important reality check that one-dimensional modeling of giant planet atmospheres, given appropriate input, satisfactorily reproduces observed properties of giant planets. Modeling extrasolar planets, however, will be more challenging: Only the incident radiation is known with certainty. Atmospheric composition, cloud properties, and thermal structure and perhaps mass and radius will all have to be inferred from comparison of models to data.

Burrows et al. (2000) reviewed the scant early work on atmospheric modeling of the cooler irradiated EGPs. Most pre-1995 investigations focused on studying the evolution of isolated objects or assumed gray atmospheres to estimate EGP detectability. *Marley* (1998) computed exploratory spectra of irradiated giant planets and found that the presence or absence of water clouds is an important spectral and albedo marker in EGP atmospheres. Particularly in the red and infrared the presence or absence of scattering clouds can change the scattered flux by a factor of 2 or more, with cloudless planets being darker.

Atmosphere models specifically of the hot Jupiters were first developed by *Seager and Sasselov* (1998). Subsequent work focusing on either specific objects or the class in general includes that by *Goukenleuque et al.* (2000), *Seager et al.* (2000, 2005), *Sudarsky et al.* (2000, 2003), *Barman et al.* (2001), *Iro et al.* (2005), *Burrows et al.* (2005), and *Fortney et al.* (2005, 2006). As with the cooler planets, the main conclusion of this body of work is that the spectra of the hot Jupiters depends sensitively on the vertical distribution and properties of condensates. Models that either postulate or predict high-altitude iron and silicate cloud decks tend to be warmer and more Planckian in thermal emission than models with deeper cloud decks. Hot-Jupiter models and observations are considered in detail in sections 3.4 and 4.

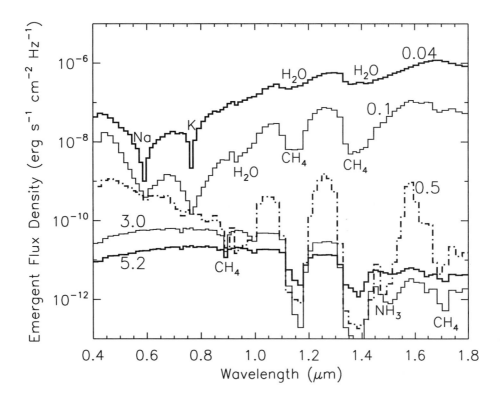

Fig. 5. Model atmosphere spectra (computed by authors J.F. and M.M.) for giant planets roughly corresponding to the atmosphere classes shown in Fig. 2. Numbers give orbital distance a from a solar type star in AU. The top model is for a hot Jupiter (a = 0.04 AU; T_{eff} = 1440 K; Class V). The atmosphere is very hot with high refractory clouds. In the second model the atmosphere is cooler (a = 0.1 AU; T_{eff} = 870 K; Class IV), the clouds are deeper, and the absorption bands are correspondingly more prominent. At a = 0.5 AU T_{eff} = 375 K; Class III) the atmosphere is cooler and relatively cloud free. Remaining two curves illustrate atmosphere with water clouds (a = 3 AU; Class II) and ammonia clouds (a = 5 AU; Class I).

The most systematic surveys of model EGP spectra include the work of *Sudarsky et al.* (2000, 2003, 2005) and *Barman et al.* (2001, 2005), who have studied model planets of a variety of masses, ages, and orbital radii (Fig. 2). *Burrows* (2005) reviews and recasts much of the former work with an eye toward detectability of EGPs. The universal conclusion of this body of work remains that molecular absorption bands and atmospheric condensates are the key diagnostics of giant planet effective temperature since giant planets cool as they age. For those planets distant enough from their stars that atmospheric temperature is primarily controlled by the loss of internal energy, not incident flux, the progression to lower atmospheric temperature with age results in a diagnostic sequence of spectroscopic changes discussed in the next section.

Planets more massive than 5 M_{Jup} may be as warm as 2000 K shortly after formation, with temperatures falling well below 1000 K by a few hundred million years. By a few billion years all planet-mass objects [<13 M_{Jup} (*Burrows et al.*, 1997)] are cooler than 500 K. The important chemical equilibrium and condensation boundaries are shown in Figs. 2 and 3. As the atmosphere cools, chemical equilibrium begins to favor first CH_4 over CO and then NH_3 over N_2. Water is present throughout this temperature range,

but the molecular bands become deeper with falling temperature. The early part of this sequence has already been well sampled by observed L- and T-type brown dwarfs, the coolest of which is about 700 K.

3.7. Spectral Signatures of Extrasolar Giant Planets

No one discussion or figure can hope to capture the range of temperature, metallicities, and cloud structures that likely define the entire suite of possible giant planets. Nevertheless, Figs. 5 and 6 help illustrate the important physical processes that control EGP spectra and give an indication of the major spectral signatures expected in atmospheres with roughly solar composition. These spectra are purposefully presented at moderate spectral resolution that will likely typify early direct detection spectra.

Figure 2 presents a set of five cloudless temperature-pressure (T-P) profiles. Also shown are the condensation curves for iron, silicate, water, and ammonia. The condensates expected in a given atmosphere depend upon the particular atmospheric temperature structure. *Sudarsky et al.* (2003) used this atmospheric characteristic to suggest that planets be categorized by which clouds form in their atmosphere. While this proposal has some drawbacks, mentioned

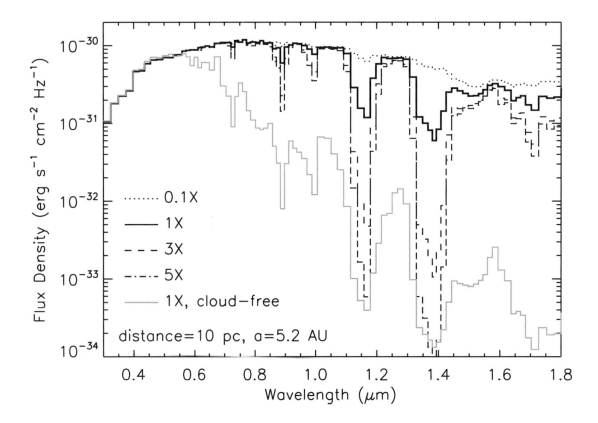

Fig. 6. Sensitivity of model Jupiter-like spectra (computed by authors J.F. and M.M.) to changes in model assumptions. Labeled curves illustrate reflected fluxes from 1 M_{Jup} planets assuming, from top to bottom, 0.1× solar abundance of heavy elements, solar abundance, 3× solar, 5× solar, and 1× solar with no cloud opacity. Note that while the continuum levels are generally unaffected by composition changes, the depths of the methane bands are highly sensitive. The cloud-free model is much darker in the red and infrared since incident photons are far more likely to be absorbed at these wavelengths than Rayleigh scattered.

above, it does nicely frame the discussion of EGP atmospheres and spectra.

The hottest EGPs orbiting most closely to their parent stars are expected to exhibit iron and silicate clouds high in their atmosphere since the atmospheric profile crosses these condensation curves at low pressures. Sudarsky et al. term such planets Class V. As seen in T-type brown dwarfs, Na and K are expected to be present in gaseous form and dominate the optical spectra, with water bands appearing in the near-infrared (Fig. 4). Thermal emission is an important contributor to the near-infrared flux, particularly between the strong water bands. Cloud scattering, however, limits the band depths.

In somewhat cooler atmospheres (Class IV) the clouds form at higher pressures in the atmosphere, which results in deeper absorption band depths. In addition, carbon is now found as CH_4 rather than CO, and thus methane features begin to appear in the near-infrared.

Somewhat cooler still, the effective temperature of a planet orbiting a G star at 0.5 AU would be about 375 K. Absorption of stellar radiation keeps the atmosphere warm enough that water clouds would not form, yet the iron and

silicate clouds lie far below the visible atmosphere. Although low abundance species like Na_2S could form low optical depth hazes, these atmospheres (Class III) will be relatively clear with a steep blue spectral slope and deep molecular bands. Like Class IV and V, thermal emission is important beyond about 1 μm.

For somewhat more distant planets, water and then ammonia condense, resulting in Class II and Class I atmospheres respectively. Continuum flux levels are controlled by the bright cloud decks and the "giant planet bands" of methane are apparent throughout the optical, particularly the strong band at 0.889 μm. An ammonia absorption feature is detectable at 1.5 μm in Class II atmospheres, but disappears in the colder Class I since the ammonia has condensed into clouds.

Figure 6 illustrates how sensitive such predictions are to atmospheric metallicity. Recall (Fig. 1) that Jupiter's atmosphere is enhanced by a factor of 3 in carbon and Uranus and Neptune by a factor of 30. The optical and near-infrared methane bands are highly sensitive to the methane abundance. Continuum levels, however, vary much less since they are controlled by the cloud structure, which is not as sensitive to abundance variations (the clouds are already optically thick). A cloud-free atmosphere, however, is much darker, again illustrating the importance of clouds.

3.8. Atmospheres of the Hot Jupiters

Despite their nickname, "hot Jupiters" likely bear little resemblance to a hotter version of our own Jupiter. Furthermore, given their extremely small orbital separations, these planets have undoubtedly experienced a very different upbringing than their frosty jovian cousins.

Early exploratory studies into the nature of hot Jupiters revealed that stellar heating leads to much shallower T-P profiles than present in isolated brown dwarfs (*Seager and Sasselov,* 1998; *Goukenleuque et al.,* 2000) (Fig. 3). Such reductions of the temperature gradient dramatically weaken the strength of otherwise prominent molecular absorption bands (e.g., due to water). Also, temperatures are high enough that the dominant carbon-based molecule is CO, unlike cooler giant planet atmospheres, which have high concentrations of CH_4. Hot-Jupiter models also indicate that, even though significant amounts of reflected optical light will be present due to Rayleigh and/or Mie scattering, Bond albedos may be well below 0.1 (*Sudarsky et al.,* 2000, 2003; *Barman et al.,* 2001).

Even though hot Jupiters are hot, their atmospheres are still cool enough that molecules, liquids, and even solids may form and many of the issues mentioned above are still relevant. As always, the expected atmospheric condensates depend on the detailed thermal structure of the atmosphere, which still varies a great deal within the hot-Jupiter class. Equilibrium chemistry suggests that high-altitude (P < 0.1 bar) Fe and silicate clouds may be present on the dayside of hot Jupiters. In general, cloud formation tends to increase the amount of scattered light and smooth out many of the spectral features. However, at these altitudes the atmosphere is purely radiative and likely fairly quiescent, which would allow condensate particles to quickly rain down to deeper levels of the atmosphere, allowing only a relatively thin haze to remain at high altitudes.

Hot-Jupiter temperature profiles can also enter a high-temperature, low-pressure domain in which the molecules TiO and VO, which have strong optical absorption bands, do not condense deeper in the atmosphere. This leads to very strong heating by incident radiation and the formation of exceptionally hot stratospheres akin to, but much hotter than, the stratospheres driven by near-infrared methane absorption in the solar system giant planets (*Hubeny et al.,* 2003; *Fortney et al.,* 2006).

4. OBSERVATIONS OF HOT JUPITERS

Since their first, surprising detection a decade ago (*Mayor and Queloz,* 1995) the hot Jupiters have received substantial attention, leading to the detection of the planets both during transit and eclipse. The chapter by Charbonneau et al.

fully explores this topic. Here we focus on the theoretical interpretation of the direct detections.

4.1. Transmission Spectra

As extrasolar planets transit their parent star, a small fraction of the stellar flux passes tangentially through the planet's limb and upper atmosphere. The absorbing properties of the planetary atmosphere (along a slant geometric path through the planet's limb) are added to the transmitted stellar absorption spectrum. There have been many published synthetic transmission spectra for hot Jupiters — sometimes presented as the wavelength-dependent planet radius that would be observed during a transit event. Some of these models assume plane-parallel slab geometry (*Seager and Sasselov,* 2000; *Hubbard et al.,* 2001; *Brown,* 2001), while others assume spherical geometry (*Barman et al.,* 2001, 2002). All these models have adopted a single one-dimensional thermal profile intended to represent an average of the planet's limb.

Seager and Sasselov (2000) predicted strong transmission absorption features due to Na and K alkali lines. *Hubbard et al.* (2001) extended the modeling of transmission spectra to near-infrared wavelengths. Their models showed that, similar to the Na and K alkali lines, H_2O bands can also imprint strong absorption feature onto the transmitted spectrum. Hubbard et al. also emphasized that while the emission spectrum of the planet may have molecular bands diminished by a reduced temperature gradient, the transmission spectrum is unaltered by such affects.

Modeling the transmission spectrum includes many potential difficulties. Transmission spectroscopy probes the low-pressure layers of the atmosphere where nonequilibrium conditions are most likely to occur. Also, the limb (or the terminator) is the transition zone between the nightside and the irradiated dayside. Consequently, the stellar radiation passes through a region that could have a steep *horizontal* temperature gradient and a correspondingly steep gradient in the chemical composition along tangent path lengths (see *Iro et al.,* 2005; *Barman et al.,* 2005). Such complications would be difficult to represent accurately using a single one-dimensional model. *Fortney* (2005) has also pointed out that, due to the relatively long tangential path lengths, trace condensates (negligible to the emission spectrum) may have a column density significant enough to impact the predicted transmission spectrum.

4.2. Thermal Emission

Impressive new datasets appeared in 2005 that placed new constraints on hot-Jupiter atmospheres and stress-tested existing hot-Jupiter atmosphere models. Those planets that transit their primary stars are eclipsed by them half an orbital period later, during which time only starlight — not planetary thermal emission — is detectable. The resulting lightcurve yields the ratio of planet to stellar flux. Observations with Spitzer have constrained this ratio for both HD 209458b [at 24 μm (*Deming et al.*, 2005)] and for TrES-1 [at 4.5 and 8 μm (*Charbonneau et al.*, 2005)]. For HD 209458b, the known stellar flux can by multiplied by the flux ratio at the same wavelength to yield the planetary flux, 55 ± 10 μJy, which can be expressed equivalently as a brightness temperature of 1130 ± 150 K. The distance to, and hence flux of, the TrES-1 star is not known. Hence only a brightness temperature can be quoted with some certainty. The TrES-1 temperature at 4.5 μm is 1010 ± 60 K and at 8 μm is 1230 ± 60 K.

Despite only three thermal emission data points for two different planets, four model interpretations (*Barman et al.*, 2005; *Burrows et al.*, 2005; *Fortney et al.*, 2005; *Seager et al.*, 2005) have already been published! Some of the published models have conflicting interpretations. For example, *Burrows et al.* (2005) claim that the model interpretation of TrES-1 suggests that the planet is presently reradiating on the dayside, while the best fit models of *Fortney et al.* (2005) are those for which the incident stellar radiation is evenly redistributed.

Given such conflicting conclusions, one might ask whether the current dataset is adequate to say anything concrete about the planetary atmospheres. Below we briefly summarize the currently published interpretations, and then provide our perspective on this question. The intense interest in HD 209458b, however, does permit comparisons between groups modeling the same object with different approaches. The range of a subset of the published models (Fig. 4) provides a measure of the uncertainty at the current state of the art.

Burrows et al. (2005) find that their predictions for the planet-to-star flux density ratios of both planets to be robust given the uncertainties in the planets' and primary stars' physical properties. They inferred the presence of CO and perhaps H_2O, and have determined that the atmospheres are hot. They suggest that the difference between the theoretical models and all three new measurements may be explained by an infrared-brighter hot dayside.

Fortney et al. (2005) find that while standard solar metallicity models can fit the single datapoint for HD 209458b, they do not for TrES-1, as the planetary spectral slope implied by the 4.5- and 8-μm observations is redder than expected. Model atmospheres that include a 3–5× metal enhancement, or energy deposition into the atmosphere from 1 to 10 mbar, lead to a redder spectral slope. With these models they find they can match the TrES-1 observations at 4.5 μm to 1σ, and at 8 μm to 2σ. Fortney et al. find that the best-fit models for both planets assume that reradiation of absorbed stellar flux occurs over the entire planet. They also note the excellent agreement to date between Spitzer ultracool dwarf infrared spectral data and models (*Roellig et al.*, 2004).

In addition to standard solar abundance f = 1/4 and 1/2 models, *Barman et al.* (2005) compute the two-dimensional vertical *and* horizontal temperature gradient over the entire dayside in the static no-redistribution (f = 1) case. For TrES-1, they find that all three models are consistent with

the 8-μm observations at the 2σ level. However, only their f = 1/4 model agrees with the 4.5-μm observation. More importantly, agreement between the f = 1 model and the 4.5-μm observation would require an unrealistic change to the atmospheric abundances. Consequently, some process must be redistributing the absorbed stellar flux at a level more comparable to an f = 1/2 or 1/4 scenario. This is in agreement with the findings of *Barman et al.* and *Fortney et al.* (2005) for HD 209458b where an f = 1/4 model shows the best agreement with the 24-μm Multiband Imaging Photometer (MIPS) observations. Figure 7 also illustrates just how red TrES-1 is compared to an typical brown dwarf spectrum of the same luminosity.

Finally, *Seager et al.* (2005) also conclude that a wide range of models fit the observational error bars. Starting with this philosophy, and including a 2.2-μm observational upper limit (*Richardson et al.*, 2003) neglected by the other groups, they rule out some models for HD 209458b at the hot and cold end of the plausible temperature range. They show that models with C/O > 1 can fit the HD 209458b data, including a paucity of H_2O (Fig. 8), and describe how the same models could fit TrES-1. They suggest that the models show an atmospheric circulation regime intermediate between pure *in situ* reradiation and very efficient heat redistribution.

All modelers agree on one point: Hot Jupiters are indeed hot. (Because the brightness temperatures of both HD 209458b and TrES-1 are over 1000 K, this conclusion does not, in fact, require a model interpretation at all.) The

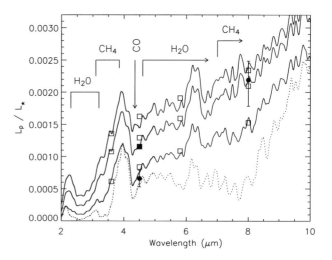

Fig. 7. The ratio of the model flux from the day side of TrES-1 to that of its parent star, from *Barman et al.* (2005), assuming no redistribution or f = 1 (top curve) and redistribution models with f = 0.5 (middle solid curve) and f = 0.25 (bottom solid curve). IRAC band fluxes for each model (found by convolving with the IRAC response curves) are indicated with open squares and filled circles show the Spitzer data with 1σ error bars. The 4.5-μm IRAC value for a 10× solar, f = 0.5 model is also shown (solid square). The lower dotted line corresponds to an isolated brown dwarf model with T_{eff} = 1150 K. Note the much redder slope of the planet. Major absorption bands are indicated.

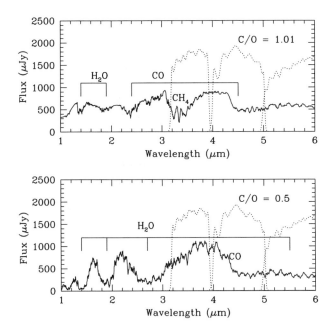

Fig. 8. Thermal emission spectrum for HD 209458b with C/O = 1.01 (and other elements in solar abundance). For C/O > 1 at this planet's temperature, the water abundance is low and CH_4 abundance is increased, compared to a solar abundance spectrum (e.g., Fig. 6). For C/O greater than solar (0.5) but less than 1, the spectrum would still show reduced H_2O and increased CH_4 but to a lesser extent. The Spitzer IRAC bandpasses are shown as dotted lines.

second point that all four modelers agree upon is that the TrES-1 4.5- and 8-μm data are not fit by a basic model with solar abundances: The model flux is too high in the 8-μm band compared to observations. These two points are probably the only concrete inferences that can be made from the observations. The detailed arguments for validity of specific atmosphere conditions (stratosphere, clouds, photochemistry, etc.) must await further Spitzer data at other wavelengths.

4.3. Nonequilibrium Effects

So far nearly all hot-Jupiter models have assumed local thermodynamic equilibrium (LTE) when solving the radiative transfer equation. Assuming LTE is tantamount to assuming that all species have level populations given by the Saha-Boltzmann distribution and that the frequency- and depth-dependent source function is simply a black body. Consequently, potentially important effects like photoionization and non-Boltzmann-like level populations are ignored. Given that a large fraction of the radiation field in a hot-Jupiter atmosphere is a nonlocal phenomena, LTE may be a rather risky assumption. *Yelle* (2004) reviews these and other issues related to understanding the likely very hot upper atmospheres of the hot Jupiters.

Shortly after the detection of Na D absorption in the atmosphere of HD 209458b (*Charbonneau et al., 2002*), *Barman et al.* (2002) and *Fortney et al.* (2003) explored the

possibility that non-LTE effects could alter the predicted strength of Na absorption in hot-Jupiter atmospheres. One possibility is that Na is ionized to pressures less than about 0.01 bar. Also, the relevant level populations of Na may be underpopulated, thereby reducing Na D line strengths.

4.4. Horizontal Gradients

Unlike the atmospheres of isolated brown dwarfs, which are heated entirely from the inside out, hot Jupiters experience significant heating by both internal (heat leftover from formation) and external (heat from the star) sources of energy. The presence of this external source of energy breaks the spherical symmetry implicitly assumed when one-dimensional model atmospheres are compared to observations of brown dwarfs and stars, for example. A lack of symmetry and dual heating sources bring a number of challenging issues to the forefront of the hot-Jupiter model atmosphere problem. Indeed, the day-night asymmetry and the need for more sophisticated modeling was recognized very early on, as discussed in section 2.1.

Barman et al. (2005) have made approximate two-dimensional static radiative-convective equilibrium models for the dayside atmosphere for HD 209458b and TrES-1. These models estimate the horizontal temperature gradients — in the absence of winds — to be quite steep (~1000 K at P = 1 bar) and can lead to a complex chemistry gradient over the dayside. For example, near the terminator, CO can potentially be replaced as the dominant carbon-bearing molecule by CH_4. Sodium condensation may also become important in this region (see also *Iro et al., 2005*).

As they are likely unstable, the steep horizontal gradients in the *Barman et al.* (2005) model strengthen the case for modeling the effects of global circulation. How significant the impact of such circulations would be on the atmospheric structure (and thus on the emergent spectrum) depends largely on the depth at which the stellar flux is absorb and the radiative and advective timescales in this region (*Seager et al., 2005*).

The strong dayside irradiation has motivated several groups to model the global atmospheric circulation currents in hot-Jupiter atmospheres. Three-dimensional simulations for HD 209458b suggest the possibility of strong zonal winds approaching or surpassing the sound speed (~1 km s⁻¹ winds) and a significant displacement of the atmospheric hot spot (*Showman and Guillot, 2002; Cooper and Showman, 2005*). Additionally, *Showman and Cooper* (2005) demonstrated the impact global winds could potentially have on the depth-dependent temperature structure. Their simulations predict low-pressure temperature inversions not seen in static one-dimensional atmosphere models. *Cho et al.* (2003), assuming a characteristic wind speed of 400 m s⁻¹ for HD 209458b, found localized jets and vortices producing hot and cold regions differing by as much as 300 K. These localized "hot spots" in the Cho et al. simulations also move about the poles with ~25-day periods. *Menou et al.* (2003) applied the results of *Cho et al.* (2003) to other short-period EGPs and concluded that these kinds of circulation patterns are likely to be common among hot Jupiters. Despite predicting very different atmospheric flows, all these simulations agree that circulation currents will most likely reposition the atmospheric hot spot(s) away from the substellar point. Consequently, maximum and minimum infrared fluxes would not necessarily coincide with orbital phases that align the substellar and antistellar points with Earth, a result that can be tested by infrared lightcurves or, more easily, by secondary eclipse diagnostics (*Williams et al.*, 2006).

While the various hydrodynamic simulations differ significantly in details, there appears to be agreement that variations on the order of ~300 to 500 K may be present at "photospheric" pressures. The studies of *Showman and Guillot* (2002) and more recently *Cooper and Showman* (2005) predict steady eastward supersonic winds producing an atmospheric hot spot that may be displaced 60° from the planet's substellar point.

5. THE FUTURE

Transiting planets and hot Jupiters are the most favorable extrasolar planets for observational study in the near future. They can be studied in the combined light of the planet-star system without direct imaging. Although groundbased observations will surely continue, the stable conditions of space make it the best place for the requisite high-precision observations.

The transiting planets orbiting the brightest stars, particularly HD 209458b (the touchstone for hot Jupiters) as well as the newly discovered HD 189733b, will certainly continue to receive great attention. Spitzer photometric and spectral observations will determine if there are phase variations in the thermal emission and will search for spectral signatures of the atmosphere. At visible wavelengths, the Canadian Microvariabilité and Oscillations Stellaires (MOST) space telescope will observe HD 209458b during secondary eclipse in scattered light and will reach a geometric albedo of 0.15. Hubble Space Telescope (HST) data will add to the variety of data on the same planet.

Other non-transiting hot Jupiters will likely be monitored with Spitzer for phase variation with both the Infrared Array Camera (IRAC) and MIPS. New transiting planets around bright stars, such as TrES-1 and HD 149026, will also be observed with Spitzer and HST. The Stratospheric Observatory for Infrared Astronomy (SOFIA) and large groundbased observatories such as Keck and the ESO Very Large Telescope (VLT) may be able to detect thermal emission at shorter wavelengths than Spitzer (2–4 μm) during secondary eclipse of transiting hot Jupiters.

The more distant future for extrasolar planet characterization looks even more promising. In the next decade Kepler will find dozens of transiting EGPs at a variety of semimajor axes out to 1 AU. Some of those planets' atmospheres may be detectable by the James Webb Space Telescope (JWST). Further to the future, in the latter part of the next decade, planned 20- to 30-m groundbased telescopes

may be able to directly image massive Jupiters in Jupiter-like orbits. Around the same time, NASA's Terrestrial Planet Finders and ESA's Darwin aim to directly detect and characterize nearby extrasolar planets ranging from giant planets down to Earth-sized planets.

As the era of EGP characterization moves from the hot Jupiters to the realm of true Jupiter analogs, the focus of characterization will change. Mass and particularly radius, which are most easily constrained by transits, will be much less well constrained for more distant planets. Instead spectra, which are especially sensitive to cloud structure, atmospheric composition, and atmospheric chemistry, will provide the primary method for planet characterization. Ultimately, obtaining the equivalent of the compositional fingerprint shown in Fig. 1 for many planets in many planetary systems will illuminate the giant-planet-formation process as a function of stellar mass, metallicity, and planetary system architecture.

Acknowledgments. This work was partially supported by the NASA Astrophysics Theory and Origins programs. We thank H. Hammel, K. Zahnle, and the anonymous referee for thoughtful suggestions that improved the manuscript, as well as J. Moses for providing advice on photochemical issues.

REFERENCES

Ackerman A. S. and Marley M. S. (2001) *Astrophys. J., 556,* 872–884.
Allende Prieto C., Lambert D. L., and Asplund M. (2002) *Astrophys. J., 573,* L137–L140.
Appleby J. F. (1986) *Icarus, 65,* 383–405.
Appleby J. F. and Hogan J. S. (1984) *Icarus, 59,* 336–366.
Asplund M. (2005) *Ann. Rev. Astron. Astrophys., 43,* 481–530.
Atreya S. K., Mahaffy P. R., Niemann H. B., Won M. H., and Owen T. C. (2003) *Planet. Space Sci., 51,* 105–112.
Baraffe I., Chabrier G., Barman T. S., Allard F., and Hauschildt P. H. (2003) *Astron. Astrophys., 402,* 701–712.
Barman T. S., Hauschildt P. H., and Allard F. (2001) *Astrophys. J., 556,* 885–895.
Barman T. S., Hauschildt P. H., Schweitzer A., Stancil P. C., Baron E., and Allard F. (2002) *Astrophys. J., 569,* L51–L54.
Barman T. S., Hauschildt P. H., and Allard F. (2005) *Astrophys. J., 632,* 1132–1139.
Bodenheimer P., Lin D. N. C., and Mardling R. A. (2001) *Astrophys. J., 548,* 466–472.
Bodenheimer P., Laughlin G., and Lin D. N. C. (2003) *Astrophys. J., 592,* 555–563.
Brown T. M. (2001) *Astrophys. J., 553,* 1006–1026.
Buriez J. C. and de Bergh C. (1981) *Astron. Astrophys., 94,* 382–390.
Burkert A., Lin D. N. C., Bodenheimer P. H., Jones C. A., and Yorke H. W. (2005) *Astrophys. J., 618,* 512–523.
Burrows A. (2005) *Nature, 433,* 261–268.
Burrows A., Marley M., Hubbard W. B., Lunine J. I., Guillot T., et al. (1997) *Astrophys. J., 491,* 856–875.
Burrows A., Hubbard W. B., Lunine J. I., Marley M. S., and Saumon D. (2000) In *Protostars and Planets IV* (V. Mannings et al., eds), pp. 1339–1361. Univ. of Arizona, Tucson.
Burrows A., Hubbard W. B., Lunine J. I., and Liebert J. (2001) *Rev. Mod. Phys., 73,* 719–765.
Burrows A., Sudarsky D., and Hubbard W. B. (2003) *Astrophys. J., 594,* 545–551.
Burrows A., Hubeny I., and Sudarsky D. (2005) *Astrophys. J., 625,* L135–L138.

Chabrier G., Barman T., Baraffe I., Allard F., and Hauschildt P. H. (2004) *Astrophys. J., 603*, L53–L56.

Charbonneau D., Brown T. M., Noyes R. W., and Gilliland R. L. (2002) *Astrophys. J., 568*, 377–384.

Charbonneau D., Allen L. E., Megeath S. T., Torres G., Alonso R., et al. (2005) *Astrophys. J., 626*, 523–529.

Cho J. Y.-K., Menou K., Hansen B., and Seager S. (2003) *Astrophys. J., 587*, L117–L120.

Cooper C. S. and Showman A. (2005) *Astrophys. J., 629*, L45–L48.

Cooper C. S., Sudarsky D., Milsom J. A., Lunine J. I., and Burrows A. (2003) *Astrophys. J., 586*, 1320–1337.

Deming D., Seager S., Richardson L. J., and Harrington J. (2005) *Nature, 434*, 740–743.

Dlugach J. M. and Yanovitskij E. G. (1974) *Icarus, 22*, 66–81.

Dyudina U. A., Sackett P. D., Bayliss D. D. R., Seager S., Porco C. C., Throop H. B., and Dones L. (2005) *Astrophys. J., 618*, 973–986.

Fegley B. Jr. and Lodders K. (1994) *Icarus, 110*, 117–154.

Fegley B. Jr. and Lodders K. (1996) *Astrophys. J., 472*, L37–L40.

Fegley B. and Prinn R. G. (1988) *Astrophys. J., 324*, 621–625.

Fegley B. J., Gautier D., Owen T., and Prinn R. G. (1991) In *Uranus* (J. T. Bergstrahl et al., eds.), pp. 147–203. Univ. of Arizona, Tucson.

Flasar F. M., Achterberg R. K., Conrath B. J., Pearl J. C., Bjoraker G. L., et al. (2005) *Science, 307*, 1247–1251.

Fortney J. J. (2005) *Mon. Not. R. Astron. Soc., 364*, 649–653.

Fortney J. J., Sudarsky D., Hubeny I., Cooper C. S., Hubbard W. B., Burrows A., and Lunine J. I. (2003) *Astrophys. J., 589*, 615–622.

Fortney J. J., Marley M. S., Lodders K., Saumon D., and Freedman R. (2005) *Astrophys. J., 627*, L69–L72.

Fortney J. J., Saumon D., Marley M. S., Lodders K., and Freedman R. S. (2006) *Astrophys. J., 642*, 495–504.

Gautier D., Conrath B., Owen T., de Pater I., and Atreya S. (1995) In *Neptune and Triton* (D. Cruikshank, ed.), pp. 547–611. Univ. of Arizona, Tucson.

Golimowski D. A., Leggett S. K., Marley M. S., Fan X., Geballe T. R., et al. (2004) *Astron. J., 127*, 3516–3536.

Goukenleuque C., Bézard B., Joguet B., Lellouch E., and Freedman R. (2000) *Icarus, 143*, 308–323.

Griffith C. A. and Yelle R. V. (1999) *Astrophys. J., 519*, L85–L88.

Guillot T. and Showman A. P. (2002) *Astron. Astrophys., 385*, 156–165.

Guillot T., Burrows A., Hubbard W. B., Lunine J. I., and Saumon D. (1996) *Astrophys. J., 459*, L35–L38.

Helling C., Klein R., Woitke P., Nowak U., and Sedlmayr E. (2004) *Astron. Astrophys., 423*, 657–675.

Hogan J. S., Rasool S. I., and Encrenaz T. (1969) *J. Atmos. Sci., 26*, 898–905.

Hubbard W. B. (1968) *Astrophys. J., 152*, 745–754.

Hubbard W. B. (1977) *Icarus, 30*, 305–310.

Hubbard W. B., Fortney J. J., Lunine J. I., Burrows A., Sudarsky D., and Pinto P. (2001) *Astrophys. J., 560*, 413–419.

Hubeny I., Burrows A., and Sudarsky D. (2003) *Astrophys. J., 594*, 1011–1018.

Iro N., Bézard B., and Guillot T. (2005) *Astron. Astrophys., 436*, 719–727.

Kuiper G. P. (1952) *The Atmospheres of the Earth and Planets.* Univ. of Chicago, Chicago.

Kurucz R. (1993) *CD-ROM 13, ATLAS9 Stellar Atmosphere Programs and 2 km/s Grid.* Smithsonian Astrophysical Observatory, Cambridge.

Laughlin G., Wolf A., Vanmunster T., Bodenheimer P., Fischer D., Marcy G., Butler P., and Vogt S. (2005) *Astrophys. J., 621*, 1072–1078.

Liang M.-C., Seager S., Parkinson C. D., Lee A. Y.-T., and Yung Y. L. (2004) *Astrophys. J., 605*, L61–L64.

Lissauer J. J. (1993) *Ann. Rev. Astron. Astrophys., 31*, 129–174.

Lodders K. (2003) *Astrophys. J., 591*, 1220–1247.

Lodders K. (2004) *Astrophys. J., 611*, 587–597.

Lodders K. and Fegley B. (2006) In *Astrophysics Update 2* (J. W. Mason, ed.), pp. 1–28. Praxis, Chichester, UK.

Marley M. S. (1998) In *Brown Dwarfs and Extrasolar Planets* (R. Rebolo et al., eds.), pp. 383–393. ASP Conf. Series 134, San Francisco.

Marley M. S. and McKay C. P. (1999) *Icarus, 138*, 268–286.

Marley M. S., Saumon D., Guillot T., Freedman R. S., Hubbard W. B., Burrows A., and Lunine J. I. (1996) *Science, 272*, 1919–1921.

Marley M. S., Gelino C., Stephens D., Lunine J. I., and Freedman R. (1999) *Astrophys. J., 513*, 879–893.

Marley M. S., Seager S., Saumon D., Lodders K., Ackerman A. S., Freedman R. S., and Fan X. (2002) *Astrophys. J., 568*, 335–342.

Marley M. S., Ackerman A. S., Burgasser A. J., Saumon D., Lodders K., and Freedman R. (2003) In *Brown Dwarfs* (E. Martin, ed.), pp. 333–344. IAU Symposium 211, ASP, San Francisco.

Mayor M. and Queloz D. (1995) *Nature, 378*, 355–359.

Menou K., Cho J. Y.-K., Seager S., and Hansen B. (2003) *Astrophys. J., 587*, L113–L116.

Owen T., Mahaffy P., Niemann H. B., Atreya S., Donahue T., Bar-Nun A., and de Pater I. (1999) *Nature, 402*, 269–270.

Pollack J. B., Rages K., Baines K. H., Bergstrahl J. T., Wenkert D., and Danielson G. E. (1986) *Icarus, 65*, 442–466.

Prinn R. G. and Barshay S. S. (1977) *Science, 198*, 1031–1034.

Richardson L. J., Deming D., Wiedemann G., Goukenleuque C., Steyert D., Harrington J., and Esposito L. W. (2003) *Astrophys. J., 584*, 1053–1062.

Roellig T. L., Van Cleve J. E., Sloan G. C., Wilson J. C., Saumon D., et al. (2004) *Astrophys. J. Suppl., 154*, 418–421.

Rossow W. B. (1978) *Icarus, 36*, 1–50.

Saumon D., Marley M., and Lodders K. (2003) *ArXiv Astrophys. e-prints* (astro-ph/0310805).

Seager S. and Sasselov D. D. (1998) *Astrophys. J., 502*, L157–L161.

Seager S., Whitney B. A., and Sasselov D. D. (2000) *Astrophys. J., 540*, 504–520.

Seager S., Richardson L. J., Hansen B. M. S., Menou K., Cho J. Y.-K., and Deming D. (2005) *Astrophys. J., 632*, 1122–1131.

Showman A. P. and Guillot T. (2002) *Astron. Astrophys., 385*, 166–180.

Sudarsky D., Burrows A., and Pinto P. (2000) *Astrophys. J., 538*, 885–903.

Sudarsky D., Burrows A., and Hubeny I. (2003) *Astrophys. J., 588*, 1121–1148.

Sudarsky D., Burrows A., Hubeny I., and Li A. (2005) *Astrophys. J., 627*, 520–533.

Trafton L. M. (1967) *Astrophys. J., 147*, 765–781.

Tsuji T. (2005) *Astrophys. J., 621*, 1033–1048.

Vidal-Madjar A., Lecavelier des Etangs A., Désert J.-M., Ballester G. E., Ferlet R., Hébrard G., and Mayor M. (2003) *Nature, 422*, 143–146.

Visscher C. and Fegley B. J. (2005) *Astrophys. J., 623*, 1221–1227.

Westphal J. A., Matthews K., and Terrile R. J. (1974) *Astrophys. J., 188*, L111–L112.

Wetherill G. W. and Stewart G. R. (1989) *Icarus, 77*, 330–357.

Williams P. K. G., Charbonneau D., Cooper C. S., Showman A. P., and Fortney J. J. (2006) *Astrophys. J., 649*, 1020–1027.

Yelle R. (2004) *Icarus, 170*, 167–179.

Young R. E. (2003) *New Astron. Rev., 47*, 1–51.

Yung Y. L., Drew W. A., Pinto J. P., and Friedl R. R. (1988) *Icarus, 73*, 516–526.

Part VIII:

Dust, Meteorites, and the Early Solar System

The Chemical Evolution of Protoplanetary Disks

Edwin A. Bergin
University of Michigan

Yuri Aikawa
Kobe University

Geoffrey A. Blake
California Institute of Technology

Ewine F. van Dishoeck
Leiden Observatory

In this review we reevaluate our observational and theoretical understanding of the chemical evolution of protoplanetary disks. We discuss how improved observational capabilities have enabled the detection of numerous molecules, exposing an active disk chemistry that appears to be in disequilibrium. We outline the primary facets of static and dynamical theoretical chemical models. Such models have demonstrated that the observed disk chemistry arises from warm surface layers that are irradiated by X-ray and FUV emission from the central accreting star. Key emphasis is placed on reviewing areas where disk chemistry and physics are linked, including deuterium chemistry, gas temperature structure, disk viscous evolution (mixing), ionization fraction, and the beginnings of planet formation.

1. INTRODUCTION

For decades, models of our own solar nebular chemical and physical evolution have been constrained by the chemical record gathered from meteorites, planetary atmospheres, and cometary comae. Such studies have provided important clues to the formation of the Sun and planets, but large questions remain regarding the structure of the solar nebula, the exact timescale of planetary formation, and the chemical evolution of nebular gas and dust. Today we are on the verge of a different approach to nebular chemical studies, one where the record gained by solar system studies is combined with observations of numerous molecular lines in a multitude of extrasolar protoplanetary disk systems tracking various evolutionary stages.

Our observational understanding of extrasolar protoplanetary disk systems is still in its infancy as the current capabilities of millimeter-wave observatories are limited by sensitivity and also by the small angular size of circumstellar disks, even in the closest star-forming regions (<3–4"). Nonetheless, numerous molecules have been detected in protoplanetary disks, exposing an active chemistry (*Dutrey et al.,* 1997; *Kastner et al.,* 1997; *Qi et al.,* 2003; *Thi et al.,* 2004). Since the last *Protostars and Planets* review (*Prinn,*

1993) these observations have led to a paradigm shift in our understanding of disk chemistry. For many years focus was placed on thermochemical models as predictors of the gaseous composition, and these models have relevance in the high-pressure, and $\geq 10^{-6}$ bar (inner) regions of the nebula (e.g., *Fegley,* 1999). However, for most of the disk mass, the observed chemistry appears to be in disequilibrium and quite similar to that seen in dense regions of the interstellar medium (ISM) that are directly exposed to radiation (*Aikawa and Herbst,* 1999; *Willacy and Langer,* 2000; *Aikawa et al.,* 2002).

In this review we focus on gains in our understanding of the chemistry that precedes and is contemporaneous with the formation of planets from a perspective guided by observations of other stellar systems whose masses are similar to the Sun. We examine both the observational and theoretical aspects of this emerging field, with an emphasis on areas where the chemistry directly relates to disk physics. Portions of this review overlap with other chapters, such as the observational summary of molecular disks, inner disk gas, and disk physical structure (see the chapters by Dutrey et al., Najita et al., and Dullemond et al., respectively). For this purpose we focus our review on the physics and chemistry of the outer disk (r > 10 AU) in systems with ages of

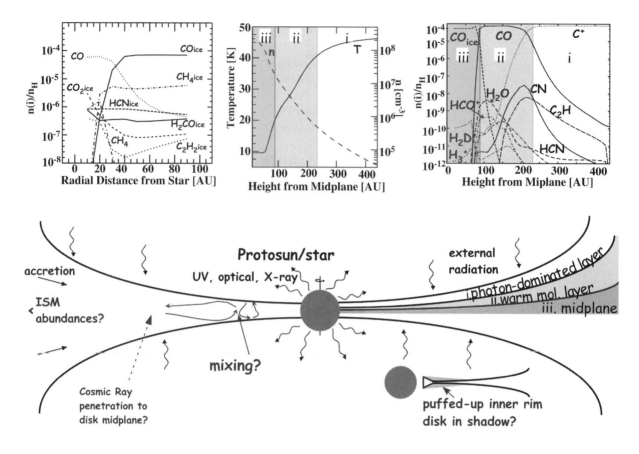

Fig. 1. Chemical structure of protoplanetary disks. Vertically the disk is schematically divided into three zones: a photon-dominated layer, a warm molecular layer, and a midplane freeze-out layer. The CO freeze-out layer disappears at r ≤ 30–60 AU as the midplane temperature increases inward. Various nonthermal inputs, cosmic rays, UV, and X-rays drive chemical reactions. Viscous accretion and turbulence will transport the disk material both vertically and radially. The upper panels show the radial and vertical distribution of molecular abundances from a typical disk model at the midplane (*Aikawa et al.*, 1999) and r ~ 300 AU (*van Zadelhoff et al.*, 2003). A sample of the hydrogen density and *dust* temperature at the same distance (*D'Alessio et al.*, 1999) is also provided. In upper layers (≥150 AU) the *gas* temperature will exceed the dust temperature by ≥25 K (*Jonkheid et al.*, 2004).

0.3–10 m.y. In support of our theory we also present an observational perspective extending from the infrared (IR) to the submillimeter to motivate the theoretical background and supplement other discussions.

2. GENERAL THEORETICAL PICTURE

2.1. Basic Physical and Chemical Structure of the Disk

Chemical abundances are determined by physical conditions such as density, temperature, and the incident radiation field. Recent years have seen significant progress in characterizing the physical structure of the disk, which aids in understanding disk chemical processes. Isolated disks can be quite extended, with r_{out} ~ 100 to a few hundred AU (*Simon et al.*, 2000), much larger than expected from comparison with the minimum mass solar nebula (MMSN) (*Hayashi*, 1981). However, it should be stated that we have an observational bias toward detecting larger disks due to our observational limitations. The radial distribution

of column density and midplane temperature have been estimated by observing thermal emission of dust; they are fitted by a power-law $\Sigma(r) \propto r^{-p}$ and $T(r) \propto r^{-q}$, with p = 0–1 and q = 0.5–0.75. The temperature at 1 AU is ~100–200 K, while the surface density at 100 AU is 0.1–10 g cm^{-2} (e.g., *Beckwith et al.*, 1990; *Kitamura et al.*, 2002).

The vertical structure is estimated by calculating the hydrostatic equilibrium for the density and radiation transfer for the dust temperature (see chapter by Dullemond et al.). Beyond several AU the disk is mainly heated by irradiation from the central star. The stellar radiation is absorbed by grains at the disk surface, which then emit thermal radiation to heat the disk interior (*Calvet et al.*, 1992; *Chiang and Goldreich*, 1997; *D'Alessio et al.*, 1998). Hence the temperature decreases toward the midplane, as seen in Fig. 1. For small radii (r < few AU), however, heating by mass accretion is not negligible and the midplane can be warmer than the disk surface. At r = 1 AU, for example, the midplane temperature can be as high as 1000 K, if the accretion rate is large (*D'Alessio et al.*, 1999; *Nomura*, 2002). The density distribution is basically Gaussian, exp –(Z/H)^2, with some deviation due to vertical temperature variations (see Fig. 1). As a whole the disk has a flared-up structure, with a geometrical thickness that increases with radius (*Kenyon and Hartmann*, 1987).

Based on such physical models, the current picture of the general disk chemical structure is schematically shown in Fig. 1. At r ≥ 100 AU, the disk can be divided into three layers: the photon-dominated region (PDR), the warm molecular layer, and the midplane freeze-out layer. The disk is irradiated by UV radiation from the central star and interstellar radiation field that ionize and dissociate molecules and atoms in the surface layer. In the midplane the temperature is mostly lower than the freeze-out temperature of CO (~20 K), one of the most abundant and volatile molecules in the ISM. Since the timescale of adsorption onto grains is short at high density [~$10(10^9 cm^{-3}/n_H)$ yr], heavy-element species are significantly depleted onto grains. At intermediate heights, the temperature is several tens of Kelvin, and the density is sufficiently high ($\geq 10^6$ cm^{-3}) to ensure the existence of molecules even if the UV radiation is not completely attenuated by the upper layer (*Aikawa and Herbst*, 1999; *Willacy and Langer*, 2000; *Aikawa et al.*, 2002). Here water is still frozen onto grains, trapping much of the oxygen in the solid state. Thus, the warm CO-rich gas layers will have C/O ~ 1, leading to a rich and extensive carbon-based chemistry.

These models provide a good match to observed abundances. *Dutrey et al.* (1997) found that in the DM Tau disk, molecular abundances are generally lower than in dense clouds, but the CN/HCN ratio is higher (see also *Thi et al.*, 2004). The low molecular abundances are caused by depletion in the midplane, and the high CN/HCN ratio originates in the surface PDR (cf. Fig. 1), as seen in PDRs in the ISM (*Rodríguez-Franco et al.*, 1998).

At r ≤ 100 AU, the midplane temperature is high enough to sublimate various ice materials that formed originally in the outer disk radius and/or parental cloud core (e.g., *Markwick et al.*, 2002). This sublimation will be species dependent with the "snowline" for a given species appearing at different radii. For example, in the solar nebula the water ice snowline appeared near 3–5 AU, while the CO snowline would appear at greater distances where the midplane dust temperatures drop below ~20 K. Within these species-selective gaseous zones, sublimated molecules will be destroyed and transformed to other molecules by gas-phase reactions. In this fashion, the chemistry is similar to the so-called "hot core" chemistry, which appears in star-forming cores surrounding protostars (e.g., *Hatchell et al.*, 1998; see chapter by Ceccarelli et al.). For example, sublimated CH$_4$ is transformed to larger and less-volatile carbon-chain species, which can then accumulate onto grains (*Aikawa et al.*, 1999).

2.2. Key Ingredients: Ultraviolet, X-Ray, and Cosmic-Ray

Thermochemistry and its consequences are described by *Fegley and Prinn* (1989) and *Prinn et al.* (1993). In recent years nonthermal events such as cosmic rays, UV, and X-

rays have been included in the disk chemistry models and are shown to play important roles.

In molecular clouds, chemical reactions are driven by cosmic-ray ionization (e.g., *Herbst and Klemperer*, 1973). Since cosmic-ray ionization has an attenuation length of 96 g cm^{-2} (*Umebayashi and Nakano*, 1981), in the disk it can be important, for r ≥ several AU, in driving ion-molecule reactions and producing radicals that undergo neutral-neutral reactions (*Aikawa et al.*, 1997). Although cosmic rays may be scattered by the magnetic fields within and around the protostar-disk system, detection of ions (e.g., HCO$^+$ and H$_2$D$^+$) and comparison with theoretical models indicate that some ionization mechanisms, perhaps cosmic rays, is available at least for r ≥ 100 AU (*Semenov et al.*, 2004; *Ceccarelli and Dominik*, 2005) (see also section 4.3).

T Tauri stars have excess UV flux that is much higher than expected from their effective temperature of ~3000 K (e.g., *Herbig and Goodrich*, 1986). It is considered to originate in the accretion shock on the stellar surface (*Calvet and Gullbring*, 1998), with potential contributions from an active chromosphere (*Alexander et al.*, 2005). For low-mass T Tauri stars, the strength of the UV field is often parameterized in terms of the local interstellar radiation field (*Habing*, 1968) (G$_0$ = 1), and have values of G$_0$ ~ 300–1000 at 100 AU (*Bergin et al.*, 2004). This radiation impinges on the flared disk surface with a shallow angle of incidence. Stellar and interstellar UV photons dissociate and ionize molecules and atoms in the flared disk surfaces and detailed two-dimensional radiative transfer models are required to quantitatively predict molecular abundances. *van Zadelhoff et al.* (2003) showed scattering of stellar UV at the disk surface significantly enhances the abundance of radical species in deeper layers, and examined the resulting chemical evolution for different wavelength dependencies of the stellar UV radiation field. *Bergin et al.* (2003) pointed out the importance of Lyα emission, which stands out in the UV spectrum of T Tauri stars (Fig. 2) and is absent from the interstellar field. In the one case where the line is unaffected by interstellar absorption, TW Hya, Lyα radiation carries ~85% of the FUV flux (*Herczeg et al.*, 2004). Since photodissociation cross sections are a function of wavelength, species that absorb Lyα, such as HCN and H$_2$O, will be selectively dissociated, while others, such as CO and H$_2$, are unaffected (*van Dishoeck et al.*, 2006).

Most stars do not form in isolation, rather 70–90% stars are born in GMCs (which contain most of the galactic molecular mass) and are found in embedded stellar clusters (*Lada and Lada*, 2003). In this light there exists growing evidence that the Sun formed in a cluster in the vicinity of a massive star (see chapter by Wadhwa et al.; *Hester and Desch*, 2005). In this case the UV radiation field can be much higher, depending on the spectral type of the OB star, the proximity of the low-mass star to the source of energetic radiation, and the dissipation timescale of the surrounding dust and gas (which can shield forming low-mass disks from radiation). After gas/dust dissipation the external radiation has greater penetrating power, because it can impinge on the disk with a greater angle of incidence. The

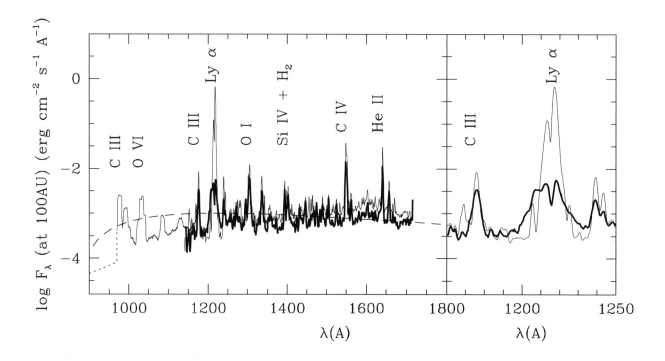

Fig. 2. UV spectra of T Tauri stars. Heavy solid lines and light solid lines represent the spectra of BP Tau and TW Hya, respectively. The spectrum of TW Hya is scaled by 3.5 to match the BP Tau continuum level. The long dashed line represents the interstellar radiation field of *Draine* (1978) scaled by a factor of 540. The region around the Lyα line is enlarged in the right panel. Taken from *Bergin et al.* (2003).

primary effects of external radiation, if it dominates the stellar contribution, will be to magnify the chemistry (e.g., CO driven photochemistry) produced by the stellar radiation and increase the size of the warm molecular layer.

Strong X-ray emission is observed toward T Tauri stars (e.g., *Kastner et al.*, 2005). It may originate in the magnetic reconnections either in the stellar magnetosphere, at the star-disk interface, or above the circumstellar disk (*Feigelson and Montmerle*, 1999, and references therein). X-rays affect the chemistry in several ways (*Maloney et al.*, 1996; *Stäuber et al.*, 2005): (1) They ionize atoms and molecules to produce high-energy photoelectrons that further ionize the gas. On the disk surface X-ray ionization produces a higher ionization rate than cosmic rays, and can even be the dominant ionization source if the cosmic rays are scattered by the magnetic field (*Glassgold et al.*, 1997; *Igea and Glassgold*, 1999). (2) High-energy photoelectrons heat the gas. For example, at r = 1 AU, the gas temperature in the uppermost layer can be as high as 5000 K due to the X-ray heating together with mechanical heating such as turbulent dissipation (*Glassgold et al.*, 2004; see chapter by Najita et al.). (3) Collision of the high-energy electrons with hydrogen atoms and molecules results in excitation of these species and then the emission of UV photons within the disk (*Maloney et al.*, 1996; *Herczeg et al.*, 2004). Recently *Bergin et al.* (2004) found H₂ FUV continuum emission caused by this mechanism. The high ionization rate and induced photodissociation of CO enhance the abundances of organic

species such as CN, HCN, and HCO⁺ (*Aikawa and Herbst*, 1999, 2001).

Nonthermal particles and radiation can also drive the desorption of molecular species from the grain surface. Because of the high densities, gaseous species collide with grains on short timescales. In the low-temperature region in the outer disk the colliding molecules are adsorbed onto grains, and thermal desorption is inefficient except for very volatile species such as CO, N₂, and CH₄, depending on the temperature. Various gaseous molecules are still observable since they are formed from CO and N₂ via gas-phase reactions, which compensates for adsorption (e.g., *Aikawa et al.*, 2002). However, observations may indicate the need for more efficient (thermal or nonthermal) desorption, especially for species mainly formed by grain-surface reactions. For example, *Dartois et al.* (2003) find evidence for gaseous CO in layers with a temperature below the CO sublimation temperature. Cosmic rays and X-rays can temporally "spot heat" the grains to enhance desorption rates (e.g., *Léger et al.*, 1985; *Hasegawa and Herbst*, 1993). UV radiation can also desorb molecules, possibly by producing radicals within the ice mantle, which react with other radicals on the grain surface to release excess energies (*Westley et al.*, 1995). It should be noted, however, that the nonthermal desorption rates are uncertain and depend on various parameters such as structure of the grain particle (*Najita et al.*, 2001), UV flux, number density of radical species in grain mantle (*Shen et al.*, 2004), and detailed desorption process

(*Bringa and Johnson*, 2004). Mixing could also play a role in moving material from warmer layers to colder ones (e.g., section 4.4).

3. OBSERVATIONS

Observational studies of the chemistry in extrasolar protoplanetary disks began only in the last decade thanks to improved sensitivity and spatial resolution at millimeter and IR wavelengths. At millimeter wavelengths, molecules other than CO have now been detected and imaged in a handful of disks with single-dish telescopes and interferometers. This technique has the advantage that molecules with very low abundances (down to 10^{-11} with respect to H_2) can be detected through their pure rotational transitions, and that the spatial distribution in the disk can be determined. With a spectral resolving power $R \approx \lambda/\Delta\lambda > 10^6$, the line profiles are fully resolved and kinematic information can be derived. Infrared spectroscopy has the main advantage that not only gas but also solid material can be probed through their vibrational transitions, including ices and silicates. Also, gas-phase molecules without dipole moments, including H_2, CH_4, C_2H_2, and CO_2, can only be observed at IR wavelengths. Finally, polycyclic aromatic hydrocarbons (PAHs) have unique IR features. For spacebased instruments, the resolving power is usually low, typically $R \approx 300$–3000, making it difficult to observe and resolve gas-phase lines.

3.1. Infrared Observations

3.1.1. Silicates, ices, and polycyclic aromatic hydrocarbons. The Infrared Space Observatory (ISO) opened up mid-IR spectroscopy of disks over the full 2–200-μm range unhindered by Earth's atmosphere, revealing a wealth of features (see *van Dishoeck*, 2004, for a review). Because of limited sensitivity, ISO could only probe the chemistry in disks around intermediate-mass Herbig Ae/Be stars. The Spitzer Space Telescope has the sensitivity to obtain 5–40-μm spectra of solar-mass T Tauri stars, while large 8–10-m optical telescopes can obtain higher spectral and spatial resolution data in atmospheric windows, most notably at 3–4, 4.6–5, and 8–13 μm. The features are usually in emission, except if the disk is viewed nearly edge-on when the bands occur in absorption against the continuum of the warm dust in the inner disk.

The amorphous broad silicate features at ~10 and 20 μm are the most prominent emission bands in disk spectra (e.g., *Przygodda et al.*, 2003; *Kessler-Silacci et al.*, 2005; 2006) (see Fig. 3). They arise in the superheated layers of the inner disk at <1–10 AU and are not representative of the outer disk. For at least half of the sources narrower features are seen as well, which can be ascribed to crystalline silicates such as forsterite, Mg_2SiO_4 (e.g., *Malfait et al.*, 1998; *Forrest et al.*, 2004). Silicates are discussed extensively in the chapter by Natta et al.; the main point for this chapter is that the shape and strength of the silicate features, together with the overall spectral energy distribution, can be used

to determine whether grain growth and settling has occurred (e.g., *van Boekel et al.*, 2005). This, in turn, affects the chemistry and heating in the disk (see section 4.1). Also, the presence of crystalline silicates may be an indication of significant radial and vertical mixing (see section 4.4).

Ices can only be present in the cold, outer parts of the disk where the temperature drops below 100 K. Thus, the strongest ice emission bands typically occur at far-IR wavelengths. Crystalline water ice has been seen in a few disks through librational features at 44 and 63 μm (e.g., *Malfait et al.*, 1998, *Chiang et al.*, 2001). The data can be reproduced in models assuming that 50% of the available oxygen is in water ice. Edge-on disks offer a special opportunity to study ices at mid-IR wavelengths in absorption. Examples are L1489 (*Boogert et al.*, 2002), DG Tau B (*Watson et al.*, 2004), and CRBR2422.8-3423 (*Thi et al.*, 2002; *Pontoppidan et al.*, 2005) (Fig. 3). Extreme care has to be taken in the interpretation of these data since a large fraction of the ice features may arise in foreground clouds. For the case of CRBR 2422.8-3423, a disk viewed at an inclination of ~70°, comparison with nearby lines of sight through the same core combined with detailed disk modeling has been used to constrain the amount of ice in the disk. H_2O ice has an average line-of-sight abundance of ~10^{-4} relative to H_2, consistent with significant freeze-out. CO_2 and CO ice are also present, the latter only in the form where it is mixed with H_2O ice. The shape of the 6.85-μm ice band — usually ascribed to NH_4^+ (e.g., *Schutte and Khanna*, 2003) — shows evidence for heating to 40–50 K, as expected in the warm intermediate layers of the disk. Future studies of a large sample of edge-on disks can provide significant insight into the abundance and distribution of ices in disks, because the ice absorption depths, band shapes, and feature ratios depend strongly on the disk temperature structure and line of sight, i.e., inclination (*Pontoppidan et al.*, 2005).

PAHs are important in the chemistry for at least three reasons: as absorbers of UV radiation, as a heating agent for the gas, and as potential sites of H_2 formation when classical grains have grown to large sizes. Since they require UV radiation for excitation, PAHs are also excellent diagnostics of the stellar radiation field and disk shape (flaring or flat). PAHs are detected in the spectra of at least 50% of Herbig Ae stars with disks through emission features at 3.3, 6.2, 7.7, 8.6, 11.2, and 12.8 μm (*Acke and van den Ancker*, 2004). For T Tauri stars, the features are weaker and more difficult to see on top of the strong continuum, but at least 8% of sources with spectral types later than F7 show the 11.2-μm PAH feature (Geers et al., in preparation). Ground-based long-slit spectroscopy and narrow-band imaging at subarcsecond resolution has demonstrated that, at least for some disks, the PAH emission comes from a region of radius 10–100 AU (e.g., *Habart et al.*, 2004). The inferred PAH abundance is typically 10^{-7} with respect to H_2, assuming that ~10% of the carbon is in PAHs with 50–100 carbon atoms. PAHs have been detected in transitional disks, where there is evidence for grain growth to micrometer size, albeit at a low abundance of 10^{-9} (*Li and Lunine*, 2003;

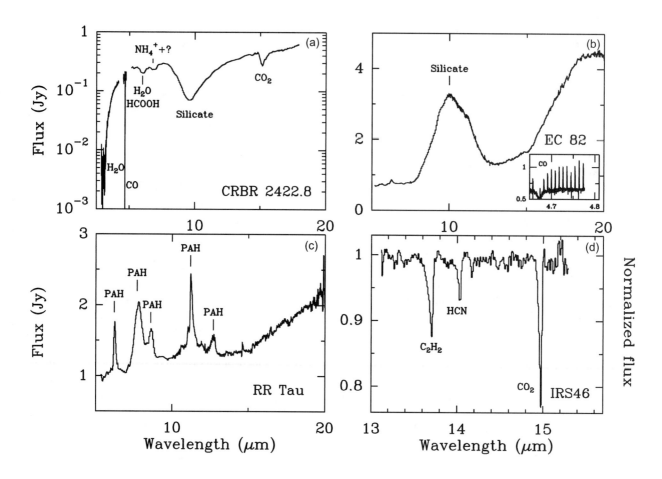

Fig. 3.   **(a)** Ice features toward edge-on disk CRBR 2422.8-3423. Some absorptions arise in the cold foreground core (*Pontoppidan et al.*, 2005). *(b)* Silicate emission at 10 and 20 μm toward EC 82 (*Kessler-Silacci et al.*, 2006); inset, gas-phase CO v = 1–0 emission (Blake and Boogert, personal communication). **(c)** PAH features toward RR Tau (Geers et al., in preparation). **(d)** Gaseous C_2H_2, HCN, and CO_2 toward IRS 46 in Oph (*Lahuis et al.*, 2006).

Jonkheid et al., 2006). This indicates that the PAHs and very small grains in the upper disk layer may be decoupled dynamically from the larger silicate grains and have a much longer lifetime toward grain growth and settling.

3.1.2. Gas-phase molecules.   Vibration-rotation emission lines of gaseous CO at 4.7 μm are detected toward a large fraction (>80%) of T Tauri and Herbig Ae stars with disks (e.g., *Najita et al.*, 2003; *Brittain and Rettig*, 2002; *Blake and Boogert*, 2004). The CO lines can be excited by collisions in the hot gas in the inner (<5–10 AU) disk as well as by IR or UV pumping in the upper layers of the outer (r > 10 AU) disk. Searches for other molecular emission lines have so far been largely unsuccessful except for the detection of hot H_2O toward one young star, presumably arising in the inner disk (*Carr et al.*, 2004). A tentative detection of H_3^+ — a potential tracer of protoplanets — has been claimed toward one young star (HD141569) (*Brittain and Rettig*, 2002), but this remains unconfirmed (*Goto et al.*, 2005).

Surprisingly, Spitzer has recently detected strong absorption from C_2H_2, HCN, and CO_2 at R = 600 toward one young star with an edge-on disk, IRS 46 in Ophiuchus (Fig. 3) (*Lahuis et al.*, 2006). The inferred abundances are high, ~10^{-5} with respect to H_2, and the gas is hot, 300–700 K, with linewidths of ~20 km s^{-1}. The most likely location for this hot gas rich in organic molecules is in the inner (<few AU) disk, with the line of sight passing through the puffed-up inner rim. The observed abundances are comparable to those found in inner disk chemistry models (e.g., *Markwick et al.*, 2002).

A novel method to probe the chemistry in the outer (r > 10 AU) disk is through observations of narrow, low-velocity [O I] ^1D–^3P lines at 6300 Å showing signs of Keplerian rotation (*Bally et al.*, 1998; *Acke et al.*, 2005). The excited O ^1D atoms most likely result from photodissociation of OH in the upper flaring layers of the disk out to at least 10 AU (*Störzer and Hollenbach*, 1998; *van Dishoeck and Dalgarno*, 1984). The OH abundance required to explain the observed [O I] fluxes is ~10^{-7}–10^{-6}, significantly larger than that found in current chemical models.

The main reservoir of the gas in disks is H_2, which has its fundamental rotational quadrapole lines at mid-IR wavelengths. Searches for these lines have been performed by *Thi et al.* (2001) with ISO, but tentative detections have not been confirmed by subsequent groundbased data (e.g., *Richter et al., 2002; Sako et al., 2005*), although H_2 vibrational lines have been detected (*Bary et al., 2003*). Constraining the amount of gas in disks is important not only for jovian planet formation, but also because the gas/dust ratio affects the chemistry and thermal balance of the disk as well as the dust dynamics.

3.2. Millimeter- and Submillimeter-Wave Spectroscopy

At long wavelengths, where the dust emission is largely optically thin, rotational line emission forms a powerful probe of the physics and chemistry in disks. Indeed, at a fiducial radius of 100 AU, with a temperature of 20–30 K, the disk radiates preferentially at (sub)millimeter wavelengths. Surveys of the millimeter-continuum and SEDs from disks (*Beckwith et al., 1990*) have confirmed the spatio-temporal properties of dust disks. However, only a *handful* of objects have been investigated by detailed imaging, with millimeter-wave interferometer observations of CO from disks around T Tauri stars providing among the earliest and most conclusive evidence for the expected ~Keplerian velocity fields (see *Koerner et al., 1993; Dutrey et al., 1994*).

From such studies a suite of disks have been identified that are many arcseconds in diameter, either because they are nearby (TW Hya) or are intrinsically large (GM Aur, LkCa 15, DM Tau). The age, large size, and masses of these disks make them important for further study since they may represent an important transitional phase in which viscous disk spreading and dispersal competes with planetary formation processes. Their large size makes them difficult, but feasible, targets for further chemical study. Work in this area began with the pioneering observations of DM Tau and GG Tau with the IRAM 30-m telescope (*Dutrey et al., 1997*) and TW Hya with the JCMT (*Kastner et al., 1997*). Further detections of the higher –J lines of high dipole moment species such as HCN and HCO⁺, along with statistical equilibrium analyses, demonstrated that the line emission arises from the warm molecular layer with $n_{H_2} \approx 10^6$–10^8 cm^{-3}, T ≥ 30 K (*van Zadelhoff et al., 2001*). The very deepest integrations have begun to reveal more complex species (*Thi et al. 2004*), although the larger organics often seen toward protostellar hot cores remain out of reach of existing single-dish telescopes.

At present, aperture synthesis observations can only sense the outer disk (r > 30–50 AU) (*Dutrey and Guilloteau, 2004;* see chapter by Dutrey et al.) for stars in the nearest molecular clouds. Thus, the *chemical* imaging of disks is rarer still, with studies concentrating on a few of the best characterized T Tauri and Herbig Ae stars. Imaging studies of LkCa 15, for example, have detected a number of isotopologues of CO along with the molecular ions HCO⁺ and N_2H^+ and the more complex organics formaldehyde and methanol (*Duvert et al., 2000; Aikawa et al., 2003; Qi et al., 2003*). For this disk at least, molecular depletion of molecules onto the icy mantles of grains near the disk midplane is found to be extensive, but the fractional abundances and ionization in the warm molecular layer are in line with those seen toward dense PDRs (section 2.1). While the lines from the less-abundant species can be detected, they were too weak to image with good signal-to-noise. Thus, while millimeter-wave rotational line emission is a good tracer of the outer disk velocity field, it is not a robust tracer of the mass unless the chemistry is very well understood.

New observational facilities are poised to change this situation dramatically, as illustrated by the recent SMA results on the TW Hya disk presented in Fig. 4 (*Qi et al., 2006,* in prep.). At a distance of only 56 pc, observations of this source provide nearly 2–3× the effective linear resolution of studies in Taurus and Ophiuchus. Thus, channel maps such as those presented can be used to derive a great deal about the physical and chemical structure of the disk — its size and inclination (*Qi et al., 2004*), the run of mass surface density and temperature with radius, the chemical abundance ratios with radius in the outer disk, etc. Ongoing improvements to existing arrays such as the eSMA, PdBI, and CARMA will enable similar studies for a large number of disks in the near future, and will push the radii over which chemical studies can be pursued down to 10–20 AU. Resolving the chemical gradients discussed in section 4.3 and 4.4 and studying the chemistry in the 1–10-AU zone of active planet formation will require even greater sensitivity and spatial resolution, and awaits ALMA.

4. CHEMICAL AND PHYSICAL LINKS

Protoplanetary disks are evolving in many different, but connected, ways. Small micrometer-sized dust grains collide, coagulate, collisionally fragment, and settle in a process that ultimately produces planets (*Weidenschilling, 1997*). At the same time, the disk is viscously evolving, with indications that the disk mass accretion rate decreases with age (*Hartmann et al., 1998*). Finally, disks evolve chemically, with the eventual result that all heavy elements are frozen on grains in the midplane and, prior to gas dissipation, chemistry consists of H_2 ionization and the deuteration sequence to H_2D^+, D_2H^+, and D_3^+ (see section 4.3 and section 5). The interconnections between these types of evolution is, at present, poorly understood. Nonetheless, some physical and chemical connections have become clear, which we review here.

4.1. Grain Evolution

The onset of grain evolution within a protoplanetary disk consists of collisional growth of submicrometer-sized particles into larger grains; the process continues until the larger grains decouple from the gas and settle to an increasingly dust-rich midplane (*Nakagawa et al., 1981; Weidenschilling and Cuzzi, 1993; Beckwith et al., 2000*). Evolving dust grains within irradiated disks reprocess stellar and accretion-gener-

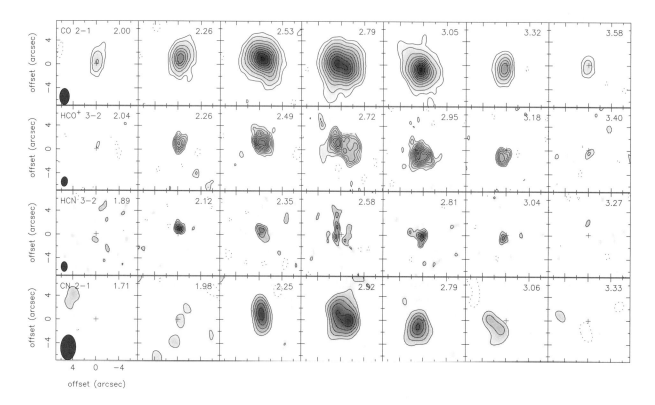

Fig. 4. SMA channel maps of the millimeter and submillimeter spectral line emission from the TW Hya circumstellar disk. The ellipses at lower left in each series of panels display the synthesized beam, which for the HCN 3–2 observations achieves an effective spatial resolution of ~60 AU. Kindly provided by C. Qi in advance of publication (Qi et al., in preparation).

ated UV and optical photons into the IR and submillimeter. Thus, the dust thermal emission spectrum bears information on the grain size distribution and spatial location. This, together with the spectral features discussed in section 3.1.1, is used to provide evidence for grain evolution in ~1-m.y. T Tauri systems (*Beckwith et al., 2000*; see chapter by Dullemond et al., and references therein).

Grain coagulation can alter the chemistry through the reduction in the total geometrical cross-section, lowering the adsorption rate and the Coulomb force for ion-electron grain recombination. Micrometer-sized grains couple to the smallest scales of turbulence (*Weidenschilling and Cuzzi, 1993*) and have a thermal, Brownian, velocity distribution. Thus, the timescale of grain-grain collisions is $\tau_{gr-gr} \propto a_d^{5/2}/(T_d^{1/2}\xi n_H)$, where a_d is the grain radius, ξ the gas-to-dust mass ratio, and T_d the dust temperature (*Aikawa et al., 1999*). In this fashion grain coagulation proceeds faster at small radii where the temperatures and densities are higher. *Aikawa et al.* (1999) note that the longer timescale for adsorption on larger grains leaves more time for gas-phase reactions to drive toward a steady-state solution; this involves more carbon trapped in CO as opposed to other, more-complex species.

Overall, the evolution of grains, both coagulation and sedimentation, can be a controlling factor for the chemistry. As grains grow the UV opacity, which is dominated by small grains, decreases, allowing greater penetration of ionizing/dissociating photons (e.g., *Dullemond and Dominik, 2005*). As an example, in the coagulation models of *Dullemond and Dominik* (2005) the integrated vertical UV optical depth at 1 AU decreases over several orders of magnitude, toward optically thin over the entire column (see also *Weidenschilling, 1997*). The chemical effects are demonstrated in Fig. 5b, where we show the C+ to CO transition as a function of vertical distance for two dust models, well-mixed and settled (*Jonkheid et al., 2004*). In the settled model the CO transition occurs at slightly smaller height, even though this model assumes that PAHs remain present in the upper atmosphere. Thus, as grains evolve, there will be a gradual shifting of the warm molecular layer deeper into the disk, eventually into the midplane. Because the grain emissivity, density, and temperature will also change, the chemical and emission characteristics of this layer may be altered (*Aikawa and Nomura, 2006*). These effects are magnified in the inner disk, where there is evidence for significant grain evolution in a few systems (*Uchida et al., 2004*; *Calvet et al., 2005*) and deeper penetration of energetic radiation (*Bergin et al., 2004*). A key question in this regard is the number of small grains (e.g., PAHs) present in the atmosphere of the disk during times when significant coagulation and settling has occurred (e.g., *Dullemond and Dominik, 2005*; *D'Alessio et al., 2006*).

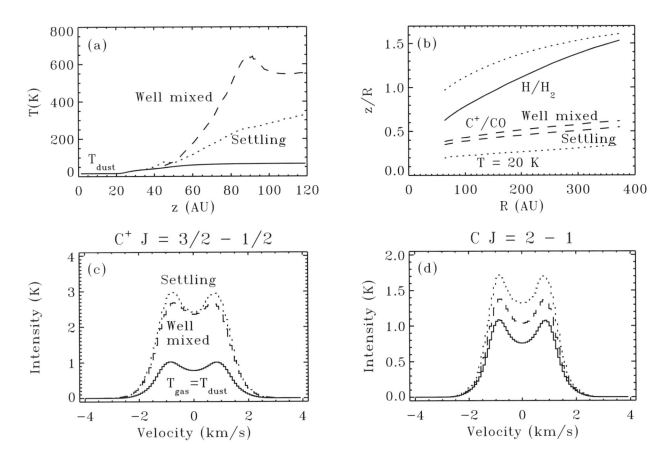

Fig. 5. **(a)** Vertical distribution of gas (well-mixed model: dashed line; settling model: dotted line) and dust temperatures (solid line). **(b)** Vertical chemical structure with the edge of the disk and 20 K isotherm as dotted lines, the H/H₂ transition as a solid line, and the C⁺/CO transitions in dashed lines (for both well-mixed and settled models). **(c)** C⁺ 158-μm emission and **(d)** C 370-μm emission lines for $T_g = T_d$ (solid), and T_g calculated independently for well-mixed (dashed) and settling (dotted) models. From *Jonkheid et al.* (2004).

It is worth noting that the penetration of X-ray photons is somewhat different than those at longer UV wavelengths. Absorption of UV radiation is dominated by the small grains, while X-rays are absorbed at the atomic scale by heavy metals predominantly trapped in the grain cores. Thus, coagulation will have a greater effect on UV photons. When the grain mass is distributed downward by settling, the X-ray penetration depth increases, eventually to the limit of total heavy element depletion where the opacity at 1 keV decreases by a factor of ~4.5 (i.e., the H and He absorption limit) (*Morrison and McCammon,* 1983).

4.2. Gas Thermal Structure

Models of disk physical structure have generally assumed that dust and gas temperatures are in equilibrium (*Chiang and Goldreich,* 1997; *D'Alessio et al.,* 1998). However, the disk vertical structure is set by the temperature of the dominant mass component, hydrogen, which under some conditions in the upper disk atmosphere is thermally decoupled from dust (*Chiang and Goldreich,* 1997).

Irradiated disk surfaces are analogs to interstellar PDRs, which have a history of detailed thermal balance calculations (see *Hollenbach and Tielens,* 1999, and references therein). In the studies of the gas thermal balance in disk atmospheres (*Jonkheid et al.,* 2004; *Kamp and Dullemond,* 2004; *Nomura and Millar,* 2005) a number of heating mechanisms have been investigated, including photoelectric heating by PAHs and large grains, UV excitation of H₂ followed by collisional deexcitation, H₂ dissociation, H₂ formation, gas-grain collisions, carbon ionization, and cosmic rays. In decoupled layers, the gas cools primarily by atomic ([O I], [C II], [C I]) and molecular (CO) emission, with the dominant mechanism a function of radial and vertical distance.

Some of these results are shown in Fig. 5; note that the gas temperature can exceed that of the dust in the upper atmosphere (Fig. 5a), which has consequences for the gas phase emission (Figs. 5c,d). The inclusion of PAHs into the models has an effect as PAHs provide additional heating power and are strong UV absorbers. Thus, grain evolution can significantly alter the thermal structure (see Fig. 5) by

TABLE 1. Disk ionization processes and vertical ion structure.

Layer/Carrier	Ionization Mechanism	$\Sigma_{\tau=1}$ (g cm^{-2})*	α_r (cm^{-3} s^{-1})[†]	x_e[‡]
Upper Surface H^+	UV photoionization of H[§] $k_{H^+} \sim 10^{-8}$ s^{-1}	6.9×10^{-4}	$\alpha_{H^+} = 2.5 \times 10^{-10} T^{-0.75}$	$>10^{-4}$
Lower Surface C^+	UV photoionization of C[¶] $k_{C^+} \sim 4 \times 10^{-8}$ s^{-1}	1.3×10^{-3}	$\alpha_{C^+} = 1.3 \times 10^{-10} T^{-0.61}$	$\sim 10^{-4}$
Warm Molecular H_3^+, HCO$^+$	Cosmic[**] and X-ray[††] ionization $\zeta_{cr} = \dfrac{\zeta_{cr,0}}{2}\left[\exp\left(-\dfrac{\Sigma_1}{\Sigma_{cr}}\right) + \exp\left(-\dfrac{\Sigma_2}{\Sigma_{cr}}\right)\right]$ $\zeta_X = \zeta_{X,0} \dfrac{\sigma(kT_X)}{\sigma(1\,\text{keV})} L_{29} J(r/\text{AU})^{-2}$	96 (CR) 0.008 (1 keV) 1.6 (10 keV)	$\alpha_{H_3^+} = -1.3 \times 10^{-8} +$ $1.27 \times 10^{-6} T^{-0.48}$ $\alpha_{HCO^+} = 3.3 \times 10^{-5} T^{-1}$	$10^{-11} \to -6$
Mid-Plane Metal$^+$/gr HCO$^+$/gr $H_3^+ - D_3^+$	Cosmic-ray[**] and Radionuclide[‡‡] $\zeta_R = 6.1 \times 10^{-19}$ s^{-1} (r < 3 AU) (3 < r < 60 AU) (r > 60 AU)		$\alpha_{Na^+} = 1.4 \times 10^{-10} T^{-0.69}$ α_{gr} (see text) $\alpha_{D_3^+} = 2.7 \times 10^{-8} T^{-0.5}$	$<10^{-12}$ $10^{-13, -12}$ $>10^{-11}$

* Effective penetration depth of radiation (e.g., $\tau = 1$ surface).

[†] Recombination rates from UMIST database (*Le Teuff et al.*, 2000) except for H_3^+, which is from *McCall et al.* (2004), and D_3^+, from *Larsson et al.* (1997).

[‡] Ion fractions estimated from *Semenov et al.* (2004) and *Sano et al.* (2000). Unless noted values are relevant for all radii.

[§] Estimated at 100 AU assuming 10^{41} s^{-1} ionizing photons (*Hollenbach et al.*, 2000) and $\sigma = 6.3 \times 10^{-18}$ cm^2 (H photoionization cross-section at threshold). This is an overestimate as we assume all ionizing photons are at the Lyman limit.

[¶] Rate at the disk surface at 100 AU using the radiation field from *Bergin et al.* (2003).

[**] Taken from *Semenov et al.* (2004). $\zeta_{cr,0} = 1.0 \times 10^{-17}$ s^{-1} and $\Sigma_1(r, z)$ is the surface density above the point with height z and radius r with $\Sigma_2(r, z)$ the surface density below the same point. $\Sigma_{cr} = 96$ g cm^{-2} as given above (*Umebayashi and Nakano*, 1981).

[††] X-ray ionization formalism from *Glassgold et al.* (2000). $\zeta_{X,0} = 1.4 \times 10^{-10}$ s^{-1}, while $L_{29} = L_X/10^{29}$ erg s^{-1} is the X-ray luminosity and J is an attenuation factor, $J = A\tau^{-a}e^{-B\tau^b}$, where A = 0.800, a = 0.570, B = 1.821, and b = 0.287 (for energies around 1 keV and solar abundances).

[‡‡] ^{26}Al decay from *Umebayashi and Nakano* (1981). If ^{26}Al is not present ^{40}K dominates with $\zeta_R = 6.9 \times 10^{-23}$ s^{-1}.

removing PAHs and small grains through coagulation and larger grains by settling, reducing photoelectric heating. In disks where grains have grown to micrometer size, and which are optically thin to UV radiation, other heating processes such as the drift velocity between the dust and gas may become important (see *Kamp and van Zadelhoff*, 2001).

One of the largest chemical influences for most of the disk mass is the freeze-out of molecular species onto grain surfaces. In general, the loss of gas coolants would produce a temperature rise, but in (midplane) layers dominated by freeze-out, the densities are high enough to thermally couple the gas to the dust (see, e.g., *Goldsmith*, 2001).

4.3. The Ionization Fraction

Over the past few years the disk fraction ionization has received a high degree of attention owing to the appreciation of the magneto-rotational instability (MRI) as a potential mechanism for disk angular momentum transport (e.g., *Hawley and Balbus*, 1991; *Stone et al.*, 2000). The chemical evolution is linked to dynamics as the presence of ions is necessary to couple the gas to the magnetic field. As we

will outline, most of the ionization processes are active at the surface and there exists the potential that accretion may only be active on the surface (*Gammie*, 1996; but see also *Klahr and Bodenheimer*, 2003; *Inutsuka and Sano*, 2005).

In equilibrium, the ion fraction, $x_e = n_e/n_H$, can be expressed by

$$x_e = \sqrt{\zeta/(\alpha_r n_H)}$$

where ζ is the ionization rate and α_r the electron recombination rate. In Table 1 we provide an overview of disk ionization. The top left panel in Fig. 6 also shows the electron abundance (equivalent to the ionization fraction) from a detailed model (*Semenov et al.*, in preparation). This figure, along with Table 1, can be used as a guide for the following discussion. For a discussion of the validity of the equilibrium assumption see *Semenov et al.* (2004), while *Ilgner and Nelson* (2005a) provide a detailed comparison of ionization and MRI for a variety of reaction networks.

The observed FUV radiation excess produces high ion fractions, but only over a small surface column, with C$^+$ as the charge carrier (the ionized hydrogen layer will be quite

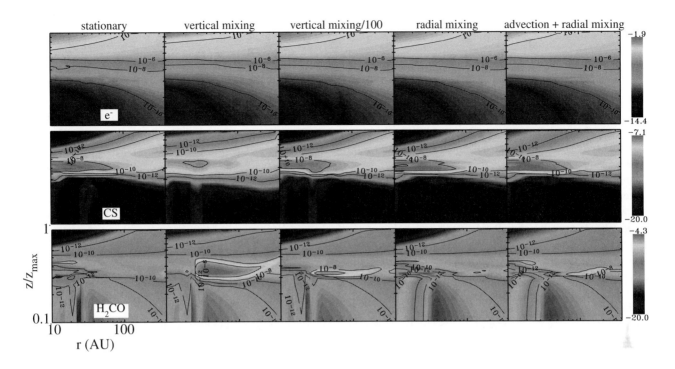

Fig. 6. Shown are molecular abundances in the two-dimensional flared disk of *D'Alessio et al.* (1999) after 10^6 yr of evolution. The x axis represents radii from 10 to 370 AU on logarithmic scale, while the y axis corresponds to the normalized vertical extent of the disk, z/z_{max}, which ranges from 0.1 to 1.0 on a logarithmic scale. The abundance values are given in respect to the total amount of hydrogen nuclei. Several models with variable mixing were considered and are labeled above each panel (vertical mixing/100 is a model with reduced mixing). Kindly provided by D. Semenov in advance of publication (Semenov et al., in preparation).

small). Deeper inside the warm molecular layer is reached, where primary charge carriers are molecular ions produced by X-ray ionization of H_2 (*Glassgold et al.*, 1997) and, when this decays, cosmic-ray ionization. It is worth noting that X-ray flares are observed in T Tauri systems (e.g., *Favata et al.*, 2005), after which there will exist a burst of ionization on the disk surface that will last for $\tau_r \sim 1/(\alpha_r n_e)$.

An important question is whether cosmic rays penetrate the inner disk. Within our own planetary system, the solar wind excludes ionizing cosmic rays. Estimates of mass loss rates from young star winds significantly exceed the solar mass loss rate (*Dupree et al.*, 2005), and may similarly exclude high-energy nuclei. In the case of cosmic-ray penetration, primary charge carriers range from metal ions and/or grains at small radii (~1 AU) to molecular ions for r > few AU with $x_e \sim 10^{-13}$ near the midplane. If ionizing cosmic rays are excluded, radionuclides can produce

$$x_e \sim 10^{-3}(T/20 \text{ K})^{-0.5}/\sqrt{n_H}$$

(assuming H_3^+ as the dominant ion and ^{26}Al is present; if ^{26}Al is not present the prefactor is 10^{-8}). The presence or absence of metal ions in the midplane can also affect MHD-driven dynamics (*Ilgner and Nelson*, 2005b). Metal ions have recombination times longer than molecular ions or the diffusive timescale and, if present, can be important charge reservoirs.

In dense protostellar cores models and observations now suggest near total freeze-out of heavy species that results in D_3^+ and other forms of deuterated H_3^+ becoming important charge carriers (*Roberts et al.*, 2004; *Walmsley et al.*, 2004). The disk midplane should present a similar environment. The recent detection of H_2D^+ by *Ceccarelli et al.* (2004) in the outer disks of TW Hya and DM Tau supports this view.

In the midplane one issue is the electron sticking coefficient (S_e) to grains. *Sano et al.* (2000) assume $S_e = 0.6$ based on the work of *Nishi et al.* (1991), while *Semenov et al.* (2004) assume a strong temperature dependence with S_e essentially zero at high temperatures. Thus in the inner disk the primary charge carrier differs between these two models, with grains dominating in the *Sano et al.* (2000) model and molecular/metal ions in the *Semenov et al.* (2004) model [for a discussion of electron sticking coefficients see *Weingartner and Draine* (2001)]. If grains are the dominant charge carrier the recombination rate α_{gr} is the grain collisional timescale with a correction for long-distance Coulomb focusing: $\alpha_{gr} = \pi a_d^2 n_{gr} v(1 + e^2/ka_d T_d)$. At $T_d = 20$ K *Draine and Sutin* (1987) show that for molecular ions, grain recombination will dominate when $n_e/n_H < 10^{-7}(a_{min}/3\text{Å})^{-3/2}$. Grains can be positive or negative and carry multiple charge: *Sano et al.* (2000) find that the total grain charge is typically negative, while the amount of charge is 1–2e^-, varying with radial and vertical distance. Since the criteria for the MRI instability include the mass of the charged particles, it is important to determine whether grains or molecules are the primary charge carriers.

4.4. Mixing

Chemical mixing within the solar nebula has a long history due to important questions regarding the potential transport of material from warm, thermochemically active regions (either inner solar nebula or jovian subnebula) into colder inactive regions (*Lewis and Prinn*, 1980; *Prinn and Fegley*, 1981). It is clear that some movement of processed material is likely; for example, the detection of crystalline silicates in comets (*Crovisier et al.*, 1997; *Bockelée-Morvan et al.*, 2002) and the chondritic refractory inclusions in meteorites (*MacPherson et al.*, 1988) imply some mixing. However, the mixing efficiency has been a matter of debate (*Stevenson*, 1990; *Prinn*, 1990, 1993).

In terms of the dynamical movement of gas within a protoplanetary disk and its chemical effects, a key question is whether the chemical timescale, τ_{chem}, is less than the relevant dynamical timescale, τ_{dyn}, in which case the chemistry will be in equilibrium and unaffected by the motion. If $\tau_{dyn} < \tau_{chem}$ then mixing will alter the anticipated composition. These two constraints are the equilibrium and disequilibrium regions (respectively) outlined in *Prinn* (1993). What is somewhat different in our current perspective is the recognition of an active gas-phase chemistry on a photon-dominated surface. This provides another potential mixing reservoir in the vertical direction, as opposed to radial, which was the previous focus.

It is common to parameterize the transfer of angular momentum in terms of the turbulent viscosity, $\nu = \alpha c_s H$, where ν is the viscosity, c_s the sound speed, H the disk scale height, and α is a dimensionless parameter (*Shakura and Sunyaev*, 1973; *Pringle*, 1981). *Hartmann et al.* (1998) empirically constrained the α-parameter to be $\leq 10^{-2}$ for a sample of T Tauri disks.

The radial disk viscous timescale is $\tau_\nu = r^2/\nu$ or

$$\tau_\nu \sim 10^4 \text{ yr} \left(\frac{\alpha}{10^{-2}}\right)^{-1} \left(\frac{T}{100 \text{ K}}\right)^{-1} \left(\frac{r}{1 \text{ AU}}\right)^{\frac{1}{2}} \left(\frac{M_*}{M_\odot}\right)^{\frac{1}{2}}$$

The diffusivity, D, is not necessarily the same as the viscosity, ν (e.g., *Stevenson*, 1990), even though some treatments equate the two (*Ilgner et al.*, 2004; *Willacy et al.*, 2006). In the case of MRI, *Carballido et al.* (2005) estimate $\nu/D \sim 11$, i.e., turbulent mixing is much less efficient than angular momentum transport [but see also *Turner et al.* (2006), where $\nu/D \sim 1$–2]. With knowledge of the ν/D ratio, the above equation for τ_ν serves as an estimate of the dynamical timescale for diffusion.

Recent models that include dynamics with chemistry generally can be grouped into two categories: ones that include only advection (*Duschl et al.*, 1996; *Finocchi et al.*, 1997; *Willacy et al.*, 1998; *Aikawa et al.*, 1999; *Markwick et al.*, 2002) and ones that investigate the effects of vertical and/or radial mixing including advection (*Wehrstedt and*

Gail, 2003; *Ilgner et al.*, 2004; *Willacy et al.*, 2006; Semenov et al., in preparation). The effects of advection on the chemical evolution are dominated by migration of icy grains toward the warmer inner disk where the volatile ices evaporate. Most of the species that desorb are processed via the active gas phase and/or gas-grain chemistry into less-volatile species that freeze out (e.g., *Aikawa et al.*, 1999). For instance, N_2 evaporates at T > 20 K (*Öberg et al.*, 2005) and is converted to HCN, and other less-volatile species, via ion-molecule reactions. This would predict a strong temporal dependence on the chemistry in the accreting material within the evaporative zones.

As an example of more complex dynamical models, Fig. 6 shows results from Semenov et al. (in preparation) (see also *Willacy et al.*, 2006). For these models, radial and vertical mixing is incorporated into a disk model that also includes all other relevant physical/chemical processes described in section 1.2. The model assumes a traditional α-disk that is fully dynamically active with $\nu/D = 1$. This model reproduces the basic structure shown in Fig. 1, with the presence of a warm molecular layer and a drop in abundance toward the midplane (most dramatically seen in CS). When vertical transport and/or radial transport is included the essential features of the basic structure are preserved, in the sense that the warm molecular layer still exists, although it may be expanded (*Willacy et al.*, 2006). This is readily understood as the chemical timescales driven by the photodissociation, $\tau_{chem} \leq 100$–$1000(r/100 \text{ AU})^2$ yr, are less than the dynamical timescale at all radii, thus the photochemical equilibrium is preserved. The effects of radial mixing and advection are mostly important for the upper disk atmosphere. Two results stand out: (1) The electron abundance structure shows little overall change, thus ionization equilibrium is preserved throughout the disk. (2) There is clear evidence for abundance enhancements of key species (such as H_2CO) in models with vertical mixing. The molecules with the largest abundance enhancements in the vertical mixing models are those that are more volatile than water, but are also major components of the grain mantle. This suggests that observations of these species may ultimately be capable of constraining disk mixing.

However, all predictions are highly uncertain. If MRI powers accretion, and cosmic rays do not penetrate to the midplane, then the bulk of the disk mass will not participate in any global mixing. Moreover, the outer disk may be actively turbulent while the inner disk may be predominantly quiescent (excluding the surface ionized by X-rays). Thus, while current models are suggestive of the importance of mixing for the disk chemical evolution, significant questions remain.

5. DEUTERATED SPECIES IN DISKS AND COMETS

Isotopic fractionations are measured in primordial materials in comets and meteorites, and considered to be good tracers of their origin and evolution. Here we review recent progress on deuterium fractionation in relation to comets

TABLE 2.   D/H ratios in comets, disks and cores.

Region Type	Object	Species	D/H ratio	Reference
Hot cores	various	HDO	3.0×10^{-4}	*Gensheimer et al.* (1996)
	various	DCN	$0.94.0 \times 10^{-3}$	*Hatchell et al.* (1998)
Low-mass protostars	IRAS 16293-2422	HDO	3.0×10^{-2}	*Parise et al.* (2005)
	IRAS 16293-2422	CH_3OD	2.0×10^{-2}	*Parise et al.* (2004)
	IRAS 16293-2422	CH_2DOH	3.0×10^{-1}	*Parise et al.* (2004)
	various	HDCO	$5.0–7.0 \times 10^{-2}$	*Roberts et al.* (2002)
	various	NH_2D	$1.0–2.8 \times 10^{-1}$	*Roueff et al.* (2005)
	various	DCO^+	1.0×10^{-2}	Class 0 average; *Jørgensen et al.* (2004)
	various	DCN	1.0×10^{-2}	Class 0 average; *Jørgensen et al.* (2004)
Dark Cores	L1544	DCO^+	4.0×10^{-2}	*Caselli et al.* (2002)
	L134N	DCO^+	1.8×10^{-1}	*Tiné et al.* (2000)
	TMC-1	DCN	2.3×10^{-2}	*van Dishoeck et al.* (1995)
Disks	TW Hya	DCO^+	3.5×10^{-2}	*van Dishoeck et al.* (2003)
	LkCa 15	HDO	6.4×10^{-2}	*Kessler et al.* (2003)
	DM Tau	HDO	1.0×10^{-2}	*Ceccarelli et al.* (2005)
	LkCa 15	DCN	<0.002	*Kessler et al.* (2003)
Comets	Halley	HDO	$(3.2 \pm 0.3) \times 10^{-4}$	*Eberhardt et al.* (1995)
	Hyakutake	HDO	$(2.9 \pm 1.0) \times 10^{-4}$	*Bockelée-Morvan et al.* (1998)
	Hale-Bopp	HDO	$(3.3 \pm 0.8) \times 10^{-4}$	*Meier et al.* (1998)
	Hale-Bopp	HDO	2×10^{-3}	*Blake et al.* (1999)
	Hale-Bopp	DCN	$(2.3 \pm 0.4) \times 10^{-3}$	*Meier et al.* (1998)

HDO has only been *tentatively* detected in disks, whereas H_2O has not. In these cases (*Ceccarelli et al.*, 2005; *Kessler et al.*, 2003) the D/H ratio is estimated by model calculation and is therefore highly uncertain.

(see the chapter by Yurimoto et al. for a discussion of oxygen fractionation). In Table 2 we list molecular D/H ratios, defined as n(XD)/n(XH), observed in protoplanetary disks, comets, and low-mass cores.

The two-dimensional (r–z) distributions of deuterated species at $r \geq 26$ AU have been calculated in models of *Aikawa and Herbst* (2001) and *Aikawa et al.* (2002). At low temperatures D/H fractionation proceeds via ion-molecule reactions; species such as H_3^+ and CH_3^+ are enriched in deuterium because of the difference in zero-point energies between isotopomers and rapid exchange reactions such as $H_3^+ + HD \rightarrow H_2D^+ + H_2$ (e.g., *Millar et al.*, 1989). Since CO is the dominant reactant with H_2D^+, CO depletion further enhances the H_2D^+/H_3^+ ratio. The deuterium enrichment propagates to other species via chemical reactions (see chapters by Ceccarelli et al. and Di Francesco et al.). Hence the D/H ratios of HCN and HCO^+ tend to increase toward the midplane with low temperature and heavy molecular depletion, while their absolute abundances reach the maximum value at some intermediate height. The column density D/H ratios of HCO^+, HCN, and H_2O integrated in the vertical direction are 10^{-2} at $r \geq 100$ AU and 10^{-3} at 26 AU $\leq r \leq 100$ AU in *Aikawa et al.* (2002). H_3^+ and its deuterated families (H_2D^+, HD_2H^+, and D_3^+), on the other hand, are abundant in the midplane, and the D/H ratio can even be higher than unity. At present, the data on D/H ratios in disks (with both deuterated and hydrogenated species observed) is limited to the detection of DCO^+ (*van Dishoeck et al.*, 2003), which is in agreement with models (*Aikawa et al.*, 2002).

In contrast to the millimeter observations of gas in the outer disk, comets carry information on ice (rather than gas) at radii of 5–30 AU (e.g., *Mumma et al.*, 1993). The similarity in molecular D/H ratios between comets and high-mass hot cores has been used to argue for an interstellar origin of cometary matter, but the D/H ratios in low-mass star-forming regions (e.g., TMC-1) are higher than those in comets, casting questions as to the interstellar origin scenario (e.g., *Irvine*, 1999). In recent years hot cores are found around low- or intermediate-mass protostars (see IRAS 16293-2422). In addition, temporal and spatial variation in molecular D/H ratios are found in low-mass dense cores, where the bulk of ices formed (e.g., *Bacmann et al.*, 2003; *Caselli*, 2002). Hence, what one means by "interstellar" is ambiguous. Eventually, molecular evolution from cores to disks and within disks should be investigated.

Aikawa and Herbst (1999) calculated molecular abundances and D/H ratios in a fluid parcel accreting from a core to the disk, and then from the outer disk radius to the comet-forming region (30 AU), showing that ratios such as DCN/HCN depend on the ionization rate in the disk, and can decrease from 0.01 to 0.002 (if the migration takes 10^6 yr and the ionization rate is 10^{-18} s^{-1}) due to chemical reactions during migration within the disk. This model, however, assumed that the fluid parcel migrates only inward within the cold ($T \leq 25$ K) midplane, which results in the survival of highly deuterated water accreted from the core. *Hersant et al.* (2001) solved the diffusion equation to obtain the D/H ratio in the disk; initially high D/H ratios of H_2O and HCN are lowered by mixing with the poorly deuterated material from the smaller radii. This model, however, considered only thermal reactions. In reality deuterium fractionation (or backward reactions) via ion-molecule reactions would proceed within the disk depending on the local ionization rate, temperature, and degree of molecular depletion, while the vertical and radial diffusion will tend to lessen the spatial gradient of D/H ratios. Inclusion of deuterated species in the recent models with nonthermal chemistry and two-dimensional diffusion is desirable.

6. OUTSTANDING ISSUES AND FUTURE PROSPECTS

Significant gains have been made in our observational and theoretical understanding of the chemical evolution of protoplanetary disks in the decade since the last review in this conference series. We now recognize the importance of the irradiated surface, which at the very least contains an active chemistry and is responsible for most observed molecular emission lines. The gross characteristics and key ingredients of this surface are roughly understood. How the varied effects (grain evolution, UV/X-ray radiation dominance, etc.) play out on the chemical evolution in terms of X-ray/UV dominance and the dependence on other evolutionary factors is one of the challenges for future models. Given the observed chemical complexity, a detailed understanding of the chemistry is a prerequisite for the interpretation of ongoing and future observations of molecular emission in protoplanetary disks.

Disk surface processes may dominate the observed chemistry, but it is not certain how much of a role this chemistry plays in altering the chemical characteristics within the primary mass reservoir, the disk midplane. Thus one of the main outstanding questions for disk chemistry remains the question of how much material remains pristine and chemically unaltered from its origin in the parent molecular cloud. We now have observational and theoretical evidence for active chemical zones; it is thus likely that the most volatile species, which are frozen on grains in the infalling material (e.g., CO, N_2), do undergo significant processing. This will trickle down to other, less-abundant molecules that form easily from the "parent species" (e.g., H_2CO, HCN, and the deuterated counterparts). Disentangling these effects will be complicated because the chemistry in the outer disk (r > 30 AU), which through advection feeds the inner disk, is quite similar to that seen in dense regions of the interstellar medium. For the least-volatile molecules, in particular water ice, sublimation and subsequent gas-phase alteration is less likely, unless there is significant radial mixing from the warmer inner disk to colder outer regions.

In part, our recognition of the warm molecular layer and the importance of photochemistry is driven by our current observational facilities, which are unable to resolve the innermost regions of the disk (e.g., planet-forming zones), coupling better to the larger surface area of the outer disk. Within r < 10–30 AU, the midplane and surface are both hot enough to sublime even the least-volatile molecules (e.g., H_2O), eventually producing an active chemistry that

is described by the earlier *Protostars and Planets III* review (*Prinn*, 1993) and in the chapter by Najita et al. The transition between these layers, the so-called "snowline," and the chemistry within the planet-forming zone will be species specific and should be readily detectable with upcoming advances in our capabilities, in particular the eagerly awaited ALMA array.

In summary, we stand on the cusp of the marriage of a rapidly emerging new field, studies of extrasolar protoplanetary disk chemical evolution, and an old one, the cosmochemical study of planets, meteorites, asteroids, and comets. In this review we have outlined broad areas where the evolving chemistry can be altered through changes induced by vertical and horizontal temperature gradients, the evolution of grain properties, and disk dynamics (mixing). Thus studies of active chemistry in extrasolar disks offers the promise and possibility to untangle long-standing questions regarding the initial conditions, chemistry, and dynamics of planet formation, the origin of cometary ices, and, ultimately, a greater understanding of the organic content of gas/solid reservoirs that produced life at least once in the galaxy.

Acknowledgments. E.A.B. thanks L. Hartmann for an initial reading. We also gratefully acknowledge receipt of unpublished material from B. Jonkheid, C. Qi, D. Semenov, N. Turner, and K. Willacy. This work was supported by in part by NASA through grants NNG04GH27G and 09374.01-A from STScI, by a Grant-in-Aid for Scientific Research (17039008) and "The 21st Century COE Program of the Origin and Evolution of Planetary Systems" from MEXT in Japan, and by a Spinoza award of NWO.

REFERENCES

Acke B. and van den Ancker M. E. (2004) *Astron. Astrophys., 426,* 151–170.
Acke B., van den Ancker M. E., and Dullemond C. P. (2005) *Astron. Astrophys., 436,* 209–230.
Aikawa Y. and Herbst E. (1999) *Astron. Astrophys., 351,* 233–246.
Aikawa Y. and Herbst E. (2001) *Astron. Astrophys., 371,* 1107–1117.
Aikawa Y. and Nomura H. (2006) *Astrophys. J., 642,* 1152–1162.
Aikawa Y., Umebayashi T., Nakano T., and Miyama S. M. (1997) *Astrophys. J., 486,* L51.
Aikawa Y., Umebayashi T., Nakano T., and Miyama S. M. (1999) *Astrophys. J., 519,* 705–725.
Aikawa Y., van Zadelhoff G. J., van Dishoeck E. F., and Herbst E. (2002) *Astron. Astrophys., 386,* 622–632.
Aikawa Y., Momose M., Thi W.-F., van Zadelhoff G.-J., et al. (2003) *Publ. Astron. Soc. Japan, 55,* 11–15.
Alexander R. D., Clarke C. J., and Pringle J. E. (2005) *Mon. Not. R. Astron. Soc., 358,* 283–290.
Bacmann A., Lefloch B., Ceccarelli C., Steinacker J., Castets A., and Loinard L. (2003) *Astrophys. J., 585,* L55–L58.
Bally J., Sutherland R. S., Devine, D., and Johnstone D. (1998) *Astron. J., 116,* 293–321.
Bary J. S., Weintraub D. A., and Kastner J. H. (2003) *Astrophys. J., 586,* 1136–1147.
Beckwith S. V. W., Sargent A. I., Chini R. S., and Guesten R. (1990) *Astron. J., 99,* 924–945.
Beckwith S. V.W., Henning T., and Nakagawa Y. (2000) In *Protostars and Planets IV* (V. Mannings et al., eds.), pp. 533–558. Univ. of Arizona, Tucson.
Bergin E., Calvet N., D'Alessio P., and Herczeg G. J. (2003) *Astrophys. J., 591,* L159–L162.

Bergin E., Calvet N., Sitko M. L., Abgrall H., D'Alessio P., Herczeg G. J., et al. (2004) *Astrophys. J., 614*, L133–L136.

Blake G. A. and Boogert A. C. A. (2004) *Astrophys. J., 606*, L73–L76.

Blake G. A., Qi C., Hogerheijde M. R., Gurwell M. A., and Muhleman D. O. (1999) *Nature, 398*, 213.

Bockelée-Morvan D., Gautier D., Lis D. C., Young K., Keene J., Phillips T., et al. (1998) *Icarus, 133*, 147–162.

Bockelée-Morvan D., Gautier D., Hersant F., Huré J.-M., and Robert F. (2002) *Astron. Astrophys., 384*, 1107–1118.

Boogert A. C. A., Hogerheijde M. R., and Blake G. A. (2002) *Astrophys. J., 568*, 761–770.

Bringa E. M. and Johnson R. E. (2004) *Astrophys. J., 603*, 159–164.

Brittain S. D. and Rettig T. W. (2002) *Nature, 418*, 57–59.

Calvet N. and Gullbring E. (1998) *Astrophys. J., 509*, 802–818.

Calvet N., Magris G. C., Patino A., and D'Alessio P. (1992) *Rev. Mex. Astron. Astrofis., 24*, 27.

Calvet N., D'Alessio P., Hartmann L., Wilner D., et al. (2002) *Astrophys. J., 568*, 1008–1016.

Calvet N., D'Alessio P., Watson D. M., Franco-Hernández R., Furlan E., et al. (2005) *Astrophys. J., 630*, L185–L188.

Carballido A., Stone J. M., and Pringle J. E. (2005) *Mon. Not. R. Astron. Soc., 358*, 1055–1060.

Carr, J. S., Tokunaga, A. T., and Najita J. (2004) *Astrophys. J., 603*, 213–220.

Caselli P. (2002) *Planet. Space Sci., 50*, 1133–1144.

Caselli P., Walmsley C. M., Zucconi A., Tafalla M., et al. (2002) *Astrophys. J., 565*, 344–358.

Ceccarelli C. and Dominik C. (2005) *Astron. Astrophys., 440*, 583–593.

Ceccarelli C., Dominik C., Lefloch B., Caselli P., and Caux E. (2004) *Astrophys. J., 607*, L51–L54.

Ceccarelli C., Dominik C., Caux E., Lefloch B., and Caselli P. (2005) *Astrophys. J., 631*, L81–L84.

Chiang E. I. and Goldreich P. (1997) *Astrophys. J., 490*, 368.

Chiang E. I., Joung M. K., Creech-Eakman M. J., Qi C., et al. (2001) *Astrophys. J., 547*, 1077–1089.

Crovisier J., Leech K., Bockelée-Morvan D., Brooke T., et. al. (1997) *Science, 275*, 1904–1907.

D'Alessio P., Canto J., Calvet N., and Lizano S. (1998) *Astrophys. J., 500*, 411.

D'Alessio P., Calvet N., Hartmann L., Lizano S., and Cantó J. (1999) *Astrophys. J., 527*, 893–909.

D'Alessio P., Calvet N., Hartmann L., Franco-Hernández R., and Servín H. (2006) *Astrophys. J., 638*, 314–335.

Dartois E., Dutrey A., and Guilloteau S. (2003) *Astron. Astrophys., 399*, 773–787.

Draine B. T. (1978) *Astrophys. J. Suppl., 36*, 595–619.

Draine B. T. and Sutin B. (1987) *Astrophys. J., 320*, 803–817.

Dullemond C. P. and Dominik C. (2005) *Astron. Astrophys., 434*, 971–986.

Dupree A. K., Brickhouse N. S., Smith G. H., and Strader J. (2005) *Astrophys. J., 625*, L131–L134.

Duschl W. J., Gail H.-P., and Tscharnuter W. M. (1996) *Astron. Astrophys., 312*, 624–642.

Dutrey A. and Guilloteau S. (2004) *Astrophys. Space Sci., 292*, 407–418.

Dutrey A., Guilloteau S., and Simon M. (1994) *Astron. Astrophys., 291*, L23–L26.

Dutrey A., Guilloteau S., and Guelin M. (1997) *Astron. Astrophys., 317*, L55–L58.

Duvert G., Guilloteau S., Ménard F., Simon M., and Dutrey A. (2000) *Astron. Astrophys., 355*, 165–170.

Eberhardt P., Reber M., Krankowsky D., and Hodges R. R. (1995) *Astron. Astrophys., 302*, 301.

Favata F., Flaccomio E., Reale F., Micela G., Sciortino S., et. al. (2005) *Astrophys. J. Suppl., 160*, 469–502.

Fegley B. J. (1999) *Space Sci. Rev., 90*, 239–252.

Fegley B. J. and Prinn R. G. (1989) In *The Formation and Evolution of Planetary Systems* (H. Weaver and L. Danly, eds.), pp. 171–205. Cambridge Univ., New York.

Feigelson E. D. and Montmerle T. (1999) *Ann. Rev. Astron. Astrophys., 37*, 363–408.

Finocchi F. and Gail H.-P. (1997) *Astron. Astrophys., 327*, 825–844.

Forrest W. J., Sargent B., Furlan E., D'Alessio P., Calvet N., et al. (2004) *Astrophys. J. Suppl., 154*, 443–447.

Gammie C. F. (1996) *Astrophys. J., 457*, 355.

Gensheimer P. D., Mauersberger R., and Wilson T. L. (1996) *Astron. Astrophys., 314*, 281–294.

Glassgold A. E., Najita J., and Igea J. (1997) *Astrophys. J., 480*, 344.

Glassgold A. E., Feigelson E. D., and Montmerle T. (2000) In *Protostars and Planets IV* (V. Mannings et al., eds.), pp. 429–456. Univ. of Arizona, Tucson.

Glassgold A. E., Najita J., and Igea J. (2004) *Astrophys. J., 615*, 972–990.

Goldsmith P. F. (2001) *Astrophys. J., 557*, 736–746.

Goto M., Geballe T. R., McCall B. J., Usuda T., et. al. (2005) *Astrophys. J., 629*, 865–872.

Habart E., Natta A., and Krügel E. (2004) *Astron. Astrophys., 427*, 179–192.

Habing H. J. (1968) *Bull. Astron. Inst. Neth., 19*, 421.

Hartmann L., Calvet N., Gullbring E., and D'Alessio P. (1998) *Astrophys. J., 495*, 385.

Hasegawa T. I. and Herbst E. (1993) *Mon. Not. R. Astron. Soc., 261*, 83–102.

Hatchell J., Thompson M. A., Millar T. J., and MacDonald G. H. (1998) *Astron. Astrophys. Suppl., 133*, 29–49.

Hawley J. F. and Balbus S. A. (1991) *Astrophys. J., 376*, 223.

Hayashi C. (1981) *Prog. Theor. Phys. Suppl., 70*, 35–53.

Herbig G. H. and Goodrich R. W. (1986) *Astrophys. J., 309*, 294–305.

Herbst E. and Klemperer W. (1973) *Astrophys. J., 185*, 505–534.

Herczeg G. J., Wood B. E., Linsky J. L., Valenti J. A., and Johns-Krull C. M. (2004) *Astrophys. J., 607*, 369–383.

Hersant F., Gautier D., and Huré J.-M. (2001) *Astrophys. J., 554*, 391–407.

Hester J. J. and Desch S. J. (2005) In *Chondrites and the Protoplanetary Disk* (A. N. Krot et al., eds.), pp. 107–131. ASP Conf. Series 341, San Francisco.

Hollenbach D. J. and Tielens A. G. G. M. (1999) *Rev. Mod. Phys., 71*, 173–230.

Hollenbach D. J., Yorke H. W., and Johnstone D. (2000) In *Protostars and Planets IV* (V. Mannings et al., eds.), pp. 401–429. Univ. of Arizona, Tucson.

Igea J. and Glassgold A. E. (1999) *Astrophys. J., 518*, 848–858.

Ilgner M. and R. P. Nelson (2005a) *Astron. Astrophys., 445*, 205–222.

Ilgner M. and R. P. Nelson (2005b) *Astron. Astrophys., 445*, 223–232.

Ilgner M., Henning T., Markwick A. J., and Millar T. J. (2004) *Astron. Astrophys., 415*, 643–659.

Inutsuka S.-i. and T. Sano (2005) *Astrophys. J., 628*, L155–L158.

Irvine W. M. (1999) *Space Sci. Rev., 90*, 203–218.

Jonkheid B., Faas F. G. A., van Zadelhoff G.-J., and van Dishoeck E. F. (2004) *Astron. Astrophys., 428*, 511–521.

Jonkheid B., Kamp I., Augereau J.-C., and van Dishoeck E. F. (2006) *Astron. Astrophys., 453*, 163–171.

Jørgensen J. K., Schöier F. L., and van Dishoeck E. F. (2004) *Astron. Astrophys., 416*, 603.

Kamp I. and Dullemond C. P. (2004) *Astrophys. J., 615*, 991–999.

Kamp I. and van Zadelhoff G.-J. (2001) *Astron. Astrophys., 373*, 641–656.

Kastner J. H., Zuckerman B., Weintraub D. A., and Forveille T. (1997) *Science, 277*, 67–71.

Kastner J. H., Franz G., Grosso N., Bally J., et. al. (2005) *Astrophys. J. Suppl., 160*, 511–529.

Kenyon S. J. and Hartmann L. (1987) *Astrophys. J., 323*, 714–733.

Kessler J. E., Blake G. A., and Qi C. (2003) In *Chemistry as a Diagnostic of Star Formation* (C. L. Curry and M. Fich, eds.), p. 188. NRC, Ottawa.

Kessler-Silacci J. E., Hillenbrand L. A., Blake G. A., and Meyer M. R. (2005) *Astrophys. J., 622*, 404–429.

Kessler-Silacci J. E., Augereau J.-C., Dullemond C. P., Geers V., Lahuis F., et al. (2006) *Astrophys. J., 639*, 275–291.

Kitamura Y., Momose M., Yokogawa S., Kawabe R., et al. (2002) *Astrophys. J., 581*, 357–380.

Klahr H. H. and Bodenheimer P. (2003) *Astrophys. J., 582,* 869–892.

Koerner D. M., Sargent A. I., and Beckwith S. V. W. (1993) *Icarus, 106,* 2–12.

Lada C. J. and Lada E. A. (2003) *Ann. Rev. Astron. Astrophys., 41,* 57–115.

Lahuis F., van Dishoeck E. F., Boogert A. C. A., Pontoppidan K. M., Blake G. A., Dullemond C. P., et al. (2006) *Astrophys. J., 636,* L145–148.

Larsson M., Danared H., Larson A., Le Padellec A., Peterson J. R., et. al. (1997) *Phys. Rev. Lett., 79,* 395–398.

Léger A., Jura M., and Omont A. (1985) *Astron. Astrophys., 144,* 147–160.

Le Teuff Y. H., Millar T. J., and Markwick A. J. (2000) *Astron. Astrophys. Suppl., 146,* 157–168.

Lewis J. S. and Prinn R. G. (1980) *Astrophys. J., 238,* 357–364.

Li A. and Lunine J. I. (2003) *Astrophys. J., 594,* 987–1010.

MacPherson G. J., Wark D. A., and Armstrong J. T. (1988) In *Meteorites and the Early Solar System* (J. F. Kerridge and M. S. Matthews, eds.), pp. 746–807. Univ. of Arizona, Tucson.

Malfait K., Waelkens C., Waters L. B. F. M., Vandenbussche B., et. al. (1998) *Astron. Astrophys., 332,* L25–L28.

Maloney P. R., Hollenbach D. J., and Tielens A. G. G. M. (1996) *Astrophys. J., 466,* 561.

Markwick A. J., Ilgner M., Millar T. J., and Henning T. (2002) *Astron. Astrophys., 385,* 632–646.

McCall B. J., Huneycutt A. J., Saykally R. J., Djuric N., Dunn G. H., et al. (2004) *Phys. Rev. A, 70,* 052716.

Meier R., Owen T. C., Jewitt D. C., Matthews H. E., et al. (1998) *Science, 279,* 1707.

Millar T. J., Bennett A., and Herbst E. (1989) *Astrophys. J., 340,* 906–920.

Morrison R. and McCammon D. (1983) *Astrophys. J., 270,* 119–122.

Mumma M. J., Weissman P. R., and Stern S. A. (1993) In *Protostars and Planets III* (E. Levy and J. Lunine, eds.), pp. 1177–1252. Univ. of Arizona, Tucson.

Najita J., Bergin E. A., and Ullom J. N. (2001) *Astrophys. J., 561,* 880–889.

Najita J., Carr J. S., and Mathieu R. D. (2003) *Astrophys. J., 589,* 931–952.

Nakagawa Y., Nakazawa K., and Hayashi C. (1981) *Icarus, 45,* 517–528.

Nishi R., Nakano T., and Umebayashi T. (1991) *Astrophys. J., 368,* 181–194.

Nomura H. (2002) *Astrophys. J., 567,* 587–595.

Nomura H. and T. J. Millar (2005) *Astron. Astrophys, 438,* 923–938.

Öbert K. I., van Broekhuizen F., Fraser H. J., Bisschop S. E., et al. (2005) *Astrophys. J., 621,* L33–L36.

Parise B., Castets A., Herbst E., Caux E., Ceccarelli C., et. al. (2004) *Astron. Astrophys., 416,* 159–163.

Parise B., Caux E., Castets A., Ceccarelli C., Loinard L., et al. (2005) *Astron. Astrophys., 431,* 547–554.

Pontoppidan K. M., Dullemond C. P., van Dishoeck E. F., Blake G. A., et al. (2005) *Astrophys. J., 622,* 463–481.

Pringle J. E. (1981) *Ann. Rev. Astron. Astrophys., 19,* 137–162.

Prinn R. G. (1990) *Astrophys. J., 348,* 725–729.

Prinn R. G. (1993) In *Protostars and Planets III* (E. Levy and J. Lunine, eds.), pp. 1005–1028. Univ. of Arizona, Tucson.

Prinn R. G. and Fegley B. (1981) *Astrophys. J., 249,* 308–317.

Przygodda F., van Boekel R., Àbrahàm P., Melnikov S. Y., et. al. (2003) *Astron. Astrophys., 412,* L43–L46.

Qi C., Ho P. T. P., Wilner D. J., Takakuwa S., Hirano N., et al. (2004) *Astrophys. J., 616,* L11–L14.

Qi C., Kessler J. E., Koerner D. W., Sargent A. I., and Blake G. A. (2003) *Astrophys. J., 597,* 986–997.

Richter M. J., Jaffe D. T., Blake G. A., and Lacy J. H. (2002) *Astrophys. J., 572,* L161–L164.

Roberts H., Fuller G. A., Millar T. J., Hatchell J., and Buckle J. V. (2002) *Astron. Astrophys., 381,* 1026–1038.

Roberts H., Herbst E., and Millar T. J. (2004) *Astron. Astrophys., 424,* 905–917.

Rodriguez-Franco A., Martín-Pintado J., and Fuente A. (1998) *Astron. Astrophys., 329,* 1097–1110.

Roueff E., Lis D. C., van der Tak F. F. S., Gerin M., and Goldsmith P. F. (2005) *Astron. Astrophys., 438,* 585–598.

Sako S., Yamashita T., Kataza H., Miyata T., et al. (2005) *Astrophys. J., 620,* 347–354.

Sano T., Miyama S. M., Umebayashi T., and Nakano T. (2000) *Astrophys. J., 543,* 486–501.

Schutte W. A. and Khanna R. K. (2003) *Astron. Astrophys., 398,* 1049–1062.

Semenov D., Wiebe D., and Henning T. (2004) *Astron. Astrophys., 417,* 93–106.

Shakura N. I. and Sunyaev R. A. (1973) *Astron. Astrophys., 24,* 337–355.

Shen C. J., Greenberg J. M., Schutte W. A., and van Dishoeck E. F. (2004) *Astron. Astrophys., 415,* 203–215.

Simon M., Dutrey A., and Guilloteau S. (2000) *Astrophys. J., 545,* 1034–1043.

Stäuber P., Doty S. D., van Dishoeck E. F., and Benz A. O. (2005) *Astron. Astrophys., 440,* 949–966.

Stevenson D. J. (1990) *Astrophys. J., 348,* 730–737.

Stone J. M., Gammie C. F., Balbus S. A., and Hawley J. F. (2000) In *Protostars and Planets IV* (V. Mannings et al., eds.), p. 589. Univ. of Arizona, Tucson.

Störzer H. and Hollenbach D. (1998) *Astrophys. J., 502,* L71.

Thi W. F., van Dishoeck E. F., Blake G. A., van Zadelhoff G. J., Horn J., et al. (2001) *Astrophys. J., 561,* 1074–1094.

Thi W. F., Pontoppidan K. M., van Dishoeck E. F., Dartois E., and d'Hendecourt L. (2002) *Astron. Astrophys., 394,* L27–L30.

Thi W.-F., van Zadelhoff G.-J., and van Dishoeck E. F. (2004) *Astron. Astrophys., 425,* 955–972.

Tiné S., Roueff E., Falgarone E., Gerin M., and Pineau des Forêts G. (2000) *Astron. Astrophys., 356,* 1039–1049.

Turner N. J., Willacy K., Bryden G., and Yorke H. W. (2006) *Astrophys. J., 639,* 1218–1226.

Uchida K. I., Calvet N., Hartmann L., Kemper F., Forrest W. J., et al. (2004) *Astrophys. J. Suppl., 154,* 439–442.

Umebayashi T. and Nakano T. (1981) *Publ. Astron. Soc. Japan, 33,* 617.

van Boekel R., Min M., Waters L. B. F. M., de Koter A., et al. (2005) *Astron. Astrophys., 437,* 189–208.

van Dishoeck E. F. (2004) *Ann. Rev. Astron. Astrophys., 42,* 119–167.

van Dishoeck E. F. and Dalgarno A. (1984) *Icarus, 59,* 305–313.

van Dishoeck E. F., Blake G. A., Jansen D. J., and Groesbeck T. D. (1995) *Astrophys. J., 447,* 760–782.

van Dishoeck E. F., Thi W.-F., and van Zadelhoff G.-J. (2003) *Astron. Astrophys., 400,* L1–L4.

van Dishoeck E. F., Jonkheid B., and van Hemert M. C. (2006) *Faraday Discussions, 133,* in press.

van Zadelhoff G. J., van Dishoeck E. F., Thi W. F., and Blake G. A. (2001) *Astron. Astrophys., 377,* 566–580.

van Zadelhoff G. J., Aikawa Y., Hogerheijde M. R., and van Dishoeck E. F. (2003) *Astron. Astrophys., 397,* 789–802.

Walmsley C. M., Flower D. R., and Pineau des Forêts G. (2004) *Astron. Astrophys., 418,* 1035–1043.

Watson D. M., Kemper F., Calvet N., Keller L. D., Furlan E., et al. (2004) *Astrophys. J. Suppl., 154,* 391–395.

Weidenschilling S. J. (1997) *Icarus, 127,* 290–306.

Weidenschilling S. J. and Cuzzi J. N. (1993) In *Protostars and Planets III* (E. Levy and J. Lunine, eds.), pp. 1031–1060. Univ. of Arizona, Tucson.

Weingartner J. C. and Draine B. T. (2001) *Astrophys. J., 563,* 842–852.

Wehrstedt M. and Gail H.-P. (2003) *Astron. Astrophys., 410,* 917–935.

Westley M. S., Baragiola R. A., Johnson R. E., and Baratta G. A. (1995) *Nature, 373,* 405.

Willacy K. and Langer W. D. (2000) *Astrophys. J., 544,* 903–920.

Willacy K., Klahr H. H., Millar T. J., and Henning T. (1998) *Astron. Astrophys., 338,* 995–1005.

Willacy K., Langer W. D., Allen M., and Bryden G. (2006) *Astrophys. J., 644,* 1202–1213.

Dust in Protoplanetary Disks: Properties and Evolution

Antonella Natta and Leonardo Testi
Osservatorio Astrofisico di Arcetri

Nuria Calvet
University of Michigan

Thomas Henning
Max Planck Institute for Astronomy, Heidelberg

Rens Waters
University of Amsterdam and Catholic University of Leuven

David Wilner
Harvard-Smithsonian Center for Astrophysics

We review the properties of dust in protoplanetary disks around optically visible pre-main-sequence stars obtained with a variety of observational techniques, from measurements of scattered light at visual and infrared wavelengths to mid-infrared spectroscopy and millimeter interferometry. A general result is that grains in disks are on average much larger than in the diffuse interstellar medium (ISM). In many disks, there is evidence that a large mass of dust is in grains with millimeter and centimeter sizes, more similar to "sand and pebbles" than to grains. Smaller grains (with micrometer sizes) exist closer to the disk surface, which also contains much smaller particles, e.g., polycyclic aromatic hydrocarbons. There is some evidence of a vertical stratification, with smaller grains closer to the surface. Another difference with ISM is the higher fraction of crystalline relative to amorphous silicates found in disk surfaces. There is a large scatter in dust properties among different sources, but no evidence of correlation with the stellar properties, for samples that include objects from intermediate to solar mass stars and brown dwarfs. There is also no apparent correlation with the age of the central object, over a range roughly between 1 and 10 m.y. This suggests a scenario in which significant grain processing may occur very early in the disk evolution, possibly when it is accreting matter from the parental molecular core. Further evolution may occur, but not necessarily rapidly, since we have evidence that large amounts of grains, from micrometer to centimeter size, can survive for periods as long as 10 m.y.

1. INTRODUCTION

Young stars are surrounded by circumstellar disks made of gas and dust. Some of these disks will form planets, and one of the current open questions is to understand which disks will do it, which will not, and why. In this chapter, we will refer to all circumstellar disks as *protoplanetary*, even if some (many) may only have the potential to evolve into a planetary system, and some (many) not even that.

The mass of protoplanetary disks is dominated by gas, but the solid component (dust grains) is of great importance. Observations of light scattered on dust grains and their thermal emission remain key diagnostic tools for detecting disks and characterizing their structure. Dust grains play an active role in determining the thermal and geometrical structure of disks because their opacity dominates over the gas opacity whenever they are present. Furthermore, grains shield

the disk midplane from energetic radiation, thereby influencing the ionization structure, possibly leading to a "dead" zone where the magneto-rotational instability cannot operate. The formation of dust grains is a phase transition that provides solid surfaces important for chemical reactions and the freeze-out of molecular components such as CO and H_2O in the colder parts of disks. Finally, the solid mass of the disks is important because dust grains are the building blocks for the formation within the disk of planetesimals and eventually planets.

Grains in protoplanetary disks are very different from grains in the diffuse interstellar medium (ISM); their properties change from object to object and we believe that in each object they evolve with time. Without a comprehensive understanding of dust properties and evolution, many disk properties cannot be understood. An immediate example is the determination of disk solid masses from millimeter

767

continuum emission, which requires knowledge of the dust absorption coefficient for a specific grain composition and structure, and of the grain temperature.

Protoplanetary disks interest us also because they are the place where planets form. Grains are thought to be the primary building blocks on the road to planet formation, and the continuing and growing interest in this field is largely motivated by the need to understand the diversity of extrasolar planetary systems and reasons for this diversity. Earlier reviews by *Weidenschilling and Cuzzi* (1993) and *Beckwith et al.* (2000) in *Protostars and Planets III* and *IV* summarize what was known at the time. Since *Protostars and Planets IV* we have seen a lot of progress in our knowledge of grain properties in disks, thanks to high-resolution data provided by infrared (IR) long-baseline interferometry, the high sensitivity of Spitzer and groundbased 10 m-class telescopes, millimeter interferometers such as PdB and OVRO, and the VLA capabilities at millimeter wavelength. Data now exist for a large number of disks, around stars with very different mass and luminosity, from intermediate-mass objects (Herbig AeBe stars, hereafter referred to as HAeBe) to T Tauri stars (TTS) and brown dwarfs (BDs). On the theoretical side, there is a revived effort in modeling grain processing in disks (via coalescence, sedimentation, fragmentation, annealing, etc.) and its relation with the chemical and dynamical evolution of the gas. Although it is still very difficult to integrate observations and theory in a quantitative description of disk evolution, we can expect major advances in the near future.

In this review, we concentrate on observational evidence for grain evolution with a special emphasis on grain growth and mineralogy. As already mentioned, grain properties and their distribution within the disk affect many observable quantities. These aspects will be covered in other chapters of this book, and we will concentrate on observations that measure directly the grain physical and chemical structure. Prior to that, we will review very briefly the most important processes that control the grain properties (section 2). We will then outline the various observational techniques and their limitations (section 3), discuss evidence for grains of different size, from polycyclic aromatic hydrocarbons (PAHs; section 4) to micrometer size (section 5) and centimeter size (section 6). We will discuss grain mineralogy in section 7. Section 8 will summarize the main conclusions one can derive from the observations and outline some of the open questions we need to address in the future.

2. GRAIN GROWTH: WHY AND HOW

Grains in the ISM are very likely a mixture of silicates and carbons, with a size distribution from ~100 Å to maximum radii of ~0.2–0.3 μm. Smaller particles, most likely PAHs and very small carbonaceous grains, are also present (e.g., *Draine*, 2003). The composition of dust in molecular clouds is similar. Models of dust evolution in collapsing cores (*Krügel and Siebenmorgen*, 1994; *Ossenkopf and Henning*, 1994; *Miyake and Nakagawa*, 1995) predict only minor changes, as confirmed by the observations of Class 0

and Class I objects (e.g., *Beckwith and Sargent*, 1991; *Bianchi et al.*, 2003; *Kessler-Silacci et al.*, 2005).

Major changes occur once dust is collected in a circumstellar disk, where the pristine interstellar grains may grow from submicrometer-sized or even smaller particles, to kilometer-sized bodies (planetesimals), and eventually planets. This process is driven by coagulation of smaller particles into larger and larger ones, and has a long history of theoretical and laboratory studies, which it is not the scope of this paper to analyze. We refer for a critical discussion to the most recent reviews of *Henning et al.* (2006) and the chapter by Dominik et al. However, the relevance of the observations of grain properties we will discuss in the following is better understood in the context of grain growth and planetesimal formation models, which we will therefore briefly summarize here.

Grain growth to centimeter and even meter sizes is mainly driven by collisional aggregation, although gravitational instabilities may play a role in overdense dust regions. The relative velocities leading to such collisions are caused by the differential coupling to the gas motion (see, e.g., *Weidenschilling and Cuzzi*, 1993; *Beckwith et al.*, 2000). This immediately demonstrates that grain growth cannot be understood without a better characterization of the gas velocity field in disks, especially the degree of turbulence. For submicrometer-sized grains this is not a real issue and their growth by Brownian motion is reasonably well understood thanks to numerical simulations and extensive laboratory experiments (see, e.g., *Blum*, 2004, and the chapter by Dominik et al.). The relative velocities of grains in the 1–100-μm size range are on the order of 10^{-3} m/s to 10^{-4} m/s, low enough for sticking. The outcome of this early growth process is a relatively narrow mass distribution function with fractal aggregates having open and chainlike structures.

While growing in mass, particles also start to sediment [due to the vertical component of the stellar gravitational field (*Schräpler and Henning*, 2004; *Dullemond and Dominik*, 2005, and references therein)] and the relative velocities may eventually become higher, leading to the compaction of aggregates. Above the compaction limit, a runaway growth can be expected, where a few large aggregates grow by collisions with smaller particles. In this regime an exponential increase of mass with time can be expected. The coupling between sedimentation, grain growth, and structure of particles remains the next challenge for protoplanetary dust models, which have to deal with the transition from open to compact aggregates. For grains of boulder size (larger than 1 m), the relative velocities remain more or less constant at about 50 m/s, challenging growth models because this velocity is above the destruction threshold (but see *Wurm et al.*, 2005). Furthermore, such boulders are rapidly transported to the central star within a few hundred years. There are quite a number of proposals to solve the problem of the growth barrier, none with complete success yet.

In spite of the many uncertainties, the models converge on predicting that grains can grow to very large sizes and sediment under the effect of the stellar gravity, leaving a population of smaller grains closer to the disk surface and of increasingly larger bodies toward the midplane. Grain growth should be much slower in regions further away from the star than in regions closer to it, which should be completely depleted of small grains. When only coagulation mechanisms are considered, small grains are removed very quickly, leading to the complete disappearance of the dusty disks on timescales much shorter than the age of observed disks. Mechanisms that replenish disks of small grains are clearly required, for example, aggregate fragmentation. Radial and meridian circulation can also play an important role in grain evolution.

The process of growth of grains from submicrometer to kilometer size we have described assumes that the original solid mass never returns to the gas phase. This may not be the case, as the temperatures in the inner disk are high enough to evaporate grains. If, over time, radial circulations carry close to the star most of the disk material, all the original solid mass is destroyed and condensed again in the disk. If so, the newly formed grains may be very different from those in the ISM, even if actual growth does not occur, or is important only above a certain (large) size. Gassification and recondensation of grains in the inner disk, coupled with strong radial drifts, has been suggested (*Gail,* 2004) as an explanation for the large fraction of crystalline grains seen in several disks (see section 7), and may also play an important role in determining the grain size distribution.

For a long time, the only test for planet formation theories was our own solar system. This has changed completely over the last decade, first with the discovery of the variety of planetary systems that can actually form, but also with a much better knowledge of the grain properties and distribution in disks that are the progenitors of planetary systems. Although our knowledge and understanding of this second aspect is still limited, and the answers to some key questions (e.g., which planetary systems will form and from which disks) elusive, we will show in the following that the observations may already provide important constraints to theory.

3. OBSERVATIONAL TECHNIQUES

Grain properties in disks can be explored with a variety of techniques and over a large range of wavelengths, from visual to centimeter. Before reviewing the results and their implication, it is useful to summarize briefly the capabilities and limitations of the different techniques.

The most severe limits to the characterization of grains in disks do not come from observational limits, such as spatial resolution or sensitivity (although these too can be a problem), but from the physical structure of disks. Their extremely high optical depth allows us to get information only on grains located in the $\tau = 1$ layer, which at optical and IR wavelengths contains a tiny fraction of the dust mass. Moreover, if one measures dust emission, the observed flux is strongly biased by the temperature dependence of the Planck function.

The only way to measure the properties of the bulk of the dust mass is to go to longer wavelengths, where an increasing fraction of the disk becomes optically thin. Interferometric observations at millimeter and centimeter wavelengths have provided the strongest evidence so far that most of the original solid mass in protoplanetary disks has grown to centimeter size by the time the central star becomes optically visible. However, one should remember that this technique is limited today by the sensitivity and resolution of existing millimeter interferometers. For example, studies of the radial dependence of the dust properties are still very difficult. Furthermore, even at these very long wavelengths, the regions of the disk closer to the star (typically up to a few AU) are optically thick, and thus this very interesting inner portion of the midplane remains inaccessible at all wavelengths.

The mid-IR spectral region, roughly between 3 and 100 μm, is rich in vibrational resonances of abundant dust species, silicates in particular. The spectral region between 3 and 20 μm also contains prominent PAH C-C and C-H resonances. The wavelength, spectral shape, and strength of the resonances are sensitive probes of the chemical composition, lattice structure, size, and shape of the particles. In disks around optically visible stars these dust features are generally seen in emission, having their origin in the optically thin surface layers, heated by the stellar radiation to temperatures higher than the disk midplane (e.g., *Calvet et al.,* 1991; *Chiang and Goldreich,* 1997; *Menshchikov and Henning,* 1997). Infrared spectroscopy roughly probes dust particles with sizes of up to a few micrometers (depending on material and wavelength) in the temperature range between 1500 and about 50 K. In protoplanetary disks, this implies that such spectra are sensitive to dust in the innermost regions, and in the surface layers of the disk only. No information is obtained about chemical and allotropic properties of large and/or cold dust grains.

The 10-μm spectral region deserves special attention because it is accessible from the ground, is available for high-spatial-resolution interferometric studies, and a fairly large body of data has been collected, covering a wide range of stellar mass and age. It contains the strongest resonances of both amorphous and crystalline silicates. The silicate emission bands have important but somewhat limited diagnostic value. First, the 10-μm spectral region probes grains of a certain temperature range, mostly between 200 and 600 K. Second, both the amorphous and crystalline silicate resonances near 10 μm only probe the presence of grains smaller than a few micrometers in size, the exact value depending on the dust species. This limits the diagnostic value of the 10-μm region to warm, small silicates mostly located on the disk surface in regions within ~10 AU for protoplanetary disks around HAe stars, ~0.5–1 AU for TTS, and ~0.1 AU for BDs.

At shorter wavelengths (visual and near-IR), scattering becomes important and one can observe scattered light extending to very large distances from the star (e.g., *McCaughrean et al.,* 2000). Multiwavelength images in scattered light can be used to derive grain properties over a large range of disk radii, but limited to a very narrow layer on the disk surface.

Interesting information on grain properties deeper in the disk can be obtained from edge-on disks (silhouette), seen as a dark lane against a luminous background. Information on dust properties can be obtained in favorable cases by studying the wavelength dependence of the dark lane absorption against the background of the scattered light at visual and near-IR wavelengths. Silhouette disks should have deep silicate features at 10 μm in absorption, which, contrary to the features in emission, sample the colder grains in the outer disk. Silicate absorption is observed in a few cases, but it is often difficult to exclude contamination from grains in the surrounding material.

Multiwavelength studies are obviously necessary to completely characterize the grain population in size, composition, and physical structure. So far, this cannot be done in disks, where observations at different wavelengths sample different regions of the disk. As we will describe in the following, there is clear evidence that grains of all sizes (from PAHs to centimeter-sized bodies at least) are present in disks. However, the information is always partial, and it is very difficult to get a global picture, for example, to measure the total mass fraction of grains of different size.

An additional caveat is that none of the techniques we have mentioned is sensitive to bodies larger than few centimeters. Kilometer-sized planetesimals can be detected through the dynamical perturbations they create, but the detection of meter-sized bodies is practically impossible.

4. THE SMALLEST PARTICLES: POLYCYCLIC AROMATIC HYDROCARBONS

Emission from transiently heated very small particles (at 3.3, 6.2, 7.7–7.9, 8.6, 11.3, and 12.7 μm, etc.) has been detected in many disk systems, mostly HAeBe stars (e.g., *Acke and van den Ancker,* 2004) and generally attributed to PAHs. For a long time, it has not been clear if the observed emission is from PAHs in the disks or from the reflection nebulosities associated with many HAeBe. There is now convincing evidence in favor of the disk origin.

Peeters et al. (2002), *van Diedenhoven et al.* (2004), and *Sloan et al.* (2005) have shown that in HAeBe stars the PAH features differ in shape and wavelength when compared to the ISM. *Meeus et al.* (2001) and *Acke and van den Ancker* (2004) found that the strength of the PAH bands correlates with the shape of the spectrum in the 10–60-μm region. Flared disks [i.e., with flux F_ν increasing with wavelength in the 10–60-μm range; group I in the *Meeus et al.* (2001) classification] have strong PAH features, while flat disks (i.e., with F_ν decreasing with wavelength in the 10–60-μm range; group II) have no or very weak emission. This trend has been analyzed by *Habart et al.* (2004a), who have computed disk models that include PAHs and have shown that the observed correlation between the strength of the PAH bands and the shape of the SED is well explained in terms of the solid angle the disk subtends as seen from the star. In a few stars, other features, at 3.43 and 3.53 μm, from

transiently heated carbonaceous materials, possibly identified as nanodiamonds (*Guillois et al.,* 1999), have also been shown to have a disk origin (*Van Kerckhoven et al.,* 2002).

Recently, PAH and nanodiamond emission has been spatially resolved in several objects (*van Boekel et al.,* 2004b; *Ressler and Barsony,* 2003; *Habart et al.,* 2004b, 2006); the emitting region has a size consistent with disk model predictions (*Habart et al.,* 2004b, 2006; see Fig. 1). The emission of such transiently heated species (which are "hot" whenever excited) is much more extended than that of the adjacent continuum, emitted by larger grains in thermal equilibrium with the radiation field that rapidly become cool as the distance from the star increases. Using images in the features of PAHs and nanodiamonds it has been possible to measure disk inclination and position angle on the sky (*Habart et al.,* 2004b) at spatial scales of a few tens of AU.

The presence of very small particles on disk surfaces has a strong impact on the gas physical properties, since they contribute a large fraction of the gas heating via the photoelectric effect and may dominate the H_2 formation on grain surfaces. Polycyclic aromatic hydrocarbons are thus an essential ingredient of disk gas models (e.g., see the chapter

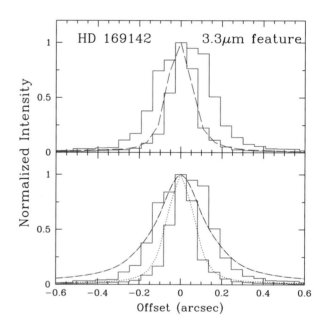

Fig. 1. Intensity profile of the 3.3-μm PAH feature in the HAe star HD 169142, observed with NAOS/CONICA on the VLT (*Habart et al.,* 2006). The top panel shows the intensity in the feature after continuum subtraction (thick solid line) and in the adjacent continuum (thin solid line); the dashed line plots the pointspread function as measured on a nearby unresolved star. The PAH feature is clearly spatially resolved, with FWHM size of 0.3 arcsec (roughly 40 AU at the distance of the star). The continuum is unresolved. The bottom panel compares the observations with disk model predictions for the feature (dot-dashed line) and the continuum (dotted line). Details of observations and calculations, including model parameters, are in *Habart et al.* (2004a, 2006).

by Dullemond et al.). However, the observations have severe limitations, as they can only probe matter at large altitudes above the disk midplane, so that we do not know if PAHs survive deeper into the disk; moreover, the intensity in the features decreases with radius, and is well below current observational capabilities for distances >40 AU (at least for the 3.3-μm feature).

There are also limitations in the current models, which need to explore a much wider range of parameters, both in disk and PAH properties.

5. FROM SUBMICROMETER- TO MICROMETER-SIZED GRAINS

5.1. Scattering and Polarization

Observations at visible and near-IR wavelengths of the dark disk silhouette seen against the background light scattered by the disk surface provide direct view of disks around young stars. The properties of the scattered radiation and the fraction of polarization can be used in principle to measure the size of the scattering grains. However, we know from laboratory work that particle sizing based on the angular distribution of scattered light or the measurement of polarization is not without problems and usually only works well for monodisperse distributions of spheres. In astronomical sources, the analysis of scattered radiation and polarization is even more complicated because it is not possible to obtain measurements for all scattering angles. In addition, polarization measurements are often technically challenging, which may explain why we only have limited data using this technique and grain size information from such observations is still scarce.

Scattering is well described by the 4 × 4 scattering matrix (Mueller matrix) transforming the original set of Stokes parameters into a new set after the scattering event. The matrix elements are angular-dependent functions of wavelength, particle size, shape, and material. For particles that are small compared with the wavelengths, the scattered light is partially polarized with a typical bell-shaped angular dependence with 100% polarization at a scattering angle of 90°. The polarization degree depends only on the scattering angle and not on particle size. This practically means that the radius of small spherical particles cannot be determined from scattering measurements.

For very small particles scattering does not vary a lot with direction. This changes drastically for larger spheres with more ripples and peaks in the scattering pattern, for which forward scattering becomes very important. In addition, the peak polarization decreases and moves to larger scattering angles. The very structured scattering pattern for large spheres is used for experimental particle sizing, but the characteristic ripples disappear if a grain size distribution is considered instead of a single size population. For very large spheres we again find smooth curves. Furthermore, nonspherical particles may behave very differently.

We should also note that an ensemble of nonidentical scattering particles leads to depolarization of scattered light, adding to the complexity of our understanding of the polarization degree (in terms of grains sizes) of radiation coming from a disk. We refer to *Voshchinnikov and Krügel* (1999) for Mie calculations, demonstrating the effect of different grain sizes.

One may conclude that solving the inverse problem of determining grain sizes from scattering and polarization observations is hopeless. This is certainly not true. A growing number of objects are studied using multiwavelength imaging and polarization techniques. For our purposes, it is sufficient to note that, in all cases, the evidence points toward grain growth on the disk surface to sizes of up to a few micrometers; often, there is also evidence of sedimentation, i.e., that larger grains are closer to the disk midplane. Among the more recent studies, we note the work of *Lucas et al.* (2004), who nicely demonstrated the power of a detailed analysis of high-resolution imaging polarization data for the case of HL Tau. They found that silicate core-ice mantle grains with the largest particles having radii slightly in excess of 1 μm best fit the data. A similar study has been performed for the ring around GG Tau by *Duchêne et al.* (2004). They found again a slight increase in the grain size toward micrometer-sized grains, and evidence for a vertical stratification of dust. The surface layers, located ~50 AU above the ring midplane, contain dust grains that are consistent with being as small as in the ISM, while the region of the ring located ~25 AU from the midplane contains significantly larger grains (>1 μm). This stratified structure is likely the result of vertical dust settling and/or preferential grain growth in the densest parts of the ring. The first scattering measurements at mid-IR wavelengths obtained for the T Tauri star HK Tau B (*McCabe et al.*, 2003) showed similar results.

Evidence of growth of grains to sizes of up to a few micrometers and of a vertical stratification of grain sizes within the disk surface has also been inferred in silhouette disks in Taurus (*D'Alessio et al.*, 2001) and Orion (*Throop et al.*, 2001; *Shuping et al.*, 2003; see also *McCaughrean et al.*, 2000, and references therein) by studying the behavior with wavelength of the translucent edges of the dark lane.

5.2. Mid-Infrared Spectroscopy

High-quality mid-IR spectra are available for an increasing number of objects. First the Infrared Space Observatory (ISO) and now the Spitzer Space Telescope are providing us with data of a quality that could not be achieved from the ground, even if the spatial resolution is in both cases too low to resolve the emission. Spitzer spectra are becoming available in increasing number at the time we write, and we expect further improvement in the statistics and quality of the data, for low mass objects in particular. In the following, we provide a summary of the silicate properties, as derived from the available observations.

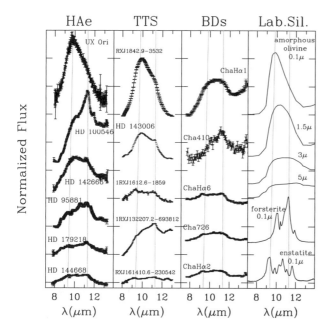

Fig. 2. Observed profiles of the 10-μm silicate feature. The first three panels (from left to right) show a selection of profiles for HAe stars (*van Boekel et al.,* 2005), TTS (Bouwman et al., in preparation), and BDs in Chamaeleon (*Apai et al.,* 2005). All the profiles have been normalized to the 8-μm flux and shifted for an easier display. In each panel, they have been ordered according to the F_{peak}/F_{cont} value. The dashed vertical lines show the location (at 9.8 μm) of the peak of small amorphous olivine (Mg_2SiO_4) and of the strongest crystalline feature of forsterite at 11.3 μm. Amorphous silicate grains of composition $Mg_{2x}Fe_{2-2x}SiO_4$, generally with x ≪ 1 have the same composition of olivines, and are often referred to as "amorphous olivines." Although this is not correct (olivine is crystalline), we will sometime use this definition. The panel to the far right shows for comparison a selection of profiles of laboratory silicates: starting at the top, amorphous olivine with radius 0.1, 1.5, 3, and 5 μm, as labeled. In all cases, we have added the same continuum (30% of the peak of the 0.1-μm amorphous olivine), and normalized as the observed profiles. The two bottom curves show the profile of small crystalline grains, forsterite (Mg_2SiO_4) and enstatite ($MgSiO_3$), respectively; larger crystalline grains have broader and weaker features (e.g., *van Boekel et al.,* 2005). Several of the features displayed by crystalline silicates can be identified in some of the observed spectra.

Most of the information is obtained from observations of the 10-μm emission feature. The spectra show a large variation in the strength and shape of this feature, for objects of all mass, from HAe stars to TTS and BDs. Figure 2 plots a representative selection of HAe stars, TTS, and BDs, which shows the variety of observed profiles. In some cases, the shape of the feature is strongly peaked at about 9.8 μm, as for small (≪1 μm) amorphous grains in the ISM. In other objects, the feature is very weak with respect to the continuum, and much less peaked. In other cases, many narrow features, typical of crystalline silicates, are clearly visible, superimposed on the smoother and broader amorphous silicate emission. These features will be discussed in detail in

section 7. In spite of these large variations, there is a general trend, valid for objects of all mass, first identified by *van Boekel et al.* (2005), who noted that the shape and strength of the silicate band are correlated, so that weaker features are also flatter. This is consistent with the growth of silicate grains from submicrometer to micrometer sizes. If grains grow further, the silicate emission disappears; there are in fact a few objects that show no silicate emission, most likely because *all* small silicates have been removed (e.g., *Meeus et al.,* 2003).

A useful way to summarize the properties of the 10-μm feature of a sample of objects is shown in Fig. 3. For each object, the shape of the 10-μm silicate emission feature is characterized by two parameters, the ratio of the flux at the peak of the feature over the continuum (F_{peak}/F_{cont}) and the ratio of the continuum subtracted 11.3 μm over the 9.8-μm flux. Amorphous silicates of increasing size have weaker and flatter features (cf. Fig. 2), with values of F_{peak}/F_{cont} decreasing from 3–4 (for sizes ≪1 μm) to ~1, for grains

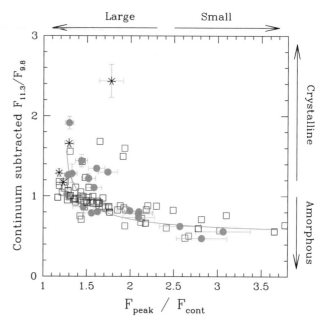

Fig. 3. The ratio of the continuum-subtracted flux at 11.3 μm over that at 9.8 μm ($F_{11.3}/F_{9.8}$) is plotted as function of flux at the peak of the 10-μm feature over the continuum, F_{peak}/F_{cont}. Open and filled symbols show the observations for HAe stars [filled dots, *van Boekel et al.* (2005)], TTS [squares, *Kessler-Silacci et al.* (2006); Bouwman et al. (in preparation); *Przygodda et al.* (2003)], BDs [stars, *Apai et al.* (2005)]. All the data but those of *Kessler-Silacci et al.* (2006) and Bouwman et al. (in preparation) have been reanalyzed by *Apai et al.* (2005) in a homogeneous way. Error bars are shown when available in the literature. The solid curve shows the result for a dust model that includes amorphous olivine of two sizes (1 and 10 μm) in proportion varying from all 1 μm (extreme right) to all 10 μm (extreme left), and an additional contribution of a 2% mass fraction of 0.1-μm forsterite grains (from J. Bouwman, personal communication, 2005); the same continuum (25% of the peak flux of the smallest grains) is added to all model spectra.

larger than few micrometers. At the same time, the ratio $F_{11.3}/F_{9.8}$ increases from ~0.5–0.6 to ~1. Larger values of the ratio $F_{11.3}/F_{9.8}$ are due to a significant contribution from forsterite (i.e., crystalline Mg_2SiO_4).

Note that the ratio F_{peak}/F_{cont} depends not only on grain cross section but also on other quantities (e.g., the disk inclination and geometry) that affect feature and continuum emission differently (e.g., *Chiang and Goldreich*, 1999). Also, changes in grain properties other than size [e.g., porosity (see *Voshchinnikov et al.*, 2006; *Min et al.*, 2006)] can reproduce part of the observed trend, and make the simple analysis in terms of grain size more uncertain. However, the interpretation of the 10-μm feature profiles in terms of growth of the grains in the disk surface is convincing, and in agreement with the results from other kinds of observations discussed in section 5.1.

Additional information will be derived in the future from the properties of the weaker 20-μm silicate band, which samples slightly larger grains, and regions on the disk surface further away from the star. Spitzer spectra are becoming available for a rapidly growing number of objects, and the first results seem to confirm the results obtained from the 10-μm feature analysis (*Kessler-Silacci et al.*, 2006; Bouwman et al., in preparation).

6. GROWTH TO CENTIMETER-SIZED GRAINS

As discussed in section 3, at (sub-)millimeter wavelengths and beyond the dust emission in disks begins to be moderately optically thin. At these wavelengths it is thus possible to probe the bulk of the dust particles, which are concentrated in the disk midplane. The spectral energy distribution of the continuum emission can be directly related to the dust emissivity, which in turn is related to the grain properties. The variation of the dust opacity coefficient per unit mass with frequency, in the millimeter wave range, can be approximated to relatively good accuracy by a power law $k \sim \lambda^{-\beta}$. The value of the β exponent is directly related to the dust properties, in particular the grain size distribution and upper size cutoff; in the limit of dust composed only of grains much larger than the observing wavelength, the opacity becomes gray and $\beta = 0$.

It has been known for many years that pre-main-sequence disks have (sub-)millimeter spectral indices α ($F_\nu \propto \lambda^{-\alpha}$) shallower than prestellar cores and young protostars (*Beckwith and Sargent*, 1991). If the disk emission is optically thin and $h\nu/kT_d \ll 1$, the observed values of α would translate very simply in opacity power law indices $\beta = \alpha - 2$; with observed typical values of $\alpha \lesssim 3$, this would imply $\beta \sim 1$ or lower. This immediately suggests that disk grains are very different from ISM grains, which have $\beta \sim 1.7$ (*Weingartner and Draine*, 2001).

This result, however, has been viewed with great caution, since the effect of large optical depth at the frequency of the observations cannot be ruled out for spatially unresolved observations. The extreme case of an optically thick

disk made of ISM grains has $\alpha = 2$, the same value as an optically thin disk with very large grains, for which $\beta = 0$ (see, e.g., *Beckwith and Sargent*, 1991; *Beckwith et al.*, 2000). The combination of optical depth and grain properties can be disentangled if the emission is resolved spatially and the disk size at the observed wavelength is measured (see, e.g., *Testi et al.*, 2001, 2003).

There are now several disks whose millimeter emission has been imaged with interferometers. A major step forward has come from the use of 7-mm VLA data, which, in combination with results at shorter wavelengths (1.3 and 2.6 mm) with PdB and OVRO, provides not only high spatial resolution and larger wavelength range (minimizing the uncertainties on α), but also the possibility of probing population of grains as large as a few centimeters. Additionally, the VLA at centimeter wavelengths has been useful to check whether the dust emission at shorter wavelength is contaminated by free-free emission (e.g., *Testi et al.*, 2001; *Wilner et al.*, 2000; *Rodmann et al.*, 2006). We summarize in Fig. 4 the results for the objects with 7-mm measurements from

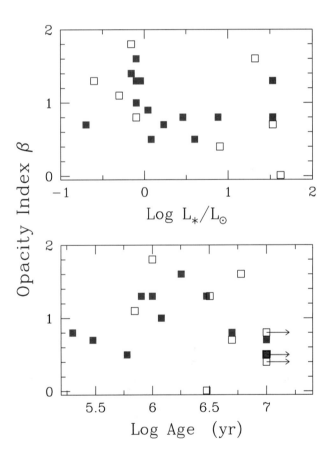

Fig. 4. The values of the millimeter opacity index β ($\kappa \propto \lambda^{-\beta}$) plotted as a function of the stellar luminosity (top panel) and of the stellar age (bottom panel); data from *Natta et al.* (2004a) and *Rodmann et al.* (2006). Filled squares are objects where the disk millimeter emission has been spatially resolved, open squares are objects that have not been observed with high spatial resolution, or that were found to be not resolved with the VLA (resolution ~0.5 arcsec).

Natta et al. (2004a) and references therein and *Rodmann et al.* (2006). For these objects, the contribution from gas emission has been subtracted from the millimeter emission, and the spectral index α has been computed over the wavelength interval 1.3–7 mm; in three cases the 7-mm flux is an upper limit, once the gas emission has been subtracted from the total. The values of β have been derived fitting disk models to the data. Note that these model-fitted β values are slightly larger than the optically thin determination β = α – 2, typically by about 0.2. This difference is due to the fact that the emission of the inner optically thick disk, albeit small, is not entirely negligible. The uncertainty on the model-derived β is of about ±0.2 (*Natta et al.,* 2004a; *Rodmann et al.,* 2006. The figure shows also the results for those objects for which no resolved maps exist, but have measurements of the integrated flux at 7 mm. For these, β has been derived from the measured α assuming that the disk is larger than ~100 AU, sufficient to make the optically thick contribution small.

If we consider only the 14 resolved disks, which have all radii larger than 100 AU, β ranges from 1.6 (i.e., very similar to the ISM value), to 0.5. Ten objects have β ≤ 1, i.e., much flatter than ISM grains. The unresolved disks behave in a similar way; the extreme case is UX Ori, with β ~ 0. Unfortunately, the UX Ori disk has not been resolved so far; at the distance of ~450 pc, and, with an integrated flux at 7 mm of only 0.8 mJy (*Testi et al.,* 2001), it remains a tantalizing object. High and low values of β are found both for HAe stars and for TTS, and there is no apparent correlation of β with the stellar luminosity or mass. Similarly, we do not observe any correlation of β with the age of the star. This last quantity, however, is often very uncertain, and the observed sample limited.

Once we have established that the grains originally in the collapsing cores, for which β ~ 1.7–2, have gone through a large degree of processing, one wants to derive from β the properties of the actual grain population, and in particular its size. The properties of the millimeter opacity of grains of increasing size have been discussed by, e.g., *Beckwith et al.* (2000) and references therein. Figure 5 illustrates the behavior of β computed between 1 and 7 mm for a population of grains with size distribution $n(a) \propto a^{-q}$ between a_{min} and a_{max}. The index β is plotted as function of a_{max} for different values of q; in all cases, $a_{min} \ll 1$ mm. One can see that, for all values of q, as a_{max} increases β is first constant at the value typical of grains of size ≪1 mm, it has a strong and rather broad peak at a_{max} ~ 1 mm and decreases below the initial value for $a_{max} \gtrsim$ few mm. However, only for q < 3 does β go to zero for large a_{max}; for q > 3, the small grains always contribute to the opacity, so that β reaches an asymptotic value that depends on q and on the β of the small grains. For q = 4, the asymptotic value is practically that of the small grains.

The results shown in Fig. 5 have been computed for compact segregated spheres of olivine, organic materials, and water ice (*Pollack et al.,* 1994). The exact values of β

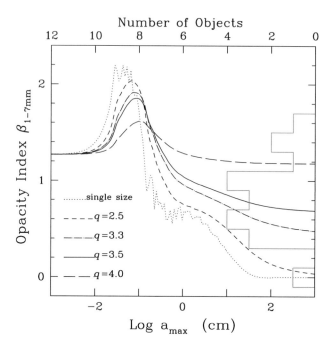

Fig. 5. Opacity index β for grains with a size distribution $n(a) \propto a^{-q}$ between a_{min} and a_{max}. The index β is computed between 1 and 7 mm for compact segregated spheres of olivine, organic materials, and water ice (*Pollack et al.,* 1994), and plotted as function of a_{max}. Different curves correspond to different values of q, as labeled. In all cases, $a_{min} \ll 1$ mm. The dotted curve shows β for grains with single size a_{max}. The histogram on the right shows the distribution of the values of β derived from millimeter observations (see Fig. 4). About 60% of the objects have β ≤ 1, and $a_{max} \gtrsim 1$ cm. Different dust models give results that are not very different (e.g., *Natta and Testi,* 2004).

depend on grain properties, such as their chemical composition, geometrical structure, and temperature (e.g., *Henning et al.,* 1995); examples for different grain models can be found, e.g., in *Miyake and Nakagawa* (1993), *Krügel and Siebenmorgen* (1994), *Calvet et al.* (2002), and *Natta et al.* (2004a). However, these differences do not undermine the general conclusion that the observed spectral indices require that grains have grown to sizes much larger than the observing wavelengths (e.g., *Beckwith et al.,* 2000, and references therein; *Draine,* 2006). To account for the observations that extend to 7 mm, objects with β ~ 0.5–0.6 need a distribution of grain sizes with maximum radii of a few centimeters, at least. The minimum grain radii are not constrained, and could be as small as in the ISM; however, the largest particles need to contribute significantly to the average opacity, which implies that the grain size distribution cannot be steeper than $n(a) \propto a^{-(3-3.5)}$; in any case, the large grains contain most of the solid mass. A detailed analysis can be found in, e.g., *Natta and Testi* (2004).

Of the objects observed so far, TW Hya is a particularly interesting case. Millimeter observations can be fitted well with models that include grain growth up to 1 cm (*Calvet et al.,* 2002; *Natta and Testi,* 2004; *Qi et al.,* 2004). However, this is probably an underestimation of the maximum grain sizes in this disk. *Wilner et al.* (2005) have demonstrated that the 3.5-cm emission of this object, contrary to expectations, is not dominated by gas emission, but rather by thermal emission from dust grains in the disk, and that a population of particles as large as several centimeters residing in the disk is needed to explain both the long-wavelength SED and the spatial distribution of the 3.5-cm emission.

The result that dust particles have grown to considerable sizes, in fact to "pebbles," is thus a solid one and verified in many systems. So far, it has been derived as a "global" property of the dust grain population of the disk. In order to constrain dust growth models, the next logical step is to check for variation of dust properties as a function of radius. A first attempt at this type of study is shown in Fig. 6. High angular resolution and high signal-to-noise 7-mm (VLA) and 1.3-mm (PdBI) maps of the HD 163296 system have been combined to reconstruct the millimeter spectral index profile as a function of radius. The result suggests that the spectral index varies as a function of radius from $\alpha \sim 2.5$ in the inner ~60 AU to $\alpha \sim 3.0$ in the far outer disk. If interpreted as differential grain growth, this result implies that the outer disk contains less-evolved particles than the inner disk.

7. MINERALOGY

The ISO and the latest generation of mid-IR spectrographs on large groundbased telescopes have revealed the presence of many new dust species, in addition to amorphous silicates, in the protoplanetary disks surrounding pre-main-sequence stars. The position and strength of the observed emission bands match well those of Mg-rich and Fe-poor crystalline silicates of the olivine ($Mg_{2x}Fe_{2-2x}SiO_4$ and pyroxene $Mg_xFe_{1-x}SiO_3$ families, in particular their Mg-rich end members forsterite and enstatite respectively (x = 1). Apart from these components, evidence for FeS and SiO_2 was also found (*Keller et al.,* 2002). Spitzer is now providing similar results for stars of lower mass, TTS and BDs.

The presence of these dust species in protoplanetary disks implies substantial processing (chemical and physical) of the dust because their abundance, relative to that of amorphous silicates, is far above limits set for these species in the interstellar medium (e.g., *Kemper et al.,* 2004). Therefore, these dust species must be formed sometime during the collapse of the molecular cloud core or in the accretion disk surrounding the young star. There are several possible mechanisms that may be responsible for grain processing, and it is quite likely that several contribute. In the innermost disk regions, where the dust temperature is above 1000 K, heat will induce a change of the grain lattice ordering (long-range ordering) from chaotic to regular (thermal annealing), leading to the transformation of amorphous silicates into crystalline ones (olivines and pyroxenes). Above about 1200–1300 K, chemical equilibrium processes will lead to vaporization and gas-phase condensation of silicates, mostly in the form of crystalline forsterite. Local high-energy processes (shocks, lightning) as well as radial mixing may increase substantially the amount of crystalline silicates in the cool outer regions of the disk. Note that thermal annealing of amorphous, presumably Fe-rich silicates would lead to Fe-rich crystalline silicates. Therefore the Fe content of crystalline silicates may be used as an indication of the chemical processing that has occurred (gas-phase condensation, annealing, local processes). In this context, we stress the importance of improving the still poorly constrained stochiometry of interstellar amorphous silicates, in particular the Fe/Mg and Mg/Si ratio.

The mineral composition of dust in protoplanetary disks can be compared directly to that of solar system comets, asteroids, and interplanetary dust particles (IDPs). Only one high-quality 2–200-μm spectrum of a solar system comet, Hale-Bopp, is available from ISO. The crystalline silicates

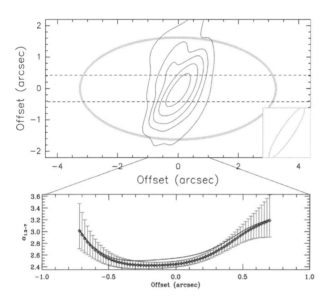

Fig. 6. The top panel shows the IRAM-PdBI contour map of the 1.3-mm emission from the disk, and the approximate shape and size of the scattered light disk as detected with HST/STIS. The images have been rotated so that the major axis of the disk is aligned with the abscissa of the plot. The spatial resolution of the millimeter observations is shown by the ellipse in the lower right box of the upper panel. In the bottom panel the diamonds with error bars show the variation along the disk major axis of the continuum spectral index measured between 1.3 and 7 mm. In the same panel, the solid line shows the values of the spectral index when a correction for the maximum possible gas contribution is applied to the 7-mm observations. The data suggest a variation of the spectral index as a function of disk radius. If interpreted as a variation of the grain properties, they indicate that larger grains are found in the inner ($r \leq 60$ AU) regions of the disk, while the grains in the outer disk may still be in a less-evolved stage.

in Hale-Bopp are very Mg-rich and Fe-poor, similar to those in protoplanetary disks. Laboratory studies of IDPs of cometary origin also reveal the presence of Mg-rich crystalline silicates. The situation for asteroids is much more diverse. Apart from Mg-rich crystalline silicates, many asteroids show Fe-containing silicates, probably related to parent body processing or to nebular processing in the inner solar nebula.

7.1.   Spatially Unresolved 10-Micrometer Spectra

Information on the abundance of crystalline silicates in a large sample of objects of different mass are derived from the shape of the 10-μm feature (mostly through the 11.3-μm forsterite peak) in spatially unresolved observations. Figure 2 shows how this feature can be prominent in stars of all mass, from HAeBe to brown dwarfs. Figure 3 shows how in many objects the flux at 11.3 μm is stronger than at 9.8 μm, where the emission of small amorphous silicates roughly peaks. A strong 11.3-μm peak is typical of profiles with a large component of crystalline silicates, in addition to the amorphous ones. This interpretation has been confirmed by the presence of crystalline features at longer wavelengths, when the data are available. Contamination from the 11.3-μm PAH feature has been ruled out in many cases, from the absence of the other PAH features at shorter wavelengths (see, e.g., *Acke and van der Ancker,* 2004).

A determination of the fractional mass abundance of the crystalline silicates has been derived by various authors fitting the observed profiles with mixtures of grains of different composition (typically, olivines and pyroxenes), size (from submicrometer to few micrometers) and allotropic state (amorphous and crystalline). In general, the models assume a single temperature for all grains and that the emission is optically thin. These assumptions, and the limited number of dust components that are included in models, make the estimates very uncertain. However, the results are interesting. For HAeBe stars, *van Boekel et al.* (2005) find that crystalline silicates are abundant only when the amorphous silicate grain population is dominated by large (≥1.5 μm) grains. In Fig. 3, there are no points with 11.3/9.8 ratio ≥ 1 and large values of F_{peak}/F_{cont}. The fraction of grains that is crystalline ranges from 5% (detection limit) to ~30%. Note that grains larger than few micrometers are not accounted for; if crystalline silicates are mostly submicrometer in size, while the amorphous species have a much broader distribution, the degree of crystallinity can be much lower than current estimates. This seems to be the case in Hale-Bopp (*Min et al.,* 2005), and may occur in protostellar disks as well. In addition, the derived crystalline/amorphous abundances are spatially averaged values; there are likely strong radial gradients in crystallinity (see below). As more data for lower-mass objects become available, we see that they show a similar trend, although relatively high crystallinity and relatively few large amorphous grains have been seen in one very low mass TTS (e.g.,

Sargent et al., 2005). Recent Spitzer data for brown dwarfs in ChaI show that objects of very low mass also have a large fraction of crystalline silicates, from 9% to about 50% (*Apai et al.,* 2005).

As far as the chemical composition of the silicates is concerned, there are some general trends, showing that crystalline pyroxene grains can be found in disks with relatively large fractions of forsterite and silica, but they are typically not found in disks with small fractions of forsterite and silica (*van Boekel et al.,* 2005; *Sargent et al.,* 2006).

7.2.   Spatially Resolved Spectroscopy

The advent of spectrally resolved mid-IR interferometry using large baselines at the Very Large Telescope Interferometer (VLTI) has made it possible for the first time to study the nature of the dust grains on AU spatial scales. The MIDI instrument at the VLTI has been used to study the innermost regions of HAeBe star disks. *van Boekel et al.* (2004a) (see Fig 7) show that the inner 1–2 AU of the disk surface surrounding three HAe stars is highly crystalline, with between 50% and 100% of the small silicates being crystalline. However, crystalline silicates are also present at larger distance, with a wide range of abundances. A similar behavior is found in other objects observed with MIDI

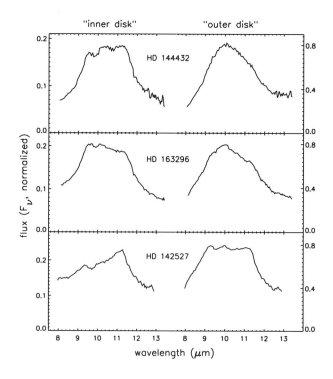

Fig. 7.   Comparison of the silicate profiles from the inner (1–2 AU) and outer (2–20 AU) disk in three HAe stars (*van Boekel et al.,* 2004a). The data have been obtained with the mid-IR interferometric instrument MIDI on the VLTI. The comparison shows that, for all three stars, the crystalline features are much stronger in the inner than in the outer disk.

(Leinert et al., in preparation). Clearly, large star-to-star variations in the amount and distribution of the crystalline silicates exist, but it seems that crystalline grains are relatively more prominent in the inner than in the outer disk. Note that, because silicate emission from cold grains become very weak, in this case "outer" refers to warm regions of the disk between 2 and 20 AU at most (*van Boekel et al., 2004a*).

In the star with the highest crystallinity, HD 142527, the observations suggest a decrease in the ratio between forsterite and enstatite with increasing distance from the star. Such a trend in the nature of the crystals with distance is predicted by chemical equilibrium and radial mixing models (*Gail, 2004*). The innermost disk is expected to contain mostly forsterite, while at larger distance from the star a conversion from forsterite to enstatite takes place. Thermal annealing produces both olivines and pyroxenes, with a ratio depending on the stochiometry of the amorphous material accreted from the parent molecular cloud. *van Boekel et al. (2004a)* also find evidence for a decrease of the average grain size with distance from the star, suggesting that grain aggregation has proceeded further in the innermost, densest disk regions.

7.3. TW Hya Systems

A number of objects with distinct signs of evolution of their inner disk has been identified in the last decade. These objects with "transitional" disks are characterized by spectral energy distributions (SEDs) with a flux deficit in the near-IR relative to the median SED of classical T Tauri stars in Taurus [a proxy for the expected emission from optically thick disks (see *D'Alessio et al., 1999*)], while at longer wavelengths fluxes are comparable or sometimes higher than the median SED of Taurus. Groundbased near- and mid-IR photometry combined with IRAS fluxes allowed the identification of objects with these properties in Taurus (*Marsh and Mahoney, 1992; Jensen and Mathieu, 1997; Bergin et al., 2004*) and in the TW Hya association (*Jayawardhana et al., 1999; Calvet et al., 2002*). However, instruments onboard Spitzer are providing a much better view of the characteristics and frequencies of these objects (*Uchida et al., 2004; Forrest et al., 2004; Muzerolle et al., 2004; Calvet et al., 2005; Sicilia-Aguilar et al., 2006*). The inner disk clearing may be related to photoevaporation of the outer disk by UV radiation (*Clarke et al., 2001*). However, transitional disks have now been found in brown dwarfs (*Muzerolle et al., 2006*), for which the UV supply of energy is negligible. An alternative explanation is that a planet has formed and carved out a gap in the disk; hydrodynamical simulations seem to support this view (*Rice et al., 2003; Quillen et al., 2004*).

The 10-m.y.-old TW Hya was the first of these objects analyzed in detail. The peculiar SED of this object (*Calvet et al., 2002; Uchida et al., 2004*) can be understood by an optically thick disk truncated at ~4 AU; the wall at the edge of this disk is illuminated directly by stellar radiation, producing the fast rise of emission at wavelengths >10 μm. The region encircled by this wall is not empty. For one thing, it has gas, because the disk is still accreting mass onto the star (*Muzerolle et al., 2000*). In addition, this inner region contains ~0.5 $M_\oplus$ of micrometer-sized particles, responsible for the small near-IR flux excess over photospheric emission. This dust is also responsible for the strong silicate feature at 10 μm, which allows us to examine the conditions of the dust in the inner disk (*Calvet et al., 2002*). The profile of the 10-μm feature, integrated over the disk, is typical of amorphous silicates with sizes of ~2 μm (*Uchida et al., 2004*), with a very low content of crystalline silicates, <2% fraction by mass (*Sargent et al., 2005*). However, interferometric observations with MIDI resolve the feature from the inner disk [within ~1 AU (Leinert et al., in preparation; see chapter by Millan-Gabet et al.)], showing clearly the 11.3-μm peak typical of forsterite, while the uncorrelated profile is very similar to that measured with single-dish instruments.

Sargent et al. (2006) have analyzed the dust content of a number of transitional disks, in addition to TW Hya. They find, averaged over the whole disk, a very low content of crystalline silicates. In contrast, there is a large spread of the crystalline silicate fraction (from 0.3% to 20% in mass) in their sample of objects with optically thick inner disks.

These results support the idea that disks accumulate crystalline silicate grains in their innermost regions and that radial mixing transports outward some fraction of this material, with an efficiency that decreases with radius. In transitional disks, a large fraction of the region within approximately a few AU is cleared out, leaving only material with low crystalline content. Spatially resolved observations of the innermost regions are needed to detect the small amount of remaining grains, with their higher crystallinity.

8. SUMMARY AND OPEN PROBLEMS

8.1. Evidence of Grain Growth

The results we have discussed come from a variety of techniques, including optical, near- and mid-IR imaging, mid-IR spectrometry, and millimeter interferometry. All the data show clear evidence that grains in protoplanetary disks differ significantly from grains in the diffuse ISM and in molecular clouds. They are much larger on average, and there is increasing support for a scenario in which grains grow and sediment toward the disk midplane. On the surface of the disk, within a few tens of AU from the star, we know that in most cases silicates have grown to sizes of few micrometers, although much smaller particles (PAHs) may be present as well, and there are hints of vertical stratification (at several scale heigths from the midplane) of these micrometer-sized grains from studies of scattered light and silhouette disks. We have evidence from millimeter interferometry that, if we consider not only grains on the disk surface but the bulk of the dust mass, in many disks grain growth has not stopped at micrometer sizes, but there is a dominant population of "sand and pebbles," millimeter and centimeter grains. In addition to vertical gradients of grain sizes, there is also evidence that radial gradients of grain properties exist, as shown by mid-IR interferometry, absorption mid-IR spectra of silhouette disks, and millimeter images at different wavelengths, as discussed before. Grains in the very outer disk seem to be less processed than grains closer to the star.

One important aspect to keep in mind is that not all the objects show evidence of processed grains: In some stars, for example, the 10-μm silicate feature has a shape very similar to that of the small silicates in the ISM; some disks have a rather steep dependence of the millimeter flux on wavelength, again typical of the small ISM grains. Some objects seem to have small silicates on the disk surface, while very large grains are implied by the millimeter spectral shape. An impressive case is that of UX Ori, which has a very peaked and strong 10-μm emission feature, typical of the small ISM amorphous silicates (Fig. 2), and an extremely flat millimeter spectrum, which may indicate the presence of very large grains in the midplane.

One should also remember that the very nature of protoplanetary disks (i.e., high optical depth, strong temperature radial gradient, etc.) prevents a complete census of the dust population, so that deriving from the observations the global properties of the solids in any given disk (such as the density of grains of different size and composition as function of radius and altitude) remains impossible for the moment, and we have to do with a piecemeal picture.

8.2. Dependence on Stellar Properties

One important step forward since *Protostars and Planets IV* has been the capability to study disks around stars of very different properties. The advent of 10-m-class telescopes on the ground, the success of Spitzer, and the improvement of the millimeter interferometers, including the VLA, have given us access to objects of increasingly lower mass. We have now millimeter spatially resolved data for TTS, and mid-IR spectra for very low mass stars and brown dwarfs, which allow us to investigate if there is a dependence of grain evolution on stellar and disk parameters. It may be worth noting that, within the millimeter-observed sample, which includes TTS and HAe stars, the stellar mass varies by a factor of 5 and the luminosity by a factor of 250; assuming typical scaling laws (*Muzerolle et al.,* 2003; *Natta et al.,* 2004b, 2006), the accretion rate in the disk is likely to vary by at least a factor of 25. The range of stellar properties is even larger for the mid-IR sample, which includes now several brown dwarfs and covers the interval between ~0.04 and 2–3 $M_\odot$ in mass, ~0.01 and 50 $L_\odot$ in luminosity, and ~10^{-10} and 10^{-6} $M_\odot$ yr in accretion rate. Over this large range of physical conditions, we do not detect *systematic* differences, i.e., grain properties do not seem to correlate with any star or disk parameter. At any given

mass, there are large variations of grain properties, although objects with unprocessed grains seem to be rare.

It seems that disks around all kind of stars have the potential of processing grains, to a degree that varies for reasons not understood so far.

8.3. Time Dependence

The time dependence of grain growth is crucial information for models, and one that we would like to derive from the observations with as much detail as possible. We should stress, first of all, that there is not at present anything close to an *unbiased* sample that can be used to study grain evolution. All the observing techniques discussed in this review have been applied to selected objects, known to be suitable for detection and analysis. In addition, ages of individual pre-main-sequence stars (our best clock so far) are measured from the location on the HR diagram and have large uncertainties, especially for HAe stars and brown dwarfs. Even more importantly, the objects for which we have observations vary in age by only about a factor of 10 (from ~1 to ~10 m.y.), and individual scatter may dominate over an underlying time dependence.

The data available so far show no evidence of a correlation of the grain properties (at any wavelength) with time: We find very processed grains in the disks of the youngest as well as the oldest pre-main-sequence stars. One possible explanation is that grains are processed efficiently in the very early stages of disk evolution, when the system star + disk is still embedded and accreting actively from the natal core [see, e.g., the theoretical models of dust dynamics and evolution during the formation of a protostellar disk by *Suttner and Yorke* (2001)]. Then, the large star-to-star variation seen in, e.g., the HAe stars would be due to differences in processing during this initial, very active phase. However, we cannot exclude that further modifications also take place in the ~10 m.y. in which the disk continues to exist.

Some information can be obtained by studying grain properties in disks in Class I sources, which are in an earlier evolutionary stage than the disks around optically visible objects discussed in this review. In Class I objects, the silicate features are seen in general in absorption, and show profiles typical of small, unprocessed amorphous silicates. It is likely, however, that the absorption is dominated by grains in the surrounding envelope, for which we expect very little growth. Only if the balance between the disk emission and the intervening absorption in the core is favorable can we have a glimpse at the properties of the disk silicates, and in a few cases there is evidence that the disk silicates are much larger than those in the core (e.g., *Kessler-Silacci et al.,* 2005). Also, the young protostellar binary SVS20 in Serpens shows evidence of the 11.3-μm forsterite feature in its spectrum (*Ciardi et al.,* 2005). For the moment, there are only a few good-quality spectra, but we can expect major progress soon, as Spitzer data will become available. One would also like to know if Class I disks have millimeter spectral energy distributions as flat as those of

Class II objects. Unfortunately, using millimeter interferometry to measure the spectral index of the emission of disks embedded in Class I cores has so far proven very difficult, as the core emission is substantial even at the smallest physical scales one can probe. The extremely flat spectra of the few objects studied so far come from unresolved central condensations, and cannot be interpreted as evidence of very large grains in disks until the emission can be spatially resolved (*Hogerheijde et al.*, 1998, 1999).

8.4. Comparison with Grain Evolution Models

Models of grain growth by collisional coagulation and sedimentation (e.g., *Dullemond and Dominik*, 2005; *Weidenschilling and Cuzzi*, 1993, and references therein) predict that grain growth will occur on very short timescales. This, at first glance, is in agreement with the idea that grain properties are already changing even in the early evolutionary phases, when disks are actively accreting matter from the parental core.

However, models also predict that the growth will not stop at micrometer or even at centimeter sizes, but will proceed very quickly (compared to the pre-main-sequence stellar life times) to form planetesimals. This is not consistent with the observations, which show us disks around stars as old as 10^7 yr with grains that, albeit very large, are far from being planetesimals. When fragmentation of the conglomerates is included, in many cases the outcome is a bimodal distribution for the solids, with most of the mass in planetesimals and a tail of smaller fragments, which can be as large as centimeters and meters, but contains only a tiny fraction of the mass in planetesimals. If these smaller bodies provide sufficient optical depth, the disk properties may look similar to what is observed. However, we think this is unlikely to be the case. The millimeter observations described in section 6 tell us not only that grains need to have very big sizes but also the mass of these grains, which turns out to be very large. Using, e.g., the same grain models as in Fig. 5, we derive dust masses, for the disks with evidence of very large grains, between 10^{-3} and 10^{-2} M$_\odot$; assuming the standard gas-to-dust ratio of 100 (which is probably close to true when disks formed), this implies disk masses between ~0.1 and 1 M$_\odot$, i.e., ratios of the disk to star masses approaching unity. If a large fraction of the original dust is not in the observed population of millimeter and centimeter grains, but in planetesimals, the disk mass should be even higher, well above the gravitational instability limit. Although we cannot exclude that a *small* fraction of the original solid mass is in planetesimals, we think it is unlikely that they have collected most of it.

The survival of a large mass of millimeter- and centimeter-sized grains over a timescale of several million years, as observed in many disks, is challenging the models on several grounds. It suggests that in many cases the process of planetesimal formation is much less efficient than predicted by theoretical models. If disks are forming planetesimals as fast as predicted, then this can only happen at the end of a long quiescent phase, where growth is limited to sizes of a few centimeters at most; otherwise, the process has to be very slow, involving a modest fraction of the dust mass for most of the pre-main-sequence life of the star. The "inefficiency" may result from fragmentation of the larger bodies. This *in situ* production of the grains we observe may take care of the other severe difficulty one has, namely that millimeter and centimeter solid bodies (in a gas-rich disk) migrate toward the star on very short timescales. This is a very open field, where progress can be made only by combining together theoretical studies, laboratory experiments, and observations.

It is possible that the disks we can study with current techniques are those that will never form planets. Disks where planets form may evolve indeed very quickly, and could be below present-day detection limits. This, however, seems unlikely, as we have objects like TW Hya and HD 100546, which have evidence of very large grains in their outer disk, and an inner gap possibly due to the action of a large planet. Similar "transitional" objects are found by Spitzer in increasing number, and a detailed characterization of their millimeter properties will be important. In any case, the properties and evolution of disks that we have studied so far and their dust content provide the only observational constraints to theoretical models of grain evolution and planet formation.

8.5. Crystalline Silicates and Mineralogy

The ISO discovery of large fractional amounts of crystalline silicates in disks around stars of all masses has been a surprise. They are enhanced in the innermost regions of disks, as shown by interferometric observations of HAe stars and by the spectra of transitional disks, but they are also present, in variable degree, further out. Fractional masses of crystalline silicates, integrated over the whole disk-emitting region, have been derived for many objects, HAe, TTS, and BDs, with values ranging from zero to about 60%. These estimates refer to grains on the disk surface, typically within 10–20 AU from the star for the HAe stars, and 0.1–0.2 AU for BDs. One should note that these values are extremely model dependent, and refer to a limited range of grain sizes only; they should be taken with the greatest care. We can expect great progress in the immediate future, as spectra in the 15–45-μm spectral region, which can detect colder (and larger) crystals, are becoming available. Further insight can also come from improved characterization of the mineralogy in disks, e.g., measuring the ratio of species such as crystalline pyroxene, forsterite, silica, etc. Self-consistent radiation transfer models in disks that can include a large number of dust species exist, and can be used to provide more reliable values for the relative abundances of the various species observed on disk surfaces.

The formation of crystalline silicates requires energetic processes, which have to be efficient not only in HAe stars, where crystalline silicates were first discovered, but also in the much less luminous TTS and brown dwarfs. If crystallization is restricted to the very inner disks, as in the *Gail* (2004) models, then strong radial mixing must occur, transporting material outwardly. Radial and meridian drifts can have an important role in the growth of grains, which needs to be addressed in the future.

8.6. Final Remarks

The observations we have discussed in this chapter show clearly that very few disks (if any) contain unprocessed grains, i.e., with properties similar to those of grains in the ISM. In general, dust has been largely processed in all objects.

However, the degree and the result of these changes vary a great deal from object to object, as we find large differences in the grain size, composition, and allotropic properties. At present, we have not been able to identify any correlation between the grains and other properties of the central star (such as its mass and luminosity) and of the disk (for example, mass, accretion rate, etc.), nor with the age of the system. This last point is particularly distressing, since planet formation theories require understanding if, when, and how grains change with time, or, in other words, how they "evolve."

The study of pre-main-sequence stars has shown us that, no matter which aspect of their rich phenomenology we are interested in, there is a large scatter between individuals, which can hide underlying trends. In discussing grain properties, it is possible that the objects observed so far do not sample the right range of parameters, age in particular, or that the available samples are still too small, or restricted to a too narrow range in the parameter space. We see the noise, and cannot identify trends that remain, if present, hidden. If this is really the case, we can expect major advances in the near future, as new space- and groundbased facilities will make it possible to study much larger samples of disks, in different star-forming regions, surrounding objects distributed over a broader range of mass, age, and multiplicity.

Acknowledgments.   We are indebted to a number of colleagues who have provided us with unpublished material and helped in preparing some of the figures: J. Bouwman, R. van Boekel, I. Pascucci, D. Apai, J. Kessler-Silacci, J. Rodmann, M. Min, and the IRS Spitzer team. During a visit to the MPA in Heidelberg, A.N. enjoyed discussions with K. Dullemond, C. Leinert, J. Bouwman, R. van Boekel, and J. Rodmann, among others. This work was partially supported by MIUR grant 2004025227/2004 to the Arcetri Observatory.

REFERENCES

Acke B. and van den Ancker M. E. (2004) *Astron. Astrophys., 426,* 151–170.

Apai D., Pascucci I., Bouwman J., Natta A., Henning Th., and Dullemond C. P. (2005) *Science, 310,* 834–836.

Beckwith S. V. W. and Sargent A. I. (1991) *Astrophys. J., 381,* 250–258.

Beckwith S. V. W., Henning Th., and Nakagawa Y. (2000) In *Protostars and Planets IV* (V. Mannings et al., eds.), pp. 533–558. Univ. of Arizona, Tucson.

Bergin E., Calvet N., Sitko M. L., Abgrall H., D'Alessio P., et al. (2004) *Astrophys. J., 614,* L133–L136.

Bianchi S., Gonçalves J., Albrecht M., Caselli P., Chini R., Galli D., and Walmsley M. (2003) *Astron. Astrophys., 399,* L43–L46.

Blum J. (2004) In *Astrophysics of Dust* (A. N. Witt et al., eds.), pp. 369–391. ASP, San Francisco.

Calvet N., Patino A., Magris G. C., and D'Alessio P. (1991) *Astrophys. J., 380,* 617–630.

Calvet N., D'Alessio P., Hartmann L., Wilner D., Walsh A., and Sitko M. (2002) *Astrophys. J., 568,* 1008–1016.

Calvet N., D'Alessio P., Watson D. M., Franco-Hernández R., Furlan E., et al. (2005) *Astrophys. J., 630,* L185–188.

Chiang E. I. and Goldreich P. (1997) *Astrophys. J., 490,* 368–376.

Chiang E. I. and Goldreich P. (1999) *Astrophys. J., 519,* 279–284.

Ciardi D. R., Telesco C. M., Packham C., Gómez Martin C., Radomski J. T., De Buizer J. M., Phillips Ch. J., and Harker D. E. (2005) *Astrophys. J., 629,* 897–902.

Clarke C. J., Gendrin A., and Sotomayor M. (2001) *Mon. Not. R. Astron. Soc., 328,* 485–491.

D'Alessio P., Calvet N., Hartmann L., Lizano S., and Cantó J. (1999) *Astrophys. J., 527,* 893–909.

D'Alessio P., Calvet N., and Hartmann L. (2001) *Astrophys. J., 553,* 321–334.

Draine B. T. (2003) *Ann. Rev. Astron. Astrophys., 41,* 241–289.

Draine B. T. (2006) *Astrophys. J., 636,* 1114–1120.

Duchêne G., McCabe C., Ghez A. M., and Macintosh B. A. (2004) *Astrophys. J., 606,* 969–982.

Dullemond C. P. and Dominik C. (2005) *Astron. Astrophys., 434,* 971–986.

Forrest W. J., Sargent B., Furlan E., D'Alessio P., Calvet N., et al. (2004) *Astrophys. J. Suppl., 154,* 443–447.

Gail H.-P. (2004) *Astron. Astrophys., 413,* 571–591.

Guillois O., Ledoux G., and Reynaud C. (1999) *Astrophys. J., 521,* L133–L136.

Habart E., Natta A., and Krügel E. (2004a) *Astron. Astrophys., 427,* 179–192.

Habart E., Testi L., Natta A., and Carbillet M. (2004b) *Astrophys. J., 214,* L129–L132.

Habart E., Natta A., Testi L., and Carbillet M. (2006) *Astron. Astrophys., 449,* 1067–1075.

Henning Th., Michel B., and Stognienko R. (1995) *Planet. Space Sci., 43,* 1333–1343.

Henning Th., Dullemond C. P., Wolf S., and Dominik C. (2006) In *Planet Formation: Theory, Observation and Experiments* (H. Klahr and W. Brandner, eds.), pp. 112–128. Cambridge Univ., Cambridge.

Hogerheijde M. R., van Dishoeck E. F., Blake G. A., and van Langevelde H. J. (1998) *Astrophys. J., 502,* 315–336.

Hogerheijde M. R., van Dishoeck E. F., Salverda J. M., and Blake G. A (1999) *Astrophys. J., 513,* 350–369.

Jayawardhana R., Hartmann L., Fazio G., Fisher R. S., Telesco C. M., and Piña R. K. (1999) *Astrophys. J., 521,* L129–L132.

Jensen E. L. N. and Mathieu R. D. (1997) *Astron. J., 114,* 301–316.

Keller L. P., Hony S., Bradley J. P., Molster F. J., Waters L. B. F. M., et al. (2002) *Nature, 417,* 148–150.

Kemper F., Vriend W. J., and Tielens A. G. G. M. (2004) *Astrophys. J., 609,* 826–837.

Kessler-Silacci J. E., Hillenbrand L. A., Blake Geoffrey A., and Meyer M. R. (2005) *Astrophys. J., 622,* 404–429.

Kessler-Silacci J. E., Augereau J.-C., Dullemond C. P., Geers V., Lahuis F., et al. (2006) *Astrophys. J., 639,* 275–291.

Krügel E. and Siebenmorgen R. (1994) *Astron. Astrophys., 288,* 929–941.

Lucas P. W., Fukagawa M., Tamura M., Beckford A. F., Itoh Y., et al. (2004) *Mon. Not. R. Astron. Soc., 352,* 1347–1364.

Marsh K. A. and Mahoney M. J. (1992) *Astrophys. J., 395,* L115–L118.

McCabe C., Duchêne G., and Ghez A. M. (2003) *Astrophys. J., 588,* L113–L116.

McCaughrean M. J., Stapelfeldt K. R., and Close L. M. (2000) In *Protostars and Planets IV* (V. Mannings et al., eds.), pp. 485–507. Univ. of Arizona, Tucson.

Meeus G., Waters L. B. F. M., Bouwman J., van den Ancker M. E., Waelkens C., and Malfait K. (2001) *Astron. Astrophys., 365,* 476–490.

Meeus G., Sterzik M., Bouwman J., and Natta A. (2003) *Astron. Astrophys., 409,* L25–L29.

Menshchikov A. B. and Henning Th. (1997) *Astron. Astrophys., 318,* 879–907.

Min M., Hovenier J. W., de Koter A., Waters L. B. F. M., and Dominik C. (2005) *Icarus, 179,* 158–173.

Min M., Dominik C., Hovenier J. W., de Koter A., and Waters L. B. F. M. (2006) *Astron. Astrophys., 445,* 1005–1014.

Miyake K. and Nakagawa Y. (1993) *Icarus, 106,* 20–41.

Miyake K. and Nakagawa Y. (1995) *Astrophys. J., 441,* 361–384.

Muzerolle J., Calvet N., Briceño C., Hartmann L., and Hillenbrand L. (2000) *Astrophys. J., 535,* L47–L50.

Muzerolle J., Hillenbrand L., Calvet N., Briceño C., and Hartmann L. (2003) *Astrophys. J., 592,* 266–281.

Muzerolle J., Megeath S. T., Gutermuth R. A., Allen L. E., Pipher J. L., et al. (2004) *Astrophys. J. Suppl., 154,* 379–384.

Muzerolle J., Adame L., D'Alessio P., Calvet N., Luhman K. L., et al. (2006) *Astrophys. J., 643,* 1003–1010.

Natta A. and Testi L. (2004) In *Star Formation in the Interstellar Medium: In Honor of David Hollenbach, Chris McKee and Frank Shu* (D. Johnstone et al., eds.), pp. 279–285. ASP, San Francisco.

Natta A., Testi L., Neri R., Shepherd D. S., and Wilner D. J. (2004a) *Astron. Astrophys., 416,* 179–186.

Natta A., Testi L., Muzerolle J., Randich S., Comerón F., and Persi P. (2004b) *Astron. Astrophys., 424,* 603–612.

Natta A., Testi L., and Randich S. (2006) *Astron. Astrophys.,* in press (astro-ph/0602618).

Ossenkopf V. and Henning Th. (1994) *Astron. Astrophys., 291,* 943–959.

Peeters E., Hony S., Van Kerckhoven C., Tielens A. G. G. M., Allamandola L. J., Hudgins D. M., and Bauschlicher C. W. (2002) *Astron. Astrophys., 390,* 1089–1113.

Pollack J. B., Hollenbach D., Beckwith S., Simonelli D. P., Roush T., and Fong W. (1994) *Astrophys. J., 421,* 615–639.

Przygodda F., van Boekel R., Ábrahám P., Melnikov S. Y., Waters L. B. F. M., and Leinert Ch. (2003) *Astron. Astrophys., 412,* L43–L46.

Qi C., Ho P. T. P., Wilner D. J., Takakuwa S., Hirano N., et al. (2004) *Astrophys. J., 616,* L11–L14.

Quillen A. C., Blackman E. G., Frank A., and Varnière P. (2004) *Astrophys. J., 612,* L137–L140.

Ressler M. E. and Barsony M. (2003) *Astrophys. J., 584,* 832–842.

Rice W. K. M., Wood K., Armitage P. J., Whitney B. A., and Bjorkman J. E. (2003) *Mon. Not. R. Astron. Soc., 342,* 79–85.

Rodmann J., Henning Th., Chandler C. J., Mundy L. G., and Wilner D. J. (2006) *Astron. Astrophys., 446,* 211–223.

Sargent B., Forrest W. J., D'Alessio P., Najita J., Li A., et al. (2005) Poster presented at IAU Symp. 231 on "Astrochemistry Throughout the Universe: Recent Successes and Current Challenges," Asilomar, California.

Sargent B., Forrest W. J., D'Alessio P., Li A., Najita J., et al. (2006) *Astrophys. J.,* in press (astro-ph/0605415).

Schräpler R. and Henning Th. (2004) *Astrophys. J., 614,* 960–978.

Shuping R. Y., Bally J., Morris M., and Throop H. (2003) *Astrophys. J., 587,* L109–L112.

Sicilia-Aguilar A., Hartmann L., Calvet N., Megeath S. T., Muzerolle J., et al. (2006) *Astrophys. J., 638,* 897–919.

Sloan G. C., Keller L. D., Forrest W. J., Leibensperger E., Sargent B., et al. (2005) *Astrophys. J., 632,* 956–963.

Suttner G. and Yorke H. W. (2001) *Astrophys. J., 551,* 461–447.

Testi L., Natta A., Shepherd D. S., and Wilner D. J. (2001) *Astrophys. J., 554,* 1087–1094.

Testi L., Natta A., Shepherd D. S., and Wilner D. J. (2003) *Astron. Astrophys., 403,* 323–328.

Throop H. B., Bally J., Esposito L. W., and McCaughrean M. J. (2001) *Science, 292,* 1686–1689.

Uchida K. I., Calvet N., Hartmann L., Kemper F., Forrest W. J., et al. (2004) *Astrophys. J. Suppl., 154,* 439–442.

van Boekel R., Min M., Leinert Ch., Waters L. B. F. M., Richichi A., et al. (2004a) *Nature, 432,* 479–482.

van Boekel R., Waters L. B. F. M., Dominik C., Dullemond C. P., Tielens A. G. G. M., and de Koter A. (2004b) *Astron. Astrophys., 418,* 177–184.

van Boekel R., Min M., Waters L. B. F. M., de Koter A., Dominik C., van den Ancker M. E., and Bouwman J. (2005) *Astron. Astrophys., 437,* 189–208.

van Diedenhoven B., Peeters E., Van Kerckhoven C., Hony S., Hudgins D. M., Allamandola L. J., and Tielens A. G. G. M. (2004) *Astrophys. J., 611,* 928–939.

Van Kerckhoven C., Tielens A. G. G. M., and Waelkens C. (2002) *Astron. Astrophys., 384,* 568–584.

Voshchinnikov N. V. and Krügel E. (1999) *Astron. Astrophys., 352,* 508–516.

Voshchinnikov N. V., Ilin V. B., Henning Th., and Dubkova D. N. (2006) *Astron. Astrophys., 445,* 167–177.

Weidenschilling S. J. and Cuzzi J. N. (1993) In *Protostars and Planets III* (E. H. Levy and J. N. Lunine, eds.), pp. 1031–1060. Univ. of Arizona, Tucson.

Weingartner J. C. and Draine B. T. (2001) *Astrophys. J., 548,* 296–309.

Wilner D. J., Ho P. T. P., Kastner J. H., and Rodríguez L. F. (2000) *Astrophys. J., 534,* L101–104.

Wilner D. J., D'Alessio P., Calvet N., Claussen M. J., and Hartmann L. (2005) *Astrophys. J., 626,* L109–L112.

Wurm G., Paraskov G., and Krauss O. (2005) *Icarus, 178,* 253–263.

Growth of Dust as the Initial Step Toward Planet Formation

Carsten Dominik
Universiteit van Amsterdam

Jürgen Blum
Technische Universität Braunschweig

Jeffrey N. Cuzzi
NASA Ames Research Center

Gerhard Wurm
Universität Münster

We discuss the results of laboratory measurements and theoretical models concerning the aggregation of dust in protoplanetary disks as the initial step toward planet formation. Small particles easily stick when they collide and form aggregates with an open, often fractal structure, depending on the growth process. Larger particles are still expected to grow at collision velocities of about 1 m/s. Experiments also show that, after an intermezzo of destructive velocities, high collision velocities above 10 m/s on porous materials again lead to net growth of the target. Considerations of dust-gas interactions show that collision velocities for particles not too different in surface-to-mass ratio remain limited up to sizes of about 1 m, and growth seems to be guaranteed to reach these sizes quickly and easily. For meter sizes, coupling to nebula turbulence makes destructive processes more likely. Global aggregation models show that in a turbulent nebula, small particles are swept up too fast to be consistent with observations of disks. An extended phase may therefore exist in the nebula during which the small particle component is kept alive through collisions driven by turbulence, which frustrates growth to planetesimals until conditions are more favorable for one or more reasons.

1. INTRODUCTION

The growth of dust particles by aggregation stands at the beginning of planet formation. Whether planetesimals form by incremental aggregation, or through gravitational instabilities in a dusty sublayer, particles have to grow and settle to the midplane regardless. On the most basic level, the physics of such growth is simple: Particles collide because relative velocities are induced by random and (size-dependent) systematic motions of grains and aggregates in the gaseous nebula surrounding a forming star. The details are, however, highly complex. The physical state of the disk, in particular the presence or absence of turbulent motions, set the boundary conditions. When particles collide with low velocities, they stick by mutual attractive forces, be it simple van der Waals attraction or stronger forces (molecular dipole interaction in polar ices, or grain-scale long-range forces due to charges or magnetic fields). While the lowest velocities create particle shapes governed by the motions

alone, larger velocities contribute to shaping the aggregates by restructuring and destruction. The ability to internally dissipate energy is critical in the growth through intermediate pebble and boulder sizes. In this review we will concentrate on the physical properties and growth characteristics of these small and intermediate sizes, but also make some comments on the formation of planetesimals.

Relative velocities between grains in a protoplanetary disk can be caused by a variety of processes. For the smallest grains, these are dominated by Brownian motions, which provide relative velocities in the millimeter per second to centimeter per second range for (sub)micrometer-sized grains. Larger grains show systematic velocities in the nebula because they decouple from the gas, settle vertically, and drift radially. At 1 AU in a solar nebula, these settling velocities reach meters per second for centimeter-sized grains. Radial drift becomes important for even larger particles and reaches tens of meters per second for meter-sized bodies. Finally, turbulent gas motions can induce relative motions between particles. For details see, e.g., *Weidenschilling* (1977, 1984), *Weidenschilling and Cuzzi* (1993), and *Cuzzi and Hogan* (2003).

The timescales of growth processes, and the density and strength of aggregates formed by them, will depend on the structure of the aggregates. A factor of overriding importance for dust-gas interactions (and therefore for the timescales and physics of aggregation), for the stability of aggregates, and for optical properties alike is the structure of aggregates as they form through the different processes.

The interaction of particles with the nebula gas is determined primarily by their gas drag stopping time t_s, which is given by

$$t_s = \frac{mv}{F_{fric}} = \frac{3}{4c_s\rho_g}\frac{m}{\sigma} \quad (1)$$

where m is the mass of a particle, v its velocity relative to the gas, σ the average projected surface area, ρ_g the gas density, c_s the sound speed, and F_{fric} the drag force. The second equal sign in equation (1) holds under the assumption that particles move at subsonic velocities and that the mean free path of a gas molecule is large compared to the size of the particle (Epstein regime). In this case, the stopping time is proportional to the ratio of mass and cross section of the particle. For spherical nonfractal (i.e., compact or porous) particles of radius a and mass density ρ_s, this can be written as $t_s = a\rho_s/c\rho_g$. Fractal particles are characterized by the fact that the average density of a particle depends on size in a powerlaw fashion, with a power (the fractal dimension D_f) smaller than 3.

$$m(a) \propto a^{D_f} \quad (2)$$

For large aggregates, this value can in principle be measured for individual particles. For small particles, it is often more convenient to measure it using sizes and masses of a distribution of particles.

Fractal particles generally have large surface-to-mass ratios; in the limiting case of long linear chains ($D_f = 1$) of grains with radii a_0, σ/m approaches the constant value $3\pi/(16a_0\rho_s)$. This value differs from the value for a single grain $3/(4a_0\rho_s)$ by just a factor $\pi/4$. Figure 1 shows how the cross section of particles varies with their mass for different fractal dimensions. It shows that for aggregates made of 10,000 monomers, the surface-to-mass ratio can easily differ by a factor of 10. An aggregate made from 0.1 μm particles with a mass equivalent to a 10 μm particle consists of 10^6 monomers and the stopping time could vary by a factor on the order of 100. Just how far the fractal growth of aggregates proceeds is really not yet known.

This review is organized as follows: In section 2 we cover the experiments and theory describing the basic growth processes of dust aggregates. In section 3 we discuss particle-gas interactions and the implications for interparticle collision velocities as well as planetesimal formation. In sec-

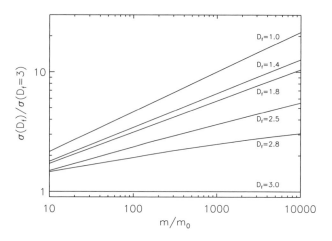

Fig. 1. Projected area of aggregates as a function of aggregate size and fractal dimension, normalized to the cross section of a compact particle with the same mass.

tion 4 we describe recent advances in the modeling of dust aggregation in protoplanetary disks and observable consequences.

2. DUST AGGREGATION EXPERIMENTS AND THEORY

2.1. Interactions Between Individual Dust Grains

2.1.1. Interparticle adhesion forces. Let us assume that the dust grains are spherical in shape and that they are electrically neutral and nonmagnetic. In that case, two grains with radii a_1 and a_2 will always experience a short-range attraction due to induced dielectric forces, e.g., van der Waals interaction. This attractive force results in an elastic deformation leading to a flattening of the grains in the contact region. An equilibrium is reached when the attractive force equals the elastic repulsion force. For small, hard grains with low surface forces, the equilibrium contact force is given by (*Derjaguin et al.*, 1975)

$$F_c = 4\pi\gamma_s R \quad (3)$$

where γ_s and R denote the specific surface energy of the grain material and the local radius of surface curvature, given by $R = a_1a_2/(a_1 + a_2)$, respectively. Measurements of the separation force between pairs of SiO_2 spheres with radii a between 0.5 μm and 2.5 μm (corresponding to reduced radii R = 0.35 . . . 1.3 μm) confirm the validity of equation (3) (*Heim et al.*, 1999).

2.1.2. Interparticle rolling-friction forces. Possibly the most important parameter influencing the structure of aggregates resulting from low-velocity collisions is the resistance to rolling motion. If this resistance is very strong, both aggregate compaction and internal energy dissipation in aggregates would be very difficult. Resistance to rolling first of all depends strongly on the geometry of the grains. If grains contain extended flat surfaces, contact made on such locations could not be moved by rolling — any attempt to roll them would inevitably lead to breaking the contact. In the contact between round surfaces, resistance to rolling must come from an asymmetric distribution of the stresses in the contact area. Without external forces, the net torque exerted on the grains should be zero. *Dominik and Tielens* (1995) showed that the pressure distribution becomes asymmetric, when the contact area is slightly shifted with respect to the axis connecting the curvature centers of the surfaces in contact. The resulting torque is

$$M = 4F_c \left(\frac{a_{contact}}{a_{contact,0}} \right)^{3/2} \xi \qquad (4)$$

where $a_{contact,0}$ is the equilibrium contact radius, $a_{contact}$ the actual contact radius due to pressure in the vertical direction, and ξ is the displacement of the contact area due to the torque. In this picture, energy dissipation, and therefore friction, occurs when the contact area suddenly readjusts after it has been displaced because of external forces acting on the grains. The friction force is proportional to the pull-off force F_c.

Heim et al. (1999) observed the reaction of a chain of dust grains using a long-distance microscope and measured the applied force with an atomic force microscope (AFM). The derived rolling-friction forces between two SiO_2 spheres with radii of $a = 0.95$ μm are $F_{roll} = (8.5 \pm 1.6) \times 10^{-10}$ N. If we recall that there are two grains involved in rolling, we get for the rolling-friction energy, defined through a displacement of an angle $\pi/2$

$$E_{roll} = \pi a F_{roll} = O(10^{-15} \text{ J}) \qquad (5)$$

Recently, the rolling of particle chains has been observed under the scanning electron microscope while the contact forces were measured simultaneously (*Heim et al.*, 2005).

2.1.3. Sticking efficiency in single-grain collisions. The dynamical interaction between small dust grains was derived by *Poppe et al.* (2000a) in an experiment in which single, micrometer-sized dust grains impacted smooth targets at various velocities (0 . . . 100 m/s) under vacuum conditions. For spherical grains, a sharp transition from sticking with an efficiency of $\beta \approx 1$ to bouncing (i.e., a sticking efficiency of $\beta = 0$) was observed. This threshold velocity is $v_s \approx 1.2$ m/s for $a = 0.6$ μm and $v_s \approx 1.9$ m/s for $a = 0.25$ μm. It decreases with increasing grain size. The target materials were either polished quartz or atomically smooth (surface-oxidized) silicon. Currently, no theoretical explanation is available for the threshold velocity for sticking. Earlier attempts to model the low-velocity impact behavior of spherical grains predicted much lower sticking velocities (*Chokshi et al.*, 1993). These models are based upon impact experiments with "softer" polystyrene grains

(*Dahneke*, 1975). The main difference becomes visible when studying the behavior of the rebound grains in nonsticking collisions. In the experiments by *Dahneke* (1975) and also in those by *Bridges et al.* (1996) using macroscopic ice grains, the behavior of grains after a bouncing collision was a unique function of the impact velocity, with a coefficient of restitution (rebound velocity divided by impact velocity) always close to unity and increasing monotonically above the threshold velocity for sticking. For harder, still spherical, SiO_2 grains (*Poppe et al.*, 2000a), the *average* coefficient of restitution decreases considerably with increasing impact velocity. In addition to that, *individual* grain impacts show considerable scatter in the coefficient of restitution.

The impact behavior of irregular dust grains is more complex. Irregular grains of various sizes and compositions show an overall decrease in the sticking probability with increasing impact velocity. The transition from $\beta = 1$ to $\beta = 0$, however, is very broad so that even impacts as fast as $v \approx 100$ m/s can lead to sticking with a moderate probability.

2.2. Dust Aggregation and Restructuring

2.2.1. Laboratory and microgravity aggregation experiments. In recent years, a number of laboratory and microgravity experiments have been carried out to derive the aggregation behavior of dust under conditions of young planetary systems. To be able to compare the experimental results to theoretical predictions and to allow numerical modelling of growth phases that are not accessible to experimental investigation, "ideal" systems were studied, in which the dust grains were monodisperse (i.e., all of the same size) and initially nonaggregated. Whenever the mean collision velocity between the dust grains or aggregates is much smaller than the sticking threshold (see section 2.1.3), the aggregates formed in the experiments are "fractal," i.e., $D_f < 3$ (*Wurm and Blum*, 1998; *Blum et al.*, 1999, 2000; *Krause and Blum*, 2004). The precise value of the fractal dimension depends on the specific aggregation process and can reach values as low as $D_f = 1.4$ for Brownian-motion-driven aggregation (*Blum et al.*, 2000; *Krause and Blum*, 2004; *Paszun and Dominik*, 2006), $D_f = 1.9$ for aggregation in a turbulent gas (*Wurm and Blum*, 1998), or $D_f = 1.8$ for aggregation by gravitationally driven sedimentation in gas (*Blum et al.*, 1999). It is inherent to a dust aggregation process in which aggregates with low fractal dimensions are formed that the mass distribution function is rather narrow (quasi-monodisperse) at any given time. In all realistic cases, the mean aggregate mass $\bar{m}$ follows either a power law with time t, i.e., $\bar{m} \propto t^\gamma$ with $\gamma > 0$ (*Krause and Blum*, 2004), or grows exponentially fast, $\bar{m} \propto \exp(\delta t)$ with $\delta > 0$ (*Wurm and Blum*, 1998), which can be verified in dust-aggregation models (see section 2.2.2).

As predicted by *Dominik and Tielens* (1995, 1996, 1997), experiments have shown that at collision velocities near the velocity threshold for sticking (of the individual dust grains), a new phenomenon occurs (*Blum and Wurm*, 2000).

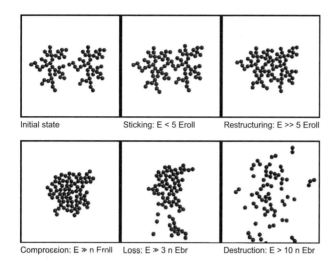

Initial state   Sticking: E < 5 Eroll   Restructuring: E >> 5 Eroll

Compression: E ≫ n Froll   Loss: E ≫ 3 n Ebr   Destruction: E > 10 n Ebr

Fig. 2. Dominating processes and associated energies in aggregate collisions, after *Dominik and Tielens* (1997) and *Wurm and Blum* (2000). E_{br} is the energy needed to break a contact, E_{roll} is the energy to roll two grains through an angle $\pi/2$, and n is the total number of contacts in the aggregates.

Whereas at low impact speeds, the aggregates' structures are preserved in collisions (the so-called "hit-and-stick" behavior), the forming aggregates are compacted at higher velocities. In even more energetic collisions, the aggregates fragment so that no net growth is observable. The different stages of compaction and fragmentation are depicted in Fig. 2.

2.2.2. Modeling of dust aggregation. The evolution of grain morphologies and masses for a system of initially monodisperse spherical grains that are subjected to Brownian motion has been studied numerically by *Kempf et al.* (1999). The mean aggregate mass increases with time following a power law (see section 2.2.1). The aggregates have fractal structures with a mean fractal dimension of $D_f = 1.8$. Analogous experiments by *Blum et al.* (2000) and *Krause and Blum* (2004), however, found that the mean fractal dimension was $D_f = 1.4$. Recent numerical work by *Paszun and Dominik* (2006) showed that this lower value is caused by Brownian rotation [neglected by *Kempf et al.* (1999)].

More chain-like dust aggregates can form if the mean free path of the colliding aggregates becomes smaller than their size, i.e., if the assumption of ballistic collisions breaks down and a random-walk must be considered for the approach of the particles. The fractal dimension of thermally aggregating dust grains is therefore dependent on gas pressure and reaches an asymptotic value of $D_f = 1.5$ for the low-density conditions prevailing through most of the presolar nebula. Only in the innermost regions are the densities high enough to cause deviations.

The experimental work reviewed in section 2.2.1 can be used to test the applicability of theoretical dust aggregation models. Most commonly, the mean-field approach by *Smoluchowski* (1916) is used for the description of the number density n(m, t) of dust aggregates with mass m as a func-

tion of time t. Smoluchowski's rate equation reads in the integral form

$$\frac{\partial n(m,t)}{\partial t} = \frac{1}{2}\int_0^m K(m', m - m')$$
$$\cdot\, n(m',t)n(m - m',t)dm' \qquad (6)$$
$$-n(m,t)\int_0^\infty K(m',m)n(m',t)dm'$$

Here, $K(m_1,m_2)$ is the reaction kernel for aggregation of the coagulation equation (6). The first term on the rhs of equation (6) describes the rate of sticking collisions between dust particles of masses m' and m – m' whose combined masses are m (gain in number density for the mass m). The second term denotes a loss in the number density for the mass m due to sticking collisions between particles of mass m and mass m'. The factor 1/2 in the first term accounts for the fact that each pair collision is counted twice in the integral. In most astrophysical applications the gas densities are so low that dust aggregates collide ballistically. In that case, the kernel in equation (6) is given by

$$K(m_1,m_2) = \beta(m_1,m_2;\, v)v(m_1,m_2)\sigma(m_1,m_2) \qquad (7)$$

where $\beta(m_1,m_2;\, v)$, $v(m_1,m_2)$, and $\sigma(m_1,m_2)$ are the sticking probability, the collision velocity, and the cross section for collisions between aggregates of masses m_1 and m_2, respectively.

A comparison between numerical predictions from equation (6) and experimental results on dust aggregation was given by *Wurm and Blum* (1998), who investigated dust aggregation in rarefied, turbulent gas. Good agreement for both the mass distribution functions and the temporal behavior of the mean mass was found when using a sticking probability of $\beta(m_1,m_2;\, v) = 1$, a mass-independent relative velocity between the dust aggregates and the expression by *Ossenkopf* (1993) for the collision cross section of fractal dust aggregates. *Blum* (2006) showed that the mass distribution of the fractal aggregates observed by *Krause and Blum* (2004) for Brownian-motion-driven aggregation can also be modeled in the transition regime between free-molecular and hydrodynamic gas flow.

Analogous to the experimental findings for the collisional behavior of fractal dust aggregates with increasing impact energy (*Blum and Wurm,* 2000) (see section 2.2.1), *Dominik and Tielens* (1997) showed in numerical experiments on aggregate collisions that with increasing collision velocity the following phases can be distinguished: hit-and-stick behavior, compaction, loss of monomer grains, and complete fragmentation (see Fig. 2). They also showed that the outcome of a collision depends on the impact energy, the rolling-friction energy (see equation (5) in section 2.1.2) and the energy for the breakup of single interparticle contacts (see section 2.1.1). The model by *Dominik and Tielens* (1997) was quantitatively confirmed by the experiments of *Blum and Wurm* (2000) (see Fig. 2).

To analyze observations of protoplanetary disks and model the radiative transfer therein, the optical properties of particles are important (*McCabe et al.*, 2003; *Ueta and Meixner*, 2003; *Wolf*, 2003). Especially for particle sizes comparable to the wavelength of the radiation, the shape and morphology of a particle are of major influence for the way the particle interacts with the radiation. With respect to this, it is important to know how dust evolution changes the morphology of a particle. As seen above, in most cases dust particles are not individual monolithic solids but rather aggregates of primary dust grains. Numerous measurements and calculations have been carried out on aggregates (e.g., *Kozasa et al.*, 1992; *Henning and Stognienko*, 1996; *Wurm and Schnaiter*, 2002; *Gustafson and Kolokolova*, 1999; *Wurm et al.*, 2004a; *Min et al.*, 2006). No simple view can be given within the frame of this paper. However, it is clear that the morphology and size of the aggregates will strongly influence the optical properties.

2.2.3. Aggregation with long-range forces. Long-range forces may play a role in the aggregation process, if grains are either electrically charged or magnetic. Small iron grains may become spontaneously magnetic if they are single domain (*Nuth et al.*, 1994; *Nuth and Wilkinson*, 1995), typically at sizes of a few tens of nanometers. Larger grains containing ferromagnetic components can be magnetized by an impulse magnetic field generated during a lightning discharge (*Túnyi et al.*, 2003). For such magnetized grains, the collisional cross section is strongly enhanced compared to the geometrical cross section (*Dominik and Nübold*, 2002). Aggregates formed from magnetic grains remain strong magnetic dipoles, if the growth process keeps the grain dipoles aligned in the aggregate (*Nübold and Glassmeier*, 2000). Laboratory experiments show the spontaneous formation of elongated, almost linear aggregates, in particular in the presence of an external magnetic field (*Nübold et al.*, 2003). The relevance of magnetic grains to the formation of macroscopic dust aggregates is, however, unclear.

Electric charges can be introduced through tribo-electric effects in collisions, through which electrons and/or ions are exchanged between the particles (*Poppe et al.*, 2000b; *Poppe and Schräpler*, 2005; *Desch and Cuzzi*, 2000). The number of separated elementary charges in a collision between a dust particle and a solid target with impact energy E_c can be expressed by (*Poppe et al.*, 2000b; *Poppe and Schräpler*, 2005)

$$N_e \approx \left(\frac{E_c}{10^{-15} \text{ J}} \right)^{0.8} \qquad (8)$$

The cumulative effect of many nonsticking collisions can lead to an accumulation of charges and to the buildup of strong electrical fields at the surface of a larger aggregate. In this way, impact charging could lead to electrostatic trapping of the impinging dust grains or aggregates (*Blum*, 2004). Moreover, impact charging and successive charge separation can cause an electric discharge in the nebula gas. For nebula lighting (*Desch and Cuzzi*, 2000) a few hundred to thousand elementary charges per dust grain are required. This corresponds to impact velocities in the range 20 . . . 100 m/s (*Poppe and Schräpler*, 2005) which seems rather high for millimeter particles.

Electrostatic attraction by dipole-dipole forces has been seen to be important for grains of several hundred micrometer radius (chondrule-sized) forming clumps that are centimeters to tens of centimeters across (*Marshall et al.*, 2005; *Ivlev et al.*, 2002). Spot charges distributed over the grain surfaces lead to a net dipole of the grains, with growth dynamics very similar to that of magnetic grains. Experiments in microgravity have shown spontaneous aggregation of particles in the several hundred micrometer size regime (*Marshall and Cuzzi*, 2001; *Marshall et al.*, 2005; *Love and Pettit*, 2004). The aggregates show greatly enhanced stability, consistent with cohesive forces increased by factors of 10^3 compared to the normal van der Waals interaction. Based on the experiments, for weakly charged dust grains, the electrostatic interaction energy at contact for the charge-dipole interaction is in most cases larger than that for the charge-charge interaction. For heavily charged particles, the mean mass of the system does not grow faster than linearly with time, i.e., even slower than in the noncharged case for Brownian motion (*Ivlev et al.*, 2002; *Konopka et al.*, 2005).

2.3. Growth and Compaction of Large Dust Aggregates

2.3.1. Physical properties of macroscopic dust aggregates. Macroscopic dust aggregates can be created in the laboratory by a process termed random ballistic deposition (RBD) (*Blum and Schräpler*, 2004). In its idealized form, RBD uses individual, spherical and monodisperse grains that are deposited randomly but unidirectionally on a semi-infinite target aggregate. The volume filling factor $\phi = 0.11$ of these aggregates, defined as the fraction of the volume filled by dust grains, is identical to ballistic particle-cluster aggregation, which occurs when a bimodal size distribution of particles (aggregates of one size and individual dust grains) is present and when the aggregation rates between the large aggregates and the small particles exceed those between all other combinations of particle sizes. When using idealized experimental parameters, i.e., monodisperse spherical SiO_2 grains with 0.75 µm radius, *Blum and Schräpler* (2004) measured a mean volume filling factor for their macroscopic (centimeter-sized) RBD dust aggregates of $\phi = 0.15$, in full agreement with numerical predictions (*Watson et al.*, 1997). Relaxing the idealized grain morphology resulted in a decrease of the volume filling factor to values of $\phi = 0.10$ for quasi-monodisperse, irregular diamond grains and $\phi = 0.07$ for polydisperse, irregular SiO_2 grains (*Blum*, 2004).

Static uniaxial compression experiments with the macroscopic RBD dust aggregates consisting of monodisperse spherical grains (*Blum and Schräpler*, 2004) showed that the volume filling factor remains constant as long as the stress on the sample is below ~500 N m^{-2}. For higher stresses, the volume filling factor monotonically increases from $\phi = 0.15$ to $\phi = 0.34$. Above ~10^5 N m^{-2}, the volume filling factor remains constant at $\phi = 0.33$. Thus, the compressive strength of the uncompressed sample is $\Sigma \approx 500$ N m^{-2}. These values differ from those derived with the models of *Greenberg et al.* (1995) and *Sirono and Greenberg* (2000) by a factor of a few. The compressive strengths of the macroscopic dust aggregates consisting of irregular and polydisperse grains was slightly lower at $\Sigma \sim 200$ N m^{-2}. The maximum compression of these bodies was reached for stresses above ~5×10^5 N m^{-2} and resulted in volume filling factors as low as $\phi = 0.20$ (*Blum,* 2004). As a maximum compressive stress of ~$10^5 \ldots 10^6$ N m^{-2} corresponds to impact velocities of ~$15 \ldots 50$ m/s, which are typical for meter-sized protoplanetary dust aggregates, we expect a maximum volume filling factor for these bodies in the solar nebula of $\phi = 0.20 \ldots 0.34$. *Blum and Schräpler* (2004) also measured the tensile strength of their aggregates and found for slightly compressed samples ($\phi = 0.23$) T = 1000 N m^{-2}. Depending on the grain shape and the size distribution, the tensile strength decreased to values of T ~ 200 N m^{-2} for the uncompressed case (*Blum,* 2004).

Sirono (2004) used the above continuum properties of macroscopic dust aggregates, i.e., compressive strength and tensile strength, to model the collisions between protoplanetary dust aggregates. For sticking to occur in an aggregate-aggregate collision, *Sirono* (2004) found that the impact velocity must follow the relation

$$v < 0.04 \sqrt{\frac{d\Sigma(\phi)}{d\rho(\phi)}} \qquad (9)$$

where $\rho(\phi) = \rho_0 \times \phi$ is the mass density of the aggregate and ρ_0 denotes the mass density of the grain material. Moreover, the conditions $\Sigma(\phi) < Y(\phi)$ and $\Sigma(\phi) < T(\phi)$ must be fulfilled. For the shear strength, *Sirono* (2004) applies $Y(\phi) = \sqrt{2\Sigma(\phi)T(\phi)/3}$. A low compressive strength of the colliding aggregates favors compaction and thus damage restoration, which can otherwise lead to a breakup of the aggregates. In addition, a large tensile strength also prevents the aggregates from being disrupted in the collision.

Blum and Schräpler (2004) found an approximate relation between compressive strength and volume filling factor

$$\Sigma(\phi) = \Sigma_s(\phi - \phi_0)^{0.8} \qquad (10)$$

which is valid in the range $\phi_0 = 0.15 \leq \phi \leq 0.21$. Such a scaling law was also found for other types of macroscopic aggregates, e.g., for jammed toner particles in fluidized bed experiments (*Valverde et al.,* 2004). For the aggregates consisting of monodisperse SiO$_2$ spheres, the scaling factor Σ_s can be determined to be $\Sigma_s = 2.9 \times 10^4$ N m^{-2}. If we apply equation (10) to equation (9) we get, with $\rho(\phi) = \rho_0 \times \phi$ and $\rho_0 = 2 \times 10^3$ kg m^{-3}, for the impact velocity of low-density dust aggregates

$$v < 0.04 \sqrt{\frac{0.8\Sigma_s}{\rho_0(\phi - \phi_0)^{0.2}}} \approx 0.14(\phi - \phi_0)^{-0.1} \text{ m/s} \qquad (11)$$

Although the function in equation (11) goes to infinity for $\phi \to \phi_0$, for all practical purposes the characteristic velocity is strongly restricted. For volume filling factors $\phi \geq 0.16$ we get $v < 0.22$ m/s. Thus, following the SPH simulations by *Sirono* (2004), we expect aggregate sticking in collisions for impact velocities $v \lesssim 0.2$ m/s.

2.3.2. Low-velocity collisions between macroscopic dust aggregates. Let us now consider recent results in the field of high-porosity aggregate collisions. Langkowski and Blum (unpublished data) performed microgravity collision experiments between 0.1–1-mm-sized (projectile) RBD aggregates and 2.5-cm-sized (target) RBD aggregates. Both aggregates consisted of monodisperse spherical SiO$_2$ grains with radii of a = 0.75 μm. In addition to that, impact experiments with high-porosity aggregates consisting of irregular and/or polydisperse grains were performed. The parameter space of the impact experiments by Langkowski and Blum encompassed collision velocities in the range $0 < v < 3$ m/s and projectile masses of 10^{-9} kg $\leq$ m $\leq 5 \times 10^{-6}$ kg for all possible impact parameters (i.e., from normal to tangential impact). Surprisingly, through most of the parameter space, the collisions did lead to sticking. The experiments with aggregates consisting of monodisperse spherical SiO$_2$ grains show, however, a steep decrease in sticking probability from $\beta = 1$ to $\beta = 0$ if the tangential component of the impact energy exceeds ~10^{-6} J (see the example of a nonsticking impact in Fig. 3a). Other materials also show the tendency toward lower sticking probabilities with increasing tangential impact energies. As these aggregates are "softer," the decline in sticking probability in the investigated parameter space is not complete. When the projectile aggregates did not stick to the target aggregate, considerable mass transfer from the target to the projectile aggregate takes place during the impact (Langkowski and Blum, unpublished data). Typically, the mass of the projectile aggregate was doubled after a nonsticking collision (see Fig. 3a).

The occurrence of sticking in aggregate-aggregate collisions at velocities ≥ 1 m/s is clearly in disagreement with the prediction by *Sirono* (2004) (see equation (9)). In addition, the evaluation of the experimental data shows that the condition for sticking, $\Sigma(\phi) < Y(\phi)$, seems not to be fulfilled for high-porosity dust aggregates. This means that the continuum aggregate model by *Sirono* (2004) is still not precise enough to fully describe the collision and sticking behavior of macroscopic dust aggregates.

2.3.3. High-velocity collisions between macroscopic dust aggregates. The experiments described above indicate that at velocities above approximately 1 m/s, collisions turns from sticking to bouncing, at least for oblique impacts. At higher velocities one would naively expect that bouncing

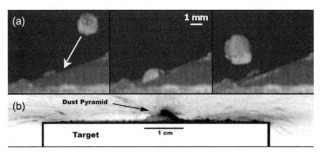

Fig. 3. (a) Nonsticking (v = 1.8 m/s) oblique impact between high-porosity dust aggregates (Langkowski and Blum, unpublished data). The three images show (left to right) the approaching projectile aggregate before, during, and after impact. The arrow in the left image denotes the impact direction of the projectile aggregate. The experiment was performed under microgravity conditions. It is clearly visible that the rebounding aggregate (right image) is more massive than before the collision. (b) Result of a high-velocity normal impact (v = 23.7 m/s) between compacted aggregates (*Wurm et al., 2005*). About half the projectile mass sticks to the target after the impact and is visible by its pyramidal structure on the flat target. Note the different size scales in (a) and (b).

and eventually erosion will continue to dominate, and this is also observed in a number of different experiments (*Colwell, 2003; Bridges et al., 1996, Kouchi et al., 2002; Blum and Münch, 1993; Blum and Wurm, 2000*).

Growth models that assume sticking at velocities ≫1 m/s are therefore often considered to be impossible (e.g., *Youdin, 2004*). As velocities ≥10 m/s clearly occur for particles that have exceeded meter size, this is a fundamental problem for the formation of planetesimals.

However, recent experiments (*Wurm et al., 2005*) have studied impacts of millimeter-sized compact dust aggregates onto centimeter-sized compact aggregate targets at impact velocities between 6 and 25 m/s. Compact aggregates can be the result of previous sticking or nonsticking collisions (see sections 2.3.1 and 2.3.2). Both projectile and target consisted of micrometer-sized dust particles. In agreement with the usual findings at lower impact velocities around a few meters per second, the projectiles just rebound, slightly fracture or even remove some parts of the target. However, as the velocity increases *above* a threshold of 13 m/s, about half the mass of the projectile rigidly sticks to the target after the collision while essentially no mass is removed from the target (see Fig. 3b). Obviously, higher collision velocities can be favorable for growth, probably by destroying the internal structure of the porous material and dissipating energy in this way.

Only about half of the impactor contributes to the growth of the target in the experiments. The other half is ejected in the form of small fragments, with the important implication that these collisions both lead to net growth of the target and return small particles to the disk. This keeps dust abundant in the disk over a long time. For the specific experiments by *Wurm et al.* (2005), the fragments were evenly distributed in size up to 0.5 mm. In a certain sense, the disk might thus quickly turn into a "debris disk" already at early times. We will get back to this point in section 4.4.

3. PARTICLE-GAS INTERACTION

Above we have seen that small solid particles grow rapidly into aggregates of quite substantial sizes, while retaining their fractal nature (in the early growth stage) or a moderate to high porosity (for later growth stages). From the properties of primitive meteorites, we have a somewhat different picture of nebula particulates — most of the solids (chondrules, CAIs, metal grains, etc.) were individually compacted as the result of unknown melting processes, and were highly size-sorted. Even the porosity of what seem to be fine-grained accretion rims on chondrules is 25% or less (*Scott et al., 1996; Cuzzi, 2004; Wasson et al., 2005*). Because age-dating of chondrules and chondrites implies a delay of a million years or more after formation of the first solids, it seems possible that, in the asteroid formation region at least, widespread accretion to parent-body sizes did not occur until after the mystery melting events that formed the chondrules began.

It may be that conditions differed between the inner and outer solar system. Chondrule formation might not have occurred at all in the outer solar system where comet nuclei formed, so some evidence of the fractal aggregate growth stage may remain in the granular structure of comet nuclei. New results from *Deep Impact* imply that Comet Tempel 1 has a porosity of 60–80% (*A'Hearn et al., 2005*)! This value is in agreement with similar porosities found in several other comets (*Davidsson, 2006*). Even in the terrestrial planet/asteroid belt region, there is little reason to doubt that growth of aggregates started well before the chondrule formation era, and continued into and (probably) throughout it. Perhaps, after chondrules formed, previously ineffective growth processes might have dominated (sections 3.2 and 3.3).

3.1. Radial and Vertical Evolution of Solids

3.1.1. Evolution prior to formation of a dense midplane layer. The nebula gas (but not the particles) experiences radial pressure gradients because of changing gas density and temperature. These pressure gradients act as small modifications to the central gravity from the star that dominates orbital motion, so that the gas and particles orbit at different speeds and a gas drag force exists between them that constantly changes their orbital energy and angular momentum. Because the overall nebula pressure gradient force is outward, it counteracts a small amount of the inward gravitational force and the gas generally orbits more slowly than the particles, so the particles experience a headwind that saps their orbital energy, and the dominant particle drift is inward. Early work on gas-drag-related drift was by *Whipple* (1972), *Adachi et al.* (1976), and *Weidenschilling* (1977).

Analytical solutions for how particles interact with a nonturbulent nebula having a typically outward pressure gradient were developed by *Nakagawa et al.* (1986). For instance, the ratio of the pressure gradient force to the dominant central gravity is $\eta \sim 2 \times 10^{-3}$, leading to a net velocity difference between the gas and particles orbiting at Keplerian velocity V_K of ηV_K (see, e.g., *Nakagawa et al.*, 1986). However, if local radial maxima in gas pressure exist, particles will drift toward their centers from both sides, possibly leading to radial bands of enhancement of solids (see section 3.4.1).

Small particles generally drift slowly inward, at perhaps a few centimeters per second; even this slow inexorable drift has generated some concern over the years as to how CAIs (early, high-temperature condensates) can survive over the apparent 1–3-m.y. period between their creation and the time they were incorporated into chondrite meteorite parent bodies. This concern, however, neglected the role of turbulent diffusion (see section 3.1.2). Particles of meter size drift inward very rapidly: 1 AU per century. It has often been assumed that these particles were "lost into the Sun," but more realistically, their inward drift first brings them into regions warm enough to evaporate their primary constituents, which then become entrained in the more slowly evolving gas and increase in relative abundance as inward migration of solids supplies material faster than it can be removed. Early models describing significant global redistribution of solids relative to the nebula gas by radial drift were presented by *Morfill and Völk* (1984) and *Stepinski and Valageas* (1996, 1997); these models either ignored midplane settling or made simplifying approximations regarding it, and did not emphasize the potential for enhancing material in the vapor phase. Indeed, however, because of the large mass fluxes involved, this "evaporation front" effect can alter the nebula composition and chemistry significantly (*Cuzzi et al.*, 2003; *Cuzzi and Zahnle*, 2004; *Yurimoto and Kuramoto*, 2004; *Krot et al.*, 2005; *Ciesla and Cuzzi*, 2006; see also *Cyr et al.*, 1999, for a discussion); however, the results of this paper are inconsistent with similar work by *Supulver and Lin* (2000) and *Ciesla and Cuzzi* (2006). This stage can occur very early in nebula history, long before formation of objects large enough to be meteorite parent bodies.

3.1.2. The role of turbulence. The presence or absence of gas turbulence plays a critical role in the evolution of nebula solids. There is currently no widespread agreement on just how the nebula gas may be maintained in a turbulent state across all regions of interest, if indeed it is (*Stone et al.*, 2000; *Cuzzi and Weidenschilling*, 2006). Therefore we discuss both turbulent and nonturbulent situations. For simplicity we will treat turbulent diffusivity $\mathcal{D}$ as equal to turbulent viscosity $\nu_T = \alpha c H$, where c and H are the nebula sound speed and vertical scale height, and $\alpha \ll 1$ is a nondimensional scaling parameter. Evolutionary timescales of observed protoplanetary nebulae suggest that $10^{-5} < \alpha < 10^{-2}$ in some global sense. The largest eddies in turbulence have scale sizes $H\sqrt{\alpha}$ and velocities $\nu_{turb} = c\sqrt{\alpha}$ (*Shakura et al.*, 1978; *Cuzzi et al.*, 2001).

Particles respond to forcing by eddies of different frequency and velocity as described by *Völk et al.* (1980) and *Markiewicz et al.* (1991), determining their relative velocities with respect to the gas and to each other. The diffusive properties of MRI turbulence, at least, seem not to differ in any significant way from the standard homogeneous, isotropic models in this regard (*Johansen and Klahr*, 2005). Analytical solutions for resulting particle velocities in these regimes were derived by *Cuzzi and Hogan* (2003). These are discussed in more detail below and by *Cuzzi and Weidenschilling* (2006).

Vertical turbulent diffusion at intensity α maintains particles of stopping time t_s in a layer of thickness $h \sim H\sqrt{\alpha/\Omega t_s}$ (*Dubrulle et al.*, 1995; *Cuzzi et al.*, 1996), or a solid density enhancement $H/h = \sqrt{\Omega t_s/\alpha}$ above the average value. For particles of 10-cm size and smaller and $\alpha > \alpha_{min} = 10^{-6}(a/1\ cm)$ (*Cuzzi and Weidenschilling*, 2006), the resulting layer is much too large and dilute for collective particle effects to dominate gas motions, so radial drift and diffusion continue unabated. Outward radial diffusion relieves the longstanding worry about "loss into the Sun" of small particles, such as CAIs, which are too small to sediment into any sort of midplane layer unless turbulence is vanishingly small ($\alpha \ll \alpha_{min}$), and allows some fraction of them to survive over one to several million years after their formation as indicated by meteoritic observations (*Cuzzi et al.*, 2003). A similar effect might help explain the presence of crystalline silicates in comets (*Bockelée-Morvan*, 2002; *Gail*, 2004).

3.1.3. Dense midplane layers. When particles *are* able to settle to the midplane, the particle density gets large enough to dominate the motions of the local gas. This is the regime of *collective effects*; that is, the behavior of a particle depends indirectly on how all other local particles combined affect the gas in which they move. In regions where collective effects are important, the mass-dominant particles can drive the entrained gas to orbit at nearly Keplerian velocities (if they are sufficiently well coupled to the gas), and thus the headwind the gas can exert upon the particles diminishes from ηV_K (section 3.1.1). This causes the headwind-driven radial drift and all other differential particle velocities caused by gas drag to diminish as well.

The particle mass loading ρ_p/ρ_g cannot increase without limit as particles settle, even if the global turbulence vanishes, and the density of settled particle layers is somewhat self-limiting. The relative velocity solutions of *Nakagawa et al.* (1986) apply in particle-laden regimes once ρ_p/ρ_g is known, but do not provide for a fully self consistent determination of ρ_p/ρ_g in the above sense; this was addressed by *Weidenschilling* (1980) and subsequently *Cuzzi et al.* (1993), *Champney et al.* (1995), and *Dobrovolskis et al.* (1999). The latter numerical models are similar in spirit to the simple analytical solutions of *Dubrulle et al.* (1995) mentioned earlier, but treat large-particle, high-mass-loading regimes in globally nonturbulent nebulae that the analytic solutions cannot address. Basically, as the midplane particle density increases, local, entrained gas is accelerated to near-Keplerian velocities by drag forces from the particles. Well above the dense midplane, the gas still orbits at its pressure-supported, sub-Keplerian rate. Thus there is a vertical shear gradient in the orbital velocity of the gas, and the velocity shear creates turbulence that stirs the particles. This is sometimes called "self-generated turbulence." Ultimately a steady-state condition arises where the particle layer thickness reaches an equilibrium between downward settling and upward diffusion. This effect acts to block a number of gravitational instability mechanisms in the midplane (section 3.3.2).

3.2. Relative Velocities and Growth in Turbulent and Nonturbulent Nebulae

In both turbulent and nonturbulent regimes, particle relative velocities drive growth to larger sizes. Below we show that relative velocities in both turbulent and nonturbulent regimes are probably small enough for accretion and growth to be commonplace and rapid, at least until particles reach meter size or so. We only present results here for particles up to a meter or so in size, because the expression for gas drag takes on a different form at larger sizes. As particles grow, their mass per unit area increases so they are less easily influenced by the gas, and "decouple" from it. Their overall drift velocities and relative velocities all diminish roughly in a linear fashion with particle radius larger than a meter or so (*Cuzzi and Weidenschilling*, 2006).

3.2.1. Relative velocities. We use particle velocities relative to the gas as derived by *Nakagawa et al.* (1986) for a range of local particle mass density relative to the gas density (their equations (2.11), (2.12), and (2.21)) to derive particle velocities relative to each other in the same environment; all relative velocities scale with ηV_K.

For simplicity we will assume particles that differ by a factor of 3 in radius; *Weidenschilling* (1997) finds mass accretion to be dominated by size spreads on this order; the results are insensitive to this factor. Relative velocities for particles of radii a and a/3, in the absence of turbulence and due only to differential, pressure-gradient-driven gas drag, are plotted in the top two panels of Fig. 4. In the top panel we show cases where collective effects are negligible (particle density $\rho_p \ll \rho_{gas}$ density ρ_g). Differential vertical settling (shown at different heights z above the midplane, as normalized by the gas vertical scale height H) dominates relative velocities and particle growth high above the midplane (z/H > 0.1), and radial relative velocities dominate at lower elevations. Except for the largest particles, relative velocities for particles with this size difference are much less than ηV_K; particles closer in mass would have even smaller relative velocities.

Moreover, in a dense midplane layer, when collective effects dominate (section 3.1.3), all these relative velocities are reduced considerably from the values shown. In the second panel we show radial and azimuthal relative velocities for several values of ρ_p/ρ_g. When the particle density

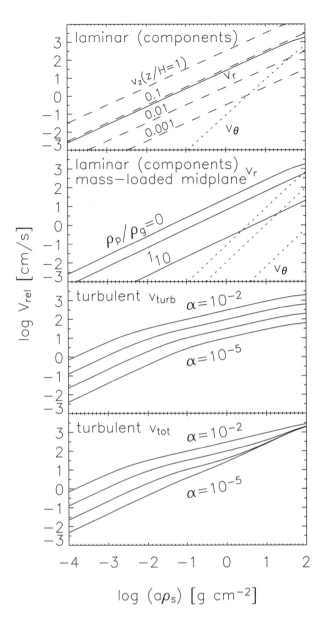

Fig. 4. Relative velocities between particles of radii a and a/3, in nebulae that are nonturbulent (top two panels) or turbulent (bottom two panels), for a minimum mass solar nebula at 2.5 AU. In the top panel, the particle density ρ_p is assumed to be much smaller than the gas density ρ_g: $\rho_p/\rho_g \ll 1$. Shown are the radial (solid line) and azimuthal (dotted line) components of the relative velocities. The dashed curves show the vertical relative velocities, which depend on height above the midplane and are shown for different values of z/H. For nonturbulent cases, particles settle into dense midplane layers (section 3.1.3), so a more realistic situation would be $\rho_p/\rho_g \geq 1$ or even $\gg 1$ (*Cuzzi et al.*, 1993); thus in the second panel we show relative radial and angular velocities for three different values of $\rho_p/\rho_g = 0$, 1, and 10. For these high mass loadings, z/H must be small, so the vertical velocities are smaller than the radial velocities. In the third panel we show relative velocities for the same particle size difference due only to turbulence, for several values of α. In the bottom panel, we show the quadrature sum of turbulent and nonturbulent velocities, assuming z/H = 0.01.

exceeds the gas density, collective effects reduce the headwind, and all relative velocities diminish.

Relative velocities in turbulence of several different intensities, as constrained by the nebula α (again for particles of radii a and a/3), are shown in the two bottom panels. In the second panel from the bottom, relative velocities are calculated as the difference of their velocities relative to the turbulent gas, neglecting systematic drifts and using analytical solutions derived by *Cuzzi and Hogan* (2003, their equation (20)) to the formalism of *Völk et al.* (1980). Here, the relative velocities are forced by turbulent eddies with a range of size scales, having eddy turnover times ranging from the orbit period (for the large eddies) to much smaller values (for the smaller eddies), and scale with $v_{turb} = c\alpha^{1/2}$.

In the bottom panel we sum the various relative velocities in quadrature to get an idea of total relative velocities in a turbulent nebula in which particles are also evolving by systematic gas-pressure-gradient-driven drift. This primarily increases the relative velocities of the larger particles in the lower α cases.

Overall, keeping in mind the critical velocities for sticking discussed in section 2 (~m/s), and that particle surfaces are surely crushy and dissipative, one sees that for particles up to a meter or so, growth by sticking is plausible even in turbulent nebulae for a wide range of α. Crushy aggregates will grow by accumulating smaller crushy aggregates as described in earlier sections [e.g., *Weidenschilling* (1997) for the laminar case]. After this burst of initial growth to roughly meter size, however, the evolution of solids is very sensitive to the presence or absence of global nebula turbulence, as described in sections 3.3 and 3.4 below. Meter-sized particles inevitably couple to the largest eddies, with $v_{turb} \gtrsim$ several meters per second, and would destroy each other if they were to collide. We refer to this as the fragmentation limit. However, if particles can somehow grow their way past 10 m in size, their survival becomes more assured because all relative velocities, such as shown in Fig. 4, decrease linearly with $a\rho_s$ for values larger than shown in the plot due to the linear decrease of the area/mass ratio.

3.2.2. Gas flow and reaccretion onto large bodies. The role of gas in protoplanetary disks is not restricted to generate relative velocities between two bodies that then collide. The gas also plays an important role *during* individual collisions. A large body that moves through the disk faces a headwind and collisions with smaller aggregates take place at its front (headwind) side. Fragments are thus ejected against the wind and can be driven back to the surface by the gas flow.

For small bodies the gas flow can be regarded as free molecular flow. Thus streamlines end on the target surface and the gas drag is always toward the surface. Whether a fragment returns to the surface depends on its gas-particle coupling time (i.e., size and density) and on the ejection speed and angle. Whether reaccretion of enough fragments for net growth occurs eventually depends on the distribution of ejecta parameters, gas density, and target size. It was shown by *Wurm et al.* (2001) that growth of a larger body due to impact of dust aggregates entrained in a headwind

is possible for collision velocities above 12 m/s. At 1 AU a 30-cm body in a disk model according to *Weidenschilling and Cuzzi* (1993) can grow in a collision with small dust aggregates even if the initial collision is rather destructive.

Sekiya and Takeda (2003) and *Künzli and Benz* (2003) showed that the mechanism of aerodynamic reaccretion might be restricted to a maximum size due to a change in the flow regime from molecular to hydrodynamic. Fragments are then transported around the target rather then back to it. *Wurm et al.* (2004b) argue that very porous targets would allow some flow going through the body, which would still allow aerodynamic reaccretion, but this strongly depends on the morphology of the body (*Sekiya and Takeda*, 2005). As the gas density decreases outward in protoplanetary disks, the maximum size for aerodynamic reaccretion increases. However, the minimum size also increases and the mechanism is only important for objects that have already grown beyond the fragmentation limit in some other way — e.g., by immediate sticking of parts of larger particles as discussed above (*Wurm et al., 2005*).

3.3. Planetesimal Formation in a Midplane Layer

3.3.1. Incremental growth. Based on relative velocity arguments such as given above, *Weidenschilling* (1988, 1997) and *Dullemond and Dominik* (2004, 2005) find that growth to meter size is rapid (100–1000 yr at 1 AU; 6–7 × 10^4 yr at 30 AU) whether the nebula is turbulent or not. Such large particles settle toward the midplane within an orbit period or so. However, in turbulence, even meter-sized particles are dispersed sufficiently that the midplane density remains low, and growth remains slow. A combination of rapid radial drift, generally erosive, high-velocity impacts with smaller particles, and occasional destructive collisions with other meter-sized particles, frustrates growth beyond meter size or so under these conditions.

In *nonturbulent* nebulae, even smaller particles can settle into fairly thin midplane layers and the total particle densities can easily become large enough for collective effects to drive the entrained midplane gas to Keplerian, diminishing both headwind-induced radial drift and relative velocities. In this situation, meter-sized particles quickly grow their way out of their troublesome tendency to drift radially (*Cuzzi et al.*, 1993); planetesimal-sized objects form in only 10^3–10^4 yr at 1 AU (*Weidenschilling*, 2000) and a few times 10^5 yr at 30 AU (*Weidenschilling*, 1997). However, such robust growth may, in fact, be too rapid to match observations of several kinds (see section 4.4 and chapters by Dullemond et al. and Natta et al.).

3.3.2. Particle-layer instabilities. While to some workers the simplicity of "incremental growth" by sticking in the dense midplane layer of a nonturbulent nebula is appealing, past uncertainty in sticking properties has led others to pursue instability mechanisms for particle growth that are insensitive to these uncertainties. Nearly all instability mechanisms discussed to date (*Safronov*, 1969, 1991; *Goldreich and Ward*, 1973; *Ward*, 1976, 2000; *Sekiya*, 1983, 1998; *Goodman and Pindor*, 2000; *Youdin and Shu*, 2002) occur *only* in nebulae where turbulence is essentially absent, and particle relative velocities are already very low. Just how low the global turbulence must be depends on the particle size involved, and the nebula α (sections 3.1.1 and 3.1.3).

Classical treatments [the best known is *Goldreich and Ward* (1973)] assume that gas pressure plays no role in gravitational instability, being replaced by an effective pressure due to particle random velocities (below we note this is not the case). Particle random velocities act to puff up a layer and reduce its density below the critical value, which is always on the order of the so-called Roche density $\rho^* \sim M_\odot/R^3$ where R is the distance to the central star; different workers give constraints that differ by factors on the order of unity (cf. *Goldreich and Ward*, 1973; *Weidenschilling*, 1980; *Safronov*, 1991; *Cuzzi et al.*, 1993). These criteria can be traced back through *Goldreich and Ward* (1973) to *Goldreich and Lynden-Bell* (1965), *Toomre* (1964), *Chandrasekhar* (1961), and *Jeans* (1928), and in parallel through *Safronov* (1960), *Bel and Schatzman* (1958), and *Gurevitch and Lebedinsky* (1950). Substituting typical values one derives a formal, nominal requirement that the local particle mass density must exceed about 10^{-7} g cm^{-3} at 2 AU from a solar-mass star even for *marginal* gravitational instability — temporary gravitational clumping of small amplitude — to occur. This is about $10^3\times$ larger than the gas density of typical minimum mass nebulae, requiring enhancement of the solids by a factor of about 10^5 for a typical average solids-to-gas ratio. From section 3.1.2 we thus require the particle layer to have a thickness h < 10^{-5} H, which in turn places constraints on the particle random velocities hΩ and on the global value of α.

Even assuming global turbulence to vanish, *Weidenschilling* (1980, 1984) noted that turbulence stirred by the very dense particle layer itself will puff it up to thicknesses h that precluded even this marginal gravitational instability. This is because turbulent eddies induced by the vertical velocity profile of the gas (section 3.1.3) excite random velocities in the particles, diffusing the layer and preventing it from settling into a sufficiently dense state. Detailed two-phase fluid models by *Cuzzi et al.* (1993), *Champney et al.* (1995), and *Dobrovolskis et al.* (1999) confirmed this behavior.

It is sometimes assumed that ongoing, but slow, particle growth to larger particles, with lower relative velocities and thus thinner layers (section 3.2), can lead to $\rho_p \sim \rho^*$ and gravitational instability can then occur. However, merely achieving the formal requirement for marginal gravitational instability does not inevitably lead to planetesimals. For particles that are large enough to settle into suitably dense layers for *marginal* instability under self-generated turbulence (*Weidenschilling*, 1980; *Cuzzi et al.*, 1993), random velocities are not damped on a collapse timescale, so incipient instabilities merely "bounce" and tidally diverge. This is like the behavior seen in Saturn's A ring, much of which is gravitationally unstable by these same criteria (*Salo*, 1992; *Karjalainen and Salo*, 2004). Direct collapse to planetesimals is much harder to achieve, requiring much

lower relative velocities, and is unlikely to have occurred this way (*Cuzzi et al.*, 1994; *Weidenschilling*, 1995; *Cuzzi and Weidenschilling*, 2006). Recent results by *Tanga et al.* (2004) assume an artificial damping by gas drag and find gravitationally bound clumps form that, while not collapsing directly to planetesimals, retain their identity for extended periods, perhaps allowing for slow shrinkage; this is worth further numerical modeling with more realistic damping physics, but still presumes a globally laminar nebula.

For very small particles (a < 1 mm; the highly relevant chondrule size), a different type of instability comes into play because the particles are firmly trapped to the gas by their short stopping times, and the combined system forms a single "one-phase" fluid that is stabilized against producing turbulence by its vertical density gradient (*Sekiya*, 1998; *Youdin and Shu*, 2002; *Youdin and Chiang*, 2004; *Garaud and Lin*, 2004). Even for midplane layers of such small particles to *approach* a suitable density for this to occur requires nebula turbulence to drop to what may be implausibly low values ($\alpha < 10^{-8}$ to 10^{-10}). Moreover, such one-phase layers, with particle stopping times t_s much less than the dynamical collapse time $(G\rho_p)^{-1/2}$, cannot become "unstable" and collapse on the dynamical timescale as normally envisioned, because of *gas* pressure support, which is usually ignored (*Sekiya*, 1983; *Safronov*, 1991). *Sekiya* (1983) finds that particle densities must exceed $10^4 \rho^*$ for such particles to undergo instability and actually collapse. While especially difficult on one-phase instabilities by definition, this obstacle should be considered for any particle with stopping time much shorter than the dynamical collapse time — that is, pretty much anything smaller than a meter for $\rho_p \sim \rho^*$.

A slower "sedimentation" from axisymmetric rings (or even localized blobs of high density, which might form through fragmentation of such dense, differentially rotating rings) has also been proposed to occur under conditions normally ascribed to marginal gravitational instability (*Sekiya*, 1983; *Safronov*, 1991; *Ward*, 2000), but this effect has only been modeled under nonturbulent conditions where, as mentioned above, growth can be quite fast by sticking alone. In a turbulent nebula, diffusion (or other complications discussed below, such as large vortices, spiral density waves, etc.) might preclude formation of all but the broadest-scale "rings" of this sort, which have radial scales comparable to H and grow only on extremely long timescales.

3.4. Planetesimal Formation in Turbulence

A case can be made that astronomical, asteroidal, and meteoritic observations require planetesimal growth to stall at sizes much smaller than several kilometers, for something like a million years (*Dullemond and Dominik*, 2005; *Cuzzi and Weidenschilling*, 2006; *Cuzzi et al.*, 2005). This is perhaps most easily explained by the presence of ubiquitous weak turbulence ($\alpha > 10^{-4}$). Once having grown to meter size, particles couple to the largest, most energetic turbulent eddies, leading to mutual collisions at relative velocities on the order of $v_{turb} \sim \sqrt{\alpha}c \sim 30$ m/s, which are probably disruptive, stalling incremental growth by sticking at around a meter in size. Astrophysical observations supporting this inference are discussed in the next section. In principle, planetesimal formation could merely await cessation of nebula turbulence and then happen all at once; pros and cons of this simple concept are discussed by *Cuzzi and Weidenschilling* (2006). The main difficulty with this concept is the very robust nature of growth in dense midplane layers of nonturbulent nebulae, compared to the very extended duration of 10^6 yr, which apparently characterized meteorite parent-body formation (see chapter by Wadhwa et al.). Furthermore, if turbulence merely ceased at the appropriate time for parent body formation to begin, particles of all sizes would settle and accrete together, leaving unexplained the very-well-characterized chondrite size distributions we observe. Alternately, several suggestions have been advanced as to how the meter-sized barrier might be overcome even in ongoing turbulence, as described below.

3.4.1. Concentration of boulders in large nebula gas structures. The speedy inward radial drift of meter-sized particles in nebulae where settling is precluded by turbulence might be slowed if they can be, even temporarily, trapped by one of several possible fluid dynamical effects. It has been proposed that such trapping concentrates them and leads to planetesimal growth as well.

Large nebula gas dynamical structures such as systematically rotating vortices (not true turbulent eddies) have the property of concentrating large boulders near their centers (*Barge and Sommeria*, 1995; *Tanga et al.*, 1996; *Bracco et al.*, 1998; *Godon and Livio*, 2000; *Klahr and Bodenheimer*, 2006). In some of these models the vortices are simply prescribed and/or there is no feedback from the particles. Moreover, there are strong vertical velocities present in realistic vortices, and the vortical flows that concentrate meter-sized particles are not found near the midplane, where the meter-sized particles reside (*Barranco and Marcus*, 2005). Finally, there may be a tendency of particle concentrations formed in modeled vortices to drift out of them and/or destroy the vortex (*Johansen et al.*, 2004).

Another possibility of interest is the buildup of solids near the peaks of nearly axisymmetric, localized radial pressure maxima, which might for instance be associated with spiral density waves (*Haghighipour and Boss*, 2003a,b; *Rice et al.*, 2004). *Johansen et al.* (2006) noted boulder concentration in radial high-pressure zones of their full simulation, but (in contrast to above suggestions about vortices), saw no concentration of meter-sized particles in the closest thing they could resolve in the nature of actual turbulent eddies. Perhaps this merely highlights the key difference between systematically rotating (and often artificially imposed) vortical fluid structures, and realistic eddies in realistic turbulence.

Overall, models of boulder concentration in large-scale fluid structures will need to assess the tendency for rapidly colliding meter-sized particles in such regions to destroy each other, in the real turbulence that will surely accompany such structures. For instance, breaking spiral density

waves are themselves potent drivers for strong turbulence (*Boley et al.*, 2005).

3.4.2. Concentration of chondrules in three-dimensional turbulence. Another suggestion for particle growth beyond a meter in turbulent nebulae is motivated by observed size sorting in chondrites. *Cuzzi et al.* (1996, 2001) have advanced the model of turbulent concentration of chondrule-sized (millimeter or smaller diameter) particles into dense zones, which ultimately become the planetesimals we observe. This effect, which occurs in genuine, three-dimensional turbulence (both in numerical models and laboratory experiments), naturally satisfies meteoritic observations in several ways under quite plausible nebula conditions. It offers the potential to leapfrog the problematic meter-size range entirely and would be applicable (to differing particle types) throughout the solar system (for reviews, see *Cuzzi and Weidenschilling*, 2006; *Cuzzi et al.*, 2005). This scenario faces the obstacle that the dense, particle-rich zones that certainly *do* form are far from solid density, and might be disrupted by gas pressure or turbulence before they can form solid planetesimals. As with dense midplane layers of small particles, gas pressure is a formidable barrier to gravitational instability on a dynamical timescale in dense zones of chondrule-sized particles formed by turbulent concentration. However, as with other small-particle scenarios, sedimentation is a possibility on longer timescales than that of dynamical collapse. It is promising that *Sekiya* (1983) found that zones of these densities, while "incompressible" on the dynamical timescale, form stable modes. Current studies are assessing whether the dense zones can survive perturbations long enough to evolve into planetesimals.

3.5. Summary of the Situation Regarding Planetesimal Formation

As of the writing of this chapter, the path to planetesimal formation remains unclear. In nonturbulent nebulae, a variety of options seem to exist for growth that, while not on dynamical collapse timescales, is rapid on cosmogonic timescales ($\ll 10^5$ yr). However, this set of conditions and growth timescales seems to be at odds with asteroidal, meteoritic, and astronomical observations of several kinds (*Russell et al.*, 2006; *Dullemond and Dominik*, 2005; *Cuzzi and Weidenschilling*, 2006; *Cuzzi et al.*, 2005; chapter by Wadhwa et al.). The alternate set of scenarios — growth beyond a meter or so in size in turbulent nebulae — are perhaps more consistent with the observations but are still incompletely developed beyond some promising directions. The challenge is to describe quantitatively the rate at which planetesimals form under these inefficient conditions.

4. GLOBAL DISK MODELS WITH SETTLING AND AGGREGATION

Globally modeling a protoplanetary disk, including dust settling, aggregation, radial drift, and mixing, along with radiative transfer solutions for the disk temperature and spectrum, forms a major numerical challenge because of the many orders of magnitude that have to be covered both in timescales (inner disk vs. outer disk, growth of small particles vs. growth of large objects) and particle sizes. Further numerical difficulties result from the fact that small particles may contribute significantly to the growth of larger bodies, and careful renormalization schemes are necessary to treat these processes correctly and in a mass-conserving manner (*Dullemond and Dominik*, 2005). Further difficulties arise from uncertainty about the strength and spatial extent of turbulence during the different evolutionary phases of a disk. A complete model covering an entire disk and the entire growth process along with all relevant disk physics is currently still out of reach. Work so far has therefore either focused on specific locations in the disk, or has used parameterized descriptions of turbulence with limited sets of physical growth processes. However, these "single-slice" models have the problem that radial drift can become so large for meter-sized objects that these leave the slice on a timescale of a few orbital times (*Weidenschilling*, 1977) (section 3.1.1). Nevertheless, important results that test underlying assumptions of the models have come forth from these efforts.

For the spectral and imaging appearance of disks, there are two main processes that should produce easily observable results: particle settling and particle growth. Particle settling is due to the vertical component of gravity acting in the disk on the pressureless dust component (section 3.1.2). Neglecting growth for the moment, settling leads to a vertical stratification and size sorting in the disk. Small particles settle slowly and should be present in the disk atmosphere for a long time, while large particles settle faster and to smaller scale heights. While in a laminar nebula this is a purely time-dependent phenomenon, this result is permanent in a turbulent nebula as each particle size is spread over its equilibrium scale height (*Dubrulle et al.*, 1995). From a pure settling model, one would therefore expect that *small dust grains* will increasingly dominate dust-emission features (cause strong feature-to-continuum ratios) as large grains disappear from the surface layers.

Grain growth may have the opposite effect. While vertical mixing and settling still should lead to a size stratification, particle growth can become so efficient that all small particles are removed from the gas. In this case, dust emission features should be characteristic for *larger particles* (i.e., no or weak features) (see chapter by Natta et al.). At the same time, the overall opacity decreases dramatically. This effect can become significant, as has been realized already early on (*Weidenschilling*, 1980, 1984; *Mizuno*, 1989). In order to keep the small particle abundance at realistic levels and the dust opacity high, *Mizuno et al.* (1988) considered a steady inflow of small particles into a disk. However, disks with signs of small particles are still observed around stars that seem to have completely removed their parental clouds, so this is not a general solution for this problem. In the following we discuss the different disk models documented in the literature. We begin with a discussion

of earlier models focusing on specific regions of the solar system.

4.1. Models Limited to Specific Regions in the Solar System

Models considering dust settling and growth in a single vertical slice have a long tradition, and have been reviewed previously in *Protostars and Planets III* (*Weidenschilling and Cuzzi*, 1993). We therefore refrain from an indepth coverage and only recall a few of the main results. The global models discussed later are basically similar calculations, with higher resolution, and for a large set of radii. Weidenschilling has studied the aggregation in laminar (*Weidenschilling*, 1980, 2000) and turbulent (*Weidenschilling*, 1984, 1988) nebulae, focusing on the region of terrestrial planet formation, in particular around 1 AU. These papers contain the basic descriptions of dust settling and growth under laminar and turbulent conditions. They show the occurrence of a rainout after particles have grown to sizes where the settling motion starts to exceed the thermal motions. *Nakagawa et al.* (1981, 1986) study settling and growth in vertical slices, also concentrating on the terrestrial planet-formation regions. They find that within 3000 yr, the midplane is populated by centimeter-sized grains.

Weidenschilling (1997) studied the formation of comets in the outer solar system with a detailed model of a nonturbulent nebula, solving the coagulation equation around 30 AU. In these calculations, growth initially proceeds by Brownian motion, without significant settling, for the first 10,000 yr. Then, particles become large enough and start to settle, so that the concentration of solids increases quickly after 5×10^4 yr. The particle layer reaches the critical density where the layer gravitational instability is often assumed to occur, but first the high-velocity dispersion prevents the collapse. Later, a transient density enhancement still occurs, but due to the small collisional cross section of the typically 1-m-sized bodies, growth must still happen in individual two-body collisions.

4.2. Dust Aggregation During Early Disk Evolution

Schmitt et al. (1997) implemented dust coagulation in an α-disk model. They considered the growth of PCA in a one-dimensional disk model, i.e., without resolving the vertical structure of the disk. The evolution of the dust size distribution is followed for only 100 yr. In this time, at a radius of 30 AU from the star, first the smallest particles disappear within 10 yr, due to Brownian motion aggregation. This is followed by a self-similar growth phase during which the volume of the particles increases by 6 orders of magnitude. Aggregation is faster in the inner disk, and the decrease in opacity followed by rapid cooling leads to a *thermal gap* in the disk around 3 AU. Using the CCA particles, aggregation stops in this model after the small grains have been removed. For such particles, longer timescales are required to continue the growth.

Global models of dust aggregation during the prestellar collapse stage and into the early disk formation stage are numerically feasible because the growth of particles is limited. *Suttner et al.* (1999) and *Suttner and Yorke* (2001) study the evolution of dust particles in protostellar envelopes, during collapse, and the first 10^4 yr of dynamical disk evolution, respectively. These very ambitious models include a radiation hydrodynamic code that can treat dust aggregation and shattering using an implicit numerical scheme. They find that during a collapse phase of 10^3 yr, dust particles grow due to Brownian motion and differential radiative forces, and can be shattered by high-velocity collisions cause by radiative forces. During early disk evolution, they find that at 30 AU from the star within the first pressure scale height from the midplane, small particles are heavily depleted because the high densities lead to frequent collisions. The largest particles grow by a factor of 100 in mass. Similar results are found for PCA particles, while CCA particles show accelerated aggregation because of the enhanced cross section in massive particles. Within 10^4 yr, most dust moves to the size grid limit of 0.2 mm. While aggregation is significant near the midplane (opacities are reduced by more than a factor of 10), the overall structure of the model is not yet affected strongly, because at the low densities far from the midplane aggregation is limited and changes in the opacity are only due to differential advection.

4.3. Global Settling Models

Settling of dust without growth goes much slower than settling that is accelerated by growth. However, even pure settling calculations show significant influence on the spectral energy distributions of disks. While the vertical optical depth is unaffected by settling alone, the height at which stellar light is intercepted by the disk surface changes. *Miyake and Nakagawa* (1995) computed the effects of dust settling on the global SED and compared these results with IRAS observations. They assume that after the initial settling and growth phase, enough small particles are left in the disk to provide an optically thick surface and follow the decrease of the height of this surface, concluding that this is consistent with the lifetimes of T Tauri disks, because the settling time of a 0.1 μm grain within a single pressure scale height is on the order of 10 m.y. However, the initial settling phase does lead to strong effects on the SED, because settling times at several pressure scale heights are much shorter.

Dullemond and Dominik (2004) show that settling from a fully mixed passive disk leads to a decrease of the surface height in 10^4–10^5 yr, and can even lead to self-shadowed disks (see chapter by Dullemond et al.).

4.4. Global Models of Dust Growth

Mizuno (1989) computes global models including evaporation and a steady-state assumption using small grains continuously raining down from the ISM. The vertical disk structure is not resolved; only a single zone in the midplane is considered. He finds that the Rosseland mean opacity decreases, but then stays steady due to the second-generation grains.

Kornet et al. (2001) model the global gas and dust disk by assuming that at a given radius, the size distribution of dust particles (or planetesimals) is exactly monodisperse, avoiding the numerical complications of a full solution of the Smoluchowski equation. They find that the distribution of solids in the disk after 10^7 yr depends strongly on the initial mass and angular momentum of the disk.

Ciesla and Cuzzi (2006) model the global disk using a four-component model: dust grains, meter-sized boulders, planetesimals, and the disk gas. This model tries to capture the main processes happening in a disk: growth of dust grains to meter-sized bodies, the migration of meter-sized bodies and the resulting creation of evaporation fronts, and the mixing of small particles and gas by turbulence. The paper focuses on the distribution of water in the disk, and the dust growth processes are handled by assuming timescales for the conversion from one size to the next. Such models are therefore mainly useful for the chemical evolution of the nebula and need detailed aggregation calculations as input.

The most complete long-term integrations of the equations for dust settling and growth are described in recent papers by *Tanaka et al.* (2005, henceforth *THI05*) and *Dullemond and Dominik* (2005, henceforth *DD05*). These papers implement dust settling and aggregation in individual vertical slices through a disk, and then use many slices to stitch together an entire disk model, with predictions for the resulting optical depth and SED from the developing disk. Both models have different limitations. *THI05* consider only laminar disk models, so that turbulent mixing and collisions between particles driven by turbulence are not considered. Their calculations are limited to compact solid particles. *DD05*'s model is incomplete in that it does not consider the contributions of radial drift and differential angular velocities between different particles. But in addition to calculations for a laminar nebula, they also introduce turbulent mixing and turbulent coagulation, as well as PCA and CCA properties for the resulting dust particles. *THI05* use a two-layer approximation for the radiative transfer solution, while *DD05* run a three-dimensional Monte Carlo radiative transfer code to compute the emerging spectrum of the disk. Both models find that aggregation proceeds more rapidly in the inner regions of the disk than in the outer regions, quickly leading to a region of low optical depth in the inner disk.

Both calculations find that a bimodal size distribution is formed, with large particles in the midplane formed by rainout (the fast settling of particles after their settling time has decreased below their growth time) and continuing to grow quickly, and smaller particles remaining higher up in the disk and then slowly trickling down. In the laminar disk, growth stops in the *DD05* calculations at centimeter sizes because radial drift was ignored. In *THI05*, particles continue to grow beyond this regime.

The settling of dust causes the surface height of the disk to decrease, reducing the overall capacity of the disk to reprocess stellar radiation. *THI05* find that at 8 AU from the star the optical depth of the disk at 10 μm reaches about unity after a bit less than 10^6 yr. In the inner disk, the surface height decreases to almost zero in less than 10^6 yr. The SED of the model shows first a strong decrease at wavelength of 100 μm and longer, within the first 10^5 yr. After that the near-IR and mid-IR radiation also decreases sharply. *THI05* consider their results to be roughly consistent with the observations of decreasing fluxes at near-IR and millimeter wavelengths in disks.

The calculations by *DD05* show a more dramatic effect, as shown in Fig. 5. In the calculations for a laminar disk, here the surface height already significantly decreases in the first 10^4 yr, then the effect on the SED is initially strongest in the mid-IR region. After 10^6 yr, the fluxes have dropped globally by at least a factor of 10, except for the millimeter regime, which is affected greatly only after a few × 10^6 yr.

In the calculations for a turbulent disk (*DD05*), the depletion of small grains in the inner disk is strongly enhanced. This result is caused by several effects. First, turbulent mix-ing keeps the particles moving even after they have settled to the midplane, allowing them to be mixed up and rain down again through a cloud of particles. Furthermore, vertical mixing in the higher disk regions mixes low-density material down to higher densities, where aggregation can proceed much faster. The material being mixed back up above the disk is then largely deprived of solids, because the large dust particles decouple from the gas and stay behind, settling down to the midplane. The changes to the SED caused by coagulation and settling in a turbulent disk are dramatic, and clearly inconsistent with the observations of disks around T Tauri stars that indicate lifetimes of up to 10^7 yr. *DD05* conclude that ongoing particle destruction must play an important role, leading to a steady-state size distribution for small particles (section 2.3.1).

4.5. The Role of Aggregate Structure

Up to now, most solutions for the aggregation equation in disks are still based on the assumption of compact particles resulting from the growth process. However, at least for the small aggregates formed initially, this assumption is certainly false. First of all, if aggregates are fluffy, with large surface-to-mass ratios, it will be much easier to keep these particles in the disk surface where they can be observed as scattering and IR-emitting grains. Observations of the 10-μm silicate features show that in many disks, the population emitting in this wavelength range is dominated by particles larger than interstellar (*van Boekel et al.*, 2005; *Kessler-Silacci et al.*, 2005). When modeled with compact grains, the typical size of such grains is several micrometers, with corresponding settling times less than 1 m.y. When modeled with aggregates, particles have to be much larger to produce similar signatures [flattened feature shapes (e.g., *Min et al.*, 2006)].

When considering the growth timescales, in particular in regions where settling is driving the relative velocities, the timescales are surprisingly similar to the case of compact particles (*Safronov*, 1969; *Weidenschilling*, 1980). While initially, fluffy particles settle and grow slowly because of small settling velocities, the larger collisional cross section soon leads to fast collection of small particles, and fluffy particles reach the midplane as fast as compact grains, and with similar masses collected.

5. SUMMARY AND FUTURE PROSPECTS

A lot has been achieved in the last few years, and our understanding of dust growth has advanced significantly. There are a number of issues where we now have clear an-

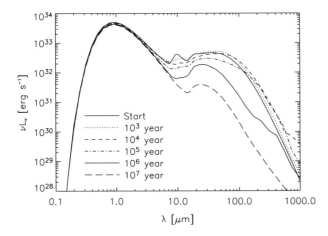

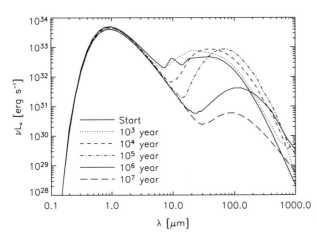

Fig. 5. Time evolution of the disk SED in the laminar case (top panel) and the turbulent case (bottom panel). From *DD05*.

TABLE 1.   Overview.

What We Really Know	Main Controversies/Questions	Future Priorities
Microphysics		
Dust particles stick in collisions with less than ~1 m/s velocity due to van der Waals force or hydrogen bonding.	At what aggregate size does compaction happen in a nebula environment?	More empirical studies in collisions between macroscopic aggregates required.
For low relative velocities ($\ll$1 m/s) a cloud of dust particles evolves into fractal aggregates (D_f < 2) with a quasi-monodisperse mass distribution.	When do collisions between macroscopic aggregates result in sticking? Some experiments show no sticking at rather low impact velocities, while others show sticking at high impact speeds.	Macroscopic model for aggregate collisions (continuum description) based on microscopic model and experimental results.
Due to the increasing collision energy, growing fractal aggregates can no longer keep their structures so that nonfractal (but very porous) aggregates form (still at v $\ll$ 1 m/s).	How important are special material properties: organics, ices, magnetic and electrically charged particles?	Develop recipes for using the microphysics in large-scale aggregation calculations.
Macroscopic aggregates have porosities >65% when collisional compaction, and not sintering or melting occurs.	What are the main physical parameters (e.g., velocity, impact angle, aggregate porosity/material/shape/mass) determining the outcome of a collision?	Develop aggregation models that treat aggregate structure as a *variable* in a self-consistent way.
Nebula Processes		
Particle velocities and relative velocities in turbulent and nonturbulent nebulae are understood; values are <1 m/s for aρ < 1–3 g cm^{-2} depending on α.	What happens to dust aggregates in highly mass-loaded regions in the solar nebula, e.g., midplane, eddies, stagnation points?	Relative velocities in highly mass-loaded regions in the solar nebula, e.g., midplane, eddies, stagnation points.
Radial drift decouples large amounts of solids from the gas and migrates it radially, changing nebula mass distribution and chemistry	Is the nebula turbulent? If so, how does the intensity vary with location and time? Can purely hydrodynamical processes produce self-sustaining turbulence in the terrestrial planet formation zone?	Improve our understanding of turbulence production processes at very high nebula Reynolds numbers.
Turbulent diffusion can offset inward drift for particles of centimeter size and smaller, relieving the "problem" about age differences between CAIs and chondrules.	Can large-scale structures (vortices, spiral density waves) remain stable long enough to concentrate boulder-sized particles?	Model effects of MRI-active upper layers on dense, nonionized gas in magnetically dead zones.
		Model the evolution of dense strengthless clumps of particles in turbulent gas.
	Can dense turbulently concentrated zones of chondrule-sized particles survive to become actual planetesimals?	Model collisional processes in boulder-rich vortices and high-pressure zones.
		Model evolution of dense clumps in turbulent gas.
Global Modeling and Comparison with Observations		
Small grains are quickly depleted by incorporation into larger grains.	What is the role of fragmentation for the small grain component?	Study the optical properties of *large* aggregates, fluffy and compact.
Growth timescales are short for small compact and fractal grains alike.	Are the "small" grains seen really large, fluffy aggregates?	Implement realistic opacities in disk models to produce predictions and compare with observations.
Vertical mixing and small grain replenishment are necessary to keep the observed disk structures (thick/flaring).	Are the millimeter-/centimeter-sized grains seen in observations compact particles, or much larger fractal aggregates?	Construct truly global models including radial transport.
	What is the global role of radial transport?	More resolved disk images at many wavelengths, to better constrain models.

swers. However, a number of major controversies remain, and future work will be needed to address these before we can come to a global picture of how dust growth in protoplanetary disks proceeds and which of the possible ways toward planetesimals are actually used by nature. Table 1 summarizes our main conclusions and questions, and notes some priorities for research in the near future.

Acknowledgments. We thank the referee (S. Weidenschilling) for valuable comments on the manuscript. This work was partially supported by J.N.C.'s grant from the Planetary Geology and Geophysics program. C.D. thanks C. P. Dullemond for many discussions and D. Paszun for preparing Fig. 1. J.N.C. thanks A. Youdin for a useful conversation regarding slowly evolving, large-scale structures.

REFERENCES

Adachi I., Hayashi C., and Nakazawa K. (1976) *Prog. Theor. Phys., 56,* 1756–1771.

A'Hearn M. A. and the Deep Impact Team (2005) *37th DPS Meeting,* Cambridge, England, Paper #35.02.

Barge P. and Sommeria J. (1995) *Astron. Astrophys., 296,* L1–L4.

Barranco J. A. and Marcus P. S. (2005) *Astrophys. J., 623,* 1157–1170.

Bel N. and Schatzman E. (1958) *Rev. Mod. Phys., 30,* 1015–1016.

Blum J. (2004) In *Astrophysics of Dust* (A. Witt et al., eds.), pp. 369–391. ASP Conf. Series 309, San Francisco.

Blum J. (2006) *Adv. Phys.,* in press.

Blum J. and Münch M. (1993) *Icarus, 106,* 151–167.

Blum J. and Schräpler R. (2004) *Phys. Rev. Lett., 93,* 115503.

Blum J. and Wurm G. (2000) *Icarus, 143,* 138–146.

Blum J., Wurm G., Poppe T., and Heim L.-O. (1999) *Earth Moon Planets, 80,* 285.

Blum J., Wurm G., Kempf S., Poppe T., Klahr H., et al. (2000) *Phys. Rev. Lett., 85,* 2426–2429.

Bockelée-Morvan D., Gautier D., Hersant F., Huré J.-M., and Robert F. (2002) *Astron. Astrophys., 384,* 1107–1118.

Boley A. C., Durisen R. H., and Pickett M. K. (2005) In *Chondrites and the Protoplanetary Disk* (A. N. Krot et al., eds.), pp. 839–848. ASP Conf. Series 34, San Francisco.

Bracco A., Provenzale A., Spiegel E. A., and Yecko P. (1998) In *Theory of Black Hole Accretion Disks* (M. A. Abramowicz et al., eds.), p. 254. Cambridge Univ., Cambridge.

Bridges F. G., Supulver K. D., and Lin D. N. C. (1996) *Icarus, 123,* 422–435.

Champney J. M., Dobrovolskis A. R., and Cuzzi J. N. (1995) *Phys. Fluids, 7,* 1703–1711.

Chandrasekhar S. (1961) *Hydrodynamic and Hydromagnetic Stability,* p. 589. Oxford Univ., New York.

Chokshi A., Tielens A. G. G. M., and Hollenbach D. (1993) *Astrophys. J., 407,* 806–819.

Ciesla F. J. and Cuzzi J. N. (2006) *Icarus, 181,* 178–204.

Colwell J. E. (2003) *Icarus, 164,* 188–196.

Cuzzi J. N. (2004) *Icarus, 168,* 484–497.

Cuzzi J. N. and Hogan R. C. (2003) *Icarus, 164,* 127–138.

Cuzzi J. N. and Weidenschilling S. J. (2006) In *Meteorites and the Early Solar System II* (D. Lauretta et al., eds.), pp. 353–381. Univ of Arizona, Tucson.

Cuzzi J. N. and Zahnle K. (2004) *Astrophys. J., 614,* 490–496.

Cuzzi J. N., Dobrovolskis A. R., and Champney J. M. (1993) *Icarus, 106,* 102–134.

Cuzzi J. N., Dobrovolskis A. R., and Hogan R. C. (1994) In *Lunar Planet. Sci. XXV,* pp. 307–308. LPI, Houston.

Cuzzi J. N., Dobrovolskis A. R., and Hogan R. C. (1996) In *Chondrules and the Protoplanetary Disk* (R. Hewins et al., eds.), pp. 35–44. Cambridge Univ., Cambridge.

Cuzzi J. N., Hogan R. C., Paque J. M., and Dobrovolskis A. R. (2001) *Astrophys. J., 546,* 496–508.

Cuzzi J. N., Davis S. S., and Dobrovolskis A. R. (2003) *Icarus, 166,* 385–402.

Cuzzi J. N., Ciesla F. J., Petaev M. I., Krot A. N., Scott E. R. D., and Weidenschilling S. J. (2005) In *Chondrites and the Protoplanetary Disk* (A. N. Krot et al., eds.), pp. 732–773. ASP Conf. Series 341, San Francisco.

Cyr K., Sharp C. M., and Lunine J. I. (1999) *J. Geophys. Res., 104,* 19003–19014.

Dahneke B. E. (1975) *J. Colloid Interf. Sci., 1,* 58–65.

Davidsson B. J. R. (2006) *Adv. Geosci.,* in press.

Derjaguin B. V., Muller V. M., and Toporov Y. P. (1975) *J. Colloid Interf. Sci., 53,* 314–326.

Desch S. J. and Cuzzi J. N. (2000) *Icarus, 143,* 87–105.

Dobrovolskis A. R., Dacles-Mariani J. M., and Cuzzi J. N. (1999) *J. Geophys. Res.–Planets, 104,* 30805–30815.

Dominik C. and Nübold H. (2002) *Icarus, 157,* 173–186.

Dominik C. and Tielens A. G. G. M. (1995) *Phil. Mag., 72,* 783–803.

Dominik C. and Tielens A. G. G. M. (1996) *Phil. Mag., 73,* 1279–1302.

Dominik C. and Tielens A. G. G. M. (1997) *Astrophys. J., 480,* 647–673.

Dubrulle B., Morfill G. E., and Sterzik M. (1995) *Icarus, 114,* 237–246.

Dullemond C. P. and Dominik C. (2004) *Astron. Astrophys., 421,* 1075–1086.

Dullemond C. P. and Dominik C. (2005) *Astron. Astrophys., 434,* 971–986.

Gail H.-P. (2004) *Astron. Astrophys., 413,* 571–591.

Garaud P. and Lin D. N. C. (2004) *Astrophys. J., 608,* 1050–1075.

Godon P. and Livio M. (2000) *Astrophys. J., 537,* 396–404.

Goldreich P. and Lynden-Bell D. (1965) *Mon. Not. R. Astron. Soc., 130,* 97–124.

Goldreich P. and Ward W. R. (1973) *Astrophys. J., 183,* 1051–1061.

Goodman J. and Pindor B. (2000) *Icarus, 148,* 537–549.

Greenberg J. M., Mizutani H., and Yamamoto T. (1995) *Astron. Astrophys., 295,* L35–L38.

Gurevich L. E. and Lebedinsky A. I. (1950) *Izv. Acad. Nauk USSR, 14,* 765.

Gustafson B. Å. S. and Kolokolova L. (1999) *J. Geophys. Res., 104,* 31711–31720.

Haghighipour N. and Boss A. P. (2003a) *Astrophys. J., 583,* 996–1003.

Haghighipour N. and Boss A. P. (2003b) *Astrophys. J., 598,* 1301–1311.

Heim L., Blum J., Preuss M., and Butt H.-J. (1999) *Phys. Rev. Lett., 83,* 3328–3331.

Heim L., Butt H.-J., Schräpler R., and Blum J. (2005) *Aus. J. Chem., 58,* 671–673.

Henning T. and Stognienko R. (1996) *Astron. Astrophys., 311,* 291–303.

Ivlev A. V., Morfill G. E., and Konopka U. (2002) *Phys. Rev. Lett., 89,* 195502.

Jeans J. H. (1928) *Astronomy and Cosmogony,* p. 337. Cambridge Univ., Cambridge.

Johansen A. and Klahr H. (2005) *Astrophys. J., 634,* 1353–1371.

Johansen A., Andersen A. C., and Brandenburg A. (2004) *Astron. Astrophys., 417,* 361–374.

Johansen A., Klahr H., and Henning Th. (2006) *Astrophys. J., 636,* 1121–1134.

Karjalainen R. and Salo H. (2004) *Icarus, 172,* 328–348.

Kempf S., Pfalzner S., and Henning Th. (1999) *Icarus, 141,* 388.

Kessler-Silacci J. E., Hillenbrand L. A., Blake G. A., and Meyer M. R. (2005) *Astrophys. J., 622,* 404–429.

Klahr H. and Bodenheimer P. (2006) *Astrophys. J., 639,* 432–440.

Konopka U., Mokler F., Ivlev A. V., Kretschmer M., Morfill G. E., et al. (2005) *New J. Phys., 7,* 227, 1–11.

Kornet K., Stepinski T. F., and Różyczka M. (2001) *Astron. Astrophys., 378,* 180–191.

Kouchi A., Kudo T., Nakano H., Arakawa M., Watanabe N., Sirono S.-I., Higa M., and Maeno N. (2002) *Astrophys. J., 566,* L121–L124.

Kozasa T., Blum J., and Mukai T. (1992) *Astron. Astrophys., 263,* 423–432.

Krause M. and Blum J. (2004) *Phys. Rev. Lett., 93,* 021103.

Krot A. N., Hutcheon I. D., Yurimoto H., Cuzzi J. N., McKeegan K. D., Scott E. R. D., Libourel G., Chaussidon M., Aleon J., and Petaev M. I. (2005) *Astrophys. J., 622,* 1333–1342.

Künzli S. and Benz W. (2003) *Meteoritics & Planet. Sci., 38,* 5083.

Love S. G. and Pettit D. R. (2004) In *Lunar Planet. Sci. XXXV,* Abstract #1119. LPI, Houston.

Markiewicz W. J., Mizuno H., and Völk H. J. (1991) *Astron. Astrophys., 242,* 286–289.

Marshall J. and Cuzzi J. (2001) In *Lunar Planet. Sci. XXXII,* Abstract #1262. LPI, Houston.

Marshall J., Sauke T. A., and Cuzzi J. N. (2005) *Geophys. Res. Lett., 32,* L11202–L11205.

McCabe C., Duchêne G., and Ghez A. M. (2003) *Astrophys. J., 588,* L113.

Min M., Dominik C., Hovenier J. W., de Koter A., and Waters L. B. F. M. (2006) *Astron. Astrophys., 445,* 1005–1014.

Miyake K. and Nakagawa Y. (1995) *Astrophys. J., 441,* 361–384.

Mizuno H. (1989) *Icarus, 80,* 189–201.

Mizuno H., Markiewicz W. J., and Voelk H. J. (1988) *Astron. Astrophys., 195,* 183–192.

Morfill G. E. and Völk H. J. (1984) *Astrophys. J., 287,* 371–395.

Nakagawa Y., Nakazawa K., and Hayashi C. (1981) *Icarus, 45,* 517–528.

Nakagawa Y., Sekiya M., and Hayashi C. (1986) *Icarus, 67,* 375–390.

Nübold H. and Glassmeier K.-H. (2000) *Icarus, 144,* 149–159.

Nübold H., Poppe T., Rost M., Dominik C., and Glassmeier K.-H. (2003) *Icarus, 165,* 195–214.

Nuth J. A. and Wilkinson G. M. (1995) *Icarus, 117,* 431–434.

Nuth J. A., Faris J., Wasilewski P., and Berg O. (1994) *Icarus, 107,* 155–163.

Ossenkopf V. (1993) *Astron. Astrophys., 280,* 617–646.

Paszun D. and Dominik C. (2006) *Icarus, 182,* 274–280.

Poppe T. and Schräpler R. (2005) *Astron. Astrophys., 438,* 1–9.

Poppe T., Blum J., and Henning Th. (2000a) *Astrophys. J., 533,* 454–471.

Poppe T., Blum J., and Henning Th. (2000b) *Astrophys. J., 533,* 472–480.

Rice W. K. M., Lodato G., Pringle J. E., Armitage P. J., and Bonnell I. A. (2004) *Mon. Not. R. Astron. Soc., 355,* 543–552.

Russell S. S., Hartmann L. A., Cuzzi J. N., Krot A. N., and Weidenschilling S. J. (2006) In *Meteorites and the Early Solar System II* (D. Lauretta et al., eds.), pp. 233–251. Univ. of Arizona, Tucson.

Safronov V. S. (1960) *Ann. Astrophys., 23,* 979–982.

Safronov V. S. (1969) *Evolution of the Protoplanetary Cloud and Formation of the Earth and the Planets.* Nauka, Moscow (translated as NASA TTF-677).

Safronov V. S. (1991) *Icarus, 94,* 260–271.

Salo H. (1992) *Nature, 359,* 619–621.

Schmitt W., Henning Th., and Mucha R. (1997) *Astron. Astrophys., 325,* 569–584.

Scott E. R. D., Love S. G., and Krot A. N. (1996) in *Chondrules and the Protoplanetary Disk* (R. Hewins et al., eds.), pp. 87–96. Cambridge Univ., Cambridge.

Sekiya M. (1983) *Prog. Theor. Phys., 69,* 1116–1130.

Sekiya M. (1998) *Icarus, 133,* 298–309.

Sekiya M. and Takeda H. (2003) *Earth Planets Space, 55,* 263–269.

Sekiya M. and Takeda H. (2005) *Icarus, 176,* 220–223.

Shakura N. I., Sunyaev R. A., and Zilitinkevich S. S. (1978) *Astron. Astrophys., 62,* 179–187.

Sirono S. (2004) *Icarus, 167,* 431–452.

Sirono S. and Greenberg J. M. (2000) *Icarus, 145,* 230–238.

Smoluchowski M. v. (1916) *Phys. Zeit., 17,* 557–585.

Stepinski T. F. and Valageas P. (1996) *Astron. Astrophys., 309,* 301–312.

Stepinski T. F. and Valageas P. (1997) *Astron. Astrophys., 319,* 1007–1019.

Stone J. M., Gammie C. F., Balbus S. A., and Hawley J. F. (2000) In *Protostars and Planets IV* (V. Mannings et al., eds.), pp. 589–599. Univ. of Arizona, Tucson.

Supulver K. and Lin D. N. C. (2000) *Icarus, 146,* 525–540.

Suttner G. and Yorke H. W. (2001) *Astrophys. J., 551,* 461–477.

Suttner G., Yorke H. W., and Lin D. N. C. (1999) *Astrophys. J., 524,* 857–866.

Tanaka H., Himeno Y., and Ida S. (2005) *Astrophys. J., 625,* 414–426.

Tanga P., Babiano A., Dubrulle B., and Provenzale A. (1996) *Icarus, 121,* 158–170.

Tanga P., Weidenschilling S. J., Michel P., and Richardson D. C. (2004) *Astron. Astrophys., 427,* 1105–1115.

Toomre A. (1964) *Astrophys. J., 139,* 1217–1238.

Túnyi I., Guba P., Roth L. E., and Timko M. (2003) *Earth Moon Planets, 93,* 65–74.

Ueta T. and Meixner M. (2003) *Astrophys. J., 586,* 1338–1355.

Valverde J. M., Quintanilla M. A. S., and Castellanos A. (2004) *Phys. Rev. Lett., 92,* 258303.

van Boekel R., Min M., Waters L. B. F. M., de Koter A., Dominik C., van den Ancker M. E., and Bouwman J. (2005) *Astron. Astrophys., 437,* 189–208.

Völk H. J., Jones F. C., Morfill G. E., and Röser S. (1980) *Astron. Astrophys., 85,* 316–325.

Ward W. R. (1976) In *Frontiers of Astrophysics* (E. H. Avrett, ed.), pp. 1–40. Harvard Univ., Cambridge.

Ward W. R. (2000) In *Origin of the Earth and Moon* (R. M. Canup and K. Righter, eds.), pp. 75–84. Univ. of Arizona, Tucson.

Wasson J. T., Trigo-Rodriguez J. M., and Rubin A. E. (2005) In *Lunar Planet. Sci. XXXVI,* Abstract #2314. LPI, Houston.

Watson P. K., Mizes H., Castellanos A., and Pérez A. (1997) In *Powders and Grains 97* (R. Behringer and J. T. Jenkins, eds), pp. 109–112. Balkema, Rotterdam.

Weidenschilling S. J. (1977) *Mon. Not. R. Astron. Soc., 180,* 57–70.

Weidenschilling S. J. (1980) *Icarus, 44,* 172–189.

Weidenschilling S. J. (1984) *Icarus, 60,* 553–567.

Weidenschilling S. J. (1988) In *Meteorites and the Early Solar System* (J. A. Kerridge and M. S. Matthews, eds), pp. 348–371. Univ. of Arizona, Tuscon.

Weidenschilling S. J. (1995) *Icarus, 116,* 433–435.

Weidenschilling S. J. (1997) *Icarus, 127,* 290–306.

Weidenschilling S. J. (2000) *Space Sci. Rev., 92,* 295–310.

Weidenschilling S. J. and Cuzzi J. N. (1993) In *Protostars and Planets III* (E. H. Levy and J. I. Lunine, eds.), pp. 1031–1060. Univ. of Arizona, Tucson.

Whipple F. (1972) In *From Plasma to Planet* (A. Elvius, ed.), pp. 211–232. Wiley, New York.

Wolf S. (2003) *Astrophys. J., 582,* 859–868.

Wurm G. and Blum J. (1998) *Icarus, 132,* 125–136.

Wurm G. and Schnaiter M. (2002) *Astrophys. J., 567,* 370–375.

Wurm G., Blum J., and Colwell J. E. (2001) *Icarus, 151,* 318–321.

Wurm G., Relke H., Dorschner J., and Krauss O. (2004a) *J. Quant. Spect. Rad. Transf., 89,* 371–384.

Wurm G., Paraskov G., and Krauss O. (2004b) *Astrophys. J., 606,* 983–987.

Wurm G., Paraskov G., and Krauss O. (2005) *Icarus, 178,* 253–263.

Youdin A. N. (2004) In *Star Formation in the Interstellar Medium* (D. Johnstone et al., eds.), pp. 319–327. ASP Conf. Series 323, San Francisco.

Youdin A. N. and Chiang E. I. (2004) *Astrophys. J., 601,* 1109–1119.

Youdin A. and Shu F. (2002) *Astrophys. J., 580,* 494–505.

Yurimoto H. and Kuramoto N. (2004) *Science, 305,* 1763–1766.

Astronomical and Meteoritic Evidence for the Nature of Interstellar Dust and its Processing in Protoplanetary Disks

C. M. O'D. Alexander and A. P. Boss
Carnegie Institution of Washington

L. P. Keller
NASA Johnson Space Center

J. A. Nuth
NASA Goddard Space Flight Center

A. Weinberger
Carnegie Institution of Washington

Here we compare the astronomical and meteoritic evidence for the nature and origin of interstellar dust, and how it is processed in protoplanetary disks. The relative abundances of circumstellar grains in meteorites and interplanetary dust particles (IDPs) are broadly consistent with most astronomical estimates of galactic dust production, although graphite/amorphous C is highly underabundant. The major carbonaceous component in meteorites and IDPs is an insoluble organic material (IOM) that probably formed in the interstellar medium, but a solar origin cannot be ruled out. GEMS (glass with embedded metal and sulfide) that are isotopically solar within error are the best candidates for interstellar silicates, but it is also possible that they are solar system condensates. No dust from young stellar objects has been identified in IDPs, but it is difficult to differentiate them from solar system material or indeed some circumstellar condensates. The crystalline silicates in IDPs are mostly solar condensates, with lesser amounts of annealed GEMS. The IOM abundances in IDPs are roughly consistent with the degree of processing indicated by their crystallinity if the processed material was ISM dust. The IOM contents of meteorites are much lower, suggesting that there was a gradient in dust processing in the solar system. The microstructure of much of the pyroxene in IDPs suggests that it formed at temperatures >1258 K and cooled relatively rapidly (~1000 K/h). This cooling rate favors shock heating rather than radial transport of material annealed in the hot inner disk as the mechanism for producing crystalline dust in comets and IDPs. Shock heating is also a likely mechanism for producing chondrules in meteorites, but the dust was probably heated at a different time and/or location to chondrules.

1. INTRODUCTION

There are two sources of information on protoplanetary disk evolution: astronomical observations, and for our solar system, primitive chondritic meteorites, interplanetary dust particles (IDPs), and comets. Astronomical observations are largely confined to the surfaces of disks and have relatively low spatial resolution. Primitive meteorites, IDPs, and comets retain a complex record of processes that occurred throughout the early solar protoplanetary disk (solar nebula). This record is still being deciphered. How complete it is and how representative it is of disk evolution in general are open questions.

Silicate dust in the interstellar medium is observed to be largely amorphous (e.g., *Mathis,* 1990; *Kemper et al.,* 2004). One of the most striking observations of protoplanetary disks is that their dust has a significant crystalline component (e.g., *Meeus et al.,* 1998; *Bouwman et al.,* 2001; *van Boekel et al.,* 2004) and it tends to be coarser grained than interstellar dust. Both observations suggest that dust has been thermally processed and has aggregated in these disks even at large radial distances from the central stars. These observations imply either extensive transport of dust from the hot inner regions of the protoplanetary disks (*Nuth et al.,* 2000b; *Gail,* 2004), or perhaps more localized heating mechanisms that operate over large portions of disks (e.g., *Chick and Cassen,* 1997; *Harker and Desch,* 2002).

Comets that formed at large radial distances from the Sun, including Halley (*Swamy et al.,* 1988) and 9P/Tempel 1 (*Harker et al.,* 2005), have a large crystalline component in their silicate dust. Thus, the solar system appears to have at least one process in common with other protoplanetary disks. Some IDPs may come from comets, and components of chondrites share some features in common with IDPs and comets. This includes presolar material inherited from the protosolar molecular cloud. Thus, it is likely that we can study in the laboratory unprocessed and processed dust that may help constrain the conditions and mechanism of thermal processing.

Here we review astronomical observations of dust in the interstellar medium (ISM) and compare them to observations of annealed dust analogs, cometary dust, chondrites, and IDPs. Components of meteorites and IDPs retain evidence of several distinct thermal processes that operated in the solar nebula, and we discuss which, if any, may have been responsible for the processing of dust observed in protoplanetary disks. Finally, we discuss the implications and challenges these observations have for the dynamics of protoplanetary disks.

2. CIRCUMSTELLAR, INTERSTELLAR, AND PROTOPLANETARY DUST

2.1. Sources of Interstellar Medium Dust: Evolved Stars and Young Stellar Objects

The relative importance of ISM dust sources is very uncertain. Most estimates find that red giant (RGB) and asymptotic giant branch (AGB) stars are the main stellar sources of O-rich and C-rich dust (e.g., *Jones*, 2001). *Kemper et al.* (2004) suggest that M supergiants may be as important sources of silicate dust as AGB stars. On the other hand, *Tielens et al.* (2005) estimate that most silicate dust comes in roughly equal quantities from O-rich AGB stars and from young stellar objects (YSO). They also suggest that supernovae could be important sources of dust, but are only able to set upper limits. The crystalline fractions of silicates from O-rich AGB stars and M supergiants are ~10–20% (*Kemper et al.*, 2004). The principle crystalline components are very Mg-rich pyroxene ($Mg_xFe_{1-x}SiO_3$) and olivine ($Mg_{2x}Fe_{2-2x}SiO_4$), with on average 3–4× as much pyroxene as olivine (*Molster et al.*, 2002). The crystallinity of the YSO dust is unknown, but it is likely to be dominated by high-temperature condensates and annealed material since it is ejected in winds/jets that are generated in the hot inner disk regions.

Carbon-rich AGB stars are the main producers of carbonaceous dust, and they produce about as much dust as O-rich AGB stars. Models of the IR spectra of C stars suggest that the dust is mostly amorphous C and SiC (*Jura*, 1994).

2.2. Interstellar Dust

In the diffuse ISM, grains are subject to modification or destruction in supernova-generated shock waves (*Jones et al.*, 1996). Shattering and vaporization, both associated with grain-grain collisions, and sputtering are the principal modes of modification/destruction. Analysis of gas phase depletions suggests that there is preferential sputtering of the elements, with Si > Mg > Fe (*Tielens*, 1998; *Jones*, 2000), possibly because they are in different phases.

Best estimates of the range of possible diffuse ISM dust compositions indicate that ~59–65 wt% is silicates and ~34–38 wt% carbonaceous (*Zubko et al.*, 2004). The silicate dust appears amorphous and dominated by ≤0.1-μm grains with olivine-like (85%) and pyroxene-like (15%) compositions (Fe/Mg ≈ 1 for both) (*Kemper et al.*, 2004). The crystalline fraction is 0.2 ± 0.2% by mass and if present at all, the crystals are probably mostly forsterite. Given the ~10–20% crystallinity of circumstellar dust being added to the ISM, to maintain such a low crystalline fraction grains must be amorphized on timescales that are ~1–2% of their diffuse ISM lifetime. The amorphization is probably the result of irradiation in supernova-generated shocks (e.g., *Demyk et al.*, 2001; *Brucato et al.*, 2004)

There is considerable uncertainty about the nature and origin of the C-rich dust in the diffuse ISM. *Zubko et al.* (2004) found that between 13% and 100% of the C-rich dust could be in refractory organics, including polycyclic aromatic hydrocarbons (PAHs), the remainder being graphite/amorphous C. Here refractory organics include all nonvolatile organic matter that remains in the dust in the diffuse ISM, cometary comas, etc. *Pendleton and Allamandola* (2002) proposed that ~80% of the C in the refractory organic material is in large PAHs (C_{20-200}), originally formed in C-star outflows, that are linked by short, branched aliphatic chains. *Dartois et al.* (2004) suggest that the refractory organics form by UV photolysis of methane ices and contain at most 15% PAHs. Nanosized carbon grains irradiated with H also reproduce the 3–7-μm IR features ascribed to the refractory organics (*Mennella et al.*, 2002).

The larger gas phase depletions of refractory elements like Mg and Si in diffuse clouds compared to the diffuse ISM (*Savage and Sembach*, 1996) requires condensation of material sputtered and/or vaporized in shocks along with uncondensed material from stellar outflows. The condensed material is probably amorphous and enriched in Si relative to Mg and Fe (*Tielens*, 1998).

In the dense molecular clouds, all but the most volatile elements condense out of the gas. The nature of the silicate dust in molecular clouds is not well understood, but spectra of embedded Class I protostars in the Taurus star-forming region show purely amorphous silicate absorption features (*Watson et al.*, 2004) arising from their molecular cloud envelopes and, possibly, their disks. Grains in molecular clouds are protected from the destructive shocks in the diffuse ISM. The amorphous nature of the molecular cloud silicates probably reflects the rapid cycling of material between the dense and diffuse phases of the ISM (*McKee*, 1989; *Draine*, 1990).

2.3. Protoplanetary Disks

Observations of classical T Tauri stars in the 10-μm region show much more complex spectra than the ISM, particularly an increased emission at 11.3 μm, resulting in a flat-topped appearance (*Przygodda et al.*, 2003; *Forrest et al.,* 2004). Individual T Tauri spectra have been fitted with mixtures of amorphous and crystalline grains (e.g., *Honda et al.,* 2003; *Ciardi et al.,* 2005). However, in general the increased flux at 11.3 μm, overall spectral shapes, and diminished total intensities are consistent with the removal of small grains (~0.1 μm) and only large grains (~2 μm) remaining (*Przygodda et al.,* 2003; *Kessler-Silacci et al.,* 2005). Observations with the Spitzer telescope have unambiguously detected crystalline silicates from T Tauri star disks (Bouwman et al., in preparation). However, at present the data do not exist for correlating crystalline content with age or distance from the star.

The picture is somewhat clearer for disks around the more massive (>2 M$_\odot$) Herbig AeBe stars. Crystalline silicates, both forsterite and enstatite, have been unambiguously detected in Herbig AeBe disks (e.g., *Meeus et al.,* 1998; *Bouwman et al.,* 2001; *van Boekel et al.,* 2005) with crystalline mass fractions of up to ~15%. Gradients in the crystallinity of three Herbig AeBe disks have been observed with much higher degrees of crystallinity in the inner 1–2 AU (40–95%) than in the outer 2–20 AU (10–40%) (*van Boekel et al.,* 2004). One of these disks shows a higher forsterite to enstatite ratio in the inner (2.1) compared to the outer (0.9) disk. The ages of Herbig AeBe stars are difficult to estimate, but those with crystalline dust, such as HD 142527 (~1 m.y.) and HD 100546 (~10 m.y.), span the range of pre-main-sequence ages.

Little is known about the organic content of protoplanetary disk dust. Gaseous PAH emission has been detected from disks (*Acke and van den Ancker,* 2004; *Habart et al.,* 2004a), but its origin is uncertain. *Li and Lunine* (2003) assume that the gaseous PAHs are released by sublimation of interstellar ices. Alternatively, PAHs might also be generated by pyrolysis of organic matter in dust during the thermal processing that produced the crystalline silicates, or by irradiation/sputtering of organics in the dust.

2.4. Comets

Crystalline silicates, predominantly forsterite, have long been observed in Oort cloud (long period) comets, such as Halley (*Swamy et al.,* 1988) and Hale-Bopp (*Hanner et al.,* 1999). These objects presumably finished their aggregation in the giant planet region (5–10 AU) and were then scattered to large distances by interactions with Jupiter and the other giant planets (*Weidenschilling,* 1997). Kuiper belt objects are thought to have formed and stayed in the outer solar system at distances of ~35–50 AU from the Sun. Evidence for crystals in Kuiper belt comets has been scarce, but new results for 9P/Tempel 1 show that crystalline forsterite composed roughly half of the silicates (*Harker et al.,* 2005). Thus, there is little evidence for a gradient in crystallinity in the solar system between the formation regions of Oort cloud and Kuiper belt comets. No mechanism has been proposed for making crystalline material beyond 5–10 AU. The high crystalline content of 9P/Tempel 1 seems to require significant radial transport in the outer solar nebula. It has been argued that some IDPs are cometary in origin (*Joswiak*

et al., 2000), and are most likely to have come from Kuiper belt comets (*Flynn,* 1989). These IDPs potentially provide much more detailed information about the mineralogy of Kuiper belt objects than is possible astronomically (see section 4.4).

3. CONDENSATION, IRRADIATION, AND ANNEALING OF DUST ANALOGS

3.1. Condensation

Models of condensation in stellar outflows and protoplanetary disks usually assume thermodynamic equilibrium and predict crystalline condensates. However, whether equilibrium is maintained will depend on the relative rates of cooling, nucleation, and growth. It is evident from observations of AGB outflows that silicate condensates are largely amorphous, and therefore presumably never achieved equilibrium. Amorphous condensates may also form in the ISM and during thermal events in protoplanetary disks.

In the laboratory, amorphous Mg-Fe-Al-Si-O condensate "smokes" (nanometer-sized particulates) have been made in a gas-flow reactor (*Hallenbeck et al.,* 1998; *Nuth et al.,* 2000a) and by laser ablation of target materials (*Fabian et al.,* 2000; *Brucato et al.,* 2002). Not surprisingly, the physical properties of the "smokes" prepared by the two techniques differ, but in both, individual particles from a single experiment exhibit a wide range of compositions even in relatively simple systems (e.g., *Rietmeijer et al.,* 2002). Thus, stellar condensates and condensates from other low-pressure and relatively fast-cooling environments are likely to be amorphous and compositionally heterogeneous at microscopic scales. Their varied compositions will also influence how they anneal since individual grains will not have ideal mineral compositions, requiring diffusion to or away from growing crystals.

3.2. Irradiation and Amorphization

All grains in the ISM, and perhaps some in protoplanetary disks, should have been heavily irradiated. *Demyk et al.* (2001, 2004) estimate that most diffuse ISM silicate grains would be amorphized in a single fast supernova shock. Intense irradiation can produce distinctive microstructures and even chemical changes in grains that can be looked for.

Irradiation of submicrometer-sized grains by ≥400 keV H and He cosmic rays does little to their crystal structure, but leaves identifiable tracks (*Jäger et al.,* 2003; *Brucato et al.,* 2004). Irradiation of olivine and enstatite by He$^+$ at energies of 0.05, 1, 2.5, 5, and 12.5 keV/amu produces amorphous layers of approximately 4, 40, 90, 200, and 400 nm (*Demyk et al.,* 2001, 2004; *Carrez et al.,* 2002a; *Brucato et al.,* 2004). These energies are equivalent to shock velocities of roughly 100, 450, 700, 1000, and 1500 km/s. Amorphization by H rather than He to equivalent depths would require even faster shocks, but other abundant more massive ions (e.g., C, N, O) will contribute to amorphization at greater depths. Most of the amorphous grains in the ISM are <100 nm in size (*Mathis,* 1990; *Kemper et al.,* 2004). This would imply that shock speeds of up to ~500 km/s are needed to produce the almost complete amorphization of ISM grains, much higher than is typically assumed (100–200 km/s) (*Jones et al.,* 1996). *Slavin et al.* (2004) have shown that grains >0.1 μm can be betatron accelerated to much higher velocities than the shock speed, but they are largely destroyed.

Irradiation can also result in preferential sputtering of elements and changes in composition of the target. However, the experimental results often seem contradictory. Irradiation of olivine (Mg/Mg + Fe = 0.9) and enstatite by 4–10 keV He$^+$ produced preferential sputtering of O and Mg (*Demyk et al.,* 2001; *Carrez et al.,* 2002a; *Joswiak et al.,* 2004). On the other hand, *Dukes et al.* (1999) found no preferential sputtering of Mg or Fe when they irradiated olivine (Mg/Mg + Fe = 0.9) with 1 keV H$^+$ and 4 keV He$^+$, although preferential sputtering of O reduced the Fe to metal and produced some elemental Si. Reduction of Fe by preferential sputtering of O also occurs in lunar soil grains irradiated at similar energies by the solar wind (*Christoffersen et al.,* 1996). *Bradley* (1994) reported preferential sputtering of Mg and Ca relative to Si, and enrichment of O and Fe, during irradiation of olivine and pyroxene with 20 keV H$^+$ ions. While *Jäger et al.* (2003) found no preferential sputtering when irradiating enstatite with 50 keV He$^+$. As a result of preferential sputtering and implantation in interstellar shocks, diffuse ISM grains are unlikely to have ideal (stoichiometric) mineral compositions even if initially they were crystalline, and if they originally contained FeO it is likely to be reduced to metal.

Since amorphization timescales in the ISM are only ~1–2% of the typical grain lifetime, if irradiation is the amorphization mechanism it seems likely that silicates in the ISM will receive ion fluences that exceed amorphization thresholds by factors of up to 50–100. In the experiments, when fluences approach that required for amorphization the samples become vesiculated (*Demyk et al.,* 2001; *Jäger et al.,* 2003; *Joswiak et al.,* 2004). The amorphous rinds on mature lunar soil grains are saturated in solar-wind-implanted gases, but they are not vesiculated (*Keller and McKay,* 1997). However, in friable lunar breccias the rinds are vesiculated (*Noble et al.,* 2005), as are the solar-wind-saturated rinds on IDPs heated during atmospheric entry (*Brownlee et al.,* 1998), suggesting that a thermal pulse is needed to produce the vesiculation in these materials. There may be localized heating at the much higher fluxes of the irradiation experiments (*Jäger et al.,* 2003). The fluxes and timescales of interstellar shocks are closer to those of the experiments than solar-wind irradiation of lunar grains (hundreds of years). Thus, one might expect interstellar grains to be vesiculated or to become vesiculated on heating.

3.3. Annealing of Glasses, Smokes, and Irradiated Grains

The mechanisms and kinetics of annealing that might be expected for dust in protoplanetary disks have been explored

using glasses, condensate "smokes," and irradiated grains. Thermal annealing of enstatite glass, produced by quenching of a melt, generated orthoenstatite, except in the lowest-temperature experiments conducted at 1000 K when there was incipient formation of forsterite and the SiO$_2$ polymorph tridymite (*Fabian et al.,* 2000). Annealing of an enstatitic glass during electron irradiation in the transmission electron microscope (TEM) also resulted in the formation of forsterite and silica (*Carrez et al.,* 2002b). *Rietmeijer et al.* (1986) previously found that for nanoscale grains tridymite and forsterite appears to be more stable than enstatite. MgSiO$_3$-Mg$_2$SiO$_4$ "smokes" predominantly recrystallize as polycrystalline forsterite and silica (both amorphous and crystalline), along with some MgO in the forsteritic "smokes" (*Hallenbeck et al.,* 1998; *Fabian et al.,* 2000).

Activation energies for annealing rates of "smokes" and glass are similar, although there is some dependence on composition and how the materials were made (*Hallenbeck et al.,* 1998; *Fabian et al.,* 2000; *Brucato et al.,* 2002). Despite the defects produced during amorphization, the activation energy associated with recrystallization of irradiated olivine is similar to that of glasses and "smokes" (*Djouadi et al.,* 2005). Some amorphous silica is produced during the annealing of irradiated olivine.

The experiments suggest that annealed ISM amorphous condensates and irradiated grains are likely to be polycrystalline and even polymineralic. The shapes of annealed grains that were heavily irradiated at low energies may also preserve their original void-rich morphologies. The measured activation energies mean that the annealing times will be a strong function of temperature. For instance, during annealing Mg-Si-O smokes prepared in a gas-flow reactor go through a stall phase. To reach this stall phase requires a few minutes at 1067 K, 23 d at 1000 K, 3900 yrs at 900 K, and roughly 10^{14} yr at 800 K (*Hallenbeck et al.,* 2000). Thus, the presence of crystalline material in disk dust indicates a lower limit on processing temperatures of 800–900 K.

4. EVIDENCE FOR THERMAL PROCESSING OF DUST IN METEORITES AND INTERPLANETARY DUST PARTICLES

4.1. Meteorites and Interplanetary Dust Particles

4.1.1. Sources. With a few exceptions, those that come from the Moon and Mars, meteorites are fragments of main-belt asteroids (2–4 AU), with a strong bias toward inner-belt asteroidal sources (*Morbidelli et al.,* 2002). While there is an overall gradient of spectral classes in the asteroid belt (*Gradie and Tedesco,* 1982), there has been considerable mixing and it is likely that the meteorite collection has sampled many of them (*Burbine et al.,* 2002).

Some IDPs almost certainly come from comets (*Joswiak et al.,* 2000), but dynamical arguments suggest that most are asteroidal (*Dermott et al.,* 2002). Asteroidal particles would have been part of the zodiacal cloud, and evolved into Earth-crossing orbits through the Poynting-Robertson effect. Thus, they should sample a broader range of parent bodies than do meteorites. Dynamical considerations suggest that cometary IDPs are much more likely to be from Kuiper belt rather than Oort cloud comets (*Flynn*, 1989). Oort cloud comets tend to have higher eccentricities and therefore higher Earth encounter velocities (50–70 km/s) than Kuiper belt comets (10–40 km/s). The great majority of particles with velocities >25 km/s are not expected to survive atmospheric entry.

4.1.2. Classification. While there are variations, the bulk compositions of the chondrites are remarkably similar to that of the solar photosphere (excluding H, He, etc.). In terms of bulk composition, the most solar-like of the chondrites are the CIs (*Lodders*, 2003). Because solar (or CI) is the starting composition from which all solar system materials evolved, the compositional variations of chondrites and their components are generally expressed as deviations relative to CI.

The compositional variations (fractionations relative to solar) among chondrites are largely controlled by the volatility of the elements (Fig. 1). The volatility of an element is normally expressed as its calculated 50% equilibrium condensation temperature from a gas of solar composition at a total pressure of 10^{-4} bar (e.g., *Lodders*, 2003).

Historically, the chondrites have been divided into three groups based on their compositions and mineralogies (ordinary, carbonaceous, and enstatite). These in turn have been subdivided (*Scott and Krot*, 2003) into a number of classes: ordinary chondrites into H, L, and LL; carbonaceous chondrites into CI, CM, CR, CV, CO, and CK; and enstatite chondrites into EH and EL. The classification scheme is still evolving as more meteorites are found — two new groups (R and K chondrites) have been identified, and a number of individual meteorites do not belong to any recognized group.

After formation, the chondrites experienced secondary modification (thermal metamorphism and aqueous alteration) on their parent bodies. A petrographic classification

scheme for secondary processes divides the chondrites into six types: types 3–6 reflect increasing thermal metamorphism, and types 3–1 reflect increasing degrees of aqueous alteration. By convention, the chemical classification is followed by the petrologic one (e.g., CI1, CM2, CV3).

Chondritic IDPs, those with roughly chondritic bulk compositions, can be divided into two broad categories (*Bradley*, 2003): compact hydrous particles and porous anhydrous particles often referred to as chondritic porous (CP)-IDPs. The hydrous particles share many mineralogical similarities with the CM and CI chondrites. The anhydrous particles do not seem to have an affinity to any known meteorite class. Based on their very fine grain size, disequilibrium assemblage of minerals and amorphous silicates, and abundant presolar materials, the CP-IDPs are thought to be the most primitive solar system objects found to date. It has been argued that most CP-IDPs are cometary because of their high inferred atmospheric entry velocities (*Joswiak et al.*, 2000). The Stardust mission that recently returned samples of Comet Wild 2 should help establish whether CP-IDPs are indeed cometary.

4.1.3. Major components of meteorites. The chondrites are principally made up of three components — chondrules, refractory inclusions, and matrix — whose relative abundances vary widely. Refractory inclusions are a diverse group of objects with sizes that range from a few micrometers to centimeters across and abundances that range from <0.1 to 13 vol% (*MacPherson*, 2003). They are enriched in the most refractory elements (e.g., Ca and Al) in the canonical condensation sequence. Refractory inclusions formed at high temperatures, some by melting of preexisting material, others by condensation from a cooling gas. Their conditions of formation are somewhat uncertain, but most probably formed at ~1700–1800 K on timescales of hours to days in a system of roughly solar composition. They are the earliest solar system objects preserved in meteorites. Most models assume that refractory inclusions formed sunward of the asteroid belt and were transported to the asteroid belt via an X-wind or through turbulent diffusion.

Chondrules are the most abundant objects in most chondrites (up to ~80 vol%) (*Connolly*, 2005). They are ~0.1–1-mm-sized silicate and metal/sulfide spherules that formed by partial or complete melting of solid precursors. They experienced peak temperatures of ~1700–2100 K and formed on timescales of minutes to hours (cooling rates of ~10–1000 K/h) in environments that were probably enriched in dust relative to gas by factors of ~100–1000 compared to solar. There is considerable debate about when and how chondrules formed. However, it seems that the chondrules in at least some chondrites formed 1–2 m.y. after refractory inclusions. Currently, the most popular mechanism for chondrule formation is in shockwaves in the asteroid belt, although other explanations such as the X-wind model have not been ruled out.

The fine-grained matrix cements the chondrites together and makes up 10–50 vol% of most chondrites (*Huss et al.*, 2005). Matrix abundance in CI chondrites may have been 90–100 vol%, but they have been so extensively aqueously

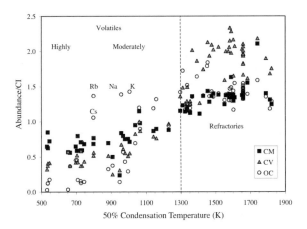

Fig. 1. The CI-normalized elemental abundances vs. 50% condensation temperature (*Lodders*, 2003) in three chondrite groups (*Wasson and Kallemeyn*, 1988) showing the volatility dependence of the elemental fractionations.

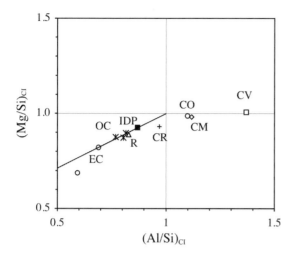

Fig. 2. The CI-normalized Mg/Si vs. Al/Si elemental ratios of bulk chondrites (*Wasson and Kallemeyn*, 1988) and anhydrous, CP-IDPs (*Schramm et al.*, 1989; *Thomas et al.*, 1993). The OC, EC, R, and CR chondrites, as well as the CP-IDPs, all form a trend that is probably linked to the loss of refractory material. The CO, CM, and CV chondrites form a distinct trend that may be due to addition of refractory inclusions.

altered that their primordial matrix abundances remain uncertain. Because of its fine grain size and relatively high volatile-element contents, matrix is generally assumed to be more primitive than chondrules and refractory inclusions, an assumption that is confirmed by the presence of presolar materials in it (see section 4.3). Matrix shares some similarities with IDPs, and it is in matrix and IDPs that we are most likely to find evidence for the thermally annealed dust.

Most of the least atmospheric entry heated CP-IDPs are enriched relative to CI in volatile elements (*Kehm et al.*, 2002; *Flynn et al.*, 2004). The average major-element compositions of IDPs are also fractionated (Fig. 2). The major- and trace-element fractionations may result either from preferential destruction of coarser/denser more refractory IDP material during atmospheric entry, or because like most chondrites there has been loss of more refractory material during processing in the nebula (*Alexander*, 2005).

4.2. Causes of Element Fractionations in Chondrites

The correlation between elemental abundances in chondrites and volatility (Fig. 1) clearly points to the role of thermal processes in generating the fractionations. The most widely accepted explanation for the fractionations has been that they reflect variations in conditions during cooling and condensation of an initially totally vaporized inner solar nebula (*Wasson and Chou*, 1974; *Bland et al.*, 2005). The continuous variation in abundance of moderately and highly volatile elements with condensation temperature in chondrites (Fig. 1) requires a more-or-less continuous process of separation of condensates from the gas (*Cassen*, 2001).

Variations in the more refractory element abundances seem to be associated with the addition of refractory inclusions to CM, CV, and CO chondrites, and loss of refractory material from most other chondrites (Figs. 1 and 2).

An alternative explanation for the elemental fractionations is the so-called two-component model (e.g., *Alexander*, 2005). In this model, the moderately and highly volatile element abundances of chondrites were largely determined by mixing of volatile-rich, primitive matrix and volatile-depleted chondrules and refractory inclusions. The presence of presolar materials in the matrix of all chondrites in roughly CI abundances shows that the chondrites did accrete a primitive matrix component. Most problematic for the two-component model is the fact that bulk matrix compositions are not CI-like, but the addition of a few tens of percent of more refractory material to matrix and secondary redistribution of elements during parent-body processes might explain this.

A third class of explanations are motivated by the rough complementarity between refractory element depletions in the gas of the diffuse ISM and the volatile-element depletions in the chondrites (*Palme*, 2002; *Huss et al.*, 2003; *Yin*, 2005). The models suggest that the fractionations in chondrites were largely the result of sublimation of volatile-rich ices and amorphous material inherited from the presolar molecular cloud. Hence, the volatile-element fractionations in meteorites should resemble those in the dust in the diffuse ISM. However, the CI chondrites are unfractionated although they are not composed of unprocessed interstellar material (see section 4.5), and these models cannot explain all the elemental fractionations seen in chondrites, or the isotopic systematics of radiogenic systems (e.g., U-Pb and Rb-Sr) in which the parent and daughter have very different volatilities (*Palme*, 2002).

4.3. Presolar Materials

While most components of chondrites and IDPs formed in the solar system, they also contain presolar materials that were inherited from the protosolar molecular cloud. These materials include circumstellar grains (formed around evolved stars) and interstellar organic matter. Interstellar grains (formed in the ISM) and grains from YSOs must be present, but it has proved to be much more difficult to definitively identify them. The presolar materials potentially retain a record of thermal processing in the solar nebula and also could act as tracers of unaltered primordial dust abundances.

4.3.1. Circumstellar grains. The major types of circumstellar grain found to date include nanodiamonds, silicates, SiC, Si_3N_4, graphitic spherules, and oxides. Based on their isotopic compositions, the circumstellar grains mostly formed around RGB and AGB stars, with a few percent from supergiants, supernovae, and, possibly, novae (e.g., *Zinner*, 2003; *Clayton and Nittler*, 2004). Circumstellar graphitic grains are highly underabundant in meteorites compared to galactic dust production rates (*Alexander,* 2001). Indeed, graphite and poorly graphitized C of any origin are rare in chondrites and IDPs, which is inconsistent with some diffuse ISM dust models. The major carbonaceous components are organic matter and nanodiamonds. Nanodiamonds have been tentatively identified in circumstellar outflows (*Hill et al.,* 1998), may be very abundant in the ISM (*Allamandola et al.,* 1993; *Jones and d'Hendecourt,* 2000), and have been observed in one protoplanetary disk (*Habart et al.,* 2004b).

To date, five presolar silicate grains have been examined in the TEM to determine their major-element compositions and mineralogies (*Messenger et al.,* 2003, 2005; *Nguyen et al.,* 2005). Contrary to astronomical expectations, two of them were crystalline forsteritic olivine, and one of these is a supernova grain (*Messenger et al.,* 2005). This compares with 0.2 ± 0.2% crystalline silicates in the diffuse ISM (see section 2.2) and the inferred high enstatite/forsterite ratio in stellar outflows (see section 2.1). Neither of the olivine grains shows evidence for having been irradiated, although the supernova grain is composed of multiple subgrains and may have been annealed.

With the exception of ~1% of highly disordered grains, the SiC is highly crystalline with little evidence for radiation damage (*Daulton et al.,* 2003; *Stroud et al.,* 2004b; *Stroud and Bernatowicz,* 2005). The larger pristine SiC grains (≥0.5 μm) show little evidence for sputtering or cratering (*Bernatowicz et al.,* 2003). Smaller SiC grains found in acid residues are more irregular in shape (*Daulton et al.,* 2003), but whether this is the result of grain growth, ISM processing, or the harsh chemical treatments used to isolate them is not known. The size distribution of SiC in meteorites is consistent with the inferred size distribution around C stars (*Russell et al.,* 1997).

Stroud et al. (2004a, 2005) reported detailed studies of two Al_2O_3 grains and two hibonite ($CaAl_{12}O_{19}$) grains. The hibonite and one of the Al_2O_3 grains were crystalline, while the other Al_2O_3 grain was amorphous. *Stroud et al.* (2004a) argued that the grains that survive in meteorites "were not significantly processed by radiation in the ISM" and that the differences in microstructure of the two Al_2O_3 grains reflect conditions during condensation in the AGB outflows.

It seems likely that the circumstellar grains found in meteorites and IDPs avoided processing in the diffuse ISM. Direct injection of dust from AGB stars, etc., into molecular clouds is rare (*Kastner and Myers,* 1994) and cannot explain the tens to hundreds of sources that are represented in the circumstellar grains (*Alexander,* 2001). There is no evidence that the circumstellar grains have protective layers, of organics for instance, that are thick enough to have protected them from grain-grain collisions in ISM shocks (*Bernatowicz et al.,* 2003). SiC grains >0.5 μm can largely survive (≤10% loss) an ISM shock provided that the shock speed lies between ~50–80 km/s (*Slavin et al.,* 2004). Graphite and silicate grains can also largely survive provided that they are >1 μm and shock speeds fall in much more restricted ranges. The much more restricted range of

conditions over which graphite and silicate grains survive shocks means that they should be underabundant relative to SiC, particularly at grain sizes of <1 μm. Graphite is underabundant (*Alexander,* 2001). Circumstellar silicate abundances are still rather uncertain, but almost all the grains found to date are <1 μm. Perhaps the best explanation is that stochastic processes result in a small fraction of grains only spending a short time in the diffuse ISM and/or only encountering shocks that do not significantly affect them.

4.3.2. Interstellar organics. The organic matter in chondrites can be divided into soluble and insoluble fractions (*Gilmour,* 2003). The soluble fraction has been intensively studied, but the insoluble organic material (IOM) makes up the majority >75% of the organic material in chondrites. The large D and ^{15}N isotopic enrichments in IOM in chondrites and IDPs are thought to be the result of ion-molecule reactions and other ISM processes (e.g., *Messenger,* 2000; *Aléon et al.,* 2001), although D enrichments in gas-phase molecules are possible in the outer solar nebula (*Aikawa et al.,* 2002). There are variations in the isotopic composition of IOM within and between chondrites and IDPs, but this may reflect parent-body processing in chondrites (*Alexander et al.,* 1998) and atmospheric entry heating of IDPs (*Keller et al.,* 2004). The most primitive meteoritic IOM at least isotopically resembles the IOM in IDPs (*Busemann et al.,* 2006), and has an elemental composition ($C_{100}H_{75}N_4O_{15}S_4$) that is similar to the average composition ($C_{100}H_{80}N_4O_{20}S_2$) of Comet Halley CHON particles (*Kissel and Krueger,* 1987). Thus the organics in meteorites, CP-IDPs, and comets appear to be related despite the very different formation conditions and locations of their parent bodies.

Further evidence in favor of the IOM's presolar origins is found in the similarity between the 3–4-μm infrared (IR) spectrum of the IOM and the refractory organic matter in the diffuse ISM (*Pendleton et al.,* 1994). The IOM may also be partly responsible for the ubiquitous 2175 Å UV absorption feature in the diffuse ISM (*Bradley et al.,* 2005). This would be consistent with the inference of *Adamson et al.* (1999) that the carriers of the 3–4-μm and 2175 Å features share many of the same characteristics.

The IOM is composed of 50–60% of small PAHs (<C_{20}) and 40–50% short, highly branched aliphatic chains (e.g., *Sephton et al.,* 2004; *Cody and Alexander,* 2005). Of the models for refractory organics in the diffuse ISM (see section 2.1), this most closely resembles that of *Pendleton and Allamandola* (2002), although their PAHs are larger (C_{20-200}) and more abundant (~80%).

4.4. Composition and Microstructure of Interplanetary Dust Particles

Presolar and primitive solar materials are preserved in the matrices of chondrites (*Huss et al.,* 2005). However, the abundance of presolar material in chondrite matrices are not as high as in CP-IDPs, and no chondrites have entirely escaped parent-body processing that will have modified or destroyed the finest grained silicates. Hence here we only consider CP-IDPs.

CP-IDPs contain Mg-rich crystalline silicates (mostly enstatite and forsterite), equilibrated aggregates (EAs), amorphous silicates largely in the form of GEMS (glass with embedded metal and sulfide), and IOM (*Bradley*, 2003). Crystalline silicates are a major (~20–50 vol%) component of anhydrous IDPs and occur as single crystals ranging in size from 0.1 μm to several micrometers, as well as in polycrystalline aggregates with constituent grain sizes <0.5 μm. They are typically enstatitic pyroxene and forsteritic olivine (Mg/(Mg + Fe) > 0.95), although they can be more Fe-rich and other minerals are present. The microstructure (multiple twins and intergrowth of ortho- and clino-polymorphs) of much of the pyroxene (Fig. 3) suggests that it formed by condensation at temperatures above 1258 K (*Bradley et al.,* 1983) and cooled relatively rapidly. There has been no systematic study of the pyroxenes with this microstructure. The cooling rate, estimated from the ~20–25 vol% of orthoenstatite in one grain (*Bradley et al.,* 1983) and the experiments of *Brearley et al.* (1993), is ~1000 K/h, but clearly more work needs to be done. A similar microstructure is seen in chondrule pyroxenes and chondrules are thought to have cooled at ~10–1000 K/h. So-called enstatite whiskers and rods seem to be metastable condensates that cannot be used for cooling rate estimates (*Bradley et al.,* 1983). The pyroxene/olivine ratio varies considerably between IDPs. The pyroxene-rich ones are generally considered to be more primitive because in a small sample they tended to be more C-rich (*Thomas et al.,* 1993), but forsterite seems to dominate in comets (see section 2.4).

Crystalline silicates also occur in EAs that are a common minor (<10 vol%) component of CP-IDPs (Fig. 4). These micrometer-sized aggregates contain numerous grains of enstatite, pyrrhotite (Fe$_{1-x}$S), minor forsterite, and an interstitial amorphous Si-rich phase. The textures, mineralogy, and mineral chemistry of EAs are consistent with the annealing of GEMS precursors at T ≥ 1000 K for hours (*Brownlee et al.,* 2005). A continuum of morphologies is

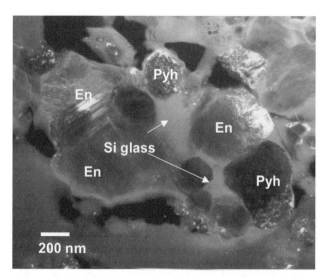

Fig. 4. An annealed aggregate in a CP-IDP composed of enstatite (En: MgSiO$_3$) and pyrrhotite (Pyr: Fe$_{1-x}$S) set in a SiO$_2$-rich glass.

observed from porous to solid GEMS to EAs that probably reflect a sequence of thermal annealing at subsolidus temperatures (below the thermal stability limit of pyrrhotite) in the nebula. The presence of the Si-rich amorphous phase as a by-product of the annealing is consistent with IR observations of protoplanetary disks that invoke a Si-rich component to fit the 10-μm silicate feature of the processed silicates (e.g., *Bouwman et al.,* 2001).

GEMS grains are <0.5 μm in diameter and consist of numerous 10–50-nm-sized Fe-Ni metal and Fe-Ni sulfide grains dispersed in a Mg-Si-Al-Fe amorphous silicate matrix (Fig. 5). The FeO contents of GEMS are very low, in contrast to what has been inferred for ISM dust (see section 2.2). *Keller et al.* (2005) have demonstrated that most GEMS grains are aggregates composed of even smaller subgrains (<100 nm) exhibiting strongly heterogeneous chemical compositions. GEMS grains are systematically subsolar (~0.6 × solar) with respect to S/Si, Mg/Si, Ca/Si, and Fe/Si (*Keller and Messenger,* 2004), although the average Al/Si ratio in GEMS is indistinguishable from solar.

Bradley (1994) proposed that GEMS are preserved interstellar silicates, based primarily on O excesses and Mg depletions at the edges of GEMS that they attributed to preferential sputtering. Other irradiation experiments have found that O is depleted along with Mg (see section 3.2). *Bradley et al.* (1999) also showed that there is a close resemblance in the IR spectra of GEMS and interstellar silicates. *Westphal and Bradley* (2004) propose that GEMS grains are the products of intense irradiation in fast (~1000 km/s) shocks of initially crystalline circumstellar grains from supergiant stars in OB associations. However, the presolar grains and most dust production estimates suggest that supergiants are minor contributors of dust to the ISM (see sections 2.2 and 4.3). Also, the O elemental enrichments in GEMs and lack of vesicles are contrary to expectations from most irradiation experiments, although the metal could be the result of irradiation-induced reduction. The model predicts that ulti-

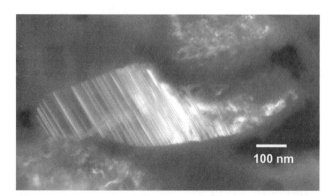

Fig. 3. A typical isolated enstatite grain from a chondritic porous IDP. The fine lamellae are the result of relatively rapid cooling through the protoenstatite-orthoenstatite (1258 K) transition (*Bradley et al.,* 1983).

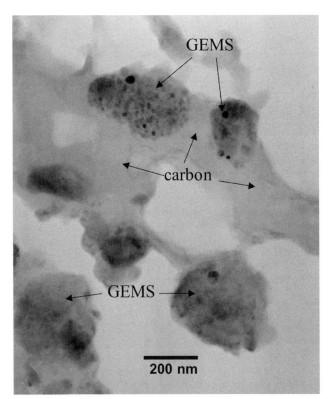

Fig. 5. GEMS from a CP-IDP set in an organic carbon matrix. The GEMS are polymineralic objects. The dark subgrains are mostly Fe-Ni metal and Fe sulfide.

mately there is almost complete replacement of the original atoms by the implanted shock gas. Both the original supergiant grains and the shock gas, which is highly enriched in supernovae ejecta, will be isotopically very anomalous in O. Yet within the measurement errors, <1–5% of GEMS have demonstrably nonsolar O-isotopic compositions, and most that are anomalous probably formed around AGB stars (see section 4.3). The only supernova silicate grain found to date is crystalline.

One explanation for the lack of vesicles and roughly solar isotopic compositions would be if GEMS were ISM condensates. *Keller and Messenger* (2004) point out that on average GEMS compositions are not consistent with average diffuse ISM dust; relative to Si their Mg, Fe, and Ca contents are too low and their S contents are too high. They estimate that only 10–20% of GEMS have roughly diffuse ISM-like dust compositions, while the remaining 80–90% are probably solar system in origin. Condensates that formed in the ISM or in YSO outflows may be present, but if their isotopic anomalies are small they would not have been recognized. Even with isotopically anomalous silicate/ oxide grains, it may be difficult to distinguish between circumstellar and YSO grains under some circumstances. *Keller et al.* (2005) also argue that the heterogeneous chemical compositions of the GEMS subunits are at odds with uniform chemical gradients expected from extensive irradiation, and that GEMS are more likely to be the result of coagulation of compositionally distinct subgrains that formed

during fractional condensation, probably in the solar nebula (*Keller et al.,* 2005).

However, there is a conflict between a solar system origin for most GEMS and an interstellar origin for the IOM. The IOM would be destroyed at high temperatures. Consequently, if most crystalline material and most GEMS are solar condensates, no more than 10–20% of the original presolar IOM should remain in IDPs. Yet, the IOM content of IDPs is ~40–70% of that expected of ISM dust, which is roughly consistent with their crystallinity (see section 4.5). Either a higher fraction of the GEMS must be interstellar, or much of the IOM is solar system in origin.

4.5. Constraints on the Thermal Processing of Dust

The crystalline material in IDPs is dominated by what appear to be solar nebula condensates, with lesser amounts of annealed GEMS heated to ≥1000 K. The microstructure of much of the pyroxene in IDPs seems to require relatively rapid cooling from above 1258 K. The IOM would be destroyed or heavily modified by temperatures of ≥1000 K. Therefore, if the IOM is interstellar, its abundance is a useful indicator of the degree of thermal processing, particularly in the chondrites where parent-body processes have modified much of the fine-grained material.

Assuming that all Mg was condensed in dust and there have not been any dust fractionations, the solar-normalized C/Mg ratios give the fractions of the total C in refractory organics in the dust formation regions. The estimated fraction of C in refractory organics in Comet Halley dust, including the quoted possible factor of 2 error, is in the range 1–0.3 (*Jessberger and Kissel,* 1991; *Schulze et al.,* 1997), in the CP-IDPs is ~0.35 (*Schramm et al.,* 1989; *Thomas et al.,* 1993), and in CIs is 0.07 (*Alexander et al.,* 1998). Roughly 15% of the original IOM in CIs may have been destroyed by aqueous alteration (*Cody and Alexander,* 2005). Halley dust has a crystallinity of ~50%. If this crystallinity reflects the fraction of material that has been thermally processed, the original fraction of C in refractory organics in dust from the presolar molecular cloud would have been ≥0.6. The abundance of refractory organics in dust in molecular clouds has not been determined directly. In the diffuse ISM, assuming that all C-rich dust is organic (including PAHs), ~50% of the total C is in the organics (*Zubko et al.,* 2004) and ~80% if the *Lodders* (2003) solar composition is adopted.

The low abundance of IOM in CI chondrites may be due to >75–85% of the original dust having been thermally processed. If it was the result of thermal processing, somehow it was done without significantly fractionating the more volatile elements because the CI and solar abundances are very similar (*Lodders,* 2003). This means that gas and dust cannot have fractionated from one another during the thermal processing (evaporation/condensation and annealing), perhaps because the dust was fine-grained enough for it to remain coupled to the gas. The difference in IOM abundance between chondrites, CP-IDPs, and comets suggests that there was a gradient in the dust processing in the solar nebula. How steep the gradient was will depend on where the CP-IDPs formed.

Scott and Krot (2005) have suggested that dust processing could have occurred during chondrule formation. However, the abundance of IOM in matrix is roughly CI-like in all primitive chondrites, and apparently unrelated to chondrule or refractory inclusion abundances. The CI chondrites contained few if any chondrules or refractory inclusions, yet they have much lower IOM contents than CP-IDPs or comets. There is not evidence for chondrule fragments in CP-IDPs, although this could reflect a bias because larger, denser particles are more intensely heated during atmospheric entry. Finally, the lack of large isotopic fractionations in chondrules suggests that they formed as stable melts (e.g., *Davis et al.,* 2005). In this case, it is unlikely that the gas would have overcome kinetic barriers to nucleation and condensed as small isolated crystals rather than condensed directly onto the already existing chondrules. At present, it seems unlikely that the crystalline dust in CP-IDPs is the direct product of chondrule formation. Also, the evidence from chondrules and CAIs is that volatile elements were lost during their formation.

5. MODELING OF GRAIN HEATING IN PROTOPLANETARY DISKS

5.1. During Accretion onto the Disk

Infalling dust from the molecular cloud envelope surrounding a forming stellar system will be heated first by radiation from the central star and disk, and subsequently at the accretion shock. *Chick and Cassen* (1997) modeled the heating grains would experience during infall onto the disk. The results depend on the model assumptions, but they conclude that grain temperatures would never exceed 1000 K beyond ~1.5 AU, and that refractory organics would survive beyond 0.5–4 AU depending on the conditions. Since most material will accrete onto the disk beyond 4 AU, infall heating is an unlikely explanation for the crystallinity of silicates beyond the terrestrial planet region or the variations in refractory organic abundances between chondrites, IDPs, and comets.

5.2. Shocks

At radial distances out to ~5–10 AU, heating in shocks generated by disk instabilities or the giant planets are sufficient to anneal grains and possibly vaporize small ones (<0.25 µm) (*Harker and Desch,* 2002). Oort cloud (long period) comets are thought to have formed at 5–10 AU and then been scattered into the Oort cloud by Jupiter. Thus, shock heating could explain the high fraction of crystalline material in Oort cloud comets without requiring vigorous radial mixing. Shock heating would also be consistent with the relatively rapid cooling required by the CP-IDP pyroxene microstructures (see section 4.4). Since small grains will remain coupled to the gas while they are being heated,

it is possible that shock heating could anneal or vaporize and recondense grains without producing volatile element fractions. However, shock heating cannot explain the high crystallinity of 9P/Tempel 1 (see section 2.4), a Kuiper belt comet, unless there was vigorous transport of dust in the disk or some Kuiper belt objects formed in the giant planet region. The same is true if CP-IDPs are mostly from Kuiper belt comets (see section 4.1.1).

5.3. Radial Transport in Disks

Radial transport of dust from the hot inner portions of a disk is another explanation for the high crystallinity of silicates in comets (*Nuth et al.,* 2000b). The distances that material must be transported will depend on the stage of disk evolution. For typical T Tauri disks with low accretion rates, the estimated midplane temperatures range from about 1500 K inside about 0.2 AU, to about 300 K at 1 AU (e.g., *D'Alessio et al.,* 2001). During periods of significantly higher mass accretion, midplane temperatures are expected to be correspondingly higher, with temperatures exceeding 1000 K at 1–5 AU under some conditions (*Boss,* 1998; *Bell et al.,* 2000).

Outward flow of material near the midplane of disks with return flow in the upper regions of the disk, turbulent diffusion, and large-scale motions associated with spiral arms could all contribute to the transport of some matter from the inner to the outer portions of a disk on reasonable timescales (*Gail,* 2001, 2004; *Bockelée-Morvan et al.,* 2002; *Boss,* 2004). This could possibly explain observations of crystallized silicates in comets, but the timescales for cooling of condensed grains in these large-scale motions are likely to be too slow to explain the microstructures of crystalline pyroxene in CP-IDPs (see section 4.3).

Gail (2002) also models the destruction of refractory organic C as primordial dust is transported within a disk. The abundance of refractory C increases with increasing radial distance, and at any given radial distance the abundance depends on the mass accretion rate. Qualitatively these results are consistent with the relatively low C content in chondrites (including CIs) compared to comets and IDPs. The models suggest that the low C contents of chondrites forming at 2–3 AU would require high accretion rates (>$10^{-6} M_\odot$/yr), but it is not clear whether the refractory C remaining has been heated significantly. The many similarities between the IOM in chondrites (~2 wt%) and CP-IDPs (5–40 wt%) suggest that the chondritic material has not been heated significantly. Analysis of the IOM by pyrolysis GC-MS is usually carried out by heating the sample to ~873 K for a few seconds. Experiments by Cody et al. (in preparation) also find that heating at 873 K for even a few seconds significantly affects the IOM.

6. SUMMARY

The relative abundances of circumstellar grains in meteorites broadly conform to astronomical estimates, with most coming from AGB stars, although the abundance of graphite/amorphous C in meteorites is lower and nanodiamonds higher than expected. The relative abundances of supernova grains are only of the order of a few percent, which is much lower than some estimates or upper limits. It seems likely that the circumstellar grains preserved in meteorites avoided processing in the ISM.

The evidence for interstellar material in meteorites is less clear. Many properties of the IOM in meteorites and IDPs are consistent with an interstellar origin, but whether synthesis of similar materials in protoplanetary disks is possible has yet to be determined. The evidence for an interstellar origin for GEMS is a matter of debate. Many of the properties of GEMS are consistent with an interstellar origin. However, their often high S contents and lack of evidence for implanted H and He seem to be inconsistent with GEMS being highly irradiated interstellar grains. Within the measurement uncertainties, <1–5% of GEMS have nonsolar O-isotopic compositions. Some GEMS may be condensates, probably of solar system origin, although at present interstellar and YSO origins cannot be excluded. If most GEMS are solar system condensates, the crystallinity of IDPs underestimates the degree of thermal processing and most IOM must be solar.

The crystalline fraction of CP-IDPs is dominated by solar system condensates, with lesser amounts of annealed GEMS. The microstructure of the pyroxene suggests that it formed above 1258 K and then cooled relatively fast (~1000 K/h). Because the IOM would be modified or destroyed at temperatures >1000 K, provided that it is interstellar its abundance is a useful additional indicator of the degree of thermal processing. In CP-IDPs, the IOM abundance is about 40–70% of estimates of the ISM dust organic content. This is roughly consistent with the 20–50% silicate crystallinity in CP-IDPs. The IOM content in CI chondrites is much lower than this. Thus, there probably was a gradient of thermal processing in the solar system, but how steep it was depends on the origin of the CP-IDPs (outer main-belt asteroids or comets). Whatever the mechanism for thermal processing of dust, it did not result in the loss of the volatile elements, perhaps because the dust was fine-grained enough to remain coupled to the gas as it cooled.

Models show that radial transport of material from the hot inner nebula out to distances of tens of AU is possible on reasonable timescales. However, the cooling rates in the inner nebula may be too slow to explain the microstructure of pyroxene in IDPs that require cooling rates after formation of ~1000 K/h. Shock heating predicts faster cooling times. Shock heating is unlikely to be strong enough to anneal/vaporize dust beyond ~10 AU. If crystalline material in Kuiper belt objects was produced by shock heating, an efficient mechanism for radial transport in the outer solar system is still required. Shock heating is also the currently most favored mechanism for making chondrules. Chondrule formation was probably not directly responsible for thermal processing of dust. However, it is possible that IDP dust processing and chondrule formation were driven by the same mechanism and occurred at the same time but in different places — dust processing occurred in regions of lower dust

density with few large grains (e.g., above the midplane in the asteroid region and/or at >3–4 AU).

REFERENCES

Acke B. and van den Ancker M. E. (2004) *Astron. Astrophys., 426,* 151–170.

Adamson A. J., Whittet D. C. B., Chrysostomou A., Hough J. H., Aitken D. K., Wright G. S., and Roche P. F. (1999) *Astrophys. J., 512,* 224–229.

Aikawa Y., van Zadelhoff G. J., van Dishoeck E. F., and Herbst E. (2002) *Astron. Astrophys., 386,* 622–632.

Aléon J., Engrand C., Robert F., and Chaussidon M. (2001) *Geochim. Cosmochim. Acta, 65,* 4399–4412.

Alexander C. M. O'D. (2001) *Philos. Trans. R. Soc. London, A359,* 1973–1988.

Alexander C. M. O'D. (2005) *Meteoritics & Planet. Sci., 40,* 943–965.

Alexander C. M. O'D., Russell S. S., Arden J. W., Ash R. D., Grady M. M., and Pillinger C. T. (1998) *Meteoritics & Planet. Sci., 33,* 603–622.

Allamandola L. J., Sandford S. A., Tielens A. G. G. M., and Herbst T. M. (1993) *Science, 260,* 64–66.

Bell K. R., Cassen P. M., Wasson J. T., and Woolum D. S. (2000) In *Protostars and Planets IV* (V. Mannings et al., eds.), pp. 897–926. Univ. of Arizona, Tucson.

Bernatowicz T. J., Messenger S., Pravdivtseva O., Swan P., and Walker R. M. (2003) *Geochim. Cosmochim. Acta, 67,* 4679–4691.

Bland P. A., Alard O., Benedix G. K., Kearsley A. T., Menzies O. N., Watt L. E., and Rogers N. W. (2005) *Proc. Natl. Acad. Sci., 102,* 13755–13760.

Bockelée-Morvan D., Gautier D., Hersant F., Huré J. M., and Robert F. (2002) *Astron. Astrophys., 384,* 1107–1118.

Boss A. P. (1998) *Ann. Rev. Earth Planet. Sci., 26,* 53–80.

Boss A. P. (2004) *Astrophys. J., 616,* 1265–1277.

Bouwman J., Meeus G., de Koter A., Hony S., Dominik C., and Waters L. B. F. M. (2001) *Astron. Astrophys., 375,* 950–962.

Bradley J. P. (1994) *Science, 265,* 925–929.

Bradley J. P. (2003) In *Treatise on Geochemistry, Vol. 1: Meteorites, Comets, and Planets* (A. M. Davis, ed.), pp. 689–712. Elsevier-Pergamon, Oxford.

Bradley J. P., Brownlee D. E., and Veblen D. R. (1983) *Nature, 301,* 473–477.

Bradley J. P., Keller L. P., Snow T. P., Hanner M. S., Flynn G. J., Gezo J. C., Clemett S. J., Brownlee D. E., and Bowey J. E. (1999) *Science, 285,* 1716–1718.

Bradley J. P., Dai Z. R., Erni R., Browning N., Graham G., et al. (2005) *Science, 307,* 244–247.

Brearley A. J., Jones R. H., and Papike J. J. (1993) In *Lunar Planet. Sci. XXIV,* pp. 185–186.

Brownlee D. E., Joswiak D. J., Bradley J. P., Schlutter D. J., and Pepin R. O. (1998) In *Lunar Planet. Sci. XXIX,* p. 1869.

Brownlee D. E., Joswiak D. J., Bradley J. P., Matrajt G., and Wooden D. H. (2005) In *Lunar Planet. Sci. XXXVI,* Abstract #2391.

Brucato J. R., Mennella V., Colangeli L., Rotundi A., and Palumbo P. (2002) *Planet. Space Sci., 50,* 829–837.

Brucato J. R., Strazzulla G., Baratta G., and Colangeli L. (2004) *Astron. Astrophys., 413,* 395–401.

Burbine T. H., McCoy T. J., Meibom A., Gladman B., and Kiel K. (2002) In *Asteroids III* (W. F. Bottke Jr. et al., eds.), pp. 653–667. Univ. of Arizona, Tucson.

Busemann H., Young A. F., Alexander C. M. O'D., Hoppe P., Mukhopadhyay S., and Nittler L. R. (2006) *Science, 314,* 727–730.

Carrez P., Demyk K., Cordier P., Gengembre L., Grimblot J., d'Hendecourt L., Jones A. P., and Leroux H. (2002a) *Meteoritics & Planet. Sci., 37,* 1599–1614.

Carrez P., Demyk K., Leroux H., Cordier P., Jones A. P., and d'Hendecourt L. (2002b) *Meteoritics & Planet. Sci., 37,* 1615–1622.

Cassen P. (2001) *Meteoritics & Planet. Sci., 36,* 671–700.

Chick K. M. and Cassen P. (1997) *Astrophys. J., 477,* 398–409.

Christoffersen R., Keller L. P., and McKay D. S. (1996) *Meteoritics & Planet. Sci., 31*, 835–848.

Ciardi D. R., Telesco C. M., Packham C., Gómez Martin C., Radomski J. T., De Buizer J. M., Phillips C. J., and Harker D. E. (2005) *Astrophys. J., 629*, 897–902.

Clayton D. D. and Nittler L. R. (2004) *Ann. Rev. Astron. Astrophys., 42*, 39–78.

Cody G. D. and Alexander C. M. O'D. (2005) *Geochim. Cosmochim. Acta, 69*, 1085–1097.

Connolly H. C. Jr. (2005) In *Chondrites and the Protoplanetary Disk* (A. N. Krot et al., eds.), pp. 215–224. ASP Conf. Ser. 341, San Francisco.

D'Alessio P., Calvet N., and Hartmann L. (2001) *Astrophys. J., 553*, 321–334.

Dartois E., Muñoz-Caro G. M., Deboffle D., and d'Hendecourt L. (2004) *Astron. Astrophys., 423*, L33–L36.

Daulton T. L., Bernatowicz T. J., Lewis R. S., Messenger S., Stadermann F. J., and Amari S. (2003) *Geochim. Cosmochim. Acta, 67*, 4743–4767.

Davis A. M., Alexander C. M. O'D., Nagahara H., and Richter F. M. (2005) In *Chondrites and the Protoplanetary Disk* (A. N. Krot et al., eds.), pp. 432–455. ASP Conf. Ser. 341, San Francisco.

Demyk K., Carrez P., Leroux H., Cordier P., Jones A. P., Borg J., Quirico E., Raynal P. I., and d'Hendecourt L. (2001) *Astron. Astrophys., 368*, L38–L41.

Demyk K., d'Hendecourt L., Leroux H., Jones A. P., and Borg J. (2004) *Astron. Astrophys., 420*, 233–243.

Dermott S. F., Durda D. D., Grogan K., and Kehoe T. J. J. (2002) In *Asteroids III* (W. F. Bottke Jr. et al., eds.), pp. 423–442. Univ. of Arizona, Tucson.

Djouadi Z., d'Hendecourt L., Leroux H., Jones A. P., Borg J., Deboffle D., and Chauvin N. (2005) *Astron. Astrophys., 440*, 179–184.

Draine B. T. (1990) In *Evolution of the Interstellar Medium* (L. Blitz, ed.), pp. 193–205. ASP Conf. Ser. 12, San Francisco.

Dukes C. A., Baragiola R. A., and McFadden L. A. (1999) *J. Geophys. Res., 104*, 1865–1872.

Fabian D., Jäger C., Henning T., Dorschner J., and Mutschke H. (2000) *Astron. Astrophys., 364*, 282–292.

Flynn G. J. (1989) *Icarus, 77*, 287–310.

Flynn G. J., Keller L. P., and Sutton S. R. (2004) In *Lunar Planet. Sci. XXXV*, Abstract #1334.

Forrest W. J., Sargent B., Furlan E., D'Alessio P., Calvet N., et al. (2004) *Astrophys. J. Suppl., 154*, 443–447.

Gail H.-P. (2001) *Astron. Astrophys., 378*, 192–213.

Gail H.-P. (2002) *Astron. Astrophys., 390*, 253–265.

Gail H.-P. (2004) *Astron. Astrophys., 413*, 571–591.

Gilmour I. (2003) In *Treatise on Geochemistry, Vol. 1: Meteorites, Comets, and Planets* (A. M. Davis, ed.), pp. 269–290. Elsevier-Pergamon, Oxford.

Gradie J. and Tedesco E. (1982) *Science, 216*, 1405–1407.

Habart E., Natta A., and Krügel E. (2004a) *Astron. Astrophys., 427*, 179–192.

Habart E., Testi L., Natta A., and Carbillet M. (2004b) *Astrophys. J., 614*, L129–L132.

Hallenbeck S. L., Nuth J. A. III, and Daukantas P. L. (1998) *Icarus, 131*, 198–209.

Hallenbeck S. L., Nuth J. A. III, and Nelson R. N. (2000) *Astrophys. J., 535*, 247–255.

Hanner M. S., Gehrz R. D., Harker D. E., Hayward T. L., Lynch D. K., et al. (1999) *Earth Moon Planets, 79*, 247–264.

Harker D. E. and Desch S. J. (2002) *Astrophys. J., 565*, L109–L112.

Harker D. E., Woodward C. E., and Wooden D. H. (2005) *Science, 310*, 278–280.

Hill H. G. M., Jones A. P., and d'Hendecourt L. B. (1998) *Astron. Astrophys., 336*, L41–L44.

Honda M., Kataza H., Okamoto Y. K., Miyata T., Yamashita T., Sako S., Takubo S., and Onaka T. (2003) *Astrophys. J., 585*, L59–L63.

Huss G. R., Meshik A. P., Smith J. B., and Hohenberg C. M. (2003) *Geochim. Cosmochim. Acta, 67*, 4823–4848.

Huss G. R., Alexander C. M. O'D., Palme H., Bland P. A., and Wasson J. T. (2005) In *Chondrites and the Protoplanetary Disk* (A. N. Krot et al., ed.), pp. 701–731. ASP Conf. Ser. 341, San Francisco.

Jäger C., Fabian D., Schrempel F., Dorschner J., Henning T., and Wesch W. (2003) *Astron. Astrophys., 401*, 57–65.

Jessberger E. K. and Kissel J. (1991) In *Comets in the Post-Halley Era* (R. L. Newburn Jr. et al., eds.), pp. 1075–1092. Kluwer, Dordrecht.

Jones A. P. (2000) *J. Geophys. Res., 105*, 10257–10268.

Jones A. P. (2001) *Philos. Trans. R. Soc. London, A359*, 1961–1972.

Jones A. P. and d'Hendecourt L. (2000) *Astron. Astrophys., 355*, 1191–1200.

Jones A. P., Tielens A. G. G. M., and Hollenbach D. J. (1996) *Astrophys. J., 469*, 740–764.

Joswiak D. J., Brownlee D. E., Pepin R. O., and Schlutter D. J. (2000) In *Lunar Planet. Sci. XXXI*, Abstract #1500.

Joswiak D. J., Brownlee D. E., Schlutter D. J., and Pepin R. O. (2004) In *Lunar Planet. Sci. XXXV*, Abstract #1919.

Jura M. (1994) *Astrophys. J., 434*, 713–718.

Kastner J. H. and Myers P. C. (1994) *Astrophys. J., 421*, 605–614.

Kehm K., Flynn G. J., Sutton S. R., and Hohenberg C. M. (2002) *Meteoritics & Planet. Sci., 37*, 1323–1335.

Keller L. P. and McKay D. S. (1997) *Geochim. Cosmochim. Acta, 61*, 2331–2341.

Keller L. P. and Messenger S. (2004) In *Lunar Planet. Sci. XXXV*, Abstract #1985.

Keller L. P., Messenger S., Flynn G.. J., Clemett S., Wirick S., and Jacobsen C. (2004) *Geochim. Cosmochim. Acta, 68*, 2577–2589.

Keller L. P., Messenger S., and Christoffersen R. (2005) In *Lunar Planet. Sci. XXXVI*, Abstract #2088.

Kemper F., Vriend W. J., and Tielens A. G. G.. M. (2004) *Astrophys. J., 609*, 826–837.

Kessler-Silacci J. E., Hillenbrand L. A., Blake G. A., and Meyer M. R. (2005) *Astrophys. J., 622*, 404–429.

Kissel J. and Krueger F. R. (1987) *Nature, 326*, 755–760.

Li A. and Lunine J. I. (2003) *Astrophys. J., 594*, 987–1010.

Lodders K. (2003) *Astrophys. J., 591*, 1220–1247.

MacPherson G. J. (2003) In *Treatise on Geochemistry, Vol. 1: Meteorites, Comets, and Planets* (A. M. Davis, ed.), pp. 201–246. Elsevier-Pergamon, Oxford.

Mathis J. S. (1990) *Ann. Rev. Astron. Astrophys., 28*, 37–70.

McKee C. F. (1989) In *Interstellar Dust* (L. J. Allamandola and A. G. G. M. Tielens, eds.), pp. 431–443. Kluwer, Dordrecht.

Meeus G., Waelkens C., and Malfait K. (1998) *Astron. Astrophys., 329*, 131–136.

Mennella V., Brucato J. R., Colangeli L., and Palumbo P. (2002) *Astrophys. J., 569*, 531–540.

Messenger S. (2000) *Nature, 404*, 968–971.

Messenger S., Keller L. P., Stadermann F. J., Walker R. M., and Zinner E. (2003) *Science, 300*, 105–108.

Messenger S., Keller L. P., and Lauretta D. S. (2005) *Science, 309*, 737–741.

Molster F. J., Waters L. B. F. M., Tielens A. G. G. M., Koike C., and Chihara H. (2002) *Astron. Astrophys., 382*, 241–255.

Morbidelli A., Bottke W. F. Jr., Froeschle C., and Michel P. (2002) In *Asteroids III* (W. F. Bottke Jr. et al., eds.), pp. 409–422. Univ. of Arizona, Tucson.

Nguyen A. N., Zinner E., and Stroud R. M. (2005) In *Lunar Planet. Sci. XXXVI*, Abstract #2196.

Noble S. K., Keller L. P., and Pieters C. M. (2005) *Meteoritics & Planet. Sci., 40*, 397–408.

Nuth J. A., Hallenbeck S. L., and Rietmeijer F. J. M. (2000a) *J. Geophys. Res., 105*, 10387–10396.

Nuth J. A., Hill H. G. M., and Kletetschka G. (2000b) *Nature, 406*, 275–276.

Palme H. (2002) In *Lunar Planet. Sci. XXXIII*, Abstract #1709.

Pendleton Y. J. and Allamandola L. J. (2002) *Astrophys. J. Suppl., 138,* 75–98.

Pendleton Y. J., Sandford S. A., Allamandola L. J., Tielens A. G. G. M., and Sellgren K. (1994) *Astrophys. J., 437,* 683–696.

Przygodda F., van Boekel R., Àbrahàm P., Melnikov S. Y., Waters L. B. F. M., and Leinert C. (2003) *Astron. Astrophys., 412,* L43–L46.

Rietmeijer F. J. M., Nuth J. A., and MacKinnon I. D. R. (1986) *Icarus, 66,* 211–222.

Rietmeijer F. J. M., Nuth J. A. III, and Karner J. M. (2002) *Phys. Chem. Chem. Phys., 4,* 546–551.

Russell S. S., Ott U., Alexander C. M. O'D., Zinner E. K., and Pillinger C. T. (1997) *Meteoritics & Planet. Sci., 32,* 719–732.

Savage B. D. and Sembach K. R. (1996) *Ann. Rev. Astron. Astrophys., 34,* 279–330.

Schramm L. S., Brownlee D. E., and Wheelock M. M. (1989) *Meteoritics, 24,* 99–112.

Schulze H., Kissel J., and Jessberger E. K. (1997) In *From Stardust to Planetesimals* (Y. J. Pendleton and A. G. G. M. Tielens, ed.), pp. 397–414. ASP Conf. Ser. 122, San Francisco.

Scott E. R. D. and Krot A. N. (2003) In *Treatise on Geochemistry, Vol. 1: Meteorites, Comets and Planets* (A. M. Davis, ed.), pp. 143–200. Elsevier-Pergamon, Oxford.

Scott E. R. D. and Krot A. N. (2005) *Astrophys. J., 623,* 571–578.

Sephton M. A., Love G. D., Watson J. S., Verchovsky A. B., Wright I. P., Snape C. E., and Gilmour I. (2004) *Geochim. Cosmochim. Acta, 68,* 1385–1393.

Slavin J. D., Jones A. P., and Tielens A. G. G. M. (2004) *Astrophys. J., 614,* 796–806.

Stroud R. M. and Bernatowicz T. J. (2005) In *Lunar Planet. Sci. XXXVI,* Abstract #2010.

Stroud R. M., Nittler L. R., and Alexander C. M. O'D. (2004a) *Science, 305,* 1455–1457.

Stroud R. M., Nittler L. R., and Hoppe P. (2004b) *Meteoritics & Planet. Sci., 39,* A101.

Stroud R. M., Nittler L. R., Alexander C. M. O'D., and Stadermann F. J. (2005) *Meteoritics & Planet. Sci., 40,* A148.

Swamy K. K. S., Sandford S. A., Allamandola L. J., Witteborn F. C., and Bregman J. D. (1988) *Icarus, 75,* 351–370.

Thomas K. L., Blanford G. E., Keller L. P., Klöck W., and McKay D. S. (1993) *Geochim. Cosmochim. Acta, 57,* 1551–1566.

Tielens A. G. G. M. (1998) *Astrophys. J., 499,* 267.

Tielens A., Waters L. B. F. M., and Bernatowicz T. (2005) In *Chondrites and the Protoplanetary Disk* (A. N. Krot et al., eds.), pp. 605–631. ASP Conf. Ser. 341, San Francisco.

van Boekel R., Min M., Leinert C., Waters L. B. F. M., Richichi A., et al. (2004) *Nature, 432,* 479–482.

van Boekel R., Min M., Waters L. B. F. M., de Koter A., Dominik C., van den Ancker M. E., and Bouwman J. (2005) *Astron. Astrophys., 437,* 189–208.

Wasson J. T. and Chou C.-L. (1974) *Meteoritics, 9,* 69–84.

Wasson J. T. and Kallemeyn G. W. (1988) *Philos. Trans. R. Soc. London, A325,* 535–544.

Watson D. M., Kemper F., Calvet N., Keller L. D., Furlan E., et al. (2004) *Astrophys. J. Suppl., 154,* 391–395.

Weidenschilling S. J. (1997) *Icarus, 127,* 290–306.

Westphal A. J. and Bradley J. P. (2004) *Astrophys. J., 617,* 1131–1141.

Yin Q. (2005) In *Chondrites and the Protoplanetary Disk* (A. N. Krot et al., eds.), pp. 632–644. ASP Conf. Ser. 341, San Francisco.

Zinner E. (2003) In *Treatise on Geochemistry, Vol. 1: Meteorites, Comets, and Planets* (A. M. Davis, eds.), pp. 17–40. Elsevier-Pergamon, Oxford.

Zubko V., Dwek E., and Arendt R. G. (2004) *Astrophys. J. Suppl., 152,* 211–249.

Comet Grains and Implications for Heating and Radial Mixing in the Protoplanetary Disk

Diane Wooden
NASA Ames Research Center

Steve Desch
Arizona State University

David Harker
University of California, San Diego

Hans-Peter Gail
Universität Heidelberg

Lindsay Keller
NASA Johnson Space Center

Observations of comets and chondritic porous interplanetary dust particles (CP IDPs, grains likely shed from comets), as well as of protoplanetary disks, show that a large fraction of the submicrometer silicate grains in these objects are Mg-rich crystalline silicates. Here we review observations of the mineralogy and crystallinity of cometary grains and anhydrous CP IDPs, including new spectroscopy of the dust liberated by the Deep Impact experiment on 9P/Tempel. Some key results of these observations are that crystalline silicates are very Mg-rich, and in most disks (including the solar system's) a gradient in the silicate crystalline fraction exists. We discuss the mechanisms by which Mg-rich crystals can be produced in protoplanetary disks, including complete evaporation followed by slow recondensation or reduction of Fe in Mg-Fe silicates (possibly facilitated by C combustion to CO or CO_2), combined with thermal annealing. Finally, we discuss how these processes might occur in protoplanetary disks. We conclude that there are three viable scenarios that may operate in protoplanetary disks to produce Mg-rich crystalline silicates with a crystallinity gradient: (1) Steady-state conditions can maintain temperatures high enough in the inner disk (<a few AU) to evaporate dust and allow it to recondense; this must be followed by moderately effective outward radial transport of the dust produced. (2) Transient heating events, probably shocks, may evaporate dust in the outer disk (<tens of AU) and allow it to recondense over timescales of hours to days, directly producing Mg-rich silicates; alternatively, it may recondense rapidly but be thermally annealed by a second event. (3) Transient heating events (shocks) may heat amorphous Mg-Fe silicates only to ≈1200–1400 K, enough to anneal the dust without destroying it; excess Fe in the silicate must be simultaneously reduced to Fe metal. On the other hand, the presence and characteristics of volatile and refractory organics in cometary materials demonstrate that a significant fraction of the outer disk mass sustained low temperatures (~30–150 K), and so did not pass through the hottest, inner regions of the disk in the early collapse phase nor was shocked. All models are as yet too incomplete for us to favor any of them: Outward radial transport by turbulent diffusion suffers from a lack of knowledge of the cause of the turbulence, and shock models have not been developed sufficiently to say where in disks shocks can heat or vaporize dust, and for how long. All processes might be simultaneously occurring in disks. What is clear is that cometary grains and anhydrous CP IDPs contain a component of dust — crystalline Mg-rich silicates — that necessarily saw very high temperatures, either by large-scale radial excursions through the solar nebula disk, or by very energetic transient heating events in the comet-forming zone.

1. INTRODUCTION

Data on comets and interplanetary dust particles of probable cometary origin show that amorphous and crystalline silicates were abundant in the protosolar disk. While amorphous silicates are abundant in the interstellar medium (ISM), crystalline silicates are rare. In interstellar clouds in our galactic disk, amorphous silicates contain magnesium (Mg) and iron (Fe). In contrast to ISM Mg-Fe amorphous silicates, crystalline silicates in comets and protoplanetary disks are Mg-rich and Fe-poor. In the solar nebula, Mg-rich crystalline silicates formed as the result of high-temperature (≥1000 K) processes.

Comet nuclei are the most primitive icy bodies in the solar system, having suffered no postaccretion aqueous or thermal alteration and minimal alteration by cosmic rays and collisions. When perturbed into orbits that pass comet nuclei close to the Sun, nuclear ices, trapped volatile gases, and dust grains are released into a cometary comae. At heliocentric distances of ~1 AU, 0.1-µm silicate grains in cometary comae reach radiative equilibrium temperatures of ≤400 K (Fig. 1) (*Harker et al.*, 2002), insufficient to anneal amorphous silicates. Therefore, Mg-rich crystalline silicates formed in the solar nebula prior to their incorporation into cometary nuclei (*Hanner et al.*, 1994).

Cometary nuclei also contain materials from the protosolar disk that never saw temperatures above ~30–150 K. For example, in cometary comae the ortho-to-para ratio of H_2O and NH_2 demonstrate that cometary nuclei contain water ice and ammonia ice that last equilibrated at ~30 K (*Kawakita et al.*, 2004b). At sufficiently large heliocentric distances where water ice survives in the coma ($r_h \geq 3$ AU), amorphous water ice has been detected in the near-infrared reflection spectra of comets C/1995 O1 (Hale-Bopp) (*Davies et al.*, 1997) and C/2004 T7 (LINEAR) (*Kawakita et al.*, 2004a). Amorphous water ice has never experienced temperatures above its crystallization temperature of ~150 K.

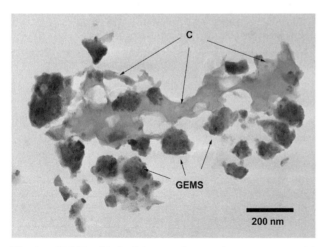

Fig. 1. GEMS with D-rich amorphous carbonaceous material (C) in an anhydrous CP IDP (L2009*E2) [Fig. 3 of *Keller et al.* (2000)].

Cometary nuclei accreted from the colder, icy regions of the solar nebula, beyond the snow line (≥5–100 AU). The incorporation into cometary nuclei of low-temperature materials, e.g., amorphous water ice, and high-temperature materials, i.e., crystalline silicates, implies that high-temperature materials had to be radially transported from hotter regions of the disk to the colder regions, prior to their accretion into cometary nuclei. If transient heating events, such as shocks, occurred and formed crystalline silicates, then there had to be mixing between the shocked and unshocked regions.

Crystalline silicates are seen in external protoplanetary disk systems, and now are thought to be tracers of thermal processing at high temperatures (≥1000 K) and of radial transport of grains (e.g., *Furlan et al.*, 2005; *Wooden et al.*, 2005). Infrared (IR) spectral features occur with remarkable similarity in comet C/1995 O1 (Hale-Bopp) and in the protoplanetary disk around Herbig Ae/Be star HD 100546 at wavelengths 11.2, 19.5, 23.5, and 33.5 µm (*Bouwman et al.*, 2003), which are attributable to Mg-rich crystalline silicates (*Koike et al.*, 2003; *Chihara et al.*, 2002). Mg-rich crystalline silicates are detected frequently in intermediate-mass pre-main-sequence Herbig Ae/Be stars (*Bouwman et al.*, 2001) and are present in some low-mass pre-main-sequence T Tauri stars (*Furlan et al.*, 2005; *Honda et al.*, 2004).

Understanding the formation conditions of Mg-rich crystalline silicates in the protosolar disk sets the context for interpreting Mg-rich crystalline silicate features in external protoplanetary disks. Mg-rich crystalline silicates are present in the most primitive solar system materials — comets, anhydrous chondritic porous (CP) interplanetary dust particles (IDPs) of probable cometary origins, and type 3.0 chondrites (*Brearley*, 1989; *Scott and Krot*, 2005; *Wooden et al.*, 2005). On the other hand, Fe-rich crystalline silicates are abundant in meteoritic materials that have experienced "secondary formation" mechanisms, such as thermal (~500 K) and aqueous alteration (*Scott and Krot*, 2005) on asteroidal parent bodies. Therefore, Mg-rich crystalline silicates are "primary formation" products of thermal processing in the protosolar disk (≥1000 K), by either condensation or annealing. Vaporization of ISM Mg-Fe amorphous silicates followed by condensation thermodynamically favors the formation of Mg-rich crystalline silicates and Fe-metal, thus transforming the reservoir in which the Fe resides from within silicates to a separate mineral phase.

Annealing, i.e., the devitrification of amorphous silicates into crystalline silicates by heating, does not change the chemical composition (stoichiometry) of the grain: Mg-rich amorphous silicates anneal to Mg-rich crystalline silicates and Fe-rich amorphous silicates anneal to Fe-rich crystalline silicates. The formation of Mg-rich crystalline silicates from the protosolar disk's reservoir of ISM Mg-Fe amorphous silicates by annealing requires a mechanism to metamorphose Mg-Fe silicates into Mg-rich silicates. When Mg-Fe silicates are heated in the presence of elemental carbon (C), the C combusts to CO or CO_2 and FeO in the silicates is reduced to Fe-metal. Reduction of Fe is one "primary formation" mechanism that may contribute to the viability of the annealing scenario for the formation of Mg-rich crystalline silicates from ISM Mg-Fe amorphous silicates.

The abundance of carbon in CP IDPs is on average ~12 wt%, which is 2–3× more than the most primitive CI chondrites (*Thomas et al.,* 1994, 1996; *Keller et al.,* 2004). In CP IDPs, carbon takes the forms of amorphous or poorly graphitized carbon and organic matter. The carbonaceous matter appears as coatings on single mineral crystals or as the matrix that "glues" the CP IDP aggregate together.

There is a potential link between the primary thermal processing of silicates and carbonaceous matter in the protosolar disk: C combusts to CO or CO_2, consuming oxygen and driving the reduction of Fe^{2+} in silicates to surface Fe-metal and silica. Surface Fe-metal is an important catalyst for the formation of organic molecules. Heating may transform organic materials to elemental carbon. Comets being more abundant in carbon than CI chondrites suggests that materials in the colder, outer protosolar disk regions probably provided carbon to the inner disk or to regions of shocks for primary thermal processing.

Understanding the origins of cometary materials provides a basis for interpreting silicate crystallinity and mineralogy and carbonceous species in terms of the physical conditions under which dust grain materials formed and aggregated, and were transported in our protosolar disk and in external protoplanetary disks. The properties of silicates and carbonaceous species in cometary comae and in anhydrous CP IDPs are presented in section 2. Plausible formation scenarios for Mg-rich crystalline silicates, including condensation, and annealing combined with Fe reduction are discussed in section 3. Models for protoplanetary disks and for shock heating in the protosolar disk are presented in the context of crystalline silicate formation scenarios in section 4. Conclusions are presented in section 5.

2. COMET GRAINS AND ANHYDROUS CHONDRITIC POROUS INTERPLANETARY DUST PARTICLES

We utilize a combination of investigative techniques, including remote sensing of cometary comae materials, *in situ* mass spectrometer measurements on flyby spacecraft, and laboratory investigations of anhydrous CP IDPs to assemble the inventory of cometary carbonaceous and siliceous grain materials. Three types of cometary grain materials are presented based on their degree of volatility: volatile organics, refractory organics, and refractory minerals. The volatile organics have such short lifetimes that they are only observable in cometary comae. The refractory organic material survives transit through cometary comae into the interplanetary medium and is present in CP IDPs. Cometary refractory organics may be a mixture of presolar organics and organics formed by "primary" processes in the early solar nebula. Cometary refractory organic matter constitutes the matrix of anhydrous CP IDPs that "glues" the refractory minerals together. Cometary refractory minerals

are the Mg-Fe amorphous and Mg-rich crystalline silicates and amorphous carbon detected in cometary comae grains, as well as FeS found in anhydrous CP IDPs, and less abundantly silica (SiO_2) in aggregate GEMS (*Keller et al.,* 2005). The refractory minerals represent relic ISM Mg-Fe amorphous silicates and Mg-rich crystalline silicates that formed in "primary" thermal processes including condensation and annealing.

To date, the best-studied laboratory samples of the most primitive solar system materials are anhydrous CP IDPs collected in Earth's stratosphere. Anhydrous CP IDPs are thought to be of cometary origin (*Hanner and Bradley,* 2004; *Wooden,* 2002; *Joswiak et al.,* 1996) because (1) they contain organic materials with deuterium-to-hydrogen (D/H) isotopic ratios that vary on submicrometer scales, ranging from subsolar deuterium (D) (≤–400‰) to enormous D-enriched materials (≥3000‰) of presolar origin (*Messenger,* 2000; *Keller et al.,* 2000); (2) they have relatively high atmospheric entrance velocities ($v \geq 16$ km s^{-1}) (*Brownlee et al.,* 1995; *Nier and Schlutter,* 1993), which imply they were released from comets on more eccentric orbits than asteroid bodies; (3) they possess anhydrous silicate mineral grains (*Bradley,* 1988); (4) they have highly porous aggregate structures (*Bradley,* 1988); (5) they are unequilibrated aggregates, i.e., aggregates of mineral species that would not exist as adjacent subgrains if heating had occurred after the grains had aggregated; and (6) their refractory mineral subgrains include some of the most primitive materials known in the solar nebula, including GEMS subgrains (nonstoichiometric Mg-Fe noncrystalline or amorphous silicates referred to as glasses with embedded metal and sulfides); submicrometer domains of partially melted silicates, Si-rich material, and FeS (*Keller and Messenger,* 2005); submicrometer domains of microcrystalline minerals of similar composition called "equilibrated aggregates"; single crystal FeS (*Keller et al.,* 2000); and single crystal silicate minerals with Mg contents higher than any other solar nebula materials (*Bradley et al.,* 1999b).

Anhydrous CP IDPs are held together by a matrix of organic materials: Anhydrous CP IDP subgrains were exposed to either high or low temperatures prior to but not subsequent to grain aggregation. The properties of anhydrous CP IDPs reveals that primary thermal processes and radial transport was happening to submicrometer-sized grain material at the time grains were aggregating, prior to their accretion into cometesimals. Within this next year, laboratory studies of grains collected from the coma of Jupiter-family comet 81P/Wild 2 by the Stardust Comet Sample Return Mission will add significantly to our knowledge of anhydrous CP IDPs.

2.1. Semirefractory Organic Grain Species Produce "Distributed" Sources in Cometary Comae

The greatest alteration to the semirefractory organic grain species in comets occurs in cometary comae (*Mann et al.,* 2005). In many but not all comets, gaseous species are observed in comae to be more radially extended than the dust (*Bockelée-Morvan et al.,* 2004; *Bockelée-Morvan and Crovisier,* 2002). These radially "extended" sources of volatile gases are thought to be "distributed" into cometary comae by dust grains: Molecular species are hypothesized to appear in the coma as a result of the desorption of an organic grain component(s) after a finite lifetime (typically hours at ≤3–4 AU) (*Bockelée-Morvan and Crovisier,* 2002; *DiSanti et al.,* 2001). The production rates from distributed sources can be similar in magnitude to the production rates of "native" volatile gases released from the nucleus. To date, the carriers of the distributed sources are a great mystery in cometary science. Different species appear as distributed sources depending on the heliocentric (r_h) distance of the comet. In comet Hale-Bopp, the molecular species that were observed to be "distributed" were at 5–14 AU, CO; at 3–5 AU, CO, H_2O, CH_3OH, and H_2CO; at ≤3 AU, H_2CO, HCN, OCS, and SO, and other species; and at ≤1.5 AU, CO, CN, and C_2. Understanding the carriers of the distributed sources is important to translating comae abundances into nuclear abundances and assessing the solar nebula reservoir out of which comets accreted.

Grain aggregates may fragment in comae when distributed sources are released as the organic materials that serve as the "glue" that binds aggregates of mineral subgrains together desorb (*Jessberger et al.,* 2001; *Harker et al.,* 2002). The inner coma of comet 1P/Halley (8900–29000 km) had ~2.4× the abundance of organic matter (CHON particles) than the outer coma (29000–40000 km), according to the Vega 1 flyby mass spectrometer measurements of ~1 ng of comae cluster grains (*Fomenkova et al.,* 1994). Distributed sources in cometary comae may be the primary tool for investigating the semirefractory organic grain species extant in the early solar nebula.

2.2. Refractory Organic Materials in Comet Grains and in Anhydrous Chondritic Porous Interplanetary Dust Particles

One-half of the carbon in anhydrous CP IDPs is in organic materials. The other half of the carbon is elemental C in the form of amorphous or poorly graphitized carbon (*Flynn et al.,* 2003; *Thomas et al.,* 1996). On average, anhydrous CP IDPs are ~6 wt% amorphous carbon. Given the extreme variability in the C wt% in an individual anhydrous CP IDP, these laboratory measurements of anhydrous CP IDPs agree well with the amorphous carbon mass fraction of 7–12 wt% deduced for comet Hale-Bopp at 0.95–2.8 AU from thermal emission model fits to observed IR spectral energy distributions (*Harker et al.,* 2002, 2004) (section 2.3).

Understanding the composition of cometary organic refractory materials is important because (1) these materials probably are the precursors of primitive meteoritic organics and by comparison reveal the effects of primary thermal processing on organics, and (2) these materials are the best representative samples of solid-phase organics formed in the ISM (*Keller et al.,* 2004). Cometary refractory or-

ganic grain species are best studied with laboratory investigations of anhydrous CP IDPs, *in situ* measurements, and comet sample return missions (Stardust). To date, there is no direct spectroscopic detection of a refractory organic grain species in cometary comae: a 3.4-μm feature seen in a few comets including Halley, C/1989 X1 (Austin), C/1990 K1 (Levy), and the Deep Impact-induced ejecta of comet 9P/Tempel (*A'Hearn et al.,* 2005). The 3.4-μm feature is partly attributable to methanol, with the remainder of the band arising from gas-phase fluorescence and not from refractory organics (*Bockelée-Morvan et al.,* 1995, 2004). An indirect assessment of the cometary refractory organic grain component comes from models for the observed visible light scattering and polarization of comet comae, which require a highly absorbing organic grain mantle (*Kimura et al.,* 2003).

In the coma of comet Halley, refractory organics were studied *in situ* by the mass spectrometers on Vega 1, Vega 2, and Puma 1, and found to contain primarily C, H, O, and N with a wide variation in relative elemental abundances. Halley's so-called "CHON" particles were more abundant in the inner coma and constituted ~25% of the total population of cluster particles, with another ~25% being CHON-free siliceous particles (*Schulze et al.,* 1997) and 50% being mixtures of CHON and silicecous materials (*Kissel,* 1999). While the majority (~70%) of the CHON clusters were kerogen-like organic compounds (*Fomenkova,* 1999), a significant minority was elemental carbon (19%) and hydrocarbons (10%) (*Fomenkova et al.,* 1994). The heterogeneity of the chemical properties of the CHON refractory organics is incompatible with their formation in the same protosolar disk environment: Halley's nucleus accreted refractory organic materials that formed in different spatial regions, e.g., the ISM, the prenatal molecular cloud, and the protosolar disk, or experienced different thermal histories (*Fomenkova et al.,* 1994).

In anhydrous CP IDPs, carbonacous material occurs in three different morphologies: (1) as thin (0.01-μm) coatings on single mineral crystals, (2) as discrete submicrometer- to micrometer-sized domains (carbonaceous units), and (3) as the matrix (~0.5-μm coatings or domains) that apparently "glue" subgrains together (*Flynn et al.,* 2003). Individual CP IDPs are found to have a wide range (2–90 wt%) of carbon abundances (*Thomas et al.,* 1994, 1996; *Flynn et al.,* 2003, 2004; *Keller et al.,* 2004). The average carbon abundance determined for 100 CP IDPs is ~12 wt%, which is a factor of 2–3× more than CI chondrites. For a sample of 19 IDPs, on average 30–50% of the carbon is in organic species with either aliphatic hydrocarbons (1–3 wt%) or carbonyl (C=O) bonds (~2 wt%) (*Flynn et al.,* 2004). In anhydrous CP IDPs the aliphatic hydrocarbon is the carrier of the D enrichments (*Keller et al.,* 2004). The aliphatic hydrocarbons, i.e., chains of CH_2 bonds with terminal ends of CH_3 groups, have approximately the same mean aliphatic chain lengths in anhydrous CP IDPs and hydrous CP IDPs, but are longer than in the Murchison CI chondrite (*Flynn et al.,* 2003). Analyses of three anhydrous CP IDPs suggests that the organic matter is more complicated (containing two C=O groups) than in hydrous CP IDPs and the Orgueil CI meteorite (*Flynn et al.,* 2004).

The large variations in the D and ^{15}N enrichments in the matrices of anhydrous CP IDPs indicate that the organic materials have experienced various degrees of thermal processing and are a mixture of protosolar disk and presolar materials (*Keller et al.,* 2004). On submicrometer size scales, D enrichments range from subchondritic (–400‰) to strongly enriched (≥+10,000) (*Keller et al.,* 2004; *Messenger et al.,* 1996). In one anhydrous CP IDP fragment, high ^{15}N/^{14}N isotopic ratios occur in a more refractory phase than the D-rich material, perhaps bonded to the relatively rare aromatic hydrocarbons (*Keller et al.,* 2004). The differences between the anhydrous CP IDP organics and organics in primitive carbonceous meteorites (e.g., *Kerridge,* 1999) indicates that thermal processing (1) converts aliphatic hydrocarbons from longer to shorter chains; (2) converts alphatic bonds to aromatic bonds; and (3) alters the organic material with C=O bonds (*Flynn et al.,* 2003).

2.3. Silicate Mineralogy and Amorphousness Versus Crystallinity in Comets and in Anhydrous Chondritic Porous Interplanetary Dust Particles

Infrared spectra of comet comae, laboratory studies of anhydrous CP IDPs, and *in situ* measurements of Halley's coma all indicate that the predominant refractory minerals are amorphous carbon (also discussed in section 2.2) and silicates with compositions similar to olivine, $(Mg_y,Fe_{(1-y)})_2 SiO_4$, and pyroxene, $(Mg_x,Fe_{(1-x)})SiO_3$. Specifically, olivines can be considered as solid solutions of Mg_2SiO_4 (y = 1) and Fe_2SiO_4 (y = 0), respectively named forsterite (Fo_{100}) and fayalite (Fo_0). In the nomenclature of mineralogists, olivine is crystalline, and Fo_{100}–Fo_{92} is "forsterite" (*Henning et al.,* 2005). Olivine crystals in cometary materials, however, have higher Mg contents (Fo_{98}) than in any other solar system materials (*Bradley et al.,* 1999b). Because of this distinction, we will denote Mg-rich for 1.0 ≥ y ≥ 0.9 and Mg-Fe for y ≈ 0.5, and in certain cases, we will use the corresponding notation Fo_{100}–Fo_{90} and Fo_{50}. Correspondingly, pyroxenes range from pure-Mg enstatite (x = 1) to pure-Fe ferrosilite (x = 0), denoted En_{100} and En_0, respectively.

By comparison with laboratory minerals, the predominant 9.7-μm silicate feature in comets is attributed to Mg-Fe amorphous olivine, and the short wavelength rise on the feature is attributed to Mg-Fe amorphous pyroxene (*Hanner et al.,* 1994; *Wooden et al.,* 1999, 2005; *Wooden,* 2002; *Harker et al.,* 2002, 2004; *Hanner and Bradley,* 2004). In amorphous olivine and pyroxene, the vibrational stretching and bending modes of the Si-O bonds produce the 10-μm and 20-μm features, respectively. At any single comet observation epoch, all coma grains are at the same heliocentric distance and the IR emission from the coma is optically thin, so the observed flux density, especially the ratio of the 10-μm/20-μm features, constrains the grain temperatures. Radiative equilibrium grain temperatures (T_{dust}) are strongly dependent on the absorptivity (Q_{abs}) of sunlight at visible and near-IR

wavelengths (Fig. 4), and weakly dependent on grain radius ($T_{dust} \propto a^{1/5}$), so grain temperatures probe the optical properties of the grains. From their grain temperatures, cometary amorphous silicates are deduced to have 0.5 ≤ x ≃ y ≤ 0.7 (*Harker et al.,* 2002), i.e., they have approximately chondritic (x = y = 0.5) Mg and Fe contents.

Most cometary silicate features are more "flat-topped" than amorphous silicate features (*Colangeli et al.,* 1995): IR spectroscopy first detected the 11.2-μm crystalline olivine feature in comet 1P/Halley (*Campins and Ryan,* 1989). The resonant features from crystalline silicates are narrower than features from amorphous silicates. Strong far-IR features at 19.5, 23.5, and 33.5 μm, and weaker features at 16 and 27.5 μm, from Mg-rich crystalline olivine ($Mg_y,Fe_{(1-y)})_2SiO_4$ with y ≥ 0.9 were detected first in the ISO SWS spectra of comet Hale-Bopp at 2.8 AU preperihelion (Fig. 3), providing irrefutable confirmation of the identification of the 11.2- and 10.0-μm features with Mg-rich crystalline olivine (*Crovisier et al.,* 1997). The mineralogical decomposition of the 10-μm silicate feature of comet Hale-Bopp is demonstrated in Fig. 2. Features from Mg-rich crystalline pyroxene ($Mg_x,Fe_{(1-x)})SiO_3$ with x ≥ 0.9 were detected at 9.3 and 10.5 μm in groundbased mid-IR spectra of Hale-Bopp at 0.93 AU, when the comet was close to perihelion (*Wooden et al.,* 1999; *Harker et al.,* 2002, 2004). The wavelengths of the features observed in comets are consistent with high-Mg-content crystals, (0.9 ≤ x ≃ y ≤ 1.0) (*Koike et al.,* 2003; *Chihara et al.,* 2002). In comets, the Mg content of the crystalline silicates is significantly greater than in the amorphous silicates.

The relative abundances of refractory mineral grains are derived by χ^2-fitting thermal emission models to observed spectral energy distributions (SEDs) (*Harker et al.,* 2002, 2004; *Wooden et al.,* 2004). The best-fit thermal emission model for comet Hale-Bopp at 2.8 AU is shown in Fig. 3. The derived relative abundances depend on the relative grain temperatures, which depend on the mineralogy and Mg contents. Mg-rich crystals are more transparent at optical and near-IR wavelengths than Mg-Fe amorphous silicates,which are more transparent than amorphous carbon and Fe-metal grains (Fig. 4). A Mg-rich crystal absorbs sunlight less efficiently than a Mg-Fe amorphous silicate of the same grain radius (a), so a Mg-rich crystal has a lower radiative equilibrium temperature and emits less flux ($F_\lambda = \pi a^2 \pi B_\lambda(T_{dust}) Q_\lambda(a)$). A Mg-rich crystal is harder to detect than a warmer Mg-Fe amorphous silicate grain of the same size. Therefore, the silicate crystalline fraction f_{cryst}, i.e., the mass ratio of submicrometer Mg-rich crystals to the sum of submicrometer amorphous and crystalline silicates, needs to be greater than ~0.5 for Mg-rich crystals to be detected (in IR spectra with a spectral resolution of R ~ 100 and signal-to-noise ratio of ~100).

Relatively high crystalline mass fractions are deduced for the submicrometer portion of the grain size distribution in Oort cloud comets: For Hale-Bopp, f_{cryst} ≃ 0.6–0.8 (*Harker et al.,* 2002, 2004) and for C/2001 Q4 (NEAT) f_{cryst} ≃ 0.7 (*Wooden et al.,* 2004); for Hale-Bopp, f_{cryst} = 0.79, 0.67, and

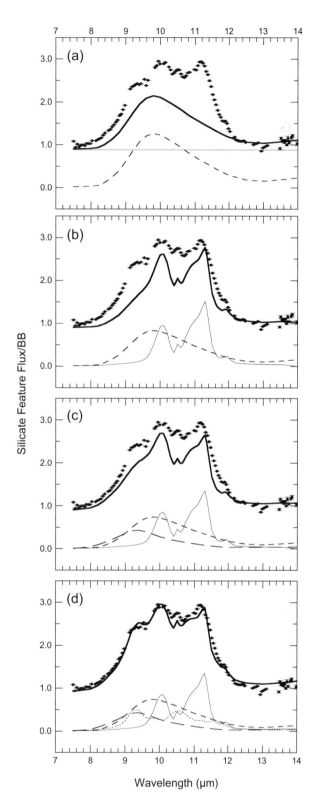

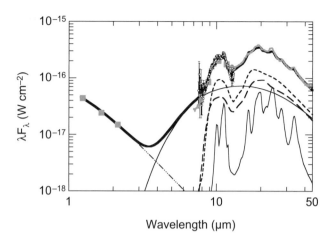

Fig. 3. SED (λF_λ vs. λ) of groundbased photometry (squares) and spectra (circles), and ISO SWS (thick gray line) data of Comet Hale-Bopp at 2.8 AU plotted with the best-fit thermal emission model (heavy line). SED is the sum of scattered sunlight (≤ 5 μm) and thermal emission (≥ 3 μm) from a size distribution [slope = –3.4, peak grain radius a_p = 0.2 μm (*Hanner*, 1983), 0.1 μm $\leq a \leq$ 100 μm] of discrete mineral grains: porous grains of amorphous carbon (solid), amorphous Mg-Fe olivine (Fo$_{50}$) (short dash), amorphous Mg-Fe pyroxene (En$_{50}$) (long dash), and 0.1–1-μm solid crystalline Mg-rich olivine (Fo$_{90}$) (solid). Adapted from *Harker et al.* (2002, 2004).

0.60 in Figs. 2, 5, and 3, respectively. In contrast, the crystalline mass fraction ($f_{cryst} \simeq 0.35$) deduced for the subsurface layers of the nucleus of Jupiter-family comet 9P/Tempel, which were ejected when the Deep Impact Mission hit the comet (Fig. 5) (*Harker et al.*, 2005b), is lower than for Oort cloud comets. Oort cloud comets and Jupiter-family comets were ejected from the transjovian and transneptunian regions of our protoplanetary disk out to the Oort cloud and Kuiper belt, respectively (*Gomes*, 2003; *Morbidelli and Levison*, 2003). Comparing f_{cryst} for Oort cloud and Jupiter-family comets implies a radial dependence to f_{cryst} possibly existed in the protosolar disk (*Harker et al.*, 2005b).

Submicrometer solid crystals produce the shapes of the observed IR resonances, so the ratio of the mass of submicrometer crystals to the total mass of submicrometer silicates defines the crystalline fraction (*Harker et al.*, 2002). Nevertheless, a size distribution of grains that includes larger (10–100-μm-radii) amorphous silicates contributes to the observed SEDs. If we try to mimic grain aggregates with a grain size distribution that contains submicrometer crystals as well as submicrometer- to 100-μm-radii amorphous silicate grains, we may underestimate the crystalline fraction. In fact, crystals may be effectively spectroscopically hidden within larger aggregates: IR spectral features of anhydrous CP IDP thin sections reveal resonances of stronger contrast than bulk particles (*Molster et al.*, 2001; *Bradley et al.*, 1999c). Given these caveats, a grain size distribution spanning radii of 0.1–10 μm, rather than just the 0.1–1-μm grains, yields lower crystalline mass fractions: $f_{cryst} \simeq 0.4$

Fig. 2. The 10-μm silicate feature of Comet Hale-Bopp at 0.93 AU on 1997 April 10 UT, shown as the flux divided by a blackbody (BB) fitted to the $\lambda \leq 8.4$ and $\lambda \geq 12.4$-μm emission, is compared with the progressive addition of **(a)** amorphous carbon (light solid line) and Mg-Fe amorphous olivine (Fo$_{50}$) (short dash); **(b)** Mg-rich crystalline olivine (Fo$_{90}$) (solid); **(c)** Mg-Fe amorphous pyroxene (En$_{50}$) (long dash); and **(d)** Mg-rich crystalline orthopyroxene (Fo$_{90}$) (dotted) [cf. Fig. 6 of *Wooden et al.* (1999)].

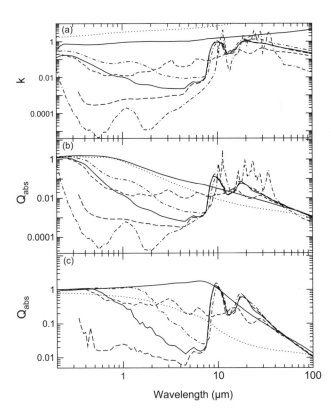

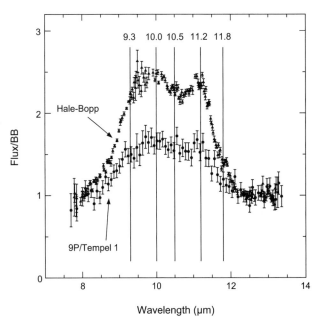

Fig. 4. Optical properties of minerals given by **(a)** the imaginary index of refraction, k, and by grain absorptivities ($Q_{abs}(\lambda)$) for grains of **(b)** 0.1-μm radii and **(c)** 1-μm radii. From the most absorptive to least absorptive minerals (top to bottom at 1 μm): iron (dotted line); amorphous carbon (solid line); Mg-Fe amorphous olivine (Fo_{50}) (dash-dot); kerogen type II organic (short dash) (*Khare et al.*, 1990); Mg-Fe amorphous pyroxene (En_{50}) (solid line); Mg-rich amorphous pyroxene (En_{95}) (long dash); and Mg-rich crystalline olivine (Fo_{100}) (dash-dash-dot). From *Wooden et al.* (2005).

Fig. 5. Crystalline silicate features in the 10-μm spectra of Oort cloud comet Hale-Bopp (top, 1.7 AU post-perihelion) and in the Deep Impact-induced ejecta of Jupiter-family comet 9P/Tempel 1 (bottom, 1.5 AU, 1 h after impact). From Fig. 2 of *Harker et al.* (2005b).

for Hale-Bopp, and $f_{cryst} \simeq 0.2$ for 9P/Tempel, which is still higher than $f_{cryst} \lesssim 0.02$ deduced for any other Jupiter-family comet (*Sugita et al.*, 2005). A significantly lower crystalline mass fraction of $f_{cryst} \simeq 0.075$ is derived for Hale-Bopp using a grain size distribution with submicrometer crystals, submicrometer- to ~100-μm-radii amorphous minerals, and nonspherical particle shapes (*Min et al.*, 2005a).

In situ mass spectrometer measurements of Halley do not reveal the amorphousness or crystallinity of the comae grains (*Kissel*, 1999), but do yield the relative mass fractions of elements (Mg, Si, S, Fe), whose relative ratios reveal the possible minerals present. Given (Mg + Fe):Si $\simeq$ 1 for pyroxene and 2 for olivine, the mass spectrometer measurements (Table 2 of *Schulze et al.*, 1997; *Schulze and Kissel*, 1992) indicate that the dominant ($\geq$50% by mass) refractory mineral is Mg-rich pyroxene ($0.8 \geq x \geq 0.9$). Also present are significant mass fractions (~25%) of Mg-Fe olivine (y $\simeq$ 0.5) and smaller mass fractions of pure-Mg olivine (~5%), FeS (~10%), and Fe-metal grains (~3%) (*Schulze et al.*, 1997). Most (70%) of the iron is in Fe or FeS grains, with

the remaining fraction (30%) being in siliceous grains; some siliceous grains also contain S. Due to the similarities in Fe content, the Mg-Fe olivine *in situ* to Halley may be the Mg-Fe amorphous silicates detected in IR spectra of comets.

Anhydrous CP IDPs are important comparison materials for fine-tuning our understanding of the refractory minerals in cometary dust. The dominant refractory mineral phases in anhydrous CP IDPs are Mg-rich crystalline silicates, GEMS (section 2) (*Bradley et al.*, 1999c), Fe-sulfides (*Keller et al.*, 2000), and amorphous or poorly graphitized carbon. Submicrometer- and micrometer-sized single crystals of Mg-rich pyroxene and olivine can constitute 20–30% of mass of an anhydrous CP IDP. For example, IDP L2009*E2 (Fig. 1) has 5–10% by mass Mg-rich crystalline silicates.

Infrared spectra of the major mineral phases in anhydrous CP IDPs have features at the same wavelengths as the IR spectra of comets (*Bradley et al.*, 1992, 1999a,c; *Wooden et al.*, 2000; *Keller and Flynn*, 2003). The 10-μm absorption spectrum derived from the average of 19 anhydrous CP IDPs (Fig. 6) has features at the same wavelengths as in the 10-μm emission spectrum of comets Hale-Bopp and 9P/Tempel (Fig. 5). Crystalline olivine and pyroxene are more Mg-rich than in other primitive solar system laboratory samples (carbonaceous chondrites, Antarctic micrometeorites) (*Bradley et al.*, 1999b). The wavelength positions of the far-IR spectral resonances in one anhydrous CP IDP thin section (Fig. 7), however, reveal a domain of microcrystalline olivine of Fo_{75} that is less than Fo_{90-100} deduced for crystalline olivine in comet Hale-Bopp.

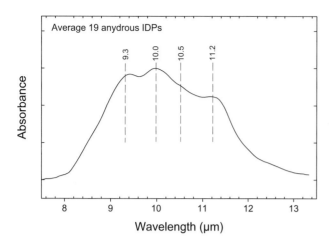

Fig. 6. Average absorption spectra of 19 anhydrous IDPs. See also Fig. 2 of *Keller and Flynn* (2003).

In anhydrous CP IDPs, iron is contained in macroscopic (micrometer-sized) FeS crystals and nanophase Fe and FeS in GEMS. Sulfur is concentrated in Fe-sulfides (*Flynn et al.,* 2004). FeS is most frequently observed toward the outer surfaces of GEMS than the centers. This gradient may be a consequence of S diffusion into the GEMS, while in the solar nebula (L. Keller, personal communication).

When detailed laboratory studies of dozens of individual anhydrous CP IDPs are compared to the properties of comae grains deduced from IR spectroscopy alone, we find that anhydrous CP IDPs are similar to comae grains in that they contain (a) GEMS and Mg-Fe amorphous silicates, respectively; (b) Mg-rich crystalline olivine and Mg-rich crystalline orthopyroxene; and (c) amorphous carbon. At the same time, anhydrous CP IDPs are different in that they contain (d) more enstatite than forsterite (*Keller et al.,* 2004; *Keller and Flynn,* 2003); (e) a matrix composed of not only amorphous carbon but also refractory organics that are not detected spectroscopically; (f) domains of Mg-rich crystalline olivine with lower Mg contents compared to that deduced from IR spectra (far-IR comparison in Fig. 7); and (g) FeS single crystals (*Keller et al.,* 2000). *In situ* measurements of a nanogram of mass of Halley's comae grains are in concurrence with studies of anhydrous CP IDPs in that they contain (d), (e), and (g), while (f) is not determinable from mass spectrometer measurements. It is important to note than pure-Mg amorphous silicates as well as discrete Fe-rich crystalline silicates are absent from anhydrous CP IDPs. In summary, IR spectra do not reveal the presence of Fe or FeS grains nor detect the refractory organic grain component that are seen in anhydrous CP IDPs and *in situ* measurements of comet comae.

2.4. Magnesium-Iron Amorphous Silicates are Interstellar

Along lines-of-sight through the ISM, silicate absorption features are well-matched by Mg-Fe amorphous olivine and pyroxene (*Kemper et al.,* 2004, 2005). Mg-rich crystal-

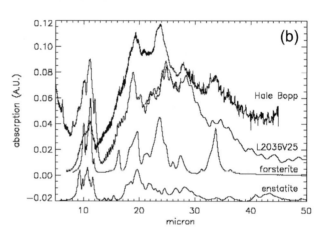

Fig. 7. (a) Atomic force microscope image of Mg-rich crystals within thin section of anhydrous CP IDP L2036V25 (10 μm × 10 μm × 3 μm). (b) 5–50-μm transmission spectrum of IDP L2036V25 thin section compared to pure-Mg crystalline olivine (forsterite) and pyroxene (enstatite), and to the emission spectrum of Comet Hale-Bopp (*Crovisier et al.,* 1997). The Mg content of the olivine $(Mg_y,Fe_{(1-y)})_2SiO_4$ crystals in Hale-Bopp (y ≥ 0.9) are higher than in this IDP (y ≥ 0.75), based on the wavelengths of the far-IR features. From *Molster et al.* (2003).

line silicates are rare (<5%) (*Li and Draine,* 2001) to absent (≤1.1 ± 1.1% toward the galactic center) (*Kemper et al.,* 2004, 2005). Detailed analysis of the ISM extinction curve implies the presence of Mg-Fe amorphous silicates (*Li and Draine,* 2001). Studies of the depletion of gas-phase atoms into dust grains in galactic diffuse ISM clouds show Fe, Mg, and Si are depleted together in grains, and in increasing amounts for denser clouds (*Jones,* 2000; *Savage and Sembach,* 1996). Through depletion studies, a less-refractory dust phase that contains Mg and Si and no Fe is seen to exist in clouds in the galactic halo above the galactic disk (*Jones,* 2000). The ISM silicate absorption feature toward the galactic center can be fitted by nonspherical amorphous pure-Mg silicates as well as by spherical Mg-Fe amorphous silicates (*Min et al.,* 2005b), demonstrating that particle shape can affect the shape of the resonances and the interpretation of the mineralogy and Mg and Fe contents. However, the depletion studies taken together with models for the shape of the 10-μm absorption feature imply that galactic disk ISM silicates are Mg-Fe amorphous olivine and pyroxene.

Mg-Fe amorphous silicates are deduced to be dominant components of comet comae grain populations (section 2.3), and are present in anhydrous CP IDPs as GEMS (*Bradley et al.*, 1999c; *Flynn et al.*, 2003). GEMS are not well-defined stoichiometric olivines and pyroxenes, but have elemental ratios suggesting a mixture of compositions spanning olivine and pyroxene compositions (*Bradley et al.*, 1992; *Brownlee et al.*, 1999; *Keller and Messenger*, 2004a). The absorption feature from GEMS-rich anhydrous CP IDP material has a similar shape as the ISM absorption feature (Fig. 8) (*Bradley*, 1994a,b). For more than a decade, GEMS have been considered as the prototypical ISM Mg-Fe amorphous silicate (*Hanner and Bradley*, 2004).

A recent hypothesis that 80–90% of GEMS are "non-equilibrium condensates that have escaped annealing and further reaction with the gas phase (except for late sulfidization of Fe)," *which formed in the solar system* (*Keller and Messenger*, 2004a,b; *Keller et al.*, 2005) implies a similar mass of more volatile ISM organics and ices would have been subjected to high temperatures in the protosolar disk. However, the large abundance of ISM volatile gases and ices (*Bockelée-Morvan et al.*, 2000; *Ehrenfreund and Charnley*, 2000; *Ehrenfreund et al.*, 2004) and organics (*Keller et al.*, 2003) in cometary materials refutes this concept.

An argument used to substantiate the solar system origin for both GEMS and Mg-rich crystalline silicates is that ~99% of the mass of more than 1000 subgrains of anhydrous CP IDPs have homogeneous solar-isotopic ratios of $^{17}O/^{16}O$, $^{18}O/^{16}O$, and $^{15}N/^{14}N$ (*Messenger*, 2000; *Messenger et al.*, 2003). Nevertheless, grains could lose the isotopic signatures of their sites of origin in AGB stars, novae, and supernovae if reformed in ISM clouds after being sputtered or destroyed by shocks (*Jones*, 2000). To date, there is no consensus as to whether the isotopic homogeneities in subgrains in anhydrous CP IDPs can be interpreted as evi-

dence for their formation in the protosolar disk or for grain homogenization in the ISM (see *Tielens*, 2003).

Sources for ISM Mg-Fe amorphous silicates include O-rich asymptotic giant branch (AGB) stars and supernovae. AGB stars primarily shed Mg-Fe amorphous silicates, but up to 4–20% of their siliceous grains may be Mg-rich crystalline silicates (*Kemper et al.*, 2001). Crystalline silicates dispersed via stellar winds into the ISM, however, probably are rapidly amorphized by the bombardment by cosmic rays. Bombardment by low-energy galactic cosmic rays (few–50 keV H^+ or He^+ ions) probably efficiently amorphizes crystalline silicates (*Carrez et al.*, 2002; *Jäger et al.*, 2003; *Brucato et al.*, 2004; *Bradley*, 1994a; *Dukes et al.*, 1999), with chemical composition alteration occurring at ≤20 keV (*Carrez et al.*, 2002; *Bradley*, 1994a). Also, low-energy (4–10 keV) cosmic-ray exposure can reduce iron from its stoichiometric inclusion in an amorphous Mg-Fe silicate mineral grain to nanophase Fe metal embedded within the Mg-rich amorphous silicate grain (*Carrez et al.*, 2002), such as seen in GEMS (*Bradley*, 1994a). Amorphization by high-energy, heavy cosmic rays (1.5 MeV Kr^+ ions) occurs at one-fourth the dose for Fe-rich crystalline olivine (Fe_2SiO_4) than for Mg-rich crystalline olivine ($Mg_{0.88}Fe_{0.12})_2SiO_4$, and the distances from supernovae shocks traveled by MeV cosmic rays (~2 pc) (*Spitzer and Jenkins*, 1975) are greater than for keV cosmic rays, so preferential amorphization of Fe-bearing crystalline silicates in the ISM may explain the preponderance of Fe-bearing amorphous silicates in astrophysical objects including comets, AGB stars, and post-AGB environments (*Jäger et al.*, 2003).

3. FORMING MAGNESIUM-RICH CRYSTALS IN DISKS

In protoplanetary disks, some fraction of the initial reservoir of interstellar Mg-Fe amorphous silicate grains accreted into the outer regions from the prenatal cloud core are converted into the Mg-rich silicate crystals. The processes that form pure-Mg or Mg-rich crystals are (1) vaporization followed by gas-phase condensation of solids under conditions that favor thermodynamically stable pure-Mg crystalline silicate minerals in chemical equilibrium; (2) vaporization followed by rapid nucleation and growth of amorphous Mg-rich silicates under kinetically controlled conditions, followed by annealing; and (3) annealing of Mg-Fe amorphous silicates (formed by rapid condensation or interstellar in origin) into crystalline Mg-Fe silicates coupled with reduction of Fe out of the grains under the typical reducing (as opposed to oxidizing) conditions in the solar nebula gas. Since neither Mg-rich amorphous silicates nor Mg-Fe crystalline silicates are a substantial component of cometary comae or anhydrous CP IDPs, so scenarios (2) and (3) imply ~100% efficiency.

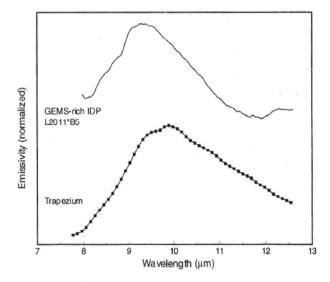

Fig. 8. The 10-μm feature of a GEMS-rich section of an anhydrous CP IDP is broad (*Bradley et al.*, 1999c), characteristic of amorphous silicates, and is compared to the ISM feature toward the Trapezium. From *Keller et al.* (2000).

3.1. Condensation from Nebular Gases

Mg-rich crystalline silicates are thermodynamically favored to condense in hot, dense inner disks. When partial pressures of condensible materials are high enough that kinetic barriers are overcome, often under conditions of "supersaturation," molecular species nucleate into molecular clusters and cluster growth ensues (e.g., *Colangeli et al.,* 2003). Rapid growth produces highly disordered particles (*Nuth et al.,* 2002; *Rietmeijer et al.,* 2002). If temperatures are high enough, clusters rearrange themselves into regular structures or crystals (*Colangeli et al.,* 2003), or clusters grow "epitaxially" as crystals on preexisting crystalline "seed" grains. Growth of single crystals from nebular gases can explain the relatively infrequent (J. Bradley, personal communication) platelet or ribbon forms of forsterite and enstatite crystals in anhydrous CP IDPs (*Bradley et al.,* 1983; *Scott and Krot,* 2005; see Fig. 2 of *Keller et al.,* 2000).

In protoplanetary disk environments where there are sustained high partial pressures of refractory elements (*Grossman,* 1972; *Pataev and Wood,* 2005), such as in the midplanes of inner disks, a high proportion of crystalline minerals are expected to be produced. The dominant mineral phases to condense as the gas cools are (Fig. 9) (*Gail,* 2003): first, oxides, e.g., corundum or Al_2O_3; second, Mg-rich crystalline silicates, i.e., forsterite (Mg_2SiO_4), enstatite ($MgSiO_3$), and Fe-metal; third, at much cooler temperatures, troilite or FeS.

In chemical equilbrium models, orthoenstatite is predicted to be the dominant silicate condensate (e.g., *Grossman,* 1972; *Petaev and Wood,* 1998). The conversion of forsterite into enstatite is governed by the diffusion of cations and occurs on fairly long timescales compared to disk evolution timescales: A 0.1-μm radii forsterite crystal is converted into an enstatite crystal in 10^6 yr at ~1350 K (*Gail,* 2004). During 10^6 yr, protoplanetary disk midplanes cool

as their mass accretion rates decline (*Wehrstedt and Gail,* 2002; *Keller and Gail,* 2004), so the relatively slow forsterite-enstatite conversion probably does not attain chemical equilibrium at all times throughout the protosolar disk. The forsterite-to-enstatite ratio is diverse in primitive chondrite matrices (*Scott and Krott,* 2005); appears high based on IR spectroscopy of comet comae (Fig. 2) but low based on Halley flyby measurements; and is high in anhydrous CP IDPs for the domains of polycrystalline "equilibrated aggregates" (Fig. 7) but low for the larger (~0.5–5 μm) single crystals (see Fig. 2 of *Keller et al.,* 2000). A larger sample of high signal-to-noise comet IR spectra as well as analyses of samples returned by the Stardust mission will provide further constraints on the degree of forsterite enstatite conversion in the protosolar disk.

3.2. Condensation of Amorphous Silicates

The conditions for equilibrium cannot be expected to persist throughout the disk, as transient heating events such as shocks are likely to occur. A higher proportion of amorphous silicates relative to crystalline silicates are expected to be produced in environments characterized by low partial pressures of condensable elements and rapid cooling, e.g., in stellar outflows (*Molster and Kemper,* 2005), and possibly in X-winds (*Shu et al.,* 2001), protostellar winds (*Kenyon et al.,* 1991), and shocks that melt grains (*Desch and Connolly,* 2002). In regions where low partial pressures of condensable elements exist, condensation will be controlled largely by kinetics, and not by thermodynamics (*Gail and Sedlmayr,* 1999; *Gail,* 2004). Rapid growth of more highly disordered clusters will ensue when gas temperatures drop and supersaturation occurs.

Experiments of vapor-phase condensation from mixtures of gases at ~1000 K form highly disordered "smokes" or fine-grained films of magnesiosilica ($MgO.SiO_2$) or ferrosilica ($Fe_2O_3 + SiO_2$) compositions but not Mg-Fe silicates (*Hallenbeck et al.,* 1998; *Rietmeijer,* 1998; *Rietmeijer et al.,* 2002; *Nuth et al.,* 2002). Rapid condensation scenarios for amorphous silicates in the chondrule-forming region are discussed by *Nuth et al.* (2005). In addition, amorphous silicate minerals are formed by rapid quenching (~2000°C/s) of a melt (*Dorschner et al.,* 1995), by gel dissication (chemical reactions forming a gelatinous precipitate) (*Thompson et al.,* 2002, 2003), or by vaporizing crystalline silicate mineral samples at ~5000 K with a pulsed laser (*Brucato et al.,* 1999, 2002; *Fabian et al.,* 2000). Amorphous enstatite prepared by gel desiccation have Si-O tetrahedonal "protoforsteritic" structures (*Thompson et al.,* 2002, 2003). Amorphous silicate condensates formed by laser ablation, including forsterite, fayalite (Fe_2SiO_4), enstatite, and silica (SiO_2), appear to have more homogeneous mineral compositions than vapor-phase amorphous condensates, possibly because at higher temperatures congruent evaporation of molecular clusters yields condensates with compositions similar to the starting minerals, whereas vapor-phase condensates must

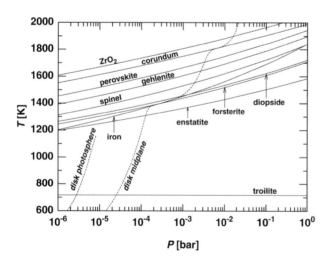

Fig. 9. The stability limits of the important solid phases in the pressure regime of interest for accretion disks. Also shown are the typical P-T-combinations in the midplane of the disk and in the disk atmosphere. As the gas cools in the midplane, at P > 10^{-3} (bar) Fe-metal grains condense out before forsterite (Mg_2SiO_4) and enstatite ($MgSiO_3$).

nucleate clusters from the gas (J. Brucato, personal communication; J. Nuth, personal communication).

3.3. Annealing of Amorphous Silicates

Amorphous silicates are prepared for laboratory annealing experiments using three different processes: vapor condensation, laser ablation, and gel desiccation (section 3.2). Sample preparation techniques impact the composition and structures of amorphous silicate materials and the temperatures and durations of heating required to anneal these samples (see *Wooden et al.*, 2005, for a detailed review).

Despite the differences between the chemical compositions and structures of the amorphous silicate samples prepared by different laboratory techniques, the experiments to anneal amorphous silicates into crystalline silicates come to some general conclusions: (1) The stoichiometry or chemical composition of the starting amorphous material is not altered by annealing, but rather revealed by annealing; Mg-rich amorphous silicates anneal to Mg-rich crystals, Fe-rich amorphous silicates anneal to Fe-rich crystals, Mg-Fe amorphous silicates anneal to moderately Fe-rich crystals, and protoforsteritic structures in amorphous enstatite samples prepared by gel desiccation anneal to forsterite. [Amorphous oxide mixtures ($MgO-SiO_2-Fe_2O_3$) prepared by laser ablation and then annealed at 1000 K for 1 h have mid-IR resonant features at the same wavelengths as annealed forsterite, but far-IR features at slightly (~0.1–0.3 μm) different wavelengths (*Brucato et al.*, 2002). This experiment has not been repeated.] (2) Amorphous magnesiosilica materials anneal at lower temperatures than amorphous ferromagnesia materials (*Hallenbeck et al.*, 1998; *Brucato et al.*, 2002; *Nuth et al.*, 2005). (3) At ~1100–1000 K, Mg-rich amorphous pyroxenes complete annealing in ~1–4 h, with the exception of vapor-condensed samples, which may take up to ~200 h. (4) At ~1100–1000 K, Mg-rich amorphous olivines complete their annealing in ~4–12 h or longer (see Table 1 of *Wooden et al.*, 2005, for a summary of activation energies). (5) Mg-rich crystalline silicate resonances arise from microcrystalline structures as well as from crystallized 0.1-μm spherules (*Brucato et al.*, 1999). (6) Domains of long-range order still may exist within amorphous silicates, even though neither IR spectroscopy nor X-ray diffraction can detect periodic structures of crystalline materials in laboratory-prepared amorphous silicates (e.g., *Colangeli et al.*, 2003). Such domains can act as nucleation sites that promote crystallization more rapidly or at lower annealing temperatures (*Thompson et al.*, 2003; *Colangeli et al.*, 2003; *Brucato et al.*, 2002). (7) The most highly disordered vapor-phase condensates take the longest to anneal, with a "stall" (*Hallenbeck et al.*, 1998) that may be analogous to rearrangement into congruent molecular clusters that exist *a priori* in amorphous silicates prepared by laser ablation (*Rietmeijer et al.*, 2002). Hence, the most conservative approach to computing annealing times is to use the silicate evolution index (SEI) (*Hallenbeck et al.*, 2000), which char-

acterizes the annealing times of the most disordered amorphous silicate samples, i.e., the vapor-phase condensates (*Rietmeijer et al.*, 2002).

Fe-rich amorphous silicates and Fe-rich crystalline silicates are not identified in the IR spectra of comets, are not abundant in anhydrous CP IDPs, and are not inferred to exist from *in situ* measurements of Halley, so the condensation and annealing of Fe-rich amorphous silicate phases appear to play an insignificant role in primary thermal processing in the protosolar disk. Even so, it is interesting to note that Fe-rich amorphous silicates anneal at such high temperatures that partial vaporization may occur, lessening the resultant radii of Fe-rich silicate crystals (*Nuth and Johnson*, 2006).

Mg-rich amorphous silicates are not an abundant component of anhydrous CP IDPs, nor are they deduced to dominate cometary comae [i.e., comet Hale-Bopp (*Harker et al.*, 2002)]. If the chemically favored formation of pure-Mg amorphous silicates by rapid condensation occurred in the protosolar disk (Fig. 11), then these Mg-rich amorphous silicates were annealed to Mg-rich crystalline silicates at ≥1000 K with nearly 100% efficiency.

Heating GEMS, the Mg-Fe amorphous silicates in anhydrous CP IDPs, produces moderately Fe-rich (Mg-Fe) olivine crystals through a subsolidus devitrification (*Brownlee et al.*, 2005). The absence of Mg-Fe crystalline silicates in anhydrous CP IDPs and in comet spectra indicate that the majority of GEMS-like materials never saw temperatures above ~1000 K (*Brownlee et al.*, 2005).

Mg-rich amorphous silicates anneal at lower temperatures or in less time than Mg-Fe amorphous silicates, so given a reservoir of Mg-rich and Mg-Fe amorphous silicates, Mg-rich crystalline silicates are more likely an outcome of the primary thermal processing of annealing. However, the transient heating events would have to be of just the right duration to anneal all the Mg-rich amorphous silicates into Mg-rich crystals but not anneal a significant fraction of the Mg-Fe amorphous silicates into Mg-Fe crystalline silicates. Given the improbability of this "fine-tuning" of transient heating events, we look to combine another physical process — Fe reduction — with annealing to metamorphose Mg-Fe amorphous silicates into Mg-rich crystalline silicates.

3.4. Removal of Iron from Magnesium-Iron Silicates by Reduction, on Magnesium-Iron Interdiffusion Timescales

Experimental heating of Mg-Fe olivines, typically $(Mg_{0.9},Fe_{0.1})_2SiO_4$, in a reducing environment causes the FeO in the mineral to be reduced to Fe-metal, increasing the Mg content of the olivine, and in many cases forming Fe-metal blebs on the surface (*Allen et al.*, 1993; *Weisberg et al.*, 1994). In these experiments, O is scavenged from the silicates in a "locally reducing" environment caused either by a low partial pressure of oxygen or by a "reducing agent" in contact with the grain surface, such as graphitic C that combusts to CO or CO_2.

Experiments demonstrating Fe-reduction have utilized a wide variety of heating temperatures and sample preparations, including ~20–100-μm-radii powders, bulk minerals, and thin films. Flash-melting (≥1873 K) chondrule precursor 20–100-μm-radii ground mineral powders mixed with ≥5 wt% C produces from $(Mg_y,Fe_{(1-y)})_2SiO_4$, y = 0.94–0.89 nearly pure-Mg olivine $(Mg_{0.99},Fe_{0.01})_2SiO_4$, Fe-metal grains, and silica (SiO_2) (*Connolly et al.*, 1994a,b; *Connolly*, 1996). Even though within the furnace the oxygen fugacity is $\log(f_{O_2}[bar]) \approx -9.5$, the oxygen fugacity in the "local environment" of the melt is lower, $\log(f_{O_2}[bar]) \approx -12$. At 1873 K, the canonical solar nebula had $\log(f_{O_2}[bar]) = -14.7$ (–2.7 dex lower than in the experiment), so Fe reduction can occur in an "oxygen-rich" environment compared to the solar nebula.

For Fe reduction to occur in the subsolidus, C must be present on grain surfaces: Heating of a bulk sample of Mg-rich olivine $(Mg_{0.90},Fe_{0.1})_2SiO_4$ for 5–48 h at 1373 K, i.e., at a lower temperature than melts, generates $\log(f_{O_2}[bar]) = -13$, well below (by more than 4 dex) the equilibrium oxygen fugacity, and causes Fe-metal blebs to form on the surface, as well as a gradient in the Mg^{2+}-Fe^{2+} concentration (less Fe^{2+}) and excess SiO_2 in the near-surface region, by the reaction

$$Fe_2SiO_4 \text{ in olivine} + 2C_{graphite} =$$

$$2Fe_{\text{in surface metal}} + SiO_2 \text{ in silicate} + 2CO_{\text{in gas}}$$

Heating at 1000 K a thin film of Mg-rich amorphous olivine, $(Mg_y,Fe_{(1-y)})_2SiO_4$ with y = 0.9, reduces the FeO to 0.002–0.05-μm Fe-metal within the amorphous olivine and generates domains of 0.05–0.25-μm-sized pure-Mg crystalline olivine (Mg_2SiO_4) (*Davoisne et al.*, 2006). In summary, under low oxygen fugacity (f_{O_2}) conditions produced on grain surfaces or in the vicinity of grains by the combustion of C into CO and CO_2, Fe reduction can produce embedded nanophase or surface Fe-metal grains. Under solar nebula conditions S is available in the gas phase, so in conjunction with Fe reduction, $Fe_{1-x}S$ grains may form (*Keller et al.*, 2000). $Fe_{1-x}S$ grains are a component of CP IDPs and GEMS subgrains (*Keller et al.*, 2005).

Carbon is an abundant component of cometary grain materials, either as elemental C in the form of amorphous carbon or poorly graphitized carbon, or in organic grain materials that can be converted to elemental C by heating (*Keller et al.*, 1996). The higher relative abundance of C in cometary materials compared with meteoritic materials, as well as the intimate contact between carbonaceous and siliceous minerals in anhydrous CP IDPs, supports the plausibility of the scenario that Fe reduction occurred in the solar nebula, fueled by abundant C in primitive grains (sections 2.2 and 2.3).

Can Fe reduction be efficient enough to explain the absence of Mg-Fe silicate crystals in cometary grains? The

rate of Fe reduction appears to be governed by the Mg^{2+}-Fe^{2+} interdiffusion rates (*Lemelle et al.*, 2001). Fe reduction generates a gradient in the Mg^{2+}-Fe^{2+} concentration because there is less Fe^{+2} beneath the surface where Fe-metal has been reduced on the surface. A "diffussivity" parameter is derived for Fe reduction that "is close to the diffusivity of Fe^{2+}-Mg^{2+} in olivine" (*Lemelle et al.*, 2001, citing *Nakamura and Schmalzried*, 1984; *Chakraborty*, 1997). Figure 10 shows the interdiffusion rates for Fe^{2+}-Mg^{2+} (data taken from figures in *Chakraborty*, 1997) and the diffusion rate derived for the Fe reduction to Fe-metal on the surface at 1373 K (*Lemelle et al.*, 2001). Also shown in Fig. 10 is the diffusion rate deduced from the heating of 0.05–0.1-μm-radii GEMS at 1000 K for "a few hours" (*Brownlee et al.*, 2005), the consequence of which was both the homogenization (complete Mg^{2+}-Fe^{2+} interdiffusion) and the annealing to a "moderately Fe-rich" (denoted herein Mg-Fe) crystalline olivine. The lower the concentration of Fe, the longer the diffusion time and longer the reduction time.

From the interdiffusion rates (Fig. 10), we compute the Fe reduction times for a 0.1-μm-radii grain, the typical size of a monomer within an aggregate CP IDP (Table 1). Diffusion is a "random walk" process, so $t(s) = a(cm)^2/D(cm^2 s^{-1})$. At 1200 K, the timescales for Fe reduction of a 0.1-μm-radii $(Mg_{0.5},Fe_{0.5})_2SiO_4$ (Fo$_{50}$) crystal is about 0.2 h, which is faster than the completion time for annealing for Fe_2SiO_4 (Fo$_0$) (1 h at 1273 K), slower than for Mg_2SiO_4 (Fo$_{100}$) (1 h at 1073 K) (*Brucato et al.*, 2002; see *Wooden et al.*, 2005, for a comparison of annealing times), and slower for GEMS (*Brownlee et al.*, 2005). A better comparison between the Fe reduction times and annealing time must await future experiments on the annealing of amorphous Mg-Fe silicates and GEMS. The best we can infer from the current laboratory data is that Fe reduction may transform submicrometer Mg-Fe silicates to Mg-rich silicates on timescales similar to or somewhat (maybe up to a factor of 10×) longer than annealing timescales.

4. GRAIN HEATING AND RADIAL MIXING IN DISKS

The metamorphism of Mg-Fe amorphous silicates into Mg-rich crystalline silicates can occur in the hotter, inner disk on long timescales under conditions of "equilibrium + radial diffusion" (Fig. 11). Large-scale radial transport subsequently is required to transport the crystals out to the colder, icy comet-forming regions. Midplane temperatures of 1400–1000 K needed to condense or anneal Mg-rich crystalline silicates only occur out to 2–4 AU at early times in protosolar disk evolution when the mass accretion rates are high (e.g., $10^{-5} M_\odot yr^{-1} \leq \dot{M} \leq 10^{-6} M_\odot yr^{-1}$) (*Bell et al.*, 2000). In the early (≤300,000 yr) collapse, rapid redistribution of angular momentum phase of the protosolar disk, a warm nebula model generates a uniform crystalline mass fraction of ~0.1 throughout the disk (*Bockelée-Morvan et al.*, 2002). A greater fraction of the collapsing mass can be

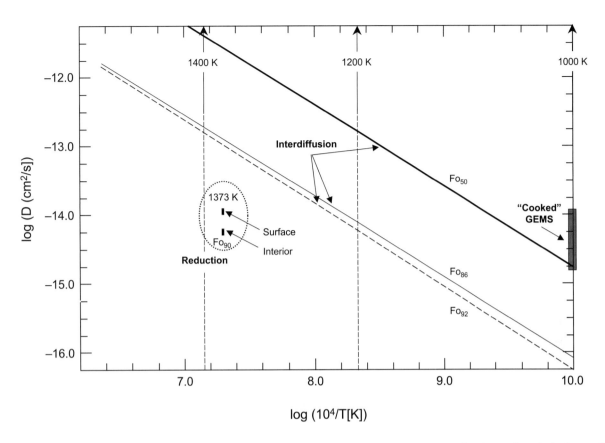

Fig. 10. Temperature-dependence of diffusion coefficients D (cm^2 s^{-1}) for the interdiffusion of Fe^{2+} with Mg^{2+} (*Chakraborty*, 1997), and for Fe reduction from crystalline olivine (Fo$_{90}$) at 1373 K. Reduction proceeds at approximately interdiffusion rates (*Lemelle et al.*, 2001). Fe^{2+} diffuses faster in Mg-Fe (Fo$_{50}$) compared to Mg-rich (Fo$_{86}$, Fo$_{92}$) crystalline olivine.

TABLE 1. Fe reduction times for a 1-μm-radius grain.

T(K)	(Mg$_{0.5}$,Fe$_{0.5}$)$_2$SiO$_4$		(Mg$_{0.92}$,Fe$_{0.08}$)$_2$SiO$_4$	
	t(s)	t(h)	t(s)	t(h)
1000	2.3E4	6.4	1.8E6	494
1200	630	0.2	1.6E4	4.4
1400	20	0.006	630	0.2

processed through this hotter, inner region of the disk for higher angular momentum cloud cores, yielding a higher uniform crystalline fraction (*Dullemond et al.*, 2006).

In contrast to these uniform crystalline fractions produced in very early stages of disk evolution, models for active disks that incorporate the advection-diffusion-reaction equations can produce radial-gradients in the crystalline fractions (f$_{cryst}$). Even as the accretion rates decline with time, f$_{cryst}$ spreads to the outer disk regions. These models for actively accreting disks produce radial gradients in the silicate crystalline fraction, with values of a few tens of percent at ~5 AU (section 4.1). These values of f$_{cryst}$ are at most factors of a few lower than the range determined for Oort cloud comets (section 2.3): Methods to assess f$_{cryst}$ use the mass of submicrometer silicate crystals ratioed to the mass of only submicrometer amorphous silicates or to the mass of submicrometer- to 10-μm-radii amorphous silicates, deriving f$_{cryst}$ ~ 0.6–0.8 or f$_{cryst}$ ~ 0.4, respectively, from the same IR spectra. A model that uses nonspherical particles, submicrometer crystals, and submicrometer- to 100-μm-radii amorphous silicates derives a comparitively very low crystalline fraction of f$_{cryst}$ ~ 0.075 (section 2.3) (*Min et al.*, 2005a).

Mg-rich crystals may form by "transient heating" events such as shocks. Larger volumes of the disk are at lower temperatures, so annealing mechanisms (≥1000 K) can act over larger volumes than can evaporation and condensation (≥1400 K). As the mass accretion rates decline as the disks evolve with time, annealing can continue to act at disk radii where temperatures are too low for evaporation and condensation. Transient heating events include shocks, X-ray flares, and temperature increases arising from increases in opacity, such as may occur when inwardly migrating planetesimals evaporate and increase the dust-to-gas ratio (*Cuzzi et al.*, 2003). At larger disk radii, shocks may not require large-scale radial transport (depending on the number of shocks and the shock speeds) to deliver Mg-rich crystals to comet-forming zones (section 4.2). To produce Mg-rich crystals by annealing, transient heating events need to remove Fe, pos-

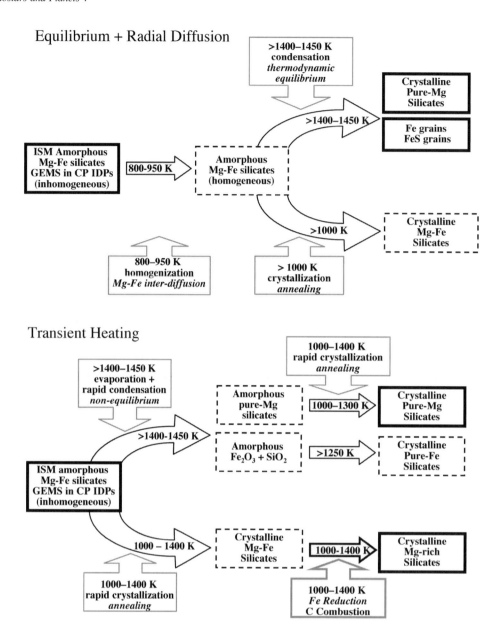

Fig. 11. Scenarios for the thermal metamorphosis of the silicate mineral complex in protoplanetary disks. Equilibrium + radial transport in accretion disks provide conditions for thermodynamically regulated condensation, which converts ISM amorphous Mg-Fe silicates into crystalline pure-Mg silicates and Fe-grains. Transient heating processes, such as in shocks, can (upper path) rapidly condense amorphous pure-Mg silicates that can subsequently be annealed into crystalline pure-Mg silicates. Transient heating also can (lower path) anneal amorphous Mg-Fe silicates to crystalline Mg-Fe silicates. Fe reduction in the subsolidus phase may remove Fe from crystalline Mg-Fe silicates, transforming them to crystalline Mg-rich silicates. Annealing and Fe reduction can occur at similar temperatures, but Fe reduction may be slower (see section 3.3). In comets and anhydrous CP IDPs, some grain materials are present (bold borders) and some are not (dashed borders).

sibly via Fe reduction (Fig. 11), from the Mg-Fe amorphous silicate grains (sections 3.1 and 3.3).

Both Deep Impact Mission results (section 2.3) and recent observations of Herbig Ae/Be stars with the MIDI 10-μm interferometric instrument indicate that radial-dependent crystalline fractions exist in protoplanetary disks. MIDI observations show crystalline fractions are enhanced in inner disks by factors of 2–6 (*van Boekel et al.,* 2004). Similar

enhancements in the inner disk crystalline fraction are deduced from χ^2-fits of passive disk models to observed Herbig Ae/Be star SEDs (*Harker et al.,* 2005a). Both "equilibrium + radial diffusion" and "transient heating" models for the formation of Mg-rich crystals out of the initial reservoir of Mg-Fe amorphous silicates have the potential to produce the silicate crystalline fractions derived from observations of comets and external protoplanetary disks.

4.1. Crystalline Radial Gradients in Accretion Disks

In recent time-dependent two-dimensional models for the mineralogical evolution of grains in active accretion disks (*Gail, 2001, 2004;* Wehrstedt and Gail, in preparation), mass accretion carries amorphous Mg-Fe silicate grains into the warm inner disk zones where high temperature equilibrium processes, e.g., homogenization, Mg-Fe diffusion, annealing, and vaporization followed by condensation (*Gail, 2004*), convert the Mg-Fe amorphous silicates, which are nonequilibrium mixtures, into a chemical equilibrium mixture of Fe-grains and Mg-rich crystalline silicates. The Mg-rich crystals are then mixed outward by radial transport mechanisms including turbulent diffusion and large-scale circulation currents. The mass accretion rate follows a somewhat rapid decline (Fig. 12), yielding a mass accretion rate of 3×10^{-8} $M_\odot$ yr^{-1} at a disk age of t = 10^6 yr for the initial set of disk parameters (e.g., angular momentum and initial mass) (Wehrstedt and Gail, in preparation). The fraction of all olivines that have been transformed by time t to Mg-rich crystalline olivine is shown in Fig 13. In the time-dependent two-dimensional model, the radial dependences of the crystalline fractions for olivines and pyroxenes practically are equivalent (Table 2 in Wehrstedt and Gail, in preparation). Specifically, in the two-dimensional disk model, at 5 AU the crystalline fraction f_{cryst} evolves from ~0.3 to 0.1 between 10^5 and 10^6 yr. These theoretical results therefore suggest that models for a less rapidly evolving protosolar disk will produce higher silicate crystalline fractions in the comet-forming zones (section 2.3).

Model calculations based on stationary one-zone α-disks for the dust evolution (*Gail, 2004*), however, overestimate the crystalline fraction (f_{cryst}) at >20 AU because mixing cannot transport matter from the innermost parts of the disk (r < 1 AU) significantly beyond 20 AU within 10^6 yr.

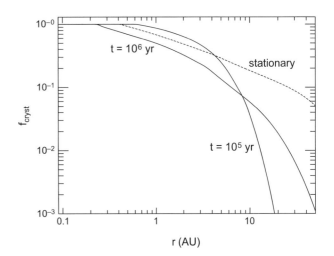

Fig. 13. From time-dependent two-dimensional models, the radial dependence of the crystalline olivine fraction (f_{cryst}) at t = 10^5 and 10^6 yr (Wehrstedt and Gail, in preparation). A stationary model (*Gail, 2004*) with an accretion rate of 3×10^{-8} $M_\odot$ yr^{-1} over estimates (f_{cryst}) at r > 10 AU.

Dust grains contribute significantly to the disk opacity. Changes in opacity due to the metamorphosis of the silicate mineralogy from Mg-Fe amorphous silicates ("Fe-rich" silicates of *Pollack et al., 1994*) to Mg-rich crystalline silicates, however, only moderately impacts the vertical optical depth and the disk's vertical structure (Fig. 6 of *Gail, 2004*).

4.2. Transient Grain Heating in Shocks

The physics of particle heating in shock waves dictates the peak temperature and duration of heating, as discussed and compared in detail in *Desch et al.* (2005). Several shock mechanisms, advanced as explanations for the melting of chondrules, could create shocks with sufficient speeds to anneal amorphous silicate grains. Accretion shocks with V_s > a few km s^{-1} are predicted in the vicinity of a protojovian nebula (*Nelson and Ruffert, 2005*), which may anneal grains in the 5–10 AU region where Oort cloud comets form. Gravitational instabilities in a disk with mass ~0.1 $M_\odot$ are predicted to generate shocks with speeds up to ~10 km s^{-1} throughout the disk (*Boss and Durisen, 2005*). Annealing by shocks could persist as long as the disk is gravitationally unstable beyond these typical radii. According to the calculations of *Bell et al.* (1997), this requires only that the mass flux through the protoplanetary disk exceed ~10^{-9} $M_\odot$ yr^{-1}, a condition that could persist for many millions of years in protoplanetary disks (*Gullbring et al., 1998*). Shock speeds ~5 km s^{-1} therefore seem achievable at heliocentric distances less than about 16 AU (where the Keplerian speed is 15 km s^{-1}) in the protosolar disk.

The fraction of silicate grains that could be processed by shocks is not known, but in principle could be a substantial fraction of all dust interior to the radii at which these

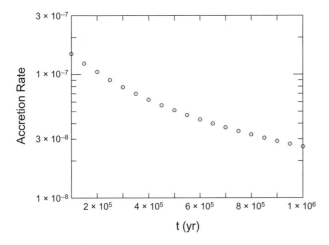

Fig. 12. Time dependence of the mass accretion rate in two-dimensional active disk models (Wehrstedt and Gail, in preparation).

shocks occur. It is important to note that shock models for water-rich regions of the nebula suggest the possible *in situ* aqueous alteration of olivine into layer lattice or phyllosilicates (*Ciesla and Cuzzi,* 2005). Phyllosilicates are missing from comet comae [<1% of montmorillonite in comet Hale-Bopp (*Wooden et al.,* 1999; *Wooden,* 2002)] and rare in anhydrous CP IDPs (*Bradley,* 1988; *Flynn et al.,* 2003); this suggests *in situ* aqueous alteration of siliceous grains was not efficient, shocks were not prevalent in the water-rich region, or radial mixing of phyllosilicates out to the comet-forming zone was extremely inefficient. An important aspect of shock scenarios is that once the transient heating has ended, the disk gas and dust must necessarily return to its ambient temperature. The presence of high- and low-temperature materials in comets implies mixing between shocked and unshocked regions.

Here, an approximation to the physics of the *Connolly and Desch* (2002) model is used to demonstrate that Mg-rich amorphous silicate grains can reach temperatures of ~1200 K for a fraction of a second, and therefore can be annealed (*Harker and Desch,* 2002). We consider a 5 km s^{-1} shock incident upon a gas of ambient density ($\rho_{amb} \approx$ 10^{-10} g cm^{-3}). The shock of Mach number 5.4 compresses the H$_2$ gas by a factor of 6. After crossing the shock front, gas is immediately ($\Delta t \ll 1$ ms) slowed to one-sixth of the incident shock speed, gas heated from T$_{gas} \approx$ 150 K $\rightarrow$ 2000 K. The gas then cools quickly as it collides with and exchanges thermal energy with dust grains; the stopping times for grains of radii a are short, 1.7 (a/1 μm) s. The grains are heated by frictional heating and by absorbing radiation from other grains in the vicinity, and are cooled by emitting radiation. Because the timescale for radiative equilibrium is so short, the peak temperature is found by balancing heating and cooling

$$4\pi a^2 \bar{Q}_{em}\sigma T^4_{dust} = 4\pi a^2 \bar{Q}_{abs}\pi J + 4\pi a^2 \frac{1}{8} \rho_d V^3_{drift}$$

Here $\bar{Q}_{em}$ is the Planck-average emissivity [$Q_{em}(\lambda) = Q_{abs}(\lambda)$] (Fig. 4) of a grain at its peak temperature; $\bar{Q}_{abs}$ is the Planck-average emissivity of the grain at the temperature of the incident radiation field, characterized by the (frequency integrated) mean intensity J_{grains} of the emission from the surrounding dust, which depends on the local (dust-dominated) opacity. Given the density of the silicate dust grain $\rho_d \approx 3$ g cm^{-3}, {$\bar{Q}_{em}, \bar{Q}_{abs}$} for Mg-Fe amorphous pyroxene (Mg$_{0.5}$,Fe$_{0.5}$)SiO$_3$, and a shock with speed V$_s$ = 5 km s^{-1}, then minimum ambient gas densities of 0.6, 2.0, 4.0, and 5.0 × 10^{-10} are required for grains of radii a = 0.10, 0.25, 0.5, and 1.0 μm to reach 1200 K for ~0.1 s and 1200 ≤ T$_{dust}$ ≤ 1150 K for ~1 s, sufficient to anneal amorphous Mg-silicates (SEI) (*Hallenbeck et al.,* 1998, 2000; *Brucato et al.,* 1999; *Wooden et al.,* 2005).

For the protosolar disk, these ambient gas densities of ~10^{-10} g cm^{-3} are typically found at ~1–10 AU, depending on the mass flux and turbulent viscosity parameter α (*Bell et al.,* 1997). For disks around 2 M$_{\odot}$ Herbig Ae/Be stars, midplane gas densities of ~10^{-10} g cm^{-3} can occur, with un-

certainty of at least a factor of 3, at ~1.5× greater distances than in the protosolar disk. Shocks driven by gravitational instabilities appear capable of annealing submicrometer silicate grains out to ~10 AU in the solar system disk, and out to perhaps 20–30 AU in Herbig Ae/Be disks.

The peak grain temperature in the shock depends on the silicate grain mineralogy. Silicate grains are bathed in the radiation field of other grains of approximately the same temperatures, so $\bar{Q}_{em}(a) \approx \bar{Q}_{abs}(a) \propto aT_{dust}$, where a is the grain radius. Thus, the peak temperature may be approximated as

$$T^4_{dust} \simeq \frac{\pi}{\sigma}J_{grains} + \frac{\rho V^3_{drift}}{8\sigma \bar{Q}_{em}}$$

From this approximate formula one sees that high peak temperatures are achieved by high gas densities, high shock speeds, or *low* wavelength-dependent emissivities [$Q_\lambda(a)$] (Fig. 4) at wavelengths near the peak of their absorption/emission (≈ 3 μm for T = 1200 K), i.e., low $Q_\lambda(a)$ by either their mineralogy or their smaller grain radii a. Our calculated dust temperatures are for amorphous Mg-Fe grains. Since Mg-rich silicates have lower emissivities at near-IR wavelengths than Mg-Fe silicates by a factor of ~30 in the near-IR (Fig. 4) (*Dorschner et al.,* 1995), one expects amorphous Mg-rich silicate grains to achieve higher temperatures (by a factor of $\approx 30^{1/4} \approx 2.3\times$) than amorphous Mg-Fe silicate grains of the same grain radius, although the exact factor of temperature increase is sensitive to the calculated radiation field J_{grains}. At near-IR wavelengths, $\bar{Q}_{em}$ is approximately linearly dependent on grain radius (Fig. 4), so smaller shock velocities are required to heat smaller grains to the same temperatures. Smaller particles also are stopped quicker ($t_{stop} \approx a\rho_p/\rho C_s \propto a$) and therefore heated for a shorter time.

While the peak temperatures of 1200–1150 K occur for only ~0.1–1 s, grain temperatures of ≥1000 K are sustained for fractions of an hour (see *Desch et al.,* 2005, for a detailed discussion of the parameters that affect this temperature plateau). If at 1200 K, a 0.1-μm Mg-Fe (Fo$_{50}$) olivine grain requires 600 s (10 min) to reduce the Fe out of the grain on diffusion timescales (Table 1), then this grain would have to be exposed to multiple shocks for it to undergo a metamorphosis to a Mg-rich crystal (section 3.3). *If a plateau of warmer temperatures (~1000 K) precedes the shock (for ≥1 h) and is sustained postshock (for more than several hours, as suggested by Fig. 1 of Desch et al., 2005), then annealing at ~1200 K and Fe reduction at ~1000 K could happen in the same shock.*

5. CONCLUSIONS: CONSTRAINTS ON TRANSIENT HEATING EVENTS PROVIDED BY COMETARY GRAINS

Cometary grains contain volatile organics, refractory organics, and refractory mineral species. Studies of anhydrous CP IDPs and *in situ* measurements show that comet grains are intimate mixtures of ISM materials and materials thermally processed in the protosolar disk. Primary thermal processing and radial transport was happening to submicrometer-sized subgrains at the time of growth of aggregate grains, prior to the incorporation into cometary nuclei. Cometary organic materials were the probable precursors of meteoritic organics and best represent solid-phase ISM organic species. Comparisons between organic species in anhydrous CP IDPs and primitive carbonaceous meteorites reveal the effects of thermal processsing. Cometary volatile organic grain species produce "distributed sources" in cometary comae and have such short lifetimes that they only can be studied in comae. The mere survival of these volatile and refractory organics demonstrates that a significant mass of the outer disk was not subjected to the temperatures required to form crystalline silicates ($\geq$1000 K).

Based on the evidence presented here, Mg-rich crystals are (1) gas-phase condensates that formed under "equilibrium conditions" (thermodynamically controlled conditions) *or* (2) Mg-rich amorphous silicate gas-phase condensates that formed during "transient heating" events (kinetically controlled conditions) and that were subsequently annealed with ~100% efficiency, *or* (3) Mg-Fe amorphous silicates that experienced both annealing and Fe reduction (and Mg-Fe interdiffusion) metamorphosis that removed the Fe from the grains. Mg-Fe amorphous silicates are abundant in cometary materials, while Mg-Fe crystalline silicates are absent. Therefore, the proposed metamorphosis mechanism (3) needs to be either efficient in converting over a sufficiently long time the entire volume of annealed grains or so efficient that it works in shocks.

Once Mg-rich silicates form, the reducing conditions [low oxygen fugacity $\log(f_{O_2})$] of the solar nebula favor the maintenance of the separation of Fe into Fe-metal or FeS (i.e., Fe not in the silicates), so condensation is a straightforward mechanism to form Mg-rich crystals. Time-dependent two-dimensional models for the protosolar disk under conditions of "equilibrium + radial diffusion" can produce radial gradients in the crystalline fraction f_{cryst}. Relatively rapidly evolving two-dimensional disks produce a crystalline fraction at ~5 AU of $f_{cryst} \approx 0.3$ in 10^5 yr, which declines at ~5 AU to $f_{cryst} \approx 0.1$ at 10^6 yr while increasing f_{cryst} at larger radii (section 4.1). These f_{cryst} are near or a factor of a few lower than the crystalline fraction deduced for the submicrometer portion of the comae grain populations of Oort cloud comets (see section 2.3 for caveats in computing the crystalline fraction using different grain size distributions).

Mass accretion rates decline with time, shrinking the regimes of the disk that are hot enough for condensation and annealing. Therefore, it is advantageous to consider the effects that "transient heating" events such as shocks can have on the production of Mg-rich crystalline silicates.

Shocks are likely to occur in protoplanetary disks (section 4.2). At 5–10 AU, shock speeds produce dust temperatures insufficient to vaporize dust grains but sufficient to anneal Mg-rich amorphous silicates into Mg-rich crystals. Grains that are shock heated may be exposed to a peak

shock temperature of ~1150–1200 K for 0.1–1 s and a plateau of higher temperatures of ~1000 K for 0.3–1 h preshock and for approximately several hours postshock. Therefore, Fe reduction at ~1000 K concurrent with annealing is a possible mechanism for the metamorphosis of Mg-Fe silicates to Mg-rich silicates during transient heating events in the solar nebula.

Fe reduction is facilitated by the relatively high carbon abundances in cometary grains, including the amorphous carbon or poorly graphitized carbon and organic materials that appear to be the "glue" that holds together the refractory mineral subgrains, primarily ($\geq$70% by mass, section 2.3) Mg-Fe amorphous silicates. When carbon is in contact with silicate mineral grain surfaces, the combustion of C into CO and CO_2 produces an even lower "local" oxygen fugacity so that Fe reduction can occur in the subsolidus, forming embedded nanophase Fe or Fe-metal blebs on grain surfaces. For submicrometer-radii grains, Fe reduction can occur in 10 min to ~6 h at temperatures of 1200–1000 K. Annealing occurs at these timescales or faster, depending on the Mg content of the initial amorphous silicate material.

Annealing plus Fe reduction implies the presence of crystalline Mg-rich silicates, e.g., Fo_{90}–Fo_{75}, whereas condensation in chemical equilibrium implies crystalline pure-Mg silicates, e.g., Fo_{100}. Mg-rich and pure-Mg crystalline silicates are found in anhydrous CP IDPs (section 2.3), and a census of their relative mass fractions is a subject of current research on anhydrous CP IDPs.

In conclusion, an efficient mechanism for the metamorphosis of Mg-Fe amorphous silicates to Mg-rich amorphous silicates is needed for annealing scenarios to succeed and for shock-annealing scenarios to prevail. Furthermore, large-scale radial transport of crystalline grains is required by scenarios (1) and (2), while it may not be for scenario (3). Future models of radial transport will contribute to our assessment of how far, at what rate, and over what timescales different radial transport mechanisms can act.

Acknowledgments.   We thank the referee E. Scott for his thoughtful comments, and H. Connolly for discussions on Fe reduction. This work was partially supported by NASA's Solar Systems Origins (RTOP 344-37-21-03) and Planetary Astronomy (RTOP 344-32-21-04) programs (D.H.W., D.E.H.), and by the National Science Foundation (AST-03-07466) (D.E.H.).

REFERENCES

A'Hearn M. F., Belton M. J. S., Delamere W. A., Kissel J., Klaasen K. P., et al. (2005) *Science, 310,* 258–264.

Allen C. C., Morris R. V., Lauer H. V. Jr., and McKay D. S. (1993) *Icarus, 104,* 291–300.

Bell K. R., Cassen P. M., Klahr H. H., and Henning T. (1997) *Astrophys. J., 486,* 372–387.

Bell K. R., Cassen P. M., Wasson J. T., and Woolum D. S. (2000) In *Protostars and Planets IV* (V. Mannings et al., eds.), pp. 897–926. Univ. of Arizona, Tucson.

Bockelée-Morvan D. and Crovisier J. (2002) *Earth Moon Planets, 89,* 5371.

Bockelée-Morvan D., Brooke T. Y., and Crovisier J. (1995) *Icarus, 116,* 18.

Bockelée-Morvan D., Lis D. C., Wink J. E., Despois D., Crovisier J., et al. (2000) *Astron. Astrophys., 353,* 1101–1114.

Bockelée-Morvan D., Gautier D., Hersant F., Huré J.-M., and Robert F. (2002) *Astron. Astrophys., 384,* 1107–1118.

Bockelée-Morvan D., Crovisier J., Mumma M. J., and Weaver H. A. (2004) In *Comets II* (M. C. Festou et al., eds.), pp. 391–423. Univ. of Arizona, Tucson.

Boss A. P. and Durisen R. H. (2005) *Astrophys. J., 621,* L137–L140.

Bouwman J., Meeus G., de Koter A., Hony S., Dominik C., and Waters L. B. F. M. (2001) *Astron. Astrophys., 375,* 950–962.

Bouwman J., de Koter A., Dominik C., and Waters L. B. F. M. (2003) *Astron. Astrophys., 401,* 577–592.

Bradley J. P. (1988) *Geochim. Cosmochim. Acta, 52,* 889–900.

Bradley J. P. (1994a) *Science, 265,* 925–929.

Bradley J. P. (1994b) *Geochim. Cosmochim. Acta, 58,* 2123–2134.

Bradley J. P., Brownlee D. E., and Veblen D. R. (1983) *Nature, 301,* 473–477.

Bradley J. P., Humecki H. J., and Germani M. S. (1992) *Astrophys. J., 394,* 643–651.

Bradley J. P., Keller L. P., Gezo J., Snow T., Flynn G. J., Brownlee D. E., and Bowey J. (1999a) In *Lunar Planet. Sci. XXX,* Abstract #1835.

Bradley J. P., Snow T. P., Brownlee D. E., and Hanner M. S. (1999b) In *Solid Interstellar Matter: The ISO Revolution* (L. d'Hendecourt et al., eds.), pp. 297–315. EDP Sciences and Springer-Verlag, Les Houches.

Bradley J. P., Keller L. P., Snow T. P., Hanner M. S., Flynn G. J., et al. (1999c) *Science, 285,* 1716–1718.

Brearley A. J. (1989) *Geochim. Cosmochim. Acta, 53,* 2395–2411.

Brownlee D. E., Joswiak D. J., Schlutter D. J., Pepin R. O., Bradley J. P., et al. (1995) In *Lunar Planet. Sci. XXVI,* pp. 183–184.

Brownlee D. E., Joswiak D. J., and Bradley J. P. (1999) In *Lunar Planet. Sci. XXX,* Abstract #2031.

Brownlee D. E., Joswiak D. J., Bradley J. P., Matrajt G., and Wooden D. H. (2005) In *Lunar Planet. Sci. XXXVI,* Abstract #2391.

Brucato J. R., Colangeli L., Mennella V., Palumbo P., and Bussoletti E. (1999) *Astron. Astrophys., 348,* 1012–1019.

Brucato J. R., Mennella V., Colangeli L., Rotundi A., and Palumbo P. (2002) *Planet. Space Sci., 50,* 829–837.

Brucato J. R., Strazzula G., Baratta G., and Colangeli L. (2004) *Astron. Astrophys., 413,* 395–401.

Campins H. and Ryan E. (1989) *Astrophys. J., 341,* 1059–1066.

Carrez P., Demyk K., Cordier P., Gengembre L., Grimblot J., d'Hendecourt L., Jones A. P., and Leroux H. (2002) *Meteoritics & Planet. Sci., 37,* 1599–1614.

Chakraborty S. (1997) *J. Geophys. Res., 102,* 12317–12332.

Chihara H., Koike C., Tsuchiyama A., Tachibana S., and Sakamoto D. (2002) *Astron. Astrophys., 391,* 267–273.

Ciesla F. J. and Cuzzi J. N. (2005) In *Lunar Planet. Sci. XXXVI,* Abstract #1479.

Colangeli L., Mennella V., Di Marino C., Rotundi A., and Bussoletti E. (1995) *Astron. Astrophys., 293,* 927–934.

Colangeli L., Henning Th., Brucato J. R., Clément D., Fabian D., et al. (2003) *Astron. Astrophys. Rev., 11,* 97–152.

Connolly H. C. Jr. (1996) *Meteoritics & Planet. Sci., 31,* A30–A31.

Connolly H. C. Jr. and Desch S. J. (2002) *Chem. Erde, 64,* 95–125.

Connolly H. C. Jr., Hewins R. H., Ash R. D., Zanda B., Lofgren G. E., and Bourot-Denise M. (1994a) *Nature, 371,* 136–139.

Connolly H. C. Jr., Hewins R. H., Ash R. D., Lofgren G. E., and Zanda B. (1994b) In *Lunar Planet. Sci. XXV,* pp. 279–280.

Crovisier J., Leech K., Bockelée-Morvan D., Brooke T. Y., Hanner M. S., et al. (1997) *Science, 275,* 1904–1907.

Cuzzi J. N., Davis S. S., and Dobrovolskis A. R. (2003) *Icarus, 166,* 385–402.

Davies J. K., Roush T. L., Cruikshank D. P., Bartholomew M. J., Geballe T. R., et al. (1997) *Icarus, 127,* 238–245.

Davoisne C., Djouadi Z., Leroux H., d'Hendecourt L., Jones A., and Debouffle D. (2006) *Astron. Astrophys., 448,* L1–L4.

Desch S. J. and Connolly H. C. (2002) *Meteoritics & Planet. Sci., 37,* 183–207.

Desch S. J., Ciesla F. J., Hood L. L., and Nakamoto T. (2005) In *Chondrites and the Protoplanetary Disk* (A. N. Krot et al., eds.), pp. 849–872. ASP Conf. Series 341, San Francisco.

DiSanti M. A., Mumma M. J., Dello Russo N., and Magee-Sauer K. (2001) *Icarus, 153,* 361–390.

Dorschner J., Begemann B., Henning T., Jaeger C., and Mutschke H. (1995) *Astron. Astrophys., 300,* 503–520.

Dukes C., Baragiola R., and McFadden L. (1999) *J. Geophys. Res., 104,* 1865–1872.

Dullemond C. P., Apai D., and Walch S. (2006) *Astrophys. J., 640,* L67–L70.

Ehrenfreund P. and Charnley S. B. (2000) *Ann. Rev. Astron. Astrophys., 38,* 427–483.

Ehrenfreund P., Charnley S. B., and Wooden D. H. (2004) In *Comets II* (M. C. Festou et al., eds.), pp. 115–133. Univ. of Arizona, Tucson.

Fabian D., Jäger C., Henning T., Dorschner J., and Mutschke H. (2000) *Astron. Astrophys., 364,* 282–292.

Flynn G. J., Keller L. P., Feser M., Wirick S., and Jacobsen C. (2003) *Geochim. Cosmochim. Acta, 67,* 4719–4806.

Flynn G. J., Keller L. P., Jacobsen C., and Wirick S. (2004) *Adv. Space Res., 33,* 57–66.

Fomenkova M. N. (1999) *Space Sci. Rev., 90,* 109–114.

Fomenkova M. N., Chang S., and Mukhin L. M. (1994) *Geochim. Cosmochim. Acta, 58,* 4503–4512.

Furlan E., Calvet N., D'Alessio P., Hartmann L., Forrest W. J., Watson D. M., et al. (2005) *Astrophys. J., 621,* L129–L132.

Gail H.-P. (2001) *Astron. Astrophys., 378,* 192–213.

Gail H.-P. (2003) In *Astromineralogy: Lecture Notes in Physics, Vol. 309* (T. K. Henning, ed.), pp. 55–120. Springer, Berlin.

Gail H.-P. (2004) *Astron. Astrophys., 413,* 571–591.

Gail H.-P. and Sedlmayr E. (1999) *Astron. Astrophys., 347,* 594–616.

Gomes R. S. (2003) *Icarus, 161,* 404–418.

Grossman L. (1972) *Geochim. Cosmochim. Acta, 38,* 47–64.

Gullbring E., Hartmann L., Briceno C., and Calvet N. (1998) *Astrophys. J., 492,* 323–341.

Hallenbeck S. L., Nuth J. A., and Daukantas P. L. (1998) *Icarus, 131,* 198–209.

Hallenbeck S. L., Nuth J. A., and Nelson R. N. (2000) *Astrophys. J., 535,* 247–255.

Hanner M. S. (1983) In *Cometary Exploration, Vol. 2* (T. I. Gombosi, ed.), pp. 1–22. Hungarian Acad. Sci., Budapest.

Hanner M. S. and Bradley J. P. (2004) In *Comets II* (M. C. Festou et al., eds.), pp. 555–564. Univ. of Arizona, Tucson.

Hanner M. S., Lynch D. K., and Russell R. W. (1994) *Astrophys. J., 425,* 274–285.

Harker D. E. and Desch S. (2002) *Astrophys. J., 565,* L109–L112.

Harker D. E., Wooden D. H., Woodward C. E., and Lisse C. M. (2002) *Astrophys. J., 580,* 579–597.

Harker D. E., Wooden D. H., Woodward C. E., and Lisse C. M. (2004) *Astrophys. J., 615,* 1081.

Harker D. E., Woodward C. E., Wooden D. H., and Temi P. (2005a) *Astrophys. J., 622,* 430–439.

Harker D. E., Woodward C. E., and Wooden D. H. (2005b) *Science, 310,* 278–280.

Henning Th., Mutschke H., and Jäger C. (2005) In *Astrochemistry: Recent Successes and Current Challenges* (D. C. Lis et al., eds.), pp. 457–468. IAU Symposium No. 231, Cambridge Univ., Cambridge.

Honda M., Kataza H., Okamota Y. K., Miyata T., Yamashita T., Sako S., et al. (2004) *Astrophys. J., 585,* L59–L63.

Jäger C., Fabian D., Schrempel F., Dorschner J., Henning, Th., and Wesch W. (2003) *Astron. Astrophys., 401,* 57–65.

Jessberger E. K., Stephan T., Rost D., Arndt P., Maetz M., et al. (2001) In *Interplanetary Dust* (E. Grün et al., eds.), pp. 253–294. Springer-Verlag, Berlin.

Jones A. P. (2000) *J. Geophys. Res., 105,* 10257–10268.

Joswiak D. J., Brownlee D. E., Bradley J. P., Schlutter D. J., and Pepin R. O. (1996) In *Lunar Planet. Sci. XXVII,* pp. 625–626.

Kawakita H., Watanabe J., Furusho R., Fuse T., Capria M. T., and De Sanctis M. C. (2004a) *Astrophys. J., 601,* 1152–1158.

Kawakita H., Watanabe J., Ootsubo T., Nakamura R., Fuse T., et al. (2004b) *Astrophys. J., 601,* L191–L194.

Keller Ch. and Gail H.-P. (2004) *Astron. Astrophys., 415,* 1177–1185.

Keller L. P. and Flynn G. J. (2003) In *Workshop on Cometary Dust in Astrophysics,* pp. 38–39. LPI Contribution No. 1182, Houston.

Keller L. P. and Messenger S. (2004a) In *Lunar Planet. Sci. XXXV,* Abstract #1985.

Keller L. P. and Messenger S. (2004b) *Meteoritics & Planet. Sci., 39,* 5186.

Keller L. P. and Messenger S. (2005) In *PPV Poster Proceedings,* www. lpi.usra.edu/meetings/ppv2005/pdf/8570.pdf.

Keller L. P., Thomas K. L., and McKay D. S. (1996) In *Physics, Chemistry, and Dynamics of Interplanetary Dust* (B. A. S. Gustafson and M. S. Hanner, eds.), pp. 295–298. ASP Conf. Series 104, San Francisco.

Keller L. P., Messenger S., and Bradley J. P. (2000) *J. Geophys. Res., 105,* 10397–10402.

Keller L. P., Messenger S., Flynn G. J., Clemett S., Wirick S., and Jacobsen C. (2004) *Geochim. Cosmochim. Acta, 68,* 2577–2589.

Keller L. P., Messenger S., and Christoffersen R. (2005) In *Lunar Planet. Sci. XXXVI,* Abstract #2088.

Kemper F., Waters L. B. F. M., de Koter A., and Tielens A. G. G. M. (2001) *Astron. Astrophys., 369,* 132–141

Kemper F., Vriend W. J., and Tielens A. G. G. M. (2004) *Astrophys. J., 609,* 826–837.

Kemper F., Vriend W. J., and Tielens A. G. G. M. (2005) *Astrophys. J., 633,* 534–534.

Kenyon S. J., Hartmann L. W., and Kolotilov E. A. (1991) *Publ. Astron. Soc. Pac., 103,* 1069–1076.

Kerridge J. F. (1999) *Space Sci. Rev., 90,* 275–288.

Khare B. N., Thompson W. R., Sagan C., Arakawa E. T., Meisse C., and Gilmour I. (1990) In *Lunar Planet. Sci. XXI,* p. 627.

Kimura H., Kolokolova L., and Mann I. (2003) *Astron. Astrophys., 407,* L5–L8.

Kissel J. (1999) In *Formulation and Evolution of Solids in Space* (J. M. Greenberg and A. Li, eds.), pp. 427–445. Kluwer, Dordrecht.

Koike C., Chihara H., Tsuchiyzma A., Suto H., Sogawa H., and Okuda H. (2003) *Astron. Astrophys., 399,* 1101–1107.

Lemelle L., Guyot F., Leroux H., and Libourel G. (2001) *Am. Mineral., 86,* 47–54.

Li A. and Draine B. T. (2001) *Astrophys. J., 550,* L213–L216.

Mann I., Czechowski A., Kimura H., Köhler M., Minato T., and Yamamoto T. (2005) In *Asteroids, Comets, and Meteors* (D. Lazzaro et al., eds.), pp. 41–65. IAU Symposium No. 229, Cambridge Univ., Cambridge.

Messenger S. (2000) *Nature, 404,* 968–971.

Messenger S., Walker R. M., Clemett S. J., and Zare R. N. (1996) In *Lunar Planet. Sci. XXVII,* pp. 867–868.

Messenger S., Keller L. P., Stadermann F. J., Walker R. M., and Zinner E. (2003) *Science, 300,* 105–108.

Min M., Hovenier J. W., de Koter A., Waters L. B. F. M., and Domonik C. (2005a) *Icarus, 179,* 158–173.

Min M., Waters L. B. F.M., Hovenier J. W., de Koter A., Keller L. P., and Markwick-Kemper F. (2005b) In *PPV Poster Proceedings,* www.lpi. usra.edu/meetings/ppv2005/pdf/8478.pdf.

Molster F. J. and Kemper C. (2005) *Space Sci. Rev., 119,* 3–28.

Molster F. J., Bradley J. P., Sitko M. L., and Nuth J. A. (2001) In *Lunar Planet. Sci. XXXII,* Abstract #1391.

Molster F. J., Demyk A., d'Hendecourt L., Bradley J. P., Bonal L., and Borg J. (2003) In *Lunar Planet. Sci. XXXIV,* Abstract #1148.

Morbidelli A. and Levison H. F. (2003) *Nature, 422,* 30–31.

Nakamura A. and Schmalzried H. (1984) *Berichte Bunsengesellsch. Physik. Chem., 88,* 140–145.

Nelson A. F. and Ruffert M. (2005) In *Chondrites and the Protoplanetary Disk* (A. N. Krot et al., eds.), pp. 903–914. ASP Conf. Series 341, San Francisco.

Nier A. O. and Schlutter D. J. (1993) *Meteoritics, 28,* 412–461.

Nuth J. A. and Johnson N. M. (2006) *Icarus, 180,* 243–250.

Nuth J. A. III, Rietmeijer J. M., and Hill H. G. M. (2002) *Meteoritics & Planet. Sci., 37,* 1579

Nuth J. A. III, Brearley A. J., and Scott E. R. D. (2005) In *Chondrites and the Protoplanetary Disk* (A. N. Krot et al., eds.), pp. 675–700. ASP Conf. Series 341, San Francisco.

Petaev M. I. and Wood J. A. (1998) *Meteoritics & Planet. Sci., 33,* 1123.

Petaev M. I. and Wood J. A. (2005) In *Chondrites and the Protoplanetary Disk* (A. N. Krot et al., eds.), pp. 373–406. ASP Conf. Series 341, San Francisco.

Pollack J. B., Hollenbach D., Beckwith S., Simoneli D. P., Roush T., and Fong W. (1994) *Astrophys. J., 421,* 615–639.

Rietmeijer F. J. M. (1998) In *Planetary Materials* (J. J. Papike, ed.), pp. 2-1 to 2-95. Reviews in Mineralogy, Vol. 36, Mineralogical Society of America.

Rietmeijer F. J. M., Hallenbeck S. L., Nuth J. A., and Karner J. M. (2002) *Icarus, 156,* 269–286.

Savage K. R. and Sembach B. D. (1996) *Ann. Rev. Astron. Astrophys., 34,* 279–329.

Schulze H. and Kissel J. (1992) *Meteoritics, 27,* 286–287.

Schulze H., Kissel J., and Jessberger E. K. (1997) In *From Stardust to Planetesimals* (Y. J. Pendleton and A. G. G. M. Tielens, eds.), pp. 397–414. ASP Conf. Series 122, San Francisco.

Scott E. R. D. and Krot A. N. (2005) *Astrophys. J., 623,* 571–578.

Shu F. H., Shang H., Gounelle M., Glassgold A. E., and Lee T. (2001) *Astrophys. J., 548,* 1029–1050.

Spitzer L. Jr. and Jenkins E. B. (1975) *Ann. Rev. Astron. Astrophys., 13,* 133–164.

Sugita S., Ootsubo T., Kadono T., Honda M., Sako S., et al. (2005) *Science, 310,* 274–278.

Thomas K. L., Keller L. P., Blanford G. E., and McKay D. S. (1994) In *Analysis of Interplanetary Dust* (M. E. Zolensky et al., eds.), pp. 165–174. AIP Conf. Proceedings 310, Springer-Verlag, Berlin.

Thomas K. L., Keller L. P., and McKay D. S. (1996) In *Physics, Chemistry, and Dynamics of Interplanetary Dust* (B. A. S. Gustafson and M. S. Hanner, eds.), pp. 283–286. ASP Conf. Series 104, San Francisco.

Thompson S. P., Fonti S., Verrienti C., Blanco A., Orofino V., and Tang C. C. (2002) *Astron. Astrophys., 395,* 705–717.

Thompson S. P., Fonti S., Verrienti C., Blanco A., Orofina V., and Tang C. C. (2003) *Meteoritics & Planet. Sci., 38,* 457–478.

Tielens A. G. G. M. (2003) *Science, 300,* 68–70.

van Boekel R., Min M., Leinert Ch., Waters L. B. F. M., Richichi A., Chesneau O., et al. (2004) *Nature, 432,* 479–482.

Wehrstedt M. and Gail H.-P. (2002) *Astron. Astrophys., 385,* 181–204.

Weisberg M. K., Prinz M., and Fogel R. A. (1994) *Meteoritics, 29,* 362–373.

Wooden D. H. (2002) *Earth Moon Planets, 89,* 247.

Wooden D. H., Harker D. H., Woodward C. E., Butner H. M., Koike C., et al. (1999) *Astrophys. J., 517,* 1034–1058.

Wooden D. H., Butner H. M., Harker D. E., and Woodward C. E. (2000) *Icarus, 143,* 126–137.

Wooden D. H., Woodward C. E., and Harker D. E. (2004) *Astrophys. J., 612,* L77–L80.

Wooden D. H., Harker D. E., and Brearley A. J. (2005) In *Chondrites and the Protoplanetary Disk* (A. N. Krot et al., eds.), pp. 774–810. ASP Conf. Series 341, San Francisco.

From Dust to Planetesimals: Implications for the Solar Protoplanetary Disk from Short-lived Radionuclides

M. Wadhwa
The Field Museum

Y. Amelin
Geological Survey of Canada

A. M. Davis
The University of Chicago

G. W. Lugmair
University of California at San Diego

B. Meyer
Clemson University

M. Gounelle
Muséum National d'Histoire Naturelle

S. J. Desch
Arizona State University

Since the publication of the *Protostars and Planets IV* volume in 2000, there have been significant advances in our understanding of the potential sources and distributions of short-lived, now extinct, radionuclides in the early solar system. Based on recent data, there is definitive evidence for the presence of two new short-lived radionuclides (^{10}Be and ^{36}Cl) and a compelling case can be made for revising the estimates of the initial solar system abundances of several others (e.g., ^{26}Al, ^{60}Fe, and ^{182}Hf). The presence of ^{10}Be, which is produced only by spallation reactions, is either the result of irradiation within the solar nebula (a process that possibly also resulted in the production of some of the other short-lived radionuclides) or of trapping of galactic cosmic rays in the protosolar molecular cloud. On the other hand, the latest estimates for the initial solar system abundance of ^{60}Fe, which is produced only by stellar nucleosynthesis, indicate that this short-lived radionuclide (and possibly significant proportions of others with mean lives ≤ 10 m.y.) was injected into the solar nebula from a nearby stellar source. As such, at least two distinct sources (e.g., irradiation and stellar nucleosynthesis) are required to account for the abundances of the short-lived radionuclides estimated to be present in the early solar system. In addition to providing constraints on the sources of material in the solar system, short-lived radionuclides also have the potential to provide fine-scale chronological information for events that occurred in the solar protoplanetary disk. An increasing number of studies are demonstrating the feasibility of applying at least some of these radionuclides as high-resolution chronometers. From these studies, it can be inferred that the millimeter- to centimeter-sized refractory calcium-aluminum-rich inclusions in chondritic meteorites are among the earliest solids to form (at 4567.2 ± 0.6 Ma). Formation of chondrules (i.e., submillimeter-sized ferromagnesian silicate spherules in chondrites) is likely to have occurred over a time span of at least ~3 m.y., with the earliest ones possibly forming contemporaneously with CAIs. Recent work also suggests that the earliest planetesimals began accreting and differentiating within a million years of CAI formation, i.e., essentially contemporaneous with chondrule formation. If so, it is likely that undifferentiated chondrite parent bodies accreted a few million years thereafter, when the short-lived radionuclides that served as the main heat sources for melting planetesimals (^{26}Al and ^{60}Fe) were nearly extinct.

1. INTRODUCTION

Short-lived radionuclides are characterized by half-lives ($T_{1/2}$) that are significantly shorter (i.e., ≤100 m.y.) than the 4.56-Ga age of the solar system. Although now extinct, their former presence at the time of solar system formation can be inferred if variations in their daughter isotopes are demonstrated to correlate with parent/daughter element ratios in meteorites and their components. These radionuclides are of particular interest since (1) an understanding of their sources and distributions in the early solar system (ESS) can provide constraints on the formation environment and astrophysical setting of the solar protoplanetary disk and (2) they have the potential for application as fine-scale chronometers (in many cases with a time resolution of ≤1 m.y.) for events occurring in the early history of the solar system.

A prerequisite for the application of a fine-scale chronometer based on a short-lived radionuclide is that the initial abundance of this radionuclide must be demonstrated to be uniform in the region of the solar system where rocky bodies were forming. Moreover, since the slope of an isochron derived from such a chronometer provides not an age but a measure of the abundance of the radionuclide at the time of last isotopic closure, comparison of the isochron slopes for two separate events can provide only a relative time difference between these events. For such high-resolution relative ages to be mapped on to an absolute timescale, they need to be pinned to a precise time "anchor," which is usually provided by the U-Pb chronometer (which is capable of providing absolute ages with a precision comparable to that of the short-lived chronometers). Further details on the application of short-lived radionuclides as chronometers and the caveats involved have been discussed in several review articles (e.g., *Wasserburg*, 1985; *Podosek and Nichols*, 1997; *Wadhwa and Russell*, 2000; *McKeegan and Davis*, 2003; *Kita et al.*, 2005; *Gounelle and Russell*, 2005).

The purpose of this review is not to provide a comprehensive overview of short-lived radionuclides and their application to the study of meteorites and their components (which may be found in several of the reviews mentioned above). Instead, we will focus on the most recent results and the advances in our understanding of the sources and distributions of these radionuclides since the publication of *Protostars and Planets IV* (PPIV). In the following sections, we will discuss their two main potential sources, i.e., stellar nucleosynthesis and local production by irradiation. Furthermore, based on our current understanding of the abundances and distributions of short-lived radionuclides in the ESS, we will discuss the implications for the astrophysical setting and for the timing of events from "dust to planetesimals" in the solar protoplanetary disk.

2. SHORT-LIVED RADIONUCLIDES IN THE EARLY SOLAR SYSTEM: THE LATEST RESULTS

Table 1 provides a listing of the short-lived radionuclides for which there is now definitive evidence of their former presence in the ESS, although the initial solar system abun-

TABLE 1. Short-lived radionuclides in the early solar system.

Parent Isotope	$T_{1/2}$*	Daughter Isotope	Solar System Initial Abundance[†]
[10]Be	1.5	[10]B	[10]Be/[9]Be ≈ 10^{-3}
[26]Al	0.72	[26]Mg	[26]Al/[27]Al ≈ 5–7 × 10^{-5}
[36]Cl	0.3	[36]Ar (98.1%)	[36]Cl/[35]Cl ≥ 1.6 × 10^{-4}
		[36]S (1.9%)	
[41]Ca	0.1	[41]K	[41]Ca/[40]Ca ≥ 1.5 × 10^{-8}
[53]Mn	3.7	[53]Cr	[53]Mn/[55]Mn ≈ 10^{-5}
[60]Fe	1.5	[60]Ni	[60]Fe/[56]Fe ≈ 3–10 × 10^{-7}
[92]Nb	36	[92]Zr	[92]Nb/[93]Nb ≈ 10^{-5}–10^{-3}
[107]Pd	6.5	[107]Ag	[107]Pd/[108]Pd ≈ 5–40 × 10^{-5}
[129]I	15.7	[129]Xe	[129]I/[129]Xe ≈ 10^{-4}
[146]Sm	103	[142]Nd	[146]Sm/[144]Sm ≈ 7 × 10^{-3}
[182]Hf	8.9	[182]W	[182]Hf/[180]Hf ≈ 10^{-4}
[244]Pu	82	Fission Xe	[244]Pu/[238]U ≈ 7 × 10^{-3}

*Half-life in millions of years.
[†]Data sources: [10]Be: *McKeegan et al.* (2000); [26]Al: *Lee et al.* (1976), *MacPherson et al.* (1995), *Bizzarro et al.* (2004), *Young et al.* (2005); [36]Cl: *Lin et al.* (2005); [41]Ca: *Srinivasan et al.* (1994, 1996); [53]Mn: *Lugmair and Shukolyukov* (1998); [60]Fe: *Tachibana and Huss* (2003), *Mostefaoui et al.* (2005), *Tachibana et al.* (2006); [92]Nb: *Harper* (1996), *Münker et al.* (2000), *Sanloup et al.* (2000), *Yin et al.* (2000), *Schönbächler et al.* (2002); [107]Pd: *Chen and Wasserburg* (1996), *Carlson and Hauri* (2001); [129]I: *Swindle and Podosek* (1988) and references therein, *Brazzle et al.* (1999); [146]Sm: *Lugmair and Galer* (1992) and references therein; [182]Hf: *Kleine et al.,* (2002, 2005), *Yin et al.* (2002); [244]Pu: *Podosek* (1970), *Hudson et al.* (1989).

dances of some of these are somewhat uncertain. Several others, such as [7]Be ($T_{1/2}$ = 53 d) (*Chaussidon et al.,* 2006), [99]Tc ($T_{1/2}$ ~ 0.2 m.y.) (*Yin et al.,* 1992), [135]Cs ($T_{1/2}$ ~ 2.3 m.y.) (*Hidaka et al.,* 2001) and [205]Pb ($T_{1/2}$ ~ 15 m.y.) (*Chen and Wasserburg,* 1987; *Nielsen et al.,* 2004), may also have been present but evidence for these is as yet suggestive rather than definitive.

Since PPIV, two new short-lived radionuclides ([10]Be and [36]Cl) have been added to the roster of those for which there is now compelling evidence for their former presence in the ESS (Table 1). In addition, the presence of [92]Nb, for which there was only suggestive evidence prior to 2000, has been confirmed by several recent studies, although its initial abundance is still debated. Also, on the basis of recent analyses of meteorites and their components, the initial abundances of several of the short-lived radionuclides listed in Table 1 have been revised. Some of the implications of these new results will be discussed in the following sections.

2.1. Beryllium-10

McKeegan et al. (2000) showed that excesses in [10]B/[11]B are correlated with [9]Be/[11]B ratios in calcium-aluminum-rich inclusions (CAI) from the Allende carbonaceous chondrite, indicating an initial [10]Be/[9]Be ratio of ~10^{-3} in the ESS (Fig. 1). Subsequently, additional studies of CAIs from other chondrite groups have confirmed this finding (*Sugiura et al.,* 2001; *Marhas et al.,* 2002; *MacPherson et al.,* 2003).

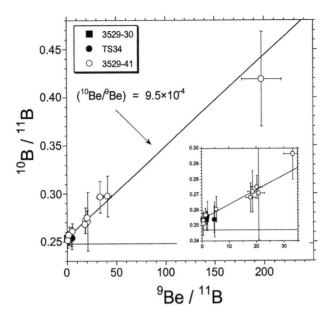

Fig. 1. Boron-isotopic composition of Allende CAIs vs. Be/B ratios (*McKeegan et al., 2000*).

2.2. Chlorine-36

Recently, *Lin et al.* (2005) presented new evidence (excesses in ^{36}S that correlated with Cl/S ratios) for the presence of live ^{36}Cl in sodalite, a chlorine-rich mineral that most likely formed from aqueous alteration on the parent body, in a CAI from the Ningqiang carbonaceous chondrite (Fig. 2). This work indicates that the $^{36}Cl/^{35}Cl$ ratio in the ESS was at least ~1.6×10^{-4}.

2.3. Niobium-92

Until just a few years ago, the only hint of the former presence of ^{92}Nb in the ESS had been a well-resolved excess of ^{92}Zr in rutile, a rare mineral with a high Nb/Zr ratio, in the Toluca iron meteorite, based upon which an initial $^{92}Nb/^{93}Nb$ ratio of ~2×10^{-5} was inferred for the solar system (*Harper, 1996*). Subsequently, several studies also reported excesses in ^{92}Zr in bulk samples and mineral separates from a variety of primitive and differentiated meteorites, but suggested a substantially higher initial $^{92}Nb/^{93}Nb$ ratio of ~10^{-3} (*Münker et al., 2000; Sanloup et al., 2000; Yin et al., 2000*). However, *Schönbächler et al.* (2002) reported internal $^{92}Nb-^{92}Zr$ isochrons for the H6 chondrite Estacado and a basaltic clast from the Vaca Muerta mesosiderite and inferred an initial solar system $^{92}Nb/^{93}Nb$ ratio of ~10^{-5}. *Yin and Jacobsen* (2002) have suggested that Estacado and the Vaca Muerta clast may record secondary events that postdated solar system formation by ~150 ± 20 m.y. If so, the lower $^{92}Nb/^{93}Nb$ ratio inferred by *Schönbächler et al.* (2002) would be compatible with the higher value of ~10^{-3} (which would then reflect the true initial value) reported by others (*Münker et al., 2000; Sanloup et al., 2000; Yin et al., 2000*). Although the $^{40}Ar-^{39}Ar$ ages for Estacado (*Flohs, 1981*) and the $^{147}Sm-^{143}Nd$ ages for Vaca Muerta

clasts (*Stewart et al., 1994*) may indeed record late disturbances >100 m.y. after the beginning of the solar system, there is no definitive indicator that the Nb-Zr system was reset in these samples at this time. As such, the initial abundance of ^{92}Nb in the ESS is as yet unclear.

2.4. Initial Abundances of Aluminum-26, Calcium-41, Iron-60, and Hafnium-182

The initial abundances of several of the radionuclides listed in Table 1 have been revised significantly since PPIV. Until recently, the initial $^{26}Al/^{27}Al$ ratio in the ESS was thought to have the canonical value of ~5×10^{-5} (*Lee et al., 1976; MacPherson et al., 1995*). However, recent high-precision Mg-isotopic analyses of CAIs indicate that the initial $^{26}Al/^{27}Al$ ratio may have been as high as $6-7 \times 10^{-5}$ (*Bizzarro et al., 2004, 2005a; Young et al., 2005; Taylor et al., 2005*). If this higher value for the initial $^{26}Al/^{27}Al$ ratio is assumed, then the initial $^{41}Ca/^{40}Ca$ ratio may also have been correspondingly higher than the previously inferred value of ~1.5×10^{-8} by at least an order of magnitude, because the initial $^{41}Ca/^{40}Ca$ ratio was measured on CAIs with internal isochrons indicating an initial $^{26}Al/^{27}Al$ ratio of ~5×10^{-5}.

Although *Birck and Lugmair* (1988) had noted excesses in ^{60}Ni in Allende CAIs, these could be attributable to nucleosynthetic anomalies and the first definitive evidence for the presence of live ^{60}Fe in the ESS came from the work of *Shukolyukov and Lugmair* (1993a,b). These authors showed that excesses in ^{60}Ni were correlated with Fe/Ni ratios in bulk samples of the eucrites Chervony Kut and Juvinas. Based on their analyses, *Shukolyukov and Lugmair*

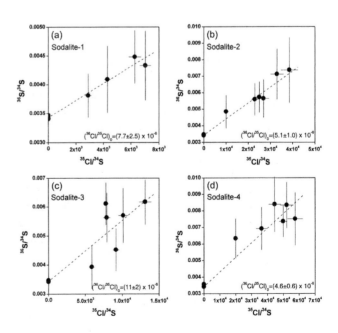

Fig. 2. Sulfur-isotopic compositions of sodalite-rich assemblages in a Ningqiang CAI (*Lin et al., 2005*). The best inferred $^{36}Cl/^{35}Cl$ ratio from these data is 5×10^{-6}. Assuming that the sodalites formed ≥1.5 m.y. after CAIs, an initial $^{36}Cl/^{35}Cl$ ratio of ≥1.6×10^{-4} is inferred.

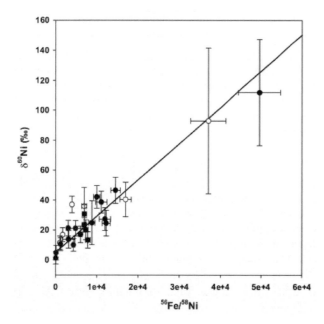

Fig. 3. Excesses in the $^{60}Ni/^{62}Ni$ ratio relative to a terrestrial standard in parts per 10^3 vs. Fe/Ni ratios in troilites from metal-free assemblages in the Semarkona unequilibrated ordinary chondrite (*Mostefaoui et al.*, 2005). The slope of the Fe-Ni isochron yields a $^{60}Fe/^{56}Fe$ ratio of $(0.92 \pm 0.24) \times 10^{-6}$.

(1993a,b) inferred an initial $^{60}Fe/^{56}Fe$ ratio of $\sim 10^{-8}$. Recently, ion microprobe analyses of components in unequilibrated chondrite meteorites indicate that the initial solar system $^{60}Fe/^{56}Fe$ ratio is likely to be as high as $\sim 10^{-6}$ (*Tachibana et al.*, 2006; *Mostefaoui et al.*, 2005) (Fig. 3). The lower initial $^{60}Fe/^{56}Fe$ ratio inferred from the eucrites (which are known to have undergone varying degrees of thermal metamorphism) is thought to be the result of partial equilibration of the Fe-Ni system.

The earliest estimates of an upper limit on the initial $^{182}Hf/^{180}Hf$ ratio for the ESS based on analyses of meteoritic material suggested that it was $\leq \sim 2 \times 10^{-4}$ (*Ireland*, 1991; *Harper and Jacobsen*, 1996). Subsequently, the work of *Lee and Halliday* (1995, 1996) indicated that the W isotope composition of the bulk silicate Earth (BSE) was identical to that of the chondrites and the initial $^{182}Hf/^{180}Hf$ ratio for the ESS was $\sim 2.5 \times 10^{-4}$. However, several recent studies demonstrated that the W isotope composition of the BSE is more radiogenic than chondrites by $\sim 2\varepsilon$ units (*Kleine et al.*, 2002; *Schoenberg et al.*, 2002; *Yin et al.*, 2002) and indicated a lower initial $^{182}Hf/^{180}Hf$ ratio of $\sim 1 \times 10^{-4}$ (note that W isotope composition in ε units or $\varepsilon^{182}W$ is defined as the $^{182}W/^{183}W$ or the $^{182}W/^{184}W$ ratio relative to the terrestrial standard in parts per 10^4). Based on the extremely unradiogenic W isotope composition of the Tlacotepec iron meteorite, *Quitté and Birck* (2004) suggested an intermediate value of $\sim 1.6 \times 10^{-4}$ for the initial $^{182}Hf/^{180}Hf$ ratio. However, the highly unradiogenic $\varepsilon^{182}W$ values reported in some iron meteorites, i.e., lower than the initial value of -3.5ε units inferred from chondrites and their components, may be due to burnout of W isotopes from long exposure to galactic cosmic rays (GCRs) (*Markowski et al.*, 2006; *Qin et al.*, 2005). Two recent Hf-W studies of meteoritic zircons (which are good candidates for Hf-W chronometry owing to their typically high Hf/W ratios) provide somewhat conflicting results. *Ireland and Bukovanská* (2003) confirmed an initial $^{182}Hf/^{180}Hf$ ratio based on Hf-W systematics in zircons from the H5 chondrite Simmern of close to $\sim 1 \times 10^{-4}$. These authors also reported Hf-W systematics in zircons from the Pomozdino eucrite that gave a substantially lower $^{182}Hf/^{180}Hf$ ratio of $\sim 2 \times 10^{-5}$, perhaps suggestive of late-metamorphic resetting in this eucrite. In contrast, *Srinivasan et al.* (2004), who analyzed Hf-W and U-Pb systematics in zircons from another eucrite, have suggested the initial $^{182}Hf/^{180}Hf$ ratio was at least $\sim 3 \times 10^{-4}$. The reason for these apparently discrepant values for the initial $^{182}Hf/^{180}Hf$ ratio is not clear. Nevertheless, the most recent work on the W isotopes in CAIs appears to support an initial $^{182}Hf/^{180}Hf$ ratio of $\sim 1 \times 10^{-4}$ (*Kleine et al.*, 2005).

3. SOURCES OF SHORT-LIVED RADIONUCLIDES AND THEIR IMPLICATIONS

3.1. Stellar Nucleosynthesis

Most of the short-lived radioactive nuclides present in the ESS could be produced and ejected from stars. In this section we briefly review the stellar synthesis of these isotopes and discuss possible implications for their distribution in the solar nebula.

3.1.1. Aluminum-26. Aluminum-26 is produced in hydrogen burning by the reactions $^{25}Mg(p,\gamma)^{26}Al$ and $^{26}Mg(p,n)^{26}Al$. Such ^{26}Al may be ejected from dredged-up hydrogen-shell-burning material in low-mass stars (stars with masses less than about 8 $M_\odot$) during their red giant branch (RGB) or asymptotic giant branch (AGB) phases or from high-mass stars (stars with masses greater than roughly 10 $M_\odot$) when they explode as core-collapse supernovae. This radioisotope is also made during carbon burning as neutrons and protons liberated in the fusion of two carbon nuclei drive capture reactions on Mg isotopes. Carbon-burning-produced ^{26}Al would only be ejected from massive stars.

3.1.2. Chlorine-36 and calcium-41. The isotopes ^{36}Cl and ^{41}Ca are synthesized in s-process nucleosynthesis, predominantly during core helium burning in massive stars. Production of ^{36}Cl and ^{41}Ca also occurs during explosive oxygen burning in supernova events.

3.1.3. Manganese-53. Manganese-53 is produced predominantly in silicon burning with some contribution from oxygen burning. While production of ^{53}Mn does occur in the presupernova evolution of a massive star, most of the ^{53}Mn ejected is synthesized during the explosive phase as the shock wave generated by the stellar collapse passes through silicon- and oxygen-rich layers of the star. These layers lie near the boundary between what escapes the star and what remains behind in the neutron star or black hole resulting from the explosion. A roughly comparable amount is made in typical Type Ia supernovae, which are thermonuclear explosions of white dwarf stars. Manganese-53 is not produced in low-mass stars.

3.1.4. Iron-60. Iron-60 cannot be produced in any significant amount during mainline *s*-processing since the short lifetime of ^{59}Fe prevents much neutron capture flow to heavier isotopes of Fe. The later stage of carbon burning achieves a higher neutron density and allows for significant production of ^{60}Fe, a fact that strongly favors massive stars as the site of production of this isotope. Most of the ^{60}Fe ejected from stars, however, is likely produced in explosive carbon burning during the supernova explosion with significant production also occurring in the neutron burst in the helium shell (e.g., *Meyer*, 2005). Significant production may occur in higher-mass AGB stars (e.g., *Gallino et al.*, 2004) and some may even occur in the rare deflagrating or detonating white dwarf stars that likely produced the bulk of the solar system's supply of ^{48}Ca (*Meyer et al.*, 1996; *Woosley*, 1997), but the yield from these events is not yet certain.

3.1.5. Palladium-107, iodine-129, hafnium-182, and plutonium-244. The isotopes ^{107}Pd, ^{129}I, ^{182}Hf, and ^{244}Pu are produced by the *r*-process of nucleosynthesis whose site is not yet determined. The most promising setting for the *r*-process is in neutrino-heated ejecta from core-collapse supernovae (e.g., *Woosley et al.*, 1994); however, tidal disruptions of neutron stars have not been ruled out (e.g., *Freiburghaus et al.*, 1999). Palladium-107 is also produced in the *s*-process of nucleosynthesis, which occurs when neutrons liberated by the reaction $^{13}C(\alpha,n)^{16}O$ and $^{22}Ne(\alpha,n)^{25}Mg$ are subsequently captured by heavier seed nuclei. This production may occur in AGB stars during shell helium burning or in massive stars during core or shell helium burning (*Gallino et al.*, 2004). Importantly, ^{129}I and ^{182}Hf are also produced in the neutron burst that occurs in the inner parts of the helium-rich shell in the massive star during a supernova explosion (e.g., *Meyer*, 2005). This neutron burst contributes only a small amount of the ^{129}I and ^{182}Hf that has ever existed in the galaxy; however, it may have contributed a dominant portion of the abundance of these isotopes present in the ESS. It is important to note, however, that a neutron burst will not produce ^{244}Pu since the seed uranium or thorium nuclei would have been burned up by *s*-processing prior to the neutron burst. Any ^{244}Pu alive in the ESS must have been a residue of galactic *r*-process nucleosynthesis.

3.1.6. Niobium-92 and samarium-146. A handful of heavy, proton-rich nuclei are bypassed by neutron capture processes. These nuclei are made in a separate process, known as the *p*-process, which is thought to occur in core-collapse supernovae as the shock wave passes through the oxygen-/neon-rich layers of a massive star during the stellar explosion. The heating due to shock passage causes first neutrons, then protons and α particles to disintegrate from preexisting seed nuclei. Once this disintegration process

freezes out, a distribution of proton-rich nuclei remains. This synthesis process is also known in the literature as the γ-process (*Woosley and Howard*, 1978; *Arnould and Goriely*, 2003). It is also possible that the γ-process occurs in the outer layers of a Type Ia supernova if the burning front is a deflagration when it reaches the surface (*Howard et al.*, 1991).

While the γ-process can account for the abundance of heavy *p*-process nuclei, including ^{146}Sm, it falls short of producing the light *p*-nuclei, particularly the *p*-process isotopes of molybdenum and ruthenium. It is therefore likely that some other process is responsible for the bulk production of those isotopes, and, concomitantly, for ^{92}Nb. The most promising process is the freeze-out from high-entropy nuclear statistical equilibrium near the mass cut of a core-collapse supernova (e.g., *Fuller and Meyer*, 1995; *Hoffman et al.*, 1996). Interactions between nuclei and the copious supply of supernova neutrinos may be necessary to account for the proper supply of *p*-process isotopes of molybdenum, ruthenium, and niobium (e.g., *Meyer*, 2003; *Frölich et al.*, 2006).

Table 2 summarizes the above discussion. The nucleosynthesis processes delineated above synthesized the short-lived isotopes over the course of the galaxy's evolution. A steady-state abundance of these isotopes developed in the interstellar medium (ISM) as the rate of production in stars

TABLE 2.   Stellar nucleosynthetic processes and sources of short-lived radionuclides.

Isotope	Nucleosynthesis Process	Site
^{26}Al	Hydrogen burning	MS, RGB, AGB
	Carbon burning	MS
^{36}Cl	*s*-process	MS, AGB
	Oxygen burning	MS
^{41}Ca	*s*-process	MS, AGB
	Oxygen burning	MS
^{53}Mn	Silicon burning	MS
	NSE	SNIa
^{60}Fe	Carbon burning	MS
	Neutron burst	MS
	s-process	AGB
	Neutron-rich NSE	Rare SNIa
^{92}Nb	*p*-process	MS, SNIa
^{107}Pd	*r*-process	MS, NS
	s-process	MS, AGB
^{129}I	*r*-process	MS, NS
	Neutron burst	MS
^{146}Sm	*p*-process	MS, SNIa
^{182}Hf	*r*-process	MS, NS
	Neutron burst	MS
^{244}Pu	*r*-process	MS, NS

MS = massive star; RGB = red giant branch star; AGB = asymptotic giant branch star; SNIa = Type Ia supernova; NS = neutron star disruptions; NSE = nuclear statistical equilibrium.

came to balance the rate of destruction by decay and astration. The solar system inherited this ISM abundance. Since we expect the dust and gas that carried the short-lived radionuclides into the solar cloud to be fairly well mixed, we would therefore expect a reasonably uniform distribution of these radionuclides in the protosolar cloud and a well-defined value for their abundance when the minerals to be dated were formed. Such results would make these isotopes valid chronometers.

These conclusions, however, rely on the assumption that the abundances in the early solar nebula are those of the steady-state ISM. This is probably true for the longer-lived isotopes such as ^{92}Nb, ^{146}Sm, and ^{244}Pu but possibly also ^{53}Mn (e.g., *Meyer and Clayton, 2000*); however, it is not for ^{26}Al, ^{36}Cl, ^{41}Ca, and ^{60}Fe. These isotopes had ESS abundances greater than those expected from a steady-state ISM. A plausible explanation for the high abundance of ^{26}Al, ^{36}Cl, ^{41}Ca, and ^{60}Fe is that the bulk of these isotopes in the early solar nebula were injected by a supernova that may have triggered the collapse of the solar cloud (e.g., *Cameron and Truran, 1977; Cameron et al., 1995; Meyer and Clayton, 2000*) or were injected directly into the protoplanetary disk (*Ouellette et al., 2005*). Remarkably, such a supernova could also have injected significant amounts of ^{129}I and ^{182}Hf as well (e.g., *Meyer, 2005*). The actual manner in which the short-lived radionuclides were injected remains to be worked out. Therefore, it is conceivable that the injection process could have given rise to an inhomogeneous distribution of these radionuclides. Such a result could cloud their use as chronometers; however, the lack of collateral anomalies in stable isotopes argues against such an inhomogeneous injection (*Nichols et al., 1999*).

3.2. Local Production

Production of the isotopes of elements lighter than Fe via irradiation of matter in the solar accretion disk by high-energy particles (>MeV) has long been considered as a promising alternative path to stellar nucleosynthesis (*Fowler et al., 1962*). The discovery that ^{26}Al was alive in the solar accretion disk (*Lee et al., 1976*) prompted studies examining its production via proton irradiation (e.g., *Heymann and Dziczkaniec, 1976; Lee, 1978*). These authors estimated that large fluences were needed to reproduce the observed ^{26}Al abundance. In the absence of experimental evidence for elevated fluences, irradiation failed to be considered as a viable means of producing short-lived radionuclides until recently revived by *Lee et al.* (1998).

At present, experimental results support the possibility that some short-lived radionuclides were produced by irradiation. First, it is now well established that virtually all low-mass stars display an enhanced X-ray activity (*Feigelson et al., 2002; Wolk et al., 2005*). Radio observations of young stellar objects (YSOs) have resulted in the direct detection of gyrosynchrotron radiation from MeV electrons (*Güdel, 2002*). YSOs also show hard X-ray spectra associated with violent magnetic reconnection flares and baryon acceleration

at energies up to hundreds of MeV/A (*Wolk et al., 2005*). A second evidence in favor of irradiation within the ESS could be the ubiquitous presence of ^{10}Be at the time of formation of CAIs (*McKeegan et al., 2000; Marhas et al., 2002; MacPherson et al., 2003*) and the possible presence of ^{7}Be at the time of formation of the Allende CAI 3529-41 (*Chaussidon et al., 2006*).

Trapping of GCRs in the protosolar molecular cloud core likely contributed at some level to the initial abundance of ^{10}Be in the ESS and possibly accounted for all of it (*Desch et al., 2004*), but GCRs are not likely to have contributed significantly to other radionuclides. For example, <0.1% of the solar system's initial abundance of ^{26}Al is attributable to GCRs (*Desch et al., 2004*). While this alternative origin is a viable one for ^{10}Be, it is also feasible that some or all of it has an irradiation origin within the solar system (*McKeegan et al., 2000; Gounelle et al., 2001; Marhas and Goswami, 2004*). The very short half-life of ^{7}Be ($T_{1/2}$ = 53 d) precludes its origin outside the solar system. Therefore, if its presence can be confirmed by further analyses, it would establish a definitive proof of irradiation within the ESS.

In any irradiation model, the main parameters are the proton flux, the irradiation duration, the abundance of heavier cosmic rays (^{3}He, ^{4}He) relative to protons, the target abundance, the nuclear cross sections, and the energy spectrum of protons (*Chaussidon and Gounelle, 2006*). What mainly distinguishes the different irradiation models are (1) the astrophysical context of irradiation, (2) the physical nature of the targets (solid or gaseous), (3) the chemistry of the targets, and (4) the location of the irradiated targets relative to the source of the cosmic rays. The proton energy spectrum is usually considered to satisfy a power law, $N(E) \sim E^{-p}$, with varying index p. Models considering irradiation in the context of the progenitor molecular cloud (e.g., *Clayton and Jin, 1995*) failed at reproducing the observed abundances of ^{26}Al, ^{41}Ca, and ^{53}Mn, and have now fallen into abeyance. As such, the most likely astrophysical context for irradiation synthesis of short-lived radionuclides is the solar accretion disk. In this context, it is recognized that the Sun's magnetic activity consists of two broad classes of events. Gradual events emitting soft X-rays are electron- and ^{3}He-poor. More frequent impulsive events emit hard X-rays and are electron- and ^{3}He-rich (*Reames, 1995*). Impulsive flares have steeper energy spectra than gradual flares.

Goswami et al. (2001), followed by *Marhas et al.* (2002) and *Marhas and Goswami* (2004), developed models examining the possibility of producing short-lived radionuclides by irradiation of solar system dust by proton and ^{4}He nuclei at asteroidal distances. In these studies, shielding of the whole solar accretion disk is neglected and accelerated solar particles are supposed to have free access to dust at ~3 AU. These authors limited their models to examining gradual events having shallow proton energy spectra (p < 3) and normalized their yields to ^{10}Be. Based on these models, *Marhas and Goswami* (2004) contended that irradiation could account for all the ^{10}Be, 10–20% of ^{41}Ca and ^{53}Mn, and none of the ^{26}Al inferred to be present in the ESS.

Lee et al. (1998) first examined the possibility that some short-lived radionuclides could be synthesized by irradiation in the context of the X-wind theory of low-mass star formation (*Shu et al., 1996*), introducing three important conceptual modifications compared to previous models: (1) Irradiation takes place close to the Sun, in a gas-poor region (i.e., the reconnection ring where magnetic lines tying the protostar and the accretion disks reconnect), providing a powerful mechanism for accelerating ^1H, ^3He, and ^4He nuclei to energies up to a few tens of MeV. The X-wind provides a natural transport mechanism from the regions close to the Sun to asteroidal distances (*Shu et al., 1996*). (2) The X-wind model has opened the ^3He channel for the production of short-lived radionuclides, enhancing the ^{26}Al production via the reaction $^{24}Mg(^3He,p)^{26}Al$. (3) It calculates absolute yields instead of yields relative to a given short-lived radionuclide (such as ^{10}Be), scaling the proton flux to observations of X-ray protostars. In this model, ^{26}Al and ^{53}Mn are produced at their observed abundance for parameters corresponding to impulsive events (p = 3.5, ^3He/^1H = 1.4). Calcium-41 is overproduced by 2 orders of magnitude relative to its observed abundance, while ^{60}Fe is underproduced by several orders of magnitude. *Gounelle et al.* (2001) refined this model, and calculated yields of the recently discovered ^{10}Be (*McKeegan et al., 2000*). The production of ^{10}Be (as well as ^{26}Al and ^{53}Mn) was found to agree with the observed value in the case of impulsive events (p = 4, ^3He/^1H = 0.3). They also proposed that the ^{41}Ca overproduction could be alleviated if proto-CAIs had a layered structure (*Shu et al., 2001*). Using the same model and a preliminary estimate of ^7Be nuclear cross sections, *Gounelle et al.* (2003) calculated a ^7Be/^9Be ratio of ~0.003, at odds with the initial claim by *Chaussidon et al.* (2002) of ^7Be/^9Be up to ~0.22 inferred in an Allende CAI. Subsequently, *Chaussidon et al.* (2006) revised their estimate of the ^7Be/^9Be ratio to ~0.005, compatible within a factor of 2 with the X-wind model prediction. The ability of the X-wind model to produce a relatively high abundance of ^7Be despite its very short half-life is due to the high flux of accelerated particles adopted since *Lee et al.* (1998). This contrasts with the otherwise similar model of *Leya et al.* (2003) that invokes special conditions for the production of ^7Be. The yields of ^{36}Cl presented in *Gounelle et al.* (2006) are slightly lower than the observed value (*Lin et al., 2005*), but still in line with it, given the model uncertainties (^{36}Cl-producing cross sections have not been experimentally determined). Furthermore, if the initial solar system abundance of ^{26}Al was indeed supercanonical (*Young et al., 2005*), the initial ^{41}Ca/^{40}Ca ratio was probably higher than 1.5×10^{-8}. Therefore, a layered structure of CAIs may no longer be required to account for the ^{41}Ca abundance (*Gounelle et al., 2006*). The decoupling of ^{10}Be and ^{26}Al observed in some hibonites (*Marhas et al., 2002*) may be accounted for by irradiation during gradual events instead of impulsive ones (*Gounelle et al., 2006*).

To summarize, it is recognized that irradiation at asteroidal distances using "normal" proton fluences and gradual events fails to produce the observed amount of short-lived radionuclides such as ^{26}Al, ^{41}Ca, and ^{53}Mn (*Goswami et al., 2001*). However, there is strong evidence that the proton fluence of protostars was $10^5\times$ higher than at present (*Feigelson et al., 2002*), and that there was intensive radial transport from the inner disk to larger heliocentric distances (*Wooden et al., 2004*), either via turbulence (*Cuzzi et al., 2003*) or by X-wind (*Shu et al., 1996*). The specific irradiation model developed in the context of the X-wind theory can reproduce the observed abundances of ^7Be, ^{10}Be, ^{26}Al, ^{36}Cl, ^{41}Ca, and ^{53}Mn within uncertainties for cosmic-ray parameters corresponding to impulsive events. X-ray observations have shown that the impulsive phase is often present in YSOs (*Wolk et al., 2005*).

Uncertainties of the model arise mainly from poor knowledge of the nuclear cross sections, especially for the ^3He channel. To reduce these uncertainties, nuclear physicists based at Orsay have undertaken the measurement of ^3He-induced cross sections (*Fitoussi et al., 2004*). They found that the experimental $^{24}Mg(^3He,p)^{26}Al$ cross section is a factor of ~2 to 3 lower than the estimate of *Lee et al.* (1998), but within the range of the reported uncertainties.

At present, it is not yet clear how much irradiation processes contributed to the inventory of short-lived radionuclides. X-ray observations of protostars and realistic models reproducing the observed initial abundances of a handful of these radionuclides call for further investigation of this possibility.

4. ASTROPHYSICAL SETTING OF THE SOLAR PROTOPLANETARY DISK

The presence of short-lived radionuclides with half-lives ≪10 m.y. in the ESS, at initial abundances noted in Table 1, provides a record of the dramatic processes occurring within a few million years of the solar system's birth. This contrasts with the case for the relatively longer-lived radionuclides such as ^{146}Sm ($T_{1/2}$ ~ 103 m.y.) and ^{244}Pu ($T_{1/2}$ ~ 82 m.y.). As mentioned in the previous section, supernovae, novae, and AGB stars maintain steady-state abundances of these radionuclides in the galaxy that are consistent with the abundances derived from meteorites, provided there is a period of free decay on the order of ~10^8 yr prior to incorporation into the solar nebula (*Schramm and Wasserburg, 1970; Harper, 1996; Jacobsen, 2005*). Ongoing galactic nucleosynthesis might possibly contribute to ^{53}Mn, ^{107}Pd, ^{129}I, and ^{182}Hf as well. However, the levels at which the shorter-lived radionuclides ^{26}Al, ^{41}Ca, and ^{60}Fe (and probably ^{36}Cl) are maintained in the galaxy are significantly lower than those inferred from meteorites in the ESS and after a delay of ~10^8 yr, essentially none of these radionuclides remains in the molecular cloud from which the solar system formed (*Harper, 1996; Wasserburg et al., 1996; Meyer and Clayton, 2000*). As such, some nearby processes were creating radionuclides within ~10^6 yr of the birth of the solar system.

It is clear that more than one process was involved since there is no proposed source that can simultaneously produce enough [10]Be and [60]Fe. Either local irradiation or trapping of GCRs might yield the observed abundance of [10]Be (see section 2.2), but both processes underproduce [60]Fe by ~3 orders of magnitude (*Lee et al.*, 1998; *Leya et al.*, 2003). Iron-60 can be produced at the levels inferred from meteorites (i.e., [60]Fe/[56]Fe ratio of up to ~10[-6]) only by stellar nucleosynthetic sources, in which beryllium is destroyed (see section 2.1.). The meteoritic data additionally suggest separate origins. Specifically, while [26]Al and [41]Ca correlate with each other in meteoritic components (*Sahijpal and Goswami*, 1998), the presence of [26]Al is not correlated with the presence of [10]Be (*Marhas et al.*, 2002). Two distinct sources are therefore required: one for [10]Be, and one for [60]Fe. However, as is evident from discussion in the previous section, some of the short-lived radionuclides, particularly [26]Al, [36]Cl, [41]Ca, and perhaps [53]Mn, could be produced by both sources. The question then is what proportions of each of these radionuclides were derived from one or the other of these sources.

The source of the [60]Fe was almost certainly a massive star that went supernova (i.e., Type II supernova). While other types of stellar sources could produce this isotope in sufficient abundance (Table 2), injection of material into the solar nebula by an AGB star (or a rare Type 1a supernova) is exceedingly improbable. In particular, *Kastner and Myers* (1994) quantified the spatial distribution of molecular clouds and AGB stars and estimated an upper limit to this probability of only 3 × 10[-6]. At any rate, an AGB star is unlikely to produce sufficient [60]Fe relative to [26]Al (*Tachibana and Huss*, 2003). Therefore, the most plausible source of [60]Fe in the ESS is a Type II supernova. When that supernova injected [60]Fe into the material that formed the solar system, it is also likely to have injected other short-lived radionuclides such as [26]Al, [36]Cl, [41]Ca, and [53]Mn (*Meyer and Clayton*, 2000; *Goswami and Vanhala*, 2000; *Meyer*, 2005).

Therefore, while the inferred initial abundance of [60]Fe in the ESS places its formation near a massive star that went supernova, the timing of this event and the distance from this supernova are uncertain. The distance may have been several parsecs from the protosolar molecular cloud core and triggered its collapse (*Cameron and Truran*, 1977; *Goswami and Vanhala*, 2000; *Vanhala and Boss*, 2002). Alternatively, it may have occurred <1 pc away from the protoplanetary disk (*Chevalier*, 2000; *Ouellette et al.*, 2005). Given the extreme spatial and chemical heterogeneities of supernova ejecta, it is difficult to definitively predict the expected abundances of short-lived radionuclides that would be incorporated into the solar nebula. Nevertheless, a supernova is capable of producing all the short-lived radionuclides in the ESS (except for [10]Be, which has plausible alternative sources). It has been shown that injection into the protoplanetary disk of selected shells of the supernova can reproduce the inferred initial abundances of various short-lived radionuclides to within a factor of ~2 (*Meyer and Clayton*, 2000; *Meyer*, 2005; *Ouelette et al.*, 2005). Future work is clearly needed to determine the relative con-

tributions of local production and stellar ejecta to the abundances of short-lived radionuclides, particularly [26]Al, [36]Cl, [41]Ca, and perhaps also [53]Mn.

5. FROM DUST TO PLANETESIMALS

5.1. Short-lived Radionuclides in Presolar Grains and Implications for Stellar Sources of Primordial Dust in the Solar Nebula

Since their discovery in the late 1980s, presolar grains in primitive meteorites have proven to be powerful tools for improving our understanding of stellar nucleosynthesis within and dust formation around a variety of types of stars (e.g., *Bernatowicz and Zinner*, 1997; *Nittler*, 2003; *Zinner*, 2003; *Clayton and Nittler*, 2004; *Mostefaoui and Hoppe*, 2004). These grains sample most of the types of stars that have been suggested as potential sources of short-lived isotopes in the ESS, including low-mass AGB stars and Type II supernovae. The most common types of presolar grains are diamond, silicates, silicon carbide, graphite, and the oxides corundum, hibonite, and spinel. The grain size of individual diamond grains is too small for individual isotopic analysis and even in aggregate samples, the concentrations of most elements are too low to measure. Presolar silicates have only recently been discovered (*Messenger et al.*, 2003; *Nguyen and Zinner*, 2004; *Nagashima et al.*, 2004). However, they are small (<1 μm) and the most common phases, olivine and pyroxene, do not readily take up trace elements. Most of the evidence regarding short-lived radionuclides in presolar grains comes from silicon carbide, with additional evidence from oxides and graphite. About 90% of presolar SiC grains come from low-mass AGB stars and are termed "mainstream"; 1–2%, the X-grains, have the isotopic signature of Type II supernovae. Similarly, most presolar oxides are from AGB stars and a few are from Type II supernovae.

In low-mass AGB stars, ^{12}C and *s*-process products, including some short-lived nuclides, are produced in the helium shell and periodically dredged up and mixed into the convective envelope, where they can be trapped in dust grains that form in stellar winds as the star loses mass (*Gallino et al.*, 1998; *Busso et al.*, 1999). Dust grains from AGB stars carry short-lived nuclei made throughout the AGB phase (lasting several hundred thousand years). In Type II supernovae, dust condenses from the envelope of a massive star, which is thrown off by the explosion that results from the core collapsing to a neutron star or black hole. Short-lived nuclei are made both by nuclear reactions prior to the explosion and due to the explosion itself (e.g., *Rauscher et al.*, 2002). Presolar grains from both low-mass AGB stars and Type II supernovae preserve evidence of nucleosynthesis of a number of short-lived radionuclides and we review the evidence here.

5.1.1. Sodium-22.   There are two neon components highly enriched in [22]Ne in meteorites, Ne-E(H) and Ne-

E(L). Ne-E(H) is believed to represent the Ne-isotopic composition of the helium shells of AGB stars, implanted into mainstream presolar SiC after loss of the stellar envelope (*Gallino et al.*, 1990). Ne-E(L) is believed to be radiogenic, from the decay of ^{22}Na ($T_{1/2}$ = 2.6 yr), for which the meteoritic carrier is presolar graphite (*Amari et al.*, 1990). A significant fraction of presolar graphite is from supernovae, accounting for the preservation of Ne-E(L).

5.1.2. Aluminum-26. The presolar grain record for this radionuclide is fairly complete (*Zinner*, 2003, and references therein), as both SiC and oxides tend to have high Al/Mg ratios as well as high initial ^{26}Al/^{27}Al ratios. Mainstream SiC grains tend to have initial ratios between 10^{-4} and 10^{-3}, but a few have ratios as high as 10^{-2}. Presolar spinel, hibonite, and corundum from low-mass AGB stars formed with somewhat high ^{26}Al/^{27}Al (*Zinner et al.*, 2005). The ^{26}Al/^{27}Al ratios in mainstream SiC are similar to those predicted by AGB stellar models, but the ratios in spinel grains require an extra production process, most likely cool-bottom processing (*Zinner et al.*, 2005; *Nollett et al.*, 2003).

Most X-grains were quite ^{26}Al-rich, with initial ^{26}Al/^{27}Al ratios of 0.1–0.4. Low-density graphite grains, also believed to come from Type II supernovae, have a somewhat wider range of initial ^{26}Al/^{27}Al, from 10^{-5} to 0.2 (*Travaglio et al.*, 1999). There is abundant isotopic evidence that SiC and graphite grains do not uniformly sample the different layers of ejecta from Type II supernovae, but the levels of ^{26}Al can be explained by mixing of different layers (*Travaglio et al.*, 1999).

5.1.3. Calcium-41. Large ^{41}K excesses attributed to ^{41}Ca decay have been reported in graphite from Type II supernovae (*Travaglio et al.*, 1999) and oxide grains from low-mass AGB stars (*Nittler et al.*, 2005). The inferred ^{41}Ca/^{40}Ca ratios are 10^{-3}–10^{-2} and 10^{-5}–5 × 10^{-4}, respectively. Both types of stars produced ^{41}Ca by neutron capture on ^{40}Ca, in amounts consistent with those found in presolar grains.

5.1.4. Titanium-44. This isotope has a half-life of only 67 yr, yet evidence of *in situ* decay has been found in graphite and SiC from supernovae. In fact, the observed excesses in daughter ^{44}Ca are among the strongest arguments in favor of a supernova origin for these types of grains (*Amari et al.*, 1992; *Hoppe et al.*, 1996; *Travaglio et al.*, 1999; *Besmehn and Hoppe*, 2003).

5.1.5. Vanadium-49. With a half-life of only 330 d, ^{49}V holds the record for the shortest-lived extinct radionuclide for which evidence has been found in presolar grains. Excesses in daughter ^{49}Ti are correlated with V/Ti ratios in several X-type SiC grains (*Hoppe and Besmehn*, 2002), and show that grain condensation occurred within a year or so of a supernova explosion, consistent with astronomical observations of dust condensation about 500 d after the explosion of supernova 1987A (*Wooden*, 1997).

5.1.6. Zircon-93. This isotope has a half-life that is long enough (2.3 m.y.) that it behaves as if it were stable during *s*-process nucleosynthesis. It decays to monoisotopic ^{93}Nb, mostly in the ejecta of low-mass stars after the AGB phase has ended. A strong correlation between the elemental abundances of zirconium and niobium in individual presolar SiC grains measured with a synchrotron X-ray fluorescence microprobe shows that the grains condensed with live ^{93}Zr (*Kashiv et al.*, 2006).

5.1.7. Technetium-99. Technetium has no stable isotopes, but was detected in spectra of red giant stars more than 50 yr ago (*Merrill*, 1952). ^{99}Tc ($T_{1/2}$ = 213 k.y.) lies along the main *s*-process path. In fact, most ^{99}Ru is made as ^{99}Tc, which subsequently decays in AGB star envelopes or ejecta. Comparison of Ru isotope compositions of individual presolar SiC grains with models of nucleosynthesis in AGB stars show that the grains condensed with live ^{99}Tc in them (*Savina et al.*, 2004).

5.1.8. Cesium-135. Cesium is a volatile element not expected to condense into grains at high temperature, whereas barium (^{135}Cs decays to ^{135}Ba) is refractory and observed in presolar SiC. Cesium-135 ($T_{1/2}$ = 2.3 m.y.) is produced in fairly high abundance in AGB stars. Comparison of high-precision Ba isotope data on aggregates of presolar SiC (*Prombo et al.*, 1993) with stellar nucleosynthesis models strongly suggests that when the grains condensed, cesium remained in the gas (*Lugaro et al.*, 2003).

5.1.9. Outlook. There are a number of other short-lived isotopes for which records of decay could potentially be found in presolar grains from meteorites, but all are more difficult than the cases given above. Manganese-53 is only made in supernovae and there are no promising host phases for manganese. Iron-60 is made in supernovae and AGB stars and could be searched for in iron-bearing presolar silicates. Palladium-107 is made at fairly high abundance in supernovae and AGB stars, but no appropriate host exists among known types of presolar grains. Samarium-146 is made in supernovae and in small amounts in AGB stars, but presolar SiC tends to have low Sm/Nd ratios (*Yin et al.*, 2005). Hafnium-182 is also made in both supernovae and AGB stars. Hafnium as well as tungsten form carbide and are likely to be present in presolar SiC, but both are refractory and variations in Hf/W ratios are unlikely. Finally, ^{205}Pb is produced in some abundance in AGB stars, but is volatile.

5.2. Formation Timescales from Dust to Planetesimals

Evolution of the protoplanetary disk from the formation of the smallest (millimeter- to centimeter-sized) solid objects to planetary-sized bodies was long thought to be a broadly sequential process: CAIs representing the earliest material are followed by chondrules and then larger objects of asteroidal to planetary sizes. However, recent chronological studies of short-lived radionuclides and U-Pb systematics in meteorites and their components are revealing a picture that is not as orderly.

5.2.1. Formation of calcium-aluminum-rich inclusions and chondrules. The state of isotopic chronology of chondrule and CAI formation as of mid-2004 has been thoroughly reviewed by *Kita et al.* (2005) and need not be discussed here in detail. To summarize briefly, Pb-Pb systematics in CAIs from the Efremovka carbonaceous chondrite give an age of 4567.2 ± 0.6 Ma (*Amelin et al., 2002*), which is consistent with, but more precise than, the previously determined Pb-Pb age of 4566 ± 2 Ma for Allende CAIs (*Göpel et al., 1994; Allègre et al., 1995*). This is also the oldest absolute age date for any solid formed in the solar system, and as such the CAIs are believed to be the earliest solids to form within the protoplanetary disk. Based primarily on ion microprobe analyses of ^{26}Al-^{26}Mg systematics in individual grains within CAIs and chondrules, a time difference of ~1–3 m.y. between CAI and chondrule formation has been suggested (e.g., *Kita et al., 2000; Huss et al., 2001; Amelin et al., 2002*). This time difference is supported by Pb-Pb ages of CAIs from the Efremovka (reduced CV3) chondrite and chondrules from the Acfer 059 (CR) chondrite (*Amelin et al., 2002*).

Detailed *in situ* studies (by ion microprobe or laser ablation multicollector inductively coupled plasma mass spectrometer) of ^{26}Al-^{26}Mg systematics in CAIs (e.g., *Hsu et al., 2000; Young et al., 2005; Taylor et al., 2005*) suggest a prolonged residence, up to ~300,000 yr, of CAIs in the protoplanetary disk. In contrast, however, Mg-isotopic analyses of "bulk" CAIs appear to indicate that they were formed within a relatively narrow time interval of ~50,000 yr (*Bizzarro et al., 2004*). This apparent discrepancy could be indicative of the possibility that the *in situ* and bulk analyses are recording different Al/Mg fractionation events in the history of CAI formation.

The existence of compound chondrule-CAI objects (*Krot and Keil, 2002; Itoh and Yurimoto, 2003; Krot et al., 2005a*) indicates that chondrule formation and remelting of CAIs overlapped in time; ^{26}Al-^{26}Mg systematics in bulk chondrules from Allende further suggest that chondrules began forming contemporaneously with CAIs, and then continued to form over a time span of at least ~2–3 m.y. (*Bizzarro et al., 2004*). Near-contemporaneous formation of at least some chondrules with CAIs is additionally supported by the Pb-Pb age of 4566.7 ± 1.0 Ma obtained for a group of Allende chondrules (*Amelin et al., 2004*). However, as suggested by *Krot et al.* (2005a), chronologic information derived from bulk chondrules may reflect the timing of formation of chondrule precursor materials rather than the time of chondrule formation.

Chondrules from metal-rich CB carbonaceous chondrites Gujba and Hammadah al Hamra 237 have the youngest absolute age (i.e., 4562.8 ± 0.9 Ma) yet reported for chondrules from any of the unequilibrated chondrites (*Krot et al., 2005b*). It is likely that these chondrules formed from a vapor-melt plume produced by a giant impact between planetary embryos after dust in the protoplanetary disk had largely dissipated. It is inferred from these results that planet-sized objects existed in the early asteroid belt ~4–5 m.y. after the formation of CAIs.

It has been recently shown that composition of chondrule minerals is inconsistent with crystallization from the melt under closed-system conditions, and that gas-melt interaction must have occurred during chondrule formation (*Libourel et al., 2005*). Formation of chondrules in open-system conditions explains their compositional and structural diversity, but it also creates an additional difficulty in dating these objects. Matching the compositional variations in chondrules with their isotopic systematics will be the matter of future studies.

5.2.2. Accretion and differentiation of planetesimals. From dating of achondrites (i.e., meteorites that formed as a result of extensive melting on their parent planetesimals) using long-lived isotope chronometers, such as ^{87}Rb-^{87}Sr ($T_{1/2}$ ~ 56 G.y.) or ^{147}Sm-^{143}Nd ($T_{1/2}$ ~ 106 G.y.), it has been known for a long time that their parent bodies had undergone planet-wide melting and differentiation early in solar system history (e.g., *Lugmair, 1974; Allègre et al., 1975; Nyquist et al., 1986; Wadhwa and Lugmair, 1995; Kumar et al., 1999*). However, the uncertainties of these absolute ages were too large — typically tens of millions of years — to really pin down the timescales at a desirable resolution. During the last decade or so significant advances have been made with the use of chronometers based on short-lived radionuclides toward obtaining high-resolution timescales of planetesimal melting and differentiation, which in turn have helped to place limits on the timescales required to accrete larger (tens to hundreds of kilometers in diameter) bodies from dust-sized particles. Here we will briefly summarize some of the more significant recent results bearing on the timescales of planetesimal accretion and differentiation. More detailed discussions on this topic may be found in *Nichols* (2006) and *Wadhwa et al.* (2006).

Using the ^{53}Mn-^{53}Cr system ($T_{1/2}$ = 3.7 m.y.), one of the first comprehensive studies on differentiated meteorites belonging to the howardite-eucrite-diogenite (HED) group, assumed to originate from the differentiated asteroid 4 Vesta, was published several years ago (*Lugmair and Shukolyukov, 1998*). It was shown that a planetesimal-wide differentiation caused the fractionation of Mn/Cr ratios in the mantle sources of the HED meteorites and that this episode had concluded 7.8 ± 0.8 m.y. before the formation of the LEW 86010 angrite (which serves as the time anchor for the short-lived ^{53}Mn-^{53}Cr chronometer). This translates to an age of 4564.8 ± 0.9 Ma for the conclusion of this Mn/Cr fractionation event on the HED parent body. This age can be compared with that of refractory inclusions (i.e., CAIs) found in primitive chondrites (4567.2 ± 0.6 Ma) (*Amelin et al., 2002*), which, as discussed earlier, are believed to be the earliest condensates from the solar nebula. Considering the time difference of 2.4 ± 1.1 m.y. and the time required to assemble and melt a body the size of 4 Vesta, this clearly demonstrates that the accretion of large objects occurred at a very early time and at a very fast pace.

While both manganese and chromium are elements that mainly reside in the silicate mantle and crust of a differentiated planetesimal, they generally are not very helpful when trying to answer questions concerning silicate-metal segregation or core formation. Here a system based on another now extinct radioisotope, ^{182}Hf, that decays to ^{182}W with a half-live of 8.9 m.y., has proven to be very useful. While both elements, hafnium and tungsten, are refractory, they are strongly fractionated during silicate-metal segregation: Hafnium remains preferentially in the silicates, while tungsten partitions mainly into the metal fraction. Measuring the radiogenic contribution to ^{182}W from the decay of ^{182}Hf in the remaining tungsten in silicate samples provides information on the timing of Hf/W fractionation in the mantle, while the main Hf/W fractionation from a chondritic value may have preceded the former during core formation.

The ^{182}Hf-^{182}W system was first applied to the HED meteorites by *Quitté et al.* (2000), followed by additional analyses by *Yin et al.* (2002) and *Kleine et al.* (2004). Using the Ste. Marguerite H chondrite as the time anchor for the ^{182}Hf-^{182}W system, *Kleine et al.* (2004) obtained an age for HED parent-body mantle differentiation of 4563.2 ± 1.4 Ma, which is in agreement with the differentiation age derived from the ^{53}Mn-^{53}Cr system as discussed above. In addition, combining the HED data with ^{182}Hf-^{182}W systematics in chondrites indicates that core formation on 4 Vesta may have preceded mantle differentiation by about 1 m.y. (*Kleine et al., 2004*).

It should be noted that the decay products of other short-lived, now extinct, radioactive isotopes have been detected in the HED meteorites and other achondrites (e.g., basaltic meteorites belonging to the angrite group) and also show their antiquity. The former presence in achondrites of live ^{26}Al ($T_{1/2}$ = 0.73 m.y.) (e.g., *Srinivasan et al., 1999; Nyquist et al., 2003; Baker et al., 2005; Bizzarro et al., 2005b; Spivak-Birndorf et al., 2005; Amelin et al., 2006*) and ^{60}Fe ($T_{1/2}$ = 1.5 Ma) (*Shukolyukov and Lugmair, 1993a,b; Quitté et al., 2005*) has been clearly demonstrated. The important aspect of these findings is that both of these nuclei can serve as potent heat sources for melting and differentiation if their abundances were sufficiently high and if the meteorite parent body had accreted at a very early time. In this context, rather tight constraints on the timing of planetesimal accretion have additionally been placed by new high-precision Pb-isotopic ages of 4566.2–4566.5 Ma of recently discovered differentiated meteorites (*Baker et al., 2005; Amelin et al., 2006*), which indicate that their parent asteroids accreted and differentiated within ~1 m.y. of the formation of CAIs, essentially contemporaneously with chondrule-forming events (*Amelin et al., 2002, 2004*). Taken together, these observations suggest that CAIs, chondrules, and differentiated asteroids formed over the same, relatively short period of time in rather complex and diverse disk environments.

One somewhat puzzling development in the last few years has resulted from a refinement of the precision of W-isotopic analyses and application to iron meteorites (some of which are thought to represent the cores of differentiated planetesimals). If differentiation and core formation on the parent planetesimals of the iron meteorites occurred during the lifetime of ^{182}Hf but after CAI formation, the expectation is that the ^{182}W/^{184}W ratios in these samples would be more radiogenic than the solar system initial value inferred

from CAIs and chondrites but less radiogenic compared to bulk chondrites (i.e., between –3.5 and –2ε units relative to BSE). The earliest data on the W-isotopic compositions of iron meteorites had shown that these samples indeed have the lowest ^{182}W/^{184}W ratios of any solar system material, with values ranging from ~–3 to –5ε units relative to BSE (e.g., *Lee and Halliday,* 1995, 1996; *Horan et al.,* 1998). However, the precision of these earliest measurements was insufficient to definitively ascertain whether any of the iron meteorites had W-isotopic compositions that were resolvably lower than the initial value inferred for the solar system. More recent, higher-precision, W-isotopic analyses of iron meteorites have shown that some iron meteorites do indeed have ^{182}W/^{184}W ratios that are resolvably lower than –3.5 (*Markowski et al.,* 2006; *Kleine et al.,* 2005; *Qin et al.,* 2005). Taken at face value, this suggests that these iron meteorites formed (and that core formation on their parent planetesimals took place) earlier than CAI formation. There are, however, incompletely understood and possibly significant effects in the ^{182}Hf-^{182}W system in iron meteorites resulting from long exposure to GCRs, which may results in an apparent lowering of the measured ^{182}W/^{184}W ratios in these samples. In fact, recent results demonstrate that GCR exposure could indeed account for the least-radiogenic W-isotopic compositions in iron meteorites, but the effect on the ^{182}W/^{184}W ratio due to irradiation is unlikely to be significantly larger than ~0.5ε units (*Markowski et al.,* 2006; *Qin et al.,* 2005). As such, this leaves us with the conclusion stated earlier that formation of CAIs and chondrules on the one hand and the accretion and differentiation of planetesimals on the other occurred within a very short time span of perhaps no more than a couple of million years.

6. OUTLOOK AND FINAL REMARKS

The emerging picture of the protoplanetary disk and the potentially complex and spatially and temporally diverse environments within it leaves many open questions, and poses challenges for astronomers, astrophysicists, disk modelers, cosmochemists, and petrologists. The following are some of the topics that need more detailed exploration.

1. Because of the growing evidence that the short-lived radionuclides in the protoplanetary disk came from multiple sources, we cannot *a priori* assume homogeneous distribution for any such radionuclide in the protoplanetary disk. The relatively longer lived of these short-lived nuclides that are produced by stellar nucleosynthesis, such as ^{146}Sm, ^{244}Pu, and possibly ^{129}I and ^{182}Hf, may be mixtures of material present in the presolar molecular cloud, and freshly synthesized material injected into the disk by a nearby supernova. The presence of ^{10}Be in the ESS may be due to irradiation from the young Sun [some other short-lived radionuclides may also have been produced in this manner (*Gounelle et al.,* 2001)] and/or from trapping of GCRs in the protosolar molecular cloud (*Desch et al.,* 2004). These sources are not mutually exclusive; for example, ^{26}Al, the most widely used short-lived chronometer nuclide, can be

a mixture of material from stellar sources and irradiation mechanisms. The extent of heterogeneity in distribution of short-lived nuclides has to be evaluated by extensive comparative studies of various ESS materials: CAIs, chondrules of various origins (nebular and asteroidal/planetary), and differentiated meteorites, with a set of short-lived and long-lived (U-Pb) isotopic systems. "Mapping" an extinct-nuclide chronometer onto the absolute timescale may work for a certain nuclide in a certain group of meteorites (e.g., ^{53}Mn-^{53}Cr in the angrites and the HED meteorites) that come from a single parent asteroid or a homogeneous population of asteroids. We cannot assume, however, that the same chronometer would give compatible results for chondrules and CAIs, because short-lived radionuclides could be heterogeneously distributed at this scale (*Gounelle and Russell, 2005a,b*).

2. A related question is the timing of injection of radionuclides into the protoplanetary disk. At what point during its evolution did the disk encounter collision with the supernova ejecta? This question could potentially be addressed by establishing a correlation between (1) isotopic anomalies of nucleosynthetic origin in meteoritic components; (2) the abundances of short-lived radionuclides, e.g., ^{26}Al, ^{60}Fe, and ^{41}Ca, in these materials; and (3) their absolute ages. This may be achieved by comparative studies of U-Pb and short-lived isotopic systems in CAIs and other refractory materials (e.g., hibonite grains from CM chondrites).

3. Based on the emerging picture of early accretion (i.e., within ~1 m.y. of CAI formation) of the differentiated planetesimals, it seems plausible that the accretion of undifferentiated chondrite parent asteroids occurred late (unless such bodies were extremely small), because otherwise large asteroids would have melted as a result of heat generated by ^{26}Al and ^{60}Fe decay. How primitive materials such as presolar grains, CAIs, and chondrules could survive a few million years in the disk (i.e., until they were accreted into chondrite parent bodies) alongside the accreting and differentiating asteroids is still unclear. A better understanding of the dynamics within a protoplanetary disk would help to clarify this.

REFERENCES

Allègre C. J., Birck J. L., Fourcade S., and Semet M. P. (1975) *Science, 187*, 436–438.

Allègre C. J., Manhès G., and Göpel C. (1995) *Geochim. Cosmochim. Acta, 59*, 1445–1456.

Amari S., Anders E., Virag A., and Zinner E. (1990) *Nature, 345*, 238–240.

Amari S., Hoppe P., Zinner E., and Lewis R. S. (1992) *Astrophys. J., 394*, L43–L46.

Amelin Y., Krot A. N., Hutcheon I. D., and Ulyanov A. A. (2002) *Science, 297*, 1678–1683.

Amelin Y., Krot A. N., and Twelker E. (2004) *Geochim. Cosmochim. Acta, 68*, E958.

Amelin Y., Wadha M., and Lugmair G. W. (2006) *Lunar Planet. Sci. XXXVII*, Abstract #1970.

Arnould M. and Goriely S. (2003) *Physics Reports, 384*, 1–84.

Baker J., Bizzarro M., Wittig M., Connelly J., and Haack H. (2005) *Nature, 436*, 1127–1131.

Bernatowicz T. J. and Zinner E., eds. (1997) *Astrophysical Implications of the Laboratory Study of Presolar Materials.* 750 pp. American Inst. of Physics, Woodbury.

Besmehn A. and Hoppe P. (2003) *Geochim. Cosmochim. Acta, 67*, 4693–4703.

Birck J. L. and Lugmair G. W. (1988) *Earth Planet. Sci. Lett., 90*, 131–143.

Bizzarro M., Baker J. A., and Haack H. (2004) *Nature, 431*, 275–278.

Bizzarro M., Baker J. A., and Haack H. (2005a) *Nature, 435*, 1280.

Bizzarro M., Baker J. A., Haack H., and Lundgaard K. L. (2005b) *Astrophys. J., 632*, L41–L44.

Brazzle R. H., Pravdivtseva O. V., Meshik A. P., and Hohenberg C. M. (1999) *Geochim. Cosmochim. Acta, 63*, 739–760.

Busso M., Gallino R., and Wasserburg G. J. (1999) *Ann. Rev. Astron. Astrophys., 37*, 239–309.

Cameron A. G. W. and Truran J. W. (1977) *Icarus, 30*, 447–461.

Cameron A. G. W., Hoeflich P., Myers P. C., and Clayton D. D. (1995) *Astrophys. J., 447*, L53–L57.

Carlson R. W. and Hauri E. H. (2001) *Geochim. Cosmochim. Acta, 65*, 1839–1848.

Chaussidon M. and Gounelle M. (2006) In *Meteorites and the Early Solar System II* (D. S. Lauretta and H. Y. McSween Jr., eds.), pp. 323–339. Univ. of Arizona, Tucson.

Chaussidon M., Robert F., and McKeegan K. D. (2002) *Meteoritics & Planet. Sci., 37*, A31.

Chaussidon M., Robert F., and McKeegan K. D. (2006) *Geochim. Cosmochim. Acta, 70*, 224–245.

Chen J. H. and Wasserburg G. J. (1987) *Lunar Planet Sci. XVIII*, pp. 165–166.

Chen J. H. and Wasserburg G. J. (1996) In *Earth Processes: Reading the Isotope Code* (A. Basu and S. R. Hart, eds.), pp. 1–20. AGU Monograph 95, Washington, DC.

Chevalier R. A. (2000) *Astrophys. J., 538*, L151–L154.

Clayton D. D. and Jin L. (1995) *Astrophys. J., 451*, 681–699.

Clayton D. D. and Nittler L. R. (2004) *Ann. Rev. Astron. Astrophys., 42*, 39–78.

Cuzzi J. N., Davis S. S., and Doubrovolskis A. R. (2003) *Icarus, 166*, 385–402.

Desch S. J., Connolly H. C. Jr., and Srinivasan G. (2004) *Astrophys. J., 602*, 528–542.

Feigelson E. D., Garmire G. P., and Pravdo S. H. (2002) *Astrophys. J., 584*, 911–930.

Fitoussi C., et al. (2004) *Lunar Planet Sci. XXXV*, Abstract #1586.

Flohs I. (1981) *Eos Trans. AGU, 62*, 17.

Fowler W. A., Greenstein J. L., and Hoyle F. (1962) *Geophys. J., 6*, 148–220.

Freiburghaus C., Rosswog S., and Thielemann F.-K. (1999) *Astrophys. J., 525*, L121–L124.

Frölich C., Hauser P., Liebendoerfer M., Martinez-Pinedo G., Thielemann F.-K., Bravo E., Zinner N. T., Hix W. R., Langanke K., Mezzacappa A., and Nomoto K. (2006) *Astrophys. J., 637*, 415–426.

Fuller G. and Meyer B. S. (1995) *Astrophys. J., 453*, 792–809.

Gallino R., Busso M., Picchio G., and Raiteri C. M. (1990) *Nature, 348*, 298–302.

Gallino R., Arlandini C., Busso M., Lugaro M., Travaglio C., Straniero O., Chieffi A., and Limongi M. (1998) *Astrophys. J., 497*, 388–403.

Gallino R., Busso M., Wasserburg G. J., and Straniero O. (2004) *New Astron. Rev., 48*, 133–138.

Göpel C., Manhès G., and Allègre C. J. (1994) *Earth Planet. Sci. Lett., 121*, 153–171.

Goswami J. N. and Vanhala H. A. T. (2000) In *Protostars and Planets IV* (V. Mannings et al., eds.), pp. 963–994. Univ. of Arizona, Tucson.

Goswami J. N., Marhas K. K., and Sahijpal S. (2001) *Astrophys. J., 549*, 1151–1159.

Gounelle M. and Russell S. S. (2005a) In *Chondrites and the Protoplanetary Disk* (A. N. Krot et al., eds.), pp. 588–601. ASP Conf. Series 341, San Francisco.

Gounelle M. and Russell S. S. (2005b) *Geochim. Cosmochim. Acta, 69*, 3129–3144.

Gounelle M., Shu F. H., Shang H., Glassgold A. E., Rehm K. E., and Lee T. (2001) *Astrophys. J., 548*, 1051–1070.

Gounelle M., Shang H., Glassgold A. E., Shu F. H., Rehm E. K., and Lee T. (2003) *Lunar Planet Sci. XXXIV,* Abstract #1833.

Gounelle M., Shu F. H., Shang H., Glassgold A. E., Rehm E. K., and Lee T. (2006) *Astrophys. J., 640*, 1163–1170.

Güdel M. (2002) *Ann. Rev. Astron. Astrophys., 40*, 217–261.

Harper C. L. Jr. (1996) *Astrophys. J., 466*, 437–456.

Harper C. L. Jr. and Jacobsen S. (1996) *Geochim. Cosmochim. Acta, 60*, 1131–1153.

Heymann D. and Dziczkaniec M. (1976) *Science, 191*, 79–81.

Hidaka H., Ohta Y., Yoneda S., and DeLaeter J. R. (2001) *Earth Planet. Sci. Lett., 193*, 459–466.

Hoffman R. D., Woosley S. E., Fuller G. M., and Meyer B. S. (1996) *Astrophys. J., 460*, 478–488.

Hoppe P. and Besmehn A. (2002) *Astrophys. J., 576*, L69–L72.

Hoppe P., Strebel R., Eberhardt P., Amari S., and Lewis R. S. (1996) *Science, 272*, 1314–1316.

Horan M. F., Smoliar M. I., and Walker R. J. (1998) *Geochim. Cosmochim. Acta, 62*, 545–554.

Howard W. M., Meyer B. S., and Woosley S. E. (1991) *Astrophys. J., 373*, L5–L8.

Hsu W., Wasserburg G. J., and Huss G. R. (2000) *Earth Planet. Sci. Lett., 182*, 15–29.

Hudson G. B., Kennedy B. M., Podosek F. A., and Hohenberg C. M. (1989) *Proc. Lunar Planet. Sci. Conf. 19th*, pp. 547–557.

Huss G. R., MacPherson G. J., Wasserburg G. J., Russell S. S., and Srinivasan G. (2001) *Meteoritics & Planet. Sci., 36*, 975–997.

Ireland T. (1991) *Lunar Planet. Sci. XXII*, pp. 609–610.

Ireland T. and Bukovanská M. (2003) *Geochim. Cosmochim. Acta, 67*, 3165–3179.

Itoh S. and Yurimoto H. (2003) *Nature, 423*, 728–731.

Jacobsen S. B. (2005) In *Chondrites and the Protoplanetary Disk* (A. N. Krot et al., eds.), pp. 548–557. ASP Conf. Series 341, San Francisco.

Kashiv Y., Davis A. M., Cai Z., Lai B., Sutton S. R., Lewis R. S., Gallino R., and Clayton R. N. (2006) *Lunar Planet. Sci. XXXVII,* Abstract #2464.

Kastner J. H. and Myers P. C. (1994) *Astrophys. J., 421*, 605–615.

Kita N. T., Nagahara H., Togashi S., and Morishita Y. (2000) *Geochim. Cosmochim. Acta, 64*, 3913–3922.

Kita N. T., Huss G. R., Tachibana S., Amelin Y., Nyquist L. E., and Hutcheon I. D. (2005) In *Chondrites and the Protoplanetary Disk* (A. N. Krot et al., eds.), pp. 558–587. ASP Conf. Series 341, San Francisco.

Kleine T., Münker C., Mezger K., and Palme H. (2002) *Nature, 418*, 952–955.

Kleine T., Mezger K., Münker C., Palme H., and Bischoff A. (2004) *Geochim. Cosmochim. Acta, 68*, 2935–2946.

Kleine T., Mezger K., Palme H., and Scherer E. (2005) *Lunar Planet. Sci. XXXVI,* Abstract #1431.

Krot A. N. and Keil K. (2002) *Meteoritics & Planet. Sci., 37*, 91–111.

Krot A. N., Yurimoto H., Hutcheon I. D., and MacPherson G. J. (2005a) *Nature, 434*, 998–1001.

Krot A. N., Amelin Y., Cassen P., and Meibom A. (2005b) *Nature, 434*, 989–992.

Kumar A., Gopalan K., and Bhandari N. (1999) *Geochim. Cosmochim. Acta, 63*, 3997–4001.

Lee D.-C. and Halliday A. N. (1995) *Nature, 378*, 771–774.

Lee D.-C. and Halliday A. N. (1996) *Science, 274*, 1876–1879.

Lee T. (1978) *Astrophys. J., 224*, 217–226.

Lee T., Papanastassiou D. A., and Wasserburg G. J. (1976) *Geophys. Res. Lett., 3*, 109–112.

Lee T., Shu F. H., Shang H., Glassgold A. E., and Rehm K. E. (1998) *Astrophys. J., 506*, 898–912.

Leya I., Halliday A. N., and Wieler R. (2003) *Astrophys. J., 594*, 605–616.

Libourel G., Krot A. N., and Tissandier L. (2005) *Lunar Planet. Sci. XXXVI,* Abstract #1877.

Lin Y., Guan Y., Leshin L. A., Ouyang Z., and Wang D. (2005) *Proc. Natl. Acad. Sci., 102*, 1306–1311.

Lugaro M., Davis A. M., Gallino R., Pellin M. J., Straniero O., and Käppeler F. (2003) *Astrophys. J., 593*, 486–508.

Lugmair G. W. (1974) *Meteoritics, 9*, 369.

Lugmair G. W. and Galer S. J. G. (1992) *Geochim. Cosmochim. Acta, 56*, 1673–1694.

Lugmair G. W. and Shukolyukov A. (1998) *Geochim. Cosmochim. Acta, 62*, 2863–2886.

MacPherson G. J., Davis A. M., and Zinner E. K. (1995) *Meteoritics, 30*, 365–386.

MacPherson G. J., Huss G. R., and Davis A. M. (2003) *Geochim. Cosmochim. Acta, 67*, 3165–3179.

Marhas K. K. and Goswami J. N. (2004) *New Astron. Rev., 48*, 139–144.

Marhas K. K., Goswami J. N., and Davis A. M. (2002) *Science, 298*, 2182–2185.

Markowski A., Quitté G., Halliday A. N., and Kleine T. (2006) *Earth Planet. Sci. Lett., 242*, 1–15.

McKeegan K. D. and Davis A. M. (2003) In *Treatise on Geochemistry, Vol. 1: Meteorites, Comets, and Planets* (A. M. Davis, ed.), pp. 431–460. Elsevier-Pergamon, Oxford.

McKeegan K. D., Chaussidon M., and Robert F. (2000) *Science, 289*, 1334–1337.

Merrill P. W. (1952) *Astrophys. J., 116*, 21–26.

Messenger S., Keller L. P., Stadermann F. J., Walker R. M., and Zinner E. (2003) *Science, 300*, 105–108.

Meyer B. S. (2003) *Nucl. Phys., A719*, 13–20.

Meyer B. S. (2005) In *Chondrites and the Protoplanetary Disk* (A. N. Krot et al., eds.), pp. 515–526. ASP Conf. Series 341, San Francisco.

Meyer B. S. and Clayton D. D. (2000) *Space Sci. Rev., 92*, 133–152.

Meyer B. S., Krishnan T. D., and Clayton D. D. (1996) *Astrophys. J., 462*, 825–838.

Mostefaoui S. and Hoppe P. (2004) *Astrophys. J., 613*, L149–L152.

Mostefaoui S., Lugmair G. W., and Hoppe P. (2005) *Astrophys. J., 625*, 271–277.

Münker C., Weyer S., Mezger K., Rehkämper M., Wombacher F., and Bischoff A. (2000) *Science, 289*, 1538–1542.

Nagashima K., Krot A. N., and Yurimoto H. (2004) *Nature, 428*, 921–924.

Nguyen A. N. and Zinner E. (2004) *Science, 303*, 1496–1499.

Nichols R. H. Jr. (2006) In *Meteorites and the Early Solar System II* (D. S. Lauretta and H. Y. McSween Jr., eds.), pp. 463–472. Univ. of Arizona, Tucson.

Nichols R. H. Jr., Podosek F. A., Meyer B. S., and Jennings C. L. (1999) *Meteoritics & Planet. Sci., 34*, 869–884.

Nielsen S. G., Rehkämper M., and Halliday A. N. (2004) *Eos Trans. AGU,* #P33A-1002.

Nittler L. R. (2003) *Earth Planet. Sci. Lett., 209*, 259–273.

Nittler L. R., Alexander C. M. O'D., Stadermann F. J., and Zinner E. K. (2005) *Lunar Planet. Sci. XXXVI,* Abstract #2200.

Nollett K. M., Busso M., and Wasserburg G. J. (2003) *Astrophys. J., 582*, 1036–1058.

Nyquist L. E., Takeda H., Bansal B. M., Shih C.-Y., Wiesmann H., and Wooden J. L. (1986) *J. Geophys. Res., 91*, 8137–8150.

Nyquist L. E., Reese Y., Wiesmann H., Shih C.-Y., and Takeda H. (2003) *Earth Planet. Sci. Lett., 214*, 11–25.

Ouellette N., Desch S. J., Hester J. J., and Leshin L. A. (2005) In *Chondrites and the Protoplanetary Disk* (A. N. Krot et al., eds.), pp. 527–538. ASP Conf. Series 341, San Francisco.

Podosek F. A. (1970) *Geochim. Cosmochim. Acta, 34*, 341–365.

Podosek F. A. and Nichols R. H. Jr. (1997) In *Astrophysical Implications of the Laboratory Study of Presolar Materials* (T. J. Bernatowicz and E. Zinner, eds.), pp. 617–647. American Inst. Physics, Woodbury.

Prombo C. A., Podosek F. A., Amari S., and Lewis R. S. (1993) *Astrophys. J., 410*, 393–399.

Qin L., Dauphas N., Janney P. E., Wadhwa M., and Davis A. M. (2005) *Meteoritics & Planet. Sci., 40*, A124.

Quitté G. and Birck J. L. (2004) *Earth Planet. Sci. Lett., 219*, 201–207.

Quitté G., Birck J. L., and Allègre C. J. (2000) *Earth Planet. Sci. Lett., 184*, 83–94.

Quitté G., Latkoczy C., Halliday A. N., Schönbächler M., and Günther D.

(2005) *Lunar Planet. Sci. XXXVI*, Abstract #1827.

Rauscher T., Heger A., Hoffman R. D., and Woosley S. E. (2002) *Astrophys. J., 576*, 323–348.

Reames D. V. (1995) *Rev. Geophys. Suppl., 33*, 585–589.

Sahijpal S. and Goswami J. N. (1998) *Astrophys. J., 509*, L137–L140.

Sanloup C., Blichert-Toft J., Télouk P., Gillet P., and Albarède F. (2000) *Earth Planet. Sci. Lett., 184*, 75–81.

Savina M. R., Davis A. M., Tripa C. E., Pellin M. J., Gallino R., Lewis R. S., and Amari S. (2004) *Science, 303*, 649–652.

Schoenberg R., Kamber B. S., Collerson K. D., and Eugster O. (2002) *Geochim. Cosmochim. Acta, 66*, 3151–3160.

Schönbächler M., Rehkämper M., Halliday A. N., Lee D.-C., Bourot-Denise M., Zanda B., Hattendorf B., and Günther D. (2002) *Science, 295*, 1705–1708.

Schramm D. N. and Wasserburg G. J. (1970) *Astrophys. J., 162*, 57–69.

Shu F. H., Shang H., and Lee T. (1996) *Science, 271*, 1545–1552.

Shu F. H., Shang S. H., Gounelle M., Glassgold A. E., and Lee T. (2001) *Astrophys. J., 548*, 1029–1050.

Shukolyukov A. and Lugmair G. W. (1993a) *Science, 259*, 1138–1142.

Shukolyukov A. and Lugmair G. W. (1993b) *Earth Planet. Sci. Lett., 119*, 159–166.

Spivak-Birndorf L., Wadhwa M., and Janney P. E. (2005) *Meteoritics & Planet. Sci., 40*, A145.

Srinivasan G., Ulyanov A. A., and Goswami J. N. (1994) *Astrophys. J., 431*, L67–L70.

Srinivasan G., Sahijpal S., Ulyanov A. A., and Goswami J. N. (1996) *Geochim. Cosmochim. Acta, 60*, 1823–1835.

Srinivasan G., Goswami J. N., and Bhandari N. (1999) *Science, 284*, 1348–1350.

Srinivasan G., Whitehouse M. J., Weber I., and Yamaguchi A. (2004) *Lunar Planet. Sci. XXXV*, Abstract #1709.

Stewart B. W., Papanastassiou D. A., and Wasserburg G. J. (1994) *Geochim. Cosmochim. Acta, 58*, 3487–3509.

Sugiura N., Shuzou Y., and Ulyanov A. (2001) *Meteoritics & Planet. Sci., 36*, 1397–1408.

Swindle T. D. and Podosek F. A. (1988) In *Meteorites and the Early Solar System* (J. F. Kerridge and M. S. Matthews, eds.), pp. 1093–1113. Univ. of Arizona, Tucson.

Tachibana S. and Huss G. R. (2003) *Astrophys. J., 588*, L41–L44.

Tachibana S., Huss G. R., Kita N. T., Shimode G., and Morishita Y. (2006) *Astrophys. J., 639*, L87–L90.

Taylor D. J., McKeegan K. D., and Krot A. N. (2005) *Lunar Planet. Sci. XXXVI*, Abstract #2121.

Travaglio C., Gallino R., Amari S., Zinner E., Woosley S., and Lewis R. S. (1999) *Astrophys. J., 510*, 325–354.

Vanhala H. A. T. and Boss A. P. (2002) *Astrophys. J., 575*, 1144–1150.

Wadhwa M. and Lugmair G. W. (1995) *Lunar Planet. Sci. XXVI*, pp. 1453–1454.

Wadhwa M. and Russell S. S. (2000) In *Protostars and Planets IV* (V. Mannings et al., eds.), pp. 995–1018. Univ. of Arizona, Tucson.

Wadhwa M., Srinivasan G., and Carlson R. W. (2006) In *Meteorites and the Early Solar System II* (D. S. Lauretta and H. Y. McSween Jr., eds.), pp. 715–731. Univ. of Arizona, Tucson.

Wasserburg G. J. (1985) In *Protostars and Planets II* (D. C. Black and M. S. Matthews, eds.), pp. 703–737. Univ. of Arizona, Tucson.

Wasserburg G. J., Busso M., and Gallino R. (1996) *Astrophys. J., 466*, L109–L113.

Wolk S. J., Hardnen F. R., Flaccomio E., Micela G., Favata F., Shang H., and Feigelson E. D. (2005) *Astrophys. J. Suppl., 160*, 423–449.

Wooden D. H. (1997) In *Astrophysical Implications of the Laboratory Study of Presolar Materials* (T. J. Bernatowicz and E. Zinner, eds.), pp. 317–376. American Inst. Physics, Woodbury.

Wooden D. H., Woodward C. E., and Harker D. E. (2004) *Astrophys. J., 612*, L77–L80.

Woosley S. E. (1997) *Astrophys. J., 476*, 801–810.

Woosley S. E. and Howard W. M. (1978) *Astrophys. J. Suppl., 36*, 285–305.

Woosley S. E., Wilson J. R., Mathews G. J., Hoffman R. D., and Meyer B. S. (1994) *Astrophys. J., 433*, 229–246.

Yin Q.-Z. and Jacobsen S. B. (2002) *Meteoritics & Planet. Sci., 37*, A152.

Yin Q.-Z., Jagoutz E., and Wänke H. (1992) *Meteoritics, 27*, 310.

Yin Q.-Z., Jacobsen S. B., McDonough W. F., Horn I., Petaev M. I., and Zipfel J. (2000) *Astrophys. J., 536*, L49–L53.

Yin Q.-Z., Jacobsen S.B., Yamashita K., Blichert-Toft J., Télouk P., and Albarède A. (2002) *Nature, 418*, 949–952.

Yin Q.-Z., Ott U., and Lee C.-Y. (2005) *Meteoritics & Planet. Sci., 40*, A171.

Young E. D., Simon J. I., Galy A., Russell S. S., Tonui E., and Lovera O. (2005) *Science, 308*, 223–227.

Zinner E. K. (2003) In *Treatise on Geochemistry, Vol. 1: Meteorites, Comets, and Planets* (A. M. Davis, ed.), pp. 17–39. Elsevier-Pergamon, Oxford.

Zinner E., Nittler L. R., Hoppe P., Gallino R., Straniero O., and Alexander C. M. O'D. (2005) *Geochim. Cosmochim. Acta, 69*, 4149–4165.

Origin and Evolution of Oxygen-Isotopic Compositions of the Solar System

Hisayoshi Yurimoto and Kiyoshi Kuramoto
Hokkaido University

Alexander N. Krot and Edward R. D. Scott
University of Hawaii at Manoa

Jeffrey N. Cuzzi
NASA Ames Research Center

Mark H. Thiemens
University of California San Diego

James R. Lyons
University of California, Los Angeles

On a three-isotope diagram oxygen-isotopic compositions of most primitive meteorites (chondrites), chondritic components (chondrules, refractory inclusions, and matrix), and differentiated meteorites from asteroids and Mars deviate from the line along which nearly all terrestrial samples plot. Three alternative mechanisms have been proposed to explain this oxygen-isotopic anomaly: nucleosynthetic effects, chemical mass-independent fractionation effects, and photochemical self-shielding effects. Presently, the latter two are the most likely candidates for production of the isotopic anomalies. Recent data on solar wind oxygen isotopes lends support to the photochemical self-shielding scenario, but additional solar-isotopic data are needed. Observations, experiments, and modeling are described that will advance our understanding of the complex history of oxygen in the solar system.

1. INTRODUCTION

Oxygen is the third most abundant element in the solar system and the most abundant element of the terrestrial planets. The presence of oxygen in both gaseous and solid phases makes O isotopes (the terrestrial abundance: $^{16}O = 99.757\%$, $^{17}O = 0.038\%$, and $^{18}O = 0.205\%$) important tracers of various fractionation processes in the solar nebula, which are essential for understanding the evolution of gaseous and solid phases in the early solar system.

Oxygen-isotopic compositions are normally expressed in δ units, which are deviations in part per thousand (permil, ‰) in the $^{17}O/^{16}O$ and $^{18}O/^{16}O$ ratios from standard mean ocean water (SMOW) with $^{17}O/^{16}O = 0.0003829$ and $^{18}O/^{16}O = 0.0020052$ (*McKeegan and Leshin,* 2001): $\delta^{17,18}O_{SMOW} = [(^{17,18}O/^{16}O)_{sample}/(^{17,18}O/^{16}O)_{SMOW} - 1] \times 1000$. On a three-isotope diagram of $\delta^{18}O$ vs. $\delta^{17}O$, compositions of nearly all terrestrial samples plot along a single line of slope 0.52 called the terrestrial fractionation line. This line reflects *mass-dependent fractionation* from a single homogeneous source during chemical and physical processes that results from differences in the masses of the O isotopes. The slope 0.52 results from changes in $^{17}O/^{16}O$ that are nearly half those in $^{18}O/^{16}O$ because of isotopic mass

differences; the precise value of the slope depends on the nature of the isotopic species or isotopologues (e.g., *Thiemens,* 2006). In contrast, O-isotopic compositions of the vast majority of extraterrestrial samples, including primitive (chondrites) and differentiated (achondrites) meteorites, deviate from the terrestrial fractionation line (Fig. 1; see section 4 for details), reflecting *mass-independent fractionation* processes that preceded accretion of these bodies in the protoplanetary disk. Samples from bodies that were largely molten and homogenized such as Mars and Vesta lie on lines that are parallel to the terrestrial fractionation line. Lunar samples show no detectable deviations from the terrestrial fractionation line, for reasons that are still debated (see section 8.4.3 for details). The deviation from the terrestrial fractionation line is commonly expressed as $\Delta^{17}O_{SMOW} = \delta^{17}O_{SMOW} - 0.52 \times \delta^{18}O_{SMOW}$.

The origin of the O-isotopic variations or anomalies in solar system materials has been a major puzzle for planetary scientists since they were discovered over 30 years ago (*Clayton et al.,* 1973). The interpretation of the mass-independent O-isotopic variations or anomalies is one of the most important outstanding problems in cosmochemistry (*McKeegan and Leshin,* 2001). Here we discuss the nature of the O-isotopic anomalies in the solar system, the evolu-

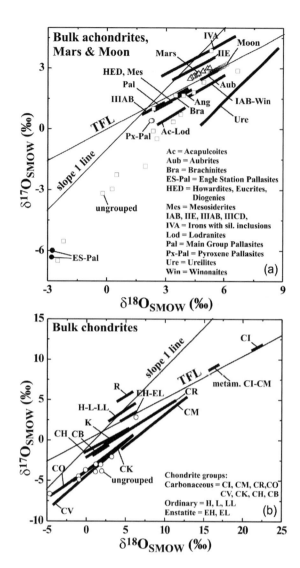

Fig. 1. Bulk O-isotopic compositions of achondrites and meteorites from **(a)** Mars and Moon and **(b)** chondrites. Each group of meteorites probably represents a single asteroidal or planetary body. Oxygen-isotopic compositions of most meteorites deviate from the terrestrial fractionation line (TFL). Data from *Clayton and Mayeda* (1996, 1999).

tion of O-isotopic compositions in the solar nebula, and possible implications for other protoplanetary disks and planetary systems.

2. CHONDRITES AND THEIR COMPONENTS

Mass-independent O-isotopic variations were discovered in chondrites as their diverse components show effects that are much larger than those shown by bulk chondrites or achondrites (*Clayton et al.*, 1973). Chondritic meteorites consist of three major components, which may have formed at separate locations and/or times in the solar nebula: refractory inclusions [Ca,Al-rich inclusions (CAIs) and amoeboid

olivine aggregates (AOAs)], chondrules, and fine-grained matrix (e.g., *Scott and Krot*, 2003). CAIs are ~1-μm- to ~1-cm-sized irregularly shaped or spheroidal objects composed mostly of oxides and silicates of calcium, aluminum, titanium, and magnesium. AOAs are physical aggregates of individual grains of forsterite (Mg_2SiO_4), Fe,Ni-metal, and small CAIs. Evaporation and condensation appear to have been the dominant processes during formation of refractory inclusions; subsequently some CAIs, called igneous CAIs, experienced extensive melting and partial evaporation (*MacPherson et al.*, 2005; *Wood*, 2004).

Chondrules are igneous, rounded objects, 0.01–10 mm in size, composed largely of crystals of ferromagnesian olivine ($Mg_{2-x}Fe_xSiO_4$) and pyroxene ($Mg_{1-x}Fe_xSiO_3$, where 1 < x < 0) and Fe,Ni-metal with interstitial glassy or microcrystalline material. Most chondrules have textures that are consistent with crystal growth from a rapidly cooling (100–1000 K hr^{-1}) silicate melts (e.g., *Hewins et al.*, 2005). Some chondrules contain relict fragments of refractory inclusions and earlier generations of chondrules, and are surrounded by igneous rims, suggesting chondrule formation was repetitive (e.g., *Jones et al.*, 2005; *Russell et al.*, 2005). Additionally, chondrules in primitive (unmetamorphosed) chondrites are often surrounded by fine-grained rims. Based on these observations, it is generally inferred that chondrules formed by varying degrees of melting of dense aggregates of ferromagnesian silicate, metal, and sulfide grains during multiple flash-heating events, possibly by shock waves, in the dusty, inner (<5 AU) solar nebula (e.g., *Desch and Connolly*, 2002).

Matrix material is an aggregate of mineral grains, 10 nm-5 μm in size, that surrounds refractory inclusions and chondrules and fills in the interstices between them. In primitive chondrites, matrix is made largely of magnesian olivine and pyroxene crystals and amorphous ferromagnesian silicate particles (e.g., *Scott and Krot*, 2003). Matrices are chemically complementary to chondrules (*Brearley*, 1996), and may have largely experienced extensive evaporation and recondensation during chondrule formation (e.g., *Bland et al.*, 2005).

The ^{207}Pb-^{206}Pb absolute ages of CAIs from CV chondrites (4567.2 ± 0.6 m.y.), and chondrules from the CV (4566.6 ± 1.0 m.y.), ordinary (4566.3 ± 1.7 m.y.), CR (4564.7 ± 0.6 m.y.), and CB (4562.7 ± 0.5 m.y.) chondrites indicate that chondrule formation started shortly after or maybe even contemporaneously with the CV CAIs and lasted for ~5 m.y. (*Amelin et al.*, 2002; *Kita et al.*, 2005). We note, however, that chondrules in CB chondrites possibly formed from a vapor-melt plume caused by a giant impact between planetary embryos after the protoplanetary disk largely dissipated (*Krot et al.*, 2005a). Based on short-lived, ^{26}Al-^{26}Mg isotope chronology (see chapter by Wadhwa et al.), it is inferred that most CAIs formed over a short period of time, perhaps only 0.3 m.y. (*Bizzarro et al.*, 2004; *Young et al.*, 2005); some CAIs experienced late-stage remelting (*Krot et al.*, 2005c; *Russell et al.*, 2005).

The location of CAI and chondrule formation remains controversial. According to the X-wind model, both components formed near the Sun and then were transported radially by the X-wind, where they accreted together with matrix into chondrite parent asteroids (e.g., *Itoh and Yurimoto*, 2003; *Shu et al.*, 1996, 2001). Alternatively, chondrules and matrix formed in closed proximity to their accretion regions (e.g., *Jones et al.*, 2005), whereas refractory inclusions originated closer to the proto-Sun (*Cuzzi et al.*, 2003; *Scott and Krot*, 2003).

3. OXYGEN-ISOTOPIC COMPOSITIONS OF CHONDRITIC COMPONENTS

CAIs, AOAs, and chondrules in primitive chondrites plot along a line of slope ~1 and show a large range of $\Delta^{17}O_{SMOW}$, from <–20‰ to +5‰ (e.g., *Aléon et al.*, 2002; *Clayton*, 1993; *Itoh et al.*, 2004). Within a chondrite group, AOAs and most CAIs are ^{16}O-rich relative to chondrules. Some igneous CAIs are $^{17,18}O$-enriched to a level observed in chondrules (Fig. 2). Although matrix minerals show variations in O-isotopic compositions (e.g., *Kunihiro et al.*, 2005), their bulk O-isotopic compositions are probably similar to average compositions of chondrules [based on bulk O-isotopic

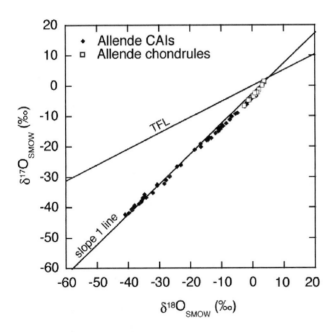

Fig. 2. Oxygen-isotopic compositions of the refractory inclusions and chondrules in the CV3 Allende carbonaceous chondrite. Most refractory inclusions and chondrules deviate from the terrestrial fractionation line (TFL) and plot close to a line with a slope of 1. Data from *Clayton* (1993). Recently many new microanalysis data using SIMS and laser-ablation-MS have been reported (e.g., references in the text). The general distribution in the diagram remains nearly unaffected by the new data although understanding of the formation processes has been greatly developed by the new data.

compositions of chondrites (Fig. 1b), average O-isotopic compositions of chondrules, and abundance of matrix].

The observed variations in O-isotopic compositions of chondritic components are commonly explained as a result of mixing of ^{16}O-rich and $^{17,18}O$-rich reservoirs in the solar nebula (e.g., *Clayton*, 1993).

The existence of an ^{16}O-rich gaseous reservoir in the inner solar nebula has been inferred from the ^{16}O-rich ($\Delta^{17}O_{SMOW}$ < –20‰) compositions of refractory solar nebula condensates, such as AOAs and fine-grained, spinel-rich CAIs with Group II rare-earth-element patterns (volatility-fractionated patterns indicating condensation from a gas from which more refractory rare earth elements have already been removed) (e.g., *Aléon et al.*, 2005a; *Krot et al.*, 2002; *Scott and Krot*, 2001).

The existence of an $^{17,18}O$-rich gaseous reservoir in the inner solar nebula can be inferred from $^{17,18}O$-rich compositions of chondrules, which probably experienced O-isotopic exchange with the nebular gas during chondrule melting events (*Krot et al.*, 2006; *Maruyama et al.*, 1999; *Yu et al.*, 1995), and from the $^{17,18}O$-rich compositions of aqueously formed secondary minerals in chondrites (*Choi et al.*, 1998).

Because most solids in the inner solar system experienced thermal processing that could have modified their initial O-isotopic compositions, the bulk O-isotopic composition of solids in the protosolar molecular cloud is unknown. Based on ^{16}O-rich compositions of the CAIs considered to be evaporative residues of the initial solar nebula dust (e.g., *Fahey et al.*, 1987; *Goswami et al.*, 2001; *Lee and Shen*, 2001; *Wood*, 1998), and from an ^{16}O-rich magnesian chondrule, which has been interpreted as a solidified melt of this dust (*Kobayashi et al.*, 2003), it is inferred that primordial solids in the protosolar molecular cloud were ^{16}O-rich ($\Delta^{17}O_{SMOW}$ < –20‰) (*Itoh and Yurimoto*, 2003; *Scott and Krot*, 2001).

The chondrite matrix contains small amounts of silicate and oxide grains with grossly anomalous O-isotopic compositions that are inferred to have formed around massive stars (*Clayton and Nittler*, 2004). Silica grains with extremely high enrichments in ^{17}O and ^{18}O might have originated in a protostellar outflow (*Aléon et al.*, 2005b). Only a small fraction of the presolar grains are ^{16}O-rich, and none of them plot along the slope-1 line (*Nittler et al.*, 1997, 1998).

4. BULK OXYGEN-ISOTOPIC COMPOSITIONS OF CHONDRITES AND ACHONDRITES

Bulk O-isotopic compositions of chondrites (Fig. 1b) show much smaller deviations from the terrestrial fractionation line than their components (Fig. 2) (*Clayton and Mayeda*, 1999). Most chondrite groups plot above or below the terrestrial fractionation line; the only exceptions are enstatite and CI carbonaceous chondrites (Fig. 1b). Chondrites in a single group define lines with slopes between 0.5 and 1 reflecting the mass-independent effects associated with their components as well as mass-dependent processes largely in their parent asteroids, such as aqueous alteration and thermal metamorphism.

Bulk O-isotopic compositions of rocks from Mars and the Moon and achondrites tend to lie closer to the terrestrial fractionation line than the chondrites (Fig. 1a). Only aubrite (enstatite achondrite) and Moon samples lie on the terrestrial fractionation line. Samples from extensively melted bodies define lines that are parallel to the terrestrial fractionation line (e.g., angrites and HED meteorites from the asteroid Vesta), whereas meteorites from partly melted asteroids like the ureilites, acapulcoites, and winonaites define steeper lines as they were not homogenized during melting (*Greenwood et al., 2005*).

5. OXYGEN-ISOTOPIC COMPOSITION OF THE SUN

The O-isotopic composition of the Sun has not yet been measured directly. Oxygen-isotopic compositions of solar wind implanted into the outermost (from two to a few hundred nanometers) surface layers of metal grains in the lunar regolith have been recently reported by *Hashizume and Chaussidon* (2005) and by *Ireland et al.* (2005). The measurements of *Hashizume and Chaussidon* (2005) revealed the presence of an ^{16}O-rich component with $\Delta^{17}O_{SMOW} < -20 \pm 4‰$ that was interpreted as the O-isotopic composition of the solar wind and, hence, of the Sun, consistent with the earlier predictions of *Clayton* (2002), *Yurimoto and Kuramoto* (2004), *Krot et al.* (2005b), and *Lyons and Young* (2005). In contrast, the O-isotopic compositions of more surficial layers of metal grains reported by *Ireland et al.* (2005) are $^{17,18}O$-enriched ($\Delta^{17}O_{SMOW} \sim +35‰$). The preservation of such anomalous O component below the surface suggests solar-wind implantation. The reasons for the discrepancy between the two datasets remain unclear. The ultimate test of both hypotheses will be provided by O-isotopic measurements of solar wind returned by the Genesis spacecraft.

6. ORIGIN OF OXYGEN-ISOTOPIC ANOMALY IN THE SOLAR SYSTEM

Three different mechanisms have been proposed to explain the origins of the mass-independent O-isotopic fractionation with a slope of ~1 observed in chondrite components (Fig. 2).

6.1. Nucleosynthetic Effects

According to the first group of hypotheses, the slope-1 line reflects an inherited O-isotopic heterogeneity in solar nebula materials (^{16}O-rich solids and $^{17,18}O$-rich gas) resulting from either nucleosynthesis in stars (*Clayton et al., 1973*) or from nuclear reactions that involve energetic particles from the proto-Sun or from galactic cosmic rays (*Lee,*

1978). However, since isotopic variations of other elements (e.g., Mg, Si, Ca, Ti) are much smaller and uncorrelated with the O-isotopic anomaly, this mechanism appears to be unlikely (*Clayton, 1993*). In addition, ^{16}O-rich presolar grains are exceptionally rare in meteorites (*Nagashima et al., 2004; Nittler, 2003*).

6.2. Chemical Mass-independent Fractionation Effects During Formation of Solids

According to the second hypothesis, the slope-1 line resulted from chemical mass-independent fractionation effects similar to those observed in gas-phase O_3 production (*Thiemens and Heidenreich*, 1983). Some solid, terrestrial nitrates and sulfates possess mass-independent isotopic effects that have survived for at least 100 m.y. (*Thiemens et al., 2001*). Sulfur is now known to have preserved mass-independent isotopic effects on Earth for at least 3.8 G.y. (*Farquhar et al., 2000*).

Because the solar nebula is too reducing for O_3 to be a significant component, *Thiemens* (1999, 2006) proposed that reactions involving symmetric components of silicate vapors such as $O + SiO \rightarrow SiO_2$ were responsible for the O-isotopic anomalies in chondrites. These molecules are vapor-phase precursors to minerals common in CAIs and chondrules. The analogous reaction $O + CO \rightarrow CO_2$ has been shown to yield a mass-independent fractionation signature. SiO_2 has its symmetry modified by substitution of ^{17}O or ^{18}O, as occurs for isotope substitution at the terminal oxygens of O_3. The same would also hold for reactions involving AlO, FeO, MgO, or CaO, all relevant nebular constituents. The chemical mechanism of mass-independent fractionation in O_3 is believed to be a preferential stabilization of the asymmetric isotopologue arising from a lack of intramolecular equilibrium in the symmetric isotopologue, referred to as an η-effect (*Gao et al., 2002*). Although an η-effect could also occur during $SiO_{2(g)}$ formation, *Marcus* (2004) has argued that $O + SiO \rightarrow SiO_2$ in the inner gas phase solar nebula will be too slow by several orders of magnitude vs. the reaction of O with H_2 to form H_2O. Instead, *Marcus* (2004) proposed that mass-independent fractionation occurred during SiO_2 formation on grain surface during CAI formation. This removes the spatial dependency and allows the reaction to occur anywhere in the nebula. It is known that O_3 formed on the walls of reaction vessels does not have a mass-independent fractionation signature, but how relevant that result is to SiO_2 formation, in which SiO is incorporated into the grain surface, is unclear. Experiments involving metal oxide reactions will be of importance in developing a model for the reactions in the nebula. The reactions of $O + CO$ and $OH + CO$ have been demonstrated to produce mass-independent isotopic effects, consistent with the proposed role in the nebula via symmetry reactions.

A past limitation in model considerations has been the inability to define the role of the Sun and photochemistry in the solar system. Recently, *Rai et al.* (2005) have reported mass-independent S-isotopic effects in four achondritic meteorite classes at the bulk level. The results appear consistent with a photochemical production, as observed in terrestrial Precambrian solids (*Farquhar et al., 2000*). As shown in this paper, the only explanation that may account for these observations requires gas phase photochemistry of S species. This requires photochemistry prior to condensation and incorporation into meteorites, directly implicating photochemistry in the nebula. This observation and photochemical requirement is consistent with both the symmetry (production of O atoms and OH radicals) and self-shielding models discussed in the following section.

6.3. Photochemical Self-shielding Effects

The third mechanism proposed for the mass-independent O-isotopic fractionation is photochemical self-shielding effects in CO. Self-shielding in O_2 was first considered as a possible solar nebular process (*Kitamura and Shimizu, 1983; Thiemens and Heidenreich, 1983*). It was shown later that trapping of photodissociated O by metal atoms, H, and O_2 is inefficient compared to O-isotopic exchange reactions in the solar nebula (*Navon and Wasserburg, 1985*). *Thiemens and Heidenreich* (1983) and *Navon and Wasserburg* (1985) proposed that self-shielding in CO, as observed in astronomical environments, was responsible for production of the meteoritic isotopic anomalies.

The isotopically selective photolysis of CO has been well known in the astronomical community for a long time (*Bally and Langer, 1982; van Dishoeck and Black, 1988*), and this mechanism has also been also suggested to explain the origin of the O-isotopic anomaly in meteorites and the evolution of O-isotopic compositions of the inner solar nebula (*Clayton, 2002; Lyons and Young, 2005; Yurimoto and Kuramoto, 2004*).

Three different astrophysical settings for CO self-shielding have been proposed to account for the solar system O-isotopic anomalies: the protosolar molecular cloud (*Yurimoto and Kuramoto, 2004*), the inner protoplanetary disk (*Clayton, 2002*), and the outer protoplanetary disk (*Lyons and Young, 2005*) (for details see sections 7 and 8). In all three cases it is assumed that the bulk O-isotopic composition of the protosolar molecular cloud, and, hence, the bulk composition of the Sun, is ^{16}O-rich; i.e., $^{17,18}O_{SMOW}$ = $-50‰$. We note that this assumption is consistent with O-isotopic measurements of solar wind by *Hashizume and Chaussidon* (2005), but is inconsistent with the results of *Ireland et al.* (2005). In the following discussion, it is convenient to use bulk composition of the molecular cloud as the new reference standard ($\delta^{17,18}O_{MC}$). In this case, $\delta^{17,18}O_{MC}$ = $\delta^{17,18}O_{SMOW} + 50‰$.

7. SELF-SHIELDING OF CARBON MONOXIDE IN MOLECULAR CLOUDS

Self-shielding effects of CO as a mechanism for O-isotopic fractionation were originally discussed in the context

of molecular clouds by *Solomon and Klemperer* (1972). The self-shielding of CO is a process of isotope-selective photodissociation that occurs at far-ultraviolet (FUV) wavelengths from 91.3 nm to 107.6 nm. Because CO first goes to a bound excited state before dissociating (predissociation), the absorption spectrum of CO consists of many narrow lines. The wavelength of each line is determined by the specific vibrational and rotational levels involved, and is shifted when the mass of the molecules is changed as a result of an isotopic substitution. The absorption spectra of the various CO isotopologues do not overlap significantly, particularly at the low temperatures of molecular clouds. In the environment of molecular clouds, predissociation due to line spectrum absorption of UV photons is the dominant mechanism for photodissociation of CO (*Bally and Langer, 1982; Chu and Watson, 1983; Glassgold et al., 1985; Marechal et al., 1997b; van Dishoeck and Black, 1988; Warin et al., 1996*). Ultraviolet intensity at the wavelengths of dissociation lines for abundant $C^{16}O$ rapidly attenuates in the surface layer of a molecular cloud, because the cloud becomes optically thick in the lines of $C^{16}O$ after only short penetration depths. For less-abundant $C^{17}O$ and $C^{18}O$, which have shifted absorption lines because of the differences in the vibrational-rotational energy levels, the attenuation is much slower. As a result, $C^{17}O$ and $C^{18}O$ are dissociated by UV photons even in a deep molecular cloud interior. This process results in selective enrichment of atomic oxygen in ^{17}O and ^{18}O.

Large mass-independent fractionation of O isotopes resulting in $^{17,18}O$ depletion, i.e., ^{16}O enrichment, in CO has been observed in a diffuse molecular cloud toward X Persei (*Sheffer et al., 2002*), supporting the CO self-shielding hypothesis (Fig. 3).

Evidence for mass-independent fractionation of O isotopes has also been observed in star-forming molecular clouds. Systematic observations of the variation of multiple

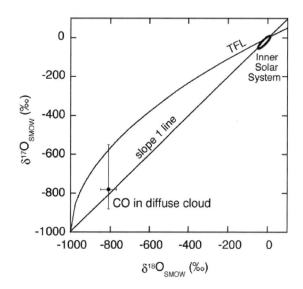

Fig. 3. Oxygen-isotopic composition of CO in a diffuse molecular cloud toward X Persei. Data from *Sheffer et al.* (2002).

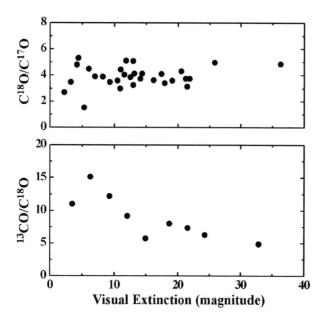

Fig. 4. Isotopic compositions of CO isotopologues in IC 5146 dark cloud. Data from *Lada et al.* (1994) and *Bergin et al.* (2001).

minor CO isotopologues with dust column density have been reported for the IC 5146 dark cloud (Fig. 4). ^{13}CO/C^{18}O ratios in IC 5146 increase with decreasing visual extinction (A_v) (Fig. 4) (*Lada et al.*, 1994). Because the C^{18}O abundance is about five times lower than the ^{13}CO abundance, this trend indicates a selective UV photodissociation of C^{18}O by self-shielding. It is also to be noted that the effect of selective photodissociation of ^{13}CO is diluted by exothermic ion molecule reaction ^{12}CO + ^{13}C$^+$ → ^{13}CO + ^{12}C$^+$, thereby ^{13}CO is expected to track fairly well the abundance of ^{12}CO, i.e., the ^{13}CO/^{12}CO ratio little fractionated in molecular clouds (e.g., *Warin et al.*, 1996). On the other hand, the C^{18}O/C^{17}O ratios are nearly constant for all A_v range from 0 to 40 mag, suggesting equal photodissociation efficiency for C^{18}O and C^{17}O (*Bergin et al.*, 2001). These results indicate that mass-independent O-isotopic fractionation with ^{17,18}O-depletion of CO occurs in IC 5146. Similar isotopic characteristics for ^{13}CO, C^{17}O, and C^{18}O isotopologues have been observed in the ρ Oph dark cloud, the nearest low-mass star-forming region (*Wouterloot et al.*, 2005), and Orion molecular cloud (OCM1), an active high-mass star-forming region near to us (*White and Sandell*, 1995).

Based on these observations, we infer that mass-independent isotopic fractionation is a common feature of the diffuse, dark, and giant molecular clouds, and that most low- and high-mass stars probably formed from molecular clouds having mass-independent fractionation of CO. However, due to large uncertainties in the O-isotopic analysis, more precise measurements are necessary to state whether the effect is the requisite equal ^{17}O/^{18}O pattern required by

current models. The isotopic fractionation factor associated with the actual photodissociation has not been measured in the laboratory. Large mass-dependent isotopic fractionations are known to occur in some photodissociation processes.

Because CO and atomic O are the dominant O-bearing gas species in molecular clouds (*Marechal et al.*, 1997a), their isotopic fractionation caused by UV self-shielding may propagate to other O-bearing species. Water ice is the dominant O-bearing species among ices in molecular clouds (Fig. 5), where it nucleates and grows on silicate dust grains by surface hydrogenation reactions between atomic oxygen and hydrogen (*Aikawa et al.*, 2003; *Greenberg*, 1998). Therefore, O-isotopic compositions of H$_2$O ice should be close to those of gaseous atomic O; i.e., enriched in ^{17}O and ^{18}O (*Yurimoto and Kuramoto*, 2004), assuming there are no large mass-fractionation effects associated with the formation process. Water ice is observed in molecular clouds with $A_v > 3.2$; its abundance increases with increasing A_v (*Whittet et al.*, 2001).

In collapsing protostellar molecular cloud cores, most of the atomic O reacts to form H$_2$O ice, and CO becomes the dominant O-bearing gas species within 10^5 yr (*Bergin et al.*, 2000). In more evolved, cold molecular cloud cores, most CO could become frozen onto dust grains (*Aikawa et al.*, 2003). Because of the low temperature, O-isotopic exchange among CO gas, CO ice, and H$_2$O ice is insignificant, and the evolved isotope fractionation of O is preserved in each phase. Thus, O-isotopic compositions of CO and H$_2$O become depleted and enriched in ^{17,18}O with time, respectively.

Low-mass stars form by collapse of individual cores in a cold, dark cloud with $5 < A_v < 25$ and temperatures as low as ~10 K (*van Dishoeck et al.*, 1993). Under these conditions, a model simulating photochemical isotopic fraction-

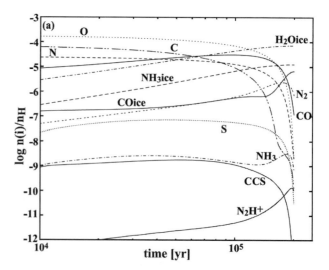

Fig. 5. Temporal variations of molecular abundances with grain-surface reactions. From *Aikawa et al.* (2003).

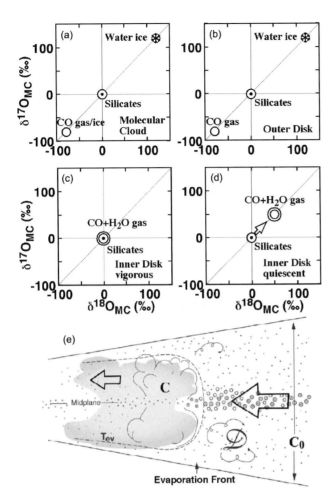

Fig. 6. (a)–(d) Schematic diagrams of the evolution of O-isotopic compositions of the solar system from (a) a protosolar molecular cloud to the protoplanetary disk at different locations and various stages of its evolution: (b) outer disk during vigorous and quiescent stages, (c) vigorous inner disk, (d) and quiescent inner disk. Arrow in (d) shows a direction of solid-gas equilibration. After *Yurimoto and Kuramoto* (2004). (e) Sketch illustrating inwardly drifting water-ice-rich material (circles) crossing water evaporation front (snow line) with midplane temperature T_{ev}; dots show the more refractory material (silicates, oxides, metal, sulfides, and organics). The large inward drift flux of water-ice-rich material cannot be offset by water removal processes until the concentration of the water vapor is much greater than its nominal solar value C_0. After *Cuzzi and Zahnle* (2004).

ation implies that the O-isotopic compositions ($\delta^{17,18}O_{MC}$) of H_2O ice and CO are in the range of +100 to +250‰ and –60 to –400‰, respectively (*Marechal et al.*, 1997b). These values are consistent with astronomical observations (*Ando et al.*, 2002; *Lada et al.*, 1994; *White and Sandell*, 1995; *Wouterloot et al.*, 2005).

For the protosolar molecular cloud, *Yurimoto and Kuramoto* (2004) assumed average $\delta^{17,18}O_{MC}$ values for silicates, H_2O ice, and CO of 0‰, +120‰, and –80‰, respectively (Fig. 6a). These values were chosen to be consistent with

the astronomical studies and the relative O abundances of silicates, H_2O ice, and CO (1:2:3) of a molecular cloud of solar bulk composition (*Greenberg*, 1998). The [17,18]O-depleted compositions of primary solids and [17,18]O-enriched nature of H_2O ice in the solar nebula are supported by meteoritic evidence (see section 3 for details).

8. EVOLUTION OF PROTOPLANETARY DISKS AND OXYGEN ISOTOPES

8.1. Protoplanetary Disk Evolution

Protoplanetary disks form as a result of gravitational collapse of dense molecular cloud cores. Protoplanetary disks are evolving objects; the surface mass density and mass accretion rate both decrease as matter is accreted onto the star or photoevaporated into interstellar space, and the disk outer edge expands (*Hartmann*, 2005). Temperatures in the inner disk regions depend on these accretion rates, but temperatures in the outer disk regions are nearly independent of accretion rate and are probably low enough that the primordial O-isotopic compositions of the molecular cloud components would have been preserved there in the absence of other effects. For instance, CO ice may sublime while preserving its O-isotopic composition (Fig. 6b). However, as material collapses from the protostellar molecular cloud into the disk, some of it passes through accretion shocks (*Chick and Cassen*, 1997) which may result in some isotopic reequilibration of CO and H_2O and with other O reservoirs (*Lyons and Young*, 2005).

During all early stages of nebula evolution, UV radiation from the central protostar and perhaps neighboring massive stars (*Hester and Desch*, 2005) will be present — strongest at the innermost edge of the disk near the protostar, but also important in the low-density outer surfaces of the disk (*Adams et al.*, 2004). This irradiation may result in isotopic self-shielding effects in either or both places, as described below (see sections 8.2 and 8.3).

Astronomical observations set timescales for the evolution of disks (*Haisch et al.*, 2001; *Hartmann*, 2005), and absolute and relative chronologies of chondrite components provide constraints on the duration of the formation and thermal processing of solids in our own disk (*Amelin et al.*, 2002; *Baker et al.*, 2005; *Bizzarro et al.*, 2004; *Kita et al.*, 2005; *Krot et al.*, 2005b). These timescales are in general accord, lying in the range of 3–5 m.y. for the duration of dust-rich disks.

Young disks undergo a *vigorously accreting stage* lasting <1 m.y., characterized by accretion rates of several times 10^{-6} $M_\odot$ per year and temperatures in the innermost nebula hot enough to vaporize most silicates (*Cuzzi et al.*, 2005; *Muzerolle et al.*, 2003; *Woolum and Cassen*, 1999); this stage might be associated with Class 0 or even some Class I protostars. Most Class II objects are in the 1–5-m.y.-old range, and are accreting at lower rates (between 10^{-7} and 10^{-9} $M_\odot$ per year), with inner nebula temperatures cool enough for chondrules and other rock-forming minerals, but not water ice, to be solid. Water ice condenses only outside the snow line, a location that varies with model and with time as the nebula cools, but lies well outside the asteroid-formation region for most of the evolution of the nebula in most models. Superimposed on this nebula gas density and temperature evolution is the transport of trace gases and solids by advection, diffusion, and gas drag migration, which changes the chemistry of the nebula over its lifetime. This evolution will depend on the intensity of turbulence, and turbulent diffusion, in the nebula — something that remains poorly understood but will be assumed here.

Particles immersed in the disk have a relative motion against ambient gas, which rotates with a slightly slower speed than the Keplerian velocity due to the generally outward radial pressure gradient. The frictional loss of angular momentum due to their relative motion causes a drift of the dust grains toward the central star (*Weidenschilling*, 1977; for recent reviews see also *Cuzzi et al.*, 2005; *Cuzzi and Weidenschilling*, 2005; see chapter by Dominik et al.). Solids probably can grow by coagulation into meter-sized particles that move rapidly inward due to gas drag. Instead of being "lost into the Sun" as sometimes assumed, particles migrate inward only until the temperature becomes high enough to evaporate them (Fig. 6e). If about 10% of the mass in solids is in the rapidly drifting meter-sized range, material is carried inward across the snow line as solids faster than it can be removed as a vapor by the advection of the disk and diffusion of its vapor. The disk immediately inside the evaporation front thus becomes enhanced above solar in its vapor concentration, and as time goes on and the process continues, this enhancement spreads throughout the entire inner disk (Fig. 6e). This evaporation front effect occurs for any volatile, but silicates and water are especially interesting (*Cuzzi et al.*, 2003; *Cuzzi and Zahnle*, 2004).

Here we focus on the water evaporation front as having direct relevance to O-isotope compositions and their variations with space and time. The most recent modeling of this process has been done by *Ciesla and Cuzzi* (2006), where a globally evolving disk is followed for several million years, for a number of parameter choices. The surface density and temperature structure change consistently over the disk lifetime. As the gas disk evolves, the radial distribution of water solids and vapor evolves differently by transport, vaporization, condensation, accretion, and collisional destruction. Prior modeling by *Stepinski and Valageas* (1996) and *Kornet et al.* (2001) first showed these decoupling effects for the solids. The *Ciesla and Cuzzi* (2006) model considers four types of nebular materials: vapor, dust, migrators, and planetesimals. Vapor and dust (solid particles smaller than a meter) are transported by diffusion, which carries them either inward or outward along concentration gradients. Migrators represent the meter-sized rubble in the disk that drifts rapidly inward (~1 AU/century) because of gas drag. Planetesimals are larger objects (>1 km) that are massive enough to be unaffected by the gas in the disk and are immobile in the simulations. *Stevenson and Lunine* (1988) proposed that formation of immobile planetesimals

outside the snow line served as a cold finger that could dry out the entire inner nebula; we now see that this is but one of several possible stages nebula water content can go through — perhaps the final stage.

Figure 7 shows a schematic of the temporal and spatial evolution of the water content of the nebula, and Fig. 8 shows recent numerical model results from *Ciesla and Cuzzi*

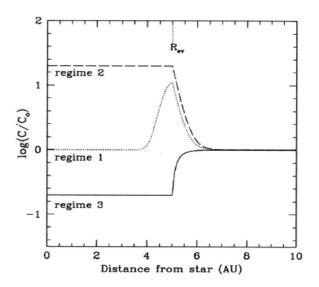

Fig. 7. Schematic of three regimes of the evolution of water vapor inside its evaporation front. An initial intense pulse is short-lived, and propagates throughout the inner solar system on a timescale of 10^5–10^6 yr depending on nebula properties. Later on, after immobile planetesimals provide a cold sink outside the snow line, the inner nebula can be dried out. From simplified analytical models by *Cuzzi and Zahnle* (2004).

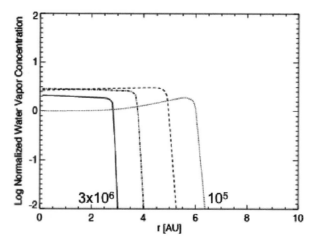

Fig. 8. Results from an evolving nebula model by *Ciesla and Cuzzi* (2006). The snow line moves to smaller radii with time as the nebula cools (water vapor abundance is shown at 10^5, 10^6, 2×10^6, and 3×10^6 yr, respectively). This is typical of a range of models, most of which show water vapor enhancements reaching only a factor of several because of limits on how much icy material can migrate in from the outer solar system. This particular model has not reached regime 3 by the end of the simulation.

(2006). The numerical models show the enhanced water abundance propagating into the inner solar system (regime 2) on a timescale that is in good accord with simple diffusion and advection arguments (*Cuzzi and Zahnle*, 2004).

The enrichment of the inner disk in water vapor decreases at later times (regime 3) because the growth of planetesimals prevents migrators from reaching the inner disk. Because vapor is not being supplied to the inner disk at such a rapid rate, diffusion of water vapor becomes the dominant transport mechanism. Vapor is thus carried outward where it condenses and is locked up on the immobile planetesimals outside the snow line. The concentration of vapor in the inner disk then decreases to less-than-solar values.

In an accretion disk, the coalescence of dust particles may be inhibited as a dust particle grows near the size comparable with the mean free path of nebular gas (e.g., *Sekiya and Takeda*, 2005) because collision velocity with smaller particles begins to rapidly increase. However, even millimeter-sized dust grains possibly migrate selectively enough to cause significant enrichment of water vapor in the inner disk when the disk mass accretion rate becomes lower than ~10^{-8} $M_\odot$ per year (*Kuramoto and Yurimoto*, 2005; *Yurimoto and Kuramoto*, 2004). In such a case, fluctuation of nebular accretion rate as expected from observation (e.g., *Calvet et al.*, 2000) would cause significant temporal variation of water vapor content and redox state in the innermost disk where refractory condensates are actively reprocessed (*Kuramoto and Yurimoto*, 2005).

8.2. Self-shielding in the Inner Disk

Because during the early stages of disk evolution, the central protostar provides a strong source of UV radiation, it was proposed that preferential photodissociation of $C^{17}O$ and $C^{18}O$, and formation of isotopically heavy atomic oxygen, occurred at the innermost edge of the disk (*Clayton*, 2002). This scenario made use of the astrophysical setting of the X-wind model (*Shu et al.*, 1996, 2001), suggesting that isotopically heavy atomic oxygen produced this way was efficiently trapped in oxides, silicates, and H_2O vapor produced by oxidation of metallic elements and H_2 near the X-point, processed near the X-point to form chondrules, and launched by an X-wind to 1–3 AU from the proto-Sun, where together with matrix and refractory inclusions they accreted into chondrite parent asteroids. According to this model, refractory inclusions formed in a gas-depleted region of the disk inside the X-point, and hence largely preserved the initial, ^{16}O-rich isotopic composition of the primordial solids ($\delta^{17,18}O_{MC} = 0‰$). This model postulates that all solids in the inner solar system were thermally processed through an X-region and attempts to unify the O-isotopic photochemistry with the chemistry formation of chondrules and CAIs (*Clayton*, 2002; *Shu et al.*, 1996). This model, however, does not account for the effect of isotopic exchange between the anomalous oxygen atom and CO, a limitation proposed by *Navon and Wasserburg* (1985).

There are several arguments against significant isotope-selective photodissociation of CO at the X-point. First, for the expected high gas temperatures of the innermost solar nebula (~1000–2000 K), H_2O is likely to be present with a partial pressure comparable to CO although the ratio would be depend on the photodissociation rates of the molecules and collision rates of the photodissociated species at the X-point. H_2O is a continuum absorber and will diminish $C^{17}O$ and $C^{18}O$ dissociation in the region of $C^{16}O$ self-shielding (*Lyons and Young*, 2003). Second, at high temperatures, band overlap occurs because of the increase in high-J rotational states, and because v > 0 ground-state vibrational levels are populated. Both of these effects increase the likelihood of line overlap, which diminishes the effectiveness of CO self-shielding (*Navon and Wasserburg*, 1985). Third, the preservation of ^{17}O and ^{18}O enrichment for H_2O near the X-point is problematic. The high temperatures at the innermost part of the disk facilitate reaction of C (either atomic or as C^+) with isotopically enriched H_2O, which erases the O-isotopic fractionation produced during CO self-shielding (*Lyons and Young*, 2003). Finally, the radial penetration depth of solar UV is extremely small, and it would be hard for chondrules to be produced sufficiently close to the proto-Sun for them to benefit from any $^{17,18}O$ so released. In fact, *Muzerolle et al.* (2003) find that vigorously accreting stars vaporize essentially all solids (certainly all chondrules) out to distances of 0.07–0.54 AU, slightly far distant for an X-point-related self-shielding effect to be relevant. These arguments against CO self-shielding at the X-point are only semiquantitative, and await a more complete assessment.

8.3. Self-shielding in the Outer Disk

During the early stages of star evolution, UV radiation from the central protostar and from neighboring massive stars (if star formation occurs in an aggregate of at least ~100 stars) is expected to produce heavily irradiated outer disk surfaces (*Adams et al.*, 2004). This irradiation may result in isotopic self-shielding effects in the surface skin layer of the disk. The isotopically heavy atomic oxygen (^{17}O and ^{18}O) is produced during photodissociation of $C^{17}O$ and $C^{18}O$, and is transferred to H_2O ice by the same process as described for molecular clouds (see section 7).

Surface self-shielding has been proposed in a photochemical model of a turbulent (α-disk) solar nebula (*Lyons and Young*, 2005). In this model FUV radiation was taken to be normal to the disk midplane, and a set of time-dependent chemical continuity equations was solved at a specified heliocentric distance and for a vertical eddy diffusion coefficient assumed to be equal to the turbulent viscosity, $v_t = \alpha c_s H$. The chemical network includes H-C-O species in a set of gas-phase, ion-molecule, and gas-grain reactions, in which all three isotopes of O are explicitly computed. CO shielding functions from *van Dishoeck and Black* (1988) were used to compute the photolysis rates of CO isotopologues. Oxygen atoms liberated during CO photolysis form H_2O enriched in ^{17}O and ^{18}O by gas-grain reactions with H atoms. Figure 9 shows model results at 30 AU (midplane temperature = 51 K) for $\alpha = 10^{-2}$ and for a range of FUV enhancements above the local ISM FUV field. The

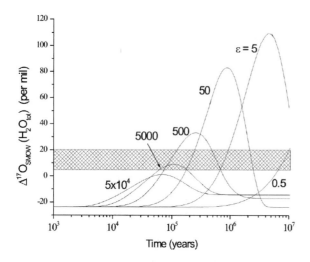

Fig. 9. Mass-independent fractionation signature ($\Delta^{17}O_{SMOW}$) predicted for total nebular water in an α-disk at 30 AU for a range of FUV fluxes. Total nebular water consists of water inherited from the protosolar molecular cloud and water produced as a result of isotope-selective photodissociation of CO in the nebula. Oxygen-isotopic composition of water in the protosolar molecular cloud is assumed to be equal to the bulk isotopic composition of the molecular cloud. Results are shown for $\alpha = 10^{-2}$ and for a FUV enhancement factor $\varepsilon = 0.5$ to 5×10^4 times the local interstellar medium FUV flux. The shaded region indicates the range of minimum $\Delta^{17}O$ inferred for nebular water from primitive meteorites (*Clayton and Mayeda*, 1984; *Young*, 2001). About 10^5 yr are required to reach meteorite water values. From *Lyons and Young* (2005).

shaded region shows the minimum initial nebular water $\Delta^{17}O_{SMOW}$ value necessary to account for the CAI slope-1 mixing line, as inferred from carbonaceous chondrites (*Clayton and Mayeda*, 1984; *Young*, 2001). A FUV flux enhancement factor of $\sim10^3$ is needed for total nebular water (i.e., water produced as a result of CO photodissociation at the disk surface plus water inherited from the parent molecular cloud) at the midplane of the model to reach this range of $\Delta^{17}O$ values within a time $\ll 1$ m.y., the approximate minimum residence time for gas in the nebula. The radial motion of disk materials mentioned in section 8.1 should be incorporated in the future study.

Figure 10 gives the time trajectory of total nebular water on a three-isotope plot for the same FUV enhancement factor. The model slope is ~5–10% greater than the measured CAI mixing line due to self-shielding by $C^{18}O$. The model slope is reduced by $\sim5\%$ when differential shielding by H_2 is accounted for in a particularly important band of CO for ^{17}O and ^{18}O formation [band 31 in *van Dishoeck and Black* (1988)], which yields better agreement with the meteorite data. Mass-dependent effects, which have not been included here, may also affect the model slope for nebular water. A distinct advantage of this model as compared to that of *Clayton* (2002) is that the low temperature kinetically quenches the isotopic exchange process. On the other hand, the high temperatures of *Clayton* (2002) reduce

the magnitude of mass-dependent fractionation to 1–2‰, i.e., into the "noise."

8.4. Stages of Disk Evolution and Oxygen Isotopes

8.4.1. Vigorous early disk stage. The very early (Class 0) stage of nebula evolution, accompanied by vigorous accretion and high inner nebula temperatures, was fairly short (perhaps 10^5 yr), so even if meter-sized particles had begun to form and drift radially, the water plume at the water evaporation front had not yet had time to propagate into the inner solar system. Thus, the mean O-isotopic composition of the gas in the inner part of the disk probably remained similar to the bulk O-isotopic composition of the protosolar molecular cloud ($\delta^{17,18}O_{MC} = 0‰$; Fig. 6c). The earliest transient heating events in the inner solar nebula — formation of refractory inclusions (*Russell et al.*, 2005) — probably caused evaporation, condensation, and melting of the primordial dust in the presence of the nebular gas with $\delta^{17,18}O_{MC} = 0‰$ during which the refractory inclusions would have formed and isotopically equilibrated with the nebular gas (*Itoh and Yurimoto*, 2003; *Krot et al.*, 2002). The similarity between the inferred ^{16}O-enriched composition of the Sun (*Hashizume and Chaussidon*, 2005) and the ^{16}O-rich compositions of the majority refractory in primitive chondrites is consistent with this scenario (*Yurimoto and Kuramoto*, 2004).

8.4.2. Toward quiescent disk stage. During the early stages of disk evolution, water transport is dominated by coagulation of submicrometer-sized dust into millimeter- to meter-sized particles that then rapidly move inward due to

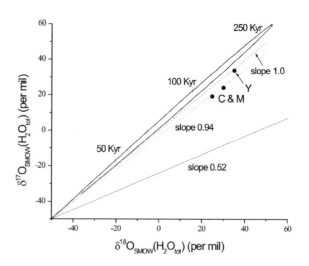

Fig. 10. Model prediction of the time trajectory of O-isotopic ratios of total nebular water for the conditions of Fig. 7 and for a FUV enhancement $\varepsilon = 500$. The points labeled "C & M" and "Y" correspond to the inferred nebular water values from *Clayton and Mayeda* (1984) and *Young* (2001), which lie on lines of slope 0.94 and 1.00, respectively. The model slope ($\delta^{17}O/\delta^{18}O$ ratio) is 5–10% higher than the slope inferred from primitive meteorites. From *Lyons and Young* (2005).

gas drag (*Weidenschilling*, 1977). This results in more icy material being carried inward across the snow line than can be removed by the advection of the disk and diffusion of water vapor (Fig. 6e) (*Ciesla and Cuzzi, 2006; Cuzzi and Zahnle*, 2004). As a result, the inner disk immediately inside the snow line becomes enhanced in its water vapor concentration (Fig. 7). This inward flux continues as the water vapor is redistributed in the inner disk, resulting in a uniform enhancement of water vapor inside of the snow line, with the maximum enhancement reaching ~10× solar (*Ciesla and Cuzzi*, 2006). If these inward migrating particles are enriched in isotopically heavy O, as predicted in the self-shielding models of *Yurimoto and Kuramoto* (2004) and *Lyons and Young* (2005), the inner disk gas becomes increasingly enriched in $^{17,18}O$ over time (Fig. 6d) (*Yurimoto and Kuramoto, 2004*).

The degree of $^{17,18}O$-enrichment of the inner disk gas relative to the bulk O-isotopic composition of the protosolar molecular cloud is determined by the enrichment factor of water vapor and O-isotopic composition of CO and H_2O. If the average $\delta^{17,18}O_{MC}$ values for silicates, H_2O ice, and CO are 0‰, +120‰, and –80‰, and these are preserved in the outer disk (outside the snow line), the enrichment of the inner disk in water vapor by a factor of 3 relative solar abundance will produce gas with $\delta^{17,18}O_{MC}$ of +50‰, which is enough to explain the O-isotopic compositions of chondrules (Fig. 6d) (*Yurimoto and Kuramoto, 2004*). The $^{17,18}O$-rich signatures shown in chondrite components results from high-temperature thermal processing in the $^{17,18}O$-rich disk gas, and of lower-temperature aqueous alteration in the disk or on the planetesimal (the arrow of Fig. 6d). We note, however, that the presence of abundant crystalline silicates in anhydrous interplanetary dust particles, possibly of cometary origin (*Keller and Messenger,* 2005), and in cometary nuclei (*Harker et al.,* 2005) may indicate extensive thermal processing of primordial dust outside the snow line, possibly by shock waves (see chapter by Alexander et al.; *Harker and Desch,* 2002). If this is the case, thermal processing of water ice and its isotopic equilibration with CO gas is expected. As a result, (1) it seems unlikely that the isotopic signature of H_2O and CO of the molecular cloud or of the outer disk will be entirely preserved, and (2) a higher enrichment factor of water vapor will be required to explain the O-isotopic composition of chondrules.

The enrichment of the inner disk in water vapor decreases over time because growth of planetesimals outside the snow line prevents fast-moving, ice-rich rubble from drifting into the inner disk. Because vapor is not being supplied to the inner disk, diffusion carries water vapor outward where it condenses and is locked up in the immobile planetesimals. The concentration of water vapor in the inner disk then decreases over time (*Ciesla and Cuzzi,* 2006; *Stevenson and Lunine,* 1988). It may result in gradual enrichment of the inner disk gas in ^{16}O. It remains unclear whether any record of this fluctuating O-isotopic composition of the inner disk gas is preserved in chondritic components.

8.4.3. Planet accretion stage.

Because the planets accrete during the quiescent disk stage (*Hayashi et al.,* 1985), each planet is expected to record the mean O-isotopic composition corresponding to the accretion regions of the disk. During the quiescent disk stage, the inner disk is probably enriched in ^{17}O and ^{18}O relative to bulk molecular cloud because of self-shielding and migration of H_2O-bearing icy particles across the snow line and thermal processing of dust in this $^{17,18}O$-enriched gas (Fig. 6d; see section 8.4.2). As a result, the inner solar system planets are expected to be rather uniformly $^{17,18}O$-rich, which is generally consistent with observations (Fig. 1a). For example, the difference in O-isotopic composition between Earth and Mars (in terms of $\Delta^{17}O_{SMOW}$) is 0.3‰ (*Clayton,* 1993). Variations in O-isotopic compositions of meteorite parent asteroids are less than 10‰ (Fig. 1). The small differences observed among the meteorite parent asteroids could be explained by variations in time of asteroid accretion, incorporation of different amounts of nonvaporized water ice and solids with different degrees of the solid-gas equilibration, etc., and the smaller variations between planets to the averaging associated with their growth.

Earth and Moon have identical (±0.016‰, $\Delta^{17}O$) O-isotopic compositions (Fig. 1a). There is a general consensus that the Moon formed from hot mantle material thrown into orbit around Earth after the planet was hit by a Mars-sized impactor (e.g., *Cameron,* 1997). Theoretical work suggests that ~80% of the Moon-forming material comes from the impactor (*Canup and Asphaug,* 2001). As a result, the similarity in O-isotope composition between Earth and the impactor has been interpreted as the formation at about the same heliocentric distances (*Wiechert et al.,* 2001). This conclusion, however, is inconsistent with the formation of the terrestrial planets by the stochastic collisions of embryos originating over much of the inner solar system (e.g., *Chambers,* 2001). It has therefore been proposed that the observed O-isotopic similarity between Earth and the Moon is due to O-isotopic exchange between Earth's mantle and the circumterrestrial vapor-melt silicate disk that produced the Moon (*Pahlevan and Stevenson,* 2005). Additional work is needed to test this hypothesis.

In contrast to the relatively uniform $^{17,18}O$-rich compositions of the inner solar system bodies at the quiescent disk stage, O-isotopic heterogeneity between H_2O and CO ices and silicates may be better preserved in the outer disk region (Fig. 6b). According to the standard core-accretion model of giant planet formation (*Pollack et al.,* 1996), runaway accretion of a solid core composed of ice and silicate dust is followed by accretion of disk gas. Because little fractionation is expected for H_2O ice and for silicate dust in the outer disk, the O-isotopic composition of the giant planet cores and icy planets would be $\delta^{17,18}O_{MC}$ ~ +70‰, assuming solar proportions of ice and silicate dust. This value is also expected for Edgeworth-Kuiper-belt objects (KBOs) and comets.

Because the O-isotopic compositions of gas planets change with the ratio between captured gas and ice-dust

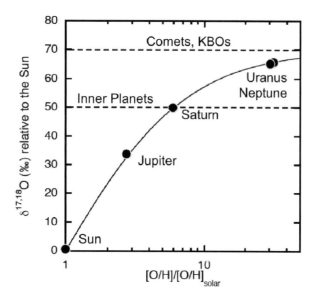

Fig. 11. Estimated O-isotopic compositions of giant planets and KBOs in the solar system.

core, observations of heavy elements in their envelopes would be an indicator of the ratio. The envelopes of gas planets are observed to be enriched in heavy elements relative to the solar composition (*Atreya et al.,* 1999; *Gautier and Owen,* 1989). Based on the observed enrichment of envelopes in heavy elements and the assumption that the O-isotopic compositions of silicate dust, H_2O ice, and CO in the outer disk are similar to those in the protosolar molecular cloud, *Kuramoto and Yurimoto* (2005) predicted a relationship between the O abundances and isotopic compositions of the giant planets (Fig. 11). We note that such a relationship would not expected if the molecular cloud O-isotopic signatures of these components were modified during the molecular cloud collapse, during transient heating events in the outer disk (e.g., *Harker and Desch,* 2002), or by self-shielding in the outer disk (*Lyons and Young,* 2005).

9. FUTURE WORK

The evolution scenario discussed in this paper implies that the O-isotopic anomaly occurs not only in the solar system but it is also ubiquitous in other protoplanetary disks and planetary systems. Therefore, a full understanding of O isotopes in the early solar system will require further astronomical observations, experiments, and modeling. A direct high-precision measurement of CO isotopologues in gas disks would confirm the presence of strong CO fractionation as predicted by self-shielding. Observation of a large O-isotopic fractionation between gas-phase H_2O and CO in a disk would demonstrate isotopic partitioning of O between these two key nebular species. ALMA may be capable of making these exceedingly difficult measurements. Key experiments needed are measurement of mass-independent fractionation in high-temperature silicate reactions (both

gas-phase and surface) to verify the occurrence of mass-independent fractionation. Measurement of the $^{17}O/^{18}O$ ratio associated with isotope-selective photodissociation of CO are required at a geochemical (i.e., mass spectrometer) level of precision. Finally, modeling of O isotopes in molecular clouds, collapsing cores, and protostellar nebulae, over temperatures ranging from 10 K to several 1000 K, will be needed to fully synthesize the meteorite and astronomical data.

Acknowledgments. This work was supported by Monkasho grants (H. Yurimoto, P.I.) and NASA grants NAG5-10610 (A. N. Krot, P.I.), NAG5-11591 (K. Keil, P. I.), and NASA's Planetary Geology and Geophysics and Origins of Solar Systems programs (J. N. Cuzzi, P.I.). J.R.L., A.N.K., and E.R.D.S. acknowledge support from the NASA Astrobiology Institute (NNA04CC08A, K. Meech, P.I.).

REFERENCES

Adams F. C., Hollenbach D., Laughlin G., and Gorti U. (2004) *Astrophys. J., 611,* 360–379.

Aikawa Y., Ohashi N., and Herbst E. (2003) *Astrophys. J., 593,* 906–924.

Aleon J., Krot A. N., and McKeegan K. D. (2002) *Meteoritics & Planet. Sci., 37,* 1729–1755.

Aleon J., Krot A. N., McKeegan K. D., MacPherson G. J., and Ulyanov A. A. (2005a) *Meteoritics & Planet. Sci., 40,* 1043–1058.

Aleon J., Robert F., Duprat J., and Derenne S. (2005b) *Nature, 437,* 385–388.

Amelin Y., Krot A. N., Hutcheon I. D., and Ulyanov A. A. (2002) *Science, 297,* 1678–1683.

Ando M., Nagata T., Sato S., Mizuno N., Mizuno A., et al. (2002) *Astrophys. J., 574,* 187–197.

Atreya S. K., Wong M. H., Owen T. C., Mahaffy P. R., Niemann H. B., et al. (1999) *Planet. Space Sci., 47,* 1243–1262.

Baker J., Bizzarro M., Wittig N., Connelly J., and Haack H. (2005) *Nature, 436,* 1127–1131.

Bally J. and Langer W. D. (1982) *Astrophys. J., 255,* 143–148.

Bergin E. A., Melnick G. J., Stauffer J. R., Ashby M. L. N., Chin G., et al. (2000) *Astrophys. J., 539,* L129–L132.

Bergin E. A., Ciardi D. R., Lada C. J., Alves J., and Lada E. A. (2001) *Astrophys. J., 557,* 209–225.

Bizzarro M., Baker J. A., and Haack H. (2004) *Nature, 431,* 275–278.

Bland P. A., Rost D., Vicenzi E. P., Stadermann F. J., Floss C., et al. (2005) In *Lunar Planet. Sci. XXXVI,* Abstract #1841.

Brearley A. J. (1996) In *Chondrules and the Protoplanetary Disk* (R. H. Hewins et al., eds.), pp. 137–151. Cambridge Univ., Cambridge.

Calvet N., Hartmann L., and Strom S. E. (2000) In *Protostars and Planets IV* (V. Mannings et al., eds.), pp. 377–399. Univ. of Arizona, Tucson.

Cameron A. G. W. (1997) *Icarus, 126,* 126–137.

Canup R. M. and Asphaug E. (2001) *Nature, 412,* 708–712.

Chambers J. (2001) *Icarus, 152,* 205–224.

Chick K. M. and Cassen P. M. (1997) *Astrophys. J., 477,* 398–409.

Choi B.-G., McKeegan K. D., Krot A. N., and Wasson J. T. (1998) *Nature, 392,* 577–579.

Chu Y.-H. and Watson W. D. (1983) *Astrophys. J., 267,* 151–155.

Ciesla F. J. and Cuzzi J. N. (2006) *Icarus, 181,* 178–204.

Clayton D. D. and Nittler L. R. (2004) *Ann. Rev. Astron. Astrophys., 42,* 39–78.

Clayton R. N. (1993) *Ann. Rev. Earth Planet. Sci., 21,* 115–149.

Clayton R. N. (2002) *Nature, 415,* 860–861.

Clayton R. N. and Mayeda T. K. (1984) *Earth Planet. Sci. Lett., 67,* 151–161.

Clayton R. N. and Mayeda T. K. (1996) *Geochim. Cosmochim. Acta, 60,* 1999–2018.

Clayton R. N. and Mayeda T. K. (1999) *Geochim. Cosmochim. Acta, 63,* 2089–2104.

Clayton R. N., Grossman L., and Mayeda T. K. (1973) *Science, 182,* 485–488.

Cuzzi J. N., Davis A., and Dobrovolskis A. (2003) *Icarus, 166,* 385–402.

Cuzzi J. N. and Weidenschilling S. J. (2005) In *Meteorites and the Early Solar System II* (D. S. Lauretta and H. Y. McSween Jr., eds.), pp. 353–381. Univ. of Arizona, Tucson.

Cuzzi J. N. and Zahnle K. J. (2004) *Astrophys. J., 614,* 490–496.

Cuzzi J. N., Petaev M., Scott E. R. D., and Ciesla F. (2005) In *Chondrites and the Protoplanetary Disk* (A. N. Krot et al., eds.), pp. 732–773. ASP Conference Series 341, San Francisco.

Desch S. J. and Connolly H. C. (2002) *Meteoritics & Planet. Sci., 37,* 183–207.

Fahey A. J., Goswami J. N., McKeegan K. D., and Zinner E. K. (1987) *Astrophys. J., 231,* L91–L95.

Farquhar J., Bao H., and Thiemens M. H. (2000) *Science, 289,* 756–758.

Gao Y. Q., Chen W.-C., and Marcus R. A. (2002) *J. Chem. Phys., 117,* 1536–1543.

Gautier D. and Owen T. (1989) In *Origin and Evolution of Planetary and Satellite Atmospheres* (S. K. Atreya et al., eds.), pp. 487–512. Univ. of Arizona, Tucson.

Glassgold A. E., Huggins P. J., and Langer W. D. (1985) *Astrophys. J., 290,* 615–626.

Goswami J. N., Marhas K. K., and Sahijpal S. (2001) *Astrophys. J., 549,* 1151–1159.

Greenberg J. M. (1998) *Astron. Astrophys., 330,* 375–380.

Greenwood R. C., Franchi I. A., Jambon A., and Buchanan P. C. (2005) *Nature, 435,* 916–918.

Haisch K. E., Greene T. P., Barsony M., and Ressler M. (2001) *Astrophys. J., 553,* L153–L156.

Harker D. E. and Desch S. J. (2002) *Astrophys. J., 565,* L109–L112.

Harker D. E., Woodward C. E., and Wooden D. H. (2005) *Science, 310,* 278–280.

Hartmann L. (2005) In *Chondrites and the Protoplanetary Disk* (A. N. Krot et al., eds.), pp. 131–144. ASP Conference Series 341, San Francisco.

Hashizume K. and Chaussidon M. (2005) *Nature, 434,* 619–622.

Hayashi C., Nakazawa K., and Nakagawa Y. (1985) In *Protostars and Planets II* (D. C. Black and M. S. Matthews, eds.), pp. 1100–1153. Univ. of Arizona, Tucson.

Hester J. J. and Desch S. J. (2005) In *Chondrites and the Protoplanetary Disk* (A. N. Krot et al., eds.), pp. 107–131. ASP Conference Series 341, San Francisco.

Hewins R. H., Connolly H. C., Lofgren G. E., and Libourel G. (2005) In *Chondrites and the Protoplanetary Disk* (A. N. Krot et al., eds.), pp. 286–317. ASP Conference Series 341, San Francisco.

Ireland T. R., Holden P., and Norman M. D. (2005) In *Lunar Planet. Sci. XXXVI,* Abstract #1572.

Itoh S. and Yurimoto H. (2003) *Nature, 423,* 728–731.

Itoh S., Kojima H., and Yurimoto H. (2004) *Geochim. Cosmochim. Acta, 68,* 183–194.

Jones R. H., Grossman G. N., and Rubin A. E. (2005) In *Chondrites and the Protoplanetary Disk* (A. N. Krot et al., eds.), pp. 251–286. ASP Conference Series 341, San Francisco.

Keller L. P. and Messenger S. (2005) In *Chondrites and the Protoplanetary Disk* (A. N. Krot et al., eds.), pp. 657–668. ASP Conference Series 341, San Francisco.

Kita N. T., Huss G. R., Tachibana S., Amelin Y., Nyquist L. E., and Hutcheon I. D. (2005) In *Chondrites and the Protoplanetary Disk* (A. N. Krot et al., eds.), pp. 558–588. ASP Conference Series 341, San Francisco.

Kitamura Y. and Shimizu M. (1983) *Moon and Planets, 29,* 199–202.

Kobayashi S., Imai H., and Yurimoto H. (2003) *Geochem. J., 37,* 663–669.

Kornet K., Stepinski T. F., and Różyczka M. (2001) *Astron. Astrophys., 378,* 180–191.

Krot A. N., McKeegan K. D., Leshin L. A., MacPherson G. J., and Scott E. R. D. (2002) *Science, 295,* 1051–1054.

Krot A. N., Amelin Y., Cassen P., and Meibom A. (2005a) *Nature, 436,* 989–992.

Krot A. N., Hutcheon I. D., Yurimoto H., Cuzzi J. N., McKeegan K. D., et al. (2005b) *Astrophys. J., 622,* 1333–1342.

Krot A. N., Yurimoto H., Hutcheon I. D., and MacPherson G. J. (2005c) *Nature, 434,* 998–1001.

Krot A. N., Libourel G., and Chaussidon M. (2006) *Geochim. Cosmochim. Acta, 70,* 767–769.

Kunihiro T., Nagashima K., and Yurimoto H. (2005) *Geochim. Cosmochim. Acta, 69,* 763–773.

Kuramoto K. and Yurimoto H. (2005) In *Chondrites and the Protoplanetary Disk* (A. N. Krot et al., eds.), pp. 181–192. ASP Conference Series 341, San Francisco.

Lada C. J., Lada E. A., Clemens D. P., and Bally J. (1994) *Astrophys. J., 429,* 694–709.

Lee T. (1978) *Astrophys. J., 224,* 217–226.

Lee T. and Shen J. J. (2001) *Meteoritics & Planet. Sci., 36,* A111.

Lyons J. R. and Young E. D. (2003) In *Lunar Planet. Sci. XXXIV,* Abstract #1981.

Lyons J. R. and Young E. D. (2005) *Nature, 435,* 317–320.

MacPherson G. J., Simon S. B., Davis A. M., Grossman L., and Krot A. N. (2005) In *Chondrites and the Protoplanetary Disk* (A. N. Krot et al., eds.), pp. 225–250. ASP Conference Series 341, San Francisco.

Marcus R. A. (2004) *J. Chem. Phys., 121,* 8201–8211.

Marechal P., Viala Y. P., and Benayoun J. J. (1997a) *Astron. Astrophys., 324,* 221–236.

Marechal P., Viala Y. P., and Pagani L. (1997b) *Astron. Astrophys., 328,* 617–627.

Maruyama S., Yurimoto H., and Sueno S. (1999) *Earth Planet. Sci. Lett., 169,* 165–171.

McKeegan K. D. and Leshin L. A. (2001) In *Stable Isotope Geochemistry* (J. W. Valley and D. R. Cole, eds.), pp. 279–318. Reviews in Mineralogy and Geochemistry, Vol. 43, Mineralogical Society of America.

Muzerolle J., Calvet N., Hartmann L., and D'Alessio P. (2003) *Astrophys. J., 597,* L149–L152.

Nagashima K., Krot A. N., and Yurimoto H. (2004) *Nature, 428,* 921–924.

Navon O. and Wasserburg G. J. (1985) *Earth Planet. Sci. Lett., 73,* 1–16.

Nittler L. R. (2003) *Earth Planet. Sci. Lett., 209,* 259–273.

Nittler L. R., Alexander C. M. O'D., Gao X., Walker R. M., and Zinner E. (1997) *Astrophys. J., 483,* 475–495.

Nittler L. R., Alexander C. M. O'D., Wang J., and Gao X. (1998) *Nature, 393,* 222.

Pahlevan K. and Stevenson D. J. (2005) In *Lunar Planet. Sci. XXXVI,* Abstract #2382.

Pollack J. B., Hubickyj O., Bodenheimer P., Lissauer J. J., Podolak M., and Greenzweig Y. (1996) *Icarus, 124,* 62–85.

Rai V. K., Jackson T. L., and Thiemens M. H. (2005) *Science, 309,* 1062–1065.

Russell S. S., Krot A. N., Huss G. R., Keil K., Itoh S., et al. (2005) In *Chondrites and the Protoplanetary Disk* (A. N. Krot et al., eds.), pp. 317–350. ASP Conference Series 341, San Francisco.

Scott E. R. D. and Krot A. N. (2001) *Meteoritics & Planet. Sci., 36,* 1307–1319.

Scott E. R. D. and Krot A. N. (2003) In *Treatise on Geochemistry, Vol. 1: Meteorites, Comets, and Planets* (A. M. Davis, ed.), pp. 143–200. Elsevier, Oxford.

Sekiya M. and Takeda H. (2005) *Icarus, 176,* 220–223.

Sheffer Y., Lambert D. L., and Federman S. R. (2002) *Astrophys. J., 574,* L171–L174.

Shu F. H., Shang H., and Lee T. (1996) *Science, 271,* 1545–1552.

Shu F. H., Shang H., Gounelle M., Glassgold A. E., and Lee T. (2001) *Astrophys. J., 548,* 1029–1050.

Solomon P. M. and Klemperer W. (1972) *Astrophys. J., 178,* 389–422.

Stepinski T. F. and Valageas P. (1996) *Astron. Astrophys., 309,* 301–312.

Stevenson D. J. and Lunine J. I. (1988) *Icarus, 75,* 146–155.

Thiemens M. H. (1999) *Science, 283,* 341–345.

Thiemens M. H. (2006) *Ann. Rev. Earth Planet Sci., 34,* 217–2620.

Thiemens M. H. and Heidenreich J. E. (1983) *Science, 219,* 1073–1075.

Thiemens M. H., Savarino J., Farquhar J., and Bao H. (2001) *Accounts Chem. Res., 34,* 645–652.

van Dishoeck E. F. and Black J. H. (1988) *Astrophys. J., 334,* 771–802.

van Dishoeck E. F., Blake G. A., Draine B. T., and Lunine J. I. (1993) In *Protostars and Planets III* (E. Levy and J. I. Lunine, eds.), pp. 163–241. Univ. of Arizona, Tucson.

Warin S., Benayoun J. J., and Viala Y. P. (1996) *Astron. Astrophys., 308,* 535–564.

Weidenschilling S. J. (1977) *Mon. Not. R. Astron. Soc., 180,* 57–70.

White G. J. and Sandell G. (1995) *Astron. Astrophys., 299,* 179–192.

Whittet D. C. B., Gerakines P. A., Hough J. H., and Shenoy S. S. (2001) *Astrophys. J., 547,* 872–884.

Wiechert U., Halliday A. N., Lee D.-C., Snyder G. A., Taylor L. A., and Rumble D. (2001) *Science, 294,* 345–348.

Wood J. A. (1998) *Astrophys. J., 503,* L101–L104.

Wood J. A. (2004) *Geochim. Cosmochim. Acta, 68,* 4007–4021.

Woolum D. S. and Cassen P. (1999) *Meteoritics & Planet. Sci., 34,* 897–907.

Wouterloot J. G. A., Brand J., and Henkel C. (2005) *Astron. Astrophys., 430,* 549–560.

Young E. D. (2001) *Philos. Tran. R. Soc. Lond., A359,* 2095–2110.

Young E. D., Simon J. I., Galy A., Russell S. S., Tonui E., and Lovera O. (2005) *Science, 308,* 223–227.

Yu Y., Hewins R. H., Clayton R. N., and Mayeda T. K. (1995) *Geochim. Cosmochim. Acta, 59,* 2095–2104.

Yurimoto H. and Kuramoto K. (2004) *Science, 305,* 1763–1766.

Water in the Small Bodies of the Solar System

David Jewitt and Lysa Chizmadia
University of Hawaii

Robert Grimm
Southwest Research Institute

Dina Prialnik
Tel Aviv University

Water is important for its obvious role as the enabler of life but more generally as the most abundant volatile molecule in the solar system, containing about half the condensible mass in solids. In its solid phase, water strongly influences the opacity of the protoplanetary disk and may determine how fast, and even whether, gas giant planets form. Water ice is found or suspected in a wide range of small-body populations, including the giant planet Trojan librators, comets, Centaurs, Kuiper belt objects, and asteroids in the outer belt. In addition to ice, there is mineralogical evidence for the past presence of liquid water in certain meteorites and, by inference, in their parent main-belt asteroids. The survival and evolution of liquid and solid water in small bodies is discussed.

Water is the driving force of all nature.
—Leonardo da Vinci

1. INTRODUCTION

Leonardo's claim is only a slight exaggeration. After hydrogen and helium, oxygen is the third most cosmically abundant element (H/O ~ 1200 by number). Helium is chemically inert, leaving hydrogen atoms free to combine to make H_2, or with oxygen to form H_2O. Water is the second most abundant molecule (after H_2) in those astrophysical environments where the temperature is below the thermal dissociation limit (~2000 K). These environments include almost the entire protoplanetary disk of the Sun. Therefore, it should be no surprise that water played an important role over a wide range of heliocentric distances in the solar system from the region of the terrestrial planets out to the Kuiper belt and beyond.

At low temperatures, water is thermodynamically stable as a solid; it condenses as frost or ice. Ice grains have a strong influence on the mean radiative opacity (*Pollack et al.*, 1994) and so can change the temperature structure in the protoplanetary disk. In turn, the disk viscosity is related to the temperature of the gas. Hence, the freeze-out and distribution of icy grains together have a surprisingly strong impact on the transport of energy, mass, and angular momentum and so influence the overall structure and evolution of the protoplanetary disk (*Ciesla and Cuzzi,* 2006).

The "snow line" (actually, it is a surface) divides the outer, cold, ice-rich region of the protoplanetary disk from the inner, steamy hot zone. Outward gas motions across the snow line result in the condensation of water into ice grains, which can collide, stick, and grow until they dynamically decouple from the flow and are left behind. In this way, the snow line defines the inner edge of a cold trap in which the density of solids may have been sufficiently enhanced as to help speed the growth of planetesimals and, ultimately, of the cores of the giant planets (*Stevenson and Lunine,* 1988). Aerodynamic drag causes 100-m-scale ice-rich planetesimals from the outer disk to spiral inward on very short timescales (perhaps 1 AU per century). Some are swept up by bodies undergoing accelerated growth just outside the snow line. Others cross the snow line to quickly sublimate, enhancing the local water vapor abundance. Coupled with ongoing dissociation from ultraviolet photons, these freeze out and sublimation processes lead to isotopic fractionation anomalies in water (*Lyons and Young,* 2005) that have already been observed in the meteorite record (*Krot et al.,* 2005). In the protoplanetary disk, opacity due to grains inhibited radiative cooling and raised the midplane kinetic temperature. Growth and migration of solids in the disk would have caused the opacity to change, so moving the snow line. For a fraction of a million years, it may have pushed in to the orbit of Mars or even closer (*Sasselov and Lecar,* 2000; *Ciesla and Cuzzi,* 2006) (cf. Fig. 1), meaning that asteroids in the main belt, between Mars and Jupiter, could have incorporated water ice upon formation. Indeed, small samples of certain asteroids, available to us in the form of meteorites, contain hydrated minerals that probably formed when buried ice melted and chemically reacted with surrounding refractory materials (*Kerridge et al.,* 1979; *McSween,* 1979). Short-lived radionuclides were the most probable heat source in the early solar system. Cosmochemical evidence for [26]Al (half-life 0.71 m.y.) is firmly established

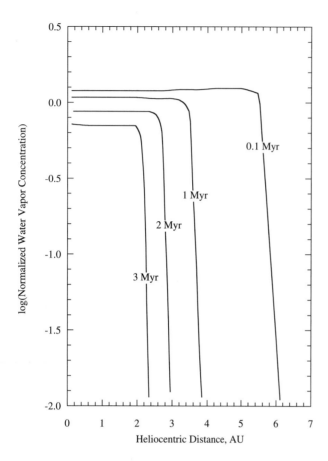

Fig. 1. Results from a model of the temporal evolution of the water vapor content in the protoplanetary disk as a function of the heliocentric distance. The curves are labeled with the time since disk formation. The steep drop on each curve is caused by condensation of the vapor into ice grains, and marks the moving snowline. These results are for one simplistic model of the protoplanetary disk and should not be taken literally. Their value is in showing the migration of the snow line in response to grain growth, gas drag, and time-dependent opacity variations in the disk. Figure replotted from *Ciesla and Cuzzi* (2006).

(*Lee et al.*, 1976; see *MacPherson et al.*, 1995, for a review). Additionally, ^{60}Fe (half-life ~1.5 m.y.) may have provided supplemental heating (*Mostefaoui et al.*, 2004).

Further from the Sun the ice trapped in solid bodies may have never melted. This is mainly because the lower disk surface densities in the outer regions lead to long growth times. Small bodies in the outer regions, especially the Kuiper belt objects and their progeny (the Centaurs and the nuclei of Jupiter-family comets), probably escaped melting because they did not form quickly enough to capture a full complement of the primary heating agent, ^{26}Al. Melting is nevertheless expected in the larger Kuiper belt objects, where the surface area to volume ratio is small and the longer-lived radioactive elements can play a role (see section 5).

As this preamble shows, the study of water in small solar system bodies involves perspectives that are astronomical,

geochemical, and physical. As is common in science, the study of one broad subject is breaking into many smaller subfields that will eventually become mutually incomprehensible. Here, we aim to provide an overview that will discuss the most significant issues concerning water in the solar system's small bodies in a language accessible to astronomers and planetary scientists, the main readers of this volume. We refer the reader to earlier reviews of the astronomical aspects by *Rivkin et al.* (2002), of the thermal models by *McSween et al.* (2002), and of the meteorite record by *Keil* (2000).

2.   EVIDENCE FROM THE METEORITES

2.1.   Evidence for Extraterrestrial Liquid Water

Meteorites provide samples of rock from asteroids in the main belt between Mars and Jupiter. Some contain evidence for the past presence of liquid water in the form of hydrated phases. Here, "hydrated" means that the mineral, or phase, contains structurally bound OH or H_2O, with chemical bonds between the OH/H_2O molecule and the other elements in the crystal. For example, in serpentine ($Mg_3Si_2O_5(OH)_4$) the OH molecule shares electrons with the SiO_4 tetrahedra and is an integral part of the mineral. Hydrated meteorites are those that contain hydrated phases and/or minerals that likely formed in the presence of an aqueous fluid. Not all minerals that formed as a consequence of liquid water contain OH or H_2O molecules. Hydrated meteorites include most groups of carbonaceous chondrites and some of the type 3 ordinary chondrites.

Evidence that the aqueous alteration of these meteorites occurred extraterrestrially includes:

1. Some of the meteorites were observed to fall and immediately recovered; not enough time elapsed for terrestrial alteration reactions to occur.

2. Radioisotopes in carbonate assemblages indicate formation ~5–10 m.y. after the solidification of the chondrules. Earth at this time was too hot to sustain liquid water.

3. The O-isotopic composition of different meteorite groups are distinct from each other and from Earth (*Clayton et al.*, 1976; *Clayton and Mayeda*, 1984; *Clayton*, 1993, 1999). Materials that form on Earth fall on the terrestrial fractionation line. The hydrated phases in chondrites do not fall on this line.

4. The Fe^{2+}/Fe^{3+} ratio in the secondary minerals is generally significantly higher than in secondary minerals that form on the surface of Earth (*Zolensky et al.*, 1993). The surface of Earth is much more oxidizing than the environment in which these reactions took place.

Several styles of aqueous alteration have been proposed (Fig. 2). Gas-phase hydration reactions and the condensation of carbonate were initially thought, on theoretical grounds, to be kinetically inhibited (*Prinn and Fegley*, 1987). However, minerals including calcite and dolomite have been found in the dust ejected from old stars (e.g., *Kemper et al.*, 2002) where gas phase reactions are now

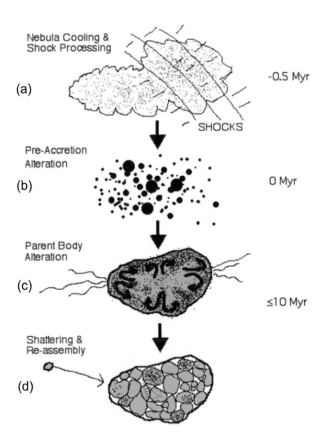

Fig. 2. Cartoon showing possible stages in the evolution of water in a growing small body. **(a)** Nebula cooling and shock processing occur as the protoplanetary nebula cools by radiation into space but is briefly shocked to higher temperatures and densities. **(b)** Preaccretion alteration occurs in small precursor bodies that later mix with others (perhaps from far away in the disk) to form the final parent body. **(c)** Parent-body alteration involves the flow of water through the formed asteroid or comet. **(d)** Later, collisional shattering and reassembly under gravity introduces macroporosity and chaotically jumbles the hydrated materials with others. Rough times are noted relative to the solidification of the CAIs.

thought possible (*Toppani et al.,* 2005). Similar reactions, as well as hydration reactions, may have occurred in the cooling solar nebula, perhaps aided by shock heating and compression (*Ciesla et al.,* 2003). Evidence consistent with gas-phase alteration comes from the fine-grained rims of chondrules. Alternatively, or in addition, alteration by liquid water could have occurred in the parent asteroids following the melting of buried ice by heat liberated by radioactive decay. Evidence for aqueous alteration involving liquid water comes from geochemical and textural evidence in the carbonaceous chondrites. For example, the systematic redistribution of elemental constituents on scales of millimeters [e.g., Mg and Ca into the matrix and Fe into the chondrules (*McSween,* 1979)] is best explained by liquid water. The consistent degree and state of alteration of chondrules with similar compositions in the same chondrites is also difficult to understand in the context of gas-phase pro-

cessing but would be naturally produced by aqueous alteration (e.g., *Hanowski and Brearley,* 2001). Furthermore, mineral veins and Fe-rich aureoles indicate soaking in a fluid for an extended period of time (*Zolensky et al.,* 1989, 1998; *Browning et al.,* 1996; *Hanowski and Brearley,* 2001). A reasonable view is that gas-phase alteration probably occurred, but postaccretion alteration by liquid water played a larger role, often overwhelming any products of gas-phase alteration (*Brearley,* 2003). The distribution of hydrated materials within an asteroid is likely to be further complicated, during and after the hydration phase, by brecciation, impact shattering, and rearrangement (Fig. 2).

2.2. Carbonaceous Chondrites

Several observations indicate contact with water. Most directly, the bulk rock contains H_2O when analyzed (e.g., carbonaceous chondrites can contain up to 12 wt% H_2O). Secondary minerals containing structurally bound OH/H_2O are also present, as are minerals that generally form in the presence of liquid H_2O. The oxygen in hydrated minerals is ^{16}O poor compared to the primary minerals, such as olivine, indicating that the water/ice must have formed at a different time or place from the primary phases. Oxygen isotopes are distinct between different bodies and groups of meteorites. Since we know that some of these bodies formed at different distances from the Sun (e.g., Earth, Mars, 4 Vesta), the O isotopes imply that different regions of the solar nebula had different O-isotopic compositions.

Water-related secondary minerals are mainly clays with related serpentine (Fig. 3), carbonates (Fig. 4), phosphates, sulfates, sulfides, and oxides. Carbonates, phosphates, sulfates, sulfides, and oxides formed under oxidizing conditions, unlike the H_2-rich, reducing solar nebula, and generally precipitate from liquid water. Some of these are found in cross-cutting relationships with other phases, indicating that they must have formed later (Fig. 5). We briefly discuss important secondary minerals that imply the action of liquid water in meteorite parent bodies.

Clay minerals and serpentines contain structurally bound OH (Fig. 3). The aqueous alteration of CM chondrites is thought to have occurred at ~20°C (*Clayton et al.,* 1976; *Clayton and Mayeda,* 1984; *Clayton,* 1993, 1999; *Zolensky et al.,* 1993).

Carbonates, such as calcite ($CaCO_3$) and dolomite ($CaMgC_2O_6$), are often present in the most aqueously altered carbonaceous chondrites, the CMs and the CIs. It is unlikely that the partial pressure of CO_2 gas was high enough in the nebula for carbonate to have formed. Origin in solution is more plausible (*Prinn and Fegley,* 1987). The ages, based on the decay of ^{53}Mn (half-life 3.7 m.y.), indicate that most carbonate assemblages in CI, CM, and some other chondrites formed well after the formation of chondrules and almost certainly after the gaseous protoplanetary nebula had dissipated and after the parent-body asteroids were assembled (*Hutcheon et al.,* 1998; *Brearley and Hutcheon,* 2002). This is all consistent with *in situ* aqueous alteration following the melting of bulk ice by radioactive decay heating.

2.3. Ordinary Chondrites

Although they are less oxidized and contain fewer volatiles than carbonaceous chondrites, ordinary chondrites also

show several pieces of evidence indicating contact with an aqueous fluid.

First, phyllosilicates have been observed among the matrix materials in the unmetamorphosed (unheated) meteorites Semarkona and Bishunpur, indicating interaction with water (*Hutchison et al.,* 1987; *Alexander et al.,* 1989; *Grossman et al.,* 2000). Calcite, probably produced by aqueous alteration, has also been identified in Semarkona (*Hutchison et al.,* 1987).

Second, the outer portions of some chondrules show a "bleached" texture. Interaction with an aqueous fluid has resulted in the replacement of chondrule glass by phyllosilicates, while soluble elements such as Ca and Na have been removed (*Grossman et al.,* 2000; *Grossman and Brearley,* 2003).

Third, halite (NaCl) and sylvite (KCl) have been discovered in the ordinary chondrites Zag and Monahans (*Zolensky et al.,* 1998, 1999; *Bridges et al.,* 2004). The freezing-

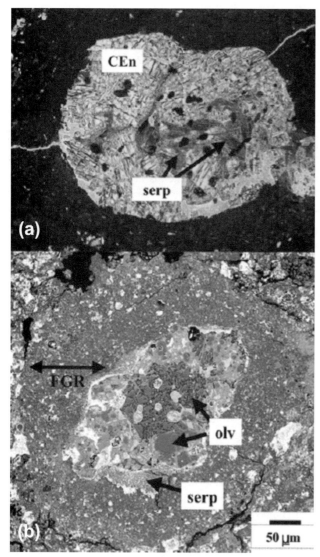

Fig. 3. Images illustrating the replacement of chondrule phases by secondary serpentine in CM2 chondrites. **(a)** Optical micrograph of a chondrule in Murchison, in which the primary chondrule glass has been replaced by serpentine (serp) probably during contact with liquid water. Image is approximately 2 mm across (from A. Brearley, personal communication, 2003). CEn is the high-temperature (~800° to 1000°C) phase clinoenstatite. Serpentine is a lower-temperature (<400°C) phase and it is unlikely that these two phases formed during the same event. Serpentine is formed by secondary, low-temperature alteration involving water. **(b)** Backscattered electron (BSE) image of a chondrule in ALH 81002, in which the primary olivine (olv) has been partially replaced by serpentine, again as a likely consequence of contact with liquid water. FGR is the fine-grained rim (from Chizmadia and Brearley, in preparation).

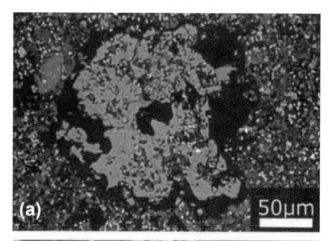

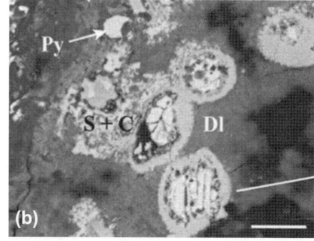

Fig. 4. Backscattered electron images of carbonate assemblages in carbonaceous chondrites. **(a)** Aragonite in the carbonaceous chondrite Tagish Lake (from *Nakamura et al.,* 2003) and **(b)** a complex intergrowth of secondary phases in the CI-like chondrite, Y 86029 (from *Tonui et al.,* 2003). In **(b)**, S is sulfide, C is calcite, Dl is dolomite, and Py is pyrrhotite (Fe_xS_{1-x}). The textural complexity seen in the figure suggests formation *in situ* from liquid water.

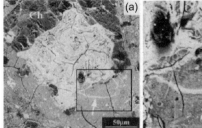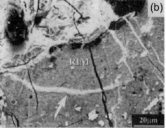

Fig. 5. Backscattered electron images of Fe-rich veins (arrows) protruding from an Fe-rich area of a chondrule (Ch) in the CM2 chondrite Murchison; **(b)** shows an enlargement of the boxed region in **(a)**. The Fe-rich area of the chondrule was originally a metal grain that has been altered by contact with liquid water. The veins cut through the fine-grained rim and therefore must have formed after it, probably in conjunction with the aqueous alteration of the metal grain. From *Hanowski and Brearley* (2001).

melting temperatures of the fluid inclusions in these salts are consistent with their having formed from a $NaCl-H_2O$ brine at <50°. These are metamorphosed meteorites and the halite was formed after metamorphism. Moreover, $^{40}Ar-^{39}Ar$ and $^{129}I-^{129}Xe$ dates show that the halite crystals formed 4.0–4.2 G.y. ago (*Burgess et al.,* 2000).

Fourth, sulfur maps of ordinary chondrites and CO3 chondrites show that, with increasing degree of alteration, sulfur mobilizes into chondrules from matrix (*Grossman and Rubin,* 1999).

2.4. Interplanetary Dust Particles

Interplanetary dust particles (IDPs) are ejected from active comets and created by collisional grinding of asteroids (*Rietmeijer,* 1998). Some include a relic presolar dust component. Many IDPs are chondritic in bulk composition and share many mineralogic similarities with chondritic meteorites (*Rietmeijer,* 1998). In addition, the dust measured from Comet Halley by the Giotto spacecraft was chondritic in composition (*Rietmeijer,* 1998). Some chondritic IDPs contain such secondary minerals as clays, serpentines, salts, carbonates, sulfides, and oxides (*Brownlee,* 1978; *Rietmeijer,* 1991), again consistent with contact with liquid water.

3. ASTRONOMICAL EVIDENCE

Astronomically, water ice and hydrated minerals are detected primarily through their near-infrared spectra. Ice has characteristic absorption bands due to overtones and combination frequencies in the thermally vibrating H_2O molecule. These occur in the near-infrared at 1.5 μm, 2.0 μm, and 3.0 μm wavelength, with the 3-μm band being the deepest but, observationally, most difficult to detect. Contamination of the water ice by absorbing matter (e.g., organics) can also reduce the depths of the characteristic bands, hiding water from spectroscopic detection. For example, laboratory experiments in which just a few percent (by weight) of absorbing matter are added to otherwise pure ice show

that the band depths can be reduced to near invisibility in all but the highest-quality spectra (*Clark and Lucey,* 1984).

Hydrated minerals show a characteristic band near 3 μm (due to structural OH and H_2O) but the shorter wavelength bands are normally absent or very weak. An alternative explanation in which the 3-μm band is due to OH radicals formed by implanted solar wind protons may be compatible with some observations, but appears unable to account for the very deep features seen in phyllosilicates and many hydrated asteroids (*Starukhina,* 2001). Because of the practical difficulties of observing at 3 μm, there is interest in using correlated optical features as proxies. In particular, many clays show a band near 0.7 μm that has been attributed to an $Fe^{2+} \rightarrow Fe^{3+}$ charge transfer transition and that, empirically, is correlated with the harder-to-observe 3-μm band (*Vilas,* 1994). The fraction of the asteroids showing this feature varies with semimajor axis, being larger beyond ~3 AU where the C-type asteroids (low albedo, neutral to slightly red) are dominant, but falling to zero in the D-type (dark and redder) Trojans beyond (Fig. 6).

3.1. Asteroids

Radiation equilibrium temperatures on the surfaces of main-belt asteroids are too high for the long-term stability of water ice. Therefore, little direct evidence exists for surface ice in the main belt (however, see discussion of Ceres below). On the other hand, spectral evidence for hydrated phases is very strong, particularly in the outer-belt (semimajor axis >2.5 AU) populations. Roughly half the C-type asteroids, for instance, show spectral evidence for hydrated silicates in their 3-μm spectra (*Jones et al.,* 1990). Conversely, hydration features are generally not observed in the D-type and P-type asteroids that dominate the jovian Trojan population at ~5 AU. One interpretation is that the C types were heated sufficiently for hydration reactions to occur but ice in the D and P types never reached the melting temperature, no hydration reactions took place, and so no hydrated materials are detected. Bulk ice could still be present in these more distant objects (*Jones et al.,* 1990; *Jewitt and Luu,* 1990).

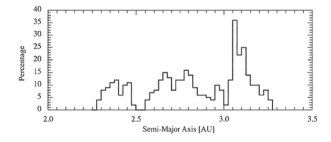

Fig. 6. Fraction of asteroids (of all spectral types) showing the 0.7-μm band attributed to the $Fe^{2+} \rightarrow Fe^{3+}$ transition in hydrated material. Given that the 0.7-μm band is a proxy for OH, the figure may show the distribution of water across the main belt. Replotted from *Carvano et al.* (2003).

3.1.1. Ceres. Main-belt asteroid 1 Ceres has a mean diameter of 952 ± 3.4 km, a density of ~2077 ± 36 kg m^{-3}, and follows a nearly circular orbit (eccentricity e = 0.08) of semimajor axis 2.77 AU (*Thomas et al.,* 2005). The surface reflectivity is flat at visible wavelengths with an albedo near 0.11. Observations in the ultraviolet have shown tentative evidence for off-limb hydroxyl (OH) emission (*A'Hearn and Feldman,* 1992). Hydroxyl radicals could be produced by the photodissociation of water, and A'Hearn and Feldman suggested that their observation might be explained by the sublimation of water ice from polar regions of Ceres. Spectroscopic evidence for surface water ice is given by an absorption band at 3.06 ± 0.02 μm and about 10% deep (Fig. 7) (see *Lebofsky et al.,* 1981; *King et al.,* 1992; *Vernazza et al.,* 2005). Observations in this wavelength region are very difficult, and interpretations of the band vary based on details of the measured band shape. *King et al.* (1992) suggest that broadening of the band relative to pure ice could be due to NH_4 in the smectite clay mineral saponite. *Vernazza et al.* (2005) suggest instead a composite model incorporating crystalline water ice at 150 K and ion-irradiated asphaltite (a complex organic material).

Water ice on the surface of Ceres must have been recently emplaced because the sublimation lifetime at 2.8 AU is short (the sublimation rate is roughly 1 m yr^{-1}, if the ice is dirty and dark). On the other hand, although short-lived on the surface, *subsurface* ice on Ceres can be stabilized indefinitely at high latitudes by as little as ~1 m of overlying refractory mantle (*Fanale and Salvail,* 1989). We may surmise that water migrated to the surface from deeper down, perhaps in response to internal heating that has driven out

volatiles (*Grimm and McSween,* 1989), perhaps creating fumarolic activity at the surface. Alternatively, ice could be excavated from beneath the mantle by impacts or deposited on the surface by impacting comets (as has been suggested to explain suspected polar ice deposits on the Moon and Mercury).

Thermal models (*McCord and Sotin,* 2005) show that, because of Ceres' considerable size, even heating by long-lived nuclei (K, Th, U) could lead to melting of deeply buried ice, followed by widespread hydration reactions with silicates in globally distributed hydrothermal systems. *Thomas et al.* (2005) report shape measurements that show that Ceres is rotationally relaxed with (weak) evidence for a nonuniform internal structure. Their models are consistent with water ice mantles containing ~15% to 25% of the total mass: Perhaps this is the ultimate source of the off-limb hydroxyl and the near-infrared absorption band.

3.1.2. Main-belt comets. Other evidence for ice in the main belt comes from observations of the newly identified class of main-belt comets (MBCs) (*Hsieh and Jewitt,* 2006). Three examples are currently known; asteroid (7968) 1996 N2 (also known as 133P/Elst-Pizarro or "EP") is shown in Fig. 8, while 118401 and P/2005 U1 (Read) are reported in *Hsieh and Jewitt* (2006). It displays a narrow, linear trail of dust morphologically reminiscent of the large-particle dust trails detected in the thermal infrared in the orbits of comets (*Sykes and Walker,* 1992). At first, the dust production was attributed to a recent impact and the likely parent of

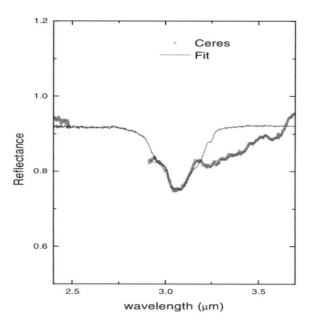

Fig. 7. Spectrum of asteroid 1 Ceres showing the 3.06-μm absorption band attributed to crystalline water ice. The thin line is a model fit to the data. An additional absorber is required to match the data at wavelengths longer than 3.15 μm. From *Vernazza et al.* (2005).

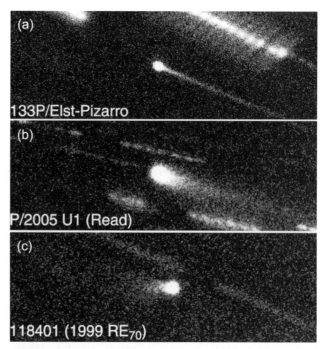

Fig. 8. Composite R-band images of three main-belt comets: **(a)** 133P/Elst-Pizarro, **(b)** P/2005 U1 (Read), and **(c)** 118401 (1999 RE70). These composites were produced by shifting and adding sets of short exposure images, causing fixed stars and galaxies to be streaked. Orbits of these three objects are identified in Fig. 9. Image from *Hsieh and Jewitt* (2006).

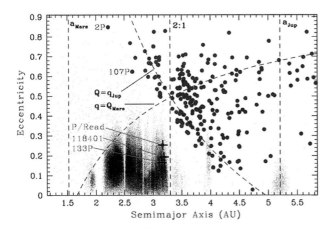

Fig. 9. Semimajor axis vs. orbital eccentricity showing main-belt asteroids (small dots), active Jupiter-family comets (large filled circles), and the three established main-belt comets (plus signs, labeled). Vertical dashed lines at 1.5 AU and 5.2 AU mark the semimajor axes of the orbits of Mars and Jupiter, respectively, while the 2:1 mean-motion resonance with Jupiter (which practically defines the outer edge of the main belt) is shown at 3.3 AU. Curved dashed lines show the loci of orbits that have perihelia inside Mars' orbit or aphelia outside Jupiter's. Objects above these lines are Mars and/or Jupiter crossing. The MBCs are far below both lines with the main-belt asteroids and distinct from the Jupiter-family comets. Figure from *Hsieh and Jewitt* (2006).

the impactor was identified as asteroid 427 Galene (*Toth*, 2000). The reappearance of the trail in 2002 makes this explanation implausible and the dust is most likely ejected by sublimating ice (*Hsieh et al.*, 2004). Temporal evolution of the trail shows that the 2×3-km nucleus of EP ejects dust particles ~10 μm in size at speeds ~1.5 m s^{-1} over periods of months.

Exposed ice at the ~3 AU heliocentric distance of the MBCs (Fig. 9) has a short lifetime to sublimation, meaning that the ice must have been recently emplaced in order for the activity to still be observed. *Hsieh et al.* (2004) considered two basic ideas concerning how these main-belt objects came to sublimate ice. The first possibility is that MBCs formed with the Jupiter-family comets in the Kuiper belt and were transported by unknown processes to orbits that coincidentally appear asteroidal. However, published models of pure dynamical transport from the Kuiper belt fail to produce asteroid-like orbits (*Fernández et al.*, 2002). It is possible that model simplifications, including the neglect of gravitational deflections by the low-mass terrestrial planets and of rocket forces due to asymmetric mass loss, might account for this failure. However, recent simulations including these extra forces do not produce MBCs with inclinations as low as that of 133P (*Fernández et al.*, 2002; *Levison et al.*, 2006).

The second idea is that MBCs formed in the main belt, incorporating a substantial ice component. The observation that the orbital elements of the MBCs are close to those of the collisionally produced Themis family adds support to

the conjecture that the MBCs are fragments from larger, parent bodies that were shattered by impact. This cannot explain why the MBCs are outgassing now, however, since the age of the Themis family, estimated from the dispersion of orbital elements of its members, is near 2 G.y. (*Marzari et al.*, 1995) and surface ice should have long since sublimated away. Another trigger for outgassing must be invoked. One possibility is that recent impacts exposed buried ice, leading to sublimation that is modulated by seasonal variations in the insolation. Preliminary calculations suggest that crater sizes of ~20 m are needed to supply the measured mass loss. Such craters would be produced by impacting boulders only a few meters across. The idea that the MBCs contain primordial ice is interesting in the context of the evidence for aqueous activity from the meteorites, discussed in section 2. What is most surprising is that the ice has survived at temperatures that are high (~150 K) by the standards of the much more distant Kuiper belt and Oort cloud. The MBCs appear to constitute an ice repository within the solar system that is distinct from the better known Kuiper belt and Oort cloud reservoirs, and to sample a region of the protoplanetary disk much closer to the Sun.

3.2. Jovian Trojans

The jovian Trojans co-orbit with the planet at 5.2 AU but librate around the L4 and L5 Lagrangian points, leading and trailing the planet by ±60° of longitude. The leading and following swarms together contain roughly 10^5 objects larger than kilometer scale (cf. *Jewitt et al.*, 2004). The largest Trojan is 624 Hektor (~370 × 195 km).

The jovian Trojan surfaces are located near the edge of the stability zone for water ice in the modern solar system. They are outside the expected locations of the snow line through most or all of the early evolution of the planetary system. If they formed *in situ*, or at larger distances from the Sun, it is quite likely that the Trojans should be ice-rich bodies much like the nuclei of comets and the Kuiper belt objects.

Paradoxically, no direct evidence for water on the Trojans has been reported. The reflection spectra in the 0.4-μm to 2.5-μm wavelength range are generally featureless and slightly redder than the Sun (*Luu et al.*, 1994; *Dumas et al.*, 1998; *Cruikshank et al.*, 2001; *Emery and Brown*, 2004). The geometric albedos are low, with a mean value of 0.041 ± 0.002 (*Fernández et al.*, 2003). In these properties, the Trojans closely resemble the nuclei of comets. The data are consistent with the hypothesis that the Trojans, like the cometary nuclei, have ice-rich interiors hidden by a protective cap or "mantle" of refractory matter, probably left behind as a sublimation lag deposit. The low density of 617 Patroclus (800^{+200}_{-100} kg m^{-3}) also suggests a porous, ice-rich composition (*Marchis et al.*, 2006).

Might water ice be detectable on the Trojans? Clean (high-albedo) water ice at 5 AU is cold enough to be stable against sublimation while dirty (absorbing, warmer) ice will sublimate away on cosmically short (<10^4 yr) timescales.

Therefore, surface ice on Trojan asteroids that has been recently excavated by impacts might remain spectroscopically detectable until surface gardening (in which micrometeorite bombardment overturns the surface layers and mixes them with dirt) lowers the albedo, whereupon the ice would promptly sublimate. Spectroscopic surveys targeted to the smaller Trojans (on which collisions are more likely to provide large fractional resurfacing) might one day detect the ice.

The origin of the Trojans is presently unknown, with two distinct source locations under discussion. The Trojans could have formed locally and been trapped in the 1:1 resonance as the planet grew (*Chiang and Lithwick*, 2005). In this case, the compositions of the jovian Trojans would reflect the ~100 to ~150 K temperature at 5 AU heliocentric distance and it would be permissible to think of the Trojans as surviving samples of the type material that formed Jupiter's high molecular weight core. Another possibility is that the Trojans formed at some remote location, possibly the Kuiper belt, and were then scattered and captured during a chaotic phase in the solar system (*Morbidelli et al.*, 2005). In this case, the Trojans would be captured Kuiper belt objects, presumably with initial ice contents equal to those of the comets. In either case, the volatile-rich interiors of the Trojans would be protected from insolation by a lag deposit consisting of refractory matter. It is interesting that the Trojans and the cometary nuclei are so similar in the color-albedo plane. This fact is not proof that Trojans are comets, but it is consistent with this inference. On the other hand, the Trojans are less red (*Jewitt*, 2002) and darker (*Fernández et al.*, 2003) than the KBOs. If they were captured from the Kuiper belt then unspecified evolutionary processes must have acted to both darken their surfaces and remove very red material.

One meteorite known as Tagish Lake shows spectral similarities to D-type (and P-type) asteroids (Fig. 10) (*Hiroi et al.*, 2001; *Hiroi and Hasegawa*, 2003) and so is a candidate for the type of material that might make up the Trojans. Similarities include a low albedo (a few percent at visual wavelengths) and reddish color, with a featureless optical to near-infrared reflection spectrum. Tagish Lake is an anomalous carbonaceous chondrite that was observed to fall and immediately recovered from frozen ground in Canada in 1999. It shows abundant evidence for aqueous alteration in the form of Mg-Fe carbonates and phyllosilicates (*Nakamura et al.*, 2003).

3.3. Cometary Nuclei

Water is well known in the nuclei of comets. It sublimates inside a critical distance near 5 AU or 6 AU to produce the distinctive coma (gravitationally unbound atmosphere) and tails that observationally define comets. The thermal response of water ice to the varying insolation on a comet moving in an eccentric orbit about the Sun formed the basis of *Whipple*'s (1950) classic comet model. Whipple thought that water might carry about 50% of the nucleus mass. Recent estimates based on long-wavelength (infrared and submillimeter) radiation show that water is less important, perhaps carrying only 20–30% of the mass in typical nuclei (*Sykes et al.*, 1986).

The water ice in comets appears to have condensed at temperatures in the range 25–50 K. Several lines of evidence lead to this temperature range, which (in the present epoch) is characteristic of heliocentric distances beyond Neptune. First, comets are sources of CO, with production rates relative to water in the ~1% to ~20% range. The CO molecule is highly volatile and difficult to trap at temperatures much above 50 K. Laboratory experiments in which CO is trapped in amorphous ice suggest accretion temperatures near 25 K (*Notesco and Bar-Nun*, 2005). The nuclear spin temperature, which is thought to be set when the water molecule forms, is near 30 K in those comets for which it has been accurately measured. Measurements of the D/H ratio in H_2O and HCN in comet C/Hale-Bopp are compatible with ion-molecule reactions at 30 ± 10 K, probably in the collapsing cloud that formed the solar system (*Meier et al.*, 1998b). Water ice formed at temperatures below ~110 K is thermodynamically stable in the amorphous (disordered) state. The ice in two comets [C/Hale-Bopp (*Davies et al.*, 1997) and C/2002 T7 (LINEAR) (*Kawakita et al.*, 2004)] has indeed been inferred to be amorphous, based on the absence of the 1.65-μm feature of crystalline ice. However, since water ice is rarely detected in comets, we do not know from observations whether the ice in comets is generally amorphous. Models assuming abundant amorphous ice have been constructed to take advantage of the exothermic transition to the crystalline phase in order to power cometary activity at distances slightly beyond the critical distance for sublimation in radiative equilibrium with the Sun (*Prialnik*, 1992; *Prialnik et al.*, 2005).

Most recently, thermal infrared observations of the ejecta produced by the impact of NASA's Deep Impact spacecraft into the nucleus of P/Tempel 1 have shown tentative evi-

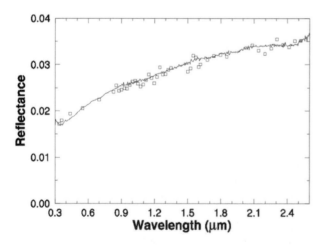

Fig. 10. A comparison of the reflectance spectra of Tagish Lake (line) with the D-type asteroid (368) Haidea (points). Figure courtesy of *Hiroi et al.* (2001).

dence for hydrated minerals (*Lisse et al.*, 2006). If confirmed (for instance by samples returned in NASA's Stardust mission) this detection would seem to require that the hydrated minerals be formed in the gas phase before incorporation into the nucleus, since it is hard to imagine how this small cometary nucleus could have been hot enough to sustain liquid water and yet retain substantial quantities of volatile ices such as CO.

3.4. Kuiper Belt Objects and Centaurs

As with the Trojan asteroids and cometary nuclei, the optical reflection spectra of most KBOs and Centaurs rise linearly with wavelength and are spectrally featureless. The proportion showing identifiable absorption bands is, however, higher and this presumably reflects the lower rates of sublimation at the greater distances of these objects. The main problem in the study of KBOs and Centaurs is a practical one: Most are very faint, technically challenging astronomical targets.

Figure 11 shows the 1.5-μm and 2.0-μm bands of water ice in the reflection spectrum of ~1200 km diameter KBO 50000 Quaoar (*Jewitt and Luu*, 2004). A third band at 1.65 μm is diagnostic of the presence of crystalline (not amorphous) water ice. This result is curious, given that Quaoar is at 43 AU and the surface temperature is near 50 K or less: Ice would be indefinitely stable in the amorphous form at this temperature. The inevitable conclusion is that the ice exposed on the surface of Quaoar, although now cold, has been heated above the critical temperature (~110 K) for transformation to the crystalline form. Furthermore, the ice on Quaoar (and all other airless bodies) is exposed to continual bombardment by energetic photons from the Sun, and particles from the solar wind and from interstellar space (the cosmic rays). These energetic photons and particles are capable of breaking the bonds in crystalline ice, converting it back to a disordered, amorphous state on a timescale that may be as short as 10^6–10^7 yr (*Mastrapa et al.*, 2005). Thus, we may also conclude that the ice surface on Quaoar is young. How could this be?

The bulk of the ice inside Quaoar could have been heated above the amorphous to crystalline phase transition temperature by internal processes. Quaoar is large and radiogenic heat is capable of raising the internal temperatures substantially above the low values on the surface and in the nebula at 43 AU (see section 5). This bulk crystalline material must be replenished at the surface in order to avoid reamorphization. One mechanism is micrometeorite "gardening," in which continual bombardment overturns the surface layers, dredging up fresh material and burying the instantaneous surface layer. Given that near-infrared photons probe to depths of only a few millimeters, the gardening rate only needs to be large compared to a few millimeters per 10^6 or 10^7 yr. We know essentially nothing about the micrometeorite flux in the Kuiper belt, but it seems reasonable that such a low rate could be sustained. Another mechanism is simple eruption of crystalline ice onto the surface via the action of cryovolcanism. In the context of the Kuiper belt, cryovolcanism need not involve the eruption of liquid water, but could be as simple as the leakage of a volatile gas, perhaps CO mobilized by heating at great depth, with the entrainment of crystalline ice particles into the flow. It is worth noting that Quaoar is not alone in showing crystalline water ice: Several satellites of Uranus as well as Pluto's satellite Charon also have crystalline ice on the surfaces. Are they all cryovolcanically active? Intuition says this is unlikely, but recent observations of Saturn's satellite Enceladus show that it is outgassing at a few hundred kilograms per second, populating the E ring with dust particles (*Hansen et al.*, 2006). Enceladus has a source of energy (tidal heating) that is unavailable to KBOs, but it is also much smaller (500 km diameter, compared to Quaoar's 1200 km) and faster to cool.

4. ISOTOPIC EVIDENCE

4.1. Hydrogen

Most models indicate that Earth formed inside the snow line and so formed dry (see Fig. 1). However, Earth is not dry (the mass of the oceans is ~3×10^{-4} $M_{\oplus}$), raising the question of how it acquired its water. Likely sources include ice-rich bodies accreted outside the snow line and transported inward to impact Earth after its formation, providing a late-stage volatile veneer (*Owen and Bar-Nun*, 1995). These could be bona fide comets from the outer regions or asteroids from the main belt (the MBCs of section 3.1). A potentially important constraint on the source population can be set from measurements of the deuterium/hydrogen

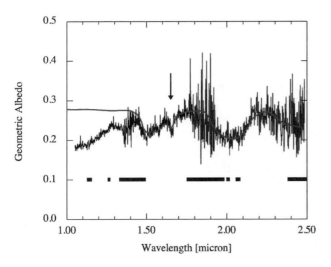

Fig. 11. Reflection spectrum of KBO 50000 Quaoar in the near-infrared, showing broad bands due to water at 1.5 μm and 2.0 μm, and a narrower feature at 1.65 μm that is specifically due to crystalline water ice (arrow). The smooth line is a water ice spectrum that is overplotted (but not fitted) to the astronomical data. Horizontal bars mark wavelengths at which Earth's atmosphere absorbs. From *Jewitt and Luu* (2004).

(D/H) ratio. The standard mean ocean water (SMOW) value is D/H = 1.6×10^{-4}, while measurements in comets give a value about twice as large, namely D/H ~3.3×10^{-4} (for descriptions, see *Meier et al., 1998a; Meier and Owen, 1999*). It is tempting to conclude on this basis that the comets cannot be the dominant suppliers of Earth's water but this conclusion is probably premature. First, the formal significance of the difference between D/H in comets and in SMOW is marginal (and also difficult to judge, given that systematic errors rival statistical ones in the remote determination of D/H). Second, the measured comets are Halley-family or long-period objects, dynamically unlike the Jupiter-family comets (JFCs) that may have dominated the volatile delivery to Earth. The JFCs and long-period comets may originate from different regions within the protoplanetary disk and at different temperatures (although this is itself far from certain). If the measured comets have D/H ratios that are not representative of the dominant impactors, then objections based on isotopic differences with SMOW would disappear. Unfortunately, most comets are too faint for astronomical determinations of D/H and *in situ* measurements from spacecraft will be needed to acquire a large, dynamically diverse sample.

Other possible sources include ice-rich "asteroids" from the outer parts of the main belt. This source is appealing because some measurements of D/H in hydrated meteorites are close to the SMOW value. Furthermore, some dynamical models purport to show that the main belt, in an early clearing phase, could have supplied a sufficient mass of water to Earth to explain the oceans (*Morbidelli et al., 2000*). Unfortunately, the mechanism by which the main belt was cleared and the timing of this event relative to the formation of Earth remain uncertain. Also, noble gas abundance patterns on Earth are unlike those in the meteorites and it is not obvious how this difference can be explained if these gases were delivered, with water, from the asteroid belt. Noble gas abundances in comets are unmeasured. Future measurements of D/H in water from main-belt comets (section 3.1.2) will be of great interest in this regard.

4.2. Oxygen

Measurements of the $^{16}O/^{17}O$ and $^{16}O/^{18}O$ ratios in primitive meteorites show mass-independent variations that are distinct from the mass-dependent fractionation produced from chemical reactions. Mass-independent variations may be caused by the mixing of ^{16}O-rich with ^{16}O-poor material in the asteroid-belt source region where the meteorite parents formed (*Krot et al., 2005*). Optical depth effects in the protoplanetary nebula, coupled with the freeze-out of water, could be responsible (*Krot et al., 2005*). Gas-phase $C^{16}O$ would be optically thick to ultraviolet photons in much of the young nebula and thus immune to photodissociation. The more rare isotopomers $C^{17}O$ and $C^{18}O$ could be optically thin, especially above the disk midplane and at large heliocentric distances. Atoms of ^{17}O and ^{18}O produced

by dissociation beyond the snow line would be picked up by water molecules and soon trapped as ice. In this way, ice-bearing solids would become progressively depleted in ^{16}O relative to the gas (which is ^{16}O enhanced), by an amount that depends upon the disk gas mass, the heliocentric distance, and the degree of mixing upward from the midplane (*Lyons and Young, 2005*). Inward drift of icy planetesimals across the snow line leading to their sublimation would return ^{16}O-poor water to the gas phase. One feature of this model is that it requires a flux of UV photons comparable to that expected from an O- or B-type star located within ~1 pc of the Sun (*Lyons and Young, 2005*). If this is the correct interpretation of the mass-independent O-isotopic variations, then it provides evidence that the Sun was part of a stellar cluster (since O- and B-type stars are rare, and on average found at much larger distances).

5. THERMAL EVOLUTION OF SMALL BODIES

5.1. Analytical Considerations

Thermal evolution in the presence of ice is an exciting topic that promises to throw much light on the interrelations between comets and asteroids (*Jewitt, 2005*) and on their water and volatile histories (*Prialnik et al., 2005*). Minimal requirements for the stability of liquid water include temperatures *and* pressures above the triple-point values 273 K and 600 N m^{-2}, respectively. Central hydrostatic pressure in a spherical body of density ρ and radius r is $P_c \sim 2\pi/3\ G\ \rho^2\ r^2$. With $\rho = 10^3$ kg m^{-3}, $P_c \geq 600$ N m^{-2} for r $\geq$ 2 km. Smaller bodies can never sustain the triple-point pressure unless they are strong (which, in the context of prevailing formation scenarios seems unlikely). In larger bodies, whether liquid water can be stable at depth depends mainly on the existence of heat sources adequate to raise the temperatures above the triple-point value. Gravitational binding energy is too small to melt or vaporize ice in any but the largest bodies considered in this chapter (see Fig. 12). Instead, in the asteroid belt the dominant heat source was likely ^{26}Al (*Lee et al., 1976; MacPherson et al., 1995*). The influence of ^{26}Al depends on the size of the body, on the mechanisms of heat loss, and on the time elapsed between the explosion of the ^{26}Al-producing supernova and its incorporation into the growing body (*Ghosh et al., 2003*).

A lower limit to the rate of loss of heat from a body of radius r is set by thermal conduction. The conduction cooling time is

$$\tau_c \sim \frac{r^2}{\kappa} \qquad (1)$$

where κ is the thermal diffusivity given by $\kappa = k/(\rho c_p)$, k = thermal conductivity, ρ = density, and c_p = specific heat

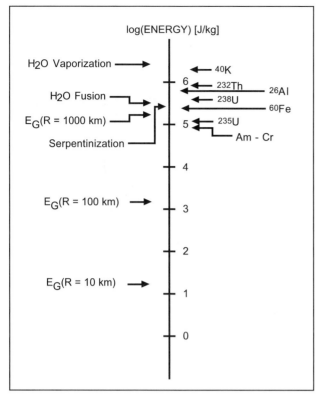

Fig. 12. Energy per unit mass for processes relevant to the thermal evolution of small bodies. Isotope symbols show the time-integrated energy production from the given element divided by the total mass of the small body, computed assuming cosmic initial abundances of the elements. Energies for the amorphous-crystalline (Am-Cr) phase transition and for H_2O fusion and vaporization are per unit mass of ice, while that for serpentinization is per unit mass of serpentine (as in section 4). E_G is the gravitational binding energy per unit mass, given for a homogeneous sphere by $E_G = 3 GM/(5 R)$ and computed assuming a density of 1000 kg m^{-3}. Another heat source due to electromagnetic induction in the magnetic field of the young sun has been discussed in the context of asteroids (cf. *McSween et al.*, 2002), but is of uncertain magnitude, declines precipitously with heliocentric distance, and is not plotted here.

capacity. Common nonporous solids have $\kappa \sim 10^{-6}$ m^2 s^{-1}, while porosity may reduce this by a factor of ~10. The heat released by ^{26}Al can only be effective when $\tau_c \gg \tau$, which corresponds to $r_c \gg (\kappa\tau)^{1/2}$, where τ is the time constant for the decay of ^{26}Al. Substituting $\kappa = 10^{-6}$ m^2 s^{-1} and $\tau \sim$ 1 m.y., we obtain $r_c \gg 5$ km for the critical size above which strong thermal effects are expected. Bodies much larger than a few kilometers in size, then, are candidates for having central hydrostatic pressures high enough, and cooling times long enough, to nurture liquid water.

A crude calculation of the effect of radioactive heating may be obtained by considering a body of radius r and initial temperature T_0, represented by a unique (average) internal temperature T(t) as a function of time t. Assuming a radioactive source of initial mass fraction X_0 and decay en-

ergy per unit mass of rock H, the energy balance equation for the body is

$$c_p \frac{dT(t)}{dt} = X_0 H \tau^{-1} e^{-t/\tau} - \frac{3kT(t)}{\rho r^2} \qquad (2)$$

where the second term on the RHS of equation (2) is the cooling flux (*Merk and Prialnik*, 2003). Assuming constant, average values c_p and k, this equation can be solved analytically to obtain the time t_1 required to reach the melting temperature $T(t_1) = T_m = 273$ K, and the amount of radioactive material still available $X_1 = X_0 e^{-t_1/\tau}$,

$$T(t) = T_0 e^{-at} + \frac{b}{a\tau - 1}[e^{-t/\tau} - e^{-at}] \qquad (3)$$

where

$$a(r) = \frac{3k}{\rho c_p r^2} = \frac{3\kappa}{r^2}, \quad b = \frac{X_0 H}{c_p} \qquad (4)$$

When the melting temperature is reached, the heat released is absorbed by the melting ice and the temperature does not rise any longer. Thus, energy balance for times later than t_1 reads

$$H_m \dot{Y}_m = X_1 H \tau^{-1} e^{-t/\tau} - \frac{3kT_m}{\rho r^2} \qquad (5)$$

where H_m = latent heat of melting, $Y_m(t)$ is the volume fraction of liquid water, and t is set to zero at the onset of melting. The solution

$$Y_m(t) = H_m^{-1}\left[X_1 H(1 - e^{-t/\tau}) - \frac{3kT_m}{\rho r^2} t\right] \qquad (6)$$

rises to a maximum and declines to zero. The timespan for which $Y_m > 0$, that is, for which liquid water is present, can be easily estimated. (Equation (6) gives a strong upper limit to the duration of the liquid phase, since energy transport by convection in liquid water will dominate the cooling flux and suppress temperatures much faster than conduction.) Representative results for the time required to reach the melting temperature, the maximal volume fraction of water attained, and the duration of the liquid phase are given in Fig. 13, considering the effect of ^{26}Al and of the most potent long-lived radioactive species, 40 K. The values of constants assumed are $\kappa = 10^{-6}$ m^2 s^{-1} (as above) and $c_p = 10^3$ J kg^{-1}; the radioactive abundances assumed correspond to a 1:1 mass ratio of ice to dust. A time delay of 2 m.y. was assumed in the case of ^{26}Al to account for the imperfect trapping of this short-lived isotope.

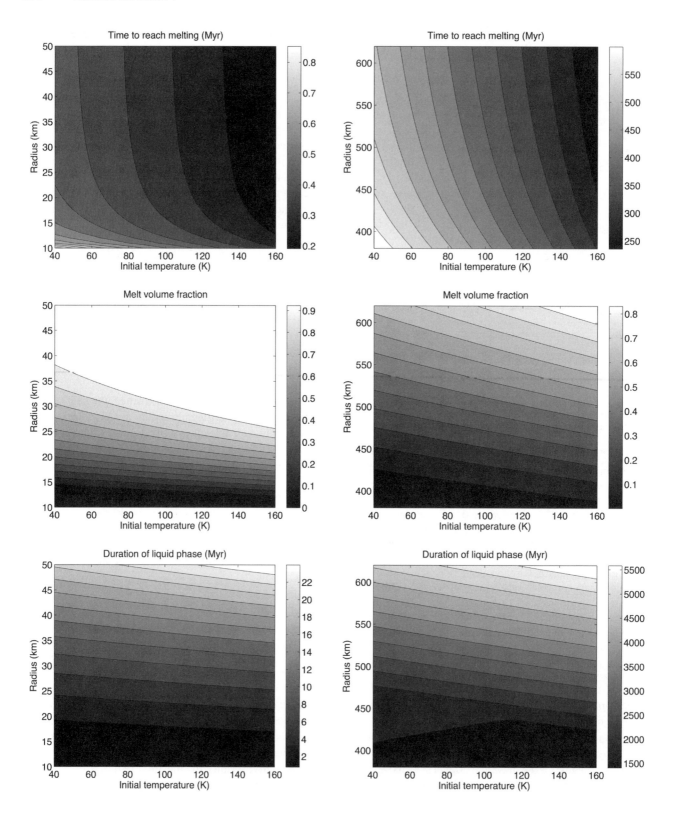

Fig. 13. Models for the thermal evolution of small bodies heated by radioactive decay as a function of their radius and initial temperature assuming a 1:1 ice:rock mass ratio. Left and right columns show models heated by the decay of [26]Al and [40]K, respectively (see text). At the top is shown the time to reach the water ice melting point, in the middle is the fractional melt volume while at the bottom is the duration of the liquid phase.

These simple models show that, while small bodies — a few kilometers in size — are completely unaffected by 40 K, bodies of a few hundred kilometers radius may have retained liquid water in their interiors up to the present. Significant amounts of liquid water may be expected in smaller bodies, heated by ^{26}Al, only for radii above ~20 km; the water is expected to refreeze on a timescale of ~20 m.y.

5.2. Numerical Modeling: Icy Bodies (Comets)

Obviously, all these quantities are only rough estimates based on a crude analytical treatment. Somewhat egregiously, we have allowed cooling only by conduction when, in fact, solid-state convection may dominate the cooling rate long before the melting temperature is reached (*Czechowski and Leliwa-Kopystynski*, 2002). Countering this, water may add new sources of heat. First, if the accreted ice is initially amorphous (unlikely in the hot inner solar system but possible in the cold Kuiper belt), then the energy of the amorphous to crystalline phase change is available. This is ~9×10^4 J kg^{-1}, sufficient to drive a wave of crystallization through the icy portions of any body. Second, if the liquid state is reached, powerful serpentinization reactions may come to dominate the heat production, at least locally (section 5.3).

On the other hand, numerical models that treat thermal evolution in detail necessarily have a large number of free parameters. Thus, models usually focus on the effect of one parameter or characteristic and the results obtained may only be taken to indicate trends of behavior. For example, *Haruyama et al.* (1993) investigate the potential effect of a very small thermal conductivity on the crystallization of amorphous ice due to radioactive heating in a H$_2$O ice and dust body. Not surprisingly, they find that for very low conductivities, a runaway increase in the internal temperature occurs and most of the ice crystallizes, while for sufficiently high conductivities, the initially amorphous ice may be almost entirely preserved. A similar study by *Yabushita* (1993) investigates the possibility of melting and formation of organic chemicals due to radioactive heating and finds that a very low diffusivity allows for melting in the interior of relatively large icy bodies (r $\gtrsim$ 200 km). However, these studies neglect heat transport by vapor flow through the porous nucleus and the possibility of internal sublimation or recondensation. Vapor flow is taken into account by *Prialnik and Podolak* (1995), who consider radioactive heating in porous nuclei, allowing for trapped CO gas in the amorphous ice, which is released upon crystallization. A large variety of results are obtained by varying the basic parameters: porosity, radius, thermal conductivity, and ^{26}Al content. Porosity and heat advection by the flowing gas are shown to affect the thermal evolution significantly. Thus, only for a very low porosity and very low conductivity is it possible to obtain an extended liquid core for a considerable period of time. However, this study does not consider ices of volatiles besides H$_2$O in the initial composi-

tion. These are considered by *Choi et al.* (2002), who show that ices of highly volatile species have a considerable effect on the early crystallization pattern of H$_2$O ice induced by radioactive heating, both by absorbing and by releasing latent heat. Thus, they find that even large bodies (r ~ 50 km) may retain the ice in amorphous form, despite radioactive heating (even including some ^{26}Al). However, the effect of advection by volatile flow is maximized in their study by assuming a very high permeability.

Despite the nonuniqueness inherent in presently available models, it is clear that heating by ^{26}Al is potentially so strong that even silicates could have been melted in small asteroids if they formed early enough to trap this isotope (e.g., *Herndon and Herndon*, 1977; *Grimm and McSween*, 1993). Conversely, delayed aggregation into large bodies would minimize the thermal impact of ^{26}Al: After 10 half-lives (~7 m.y.), the available heat from ^{26}Al is reduced by a factor of 2×10^4 and can be ignored. In bodies that formed quickly enough relative to the ^{26}Al decay, radioactive heating may drive melting in the central region of a relatively small icy body; the question remains of the consequences of a liquid core, considering that water is much denser than ice. So far, this problem has not been considered in detailed numerical simulations — largely due to the fact that structural parameters of cometary material are poorly known — and not much progress has been made beyond the early analytical work of *Wallis* (1980), who treated a vapor-droplet mixture within an icy shell.

The formation timescale, t$_f$, is crucially important in determining the thermal evolution of a body. For bodies of a given size, t$_f$ varies in inverse proportion to the surface density in the accretion disk of the Sun (the latter may have varied as $\propto$ R$^{-3/2}$) and in proportion to the Keplerian orbit period ($\propto$ R$^{3/2}$). When multiplied together, these give t$_f \propto$ R^3. All else being equal, an object taking 2 m.y. to accrete in the main belt at 2.5 AU would take ~20 m.y. at the 5 AU distance of the jovian Trojans. Twenty million years corresponds to about 28 half-lives of ^{26}Al, precluding any heating from this source in the Trojans, whereas heating of the main-belt objects could be strong. The strong heliocentric distance dependence of the growth time is consistent with a large radial gradient in their hydration properties across rather modest radial distance differences (*Grimm and McSween*, 1993). Taken out to the Kuiper belt (30 AU and beyond) the formation times predicted by this simple scaling law approach the age of the solar system, and are unreasonable. Detailed models of Kuiper belt object accretion show that such objects can form on much more reasonable (10–100 m.y.) timescales provided very low velocity dispersions are maintained (*Kenyon*, 2002), but reinforce the point that the KBOs are probably much less heated by trapped ^{26}Al or other short-lived nuclei compared to inner disk objects of comparable size. The primitive meteorites, especially those that are hydrated, show that early melting of silicates was not a universal fate even in the asteroid belt, and in the parents of these meteorites we seek to examine the effects of heating on buried water ice. We should bear in mind, however, that heating by [26]Al may well take place during the accretion process itself; moreover, the accretion rate is not linear in time. Thus, the core of an accreting body may be significantly heated before the body reaches its final size long after the decay of [26]Al. The effect of combined growth and internal heating by [26]Al decay was investigated by *Merk and Prialnik* (2003) for an amorphous ice and dust composition, without other volatiles. They found that while it still remains true that small objects remain almost unaffected by radioactive heating, the effect on larger bodies is not linear with size. There is an intermediate size range (around 25 km) where the melt fraction and duration of liquid water are maximal, and this range depends strongly on formation distance (ambient temperature). The corresponding calculation for an astcroid is by *Ghosh et al.* (2003).

5.3. Small Bodies and Serpentinization

If the water triple point is reached, exothermic hydration reactions with rock (so-called "serpentinization" reactions) can generate their own heat. The heat produced varies depending on the specific hydration reaction under consideration. For the (idealized) serpentine-producing reaction

$$Mg_2SiO_4 + MgSiO_3 + 2H_2O \rightleftharpoons Mg_3Si_2O_5(OH)_4 \quad (7)$$

the energy released per unit mass of serpentine produced is 2.4×10^5 J kg^{-1} (*Grimm and McSween*, 1989; *Cohen and Coker*, 2000) [see also *Jakosky and Shock* (1998) for energetics of other reactions]. In this reaction, two moles (about 36 g) of water react with rock to produce one mole (about 276 g) of serpentine. Enough energy is produced to melt about 40 moles (730 g) of water ice (the latent heat of fusion is 3.3×10^5 J kg^{-1}). Therefore, depending on the details of the heat transport within the parent body, there is the possibility for a thermal runaway in which heat from serpentinization in a localized region of the asteroid triggers the melting of ice throughout the body. An everyday example of this type of chemical energy release is commercial concrete, which, when mixed with water, liberates heat. A natural, terrestrial example may have been found in the mid-Atlantic, Lost City hydrothermal complex (Fig. 14). Lost City is located far from the midocean ridge heat source and emits fluids with distinctive chemical properties different from those at the midocean ridge, suggesting the action of serpentinization reactions (*Kelley et al.*, 2001) [note that this interpretation is contested by *Allen and Seyfried* (2004) on geochemical grounds]. Lost City has been active for 30,000 yr, produces a few MW of power (corresponding to serpentinization at a rate ~10 kg s^{-1}), and has enough reactants to continue for ~10[6] yr (*Fruh-Green et al.*, 2003).

Models of runaway serpentinization in main-belt asteroids have been computed by different groups, leading to some spectacular, if arguable, conclusions. In these models, ice melting occurs first as a result of radioactive decay heating by short-lived isotopes. Serpentinization then re-

Fig. 14. Carbonate chimneys at the Lost City field rise up to 60 m above the surrounding terrain and heat sea water to 40° to 75°C above ambient. They may be entirely powered by serpentinization. Figure courtesy of University of Washington.

leases a tremendous amount of heat that dominates the subsequent thermal evolution. Such heating may be moderated in reality by the rate at which water can be introduced to reaction sites or by efficient heat loss due to hydrothermal convection. Bodies of asteroidal porosity must be a few hundred kilometers in diameter, however, to initiate hydrothermal convection (*Grimm and McSween*, 1989; *Young et al.*, 2003). If heating is too rapid or hydrothermal circulation does not otherwise develop, high steam pressures could cause parent-body rupture under conditions of low permeability (*Grimm and McSween*, 1989; *Wilson et al.*, 1999).

Bodies that are too small, or those that form too late, are unable to trap much [26]Al and never attain temperatures sufficient to melt ice. In these, the serpentinization reactions are never triggered, except perhaps sporadically by heating events caused by large impacts (*Abramov and Kring*, 2004). For this reason, we expect that the effects of serpentinization reactions in the KBOs are also much diminished, except perhaps in the larger objects where longer-lived unstable nuclei ([40]K, [235]U, [238]U, and [232]Th) might drive water above the melting temperature, at least locally, at later times (*Busarev et al.*, 2003). Serpentinization also releases hydrogen, which, following Fischer-Tropsch-type reactions, could lead to the production of methane (CH_4) from CO or CO_2. Methane is known on the surface of Pluto and has been recently detected on other large Kuiper belt objects (*Brown et al.*, 2005) but is not expected to have been abundant in the solar nebula from which these objects formed (*Prinn and Fegley*, 1981). Serpentinization could be the ultimate source. Finally, it should not go unremarked that, on Earth, chemical reactions driven by deep-ocean hydrothermal systems are the foundation for richly populated abodes for life. Detailed investigations of the water-rich small bodies of the middle and outer solar system promise to be extremely interesting from an astrobiological perspective.

6. SUMMARY

Water is the signature volatile dividing the hot, dry inner regions of the protoplanetary disk from the frigid abyss beyond. Observations show that water has survived as ice in many of the solar system's small-body populations, including the nuclei of comets, the Kuiper belt objects, the Centaurs, and even in main-belt asteroids beyond about 3 AU.

Evidence from the carbonaceous meteorites shows that *liquid* water was present within a few million years to ~10 m.y. after the accretion of the first macroscopic solid bodies between Mars and Jupiter. The energy sources for melting ice are unknown but almost certainly included the radioactively decaying elements ^{26}Al and ^{60}Fe, both of which have very short half-lives (~1 m.y.). The thermal impact of these elements in small bodies is a function of their initial abundances but also of the solid body accretion timescales relative to the radioactive half-lives. Inner main-belt asteroids formed quickly in (what was) a high-density region of the protoplanetary disk and show the results of heating by trapped, short-lived nuclei. Ice was either not incorporated, or was melted in most of the asteroids of the inner belt but survives in the outer belt (beyond about 3 AU) where lower densities and longer growth times reduced the peak temperatures caused by radioisotope decay. Still more extreme conditions in the Kuiper belt resulted in growth times most likely larger than the ^{26}Al and ^{60}Fe half-lives, minimizing their thermal impact. On the other hand, large Kuiper belt objects might have been heated to the ice melting point by longer-lived radioactive decays (especially ^{40}K, ^{232}Th, and ^{238}U). Once formed, liquid water provides an independent source of heat through exothermic (and hydrogen-producing) "serpentinization" reactions with rock that, under some circumstances, could lead to runaway melting of ice throughout the body. Byproducts of serpentinization (including CH_4) may be found on some of the large KBOs.

Acknowledgments. We thank the anonymous referee for a review and A. Delsanti, H. Hsieh, P. Lacerda, J. Luu, T. Owen, and B. Yang for helpful comments. This work was supported by NASA grant NNG05GF76G to D.J., by the NASA Astrobiology Institute under Cooperative Agreement No. NNA04CC08A issued through the Office of Space Science, by NASA cosmochemistry grant NAG5-11591 to K. Keil, by Israel Science Foundation grant 942/04 to D.P., and by NASA's Outer Planets Research grant NNG05GG60G to R.G.

REFERENCES

Abramov O. and Kring D. (2004) *J. Geophys. Res.–Planets, 109,* E10007.

A'Hearn M. and Feldman P. (1992) *Icarus, 98,* 54–60.

Alexander C. M. O'D., Barber D. J., and Hutchison R. (1989) *Geochim. Cosmochim. Acta, 53,* 3045–3057.

Allen D. E. and Seyfried W. E. (2004) *Geochim. Cosmochim. Acta, 68,* 1347–1354.

Brearley A. J. (2003) In *Treatise on Geochemistry* (H. D. Holland and K. K. Turekian, eds.), pp. 247–268. Elsevier, Oxford.

Brearley A. J. and Hutcheon I. D. (2002) *Meteoritics & Planet. Sci., 37,* A23.

Bridges J. C., Banks D. A., Smith M., and Grady M. M. (2004) *Meteoritics & Planet. Sci., 39,* 657–666.

Brown M. E., Trujillo C. A., and Rabinowitz D. L. (2005) *Astrophys. J., 635,* L97–L100.

Browning L. B., McSween H. Y., and Zolensky M. E. (1996) *Geochim. Cosmochim. Acta, 60,* 2621–2633.

Brownlee D. E. (1978) In *Protostars and Planets* (T. Gehrels, ed.), pp. 134–150. Univ. of Arizona, Tucson.

Burgess R., Whitby J. A., Turner G., Gilmour J. D., and Bridges J. C. (2000) In *Lunar Planet. Sci. XXXI,* Abstract #1330.

Busarev V. V., Dorofeeva V. A., and Makalkin A. B. (2003) *Earth Moon Planets, 92,* 345–375.

Carvano J. M., Mothé-Diniz T., and Lazzaro D. (2003) *Icarus, 161,* 356–382.

Chiang E. I. and Lithwick Y. (2005) *Astrophys. J., 628,* 520–532.

Choi Y.-J., Cohen M., Merk R., and Prialnik D. (2002) *Icarus, 160,* 300–312.

Ciesla F. J. and Cuzzi J. (2006) *Icarus, 181,* 178–204.

Ciesla F. J., Lauretta D. S., Cohen B. A., and Hood L. L. (2003) *Science, 299,* 549–552.

Clark R. N. and Lucey P. G. (1984) *J. Geophys. Res., 89,* 6341–6348.

Clayton R. N. (1993) *Ann. Rev. Earth Sci., 21,* 115–149.

Clayton R. N. (1999) *Geochim. Cosmochim. Acta, 63,* 2089–2104.

Clayton R. N. and Mayeda T. K. (1984) *Earth Planet. Sci. Lett., 67,* L151–L161.

Clayton R. N., Mayeda T. K., and Onuma N. (1976) *Earth Planet. Sci. Lett., 30,* L10–L18.

Cohen B. and Coker R (2000) *Icarus, 145,* 369–381.

Cruikshank D., Dalle Ore C., Roush T. L., Geballe T., Owen T., et al. (2001) *Icarus, 153,* 348–360.

Czechowski L. and Leliwa-Kopystyski J. (2002) *Adv. Space Res., 29,* 751–756.

Davies J. K., Roush T. L., Cruikshank D. P., Bartholomew M. J., Geballe T. R., Owen T., and de Bergh C. (1997) *Icarus, 127,* 238–245.

Dumas C., Owen T., and Barucci M. A. (1998) *Icarus, 133,* 221–232.

Emery J. and Brown R. (2004) *Icarus, 170,* 131–152.

Fanale F. P. and Salvail J. R. 1989, *Icarus, 82,* 97–110.

Fernández J. A., Gallardo T., and Brunini A. (2002) *Icarus, 159,* 358–368.

Fernández Y., Sheppard S., and Jewitt D. (2003) *Astron. J., 126,* 1563–1574.

Fruh-Green G., Kelley D., Bernasconi S., Karson J., Ludwig K., et al. (2003) *Science, 301,* 495–498.

Ghosh A., Weidenschilling S., and McSween H. (2003) *Meteoritics & Planet. Sci., 38,* 711–724.

Grimm R. E. and McSween H. Y. (1989) *Icarus, 82,* 244–280.

Grimm R. E. and McSween H. Y. (1993) *Science, 259,* 653–655.

Grossman J. and Brearley A. J. (2003) In *Lunar Planet. Sci. XXXIV,* Abstract #1584.

Grossman J. and Rubin A. (1999) In *Lunar Planet. Sci. XXX,* Abstract #1639.

Grossman J., Alexander C. M. O'D., Wang J., and Brearley A. J. (2000) *Meteoritics & Planet. Sci., 35,* 467–486.

Haruyama J., Yamamoto T., Mizutani H., and Greenberg J. M. (1993) *J. Geophys. Res., 98,* 15079.

Hanowski N. and Brearley A. J. (2001) *Meteoritics & Planet. Sci., 35,* 1291–1308.

Hansen C., Esposito L., Stewart A., Colwell J., Hendrix A., Pryor W., Shemansky D., and West R. (2006) *Science, 311,* 1422–1425.

Herndon J. M. and Herndon M. A. (1977) *Meteoritics, 12,* 459–465.

Hiroi T. and Hasegawa S. (2003) *Antarc. Meteorite Res., 16,* 176–184.

Hiroi T., Zolensky M. E., and Pieters C. M. (2001) *Science, 293,* 2234–2236.

Hsieh H. and Jewitt D. (2006) *Science, 312,* 561–563.

Hsieh H., Jewitt D., and Fernández Y. (2004) *Astron. J., 127,* 2997–3017.

Hutcheon I., Krot A., Keil K., Phinney D., and Scott E. (1998) *Science, 282,* 1865–1867.

Hutchison R., Alexander C. M. O., and Barber D. J. (1987) *Geochim. Cosmochim. Acta, 51,* 1875–1882.

Jakosky B. M. and Shock E. L. (1998) *J. Geophys. Res., 103,* 19359–19364.

Jewitt D. (2002) *Astron. J., 123,* 1039–1049.

Jewitt D. (2005) In *Comets II* (M. C. Festou et al., eds.), pp. 659–676. Univ. of Arizona, Tucson.

Jewitt D. and Luu J. (1990) *Astron. J., 100,* 933–944.

Jewitt D. and Luu J. (2004) *Nature, 432,* 731–733.

Jewitt D., Sheppard S., and Porco C. (2004) In *Jupiter: The Planet, Satellites and Magnetosphere* (F. Bagenal and W. McKinnon, eds.), pp. 263–280. Univ. of Arizona, Tucson.

Jones T., Lebofsky L., Lewis J., and Marley M. (1990) *Icarus, 88,* 172–192.

Kawakita H., Watanabe J.-I., Ootsubo T., Nakamura R., Fuse T., Takato N., Sasaki S., and Sasaki T. (2004) *Astrophys. J., 601,* L191–L194.

Keil K. (2000) *Planet. Space Sci., 48,* 887–903.

Kelley D., Karson J., Blackman D., Fruh-Green G., Butterfield D., et al. (2001) *Nature, 412,* 145–149.

Kemper F., Jäger C., Waters L, Henning T., Molster F., Barlow M., Lim T., and de Koter A. (2002) *Nature, 415,* 295–297.

Kenyon S. J. (2002) *Publ. Astron. Soc. Pac., 114,* 265–283.

Kerridge J. F., Mackay A. L., and Boynton W. V. (1979) *Science, 205,* 395–397.

King T., Clark R., Calvin W., Sherman D., and Brown R. (1992) *Science, 255,* 1551–1553.

Krot A. N., Hutcheon I., Yurimoto H., Cuzzi J., McKeegan K., et al. (2005) *Astrophys. J., 622,* 1333–1342.

Lebofsky L., Feierberg M., Tokunaga A., Larson H., and Johnson J. (1981) *Icarus, 48,* 453–459.

Lee J., Papanastassiou D. A., and Wasserburg G. J. (1976) *Geophys. Res. Lett., 3,* 41–44.

Levison H., Terrell D., Wiegert P., Dones L., and Duncan M. (2006) *Icarus, 182,* 161–168.

Lisse C., VanCleve J., Adams A., A'Hearn M., Fernández Y. R., et al. (2006) *Science, 313,* 635–640.

Luu J., Jewitt D., and Cloutis E. (1994) *Icarus, 109,* 133–144.

Lyons J. R. and Young E. D. (2005) *Nature, 435,* 317–320.

MacPherson G. J., Davis A. M., and Zinner E. K. (1995) *Meteoritics, 30,* 365.

Marchis F., et al. (2006) *Nature, 439,* 565–567.

Marzari F., Davis, D., and Vanzani V. (1995) *Icarus, 113,* 168–187.

Mastrapa R., Moore M., Hudson R., Ferrante R., and Brown R. (2005) *AAS/Division for Planetary Sciences Meeting Abstracts, 37.*

McCord T. B. and Sotin C. (2005) *J. Geophys. Res.–Planets, 110,* E05009.

McSween H. Y. (1979) *Geochim. Cosmochim. Acta, 43,* 1761–1770.

McSween H., Ghosh A., Grimm R., Wilson L., and Young E. (2002) In *Asteroids III* (W. F. Bottke Jr. et al., eds.), pp. 559–571. Univ. of Arizona, Tucson.

Meier R. and Owen T. C. (1999) *Space Sci. Rev., 90,* 33–44.

Meier R., Owen T., Matthews H., Jewitt D., Bockelée-Morvan D., et al. (1998a) *Science, 279,* 842–844.

Meier R., Owen T., Jewitt D., Matthews H., Senay M., et al. (1998b) *Science, 279,* 1707–1709.

Merk R. and Prialnik D. (2003) *Earth Moon Planets, 92,* 359–374.

Morbidelli A., Chambers J., Lunine J. I., Petit J. M., Robert F., et al. (2000) *Meteoritics & Planet. Sci., 35,* 1309–1320.

Morbidelli A., Levison H., Tsiganis K., and Gomes R. (2005) *Nature, 435,* 462–465.

Mostefaoui S., Lugmair G. W., Hoppe P., and El Goresy A. (2004) *New Astron. Rev., 48,* 155–159.

Nakamura T., Noguchi T., Zolensky M., and Tanaka M. (2003) *Earth Planet. Sci. Lett., 207,* L83–L101.

Notesco G. and Bar-Nun A. (2005) *Icarus, 175,* 546–550.

Owen T. and Bar-Nun A. (1995) *Icarus, 116,* 215–226.

Pollack J., Hollenbach D., Beckwith S., Simonelli D., Roush T., and Fong W. (1994) *Astrophys. J., 421,* 615–639.

Prialnik D. (1992) *Astrophys. J., 388,* 196–202.

Prialnik D. and Podolak M. (1995) *Icarus, 117,* 420–430.

Prialnik D., Benkhoff J., and Podolak M. (2005) In *Comets II* (M. C. Festou et al., eds.), pp. 359–387. Univ. of Arizona, Tucson.

Prinn R. G. and Fegley B. (1981) *Astrophys. J., 249,* 308–317.

Prinn R. G. and Fegley B. (1987) *Ann. Rev. Earth Planet. Sci., 15,* 171–212.

Rietmeijer F. (1991) *Earth Planet. Sci. Lett., 102,* L148–L157.

Rietmeijer F. (1998) In *Planetary Materials* (J. J. Papike, ed.), pp. 201–287. Reviews in Mineralovy, Vol. 36, Mineralogical Society of America.

Rivkin A., Howell E., Vilas F., and Lebofsky L. (2002) In *Asteroids III* (W. F. Bottke Jr. et al., eds.), pp. 235–253. Univ. of Arizona, Tucson.

Sasselov D. D. and Lecar M. (2000) *Astrophys. J., 528,* 995–998.

Starukhina L. (2001) *J. Geophys. Res., 106,* 14701–14710.

Stevenson D. J. and Lunine J. I. (1988) *Icarus, 75,* 146–155.

Sykes M. V. and Walker R. G. (1992) *Icarus, 95,* 180–210.

Sykes M. V., Lebofsky L. A., Hunten D. M., and Low F. (1986) *Science, 232,* 1115–1117.

Thomas P., Parker J., McFadden L., Russell C., Stern S., Sykes M., and Young E. (2005) *Nature, 437,* 224–227.

Tonui E. K., Zolenksy M. E., Lipschutz M. E., Wang M.-S., and Nakamura T. (2003) *Meteoritics & Planet. Sci., 38,* 269–292.

Toppani A., Robert F., Libeourel G., de Donato P., Barres O., d'Hendecourt L., and Ghanbaja J. (2005) *Nature, 437,* 1121–1124.

Toth I. (2000) *Astron. Astrophys., 360,* 375–380.

Vernazza P., et al. (2005) *Astron. Astrophys., 436,* 1113–1121.

Vilas F. (1994) *Icarus, 111,* 456–467.

Wallis M. K. (1980) *Nature 284,* 431–433.

Whipple F. L. 1950, *Astrophys. J., 111,* 375–394.

Wilson L., Keil K., Browning L., Krot A., and Bourcier W. (1999) *Meteoritics & Planet. Sci., 34,* 541–557.

Yabushita S. (1993) *Mon. Not. R. Astron. Soc., 260,* 819–825.

Young E. D., Zhang K. K., and Schubert G. (2003) *Earth Planet. Sci. Lett., 213,* L249–L259.

Zolensky M., Bourcier W., and Gooding J. (1989) *Icarus, 78,* 411–425.

Zolensky M. E., Barrett R. A., and Browning L. B. (1993) *Geochim. Cosmochim. Acta, 57,* 3123–3148.

Zolensky M. E., Gibson K. E. Jr., Lofgren E. G., Morris V. R., and Yang V. S. (1998) *Antarctic Meteorites XXIII, 23,* 189–191.

Zolensky M. E., Bodnar R. J., and Rubin A. E. (1999) *Meteoritics & Planet. Sci., 34,* A124.

Physical Properties of Transneptunian Objects

D. P. Cruikshank
NASA Ames Research Center

M. A. Barucci
Observatoire de Paris, Meudon

J. P. Emery
SETI Institute and NASA Ames Research Center

Y. R. Fernández
University of Central Florida

W. M. Grundy
Lowell Observatory

K. S. Noll
Space Telescope Science Institute

J. A. Stansberry
University of Arizona

In 1992, the first body beyond Neptune since the discovery of Pluto in 1930 was found. Since then, nearly a thousand solid bodies, including some of planetary size, have been discovered in the outer solar system, largely beyond Neptune. Observational studies of an expanding number of these objects with space- and groundbased telescopes are revealing an unexpected diversity in their physical characteristics. Their colors range from neutral to very red, revealing diversity in their intrinsic surface compositions and/or different degrees of processing that they have endured. While some show no diagnostic spectral bands, others have surface deposits of ices of H_2O, CH_4, and N_2, sharing these properties with Pluto and Triton. Thermal emission spectra of some suggest the presence of silicate minerals. Measurements of thermal emission allow determinations of the dimensions and surface albedos of the larger (diameter > ~75 km) members of the known population; geometric albedos range widely from 2.5% to >60%. Some 22 transneptunian objects (including Pluto) are multiple systems. Pluto has three satellites, while 21 other bodies, representing about 11% of the sample investigated, are binary systems. In one binary system where both the mass and radius are reliably known, the mean density of the primary is ~500 kg/m³, comparable to some comets [e.g., Comet 1P/Halley (*Keller et al.,* 2004)].

1. INTRODUCTION

The many objects found orbiting the Sun beyond Neptune, beginning with the first discovery in 1992 (*Jewitt and Luu,* 1993), open a new window on the content, dynamics, origin, and evolution of our own solar system, and give rich insights into those same aspects of protoplanetary disks and extrasolar planetary systems. Our present understanding of the population of transneptunian objects is significantly greater than our understanding of the asteroid belt was at the time when a comparable number of asteroids were known (early 1920s). This improvement arises from the fact that, in contrast with the development of asteroid science, we are studying the distant transneptunian objects with the full range of available observational techniques in parallel rather than in a serial way. By bringing together dynamical studies of their orbits and photometric, spectroscopic, and radiometric observations, we are rapidly gaining insight into the origin of transneptunian objects and their relationship to the solar nebula and protoplanetary disks around young or forming stars.

This chapter is focused on the physical properties of the objects beyond Neptune derived from observational studies with Earth- and spacebased telescopes. Those physical properties include dimensions, surface compositions, colors, and a taxonomy derived from that information, the occurrence of binary systems, and in exceptional cases masses and mean densities.

The objects with perihelia beyond Neptune, which as of early 2006 some 900 were known, have received many designations and names, both formal and informal. Three distinct dynamical populations have become apparent as the discoveries continue. The first grouping, which includes the majority of the objects discovered to date, is known as the *classical Kuiper belt*. It consists of objects in near circular orbits with semimajor axes around 45 AU. These orbits are stable against Neptune's perturbations over the age of the solar system. The second population consists of objects in orbits with 2:3 resonance with Neptune, as in the case of Pluto. These objects are informally referenced by their etymologically grotesque name, *"Plutinos."* Taken together, these two populations are frequently referred to as Kuiper belt objects (KBOs), a term that is used in this paper. The third population consists of objects in highly eccentric orbits with perihelia generally within the classical Kuiper belt (although some are inside Neptune's orbit), but with aphelia far outside. These bodies have been dynamically excited by Neptune; they are called *scattered disk* objects. In this chapter we refer to all of them with the most general term, *transneptunian objects*, or TNOs, but where it is useful to discuss dynamical categories, we use the generally accepted terms, resonant and nonresonant Kuiper belt objects (KBOs) and scattered disk objects (see chapter by Chiang et al.). An additional class, the Centaurs, consists of objects derived from the TNO population and occupying temporary orbits that cross those of the major planets. About 80 such objects having perihelia <30 AU are known as of early 2006.

In this review we will find it useful to include the planet Pluto, Neptune's satellite Triton, and the Centaurs; in discussing spectroscopy in the thermal spectral region ($\lambda >$ 10 µm), we also include the jovian Trojan asteroids for comparison. Pluto shares physical characteristics with some of the more recently discovered objects in the transneptunian region (e.g., *Owen et al.*, 1993; *Douté et al.*, 1999), while Triton has similar physical characteristics to Pluto (e.g., *Cruikshank et al.*, 1993; *Quirico et al.*, 1999) and is widely thought to have been captured into a retrograde orbit by Neptune from the TNO population (e.g., *McKinnon et al.*, 1995). The jovian Trojans are a family of small bodies that share more characteristics with comets and Centaurs than with main-belt asteroids (*Emery et al.*, 2006), and probably originated in the outer solar system.

2. COLORS AND TAXONOMY

Observations of brightness and color in the extended visible spectral region (0.3–1.0 µm) have been obtained for a large sampling of TNOs, and photometric observations in several colors obtained in recent years have yielded high quality B, V, R, and I measurements for more than 130 objects (see *Barucci et al.*, 2005a, for complete references). These observations reveal a clear diversity in color among the TNO population, noted early in the investigation of these objects (*Luu and Jewitt*, 1996). Relevant statistical analyses have been performed and a wide range of correlations

between optical colors and orbital parameters have been analyzed (*Tegler and Romanishin*, 2000, 2003; *Doressoundiram et al.*, 2002; *Trujillo and Brown*, 2002; *Tegler et al.*, 2003; *Peixhino et al.*, 2004; *Doressoundiram et al.*, 2005a). Color variation is an important diagnostic of diversity in surface composition and the evolution of surface composition and microstructure through processing by various physical processes, a subject to which we return below.

When dealing with a large number of objects it is useful to distinguish groups of objects with similar surface properties. *Barucci et al.* (2005a) applied the multivariate statistic analysis (G mode) (*Barucci et al.*, 1987) to the TNO and Centaur populations; this and the principal component analysis are the same techniques used in the 1980s to classify the asteroid population (*Tholen*, 1984; *Tholen and Barucci*, 1989).

Barucci et al. (2005a), considering all the high-quality available colors on TNOs and Centaurs published after 1996, analyzed (1) a set of data for 135 objects observed in the B, V, R, and I bands; (2) a set of 51 objects observed in the B, V, R, I, and J bands with high-quality homogeneous data; and (3) a subsample of 37 objects that also included H-band measurements. They selected as a primary sample for the analysis a complete and homogeneous set of 51 objects observed in five filters (B, V, R, I, J), adopting the mean values weighted with the inverse of the error of individual measurement when multiple observations of an object were available.

The results of the analyses show the presence of four groups where J color plays the main role in discriminating the groups (J weight is 46%). The four homogeneous groups showing a high confidence level (corresponding to 3s) have been named BB, BR, IR, and RR. The same well-determined four groups were found when the analysis was applied to a subset of 37 objects for which the H color was also available.

In Table 1, the color average value and the standard deviation for each group are given. The BB (blue) group consists of objects having neutral colors with respect to the Sun, while the RR group has a very strong red color. The BR group includes objects with an intermediate blue-red color, while the IR group includes moderately red objects.

The G mode can be extended (*Fulchignoni et al.*, 2000) to assign objects for which the same set of variables become available to one of the already defined taxonomic groups, even if only a subset of variables used in the initial taxonomy is known for new objects. Such an extension has been applied to the 84 other TNOs for which only B, V, R, and I colors were available, and an indication of the classification has been obtained for 69 objects.

The investigation of the diversity among the TNOs can help in understanding their characteristics and the different physical processes that have affected their surfaces. The four groups noted here are well defined and homogeneous in color properties. To better understand the surface composition properties of each group, we have superimposed representative compositional surface models calculated for some of the well-studied objects (Fig. 1). We return below

TABLE 1. Average colors and the relative standard deviation for the four taxonomic groups obtained by G-mode analysis on the sample of 51 objects (V-H has been computed for a subset of 37 objects).

Class	B-V	V-R	V-I	V-J	V-H
BB	0.70 ± 0.04	0.39 ± 0.03	0.77 ± 0.05	1.16 ± 1.17	1.21 ± 0.52
BR	0.76 ± 0.06	0.49 ± 0.03	0.9 ± 0.07	1.67 ± 0.19	2.04 ± 0.24
IR	0.92 ± 0.03	0.61 ± 0.03	1.20 ± 0.04	1.88 ± 0.09	2.21 ± 0.06
RR	1.08 ± 0.08	0.71 ± 0.04	1.37 ± 0.09	2.27 ± 0.20	2.70 ± 0.24

to questions of the efficacy and uniqueness of compositional models of planetary bodies.

3. SPECTROSCOPY (0.3–5 MICROMETERS)

3.1. Visible and Near-Infrared Spectral Region

While broadband colors provide useful information on the surfaces of these objects, spectroscopy is the only way to investigate their surface compositions in detail. Although the faintness of these objects limits spectroscopic observations, the use of some of the world's largest telescopes has yielded combined visible region and infrared spectra for about 20 objects.

The visible region spectra are generally featureless, but show differences in the spectral slope consistent with multicolor photometry. Three objects show possible broad absorptions in their spectra; 47932 (2000 GN$_{171}$) shows an absorption band centered near 0.7 μm, while the spectrum of 38628 (2000 EB$_{173}$) suggests two weak features centered at 0.6 μm and at 0.745 μm (*Lazzarin et al.*, 2003; *de Bergh et al.*, 2004). The spectrum of 2003 AZ$_{84}$ (*Fornasier et al.*, 2004a) also seems to show a weak absorption centered near 0.7 μm. These features are very similar to those due to aqueously altered minerals, as found in some main-belt asteroids (*Vilas and Gaffey*, 1989). We note further that hydrated silicates are a dominant component of carbonaceous meteorites. *Jewitt and Luu* (2001) have reported absorption bands (around 1.4, 1.9 μm) of phyllosilicates in the spectrum of the Centaur 26375 1999 DE$_9$. All these features are rather weak, and require confirmation.

Finding aqueous altered materials in TNOs would not be too surprising (*de Bergh et al.*, 2004) since hydrous silicates are detected in interplanetary dust particles (IDPs) and micrometeorites, and by inference will be found in comets. In experiments with the condensation of magnesiosilica smokes in the presence of water, *Reitmeijer et al.* (2004) produced protophyllosilicates in an environment that may be similar to that in stellar outflows and the inner regions of the solar nebula. Their results may have implications for the pervasiveness of hydrated amorphous protophyllosilicates in comets, and by association, icy protoplanets like TNOs.

While broad spectral coverage of the visible and near-infrared region is a necessary condition to allow a diagnosis for silicate minerals, carbonaceous assemblages, organics, and ices and/or hydrocarbons, in some cases it is not a sufficient condition. Some objects exhibit no discrete absorption bands, in which case the analysis can use only color

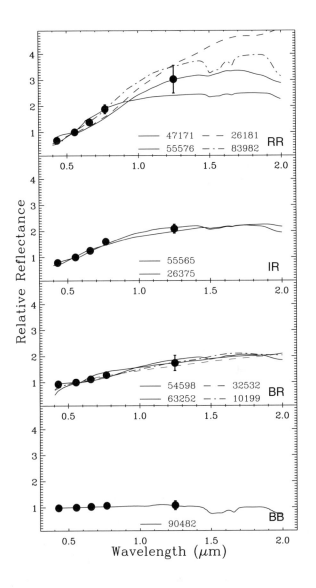

Fig. 1. Each of the four color classes is shown with the average broadband reflectance data (solid circles) and their relative error bars, normalized to the V band (0.55 μm). Overlain on each group is a compositional model of one or more well-observed members of the group. For RR the spectral models of (47171) 1999 TC$_{36}$ (*Dotto et al.*, 2003a), (26181) 1996 GQ$_{21}$ (*Doressoundiram et al.*, 2003), (55576) 2002 GB$_{10}$, and (83982) 2002 GO$_9$ (*Doressoundiram et al.*, 2005a,b) are shown; for BB the model of 90482 Orcus (*Fornasier et al.*, 2004a); for IR the models of (26375) 1999 DE$_9$ (*Doressoundiram et al.*, 2003) and (55565) 2002 AW$_{197}$ (*Doressoundiram et al.*, 2005a,b); for BR the models of (63252) 2001 BL$_{41}$ (*Doressoundiram et al.*, 2005a,b), 10199 Chariklo (*Dotto et al.*, 2003b), 54598 Bienor (*Dotto et al.*, 2003a), and 32532 Thereus (*Barucci et al.*, 2002).

properties and albedo. As noted above, only about 20 bright members of the TNO and Centaur populations have been well studied in both visible and near-infrared wavelengths and rigorously modeled. The spectra of TNOs and Centaurs have common behavior and their surface characteristics seem to show wide diversity. Several objects seem to show heterogeneity in their surfaces, such as 31824 Elatus (*Bauer et al., 2003a*), 19308 1996 TO₆₆ (*Brown et al., 1999*), and 32532 Thereus (*Barucci et al., 2002; Merlin et al., 2005*). In some cases, observed surface variations may be attributed to different viewing geometry, and possibly the low signal precision of some of the observational data.

Several large and relatively bright objects have been discovered in the last few years, thus enabling spectroscopic observations with large groundbased telescopes. The spectral reflectance of 90482 Orcus can be matched with a model consisting of kerogen (a red-colored complex and relatively refractory organic solid), amorphous carbon, and H_2O ice (*Fornasier et al., 2004a; de Bergh et al., 2005*). *Jewitt and Luu* (2004) detected crystalline water on 50000 Quaoar; its presence may imply melting by resurfacing events, since the expected low temperature of formation suggests that amorphous ice should dominate. TNO 90377 Sedna resembles Triton (*Barucci et al., 2005b*), while the largest presently known object of the TNOs population, 2003 UB₃₁₃, has a near-infrared spectrum very similar to that of Pluto (*Brown et al., 2005*) (see below).

3.2. Groupings of Transneptunian Objects by Spectroscopic Characteristics

A sufficient number of TNOs have been observed spectroscopically in the region 0.3–2.5 μm to see categories into which they fall. We do not propose the following as a taxonomy, but note the basic characteristics of these groups.

1. Objects having absorption bands of H_2O ice that, when sufficiently well defined, reveal the crystalline rather than amorphous phase. These objects exhibit a range of redness and geometric albedo similar to the previous class.

2. Objects having absorption bands of solid CH_4 with a wide range of band strengths from weak to heavily saturated. Two subclasses are seen:

2a. Absorption bands of both CH_4 and N_2; in this situation the CH_4 is often dissolved in N_2 ice. Pluto and (probably) Sedna share these properties, together with Neptune's satellite Triton.

2b. Absorption bands of CH_4 in the absence of N_2, and a clear indication that the CH_4 ice is pure and not dissolved in another ice. Examples are 2003 UB₃₁₃ and 2005 FY₉.

3. Objects having no spectral features (0.3–2.5 μm), but a wide range of colors, represented by degrees of positive spectral slope toward longer wavelengths (red color). Some objects are essentially neutral (gray) over this spectral interval, while others are among the reddest objects in the solar system. Geometric albedos appear to lie in the range 0.03–0.2.

3.3.   90377 Sedna and 2003 UB₃₁₃: Two Objects of Special Interest

The object 90377 Sedna belongs to the dynamical class called the extended scattered disk, with the most distant perihelion at 76 AU presently known. *Barucci et al.* (2005b) reported visible and near-infrared observations showing a similarity to the spectrum of Triton, in particular in terms of the presence of absorption bands of N_2 and CH_4 in the 2.0–2.5-μm range.

Cross-correlation investigation gives a high level of confidence that the spectrum of Sedna has the same spectral bands as that of Triton (*Cruikshank et al., 2000*) in the 2.0–2.45-μm wavelength region. A model of the Sedna spectrum (Fig. 2) calculated using the radiative transfer theory developed by *Shkuratov et al.* (1999) reproduces the spectrum well with an intimate mixture of 24% Triton tholin, 7% amorphous carbon, 10% N_2 ice, 26% CH_3OH ice, and 33% CH_4 ice contaminated by small inclusions of Titan tholin. Its composition could imply that during its perihelion passage, Sedna may have a temporary and thin N_2 atmosphere.

Brown et al. (2005) discovered an object larger than Pluto in a highly inclined orbit in the scattered Kuiper belt; it is currently designated 2003 UB₃₁₃. While this object is nearly neutral in color in the extended visible wavelength region, its near-infrared spectrum shows a complex of methane-ice absorptions very similar to Pluto (Fig. 3). The solid N_2 band at 2.15 μm that is visible in the spectra of Pluto and Triton cannot be reliably detected in the existing data. An analysis of the central wavelengths of the CH_4 bands does not show the shift in wavelength (the matrix shift) seen in the spectra of Pluto and Triton and attributed to the fact that most (Triton) or some (Pluto) of the CH_4 is dissolved in solid N_2. The CH_4 on 2003 UB₃₁₃ is thus thought to be present largely in its pure form (*Brown et al., 2005*).

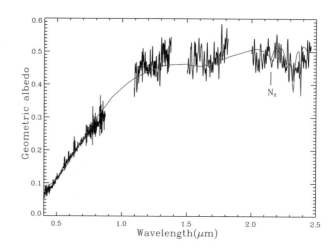

Fig. 2.   The spectrum of 90377 Sedna, 0.4–2.5 μm. The best-fitting model is overlain as a continuous line. The geometric albedo is 0.15 at 0.55 μm (*Barucci et al., 2005b*).

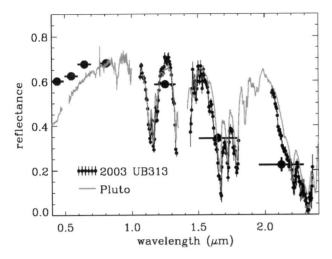

Fig. 3. Relative reflectance of 2003 UB$_{313}$ (filled circles with error bars) and absolute reflectance of Pluto (gray line). The reflectance derived from BVRIJHK photometry is shown by the large points. The UB$_{313}$ spectrum is scaled to match Pluto at 0.8 μm. From *Brown et al.* (2005), reproduced courtesy of *The Astrophysical Journal.*

4. THERMAL INFRARED SPECTROSCOPY

The Infrared Spectrograph (IRS) on the Spitzer Space Telescope brings thermal emission spectroscopy of the brightest Centaurs and KBOs within reach. The mid-infrared (5–38 μm) is well suited to investigating the silicate fraction of the surface layer of these objects because it contains the Si-O stretch and bend fundamental molecular vibration bands (typically 9–12 and 14–25 μm, respectively). Interplay between surface and volume scattering around these bands creates complex patterns of emissivity highs and lows that are very sensitive to, and therefore diagnostic of, mineralogy (e.g., *Salisbury et al.,* 1992; *Hapke,* 1996; *Cooper et al.,* 2002). We include Trojan asteroids in this section because they may be genetically related to Centaurs, they have similar emissivity spectra to at least one Centaur, and they are more accessible than the Centaurs and KBOs.

4.1. Trojan Asteroids

Trojan asteroids are located beyond the main belt, trapped in Jupiter's stable Lagrange points at 5.2 AU (leading and trailing the planet by 60° in its orbit). They typically have albedos of a few to several percent (e.g., *Cruikshank,* 1977; *Fernandez et al.,* 2003). Their reflectance spectra in the visible and near-infrared (0.3–4.0 μm) generally are featureless with moderately red spectral slopes, comparable to class BB and BR Centaurs and KBOs (e.g., *Jewitt and Luu,* 1990; *Luu et al.,* 1994; *Dumas et al.,* 1998; *Emery and Brown,* 2003; *Fornasier et al.,* 2004b) (see also section 2.1). Emissivity spectra of three Trojan asteroids have been measured with Spitzer: 624 Hektor, 911 Agamemnon, and 1172 Aneas

(*Emery et al.,* 2006) (Fig. 4). The emissivity spectrum is defined as the measured flux spectrum divided by modeled thermal continuum.

The emissivity spectra of all three Trojans exhibit emission bands with strong spectral contrast after the underlying thermal continuum throughout the 5–38-μm region is removed (Fig. 4). *Emery et al.* (2006) interpret the strong emissivity plateau near 10 μm and the broader emissivity high near 20–25 μm to indicate the presence of fine-grained silicates (Fig. 5); large grains show emissivity lows at these wavelengths (e.g., *Christensen et al.,* 2000). In addition, the emissivity bands in Trojans also do not exactly match those expected for regolith surfaces; the 10-μm plateau is narrower and the spectra do not rise as rapidly near 15 μm. Surface structure and grain size may be playing an important role; a surface composed of silicate grains that are tens or hundreds of micrometers in size show a 10-μm absorption band, while grains smaller than about 10 μm (dispersed, as in a comet's coma) show the characteristic band in emission. On a solid surface, in the absence of a suspending medium, emission from fine-grained silicate particles imbedded in a matrix that is fairly transparent at these wavelengths (e.g., macromolecular organic solids) may explain the occurrence of the emissivity peak attributed to silicates (*Emery et al.,* 2006). Grain size and surface microstructure are relevant because they are indicators of the mechanisms of origin and evolution of the surfaces of airless bodies exposed to the local environment of collisional regolith gardening, and bombardment by micrometeoroids and atomic particles over the age of the solar system. Although we are only beginning to understand these processes in various regions of the solar system, their importance in deciphering surface evolution is generally recognized (e.g., *Clark et al.,* 2002).

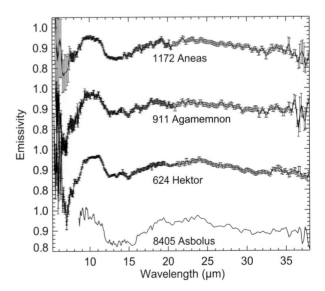

Fig. 4. Spectra of three Trojan asteroids measured with the Spitzer Space Telescope.

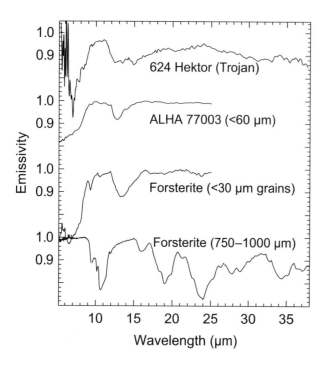

Fig. 5. The emissivity spectrum of Hektor is compared with laboratory spectra of a meteorite and minerals. The fine-grained forsterite (Fo_{88}) and the carbonaceous (CO3) meteorite ALHA 77003 are from the ASTER database (*speclib.jpl.nasa.gov*) (*Salisbury et al., 1992*), and the large-grained forsterite is from the spectral library at Arizona State University (*tes.asu.edu/speclib*) (*Christensen et al., 2000*).

4.2. Centaurs and Kuiper Belt Objects

Since these objects are at larger heliocentric distances than Trojans, they are cooler and thus observable only at the longer wavelengths of the IRS spectral range. An early program on Spitzer included two Centaurs that were observed from 7.5 to 38 μm, and four Centaurs in the range 14.2–38 μm. An additional 11 Centaurs and KBOs were observed only in the region 20–38 μm. The Centaur 8405 Asbolus was bright enough to be observed from 7.5 to 38 μm; in this region its spectrum is very similar to those of the Trojan asteroids.

Because the other Centaurs and KBOs are significantly fainter, the quality of the spectra is much lower. Longward of 20 μm, the spectrum of Pholus displays no diagnostic bands, despite having a feature-rich near-infrared reflectance spectrum (*Cruikshank et al., 1998*).

5. MODELS OF PLANETARY SURFACES

5.1. General Remarks

Before describing the spectral models for each of the four taxa presented in section 2 and for the spectra of individual objects, it is useful to review the state of modeling of diffuse reflectance from the surfaces of airless bodies, and to note some of the directions this work is currently taking.

For the spectral region in which reflected sunlight dominates the observed flux from an airless body, quantitative models of diffuse reflectance require the computation of the full spectral reflectance properties, which include not only the spectral shape, but the absolute reflectance (characterized by geometric albedo) across the full range of wavelengths for which observations exist. In these calculations, the formulations of *Hapke* (1981, 1993) and *Shkuratov et al.* (1999) are often used. In addition to the original literature sources, the discussion of the practical application of Hapke theory to reflectance spectroscopy of solid surfaces by *Verbiscer and Helfenstein* (1998) is particularly useful (see also *McEwen,* 1991). *Poulet et al.* (2002) have compared the Hapke and Shkuratov theories in the context of modeling solid surfaces, and *Cruikshank et al.* (2004) have described the computational procedures in more general terms.

The naturally occurring materials on a planetary surface can be segregated from one another or can be mixed in various combinations in a number of ways that affect the scattering properties and the abundances derived from the synthetic spectra calculated from scattering theories to match the observed spectra. These mixing scenarios have been discussed in various publications, including those cited in the previous paragraph.

5.2. Materials that May Impart Red Color

While a modest degree of red color, such as that of Trojan asteroid 624 Hektor, can be explained by the presence of the mineral Mg-rich pyroxene (*Cruikshank et al., 2001; Emery and Brown, 2003*), redder objects require other materials that are not minerals. *Gradie and Veverka* (1980) introduced the concept of *organic solids* to explain the red color of the D-type asteroids, suggesting the presence of " . . . very opaque, very red, polymer-type organic compounds, which are structurally similar to aromatic-type kerogen." They compared the reflectance spectra of the D-type asteroids to mixtures of montmorillonite (clay), magnetite (iron oxide), carbon black, and a coal-tar residue (kerogen) that was insoluble in organic solvents, and found a satisfactory match for both the low albedo and the red color of this specific class of asteroids. Organic kerogen-like structures consisting of small aromatic moieties connected by short aliphatic bridging units are found in carbonaceous meteorites (*Kerridge et al., 1987*), and the presence of similar structures in interstellar dust grains has also been deduced from infrared spectra of dusty diffuse interstellar clouds (*Pendleton and Allamandola,* 2002).

Tholins, which are the refractory residues from the irradiation of gases and ices containing hydrocarbons, have color properties that make them reasonable candidates for comparison to the spectra of solar system bodies. A number of tholins have been prepared in the context of the photochemical aerosols in Titan's atmosphere (e.g., *Khare et al.,* 1984; *Coll et al.,* 1999; *Ramírez et al.,* 2002; *Tran et al.,*

2003; *Imanaka et al.,* 2004), and while they are optically (spectrally) quite similar to Titan, they have until recently proven difficult to analyze and characterize fully from a chemical point of view. Recent work by *Tran et al.* (2003) and *Imanaka et al.* (2004), for example, has greatly improved the analysis and characterization of a class of tholins produced by photolysis and cold plasma irradiation of gaseous mixtures. Optically, the tholins are characterized by strong absorption in the ultraviolet and visible spectral regions giving them strong yellow, orange, and red colors, high reflectance at longer wavelengths, and (in some cases) absorption bands characteristic of aliphatic and aromatic hydrocarbons with varying amounts of substituted nitrogen (*Cruikshank et al.,* 2005a).

5.3. Color Classes Compared with Models

Groups BB and BR defined above have color spectra very similar to those of C-type and D-type asteroids. Going from neutral (BB group) to very red (RR group) spectra requires a higher content of organic materials to fit approximately the characteristic spectrum of the each group. Groups BB and BR in general, as discussed by *Cruikshank and Dalle Ore* (2003), do not require the presence of organic materials to reproduce their behavior, while IR and RR groups require large amounts of tholins on the surface to reproduce their strong colors [see *Roush and Cruikshank* (2004) and *Cruikshank et al.* (2005a) for additional discussion of tholins in the solar system]. Water ice seems to be present in some spectra of all the groups.

The RR group contains the reddest objects of the solar system. Some well-observed objects are members of this group, such as 5145 Pholus, 47171 (1999 TC$_{36}$), 55576 (2002 GB$_{10}$), 83982 (2002 GO$_9$), and 90377 Sedna. All these objects appear to contain at least few percent of H$_2$O ice on the surface, while 5145 Pholus has absorption bands of methanol ice (*Cruikshank et al.,* 1998) in addition to H$_2$O. No known material other than tholins can simultaneously reproduce their redness and low albedo.

The BB group contains objects having neutral reflectance spectra. Typical objects in the group are 2060 Chiron, 90482 Orcus, 19308 (1996 TO$_{66}$), and 15874 (1996 TL$_{66}$). The spectra are in general flat, sometime bluish in the NIR. The H$_2$O ice absorption bands seem generally stronger than in the other groups, although the spectrum of 1996 TL$_{66}$ is completely flat and the Chiron spectrum shows H$_2$O ice only at some times (it exhibits episodic cometary activity). Large amounts of amorphous carbon must be common to the members of this group.

The IR group is less red than the RR group. Typical members of this class are 20000 Varuna, 38628 Huya, 47932 (2000 GN$_{171}$), 26375 (1999 DE$_9$), and 55565 (2002 AW$_{197}$). Three of these objects show hydrous silicates on the surface.

The BR group is an intermediate group between BB and IR with colors closer to those of the IR group. Typical members of this class are 8405 Asbolus, 10199 Chariklo,

54598 Bienor, and 32532 Thereus. A few percent of H$_2$O is present on the surface of these objects, except Asbolus, for which *Barucci et al.* (2000) and *Romon-Martin et al.* (2002) found no ice absorption features during the object's complete rotational cycle.

The robust multivariate statistical analysis of broadband colors provides a strong indication for differences in the surface characteristics of the TNOs and Centaurs, while the classification scheme clearly indicates groupings with different physicochemical surface properties.

5.4. Nature Versus Nurture: What Do the Colors and Spectra of Transneptunian Objects Tell Us?

The color diversity at visible wavelengths of Kuiper belt objects and Centaurs has been recognized for a decade (e.g., *Luu and Jewitt,* 1996; *Tegler and Romanishin,* 1998; *Jewitt,* 2002). Investigators have noted many possible correlations between colors and dynamical characteristics, fueling what has become the TNO version of the classic "nature vs. nurture" debate in biology: Do the diverse colors reflect diverse primordial compositions, or do they result from diverse and different degrees of surface processing? Compelling arguments have been made on both sides, suggesting that both factors may influence colors.

The most robust color pattern reported to date is the redder visual colors of classical KBOs, first described by *Tegler and Romanishin* (2000). Classical KBOs are those in low-inclination, low-eccentricity, nonresonant orbits. Although the numbers vary, depending on which color index is used, it is clear that the color statistics of this subpopulation are significantly different from those of other TNO dynamical classes (e.g., *Boehnhardt et al.,* 2003), despite considerable diversity of opinions about how (or indeed, if) classical objects differ dynamically from other nonresonant dynamical classes.

Another color pattern that has been widely confirmed is a tendency for lower inclination objects to have redder colors (e.g., *Trujillo and Brown,* 2002; *Stern,* 2002a; *Doressoundiram et al.,* 2002; *Boehnhardt et al.,* 2002). However, since classical objects generally have lower inclinations and redder colors, and scattered objects have more neutral average colors and higher inclinations, much, if not all, of this trend could be due to the presence of red classical objects dominating the low-inclination end of the sample, and/or similar contamination from the more gray scattered population at the high-inclination end of the sample (e.g., *Peixinho et al.,* 2004; chapter by Chiang et al.).

Many additional trends have been proposed, and some will probably be confirmed as more data accumulate, but we will concentrate on the one (or two) described above. Part of our reasoning for restricting the focus at present is the risk of observational biases. We have already noted possible contamination of samples resulting from different dynamical classification schemes. Biases may also enter from inefficiencies in discovering particular types of objects. *Trujillo and Brown* (2002) noted an apparent paucity of gray objects with perihelion distances q > 40, but showed that a sampling bias was probably at work. *Peixinho et al.* (2004) explored the dependence of color trends with q on absolute magnitude, and found the trend only seemed to affect the fainter objects, those most likely to be afflicted by observational biases.

5.4.1. Environmental influences. Many environmental influences have been hypothesized. Radiolysis and photolysis are thought to progressively darken the surfaces of icy outer solar system materials over time, initially at blue wavelengths, leading to very red materials. As higher doses accumulate, red wavelengths darken and blue wavelengths may even brighten somewhat, culminating in spectrally neutral coatings (e.g., *Strazzulla et al.,* 1998, 2003; *Moroz et al.,* 2003, 2004). Irradiation of ices found in TNOs also produces distinct chemical changes that have been studied in the laboratory (e.g., *Moore et al.,* 2003; *Gerakines et al.,* 2004). Cratering by large impactors should excavate more pristine material from deep below the radiation-processed surface zone. *Luu and Jewitt* (1996) proposed a collisional resurfacing model in which continuous radiolytic reddening competes with sporadic impacts bringing grayer material to the surface. They were able to reproduce the observed color diversity, but the model predicted a strong size-dependence of color as well as significant color variations as objects rotate, neither of which has been observed (*Jewitt and Luu,* 2001).

Delsanti at al. (2004) folded cometary activity into the model, showing that activity could mask regional color variations on larger objects, which can reaccrete the bulk of the dust lofted during an active episode, and thus change colors in a globally uniform way. Since larger objects are generally the ones that are observed, this may offer a way around the absence of color variation with rotation. While some Centaurs are known to show continuous or episodic cometary activity (e.g., *Luu and Jewitt,* 1990; *Bauer et al.,* 2003b; *Choi and Weissman,* 2006), among the TNOs there are no reported cases of volatile activity in the current epoch.

Gil-Hutton (2002) and *Cooper et al.* (2003) applied a more realistic approach to radiolysis, noting that energetic particle fluxes vary considerably through the transneptunian region. Closer to the Sun, solar wind protons and solar UV photons drive increasingly rapid chemical processing, while further out, charged particles diffuse in from the heliosphere termination shock region, resulting in higher fluxes at large heliocentric distances as well. In high-flux regions, radiolysis could progress so rapidly that it makes surfaces gray, rather than red. Minimum radiolysis rates probably occur somewhere between 40 and 80 AU, where the classical KBOs reside. *Cooper et al.* (2003) also note the potential importance of high-velocity interstellar dust grains and sputtering processes, both of which erode surfaces, exposing much shallower materials than large impacts do. These shallow materials could be radiolytically reddened even before being exposed at the surface, offering a mechanism for maintaining red surfaces, at least in regions experiencing lower radiation doses, such as the classical belt.

Stern (2002a) recast the apparent trend of color with inclination among nonresonant objects in terms of mean random collision speed. Higher-inclination objects tend to suffer higher-speed collisions with cold disk objects, since they pass through the cold disk with higher velocities relative to the cold disk. This scenario favored a collisional influence on color, although the absence of convincing color trends with mean collision speed among the objects of any individual dynamical class argues against it. Also, *Thébault and Doressoundiram* (2003) have noted that collision frequencies should depend strongly on eccentricity, but compelling correlations between color and eccentricity have not been seen.

5.4.2. Compositional influences. The Kuiper belt may have experienced considerable dynamical churning since its objects accreted (e.g., chapter by Chiang et al.). The mixing that ensued frustrates efforts to search for the signature of compositional trends in the protosolar nebula in TNO colors, as was done so successfully in the main asteroid belt. However, compositionally distinct source regions may have existed, and could have resulted in compositionally distinct classes of TNOs.

Tegler and Romanishin (1998) presented perhaps the first evidence for a primordial compositional influence on TNO colors with their report of a bimodal color distribution. It has since emerged that the bimodal signature came from the Centaurs in their sample (*Peixinho et al.,* 2003; *Tegler and Romanishin,* 2003). Bimodal Centaur colors do not necessarily challenge collisional resurfacing scenarios, if there are two distinct dynamical sources feeding Centaurs into the region of the major planets. For instance, erosion of the inner edge of the classical belt and of the inner edge of the scattered disk could sample two distinct color distributions, both explainable by environmental factors.

Barucci et al. (2005a) have argued that even the non-Centaurs exhibit subtle clustering in their colors, rather than a continuum between the gray and red ends of the color distribution. These sorts of clusters are difficult to reconcile with collisional resurfacing scenarios, which generally predict continuous color distributions.

Comparing the *Barucci et al.* (2001) color classes with the dynamical classification scheme of Chiang et al. (this volume), we find several interesting patterns. First, classical objects almost all have RR-type colors, consistent with their well-known redder color distribution (e.g., *Tegler and Romanishin,* 2000). Second, Centaurs and resonant objects show remarkably similar G-mode color distributions, both groups being dominated by BR and RR colors. Third, scattered objects have much higher frequencies of BB objects than other dynamical classes, but otherwise look similar to the resonant and Centaur color distributions. Some caution is required here, since no effort was made to control for biases, which could play a significant role, since good color data are generally only available for brighter TNOs and discovery and recovery rates strongly favor intrinsically brighter objects, those that come closer to the Sun, and those having more "normal" types of orbits.

6. SIZES AND ALBEDOS OF TRANSNEPTUNIAN OBJECTS

The sizes of TNOs are of interest for several reasons: (1) Our estimates of the mass distribution in the outer solar nebula depend on the third power of TNO sizes. (2) Their size in relation to the sizes of Jupiter-family comets reveals something about both the size distribution of TNOs, from which these comets are derived (*Levison and Duncan,* 1997), and potentially about the mechanisms by which TNOs are perturbed onto cometary orbits. (3) Once the size of an object is known, its albedo is also known, providing an important constraint for interpreting colors and spectra (e.g., *Grundy and Stansberry,* 2004). Albedo diversity, to the extent that it is not correlated with color and spectral diversity, provides additional insight into the mechanisms underlying the remarkable spectral reflectance diversity of the TNO population.

The sizes of some outer solar system objects have been determined by a variety of methods. The sizes of Pluto and Charon have been measured via stellar occultation (e.g., *Elliot and Young,* 1992), and they have also been mapped using the Hubble Space Telescope (HST) (*Buie et al.,* 1997; *Young et al.,* 1999). Using HST, *Brown and Trujillo* (2004) resolved the TNO Quaoar, and placed an upper limit on the size of Sedna (*Brown et al.,* 2004). Recently, *Rabinowitz et al.* (2006) placed constraints on the size and albedo of 2003 EL$_{61}$ based on its short rotation period (3.9 h), the mass determined from the orbit of a satellite, and an analysis of the stability of a rapidly rotating ellipsoid. *Grundy et al.* (2005) used the masses of binary TNO systems and an assumed range of mass densities to place plausible limits on their sizes and albedos.

Measurements of thermal emission can also be used to also constrain the sizes, and thereby albedos, of unresolved targets. *Tedesco et al.* (2002) used Infrared Astronomical Satellite (IRAS) thermal detections of asteroids to build a catalog of albedos and diameters. IRAS also detected thermal emission from the Centaur object Chiron and the Pluto-Charon system, and those data were used to determine albedos and sizes for those objects (*Sykes et al.,* 1987, 1991, 1999). Advances in the sensitivity of far-infrared and submillimeter observatories have recently allowed the detection of thermal emission from a small sample of TNOs, providing the first meaningful constraints on their sizes and albedos (e.g., *Jewitt et al.,* 2001; *Lellouch et al.,* 2002; *Altenhoff et al.,* 2004).

6.1. The Radiometric Method

The radiometric method for determining albedos and sizes typically utilizes measurements of both the visible and thermal-infrared brightness of an object. The visible brightness is proportional to the product of an object's visible geometric albedo, p_V, and cross-sectional area, πr^2, while the thermal brightness depends on the bolometric albedo, A (which determines the temperature), and the cross-sec-

tional area. Given knowledge or an assumption for the phase integral, q (A = qp_V), measurements of the visible and thermal brightness can in principle be combined to solve directly for both the size of the object and its albedo.

In practice the thermal flux depends sensitively on the temperature distribution across the surface of the object: Warm regions near the subsolar point dominate the thermal emission, and their temperature depends on the thermal inertial of the surface and the rotation vector of the object as well as the albedo. Thermal inertia is a measure of the resistance of surface materials to a change in temperature; it is related to the thermal conductivity, heat capacity, and density. There are two commonly employed end-member models for the distribution of temperature: the slow-rotator, or standard thermal model (STM), and the fast-rotator, or isothermal latitude model (ILM) (see *Lebofsky and Spencer,* 1989). The STM is based on the assumption of a nonrotating (or equivalently, zero thermal inertia) spherical object, and the temperature distribution is specified by $T(\theta) = T_0 \cos^{1/4}\theta$, where $T_0 = [(1-A)S/(\eta\varepsilon\sigma)]^{1/4}$ is the subsolar point temperature. Here θ is the angular distance from the subsolar point, S is the solar constant at the distance of the object, ε the bolometric emissivity, and σ is the Stefan-Boltzmann constant. The beaming parameter, η, is an assignable factor that accounts for the nonuniform angular distribution of thermal emission (infrared beaming). The ILM incorporates the assumption that the object is illuminated at the equator and rotating very quickly (or equivalently, with infinite thermal inertia). The resulting temperature distribution is $T(\phi) = T_0 \cos^{1/4}\phi$, where in this case the subsolar point temperature is given by $T_0 = [(1-A)S/(\pi\eta\varepsilon\sigma)]^{1/4}$ (note the factor of π).

While more sophisticated extensions to the STM and ILM include the effects of surface roughness and thermal inertia (*Spencer,* 1990), as well as viewing geometry (*Harris,* 1998), their effects are semiquantitatively captured by the parameter η (e.g., *Spencer et al.,* 1989; *Stansberry et al.,* 2006). Furthermore, thermal measurements at multiple wavelengths near the blackbody peak directly give the color-temperature of the surface, in which case the systematic uncertainties between STM- and ILM-derived albedos and sizes (or those from thermophysical models) largely disappear (see Fig. 6). Here we only present results for objects that we have detected at both 24 and 70 μm with Spitzer. For all these reasons, we restrict our discussion of our Spitzer observations of the thermal emission from KBOs to albedos and diameters based on the STM.

6.2. Results of the Radiometric Method

Radiometric detections of TNOs have been made using the Infrared Space Observatory (ISO), the James Clerk Maxwell Telescope (JCMT) in Hawaii, the 30-m IRAM submillimeter telescope in Spain, and the Spitzer Space Telescope. *Thomas et al.* (2000) reported the first thermal detection of a TNO (excepting Pluto/Charon) based on ISO observations at a wavelength of 90 μm. *Altenhoff et al.*

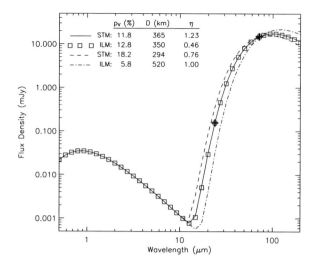

	pv (%)	D (km)	η
STM:	11.8	365	1.23
ILM:	12.8	350	0.46
STM:	18.2	294	0.76
ILM:	5.8	520	1.00

Fig. 6. Thermal models fitted to Spitzer data for (55565) 2002 AW$_{197}$ (*Cruikshank et al.*, 2005b). All the models fit the visual magnitude. The heavy solid line and circles show the STM and ILM models fitted to both the 24 and 70 μm data, where η was allowed to be a free parameter. The resulting diameters and albedos are very similar. The dashed and dash-dot lines show STM and ILM models with the canonical values for η, and are fitted to just the 70-μm data point. The resulting diameters and albedos are highly discrepant, demonstrating the power of measuring thermal emission in two bands for determining albedos and sizes via the radiometric method.

(2004) report submillimeter measurements or limits for six TNOs, and review the observations of Varuna (*Jewitt et al.*, 2001; *Lellouch et al.*, 2002) and 2002 AW$_{197}$ (*Margot et al.*, 2004). The IRAM submillimeter data were all taken at a wavelength of 1.2 mm. *Grundy et al.* (2005) also reanalyzed the submillimeter data from all four of those studies using a consistent thermal-modeling approach. *Cruikshank et al.* (2005b) and *Stansberry et al.* (2005, 2006) have reported Spitzer observations of six TNOs at wavelengths of 24 and 70 μm.

Table 2 summarizes the albedos and diameters of TNOs from all these studies, as well as providing values for the absolute magnitude and the spectral slope in the visible. In many cases *Grundy et al.* (2005) derived values from submillimeter observations that were in basic agreement with those found in the original studies. In these cases we have only included the original values. In a number of instances albedos and diameters are constrained by more than one measurement or method. In particular, a number of targets have been observed by both Spitzer and in the submillimeter. In general the submillimeter albedos are lower than those from Spitzer, and the submillimeter diameters correspondingly higher. While this discrepancy is not large, it may be systematic, and is significant at about the 2σ level (relative to flux uncertainties given in the studies) for individual objects. Another interesting case is 47171 (1999 TC$_{36}$). Both the Spitzer and submillimeter data indicate that

it is considerably larger than was deduced by *Grundy et al.* (2005) by assuming a minimum plausible density of 0.5 g cm^{-3} and using the binary mass to deduce the size. *Stansberry et al.* (2006) derive a density of 0.5 (0.3–0.8) g cm^{-3} using their Spitzer size, with the density being about half as large using the size from *Altenhoff et al.* (2004).

Excluding objects with upper limits on their thermal fluxes and those with albedos ≥25%, the average geometric albedo of TNOs in Table 2 is 9 ± 5%, with the extremes being 2.5% (2001 QC$_{298}$) and 19% (28978 Ixion). Of the objects with albedos ≥25%, three are binaries [(47171) 1999 TC$_{36}$, (58534) 1997 CQ$_{29}$, and (66652) 1999 RZ$_{253}$] with albedos and sizes derived from assumed densities, and one is the Pluto-class object 2003 EL$_{61}$. If these binaries have densities <0.5 g cm^{-3} [the lower limit assumed by *Grundy et al.* (2005)], their albedos will be lower than shown in Table 2, and more in accord with the measured albedos. This may be a hint of the existence of many more TNOs of very low density. Object 2003 EL$_{61}$, with an albedo of 65%, could have high-albedo deposits of volatile ices by reason of its extreme size and very distant orbit. Its spectrum shows strong absorptions due to water ice (*Trujillo et al.*, 2005), and places fairly strong limits on the amount of CH$_4$ ice that could be present unless the grain size is miniscule. However, solid N$_2$ is very difficult to detect (particularly in the α-phase), and could easily produce the high albedo. It seems likely that its albedo is representative of a class of super-TNOs (Pluto/Charon, 2003 UB$_{313}$, Sedna, 2005 FY$_9$, 2003 EL$_{61}$) rather than of more typical TNOs.

6.3. Size and Albedo Determinations from Thermal Spectra

Just as direct radiometric observations yield information on the sizes and albedos of outer solar system bodies, spectra in the thermal region (described in section 4) can be similarly used. The emissivity spectra shown in section 4 were created by dividing the measured spectral energy distribution (SED) by a model of the thermal continuum. An estimate of the size and albedo of a body can be obtained by allowing the radius and albedo to vary in the model in order to find the best thermal continuum fit to the SED, just as was done with the Spitzer MIPS radiometry in section 5, but with a different dataset. The absolute calibration of IRS has an uncertainty of ~20%, which propagates to uncertainties of ~5% in the size estimate (see *Emery et al.*, 2006, for further discussion). The derived effective radii and albedos using the standard thermal model (STM) are listed in Table 3; close agreement with the MIPS radiometry is found in most cases.

7.   MULTIPLICITY IN THE KUIPER BELT

The existence of numerous Kuiper belt binaries (hereafter abbreviated as KBB) has been revealed by a string of discoveries starting with the detection of the companion to 1998 WW$_{31}$ in 2001 (*Veillet et al.*, 2002). Since then, the dis-

TABLE 2. Albedos and diameters of TNOs.

Object	H_v	S (%/100 nm)	Spitzer and ISO p_v (%)	D (km)	STM η	Submillimeter p_v (%)	D (km)	Ref.
(15789) 1993 SC		27.7	3.5 (2.2–5.1)	398 (227–508)				t
(15874) 1996 TL$_{66}$		0.74	>1.8	<958				t
(47171) 1999 TC$_{36}$	5.37	23.3	7.9 (5.8–11)	405 (350–470)*	1.2 (1.0–1.4)	5 (4–6)	609 (562–702)	s,a
(29981) 1999 TD$_{10}$	9.1	8.4	5.3 (3.8–7.8)	88 (73–104)	1.3 (1.0–1.6)			s
(55565) 2002 AW$_{197}$	3.62	16.5	12 (8.8–18)	734 (599–857)	1.2 (1.0–1.4)	9 (7–11)	977 (890–1152)	s,m2
(28978) Ixion	4.04	16.8	19 (11–37)	480 (344–632)	0.6 (0.4–0.9)	>15	<804	s,a
(38628) Huya	5.1	17.7	6.6 (5.0–8.9)	500 (431–575)	1 (0.9–1.2)	>8	<540	s,a
(20000) Varuna	3.94	17.9	14 (8.0–23)	586 (457–776)	1.6 (1.2–2.3)	6 (4–8)	1016 (915–1218)	s,j
(20000) Varuna						7 (4–10)	914 (810–1122)	l
			Other Approaches		Method			
(19308) 1996 TO$_{66}$	4.77	1.382	>3.8	<902	submillimeter limit			a,g
(19521) Chaos	4.95	18.958	>6.6	<748	submillimeter limit			a,g
(24835) 1995 SM$_{55}$	4.58	2.097	>7.6	<705	submillimeter limit			a,g
(26308) 1998 SM$_{165}$	6.38	20.961	17 (9.5–24)	238 (184–292)	binary			m4,g
(47171) 1999 TC$_{36}$	5.39	23.322	25 (14–35)	309 (239–379)	binary			m4,g
(50000) Quaoar	2.74	20.392	10 (6.9–13)	1260 (1070–1450)	imaging			b2
(55636) 2002 TX$_{300}$	3.47	−0.471	>22	<712	submillimeter limit			or,g
(58534) 1997 CQ$_{29}$	7.38	18.401	44 (25–63)	77 (60–95)	binary			m4,n,g
(66652) 1999 RZ$_{253}$	6.03	23.094	31 (18–44)	170 (131–208)	binary			n,g
(84522) 2002 TC$_{302}$	4.94	0	>5.1	<1211	submillimeter limit			a,g
(88611) 2001 QT$_{297}$	7.01	20.346	9.8 (5.6–14)	168 (130–206)	binary			os,g
(90377) Sedna	1.2	35.924	>18	<1800	imaging			b4
1998 WW$_{31}$	7.76	0.477	6.0 (3.4–8.6)	152 (117–186)	binary			v,g
2001 QC$_{298}$	7.69	4.249	2.5 (1.4–3.5)	244 (189–299)	binary			m4,g
2003 EL$_{61}$	−0.26	0.36	65 (60–73)	2050 (1960–2500)	shape, rotation			

*Implies a density of 500 (300–800) kg m^{-3} (*Stansberry et al.,* 2006).

References: [a] *Altenhoff et al.* (2004); [b2] *Brown and Trujillo* (2004); [b4] *Brown and Trujillo* (2004); [g] *Grundy et al.* (2005); [j] *Jewitt et al.* (2001); [l] *Lellouch et al.* (2002); [m2] *Margot et al.* (2002); [m4] *Margot et al.* (2004); [n] *Noll et al.* (2004b); [os] *Osip et al.* (2003); [or] *Ortiz et al.* (2004); [v] *Veillet et al.* (2002).

TABLE 3. Best fit parameters for the standard thermal model (STM).

Object	H_v*	R (km)	p_v	η	T_{ss} (K)
Hektor	7.49	110.0	0.038	0.95	179.4
Agamemnon	7.89	71.5	0.062	0.89	175.3
Aneas	8.33	69.1	0.044	0.96	170.4
Pholus	7.65	74.2	0.072	1.42	85.4
Asbolus	8.96	50.7	0.046	0.91	148.3
1999 RG$_{33}$	12.10	8.1	0.100	0.42	143.3
1999 TD$_{10}$	8.64	49.4	0.065	1.48	97.0
Elatus	10.98	17.9	0.057	0.89	125.0

*H_v is from the IAU Minor Planet Center. We assume G = 0.15 for all objects except Pholus and Elatus, for which G = 0.16 and 0.13, respectively, have been derived.

coveries have accumulated rapidly so that now a total of 25 KBBs are known in 22 systems (Table 4).

Gravitationally bound systems are important in three ways. First, they are tools for measuring the masses of Kuiper belt objects by observing their mutual orbits. Second, the frequency of binaries in separate dynamical populations may provide important clues on the origin and survival of these bound systems. Third, they are an important detail in understanding the general nature of debris disks.

7.1. Orbital Properties of Kuiper Belt Binaries

Of the 22 known systems, orbits have been observed for a subset of nine. For each of these nine systems, observation of the relative separation of the components and their orientation at a number of different epochs forms the basis for determining the binary orbit. The steps involved in orbit determination, while conceptually simple, are often difficult in practice because of the observational challenges and consequent sparseness of most datasets. The large range of uncertainties shown in Table 4 reflects these observational challenges.

As with many other measurable quantities in the Kuiper belt, the binary orbits show a large diversity in semimajor axis, eccentricity, and period. The shortest period so far determined is for the Pluto/Charon system at 6.4 d while the longest period (2001 QT$_{297}$) is roughly 100× greater. The most widely separated system, 2001 QW$_{322}$, had components separated by ~4 arcsec at the time of discovery (*Kavelaars et al.,* 2001). The lower limit is set by observational limitations with the closest reported binary separation being the 15 milliarcsec (0.2 pixel) separation of 1995 TL$_8$, an unresolved pair identified through PSF-fitting of data obtained with HST (*Stephens and Noll,* 2006). Theoretical studies (*Goldreich et al.,* 2002) do not indicate any reason

TABLE 4.   Binary Kuiper belt objects.

Object	Class*	a (km)[†]	e[†]	P (d)[†]	M System (Zg)[†]	Ref.
Pluto (Charon)	3:2	19,636(8)	0.0076(5)	6.38722(2)	14,600(100)	[1]
(P1)						[2],[3]
(P2)						[2],[3]
(26308) 1998 SM$_{165}$	2:1	11,310(110)		130(1)	6.8(2)	[4],[5]
(47171) 1999 TC$_{36}$	3:2	7,640(460)		50.4(5)	14(3)	[4],[5]
2003 EL$_{61}$	S	49,500(400)	0.050(3)	49.12(3)	4,200(100)	[6]
					[7]	
2001 QC$_{298}$	S	3,690(70)		19.2(2)	10.8(7)	[4],[5]
(82075) 2000 YW$_{134}$	S					[8]
(48639) 1995 TL$_{8}$	S					[8]
2003 UB$_{313}$	S					[7]
(88611) 2001 QT$_{297}$	C	31,409(2500)	0.31(8)	876(227)	3.2(3/2)	[9],[10]
1998 WW$_{31}$	C	22,300(800)	0.82(5)	574(10)	2.7	[11]
(58534) 1997 CQ$_{29}$	C	8,010(80)	0.45(3)	312(3)	0.42(2)	[12]
(66652) 1999 RZ$_{253}$	C	4,660(170)	0.460(13)	46.263(6/74)	3.7(4)	[13]
2001 QW$_{322}$	C					[14]
2000 CF$_{105}$	C					[15]
2000 CQ$_{114}$	C					[16]
2003 UN$_{284}$	C					[17]
2003 QY$_{90}$	C					[18]
2005 EO$_{304}$	C					[19]
(80806) 2000 CM$_{105}$	C					[8]
2000 OJ$_{67}$	C					[8]
(79360) 1997 CS$_{29}$	C					[8]
1999 OJ$_{4}$	C					[8]

*Dynamical classes: S = scattered disk, C = cold classical, n:m = resonant.
[†]Uncertainty in last significant digit(s) indicated in parentheses.

References: [1] *Tholen and Buie* (1997); [2] *Weaver et al.* (2006); [3] *Buie et al.* (2006); [4] *Margot et al.* (2004); [5] Grundy and Noll (personal communication, 2004); [6] *Brown et al.* (2005); [7] *Brown et al.* (2006); [8] *Stephens and Noll* (2006); [9] *Osip et al.* (2003); [10] *Kern* (2005); [11] *Veillet et al.* (2002); [12] *Noll et al.* (2004a); [13] *Noll et al.* (2004b); [14] *Kavelaars et al.* (2001); [15] *Noll et al.* (2002); [16] *Stephens et al.* (2004); [17] *Millis and Clancy* (2003); [18] *Elliot et al.* (2003); [19] *Kern and Elliot* (2005).

to suspect a lower limit to binary separation, and the large-amplitude, long-period lightcurve of 2001 QG$_{298}$ suggests it as a candidate member of a potentially significant population (~10–20%) of near-contact binaries (*Sheppard and Jewitt,* 2004).

So far, none of the nine measured systems have orbits that are known well enough that the timing of mutual transits and occultations can be predicted with certainty. Because the KBB components typically have diameters of hundreds of kilometers, mutual event observability is strongly biased toward systems with shorter periods (smaller semimajor axes). These are among the most recently discovered and least well characterized in terms of orbits. Several ongoing observational programs are measuring the orbits of additional KBBs, so the list is likely to expand and uncertainties are likely to decrease on relatively short timescales.

7.2. Physical Properties of Kuiper Belt Binaries and Multiples

The determination of mutual orbital period and semimajor axis allows the measurement of the total system mass following Kepler's second law. For six of the nine systems the mass is determined to an uncertainty of 10% or better.

Masses range from a low of 0.42 Zg (1 Zg = 10^{18} kg) for (58534) 1997 CQ$_{29}$ to 14,600 Zg for Pluto.

Even in the absence of size information, the availability of mass information makes it possible to constrain two other physical parameters of interest, albedo and density. If photometric data at optical or near-infrared wavelengths are available (i.e., the reflected sunlight portion of the spectrum) it is possible to constrain albedo and density to lie along a well-defined function with albedo varying as the 2/3 power of the density. This derived information can be usefully reported as the geometric albedo in a given wavelength band at a reference density; a typical choice is to report the Cousins R band geometric albedo for an assumed density of 1000 kg/m^3. Derived albedos using this method show a very broad range (*Noll et al.,* 2004a,b; J.-L. Margot, personal communication), suggesting significant differences among the surfaces of KBBs. No strong trends of albedo of KBOs with other measurable quantities appear in the current set of available data, which includes both binaries and singles with measured albedos (*Grundy et al.,* 2005). The calculation of geometric albedo is subject to an uncertain phase correction. Most authors have used β = 0.15 mag/degree (G = –0.21) found by *Sheppard and Jewitt* (2002). The binary 1997 CQ$_{29}$ has a surprisingly high albedo of p_R = 0.37

($\rho = 1000$ kg/m^3)$^{2/3}$ for a primary that would be only ~40 km in diameter (*Noll et al.*, 2004a).

When the diameter of at least one component of a binary is known, either from direct measurements or indirectly from thermal emission observations and modeling, it is possible to uniquely constrain the density of the object. This is useful for removing the density dependence of albedo, as discussed above. Even more importantly, the density provides constraints on the internal structure of an object including constraints on mixtures of cosmogonically plausible compositions and internal macro- and microporosity. Beyond the Pluto-Charon binary (*Buie et al.*, 2006), this information is currently published for only two objects, (47171) 1999 TC$_{36}$ (*Stansberry et al.*, 2005) and 2003 EL$_{61}$ (*Brown et al.*, 2005). *Stansberry et al.* (2005) find a density of 550–800 kg/m^3 for (47171) 1999 TC$_{36}$. The extremely low density requires a very high porosity for plausible compositions and at the lowest densities, highest porosities raises the question of whether it can be consistent with compaction expected in a self-supported sphere of diameter ~325–425 km. One possible explanation suggested by Stansberry et al. is that the primary of (47171) 1999 TC$_{36}$ is itself a close binary. By decreasing the enclosed volume for a given surface area, the apparently anomalously low density is partly mitigated.

7.3. Formation and Destruction of Kuiper Belt Binaries

Objects in the Kuiper belt are grouped by their heliocentric orbital properties into dynamical families. Binaries detectable with current data appear to be significantly more common in the so-called cold classical Kuiper belt than in other populations (*Stephens and Noll*, 2006). The reason for the prevalence of binaries among the dynamically cold objects in the Kuiper belt may be related to formation mechanisms and/or to subsequent events that may have destroyed preexisting binaries.

None of the formation models proposed to date can form binaries in the current Kuiper belt (*Stern*, 2002b; *Weidenschilling*, 2002; *Goldreich et al.*, 2002; *Funato et al.*, 2004; *Astakhov et al.*, 2005); all require a 100-fold or more higher density of objects. This points to a primordial origin for KBBs and is consistent with models of early solar system evolution that predict 2 or more orders of magnitude reduction in the surface mass density in the Kuiper belt from the orbital migration of Neptune (*Levison and Morbidelli*, 2003). The observed prevalence of binary systems with similar-mass components (i.e., m_2/m_1 on the order of 1) suggests chaos-assisted capture as the most likely mode of formation for most of the known KBBs (*Astakhov et al.*, 2005). Models of KBB survival predict the loss of roughly 10× the current number of widely separated KBBs, although most of the smaller-separation binaries can survive for 4.5 G.y. (*Petit and Mousis*, 2004).

Two multiple systems are now known in the transneptunian region. The satellites in these systems have small masses relative to their primaries, which together with their orbital properties and the large masses of the primaries

suggests a mode of formation other than capture. Preliminary orbits determined for the two newly discovered satellites of Pluto are consistent with zero or low eccentricity (P2 and P1 respectively) (*Weaver et al.*, 2006; *Buie et al.*, 2006), suggestive of formation in a post-giant-impact accretion disk (*Stern et al.*, 2006) that would also be consistent with the proposed impact origin for Charon (*Canup*, 2005). Object 2003 EL$_{61}$ has two reported satellites (*Brown et al.*, 2005, 2006) and the authors speculate that these may also be compatible with an impact origin. It appears that, as is the case with other physical properties of KBOs, bound systems in the Kuiper belt come in a fascinatingly diverse variety, begging for further study.

8. SUMMARY AND CONCLUSIONS

TNOs are remnants of the protoplanetary disk in which the accretion of the planetesimals that built the giant planets occurred. TNO population statistics and dynamics demonstrate that the region beyond Neptune has not been stable since the planets formed, having been dynamically disrupted by the outward migration of Uranus and Neptune, with the possibility of significant mass loss resulting from mutual collisions.

In this chapter we have shown that the physical properties of TNOs are highly varied in every respect: size, color, reflectance (albedo), surface composition, mean density, and multiplicity. The diversity in all these parameters suggests that the TNOs represent a substantial range in original bulk composition, with differing fractions of ices, silicates, and organic solids; all these components are found in interstellar dust in giant molecular clouds. An alternative view suggests that varying degrees of postaccretional processing of bodies of approximately uniform composition have produced the colorful tableau of physical properties that has emerged from just over a decade of physical studies of these objects.

REFERENCES

Altenhoff W. J., Bertoldi F., and Menten K. M. (2004) *Astron. Astrophys., 415,* 771–775.
Astakhov S. A., Lee E. A., and Farrelly D. (2005) *Mon. Not. R. Astron. Soc., 360,* 401–415.
Barucci M. A., Capria M. T., Coradini A., and Fulchignoni M. (1987) *Icarus, 72,* 304–324.
Barucci M. A., de Bergh C., Cuby J.-G., Le Bras A., Schmitt B., and Romon J. (2000) *Astron. Astrophys., 357,* L53–L56.
Barucci M. A., Fulchignoni M., Birlan M., Doressoundiram A., Romon J., and Boehnhardt H. (2001) *Astron. Astrophys., 371,* 1150–1154.
Barucci M. A., Boehnhardt H., Dotto E., Doressoundiram A., Romon J., et al. (2002) *Astron. Astrophys., 392,* 335–339.
Barucci M. A., Belskaya I. N., Fulchignoni M., and Birlan M. (2005a) *Astron. J., 130,* 1291–1298.
Barucci M. A., Cruikshank D. P., Dotto E., Merlin F., Poulet F., et al. (2005b) *Astron. Astrophys., 439,* L1–L4.
Bauer J. M., Meech K. J., Fernández Y. R., Pittichova J., Hainaut O. R., Boehnhardt H., and Delsanti A. C. (2003a) *Icarus, 166,* 195–211.
Bauer J. M., Fernandez Y. R., and Meech K. J. (2003b) *Publ. Astron. Soc. Pac., 115,* 981–989.
Boehnhardt H., Delsanti A., Barucci A., Hainaut O., Doressoundiram A., et al. (2002) *Astron. Astrophys., 395,* 297–303.

Boehnhardt H., Barucci M. A., Delsanti A., De Bergh C., Doressoundiram A., et al. (2003) *Earth Moon Planets, 92,* 145–156.

Brown M. E. and Trujillo C. A. (2004) *Astron. J., 127,* 2413–2417.

Brown M. E., Trujillo C. A., and Rabinowitz D. (2004) *Astrophys. J., 617,* 645–649.

Brown M. E., Trujillo C. A., and Rabinowitz D. L. (2005) *Astrophys. J., 635,* L97–L100.

Brown M. E., van Dam M. A., Bouchez A. H., Le Mignant D., Campbell R. D., et al. (2006) *Astrophys. J., 639,* L43–L46.

Brown R. H., Cruikshank D. P, and Pendleton Y. (1999) *Astrophys. J., 519,* L101–L104.

Buie M. W., Tholen D. J., and Wasserman L. H. (1997) *Icarus, 125,* 233–244.

Buie M. W., Grundy W. M., Young E. F., Young L. A., and Stern S. A. (2006) *Astron. J., 132,* 290–298.

Canup R. M. (2005) *Science, 307,* 546–550.

Choi Y. J. and Weissman P. R. (2006) *IAU Circ. 8656.*

Christensen P. R., Bandfield J. L., Hamilton V. E., Howard D. A., Lane M. D., Piatek J. L., Ruff S. W., and Stefanov W. L. (2000) *J. Geophys. Res., 105(E4),* 9735–9739.

Clark B. E., Hapke B., Pieters C., and Britt D. (2002) In *Asteroids III* (W. F. Bottke Jr. et al., eds.), pp. 585–599. Univ. of Arizona, Tucson.

Coll P., Coscia D., Smith N., Gazeau M.-C., Ramírez S. I., et al. (1999) *Planet Space Sci., 47,* 1331–1340.

Cooper B. L., Salisbury J. W., Killen R. M., and Potter A. E. (2002) *J. Geophys. Res., 107(E4),* 1–12.

Cooper J. F., Christian E. R., Richardson J. D., and Wang C. (2003) *Earth Moon Planets, 92,* 261–277.

Cruikshank D. P. (1977) *Icarus, 30,* 224–230.

Cruikshank D. P. and Dalle Ore C. M. (2003) *Earth Moon Planets, 92,* 315–330.

Cruikshank D. P., Roush T. L., Owen T. C., Geballe T. R., de Bergh C., et al. (1993) *Science, 261,* 742–745.

Cruikshank D. P., Roush T. L., Bartholomew M. J., Geballe T. R., Pendleton Y. J., et al. (1998) *Icarus, 135,* 389–407.

Cruikshank D. P., Schmitt B., Roush T. L., Owen T. C., Quirico E., et al. (2000) *Icarus, 147,* 309–316.

Cruikshank D. P., Dalle Ore C. M., Roush T. L., Geballe T. R., Owen T. C., et al. (2001) *Icarus, 153,* 348–360.

Cruikshank D. P., Roush T. L., and Poulet F. (2004) *C. R. Physique, 4,* 783–789.

Cruikshank D. P., Imanaka H., and Dalle Ore C. M. (2005a) *Adv. Space Res., 36,* 178–183.

Cruikshank D. P., Stansberry J. A., Emery J. P., Fernández Y. R., Werner M. W., Trilling D. E., and Rieke G. H. (2005b) *Astrophys. J., 624,* L53–L56.

de Bergh C., Boehnhardt H., Barucci M. A., Lazzarin M., Fornasier S., et al. (2004) *Astron. Astrophys., 416,* 791–798.

de Bergh C., Delsanti A., Tozzi G. P., Dotto E., Doressoundiram A., and Barucci M. A. (2005) *Astron. Astrophys., 437,* 1115–1120.

Delsanti A., Hainaut O., Jourdeuil E., Meech K. J., Boehnhardt H., and Barrera L. (2004) *Astron. Astrophys., 417,* 1145–1158.

Doressoundiram A., Peixinho N., de Bergh C., Fornasier S., Thébault P., Barucci M. A., and Veillet C. (2002) *Astron. J., 124,* 2279–2296.

Doressoundiram A., Tozzi G. P., Barucci M. A., Boehnhardt H., Fornasier S., and Romon J. (2003) *Astron. J., 125,* 2721–2727.

Doressoundiram A., Barucci M. A., Boehnhardt H., Tozzi G. P., Poulet F., de Bergh C., and Peixinho N. (2005a) *Planet. Space Sci., 53,* 1501–1509.

Doressoundiram A., Peixinho N., Doucet C., Mousis O., Barucci M. A., Petit J. M., and Veillet C. (2005b) *Icarus, 174,* 90–104.

Dotto E., Barucci M. A., Boehnhardt H., Romon J., Doressoundiram A., Peixinho N., de Bergh C., and Lazzarin M. (2003a) *Icarus, 164,* 408–414.

Dotto E., Barucci M. A., Leyrat C., Romon J., de Bergh C., and Licandro J.-M. (2003b) *Icarus, 164,* 122–126.

Douté S., Schmitt B., Quirico E., Owen T. C., Cruikshank D. P., et al. (1999) *Icarus, 142,* 421–444.

Dumas C., Owen T. C., and Barucci M. A. (1998) *Icarus, 133,* 221–232.

Elliot J. L. and Young L. A. (1992) *Icarus, 103,* 991–1015.

Elliot J. L., Kern S. D., and Clancy K. B. (2003) *IAU Circ. 8235.*

Emery J. P. and Brown R. H. (2003) *Icarus, 164,* 104–121.

Emery J. P., Cruikshank D. P., and Van Cleve J. (2006) *Icarus, 182,* 496–512.

Fernández Y. R., Sheppard S. S., and Jewitt D. C. (2003) *Astron. J., 126,* 1563–1574.

Fornasier S., Dotto E., Barucci M. A., and Barbieri C. (2004a) *Astron. Astrophys., 422,* L43–L46.

Fornasier S., Dotto E., Marzari F., Barucci M. A., Boehnhardt H., Hainaut O., and de Bergh C. (2004b) *Icarus, 172,* 221–232.

Fulchignoni M., Birlan M., and Barucci M. A. (2000) *Icarus, 146,* 204–212.

Funato Y., Makino J., Hut P., Kokubo E., and Kinoshita D. (2004) *Nature, 427,* 518–520.

Gerakines P. A., Moore M. H., and Hudson R. L. (2004) *Icarus, 170,* 202–213.

Gil-Hutton R. (2002) *Planet. Space Sci., 50,* 57–62.

Goldreich P., Lithwick Y., and Sari R. (2002) *Nature, 420,* 643–646.

Gradie J. and Veverka J. (1980) *Nature, 283,* 840–842.

Grundy W. M. and Stansberry J. A. (2004) *Earth Moon Planets, 92,* 331–336.

Grundy W. M., Noll K. S., and Stephens D. C. (2005) *Icarus, 176,* 184–191.

Hapke B. (1981) *J. Geophys. Res., 96,* 3039–3054.

Hapke B. (1993) *Theory of Reflectance and Emittance Spectroscopy.* Cambridge Univ., New York. 455 pp.

Hapke B. (1996) *J. Geophys. Res., 101,* 16833–16840.

Harris A. W. (1998) *Icarus, 131,* 291–301.

Imanaka H., Khare B. N., Elsila J. E., Bakes E. L. O., McKay C. P., et al. (2004) *Icarus, 168,* 344–366.

Jewitt D. C. (2002) *Astron. J., 123,* 1039–1049.

Jewitt D. C. and Luu J. X. (1990) *Astron. J., 100,* 933–944.

Jewitt D. C. and Luu J. X. (1993) *Nature, 362,* 730–732.

Jewitt D. C. and Luu J. X. (2001) *Astron. J., 122,* 2099–2114.

Jewitt D. C. and Luu J. X. (2004) *Nature, 432,* 731–733.

Jewitt D. C., Aussel H., and Evans A. (2001) *Nature, 411,* 446–447.

Kavelaars J. J., Petit J.-M., Gladman B., and Holman M. (2001) *IAU Circ. 7749.*

Keller H. U., Britt D., Buratti B. J., and Thomas N. (2004) In *Comets II* (M. C. Festou et al., eds.), pp. 211–222. Univ. of Arizona, Tucson.

Kern S. D. (2005) Ph.D. thesis, Massachusetts Institute of Technology.

Kern S. D. and Elliot J. L. (2005) *IAU Circ. 8526.*

Kerridge J., Chang S., and Shipp R. (1987) *Geochim. Cosmochim. Acta, 51,* 2527–2540.

Khare B. N., Sagan C., Arakawa E. T., Suits R., Callcot T. A., and Williams M. W. (1984) *Icarus, 60,* 127–137.

Lazzarin M., Barucci M. A., Boehnhardt H., Tozzi G. P., de Bergh C., and Dotto E. (2003) *Astrophys. J., 125,* 1554–1558.

Lebofsky L. A. and Spencer J. R. (1989) In *Asteroids II* (R. P. Binzel et al., eds.), pp. 128–147. Univ. of Arizona, Tucson.

Lellouch E., Moreno R., Ortiz J. L., Paubert G., Doressoundiram A., and Peixinho N. (2002) *Astron. Astrophys., 391,* 1133–1139.

Levison H. F. and Duncan M. J (1997) *Icarus, 127,* 13–32.

Levison H. F. and Morbidelli A. (2003) *Nature, 426,* 419–421.

Luu J. and Jewitt D. (1990) *Astron. J., 100,* 913–932.

Luu J. and Jewitt D. (1996) *Astron. J., 112,* 2310–2318.

Luu J. X., Jewitt D. C., and Cloutis E. (1994) *Icarus, 109,* 133–144.

Margot J. L., Trujillo C., Brown M. E., and Bertoldi F. (2002) *Bull. Am. Astron. Soc., 34,* 871 (abstract).

Margot J. L., Brown M. E., Trujillo C. A., and Sari R. (2004) *Bull. Amer. Astron. Soc., 36,* 1081 (abstract).

McEwen A. S. (1991) *Icarus, 92,* 298–311.

McKinnon W. B, Lunine J. I., and Banfield D. (1995) In *Neptune and Triton* (D. P. Cruikshank, ed.), pp. 807–877. Univ. of Arizona, Tucson.

Merlin F., Barucci M. A., Dotto E., de Bergh C., and Lo Curto G. (2005) *Astron. Astrophys., 444,* 977–982.

Millis R. L. and Clancy K. B. (2003) *IAU Circ. 8251.*

Moore M. H., Hudson, R. L., and Ferrante R. F (2003) *Earth Moon Planets, 92,* 291–306.

Moroz L. V., Baratta G., Distefano E., Strazzulla G., Starukhina L. V., Dotto E., and Barucci M. A. (2003) *Earth Moon Planets, 92,* 279–289.

Moroz L. V., Baratta G., Strazzulla G., Starukhina L., Dotto E., Barucci M. A., Arnold G., and Distefano E. (2004) *Icarus, 170,* 214–228.

Noll K., Stephens D., Grundy W., Spencer J., Millis R., et al. (2002) *IAU Circ. 7857.*

Noll K. S., Stephens D. C., Grundy W. M., Osip D. J., and Griffin I. (2004a) *Astron. J., 128,* 2547–2552.

Noll K. S., Stephens D. C., Grundy W. M., and Griffin I. (2004b) *Icarus, 172,* 402–407.

Ortiz J. L., Sota A., Moreno R., Lellouch E., Biver N., et al. (2004) *Astron. Astrophys., 420,* 383–388.

Osip D. J., Kern S. D., and Elliot J. L. (2003) *Earth Moon Planets, 92,* 409–421.

Owen T. C., Cruikshank D. P., Roush T., DeBergh C., Brown R. H., et al. (1993) *Science, 261,* 745–748.

Peixinho N., Doressoundiram A., Delsanti A., Boehnhardt H., Barucci M. A., and Belskaya I. (2003) *Astron. Astrophys., 410,* L29–L32.

Peixinho N., Boehnhardt H., Belskaya I., Doressoundiram A., Barucci M. A., and Delsanti A. (2004) *Icarus, 170,* 153–166.

Pendleton Y. J. and Allamandola L. J. (2002) *Astrophys. J. Suppl., 128,* 75–98.

Petit J.-M. and Mousis O. (2004) *Icarus, 168,* 409–419.

Poulet F., Cuzzi J. N., Cruikshank D. P., Roush T., and Dalle Ore C. M. (2002) *Icarus, 160,* 313–324.

Quirico E., Douté S., Schmitt B., de Bergh C., Cruikshank D. P., et al. (1999) *Icarus, 139,* 159–178.

Rabinowitz D. L., Barkume K., Brown M. E., Roe H., Schwartz M., Tourtellotte S., and Trujillo C. (2006) *Astrophys. J., 639,* 1238–1251.

Ramírez S. I., Coll P., da Silva A., Navarro-González R., Lafait J., and Raulin F. (2002) *Icarus, 156,* 515–529.

Rietmeijer F. J. M., Nuth J. A. III, and Nelson R. N. (2004) *Meteoritics & Planet. Sci., 39,* 723–746.

Romon-Martin J., Barucci M. A., de Bergh C., Doressoundiram A., Peixinho N., and Poulet F. (2002) *Icarus, 160,* 59–65.

Roush T. L. and Cruikshank D. (2004) In *Astrobiology: Future Prospectives* (P. Ehrenfreund, ed.), pp. 149–177. Kluwer, Dordrecht.

Salisbury J. W., Walter L. S., Vergo N., and D'Aria D. M. (1992) *Mid-Infrared (2.1–25 µm) Spectra of Minerals.* Johns Hopkins Univ., Baltimore. 346 pp.

Sheppard S. S. and Jewitt D. C. (2002) *Astron. J., 124,* 1757–1775.

Sheppard S. S. and Jewitt D. (2004) *Astron. J., 127,* 3023–3033.

Shkuratov Y., Starukhina L., Hoffmann H., and Arnold G. (1999) *Icarus, 137,* 235–246.

Spencer J. R. (1990) *Icarus, 83,* 27–38.

Spencer J. R., Lebofsky L. A., and Sykes M. V. (1989) *Icarus, 78,* 337–354.

Stansberry J. A., Cruikshank D. P., Grundy W. G., Margot J. L., Emery J. P., Fernández Y. R., and Rieke G. H. (2005) *Bull. Am. Astron. Soc., 37,* 737 (abstract).

Stansberry J. A., Grundy W. M., Margot J .L., Cruikshank D. P., Emery J. P., Rieke G. H., and Trilling D. E. (2006) *Astrophys. J., 643,* 556–566.

Stephens D. C. and Noll K. S. (2006) *Astron. J., 131,* 1142–1148.

Stephens D. C., Noll K. S., and Grundy W. (2004) *IAU Circ. 8289.*

Stern S. A. (2002a) *Astron. J., 124,* 2297–2299.

Stern S. A. (2002b) *Astron. J., 124,* 2300–2304.

Stern S. A., Weaver H. A., Steffl A. J., Mutchler M. J., Merline W. J., et al. (2006) *Nature, 439,* 946–948.

Strazzulla G. (1998) In *Solar System Ices* (B. Schmitt et al., eds.), pp. 281–301. Kluwer, Dordrecht.

Strazzulla G. (2003) *C. R. Physique, 4,* 791–801.

Sykes M. V. (1999) *Icarus, 142,* 155–159.

Sykes M. V. and Walker R. G. (1991) *Science, 251,* 777–780.

Sykes M. V., Cutri R. M., Lebofsky L. A., and Binzel R. P. (1987) *Science, 237,* 1336–1340.

Tedesco E. F., Noah P. V., Noah M., and Price S. D. (2002) *Astron. J., 123,* 1056–1085.

Tegler S. C. and Romanishin W. (1998) *Nature, 392,* 49–51.

Tegler S. C. and Romanishin W. (2000) *Nature, 407,* 979–981.

Tegler S. C. and Romanishin W. (2003) *Icarus, 161,* 181–191.

Tegler S. C., Romanishin W., and Consolmagno G. J. S. J. (2003) *Astrophys. J., 599,* L49–L52.

Thomas N., Eggers S., Ip W.-H., Lichtenberg G., Fitzsimmons A., Jorda L., et al. (2000) *Astrophys. J., 534,* 446–455.

Thébault P. and Doressoundiram A. (2003) *Icarus, 162,* 27–37.

Tholen D. J. (1984) Ph.D. thesis, Univ. of Arizona.

Tholen D. J. and Barucci M. A. (1989) In *Asteroids II* (R. P. Binzel et al., eds.), pp. 298–315. Univ. of Arizona, Tucson.

Tholen D. J. and Buie M. W. (1997) In *Pluto and Charon* (S. A. Stern and D. J. Tholen, eds.), pp. 298–315. Univ. of Arizona, Tucson.

Tran B. N., Ferris J. P., and Chera J. J. (2003) *Icarus, 162,* 114–124.

Trujillo C. A. and Brown M. E. (2002) *Astrophys. J., 566,* L125–L128.

Trujillo C. A., Brown M. E., Rabinowitz D. L., and Geballe T. R. (2005) *Astrophys. J., 627,* 1057–1065.

Veillet C., Parker J. W., Griffin I., Marsden B., Doressoundiram A., et al. (2002) *Nature, 416,* 711–713.

Vilas F. and Gaffey M. J. (1989) *Science, 246,* 790.

Virbiscer A. and Helfenstein P. (1998) In *Solar System Ices* (B. Schmitt et al., eds.), pp. 157–197. Kluwer, Dordrecht.

Weaver H. A., Stern S. A., Mutchler M. J., Steffl A. J., Buie M. W., et al. (2006) *Nature, 439,* 943–945.

Weidenschilling S. J. (2002) *Icarus, 160,* 212–215.

Young E. F., Galdamez K., Buie M. W., Binzel R. P., and Tholen D. J. (1999) *Astron. J., 117,* 1063–1076.

A Brief History of Transneptunian Space

Eugene Chiang, Yoram Lithwick, and Ruth Murray-Clay
University of California at Berkeley

Marc Buie and Will Grundy
Lowell Observatory

Matthew Holman
Harvard-Smithsonian Center for Astrophysics

The Edgeworth-Kuiper belt encodes the dynamical history of the outer solar system. Kuiper belt objects (KBOs) bear witness to coagulation physics, the evolution of planetary orbits, and external perturbations from the solar neighborhood. We critically review the present-day belt's observed properties and the theories designed to explain them. Theories are organized according to a possible timeline of events. In chronological order, epochs described include (1) coagulation of KBOs in a dynamically cold disk, (2) formation of binary KBOs by fragmentary collisions and gravitational captures, (3) stirring of KBOs by Neptune-mass planets ("oligarchs"), (4) eviction of excess oligarchs, (5) continued stirring of KBOs by remaining planets whose orbits circularize by dynamical friction, (6) planetary migration and capture of resonant KBOs, (7) creation of the inner Oort cloud by passing stars in an open stellar cluster, and (8) collisional comminution of the smallest KBOs. Recent work underscores how small, collisional, primordial planetesimals having low velocity dispersion permit the rapid assembly of ~5 Neptune-mass oligarchs at distances of 15–25 AU. We explore the consequences of such a picture. We propose that Neptune-mass planets whose orbits cross into the Kuiper belt for up to ~20 m.y. help generate the high-perihelion members of the hot classical disk and scattered belt. By contrast, raising perihelia by sweeping secular resonances during Neptune's migration might fill these reservoirs too inefficiently when account is made of how little primordial mass might reside in bodies having sizes on the order of 100 km. These and other frontier issues in transneptunian space are discussed quantitatively.

1. INTRODUCTION

The discovery by *Jewitt and Luu* (1993) of what many now regard as the third Kuiper belt object opened a new frontier in planetary astrophysics: the direct study of transneptunian space, that great expanse extending beyond the orbit of the last known giant planet in our solar system. This space is strewn with icy, rocky bodies having diameters ranging up to a few thousand kilometers and occupying orbits of a formerly unimagined variety.

Kuiper belt objects (KBOs) afford insight into processes that form and shape planetary systems. In contrast to main-belt asteroids, the largest KBOs today have lifetimes against collisional disruption that well exceed the age of the universe. Therefore their size spectrum may preserve a record, unweathered by erosive collisions, of the process by which planetesimals and planets coagulated. At the same time, KBOs can be considered test particles whose trajectories have been evolving for billions of years in a time-dependent gravitational potential. They provide intimate testimony of how the giant planets — and perhaps even planets that once resided within our system but have since been ejected — had their orbits sculpted. The richness of structure revealed by studies of our homegrown debris disk is unmatched by more distant, extrasolar analogs.

Section 2 summarizes observed properties of the Kuiper belt. Some of the data and analyses concerning orbital elements and spectral properties of KBOs are new and have not been published elsewhere. Section 3 is devoted to theoretical interpretation. Topics are treated in order of a possible chronology of events in the outer solar system. Parts of the story that remain missing or that are contradictory are identified. Section 4 recapitulates a few of the bigger puzzles.

Our review is packed with simple and hopefully illuminating order-of-magnitude calculations that readers are encouraged to reproduce or challenge. Some of these confirm claims made in the literature that would otherwise find no support apart from numerical simulations. Many estimates are new, concerning all the ways in which Neptune-sized planets might have dynamically excited the Kuiper belt. While we outline many derivations, space limitations prevent us from spelling out all details. For additional guid-

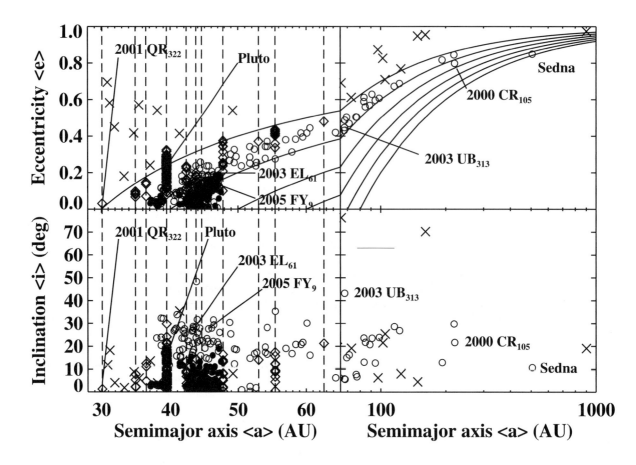

Fig. 1. Orbital elements, time-averaged over 10 m.y., of 529 securely classified outer solar system objects as of October 8, 2005. Symbols represent dynamical classes: Centaurs (×), resonant KBOs (◇), classical KBOs (●), and scattered KBOs (○). Dashed vertical lines indicate occupied mean-motion resonances; in order of increasing heliocentric distance, these include the 1:1, 5:4, 4:3, 3:2, 5:3, 7:4, 9:5, 2:1, 7:3, 5:2, and 3:1 (see Table 1). Solid curves trace loci of constant perihelion q = a(1 − e). Especially large [2003 UB$_{313}$, Pluto, 2003 EL$_{61}$, 2005 FY$_9$ (*Brown et al.*, 2005a,b)] and dynamically unusual [2001 QR$_{322}$ (Trojan) (*Chiang et al.*, 2003; *Chiang and Lithwick*, 2005), 2000 CR$_{105}$ (high q) (*Millis et al.*, 2002; *Gladman et al.*, 2002), Sedna (high q) (*Brown et al.*, 2004)] KBOs are labeled. For a zoomed-in view, see Fig. 2.

ance, see the pedagogical review of planet formation by *Goldreich et al.* (2004a, hereafter *G04*), from which our work draws liberally.

2. THE KUIPER BELT OBSERVED TODAY

2.1. Dynamical Classes

Outer solar system objects divide into dynamical classes based on how their trajectories evolve. Figure 1 displays orbital elements, time-averaged over 10 m.y. in a numerical orbit integration that accounts for the masses of the four giant planets, of 529 objects. Dynamical classifications of these objects are secure according to criteria developed by the Deep Ecliptic Survey (DES) team (*Elliot et al.*, 2005, hereafter *E05*). Figure 2 provides a close-up view of a portion of the Kuiper belt. We distinguish four classes:

1. *Resonant KBOs* (122/529) exhibit one or more mean-motion commensurabilities with Neptune, as judged by steady libration of the appropriate resonance angle(s)

(*Chiang et al.*, 2003, hereafter *C03*). Resonances most heavily populated include the exterior 3:2 (Plutino), 2:1, and 5:2 (see Table 1). Of special interest is the first discovered Neptune Trojan (1:1). All resonant KBOs (except the Trojan) are found to occupy e-type resonances; the ability of an e-type resonance to retain a KBO tends to increase with the KBO's eccentricity e (e.g., *Murray and Dermott*, 1999). Unless otherwise stated, orbital elements are heliocentric and referred to the invariable plane. Several (9/122) also inhabit inclination-type (i^2) or mixed-type (ei^2) resonances. None inhabits an e$_N$-type resonance whose stability depends on the (small) eccentricity of Neptune. The latter observation is consistent with numerical experiments that suggest e$_N$-type resonances are rendered unstable by adjacent e-type resonances.

2. *Centaurs* (55/529) are nonresonant objects whose perihelia penetrate inside the orbit of Neptune. Most Centaurs cross the Hill sphere of a planet within 10 m.y. Centaurs are likely descendants of the other three classes, recently dislodged from the Kuiper belt by planetary per-

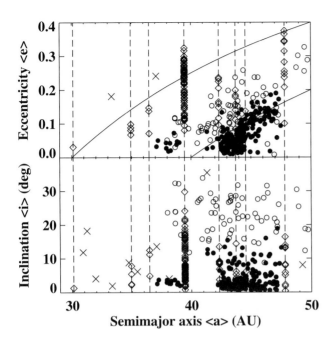

Fig. 2. Same as Fig. 1, zoomed in.

TABLE 1. Observed populations of Neptune resonances (securely identified by the DES team as of October 8, 2005).

Order 0		Order 1		Order 2		Order 3		Order 4	
m:n	#	m:n	#	m:n	#	m:n	#	m:n	#
1:1	1	5:4	4	5:3	9	7:4	8	9:5	2
		4:3	3	3:1	1	5:2	10	7:3	1
		3:2	72						
		2:1	11						

turbations (*Holman and Wisdom*, 1993, hereafter *HW93*; *Tiscareno and Malhotra*, 2003). They will not be discussed further.

3. *Classical KBOs* (246/529) are nonresonant, nonplanet-crossing objects whose time-averaged $\langle e \rangle \leq 0.2$ and whose time-averaged Tisserand parameters

$$\langle T \rangle = \langle (a_N/a) + 2\sqrt{(a/a_N)(1 - e^2)} \cos \Delta i \rangle \quad (1)$$

exceed 3. Here Δi is the mutual inclination between the orbit planes of Neptune and the KBO, a is the semimajor axis of the KBO, and a_N is the semimajor axis of Neptune. In the circular, restricted, three-body problem, test particles with T > 3 and a > a_N cannot cross the orbit of the planet [i.e., their perihelia q = a(1 − e) remain greater than a_N]. Thus, classical KBOs can be argued to have never undergone close encounters with Neptune in its current nearly circular orbit and to be relatively pristine dynamically. Indeed, many classical KBOs as identified by our scheme have low inclinations $\langle i \rangle < 5°$ ("cold classicals"), though some do not ("hot classicals"). Our defining threshold for $\langle e \rangle$ is arbitrary; like our threshold for $\langle T \rangle$, it is imposed to

suggest — perhaps incorrectly — which KBOs might have formed and evolved *in situ*.

Classical KBOs have spectral properties distinct from those of other dynamical classes: Their colors are more uniformly red (Fig. 3) (see chapter by Cruikshank et al.). According to the Kolmogorov-Smirnov test (*Press et al.*, 1992; *Peixinho et al.*, 2004), the probabilities that classical KBOs have B-V colors and Boehnhardt-S slopes (*Boehnhardt et al.*, 2001) drawn from distributions identical to those for resonant KBOs are 10^{-2} and 10^{-3}, respectively. When classical KBOs are compared to scattered KBOs (see below), the corresponding probabilities are 10^{-6} and 10^{-4}. An alternative interpretation is that low-i KBOs are redder than high-i KBOs (*Trujillo and Brown*, 2002; *Peixinho et al.*, 2004). This last claim is statistically significant when classical, scattered, and resonant KBOs are combined and analyzed as one set (Fig. 4). However, no correlation between physical properties and i (or any other measure of excitation) has proven significant within any individual class.

Both the inner edge of the classical belt at a ≈ 37 AU, and the gap in the classical belt at a ≈ 40–42 AU and $\langle i \rangle \leq 10°$ (see Fig. 2), reflect ongoing sculpting by the present-day planets. The inner edge marks the distance out to which the planets have eroded the Kuiper belt over the last few billion years (*HW93*; *Duncan et al.*, 1995, hereafter *D95*). The gap is carved by the ν_{18}, ν_{17}, and ν_8 secular resonances (*HW93*; *D95*; *Knežević et al.*, 1991). At a secular resonance denoted by ν_j, the orbital precession frequency of a test particle — apsidal if j < 10 and nodal if j > 10 — matches one of the precession eigenfrequencies of the planets (see Chapter 6 of *Murray and Dermott*, 1999). For example, at

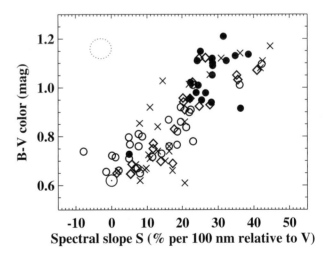

Fig. 3. Visual colors of KBOs and Centaurs calculated from published photometry, with the average uncertainty indicated by the upper left oval. The spectral slope S is calculated for wavelengths in the range of Johnson V through I. Neutral (solar) colors are indicated by ○. Symbols for dynamical classes are the same as those for Fig. 1. Classical KBOs constitute a distinct red cluster, except for (35671), which also has a small semimajor axis of 38 AU. Other classes are widely dispersed in color.

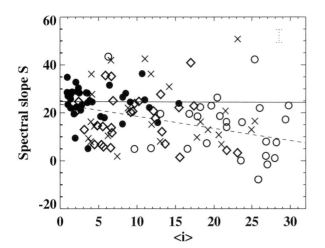

Fig. 4. Spectral slope S vs. time-averaged inclination. The typical uncertainty in S is indicated by the dotted bar. Classical KBOs evince no trend of color with $\langle i \rangle$. The solid line is fitted to classicals only; statistical tests using Spearman's rank-order coefficient and Kendall's tau (*Press et al.*, 1992) show that no significant correlation exists. When classical, resonant, and scattered KBOs are combined, S and $\langle i \rangle$ correlate significantly (with a false alarm probability of 10^{-5}); the dashed line is a fit to all three classes. The two most neutral classicals are (35671) and 1998 WV$_{24}$, having semimajor axes of 38 and 39 AU, respectively.

low i, the ν_8 resonance drives e to Neptune-crossing values in $\sim 10^6$ yr. Particles having large i, however, can elude the ν_8 (*Knežević et al.*, 1991). Indeed, 18 KBOs of various classes and all having large $\langle i \rangle$ reside within the gap.

By contrast, the outer edge of the classical belt at a $\approx$ 47 AU is likely primordial. Numerous surveys (e.g., *E05*; *Bernstein et al.*, 2004; and references therein) carried out after an edge was initially suspected (*Jewitt et al.*, 1998) all failed to find a single object moving on a low-e orbit outside 47 AU. The reality of the "Kuiper Cliff" is perhaps most convincingly demonstrated by *Trujillo and Brown* (2001), who simply plot the distribution of heliocentric discovery distances of (mostly classical) KBOs after correcting for the bias against finding more distant, fainter objects. This distribution peaks at 44 AU and plummets to a value 10× smaller at 50 AU. The statistical significance of the Cliff hinges upon the fact that the bias changes less dramatically — only by a factor of 2.2–2.4 for reasonable parameterizations of the size distribution — between 44 and 50 AU. The possibility remains that predominantly small objects having radii R < 50 km reside beyond 47 AU, or that the Cliff marks the inner edge of an annular gap having radial width ≥30 AU (*Trujillo and Brown*, 2001).

4. *Scattered KBOs* (106/529) comprise nonclassical, nonresonant objects whose perihelion distances q remain outside the orbit of Neptune. [The "scattered-near" and "scattered-extended" classes defined in *E05* — see also *Gladman et al.* (2002) — are combined to simplify discussion. Also, while we do not formally introduce Oort cloud objects as a class, we make connections to that population through-

out this review.] How were scattered KBOs emplaced onto their highly elongated and inclined orbits? Appealing to perturbations exerted by the giant planets in their current orbital configuration is feasible only for some scattered objects. A rule of thumb derived from numerical experiments for the extent of the planets' collective reach is q ≲ 37 AU (*D95*; *Gladman et al.*, 2002). Figure 1 reveals that many scattered objects possess q > 37 AU and are therefore problematic. Outstanding examples include 2000 CR$_{105}$ (q = 44 AU) (*Millis et al.*, 2002; *Gladman et al.*, 2002) and (90377) Sedna (q = 76 AU) (*Brown et al.*, 2004).

These classifications are intended to sharpen analyses and initiate discussion. The danger lies in allowing them to unduly color our thinking about origins. For example, although Sedna is classified above as a scattered KBO, the history of its orbit may be distinct from those of other scattered KBOs. We make this distinction explicit below.

2.2. Sky Density and Mass

We provide estimates for the masses of the Kuiper belt (comprising objects having q ≲ 60 AU and a > 30 AU; section 2.2.1); the inner Oort cloud (composed of Sedna-like objects; section 2.2.2); and Neptune Trojans (a ≈ 30 AU; section 2.2.3).

2.2.1. Main Kuiper belt. *Bernstein et al.* (2004, hereafter *B04*) compile data from published surveys in addition to their own unprecedentedly deep Hubble Space Telescope (HST) survey to compute the cumulative sky density of KBOs vs. apparent red magnitude m_R ("luminosity function"), shown in Fig. 5. Sky densities are evaluated near

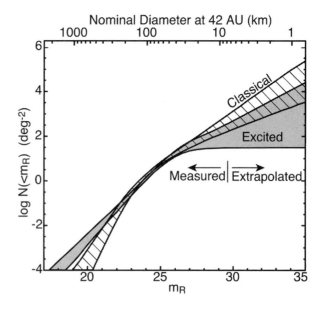

Fig. 5. Cumulative sky density vs. apparent red magnitude for CKBOs and excited KBOs, from *B04*. If $N(<m_R) \propto 10^{\alpha m_R}$, then the size distribution index $\tilde{q} = 5\alpha + 1$ (see section 2.2.1). Envelopes enclose 95% confidence intervals. The top abscissa is modified from *B04*; here it assumes a visual albedo of 10%. Figure courtesy of G. Bernstein.

the ecliptic plane. Objects are divided into two groups: "CKBOs" (similar to our classical population) having heliocentric distances 38 AU < d < 55 AU and ecliptic inclinations i ≤ 5°, and "excited" KBOs (similar to our combined resonant and scattered classes) having 25 AU < d < 60 AU and i > 5°. Given these definitions, their analysis excludes objects with ultrahigh perihelia such as Sedna. With 96% confidence, *B04* determine that CKBOs and excited KBOs have different luminosity functions. Moreover, neither function conforms to a single power law from $m_R = 18$ to 29; instead, each is well-fitted by a double power law that flattens toward fainter magnitudes. The flattening occurs near $m_R \approx 24$ for both groups. To the extent that all objects in a group have the same albedo and are currently located at the same heliocentric distance, the luminosity function is equivalent to the size distribution. We define $\tilde{q}$ as the slope of the differential size distribution, where $dN \propto R^{-\tilde{q}}dR$ equals the number of objects having radii between R and R + dR. As judged from Fig. 5, for CKBOs, $\tilde{q}$ flattens from 5.5–7.7 (95% confidence interval) to 1.8–2.8 as R decreases. For excited KBOs, $\tilde{q}$ flattens from 4.0–4.6 to 1.0–3.1. Most large objects are excited (see also Fig. 1).

By integrating the luminosity function over all magnitudes, *B04* estimate the total mass in CKBOs to be

$$M_{CKBO} \approx 0.005 \left(\frac{p}{0.10} \right)^{-3/2} \left(\frac{d}{42 \text{ AU}} \right)^6 \times \left(\frac{\rho}{2 \text{ g cm}^{-3}} \right) \left(\frac{A}{360° \times 6°} \right) M_{\oplus} \tag{2}$$

where all CKBOs are assumed to have the same albedo p, heliocentric distance d, and internal density ρ. The solid angle subtended by the belt of CKBOs is A. Given uncertainties in the scaling variables — principally p and ρ (see chapter by Cruikshank et al. for recent estimates) — this mass is good to within a factor of several. The mass is concentrated in objects having radii R ~ 50 km, near the break in the luminosity function.

The mass in excited KBOs cannot be as reliably calculated. This is because the sample is heterogeneous — comprising both scattered and resonant KBOs having a wide dispersion in d — and because corrections need to be made for the observational bias against finding objects near the aphelia of their eccentric orbits. The latter bias can be crudely quantified as $(Q/q)^{3/2}$: the ratio of time an object spends near its aphelion distance Q (where it is undetectable) vs. its perihelion distance q (where it is usually discovered). An order-of-magnitude estimate that accounts for how much larger A, d^6, and $(Q/q)^{3/2}$ are for excited KBOs than for CKBOs suggests that the former population might weigh ~10× as much as the latter, or ~0.05 $M_{\oplus}$. This estimate assumes that $\tilde{q}$ for excited KBOs is such that most of the mass is concentrated near $m_R \approx 24$, as is the case for CKBOs. If $\tilde{q}$ for the largest excited KBOs is as small as 4, then our mass estimate increases by a logarithm to ~0.15 $M_{\oplus}$.

2.2.2. Inner Oort cloud (Sedna-like objects).

What about objects with unusually high perihelia such as Sedna, whose mass is $M_S \sim 6 \times 10^{-4}(R/750 \text{ km})^3 M_{\oplus}$? The Caltech Palomar Survey searched f ~ 1/5 of the celestial sphere to discover one such object (*Brown et al.,* 2004). By assuming Sedna-like objects are distributed isotropically (not a forgone conclusion; see section 3.7), we derive an upper limit to their total mass of $M_S(Q/q)^{3/2}f^{-1} \sim 0.1 M_{\oplus}$. If all objects on Sedna-like orbits obey a size spectrum resembling that of excited KBOs (*B04*), then we revise the upper limit to ~0.3 $M_{\oplus}$. The latter value is 20× smaller than the estimate by *Brown et al.* (2004); the difference arises from our use of a more realistic size distribution.

2.2.3. Neptune Trojans.

The first Neptune Trojan, 2001 QR$_{322}$ (hereafter "QR"), was discovered by the DES (*C03*). The distribution of DES search fields on the sky, coupled with theoretical maps of the sky density of Neptune Trojans (*Nesvorný and Dones,* 2002), indicate that N ~ 10–30 objects resembling QR librate on tadpole orbits about Neptune's forward Lagrange (L4) point (*C03*). Presumably a similar population exists at L5. An assumed albedo of 12–4% yields a radius for QR of R ~ 65–115 km. Spreading the inferred population of QR-like objects over the area swept out by tadpole orbits gives a surface mass density in a single Neptune Trojan cloud that approaches that of the main Kuiper belt to within factors of a few (*Chiang and Lithwick,* 2005). Large Neptune Trojans are at least comparable in number to large Jupiter Trojans and may outnumber them by a factor of ~10 (*C03*).

2.3. Binarity

Veillet et al. (2002) optically resolved the first binary (1998 WW$_{31}$) among KBOs having sizes R ~ 100 km. Over 20 binaries having components in this size range are now resolved. Components typically have comparable brightnesses and are separated (in projection) by 300–10^5 km (e.g., *Stephens and Noll,* 2006). These properties reflect observational biases against resolving binaries that are separated by ≤0.1" and that contain faint secondaries.

Despite these selection biases, *Stephens and Noll* (2006) resolved as many as 9 out of 81 KBOs (~10%) with HST. They further report that the incidence of binarity appears 4× higher in the classical disk than in other dynamical populations. It is surprising that so many binaries exist with components widely separated and comparably sized. A typical binary in the asteroid belt, by contrast, comprises dissimilar masses separated by distances only slightly larger than the primary's radius. Another peculiarity of KBO binaries is that components orbit one other with eccentricities on the order of unity. In addition, binary orbits are inclined relative to their heliocentric orbits at seemingly random angles. See *Noll* (2003) for more quantitative details.

Although close binaries cannot be resolved, their components can eclipse each other. *Sheppard and Jewitt* (2004) highlight a system whose lightcurve varies with large amplitude and little variation in color, suggesting that it is a near-contact binary. They infer that at least 10% of KBOs are members of similarly close binaries.

Among the four largest KBOs having R ≈ 1000 km — 2003 UB$_{313}$, Pluto, 2005 FY$_9$, and 2003 EL$_{61}$ — three are known to possess satellites. Secondaries for 2003 EL$_{61}$ and 2003 UB$_{313}$ are 5% and 2% as bright as their primaries, and separated by 49,500 and 36,000 km, respectively (*Brown et al.*, 2005b). In addition to harboring Charon (*Christy and Harrington*, 1978), Pluto possesses two, more distant companions having R ~ 50 km (*Weaver et al.*, 2006). The three satellites' orbits are nearly co-planar; their semimajor axes are about 19,600, 48,700, and 64,800 km; and their eccentricities are less than 1% (*Buie et al.*, 2006).

3. THEORETICAL TIMELINE

We now recount a possible history of transneptunian space. Throughout our narration, it is helpful to remember that the timescale for an object of size R and mass M to collide into its own mass in smaller objects is

$$t_{col} \sim \frac{M}{\dot{M}} \sim \frac{\rho R^3}{(\sigma/h)R^2 v_{rel}} \sim \frac{\rho R}{\sigma \Omega} \qquad (3)$$

where $\dot{M}$ is the rate at which mass from the surrounding disk impacts the object, σ is the disk's surface mass density (mass per unit face-on area) in smaller objects, v_{rel} is the relative collision velocity, $h \sim v_{rel}/\Omega$ is the effective vertical scale height occupied by colliders, and Ω is the orbital angular frequency. Relative velocities v_{rel} depend on how e and i are distributed. Equation (3) requires that e's and i's be comparably distributed and large enough that gravitational focusing is ignorable. While these conditions are largely met by currently observed KBOs, they were not during the primordial era. Our expressions below represent appropriate modifications of equation (3).

3.1. Coagulation

The mass inferred for the present-day Kuiper belt, ~0.05–0.3 M$_\oplus$ (section 2.2), is well below that thought to have been present while KBOs coagulated. *Kenyon and Luu* (1998, 1999) and *Kenyon* (2002), in a series of particle-in-a-box accretion simulations, find that ~3–30 M$_\oplus$ of primordial solids, spread over an annulus extending from 32 to 38 AU, are required to (1) coagulate at least one object as large as Pluto and (2) coagulate ~10^5 objects having R > 50 km. The required initial surface density, $\sigma \sim 0.06$–0.6 g cm^{-2}, is on the order of that of the condensible portion of the minimum-mass solar nebula (MMSN) at 35 AU: $\sigma_{MMSN} \sim 0.2$ g cm^{-2}.

3.1.1. The missing-mass problem. That primordial and present-day masses differ by 2 orders of magnitude is referred to as the "missing-mass" problem. The same accretion simulations point to a possible resolution: Only ~1–2% of the primordial mass accretes to sizes exceeding

~100 km. The remainder stalls at comet-like sizes of ~0.1–10 km. Stunting of accretion is attributed to the formation of several Pluto-sized objects whose gravity amplifies velocity dispersions so much that collisions between planetesimals are erosive rather than accretionary. Thus, accretion in the Kuiper belt may be self-limiting (*Kenyon and Luu*, 1999). The bulk of the primordial mass, stalled at cometary sizes, is assumed by these authors to erode away by destructive collisions over gigayear timescales.

We can verify analytically some of Kenyon and Luu's results by exercising the "two-groups method" (*G04*), whereby the spectrum of planetesimal masses is approximated as bimodal. "Big" bodies each of size R, mass M, Hill radius R$_H$, and surface escape velocity v_{esc} comprise a disk of surface density Σ. They are held primarily responsible for stirring and accreting "small" bodies of size s, surface density σ, and random velocity dispersion u. By random velocity we mean the noncircular or nonplanar component of the orbital velocity. As such, u is proportional to the root-mean-squared dispersion in e and i.

To grow a big body takes time

$$t_{acc} \equiv \frac{R}{\dot{R}} \sim \frac{\rho R}{\sigma \Omega} \left(\frac{u}{v_{esc}} \right)^2 \qquad (4)$$

where the term in parentheses is the usual gravitational focussing factor (assumed <1). Gravitational stirring of small bodies by big ones balances damping of relative velocities by inelastic collisions among small bodies. This balance sets the equilibrium velocity dispersion u

$$\frac{\rho R}{\Sigma \Omega} \left(\frac{u}{v_{esc}} \right)^4 \sim \frac{\rho s}{\sigma \Omega} \qquad (5)$$

Combining equations (4) and (5) implies

$$t_{acc} \sim 10 \left(\frac{R}{100 \text{ km}} \frac{s}{100 \text{ m}} \frac{\Sigma/\sigma}{0.01} \right)^{1/2} \left(\frac{\sigma_{MMSN}}{\sigma} \right) \text{ m.y.} \qquad (6)$$

$$u \sim 6 \left(\frac{s}{100 \text{ m}} \frac{\Sigma/\sigma}{0.01} \right)^{1/4} \left(\frac{R}{100 \text{ km}} \right)^{3/4} \text{ m s}^{-1} \qquad (7)$$

at a distance of 35 AU. All bodies reside in a remarkably thin, dynamically cold disk: Eccentricities and inclinations are at most on the order of u/(Ωa) ~ 0.001. Our nominal choices for σ, Σ, and s are informed by Kenyon and Luu's proposed solution to the missing-mass problem. Had we chosen values resembling those of the Kuiper belt today — $\Sigma \sim \sigma \sim 0.01\sigma_{MMSN}$ — coagulation times would exceed the age of the solar system.

The above framework for understanding the missing-mass problem, while promising, requires development. First, account needs to be made for how the formation of Neptune — and possibly other planet-sized bodies — influences the coagulation of KBOs. None of the simulations cited above succeeds in producing Neptune-mass objects. Yet minimum-mass disks may be capable, in theory if not yet in simulation, of producing several planets having masses approaching that of Neptune at distances of 15–25 AU on timescales much shorter than the age of the solar system (*G04*) (see sections 3.4–3.5). The inability of simulations to produce ice giants may arise from their neglect of small-s, low-u particles that can be efficiently accreted (*G04*). Sizes s as small as centimeters seem possible. How would their inclusion, and the consequent formation of Neptune-mass planets near the Kuiper belt, change our understanding of the missing-mass problem? Second, how does the outer solar system shed ~99% of its primordial solids? The missing-mass problem translates to a "clean-up" problem, the solution to which will involve some as yet unknown combination of collisional comminution, diffusive transport by interparticle collisions, gravitational ejection by planets, and removal by gas and/or radiation drag.

3.1.2. The outer edge of the primordial planetesimal disk. How far from the Sun did planetesimals coagulate? The outer edge of the classical disk at 47 AU (section 2) suggests that planetesimals failed to form outside this distance. Extrasolar disks are also observed to have well-defined boundaries. The debris disks encircling β Pictoris and AU Microscopii exhibit distinct changes ("breaks") in the slopes of their surface brightness profiles at stellocentric distances of 100 AU and 43 AU, respectively (e.g., *Kalas and Jewitt*, 1995; *Krist et al.*, 2005). This behavior can be explained by having dust-producing parent bodies reside only at distances interior to break radii (*Strubbe and Chiang*, 2006).

We cannot predict with confidence how planetesimal disks truncate. Our understanding of how micrometer-sized dust assembles into the "small," super-meter-sized bodies that coagulation calculations presume as input is too poor. Recent work discusses how solid particles might drain toward their central stars by gas drag, and how the accumulation of such solids at small stellocentric distances triggers self-gravitational collapse and the formation of larger bodies (*Youdin and Shu*, 2002; *Youdin and Chiang*, 2004). These ideas promise to explain why planetesimal disks have sharp outer edges, but are subject to uncertainties regarding the viability of gravitational instability in a turbulent gas. To sample progress on planetesimal formation, see *Garaud and Lin* (2004), *Youdin and Goodman* (2005), and *Gómez and Ostriker* (2005). In what follows, we assume that objects having R ~ 100 km coagulated only inside 47 AU.

3.2. Formation of Binaries

To have formed from a fragmentary collision, binary components observed today cannot have too much angular momentum. Consider two big bodies undergoing a gravi-

tationally focused collision. Each body has radius R, mass M, and surface escape velocity v_{esc}. Prior to the collision, their angular momentum is at most $L_{max} \sim M v_{esc} R$. After the collision, the resultant binary must have angular momentum $L < L_{max}$. Unless significant mass is lost from the collision, components can be comparably sized only if their separation is comparable to their radii. Pluto and Charon meet this constraint. Their mass ratio is ~1/10, their separation is ~20 R_{Pluto}, and hence their angular momentum obeys $L/L_{max} \sim \sqrt{20}/10 \lesssim 1$. *Canup* (2005) explains how Charon might have formed by a collision. The remaining satellites of Pluto (*Stern et al.*, 2006), the satellites of Pluto-sized KBOs 2003 EL_{61} and 2003 UB_{313}, and the candidate near-contact binaries discovered by *Sheppard and Jewitt* (2004) might also have formed by collisions.

By contrast, binary components having wide separations and comparable masses have too much angular momentum to have formed by gravitationally focused collisions. And if collisions were unfocused, collision times would exceed the age of the solar system — assuming, as we do throughout this review, that the surface density of big (R ~ 100 km) bodies was the same then as now (section 1; section 3.9).

Big bodies can instead become bound ("fuse") by purely gravitational means while they are still dynamically cold. Indeed, such binaries testify powerfully to the cold state of the primordial disk. To derive our expressions below, recall that binary components separated by the Hill radius ~R_H orbit each other with the same period that the binary's center of mass orbits the Sun, ~Ω^{-1}. Furthermore, we assume that the velocity dispersion v of big bodies is less than their Hill velocity $v_H \equiv \Omega R_H$. Then big bodies undergo runaway cooling by dynamical friction with small bodies and settle into an effectively two-dimensional disk (*G04*). Reaction rates between big bodies must be calculated in a two-dimensional geometry. Because u > v_H, reaction rates involving small bodies take their usual forms appropriate for three dimensions.

Goldreich et al. (2002, hereafter *G02*) describe two collisionless formation scenarios, dubbed L^3 and L^2s. Both begin when one big body (L) enters a second big body's (L) Hill sphere. Per big body, the entry rate is

$$\dot{N}_H \sim \frac{\Sigma\Omega}{\rho R}\left(\frac{R_H}{R}\right)^2 \sim \frac{\Sigma\Omega}{\rho R}\alpha^{-2} \qquad (8)$$

where $\alpha \equiv R/R_H \approx 1.5 \times 10^{-4}(35\,\text{AU}/a)$. If no other body participates in the interaction, the two big bodies would pass through their Hill spheres in a time Ω^{-1} (assuming they do not collide). The two bodies fuse if they transfer enough energy to other participants during the encounter. In L^3, transfer is to a third big body: $L + L + L \rightarrow L^2 + L$. To just bind the original pair, the third body must come within R_H of the pair. The probability for this to happen in time Ω^{-1} is $P_{L^3} \sim \dot{N}_H \Omega^{-1}$. If the third body succeeds in approaching this close, the probability that two bodies fuse is on the order of unity. Therefore the timescale for a given big body to fuse to another by L^3 is

$$t_{fuse,L^3} \sim \frac{1}{\dot{N}_H P_{L^3}} \sim \left(\frac{\rho R}{\Sigma}\right)^2 \frac{\alpha^4}{\Omega} \sim 2 \text{ m.y.} \qquad (9)$$

where the numerical estimate assumes R = 100 km, Σ = $0.01\sigma_{MMSN}$, and a = 35 AU.

In L^2s, energy transfer is to small bodies by dynamical friction: $L + L + s^\infty \rightarrow L^2 + s^\infty$. In time $\sim\Omega^{-1}$, the pair of big bodies undergoing an encounter lose a fraction $(\sigma\Omega/\rho R)(v_{esc}/u)^4\Omega^{-1}$ of their energy, under the assumption $v_{esc} > u > v_H$ (*G04*). This fraction is on the order of the probability P_{L^2s} that they fuse, whence

$$t_{fuse,L^2s} \sim \frac{1}{\dot{N}_H P_{L^2s}} \sim \left(\frac{\rho R}{\sigma}\right)^2 \frac{s}{R} \frac{\alpha^2}{\Omega} \sim 7 \text{ m.y.} \qquad (10)$$

where we have used equation (5) and set s = 100 m.

Having formed with semimajor axis $x \sim R_H \sim 7000$ R, a binary's orbit shrinks by further energy transfer. If L^3 is the more efficient formation process, passing big bodies predominantly harden the binary; if L^2s is more efficient, dynamical friction dominates hardening. The probability P per orbit that x shrinks from $\sim R_H$ to $\sim R_H/2$ is on the order of either P_{L^3} or P_{L^2s}. We equate the formation rate of binaries, N_{all}/t_{fuse}, with the shrinkage rate, $\Omega P N_{bin}|_{x \sim R_H}$, to conclude that the steady-state fraction of KBOs that are binaries with separation R_H is

$$f_{bin}(x \sim R_H) \equiv \left.\frac{N_{bin}}{N_{all}}\right|_{x \sim R_H} \sim \frac{\Sigma}{\rho R}\alpha^{-2} \sim 0.4\% \qquad (11)$$

As x decreases, shrinkage slows. Therefore f_{bin} increases with decreasing x. Scaling relations can be derived by arguments similar to those above. If L^2s dominates, $f_{bin} \propto x^0$ for $x > R_H(v_H/u)^2$ and $f_{bin} \propto x^{-1}$ for $x < R_H(v_H/u)^2$ (*G02*). If L^3 dominates, $f_{bin} \propto x^{-1/2}$.

Alternative formation scenarios require, in addition to gravitational scatterings, physical collisions. *Weidenschilling* (2002) suggests a variant of L^3 in which the third big body collides with one member of the scattering pair. Since physical collisions have smaller cross-sections than gravitational interactions, this mechanism requires $\sim 10^2$ more big (R $\sim$ 100 km) bodies than are currently observed to produce the same rate that is cited above for L^3 (*Weidenschilling*, 2002). *Funato et al.* (2004) propose that observed binaries form by the exchange reaction $Ls + L \rightarrow L^2 + s$: A small body of mass m, originally orbiting a big body of mass M, is ejected by a second big body. In the majority of ejections, the small body's energy increases by its orbital binding energy $\sim mv_{esc}^2/2$, leaving the big bodies bound to each other with separation $x \sim (M/m)R$. The rate-limiting step is the formation of

the preexisting (Ls) binary, which requires (as in the asteroid belt) two big bodies to collide and fragment. Hence

$$t_{fuse,exchange} \sim \frac{\rho R}{\Sigma\Omega}\alpha^{3/2} \sim 0.6 \text{ m.y.} \qquad (12)$$

Estimating f_{bin} as a function of x under the exchange hypothesis requires knowing the distribution of fragment masses m. Whether L^2s, L^3, or exchange reactions dominate depends on the uncertain parameters Σ, σ, and s.

As depicted above, newly formed binaries should be nearly co-planar with the two-dimensional disk of big bodies, i.e., binary orbit normals should be nearly parallel. Observations contradict this picture (section 2.3). How dynamical stirring of the Kuiper belt subsequent to binary formation affects binary inclinations and eccentricities has not been investigated.

3.3. Early Stirring by Growing Planetary Oligarchs

Coagulation of KBOs and fusing of binaries cannot proceed today, in part because velocity dispersions are now so large that gravitational focusing is defeated on a wide range of length scales. What stirred the Kuiper belt? There is no shortage of proposed answers. Much of the remaining review (sections 3.3–3.7) explores the multitude of nonexclusive possibilities. We focus on stirring "large" KBOs like those currently observed, having R $\sim$ 100 km. Our setting remains the primordial disk, of whose mass large KBOs constitute only a small fraction (1–2%; section 3.1.1).

Neptune and Uranus are thought to accrete as oligarchs, each dominating their own annulus of full-width $\sim$5 Hill radii (*G04*; *Ida and Makino*, 1993; *Greenberg et al.*, 1991). (The coefficient of 5 presumes that oligarchs feed in a shear-dominated disk in which planetesimals have random velocities u that are less than the oligarch's Hill velocity $v_H = \Omega R_H$. If $u > v_H$, oligarchs' feeding annuli are wider by $\sim u/v_H$. In practice, u/v_H does not greatly exceed unity since it scales weakly with input parameters.) Each oligarch grows until its mass equals the isolation mass

$$M_p \sim 2\pi a \times 5R_{H,p} \times \sigma \qquad (13)$$

where $R_{H,p}$ is the oligarch's Hill radius. For a = 25 AU and M_p equal to Neptune's mass $M_N = 17 M_\oplus$, equation (13) implies $\sigma \sim 0.9$ g cm^{-2} $\sim 3\sigma_{MMSN}$. About 5 Neptune-mass oligarchs can form in nested annuli between 15 and 25 AU. Inspired by *G04* who point out the ease with which ice giants coagulate when the bulk of the disk mass comprises very small particles (section 3.1.1), we assume that all five do form in a disk that is a few times more massive than the MMSN and explore the consequences of such an initially packed system.

While oligarchs grow, they stir large KBOs in their immediate vicinity. A KBO that comes within distance b of mass M_p has its random velocity excited to $v_K \sim (GM_p/b)^{1/2}$. Take the surface density of perturbers to be Σ_p. Over time t, a KBO comes within distance b ~ $[M_p/(\Sigma_p\Omega t)]^{1/2}$ of a perturber. Therefore

$$v_K \sim G^{1/2}(M_p\Sigma_p\Omega t)^{1/4} \qquad (14)$$

Since Neptune and Uranus contain more hydrogen than can be explained by accretion of icy solids alone, they must complete their growth within $t_{acc,p} \sim$ 1–10 m.y., before all hydrogen gas in the MMSN photoevaporates (e.g., *Matsuyama et al.,* 2003, and references therein). For $t = t_{acc,p} =$ 10 m.y., $M_p = M_N$, $\Sigma_p = 0.9$ g cm^{-2}, and $\Omega = 2\pi/(100$ yr), equation (14) implies $v_K \sim 1$ km s^{-1} or $e_K \sim 0.2$. It is safe to neglect damping of v_K for large KBOs, which occurs by inelastic collisions over a timescale $t_{col} \sim 400$ (0.9 g cm$^{-2}/\sigma$) m.y. $\gg t_{acc,p}$.

3.4. Velocity Instability and Ejection of Planets

Once the cohort of Neptune-mass oligarchs consumes approximately one-half the mass of the parent disk, they scatter one another onto highly elliptical and inclined orbits (equation (111) of *G04*; *Kenyon and Bromley,* 2006). This velocity instability occurs because damping of planetary random velocities by dynamical friction with the disk can no longer compete with excitation by neighboring, crowded oligarchs.

The epoch of large planetary eccentricities lasts until enough oligarchs are ejected from the system. We can estimate the ejection time by following the same reasoning that led to equation (14). Replace v_K with the system escape velocity $v_{esc,sys} \sim \Omega a$, and replace Σ_p with the surface density of oligarchs $\sim M_p/a^2$ (see equation (13)). Then solve for

$$t = t_{eject} \sim \left(\frac{M_\odot}{M_p}\right)^2 \frac{0.1}{\Omega} \qquad (15)$$

The coefficient of 0.1 is attributed to more careful accounting of encounter geometries; equation (15) gives ejection times similar to those found in numerical simulations (*G04*). Neptune-mass oligarchs at a $\approx$ 20 AU kick their excess brethren out over $t_{eject} \sim 600$ m.y. Removal is faster if excess oligarchs are passed inward to Jupiter and Saturn.

Oligarchs moving on eccentric orbits likely traverse distances beyond 30 AU and stir KBOs. We expect more members are added to the scattered KBO disk during this stage.

We have painted a picture of dynamically hot oligarchs similar to that drawn by *Thommes et al.* (1999) (see also *Tsiganis et al.,* 2005), who hypothesize that Neptune and Uranus form as oligarchs situated between the cores of Jupiter and Saturn at 5–10 AU. The nascent ice giants are scattered outward onto eccentric orbits once the gas giant cores amass their envelopes. While Neptune and Uranus reside on eccentric orbits, they can stir KBOs in much the same way as we have described above (*Thommes et al.,* 2002). Despite the similarity of implications for the stirring of

KBOs, the underlying motivation of the cosmogony proposed by *Thommes et al.* (1999) is the belief that Neptune-mass bodies do not form readily at distances of ~30 AU. Recent work highlighting the importance of inelastic collisions among very small bodies challenges this belief (*G04*) (see section 3.1).

3.5. Dynamical Friction Cooling of Surviving Planets

Planetary oligarchs that survive ejection — i.e., Uranus and Neptune — have their e's and i's restored to small values by dynamical friction with the remnant disk (comprising predominantly small KBOs of surface density σ and velocity dispersion u) over time

$$t_{df,cool} = \frac{v_p}{\dot{v}_p} \sim \frac{\rho R_p}{\sigma\Omega}\left(\frac{v_p}{v_{esc,p}}\right)^4 \qquad (16)$$

where R_p, $v_{esc,p}$, and $v_p \gg u$ are the planet's radius, surface escape velocity, and random velocity, respectively. For $v_p = \Omega a/2$ (planetary eccentricity $e_p \sim 0.5$), a = 25 AU, $R_p =$ 25000 km, $v_{esc,p} = 24$ km s^{-1}, and $\sigma = \Sigma_p = 0.9$ g cm^{-2} (since the velocity instability occurs when the surface density of oligarchs equals that of the parent disk; section 3.4), we find $t_{df,cool} \sim 20$ m.y.

While Neptune's orbit is eccentric, the planet might repeatedly invade the Kuiper belt at a $\approx$ 40–45 AU and stir KBOs. Neptune would have its orbit circularized by transferring energy to both small and large KBOs. Unlike small KBOs, large ones cannot shed this energy because they cool too inefficiently by inelastic collisions (see the end of section 3.3). Insert equation (16) into equation (14) and set $\Sigma_p = \sigma$ to estimate the random velocity to which large KBOs are excited by a cooling Neptune

$$v_K \sim v_p \qquad (17)$$

Thus large KBOs are stirred to the same random velocity that Neptune had when the latter began to cool, regardless of the numerical value of $t_{df,cool}$. Large KBOs effectively record the eccentricity of Neptune just prior to its cooling phase. Final eccentricities e_K might range from ~0.1 to nearly 1. During this phase, the population of the scattered KBO disk would increase, perhaps dramatically so. If all large KBOs are stirred to $e_K \gg 0.1$, new large KBOs must coagulate afterward from the remnant disk of small, dynamically cold bodies to reconstitute the cold classical disk. Cold classicals might therefore postdate hot KBOs.

3.6. Planetary Migration

Having seen a few of its siblings evicted, and having settled onto a near-circular, flattened orbit, Neptune remains immersed in a disk of small bodies. The total mass of the disk is still a few times that of the planet because the prior velocity instability occurred when the surface density of oligarchs was comparable to that of the disk. By continuing to scatter small bodies, Neptune migrates: Its semimajor axis changes while its eccentricity is kept small by dynamical friction. Absent other planets, migration would be sunward on average. Planetesimals repeatedly scattered by Neptune would exchange angular momentum with the planet in a random-walk fashion. Upon gaining specific angular momentum $\sim(\sqrt{2} - 1)\Omega a^2$, where Ω and a are appropriate to Neptune's orbit, a planetesimal initially near Neptune would finally escape. Having lost angular momentum to the ejected planetesimal, Neptune would migrate inward. [A single planet can still migrate outward if it scatters material having predominantly higher specific angular momentum. *Gomes et al.* (2004) achieve this situation by embedding Neptune in a disk whose mass is at least 100 $M_\oplus$ and is weighted toward large distances ($\sigma a^2 \propto$ a); see also the chapter by Morbidelli et al.).]

Other planets complicate this process. Numerical simulations by *Fernández and Ip* (1984) and *Hahn and Malhotra* (1999) incorporating all four giant planets reveal that planetesimals that originate near Neptune are more likely ejected by Jupiter. Over the course of their random walks, planetesimals lose angular momentum to Neptune and thereby cross Jupiter's orbit. Jupiter summarily ejects them (see equation (15) and related discussion). Thus, on average, Neptune gains angular momentum and migrates outward, as do Saturn and Uranus, while Jupiter's orbit shrinks.

An outward-bound Neptune passes objects to the interior planets for eventual ejection and seeding of the Oort cloud. We refer to this process as "scouring" the transneptunian disk. Scouring and migration go hand in hand; the fraction by which Neptune's semimajor axis increases is on the order of the fraction that the disk mass is scoured. Scouring is likely a key part of the solution to the clean-up (a.k.a. missing-mass) problem. If clean-up is not achieved by the end of Neptune's migration, one must explain how to transport the bulk of the transneptunian disk to other locales while keeping Neptune in place (*Gomes et al.*, 2004). Scouring has only been treated in collisionless N-body simulations. How scouring and migration proceed in a highly collisional disk of small bodies is unknown (section 3.6.4). In addition to scouring the disk, Neptune's migration has been proposed to sculpt the disk in other ways — by capturing bodies into mean-motion resonances (section 3.6.1), redistributing the classical disk by resonance capture and release (section 3.6.2), and deflecting objects onto scattered orbits (section 3.6.3). We critically examine these proposals below.

3.6.1. Capture and excitation of resonant Kuiper belt objects. As Neptune migrates outward, its exterior mean-motion resonances (MMRs) sweep across transneptunian space. Provided the migration is sufficiently slow and smooth, MMRs may trap KBOs and amplify their orbital eccentricities and, to a lesser extent, their inclinations. The eccentric orbits of Pluto and the Plutinos — objects that all inhabit Neptune's 3:2 resonance — may have resulted from resonance capture and excitation by a migrating Neptune (*Mal-*

hotra, 1993, 1995; *Jewitt and Luu,* 2000). The observed occupation of other low-order resonances — e.g., the 4:3, 5:3, and 2:1 MMRs — by KBOs on eccentric orbits (see Fig. 2 and Table 1) further support the migration hypothesis (*C03*). In this section, we review the basic mechanism of resonant excitation of eccentricity, examine how the migration hypothesis must change in light of the unexpected occupation of high-order (e.g., the 7:4, 5:2, and 3:1) MMRs, and discuss how m:1 resonances serve as speedometers for Neptune's migration.

Consider the interaction between a test particle (KBO) and a planet on an expanding circular orbit. In a frame of reference centered on the Sun and rotating with the planet's angular velocity $\Omega_p(t)$, the particle's Hamiltonian is

$$\mathcal{H} - \mathcal{E} - \Omega_p(t)\mathcal{L} - \mathcal{R}(t) \qquad (18)$$

where $\mathcal{E} = -GM_\odot/2a$, $\mathcal{L} = [GM_\odot a(1 - e^2)]^{1/2}$, and $\mathcal{R}$ is the disturbing potential due to the planet (these quantities should be expressed in canonical coordinates). From Hamiltonian mechanics, $d\mathcal{H}/dt = \partial\mathcal{H}/\partial t = -\dot{\Omega}_p\mathcal{L} - \partial\mathcal{R}/\partial t$. Therefore

$$\frac{d\mathcal{E}}{dt}(1 - \varepsilon) - \Omega_p\frac{d\mathcal{L}}{dt} = 0 \qquad (19)$$

where $\varepsilon \equiv (d\mathcal{R}/dt - \partial\mathcal{R}/\partial t)/(d\mathcal{E}/dt)$. We rewrite equation (19) as

$$\frac{de^2}{dt} = \frac{(1 - e^2)^{1/2}}{a}[(1 - e^2)^{1/2} - \Omega/\Omega_p(1 - \varepsilon)]\frac{da}{dt} \qquad (20)$$

where Ω is the particle's angular frequency.

For a particle trapped in m:n resonance (where m and n are positive, relatively prime integers), a, e, and the resonance angle change little over the particle's orbital period. If the synodic period is not much longer than the orbital period, we may average the Hamiltonian over the former (we may do this by choosing appropriate terms in the expansion of $\mathcal{R}$). This yields $\Omega/\Omega_p(1 - \varepsilon) = n/m$. For a particle in resonance, $|\varepsilon| \ll 1$. By change of variable to $x \equiv (1 - e^2)^{1/2}$, equation (20) integrates to

$$[(1 - e^2)^{1/2} - n/m]^2 a = \text{constant} \qquad (21)$$

which relates changes in a to changes in e for any resonance — exterior m > n, interior m < n, or Trojan m = n. In the case of a planet that migrates toward a particle in exterior resonance, a increases to maintain resonant lock (*Goldreich*, 1965; *Peale*, 1986). Then by equation (21), e also tends to increase, toward a maximum value $[1 - (n/m)^2]^{1/2}$. Particles inhabiting either an exterior or interior resonance have their eccentricities amplified from 0 because they are perturbed by a force pattern whose angular speed Ω_p does not equal their orbital angular speed Ω. Particles receive energy and angular momentum from the planet in a ratio that cannot maintain circularity of orbits.

Among observed 2:1 resonant KBOs, max(e) ≈ 0.38 (Fig. 2). If 2:1 resonant KBOs had their eccentricities amplified purely by migration, they must have migrated by $\Delta a \approx 13$ AU (equation (21)). Neptune must have migrated correspondingly by $\Delta a_p \approx 8$ AU. This is an upper bound on Δa_p because it does not account for nonzero initial eccentricities prior to capture.

In early simulations (*Malhotra*, 1993, 1995) of resonance capture by a migrating Neptune, resonances swept across KBOs having initially small e's and i's. These models predicted that if Neptune's orbit expanded by $\Delta a_p \approx 8$ AU, low-order resonances such as the 4:3, 3:2, 5:3, and 2:1 MMRs would be occupied by objects having $0.1 \lesssim e \lesssim 0.4$ and $i \lesssim 10°$. Eccentric KBOs indeed inhabit these resonances (Fig. 2). Two observations were not anticipated: (1) Resonant KBOs are inclined by up to $i \approx 30°$, and (2) high-order resonances — e.g., the 5:2, 7:4, and 3:1 — enjoy occupation. These observations suggest that Neptune's MMRs swept across not only initially dynamically cold objects, but also initially hot ones: The belt was preheated. For example, to capture KBOs into the 5:2 MMR, preheated eccentricities must be ≥ 0.1 (*C03*). Neptune-sized perturbers (sections 3.3–3.5) might have provided the requisite preheating in e and i.

To understand why capture into high-order resonances favors particles having larger initial e, recognize that capture is only possible if, over the time the planet takes to migrate across the maximum possible libration width $\max(\delta a_{lib})$, the particle completes at least one libration

$$\frac{\max(\delta a_{lib})}{|\dot{a}_p|} > T_{lib} \qquad (22)$$

where T_{lib} is the libration period (*Dermott et al.*, 1988). Otherwise, the particle would hardly feel the resonant perturbation as the planet races toward it. Since $\max(\delta a_{lib}) \sim (T_{orb}/T_{lib})a_p$ and $T_{lib} \sim T_{orb}(M_\odot e^{-|m-n|}/M_p)^{1/2}$ (*Murray and Dermott*, 1999), where T_{orb} is the orbital period of the particle and M_p is the mass of the planet, we rewrite equation (22) as

$$\frac{T_{lib}^2}{T_{orb}T_{mig}} \sim \frac{T_{orb}}{T_{mig}}\frac{M_\odot}{M_p}\frac{1}{e^{|m-n|}} < 1 \qquad (23)$$

where the migration timescale $T_{mig} \equiv a_p/|\dot{a}_p|$. The higher the order $|m - n|$ of the resonance, the greater e must be to satisfy equation (23) (*C03*; *Hahn and Malhotra*, 2005).

Asymmetric (m:1) resonances afford a way to estimate the migration timescale observationally. An asymmetric MMR furnishes multiple islands of libration. At the fixed point of each island, a particle's direct acceleration by Neptune balances its indirect acceleration by the Sun due to the Sun's reflex motion (*Pan and Sari*, 2004; *Murray-Clay and Chiang*, 2005, hereafter *MC05*). The multiplicity of islands translates into a multiplicity of orbital longitudes, measured relative to Neptune's, where resonant KBOs cluster on the sky. The pattern of clustering varies systematically with migration speed at the time of capture (*Chiang and Jordan*, 2002). For example, when migration is fast—occurring on timescales $T_{mig} \lesssim 20$ m.y. — objects are caught into 2:1 resonance such that more appear at longitudes trailing, rather than leading, Neptune's. The degree of asymmetry can be as large as 300%. When migration is slow, the distribution of captured 2:1 objects is symmetric about the Sun-Neptune line. The preference for trailing vs. leading longitudes arises from migration-induced shifts in the stable and unstable equilibria of the resonant potential. Shifts in the equilibrium values of the resonance angle are given in radians by equation (23) and are analogous to the shift in the equilibrium position of a spring in a gravitational field (*MC05*). The observation that trailing 2:1 KBOs do not outnumber leading ones constrains $T_{mig} > 20$ m.y. with nearly 3σ confidence (*MC05*). This measurement accords with numerical simulations of the migration process itself by *Hahn and Malhotra* (1999) and by *Gomes et al.* (2004, their Fig. 10); in these simulations, $T_{mig} \geq 40$ m.y.

3.6.2. Stochastic migration and resonance retainment. Finite sizes of planetesimals render planetary migration stochastic ("noisy"). The numbers of high- and low-momentum objects that Neptune encounters over fixed time intervals fluctuate randomly. These fluctuations sporadically hasten and slow — and might occasionally even reverse — the planet's migration. Apportioning a fixed disk mass to larger (fewer) planetesimals generates more noise. Extreme noise defeats resonance capture. Therefore the existence of resonant KBOs — which we take to imply capture efficiencies on the order of unity — sets an upper limit on the sizes of planetesimals (small bodies) comprising the bulk of the mass of the disk. *Murray-Clay and Chiang* (2006, hereafter *MC06*) estimate this upper limit to be $s_{max} \sim O(100)$ km; a shortened derivation of their result reads as follows.

For a given planetesimal size, most noise is generated per unit mass disk by planetesimals having sub-Hill ($u < v_{H,p} = \Omega R_p/\alpha$) velocity dispersions and semimajor axes displaced $\pm R_{H,p}$ from the planet's (*MC06*). A single such planetesimal of mass μ, after undergoing a close encounter with the planet, changes the planet's semimajor axis by $\Delta a_1 \sim \pm(\mu/M_p)R_{H,p}$. The planet encounters such planetesimals at a rate $\dot{N} \sim \sigma R_{H,p}^2 \Omega_p/\mu$. Over the duration of migration $\sim(\Delta a_p/a_p)T_{mig}$, the planet's semimajor axis random walks away from its nominal (zero-noise) value by $\Delta a_{rnd} \sim \pm(\dot{N}\Delta a_p T_{mig}/a_p)^{1/2}|\Delta a_1|$. The libration amplitude in a of any resonant KBO increases by about this same $|\Delta a_{rnd}|$. Then stochasticity does not defeat resonance capture if $|\Delta a_{rnd}| < \max(\delta a_{lib})$; that is, if

$$s \lesssim \left(\frac{M_p}{M_\odot}\right)^{1/9}\left(\frac{\rho R_p e a_p}{\sigma \Omega_p T_{mig}\Delta a_p}\right)^{1/3}\alpha^{2/3}R_p \qquad (24)$$

which evaluates to $s \lesssim O(100)$ km for $a_p = 30$ AU, $\Delta a_p = 8$ AU, $T_{mig} \approx 40$ m.y., $\sigma = 0.2$ g cm^{-2}, and e = 0.2.

The above constraint on size applies to those planetesimals that comprise the bulk of the disk mass. Noise is also introduced by especially large objects that constitute a small fraction of the disk mass. The latter source of noise has been invoked to explain the curious near-coincidence between the edge of the classical disk (a = 47 AU) and Neptune's 2:1 resonance (a = 47.8 AU). *Levison and Morbidelli (2003)* suggest that the sweeping 2:1 MMR captures KBOs only to release them en route because of close encounters between Neptune and objects having ~10× the mass of Pluto ("super-Plutos"). Dynamically cold KBOs, assumed to coagulate wholly inside 35 AU (section 3.1.2), are thereby combed outward to fill the space interior to the final location of the 2:1 MMR. Why the super-Plutos that are invoked to generate stochasticity have not been detected by wide-field surveys is unclear (*Morbidelli et al., 2002*). The scenario further requires that ~3 $M_\oplus$ be trapped within the 2:1 MMR so that a secular resonance maintains a population of 2:1 resonant KBOs on low-e orbits during transport.

3.6.3. Contribution of migration to scattered Kuiper belt objects. Neptune migrates by scattering planetesimals. What fraction of these still reside today in the scattered belt? Do hot classicals (having i ≳ 5°) owe their excitation to a migratory Neptune? Many scattered and hot classical KBOs observed today have q > 37 AU. This fact is difficult to explain by appealing to perturbers that reside entirely inside 30 AU. Insofar as a close encounter between a perturber and a particle can be modeled as a discontinous change in the particle's velocity at fixed position, the particle (assuming it remains bound to the Sun) tends to return to the same location at which it underwent the encounter.

Gomes (2003a,b) proposes that despite this difficulty, objects scattered by Neptune during its migration from ~20 to 30 AU can evolve into today's scattered and hot classical KBOs by having their perihelia raised by a variety of sweeping secular resonances (SRs; see section 2.1). As the outer planets migrate, SRs sweep across transneptunian space. After having its e and i amplified by close encounters with Neptune, a planetesimal may be swept over by an SR. Unlike MMRs, SRs cannot alter particle semimajor axes and therefore do not permanently trap particles. However, a particle that is swept over by an apsidal-type SR can have its eccentricity increased or decreased. A particle swept over by a secular resonance is analogous to an ideal spring of natural frequency ω_0, driven by a force whose time-variable frequency $\omega(t)$ sweeps past ω_0. Sweeping ω past ω_0 can increase or decrease the amplitude of the spring's free oscillation (the component of the spring's displacement that varies with frequency ω_0), depending on the relative phasing between driver and spring near the moment of resonance crossing when $\omega \approx \omega_0$.

Lowering e at fixed a raises q. *Gomes (2003ab)* and *Gomes et al. (2005)* find in numerical simulations of planetary migration that Neptune-scattered planetesimals originating on orbits inside 28 AU can have their perihelia raised up to 69 AU by a combination of sweeping SRs, MMRs, and Kozai-type resonances (which are a kind of SR). In addition to offering an explanation for the origin of high-q, high-i KBOs, this scenario also suggests a framework for

understanding differences in physical properties between dynamical classes. Compared to classical KBOs, which are held to coagulate and evolve largely *in situ*, scattered KBOs originate from smaller heliocentric distances d. To the (unquantified) extents that coagulation rates and chemical environments vary from d ≈ 20–50 AU, we can hope to understand why a large dispersion in i — which in the proposed scenario reflects a large dispersion in birth distance d — implies a large dispersion in color/size.

The main difficulty with this perihelion-raising mechanism is its low efficiency: Only ~0.1% of all objects that undergo close encounters with a migratory Neptune have their perihelia raised to avoid further close encounters over the age of the solar system (*Gomes, 2003a,b*). Based on this mechanism alone, a disk weighing ~50 $M_\oplus$ prior to migration would have ~0.05 $M_\oplus$ deposited into the scattered and hot classical belts for long-term storage. But only ~1–2% of this mass would be in bodies having sizes R ≳ 100 km (*Kenyon and Luu*, 1998, 1999; *Kenyon*, 2002) (section 3.1). Therefore this scenario predicts that scattered and hot classical KBOs having R ≳ 100 km would weigh, in total, ~10^{-3} $M_\oplus$ — about 50–150× below what is observed (section 2.2). This discrepancy is missed by analyses that neglect consideration of the KBO size distribution. A secondary concern is that current numerical simulations of this mechanism account for the gravitational effects of disk particles on planets but not on other disk particles. Proper calculation of the locations of secular resonances requires, however, a full accounting of the mass distribution.

Given the low efficiency of the mechanism, we submit that the high-q orbits of hot classical and scattered KBOs did not arise from Neptune's migration. Instead, these orbits may have been generated by Neptune-mass oligarchs whose trajectories passed through the Kuiper belt. While a numerical simulation is necessary to test this hypothesis, our order-of-magnitude estimates (sections 3.3–3.5) for the degree to which oligarchs stir the belt by simple close encounters are encouraging. No simulation has yet been performed in which the Kuiper belt is directly perturbed by a mass as large as Neptune's for a time as long as $t_{df,cool}$ ~ 20 m.y. Differences in physical properties between classical and scattered/resonant KBOs might still be explained along the same lines as described above: Scattered/resonant KBOs were displaced by large distances from their coagulation zones and so might be expected to exhibit a large dispersion in color and size, while classical KBOs were not so displaced. Even if all KBOs having R ≳ 100 km were heated to large e or i by planetary oligarchs, the cold classical disk might have regenerated itself in a second wave of coagulation from a collisional disk of small bodies.

3.6.4. Problems regarding migration. The analyses of migration cited above share a common shortcoming: They assume that planetesimals are collisionless. But coagulation studies (section 3.1) indicate that much of the primordial mass remains locked in small bodies for which collision times threaten to be shorter than the duration of planetary migration. By equation (3), planetesimals having sizes ≪1 km in a minimum-mass disk have collision times ≪20 m.y. How Neptune's migration unfolds when most of the disk comprises highly collisional bodies has not been well explored. Neptune may open a gap in the disk (in the same way that moons open gaps in collisional planetary rings) and the planet's migration may be tied to how the disk spreads by collisional diffusion (*Goldreich et al.*, 2004b).

How does the classical belt shed 99% of its primordial mass? Situated at 40–47 AU, it may be too distant for Neptune to scour directly. Perhaps the small bodies of the classical belt are first transported inward, either by gas drag or collisional diffusion, and subsequently scoured. Clean-up and migration are intertwined, but the processes are often not discussed together (but see *Gomes et al.*, 2004).

Are there alternatives to migration for the capture of resonant KBOs? Perhaps resonant KBOs are captured as Neptune's orbit cools by dynamical friction (section 3.5). Before capture, many belt members would already be stirred to large e and i, not only by unstable oligarchs (section 3.4), but also by Neptune while it cools. Cooling accelerates as it proceeds (equation (16)). A rapid change in the planet's semimajor axis toward the end of cooling might trap KBOs into resonance by serendipity. Just after Neptune's semimajor axis changes, objects having orbital elements (including longitudes) suitable for libration would be trapped. This speculative "freeze-in" mechanism might be too inefficient, since it requires that the fraction of phase-space volume occupied by resonances equal the fraction of KBOs that are resonant. Taken at face value, observations suggest the latter fraction is not much smaller than on the order of unity (section 2.1).

3.7. Stellar Encounters

A passing star may have emplaced Sedna onto its high-perihelion orbit. For the last $t \sim 4$ G.y., solar-mass stars in the solar neighborhood have had an average density $n_* \sim 0.04$ stars pc^{-3} and a velocity dispersion $\langle v_*^2 \rangle^{1/2} \sim 30$ km s^{-1}. If we assume that the Sun once resided within a "typical" open cluster, then $n_* \sim 4$ stars pc^{-3} and $\langle v_*^2 \rangle^{1/2} \sim 1$ km s^{-1} over $t \sim 200$ m.y. Over $t \gtrsim 200$ m.y., open clusters dissolve by encounters with molecular clouds (*Binney and Tremaine*, 1987). The number of stars that fly by the Sun within a distance q_* large enough that gravitational focussing is negligible ($q_* \gtrsim GM_\odot / \langle v_*^2 \rangle \sim 900$ AU for $\langle v_*^2 \rangle^{1/2} \sim 1$ km s^{-1}) increases as $\int^t n_* \langle v_*^2 \rangle^{1/2} dt$. Therefore flybys during the current low-density era outnumber those during the cluster era by a factor of ~6. Nonetheless, intracluster encounters can be more effective at perturbing KBO trajectories because encounter velocities are 30× lower.

Fernández and Brunini (2000) simulate the formation of the Oort cloud within an open cluster having parameters similar to those cited above. They find that passing stars create an "inner Oort cloud" of objects having $35 \leq q(AU) \leq 1000$, $300 \leq a(AU) \leq 10^4$, $\langle e \rangle \sim 0.8$, and $\langle i^2 \rangle^{1/2} \sim 1$. Sedna may be the first discovered member of this inner Oort cloud (*Brown et al.*, 2004). Such objects coagulate in the vicinity

of the giant planets and are scattered first by them. Since a scattering event changes velocities more effectively than it does positions, objects' perihelia remain at heliocentric distances of ~5–30 AU while aphelia diffuse outward. Aphelia grow so distant that objects are scattered next by cluster stars. These stars raise objects' perihelia beyond the reach of the giant planets.

We confirm the ability of cluster stars to raise the perihelion of Sedna with an order-of-magnitude calculation. During the open cluster phase, the number of stars that pass within distance q_* of the Sun is

$$N_* \sim 1 \left(\frac{q_*}{4000 \text{ AU}} \right)^2 \left(\frac{n_*}{4 \text{ pc}^{-3}} \frac{\langle v_*^2 \rangle^{1/2}}{1 \text{ km s}^{-1}} \frac{t}{200 \text{ m.y.}} \right) \quad (25)$$

A star of mass M_* having perihelion distance q_* much greater than a planetesimal's aphelion distance ($Q \approx 2a$) perturbs that object's specific angular momentum by

$$\delta h = \pm C \frac{GM_*}{\langle v_*^2 \rangle^{1/2}} \left(\frac{a}{q_*} \right)^2 \quad (26)$$

where the numerical coefficient C depends on the encounter geometry (*Yabushita*, 1972). We can derive the form of equation (26) by noting that $\delta h \sim Q \delta v$, where δv is the perturbation to the object's velocity relative to the Sun. We write δv as the tidal acceleration $GM_* Q/q_*^3$ induced by the star, multiplied by the duration $q_*/\langle v_*^2 \rangle^{1/2}$ of the encounter, to arrive at equation (26). For highly eccentric orbits $\delta q = h \delta h/(GM_\odot)$, whence

$$\frac{\delta q}{q} \sim \pm C \frac{M_*}{M_\odot} \left(\frac{a}{q_*} \right)^2 \left(\frac{2GM_\odot}{q \langle v_*^2 \rangle} \right)^{1/2} \quad (27)$$

For $M_* = M_\odot$, $q_* = 4000$ AU, $C \approx 6$ [see equation (3.17) of *Yabushita* (1972)], $\langle v_*^2 \rangle^{1/2} = 1$ km s^{-1}, and preencounter values of $q = 35$ AU and $a = 600$ AU, $\delta q/q \sim \pm 1$. Thus, Sedna's perihelion could have doubled to near its current value, $q \approx 76$ AU, by a single slow-moving cluster star. Multiple encounters at larger q_* cause q to random walk and change its value less effectively: $\langle (\delta q)^2 \rangle^{1/2} \propto (N_*)^{1/2} q_*^{-2} \propto q_*^{-1}$.

Had we performed this calculation for parameters appropriate to the present-day stellar environment, we would have found $\delta q/q \approx \pm 0.2$. The reduction in efficacy is due to the larger $\langle v_*^2 \rangle$ today.

The cluster properties cited above are averaged over a half-light radius of 2 pc (*Binney and Tremaine*, 1987). For comparison, the Hyades cluster has 4× lower n_*, 3× lower $\langle v_*^2 \rangle^{1/2}$, and 6× longer lifetime t (*Binney and Merrifield*, 1998; *Perryman et al.*, 1998); the Hyades therefore generates 2× fewer encounters than does our canonical cluster. Younger clusters like the Orion Trapezium maintain 15× higher n_* and similar $\langle v_*^2 \rangle^{1/2}$ over 200× shorter t (*Hillenbrand and Hartmann*, 1998), and therefore yield even fewer encounters. Scenarios that invoke stellar encounters for which $q_* \ll 1000$ AU to explain such features as the edge of the classical belt require that the Sun have resided in a cluster having atypical properties, i.e., dissimilar from those of the Orion Trapezium, the Hyades, and all open clusters documented by *Binney and Merrifield* (1998). That parent bodies in extrasolar debris disks also do not extend beyond ~40–100 AU (section 3.1.2) argues against explanations that rely on unusually dense environments.

3.8. Coagulation of Neptune Trojans

Planetesimal collisions that occur near Neptune's Lagrange points insert debris into 1:1 resonance. This debris can coagulate into larger bodies. The problem of accretion in the Trojan resonance is akin to the standard problem of planet formation, transplanted from a star-centered disk to a disk centered on the Lagrange point. As with other kinds of transplant operations, there are complications: Additional timescales not present in the standard problem, such as the libration period T_{lib} about the Lagrange point, require juggling. *Chiang and Lithwick* (2005, hereafter *CL05*) account for these complications to conclude that QR-sized Trojans may form as miniature oligarchs, each dominating its own tadpole-shaped annulus in the ancient Trojan subdisk. Alternative formation scenarios for Trojans such as pull-down capture and direct collisional emplacement of QR-sized objects into resonance are considered by *CL05* and deemed unlikely. Also, the mechanism proposed by *Morbidelli et al.* (2005) to capture Jupiter Trojans cannot be applied to Neptune Trojans since Uranus and Neptune today lie inside their 1:2 MMR and therefore could not have divergently migrated across it (A. Morbidelli, personal communication). We focus on *in situ* accretion, but acknowledge that a collisionless capture scenario might still be feasible and even favored by late-breaking data; see the end of this subsection.

In the theory of oligarchic planet formation (e.g., *G04*), each annulus is on the order of $5\,R_H$ in radial width; the number of QR-sized oligarchs that can be fitted into the tadpole libration region is

$$N_{Trojan} \sim \frac{(8M_N/3M_\odot)^{1/2}a_N}{5\,R_H} \sim 20 \qquad (28)$$

attractively close to the number of QR-sized Neptune Trojans inferred to exist today (section 2.2.3). The numerator in equation (28) equals the maximum width of the 1:1 MMR, $a_N \approx 30$ AU is Neptune's current semimajor axis, $R_H = R/\alpha$ is the Trojan's Hill radius, and $R \approx 90$ km is the radius of QR.

The input parameters of the coagulation model are the surface density σ and sizes s of small bodies in 1:1 resonance. Big bodies grow by consuming small bodies, but growth is limited because small bodies diffuse out of resonance by colliding with other small bodies. The time for a

small body to random walk out of the Trojan subdisk is

$$t_{esc} \sim \frac{\rho s}{\sigma\Omega}\left[\frac{(M_N/M_\odot)^{1/2}a_N}{u/\Omega}\right]^2 \qquad (29)$$

The term in square brackets follows from noting that a small body shifts its orbital guiding center by of order its epicyclic amplitude $\sim\pm u/\Omega$ every time it collides with another small body in an optically thin disk. To escape resonance, the small body must random walk the maximum libration width. We equate t_{esc} to the growth time of a big body t_{acc} (equation (4)) to solve for the maximum size to which a large body coagulates

$$R = R_{final} \sim 100\left(\frac{2}{u/v_H}\right)^{4/3}\left(\frac{s}{20\,\text{cm}}\right)^{1/3}\text{km} \qquad (30)$$

Our normalization of $u/v_H \approx 2$ is derived from $s \sim 20$ cm and $\sigma \sim 4 \times 10^{-4}$ g cm^{-2} $\sim 10\times$ the surface density inferred in QR-sized objects today; we derive u/v_H by balancing gravitational stirring by big bodies with damping by inelastic collisions between small bodies (*CL05*). For these parameter values, $t_{esc} \sim t_{acc} \sim 1 \times 10^9$ yr. Unlike Neptune-sized oligarchs that may have been ejected out of the solar system (section 3.4), all ~10–30 Trojan oligarchs in a single cloud should be present and eventually accounted for.

As speculated by *CL05*, orbital inclinations of Trojans with respect to Neptune's orbit plane might be small; perhaps $\langle i^2\rangle^{1/2} \lesssim 10°$. A thin disk of Neptune Trojans would contrast with the thick disks occupied by Jupiter Trojans, main-belt asteroids, and nonclassical KBOs, and would reflect a collisional, dissipative birth environment. Three other Neptune Trojans have since been announced after the discovery of QR, having inclinations of 1.4°, 25.1°, and 5.3° (*Sheppard and Trujillo*, 2006). If a large fraction of Neptune Trojans have high i, we might look to the ν_{18} secular resonance, unmodeled by *CL05*, to amplify inclinations. See also *Tsiganis et al.* (2005), who find that Neptune Trojans can be captured collisionlessly; the capture process is related to "freeze-in" as described in section 3.6.4.

3.9. Collisional Comminution

Over the last few billion years, sufficiently small and numerous bodies in the Kuiper belt suffer collisional attrition. As interpreted by *Pan and Sari* (2005, hereafter *PS05*), the break in the size distribution of KBOs at $R \approx 50$ km as measured by *Bernstein et al.* (2004) (section 2.2.1) divides the collisional spectrum at small R from the primordial coagulation spectrum at large R. For the remainder of this subsection, we do not distinguish between the various dynamical classes but instead analyze all KBOs together as a single group. At $R > R_{break}$, the size spectrum $dN/dR \propto R^{-\bar{q}_0}$, where dN is the number of objects per unit face-on area of the belt having sizes between R and R + dR (the differential surface number density). The slope $\tilde{q}_0 \sim 5$ (see section 2.2.1 for more precise values) presumably represents the unadulterated outcome of coagulation. Bodies at this large-R end of the spectrum are insufficiently numerous to collide among themselves and undergo attrition. At $R < R_{break}$, $dN/dR \propto R^{-\tilde{q}}$, where $\tilde{q}$ derives from a quasisteady collisional cascade (*Dohnanyi*, 1969; *PS05*). By definition of R_{break}, the time for a body of radius R_{break} to be catastrophically dispersed equals the time elapsed

$$\frac{1}{N_{proj} \times \pi R_{break}^2 \times \Omega} \sim t \qquad (31)$$

where πR_{break}^2 is the collision cross-section and N_{proj} is the surface number density of projectiles that are just large enough to disperse R_{break}-sized targets (catastrophic dispersal implies that the mass of the largest postcollision fragment is no greater than half the mass of the original target and that collision fragments disperse without gravitational reassembly). This expression is valid for the same assumptions underlying equation (3), i.e., for today's dynamically hot belt.

We proceed to estimate R_{break} given the parameters of the present-day Kuiper belt. For $R > R_{break}$, $N = N_0(R/R_0)^{1-\tilde{q}_0}$, where N is the surface number density of objects having sizes between R and 2 R. We estimate that for fiducial radius $R_0 = 100$ km, $N_0 \approx 20$ AU^{-2} at $a \approx 43$ AU. The minimum radius R_{proj} of the projectile that can catastrophically disperse a target of radius R_{break} is given by

$$\frac{1}{2} R_{proj}^3 v_{rel}^2 = R_{break}^3 Q^* \qquad (32)$$

where

$$Q^* = Q_0^* \left(\frac{R}{R_0} \right)^y \qquad (33)$$

is the collisional specific energy (*Greenberg et al.*, 1978; *Fujiwara et al.*, 1989) and v_{rel} is the relative collision velocity. Since for $R < R_{break}$ as much mass is ground into every logarithmic interval in R as is ground out (e.g., *PS05*),

$$\tilde{q} = \frac{21 + y}{6 + y} \qquad (34)$$

We assume (and can check afterward) that $R_{proj} < R_{break} < R_0$ to write

$$N_{proj} = N_0 \left(\frac{R_{break}}{R_0} \right)^{1-\tilde{q}_0} \left(\frac{R_{proj}}{R_{break}} \right)^{1-\tilde{q}} \qquad (35)$$

Combining the above relations yields

$$\frac{R_{break}}{R_0} \sim (\pi N_0 R_0^2 \Omega t)^{z_1} \left(\frac{v_{rel}^2}{2Q_0^*} \right)^{z_2} \qquad (36)$$

where $z_1 = (6 + y)/[5y + (6 + y)(\tilde{q}_0 - 3)]$ and $z_2 = 5/[5y + (6 + y)(\tilde{q}_0 - 3)]$. For targets held together by self-gravity, $Q^* \approx 3v_{esc}^2/10$ and $y = 2$. If we insert these values into equation (36), together with $v_{rel} = 1$ km s^{-1}, $\tilde{q}_0 = 5$, $\Omega = 2\pi/(300$ yr), and $t = 3 \times 10^9$ yr, we find that $R_{break} \approx 0.4 R_0 \approx 40$ km, in good agreement with the observed break in the luminosity function (Fig. 5) (*PS05*). The small-R end of the KBO size spectrum as observed today reflects the catastrophic comminution of bodies that derive their strength from self-gravity ("rubble piles"). Furthermore, the Kuiper belt has been dynamically hot for the last few billion years (*PS05*).

4. DIRECTIONS FOR FUTURE WORK

1. *Collisional vs. collisionless:* Most explorations of planetary migration and of how the Kuiper belt was stirred utilize collisionless gravitational simulations. But the overwhelming bulk of the primordial mass may have resided in small, collisional bodies. Simultaneously accounting for collisions and gravity might revolutionize our understanding of the clean-up (a.k.a. missing-mass) problem. Insights from the study of planetary rings will be helpful.

2. *Classical Kuiper belt object colors vs. heliocentric distance:* Do classical KBOs exhibit a trend in color from neutral to red with increasing heliocentric distance d? The two neutral classicals at $d \approx 38$ AU, contrasted with the predominantly red classicals at $d \approx 42$ AU, suggest the answer is yes (Figs. 3 and 4). Confirmation would support ideas that classicals coagulated *in situ*, and that neutrally colored resonant/scattered KBOs coagulated from small d and were transported outward. We must also ask why trends in color with birth distance d would exist in the first place.

3. *Formation of the scattered belt by Neptune-mass oligarchs:* We argue that Neptune's migration and the concomitant sweeping of secular resonances do not populate the scattered and hot classical belts with enough objects to explain observations. When account is made of the primordial size distribution of planetesimals — a distribution that should be preserved today at large sizes (sections 1 and 3.9) — the expected population of scattered/hot classical objects having sizes above 100 km is less than that observed by a factor of 50–150. We propose instead that planetesimals were deflected onto scattered/hot classical orbits by simple close encounters with marauding Neptune-mass oligarchs that have since been ejected from the solar system, and by Neptune while its orbit circularized by dynamical friction. These contentions are supported by order-of-magnitude estimates but require numerical simulations to verify.

4. *Kuiper Cliff:* Why do planetesimal disks have sharp outer edges?

5. *Binaries*: Kuiper belt binaries might prove the most informative witnesses we have to the history of transneptunian space. They hearken back to a primordially dense and cold disk in which collisions and multiple-body encounters were orders of magnitude more frequent than they are today. Binary orbit properties must also reflect how the Kuiper belt was stirred as a whole. How binary inclinations, eccentricities, and component mass ratios are distributed, and how/why the incidence of binarity correlates with dynamical class, are open issues for observer and theorist alike.

Acknowledgments. This work was supported by the Alfred P. Sloan Foundation, the National Science Foundation, and NASA. We acknowledge helpful exchanges with G. Bernstein, P. Goldreich, R. Gomes, D. Jewitt, S. Kenyon, A. Morbidelli, D. Nesvorný, R. Sari, L. Strubbe, A. Youdin, and an anonymous referee. We are grateful to the Deep Ecliptic Survey (DES) team for their unstinting support.

REFERENCES

Bernstein G. M., Trilling D. E., Allen R. L., Brown M. E., Holman M., and Malhotra R. (B04) (2004) *Astron. J., 128*, 1364–1390.

Binney J. and Merrifield M. (1998) *Galactic Astronomy*, pp. 377–386. Princeton Univ., Princeton.

Binney J. and Tremaine S. (1987) *Galactic Dynamics*, pp. 26, 440–443. Princeton Univ., Princeton.

Boehnhardt H., Tozzi G. P., Birkle K., Hainaut O., Sekiguchi T., et al. (2001) *Astron. Astrophys., 378*, 653–667.

Brown M. E., Trujillo C., and Rabinowitz D. (2004) *Astrophys. J., 617*, 645–649.

Brown M. E., Trujillo C. A., and Rabinowitz D. L. (2005a) *Astrophys. J., 635*, L97–L100.

Brown M. E., van Dam M. A., Bouchez A. H., Le Mignant D., Campbell R. D., et al. (2005b) *Astrophys. J., 632*, L45–L48.

Buie M. W., Grundy W. M., Young E. F., Young L. A., and Stern S. A. (2006) *Astron. J., 132*, 290–298.

Canup R. M. (2005) *Science, 307*, 546–550.

Chiang E. I. and Jordan A. B. (2002) *Astron. J., 124*, 3430.

Chiang E. I. and Lithwick Y. (CL05) (2005) *Astrophys. J., 628*, 520–532.

Chiang E. I., Jordan A. B., Millis R. L., Buie M. W., Wasserman L. H., et al. (C03) (2003) *Astron. J., 126*, 430–443.

Christy J. W. and Harrington R. S. (1978) *Astron. J., 83*, 1005–1008.

Dermott S. F., Malhotra R., and Murray C. D. (1988) *Icarus, 76*, 295–334.

Dohnanyi J. W. (1969) *J. Geophys. Res., 74*, 2531–2554.

Duncan M. J., Levison H. F., and Budd S. M. (D95) (1995) *Astron. J., 110*, 3073–3081.

Elliot J. L., Kern S. D., Clancy K. B., Gulbis A. A. S., Millis R. L., et al. (E05) (2005) *Astron. J., 129*, 1117–1162.

Fernández J. A. and Brunini A. (2000) *Icarus, 145*, 580–590.

Fernández J. A. and Ip W.-H. (1984) *Icarus, 58*, 109–120.

Fujiwara A., Cerroni P., Davis D., Ryan E., and di Martino M. (1989) In *Asteroids II* (R. P. Binzel et al., eds.), pp. 240–265. Univ. of Arizona, Tucson.

Funato Y., Makino J., Hut P., Kokubo E., and Kinoshita D. (2004) *Nature, 427*, 518–520.

Garaud P. and Lin D. N. C. (2004) *Astrophys. J., 608*, 1050–1075.

Gladman B., Holman M., Grav T., Kavelaars J., Nicholson P., et al. (2002) *Icarus, 157*, 269–279.

Goldreich P. (1965) *Mon. Not. R. Astron. Soc., 130*, 159–181.

Goldreich P., Lithwick Y., and Sari R. (G02) (2002) *Nature, 420*, 643–646.

Goldreich P., Lithwick Y., and Sari R. (G04) (2004a) *Ann. Rev. Astron. Astrophys., 42*, 549–601.

Goldreich P., Lithwick Y., and Sari R. (2004b) *Astrophys. J., 614*, 497–507.

Gomes R. S. (2003a) *Icarus, 161*, 404–418.

Gomes R. S. (2003b) *Earth Moon Planets, 92*, 29–42.

Gomes R. S., Morbidelli A., and Levison H. F. (2004) *Icarus, 170*, 492–507.

Gomes R. S., Gallardo T., Fernández J. A., and Brunini A. (2005) *Cel. Mech. Dyn. Astron., 91*, 109–129.

Gómez G. C. and Ostriker E. C. (2005) *Astrophys. J., 630*, 1093–1106.

Greenberg R., Hartmann W. K., Chapman C. R., and Wacker J. F. (1978) *Icarus, 35*, 1–26.

Greenberg R., Bottke W. F., Carusi A., and Valsecchi G. B. (1991) *Icarus, 94*, 98–111.

Hahn J. M. and Malhotra R. (1999) *Astron. J., 117*, 3041–3053.

Hahn J. M. and Malhotra R. (2005) *Astron. J., 130*, 2392–2414.

Hillenbrand L. A. and Hartmann L. W. (1998) *Astrophys. J., 492*, 540–553.

Holman M. J. and Wisdom J. (HW93) (1993) *Astron. J., 105*, 1987–1999.

Ida S. and Makino J. (1993) *Icarus, 106*, 210–217.

Jewitt D. and Luu J. (1993) *Nature, 362*, 730–732.

Jewitt D. and Luu J. (2000) In *Protostars and Planets IV* (V. Mannings et al., eds.), pp. 1201–1229. Univ. of Arizona, Tucson.

Jewitt D., Luu J., and Trujillo C. (1998) *Astron. J., 115*, 2125–2135.

Kalas P. and Jewitt D. (1995) *Astron. J., 110*, 794–804.

Kenyon S. J. (2002) *Publ. Astron. Soc. Pac., 793*, 265–283.

Kenyon S. J. and Bromley B. C. (2006) *Astron. J., 131*, 1837–1850.

Kenyon S. J. and Luu J. X. (1998) *Astron. J., 115*, 2136–2160.

Kenyon S. J. and Luu J. X. (1999) *Astron. J., 118*, 1101–1119.

Knežević Z., Milani A., Farinella P., Froeschle Ch., and Froeschle Cl. (1991) *Icarus, 93*, 315–330.

Krist J. E., Ardila D. R., Golimowski D. A., Clampin M., Ford H. C., et al. (2005) *Astron. J., 129*, 1008–1017.

Levison H. F. and Morbidelli A. (2003) *Nature, 426*, 419–421.

Malhotra R. (1993) *Nature, 365*, 819–821.

Malhotra R. (1995) *Astron. J., 110*, 420–429.

Matsuyama I., Johnston D., and Hartmann L. (2003) *Astrophys. J., 582*, 893–904.

Millis R. L., Buie M. W., Wasserman L. H., Elliot J. L., Kern S. D., and Wagner R. M. (2002) *Astron. J., 123*, 2083–2109.

Morbidelli A., Jacob C., and Petit J.-M. (2002) *Icarus, 157*, 241–248.

Morbidelli A., Levison H. F., Tsiganis K., and Gomes R. (2005) *Nature, 435*, 462–465.

Murray C. D. and Dermott S. F. (1999) In *Solar System Dynamics*, pp. 63–128, 225–406. Cambridge Univ., Cambridge.

Murray-Clay R. A. and Chiang E. I. (MC05) (2005) *Astrophys. J., 619*, 623–638.

Murray-Clay R. A. and Chiang E. I. (MC06) (2006) *Astrophys. J., in press* (astro-ph/0607203).

Nesvorný D. and Dones L. (2002) *Icarus, 160*, 271–288.

Noll K. (2003) *Earth Moon Planets, 92*, 395–407.

Pan M. and Sari R. (2004) *Astron. J., 128*, 1418–1429.

Pan M. and Sari R. (PS05) (2005) *Icarus, 173*, 342–348.

Peale S. J. (1986) In *Satellites* (J. A. Burns and M. S. Matthews, eds.), pp. 159–223. Univ. of Arizona, Tucson.

Peixinho N., Boehnhardt H., Belskaya I., Doressoundiram A., Barucci M. A., and Delsanti A. (2004) *Icarus, 170*, 153–166.

Perryman M. A. C., Brown A. G. A., Lebreton Y., Gomez A., Turon C., et al. (1998) *Astron. Astrophys., 331*, 81–120.

Press W. H., Teukolsky S. A., Vetterling W. T., and Flannery B. P. (1992) In *Numerical Recipes in C: The Art of Scientific Computing*, pp. 609–639. Cambridge Univ., Cambridge.

Sheppard S. S. and Jewitt D. (2004) *Astron. J., 127*, 3023–3033.

Sheppard S. S. and Trujillo C. A. (2006) *Science, 313*, 511–514.

Stephens D. C. and Noll K. S. (2006) *Astron. J., 131*, 1142–1148.

Stern S. A., Weaver H. A., Steffl A. J., Mutchler M. J., Merline W. J., et al. (2006) *Nature, 439*, 946–948.

Strubbe L. E. and Chiang E. I. (2006) *Astrophys. J., 648*, 652–665.

Thommes E. W., Duncan M. J., and Levison H. F. (1999) *Nature, 402,* 635–638.

Thommes E. W., Duncan M. J., and Levison H. F. (2002) *Astron. J., 123,* 2862–2883.

Tiscareno M. S. and Malhotra R. (2003) *Astron. J., 126,* 3122–3131.

Trujillo C. A. and Brown M. E. (2001) *Astrophys. J., 554,* L95–L98.

Trujillo C. A. and Brown M. E. (2002) *Astrophys. J., 566,* L125–L128.

Tsiganis K., Gomes R., Morbidelli A., and Levison H. F. (2005) *Nature, 435,* 459–461.

Veillet C., Parker J. W., Griffin I., Marsden B., Doressoudiram A., et al. (2002) *Nature, 416,* 711–713.

Weaver H. A., Stern S. A., Mutchler M. J., Steffl A. J., Buie M. W., et al. (2006) *Nature, 439,* 943–945.

Weidenschilling S. J. (2002) *Icarus, 160,* 212–215.

Yabushita S. (1972) *Astron. Astrophys., 16,* 395–403.

Youdin A. N. and Shu F. H. (2002) *Astrophys. J., 580,* 494–505.

Youdin A. N. and Chiang E. I. (2004) *Astrophys. J., 601,* 1109–1119.

Youdin A. N. and Goodman J. (2005) *Astrophys. J., 620,* 459–469.

Part IX:

Life

Comparative Planetology and the Search for Life Beyond the Solar System

Charles A. Beichman
California Institute of Technology

Malcolm Fridlund
European Space Agency

Wesley A. Traub and Karl R. Stapelfeldt
Jet Propulsion Laboratory

Andreas Quirrenbach
University of Leiden

Sara Seager
Carnegie Institute of Washington

The study of planets beyond the solar system and the search for other habitable planets and life is just beginning. Groundbased (radial velocity and transits) and spacebased surveys (transits and astrometry) will identify planets spanning a wide range of size and orbital location, from Earth-sized objects within 1 AU to giant planets beyond 5 AU, orbiting stars as near as a few parsec and as far as a kiloparsec. After this initial reconnaissance, the next generation of space observatories will directly detect photons from planets in the habitable zones of nearby stars. The synergistic combination of measurements of mass from astrometry and radial velocity, of radius and composition from transits, and the wealth of information from the direct detection of visible and mid-IR photons will create a rich field of comparative planetology. Information on protoplanetary and debris disks will complete our understanding of the evolution of habitable environments from the earliest stages of planet formation to the transport into the inner solar system of the volatiles necessary for life. The suite of missions necessary to carry out the search for nearby, habitable planets and life requires a "Great Observatories" program for planet finding (SIM PlanetQuest, Terrestrial Planet Finder-Coronagraph, and Terrestrial Planet Finder-Interferometer/Darwin), analogous to the highly successful "Great Observatories Program" for astrophysics. With these new Great Observatories, plus the James Webb Space Telescope, we will extend planetology far beyond the solar system, and possibly even begin the new field of comparative evolutionary biology with the discovery of life itself in different astronomical settings.

1. STUDIES OF PLANETARY SYSTEMS AND THE SEARCH FOR LIFE

The search for habitable planets and life beyond Earth represents one of the oldest questions in natural philosophy, but is one of the youngest fields in astronomy. This new area of research derives its support among the scientific community and the general public from the fact that we are using twenty-first century technology to address questions that were first raised by inquiring minds almost 2500 years ago. Starting in 1995, radial velocity and, most recently, transit and microlensing studies have added more than 168 planets to the 9 previously known in our own solar system. With steadily improving instrumentation, the mass limit continues to drop while the semimajor axis limit continues to grow: a 7.5 $M_\oplus$ planet orbiting the M star GL 876 at 0.02 AU (*Rivera et al.*, 2005) and a 4 M_J planet orbiting 55 Cnc at 5.2 AU (*Marcy et al.*, 2002) define boundaries that are certain to be eclipsed by newer discoveries. While we do not yet have the tools to find true solar system analogs or an Earth in the habitable zone of its parent star, evidence continues to accumulate (*Marcy et al.*, 2005) that the number of potential Earths is large and that some could be detected nearby, if only we had the tools. In this chapter, we assess the prospects for the discovery and eventual characterization of planets of all sizes — from gas giants to habitable terrestrial analogs. Considerations of length necessarily make this discussion incomplete and we have omitted discussion of valuable techniques that do not by their nature (e.g. microlensing) lend themselves to follow-up observations relevant to physical characterization of planets and the search for life.

TABLE 1.   Space projects presently under consideration.

	Approximate Sensitivity Planet Size, Orbit, Distribution	Planet Yield*	Approximate Launch Date
Survey for Distant Planets (Transits)			
COROT	2 $R_\oplus$ at 0.05 AU (500 pc)	10s–100	2006
Kepler	$R_\oplus$ at 1 AU (500 pc)	100s	2008
GAIA	10 $R_\oplus$ at <0.1 AU	4,000	2012
JWST	2 $R_\oplus$ (100–500 pc)	250[†]	2013
Determine Masses/Orbits (Astrometry)			
SIM	3 $R_\oplus$ at 1 AU (10 pc)	250	2012[§]
GAIA	30 $R_\oplus$ at 1 AU (<200 pc)	10,000	2012
Characterize Planets and Search for Life (Direct Detection)			
JWST	1 R_J at >30 AU (50–150 pc)	250 young stars	2013
TPF-Coronagraph[‡]	1 $R_\oplus$ at 1 AU (15 pc)	250	2018[§]
TPF-Interferometer/Darwin[‡]	1 $R_\oplus$ at 1 AU (15 pc)	250	2018[§]

*Yield is highly approximate and assumes roughly one planet orbiting each star surveyed.
[†]Approximate number of follow-ups of groundbased, COROT, and Kepler transit targets.
[‡]Parameters for TPF-C and TPF-I/Darwin are still being developed.
[§]Launch dates are uncertain in light of recent NASA budget submissions for 2007 and beyond.

Papers in this volume describe many results from planet-finding experiments. With the exception of a few (very exciting) HST and Spitzer observations, these are drawn primarily from groundbased observations. The discussion of space activities necessarily focuses on future activities. While the space environment is highly stable and offers low backgrounds and an unobscured spectral range, the technology needed to take full advantage of the environment will take many years to develop and the missions to exploit the technology and the environment will be expensive to construct, launch, and operate. But when the missions described here are completed, they will revolutionize our conception of our place in the universe.

Table 1 identifies the major spacebased projects presently under consideration grouped by observing technique: transit photometry, astrometry, and direct detection. Plate 7 summarizes the discovery space of these projects as well as some groundbased activities in the mass-semimajor axis plane. COROT, Kepler, and SIM will provide the many order-of-magnitude improvements in sensitivity and resolution relative to groundbased efforts needed to detect other Earths in the habitable zones of their parent stars using indirect techniques of transits, radial velocity, and astrometry.

Once we have detected these planets via indirect means (section 2 and section 3), we argue below that we will need a number of missions to characterize these planets physically and to search for evidence of life in any atmospheres these planets may possess. Analogously to the improved knowledge gained about astrophysical phenomena from using all of NASA's four Great Observatories (the Hubble Space Telescope, the Compton Gamma Ray Observer, the Chandra X-ray Observatory, and the Spitzer infrared telescope) and the cornerstone missions of the European Space Agency (ESA) (the Infrared Space Observatory — ISO — and the X-ray Multi-Mirror satellite — XMM), a compa-

rable "Great Observatories" program for planet finding will yield an understanding of planets and the search for life that will greatly exceed the contributions of the individual missions.

As discussed in section 2, the combination of transit photometry with COROT and Kepler, follow-up spectrophotometry from the James Webb Space Telescope (JWST), and follow-up radial velocity data gives unique information on planetary mass, radius, density, orbital location, and, in favorable cases, composition of the upper atmosphere. These data, available for large numbers of planets, will revolutionize our understanding of gas-giant and icy planets. While less information will be available for smaller, rocky planets, a critical result of the transit surveys will be the frequency of Earth-sized planets in the habitable zone, $\eta_\oplus$ (*Beichman*, 2000). Since the angular resolution and collecting area needed for the direct detection of nearby planets are directly related to the distance to the closest host stars, the value of $\eta_\oplus$ will determine the scale and cost of missions to find and characterize those planets.

Subsequent to the transit surveys, we will embark on the search for and the characterization of nearby planets, and the search for a variety of signposts of life. We will ultimately require three complementary datasets: masses via astrometry (SIM PlanetQuest, section 3); optical photons (TPF-Coronagraph, section 4); and mid-IR photons (TPF-Interferometer/Darwin, section 5). JWST will play an important role in follow-up activities looking at groundbased, COROT, and Kepler transits; making coronagraphic searches for hot, young Jupiters; and studying protoplanetary and debris disks. The synergy between the planet-finding missions, with an emphasis on studies of nearby, terrestrial planets and the search for life, is addressed in section 6. Studies of potential target stars are ongoing but will need to be intensified and their results collated to select the best targets (section 7). The ordering of these missions will be the result of the optimization of a highly nonlinear function incorporating technical readiness, cost, and political and scientific support on two or more continents (section 8).

2. TRANSITING PLANETS

The age of comparative planetology is upon us with a growing number of transiting giant planets — nine and counting. As they pass in front of and behind their parent stars, the transiting planets' size — and hence density, transmission spectrum, and thermal emission and albedo — can potentially be measured. This makes the group of transiting planets the ones that can best be physically characterized before direct imaging is available (see chapters by Udry et al., Charbonneau et al., and Marley et al. for details on the recent planet transit discoveries, observations, and interpretations).

The dedicated, spacebased, transit survey missions — Kepler and COROT — will build upon the exciting ground-based transit detections of giant planets. With very high-precision photometry enabled by the stable space environment and lack of day/night and weather interruptions, COROT and Kepler will push to planetary sizes as small as the Earth's. With long-duration observing campaigns they will extend transit planet discoveries to larger semimajor axes. JWST will similarly build on the pioneering HST and Spitzer measurements of planetary atmospheres. The legacy of Kepler and COROT combined with JWST will be to enable comparative planetology on a wide range of planet types, encompassing a range of planet masses, temperatures, and host stars, before direct imaging of solar-system-aged planets is possible.

2.1. Prospects for Planet Transit Discoveries

2.1.1. HST and MOST. The Hubble Space Telescope (HST) and the MOST (Microvariability and OScillations of STars) microsatellite of the Canadian Space Agency have shown the promise of spacebased transit studies. HST monitored 34,000 stars in the globular cluster 47 Tuc (*Gilliand et al., 2000*) continuously for 8.3 days. While 17 short-period giant planets were expected, none were found, suggesting that either low-metallicity or high stellar density interfere with planet formation or migration. More recently, HST ACS/WFC monitored a field of 160,000 main-sequence stars in the galactic bulge field for 7 days (*Sahu et al., 2005*). Over 100 transiting planets were expected if the frequency of hot Jupiters in the galactic bulge is similar to that in the solar neighborhood. MOST, launched in June 2003, is a 15-cm telescope with a 350–700-nm broadband filter and a part-per-million photometric accuracy capability for bright stars monitored for one month or more. MOST is not a transit survey instrument, but has monitored four stars hosting known hot Jupiters for 10–30 days. Of relevance to COROT and Kepler, MOST has put an upper limit on the albedo of HD 209458b of 0.15 (1σ) (*Rowe et al., 2006*) and is finding hints that host stars of the short-period planets (hot Jupiters)

may be too variable to detect the illumination phase curve (*Walker et al., 2005*).

2.1.2. COROT and Kepler. COROT (COnvection, ROtation, and planetary Transit) and Kepler are wide-field-survey space telescopes designed to detect small transiting planets via extremely high-precision photometry. These telescopes will initiate the next generation of exoplanetary science by uncovering Neptune- (17 $M_{\oplus}$) to Earth-sized planets around a range of stellar types. A large pool of this as-yet-unknown class of low-mass planets will provide the planet frequency and orbital distribution for insight into their formation and migration. The same group of planets will yield many objects suitable for follow-up physical characterization.

COROT is a CNES/ESA mission to be launched in October 2006 (*Baglin et al., 2003*). COROT is a 27-cm telescope with a 3.5 deg^2 field of view. For the planet survey part of the program, five fields containing approximately 12,000 dwarf stars in the range $11 < V < 16.5$ mag will be continuously monitored for 150 days. Assuming all stars have a 2 $R_{\oplus}$ radius planet and assuming that these planets are uniformly distributed in semimajor axis, then COROT will detect about 100 of these planets (*Bordé et al., 2003*). [See *Gillon et al.* (2005) and *Moutou et al.* (2005) for details including the radius and semimajor axis distribution of expected transiting planets around different star types.]

Kepler is a NASA Discovery mission (*Borucki et al., 2003*) to be launched in June 2008. Kepler has a 0.95-m diameter mirror and an extremely wide field of view — 105 deg^2. Kepler will simultaneously monitor more than 100,000 main-sequence stars ($V < 14$ mag) for its four-year mission duration. Kepler will find 50 transiting Earth-sized planets in the 0.5–1.5-AU range, if every star has two terrestrial planets (as the Sun does). This number increases to 650 planets if most terrestrial planets have a size of 2 $R_{\oplus}$. If Kepler finds few Earth-sized planets it will come to the surprising and significant conclusion that Earth-sized planets in Earth-like orbits are rare. Aside from detecting Earth-sized planets in the habitable zone, Kepler will advance the hot Neptune and hot Earth studies started by COROT, detecting up to hundreds of them down to a size as small as that of Mercury.

Both Kepler and COROT will produce exciting extra-solar giant planet science with tens of transiting giant planets with semimajor axes from 0.02 to 1 AU. Even giant planets in outer orbits — beyond 1 AU — can be detected with Kepler from single transit events at 8σ. Follow-up radial velocity observations are required to confirm that the photometric dips are really due to planetary transits, as well as to measure the planetary masses and in some cases to determine the orbital period.

2.2. Physical Characterization of Transiting Planets

The prospects for physical characterization of transiting planets first with Spitzer and then with JWST are truly astonishing. By the time JWST launches, 100 or more transiting planets from Jupiter down to Earth-sizes should be available for observation. While the most favorable planets of all sizes will be at short semimajor axes, a decent number of larger planets out to 1 AU will also be suitable for density and atmosphere measurements. For transiting giant planets, their albedo, moons, and rings (*Brown et al., 2001*), and even the oblateness and hence rotation rate (*Seager and Hui, 2002; Barnes and Fortney, 2003*), can potentially be measured. Some specific possibilities of physical characterization are given below.

2.2.1. Photometry. JWST's NIRSpec is a high-resolution spectrograph operating from 0.7 to 5 μm. With its spectral dispersion and high cadence observing NIRSpec will be capable of high-precision spectrophotometry on bright stars. NIRSpec data can then be used in the same way that the HST STIS spectral data for HD 209458 was rebinned for photometry (*Brown et al., 2001*). For example, at 0.7 μm JWST can obtain a 35σ transit detection for two interesting cases: an Earth-sized moon orbiting HD 209458b (3-hour transit time at 47 pc) and a 1 AU Earth-sized planet orbiting a Sun-like star at 300 pc (*Gilliland, 2005*). Kepler stars are about 300 pc distant — meaning JWST will be capable of confirming Kepler's Earth-sized planet candidates.

Planetary density is the key to the planetary bulk composition. With precise radii from JWST and complementary radial velocity mass measurements, densities of many planets can be determined, even for super-Earth-mass planets close to the star. This will identify the nature of many Neptune-mass and super-Earth-mass planets. Are they ocean planets? Carbon planets? Small gaseous planets? Remnant cores of evaporated giant planets? Or some of each? In this way JWST + Kepler/COROT will be able to provide insight into planet formation and migration of the low-mass planets.

2.2.2. Spectroscopy. Planetary temperature is important for understanding planetary atmospheres and composition. Spitzer has initiated comparative exoplanetology by measuring transiting hot Jupiters in the thermal infrared using secondary eclipse (*Deming et al., 2005; Charbonneau et al., 2005*). Four transiting planets were observed in 2005. Spitzer's broadband photometry from 3 to 8 μm, together with photometry at 14 and 24 μm, will help constrain the temperatures and compositions of hot Jupiters, including possibly their metallicity. The thermal infrared phase variation of seven hot Jupiters may also be detected this year, providing clues about planetary atmospheric circulation in the intense irradiation environment.

The JWST thermal-IR detection capability can be explored by scaling the Spitzer results. The 5–8-μm region is ideal for solar-type stars because the planet-star contrast is high and the exozodiacal background is low. For an estimate we can scale the TrES-1 5σ detection at 4.5 μm, taking into account that JWST has 45× the collecting area of Spitzer, and assuming that the overall efficiency is almost 2× times lower, giving an effective collecting area improvement of ~25×. JWST will be therefore be able to detect hot Jupiter thermal emission at an SNR of 25 around stars at TrES-1's distance (~150 pc; a distance that includes most stars from shallow groundbased transit surveys). Similarly,

JWST can detect a hot planet five times smaller than TrES-1, or down to 2 Earth radii, for the same set of stars *assuming instrument systematics are not a limiting factor*. Scaling with distance, JWST can detect hot Jupiters around stars five times more distant than TrES-1 to SNR of 5, which includes all the Kepler and COROT target stars. Beyond photometry, JWST can obtain thermal emission spectra (albeit at a lower SNR than for photometry for the same planet). Rebinning the R = 3000 NIRSpec data to low-resolution spectra will enable detection of H_2O, CO, CH_4, and CO_2.

Transiting planets too cold to be observed in thermal emission can still be observed via transmission spectroscopy during primary planet transit. Such planets include giant planets at all semimajor axes from their stars. In visible and near-IR wavelengths, NIRSpec might detect H_2O, CO, CH_4, Na, K, O_2, and CO_2. With NIRSpec capabilities to a wavelength as short as 0.7 μm, JWST has the potential to identify molecular oxygen at 0.76 μm (a sign of life as we know it) in the outer atmosphere of a super-Earth-mass planet.

3. ASTROMETRY

3.1. Why Astrometry?

The principle of planet detection with astrometry is similar to that behind the Doppler technique: The presence of a planet is inferred from the motion of its parent star around the common center of gravity. In the case of astrometry one observes the two components of this motion in the plane of the sky; this gives sufficient information to solve for the orbital elements without the sin i ambiguity plaguing Doppler measurements. In addition, the astrometric method can be applied to all types of stars (independently of their spectral characteristics), is less susceptible to noise induced by the stellar atmosphere, and is more sensitive to planets with large orbital semimajor axes. From simple geometry and Kepler's Laws it follows immediately that the astrometric signal θ of a planet with mass m_p orbiting a star with mass m_* at a distance d in a circular orbit of radius a is given by

$$\theta = \frac{m_p}{m_*} \frac{a}{d} = \left(\frac{G}{4\pi^2}\right)^{1/3} \frac{m_p}{m_*^{2/3}} \frac{P^{2/3}}{d} =$$

$$3 \, \mu as \cdot \frac{m_p}{M_\oplus} \cdot \left(\frac{m_*}{M_\odot}\right)^{-2/3} \left(\frac{P}{yr}\right)^{2/3} \left(\frac{d}{pc}\right)^{-1} \tag{1}$$

This signature is represented in Plate 7, which shows astrometric detection limits for groundbased and spacebased programs.

3.2. Astrometry from the Ground

The best prospects for astrometric planet detection from the ground will be offered by the development of narrow-angle dual-star interferometry (*Shao and Colavita, 1992; Quirrenbach et al., 1998; Traub et al., 1996*), which is being pursued at the Palomar Testbed Interferometer (PTI) (*Mu-*

terspaugh et al., 2005), the Keck Interferometer (KI), and at ESO's Very Large Telescope Interferometer (VLTI) (*Quirrenbach et al.,* 2000). While Earth's atmosphere imposes limits on groundbased interferometers, in favorable cases where a suitable reference star is available within 10", a precision of 10 μas can be reached. While adequate for gas giant planets, this limit excludes the use of groundbased facilities for searches for Earth analogs (cf. Plate 7). The scientific goals of groundbased projects thus include determining orbital inclinations and masses for planets already known from radial-velocity surveys, searches for giant planets around stars that are not amenable to high-precision radial-velocity observations, and a search for large rocky planets around nearby low-mass stars.

3.3. SIM PlanetQuest: Nearby Terrestrial Planets

NASA's Space Interferometry Mission (SIM PlanetQuest) will push the precision of astrometric measurements far beyond the capabilities of any other project currently in existence or under development. SIM, to be launched in 2012, will exploit the advantages of space to perform a diverse astrometric observing program (e.g., *Unwin,* 2005; *Quirrenbach,* 2002). SIM consists of a single-baseline interferometer with 30-cm telescopes on a 9-m baseline. SIM is a pointed mission so that targets can be observed whenever there is a scientific need, subject only to scheduling and solar exclusion angle constraints. Additionally, the integration time can be matched to the desired signal-to-noise ratio, enabling observation of very faint systems.

In its "narrow-angle" mode (i.e., over a field of about 1°), SIM will provide an accuracy of ~1 μas for each measurement. SIM PlanetQuest will carry out a high-precision survey of 250 nearby stars reaching down to 1–3 $M_\oplus$ (depending on stellar mass and distance), and a less-sensitive survey of some 2000 stars, establishing better statistics on massive planets in the solar neighborhood. In addition, SIM will observe a sample of pre-main-sequence stars to investigate the epoch of planet formation.

It is very likely that SIM PlanetQuest will discover the first planets in the habitable zone around nearby stars. True Earth analogs are just within reach. With 200 visits at a single-measurement precision of 1 μas, astrometric signatures just below 1 μas constitute secure planet detections, with a false-alarm probability of only 1%. This means that planets with 1 $M_\oplus$ in 1 AU orbits can be discovered around seven nearby G and K dwarfs; planets twice as massive would be found around 28 G and K stars. Assuming $\eta_\oplus$ = 0.1, SIM would have a ~50% chance of finding at least one 1 $M_\oplus$/1 AU planet, and a ~95% chance of discovering at least one 2 $M_\oplus$ planet in a 1 AU orbit. Figure 1 (cf. *Catanzarite et al.,* 2006) summarizes the number of terrestrial planets of various masses that SIM might find in the habitable zones [~1 AU(L/L$_\odot$)$^{0.5}$] surrounding 250 nearby solar-type stars (sections 6 and 7).

One should note that the astrometric signature of Earth, 450 km or 1/1500 $R_\odot$, is several times larger than the mo-

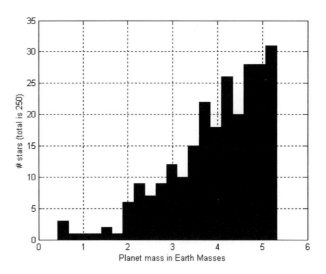

Fig. 1. Histogram showing the number of terrestrial planets in different mass ranges that SIM PlanetQuest could find in the habitable zones surrounding 250 nearby solar-type stars assuming 100 visits with 1-μas accuracy over 5 yr (cf. *Catanzarite et al.,* 2006).

tion of the photocenter of the Sun induced by spots. Starspots are not expected to contribute significantly to the noise for planet searches around Sun-like stars. Although starspots are a cause for concern for more active types of star, their effects are less than the radial velocity noise associated with young stars. SIM astrometry will be able to find gas giant planets within a few AU of T Tauri stars for the first time.

3.4. GAIA: A Census of Giant Planets

The European Space Agency (ESA) is planning to launch an astrometric satellite, GAIA, in roughly the same time frame as SIM. GAIA's architecture builds on the successful Hipparcos mission (*Lindegren and Perryman,* 1996; *Perryman et al.,* 2001). Unlike SIM, GAIA will be a continuously scanning survey instrument with a large field of view, which will cover the whole sky quite uniformly, observing each star hundreds of times over its five-year mission. Among the many scientific results of the GAIA mission will thus be a complete census of stellar and substellar companions down to the accuracy limit of the mission. This limit depends on the magnitude and color of each star: For a G2V star the expected accuracy, expressed as the parallax error at the end of the mission, is 7 μas at V = 10 mag and 25 μas at V = 15 mag. Very roughly, the corresponding *single measurement accuracy* relevant to planet detection will be 70 μas at V = 10 mag and 250 μas at V = 15 mag, assuming 100 measurements per star over the course of the mission. GAIA is thus expected to detect some 10,000 Jupiter-like planets in orbits with periods ranging from 0.2 to 10 years, out to a typical distance of 200 pc. Around nearby stars, the detection limit will be about 30 $M_\oplus$. In addition, over 4000 transiting "hot Jupiters" will be detected in the GAIA photometry. GAIA will thus provide complementary information to SIM on the incidence of planetary systems as a function of metallicity, mass, and other stellar properties.

4. CORONAGRAPHY

4.1. Groundbased Coronagraphy

The search for faint planets located close to a bright host star requires specialized instrumentation capable of achieving contrast ratios of 10^{-9}–10^{-10} at subarcsecond separations. The most common approach is a Lyot coronagraph, where direct starlight is occulted at a first focal plane; diffraction from sharp edges in the entrance pupil is then suppressed with a Lyot mask in an intermediate pupil plane; and finally, imaging at a second focal plane occurs with greatly reduced diffraction artifacts. In addition to diffraction control, good wavefront quality (Strehl ratio) and wavefront stability are needed to achieve high image contrast.

Future extremely large groundbased telescopes will complement space coronagraphy through studies of young or massive jovian planets. Contrast levels of 10^{-7} or perhaps 10^{-8} may be achievable at subarcsecond separations through future developments in extreme adaptive optics. However, achieving the higher contrasts needed to detect Earths would require working at levels of 10^{-3} and 10^{-4} below the level of the residual background, many orders of magnitude greater than has been demonstrated to date (*Stapelfeldt*, 2006; *Chelli*, 2005; see chapter by Beuzit et al.). As discussed below, the TPF-coronagraph will greatly reduce the residual stellar background by taking advantage of a stable space platform and the absence of a variable atmosphere. Furthermore, the study of weak atmospheric features (including key biomarkers) on distant planets will be straightforward from space, but could be compromised by observing at relatively low spectral resolution through Earth's atmosphere.

4.2. High-Contrast Imaging from Space

Groundbased adaptive optics systems and three Hubble Space Telescope (HST) instruments using simple coronagraphs have achieved valuable results on circumstellar disks and substellar companions at moderate image contrast (see chapters by Beuzit et al. and Ménard et al.; *Schneider et al.*, 2002; *Krist et al.*, 2004, and references therein). However, only by taking full advantage of stable, spacebased platforms can the necessary starlight rejection ratios be achieved. To date, no space astronomy mission has been designed around the central goal of very high contrast imaging to enable detection of extrasolar planets. The HST coronagraphs were low-priority add-ons to general-purpose instruments, and achieved only partial diffraction control and image contrasts of 10^{-4}–10^{-5}.

JWST, expected to launch in 2013, will offer coronagraph modes for its two infrared imaging instruments. Even though JWST's segmented primary mirror is poorly suited to high-contrast coronagraphy, JWST's coronagraphs can access the bright 4.8-μm emission feature expected in the spectra of giant planets and brown dwarfs, enabling the detection of substellar companions at contrast ratios of 10^{-5}–10^{-6}. Detailed performance estimates show that JWST should be able to detect warm planets in nearby young stellar associations at radii >0.7", and perhaps even a few old (5 G.y.) Jupiter-mass planets around nearby, late-M-type stars (*Green et al.*, 2005).

The key to achieving such high contrasts is precision wavefront sensing and control. After a coronagraph suppresses the diffracted light in a telescope, the detection limits are governed by how much light is scattered off surface imperfections on the telescope primary mirror. In HST, these imperfections produce a background field of image speckles that is more than 1000× greater than the expected brightness of a planet in visible light. At the time HST was designed, the only way to reduce these imperfections would have been to polish the primary mirror to 30× better surface accuracy — infeasible then and now. But during the 1990s, another solution became available: Use deformable mirrors (DMs) to actively correct the wavefront. The needed wavefront quality can be achieved by canceling out primary mirror surface figure errors with a DM — exactly as Hubble's spherical aberration was corrected, but now at higher precision and higher spatial frequencies.

Over the past five years, the community has been developing deformable mirror technology, innovative pupil masks, and instrument concepts to enable very-high-contrast imaging (*Green et al.*, 2003; *Kuchner and Traub*, 2002; *Vanderbei et al.*, 2004; *Guyon et al.*, 2005; *Traub and Vanderbei*, 2005; *Shao et al.*, 2004). At the Jet Propulsion Laboratory, the High Contrast Imaging Testbed (HCIT) has been developed and consists of a vacuum-operated optical bench with a Lyot coronagraph and a 64 × 64 format deformable mirror. New wavefront control algorithms have been developed. As of mid-2005, HCIT had *demonstrated* 10^{-9} contrast in 785-nm narrowband light, at a distance of 4λ/D from a simulated star image (*Trauger et al.*, 2004) — a huge improvement over the previous state of the art. While additional progress is needed to achieve the requirement of 10^{-10} contrast in broadband light, these results are extremely encouraging evidence that a coronagraphic version of the Terrestrial Planet Finder mission (TPF-C) is feasible. NASA is now defining a TPF-C science program and mission design concept.

4.3. TPF-C Configuration

The configuration of the TPF-C observatory is driven by the specialized requirements of its high-contrast imaging mission (Plate 8). To enable searches of the habitable zone in at least 100 nearby FGK stars an 8-m aperture is needed. A monolithic primary mirror is essential to achieving the needed wavefront quality and maximizing throughput, but a conventional circular 8-m primary cannot be accommodated in existing rocket launcher shrouds. The unconventional solution is an 3.5 m × 8 m elliptical primary, tilted on-end within the rocket shroud for launch, with a deployed secondary mirror. The telescope has an off-axis (unobscured) design, to minimize diffraction effects and maximize coronagraph throughput. To provide very high thermal stability, the telescope is enclosed in a multilayer V-groove sunshade and will be operated in L2 orbit. The large sunshade dictates a solar sail to counterbalance radiation pressure torques on the spacecraft. Milliarcsecond-level pointing will be required to maintain alignment of the bright tar-get star on the coronagraph occulting spot.

TPF-C's main instrument will be a coronagraphic camera/spectrometer operating over the wavelength range $0.4 < \lambda < 1.1$ μm. It will have a very modest field of view, perhaps 2" in diameter, with a corrected high contrast dark field extending to a radius of ~1" from the bright central star. The science camera will also serve as a wavefront sensor to derive the adaptive correction settings for the DM, which must be set and maintained to subangstrom accuracy to achieve the required contrast. The camera will provide multispectral imaging, at resolutions from 5 to 70, in order to separate planets from residual stellar speckles, and to characterize their atmospheres. Spectral features of particular interest for habitability are the O_2 A band at 0.76 μm, an H_2O band at 0.81 μm, and the chlorophyll red edge beyond 0.7 μm.

Terrestrial exoplanets are very faint targets, with typical V > 29 mag. Even with a collecting area five times greater than HST's, exposure times on the order of a day will be needed to detect them. Repeat observations at multiple epochs will be needed to mitigate against unfavorable planetary elongations or illumination phases. Each of the target stars must therefore be searched for roughly a week of cumulative integration time. In systems where candidate planets are detected, follow-up observations to establish common proper motion, measure the planet's orbital elements, and spectroscopic characterization would add weeks of additional integration time per target. Accompanying giant planets and zodiacal dust should be readily detected in the systems selected for terrestrial planet searches.

5. INTERFEROMETRY

While COROT and Kepler will tell us about the general prevalence of terrestrial planets, we need to study planets in their habitable zones around nearby (25 pc) solar-type stars in a large enough sample (150 to over 500) to make statistically significant statements about habitable planets and the incidence of life. In order to cover a range of spectral types, metallicities, and other stellar properties, we would like to include many F, G, K, and some M dwarfs. As described in section 4, the angular resolution of a space-based coronagraph is limited to about 60 mas ($4\lambda/D$ at 0.7 μm) by the size of the largest monolithic telescope one can launch (roughly an ~8-m major axis). Beyond about 15 pc such a coronagraph will be limited to searches of the habitable zones around more luminous F-type stars. An interferometer with each telescope on a separate spacecraft suffers no serious

constraints on angular resolution and would be able to study stars over a large span of distances and luminosities with a consequently larger number and greater variety of potential targets.

Nulling (or destructive) interferometry uses an adaptation of the classical Michelson interferometry currently being developed at groundbased sites like the Keck Interferometer and ESO's Very Large Telescope Interferometer. Interferometers operating in the mid-infrared offer a number of advantages over other systems: (1) inherent flexibility that allows the observer to optimize the angular resolution to suit a particular target star; (2) a star-planet contrast ratio that is only 10^{-6}–10^{-7} at 10 μm, roughly a factor of 1000 more favorable than at visible wavelengths; (3) a number of temporal and spatial chopping techniques to filter out optical and mechanical instabilities, thereby relaxing some difficult requirements on the optical system; and (4) the presence of deep, broad spectral lines of key atmospheric tracers that can be observed with low spectral resolution.

There are, of course, disadvantages to a mid-IR system, including the need for cryogenic telescopes to take advantage of low space background; the complexity of multi-spacecraft-formation flying; and the complexity of signal extraction from interferometric data compared to more direct coronagraphic imaging.

ESA and NASA are investigating interferometric planet-finding missions and investing heavily in key technologies to understand the tradeoffs between different versions of the interferometer and, more generally, with coronagraphs.

5.1. Nulling Interferometry

In a nulling interferometer the outputs of the individual telescopes are combined after injecting suitable phase differences (most simply a half-wavelength) so that the on-axis light is extinguished while, at the same time, slightly off-axis light will be transmitted. By rotating the interferometer or by using more than two telescopes in the system one can sweep around the optical axis with high transmission while constantly obscuring the central object. In this manner, first proposed by *Bracewell* (1978), one can achieve the very high contrast ratios needed to detect a planet in the presence of its parent star. The depth and the shape of the null in the center depend critically on the number of telescopes and the actual configuration (*Angel and Woolf*, 1997). While better angular resolution is desirable to probe closer to the star, at high enough resolution the central stellar disk becomes resolved and light leaks out of the central null. This leakage is one of the noise sources in a nulling interferometer designed to search for terrestrial planets. When full account is taken of the leakage and other noise sources including local zodiacal emission, telescope background, and zodiacal light from dust orbiting the target star, the performance of a nulling interferometer using 3–4-m telescopes is well matched to the study of terrestrial planets around hundreds of stars (*Beichman et al.*, 1998; *Mennesson et al.*, 2004).

TABLE 2.   Measurement synergy for TPF-C, TPF-I/Darwin, and SIM.

	SIM	TPF-C	TPF-I
Orbital Parameters			
Stable orbit in habitable zone	**Measurement**	**Measurement**	**Measurement**
Characteristics for Habitability			
Planet temperature	Estimate	Estimate	**Measurement**
Temperature variability due to eccentricity	**Measurement**	**Measurement**	**Measurement**
Planet radius	*Cooperative*	*Cooperative*	**Measurement**
Planet albedo	*Cooperative*	*Cooperative*	*Cooperative*
Planet mass	**Measurement**	Estimate	Estimate
Surface gravity	*Cooperative*	*Cooperative*	*Cooperative*
Atmospheric and surface composition	*Cooperative*	**Measurement**	**Measurement**
Time-variability of composition		**Measurement**	**Measurement**
Presence of water		**Measurement**	**Measurement**
Solar System Characteristics			
Influence of other planets, orbit coplanarity	**Measurement**	Estimate	Estimate
Comets, asteroids, and zodiacal dust		**Measurement**	**Measurement**
Indicators of Life			
Atmospheric biomarkers		**Measurement**	**Measurement**
Surface biosignatures (red edge of vegetation)		**Measurement**	

"Measurement" indicates a directly measured quantity from a mission; "Estimate" indicates that a quantity that can be estimated from a single mission; and *"Cooperative"* indicates a quantity that is best determined cooperatively using data from several missions.

Excellent technical progress has been made in both the U.S. and Europe on the key "physics" experiment of producing a broadband null. Depths less than 10^{-6} have been achieved in a realistic four-beam configuration in the laboratory (*Martin et al.,* 2003). Operational nulling systems are being deployed on both the Keck and Large Binocular Telescope Interferometers (*Serabyn et al.,* 2004; *Herbst et al.,* 2004).

5.2.   Darwin and the Terrestrial Planet Finder

ESA's Darwin and NASA's TPF-I are currently foreseen to be implemented on four to five spacecraft flying in a precision formation. The system would consist of three or four telescopes each of 3–4 m diameter and flying on its own spacecraft. An additional spacecraft would serve as a beam combiner. When searching nearby stars out to 25 pc, the distance between the outermost telescopes would be roughly 100 m. The operating wavelength would be 6–17 µm (possibly as long as 20 µm) where the contrast between parental star and Earth-like planet is most favorable and where there are important spectral signatures characterizing the planets. ESA and NASA are collaborating on TPF-I/Darwin under a Letter of Agreement that calls for joint science team members, conferences and workshops, as well as periodic discussions on technology and the various configurations under study individually by the two space agencies. The common goal is to implement one interferometric mission, since complexity and cost indicate the necessity for a collaborative approach. Systems utilizing 1–2-m telescopes could also be considered if it were known

in advance, e.g., from COROT and Kepler, that Earths in the habitable zone were common so that one could limit the search to 10 pc instead of 25 pc to detect a suitable number of planets.

6.   SYNERGY AMONG TECHNIQUES FOR NEARBY PLANETS

The major missions discussed in this chapter approach the search for terrestrial planets from different perspectives: COROT and Kepler for transits (section 2); SIM using astrometry (section 3); TPF-C directly detecting visible photons (section 4); and TPF-I/Darwin directly detecting mid-infrared photons (section 5). In this section we argue that all these perspectives are needed to determine habitability and search for signs of life. The synergy between the transit missions and JWST has already been discussed (section 2). In this section we focus on investigations of nearby stars and the potential for SIM, TPF-C, and TPF-I/Darwin to determine important physical parameters either individually or in combination (Table 2). *The cooperative aspects are discussed below in italics.* We anticipate that the strongest, most robust statements about the characteristics of extrasolar planets will come from these cooperative efforts.

6.1.   Stable Orbit In Habitable Zone

Each mission can measure an orbit and determine if it lies within the habitable zone (where the temperature permits liquid water on the surface of the planet). SIM does this by observing the wobble of the star and calculating where the planet must be to cause that wobble. TPF-C and TPF-I/Darwin do this by directly imaging the planet and noting how far it appears to be from the star. The missions work together and separately to determine orbital information:

1. SIM detections of planets of a few Earth masses would provide TPF-C and TPF-I/Darwin with targets to be characterized and the optimum times for observing them, thus increasing the early-mission characterization yield of TPF-C or TPF-I/Darwin.

2. Where SIM finds a planet, of any mass, in almost any orbit, TPF-C and TPF-I/Darwin will want to search as well, because we expect that planetary multiplicity may well be the rule (as in our solar system). Thus SIM will help TPFC and TPF-I/Darwin to prioritize likely target stars early in their missions.

3. For stars where SIM data suggest that planets exist below SIM's formal detection threshold, TPF-C or TPF-I/Darwin could concentrate on those stars to either verify or reject the detection. Such verification would lower the effective mass-detection threshold for planets with SIM data.

4. All three missions can detect several planets around a star, within their ranges of sensitivity. Thus there may be a planet close to the star that SIM can detect, but is hidden from TPF-C. Likewise there may be a distant planet that TPF-C or TPF-I/Darwin can detect, but has a period that is too long for SIM. For the more subtle issue of whether the planets have orbits in or out of the same plane, SIM will do the best job. *In general, each of the three missions will detect some but not necessarily all of the planets that might be present in a system, so the combination will deliver a complete picture of what planets are present, their masses, their orbits, and how they are likely to influence each other over the age of the system, including co-planarity.*

6.2. Gross Physical Properties of Planets

1. Planet temperature: A planet's effective temperature can be roughly estimated by noting its distance from its star and assuming a value for the albedo. TPF-C can estimate the temperature by noting the distance and using planet color to infer its albedo by analogy with planets in our solar system. TPF-I/Darwin can observe directly the thermal-infrared emission continuum at several wavelengths (i.e., infrared color) and use Planck's law to calculate the effective temperature. For a planet like Venus with a thick or cloudy atmosphere, the surface temperature is different from the effective temperature, but might still be inferred from a model of the atmosphere. *With all three missions combined, the orbit, albedo, and greenhouse effect can be estimated, and the surface temperature as well as temperature fall-off with altitude can be determined cooperatively and more accurately than with any one mission alone.*

2. Temperature variability due to distance changes: Each mission alone can observe the degree to which the orbit is circular or elliptical, and thereby determine if the temperature is constant or varying. In principle TPF-C and TPF-I/Darwin can tell whether there is a variation in color or spectrum at different points in the planet's orbit, perhaps due to a tilt of the planet's axis, which would lead to a seasonal temperature variability. *The measurement of a terrestrial planet's orbital eccentricity using combined missions (SIM plus TPF-C and/or TPF-I) can be much more accurate than from any one mission alone, because complementary sensitivity ranges in planet mass and distance from star combine favorably. SIM gives eccentricity data that aids TPF-C and TPF-I/Darwin in selecting optimum observation times for measuring planet temperature, clouds, and atmospheric composition.*

3. Planet radius: SIM measures planet mass from which we can estimate radius to within a factor of 2 if we assume a value for the density (which in the solar system spans a factor of 8). TPF-C measures visible brightness, which along with an estimate of albedo, can give a similarly rough estimate of radius. A TPF-C color-based estimate of planet type can give a better estimate of radius. TPF-I/Darwin measures infrared brightness and color temperature, which, using Planck's law, gives a more accurate planet radius. *Planet radius and mass, or equivalently density, is very important for determining the type of planet (rocks, gas, ice, or combination), its habitability (solid surface or not; plate tectonics likely or not), and its history (formed inside or outside the ice line). With SIM's mass, and one or both TPF brightness measurements, we can dramatically improve the estimate of planet radius.*

4. Planet albedo: The albedo controls the planet's effective temperature, which is closely related to its habitability. SIM and TPF-C combined can estimate possible pairs of values of radius and albedo, but cannot pick which pair is best (see above). We can make a reasonable estimate of albedo by using TPF-C to measure the planet's color, then appealing to the planets in our solar system to convert a color to an absolute albedo. By adding TPF-I/Darwin measurements we can determine radius (above), then with brightness from TPF-C we can compute an accurate albedo. *SIM and TPF-C together give a first estimate of planet albedo. Adding TPF-I/Darwin gives a conclusive value of albedo, and therefore effective temperature and potential habitability.*

5. Planet mass: SIM measures planet mass directly and accurately. TPF-C and TPF-I/Darwin depend entirely on SIM for a true measurement of planet mass. If TPF-C and TPF-I/Darwin do not have a SIM value for planet mass, then they will use theory and examples from our solar system to estimate masses (see above). *SIM plus TPF-C and TPF-I/Darwin are needed to distinguish among rock-, ice-, and gas-dominated planet models, and to determine with confidence whether the planet could be habitable.*

6. Surface gravity: The planet's surface gravity is calculated directly using mass from SIM and radius from TPFC and TPF-I/Darwin (see above). *Surface gravity governs whether a planet can retain an atmosphere or have plate tectonics (a crucial factor in Earth's evolution). Cooperative measurements are the only way to obtain these data.*

7. Atmosphere and surface composition: The TPF missions are designed to measure a planet's color and spectra, from which we can determine the composition of the atmos-

phere and surface. For the atmosphere, TPF-C can measure water, molecular oxygen, ozone, and the presence of clouds for a planet like the present Earth; in addition, it can measure carbon dioxide and methane for a planet like the early Earth or a giant planet. For the surface, TPF-C can measure vegetation using the red edge effect (see below). TPF-I/Darwin will add to this suite of observations by measuring carbon dioxide, ozone, water, methane, and nitrous oxide using different spectroscopic features, and in general probing a different altitude range in the atmosphere. SIM is important to this interpretation because it provides planet mass, crucial to interpreting atmospheric measurements.

Both TPF-C and TPF-I/Darwin are needed in order to determine whether a planet is habitable, because they make complementary observations, as follows (assuming an Earth-like planet). Ozone has a very strong infrared (TPF-I) feature, and a weak visible (TPF-C) one, so if ozone is abundant, both can be used to extract the abundance. If ozone is only weakly present, then only the TPF-I/Darwin feature will be useable. Water as seen by TPF-C will be in the lower atmosphere of the planet, but as seen by TPF-I/Darwin it will be in the upper atmosphere; together both give a more complete picture of the atmosphere. Methane and carbon dioxide could be detected by TPF-I/Darwin at levels similar to present-day concentrations in Earth's atmosphere. Methane and carbon dioxide, in large amounts (as for the early Earth), can be detected by TPF-C. For large amounts of methane or carbon dioxide, TPF-I/Darwin will see mainly the amount in the upper atmosphere, but TPF-C will see mainly the amount in the lower atmosphere, so both are needed for a complete picture. In addition to these overlap topics, only TPF-C can potentially measure oxygen, vegetation, and the total column of air (Rayleigh scattering); likewise only TPF-I/Darwin can measure the effective temperature. TPF-I's wavelength coverage includes a spectral line of nitrous oxide, a molecule strongly indicative of the presence of life. Unfortunately, there is only a small hope of detecting this biomarker at low spectral resolution. *In short, SIM is needed for planet mass, TFP-C and TPF-I/Darwin are needed to characterize the atmosphere for habitability, and all three are needed to fully characterize the planet.*

8. Temporal variability of composition: Both TPF-C and TPF-I/Darwin potentially can measure changes in color and the strengths of spectral features as the planet rotates. These changes can tell us the length of day on the planet, and can indicate the presence of large oceans or land masses (with different reflectivities or emissivities, by analogy to Earth). Superposed on this time series of data could be random changes from weather patterns, possibly allowing the degree of variability of weather to be measured. *The TPF missions can potentially measure variability of composition over time, which we know from our Earth to be an indicator of habitability.*

9. Presence of water: Both TPF-C and TPF-I/Darwin have water absorption features in their spectra, so if water vapor is present in the atmosphere, we will be able to measure it. However, habitability requires liquid water on the surface, which in turn requires a solid surface as well as a temperature that permits the liquid state; only with the help of a value of mass from SIM will we be able to know the radius, and when TPF-I/Darwin is launched, the temperature. *To know whether liquid water is present on the surface of a planet, we need mass data from SIM and spectroscopic data from TPF-C or TPF-I/Darwin.*

6.3. Biomarkers

The simultaneous presence of an oxidized species (like oxygen or ozone) and a reduced species (like methane) is considered to be a sign of nonequilibrium that can indirectly indicate the presence of life on a planet. The presence of a large amount of molecular oxygen, as on the present Earth, may also be an indirect sign of life. In addition, since water is a prerequisite for life, as we consider it here, the presence of liquid water (indicated by water vapor and an appropriate temperature) is needed. Together these spectroscopically detectable species are our best current set of indicators of life on a planet. *These markers will be measured exclusively by TPF-C and TPF-I, but to know that we are observing an Earth-like planet will require SIM data on mass. If we do (or do not) find biomarkers, we will certainly want to know how this is correlated with planet mass.*

The "red edge" of vegetation is a property of land plants and trees whereby they are very good reflectors of red light just beyond the long-wavelength limit of our eyes. This is a useful feature for measuring plant cover on Earth. If extrasolar planets have developed plant life like that on Earth, and if the planet is bright, has few clouds, and a lot of vegetated land area, then we may use this feature to detect living vegetation. *As for other biomarkers (above), we will want to correlate the presence of vegetation with the planet mass, requiring SIM as well as TPF-C.*

7. TARGET STARS

7.1. Stars Suitable for SIM, TPF-C, and TPF-I/Darwin

Table 3 suggests just a few of the considerations that will go into choosing the best targets for SIM, TPF-C, and TPF-I/Darwin. Some of these are of a scientific nature, e.g., age, spectral type, and metallicity (which seems to be highly correlated with the presence of gas giant planets), while other constraints are more of an engineering nature: zodiacal emission, ecliptic latitude, binarity, variability, or the presence of confusing background objects. An extensive program of observation and data gathering must be an essential part of preparation for the planet-finding program. A report entitled *Terrestrial Planet Finder Precursor Science* (*Lawson et al.*, 2004) describes a complete roadmap for the acquisition and assessment of relevant data. NASA has begun development of the Stellar Archive and Retrieval

TABLE 3. Key target star properties.

Stellar age	Evolutionary phase
Spectral type	Mass
Variability	Metallicity
Distance	Galactic kinematics
Multiplicity	Giant planet companions
Exozodiacal emission	Background confusion
Position in ecliptic	Position in galaxy

System (StARs) to provide a long-term repository and network accessible database for this effort. A parallel and complementary effort is currently being developed in Europe.

The science teams for SIM, TPF-C, and TPF-I/Darwin have developed preferred lists of 100–250 stars optimized for their particular instrumental capabilities as described in *Traub et al.* (2006):

1. While closer, lower-mass stars maximize SIM's astrometric signal (cf. equation (1)), more luminous stars have a larger habitable zone (scaling as $L_*^{0.5}$) that offsets their higher

mass and typically greater distance and eases the astrometric search for habitable planets in such systems. Assuming a 1-μas sensitivity limit, the minimum mass planet in the habitable zone detectable by SIM is given by M_{min}(SIM) = 0.33 $M_\oplus$d$L^{-0.26}$ for a planet at d pc of orbiting a star of luminosity L ($L_\oplus$) (*Traub et al.*, 2006).

2. With its limited angular resolution, TPF-C favors the closest stars with the larger habitable zones. Assuming a limiting contrast ratio of 10^{-10} or 25 mag, the minimum-mass planet in the habitable zone detectable by TPF-C is given by M_{min}(TPF-C) = 0.81 $M_\oplus$$L^{1.5}$ (*Traub et al.*, 2006).

3. With its nearly unlimited angular resolution but more limited sensitivity, TPF-I/Darwin is best suited to the study of later spectral types with smaller habitable zones.

Table 4 lists the top 25 stars suitable for joint observation by TPF-I/Darwin, TPF-C, and SIM based on information gathered from the science groups of each mission. These stars have no known companions within 10" and an inner edge to their habitable zones (roughly the orbit of Venus) larger than 62 mas. Stars with ecliptic latitudes in excess of 45° are excluded due to the need to shade the

TABLE 4. Likely targets for TPF-C, TPF-I/Darwin, and SIM.

Hip	HD	Name	Spectral Type	Distribution (pc)	Inner HZ (mas)	SIM Mass Limit ($M_\oplus$)*	Zodiacal Emission
8102	10700	τ Ceti	G8V	3.65	120	0.97	Yes, IRAS
19849	26965	DY Eri	K0/1V	5.04	75	1.43	No 24/70, Nearby[†]
99240	190248	δ Pav	G6/8IV	6.11	116	1.29	No 24/70, FGK[†]
64924	115617	61 Vir	G5V	8.53	68	1.99	Strong 70, FGK
64394	114710	β Com	G0V	9.15	88	1.81	No 24/70, FGK
15457	20630	κ Cet	G5V	9.16	65	2.1	N/A, FGK
108870	209100	ϵ Ind	K4/5V	3.63	60	1.35	N/A, FGK
57443	102365	GL 442A	G3/5V	9.24	64	2.13	No 24/70, SIMTPF[†]
14632	19373	ι Per	G0V	10.53	96	1.88	N/A, FGK
12777	16895	13 Per	F7V	11.23	96	1.95	No 24/70, SIMTPF
53721	95128	47 UMa	G0V	14.08	61	2.72	No 24/70, FGK
47592	84117	GL 364	G0V	14.88	65	2.72	No 24/70, FGK
56997	101501	61 Uma	G8V	9.54	52	2.4	No 24/70, FGK
22449	30652	π^3 Ori	F6V	8.03	147	1.34	N/A, FGK
78072	142860	γ Ser	F6V	11.12	108	1.83	No 24/70, FGK
25278	35296	GL 202	F8V	14.66	63	2.74	No 24/70, FGK
16852	22484	GL147	F8V	13.72	87	2.27	N/A, FGK
80337	147513	GL 620.1A	G3/5V	12.87	52	2.8	No 24/70, SIMTPF
57757	102870	β Vir	F9V	10.9	120	1.73	No 24/70, FGK
7513	9826	υ And	F8V	13.47	95	2.15	N/A, FGK
3909	4813	GL 37	F7V	15.46	59	2.91	No 24/70, SIMTPF
116771	222368	ι Psc	F7V	13.79	95	2.19	N/A, FGK
71284	128167	σ Boo	F3V	15.47	83	2.49	N/A, Dirty Dozen[†]
86796	160691	μ Ara	G3IV/V	15.28	57	2.93	No 24/70, FGK
40843	69897	χ Cnc	F6V	18.13	60	3.13	N/A, FGK

*Assumes 125 2-Dimensional Visits.

[†]Zodiacal emission at 24 or 70 μm in Nearby Stars MIPS Survey, *Gautier et al.* (2006) in preparation; FGK survey (*Beichman et al.*, 2005a, 2006; *Bryden et al.*, 2006); or the SIMTPF Comparative Planetology survey (*Beichman et al.* 2006, in preparation; MIPS/IRS survey of the "Dirty Dozen," bright, potentially resolvable disks, *Stapelfeldt et al.* (2006) in preparation. N/A denotes data not yet available in a given survey.

TPF-I/Darwin telescopes from the Sun. Also included in the table is the angular extent of the inner edge of the habitable zone (the orbit of Venus scaled by the square root of the stellar luminosity) and any indication of an infrared excess from exozodiacal dust as determined by IRAS or Spitzer. This information is also portrayed graphically in Plate 9, which shows stars observable in common between SIM, TPF-C, and TPF-I/Darwin.

7.2. Zodiacal Dust and Planet Detection

Planetary systems include many constituents: gas-giant planets, ice-giant planets, and rocky planets, as well as comet and asteroid belts. Understanding the interrelated evolution of all these constituents is critical to understanding the astronomical context of habitable planets and life. For example, the presence of a large amount of zodiacal emission from the debris associated with either a Kuiper belt of comets or a rocky zone of asteroids may indicate conditions hostile to the habitable planets due to a potentially high rate of bombardment. At the same time, the transfer of water and other volatile (organic) species from the outer to the rocky planets in the habitable zone may be an essential step in the formation of life. Thus, from a scientific standpoint, we want to gather information on disks at all stages in the evolution of planetary systems, including debris disks surrounding nearby stars. TPF-C, TPF-I/Darwin, and JWST will work in conjunction to make images and spectra of scattered light and thermally emitted radiation from a large number of targets, spanning distant stars (25–150 pc) with bright disks where planets may still be forming to nearby systems with zodiacal clouds no brighter than our own.

From an engineering standpoint, zodiacal dust is a critical factor for direct detection of planets due to increased photon shot noise and potential confusion with zodiacal structures. Sensitivity calculations for various TPF-I/Darwin and TPF-C designs suggest that the integration time needed to reach a certain level increases by a factor of 2–3 for zodiacal levels roughly 10× the solar system's. Spitzer is carrying out a number of programs to assess the level of exozodiacal emission. Initial results suggest that only 15% of solar type stars have more than 50× the solar system's level of zodiacal emission at 30 μm, corresponding to material just outside the habitable zone, beyond about 5 AU (*Beichman et al.*, 2006; *Bryden et al.*, 2006). This result is encouraging, but must be expanded to more TPF target stars using Spitzer and Herschel, and to lower levels of zodiacal emission using interferometric nulling at 10 μm on the Keck and LBT interferometers. The combination of these results will yield the "luminosity function" of disks for statistical purposes and allow us to screen potential targets. As an example of the sort of problem that can arise is the remarkable star HD 69830, a 2–4-G.y.-old K0 star at 14 pc that might be a TPF target except for a zodiacal dust level in the habitable zone that is 1400× higher than seen in the solar system (*Beichman et al.*, 2005b). While SIM may be able

to identify planets around this star astrometrically, no direct searches of the habitable zone will be possible.

8. DISCUSSION AND CONCLUSIONS

The exploration of extrasolar planetary systems is a rich and diverse field. It calls for measurements with many kinds of instruments, as well as theoretical studies and numerical modeling. To discover and characterize extrasolar planets that are habitable and to be sure beyond a reasonable doubt that we can detect life, we need to measure the statistical distribution of planet diameters, the masses of nearby planets, and the spectra at visible and infrared wavelengths.

Each of the missions listed in Table 1 is a vital element of the program. Not only does each mission by itself produce its own compelling science, but the ensemble will provide a coherent set of data that will advance our understanding better than could any single mission.

The exciting scientific promise described in preceding sections will not happen cheaply or overnight. The Great Observatories program for astrophysics spanned more than a decade between the first and last launches — HST in 1990 and Spitzer in 2003 — and was still longer in gestation. While the transit missions COROT and Kepler are being readied for flight, the "Great Observatories" for planet finding will take a generation of scientific and political advocacy: SIM PlanetQuest has completed its technology development, has been endorsed repeatedly by the U.S. science community, and awaits final NASA approval; TPF-C and TPF-I/Darwin are well along in their programs of technology development and will need strong endorsement by U.S. and European scientific communities in coming years. Interest is TPF goals is growing in Japan as well (*Tamura and Abe*, 2006). JWST is moving into its construction phase and will provide many observations useful to the planet finding endeavor.

While the study of extrasolar planets is a new field of research with a relatively small number of (young) practitioners, our field is growing rapidly. We will fare well in any assessment of the importance of our field to progress in astronomy and of the great interest our program holds for the general public. We will also fare well in any assessment of the value of these planet-finding facilities for general astrophysical investigations: SIM has already allocated a significant amount of observing time to a broad suite of general astrophysics; TPF-C will provide wide-field optical imaging with unprecedented sensitivity and angular resolution to complement JWSTs mid- and near-IR capabilities; TPF-I/Darwin will break new ground with milliarcsecond mid-IR imaging and micro-Jy sensitivity.

Near-term political considerations must not discourage us as we plan this new era of planetary exploration. *While this volume was going to press, information about the NASA budget suggested the possibility of long delays for SIM, the possible cancellation of TPF-C/I, and a reduction in a variety of grants programs. Yet Kepler and COROT are still going ahead, so that some of the exciting new data dis-*

cussed in this chapter will become available in the next few years. Despite delays and reverses, the remainder of the program outlined above will one day be carried out since it is grounded in fundamental scientific principles. To accomplish our goals, we must demand from the funding agencies the budgets needed to nurture young scientists and senior researchers, to prepare the difficult technologies of nulling, large space optics, and formation flying, and eventually to build these "Great Observatories." With our colleagues, we must argue forcefully for a balanced program based on scientific priorities free from parochial considerations of individual facilities or institutions. Twentieth-century cosmologists expanded our conception of the universe with the discovery of galaxies, the expanding universe, and dark matter; twenty-first-century planet finders will expand our conception of humanity's place in the universe with the discoveries of other habitable worlds and possibly of life itself.

Acknowledgments. We would like to thank P. Lawson for creating Plate 7 and O. Lay for his work on putting together Plate 9 and Table 4. The science teams of SIM, TPF-C, and TPF-I, and in particular J. Kasting, M. Shao, and K. Johnston, helped develop the synergy arguments presented in section 6. S. Unwin, who has been unstinting in his efforts on behalf of SIM and TPF over many years, helped to organize the special session on planet finding from space. We are grateful to B. Reipurth for his heroic efforts in putting together the wonderful conference on which this volume is based. Some of the research described in this publication was carried out at the Jet Propulsion Laboratory, California Institute of Technology, under a contract with the National Aeronautics and Space Administration.

REFERENCES

Angel J. R. P. and Woolf N. (1997) *Astrophys. J., 475,* 373.

Baglin A. (2003) *Adv. Space Res, 31,* 345.

Barnes J. W. and Fortney J. J. (2003) *Astrophys. J., 588,* 545.

Beichman C. A. (1998) In *Exozodiacal Dust Workshop* (D. Backman and L. Caroff, eds.), pp. 149–159. NASA CP-1998-10155.

Beichman C. A. (2000) In *Planetary Systems in the Universe: Observation, Formation and Evolution* (A. Penny et al., eds.), p. 432. IAU Symposium 202, ASP, San Francisco.

Beichman C. A., Bryden G., Rieke G. H., Stansberry J. A., Trilling D. E., et al. (2005a) *Astrophys. J., 622,* 1160–1170.

Beichman C. A., Bryden G., Gautier T. N., Stapelfeldt K. R., Werner M. W., et al. (2005b) *Astrophys. J., 626,* 1061–1069.

Beichman C. A., Tanner A., Bryden G., Stapelfeldt K. R., Werner M. W., Rieke G. H., et al. (2006) *Astrophys. J., 639,* 1166–1176.

Bordé P., Rouan D., and Léger A. (2003) *Astron. Astrophys., 405,* 1137–1144.

Borucki W. J., Koch D. G., Lissauer J. J., Basri G. B., Caldwell J. F., et al. (2003) *Proc. SPIE, 4854,* 129–140.

Bracewell R. (1978) *Nature, 274,* 780–781.

Brown R. A. (2005) *Astrophys. J., 624,* 1010–1024.

Brown T. M., Charbonneau D., Gilliland R. L., Noyes R. W., and Burrows A. (2001) *Astrophys. J., 552,* 699–709.

Bryden G., Beichman C. A., Trilling D. E., Rieke G. H., Holmes E. K., et al. (2006) *Astrophys. J., 636,* 1098–1113.

Catanzarite J., Shao M., Tanner A., Unwin S., and Yu J. (2006) *Publ. Astron. Soc. Pac., 118,* in press.

Charbonneau D., Allen L. E., Megeath S. T., Torres G., Alonso R., Brown T. M., et al. (2005) *Astrophys. J., 626,* 523–529.

Chelli A. (2005) *Astron. Astrophys., 441,* 1205–1210.

Deming D., Seager S., Richardson L. J., and Harrington J. (2005) *Nature, 434,* 740–743.

Gilliland R. (2005) In *Astrobiology and JWST. A Report to NASA.* Available on line at http://www.dtm.ciw.edu/seager/NAIAFG/JWST.pdf.

Gilliland R. L., Brown T. M., Guhathakurta P., Sarajedini A., Milone E. F., et al. (2000) *Astrophys. J., 545,* L47–51.

Gillon M., Courbin F., Magain P., and Borguet B. (2005) *Astron. Astrophys., 442,* 731–744.

Green J. J., Basinger S. A., Cohen D., Niessner A. F., Redding D. C., Shaklan S. B., and Trauger J. T. (2003) *Proc. SPIE, 5170,* 38–48.

Green J. J., Beichman C. A., Basinger S. A., Horner S., Meyer M., Redding D. C., Rieke M., and Trauger J. T. (2005) *Proc. SPIE, 5905,* 185–195.

Guyon O., Pluzhnik E. A., Galicher R., Martinache F., Ridgway S. T., and Woodruff R. A. (2005) *Astrophys. J., 622,* 744–758.

Herbst T. M. and Hinz P. M. (2004) *Proc. SPIE, 5491,* 383–384.

Krist J. E. (2004) *Proc. SPIE, 5487,* 1284–1295.

Kuchner M. J. and Traub W. A. (2002) *Astrophys. J., 570,* 900–908.

Lawson P., Unwin S., and Beichman C. A. (2004) *Terrestrial Planet Finder Precursor Science.* JPL Tech. Rpt. #04-014, available on line at http://planetquest.jpl.nasa.gov/documents/RdMp273.pdf.

Lindegren L. and Perryman M. A. C. (1996) *Astron. Astrophys. Suppl., 116,* 579–595.

Marcy G. W., Butler R. P., Fischer D. A., Laughlin G., Vogt S. S., Henry G., and Pourbaix D. (2002) *Astrophys. J., 581,* 1375–1388.

Marcy G. W., Butler R. P., Fischer D., Vogt S., Wright J. T., Tinney C. G., and Jones H. R. A. (2005) *Progr. Theor. Phys. Suppl., 158,* 1.

Martin S. R., Gappinger R. O., Loya F. M., Mennesson B. P., Crawford S. L., and Serabyn E. (2003) *Proc. SPIE, 5170,* 144–154.

Mennesson B. P., Johnston K. J., and Serabyn E. (2004) *Proc. SPIE, 5491,* 136–137.

Moutou C., Pont F., Barge P., Aigrain S., Auvergne M., Blouin D., et al. (2005) *Astron. Astrophys., 437,* 355–368.

Muterspaugh M. W., Lane B. F., Konacki M., Burke B. F., Colavita M. M., Kulkarni S. R., and Shao M. (2005) *Astron. J., 130,* 2866–2875.

Perryman M. A. C., de Boer K. S., Gilmore G., Høg E., Lattanzi M. G., et al. (2001) *Astron. Astrophys., 369,* 339–363.

Quirrenbach A. (2000) In *From Extrasolar Planets to Cosmology: The VLT Opening Symposium* (J. Bergeron and A. Renzini, eds.), pp. 462–467. Springer-Verlag, Berlin.

Quirrenbach A. (2002) In *From Optical to Millimetric Interferometry: Scientific and Technological Challenges* (J. Surdej et al., eds.), pp. 51–67. Univ. de Liège.

Quirrenbach A., Coudé du Foresto V., Daigne G., Hofmann K. H., Hofmann R., et al. (1998) *Proc. SPIE, 3350,* 807–817.

Rivera E., Lissauer J., Butler R. P., Marcy G. W., Vogt S., Fischer D. A., Brown T., and Laughlin G. (2005) *Astrophys. J., 634,* 625–640.

Rowe J. F., Matthews J. M., Seager S., Kuschnig R., and Guenther D. B. (2006) *Bull. Am. Astron. Soc., 207,* 110.07.

Sahu K., et al. (2005) HST Proposal ID#10466. Available on line at http://adsabs.harvard.edu/cgi-bin/nph-bib_query?bibcode=2005hst..prop.6787S.

Schneider G. (2002) *Domains of Observability in the Near- Infrared With HST NICMOS and Large Ground-based Telescopes.* Available on line at http://nicmos.as.arizona.edu:8000/REPORTS/.

Seager S. and Hui L. (2002) *Astrophys. J., 574,* 1004–1010.

Serabyn E., Booth A. J., Colavita M. M., Creech-Eakman M. J., et al. (2004) *Proc. SPIE, 5491,* 806–807.

Shao M. and Colavita M. M. (1992) *Astron. Astrophys., 262,* 353–358.

Shao M., Wallace J. K. Levine, B. M., and Liu D. T. (2004) *Proc. SPIE, 5487,* 1296–1303.

Stapelfeldt K. R. (2006) In *The Scientific Requirements for Extremely Large Telescopes* (P. A. Whitelock et al., eds.), pp. 149–158. Cambridge Univ., Cambridge.

Tamura M. and Abe L. (2006) In *Direct Imaging of Exoplanets: Science and Techniques* (C. Aime and F. Vakili, eds.), p. 323. IAU Colloquium #200, Cambridge Univ., Cambridge.

Traub W. A. and Vanderbei R. J. (2005) *Astrophys. J., 626,* 1079–1090.

Traub W. A., Carleton N. P., and Porro I. L. (1996) *J. Geophys. Res., 101,* 9291–9296.

Traub W. A, Kasting J., Shao M., Johnston K. J., and Beichman C. A. (2006) In *Direct Imaging of Exoplanets: Science and Techniques* (C. Aime and F. Vakili, eds.), p. 399. IAU Colloquium #200, Cambridge Univ., Cambridge.

Trauger J. T., Burrows C., Gordon B., and Green J. J., et al. (2004) *Proc. SPIE, 5487,* 1330–1336.

Unwin S. C. (2005) In *Astrometry in the Age of the Next Generation of Large Telescopes* (P. K. Seidelmann and A. K. B. Monet, eds.), pp. 37–42. ASP Conf. Ser. 338, San Francisco.

Vanderbei R. J., Kasdin N. J., and Spergel D. N. (2004) *Astrophys. J., 615,* 555–561.

Walker G. A. H., Kuschnig R., Matthews J. M., Cameron C., Saio H., et al. (2005) *Astrophys. J., 635,* L77–L80.

From Protoplanets to Protolife:
The Emergence and Maintenance of Life

Eric Gaidos
University of Hawai'i

Franck Selsis
Centre de Recherche Astronomique de Lyon

Despite great advances in our understanding of the formation of the solar system, the evolution of Earth, and the chemical basis for life, we are not much closer than the ancient Greeks to an answer of whether life has arisen and persisted on any other planet. The origin of life as a planetary phenomenon will probably resist successful explanation as long as we lack an early record of its evolution and additional examples. Plausible but meagerly investigated scenarios for the origin of important prebiotic molecules and their polymers on Earth involving atmospheric chemistry, meteorites, deep-sea hot springs, and tidal flat sediments have been developed. Our view of the diversity of extant life, from which properties of a last universal common ancestor (LUCA) can be inferred, has also improved in scope and resolution. It is widely thought that the geologic record shows that life emerged quickly after the end of prolonged bombardment of Earth. New data and simulations contradict that view and suggest that more than half a billion years of unrecorded Earth history may have elapsed between the origin of life and LUCA. The impact-driven exchange of material between the inner planets may have allowed earliest life to be more cosmopolitan. Indeed, terrestrial life may not have originated on Earth, or even on any planet. Smaller bodies, e.g., the parent bodies of primitive meteorites, in which organic carbon molecules and catalytic transition metals were abundant, and in which hydrothermal circulation persisted for millions of years, offer alternative environments for the origin of life in the solar system. However, only planet-sized bodies offer the stable physiochemical conditions necessary for the persistence of life. The search for past or present life on Mars is an obvious path to greater enlightenment. The absence of intense geologic activity on Mars, which contributes to its inhospitable state today, has also preserved its ancient history. If life did emerge on Mars or was transferred from Earth, the lack of sterilizing impacts (due to a low gravity and no oceans) means that a more diverse biota may have thrived than is represented by extant life on Earth. On the other hand, a habitable but still lifeless early Mars would be strong evidence against efficient transfer of life between planets. The subsurface oceans of some icy satellites of the outer planets represent the best locales to search for an independent origin of life in the solar system because of the high dynamical barriers for transfer, intense radiation at their surfaces, and thick ice crusts. These also present equally formidable barriers to our technology. The "ultimate" answer to the abundance of life in the cosmos will remain the domain of speculation until we develop observatories capable of detecting habitable planets — and signs of life — around the nearest million or so stars.

1. INTRODUCTION

This contribution's place as the final chapter in *Protostars and Planets V* may betray a subtle conceit in how we view our place in a cosmic order that runs from the interstellar medium to planetary bodies. (Read in reverse order, the chapters would suggest a more humble search for our origins among wisps of interstellar gas and dust.) Nevertheless, this sequence makes sense, both in a temporal and also a physical order: It describes a gradation in phenomena in which physical and chemical inevitability (the laws of gravity, classical and quantum mechanics, and electromagnetism) that govern the collapse of the interstellar medium and the formation of stars are replaced by more stochastic processes such as accretion during planet formation and evolution. For example, it may be inevitable that a cooling molecular cloud collapses, a disk forms, and that runaway growth of planetesimals occurs in that disk, but the final masses, orbits, and surface environments of planets may not be predictable in more than a statistical sense. Ultimately it is no longer sufficient to describe what could happen; one must also describe what *did* happen. Whereas stars can be described by a relatively small number of variables (age, rotation, and heavy element abundance, for example), planets, particularly terrestrial planets, cannot. In that context, the origin and survival of life might be the ultimate contingency.

On the other hand, what little we know about the origin of life seems to suggests some element of inevitability. The

929

primary constituents of life (C, H, N, and O) are four of the five most abundant elements in the universe. Some of the monomeric molecules of life (amino acids, sugars, etc.) are found everywhere. Laboratory experiments have suggested possible pathways along which those monomers might become polymers, make copies of themselves, and interact in complex ways on which Darwinian selection can act. Evidence for life appears in Earth's rock record as soon as there is any geologic record at all.

The dichotomy between chemical inevitability and historical contingency infuses studies of the origin and propensity of life in the universe (not to mention the question of what life is), and it has spawned numerous popular books on the subject. We leave resolution of that problem to scientists-*cum*-philosophers. In this review we concentrate on those lines of inquiry that have experienced especially fruitful development since the review of this subject by *Chyba et al.* (2000) for *Protostars and Planets IV*, including new age constraints on the appearance of clement environments and life on Earth, a reassessment of predictions for the chemistry of the prebiotic atmosphere and oceans, the formulation of a dynamical scenario for a "late" cataclysmic bombardment that may have profoundly influenced the emergence of life, and the development of new theories for the origin of Earth's water. Because science knows so little about the origin of life on Earth and the potential environments for its origin elsewhere, we feel it is important to be open-minded — and even provocative — in the scenarios that we consider. Our review is structured as follows: We consider the timing and environment of the origin of terrestrial life (section 2) and our understanding of the combination of factors that permit Earth-like life to persist on a planet for an astronomically interesting period of time (section 3). Finally, we address how the search for life elsewhere in the solar system, and particularly for life-bearing planets around other stars, promises to ultimately inform us about the evolution of our own habitable planet and the possibility of other origins elsewhere in the cosmos (section 4). Other relevant reviews since that of *Chyba et al.* (2000) include *Shock et al.* (2000), *Kasting and Catling* (2003), *Gaidos et al.* (2005), and *Chyba and Hand* (2005).

2. WHEN AND WHERE DID LIFE EMERGE?

2.1. Origin of a Theory of Origin

Recorded speculation on the setting of the origin of life goes back at least to ancient Greek civilization. Thales of Miletus (640–546 BCE) presciently suggested that all life, including humans, arose from the single "element" water — i.e., the sea. His student Anaximander (611–545 BCE) slightly modified his master's idea, substituting mud for water and thus proposing the first primordial "soup" hypothesis. Empedocles (490–435 BCE) further elaborated (or obfuscated) the theory, proposing that life emerged in a random fashion from a combination of the four classic Greek "elements." The concept of "spontaneous generation" of life from nonliving matter relied on unsupported anecdote and uncontrolled experiment for two full millenia, but was doomed by the invention of the compound microscope ca. 1590, the discovery of ubiquitous microorganisms by Antonie von Leeuwenhoek a century later, and the *coup de grace* delivered by Louis Pasteur's irrefutable 1864 demonstration of microbial contamination in all previous origin-of-life experiments. Modern inquiry into the origin of life began once science had discovered aspects of the chemical basis for life, described the theory of evolution by natural selection, and appreciated the age of Earth: In the 1920s Oparin and Haldane independently described a new theory in which life emerged from "prebiotic" chemistry driven by electricity or solar ultraviolet radiation in a reducing atmosphere of the early Earth.

By necessity, tests of such theories have been limited to demonstrations of plausibility by laboratory experiments. This is because the same geologic activity (volcanism, plate tectonics, and metamorphism) that sustains geochemical cycles and life on Earth today has destroyed nearly all of the earliest record of surface conditions and possible life that could be used to test such theories. Earth formed 4.56 b.y. ago (Ga) but the paltry record of the first 600 m.y. consists of a handful of zircon crystals as old as 4.4 Ga (*Wilde et al.*, 2001) and a single outcrop of heavily metamorphosed gneiss dated at 4.0 Ga (*Bowring and Williams*, 1999). The oldest putative evidence for life on Earth is isotopically fractionated C in 3.85 Ga rocks from Akila Island and the 3.8–3.7 Ga Isua formation in Greenland (*Schidlowski*, 1988; *Mozjsis et al.*, 1996; *Rosing*, 1999). However, some of this evidence has recently been challenged (*van Zuilen et al.*, 2002; *Fedo and Whitehouse*, 2002; *Mojzsis et al.*, 2002). Likewise, the origin and provenance of the oldest (3.46-Ga) putative microfossils, from the Apex chert in Australia (*Schopf and Packer*, 1987), have been disputed (*Brazier et al.*, 2002, 2004). The biological nature of even the 3.5–3.4 Ga fossil stromatolites, laminated microbial mats, in Australia and South Africa (*Walter et al.*, 1980) has been questioned (*Grotzinger and Rothman*, 1996). Despite the controversy, it seems likely that at least some of the evidence for life by 3.5 Ga will withstand scrutiny and new kinds of evidence may emerge (*Furnes et al.*, 2004). However, the geologic record of the origin and evolution of earlier, more primitive life seems irretrievably lost.

Any successful theory of biogenesis must provide a prebiotic source of the organic monomers (e.g., amino acids and nucleotides) as a starting point, and one or more mechanisms of chemical condensation of these monomers into polymers and more complex molecules. The Oparin-Haldane conjecture of an atmospheric source assumed a reducing primordial atmosphere containing abundant CH_4, NH_3, and H_2. This mechanism was brilliantly supported by Stanley Miller's experiment (*Miller*, 1953). However, this scenario fell into disfavor upon the development of models predicting that planetary core formation was contemporaneous with homogeneous accretion (*Stevenson*, 1980), leaving the mantle depleted of metallic Fe, and volcanic gases relatively oxidized (N_2, CO_2, and H_2O). Discharge experiments with such gas mixtures fail to produce significant quantities of organic molecules and underscore the particular importance of CH_4 and H_2 (*Miller and Schlesinger,* 1983; *Sleep et al.,* 2004).

New models of Earth's earliest atmosphere predict chemically significant concentrations of H_2 and CH_4: Although most of the Fe in Earth would have been sequestered into the core, degassing during impact of material with a carbonaceous chondrite composition would have created a reducing atmosphere composed of CH_4, N_2, NH_3, H_2, and H_2O (*Schaefer and Fegley,* 2005). The isotopic and elemental abundances of rare gases suggest that this primordial atmosphere was lost: Massive H escape was probably complete by 4.47 Ga (*Podesek and Ozima,* 2000) and the atmosphere was closed to all elements except H and He by 4.3 Ga (*Tolstikhin and O'Nions,* 1994). However, this still leaves a period of between 30 and 200 m.y. after core formation in which a Urey-Miller atmosphere could have existed, perhaps plenty of time for biogenesis to occur. Furthermore, H outgassing later from volcanos may have been more strongly retained by an anoxic atmosphere where the upper atmosphere did not contain singlet O that absorbs extreme ultraviolet radiation from the Sun (*Tian et al.,* 2005), although this is not conclusive (*Catling,* 2006). Nevertheless, alternative sources of organic monomers are available: One appeared serendipitously in the form of a meteorite that fell near the town of Murchison, Australia, in 1969. The archetype CM meteorite was found to contain a suite of organic molecules including many of the biotic amino acids (see review by *Ehrenfreund and Charnley,* 2000). Both meteorites and comets might have provided organics to the early Earth (*Chyba et al.,* 1990).

A decade after the Murchison meteorite fell, the first deep-sea hot spring chemotrophic ecosystem supported by the mixing of sulfidic hydrothermal fluids with oxygenated seawater was discovered (*Corliss et al.,* 1979). The appreciation that microorganisms could have colonized such high-temperature settings and exploited chemical energy sources before the advent of photosynthesis led to interest in their potential role in the origin and early evolution of life. Currently, the hypothesis of a hydrothermal origin of life draws support from three observations: First, hydrothermal systems are sites where organic synthesis is thermodynamically favored (*Shock and Schulte,* 1988; *Shock et al.,* 2000). Second, these environments contain abundant Fe and Ni sulfides that may catalyze reactions of potential prebiotic importance (*Huber and Wächtershäuser,* 1997) and are present as co-factors in many enzymes (*Johnson et al.,* 2005). *Cody et al.* (2001) showed that reaction of iron sulfide (FeS) with alkyl thiols (RSH), where R is an alkane group, produces carbonylated Fe-S compounds via

$$2FeS + 6CO + 2RSH \rightarrow Fe_2(RS)_2(CO)_6 + 2S^0 + H_2 \quad (1)$$

which they suggest to be responsible for catalysis, in lieu of mineral surfaces themselves. [The possible role of metal sulfides in prebiotic chemistry and subsequent incorporation into central metabolic pathways has been recently reviewed by *Cody* (2004). *Holm and Andersson* (2005) dis-

cuss the challenges of conducting hydrothermal chemistry under geologically relevant conditions.] Third, many thermophilic and hyperthermophilic archaea and bacteria are located near the root of phylogenetic trees constructed from small subunit ribosomal RNA gene sequences. This has been taken to suggest that a primitive character of the last universal common ancestor of all life was adaptation to high temperature, as originally suggested by *Woese* (1987), an inference widely, but not completely, accepted (*Galtier et al.,* 1999; *Brochier and Philippe,* 2002; *Di Giulio,* 2003). (See the next section for an alternative explanation of thermophily.)

Another successful conjecture in origin of life studies is the idea of an "RNA world" in which ribonucleic acid (RNA) played the role of both DNA and protein in primitive organisms by carrying information and catalyzing chemistry (*Orgel,* 1968; *Crick,* 1968; *Gilbert,* 1986). This conjecture is supported by the appearance of RNA in ubiquitous and highly conserved — and thus evolutionarily ancient — parts of the cellular machinery such as the ribosome, the demonstration that ribonucleotides are catalytically active (*Cech,* 1986), and by the success of evolving catalytically active RNA molecules in the laboratory (*Joyce,* 2004). In contrast to the hypothetical high-temperature origin of life described above, the phosphodiester backbone of RNA and the nucleobases themselves are unstable under high-temperature aqueous conditions (e.g., *Levy and Miller,* 1998). One scenario is for an RNA world to evolve under near-freezing conditions, perhaps in pockets of eutectic brine within ice where components were cyclically frozen and rehydrated (*Orgel,* 2004; *Vlassov et al.,* 2005). Brines have also been suggested as the site of prebiotic purine and pyrimidine synthesis and polymerization (*Bada et al.,* 1994, *Miyakawa et al.,* 2002a,b).

Recently, investigators have turned to wet-dry cyclic chemistry at clement temperatures, perhaps driven in the sediments of intertidal flats. *Commeyras et al.* (2004) describe a mechanism of prebiotic polypeptide synthesis through cyclic condensation of N-carbonyl amino acids under alternating pH conditions in the presence of significant N oxides in the atmosphere. [See also *Lathe* (2004) for a speculative scenario based on salt concentrations.] Alternatively, a more stable predecessor to RNA such as a peptidal molecule has been posited. *Russell and Arndt* (2005) argue for biogenesis at low-temperature, alkaline submarine seeps. These seeps form mounds containing precipitated Fe-Ni sulfides through which strong chemical gradients are maintained between the H_2-rich, reducing fluids and more oxidizing oceans, driving the reduction of CO_2 or HCO_3^- to acetate (COOH).

If core metabolism reflects a hydrothermal environment, and RNA evolved before protein, then the thermal instability of RNA suggests that it in turn was preceded by an unknown protobiotic world that functioned at higher temperatures, and therefore the thermophilic character of a LUCA is unrelated to a high-temperature origin of metabolism. Alternatively, RNA and the core metabolism of extant organisms appeared in different lineages. These considerations suggest a substantial evolutionary history preceding LUCA. Such a history may have involved the extensive chimerism of lineages that evolved from different environments. An analogous history is recorded in the complex organelle structure of eukaryotic microalgae that have experienced engulfment and endosymbiosis of independent unicellular lineages (*McFadden,* 2001). *Woese* (2000) has suggested that the earliest history of RNA/DNA-based life was marked by the rampant "horizontal" transfer of genes between organisms, absence of distinct lineages, and communal evolution of the gene pool. Less efficient and redundant components would have been discarded (e.g., the information-carrying molecules in the high-temperature contributor, the metabolic machinery in the low-temperature contributor), leaving an organism whose chemical ancestry derives from very different environments.

Furthermore, the environment(s) in which the origin of life took place need not resemble any environment on the modern Earth, and indeed may not be habitable by the standards of modern organisms. The evolution of life may have involved "frozen accidents" in which universal biological attributes selected for in an archaic environment are retained, even in the face of maladaptation in a new environment, because any changes in them would be too costly to the fitness of organisms. For example, while the eukaryotic cell may have arisen from a chimeric fusion of representatives of the Bacteria and Archaea, both domains of life that contain species that thrive at temperatures near 100°C, no eukaryote has been found that grows at temperatures above ~60°C, probably because the incorporation of membrane-surrounded organelles such as the nucleus requires permeability that renders the membrane susceptible to destruction at high temperatures. It is conceivable that life arose at temperature exceeding 120°C, but that the universal use of lipid membranes for structure and triphosphates for energy has rendered those environments forever inaccessible to life.

Darwin's proposal that all life on Earth shares a common ancestry is supported by vast amounts of molecular work. Yet, much of the microscopic world is classified only by molecular techniques such as the polymerase chain reaction (PCR) and it is conceivable that completely "alien" organisms based on different molecules flourish undetected under our feet (*Davies and Lineweaver,* 2005). If all Earthly life does have a single origin this might mean that the origin of life is sufficiently infrequent that the probability of it happening more than once on the same planet is low. Alternatively, it might mean that sometime in Earth history all other forms of life went extinct. Although it may be chauvinistically satisfying to think that other forms of life were out-competed by our common ancestor, nature tolerates the competitive or noncompetitive coexistence of countless forms of life, often within the same ecological niche (e.g., there are 300,000 known species of plants). Although there is no evidence that independent forms of life ever existed, it is difficult to exclude them from the first billion years of history in the absence of morphological fossils, and impossible to exclude them from the first 600 m.y.

as there is no record at all. Such a loss in diversity would not be the first to be inferred in the history of life. For example, the diversity of animal body plans recorded in fossil deposits of exceptional preservation such as the Burgess Shale is thought to greatly exceed later body plan diversity. Instead of competition, perhaps a uniquely catastrophic event extinguished all but a few, related forms of life that occupied some refuge.

2.2. Impacts, Bottlenecks, and Frozen Accidents

Giant impacts capable of vaporizing the oceans may have provided such an extinction event. A "late" (3.9-Ga) episode of impacts is recorded on the Moon and in the martian meteorite ALH 84001 (*Turner et al.,* 1997). Sterilizing impacts may have limited the emergence of life (*Maher and Stevenson,* 1988) and imposed a high-temperature "bottleneck" through which only adapted organisms could have passed, thus explaining the inference that LUCA was a thermophile (*Sleep and Zahnle,* 1998; *Nisbet and Sleep,* 2001). Giant impacts may also have contributed to the destruction of the rock record itself. One model for this "late heavy bombardment" involves the decay of a long-lived reservoir of impactors somewhere in the outer solar system (*Fernández and Ip,* 1983). However, searches for geochemical evidence for an extraterrestrial input to the Earth system at 3.8–3.7 Ga have yielded ambiguous results (*Anbar et al.,* 2001; *Schoenberg et al.,* 2002; *Frei and Rosing,* 2005). A null result from such searches supports an alternative scenario in which the impacts occurred in a single cataclysm ca. 3.9 Ga (*Dalrymple and Ryder,* 1993; *Cohen et al.,* 2000). Such an event can be produced by a 1:2 mean-motion resonance crossing of Jupiter and Saturn (*Tsiganis et al.,* 2005) during an early period of giant planet migration driven by planetesimal scattering (*Hahn and Malhotra,* 1999). This scenario is consistent with evidence for an asteroidal origin of the impacts (*Strom et al.,* 2005).

Previously, the earliest evidence for life in the rock record, apparently at the tail end of a continuous period of sterilizing giant impacts, was taken to suggest that the origin of life was geologically instantaneous and would occur just as quickly on other planets were conditions correct (e.g., *Lineweaver and Davis,* 2002). If the scenario of a "brief" cataclysm is correct, life may have emerged during the previous 600-m.y. period that followed a magma ocean (*Boyet and Carlson,* 2005) and Moon-forming impact (*Lee et al.,* 2002) at around 4.5 Ga. During that time the impact rate may have been permissible for life, and considerable prebiotic and biological evolution could have taken place of which we have no record. Or do we? Assuming that life emerged prior to 3.9 Ga and survived the impact bottleneck in deep refugia, the genetic information carried in the LUCA(s) might tell us something about that early environment. For example, molecular oxygen in a pre-3.9-Ga atmosphere would explain the paradox of the presence of cytochrome *c* terminal oxidases in many species of both bacteria and archaea, and thus presumably in a LUCA, and before the origin of oxygenic photosynthetic cyanobacteria (*Castresana et al.,* 1994). A giant ocean-vaporizing impact would extinguish photosynthetic life, but perhaps not deeper-living organisms that had profited from that O_2 (such as those that exist in modern vent systems). A narrow bottleneck would be a convenient explanation for why only one form of life exists on modern Earth. The requirement of giant planets near a resonance suggests that such cataclysms may not occur (or may occur at a different time) in extrasolar planetary systems with different giant planet architectures.

Impacts also provide a mechanism by which life might be transferred from one planet to another. Interest in the interplanetary transfer of life ("lithopanspermia"; related to, but to be distinguished from conjectures of cosmological "panspermia") was catalyzed by the discovery of meteorites from Mars, the elaboration of the spallation mechanism of impact ejection (*Melosh,* 1984), and dynamical simulations showing small but finite probabilities that such ejecta could be transferred between the inner planets on timescales of thousands of years or less (*Gladman and Burns,* 1996). Magnetic constraints on the thermal history of the ALH 84001 meteorite during its ~17-m.y. transit (*Goswami et al.,* 1997) are permissive of life (*Weiss et al.,* 2000a). Laboratory experiments indicate that bacteria and their spores can survive the shock pressures and acceleration associated with impact ejection (*Mastrapa et al.,* 2001; *Burchell et al.,* 2001, 2003, 2004) and can find sufficient protection from radiation within rock fragments a few centimeters in size (*Horneck et al.,* 2001).

Transfer between the inner planets may have been a ubiquitous process. Simulations by *Gladman et al.* (2005) show that 1%, 0.1%, and 0.001% of ejecta from Earth reach Earth, Venus, and Mars in 30,000 years. In the first case, this suggests that ejecta may have been a refugia for microorganisms during a giant impact event in which sterilizing conditions existed for thousands of years (*Wells et al.,* 2003). Alternatively, ejecta on "express" trajectories (a few years) could have reseeded planets after giant impact extinction events, provided there was a second, life-bearing planet. Climate models suggest that Venus, if it started out with an Earth-like inventory of water, could have experienced clement surface temperatures (*Kasting et al.,* 1993) and there is geomorphological evidence for a very early warm, wet Mars (*Jakosky and Phillips,* 2001). Even if a sterilizing impact was inevitable on each planet, the probability of simultaneous events (within a few thousand years) on the two planets would be vanishingly small. This could mean a novel requirement for planetary habitability, that of a second habitable planet.

If life can be transferred between planets then it is not too great a leap of logic to suppose that it arose on another planet and was later transferred to Earth. [Although it appears unlikely that meteorites could be exchanged between planetary systems (*Melosh,* 2003; *Wallis and Wickramasinghe,* 2004) it was more likely for stars (possibly like the Sun) formed in a dense cluster (*Adams and Spergel,* 2005).] Mars is *a priori* the favorite alternate planet of origin because of its lower escape velocity and because there is evidence for

at least episodic Earth-like conditions in the past — although the exact conditions are controversial (*Carr,* 1999; *Craddock and Howard,* 2002; *Bhattacharya et al.,* 2005). There is no such evidence (one way or another) for Venus and it has a deeper gravity well. *Sleep and Zahnle* (1998) have also found that any organisms on Mars would have been more likely to survive giant impacts in the past, again because the kinetic energy of the impact is smaller, and because of the absence of the latent heat of fusion of a vaporized global ocean that would delay cooling (assuming Mars had no such ocean). However speculative such theories may seem, the absence of any record of early life on Earth suggests that we keep an open mind on such matters.

2.3. Life First, Planets Second?

Indeed, planetary bodies much smaller than Mars represent a potential site for the origin (but not maintenance) of Earth life. Many carbonaceous chondrite meteorites record geochemical alteration by liquid water, and it is presumed that they originate from parent bodies a few tens of kilometers across, i.e., large enough to have maintained temperatures above the freezing point of water for millions of years, but too small to have experienced differentiation and high-temperature metamorphism (*Keil,* 2000). The main asteroid belt presently contains more than 300 asteroids with diameters larger than 50 km and the primordial belt may have contained 10^3–10^4 times as many (*Weidenschilling,* 1977). A scenario for the origin of life in a primitive planetesimal and its subsequent transfer to Earth would involve biogenesis while liquid water was present, transfer of protoorganisms to Earth after the Moon-forming impact approximately 30 m.y. into solar system history (*Jacobsen,* 2005), and preservation of the organisms during any intervening period. This scenario is distinct from the survivability of organisms in asteroids to the present day, which *Clark et al.* (1999) have dismissed based on thermal, radiation, and energetic arguments.

Carbonaceous chondrite meteorites contain abundant (up to a few weight percent) water. Masses of several main-belt asteroids determined by the orbits of satellites give low densities suggestive of high water ice content and/or high void space (*Marchis et al.,* 2005) and consistent with a picture of an asteroid as an icy "rubble pile" (*Weidenschilling,* 1981). Highly permeable, water-rich asteroids would have been sites of hydrothermal circulation early in their history. Water in the interior of parent bodies would be liquefied and mobilized by the heat from decaying ^{26}Al and ^{60}Fe while protected by an ice-filled impermeable crust a few kilometers thick. Additional internal heat may come from exothermic serpentinization reactions (see the chapter by Jewitt et al.) and possibly impacts. Detailed three-dimensional simulations of hydrothermal convection in a 40-km body show interior temperatures remain well above the freezing point for millions of years (*Travis and Schubert,* 2005).

Carbonaceous chondrites (and by inference their parent bodies) also contain organic molecules, including amino acids (*Kvenvolden et al.,* 1970) and polyhydroxylated compounds (e.g., sugars) (*Cooper et al., 2001*), and their possible role as a source of important biotic precursor molecules has long been scrutinized. The stable isotopes of C and N in this organic matter suggests an origin in the interstellar medium (*Alexander et al., 1998*), but significant processing could have occurred in the solar nebula and in meteorite parent bodies. Although aqueous alteration in many parent bodies involved relatively oxidizing conditions and thus led to loss of organic material (e.g., conversion to CO_2 and carbonates) (*Naraoka et al., 2004*), a few meteorites, particularly CM meteorites like Murchison, seemed to have been altered by reducing fluids (*Browning and Bourcier, 1996*). Moreover, *Shock and Schulte* (1990) make thermodynamic arguments for amino acid synthesis by aqueous alteration of polycyclic aromatic hydrocarbons (PAHs), a common organic in the interstellar medium and primitive meteorites, and Strecker synthesis by reaction of ketones or aldehydes with HCN and NH_3 (*Schulte and Shock, 1992*).

Clark et al. (1999) argue that the emergence of endogenous organisms is *a priori* less likely in an asteroid than on a planet because the former are smaller, and because they supposedly comprise less diverse environments. However, the macroscopic scale of an environment is unlikely to affect its potential to host microscopic prebiotic chemistry. First-order chemical kinetics depends on the *concentration* of reactants rather than the total molar quantity and high concentrations of reactants (a "soup") are more plausibly produced in small environments ("puddles") than in large ones. If the first steps in the origin of life consist of prebiotic chemistry, it is chemical diversity rather than physical or geologic diversity that is important. Melting and high-temperature metamorphism associated with the accretion and differentiation of planetary embryos and planets results in chemical equilibrium and the destruction of chemical diversity. Besides many of the important terrestrial minerals such as olivines, pyroxenes, and clays, meteorites contain a diverse suite of minerals that have not been found on Earth, including various metal sulfides and phosphates (Table 1). Carbonaceous chondrite meteorites also contain abundant metallic Fe-Ni grains, in contrast to the surface of Earth where such metal alloys are extremely rare and found only associated with ophiolites (preserved pieces of oceanic crust that have been heavily altered by the reducing fluids associated with serpentinization). As discussed above, metal sulfides and metals may have played an important catalytic role in prebiotic chemistry.

Although these parent bodies were small, they were extremely numerous and diverse. Each of these bodies would have differed because of chemical gradients in the solar nebula, their precise accretion history, and their final size. The simulations of *Travis and Schubert* (2005) also show that within a single (undifferentiated) body there is a diversity of hydraulic histories and presumably, degrees of chemical alteration. Individual impacts at speeds low enough to be nonsterilizing would induce additional heterogeneity in physiochemical conditions. Essentially, each of these bodies would represent a different "experiment" in low-temperature

TABLE 1.   Uniquely extraterrestrial minerals.

Name	Chemical Formula
barringerite	$Fe_{2-x}Ni_xP$
brezinaite	Cr_3S_4
brianite	$Na_2CaMg(PO_4)_2$
carlsbergite	CrN
daubreelite	$FeCr_2S_4$
farringtonite	$Mg_3(PO_4)_2$
gentnerite	$Cu_8Fe_3Cr_1S_{18}$
haxonite	$Fe_{20}Ni_3C_6$
heideite	$(Fe,Cr)_{1+x}(Ti,Fe)_2S_4$
krinovite	$NaMg_2CrSi_3O_{10}$
lawrencite	$(Fe,Ni)Cl_2$
majorite	$Mg_3(Fe,Al,Si)_2(SiO_4)_3$
merrihueite	$(K,Na)_2(Fe,Mg)_5Si_{12}O_{30}$

inorganic and organic chemistry. Many or most of these experiments would be cut short by accretion onto large embryos where melting and differentiation would occur. However enough bodies might have survived the 30 m.y. during which accretion of Earth was completed. Disruption of these bodies by mutual collisions induced by the gravitational perturbations of Jupiter might have produced frozen fragments containing protolife that successfully transited the thick atmosphere of an abiotic Earth to thaw on and colonize its surface.

Could some form of protolife have emerged in a primordial asteroid and then persisted long enough (perhaps in a frozen state) to await collisional disruption of the body into fragments small enough for a relatively gentle arrival on Earth? Such a scenario requires that (1) life evolved "very quickly" (within a few to tens of millions of years); (2) it was preserved in the parent body or fragments of the parent body during the period of the formation and cooling of the terrestrial planets (perhaps 30–100 m.y.), (3) it was successfully transferred to Earth (or Mars) intact, perhaps in a small fragment; and (4) it arrived in an environment in which it could thrive.

The unsuccessful (or overly successful) search for fossil life in meteorites has been well documented (e.g., *Anders et al.*, 1964). If life did emerge in the interior of primitive planetesimals, why has it or evidence for biological activity not been found in a collected primitive meteorite? One possibility is that any organisms or biomarkers have been degraded by radiation or impacts over the intervening 4.5 b.y. since these bodies were warm. Furthermore, only the small fraction of organics that are soluble have been thoroughly studied. The remainder is thought to be dominated by complex (poly)aromatic hydrocarbons (*Cody et al.*, 2002; *Sephton et al.*, 2003). There are controversial measurements of L-excess chirality of meteoritic amino acids (*Engel and Nagy*, 1982; *Pizzarello and Cronin*, 2000). Another explanation is that the world's meteorite collection probably samples only ~100 parent bodies in the present asteroid belt. Finally, the population of bodies that could have seeded

Earth within a few tens of millions of years has been completely depleted over the age of the solar system. In other words, if terrestrial life did emerge in a planetesimal, then we do not find it in our meteorites because that body or its fragments already arrived long ago, and we, and all life on Earth, are the result.

The scenario that life arose in the interior of an undifferentiated, primitive body and subsequently found a permanent home on a differentiated planet requires a population of small bodies with a dynamical lifetime longer than (but not much longer than) the accretion timescale of a potentially habitable planet. Terrestrial planet formation is a relatively efficient process, i.e., most planetesimals are accreted into large embryos (which differentiate and melt) rather than small bodies; nevertheless, final clearing may take well over 100 m.y. (*Goldreich et al.,* 2004). In addition, the gravitational perturbation of a gas giant planet such as Jupiter inhibits planet formation and scatters bodies at large distances. Thus, the formation of a giant planet and the equivalent of an asteroid belt may be a prerequisite for the emergence of life in a planetary system.

3. ELEMENTS OF HABITABILITY

3.1. The Habitable Zone

Once life is established on a planet, and assuming it survives catastrophes such as giant impacts, what factors are important to its persistence over a significant (i.e., observable) period of time? The range of orbital semimajor axes for which the surface temperatures on Earth-like planets would permit liquid water describes a "habitable zone" around a star (*Huang,* 1959). This will change with stellar luminosity evolution (*Hart,* 1979) and will depend on the concentration of greenhouse gases in the atmosphere and therefore on geochemical feedbacks (*Kasting et al.,* 1993) and rates of geologic activity such as volcanism (*Franck et al.,* 2000). That region of space in which a planet on a stable orbit will remain in the habitable zone over an extended period of time is known as the continuously habitable zone. Earth's orbit is relatively stable against the perturbations of the other planets over billion-year timescales (*Laskar,* 1994). It will remain in the habitable zone for another 1–2 b.y. before experiencing a runaway greenhouse (*Caldeira and Kasting,* 1992).

However, the known systems of extrasolar planets have giant planet configurations quite unlike that of our solar system. Yet unseen terrestrial planets in the habitable zones of these stars may have orbits that are dynamically unstable against gravitational perturbation by the detected giant planets. The criterion of dynamical habitability has motivated a host of publications that explore the stability of small (i.e., massless) planets within known giant planet systems (*Érdi et al.,* 2004; *Asghari et al.,* 2004; *Ji et al.,* 2005; *Jones et al.,* 2005, see also references in *Gaidos et al.,* 2005). These show that small planets could persist in the habitable zone of some, but not all these systems for the duration of the

simulations (which tend to be limited to millions of years). The kinematics of hypothetical extrasolar planets and the implications for habitability have been less explored: In the presence of at least two other planets, planets may experience chaotic obliquity fluctuations. The presence of oceans would moderate surface temperatures, however, making them habitable at least for simple life (*Williams and Pollard,* 2003). A similar conclusion is reached for planets on eccentric orbits (*Williams and Pollard,* 2002). Planets in the close-in habitable zones around much fainter M stars will be subject to tidal locking; however, even in this case, sufficient convective heat transport to the dark side can maintain atmospheres against collapse (*Joshi et al.,* 1997). Although we may have a quantitative understanding of the allowed ranges of orbital and rotation necessary for the habitability of an Earth *twin,* many other factors determine whether a planet can support life (*Taylor,* 1999). Some of these, including the frequency of supernovae and giant impacts, have been explored by *Gonzalez et al.* (2001).

3.2. Planetary Water

Water is an indisputably indispensable commodity of planetary habitability and a defining constituent of Earth's surface. Any model of terrestrial planet habitability must include a component that addresses the abundance of water, and any such component must satisfactorily explain the origin of Earth's water. The inner regions (~1 AU) of model primordial solar nebulae are devoid of water, as a consequence of diffusion of water vapor outward along a thermal gradient and condensation at a "snow line," and in apparent agreement with the correlation between the water content and the orbital distance of asteroids (assumed to be their formation distance). It is also thought that retention of water against loss to space is efficient only when a planet had grown to a certain mass. Compared to the abundance of water in primitive materials such as CI chondrites (1–10%), the bulk Earth is indeed dry; roughly 0.023% by weight for the oceans and an uncertain but probably similar amount for the water in the hydrous mantle (*Lécuyer,* 1998). Rare gas isotopic and elemental abundances also indicate the loss of copious H to space (*Pepin,* 1991) and since water is the major reservoir of H (at least on the modern Earth), this must be accounted for as well (see below).

The accretion of a late "veneer" of water-rich material has been postulated as the source of Earth's water. Water-rich carbonaceous chondrite meteorites were early suspects (*Boato,* 1954). Observations and models of the solar nebula suggest that bodies beyond 2.5 AU may be water rich and the source of carbonaceous chondrites. The relative abundance of D to H of H_2O in these meteorites spans the value of seawater (1.53×10^{-4}). (In these discussion, it should be kept in mind that the material that was the source of Earth's water may not have any representatives in our meteorite collections or indeed in the solar system; terrestrial planet accretion is a relatively efficient process!) A major contribution by comets (*Chyba,* 1987), is not consistent with the

D/H values nor the abundances of rare gases (*Dauphas and Marty*, 2002) and is dynamically difficult. Another mechanism of inward water transport is the condensation of ice grains beyond the "snow lines" where temperatures are below 160 K, inward migration by gas drag, and sublimation (*Cyr et al.,* 1998; *Cuzzi and Zahnle,* 2004; *Mousis and Alibert,* 2005).

New developments in isotopic geochemistry and numerical dynamics calculations have added substance to investigations of the source and timescales of delivery of Earth's water. Investigators have sought to use the abundance of siderophilic elements (Ni, Co, Ge, and the platinum group elements) in Earth's crust as a constraint on any "late" (post-core-formation) accretion of primitive material onto Earth (*Chyba et al.,* 1990). *Righter and Drake* (1997) has proposed that the high abundances are instead controlled by equilibrium with metallic Fe at the base of an early magma ocean. New results illuminate, but do not resolve, this controversy; Nd-isotopic data support the existence of a magma ocean (*Boyet and Carlson,* 2005) but new high-pressure experiments for some elements have not supported Righter's explanation for crustal siderophile abundances (*Holzheid et al.,* 2000; *Righter,* 2003; *Kegler et al.,* 2005). Based on analysis of the Hf-W and Sm-Nd isotope systems, the bulk of Earth is now thought to have accreted in about 10 m.y., and was essentially complete at 30 m.y. (*Jacobsen,* 2005; *Boyet and Carlson,* 2005). Rapid accretion of Earth makes the delivery of siderophilic elements more dynamically plausible since complete clearing of planetesimals may have taken as long as 300 m.y. (*Goldreich et al.,* 2004). This implies that dehydrated but undifferentiated material near Earth's orbit supplied the siderophilic elements — but no water. (Of course, those same simulations fail to produce Earth in the required 30 m.y.!)

Numerical simulations have been employed to investigate mechanisms by which water-bearing material beyond 2.5 AU might be transported inward to the orbit of Earth. The late impactor cataclysm scenario described in *Gomes et al.* (2005) is not a contender as the event occurs long after the earliest evidence for water on the planet, i.e., the isotopic composition of O in 4.4–4.3-Ga zircons (*Mojzsis et al.,* 2001). [Zircons are abundant in granitic rocks produced by partial melting in the presence of water, but zircons have also been found in lunar igneous rocks (*Meyer et al.,* 1996)]. Also, the estimated total accreted mass is too low to supply the water. An alternative mechanism is that self-scattering of planetary embryos (and their water) in the late stages of planetary accretion moved water inward (*Morbidelli et al.,* 2000). N-body simulations (*Chambers and Wetherill,* 1998; *Chambers,* 2001) suggest that Earth is the result of the fusion of tens of individual planetary embryos that formed within a broad range of orbital distances. Some of them may originate from regions at or beyond 2.5 AU where hydrated minerals or even ices were stable. Only a small number of these volatile-rich embryos are expected to contribute to the formation of an Earth at 1 AU, but a single Moon-sized embryo formed at 3 AU and made of 10% water by mass

would give Earth the equivalent of five modern oceans. In this scenario, the delivery of water to the telluric planets by "wet" embryos from more distant parts of the primordial solar system is a stochastic process relying on a small number of collisions. As a consequence, the water content of terrestrial planets is expected to be variable, even within a single planetary system. *Raymond et al.* (2004) carried out simulations of embryo scattering and accretion terrestrial planet formation with different nebular solid densities, position of the "snow line," and orbit of an outer giant planet. The vast majority of planets that formed in the "habitable zone" (0.8–1.5 AU) had water inventories equal to or greater than that of Earth. They found that the terrestrial planets in their simulations ended with an average water abundance about that of Earth, as long as the giant planet configuration was not too different from the one in the solar system. They showed that dry planets and extremely water-rich planets can also be expected.

This mechanism of water delivery can explain the difference in the water inventories of Earth and Mars: At the orbital distance of Mars, planetary formation is less efficient because of the influence of Jupiter, and Mars can be a remaining dry embryo (or the result of a very small number of dry embryos) formed locally and to which water was only brought by the late bombardment (*Lunine et al.,* 2003). However, some discrepancies between N-body simulations and observations still need to be explained. *Wiechert et al.* (2001) pointed out that the identical isotope fractionation of O on Earth and the Moon implies a similar composition of the Moon-forming impactor "Theia" and the proto-Earth. Oxygen-isotopic fractionation is a signature of the heliocentric distance of formation. Even if Earth and Theia formed at the same distance from the Sun (*Belbruno and Gott,* 2005) it is difficult to explain how Theia and the proto-Earth could have shared the same isotopic signature. Although O isotopes might have been homogenized in the circumterrestrial disk in the aftermath of the giant impact (*Pahlevan and Stevenson,* 2005), this would not explain its terrestrial-like superchondritic $^{142}Nd/^{144}Nd$ (*Boyet and Carlson,* 2005).

Another potential issue with the delivery of water by embryos is its escape from the embryos themselves. "Wet" embryos formed from kilometer-sized objects in $\sim10^4$ yr but were unable to radiate away the energy of accretion ($>3GM^2/5R$) in this period because the required cooling rate exceeds (by orders of magnitude) the ~300 W m^{-2} runaway greenhouse limit. This created a "magma ocean" phase, during which a dense steam atmosphere equilibrated with a molten rocky surface (*Zahnle,* 1998). For embryos with masses between 0.01 and 0.1 M$_\oplus$, this phase lasted 0.5 to 4 m.y., which is comparable to the typical lifetime for protoplanetary gas disks (*Lyo et al.,* 2003; *Armitage et al.,* 2003). While the disk was present, its opacity screened the embryos from intense UV radiation from the young star (*Ribas et al.,* 2005). Once the disk is absent, however, this radiation can drive photolysis of water in the upper atmospheres of water and escape of H to space. Furthermore, if core formation in these embryos is incomplete, water reacts with Fe in the mantle, releasing large amounts of molecular H (*Zahnle*, 1998). Escape to space of H from the relatively low gravitational potential of lunar-sized embryos would be efficient. The history of water may be very different in the inner regions of planetary systems that hosted different-sized embryos (due to a different mass surface density and isolation mass, for example) or had a different disk lifetime than that of our solar system.

How much water is "enough," and where does it end up? *Matsui and Abe* (1986) showed that the amount of water at Earth's surface is roughly what would be expected were it controlled by the solubility of water in silicate melt, i.e., an early magma ocean. Besides the reservoir of the global ocean, a significant amount of water may be sequestered in the mantle. The concentration of water in Earth's mantle is a subject of active research (*Tarits et al.*, 2004) but it may be the equivalent of several oceans (*Litasov et al.*, 2003). A significant amount of water could have been lost as the hydrous silicates reacted with metallic Fe during core formation to form Fe hydrides (FeH_x) that would be sequestered into core. The residual O then reacted with ferrous Fe in the mantle. *Hirao et al.* (2004) estimates that the core could contain H that is the equivalent of 8–24 oceans of water. Water may also have been lost by erosion of the atmosphere by giant impacts, and (as H) by continued hydrodynamic escape from the growing planet (*Pepin*, 1991). *Chen and Ahrens* (1997) estimated that such impacts produce ground velocities above the escape velocity, resulting in the escape of almost all the atmosphere. However, the question was revisited by *Genda and Abe* (2003): They found that, even in a collision the size of the Moon-forming impact, less than 30% of the atmosphere of both bodies is lost to space. Therefore, giant impacts can actually result in a net delivery of water to the growing protoplanet.

There may be other, important mechanisms for the removal of volatiles, including water from the surfaces of otherwise "habitable" planets. The habitable zone of M stars is very close to the star. Because M stars tend to have a higher ratio of X-ray and ultraviolet to bolometric luminosity, radiation and stellar wind-driven escape of planetary atmospheres may be important. Exospheric temperatures between 10,000 and 30,000 K are expected. It is within this range of temperature that Jeans (thermal) escape of the atmosphere is significant. Figure 1 shows the mass loss from a terrestrial planet for O (solid), N (dotted), and C (dashed) for a CO_2-rich atmosphere with 10% of N, as a function of the planetary mass. (Planets with high CO_2 levels are attractive in this context because the diurnal temperature difference on the tidally locked planet is damped.) The mass loss is given in mass of Earth atmosphere per billion year. Thin lines are for T_{exo} = 10,000 K and thick lines for 30,000 K. At these temperatures, H loss is of course diffusion-limited.

But around G stars, terrestrial planets may have water abundances much larger than that of the Earth. *Kuchner* (2003) described another mechanism of forming water-rich worlds; migration of entire icy planets inward by interaction with a gas or planetesimal disk. Such "ocean planets" have

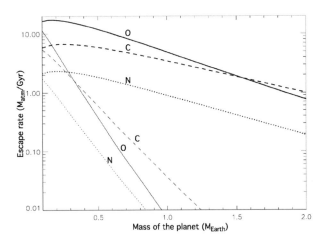

Fig. 1. Mass loss from a terrestrial planet in the habitable zone of an M stars for O (solid), N (dotted), and C (dashed) for a CO_2-rich atmosphere with 10% N, as a function of the planetary mass. The mass loss is given in units of Earth's present atmosphere per billion years. Thin lines are for an exosphere temperature of 10,000 K and thick lines are for 30,000 K. At these temperatures, H loss is diffusion limited (*Kulikov et al.*, 2006).

also been described by *Léger et al.* (2004). The abundance of water in a planet-forming nebula may have other secondary but potentially important implications for habitability, namely the presence of a giant planet and its dynamical effects. For example, the leading theory for the formation of Jupiter (and some of the habitability properties that it may confer to Earth) involves the rapid accretion of a core before depletion of nebular gas, an accretion accelerated by condensation of water beyond the "snow line" (*Stevenson and Lunine*, 1988). Nebulae with varying water abundances would presumably be more or less likely to form gas-accreting cores.

3.3. Planetary Composition and Diversity

As proposed by *Kuchner* (2003) and *Léger et al.* (2004), Earth-sized planets around other stars may have very different bulk compositions than that of our planet. Even seemingly minor differences in planetary composition could affect — perhaps dramatically — geologic activity and geochemical cycles at the planet's surface. Just as distance from the Sun, accretion history, and incorporation of varying amounts of nebular gas have produced a diversity of planets in our solar system, we should expect no less diversity, or probably much more, among a collection of planetary systems with different cosmochemical inheritances and formation histories. For example, two abundant planet-forming elements are Si and Fe. Silicon is an α-chain element and produced in massive stars, whereas Fe is produced primarily in type I SN from intermediate stars. As a consequence the ratio Fe/Si has increased with time. This will influence the size of planetary cores relative to the mantle as well as the abundance of radiogenic ^{60}Fe, an important heat source in the early nebula. Even the relative abundances of the major silicate mineral-forming elements (which controls such properties as melting temperature) vary more from star to star than they do within the solar system (Fig. 2). Some potential relationships between cosmochemistry, planetary composition, and habitability have been discussed by *Gonzalez et al.* (2001) and *Gaidos et al.* (2005). Gaidos (in preparation) calculated the relative rates of geologic activity on an Earth whose bulk mantle composition was that of CI chondrites (perhaps not far from the actual Earth) and a planet of identical size whose composition was that of enstatite EH chondrite after the metal has been removed. The latter has a significantly higher concentration of the long-lived radioisotopes ^{40}K, ^{232}Th, ^{235}U, and ^{238}U (*Anders and Grevesse*, 1989; *Newsom*, 1995) and such a body would have significantly enhanced rates of geologic activity, and would remain active for a longer period of time.

A major parameter that controls the composition of planets is the ratio of C to O (C/O) in the primordial nebula. Carbon and O are the two most abundant elements in the interstellar medium after H and He; their predominant form in the interstellar medium is thermodynamically stable CO molecules. Collapse of molecular cloud gas leads to higher pressures that favor the formation of water and methane

$$CO + 3H_2 \rightarrow CH_4 + H_2O \qquad (2)$$

However, this reaction is kinetically inhibited on formation timescales (millions of years) and requires a catalyst such

as free Fe (*Lewis and Prinn*, 1980). If O is more abundant than C, then nearly all C is bound in CO and remaining O is available for the formation of H_2O. Conversely, excess C results in all O being bound in CO, absence of H_2O, and the formation of graphite and organic molecules.

The solar photosphere has a C/O of 0.5 ± 0.07 (*Allendo Prieto et al.*, 2002), and presumably the primordial nebula was oxidizing and water-rich. Measurements of C and O abundances in nearby solar-type stars both with and without planets suggest a significant scatter in C/O (Gaidos, in preparation) with the Sun occupying a relatively C-poor, "water-rich" region of the distribution and some stars with C/O > 1. Solar-mass stars do not themselves produce significant C or O, and therefore these abundances reflect that of the gas and dust (ISM and molecular clouds) from which the stars formed. Stellar nucleosynthesis theory predicts that the relative production and ejection of C and O from massive stars (in winds and supernova ejecta) depend on stellar mass, metallicity, and the amount of "dredge-up" from the C-rich interior (*Woosley and Weaver*, 1995). About 57% of the C returned to the ISM from a solar-metallicity stellar population is via the winds of massive stars: 33% is produced in intermediate-mass stars and the remainder in high-mass star ejecta. Oxygen is almost entirely (87%) derived from supernovae and the rest is from their winds. Molecular clouds and their offspring can have different C/O because of local supernovae. Thus stars and disks that form from the chemically heterogeneous and evolving interstellar medium will start with different C/O ratios. The mean C/O of stella ejecta increases with galactic radius such that the older bulge should be more O-rich than the younger disk. As the galaxy ages, the C/O ratio of the ISM and the stars that form from it increases (Fig. 3). This picture is consistent with observations of dwarf galaxies (*Garnett et al.*, 1995).

Within a single star-forming region, the C/O can vary bcause of condensation and sedimentation of grains (*Lattanzio*, 1984) or contamination by very massive, short-lived stars within the same generation. In fact, the primordial chemistry of the solar system may have been influenced by mass loss from nearby massive stars. *Olive and Schramm* (1982), among others, have suggested that anomalous Al-, Pd-, and O-isotopic ratios in the solar system can be explained if the primordial nebula was contaminated with ejected from supernovae, possibly from short-lived massive stars formed in association with the Sun. Local C/O in a planet-forming disk will also be altered by diffusion of water outward along the thermal gradient (*Cyr et al.*, 1999) and pile-up of C-rich interstellar dust in the inner regions of a disk.

The condensation sequence in a nebula with C/O ~ 1 will be markedly different than that proposed for solar conditions, namely carbides will replace silicates and C will precipitate as graphite (*Larimer*, 1975; *Sharp*, 1990). *Gaidos* (2000) suggested that terrestrial planets would be composed of SiC, a ceramic with melting temperatures exceeding 3000 K, as well as other carbides. *Kuchner and Seager* (2006) discuss the properties of potential C/O ≫ 1 planets and cal-

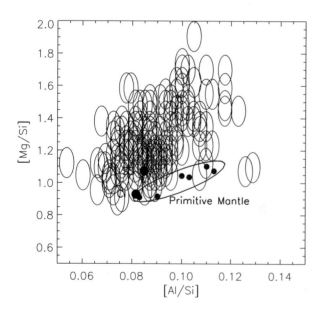

Fig. 2. Plot of Mg/Si vs. Al/Si on hypothetical planet-forming nebula based on the solar-type star photosphere data of *Edvardsson et al.* (1993). Circles represent the approximate range due to measurement errors. Edvardsson et al. measurements are compared with solar system (SS), chondritic (CI), and several primitive terrestrial mantle models. From Gaidos, in preparation.

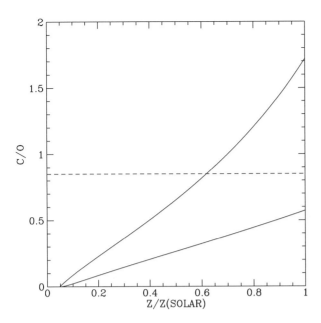

Fig. 3. Calculated evolution of the C/O of stellar wind and supernova ejecta (top line) and the average interstellar medium (bottom line) in the disk as a function of the abundance of heavy elements normalized to the solar value. The dashed line is the approximate threshold above which reducing, rather than oxidizing, conditions are expected. The solar photosphere has C/O of 0.5. From Gaidos, in preparation.

culated an atmospheric spectrum. They proposed that the surface of these planets will be covered with organics. A "ceramic planet" will have a Fe-Ni core containing 5–7% of dissolved C. Because of the high melting temperature of SiC, the planet will heat up by a corresponding amount until mantle convection can remove the heat produced by radiogenic elements. The core will be entirely molten and this may mean that such a planet will lack a magnetic field (Gaidos, in preparation). Excess C in the mantle will exist as either graphite, diamond, or liquid C, depending on conditions. The last will be extremely buoyant and may erupt to the surface. Because of the high thermal conductivity of SiC (2–3 times that of silicates), a thick, rigid lithosphere will develop and plate tectonics will be less likely. This example shows that future searches for other Earths may find instead rather exotic planets. There is really only one way to find out.

4. EXTRASOLAR EARTHS AND OTHER ORIGINS

4.1. Prospects in the Solar System

It is difficult to test theories of the origin of life when we are limited to a single example and when all of the early record of that life is lost. Thus searches for a second origin of life outside the Earth are paramount to understanding our own origins. Historically, Mars has been the favorite target in the solar system; it is the nearest planet with an accessible surface, and has an atmosphere and evidence for past geological processes and water. Initial disappointment that the Viking missions did not turn up unambiguous evidence for even simple life forms and that the surface proved chemically inhospitable directed subsequent searches for habitable conditions (i.e., liquid water) into Mars' past (or most recently with the MARSIS radar, deep beneath its surface). Geomorphological evidence from orbit in the form of outflow channels, valley networks, and possible playa lakes has now been complemented by more direct geological evidence in the form of aqueous alteration and evaporite deposition (*Squyres et al.,* 2004; *Herkenhoff et al.,* 2004; *Klingelhöfer et al.,* 2004; *Rieder et al.,* 2004; *Haskin et al.,* 2005; *Hynek,* 2004). A picture is emerging of a very early period (of uncertain duration, but perhaps a few hundred million years) of a warm, wet Mars, and a cold Mars in the intervening time (*Jakosky and Phillips,* 2001; *Gaidos and Marion,* 2003; *Solomon et al.,* 2005). A very exciting possibility is that, due to a cold climate regime and the absence of plate tectonics, Mars has preserved information about early prebiotic conditions that has been lost on Earth. The oldest rock on Earth is a meteorite from Mars (4.5 Ga).

Recent discoveries have also rejuvenated the possibility of habitable environments on current Mars, albeit at isolated locations in the subsurface. These include the presence of abundant regolith ice, the discovery of "young" gully-like formations, and the detection of atmospheric methane (*Mumma et al.,* 2004; *Krasnopolsky et al.,* 2004; *Formisano et al.,* 2004). Methane can be produced from the high-temperature reduction of CO_2 by H_2 during hydrothermal serpentinization of mafic rock (*Oze and Sharma,* 2005; *Lyons et al.,* 2005). While the possibility of biogenic methane cannot yet be ruled out, the estimated atmospheric concentration of a few tens of parts per billion and the lifetime in the atmosphere (~300 yr) suggest a source flux much weaker than the estimated abiogenic flux of methane on Earth. If reports of latitudinal variation in methane abundance are correct (M. Mumma, personal communication), the lifetime must be much shorter (~1 yr) and the flux commensurately higher. Combined with an upper limit for SO_2 (*Krasnopolsky,* 2005) this might disfavor an abiotic seepage source. However, martian geochemistry might be more reducing, thereby favoring a higher CH_4/SO_2 ratio, and SO_2 disproportionates in water to sulfate (which is soluble) and hydrogen sulfide (which will rapidly oxidize to sulfuric acid in the Mars atmosphere). Regardless, Mars CH_4 gives future astrobiological investigations a focus, e.g., measurement of the ratio of stable C isotopes to search for biogenic fractionation. If life is found on Mars, one possibility is that it will be unexpectedly familiar. Efficient ejection and transfer of material between the planets may have produced a common ancestry between the planets. However, if Mars was once habitable and no evidence for past or present life is found, this constrains models of lithopanspermia.

Beyond Mars, there are prospects for habitable environments in the water-rich interiors of the icy satellites of Jupiter, including Europa and Callisto, and the satellites of Saturn, Titan and Enceladus. The debate on the suitability of these objects to support life centers around the potential energy sources available; while plausible energy sources are many orders of magnitude lower than the potential energy from sunlight on Earth (*Gaidos et al.,* 1999), there are several mechanisms by which very low energy fluxes might be generated in the form of a redox gradient between the atmosphere and the surface, or between the crust and an interior ocean (*Gaidos et al.,* 1999; *Chyba and Phillips,* 2002). At the minimum, these bodies offer examples of possible prebiotic chemistries in the solar system that might be literally frozen in time. However, the same dynamical barriers, radiation environment, and thick crust that have isolated these bodies from contamination by interplanetary transfer of Earth material also challenge the technologies of humans that choose to investigate these intriguing environments.

4.2. Extrasolar Planets

Because the objects in our solar system are likely to represent a meager sample of the cosmic diversity of possible habitats for life, a more complete understanding of the potential abundance and distribution of life depends on the successful exploration of other planetary systems. The Kepler (*Borucki et al.,* 2003) and Corot (*Bordé et al.,* 2003) observatories will be capable of detecting Earth-sized planets as they transit the parent star and will foreshadow the eventual deployment of far more advanced telescopes that can directly detect the emitted or reflected light from such planets. As spatial resolution of such planets is beyond foreseeable technology and sources of funding, such characterization will rely on spectroscopy of their surfaces and atmosphere. Life manifests itself by *biosignatures,* in this case spectral features of the surface or atmosphere that reflect its biogeochemical activity and cannot be found in the absence of life. However, it is possible that abiotic mechanisms that are not known in the solar system might reproduce what was thought to be a reliable biomarker. In fact, many features once claimed to be biosignatures now have convincing abiotic explanations, e.g., martian vegetation (*Sinton,* 1957) and "nanobacteria" in the ALH 84001 meteorite (*McKay et al.,* 1996). The reliability of a biosignature depends strongly on contextual information. For instance, the detection of an O_2-containing atmosphere does not have the same implications on the icy moon of a giant planet compared to a terrestrial planet in the habitable zone of its star (*Selsis et al.,* 2002). This is because on the latter the weathering of minerals will consume O_2 and the only source of comparable intensity is oxygenic photosynthesis. Conversely, the detection of O_2 or O_3 is certainly a better biomarker when associated with a reducing compound such as CH_4 or NH_3 (*Lovelock,* 1975).

Moreover, the absence of a biosignature may not be evidence that a planet is lifeless, just that a particular metabolism is not present, that the activity is below detectable limits, or that differences in the planet's abiotic chemistry mask

the biological effect. Let us consider that a metabolism M (for instance, oxygenic photosynthesis) produces a biogenic species S (O_2) that, upon accumulation in the atmosphere, can result in a spectral signature B (the 760-nm band of O_2 or the 9.6 μm of O_3). The nondetection of B could have multiple explanations: (1) Life forms based on M do not exist on this planet. (2) Life forms based on M do exist but S does not reach detectable concentrations. This was probably the case on Earth between the emergence of O_2 producers and the rise of O_2, a period that could have lasted 500–1500 m.y. (*Catling and Claire,* 2005). (3) S reaches levels that would be detectable alone but B is masked by other spectral features; e.g., the 9.6-μm O_3 band would be masked by the high CO_2 level required for greenhouse warming in most of the habitable zone (*Selsis et al.,* 2002). The Lovelock example is another case in point. The thermodynamic disequilibrium that Lovelock advocated as a biosignature is a result of photosynthesis and the conversion of electromagnetic energy into potential chemical energy in the atmosphere. In the absence of photosynthesis, one biosignature might be the *absence* of such a disequilibrium, as this represents an unused source of energy for microorganisms (*Weiss et al.,* 2000b).

The spectrum of Earth exhibits biosignatures, including the presence of O_2 (and O_3) simultaneously with CH_4, that are detectable from space (*Sagan et al.,* 1993). Table 2 gives some groups of atmospheric molecular bands that could serve as biosignatures for future missions. Any biomarker should include the signature of H_2O, water being considered as a requisite for life as we know it. Some of the listed features are not observable in the spectrum of present Earth but may have been present in the past. Some other biogenic compounds, such as N_2O, were probably never observable in a low-resolution spectrum of Earth but would be at slightly higher concentrations. In addition to atmospheric molecules, the vegetation "red edge" (the increase of plant reflectivity between 700 and 800 nm) may be another promising way to detect complex extraterrestrial life (*Arnold et al.,* 2002, *Seager et al.,* 2005). However, the red edge results from photosynthetic pigments like chlorophyll that are much more complex than simple gases such as O_2. A life form able to use H_2O as an electron donor to reduce CO_2 will produce O_2 whatever the pigments or the energy source. On the other hand, evolution could select other pigments, characterized by different radiative properties. Moreover, detecting the red-edge on a distant Earth replica requires a level of resolution and sensitivity that will not be reached by the next generation of telescopes. There may be other, more readily obtainable pieces of information contained in the time-variability of emitted or reflected radiation from a planet about its ability to support life, e.g., period of rotation and the presence of an ocean or thick atmosphere (*Ford et al.,* 2001; *Gaidos and Williams,* 2004; *Williams and Gaidos,* 2005).

Our ignorance of when and where life emerged in our solar system, as well as the complexities associated with the maintenance of life on planetary bodies, means that this area of scientific inquiry will be driven by observations into the

TABLE 2. Atmospheric biomarkers: Molecular bands detectable by future space observatories at infrared (6–20 μm) and optical (0.5–0.8 μm) wavelengths with plausible spectral resolution.

	IR ($\lambda/\Delta\lambda = 25$)						Visible ($\lambda/\Delta\lambda = 70$)				
	H_2O	CO_2	O_3	CH_4	N_2O	NX	H_2O	O_2	O_3	CH_4	
λ (μm)	<8 >18	15	9.6	7.5	7.8 17	*	0.72	0.76 0.82	0.6 ±0.1	0.73 0.79	
Level†	<1	<1	<10	‡	>10	>100	≤1	<10	≥1	>50	
IR alone							**Visible alone**				
	•	•	•				○	○	(○)		§
	•		•	•			•	•	(•)	•	§
	•				•						
	•		•								
	•				•						
	○						○				
Examples of Biosignatures Requiring both IR and Visible											
	(○)	○					○	○	(○)		¶
	•	(•)	•				•			•	**

A planet exhibits a biosignature if all the marked bands from a same line are detected in its spectrum. Bands in parentheses are conditional. Empty circles indicate an unlikely but known possible abiotic origin (*Owen*, 1980; *Léger et al.*, 1993, 1999; *Léger*, 2000; *Des Marais et al.*, 2002; *Selsis et al.*, 2002; *Segura et al.*, 2003; *Kasting and Catling*, 2003; *Selsis et al.*, 2005).

* NX = NO, NO_2, or NH — See *Selsis et al.* (2005) for wavelengths and required abundance.

† Levels in present atmospheric level (PAL) required for detection at the expected resolution.

‡ 1 PAL without H_2O, >20 PAL with H_2O.

§ O_3 conditional: tracer of O_2.

¶ Dense CO_2 atmosphere: the IR band of O_3 is hidden.

** O_2 and O_3 too low for visible but O_3 detected in IR, CH_4 hidden by H_2O in IR.

foreseeable future. As a consequence, the first planet-characterizing missions must be designed for broader objectives than the search for a specific biomarker. Perhaps the best approach is to "expect the unexpected" and to design instruments not on the basis of a specific biosignature, but to maximize the potential for characterization of the physical and chemical properties of the planet. Inference of biological activity on a planet could emerge from a more general understanding of its spectrum, even if none of the expected biosignatures are found. For the foreseeable future, a working definition for the biosignatures of remote, inaccessible planetary life might remain *chemical phenomena that cannot be explained by all known abiotic chemistry*. This is ultimately an unsatisfactory state of affairs but we should not despair too quickly: Not quite four centuries have elapsed since Galileo turned his telescope to the other planets in our solar system, and it has been a mere decade since the discovery of the first extrasolar planet around a main-sequence star. Should our species desist from threatening the life and habitability of this world, our progeny will have the fullness of time to answer the question of whether other planets host living beings and whether any of them also ponder the same questions.

Acknowledgments. E.G. thanks the Centre Recherche Astronomique de Lyon, the NASA Terrestrial Planet Finder Foundation Science program, and the NASA Astrobiology Institute for travel support during this chapter's writing. B. Jakosky provided a helpful review.

REFERENCES

Adams F. C. and Spergel D. N. (2005) *Astrobiol., 5,* 497–514.

Alexander C. M. O'D., Russell S. S., Arden J. W., Ash R. D., Grady M. M., and Pillinger C. T. (1998) *Meteoritics & Planet. Sci., 33,* 603–622.

Allende-Prieto C., Lambert D. L., and Asplund M. (2002) *Astrophys. J., 573,* L137–L140.

Anbar A. D., Zahnle K. J., Arnold G. L., and Mojzsis S. J. (2001) *J. Geophys. Res., 106,* 3219–3236.

Anders E. and Grevesse N. (1989) *Geochim. Cosmochim. Acta, 53,* 197–214.

Anders E., DuFresne A., Fitch F. W., Cavaillé A., Dufresne E. R., and Hayatsu R. (1964) *Science, 146,* 1157–1161.

Armitage P. J., Clarke C. J., and Palla F. (2003) *Mon. Not. R. Astron. Soc., 342,* 1139–1146.

Arnold L., Gillet S., Lardiére O., Riaud P., and Schneider J. (2002) *Astron. Astrophys., 392,* 231–237.

Asghari N., Broeg C., Carone L., Casa-Miranda R., Castro Palacio J. C., et al. (2004) *Astron. Astrophys., 426,* 353–365.

Bada J. L., Bigham C., and Miller S. L. (1994) *Proc. Natl. Acad. Sci. USA, 91,* 1248–1250.

Belbruno E. and Gott J. R. III (2005) *Astron. J., 129,* 1724–1745.

Bhattacharya J. P., Payenberg T. H. D., Lang S. C., and Bourke M. (2005) *Geophys. Res. Lett., 32,* L10201.

Boato G. (1954) *Geochim. Cosmochim. Acta, 6,* 209–220.

Bordé P., Rouan D., and Léger A. (2003) *Astron. Astrophys., 405,* 1137–1144.

Borucki W. J., Koch D. G., Lissauer J. J., Basri G. B., Caldwell J. F., et al. (2003) *Proc. SPIE Conf., 4854,* 129–140.

Bowring S. A. and Williams I. S. (1999) *Contrib. Mineral. Petrol., 134,* 3–16.

Boyet M. and Carlson R. W. (2005) *Science, 309,* 576–581.

Brasier M. D., Green O. R., Jephcoat A. P., Kleppe A. K., van Krenedonk M. J., et al. (2002) *Nature, 416,* 76–81.

Brasier M., Green O., Lindsay J., and Steele A. (2004) *Orig. Life Evol. Biosph., 34,* 257–269.

Brochier C. and Philippe H. (2002) *Nature, 417,* 244.

Browning L. and Bourcier W. (1996) *Meteoritics & Planet. Sci., 31,* A22.

Burchell M. J., Shrine N. R. G., Mann J., Bunch A. W., Brandão P., et al. (2001) *Adv. Space Res., 28,* 707–712.

Burchell M. J., Galloway J. A., Bunch A. W., and Brandão P. F. B. (2003) *Orig. Life Evol. Biosph., 33,* 53–74.

Burchell M. J., Mann J. R., and Bunch A. W. (2004) *Mon. Not. R. Astron. Soc., 352,* 1273–1278.

Caldeira K. and Kasting J. F. (1992) *Nature, 360,* 721–723.

Carr M. H. (1999) *J. Geophys. Res., 104,* 21897–21910.

Castresana J., Lübben M., Saraste M., and Higgins D. G. (1994) *EMBO J., 13,* 2516–2525.

Catling D. C. (2006) *Science, 311,* 38a.

Catling D. C. and Claire M. W. (2005) *Earth Planet. Sci. Lett., 237,* 1–20.

Cech T. R. (1986) *Cell, 44,* 207–210.

Chambers J. E. (2001) *Icarus, 152,* 205–224

Chambers J. E. and Wetherill G. W. (1998) *Icarus, 136,* 304–327.

Chen G. Q. and Ahrens T. J. (1997) *Phys. Earth Planet. Inter., 100,* 21–26.

Chyba C. F. (1987) *Nature, 330,* 632–635.

Chyba C. F. and Hand K. P. (2005) *Ann. Rev. Astron. Astrophys., 43,* 31–74.

Chyba C. F. and Phillips C. B. (2002) *Origins Life Evol. Biosph., 32,* 47–67.

Chyba C. F., Thomas P. J., Brookshaw L., and Sagan C. (1990) *Science, 249,* 366–373.

Chyba C. F., Whitmire D. P., and Reynolds R. (2000) In *Protostars and Planets IV* (V. Mannings et al., eds.), pp. 1365–1393, Univ. of Arizona, Tucson.

Clark B. C., Baker A. L., Cheng A. F., Clement S. J., McKay D., et al. (1999) *Orig. Life Evol. Biosph., 29,* 521–545.

Cody G. D. (2004) *Ann. Rev. Earth Planet. Sci., 32,* 569–599.

Cody G. D., Boctor N., Filley T. R., Hazen R. M., Scott J. H., Sharma A., and Yoder H. S. Jr. (2001) *Science, 289,* 1337–1340.

Cody G. D., Alexander C. M. O., and Tera F. (2002) *Geochim. Cosmochim. Acta, 66,* 1851–1865.

Cohen B. A., Swindle T. D., and Kring D. A. (2000) *Science, 290,* 1754–1755.

Commeyras A., Taillades J., Collet H., Boiteau L., Vandenbeele-Trambouze O., et al. (2004) *Orig. Life Evol. Biosph., 34,* 35–55.

Cooper G., Kimmich N., Bellsle W., Sarinana J., Brabham K., and Garrel L.(2001) *Nature, 414,* 879–883.

Corliss J. B., Dynmond J., Gordon L. I., Edmont J. M., von Herzen R. P., et al. (1979) *Science, 203,* 1073–1083.

Craddock R. A. and Howard A. D. (2002) *J. Geophys. Res., 107,* 5111.

Crick F. H. C. (1968) *J. Mol. Biol., 38,* 367–379.

Cuzzi J. N. and Zahnle K. J. (2004) *Astrophys. J., 614,* 490–496.

Cyr K. E., Sears W. D., and Lunine J. I. (1998) *Icarus, 135,* 537–548.

Cyr K. E., Sharp C. M., and Lunine J. I. (1999) *J. Geophys. Res., 104,* 19,003–19,014.

Dalrymple G. B. and Ryder G. (1993) *J. Geophys. Res., 98,* 13,085–13,095.

Dauphas N. and Marty B. (2002) *J. Geophys. Res., 107,* 5129, doi:10.1029/2001JE001617.

Davies P. C. W. and Lineweaver C. H. (2005) *Astrobiol., 5,* 154–163.

Des Marais D. J., Harwit M. O., Jucks K. W., Kasting J. F., Lin D. N. C., et al. (2002) *Astrobiol., 2,* 153–181.

Di Giulio M. (2003) *J. Mol. Evol., 57,* 721–730.

Edvardsson B., Andersen J., Gustafsson B., Lambert D. L., Nissen P. E., and Tomkin J. (1993) *Astron. Astrophys., 275,* 101–152

Ehrenfreund P. and Charnley S. B. (2000) *Ann. Rev. Earth Planet. Sci., 38,* 427–483.

Engel M. H. and Nagy B. (1982) *Nature, 296,* 837–840.

Érdi B., Dvorak R., Sándor Z., Pilat-Lohinger E., and Funk B. (2004) *Mon. Not. R. Astron. Soc., 351,* 1043–1048.

Fedo C. M. and Whitehouse M. J. (2002) *Science, 296,* 1448–1452.

Fernández J. A. and Ip W. H. (1983) *Icarus, 54,* 377

Ford E. B., Seager S., and Turner E. L. (2001) *Nature, 412,* 885–887.

Formisano V., Atreya S., Encrenaz T., Ignatiev N., and Giuranna M. (2004) *Science, 306,* 1758–1761.

Franck S., Block A., von Bloh W., Bounama C., Schellnhuber H.-J., and Svirezhev Y. (2000) *Planet. Space Sci., 48,* 1099–1105.

Frei R. and Rosing M. T. (2005) *Earth Planet. Sci. Lett., 236,* 28–40.

Furnes H., Banerjee N. R., Muehlenbachs K., Staudigel H., and de Wit M. (2004) *Science, 304,* 578–581.

Gaidos E. J. (2000) *Icarus, 145,* 637–640.

Gaidos E. J. and Marion G. (2003) *J. Geophys. Res., 108,* doi:10.1029/2002JE002000.

Gaidos E. and Williams D. M. (2004) *New Astron., 10,* 67–77.

Gaidos E. J., Nealson K. H., and Kirschvink J. L. (1999) *Science, 284,* 1631–1633.

Gaidos E., Deschenes B., Dundon L., Fagan K., McNaughton C., Menviel-Hessler L., Moskovitz N., and Workman M. (2005) *Astrobiol., 5,* 100–126.

Galtier N., Tourasse N., and Gouy M. (1999) *Science, 283,* 220–221.

Garnett D. R., Skillman E. D., Dufour R. J., Peimbert M., Torres-Peimbert S., et al. (1995) *Astrophys. J., 443,* 64–76.

Genda H. and Abe Y. (2003) *Icarus, 165,* 149–162.

Gilbert W. (1986) *Nature, 319,* 618.

Gladman B. and Burns J. A. (1996) *Science, 274,* 161–162.

Gladman B., Dones L., Levison H. F., and Burns J. A. (2005) *Astrobiol., 5,* 483–496.

Goldreich P., Lithwick Y., and Sari R. (2004) *Astrophys. J., 614,* 497–507.

Gomes R., Levison H. F., Tsiganis K., and Morbidelli A. (2005) *Nature, 435,* 466–469.

Gonzalez G., Brownlee D., and Ward P. (2001) *Icarus, 152,* 185–200.

Goswami J. N., Sinha N., Murty S. V. S., Mohapatra R. K., and Clement C. J. (1997) *Meteoritics & Planet. Sci., 32,* 91–96.

Grotzinger J. P. and Rothman D. H. (1996) *Nature, 383,* 423–425.

Hahn J. M. and Malhotra R. (1999) *Astron. J., 117,* 3041–3053.

Hart M. H. (1979) *Icarus, 37,* 351–357.

Haskin L. A., Wang A., Jolliff B. L., McSween H. Y., Clark B. C., et al. (2005) *Nature, 436,* 66–69.

Herkenhoff K. E., Squyres S. W., Arvidson R., Bass D. S., Bell J. F. III, et al. (2004) *Science, 306,* 1727–1730.

Hirao N., Kondo T., Ohtani E., Takemura K., and Kikegawa T. (2004) *Geophys. Res. Lett., 31,* L06616.

Holm N. G. and Andersson E. (2005) *Astrobiol., 5,* 444–460.

Holzheid A., Sylvester P., O'Neill H. S. C., Rubie D. C., and Palme H. (2000) *Nature, 406,* 396–399.

Horneck G., Rettberg P., Reitz G., Wehner J., Eschweiler U., et al. (2001) *Orig. Life Evol. Biosph., 31,* 527–546.

Huang S.-S. (1959) *Am. Sci., 47,* 392–402.

Huber C. and Wächtershäuser G. (1997) *Science, 276,* 245–247.

Hynek B. M. (2004) *Nature, 431,* 156–159

Jacobsen S. B. (2005) *Ann. Rev. Earth Planet. Sci., 33,* 531–570.

Jakosky B. M. and Phillips R. (2001) *Nature, 412,* 237–244.

Ji J., Liu L., Kinoshita H., and Li G. (2005) *Astrophys. J., 631,* 1191–1197.

Johnson D. C., Dean D. R., Smith A. D., and Johnson M. K. (2005) *Ann. Rev. Biochem., 74,* 247–281.

Jones B. W., Underwood D. R., and Sleep P. N. (2005) *Astrophys. J., 622,* 1091–1101.

Joshi M. M., Haberle R. M., and Reynolds R. T. (1997) *Icarus, 129,* 450–465.

Joyce G. (2004) *Ann. Rev. Biochem., 73,* 791–836.

Kasting J. F. and Catling D. (2003) *Ann. Rev. Astron. Astrophys., 41,* 429–463.

Kasting J. F., Whitmire D. P., and Reynolds R. T. (1993) *Icarus, 101,* 108–128.

Kegler P., Holzheid A., Rubie D. C., Frost D., and Palme H. (2005) In *Lunar Planet. Sci. Conf. XXXVI,* Abstract #2030.

Keil K. (2000) *Planet. Space Sci., 48,* 887–903.

Klingelhöfer G., Morris R. V., Bernhardt B., Schröder C., Rodionov D. S., et al. (2004) *Science, 306,* 1740–1745.

Krasnopolsky V A. (2005) *Icarus, 178,* 487–492.

Krasnopolsky V. A., Maillard J. P., and Owen T. C. (2004) *Icarus, 172,* 537–547.

Kuchner M. J. (2003) *Astrophys. J., 596,* L105–L108.

Kuchner M. J. and Seager S. (2006) *Astrophys. J.,* in press.

Kulikov Y. N., Lammer H., Lichtenegger H. I. M., Terada N., Ribas I., et al. (2006) *Icarus,* in press.

Kvenvolden K., Lawless J., Pering K., Peterson E., Flores J., et al. (1970) *Nature, 228,* 923–926.

Larimer J. W. (1975) *Geochim. Cosmochim. Acta, 39,* 389–392.

Laskar J. (1994) *Astron. Astrophys., 287,* L9–L12.

Lathe R. (2004) *Icarus, 168,* 18–22.

Lattanzio J. C. (1984) *Mon. Not. R. Astron. Soc., 207,* 309–322.

Lécuyer C. (1998) *Chem. Geol., 145,* 249–261.

Lee D.-C., Halliday A. N., Leya I., Wieler R., and Wiechert U. (2002) *Earth Planet. Sci. Lett., 198,* 267–274.

Léger A. (2000) *Adv. Space Res., 25,* 2209–2223.

Léger A., Pirre M., and Marceau F. J. (1993) *Astron. Astrophys., 277,* 309–313.

Léger A., Ollivier M., Altwegg K., and Woolf N. J. (1999) *Astron. Astrophys., 341,* 304–311.

Léger A., Selsis F., Sotin C., Guillot T., Despois D., Mawet D., Ollivier M., Labèque A., Valette C., Brachet F., Chazelas B., and Lammer H. (2004) *Icarus, 169,* 499–504.

Levy M. and Miller S. L. (1998) *Proc. Natl. Acad. Sci. USA, 95,* 7933–7938.

Lewis J. S. and Prinn R. G. (1980) *Astrophys. J., 238,* 357–364.

Lineweaver C. H. and Davis T. (2002) *Astrobiol., 2,* 293–304.

Litasov K., Ohtani E., Langenhorst F., Yurimoto H., Kubo T., and Kondo T. (2003) *Earth Planet. Sci. Lett., 211,* 189–203.

Lovelock J. E. (1975) *Proc. R. Soc. London, B189,* 167–180.

Lunine J., Chambers J., Morbidelli A., and Leshin L. (2003) *Icarus, 165,* 1–8.

Lyo A.-R., Lawson W. A., Mamajek E. E., Feigelson E. D., Sung E.-C., and Crause L. A. (2003) *Mon. Not. R. Astron. Soc., 338,* 616–622.

Lyons J. R., Manning C., and Nimmo F. (2005) *Geophys. Res. Lett., 32,* L13201.

Maher K. A. and Stevenson D. J. (1988) *Nature, 331,* 612–614.

Marchis F., Descamps P., Hestroffer D., Berthier J., and de Pater I. (2005) *Bull Am. Astron. Soc., 36,* Abstract #46.02.

Mastrapa R. M. E., Glanzberg H., Head J. N., Melosh H. J., and Nicholson W. L. (2001) *Earth Planet Sci. Lett., 189,* 1–8.

Matsui T. and Abe Y. (1986) *Nature, 322,* 526–528.

McFadden G. I. (2001) *J. Phycol., 37,* 951–959.

McKay D. S., Gibson E. K. Jr., Thomas-Keprta K. L., Hojatollah V., Romanek C. S., et al. (1996) *Science, 273,* 924–930.

Melosh H. J. (1984) *Icarus, 59,* 234–260.

Melosh H. J. (2003) *Astrobiol., 3,* 207–215.

Meyer C., Williams I. S., and Compston W. (1996) *Meteoritics & Planet. Sci., 31,* 370–387.

Miller S. L. (1953) *Science, 117,* 528–529.

Miller S. L. and Schlesinger G. (1983) *Adv. Space Res., 3,* 47–53.

Miyakawa S., Cleaves H. J., and Miller S. L. (2002a) *Orig. Life Evol. Biosph., 32,* 195–208.

Miyakawa S., Cleaves H. J., and Miller S. L. (2002b) *Orig. Life Evol. Biosph., 32,* 209–218.

Mojzsis S. J., Arrhenius G., McKeegan K. D., Harrison T. M., Nutman A. P., and Friend C. R. L. (1996) *Nature, 384,* 55–59.

Mojzsis S. J., Harrison T. M., and Pidgeon R. T. (2001) *Nature, 409,* 178–181.

Mojzsis S. J., Harrison T. M., Friend C. R. L., Nutman A. P., Bennett V. C., Fedo C. M., and Whitehouse M. J. (2002) *Science, 298,* 917a.

Morbidelli A., Chambers J., Lunine J. I., Petit J. M., Robert F., Valsecchi G. B., and Cyr K. E. (2000) *Meteoritics & Planet. Sci., 35,* 1309–1320.

Mousis O. and Alibert Y. (2005) *Mon. Not. R. Astron. Soc., 358,* 188–192.

Mumma M. J., Novak R. E., DiSanti M. A., Bonev B. P., and Dello Russo N. (2004) *Bull. Am. Astron. Soc., 36,* 26.02

Naraoka H., Mita H., Komiya M., Yoneda S., Kojima H., and Shimoyama A. (2004) *Meteoritics & Planet. Sci., 39,* 401–406.

Newsom H. E. (1995) In *Global Earth Physics: A Handbook of Physical Constants* (T. J. Ahrends, ed.), pp. 159–189. AGU, Washington, DC.

Nisbet E. G. and Sleep N. H. (2001) *Nature, 409,* 1083–1091.

Olive K. A. and Schramm D. N. (1982) *Astrophys. J., 257,* 276–282.

Orgel L. E. (1968) *J. Mol. Biol., 38,* 381–393.

Orgel L. E. (2004) *Orig. Life Evol. Biosph., 34,* 361–369.

Owen T. (1980) In *Strategies for the Search for Life in the Universe* (M. Papagiannis, ed.), pp. 177–188. Reidel, Dordrecht.

Oze C. and Sharama M. (2005) *Geophys. Res. Lett., 32,* L10203.

Pahlevan K. and Stevenson D. J. (2005) In *Lunar Planet. Sci. Conf. XXXVI,* Abstract #2382.

Pepin R. O. (1991) *Icarus, 92,* 2–79.

Pizzarello S. and Cronin J. R. (2000) *Geochim. Cosmochim. Acta, 64,* 329–338.

Podesek F. A. and Ozima M. (2000) In *Origin of the Earth and Moon* (R. M. Canup and K. Righter, eds.), pp. 63–72. Univ. of Arizona, Tucson.

Raymond S. N., Quinn T., and Lunine J. I. (2004) *Icarus, 168,* 1–17.

Ribas I., Guinan E. F., Güdel M., and Audard M. (2005) *Astrophys. J., 622,* 680–694.

Rieder R., Gellert R., Anderson R. C., Brückner J., Clark B. C., et al. (2004) *Science, 306,* 1746–1749.

Righter K. (2003) *Ann. Rev. Earth Planet. Sci., 31,* 135–174.

Righter K. and Drake M. J. (1997) *Earth Planet. Sci. Lett., 146,* 541–553.

Rosing M. T. (1999) *Science, 283,* 674–676.

Russell M. J. and Arndt N. T. (2005) *Biogeosci., 2,* 97–111.

Sagan C., Thompson W. R., Carlson R., Gurnett D., and Hord C. (1993) *Nature, 365,* 715–721.

Schidlowski M. A. (1988) *Nature, 333,* 313–318.

Schaefer L. and Fegley B. Jr. (2005) *Bull. Am. Astron. Soc., 37,* 29.15.

Schoenberg R., Kamber B. S., Collerson K. D., and Eugster O. (2002) *Geochim. Cosmochim. Acta, 66,* 3151–3160.

Schopf J. W. and Packer B. M. (1987) *Science, 237,* 70–73.

Schulte M. D. and Shock E. L. (1992) *Meteoritics, 27,* 286.

Seager S., Turner E. L., Schafer J. and Ford E. B. (2005) *Astrobiol., 5,* 372–390.

Segura A., Krelove K., Kasting J. F., Sommerlatt D., Meadows V., et al. (2003) *Astrobiol., 3,* 689–708.

Selsis F. (2002) In *The Evolving Sun and Its Influence on Planetary Environments* (B. Montesinos et al., eds.), pp. 273–281. ASP Conf. Series 269, San Francisco.

Selsis F. and Ollivier M. (2005) In *Lectures in Astrobiology, Vol. 1* (M. Gargaud et al., eds.), pp. 385–419. Springer, Berlin.

Selsis F., Despois D., and Parisot J.-P. (2002) *Astron. Astrophys., 388,* 985–1003.

Sephton M. A., Verchovsky A. B., Bland P. A., Gilmour I., Grady M. M., and Wright I. P. (2003) *Geochim. Cosmochim. Acta, 67,* 2093–2108.

Sharp C. M. (1990) *Astrophys. Space Sci., 171,* 185–188.

Shock E. L. and Schulte M. D. (1998) *J. Geophys., 103,* 28513–28528.

Shock E. L. and Schulte M. D. (1990) *Nature, 343,* 728–731.

Shock E. L., Amend J. P., and Zolotov M. Y. (2000) In *The Origin of the Earth and Moon* (R. Canup and K. Righter, eds.), pp. 527–543. Univ. of Arizona, Tucson.

Sinton W. (1957) *Astrophys. J., 126,* 231–239.

Sleep N. H. and Zahnle K. (1998) *J. Geophys. Res., 103,* 28529–28544.

Sleep N. H., Meibom A., Fridriksson Th., Coleman R. G., and Bird D. K. (2004) *Proc. Natl. Acad. Sci. USA, 101,* 12,818–12,823.

Solomon S. C., Aharonson O., Aurnou J. M., Banerdt W. B., Carr M. H., et al. (2005) *Science, 307,* 1214–1220.

Squyres S. W., Arvidson R. E., Bell J. F. III, Brückner J., Cabrol N. A., et al. (2004) *Science, 306,* 1709–1714.

Stevenson D. J. (1980) In *Lunar Planet. Sci. Conf. XI,* pp. 1088–1090.

Stevenson D. J. and Lunine J. I. (1988) *Icarus, 75,* 146–155.

Strom R. G., Malhotra R., Ito T., Yoshida F., and Kring D. A. (2005) *Science, 309,* 1847–1850.

Tarits P., Hautot P., and Perrier F. (2004) *Geophys. Res. Lett., 31,* L06612.

Taylor S. R. (1999) *Meteoritics & Planet. Sci., 34,* 317–239.

Tian F., Toon O. B., Pavlov A. A., and De Sterck H. (2005) *Science, 538,* 1014–1017.

Tolstikhin I. N. and O'Nions R. K. (1994) *Chem. Geol., 115,* 1–6.

Travis B. J. and Schubert G. (2005) *Earth Planet. Sci. Lett., 240,* 234–250.

Tsiganis K., Gomes R., Morbidelli A., and Levison H. G. (2005) *Nature, 135,* 459–461.

Turner G., Knott S. F., Ash R. D., and Gilmour J. D. (1997) *Geochim. Cosmochim. Acta, 61,* 3835–3850.

van Zuilen M. A., Lepland A., and Arrhenius G. (2002) *Nature, 418,* 627–630.

Vlassov A. V., Kazakov S. A., Johnston B. H., and Landweber L. F. (2005) *J. Mol. Evol., 61,* 264–273.

Wallis M. K. and Wickramasinghe N. C. (2004) *Mon. Not. R. Astron. Soc., 348,* 52–61.

Walter M. R., Buick R., and Dunlop J. S. R. (1980) *Nature, 284,* 443–445.

Weidenschilling S. J. (1977) *Astrophys. Space Sci., 51,* 153–158.

Weidenschilling S. J. (1981) *Icarus, 46,* 124–126.

Weiss B. P., Kirschvink J. L., Baudenbacher F. J., Vali H., Peters N., Macdonald F. A., and Wikswo J. P (2000a) *Science, 290,* 791–795.

Weiss B. P., Yung Y. L., and Nealson K. H. (2000b) *Proc. Natl. Acad. Sci. USA, 97,* 1395–1399.

Wells L. E., Armstrong J. C., and Gonzalez G. (2003) *Icarus, 162,* 38–46.

Wiechert U., Halliday A. N., Lee D.-C., Snyder G. A., Taylor L. A., and Rumble D. (2001) *Science, 294,* 345–348.

Wilde S. A., Valley J. W., Peck W. H., and Graham C. M. (2001) *Nature, 409,* 175–178.

Williams D. M. and Gaidos E. (2005) *Bull. Am. Astron. Soc., 37,* 31.13.

Williams D. M. and Pollard D. (2002) *Intl. J. Astrobiol., 1,* 61–69.

Williams D. M. and Pollard D. (2003) *Intl. J. Astrobiol., 2,* 1–19.

Woese C. R. (1987) *Microbiol. Rev., 51,* 221–271.

Woese C. R. (2000) *Proc. Natl. Acad. Sci. USA, 97,* 8392–8396.

Woosley S. E. and Weaver T. A. (1995) *Astrophys. J. Suppl., 101,* 181–234.

Zahnle K. (1998) In *Origins* (C. E. Woodward et al., eds.), pp. 364–391. ASP Conf. Series 148, San Francisco.

Color Section

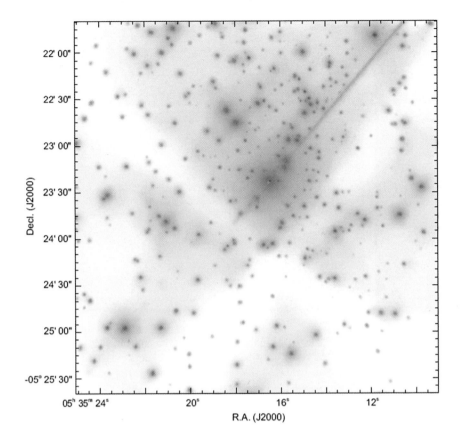

Plate 1. Inner region of the Orion Nebula viewed by the Chandra Orion Ultradeep Project (COUP). The image is smoothed and colors represent soft (red, 0.5–2 keV) and hard (blue, 2–8 keV) X-rays. The brightest source is θ^1C Orion (O7), and a group of embedded sources in the Becklin-Neugebauer region can be seen 1' to the northwest. From *Getman et al.* (2005a).

Accompanies chapter by Feigelson et al. (pp. 313–328).

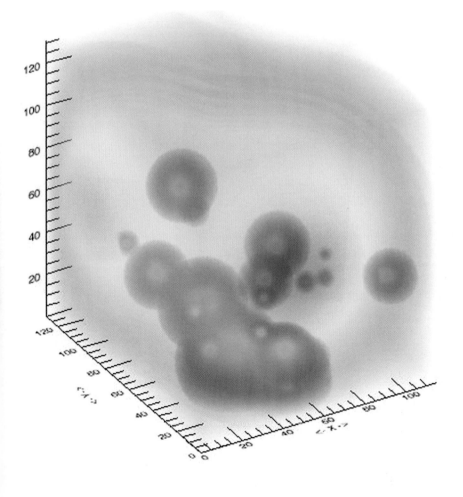

Plate 2. X-ray dissociation regions from X-ray stars embedded in the Orion Becklin-Neugebauer cloud core. Yellow-to-red shows assumed three-dimensional structure of the molecular gas. Blue-to-green shows the inferred XDR structures. From Lorenzani et al. (in preparation).

Accompanies chapter by Feigelson et al. (pp. 313–328).

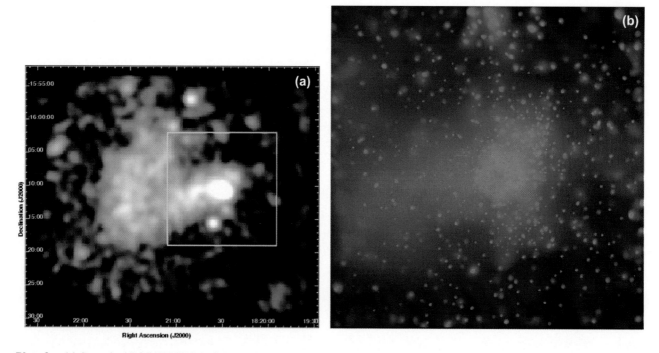

Plate 3. (a) Smoothed ROSAT PSPC (soft X-ray) image of M 17, ~39' × 42', with the outline of the Chandra pointing overlaid in blue. (b) The Chandra observation of M 17 showing hundreds of PMS stars and the soft X-ray flow from shocked O star winds. Red intensity is scaled to the soft (0.5–2 keV) emission and blue intensity is scaled to the hard (2–8 keV) emission. From *Townsley et al.* (2003).

Accompanies chapter by Feigelson et al. (pp. 313–328).

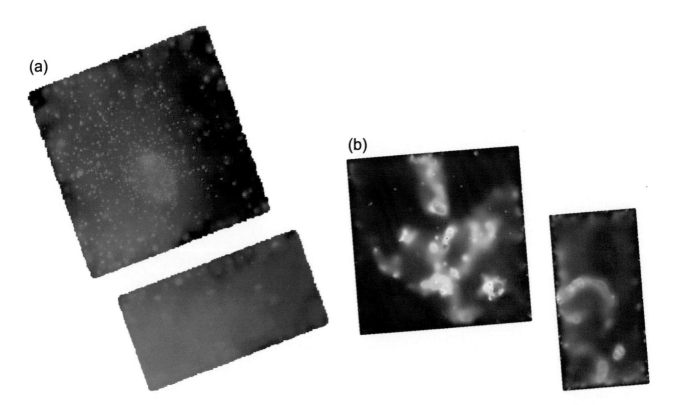

Plate 4. (a) A smoothed Chandra image (red = 0.5–2 keV, blue = 2–8 keV) of Trumpler 14 in the Carina Nebula. The field is 17' × 17', or ~14 × 14 pc. About 1600 point sources plus extensive diffuse emission is seen. From Townsley et al. (in preparation). (b) Smoothed soft-band image (red = 0.5–0.7 keV, green = 0.7–1.1 keV, blue = 1.1–2.3 keV) of the 30 Dor starburst in the Large Magellanic Cloud. The image covers ~250 pc on a side. R136a lies at the center; the plerionic SNR N157B lies to the southwest; the superbubble 30 Dor C and the Honeycomb and SN1987A SNRs are seen in the two off-axis CCDs. The colorful inhomogeneous diffuse structures are the superbubbles produced by past generations of OB star winds and supernovae. From *Townsley et al.* (2006a).

Accompanies chapter by Feigelson et al. (pp. 313–328).

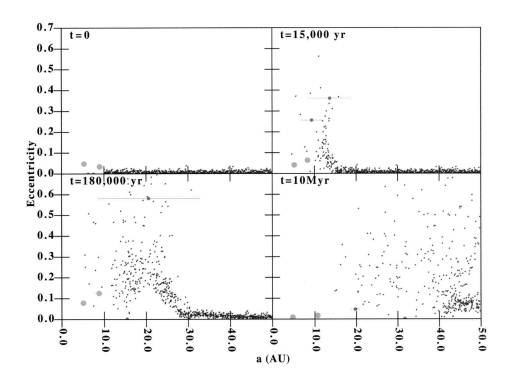

Plate 5. Four snapshots of the dynamical evolution of a system where Uranus and Neptune (red) are originally between Jupiter and Saturn (green), and there originally is a 200 $M_\oplus$ planetesimal disk (black) beyond Saturn. The run was taken from *Thommes et al.* (1999). For the planets, the error bars show the range of heliocentric distances that their orbits cover, and thus are a function of eccentricity.

Accompanies chapter by Levison et al. (pp. 669–684).

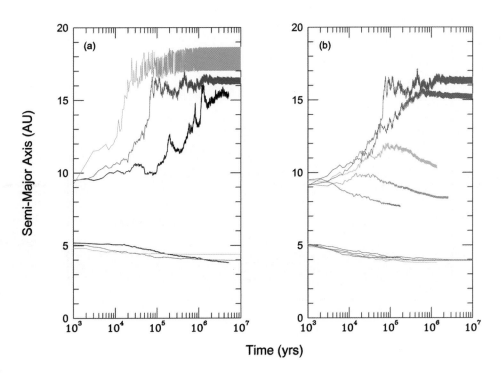

Plate 6. The temporal evolution of the semimajor axes of the two giant planets in our two planet simulations. Each color represents a different run. **(a)** Runs where the masses of the planets and disk are scaled by the same value. The black curve shows a run with Jupiter and Saturn at their current masses. The red and blue curves shows runs where the masses were scaled by a factor of 3 and 10, respectively. **(b)** Runs where the total mass of the planets remained fixed, but the mass ratio of the planets varied from run to run. Red, purple, green, orange, and brown refer to runs where the inner to outer mass ratio was 3.3, 2, 1.5, 1, and 0.5, respectively.

Accompanies chapter by Levison et al. (pp. 669–684).

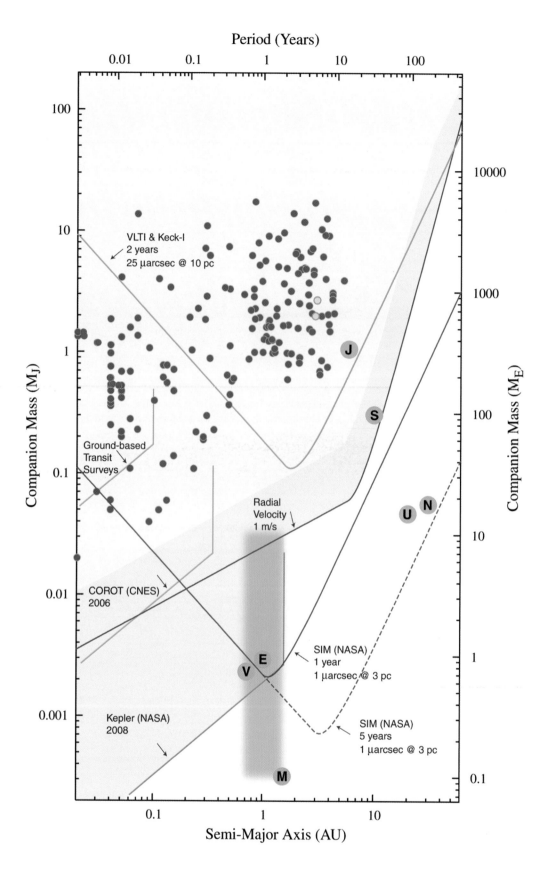

Plate 7. A wide variety of space- and groundbased capabilities showing the detectability of planets of varying sizes and orbital locations. Spacebased techniques become critical as one goes after terrestrial planets in the habitable zone. For a detailed description of this figure, see *Lawson et al.* (2004).

Accompanies chapter by Beichman et al. (pp. 915–928).

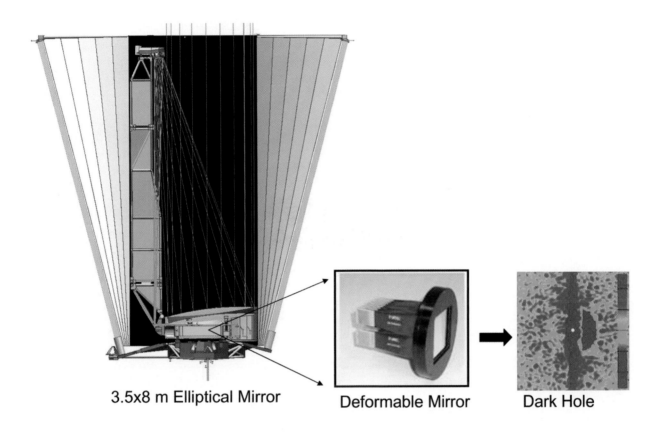

3.5x8 m Elliptical Mirror   **Deformable Mirror**   **Dark Hole**

Plate 8. (*Left*) An artist's concept for TPF showing the sunshield surrounding the 3.5 × 8 m primary mirror; (*middle*) picture of the deformable mirror, which is the key development for wavefront control; (*right*) a "dark hole" is created when the deformable mirror is adjusted to take out wavefront errors in the optical system. In this rectangular area the contrast demonstrated in the laboratory has reached 10^{-9} starting at a field angle of $4\lambda/D$, within a factor of 10 of that needed to detect planets.

Accompanies chapter by Beichman et al. (pp. 915–928).

Overlap between TPF-C, TPF-I and SIM capabilities

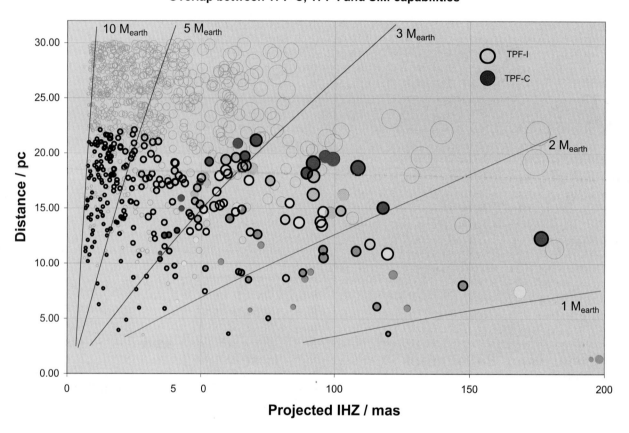

Plate 9. Potential targets for SIM, TPF-C, and TPF-I/Darwin are shown in terms of stellar distance vs. the angular extent of the inner edge of the habitable zone (IHZ, which is set to be the orbit of Venus scaled by the square root of the stellar luminosity). Stellar diameter is indicated by the size of the symbol. High-priority TPF-C targets are shown with filled symbols shaded to denote different probabilities of completeness (*Brown,* 2005) in a survey for Earth-sized planets in the center of the habitable zone [green denotes highest completeness (>75%); purple is lowest (<25%); yellow is intermediate]. TPF-I/Darwin targets are shown as black open circles. The loci of minimum masses detectable by SIM in 125 two-dimensional visits are also shown. Table 4 gives a subset of the most favorable targets.

Accompanies chapter by Beichman et al. (pp. 915–928).

Index

2MASS 10, 366, 373

Accretion 119ff, 156ff, 448ff, 485ff, 631
    diagnostics 448, 485ff
    models 489ff
    observations 487ff
    rates 125, 449, 519, 631ff
    time variable 488
    X-rays 316
Adaptive mesh refinement (AMR) 105ff, 141, 144
Adaptive optics 721
Advanced Camera for Surveys (ACS) 223
Ambipolar diffusion 9, 35, 69, 159, 263, 356
Amoeboid olivine aggregates (AOAs) 850
Amorphous silicates 772, 819ff, 822ff
Angular momentum 279, 300ff
Angular momentum transport 614
Anhydrous chondritic porous IDPs 817ff
Annealing 804, 816, 825
Asteroids 863ff, 867ff
    hydrated minerals 867ff, 881
    jovian Trojans 677ff, 869ff, 883ff
Asteroids, individual
    1 Ceres 868
    4 Vesta 844ff
    624 Hektor 884
Astrometry 918
Asymptotic giant branch (AGB) stars 802, 806, 823
Atacama Large Millimeter Array (ALMA) 45, 94, 499, 506
Atmospheric escape 937

Berkeley Illinois Maryland Association (BIMA) interferometer 82
Bernoulli flow 567
Binary systems 133ff, 151, 379ff, 395ff, 411ff, 427ff
    disk diagnostics 396
    disk evolution 401
    disk orientation 398
    disk resupply 405
    dynamical mass 380, 412ff, 416ff
    eclipsing 412
    fraction 134, 429ff
    mass ratio 134, 428, 432
    massive stars 209
    orbital motion 387
    planet formation 405
    separation 134, 402ff, 431ff
    spectroscopic 381, 413
    statistics 433ff, 460ff
    very low mass 429ff, 446, 465
Binary systems, individual
    2M0535-05 414
    AK Sco 401
    DQ Tau 401
    FU Ori 548
    GG Tau 400, 534
    GW Ori 401
    HD 98800B 415, 548
    ISO-Cha I 97 386
    SR 24N 400
    SVS 20 386
    V1174 Ori 418ff
    V4046 Sgr 401

Binary systems, individual (continued)
    WL 20 386
    YLW 15 386
Biosignatures 940
Bipolar outflows, *see* Outflows
Birth binary population 137, 138
Bok globules 8, 292
Bondi-Hoyle accretion 109ff, 152, 157, 159, 172, 631
Bonnor-Ebert sphere 7, 22, 43, 48, 68, 103, 153, 291, 629
Brown dwarf desert 428, 446, 592
Brown dwarfs 443ff, 459ff, 623ff
    accretion 338ff, 423, 449, 461, 633ff
    atmospheres 420
    disks 338ff, 450ff, 461
    ejection 136, 160, 337, 339, 437, 470ff
    evolution 420
    kinematics 445
    multiplicity 445ff, 460, 465
    outflows 450, 461
    photoevaporation 438, 472
    planet formation 454
    rotation 303ff, 461
    variability 339
    vs. planets 634ff
    X-rays 336, 338, 461
Brown dwarfs, individual
    2M0535-05 414

Calcium-aluminum-rich inclusions (CAIs) 325, 644, 789, 843ff, 850
    formation 844ff
Canada-France-Hawaii Telescope 329ff, 334ff
Carbide planets 939
Carina Nebula 224, 225, 322
Centaurs 871, 884, 896
Centaurus A 13
Center for High Angular Resolution Astronomy (CHARA) 551, 728ff
Champagne flow 75, 186ff
Chandra 217, 313ff
Chandra Orion Ultradeep Project (COUP) 217, 313ff
Chondrules 325, 618, 644, 789, 794, 805, 843ff, 850, 866
    formation 789, 829, 844ff
    shock models 829
Circumbinary disks 400
Circumstellar disks, *see* Disks
Class 0 sources 50, 53, 118
Class I sources 118, 778, 802
Class II sources 118
Class III sources 118
Classical T Tauri stars (CTTSs) 336, 479ff
    accretion 487ff
    spin-down 490
    winds 488
    X-rays 340, 754
Clustered star formation 151, 176
Clusters 361ff
    evolution 372
    lifetime 370
    mid-infrared 362
    millimeter 364
    near-infrared 362
    protostars 370
    star counts 364

Page numbers refer to specific pages on which an index term or concept is discussed.
"ff" indicates that the term is also discussed on the following pages.

945